Proceedings of the
Twenty First Annual Conference
of the
Cognitive Science Society

Martin Hahn and Scott C. Stoness
Editors

August 19-21, 1999
Simon Fraser University
Vancouver, British Columbia

LEA

LAWRENCE ERLBAUM ASSOCIATES, PUBLISHERS

1999   Mahwah, New Jersey   London

Distributed by
Lawrence Erlbaum Associates, Inc.
10 Industrial Avenue
Mahwah, New Jersey 07430

ISBN 0-8058-3581-4

ISSN 1047-1316

Printed in the United States of America

Table of Contents

Symposia

Long Papers

Abstract Posters

Twenty First Annual Conference of the Cognitive Science Society

August 19-21, 1999
Simon Fraser University
Vancouver, British Columbia

Conference Chair

Martin Hahn, Simon Fraser Unversity

Conference/Program Committee

Kathleen Akins	Simon Fraser University
Veronica Dahl	Simon Fraser University
Steven Davis	Simon Fraser University
Brian Fisher	Simon Fraser University, University of B.C.
Robert Hadley	Simon Fraser University
Nancy Hedberg	Simon Fraser University
Alan Lesgold	University. of Pittsburgh
Fred Popopowich	Simon Fraser University
Colleen Siefert	University of Michigan
Paul Thagard	University of Waterloo

Administrative Assistant and Volume Co-editor

Scott C. Stoness

Web Site

Centre for System Science, SFU

1999 Marr Prize

Lera Boroditsky, Department of Psychology, Stanford University
*First-Language thinking for Second Language Understanding:
Mandarin and English Speakers' Conceptions of Time.*

This conference was supported by the Cognitive Science Society, the Office of Vice-President-Academic, the Department of Philosophy and the School of Computing Science at Simon Fraser University, and the Media and Graphics Interdisciplinary Centre at the University of British Columbia.

The Cognitive Science Society

The Cognitive Science Society was founded in 1979 to promote interchange among workers in the various disciplines that comprise cognitive science. The Society sponsors an annual conference and publishes the journal *Cognitive Science*. Members of the Society are persons qualified to conduct or supervise research in cognitive science or allied sciences. Anyone is eligible to become an Associate Member of the Society, including students of cognitive science.

Cognitive Science Society, Univ. of Michigan, 525 East University, Ann Arbor, MI, 48109-1109; cogsci@umich.edu; (734) 429-4286; fax (734) 763-7480.

Tutorial Program
August 18, 1999

Introduction to the ACT-R cognitive architecture
John R. Anderson & Christian Lebiere

Psychological Soar Tutorial
Richard M. Young & Frank E. Ritter

An introduction to the COGENT Cognitive Modelling Environment
Richard Cooper & Peter Yule

Computational Cognitive Neuroscience Modeling using PDP++
Randy O'Reilly

Tutorial Progam Committee Co-Chairs

Frank Ritter   University of Nottingham
Richard Young   University of Hertfordshire,

Tutorial Progam Committee

Randy Jones , University of. Michigan   Kevin Korb, Monash Univesity
John Anderson ,Carnegie Mellon University   Michail Lagoudakis, Duke University
Christian Lebiere, Carnegie Mellon University Christine Manning, University of Minnesota
Todd Johnson, University of Texas, Houston   Brian Fisher, Simon Fraser University
Vasant Honavar, Iowa StateUniversity

The Office of Navy Research has helped support this tutorial series.

Workshops on Teaching Cognitive Science
August 18, 1999

Different models for undergraduate programs in cognitive science.
David Anderson (Illinois State), Jan Andrews (Vassar), Mark Rollins (Washington U.), Lon Shapiro and Johnna Shapiro (Illinois Wesleyan), Neil Stillings (Hampshire), Tom Wasow (Stanford)

Tools for teaching cognitive science courses, I.
Christine Diehl and Todd Shimoda (Berkeley), Ken Livingston (Vassar)

Tools for teaching cognitive science courses, II.
John Barker (Southern Illinois), Steven Weisler (Hampshire), David Anderson (Illinois State)

Graduate education in cognitive science and issues of cross-disciplinary communication. Andy
Brook (Carleton U.), Stellan Ohlsson (U. Illinois at Chicago), TBA

Tutorial Program Co-Chairs

Kenneth R. Livingston   Vassar College
Janet K. Andrews   Vassar College

Invited Symposia

Conceptual Foundations of Neuroscience
Convenor: Ian Gold, McGill University

Ralph Adolphs, University of Iowa
Earl Miller, Massachusetts Institute of Techonology
Steve Quartz, Salk Institute
Cathy Rankin, University of British Columbia

The object-based nature of visual attention and cognition
Convenor: Zenon Pylyshin

Object continuity under occlusion
Steve Yantis, Johns Hopkins University
Barry Vaughan, Johns Hopkins University

Competition for consciousness among visual events
Iterative reentrant processing in object perception
Vince Di Lollo & Jim Enns, UBC

Reflexive joint attention depends on lateralized cortical connections
Alan Kingstone, UBC

Clusters precede shapes in perceptual organization
Lana Trick, Kwantlen College, & Jim Enns, UBC

Two ways of asking 'What is a visual object?' (and some answers).
Brian Scholl, Rutgers University

.Connecting vision, cognition and the worldTracking as the basic paradigm.
Zenon Pylyshyn, Rutgers Center for Cognitive Science

Constraints, rules, and defaults: Optimality Theory and Cognitive Science
Convenor: Joe Stemberger, University of Minnesota

Who Rules LanguageShannon or Chomsky
Convenor: Fernando Pereira , AT&T Research

Hinrich Schuetze, Xerox PARC, Palo Alto, CA
Mark Johnson, Brown University

Sensation, Perception, and Sensory Quality
Convenor: David Rosenthal, CUNY Graduate School and University Center

Robert Schwartz, University of Wisconsin, Milwaukee
Rainer Mausfeld University of Kiel

Reviewers for the Twenty-First Annual Conference of the Cognitive Science Society

Agnar Aamodt
Clifford Abbott
Ralph Adolphs
Jeanette Altarriba
Rick Alterman
Mike Anderson
Richard N. Aslin
Bernard Baars
Bruno Bara
John Barnden
Lawrence W. Barsalou
William Bechtel
Robert C. Berwick
John Best
Dorrit Billman
D. S. Bree
Sarah Brem
Bruce Bridgeman
Bruce K. Britton
Paul Brna
Andrew Brook
Curtis Brown
Curt Burgess
Michael D. Byrne
Antonio Fernandez Caballero
Richard Carlson
Richard Catrambone
Valerie M. Chase
Nick Chater
Peter Cheng
Daniel Chester
Michelene T.H. Chi
Clark A. Chinn
Todd L. Chmielewski
Martin Chodorow
Marten H. Christiansen
James I. Chumbley
Axel Cleeremans
Catherine Clement
Chuck Clifton
Frederick Conrad
Alain Content
Andrew R.A. Conway
Richard Paul Cooper
Gary Cottrell
Neilson Cowan
Nils Dahlbäck
Kathleen Dahlgren
Edward L. DeLosh
Guy Denhiere
Karl Diller
Stephanie Doane
Danièle Dubois

Renee Elio
Eileen B. Entin
Debra C. Evans
Marte Fallshore
Martha Farah
Gilles Fauconnier
Rodolfo Fiorini
Ute Marie Fischer
Brian Fisher
Charles R. Fletcher
Kenneth D. Forbus
Carl H. Frederiksen
Michael Freed
Eric Freedman
Reva Freedman
Christian Freksa
Alinda Friedman
Joachim Funke
Jennifer B. Ganger
Michael Gasser
Dedre Gentner
Ted Gibson
Helen M. Gigley
Fernand Gobet
John A. Goldsmith
Barbara L. Gonzalez
Adrian Gordon
Peter C. Gordon
Hank Gorman
Wayne D. Gray
Sami Gulgoz
Karl Haberlandt
Robert F. Hadley
Udo Hahn
Henry M. Halff
Graeme S. Halford
James Hampton
Trevor A. Harley
Gilbert Harman
Gary Hatfield
Barbara Hemforth
Randall Hendrick
Petra Hendriks
Friedrich W. Hesse
Graeme Hirst
Edward Hoenkamp
Jim Hollan
Keith J. Holyoak
Vasant Honavar
Harry Howard
Lumei Hui
John E. Hummel
Gavin Huntley-Fenner

Kiyoto Ishimaru
Anthony Jameson
Dietmar Janetzko
Bonnie John
Philip Nicholas Johnson-Laird
Arne Jonsson
Bob Kachelski
James Kahn
Demetrios Karis
David Kaufman
Mark T. Keane
Brian L. Keeley
Alan W. Kersten
Jinwoo Kim
Thomas King
Sheldon Klein
Markus Knauff
Janet L. Kolodner
Hyung Joon Kook
Timothy Koschmann
Gilbert Krulee
Kenneth Kurtz
Howard S. Kurtzman
Yannick Lallement
William Langston
David Leake
Sylvie Leclerc
Adrienne Y. Lee
Jung-Mo Lee
Yuh-shiow Lee
F. Javier Lerch
Leonardo Lesmo
James Lester
Ping Li
Ken Livingston
Vincenzo Lombardo
Paul A. Luce
Sten R. Ludvigsen
George Luger
Juan Magarinos de Morentin
Lorenzo Magnani
Margo Malakoff
Barbara Malt
Arthur Brian Markman
Gail Martino
Dominic W. Massaro
Yuji Matsumoto
Mark Mattson
Devin J. McAuley
James L. McClelland
Gary McGraw
Jean McKendree
Danielle S. McNamara

Ken McRae
Lise Menn
Craig Miller
George Miller
Toby Mintz
Naomi Miyake
Fumio Mizoguchi
Joyce L. Moore
Kenneth Moorman
Erik T. Mueller
Paul Munro
Gregory L. Murphy
Kumiyo Nakakoji
N. Hari Narayanan
Aldo Nemesio
Josef Nerb
Nancy J. Nersessian
Tim Norman
Angela O'Donnell
Padraig O'Seaghdha
Ruediger Oehlmann
Stellan Ohlsson
Clark G. Ohnesorge
Klaus Opwis
Thomas C. Ormerod
Yuriko Oshima-Takane
Vimla Patel
T. Pattabhiraman
Jim Peters
Steven Phillips
John L. Pollock
Ian Pratt
Sadhana Puntambekar
Clark Quinn
Paul Quinn
Mitch Rabinowitz
Satyajit Rao
Stephen John Read
Mike Redmond
Terry Regier
Daniel Reisberg
Ehud Reiter
Valerie F. Reyna
Juliet Richardson
Chris Riesbeck
Robert Rist
Frank Ritter
Paul Rosenbloom
J. Edward Russo
Ala Samarapungavan
Eric Saund
Patricia Schank
Christian Schunn
Dan Schwartz
Julie C. Sedivy
Erwin Segal

Mark S. Seidenberg
Michael Shafto
Stuart C. Shapiro
Bruce Sherin
Atsushi Shimojima
Thomas R. Shultz
Richard Allan Shweder
Daniel Silverman
Judith E. Sims-Knight
Peter Slezak
Vladimir M. Sloutsky
Linda B. Smith
Steve M. Smith
Hans Spada
Michael Spivey
Mark St. John
James J. Staszewski
Nancy L. Stein
Keith Stenning
Suzanne Stevenson
Robert Stufflebeam
Ron Sun
Dan Suthers
David Swinney
Niels Taatgen
Heike Tappe
Roman Taraban
Paul Thagard
Jean-Pierre Thibaut
Charles Tijus
Maurizio Tirassa
Shingo Tokimoto
Elise Turner
Roy M. Turner
Barbara Tversky
Ryan D. Tweney
Jody Underwood
Erminia Vaccari
Maria Dolores Valita
Frederic Vallee-Tourangeau
Frank Van Overwalle
Mark Van Selst
Jon Vaughan
Alonso Vera
M. Felisa Verdejo
Greg Vesonder
Regina Vollmeyer
Michael R. Waldmann
J. G. Wallace
Arnold D. Well
John Whalen
Bob Widner
Giovanna Winchkler
Stanton Wortham
Yan Xiao
Fei Xu

Yingrui Yang
Joseph L. Young
Richard M. Young
Jiajie Zhang
Alf Zimmer

Integrated Models of Perception, Cognition, and Action

Michael D. Byrne (byrne@acm.org), Organizer
Department of Psychology, Rice University
Houston, TX 77005-1892

Ronald S. Chong (rchong@soartech.com)
Soar Technology Incorporated
Ann Arbor, MI 48105

Michael Freed (mfreed@mail.arc.nasa.gov)
NASA Ames Research Center
Moffett Field, CA 94035

Frank E. Ritter (ritter@psychology.nottingham.ac.uk)
School of Psychology, University of Nottingham
Nottingham, UK NG7 2RD

Wayne D. Gray (gray@gmu.edu), Discussant
Department of Psychology, George Mason University
Fairfax, VA 22030-4444

"One thing wrong with much theorizing about cognition is that it does not pay much attention to perception on the one side or motor behavior on the other... The result is that the theory gives up the constraint on...cognition that these systems could provide. The loss is serious--it assures us that theories will never cover the complete arc from stimulus to response, which is to say, will never be able to tell the full story about any particular behavior." - Allan Newell, *Unified Theories of Cognition,* pp. 159-160.

When that quote was delivered to the audience of William James lectures more than a decade ago, Cognitive Science as a field was guilty as charged of neglecting the integration of cognition, perception, and action. However, in recent years serious efforts have been made to construct models that encompass all three systems and take seriously the mutual constraints they impose on one another. These include: EPIC-Soar, a system based on integrating Soar and the EPIC perceptual-motor architecture (Chong); ACT-R/PM, which integrates the ACT-R cognitive architecture with a system of EPIC-like perceptual-motor modules (Byrne); and APEX, a system inspired by the Model Human Processor (MHP) designed to model performance in complex, dynamic environments (Freed). A more generic perceptual-motor system, able to interact with multiple cognitive architectures and designed to interact with multiple real-world tasks, has also been proposed (Ritter).

Research on these systems has been fruitful practically, empirically, and theoretically. Each symposium participant will be asked to discuss the issues involved and the hurdles they have overcome in constructing systems that coordinate cognition, perception, and action. Examples include:

• What new empirical questions have been raised by broadening the scope of research from cognition to include perception and action?

• How has the inclusion of perception and action capabilities constrained or informed the development of the cognitive aspects of models developed with your system(s)?

• Conversely, have the cognitive capabilities of the system constrained or influenced the design of the perceptual-motor systems?

• What kinds of tasks and domains are you able to model that you could not have modeled successfully without serious consideration of perceptual-motor capabilities?

• Working at "lower" levels of analysis required to model perception and action in detail may also have costs. For example, has it become more difficult to model higher-level cognition such as problem solving and reasoning?

• How is learning affected by perception and action? Conversely, how are perception and action affected by changes in cognition?

• How is communication between the three subsystems managed?

• What technical issues have you had to overcome to develop a more integrated approach?

• What model of visual attention is used in your system? What is your system's perspective on the relationship between gaze position and attention?

Productive Interdisciplinarity: The Challenge that Human Learning Poses to Machine Learning

Helen M. Gigley, Ph.D (gigley@itd.nrl.navy.mil)
Office of Naval Research
800 N Quincy Street (Code 342)
Arlington, VA 22217-5660

Susan F. Chipman, Ph.D. (chipmas@onr.navy.mil)
Office of Naval Research
800 N. Quincy Street (Code 342)
Arlington, VA 22217-5660

Overview

Recent efforts in the Hybrid Architectures for Learning Program sponsored by the Office of Naval Research were based on applying general computational hybrid models of learning to three human learning tasks. Each task had learning performance data available. The issue was to run the basic hybrid model on a selected task to verify the model's performance relative to the actual human data and to evolve the model, increasing the match between the learned performances, to obtain a better predictive/explanatory model of the human process.

There were three tasks used, an Air Traffic Controller simulation task (The Kanfer-Akerman ATC-Task ™), an Obstacle Avoidance or Navigation simulation task, and a Command Information Center(CIC) decision making task. Eight groups participated, selecting one of the three tasks. This symposium will present results from three groups on two of the tasks. These projects raise exciting new questions and issues specifically for machine learning approaches to the study of learning. Human learning presents challenging performance for current machine learning approaches to meet. Thus, cognitive modeling applications contribute to the understanding of computational approaches as much as to the understanding of human cognition.

Format for the Symposium

Dr. Helen Gigley Introduction to the Hybrid Architectures Program

Dr. Devika Subramanian* (devika@cs.rice.edu)
Rice University -- Learning the NRL Navigation Task

Dr. Bonnie John (Bonnie_John@cs.cmu.edu)
Carnegie Mellon University -- Short Introduction to the ATC-Task

Dr.PrasadTadepalli** (tadepall@cs.orst.edu)
Oregon State University -- Learning Hierarchical Control: Humans vs. Machines

Dr. Bonnie John *** -- Strategy use and its implications for computational models of learning

Dr. Susan Chipman -- Summary

Open discussion

---

* Joint work Dr.Diana Gordon at Naval Research Laboratory and Dr. Sandra Marshall at San Diego State University.
** Joint work with Thomas G. Dietterich and Chandra Reddy at Oregon State University.
*** Joint work with Dr. Yannick Lallement, Novator Systems, Toronto, Ontario, Canada.

Connectionism: What's structure got to do with it?

Gary F. Marcus (gary.marcus@nyu.edu)
Department of Psychology, New York University, 6 Washington Place, New York, NY 10012

John Hummel (jhummel@psych.ucla.edu)
Department of Psychology, 405 Hilgard Ave, UCLA., Los Angeles, CA 90095

Risto Miikkulainen (risto@cs.utexas.edu)
Department of Computer Science, The University of Texas at Austin, Austin, TX 78712

Lokendra Shastri (shastri@ICSI.Berkeley.EDU)
International Computer Science Institute and UC Berkeley, 1947 Center Street Suite 600, Berkeley CA 94704

Most discussions of connectionism in cognitive science have focused on whether or not simple unstructured networks such as multilayer perceptrons suffice for cognition. Other types of connectionist models have received relatively less attention. The purpose of this symposium is to increase awareness of these other kinds of models, situating them in a context that allows a greater understanding of what kinds of models are good for what kinds of tasks. Our assumption is that in any given domain it is possible build some adequate model, but that the interesting question is how. The emphasis, then, will be on the strengths and weaknesses of competing connectionist architectures and how they can be usefully applied to models of cognition.

The symposium will start with a discussion by Marcus of some important cognitive phenomena that are difficult to handle with multilayer perceptrons, and then continue with discussions Hummel, Shastri, and Miikkulainen of how other kinds of connectionist models can account for some of those phenomena.

Representations: New approaches to old problems

Arthur B. Markman (markman@psy.utexas.edu)
Department of Psychology; University of Texas, Mezes Hall 330
Austin, TX 78712 USA

William Bechtel (bill@twinearth.wustl.edu)
Department of Philosophy, Washington University in St. Louis,
St. Louis, MO 63130-4899

Overview of the Symposium

Modern cognitive science is built on a foundation of representation. All cognitive scientists assume that there are internal information carrying states that mediate thought and behavior. Despite this general agreement, there is significant disagreement about the nature of these internal states.

The classical view in cognitive science is that representations consist of abstract symbols that cut across modalities. This view takes the data structures of a computer as a model for cognitive representations. This view has come under attack from a number of directions because of observed limitations with this approach. For example, connectionists have pointed out that symbol systems are often too rigid and brittle to capture the context sensitivity of behavior. Advocates of perceptual representation have pointed out that there is no reason to believe that cognitive representations are divorced from perceptual modalities. It has also proven difficult to reconcile perceptual representations with these amodal symbols. Researchers using dynamic systems have focused on the self-organizing properties of behavior and have argued that representational theories lead to models of cognition that fail to capture the transitions between stable states of behavior. That is, they argue that models based on classical representations provide coarse descriptions of behavior, but fail to capture the fine details of individual performance.

The attacks on the standard view of representation typically begin with a core example that highlights an important problem with representation. From there, an argument is developed that cognitive science should dispense with the classical model of representation in favor of the approach that handles the example presented. Unfortunately, there is a tendency for researchers to sketch the way their new proposal for representation will handle cases beyond the example that initially posed a problem for the classical view of representation. Thus, debates over representation often degenerate into rhetorical battles.

In this symposium, we bring together researchers from different perspectives with the goal of having them talk explicitly about how to create a broader theory of representation. The symposium begins with an introduction presented by Arthur Markman, who will discuss the importance of having multiple approaches to representation, and will give an overview of the talks to be presented.

After the introduction, there are three speakers. The first speaker is William Bechtel. In his previous work, he has analyzed the steam engine governor that has been used by advocates of dynamic systems as an example of how cognition might take place without representation. He argues that the steam engine governor does indeed have representations, and describes the representational aspects of the governor. He gives a brief overview of his analysis of the steam engine governor and then focuses on whether cognitive science needs representations above and beyond those he suggests for the governor.

The second speaker is Lawrence W. Barsalou, whose recent work focuses on perceptual symbol systems. On his view, the classical construct of representation went astray by assuming that cognitive and sensory-motor representations are separate. After presenting initial arguments for why the brain evolved to be a representational device, he shows how perceptual symbol systems implement classical representational phenomena such as productivity and propositions. He argues that it is possible for a theory of structured representations to be dynamical, context-sensitive, and embodied, integrating insights of connectionism, dynamic systems, and classical representation.

The third speaker is John Hummel. He talks about the coordination of context sensitivity and structured symbol processing in connectionist modeling. A key strength of connectionist models is that they allow processing to vary with context. The problem with connectionist systems is that they often fail to represent relational (symbolic) structures. Dr. Hummel's work on connectionist models of object recognition and analogy has tried to bridge the gap by incorporating techniques for building relational representations within context-sensitive connectionist architectures.

Following the talks, Eric Dietrich will serve as a discussant. His presentation ties together the central themes raised by the speakers. In addition, he highlights the problems that remain. Finally, he explores the likelihood that researchers from different perspectives will be able to merge their perspectives. The session ends with a 20 minute discussion period moderated by Arthur Markman.

In sum, the goal of this symposium is to introduce a variety of approaches to representation. Furthermore, this session aims to go beyond the simple examples that motivate alternative approaches to representation, and to begin to attack the difficult problems that lie ahead.

Symposium: Dynamic Decisions, Conflict Resolution, and Real-Time Diagnosis in Complex Domains

Chair and Organizer: Vimla L. Patel

Presenters: **Vimla L. Patel** (patel@hebb.psych.mcgill.ca)
Cognitive Studies in Medicine, Centre for Medical Education, McGill University, Montreal Canada

Guy Boy (boy@onecert.fr)
European Institute of Cognitive Sciences and Engineering (EURISCO)
Toulouse, France

Kim J. Vicente (benfica@mie.utoronto.ca)
Cognitive Engineering Laboratory
Department of Mechanical & Industrial Engineering, University of Toronto, Canada

Alan Lesgold (al+@pitt.edu)
Learning, Research and Development Center, University of Pittsburgh

Studies of cognition beyond laboratory walls have flourished in recent years. These developments have been enabled by both methodological and theoretical advances that have facilitated investigations of "cognition in the wild". These endeavors promise to profoundly impact the discipline of cognitive science. This symposium presents research pertaining to the development of decision-making expertise in complex and diverse environments.

An emerging area of research concerns investigations of cognition in dynamic "real-world" work environments. Dynamic environments are characterized by high levels of urgency, uncertainty, and shifting, ill-defined, and competing goals. Decisions are often part of an ongoing process, embedded in the flow of work, and jointly determined by teams of individuals with complementary spheres of expertise. Emergency and intensive care medicine are two exemplar disciplines characterized by high velocity decision making. Vimla Patel and David Kaufman present work related to understanding dynamic decision making in high velocity medical environments, namely intensive care and medical emergency units. This presentation focuses on a series of studies conducted by Patel and colleagues in these dynamic medical environments. The presentation examines the differential and selective use of evidence and conflict resolution in negotiating decisions in situations of varying urgency. Guy Boy discusses sources and models of conflict resolution between agents, including humans and machines, in airplane cockpits This presentation focuses on major causes of conflict in an aircraft cockpit, such as lack of knowledge, lack of training, forgetting, role confusion, workload, human errors, inability to delegate, imprecise or incomplete perception, quid pro quo, lack of power sharing, or misunderstanding. An example of accident analysis is presented to highlight how a conflict between a human and a machine is generated and evolves toward an unrecoverable situation. Boy provides some recommendations in the form of usability criteria for human-centered design of artificial agents. Kim Vicente and colleagues similarly address conflict resolution by nuclear power plant (NPP) operators. They present findings into how operators deal with these conflict situations, drawing on a number of field studies. Conflicts arise in some cases arises because their expectations are inconsistent with one or more indications provided by the control room displays and other times because the control room indications themselves contradict each other. Our final speaker, Alan Lesgold, presents studies related to the development of expertise in the diagnosis and repair of complex equipment in microchip manufacturing. Many areas of technical expertise involve complex mixtures of partial conceptual knowledge and rules of thumb that are often not well anchored in basic scientific principles. Existing theories have not adequately addressed this. This research draws on recent work on building coached apprenticeship environments. Their experience has been that carefully designed intermediate representations that are grounded in science but not necessarily fully explained or understood, can be very useful in promoting technical expertise, even when the technicians do not have much science background.

There is a gap between decision making research and education in the professions. In addition, technical skill domains have not adequately addressed the role that conceptual knowledge plays in supporting acquisition of expertise. The presentations in this symposium address some of these common concerns in diverse everyday work environments, using different methodological and theoretical approaches.

Symposium on reference axes in language and space

Emile van der Zee (evanderzee@lincoln.ac.uk)
Department of Psychology; Brayford Pool
Lincoln, LN1 1LS UK

In order to be able to talk about what we see we may use both linguistic and spatial knowledge (Landau & Jackendoff, 1993). It is generally assumed that the use of directional terms like *front* and *back* not only involves linguistic knowledge, but also knowledge of reference axis representation and categorization (see Bloom et al., 1996). Although there seems to be agreement on the possible involvement of reference axes in spatial language use, there is no agreement about the way in which reference axes are represented, which axes are cognitively more prominent than other ones, how reference axes are categorized, and whether the representation and categorization of such axes is universal or not (see, e.g., Levinson, 1996). This symposium brings together four papers that take a different standpoint on these issues. The paper presentations are followed by a half hour discussion in which the paper presenters discuss their views. Other symposium participants are also invited to take part in the discussion. The purpose of the symposium is to bring more clarity in the factors that determine reference axis representation and categorization for the purpose of language.

Edward Munnich, Barbara Landau and Barbara Dosher (University of California) present two experiments with native English, Japanese and Korean speakers. These experiments show that subjects from these languages performed the same on language tasks and memory tasks for the representation and categorization of axes, that subjects from these languages performed in a similar way with respect to the notions of contact and support in the memory tasks, but also that in relation to contact and support subjects performed differently in the language tasks. On the basis of these results the presenters suggest that knowledge of axis representation and knowledge of contact and support are non-linguistic universals, that knowledge of axis categorization may be a linguistic universal, but that knowledge of the categorization of contact and support is different across languages.

Emile van der Zee (University of Lincoln) and Rik Eshuis (Graduate Program in Cognitive Science in Hamburg) present two separate theories on reference axis representation and categorization. They assume that both reference axis representation and categorization may be based on the spatial features of an object. In addition, they assume that the mechanisms for axis representation and categorization are universal, although reference axis labeling is not. Their theory on reference axis categorization assumes that the top-down axis is the most basic axis in spatial processing, followed by the front-back and the left-right axis (see, e.g., Tversky, 1996). Three experiments with native Dutch speakers support these theories.

Urpo Nikanne (University of Oslo) presents linguistic research on the semantic representation of reference axes in Finnish and English. Drawing on examples of prepositional phrases that are used in motion expressions Nikanne argues that reference axes are represented in a semantic hierarchy. This hierarchy assumes that the front-back axis is the most basic axis, and that this axis is neutral with respect to verticality/horizontality, as long as no other axes are introduced. The next axis in the hierarchy is the left-right axis, projecting 2D entities in the horizontal plane. The top node in the hierarchy is the top-down axis, which projects 3D entities along the vertical.

Laura Carlson-Radvansky (University of Notre Dame) presents work that focuses on how the identity of verbally related objects influences the orientation and origin of a reference frame in one of these objects. One experiment shows that the functional relation of the objects involved influences how the axes of a reference frame are oriented. This experiment demonstrates a preference for defining axes with respect to an object when the objects are functionally related. Two additional experiments demonstrate that the identity and function of the objects impact where the reference frame is placed on an object for the purpose of describing the objects' spatial relation. These data implicate an additive combination of perceptual and conceptual factors for determining how reference frames are situated for language use (Carlson-Radvansky, in preparation).

References

Bloom, P., Peterson, M., Nadel, L. & Garrett, M. (1996). *Language and space.* Cambridge, MA: MIT Press.

Carlson-Radvansky, L. (in preparation). Object use and object location: The effect of function on spatial relations. In E. van der Zee and U. Nikanne (Eds.), *Cognitive interfaces: Constraints on cognitive information.* Oxford: Oxford University Press.

Landau, B & Jackendoff, R. (1996). 'What' and 'where' in spatial language and spatial cognition. *Behavioral and Brain Sciences*, 16, 217-238.

Levinson, S. (1996). Frames of reference and Molyneux's Question: Crosslinguistic evidence. In P. Bloom, M. Peterson, L. Nadel and M. Garrett (Eds.), *Language and space.* Cambridge, MA: MIT Press.

Tversky, B. (1996). Spatial perspective in descriptions. In P. Bloom, M. Peterson, L. Nadel and M. Garrett (Eds.), *Language and space.* Cambridge, MA: MIT Press.

The Role of Theory of Mind and Deontic Reasoning in the Evolution of Deception

Mauro Adenzato (adenzato@psych.unito.it)
Rita B. Ardito (ardito@psych.unito.it)

Center for Cognitive Science
University of Turin
via Lagrange, 3   10123 Torino, Italy

Abstract

Modern Darwinist perspective enables to deal with the study of several human phenomena, one of which is deception, that we define as a behaviour unfolded with the deliberate intention of producing or sustaining a state of ignorance or false belief in another person. Evolutionary Psychology, an emerging area inside Cognitive Science, represents a promising conceptual approach to the study of deception. According to it, knowledge on human mind can be improved by understanding the processes which, during evolution, shaped its architecture. This work traces back to the Evolutionary Psychology arguments (for a review see Cosmides & Tooby, 1987; Barkow, Cosmides & Tooby, 1992; Buss, 1995; 1999) and develops the hypothesis that deception is a behaviour underpinned by two psychological mechanisms that evolved in response to problems posed by group living: the theory of mind and deontic reasoning.

1. Basic Assumptions of Evolutionary Psychology

According to evolutionary psychologists, it is possible to increase our knowledge of the human mind through an understanding of the processes that in the course of phylogenesis have modelled its architecture. This in fact means to construct theories regarding the selective pressures that have recurringly acted throughout the history of our evolution, so as to be able to formulate hypotheses for the architecture of the human mind, considering it as the result of those pressures. The selective pressures that have accompanied our evolution can be seen as adaptive problems, that favourably select those individuals that have developed mechanisms capable of generating responses to them. Among the adaptive problems that have been necessary to confront are, for example, the choice of sexual partner, communication with other members of the group, the defence of progeny against predators and the recognition of deception in social exchanges.

A central assumption of Evolutionary Psychology is that there is a universal human nature to be sought among the body of psychological mechanisms that shape our cognitive architecture and that these mechanisms constitute the basis of our behaviour. One of the goals pursued by Evolutionary Psychology is to furnish a functional explanation of these psychological mechanisms, by seeking to comprehend the selective pressures encountered by our ancestors, to which they are a response. By uniting Cognitive Psychology with Evolutionary Biology, Evolutionary Psychology attempts to demonstrate that the human mind is a complex system composed of a finite number of mechanisms, each of which having been shaped by natural selection to favour individual survival through the exercise of a particular function (Symons, 1992). Evolutionary Psychology however goes well beyond the notion of the innate and acquired patrimonies as reciprocally irreducible ontogenetic dimensions, to focus its particular attention on the complex causal relations extant between selective pressures and psychological mechanisms, and between these latter and behaviour.

What the evolutionary psychologists maintain is that the few tens of thousands of years which have passed since the appearance of man in his modern form, which appearance has been traced to a time between 100 and 200 thousand years ago (Horai, Hayasaka, Kondo et al., 1995), are almost irrelevant with respect to the more than two million years that individuals of genus *Homo* lived with a social organization and a lifestyle very different from those current today. Consequently, the hypothesis can be advanced that it is not possible to fully comprehend the nature of any given psychological mechanism without referring to the type of life the individuals of our genus conducted during the Pleistocene, the life of hunter-gatherers of the savannah and the prairie.

Since human mind is the product of a slow process, it would seem reasonable to exclude the possibility of its having evolved in response to the conditions of life and environment that man has had to confront in recent times. In fact, these conditions represent only a fraction of our evolutionary history and the conditions prevailing during the course of our phylogenesis were very different.

Regarding social organization, for example, we know that genus *Homo* spent more than 99% of his evolution in groups made up of a number of members varying from 30-50 to 200-300. These groups of individuals, organized into true bands of hunter-gatherers, were the prevalent type of social organization until about 10,000 years ago (well after the appearance of modern man) when we see the beginnings of the progressive propagation of a new relationship with the environment. This new relationship consolidated itself only within the last 5,000 years (Diamond, 1997), leading to an economy based on agriculture and animal raising, and to a social organization evermore characterized by the creation of stable and populous urban nuclei.

These considerations aid in understanding the reasons why we cannot expect our minds to have evolved mechanisms capable of confronting the problems which arose following the appearance of agriculture, let alone those arising as a consequence of industrialization, and they clearly signal the necessity for research into the style of life and the selective pressures that accompanied the evolution of our species for over two million years.

2. Deception in the Perspective of Evolutionary Psychology

Evolutionary Psychology suggests that a series of psychological mechanisms underpins human behaviour. Since this is valid as well for deception, a satisfactory explanation of this phenomenon must be able to generate falsifiable hypotheses as to nature of the psychological mechanisms at its base. As has been previously explained, psychological mechanisms can be interpreted as structures that evolved to resolve the adaptive problems faced by our ancestral forebears. As such, in order to identify the mechanisms underpinning deception we must first ask under what selective pressures they evolved. In other words, we must single out the adaptive problem to which deception -or more precisely the mechanisms permitting its actuation- is a response.

Our hypothesis is twofold: (a), that deception is a behaviour underpinned by two psychological mechanisms that evolved in response to problems posed by group living, the theory of mind and deontic reasoning; and (b), that deception is a behaviour able to confront one specific problem among others: the problem of the constraints imposed by the group on the individual. This constraints limit individual possibilities of achieving personal goals. The hypothesis thus presented underscores how the correct interpretation of deception behaviour can emerge clearly only through consideration of the complex social organization that characterized the evolutionary history of genus *Homo* (Adenzato, 1998; Adenzato & Ardito, 1998; Adenzato & Bara, 1999).

The hypothesis that some aspects of human cognitive architecture can have evolved in response to problems posed by sociality has been yet authoritatively sustained by other researchers. According to the hypothesis of the *Social Origin of the Mind*, the increase in cerebral mass and the consequent development of cognitive capacity are adaptive traits that primates evolved in response to the complexities of social life. At the base of this hypothesis, advanced in its most explicit form by Humphrey (1976), but already delineated years before by Chance and Mead (1953) and Jolly (1966), is the observation that the social world, for the challenges it poses to the individual, is more complex than the physical one, which instead is normally more predictable. During the course of evolution there would therefore have been stronger selective pressures for mechanisms capable of resolving problems posed by group living than for those operating in response to the physical world.

One of the most interesting developments of the hypothesis of the social organization of intelligence is the concept of "Machiavellian intelligence", proposed by Byrne and Whiten (1988). This term, inspired by the Florentine tutor of deceitful politicians, refers to the fact that among social primates intelligence is often used to deliberately manipulate and exploit other members of the group. A social primate thus demonstrates its possession of profound knowledge of both the complex network of relationship linking the members of the group and the particular characteristics of each individual (de Waal, 1982; Cheney & Seyfarth, 1990). Machiavellian intelligence is manifested in its clearest form in the capability it confers upon individual primates to utilize such knowledge in order to increase their reproductive success, or to form alliances with certain individuals to obtain advantages that are to the detriment of others.

We maintain that the selective pressures arising from the complex social organization that accompanied the development of our ancestral forebears during the course of their evolutionary adaptation are the basis for the first psychological mechanism that underpins the human capacity to deceive, the theory of mind.

Having a theory of mind signifies to comprehend that human beings are entities gifted with mental states such as beliefs, desires, intentions, thoughts and that these mental states are in casual relationship with the events of the physical world, i.e., capable of being the causes as well as the effects of these events. Moreover it signifies being able to refer to one's own and to others' minds for the explanation and prediction of individual behaviour (Leslie, 1987; Astington, Harris & Olson, 1988; Wellman, 1990; Povinelli & Preuss, 1995). A disturbed development of the theory of mind has been associated with the syndrome of autism (Baron-Cohen, Leslie & U. Frith, 1985; Baron-Cohen, 1995) while its deterioration at an advanced stage has been connected with certain schizophrenic manifestations (C.D. Frith, 1992; Corcoran, Mercer & C.D. Frith, 1995).

It has been maintained that the appearance of the theory of mind during the childhood of our species coincides with the attainment of comprehension of false belief. Comprehension of a false belief concerns the capacity of a subject to consider that another person can have a belief that he or she retains to be true but which the subject knows to be false. It requires therefore an ability to represent to oneself how another's representation coincides with reality. Experimental results in the literature seem to show that comprehension of the false belief appears at the age of three or four. Children reaching this stage of development are able not only to understand how another person can form an erroneous belief about something, but also to extrapolate what the erroneous belief will be and what effect it will have on the behaviour of its holder (Wimmer & Perner, 1983; Perner, 1991). The most important result of this acquisition from the adaptive point of view is that by becoming part of the casual network of the world, mental states become inferable and reliable, and as such can be explained and predicted, on the basis of such clues as personal behaviour and the elements of a given situation.

The theory of mind is a mechanism without which deception is impossible, for two reasons at least. First, because lacking such a mechanism an individual is unable to create a state of false belief in another, that is, he is unable to induce the other to believe in something false because he cannot create for himself a satisfactory representation of the other's beliefs, which is an obvious impediment to the manipulation of that person with the aim of deceiving him. Moreover, an individual that wants to deceive, before concerning himself with the creation of a false belief in another, must first worry about convincing that person to interact with him. Without a theory of mind however, it is not possible to make plausible inferences as to the desires, the beliefs and the motivations that could induce someone to participate in an interaction that will reveal itself to be a deception.

In the existing literature on deception, attention has up to now been concentrated on the role played by the theory of mind. According to the previously presented hypothesis however, the theory of mind is not the only mechanism to underpin deception behaviour. An evolutionary reading of the phenomenon leads us to suppose that beside this psychological mechanism stands another, that of deontic reasoning. The evidence for this comes from a consideration of the type of social organization that accompanied the evolution of genus *Homo*, and of the kinds of selective pressures that such organization brought to bear.

It is well-known that group living, which has characterized the course of development of genus *Homo*, guarantees a series of noteworthy advantages: a joint defence against predators is more effective, obtaining otherwise inaccessible food from larger prey is facilitated, and the young are more highly safeguarded (Alexander, 1974). It must be borne in mind however, that although

these benefits are of clear importance to the survival of individual members of the group, only very rarely are they equitably distributed. What is normally observed in animal societies is that access to resources, whether they be food, or safe places to sleep, or access to sexual partners, is regulated by a hierarchy of domination, which may be more or less rigid according to the species.

Domination hierarchies determine the social status of each individual belonging to the group, and they can be seen as a scale of ranks, within which the individuals arrange themselves after having confronted each other in aggressive or ritualized interactions. The individual who holds the highest rank is the one to whom falls the right of access to the best of the available resources. He will be able, for example, to choose the richest places for eating, the safest for sleeping, and have a greater number of sexual partners (Clutton-Brock & Harvey, 1976). The other individuals, in their turn, will manage the remaining resources in relation to their respective positions in the hierarchy, with the consequence that those occupying the lower positions must adapt to an uncomfortable life in which their prospects of reproductive success have been greatly reduced. As such, the hierarchy of domination can be viewed as a structure capable of imposing rules of social conduct on individuals and influencing individual reproductive success, according to the rank attained by the individual in question (Fedigan, 1983; Clutton-Brock, 1988; Cheney & Seyfarth, 1990).

Among social primates, the domination hierarchy is a structure which emerges from the cooperative and competitive relationships that develop among members of the group. Managing these relationships adequately is a primary necessity for the individual members of a social species, and surely during the course of evolution there was a strong selective pressure that favoured those individuals who, as a result of the processes of genetic recombination, presented cognitive structures capable of drawing the maximum advantages from the complex cooperative and competitive interactions. Cummins (1996a; 1996b; 1996c) has proposed that the capability of reasoning deontically emerged precisely in such a way, and that deontic reasoning can be considered as an innate mechanism of human cognitive architecture.

To reason deontically means reasoning regards to what is permitted, forbidden, or alternatively obligatory with regards to other individuals. Life within a hierarchy of domination requires individuals to enlarge continuously in deontic considerations, since lower ranked individuals must decide whether to commit themselves to forbidden activities with the aims of fraudulently procuring resources, and higher ranked individuals must continually defend their positions of privileged access to resources, while recognizing and punishing others' attempts at deception. Deontic reasoning also plays an important role in establishing alliances. If among primates the individual's social status were determined exclusively by corporeal mass, as it is for example among elephant seals were the largest individuals invariably occupy the highest places in the hierarchy (Le Boeuf, 1974), then there would have been no need to develop a capacity for deontic reasoning, and pure physical force would have been enough. But since among primates the rank of an individual is determined to a great degree by his ability to form alliances (de Waal, 1982; Harcourt, 1988; Harcourt & de Waal, 1992), deontic reasoning has been selected, because it permits control over how the contracted obligations of the members of the coalition are respected.

According to the hypothesis here presented, deontic reasoning plays a determining role in deception, for without it deception behaviour could not effectively articulate itself. The importance of deontic reasoning to the capacity to deceive emerges clearly upon consideration of the relational dimension of deception. In fact, when in a given situation an individual decides that deception is the best means of achieving a given objective, he commits itself to confronting a situation of social interaction with one or more persons, and for the deception to succeed the deceiver must be able to manage such a situation. The deceiver must know what social bonds and rules tie him to other individuals, what he can do and what he cannot, what the others expect of him and what he can expect of them, what obligations he must keep and where he has freedom of action. Without this body of knowledge it would not be possible to deceive.

Only when a rule is known it is possible intentionally violate it, and if an individual is unaware of the existence of a rule and he violates it, he is not being deceitful, but he is exposing himself to a situation with consequences that are unpredictable and not open to active influence. If, for example, I do not realize that the woman I am courting belongs to the chief, then I will be vulnerable to retaliation without knowing the reason; if instead I am conscious of breaking a rule then it is possible for me to manipulate the situation in such a way that my fraudulent comportment in not discovered. Deception behaviour as such, basing itself on an individual's social knowledge, abilities and alliances, gives him or her the possibility to achieve personal goals, and therefore to directly or indirectly influence his or her probabilities of reproductive success.

3. Some Testable Hypotheses

The analysis of deception presented in this work lends itself to series of falsifiable predictions. Firstly, if deception is effectively a behaviour able to affront the problems of social constraints that group living imposes upon the individual, then it is possible to predict that its manifestation will be more probable in situations where social obligations, status, and prohibitions are obstacles to the achievement of personal goals. We are currently engaged in a detailed study to validate this hypothesis on the basis of an extensive data base of incidents of deception behaviour drawn from natural situations; the preliminary results are particularly promising.

Secondly, if it is correct to affirm that deception behaviour is based on the theory of mind and deontic reasoning, then the specific lack of these psychological mechanism should clearly compromise an individual's capacity to deceive. As regards the theory of mind there are already studies in the literature that demonstrate how autistic subjects who are characterized by the lack of this particular mechanism, have a significantly diminished capacity do deceive (Oswald & Ollendick, 1989; Baron-Cohen, 1992; Sodian & U. Firth, 1992; 1993). Regarding deontic reasoning, Brothers (1994) and Damasio (1994) cite several cases in which a specific damage to frontal lobes is associated to a specific impairments of social reasoning, but not with others areas of cognition. The considerations made in the course of this work induce us to predict that a subject suffering of a frontal syndrome will be seen to have difficulty demonstrating an adequate capacity to deceive in the course of normal social interactions. We have already construed a specific test for to explore this hypothesis in that we are sure that studies of such kind cannot help but enrich the field of our knowledge, and the future will see the clinical study become an ever more useful test bench for hypotheses developed according an evolutionary perspective.

4. Conclusions

The present work would emphasize the impossibility of studying deception behaviour without giving due consideration to the social organization that has accompanied the evolution of genus *Homo* for over two millions years. Only thanks to such considerations it is possible to recognize in deontic reasoning a psychological mechanism indispensable to deception. The role of this mechanism in the generation of deception behaviour should be studied, with an attention equal to that which the literature has until now justly dedicated to the theory of mind.

Acknowledgments

We are grateful to Bruno G. Bara for helpful suggestions and discussions.

This research has been supported by the Ministero dell'Università e della Ricerca Scientifica e Tecnologica (M.U.R.S.T.) of Italy, Coordinate Project 60%, 1997 "Sviluppo della competenza pragmatica".

The names of the authors are in alphabetic order.

References

Adenzato M. 1998. The psychological bases of deception. *10th Annual Meeting of the Human Behavior and Evolution Society*, Davis, California.

Adenzato M. & Ardito R.B. 1998. The role of social complexity in the evolution of human communication. *6th International Pragmatics Conference*, Reims, France.

Adenzato M. & Bara B.G. 1999. Deceiving: implications for primates. *Folia Primatologica*, in press.

Alexander R.D. 1974. The evolution of social behavior. *Annual Review of Ecology and Systematics*, 5, 325-383.

Astington J.W., Harris P.L. & Olson D.R., eds. 1988. *Developing theories of mind*. Cambridge University Press, Cambridge.

Barkow J.H., Cosmides L. & Tooby J., eds. 1992. *The adapted mind. Evolutionary psychology and the generation of culture*. Oxford University Press, New York.

Baron-Cohen S. 1992. Out of sight or out of mind: another look at deception in autism. *Journal of Child Psychology and Psychiatry*, 33, 1141-1155.

Baron-Cohen S. 1995. *Mindblindness. An essay on autism and Theory of Mind*. MIT Press, Cambridge, MA.

Baron-Cohen S., Leslie A.M. & Frith U. 1985. Does the autistic child have a "theory of mind"? *Cognition*, 21, 37-46.

Brothers L. 1994. Neurophysiology of social interactions. In: Gazzaniga M., ed. *The cognitive neurosciences*. MIT Press, Boston, MA.

Buss D.M. 1995. Evolutionary psychology: a new paradigm for psychological science. *Psychological Inquiry*, 6, 1-30.

Buss D.M. 1999. Evolutionary psychology: the new science of the mind, Allyn and Bacon, Needham Heights, MA.

Byrne R.W. & Whiten A., eds. 1988. *Machiavellian Intelligence: social expertise and the evolution of intellect in monkeys, apes, and humans*. Oxford University Press, Oxford.

Chance M.R.A. & Mead A.P. 1953. Social behavior and primate evolution. *Symposia of the Society for Experimental Biology, Evolution*, 7, 395-439.

Cheney D.L. & Seyfarth R.M. 1990. *How monkeys see the world*. University of Chicago Press, Chicago.

Clutton-Brock T.H. 1988. Reproductive success. In: Clutton-brock T.H., ed., *Reproductive success*. University of Chicago Press, Chicago.

Clutton-Brock T.H. & Haevey P.H. 1976. Evolutionary rules and primates societies. In: Bateson P.P.G. & Hinde R.A., eds., *Growing points in ethology*. Cambridge University Press, Cambridge.

Corcoran R., Mercer G. & Frith C.D. 1995. Schizophrenia, symptomatology and social inference: investingating "theory of mind" in people with schizophrenia. *Schizophrenia Research*, 17, 5-13.

Cosmides L. & Tooby J. 1987. From evolution to behavior: evolutionary psychology as the missing link. In: Dupre J., ed., *The latest on the best: essays on evolution and optimality*. MIT Press, Cambridge, MA.

Cummins D.D. 1996a. Dominance hierarchies and the evolution of human reasoning. *Minds and Machines*, 6, 4, 463-480.

Cummins D.D. 1996b. Evidence for the innateness of deontic reasoning. *Mind and Language*, 11, 2, 160-190.

Cummins D.D. 1996c. Evidence of deontic reasoning in 3- and 4-year-old children. *Memory & Cognition*, 24, 6, 823-829.

Damasio A.R. 1994. *Descartes' Error. Emotion, Reason, and the Human Brain*. Grosset/Putnam, New York.

de Waal F. 1982. *Chimpanzee politics: power and sex among apes*. Jonathan Cape, London.

Diamond J. 1997. *Guns, germs and steel. The fates of human societies*. W.W. Norton & Company, New York-London.

Fedigan L. 1983. Dominance and reproductive success in primates. *Yearbook of physical anthropology*, 26, 91-129.

Frith C.D. 1992. *The cognitive neuropsychology of schizophrenia*. Lawrence Erlbaum Associates, Hove, UK and Hillsdale, NJ.

Harcourt A.H. 1988. Alliances in contests and social intelligence. In: Byrne R.W. & Whiten A., eds., *Machiavellian Intelligence: social expertise and the evolution of intellect in monkeys, apes, and humans*. Oxford University Press, Oxford.

Harcourt A.H. & de Waal F., eds. 1992. *Coalitions and alliances in humans and other animals*. Oxford University Press, Oxford.

Horai S., Hayasaka K., Kondo R., Tsugane K. & Takahata N. 1995. Recent african origin of modern humans revealed by complete sequences of hominid mitochondrial DNAs. *Proceedings of the National academy of science*, 92, 532-536.

Humphrey N.K. 1976. The social function of intellect. In: Bateson P.P.G. & Hinde R.A., eds., *Growing points in ethology*. Cambridge University Press, Cambridge.

Jolly A. 1966. Lemur social behaviour and primate intelligence. *Science*, 153, 501-506.

Le Boeuf B.J. 1974. Male-male competition and reproductive success in elephant seals. *American Zoologist*, 14, 163-176.

Leslie A.M. 1987. Pretense and representation: the origins of "theory of mind". *Psychological Review*, 94, 412-426.

Oswald D. & Ollendick T. 1989. Role taking and social competence in autism and mental retardation. *Journal of Autism and Developmental Disorders*, 19, 119-128.

Perner J. 1991. *Understanding the representational mind*. MIT Press, Cambridge, MA.

Povinelli D.J. & Preuss T.M. 1995. Theory of mind: evolutionaru history of a cognitive specialization. *Trends in Neurosciences*, 18, 418-424.

Sodian B. & Frith U. 1992. Deception and sabotage in autistic, retarded and normal children. *Journal of Child Psychology and Psychiatry*, 33, 591-606.

Sodian B. & Frith U. 1993. The theory of mind deficit in autism: evidence from deception. In: Baron-Cohen S., Tager-Flusberg H. & Cohen D.J., eds., *Understanding others minds: perspectives from autism.* Oxford University Press, Oxford.

Symons D. 1992. On the use and misuse of darwinism in the study of human behavior. In: Barkow J.H., Cosmides L. & Tooby J., eds. *The adapted mind. Evolutionary psychology and the generation of culture.* Oxford University Press, New York.

Wellman H.M. 1990. *The Child's Theory of Mind.* MIT Press, Cambridge, Mass.

Wimmer H. & Perner J. 1983. Belief about beliefs. Representation and constraining functions of wrong beliefs in young children's understanding of deception. *Cognition*, 13, 103-128.

Verbal and embodied priming in schema mapping tasks

Tracy Packiam Alloway[1], Michael Ramscar[2,3] and Martin Corley[1,3]
({tracyp,michael,martinco}@dai.ed.ac.uk)
[1]Department of Psychology
[2]Institute for Communicating and Collaborative Systems, Division of Informatics
[3]Human Communication Research Centre
University of Edinburgh, 80 South Bridge, Edinburgh EH1 1HN, Scotland

Abstract

The question of whether language influences thought or not has been much discussed and disputed in the cognitive science literature. A recent proposal by Lakoff and Johnson (1999) adds an interesting slant to this debate by arguing that although language can influence thought via conceptual metaphors, the overall shape of the human conceptual system is determined by its embodied, perceptual nature. In this way, language is ultimately the slave of thought.

We present an experiment aimed at exploring this question empirically. Exploiting evidence that has shown that schema consistent priming can bias the outcome of reasoning tasks, we performed a study in a well mapped conceptual domain in order to examine whether embodied experience or language is the greater determinant of conceptual inferences. In this study, we found that language, rather than thought, is maybe what counts.

Introduction

Concepts are an essential part of cognition. The ability to group things together - whether as edible, dangerous, or even friendly - confers many benefits, both in terms of cognitive economy and, perhaps ultimately, evolutionary advantage. The relationship between concepts, the cognitive capabilities that facilitate grouping, and words, the labels that are the primary manifestation of categorisation is close, somewhat controversial, and goes to the heart of cognitive science. The question of whether thought influences language or language influences thought is an old one, with powerful adherents on either side of the argument. In this paper, we explore a recent proposal by Lakoff and Johnson (1999) which adds an interesting slant to this debate by arguing that although language can influence thought via conceptual metaphors, the overall shape of the human conceptual system is determined by its embodied, perceptual nature. In this way, language is ultimately the slave of thought. We present an experiment aimed at exploring this question empirically. Exploiting evidence that has shown that schema consistent priming can bias the outcome of reasoning tasks, we describe a study in a well mapped conceptual domain in order to examine whether embodied experience or language is the greater determinant of conceptual inferences.

Space And Time - A "Conceptual Domain"

There is a great deal of overlap in the lexical terms we use in talking about space and time. A number of researchers have noted systematic correspondences between the words we use in talking about space and time (McTaggart, 1908; Clark, 1973; Traugott, 1978; Lakoff and Johnson, 1980; Boroditsky, 1998). We often use phrases like "Christmas is *coming*" or "Our vacation is *ahead* of us," or "The honeymoon *followed* the wedding" without being aware of the spatial metaphors that appear to underpin our temporal speech. We employ this type of metaphor with such frequency that they have acquired a ubiquity that tends to hide their origins.

As Gentner and Imai (1992) note time is usually seen as unidirectional and unidimensional because it moves in one direction and in a linear form. For this reason the terms that are borrowed from the domain of space and used to express time are also unidimensional, such as forward/backward, and front/back rather than multidimensional terms like shallow/deep, narrow/wide.

The *motion of time* represents one framework for how spatio-temporal metaphors are comprehended and is determined by the future moving to the past. This is explained by a simple example. In the month of February, Christmas is now in the future; in a few months it will soon be moved to the present and then to the past. The individual is a stationary observer as time "flows" past him, as in the example *The party is after the seminar*. This system is known as the *Time Moving* metaphor (in this metaphor, temporal events are seen as moving past an observer like "objects", hence its spatial equivalent is the *Object Moving* metaphor).

The second system is the *Ego-Moving* metaphor where the ego or the individual moves from the past to the future such as the sentence *His vacation at the beach lies before him* (in this metaphor, the observer is seen as moving forward through time, passing temporal events which are seen as stationary points, hence it is the temporal equivalent of the spatial *Ego Moving* system, where the observer moves forward through space).

These spatial metaphors for understanding time appear to represent an instance of the kind of conceptual scheme proposed by Lakoff and Johnson (1980) in their Conceptual Metaphor hypothesis. According to this hypothesis, metaphors are not just a manner of speaking but a deeper reflection of human thought processes. Metaphoric speaking is reflective, say Lakoff and Johnson, of deeper conceptual mappings that occur in our thinking and is depicted as an over-arching and general metaphor termed as the Conceptual Metaphor (see also Gibbs, 1992). Consider the following statements:

Your claims are *indefensible*.

He *attacked every weak point* in my argument.

He *shot down* all of my arguments.

According to the Conceptual Metaphor (metaphoric representation) hypothesis when we use statements such as these we are making use of a larger conglomerate metaphor, in this instance, ARGUMENT IS WAR.[1]

The thrust of the Conceptual Metaphor argument is as follows: arguments are similar to wars in that there are winners and losers, positions are attacked and defended, and one can gain or lose ground. The theory of Conceptual Metaphor suggests that we process metaphors by mapping from a base domain to a target domain. In this particular example, the base domain is ARGUMENT IS WAR and the target domain is a subordinate metaphor such as *Your claims are indefensible.*

Lakoff and Johnson extend the idea of Conceptual Metaphor to spatio-temporal metaphors by invoking the locative terms of FRONT/BACK to represent how we view time and space. FRONT is assigned on the assumption of motion (Fillmore, 1978). According to this theory, in the *ego-moving* system, FRONT is used to designate a future event because the ego is moving forward and encounters the future event in front of him. In the *time-moving* system the FRONT term denotes a past event where the ego or the individual is stationary but the events are moving. (for a critique of this view see McGlone, 1996; Murphy, 1996).

Embodiment Theory

The notion of Conceptual Metaphor is part of a deeper theory concerning the way we process and categorise objects around us (Lakoff and Johnson, 1999). Lakoff and Johnson introduce the idea of *embodiment*, which incorporates our experiences as an integral part in the formation of concepts (see also Barsalou, in press). They claim that categorisation is not a product of conscious reasoning or the intellect but results instead as a product of our embodied experiences. It is our interaction with the circumstances in which we are immersed that, according to this view, helps us to formulate the structures that enable us to function in, and comprehend the everyday situations in which we find ourselves.

The embodiment theory can be summed up in the following statement: "An embodied concept is a neural structure that is actually part of, or makes use of, the sensorimotor system of our brains. Much of conceptual

inference is, therefore, sensorimotor inference" (Lakoff and Johnson, 1999, p. 20).

Although the embodiment theory makes much reference to neural structure, Lakoff & Johnson cite no neurophysical evidence to confirm their theory (but c.f. Pulvermüller, in press). Instead they make exclusive reference to language to support the embodiment theory. For example, our use of words like *front, back, forward* are all contingent on our bodies and its interaction with things around us. Because of our dependence on our bodily projections to conceptualise objects, this theory is labelled as a "phenomenological embodiment" (Lakoff and Johnson, 1999, p.36).

Lakoff and Johnson go further to claim that this notion of embodiment blurs the distinction between perception and conception. It has previously been assumed that the formulation of concepts is based purely on reason and that, while perception may influence reason and cause motion, neither perception or movement is considered part of reason. On the other hand, perception has been associated with movement and separate from conception or mental processes. However, according to the embodiment theory, perception and movement are fundamental to conception as well, because of the important role embodiment plays in categorisation.

Our spatial-relation words (*ahead, under, forward*) depend on our embodied perception and movement, which allow us to conceptualise actions or events. Thus the theory of Embodiment is intimately connected to the theory of Conceptual Metaphor because it is our experiences that drive our formulation of Complex Metaphors such as LOVE IS A JOURNEY or ARGUMENT IS WAR. From our daily experiences, we form an understanding of events and cluster them in a category that serves to allow us to function more effectively in daily life.

The Schema-Mapping Hypothesis

Gentner and Boronat (1991) have observed that since metaphors are processed from a common base schema to a common target schema, such processing should be fluent because of important similarities in the underlying metaphoric schemas. On the other hand, if metaphors from different schemas are presented, the processing time should increase because the individual has to shift between different perspectives.

Gentner and Boronat tested this idea by presenting participants with consistent and inconsistent metaphors. They discovered that there was a significant decrease in reading time when an inconsistent metaphor was presented after a series of consistent metaphors. Gentner and Boronat suggested that this time decrease was a result of remapping because the metaphors were processed as schema consistent mappings. Thus the schema consistency paradigm suggests that when an individual is presented with a metaphor from a different schema, they have to make a shift from one schema to the other and this shift causes a lapse or disturbance in processing.

The idea that schema consistency could increase mapping efficiency has been extended to encompass spatio-temporal metaphors. Gentner and Imai (1992)

---

[1] Following Lakoff and Johnson's convention (1980), all Conceptual Metaphors are typed in the uppercase to distinguish them from the subordinate metaphors

propose that these particular types of metaphors are processed via two distinct internally consistent systems. Gentner and Imai carried out several experiments to test this hypothesis. Participants were presented with either *ego-moving* or *time-moving* metaphor materials that used words like *before, ahead,* or *behind* to serve as locative prepositions. Subsequently, participants were asked to respond to questions that were either consistent or inconsistent with the type of metaphor embodied in these priming materials.

Gentner and Imai found that participants responded faster to questions that were consistent with the priming than to questions that were inconsistent with their primes. Gentner and Imai argue that this supports the theory that metaphors are mapped in distinct schemas: the shift from one schema to another causes a disruption in the processing, reflected in increased processing time. They argue that their study indicates that the relations between space and time are reflective of a psychologically real conceptual system as opposed to an etymological relic.[2]

A study by McGlone and Harding (1998) involved participants answering questions about days of the week - relative to Wednesday - which were posed in either the *ego-moving* or the *time-moving* metaphor. *Ego-moving* metaphor trials comprised statements such as "We passed the deadline two days ago", whilst *time-moving* metaphor trials involved statements such as "The deadline was passed two days ago"; in each case, participants read the statements and were then asked to indicate the day of the week that a given event had occurred or was going to occur. At the end of each block of such priming statements, participants read an ambiguous statement, such as "The reception scheduled for next Wednesday has been moved forward two days"[3] and then were asked to indicate the day of the week on which this event was now going to occur. Participants who had answered blocks of priming questions about statements phrased in a way consistent with the *ego-moving* metaphor tended to disambiguate "moved forward" in a manner consistent with the *ego-moving* system (they assigned 'forward' - the front - to the future, and hence thought the meeting had been re-scheduled for Friday), whereas participants who had answered blocks of questions about statements phrased a way consistent with the *time-moving* metaphor tended to disambiguate "moved forward" in a manner consistent with the *time-moving* system (they assigned 'forward' - the front - to the past, and hence thought the meeting had been re-scheduled for Monday).

These experiments offer some support to the embodiment theory, with its concomitant claim that embodiment in the world affects conceptualisation. They appear to show that participants ' perception of space has a direct effect on their conceptualisation of time.

Lakoff and Johnson (1999) cite experiments carried out by Boroditsky (1998) in support of the embodiment theory. Boroditsky (1998) suggests that there is an explicit

analogy between two schemas for organising space and time. On this analogy, *ego-moving* schemas are defined - for both space and time - in respect to an observer's direction of motion. The 'front' is assigned as the furthest forward point in the observer's direction of motion: thus in time, 'front' is assigned to the future, and in space, if objects are conceived of in linear fashion along a path, then 'front' is assigned to the objects that are furthest forward - relative to the observer's direction of motion - along the path. For *time-* and *object-moving* schemas, front is set to the furthest forward point in the direction of the movement of time or objects. Since time is usually conceived of as moving from future to past, 'front' is assigned to past, or earlier events. By analogy, in space, if two objects are moving (whether they have intrinsic 'fronts' or not),[4] then front is assigned to the leading part of the leading object.

Boroditsky presented participants with the phrase *Next Wednesday's meeting has been moved forward two days* together following either *ego-moving* metaphor or *time-moving* metaphor primes and asked them what day the meeting would be on. Her findings suggest that participants were not influenced by primed temporal schemas in responding to a problem about space; but spatial schemas had an effect in their responses to temporal questions.

Of central concern to us in this paper is the assertion by Lakoff and Johnson that Boroditsky's findings (and those of Gentner and Imai, 1992, and McGlone and Harding, 1998) should be accounted for by an *embodied* theory of conceptual understanding. In arguing for embodiment as the basis for our concepts, Lakoff and Johnson (1980; 1999) argue that language - through metaphoric representation, or Conceptual Metaphors - influences thought; and equally, they imply that ultimately it is thought - manifest in embodied perception - that influences language.

The Saphir-Whorf hypothesis

A different perspective on the language-thought debate - the theory that language is primary - is put forward in the Saphir-Whorf hypothesis. This can take numerous forms: the strong version of the Saphir-Whorf hypothesis claims that language *determines* thought, whilst weaker versions suggests that language *affects* perception (and hence thought). Other theorists have taken even stronger positions: Wittgenstein (1953) explicitly questions and rejects the idea that "thought" - as we understand it as linguistic animals - can exist independently of language at all.

A significant portion of the evidence for the Saphir-Whorf hypothesis is anthropological. For example, it has been claimed that languages indigenous to several different Native American Indian groups (such as Hopi, Nootka, Apache, and Aztec) each have a unique vocabulary that allows them to express events or spatial movement differently to what is possible in English. According to

---

[2] Although McGlone and Harding (1998) criticise some aspects of Gentner and Imai's methodology, their corrected replication of the original study confirms its findings.
[3] All trials were conducted on a Wednesday.

---

[4] Thus, when a car reverses, for instance, the back of the car is in front.

the Saphir-Whorf hypothesis this difference in vocabulary suggests that these cultures perceive events and space differently.

Proponents of the Saphir-Whorf hypothesis would also claim that grammatical differences document a pattern of how attention is attributed to particular objects. For example, in Navaho the endings of verbs correspond with the shape and rigidity of the object in mention. This grammatical distinction may well be a consequence of the attention that Navahos give to the properties of objects; it is claimed that the prominence of these forms of verb and verb endings are is indicative of deeper ways of thinking among the Navaho Indians. Fillmore (1971) suggests that our understanding of space is indicated in a similar way by language. He cites several different cultural and language groups to support the theory that language reveals our categorisation of time (or systems of spatial-relations concepts as Lakoff and Johnson refer to it). In English the different uses of the prepositions *on* and *in* are indicative of different spatial features. Fillmore claims that the use of *on* is reserved for surface words like *on the lawn*, and when used as a phrase like *on the earth*, it denotes the surface of a three-dimensional object. The word *in* applies to three-dimensional spaces (e.g., *in the yard*); so *in the earth* refers to the three-dimensional area of the earth as opposed to just the surface area. In other languages such as Samal (spoken in the Philippines), there are single terms that refer to concepts such as *near me, near you* or *away from all of the above*. The prevalence of these expressions reveals how speakers of languages in this particular group locate people spatially, and thus, according to Fillmore, how they perceive the spatial world.

Thought and Language

On the one hand, the Saphir-Whorf hypothesis suggests that by studying the language of a culture, we can begin to understand that culture's cognitive processes. On the other, the embodiment theory presents a very different notion. Lakoff and Johnson (1999) suggest that in the final analysis, thought influences language - language is ultimately the slave of our (universal) embodied thought.

Clearly the results of the experiments described above, which are based upon the schema consistency paradigm, (Gentner and Imai, 1992; McGlone and Harding, 1998; Boroditsky, 1998) are consistent with either view. It could be the case - as the embodiment theory would predict - that our understanding of space and time is causally determined by the embodied shape of our perceptions. On the other hand, it may be that language is the final arbiter of our concepts, and whatever words and metaphors we choose to employ in talking about space and time is by far the greater determinant of the way we conceptualise them. In either case, schema consistency would be promoted by the experimental conditions that obtained in the studies reviewed above.

Unfortunately, as their mutual compatibility with the evidence shows, a major problem with speculations about the ontogeny of temporal and spatial concepts is finding any way of empirically distinguishing between them. The following experiment was designed as an attempt to disentangle these competing accounts. We aimed to determine what is more important in ensuring schema consistent reasoning - embodied thought, or explicit language.

Experiment 1

Participants in Boroditsky's (1998) experiment were given visual primes that also included an explicit linguistic element. Participants were presented with a diagram of an observer moving towards some objects, and had to answer the question

[OBJECT] IS AHEAD OF ME   TRUE/FALSE

Thus it is hard to distinguish whether it was participants' embodied perception of the movement in the stimuli, or their reaction to the explicit linguistic elements that played the predominant role in determining their schema consistent inferences.

By fully involving participants in a non-linguistic task that was heavily biased towards a particular space metaphor schema, we aimed to explore the respective influences of simple immersion in such a task, and immersion with explicit linguistic priming, on schema consistency.

Participants

Sixty-eight undergraduate students of the University of Edinburgh, all native English speakers, served as volunteers. They were not aware of the nature of the experiment.

Materials

In order to immerse participants in a task that embodied a particular spatial schema we utilised a video game where the individual was stationary and had to defend himself against approaching attacks. Thus the game embodied the *object-moving* metaphor. The game was mouse-controlled and was uncomplicated. The participants controlled a stationary anti-aircraft gun at the bottom of the computer screen and had to shoot parachuters that came from the sky. At the end of each level, participants were able to replenish their ammunition The levels did not differ in their objective.

Movement in this game was limited, in that the parachutists that came towards the stationary subject were the only moving objects. Thus consistent with the moving-object schema, participants were stationary, with FRONT being assigned to the furthest point forward in the direction of the movement of objects (so that the parachutist that had travelled furthest towards the gun was "in front").

Next to the computer was a box with a three-way switch with attached light bulbs. No attention was directed to this box until the middle of the game.

Procedure

All participants were tested individually and - apart from the control group - were required to play the video game for five minutes. (Instructions given to the participants were that they were required to progress to level 4 without being eliminated.) There were three groups of participants.

The first group acted as a control group. They were simply presented with the target task - the switch box - whilst performing an unrelated task, and then asked to

"move the switch forwards"

The second group played the game for the requisite five minutes. Once they had completed their session, they were instructed to:

"move the switch forwards"

to indicate that they had completed playing. Two further questions were designed to check the effectiveness of the prime. At the end of the computer game, participants were asked an open-ended question, which required them to describe which parachuters (as described by screen location) they perceived to be the biggest threat.

As a final priming check, they were then asked to determine if the following statement was true or false:

During the game, it was more important to first shoot the parachuters in the front.

These questions evaluated whether participants had represented the parachuters that were closer to the ground as being in front. This would be consistent with an *object-moving* schema and in a schema-consistent mapping should determine their assignment of forward in moving the switch as well. Since participants were mapping FRONT to objects moving towards themselves (in their role as the gun controller), they should move the switch towards themselves when moving it *"forward"*.

The third group of participants also played the game for the requisite five minutes, but they received an explicit linguistic assignment of FRONT during their session. Instead of being asked about the game after they had finished playing, as in condition 2, participants were posed the priming check questions (identical to those for condition 2) during the course of their game playing session. That is, participants were required to explicitly verbalise their priming whilst they were immersed in the task.

Once again, at the end of level four in the game, these participants were instructed to *Move the switch forward* on the switch board.

In all conditions, a light bulb indicated which direction the switch had been moved.

Hypotheses

Since the switch moving task was inherently ego-centric (the actor in the task being the subject) - we expected in the control condition that participants, when asked to perform the task of moving a switch forward with no additional priming, would assign FRONT to the direction they were facing, and hence *"move the switch forward"* would map to FRONT in the *ego-moving* system, with *forward* being assigned the direction away from the subject. Since we expected the natural bias of interpretation in the switch task to be against the bias of

the priming, we expected to be able to measure the relative strength of embodied versus verbal priming by gauging the extent to which this natural bias could be defeated.

In the two game playing conditions, we hypothesised the following:

• If embodiment was the most significant determinant of thought, then participants primed in the *ego-moving* system by the task should tend to prefer schema consistency and hence reassign FRONT to objects coming towards them, mapping *"move the switch forward"* to FRONT in the *object-moving* system, with *forward* being the direction towards the subject

• If language was a more significant determinant of thought, then participants primed in the *ego-moving* system by the task *with* verbal priming should tend to prefer schema consistency and reassign FRONT to objects coming towards them, and hence map *"move the switch forward"* to FRONT in the *object-moving* system, with *forward* being the direction towards the subject. To the extent which verbal priming contributes to the adoption of a particular schema, participants in this condition should be more likely to be primed, compared to those primed only by the embodied task.[5]

Results

Data from all of the participants that responded incorrectly to the prime testing questions in the primed conditions was rejected. The overall error rate was 9% and it was equally distributed between the embodied prime and embodied-plus-explicit-verbal prime conditions.

As predicted, in the control condition, 100% of participants mapped FORWARD in *"move the switch forward"* to FRONT in the *ego-moving* system, and moved the switch on the switch-board away from themselves.

In the embodied prime condition, only 16% of participants responded in a prime consistent manner (assigning front according to the *object-moving* system), whereas 84% of participants continued to respond in a way consistent with the *ego-moving* system. This change was not significant when considered together with the result from the control group.

In the linguistic prime condition, 50% of participants responded in the embodied prime consistent manner (assigning front according to the *object-moving* system), whilst 50% of participants continued to respond in a way consistent with the *ego-moving* system. A chi-squared[6] analysis showed this to be significantly different from the distribution found in the control group χ (1, N=43) = 12.314, $p<.0.001$.

---

[5] Since Lakoff and Johnson allow for both linguistic and embodied influences on concepts, the experiment was designed to separate the effects of embodied priming against linguistic priming, but not vice versa.

[6] Because of the low numbers (0) in the cells in the control, this analysis used Yates' corrected chi.

Discussion

Our results showed that language - in the form of explicit verbal acknowledgements of primes in the *object-moving* metaphoric system - could significantly reverse participants' natural bias to assign FORWARD in an *ego-moving* manner. However, no such reversal was evident when the primes were solely of an embodied nature, despite the fact that subsequent tests showed that participants had been affected by that priming.

If, as Lakoff and Johnson (1999) suggest, thought influences language - and therefore language is ultimately the slave of our (universal) embodied thought - then we would have expected pure embodied priming to have at least as much an influence as embodied plus verbal priming. However, the effects of pure embodied priming were negligible.

The Saphir-Whorf hypothesis, on the other hand, maintains that language influences thought. Our finding that participants that were required to explicitly verbalise the concept of FRONT they had assigned as a result of the *object-moving* metaphor were more likely to be primed is consistent with this. Our experiment suggests that linguistic priming might be the key factor in overriding the natural bias of an individual to assign FRONT based on himself. This is an interesting twist in the relationship between language and thought and we look forward to pursuing further research in this area.

References

Barsalou, L. (in press). *Perceptual symbol systems.* Behavioural and Brain Sciences.

Boroditsky, L. (1998). Evidence for metaphoric representation: Understanding time. In K.Holyoak, D. Gentner and B. Kokinov (Eds). *Advances in Analogy Research*, New Bulgarian University Press, Sofia, Bulgaria.

Clark, H. H. (1973). Space, time semantics, and the child. In T. E. Moore (Ed.), *Cognitive development and the acquisition of language.* New York, NY: Academic Press.

Fillmore, C.J. (1971). *The Santa Cruz lectures on deixis.* Bloomington, IN:Indiana University Linguistics Club

Gentner, D., & Boronat , C. (1991). Metaphors are (sometimes) represented as domain mappings. Paper presented at the symposium on Metaphor and Conceptual Change, Meeting of the Cognitive Science Society, Chicago, IL.

Gentner, D., & Imai, M. (1992). Is the future always ahead? Evidence for system mappings in understanding space-time metaphors. In *Proceedings of the Fourteenth Annual Conference of the Cognitive Science Society*,

Gibbs, R. (1992). Categorization and Metaphor Understanding. *Psychological Review, 99(3)*, 572-577.

Glucksberg, S., & Keysar, B. (1990). Understanding Metaphorical Comparisons: Beyond Similarity. *Psychological Review, 97(1)*, 3-18.

Glucksberg, S., McGlone, M., & Manfredi, D. (1997). Property Attribution in Metaphor Comprehension. *Journal of Memory and Language, 36,* 50-67.

Lakoff, G., & Johnson, M. (1980). *Metaphors we live by.* Chicago: University of Chicago Press.

Lakoff, G., & Johnson, M. (1999). *Philosophy in the Flesh.* New York:Basic Books.

McGlone, M. (1996). Conceptual Metaphors and Figurative Language Interpretation: Food for Thought? *Journal of Memory and Language, 35,* 544-565.

McGlone, M., & Harding, J. (1998). Back (or Forward?) to the Future: The Role of Perspective in Temporal Language Comprehension. *Journal of Experimental Psychology, 24,* 1211-1223.

McTaggart, J.E. (1908). The unreality of time. *Mind,* 17, 457-474.

Murphy, G. (1996). On metaphoric representation. *Cognition, 60,* 173-204.

Murphy, G. (1997). Reasons to doubt the present evidence for metaphoric representation. *Cognition, 62,* 99-108.

Nayak, N., & Gibbs, R. (1990). Conceptual Knowledge in the Interpretation of Idioms. *Journal of Experimental Psychology: General, 119(3),* 315-330.

Pulvermüller, F. (in press). Words in the brain's language. *Behavioral and Brain Sciences.*

Traugott, E. (1978). On the expression of spatio-temporal relations in language. In J.H. Greenberg (Ed.), *Universals of human language: Vol. 3. Word structure.* Stanford, CA: Stanford University Press.

Wittgenstein, L. trans. Anscombe, E. (1953). *Philosophical Investigations* Blackwell, Oxford.

Memory for Goals: An Architectural Perspective

Erik M. Altmann (altmann@gmu.edu)
Human Factors & Applied Cognition
George Mason University
Fairfax, VA 22030

J. Gregory Trafton (trafton@itd.nrl.navy.mil)
Naval Research Laboratory, Code 5513
4555 Overlook Ave., SW
Washington, DC 20375-5337

Abstract

The notion that memory for goals is organized as a stack is central in cognitive theory in that stacks are core constructs leading cognitive architectures. However, the stack over-predicts the strength of goal memory and the precision of goal selection order, while under-predicting the maintenance cost of both. A better way to study memory for goals is to treat them like any other kind of memory element. This approach makes accurate and well-constrained predictions and reveals the nature of goal encoding and retrieval processes. The approach is demonstrated in an ACT-R model of human performance on a canonical goal-based task, the Tower of Hanoi. The model and other considerations suggest that cognitive architectures should enforce a two-element limit on the depth of the stack to deter its use for storing task goals while preserving its use for attention and learning.

Introduction

The ability to decompose a complex problem into subgoals is to complex cognition roughly as the opposable thumb is to complex action. However, despite a generation of research into cognitive goal-based processing strategies like means-ends analysis, and into goal-based processing in artificial intelligence, we lack an adequate theory of how cognition manages the fine-grain goals of everyday tasks. The most common description we have is the stack — that is, a data type taken from computer science and interpreted as a description of cognitive processing.

The stack is an appealing descriptive tool because it elegantly captures the structure of everyday tasks that have to be decomposed to become achievable (Miller, Galanter, & Pribram, 1960). For example, traveling to a conference may entail taking an airplane, and whereas getting to the conference is in some sense the top-level goal, getting on the airplane has to happen first. Thus the order in which steps are planned (conference, then airplane) is opposite the order in which steps are executed (airplane, then conference). This is precisely the kind of reversal supplied by a stack, if goals are pushed in the order they are planned and popped in the order they are achieved (Figure 1). The theoretical inference has been that, because tasks are often decomposable in this way, the human cognitive architecture incorporates a stack to support such processing.

However, from a theoretical and empirical perspective, the stack is an idealized model of memory for goals that raises more questions than it answers. For example, the stack implemented in the ACT-R architecture invites questions about reactivity, dual-tasking, and memory for old goals (Anderson & Lebiere, 1998). The stack implemented in Soar (Newell, 1990) invites similar questions about similar issues (John, 1996; Rosenbloom, Laird, & Newell, 1988; Young & Lewis, in press).

We offer a critical analysis of the stack as cognitive theory. The supply of such theory is limited essentially to the ACT-R and Soar cognitive architectures, but in both cases the stack has been used to model a broad range of goal-based behavior. The role of the stack is typically to store goals associated with the task — *task goals* — like taking a flight or attending a conference. In both architectures, stacked task goals are immediately available for error-free retrieval with their order preserved regardless of when they were placed on the stack and regardless of how much or how little they were used since. These properties make the stack as poor a cognitive theory as it is useful a programming construct. As a theory it masks strategic adaptation and variation in goal management across tasks and individuals, and its usefulness as a programming tool simply adds to the problem by tempting the analyst. A better way to represent goals is to treat them like any other kind of memory element and then ask what (if any) supporting processes are necessary to account for how people remember them.

We start by comparing predictions of the task-goal stack to evidence from the literature. We then present an ACT-R model of the Tower of Hanoi, a task often taken to reveal the cognitive reality of the task-goal stack (Anderson & Lebiere, 1998; Anderson, Kushmerick, & Lebiere, 1993; Egan & Greeno, 1973). The model conforms to our proposal to treat task goals like any other kind of memory element, and shows that stacking them is not necessary to account for human performance. The model uses ACT-R's stack in prin-

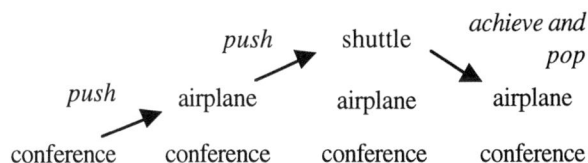

Figure 1: Pushing and popping task goals on a stack. Getting to the conference means flying, which in turn means shuttling to the airport.

cipled ways to support encoding and retrieval processes, and goes beyond a previous ACT-R model to predict errors as well as response-time data. We analyze this and other uses of the stack in ACT-R and Soar to propose a limit of two elements as an architectural constraint on stack depth.

The Stack: Predictions and Contradictions

A stack is an ordered set of elements, with the element at one end designated the top. The only operations defined on a stack affect this top element. A new element can be pushed onto the stack, covering the top element and itself becoming the top, or the top element can be popped off the stack, uncovering the element underneath it as the new top element (Figure 1).[1] Thus stack elements are ordered by age, with the newest element always on top.

One property of the stack is that once an element is pushed, it remains on the stack until it is popped. If the element is a task goal being pushed onto a stack as part of a planning process, then that goal remains on the stack as long as it takes to accomplish all task subgoals pushed on top of it. This is how task goals are stored on the ACT-R stack. The top goal controls the system's behavior, and when that goal is popped, the next older one takes over, no matter how old it is. Thus, ACT-R's stack predicts that memory for old goals is perfect.

Empirically, however, memory for pending goals (often referred to as prospective memory) is generally variable and strategic. Subjectively it seems quite common to embark on a simple errand or chore and midway through forget the purpose. In one study of intentions, actions to be carried out directly were remembered better than actions to be verified as carried out by someone else (Goschke & Kuhl, 1993). A stack-based account, in which memory for goals is perfect, predicts that memory should be at ceiling in either case.

A second property of the stack is that its elements are ordered last-in first-out, or LIFO. If these elements are again task goals pushed as they arise during planning, then the stack preserves that order perfectly, in addition to the elements themselves.

LIFO goal selection is not as pervasive as an architectural stack might predict. For example, goal selection in arithmetic can be highly variable both within and between subjects, and the variance is better explained by idiosyncratic goal selection strategies than by uniform LIFO ordering (Van-Lehn, Ball, & Kowalski, 1989). Similarly, goal selection in mundane tasks like VCR programming is guided by background knowledge and by simple difference-reduction strategies that make extensive use of perceptual information (Gray, in press). Finally, a phenomenon that defies LIFO goal selection is post-completion error, for example leaving the originals in the photocopier after taking the copies (Byrne & Bovair, 1997). Making copies is the top level goal and hence should be the first goal pushed and the last goal popped, with no stragglers. Thus if cognition had a task-goal stack, post-completion error would not be the common procedural error that it is.

A third property of the stack, at least as implemented in Soar and ACT-R, is that it depletes no resources directly affecting declarative memory. An element on the stack requires neither activation nor active maintenance to stay available and maintain its rank. Thus, memory for goals is not only perfect but free.

Several studies show that maintaining goals in memory does involve cognitive opportunity cost. For example, goal management strategies that reduce memory load have been linked to performance on Raven's Progressive Matrices in that better strategies allow more activation to be focussed on the underlying inference task (Carpenter, Just, & Shell, 1990). Similarly, working memory capacity has been linked to goal-selection errors in the Tower of Hanoi (Just, Carpenter, & Hemphill, 1996). Finally, fewer post-completion errors occurred when completed goals were allowed to decay, suggesting that activation is a limited resource that can shift among goals (Byrne & Bovair, 1997). The traditional task-goal stack, for example as found in ACT-R and Soar, simply cannot account for these results.

In sum, a variety of evidence suggests that memory for pending goals is strategic and effortful rather than perfect and automatic, contradicting the stack's fundamental predictions.

Storing Task Goals in Memory

If the analyst must do without the stack as a goal store, then other storage mechanisms must provide whatever critical functionality is lost. The natural stores to consider as substitutes are those that hold other kinds of declarative information, namely memory and the environment. These stores, together with supporting cognitive processes, would have to provide memory for goals and guidance for goal selection, and do so at reasonable cognitive cost.

To examine this possibility, we modeled human performance on a canonical goal-based task, the Tower of Hanoi. The structure of this task is such that the overall goal (of relocating a tower of disks) has to be recursively decomposed into subgoals (of moving individual disks). The task has been widely studied and modeled, generally on the assumption that people manage this decomposition using a stack.

Our starting point was a model by Anderson and Lebiere (1998) that uses ACT-R's stack in traditional fashion to store task goals. Their model (Traditional Goal Stack, or TGS) fits their data set (Anderson et al., 1993) remarkably well, with $R^2 = .99$ for response times on 4-disk trials and $R^2 = .95$ for 5-disk trials.

Our model (Memory as Goal Store, or MAGS) fits the same data set equally well, with $R^2 = .99$ for 4-disk trials and $R^2 = .95$ for 5-disk trials.

MAGS also goes beyond TGS to predict errors. TGS performs every trial perfectly in the fewest possible moves, which is 15 moves for 4-disk trials and 31 moves for 5-disk trials. In contrast, MAGS veers off the optimal path due to noisy retrieval of task goals from memory. In Monte Carlo simulations MAGS predicted a mean of 18.0 moves per 4-disk trial (3.0 errors) and 50.4 moves per 5-disk trial (19.4 errors). In the Anderson et al. (1993) data, empirical means

---

[1] The pop operation in Soar is generalized to allow popping any stack element along with all newer elements. This lets Soar react to events that, for example, achieve an older task goal and thus make all its subgoals on the stack moot. However, this more flexible pop operation does not materially alter the predictions of perfect, zero-cost memory for goals and goal order.

are 19.2 moves per 4-disk trial and 53.5 moves per 5-disk trial, both within 10% of our predictions.

Instead of a task-goal stack, MAGS incorporates commonly-recognized limitations on memory like decay and noise. Task goals are ordinary memory elements that have to be encoded and retrieved in ways that overcome these limitations. The processes that accomplish this goal management, adopted from an independently-constructed memory model (Altmann & Gray, 1999), account for response-time patterns in the data set. Moreover, instead of goal order being retained internally, goal selection is guided by cues from the environment. Finally, in addition to functioning without the stack, MAGS uses fewer free parameters than TGS.[2]

Figure 2 shows response-time data from 4-disk trials solved in the fewest possible moves. The empirical data (solid ink) are from Anderson et al. (1993), and the simulation data (dashed ink) are from MAGS. There are several patterns in the data that both MAGS and TGS account for, but we focus on how MAGS accounts for the large latency peaks at moves 1, 9, and 13, the latency valleys at even-numbered moves, and the small peaks at remaining moves.

Task Goals in the Tower of Hanoi

The algorithm used by the Anderson et al. (1993) participants combines the goal-recursive and perceptual strategies described by Simon (1975). The algorithm, shown in Figure 3, starts by focusing on the largest out-of-place disk, which we refer to as the *LOOP* disk. The LOOP disk is initially disk 4 and rests on peg A with target peg C. The algorithm first checks whether disk 3 blocks disk 4. (One disk *blocks* another if it is smaller than the other and rests on the other's

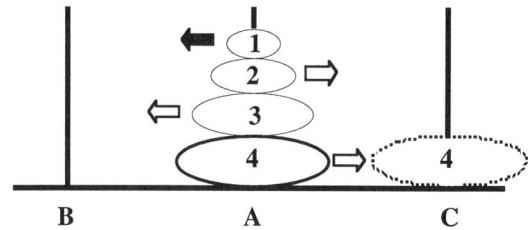

Figure 3: Planning a move in the Tower of Hanoi. Disk 4 must move to peg C (4:C), which entails 3:B, which entails 2:C, which entails 1:B as the move.

source or target pegs.) Disk 3 blocks disk 4 so it must be moved out of the way to peg B. This move is blocked by disk 2, so disk 2 must be moved to peg C. This move is blocked by disk 1, which must be moved to peg B (the filled arrow in Figure 3). Thus, moving disk 4 to peg C (which we designate *4:C*) entails a plan whose first step is 1:B.

A task goal in this context is an association between a disk and a target peg. The first task goal produced by the algorithm is 4:C, the second 3:B, and the third 2:C. The first of these (4:C) is readily inferred from the display by comparing the current state of the puzzle to the end state. However, the recursive task goals formulated in service of 4:C are not as easily available. These must be re-inferred after every move using the algorithm or they must be stored in memory and retrieved at the right time. Thus, for example, once 1:B has been made, the next move could be inferred anew by focussing on disk 4 again, then on disk 3, and then on disk 2. Alternatively, if a task goal for 2:C had been stored in memory during the first pass of the algorithm, and if that task goal could be retrieved now, then a second pass could be spared; memory would indicate 2:C.

The Anderson and Lebiere TGS model stores task goals on the ACT-R stack. For example, the model pushes 4:C, 3:B, and 2:C as it infers 1:B. After making 1:B the model examines the stack to see what move is on top. The top move is 2:C, which can now be made. If the top move had been blocked instead, the model would have pushed a new goal to move the blocking disk. For example, once 2:C is made, 3:B (underneath 2:C on the stack) is blocked because disk 1 is at peg B, so the model would push a goal for 1:A.[3]

The MAGS model, in contrast, treats a task goal like any other memory element. This poses two functional challenges. First, declarative memory elements in ACT-R, or *chunks*, decay over time. Thus an old goal, like 3:B in the example above, could be difficult to retrieve. Second, when one goal is achieved another must be selected. LIFO ordering is optimal for the Tower of Hanoi, so without the stack a useful selection order must come from some other source.

One of the patterns in the data in Figure 2 is that the peaks at moves 1, 9, and 13 decrease in size. These peaks correspond to the planning phase in the MAGS model, and the decrease is caused by successive shortening of the plan-

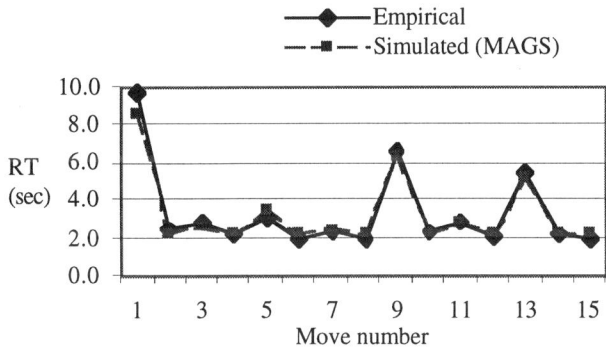

Figure 2: Tower of Hanoi response times (RTs) for 4-disk problems. Empirical RTs are from Anderson and Lebiere (1998) and simulated RTs are from the MAGS model.

---

[2] Default values are used for the ACT-R parameters of goal activation ($W = 1.0$), base-level learning ($d = 0.5$), and latency factor ($F = 1.0$). Transient activation noise ($s = 0.3$) and retrieval threshold ($\tau = 4.0$) are taken from a model of a different but related goal-management task (Altmann & Gray, 1999), along with the mechanisms that depend on them. Encoding time (185 msec) is taken from ACT-R models of menu scanning and the Sperling task (Anderson, Matessa, & Lebiere, 1998). The one free parameter is move time, which we set to the same value as the Anderson and Lebiere model (2.15 sec). The model code is available for downloading at hfac.gmu.edu/people/altmann/toh.

---

[3] Note that the retention interval is longer for 3:B than it is for 2:C, at both ends: 3:B is both encoded earlier and retrieved later. Despite this, both goals are equally available at retrieval time because the stack preserves them perfectly.

ning phase. For example, planning move 1 means starting with disk 4, whereas planning move 9 means starting with disk 3, because disk 4 is in place and can be ignored for the rest of the trial. With one fewer task goal to encode, planning move 9 is faster than planning move 1.

Encoding and Retrieving Task Goals

The first challenge in storing task goals in memory is to encode them so as to resist decay. In MAGS the process that accomplishes this is interleaved with the planning algorithm. Whenever the model recurses to focus on a blocking disk, it encodes a task goal for where it wanted to move the blocked disk. The encoding process strengthens a goal using cumulative mental rehearsal to a criterion that anticipates the retention interval (Altmann & Gray, 1999). This process accounts for the large latency peaks at moves 1, 9, and 13 (Figure 2). On these moves the model focuses on the LOOP disk and encodes a task goal for that disk, then focuses on the blocking disk and encodes a task goal for it, and so on, until it reaches a disk it can move. These task goals represent a plan that was time-consuming to encode in memory.

The second challenge to storing task goals in memory is to retrieve them in a useful order. In the Tower of Hanoi, perceptual cues afford the same ordering information as the stack does, and the MAGS model makes use of this information. For example, when the model uncovers a disk it asks itself, Where did I want to move the disk I just uncovered? The nature of the task is such that simple heuristics like this suffice to generate stack-like behavior.

Three simple heuristics produce optimal retrieval cues from perceptual information.[4] The first heuristic, *retrieve-uncovered*, applies when a move uncovers a disk on the source peg, and says simply to use the uncovered disk as a cue (as described above). The second heuristic, *retrieve-larger*, applies when a move empties a peg instead of uncovering a disk. It says to use the next larger disk as a cue, wherever that disk is now.

The third rule, *retrieve-next*, is needed because task goals can become "stale". Some disks move more than once while the LOOP disk is being unblocked, so an old task goal may indicate an obsolete target. For example, when the LOOP disk is 4, disks 4 and 3 each move once, disk 2 moves twice, and disk 1 moves four times before disk 4 is unblocked. The model encodes no task goals for disk 1, because disk 1 can always be moved. The first time disk 2 is

---

[4] Pseudo-productions for the three heuristics are:

retrieve-uncovered:
 if the last move uncovered a disk,
 then retrieve a task goal for the uncovered disk.

retrieve-larger:
 if the last move emptied the source peg,
 and there is a next larger disk,
 then retrieve a task goal for the larger disk.

retrieve-next:
 if a task goal was retrieved for a disk,
 and it says to move that disk to a peg,
 and the disk is already on that peg,
 and there is a next larger disk,
 then retrieve a task goal for the larger disk.

used as a cue, its task goal indicates the correct target. The second time, however, its task goal incorrectly indicates the peg to which disk 2 was moved before. Retrieve-next recognizes this situation and tries to retrieve a task goal for the next larger disk.

When the model retrieves a task goal, it makes that move if it can. If instead the move is blocked, the model invokes the planning algorithm and focuses recursively on the blocking disk. However, the algorithm takes less time than it would have otherwise, because the cue disk is a better starting point than the LOOP disk for inferring the next move.

Some moves can be selected efficiently without retrieving goals from memory. An admissible heuristic in the Tower of Hanoi is *don't-undo*, which produces a single candidate for all even-numbered moves. A second heuristic, *one-follows-two*, is that when disk 2 moves, disk 1 should always be moved on top of it. In one closely-studied Tower of Hanoi session, the participant used both heuristics throughout despite being a novice (VanLehn, 1991). On these grounds we assume that the Anderson et al. (1993) subjects also used both heuristics, and incorporating them in our model improved its fit to the data. Don't-undo is responsible for the latency valleys at even-numbered moves, and one-follows-two reduces the small peaks at moves 3, 7, and 11 to less than the small peak at move 5 (Figure 2). Thus many moves in the Tower of Hanoi can be governed by simple perceptual information, reducing the burden on memory for goals and the need to posit a mechanism like the stack.

Roles for the stack

Our analysis suggests that a stack is not necessary to account for human performance in a canonical goal-based task. This undermines the traditional rationale for the architectural stack in theories like Soar and ACT-R, raising more general questions about the stack's role in cognitive theory. We suggest that the stack has several roles to play, but argue that it can and should be constrained to grow no more than two elements high at run time. The argument is based on a closer examination of our model and of learning mechanisms in both architectures.

Focusing Attention

MAGS uses the stack to focus attention on task goals. In ACT-R, the construct of attention is represented in part as activation, which affects the availabilty of chunks in memory. The more active a chunk, the greater the speed and likelihood of its retrieval (Anderson & Lebiere, 1998).

Activation in ACT-R has two components, source activation and base-level activation. *Source activation* captures the effect of context. Metaphorically, source activation acts as a spotlight on one chunk in memory. This focussed-on chunk, always the top chunk on the ACT-R stack, then radiates source activation to other, related chunks in memory. The link through which source activation spreads, which we refer to as a *cue*, is itself a chunk contained in the focussed-on and related chunks. Should the related chunk be the target of a retrieval, source activation is one component of its total activation. The other component is the target's *base-level activation*, which captures the effect of history. Base-level activation increases (strengthens) with use and decreases (decays) without. Noise in memory is represented as transient fluctuation in total activation.

The need to focus attention arises because the amount of source activation is fixed and divided equally among the cues in the focussed-on chunk. Thus the more cues there are, the less effective is any one of them in activating a given target. The stack enables strategic control over attention by allowing the system to push a subgoal containing cues selected so that all available source activation reaches the target. When retrieval succeeds (or the attempt is abandoned), the subgoal pops and the architecture restores the greater context by refocussing on the older goal. The hypothesis is that with preparation (creating and pushing a subgoal), cognition can maintain context through intervals of focussed attention.

In MAGS, the need to focus attention arises because the model keeps a variety of state information in the focussed-on chunk. This information includes the source and target pegs of the previous move and the disks that were covered and uncovered. This information interferes with task-goal retrieval by drawing source activation away from the cue disk. To overcome this interference, the model pushes a subgoal containing only the cue disk, which then directs all available source activation to the target task goal. After retrieving the task goal the model pops the retrieval subgoal, causing ACT-R to refocus on the chunk below it on the stack. This older chunk contains the same state information it did before, now augmented with the target peg from the task goal.

The encoding process also needs to focus attention, because it needs to simulate the retrieval context to know when to stop. The encoding process functions by strengthening the task goal's base-level activation. To determine when this activation is high enough, the encoding process pushes a subgoal containing only the retrieval cue and attempts a retrieval.[5] If this retrieval succeeds, the encoding process exits because it has verified that the task goal's retrieval demands are likely to be met.

In sum, ACT-R's stack plays a key role in MAGS, not directly by storing task goals but indirectly by supporting the processes that encode and retrieve them. Thus the stack remains a core construct, but at a finer grain of analysis than its traditional interpretation as a theory of memory for goals.

Learning

The stack plays other roles in both ACT-R and Soar, one of which is learning of productions. Productions specify a conditional transformation on the current focus of attention, or *state*. Half the production (the condition) specifies the input state to the transformation, and the other half (the action) specifies the output state. Learning a new production therefore entails retaining a memory of the input state while the output state is generated. The stack is a natural way to implement this learning. For example, Soar models often learn by pushing a duplicate of the input state onto the stack and transforming the duplicate into the output state. When the transformation is done, both states are available as blueprints for the new production, whose conditions are patterned on the input state and whose actions are patterned on the output state. The new production caches in one step the sequence of steps that transformed the state

Several learning mechanisms in ACT-R also make use of the stack. First, in declarative learning a chunk enters long-term memory when popped off the stack, on the premise that any information deliberately focussed on leaves a trace in memory. Second, production learning is also deliberate in that it requires pushing a special kind of subgoal to create a new production. In both cases, the stack maintains a previous context while the system focuses temporarily on learning, much as the stack maintains an older context while MAGS focuses attention. Third, in learning the utility of productions ACT-R credits a production pending the outcome of the goal created by that production. This credit assignment requires that the old goal stay available over time. Fourth, the stack has been used to control the associations learned between cues and facts (Anderson & Lebiere, 1998, Ch. 9) in a manner also related to attention focussing.

A two-high stack suffices for the architectural learning processes outlined above. Two states support symbolic learning, and support subsymbolic learning for one unit of procedural knowledge or one declarative association at a time. A two-high stack is also sufficient for modeling complex, goal-based behavior in a range of domains, including programming (Altmann & John, in press), exploratory learning (Howes & Young, 1996; Rieman, Young, & Howes, 1996), and serial attention (Altmann & Gray, 1999), as well as the Tower of Hanoi. Indeed, representation decisions concerning goals and goal selection are simpler when there is no open-ended choice for the analyst to make about what stack depth best accounts for the participant's state of mind at a given instant.

Conclusion

The stack captures the means-ends structure of many tasks, but, we argue, is a poor explanation of cognitive performance on those tasks. It over-predicts memory for goals and goal order and under-predicts the cost of encoding and retrieval. We made this case in part with reference to data on prospective memory, goal selection order, and the relationship between goals and working memory.

We also made the case with an analysis of the Tower of Hanoi. This task is traditionally assumed to induce cognitive goal stacking, and the TGS model of Anderson and Lebiere (1998) incorporates this premise. However, our MAGS model improves on TGS in several ways. First, it dispenses with the task-goal stack, which we argue is cognitively implausible, while achieving an equally good fit to response-time data. Second, in place of the task-goal stack MAGS incorporates independently-constrained memory mechanisms (Altmann & Gray, 1999). Third, MAGS goes beyond TGS to predict errors as well as response times. Fourth, MAGS uses fewer free parameters at the subsymbolic level. MAGS thus offers an accurate, well-constrained, and broad account of empirical data, increasing our confidence that performance in the Tower of Hanoi reflects the reality of memory and attention rather than the reality of the task-goal stack.

The use of perceptual cues in MAGS is congruent with routine use of such cues in behavior that appears plan-like to

---

[5] This test retrieval is scaled to the size of the disk, to model our assumption that practiced subjects will adapt to the fact that task goals for larger disks have longer retention intervals.

the observer (Suchman, 1987; Vera, in press). It also predicts that tasks will be much more difficult when goal order is relevant but perceptual cues to order are not available. In such situations goal order must be explicitly encoded in memory, requiring more time and possibly the knowledge and techniques underlying skilled memory (Ericsson & Kintsch, 1995). If selection order is not relevant, then selection should be guided by factors like the recency and frequency with which pending goals were checked or rehearsed.

Whereas the stack is a poor theory of memory for task goals, a limited stack is a natural basis for sub-theories of attention and learning. Moreover, a limit of two elements has proven not only workable but parsimonious in several models of goal-based behavior, and thus may be appropriate to enforce in cognitive architectures like ACT-R and Soar.

Cognition may of course still provide specialized, stack-like support for means-ends processing under circumstances that we have not anticipated. In favor of this view, one could argue that memory has likely adapted to the means-ends structure of the environment as it has adapted to the structure of the environment in other ways (Anderson, 1990). It remains for future research to identify control processes or memory subsystems that function like a stack under appropriate circumstances, or to add to the evidence that pending goals are stored in memory like any other information.

Acknowledgments

The first author is supported by Air Force Office of Scientific Research grant F49620-97-1-0353, and the second author by funds from the Office of Naval Research. We thank B. D. Ehret, W. T. Fu, W. D. Gray, I. R. Katz, A. W. Lipps, S. L. Miller, L. D. Saner, M. J. Schoelles, C. D. Schunn, S. B. Trickett, S. Varma, and an anonymous reviewer for their comments.

References

Altmann, E. M. & John, B. E. (in press). Episodic indexing: A model of memory for attention events. *Cognitive Science*.

Altmann, E. M. & Gray, W. D. (1999). Serial attention as strategic memory. *Proceedings of the twenty-first annual conference of the Cognitive Science Society*, to appear.

Anderson, J. R. (1990). *The adaptive character of thought*. Hillsdale, NJ: Erlbaum.

Anderson, J. R. & Lebiere, C. (1998). *The atomic components of thought*. Hillsdale, NJ: L. Erlbaum.

Anderson, J. R., Kushmerick, N., & Lebiere, C. (1993). The Tower of Hanoi and goal structures. In J. R. Anderson (Ed.), *Rules of the mind*. Hillsdale, NJ: L. Erlbaum. 121-142.

Anderson, J. R., Matessa, M., & Lebiere, C. (1998). ACT-R: A theory of higher-level cognition and its relation to visual attention. *Human-Computer Interaction*, **12**, 439-462.

Byrne, M. D. & Bovair, S. (1997). A working memory model of a common procedural error. *Cognitive Science*, **21**, 31-61.

Carpenter, P. A., Just, M. A., & Shell, P. (1990). What one intelligence test measures: A theoretical account of the processing in the Raven Progressive Matrices test. *Psychological Review*, **97**, 404-431.

Just, M. A., Carpenter, P. A., & Hemphill, D. D. (1996). Constraints on processing capacity: Architectural or implementational? In D. M. Steier & T. M. Mitchell (Eds.), *Mind matters: A tribute to Allen Newell*. Hillsdale, NJ: L. Erlbaum. 141-178.

Egan, D. S. & Greeno, J. G. (1973). Theory of rule induction: Knowledge acquired in concept learning, serial pattern learning, and problem solving. In L. W. Gregg (Ed.), *Knowledge and cognition*. Hillsdale, NJ: Erlbaum. 43-103.

Goschke, T. & Kuhl, J. (1993). Representation of intentions: Persisting activation in memory. *Journal of Experimental Psychology: Learning, memory, and cognition*, **19**, 1211-1226.

Gray, W. D. (in press). The nature and processing of errors in interactive behavior. *Cognitive Science*.

Ericsson, K. A. & Kintsch, W. (1995). Long-term working memory. *Psychological Review*, **102**, 211-245.

Howes, A. & Young, R. M. (1996). Learning consistent, interactive and meaningful task-action mappings: A computational model. *Cognitive Science*, **20**, 301-356.

John, B. E. (1996). Task matters. In D. M. Steier & T. M. Mitchell (Eds.), *Mind matters: A tribute to Allen Newell*. Hillsdale, NJ: L. Erlbaum. 313-324.

Miller, G. A., Galanter, E., & Pribram, K. H. (1960). *Plans and the structure of behavior*. New York: Holt, Rinehart, & Winston.

Newell, A. (1990). *Unified theories of cognition*. New York: Harvard.

Rieman, J., Young, R. M., & Howes, A. (1996). A dual-space model of iteratively deepening exploratory learning. *International Journal of Human-Computer Studies*, **44**, 743-775.

Rosenbloom, P. S., Laird, J. E., & Newell, A. (1988). Meta-levels in Soar. In P. Maes & D. Nardi (Eds.), *Meta-level architectures and reflection*. Amsterdam, The Netherlands: Elsevier Science Publishers B.V. 227-240.

Simon, H. A. (1975). The functional equivalence of problem solving skills. *Cognitive Psychology*, **7**, 268-288.

Suchman, L. A. (1987). *Plans and situated action: The problem of human-machine communication*. New York: Cambridge University Press.

VanLehn, K. (1991). Rule acquisition events in the discovery of problem solving strategies. *Cognitive Science*, **15**, 1-47.

VanLehn, K., Ball, W., & Kowalski, B. (1989). Non-LIFO execution of cognitive procedures. *Cognitive Science*, **13**, 415-465.

Vera, A. H. (in press). By the seat of our pants: The evolution of research on cognition and action - Review of *Plans and situated action*. *Journal of the Learning Sciences*.

Young, R. M. & Lewis, R. L. (in press). The Soar cognitive architecture and human working memory. In A. Miyake & P. Shah (Eds.), *Models of working memory: Mechanisms of active maintenance and executive control*. New York: Cambridge University Press.

Serial Attention as Strategic Memory

Erik M. Altmann (altmann@gmu.edu)
Wayne D. Gray (gray@gmu.edu)
Human Factors & Applied Cognition
George Mason University
Fairfax, VA 22030

Abstract

Serial attention is the process of focussing mentally on one item at a time. This process has two phases: attention switching and attention maintenance. Attention switching involves rapidly building up the activation of a new item to dominate old items. Attention maintenance involves letting the current item decay while in use to prevent it from intruding on the next item later on. SASM, a model based on this analysis, suggests that this balance of high initial activation followed by gradual decay reflects a strategic adaptation to task demands on one hand and principles of memory on the other. The model makes novel and accurate predictions about response times and error rates, integrates past use and current context as memory activation sources, and integrates attention switching and attention maintenance into one unified account.

Introduction

To think about mental attention, we can adopt a metaphor from visual attention (e.g., Posner, 1980) and imagine a spotlight directed internally at memory and focussed on the current thought. A sequence of thoughts, or *serial attention*, would then involve moving the spotlight around — maintaining it at one position for a time, then switching it to the next, and so forth. Serial attention is basic to many cognitive activities, from searching a memory set for a probe, to achieving one goal and switching to another during problem solving.

Understanding the *costs* of paying attention and where they occur is crucial to understanding serial attention as a whole. For example, the cost of switching attention has often been interpreted as the time needed to pull a mental lever that switches attention from one task to another (e.g., Garavan, 1998; Gopher, Greenshpan, & Armony, 1996; Rogers & Monsell, 1995). Under additive-factors logic, this switch cost would be reflected in the first measurement taken after the lever is pulled. By extension, this switching action, once complete, should have no effect on the train of thought, which simply continues down the new track. Thus the mental-lever view suggests that attention switching is an active process but attention maintenance is essentially a passive system state and should produce stable performance.

Unfortunately for the mental-lever view, attention maintenance has its own cost, measured as a gradual increase in response time between attention switches (Altmann & Gray, 1999). Our initial explanation for this *maintenance cost* was in terms of interference in memory (Altmann & Gray, 1998). In brief, our proposal was that interference among trials made attention maintenance more difficult over time, and that this interference was "released" by the act of switching attention. However, the functional role of this interference was not clear. Did it satisfy some constraint other than fitting the data? Also, though our explanation was grounded in a cognitive theory (ACT-R; see Anderson & Lebiére, 1998), it ignored basic operational principles of that theory, including strengthening, decay, and noise in memory. The mechanisms representing these principles were simply "turned off" in our computational ACT-R model.

Here we present an expanded model of serial attention that explains maintenance cost and offers a preliminary account of switch cost. From the previous model we carry over the premise that serial attention is essentially a memory phenomenon. We now also adopt the premise that memory in serial attention acts like memory in general in that it strengthens with use and decays without, and is subject to noise like any other data channel. The implication is that cognition must employ active processes or *strategies* that manipulate memory strength to cope with these constraints. These memory-manipulation strategies are responsible for the costs of maintaining and switching attention. We thus characterize *serial attention as strategic memory*, or *SASM*.

We first briefly review our serial attention paradigm and the evidence for maintenance cost. Next, we argue that maintenance costs are inevitable given our premises. We then develop the parameters of the SASM model in geometric terms, and develop and test novel predictions against empirical data. We end by discussing implications of the model for such questions as cognitive workload. The appendix presents an algebraic derivation of the model. We have also implemented SASM as a computational ACT-R model and fit Monte Carlo simulations to our empirical data, but this work is not reported here.

The Paradigm and Sample Data

Our serial attention paradigm involves giving participants two simple tasks and periodically issuing an instruction to switch from one task to the other. For example, the tasks might be to judge a single digit (from the set 1, 2, 3, 4, 6, 7, 8, 9) as Odd or Even or as High (> 5) or Low (< 5). In a typical experiment, trials are presented in blocks. Each block begins with an *instruction trial* saying which task to do, for example "Odd Even." This trial is followed by a sequence of classification trials. These are single digits and provide no clue to the current task. The run of classification trials is interrupted by a second instruction trial, which may or may not indicate the same task as the first instruction trial. The second instruction trial is followed by a second run of classification trials, followed by feedback on accuracy and response time for that block. A block contains 20 classification trials, and the second instruction trial occurs randomly between the 7th and 14th classification trials. In a typical experiment, participants receive 192 blocks, for a total of 384 instruction trials (192 for each task). Responses to all trials are self-paced and there is no calibrated inter-trial interval. All stimuli appear at the same location in the center of the screen.

Data from this paradigm appear in Figure 1 (from Altmann & Gray, 1998). The abscissa shows the first seven classification trials after the second instruction trial in a block, with *trial position* meaning position relative to the instruction. Switch cost occurs on P1, in that response time (RT) is substantially slower than on P2. Maintenance cost is the gradual slowing from P2 to P7.

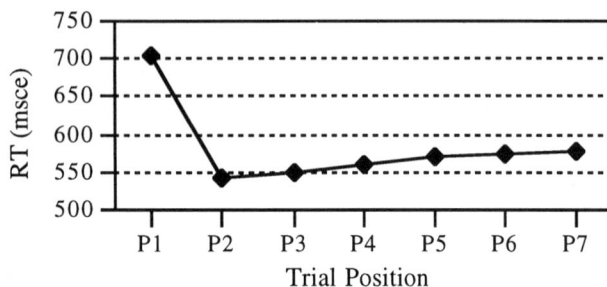

Figure 1: Response time (RT) on post-instruction trials. Maintenance cost is the slowing trend from P2 to P7.

Memory for Instructions

An important distinction is that between a *task* and an *instruction*. A task is semantic and in our paradigm there are only two (e.g., OddEven and HighLow). An instruction is semantic and episodic – there are as many instructions as there are instruction trials (384 in a typical experiment). An instruction specifies what task to do *now*, superseding all previous instructions.

An assumption central to SASM is that each instruction is encoded as a distinct trace in memory. No instruction is instantaneously forgotten or deleted. Rather, old instructions

linger and may interfere with the current one. Cognition must cope with this potential interference by encoding each new instruction to resist intrusions from old ones.

The mechanism for coping with this interference is grounded in basic laws of memory. Under the law of exercise a memory element becomes stronger with use, and under the law of forgetting a memory element becomes weaker when unused. Both laws are implemented in ACT-R in the functions governing the activation of declarative memory elements (*chunks*). A chunk *use* constitutes either a new *encoding* of that chunk in memory or a *retrieval* of that chunk from memory.

When cognition attempts a retrieval, ACT-R's memory system returns the most active chunk. This reflects the *rational memory* assumption, in which activation represents the memory system's best guess at the chunk most likely to be needed now (Anderson, 1990). Activation in ACT-R has two components, one representing a chunk's history of use and the other representing the chunk's relevance to the current context. *Base-level activation* represents history of use. For example, a period of concentrated rehearsal or encoding makes a chunk very active. *Associative activation*, which we refer to as *priming*, represents relevance to the current context. Priming accounted for maintenance cost in our previous model (Altmann & Gray, 1998); in the current model it complements base-level activation to improve memory accuracy, as we discuss later.

The base-level activation of instructions is a critical factor in serial attention, as illustrated in Figure 2. The abscissa shows two contiguous *runs* of trials, with the *previous* run on the left and the *current* run on the right. Each run is governed by an instruction (I_P and I_C, respectively). The ordinate shows instruction activation. The top two curves (in solid ink) show each instruction at peak activation initially (at $P1_P$ and $P1_C$) then decaying throughout its run. When I_P gives way to I_C (at $P1_C$), it decays somewhat faster because it is no longer being used.

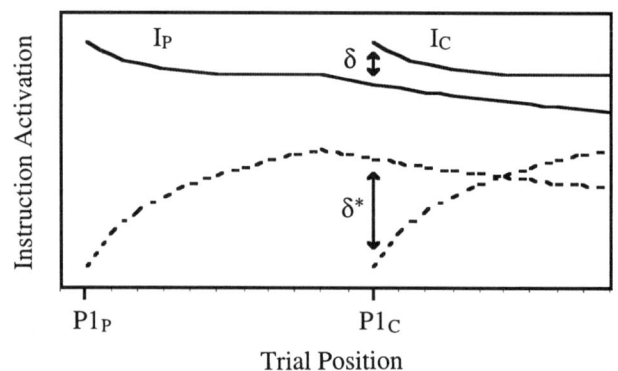

Figure 2: When instructions decay (solid curves), the current instruction (I_C) is always stronger than the previous instruction (I_P), by amount δ at $P1_C$. Were instructions to strengthen (dashed curves), serial attention would fail because I_C would be weaker than I_P, by amount δ^* at $P1_C$.

The important relationship in Figure 2 is between the activation of I_C and I_P. Once I_C is encoded, both instructions coexist in memory because I_P is not completely forgotten. However, the top two curves show I_C always being more active than I_P. Under the rational memory assumption, this ensures that the memory system returns I_C on each trial during the current run, producing correct performance. I_C dominates I_P because both activation curves slope downwards — if all instructions decay from a high initial level, the newest one will always be the most active.

The bottom two curves in Figure 2 (in dashed ink) show an intuitive but problematic interpretation of the law of exercise in this paradigm. The intuition is that if an instruction is retrieved on each trial, it should *gain* activation over trials. However, this would mean that it ends up more active than it begins. At $P1_C$, therefore, I_C would be *less* active than I_P by amount $\delta*$. Under the rational memory assumption, this would preclude correct performance. Hence, instruction activation must start high and end low.

Thus decay in episodic memory is a necessary condition for serial attention and this decay implies maintenance cost. Time to retrieve a memory element, in ACT-R as in other memory models, is a function of its activation, with higher activation implying faster retrieval. If instructions decay from when they are first used, retrieval time will increase on each subsequent trial within a run.

Parameters of Instruction Memory

We have argued that instructions must decay *ab initio* for serial attention to be possible. We next examine four parameters that determine what initial level of activation is necessary to produce such decay. Here we use a geometric notation, leaving the algebra to the Appendix.

One parameter is noise, which we assume affects memory much as it affects any data channel. Following ACT-R, we take noise to be manifested as transient increases or decreases in chunk activation. Each chunk has an expected level of activation, but its actual level on a given retrieval cycle varies according to a logistic (roughly normal) distribution (Anderson & Lebiére, 1998, ch. 3). This activation variance can cause memory errors and in turn performance errors.

A memory error occurs when noise makes the target less active than some other chunk on a given retrieval cycle. The likelihood of such an error depends both on the amount of noise in the system and on how active the target is, on average, compared to other chunks. This is illustrated in Figure 3, which shows activation distributions for I_C (the target) and I_P. Activation is now on the abscissa and the probability of an instruction having a given activation is on the ordinate. Noise is reflected in the dispersion of each distribution. This dispersion is one factor determining the overlap of the activation distributions. The greater the overlap, the more likely I_P will be retrieved in place of I_C, and hence the greater the number of memory errors.

The other factor affecting memory error is δ, the difference in expected activation between I_C and I_P (Figure 3). This difference, resembling the d' of signal detection theory, is itself a function of three parameters, two affecting base-level activation and one affecting associative activation.

The first parameter affecting δ is the amount of time spent encoding the instruction while it is visible on the display. We assume that more time spent encoding the instruction means more base-level activation for the instruction chunk, consistent with memory paradigms in which stimulus exposure and trace strength are taken to be synonymous. With respect to Figure 3, a longer encoding time would shift I_C to the right, thereby increasing δ. Because instruction trials are self-paced, participants can wait to dismiss the instruction until it is sufficiently encoded. Thus, encoding time is under strategic control.

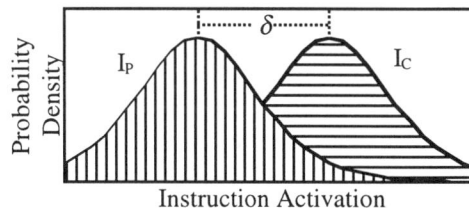

Figure 3: Activation distributions of the previous (I_P) and current (I_C) instructions. The less overlap between them the more likely I_C will be retrieved.

The second parameter affecting δ is run length. As this grows, all else being equal, so does the amount of base-level activation coming from trial-by-trial retrieval as opposed to initial encoding. In Figure 2, each curve decays at first and then flattens out; with more trials per run but no extra encoding, the curve would inflect before the end of the run and begin to slope upwards. In Figure 3, greater run length (all else being equal) would shift I_P to the right, decreasing δ and thus increasing the chance of memory error.

The third parameter affecting δ is associative activation, or priming from cues in the cognitive context. For priming to contribute to δ, some cue must prime I_C more than it primes I_P. This might occur, for example, were the trial stimulus to cue the task. In a variant on our paradigm, the two tasks might be OddEven and ConsonantVowel (instead of OddEven and HighLow), each with a different stimulus set (i.e., numbers vs. letters). In this case, the stimulus itself would be an effective cue for I_C. In contrast, in our paradigm the stimulus (e.g., always a number) affords either task, making it unhelpful as a cue.

However, not all cues need be external. One likely *internal* cue is a residual memory for the previous trial. Although this may be weak, it may play a role in priming the current trial, producing a kind of repetition effect. Within a run, on trial positions after P1, the task performed on the previous trial primes I_C but does not I_P, which specifies the other task. Thus, the previous trial increases δ for the current trial by contributing to the expected activation of I_C.

In sum, we have identified four parameters of memory for the current instruction, or more generally of memory for the current item in serial attention. Memory noise determines activation dispersion; encoding time, run length, and priming affect an instruction's expected total activation. We next examine predictions of a closed-form model built on these parameters.

Predictions of the Model

The parameters described above are related to each other and to empirical measures in a system of mutual constraint. For example, increased accuracy might require increased encoding time. The model that captures this system is formalized in the Appendix; here we derive two predictions from it, one empirical and one theoretical.

Errors Increase Within a Run

One possible interpretation of maintenance cost is that it reflects increasingly careful processing across trial positions. People might be shifting their speed-accuracy criterion gradually toward accuracy. Such a shift would imply a constant or decreasing error rate across trials.

In contrast, SASM predicts that error rates will increase within a run for two reasons. First, over the first few trials of a run the activation curves for I_C and I_P approach each other (Figure 2). This decreases δ which increases memory errors and, hence, performance errors. Second, errors early in a run beget more errors later in that run. An error occurs when I_P is used in place of I_C. This use causes I_P to gain in base-level activation at the expense of I_C. Thus, an error decreases δ for all subsequent trials in that run.

Error data from the experiment described earlier are shown in Figure 4. The ordinate shows total errors out of 176 trials and the abscissa shows trial position. Rather than a constant or decreasing error rate, errors increase throughout the run, as SASM predicts. The effect of trial position is significant, $F(6, 114) = 4.9$, $p < .001$, as is the linear trend, $F(1, 114) = 24.2$, $p < .001$.

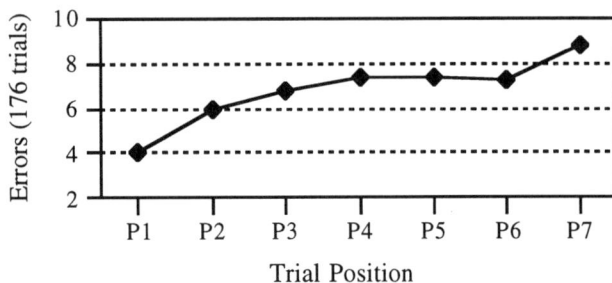

Figure 4: Errors error on post-instruction trials, out of 176 trials. Error maintenance cost spans P1 to P7.

This correct prediction strongly supports our model. We assumed that each instruction is encoded distinctly in episodic memory and decays gradually rather than instantaneously when superseded. These assumptions, together with ACT-R's memory theory, imply the decay trajectory in Figure 2, which in turn predicts the observed error pattern.

Base-Level Activation and Priming are Irreducible

Under the rational memory assumption, activation is composed of two terms: base-level activation (from past use) and priming (from the current cognitive context). This decomposition is based on a Bayesian characterization of the statistical structure of the environment. However, to our knowledge, there has been no analysis of whether both components of activation are functionally necessary. Is it possible that one can be reduced to the other?

By binding the model's parameters we can determine what combinations of base-level and associative activation yield a given error rate. The first parameter, noise, has been estimated many times and typically falls in a limited range (0.30 to 0.85 in terms of the logistic parameter s; Anderson & Lebiére, 1998, p. 217). Hence, although noise is not fixed, it is constrained enough for a sensitivity analysis. As a measure of encoding time we take mean instruction response time, which in our data is 0.97 sec. Run length in our paradigm is 10 trials on average. Finally, the dependent variable, error rate, is 0.023 on trial position P1.

The remaining parameter is priming. Because base-level activation and priming are the only two sources of activation, we can express one as a combination of the other and the δ needed for a given level of accuracy. We designate P_B as the proportion of δ due to base-level activation. P_B is thus nominally defined on an interval of 0 (δ entirely due to priming) to 1 (δ entirely due to base-level activation).

Figure 5 shows a sensitivity analysis of SASM for noise and priming, the two parameters constrained by boundary values. The abscissa shows P_B and the curves predict instruction response time for boundary values of the noise parameter s. Thus the predicted times are those required to achieve δ for given values of P_B and s.

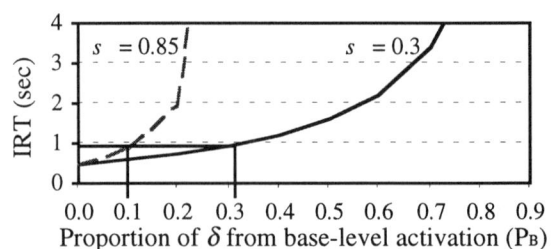

Figure 5: Instruction response time (IRT) as predicted by P_B and upper and lower bounds on activation noise s (see Appendix, Eqn. 4). The empirical IRT of 0.97 is predicted by $P_B = 0.1$ for $s = 0.85$ and $P_B = 0.32$ for $s = 0.30$.

Figure 5 shows that neither base-level activation nor priming alone can achieve the δ needed for high-accuracy serial attention. A P_B of zero predicts a minimum encoding time of roughly 500 msec (regardless of s). Even if δ is entirely due to priming, this amount of initial encoding is needed to bring the initial base-level activation of I_C up to the final base-level activation of I_P.

For large values of P_B, predicted instruction response times go off the scale. That is, even large amounts of initial encoding cannot completely replace priming. As encoding time increases, δ due to base-level activation approaches a limit (see Appendix, Eqn. 3). Performance accuracy requires δ to be above this limit, so priming must supply the balance of the needed activation. In sum, high-accuracy serial attention depends on both base-level activation and priming.

If some amount of δ must come from priming, what cues could provide this? The environment offers no effective cues; the classification-trial stimulus (a number) equally primes both kinds of instruction (OddEven and HighLow) and thus fails to contribute to δ. Therefore, any effective cues must be internal. We proposed earlier that residual memory for the previous trial is a likely cue. Indeed, it is unclear what other internal cues there could be that affect I_C and I_P differentially. Thus, SASM predicts an inherent inertia to serial-attention performance. Priming by the previous trial implies an architectural propensity to do the same task over again. However, because priming alone cannot produce the needed δ, perseveration is not a danger.

Upper and lower bounds on P_B can be estimated from our data. Figure 5 shows that P_B between 0.1 and 0.32 predicts the empirical instruction response time of 0.97 sec. We interpret this to mean that inertia from trial-to-trial priming substantially facilitates serial attention.

In sum, serial attention depends on both base-level and associative activation — neither is reducible to the other. The general implication is that memory retrieval relies not on one but on two sources of information about the target. This is an axiom of Bayesian analyses in general and ACT-R in particular, but SASM suggests that two sources of information are *required* for reliable retrieval in a dynamic environment. To the extent that serial attention is a building block of higher-level cognition, base-level and associative activation are building blocks as well and earn their designation as atomic components of thought.

Discussion

The SASM model reduces serial attention to a set of memory phenomena, going some way toward banishing the homunculous of mental attention. Switching attention involves rapidly strengthening a new item to be temporarily dominant, and maintaining attention involves letting the current item decay slightly to prevent it from intruding later.

The signature evidence for the model is maintenance cost, as measured by the gradual increase in response times across trials in a run. Although to our knowledge this effect is novel, we have replicated it under a variety of situations (Altmann & Gray, 1999). We explain this effect in terms of short-term, trial-by-trial decay of instructions encoded in episodic memory. This decay is a feature, not a bug, in that

it contributes to δ, the activation difference that makes the current instruction always the most active.

The complement to maintenance cost is switch cost. This is typically measured on the first classification trial governed by the new task (Allport, Styles, & Hsieh, 1994; Garavan, 1998; Gopher et al., 1996; Rogers & Monsell, 1995). Switch cost is often (but not always; Allport et al., 1994) interpreted as the cost of moving an attentional spotlight from one location to another, with no related account of maintenance cost. SASM suggests a broader interpretation of switch cost as the cost of processes that increase δ. These processes may occur on classification trials, instruction trials, or elsewhere. On this view, the largest switch cost in our paradigm is instruction response time. On instruction trials, participants strategically encode the displayed instruction to decay through use.

From an applied perspective, SASM may help operationalize cognitive workload in real-time dynamic tasks. Excess workload in a dynamic environment means essentially that too much is happening too fast, causing the operator's performance to suffer. SASM provides a way to analyze such excess workload as a memory phenomenon. For example, in modeling sustained operations, fatigue may be instantiated as increased noise in memory, and accuracy and run length may map directly to measures of task performance. Thus SASM might be used, for example, to predict need for external memory aids as a function of task complexity and opacity (Brehmer & Dorner, 1993).

An open question concerns the first trial position of a run (P1). This position appears not to benefit from high initial instruction strength, in that RT is substantially slower than on later trial positions (Figure 1). One possible explanation is that this slowdown reflects a final phase of the encoding process. If initial encoding stops when activation reaches criterion, then transient noise may boost activation to this criterion prematurely and bring an early end to the instruction trial. However, if the instruction persists in iconic memory, a final encoding phase would be possible as P1 begins. In our computational ACT-R model, this final encoding phase regresses activation toward its criterion value, because instruction chunks not made active enough during the instruction trial get a second chance. In future research it will be important to investigate empirically the extra processing that seems to take place on P1, and the role of this processing in the general scheme of serial attention.

Acknowledgments

This work was supported by a grant from the Air Force Office of Scientific Research (F49620-97-1-0353) to Wayne D. Gray. We thank Wai-Tat Fu, Michael J. Schoelles, Brian D. Ehret, Willard Larkin, and an anonymous reviewer for their comments.

References

Allport, A., Styles, E. A., & Hsieh, S. (1994). Shifting intentional set: Exploring the dynamic control of tasks. In C. Umilta & M. Moscovitch (Eds.), *Attention and performance IV*. Cambridge, MA: MIT Press. 421-452.

Altmann, E. M., & Gray, W. D. (1998). Pervasive episodic memory: Evidence from a control-of-attention paradigm. In M. A. Gernsbacher & S. J. Derry (Eds.), *Proceedings of the twentieth annual conference of the Cognitive Science Society* (pp. 42-47). Hillsdale, NJ: Erlbaum.

Altmann, E. M., & Gray, W. D. (1999). The within-run slowing effect in serial attention. *Manuscript in preparation*.

Anderson, J. R. (1990). *The adaptive character of thought*. Hillsdale, NJ: Erlbaum

Anderson, J. R., & Lebiére, C. (1998). *The atomic components of thought*. Hillsdale, NJ: Erlbaum.

Brehmer, B. & Dorner, D. (1993). Experiments with computer-simulated microworlds: Escaping both the narrow straits of the laboratory and the deep blue sea of the field study. *Computers in Human Behavior, 9*, 171-184.

Garavan, H. (1998). Serial attention within working memory. *Memory and Cognition, 26*, 263-276.

Gopher, D., Greenshpan, Y., & Armony, L. (1996). Switching attention between tasks: Exploration of the components of executive control and their development with training. *Proc. HFES 40th Annual Meeting*. Philadelphia: HFES. 1060-1064.

Posner, M. I. (1980). Orienting of attention. *Quarterly Journal of Experimental Psychology, 32*, 3-25.

Rogers, R. D., & Monsell, S. (1995). Costs of a predictable switch between simple cognitive tasks. *Journal of Experimental Psychology: General, 124*, 207-231.

Appendix

SASM is predicated on the notion that, to be reliably retrieved, the current item (e.g., an instruction) must be more active than the previous item by some amount δ:

$$B_C = B_P + \delta \quad \text{Serial Attention Equation} \quad (1)$$

δ depends on the probability of retrieving the current instruction, $P(I_C)$. This is given by the Chunk Choice Equation (Anderson & Lebiére, 1998, p. 77):

$$P(i) = e^{m_i/t} \left[\sum_j e^{m_j/t} \right]^{-1}$$

$P(i)$ is the probability of retrieving chunk i given noise t and given j chunks in memory each with expected activation m_j. We assume an infinite number of previous instructions of which the most recent few materially affect the probability of a previous instruction intruding on the current one. The expected activations of these few previous instructions are roughly δ apart, so we estimate $m_j \approx m_i - j\delta$. This allows

for a closed-form solution to the Chunk Choice Equation, $P(I_C) = 1 - e^{-\delta/t}$. Rearranging, we get $\delta = -t \ln[1 - P(I_C)]$.

$P(I_C)$ is also constrained by performance accuracy, A. We assume an accurate response if the retrieved instruction is (a) I_C, which always specifies the appropriate task; (b) a previous instruction I_{PA} that specifies the appropriate task; or (c) a previous instruction I_{PN} that specifies the not-appropriate task but whose response for the current stimulus is the same as that of the appropriate task. Thus, $A = P(I_C) + P(I_{PA}) + P(I_{PN})$. The paradigm is structured such that $P(I_{PA}) = 2P(I_{PN})$ and, assuming that an instruction is always retrieved when retrieval is attempted, $P(I_C) = 1 - [P(I_{PA}) + 2P(I_{PN})]$. These constraints together imply that $P(I_C) = 4A - 3$. Substituting this for $P(I_C)$ in the equation for δ, and using the error rate $E = 1 - A$ instead of A, yields the equation below, where P_B, defined on $[0...1]$, limits δ to the proportion due to base-level activation.

$$\delta = -t P_B \ln(4E) \quad (2)$$

With δ bound by E, we can find how many initial uses are needed to achieve that δ. Assuming one use per trial, average run length R, and N initial uses, we can express the age of an instruction in terms of uses. At $P1_C$, the age of I_P is one instruction trial and R classification trials from the previous run, plus another instruction trial and one classification trial in the current run, or $R+3$. The age of I_C is only 2. The number of uses of I_P is $N+R$ and of I_C is $N+1$. We can now expand Eqn. 1 using the Base-Level Learning Equation (Anderson & Lebiére, 1998, p. 124), which defaults to $B = \ln(2nT^{-0.5})$ for n uses over chunk age T. Eqn. 1 in terms of N, R, and δ (and exponentiated) is then:

$$(N+1)2^{-0.5} = (N+R)(R+3)^{-0.5} e^{\delta} \quad (3)$$

To estimate the maximum contribution of initial use to δ, we can solve Eqn. 3 for δ and take the limit as N goes to infinity. This produces $\ln\sqrt{0.5(R+3)}$, or 0.94 for $R=10$. However, E with minimal noise entails a δ of at least 1.02 (via Eqn. 2 with $P_B=1$; E and t are bound below). Therefore, some δ must come from differential priming of I_C over I_P.

To estimate encoding time as a function of P_B, we can solve Eqn. 3 for N and substitute for δ from Eqn. 2. Two ACT-R production firings serve to encode a chunk once, one to create the chunk and push it onto the goal stack and one to pop it into memory. Default firing time is 50 msec, so time per use is 100 msec. Thus predicted encoding time, as measured by instruction response time (IRT), is:

$$IRT = 100 \frac{\sqrt{R+3} - R\sqrt{2}(4E)^{-tP_B}}{\sqrt{2}(4E)^{-tP_B} - \sqrt{R+3}}, \text{ with } t = \sqrt{2}s \quad (4)$$

Eqn. 4, with $E=0.023$ (bound empirically), $s=0.30$ and 0.85 (Anderson & Lebiére, 1998, ch. 7), $R=10$ (task-specific), and $P_B = [0...1]$, produces the curves in Figure 5.

30

The effects of Referent Specificity and Utterance Contribution on pronoun resolution

Jennifer E. Arnold (jarnold@linc.cis.upenn.edu)
Institute for Research in Cognitive Science, University of Pennsylvania
3401 Walnut St. Suite 400A, Philadelphia, PA 19104

Maryellen C. MacDonald (mcm@gizmo.usc.edu)
Department of Psychology, Hedco Neuroscience Building
University of Southern California, Los Angeles, CA 90089-2520

Abstract

Two experiments explore how pronoun resolution is influenced by a) properties of discourse referents, specifically whether they are underspecified and in need of description, and b) the contribution of the pronoun-containing utterance, specifically whether it provides a description or specifies an event. We find that these factors interact, such that when an underspecified referent is in focus, reading is facilitated for description continuations, but when a specified referent is in focus, reading is facilitated in event continuations when the specified referent continues as the topic. This study reveals one of the complex interactions that underlies pronoun resolution.

Introduction

How do readers interpret pronouns? Research has identified numerous relevant factors, many of which are claimed to affect resolution only at the point where the pronoun is encountered. For example, the interpretation of the pronoun in (1) is guided by the roles of the potential antecedents, "Mary" and "Sarah", and the fact that one character is the more likely cause of the blaming event (e.g., Garvey and Caramazza, 1974).

> 1. Mary blamed Sarah because she...

It has been further suggested that verb biases only come into play at the moment that the reader encounters the pronoun, and that they do not lead the implicit cause to be more generally accessible beforehand (McDonald and MacWhinney, 1995; Garnham et al., 1996).

By contrast, Arnold (1998) proposed that reference processing is influenced by the likelihood that a given entity will be important to the following discourse, which is construed dynamically and is not localized to the referring form itself. If the information available to the comprehender suggests that the speaker is more likely to refer to one entity, comprehension is facilitated when such a reference occurs. On this view, referent activation is linked to the probability that the entity will be referred to, and in some cases activation can be anticipatory.

This approach to discourse processing is inspired by research on syntactic ambiguity resolution, which has recently come to focus on how various aspects of the context make it more likely for the speaker to provide certain types of information, well before an ambiguity is encountered. For example, research on modifier ambiguities has found that NP-modifiers are easier to comprehend if the referential context makes the noun need modification. For example, a context containing a set of books makes it is easier to parse "Put the book on the table on the floor", since without modification the bare NP "the book" is ambiguous (e.g., Altmann, Garnham, and Dennis, 1992; Crain and Steedman, 1985; Tanenhaus et al., 1995). In other cases, the need for modification is determined by properties of the referring expression itself. Thornton, MacDonald, and Gil (in press) found that a non-specific NP like "a house" was more modifiable than a more specific NP like "my house" ("a house with shutters" vs. "my house with shutters"), and that it was easier for readers to attach PPs to non-specific NPs than to specific NPs.

The approach in this line of work is fundamentally referential: these studies show that comprehenders attempt to find referents for referring expressions immediately and incrementally, and when a bare NP is not sufficiently informative, they search for further information in the linguistic input.

Referent Specificity. Our study applies the preceding logic to local discourse comprehension, and investigates the role of referent specificity during pronoun comprehension. We hypothesize that readers may find it likely that an underspecified character will be mentioned again soon, because they may expect the speaker to justify having introduced this character to the story. For example, in (2) readers may focus on "a student" as a likely topic of the following utterance.

> 2. On the first day of class, I saw the professor talking to a student in the front row. She...

If the underspecified character is indeed likely to be mentioned again, it should be easier to interpret a subsequent pronoun referring to this character.

At the same time, other aspects of referent specificity make contradictory predictions. One character, "the professor", has been introduced with a specific, definite NP. Definite NPs are often used for given, topical entities in a discourse (Prince, 1992), and although "the professor" is not given, it is inferrable from the context of a class. If the speaker chooses a definite NP for this character, the comprehender may assume that it is meant to be a central character in the story. Thus, the definite, specific nature of the professor character may make it a probable topic of the following utterance.

Because of these contrasting predictions, we hypothesize that the comprehender's tendency to focus on one character or the other will be influenced by another factor: the comprehender's perception of how the following utterance relates to the story.

Utterance Contribution. A crucial part of utterance comprehension is interpreting how an utterance contributes to the task at hand (e.g., Clark, 1996; Grosz and Sidner, 1986). When the task is primarily linguistic, comprehension is driven by how the listener perceives the relation between the incoming utterance and the previous discourse (e.g., Garvey and Caramazza, 1974; Garnham et al., 1996; McDonald and MacWhinney, 1995; Stevenson, Crawley, and Kleinman, 1994). With respect to causal relations, as in (1), the interpretation of the pronoun depends on the comprehender knowing that the second clause is specifying the cause of the event described in the first clause. In this example, the connector "because" provides strongly constraining information about this relationship.

When the beginning of an utterance signals what its role is with respect to the previous utterance, it probabilistically influences the comprehender's expectations about where the discourse is going, which in turn impacts the likelihood that a given entity will be mentioned. For example, if the comprehender infers that an utterance will provide descriptive information, underspecified referents will be more likely to be mentioned. We hypothesized that if readers know a description is coming, they are likely to focus on things that need to be described, like "the student" in 2. In contrast, if the utterance appears to specify a subsequent event, readers will focus on characters they perceive as more topical, such as the more specified referent in 2, "the professor".

Hypothesis. We hypothesized that Referent Specificity and Utterance Contribution would interact to make underspecified referents more likely discourse continuations in descriptive contexts, and specified referents more likely continuations in event contexts, and that this would influence pronoun comprehension in the second utterance. Experiment 1 investigated which character was more likely to be mentioned in the continuation of a story. Experiment 2 looked at the comprehension of pronouns under different

conditions of Utterance Contribution and the specificity of the pronoun referent.

Experiment 1: Story-continuation

Methods and Participants. This experiment investigated whether specific or unspecific characters would be considered more likely continuations of a story, depending on whether the following utterance was perceived to be a description or an event. Participants were asked to read short "stories" like (3) and add a natural continuation to the end.

3a. SPECIFIC FIRST: I arrived at the café and discovered the waitress talking to a little boy.
a'. UNSPECIFIC FIRST: I arrived at the café and discovered a little boy talking to the waitress.

b. DESCRIPTION CONDITION: It looked like...
b'. EVENT CONDITION: Right then...

The sentence began with a scene-setting phrase, presented from the perspective of an observer (usually "I" or "we"). Each stimulus item included two characters, of different genders, denoted by NPs typically associated with only one gender (e.g., man, woman, actress, sailor). One character was specific and the other unspecific. Referent Specificity was manipulated by both NP definiteness and role specificity. Specific characters were identified by their roles, e.g. "the waitress" or "the mailman", and were consistent with the scene described in the first part of the sentence. All unspecific characters were either "a man", "a woman", "a (little) boy", or "a (little) girl".

All characters were human and animate. This was important, because our hypothesis was that comprehenders have some expectation for underspecified characters to be described under certain conditions. However, the perceived importance of an unspecific character is probably determined by many factors, one of which may be animacy. For example, some inanimates may be unimportant to the story, like "a beer" in "John drank a beer".

In addition, we manipulated two factors: a) Utterance Contribution (DESCRIPTION vs. EVENT), as described above, and b) Order of Mention (specific first vs. unspecific first).

Order of Mention is one of the strongest known factors affecting pronoun resolution. First-mentioned characters are more likely to be pronominalized in subsequent references, and pronouns are easier to understand if the referent is a first-mentioned character (e.g., Gordon et al., 1993; Stevenson et al., 1994; among others). The first character is the "starting-point" of the utterance (Chafe, 1994), it is the basis by which readers lay the foundation for the rest of the discourse (Gernsbacher, 1990), and it is the most likely character to be mentioned in the following discourse (Arnold, 1998). Because of the demonstrated strength of this factor, we hypothesized that it might interact with Referent Specificity and Utterance Contribution.

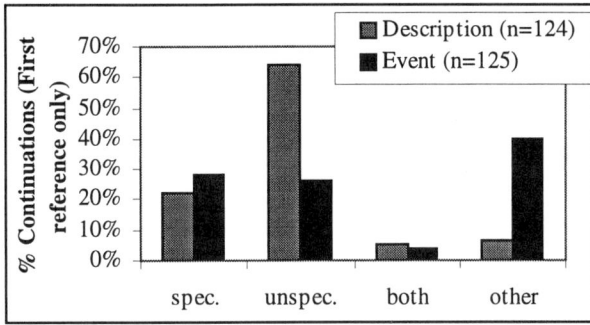

Figure 1: Percentage of participant completions beginning with reference to the specific character, unspecific character, both (as a compound NP or "they"), or other referent.

Table 1: Example continuations in Experiment 1, corresponding to categories in Figure 1.

STIMULUS (first sentence):
The first scene of the movie was the cowboy talking to a woman.

COMPLETION TYPE	EXAMPLE (with relevant referring form underlined)
specific	After that ... the cowboy got shot and the woman cried.
unspecific	It seemed like ... she was about to swoon ove- . all over him.
both	After that ... the woman and the cowboy drove off . in the wagon.
other	It seemed like ... one of those hokey old Westerns that Jimmy Stewart was in.

Each of the 12 experimental items was rotated through the four conditions that resulted from crossing the two factors (Utterance Contribution and Order of Mention). These were presented in 4 lists to 24 members of the Stanford University community,[1] along with 24 items from another experiment and 36 fillers.

The experiment was conducted using an oral story completion method, where participants read the stimuli sentences out loud into a tape recorder, and provided their continuation orally. This method has the advantage that people respond more quickly, which means that their responses reflect the on-line processes occurring as they reach the end of the stimulus. In addition, they do not restrict themselves to extremely short responses, as can be the case with written sentence-completion.

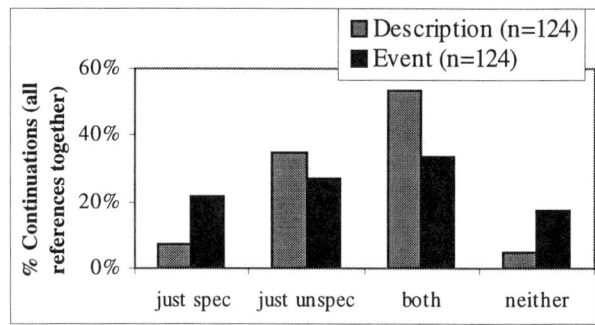

Figure 2: Percentage of participant completions that included references to both unspecific and specific characters, just one or the other, or neither (only some other referent).

Predictions. The goal of this experiment was to see which character participants referred to in their continuations more frequently. We expected that unspecific characters would be relatively more frequent in the DESCRIPTION condition, and specific characters more frequent in the EVENT condition. We also expected that these factors would interact with a tendency for story continuations to refer more often to first-mentioned than second-mentioned characters.

Results. The participant completions were tape recorded, transcribed, and analyzed to answer two questions: a) which character did the participant begin the continuation with? (Figure 1), and b) considering all references in the continuation, how often did the participant refer to both characters or just one or the other? (Figure 2).

Figure 1 shows that in the DESCRIPTION condition, speakers were most likely to begin their continuation with a reference to the unspecific character. In the EVENT condition, by contrast, both specific and unspecific characters were likely beginnings for the continuation (comparing specific and unspecific characters only, $\chi^2(1)=12.9$, p<.001). Counter to our expectations, there was no effect of Order of Mention nor any interaction between it and the other factors. Examples of each type of completion are listed in Table 1. The data in Figure 1 suggest that as predicted, in the DESCRIPTION condition the unspecific character became more accessible, and speakers began their continuations with this character.

However, Figure 2 shows a further difference between DESCRIPTION and EVENT conditions. These data consider all responses in the entire continuation, which show that participants were more likely to refer to both characters in the DESCRIPTION than in the EVENT condition (Z for two proportions=3.1, p<.002). This suggests that when comprehenders perceive that a description is coming, they are most likely to produce a description that describes the relationship between the two characters.

By contrast, the EVENT condition led to more varied responses. The introductory phrase in this condition often signaled a change in time or place. For this reason, participants were most likely to begin their response with a reference to something other than the specific or unspecific

---

[1] One subject was excluded because he focused on the question of who he referred to in his continuations, and two subjects were replaced because they were non-native speakers of English. One item was excluded from the analysis due to experimenter error in stimulus construction. Four continuations were excluded because the participant produced an unintelligible response or repeated the stimulus.

character (e.g., "All of a sudden...the lights went out.") Taking the entire continuation into account, responses were relatively evenly split between those that referred to both characters, just the specific, just the unspecific, or neither. Contrary to expectations, event continuations did not focus primarily on the specific character, perhaps because specific characters are not strongly marked as likely topics of the next utterance, in the absence of other discourse cues like repeated reference. However, Figure (2) shows that continuations referring **just** to the specific character were more common in the Event than in the DESCRIPTION condition (Z=3.2, p<.002), suggesting that the EVENT condition does promote the accessibility of specific characters to a certain extent.

One limitation of the oral story-continuation methodology is that participants tend to focus on the second-mentioned character more than usual. In naturally occurring language, first-mentioned entities tend to be discourse-given, tend to be continued in the following discourse, and when they are referred to, are often pronominalized. However, it has been observed that task demands of the story-completion task lead to more frequent mention of the second-mentioned character, possibly reflecting a recency effect (see Arnold, 1998 for a discussion of this methodology). This pattern also emerged in our data here, in that participants referred equally often to the first-mentioned (n=92) and second-mentioned (n=86) characters (Z=.39, p>.6). This may explain why Order of Mention did not interact with the other variables of interest, Referent Specificity and Utterance Contribution.

In sum, Experiment 1 confirmed that the need for specification of some discourse characters interacts with the comprehender's perception of how a given utterance relates to the previous discourse. When the beginning of the utterance signaled a description, people began their continuations more often with the unspecific character, and they were more likely to mention both characters during the continuation than in the EVENT condition. The EVENT condition produced more varied responses, including a higher tendency to focus exclusively on the specific character than in the DESCRIPTION condition.

Our next question was how these patterns of probable story continuation relate to the on-line comprehension of pronominal references.

Experiment 2: On-line pronoun resolution

Methods and Participants. We used a self-paced moving window paradigm to present 16 two-sentence stories to 40 USC undergraduates, one word at a time. These items were combined with 40 items from two other experiments, 9 practice items, and 40 fillers, which were randomized in 8 lists.

The stimuli followed the same structure as those in Experiment 1. Each sentence contained one specific and one unspecific character. We also manipulated three variables: a) Order of Mention (specific first vs. unspecific first), b) Utterance Contribution (DESCRIPTION vs. EVENT

continuation), and c) Pronoun Referent (specific vs. unspecific character). Sample stimuli are in (4).

4a. SPECIFIC FIRST: When I got to the kitchen, I saw the maid yelling at a man.
a'. UNSPECIFIC FIRST: When I got to the kitchen, I saw a man yelling at the maid.

b. DESCRIPTION CONDITION: It seemed that {he/she} had spilled milk all over the floor.
b'. EVENT CONDITION: Shortly after that {he/she} stormed out the door.

The data from 10 participants were excluded from the analysis due to errors on more than 15% of the comprehension questions for this experiment (n=8) or extremely long reading times on (n=2). Data were trimmed at 2 standard deviations above and below the cell means.

We divided the stimulus items into nine regions, and analyzed the residual reading times for each region (Ferreira and Clifton, 1986). The scheme for regionization is detailed in Table 2.

Table 2. Regions analyzed in Experiment 2

Region	Example
intro to sentence 1	I walked into the room and saw
NP 1	a man
verb region	talking with
NP 2	the nanny.
intro to sentence 2	It seemed like
pronoun	she / he
next word (1)	was
next word (2)	very
end region[2]	angry.

Predictions. We predicted that the specificity of the characters would interact with the contribution of the second utterance. We expected that in the DESCRIPTION condition, reading times would be shorter when the pronoun referred to the unspecific character, and in the EVENT condition, reading times would be shorter when the pronoun referred to the specific character. We predicted a possible interaction of these variables with Order-of-Mention, since this factor has been shown to be significant in other studies, despite its lack of influence in Experiment 1.

We also expected these results to occur in the region(s) immediately following the pronoun. Past work using the moving-window paradigm has established that the processing load for a given word is often observed one or two words later.

Results. The major finding was that Utterance Contribution produced different patterns of facilitation, depending on which referent was in focus: 1) When the unspecific

---

[2] The reading times for the last region are shown but were not analyzed, because this region contained a different number of words in each item.

character mentioned first (and therefore was in focus), the DESCRIPTION continuations were facilitated, and 2) when the specific character was mentioned first, the EVENT continuation was facilitated, but only when the pronoun referred to the specific character.

The first difference among the stimuli occurred during the first sentence, where half the items mentioned the specific character first, and half mentioned the unspecific character first. This distinction yielded two results. The more relevant result[3] was that it influenced the way the rest of the item was read. Readers focused on the first-mentioned character, which determined whether facilitation occurred in EVENT or DESCRIPTION conditions.

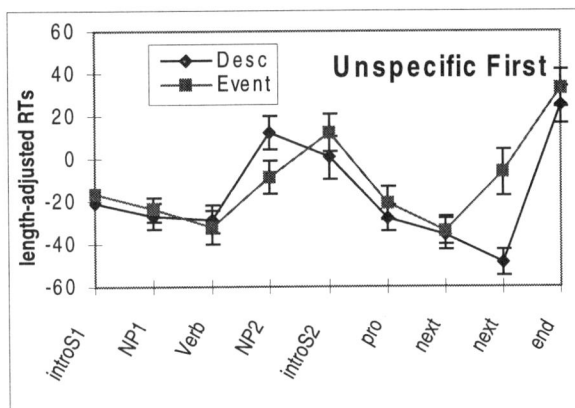

Figure 3: Reading times for each region for items where the unspecific character came first.

Result 1: Unspecific first/Description facilitation. When the unspecific character was mentioned first, participants were focused on an underspecified character in need of description. This need for specification was fulfilled in the DESCRIPTION condition, which resulted in facilitation at the second word after the pronoun. Figure (3) shows the contrast between DESCRIPTION and EVENT conditions for items where the unspecific character came first. An analysis of each region indicates that the only reliable difference between DESCRIPTION and EVENT conditions occurred at the second word after the pronoun ($F1(1,29)=17.0$; $F2(1,15)=12.5$; p's $<.005$), where reading times were shorter in the DESCRIPTION condition.

Note that the facilitation in the DESCRIPTION condition occurred equally for items with pronouns referring to specific and unspecific characters. That is, if readers were focussing on the unspecific character, reading was facilitated when they got a description, but it didn't matter if this description referred to the specific or unspecific

---

[3] The less relevant result was that the reading times for the region following the noun phrases were longer for the specific characters than the unspecific characters (for NP1, $F1(1,29)=7.6$, $F2(1,15)=5.7$; for NP2, $F1(1,29)=5.5$ $F2(1,15)=5.4$; p's $<.05$). This may reflect one of two things: a) the infelicity of using a definite NP for introducing a new character, or b) the simple fact that our definite NPs (e.g., the fire chief, the nanny) were lower frequency words than our indefinite NPs (e.g., a woman, a boy).

character. This result is consistent with the findings from Experiment 1, where the DESCRIPTION condition made participants most likely to mention both characters during their continuation. We hypothesized that this was because describing the unspecific character was usually accomplished through a description of the relationship between the two characters. This is consistent with the idea that when readers are focused on the character that needs description, they accept all descriptive continuations as informative, regardless of which character the pronoun refers to.

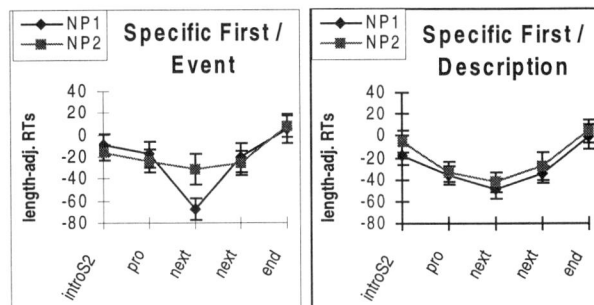

Figures 4a and b: Reading times for Sentence 2 for items where the Specific character came first in Sentence 1. The contrast between items with pronouns co-referring with NP1 or NP2 is shown separately for DESCRIPTION and EVENT continuations.

Result 2: Specific First/Event facilitation. In contrast, when the specific character appeared first, the facilitation occurred in the EVENT continuation condition. Here, however, the facilitation only occurred in cases where the pronoun referred to the specific character. We conducted ANOVAs at each word in the second sentence, looking separately at the EVENT and DESCRIPTION conditions when the Specific character came first, comparing conditions where the pronoun co-referred with the first-mentioned NP (i.e., the specific character) or the second-mentioned NP. The only reliable difference occurred at the word after the pronoun in the EVENT condition ($F1(1,29)=5.7$; $F2(1,15)=4.6$; p's$<.05$).

Discussion

The major result of these studies was that comprehension was influenced by an interaction between character specificity and the perceived relationship between the two utterances. Experiment 1 showed that whether a specific or unspecific character was considered a more likely continuation depended on the perceived role of the next utterance. Experiment 2 showed that these factors influenced on-line reading times, and further that they interacted with Order-of-Mention. When the unspecific character was mentioned first, participants found a DESCRIPTION continuation easier to read in the region following the pronoun. In contrast, when the specific character was mentioned first, reading was facilitated if the pronoun referred to the specific character in an Event continuation.

These results support a view in which reading comprehension is influenced by the reader's estimation of where the discourse is going. An important feature of this view is that this estimation is built up dynamically, and is influenced by both properties of focused referents (e.g., whether they are specific or unspecific), and other information that signals how the following utterance will relate to the story.

This study also shows that these factors influence how the pronoun is resolved and integrated with the predicate, as indicated by the fact that the observed effects occurred immediately following the pronoun. This suggests that pronoun resolution is not guided by simple rules like "pronoun refers to focused character". Rather than a general first-mentioned advantage, Experiment 2 showed that the features of the focussed referent determined how the contribution of the next utterance impacted comprehension. These data are consistent with a view that information relevant to pronoun resolution accrues from information throughout the discourse, and is not localized to either the introduction of the discourse entities or to the pronoun itself. This study manipulated the introduction to the second sentence as a way of signaling its role, but other factors like the tense of a phrase, discourse genre, or task demands may also influence the perception of utterance contribution and pronoun resolution.

These two experiments have begun to unravel some of the complex interactions that affect language comprehension. However, there are many unanswered questions. For example, why did Order of Mention interact with Referent Specificity and Utterance Contribution in Experiment 2 but not in Experiment 1? We suspect that this occurred because of task differences between the experiments. However, this and other questions need to be explored in future studies.

In sum, language comprehension is a referentially driven process. Speakers and writers establish discourse entities and predicate information about them, and comprehenders need to identify these referents. We knew that this influences syntactic ambiguity resolution; this study shows that a similar factor affects reading and pronoun resolution.

Acknowledgements

Thank you to Robert Thornton for his help with the experiment and discussions of the data. This research was partially funded by a Stanford University Graduate Research Opportunity grant, and partially funded by NSF Grant SBR-95-11270.

References

Altmann, G. T. M., Garnham, A., and Dennis, Y. (1992). Avoiding the Garden Path: Eye Movements in Context. *Journal of Memory and Language*, 31, 685-712.

Arnold, J. (1998). *Reference Form and Discourse Patterns*. Ph.D. Dissertation, Stanford University.

Chafe, W. (1994). *Discourse, consciousness, and time.* Chicago: Chicago University Press.

Chafe, W. L. (1976). Givenness, contrastiveness, definiteness, subjects, topics, and point of view. *Subject and topic*, C. N. Li (Ed.), 25-56. New York: Academic Press, Inc.

Clark, H. H. (1996). *Using language.* Cambridge: Cambridge University Press.

Crain, S., & Steedman, M. (1995). On not being led up the garden path: The use of context by the psychological syntax processor. In D. Dowty, L. Kartunnen, & A. M. Zwicky (Eds.), *Natural language parsing: Psychological, computational, and theoretical perspectives* (pp. 320-358). Cambridge, UK: Cambridge University Press.

Ferreira, F., & Clifton, C., Jr. (1986). The independence of syntactic processing. *Journal of Memory and Language*, 25, 248-368.

Garnham, A., Traxler, M., Oakhill, J. & Gernsbacher, M. A. (1996). The locus of implicit causality effects in comprehension. *Journal of Memory and Language*, 35.517-543.

Garvey, C. & Caramazza, A. (1974). Implicit causality in verbs. *Linguistic Inquiry*, 5.459-64.

Gernsbacher, M. A. (1990). Language comprehension as structure building. Hillsdale, NJ: Erlbaum.

Gordon, P. C., Grosz, B. J. & Gilliom, L. A. (1993). Pronouns, names, and the centering of attention in discourse. Cognitive Science, 17.311-357.

Grosz, B. & Sidner, C. (1986). Attention, intentions, and the structure of discourse. *Computational Linguistics*, 12.175-204.

McDonald, J. & MacWhinney, B. (1995). The time course of anaphor resolution: Effects of implicit verb causality and gender. *Journal of Memory and Language*, 34.543-566.

Prince, E. F. (1992). The ZPG letter: Subjects, definiteness, and Information-status. *Discourse description: diverse linguistic analyses of a fund-raising text.*, W. C. Mann & S. A. Thompson (Ed.), 295-325. Amsterdam: John Benjamins.

Stevenson, R. J., Crawley, R. A. & Kleinman, D. (1994). Thematic roles, focus and the representation of events. *Language and Cognitive Processes*, 94.473-592.

Tanenhaus, M. K., Spivey-Knowlton, M. J., Eberhard, K. M., and Sedivy, J. C. (1995). Integration of Visual and Linguistic Information in Spoken Language Comprehension. *Science*, 268, 1632-1634.

Thornton, R., MacDonald, M. C., and Gil, M. (in press). Pragmatic Constraint on the Interpretation of Complex Noun Phrases in Spanish and English. Journal of Experimental Psychology: Language, Memory, and Cognition.

Using a High-dimensional Memory Model to Evaluate the Properties of Abstract and Concrete Words

Chad Audet (chad@cassandra.ucr.edu)
Department of Psychology; University of California, Riverside
and Language Machines, Inc.
Riverside, CA 92521

Curt Burgess (curt@doumi.ucr.edu)
Department of Psychology; University of California, Riverside
and Language Machines, Inc.
Riverside, CA 92521

Abstract

The evidence that the comprehension of abstract and concrete words differ prompts one to consider how the lexical representations for these word types differ. The context-availability model (Schwanenflugel & Shoben, 1983) suggests that abstract words are more difficult to process because associated contextual information stored in memory for these words is more difficult to retrieve than for concrete words. Schwanenflugel (1991) provides two hypotheses regarding how these differences in retrieval of contextual information may come about. Three simulations using context representations from the Hyperspace Analogue to Language (HAL) model of memory (Burgess & Lund, 1997; Lund & Burgess, 1996) are used to evaluate Schwanenflugel's hypotheses, as well as to provide insight into the representational differences between abstract and concrete words.

While most empirical results have consistently demonstrated that abstract words like "permission" and "issue" are more difficult to comprehend than concrete words like "newspaper" and "apple," the source of this relative difficulty for understanding abstract words is unclear. The finding that concrete words are more readily processed by language users--the "concreteness effect"--has been found in a number of experiments using a variety of tasks (for reviews, see Balota, Ferraro, & Conner, 1991; and Schwanenflugel, 1991; for representative studies that have failed to find concreteness effects, see these same reviews). Concreteness effects have been shown for words presented in lexical decision (Bleasdale, 1987; Chiarello, Senehi, & Nuding, 1987; Ransdell & Fischler, 1987; Schwanenflugel, Harnishfeger, & Stowe, 1988; Schwanenflugel & Shoben, 1983), naming (Bleasdale, 1987; de Groot, 1989; Schwanenflugel & Stowe, 1989), free recall (Paivio, 1986; Ransdell & Fischler, 1987; Schwanenflugel, Akin, & Luh, 1992), and word association (de Groot, 1989). Concreteness effects have also been seen in experimental tasks that involved comprehension or recall of sentences or paragraphs that differed on concreteness (Belmore, Yates, Bellack, Jones, & Rosenquist, 1982; Holmes & Langford, 1976; Ransdell & Fischler, 1989; Schwanenflugel & Shoben, 1983).

Given the wealth of evidence that suggests that the processing of concrete and abstract words differs, an explanation of the representational differences between these word types would seem to be essential for any robust theory of word meaning. One potential explanation of this difference is extended by the context-availability model discussed by Schwanenflugel and Shoben (1983; and Schwanenflugel, 1991). This model posits that language comprehension is aided by the retrieval from memory of contextual information associated with the material being processed. If the appropriate contextual information can not be retrieved from memory (and is not provided by some external source, such as a conversation), then comprehension is difficult. As for the concreteness effect in language comprehension, the context-availability model assumes that retrieving associated contextual information for abstract words is more difficult than for concrete items because abstract word representations presumably have weaker connections to contextual information than do representations for concrete words.

What follows is a series of simulations using the Hyperspace Analogue to Language (HAL) model (Burgess & Lund, 1997; Lund & Burgess, 1996) to examine cognitively-relevant differences between the representations for abstract and concrete words. One particular goal is to determine whether abstract and concrete word representations differ in the availability of associated contextual information stored in memory (as suggested by Schwanenflugel & Shoben). HAL is a context model that develops word meaning from global co-occurrence statistics extracted from human on-line language use. A ~320 million word corpus of Usenet text is the input stream from which HAL records weighted co-occurrence information for the 70,000 most frequent vocabulary items. The process of recording these co-occurrences allows for the formation of a co-occurrence matrix from which word vectors are derived. Mathematically, these vectors represent points in a

37

high-dimensional space. The similarity between words corresponds inversely to inter-point distances, with the assumption that the more similar two words are, the closer their points in the high-dimensional space (see Burgess & Lund, 1997, for further discussion of the HAL methodology). Conceptually, each vector represents the entire learning history of a given word in the context of other words. We claim that these context vectors provide robust representations of a number of important aspects of word meaning (Burgess, Livesay, & Lund, 1998). HAL has been used to account for grammatical class distinctions and semantic affects on syntactic processing (Burgess & Lund, 1997), several semantic and associative priming effects (Lund & Burgess, 1996; Lund, Burgess, & Audet, 1996), the sort of semantic errors made by deep dyslexia patients (Buchanan, Burgess, Lund, 1996), and cerebral asymmetries in semantic memory processing (Burgess & Lund, 1998).

Experiment 1: Demonstrating Separation of Abstract and Concrete Words in the HAL Model

Semantic differentiation of a small set of dissociable abstract words (five emotional words, e.g., *love*, *sorrow*; five legal terms, e.g., *judge*, *law*) has been demonstrated before using HAL context vectors (Burgess & Lund, 1997). However, what remains unclear is whether these context vectors can be used to examine more systematically the proposed distinctions between abstract and concrete words. The goal of this first simulation is to provide new evidence that contextual representations extracted from HAL can be categorized along the concreteness dimension. In this simulation three larger sets of abstract and concrete words are subjected to multidimensional scaling (MDS) in order to show that the interword distances in the high-dimensional space can provide a basis for this categorization, thus providing evidence that HAL representations are relevant to the issues presented in the Introduction.

Method

Materials. Bleasdale (1987) presented a list of 80 concrete and 80 abstract words arranged in prime-target pairs (see his Appendix A). These words had been rated by undergraduate students for concreteness and imageability, with concrete primes and targets rated reliably higher than abstract primes and targets on both characteristics. Without regard to prime/target status, 159 of these words (the word "glutton" did not occur in the HAL matrix, and was not included in any simulations or analyses presented herein) were placed into a stimulus pool for possible inclusion into the following simulations and subsequent analyses. As the amount of information that can be displayed in a two-dimensional MDS in an interpretable manner is somewhat limited, we chose not to include all 159 words in a single MDS. Rather, three separate sets of 20 concrete and 20 abstract words were pseudo-randomly sampled (without

replacement) from the larger pool, and these individual subsets were treated as representative of the entire pool of words. The only restriction placed on this sampling procedure was an attempt to avoid including highly related words in the same MDS as highly associated and/or similar words have a tendency to "pair off" and increase apparent dissociations between word categories within an MDS (e.g., *truth* and *false*, *son* and *daughter*, *democracy* and *government*) that might tend to exaggerate the examined effects.

Procedure. Global co-occurrence vectors were extracted from the HAL model for each set of 40 words. Each vector was treated as a set of coordinates in a high-dimensional Euclidean space, and for each set of words a MDS solution was computed. The hypothesis was that these word vectors, representing the interword distances for the chosen set of words, would operate as a similarity matrix (Lund & Burgess, 1996).

Results

Each similarity matrix was analyzed by a MDS algorithm that projects points from a high-dimensional space into a lower-dimensional space in a nonlinear fashion that attempts to preserve the distances between points. The lower-dimensional projection allows for the visualization of the spatial relationships between the co-occurrence vectors. The two-dimensional MDS solutions for the three subsets of words are shown in Figures 1a-1c. To further simplify interpretation of the information present in the MDS solutions, concrete and abstract words are represented by Cs or As, respectively.

Visual inspection of Figures 1a-1c suggests that concrete and abstract words were differentiated in the MDS solutions. In each case the abstract words appear to occupy a space separate from the concrete words. However, given the nature of the MDS procedure, in which an extreme reduction of dimensionality occurs when projecting data from a high-dimensional space down to only two dimensions, it was necessary to perform appropriate inferential statistics. In order to determine whether abstract words do exist in a separate (high-dimensional) space than concrete words an analysis of variance was performed separately for each set of 40 words which compared intragroup distances between these words with intergroup distances between these items. Distances between all combinations of word pairs within a group (i.e., concrete or abstract words) were calculated and compared to the distances between all combinations of word pairs between groups. In each of the three analyses concrete words were differentiated from abstract words, $F_{set1}(1, 778) = 5.59$, $p = .018$; $F_{set2}(1, 778) = 11.7$, $p = .0007$; $F_{set3}(1, 778) = 8.13$, $p = .01$. As well, abstract words were differentiated from concrete words, $F_{set1}(1, 778) = 143.7$, $p < .0001$; $F_{set2}(1, 778) = 74.23$, $p < .0001$; $F_{set3}(1, 778) = 116.1$, $p < .0001$.

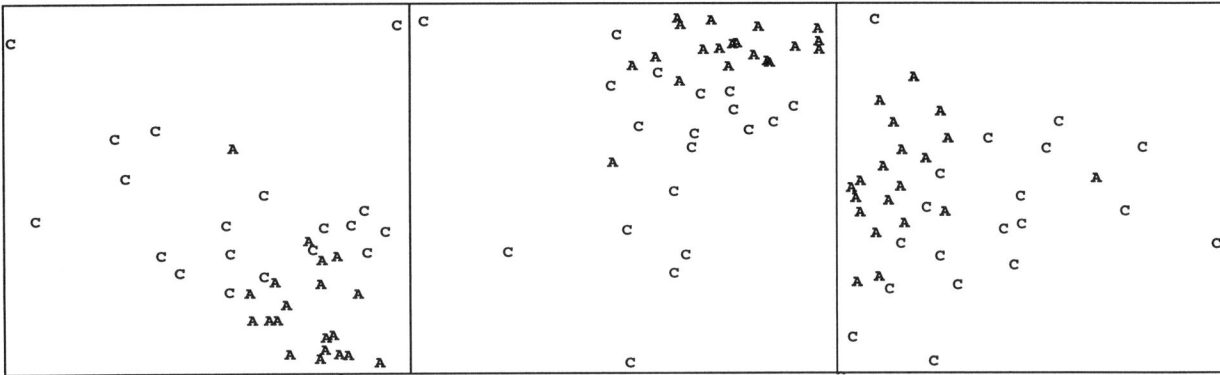

Figures 1a-1c: Two-dimensional multidimensional scaling solutions for word vectors for abstract (A) and concrete (C) words from Bleasdale (1987).

Discussion

As seen in Figures 1a-1c, it is clear that the information carried in HAL context vectors is sufficient to distinguish concrete from abstract words. This effect supports the range of theoretical and empirical evidence that suggests important differences in the processing and comprehension of these two word types. However, what remains unclear is the nature of the information contained in the HAL vectors that contributes to the present results. Moreover, it is also unclear what representational differences exist between abstract and concrete words for language users, differences that, presumably, play some important role in the language comprehension process. A number of possible explanations for the representational and processing differences between abstract and concrete words have been proposed (e.g., Paivio's dual-coding model, Schwanenflugel & Shoben's context-availability model; see Schwanenflugel, 1991 for a review of these proposals). Experiment 2 is an evaluation of one such theoretical model, the context-availability model, which attributes variations in word comprehension difficulty to differences in the retrieval of contextual information from memory. Given that HAL representations are derived from local and global contextual information it is expected that the context vectors extracted from this model will provide an avenue for evaluation of the context-availability model.

Experiment 2: Evaluating the Frequency and Context Diversity Hypotheses

Schwanenflugel and Shoben (1983; Schwanenflugel, 1991) suggest that concreteness effects in language comprehension can be explained with the context-availability model. As discussed in the Introduction, at the heart of this model is the idea that abstract words are more difficult to understand than concrete items because a person tends to have greater difficulty in retrieving associated contextual information from the "knowledge base" (i.e., the mental lexicon) for abstract words. In her discussion of this model Schwanenflugel (1991, p. 243) provided two distinct hypotheses regarding how abstract words might come to have weaker connections to contextual information.

Hypothesis 1: Abstract words occur so infrequently in language that representations for these words have relatively few opportunities to develop strong connections to contextual information stored in memory.

Hypothesis 2: Abstract words occur frequently, but in such a diversity of contexts that the opportunity to develop strong connections to one or a few particular contexts does not arise.

A corollary to the latter hypothesis is that although concrete words might occur in a relatively few number of contexts, these words are presumably well-grounded in one or a few contexts (perhaps due to strong connections to the visual environment, as Paivio would suggest), thus providing for secure links between concrete word representations and stored contextual information. As for the former hypothesis, it may be that more frequent words simply have more opportunities to establish stronger ties to stored contextual information, and concrete words may, in fact, be more frequent.

Materials. As Schwanenflugel did not provide a list of stimulus words in her articles, we chose to use the 159 words provided by Bleasdale (1987), along with an additional set of abstract and concrete words borrowed from Chiarello, Senehi and Nuding (1987) that were added to increase the scope of our investigation. Both Bleasdale and Chiarello et al. had subjects norm their stimuli for concreteness and imageability, demonstrating that their concrete words were rated reliably higher than abstract words on both characteristics. After removing duplicate words we were left with a list of 119 abstract and 129 concrete words.

Testing Hypothesis 1: Testing Schwanenflugel's first hypothesis regarding the possibly infrequent occurrence of abstract words in language was a simple matter of comparing raw frequency counts for the abstract and concrete items in the 320 million word HAL corpus (see Burgess & Livesay, 1998). Contrary to Hypothesis 1, abstract words were more frequent (228 occurrences per million words) than concrete words (136 occurrences per million), $t(246) = 2.84$, *p* = .0049. This result shows that language users (specifically, English speakers) encounter abstract words much more frequently than concrete words, thus providing evidence that language users have more opportunities to develop strong connections between abstract words and associated contextual information, as compared to concrete words.

Testing Hypothesis 2: The disconfirmation of Hypothesis 1 provides some evidence for Schwanenflugel's second hypothesis: abstract words do occur more frequently than concrete words. However, this is not a direct indication that abstract words occur in a greater diversity of contexts. Examining this notion requires a measure that moves beyond raw word frequencies, and provides a metric of the different contexts in which a word was experienced by the HAL model. We believe that decomposing HAL's context vectors provides such a metric, in that each element in a word's context vector is a weighted co-occurrence count of that word with some other word in the text stream input to the HAL model. Considered more simply, each of the vector elements is a direct record of each of the contexts in which a target word occurred. If a target word never co-occurred with a particular word in the input stream--thus indicating that the target word was never experienced by the HAL model in that particular context--the vector element recording that potential co-occurrence would have a value of zero. On the other hand, any actual co-occurrences between the target word and some other word would result in the element representing that co-occurrence having a value greater than zero, thus representing the model's experience with that word in that context.

With the above consideration in mind, it is a simple matter to calculate the proportion of non-zero elements in a word's context vector (i.e., across the entire 70,000-element vector). This proportion--which we have labeled "context density"--directly represents the number of contexts in which a given word has been encountered by the HAL model across its entire learning history for that word. Analysis of the abstract and concrete words shows that the context density for the abstract words (16.8%) is greater than for concrete words (12.7%), *t*(246) = 2.12, *p* = .035, thus supporting Schwanenflugel's notion that abstract words occur in a greater diversity of contexts. In fact, the roughly 4% difference between the context densities for abstract and concrete words represents a difference of approximately 2,800 contexts--a rather substantial difference.

Discussion

Abstract words were shown to be more frequent than concrete words in a large sampling of human language use. This finding disconfirms Schwanenflugel's first hypothesis, and suggests that whatever differences between abstract and concrete word representations contribute to the concreteness effect in language comprehension, these differences are not attributable to a lack of experience with abstract words. Not only are abstract words more frequent than concrete words, abstract words also appear in a greater number of contexts. This latter finding supports Schwanenflugel's second hypothesis, and suggests that abstract word representations might have more diffuse connections to associated contextual information. To understand this conclusion, one might consider that a frequently occurring abstract word, such as "vacation," appears in such a variety of contexts that the representation for that word has few significant ties to any particular context. In contrast, a less frequent concrete word like "pyramid" is strongly associated with only a few distinct contexts (e.g., Egypt and pharaohs). We are aware that the *diversity* of contexts in which a word appears is not reflected only in the raw number of contexts for that word. Contextual diversity is also a function of the differences between the number of occurrences of a word in each of the different contexts in which that word was experienced. These different patterns of co-occurrence translate into different variances across the elements in HAL context vectors (Lund & Burgess, 1996). Put more simply, two words might occur in roughly the same number of contexts, but vary in the overall pattern of occurrences across these contexts, thus leading to quantitatively and qualitatively different representations.

Though the results from Experiment 2 do not fully address the veracity of the context-availability model as an adequate explanation of concreteness effects, these findings do support the basic assumption that abstract and concrete word representations differ on context availability (i.e., context density). We argue that the representational differences between abstract and concrete words are largely due to differences in how these words are used in language, and the context vectors extracted from the Hal model are transparently sensitive to such differences in word use. In fact, these results provide the first quantitative evidence that abstract and concrete words differ in the number of contexts in which these word types occur in natural language. What remains to be seen is whether differences in context density relate to the processing differences between abstract and concrete words seen in human studies. Experiment 3 involves a consideration of the concreteness effect as a function of the distinction between automatic and controlled processing.

Experiment 3: The Relationship Between Priming and Semantic Distance

As discussed in the Introduction, concreteness effects have been found using a number of experimental paradigms including both lower-level (e.g., lexical decision and naming) and higher-level tasks (e.g., sentence verification). However, few studies have examined concreteness effects along with an explicit consideration of the difference between lower-level and higher-level *processing*, in other words, the distinction between automatic and controlled processing. Two exceptions are the semantic priming studies presented by Bleasdale (1987) and Chiarello et al. (1987) in which the authors explicitly manipulated both word concreteness and the degree of automatic versus controlled processing. In a controlled priming paradigm using lexical decision (Exp. 1: 75% proportion of related prime-target pairs, and subject instructions designed to focus attention on prime word identities) Chiarello et al. showed greater priming for targets when preceded by a concrete prime, as compared to an abstract prime. This effect was not found in an automatic priming paradigm (Exp. 3: 25% proportion of related prime-target pairs; no instruction focusing attention on prime word identities) in which priming was equivalent for targets preceded by concrete or abstract primes. Bleasdale showed a similar pattern of results for both automatic and controlled priming. Bleasdale presented data from a naming (Exp. 1) and lexical decision task (Exp. 2), each involving controlled processing (i.e., a long stimulus-onset asynchrony between primes and targets, and a 50% proportion of related prime-target pairs), in which targets preceded by concrete primes showed greater priming than when preceded by abstract primes. In contrast, in a lexical decision task (Exp. 3) using an automatic priming procedure (i.e., a brief stimulus-onset asynchrony between primes and targets; a 50% proportion of related prime-target pairs) Bleasdale showed equivalent priming for targets following either a concrete or abstract prime.

In the following simulation we examine Bleasdale's prime-target pairs with the intent to show that the context vectors extracted from the HAL model mimic the initial bottom-up activation of semantic representations in memory. We have argued elsewhere (Burgess & Lund, in press) that the information carried by the context vectors best represents the sorts of information that are activated in memory early on in language processing (e.g., in word recognition). In fact, HAL context vectors are removed from any attentional or strategic effects that appear in controlled processing situations. Based on the results of Bleasdale and Chiarello et al. in their automatic processing paradigms, we expect to find equivalent context distance (i.e., semantic) priming with our vector representations when targets are paired with abstract or concrete primes.

Method

Materials. For our related condition we borrowed 79 prime-target pairs listed in Bleasdale's (1987) Appendix A (one pair was removed because the target did not occur in the HAL matrix). These stimuli consisted of 20 pairs with concrete primes and targets, 20 pairs with abstract primes and targets, 20 pairs with concrete primes and abstract targets, and 19 pairs with abstract primes and concrete targets. From these related pairs we generated 79 unrelated pairs by pseudo-randomly pairing each target with an unrelated prime. Though we have no way of directly comparing our unrelated pairs to those used by Bleasdale, when producing these pairs we did explicitly avoid

producing unrelated pairs in which the prime and target shared any obvious relationship.

Procedure. Context vectors for all primes and targets were extracted from the HAL model. For each prime-target pair the Euclidean distance (in the high-dimensional context space represented by the HAL global co-occurrence matrix) between the two words was calculated. These distances are based on context vectors which have been normalized simply to provide a more easily interpretable set of values that map onto somewhat realistic word recognition reaction times. Analogous to the procedure used in human priming studies, in this simulation each target acted as its own control, with the context distance between a target and its related prime being compared to the distance between the target and its unrelated prime. A priming effect was then calculated for each word pair by subtracting the distance for the related pair from the distance for the unrelated pair.

Results

The comparison of primary interest was the difference in distance priming between targets that were paired with concrete primes compared to those paired with abstract primes. While an analysis of variance indicated an overall priming effect with related prime-target pairs displaying smaller context distances (603 distance units) than unrelated prime-target pairs (661 units), $F(1, 154) = 8.59$, $p = .004$, this effect did not interact with prime concreteness, $F < 1$. Thus, the distance priming seen for targets paired with abstract primes (53 units) did not differ from the distance priming found for targets paired with concrete primes (62 units).

Discussion

The pattern of results provided by Bleasdale and Chiarello et al. suggests that concreteness effects tend to appear when the experimental task promotes the use of higher-level, controlled processing of verbal stimuli, such as when subjects have an opportunity and the motivation to develop expectancies and response strategies in word recognition (Neely, 1991). The results of Experiment 3 using Bleasdale's stimuli replicate previous failures to find an effect of word concreteness in automatic processing paradigms. This supports our previous findings that HAL's context vectors provide a good match to empirical semantic priming results (Lund, Burgess, & Audet, 1995).

General Discussion

While it appears that processing differences do exist between abstract and concrete words, the nature of the representational differences between these word types that presumably underlie such processing differences is not well-understood. An evaluation of one proposed explanation of these representational and processing differences--the context-availability model--using word meaning representations provided by the HAL memory model suggests that important differences may exist between the diversity of contexts in which abstract and concrete words are experienced by language users. Our findings indicate that context diversity, as measured by context density in HAL's word representations (and perhaps context variance, as well), may have some utility in providing an explanation of concreteness effects in higher-level processing tasks. As for the issue of how context density might relate to the lack of concreteness effects in automatic processing paradigms, the predictions provided by the context-availability model are not motivated by differences in the time-course of word meaning activation, per se, but rather by the notion that contextual information associated with abstract words is simply more difficult to access and retrieve. We propose that the context distances between HAL context vectors are better predictors of early time-course effects in semantic memory activation than are variances in context diversity.

Acknowledgements

This research was supported by NSF Presidential Faculty Fellow award SBR-9453406 to Curt Burgess.

References

Balota, D. A., Ferraro, F. R., & Connor, L. T. (1991). On the early influence of meaning in word recognition: A review of the literature. In P. J. Schwanenflugel (Ed.), *The psychology of word meanings* (pp. 223-250). Hillsdale, NJ: Lawrence Erlbaum Associates.

Belmore, S. M., Yates, J. M., Bellack, D. R., Jones, S. N., & Rosenquist, S. E. (1982). Drawing inferences from concrete and abstract sentences. *Journal of Verbal Learning and Verbal Behavior, 21,* 338-351.

Bleasdale, F. A. (1987). Concreteness-dependent associative priming: Separate lexical organization for concrete and abstract words. *Journal of Experimental Psychology: Learning, Memory, and Cognition, 13,* 582-594.

Buchanan, L., Burgess, C., & Lund, K. (1996). Overcrowding in semantic neighborhoods: Modeling deep dyslexia. *Brain and Cognition, 32,* 111-114.

Burgess, C., & Livesay, K. (1998). The effect of corpus size in predicting reaction time in a basic word recognition task: Moving on from Kucera and Francis. *Behavior Research Methods, Instruments, and Computers, 30,* 272-277.

Burgess, C., & Lund, K. (1997). Modelling parsing constraints with high-dimensional context space. *Language and Cognitive Processes, 12,* 1-34.

Burgess, C., & Lund, K. (1998). Modeling cerebral asymmetries in high-dimensional semantic space. In M. Beeman & C. Chiarello (Eds.), *Right hemisphere language comprehension: Perspectives from cognitive neuroscience.* Hillsdale, NJ: Lawrence Erlbaum Associates.

Burgess, C., & Lund, K. (In press). The dynamics of meaning in memory. In E. Dietrich & A. Markman (Eds.) *Cognitive dynamics: Conceptual and representational change in humans and machines.*

Chiarello, C., Senehi, J., & Nuding, S. (1987). Semantic priming with abstract and concrete words: Differential asymmetry may be postlexical. *Brain and Language, 31,* 43-60.

Collins, A. M. & Loftus, E. F. (1975). A spreading-activation theory of semantic processing. *Psychological Review, 82,* 407-28.

de Groot, A. M. B. (1989). Representational aspects of word imageability and word frequency assessed through word association. *Journal of Experimental Psychology: Learning, Memory, and Cognition, 15,* 824-845.

Holmes, V. M., & Langford, J. (1976). Comprehension and recall of abstract and concrete sentences. *Journal of Verbal Learning and Verbal Behavior, 15,* 559-566.

Lund, K., & Burgess, C. (1996). Producing high-dimensional semantic spaces from lexical co-occurrence. *Behavior Research Methods, Instrumentation, and Computers, 28,* 203-208.

Lund, K., Burgess, C., & Audet, C. (1996). Dissociating semantic and associative word relationships using high-dimensional semantic space. *Proceedings of the Cognitive Science Society* (pp. 603-608). Hillsdale, NJ: Lawrence Erlbaum Associates.

Neely, J. (1991). Semantic priming effects in visual word recognition: A selective review of current findings and theories. In D. Besner & G.W. Humphreys (Eds.), *Basic processes in reading: Visual word recognition* (pp. 264-336). Hillsdale, NJ: Lawrence Erlbaum Associates.

Paivio, A. (1986). *Mental representations: A dual coding approach.* Oxford, England: Oxford University Press.

Ransdell, S. E., & Fischler, I. (1987). Memory in a monolingual mode: When are bilinguals at a disadvantage? *Journal of Memory and Language, 26,* 392-405.

Schwanenflugel, P. J. (1991). Why are abstract concepts hard to understand? In P. J. Schwanenflugel (Ed.), *The psychology of word meanings* (pp. 223-250). Hillsdale, NJ: Lawrence Erlbaum Associates.

Schwanenflugel, P. J., Akin, C., & Luh, W. (1992). Context availability and the recall of abstract and concrete words. *Memory and Cognition, 20,* 96-104.

Schwanenflugel, P. J., Harnishfeger, K. K., & Stowe, R. W. (1988). Context availability and lexical decisions for abstract and concrete words. *Journal of Memory and Language, 27,* 499-520.

Schwanenflugel, P. J., & Shoben, E. J. (1983). Differential context effects in the comprehension of abstract and concrete materials. *Journal of Experimental Psychology: Learning, Memory, and Cognition, 9,* 82-102.

Schwanenflugel, P. J., & Stowe, R. W. (1989). Context availability and the processing of abstract and concrete words. *Reading Research Quarterly, 24,* 114-126.

Causal Relationships and Relationships between Levels: The Modes of Description Perspective

Bram Bakker (bbakker@fsw.leidenuniv.nl)
Unit of Experimental and Theoretical Psychology; Leiden University
P.O. Box 9555; 2300 RB, Leiden, The Netherlands

Paul den Dulk (dendulk@psy.uva.nl)
Psychonomics department; University of Amsterdam
Roetersstraat 15; 1018 WB, Amsterdam, The Netherlands

Abstract

Many researchers have argued for a description of nature using multiple levels, or modes of description, as we call them. This paper focuses on a confusion that follows from the multiple-mode approach, a confusion due to the notion of causation between modes. Causation between modes is reinterpreted as ordinary causation but with cause and effect described in different modes. In the first part of the paper the framework of modes of description is presented. In the second part it is applied to examples from cognitive science, which are taken from debates on the mind-brain issue and the dynamical systems approach to cognition.

Introduction

Many researchers have argued for scientific analysis at different levels when studying complex phenomena (e.g. Pylyshyn, 1984; Marr, 1982; Newell, 1990; Arbib, 1989; Hofstadter, 1979; Kelso, 1995; Dennett, 1987; Salthe, 1985; Bechtel and Richardson, 1993). Multiple levels— or *modes of description*, as we call them—may be useful to provide a richer picture of a complex phenomenon, and the framework of levels may help thinking about how different descriptions of the same phenomenon relate to each other.

However, the main focus in this paper is on confusions that sometimes arise within this framework, confusions due to the inappropriately mixing up of different modes or levels of description. In particular, we focus on confusions involving *causal relationships* within a multi-level approach.

The next section introduces the framework of modes of description. It is based on a few simple principles, such as the explicit distinction between reality and knowledge and a view of knowledge as many descriptions of a single reality. Our purpose is not to introduce a new, complete philosophical theory of knowledge or to provide a comprehensive overview and rebuttal of all alternative views, but rather to shape and restructure existing (although sometimes disputed) ideas into a simple and coherent framework that is applicable to issues in modern empirical science. Its usefulness when applied to those issues should be the test of its value.

For this reason, we apply the framework to examples of ongoing scientific debates in section 3. The examples concern the mind-brain issue and the dynamical systems approach to cognition.

Theory

One Reality, Multiple Descriptions

Central to the arguments presented in this paper is the straightforward notion that reality and knowledge are different things. Scientific knowledge can be thought of as *descriptions of* reality, but it is not reality itself. Pragmatists (e.g. James, 1907) have argued that reference to reality is unfruitful and unnecessary, but we feel that it is actually more pragmatic, at least for the purposes of this paper, to explicitly assume the existence of a single reality. We hope the usefulness of this assumption will become clear in the remainder of this paper. Thus, we assume that there is only one reality; but many different descriptions of this single reality are possible.

The importance of using multiple descriptions is commonly accepted in cognitive science (see Pylyshyn, 1984; Marr, 1982; Newell, 1990; Arbib, 1989; Hofstadter, 1979; Kelso, 1995; Dennett, 1987). Marr's (1982) implementational, algorithmic, and computational levels of analysis, and Dennett's (1987) physical, design, and intentional stance are particularly well-known examples. In practice, a distinction is often made between the functional level and the implementational level. "Classical", functionalist, symbolic models are viewed by many as descriptions at the abstract, functional level, while connectionist models are viewed as descriptions at a more hardware-oriented implementational level. Neurophysiologists zoom in to low-level physical processes even more when they study chemical processes within a single neuron.

In some cases theories at different levels can be compared as competing theories. For instance, a symbolic model and a connectionist model can compete for the best predictions on the same psychological experiment. They are in competition because the assumption of one reality implies that if two theories make contradictory predictions of data, they cannot both be right. Furthermore, descriptions that explain and predict more data than other descriptions are better descriptions of reality, even if neither of them predicts perfectly.

For such comparisons between different descriptions to make sense, the descriptions must be competing for the same turf, in the sense that they aim to predict the same data. In contrast, very often different levels are intended to predict different data. Functionalist models answer questions about overt behavior, but none about the neural substrate. Neurophysiological models answer questions about the effect of neurotransmitters, but they usually do not predict overt behavior.

Finally, sometimes multiple descriptions need to be considered at the same time in order to make good predictions of data. In that case, they can be said to complement each other (this is an important topic in the paragraph about dynamical systems).

Modes of Description

Some authors identify a level of description more or less directly with physical size or scale (e.g. Salthe, 1985; Yates, 1993). However, sometimes descriptions cannot be neatly arranged within such a hierarchy, and one description cannot easily be thought of as being at a higher level or a lower level than the other. A neuron can be described focusing on electrical properties, or it can be described focusing on chemical processes. Which one is the higher level description? The notion of size is even less of a help when we give a description of the function of a neuron. The "electrical", "chemical" and "function" descriptions are just *different approaches* to understanding reality, different ways of looking at the neuron, all aimed at roughly the same scale. The reason for maintaining them all is that each has its own value for answering different questions.

Conversely, a single approach to understanding reality may be used to describe things at very different scales. Molecules, billiard balls, and planets can often be viewed simply as point masses and be described accordingly, using Newtonian mechanics. The scales are very different, but the description is very much the same and the predictions that follow from this single description are very accurate.

The notion of "levels" suggests a clear-cut hierarchy in which descriptions have an unambiguously defined position that is higher or lower relative to other descriptions. In practice, such a neat ordering seems difficult if not impossible to attain, and we see no reason to search for it. The important point is that there may be different descriptions of the same phenomenon, sometimes differing in scale, sometimes differing in abstractness, sometimes differing in what aspects of the phenomenon the focus is on. This is a more generic outlook on different descriptions than the notion of level can convey, and for that reason we prefer to speak of different *modes of description*.

Concepts

If a theoretical concept plays a pivotal role in a successful description—successful in the sense that it explains and predicts a lot of data—and without this concept the description falls apart completely, we can assume some sort of concurrent *regularity* in reality—a "real pattern", Dennett might say (Dennett, 1991b). All this means is that apparently reality is such that it can be aptly described using that concept. Put the other way around, if there were no such concurrent real regularity, the description could not possibly be as successful as it is. For instance, the concept of atom must have some concurrent regularity in reality, simply because predictions made with theories in which atoms are indispensable are usually extremely accurate. However, arguing that there must be a concurrent regularity in reality is not the same as arguing that there is an *absolute thing* in reality corresponding to the concept in a straightforward way. The exact nature of the regularity cannot be further specified. After all, we only have descriptions, featuring those very concepts, to tell us what reality is like.

Because there are better and worse descriptions, there are better and worse concepts. The concept of atom is better than the concept of phlogiston, because the theory involving atoms is better than the theory involving phlogiston. *Apparently* the concept of atom better corresponds to regularities in reality than the concept of phlogiston does. At the other end, of course, we should also simply discard a concept when it turns out that its corresponding theory is not successful at all, rather than trying to look for the "true meaning" of the concept. After it was discovered that many substances gain weight after combustion rather than lose weight, it was briefly suggested that phlogiston has negative weight (Chalmers, 1982), in an ad hoc attempt to save a concept that had completely lost its value.

Translation of Modes

Even though different modes of descriptions are *separate* ways of looking at reality, they are not *free-floating* ways of looking at reality. Because they are all about one reality, at the very least they cannot contradict each other, in terms of prediction of data. In that sense, different modes must be compatible. What is more, it is often possible to "map" concepts in one mode directly onto concepts in another mode. For instance, the concept of gas—as used in the so-called thermodynamic mode of description—can be translated relatively straightforwardly into concepts used in a molecular mode, when we describe gas as a large number of molecules interacting in a particular way. A molecule (molecular mode), in turn, can often be viewed as a particular threedimensional configuration of atoms (atomic mode).

Sometimes, however, concepts in one mode cannot be translated as easily to concepts in other modes. Consider the concept of life. We cannot map life straightforwardly onto concepts in an atomic or molecular mode of description. If we use the atomic mode of description, we see that atoms and structures of atoms making up living things are basically the same as those making up non-living things. This does not logically imply we have overlooked an important thing or property in the atomic mode that is somehow responsible for life. Nor does it imply that in a life-mode of description something is added to reality that is absent in the atomic mode—we assume there is only one reality. A new mode does not suddenly add things to reality in a mystical way; it only describes this single reality differently. The concept of life in the life-mode corresponds to collective properties and mechanisms of large groups of atoms that are just *not easily described* in the atomic mode. Even though properties described in the life-mode must somehow be realized by mechanisms that can be described in the atom-mode, there is no simple one-to-one mapping from the concept of life to single concepts in the atom-mode. In that sense, not every concept can be "reduced" straightforwardly to concepts in the atomic mode. And for this reason, not everything can or should be predicted from the atom-mode.

Causation

Can we say, at least, that life is *caused* by the (interaction of) atoms? No, we are talking about multiple descriptions—in which the concepts of life and atoms have roles—of one single reality. Different descriptions of the same thing at the same time cannot be said to "cause" each other. Atoms do not cause life, nor do they cause molecules or planets; and life does not cause atoms.

Causation in this sense of levels causing each other tends to get confused with causation in the regular sense, as we shall see in the Applications section. Causation in the regular sense refers to *processes over time, with causes and effects that are not just different descriptions of the same thing.* Suppose we are looking at one neuron in a brain, and the neuron fires. What is the cause of the effect "the neuron fires"?[1] One valid answer is that the presynaptic neurons fired, thus causing our neuron to fire as well. Another valid answer to the same question is that the neuron's membrane potential increased beyond a threshold value, causing the voltage-gated sodium channels to open, causing a massive influx of sodium, leading to a fast and large increase in potential: a spike. Still another valid answer is that an object is present in the visual field that this neuron is sensitive to, so the neuron becomes active.

The first causal story identifies "presynaptic neurons firing" as the cause, and thus stays in the same neural network mode of description as the effect ("the neuron fires"). The other two causal stories identify causes described in very different modes from the neural network mode in which the effect is described: "the membrane potential threshold was crossed" and "the object was present", respectively. Thus, causal explanations may involve *mode switches*: cause and effect need not be described in the same mode. But what makes these examples regular causal explanations is that they all refer to a process over time, and the cause and the effect are not different descriptions of the same thing. In these examples, causation in the other, inappropriate sense would be: "the fast and large increase in potential causes the spike", or: "the feature detection by the neuron causes it to fire". The spike and the fast, large increase in potential are the same thing at the same time, but described in different modes. Likewise, the neuron firing and the detection of the feature are the same thing, but described in different modes.

---

[1]This example is based on Hofstadter (1981, p. 193–196).

Applications

Mind and Brain

The relationship between mind and brain is one of the classical issues of cognitive science. Descartes (1662) argued that mind and brain are two fundamentally different substances, and although this dualism has lost a lot of ground, there are still many who adhere to it (e.g. Popper & Eccles, 1977; Koestler, 1967). At the opposite end of the spectrum, eliminative materialists argue that the mind is really the brain (Hobbes, 1651, is an early example), and that the mind is at best a fair but limited abstraction of the workings of the brain (e.g. Churchland, 1986). As Armstrong (1968) puts it, "The mind is nothing but the brain.... We can give a complete account of man in purely physio-chemical terms".

Within the modes of description perspective, the crucial step is to consider talk of the mind and talk of the brain as two different modes of description. On the one hand, our earlier assumption of one reality precludes *a priori* a Cartesian duality of mind and brain in reality. On the other hand, this does not mean that we can or need to give "a complete account of man in purely physio-chemical terms". First of all, *all* modes of description are "at best fair but limited abstractions" of reality: the brain mode is still just a description, and not intrinsically "real" (as opposed to the mind mode) because it is "lower" (see Russell, 1962).

The brain mode of description talks about regions in the brain, spiking patterns in networks of neurons, and motor neurons activating muscles. It is used when, for instance, the question is: "How does a person coordinate a hitting movement?". The mind mode of description is a description of reality from "the intentional stance" (Dennett, 1987), and it involves concepts such as intentions, beliefs, desires, the self, and free will. It is used—for now, for a long time to come, and possibly for ever—for answering questions of the kind: "Why does this person hit that other person?". We cannot and need not answer all questions about human behavior using just the brain-mode (or just the mind-mode). Both modes are useful for answering *different* sets of questions, so they can coexist.

Given that both the brain and the mind are valuable modes, what is their relationship to each other? A frequently encountered intuition is that the brain somehow causes the mind. The mind may then be conceived as some sort of "epiphenomenon" of the brain (e.g. Jackson, 1982), an effect of the workings of the brain, but without any real causal role to play in the person's behavior: the mind does not cause muscle fibers to contract, the brain does. That view implies one-way causation from brain to mind. Others—so-called emergent-interactionists—suggest two-way causation: the brain and the mind interact, so that while brain activities give rise to the mind, the mind has "emergent top-down control" over the brain (Sperry, 1991, p. 221). However, just as in the earlier example of atoms and life, the brain mode and the mind mode are two ways of describing one reality, so they do not cause each other, neither one-way nor two-way. Neurons firing do not cause the belief, neurons firing and the belief are the same thing, but described in a different mode.

If relationships between modes are not clearly distinguished from causal relationships, confusions arise. Libet (1985) performed a well-known series of experiments in which he investigated the question whether neural activity caused mental activity or the other way around. Although his research received much criticism, most of it focused on methodological issues. From the modes of description perspective, the question should not even have been asked (see also MacKay, 1985).

Another type of confusion can be found in Hauser (1997). He concludes from his experiments on animal communication (ibid., p. 586):

> These experiments suggest that physiological changes may largely dictate the conditions for call production and suppression, and that an intentional decision need not enter into the equation.

This suggests that the causes for call production and suppression must be mental or physical or possibly a combination of both, and that these options refer to different processes in reality. But because the mental and the physical are just different descriptions of the same phenomenon, talking about deciding between these options makes no sense; an intentional decision in the mind mode must correspond to some physiological changes in the brain mode (even if we don't know what they are).

As a final example of confusion, Sperry (1991, p. 226) vehemently denies Cartesian dualism, but his talk of "emergent interaction" and "downward causation" in mind and brain has led many to reject him as a dualist (e.g. Dennett, 1991a; Vandervert, 1991) or to conclude that even established neuroscience suggests that dualism is true (e.g. Popper & Eccles, 1977, p. 209; Zimbardo, McDermott, Jansz, & Metaal, 1995, p. 90).

If the brain does not cause the mind and the mind does not cause the brain, does this mean that the *mental* event of believing to see a familiar face is not caused by appropriate *physical* stimulation of the retina? Does it mean that a person cannot be said to contract her *physical* muscles because of her own *mental* voluntary decision? No, it doesn't: a causal explanation may contain mode switches, as long as cause and effect do not refer to the same thing at the same time. In this case, description of the different events in the causal chain may be done best in different modes. Stimulation of the retina is an event described in the brain mode which causes, later in time, an event described in the mind mode: believing to see a familiar face. A voluntary decision is an event described in the mind mode, and it causes an event later in time described in the brain mode: motor neurons making muscle fibers contract.

Thus, causal explanations involving voluntary decisions are allowed within the modes of description perspective. Keeping in mind our emphasis on compatibility between modes, that notion may seem incompatible with the basic insight of *determinism*[2] in the physical sciences, but it is not. Concepts from different modes may be *very* different, and the meaning of a concept in one mode does not directly "transfer" to another mode. The notion of voluntary decisions works well in the mind mode of description: people choose freely whether or not to move a limb. In the brain mode or the atom mode there is no such thing as personal choice. But that is a normal state of affairs; in the atom mode there is no such thing as life either. The voluntary decision, described in the mind mode, corresponds to deterministic processes in the brain mode, just as lifelike properties in the life mode correspond to lifeless properties in the atom mode. In other words, demanding compatibility between modes does not mean that concepts from different modes are necessarily mapped onto each other *straightforwardly*. We do not have to be able to identify beliefs with clear-cut brain states in some obvious way in order for us to accept beliefs as good descriptions of cognitive functioning (see Fodor, 1975; Dennett, 1991b). Similarly, we do not have to find voluntary decisions somewhere in the concepts of the brain mode in order for us to accept voluntary decisions as good descriptions of cognitive functioning. Just as in the case of life and atoms, concepts in the mind mode may correspond to collective properties and mechanisms of large groups of neurons—or even the whole person, if we talk about broad concepts such as the self and voluntary decisions (Dennett, 1991a)—that are not easily described from within the brain mode.

The Dynamical Systems Approach to Cognition

One of the most promising and exciting new perspectives in cognitive science is the dynamical systems (DS) approach. It challenges the still dominant idea that the best abstraction of cognitive systems is in terms of "classical", discrete computation and distinct functional modules. Instead of static modules, symbols, propositional logic, and rule-based reasoning, it puts forward the language of dynamical systems, featuring concepts like continuous state spaces, attractors, and bifurcations (e.g. Port & Van Gelder, 1995; Kelso, 1995; Haken, 1983, 1995; Thelen & Smith, 1994; Beer, 1995).

However, there are some confusions in the interpretations of DS models that distract from the important contributions and core issues in the debate. In the DS approach, great emphasis is put on how, in a process called "self-organization", a complex pattern that can be described using the "order parameter" can "emerge" spontaneously when simple units interact. This is a valid and fascinating point, showing that a great deal of regularity and functionality can arise in another, "cheaper", and possibly more robust way than through explicit, detailed instructions that code for the pattern (see Kauffman, 1995).

---

[2]For argument's sake, we ignore quantum mechanics and deterministic chaos. Deterministic should be read here as "subject to the impersonal laws of nature", being the opposite of "controlled by one's free will".

The relationship between the emerged pattern and the simple units is often referred to as "circular causality" (e.g. Haken, 1995; Kelso, 1995; Keijzer, 1997). Kelso (1995) puts it like this (p. 8–9):

> ...the order parameter is created by the cooperation of the individual parts of the system... Conversely, it governs or constrains the behavior of the individual parts. This is a strange kind of circular causality (which is the chicken and which is the egg?), but we will see that it is typical of all self-organizing systems.

From the modes of description perspective, emergence means that a new mode of description becomes applicable (applicable in the sense that it has predictive and explanatory power) to reality where before it was not applicable, and the order parameter is a concept within this new mode of description. It is basically the same situation as the molecular mode becoming applicable in certain cases, where before only the atom mode was applicable. The order parameter description can be said to *summarize* the behavior of the individual parts (Haken, 1995, p. 26):

> While a huge amount of information is required if we have to describe the behavior of the system by means of the variables q_j of the components, an enormous information compression is achieved by means of the order parameter.

This is one of many cases where a single system ("the system") is described using different modes: one mode focusing in detail on the components q_j, and the other mode providing a compressed description. Therefore, at any one point in time, the emerged pattern cannot be said to be caused by the individual parts or vice versa.

However, causal explanations of the behavior of the individual parts or the order parameter over time may contain two mode switches: change in the order parameter is caused by change in the individual parts, and change in the individual parts in turn is caused by change in the order parameter. In that sense, the total causal story could be called "circular". But it should be clear that this, however interesting, is not a truly new type of causality.

If circular causality is perceived as a truly new type of causality it may give rise to questions such as "how are the causal powers divided between the levels?". It is this type of question that leads to problems, as we saw in the discussion of mind and brain. Kelso sees a strong contrast between circular causality and traditional causality (ibid., p. 9):

> What we have here is one of the main conceptual differences between the circularly causal underpinnings of pattern formation in nonequilibrium systems and the linear causality that underlies most of modern physiology and psychology, with its inputs and outputs, stimuli and responses. Some might argue that the concept of feedback closes the loop, as it were, between input and output. This works fine in simple systems that have only two parts to be joined, each of which affects the other. But add a few more parts interlaced together and very quickly it becomes impossible to treat the system in terms of feedback circuits.

This passage may be read in two ways. If the difference between circular causality and "the linear causality that underlies most of modern physiology and psychology" boils down to the difference between "relationships between modes" on the one hand and "regular causation" on the other, then a meaningless comparison is made between these two entirely different types of relationships. If, alternatively, circular causality is different only because the circularly causal story happens to involve two mode switches, then there is no real "conceptual difference": circular causality and "linear" causality are both just regular causality.

In either case, there seems to be no reason to choose between circular causality on the one hand and concepts such as "inputs" and "outputs" on the other. All systems have input and output. Moreover, there seems to be no reason to discuss the problems of "feedback", because those problems, if any, apply to all systems, including the self-organizing ones.

The unwarranted idea of causation between modes is similarly suggested by phrases like "interaction between levels" (Kelso, 1995; Keijzer, 1997; Salthe, 1985), "leaky levels" (Saunders, Kolen, & Pollack, 1994), "organization across levels" (Keijzer, 1997), "interference between levels" (Salthe, 1985; Keijzer, 1997), and "transactions between levels" (Salthe, 1985). It has led some authors to describe a contrast between systems following a "Newtonian trajectory" and systems following a "regular, biological trajectory" (Yates, 1993; Keijzer, 1997), the latter of which allegedly moves up and down the level hierarchy.

Explaining his idea of levels, Salthe (1985, p.47) says:

> We have seen that current interpretations of complexity tend to lead into matters of scale... we might get into the idea of scale by considering maps. Maps can be drawn to different scales, arbitrarily chosen. A large-scale map might show all of North America... we now decrease the scale of the map, so that we have only the New England coastline showing on a map of the same size...

A map is as clear an example of a description as one can think of: each level is construed as a separate description at a different scale. But later he talks about how lower levels and higher levels "constrain" a certain level of interest, the "focal" level (ibid., p. 82–85):

> Like lower-level constraints, they [higher-level constraints] inform and influence focal-level processes without participating in them dynamically. Where do they come from? In general, from the environment of a process... For some diverse concrete examples: *in vitro* synthesis of sugars and amino acids results in a mixture of racemates while *in vivo* synthesis results only in dextro sugars and levo amino acids; sex determination in turtles depends on environmental temperature; the particular stages a fluid system goes through on its way to turbulence depends upon the width, etc., of the observation chamber.

Here Salthe is talking about how differences in a system's environment cause differences in the system's behavior: causal relationships. Causal relationships and relationships

47

between modes (levels) are explicitly confused. This confusion leads to many peculiar conclusions, such as the conclusion that the interaction between levels must be "limited", because "interactional complexity, of course, would destroy neatly organized level structures" (ibid., p. 53; see also ibid., p.74; Keijzer, 1997, p. 22, 210).

Conclusion

In this paper we have argued against the notion of causation between levels or modes. That notion prompts the question whether a mode is cause, effect, or is interacting with other modes. Actual examples from cognitive science were given to illustrate how this results in confusions and mistaken conclusions. In the framework of modes of descriptions, both cause and effect can be described in many different modes; the issue whether a causal relationship is from one mode to another, the other way around, or interactive, does not even arise. The remaining question is: "In which—possibly different—modes can cause and effect best be described?".

References

Arbib, M.A. (1989). *The Metaphorical Brain II: Neural Networks and Beyond.* New York: John Wiley & Sons.

Armstrong, D.M. (1968). *A Materialist Theory of the Mind.* London: Routledge & K. Paul.

Beer, R.D. (1995). A Dynamical Systems Perspective on Agent-Environment Interaction. *Artificial Intelligence, 72,* 173–215.

Bechtel, W., and Richardson, R.C. (1993). *Discovering Complexity: Decomposition and Localization as Strategies in Scientific Research.* Princeton, NJ: Princeton University Press.

Chalmers, A.F. (1982). *What is this Thing Called Science?* Milton Keynes: Open University Press.

Churchland, P.S. (1986). *Neurophilosophy: Toward a Unified Science of Mind-Brain.* Cambridge, MA: MIT Press.

Dennett, D.C. (1987). *The Intentional Stance.* Cambridge, MA: MIT Press.

Dennett, D.C. (1991a). *Consciousness Explained.* Boston: Little, Brown.

Dennett, D.C. (1991b). Real patterns. *Journal of Philosophy, 87,* 27–51.

Descartes, R. (1662). *The Treatise on Man,* translation from French and commentary by Thomas S. Hall, Cambridge, MA: Harvard Univ. Press, 1972.

Fodor, J.A. (1975). *The Language of Thought.* New York: Thomas Y. Crowell.

Haken, H. (1983). *Synergetics, An Introduction.* Berlin: Springer.

Haken, H. (1995). Some basic concepts of synergetics with respect to multistability in perception, phase transitions and formation of meaning. In: Kruse and Stadler (Eds.) *Ambiguity in Mind and Nature.* Berlin: Springer Verlag.

Hauser, M.D. (1995). *The Evolution of Communication.* Cambridge, MA: MIT Press.

Hobbes, T. (1651). *Leviathan.* London: Crooke.

Hofstadter, D. (1979). *Godel, Escher, Bach.* London: Harvester Press.

Hofstadter, D. (1981) Reflections on: Ant Fugue. In: D. Hofstadter en D.C. Dennett (Eds.) *The Mind's I.* New York: Basic Books.

Jackson, F. (1982). Epiphenomenal qualia. *Philosophical Quarterly, 32,* 127–136.

James, W. (1907). *Pragmatism: A New Name for Some Old Ways of Thinking.* New York: Longmans, Green and Company.

Kauffman, S.A. (1995). *At Home in the Universe.* London: Viking.

Keijzer, F.A. (1997). *The Generation of Behavior.* PhD Thesis, Dept. of Psychology, Leiden University.

Kelso, J.A.S. (1995). *Dynamic Patterns.* Cambridge, MA: MIT Press.

Koestler, A. (1967). *The Ghost in the Machine.* New York: Macmillan.

Libet, B. (1985). Unconscious cerebral initiative and the role of conscious will in voluntary action. *Behavioral and Brain Sciences, 8,* 529–566.

MacKay, D.M. (1985). Do we control our brains. *Behavioral and Brain Sciences, 8,* 546–546.

Marr, D. (1982). *Vision.* New York: Freeman.

Maturana H.M., and Varela J.V. (1984). *The Tree of Knowledge,* Bern: Sherz Verlag.

Newell, A. (1990). *Unified theories of cognition.* Cambridge, MA: Harvard University Press.

Popper, K.R., and Eccles, J.C. (1977). *The self and its brain.* New York: Springer International.

Port, R.F. and Van Gelder T. (Eds.) (1995). *Mind as Motion.* Cambridge, MA: MIT Press.

Pylyshyn, Z.W. (1984). *Computation and Cognition.* Cambridge, MA: MIT Press.

Russell, B.A.W. (1962). *The basic writings of Bertrand Russell.* London: Allen and Unwin.

Salthe, S.N. (1985). *Evolving Hierarchical Systems.* New York: Colombia University Press.

Saunders, G., Kolen, J.F. and Pollack, J.B. (1994). The importance of leaky levels for behavior based AI. *Proceedings of the Third International Conference on Simulation of Adaptive Behavior.*

Sperry, R.W. (1991). In defence of mentalism and emergent interaction. *The Journal of Mind and Behavior, 12,* 221–246.

Thelen, E. and Smith, L.B. (1994). *A dynamic systems approach to the development of cognition and action.* Cambridge, MA: MIT Press.

Vandervert, L. (1991). A measurable and testable brain-based emergent interactionism: An alternative to Sperry's mentalist emergent interactionism. *The Journal of Mind and Behavior, 12,* 201–220.

Yates, F.E. (1993). The Logic of Life. In: C.A.R. Boyd and D. Noble (Eds.) *The Challenge of Integrative Physiology.* Oxford: Oxford University Press.

Zimbardo, P., McDermott, M., Jansz, J., and Metaal. N. (1995). *Psychology: A European Text,* London: Harper-Collins.

The Effects of Belief and Logic in Syllogistic Reasoning: Evidence from an Eye-Tracking Analysis

Linden J. Ball (l.j.ball@derby.ac.uk)
Cognitive and Behavioral Sciences Research Group,
Institute of Behavioural Sciences, University of Derby,
Mickleover, Derby, DE3 5GX, UK

Jeremy D. Quayle (j.d.quayle@derby.ac.uk)
Cognitive and Behavioral Sciences Research Group,
Institute of Behavioural Sciences, University of Derby,
Mickleover, Derby, DE3 5GX, UK

Abstract

Studies of syllogistic reasoning report a strong non-logical tendency to endorse more believable conclusions than unbelievable conclusions. This *belief bias* effect is found to be stronger with invalid arguments than with valid ones. An experiment is reported in which participants' eye-movements were recorded in order to gain insight into the nature and time course of the reasoning processes associated with experimental manipulations of logical validity and believability. Results are considered in relation to predictions derivable from contemporary accounts of belief bias. The logical status of conclusions was found to influence the duration of gazes, supporting the view that invalid conclusions are more demanding to evaluate than valid ones and the idea that a valid-invalid processing distinction underpins the interaction that is observed between logic and belief. Predictions concerning effects of believability upon gaze behaviour that were derivable from the *mental models* account (e.g., Oakhill & Johnson-Laird, 1985) gained little support. The paper argues for the value of eye-movement analyses in reasoning research as an important adjunct to existing process-tracing techniques.

Introduction

The syllogism is a deductive reasoning problem comprising two premises and a conclusion (see example given below). The premises feature three terms: two end terms (one in each premise) and a middle term (featured in both premises). A logically valid conclusion to a syllogism is a statement that describes the relationship between the classes of items or individuals referred to by the end terms in a way that is necessarily true. Statements that are simply consistent with the premises but not necessitated by them are invalid. It should be noted that a logical argument is valid by virtue of its *form*, and not because of its content. That is, the actual words or other symbols that could be used as terms within a syllogism are irrelevant when considering validity.

> Some artists are beekeepers
> No beekeepers are carpenters
> *Therefore,* Some artists are not carpenters

Participants in syllogistic reasoning experiments are required either to produce their own logically valid conclusions from given premises, or to evaluate the validity of a presented conclusion (or conclusions) from given premises. It has been found that participants' responses in such experiments vary systematically according to three main factors. Two factors relate to the logical form of the syllogism and are termed *figure* and *mood*. Figure refers to the arrangement of the terms within the premises. There are four possible figures: A-B, B-C; A-B, C-B; B-A, C-B; and B-A, B-C (where A refers to the end-term in the first premise, C refers to the end-term in the second premise, and B refers to the middle term). Mood refers to the different combinations of logical quantifiers featured within the premises and conclusion. Four different quantifiers are used in standard English language syllogisms. These are commonly referred to by letters of the alphabet: A = *all*, E = *no*, I = *some*, and O = *some . . . are not*. The syllogism in the above example, therefore, can be said to be in the A-B, B-C figure, and in the IEO mood. Since there are four different figures and each of the two premises can feature one of four standard quantifiers, there or 64 standard syllogisms -- only 27 of these, however, yield valid conclusions. Studies have found that different combinations of figure and mood have a marked effect on reasoning performance. Indeed, some forms of syllogism are so easy that nearly all participants are able to give correct responses. Others are so difficult that few individuals respond without error (e.g., see Johnson-Laird & Bara, 1984; Johnson-Laird & Byrne, 1991).

In addition to the effects of figure and mood, participants' prior knowledge and beliefs have been found to bias responses. There are three basic findings that derive from studies of *belief bias* in syllogistic reasoning in which the validity of logical arguments has been manipulated alongside the prior believability of presented conclusions (see Garnham & Oakhill, 1994). First, believable conclusions such as "Some addictive things are not cigarettes" are more readily endorsed than unbelievable ones such as "Some millionaires are not rich people" (these examples are taken from Evans, Barston and Pollard, 1983). Second, logically valid conclusions are more readily endorsed than invalid ones. Third, there is an interaction between logic and belief such that the effects of believability are stronger on invalid problems than valid ones.

Few studies have directly attempted to investigate the processes underlying belief bias effects in syllogistic reasoning. One notable exception is the study carried out by Evans et al. (1983) who recorded and analysed the think-aloud protocols of participants evaluating the validity of believable and unbelievable presented conclusions. The majority of the protocols were classifiable under one of three headings: (a) *conclusion-only* protocols in which participants referred to a syllogism's conclusion without mentioning the premises; (b) *premises-to-conclusion* protocols in which participants referred to the premises of the syllogism and subsequently to the conclusion; and (c) *conclusion-to-premises* protocols in which participants referred to the conclusion and subsequently to the premises. A clear relationship was found between the type of protocol and the level of belief bias observed: belief bias was found to be strongest on problems where conclusion-only protocols were observed and weakest on problems where premises-to-conclusion protocols were observed. On this basis, the verbal protocols were taken to indicate distinct reasoning strategies (cf. Evans, Newstead & Byrne, 1993).

The analysis of concurrent verbal protocols is an established method in problem solving research (see Ericsson & Simon, 1993) and, to a lesser extent, in reasoning research (e.g., Beattie & Baron, 1988). There has, however, been a longstanding debate over the nature of concurrent verbal protocols and what they can reveal about cognitive processes and strategies. On the latter issue, Ericsson and Simon acknowledge that whilst many forms of verbalisations may not impact upon the *structure* of reasoning processes, such verbalisation may impact upon the *completeness* of the reports produced. This is because task-oriented processes will tend to have priority over verbalistion processes when such processes are in competition. As a consequence, participants may at times stop verbalising when the cognitive demands of the primary task are high, thus producing incomplete reports of the products of reasoning processes. Concerns over the completeness of verbal reports during reasoning encouraged us to explore an alternative process-tracing technique to investigate the processes underlying belief bias in syllogistic reasoning. The technique chosen was eye-movement analysis.

Eye-movement analysis is a technique that has been used extensively in reading research. Experimenters in this field assume a close association between patterns of eye movements and the thought processes underlying the understanding of written text (cf. Liversedge, Paterson & Pickering, 1998). That is, the position of a visual fixation is taken to indicate the piece of text that is currently being processed by the reader, and the length of a fixation or a gaze (which may include two or more fixations) is taken to indicate the ease with which a piece of text is processed. Similarly, the number of return fixations (or regressions) provides a further index of understanding (i.e., participants may need to return to an item of text in order to resolve ambiguities in meaning). The validity of a proposed association between thought processes and eye movements is supported by a large body of research which shows that the linguistic properties of text have a direct influence on readers' fixations and gazes (e.g., Rayner & Pollatsek, 1989).

Evidence from studies of reading would suggest that monitoring participants' eye movements whilst they are engaged in a text-based reasoning task could provide insights into the nature and organisation of the underlying processes associated with different types of syllogism. Indeed, eye-movement analysis may afford distinct advantages as a tracing technique over think aloud protocol analysis. Since eye movements are typically quite fast and spontaneous, an analysis of eye movements may provide more detailed records of the sequence and organisation of processing than think aloud protocols. Furthermore, since eye movements do not place additional processing demands on working memory in the manner that verbalisations might, working memory is left free for primary task-oriented processing, and eye movements associated with task-oriented processing should not cease when the processing demands of the primary task are high. Indeed, eye-movement investigations of reading would suggest quite the reverse, that is, when the processing demands of the primary task are high, lengthier fixations or a greater number of fixations upon the relevant text may be recorded.

The question remains, however, what can an analysis of participants' eye movements during a syllogistic reasoning task tell us about the cognitive processes underlying the task? This question can be addressed in the following way. Since working memory is the cognitive system within which many aspects of complex tasks such as deductive reasoning are carried out (cf. Gilhooly, 1998; Johnson-Laird & Byrne, 1991), and limited capacity and fragility of storage are key characteristics of this system, evidence from studies of reading would suggest that participants will need to read and re-read parts of the problem information which they are processing in order to construct, refresh or flesh out their mental representations when necessary. Thus, the high processing demands associated with some syllogisms may cause participants to gaze for longer upon elements of the problem information than with other less demanding problems. Evidence for the application of proposed heuristics which motivate reasoners to scrutinize the logical validity of some arguments more than others may also be detected in eye movements, such that longer gazes upon problem information may be evident with syllogisms where consideration is given to the logical argument than with other syllogisms where less logical scrutiny is applied. Based on these assumptions, predictions can be derived from contemporary theories of belief bias.

The mental models account of syllogistic reasoning (e.g., Johnson-Laird & Byrne, 1991) assumes that people begin reasoning by constructing a mental model in which a minimal amount of information concerning the logical relationships between the terms within the premises is made explicit. If a putative conclusion is true in this initial model, then it is tested against *fleshed out* models which make explicit more of the information within the premises. If a conclusion is found not to be consistent with a mental model, then it is rejected, otherwise it is accepted. Some syllogisms (termed *one-model* problems) are said to be relatively easy because they require the construction of a single mental model, and thus place a minimal load on working memory. Others are said to be more difficult because they place less manageable loads on working memory, requiring the construction of multiple models. The idea that the difficulty of syllogisms is closely associated with the number of models that need to be constructed has received strong support from empirical studies (e.g., Johnson-Laird & Bara, 1984; Bara, Bucciarelli, & Johnson-Laird, 1995).

The mental models account of belief bias (e.g., Oakhill & Johnson-Laird, 1985; Oakhill, Johnson-Laird & Garnham, 1989) assumes that prior beliefs determine whether participants will flesh out an initial model of the premises when testing the logical validity of a putative conclusion. Conclusions that are true in an initial model are accepted if they are believable, but tested against alternative and potentially falsifying models if they are unbelievable. In this way, the logic by belief interaction is explained. This account seems to predict that there will be an overall effect of belief on gazes, since greater consideration is given to the information within the premises when evaluating a conclusion that is unbelievable than one that is believable.

The mental models account assumes that two stages of mental models construction take place with syllogisms that lead to unbelievable conclusions: (a) a *pre-conclusion-gaze* stage, and (b) a *post-conclusion-gaze* stage. Only one stage of mental models construction, however, would occur with syllogisms leading to believable conclusions: a pre-conclusion-gaze stage. If two clear stages such as these can be identified from eye-movement analyses, then an interaction between reasoning stage (pre-conclusion / post-conclusion-gaze) and believability might be expected. This interaction would be such that the effect of belief upon gazes - whether measured in terms of duration or number of gazes - would be greater in the post-conclusion-gaze stage than in the pre-conclusion-gaze stage.

Although much support has been claimed for the mental models account of belief bias, it has been criticised for failing to explain some key findings of belief bias research (e.g., Evans et al., 1993). For example, the account predicts that no logic by belief interaction should be observed with one-model syllogisms, since valid conclusions will be accepted and invalid conclusions rejected with such problems irrespective of their believability status. Although support for this prediction was found by Newstead et al. (1992), without ad hoc modifications (e.g., the addition of a *conclusion filter*) the account has difficulty in explaining the unpredicted finding of an effect of belief with one-model syllogisms (see Oakhill, et al., 1989).

In an attempt to explain belief bias findings, Quayle and Ball (1997) have proposed an account of belief bias based around the notion of *metacognitive uncertainty*. Set within the general framework of the mental models theory, this account assumes that the tendency to respond in accordance with belief is determined by the relative loads placed on limited working memory resources by different types of syllogism. In accordance with the mental models approach, it is assumed that some syllogisms place greater and less manageable loads on working memory than others, and that when working memory is overloaded participants are no longer able to test the truth of putative conclusions against models of the premises. In this instance, it is argued that

being uncertain of a conclusion's logical status, participants fall back on a belief-based response as a *second-best* option.

The metacognitive uncertainty account's explanation of the logic by belief interaction hinges on the observation that with valid multiple-model syllogisms, correct responses can be given after the construction of a single mental model, whilst invalid multiple-model syllogisms require the consideration of more than one model (cf. Garnham, 1993; Hardman and Payne, 1995). In order to illustrate this idea, using Johnson-Laird & Byrne's (1991) notation the mental models that would be constructed for the syllogism given as an example earlier are shown below.

a	[b]		a	[b]		a	[b]	
a	[b]		a	[b]		a	[b]	
		[c]	a		[c]	a		[c]
		[c]			[c]	a		[c]
. . .			. . .			. . .		

The As, Bs and Cs in each model represent the classes "artists", "beekeepers" and "carpenters" respectively. Each horizontal line represents the relationship between the three terms (the number of individual class members represented in each model is arbitrary). For example, the top two lines of the model on the far right show that there are As that are Bs that are not Cs, whilst the bottom two lines depict As that are not Bs but are Cs. The square brackets denote *exhaustive* representation. The class of Bs in each model is shown to be exhaustively represented in relation to the class of As -- that is, there can be no Bs that are not As. The three dots below each model indicate that there is premise information not yet made explicit in the model.

The valid conclusion to this three-model syllogism is "Some A are not C". As the second term in this conclusion (the C term) is represented exhaustively in relation to the B term in the initial mental model (on the far left), the relationship between the end terms in the model that shows the conclusion to be true (the top two lines) remains unchanged when the model is fleshed out. Hence, the consideration of more than one model is unnecessary. With valid problems, therefore, participants may feel certain of the correctness of their responses after the construction of a single model. Now let's consider the indeterminately invalid conclusion "Some C are not A". The bottom two lines of the initial model show this conclusion to be true. However, since the second term in the conclusion (the A term) is not represented exhaustively in this model, it is necessary to flesh out the model to be certain of the conclusion's logical status.

In the present study, and in earlier studies of belief bias that have employed a conclusion evaluation methodology, both the valid and the invalid problems that were used had determinate premises - that is, ones that yield a valid conclusion. So long as figure is kept constant, the use of such materials means that there should be little difference in the gaze behaviour between valid and invalid syllogisms prior to the point at which the conclusion is first gazed upon. However, the idea of a valid-invalid processing distinction suggests that after viewing the conclusion participants will be likely to gaze upon the premises for longer with invalid problems than with valid ones. For this reason, the metacognitive uncertainty account predicts that an interaction between reasoning stage (pre-conclusion-gaze/post-conclusion-gaze) and logic will be evident in participants' gazes.

The main aim of this study was to test the different eye-movement predictions made by the two theories described above, and in this way, to arbitrate between these accounts of belief bias. To this end, participants' eye movements were recorded using an eye-tracking device whilst they evaluated logical arguments whose presented conclusions varied in validity and prior believability.

Method

Participants

Twenty undergraduate psychology students at the University of Derby took part in the experiment. None of the participants had taken formal instruction in logic and all were tested individually.

Materials

Whilst the participants carried out the reasoning task their eye movements were monitored using an ASL (Applied Science Laboratories) 4200R Eye Tracking System. This eye-tracking device operates on the 'Double Purkinje image' method. Measurements are taken of the relative changes in angle of infra-red light reflections from the front of the cornea and rear of the lens. This method allows for a degree of free head movement whilst monitoring is taking place (Megaw, 1990).

Two forms of three-model syllogism (e.g., as classified by Johnson-Laird and Byrne, 1991) were used. One form was valid and one form was invalid (i.e., the conclusions did not follow logically from the premises). Both the valid and invalid syllogisms were in the same A-B, B-C figure with conclusions of the form A-C. The valid problem was in the IEO mood, whilst the invalid problem was in the EIO mood. The invalid conclusions were *indeterminately* invalid (i.e., the conclusions were consistent with the premises but they were not *necessitated* by them).

A set of potential conclusions which were false by definition, (e.g., "Some kings are not men") was chosen, together with a set of believable conclusions, for example, "Some animals are not cats". The conclusions were devised so as to appear believable when the terms were presented in one order, but unbelievable when the terms were reversed. In order to assess believability, the potential conclusions were pre-rated by a group of 30 participants on a seven point scale ranging from -3 (*totally unbelievable*) to +3 (*totally believable*). Those conclusions which received the most extreme and consistent ratings were used in this study. Half of the valid and invalid syllogisms were presented with conclusions which were believable and half were presented with conclusions which were unbelievable by definition. In addition to the four types of three-model syllogism there were three one-model filler syllogisms which were used to distract the participants from the forms of the syllogisms of interest.

The syllogisms were presented individually on display cards in *times new roman* font size 36 (lower case = 6mm high, upper case = 8.5mm). The two premises were printed at the top of each card (the distance between the bottom of the letters in the first premise and the top of the lower case letters in the second premise was 23.5mm), and the conclusions were printed approximately two thirds of the way down the page (the distance between the bottom of the letters in the second premise and the top of the lower case letters in the conclusions was 112mm). The response words "yes" and "no" were printed on the left and right sides at the bottom of each page.

Design

A within participants design was used, with all of the participants receiving the four three-model syllogisms together with the three filler syllogisms. These were preceded by 3 practice syllogisms (10 syllogisms in total). The order of the problem types was varied using a four by four balanced Latin square design; with the restriction that the filler items appeared in the same position in each booklet: in 2nd, 4th and 6th places. The thematic contents of the syllogisms were rotated over the different problem types, producing four different sets of materials. The four sets of materials were distributed evenly and randomly amongst the participants (i.e., five participants per set of materials). At the beginning of each booklet three practice syllogisms were given in order to familiarise the participants with the task. The participants were unaware that these were practice syllogisms.

Procedure

Each participant was seated in a chair with a card display rack in front of them. The distance between the display rack and the participants' eyes was approximately 60cm. Once that the eye-tracking device had been calibrated, the following instructions were presented:

"This is an experiment to test people's reasoning ability. You will be given ten problems. You will be shown two statements and you are asked if certain conclusions (given below the statements) may be logically deduced from them. You should answer this question on the assumption that the two statements are, in fact, true. If, and only if, you judge that the conclusion necessarily follows from the statements, you should point to the word "yes", otherwise the word "no".

Please take your time and be sure that you have the right answer before giving your response. After you have given your response the next problem will be presented. Thank you very much for participating."

The 10 problem cards were placed in an upright pile upon the card display rack. After the participant had indicated their response to a syllogism (by pointing to either the "yes" or the "no") that problem card was removed to reveal the next problem card. The participants were allowed as much time as they required to complete the reasoning task.

A video camera mounted on the ceiling above and behind the participant recorded an image of each problem card as well as the participant's pointing responses. Horizontal and vertical eye movements were recorded using the infra-red tracking device. A small black square indicating eye movements was superimposed onto the video image by the tracking device. A digital timer (displaying hundredths of seconds) was also superimposed onto the bottom right hand corner of the video image.

Results

Conclusion Acceptances

An analysis of the conclusion-acceptance data revealed a significant effect of Belief, $p < .01$, one-tailed sign test, with participants accepting more believable conclusions than unbelievable conclusions (70% - 40% = 30%). The effect of Logic was in the standard direction, with more valid arguments accepted than invalid ones (63% - 48% = 15%), but fell short significance. There was a significant interaction between Logic and Belief, $p < .01$, one-tailed sign test, such that the effect of belief was greater on syllogisms leading to invalid conclusions than on those leading to valid conclusions.

Eye-movement Analysis

For the eye-movement analyses four fixation areas were identified: premises; conclusion; yes-no response words; and blank areas of display cards. The eye-movement video recordings were analysed frame by frame (since there were 25 frames per second, the shortest fixation that could be identified was 40 msec in duration) in order to establish the time duration of each gaze. A gaze typically included more than one separate fixation, and was defined as any time spent viewing a fixation area that was greater than 200 msec in duration. Gazes on the 'yes' and 'no' response words and blank areas of problem cards are not included in the following analyses. Due to refractive or pathological eye defects the eye-movement video recordings for three out of the 20 participants were uninterpretable. The recordings made for these participants have, therefore, been excluded from the following eye-movement analyses

In order to establish whether viewing the conclusion had an effect on subsequent premise gazes, the premise gaze times data for each type of syllogism were divided into pre-conclusion-viewing and post-conclusion-viewing times. These data were subjected to a multi-factorial analysis of variance. The factors were Stage (two levels: pre-conclusion-viewing / post-conclusion-viewing), Logic and Believability. None of the three factors was significant. However, Stage interacted with Logic, $F(1,16) = 5.06$, $p < .05$, such that with the invalid problems participants gazed upon the premises for a greater amount of time after viewing the conclusion than before. With the valid problem, however, this effect of stage was somewhat smaller and in the opposite direction. Other interactions were not significant.

Discussion

This study assumed that the lengths of gazes identified from an analysis of participants' eye movements would provide some direct insight into the processes associated with the experimental manipulations of logical validity and believability. The finding of an effect of belief together with an interaction between logic and belief in the conclusion acceptances data provided a good opportunity to investigate the processes and strategies underlying these standard effects.

The mental models account claims that there will be two stages of mental models construction when multiple-model syllogisms are presented with *unbelievable* conclusions: a pre-conclusion-gaze stage in which participants construct an initial model of the premises; and a post-conclusion-gaze stage in which participants flesh out the initial model in a search for counter-examples which might show the conclusion to be false. On the other hand, when syllogisms are presented with *believable* conclusions the mental models account claims there will be only one stage of mental models construction: a pre-conclusion-gaze stage. The mental models account, therefore, appears to predict that there should be an interaction between stage (pre- / post-conclusion-gaze) and believability. No such interaction was evident in the data. The finding of an interaction between stage and logic, however, is consistent with the idea of a valid-invalid processing distinction that lies at the heart of the metacognitive uncertainty account. In the present study, and in earlier studies of belief bias that have employed a conclusion evaluation methodology, both the valid and the invalid problems used had determinate premises - that is, ones that yield a valid conclusion. So long as figure is kept constant, the use of such materials means that there should be little difference in gaze behaviour between valid and invalid syllogisms prior to the point at which the conclusion is viewed. However, the idea of a valid-invalid processing distinction incorporated within the metacognitive uncertainty account suggests that after viewing a presented conclusion participants will gaze upon the premises for longer with invalid problems than with valid ones.

Whilst the pattern of eye-movement behaviour observed in the present study was predicted by the metacognitive uncertainty account, the conclusion-centred reasoning strategies identified from verbal protocol analysis by Evans et al. would predict quite a different pattern of eye movements. This observation supports the claim that think aloud protocols given during syllogistic reasoning tasks may be incomplete. We suspect that the processes involved in verbalisation may compete with task-oriented processing such that participants may stop verbalising at times when the processing demands of the primary reasoning task are high. Evans et al.'s identification of distinct reasoning strategies in studies of belief bias, therefore, may be a methodological artifact attributable to the employment of a think-aloud verbal protocol analysis methodology.

In conclusion, the study has demonstrated how an analysis of participants' eye-movements during reasoning can provide direct and detailed insights into the nature and time course of reasoning processes, which in turn can allow the researcher to arbitrate between conflicting accounts of reasoning phenomena. We maintain that the employment of eye-movement analysis in other reasoning paradigms alongside existing methodologies may enable researchers to gain a more detailed understanding of human deductive competence and performance than currently exists.

Acknowledgments

We thank Kevin Purdy, Mark Mugglestone, Tim Horberry and Alastair Gale of the Applied Vision Research Unit at the University of Derby for their technical support, and Kevin Paterson for helpful comments on an earlier version of this paper.

References

Bara, B.G., Bucciarelli, M., & Johnson-Laird, P.N. (1995). Development of syllogistic reasoning. *The American Journal of Psychology, 108,* 157-193.

Beattie, J., & Baron, J. (1988). Confirmation and matching biases in hypothesis testing. *Quarterly Journal of Experimental Psychology, 40A,* 269-297.

Ericsson, K.A., & Simon, H.A. (1993*). Protocol analysis: Verbal reports as data (2nd ed.).* Cambridge, MA: MIT Press.

Evans, J.St.B.T., Barston, J.L., & Pollard, P. (1983). On the conflict between logic and belief in syllogistic reasoning. *Memory and Cognition, 11,* 295-306.

Evans, J.St.B.T. Newstead, S.E., & Byrne, R.M.J. (1993). *Human reasoning: The psychology of deduction.* Hove: Lawrence Erlbaum Associates.

Garnham, A. (1993). A number of questions about a question of number. *Behavioural and Brain Sciences, 16,* 350-351.

Garnham, A., & Oakhill, J. (1994). *Thinking and Reasoning.* Oxford: Basil Blackwell.

Gilhooly, K.J. (1998). Working memory, strategies, and reasoning tasks. In R.H. Logie & K.J. Gilhooly (Eds.), *Working memory and thinking.* Hove: Psychology Press.

Hardman, D.K., & Payne, S.J. (1995). Problem difficulty and response format in syllogistic reasoning. *The Quarterly Journal of Experimental Psychology, 48A,* 945-975.

Johnson-Laird, P.N., & Bara, B.G. (1984). Syllogistic inference. *Cognition, 16,* 1-62.

Johnson-Laird, P.N., & Byrne, R.M.J. (1991). *Deduction.* Hove: Lawrence Erlbaum Associates.

Liversedge, S.P., Paterson, K.B., & Pickering, M.J. (1998). Eye movements and measures of reading time. In G. Underwood (Ed.), *Eye movement control.* Oxford: Elsevier Science.

Megaw, T. (1990). The definition and measurement of visual fatigue. In J.R. Wilson & E.N. Corlett (Eds.), *Evaluation of Human Work: A Practical Ergonomic Methodology.* London: Taylor & Francis.

Newstead, S.E. Pollard, P. Evans, J.St.B.T., & Allen, J.L. (1992). The source of belief bias effects in syllogistic reasoning. *Cognition, 45,* 257-284.

Oakhill, J., & Johnson-Laird, P.N. (1985). The effect of belief on the spontaneous production of syllogistic conclusions. *Quarterly Journal of Experimental Psychology, 37A,* 553-570.

Oakhill, J., Johnson-Laird, P.N., & Garnham, A. (1989). Believability and syllogistic reasoning. *Cognition, 31,* 117-140.

Quayle, J.D., & Ball, L.J. (1997). Subjective confidence and the belief bias effect in syllogistic reasoning. *Proceedings of the Nineteenth Annual Conference of the Cognitive Science Society.* (pp. 626-631). Mahwah, New Jersey: Lawrence Erlbaum Associates.

Rayner, K., & Pollatsek, A. (1989). *The psychology of reading.* New Jersey: Englewood Cliffs.

Simple and Complex Speech Acts: What Makes the Difference within a Developmental Perspective

Bruno G. Bara (bara@psych.unito.it)
Francesca M. Bosco (bosco@psych.unito.it)
Monica Bucciarelli (monica@psych.unito.it)

Centro di Scienza Cognitiva, Università di Torino
via Lagrange, 3 - 10123 Torino, Italy

Abstract

In the linguistic psychological literature, there is a classical distinction between direct and indirect speech acts. In particular, some theories claim that the latter are more difficult to produce and comprehend than the former. We propose to abandon such a distinction in favour of a novel one between simple and complex speech acts. This distinction applies to any kind of pragmatic phenomena, from standard speech acts to non standard ones, like irony and deceit. Our proposal is based on the types of mental representations and mental operations involved in speech acts production and comprehension.

1. Introduction

In the classical philosophy of language, a well-known distinction is drawn between direct and indirect speech acts. Searle (1975) claims that to comprehend an indirect speech act means to realize that an illocutionary act is (indirectly) being performed *via* the execution of a different, literal illocutionary act. Direct speech acts are instead those where a speaker utters a sentence and means exactly and literally what she is saying, as in:

[1] What time is it?

In indirect speech acts the speaker communicates to the hearer more than she is actually saying, by relying on the background information they mutually share, and on the hearer's general powers of rationality and inference. Examples of indirect speech acts are:

[2] a. Can you please tell me what time it is?
    b. Do you mind telling me what time it is ?
    c. I wonder if you'd be so kind as to tell me what time it is.
    d. I don't have my watch.

Searle claims that understanding [1] is straightforward, that is, does not require inferences, while understanding [2] relies on some kind of common knowledge. However, the length of the inferential path is not the same for each utterance in [2]. For example, [2d] requires a greater number of inferences than [2a].

Searle claims that the primary illocutionary force of an indirect speech act is derived from the literal one *via* a series of inferential steps. The hearer's inferential process is triggered by the assumption that the speaker is following the Principle of Cooperation (Grice, 1978), together with the evidence of an inconsistency between the utterance and the context of pronunciation. The hearer tries first to interpret the utterance literally, and only after the failure of this attempt, due to the irrelevance of the literal meaning, looks for a different meaning, which conveys the primary illocutionary force. According to the classical theory, an indirect speech act is intrinsically harder to comprehend than a direct one.

Some authors have criticized this position (cf. Clark, 1979; Recanati, 1995; Sperber & Wilson, 1986). In particular, Gibbs (1994) states that indirect speech acts with a conventionalized meaning are simpler to understand than nonconventional ones; the context specifies the necessity of using a conventional indirect and thus helps the hearer to understand the intended meaning more quickly. Gibbs (1986) claims that a speaker can use an indirect act when she thinks that there might be obstacles against the request she intends to formulate: for example, when the speaker does not know whether the hearer owns the object she desires, she can use a conventional indirect request. Gibbs suggests that the partner infers the meaning of a conventional indirect speech act *via* a habitual shortcut that facilitates its comprehension.

An alternative proposal is the theory of Cognitive Pragmatics by Airenti, Bara and Colombetti (1993a).

2. Cognitive Pragmatics Theory

A major assumption of Cognitive Pragmatics is that intentional communication requires behavioral cooperation between two agents; this means that when two agents communicate they are acting on the basis of a plan that is at least partially shared. Airenti, Bara and Colombetti (1984) call this plan a *behavior game*. Each communicative action performed by the agents realizes the moves of the behavior game they are playing. The meaning of a communicative act (either linguistic or extra-linguistic or a mix of the two) is fully understood only when it is clear what move of what behavior game it realizes.

Thus, the comprehension of any kind of speech act depends on the comprehension of the behavioral game bid by the actor. Unless a communicative failure occurs, each participant in a dialogue interprets the utterances of the interlocutors on a ground she gives as shared with them. The only distinction that can be drawn concern the chain of inferences required to pass from the utterance to the game it refers to. Direct and conventional indirect speech acts immediately make reference to the game, and thus we shall call them *simple* speech acts. On the contrary, non conventional indirect speech acts can be referred to as *complex* in that they require a chain of inferential steps because the specific behavior game of which they are a move is not immediately identifiable. For example, to understand [1] it is sufficient for the partner to refer to the game [GIVE-INFORMATION]. In order to understand [2d], a more complex inferential process is necessary: the partner, needs to share with the speaker the beliefs that if one has not a watch, she cannot know what time it is, and that when somebody looks at her watch it is because she wants to know the time. Only then, the partner can attribute to the utterance the value of a move of the game [GIVE-INFORMATION]. Thus, if the problem is how to access the game, the distinction between direct and indirect speech acts is inexistent. It is the complexity of the inferential steps necessary to refer the utterance to the game bid by the actor to account for the difficulty of speech acts comprehension.

Airenti, *et al.* (1993a) consider as standard the path of communication where default rules of inference are used to understand each other's mental states. Default rules are always valid unless their consequent is explicitly denied (cf. Reiter, 1980). Thus, according to Cognitive Pragmatics' proposal, the meaning of direct and indirect speech acts can be straightforward inferred by referring them to the game bid by the speaker, *via* default rules of inference. Non standard communication, on the contrary, involves comprehension and production of speech acts *via* the block of default rules and the occurrence of more complex inferential processes: an examples are ironic and deceitful speech acts. Cognitive Pragmatics claims that, in order to refer a non standard speech act to the game bid by the speaker, the partner has to draw a chain of inferences which can not be based on default rules.

3. Simple and Complex Speech Acts

From an empirical point of view, Searle's indirect speech acts, as well as Gibbs' indirect speech acts without conventional use, and Airenti *et al.*'s complex speech acts, are equivalent. Nonetheless, in our view the latter definition - contrary to those of Searle and Gibbs - applies to any kind of communicative act. Furthermore, in our view the comprehension of conventional speech acts does not rely on shortcuts - as Gibbs suggests -, but on a game that is immediately accessible. In other words, the simple/complex distinction is grounded in the sort of mental representations

and mental processes necessary to refer the act to the game bid by the actor.

In order to try to falsify the mentioned theories it is useful to take developmental pragmatics into account. Indeed, adult performance is almost always correct, and it does not allow to test predictions about difference in difficulty of comprehension. On the contrary, the predicted mistakes of children's performance can be considered as evidence in favour of one of the theories. The developmental perspective in Cognitive Science reminds us that to reach a better comprehension of the mind's functionning we have to consider not only the adult's steady states, but also the development from childhood, trougth adolescence, to mental maturation (Bara, 1995).

If Searle's theory is correct, then indirect speech acts would always be harder to deal with than direct ones (indirect > direct). If Gibbs is correct, then non-conventional indirect speech acts should be harder than the conventional indirect speech acts, which in turn should be equivalent to direct ones (non conventional indirect > conventional indirect or direct). If Airenti *et al.* are correct, then complex indirect speech acts should be more difficult than simple speech acts, which may indifferently be either direct or indirect acts (complex > simple: indirect ≡ direct).

In support of Searle's proposal, Garvey (1984) finds that children under 3 years have some difficulties in understanding conventional indirect requests made by an adult. The explanation she gives is that such requests are ambiguous in that, as claimed by Searle, they have simultaneously a literal meaning and a directive implicit force. In particular, Garvey reports an interaction between a mother and her 32-month-old child. The mother points at a picture in a book and asks the child, 'You see what this little boy is doing?' As the child fails to volunteer the information about what he sees and limits his response to 'Yeah', Garvey concludes that indirection is difficult for children.

In support of Gibbs' proposal, Shatz (1978) observes children between 1;7 and 2;4 playing with their mothers at home and finding that they understood conventional indirect requests like 'Can you shut the door?' or 'Are there any more suitcases?'. Shatz concludes that very young children are able to map the language they hear on to the familiar non-linguistic world of action and objects.

In line with Airenti *et al*, Reeder (1980) finds, that children between 2,6-3 comprehend that, in an adequate context, utterances like 'I want you to do that' or 'Would you mind do that' have the same illocutionary force (see also Bernicot & Legros, 1987). Also, Becker (1990) and Ervin-Tripp and Gordon (1986) find evidence that 2,6 year olds already produce different kinds of indirect speech acts. Finally, Bara and Bucciarelli (1998) show that 2;6-3 year old easily comprehend simple directives (conventional indirects) like, 'Would you like to sit down?'. On the contrary, they have difficulties with complex directives (non-conventional indirects) like, to understand that answering 'It's raining' to the proposal 'Let's go out and play' corresponds to a refusal.

Apparently, the experimental literature on indirect speech acts comprehension in children is inconsistent and does not allow to choose between the three proposals. However, the experimental results of Garvey, which support Searle's theory, deserve some considerations. There is no reason for the child observed to describe something that his mother can perfectly see; furthermore, Garvey herself admits that she does not really know the actual communicative intention of the mother.

To sum up, while we are waiting for clearer experimental data, we would prize the generality of the distinction between simple and complex speech acts as due to a difference in the complexity of the mental representations and of the chain of inferences involved. A critical analysis of the relevant developmental litterature is in Bara, Bosco & Bucciarelli (1999).

4. Simple and Complex Speech Acts in Non-Standard Communication

The distinction between simple and complex speech acts holds also for non standard speech acts, like ironies and deceits. Indeed, as we shall see, there are simple and complex ironies and simple and complex deceits.

4.1 Simple and Complex Ironies

Grice advances the so-called traditional theory of irony (1978; 1989). He claims that, in order to comprehend an ironic utterance, the hearer assigns to it a meaning opposite to that literally expressed by the speaker. In particular, Grice claims that an ironic intention can be detected when the literal interpretation does not fit with the context. Unfortunately, however, ironic utterances may consist in something different from the expression of a meaning opposite to the intended one. Further, Grice's account leaves unclear why p should be interpreted as an ironic $not-p$, and not as a lie (Morgan, 1990).

In a completely different perspective, some theories assume - more or less implicitly - that irony involves the ability to meta-represent and, as a consequence, the necessity to draw more or less complex inferential chains so to relate the utterance to its intended meaning. In particular, Relevance theory claims that an ironic utterance is intended and interpreted as an echo of a past utterance (Sperber & Wilson, 1981). Its interpretation does not require the attribution to the speaker of a precise thought, since it echoes the thought of a person or of people in general. The ironic utterance is an echoic mention where the ironist expresses her attitude toward the proposition she is echoing (see also Jorgensen, Miller & Sperber, 1984).

According to Clark and Gerrig (1984) and Morgan (1990), a listener's understanding of an ironic utterance crucially depends on the common ground he takes as shared by the ironist and the audience, their mutual beliefs, mutual knowledge and mutual presuppositions. In case there is not such a common ground, the authors show that the hearer has no way to recognize the pretense. Thus, the ironist is not using one proposition in order to get across its contradictory: rather, in saying p, the ironist is pretending to believe it (Pretense Theory of Irony).

Consistently with this theory, Kumon-Nakamura, Glucksberg and Brown (1995) claim that ironic remarks have their effects by alluding to a failed expectation. According to their proposal (Allusional Pretense Theory of

Irony), ironic utterances have two main features: the speaker expresses a certain attitude, and she is patently insincere. Irony is used to direct the hearer's attention toward a discrepancy between what is and what should have been.

Airenti, Bara and Colombetti (1993b) explain irony on the ground of agent's shared knowledge and the contrasting meaning uttered by the actor. A statement uttered by an actor becomes ironic when compared with the scenario provided by the knowledge she shares with the partner. The partner has to infer a further meaning which contrasts with the background against which the ironic utterance stands out.

Grice's proposal, according to which an ironic utterance expresses the opposite of what is meant by the speaker, is consistent with some results on very young children. For instance, Reddy (1991) found that humour in young infants comes from the violation of the expectation ($not-p$) that the canonical outcome (p) of an interactive event such as giving and taking will occur. Dunn (1991) has analyzed children jokes, finding that 2 and 3-year-olds have a remarkable and differentiated understanding of what familiar others will find funny. These results are inconsistent with the assumption that irony requires a metarepresentational ability, because infants of that age lack such ability (Hogrefe, Wimmer & Perner, 1986; Perner, 1991; Wellman & Wooley, 1990).

However, the proposal that irony involves a metarepresentational and a sophisticated inferential ability is consistent with some experiments on children older than those studied by Reddy. A study on the ability of 6- and 8-year-olds to provide ironic endings to unfinished stories has been carried out by Lucariello and Mindolovich (1995). The authors claim that the recognition and the construction of ironic events involves the metarepresentational skill of manipulating the representations of events. These representations are to be transcended, critically viewed, and disassembled in order to create new and different (and ironic) event structures. According to their model, it is possible to make a distinction between simple and complex forms of ironies; their results show that elder children construct more complex ironic derivations from the representational base than younger children do.

Consistently with this idea, Dews et al. (1996) claim that an ironic comment can either explicitly state the opposite of what is meant (direct irony), or imply something that is the opposite of what is said (indirect irony). In the former case, the speaker's meaning simply is the opposite of what is meanted and there is no echoic mention of a previous statement, while in the latter, it follows from the opposite of what is said. Their results show that adults more often rank indirect ironies as the funniest, while children more often rank direct ironies as the funniest. Dews an colleagues conclude that indirect irony is more subtle.

In conclusion, the apparently divergent data in the literature are not reconcilable, unless a theory is available which allows to cover both simple and complex ironies. Metarepresentational ability might be involved only in the latter.

Our hipothesis is that, the capacity of understanding and producing ironic speech acts develops in two stages. In the first, children start mastering simple irony *á la* Grice: A utters p to mean $not-p$ (Figure 1). Thus, simple ironies

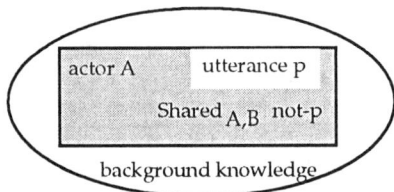

Figure 1. Actor A expresses an ironic utterance *p* which overtly contrasts with the belief *not-p*, shared between A and B.

immediately contrast with a belief shared between the agents.

In the second stage, children learn to perform more subtle inferences, until they reach the levels of indirect irony (complex irony) revealed by experimental data (see Figure 2). Complex ironies require a series of inferences to detect their contrast with the belief shared by the agents.

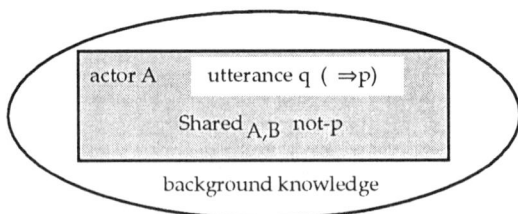

Figure 2. Actor A expresses the ironic utterance *q* which implies the belief *p*, which contrasts with the belief *not-p*, shared between A and B.

4.2 Simple and Complex Deceits

Perner (1991) claims that a deceit is an actor's intentional attempt to manipulate a partner's mental state: the actor's goal is to induce the partner to believe something wrong about reality. He calls instead *pseudo-lies* interactions like the following one:

[3] Mother: Did you finish the chocolate?
     Child: No, it has been Ben.

In Perner's view, in this case, what the child is really aiming at is not to manipulate the mother's beliefs, but to avoid a disagreeable consequence, i.e., to be rebuked. Bussey (1992) and Lewis, Stanger and Sullivan (1989) found that children start to use lies as a means to escape a disagreeable consequence from 3 years of age on. Leekman (1992) states that the liar aims at the achievement of some goal by saying something that she knows or believes is false. In her view, there are progressive steps in the structure of a lie/deceit. At the first one, the actor's intention is to affect the listener's behavior, and only at the following ones her intention is to affect the listener's beliefs.

Peskin (1996) claims that, in order to plan or understand a deceit, it is necessary that the speaker takes as shared something she does not really believe, and that the hearer comes thus to hold a false belief. He concludes that while 3-year-olds understand the former, only 4 years old understand the latter, i.e. the deceptive purpose of the actor.

Airenti *et al.* (1993b) define a deceit as a premeditated rupture of the rules governing sincerity in the behavior game at play. Deceiving requires the actor to break the rule of sincerity and to construct a suitable strategy to successfully modify the partner's knowledge. For example, the actor, while privately believing that *p* is false, tries to convince the partner that *p* is true. If this attempt succeeds, the partner will believe *p* to be shared with the actor.

In our view any kind of deceit, lies included, are attempts to modify a partner's mental state. Their difficulty of comprehension, and production, can vary according to the complexity of the agent's mental states involved into the representation of the deceit. Some deceitful speech acts are simple because they consist in an utterance (*p*) which denies something (*not-p*), that would allow the partner to immediately refer to the game that the actor wishes to conceal from the partner (Figure 3).

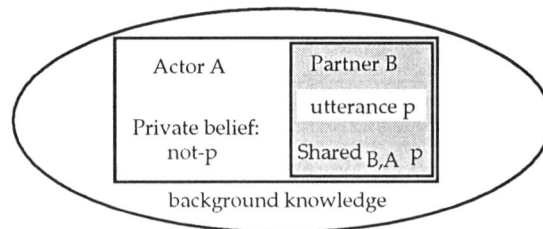

Figure 3. Actor A plans to deceive the partner B. While believing *not-p*, A tries to induce B to consider *p* as shared with her.

A complex deceitful speech act consists instead in a communicative act (*q*) which implies a belief (*p*), that leads the partner to a different game from the one he would reach, if he had access to the actor's private belief (*not-p*), (Figure 4). Thus, all deceits do not have the same complexity; their difficulty, both in production and in comprehension, depends on the number of inferences necessary to refer the utterance to the game bid by the actor. Indeed, in order to perform or to discover a complex deceit, the partner has to consider further elements besides the truthfulness of the utterance. Although there is no theoretical limit to the complexity of a deceitful situation, people's working memory can handling only a limited number of boxings.

What makes ironies different from deceits? Why should the utterance that *not p* be considered either ironic or

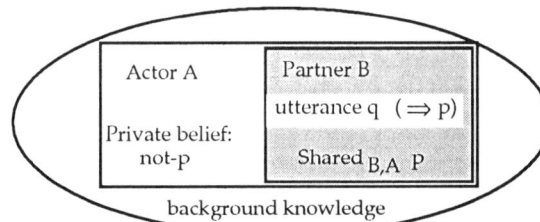

Figure 4. Actor A plans to deceive the partner B. While believing *not-p*, A tries to induce B to believe *q*, which implies *p*. The goal is to induce the partner to consider *p* as shared with A.

deceitful, given the actor's belief that p? Irony differs from deceit because the actor takes as shared with the partner a belief that contrasts with the ironic utterance (see also Sullivan, Winner & Hopfield, 1995), while in the deceit the actor does not share with the partner her private belief. Thus, the same utterance can be considered at the same time an irony or a deceit: it depends on what the actor is sharing with the partner.

According to our proposal lies are simple deceits; they are intentional messages aimed at deceiving (see also Bok, 1978). Thus, we agree with Sodian (1991) when, in his terminology, he considers lies as easier than deceits by definition. However, lies are easier than deceit for a different reason than that hypothesized by Leekman (1992) or Perner (1991). We assume a single cathegory of deceit, whitin which there are lies, whose goal is the modification of the partner's mental state as well. The crucial point is that not all deceits have the same complexity: lies are the simplest ones.

Thus, if the increasing capacity to construct and manipulate complex representations is involved in the emergence of complex deceits, we would predict that a deceptive task can be made easier by reducing the number of characters, episodes, and scenes and by including a context of deception. Actually, Sullivan, Zaitchik and Tager-Flusberg (1994) carry out an experiment on preschoolers and kindergartens and confirm this prediction. Moreover, Russell, Jarrold and Potel (1995) find that executive requirements play a major role in making complex deception hard for children as young as 3 years: when the opponent is removed from a test of complex deception the divergence between the performance of 3-year-olds and 4-year-olds remains essentially unaffected. The authors claim that 3-year-olds' difficulty with complex deception is not caused by an inability to conceive of implanting false beliefs into another person's mind: the reason for the poor performance should be the cognitive load required by complex deceits.

Also, as the ability to conceive of complex representations does increase with the age, we would also expect that children become better mendacious as they grow up. This prediction is confirmed by Leekman (1992) and Peskin (1996); they find that only 7 years olds are good mendacious.

5. Conclusions

We have suggested that the classical distinction made in literature between direct and indirect speech acts should be abandoned in favour of the more general distinction between simple and complex speech acts. The latter distinction applies not only to standard communicative acts, but also to non standard acts such as ironies and deceits. The experimental evidence in the psychological literature is in favour of the existence of simple and complex speech acts within different pragmatic phenomena.

Acknowledgments

We would like to thank our colleague Maurizio Tirassa who read and criticized earlier versions of this paper. This research was supported by the Ministero Italiano dell'Università e della Ricerca Scientifica (MURST, ex-40% for the year 1999).

The names of the authors are in alphabetic order.

References

Airenti, G., Bara, B.G. & Colombetti, M. 1984. Planning and understanding speech acts by interpersonal games. In: *Computational models of natural language processing*, eds. B.G. Bara & G. Guida. Amsterdam: North-Holland.

Airenti, G., Bara, B. G. & Colombetti, M. 1993a. Conversation and behavior games in the pragmatics of dialogue. *Cognitive Science, 17*, 197-256.

Airenti, G., Bara, B. G. & Colombetti, M. 1993b. Failures, exploitations and deceits in communication. *Journal of Pragmatics, 20*, 303-326.

Bara, B. G. 1995. *Cognitive Science: A developmental approach to the simulation of the mind*. Hillsdale, NJ: Lawrence Erlbaum Associates.

Bara, B. G., Bosco F. M. & Bucciarelli, M. 1999. *Developmental Pragmatics in normal and abnormal children*. Brain and Language, in press.

Bara, B. G. & Bucciarelli, M. 1998. Language in context: The emergence of pragmatic competence. In Quelhas, A. C. & Pereira, F. (Eds.), *Cognition and Context*, 317-345. Istituto Superior de Psicologia Aplicada, Lisbon.

Becker, J. A. 1990. Processes in the acquisition of pragmatic competence. In Conti-Ramsden, G. & Snow, C. E. (Ed.). *Children's language, Vol. 7*, Hillsdale, NJ.

Bernicot J. & Legros S. 1987. Direct and Indirect Directives: What Do Young Children Understand? *Journal of Experimental Child Psychology, 43*, 346-358.

Bok, S. 1978. *Lying: Moral choices in public and private life*. New York: Pantheon.

Bussey, K. 1992. Lying and truthfulness: Children's definitions, standards, and evaluation reactions. *Child Development, 63*, 129-137.

Clark, H. H. 1979. Responding to indirect speech acts. *Cognitive Psychology, 11*, 430-477.

Clark, H. H. & Gerrig, R. J. 1984. On the pretense theory of irony. *Journal of Experimental Psychology: General, 113*, 121-126.

Dews, S., Winner, E., Kaplan, J., Rosenblatt, E., Hunt, M., Lim, K., McGovern, A., Qaulter, A. & Smarsh, B. 1996. Children's understanding of the meaning and functions of verbal irony. *Child Development, 67*, 3071-3085.

Dunn, J. 1991. Young children's understanding of other people: Evidence from observations within the family. In Frye, D. & Moore, C. (ed.), *Children's theories of mind: mental states and social understanding*, 97-114. Hillsdale, New Jersey.

Ervin-Tripp S. & Gordon, D. 1986. The development of requests. In *Language competence*, ed. R. L. Schiefelbusch (San Diego: College Hill).

Garvey, C. 1984. *Children's talk*. New York, Fontana.

Gibbs, R. 1986. What makes some indirect speech acts conventional? *Journal of Memory and Language, 25,* 181-196.

Gibbs, R. 1994. *The poetics of mind.* Cambridge University Press, Cambridge.

Grice, H. P. 1978. Further notes on logic and conversation. In P. Cole (Ed.), *Syntax and semantics: Vol. 9. Pragmatics.* New York: Academic.

Grice, H. P. 1989. *Studies in the way of words.* Cambridge, MA, & London: Harvard University Press.

Hogrefe, G.J., Wimmer, H. & Perner, J. 1986. Ignorance versus false belief: A developmental lag in attribution of epistemic states. *Child Development, 57,* 567-582.

Jorgensen, J., Miller, G. A. & Sperber, D. 1984. Test of the mention theory of irony. *Journal of Experimental Psychology: General, 113,* 112-120.

Kumon-Nakamura, S., Glucksberg, S. & Brown, M. 1995. How about another piece of pie: The allusional pretense theory of discourse irony. *Journal of Experimental Psychology: General, Vol. 124,* 1, 3-21.

Lewis, M., Stanger, C. & Sullivan, M. 1989. Deception in 3-year old. *Developmental Psychology, 25,* 439-443.

Leekman, S. R. 1992. Believing and Deceiving: Step to becoming a good liar. In *Cognitive and Social Factor in Early Deception.* S.J., Ceci, M., DeSimone Leichtman & M., Putnick (Eds.). Hillsdale, NJ: Erlbaum.

Lucariello, J. & Mindolovich, C. 1995. The development of complex meta-representational reasoning: The case of situational irony. *Cognitive Development, 10,* 551-576.

Morgan, J. 1990. Comments on Jones and on Perrault. In: P. R. Cohen, J. Morgan and M. E. Pollack, eds., *Intentions in communication,* 187-193. Cambridge, MA: MIT press.

Perner, J. 1991. *Understanding the representational mind.* MIT press, Cambridge, MA.

Peskin, J. 1996. Guise and Guile: Children's understanding of narratives in which the purpose of pretense is deception. *Child Development, 67,* 1735-1751.

Recanati, F. 1995. The alleged priority of literal interpretation. *Cognitive Science, 19,* 207-232.

Reddy, V. 1991. Playing with others' expectations: Teasing and sucking about in the first year. In A. Whitten (ed.), *Natural Theories of Mind: Evolution, Development and Simulation of Everyday Mindreading.* Blackwell.

Reeder K. 1980. The emergence of illocutionary skills. *Journal of Child Language, 7,* 13-28.

Reiter, R. 1980. A logic for default reasoning. *Artificial Intelligence, 13,* 81-132.

Russel J., Jarrold C. & Potel D. 1995. What makes strategic deception difficult for children: the deception or the strategy. *British Journal of Developmental Psychology, 12,* 301-314.

Searle, J.R. 1975. Indirect speech acts. In: *Syntax and semantics,* vol, 3: *Speech acts,* eds. P. Cole & J.L. Morgan. New York: Academic Press.

Shatz, M. 1978. Children's comprehension of their mothers' question-directives. *Journal of Child Language, 5,* 39-46.

Sodian, B. 1991. The development of deception in children. *British Journal of Developmental Psychology, 9,* 173-188.

Sperber, D. & Wilson, D. 1981. Irony and the use-mention distinction. In Cole, P. (Ed.), *Radical pragmatics.* Academic Press, New York, 295-318.

Sperber, D. & Wilson, D. 1986. *Relevance.* Cambridge, MA: Harvard University Press.

Sullivan, K., Winner, E. & Hopfield, N. 1995. How children tell a lie from a joke: The role of second-order mental state attributions. *British Journal of Developmental Psychology, 13,* 191-204.

Sullivan, K., Zaitchik, D. & Tager-Flusberg, H. 1994. Preschoolers can attribute second-order beliefs. *Developmental Psychology, 30 (3),* 395-402.

Wellman, H.M. & Wooley, J. D. 1990. From simple desire to ordinary belief: The early development of everyday psychology. *Cognition, 35,* 245-275.

Towards the Relation between Language and Thinking – the Influence of Language on Problem-Solving and Memory Capacities in Working on a Non-Verbal Complex Task

Christina Bartl (christina.bartl@ppp.uni-bamberg.de)
Lehrstuhl Psychologie II; Otto-Friedrich-Universität
D-96045 Bamberg, Germany

Abstract

This study focuses on the „classical" topic of the relation between language and thinking. Empirical studies investigating the interaction of verbalization and problem solving show inconsistent results. Studies differ with respect to the instruction of verbalization and the characteristics of the task. The aim of the study is to compare the performance in a non-verbal problem with and without language. For this purpose we investigate the performance of six groups of subjects working under different conditions: some of them were disturbed in their language behavior, others were encouraged to verbalize. It could be shown that though they had to work on a non-verbal problem, subjects disturbed in their linguistic behavior showed a worse performance than controls. Furthermore it can be shown that „thinking aloud" in itself does not guarantee an improvement of performance. Moreover there are specific aspects of thinking aloud supporting problem solving. Case studies reveal interesting results with respect to the specific structure of „helpful" verbalization. The differences found cannot be explained by different memory loads or by the degree of distraction.

Introduction

When children work on a difficult task you can observe them talking to themselves. Verbalization seems to help to carry out activities which are close to the maximum a child can achieve according to his or her developmental state (Diaz & Berk, 1992). Several methods in psychotherapy make use of the positive effect of self-verbalization on action regulation. Meichenbaum & Goodman (1971) developed a program for impulsive children. They trained them to talk to themselves while working on cognitive or motoric tasks. The verbalization, first aloud, later whispering and finally tacitly can improve problem solving performance and help impulsive children to control themselves. Similar observations can be made when subjects work on complex problems: being in a crucial situation, they spontaneously begin to verbalize aspects of the problem or their own behavior. Although we are not yet really sure which specific aspects of verbalization facilitate the organization of cognitive processes, the need to "talk to oneself" while solving problems seems to be almost irresistable.

Experiments concerning the treatment of "thinking aloud" and "self-verbalization" however do not give clear evidence. On the one hand there are a lot of studies which support the hypothesis that thinking aloud can improve problem solving behavior in non-verbal tasks, such as the Tower of Hanoi, Raven's Standard Progressive Matrices or other non-verbal intelligence tasks (see Berry & Broadbent, 1984; Franzen & Merz, 1976; Gagné & Smith, 1963; Hussy, 1987). On the other hand, studies give indication for a contrary assumption. Phelan (1965) forced subjects to think aloud while working on concept learning tasks and found negative effects on their performance. Several studies indicate no differences in performance between a verbalizing group and controls (e.g. Deffner, 1989; Lass, Klettke, Lüer & Ruhlender, 1991; Rhenius & Heydemann, 1984).

How can we explain such inconsistencies? Looking exactly upon the experimental designs of the cited studies, the hypothesis arises that differences in the instruction of "thinking aloud" lead to contradictory results. According to the model of Ericsson & Simon (1980, 1984), verbalization can only improve performance when it contents aspects not included in working memory before. So meta-cognitive processes such as self-explanation or self-reflection could modify and improve problem solving strategies. The study of Chi et al. (1989) could demonstrate that better achievers have more spontaneous utterances reflecting and evaluating their own behavior. Other studies show that treatments enforcing meta-cognitive processes have positive effects on subjects' problem solving capacities (Beradi-Colletta et. al., 1995; Chi et al., 1994; Tisdale, 1998).

Another reason for the different results could be found in the experimental designs. Subjects instructed to think aloud were compared with controls without any specific instruction. One could argue that the controls also verbalized, either loudly or silently. Obviously it is necessary to control, whether the subjects indeed refrained from verbalizing. One method to suppress verbalization is the technique of "articulatory suppression". In this treatment subjects have to carry out speaking routine tasks, so that their motoric speech system is permanently busy. Yet studies in the context of Baddeley's phonological loop model (Baddeley, 1990) show, that articulatory suppression has negative effects on subjects' memory loads. The influence of articulatory suppression on problem solving processes has not been investigated yet.

The aim of this study is to explain the reasons for the different findings. Firstly we investigate if an instruction of self-explication and self-evaluation has stronger positive effects on problem solving than the unspecific instruction to think aloud. Secondly we test the hypothesis that suppression of the internal dialogue leads to a worse problem solving performance.

Besides the impacts of different language treatments on problem solving performance, we will have a look at possible influence of the treatments on attention and on memory for elements of the problem space.

Method

Subjects

The experiment was carried out with 60 subjects, 32 females and 28 males, who were distributed to six experimental groups by random. The age varied between 14 and 40 years with a mean of 23 years. Most of the subjects (n = 54) were students of different fields of the universities Bamberg, Erlangen and Trier. Furthermore one pupil and five postgraduated persons joined the experiment.

Task

In this study we used the computer-aided scenario "Bio-Beetle", which can be classified as an interpolation problem (Dörner, 1976) in the tradition of the Tower of Hanoi. In the "BioBeetle" scenario, subjects have to transform a given initial state into a target state by using given operators in a specific sequence. The scenario is semantically embedded as a laboratory for breeding beetles where the morphology of beetles can be transformed by radiation. The beetles differ in six characteristics (color, shape, form of feet, eyes, tentacles and mouth). Each of the characteristics can have two values (p.e. the color can be red or green), so that 26 animals can be distinguished. The morphology of the beetles can be transformed by twelve kinds of radiation (= operators). Several radiations only change one characteristic of the beetle, others affect several morphological attributes. Furthermore some radiations need certain conditions to work, so that the subjects first have to achieve sub-ordinate targets for using them.

All information necessary to solve the breeding problems is given in pictures. Starting the simulation, the subjects see on the left side of the screen the initial beetle and on the right side the target state (see figure 1).

In the left upper corner the actual state ("aktueller Zustand") of the beetle is visualized. In the right corner there is the target state ("Ziel"). In order to transform the beetle, subjects have to press the buttons named with Greek letters. Clicking on the buttons named with Greek letters, the subjects get information about the prerequisits and the effects of the chosen radiation. Then they can decide whether or not they want to apply it. If the necessary conditions for a radiation are not given, nothing will happen. After a successful modification, the actual state of the beetle appears on the screen.

The information about the operators is also given in pictures (see figure 2). As a result, the subjects always know the actual state of the beetle, the target state and the impact of any possible measure. Their problem is to apply the operators in the right sequence in order to attain all the characteristics of the target state. Looking at figure 2 the first line ("IN") represents the prerequisits. The radiation is not ap-

pliable unless the actual state of the beetle has the characteristics shown in the first line. When all prerequisits were given and the radiation is applied the beetle will change its morphology the way shown in the line below ("OUT"). In this case the radiation changes the color, if the beetle has an oval shape and long tentacles.

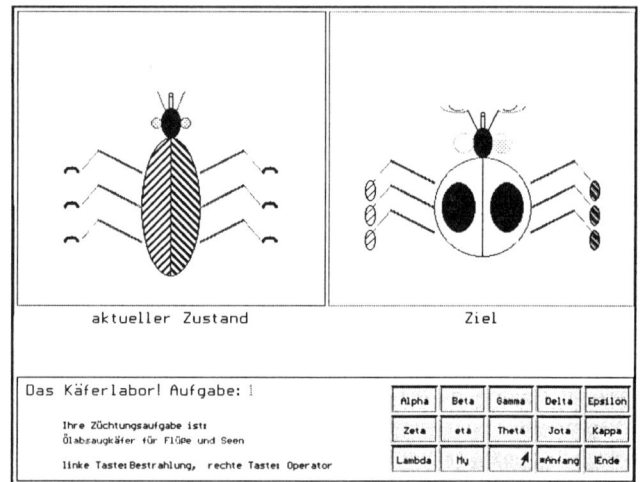

Figure 1: Surface of the computer-aided task "BioBettle".

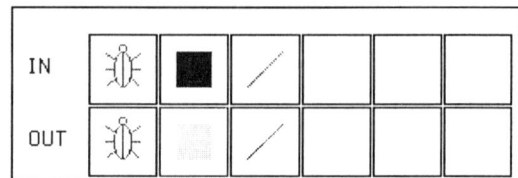

Figure 2: Pictoral information about the radiations effect.

Treatment

The subjects worked on the "BioBeetle" problem under six different conditions (10 subjects in each group). One part of the sample was encouraged in their verbal behavior by different treatments (treatment 1: "thinking aloud" and treatment 2: "self-reflection"). Others were disturbed in their inner dialogue (treatment 3: "no verbalization" and treatment 4: "articulatory suppression"). In order to determine whether an influence of articulatory suppression can be attributed to the specific verbal interference or to an unspecific impairment of attention, another group of subjects carried out a comparable motoric, but non-verbal routine task (treatment 5: motoric interference) while solving the problem. A group of controls completes the experimental design. The experimental treatments are described below:

Treatment 1 ("thinking aloud"): Before they started to work on the BioBeetle, these subjects got the instruction to verbalize their thoughts, that is to vocalize everything that would come into their mind. They were reminded of the instruction after each task.

Treatment 2 ("self-reflection"): The subjects were interrupted in intervals of four minutes. They were asked to describe what they were doing, to evaluate their strategy and to give themselves some instructions for their future behavior. In the rest of the time the subjects could verbalize aloud or silently as they liked.

Treatment 3 ("no verbalization"): In this group the subjects got the instruction to refuse verbalization, loud and tacit, as completely as possible. They should not ask themselves questions nor give themselves instructions.

Treatment 4 ("articulatory suppression"): Subjects of this group had to carry out a verbal routine task while solving the problem. They were instructed to speak two figures (32 and 43) in change again and again. We supposed that while being busy with speaking continuously, they wouldn't be able to verbalize problem-relevant topics as well as without this distraction.

Treatment 5 ("motoric interference"): Subjects were busy with a motoric routine task while solving the problem. In contrary to treatment 4, they did not carry out a task requiring the motoric system of speech, but knocked a rhythm on the table.

The sixth group ("controls") got no specific instruction.

Procedure

At the beginning of the experiment, subjects worked on three easy tasks, so that they could get used to the handling of the computer-aided scenario. Then they had to work on three experimental tasks of increasing difficulty. The first task requires the application of seven radiations in the right sequence, the second eight and the third twelve. The subjects had the possibility to restart each task and to go back to the initial state of the beetle, when they believed their sequence of operations to be misleading. Each task was aborted when the subject did not reach the target state within 30 minutes.

At the end of the experiment we tested the subjects' memory load. First the subjects had to memorize the specific effects of the radiations. Then they got a list of the pictoral codes of the twelve radiations (see figure 2) and had to assign the pictures to the labels "Alpha" to "My". In the second part of the memory test the subjects' capability to recognize the operators was evaluated.

Results

Effects of Verbalization on Problem-Solving Performance

A differentiated operationalization of problem solving performance was developed by analyzing the solution times. When a task was not solved the subject got no points for it. The maximum score of five points was achieved, when the solution time was among the shortest 20% of the sample (solution time $\leq$ 20th percentile). Four points were achieved when the solution time was between the 20th to 40th percentile and so on. In this way a maximum score of 15 points could be achieved in the three experimental tasks. The Kruskal-Wallis analysis shows significant differences between the six experimental treatments ($\chi^2 = 14.10$, df = 5, p

< 0.05). The highest scores were achieved in the treatment "self-reflection", followed by "motoric interference" and controls (see figure 3). In figure 3 black lines in the boxes mark the median of the group. Within the boxes there are all cases from the 25th to the 75th percentile. The "whiskers" show the smallest resp. the highest value.

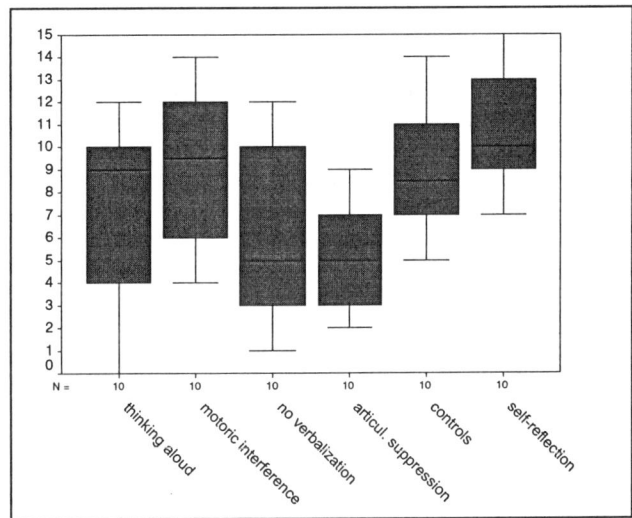

Figure 3: Box-plots of the number of points achieved in the experiment.

The "thinking aloud"-subjects showed a worse performance that the groups mentioned. This supports the hypothesis that self-explication and self-evaluation has stronger positive effects on problem solving than the unspecific instruction to think aloud. Moreover the better results of the controls indicate that the demand of thinking aloud could even disturb subjects, so that the results are even worse than without any treatment. Whereas the "self-reflection" treatment obviously improves problem solving performance, unspecific thinking aloud does not guarantee an improvement.

The two groups which were disturbed in their linguistic behavior show the worst performance. Articulatory suppression shows even stronger negative effects on problem solving that the mere instruction not to verbalize. The variances within the "no verbalization" group is significantly higher than within the "articulatory suppression" group (F = 6.28, p < .05). This result indicates that the subjects were more or less successful in suppressing their "inner dialogue" or that they followed the instruction more or less carefully. Obviously, the articulatory suppression treatment allows better control.

Besides the solution times, another indicator for problem solving performance is the number of operator applications. Some subjects find the solution directly, others go a long way round or restart the task several times, until they reach the target state. "Economy of radiation use" was one criterium explicitly mentioned in the instruction. For the following evaluation cases of subjects were excluded when they gave up before the maximum time for the task was exceeded. Looking at the number of radiations used, the six groups differ significantly (F = 3.4311, df = 5, p < 0.05). Comparing single treatments, there are significant differences between the group "articulatory suppression" and the groups "thinking aloud", "self-reflection" and controls. When solving the problem under "articulatory suppression", subjects needed more radiations to reach the goal. In contrast, subjects working under the treatments "self-reflection", "thinking aloud", and the controls made use of the radiations more carefully and economically.

Effects of Verbalization on Memory Load

Investigating the influence of verbalization on memory load, we looked at the active reproduction of elements of the problem space. A reproduction of the effects of the twelve radiations required the following steps: the subjects had to memorize how many of the six characteristics were affected by the operator. There were radiations which changed only one aspect of the beetle and others which required a specific value of all the six characteristics to work. Secondly the subjects had to memorize which characteristics were affected and which not. Thirdly they were asked to differentiate which characteristics were requirements and which of them were changed by the radiation. Finally they had to remember in which way the values changed (for example a change of color from red to green or vice versa). These aspects were summed up and for any correct answer subjects gained one point. A maximum of 160 points could be achieved.

In the recognition test, subjects had to assign pictures of the radiation effects (see figure 2) to their label (Alpha, Beta, Gamma, ..., My). For each correct answer they got twelve points, so that they could achieve a maximum of 144 points. The subjects were given the opportunity to name several alternatives. When they named two alternatives and one of them was right, they got six points, when they named three alternatives four points could be achieved and so on.

Figure 4 shows the scores achieved by the six experimental groups in the memory tests. The scores are counted in percentages of the maximum score. In general all subjects showed a bad performance. They achieved an average of 32% of the maximum score in the active reproduction and 36% of the maximum in the recognition task.

One explanation of the bad performance is the difficulty of the task. A maximum score in the active reproduction requires remembering the exact effects of twelve partly complex operators. But in solving the problem, it is not necessary to remember the effects of the operators in detail, because the pictures representing the effects can be activated on the screen at any time. Therefore it is not necessary for the subjects to memorize the operators in detail.

As to the number of points achieved in the active reproduction, analysis of variances shows significant differences between the groups (F = 4.1076, df = 5/54, p < .05). In the recognition task, the experimental groups do not differ. Single tests (Tukey-b) show significant differences between the controls, who achieved the highest scores, and the groups "no verbalization", "articulatory suppression" and "motoric interference".

The results show that experimental treatments handicapping the verbalization of subjects lead to a worse memory performance. However, the verbal motoric task ("articulatory suppression") does not show stronger effects than a non-

verbal motoric interference. As a result, one cannot be sure whether the effects on memory load are results of an unspecific impairment of attention or of the specific verbal impairment.

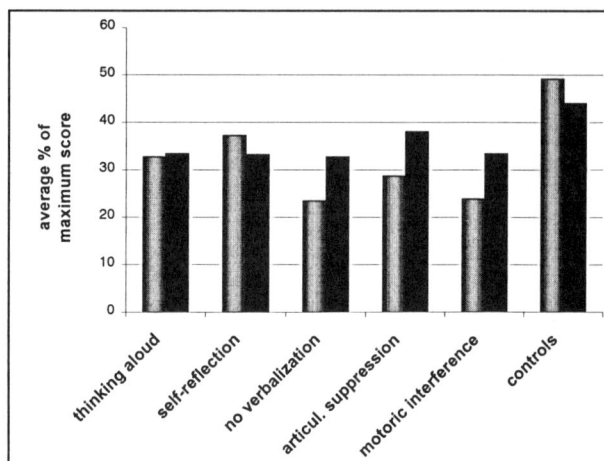

Figure 4: Average percentage of maximum score in the memory tasks. The light bars represent the active reproduction, the dark bars the recognition task.

One interesting result is that the group "motoric interference" shows comparably bad scores in the memory tests, though showing a good problem solving performance. In contrast, the group "articulatory suppression" shows the worst problem solving performance, whereas their memory loads do not differ significantly from other groups with the exception of controls.

In general, there is no correlation between the scores in the memory test and the points achieved in the problem (Kendall-Tau-b = .11, n = 60, p < .05). An interaction of problem solving performance and memory load can only be observed in the first experimental task (Kendall-Tau-b = .22, n = 60, p < .05). Subjects who can remember the operator's effects in more detail, take advantage at the beginning of the experiment, later on this advantage is not decisive any more. However, a negative correlation between the results in the memory test and the number of false clicks can be observed (Kendall-Tau-b = -.1990, n = 50, p < .05). Subjects who have worse memory performance also make more mistakes in handling the mouse. This supports the hypothesis that differences in memory loads are caused by an impairment of attention.

Case Studies

The results comparing the six experimental treatments did not fulfil the expectations in the whole. It could be shown that a disturbance of the "inner dialogue" has negative effects on problem solving performance and that "self-reflection" can improve performance. Mere thinking aloud however does not lead to the expected improvement. Furthermore the variances within the "thinking aloud"-group are high. The performance within the group is heterogenous: some subjects seemed to take advantage of verbalization, others seem to be handicapped by this instruction. The span between the performance of the "thinking aloud"-subjects varies between 0 and 12 of a maximum of 15 points.

There seem to be certain aspects of verbalization helping to solve problems more efficiently. The instruction to recapitulate one's own actions regularly, to criticize it and to give self-instructions for future strategies, leads to a clear improvement of performance.

In the following section two corner cases of thinking aloud subjects will be compared regarding to their verbalization. The comparing method refers to the method of "comparative casuistics" proposed by Jüttemann (1990). An exact analysis of the verbal data can on the one hand illustrate how subjects differ in following the instruction of thinking aloud. On the other hand, a comparative case study can show, which aspects of verbalization could improve problem solving. For the comparison the verbal data of the first task was transcribed and evaluated as to linguistic categories.

For the corner-case analysis two subjects were elected, who were in the same group, but showed different performance. One subject with a very good performance ("Logobene") will be compared with the worst subject of the "thinking aloud"-group ("Logomale"). Both subjects are female and students of psychology of the university of Bamberg.

Let us first have a look at the scores achieved by the two corner cases: The subject "Logomale" could not solve the task. After she worked on the problem for 30 minutes" the task was given up. In contrary "Logobene" solved the first task within 10 minutes and got the maximum score of 5 points. While "Logomale" applied 44 radiations, "Logobene" only needed 10 to reach the target state of the beetle.

Questions, answers and self-instructions: As with the number of utterances, the two subjects show clear differences. "Logobene" shows 126 utterances in ten minutes, whereas "Logomale" makes not more than 71 utterances while working 30 minutes on the solution. Their verbal data were analyzed with respect to the three "classical" linguistic categories of questions, answers/statements and requests (in this case: self-requests resp. self-instructions). The category sequences make the differences of the verbal data of the two corner cases obvious (see figures 5 and 6).

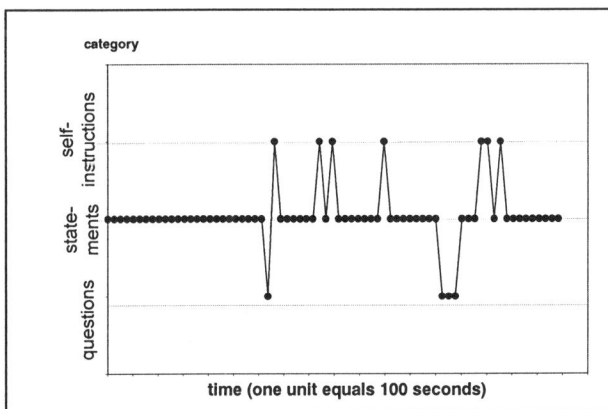

Figure 5: "Logomale": Sequence of questions, statements and self-instructions in the thinking aloud protocol.

"Logomale's" whole thinking aloud protocol shows only four questions. Moreover, there are only seven self-instructions. The main part of "Logomale's" verbalization are statements, e.g. "here the eyes can be changed", "here the color will not be changed" or "the beetle is green now and has small eyes". In the first ten minutes of the problem solving procedure, "Logomale" exclusively makes statements. Afterwards there are some questions and self-instructions, when "Logomale" deliberates whether she should start the task once more or not. until the end, however, statements dominate her thinking aloud.

The sequence of "Logobene's" verbal data shows a completely different pattern (see figure 6). She formulates hypothesis and asks herself questions from the beginning. The questions are often followed by answering statements and that leads to self-instructions. In the middle of the solution time, there is a vivid change of these categories in the sequence described. A dominance of statements can only be found at the end, when the solution is almost found and the problem turned to a solvable task.

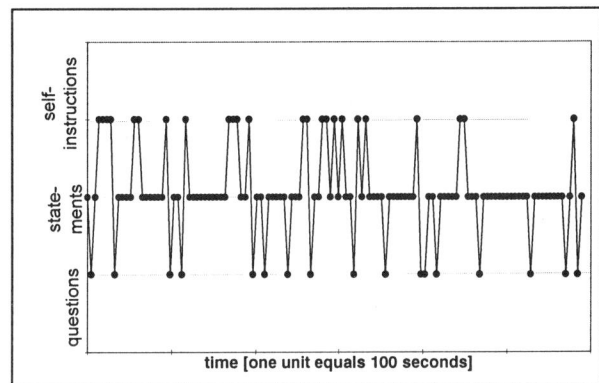

Figure 6: "Logobene's" vivid soliloquy: Sequence of questions, following statements and self-instructions.

In summary it can be shown that the thinking aloud of "Logomale" is no inner dialogue, but more a commentation of her activities. In contrary, "Logobene" talks vividly to herself. She asks herself questions and gives answers. Frequently a question-answer-sequence is followed by a self-instruction for future behavior.

Discussion

Summarizing the findings, the different treatments lead to striking effects on the performance of the "BioBeetle" tasks. The two experimental groups which were disturbed in their "inner dialogue" achieved the worst performance. Subjects who were encouraged to self-reflection showed the best performance. The other groups, "motoric interference", "thinking aloud" and controls show an average performance. Moreover it can be shown that there are strong differences in memory load. The controls show the best memory performance. The group "motoric interference", though belonging to the best problem solving groups achieve significantly lower scores in the reproduction task, as well as the groups "no verbalization" and "articulatory suppression". There is no correlation between memory load and problem solving performance, but between memory load and deficits in concentration.

Differences in problem solving performance can therefore not be explained by different memory loads, nor by different degrees of distraction. One conclusion of this study is, that a specific instruction to describe and evaluate the own behavior, which includes meta-cognitive processes leads to an improvement of problem solving. Thinking aloud in itself can not guarantee an improvement. Moreover there a some indications that an excessive commentation of action even has negative effects. The case studies could illustrate individual differences in the pattern of verbalization, both invoked by the same instruction to "think aloud". This gives some clues for explaining the high variances within the group of "loud thinkers". Some of them vividly talk to themselves, others are busy with describing the actual state and the actions carried out. The results of the case analysis led to a following study in which the patterns found ("commentation" vs. "questions, answers and self-instruction") are used as experimental treatments. Preliminary results of this experiment, in which a different scenario was used, show that this effect is robust over different settings (Bartl, in preparation).

The study has a second central finding: A disturbance of the "inner dialogue" leads to negative effects on problem solving performance. Both experimental groups that were disturbed in their linguistic behavior showed a comparably bad performance. "Articulatory suppression" has stronger effects than the instruction "not to verbalize". The high variances in the latter treatment indicate that the subjects followed this instruction more or less carefully. This experimental treatment produces the same problems as the "thinking aloud" instruction: the subjects differ in the way of following them, which leads to heterogeneous groups with high variances. In summary the results indicate a close relationship between language and thinking, whereby a soliloquy, including meta-cognitive processes, seems to improve problem solving capacities.

Acknowledgements

The experiments were conducted within the research project "Autonomie", Az. Do 200/12 supported by the Deutsche Forschungsgemeinschaft (DFG). The "BioBeetle" problem was designed by Dietrich Dörner. Special thanks to Patricia Cammarata for carrying out the experiments.

References

Baddeley, A. D. (1990). Human Memory. Theory and Practice. Hove: Lawrence Erlbaum Associates.

Bartl, C. (in preparation). Exploring an Island. The influence of commentations and meta-cognitive processes on complex problem solving. Pre-print available at Lehrstuhl Psychologie II, University of Bamberg, Markusplatz 3, D-96050 Bamberg.

Beradi-Coletta, B., Buyer, L.S., Dominowski, R.L. & Rellinger, E.R. (1995). Metacognition and Problem Solving: A Process-Oriented Approach. Journal of Experimental Psychology: Learning, Memory and Cognition, 21 (1), 205-223.

Berry, D.C. & Broadbent, D. (1984). On the relationship between task performance and associated verbalizable knowledge. Quarterly Journal of Experimental Psychology, 36A, 209-231.

Chi, M.T.H., Bassok, M., Lewis, M.W., Reimann, P. & Glaser, R. (1989). Self-Explanations: How Students Study and Use Examples in Learning to Solve Problems. Cognitive Science, 13, 145-182.

Chi, M.T.H., de Leeuw, N., Chiu, M.-H. & LaVancher, Ch. (1994). Eliciting Self-Explanations Improves Understanding. Cognitive Science, 18, 439-477.

Deffner, G. (1989). Interaktion zwischen Lautem Denken, Bearbeitungsstrategien und Aufgabenmerkmalen? Eine experimentelle Prüfung des Modells von Ericsson und Simon. (Towards the interaction of thinking aloud, solution strategies and task characteristics?). Sprache & Kognition, 8, 98-111.

Diaz, R.M. & Berk, L.E. (Eds.) (1992). Private Speech: From Social Interaction to Self-Regulation. London: Erlbaum.

Ericsson, K.A. & Simon, H.A. (1980). Verbal reports as data. Psychological Review, 87, 215-251.

Ericsson, K.A. & Simon, H.A. (1984). Protocol Analysis: Verbal Reports as Data. Cambridge, Ma: M.I.T. Press.

Franzen, U. & Merz, F. (1976). Der Einfluß des Verbalisierens auf die Leistung bei Intelligenzprüfungen: Neue Untersuchungen. (The influence of verbalization on the performance of intelligence tasks: new investigations). Zeitschrift für Entwicklungspsychologie und pädagogische Psychologie 8, 117-139.

Gagné, R.M. & Smith, E.C. (1963). A study of the effects of verbalization on problem solving. Journal of Experimental Psychology, 68, 12-18.

Hussy, W. (1987). Zur Steuerfunktion der Sprache beim Problemlösen. (The controlling function of speech in problem solving). Sprache und Kognition, 6 (1), 14-22.

Jüttemann, G. (1990). Komparative Kasuistik. (Comparative casuistics). Heidelberg: Asanger.

Lass, U., Klettke, W., Lüer, g. & Ruhlender, P. (1991). Does thinking aloud influence the structure of cognitive processes? In: R. Schmid & D. Zambarbieri (Eds.). Oculomotor Control and Cognitive Processes. Amsterdam: Elsevier.

Meichenbaum, D.H. & Goodman, J. (1971). Training impulsive children to talk to themselves: A means of developing self-control. Journal of Abnormal Psychology, 77, 115-126.

Phelan, J.G. (1965). A replication of a study on the effects of attempts to verbalize on the process of concept attainment. Journal of Psychology, 59, 282-293.

Rhenius, D. & Heydemann, M. (1984). Lautes Denken beim Bearbeiten von RAVEN-Aufgaben. (Thinking aloud while working on RAVEN matrices). Zeitschrift für Experimentelle und Angewandte Psychologie, 31, 308-327.

Tisdale, T. (1998). Selbstreflexion, Bewußtsein und Handlungsregulation. (Self-reflection, consciousness and action regulation). Weinheim: Psychologie Verlags Union.

Heuristic Identity Theory (or Back to the Future): The Mind-Body Problem Against the Background of Research Strategies in Cognitive Neuroscience

William Bechtel (bechtel@twinearth.wustl.edu)
Philosophy-Neuroscience-Psychology Program, Washington University
Campus Box 1073, St. Louis, MO 63130 USA
Robert N. McCauley (philrnm@emory.edu)
Department of Philosophy, Emory University, Atlanta, GA 30322 USA

Abstract

Functionalists in philosophy of mind traditionally raise two major arguments against the type identity theory: (1) psychological states are *multiply realizable* so that there are no one-to-one mappings of psychological states onto neural states and (2) the most that evidence could ever establish is the *correlation* of psychological and neural states, not their identity. We defend a variant on the traditional type identity theory which we call *heuristic identity theory* (HIT) against both of these objections. Drawing its inspiration from scientific practice, heuristic identity theory construes identity claims as hypotheses that guide subsequent inquiry, not as conclusions of the research.

Introduction

Functionalists in philosophy argue that the type identity theory advances an unjustifiably strong account of the metaphysics of mind. Ironically, one of the first proponents of using functional criteria to identify mental states, David Armstrong, viewed functional analysis as a means for *supporting* the identity theory. The dominant versions of functionalism, though, reject the type identity of mental and physical states, since their relations are many-to-many or at least one-to-many, not one-to-one. This is known as the *multiple realizability* objection to the identity theory.

The identity theory faces another objection to the effect that empirical investigations can never establish anything more than a correlation between mental events and physical events. We shall call this the *correlation* objection. Recent discussions of consciousness (Chalmers, 1996) have pressed this objection anew. The general argument is that, at best, neurophysiological approaches isolate brain states that correlate with conscious states. They cannot justify identifying these neural states with the conscious states (especially in light of their disparate properties). They can only establish their correlation.

A richer appreciation of the course of scientific research over time and of the thoroughly hypothetical character of all identity claims in science argues for a heuristic conception of the identity theory. Identity claims typically play a *heuristic* role in science. Scientists adopt them as hypotheses in the course of empirical investigation to guide subsequent inquiry--rather than settling on them merely as the results of such inquiry. Defending *heuristic identity theory* (HIT) from both the multiple realizability and correlation objections, we will argue that mapping at least some mental states (viz., many that figure in scientific psychology) one-to-one with physical states is a perfectly normal part of research in cognitive neuroscience and that the results often provide ample support for these hypotheses.

HIT versus the multiple realizability objection: The role of comparative studies

Underlying the multiple realizability objection is the assumption that looking across species will yield type differences in brains despite the type identities of mental states (Putnam, 1967). Perusing research in cognitive neuroscience, though, casts doubts on that assumption.

It seems obvious that when individuals from different species are in the same mental state, their neural states will differ. After all, even within the mammalian order, brains from different species clearly look different. Thus, it may prove surprising to learn that the neurobiological practice of identifying brain areas and brain processes[1] is and historically has been a *comparative* endeavor (Bechtel & Mundale, 199). To appreciate how a comparative approach informs neurobiological proposals, consider the examples that follow (two historical and one contemporary).

The first involves research on mapping the brain into functionally relevant areas by using cytoarchitectural tools, a project now largely associated with Korbinian Brodmann, but pursued at the turn of the century by many others (e.g., Oskar and Cécile Vogt, Constantin von Economo). A key foundation for Brodmann's demarcation of areas was his demonstration that cortex generally consists of six layers, for which he reported comparative studies involving fifty-five species. He distinguished areas on the basis of the relative thickness of these layers (e.g., layer 4 was very thick in areas 1, 2, and 3, but much thinner in area 4) and the particular types of neurons found (e.g., pyramidal cells). In identifying brain areas, Brodmann worked comparatively; besides his well-known map of the human cortex (figure 1), he generated maps for the lemur, flying fox, rabbit,

---

[1] Although philosophers often speak of brain states, neuroscientists are not generally interested in states but in brain areas and brain processes.

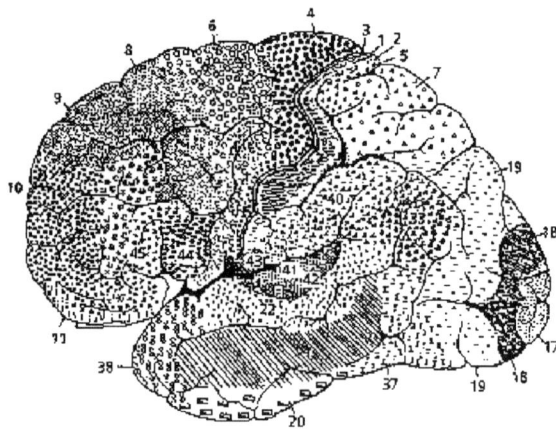

Figure 1: Brodmann's (1909) cytoarchitectural map of areas of the human cerebral cortex.

Table 1. Areas Ferrier Found to Respond to Mild Stimulation	
1.	Opposite hind limb is advanced as in walking
2.	Flexion with outward rotation of the thigh, rotation inwards of the leg, and flexion of the toes
3.	Movements comparable to 1 and 2, plus movements of the tail
4.	Opposite arm is adducted, extended, and retracted, the hand pronated
5.	Extension forward of the opposite arm (as if reaching or touching something in front).
a,b,c,d	Clenching of the fist
6.	Flexion and supination of the forearm
7.	Retraction and elevation of the angle of the mouth
8.	Elevation of the ala of the nose and upper lip
9.	Opening of the mouth, with protrusion of the tongue
10.	Opening of the mouth, with retraction of the tongue.
11.	Retraction of the angle of the mouth.
12.	Eyes open widely, pupils dilate, and head and eyes turn to the opposite side.
13,13'	Eyes move to the opposite side
14.	Pricking of the opposite ear, head and eyes turn to the opposite side, pupils dilate widely.
15.	Torsion of the lip and semiclosure of the nostril on the same side

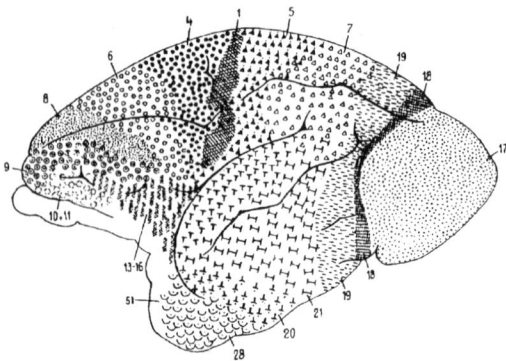

Figure 2: Brodmann's (1909) cytoarchitectural map of the lemur.

and others. (See figure 2 for an example.) For Brodmann, success in finding comparable areas in different species despite differences in brain shapes and in the relative location of areas was pivotal in establishing the reality of distinct functionally relevant areas in the brain. (Brodmann, 1909/1994).

Although Brodmann's goal was to identify functionally relevant brain areas, neuroanatomical techniques generally do not suffice to do so. Before the rise of functional brain imaging, neuroscientists primarily relied on lesion and electrophysiological techniques. David Ferrier (1876) used electrophysiological techniques in the 1870s, employing mild electrical stimulation to map brain areas in a large number of species, including monkey, dog, jackal, cat, rabbit, guinea pig, and rat. He utilized a numbering scheme shown in Table 1 to record the functional character of the responses elicited (figure 3). Although he was unable to use this technique on humans, Ferrier's goal was to extrapolate his results to humans and in his final chapter he proposed how areas found in other species project onto the human cortex (figure 4).

Historically, neuroscientific practice routinely involved identifying brain areas and processes across broad range of species as belonging to the same type. These practices continue.

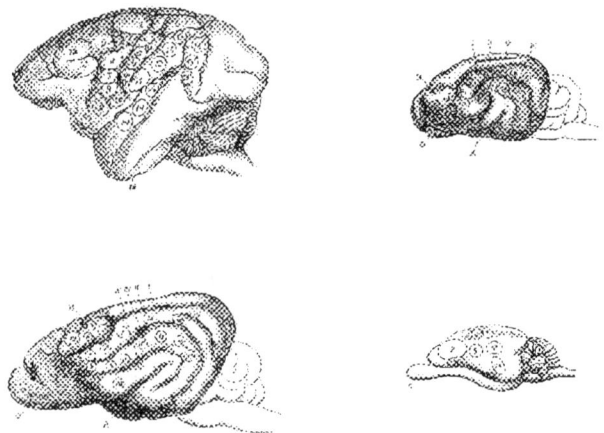

Figure 3: Areas on the cortex of monkey (upper left), dog (lower left), cat (upper right), and rabbit (lower right) where David Ferrier was able to elicit responses to mild stimulation. the common numbering system is shown in Table 1.

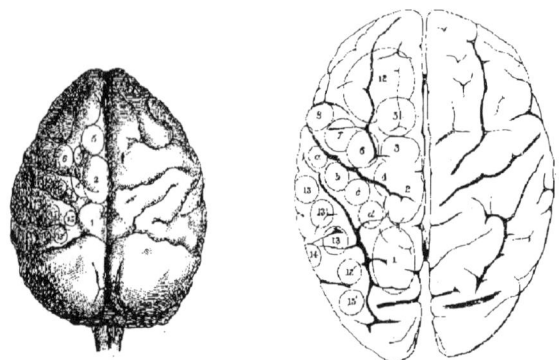

Figure 4: Ferrier's projection of areas responsive to electrical stimulation on the macaque cortex (left) onto the human cortex (right).

Maps of cortical areas have become more refined as neuroscientists have developed additional tools, such as connectivity analysis, to identify brain areas. Still, maps of, for example, visual processing areas in the brain--developed by Ungerleider and Mishkin (1982) and van Essen and Gallant (1994)--are based principally on studies of macaque monkeys.

Oddly, when they consider theories of mind-brain relations, philosophers seem to forget that the overwhelming majority of studies have been on non-human brains. Experimentally induced lesions and cell recording are two of the principal tools for unraveling the functional significance of different areas, but for obvious ethical reasons these are largely restricted to non-human animals. Although the ultimate objective is to understand the structure and function of the human brain, neuroscientists depend upon indirect, comparative procedures to apply the information from studies with non-human animals to the study of the human brain. For example, they determine the location of areas such as V4 and MT in the human brain by using neuroimaging techniques to find where tasks that would drive cells in these areas in macaques result in increased blood flow in humans (Zeki et al., 1991).

Why do the differences between the brains of different organisms, which so exercise philosophers, *not* impede neurobiological research? Part of the reason seems to be that neurobiologists often employ criteria for type identities in brains that are more coarse-grained than most philosophers have envisaged. Of course, the philosophical practice of comparing *mental* states across species is also rather coarse-grained. But no one should begrudge that. When pondering hypotheses that identify the psychological structures and processes of minds with the biological structures and processes of brains, surely one crucial issue is insuring that we compare analyses with compatible grains. Accordingly, neurobiologists do treat psychological processes (albeit not those of folk psychology, but ones that figure in information processing accounts of psychological function) as comparable across species, but they largely elude the problem of multiple realization by working with analyses from the two pertinent levels that have at least roughly similar grains (Bechtel & Mundale, 1999).[2] Ascertaining the compatibility of "grain" between research at two different levels is one of the most basic examples of the co-evolution of sciences.

---

[2]There are many occasions when neurobiologists employ a much finer grain. A major issue in recent years has been the plasticity of cortex, often demonstrated by the rewiring of sensory processing areas that occurs in response to altered sensory input. Even in the context of comparative studies, there are times when neurobiologists are concerned with micro-details (e.g., in measuring allometry or analyzing how connectivity changes between species). When neurobiologists move to this grain size for brain areas, though, they usually change the grain-size of their behavioral measures as well and attend to *differences* in behavior between animals or across species.

When Putnam (1967) employed for his example of common psychological states hunger in humans and octopi, his grain for type identifying psychological states was not especially fine. Such a broad extension of psychological types poses problems for the functional identification of psychological states, since the links to other mental states and to behaviors that are central to functional analyses differ profoundly between such radically different species. Still, given that evolution tends to conserve and extend existing mechanisms rather than create new ones, researchers could well end up type identifying even the neural mechanisms involved in hunger in the octopus and human, which would substantially defuse Putnam's intuitively plausible example. This is not to rule out the possibility of radically different ways of performing similar functions emerging in evolution. However, when researchers discover multiple mechanisms for performing similar functions, such as alternative pathways for processing visual input in invertebrates and vertebrates, it provides an impetus for psychologists to search for functional (behavioral) differences that motivate the differentiation of types at the psychological level as well. Acknowledging the possibility of different mechanisms performing similar functions does not preclude maintaining type distinctions that preserve one-to-one mapping between neural and psychological types. With just such lessons in mind, HIT looks to the comparative practices of neurobiology to dodge the multiple realizability objection.

HIT versus the correlation objection: Hypothesized identities as discovery strategies

Champions of the correlation objection, i.e., the objection that identity theorists can never establish the actual identity of neural and psychological states but only their regular correlation, assume (correctly) that identity theorists bear the burden of evidence in this debate. In a perversely Humean spirit, though, they set the bar impossibly high, requiring identity theorists to establish each identity claim's truth—in effect—beyond a shadow of a doubt. Discredited in the philosophy of science, verificationism, oddly, enjoys new life in the philosophy of mind.

Neurobiological practice provides direction for answering this objection too. Scientists often propose identities during the early stages of their inquiries. These hypothetical identities are not the conclusions of scientific research but the premises. They serve as heuristics for guiding scientific discovery. (McCauley, 1981) Instead of appealing to Leibniz's law of the identity of indiscernables as a metaphysical principle for settling things *a priori*, they opportunistically exploit its converse, the indiscernability of identicals, to guide subsequent empirical research. This formulation of Leibniz's law entails that what we learn about an entity or process under one description must apply to it under its other descriptions. Scientists propose these identities, in part, precisely because the two accounts do not mirror one another perfectly. They use each to guide discovery in the other.

This involves employing what we learn through psychological research to guide the discovery and elaboration of neural mechanisms and what we learn about neural mechanisms to develop more sophisticated psychological models. We will sketch the case of visual processing, which has involved a set of related hypothetical identities that have linked neural and psychological investigation for over a hundred years in an on-going story of progressive theoretical revision at both levels of analysis. Researchers revised their initial identification of cortical visual processing with processing in V1 as they recognized, with the help of increasingly sophisticated neurobiological accounts, that a much larger part of cortex subserves vision; these revisions in the neurobiological account are now inspiring revisions in the psychological account of vision. For practitioners in these fields at the end of a century of research, both the comparable complexity and the general compatibility of models from the psychological and neural sides of the divide render philosophers' disquiet about the incompleteness of the evidence a needlessly fastidious extravagance. Few researchers would contest identifying visual processing with processing in the areas denoted in the figure of the flattened cortex of the macaque by van Essen and Gallant (figure 5).

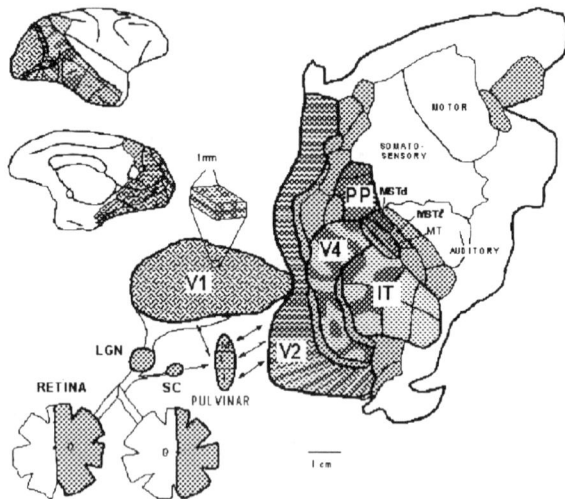

Figure 5: Van Essen and Gallant's (1994) map of distinct processing areas which figure in visual processing in a flattened macaque cortex.

Research on the neural mechanisms of vision began in the last half of the nineteenth century with efforts to locate a visual center in the brain. Based upon neuroanatomical studies indicating that the optic tract, after projecting to a part of the thalamus known as the lateral geniculate nucleus (LGN), subsequently projects to the occipital lobe (Meynert, 1870) and upon clinical evidence concerning visual deficits following stroke and other damage to the occipital lobe, most researchers identified it as *the* locus of visual processing. Ferrier, however, dissented, arguing on the basis of lesion studies in monkeys and his stimulation studies that the angular gyrus in the parietal cortex was the locus of vision. One critical piece of evidence that suggested that Ferrier was

wrong was the discovery that the organization of the occipital lobe reflected topographical layout of the visual field. Early evidence for this came from Salomen Henschen's (1893) attempt to map lesions in the occipital cortex and corresponding deficits in the visual field in humans, but the map he offered reversed the mapping contemporary scientists accept. During the Russo-Japanese War Tatsuji Inouye and during World War I Gordon Holmes and William Tindall Lister developed the modern account of the topographical arrangement of occipital cortex from their studies of wounded soldiers (Glickstein, 1988). Using single cell recording in cat and monkey Talbot and Marshall (1941) corroborated their proposals (figure 6).

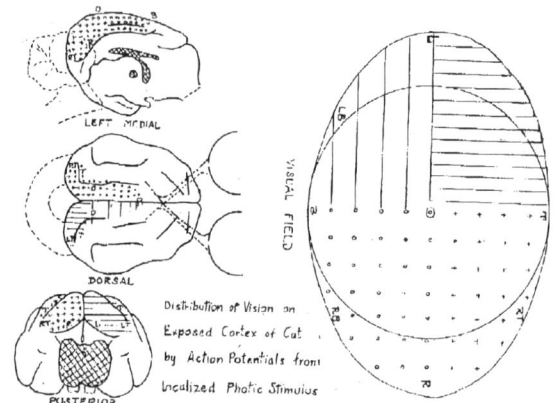

Figure 6: Talbot and Marshall's (1941) mapping of the visual field onto the primary visual cortex of the cat.

Discovering this organization in the occipital cortex supported the hypothesis that it was the location for visual processing in the brain, but it left most of the questions about how the brain processes visual information unanswered. Stephen Kuffler's (1953) research on the retina and LGN had revealed the distinctive center-surround response of cells in those areas (i.e., some cells would respond to a stimulus when it was in the center of their visual field but be suppressed when it was in their surround, while others would respond to a stimulus in the surround but not the center). Two researchers in his laboratory, David Hubel and Torsten Wiesel, set out to find similar response patterns in the occipital cortex of cats and monkeys but discovered that cells there were responsive to bars instead (Hubel & Wiesel, 1962, 1968). What they termed *simple cells* responded to bars of specific orientation at specific locations, while what they termed *complex cells* responded to bars of specific orientation at any location in a cell's receptive field and might show selective responses to bars moving in some particular direction. While Hubel and Wiesel's demonstration of specific visual function in V1 provided further support for its identification as a visual area and important details about the character of visual processing, it also showed that visual processing could not be identified with V1 alone. They ended their 1968 paper with a prophetic remark:

Specialized as the cells of 17 are, compared with rods and cones, they must, nevertheless, still represent a very elementary stage in the handling of complex forms, occupied as they are with a relatively simple region-by-region analysis of retinal contours. How this information is used at later stages in the visual path is far from clear, and represents one of the most tantalizing problems for the future. (Hubel and Wiesel, 1968, p. 242)

Although Karl Lashley (1950) had strongly resisted proposing specialized visual processing areas outside of V1, Alan Cowey (1964), relying on single cell recording, demonstrated that V2 also contained a systematic map of the topographical organization of the visual field. In 1965 Hubel and Wiesel (1965) showed that yet a third visual area, V3, preserved the topographical organization. Semir Zeki (1969) offered further evidence of the systematic nature of the maps by showing that small lesions in V1 resulted in deterioration of cells in corresponding parts of V2 and V3. In 1971 he repeated the approach by making lesions in V2 and V3 and tracing their effects into areas on the anterior bank of the lunate sulcus which he labeled V4 and V4a. Turning to cell recording, Zeki established that cells in V4 responded to the wave length of stimuli, while cells on the posterior bank of the superior temporal sulcus (an area he labeled V5 but others have designated MT) responded to motions of stimuli in specific directions (Zeki, 1973, 1974).

Various research from the 1950s to the early 1970s identified specific responses to visual stimuli in areas of the temporal cortex and of the posterior parietal cortex. Within the former, areas TE and TEO in the inferotemporal cortex responded to specific shapes (Gross, Rocha-Miranda, & Bender, 1972). In posterior parietal cortex cells responded differentially to the locations of stimuli (Goldberg & Robinson, 1980). In a relatively brief period, such research defeated the hypothesis identifying visual processing exclusively with processing in V1. Rather than undercutting the strategy of hypothesizing identities, though, determining visual function in these other areas led to more identity claims that were even more detailed and that identified various aspects of visual processing with neural processes in additional brain areas.

In 1982 Mishkin and Ungerleider proposed that visual processing in cortex followed two pathways beyond V1, a ventral pathway into inferior temporal cortex, which processed information about the identity of stimuli, and a dorsal pathway into parietal cortex that processed information about the location of stimuli (figure 7). Livingstone and Hubel (1984) extended this proposal back to the retina. Some (Milner & Goodale, 1995) have challenged the precise characterization of the processing in the two pathways, but most research (van Essen & Gallant, 1994) supports the general conception of two partially segregated processing streams.

The discovery of multiple brain areas that seem to be processing different visual information has proved the principal guide to detailed characterization of visual processing in the brain, not top-down analyses in psychology or artificial intelligence (e.g., motivated by Marr, 1982). Subsequently,

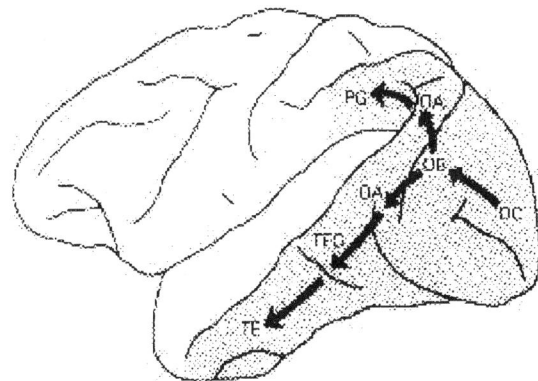

Figure 7: Ungerleider and Mishkin's (1982) proposal of two pathways in visual processing.

though, these neurobiological accounts have played a heuristic role in developing higher-level analyses of vision. Ulric Neisser (1989) was one of the first cognitive theorists to draw upon the two pathway account proposed by Mishkin and Ungerleider. Neisser construed the dorsal pathway as embodying Gibson's notion that we directly see the layout of the environment and the ventral pathway as responsible for more inferential cognitive processing (see also Milner and Goodale, 1995). Also, in the 1980s two groups of connectionist modelers developed modularized networks that separately performed *what* and *where* analyses of images on a simple retina in order to determine the computational advantages of separate pathways (Jacobs, Jordan, & Barto, 1991; Rueckl, Cave, & Kosslyn, 1989). Finally, psychologists have recently developed behavioral measurers (e.g., speed of processing) capable of demonstrating the difference in pathways in normal behaving humans (Hale, 1996).

Conclusion

HIT (Heuristic Identity Theory) proposes that identity claims between psychological processes and neural mechanisms are advanced as heuristics that serve to guide further research. Emphasizing the thoroughly hypothetical character of identity claims in science, HIT focusses on the way that proposed identifications of psychological and neural processes and structures contribute to the integration and improvement of our neurobiological and psychological knowledge. Hypothesized identities advance research by suggesting new avenues for the empirical investigation of both mind and brain. The resulting empirical findings motivate scientists at both levels to tinker with their conceptions of the pertinent processes and structures. As even the brief discussion of visual processing demonstrates, these hypothetical identities evolve in response to on-going research. Explanatory and predictive successes are what justify these identity claim and what make additional theoretical and evidential resources available in future research.

In response to both the correlation objection and the multiple realizability objection, HIT stresses the importance of attending to the contributions psychophysical identity claims have made over time to progressive programs of research in neuroscience and psychology. It is difficult to imagine that at the turn of the millennium any philosophers would regard these considerations as even secondary, let alone irrelevant, to evaluating the identity theory. We can think of no more reasonable grounds for adjudicating these matters.

References

Bechtel, W. and Mundale, J. (1999). Multiple realizability revisited: Linking cognitive and neural states. *Philosophy of Science*, in press.

Brodmann, K. (1909/1994). *Vergleichende Lokalisationslehre der Grosshirnrinde* (L. J. Garvey, Trans.). Leipzig: J. A. Barth.

Chalmers, D. (1996). *The conscious mind*. Oxford: Oxford University Press.

Cowey, A. (1964). Projection of the retina on to striate and prestriate cortex in the squirrel monkey *Saimiri Sciureus. Journal of Neurophysiology, 27*(1964), 366-393.

Ferrier, D. (1876). *The functions of the brain*. London: Smith, Elder, and Company.

Glickstein, M. (1988). The discovery of the visual cortex. *Scientific American, 259*(3), 118-127.

Goldberg, M. E., & Robinson, D. L. (1980). The significance of enhanced visual responses in posterior parietal cortex. *Behavioral and Brain Sciences, 3*, 503-505.

Gross, C. G., Rocha-Miranda, C. E., & Bender, D. B. (1972). Visual properties of neurons in inferotemporal cortex of the macaque. *Journal of Neurophysiology, 35*, 96-111.

Hale, S., Chen, J, Myerson, J. and Simon, A. (1996). Behavioral Evidence for Brain-based Ability Factors in Visuospatial Information Processing. Presentation at the Psychonomics Society.

Henschen, S. E. (1893). On the visual path and centre. *Brain, 16*, 170-180.

Hubel, D. H., & Wiesel, T. N. (1962). Receptive fields, binocular interaction and functional architecture in the cat's visual cortex. *Journal of Physiology, 28*, 229-289.

Hubel, D. H. & Wiesel, T. N. (1965). Receptive fields and functional architecture in two non-striate visual areas (18 and 19) of the cat. *Journal of Physiology, 195*, 229-289.

Hubel, D. H., & Wiesel, T. N. (1968). Receptive fields and functional architecture of monkey striate cortex. *Journal of Physiology (London), 195*, 215-243.

Jacobs, R. A., Jordan, M. I., & Barto, A. G. (1991). Task decomposition through competition in a modular connectionist architecture: The what and where vision tasks. *Cognitive Science, 15*, 219-250.

Kuffler, S. W. (1953). Discharge patterns and functional organization of mammalian retina. *Journal of Neurophysiology, 16*, 37-68.

Lashley, K. S. (1950). In search of an engram. *Symposium on Experimental Biology, 4*, 45-48.

Livingstone, M. S., & Hubel, D. H. (1984). Anatomy and physiology of a color system in the primate visual cortex. *Journal of Neuroscience, 4*, 309-356.

Marr, D. C. (1982). *Vision: A computation investigation into the human representational system and processing of visual information*. San Francisco: Freeman.

McCauley, R. N. (1981). Hypothetical identities and ontological economizing: Comments on Causey's program for the unity of science. *Philosophy of Science, 48*, 218-227.

Meynert, T. (1870). Beiträgre zur Kenntniss der centralen Projection der Sinnesoberflächen. *Sitzungberichte der Kaiserlichten Akademie der Wissenshaften, Wien. Mathematish-Naturwissenschaftliche Classe, 60*, 547-562.

Milner, A. D., & Goodale, M. G. (1995). *The visual brain in action*. Oxford: Oxford University Press.

Neisser, U. (1989). *Direct perception and recognition as distinct perceptual systems*. Paper presented to the Cognitive Science Society 1989.

Putnam, H. (1967). Psychological Predicates. In W. H. Capitan & D. D. Merrill (Eds.), *Art, Mind and Religion*. Pittsburgh: University of Pittsburgh Press.

Rueckl, J. G., Cave, K. R., & Kosslyn, S. M. (1989). Why are "what" and "where" processed by separate cortical visual systems? A computational investigation. *Journal of Cognitive Neuroscience, 1*, 171-186.

Talbot, S. A., & Marshall, W. H. (1941). Physiological studies on neural mechanisms of visual localization and discrimination. *American Journal of Ophthalmology, 24*, 1255-1263.

Ungerleider, L. G. Mishkin, M. (1982). Two cortical visual systems. In D. J. Ingle, M. A. Goodale, and R. J. W. Mansfield (Ed.), *Analysis of Visual Behavior*. Cambridge, MA: MIT Press.

van Essen, D. C., & Gallant, J. L. (1994). Neural mechanisms of form and motion processing in the primate visual system. *Neuron, 13*, 1-10.

Zeki, S. M. (1969). Representation of central visual fields in prestriate cortex monkey. *Brain Research, 14*, 271-291.

Zeki, S. M. (1973). Colour coding of the rhesus monkey prestriate cortex. *Brain Research*, 422-427.

Zeki, S. M. (1974). Functional organization of a visual area in the posterior bank of the superior temporal sulcus of the rhesus monkey. *Journal of Physiology, 236*, 549-573.

Zeki, S. M., Watson, J. D. G., Lueck, C. J., Friston, K. J., Kennard, C., Frackowiak, R. S. J. (1991). A direct demonstration of functional specialization in human visual cortex. *Journal of Neuroscience, 11*, 641-649.

Memory for Analogies and Analogical Inferences

Isabelle Blanchette (`isablan@psych.mcgill.ca`)
Department of Psychology; McGill University, 1205 Dr. Penfield Avenue,
Montréal, Québec, Canada, H3A 1B1

Kevin Dunbar (`dunbar@ego.psych.mcgill.ca`)
Department of Psychology; McGill University, 1205 Dr. Penfield Avenue,
Montréal, Québec, Canada, H3A 1B1

Abstract

An important property of analogical reasoning is that resulting inferences can be used to acquire new knowledge in a target domain. However, little is known about what happens to memory for these inferences. In this study, we explore the link between analogical reasoning, inferences, and memory. We gave participants information on a political debate. Some subjects were given a short text and other subjects were given a long text to read. In addition, half the subjects were given an analogy at the end of the text. A week later, subjects were brought back and asked to recall the information. We were particularly interested in whether subjects would (a) remember the analogy, and (b) incorporate analogical inferences into their memory for the text. We found that when they were given more information, subjects did not report the analogy, but falsely included analogical inferences in their recall. Results were different when subjects were given a lesser amount of information - they remembered the analogy and did not erroneously recall analogical inferences. Overall, the results indicate that memory for analogical inferences is highly related to the amount of information that people are given.

Introduction

While much is known on the processes underlying analogical reasoning, relatively little is known about the effects of analogy on memory. Specifically, many accounts of analogical reasoning stress the idea that analogy can be used to fill in gaps in existing knowledge about new topics (Gentner & Holyoak, 1997; Holyoak & Thagard, 1995; Vosniadou & Ortony, 1989). Thus, an important component of analogical reasoning is drawing inferences. Surprisingly, little is known about the consequences of this process; how analogies and analogical inferences are remembered. The goal of the research reported in this paper is to examine the link between analogy and memory.

Much research has been conducted on the mechanisms underlying the process of making analogical inferences. When people engage in analogical reasoning, their representations of the source and target are aligned and elements of the source are matched to those of the target (Gentner & Markman, 1997; Markman, 1997). Missing information about the target can be filled by importing knowledge from the source domain, making an analogical inference. This is a powerful mechanism for the acquisition of new knowledge. Many studies have been conducted to identify the constraints placed on analogical mapping. The process of matching the two representations and drawing inferences is influenced by structural constraints, such as isomorphism and systematicity (Clement & Gentner, 1991; Markman, 1997), pragmatic considerations (Spellman & Holyoak, 1996), task given to the subject (Blanchette & Dunbar, in press; Dunbar, in press), and the semantic content of the source and target (Bassok, Chase, & Martin, 1998). While much is known about the constraints placed on this process, we know little about the consequences of making an analogical inference, and particularly how this will alter memory for the information presented.

One aspect of the link between analogy and memory that has been explored is whether an analogy can facilitate recall of specific concepts. Educational research has shown that, under some circumstances, a relevant analogy presented with other information can enhance memory for that information (e.g.; Halpern, Hansen & Riefer, 1990; Stepich & Newby, 1988; Vosniadou & Schommer, 1988). All this research suggests that the link between analogy and memory is highly complex. However, one thing that is clear is that analogies can affect how much information will be retained. When presenting information, relating it to a better-known domain (the source) through an analogy probably increases the amount of elaboration on the new information. This, in turn, can increase memory for that information.

Although we know that analogies can impact memory for other information, little is known about memory for analogies themselves and memory for analogical inferences that can be derived from the analogy but that were not explicitly present in the information supplied. If people engage in analogical reasoning when they are presented with both a source and a target, we can expect that, through mapping elements of the source onto the target, they will draw analogical inferences. What happens to the analogical inference? Does the inference become part of the underlying representation? Research in text comprehension has shown that people often cannot differentiate between information they actually read, and inferences they drew from this information (van den Broek, 1994). Although this process does not involve analogical reasoning, the same thing could occur with analogical inferences. Thus, we would predict that when people are presented with a source and target analog, they will draw analogical inferences that will be incorporated into their representation of the information. Furthermore, it will be interesting to see whether this is related to an explicit memory for the analogical source itself.

These issues are important as in our previous research (Blanchette & Dunbar, 1997; Dunbar, 1995, 1997), we have found that analogy is frequently used in many naturalistic settings such as science and politics. We have also found that scientists often forgot the analogies that they had used.

We decided to investigate memory for analogies and analogical inferences using political debates. We used two debates that are often mentioned in the media -legalization of marijuana and funding of sports stadiums. Our choice was based on the fact that analogies used in these debates have a number of interesting properties that are common to many naturalistic uses of analogy. In naturalistic settings, analogies are usually not explicitly mapped out for the audience (see Blanchette & Dunbar, 1997). In most cases, the analogical source is described but the explicit mappings are left up to the reasoner to infer. This makes it possible to look at memory for inferences derived through analogical reasoning, information that was not explicitly presented.

One important issue that we were concerned with is whether the way people use analogy will vary with the complexity of the materials provided. Most researchers use very simple stimuli, yet in our pilot studies we found that giving subjects analogies for simple texts, containing little information, had little effect on their understanding of an issue. Thus, we created two conditions, varying the amount of information on the target problem that we provided to subjects. We used both a simple text and a more complex text, providing more information. We expected that recall of analogical inferences would be more important in the condition where people have a lot of information about the target problem, simply because these inferences would be more readily drawn when the analogy is initially presented.

Overview of study

In this experiment, subjects received information on a social/political issue. We manipulated two variables: the presence of an analogy and the amount of information presented. Participants read a text corresponding to one of the four conditions resulting from the crossing of these two variables. One week later, subjects came back and we tested their memory for the information through a free recall task. We coded their recall for inclusion of analogical inferences and the analogy itself. In addition, we measured the total number of facts included in participants' recall. We also measured participants' opinion on the target issue to examine its possible influence.

Method

Participants and procedure

Forty-eight undergraduate students participated in this study. Participants were told the goal of the experiment was to investigate reasoning about complex issues. The experiment took place in two sessions. In the first session, participants read a text about the target issue and answered a few reasoning questions. Participants from all four groups answered the same questions. One asked them to list arguments for and against a specific position on the target issue and the other asked them to state and justify their opinion. The second session was held exactly one week later. Participants answered, by writing, a free recall question. This question instructed subjects to write down, as precisely as possible, all the information they could recall from the text they had read the previous week. Four participants had to be eliminated because they did not participate in the second session. Of these, one was in the complex/analogy condition, two were in the simple/no analogy, and one was in the simple/analogy condition.

Materials

We used two different issues to ensure replication. Each participant got only one of the two. The first issue was the debate over the legalization of marijuana. In this case, the source analog was the period of the prohibition of alcohol. It described how people continued to use alcohol even though it was illegal and how a black market for alcohol products developed. The second issue was whether public funds should be used to help professional sport teams build new infrastructures such as stadiums. In this case, the source analog referred to a situation people in Quebec (where the experiment was conducted) had experienced the previous year. During an ice storm which caused massive power failure, some people selling generators were abusively raising prices because the demand was very high. This could be related to sport teams asking money from cities knowing that the demand for a team is very high. In both cases, the analogy was inserted at the end of the text by using the sentence "The situation with [marijuana/sports teams] can be compared to...". A short paragraph then described the source analog. It is important to note that there were no explicit analogical inferences in the text. Only the source analog was described and the link between source and target was only made by the first sentence described above. In the control condition, participants read the exact same text apart form the analogy. The last paragraph containing the analogy was omitted and there was no mention of the source analog.

In order to vary the amount of information presented, two texts were prepared for each issue. In the "Complex" condition, the text was approximately three pages long. The text gave a lot of detailed information on the issue, often contradictory information, and it presented arguments for both sides. In the "Simple" condition, the text contained only one paragraph stating basic facts about the issue.

Measures

Participants' answers to the free recall question were coded for three things: explicit recall of the analogy, memory for analogical inferences, and total number of facts recalled. We used the following criterion to code recall of the analogy: if the subjects mentioned anything about the analogical source, or about the fact that there was an analogy, we coded it as positive, otherwise it was coded as negative.

We coded subject's recall for inclusion of analogical inferences. The list of possible inferences was determined as follows. The description of the source analog contained a number of statements. The list of possible inferences was established by listing all these statements as they would apply to the target. We took the paragraph describing the source and simply replaced words/concepts identified to the source by the equivalent for the target. For example, the description of the prohibition of alcohol source included the statement that making alcohol illegal gave rise to elaborate criminal organizations that took over distribution and production of the substance. Here, the analogical inference would be that having marijuana illegal results in criminal organizations taking care of the distribution and production of the substance. We coded subjects' answers for all such analogical inferences.

We calculated the total number of facts contained in each participant's recall. This measure was included to control for the possibility that differences found on the other dependent measures (memory for the analogy and for analogical inferences) were actually confounded with the overall amount of information recalled by participants in the different conditions. For each subject, we counted the number of different and mutually exclusive facts that were recalled.

In order to examine its possible mediating influence, we also measured participants' opinion on the target issue - legalization of marijuana or public funding for stadiums. During session one, after they had read the text, subjects indicated their opinion on a seven-point scale, ranging from 'totally in favor' to 'totally opposed'.

Results

Memory for analogical inferences

Because most participants who included an inference in their recall only included one, we coded this measure as either yes or no. Overall, 18 out of 44 subjects (41%) included at least one analogical inference in their recall. To determine that inclusion of analogical inferences in the recall was actually related to the presence of an analogy in the text, we compared the analogy conditions to the no-analogy conditions using a chi square test. We needed to compare the two conditions as the analogical inferences were propositions that could be common knowledge and could just be intrusion errors not related to the presence of the analogy. There was a significant difference in the expected direction, χ^2 (1, N=44) = 4.86, p<.05. In the analogy conditions, 13 out of 23 subjects (57%) included at least one analogical inference in their recall whereas this was the case for only 5 of the 21 participants (24%) in the non-analogy conditions.

There were different patterns in the simple and complex conditions, as can be seen in Figure 1. There was a marked difference between the analogy and no-analogy conditions when participants read a large amount of information (complex condition). In this condition, 9 out of 12 subjects in the analogy group (75%) mentioned an inference in their recall, compared to 3 out of 12 (25%) in the no-analogy group. This difference is significant, χ^2(1, N=23) = 5.24, p<.05. However, there was no difference between the analogy and no-analogy groups in the "Simple" condition, χ^2 (1, N=21)=0.69, p>.05. When participants read a smaller amount of information (one paragraph), being presented with

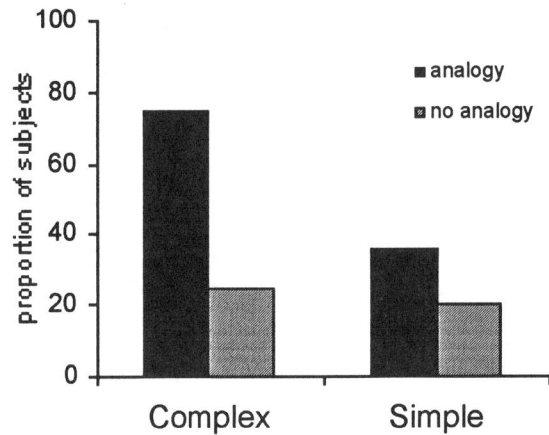

Figure 1: Inclusion of analogical inferences in recall as a function of condition and complexity

an analogy did not lead them to include inferences in their recall (4/11 in the analogy condition compared to 2/10 in the no-analogy group).

Memory for the analogy

Participants' recall of the analogies followed the opposite pattern as memory for analogical inferences. Overall, 9 out of the 23 subjects in the analogy conditions (39%) explicitly mentioned the analogy in their recall. However, there was a significant difference between the complex and simple conditions, χ^2 (1, N=23) = 5.32, p<.05 (see Figure 2). In the complex condition, only 2 out of 12 subjects (17%) reported the analogy whereas 7 out of 11 participants (64%) in the simple condition did. Thus, subjects in the simple conditions reported the analogy in their recall whereas most subjects in the complex condition did not.

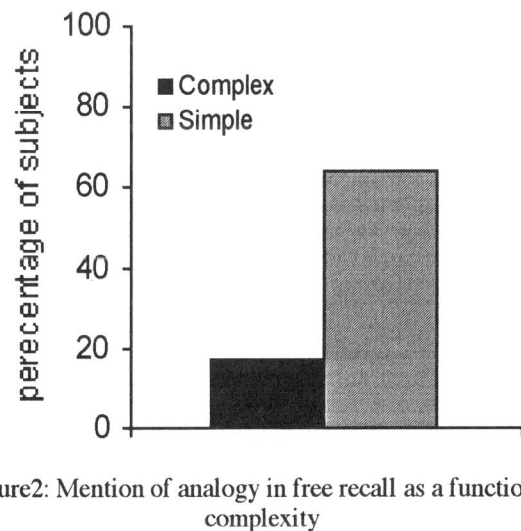

Figure2: Mention of analogy in free recall as a function of complexity

Total number of facts

We wanted to examine whether the differences in memory for the analogy and memory for analogical inferences were due to a difference in the overall amount of information recalled by participants in the different conditions. We performed a 2 by 2 (Analogy and Complexity) ANOVA on the total number of facts in the recall. There was no main effect of analogy, $F(1, 40) = 1.11$, $p > .05$. Participants in the analogy and no-analogy groups did not differ on the total number of facts that they recalled ($M = 3.96$, $M = 3.48$ respectively). The difference found for inclusion of analogical inferences in recall therefore cannot be the result of a general effect of the analogy on memory. There was of course a significant main effect of complexity, $F(1, 40) = 88.64$, $p < .05$. Participants in the Complex condition recalled more facts ($M = 5.74$) than participants in the Simple condition ($M = 1.52$), simply because there was more information to be remembered. However the two-way interaction was not significant, $F(1, 40) = 0.91$, $p > .05$, showing that there was no more difference in number of facts recalled between the analogy and no-analogy groups in either the simple or complex conditions. This analysis provides evidence that our results are not a simple artifact resulting from a widespread impact of analogy on memory.

Influence of opinion

In order to explore possible mediating effects of prior opinion, we analyzed recall for inferences and recall for the analogy in terms of participants' opinion (measured during session one). First, we needed to determine whether there was a difference in opinion between the four different groups. We performed a 2 by 2 ANOVA on the mean opinion score as a function of the two variables: presence of an analogy and complexity. There were no significant differences between the groups: there was no effect of the analogy manipulation, $F(1, 40) = 3.14$, $p > .05$, and no effect of the information manipulation, $F(1, 40) = 2.78$, $p > .05$. To see if prior opinion had an impact on the recall for the analogy, we compared, for participants in the analogy condition, the mean opinion of those who had mentioned the analogy in their recall to those who hadn't. This was done through a simple unpaired t-test. This test indicated no significant difference, $t(21) = 1.28$, $p > .05$. We performed the same analysis for analogical inferences, comparing the mean opinion of those who included analogical inferences in their recall to those who didn't. Again, a t-test revealed no significant difference, $t(42) = .35$, $p > .05$. The mean opinion scores are presented in Table 2. Overall, it appears that the pattern of results we observed is not altered by prior opinion.

Table 1: Opinion score and inclusion of inferences in recall in the simple and complex conditions

	Included analogical inference in recall		No analogical inferences in recall	
	Mean	SD	Mean	SD
Complex	4.33	1.66	5.33	1.15
Simple	3.75	2.06	3.43	2.51
Total	4.04	1.86	4.38	1.83

Discussion

In this study, we looked at people's memory for analogies and analogical inferences. We manipulated the presence/absence of an analogy in a text that subjects read, and the amount of information that was provided.

Two important findings emerged from this study. First, the fact that analogical inferences can be falsely remembered as being part of information presented. Although analogical inferences were not part of the information presented, under some conditions, participants did not report the analogy itself but reported inferences drawn from it. The second important finding is that the amount of information people have on a topic will determine how the analogy will influence their memory. Recall in the simple and complex conditions followed opposite patterns. In the complex condition, people reported analogical inferences but not the analogy itself. In the simple condition, people did recall the analogy and did not report analogical inferences. Our results also allow us to rule out the possibility that the analogy simply influenced the overall amount of information recalled by participants.

The results of this study indicate that analogy can have a powerful influence on peoples' memory for a text. When participants are given complex information, this allows inferences to be drawn from the analogy and these inferences appear to become part of the underlying representation of the text as the analogy is forgotten. When the information provided is simple, the analogy does not change the underlying representation of the text and the analogy is correctly remembered. Of course, in the case where participants received less information, the description of the analogical source represented a greater proportion of the total amount of information presented. As such, it might be easier to recall the analogy. Interestingly, when they had a lot of information, participants reported the outcome of analogical reasoning processes, without reporting the source of this reasoning, which was explicitly present in the text.

A similar phenomenon has been observed in a real-world study of scientific reasoning. Dunbar (1995, 1997) studied different scientific laboratories over extended periods. He was able to follow the unfolding of a number of scientific discoveries. In many cases, analogies were used in reasoning about data that led to a discovery. Dunbar went back to these laboratories some time later and interviewed the scientists. The researchers' memory of the events included the discovery that had been made but often kept no trace of the analogies used.

The results of this experiment are similar to those on text comprehension. Many studies have shown that people spontaneously generate inferences based on the materials they read and later cannot distinguish between these inferences and information actually presented in the text (Graesser, Singer, & Trabasso, 1994; Lorch & van den Broek, 1997; van den Broek, 1994). Similarly, research on memory has shown that people can incorporate false information, information provided afterwards, into their memory for an event (Ayers & Reder, 1998; Belli & Loftus, 1996; Loftus, 1992). Thus in both our experiment and other experiments, people incorporated new information into their underlying representation either by making inferences or being told new information. This drawing of inferences does not appear to be a totally explicit process (Schunn & Dunbar, 1997; Schacter, 1995), as people were unable to recall the source analog, or know that these inferences were drawn from an analogy. These findings thus suggest that drawing an inference from an analogy can alter the underlying representation for information just as other types of processes can.

Our results show that in order for participants' memory trace to incorporate analogical inferences, they must have a sufficient knowledge base that will allow them to draw the inference. Only participants who were provided with a greater amount of information on the target domain "recalled" the inferences. It must be emphasized that in our experiment, participants were not asked to draw inferences or to reason about the analogy. The inferences appear to have been drawn as part of the processing of the information presented, without any specific prompting being required

The results of this experiment demonstrate that people do use analogies to make inferences about a target problem and that these analogies can alter their underlying representation. Furthermore, when there is a lot of information present, people are unable to recall the analogies and incorporate inferences into their underlying representation. We suspect that one of the reasons that politicians so frequently use analogies when describing complex situations is that by making an analogy, they can deliver information covertly, without explicitly providing it.

Acknowledgments

This research was supported by a SSHRC graduate fellowship to the first author and NSERC grant OGP0037356 to the second author. The authors thank Sylvain Sirois and Tanja Rapus for their helpful comments.

References

Ayers, M.S., & Reder, L.M. (1998). A theoretical review of the misinformation effect: Predictions from an activation-based memory model. *Psychonomic Bulletin and Review, 5*, 1-21.

Bassok. M., Chase, V.M., & Martin, S.A. (1998). Adding apples and oranges: Alignment of semantic and formal knowledge. *Cognitive Psychology, 35*, 99-134.

Belli, R.F., & Loftus, E.F. (1996). The pliability of autobiographical memory. Misinformation and the false memory problem. In: D.C. Rubin (Ed). *Remembering our past: Studies in Autobiographical Memory*. New York NY: Cambridge University Press.

Blanchette, I., & Dunbar, K. (1997). Constraints underlying analogy use in a real-world context: Politics. In.: M.G. Shafto and P. Langley (Eds). *Proceedings of the Nineteenth Annual Conference of the Cognitive Science Society*, Lawrence Erlbaum Associates, Stanford, CA.

Blanchette, I., & Dunbar, K. (In Press). How Analogies are Generated: The Roles of Structural and Superficial Similarity. *Memory & Cognition*.

Clement, C.A., & Gentner, D. (1991). Systematicity as a selection constraint in analogical mapping. *Cognitive Science, 15*, 89-132.

Dunbar, K. (1995). How scientists really reason: Scientific reasoning in real-world laboratories. In: R.J. Sternberg &

J.E. Davidson (Eds). *The Nature of Insight*. Cambridge MA: MIT Press.

Dunbar, K. (1997). How scientists think: Online creativity and conceptual change in science. In: T.B. Ward, S.M. Smith, & S. Vaid (Eds). *Conceptual Structures and Processes: Emergence, Discovery, and Change*. Washington DC: APA Press.

Dunbar, K. (In Press). The analogical paradox: Why analogy is so easy in naturalistic settings, yet so difficult in the psychology laboratory. In Koichiv, B., Holyoak, K.,& Gentner, D. (Eds.) *Analogy 2000*. MIT press. Cambridge: MA.

Gentner, D., & Holyoak, K.J. (1997). Reasoning and learning by analogy: Introduction. *American Psychologist, 52*, 32-34.

Gentner, D., & Markman, A.B. (1997). Structure mapping in analogy and similarity. *American Psychologist, 52*, 45-56.

Graesser, A.C., Singer, M., & Trabasso, T. (1994). Constructing inferences during narrative text comprehension. *Psychological Review, 101*, 371-395.

Halpern, D.F., Hansen, C., & Riefer, D. (1990). Analogies as an aid to understanding and memory. *Journal of Educational Psychology, 82*, 298-305.

Holyoak, K.J., & Thagard, P. (1995). *Mental leaps: Analogy in creative thought*. MIT Press, Cambridge, US.

Loftus, E.F. (1992). When a lie becomes memory's truth: Memory distortion after exposure to misinformation. *Current Directions in Psychological Science, 1*, 121-123.

Lorch, R.F., & van den Broek, P. (1997). Understanding reading comprehension: Current and future contribution of cognitive science. *Contemporary Educational Psychology, 22*, 213-246.

Markman, A.B. (1997). Constraints on analogical inferences. *Cognitive Science, 21*, 373-418.

Schacter, D.L. (1995). *Memory distortions: How minds, brains, and societies reconstruct the past*. Harvard University Press, Cambridge, US.

Schunn, C., & Dunbar, K. (1996). Priming, analogy, and awareness in complex reasoning. *Memory and Cognition, 24*, 271-284.

Spellman, B.A., & Holyoak, K.J. (1996). Pragmatics in analogical mapping. *Cognitive Psychology, 31*, 307-346.

Stepich, D.A., & Newby, T.J. (1988). Analogical instruction within the information processing paradigm: effective means to facilitate learning. *Instructional Science, 17*, 129-144.

van den Broek, P. (1994). Comprehension and memory of narrative texts: Inferences and coherence. In: M.A. Gernsbacher (Ed.). *Handbook of Psycholinguistics*. San Diego, CA: Academic Press.

Vosniadou, S., & Ortony, A. (1989). *Similarity and analogical reasoning*. Cambridge University Press, NY, US.

Vosniadou, S. & Schommer, M. (1988). Explanatory analogies can help children acquire information from expository text. *Journal of Educational Psychology, 80*, 524-536.

Mental models and pragmatics: the case of presuppositions

Guido Boella, Rossana Damiano and Leonardo Lesmo
Dipartimento di Informatica e Centro di Scienza Cognitiva
Università di Torino
{guido,rossana,lesmo}@di.unito.it

Keywords: Mental models, linguistics, pragmatics and presuppositions

Abstract

We claim mental models are a framework that allows to shed light on the phenomenon of presuppositions. A plan-based lexical representation for verbs, together with the effect of conversational implicatures that discharge possible mental models, are the key features of this proposal.

Introduction

Presuppositions are what the utterance of a sentence assumes to be true (or takes for granted). They can be triggered by different elements in sentences, going from the use of definite NP's (*the King of France is bald* presupposes *there is a (unique) King of France*) to the presence of factive verbs (*he regrets that he has been impolite* presupposes *he has been impolite*). These inferences are characterized by resistance to negation (e.g. *he does not regret that he has been impolite* presupposes *he has been impolite*), and by cancellability in the presence of contextual information.

This paper addresses what may be called "lexical" presuppositions. In particular, the event structure, as described by certain verbs, allows to draw some inferences about the described situation. These verbs are classified in (Beaver, 1997) as "signifiers of actions and temporal/aspectual modifiers": "most verbs signifying actions carry presuppositions that the preconditions for the action are met" (Beaver, 1997) page 944.

We think that what accounts for "preconditions for the action" still requires a more precise characterization. Where do preconditions come from? It seems rather vague to state that *leaving* is a precondition for *arriving* and *climbing* is a precondition for *reaching the top*. We claim that these presuppositions arise naturally from a representation of the semantics of verbs in terms of actions and plans describing the steps required to carry out the activity referred to by the verb. It may be observed that a plan-based representation is complex, and rather expensive to build. However, we have noticed elsewhere that it is independently required both for accounting for the meaning of communication verbs (Goy and Lesmo, 1997) and for the maintenance of coherence in dialogues (Ardissono et al., 1998).

The basic idea is that, in order to understand an utterance, one must build a mental representation of the described situation: it is clear, e.g., that any representation of *arrive* must include that of *leave*. The role played by *mental models* (Johnson-Laird, 1983) is to provide a reasoning framework for these representations, that allows to explain why presuppositions survive in certain contexts. For instance, these representations make it possible to account for the fact that *John didn't arrive* presupposes *John left*, as predicted by the projection of presuppositions across negation; moreover, this inference is correctly modeled as a defeasible one, since it is possible to say *John didn't arrive, he didn't even leave*.

The interpretation is carried out in the following way: first of all, the aspectual features are accounted for by means of a plan-based representation of the lexical meaning; moreover, the temporal features represented by mental models in the style of (Schaeken et al., 1996), as well as the interpretation of the negation and of the context; this amount to building one of more *mental models* representing the described event. The information shared by all models is what the sentence entails.

At this point, potential conversational implicatures are applied: their effect is to discharge those mental models of the event that could have been more precisely described by alternative linguistic expressions. In this way, we gain more information, since we are left with the smaller set of mental models: presuppositions are the information that becomes shared by all these remaining models.

However, implicature can be defeated in the presence of further contextual information, thus blocking the discharge of models: presuppositions cannot arise anymore, because they do not occur in all event models, therefore appearing defeasible too.

In the following, we will describe a plan-based semantic representation of action verbs and our treatment of negation and of the temporal expression *before*; then, the conversational implicature mechanism will be introduced. Next, we face the problem of the projection of presuppositions. Comparison with related approaches and conclusions close the paper.

```
st                              et

━━━━━━━━━━━━━━━━━━━━━━━━━━━━━━━━━━━
at(a) ◄─prec── walk(j,a,n) ──eff─► at(n)
                  ╱  ╱  ╲  ╲
      ...   ... step(i,l)  step(m,n)
                          reference time↑
```

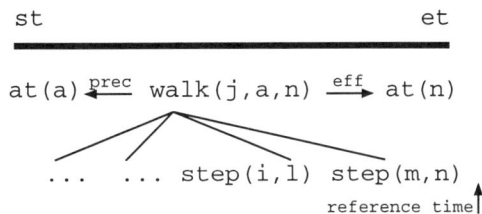

Figure 1: The mental model of *John arrived to* n. The st-et line represents the event time of the action.

A plan-based representation of verbs

We will mainly consider verbs denoting actions that are intentionally executed by an agent. The meaning of these verbs is represented by action schemata, that describe how actions are carried out by a sequence of steps. When a sentence is interpreted, an action instance is built on the basis of the schema corresponding to the main verb; then, a set of constraints expressing the temporal and aspectual information conveyed by the other linguistic elements (aspectual predicates, adverbials, verb arguments and so on) is added to this representation, by means of condition-action rules (Boella and Damiano, 1999). Temporal constraints refer to the occurrence of the action instance with respect to speech time and reference time, following a reichenbachian temporal reference schema.

An action schema Act is composed of arguments, among which the agent (agt (Act)) and the start and end time of the action (st(Act) and et(Act), respectively, denoting temporal points), the preconditions and effects of the action, and the action decomposition (body(Act)), composed of steps; the start and end time of the steps can be specified too.

Usually, a mental model representing an action does not contain all the step instances (tokens) of the plan but only a subset of them. Since steps express the focused part of the action and its temporal placement, only the steps must be included that are needed to represent the constraints resulting from the interpretation. The remaining steps can be later inferred and added to the representation, if they become necessary for reasoning purposes. Moreover, action schemata can be only partially instantiated, to account for the fact that linguistic expressions describe an event by highlighting only certain features of its, that the speaker considers relevant to his goals. In our approach, this corresponds to building models in which only the currently relevant steps are represented, in order to focus on a specific phase of the action or to represent the fact that the action has been only partially executed.

For example, the interpretation of *John walked to the store* contains only the first and the last steps of the plan and constrains them to precede the reference time (here coinciding with the speech time); on the other hand,

John arrived is not represented as a punctual event but as an instance of the action *move* (that, in this context, specializes into *walk*) containing in its decomposition only the last steps of the plan, plus the temporal constraints specifying that both the whole action of walking and its last steps happened before the reference time (in Figure 1 the decomposition links are denoted by the lines connecting the walk(j,a,n) token to the tokens representing the steps). On the contrary, if the sentence to be interpreted were *John was arriving*, an instance would be built where the reference time occurs within the sequence of steps that conclude the action. In both cases, the steps preceding the final sequence are not represented, but are assumed to exist on the basis of the action instance inner relations and can be consequently added to the representation if they become relevant. According to this representation, the mental model of *John arrived* entails the model of *John left*, where, in turn, the initial steps of the action *walk* are constrained to precede the reference time.

Therefore, the presupposition *John left* is not confined to a separate set of propositions that must be accommodated in the representation as in (Van Der Sandt, 1992); moreover, the requirement is satisfied that presuppositions are in some sense inserted at the beginning of the sentence interpretation (Beaver, 1997) and not added or calculated afterwards, as in (Karttunen, 1974; Gazdar, 1979b).

On the other hand, we do not face here the problem of the anaphorical character of presuppositions, exemplified in *he left an hour ago but he didn't arrive*, where the two clauses refer to two phases of the same underlying action of walking.

The representation of negation

We have adopted a peculiar treatment of negation, in order to account for the fact that the negated event was somehow expected to happen. We interpret such expectations by representing them in terms of intentions attributed to the involved agent; mental models containing unexecuted action instances can be readily interpreted as an agent's mental description of another agent's intentional state (Bratman et al., 1988).

The representation of *John did not arrive* includes the related plan instance of walking, and its decomposition into steps; the negation is not represented by labeling the last steps in the plan instance as denied (as mental model theory prescribes (Johnson-Laird, 1983)). On the contrary, starting from the premise that the walking action did not end before the reference time R[1], all the representations that include the plan instance (act) and satisfy the constraint $et(act) \geq R$ are allowed. Note that

---

[1] Otherwise we would have the representation of *John arrived*.

he didn't leave   he left

1   walk / step ..step ...**step**   ↑ ref time

2   start →   walk / step ..step ...**step**   ↑ ref time

3   start →   walk / step ..step ...**step**   ref time ↑

4   start   end   walk / step ..step ...**step**   ref time ↑

he didn't arrive   he arrived

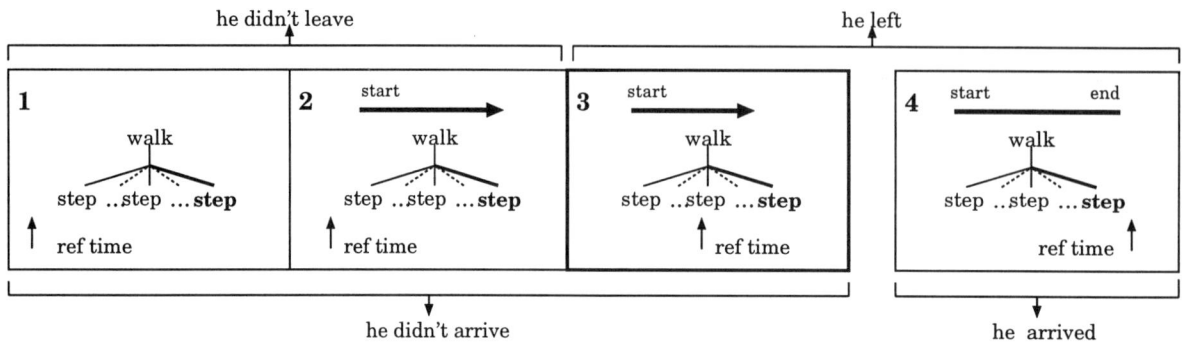

Figure 2: The relations among the mental models concerning *walk* (the thick line show whether the steps have been actually performed).

this does not imply that the *walk* action will necessarily be completed, i.e. that John will arrive later.

By using mental models, it immediately becomes apparent that two pieces of information have to be included in any event description: the fact that the end of the action necessarily follows the reference time and, given the internal constraints of the action schema, the fact that the beginning of the action precedes its conclusion. Given these constraints, two models are possible (Schaeken et al., 1996): in the former the whole action follows the reference time (R), in the latter some steps of the action precede it:

premises   models

R   et(act) ⎫   st(act)   R   et(act)
st(act)   et(act) ⎭   R   st(act)   et(act)

Actually, we have to add another variable to the model: whether the action is conceived as (partially) executed or not. This question concerns only the second model, since the part of an action preceding the reference time has certainly been executed: either the action has been or will be executed after the reference time or it hasn't and it remains as a representation of the intention of the agent.

So, the interpretation of *John did not arrive* consists of the first three models in Figure 2.

Our representation of negative sentences is strictly related to the meaning of the conjunction *before*. *Before* does not imply that the action in the subordinate clause actually happened, as shown by *Mozart died before finishing his Requiem*. As in case of negative sentences, it simply states an expectation (again represented as intentions): that the *finish* event was expected to occur at a time which follows the death (d):

d   st(act)   et(act)
st(act)   d   et(act)

Note that, for this reason, *before* is not symmetric with respect to *after*, contrarily from what (Asher and Lascarides, 1998) claim, since the latter entails the execution of both related actions: *Salieri finished Mozart's Requiem after he died in 1791*.

The parallel between negation and *before* for what concerns presuppositions will be examined below.

Conversational implicature

The gricean notion of conversational implicature is exploited to explain why some of the models are later discharged. In particular, we will consider a particular inference licensed by the maxim of quantity, the scalar implicature (Gazdar, 1979a).

When an item belongs to a salience order (i.e. a scale), the fact that the speaker has used this item instead of a different one in the scale causes the hearer to draw the inference that the speaker was not in the position to use any of the higher elements of the scale (see the model of (Hirschberg, 1985)).

For example, from *some composers died young* it is possible to infer *not all composers died young*, even if *some* per se does not entail *not all*. But *some* belongs to the scale *one, a few, some, many, most of, all*, where higher elements entail the lower ones. If the speaker has used *some* and, at the same time, he is assumed to respect the gricean maxim of quantity, this means that he cannot use the stronger element *all*, and, therefore, he intended to say that some composers last longer.

In our framework, this means that an expression (e.g. *some*) is initially interpreted by means of a set of mental models and, if some of them correspond exactly to the interpretation of another expression (e.g. *all*), then they are discharged:

Some A are B   All A are B
A
A – B
A – B
     B

A – B   A – B ⎫
A – B   A – B ⎬ Models discharged by
     B   B ⎪ the scalar implica-
A – B   A – B ⎪ ture
A – B   A – B ⎭

Arrive belongs to a scale with respect to *leave* and the same holds for *finish* and *begin*, since in both cases the former items entail the latter ones; in fact, the beginning of an action is present in every model representing its conclusion.

By reasoning with mental models it is apparent why, in case of negative sentences, scales (e.g. *leave, arrive*) can be reversed (*not arrive, not leave*). The interpretation of *John did not arrive* produces three mental models (see Figure 2); two of them (1 and 2) also represent the interpretation of *John did not leave*, as only in (3) st(act) ≤ R.

But, to describe them, the speaker would have more appropriately used *not leave*, a higher item in the scale, therefore, we can discharge the first two models from the interpretation, and keep the last one.

The presupposition *John left* now emerges since it is contained in all the remaining mental models (3).

Note that we are assuming that the speaker knows whether the higher elements of the scale (*not leave*) hold or not, otherwise this inference would not be possible. Anyway, the hearer usually knows whether the speaker has this kind of knowledge, thanks to the context in which the sentence is uttered.

In this way, the conversational implicature mechanism produces the presupposition, though in a rather indirect way, by deleting the mental models that can be better described by other sentences. Conversational implicatures are a defeasible kind of inferences: in presence of contextual information they may disappear. In our case, the cancellation of the implicature prevents the removal of the two mental models representing the fact that John has not left; therefore the presupposition cannot be drawn anymore. In *John did not arrive since he did not leave* the second clause re-asserts the first two models of Figure 2, while it negates the third one (3) - that implies that John left. In this way, the basis for the conversational implicature does not hold anymore since both items *arrive* and *leave* are denied.

In case of positive sentences, like *John arrived*, the presupposition becomes an entailment, since it is contained from the beginning of the interpretation process in the only mental model representing the sentence interpretation (4 in Figure 2), and cannot be cancelled (**John arrived but he didn't leave*).

Sentences involving *before* undergo the same reasoning based on conversational implicatures. From *Mozart met Casanova before finishing his Don Giovanni*, one can infer that he met Casanova in the period in which he was writing his opera. As stated above, *before* allows the construction of two models. One model represents the interpretation of *before he started writing his Don Giovanni*, but, since the speaker didn't use this sentence, this model is discharged, and the only model left is the one where the writing action contains the meeting event.

Furthermore, from this model it is possible to draw a further inference: Mozart actually finished his work. In the mental model we have no explicit information about the conclusion of the action of writing the composition. The inference is then licensed by another kind of motivation; as we stated above, plan instances constitute a description of an agents' intentions and the distinguishing feature of intentions is their persistency (Bratman et al., 1988): if nothing prevents him, the agent will carry out his current intentions. However, this is a defeasible kind of reasoning, and, if more information is added, the inference will be canceled: see *Mozart died before finishing his Requiem*.

The projection problem

The projection problem consists in explaining when and why the presuppositions of a clause become the presuppositions of the whole sentence where it occurs.

We start from the projection problem in disjunctive sentences. *John has stopped smoking* not only presupposes *John smoked* but, actually, implies it. In fact, the interpretation of the aspectual predicate *stop* consists of a process of smoking to which it is added the fact that the agent has not been smoking for a given period of time (see (Boella and Damiano, 1999)).

Nevertheless, in the sentence *either John stopped smoking or he never smoked* this presupposition does not hold anymore. However, since *John smoked* is implied by the first clause, we cannot resort to the cancellability of presuppositions.

A disjunction of clauses A ∨ B is represented by three mental models:[2]

A B
¬A B
A ¬B

By substituting A with the interpretation of *John stopped smoking* and B for that of *John never smoked*, we obtain a set of potential situations. The interpretation of the negated clause ¬A consists of some mental models, to which the interpretations of the clause *B* are added, resulting in a set of integrated models; then, the conversational implicatures are applied; finally, the inconsistent models are discharged (e.g. those in which is asserted that John smoked and never smoked).

John stopped smoking and *never smoked*
John did not stop smoking and *never smoked*
John stopped smoking and *smoked*

Now we add the entailments of the asserted first clause and the presuppositions of the negated one (underlined):

John stopped smoking and smoked and *never smoked*
John did not stop smoking and <u>smoked</u> and *never smoked*

---

[2] We directly flesh out the explicit models of the disjunction: in principle one should start from the implicit model that contains only the positive information:

A
  B

But in our example the negation conveys positive information, i.e. the expectation that the action happens.

John stopped smoking and smoked and *smoked*

The first model is clearly contradictory and is discharged. On the contrary, in the model ¬A B the presupposition arising from a the negated disjunct has a defeasible character, so it is canceled and the model kept. The third model is fine. At the end of the interpretation process, we have the following mental models:

John did not stop smoking and *never smoked*
John stopped smoking and *smoked*

Does this representation imply *John smoked*? Certainly not, as the two models contain opposite information.

On the contrary, in a sentence like *either John stopped smoking or he is now ill* the presupposition that John smoked is projected from the first clause to the whole sentence. In fact, the interpretation produces the following consistent models:

John stopped smoking and smoked and *is ill*
John did not stop smoking and smoked and *is ill*
John stopped smoking and smoked and *is not ill*

Now, all the models contain the information that John smoked, so, from the disjunctive sentence, it is possible to draw the inference that John smoked.

Note that the presuppositions of the two clauses are independently added to the interpretation of the whole sentence and the single models containing them are discharged if inconsistent. Therefore, differently from (Karttunen, 1974)'s approach, we are able to cope with cases in which the two clauses convey contradictory presuppositions as in: *either Fred knows he's won or he's upset that he hasn't* (Beaver, 1997).

A linguistic context where presuppositions do not survive is represented by verbs like *say* and *tell*; they prevent the projection of the presuppositions of the sentential objects: *Bill says he is not guilty* does not presuppose he is innocent. (Karttunen, 1974) has classified these verbs as *plugs*, in order to account for their behavior. In our model, such verbs are semantically interpreted as instances of the action schema of the corresponding speech acts (see (Ardissono et al., 1998)): from the precondition of the action of informing, and under the sincerity assumption, it is possible to infer only that Bill believes the proposition he uttered, while no information is given about the speaker's beliefs. Therefore, the semantic representation consists in a mental model of the action that contains an embedded mental model representing Bill's belief that he is not guilty.

Similarly, a question performed by a speaker provides a context in which the presuppositions may be cancelled, even if there is no negation: in fact, the representation of *Did John arrive?* contains an instance of the linguistic action representing questions: since it has the precondition that the questioner does not know whether the propositional content is true, two mental models are possible: one in which John arrived and another representing *John didn't arrive*.

Finally, we want to highlight how some inferences, traditionally classified as presuppositions because of their resistance to negation and cancellability, can receive a more accurate explanation than as "preconditions for actions". (Soames, 1989) noticed the different behavior of the factive verbs *regret* and *realize*: in hypothetical contexts, the former maintains its presupposition that the speaker of the utterance believes that the content of the subordinate clause holds, while the latter does not:

If I regret that I told the false, I will confess it.
If I realize that I told the false, I will confess it.

The difference emerges when the corresponding action definitions are examined. In fact, the precondition for uttering the verb *regret* is that both the speaker and the described agent, at the event time, believe that the subordinate clause p is true.

However, on the contrary, *realize* has the precondition that p is true (according to speaker's beliefs) and that the agent who realizes does not know it is: rather, the agent's knowledge that p results from the action effect. When these verbs are asserted in past tense, they share the presupposition that the speaker believes p: in fact, the action preconditions must be true. Moreover, if *realize* is asserted in the first person, the speaker and the described agent coincide. In the past tense, this means that only after the *realize* event happened, the speaker came to know that the event preconditions were true (i.e. p was true and he didn't know p). Were the sentence uttered in a hypothetical context, some mental models of the description would represent the *realize* event as not happened: in some models, the agent does not come to know that p has been true from the start, and that he was not aware of it, thus blocking the conclusion that in all models he is currently aware of p's truth.

Comparison with related works

Many approaches to presuppositions have a logical bias: the presupposition is interpreted as a function transforming contexts represented as set of sets of possible words (Beaver, 1997). However, as (Johnson-Laird, 1983) claims, logic is not a candidate for building cognitively plausible solutions to reasoning. In this work we have shown how mental models can be exploited to give an explicative solution to the problem of presuppositions that be also cognitively plausible.

(Marcu and Hirst, 1996) propose a treatment of pragmatic inferences which aims at accounting for both conversational implicatures and presuppositions in a single way. They introduce two different notions of satisfaction of a formula, where the first one is preferred in case of conflicts. This mechanism allows to distinguish between pragmatic inferences that can be canceled (i.e. conversational implicatures and presuppositions from negative sentences) and those that cannot be removed felicitously (presuppositions from positive sentences). For these reasons, different rules referring to different satisfiability notions are needed to express the fact that the same presupposition is triggered by the same lexical item in positive and negative sentences.

Moreover, they argue that a default based formalism cannot explain pragmatic inferences because they are not always cancellable. On the contrary, we keep apart the treatment of conversational implicatures from that of presuppositions. We exploit a single nonmonotonic form of reasoning for modeling implicatures, while presuppositions are explained by the interaction of mental models with scalar implicatures. Furthermore, we need no rules for deriving presuppositions, even in positive sentences, since presuppositions emerge from the plan-based representation of action verbs.

Many approaches relate presuppositions and anaphora, exploiting the DRT formalism (Van Der Sandt, 1992; Asher and Lascarides, 1998). However, the cancellability of presuppositions is not explained on the basis of a non-monotonic framework, but is based on the notion of global vs. local accommodation of the presupposed information; i.e if a presupposition in a subordinate clause is globally accommodated it becomes a presupposition of the whole sentence.

This solution has two shortcomes: presuppositions are kept apart from the asserted content and they are first introduced in the local context of the trigger: they can be removed later if they can be accommodated in a wider scope; second, defeated presuppositions (i.e. locally accommodated ones) are still present in the local context, representing information that is not true even at the local level (consider *he didn't arrived since he didn't leave*).

As a consequence, when they face the projection problem, DRT approaches have to resort on further mechanisms like discharging the global accommodation due to the lack of the informativeness of the interpretation.

In these models, the presuppositions are triggered by lexical items in an unexplained way, instead of arising from the lexical representation. Furthermore they do not take into account the differences between positive and negative contexts, and different explanations are needed for the phenomena of implicature and presuppositions.

Finally, our solution does not incur in the problem highlighted by (Zeevat, 1992): lexically triggered presuppositions must be accommodated not only globally as in (Van Der Sandt, 1992)'s approach but also locally: in our case the presupposition is not kept apart from the verb interpretation but is related to the preconditions and effects of the action.

Conclusions

Mental models were introduced by (Johnson-Laird, 1983) as a reasoning framework that is endowed with a cognitive plausibility. Here, we tried to show how mental models can be exploited to provide natural and general explanations for linguistic phenomena.

Action verb presuppositions are ruled out as an independent phenomenon, to reappear as an opportunistic phenomenon, stemming from the interaction of many other factors. In the first place, conversational implicature, and, in the second place, the reasoning on a mental model representation. This representation, in turn, relies on a plan formalism to represent actions: mental models offer a natural way to exploit action plans, and to reason on them. Defeasibility of implicatures completes the framework, causing presuppositions to look cancelled, under certain circumstances.

References

Ardissono, L., Boella, G., and Lesmo, L. (1998). An agent architecture for NL dialog modeling. In *Proc. Second Workshop on Human-Computer Conversation*, Bellagio, Italy.

Asher, N. and Lascarides, A. (1998). The semantics and pragmatics of presupposition. *Journal of Semantics*, in press.

Beaver, D. (1997). Handbook of logic and language. In van Benthem, J. and Meulen, A. T., editors, *Presuppositions*, pages 939–1008. Elsevier, Amsterdam.

Boella, G. and Damiano, R. (1999). Plan-based event representations for the analysis of tense and aspect. Submitted to conference review.

Bratman, M., Israel, D., and Pollack, M. (1988). Plans and resource-bounded practical reasoning. *Computational Intelligence*, 4:349–355.

Gazdar, G. (1979a). *Pragmatics: Implicature, Presupposition and Logical Form*. Academic Press, New York.

Gazdar, G. (1979b). A solution to the projection problem. In Oh, C. and Dineen, D., editors, *Syntax and semantics 11: Presupposition*. Academic Press, New york.

Goy, A. and Lesmo, L. (1997). Integrating lexical semantics and pragmatics: the case of italian communication verbs. In *Proc. of Int. Workshop on Computational Semantics*, Tilburg.

Hirschberg, J. (1985). *A theory of scalar implicature*. PhD thesis, University of Pennsylvania.

Johnson-Laird, P. (1983). *Mental Models*. Cambridge University Press, Cambridge.

Karttunen, L. (1974). Presuppositions and linguistic context. *Theoretical Linguistics*, 1:181–194.

Marcu, D. and Hirst, G. (1996). A formal and computational characterization of pragmatic infelicities. In *Proc. of ECAI-96*, pages 587–591, Budapest.

Schaeken, W., Johnson-Laird, P., and d'Ydewalle, G. (1996). Tense, aspect, and temporal reasoning. *Thinking & reasoning*, 2:309–327.

Soames, S. (1989). Presupposition. In Gabbay, D. and Guenther, F., editors, *Handbook of Philosophical Logic*, pages 553–616. Reidel, Dordrecht.

Van Der Sandt, R. (1992). Presupposition projection as anaphora resolution. *Journal of Semantics*, 9:333–377.

Zeevat, H. (1992). Presupposition and accomodation in update semantics. *Journal of semantics*, 9:379–412.

First-Language Thinking for Second-Language Understanding: Mandarin and English Speakers' Conceptions of Time

Lera Boroditsky (lera@psych.stanford.edu)
Department of Psychology, Bldg 420
Stanford, CA 94305-2130

Abstract

Does the language you speak affect how you think about the world? English and Mandarin speakers talk about time differently. Is this difference between the two languages reflected in the way their speakers think about time? The findings of two RT experiments show that different ways of talking about time lead to different ways of thinking. In Experiment 1, Mandarin-English bilinguals were compared to native English speakers. The results suggested that Mandarin speakers used a "Mandarin way of thinking" even when they were "thinking for English". In Experiment 2, native English speakers were trained to talk about time in "a Mandarin way". Results showed that even after a short training, native English speakers behaved more like Mandarin speakers than like untrained English speakers. It is concluded that language is a powerful tool in shaping thought.

Introduction

Does the language you speak shape the way you understand the world? Linguists, philosophers, anthropologists, and psychologists have long been interested in the role that languages might play in shaping their speakers' ways of thinking. This interest has been fueled in large part by the observation that different languages talk about the world differently. Does the fact that languages differ mean that people who speak different languages think about the world differently? Does learning new languages change the way one thinks? Do polyglots think differently when speaking different languages? Although all of these questions have long been issues of interest and controversy, definitive answers are scarce. This paper briefly reviews the empirical history of these questions and describes two experiments that demonstrate the role of language in shaping habitual thought.

The strong Whorfian view that thought is entirely determined by language has long been abandoned in the field. Rosch's (1972, 1975, 1978) work on color perception demonstrating that the Dani, despite having only two words for colors, had little trouble learning the English set of color categories was particularly effective in undermining the strong view (but see Lucy & Shweder, 1979). Although the strong linguistic determinism view seems untenable, many weaker (but still interesting) formulations can be entertained. For example, Slobin (1987, 1996) suggested that language may influence categorization during "thinking for speaking." In a similar vein, Hunt and Agnoli (1991) reviewed evidence that language may influence thought by making habitual distinctions more fluent. Recently, several lines of research have explored domains that appear more likely to reveal linguistic influences than such lower level domains as color perception. Among the new evidence have been studies of cross-linguistic differences in the object-substance distinction in Yucatec Mayan and Japanese (e.g. Imai & Gentner, 1997; Lucy, 1992), differences in thinking about spatial relations in Dutch, Korean, and Tzeltal (e.g. Bowerman, 1996; Levinson, 1996), and evidence suggesting that language influences conceptual development (e.g. Markman & Hutchinson, 1984; Waxman & Kosowski, 1990). It is possible that language is most powerful in determining thought for domains that are more abstract, i.e. ones that are not so closely tied to sensory experience (see Gentner & Boroditsky, in press for related discussion).

Although these new lines of evidence are suggestive, there are several limitations common to most of the studies. First, speakers of different languages are usually tested only in their native language. Any differences found in these kinds of comparisons can only show the effect of a language on thinking for that particular language. These differences cannot tell us whether experience with a language affects language-independent thought such as thought for other languages, or thought in non-linguistic tasks. Further, comparing results from studies conducted in different languages poses a deeper problem: there is simply no way to be certain that the stimuli and instructions are truly the same in both languages[1]. Since there is no way to know that subjects in different languages are performing the same task, it is difficult to deem the comparisons meaningful.

---

[1] This is a problem even if the verbal instructions are minimal. For example, even if the task is non-linguistic, and the instructions are simply "which one is the same?", one cannot be sure that the words used for "same" mean the same thing in both languages. If in one language the word for "same" is closer in meaning to "identical," while in the other language it's closer to "similar", the subjects in different languages may behave very differently, but due only to the difference in instructions - not because of any interesting differences in thought. There is no sure way to guard against this possibility when tasks are translated into different languages.

A second limitation is that even when non-linguistic tasks (such as sorting into categories, or making similarity judgments) are used, the tasks themselves are quite explicit. Sorting and similarity judgment tasks require a subject to decide on a strategy for completing the task. How should I divide these things into two categories? What am I supposed to base my similarity judgments on? It is quite possible that when figuring out how to perform a task, subjects simply make a conscious decision to follow the distinctions reinforced by their language. For this reason, evidence collected using such explicit dependent measures as sorting preferences or similarity judgments is not convincing as non-linguistic evidence.

Showing that experience with a language affects thought in some broader sense (other than thinking for that particular language) would require observing a cross-linguistic difference on some implicit dependent measure (e.g. reaction time) in a non-language-specific task. The set of studies described in this paper does just that. The findings show an effect of first-language thinking on second-language understanding using the implicit measure of reaction time. In particular, the studies investigate whether speakers of English and Mandarin Chinese think differently about the domain of time even when both groups are "thinking for English."

Time

Across languages people use spatial metaphors to talk about time. Whether we are looking *forward* to a brighter tomorrow, proposing theories *ahead* of our time, or falling *behind* schedule, we are relying on terms from the domain of space to talk about time. Many researchers have noted an orderly and systematic correspondence between the domains of space and time in language (Clark, 1973; Lehrer, 1990; Traugott, 1978).

This paper will focus on the event-sequencing aspect of conceptual time, that is, the way events are temporally ordered with respect to each other and to the speaker (e.g. "The worst is *behind* us" or "Thursday is *before* Saturday.") There are many striking similarities in the types of spatial terms used to talk about time across languages. In order to capture the sequential order of events, time is generally conceived as a one-dimensional, directional entity. The spatial terms imported to talk about time are one-dimensional, directional terms such as *ahead/behind*, or *up/down*, rather than multi-dimensional or symmetric terms such as *shallow/deep*, or *left/right* (Clark, 1973). Although all languages use spatial terms to talk about time, there are some interesting differences in the types of spatial terms used.

Time in English In English, we predominantly use front/back terms to talk about time. We can talk about the good times *ahead* of us, or the hardships *behind* us. We can move meetings *forward*, push deadlines *back*, and eat dessert *before* we're done with our vegetables. On the whole, the terms used to talk about the order of events in English are the same terms that are used to describe asymmetric horizontal spatial relations (e.g. "he took three steps *forward*" or "the dumpster *behind* the store").

Time in Mandarin Chinese In Mandarin, front/back spatial metaphors for time are also common (Scott, 1989). Mandarin speakers use the spatial morphemes *qián* - "front" and *hòu* - "back" to talk about time. Figure 1 shows parallel uses of *qián* and *hòu* in spatial and temporal orderings in Mandarin and their English glosses. Examples were taken from Scott (1989).

```
(1) SPACE
        zài zhuōzi qián-bian zhàn-zhe yī ge xuésheng
        there is a student standing in front of the desk

    TIME
        hǔ nián de qián yī nián shì shénme nián?
        what is the year before the year of the tiger?

(2) SPACE
        zài zhuōzi hòu-bian zhàn-zhe yī ge lǎoshi
        there is a teacher standing behind the desk

    TIME
        dàxué bìyè yǐ-hòu wǒ yòu jìn le yánjiūyuàn
        after graduating from university, I entered graduate school
```

Figure 1: Examples of spatial and temporal uses of horizontal terms *qián* and *hòu* in Mandarin given with their English glosses

What makes Mandarin interesting for our purposes, is that in addition to these front/back or horizontal metaphors, Mandarin speakers also systematically use up/down or vertical metaphors to talk about time (Scott, 1989). The spatial morphemes *shàng* - "up" and *xià* - "down" are frequently used to talk about the order of events, weeks, months, semesters, and more. Earlier events are said to be *shàng* or "up", and later events are said to be *xià* or "down". Figure 2 shows parallel uses of *shàng* and *xià* to describe spatial and temporal relations (examples taken from Scott, 1989).

```
(1) SPACE
        māo shàng shù
        cats climb trees

    TIME
        shàng ge yuè
        last (or previous) month

(2) SPACE
        tā xià le shān méi yǒu
        has she descended the mountain or not?

    TIME
        xià ge yuè
        next (or following) month
```

Figure 2: Examples of spatial and temporal uses of vertical terms *shàng* and *xià* in Mandarin given with their English glosses

Although in English vertical spatial terms can also be used to talk about time (e.g. "hand *down* knowledge from generation to generation" or "the meeting was coming *up*"), these uses are not nearly as common or systematic as is the use of *shàng* and *xià* in Mandarin (Scott, 1989). The closest English counterparts to the Mandarin uses of *shàng* and *xià* are the purely temporal terms earlier and later. Earlier and later are similar to *shàng* and *xià* in that they use an absolute framework to determine the order of events.

Relative versus absolute terms for time In Mandarin, *shàng* always refers to events closer to the past, and *xià* always refers to events closer to the future. The same is true in English for *earlier* and *later* terms respectively. This is not true, however, for the other English terms used to talk about time. Terms like *before/after*, *ahead/behind*, and *forward/back* can be used not only to order events relative to the direction of motion of time, but also relative to the observer. When ordering events relative to the direction of motion of time, we can say that Thursday is *before* Friday. Here, *before* refers to an event that's closer to the past. However, we can also order events relative to the observer as in "The best is *before* us." Here, *before* refers to an event closer to the future. The same is true for *ahead/behind* and *forward/back*. *Qián* and *hòu*, the horizontal terms used in Mandarin to talk about time, also share this flexibility. Unlike *before/after*, *ahead/behind*, and *qián/hòu*, terms like *earlier/later* and *shàng/xià* are not used to order events relative to the observer. For example, one cannot say that "the meeting is *earlier* than us" to mean that it is further in the future. *Earlier/later* and *shàng/xià* are absolute terms.

In summary, both Mandarin and English speakers use horizontal relative terms to talk about time. In addition, Mandarin speakers use the vertical absolute terms *shàng* and *xià* to talk about time. This way of talking about time is most conceptually similar to the English use of the purely temporal terms *earlier* and *later*.

Of interest to this study is whether this difference between the English and Mandarin ways of <u>talking</u> about time also leads to a difference in how the two groups <u>think</u> about time. The particular question is whether Native Mandarin speakers are more likely to rely on vertical spatial schemas to think about time in absolute terms, while Native English speakers are always more likely to rely on horizontal spatial schemas.

In order to answer the above questions, we will need a way of determining which spatial schemas Mandarin and English speakers are using when they are thinking about time. The schema-consistency paradigm developed by Boroditsky (1998) allows us to do just that. The basic rationale of the schema-consistency paradigm is as follows. If an appropriate spatial schema is primed, people should be faster to understand statements about time that employ that same schema. Therefore, if English speakers are thinking about time horizontally, then asking them to think about horizontal spatial relations right before they read a sentence about time should make them faster to understand that sentence than if they had just been thinking about vertical spatial relations. The reverse should be true for Mandarin

speakers. When thinking about time in absolute terms, Mandarin speakers should be faster after vertical primes than after horizontal primes.

In Experiment 1, Mandarin and English speakers solved sets of spatial priming questions followed by a target question about time. The spatial primes were either about horizontal spatial relations between two objects (see Figure 3), or about vertical relations (see Figure 4). After solving a set of 2 primes, participants answered a TRUE/FALSE target question about time that either used relative terms (e.g. "March comes *before* April.") or absolute terms (e.g. "March comes *earlier* than April"). Of interest was whether the prime questions had a different effect on how long it took English speakers to answer the different target questions as compared to the Mandarin speakers.

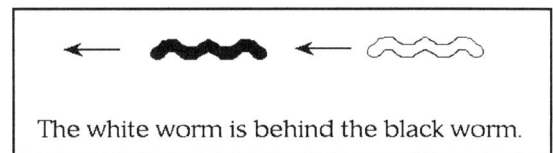

Figure 3: Example of a horizontal spatial prime used in Experiments 1 & 2

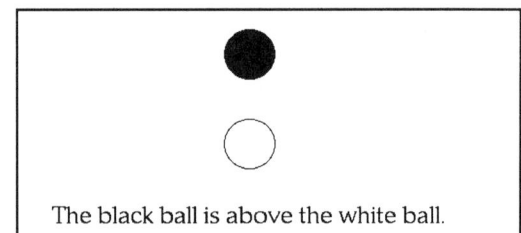

Figure 4: Example of a vertical spatial prime used in Experiments 1 & 2

If one's native language does affect how one thinks about time, then Mandarin speakers should be faster to answer absolute target questions about time after solving the vertical spatial primes than after solving the horizontal spatial primes. English speakers, on the other hand, should always be faster after horizontal primes because horizontal terms are predominantly used in English. Since both English and Mandarin speakers completed the task in English, this is a particularly strong test of the effect of one's native language on thought. If Mandarin speakers do show a vertical bias in thinking about time even when they are "thinking for English," then language must play an important role in shaping speakers' thinking habits.

Experiment 1

Method

Participants 26 native English speakers, and 20 native Mandarin speakers participated in this study. All participants were graduate or undergraduate students at Stanford University, and received either payment or course credit in return for their participation.

Design The experiment had a fully crossed within-subject 2 (prime-type) X 2 (target-type) design. Targets were statements about time: either *before/after* statements (e.g. "March comes *before* April"), or *earlier/later* statements (e.g. "March comes *earlier* than April"). Primes were spatial scenarios accompanied by a sentence description and were either horizontal (see Figure 3) or vertical (see Figure 4). Each participant completed a set of 6 practice questions and 64 experimental trials. Each experimental trial consisted of two spatial prime questions (both horizontal or both vertical) followed by one target question about time. Participants were not told that the experiment was arranged into such trials, nor did they figure it out in the course of the experiment. Participants answered each target question twice - once after each type of prime. The order of presentation of all trials was randomized for each participant.

Materials A set of 128 primes and 32 targets, all TRUE/FALSE questions, was constructed.

Primes: 128 spatial scenarios were used as primes. Each scenario consisted of a picture and sentence below the picture. Half of these scenarios were about horizontal spatial relations (see Figure 3), and the other half were about vertical spatial relations (see Figure 4). Half of the horizontal primes used the "X is *ahead* of Y" phrasing and half used the "X is *behind* Y" phrasing. Likewise, half of the vertical primes used the "X is *above* Y" phrasing and have used the "X is *below* Y" phrasing. Primes were equally often TRUE and FALSE. All of these variations were crossed into eight types of primes. In addition, the left/right orientation of the horizontal primes was counterbalanced across variations.

Targets: 16 statements about the order of the months of the year were constructed. Half of these statements used the relative terms *before* and *after*, and half used the absolute terms *earlier* and *later*. All four terms were used equally often. All target questions were "TRUE".

Fillers: 16 additional statements about months of the year were used as fillers. These statements were similar in all respects to the target questions except that all of the fillers were "FALSE". This was done to insure that subjects were alert and did not simply learn to answer "TRUE" to all questions about time. Filler time questions (along with filler spatial scenarios drawn randomly from the list of all spatial primes) were inserted randomly in-between experimental trials to ensure that participants did not deduce the trial structure of the experiment. Responses to filler trials were not analyzed.

Procedure Participants were tested individually. All participants were tested in English with English instructions. Stimuli were presented on a computer screen, one question at a time. For each question, participants were asked to make a TRUE/FALSE response as quickly as possible by pressing one of two keys on a keyboard. Response times were measured and recorded by the computer. There was a response deadline of 5 seconds. If a participant did not provide a response within 5 seconds, the computer simply went on to the next question. There was no feedback for the experimental trials.

Results

Separate 2 (prime type) X 2 (target type) repeated measures ANOVAs were conducted for data collected from English and Mandarin speakers. Response times exceeding the deadline, incorrect responses, and those following an incorrect response to a priming question were omitted from all analyses.

As expected, native English speakers were faster to solve time questions after horizontal primes ($\underline{M}$ = 2128 msecs, $\underline{SD}$ = 545 msecs) than after vertical primes ($\underline{M}$ = 2300 msecs, $\underline{SD}$ = 682 msecs). The main effect of the prime was significant, F (1, 25) = 13.76, $\underline{p}$ = 0.001. There was no interaction between prime-type and target-type, F (1, 25) = .75, $\underline{p}$ = 0.40. English speakers were faster to solve all questions about time if they followed horizontal primes than if they followed vertical primes.

The data from Mandarin speakers looked very different. There was no main effect of prime, F (1, 19) = .01, $\underline{p}$=.92. Overall, Mandarin speakers answered time questions just as fast after horizontal primes ($\underline{M}$ = 2422 msecs, $\underline{SD}$ = 493 msecs) as after vertical primes ($\underline{M}$ = 2428 msecs, $\underline{SD}$ = 443 msecs). However, there was a significant interaction between prime-type and target-type, F (1, 19) = 4.55, $\underline{p}$<.05. Like the English speakers, Mandarin speakers were faster to answer the relative *before/after* target questions after horizontal primes ($\underline{M}$ = 2340 msecs) than after vertical primes ($\underline{M}$ = 2509 msecs). As predicted, the pattern was exactly reversed for the absolute *earlier/later* targets. Unlike the English speakers, Mandarin speakers were faster to solve *earlier/later* targets after vertical primes ($\underline{M}$ = 2347 msecs) than after horizontal primes ($\underline{M}$ = 2503 msecs).

Discussion

Native English and native Mandarin speakers were found to think differently about time. This was true even though both groups performed the task in English. In Experiment 1, English speakers were always faster to answer questions about time after horizontal primes than after vertical primes. Mandarin speakers showed a very different pattern. Like the English speakers, they were faster to answer the relative *before/after* targets after horizontal primes than after vertical primes. The reverse was true for the absolute *earlier/later* targets; unlike the English speakers, Mandarin speakers were faster to answer the absolute *earlier/later* targets after vertical primes than after horizontal primes. Overall these findings suggest that while English speakers relied on horizontal spatial schemas to think about both types of targets, Mandarin speakers relied on horizontal schemas only when thinking about the relative terms *before* and *after*. For the absolute *earlier/later* targets, Mandarin speakers relied on vertical schemas. This is exactly the pattern we would predict from the way Mandarin speakers talk about time since Mandarin uses vertical metaphors (*shàng* and *xià*) to talk about time in absolute terms, and horizontal metaphors (*qián* and *hòu*) to talk about time in relative terms. In short, Mandarin speakers in our stud·

showed a pattern of first-language thinking in second-language understanding.

Although these results are highly suggestive of an effect of language on thought, there are some concerns. First, the difference in the time metaphors used in English and Mandarin is clearly not the only difference between the two languages. Many other factors could conceivably have led to the differences we observed. One important factor to consider is that of writing direction. Whereas English is written horizontally from left to right, Mandarin was traditionally written in vertical columns that ran from right to left. Although this difference in writing direction is interesting, it cannot explain the results obtained in Experiment 1. The writing direction explanation would predict a main effect of prime - since Mandarin is written vertically, Mandarin speakers should always be faster to answer time questions after vertical rather than after horizontal primes. This prediction was not borne out in data. Mandarin speakers showed an interaction between target and prime, and not the main effect predicted by writing direction. Therefore, writing direction cannot be responsible for the differences observed in this experiment.

Beyond differences in language, there may be many cultural differences between native English and native Mandarin speakers that may have lead to differences in the response patterns. Although a clear non-language-based explanation that would predict the observed pattern of results is not readily apparent, we can not *a priori* discount the possible effects of cultural differences. Experiment 2 was designed to minimize differences in non-linguistic cultural factors while preserving the interesting difference in language.

In Experiment 2, native English speakers were trained to talk about time using vertical terms. For example, they learned to say that "cars were invented *above* fax machines" and that "Wednesday is *below* Tuesday." The use of vertical terms *above* and *below* in this training was maximally similar to the use of *shàng* and *xià* in Mandarin. *Above* and *below*, like *shàng* and *xià*, were used as absolute terms. Earlier events were always said to be *above*, and later events were always said to be *below*. After the training, participants completed exactly the same experiment as in Experiment 1. If it is indeed language (and not other cultural factors) that led to the differences between English and Mandarin speakers in Experiment 1, then the "Mandarin" linguistic training given to English speakers in Experiment 2 should make their pattern of results look more like that of Mandarin speakers than that of English speakers.

Experiment 2

Method

Participants 32 Stanford University undergraduates, all native English speakers, participated in this study in exchange for course credit.

Materials and Design Participants were told they would learn "a new way to talk about time." They were given a set of 5 example sentences that "used this new system" (e.g. "Monday is *above* Tuesday") and were asked to figure out on their own how the system worked. The new system used *above* and *below* as absolute terms. Events closer to the past were always said to be *above*, and events closer to the future were always said to be *below*. Participants were then tested on a set of 90 questions that used these vertical terms to talk about time (e.g. "Nixon was president *above* Clinton"). These test questions were presented on a computer screen one at a time, and participants responded TRUE or FALSE to each statement by pressing one of two keys on the keyboard. Immediately after the training, participants went on to complete the experiment described in Experiment 1. After the initial training, all materials, instructions, and procedures were identical to those used in Experiment 1.

Results and Discussion

As in Experiment 1, response times exceeding the deadline, incorrect responses, and those following an incorrect response to a priming question were omitted from all analyses. A 2 (prime type) X 2 (target type) repeated measures ANOVA was conducted.

The pattern of results in this experiment was very similar to that obtained for Mandarin speakers in Experiment 1. There was no main effect of prime, F (1, 31) = 1.29, p=.27. Unlike untrained English speakers, participants trained in the new system for talking about time were not significantly faster to answer time questions after horizontal primes (M = 2235 msecs, SD = 599 msecs) than after vertical primes (M = 2300 msecs, SD = 588 msecs). However, just as was the case with Mandarin speakers, there was a significant interaction between prime-type and target-type, F (1, 31) = 5.40, p <.05. Just like Mandarin speakers and untrained English speakers, trained participants were faster to answer the relative *before/after* target questions after horizontal primes (M = 2141 msecs) than after vertical primes (M= 2334 msecs). As predicted, the pattern was exactly reversed for the absolute *earlier/later* targets. Like the Mandarin speakers, and unlike the untrained English speakers, trained participants were faster to solve *earlier/later* targets after vertical primes (M= 2266 msecs) than after horizontal primes (M = 2330 msecs). Overall, English speakers who were trained to talk about time using vertical *above/below* terms, showed a pattern of results very similar to that of Mandarin speakers. These results confirm that, even in the absence of other cultural differences, differences in talking can indeed lead to differences in thinking.

Conclusions

In the realm of abstract domains like time, one's native language appears to exert a strong influence on how one thinks about the world. In Experiment 1, Mandarin speakers relied on a "Mandarin" way of thinking about time even when they were thinking about English sentences. Mandarin speakers were more likely to think about time in vertical terms when an absolute reference frame was used, but not when a relative reference frame was used. This pattern of results is predicted by the way Mandarin talks about time; vertical terms are used in an absolute reference frame, and horizontal terms are used in relative reference frames. English speakers were always more likely to think about time horizontally because horizontal spatial terms predominate in English temporal descriptions. In Experiment 2, native English speakers who had just been trained to talk about time using vertical terms showed a pattern of results very similar to that of Mandarin speakers in Experiment 1. This finding confirms that the effect observed in Experiment 1 is driven by differences in language, and not by other non-linguistic cultural differences. Further, these results show that learning a new way to talk about a familiar domain, can change the way one thinks about that domain. Language can be a powerful tool in shaping abstract thought. When sensory information is scarce or inconclusive (as is the case with the domain of time), languages appear to play an important role in determining how their speakers think.

Acknowledgments

This research was funded by an NSF Graduate Research Fellowship to the author. I would like to thank Barbara Tversky, Gordon Bower, and Herb Clark for many insightful discussions of this research, and Michael Ramscar for comments on an earlier draft of this paper. Foremost, I would like to thank Jennifer Y. Lee who has made countless contributions to this research and was an invaluable source of information about the Mandarin language.

References

Boroditsky, L. (1998). Evidence for metaphoric representation: Understanding time. In K. Holyoak, D. Gentner, and B. Kokinov (Eds.), *Advances in analogy research: Integration of theory and data from the cognitive, computational, and neural sciences.* Sofia, Bulgaria: New Bulgarian University Press.

Bowerman, M. (1996). The origins of children's spatial semantic categories: cognitive versus linguistic determinants. In J. Gumperz & S. Levinson (Eds.), *Rethinking linguistic relativity.* Cambridge, MA: Cambridge University Press.

Clark, H. (1973). Space, time, semantics, and the child. In T.E. Moore (Ed.), *Cognitive Development and the acquisition of language.* New York: Academic Press.

Gentner, D., & Boroditsky, L. (in press). Individuation, relational relativity and early word learning. In M. Bowerman & S. Levinson (Eds.), *Language acquisition and conceptual development.* Cambridge, England: Cambridge University Press.

Gentner, D. & Imai, M. (1997). A cross-linguistic study of early word meaning: Universal ontology and linguistic influence. *Cognition 62 (2),* 169-200.

Heider, E.R. (1972). Universals in color naming and memory. *Journal of Experimental Psychology, 93,* 10-20.

Hunt, E., & Agnoli, F. (1991). The Whorfian hypothesis: A cognitive psychology perspective. *Psychological Review, 98,* 377-389.

Lehrer, A. (1990). Polysemy, conventionality, and the structure of the lexicon. *Cognitive Linguistics 1,* 207-246.

Levinson, S. (1996). Relativity in spatial conception and description. In J. Gumperz & S. Levinson (Eds.), *Rethinking linguistic relativity.* Cambridge, MA: Cambridge University Press.

Lucy, J.A. (1992). *Grammatical categories and cognition: a case study of the linguistic relativity hypothesis.* Cambridge, England: Cambridge University Press.

Lucy, J. A., & Shweder, R. A. (1979). Whorf and his critics: Linguistic and nonlinguistic influences on color memory. *American Anthropologist, 81,* 581-618.

Markman, E. M., & Hutchinson, J. E. (1984). Children's sensitivity to constraints on word meaning: Taxonomic versus thematic relations. *Cognitive Psychology, 16,* 1-27.

Rosch, E. (1975). Cognitive representations of semantic categories. *Journal of Experimental Psychology: General, 104,* 192-233.

Rosch, E. (1978). Principles of categorization. In R. Rosch & B. B. Lloyd (Eds.), *Cognition and categorization.* Hillsdale, NJ: Erlbaum.

Scott, Amanda. (1989). The vertical dimension and time in Mandarin. *Australian Journal of Linguistics, 9,* 295-314.

Slobin, D. (1987). Thinking for speaking. *Proceedings of the Berkeley Linguistic Society, 13.*

Slobin, D. (1996). From "thought and language" to "thinking for speaking." In J. Gumperz & S. Levinson (Eds.), *Rethinking linguistic relativity.* Cambridge, MA: Cambridge University Press.

Traugott, E. C. (1978). On the expression of spatiotemporal relations in language. In J. H. Greenberg (Ed.), *Universals of human language: Vol. 3. Word structure* (pp.369-400). Stanford, California: Stanford University Press.

Waxman, S. R., & Kosowski, T. (1990). Nouns mark category relations: Toddlers' and Preschoolers' word-learning biases. *Child Development, 61,* 1461-1473.

Metaphor Comprehension:
From Comparison to Categorization

Brian F. Bowdle
Department of Psychology
Indiana University
1101 East 10th Street
Bloomington, IN 47405
(bbowdle@indiana.edu)

Dedre Gentner
Department of Psychology
Northwestern University
2029 Sheridan Road
Evanston, IL 60208
(gentner@nwu.edu)

Abstract

In this paper, we explore the relationship between metaphor and polysemy. We begin by discussing how novel metaphoric mappings can create new word meanings in the form of domain-general representations. Turning next to consider the implications of this view for the on-line comprehension of figurative language, we suggest that there is a shift from comparison processing to categorization processing as metaphors are conventionalized. Finally, we describe a series of experimental findings that support the proposed account.

Introduction

Metaphors establish mappings between concepts from disparate domains of knowledge. For example, in the metaphor *The mind is a computer*, an abstract entity is described in terms of a complex electronic device. It is widely believed that metaphors are a major source of knowledge change, and a great deal of research has examined how metaphors can enrich and illuminate concepts that would otherwise remain vague or ambiguous. However, there have been far fewer explorations of a second generative function of metaphors – namely, lexical extension. In this paper, we will discuss (1) how novel metaphoric mappings can create new word meanings in the form of domain-general representations, and (2) how these new meanings may be applied during the comprehension of conventional metaphors. Before turning to these issues, however, it is necessary to consider the nature of metaphoric mappings in greater depth.

Metaphor and Analogy

Metaphors are traditionally viewed as comparisons between the target (a-term) and the base (b-term). According to many early models, metaphors are understood by means of a simple feature-matching process (e.g., Miller, 1979; Or-

tony, 1979; Tversky, 1977). However, more recent versions of the comparison view have assumed that metaphors act to set up correspondences between partially isomorphic conceptual structures rather than between sets of independent properties (e.g., Gentner, 1983; Indurkhya, 1987; Kittay & Lehrer, 1981; Lakoff & Johnson, 1980; Verbrugge & McCarrell, 1977). In other words, metaphor can be seen as a species of analogy.

We will use Gentner's (1983) *structure-mapping theory* to articulate the processes that may take place during metaphor comprehension. Structure-mapping theory assumes that interpreting a metaphor involves two interrelated mechanisms: alignment and projection. The alignment process operates in a local-to-global fashion to create a maximal structurally consistent match between two representations that observes *one-to-one mapping* and *parallel connectivity* (Falkenhainer, Forbus, & Gentner, 1989). That is, each object of one representation can be placed in correspondence with at most one object of the other representation, and arguments of aligned relations are themselves aligned. A further constraint on the alignment process is *systematicity*: Alignments that form deeply interconnected structures, in which higher-order relations constrain lower-order relations, are preferred over less systematic sets of commonalities. Once a structurally consistent match between the target and base domains has been found, further predicates from the base that are connected to the common system can be projected to the target as *candidate inferences*.

According to structure-mapping theory, metaphors often convey that a system of relations holding among the base objects also holds among the target objects, regardless of whether the objects themselves are intrinsically similar. Thus, the metaphor *Socrates was a midwife* highlights certain relational similarities between the individuals – both help others produce something – despite the fact that the arguments of these relations are quite different in the target and base domains: Socrates helped his *students* produce *ideas*, whereas a midwife helps a *mother* produce a *baby*. The centrality of relations during metaphor comprehension has been confirmed by a number of studies. For example, people's interpretations of metaphors tend to include more relations than simple attributes, even for statements that suggest both types of commonalities (e.g., Gentner & Clement, 1988; Shen, 1992; Tourangeau & Rips, 1991). Further, Gentner & Clement (1988) found that the relationality of people's interpretations of metaphors was positively related to the judged aptness of these same metaphors.

Metaphor and Polysemy

Like analogies, metaphors can lend additional structure to problematic target concepts, thereby making these concepts more coherent. However, this is not the only way in which metaphors can lead to knowledge change. Metaphors are also a primary source of polysemy – they allow words with specific meanings to take on additional, related meanings (e.g., Lakoff, 1987; Lehrer, 1990; Miller, 1979; Nunberg, 1979; Sweetser, 1990). For example, consider the word *roadblock*. There was presumably a time when this word referred only to a barricade set up in a road. With repeated metaphoric use, however, *roadblock* has also come to refer to any obstacle to meeting a goal (as in *Fear is a roadblock to success*).

How do metaphors create new word meanings? One recent and influential proposal is that such lexical extensions are due to stable projections of conceptual structures and corresponding vocabulary items from one (typically concrete) domain of experience to another (typically abstract) domain of experience (e.g., Lakoff, 1987; Lehrer, 1990; Sweetser, 1990). On this view, the metaphoric meaning of a polysemous word is understood directly in terms of the word's literal meaning.

We wish to consider an alternative account of the relationship between metaphor and polysemy – one that follows naturally from viewing metaphor as a species of analogy. Research on analogical problem solving has shown that the alignment of two relationally similar situations can lead to the induction of domain-general problem schemas that can be applied to future situations (e.g., Gick & Holyoak, 1983; Novick & Holyoak, 1991; Ross & Kennedy, 1990). We believe that similar forces are at work during metaphor comprehension. The central idea is that the process of structural alignment allows for the induction of metaphoric categories, which may in turn be lexicalized as secondary senses of metaphor base terms (Bowdle, 1998; Bowdle & Gentner, 1995, in preparation; Gentner & Wolff, 1997).

When a metaphor is first encountered, both the target and base terms refer to specific concepts from different semantic domains, and the metaphor is interpreted by (1) aligning the two representations, and (2) importing further predicates from the base to the target, which can serve to amplify the target representation. As a result of this mapping, the common relational structure that forms the basis of the metaphor interpretation will increase in salience relative to nonalignable aspects of the two representations. If the same base term is repeatedly aligned with different targets so as to yield the same basic interpretation, then the highlighted system may become conventionally associated with the base as an abstract metaphoric category. At this point, the base term will be polysemous, having both a domain-specific meaning and a related domain-general meaning. We will refer to this proposed evolution as *the career of metaphor hypothesis*. (For related proposals, see Holyoak & Thagard, 1995; Murphy, 1996).

Implications for Metaphor Comprehension

Research on metaphor comprehension often treats metaphor as an undifferentiated type of figurative language. However, a number of theorists have recently argued that metaphor is pluralistic, and that the manner in which a metaphor is comprehended may depend on its level of conventionality (e.g., Blank, 1988; Blasko & Connine, 1993; Giora, 1997; Turner & Katz, 1997). Our account of the relationship between metaphor and polysemy is in line with these claims. Specifically, we believe (1) that the process of conventionalization is essentially one of a base term acquiring a domain-general meaning, and (2) that this representational shift will be accompanied by a shift in mode of processing.

These ideas are illustrated in Figure 1, which shows how novel and conventional metaphors differ on the career of metaphor view. *Novel metaphors* involve base terms that refer to a domain-specific concept, but are not (yet) associated with a domain-general category. For example, the novel base term *glacier* (as in *Science is a glacier*) has a literal sense – "a large body of ice spreading outward over a land surface" – but no related metaphoric sense (e.g., "anything that progresses slowly but steadily"). Novel metaphors are therefore interpreted as comparisons, in which the target concept is structurally aligned with the literal base concept. However, metaphoric categories may arise as a byproduct of this comparison process.

In contrast to novel metaphors, *conventional metaphors* involve base terms that refer both to a literal concept and to an associated metaphoric category. For example, the conventional base term *blueprint* (as in *A gene is a blueprint*) has two closely related senses: "a blue and white photographic print in showing an architect's plan" and "anything that provides a plan." Conventional base terms are polysemous, and the literal and metaphoric meanings are semantically linked due to their similarity. Conventional metaphors may therefore be interpreted either as comparisons, by matching the target concept with the literal base concept, or as categorizations, by seeing the target concept as a member of the superordinate metaphoric category named by the base term.

There is, however, reason to expect that comparison and categorization processing will not be favored equally for conventional metaphors. Let us assume that both meanings of a conventional base term are activated simultaneously during comprehension, and that attempts to map each representation to the target concept are made in parallel. Which of these mappings wins will depend on a number of factors, including the context of the metaphor and the

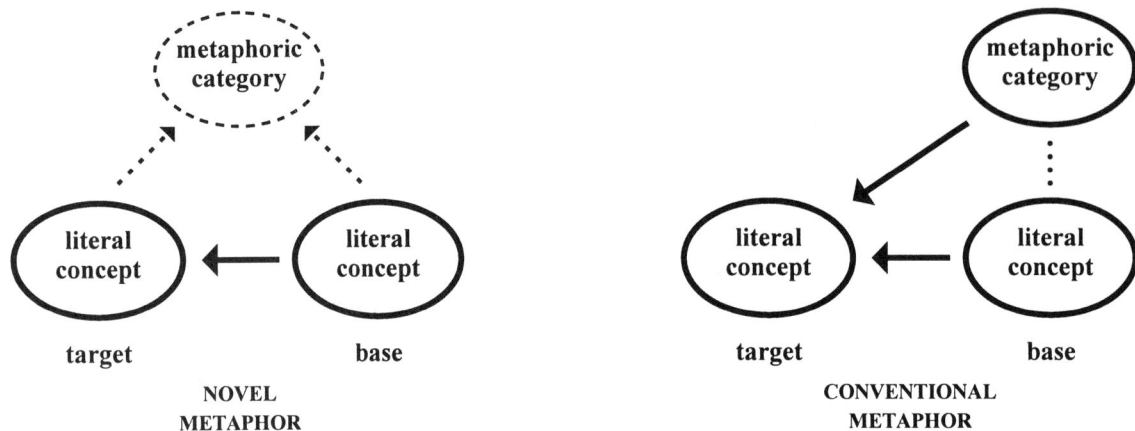

Figure 1. Novel and conventional metaphors.

relative salience of each meaning of the base term (Giora, 1997; Williams, 1992). All else being equal, however, aligning a target with a metaphoric category should be computationally less costly than aligning a target with a literal base concept. This is because metaphoric categories will be informationally sparser than the literal concepts they were derived from. Thus, the domain-general meaning of a conventional base term should be applied more rapidly than the domain-specific meaning, and conventional metaphors will more likely be interpreted as categorizations than as comparisons.

In sum, the career of metaphor hypothesis predicts that as metaphors become increasingly conventional, there is a shift in mode of processing from comparison to categorization (Bowdle, 1998; Bowdle & Gentner, 1995, in preparation; Gentner & Wolff, 1997). This is consistent with a number of recent proposals, according to which the interpretation of novel metaphors involves sense creation, but the interpretation of conventional metaphors involves sense retrieval (e.g., Blank, 1988; Blasko & Connine, 1993; Giora, 1997; Turner & Katz, 1997). On the present view, the senses retrieved during conventional metaphor comprehension are abstract metaphoric categories.

As described above, the career of metaphor hypothesis is related to an emerging alternative to comparison models of metaphor – namely, the position that metaphor is a species of categorization (e.g., Glucksberg & Keysar, 1990; Glucksberg, McGlone, & Manfredi, 1997; Honeck, Kibler, & Firment, 1987; Kennedy, 1990). On this view, the literal target and base concepts of a metaphor are never directly compared. Rather, the base concept is used to access or derive an abstract metaphoric category of which it represents a prototypical member, and the target concept is then assigned to that category. We suggest that this account is reasonably apt for conventional metaphors, but is incorrect for novel metaphors. Novel metaphors can only give rise to metaphoric categories once the original target and base concepts have been structurally aligned, and such categories do not initially contribute to the meaning of these metaphors.

Evidence: The Metaphor/Simile Distinction

We now review some recent studies we have conducted that offer support for the processing claims made by the career of metaphor hypothesis. Central to the logic of these studies was the distinction between metaphors and similes.

Nominal metaphors (figurative statements of the form *X is Y*) can often be paraphrased as similes (figurative statements of the form *X is like Y*). For example, one can say both *The mind is a computer* and *The mind is like a computer*. This linguistic alternation is interesting because metaphors are grammatically identical to literal categorization statements (e.g., *A sparrow is a bird*), and similes are grammatically identical to literal comparison statements (e.g., *A sparrow is like a robin*). Assuming that form typically follows function in both literal and figurative language, metaphors and similes may tend to promote different comprehension processes. Specifically, metaphors should invite classifying the target as a member of a category named by the base, whereas similes should invite comparing the target to the base. This makes the metaphor-simile distinction a valuable tool for examining the use of comparison and categorization processing during figurative language comprehension.

In one set of experiments, we gave subjects novel and conventional figuratives phrased as both metaphors and similes, and asked them which form they preferred for each statement (Bowdle, 1998; Bowdle & Gentner, 1995, in preparation). We consistently found that the simile form was overwhelmingly preferred for novel figuratives, but that there was a move towards the metaphor form for conventional figuratives. This supports the career of metaphor hypothesis – if conventionalization results in a processing shift from comparison to categorization, then there should be a corresponding shift at the linguistic level from the comparison (simile) form to the categorization (metaphor) form.

In a second set of experiments, we collected subjects' comprehension times for novel and conventional figuratives phrased as either metaphors or similes (Bowdle, 1998;

92

Bowdle & Gentner, 1995, in preparation). We consistently found an interaction between conventionality and grammatical form – novel figuratives were comprehended faster as similes than as metaphors, whereas conventional figuratives were comprehended faster as metaphors than as similes. Again, this supports the career of metaphor hypothesis. If novel figuratives are processed strictly as comparisons, then novel similes should be easier to comprehend than novel metaphors. This is because only the simile form directly invites comparison. At the same time, if conventional figuratives can be processed either as comparisons or as categorizations due to the polysemy of their base terms, then conventional metaphors should be easier to comprehend than conventional similes. The metaphor form invites categorization, and will therefore promote a relatively simple alignment between the target and the abstract metaphoric category named by the base. The simile form invites comparison, and will therefore promote a more complex alignment between the target and the literal base concept.

Experiment: In Vitro Conventionalization

The studies reviewed above support the claim that there is a processing shift from comparison to categorization as metaphors are conventionalized. However, because these experiments simply contrasted novel and conventional figurative statements, they do not address one of the central tenets of the career of metaphor hypothesis – namely, that it is the initial process of comparison that brings about this shift. According to the career of metaphor hypothesis, a metaphoric category is derived as a result of highlighting the common relational structure of the target and the literal base concept. If the same abstraction is derived repeatedly in the context of a given base term, then it will become lexicalized as a secondary sense of that term. In the present experiment, we directly tested these ideas. We examined whether subjects who saw multiple examples of novel similes using the same base term would derive an abstract schema and associate it with the base term. In essence, we aimed to speed up the process of conventionalization from years to minutes. We expected that after repeated comparisons involving a novel base, further figurative statements using the base will behave less like comparisons and more like categorizations. Specifically, we predicted a shift in preference from the simile form to the metaphor form.

The experiment was divided into two phases. In the study phase, subjects received triads of novel similes using the same base term. The first two similes in each triad contained different target terms but were similar in meaning. The third simile had a blank line in place of a target term. For example, a subject might receive the following set of novel similes:

(a) *An acrobat is like a butterfly.*
(b) *A figure skater is like a butterfly.*
(c) _____ *is like a butterfly.*

Subjects were asked to consider the meaning of the first two statements carefully, and then to provide a target for the third statement that would make it similar in meaning to the

first two. We hypothesized that this procedure would promote conventionalization of the novel base terms.

In the test phase, subjects received novel and conventional figuratives in both the comparison (simile) form and the categorization (metaphor) form, and were asked to indicate the strength of their preference for one form versus the other. The key manipulation was that some of the novel figuratives in the test phase used base terms previously seen in the triads of novel similes, along with a new target term (e.g., *A ballerina is (like) a butterfly*). Our prediction was that subjects' preference for the metaphor form should be stronger when the novel base term had received the conventionalization manipulation than when it had not. On the surface, this prediction is counterintuitive – seeing a given base term in two similes might be expected to result in an increased preference for the simile form. Thus, the predicted shift from simile to metaphor would constitute strong support for the career of metaphor claim that metaphoric categories are created by the initial comparison process.

Such a shift could, however, occur for reasons other than schema abstraction. Having encountered a base term in one grammatical frame, subjects might simply prefer to see it in a different grammatical frame. To control for this possibility, some of the novel figuratives in the test phase contained base terms from triads of literal comparisons previously seen in the study phase. For example, subjects might see *a ballerina is (like) a butterfly* having previously seen the following set of literal comparisons:

(a) *A bee is like a butterfly.*
(b) *A moth is like a butterfly.*
(c) _____ *is like a butterfly.*

If subjects simply prefer placing old base terms in new grammatical frames, then their preference for expressing novel figuratives as metaphors should be stronger if they use base terms that have previously been seen in literal comparisons. If not, then seeing the same base term in a triad of literal comparisons should have little or no effect on subjects' subsequent grammatical form preferences.

To ensure the generality of our results, we varied the degree of target concreteness for the figurative statements. Although most metaphors and similes involve relatively concrete base terms (e.g., Katz, 1989; Lakoff & Johnson, 1980), their target terms may be either abstract, as in *Time is (like) a river*, or concrete, as in *A soldier is (like) a pawn*. Subjects received both abstract and concrete targets paired with novel and conventional bases.

Method

Subjects
Forty-eight Northwestern University undergraduates participated in partial fulfillment of a course requirement.

Materials and Design
Twenty-four novel figurative statements and 24 conventional figurative statements were used for the test phase. Each of these sets was further divided into 12 abstract and 12 concrete statements. During the test phase, each subject received all 48 figuratives in

both the comparison (simile) form and the categorization (metaphor) form.

The key manipulation occurred during the study phase, in which the 24 novel figuratives were assigned to one of three study conditions. In the *simile* condition, the original base term was paired with two new target terms to create two new similes (e.g., *Doubt is like a tumor, A grudge is like a tumor*). The two new similes were similar in meaning to one another as well as to the novel statement seen during the subsequent test phase (e.g., *An obsession is (like) a tumor*). Half the pairs of similes contained abstract targets, and half contained concrete targets, to match the concreteness of the corresponding test-phase statements. In the *literal comparison* condition, the original base term was paired with two new target terms to create two literal comparisons (e.g., *A blister is like a tumor, An ulcer is like a tumor*). The two literal comparisons were similar in meaning to one another, but different in meaning from the test-phase statement. Finally, in the *no prior exposure* condition, subjects did not receive any statements using the original base term. The study condition assignment of the novel figuratives was counterbalanced within and between subjects. Thus, each subject saw eight pairs of novel similes, and eight pairs of literal comparisons. In addition, each subject saw eight pairs of conventional metaphors (unrelated to the conventional figuratives used in the test phase) and eight pairs of literal categorizations as filler items. The filler items were like the experimental items in that the statements in each pair used the same base term, and were similar in meaning to one another. All pairs of statements were followed by a third statement with the same base term and grammatical form as the first two, but with a blank line in place of a target term.

Procedure

For the study phase, each subject was given a booklet containing the 32 statement triads (two complete statements plus one incomplete statement) in a random order. Subjects were instructed that for each triad, they should read the first two statements carefully and then complete the third statement by writing a target term that would make it "similar in meaning to the first two". After subjects had completed the study phase, the booklets were removed and a 20-minute filler task was administered.

For the test phase, each subject was given a new booklet containing the 48 figurative statements in a random order. The statements were presented in both the comparison (simile) form and the categorization (metaphor) form, with the two grammatical forms separated by a 10-point numerical scale. Half the subjects received the comparison forms on the left and the categorization forms on the right, and half received the statements in the reverse order. Subjects indicated which form – comparison or categorization – they felt was more natural or sensible for each pair by circling a number on the 10-point scale. They were told that the stronger their preference for the form on the left, the closer their answer should be to 1, and the stronger their preference for the form on the right, the closer their answer should be to 10.

Results and Discussion

Table 1 shows the mean grammatical form preference ratings from the test phase, transformed so that higher numbers indicate a preference for the categorization (metaphor) form over the comparison (simile) form. Focusing solely on the novel figuratives, a 3 (study condition: simile, literal comparison, no prior exposure) x 2 (concreteness: abstract, concrete) repeated measures analysis of variance (ANOVA) was conducted on the subject means. There was a main effect of study condition, $F(2, 94) = 3.87, p < .05$. As predicted, the preference for the categorization form was significantly higher when the base terms had previously been seen in novel similes than when there had been no prior exposure to the base terms ($M = 3.87$ versus $M = 3.52$), $t(47) = 2.67, p < .025$. In contrast, when the base terms had been previously seen in literal comparisons, the grammatical form preference rating ($M = 3.62$) did not differ from that of the baseline condition. There was no main effect of concreteness, and no interaction between these two factors.

TABLE 1

Mean Preferences for the Categorization Form (and Standard Deviations) as a Function of Conventionality, Concreteness, and Study Condition

Conventionality Study Condition	Concreteness	
	Abstract	Concrete
Novel	3.69 (1.14)	3.65 (1.23)
Simile	3.84 (1.44)	3.90 (1.66)
Literal Comparison	3.66 (1.33)	3.58 (1.27)
No Prior Exposure	3.57 (1.41)	3.47 (1.26)
Conventional	6.16 (1.28)	6.10 (1.26)

These results are consistent with the career of metaphor claim that metaphoric categories are derived as a consequence of comparing the target and base of a novel figurative statement, which in turn allows for a shift towards categorization processing as the statement is conventionalized. Encountering a set of novel similes using the same base term encouraged the creation of an abstract schema as a kind of incipient secondary meaning of the base term, and led to a greater preference for the metaphor form of subsequent figurative statements involving that term. Indeed, this finding is particularly striking when one considers that subjects only received three novel similes for any given base term in the study phase. Thus, although novel metaphor bases may typically take years to be conventionalized, the evolutionary path described by the career of metaphor hypothesis can be sped up if the base is consistently aligned with a number of different targets within a short period of time.

Turning now to consider the entire set of data, a 2 (conventionality: novel, conventional) x 2 (concreteness: abstract, concrete) repeated measures ANOVA was conducted on the subject means. The preference for the categorization form was much higher for the conventional figuratives ($M = 6.13$) than for the novel figuratives ($M = 3.67$), $F(1, 47) = 214.51, p < .001$. That is, the move from novel to conventional figuratives was accompanied by a shift from similes to metaphors. This is as predicted by the career of metaphor hypothesis, and replicates the findings of the grammatical form preference experiments reviewed earlier. There was no main effect of concreteness, and no interaction between these two factors.

Conclusions

By viewing metaphor as a species of analogy, two generative functions of metaphors can be explained – namely, the structural enhancement of target concepts, and the lexical extension of base terms. In this paper, we have focused on the latter of these two functions, and have discussed the relationship between polysemy and conventionality in metaphors. The career of metaphor hypothesis outlined here seeks to offer a more complete theoretical framework for metaphor comprehension by describing the kinds of representational and processing changes that occur as metaphors are conventionalized.

Acknowledgments

This research was supported in part by a Northwestern University Dissertation Year Fellowship, awarded to the first author, and by NSF Grant SBR-95-11757 and ONR Grant N00014-92-J-1098, awarded to the second author.

References

Blank, G. D. (1988). Metaphors in the lexicon. *Metaphor and Symbolic Activity*, 3, 21-36.

Blasko, D. G., & Connine, C. M. (1993). Effects of familiarity and aptness on metaphor processing. *Journal of Experimental Psychology: Learning, Memory, and Cognition*, 12, 295-308.

Bowdle, B. F. (1998). *Conventionality, polysemy, and metaphor comprehension*. Unpublished doctoral dissertation, Northwestern University.

Bowdle, B. F., & Gentner, D. (1995). *The career of metaphor*. Poster given at the Thirty-Sixth Annual Meeting of the Psychonomic Society, Los Angeles, CA.

Bowdle, B. F., & Gentner, D. (in preparation). The career of metaphor.

Falkenhainer, B., Forbus, K. D., & Gentner, D. (1989). The structure-mapping engine: Algorithm and examples. *Artificial Intelligence*, 41, 1-63.

Gentner, D. (1983). Structure-mapping: A theoretical framework for analogy. *Cognitive Science*, 7, 155-170.

Gentner, D., & Clement, C. A. (1988). Evidence for relational selectivity in interpreting analogy and metaphor. In G. H. Bower (Ed.), *The psychology of learning and motivation*. New York: Academic Press.

Gentner, D., & Wolff, P. (1997). Alignment in the processing of metaphor. *Journal of Memory and Language*, 37, 331-355.

Gick, M. L., & Holyoak, K. J. (1983). Schema induction and analogical transfer. *Cognitive Psychology*, 15, 1-38.

Giora, R. (1997). Understanding figurative and literal language: The graded salience hypothesis. *Cognitive Linguistics*, 8, 183-206.

Glucksberg, S., & Keysar, B. (1990). Understanding metaphorical comparisons: Beyond similarity. *Psychological Review*, 97, 3-18.

Glucksberg, S., McGlone, M. S., & Manfredi, D. (1997). Property attribution in metaphor comprehension. *Journal of Memory and Language*, 36, 50-67.

Holyoak, K. J., & Thagard, P. (1995). *Mental leaps*. Cambridge, MA: MIT Press.

Honeck, R. P., Kibler, C. T., & Firment, M. J. (1987). Figurative language and psychological views of categorization: Two ships in the night? In R. E. Haskell (Ed.), *Cognition and symbolic structures*. Norwood, NJ: Ablex Publishing.

Indurkhya, B. (1987). Approximate semantic transference: A computational theory of metaphor and analogy. *Cognitive Science*, 11, 445-480.

Katz, A. N. (1989). On choosing the vehicles of metaphors: Referential concreteness, semantic distances, and individual differences. *Journal of Memory and Language*, 28, 486-499.

Kennedy, J. M. (1990). Metaphor – its intellectual basis. *Metaphor and Symbolic Activity*, 5, 115-123.

Kittay, E. F., & Lehrer, A. (1981). Semantic fields and the structure of metaphor. *Studies in Language*, 5, 31-63.

Lakoff, G. (1987). *Women, fire, and dangerous things*. Chicago: University of Chicago Press.

Lakoff, G., & Johnson, M. (1980). *Metaphors we live by*. Chicago: University of Chicago Press.

Lakoff, G., & Turner, M. (1989). *More than cool reason*. Chicago: University of Chicago Press.

Lehrer, A. (1990). Polysemy, conventionality, and the structure of the lexicon. *Cognitive Linguistics*, 1, 207-246.

Miller, G. A. (1979). Images and models, similes and metaphors. In A. Ortony (Ed.), *Metaphor and thought*. New York: Cambridge University Press.

Murphy, G. L. (1996). On metaphoric representation. *Cognition*, 60, 173-204.

Novick, L. R., & Holyoak, K. J. (1991). Mathematical problem solving by analogy. *Journal of Experimental Psychology: Learning, Memory, and Cognition*, 17, 398-415.

Nunberg, G. (1979). The non-uniqueness of semantic solutions: Polysemy. *Linguistics and Philosophy*, 3, 143-184.

Ortony, A. (1979). Beyond literal similarity. *Psychological Review*, 86, 161-180.

Ross, B. H., & Kennedy, P. T. (1990). Generalizing from the use of earlier examples in problem solving. *Journal of Experimental Psychology: Learning, Memory, and Cognition*, 16, 42-55.

Shen, Y. (1992). Metaphors and categories. *Poetics Today*, 13, 771-794.

Sweetser, E. (1990). *From etymology to pragmatics*. New York: Cambridge University Press.

Tourangeau, R., & Rips, L. (1991). Interpreting and evaluating metaphors. *Journal of Memory and Language*, 30, 452-472.

Turner, N. E., & Katz, A. N. (1997). The availability of conventional and of literal meaning during the comprehension of proverbs. *Pragmatics and Cognition*, 5, 199-233.

Tversky, A. (1977). Features of similarity. *Psychological Review*, 84, 327-352.

Verbrugge, R. R., & McCarrell, N. S. (1977). Metaphoric comprehension: Studies in reminding and resembling. *Cognitive Psychology*, 9, 494-533.

Williams, J. (1992). Processing polysemous words in context: Evidence for interrelated meanings. *Journal of Psycholinguistic Research*, 21, 193-218.

Conceptual Accessibility and Serial Order in Greek Speech Production

Holly P. Branigan (holly@psy.gla.ac.uk)
Department of Psychology; 53 Hillhead Street
Glasgow G12 8QF, UK

Eleonora Feleki
Centre for Cognitive Science; 2 Buccleuch Place
Edinburgh EH8 9LW, UK

Abstract

Current theories of language production disagree about the way in which conceptual accessibility influences syntactic processing (e.g. Bock, 1987; De Smedt, 1990). We present theoretical arguments that the assumption of highly incremental processing can only be reconciled with theories in which conceptual accessibility influences word order. We report a sentence recall experiment in Modern Greek that provides empirical support for this position. Our results demonstrate that Greek speakers prefer to place conceptually accessible entities in early word order positions, irrespective of grammatical function, contrary to previous findings for English (Bock & Warren, 1985; McDonald, Bock & Kelly, 1993). We interpret our results as evidence for highly incremental processing.

Introduction

Speakers are faced with the task of producing fluent, well-formed speech under time constraints. Many researchers have suggested that speakers achieve this by processing different aspects of the utterance incrementally and in parallel (e.g., De Smedt, 1990; Ferreira, 1996; Kempen & Hoenkamp, 1987; Levelt, 1989). In this way speakers can start generating an utterance as soon as a minimal amount of input is available, rather than having to wait until all elements of the utterance have been retrieved; and can begin to articulate an utterance before all aspects of its structure have been processed. If the human production system were not incremental, conversation would consist of bursts of speech punctuated by silences as the speaker planned the next utterance. This emphasis on processing partial information represents an important point of contact between language production and other aspects of human cognition (e.g. Marslen-Wilson, 1973).

Incremental accounts of language production predict an important role for information accessibility: Readily accessible information will undergo processing more quickly than less readily accessible information. Thus variations in information accessibility are hypothesized to play an important part in determining the characteristics of the utterance that is ultimately produced (Bock, 1982; De Smedt, 1990; Levelt, 1989). This paper examines that hypothesis with reference to syntactic structure. We begin by examining evidence that the conceptual features of a message influence its eventual syntactic realization. We then assess critically how current models of production account for such effects. We will argue that models in which conceptual accessibility is primarily associated with the assignment of grammatical functions are incompatible with the assumption of highly incremental processing. Instead we argue for models in which conceptual accessibility has a direct influence on word order. In the remainder of the paper, we report an experimental investigation of Modern Greek that provides empirical support for a link between word order and conceptual accessibility, contrary to previous findings for English (Bock & Warren, 1985; McDonald, Bock & Kelly, 1993). We interpret our results as support for highly incremental models of language production.

Determinants of Syntactic Processing in Production

It is generally accepted that language production begins with the speaker deciding to express a meaning. This pre-linguistic message triggers the retrieval of appropriate lexical concepts and their associated lemmas (the syntactic component of lexical entries; Levelt, Meyer & Roelofs, in press; see also Kempen & Huijbers, 1983). Syntactic structure is then generated from the syntactic information contained within the lemmas (Kempen & Hoenkamp, 1987). Under these assumptions, syntactic processing should be affected by two factors: the relative accessibility of the lemmas themselves; and the relative accessibility of the syntactic information contained within them. Evidence from syntactic priming effects in production (Bock, 1986; Bock & Loebell, 1990; Hartsuiker & Kolk, 1998; Pickering & Branigan, 1998) supports the second hypothesis; what evidence is there to support the first?

In fact, there is considerable evidence that syntactic processing is influenced by conceptual features that are plausibly associated with lemma accessibility. For example, many researchers have found a tendency for speakers of English to produce passive sentences when the patient of an action is animate and/or human (Cooper & Ross, 1975; Ferreira, 1994; McDonald et al, 1993; Sridhar, 1988). Bock and Warren (1985) (see also Bock, 1987) proposed an explicit link between variations in syntactic structure and variations in what they termed conceptual accessibility, or 'the ease with which the mental representation of some potential referent can be activated in or retrieved from memory' (p.50). We interpret this as the accessibility of a lexical concept and its associated lemma. Bock and Warren suggested that some entities are conceptually more accessible than others because they take part in more conceptual relations, and hence can be retrieved through more routes. For example, entities that are animate, concrete or prototypical are more predicable than items that are inanimate, abstract or non-prototypical (Keil, 1979).

How are variations in conceptual accessibility realized as variations in syntactic structure? We can identify two possibilities. First, conceptually accessible items might be associated with higher grammatical functions, such that easily retrieved items tend to become subjects. In that case, the preference for passive structures with animate patients would reflect an association between animacy and subjecthood. Alternatively, conceptually accessible items might be associated with early word order positions, such that easily retrieved items tend to precede less easily retrieved items. In that case, the preference for passive structures with animate patients would reflect an association between animacy and first position in the sentence.

Conceptual Accessibility and Grammatical Function Assignment

Bock (1987; see also Bock & Warren, 1985) argued for the first of these alternatives. She proposed that conceptual accessibility influences an initial stage of syntactic processing. During this stage, grammatical functions are assigned following Keenan and Comrie's (1977) NP accessibility hierarchy. The subject function is assigned first, then the direct object function, and so on. Because the lemmas associated with conceptually accessible items are retrieved more quickly, they tend to claim higher grammatical functions. Thus conceptually accessible items prefer to appear as subjects. However, conceptual accessibility does not directly influence word order. Instead, word order is determined at a subsequent stage of processing, and is influenced by the accessibility of the relevant wordforms (morpho-phonological content of a lexical entry). Bock suggested that any apparent link between conceptual accessibility and word order arises from the fact that - in English at least - higher grammatical functions tend to precede lower grammatical functions. For example, the subject of an English sentence appears at the beginning of the sentence, preceding the direct object. This means that grammatical function effects can easily be misinterpreted as word order effects. Bock argued that when the effects of grammatical function are excluded, there is no independent preference to place conceptually accessible items in early word order positions. We will term Bock's (1987) model the grammatical function model.

Two sentence recall studies provided empirical support for the grammatical function model. In this task, participants are presented with sentences that they subsequently attempt to recall. Many studies have shown that the form in which participants recall the original sentences reflects the normal biases of production (see Bock & Irwin, 1980). By manipulating the features of the sentences that are presented, and examining how this affects participants' recall of the sentences, it is possible to draw inferences about the

nature of the production system. In both of the relevant studies, the experimental manipulation was to present sentences containing pairs of nouns that varied in conceptual accessibility. Bock and Warren (1985) employed concreteness as an index of conceptual accessibility, whilst McDonald et al (1993) employed animacy. Both studies found that participants tended to recall sentences in a form that allowed the more conceptually accessible entity to appear in a higher grammatical function than the less accessible entity. For example, participants recalled an active sentence as a passive sentence when the patient was the more accessible entity, thereby promoting the accessible entity to subject. However, in NP conjunctions, where both nouns had the same grammatical function, there was no tendency for recall in a form that allowed the more accessible entity to precede the less accessible entity. These findings appear to support the hypothesis that variations in conceptual accessibility are associated with variations in grammatical function assignment but do not influence word order.

The Grammatical Function Model: Restricted Incrementality

However, both the grammatical function model and the empirical findings on which it is based appear to sit uneasily with the assumption of highly incremental processing. First, it is unclear how the model can account for the systematic variations in word order found in many languages, in which lower grammatical functions may precede higher grammatical functions. For example, Modern Greek allows an Object-Verb-Subject (OVS) ordering. Under the grammatical function model, a phrase must receive a grammatical function before it receives a serial position, and the first function to be assigned is always the subject function. There are thus only two ways in this model to account for an object's appearance preceding the subject. First, the subject's wordform might be less accessible than that of the object; in that case, the object might 'overtake' the subject at this stage of processing. But in an extensive series of studies, McDonald et al (1993) found no evidence that wordform accessibility influences serial ordering. The alternative is to sacrifice incrementality, so that in some circumstances the processor assigns the subject function but does not then place the subject in the earliest serial position. Instead, it waits until the object function has been assigned; it then assigns the object phrase the earliest position instead. Such a position is certainly possible. But it would mean that the speaker, despite having the subject phrase ready to articulate, would have to buffer it until the object phrase subsequently became ready. Thus there is no way in the grammatical function model for speakers to promote fluency by exploiting the word order variations available in their language to produce an accessible item while they are retrieving or processing a less accessible item. Rather, the production of such non-canonical sentences would seem in the grammatical function model to inherently entail disfluency.

So far, we have argued that the actual architecture of the

grammatical function model entails restricted incrementality under some circumstances. In addition, one aspect of the empirical findings that underpin the model implies restricted incrementality. This is the failure to find serial ordering effects associated with conceptual accessibility in NP conjunctions. Recall that the absence of such effects was the crucial evidence for the model's linkage between conceptual accessibility and grammatical function. This finding poses a problem for incrementality because, in an incremental processor, the first element to complete processing at one level should be the first to undergo processing at the next stage. Hence, in an NP conjunction, the conceptually more accessible entity should undergo lemma retrieval before the less accessible entity. It should therefore also undergo and complete wordform processing first (assuming that both conjuncts are of equal wordform accessibility, which was controlled for in the experiments under discussion). Under the grammatical function model, it should therefore consistently claim the earliest serial position. The failure to find such ordering effects can only be explained by assuming restricted incrementality, in which order of access does not determine order of processing.

From this conclusion, we can draw one of two alternative implications for the grammatical function model. NP conjunctions could be 'normal' structures that are processed like any other structure. In that case, evidence about the way in which they are processed is good evidence about the normal processes of production, but we must conclude that the production system is only restrictively incremental. Alternatively, NP conjunctions might be abnormal structures that are processed quite differently from other structures. In that case, failure to find serial ordering effects in these structures does not necessarily mean that the production system is restrictively incremental under normal circumstances, merely that it is restrictively incremental when processing NP conjunctions; but then evidence from such unusual structures cannot be taken as evidence about the normal workings of the system.

Conceptual Accessibility, Serial Order and Incremental Processing

Although we are not aware that the specific problems described in the previous section have been previously identified, a number of researchers have proposed alternative models of production which avoid the restricted incrementality entailed by the grammatical function model. These models differ in details; however, they all allow conceptually more accessible entities to claim early serial positions, irrespective of grammatical function (e.g. De Smedt, 1990; Kempen & Hoenkamp, 1987; Levelt, 1989). For example, in Kempen and Hoenkamp's (1987) model, lemmas are assigned grammatical functions before they claim a serial position; but functions are not necessarily assigned according to a hierarchy, and whichever lemma receives a grammatical function first claims the earliest available position. Thus an object can claim an early serial position before the subject function has been assigned. In this model, early serial positions are thus mediated via an initial stage of grammatical function assignment. By contrast,

De Smedt (1996) discussed a model in which a lemma can claim a serial position before it has been assigned a grammatical function. For example, the first lemma to be retrieved can claim the earliest serial position before the processor has committed to which grammatical function it will fulfil in the utterance. As in Kempen and Hoenkamp's model, an object can thus claim an early serial position before the subject function has been assigned. But in this model, early serial positions are directly associated with conceptual accessibility. Although these models differ in their architectural details, the important point is that they all propose that conceptual accessibility influences serial order. We therefore term them <u>serial order models</u>. In these models, variant word orders such as OVS promote fluency through incremental production, by allowing speakers to encode a readily accessible object while they are retrieving a less accessible subject. This contrasts sharply with the grammatical function model, where such variant word orders promote disfluency.

Conceptual Accessibility Effects on Syntactic Structure in Greek

Clearly, there are theoretical reasons for preferring the serial order models: They allow highly incremental processing, which we have argued to be greatly advantageous for speakers. In particular, they provide an account of how speakers of languages with flexible word orders might exploit variant orders as a means of promoting fluency. But there is limited empirical evidence to support the serial order models. Kelly, Bock and Keil (1985) found evidence in English for a link between prototypicality and serial order, using the same recall task as Bock and Warren (1985) and McDonald et al (1993). Sridhar (1988) found cross-linguistic evidence using a `simply describe' paradigm (cf. Osgood, 1971) of a tendency to produce word orders that allowed conceptually accessible entities (in his terms, more salient entities) to appear first. Prat-Sala (1997) employed a picture description task and found a tendency in Spanish and Catalan for conceptually accessible entities to precede less accessible entities. However, these experiments have other possible explanations (e.g. lexical confounds).

We therefore set out to examine whether there is evidence for a link between conceptual accessibility and word order when such alternative explanations can be excluded. Our experiment employed Modern Greek, a language allowing a wide range of structures that separate serial order from grammatical function. Bock and her colleagues' work relied crucially on NP conjunctions, since these are almost the only structure in English where grammatical function and serial order are separable. However, conjunctions are unusual structures (e.g. Chomsky, 1965). For example, they may be multiply-headed structures (Gazdar, Klein, Pullum & Sag, 1985). Thus it is possible that they are processed in unusual ways. In Modern Greek, by contrast, word order variations can be found for normal declarative sentences. For example, the subject of a sentence can appear preceding or following the verb, and preceding or following the direct object. Our experiment focused on the subject-verb-object (SVO) and object-verb-subject (OVS) orders. As in Bock and colleagues' experiments, participants heard and attempted to recall sentences like those in (1) in which the animacy of the two nouns fulfilling the subject and direct object functions was systematically manipulated:

(1a) Sta dimokratika politevmata, o politis sevete to sindagma.
in democratic regimes the citizen$_{NOM}$ respects the law$_{ACC}$
'In democratic regimes, the citizen respects the law'

(1b) Sta dimokratika politevmata, to sindagma sevete o politis.
in democratic regimes the law$_{ACC}$ respects the citizen$_{NOM}$
'In democratic regimes, the citizen respects the law'

(1c) Sta dimokratika politevmata, to sindagma sevete ton politi.
in democratic regimes the law$_{NOM}$ respects the citizen$_{ACC}$
'In democratic regimes, the law respects the citizen'

(1d) Sta dimokratika politevmata, ton politi sevete to sindagma.
in democratic regimes the citizen$_{ACC}$ respects the law$_{NOM}$
'In democratic regimes, the law respects the citizen'

The logic of the experiment was simple: If conceptual accessibility affects serial order, then we would expect participants to recall sentences in a form that allowed the conceptually more accessible entity to appear first, irrespective of grammatical function. Thus sentences like (1b) should be recalled as (1a), and sentences like (1c) should be recalled as (1d).

Method
Participants and Materials Our participants were thirty-two native Greek speakers. We constructed thirty-two items like those in (1a-d), each comprising a preposed adverbial phrase and a main clause that contained a transitive verb, an animate noun and an inanimate noun. The animate and inanimate nouns were matched for frequency over the item set as a whole. (Because frequency tables are unavailable for Modern Greek, we compared frequencies for the English translation equivalents of the target nouns, using the CELEX database [Baayen, Piepenbronck & Gulliver, 1995].) For animate nouns, the mean frequency was 39.22 per million words; for inanimate nouns, it was 38.66 per million words. There was no significant difference in frequency on a paired samples T-test (t(31) = 0.054, p = .95).

Each item had four versions, each containing the same adverbial phrase, verb and noun phrases, as in (1a-d). The four versions of each item were constructed by crossing Subjecthood (animate vs. inanimate subject) with Word Order (SVO vs. OVS). Hence the animate noun appeared as the subject in two versions (1a and 1b), and as the direct object in two versions (1c and 1d); and in sentence-initial position in two versions (1a and 1d), and in sentence-final position in two versions (1b and 1c). In addition we constructed 16 filler sentences, comprising a preposed adver-

bial phrase and a main clause of various types.

Procedure The sentences were presented on audiotape in eight blocks, each containing four experimental sentences and two filler sentences. The order of sentences within each block was randomized; the resulting order was held constant across all four lists. A three-second pause separated each recorded sentence in each block. The number of sentences presented in each block, and the duration of the pause between each sentence, were determined on the basis of a pilot study involving eight participants who did not take part in the main study. After each block of sentences, participants were prompted for oral recall using the introductory adverbial phrases; within each prompt block, the prompts appeared in a different order from the corresponding sentences in the sentence block. A block of six practice sentences was presented before the main experiment.

Scoring Participants' responses were recorded on audiotape. This represents a minor methodological change from previous sentence recall experiments, where participants wrote their responses. We asked participants to recall sentences orally because we felt that this would be more informative about spoken language production. Participants' responses were scored as correct (same function assignment and word order as the original sentence); inversion (same function assignment as the original sentence but the alternative word order); or error (anything else). As in Bock and colleagues' experiments, we used a measure that expressed the strength of the tendency to recall a sentence in a different form to that presented, when the semantic content was correctly remembered. Thus we performed analyses of variance on proportions representing the number of inversions, i.e. sentences recalled in their alternative form ((1a) recalled as (1b) and vice versa, (1c) recalled as (1d) and vice versa), relative to the total number of corrects plus inversions in each condition for each subject and item.

Results
The mean proportion of inversions in each condition is shown in Figure 1. Analyses of variance treating both participants and items as random effects revealed a main effect of Word Order ($F_1(1,31) = 40.92$, $p < .001$; $F_2(1,31) = 96.87$, $p < .001$): Participants were more likely to recall sentences in the alternative form to that originally presented when this resulted in the preferred SVO order than when it resulted in OVS order (41% vs. 6%).

More importantly, however, this tendency interacted with Subjecthood (F_1 (1,31) = 8.75, $p < .006$; F_2 (1,31) = 7.38, $p < .01$). Inspection of the results showed that participants were more likely to recall SVO sentences as OVS sentences when the subject was inanimate and such an inversion would result in the animate entity appearing first, than when the subject was animate and such an inversion would result in the inanimate entity appearing first (10% vs. 2%). Equally, participants were more likely to recall OVS sentences as SVO sentences when the subject was animate and such an inversion would result in the animate entity appearing first, than when the subject was inanimate and such an inversion would result in the inanimate entity appearing first (47% vs. 36%). Planned comparisons confirmed that there were significant effects for both SVO and OVS sentences (all $p < .001$). No other effects achieved significance.

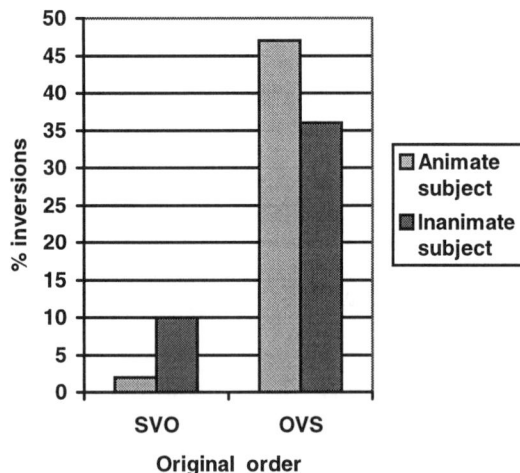

Figure 1: Percentage of sentences recalled with inverted word order, by condition.
(Columns represent original [uninverted] word order)

Discussion

These results provide strong evidence that conceptual accessibility can affect word order: Participants preferred to recall sentences in a form that allowed the conceptually more accessible entity to precede the less accessible entity, irrespective of grammatical function. This tendency was found both for the preferred SVO order and for the dispreferred OVS order. Thus animate entities tended to appear first even when they fulfilled the grammatical function of object. These results argue against the grammatical function model of conceptual influences on syntactic processing, which predicted that participants should prefer to recall animate entities as subjects but that animate entities should exhibit no independent tendency to appear in early serial positions. Instead, they provide strong support for what we have termed serial order models, in which the relative accessibility of lemmas exerts a relatively direct influence on word order decisions (e.g. De Smedt, 1990). More importantly, by disconfirming the grammatical function model, our results argue against the restrictive incrementality that we have shown that model to entail. Rather, our results argue for highly incremental syntactic structure generation, in which the processor achieves fluency by working on information as and when it becomes available.

Of course, we cannot be sure from these results whether the influence of conceptual accessibility is mediated by grammatical function in some way. There are two aspects to this point. Firstly, conceptually accessible entities might have an affinity for higher grammatical functions, in addition to early serial positions. Clearly, our experiment excludes the possibility that conceptually accessible entities are advantaged only with respect to grammatical functions; but it cannot determine whether there is a grammatical function effect of some sort. Future work could address this problem by using structures which involve variations in grammatical function that are independent of early word order positions. For example, if there is indeed a preference for conceptually accessible entities to claim higher grammatical functions, independent of serial order, then speakers should prefer to recall a sentence-final oblique object in a passive sentence as a sentence-final subject in a semantically equivalent OVS sentence. Secondly, our experiment does not allow us to distinguish between two possible architectures for syntactic processing in language production, one in which serial order can be assigned immediately a lemma has been retrieved, possibly before it has received a grammatical function, as discussed by De Smedt (1996); versus an architecture in which a lemma must be assigned a grammatical function before it claims a serial position, as in Kempen and Hoenkamp (1987). In the first architecture, conceptual accessibility would have a very direct impact upon serial order; in the second architecture, its impact would be mediated by an initial stage of function assignment. In theory, such an intermediate stage of processing could muddy the ultimate relationship between conceptual accessibility and serial order. However, the very incrementality of the processor makes it difficult to distinguish empirically between the two possibilities, at least using this experimental method. We therefore leave this as a question for future research.

A final important question concerns the differences between our findings and those of Bock and colleagues, who failed to find a link between conceptual accessibility and serial order. Do these differences reflect different processing architectures for English versus Greek, or is there some other explanation? It is obviously possible that strict word order and more flexible word order languages are associated with different processing architectures, and that the latter allow more highly incremental processing; but this seems unlikely. Rather, it seems plausible that the differences can be attributed to the structures studied. Recall that Bock and colleagues' experiments crucially relied upon NP conjunctions. These structures differ in an important way from those that we studied. Specifically, non-conjunctive NPs involve retrieval of a single noun lemma, which controls syntactic elaboration. Hence as soon as the processor has retrieved just the noun lemma, it can commence syntactic processing. In contrast, NP conjunctions require the processor to retrieve two noun lemmas, one for each conjunct. Crucially, the syntactic elaboration of the conjunctive phrase is determined by the syntactic features of both conjuncts. For example, agreement is determined with reference to both conjuncts. In syntactically elaborating an NP conjunction, therefore, the processor needs to make reference to the syntactic information contained in both lemmas. As such, it seems plausible that conjunctions are not processed incrementally like other phrases. We suggest that when processing a conjunctive phrase, the processor temporarily suspends fully incremental processing, and delays some syntactic processing until the lemmas associated with both conjuncts have been successfully retrieved and the information that they contain can be used to constrain the syntactic structure that is generated. If NP conjunctions are not processed incrementally like other phrases, it is not surprising that they do not exhibit ordering effects related to the accessibility of each conjunct.

To conclude, our results confirm a hypothesized role for conceptual accessibility in determining serial order. We interpret our results as evidence for highly incremental syntactic structure generation, in which information accessibility strongly influences the behavior of the processor.

Acknowledgments

Order of authorship is arbitrary. We thank Martin Pickering, Mercè Prat-Sala and Christoph Scheepers. The first author was supported by a British Academy Postdoctoral Fellowship.

References

Bock, J.K. (1982). Toward a cognitive psychology of syntax: Information processing contributions to sentence formulation. *Psychological Review, 89*, 1-47.

Bock, J.K. (1986). Syntactic persistence in language production. *Cognitive Psychology, 18*, 575-586.

Bock, J.K. & Loebell, H. (1990). Framing sentences. Cognitive Psychology, 35, 1-39.

Bock, J.K. (1987). Coordinating words and syntax in speech plans. In A. Ellis (ed) *Progress in the pathology of language*. London: Erlbaum.

Bock, J.K. & Irwin, D. (1980). Syntactic effects of information availability in sentence production. *Journal of Verbal Learning and Verbal Behavior, 19*, 467-484.

Bock, J.K. & Warren, R. (1985). Conceptual accessibility and syntactic structure in sentence formulation. *Cognition, 21*, 47-67.

Baayen, R. H., Piepenbronck, R., & Gulliver, L. (1995). *The CELEX Lexical Database (CD Rom)*. The Linguistic Data Consortium, University of Pennsylvania, Philadelphia.

Chomsky, N. (1965). *Aspects of the structure of syntax*. Cambridge: MIT Press.

Cooper, W.E. & Ross, J.R. (1975). World order. In R.E. Grossmann, L.J. San, and T.J. Vance (eds) *Papers from the parasession on functionalism*. Chicago Linguistic Society.

De Smedt, K. (1990). IPF: An incremental parallel formulator. In R. Dale, C. Mellish, and M. Zock (eds) *Current research in natural language generation*. London: Academic Press.

De Smedt, K. (1996). Computational models of incremental grammatical encoding. In T. Dijkstra and K. De Smedt (eds) *Computational Psycholinguistics*. London: Taylor & Francis.

Ferreira, F. (1994). Choice of passive voice is affected by verb type and animacy. *Journal of Memory and Language, 33*, 715-736.

Ferreira, V.S. (1996). Is it better to give than to donate? Syntactic flexibility in language production. *Journal of Memory and Language, 35*, 724-755.

Gazdar, G., Klein, E., Pullum, G. & Sag, I. *Generalised Phrase Structure Grammar*. Oxford: Blackwell.

Hartsuiker, R. & Kolk, H. (1998). Syntactic persistence in Dutch. *Language and Speech, 41*, 143-184.

Keenan, E. & Comrie, B. Noun phrase accessibility and universal grammar. *Linguistic Inquiry, 8*, 63-99.

Keil, F.C. (1979). *Semantic and conceptual development: An ontological perspective*. Harvard University Press.

Kelly, M.H., Bock, J.K. & Keil, F.C. (1985). Prototypicality in a linguistic context: Effects on sentence structure. *Journal of Memory and Language, 25*, 59-74.

Kempen, G. & Hoenkamp, E. (1987). An incremental procedural grammar for sentence formulation. *Cognitive Science, 11*, 201-288.

Kempen, G. & Huijbers, P. (1983). The lexicalization process in sentence production and naming: Indirect election of words. *Cognition, 14*, 185-209.

Levelt, W.J.M. (1989). *Speaking: From intention to articulation*. Cambridge: MIT Press.

Levelt, W.J.M., Meyer, A. & Roelofs, A.(in press). A theory of lexical access in speech production. *Behavioral and Brain Sciences*.

Marslen-Wilson, W. (1973). Linguistic structure and speech shadowing at very short latencies. *Nature, 244*, 522-523.

McDonald, J., Bock, J.K. & Kelly, M.H. (1993). Word and world order: Semantic, phonological and metrical determinants of serial position. *Cognitive Psychology, 25*, 188-230.

Osgood, C.E. (1971). Where do sentences come from? In D. Steinberg and L. Jakobovits (eds) *Semantics: An interdisciplinary reader in philosophy, linguistics and psychology*. Cambridge: Cambridge University Press.

Pickering, M.J. & Branigan, H.P. (1998). The representation of verbs: Evidence from syntactic priming in written production. *Journal of Memory and Language, 39*, 633-655.

Prat-Sala, M. (1997). *The production of different word orders: A psycholinguistic and developmental approach*. Unpublished PhD thesis, University of Edinburgh.

Sridhar, S.N. (1988). *Cognition and sentence production: A cross-linguistic study*. New York: Springer.

Does philosophy offer cognitive science distinctive methods?

Andrew Brook
Cognitive Science Programme
Carleton University
Ottawa ON   K1S 5B6

Abstract
Philosophy has never settled into a stable position in cognitive science and its role is not well understood. One reason for this is that the methods philosophers use to study cognition look quite peculiar to other cognitive scientists. This paper explores the methods of philosophy, laying out some of the main kinds and looking at some examples, and makes some remarks about their value to cognitive science.

Does philosophy have distinctive methods to offer cognitive science? Once upon a time, philosophers agreed at least that philosophy *has* distinctive methods. At various times, analysis of the conceptual framework of cognition or analysis of concepts or assembling reminders of how we use key terms of ordinary language or .... or ... or ... would have been presented as the method(s) in question. More recently, even this has become a matter of controversy. Starting from the heavy pressure that Quine (1952) put on the analytic/synthetic distinction and the concept/fact distinction and that Davidson(1973) put on the idea that there is a serious distinction even between facts and conceptual framework, many philosophers now deny that there is anything distinctive about the methods they use. On this view, philosophy is merely the most abstract kind of empirical theory-building and uses the same methods as other high-level theory-builders (Castañeda, 1980).

This cheery ecumenism is a bit strange. The methods of philosophy certainly look distinctive and even quite peculiar to the rest of cognitive science, especially to the real experimental theory-builders in the discipline. Maybe a few babies are in danger of being thrown out with the conceptual bathwater. There is more to philosophy than the parts of it that contribute to cognitive science but I will stick to the parts that do.

The first question I will examine is thus:

In the parts of philosophy that contribute to cognitive science, can we identify distinctive methods?

To be distinctive to philosophy as I will understand the term, a method must have three features. It must be:

(a) significantly different from hands-on experimentation, model-building, etc.,

(b) pervasive in and central to the work of philosophers who contribute to cognitive science,

and,

(c) not common elsewhere in cognitive science.

To start, notice a couple of methods would not meet one or more of these requirements:

(i) *Deductive and inductive inferences from premises.* This activity (I am not sure that it should be called a method) is quite different from hands-on experimental work, model-building, etc., and is central to philosophy but it is not at all *distinctive* to philosophy – all rationally structured investigation engages in it.

(ii) *Building cognitive or computational models.* This method is also quite distinct from experimentation and hypothesis-testing but it is not central to most of the philosophy that we find in cognitive science.

Are there any methods that would satisfy all of (a), (b) and (c)? In Section I, I will sketch some of the diverse methods that philosophers in cognitive science in fact use. In Section II, I will focus on thought experiments, one of the most characteristic activities of philosophers in cognitive science, and begin an examination of whether they satisfy (a), (b) and (c). In Section III, I will examine how thought experiments function vis-à-vis experimental science and complete the examination of whether they satisfy (a), (b) and (c).

1. The methods of philosophy

First, a preliminary point about philosophy. There is one philosophical activity that is *highly* distinctive to it: the exploration of norms – ethical norms, political norms, epistemic norms, and so on. To be more precise, philosophers investigate what our norms should be, what norms are justified. Other disciplines investigate what norms people do *in fact* accept, but only philosophy tries to determine what norms we *ought* to accept.

Moreover, philosophy does significant work on norms in cognitive science, epistemic norms in particular. (This might be thought of as a place where ethics [or metaethics] meets the philosophy of science.) All science, indeed virtually all human activity, is governed by norms. How to justify norms is a huge issue and fierce debates rage. How does the normative relate to the natural (Hatfield,1990)? Is there any source of norms independent of what some part of our past has in fact induced into us, evolution, for example, or past scientific practice? (A concept of truth independent of human interests would be an example.) Whatever the answer to the question about sources, is there any way to *justify* norms independently of evolution or past practise? –These are some of the issues in normative epistemology.

The investigation of them is highly distinctive to philosophy; other disciplines take no more than a passing interest in such questions.

It is not obvious, however, that the *methods* of normative investigation differ from the methods of philosophy in general. Even if the *subject-matter* of normative investigations is different from the subject-matter of other parts of philosophy, the former may use much the same *methods* as the rest of philosophy. At any rate, in this paper I will focus on methods used throughout philosophy's contributions to cognitive science, both normative and (largely) nonnormative.[1]

Philosophers use at least four methods in the work they do in cognitive science:

(1) *Investigation of the meanings of words:* what words do mean and what we should take them to mean. This activity, which could be extended to include such activities as the investigation of the semantic properties of different types of explanation, investigation of how various scientific (and perhaps other) activities use a given word, and so on, is conceptual analysis. Thirty years ago it was taken to be the core of philosophical investigation.

(2) *Straightforward scientific method,* but applied to more general and abstract questions than are common in (the rest of) science. Castañeda (1980) is a keen exponent of the idea that this is how analytic philosophy works; he cites Quine, Sellars and Chisholm as exemplars.

If (2) is what philosophers are doing when they do philosophy, they are straightforward empirical theory-builders, merely interested in more abstract and general questions than (other) scientists: the nature of a number rather than the nature of a neuron; how many kinds of thing exist in general rather than how many kinds of elementary particles there are; the relation of object and property rather than the relation of an ecology and its occupants; and so on.

Note that (1) and (2) might not be as different from one another as they first appear. Sorting out and the rational reconstruction of concepts is often part of what is needed to build a theory – indeed, as Quine showed, there is often no clear line between changing beliefs and changing concepts.[2]

The methods of (2) are simply the methods of good theory-building in general and when philosophy uses them, there is nothing distinctive to its methods. If science makes use of conceptual analysis in the way just suggested, then (1) is not distinctive to philosophy either. It too is merely part of the methodology of good theory-building in general. In connection with (1), the most that could be said is that philosophers *pay more attention* to the conceptual toolbox of science than (other) scientists who focus on experimentation, modelling, etc.

Note too that conceptual investigation is itself a *form* of empirical investigation, at least in part. It is not experimental but it is still empirical (experiments are only one kind of empirical investigation).[3] Conceptual analysis of the symbolic toolbox of science involves lexical semantics, the psychology of cognition (how we use concepts and why), and the exploration of what is distinctive to, perhaps even necessary for or of the "essence" of, kinds of things. Lexical semantics and psychology, at least, are straightforwardly empirical.

To be sure, analysis of concepts and other symbolic structures is not *entirely* empirical. When we go to work on the concepts, styles of reasoning, etc., used to investigate some domain, we want to generate *good* concepts, *good* styles of reasoning, etc. That is, we are at least in part interested in *normatively reconstructing* the symbolic structure so that it serves our epistemic interests better, not just in *finding out* what is built into the structure as it now exists. Rational reconstruction is at least as much normative justification as it is any kind of empirical investigation.

What do philosophers use to analyse/reconstruct concepts? This brings us to a third method:

(3) *Thought experiments* The natural sciences make some use of thought experiments, physics in particular (Brown 1991). Famous thought experiments in physics include Schrodinger's cat and Galileo's invitation to imagine a smaller and a bigger piece of matter tied together.[4] The social sciences make some use of thought experiments, too, though less use than physics. In philosophy, they are absolutely central.

In a thought experiment, we imagine some scenario (the contrast is with hands-on experiments using equipment, etc., in which we manipulate a scenario itself, rather than a representation of it; I will call these 'hands-on

---

[3.] I owe my sense of the importance of this distinction to Rob Stainton. Computational modelling is an empirical investigation that is not experimental, for one.

[4.] Schrodinger argued that a cat in a box had to be in some determinate state even if we did not know what it was. Galileo's thought experiment was aimed at Aristotle's idea that smaller mass fall slower than bigger ones. If a smaller mass falls slower than a bigger one, then if we tie the two together, the resultant mass ought to fall both faster and slower than the bigger of its components by itself. On thought experiments in science, see Brown 1991, Horowitz and Massey 1991, and Sorensen 1992.

[1.] Are any philosophical contributions entirely nonnormative? I doubt it. Virtually all philosophy has a normative element. Even an analysis of a concept is often in part a recommendation as to what we *should* take the concept to mean.

[2.] For discussion of these issues, see Kripke 1972, Nozick 1981, Cohen 1986, and Dummett 1993.

experiments'[5]). In cognitive science, some of the most famous philosophers' thought experiments are Putnam's twin earth, Jackson's Mary the colourblind colour scientist, Dennett's qualia impasses, and Searle's Chinese room. We will examine each of them shortly but first I want to ask some questions about thought experiments in general.

Like analysis of concepts, are thought experiments also empirical, at least in part? Yes; they are merely a particular way of manipulating material stored in memory, material originally gained from experience. Thought experiments are acts of imagination but there is nothing distinctively *a priori* (independent of experience) about them. Thought experiments may be empirical in another way, too. The motive for mounting them is just as much a desire to see how things hang together as when we do hands-on experiments. There are differences between the two, of course, but they do not appear in anything as general as being or not being empirical.

Thought experiments may be central to philosophy's contribution to cognitive science but they play a role in other parts of cognitive science, too. Intuitions of grammaticality are a kind of thought experiment; they play a central role in linguistics, Likewise, some empirical research into reasoning starts with subjects doing thought experiments. The subject is asked to determine whether more words begin with 'r' than end with it, for example, or to determine which is more probable in some situation, A or B, or whatever. Of course, in both these cases the thought experiments are done by the subjects, not by the researchers. Nonetheless, thought experiments are not unique to philosophy in cognitive science.

Again we must be careful. If thought experiments in the hands of linguists and psychologists play a role in finding (or displaying, or limiting) the facts, in the hands of philosophers they also play a normative role. Compare thought experiments in hands-on research into reasoning and thought experiments in a philosophical investigation of reasoning. Psychologists get subjects to do thought experiments to find out how we actually reason: what mistakes we make, what produces these mistakes, etc. When a philosopher runs a thought experiment about reasoning, her interest is different. Her interest is in finding out what *good* reasoning consists in.

So we can say this about thought experiments. In no discipline other than philosophy do they play the central role that they play in philosophy and no discipline other than philosophy uses them to study normative issues. If so, the use of thought experiments as a method is in some measure distinctive to philosophy. In what measure we will discover shortly.

(4) *Philosophical 'therapy'* Another method for doing philosophy in cognitive science is philosophical 'therapy'. This method is associated with Wittgenstein, of course. Here the theorist attempts to display that something taken to be sensible is in fact disguised nonsense. Philosophical 'therapy' and thought experiments are sometimes linked. Sometimes the point of a thought experiment is to show that something could not be as we take it to be. In such cases, philosophical 'therapy' or something very much like it is often the goal. Though some opponents of the computational model of the mind urge that it would be good idea for cognitive scientists to do a great deal more philosophical 'therapy' than they do, the method has not in fact played much of a role in philosophy's contributions to cognitive science and I won't explore it further.

Of the four methods just sketched, only (4) and perhaps (3) in some forms and some applications could satisfy (a) to (c), i.e., could be genuinely *distinctive* to philosophy. (1), (2) and some variants of (3) are also used by other branches of cognitive science, as we saw.

2. How do thought experiments work?

If thought experiments are central to philosophy's contribution to cognitive science, we need to understand how they work. Let's look into some specific examples. We introduced four thought experiments earlier, twin earth (Putnam), Mary the colourblind colour scientist (Jackson), impasses over differences in qualia (Dennett), and the Chinese room (Searle). They are good examples of the genre and have all been widely discussed.

Twin earth experiments go like this. Imagine a person here on earth, Adam, and his completely identical twin, Twadam, on twin earth. Adam and Twadam both use the word 'water' and they use it in situations that are experientially indistinguishable. Yet on earth what is called 'water' is H_2O, on twin earth it is XYZ. Does the word 'water' as used by Adam and by Twadam have the same meaning? Evidently not. Yet everything in their heads is the same. Hence, in Putnam's memorable phrase, "meanings just ain't in the head" (Putnam, 1975).

Here is the story of Mary the colourblind colour scientist. Mary is a wonderful colour scientist. Indeed, she knows absolutely everything there is to know about colour experience. Yet she has never seen a colour. One day the door is opened (or whatever) and she sees colour for the first time. It would seem that she would gain a new item of knowledge: what it is like to *experience* colour. Hence experience is not ... (draw your favourite moral).

A qualia impasse goes like this. Chase and Sanborn both notice that they don't like their favourite coffee as much any more. Chase says that the coffee tastes the same but he doesn't like that taste as much as he used to. Sanborn says, no, he would still like *that* taste as much but the coffee no

---

[5.] It is not in fact easy to find a short yet adequate way to mark the distinction. Question-begging options that don't work include: 'real experiment', 'physical experiment', ...

longer tastes the way it used to. Since there would seem to be no way in which this putative difference could make a difference to anything, we are invited to ask ourselves whether there is a real difference here.

Searle's Chinese room is probably too well known by this audience to need describing but I will describe it anyway. Someone who knows no Chinese is in a room. Sheets of paper with shapes on them come in through a slot. The person has a huge rulebook linking shapes to shapes. Every time the person finds the shape that has just come in, she moves to the linked shape and finds it on a sheet of paper. She then shoves the new sheet out a second slot. Unbeknowst to her, the shapes going in encode serious questions in Chinese and the shapes coming out are answers to these questions. Moral of the story? What the person who knows no Chinese does in the room is supposedly all that a computer processing physical symbols could do.

The philosophical contributions to contemporary cognitive science contain hundreds of such thought experiments. How do thought experiments differ from hands-on experiments?

What a thought experiment seems to do, in general, is to show that something is more than or different from the way we are inclined to think that it is. Thus, the twin earth example tries to show that meanings ain't in the head. The Mary experiment tries to show that there is something about sensible experience that is more than descriptions. Chase and Sanborne is meant to show that many distinctions that we are inclined to make with respect to conscious states do not reflect any fact of the matter. The Chinese Room is supposed to show that meanings and intentional contents are more than assemblies of physical symbols. And so on.

The general structure of the move seems to be something like this:

'If this [the object of the thought-experiment] is possible [or the case], then X [the target phenomenon] must be like *abc* and/or cannot be like *mno*.'

If so, then the next question is: What contribution do thought experiments make to cognitive science?

3. Thought experiments in cognitive science

Thought experiments play at least four different roles in cognitive science.

1. *Thought experiments isolate crisp examples of a phenomenon under investigation.* The way the Mary thought experiment isolates and displays what it is like to experience something is a good example. The utility of this activity to hands-on experiments is obvious.

2. *Thought experiments tell us what we take something to be like, what some term refers to.* Determining what we mean by a term may seem like quite a modest contribution but it is also a vital one. Studies of empirical values research a while

ago revealed that researchers were using the word 'value' to refer to up to twenty different things (Baier, 1969; Brook, 1975). No wonder everybody was talking past everybody else! We see the same thing today with notions like representation and information. It is important that everyone investigating a phenomenon have some reliable way of identifying and reidentifying examples of the phenomenon they are investigating. Otherwise they do no know what they are talking or that they are all talking about the same thing.

Thought experiments cannot give us a *final* answer about the properties of anything, of course, but they can give us a *first* answer – they can help us figure out what we have in mind when we talk about attention, or consciousness, or recognizing a word, or parsing a sentence correctly, or ... or ... – a first rough account of what we mean by a concept (Flanagan, 1992). We may revise this notion as we come to understand the phenomenon better but we have to start off with some common notion or we will get nowhere.

This function of thought experiments explains something about philosophy that often baffles the rest of the cognitive community: philosophers often care as much about possibilities as actualities. So long as something is possible, often philosophers flatly don't care whether it exists or not. Why? Because philosophers are trying to figure out what we currently mean by some concept, what we take to be the general and characteristic features of things of that kind and what the general and characteristic features *cannot* be: all the thought experiments, (We saw this negative aim at work in the thought experiments we examined.)).

For this concept-fixing task, determining what can and cannot be imagined with respect to things of kind F is not just as good as determining what is actually the case with respect to F's; it is actually better. Better because studying what is actually the case will not tell us what is characteristic of things of that kind. Only determining what can and cannot *be imagined* with respect to things of that kind can tell us that, a point that Kant made over 200 years ago (1781/7, B3).

When thought experiments help us fix what we take something to be like in this way, they are doing something very much like old-fashioned conceptual analysis. The difference is that nowadays we don't expect that investigating concepts in the imagination will give us the final word on what some concept means, just the first word: it reveals the common understanding of the concept from which we are all starting.

Earlier we saw that some thought experiments have something like Wittgensteinian therapy as their goal – weaning us away from tempting but ultimately incoherent ways of viewing something. We can now specify this connection more precisely. As we saw, some thought experiments have a negative thrust, i.e, they aim to show what something is *not* and could *not* be like, not what things of that kind *are* like. But this would be to show that there is something wrong with how we picture or think about the thing in question – precisely the aim or at least one aim of Wittgensteinian therapy.

Thought experiments play a role both in (3) hypothesis generation and (4) hypothesis elimination. Recall the Popperian distinction between the context of discovery and the context of justification. On the abductive (Piercean/ Popperian) picture of science that goes with this distinction, science proceeds by generate-and-test. First we generate hypotheses; this is the context of discovery. Then we eliminate as many of them as we can by testing; this is the context of justification. At the end, only a few or, in the ideal situation, only one hypothesis is left standing. Thought experiments play a role in both contexts.

3. *Hypothesis-generation*

The 'generate' side of generate-and-test is highly independent of data collecting, hands-on experiments, or anything else that involves direct observation and/or manipulation of the world. Indeed, hypothesis generation is pretty much a pure act of the imagination. If so, some thought experiments are a way – usually a fairly abstract and sketchy way – of generating hypotheses.

The hypothesis generating function of thought experiments displays more clearly something we saw in connection with analysis of concepts earlier: how thought experiments are different from hands-on experiments. Think of a possibility-space. Hands-on experiments are usually designed to uncover which of the possibilities in that space are actually the case. By contrast, thought experiments do two things. In their conceptual analytic role, they help us forge conceptual tools for identifying and reidentifying items in the possibility-space. And in their hypothesis generation role, they help us generate ideas for how things might hang together, for what might actually be going on among the items in the space that are actual, a new way of picturing or thinking about the phenomena under investigation.

From all this, we can conclude two things. On the one hand, thought experiments are quite different from hands-on experiments. On the other, they play a key role in generating concepts for hands-on experiments to use and hypotheses for hands-on experiments to test. We are not at the end of the matter yet, however. Thought experiments also play a role in the context of justification in cognitive science.

4. *Hypothesis elimination.* Thought experiments not only aim to help us develop tools to structure a search space and to identify possibilities. They also aim to eliminate ersatz possibilities, to downsize the search space to the genuine possibilities. If so, they play a role in the context of justification, though one quite different from the role that hands-on experiments play. All four of the thought experiments we examined had such an aim.

Earlier we saw examples of how linguistic intuitions

play a role in the context of justification in linguistics and how carrying out tasks in the imagination plays a role in the context of justification in psychology. Here is an example from physics, the famous thought experiment of Galileo's against Aristotle. Said Galileo, if a smaller mass A falls slower than larger mass B, as Aristotle says, then it would follow that an object C made up by joining A and B together will fall both faster and slower than B. This, Galileo thought, eliminated Aristotle's hypothesis and eliminated it as even a possibility. Or Schrodinger's cat. This thought experiment was meant to eliminate a central hypothesis of quantum indeterminacy theory as even a possibility.

What is interesting about thought experiments is that they do what they do without using observation. We simply imagine a situation or check to see what sounds right. When we have done so, there is meant to be nothing left for observation to do. To reiterate a point made earlier, this is not say that thought experiments are not themselves empirical. Intuitions about what sounds right and wrong are clearly empirical, and so is imagining a scenario. It is just that their empirical source is something other than (current) observation.

Compare imagining a scenario to a proof in mathematics. Justification in mathematics is (often at least) derivational: to justify a proposition, you show that it can be derived from well-accepted axioms (etc.) using well-accepted rules (etc.). Thought experiments do not work like this at all. They *display* something to us, something that was obscure or hidden and is supposed to become clear when we imagine the indicated situation. Since the materials and at least most of the relationships of the imagined situation are derived from experience, thought experiments are thus a *kind* of empirical investigation.

If thought experiments play a role in both the context of discovery and the context of justification in normal science, then use of them is not distinctive to philosophy. In both linguistics and physics, thought experiments play a role in the context of justification. Philosophers' use of thought experiments in normative investigations could have something distinctive to it. (that is the question that we set aside earlier), but when they are used in hypothesis generation and possibility elimination, they seem to function very much the way that thought experiments in linguistics or physics do.

Think of Galileo's thought experiment against Aristotle: if a smaller mass A falls slower than larger mass B, then it follows that an object C made up joining A and B together will fall both faster and slower than B. With this Galileo meant to eliminate an hypothesis as even a possibility, namely, Aristotle's hypothesis. Or Schrodinger's cat. This thought experiment was meant to eliminate a central hypothesis of quantum indeterminacy theory as even a possibility. In a roughly parallel way, intuitions of grammatically are meant to show us what the rules of our language are, not just help us generate hypotheses about them.

What is interesting about thought experiments is that they do what they do without using observation. We simply imagine a situation or check to see what sounds right. When we have done so, there is meant to be nothing left for observation to do. To reiterate a point made earlier, this is not say that thought experiments are not themselves empirical. Intuitions about what sounds right and wrong are clearly empirical, and so is imagining a scenario.

Compare the latter to proofs in mathematics. Justification in mathematics is (often at least) derivational: to justify a proposition, you show that it can be derived from well-accepted axioms (etc.) using well-accepted rules (etc.). Thought experiments do not work like this at all. They *display* something to us, something that was obscure or hidden and something that becomes clear when we imagine the right situation. Since the materials and at least most of the relationships of the imagined situation are derived from experience, thought experiments of this type are a *kind* of empirical investigation.

Concluding remarks

What have we shown? We set out to see what is distinctive to the methods of inquiry that philosophers use and how these methods contribute to cognitive research. We have discovered two things:

1. Many of the methods that philosophers have claimed as their own are clearly different from hands-on experimental methods, conceptual analysis and thought experiments being the central examples.

2. These methods are not *distinctive* to philosophy, however, because they are used in hands-on experimental science, too.

We have not exhausted the interesting questions about thought experiments and the other methods of philosophy, of course. One additional role that thought experiments play in science is in the context of *interpretation*. Findings in science have to be interpreted – what does a finding *mean*? Thought experiments play a role in this third context, too.

Of course, even if philosophy has no distinctive methods, it does not follow that philosophy does not play any distinctive role in cognitive science at all. In particular, philosophy has some distinctive preoccupations.[6] For example, experimentalists are mainly interested in whether a claim is true. Philosophers are much more interested in what terms mean, in what the possibilities are, etc. (I am reminded of the 50's television show *Dragnet*. As Jack Webb used to say, "just the facts, ma'am, nothing but the facts". A philosopher would be much more likely to ask, "Wha'd'ya mean, ma'am, wha'd'ya mean?") Second, if an

experimentalist does look at her symbolic toolkit, it is either *in extremis* because the apparatus is letting her down, or an activity of an idle moment. For philosophers, investigating symbolic structures is their main occupation.[7]

References

Baier, K. 1969. What is value? An analysis of the concept. In K. Baier and N. Rescher, eds. *Values and the Future*. Free Press.

Brook, A. 1975. *The Needs and Values Study*. Privy Council of Canada.

Brown, J. 1991. *The Laboratory of the Mind*, Routledge

Castañeda, H.-N. 1980. *On Philosophical Method*. University of Indiana Press

Cohen, L. J. 1986. *The Dialogue of Reason*. Clarendon Press

Davidson, D. The very idea of a conceptual scheme. Proc. Amer. Phil. Assoc. 47, 5-20,

Dummett, M. 1993. *The Origins of Analytical Philosophy*. Duckworth

Flanagan, O. 1992. *Consciousness Reconsidered*. MIT Press.

Hatfield, G. 1990. *The Normative and the Natural*. MIT Press

Horowitz, T. and G. Massey. 1991. *Thought Experiments in Science and Philosophy*, Rowman and Littlefield.

Kant, I. 1781/7. *Critique of Pure Reason*, trans. Norman Kemp Smith. Macmillan, 1927

Kripke, S. 1972. *Naming And Necessity*. In: D. Davidson and G. Harman, eds., *The Semantics of Natural Languages*. Reidel

Nozick, R. 1981. *Philosophical Explanations*. Harvard University Press.

Putnam, H. 1975. The meaning of meaning. In: *Mind, Language and Reality: Philosophical Essays*. Cambridge University Press.

Quine, W. v. O. 1953. Two dogmas of empiricism. In: *From a Logical Point of View*. Harvard University Press, 1961, pp. 20-46.

Sorensen, R. 1992. *Thought Experiments*, Oxford University Press.

[7] Thanks to Rob Stainton, Philosophy and Linguistics, and to Jerzy Jarmasz, Kamilla Run Johannsdottir, Ronald Boring, and Zoltan Jakab, PhD Programme in Cognitive Science, Carleton University, for helpful comments. Since none of them agrees with everything I say, none of them can be held in any way responsible for the errors therein.

[6] Jerzy Jarmasz crystallized this point for me.

Constructed vs. Received Representations for Learning about Scientific Controversy: Implications for Learning and Coaching

Violetta Cavalli-Sforza (violetta@cs.cmu.edu)

Language Technologies Institute, Carnegie Mellon University

5000 Forbes Avenue

Pittsburgh, PA 15213 USA

Abstract

The development of a graphical representation for performing a task can potentially yield a greater understanding of the task domain, but it is itself a demanding task that can distract from the primary one of learning the domain. In this research, we investigated the impact of constructing versus receiving a graphical representation on learning and coaching the analysis of scientific arguments. Subjects studied instructional materials and used the Belvedere graphical interface[1] to analyze texts drawn from an actual scientific debate. One group of subjects used a box-and-arrow representation, augmented with text, whose primitive elements had preassigned meanings tailored to the domain of instruction. In the other group, subjects used the graphical elements as they wished, thereby creating their own representation.

Our results support the following conclusions. From the perspective of learning target concepts, developing one's own representation may not hurt those students who gain a sufficient understanding of the possibilities of abstract representation, although there are costs in time on task and in the quality of the diagrams produced. The risks are much greater for less able students because, if they develop a representation that is inadequate for expressing the concepts targeted by instruction, they will use those concepts less or not at all. From the perspective of coaching students, a predefined representation has a significant advantage. If it is appropriately expressive for the concepts it is designed to represent, it provides a common language and clearer shared meaning between the student and the coach, enabling the coach to understand students' analysis more easily and to evaluate it more effectively against a model of the ideal analysis.

Background

The initial goal of our research was to develop tools for helping students in middle school through early university to engage in critical thinking. We chose scientific controversies as a domain of application, because understanding the various assumptions and reasons underlying scientific debate and change is an important part of appreciating and communicating the nature of the scientific enterprise, even to non-scientists. In contrast, until recently, school science curricula have exhibited a strong bias towards portraying scientific endeavor solely as hands-on *inquiry* and on *final-form* science, that is, the knowledge currently accepted by the scientific community (Duschl, 1990).

As part of our tool base, we developed graphical argumentation formalisms that would enable students to determine and represent how the variety of information brought to bear

---

[1]This system was a conceptual predecessor of the system described by Suthers *et al.* (1997).

in a scientific debate fits together. We investigated the use of coaching strategies to help students develop correct analyses of textual arguments and to generate their own arguments (Paolucci *et al.*, 1996; Toth97 *et al.*; Cavalli-Sforza, 1998). Some of the resulting tools and materials were made available in selected schools and over the Internet (Suthers *et al.*, 1997). Related work, focusing on belief change more than on argument structure, is described by Schank (1995).

Constructed vs. Received Graphical Representations

Although there have been several efforts to develop environments for supporting policy and design discussions (Conklin & Begemann, 1988; Fischer, & McCall, 1989; Stefik *et al*, 1987; Tatar *et al.*, 1991), authoring of arguments and argumentative text (Neuwirth & Kaufer, 1989; Schuler & Smith, 1990; Smolensky *et al.*, 1987; Streitz *et al.*, 1989), and exploration of the structure of information (Lowe, 1985; Marshall *et al.*, 1991), to our knowledge, there has been little empirical investigation of the use of diagrammatic representations in argument-like tasks and its effect on learning and coaching.

Subjects working with ConvinceMe (Schank, 1995), benefited from comparing their own beliefs and argument networks to the program's, relative to subjects working exclusively with text. The active construction, rather than the passive examination, of diagrams improves performance for verifying the validity of syllogisms when combined with a precise method for developing the diagram from the problem description (Grossen & Carnine, 1990).

In contrast, developing one's own procedure for problem solving may lead to learning that generalizes more easily to different problem types, but may also lead to "buggy" algorithms. Diagrams help solve analytical reasoning problems similar to those in the GRE, but only for subjects who show an initial aptitude for that type of reasoning (Cox *et al.*, 1994), indicating that there may be individual differences in the ability to use diagrams.

Finally, diagrammatic representations can assist in understanding computer programs, but only if the diagram supports the type of reasoning required by the task; even so, performance may be slower than with text representations (Green & Petre, 1992).

Although these results don't unanimously support the advantages of graphical representations, we believed that a diagrammatic representation of argument structure was an important part of any set of tools for helping students understand and engage in scientific argument, since, by virtue of its ex-

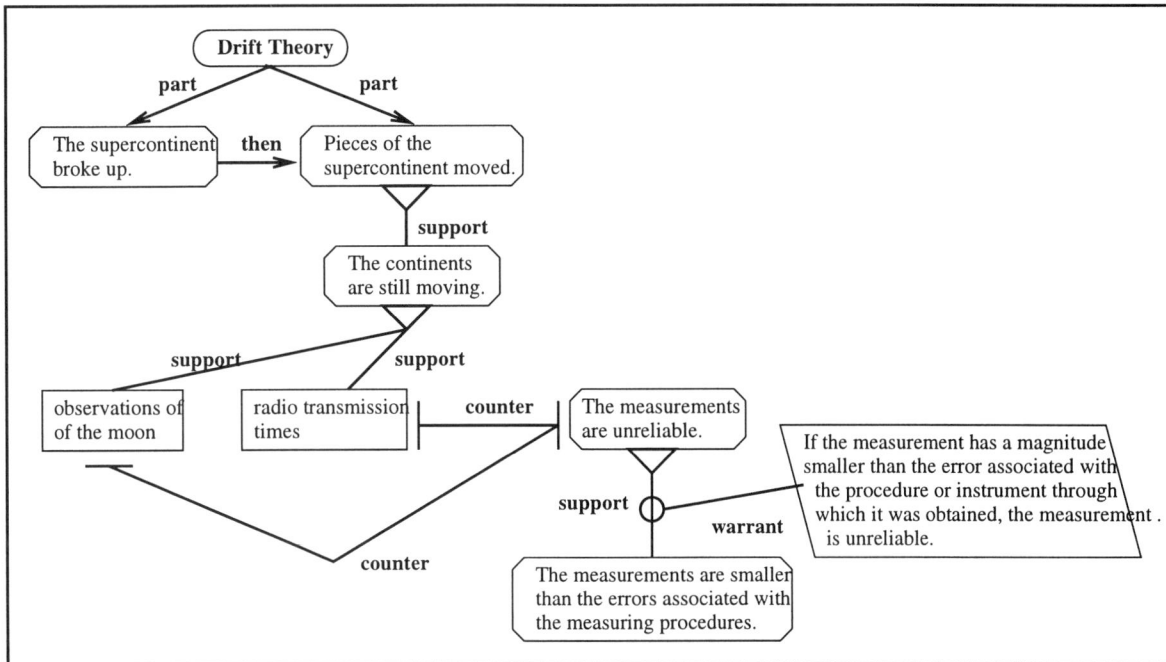

Figure 1: Graphical representation of an argument and an attack on its evidence

plicitness, it would help students ferret out from the texts the information and assumptions required by the argument.

The advantages of a box-and-arrow representation for displaying arguments are discussed in Cavalli-Sforza (1998). Very briefly, in a representation such as the one shown in Figure 1 (loosely based on Toulmin's (Toulmin *et al.*, 1984) model of argument steps), the boxes and arrows correspond to the main components of argument structure: different types of statements (positions and data) and relationships between them (supporting and contradictory, or other negative relationships). The graphical depiction of argument structure makes clear how different information is used to create an argument in favor of a claim (a rule, hypothesis or theory) or to refute that claim or the argumentation surrounding it. The presence or absence of relationships between statements also indicates opportunities for providing further support or refutation. For example, it is possible to refute a claim by arguing directly against it (making or strengthening an argument in favor of a contradictory claim), by arguing against any of the data or warrants used in supporting a claim, or by "undercutting" the support, that is, arguing that the data does not really support that claim.

The predefined graphical representation shown in Figure 1 was designed to have a good "cognitive fit" with the task, in the sense of focusing subjects' attention on the basic concepts targeted by the instructional materials and providing the kind of processing advantages that might be expected from graphical representations. There was no clearly comparable alternative representation. We tried, for example, tabular formats for representing arguments, but they lacked the richness required to represent the range of information present in an argument. In the end, we decided to let some subjects develop their own representation after providing two suggestions: a

box-and-arrow type representation for which they could determine the meaning of the shapes and links provided, and labeled-text schemata (see Figure 2). We hypothesized that the effort of developing or extending a representation might lead to carry out deeper processing of the target concepts of argument analysis.

Coaching

A further reason for using graphical representations was to provide an explicit and shared language for describing argument structure. Such an external representation facilitates interaction between student and coach by providing concrete targets of discussion during a coaching interaction. We hypothesized that the a box-and-arrow graphical representation whose primitives encode key concepts of argumentation provides a more effective channel for coaching subjects in applying those concepts than a graphical representation developed by the subjects themselves, especially for computer coaches (Paolucci *et al.*, 1996; Toth *et al.*, 1997). In this study, however, the coach was the experimenter and coaching advice was provided based on a general mental model of the argument contained in the texts being analyzed. Because there was usually more than one way of representing, or even interpreting, the content of the text, this approach was deemed more flexible and appropriate than comparing a subjects's work to a specific "expert" diagram. For similar reasons, we applied a generally non-interventionist coaching philosophy. The coach waited for the subject to produce at least a partial draft of the analysis before beginning to comment on it, unless the subject appeared to need or specifically requested assistance.

Two classes of strategies were used in providing assistance. *Scaffolding strategies* were applied when the subject seemed

PREMISES:	Space is very cold.		CLAIM:	The inside of the earth is still very hot.
	The earth was very hot when it formed.		WARRANT:	The temperature throughout an object is at least as great as that of material exiting from the object.
WARRANT:	The laws of heat transfer: An object at a different temperature from its surroundings will eventually come to a common temperature with them. The rate of cooling or warming of the object is approximately proportional to the difference in temperature.		REASONS:	Hot lava comes out of volcanoes.
CONCLUSION:	The young earth began to cool by losing heat to space.			

Figure 2: Two examples of a labeled-text schema used for a causal inference and an argument step

unable to proceed with a task and included: 1) reminding the subject of the goal and progress to date; 2) asking a question about a step in the solution; 3) suggesting a high-level plan for the subject to follow; 4) providing a limited number of choices for the subject to explore; 5) giving specific hints about what path(s) to follow; 6) suggesting (some of) the crucial steps in a solution; 7) performing some parts of the plan; 8) leading the subject through each step of a plan by asking a series of questions; 9) performing the task for the subject. *Coaching strategies* were applied to comment on the subject's work after a (partial) analysis had been produced; they included: 1) signaling a potential problem; 2) suggesting information to consider; 3) criticizing; 4) correcting; 5) explaining; 6) arguing.

The application of scaffolding and coaching strategies was interwoven. For example, a coaching strategy might be used to evaluate the subject's analysis, which would lead to the subject attempting to improve the analysis. This might call for a scaffolding strategy to help the subject carry out the improvement. Within the two general classes of strategies, individual strategies differ in the amount and type of knowledge required, as well as in the degree of coach engagement in the task. The coach always attempted to begin with low knowledge and low engagement strategies, which require the subject to do more work, and to progress to more demanding strategies only when the subject needed more assistance or feedback.

The Experiment

Four subjects, non-science majors at the University of Pittsburgh, participated in an extended experiment which included the following sessions.

- **Pre-Test and Post-Test Sessions.** In both sessions, subjects gave definitions of terms relevant to scientific argument and answered questions on an article concerning the debate about the Cretaceous-Tertiary ("dinosaur") extinctions.

- **Belvedere Sessions.** Subjects worked with the Belvedere environment for six sessions. The first four were training sessions. Subjects began with a tutorial in which they learned to use the Belvedere environment's graphical and text capabilities. Then they read instructional materials discussing concepts of scientific argumentation, and short domain texts describing two competing theories about the formation of mountains continents and ocean basins. These texts were drawn from the debate between the supporters of Continental Drift theory and the Contraction theory. The texts were chosen to exemplify different aspects of scientific argument. Some texts described the theories and showed how they explained the data, others provided different types of arguments in favor of each theory (e.g. based on data, based on analogy), others yet were dialectical texts debating the merits of the two theories relative to the data and the characteristics of the explanation (e.g. uniqueness, breadth, parsimony, internal consistency, etc.). In the last two sessions on Belvedere, subjects were given similar texts concerning a third theory, the Expansion theory, and were asked to perform comparable analyses.

Two subjects worked in each of two experimental conditions. The instructional materials, which contained the same conceptual content, introduced a related set of concepts and showed how they could be applied in analyzing a domain text. In the **FIXED** condition, subjects saw a box-and-arrow diagram representing the sample text analysis (Figure 1); in the **FREE** condition, the same analysis was carried out via one or more labeled-text schemata such as those in Figure 2. Subsequently, subjects performed similar analysis on parallel texts, equivalent in structure to the ones used in the instruction, and were coached by the experimenter according to the approach previously described. Coaching remained available to subjects in the last two sessions as well, but they no longer received the explicit guidance of the instructional materials.

- **Review Sessions.** Three sessions were spent reviewing argument concepts through a comparison of the theories relative to different criteria, further oral analysis of domain texts, and a summary of the debate.

Altogether, each subject spent between 28 and 38.5 hours in the experiment. The Belvedere sessions, ranging between 16.5 and 19.5 hours per subject, were videotaped. The videotapes were transcribed and indexed to the evolution of the diagrams produced by subjects. The review sessions were audiotaped.

Data Analysis Methodology

Questionnaire Data

The pre-test and post-test questionnaires required subjects to provide definitions for several of the terms introduced in the instructional materials. These included terms describing the status of scientific statements (e.g. theory, analogy, model), argument terms (e.g. claim, warrant, counterargument, undercut), and criteria for evaluating scientific explanations (e.g. explanatory breadth, explanatory parsimony, inconsistency and conservativeness of an explanation, sufficiency of the proposed mechanism). In total there were 37 terms, of which 27 appeared in both pre-test and post-test questionnaires; 10 appeared only in the post-test questionnaire because subjects were not expected to be familiar with these terms prior to the experiment.

For each term in the questionnaire, subjects' answers were compared to a target definition, such as was used in the instructional materials. Two raters assigned scores independently. The scores were then discussed and occasionally modified. The resulting post-discussion inter-rater correlations were 0.83 (N = 108) for items on both pre-test and post-test, and 0.86 (N = 40) for items only on the post-test.

Interaction Data

The protocols for roughly one third of the many graphs that subjects drew were segmented into units of interaction between the subject and the coach, and were analyzed quantitatively and qualitatively as follows.

For each diagram, the raw protocol of the interaction was synthesized into an *interaction summary*. First, the protocol was segmented into groups of related utterances by the subject and/or the coach. Each *utterance group* represents a self-contained thought by one person and abstracts away interruptions and simultaneous speech by the other person that were not extending the original thought in a significant way. The utterance group was then summarized into a *summary utterance*, a single statement that reflects the essential content of the utterance group. One or two summary utterances, taken together, form one *interaction unit*. An interaction unit can include:

- a summary utterance by the subject and one that represents the coach's response,

- a summary utterance by the coach with a verbal response or action by the subject, or

- a summary utterance by the coach alone, for example a comment praising the subject's work.

Several interaction units may be grouped into a single *interaction group* to indicate that they all pertain to a specific issue or topic of discussion. Interaction groups may be nested to show the overall structure of the interaction. The top-level group has a heading describing the main topic of the interaction, with nested interaction group headings giving the various subtopics addressed.

Each interaction unit was also classified on the basis of the primary **content** as well as the communicative **form** of the interaction. The content categories include: rhetorical structure, representation, scientific domain, graphical interface, and other. An interaction unit dealing primarily with the interrelationship of two statements in the domain text was coded as rhetorical structure; an interaction unit dealing primarily with the appropriate link, link name, or linkage pattern to use to represent that relationship in the graph was coded as representation.

The communicative form categories represent different communicative actions used by the coach to help the subject improve his or her graphical analysis of a text. They included, among others: review, remind, check subject's understanding of text, explain, give an example, ask for justification, check coach's understanding of subject's intentions, check coach's understanding of subject's work, ask a leading question about the text/diagram, hint at the characteristics of a solution, suggest possible actions, tell what to do, correct by giving the solution, argue, draw attention to a problem, criticize, praise.

An example of an interaction unit composed of two summary utterances is the following, which was analyzed as having content "rhetorical structure" and form "hint". Although the subject is speaking in terms of graphical links (e.g. *and*), the coach is hinting that simple conjunction is not the correct relationship among the statements.

```
S: I'll use an *and* to show the
   relation between items 1, 2, 3.

C: You should be more specific than
   *and*.  It has to do with time
   sequence.
```

For a subset of the protocols, where complex or problematic interactions had occurred, we also examined the interaction qualitatively. We analyzed the application of coaching strategies as embodied by the specific coaching actions, and their degree of success or failure in terms of bringing about the desired results.

Results and Discussion

Use of Graphical Representation

A detailed analysis of the diagrams produced by FIXED condition subjects showed that the predefined argument-specific graphical representation was an appropriate representation for the task. Both subjects used most links and shapes without difficulty; errors were usually due to an error in analyzing the domain content or the argument structure of the text. The representational problems subjects encountered were often localized, for example, confusion over the direction of a link due to naming ambiguities, and omission of links, which subjects created when needed.[2] Subjects were also able to combine primitive links in correct and creative ways. Finally, subject diagrams were often quite good even before the experimenter intervened, and were relatively easy to read even when they became crowded.

Whereas in the FIXED representation condition both subjects appeared to be largely successful in applying the concepts they had learned, subjects in the FREE representation condition differed significantly in this regard. One subject used a box-and-arrow type formalism to encode many

---

[2]There were, however, some interesting misuses of devices for representing conjunction.

of the target relational concepts, which enabled him to analyze domain texts at an appropriate level of detail and to represent complex structural relationships, especially for argument diagrams. For causal diagrams (theory description diagrams), the subject had trouble abstracting away from domain-specific relationships towards more more general temporal, causal, and explanatory relationships; consequently these diagrams were difficult to understand. Graphical primitives were often used inconsistently across both types of diagrams. The other subject did not find a way of using the box-and-arrow graphics to represent either type of content abstractly. This subject's causal diagrams were essentially analogical renderings of the physical situation; the argument diagrams exclusively employed labeled-text schemata, which emphasized the role played by different statements over their relationship, and displayed little structural detail. Since this subject did not extend the schemata to represent dialectical relationships, the diagrams did not show the relationships of the contents of the text to other parts of the debate any more explicitly than the original text.

Learning of Target Concepts

We had hypothesized that, by visually reinforcing the terminology and the target concepts, the FIXED representation would help subjects learn those concepts better than subjects in the FREE condition. However, subjects' scores for the 37 questionnaire items revealed little or no difference. Based on absolute performance scores in the Post-Test, subjects fell into a stronger and a weaker group, each containing a subject from each condition. Within the same group the FIXED condition subject was slightly ahead. Improvement scores showed the weaker subject of the FREE condition (Free-2), who started with the lowest scores of any subject, consistently improving more than the weaker subject in the FIXED condition (Fixed-2). Of the stronger subjects, Fixed-1 improved substantially more than the Free-1.

Aside from the small number of subjects, several factors may have contributed to the inconclusiveness of these results. The questionnaire was not an adequate means of assessing learning, since it tested subjects' ability to verbalize concepts, whereas the training was geared towards applying them. Examining the diagrams subjects drew, especially diagrams drawn in later sessions, we found that representation was related to subjects' ability to use the target concepts in their diagrams, if not always to their ability to supply proper definitions. Subjects in the FIXED condition consistently distinguished the status of statements in their diagrams (as observation, explanation, plain statement, etc.) because the representation required them to do so. FREE condition subjects seldom, if ever, did. Only subjects in the FIXED condition and Free-1, who developed a sufficiently expressive set of graphical primitives, consistently represented the relational concepts present in the text, especially multi-step support and dialectical relationships. For the more complex concepts that were realized in box-and-arrow diagrams with distinctive linkage patterns (e.g. analogy/model, parsimony of an explanation), there was a link between subjects' use of the concept in the diagram and their ability to give good definitions. The crucial factor underlying these results appears to be whether the representation is relation-centered, as box-and-

arrow representations are, or role-centered as the labeled-text schema adopted by Subject Free-2 is.

Coaching

The results of the analysis of interaction data supports the hypothesis that the FIXED representation condition was a more effective vehicle for coaching in several ways.

Amount of Coaching. Subjects in the FIXED condition required less coaching overall, as measured by the number of interactions units with coaching content.

Target of Coaching. Subjects in the FREE condition required a significantly greater proportion of coaching about representation itself, at the expense of verifying that the diagram encoded the text correctly and completely. This interaction, negotiating the meaning and assessing the adequacy of the representation, did not clearly benefit subjects' understanding of the structure and contents of the text.

Directiveness of Coaching. Subjects in the FIXED condition could be coached using less directive strategies, allowing the subjects to do more of the work. Most problems in FIXED condition diagrams were localized and could be fixed with a brief interaction; often the coach only needed to point out that there was a problem or to hint at the nature of the problem. In contrast, the experimenter often needed to coach FREE condition subjects in greater detail.

Ease and Acceptance of Coaching. From the coach's perspective, interaction with FIXED condition subjects was easier, more systematic and more effective than with FREE condition subjects. FIXED condition subject diagrams, especially causal diagrams, were easier for the coach to understand and to compare to her own model(s) of the correct analysis, allowing problems areas to be identified and an agenda for coaching to be quickly established. Problems requiring some global restructuring of the diagram could be addressed by applying coaching and scaffolding strategies more systematically, reducing the problem to subproblems that were smaller and easier to correct. This was partially attributable to subjects having a stock of diagramming primitives that enabled them to put the advice into effect more readily. In contrast, subjects in the FREE representation condition experienced significantly more difficulty in implementing the experimenter's advice. Subject Free-1 actually resisted accepting advice, especially on problems related to representation, preferring instead to keep on experimenting with alternative representations of his own invention. The limitations of the representations used/developed by the other subject contributed to this subject's difficulty in understanding and implementing coaching advice.

Summary and Conclusions

We investigated whether a predefined graphical representation whose primitives encode key concepts of scientific argument is more effective for learning those concepts than a graphical representation learners develop themselves. Although, we did not find evidence that such a predefined representation improves ability to *define* target concepts, there was evidence that the representation subjects ultimately adopt impacts whether they are able to *use* those concepts in their diagrams. The risk of allowing learners to construct their own graphical representation is that they may not succeed in developing a representation that is sufficiently expressive, making the diagram construction task an ineffective vehicle for learning and coaching. Even if it is sufficiently expressive, there is a cost in time and clarity of the diagrams and coaching is considerably more effortful. In contrast, a carefully designed task-appropriate representation is easily adopted and produces diagrams that encode the concepts targeted by instruction clearly and correctly. Moreover, such a representation facilitates coaching interaction over the instructional content of the task. Unlike a representation under development, which requires constant negotiation over its meaning, it provides a well-defined language for communication between the learner and the coach.

Acknowledgments

This research was funded under the National Science Foundation's Applications of Advanced Technology program under the title "Tools for Thinking About Complex Issues in Science and Public Policy," Grant MDR-9155715.

We gratefully acknowledge the assistance of Dr. Alan Lesgold, who supervised this research, and of other members of the research group at the Learning Research and Development center, University of Pittsburgh: John Connelly, Massimo Paolucci, Dan Suthers, and Arlene Weiner.

References

Cavalli-Sforza, V. (1998). *Constructed vs. received graphical representations for learning about scientific controversy: Implications for learning and coaching.* Doctoral dissertation, Intelligent Systems Program, University of Pittsburgh.

Conklin, J., & Begeman, M. L. (1988). gIBIS: A hypertext tool for exploratory policy discussion. *ACM Transactions on Office Information Systems, 6*(4), 303–331.

Cox, R., Stenning, K., & Oberlander, J. (1994). Graphical effects in learning logic: reasoning, representation and individual differences. *Proceedings of the Sixteenth Annual Conference of the Cognitive Science Society* (pp. 237–242). Hillsdale, NJ: Lawrence Erlbaum Associates.

Duschl, R. A. (1990). *Restructuring science education. The importance of theories and their development.* New York, NY: Teachers' College Press, Columbia University.

Fischer, G., McCall, R., & Morch, A. (1989). JANUS: Integrating hypertext with a knowledge-based design environment. *Hypertext '89 Proceedings* (pp. 105–117). Baltimore, MD: Association for Computing Machinery.

Green, T. R. G., & Petre, M. (1992). When visual programs are harder to read than textual programs. In G. C. van der Veer, M.J. Tauber, S. Bagnarola, & M. Antavolits (Eds.), *Human-Computer Interaction: Tasks and Organization.* Rome: CUD.

Grossen, B., & Carnine D.(1990). Diagramming a logic strategy: Effects on difficult problem types and transfer. *Learning Disability Quarterly, 13*, 168–182.

Lowe, D.G. (1985). Co-operative structuring of information: the representation of reasoning and debate. *International Journal of Man-Machine Studies, 23*, 97–111.

Marshall, C. C., Halasz, F. G., Rogers, R. A., & Janssen Jr., W. C. (1991). Aquanet: A hypertext tool to hold your knowledge in place. *Hypertext '91 Proceedings.* Baltimore, MD: Association for Computing Machinery.

Neuwirth, C. M. & Kaufer, D. S. (1989). The role of external representations in the writing process: implications for the design of hypertext-based writing tools. *Hypertext '89 Proceedings* (pp. 319–342). Baltimore, MD: Association for Computing Machinery.

Paolucci, M., Suthers, D., & Weiner, A. (1996). Automated advice-giving strategies for scientific inquiry. In C. Frasson, G. Gauthier, & A. Lesgold (Eds.) *Intelligent Tutoring Systems, Third International Conference, ITS '96. Lecture Notes in Computer Science.* New York, NY: Springer.

Schank, P. (1995). *Computational tools for modeling and aiding reasoning: Assessing and applying the Theory of Explanatory Coherence.* Doctoral dissertation, University of California, Berkeley.

Schuler, W. & Smith, J. B. (1990). Author's argumentation assistant (AAA): A hypertext-based authoring tool for argumentative tasks. In A. Rizk, N. Streitz, and J. Andrè (Eds.), *Hypertext: Concepts, systems, and applications.* Cambridge University Press, Electronic Publishing Series.

Smolensky, P. *et al.* (1987). *Computer-aided reasoned discourse, or, how to argue with a computer.* (Tech. Rep. CU-CS-358 87). University of Colorado, Department of Computer Science.

Stefik, M., Foster, G., Bobrow, D. G., Kahn, K. Lanning S., & Suchman L. (1987). Beyond the chalkboard: Computer support for collaboration and problem solving in meetings. *Communications of the ACM, 30*(1), 32–47.

Streitz, N. A., Hannemann, J., & Thüring, M., 1989. From ideas and arguments to hyperdocuments: travelling through activity spaces. *Hypertext '89 Proceedings* (pp. 343–364).

Suthers, D., Erdosne Toth, E., & Weiner, A. (1997). An Integrated Approach to Implementing Collaborative Inquiry in the Classroom. *Proceedings of Computer Supported Collaborative Learning.*

Tatar, D. G., Foster, G., & Bobrow, D. G., (1991). Design for conversation: Lessons from Cognoter. *International Journal of Man-Machine Studies, 34*, 185–209.

Toth, J. A., Suthers, D., & Weiner, A. (1997). Providing expert advice in the domain of collaborative scientific inquiry. *Proceedings of Artificial Intelligence in Education* (pp. 302–308).

Toulmin, S, Rieke, R., & Janik A. (1984). *An introduction to reasoning*, 2nd edition. New York, NY: Macmillan.

The power of statistical learning: No need for algebraic rules

Morten H. Christiansen (MORTEN@SIU.EDU)
Department of Psychology; Southern Illinois University
Carbondale, IL 62901-6502 USA

Suzanne L. Curtin (CURTIN@GIZMO.USC.EDU)
Department of Linguistics; University of Southern California
Los Angeles, CA 90089-1693 USA

Abstract

Traditionally, it has been assumed that rules are necessary to explain language acquisition. Recently, Marcus, Vijayan, Rao, & Vishton (1999) have provided behavioral evidence which they claim can only be explained by invoking algebraic rules. In the first part of this paper, we show that contrary to these claims an existing simple recurrent network model of word segmentation can fit the relevant data without invoking any rules. Importantly, the model closely replicates the experimental conditions, and no changes are made to the model to accommodate the data. The second part provides a corpus analysis inspired by this model, demonstrating that lexical stress changes the basic representational landscape over which statistical learning takes place. This change makes the task of word segmentation easier for statistical learning models, and further obviates the need for lexical stress rules to explain the bias towards trochaic stress patterns in English. Together the connectionist simulations and the corpus analysis show that statistical learning devices are sufficiently powerful to eliminate the need for rules in an important part of language acquisition.

Introduction

One of the basic questions in cognitive science pertains to whether or not explicit rules are necessary to account for complex behavior. Nowhere has the debate over rules been more heated than within the study of language acquisition. Traditionally, generative grammarians have postulated the need for rules in order to account for the patterns found in natural languages (Chomsky & Halle, 1968). In addition, much of the acquisition literature within this framework requires the child to map underlying representations to a surface realization via rules (Smith, 1973; Macken, 1980). On this account, statistical learning is assumed to play little or no role in the acquisition process; instead, abstract rules have been claimed to constitute the fundamental basis of language acquisition and processing. Recently, an alternative approach has emerged emphasizing the role of statistical learning in both the acquisition and processing of language. A growing body of research have explored the power of statistical learning in infancy from both behavioral (e.g., Saffran, Aslin, Newport, 1996) and computational perspectives (e.g., Brent & Cartwright, 1996; Christiansen, Allen & Seidenberg, 1998). This line of research has demonstrated the viability of statistical learning; including cases that were previously thought to require the acquisition of rules and cases for which the input was thought to be too impoverished for learning to take place. In this paper, we extend this research within the area of early infant speech segmentation, providing further evidence against the need for algebraic rules in language acquisition.

Within the traditional rule-based approach Marcus, Vijayan, Rao, & Vishton (1999) have recently presented results from experiments with 7-month-old infants apparently showing that they acquire abstract algebraic rules after two minutes of exposure to habituation stimuli. Marcus et al. further claim that statistical learning models—including the simple recurrent network (SRN; Elman, 1990)—are unable to fit their experimental data. In the first part of this paper we show that knowledge acquired in the service of learning to segment the speech stream can be recruited to carry out the kind of classification task used in the experiment by Marcus et al. For this purpose we took an existing model of early infant speech segmentation (Christiansen et al., 1998) and used it to simulate the results obtained by Marcus et al. Crucially, our simulations do not focus on the phonological output of the network, but rather seek to determine whether the network develops on-line internal representations—that is, transient hidden unit patterns—which can form the basis for reliable classification of input patterns. Stimulus classification then becomes a signal detection problem based on the internal representation, and the preference for one type of stimuli over another is explained in terms of differential segmentation performance. Thus, no rules are needed to account for the data; rather, statistical knowledge related to word segmentation can explain the rule-like behavior of the infants in the Marcus et al. study.

In the second part of the paper we turn our attention to another claim about the necessity of rules in language acquisition. Within the area of the acquisition of lexical stress researchers have debated whether children learn stress by rule or lexically (Hochberg 1988; Klein, 1984). The evidence so far appears to support the claim that children learn stress by rule (Hochberg, 1988) or by setting a parameter in an abstract rule-based system (Fikkert, 1994). In contrast, the segmentation model of Christiansen et al. (1998) acquires lexical stress through statistical learning. The superior performance of the model when provided with lexical stress information, suggests that lexical stress may change the basic representational landscape from which the SRN acquires the statistical regularities relevant for the word segmentation task. We investigate this suggestion through the means of a corpus analysis. The results demonstrate that representational changes caused by lexical stress facilitate learning and obviate the need for rules to explain lexical stress acquisition. Together the results from the corpus analysis and the connectionist simulations suggest that statistical learning is sufficiently powerful to avoid the postulation of abstract rules—at least within the area of speech segmentation.

Rule-Like Behavior without Rules

Marcus et al. (1999) used an artificial language learning paradigm to test their claim that the infant has two mechanisms for learning language, one that uses statistical information and another which uses algebraic rules. They conducted three experiments which tested infants' ability to generalize to items not presented in the familiarization phase of the experiment. They claim that because none of the test items appeared in the habituation part of the experiment the infants would not be able to use statistical information.

The subjects in Marcus et al. (1999) were seven-month old infants randomly placed in an experimental condition. In the first two experiments, the conditions were ABA or ABB. Each word in the sentence frame ABA or ABB consisted of a consonant and vowel sequence (e.g., "li wi li" or "li wi wi"). During the two-minute long familiarization phase the infants were exposed to three repetitions of each of 16 three-word sentences. The test phase in both experiments consisted of 12 sentences made up of words the infants had not previously been exposed to. The test items were broken into 2 groups for both experiments: consistent (items constructed with the same grammar as the familiarization phase) and inconsistent (constructed from the grammar the infants were not trained on). In the second experiment the test items were altered in order to control for an overlap of phonetic features found in the first experiment. This was to prevent the infants from using this type of statistical information. The results of the first and second experiments showed that the infants preferred the inconsistent test items over the consistent ones. In the third experiment, which we focus on in this paper, the ABA grammar was replaced with an AAB grammar. The rationale was to ensure that infants could not distinguish between grammars based solely on reduplication information. Once again, the infants preferred the inconsistent items over the consistent items.

The conclusion drawn by Marcus et al. (1999) was that a system which relied on statistical information alone could not account for the results. In addition, they claimed that a SRN would not be able to model their data because of the lack of phonological overlap between habituation and test items. Specifically, they state,

> Such networks can simulate knowledge of grammatical rules only by being trained on all items to which they apply; consequently, such mechanisms cannot account for how humans generalize rules to new items that do not overlap with the items that appeared in training (p. 79).

We demonstrate that SRNs can indeed fit the data from Marcus et al. Crucially, we do *not* build a new model to accommodate the results (see Elman, 1999, for a simulation of experiment 2[1]), but take an existing SRN model of speech segmentation (Christiansen et al., 1998) and show how this model— *without additional modification*—provides an explanation for the results.

---

[1] It is not clear that these simulation results can be extended to Experiment 3 because this SRN was trained to activate a unit when reduplication occurred. In Experiment 3, however, both conditions, and therefore both types of test items, contain reduplication and hence cannot be distinguished on the basis of reduplication alone.

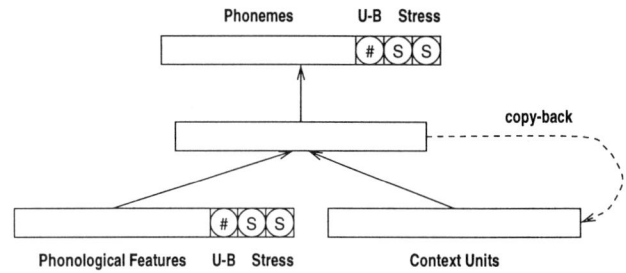

Figure 1: Illustration of the SRN used in Christiansen et al. (1998). Solid lines indicate trainable weights, whereas the dashed line denotes the copy-back weights (which are always 1). U-B refers to the unit coding for the presence of an utterance boundary.

Simulations

The model by Christiansen et al. (1998) was developed as an account of early word segmentation. An SRN was trained on a *single* pass through a corpus consisting of 8181 utterances of child directed speech. These utterances were extracted from the Korman (1984) corpus of British English speech directed at pre-verbal infants aged 6-16 weeks (a part of the CHILDES database, MacWhinney, 1991). The training corpus consisted of 24,648 words distributed over 814 types (type-token ratio = .03) and had an average utterance length of 3.0 words (see Christiansen et al. for further details). A separate corpus consisting of 927 utterances and with the same statistical properties as the training corpus was used for testing. Each word in the utterances was transformed from its orthographic format into a phonological form and lexical stress assigned using a dictionary compiled from the MRC Psycholinguistic Database available from the Oxford Text Archive[2].

As input the network was provided with different combinations of three cues dependent on the training condition. The cues were (a) phonology represented in terms of 11 features on the input and 36 phonemes on the output[3], (b) utterance boundary information represented as an extra feature marking utterance endings, and (c) lexical stress coded over two units as either no stress, secondary or primary stress. Figure 1 provides an illustration of the network.

The network was trained on the task of predicting the next phoneme in a sequence as well as the appropriate values for the utterance boundary and stress units. In learning to perform this task it was expected that the network would also learn to integrate the cues such that it could carry out the task of segmenting the input into words.

With respect to the network, the logic behind the segmentation task is that the end of an utterance is also the end of a word. If the network is able to integrate the provided cues in

---

[2] Note that these phonological *citation forms* were unreduced (i.e., they did not include the reduced vowel *schwa*). The stress cue therefore provided additional information not available in the phonological input.

[3] Phonemes were used as output in order to facilitate subsequent analyses of how much knowledge of phonotactics the net had acquired.

order to activate the boundary unit at the ends of words occurring at the end of an utterance, it should also be able to generalize this knowledge so as to activate the boundary unit at the ends of words which occur *inside* an utterance (Aslin, Woodward, LaMendola & Bever, 1996).

Classification as a Secondary Signal Detection Task

The Christiansen et al. (1998) model acquired distributional knowledge about sequences of phonemes and the associated stress patterns. This knowledge allowed it to perform well on the task of segmenting the speech stream into words. We suggest that this knowledge can be put to use in secondary tasks not directly related to speech segmentation—including artificial tasks used in psychological experiments such as Marcus et al. (1999). This suggestion resonates with similar perspectives in the word recognition literature (Seidenberg, 1995) where knowledge acquired for the primary task of learning to read can be used to perform other secondary tasks such as lexical decision.

Marcus et al. (1999) state that they conducted simulations in which SRNs were unable to fit the experimental data. As they do not provide any details of the simulations, we assume (based on other simulations reported by Marcus, 1998) that these focused on some kind of phonological output that the SRNs produced. Given our characterization of the experimental task as a secondary task, we do not think that the basis for the infants' differentiation between consistent and inconsistent stimuli should be modeled using the phonological output of an SRN. Instead, it should primarily be based on the internal representations generated during the processing of a sentence. On our account, the differentiation of the two stimulus types becomes a signal detection task involving the internal representation of the SRN (though we shall see below that a part of the *non*-phonological output can explain why the inconsistent items elicited longer looking-times).

Method *Network.* We used the SRN from Christiansen et al. (1998) trained on all three cues.

Materials. The materials from Experiment 3 in Marcus et al. (1999) were transformed into the phoneme representation used by Christiansen et al. Two habituation sets were created in this manner: one for AAB items and one for ABB items. The habituation sets used here, and in Marcus et al., consisted of 3 blocks of 16 sentences in random order, yielding a total of 48 sentences. Each sentence contained 3 monosyllabic nonsense words. As in Marcus et al. there were four different test trials: "ba ba po", "ko ko ga" (consistent with AAB), "ba po po" and "ko ga ga" (consistent with ABB). The test set consisted of three blocks of randomly ordered test trials, totaling 12 test sentences. Both the habituation and test sentences were treated as a single utterance with no explicit word boundaries marked between the individual words. The end of the utterance was marked by activating the utterance boundary unit.

Procedure. The network was habituated by providing it with a *single* pass through the habituation corpus—one phoneme at a time—with learning parameters identical to the ones used originally in Christiansen et al. (1998) (i.e., learning rate = .1 and momentum = .95). The test set was presented to the network (with the weights "frozen") and the hidden unit activation for the final input phoneme in each test sentence was recorded. Given the processing architecture of the SRN, the activation pattern over the hidden units at this point provides a representation of the sentence as a whole; that is, a compressed version of the sequence of hidden unit states that the SRN has gone through during the processing of the sentence. Each hidden unit representation constitutes an 80-dimensional vector.

Result and Discussion We used discriminant analysis (Cliff, 1987; see Christiansen & Chater, in press, for an earlier application to SRNs) to determine whether the hidden unit representations contained sufficient information to distinguish between the consistent and inconsistent items for a given habituation condition. The 12 vectors were divided into two groups depending on whether they were recorded for an AAB or ABB test item. The vectors were entered into a discriminant analysis to determine whether they contained sufficient information to be linearly separated into the relevant two groups. As a control, we randomly re-assigned three vectors from each group to the other group such that our random controls cut across the two original groupings (i.e., both random groups contained three AAB and three ABB vectors).

The results from both the AAB and ABB habituation conditions showed significant separation of the correct vectors ($df = 5, p < .001; df = 6, p < .001$), but not for the random controls ($df = 6, p = .3589; df = 6, p = .4611$). Consequently, it was possible on the basis of the hidden unit representation derived from the model to correctly predict the appropriate group membership of the test items at 100% accuracy in both conditions. However, for the random control items in both conditions the accuracy (83.3%) was not significantly different from chance.

The superficially high classification of the random vectors is due to the high number of hidden units (80) and the low number of test items (6) in each group. This increases the probability that a random variable may provide information that can distinguish between the two random groups by chance. Nonetheless, the significance statistics suggest that only the original correct grouping of hidden unit patterns contain sufficient information for the reliable categorization of the items. This information can be used by the network to distinguish between the consistent and inconsistent test items. Similarly, we argue that infants may have access to same type of information on which they can classify the test items presented to them in the Marcus et al. (1999) study.

Explaining the Preference for Inconsistent Items

The results from the discriminant analyses demonstrate that no algebraic rules are necessary to account for the differential classification of consistent and inconsistent items in Experiment 3 of Marcus et al. (1999). However, the question remains as to why the infants looked longer at the inconsistent items compared to the consistent items. To address this question we looked at the activation of the non-phonological output unit coding for utterance boundaries. Christiansen et al. (1998) used the activation of this unit as an indication of predicted word boundaries. Our prediction for the current simulations was that the SRN should show a differential ability to predict word boundaries for the words in the two test conditions. As in Christiansen et al., we used accuracy and completeness scores (Brent & Cartwright, 1996) as a quanti-

116

tative measure of segmentation performance.

$$\text{Accuracy} = \frac{\text{Hits}}{\text{Hits} + \text{False Alarms}} \quad (1)$$

$$\text{Completeness} = \frac{\text{Hits}}{\text{Hits} + \text{Misses}} \quad (2)$$

Accuracy provides a measure of how many of the words that the network postulated were actual words, whereas completeness provides a measure of how many of the actual words in the test sets that the net discovered. Consider the following hypothetical utterance example:

#t h e#d o g#s#c h a s e#t h e c#a t#

where # corresponds to a predicted word boundary. Here the hypothetical learner correctly segmented out two words, *the* and *chase*, but also falsely segmented out *dog*, *s*, *thec*, and *at*, thus missing the words *dogs*, *the*, and *cat*. This results in an accuracy of 2/(2+4) = 33.3% and a completeness of 2/(2+3) = 40.0%.

Given these performance measures, Christiansen et al. (1998) found that the network trained with all three cues (phonology, stress and utterance boundary information) achieved an accuracy of 42.71% and a completeness of 44.87%. So, nearly 43% of the words the network segmented out were actual words and it segmented out nearly 45% of the words in the test corpus. We used the same method to compare how well the network segmented the words in the test sentences from Marcus et al. (1999).

Method *Network and Materials.* Same as in the previous simulation.

Procedure. The network habituated in the previous simulation was retested on the test set (with the weights "frozen") and the output for the utterance boundary unit was recorded for every phoneme input. For each habituation condition, the output was divided into two groups dependent on whether the trials were consistent or inconsistent with the habituation. For each habituation condition, the activation of the boundary unit was recorded across all items and the mean activation was calculated. For a given habituation condition, the network was said to have postulated a word boundary whenever the boundary unit activation was above the mean.

Results and Discussion Word boundaries were posited more accurately for the inconsistent items across both conditions (80.00% and 75.00%) than for the consistent items. The scores for word completeness were also higher for the inconsistent items (see Table 1). The results indicate that overall there was better segmentation of the inconsistent items. This suggests that the inconsistent items would stand out more clearly and thus may explain why the infants looked longer towards the speaker playing the inconsistent items in the Marcus et al. (1999) study.

There was a clear effect of habituation on the segmentation performance of the model in the present study compared to the model's performance in Christiansen et al. (1998) where scores were generally lower on both measures. However, in Christiansen et al. the average number of phonemes per word was three, whereas the average number in the current study was only two phonemes per word, thus making the present task easier.

Table 1: Word completeness and accuracy for consistent and inconsistent items in the two habituation conditions.

	AAB Condition		ABB Condition	
	Con.	Incon.	Con.	Incon.
Accuracy	75.00%	80.00%	66.67%	75.00%
Completeness	50.00%	66.67%	44.44%	50.00%

Note. Con. = Consistent items; Incon. = Inconsistent items.

The simulations show how an existing SRN model of word segmentation can fit the data from Marcus et al. (1999) without invoking explicit rules. The SRN had learned to integrate the regularities governing the phonological, lexical stress, and utterance boundary information in child-directed speech. This form of statistical learning enabled it to fit the infant data. In this context, the positive impact of lexical stress information on network performance (as reported in Christiansen et al. 1998) suggests that lexical stress changes the representational landscape over which statistical learning takes place. As we shall see next, this removes the need for lexical stress rules to explain the strong/weak (trochaic) bias in English over weak/strong (iambic).

Taking Advantage of Lexical Stress without Rules

Evidence from infant research has shown that infants between one and four months are sensitive to changes in stress patterns (Jusczyk & Thompson, 1978). Additionally, researchers have found that English infants have a trochaic bias at nine-months of age yet this preference does not appear to exist at six-months (Jusczyk, Cutler & Redanz, 1993). This suggests that at some time between 6 and 9 months of age, infants begin to orient to the predominant stress pattern of the language. One might then assume that if the infant does not have a rule-like representation of stress that assigns a trochaic pattern to syllables, then he/she cannot take advantage of lexical stress information in the segmenting of speech.

The arguments put forth in the literature for rules are based on the production data of children, and based on these productions, it has been shown that word-level (lexical) stress is acquired through systematic stages of development across languages and children (Fikkert, 1994; Demuth & Fee, 1995). If children are learning stress without the use of rules, then systematic stages would not be expected. In other words, due to the consistent patterns of children's productions, a rule must be postulated in order to account for the data (Hochberg, 1988). However, we believe that this conclusion is premature. Drawing on research on the perceptual and distributional learning abilities of infants, we present a corpus analysis investigating how lexical stress may contribute to statistical learning and how this information can help infants group syllables into coherent word units. The results suggest that infants need not posit rules to perform these tasks.

117

Stress Changes the Representational Landscape: A Corpus Analysis

Infants are sensitive to the distributional (Saffran et al., 1996) and stress related (Jusczyk & Thompson, 1978) properties of language. We suggest that infants' perceptual differentiation of stressed and unstressed syllables result in a *representational* differentiation of the two types of syllables. The same syllable is represented differently depending on whether it is stressed or unstressed. This changes the representational landscape, and we employ a corpus analysis to demonstrate how this facilitates the task of speech segmentation.

Method *Materials.* For the corpus analysis we used the Korman (1984) corpus that Christiansen et al. (1998) had transformed into a phonologically transcribed corpus with indications of lexical stress. Their training corpus forms the basis for our analyses. We note that in child-directed speech there appears to be little differentiation in lexical stress between function and content words (at least at the level of abstraction we are representing here; Bernstein-Ratner, 1987; see Christiansen et al. for a discussion). Function words were therefore encoded as having primary stress. We further used a whole syllable representation to simplify our analysis, whereas Christiansen et al. used single phoneme representations.

Procedure. All 258 bisyllabic words were extracted from the corpus. For each bisyllabic word we recorded two bisyllabic nonwords. One consisted of the last syllable of the previous word (which could be a monosyllabic word) and the first syllable of the bisyllabic word, and one of the second syllable of the bisyllabic word and the first syllable of the following word (which could be a monosyllabic word). For example, for the bisyllabic word /slipI/ in /A ju eI slipI hed/ we would record the bisyllables /eIsli/ and /pIhed/. We did not record bisyllabic nonwords that straddled an utterance boundary as they are not likely to be perceived as a unit. Three bisyllabic words only occurred as single word utterances, and, as a consequence, had no corresponding nonwords. These were therefore omitted from further analysis. For each of the remaining 255 bisyllabic words we randomly chose a single bisyllabic nonword for a pairwise comparison with the bisyllabic word. Two versions of the 255 word-nonword pairs were created. In one version, the *stress condition*, lexical stress was encoded by adding the level of stress (0-2) to the representation of a syllable (e.g., /sli/ → /sli2/). This allows for differences in the representations of stressed and unstressed syllables consisting of the same phonemes. In the second version, the *no-stress condition*, no indication of stress was included in the syllable representations.

Our hypothesis suggests that lexical stress changes the basic representational landscape over which infants carry out their statistical analyses in early speech segmentation. To operationalize this suggestion we have chosen to use mutual information (MI) as the dependent measure in our analyses. MI is calculated as:

$$\text{MI} = \log\left(\frac{P(X,Y)}{P(X)P(Y)}\right) \qquad (3)$$

and provides an information theoretical measure of how significant it is that two elements, X and Y, occur together

Table 2: Mutual information means for words and nonwords in the two stress conditions.

Condition	Words	Nonwords
Stress	4.42	-0.11
No-stress	3.79	-0.46

Table 3: Mutual information means for words and nonwords from the stress condition as a function of stress pattern.

Stress Pattern	Words	Nonwords	No. of Words
Trochaic	4.53	-0.11	209
Iambic	4.28	-0.04	40
Dual	1.30	-1.02	6

given their individual probabilities of occurrence. Simplifying somewhat, we can use MI to provide a measure of how strongly two syllables form a bisyllabic unit. If MI is positive, the two syllables form a strong unit: a good candidate for a bisyllabic word. If, on the other hand, MI is negative, the two syllables form an improbable candidate for bisyllabic word. Such information could be used by a learner to inform the process of deciding which syllables form coherent units in the speech stream.

Results and Discussion The first analysis aimed at investigating whether the addition of lexical stress significantly alters the representational landscape. A pairwise comparison between the bisyllabic words in the two conditions showed that the addition of stress resulted in a significantly higher MI mean for the stress condition ($t(508) = 2.41, p < .02$)—see Table 2. Although the lack of stress in the no-stress condition resulted in a lower MI mean for the no-stress condition than for the stress condition, this trend was not significant ($t(508) = 1.29, p > .19$). This analysis thus confirms our hypothesis that lexical stress benefits the learner by changing the representational landscape in such away as to provide more information that the learner can use in the task of segmenting speech.

The second analysis investigated whether the trochaic stress pattern provided any advantage over other stress patterns—in particular, the iambic stress pattern. Table 3 provides the MI means for words and nonwords for the bisyllabic items in the stress condition as a function of stress pattern. The trochaic stress pattern provides for the best separation of words from nonwords as indicated by the fact that this stress pattern has the largest difference between the MI means for words and nonwords. Although none of the differences were significant (save for the comparison between trochaic and dual[4] stressed words: ($t(213) = 2.85, p < .006$), the results suggest that a system without any built-in bias towards trochaic stress nevertheless benefits from the existence of the abundance of such stress patterns in languages like English. In other words, the results indicate that no prior bias is needed

---

[4]According to the Oxford Text Archive, the following words were coded as having two equally stressed syllables: *upstairs, inside, outside, downstairs, hello,* and *seaside.*

toward a trochaic stress patterns because the presence of lexical stress alters the representational landscape over which statistical analyses are done such that simple distributional learning devices end up finding trochaic words easier to segment.

The segmentation model of Christiansen et al. (1998) developed a bias towards trochaic patterns, such that, when segmenting test corpora with either iambic or trochaic syllable groupings, the model was better at segmenting out words that followed a trochaic pattern. Thus, the SRN acquired the trochaic bias given the change in the distributional landscape that stress provides.

Conclusion

In this paper, we have demonstrated the power of statistical learning in two areas of language acquisition in which abstract rules have been deemed necessary for the explanation of the data. Using an existing model of infant speech segmentation (Christiansen et al., 1998), we first presented simulation results fitting the behavioral data from Marcus et al. (1999). The SRN's internal representations incorporated sufficient information for a correct classification of the test items; and the differential segmentation performance on the stimuli words in the consistent and inconsistent conditions provided an explanation for the inconsistent item preference: They are more salient. No rules are needed to explain these data. We then used a corpus analysis to test predictions from the same model concerning the way lexical stress changes the representational landscape over which statistical analyses are done. These changes result in more information being available to a statistical learner, and provide the basis for the trochaic stress bias in English. Again, no rules are needed to explain these data.

There are, of course, other aspects of language for which we have not shown that rules are not needed. Future research will have to determine whether rules may be needed outside the domain of speech segmentation. Some of our other work (Christiansen & Chater, in press) suggests that rules may not be needed to account for one of the supposedly basic rule-based properties of language: Recursion. But why is statistical learning often dismissed as a plausible explanation of language phenomena? We suggest that this may stem from an impoverished view of statistical learning. For example, Pinker (1999) in his commentary on Marcus et al. (1999) forces statistical learning, and connectionist models in particular, into a behavioristic mold: Only input-output relations are said to matter. However, connectionists have also taken part in the cognitive revolution and therefore posit internal representations mediating between input and output. As we demonstrated in the first part of the paper, hidden unit representations provide an important source of information for the modeling of rule-like behavior. Another oversight relates to the significance of combining several kinds of information within a single statistical learning device. The second part of the paper showed how the addition of lexical stress information to the phonological representations resulted in more information being available for the learner. Thus, a more sophisticated approach to statistical learning is likely to reveal its true power, and may obviate the need for algebraic rules.

References

Aslin, R.N., Woodard, J.Z., LaMendola, N.P., & Bever, T.G. (1996). Models of word segmentation in fluent maternal speech to infants. In J.L. Morgan & K. Demuth (Eds.), *Signal to syntax*. Mahwah, NJ: Lawrence Erlbaum Associates.

Bernstein-Ratner, N. (1987). The phonology of parent-child speech. In K. Nelson & A. van Kleeck (Eds.), *Children's Language, 6*. Hillsdale, NJ: Lawrence Erlbaum Associates.

Brent, M.R. & Cartwright, T.A. (1996). Distributional regularity and phonotactic constraints are useful for segmentation. *Cognition, 61*, 93–120.

Christiansen, M.H., Allen, J., & Seidenberg, M.S. (1998). Learning to segment using multiple cues: A connectionist model. *Language and Cognitive Processes, 13*, 221-268.

Christiansen, M.H. & Chater, N. (in press). Toward a connectionist model of recursion in human linguistic performance. *Cognitive Science*.

Chomsky, N. & Halle, M. (1968). *The Sound Pattern of English*. New York: Harper and Row.

Cliff, N. (1987). *Analyzing Multivariate Data*. Orlando, FL: Harcourt Brace Jovanovich.

Demuth, K & Fee, E.J. (1995). *Minimal words in early phonological development*. Ms., Brown University and Dalhousie University.

Elman, J. (1999). *Generalization, rules, and neural networks: A simulation of Marcus et. al, (1999)*. Ms., University of California, San Diego.

Fikkert, P. (1994). *On the acquisition of prosodic structure*. Holland Institute of Generative Linguistics.

Hochberg, J.A. (1988). Learning Spanish stress. *Language, 64*, 683–706.

Jusczyk, P., Cutler, A., & Redanz, N. (1993). Preference for the predominant stress patterns of English words. *Child Development, 64*, 675–687.

Jusczyk, P, & Thompson, E. (1978). Perception of a phonetic contrast in multisyllabic utterances by two-month-old infants. *Perception & Psychophysics, 23*, 105–109.

Klein, H. (1984). Learning to stress: A case study. *Journal of Child Language, 11*, 375–390.

Korman, M. (1984). Adaptive aspects of maternal vocalizations in differing contexts at ten weeks. *First Language, 5*, 44–45.

Macken, M.A. (1980). The child's lexical representation: The "puzzle-puddle-pickle" evidence. *Journal of Linguistics, 16*, 1–17.

MacWhinney, B. (1991). *The CHILDES Project*. Hillsdale, NJ: Lawrence Erlbaum Associates.

Marcus, G.F. (1998). Rethinking eliminative connectionism. *Cognitive Psychology, 37*, 243–282.

Marcus, G.F., Vijayan, S., Rao, S.B., & Vishton, P.M. (1999). Rule learning in seven month-old infants. *Science, 283*, 77–80.

Pinker, S. (1999). Out of the minds of babes. *Science, 283*, 40–41.

Saffran, J.R., Aslin, R.N., & Newport, E.L. (1996). Statistical learning by 8-month olds. *Science, 274*, 1926–1928.

Seidenberg, M.S. (1995). Visual word recognition: An overview. In Peter D. Eimas & Joanne L.Miller (Eds.), *Speech, language, and communication. Handbook of perception and cognition, 2nd ed., Vol. 11*. San Diego: Academic Press.

Smith, N.V. (1973). *The Acquisition of Phonology: A case study*. Cambridge: Cambridge University Press.

Comparative Modelling of Learning in a Decision Making Task

Richard Cooper (R.Cooper@psyc.bbk.ac.uk)
Department of Psychology, Birkbeck College, University of London, Malet St.,
London, WC1E 7HX

Peter Yule (P.Yule@psyc.bbk.ac.uk)
Department of Psychology, Birkbeck College, University of London, Malet St.,
London, WC1E 7HX

Abstract

In this paper we compare the behaviour of three competing accounts of decision making under uncertainty (a Bayesian account, an associationist account, and a hypothesis testing account) with subject performance in a medical diagnosis task. The task requires that subjects first learn a set of symptom/disease associations. Later, subjects are required to form diagnoses based on limited symptom information. The competing theoretical accounts are embodied in three computational models, each with a single parameter governing the learning rate. Subjects' diagnostic accuracy was used to calibrate the learning rates of the models. The resulting parameter-free models were then used to predict subjects' symptom querying behaviour in a subsequent task. The fit between the Associationist model's predictions and subject behaviour was poor. The fit was slightly better in the case of the Bayesian model, but the hypothesis testing account proved to provide the most adequate account of the data.

Introduction

Many decisions in real life are made under conditions of uncertain or incomplete information. Bayesian probability theory provides an optimal approach to such decisions when the uncertainty in the evidence can be quantified in the form of probabilities. In many cases, however, such quantitative information is not available, and even when it is, people frequently fail to make correct use of it, as in the well-known cases of base rate neglect (Kahneman & Tversky, 1973).

Although the Bayesian approach to decision making under uncertainty may be optimal, a number of other, sub-optimal, approaches yield plausible decision making behaviour. Gluck & Bower (1988), for example, have demonstrated that an associative network employing a Rescorla-Wagner (1972) learning rule can learn a disease categorisation task in which symptoms are probabilistically associated with diseases. The task is effectively a decision making task in which symptoms are unreliable indicators of possible diseases. Gluck & Bower compared the behaviour of their model with that predicted by a Bayesian account, and with that of human subjects. They report a correlation of 0.94 between subject performance and the model's performance, suggesting that non-Bayesian accounts are indeed capable of producing human-like performance.

Further concerns about the relevance of Bayesian approaches to human decision making under uncertainty are raised by Gigerenzer & Goldstein (1996), who suggest that the cognitive plausibility of such accounts is undermined by human performance limitations. They suggest that in real-life people rely on "fast and frugal" heuristics which approximate optimal behaviour. In support of this position Gigerenzer & Goldstein (1996) compared the decision making behaviour of four algorithms employing different forms of bounded computation with an optimal regression model. Several of the bounded algorithms achieved levels of performance equivalent to the mathematically optimal model. This work demonstrates that, at least on certain tasks, fast and frugal approaches are capable of producing near optimal decision making performance, and hence that viable alternatives to Bayesian decision algorithms do exist.

It would be wrong to suggest, however, that the results of Gluck & Bower (1988) and Gigerenzer & Goldstein (1996) provide unequivocal evidence against Bayesian processes in human decision making. Thus, although the correlation obtained by Gluck & Bower (1988) between subject performance and the performance of their associative network was high (0.94), it was not as high as the correlation between subject performance and that of a Bayesian approach (0.99). More critically, the root mean square error between disease probabilities as predicted by the associative network and the subjects was more than twice that between disease probabilities as predicted by the Bayesian account and the subjects. The correlation of 0.94 (obtained by choosing an appropriate value for the learning rate, a free parameter in the associative network) is less convincing when considered in this light.

The import of the results of Gigerenzer & Goldstein (1996) is similarly open to question. The difficulty here lies in the fact that their evaluation of non-Bayesian approaches did not involve comparison with human data. Fast and frugal algorithms were found to perform as well as a mathematically optimal account — one based on multiple regression, and, in fact, formally equivalent to an Associationist model — but human data was not collected on the task which they investigated.

In previous work (e.g., Fox, 1980; Cooper & Fox, 1997; Yule, Cooper, & Fox, 1998) we have compared the normative, quantitative, Bayesian account of decision making under uncertainty, and a qualitative, hypothesis testing account, with human performance on a medical diagnosis task. Subjects

Table 1: Conditional probabilities of symptom given disease

		Symptoms				
		sym0	*sym1*	*sym2*	*sym3*	*sym4*
Diseases	*dis0*	1.00	0.00	0.25	0.00	0.25
	dis1	0.00	0.50	1.00	0.50	1.00
	dis2	0.50	1.00	0.00	0.00	0.25
	dis3	0.00	0.00	0.25	1.00	0.00

rarely achieved the levels of performance suggested by either computational account, but the general pattern of subject performance more closely resembled the hypothesis testing account. In this paper we extend that work by 1) adding reasonable performance limitations to the Bayesian and hypothesis testing accounts (thus bringing baseline performance into line with subjects); 2) extending the comparison by including an associationist (two-layer feed-forward network) model in the set of competing computational accounts; 3) modifying the experimental task such that subject performance on one dependent measure can be used to calibrate the various models, which may in turn be used to predict subject performance on a second dependent measure; and 4) examining the effect on human and predicted behaviour of different training histories.

We begin by describing the diagnosis task in detail and outlining the mechanisms behind the three models. We then report an experiment in which human subjects learned to perform the diagnosis task. This performance is then used firstly to fix the learning rate in each model, and then to evaluate the resulting parameter-free models.

The Diagnosis Task

The task of diagnosis is essentially one of categorising a set of features or symptoms as corresponding to one of a set of known diseases. There are numerous variations on the task. Fox (1980) employed five symptoms and five diseases. All diseases were equally likely and subjects were required to query symptoms in sequence before offering a diagnosis. Gluck & Bower (1986) employed four symptoms and two diseases, but the occurrence of one disease was rare in comparison to the other. Here, subjects were allowed complete symptom information when making their diagnoses. The version of the task employed in the current work is derived from that of Fox (1980). Four diseases (hypothetical strains of 'flu) and five symptoms were employed. All diseases were equally likely, and symptoms were unreliable indicators of diseases. Table 1 shows the probability of each symptom occurring for each disease. Thus, one in four patients suffering from *dis0* would have *sym2*.

The diagnosis task can be presented in two forms. In the simplest form, the subject makes a diagnosis based on full symptom information. That is, the presence/absence of each symptom is known, and the subject must select which of the four diseases is most likely. With the symptom/disease associations employed here, it is always possible to discriminate between diseases based on full symptom information.

A more challenging form of the diagnosis task involves presenting subjects initially with one symptom (the presenting symptom), and requiring them to query just those symptoms necessary to make a diagnosis (and then to make a diagnosis when appropriate). This version of the task is naturalistic in that it corresponds closely to the task of a General Practitioner. It also yields rich data in the form of symptom querying strategies. However, the data are difficult to interpret because different subjects appear to employ different "diagnostic thresholds" — some subjects are willing to offer a diagnosis on the basis of few symptoms, whereas others query most or all symptoms before offering a diagnosis.

The freedom allowed to subjects in this more challenging version of the task, and its manifestation in the form of a diagnostic threshold, also poses methodological problems for evaluating computational accounts of subject behaviour. The diagnostic threshold effectively introduces a free parameter into the model of a subject, allowing the modeller an additional degree of freedom with which to account for subject behaviour.

The experiment reported below yields data on symptom querying strategies whilst overcoming the difficulty of diagnostic thresholds by presenting subjects with one symptom, and then requiring them to query exactly one further symptom before offering the most likely diagnosis. There is no scope for a diagnostic threshold in this form of the task. Whilst individual differences may still exist, such differences must be attributed to other factors (such as learning rate, or strategic elements).

In fact, previous research has shown large between-subject differences on diagnosis tasks (e.g. Yule, Cooper & Fox, 1998), even with little apparent variation in diagnostic threshold. The task can be taxing, and motivational factors are likely to play a role. However, between-subject differences may also be attributable to differences in training history. That is, if during training subjects are exposed to randomly generated sequences of cases of diseases, then their diagnostic behaviour may be influenced by idiosyncratic features of their training materials. This is especially likely to be true if training is limited. Training history is therefore another factor considered in the experiment reported below.

Theoretical Accounts of Diagnosis
The Bayesian Approach

The Bayesian approach to the categorisation element of diagnosis is straightforward and well documented (see, for example, Fox, 1980). The probability of each disease can be calculated from symptom information provided that disease base rates and the conditional probabilities of each symptom given each disease are known (assuming independence of symptoms). Approximations to each of these probabilities can be computed from frequency information, acquired as part of the learning of symptom/disease associations.

Symptom selection, when incomplete symptom information is provided, may be determined through calculation of the informativeness of each indeterminate symptom. Oaksford & Chater (1994) suggest that cue selection on the basis of informativeness (in conjunction with assumptions about the distribution of cues) provides a good account of subject behaviour in related tasks, and point to appropriate information theoretic measures of informativeness (Shannon & Weaver, 1949; Wiener, 1948).

Previous research has shown that this Bayesian approach significantly outperforms subjects when learning the diagnosis task (Yule, Cooper & Fox, 1998). In order to reduce performance to human levels we suggest imperfect recording of frequency information relating to both base rates of diseases and disease/symptom co-occurrences. In particular, the model that generated the data reported below includes a parameter, the learning rate, which specifies the probability that frequencies will be updated on any given trial. Thus, when this parameter is set to 0.10 (a value that yields behaviour at levels comparable to human subjects), frequency information will be recorded (and hence employed in determining the informativeness of symptoms and the probabilities of possible diagnoses) on 1 in 10 trials on average.

The Hypothesis Testing Approach

Fox (1980) described an approach to the diagnosis task which generated propositional hypotheses about possible diseases and reasoned over those hypotheses when determining appropriate symptoms to query or diagnoses to offer. Cooper & Fox (1997) extended that model to account for learning during the diagnosis task, and Yule, Cooper & Fox (1998) compared the performance of that model and a Bayesian model with subject performance on a version of the diagnosis task.

We do not describe the model again here, except to say that the presence or absence of symptoms triggers the model into forming hypotheses about possible diseases. These hypotheses lead the model to expect further symptoms, which form the basis of the model's querying strategy: the model will ask about symptoms if it expects them to be present given any of the hypothesised diseases, in an order determined by recency in its knowledge base. When provided with diagnostic feedback, the model adjusts its beliefs about the relations between symptoms and diseases, and about the symptom patterns associated with diseases.

In the model employed in the current work this updating is not performed on all trials — diagnostic feedback is sporadically ignored. The probability of updating on a given trial is determined by a learning rate parameter analogous to that in the Bayesian model. As in the Bayesian account, this probabilistic learning allows the model's diagnostic performance to be brought approximately into line with that of subjects.

The Associationist Approach

The diagnosis task may also be performed by an associative network employing a Rescorla-Wagner (1972) learning rule. As noted above, Gluck & Bower (1988) argue that this learning rule, which is formally equivalent to the Delta rule, ac-

counts well for human performance in their version of the diagnosis task.

Our implementation of the associationist approach follows traditional lines: a two-layered network maps five input nodes (corresponding to symptoms, and set equal to +1.00 when a symptom is present, −1.00 when a symptom is absent and 0.00 when a symptom's status is unknown) to four output nodes (corresponding to the four diseases). On testing cycles the symptom vector is fed to the network and the most active disease node is offered as the diagnosis. On training cycles the network's weights are adjusted according to the standard Delta rule. The learning rate constitutes a parameter of the model (again allowing the model's diagnostic accuracy to be calibrated with that of human subjects) that scales weight adjustments within the network, with weights changing by an amount equal to the learning rate times the difference between the network's output and the training signal.

Standard associative network approaches to the diagnosis task do not obviously generalise to the version of the task in which symptom selection is required. In our implementation, a second associative network is trained on instances of the identity map between symptom vectors corresponding to those symptom patterns actually presented to the main symptom/disease network. The rationale for this is that, after training, incomplete symptom information will be mapped by this network to a symptom vector resembling a previously seen pattern. The most active symptom in this vector whose presence is indeterminate is the symptom that is queried. This approach is far from optimal, but it is closer to associative principles than the most obvious alternative: to test the symptom/disease network on all possible extensions of the current symptom information and then select the symptom corresponding to the extension yielding the strongest diagnosis.

Experiment: Rationale and Method

52 second year psychology students from Birkbeck College took part in the experiment. There were four conditions, with 13 subjects in each condition. The conditions differed only in the sequences of cases presented, as described below.

Subjects in all conditions completed 5 blocks of trials. Blocks 1 and 3 were training trials in which full symptom and disease information was presented to all subjects. Subjects were required to step through trials in these blocks at their own pace, whilst attempting to learn the symptom/disease associations. These blocks comprised 12 trials each. The degree of learning in the training blocks was assessed in blocks 2 and 4. Here, subjects were required to make diagnoses based on full symptom information. Feedback (in the form of the actual underlying disease) was given on these trials, allowing subjects to further improve their diagnostic accuracy. These testing blocks also consisted of 12 trials. In the fifth and final block subjects were presented with a single presenting symptom. They were required to query the presence of exactly one further symptom and then make a "best guess" diagnosis. Diagnostic feedback (i.e., the actual underlying disease) was given in the case of error. This block consisted

Table 2: Mean diagnostic accuracy (%) in each block and training history condition (Human data).

Training History	N	Block 2		Block 4		Block 5	
		Mean	SD	Mean	SD	Mean	SD
1	13	50.6	25.1	55.1	24.7	49.6	21.0
2	13	50.6	19.7	62.8	20.3	48.5	20.2
3	13	60.9	22.7	59.6	22.5	53.1	21.9
4	13	50.0	18.0	59.6	14.8	49.2	11.5
Total	52	53.0	21.4	59.3	20.5	50.1	18.6

Table 3: Overall mean diagnostic accuracy (%) in each block for each model ($N = 52$ each).

Model	L.R.	Block 2		Block 4		Block 5	
		Mean	SD	Mean	SD	Mean	SD
Bayes.	0.10	39.3	16.3	60.3	17.4	74.8	15.1
Hypot.	0.25	41.0	15.6	57.8	13.9	56.3	16.1
Assoc.	0.015	36.3	17.2	58.0	20.3	57.0	14.6

of 20 trials. Subjects were instructed of the block structure before commencing the task, and knew that they would be required to identify diseases based on minimal information in the final block. They were instructed that in this final block they should query the symptom that would be "most helpful to them in making their diagnosis".

Symptoms and diseases were related according to the probabilities given in Table 1, with four strains of 'flu (Austrian 'flu, Belgian 'flu, Greek 'flu, and Danish 'flu) being mapped onto the disease names and five common 'flu symptoms (headache, shivering, sore throat, coughing, and sneezing) being mapped onto the symptom names. This mapping was randomised across subjects. Three cases of each disease occurred in each of the first four blocks, with five cases of each disease in the final block. In order to examine the effect of training history the sequence of cases presented to each subject was generated from one of four random seeds. Subjects were allocated at random to a training history group. (Training history was thus a between-subject independent variable.)

The experiment was administered by networked software running over the department's intranet. The software, written in JavaScript for use with web browsers, randomly assigned subjects to each of the four training conditions and collected subject responses at the end of each block.[1]

Results

Diagnostic accuracy Table 2 shows mean diagnostic accuracy for each training history condition, in each of the blocks in which a diagnosis was required, namely 2, 4 and 5. A two-factor, mixed-model ANOVA shows a significant effect of block ($F(2, 96) = 7.70$, $p < 0.0008$), but no effect of training history ($F(3, 48) = 0.28$) and no interaction ($F(6, 96) = 0.91$). The increase in diagnostic accuracy between blocks 2 and 4 ($F(1, 48) = 5.32$, $p < 0.0254$), and the decrease between blocks 4 and 5 ($F(1, 48) = 20.89$, $p < 0.0001$), are both significant. So there is evidence of learning during the training phase of the experiment, and of disruption of performance by the different task in the final block, but no evidence of any effects of training history on diagnostic accuracy.

[1]A demonstration of the client system is available at http://redback.psyc.bbk.ac.uk/expts/jdm4/demo/

Model calibration Each model contains one free parameter, a learning rate, that determines the speed and accuracy of learning. Human diagnostic accuracy data for block 4 was used to calibrate learning in all models as follows. For each model, simulations (comprising 52 virtual subjects, 13 in each training history condition) were conducted for a range of learning rates. Learning rates in each model were then fixed at values leading to diagnostic accuracy on block 4 which was most commensurate with the human data. This approach resulted in a learning rate of 0.10 for the Bayesian model, 0.015 for the Associationist model, and 0.25 for the Hypothesis Testing model. (These rates are not comparable because of the different learning mechanisms within each model.) Once calibrated, all models make parameter-free predictions for symptom query patterns and diagnostic accuracy on the final block. The experiment was deliberately designed to allow this approach to model testing.

Table 3 shows mean diagnostic accuracy for each calibrated model. It is clear that all the models show larger differences between blocks 2 and 4 than do human subjects. Also, the Bayesian model's diagnostic performance improves markedly on the final block, whereas the other models' performances do not, but none of the models show the significant decrease in performance observed in the human data.

Human symptom queries Table 4 shows frequencies of each possible symptom query for the final instance of each presenting symptom in the final block. For each row, there is a χ^2 test for nonrandom distribution of queries (d.f.=3). From the table, there are significant departures from random distribution for queries following *sym1* presenting and *sym4* presenting. By inspection of the peaks in each row, we can see that given *sym1* presenting, human Ss tend to query *sym0*, and given *sym4* presenting, Ss tend to query *sym2*.

Bayesian Model Table 5 shows the final symptom query frequency table for the Bayesian model. There are only two significant query biases, for *sym0* and *sym3* presenting. These do not correspond to either of the significant human biases. However, given *sym0* presenting, the Bayesian model tends to query *sym1*, and although the human effect does not reach significance, its maximum is also *sym1*. But the Bayesian tendency to query *sym4* given *sym3* is not reflected in the Human data.

Moreover, the significant human effects are not paralleled

Table 4: Final symptom query frequencies for each presenting symptom (Human data, row $N = 52$).

Pres. Sym.	Query					$\chi^2(3)$	p
	sym0	sym1	sym2	sym3	sym4		
sym0		19	9	14	10	4.77	
sym1	19		9	14	10	9.69	.021
sym2	13	8		14	17	3.23	
sym3	11	17	13		11	1.85	
sym4	14	6	21	11		9.08	.028

Table 5: Final symptom query frequencies for each presenting symptom (Bayesian model, row $N = 52$).

Pres. Sym.	Query					$\chi^2(3)$	p
	sym0	sym1	sym2	sym3	sym4		
sym0		26	7	9	10	17.69	.001
sym1	15		18	8	11	4.46	
sym2	13	14		10	15	1.08	
sym3	6	10	12		24	13.85	.003
sym4	14	15	12	11		0.77	

Table 6: Final symptom query frequencies for each presenting symptom (Hyp. Testing model, row $N = 52$).

Pres. Sym.	Query					$\chi^2(3)$	p
	sym0	sym1	sym2	sym3	sym4		
sym0		25	12	6	9	16.15	.001
sym1	27		7	3	15	25.85	.001
sym2	5	2		5	40	75.23	.001
sym3	4	7	18		23	18.61	.001
sym4	1	9	33	9		44.30	.001

Table 7: Final symptom query frequencies for each presenting symptom (Associationist model, row $N = 120$).

Pres. Sym.	Query					$\chi^2(3)$	p
	sym0	sym1	sym2	sym3	sym4		
sym0		33	35	22	30	3.27	
sym1	25		25	42	28	6.60	(.086)
sym2	35	24		31	30	2.07	
sym3	28	25	23		44	9.13	.028
sym4	32	29	29	30		0.20	

by the corresponding non-significant Bayesian effects; given *sym1* the Bayesian model tends to query *sym2*, unlike the human preference for *sym0*, and given *sym4* the Bayesian model queries *sym1*, not *sym2*.

Hypothesis Testing Model Table 6 shows the final symptom query frequency table for the Hypothesis Testing model. This exhibits strong, highly significant biases for each presenting symptom. With regard to the presenting symptoms which give significant querying biases in the human data, *sym1* and *sym4*, the Hypothesis Testing model generates maxima in the same places as do human subjects; it tends to query *sym0* given *sym1* (cf. Bayesian model), and *sym2* given *sym4*. But also, even where the human bias is non-significant, the model still predicts the most frequent query correctly in two of three cases, with *sym0* and *sym2* presenting, and fails to predict the human bias only with *sym3* presenting.

Associationist Model Unfortunately the Associationist model produced no significant symptom querying biases at all when calibrated to the human level of diagnostic accuracy and subject numbers. Consequently these data are not presented; instead, the model was rerun with a larger number of virtual subjects (120), at the same learning rate, resulting in Table 7.

As Table 7 shows, there is only one significant symptom query bias, for *sym3* presenting, when the model tends to query *sym4*. (Curiously, all the models predict the same bias for *sym3*, but the human bias is non-significant and in a different direction.) The Associationist model also shows an almost-significant bias to query *sym3* given *sym1*, unlike the

other models and unlike the significant human tendency to query *sym0*.

Discussion of results

The relative absence of significant biases in the Bayesian model symptom query data is attributable to the large amount of random variance in the model's behaviour, a consequence of its low learning rate. With higher learning rates, or with larger numbers of virtual subjects at the same learning rate, the model's predictions are quite clear for all presenting symptoms. But as things stand, such predictions as there are from the Bayesian model are not very well borne out in the human data.

Unfortunately, even with large numbers of virtual subjects the predictions of the Associationist model are minimal, and do not correspond to human query biases. So we can conclude that of the three, the Associationist model gives the poorest account of the human data.

The Hypothesis Testing model successfully predicts the significant symptom querying biases in the human data, as well as the directions of most of the non-significant ones, so it easily fares best of the three models. We are in the process of collecting more human data, in order to determine if more of the human query biases are significant. With more human data, we also expect to be able to investigate possible effects of training history on symptom querying strategies.

The assumption that human levels of performance can be simulated by manipulating models' learning rates was reasonably successful, in that it produced a good fit between human symptom querying patterns and the predictions of one of the models. However, none of the models generated learning curves of the same shape as the human curve, since the calibrated models performed more poorly than humans on block 2, and better than humans on block 5. It seems that humans reach peak performance quite early in the experiment, but find it hard to improve much thereafter. Their performance is then significantly disrupted on the final block, where the task is somewhat different.

General Discussion

The observed superiority of the Hypothesis Testing model over the Bayesian in fitting human symptom querying behaviour replicates previous findings (Yule *et al.*, 1998; Fox, 1980), despite substantial variations in task, materials, experimental interface and methods of analysis. The Hypothesis Testing model owes its success to two factors: it only queries symptoms expected to be present given any of the hypothesised diseases (i.e., it contains a confirmation bias), and queries are ordered according to a recency principle in memory (such that behaviour is determined more by recent events than by those in the more distant past). It is reasonable to ask if incorporation of these biases in the other models would improve their fit with the human data.

With respect to the Bayesian model, symptom queries are selected on the basis of expected information gain, and as in previous studies, while the model can predict human behaviour in a few cases, it does not yield a good overall fit with human questioning patterns. Oaksford & Chater (1994) have argued that Bayesian approaches in similar information seeking tasks can give a good account of human performance when they are supplemented with a "rarity assumption". In the current task, such an assumption would have the effect of restricting the search for evidence to symptoms expected to be present given the most likely diseases. In other words, in the current task such an assumption would amount to a Bayesian implementation of a confirmation bias. Incorporation of the rarity assumption into the Bayesian model is therefore of some importance.

The second factor contributing to the Hypothesis Testing model's superior performance, recency, might also be incorporated into a Bayesian model by weighting recent events in the estimation of event frequencies used to determine the various numerical factors required by Bayes' theorem. Such a weighting is appealing given recency effects, but its incorporation would add an extra parameter to the Bayesian model, thus raising further difficulties in model evaluation.

The Associationist model could also benefit from a re-evaluation of its approach to symptom querying. The difficulty here is that in the first four blocks the model is given full symptom/disease information. There is no obvious way in which an associative network can produce sequential symptom querying behaviour from this static information. Standard associative network approaches to sequencing (recurrent networks) offer little assistance with this problem. Network models employing competitive activation may be appropriate, but such models have little in common with the associative framework from where we started.

A final methodological point is in order. The precise form of the experiment was dictated by the requirements of model testing. We have not simply attempted to fit models to the data. Rather, the experiment was designed to yield two dependent measures: diagnostic accuracy and symptom query strategy. The first measure was used to fix the single free parameter in each model. The result was a set of predictive models. Tables 5, 6, and 7 are model *predictions* — generable (in principle) before subjects begin the final block of the experiment. Few cognitive models are parameter free. For those that are not, we strongly advocate a methodology such as ours where the requirements of model evaluation determine aspects of subsequent empirical work. This methodology, we aver, is far more sound than the more common approach of data fitting via the adjustment of parameter values.

In sum, the comparative evaluation of three very different models of decision making under uncertainty, as applied to the medical diagnosis task, leaves us cautiously optimistic with respect to fast and frugal alternatives to optimal Bayesian accounts. The evidence for purely associative processes, however, appears weak.

References

Cooper, R., & Fox, J. (1997). Learning to make decisions under uncertainty: The contribution of qualitative reasoning. In Langley, P., & Shafto, M. (Eds.), *Proceedings of the 19th Annual Conference of the Cognitive Science Society*, pp. 125–130.

Fox, J. (1980). Making decisions under the influence of memory. *Psychological Review*, 87, 190–211.

Gigerenzer, G., & Goldstein, D. G. (1996). Reasoning the fast and frugal way: Models of bounded rationality. *Psychological Review*, 103(4), 650–669.

Gluck, M. A., & Bower, G. H. (1988). From conditioning to category learning: an adaptive network model. *Journal of Experimental Psychology: General*, 117(3), 227–247.

Kahneman, D., & Tversky, A. (1973). On the psychology of prediction. *Psychological Review*, 80, 237–251.

Oaksford, M., & Chater, N. (1994). A rational analysis of the selection task as optimal data selection. *Psychological Review*, 101, 608–631.

Rescorla, R. A., & Wagner, A. R. (1972). A theory of Pavlovian conditioning: Variations in the effectiveness of reinforcement and non-reinforcement. In Black, A. H., & Prokasy, W. F. (Eds.), *Classical Conditioning II: Current Research and Theory*. Appleton-Century-Crofts, New York, NY.

Shannon, C. E., & Weaver, W. (1949). *The Mathematical Theory of Communication*. University of Illinois Press, Urbana, IL.

Wiener, N. (1948). *Cybernetics*. Wiley, New York, NY.

Yule, P., Cooper, R., & Fox, J. (1998). Normative and information processing accounts of medical diagnosis. In Gernsbacher, M. A., & Derry, S. J. (Eds.), *Proceedings of the 20th Annual Conference of the Cognitive Science Society*, pp. 1176–1181. Madison, WI.

Parsing Modifiers: The Case of Bare NP Adverbs

Martin Corley (Martin.Corley@ed.ac.uk)
Department of Psychology and Human Communication Research Centre
University of Edinburgh, Edinburgh EH8 9JZ, UK

Sarah Haywood (S.Haywood@psych.york.ac.uk)
Department of Psychology
University of York, York YO10 5DD, UK

Abstract

Current models of Human Sentence Processing fall into two broad categories: *Constraint Satisfaction* accounts, which emphasise the immediate access of the comprehension processes to detailed linguistic information as parsing progresses (e.g., MacDonald et al., 1994), and *Syntax First* accounts, which hold that parsing is essentially a two-stage process, with initial decisions being made on the basis of a subset of available information (see, e.g., Frazier, 1995). In this paper, we examine evidence from Mitchell (1987) which seems strongly to favour a syntax first position, suggesting that basic lexical information about verbs may have little influence on the early stages of sentence processing. We provide experimental evidence to show (a) that detailed linguistic information is available early, but (b) that bare NP adverbs (a type of modifier) are read surprisingly fast, a finding which appears difficult to reconcile with many current accounts of sentence processing.

Constraint Satisfaction models of Human Sentence Processing rely heavily on detailed information about the combinatorial possibilities and probabilities associated with each word of a sentence. For example, MacDonald et al. (1994) suggest that the reading of a sentence such as (1a) is affected by lexico-syntactic factors including the frequencies with which the word *raced* is used in each of its senses (for example, as a past participle or as a past tense verb), as well as thematic factors such as how good an agent or patient *horse* is of *raced*.

(1a) The horse *raced past the barn* **fell**.[1]

(1b) The landmine *buried in the sand* **exploded**.

This type of account can be contrasted with *Syntax First* positions, according to which initial parsing decisions are made on the basis of a subset of available linguistic information (generally comprising category information and phrase structure rules, e.g., Frazier, 1979). These decisions may later be revised in the light of further (syntactic, thematic or pragmatic) evidence, but these revisions have an associated processing cost. According to such accounts, (1b) should be as hard to understand as (1a), since they are syntactic homomorphs. The constraint satisfaction view, on the other hand, predicts the intuitively observable difference between the two, by pointing out that (thematically) *landmine* makes a poor agent (but a good patient) of *buried*, as well as (syntactically) the statistical facts that *buried* is typically used as a passive past participle (in comparison to *raced* which is typically used as an active past tense verb).

However, evidence from a study by Mitchell (1987) appears highly incompatible with constraint-based accounts. According to Mitchell, the most basic form of information associated with a verb—that of whether a verb is transitive or intransitive, or in other words the verb's subcategorisation information—may not be available during the initial stages of sentence comprehension. Using materials like those in (2), he found that readers of (2a) were typically slow (relative to a control with a disambiguating adverbial phrase such as *during surgery* after *visited* or *sneezed*) at reading the word *wrote*. This is to be expected within almost any framework, since it is syntactically permissible (as well as thematically acceptable) for *the doctor* to be initially interpreted as the direct object of *visited*.

(2a) After the child visited *the doctor* **wrote** him a prescription.

(2b) After the child sneezed *the doctor* **wrote** him a prescription.

Crucially, Mitchell also showed that readers were slow to read *the doctor* in (2b). Since *sneezed* is a (typically) intransitive verb,[2] Mitchell interpreted these findings as suggesting that information about the subcategorisation of *sneezed* was *not* initially available to the parsing process (but was made available as soon as later processes could check the plausibility of *the doctor* as an object of *sneezed*).

Mitchell's findings are clearly problematic for constraint-based accounts of sentence processing. If the differences between (1a) and (1b) are to be accounted for in terms of information about the (probabilistic) 'goodness of fit' between a verb and its arguments (including the relative frequencies with which particular subcategorisation frames of a given verb are used), an account must be made of the apparent *in*sensitivity of the human parser to the fact that *sneezed* is rarely, if ever, used as a transitive verb.

Less often remarked upon is the difficulty that Mitchell's results may pose for many syntax first theories. Although the

---

[1] Regions of example sentences where more than one interpretation is (potentially) available are typeset in *italics*, and disambiguating regions are set in **bold**.

[2] Almost every 'intransitive' verb, as Adams et al. (1998) note, can be used in some transitive senses, if only in highly stylised phrases such as *He sneezed a tiny sneeze*, *She yawned a big yawn*, etc.

details of models differ, a number of current positions converge on the importance of (potential) argumenthood, with incoming constituents being preferentially attached to the current phrase marker as arguments rather than as modifiers or adjuncts (e.g., Crocker, 1992; Frazier & Clifton, 1996). In order to adjudicate between the attachment of a constituent as a modifier or as an argument, the human parser *must be aware* that a potential site for argument attachment exists. If this information is unavailable at the time a verb is encountered, then an argument attachment strategy would provide a poor account, at least at the explanatory level, of the parser's behaviour.

Interpretations of Mitchell's findings tend to assume that detailed information about the verb *is* initially available, but that either linguistic or experimental factors give rise to the results obtained. For example, Tanenhaus and Trueswell (1995) observe that there may be a residual bias to interpret a noun phrase following a verb as its object, lexical biases notwithstanding. This has the advantage of providing a straightforward account of Mitchell's findings within a constraint satisfaction framework, but the disadvantage that it damages the predictive power of the theory (if difficulty in reading (1a) is accounted for in terms of information associated with a particular verb, but Mitchell's findings are interpreted in the light of the behaviour of verbs in general, how are we to know which factors are likely to influence the interpretations of hitherto untested sentences?).

Critics of Mitchell's experimental design have noted that he used a self-paced reading paradigm in his initial study. Due to the way that the materials were segmented (such that participants saw the words ... *sneezed the doctor* in a single display before pressing a key to view the matrix verb of which *the doctor* was the subject), participants may have been misled into 'strategically' attempting to interpret *the doctor* as the object of *sneezed*. This criticism (effectively of the ecological validity of the self-paced reading paradigm) was addressed by Adams, Clifton, and Mitchell (1998), who replicated Mitchell's experiment using an eyetracker, so that materials did not have to be artificially segmented. Although their findings contradict Mitchell's, they remain unconvinced that subcategorisation information *is* initially available. One argument that they make is that Mitchell's findings must still be accounted for: even if segmentation affects the way in which subjects read Mitchell's materials, it is clear that it does not cause subjects to violate phrase-structure rules (when attaching adverbials in control materials), whereas subcategorisation information appears to be easily overridden.

In this paper, we offer a different interpretation of Mitchell's findings. This rests on the observation that although intransitive verbs such as *sneezed* do not subcategorise for *object* NPs, there is a class of NPs which can legitimately follow them: namely, bare NP adverbs (Larson, 1985). For example, the sentence in (3) is a perfectly acceptable sentence in English.

(3)　　After the child sneezed the other day, the doctor wrote him a prescription.

The other day serves an adverbial function; that is, it seems to modify, rather than serve as an argument of, the verb *sneezed*. If participants *are* initially aware of the subcategorisation properties of verbs, the delay in reading *the doctor* in (2b) relative to its control might be accounted for by a system which was working on the assumption that any NP following an intransitive verb such as *sneezed* would be a bare NP adverb; the delay reflects the fact that *the doctor* must be rejected in this case.

Parsing Bare NPs

The primary aim of the first study which follows is to test this interpretation of Mitchell's findings, by explicitly including bare NP adverbs in a set of experimental materials modelled on those used by Adams et al. (1998). We have chosen to use a self-paced reading paradigm, precisely because we are interested in what happens when subjects are 'forced' to attach an NP to a verb. If our assumptions are right, then bare NP adverbs (henceforth 'bNPs') should be read faster than 'normal' NPs ('nNPs') following intransitive verbs in the subordinate clause. There should also be a cost of revision, reflected in the time taken to read the disambiguating matrix verb, in all conditions except that where an nNP follows an intransitive (in this case the revision should already have been made, in line with Mitchell's findings). The study also allows us to explore the parsing of modifiers vs. arguments. According to many 'syntax first' accounts, both bNPs and nNPs should initially be attached as arguments of the verbs, assuming that this is permissible: exact predictions are dependent on whether it is believed that subcategorisation information is initially available.

Experiment 1

Participants　24 volunteers from the University of Edinburgh undergraduate population took part in this study. All were native speakers of English, had normal or corrected-to-normal vision, and had no professed reading difficulties.

Materials　The experimental materials were adapted from those used by Adams et al. (1998). 24 materials were created, where each material had four versions similar to those in (4). Examples like (4a) and (4b) had optionally transitive verbs in the preposed subordinate clause, whereas in (4c) and (4d) these verbs were strictly intransitive. Transitive and intransitive verbs were matched for length, and for frequency according to norms from Kučera and Francis (1967).

(4a)　　Although the dog scratched *the old vet /* **seemed** */ very relaxed.*

(4b)　　Although the dog scratched *the whole day /* **seemed** */ very relaxed.*

(4c)　　Although the dog struggled *the old vet /* **seemed** */ very relaxed.*

(4d)　　Although the dog struggled *the whole day /* **seemed** */ very relaxed.*

Each material contained a (potential) ambiguity[3] in that the NP following the subordinate verb could be interpreted as an argument, or modifier, of that verb; in each case, the second (matrix) verb resolved the ambiguity in favour of an interpretation where the NP was the subject of the matrix clause. In (4a) and (4c) the NP in question was an 'object-like' NP (nNP), which might be interpreted as an argument of the subordinate verb; in (4b) and (4d) the NP was a bare NP (bNP), which might be interpreted as a modifier.

The 24 sets of four materials were sorted into four experimental packages, such that each package contained an equal number of materials in each of the four conditions in (4). Each material appeared in exactly one version in each of the experimental packages. To counterbalance the experimental materials (and dissuade subjects from assuming that all subordinate clauses read during the experiment ended with a verb) 32 control sentences were constructed, in which the NP following the first verb always belonged to the subordinate clause (e.g., *Because the waitress spat the chewing gum we demanded a refund*). 16 of these contained nNPs, and 16 contained bNPs. Finally, 46 filler sentences, of unrelated syntactic structures, were created. The 78 control and filler sentences were added to each experimental package, resulting in four experimental packages which consisted of 102 sentences each.

Materials were deliberately segmented into presentation regions in such a way that subjects would be encouraged to treat the ambiguous NP as a part of the subordinate clause (segmentation points are marked with "/" in (4)). Of critical interest are the times taken to read the first segment, which includes the ambiguous NP, and the second segment, which provides a disambiguation. Control materials were segmented analogously to experimental items, and filler materials were segmented arbitrarily.

44 of the materials (about 43%) were followed by yes/no questions referring to the content of the materials. These were used to ensure that participants read each sentence for comprehension; each participant answered 80% or more of the questions correctly (mean 92.2%), and thus all participants were included in the analysis.

Procedure Materials were presented and response times were recorded using the EXMORE experimentation package (Mitchell & Barchan, 1984) on an Acorn RISC OS computer. A five-trial practice session preceded the experiment proper, to clarify the experimental procedure. The practice and experimental sessions were identical in all respects except randomisation and content. Before both practice and experimental sessions, instructions appeared on the screen, informing participants that they would be reading sentences split into 'chunks', followed in some cases by yes-no questions.

Each trial consisted of the presentation of one material, in successive displays, followed on occasion by a related question. The trial began with the words 'Press space-bar'; this caused the first display to appear. Each successive press of the space-bar caused the display currently in view to be replaced with the next display of the material. After the final display,

one of two things could happen: if there was no question, the words 'Press space-bar' indicated the beginning of the following trial; otherwise the word 'Question' was displayed for two seconds, followed by the question itself (to which participants responded using the 'Y' and 'N' keys). Once the question had been answered, participants were instructed to press the space-bar, as usual, indicating the beginning of the subsequent trial.

For the experimental session proper, materials were randomised prior to being displayed, subject to the constraint that no more than two items which were not fillers could occur in sequence. Additionally, the first four items were always filler materials. Records were kept of the key pressed and the time to respond (in milliseconds) for each display. Subjects were informed by the computer when the experimental session had ended.

Analysis All analyses were carried out using within-subjects analysis of variance, taking into account participants (F1) and experimental items (F2) as random factors.

Results and Discussion

Figure 1 shows the mean reading times in milliseconds per character for the ambiguous (1a) and disambigiuating (1b) regions respectively. Taking the disambiguating region first, it appears from figure 1b that the matrix verb is read faster when the ambiguous region preceding it contains an intransitive verb (181.8 and 215.5 ms vs. 226.1 and 233.2 ms).[4] This impression is borne out by statistical analysis: There is a main effect of type of verb ($F1(1,23) = 5.86$, $p = .024$; $F2(1,23) = 4.96$, $p = .037$). There is also a significant effect (by participants; marginally significant by items) of type of NP ($F1(1,23) = 5.32$, $p = .030$; $F2(1,23) = 2.96$, $p = .099$). From the figure, this appears to reflect the fact that once an intransitive verb has been followed by an nNP, which could not be attached into the VP, the disambiguating information is fastest to read, although the interaction is not significant ($F1 < 1$; $F2(1,23) = 1.28$).

At first glance, these findings appear to support the view that human sentence processing is sensitive to subcategorisation information about verbs, but not to properties of NPs (this would be compatible with 'verb-driven' accounts such as that of Ford, Bresnan, & Kaplan, 1982). However, further analysis reveals that the time taken to read a disambiguating verb after an intransitive verb + nNP is faster than for any other condition, and crucially, faster than for an intransitive verb + bNP ($F1(1,23) = 4.48$, $p = .045$; $F2(1,23) = 5.11$, $p = .034$). We take this to support the view that parsing *is* sensitive to properties of NPs as well as of verbs: a revision is more likely to be necessary if a bNP follows an intransitive

---

[3]Clearly, whether each material is *actually* ambiguous depends on whether subcategorisation information from the first verb (*sneezed* or *scratched* in (4)) is initially available to the sentence comprehension processes.

[4]The fact that per-character reading times are higher in this region than for the longer ambiguous region may be accounted for by 'overspill' from a long to a short region (cf. Mitchell & Green, 1978).

128

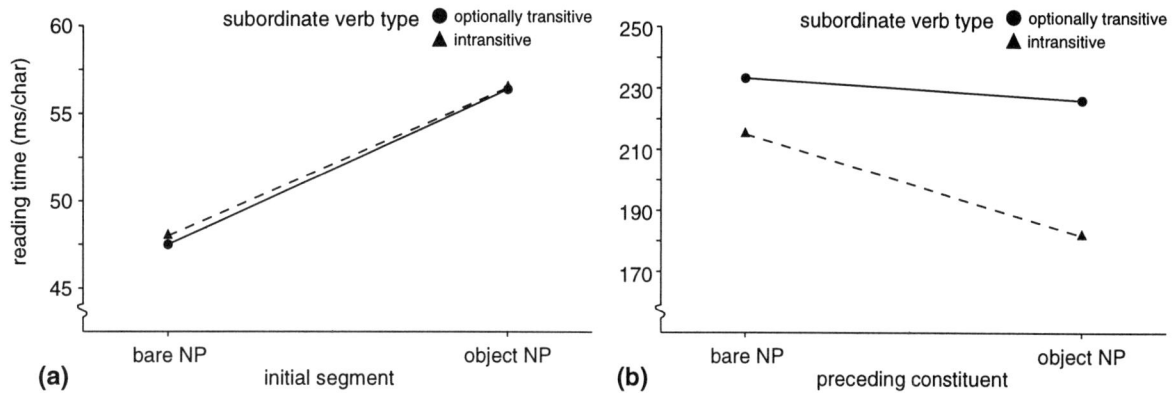

Figure 1: Mean reading times (in ms/char) for: (a) ambiguous region, including nNP or bNP; and (b) disambiguating region, consisting of matrix verb.

verb because that bNP can be incorporated into the current phrase marker as a modifier.

Figure 1a shows the reading times for the ambiguous region, containing a subordinate verb followed by either a bNP or nNP. As is clearly visible from the figure, the time to read bNPs (mean = 47.79ms) is lower than that for nNPs (56.50ms). This difference is statistically significant, in that there is a clear main effect of NP type (F1(1,23) = 20.35, p < .001; F2(1,23) = 18.30; p < .001). There are no other significant effects in the ambiguous region (all F's < 1).

The finding that bare NPs are read *faster* than nNPs is difficult to account for within a constraint-based framework, since, at least following optionally transitive verbs, nNPs should be significantly more frequently encountered than bNPs.[5] Equally, this finding is problematic for 'syntax-first' accounts which suggest that arguments should be preferred over modifiers. However, experiment 1 has the potential confound that the head nouns of the bNPs chosen are significantly more frequent than those of the nNPs (mean head noun frequency for bNPs: 583; for nNPs: 60. Figures from Kučera & Francis, 1967).

Experiment 1 has clearly demonstrated that detailed linguistic information is available during the early stages of sentence processing: by modifying Mitchell's (1987) experiment to include bare NPs we have provided the basis of an account of the differences between his findings and those of Adams et al. (1998). In experiment 2, we aim to remove the confounds from experiment 1 and explore the reading of bNPs in more detail, since confirmation that bNPs are read faster than nNPs, contrary to what would be predicted both by constraint-based and argument-favouring accounts, would have serious consequences for models of sentence processing.

Experiment 2

Experiment 2 was designed to investigate two aspects of the processing of bNPs. As well as attempting to confirm that they were read faster than nNPs, we were interested in whether they were attached into VPs in a similar way to nNPs. One potential account of bNPs (derived from Fra-

zier & Clifton, 1996) would be that, as modifiers, they are 'construed' as being associated with a VP rather than being explicitly connected to the current phrase marker. If the processes by which they are attached to VPs differ from those for nNPs, it might be possible to construct an account in which reading times reflect less 'structural work'. To investigate this possibility, we omitted intransitive verbs from experiment 2 and instead included an explicit control condition in which the ambiguous NPs remained associated with the subordinate clause. We predicted (a) that bNPs would be read faster than nNPs, and (b) that there would be a cost in encountering a matrix verb disambiguation, relative to a control, for both types of NP, reflecting the fact that both types of NP are similarly attached into the VP.

Participants A further 12 male and 12 female undergraduates volunteered to take part in the experiment (other details as for experiment 1).

Materials 24 new materials, each having four versions, were created for experiment 2 (see (5)). Materials like (5a) and (5b) were analogous to (4a) and (4b) from experiment 1; (5c) and (5d) were explicitly disambiguated (at the word *the*) in favour of an interpretation where the ambiguous NP remained associated with the subordinate clause.

(5a)   After the dogs / scratched *the whole home* / **was** / ruined / and the family were upset.

(5b)   After the dogs / scratched *the whole day* / **was** / ruined / and the family were upset.

(5c)   After the dogs / scratched *the whole home* / **the** / atmosphere / was ruined and the family were upset.

(5d)   After the dogs / scratched *the whole day* / **the** / atmosphere / was ruined and the family were upset.

---

[5]A random sample of 176 sentences containing the structure 'V+NP' from the British National Corpus yielded only 6 examples judged by the authors to contain bare NPs.

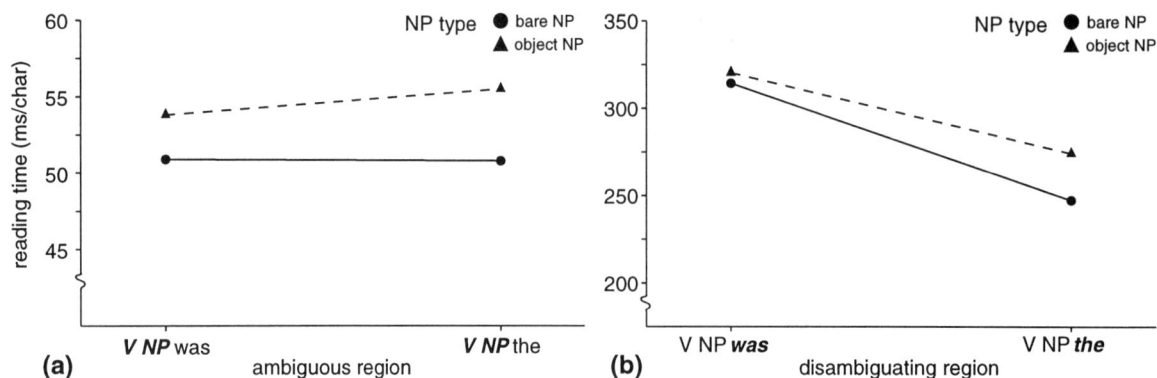

Figure 2: Mean reading times (in ms/char) for: (a) ambiguous region, including nNP or bNP; and (b) disambiguating region, consisting of *was* or *the*.

The head nouns of the nNPs and bNPs were carefully matched for frequency (using figures from Kučera & Francis, 1967)[6] and length (to ± 1 character); other words in the ambiguous NPs were identical in all conditions. The disambiguating region always consisted of the single word *was* or *the*. A further four variants were added to those in (5); these had explicit disambiguating commas before or after the ambiguous NP, as appropriate (for reasons of space, reading times for these sentences are not considered below, as they do not affect the general pattern of results).

The 24 sets of 8 materials were sorted into 8 experimental packages, in a similar way to that for experiment 1. Since each experimental package contained equal numbers of disambiguations favouring attachment of the ambiguous NP to the matrix and to the subordinate clause, additional control items were not used in this experiment, but 43 of the filler items from experiment one were added to each package file, so that each consisted of 77 sentences.

Segmentation points in the experimental materials are indicated with "/" in (5). The shorter ambiguous segment, consisting of just the verb and following NP, was chosen to reduce the variance between items. 24 Materials (about 31%) were followed by yes/no questions; one participant scored less than 80% correct and was replaced. Mean question-answering accuracy for included participants was 90.7%.

Pretest Twenty Open University students volunteered to rate the experimental materials for plausibility and acceptability. Each volunteer was given a pamphlet in which the subordinate verb + NP sections of (5) (e.g., *The dog struggled the whole day*) were followed by two seven-point scales, one for (semantic) "plausibility", and the other for (syntactic) "acceptability". The questionnaire was administered in two versions, each of which had half nNP and half bNP versions of the materials, randomised together with 24 filler materials.

Two of the 24 materials were given mean plausibilities of less than 2, and were discarded from all analyses. Mean ratings for the remaining 22 materials were 4.86 and 4.80 (plausibility and acceptability respectively) for bNPs, and 5.42 and 5.74 for nNPs. As nNPs were rated more highly than

bNPs, any processing advantage for bNPs due to plausibility/acceptability can be ruled out.

Procedure and Analysis The procedure was identical to that used for experiment 1. As for experiment 1, within-subjects analyses of variance were used to analyse the data.

Results and Discussion

Figure 2 shows the mean reading times for the ambiguous (2a) and disambiguating (2b) regions respectively. Taking the disambiguating region first, faster reading appears to occur when the ambiguous NP must be part of the subordinate clause than when it is forced to be the subject of the matrix clause; different NP types do not appear to have different effects. Statistical analyses confirm this interpretation: there is a main effect of disambiguation type ($F1(1,23) = 8.74$, $p = .004$; $F2(1,21) = 4.93$, $p = .038$); no other effects are significant (all F's ≤ 1.20). The findings for the disambiguating region confirm that there is a cost of revision for both bNPs and nNPs where they must be interpreted as the subject of the matrix clause: there is no evidence that bNPs are "less strongly" attached to the subordinate VP than are nNPs.

Turning to the ambiguous region, there appears to be a small advantage for bNPs over nNPs, confirming the results of experiment 1. This advantage, however, is only marginally significant by subjects ($F1(1,23) = 3.59$, $p = .071$) and is not significant by items ($F2(1,21) = 2.15$, $p = .158$). No other effects are significant (all F's < 1). Although it would be an over-interpretation to claim that bNPs are read more quickly than nNPs on the basis of this evidence, it is apparent that there is *no* evidence that they are read more slowly, despite being frequency-matched to, and less plausible than, their nNP counterparts. This finding remains difficult to account for within either a constraint-based or an argument-favouring framework.

General Discussion

Experiment 1 provides evidence which appears to contradict Mitchell's (1987) findings. The fact that an object NP can

---

[6]Mean frequency = 288 (bNP), 277 (nNP), F < 1.

be easily reinterpreted as the subject of the matrix clause in sentences like (4) (as witnessed by the fact that the disambiguating matrix verb is read faster for (4c) than for any other condition), coupled with the fact that there is no difference in the time taken to read nNPs following transitive or intransitive subordinate verbs, strongly suggests that detailed linguistic information *is* available initially to comprehension processes.

Experiment 1 also showed that bare NP adverbs are read more quickly than nNPs, following any kind of verb. However, the bNPs used had more frequent head nouns than their nNP counterparts, and may also have been more plausible, either of which would provide the basis for an explanation of this finding. Alternatively, it may be the case that bare NPs are not explicitly attached to VPs, perhaps resulting in a lessened cognitive load as a result of not having to alter the current phrase marker. Experiment 2 controlled for frequency and plausibility, and attempted to demonstrate that bNPs were at least as difficult to *de*tach from their host VPs as were nNPs. This latter hypothesis was confirmed by the differences in reading times for *was* and *the* in examples like (5). If there *is* a differential cost of attachment, the end results appear to be equally difficult to undo. In the absence of further evidence we choose parsimony in assuming that the attachment processes for different types of NP do not differ substantially.

Whereas experiment 1 showed a clear advantage for bare NPs, in experiment 2 the times taken to read the two types of NP did not differ. Note, however, that both constraint satisfaction and argument-favouring syntax first accounts would appear to predict that nNPs should be read more quickly than bNPs (either because it is more frequently the case that NPs following verbs are object NPs, or because attachment as an argument is a preferred strategy).[7] Constraint based accounts might further predict that (other things being equal) there should be an interaction between verb type and NP type, since NPs following intransitive verbs are much more likely to be bNPs.

One possible interpretation of these results comes from recent suggestions of a *rational analysis* of parsing (Chater, Crocker, & Pickering, 1998). Chater et al. stress the need to take into account the information gained from making a parsing decision, as well as the cost of making a revision and the probability of making a recovery. Choosing to interpret incoming NPs as bare NPs *regardless* of the type of verb that precedes them may have advantages, since (given a set of 40 or so potential head nouns for bNPs; Larson, 1985) the correctness of the parse can be quickly assessed (compare this with the situation for nNP objects, which have a far greater number of potential head nouns). Being able to quickly assess and reject a particular analysis may confer advantages on the processing mechanism, compared to pursuing a strategy where a potentially wrong analysis may be entertained for longer than necessary and may subsequently be difficult to revise. Whether this provides a useful account of the ease with which bare NP adverbs are read remains a question for future research; what seems clear from the evidence provided in this paper is that current accounts of sentence processing

have difficulties in accounting for cases where modifiers do not suffer at the expense of arguments.

References

Adams, B. C., Clifton, C., & Mitchell, D. C. (1998). Lexical guidance in sentence processing? *Psychonomic Bulletin & Review, 5*, 265–270.

Chater, N., Crocker, M. W., & Pickering, M. J. (1998). The rational analysis of inquiry: The case of parsing. In M. Oaksford & N. Chater (Eds.), *Rational models of cognition*. Oxford, UK: Oxford University Press.

Crocker, M. W. (1992). *A logical model of competence and performance in the human sentence processor.* Unpublished doctoral dissertation, Department of Cognitive Science, University of Edinburgh. (Available as research paper HCRC/RP–34)

Ford, M., Bresnan, J., & Kaplan, R. N. (1982). A competence based theory of syntactic closure. In J. Bresnan (Ed.), *The mental representation of grammatical relations*. Cambridge, MA: MIT Press.

Frazier, L. (1979). *On comprehending sentences: Syntactic parsing strategies.* Unpublished doctoral dissertation, University of Connecticut. (Indiana University Linguistics Club)

Frazier, L. (1995). Constraint satisfaction as a theory of sentence processing. *Journal of Psycholinguistic Research, 24*, 437–468.

Frazier, L., & Clifton, C. (1996). *Construal.* Cambridge, MA: MIT Press.

Kučera, H., & Francis, W. N. (1967). *Computational analysis of present-day American English.* Providence, RI: Brown University Press.

Larson, R. K. (1985). Bare-NP adverbs. *Linguistic Inquiry, 16*, 595–621.

MacDonald, M. C., Pearlmutter, N. J., & Seidenberg, M. S. (1994). The lexical nature of syntactic ambiguity resolution. *Psychological Review, 10*, 676–703.

Mitchell, D. C. (1987). Lexical guidance in human parsing: Locus and processing characteristics. In M. Coltheart (Ed.), *Attention and performance XII*. Hillsdale, NJ: Erlbaum.

Mitchell, D. C., & Barchan, J. (1984). *EXMORE: Exeter module for on-line reading experiments.* Duplicated user's manual. University of Exeter.

Mitchell, D. C., & Green, D. W. (1978). The effects of context and content on immediate processing in reading. *Quarterly Journal of Experimental Psychology, 30*, 609–636.

Tanenhaus, M. K., & Trueswell, J. C. (1995). Sentence comprehension. In J. Miller & P. Eimas (Eds.), *Speech, language and communication* Vol. 11, 2nd ed.. San Diego, CA: Academic Press.

---

[7]It is also difficult to argue that bNPs can constitute a purely syntactic category, since the same sequence of words can serve as an nNP or a bNP (and might be differentiated by, e.g., prosody).

131

Similarity & Structural Alignment:
You Can Have One Without the Other

Jodi Davenport & Mark T. Keane

Department of Computer Science,
University College Dublin, NUI Dublin,
Belfield, Dublin 4, IRELAND
`jodi.davenport@ucd.ie`
`mark.keane@ucd.ie`

Abstract

Several studies have shown that similarity judgements involve a process of structural alignment akin to analogical mapping. In particular, it has been shown that people appear to rely more on the relational structure of scenes involving cross-mappings, if they have previously carried out a similarity judgement task on these scenes (e.g., Markman & Gentner, 1993b). We report a study which shows that similarity judgements do not necessarily invoke structural alignment but that other task demands and the materials presented are more critical in selecting the comparison mechanism used in a given situation. The wider implications of these results for models of similarity and comparison are considered.

Introduction

A considerable body of recent research has shown that similarity comparisons can involve a process of structural alignment (see e.g., Goldstone, 1994; Goldstone & Medin, 1994; Goldstone, Medin & Gentner, 1991; Markman & Gentner 1993a, 1993b, 1997; Medin, Goldstone & Gentner, 1993). Representationally, this view characterises knowledge as structured hierarchies encoding objects, object attributes, relations between objects and relations between relations. Given these representations it is assumed that similarity comparisons involve the alignment of relational structure to find the most structurally consistent match between two systems of concepts, that satisfies the constraints of parallel connectivity (if two relations match, their arguments must match) and one-to-one mapping (that each item in one structure may only be mapped to one other item). Computationally, these ideas have been realised by a family of models that simulate analogical mapping (see e.g., Falkenhainer, Forbus & Gentner, 1989; Gentner, 1983, 1989; Holyoak & Thagard, 1989; Hummel & Holyoak, 1998; Keane, 1988, 1997; Keane & Brayshaw, 1988; Keane, Ledgeway & Duff, 1994; Veale & Keane, 1994, 1997, 1998). Indeed, structural alignment is fast emerging as a unified account of a diverse range of phenomena including similarity, analogy, metaphor and concept combination (see Keane & Costello, 1998).

Markman & Gentner (1993b) provided one of the key pieces of evidence supporting the role of structural alignment in similarity judgements. They used a one-shot mapping task in which subjects had to identify a cross-mapped object between two drawn scenes (see Appendix A). A cross-mapped object was defined as an object in one drawing that was perceptually similar to an object in a different relational role in the other drawing. So, for example, in the baseball scenes shown in Appendix A, the cross-mapped object would be the pitcher with the "C" on his uniform, because he is pitching in the upper scene and being pitched to in the other scene. Markman & Gentner have proposed that structural alignment is reflected in this task when subjects make relational responses (i.e., choosing the object in the same role) as opposed to object responses based on perceptual, feature similarity (i.e., choosing the perceptually similar object in a different role). The key manipulation asked participants to perform a similarity judgement task on the picture-pairs either before or after the mapping task. They found that when participants made the similarity judgement *before* the mapping task they made more relational responses than when it was presented *after* the mapping task. Thus, the result strongly suggested that the similarity judgement task invoked a structural alignment process which then carried over to the mapping task increasing the proportion of relational responses (significantly, when an aesthetic-appreciation task was given before the mapping task no facilitation in relational responding was found). However, we believe that this conclusion is unwarranted given the nature of the materials used and task demands. We argue that the similarity judgement task does not necessarily invoke structural alignment.

First, Markman & Gentner's materials may have contained unintended cues that promoted relational responding in participants. While the pictures used were designed to be understood without introducing linguistic factors, we believe that to understand the scenes subjects has to consider word-labels in the drawings. For example, in the feeding-pair, to understand that the woman is *receiving* food rather than *giving* food, one needs to use the written dialogue of the woman saying "Thank You" (see Appendix A). This dependence on linguistic factors in the picture may have promoted relational responding over a more perceptually-based response. More seriously, in some of their materials, the critical relations underlying the relational response are named in the picture (see e.g., the baseball pair in which "Pitch" is written) thereby drawing attention to responses using this relation.

Second, the task demands governing the way in which the pictures were presented may also have promoted relational responding. Participants were given 8 stimuli all of which portrayed picture pairs with relational similarities. So, independent of any effects of the similarity judgement task, participants may just have "guessed the game" and responded appropriately: that is, during the course of the similarity task subjects may have decided from the predominantly relational nature of the pictures that they were meant to map on a relational basis. Additionally, the fact that subjects were asked to make comparisons on a series of pictures of approximately equivalent similarity may also have been an added extraneous variable.

To remedy these possible deficiencies we constructed a set of materials that involved cross-mappings as defined by Markman & Gentner, but omitted any linguistic cues and names of key relations. We then expanded the set of presented materials to include 16 fillers that lacked relational similarities to balance the 8 pairs designed to have relational similarities (akin to Markman & Gentner's set). As in Markman & Gentner's study, participants either performed the similarity judgement task before or after the mapping task. The new variable introduced was the presentation order of the materials (relational-first versus relational-distributed). In the *relational-first* condition the 8 relational materials were blocked at the beginning of the booklet followed by the fillers (akin to the way in which Markman & Gentner's participants would have encountered them). In the *relational-distributed* condition, the 8 relational materials were randomly distributed among the fillers.

As such, we had a 2 x 2 between-subject design where the variables were task-order (similarity task before or after mapping task) and stimulus-order (relational materials blocked at beginning of stimuli set or randomly distributed throughout the set). As the relational stimuli meet the constraints set down by Markman & Gentner, they would predict that task-order should affect relational responding, with more relational responses being made when the similarity-judgement task is before as opposed to after the mapping task. Taking the opposing view, we do not believe that similarity judgements necessarily induce a structural alignment process and would argue that the nature of the material set is more important, that structural alignment is used when stimulus conditions appear to

require it. Hence, we predict that stimulus-order should have the dominant effect on relational responding, with more relational responses being induced when the relational materials are blocked to the front of the stimulus set as opposed to distributed throughout the set.

Method

Subjects. Forty-eight undergraduate students and staff members at University College Dublin took part voluntarily in the experiment and were randomly assigned to one of the four between-subjects conditions.

Stimuli. The stimuli for this experiment consisted of 8 pairs of pictures depicting causal scenes with matching relational structure and 16 filler pairs. Each of these 8 pairs contained a cross-mapping as operationalized by Markman and Gentner (1993b) in which a pair of perceptually-similar objects were shown which played different roles in the matching relational structure of the two scenes (see Appendix B for an example). The pictures were designed so that in half of the pairs the perceptually-similar objects were in approximately the same spatial position while in the other half the objects in the same relational roles were in the same position. Additionally, half of the pairs had relations moving in the same direction, while the other half had relations moving in opposite directions.

Eight of the filler pairs depicted comparable scenarios without matching relational structure (e.g. two beach scenes, one with a man surfing another with a child is building a sand castle) and the other 8 pairs did not match in either scenario or relational structure (e.g. a scene of an artist and a scene of a man in a grocery store).

The stimuli were presented in booklet form with one pair on each page (one picture above the other). The stimuli for the mapping task had an arrow placed above an object in the top scene. For the 8 relational pairs this was the cross-mapped object, otherwise it was an object which appeared in both scenes. The stimuli used for the similarity rating task had a scale with the numbers 1 through 9 at the bottom of the page. The words Low Similarity appeared under the 1 and the words High Similarity appeared under the 9.

Booklets in the relational-distributed condition had a completely randomised presentation of the 24 pairs for both the mapping and the similarity tasks. Booklets in the relational-first condition had a randomised block of the 8 matching relational pairs to the front of the booklet followed by the filler pairs.

Procedure. As in Gentner & Markman's study, the first page of the mapping section of the booklet instructed subjects to draw a line from the object under the arrow to the object in the bottom scene that "best went with that object". The first page of the similarity judgement section instructed subjects to rate the similarity of the two scenes by circling a number on the scale at the bottom of the page.

Subjects in the similarity-after conditions received a booklet with the mapping task followed by the similarity judgement task while subjects in the similarity-before condition received a booklet with the similarity judgement task first.

Subjects were tested in small groups of varying sizes and each experimental session took between 10 and 15 minutes.

Scoring. As in Gentner & Markman's study, participants' responses to the 8 relational pairs in the mapping task were determined as an *object mapping* if a line was drawn from the cross-mapped object to the featurally-similar object in the bottom scene; a *relational mapping*, if a line was drawn to the object in the same relational role in the bottom scene; or a *spurious mapping* if a line was drawn to another, unrelated object. As spurious mappings occurred in less than 1% of the responses, they were not considered.

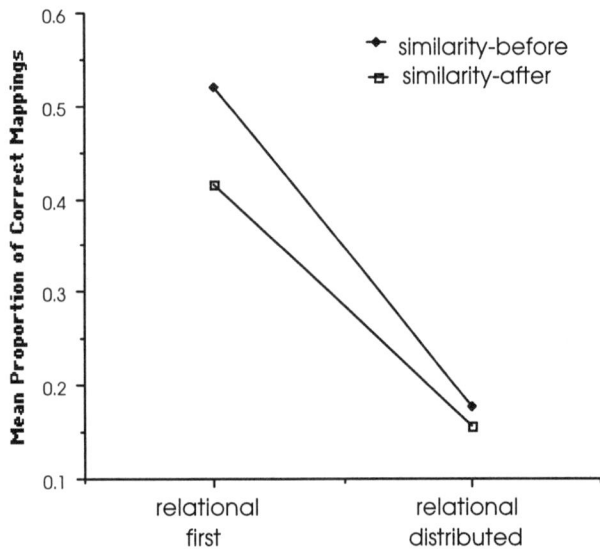

Figure 1. The Proportion of Relational Responses Made

Results & Discussion

A two-way, between-subjects ANOVA found a reliable effect of stimulus-order with a higher proportion of relational responses in the relational-first conditions ($M = 0.52$ and $M=.42$) than in the relational-distributed conditions ($M = 0.18$ and $M = .16$), $F(1,44) = 10.51$, $p<.01$, $MSError = 0.104$ (see Figure 1). This effect demonstrates that relational responding increases when the relational pictures are presented in a block before the fillers, and decreases markedly when the relational pictures are distributed among filler materials. Contrary to Markman and Gentner's predictions, no reliable effect was found for the task-order variable, $F(1,44) = .450$, $p >.10$. Finally there was no reliable interaction between task-order and materials-order, $F(1,44) = .200$, $p >.10$.

General Discussion

Markman & Gentner suggested that similarity judgements invoked structural alignment and that this promoted relational responding in a subsequent one-shot, cross-mapping task; that in this task you can't have a similarity without structural alignment. We have found that when the possible extraneous influences on the materials are ruled out and they are presented in the context of fillers, markedly different results are found. First, we found that the similarity-task produced had no sole effect or interaction effects on relational responses. So, following the logic of Markman & Gentner's study similarity must not necessarily involve structural alignment. Second, the marked increase in relational responses in the relational-first condition demonstrates that the way the materials are presented is more important in determining whether structural alignment is used or not. This suggests a more contingent interaction between people and the task situation which is not well captured by current models of structural alignment and analogy. It also shifts the focus of research in this field to the issue of the "calling conditions" for the use of one similarity mechanism rather than another.

There is one remaining mystery about the relationship of these results to those of Markman & Gentner; namely, why was it that a preceding similarity judgement task increased relational responding when an aesthetic-appreciation task with the same materials did not ? We would argue that it was not the structural alignment *per se* that produced this effect, but rather that the materials drew attention to the fact that every picture was similar to its partner in sharing the same relation. The similarity judgement task helped in "guessing the game", the aesthetic-appreciation task did not. Thus when asked to perform the relatively ambiguous mapping task, subjects responded accordingly.

The current research reveals a lack of featural or relational dominance in similarity judgements, suggesting that similarity is a pluralistic process, that is very sensitive to the conditions at hand (as has been partly argued by Medin, Goldstone & Gentner, 1993). Similarity does not *necessarily* invoke an analogical structural-alignment mechanism, but involves dynamic switching between a structural alignment and a feature-comparison mechanism. More broadly, as similarity does not appear to be governed by a process of structural alignment in every instance, a unified account of comparison in general (e.g., similarity, analogy, metaphor and concept combination) as being based on structural alignment becomes less tenable (see Keane & Costello, 1998; Costello & Keane, in press).

References

Costello, F. & Keane, M.T. (in press). Efficient creativity: Constraints on conceptual combination.*Cognitive Science*.

Falkenhainer, B., Forbus, K.D., & Gentner, D. (1989). Structure-mapping engine. *Artificial Intelligence, 41,* 1-63.

Gentner, D. (1983). Structure-mapping: A theoretical framework for analogy. *Cognitive Science, 7* , 155-170.

Gentner, D. (1989). Mechanisms of analogical learning. In S. Vosniadou & A. Ortony (Eds.), *Similarity and analogical reasoning.* Cambridge: CUP.

Goldstone, R.L. (1994). Similarity, interactive activation, and mapping. *Journal of Experimental Psychology: Memory and Cognition, 20,* 3-28.

Goldstone, R.L. & Medin, D.L. (1994). Time course of comparison. *Journal of Experimental Psychology: Language, Memory & Cognition, 20,* 29-50.

Goldstone, R.L., Medin, D.L., & Gentner, D. (1991). Relational similarity and the non-independence of features in similarity judgments. *Cognitive Psychology, 23,* 222-262.

Holyoak, K.J., & Thagard, P. (1989). Analogical mapping by constraint satisfaction. *Cognitive Science, 13,* 295-355.

Hummel, J.E. & Holyoak, K.J. (1998). Distributed representations of structure: A theory of analogical access and mapping. *Psychological Review. 105, xx-xx.*

Keane, M.T. (1988). *Analogical Problem Solving.* Chichester, England: Ellis Horwood (Cognitive Science Series). [Simon & Schuster in N. America]

Keane, M.T. (1997). What makes an analogy difficult ?: The effects of order and causal structure in analogical mapping. *Journal of Experimental Psychology: Language, Memory & Cognition, 23,* 946-967.

Keane, M.T. & Costello, F. (1998). Why Conceptual Combination is Seldom Analogy. In K.J. Holyoak, D. Gentner, & B. Kokinov (Eds.), *Proceedings of Analogy '98.* New University of Bulgaria Press: Bulgaria.

Keane, M.T., & Brayshaw, M. (1988). The Incremental Analogical Machine: A computational model of analogy. In D. Sleeman (Ed.), *European Working Session on Machine Learning.* London: Pitman.

Keane, M.T., Ledgeway, T. & Duff, S. (1994). Constraints on analogical mapping: A comparison of three models. *Cognitive Science, 18,* 387-438.

Markman, A.B. & Gentner, D. (1993a). Splitting the differences: A structural alignment view of similarity. *Journal of Memory and Language, 32,* 517-535.

Markman, A.B. & Gentner, D. (1993b). Structural alignment during similarity comparisons. *Cognitive Psychology, 25,* 431-467.

Markman, A.B. & Gentner, D. (1997). The effects of alignability on memory. *Psychological Science, 5,* 363-367.

Medin, D.L., Goldstone, R.L., & Gentner, D. (1993). Respects for similarity. *Psychological Review, 100,* 254-278.

Veale, T. & Keane, M.T. (1994). Belief modelling, intentionality and perlocution in metaphor comprehension. *Proceedings of the Sixteenth Annual Meeting of the Cognitive Science Society.* Hillsdale, NJ: Erlbaum.

Veale, T. & Keane, M.T. (1997). The competence of sub-optimal structure mapping on 'hard' analogies. *IJCAI'97: The 15th International Joint Conference on Artificial Intelligence.* Morgan Kaufmann.

Veale, T. & Keane, M.T. (1998). Principle Differences in Structure Mapping. In K.J. Holyoak, D. Gentner, & B. Kokinov (Eds.), *Proceedings of Analogy '98.* New University of Bulgaria Press: Bulgaria.

Appendix A. Two of the Materials. -- the Baseball and Giving Materials -- Used by Markman & Gentner (1993b)

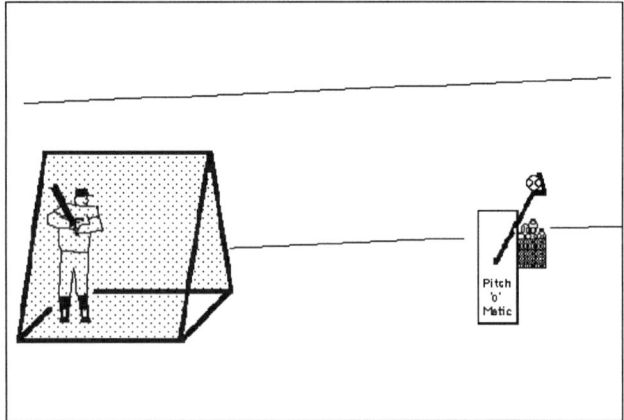 

Appendix B. An example of the materials used in the present study showing (a) a relational pair, (b) a filler with featural rather than relational overlap and (c) a filler with little featural or relational overlap.

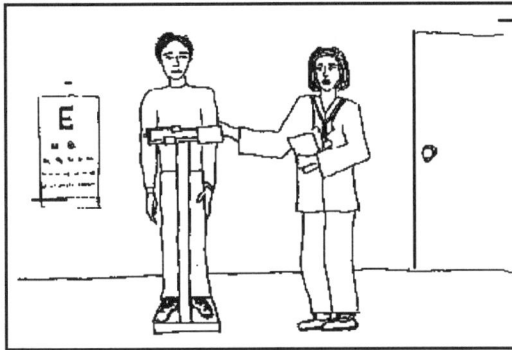

a.   b.

c.

Recognition of Exceptions and Rule-Consistent Items in the Function Learning Domain

Edward L. DeLosh (delosh@lamar.colostate.edu)
Department of Psychology, Colorado State University
Fort Collins, CO 80523 USA

Abstract

Recent studies suggest that participants commonly abstract rules when learning concepts, but a remaining question is whether they retain and apply knowledge of individual instances subsequent to rule abstraction. Research in the category learning domain indicates that exemplar information is retained and that exceptions to a category rule have special status in memory (Palmeri & Nosofsky, 1995). The present experiment examines whether these findings extend to function learning. Participants learned associations between stimulus and response magnitudes that were related according to a negative linear function. Twelve stimulus-response pairs were given, some consistent with the negative linear rule, others exceptions to the rule. After each of six training sessions, previously studied stimulus magnitudes were presented as tests of learning accuracy. Participants were also given extrapolation trials followed by a final recognition test that included old and new rule-congruent and rule-incongruent items. Extrapolation was extensive. In addition, analyses revealed poorer learning and recognition for exceptions than for rule-congruent items, plus a high rate of false alarms for new rule-congruent items. These findings suggest that although the conceptual knowledge acquired in function learning tasks centers on rules, exceptions to these rules do not have special status in memory.

Introduction

In the spirit of classic hypothesis-testing models of classification learning (e.g., Bower & Trabasso, 1963; Levine, 1975; Restle, 1962), contemporary theories have revitalized the idea that conceptual behavior is based on the abstraction and application of rules. Recent rule-based models developed by Nosofsky, Palmeri, and McKinley (1994) and DeLosh, Busemeyer, and McDaniel (1997) have been successful in accounting for a variety of data in the category and function learning domains, respectively. Both models propose that conceptual behavior reflects the joint influence of exemplar and rule-based processes. This emerging theoretical approach begs the following empirical questions: In what way do rules and exemplars jointly contribute to conceptual behavior? Are individual instances learned and retrieved subsequent to rule abstraction? Is conceptual behavior characterized by individual differences in the use of rules versus exemplars? The present experiment considers these issues as they apply to the function learning domain.

Rule Abstraction in Function Learning

Functions are abstract concepts that characterize the relationship between two causal variables. A function maps a set of input values on a stimulus continuum into a set of output values on a response continuum such that each input value is assigned only one output value. In a typical function learning task, input and output dimensions are related according to a simple mathematical function. Learning occurs on a trial-by-trial basis through experience with individual input-output pairs. In DeLosh et al. (1997), for example, participants learned associations between drug dosages and the magnitude of clinical effect caused by those dosages. The dosage-effect relationship was either linear, exponential, or quadratic. On each learning trial, a drug dosage was represented on a computer monitor as a bar length. Participants then predicted the magnitude of effect for that dosage by changing the length of a second bar. Then they were shown the "correct" magnitude of effect (represented by a third bar) as defined by the objective function. Numerous trials of this type were given, such that each of many dosage-effect pairs was presented several times.

Learning in this type of task potentially involves memory for specific input-output pairs, abstraction of relational information pertaining to the input and output dimensions, or some combination of these processes. DeLosh et al. (1997) examined these possibilities by presenting a series of extrapolation tests. Participants were given new dosage values outside the range of those given during learning. They responded to these extrapolation stimuli by generating outputs beyond the range of learned responses, and did so in a manner consistent with the form of the assigned function. A pure exemplar-based model of function learning (i.e., an extension of ALCOVE; cf. Kruschke, 1992) was unable to extrapolate to the extent observed with participants, revealing the necessity for rule learning (i.e., the abstraction of relational information during acquisition) or rule-based responding (i.e., the abstraction of relational information during retrieval) instead of or in addition to exemplar learning.

In a second line of research, DeLosh (1994) observed discontinuous patterns of responding during function learning, and these discontinuities were similar for a condition with explicit hypothesis-testing instructions and a condition with standard free-strategy instructions. This observation lends support to the idea that function learning involves the systematic sampling and testing of global input-output rules. It appears, then, that rule abstraction plays a central role in the learning and application of function-based concepts. But to what extent is exemplar information retained and used subsequent to rule abstraction?

RULEX Model of Category Learning

This issue has recently been examined as it applies to category learning. Nosofsky et al. (1994) proposed a rule-based model of classification (RULEX) in which participants abstract simple logical rules based on single dimensions or conjunctions of dimensions, supplemented by memory for exceptions to those rules. Because members of ill-defined categories can not be classified based solely on the application of logical rules, information pertaining to exceptions is central to the success of the model. The model therefore assumes that there is residual memory for old exemplars and that old exceptions have special status in memory. Consistent with this assumption, Palmeri and Nosofsky (1994) observed intact memory for old exemplars, and better recognition memory for old exceptions than for old rule-congruent items.

Note, however, that there are several differences between the category learning tasks examined in the above studies and the function learning task considered in the present experiment. In *category learning*, responses consist of discrete and nominal categories that do not have any numerical status. Rules learned in a category learning task are logical rules for mapping stimuli onto arbitrary response categories (e.g., red and square stimuli belong to Category A). In *function learning*, responses lie on a continuum and are numerically related to one another and to stimuli. Therefore the rules abstracted in a function learning task may reflect the numerical relationship between stimuli and responses (e.g., drug dosage is positively correlated with heart rate).

Despite these differences, a plausible explanation of function learning is that participants abstract a functional rule and memorize exceptions to that rule, comparable to the processing assumptions of RULEX. One might ask, then, do the findings of Palmeri and Nosofsky (1994) generalize to function learning? Is exemplar information retained subsequent to rule abstraction? Do exceptions to function-based rules have special status in memory?

Overview of the Current Experiment

The current experiment examines these questions by extending on the method used by DeLosh et al. (1997). Participants learned associations between stimulus and response magnitudes (i.e., bar heights) that were related according to a negative linear function. Twelve stimulus-response pairs were given, some consistent with the negative linear rule, others exceptions to the rule. After each of six training sessions, previously studied stimulus magnitudes were presented as tests of learning accuracy. During the final test session, participants were also given extrapolation trials to test for rule abstraction. To examine memory for individual stimulus-response pairs, participants were then given a final recognition test that consisted of old rule-congruent items, old exceptions, new rule-congruent items, and new rule-incongruent items. A random-mapping condition was included to examine performance when learning is strictly based on memory for individual input-output pairs.

Method

Participants and Apparatus

Sixty-eight Colorado State University undergraduates participated in partial fulfillment of a requirement for an introductory psychology course. Participants were tested in pairs

or groups of three in a laboratory room equipped with three computer workstations. Stimuli were presented on a 14" color monitor at a distance of approximately 60 cm and responses were collected using a standard computer keyboard placed on the desk in front of the monitor. A computer program controlled the presentation of the instructions and stimuli as well as the collection of participants' responses.

Design

The experiment included two conditions based on the mapping between stimulus and response magnitudes. In a functional-mapping condition, stimulus and response magnitudes were related according to the negative linear function $y = 200 - 1.7x$. The random-mapping condition included the same stimulus and response magnitudes used in the functional-mapping condition, but stimuli were randomly paired with responses. Mapping was manipulated between participants, with 35 participants randomly assigned to the functional-mapping condition and 33 to the random-mapping condition.

Stimuli and Responses

All stimuli and responses were presented in the form of vertical bars with the height of each bar proportional to the assigned stimulus or response magnitude. The range of possible stimulus magnitudes was 0 to 100 as indicated by an unfilled vertical bar labeled 0 to 100. The range of possible responses was 0 to 200 as indicated by an unfilled vertical bar labeled 0 to 200. Due to limited resolution of the computer monitor, response magnitudes were constrained to integer values.

A total of 12 stimulus-response pairs were given during training. For the functional-mapping condition, 10 of the stimulus-response pairs corresponded to the negative linear rule. Two were exceptions to the rule (Pairs 4 and 9; see Figure 1), with the learned response deviating from the rule-defined response by 42 units. For the random-mapping condition, the stimulus and response magnitudes used for rule-congruent items in the functional-mapping condition were randomly paired. Two random mapping sets were generated: 16 participants received Set 1 and 17 received Set 2. Pairs 4 and 9 of the random-mapping sets were identical to the exceptions used the functional-mapping condition. Figure 1 provides a graphical representation of the stimulus-response pairs used in the functional-mapping set and one of the random-mapping sets.

Note that the specific stimulus magnitudes given during training ranged from 22.5 to 77.5 on the 0 to 100 stimulus scale, with corresponding response magnitudes ranging from 68 to 162 on the 0 to 200 response scale. Values beyond this range were given after the final learning session to examine extrapolation. The following stimulus magnitudes were used on extrapolation trials: 2.5, 7.5, 12.5, 17.5, 82.5, 87.5, 92.5, and 97.5.

Procedure

At the beginning of the experiment, participants read a set of instructions on the computer monitor. In the instructions, participants were told that they would observe a hypothetical pharmacology experiment in which dosages of an

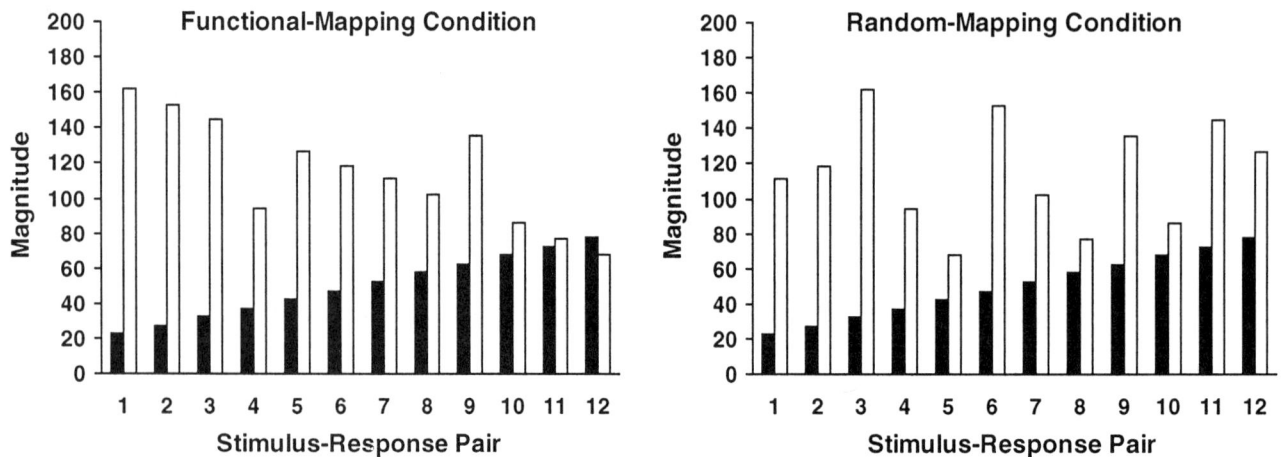

Figure 1: Graphical representation of the stimulus-response pairs given in the functional- and random-mapping conditions. Pairs 4 and 9 are exceptions.

unknown drug are given to subjects and the level of arousal produced by each dosage is measured. They were instructed to predict the level of arousal produced by each drug dosage given, and when given feedback, to remember the level of arousal associated with each dosage. They were never told to figure out the relationship between dosage and arousal levels, or that a systematic relationship might exist. With regard to the learning task itself, the instructions described the format of the presentation screen and the appropriate keys for making a prediction. Once these instructions were understood, a sample trial was given in order to familiarize the participant with the presentation screen and response procedure.

After the sample trial, participants proceeded through alternating training and test sessions, with each of six training sessions followed by a test session. During training, the stimulus magnitudes that constituted the stimulus set were presented one at a time in random order. For each trial, three unfilled vertical bars were presented simultaneously on the monitor. The leftmost bar was titled Drug Dosage and had tick marks and value labels every twenty units from 0 to 100, and the remaining two bars were titled Predicted Level of Arousal and Observed Level of Arousal, respectively, with tick marks and value labels every twenty units from 0 to 200. The relative lengths of these unfilled bars on the screen were proportional to the number of units they represented.

On a given trial, the left bar was filled in from the zero point (at the bottom of the bar) to the input value representing the amount of drug administered. Participants then used the arrow keys on their keyboard to fill in the second vertical bar from the zero point to the desired prediction value, and pressed the space bar when finished. Participants were allowed as much time as needed to make their prediction. Once the space bar was pressed, the correct level of arousal (i.e., the response value assigned to the stimulus according to the mapping condition) was shown on the rightmost vertical bar, along with an accuracy score of 0 to 100, computed as 100 minus the square of the participant's prediction

error. This correct-response feedback was displayed for 6 s. The next trial was initiated by pressing the enter key.

After participants completed a training session, they were given 12 test trials consisting of the exact stimulus values shown during training. The test stimuli were presented in random order. Individual trials proceeded in exactly the same fashion as training trials, except the rightmost bar (for presenting the correct response magnitude) was not included and no other feedback was provided. During the sixth and final test session, a sequence of eight extrapolation trials was given after the standard test trials. These extrapolation trials were given in random order, and like test trials, feedback was not provided.

The experiment concluded with a final yes-no recognition test. A series of 24 stimulus-response pairs was shown on the computer monitor. For each item participants were instructed to respond yes (press the "y" key) if they believed the pair was previously given during the experiment or no (press the "n" key) if they believed the pair was not given during the experiment. The recognition test consisted of 12 old items (each shown a total of 12 times during the alternating training and test sessions) and 12 new items. For the functional-mapping condition, these items can be grouped into four types: *old rule-congruent items* (the 10 rule-generated pairs given during training), *old rule-incongruent items* (the 2 exceptions given during training), *new rule-congruent items* (2 lures with the same stimulus values as exceptions, but paired with the appropriate rule-generated response), and *new rule-incongruent items* (10 completely new pairs inconsistent with the negative linear rule). The 24 recognition trials were presented in random order.

Results

Learning

In order to compare learning performance for exceptions versus non-exceptions, the average absolute prediction error on test trials was computed as a function of item type and test session for each participant. These averages were submitted to a 2 x 2 x 6 (Mapping x Item Type x Test Session) mixed analysis of variance (ANOVA). The rejection level was set at .05 for this and all other analyses reported in the current study. Main effects of mapping [$F(1,66) = 16.49$, $MSE = 428.41$], item type [$F(1,66) = 10.87$, $MSE = 257.86$], and test session [$F(5,330) = 15.66$, $MSE = 150.89$] were obtained. As observed in previous experiments (e.g., Carroll, 1963; DeLosh, 1996), performance was better in the functional-mapping condition than in the random-mapping condition, and improved from test session to test session. In addition, prediction accuracy was better for non-exceptions than for exceptions.

A significant interaction between mapping and item type [$F(1,66) = 70.44$, $MSE = 150.89$] was also observed. In the random-mapping condition, performance was better for exceptions (*Mean Prediction Error* = 24.85) than for randomly paired stimulus-response values (*M* = 30.58). In the functional-mapping condition, performance was better for rule-congruent items (*M* = 15.25) than for exceptions (*M* = 28.40). Therefore, the specific stimulus-response pairings used as exceptions were easier to learn than random pairings, but despite this, participants were less accurate with exceptions than with rule-congruent items in the functional-mapping condition (see Figure 2; also see Figure 3).

The analysis also yielded a significant interaction between mapping and test session [$F(5,330) = 2.36$, $MSE = 150.89$], revealing greater improvement over test sessions for the functional-mapping condition than for the random-mapping condition. None of the remaining interactions were statistically reliable (*ps* > .10).

Recognition

In order to analyze recognition performance, hit and false alarm rates were computed for each participant in the functional- and random-mapping conditions. One participant in the functional-mapping condition terminated the experiment prior to completing the recognition test, therefore the following analyses are based on the remaining 67 participants. Independent-sample *t* tests revealed a higher hit rate for items in the functional-mapping condition (*M* = .80) than for items in the random-mapping condition (*M* = .73). The false alarm rate did not significantly differ across conditions (*p* > .10; *M* = .24 and .30, respectively).

A more detailed analysis of the recognition data was then conducted for the functional-mapping condition. Hit and false alarm rates were computed for the four types of items given on the recognition test, yielding the means given in Table 1. The hit rate for old rule-congruent items significantly differed from that of old rule-incongruent items [*t*(33) = 4.29], such that recognition memory for rule-based items

Table 1: Hit and false alarm rates for each type of item in the functional-mapping condition.

Item type	Proportion of yes responses
Old rule-congruent items	.89
Old rule-incongruent items	.59
New rule-congruent items	.59
New rule-incongruent items	.17

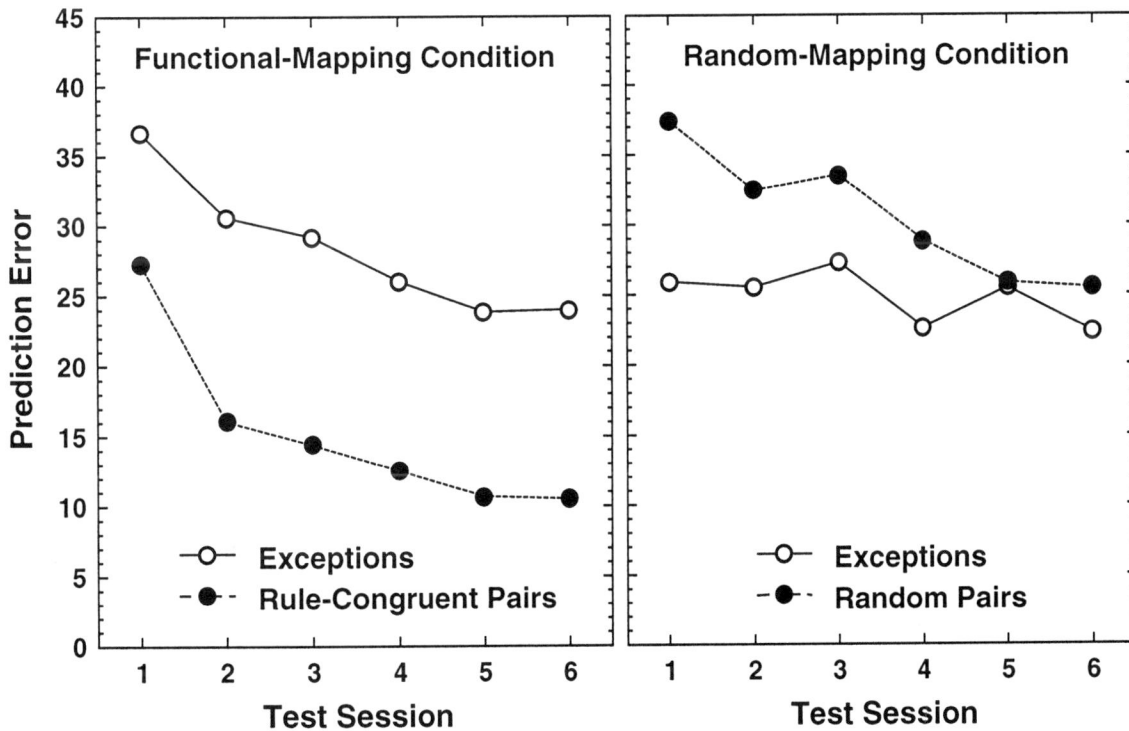

Figure 2: Accuracy of predictions across the six test sessions for exceptions and non-exceptions in the functional- and random-mapping conditions.

was better than recognition memory for exceptions.[1] In addition, the false alarm rate differed for new rule-congruent items and new rule-incongruent items, $t(33) = 7.00$. Items consistent with the negative linear rule produced a higher rate of false recognition than was observed with other new items. In fact, the false alarm rate for the rule-congruent lures did not differ from the hit rate for exceptions ($p > .10$).

Extrapolation

Performance on extrapolation trials was examined to further assess whether learning in the functional-mapping condition was based on rule abstraction. One participant did not complete the extrapolation trials, so the following data are based on the remaining 34 participants. Figure 3 shows the average of participants' predictions as compared to the objective responses across all stimulus values from the last test session. The stimulus magnitudes within the dotted lines correspond to stimuli given during learning; those outside the dotted lines are extrapolation stimuli.

The figure reveals that participants extrapolated well beyond the range of learned responses in both extrapolation regions, extending 16 and 24 units beyond learned responses in the low and high extrapolation regions, respectively. Moreover, the extrapolation responses were highly systematic, closely approximating the objective function. Participants' predictions were, in fact, closer to the objective values for extrapolation stimuli ($M = 12.19$) than for exceptions ($M = 19.49$). This extensive extrapolation replicates past findings (DeLosh, 1994, 1996) and provides strong evidence for rule abstraction (see Discussion).

It is also noteworthy that extrapolation responses in the random-mapping condition were positively correlated with stimulus magnitudes ($M = 43.85, 62.82, 83.12, 79.33, 106.42, 109.94, 114.94,$ and 117.42 for extrapolation trials 1 through 8, respectively). This suggests that participants attempted to apply a rule even in the random-mapping condition. Moreover, the particular pattern of extrapolation is consistent with findings in the function-learning literature that reveal biases toward increasing monotonic functions (Brehmer, 1974; Busemeyer et al., 1997).

Individual Differences

To determine whether the average data described above is representative of individual learners, extrapolation performance was examined for each of the 34 participants in the functional-mapping condition. Five of these participants deviated from the group data, failing to extrapolate in the two extrapolation regions. Contrary to the large advantage for rule-congruent items observed in the group data, these participants also showed: (a) similar learning performance for rule-congruent items (*Mean Prediction Error* = 20.95) and exceptions ($M = 26.57$); (b) a similar hit rate for old rule-congruent items ($M = .84$) and exceptions ($M = .80$); and (c) a similar false alarm rate for new rule-congruent ($M = .40$) and rule-incongruent items ($M = .34$).

---

[1] For comparison, there was no difference between the hit rate for exception pairs ($M = .75$) and random pairs ($M = .70$) in the random-mapping condition ($p > .10$).

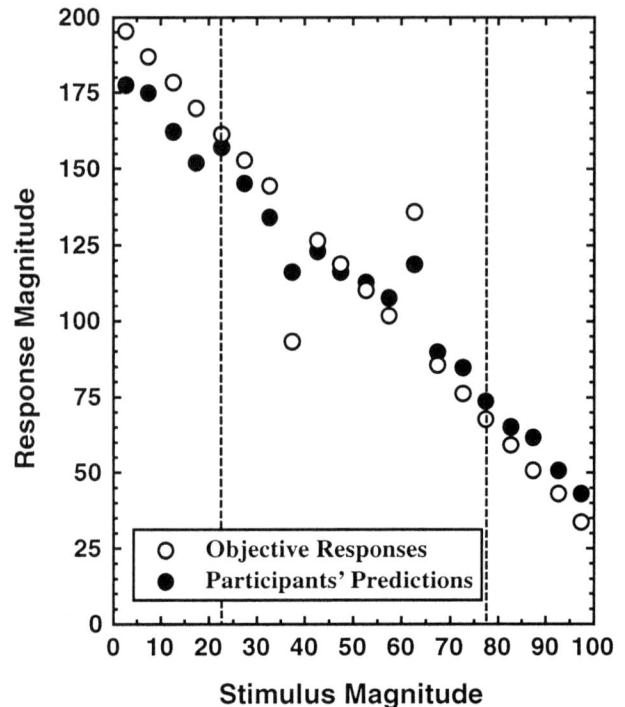

Figure 3: Participants' final test predictions for the functional-mapping condition.

Discussion

Rule Abstraction in Function Learning

The present findings support the view that participants often abstract and apply rules when learning function-based concepts, at least for simple functional relations. First consider the extrapolation results. The observed pattern of extrapolation responses approximates the objective negative linear function and therefore suggests that participants abstracted and applied information about the stimulus-response relationship. In support of this interpretation, DeLosh et al. (1997) formally tested a pure exemplar-based model of function learning (an extension of ALCOVE; see Kruschke, 1992) and showed that the model can not account for extensive extrapolation, as observed here. In order to produce extensive extrapolation, it is necessary to include a rule-based mechanism in which relational information is abstracted during the learning or retrieval of individual instances (cf. Busemeyer et al., 1997; DeLosh et al., 1997).

The present experiment also provides new corroborative support for rule abstraction. If participants only learn and remember individual stimulus-response pairs, one might not expect differences between the functional- and random-mapping conditions. There was, however, an advantage for the functional-mapping condition in both learning and recognition. Similarly, if participants rely on exemplar learning even in the functional-mapping condition, one might not expect differences between rule-congruent items and exceptions. In the random-mapping condition in which participants were required to memorize individual instances, there was no advantage for rule-congruent items. However, in the functional-mapping condition there was an advantage for rule-congruent items over exceptions in both learning and recognition. In addition, the experiment yielded a high rate of false alarms for new rule-congruent items. Participants often judged rule-congruent lures as having occurred before, and did so at a rate equivalent to that of exceptions that were shown 12 times during acquisition.

Note, however, that a few learners deviated from the group averages described above. These participants did not extrapolate beyond the range of learned responses in the two extrapolation regions. As discussed by DeLosh et al. (1997), this failure to extrapolate may reflect a strategy that centers on exemplar learning (also see DeLosh 1994, 1996). In any case, responding does not appear to be based on rules for this subset of participants. If these participants do not abstract and use rules, one would expect similar learning and recognition performance for rule-congruent and rule-incongruent items (in contrast to the large advantage for rule-congruent items observed in the group data). This is precisely what was found. It appears, then, that the large majority of participants abstract and apply rules, but a few may rely exclusively on memory for individual instances.

Memory for Instances in Function Learning

Although rule abstraction appears to play a central role in function learning for most participants, results show that these participants also have residual memory for individual instances. Within the functional-mapping condition, old rule-congruent items were more likely to be judged as having occurred before than were new rule-congruent items. Likewise, old rule-incongruent items (i.e., exceptions) were more likely to be judged as having occurred before than were new rule-incongruent items. This indicates that participants do not simply make recognition decisions by judging whether an item is consistent or inconsistent with the abstracted rule. Rather, recognition judgments appear to be based, at least in part, on familiarity with or recollection of individual stimulus-response pairs.

Memory for Exceptions to Function-Based Rules

Even though participants in function-learning tasks do appear to retain exemplar information, exceptions do not seem to have special status in memory. Contrary to findings from category learning experiments (see Palmeri & Nosofsky, 1995), rule-congruent items were better learned and better recognized than exceptions. In fact, performance for stimulus-response pairs that were used as exceptions was much worse when those pairs were learned in conjunction with rule-generated pairs (the functional-mapping condition) than when learned in conjunction with random pairs (the random-mapping condition). It therefore appears that learning a function-based rule interferes with learning and memory for specific instances that are exceptions to that rule.

Conclusions

In sum, the current study supports the view that the learning and application of function-based concepts involves rule abstraction as well as memory for specific instances. One possible instantiation of this hybrid approach, following from the RULEX model of category learning, is that participants learn functional rules and remember exceptions to those rules. However, this particular rule-plus-exemplar account is not supported by the present experiment. Unlike findings from the category learning literature, exceptions to function-based rules do not appear to have special status in memory.

Acknowledgments

The author would like to thank Jayson Johns and Jason Lickel for their assistance in participant testing, data scoring, and preliminary data analysis.

References

Bower, G., & Trabasso, T. (1963). Reversals prior to solution in concept identification. *Journal of Experimental Psychology*, 66, 409-418.

Brehmer, B. (1974). Hypotheses about relations between scaled variables in the learning of probabilistic inference tasks. *Organizational Behavior and Human Performance*, 11, 1-27.

Busemeyer, J. R., Byun, E., DeLosh, E. L, & McDaniel, M. A. (1997). Learning functional relations based on experience with input-output pairs by humans and artificial neural networks. In K. Lamberts & D. Shanks (Eds.), *Knowledge, concepts, and categories*. East Sussex, UK: Psychology Press.

Carroll, J. D. (1963). *Functional learning: The learning of continuous functional maps relating stimulus and response continua* (ETS RB 63-6). Princeton, NJ: Educational Testing Service.

DeLosh, E. L. (1994). *Rule abstraction and hypothesis testing in the learning of functional concepts*. Master's thesis, Department of Psychology, Purdue University, West Lafayette.

DeLosh, E. L. (1996). *Effects of mnemonic variables on function and category learning*. Doctoral dissertation, Department of Psychology, Purdue University, West Lafayette.

DeLosh, E. L., Busemeyer, J. R., & McDaniel, M. A. (1997). Extrapolation: The sine qua non for abstraction in function learning. *Journal of Experimental Psychology: Learning, Memory, and Cognition*, 23, 968-986.

Kruschke, J. K. (1992). ALCOVE: An exemplar-based connectionist model of category learning. *Psychological Review*, 99, 22-44.

Levine, M. (1975). *A Cognitive Theory of Learning: Research on Hypothesis Testing*. Hillsdale, NJ: Lawrence Erlbaum Associates.

Nosofsky, R. M., Palmeri, T. J, & McKinley, S. C. (1994). Rule-plus-exception model of classification learning. *Psychological Review*, 101, 53-79.

Palmeri, T. J., & Nosofsky, R. M. (1995). Recognition memory for exceptions to the category rule. *Journal of Experimental Psychology: Learning, Memory, and Cognition*, 21, 548-568.

Restle, F. (1962). The selection of strategies in cue learning. *Psychological Review*, 69, 329-343.

Selective activation as an explanation for hindsight bias

Markus Eisenhauer (markus.eisenhauer@psychol.uni-giessen.de)
FB 06 – Psychology, Justus-Liebig-University Gießen;
Otto-Behaghel-Str. 10, D-35394 Gießen, Germany

Rüdiger F. Pohl (ruediger.pohl@psychol.uni-giessen.de)
FB 06 – Psychology, Justus-Liebig-University Gießen;
Otto-Behaghel-Str. 10, D-35394 Gießen, Germany

Abstract

In hindsight, people often claim to have known more in foresight than they actually did. For example, the confidence for one of several possible outcomes is larger when it is known that this particular outcome occurred. A widespread explanation of hindsight bias assumes that the feedback serves as an anchor. How precisely this anchor takes effect and why it leads to a bias towards the anchor value has not been satisfactorily answered yet. One possible mechanism to explain hindsight bias assumes that the encoding of the feedback leads to a selective activation of the item-specific knowledge base. As a result, specific information units are strengthened and are thus more likely to be recalled when a person tries to reconstruct his or her original judgment. We tested the effect of selective activation in two hindsight experiments. The results showed a clear hindsight bias in that the recalled confidence ratings were distorted towards the feedback. Moreover, the consequences of selective activation were evident in that more information favoring the feedback was recalled

Introduction

Hindsight bias or the "Knew-it-all-along-effect" (Fischhoff, 1975) is a well-known systematic phenomenon that is of special interest for the insight it provides into the processes of judgment and recall. But how is our recall of previous knowledge states influenced by supplying new information (e.g., the outcome)? In the face of the outcome, we often seem to overestimate the quality of our previous knowledge, thus leading to a distortion towards the provided information. Suppose, for example, that a group of participants is being asked for the plausibility of absinthe being (a) a precious stone or (b) a liqueur? A second group of participants first receives the correct answer and is then being asked for the plausibility rating of the two alternatives with the instruction to ignore the solution. In comparison to the first group (without solution) the plausibility rating of the second group reveals a higher confidence in the correct alternative as suggested by the solution. That is, subjects of the second group seem to "guess better" (e.g., Hoch & Loewenstein, 1989).

The hindsight bias is even more intriguing if the same subjects are asked for the plausibility of absinthe being (a) a precious stone or (b) a liqueur and, then - usually after some time has elapsed - receive the correct answer, and finally are to remember their original plausibility rating. Now the remembered plausibility ratings are closer to the correct solu-

tion than the original ratings were (e.g., Fischhoff, 1977; Wood, 1978). The main difference between these two designs is the task that the subject has to perform, being a hypothetical judgment in the first case and a memory recollection in the second.

Of special interest` in the memory design is the stage at which the memory distortion actually occurs. Some researchers (e.g., Fischhoff, 1977; Loftus & Loftus, 1980) favor early stages, that is, they believe in destructive updating of the original information at the time of encoding the outcome information. Fischhoff (1975) used the term "creeping determinism" to point out that it is completely natural to assimilate outcome knowledge with the original information to create a coherent whole out of all the relevant knowledge. This process depicts learning from the outcome. Other experiments, though, suggest that the distortion takes place at a later stage. This can be inferred from studies showing post-outcome manipulations to be effective. Davies (1987, Experiment 1) found that supplying subjects with notes they had written in the first judgment session considerably reduced the hindsight bias. Equally effective was the post-feedback generation of reasons for all possible outcomes (Davies, 1987, Experiment 3). Hasher, Attig, and Alba (1981, Experiment 2) provided one of the rare examples in which subjects' recollections showed no hindsight bias. The critical debiasing manipulation was to warn subjects that they accidentally received false outcome information.

Conversely, Fischhoff (1977) found that informing the subjects about the bias did not reduce hindsight bias. However, his result was observed in a hypothetical design, in which the correct information was given before the first attempt to respond. Pohl and Hell (1996) found no effect of reducing hindsight bias in a memory design. Neither informing subjects in advance nor individual feedback about their recall performance reduced hindsight bias. The results showed that knowledge about the bias phenomenon did not help subjects to avoid the bias. Findings like these support automatic processes as an explanation for the observed bias and dismiss motivational accounts.

Explanations favoring the final rejudgment process as the point where biasing occurs might be labeled "cognitive-reconstruction" theories (Hawkins & Hastie, 1990). According to these, the hindsight bias is a necessary and unavoidable by-product of collecting evidence in the judgment

144

process. Hindsight bias is an automatic memory distortion that arises whenever the original response (that is being looked for) has been forgotten or - as in the case of hypothetical designs - has never been encoded. The systematic memory distortion occurs because subjects are apparently unable to ignore outcome knowledge during the rejudgment process (cf. Kahneman & Tversky, 1974).

Recently, Pohl (1998) found that when the data were separated according to whether participants considered the feedback value plausible or not, cases of unbiased recollections did emerge: feedback values that were labeled as estimates of another person and found to be implausible did not lead to hindsight bias. This finding argues against the view that hindsight bias is an automatic and unavoidable effect of feedback presentation. There are at least specific circumstances under which it is possible to avoid the influence.

In conclusion, most of the empirical evidence favors cognitive accounts, while motivational manipulations showed only minor effects. The same conclusion was drawn in a meta-analysis, covering 122 hindsight bias studies (Christensen-Szalanski & Willham, 1991). However the findings from Pohl (1998) point out that bias is not always as automatic and unavoidable as has been presumed

Previously proposed cognitive explanations of the hindsight bias are unfortunately not very satisfying. For example the anchoring and adjustment heuristic (Tversky and Kahneman, 1974) originally proposed to explain anchoring effects is also being discussed as an explanation for hindsight bias.

In a typical anchoring study (e.g., Tversky & Kahneman, 1974), participants are first asked whether the answer to a question is above or below a certain number. This number acts like an anchor because it distorts subsequent estimates towards it. Thus, mean estimates following a high anchor are higher than those following a low anchor are.

To explain such anchoring effects, Tversky and Kahneman (1974) proposed that participants start their estimation from the anchor and adjust the value in the direction they think plausible (i.e., higher or lower than the anchor). They stop at the first plausible value, thus leading to estimates that are biased towards the anchor (Jacowitz & Kahneman, 1995). Although plausible in its assumptions, it remains unclear how the anchor produces this restriction. Besides, it has been shown that highly implausible anchors lack any effects of anchoring and highly plausible anchors lead to anchoring (Pohl, 1988) albeit participants should respond with the anchor value in this case according to Jacowitz and Kahneman.

Pohl and Eisenhauer (1997) developed a detailed cognitive model that allows explaining anchoring and hindsight bias on a deeper level and that, moreover, can be used as a simulation model. As basic explanation for distorted judgment or recall, the model assumes a selective activation process of one's item-specific knowledge base. In order to reflect this focus, the model was termed SARA which stands for "Selective Activation, Reconstruction, and Anchoring" (Pohl & Eisenhauer, 1997). All processes (i.e., generating, encoding, forgetting, and reconstructing) change the associative pattern between the elements of one's knowledge base and possible retrieval cues, thus leading to

a different probability of retrieval. SARA's general architecture is based on "SAM"--the *Search of Associative Memory* model (Raaijmakers & Shiffrin, 1980).

The subsequent part of this paper describes selective activation, the central assumption of SARA, in more detail and presents two experiments that support this explanation of hindsight bias.

In a typical hindsight experiment within the memory design, participants are asked to answer difficult almanac questions. Suppose for example, that you are asked for the height of the Eiffel tower? If you don't know the correct answer, there are two options that could lead to an answer: you could guess, or activate knowledge. In the second case, you are probably neither able nor willing to access all knowledge theoretically available to answer the question. The basic idea is that the representation of the information units in memory could be described as an associative network comprising all information dealing with the specific question: the knowledge base. Thus the task to give an estimate to an almanac question leads to the attempt to recall some of the information units of one's knowledge base, depending on their level of association. In other words, only strongly associated informations are likely of being activated. Depending on the time available and on the motivation, you would probably generate not more than two or three information units. This is a reasonable assumption especially if there are 50 or more almanac questions to be answered. For example, the mean height of buildings at the turn of the century, or that the Eiffel tower is a steel construction, could come to your mind. Those informations would be translated into numerical values and summarized in one value, the estimate for the question (e.g., "250 meters").

After some time has elapsed, the solution is provided (e.g., "300 meters"). Because it is the answer to the original question you will probably try to encode the solution. The process of encoding information shares many features with the generation of an estimate. It may be seen as a reverse retrieving process. The solution will be associated with information units in the knowledge base, systematically increasing the strength of association of information units close to the solution (e.g., "steel construction"; "build for a world's fare"). The central process of SARA is that the encoding of the solution leads to a selective activation of associated units of the knowledge base. The result is that information units strongly associated to the solution are increased in their associative-strength level. Finally, the solution is added to the knowledge base, thus completing its encoding.

In the last phase of a typical hindsight experiment, when you are asked to recall your original estimate the process will be exactly the same as in the generation phase. SARA assumes that the process should be based only on some of the information units of the knowledge base. The probability to access information units varies with their associative strength towards the currently present retrieval cues: again, only strongly associated information is likely of being activated. But the pattern of association strength, has changed because of the encoding of the solution with the result that certain information units are strengthened and more likely to be activated when a person tries to reconstruct his original estimate. The probability to retrieve units closely associated to the solution in a following task should thus be increased. Consequently the recollection will most probably be systematically biased towards the solution (e.g., "250 meters"). In the remainder, we present two experiments that examine selective activation more closely.

Experiment 1

Method

Material It is next to impossible to lay out the specific information units that are potentially used to generate an estimate to a specific question. Therefore, we decided to supply the specific knowledge bases in our experiments.

We used a confidence task: Participants had to give ratings about how confident they were whether a certain quantity increased or decreased in value (e.g., increase or decrease in sales of a fictitious corporation). In order to judge how the quantity may have changed, participants received four arguments favoring an increase and four favoring a decrease of the quantity.

One example was following question:
Question:
"In 1988 88 % of the American adults believed in the right to beat their children. The percentage changed up to now."
Arguments favoring decrease:
• "authoritarian education being criticized"
• "TV-advertisements about serious consequences of violence in the family"
• "reporting in the media of abuse in the family"
• violence in the family leading theme of the universal day of the child"
Arguments favoring increase:
• "deficiency of antiauthoritarian education"
• "growth of authoritarian religious communities"
• "popularity of rigorous conservative colleges"
• "association of adolescent violence with missing limits in education"
Task:
"Please report how confident you are that the percentage increased"
Probability of increase (in percent):   e.g. 40 %

Participants, design and procedure One hundred and eight students (80 female, 28 male; between 18 and 52 years old with a mean age of 23.6 years) of different faculties of the University of Trier took part in the experiment.

The experiment consisted of two sessions: In Session 1, participants had to fill out a questionnaire with 24 verification tasks and to indicate their confidence whether the quantity increased or decreased. In Session 2 (one week later) they had to recollect their confidence ratings of the first session. The questionnaire in the second session however was presented with some of the solutions, indicating in 8 cases that the quantity had increased and in other 8 cases that it had decreased. The remaining 8 cases contained no

feedback and served as control cases. The selection of experimental items was counterbalanced across participants, so that all questions served equally often as experimental and as control items. Participants were asked to recall their own estimates given one week ago. The instructions stressed that there was no interest in the memory of the solutions but rather in the memory of the participants' first confidence ratings. The order of questions was identical in both sessions and there was no time limit. After the attempt to recollect the first confidence rating, participants had to remember in a free recall test as many arguments to each problem case as possible. At the end, participants were debriefed about hindsight bias and the goal of the experiment. The total experiment lasted about 60 minutes per person.

The dependent variable to measure hindsight bias (labeled "• %") was defined as the difference in the confidence rating for a decrease in the first session minus the confidence rating for a decrease in the second session.

•% = (confidence decrease (t1) - confidence decrease (t2))

A positive value of •% indicates a greater confidence for increase in Session 2 in comparison to Session 1, whereas a negative value indicates a greater confidence for increase. A shift towards the feedback indicates hindsight bias. The feedback that in fact "it increased" should lead to a positive value of •%. Conversely, the feedback "it decreased" should lead to a negative value of •%.

The number of arguments recalled favoring increase or decrease measured the assumed effects of selective activation. The level of significance was set to $\alpha = .05$ for all analyses.

Results

Confidence Ratings A repeated measures ANOVA for the feedback (increase – no feedback – decrease) revealed a distinct effect of the feedback. The feedback "increase" resulted in a positive value of • % (2.8), no feedback in a minor positive value (1.0) and the feedback "decrease" led to a negative value of • % (-3.3). The results showed a clear shift of the confidence ratings in the second session towards the feedback ($F_{(1,98)} = 30.7$).

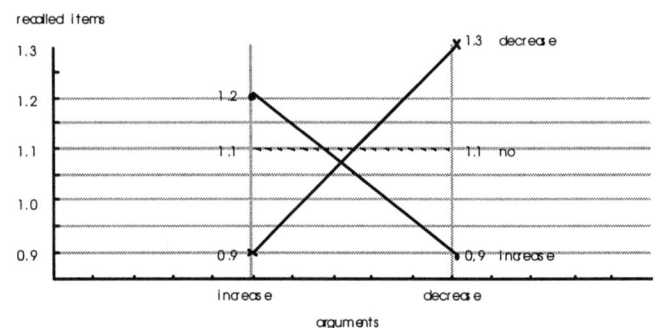

Figure 1: Mean number of recalled arguments in each experimental condition of Experiment 1.

146

Recalled arguments The number of correctly recalled arguments was analyzed in a 3 x 2 MANOVA with the factors feedback (increase – no feedback – decrease) and type of argument (favoring increase or decrease). The interaction showed a clear and distinct effect of the feedback ($F_{(2,208)} = 33.5$; see Fig. 1). The feedback that "the quantity increased" led to a better recall of arguments favoring increase (1.2 vs. 0.9) whereas the feedback that "the quantity decreased" led to a better recall of arguments favoring decrease (1.3 vs. 0.9). In the condition without feedback no difference between arguments favoring increase or decrease could be observed.

Discussion

The significant shift • % in confidence ratings depending on the feedback denotes hindsight bias. The feedback that a quantity increased augmented the confidence for increase (or lowered the confidence for decrease) in the second session as compared to the first. Whereas the feedback that a quantity decreased augmented the confidence for decrease (or lowered the confidence for increase) in the second session as compared to the first. Without feedback confidence for increase augmented slightly indicating a minor positivity bias.

The analysis of the number of recollected arguments in the free recall showed a significant interaction between type of feedback and type of argument. Whenever the feedback indicated that the fact increased, significantly more arguments were recollected favoring increase. Accordingly, a feedback of decrease led to more recollected arguments favoring decrease. Thus, significantly more arguments favoring the feedback were recollected implying a selective activation of arguments favoring the feedback. The result of this experiment can be taken as a first confirmation of selective activation as a promising explanation for hindsight bias. In a second experiment, we tried to find more evidence for selective activation. Unlike Experiment 1, we used a hypothetical design with only one session.

Experiment 2

Method

Material and design The material was the same as in the questionnaire of Experiment 1 with the only difference that the experiment took part on a computer. Participants had to answer 24 verification tasks and to indicate their confidence weather the quantity increased or decreased. Eight cases were presented with solutions indicating that the fact had increased and eight cases that it had decreased. The remaining eight cases contained no feedback and served as control cases. The design of Experiment 2 corresponded with that of Experiment 1 with the exception that the experiment took place in one session. The dependent variable was the confidence in increase dependent upon the feedback. The number of arguments in free recall favoring increase or decrease was taken to indicate selective activation.

Participants and procedure One hundred and four students (69 female, 34 male; between 18 and 40 years old with a mean age of 22.9 years) of different faculties of the University of Trier took part in the experiment. The experiment consisted of one session. Similar to the task in Experiment 1, participants received a fact on the computer screen, but this time the solution was provided immediately (except for eight control-cases without solution). The arguments favoring increase and decrease of the quantity followed subsequently. Participants were then asked to report their own confidence independent from the feedback received. There was no time limit. At the end, participants were debriefed about hindsight bias and the goal of the experiment. The experiment lasted about 30 minutes.

Results

Confidence Ratings A repeated measures ANOVA for the feedback (increase – no feedback – decrease) revealed a distinct effect of the feedback. The feedback "increase" resulted in a greater confidence for increase (64.2 %), no feedback in a medium confidence rating (55.1 %) and the feedback "decrease" led to a lesser confidence for increase (39.8 %). The results showed a clear dependency of the confidence to the feedback ($F_{(2,204)} = 58.8$).

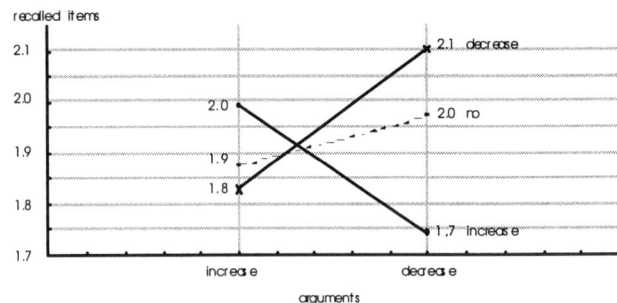

Figure 2: Mean number of recalled arguments in each experimental condition of Experiment 2.

Recalled arguments The number of correctly recalled arguments was analyzed in a 3 x 2 MANOVA with the factors feedback (increase – no feedback – decrease) and type of argument (favoring increase or decrease). The interaction showed an effect of the feedback ($F_{(2,204)} = 12.4$; see Fig. 2). The feedback that "the quantity increased" led to a better recall of arguments favoring increase (2.0 vs. 1.7) whereas the feedback that "the quantity decreased" led to a better recall of arguments favoring decrease (2.1 vs. 1.8). In the condition without feedback no noticeable difference between arguments favoring increase or decrease (1.9 vs. 2.0) could be observed.

Discussion

The difference in confidence ratings in the different feedback conditions revealed hindsight bias. The feedback that a quantity increased augmented the confidence for increase as compared to no feedback. The feedback that a fact decreased lowered the confidence for increase compared to no feedback.

The analysis of the number of recollected items in the free recall showed again a significant interaction between type of feedback and type of argument. Whenever the feedback indicated that the quantity increased significantly more arguments were recollected favoring increase. Correspondingly a feedback of decrease led to more recollected arguments favoring decrease. Significantly more arguments favoring the feedback were recollected implying again a selective activation of arguments favoring the feedback. The result of this experiment substantiates those from Experiment 1 and consolidates selective activation as a promising explanation for the hindsight bias.

Conclusion

Both experiments successfully demonstrated the existence of selective activation. Selective activation thus appears to be a promising explanation of hindsight bias. The solution or anchor proved to have a distinctive influence: Items favoring the anchor were recollected more often compared to those supporting the opposite. The central concept of SARA (Pohl & Eisenhauer, 1997), namely selective activation of the item specific knowledge base, was confirmed in the two reported experiments. SARA makes detailed assumptions about a person's pre-experimental knowledge base and how it is altered in the course of the experiment. All processes (i.e., generating, encoding, forgetting, and reconstructing) change the associative pattern between the elements of one's knowledge base and possible retrieval cues, thus leading to a different probability of retrieval. According to SARA, anchored reconstruction results from a *selective activation* of one's item-specific knowledge base (Hawkins & Hastie, 1990; Strack & Mussweiler, 1997). This activation is governed by the anchor value and is considered being selective, because information that is more similar to (or consistent with) the anchor will receive more activation than other information (Kahneman & Miller, 1986). After selective activation, the probability of retrieving a certain piece of information from one's knowledge base has changed. As a consequence, any attempt to generate or to reconstruct an "unbiased" estimate is bound to fail. Most probably, the resulting estimate will be biased towards the anchor value.

Selective activation is able to explain most of the findings in the field of anchoring. For example, Fischhoff's experiments (1975) on "creeping determinism" share many features with the reported experiments. Its participants received passages describing an unfamiliar historical event and had to evaluate the probability of four possible outcomes in the light of a solution. As in our experiments, a shift towards the solution was observed. Contrary to Fischhoff, however we don't assume an irreversible and immediate assimilation of the solution. In our opinion the shift towards the solution reflects a selective activation of the knowledge base promoting arguments in favor of the provided solution.

The results of the reported experiments support the basic idea to explain and to model distortions in judgment and memory through selective activation of one's item-specific knowledge base.

References

Christensen-Szalanski, J. J. J., & Fobian Willham, C. (1991). The hindsight bias: A meta-analysis. *Organizational Behavior and Human Decision Processes, 48*, 147–168.

Davies, M. F. (1987). Reduction of hindsight bias by restoration of foresight perspective: Effectiveness of foresight-encoding and hindsight retrieval strategies. *Organizational Behavior and Human Decision Processes, 40*, 50–68.

Fischhoff, B. (1975). Hindsight = foresight: The effect of outcome knowledge on judgment under uncertainty. *Journal of Experimental Psychology: Human Perception and Performance, 1*, 288–299.

Fischhoff, B. (1977). Perceived informativeness of facts. *Journal of Experimental Psychology: Human Performance and Perception, 3*, 349–358.

Hasher, L., Attig, M. S., & Alba, J. W. (1981). "I knew-it-all-along: Or, did I?". *Journal of Verbal Learning and Verbal Behavior, 20*, 86–96.

Hawkins, S. A., & Hastie, R. (1990). Hindsight: Biased judgments of past events after the outcomes are known. *Psychological Bulletin, 107*, 311–327.

Hell, W., Gigerenzer, G., Gauggel, S., Mall, M., & Müller, M. (1988). Hindsight bias: An interaction of automatic and motivational factors? *Memory and Cognition, 16*, 533–538.

Hoch, S. J., & Loewenstein, G. F. (1989). Outcome feedback: Hindsight and information. *Journal of Experimental Psychology: Learning, Memory, and Cognition, 15*, 605–619.

Jacowitz, K. E. & Kahneman, D. (1995). Measures of anchoring in estimation tasks. *Personality and Social Psychology Bulletin, 21*, 1161–1166.

Kahneman, D., & Miller, D. T. (1986). Norm theory: Comparing reality to its alternatives. *Psychological Review, 93*, 136–153.

Loftus, E. F., & Loftus, G. R. (1980). On the permanence of stored information in the human brain. *American Psychologist, 35*, 409–420.

Pohl, R.-F. (1998). The effects of feedback source and plausibility of hindsight bias. *European-Journal-of-Cognitive-Psychology, 10(2)*, 191–212

Pohl, R. F., & Eisenhauer, M. (1997). SARA: An associative model for anchoring and hindsight bias. In M. G. Shafto & P. Langley (Eds.), *Proceedings of the Nineteenth Annual Conference of the Cognitive Science Society* (p. 1103). Mahwah, NJ: Erlbaum.

Pohl, R.-F., & Hell,-W. (1996). No reduction in hindsight bias after complete information and repeated testing. *Organizational-Behavior-and-Human-Decision-Processes, 67(1)*, 49–58.

Raaijmakers, J. G. W., & Shiffrin, R. M. (1980). SAM: A theory of probabilistic search of associative memory. In G. H. Bower (Ed.), *The psychology of learning and motivation* (Vol. 14; pp. 207–262). San Diego, CA: Academic Press.

Strack, F., & Mussweiler, T. (1997). Explaining the enigmatic anchoring effect: Mechanisms of selective accessibility. *Journal of Personality and Social Psychology, 73*, 437–446.

Tversky, A., & Kahneman, D. (1974). Judgment under uncertainty: Heuristics and biases. *Science, 185*, 1124–1131.

Wood, G. (1978). The "knew-it-all-along" effect. *Journal of Experimental Psychology: Human Perception and Performance, 4*, 345–353.

Relevance and Feature Accessibility in Combined Concepts

Zachary Estes (zcestes@princeton.edu)
Sam Glucksberg (samg@princeton.edu)
Princeton University
Department of Psychology; Green Hall
Princeton, NJ 08544-1010 USA

Abstract

When comprehending combined concepts (e.g., 'peeled apples'), two kinds of features are potentially accessible. *Phrase features* are true only of the phrase (e.g., "white"), while *noun features* are true of both the phrase and the head noun (e.g., "round"). Phrase features are verified more quickly and more accurately than noun features. No satisfactory account of this phrase feature priority has been put forth. We propose that relevance can explain the phrase feature priority. In Experiment 1, the differential accessibility of noun and phrase features was reversed by context paragraphs that made noun features relevant. Experiment 2 more subtly replicated this effect using a single-word context. We conclude that the phrase feature priority is attributable to the discourse strategy of assigning relevance to modifiers of combined concepts.

Introduction

Meaning is an unstable phenomenon: The particular features accessed for a given concept may differ greatly across various occasions of use. For instance, Johnson-Laird (1975) noted that the sentence "The tomato rolled across the floor" emphasizes the round feature of tomatoes, while the red feature is accessed in "The sun was a ripe tomato." This same idea is captured by Barsalou's (1982) context-dependent features, which are accessed only in appropriate contexts (as opposed to context-independent features, which are accessed regardless of context). This differential feature accessibility has implications not only for semantics, but also for theories of concept representation, natural language comprehension, and referential communication.

Combined concepts are particularly rich for investigating feature accessibility because certain features emerge only when concepts are combined. For instance, 'peeled apples' are white, though neither apples nor peeled things in general are white. Rather, this feature emerges from the combination of 'peeled apples'. We refer to such features as *phrase features*, because they are true of the phrase but are not true of either constituent concept in isolation. *Noun features*, on the other hand, are true of both the combined concept and the head noun. For instance, "round" is a noun feature of 'peeled apples' because both peeled apples and apples in general are round.

Hampton & Springer (1989) and Springer & Murphy (1992) investigated the relative accessibility of noun and phrase features of combined concepts. In a typical experiment, participants indicated whether sentences such as

"Peeled apples are white" and "Peeled apples are round" are true or false (i.e., the sentence verification paradigm). Phrase features were verified more quickly and more accurately than noun features. This *phrase feature priority* has been found by several researchers (Estes & Glucksberg, 1998; Gagne & Murphy, 1996; Hampton & Springer, 1989; Springer & Murphy, 1992).

What accounts for the differential accessibility of phrase and noun features? Gagne and Murphy (1996) suggested that the given-new convention might explain the phrase feature priority. The given-new convention states that information is differentially processed according to whether it is 'given' or 'new' information (Haviland & Clark, 1974). More specifically, new information is processed prior to given information (Hornby, 1974; Singer, 1976).

To assess this given-new hypothesis, Gagne and Murphy embedded combined concepts in discourse contexts that were designed to assign 'new' information status to either the modifier or the head noun. For example, if the modifier 'peeled' is repeated twice in a paragraph but the noun 'apples' appears only once, then the repeated modifier might become the given information and the noun would be the new information. However, this and other manipulations failed to eliminate the phrase feature priority. Phrase feature statements such as "peeled apples are white" were still verified more quickly than were noun feature statements such as "peeled apples are round". Gagne and Murphy concluded that the given-new convention was not responsible for the phrase feature priority.

We propose that relevance can explain the phrase feature priority. By relevance we mean the classic sense used by Grice (1975). Dale and Reiter (1995) paraphrase Grice's maxim of relevance as follows: "A referring expression should not mention attributes that have no discriminatory power and, hence, do not help distinguish the intended referent from the members of the contrast set" (p. 240). In accordance with this, we propose that people make a discourse processing assumption that a concept has been modified because that modifier provides relevant information. That is, 'peeled apples' are mentioned because it is relevant for the comprehender to know that they are peeled rather than ordinary apples. This information is relevant because it has discriminatory power. It helps distinguish the referent from other members of the head noun category. For instance, 'peeled apples' differ from other apples in that they are white and sticky. Thus phrase features, which by definition distinguish the combined concept from the head noun category, are assumed to be relevant. This assumption of relevance, we suggest, results in the phrase feature priority.

If relevance is responsible for the phrase feature priority, then this preferential accessibility should be reversed by contexts that make noun features relevant and phrase features irrelevant. Experiment 1 tests this hypothesis by explicitly making either noun or phrase features relevant. Experiment 2 uses a more subtle, single-word context to implicitly alter feature accessibility.

Experiment 1

To test our relevance hypothesis, we constructed context paragraphs for which either a noun or a phrase feature was relevant (see Table 1). The relevance hypothesis predicts that the feature relevant to the preceding context will be more accessible, regardless of whether it is a noun or a phrase feature. That is, the phrase feature priority will be reversed by contextual relevance.

Table 1: Examples of stimuli, Experiment 1.

Noun-relevant contex: Alan and Susan were bored one Sunday afternoon, and they decided to play lawn bowling in their backyard. But they didn't have any lawn balls, so they searched around the house. The first things they found were a pair of peeled apples that were going to be used with dinner. They were a little sticky, but they worked just fine.

Phrase-relevant context: Alan was a famous French chef who used fresh fruit to garnish his meals. Each night, he spent half an hour selecting the perfect fruit for the centerpiece. Last night, Alan decided to make a colorful centerpiece. He used orange slices, kiwi and peeled apples. The centerpiece was gorgeous, until the guests began to eat it.

Noun feature verification: Peeled apples are round.
Phrase feature verification: Peeled apples are white.

Many studies have demonstrated that relevant contexts facilitate access to the features of noun concepts (Hess, Foss, & Carroll, 1995; McKoon & Ratcliff, 1988; Tabossi, 1982; 1988; Tabossi & Johnson-Laird, 1980). Thus, our hypothesis may seem obvious. However, there are two reasons to test this hypothesis directly. First, Gagne & Murphy (1996) failed to consistently affect the differential accessibility of noun and phrase features with their contexts. In fact, they concluded just the opposite of our relevance prediction: "Using a context that emphasizes a particular feature makes that feature more difficult to verify than when the feature has not been emphasized in the preceding context" (Gagne & Murphy, 1996, p. 96).

And second, there are important differences in stimuli between the present investigation and past investigations. The earlier work used simple concepts (e.g., apples), while the present experiment uses combined concepts (e.g., peeled apples). The combination of concepts brings about a host of issues that are not involved in the comprehension of simple concepts. For instance, the modifying concept may act as a local context for the head concept, possibly competing with the more global context paragraph (see Hess, Foss, & Carroll, 1995). Also, the modifier of a combined concept is often idiosyncratically construed (Wisniewski, 1996) and may be represented as only a single of its features (Estes & Glucksberg, in press), with some features emerging and others being canceled from the combination (Hampton, 1987, 1988; Medin & Shoben, 1988; Murphy, 1988). These differences in stimuli make it advisable to test any generalization from simple to combined concepts.

Materials and Design

The experiment was a 2 (context) X 2 (feature) within-subjects design, with response time and accuracy as dependent measures in a sentence verification paradigm. Feature-types were noun and phrase features, as described above. Noun and phrase features were matched for number of syllables. Contexts were brief (typically 3 or 4 sentences), and included the critical combined concept (e.g., peeled apples) once. Forty experimental contexts emphasized either the noun or phrase feature without explicitly stating the feature. Most contexts and verification sentences were taken from Gagne and Murphy (1996, Experiment 4), although the contexts were edited to make them consistent with our purposes. Forty experimental target sentences (one noun feature and one phrase feature for each combined concept) were true. Twenty filler contexts concluded with false target sentences (e.g., Pepperoni pizza is vegetarian), also taken from Gagne and Murphy. To encourage attention to the context paragraphs, each context was followed by a comprehension question. For example, the comprehension questions for the two 'peeled apples' scenarios were "Did they bowl in their front yard?" and "Did Alan have his assistant prepare the centerpiece?". The correct answer to half of these questions was "yes". These questions were fully counterbalanced across conditions. Four lists were constructed such that each consisted of 5 true items in each of the 4 experimental conditions and 20 false filler items for a total of 40 items per list. Item order was randomized for each participant.

Participants and Procedure

Thirty-seven Princeton University undergraduates participated for partial course credit or for pay. All were native speakers of American English. The procedure followed that of Gagne and Murphy (1996, Experiment 4). Participants read a context paragraph on a computer monitor and pressed the space bar upon completion (self-paced). A probe sentence was presented in the center of the screen immediately thereafter. Participants pressed one of two labeled keys to indicate whether the sentence was true or false. After this response, a comprehension question was presented in the center of the screen, and again participants responded by pressing the appropriate key. This sequence was repeated for all 40 items. Participants were instructed to read the paragraphs at their own pace, but to respond to the sentences as quickly as possible without making errors. The task lasted approximately 20 minutes.

Results and Discussion

The data of one error-prone participant were replaced by data from another participant, providing data from 36 participants in all. Two repeated measures ANOVAS were performed, one using participants as a random factor (F_s) and one using items as a random factor (F_i). Response times of less then 500ms or more than 5000ms (2.6% of data) were removed from analyses, as were incorrect responses (12.5%). The comprehension questions were answered with equivalent accuracy rates (90%) across the conditions, all $p > .15$. Thus, any differences in verification time across experimental conditions cannot be attributed to differences in comprehension or attention in the different conditions. Mean response times and percent correct as a function of condition are presented in Table 2.

Table 2: Mean response times in milliseconds (and accuracy) by condition, Experiment 1.

| | Feature-type | |
Context-type	Noun feature	Phrase feature
Noun-relevant	1980 (.89)	2117 (.82)
Phrase-relevant	2222 (.81)	1921 (.89)

As expected, both response time and accuracy were best in the target-relevant conditions. When phrase features were relevant, they were verified more quickly and more accurately than noun features. But when noun features were relevant, they were verified more quickly and accurately than phrase features. As shown in Figures 1 and 2, this predicted context by feature-type interaction was reliable for both response time [F_s (1, 35) = 21.41, $p < .01$ and F_i (1, 19) = 9.95, $p < .01$] and accuracy [F_s (1, 35) = 6.67, $p < .05$ and F_i (1, 19) = 8.12, $p = .01$]. There were no reliable main effects, including no reliable overall phrase feature priority, all $p > .10$. These results are clear: Relevant information is more accessible than irrelevant information, irrespective of whether it is a phrase feature or a noun feature.

To recapitulate, according to our relevance hypothesis, comprehenders assume that the modifier of a combined concept is relevant because it has "discriminatory power and, hence...help[s] distinguish the intended referent from the members of the contrast set" (Dale & Reiter, 1995, p. 240). That is, the phrase features added by the modifier are assumed to be relevant because they differentiate the combined concept from other members of the head noun category. Thus, in the absence of an informative context, phrase

features are more accessible than noun features—the phrase feature priority—because they are more relevant.

In this experiment we made either noun features or phrase features relevant to a context paragraph. The relevance of a feature predicted its accessibility, regardless of feature-type. Thus relevance, rather than the given-new manipulation, determines feature accessibility and can explain the phrase feature priority.

Figure 1: Mean response time by condition, Experiment 1.

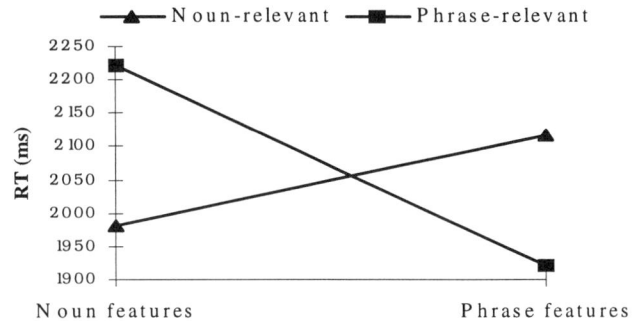

Figure 2: Mean accuracy rate by condition, Experiment 1.

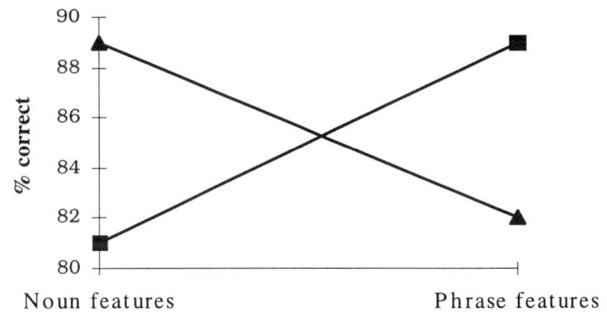

Experiment 2

In Experiment 1 we found powerful context effects using one-paragraph scenarios that explicitly rendered one or another feature-type relevant. Would minimal contexts also effectively alter the accessibility of noun and phrase features? Perhaps the most minimal context possible is a single word that changes the meaning of a sentence. The word "even" may have this effect. To see how, consider these four sentences:

(1) Peeled apples are round.
(2) Peeled apples are white.
(3) Even peeled apples are round.
(4) Even peeled apples are white.

Sentence (2) is verified more quickly and more accurately than (1) (Hampton & Springer, 1989; Springer & Murphy, 1992). From this you might expect (4) to be more easily verified than (3). However, the addition of the word "even" may alter feature accessibility. "Even" may signify to the reader that the upcoming information is somewhat obvious. That is, "even" leads the reader to expect 'given' information, or information that is not new. For instance, when we encounter the phrase "Even peeled apples...", we may expect it to be completed with information that the combined concept has in common with other members of the head noun category, such as the noun feature "round" in (3). Completing this sentence with 'new' information, such as the phrase feature "white" in (4), violates our expectations of efficient communication (cf. Grice, 1975). Thus, (4) will be difficult to verify. Experiment 2 employs such expressions to determine whether a minimal contextual manipulation can replicate the interaction found in Experiment 1.

Materials and Design

The design was a 2 (sentence-type) X 2 (feature-type) within-subjects design, with response time and accuracy of verification as dependent measures. Twenty-four combined concepts (e.g., peeled apples), each with one noun feature (e.g., round) and one phrase feature (e.g., white), were selected as experimental items. Each of these 48 experimental sentences also appeared preceded by the word "even". Thus, there were 96 total experimental items. See (1) through (4) above as an example of one of the 24 sets of stimuli. Four lists were created such that each list contained 6 unmodified noun features (e.g., (1) above), 6 even-modified noun features (e.g., (3) above), 6 unmodified phrase features (e.g., (2) above), and 6 even-modified phrase features (e.g., (4) above). No combined concept appeared in the same list more than once. In addition, 24 false filler items (e.g., Pepperoni pizza is vegetarian) were randomly interspersed in the lists. Half of the false fillers began with the word "even" (e.g., Even elevator buttons are sewn on.). Noun and phrase features were matched for number of syllables.

Participants and Procedure

Sixteen Princeton University undergraduates participated for course credit. All were native American English speakers, and none had participated in similar experiments before. Sentences were presented one at a time in the center of a computer screen, and participants indicated whether the sentences were true or false by pressing the appropriate labeled key. Each sentence was preceded by a 1000ms intertrial interval and a 500ms fixation cross. This procedure continued uninterrupted for all 48 sentences. In addition, participants were given 8 practice trials prior to the experiment proper. Participants were instructed to indicate as quickly as possible without making errors whether presented sentences were true or false. Item order was randomized for each participant.

Results and Discussion

Two repeated measures ANOVAS were performed, one using participants as a random factor (F_s) and one using items as a random factor (F_i). Response times of less than 500ms or more than 5000ms (2.6% of data) were removed from analyses, as were incorrect responses (9.4%). Results are presented in Table 3.

Table 3: Mean response times in milliseconds (and accuracy) by condition, Experiment 2.

	Feature-type	
Context-type	Noun feature	Phrase feature
Unmodified	2137 (.89)	1835 (.93)
Even-modified	2052 (.95)	2124 (.85)

The predicted interaction obtained: The word "even", when used as a single-word context, hinders verification of phrase features but not noun features. This interaction was significant in both response time [F_s (1, 15) = 5.67, p = .03 and F_i (1, 23) = 4.23, p = .05] and accuracy [F_s (1, 15) = 5.16, p = .04 and F_i (1, 23) = 4.72, p = .04]. See Figures 3 and 4 below. Sentence-type had a reliable main effect on response time in the item analysis, F_i (1, 23) = 5.93, p = .02, indicating that the word "even" generally slowed comprehension. This is not at all surprising, given that the sentences containing "even" are longer than those not containing this additional word. What is more surprising is that despite the fact that these sentences are longer, they do not slow comprehension of noun features. Only phrase feature verification is slowed by this additional word. No other main effect approached significance, all p > .15.

Figure 3: Mean response time by condition, Experiment 2.

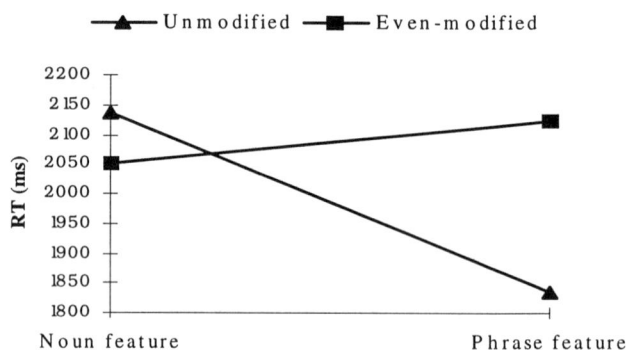

Figure 4: Mean accuracy rate by condition, Experiment 2.

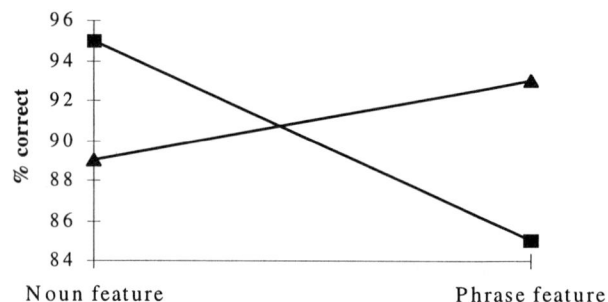

We may also look to the results of the two unmodified conditions for validation of our materials. When our experimental items are not modified by "even", noun feature verification is slower than phrase feature verification. This is a simple replication of the phrase feature priority (Hampton & Springer, 1989; Springer & Murphy, 1992), thus indicating that our effect in the present experiment is not due to the use of inappropriate items. Instead, it is attributable to the addition of a single-word context, namely, the word "even".

The given-new convention does not predict feature accessibility in combined concepts (Gagne & Murphy, 1996). Relevance does (Experiment 1 above). However, we do not intend to imply that the given-new convention plays no role in feature accessibility. Experiment 2 demonstrates that violating the given-new convention does affect feature accessibility. The addition of the word "even" apparently leads us not to expect phrase features, which are 'new' information to the combined concept. As a consequence, phrase features are more difficult to verify if they are preceded by the word "even".

Another, more speculative, claim is that Experiment 2 is further support for our relevance hypothesis. The claim here would be that "even" not only leads us not to expect phrase features, but that it also consequently makes them less relevant. However, more work needs to be done to determine whether the effect of Experiment 2 is due simply to expectancies, or if it can be explained by relevance.

General Discussion

We suggest that relevance accounts for the differential accessibility of noun and phrase features of combined concepts. Phrase features are verified more quickly and more accurately than noun features in the absence of an appropriate context. Our claim is that this phrase feature priority is attributable to a default discourse strategy of implicitly assigning relevance to the modifier of a combined concept. This default strategy is to assume, in accordance with Grice's (1975) conversational maxims, that the modifier of a combined concept serves a communicative purpose, namely, to add relevant information about the head noun concept. What information may a modifier contribute that is not already evident in the head noun concept? In short, the modification of a concept provides the addition of phrase features. Now because we have assigned relevance to the modifier, we therefore also assign relevance to those phrase features, resulting in the phrase feature priority.

If relevance is indeed responsible for the phrase feature priority, then manipulating the relevance of noun and phrase features should affect the accessibility of those features. Explicitly making noun features relevant should overturn the default strategy of assigning relevance to phrase features. Experiment 1 tested this hypothesis, using context paragraphs that emphasized either noun or phrase features. As predicted, noun features were verified faster and more accurately following contexts relevant to those noun features. Conversely, phrase features were more accessible after contexts relevant to those phrase features.

In Experiment 2 we sought to more subtly demonstrate this context- by feature-type interaction. Instead of a lengthy context, we used a single-word context. We supposed that the word "even", when preceding a combined concept (e.g., "Even peeled apples..."), indicates that the upcoming information is shared by the combined concept and the head noun in isolation. In other words, "even" leads the comprehender to expect noun features, which are 'given' information, rather than phrase features, which are 'new' information. The predicted interaction did obtain. This single-word context slowed comprehension of phrase features. However, the addition of "even" did not hinder verification of noun features, despite the fact that this single-word context made those sentences longer than others not preceded by "even". From Experiments 1 and 2 we conclude that the phrase feature priority is attributable to the discourse strategy of assigning relevance to modifiers of combined concepts.

Note that noun features may be considered *redundant* because they are true of both the combined concept and the head noun category alone. For instance, the feature "round" of 'peeled apples' is redundant because apples in general are also round. Phrase features, on the contrary, may be considered *diagnostic* because they distinguish the combined concept from other members of the head noun category. In other words, one way that 'peeled apples' differ from other apples is that they are white instead of red.

Hence, when combined concepts are encountered in the absence of an informative context, phrase features differ from noun features in two respects: (1) Whereas phrase features are diagnostic, noun features are redundant, and (2) phrase features are more relevant than noun features. Given that we observe a phrase feature priority over noun features in the lack of a context, the question then becomes whether phrase features are verified faster because they are diagnostic, or because they are relevant.

Our data directly address this question. If it is diagnosticity that produces the phrase feature priority, then (diagnostic) phrase features should be verified faster than (redundant) noun features, regardless of contextual relevance. This did not happen. Instead, we found that when those diagnostic features were made irrelevant, they were no longer preferentially processed. Rather, the relevance of the context determined relative speed of verification, regardless of the diagnosticity of the target feature. Essentially, phrase features are preferentially processed as a consequence of their relevance, not their diagnosticity.

Our results also parallel the notion of encoding specificity (Tulving & Thomson, 1973) in the memory literature. Recall is facilitated for items cued by features relevant to their original study context, in comparison to items cued by features irrelevant to the original study context (Anderson & Ortony, 1975; Barclay, Bransford, Franks, McCarrell, & Nitsch, 1974). For instance, the context for the target item 'piano' concerned either lifting a piano, for which the cue "heavy" was relevant, or tuning a piano, for which the cue "nice sound" was relevant. Cues relevant to the study context were more effective than cues that were not relevant. That is, "heavy" was a better cue for 'piano' after the lifting context than after the tuning context, while "nice sound" was a more effective cue following the tuning context than the lifting context.

We may thus frame our results as a special case of encoding specificity. When combined concepts appear in isolation or in unconstraining contexts, phrase features are verified prior to noun features as a result of the assumption of relevance described above. However, when combined concepts appear in contexts for which one feature or another is sufficiently relevant, people no longer make this default assumption, but instead selectively encode those features that are relevant in the context. In this way, the default phrase feature priority may be overturned, as the differential accessibility of noun and phrase features is governed by contextual relevance.

Acknowledgments

This research was supported by a National Science Foundation graduate research fellowship to the first author and by Grant #SBR-9712601 from the National Science Foundation to the second author. We are especially grateful to Christina Gagne for some of the materials for Experiment 1. We also thank Yevgeniya Goldvarg, Matt McGlone, Sahan Mukherji, and Mary Newsome for helpful comments.

References

Anderson, R. & Ortony, A. (1975). On putting apples into bottles: A problem of polysemy. Cognitive Psychology, 7, p. 167-180.

Barclay, J., Bransford, J., Franks, J., McCarrell, N., & Nitsch, K. (1974). Comprehension and semantic flexibility. Journal of Verbal Learning and Verbal Behavior, 13, p. 471-481.

Barsalou, L.W. (1982). Context-independent and context-dependent information in concepts. Memory & Cognition, 10, 82-93.

Estes, Z. & Glucksberg, S. (1998). Contextual activation of features of combined concepts. In Proceedings of the Twentieth Annual Conference of the Cognitive Science Society, p. 333-338. Mahwah, New Jersey: Erlbaum.

Estes, Z. & Glucksberg, S. (in press). Interactive property attribution in concept combination. Memory & Cognition.

Gagne, C.L. & Murphy, G.L. (1996). Influence of discourse context on feature availability in conceptual combination. Discourse Processes, 22, 79-101.

Grice, H.P. (1975). Logic and conversation. In P. Cole & J. Morgan (Eds.), Syntax and Semantics: Vol. 3, Speech Acts, p. 41-58. New York: Academic.

Hampton, J.A. (1987). Inheritance of attributes in natural concept conjunctions. Memory & Cognition, 15, 55-71.

Hampton, J.A. (1988). Overextension of conjunctive concepts: Evidence for a unitary model of concept typicality and class inclusion. Journal of Experimental Psychology: Learning, Memory and Cognition, 14,12-32.

Hampton, J.A. & Springer, K. (1989). Long speeches are boring: Verifying properties of conjunctive concepts. Paper presented at the 30th meeting of the Psychonomic Society, Atlanta.

Haviland, S.E. & Clark, H.H. (1974). What's new? Acquiring new information as a process in comprehension. Journal of Verbal Learning and Verbal Behavior, 13, 512-521.

Hess, D.J., Foss, D.J., & Carroll, P. (1995). Effects of global and local context on lexical processing during language comprehension. Journal of Experimental Psychology: General, 124, p. 62-82.

Hornby, P. (1974). Surface structure and presupposition. Journal of Verbal Learning and Verbal Behavior, 13, 530-538.

Johnson-Laird, P.N. (1975). Meaning and the mental lexicon. In A. Kennedy & A. Wilkes (Eds.), Studies in long-term memory, p. 123-142. London: Wiley.

McKoon, G. & Ratcliff, R. (1988). Contextually relevant aspects of meaning. Journal of Experimental Psychology: Learning, Memory, and Cognition, 14, p. 331-43.

Medin, D.L., & Shoben, E.J. (1988). Context and structure in conceptual combination. Cognitive Psychology, 20, 158-190.

Murphy, G.L. (1988). Comprehending complex concepts. Cognitive Science, 12, 529-562.

Murphy, G.L. (1990). Noun phrase interpretation and conceptual combination. Journal of Memory and Language, 29, 259-288.

Singer, M. (1976). Thematic structure and the integration of linguistic information. Journal of Verbal Learning and Verbal Behavior, 15, 549-558.

Springer, K. & Murphy, G.L. (1992). Feature availability in conceptual combination. Psychological Science, 3, 111-117.

Tabossi, P. (1982). Sentential context and the interpretation of unambiguous words. Quarterly Journal of Experimental Psychology, 34A, 79-90.

Tabossi, P. (1988). Effects of context on the immediate interpretation of unambiguous words. Journal of Experimental Psychology: Learning, Memory, and Cognition, 14, 153-162.

Tabossi, P. & Johnson-Laird, P.N. (1980). Linguistic context and the priming of semantic information. Quarterly Journal of Experimental Psychology, 32, 595-603.

Tulving, E. & Thomson, D.N. (1973). Encoding specificity and retrieval processes in semantic memory. Psychological Review, 80, 352-373.

Wisniewski, E.J. (1996). Construal and similarity in conceptual combination. Journal of Memory and Language, 35, 434-453.

Generating Support: The Influence of Perceived Category Size on Probability Judgments

Kevin W. Eva (evakw@McMaster.ca)
Department of Psychology; McMaster University
Hamilton, ON; L8S 4K1; CANADA

Lee R. Brooks (brookslr@McMaster.ca)
Department of Psychology; McMaster University
Hamilton, ON; L8S 4K1; CANADA

Abstract

When assessing the likelihood of an event, human judgment is often inconsistent with the rules inherent in standard probability theory. For example, the judged probability of an event can be heavily influenced by the alternatives that are explicitly presented. Tversky and Koehler (1994) attempted to account for this phenomenon by arguing that probability judgments are made by comparing the amount of cognitive support one holds in favour of the event in question relative to all other possibilities. They suggested that different descriptions of the same event elicit different amounts of support resulting in different probability ratings. In addition to the role played by explicitly considered alternatives, the present paper suggests that people are also sensitive to the influence of alternatives that are not considered explicitly. We present the term 'implied numerosity' in an attempt to indicate that probability ratings are influenced by a general impression of the number of potential alternatives that exist. Systematic differences in probability estimations may result from systematic changes in the perceived size of the category being evaluated.

Introduction

When judging probability or estimating frequency, the numbers raters assign often fail to follow the rules inherent in standard probability theory. For example, Redelmeier, Koehler, Liberman, and Tversky (1995) showed physicians medical case histories and asked them to estimate probabilities for specific diagnoses. Half of the physicians were asked about two diagnoses ("gastroenteritis" and "ectopic pregnancy") and the residual category ("none of the above") while the other half were asked to estimate probabilities for a list of five diagnoses (including "gastroenteritis" and "ectopic pregnancy") and a residual ("none of the above"). Although the case histories shown to both groups were identical, the average probability assigned to the residual in the short list was smaller than the sum of the corresponding probabilities in the long list. In that same paper the authors showed that, not only does the estimated probability of a particular medical diagnosis vary systematically as a function of the number of alternative diagnoses explicitly presented, but that physicians' treatment decisions were also influenced. When sinusitis was the only diagnostic hypothesis given explicitly, fewer respondents recommended a CAT scan than when a longer list of potential diagnoses was offered. Tversky and Koehler (1994) and Rottenstreich and Tversky (1997) have shown this effect using a wide variety of questions and by asking subjects to respond with either probability ratings or frequency estimations.

While these authors have described and modeled these phenomena in impressive detail, much of their focus has been on the influence of alternatives they explicitly gave the raters or alternatives raised by the raters themselves. We now propose that the alternatives that are not considered explicitly may also influence our probability ratings. The present paper introduces the idea of implied numerosity to suggest that one component of probability evaluation involves an assessment of the size of the category to which an event belongs. That is, the context of the questions, including the alternatives offered explicitly, may alter a judge's impression of the number of possible alternatives that have not been mentioned. Although descriptions of the phenomena described in the preceding paragraph might cause humans to appear to be illogical decision makers, the current framework suggests that some part of that appearance might be the by-product of a rational judgment process. We argue that experimental manipulations shown to alter probability ratings may have done so, at least partially, by changing the perceived size of the category being evaluated. This proposal will be argued for in two ways: By an assessment of its ability to explain results already present in the literature and by examining novel predictions that have been tested and are presented here. First, however, we will briefly discuss the need for such a proposal.

I. Implied Numerosity: Seeing the Forest or Counting the Trees?

Support Theory, as presented by Tversky and Koehler (1994) was first developed to explain the finding that different descriptors of the same event give rise to different probability judgments. For example, subjects typically assign a larger probability estimate when asked to evaluate the probability that a randomly selected person will die of natural causes including heart disease, pneumonia, or lung cancer, relative to when the same question is asked without the explicit mention of exemplary instances. Tversky and Koehler postulated that this is so because such judgments are made on the basis of cognitive extensions (hypotheses) rather than physical extensions (events) in the real world. In other words, rather than performing a memorial count of the ways in which people might die of natural causes relative to all other types of death, Tversky and Koehler argued that probability judgments are developed through the consideration of a ratio of cognitive 'support' for the alternative in question (dying of natural causes) relative to all other alternatives (dying of unnatural causes):

$$P(A, B) = \frac{s(A)}{s(A) + s(B)} \qquad (1)$$

where P(A, B) = probability of 'A' rather than 'B' and s(A) = support for A. Probability estimates, they suggested, are greater when specific alternatives are "unpacked" (i.e., when exemplary instances are explicitly mentioned) relative to when they remain implicit, because unpacking increases the amount of support in favour of the event in question. They modeled this suggestion in the first two terms in Equation 2:

$$s(A) \leq s(B \vee C) \leq s(B) + s(C) \qquad (2)$$

where (B ∨ C) is an explicit disjunction of the implicit disjunction 'A,' and as will be explained shortly, s (B) + s (C) are the ratings of each component made independently. Using the above example, 'A' = death due to natural causes, (B ∨ C) = heart disease, lung cancer, pneumonia, or some other natural cause of death. This effect of unpacking has been labeled "implicit subadditivity" since implicitly evaluated diagnoses typically result in a lower probability rating relative to those that are presented as unpacked, explicit possibilities. The latter two terms refer to a phenomenon labeled "explicit subadditivity" by Rottenstreich and Tversky (1997). It is included to illustrate that when probability ratings are assigned to alternatives one at a time by independent subjects, their sum is typically greater than when the same alternatives are explicitly presented, but evaluated as a whole. Again, Rottenstreich and Tversky argued that this is so because the amount of support generated in favour of those alternatives increases when they are evaluated independently.

All in all, Support Theory has had tremendous success; the original formulation as well as the extension published by Rottenstreich and Tversky have provided good explanations for (as well as predictions of) many of the inconsistencies found between subjective probability judgments and standard probability theory. In addition, Support Theory has been used as a tool in attempts to further our understanding of decisions made during the diagnostic process (Eva, Brooks, Cunnington, and Norman, Under Review; Redelmeier, et al., 1995).

There remains, however, another aspect of unpacking which, to our knowledge, has not been addressed by any of the previous writings: the influence held by the alternatives that are not explicitly evaluated. As is evidenced by Equations 1 and 2, much of the work performed in this area focuses on the effect of packing and unpacking specific alternatives. Tversky and Koehler (1994) speculated that

the amount of support in favour of any one alternative is constrained by both memory limitations (i.e., the ability to recall possible alternatives) and attentional capture (i.e., an increased salience as a result of an alternative's explicit presentation). Maintaining such a focus on the consideration of specific alternatives leaves one susceptible to the default assumption that only explicitly considered alternatives are influential. Surely, however, there is a more generic, exemplar-free way of performing frequency estimations that should not be ignored. Just as one can estimate the size of a choir by trying to pick out multiple specific voices, the volume of the chorus as a whole must also provide valuable information. The alternatives that are explicitly presented could possibly provide information that can alter a judge's perception regarding the number of potential alternatives that might be generated even without their actual generation. That is, the alternatives themselves imply numerosity.

This possibility is consistent with personal experience as well as with discussion held with experimental participants. Ask yourself the question "How probable is it that Russia hosts the world's largest prison?" While highly capable of generating a long list of countries, many of which would be reasonable alternatives (including China which is, incidentally, the correct answer), most people do not attempt to do so spontaneously. In fact, Eva et al. (Under Review) have found evidence that even the most likely alternatives in a diagnostic decision task may not be generated (or at the very least, are under-appreciated) unless explicitly mentioned. Rather, a more commonly observed pattern is the automatic consideration of the explicitly mentioned alternatives (or at most, 1 or 2 alternatives that are included only implicitly yet come to mind quickly) followed by some vague consideration (i.e., a general impression) of the question as a whole. It seems unlikely that this discounting of alternatives not explicitly mentioned is due to low motivation during the experimental situation given that this strategy has been observed in the most dedicated subjects as well as during natural conversation with other individuals. Rather, when asked to evaluate the likelihood that a given answer is correct, or that a given event might occur, there seems to be a natural tendency to make a decision on the basis of some general impression of its probability rather than through the use of specific comparisons of a number of possible alternatives.

What creates this "general impression" that seems to be driving the responses given by subjects? Given that people do not expend a lot of effort generating additional alternatives to compare with the explicitly given alternatives, what determines the magnitude of the probabilities assigned? We propose that an evaluation of the amount of 'support' one holds in favour of particular alternatives is influenced by a general impression of the size of the category judges are being asked to evaluate. We turn our attention now to an evaluation of whether this framework might still enable the explanation of known phenomena. As implied numerosity is being proposed as one mechanism by which 'support' can be generated, it maintains much of the explanatory power of Support Theory.

II.   Explaining Known Phenomena

Subadditivity

As described earlier, Support Theory predicts that unpacking the Focal hypothesis will result in an increase in its judged probability whereas unpacking the alternative hypothesis should result in a decrease in the judged probability of the Focal. That is, implicit subadditivity requires that the explicit presentation of "pneumonia, cancer, and myocardial infarction (MI)" cause the Focal hypothesis 'natural causes of death' to be assigned a higher probability rating. An implied numerosity framework would make the same predictions, but for a different reason. The argument would be that unpacking the Focal hypothesis causes the category 'natural causes of death' to seem larger within the superordinate category 'causes of death' relative to when the disorders remain implicit. If a category is perceived as being larger, then assigning it a greater probability as a whole is a perfectly rational action; an action that is consistent with standard probability theory as well as with the empirical results found throughout the literature. The converse of this, and the latter half of the above prediction, is that unpacking the alternative hypothesis by explicitly naming auto accidents, fires, or drowning as unnatural causes of death should reduce the probability assigned to 'natural causes of death' (Tversky and Koehler, 1994). Again, we would argue that such an unpacking causes the category 'unnatural causes of death' to appear larger within the superordinate category 'causes of death' relative to when these alternatives remain implicit. As a result, the probability assigned to 'natural causes of death' as a whole should be smaller.

The second type of unpacking, explicit subadditivity, suggests that the sum of disjoint components, when judged independently, receive a probability rating greater than or equal to the rating assigned to all components when evaluated as an explicit disjunction. To extend the above example, being asked to assign a probability to dying of pneumonia, cancer or MI, should result in a lower probability rating than the sum of pneumonia, cancer and MI when each are evaluated independently. Supportive empirical evidence was presented by Rottenstreich and Tversky (1997). Using the current framework, the explicit presentation of pneumonia, cancer, and MI together should cause the category 'natural causes of death' to appear larger relative to when pneumonia is presented by itself. So, any one alternative within the category 'natural causes of death' should receive a lower probability rating in the former case relative to the latter. As a result, we would expect pneumonia, cancer, and MI to receive larger probability ratings when presented independently and summed, relative to when all are presented together.

Binary Complementarity

Tversky and Koehler (1994) argued that the sum of probabilities assigned to alternatives that are judged in a binary manner (i.e., when a given alternative is judged in relation to its complement) should sum to one. That is, these types of judgments should be additive rather than

subadditive. For example, when asked the week before Super Bowl XXXIII if Denver will win or if Atlanta will win, independent judgments should sum approximately to 1.0. This would also be predicted using the implied numerosity framework outlined here as every instance of the category being evaluated is explicitly known regardless of whether one is asked to assign a probability to Denver or Atlanta. Evaluating one or the other can not change the perception of category size and so estimates are fairly stable and consistent with standard probability theory. Preliminary evidence has been gathered in support of this claim and will be described briefly in section III.

Strength of Alternatives Effect

An issue not addressed by any of the preceding work on subjective probability is the role played by the prevalence / plausibility of the unpacked alternatives themselves. While Support Theory does not make any specific predictions regarding this factor, manipulating the probability of the alternatives has the potential to yield three very different results. (1) If memory and salience are the sole causes of the subadditivity phenomenon, one might expect that the more likely the alternatives, the least effect they should have on people's probability judgments. Highly probable alternatives are the ones that should come to mind most readily and that should be considered with the greatest care. As a result, the most likely alternatives should influence probability judgments even when not mentioned explicitly. (2) If there is relative constancy in the degree to which the salience of an item increases upon its explicit presentation, then we would expect the same amount of subadditivity regardless of the unpacked item's prior probability. (3) An inversely proportional relationship might be found in which the decrease in probability assigned to the Focal diagnosis (i.e., the size of the unpacking effect) increases with the plausibility of the alternatives. The latter possibility would be predicted if implied numerosity is playing a role in our probability judgments; the more likely any given alternative is, the larger influence it should have on perception of category size (i.e., making it seem that a greater number of potential alternatives could be plausible) and hence the smaller the probability should be assigned to any single alternative.

This latter pattern is indeed what was observed by Eva, Brooks, Cunnington, and Norman (Under Review). Medical students and Motive Technology Students were shown brief case histories relevant to their field of study and asked to evaluate the probability of that case being representative of each diagnosis in the presented list. On average, the rating of the Focal hypothesis decreased when the Residual alternative was unpacked even when the unpacked alternatives were considered Implausible by expert raters - a finding consistent with Support Theory. Also consistent, but not predicted by Support Theory was the finding of the "strength of alternatives effect." It was found that the magnitude of the probability assigned to the Focal diagnosis was inversely proportional to the probability of the alternatives explicitly presented.

157

Other Effects of Context on Judgments

Within this framework, biases in probability judgments become consistent with numerous other findings in Psychology - those that have revealed that a change of context can alter the way in which we perceive and evaluate a problem. Studies of hindsight in both psychology (see Hawkins and Hastie, 1990 for a review) and medicine (Arkes et. al., 1981) have revealed that people who know an event occurred tend to believe falsely that they would have predicted the reported event. Teigen (1983) presented subjects with mystery stories and varied the number, and the role, of characters who may have committed the murder. His results suggested that the degree of suspicion drawn against one person had profound implications for the evaluation of the other characters - probability estimation was seemingly not done in isolation from the alternatives themselves. Similarly, Norman, LeBlanc, and Brooks (In Press) have shown that the features noticed by medical students in classic patient photographs are highly dependent on the diagnosis students have in mind. With knowledge of these prior studies, it seems reasonable to believe that context would play a role in our perception of category size as well. This being the case, implied numerosity is consistent with the unpacking effects described by Support Theory. In addition though, this framework makes novel predictions regarding the magnitude of the unpacking effect - predictions that are supported by the data presented here as well as another data set that will be outlined cursorily, but presented in detail elsewhere.

III.   Testing Novel Predictions

Magnitude of the Unpacking Effect

The first prediction tested by the current study is that the size of the category required to include all unpacked alternatives should systematically alter the magnitude of the unpacking effect. This will be illustrated by the use of an example. Consider again the question "what country hosts the largest prison in the world?" If subjects are sensitive to implied numerosity, then the decrease in the probability assigned to the Focal alternative (Russia) should be greater if the Residual is unpacked into alternatives which include countries from all over the world (United States and Australia) relative to when the unpacked alternatives are all Asian countries (China and Japan). The category size (i.e., the number of possible alternatives) in the latter is smaller, so any individual country should receive a larger probability rating, creating less of a drop in the rating of the Focal hypothesis relative to when the Focal hypothesis is presented on its own (i.e., the packed condition).

The second prediction is that if the alternatives which are unpacked are held constant, then there should be a greater unpacking effect the smaller the packed category is perceived to be. The smaller a category is perceived to be, the greater should be the probability assigned to any one alternative, and so the greater the effect should be upon unpacking additional alternatives. Providing a hint that the country which hosts the largest prison in the world is in Asia should serve to reduce the size of the category of focus in the

packed condition. Therefore, when a hint is given, Russia should receive a larger probability rating -- and unpacking into China and Japan should result in a greater difference in probability ratings -- relative to when no hint is given.

Methodology 25 undergraduate Psychology students were presented with 50 trivia questions and asked to evaluate what percentage of people they thought would generate particular answers in response to being asked each question. Participants were asked to assign a number between 0 and 100 to each alternative and told that the inclusion of an "all other alternatives" option should result in the numbers summing to 100 for each question. They were also told that the correct answer was not necessarily presented, so not to make that assumption. Finally, for the purpose of a manipulation check, subjects were asked on each question to assign a number estimating how many potential alternatives might possibly be generated.

Table 1 illustrates, using an example, the 5 experimental conditions described below. Trivia questions were presented in 1 of 5 conditions all of which had a Focal hypothesis consistently presented: (1) a Large Category (i.e., no hint) Packed condition (LP), (2) A Small Category (i.e., hint given) Packed Condition (SP), (3) a Large Category Unpacked Similar Condition (LUS) in which the alternatives that were unpacked were all from a relatively small category, (4) a Large Category Unpacked Dissimilar Condition (LUD) in which the unpacked alternatives came from a wide range of possible alternatives, and finally, (5) a Small Category Unpacked Condition (SU) in which the alternatives were the same as in the LUS condition. Only the Focal alternative was presented in the Packed conditions while three alternatives, including the Focal, were presented in all unpacked cases.

Table 1: Example question illustrating five conditions

"Which country saw the invention of the bicycle?"

Condition	Hint	Alternatives
Large Packed		France, "All other countries"
Small Packed	It's in Europe	France, "All other European countries"
Large Unpacked Similar		France, England, Germany, "All other countries"
Large Unpacked Dissimilar		France, U.S.A., Taiwan, "All other countries"
Small Unpacked	It's in Europe	France, England, Germany, "All other European countries"

Results Figures 1 and 2 illustrate the scores assigned to the Focal diagnosis in each condition averaged across 49 questions - one question was dropped from analysis as an error was found in the alternatives list presented.[1] Figure 1 shows that, as predicted by Support Theory, the probability

---

[1] Error bars represent Standard Error of the Mean

assigned to the Focal diagnosis was higher in the Packed Condition than in either Unpacked conditions. Furthermore, as the implied numerosity framework predicted, the decrease in the probability assigned to the Focal is greater when the alternatives are Dissimilar (i.e., taken from a large category) relative to when the unpacked alternatives are Similar (i.e., from a smaller category). The difference scores (LP - LUS vs. LP - LUD) are statistically significant ($p < 0.03$) using a repeated measures, two-tailed t-test with question as the unit of analysis.

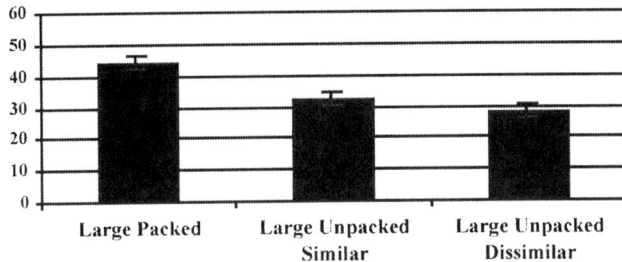

Figure 1: Mean estimated Pr (Focal alternative)

Figure 2 reveals that the second prediction made by the current framework was also observed. Keeping the unpacked alternatives constant, but varying the size of the category of focus in the Packed condition resulted in a larger unpacking effect when the category size in the Packed condition was small (SP – SU) relative to when the category size of the Packed condition was large (LP – LUS). This difference was significant ($p < 0.05$) using the same analysis as above. It can also be observed in Figure 2 that the difference in the two unpacking effects resulted from an increase in the probability assigned to the Focal diagnosis in the Small category Packed condition. This was also predicted by the impled numerosity framework, because the similarity of scores in the two Unpacked conditions is to be expected if the manipulation of unpacking into similar alternatives had the same effect as providing a hint while asking the question. That is, it appears to be true that the alternatives presented lead the judge to focus on a category of sufficient size to encompass all explicitly mentioned alternatives and nothing more. In addition to this being viewed as a manipulation check, the similar ratings in these two conditions also rules out the possibility that the higher probability assigned to the Focal diagnoses in the Small Packed condition relative to the Large Packed condition resulted solely from an increase in confidence as a result of having been given a hint.

The Ratio Rule

As described in the section entitled "Binary complementarity," the implied numerosity framework suggests that probability judgments will show additivity when every alternative is known before estimates are provided. If all alternatives are known, then changing the alternative presented can not influence the contextually determined perception of number of potential alternatives.

Although the results are preliminary and as such will not be presented in any detail, the authors have used the Ratio Rule espoused by Tversky and Koehler (1994) and elaborated on by Rottenstreich and Tversky (1997) as a measure of the consistency of subjects' probability ratings. As predicted by the current framework, it appears that the critical determinant in whether or not the behavioural data satisfies the rule is whether or not judges have an appreciation of the size of the entire sample set before evaluating the probability of any pair of alternatives. That is, only when there is no change in the number of potential alternatives are probability ratings consistent with the rules inherent in standard probability theory. This will be expounded upon at a later date - it is mentioned here simply to foreshadow further support for the implied numerosity framework.

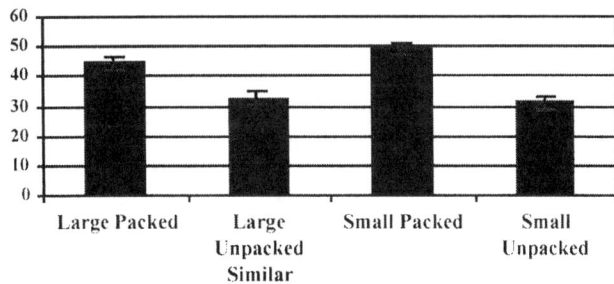

Figure 2: Mean estimated Pr (Focal alternative)

Discussion

Like Cosmides and Tooby (1995), we propose that humans may be good intuitive statisticians after all. It is a rational act to assign a probability on the basis of the number of potential alternatives -- all else being equal, the more alternatives there are, the lower the likelihood that any one alternative will be correct. Unlike Cosmides and Tooby, we do not feel that the phenomena observed in much of the judgment under uncertainty literature is simply a numbers game that results from a poor conceptualization of probabilities. Rather, we argue that the systematic variations found in probability judgments arise as a result of conceptions being altered by the way in which questions are asked or by the number of alternatives that one is explicitly asked to evaluate. The presentation of specific alternatives not only influences the support in favour of those alternatives, but can also drive people's impressions of the number of potential alternatives that they have not considered explicitly. A physician who is asked to evaluate a medical case for the presence of two specific infectious disorders might perceive the problem differently than one who is asked to evaluate the same problem via the consideration of an infectious disorder and a genetic defect. Even if additional alternatives are not explicitly considered, the two physicians' senses as to the number of plausible alternative diagnoses that exist is likely very different, but nonetheless potentially influential in both cases.

As has hopefully been made clear, we are not arguing that implied numerosity is the only factor influencing our probability judgments. On the contrary, Support Theorists have shown that a more explicit consideration of specific alternatives is very influential. In addition, using a Process Dissociation Procedure, Begg, Faulkner and Jacoby (Unpublished Manuscript) have demonstrated to our satisfaction that frequency discriminations are affected by both automatic and controlled processes such as availability and memory for frequency, respectively. We speculate that implied numerosity acts in a more automatic way in that it is not expected that judges consciously evaluate the number of alternatives that might be considered plausible before transforming the result into a probability judgment or frequency estimation. Regardless of whether or not this is the case, there remain at least two possible mechanisms through which judges might be sensitive to implied numerosity.

It has been argued throughout this paper that the alternatives explicitly mentioned create a context through which judges gain a general impression of the number of possible alternatives that exist. A second possible interpretation of the results presented is that the explicit mention of examples that constitute a relatively large category cue more additional alternatives than do examples that constitute a smaller category. The latter possibility seems unlikely for two reasons. First, as mentioned earlier, Eva et al. have presented evidence which suggests that even judges who have been trained to generate lists of differential diagnoses may fail to bring to mind even highly likely alternatives if they are not explicitly mentioned. Second, researchers examining concept formation (e.g., McRae, de Sa, and Seidenberg, 1997) often opt to use larger category decision tasks (e.g., superordinate descriptors such as "is it a living thing") so as to avoid cueing specific exemplars. Jared and Seidenburg (1992) found that narrower decision tasks (e.g., "is it a bird") were inappropriate, because they were more likely to cue specific exemplars. If this is the case, then a greater amount of spontaneous unpacking would have been expected in the "small category" condition relative to the "large category" condition, thereby resulting in predictions opposite to the ones verified in this paper. Furthermore, in observing experimental participants, it appears unlikely that frequency estimations consist of a systematic search through multiple possible alternatives. Rather, we argue that the information gained through implied numerosity consists of the creation of a general impression of the size of the category under consideration, thereby modulating the perceived probability.

Acknowledgements

This research was carried out under the support of the Natural Sciences and Engineering Research Council of Canada through grants to both authors.

We would like to express thanks to Bruce Whittlesea and Larry Jacoby for useful comments, discussion and motivation.

References

Arkes, H.R., Saville, P.D., Wortmann, R.L., & Harkness, A.R. (1981). Hindsight bias among physicians weighing the likelihood of diagnoses. *Journal of Applied Psychology, 66*, 252-254.

Begg, I.M., Faulkner, H.J. and Jacoby, L.L. (Unpublished manuscript). Dissociation of processes in frequency discrimination: The frequency attribute is controlled and intentional but the availability heuristic is automatic.

Cosmides, L., and Tooby, J. (1996). Are humans good intuitive statisticians after all? Rethinking some conclusions of the literature on judgment under uncertainty. *Cognition, 58*, 1-73.

Eva, K.W., Brooks, L.R., Cunnington, J.P.W., and Norman, G.R. (Under Review). Forgetting the unforgettable? The role of alternatives in diagnostic decision-making and probability judgments.

Hawkins, S.A., & Hastie, R. (1990). Hindsight: Biased judgments of past events after the outcomes are known. *Psychological Bulletin, 107*, 311-327.

Jared, D., & Seidenberg, M.S. (1992). Does word identification proceed from spelling to sound to meaning? *Journal of Experimental Psychology: General, 120*, 358-394.

McRae, K., de Sa, V.R., & Seidenber, M.S. (1997). On the nature and scope of featural representations of word meaning. *Journal of Experimental Psychology: General, 126*, 99-130.

Norman, G.R., LeBlanc, V.R., & Brooks, L.R. (In Press). On the difficulty of noticing the obvious. Psychological Science.

Redelmeier, D.A., Koehler, D.J., Liberman, V., & Tverksy, A. (1995). Probability judgment in medicine: Discounting unspecified possibilities. *Medical Decision Making, 15*, 227-230.

Rottenstreich, Y., & Tversky, A. (1997). Unpacking, repacking, and anchoring: Advances in Support Theory. *Psychological Review, 104*, 406-415.

Teigen, K.H. (1982). Studies in subjective probability III: The unimportance of alternatives. *Scandinavian Journal of Psychology, 24*, 97-105.

Tversky, A., & Koehler, D.J. (1994). Support Theory: A nonextensional representation of subjective probability. *Psychological Review, 101*, 547-567.

Modeling the Role of Plausibility and Verb-bias in the Direct Object/Sentence Complement Ambiguity

Todd R. Ferretti (todd@sunrae.sscl.uwo.ca)
Ken McRae (mcrae@uwo.ca)

Department of Psychology, Social Science Centre, University of Western Ontario
London, Ontario, Canada N6A 5C2

Abstract

We provide a computational account of the integration of various constraints proposed to be involved in the resolution of the direct object/sentence complement ambiguity. In the first part, competition-integration simulations show that a constraint-based model accounts for the results of Garnsey, Pearlmutter, Myers, and Lotocky (1997) at least as well as the garden-path model. In the second part, we compare the efficacy of norming techniques for capturing plausibility effects. Simulations show that norms designed to tap people's conceptual knowledge of events better capture plausibility effects than do norms that are biased toward tapping linguistic knowledge. We conclude that local information concerning event plausibility is an important constraint for understanding ambiguity resolution.

Part 1: Distinguishing Between Theories of Sentence Processing

The human language comprehension system is impressive in its ability to integrate multiple sources of information when comprehending sentences (Marslen-Wilson, 1975). Most theories of sentence comprehension include the idea that understanding sentences involves the rapid integration of general syntactic information (Frazier & Rayner, 1982), lexically-specific syntactic information (MacDonald, Pearlmutter, & Seidenberg, 1994; Garnsey et al., 1997), discourse information (Spivey & Tanenhaus, 1998), and knowledge of thematic roles (McRae, Ferretti, & Amyote, 1997; Trueswell, Tanenhaus, & Garnsey, 1994). They differ, however, in their claims about precisely when the comprehension system exploits various types of information. Two prominent theories of sentence processing have emerged in this debate.

The garden-path model (Frazier, 1987; Frazier & Rayner, 1982) claims the comprehension system naively constrains initial interpretation by using a limited subset of the relevant information, reserving other information sources for evaluating and, if necessary, revising the initial interpretation. The first stage of comprehension uses only the major syntactic category of each word (noun, verb, etc.), phrase-structure rules, and a small set of syntactic decision principles. One of these principles, minimal attachment, states that the initial interpretation *always* corresponds to the simplest structure that can be built. The second stage of comprehension temporally lags behind the first, with all potentially relevant sources of knowledge being used to evaluate, and if necessary, revise the initial structure.

In contrast, the constraint-based approach views syntactic ambiguity resolution as a continuous process in which the most likely syntactic alternatives are evaluated with respect to all available evidence (MacDonald et al., 1994; McRae, Spivey-Knowlton, & Tanenhaus, 1998). Thus, the distinguishing features of this model are that multiple constraints are combined to compute alternative interpretations in parallel, and that these alternatives compete with one another during processing. Unlike the garden-path model, lexically-specific knowledge and relevant discourse information is available to guide initial interpretations rather than solely to revise the initial interpretation if it is incorrect.

The Direct Object/Sentence Complement Ambiguity

The direct object/sentence complement ambiguity occurs when a noun phrase (NP) following a main verb is temporarily ambiguous with respect to its relationship to the verb. Consider the following examples:

1a)   The gossipy neighbor heard (that) the story had never
         actually been true.
1b)   The gossipy neighbor heard (that) the house would
         never be flooded again.

In (1), the NPs "the story" and "the house" are ambiguous as to whether they are the direct object of the verb (i.e., something that someone has heard), or the subject of a sentence complement (something that someone has heard *about*). In (1a), for example, when the complementizer "that" is removed from these sentences, readers receive confirmation that the gossipy neighbor has not directly heard the story when they read the auxiliary verb "had" or "would". Readers' use of verb-bias and plausibility information to constrain resolution of the direct object/sentence complement ambiguity has recently been investigated by Garnsey et al. (1997). They manipulated the plausibility of the post-verbal noun as a direct object. For example, in (1a) "the story" is a plausible direct object of the verb "heard" (i.e., people tend to hear stories directly), whereas "the house" is not as plausible (people do not commonly hear houses directly). Of particular interest to Garnsey et al. was how this type of plausibility

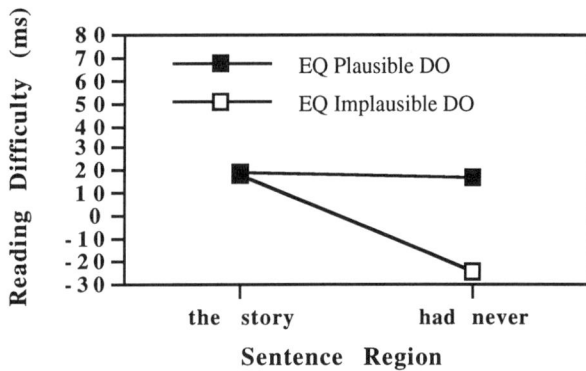

Figure 1: Reading difficulty (ambiguous - unambiguous reading times) for equi-bias verbs in the two critical regions of the sentences for both implausible and plausible direct objects.

information influences ambiguity resolution when the main verb of the sentence frequently occurs with either direct objects (e.g., heard), sentence complements (indicated), or both structures with about equal probability (felt). The main prediction of this study was that direct object plausibility would have its strongest influence when verbs appeared with the alternative syntactic structures with about equal probability. This prediction was supported by their first-pass eyetracking results, illustrated in Figure 1. When a plausible NP ("the story") followed an equi-bias verb, readers had difficulty at the disambiguating region ("had never"), but not when the NP was implausible ("the house"). This pattern of results appears to be most compatible with the constraint-based model. However, because the possibility remains that revision processes may have accounted for the data, the results may be captured by the garden-path theory as well.

Competition-Integration Modeling

To distinguish between these alternatives, two competition-integration models were constructed to implement the predictions of the opposing theories. Figure 2a presents a schematic of the garden-path version of the model, whereas Figure 2b is the constraint-based version. In these models, the syntactic alternatives compete with one another during processing until one alternative reaches a criterion value. Competition arises because multiple constraints provide support for various interpretations. A major assumption of this model is that the amount of competition produced in a region of a sentence is proportional to the reading difficulty people experience at the region, with reading difficulty measured as the reading-time difference between ambiguous (no "that") and unambiguous ("that" present) versions of the sentence.

The Relevant Constraints

The two versions of the model differ most significantly in that the garden-path version includes an additional constraint, labeled general verb bias, that represents the minimal

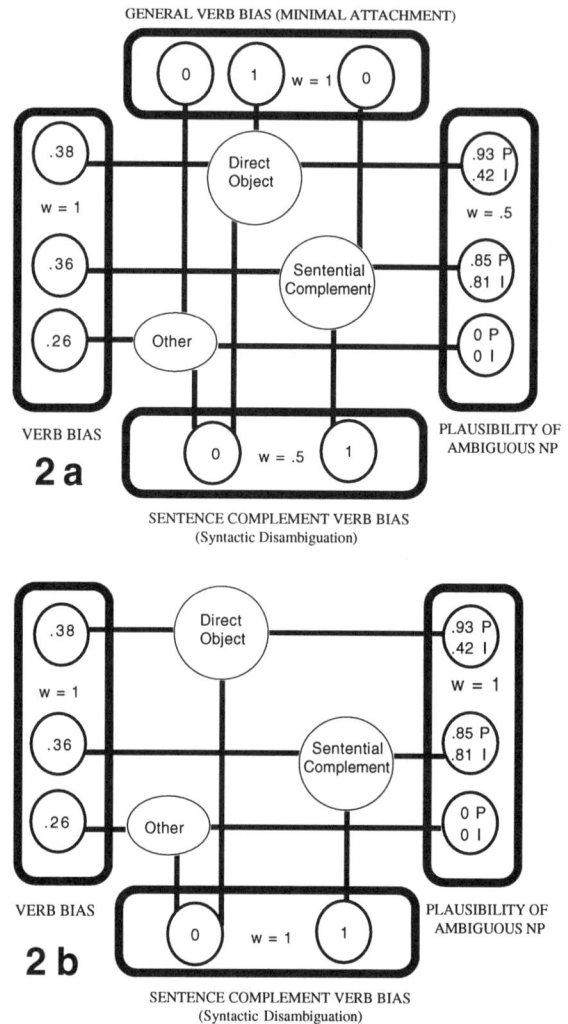

Figure 2: 2a is a schematic of the garden-path version of the model, whereas 2b is the constraint-based version.

attachment parsing principle. This constraint provides full support for the direct object interpretation in the main verb region because this is the simplest syntactic structure that can be constructed from the input to that point. In addition, there are three constraints common to both versions: verb-bias, the plausibility of the postverbal NP, and the sentence complement bias provided by the auxiliary verb. The verb-bias constraint captures the lexically-specific bias of the main verb in terms of the frequency with which it is followed by a direct object, a sentence complement, or some "other" structure such as a prepositional phrase or infinitive. The values for this constraint were taken from sentence completions collected by Garnsey et al. and the means are listed in the small circles inside the rectangle representing verb-bias.). The "plausibility of ambiguous NP" constraint captures plausibility information provided by the noun appearing in the ambiguous NP region. Note that the "P" and "I" values in the small circles represent the mean plausibility of the nouns (divided by the highest rating possible) labeled as plausible and implausible by Garnsey et al. These values

162

were collected in a norming study which is described in more detail in Part 2. Finally, the "sentence complement verb bias" constraint fully supports the sentence complement structure because it reflects the fact that the auxiliary verb disambiguates the sentence toward the complement interpretation.

There were no free parameters in the constraint-based version of the model. Each constraint was given a weight of 1 because the verb-bias entered into competition at the initial verb region, plausibility entered at the NP region ("the story") and the sentence complement verb-bias entered at the "had never" region. In the garden-path version, both the lexically-specific verb-bias and plausibility were delayed by one word. Thus, minimal attachment entered competition alone at the initial verbs and the verb-bias entered alone at the NP region, so both were given a weight of 1. Because the remaining two constraints entered at the "had never" region, the most straightforward way to weight them was .5 each. Note however that similar results were obtained using weights in the range of .2/.8 to .8/.2.

Processing

As in McRae et al. (1998), constraints were integrated at each region using a three-step normalized recurrence mechanism. First, each of the c informational constraints was normalized.

$$S_{c,a}(\text{norm}) = S_{c,a} / \sum S_{c,a} \qquad (1)$$

$S_{c,a}$ represents the activation of each constraint node (i.e., the cth constraint that is connected to the ath interpretation node). $S_{c,a}(\text{norm})$ is the normalized constraint activation. Constraints were then integrated at each interpretation node.

$$I_a = \sum [w_c * S_{c,a}(\text{norm})] \qquad (2)$$

I_a represents the activation of the ath interpretation node. The weight linking the cth constraint node to interpretation node I_a is represented by w_c. Finally, the interpretation nodes sent positive feedback to the constraints based on Equation 3. Note that the weights were equal in both directions.

$$S_{c,a} = S_{c,a}(\text{norm}) + I_a * w_c * S_{c,a}(\text{norm}) \qquad (3)$$

These three steps were computed in sequence for each cycle of competition. The network iterates at each region, with the difference between the interpretation nodes gradually increasing until a criterion is reached. This criterion changes over the iterations within a region so that it becomes more lax as competition ensues. A dynamic criterion is necessary for simulating reading across multiple regions of a sentence due to the fact that fixation durations are partially determined by a preset timing program (Rayner & Pollatsek, 1989); that is, a reader will spend only so long on a fixation before making a saccade. As in Spivey and Tanenhaus' (1998) eyetracking simulations, we used $\Delta crit = .01$.

$$dynamic\ criterion = 1 - \Delta crit * cycle \qquad (4)$$

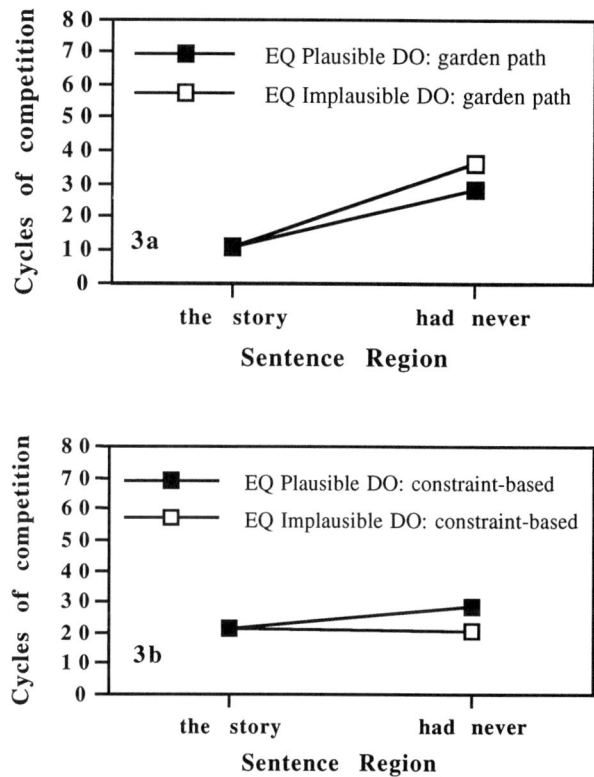

Figure 3: Cycles of competition to reach the dynamic criterion for plausible and implausible nouns in the two critical regions for the sentences containing equi-bias verbs.

Simulation Results

As illustrated in Figure 3a, the garden-path simulation predicts a different pattern of ambiguity effects than those found in the human reading-time data shown in Figure 1. The main difference between the simulations can be seen in the pattern of competition at the auxiliary verb region. The major discrepancy between the human data and the garden-path simulation is that the implausible nouns produced more competition than the plausible nouns in this region. This pattern of results is opposite to that found in the Garnsey et al.'s eyetracking data. Basically, the delay of the verb-bias and plausibility information could not be overcome. In contrast, the constraint-based simulation produced less competition for implausible nouns than for plausible nouns in the auxiliary verb region, as was found in the reading-time experiment. These simulations add to those of McRae et al. (1998) and Spivey and Tanenhaus (1998) in suggesting that architecturally-determined delays of lexically-specific syntactic and plausibility information are not present in the human sentence-comprehension system.

A look at the results of the constraint-based simulation, however, shows that although it captured the human data better than did the garden-path model, it does not completely capture the pattern of results. In particular, the pattern of reading difficulty at the verb versus at the auxiliary region was not mimicked. This is addressed in Part 2.

163

Table 1: Mean (and standard error) of the ratings for the items in the Equi-bias conditions of Garnsey et al. for each norming technique. Values represent the actual rating divided by the highest rating possible for each technique. "Plausible DO" and "Implausible DO" signify nouns that were classified as plausible or implausible direct objects by Garnsey et al.

	Garnsey et al. (1997)		Role/Filler (no subject)		Role/Filler (with subject)	
	Plausible DO	Implausible DO	Plausible DO	Implausible DO	Plausible DO	Implausible DO
Rating As Direct Object						
	.93 (.01)	.42 (.03)	.75 (.04)	.33 (.01)	.74 (.04)	.34 (.03)
Rating As Head of Sentence Complement						
	.85 (.02)	.81 (.02)	.74 (.03)	.42 (.03)	.83 (.04)	.52 (.05)

Part 2: Estimating Plausibility Constraints

One possible reason that the constraint-based simulation did not simulate the human data more closely is that the measure used to index plausibility did not properly capture it. Garnsey et al. indexed plausibility by asking participants to rate sentences or sentence fragments. To estimate the plausibility of the postverbal nouns as direct objects, participants rated sentences similar to (2a) on a 7-point scale (7=very plausible). To estimate the plausibility of postverbal nouns as the subject of a sentence complement, participants rated sentence fragments such as (2b).

(2a) The account executive concluded the speech.
(2b) The account executive concluded that the bank had

One potential problem with this type of norming is that rating sentences and sentence fragments for overall plausibility may not be the best indicator of local plausibility contingencies between lexical concepts in sentences. An alternative method for capturing the plausibility that specific objects or entities play various roles in specific events has been developed by McRae and colleagues (McRae et al, 1997; McRae et al., 1998). The underlying assumption of this norming method is that noun and verb combinations denote situations that happen in the world. Some of these situations tend to be common and others less common. Role/filler typicality norms index the plausibility of nouns as fillers for thematic roles of verbs by measuring how commonly the denoted situation occurs in the world. These norms differ from sentence ratings in that they abstract away from the linguistic structures and focus directly on knowledge about events or situations. McRae et al. (1998) showed that this world knowledge is computed and used immediately.

The following experiments and simulation examined the possibility that using a situation-based measure of plausibility might shed some light on the pattern of plausibility effects found by Garnsey et al.

Method

Participants

Sixty-eight native English-speaking psychology undergraduates from the University of Western Ontario received course credit for their participation.

Materials and Procedure

The 16 equi-bias verbs and their corresponding plausible and implausible nouns were rated in two norming studies. The first indexed the plausibility of each noun as either the direct object or subject of a sentence complement with the initial noun phrase left unspecified, thus indexing more locally the plausibility of the verb-noun combination. The second norming study did the same, but the initial noun phrase was included. This study tapped how plausibility changes as a result of certain entities being involved in the situations denoted by the verbs (i.e., noun-verb-noun combinations).

Ratings for Verb-Noun Combinations

To estimate plausibility of the post-verbal noun as a direct object, participants were asked questions such as:

How common is it for a
    story   _____
    house   _____
to be heard by someone?

Note that in the study the target nouns were placed among several filler nouns that varied in plausibility.

Similarly, to estimate plausibility of the post-verbal nouns as the subject of a sentence complement, participants rated the same nouns following questions such as:

How common is it for someone to hear something about a ?

Participants rated each combination on a 9-point scale, where 1 corresponded to very uncommon and 9 to very common. For each question type, the 16 items were placed in two separate booklets, with the second booklet containing the same question format as the first but with the items in reversed order. Participants were not under time pressure but were asked not to spend too much time on any one item, and to write down the number that corresponded to their first impression. Participants rated nouns as a direct object or as a head of a sentence complement, but not both. The booklets took approximately 30 minutes to complete.

Ratings for Noun-Verb-Noun combinations

To index plausibility when the initial noun phrases were included, participants were asked questions such as:

How common is it for a
     story   ———
     house   ———
to be heard by a gossipy neighbor?

or

How common is it for a gossipy neighbor to hear something about a ?

All other aspects were identical to the experiment performed without the initial NPs.

Results and Discussion

The results for both direct object and sentence complement plausibility are shown in Table 1. Separate analyses were performed on the direct object and sentence complement ratings. Each set of ratings were entered into an analysis of variance using items as the random variable. The dependent variable was the mean rating (by item) divided by the maximum rating, and the independent variables were norming technique (3 levels: Garnsey et al. sentence norms vs role/filler without subject NP vs role/filler with subject NP) and plausibility (2 levels: plausible DO vs implausible DO). Norming technique and plausibility were within-items.

Ratings as Direct Objects

The interaction between plausibility and norming technique did not reach significance, $F(2,45) = 2.37$, $p<.11$. However, planned comparisons revealed that the nouns labeled as plausible were rated as more plausible in the Garnsey et al. norms than in either the role/filler norms without the subject included, $F(1,85) = 10.42$, $p<.01$, or the role/filler norms with the subject included, $F(1, 85) = 10.91$, $p<.01$. There was no difference between the latter two, $F<1$. The nouns labeled as implausible were rated numerically but not significantly more plausible in the Garnsey et al. norms than in the role/filler norms without the subject, $F(1,85) = 1.45$, $p>.1$, and with the subject, $F(1,85) = 1.83$, $p>.1$. Finally, the two role/filler norms did not differ, $F<1$.

There was a main effect of norming technique, $F(2,45) = 14.08$, $p<.001$. There was also a main effect of plausibility such that nouns classified as plausible direct objects were indeed rated as more plausible, $F(1,45) = 449.82$, $p<.001$.

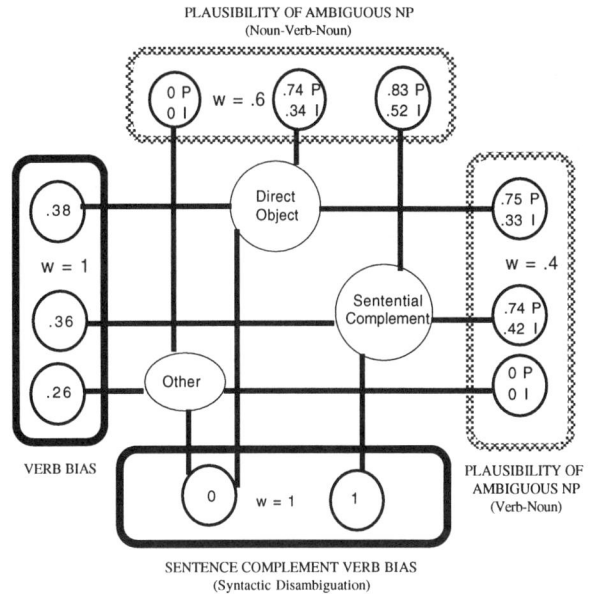

Figure 3: The model used to simulate the role/filler norms.

Ratings as Subject of Sentence Complement

Norming technique interacted with plausibility, $F(2,45) = 12.02$, $p<.001$. Planned comparisons revealed that this interaction occurred because there were no differences among the ratings for plausible nouns: Garnsey et al. versus role/filler with subjects included, $F(1,89) = 2.21$, $p>.1$; Garnsey et al. versus role/filler without subjects, $F<1$; and the two role/filler norms, $F(1,89) = 1.42$, $p>.1$. In contrast, the nouns labeled as implausible DOs were rated as more plausible in the Garnsey et al. norms than in the role/filler norms without subjects, $F(1,89) = 32.68$, $p<.001$, and with subjects, $F(1,89) = 17.58$, $p<.001$. Finally the two role filler norms did not differ, $F(1,89) = 2.32$, $p>.1$.

As above, there were main effects of norming technique, $F(2,45) = 14.08$, $p<.001$, and plausibility, $F(1,45) = 449.82$, $p<.001$.

The role/filler norming results indicate why Garnsey et al.'s plausibility manipulation produced relatively weak results. The strongest effect of this variable would be expected if the ratings had shown a high-low versus low-high pattern. Garnsey et al.'s norms suggest at least a high-high (plausible DOs) low-high (implausible DOs) pattern. However, the role/filler norms suggest a weaker manipulation of plausibility was achieved (high-high vs low-medium). In the following simulations, we examine whether these differences account for the inability of the constraint-based simulation in Part 1 to simulate the human data. The values of the role filler norming studies were entered into the constraint-based model shown in Figure 4, and the results were contrasted with the constraint-based simulation that used the norms collected by Garnsey et al.

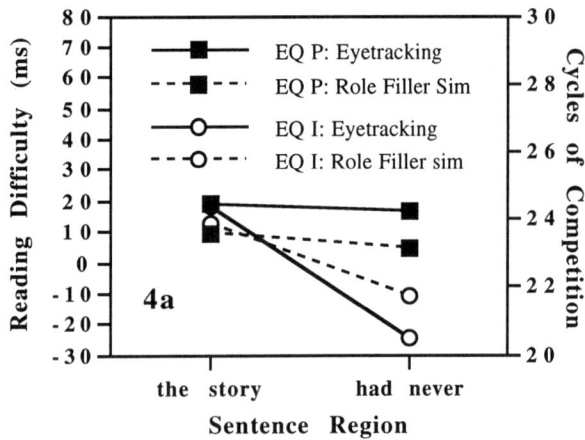

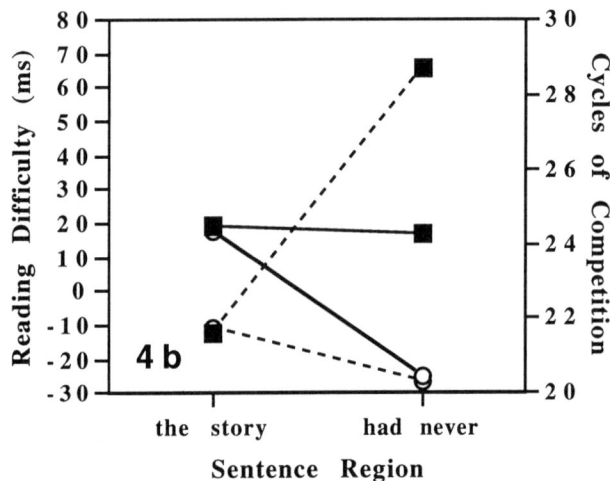

Figure 4: The mean cycles of competition produced by the constraint-based simulations are plotted with the eyetracking results of Garnsey et al. 4a shows the simulation that used plausibility as measured by the role/filler norms, whereas 4b shows the simulation that used the Garnsey et al. norms.

Simulation Results

As shown in Figure 4, the simulation using the role/filler norms to index plausibility more closely captured the eyetracking data. The main reason for the discrepancy is that the Garnsey et al. norms overestimate the difference between the plausible and implausible DOs, thus resulting in too large of a difference between these types of items at the "had never" disambiguating region. However, the norming techniques clearly make different predictions about the amount of difficulty readers should have for plausible versus implausible nouns at the disambiguation. Specifically, for the ratings collected by Garnsey et al., the model is producing more competition for the plausible nouns at the disambiguation than at the ambiguous NP. For both the role filler norms and the human data, slightly less competition (or reading difficulty) occurred for plausible nouns at the disambiguation than at the ambiguous NP. This suggests that the role/filler norms are a better measure of plausibility.

Conclusion

The simulations contrasting the garden-path and constraint-based approaches suggest that the eyetracking data collected by Garnsey et al. is better captured by a model in which all sources of information are used immediately. Moreover, the human eyetracking results were best simulated by a model that used a plausibility measure based on world knowledge abstracted away from linguistic knowledge. Understanding plausibility constraints that operate on a moment-by-moment basis during on-line sentence comprehension is best indexed by norming methods that tap knowledge of situations/events, and by an architecture in which all sources of information are immediately evaluated.

Acknowledgements

This research was supported by an NSERC Doctoral Award to the first author, and NSERC Grant 155704-98 to the second. We thank Susan Garnsey for graciously providing us with all the data necessary to conduct this research.

References

Frazier, L. (1987). Theories of syntactic processing. In J. Garfield (Ed.), *Modularity in Knowledge Representation and Natural Language Processing*. Cambridge, MA: MIT Press.

Frazier, L. & Rayner, K. (1982). Making and correcting errors during sentence comprehension: Eye movements in the analysis of structurally ambiguous sentences. *Cognitive Psychology, 14*, 178-210.

Garnsey, S. M., Pearlmutter, N. J., Myers, E., & Lotocky, M. A. (1997). The contributions of verb bias and plausibility to the comprehension of temporarily ambiguous sentences. *Journal of Memory and Language, 37*, 58-93.

MacDonald, M. Pearlmutter N. & Seidenberg, M. (1994). The lexical nature of syntactic ambiguity resolution. *Psychological Review, 101*, 676-703.

Marslen-Wilson, W. D. (1975). Sentence perception as an interactive parallel process. *Science, 189*, 226-228.

McRae, K., Ferretti, T. R., & Amyote, L. (1997). Thematic roles as verb-specific concepts. *Language and Cognitive Processes, 12*, 137-176.

McRae, K., Spivey-Knowlton, M. J., & Tanenhaus, M. K. (1998). Modeling the influence of thematic fit (and other constraints) in on-line sentence comprehension. *Journal of Memory and Language, 38*, 283-312.

Rayner, K. & Pollatsek, A. (1989). The psychology of reading. Englewood Cliffs, NJ: Prentice Hall.

Spivey, M. J., & Tanenhaus, M. K. (1998). Syntactic ambiguity resolution in discourse: Modeling the effects of referential context and lexical frequency. *Journal of Experimental Psychology: Learning, Memory, & Cognition, 24*, 1521-1543.

Trueswell, J. C., Tanenhaus, M. K., & Garnsey, S. M. (1994). Semantic influences on parsing: Use of thematic role information in syntactic ambiguity resolution. *Journal of Memory and Language, 33*, 285-318.

The Roles of Modeling, Microanalysis and Response Strategy in a Skill Acquisition Task

Jon M. Fincham (fincham@cmu.edu)
Psychology Department
Carnegie Mellon University
Pittsburgh, PA 15213-3890

John R. Anderson (ja@cmu.edu)
Psychology Department
Carnegie Mellon University
Pittsburgh, PA 15213-3890

Abstract

Researchers (see Siegler, 1987; Newell, 1973) have demonstrated the dangers of aggregating data over strategies. In this paper, we provide a current demonstration of this point using our recent work in the study of cognitive skill acquisition as a case study. Moreover, we call particular attention to the relation between cognitive modeling and microanalysis as driving forces toward a more thorough understanding of the role of strategies in cognitive skill acquisition.

Overview

We begin by reviewing the general results of our prior skill acquisition research. Following this summary, we put forth the rationale for performing a microanalysis of these data, which describes our latest effort at modeling skill learning in the experimental paradigm introduced in Anderson and Fincham (1994) and extended in Anderson, Fincham and Douglass (1997). We describe the main results of our microanalysis and finally conclude with a discussion of the importance of performing such analyses in this domain, particularly with regard to the research problem that emerges when multiple strategies, both within and between subjects, can be employed in a particular task.

Background

Anderson & Fincham (1994) and Anderson, Fincham & Douglass (1997) introduced a paradigm designed to understand the process and time course of skill acquisition. At issue in this work were two seemingly opposing theories of skill acquisition. One view was that the transition to skilled performance in a task progressed from the use of abstract rules to a reliance on retrieval of specific instances (Logan, 1988; Logan & Klapp, 1991). The other view characterized the development of skill as progressing from initially using examples to the development and use of abstract rules (Anderson, 1993). Anderson, et al. (1997), proposed a theory that is essentially a melding of these two views. The proposal describes a four stage model of skill acquisition. These (unordered, possibly overlapping) stages, ordered by increasing efficiency, characterize skill acquisition as attributable to the use of (a) analogy to examples, (b) declarative abstractions, (c) procedural rules and (d) retrieval of examples. While this four stage theory seems to be a reasonable qualitative account of the data, our goal is to elaborate the theory with a more quantitative account.

Modeling and Microanalysis Motivation

The goal of our current work is to generate a well specified, mechanistic account of the transition that occurs when moving from the novice level toward the skilled level of performance in our skill acquisition task. In particular, our intent is to develop a simulation model for the above described paradigm using ACT-R (Anderson & Lebiere, 1998; Anderson, 1993), a general cognitive modeling architecture. Developing a mechanistic account of this transition process serves several purposes. First, it is proof in principal that the qualitative account outlined above can be described by and achieved through a more formally specified process. Second, it allows us to quantify the theory in such a way that we can then use the model to make specific, testable predictions about behavior in novel tasks. Third, we hope that the modeling effort will provide additional insight into the underlying mental processes that are involved in the skill acquisition process. Finally, the act of generating the model should provide insights into subtle nuances that may exist in this specific task or within the ACT-R modeling architecture. In essence, the devil is in the details.

Serendipitously, we had recorded detailed individual mouse clicking data from Experiment 2 in Anderson, et al. (1997). In order to facilitate modeling our skill acquisition task at the atomic level of granularity outlined above, the heretofore dormant mouse clicking data were analyzed. In what follows, we will outline the general method for our task and present new results that this phase of our efforts has yielded. To foreshadow, we have found that participants employ multiple strategies when performing this task. Upon doing separate analyses for the two most common strategies, we have managed to (1) provide more compelling empirical evidence for two components of the proposed stage theory of skill acquisition and (2) identify at the mouse click level what we believe to be the atomic components involved in the execution of the task.

The Task

In the first part of these experiments participants commit to memory 8 specific facts such as "Skydiving was practiced on Saturday at 5 PM and Monday at 4 PM." Although participants were not informed of it at the time, they were learning examples of eight different rules about the time relationship between the two events for each sport. In the current example, the rule is that the second skydiving event always occurs two days later and one hour earlier than the first skydiving event. We denote this rule as "+2,-1". After successfully memorizing these eight examples, participants were told of the nature of the underlying rules associated with each sport. Participants were then tested with novel problems over a period of five days in an interface like that

illustrated in Figure 1. Participants were given either the first or second time (day and hour) and had to compute the other time using the rule associated with the given sport. Figure 1 shows a training trial where the skydiving stimulus is the first time (Friday at 3) and they would have to predict that the target was Sunday at 2. They made their prediction by clicking the relevant buttons in the boxes below. We were interested in the speed and accuracy with which they could do this. The example in Figure 1 involved going from the first time to the second time but Anderson and Fincham (1994) trained participants on 8 sports and half of them involved going from the second time to first time. On each of the five days, participants practiced 8 rules 32 times each in one direction (four with the times presented on the left hand side of the display and four presented with the times on the right hand side). Two of these rules were then tested in the other ("unpracticed") direction (also 32 times each) beginning on day 2, two more starting with day 3, two more starting with day 4, and the final two starting on day 5. We were interested in whether participants would be faster in the more practiced direction. We found more asymmetry (a greater latency advantage for rules in the more practiced direction) for rules that were reversed after more practice in one direction.

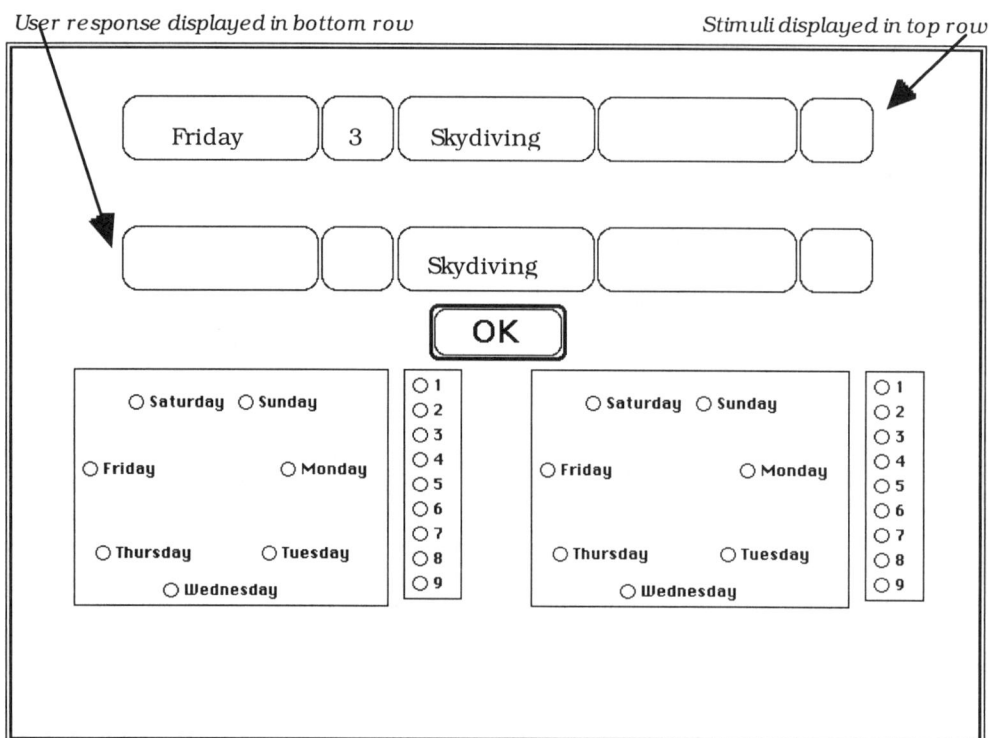

Figure 1. User interface for Anderson & Fincham skill acquisition task.

Table 1. Strategy counts as determined by response mouse click sequence for stimuli presented on the left side and right side of the display. The labels A, B, C and D represent responses of the leftmost day, leftmost hour, rightmost day and rightmost hour in the experimental interface, respectively.

Response Strategy	Ss	Left Side (A B S x x)	Right Side (x x S C D)
ST (stimulus then target)	16	ABCD	CDAB
LR (left side then right side always)	8	ABCD	ABCD
DH(st) (days then hours, stim side first)	1	ACBD	CADB
DH(lr) (days then hours, left side first)	4	ACBD	ACBD
HD(lr) (hours then days, left side first)	1	BDAC	BDAC
ST(fast) (stimulus then target, efficient mousing)	2	ABCD	DCBA
Other	15	----	----
Total	47		

Response Strategies

Using their detailed mouse clicking data, participants were classified according to the predominant response strategy they employed. The classification was performed by examining the mouse clicking patterns they exhibited on the final day of the experiment. The classification criterion for a particular strategy was that the mousing sequence was consistent for at least 80% (Siegler & Taraban, 1986) of the trials for each of the eight rules in both directions. Table 1 shows the result of this classification scheme. Remember that participants were required to fill in all four cells in the user response row (see Figure 1). Consider, for example, the response sequence we refer to as the Stimulus to Target (ST) strategy in Table 1. When employing this response strategy, participants move the mouse to the location of the stimulus, click the radio-buttons to copy the givens and then move to the opposite side of the display and enter the target transformation. Thus, when the stimulus was presented on the left, they would respond by clicking the leftmost response buttons to copy the givens followed by movement to the right side of the display and clicking the rightmost response buttons to indicate the target transformation (click sequence ABCD in the table). Conversely, when the stimulus was presented on the right, they would respond by clicking the rightmost response buttons to copy the givens followed by movement to the left side of the display and clicking the leftmost response buttons to indicate the target transformation (click sequence CDAB in the table).

The predominant response strategy was the above described ST strategy with 16 of the 47 participants using this method. The second most popular strategy (8 Ss) was the Left to Right (LR) strategy whereby participants consistently entered responses in a strictly left to right order across the display, independent of the location of the stimulus.

In order to demonstrate the importance of discovering varied strategy use by our participants, we will now present a small subset of the analyses of the two predominant strategies within the context of specific issues raised in our earlier work.

Directional Asymmetry

We have previously argued (Anderson, et al., 1997) that the emergence of a directional asymmetry in response latency is evidence for the formation of production rules, stage (c) in our skill acquisition model. In our previous work, we have shown that indeed there exists a directional asymmetry between practiced and unpracticed rules. Overall, sports tested in the more practiced direction were on average slightly (250ms) faster than when tested in the less practiced direction. While this was a significant effect, we noted that indeed it seemed relatively small.

However, we have in the past ignored the possibility of the existence of the multiple response strategies that we have currently identified. Thus, we inadvertently averaged over these strategies when performing our analyses. To correct for this problem, we have reanalyzed the latency data of the current experiment.

A repeated measures ANOVA with a single between factor of strategy (ST or LR) and within factors of day (4) and practice (practiced and unpracticed direction) was performed. There was no main effect of strategy, $F(1,19) = 1.47$, MSE = 62.95, indicating there is no overall latency advantage for one strategy over another. More interesting is the fact that strategy interacted only with practice, $F(1,19) = 5.05$, MSE = 2.97, $p<0.05$. As can be seen in Figure 2, the LR strategy shows no latency difference when applying rules in the practiced versus unpracticed direction. On the other hand, the ST strategy shows a clear asymmetry between the practiced and unpracticed directions, with the rules in the practiced direction showing about a 470ms advantage over rules in the unpracticed direction.

At first blush, it is curious that the LR strategy did not exhibit the predicted asymmetry between practiced and unpracticed rules. Yet another examination of the data sheds some light on this issue. Of the eight LR participants, only two of them adopted the LR strategy consistently from the onset of the experiment, while all ST participants

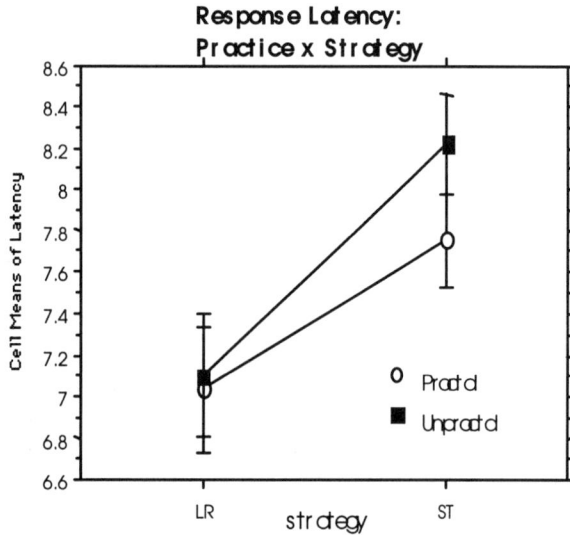

Figure 2: Response latency as a function of practice and strategy.

consistently applied their strategy throughout the experiment. In fact, it was not until the last day of the experiment that the remaining six LR participants consistently applied the strategy. Over the course of the experiment, these participants gradually changed their strategy choice from an initial ST strategy to a mixture of both ST and LR and finally entirely to the LR method. We take this then as further evidence that proceduralization goes hand in hand with asymmetry of access. Given that most of the LR group were inconsistent in their strategy choice, they were unable to develop proceduralized embodiments of the declarative versions of the rules they were applying and hence the corresponding lack of asymmetry exhibited in the latency data.

It seems clear that by taking the role of strategy into account we have managed to obtain much stronger evidence for, given substantial practice, the formation of production rules that encapsulate specific, directional transformations, stage (c) of our skill acquisition model.

Localization of Learning

To get to the heart of the atomic components involved in this task, we performed a repeated measures ANOVA of the latency data for the individual mouse clicks of user responses. This analysis allows us to examine the mouse click response profiles within strategy. Factors are day (4), practice (practiced versus unpracticed direction), mouse click sequence (ABCD or CDAB, see Table 1) and click position (first through fourth). We will only consider the data from the ST subjects here.

As we have noted, the mouse click response profile is of particular interest for the present discussion. There was a significant main effect of click position, $F(3,39) = 43.57$, $MSE = 165.95$, $p<0.0001$. Figure 3 displays this result.

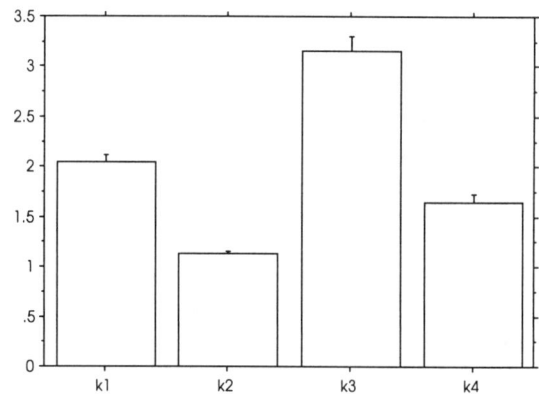

Figure 3: Mouse click latency as a function of position for the ST strategy.

The first mouse click latency corresponds to the time it takes for ST participants to orient to the stimulus and copy the given day. The second item corresponds to the copy hour operation. We see that the third item in the response carries with it the greatest latency. This is where the participant must compute and respond with the appropriate day transformation. Finally, the fourth item is where the participant must compute and respond with the appropriate hour transformation.

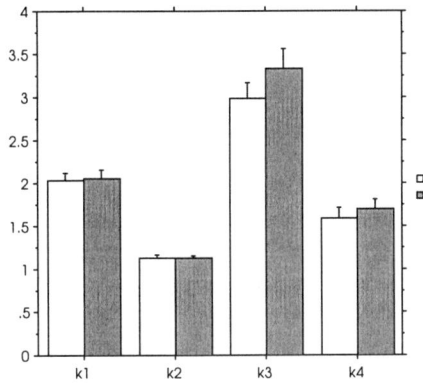

Figure 4: Mouse click latency as a function of mouse click position and practice for the ST strategy.

Note there is a subtlety here that is not necessarily obvious. Because there are two distinct mousing sequences within the ST strategy (ABCD and CDAB), we aggregated the mouse clicking data into these categories. Had we not done so, we would have missed the clear effects we have demonstrated in the mouse click response profile.

Where is the asymmetry?

There were significant interactions between practice and mouse click position, $F(3,39) = 8.87$, $MSE = 0.36$, $p<0.0001$, and between mouse click sequence and click position, $F(3,39) = 12.98$, $MSE = 1.20$, $p<0.0001$. These are shown in Figure 4 and Figure 5, respectively.

Figure 4 shows that the latency advantage for practiced over unpracticed rules reported earlier (470ms) is almost entirely driven by the third mouse click in the response sequence. Given that the third mouse click corresponds to the computation required for performing the target transformation of the day (and possibly the hour as well), we have further converging evidence that this computation has been proceduralized as per stage (c) of our skill acquisition model.

Figure 5 shows that a speed advantage for moving left to right in the response sequence is also almost entirely driven by the third mouse click in the response sequence. This is due the fact that when processing the right side of display first, the mouse must be moved from the far right of the display to the far left of the display to enter the target response.

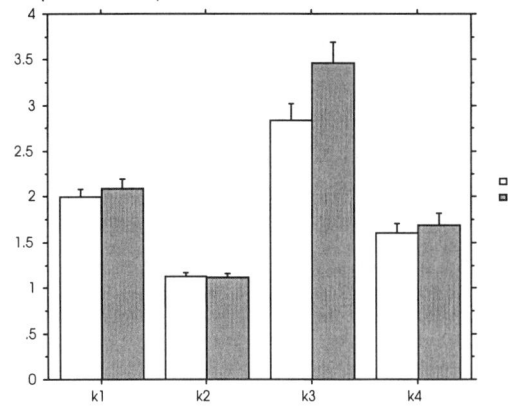

Figure 5: Mouse click latency as a function of mouse click position and response sequence (ABCD vs. CDAB) for the ST strategy.

By virtue of examining the mouse clicking data, we have discovered several previously unknown features of this particular task. The general response profile is consistent with a task analysis enumerating the procedure for solving these problems using the ST strategy, thus constraining potential models of the task. Further, we have identified the third mouse click as the predominant source of the procedural asymmetry result. Finally, we have also identified a potential problem with the current user interface employed in our skill acquisition paradigm. It simply takes longer responding from right to left across the display when compared to moving in the left to right direction when using the ST strategy.

Conclusions

We have provided here only a very small window into our efforts at generating an ACT-R model of learning in our skill acquisition task. Through this window, we hope to have shown that there is a symbiotic relationship between modeling, microanalysis, theory and empirical research methods. Indeed, the importance of this interrelationship has been characterized by others as well (Carpenter & Just, 1999).

The goal of constructing a quantitative model served as the impetus to perform a microanalysis of the task at hand. In so doing, we discovered a plethora of previously unconsidered strategies. Because we were inadvertently averaging over these strategies, we attenuated effects in support of our general skill acquisition model. Finally, we have uncovered a potential problem with the interface used in our studies that also serves to attenuate the effects in which we are most interested.

As a result of this effort, we have been able to constrain our ACT-R model of this task. Further, this research has spawned another study in which we control for response strategy and eliminate potential problematic interface problems.

Acknowledgments
This research was supported in part by grant SBR 94-21332 from the National Science Foundation. We would like to thank Robert Siegler for comments on earlier drafts of this paper.

All correspondence should be addressed to Jon M. Fincham at the Department of Psychology, Carnegie Mellon University, Pittsburgh, PA 15213-3890

References

Anderson, J. R. (1993). *Rules of the mind*. Mahwah, NJ: Erlbaum.

Anderson, J. R. & Fincham, J. M. (1994). Acquisition of procedural skills from examples. *Journal of Experimental Psychology: Learning, Memory and Cognition, 20,* 1322-1340.

Anderson, J. R., Fincham, J. M. & Douglass, S. (1997). The role of examples and rules in the acquisition of a cognitive skill. *Journal of Experimental Psychology: Learning, Memory and Cognition, 23,* 932-945.

Anderson, J. R. & Lebiere, C. (1998). *The atomic components of thought*. Mahwah, NJ: Erlbaum.

Carpenter, P. A. & Just, M. A. (1999). Computational modeling of high-level cognition vs. hypothesis testing. In R. J. Sternberg (Ed.), *The nature of cognition*. Cambridge: MIT Press.

Logan, G. D. (1988). Toward an instance theory of automatization. *Psychological Review, 95,* 492-527

Logan, G. D., & Klapp, S. T. (1991). Automatizing alphabet arithmetic: I. Is extended practice necessary to produce automaticity? *Journal of Experimental Psychology: Learning, Memory and Cognition, 17,* 179-195.

Newell, A. (1973). You can't play 20-questions with nature and win: projective comments on the papers of the symposium. In W. G. Chase (Ed.), *Visual information processing*. New York: Academic Press.

Siegler, R. S. (1976). Three Aspects of Cognitive Development. *Cognitive Psychology, 8,* 481-520.

Siegler, R. S. (1987). The perils of averaging data over strategies: an example from children's addition. *Journal of Experimental Psychology: General, 116,* 250-264.

Siegler, R. S. & Taraban, R. (1986). Conditions of Applicability of a Strategy Choice Model. *Cognitive Development, 1,* 31-51.

Modeling time perception in rats:
Evidence for catastrophic interference in animal learning

Robert M. French and André Ferrara
Department of Psychology (B32)
Univeristy of Liège, Liège, Belgium
email: rfrench@lg.ac.be; a.ferrara@ulg.ac.be

Abstract

For all intents and purposes, catastrophic interference, the sudden and complete forgetting of previously stored information upon learning new information, does not exist in healthy adult humans. But does it exist other animals? In light of recent research done by McClelland, McNaughton, & O'Reilly (1995) and McClelland & Goddard (1996) on the role of the hippocampal-neocortical interaction in alleviating catastrophic interference, it is of particular interest to ascertain whether catastrophic interference occurs in non-human higher animals, especially in those animals with a hippocampus and a neocortex, such as the rat. In this paper, we describe experimental evidence to support our claim that this type of radical forgetting does, in fact, exist for certain types of learning in some higher animals, specifically, in the rat's learning of time-durations. We develop a connectionist model that could provide an insight into how the rat might be encoding time-duration information.

Introduction

Catastrophic forgetting is a well-known problem in connectionist modeling in which the learning of new information causes the sudden and complete disappearance of previously-stored information. (For a review, see French, 1999). The severity of this problem first came to light at the end of the 1980's (McCloskey & Cohen, 1989; Ratcliff, 1990) and has been the subject of on-going research since. In healthy adult humans, however, there is no evidence of catastrophic forgetting. Humans, it seems, learn — and unlearn — gradually (Barnes & Underwood, 1959). But does catastrophic forgetting exist in non-human animals? We will present experimental results to show that catastrophic interference *does* seem to exist for certain types of learning in higher animals, specifically, in the rat's learning of time durations. We then develop a connectionist model that could provide an insight into how the rat might be encoding time information.

Catastrophic interference in connectionist networks is due, at least in part, to the overlapping nature of the network's distributed internal representations. The smaller this overlap, the less the amount of catastrophic interference (French, 1991). Various algorithms have been proposed to reduce internal representational overlap in order to decrease catastrophic interference. These algorithms generally rely on explicitly manipulating the hidden-layer representations (e.g., French, 1991, 1994; Murre, 1992; Krushke, 1992) or on orthogonalizing the input representations with the expectation that this will decrease internal representational overlap (e.g. Lewandowsky, 1991; Lewandowsky & Su-

Chen, 1993). These techniques do, in fact, produce significantly reduced catastrophic interference.

The idea of reducing the overlap of internal representations was carried to its logical conclusion in McClelland, McNaughton, and O'Reilly (1995) and McClelland & Goddard (1996). They suggested that the reason humans show no signs of catastrophic interference was because of our two complementary learning systems: the hippocampus and the neocortex. Their idea was that new information is initially learned in the hippocampus, where it cannot not adversely affect information that has been previously consolidated in the neocortex. Once the new information was learned in the hippocampus, it was transferred very gradually to neocortical long-term storage. In this way, previously learned information could be kept "out of the way" of newly arriving information. This, they claimed, was how the brain overcame catastrophic forgetting.

Since their theory is based on the key notion of a dual hippocampal/neocortical mechanism of early-processing and subsequent storage, we wondered whether there was evidence of catastrophic forgetting in animals that, like humans, had both a neocortex and a hippocampus. We believe we have discovered a likely candidate for catastrophic forgetting in animals having both hippocampus and neocortex — namely, time-duration learning (and forgetting) in the rat. Our results suggest that the hippocampal/neocortical loop may not be involved in time learning in the rat, or if it is, it may not function as it does for types of learning that are not subject to catastrophic interference. We show that a simple single-store connectionist network can provide a relatively good model this type of time-duration learning. We will conclude by suggesting that the greater amount of overlap of the network's internal representations of time-durations acquired during sequential learning compared to concurrent learning may be why catastrophic forgetting occurs after sequential learning and is absent in the case of concurrent learning. We suggest that there might be a counterpart to these differences in the internal time-duration representations in the rat.

Perception of time-duration in the rat

In their natural environment, animals are, of necessity, good at predicting significant events such as periodical food availability. This seems to be a clear indication of an ability to represent time. Laboratory researchers studying time-learning in animals have developed a number of techniques to study timing processes. One of these is the *peak procedure* (Catania, 1970) in which rats learn to press a lever to receive food after a certain fixed duration. Each trial in this procedure begins with the simultaneous onset of a sound stimulus and the insertion of a lever into the Skinner box. There are two different types randomly mixed of trials. During *reinforced (or "food") trials*, only the first lever press after the critical duration is rewarded with a little food pellet. Immediately following the reinforced lever-press, the sound is switched off and the lever is withdrawn from the box. *Test trials* begin exactly like reinforced trials, but the animal receives no reward. Test trials are necessary because in the reinforced trials, the animal's lever-pressing stops as soon as the food pellet drops into its box. These trials end independently of the lever presses made by the animal and typically last at least twice as long the duration reinforced in the rewarded trials. The number of lever presses are recorded for each one-second interval and averaged across test trials. This produces the characteristic bell-shaped response-rate function observed with this procedure (Fig. 2). The moment of maximum lever-pressing is called the *peak time* and reflects the moment of maximal food expectation by the rat.

The observation of steady-state behavior following training has long been used to understand the mechanisms underlying timing abilities (Roberts,1981; Gibbon, Church, & Meck, 1984). More recently, it has also been used to study the acquisition of a new temporal representation (Meck, Komeily-Zadeh, & Church, 1984; Ferrara, 1999).

Time perception experiment

Sequential time-duration learning

A group of 15 rats was used in a learning experiment that studied behavioral adaptations to changes in rewarded time-durations. How rats adapt their behavior when learning time-durations is assumed to reflect properties of its temporal representation. The present experiment was divided in several phases. In the first phase the animal was trained on a 40-sec. duration (14 days, with 90 trials per day of which 85% were reinforced trails and 15 % test trials). In the next phase, the animal learned an 8-sec. duration (10 days, 90 trials per day with the same proportions of fixed and test intervals).

The moment when the animals were switched from 40-sec. learning to 8-sec. learning is referred to as Transition 40-8 (1) and can be seen in the upper left graph of Figure 2. After the 8-sec. duration had been learned, the reward was switched back to the original 40-sec. duration (Transition 8-40 (2) in the lower left graph of Figure 2).

The rate of lever-pressing was recorded during three different sessions: just before a transition, the two sessions immediately following a transition. The final session of Phases 1, 2, 5 and 6 (i.e., the last session before switching to a new time-duration) is referred to as a *reference session* (indicated by a (1) in Fig. 1). The peak time of the reference session reflects the moment of maximum expectancy for

receiving a food pellet. The *transition session* (indicated by a (2) in Fig. 1) refers to the first session after switching the animal to a new time-duration. It is interesting to notice how the peak time shifts during this session compared to the reference session, since, presumably, any changes in peak time reflect modifications of the animal's internal representation of the reinforced duration. The *transition+1 session* (indicated by a (3) in Fig. 1) refers to the second session after the transition. Learning a new time-duration can be described in terms of a moving peak time. In other words, during Transition 40-8 (1), the peak time shifts from the previously learned 40-sec. duration and stabilizes around the new 8-sec. duration which is being reinforced (Fig. 2) Similarly, in Transition 8-40 (2), the peak time will gradually move from 8 to 40 seconds. It has been shown elsewhere (Lejeune, Ferrara, Simons & Wearden, 1997) that, even if the reinforcement is repeatedly switched back and forth from one time-duration to the other, there is no real improvement in the speed of the peak time adaptation. In fact, re-learning a previously learned time-duration requires as much time as learning an entirely new one. This suggests that at each transition, the previous representation of the reinforced time is "overwritten" as the new one is built. In other words, *there is no evidence of any memory savings from the previously learned time-durations.* One reasonable interpretation of this result is that new time-duration learning completely (catastrophically) wiped out the originally learned time-duration.

Concurrent learning

In the next phase (Phase 4), the rats were *concurrently* trained for 25 consecutive sessions on both 8- and 40-sec. durations. This was done by means of a random mixture of 42.5% 8-sec. trials, 42.5% 40-sec. trials and 15% test trials (each session still contains 90 trials). At the end of this learning period, the response curves (not shown here) became clearly bimodal, with a first peak located around 8 seconds and a second peak located around 40 seconds. This means that during concurrent learning of the two time-durations, the animal had developed a representation for both durations.

Phases 5, 6 and 7 were identical to phases 1-3 (except for the number of sessions). Transition 40-8 (3) and transition 8-40 (4) show the curves for lever-pressing rates for the reference, transition and transistion+1 sessions after the animals have learned 40- and 8-sec. durations concurrently as above.

Unlike the previous case of sequential learning in which there was no savings of prior learning, the animal, having now learned the two durations concurrently, can rapidly shift from the 40-sec. duration back to the 8-sec. duration. In this case, while there is still a small amount of forgetting, *there is no catastrophic forgetting of the originally learned 8-sec. duration.* This seems to imply that time-

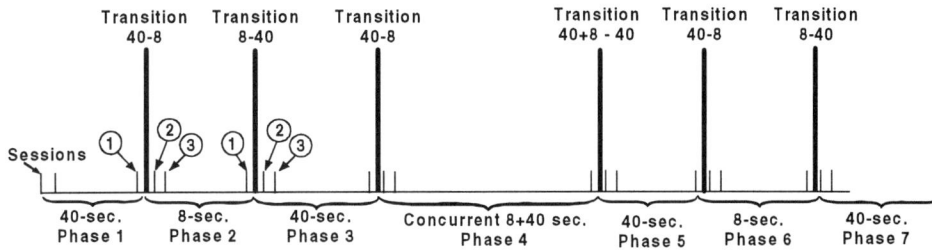

Figure 1. Lever-pressing rates were recorded during the *reference session* (1) at the end of one learning period and during the *transition* (2) *and transition+1* (3) *sessions*, i.e., the first and second sessions after switching to a new learning period.

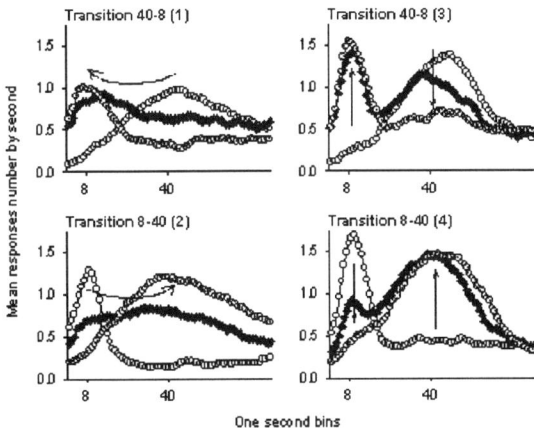

Figure 2. Average rates of lever-presses/second for the reference (white), transition (black) and transition+1 (gray) sessions.

duration representations developed during concurrent learning are significantly different than those developed during sequential learning.

Simulation

To simulate these results, we used an 11-18-2 backpropagation network with binary input coding. The learning rate was set at 0.1, momentum 0.9. The learning criterion was set at 0.001 for all output nodes. An 11-unit input layer was used because this corresponds to the oscillation periods used in the Church and Broadbent (1990) model. Eleven oscillation periods were chosen by these authors because this is enough to represent the full range of relevant short-term time-durations experienced by the animal. The oscillation periods used in the following simulation were as follows: 0.2, 0.4, 0.8, 1.6, 3.2, 6.4, 12.8, 25.6, 51.2, 102.4, 204.8 seconds.

Each time duration was translated into a binary pattern of these 11 oscillators, with each oscillator either being activated (= 1) or not (= 0). On each training session the input patterns were modified by the addition of gaussian noise. The noise added to each oscillator signal was proportional to its mean oscillation period (i.e., the larger the oscillator period, the larger the normal curve around its mean). Target outputs of 00 indicated no reinforcement; whereas 11 indicated a positive reinforcement.

Each "session" consisted of a grouped presentation of 20 patterns, P_i, learned for 16 epochs (or until the error for all patterns was below 0.001). Each pattern consisted of an input, which was an encoding for a particular time-duration and a desired output, corresponding to whether or not that particular time-duration was reinforced. For example, an "8-sec. only" session is:

P_1: Input: 2-sec.   Output: 00
P_2: Input: 4-sec.   Output: 00
P_3: Input: 6-sec.   Output: 00
P_4: Input: 8-sec.   Output: 11 ("reinforced")

. . .

P_{20}: Input: 40-sec.   Output: 00

There were six distinct phases for sequential-learning, five for concurrent-learning simulations (Figs. 3, 4).

Phase 1 and 2 were identical for both simulations. The first consisted of 20 "8-sec. only" sessions (i.e., only the 8-sec. duration was reinforced); the second by 20 "40-sec. only" sessions (i.e., only the 40-sec. duration was reinforced).

The next phases were critically different for the two simulations. For sequential learning, Phases 3 consisted forty "8-sec. only" sessions followed by Phase 4 which had forty "40-sec. only" sessions. For the concurrent learning simulation, however, these two phases were combined into a single 80-session phase during which *both* 8-sec. and 40-sec. durations were reinforced.

The test phases (i.e., Phases 5 and 6 for sequential learning and Phases 4 and 5 for concurrent-learning) in both simulations were identical and consisted of 40 sessions. The first of the two test phases consisted of eight "8-sec. only" sessions. This was long enough for the previously learned 40-sec. duration to be "forgotten" by the network, after having learned it in Phase 4 in the sequential simulation and, along with the 8-sec. duration, in Phase 3 of the concurrent simulation. In other words, "unlearning" the 40-sec. duration, whether in the sequential simulation or the concurrent simulation, meant that when 40 seconds was input to the network, it correctly responded with 00. This took eight sessions. The final 32 sessions of this phase were "40-sec. only" sessions. The critical observation was how quickly the network recovered its knowledge of the 40-sec. time duration.

All results reported in this paper were averaged over 60 independent runs of the network..

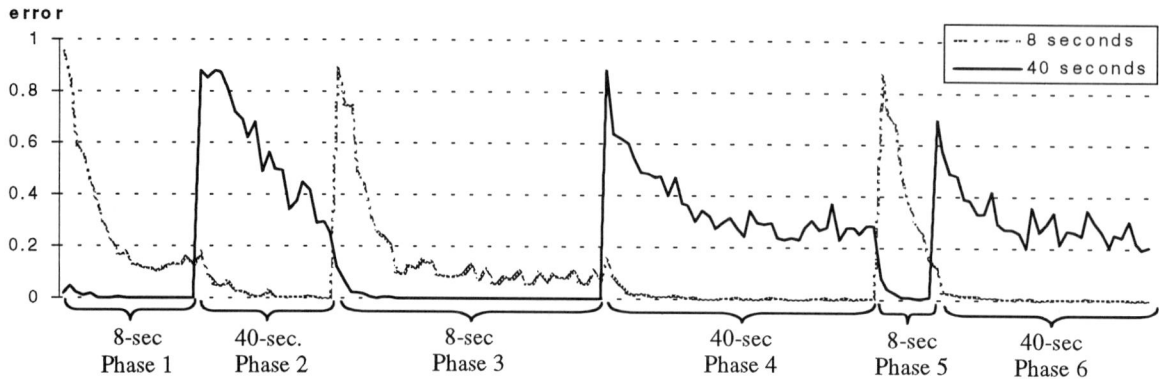

Figure 3. Phases of 8-sec. and 40-sec. learning when these durations are learned sequentially.

Results

Sequential learning

We began by running the network in "sequential mode" (Fig. 3). At the beginning of Phase 3, when the network returns to learning the 8-sec. duration, its error curve is virtually identical to the first time it ever encountered the 8-sec. duration in Phase 1. In other words, learning the 40-sec. duration in Phase 2 seems to have completely erased any memory trace of the initial 8-sec. learning. There is no evidence of any "savings" of the initial 8-sec. learning. (Ebbinghaus, 1887; see Hetherington & Seidenberg, 1989, for a discussion of this as a measure of catastrophic forgetting)..

In short, the network is responding to learning new time durations in much the same way as the rat: learning a new duration seems to completely erase (or overwrite) the memory trace of the prior time-duration learning.

Concurrent learning

In the second simulation we explore concurrent time-duration learning and find that it produces considerably different results compared to learning time-durations sequentially (Fig. 4). Phases 1 and 2 in the concurrent-learning simulation are identical to Phases 1 and 2 in the sequential-learning simulation. However, now in Phase 3 the network learned both 8-sec. and 40-sec. durations concurrently for 80 sessions (instead of 40 8-sec. sessions

followed by 40 40-sec. sessions as in the previous simulation). After this concurrent learning phase, the network is tested as in the previous simulation. In other words, in Phase 4, the network is trained for eight sessions on 8-sec. durations only. As before, this was long enough for its performance on 40-sec. durations to return to a zero-error baseline. Reinforcement was then switched back to a 40 seconds (Phase 5, Fig. 4)

Now let us compare the test phases in the two simulations. In both simulations, the eight sessions of 8-sec. learning only, allowed the network to return to the unreinforced baseline for 40-sec. learning (see the 40-sec. error rate in Phase 5 in the sequential-learning simulation, and in Phase 4 in the concurrent-learning simulation in Figures 3 and 4, respectively).

The critical difference can be seen in the error-rate just after this eight-session phase of 8-sec. learning. In the sequential learning simulation (Fig. 3), when the network is switched back to training on the 40-sec. duration, the error rate shoots up to 0.7 and remains above 0.4 for a number of sessions. On the other hand, when the 8- and 40-sec. durations were learned concurrently (Fig. 4), there is no such jump in error-rate. The error-rate for 40-sec. learning after the eight-session 8-sec. training is essentially what it was at the end of the concurrent learning phase. In short, the network now has no trouble reviving its prior memory of the 40-sec. duration.

This is very similar to what was observed in the peak-time experiments with the rat. Concurrent learning

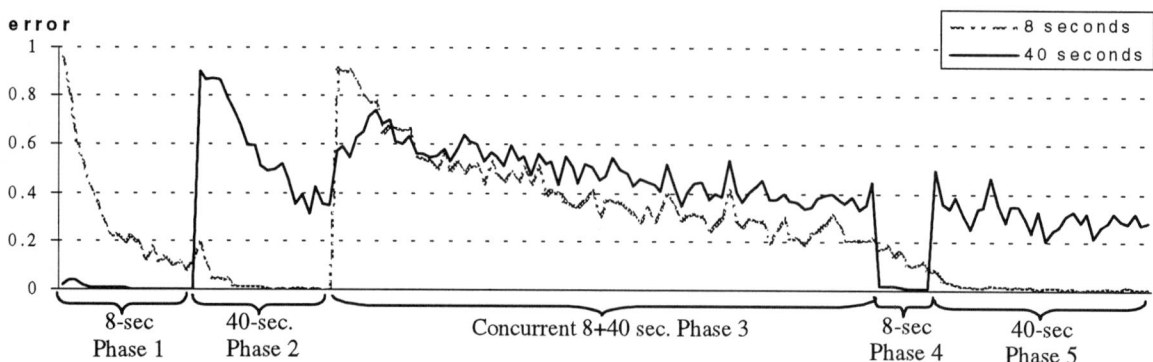

Figure 4. Concurrent learning of 8- and 40-sec. durations in Phase 3.

176

allows the rat to rapidly "revive" its memory trace of the 40-sec. duration; sequential learning does not. In the latter case, both in the case of the simulation and the rat, the memory trace seems to have been catastrophically erased by new learning.

Internal representations developed by sequential versus concurrent learning

Given the results of these simulations, we wondered whether the differences in forgetting observed in sequential and concurrent learning were related to the network's internal representations of the time information it had learned. This suggests the corresponding question for the rat — namely, are the differences observed for the rat's sequential and concurrent time-duration learning based on its development of different internal representations of these durations depending on how it learned them?

To study the internal representations of 8- and 40-sec. durations in our network, we presented "pure" (i.e., without noise) 8-sec. and 40-sec. time patterns to the network just before the eight-session 8-sec. only test phase began. This allowed us to record the hidden-unit activation patterns corresponding to 8-sec. and the 40-sec. inputs for the sequential-learning simulation and for the concurrent-learning simulation. We then calculated the hidden-unit representation overlap of these two patterns for both simulations. We then compared the differences in overlap of the hidden-unit encodings of 8-sec. and 40-sec. durations for the two different learning scenarios.

Our prediction was that, since there was considerably less forgetting of the 40-sec. time-duration in the concurrent-learning simulation, that the representations developed during Phase 3 of the concurrent-learning simulation would overlap less and be more sparse than the corresponding representations developed during sequential learning. This prediction was confirmed by the data (Fig. 5).

Hidden-layer representational overlap and sparseness

We calculated representational overlap by means of an inner-product measure of the hidden-unit vectors corresponding to 8-sec. and 40-sec. input. These values were averaged over 60 runs of the program. As predicted, in the case of case of concurrent learning, this value (0.61) was considerably lower than when the representations had developed during sequential learning (1.49). In other words, the internal representational overlap of the two time-duration patterns was *2.4 times higher for sequential learning*, accounting, at least in part, for the greater interference observed in sequential learning.

Representational sparseness can be measured by considering the total amount of activation used by a representation and by its dispersion. In general, the more sparse the representation, the less activation it will use and the lower its dispersion. For sequential learning, the average 8-sec. hidden-unit representation used 5.5 units of activation and the 40-sec. representation, 2.4 units of activation. For concurrent learning, these figures dropped to 2.2 and 2.0

units of activation, respectively. A simple measure of dispersion (standard deviation, SD) indicated how dispersed the representations were over the 18 hidden-units. The spread of the 8- and 40-sec. representations in the case of sequential learning (0.41 and 0.12 SD) is considerably greater than the spread of the same representations in the case of concurrent learning (0.27 and 0.10 SD). Figure 5 shows graphically the degree to which the sparseness of these representations differs (the activation levels of the representations have been ordered in order to facilitate comparisons between them).

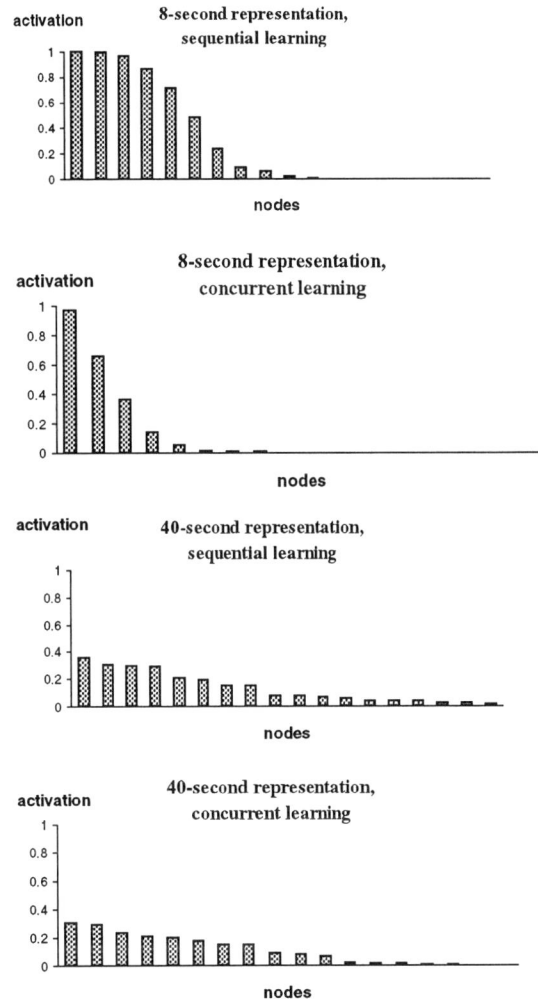

Figure 5. Differences in the sparseness of coding at the hidden layer of the representations for 8-sec. and 40-sec. durations in the case of sequential learning and concurrent learning.

In other words, the network's internal representations of 8- and 40-sec. durations when learned concurrently, were significantly sparser and less overlapping than when they were learned sequentially. It is well established that the amount of internal representational overlap is contributes significantly to the amount of catastrophic interference produced. It is therefore reasonable to conclude that forgetting is far less catastrophic for concurrent learning than for sequential learning because of, at least in part, the

177

smaller amount of interference of the internal representations of the two time-durations in concurrent learning compared to sequential learning.

Conclusion

Two suggestions emerge from this research. The first is that the rat may not store time-duration information in the same way it stores other types of information that are less susceptible to catastrophic interference. In other words, the complementary hippocampal-neocortical system proposed by McClelland, McNaughton, and O'Reilly (1995) to avoid catastrophic interference may not be used by the rat for long-term storage of time information. The fact that a single connectionist network seems to produce effects quite similar to those actually observed in rats for both sequential and concurrent time-learning would argue for the possibility of a unitary time storage area in the rat, rather than a dual hippocampal-neocortical mechanism.

Secondly, we believe these results support for the claim that in the rat there may be a distinctly different internal coding of time durations when they are learned concurrently, as opposed to when they are learned sequentially. In our connectionist simulation, we obtain forgetting results similar to those observed in the rat. These differences in the network correspond to significant differences in the internal coding of the two time-durations depending on how the network learned them. The internal representations in the case of concurrent learning are more sparse and overlap less than in the case of sequential learning. We suggest that this might imply a similar type of coding in the rat.

Acknowledgments

This work was supported in part by a grant IUAP 4-19 from the Belgian National Science Foundation.

References

Barnes, J. and Underwood, B. (1959). "Fate" of first-learned associations in transfer theory. *Journal of Experimental Psychology, 58*, 97-105

Catania, A. (1970). Reinforcement schedules and psychophysical judgements, in W. Schoenfeld (ed.). *Theory of Reinforcement Schedules*, 1-42, NY: Appleton-Century-Crofts.

Church, R. & Broadbent, H. (1990). Alternative representations of time, number and rate. *Cognition, 37*, 55-81.

Ebbinghaus, H. (1885) Über das Gedächtnis: Untersuchen zur Experimentellen Psychologie ("On memory") (H. A. Ruger and C. E. Bussenius, trans) New York: Dover, 1964

Ferrara, A. (1999). Unpublished doctoral dissertation. Psychology Department, University of Liège, Belgium.

French, R. M. (1991) Using Semi-distributed Representations to Overcome Catastrophic Forgetting in Connectionist Networks. *Proceedings of the Thirteenth Annual Cognitive Science Society Conference*, 173-178, Hillsdale, NJ: LEA

French, R. M. (1994) Dynamically constraining connectionist networks to produce distributed, orthogonal representations to reduce catastrophic interference. *Proceedings of the*

Sixteenth Annual Conference of the Cognitive Science Society, 335-340, NJ: LEA

French, R. M. (1999). Catastrophic interference in connectionist networks: Causes, consequences and solutions. *Trends in Cognitive Sciences.* (to appear).

Gibbon, J., Church, R. and Meck, W. (1984). Scalar timing in memory. In J. Gibbon and L. Allan (eds.), Annals of the New York Academy of Sciences: Vol. 423. *Timing and time perception,* 52-77, New York: New York Academy of Sciences

Hetherington, P. and Seidenberg, M., (1989), Is there "catastrophic interference" in connectionist networks?, In *Proceedings of the 11th Annual Conference of the Cognitive Science Society*, 26-33, Hillsdale, NJ: LEA

Krushke, J. (1992) ALCOVE: An exemplar-based model of category learning. *Psychological Review, 99,* 22-44

Lejeune, H., Ferrara, A., Simons, F., and Wearden, J. H. (1997). Adjusting to changes in the time of reinforcement: Peak interval transitions in rats. *Journal of Experimental Psychology: Animal Behavior Processes, 23,* 211-231.

Lewandowsky S. (1991) Gradual unlearning and catastrophic interference: a comparison of distributed architectures, in W. Hockley and S. Lewandowsky (eds.) *Relating Theory and Data: Essays on Human Memory in Honor of Bennet B. Murdock,* 445-476, NJ: LEA

Lewandowsky, S. and Shu-Chen Li (1993) Catastrophic interference in neural networks: causes, solutions, and data, in F.N. Dempster and C. Brainerd (eds.) *New Perspectives on Interference and Inhibition in Cognition.* New York, NY: Academic Press.

McClelland, J. and Goddard, N. (1996). Considerations arising from a complementary learning systems perspective on hippocampus and neocortex. *Hippocampus, 6,* 654-665.

McClelland, ,J., McNaughton, B. and O'Reilly, R. (1995) Why there are complementary learning systems in the hippocampus and neocortex: Insights from the successes and failures of connectionist models of learning and memory. *Psychological Review, 102,* 419-457

McCloskey, M. and Cohen, N. (1989) Catastrophic interference in connectionist networks: The sequential learning problem. In G. H. Bower (ed.) *The Psychology of Learning and Motivation: Vol. 24,* 109-164, NY: Academic Press

Meck, W., Komeily-Zadeh, F. and Church, R. (1984). Two-step acquisition : Modification of an internal clock's criterion. *Journal of Experimental Psychology : Animal Behavior Processes, 10,* 297-306.

Murre, J. (1992b) *Learning and Categorization in Modular Neural Networks.* Hillsdale, NJ: LEA

Ratcliff (1990) Connectionist models of recognition memory: Constraints imposed by learning and forgetting functions. *Psychological Review, 97,* 285-308

Roberts, S. (1981). Isolation of an internal clock. Journal of Experimental Psychology: *Animal Behavior Processes, 7,* 242-268

Is Snow Really Like a Shovel?

Distinguishing Similarity from Thematic Relatedness

Dedre Gentner (gentner@nwu.edu)
Northwestern University
Department of Psychology
2029 Sheridan Rd.
Evanston, IL 60208

Sarah K. Brem (sbrem@soe.berkeley.edu)
University of California, Berkeley
Graduate School of Education
4533 Tolman Hall, #1670
Berkeley, CA 94720-1670

Abstract

Traditionally, thematic relatedness (*chicken* and *egg*) and similarity (*chicken* and *turkey*) have been thought of as distinct phenomena, the former the result of associative processes, and the latter reflecting comparison processes. However, recent studies (Bassok & Medin, 1996; Wisniewski & Bassok, 1996) suggest that similarity is a result of both association and comparison. This could call for a radical redefinition of similarity as inherently fused with association. We term this view the *integration* account. We consider an alternative, the *confusability* account, under which thematic influences intrude upon assessments of similarity but are not an essential part of the similarity process. We present two experiments supporting the confusability account. The first indicates that comparison and association are independent processes. The second shows that thematic influences rise with increased cognitive load. We believe that while a redefinition of similarity is not warranted, similarity is more vulnerable to error and intrusion than is generally thought.

Similarity is central to cognitive science. It plays a role in psychological and computational models of analogical reasoning (Forbus, Gentner & Law, 1995; Gentner, 1988; Holyoak & Thagard, 1989), problem solving (Ross, 1989; Novick, 1988), categorization (Markman & Wisniewski, 1997; Medin & Heit, in press; Murphy & Allopenna, 1994) and inference (Clement & Gentner, 1991; Lassaline, 1996). Much research in cognitive development has been predicated on this distinction: e.g., research that seeks the causes of the developmental shift from early reliance on thematic associations to later reliance on common-category relations (e.g., Markman, 1989; Vygotsky, 1978). Even models that question the degree to which similarity plays a role in cognitive processes base their criticisms on the assumption that similarity is a well-defined phenomenon (e.g., Murphy & Medin, 1985; Sloman & Rips, 1998). Therefore, when the possibility arises that our basic conceptualization of similarity is in need of radical restructuring, the implications are considerable. Recently, two studies have suggested just that.

Traditionally, the similarity between two items is thought to be a function of their distance in mental space, the overlap of their features (Tversky, 1977) or of their shared structure (Gentner, 1983, 1988; Gentner & Markman, 1997; Medin, Goldstone, & Gentner, 1993). All of these approaches assume that similarity results from a process of comparison, while thematic relatedness stems from association. The difference between these two types of relationships is clear when one examines their predicate-argument structures. For example, according to Gentner (1988), a *shovel* and a *spade* are similar because they share **relations** (e.g., MOVE [shovel, here, there, stuff]/MOVE [spade, here, there, stuff]) and **attributes** (e.g., HANDLE [shovel]/HANDLE [spade]). In contrast, thematic associates like *snow* and *shovel* generally do not share attributes, and take different roles in relational predicates (e.g., MOVE [**shovel**, here, there, **snow**]). Because *snow* and *shovel* do not share features of appearance or use, both Gentner's and Tversky's models predict low similarity. Snow and shovel are related, though; experience informs people that snow and shovels interact, appear in the same propositions, and commonly co-occur. But this hinges upon noting relations that *associate* snow and shovel, rather than on *comparing* the two and noting common structure.

Perhaps because the distinction between thematic association and similarity seems obvious in many contexts its validity has not been the focus of much research. It seems clear that to base categorizations, word extensions, and inferences on thematic associations would frequently cause error. For example, whales are thematically related to the plankton they eat and to the harpoons with which whalers hunt them. However, to conclude that plankton or harpoons are warm-blooded like whales would be incorrect. Likewise, thematic associations play roles that similarity cannot. For example, associations allow us to detect and make predictions on the basis of covariation (Kelley, 1973;

Novick & Cheng, 1990). The association between smoke and fire leads us to sound the alarm when we see dark clouds billowing from an apartment window. But, looking for smoke when we see fog because they are perceptually similar would be a mistake. Differences such as these have provided a strong argument for treating similarity and thematic association as separate processes.

Beyond these intuitions, much of cognitive research depends on the theoretical assumption that thematic association and similarity relationships are distinct. Therefore, the finding that the two are not psychologically separable would radically alter our understanding of cognition. Yet, some recent theories suggest a strong link between similarity and association. Sloman (1996) proposes that that similarity and association are processed by a single system, with abstract rules reposing in a separate system. Further, Bassok and Medin (1996) have recently found that when judging the similarity of two sentences, participants are influenced by thematic relationships between the sentences. For example. participants judged the sentences *The carpenter fixed the chair* and *The carpenter sat on the chair* to be similar because "the carpenter sat on the chair to see if the repair would hold." Likewise, Wisniewski and Bassok (1996) compared the similarity ratings assigned to similar pairs (*milk-lemonade*), thematic pairs (*milk-cow*) and pairs sharing both similarity and thematic relationships (*milk-coffee*). Thematic relationships significantly increased similarity ratings: e.g., *milk* and *coffee* were rated as more similar than *milk* and *lemonade*. Although previous research has shown that similarity and thematic relationships are confusable by children (e.g., Bauer & Mandler, 1989), these experiments go further in suggesting that these confusions occur for adults as well.

These results force us to consider the possibility that traditional models oversimplify the concept of similarity. Under Medin and Bassok's (1996) view, similarity is an integration of independent processes of comparison and association. However, an even more radical position can be taken, consistent with Sloman's position. Under this view, which we term the *integration account*, thematic associations and comparisons are the result of a unitary process. Similarity is thus inherently intermixed with thematic relatedness.

Another possibility is that the comparison process is distinct from association, and thematic influences arise as the result of thematic intrusions that interfere with similarity judgments. According to this *confusability account*, similarity is the result of the comparison process, but this process can be derailed by other factors. Such derailment might occur if participants have difficulty distinguishing between the mental output that arises from accessing associations as opposed to the output of a separate, independent comparison process.

Several lines of reasoning lead us to entertain the confusability account. First, Markman (1989) argues that the developmental shift towards a preference for taxonomic groupings does not indicate a loss of salience for thematic relationships, but a rise in the salience of taxonomic relationships. Likewise, Smiley and Brown (1979) argue that the shift is one of preference, rather than a radical restructuring of knowledge. If adults remain sensitive to thematic associations, thematic relationships could interfere with other processes.

A second line of reasoning concerns individual differences. We have found evidence of substantial variation in people's ability to distinguish and identify similarity relations. In a screening task (further described below), participants were given a standard (e.g., **dart**) and had to choose which alternative was most similar: **bullseye** or **javelin**. Eleven percent of the participants were consistently unable to distinguish similarity from thematic association (Undifferentiating). Another 41% showed varying degrees of differentiation. If such confusion were universal, it would suggest a basic fusion of the processes of comparison and association, supporting the integration account. However, 48% of participants show no confusion of similarity and thematic association (Differentiating).

Our proposal is that similarity and thematic relatedness result from two separate processes. However, the output of these processes is sometimes difficult to differentiate through direct introspection, particularly when similarity is very low (the conditions in which thematic intrusions have been found in similarity tasks). We conjecture that the ability to make the distinction reliably and explicitly is learned as part of the development of metacognitive skills. There is precedent for the view that internal cognitive distinctions must be learned. Markman (1979) found that 8-11 year-old children failed to report any comprehension difficulties when reading stories that contained blatant inconsistencies (e.g., that ants rely on smell, but have no noses and cannot smell). Nevertheless, the children were slower to read sentences that led to strong inconsistencies. This suggests that they engaged in some kind of inconsistency processing but could not *label* it as such. Elsewhere, we see evidence for people's inability to accurately distinguish between the cognitive and emotional states produced by real and implanted memories (e.g., Loftus, 1997), and an inability to correctly identify what they know or do not know (e.g., Koriat, 1993). Additionally, there is evidence of developmental shifts in the ability to reflect upon the products of cognition (e.g., Flavell, Green & Flavell, 1990; Kitchener & King, 1994). Going further, there is evidence for historical shifts in the use of comparison and association. Medieval alchemy treated similarity and thematic relatedness as interchangeable to a greater degree than in modern science (Gentner & Jeziorski, 1993). In short, we suggest that reliably distinguishing the sensation of relatedness via commonalities from the sensation of relatedness via thematic associations may require a degree of metacognitive sophistication.

On the confusability account, the findings of Bassok & Medin (1996) and Wisniewski & Bassok (1996) result from people's failure to explicitly distinguish the results of an associative retrieval process and the results of a comparison process. But if these processes are indeed separable, it should be possible to find a task that draws only on comparison. In Experiment 1, we do this by utilizing a word learning task.

Experiment 1

To find out whether people can focus purely on similarity, we needed a task that naturally promotes a strong focus on commonalities. We drew on findings from cognitive development. Across many tasks researchers have noted a thematic preference in young children (Gelman & Baillargeon, 1983), and a shift towards taxonomic preferences with age and experience (Smiley & Brown, 1979). However, one task in particular has been found to elicit a reliable shift towards commonality, even among preschool children: namely, the task of extending a novel word to new items. In the classic word extension task, a child is shown a **dog** as standard, together with a **cat** (perceptually and taxonomically similar), and a **bone** (thematically related). If the child is simply asked to choose one that 'goes with' the dog, or even to choose 'another one' s/he will typically choose the bone. But if the child is told that dogs are called "feps" in a foreign language and asked to find the other "fep," s/he chooses the cat (Markman & Hutchinson, 1984; Waxman & Gelman, 1986).

The word learning task appears to invite a focus on common structure. Thus it can provide a test of whether people possess an internally separable comparison process. In this experiment we ask whether adults will show a strong focus on commonality in a word-extension task. If adults are like preschoolers in that they show only a weak preference for common structure, the results would be consistent with the integration account. To the extent that adults show a strong preference for common structure in the word extension condition, this would support the separability of comparison and association.

We showed participants triads of objects of the form:

$$\textbf{dog}$$
$$\textbf{bone} \qquad\qquad \textbf{cat}$$

So as not to bias the experiment in favor of independent processes, we designed the triads to have strong thematic associates and similarity choices that were clear but not extremely close to the standard (e.g., not in the same basic level category). In the baseline condition we asked subjects to choose the item that "goes with' the standard. In the word condition, we told them the name of the standard (e.g., 'blicket') in a foreign language, and asked them to say which alternative was also a 'blicket'. In the baseline task, since both similar and thematic items are related to the base, either could in principle be an appropriate choice. However, because we chose highly salient thematic relations, we expect to see an advantage for the thematic alternative.

The key prediction concerns the word task. If comparison is indeed a separable process, then there should be a strong similarity focus in the word task, even if baseline responding is strongly thematic. In contrast, an integration account that postulates a single process combining comparison and association, would fail to predict a difference in the mental output generated by the two tasks.

Method

Materials. Participants saw triads consisting of a standard and three alternatives: a comparison alternative, an association alternative, and an unrelated foil. There were 12 such item sets. (See Table 1 for examples).

Procedure & Design. There were two levels of Task Type (between subjects): Word extension and Baseline. In the Word condition, subjects saw a sentence of the following format: "In a foreign country, 'X,' the people call spoons 'blicks.' Which of these is also a blick?" In the Baseline condition, subjects saw a sentence of the following format: "Which of the following goes with spoon?" In both tasks, the three alternatives were presented below the sentence, and subjects circled their choice.

Participants. Twenty-two Northwestern undergraduates participated, 11 in each condition.

Table 1. Examples of items used in Experiments 1 and 2

Base item	Comparison alternative	Association alternative	Unrelated foil
spoon	ladle	cereal	shirt
rocket	missile	astronaut	belt
garlic	onion	vampire	cement

Results & Discussion

The results of Experiment 1 are presented in Table 2. As predicted, participants' choices depended overwhelmingly on Task Type. In the Baseline task, participants chose the similar item in 2.2% of the cases. In the Word task, they chose the similar item in 96.9% of the cases. This difference is significant (by participants, $\underline{t}(11) = 33.41$, $\underline{p} < 0.0001$; by items, $\underline{t}(11) = 45.51$, $\underline{p} < 0.0001$).

Table 2. Experiment 1: Percentage of similarity responses as a function of Task Type.

Task Type	% Similarity Responses
Baseline	2.2%
Word extension	96.9%

These results suggest that similarity and thematic relatedness are distinct, separable processes for adults. The strong form of the integration account is thus seriously undermined by these findings, as they cannot be accounted for by a unitary process fusing comparison and association.

Interestingly, we found no individual differences in Experiment 1. In the Baseline task, only two participants chose any similar alternatives, doing so 8% and 16% of the time. In the Word task, only two participants chose thematic alternatives, doing so 8% and 25% of the time. This striking contrast with the screening results described above suggests that the two processes may be implicitly called forth by different cognitive tasks. Just as task support helps children focus on different types of relationships, task support may aid adults in distinguishing similarity from association.

The results of Experiment 1 support the confusability account, and are not compatible with conceptualizing comparison and association as the result of a single, undifferentiating process. Participants can focus on similarity relations even in the fact of very salient thematic relations when a naturalistic task requires them to do so.

Experiment 2

While the results of Experiment 1 support the proposal that comparison and association are independent processes producing distinct output, we now need to address the confusion of these outputs. As noted above, these confusions occurred not only in Bassok & Medin (1996) and Wisniewski & Bassok (1996), but in our own screening studies. Under the confusability account, these intrusions of thematic relatedness into similarity judgments arise because when participants experience difficulty in introspectively separating the outputs of comparison and association. This account implies that the more difficult it becomes to engage in introspection, the greater should be the confusion between similarity and thematic relatedness.

In Experiment 2, we asked subjects to make a similarity choice under time pressure, and varied whether there was a competing thematic relationship present. We used triads of words containing a standard item, an item similar to the standard, and either a thematic associate of the standard, or an unrelated foil.

Participants were to ignore thematic associations, and choose only items that were similar to the standard. We provided examples and substantial practice with feedback. To manipulate cognitive load, and thus vary particpants' difficulty in examining the output of their mental processes, we employed two deadlines, one at 1000 ms and one at 2000 ms.

Any account would predict more errors at the shorter deadline. But the integration and confusability accounts make very different predictions regarding errors at the longer deadline. The confusability account predicts that shorter deadlines will differentially increase the error rate in the presence of thematic distractors. Thus there will be an interaction between triad type and deadline. In contrast, in the integration account, the inability to set aside thematic influences arises from fusion at the process level. Such fusion should compromise the separation of comparison and association at both the shorter and the longer deadlines. Thus, both accounts predict more errors at the shorter deadline, but only the 'separate but confusable processes' account predicts an interaction between deadline and the type of triad.

Method

Screening task. We tested 702 participants as part of a group testing session, measuring their ability to distinguish similarity from thematic association using a triads similarity task. Given a standard (e.g., **dart**), they chose which alternative was most similar: e.g., **bullseye** or **javelin.** We selected two extreme groups of participants. The Undifferentiating group (11% of the participants) were unable to distinguish similarity from association in over 90% of cases. The Differentiating group (48% of participants) made the distinction correctly in over 90% of cases. The remaining 41%, who were intermediate in their performance, were omitted.

The screening task was completed an average of six weeks prior to participation in this experiment, and no less than three weeks prior to participation. No mention of the screening task was made at the time of the actual experiment.

Participants. Eighty-four Northwestern undergraduates participated. From the initial screening group, 54 Differentiating and 30 Undifferentiating participants were drawn. All were fluent speakers of English.

Materials. Participants saw 64 triads consisting of a standard, a similar item, and either a thematically related alternative or an unrelated alternative. (See Table 1.)

Design and Procedure. We manipulated two variables within-subjects: Deadline (1000ms vs. 2000ms) and Alternative Type (Thematic vs. Unrelated). Deadline was blocked and counterbalanced across participants. Within each block, both Thematic and Unrelated triads were seen. Across participants, all items appeared in all conditions. We also included the categorical variable, Group, with two levels (Differentiated and Undifferentiated).

Participants were instructed to choose the similar alternative in each triad. The two types of relationships were described, and examples were given as illustrations. To allow participants to become accustomed to the procedure, they saw 24 practice items at the new deadline at the beginning of each block. They received feedback as to the accuracy of their responses during this practice session, but not during the test blocks.

The stimuli were presented by computer. For each item, participants saw lines of asterisks, which were replaced by the three words composing the triad. The triad remained on screen until the participants pressed the left or right cursor key to choose the left or right alternative. If participants failed to respond within the deadline, a buzzer sounded and the words "Too slow" flashed on the screen. Once participants made their choice, the next item was presented after a 1000 ms delay.

Results & Discussion

All responses exceeding the deadline were discarded. The percentage of similarity responses is given in Table 3.

Table 3: Percentage of Similar responses in Experiment 2.

Differentiating Participants

	1000ms	2000ms
Alternative Type		
Thematic	72.9%	86.9%
Unrelated	88.3%	96.8%

Undifferentiating Participants

	1000ms	2000ms
Alternative Type		
Thematic	68.4%	84.6%
Unrelated	84.0%	95.1%

There is a main effect of Alternative Type; this is consistent with both accounts. Participants' performance when Similar alternatives were paired with Unrelated foils was superior to their performance when Thematic alternatives were present (by item, $F(1,63) = 49.24$, $p < 0.001$; by subject, $F(1, 82) = 72.78$, $p < 0.001$). Also, as would be expected under either account, overall performance was better at the 2000ms deadline than at the 1000ms deadline (by item, $F(1, 63) = 69.68$, $p < 0.001$; by subject, $F(1,82) = 44.36$, $p < 0.001$).

However, consistent only with the confusability account, there was an interaction between Deadline and Alternative Type. Participants' ability to identify the Similar alternative was more vulnerable to time pressure in the presence of an Thematic distractor than in the presence of an Unrelated foil. The interaction is significant by subject, though marginal by item (by subject, $F(1, 82) = 6.83$, $p < 0.01$; by item, $F(1,63) = 3.16$, $p = 0.08$). The integration account predicts no such improvement in performance at the longer deadline.

With respect to individual differences, the performance of the Undifferentiating group was slightly lower in all conditions (by item, $F(1, 63) = 4.80$, $p < 0.05$; by subject, $F(1,82) = 4.37$, $p < 0.05$). The interaction of Deadline by Group was not significant (by item, $F(1,63) = 1.98$, $p > 0.10$, by subject, $F(1,82) = 1.03$, $p > 0.10$), nor was any other interaction involving Group (all $Fs < 1.1$). It appears that both Undifferentiating and Differentiating participants are challenged by shorter deadlines. This is consistent with a metacognitive process distinguishing comparison from association that requires some time to complete. Individual differences could arise because of differences in the speed and reliability of this process. Consistent with this possibility, the group differences were not specific to the similarity-thematic distinction. Rather, there was simply an overall reduction in accuracy for the Undifferentiating group.

It is also noteworthy that the Undifferentiating group did far better in this task than in the screening task. They made over 90% thematic false alarms on the screening task, but in Experiment 2 they were correct (i.e., able to ignore the association) 68% of the time at the shorter deadline, and 84% of the time at the longer deadline. This is a striking improvement. We conjecture that the initial practice session may have sharpened their ability to discern the thematic-similarity distinction. If so, this would be further evidence for separable processes with output that requires experiential practice to label reliably.

General Discussion

The results presented here are most consistent with the confusability account: thematic association and similarity are distinct processes, and thematic associations intrude upon similarity judgments. Experiment 1 provides evidence that association and similarity are distinct processes, in that participants could set aside association based on the nature of the task, choosing similar alternatives in the Word task despite the dominance of thematic alternatives in the Baseline task. However, Experiment 2 suggests that even people who make this distinction easily and consistently under normal conditions can falter under strict deadlines. This is in line with the confusability account; it is harder for participants to set aside thematic associations when they have to make quick decisions.

Although the strong form of the integration account can probably be dismissed, intermediate positions remain viable. For example, perhaps comparison and association are independent processes that are often (or perhaps typically) combined in judgments of similarity (Medin & Bassok, 1996). We also note that very common similarity relationships may be cached, rather than being recomputed each time they are encountered. Under these circumstances, similarity and thematic relatedness would often coincide.

Our conjecture is that the major culprit in the thematic intrusion phenomenon is a lack of introspective awareness of cognitive states. This is evidenced by the fact that even people who easily distinguished between similarity and association in the screening task experienced difficulty with increased cognitive load. As we noted in the introduction, there is a considerable body of evidence suggesting that people are fallible in their ability to reflect accurately upon their own mental states. For example, Goldstone (1994) comments on the fact that people can be highly variable in their interpretation of a similarity rating task, sometimes equating the task of judging similarity with that of judging purely perceptual similarity. (See Goldstone, Medin, & Gentner, 1993, for other instances of variability in rated similarity). The results of Experiment 2 as well as the screening task suggest that people experience difficulties in distinguishing and internally labeling the mental states produced by association on the one hand and comparison on the other.

Such confusion could significantly affect reasoning tasks, as well as a person's insight into their own responses. For example, in a speeded version of the word-learning task used in Experiment 1, would participants make errors of the sort seen in Experiment 2? Another intriguing question is whether the contextual tasks that support people's ability to focus on similarity also allow them to introspectively access the reason for their actions (see Markman's (1979) studies of inconsistency detection). Answers to these questions will not only help us to understand how similarity and association are cognitively represented and used, but may also shed light on individual differences in reasoning and problem solving. In the end, we believe that that these new possibilities will lead to constructive debates over the concept of similarity, a concept so pervasive in cognitive science research that it is often taken for granted.

Acknowledgments

The authors thank Philip Wolff, Andrea Boyes, and an anonymous reviewer for comments on this version and previous drafts. We also thank Kathleen Braun for her help in analyzing Experiment 1. This research was supported in part by the National Science Foundation grant SBR-9511757 awarded to the first author.

References

Bassok, M. & Medin, D. L. (1996). Birds of a feather flock together: Similarity judgments with semantically rich stimuli. Journal of Memory and Language, 36, 311-336.

Bauer, P. J., & Mandler, J. M. (1989). Taxonomies and triads: Conceptual organization in one-to-two-year-olds. Cognitive Psychology, 21, 156-184.

Cheng, P. W. & Novick, L. R. (1990). A probabilistic contrast model of causal induction. Journal of Personality & Social Psychology, 58, 545-567.

Clement, C. A. & Gentner, D. (1991). Systematicity as a selection constraint in analogical mapping. Cognitive Science, 15, 89-132.

Flavell, J. H., Green, F. L., & Flavell, E. R. (1990). Developmental changes in young children's knowledge about the mind. Cognitive Development, 5, 1-27.

Gentner, D. (1983). Structure-mapping: A theoretical framework for analogy. Cognitive Science, 7, 155-170.

Gentner, D. (1989). The mechanisms of analogical learning. In S. Vosniadou and A. Ortony (Eds.) Similarity, analogy, and thought (pp. 189-241). Cambridge, England: Cambridge University Press.

Gentner, D. & Jeziorski, M. (1993). The shift from metaphor to analogy in Western science. In A. Ortony (Ed.) Metaphor and thought. (pp. 447-480). New York: Cambridge University Press.

Gentner, D. & Markman, A. B. (1997). Structure mapping in analogy and similarity. American Psychologist, 52, 45-56.

Goldstone, R. L. (1994). The role of similarity in categorization: Providing a groundwork. Cognition 52(2), 125-157.

Holyoak, K. J. & Thagard, P. (1989). Analogical mapping by constraint satisfaction. Cognitive Science, 13, 295-355.

Kelley, H. H. (1967). Attribution theory in social psychology. Nebraska Symposium on Motivation, 15, 192-238.

King, P. M. & Kitchener, K. S. (1994). Developing Reflective Judgment : Understanding and Promoting Intellectual Growth and Critical Thinking in Adolescents and Adults. San Francisco: Jossey-Bass.

Koriat, A. (1993). How do we know what we know? The accessibility model of the feeling of knowing. Psychological Review, 100, 609-139.

Loftus, E. F. (1997). Creating childhood memories. Applied Cognitive Psychology, 11, S75-S86.

Markman, A. B. (1997). Constraints on analogical inference. Cognitive Science, 21(4), 373-418.

Markman, A. B. & Wisniewski, E. (1997). Similar and different: The differentiation of basic-level categories. Journal of Experimental Psychology: Learning, Memory, and Cognition, 23(1), 54-70.

Markman, E. M. (1979). Realizing that you don't understand: Elementary school children's awareness of inconsistencies. Child Development, 50, 643-655.

Markman, E. M. (1989). Categorization and naming in children. Cambridge, MA: MIT Press.

Markman, E. M. & Hutchinson, J. E. (1984). Children's sensitivity to constraints on word meaning: Taxonomic versus thematic relations. Cognitive Psychology, 16, 1-27.

Medin, D. L., Goldstone, R. L., & Gentner, D. (1993). Respects for similarity. Psychological Review, 100, 254-278.

Medin, D. L. & Heit, E. J. (in press). Categorization. In D. Rumelhart & B. Martin (Eds.), Handbook of cognition and perception. Hillsdale, NJ: Erlbaum.

Murphy, G. L. & Allopenna, P. D. (1994). The locus of knowledge effects in concept learning. Journal of Experimental Psychology: Learning, Memory and Cognition, 20, 904-919.

Murphy, G. L. & Medin, D. L. (1985). The role of theories in conceptual coherence. Psychological Review, 92, 289-316.

Sloman, S. A. (1996). The empirical case for two systems of reasoning. Psychological Bulletin, 119 (1), 3-22.

Tversky, A. (1977). Features of similarity. Psychological Review, 84, 327-352.

Sloman, S. A. & Rips, L. J. (1998). Similarity as an explanatory construct. Cognition, 65, 87-101.

Vygotsky, L. S. (1978). Mind in society: The development of higher psychological processes. M. Cole, V. John-Steiner, S. Scribner, & E. Souberman (Eds.), Cambridge, MA: Harvard University Press.

Waxman, S. R. & Gelman, R. (1986). Preschoolers' use of superordinate relations in classification and language. Cognitive Development, 1, 139-156.

Wisniewski, E. J. & Bassok, M. (1996). On putting milk in coffee: The effect of thematic relations on similarity judgments. Proceedings of the Eighteenth Annual Conference of the Cognitive Science Society. (pp. 464-468). Hillsdale, NJ: Erlbaum.

Understanding probability words by constructing concrete mental models

David W. Glasspool (dg@acl.icnet.uk) and John Fox (jf@acl.icnet.uk)
Advanced Computation Laboratory, Imperial Cancer Research Fund,
61 Linclon's Inn Fields, London, England.

Abstract

We propose a model of the representation and processing of uncertainty and use it to account for data from an experimental study of the use of probability words. Given two sentences, one using a probability word and the other phrased in terms of reasons-to-believe, subjects were asked to judge if the second was an acceptable paraphrase for the first. For certain word/paraphrase pairs there was a high degree of consensus about acceptability, for others the subjects were divided. We model the decision process as involving two stages. First, a concrete "mental" model is constructed which is consistent with the first phrase. The second phrase is then tested for compatibility with this model. In simulations two different representations for the meanings of phrases were tested, one based on probability intervals, and one based on qualitative argument structures. Both versions of the model give a good account for the data, both in terms of which paraphrases are judged to be acceptable and the relative proportions of subjects agreeing or disagreeing.

Introduction

What is the meaning of probability terms (such as "probable" and "possible") as used in everyday language, and how are such concepts used in cognitive processing? The second of these questions - how such terms are used in processes such as decision making - has generally been seen as closely related to the first - their underlying cognitive representation. Historically, such terms have often been taken as conveying intervals of confidence or probability over some analogue scale, such as a probability scale or fuzzy membership functions (Wallsten & Budescu 1995). An alternative possibility is based on the view that human reasoning under uncertainty involves a process of logical argumentation, in which qualitative arguments for or against (or reasons to believe or doubt) a proposition are as important, or possibly more so, than representation in terms of quantitative values (Fox, 1994). On this view probability words may convey qualitative structures of such arguments rather than numerical degrees of belief. For example, the word "probable" might mean something more like "there are better reasons to believe this is true than to doubt it" than "the probability of this being true is greater than 0.5".

In this paper we present a model of a decision process which is applied to decisions about paraphrases for common probability words. On the model, probability terms are used by constructing a concrete (internal, or "mental") model of the world that is compatible with the term. Some more abstract representation of the meaning of the phrase - be it in terms of probability intervals, argument structures or some

other formalism - is used to construct this world model, but it is the model that is the basis for the decision itself.

Before discussing the model and simulation work in more detail we first describe the experimental data on which it is based.

Experiment

The experiment reported here is designed to investigate the relationship between probability words and sets of arguments for or against propositions, with the aim of establishing a consistent set of terms for reports from risk-assessment software.

Subjects were presented with stimuli of the form "If (statement 1) then (statement 2)", as in the following examples:

If
    it is TRUE that smoking causes cancer
then
        There are better reasons to believe that
        smoking causes cancer than to deny it

If
    It can be ruled out that
    benzoate derivatives are carcinogenic
then
        it is PROBABLE that benzoate derivatives
        are carcinogenic

They were asked to judge in each case if they agreed with the "then" statement given the "if" statement. Stimuli were presented on a computer display, and subjects responded by pressing one of three buttons marked "Agree", "Disagree" and "Unclear". They were asked to disregard any opinions they might have on the truth of the statements and to concentrate only on the consistency of the second with the first. In every stimulus one of the statements used a probability word, and the other was phrased in terms of "reasons to believe". Both words and phrases were selected from a set of five possibilities (Table 1), giving 25 possible combinations. Additionally each pair of statements was presented in both orders - with the probability word first and "reasons to believe" phrase second, and vice-versa, yielding a total of 50 stimuli. These were all presented, in random order, to 33 undergraduate students who participated for credit on a psychology course at

Table 1: The probability words and "reasons to believe" phrases used in the experiment. The latter are followed by the acronyms used to identify them hereafter.

Probability Words	"Reasons to believe"	
True	The reasons to believe (p) are totally convincing	(RBTC)
Probable	There are better reasons to believe (p) than to doubt it	(BRTB)
Possible	There is no reason to doubt (p)	(NRTD)
Improbable	There are better reasons to doubt (p) than to believe it	(BRTD)
False	It can be ruled out that (p)	(CBRO)

City University, London.

Results

Of the 1650 responses of 33 subjects on 50 stimuli, only 111 (6.7%) were "unclear". In the current work we focus on the "agree" and "disagree" responses only. Table 2 shows the proportion of such responses to each stimulus which were "agree" rather than "disagree".

Table 2: Experimental results: Proportion of subjects responding "agree" in (a) the "Term then Phrase" condition and (b) the "Phrase then Term" condition. Entries show the quantity (agree responses) / (agree + disagree responses). Any "unclear" responses are not counted.

(a)

	RBTC	BRTB	NRTD	BRTD	CBRO
True	0.94	0.88	0.88	0.06	0.03
Probable	0.26	0.84	0.29	0.09	0.03
Possible	0.14	0.71	0.17	0.2	0
Improbable	0	0.22	0.03	0.87	0.19
False	0.06	0.07	0.09	0.84	0.84

(b)

	RBTC	BRTB	NRTD	BRTD	CBRO
True	0.9	0.37	0.91	0	0.12
Probable	0.79	0.97	0.79	0.26	0.03
Possible	0.69	0.96	0.68	0.66	0.13
Improbable	0.03	0.17	0.19	0.81	0.57
False	0	0.06	0.15	0.34	0.93

A number of interesting features emerge. On some stimuli the subjects were quite consistent in their responses, on others they were clearly divided with as much as a 60%:40% split in opinion. The ordering of the statements in the stimuli is important - for example, 20% agreed that if something is "possible" then there are better reasons to believe it than to doubt it, whereas 66% agreed that if there are better reasons to believe than to doubt then it is still "possible". This asymmetry makes sense given the usual intuitive meanings for these phrases, but it is interesting to note that those stimuli which show such asymmetry do so to differing degrees.

Finally, we note that the phrase "There is no reason to doubt" is treated as very similar to "The reasons to believe are totally convincing". It seems that the experiment is not sensitive to the differences in meaning between the two phrases,

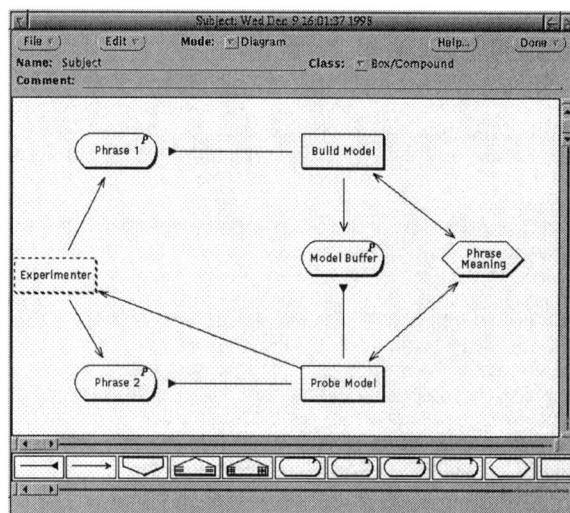

Figure 1: Overall structure of the decision model, as simulated in the COGENT modelling package.

which intuitively seem clear. We hope to probe such differences further in future experiments.

A computational model

How might one characterise the cognitive processes involved in deciding whether one phrase is an acceptable paraphrase of another? Our approach is based on two central hypotheses:

1. The task is carried out by forming a symbolic internal ("mental") model of the first phrase, then testing the second phrase against it. If the second phrase is consistent with the established model then the paraphrase is accepted.

2. The model of a phrase employed in this process takes the form of a set of alternative possible situations in which the phrase would hold.

These assumptions are embodied in the decision-making model shown in Figure 1. The model is implemented using the COGENT cognitive modelling system (Cooper & Fox, 1998), which allows the components of the model to be fully specified so that its operation can be simulated.

The overall simulation contains models of both the task environment (labelled "experimenter") and the subject, although only the subject model is shown here. The task environment model is responsible for presenting pairs of phras-

Table 3: The set of possible situation models which could be produced under each representational scheme.

Probabilistic	Reason-based
0.0	
0.1	Confirm
0.2	Exclude
0.3	Support
0.4	Oppose
0.5	(Support, Oppose)
0.6	(Support, Support, Oppose)
0.7	(Support, Oppose, Oppose)
0.8	
0.9	
1.0	

Table 4: Meanings assigned to each phrase under the different representational schemes. * "NRTD" is equated with "True" here.

Phrase	Probabilistic	Reason-based
True	$p = 1.0$	Confirmed
Probable	$p > 0.5$	Support > Opposition
Possible	$p > 0$	Not excluded
Improbable	$p < 0.5$	Opposition > Support
False	$p = 0.0$	Excluded
RBTC	$p = 1.0$	Confirmed
BRTB	$p > 0.5$	Support > Opposition
NRTD	$p = 1.0$*	Confirmed*
BRTD	$p < 0.5$	Opposition > Support
CBRO	$p = 0.0$	Excluded

es equivalent to those used in the real experiment, and for recording results.

The layout of the subject model is intended to reflect hypothesis 1. The pair of phrases presented on a particular trial are placed in two storage buffers. Phrase 1 is the operative phrase from the first statement presented on that trial (for example *probable* in the sentence "It is probable that benzoate derivatives are toxic"). Phrase 2 is the operative phrase from the second statement. The process "Build Model" implements hypothesis 2 by constructing a set of situational models, all compatible with Phrase 1. To do this, "Build Model" communicates with another process, "Phrase Meaning", to check whether each of a standard set of candidate models is compatible with the phrase. Any which are compatible accumulate in a "Model Buffer". The final set of models in this buffer is taken to conceptualise Phrase 1. The approach has similarities with the "mental models" approach to reasoning (Johnson-Laird, 1983) to the extent that a formal proposition is represented by a set of concrete world models with which it is compatible.

The process "Probe Model" checks each situational model in the buffer for compatibility with Phrase 2. Again the "Phrase Meaning" process is used to perform the compatibility check. "Probe Model" responds to the experimenter according to the number of models in the buffer that are compatible with Phrase 2. If all models are compatible the paraphrase is accepted (the response "agree" is sent to the experimenter). If all are incompatible it is rejected (the response is "disagree"). If some are compatible and some incompatible the situation is unclear. Under either of the representational schemes discussed below 20% of stimuli result in this situation, whereas only 6.7% of subjects' responses are "unclear". It is therefore not obvious that such cases should be interpreted as "unclear" responses. We return to this point below.

To implement the situational models themselves two different representational schemes were investigated, which we label "probabilistic" and "reason-based". Table 3 shows the set of candidate situational models which could be considered under each representational scheme. Each situational model is intended to represent a single possible state of the world (we refer to these as "possible worlds", although the term is not used in its formal sense). The set of models accumulat-

ed in the Model Buffer represents a set of possible worlds in each of which Phrase 1 would be true.

We will first consider the probabilistic scheme, which is perhaps the more compatible with traditional ideas about the meaning of probability words. Here each situational model comprises a single quantity, which is taken to represent the probability that an event will occur in a particular possible world. We will describe the operation of the model and the results obtained using this representational scheme, returning to the reason-based scheme later.

The "Phrase Meaning" process checks the compatibility of a particular situational model with a particular phrase. It contains a definition of the meaning of each phrase under the appropriate representational scheme, as shown in Table 4. For the probabilistic scheme the meanings of phrases are defined in terms of probability intervals, and a straightforward set of intervals is chosen for the five probability words. For consistency the "reasons-to-believe" phrases are also assigned probability interval meanings. The meaning assigned to "No Reason to Doubt" is the same as that for "Reasons to Believe are Totally Convincing", in response to the clear tendency of subjects to treat both phrases as equivalent in the experiment. We return to this point in the discussion section.

Suppose Phrase 1 is "Probable". Under the probabilistic representational scheme the "Build Model" process would build a concept for the phrase "Probable" comprising the set of probability values {0.6, 0.7, 0.8, 0.9, 1.0}. That is, all candidate models with value > 0.5. If Phrase 2 is "Possible" then the "Probe Model" process will test each of these values against the meaning for "Possible" (> 0), and find them all to be compatible. The response "agree" will be sent to the "experimenter".

Study 1

Using the probabilistic representational scheme the full set of 50 experimental stimuli was presented to the model. For an initial comparison with the subject data we examine those stimuli for which 90% or more of the subjects either agreed or disagreed with the paraphrase. There are 21 such stimuli (6 with most subjects agreeing and 15 with most disagreeing). In every one of these cases the model gives an agree response (where 90% or more of subjects agree) or a disagree response (where 90% or more disagree). This confirms that the model predicts the decisions made by subjects in those cases where the subjects themselves agree on the response.

Many of the remaining stimuli result in a mixture of compatible and incompatible situation models in the model buffer. As mentioned earlier it is not obvious how these cases should be handled. In order to investigate this further the criteria for "agree" or "disagree" responses from the model were relaxed. An "agree" response was made if 50% or more of the situational models in the model buffer were compatible with Phrase 2, a "disagree" response otherwise. This allows all stimuli to produce a response. Under these conditions, every "agree" response from the model corresponds to a stimulus for which more than 50% of subjects agree, and every "disagree" response to a stimulus for which more than 50% disagree (In only one case does the model produce exactly 50% compatible models. This is in response to a stimulus to which more than 50% of subjects responded "agree").

Clearly the proportion of "compatible" models is an excellent predictor of subjects' responses at this coarse level of analysis. This result suggests a possible interpretation for cases with both compatible and incompatible models in the Model Buffer: The ratio of compatible to incompatible models may correspond to the ratio of subjects agreeing to those disagreeing. To test this idea Table 5 shows, for each of the 50 stimuli, the proportion of "compatible" situational models produced by the simulation. The proportions do indeed correlate strongly with the proportions of "agree" responses in the experimental data of Table 2 (Spearman's rho = 0.91, p<.001, one-tailed).

The average absolute difference between simulated and actual proportions of "agree" responses over the table as a whole is 0.12, and the maximum is 0.43. If we look only at stimuli which result in both compatible and incompatible models ("mixed" responses) in the simulation, the fit appears more uniform, with an average is 0.13 and a maximum of 0.2.

Study 2

Study 1 used a representational scheme based on simple probability intervals, which is compatible with established ideas about the meaning of probability phrases. To what extent are the results of the simulation dependent on the use of a quantitative representational scheme? In this study we adopt a different approach, based on qualitative "reasons to believe" or arguments for and against a proposition. This approach is based on the idea of logical *argumentation* as a model for reasoning under uncertainty (Fox et al 1992; Fox, 1994; Krause et al, 1994), a process in which qualitative arguments for or against (or reasons to believe or doubt) a proposition are weighed up in order to make a decision.

We classify the reasons one might have for believing or disbelieving a proposition into four classes, following the type of classification common in argumentation theory (Fox, 1994). *Confirming* and *excluding* reasons are those which establish beyond doubt that a proposition is true or false, respectively. *Supporting* and *opposing* reasons provide qualitative but inconclusive evidence for or against the proposition, respectively. Table 3 shows the candidate situational models which can be chosen from by the "Build Model" process using the reason-based representational scheme. This set of candidates

Table 5: Proportion of "compatible" models produced for each stimulus under the "probabilistic" representational scheme.

(a) "Term then Phrase" condition.

	RBTC	BRTB	NRTD	BRTD	CBRO
True	1	1	1	0	0
Probable	0.2	1	0.2	0	0
Possible	0.1	0.5	0.1	0.4	0
Improbable	0	0	0	1	0.2
False	0	0	0	1	1

(b) "Phrase then Term" condition.

	RBTC	BRTB	NRTD	BRTD	CBRO
True	1	0.2	1	0	0
Probable	1	1	1	0	0
Possible	1	1	1	0.8	0
Improbable	0	0	0	1	1
False	0	0	0	0.2	1

was chosen to give the full range of qualitatively different structures using the four classes of reason - those which simply confirm, exclude, support or oppose the proposition, those which offer qualitatively balancing support and opposition, and those with more supporting than opposing arguments or vice-versa. Table 4 shows the meanings assigned to phrases under this scheme. In this case the reason-based phrases are more obvious in their meaning than the probability words, but note again that "NRTD" is assigned the same meaning as "RBTC". We have attempted to assign reasonable meanings to probability words.

Table 6 shows predicted proportion of "agree" responses using this representational scheme. The correlation (Spearman's rho) with the experimental data (Table 2) is again 0.91 (p<.001, one-tailed). The table differs from Table 5 only for the 10 "mixed" responses, for which the average absolute difference when compared with the experimental data is lower than for study 1, at 0.07, still with a maximum of 0.21. The overall average difference is 0.11.

Discussion

How are we to interpret the fit between the proportion of "compatible" models produced by the simulation and the proportion of "agree" responses from subjects? This would make sense under the assumption that the concept for phrase 1 comprises not the full set of "possible worlds" in which that phrase would be true, but only one such world model (the actual choice might be influenced by factors such as the availability, simplicity or concreteness of models, for example). In other words subjects tend to represent a probability phrase with the first appropriate situational model which comes to mind - a process of "satisficing" consistent with ideas of bounded rationality (Simon, 1956, 1982, Gigerenzer and Goldstein, 1996). The important feature of the model which allows this fit to the data seems to be the representation of the first phrase as a *concrete example* of a situation compatible with the phrase, rather than a more abstract representation capable of capturing the full range of meaning of

Table 6: Proportion of "compatible" models produced for each stimulus under the "reason-based" representational scheme.

(a) "Term then Phrase" condition.

	RBTC	BRTB	NRTD	BRTD	CBRO
True	1	1	1	0	0
Probable	0.33	1	0.33	0	0
Possible	0.17	0.5	0.17	0.33	0
Improbable	0	0	0	1	0.33
False	0	0	0	1	1

(b) "Phrase then Term" condition.

	RBTC	BRTB	NRTD	BRTD	CBRO
True	1	0.33	1	0	0
Probable	1	1	1	0	0
Possible	1	1	1	0.67	0
Improbable	0	0	0	1	1
False	0	0	0	0.33	1

the phrase. We assume that some such abstract meaning is nonetheless available at some level to allow the selection of a representative situation in the first place, and to test the second phrase against it.

An important issue in the simulation is the choice of candidate models and meanings assigned to phrases. Both clearly influence the number of compatible and incompatible models accumulated in the model buffer, which in turn determines the predicted proportions of "agree" and "disagree" responses from the simulation. As far as candidate models are concerned, the choice for the probabilistic representational scheme appears reasonable and the only real degree of freedom here would be to alter the grain size of the point probabilities available (giving a distribution in the limit), not their range. This should have no effect on the predicted proportions. The choice of candidates for the reason-based scheme was intended to capture the minimum set of qualitatively different argument structures. Various alternatives could be considered, for example including *excluding* or *confirming* arguments in the same models as *supporting* or *opposing* arguments, and this might alter the predicted proportions. In the absence of a principled procedure for selecting argument structures, however, it seems unreasonable to depart from the minimal set we have used.

For the probabilistic scheme the set of phrase meanings we have used are very simplistic probability intervals. These were chosen to give a neutral first approximation to the intended meanings of the phrases rather than with any empirical evidence in mind. There are however empirical results concerning subjects' willingness to assign particular probability values or intervals to various phrases, and this evidence could be used in a more principled version of the model. Changing the intervals would undoubtedly change the resulting predicted proportions. There would appear to be far less latitude possible in the selection of reason-based meaning definitions, which are essentially qualitative in nature. The results from study 2 can accordingly be considered more robust than those from study 1.

Table 2 suggests that subjects treat the phrase "No reason to doubt" as substantially equivalent to "The reasons to believe are totally convincing", despite the fact that intuitively the phrases do have different meanings. In both simulation studies the meanings of the two phrases have thus been made identical. It is not obvious how "no reason to doubt" would otherwise be represented on the probabilistic scheme, but there is a clear candidate meaning for the phrase under the reason-based scheme - intuitively it should correspond to an absence of opposing (and excluding) arguments. We assume that the current experimental task is insufficiently demanding to bring out any differences between the phrases, and we intend to investigate this anomaly further in subsequent studies. Another area which we will follow up in further work is the incidence of "unclear" responses from subjects. A larger study should provide a more reasonable number of these for analysis.

Study 2 shows that a representational formalism based on qualitative "argument" structures gives at least as good a fit to the data as one using quantitative probability values. This parallels findings from other modelling work in human decision making (Fox and Cooper, 1997; Cooper and Fox, 1997; Yule, Cooper and Fox, 1998) and is interesting in the light of claims that a formal theory of decision making under uncertainty based on a logic of argumentation (Fox et al 1992; Fox, 1994; Krause et al, 1994) may provide a more natural basis for understanding human decision making than traditional normative statistical approaches.

Conclusions

Both versions of the model give a good account of the data, both in terms of which paraphrases are judged to be "correct" (including the effect of order of presentation), and the relative proportions of subjects agreeing or disagreeing. The effect of the order of the two statements in the stimulus (phrases 1 and 2) can be qualitatively understood from the point of view of their logical interdependencies, but the strength of the approach presented here is that it gives a quantitative prediction for the size of the effect for different stimuli, based on the proportion of "compatible" and "incompatible" situational models generated. We conclude that the two hypotheses on which the model is based are appropriate, and we take our results as suggestive that subjects use a single concrete example to represent a probability phrase for the purposes of comparison (a result consistent with ideas of satisficing in "mental models" approaches to reasoning; Evans & Over, 1996).

The use of qualitative "argument" structures in the simulation provides at least as good a fit to the data as the use of more traditional quantitative probability values. The model is thus compatible with a view of of reasoning and decision making as involving qualitative argumentation, which we believe may provide a more natural basis for understanding these cognitive processes than quantitative statistical approaches.

Acknowledgments

The authors would like to thank David Hardman and Peter Ayton for their help in carrying out the experiment, and Andrew Coulson and Richard Cooper for their comments on the manuscript. This work was supported by an award from the UK Economic and Social Research Council (award no. L127251011) under the Cognitive Engineering Programme.

References

Cooper, R. & Fox, J. (1997). Learning to make decisions under uncertainty: The contribution of qualitative reasoning. In Langley, P. & Shafto, M. (Eds.), *Proceedings of the 19th International Conference of the Cognitive Science Society.* Madison, WI. Cognitive Science Society Incorporated.

Cooper, R. & Fox, J. (1998). COGENT: A visual design environment for cognitive modeling. *Behavior Research Methods, Instruments & Computers, 30*, 553-564.

Evans, J.St.B.T. & Over, D.E. (1996) *Rationality and Reasoning.* Hove: Psychology Press.

Fox, J. (1994). On the necessity of probability, reasons to believe and grounds for doubt. In G Wright and P Ayton (eds) *Subjective Probability.* Chichester: John Wiley.

Fox, J. & Cooper, R. (1997): Cognitive processing and knowledge representation in decision making under uncertainty. In Scholz, R. W. & Zimmer, A. C. (eds.), *Qualitative Aspects of Decision Making.* Pabst, Lengerich, Germany.

Fox, J., Krause, P. & Ambler S. (1992). Arguments, Contradictions and Practical Reasoning. In: Neumann B (Ed) ECAI92, Vienna, Austria. *Proceedings of the 10th European Conference on AI.*

Gigerenzer, G. & Goldstein, D. G. (1996). Reasoning the fast and frugal way: Models of bounded rationality. *Psychological Review, 103*, 650-669.

Johnson-Laird, P. N. (1983). *Mental Models.* Cambridge: Cambridge University Press.

Krause, P., Fox, J. & Judson P. (1994). An Argumentation Based Approach to Risk Assessment. Presented at IMA conference on Risk: Analysis and Assessment, 14-15 April '94. In *IMA Journal of Mathematics Applied to Business and Industry, 5*, 249-263.

Simon, H. A. (1956). Rational choice and the structure of the environment. *Psychological Review, 63*, 129-138.

Simon, H. A. (1982). *Models of bounded rationality.* Cambridge, MA.: MIT Press.

Wallsten, T. S. & Budescu, D. V. (1995). A review of human linguistic probability processing: General principles and empirical evidence. *Knowledge Engineering Review, 10*, 43-62.

Yule, P., Cooper, R. & Fox, J. (1998). Normative and Information Processing Accounts of Decision Making. In Gernsbacher, M. a. & Derry, S. J. (Eds.), *Proceedings of the 20th Annual Conference of the Cognitive Science Society.* Madison, WI. Cognitive Science Society Incorporated.

Reasoning with Causal Relations

Yevgeniya Goldvarg (goldvarg@phoenix.princeton.edu)
Department of Psychology
Princeton University
Green Hall
Princeton, NJ 08544

Philip N. Johnson-Laird (phil@clarity.princeton.edu)
Department of Psychology
Princeton University
Green Hall
Princeton, NJ 08544

Abstract

The mental model theory postulates that reasoners build models of the situations described in premises, and that each model represents a possibility. The present paper proposes that causal relations, such as A causes B and A allows B, have meanings that concern only possibilities and a temporal constraint that B cannot precede A. This theory predicts that causes and enabling conditions differ in meanings, contrary to a long tradition in philosophy and psychology that they are logically indistinguishable. It also predicts that individuals should reason about causation on the basis of mental models rather than on fully explicit models. Three experiments corroborated these predictions.

Introduction

People reason about causal relations in order to predict what will happen. For example, given the following inference:

A prevents B.
B causes C.
What follows?

most people respond: A prevents C, but this conclusion is invalid. Our goal in the present paper is to present a theory of causation that predicts this phenomenon and that gives a general account of both the meaning of causal relations and of how people reason from them. The theory is based on mental models. Our plan in what follows is, first, to outline the theory of mental models and its extension to causal relations; second, to explain how enabling conditions and causes differ; and, third, to describe how mental models underlie causal deductions.

The Mental Model Theory of Causal Relations

The mental model theory postulates that reasoning is a semantic process, which depends on understanding the meaning of premises and using this meaning to envisage the situations that are possible given the truth of the premises (Johnson-Laird and Byrne, 1991). A mental model accordingly represents a possibility, and its structure and content capture what is common to the different ways in which the possibility may occur. A conclusion is necessary -

- it <u>must</u> be true – if it holds in all the models of the premises. It is possible – it <u>may</u> be true – if it holds in at least one model of the premises (Bell and Johnson-Laird, 1998). And its probability – its <u>likelihood</u> of being true – depends on the proportion of models of the premises in which it is true (Johnson-Laird, Legrenzi, Girotto, Legrenzi, and Caverni, 1998). The principle of truth is a fundamental assumption of the theory: in order to minimize the load on working memory, mental models normally represent only what is true. This principle applies at two levels: individuals represent only true possibilities; and within those possibilities they represent the literal propositions in the premises only when they are true. Consider the following exclusive disjunction, for example:

There is a not a circle or else there is a triangle.

The mental models of the disjunction represent only the true possibilities, and within them, they represent only their true components:

¬o

•

where '¬' denotes negation, 'o' denotes a model of the circle, '•' denotes a model of the triangle, and each row denotes a model of a separate possibility. Hence, the first model does not represent explicitly that it is false that there is a triangle in this case; and the second model does not represent explicitly that it is false that there is not a circle in this case. Reasoners make a 'mental footnote' to keep track of this false information, but these footnotes are soon likely to be forgotten. Indeed, the failure to cope with falsity gives rise to illusory inferences about modal conclusions, i.e. inferences that nearly everyone makes, but that are wrong (Johnson-Laird & Goldvarg, 1997). Only <u>fully explicit</u> models of what is possible given the exclusive disjunction represent the false components in each model:

¬o   ¬•
o   •

where a false affirmative (there is a triangle) is represented by a true negation, and a false negative (there is not a circle) is represented by a true affirmative.

The theory deals with causal relations in a natural way. Its first assumption is that causal relations concern physical possibilities and place a temporal constraint on them. The relation A causes B means that there are three possibilities:

```
    a        b
  ¬ a        b
  ¬ a      ¬ b
```

and that B cannot precede A in time. Likewise, A allows B means that there are three possibilities:

```
    a        b
    a      ¬ b
  ¬ a      ¬ b
```

and that B cannot precede A in time. These possibilities correspond to fully explicit models of the premises, but the theory postulates that logically-untrained individuals normally represent only those possibilities in which A and B are true. Thus, their mental models of A causes B are as follows:

```
    a        b
       . . .
```

where the ellipsis represents those possibilities in which A is false. Table 1 presents the mental models and the fully explicit models of the four principal causal relations.

Table 1: The mental models and the fully explicit models of the four main causal relations.

Causal Relation	Mental Models	Fully Explicit Models
A causes B	a b . . .	a b ¬ a b ¬ a ¬ b
A allows B	a b . . .	a b a ¬ b ¬ a ¬ b
A prevents B	a ¬ b . . .	a ¬ b ¬ a b ¬ a ¬ b
A allows not-B	a ¬ b . . .	a ¬ b a b ¬ a b

The fully explicit models of A causes B correspond to A being sufficient for B. Similarly, the fully explicit models of A allows B correspond to A being necessary for B. In both cases, however, the causal models also embody a temporal constraint that is not essential for necessary or sufficient conditions. There are, of course, many other ways to describe each of the causal relations. If both A causes B and A allows B, then a strong causal relation holds between them corresponding to the following models:

```
    a        b
  ¬ a      ¬ b
```

Likewise, there is a strong relation that combines A prevents B and A allows not-B:

```
    a      ¬ b
  ¬ a        b
```

Experiment 1:

Causal Possibilities

Our first experiment was designed to test the theory embodied in Table 1. The participants' task was to list the true possibilities and the false possibilities for five causal assertions. They were the three relations: A causes B, A allows B, and A prevents B; and two different ways of paraphrasing cause: A prevents not-B, and not-A allows not-B. We devised five lexical contents that were rotated over the relations for different participants, so that each participant saw each content just once, but in the experiment as a whole each content occurred equally often with the five relations. The contents are illustrated by the following examples:

Exercise that is excessive causes the development of angina.

Use of solar energy prevents the occurrence of global warming.

Twenty Princeton undergraduates carried out the experiment. They were told to list all possibilities that would make an assertion true and all possibilities that would make it false. They could list them in any order. The model theory makes three main predictions. First, as Table 1 implies, the participants should list both true cases and false cases. Second, they should start with the possibilities corresponding to the mental models of the relations. Third, they should tend to confuse allows with causes, because they have the same mental models.

Experiment 2:

Enabling Conditions and Causes

Many philosophers and psychologists have argued that there is no logical distinction between the meaning of enabling conditions (as expressed by allows) and causes (see e.g. Mill, 1843; Mackie, 1980; Hart and Honoré, 1985). What then does distinguish them? A wide variety of answers is to be found in the literature: enabling conditions occur early but causes immediately precipitate the effect (Mill, 1843); enabling conditions are common but causes are rare (Hart and Honoré, 1985); enabling conditions are the norm but causes violate the norm (see e.g. Kahneman and Tversky, 1982; Kahneman and Miller, 1986; Einhorn and Hogarth, 1986); enabling conditions are constant but causes are inconstant (Cheng and Novick, 1991); enabling conditions are irrelevant to explanations but causes are relevant(e.g. Mackie, 1980; Turnbull and Slugoski, 1988; Hilton and Erb , 1996). And there are still other views (see Hesslow, 1988, for a review).

All of these hypotheses could be true, yet, as Experiment 1 showed, people can draw a distinction between the meaning of enabling conditions and causes. The two are logically distinct. We propose that causal interpretation depends crucially on how people conceive the circumstances of events, that is, on the particular states that they consider to be possible, whether real, hypothetical, or counterfactual. Consider, for example, the following scenario:

1. Given that there is good sunlight, if a certain new fertilizer is used on poor flowers, then they grow remarkably well. However, if there is not good sunlight, poor flowers do not grow well even if the fertilizer is used on them.

The circumstances described here correspond to the fully explicit possibilities:

Sunlight	Fertilizer	Grow-well
Sunlight	¬ Fertilizer	Grow-well
Sunlight	¬ Fertilizer	¬ Grow-well
¬ Sunlight	Fertilizer	¬ Grow-well
¬ Sunlight	¬ Fertilizer	¬ Grow-well

In these circumstances, sunlight is an enabling condition for the flowers to grow well:

Sunlight	Grow-well
Sunlight	¬ Grow-well
¬ Sunlight	¬ Grow-well

In contrast, all four possible contingencies occur concerning the fertilizer and the flowers growing well. But, <u>the sunlight enables the fertilizer to cause the flowers to grow well</u>.

Now, consider the following scenario:

2. Given the use of a certain new fertilizer on poor flowers, if there is good sunlight then the flowers grow remarkably well. However, if the new fertilizer is not used on poor flowers, they do not grow well even if there is good sunlight.

These circumstances correspond to the possibilities:

Sunlight	Fertilizer	Grow-well
¬ Sunlight	Fertilizer	Grow-well
¬ Sunlight	Fertilizer	¬ Grow-well
Sunlight	¬ Fertilizer	¬ Grow-well
¬ Sunlight	¬ Fertilizer	¬ Grow-well

In these circumstances, the respective causal roles have been swapped around: <u>the fertilizer enables the sunlight to cause the flowers to grow well</u>. In both these cases, the cause and the enabling condition have different meanings and neither of them is constant in the circumstances .

Experiment 2 tested whether 20 Princeton Undergraduates could distinguish causes and enabling conditions. Cheng and Novick (1991) showed that individuals could distinguish them in cases where the enabling conditions were constant and the causes were inconstant. Our aim was to show that they could do so when neither enabling conditions nor causes were constant. We prepared eight pairs of scenarios such as the examples above, in which there were two precursors to the effect and their respective roles as enabling condition and cause were counterbalanced in the two scenarios. We also prepared versions of the pairs of scenarios in which we reversed the order of mention of cause and enabling condition. Thus for the previous examples, there were scenarios as follows:

1'. If a certain new fertilizer is used on poor flowers, then given that there is good sunlight, they grow remarkably well. However, if there is not good sunlight, poor flowers do not grow well even if the fertilizer is used on them.

and:

2'. If there is good sunlight then poor flowers grow remarkably well given the use of a certain new fertilizer. However, if the new fertilizer is not used on poor flowers, they do not grow well even if there is good sunlight.

Each participant encountered just one version of a particular scenario, but an equal number of the four sorts of scenario in the experiment as a whole. The order of presentation was randomized for each participant. The task was to identify the enabling condition and the cause in each scenario, and the experiment included two filler items – one in which there were two joint causes and one in which there were no causes. The participants correctly identified the enabling conditions and causes on 85% of trials, and every participant was correct more often than not ($p = 0.5^{20}$). Hence, individuals can distinguish enabling conditions from causes even in scenarios where neither is constant.

Experiment 3:

Mental Models in Causal Deductions

The previous experiments corroborate the model theory of naïve causation, but they may be consistent with other accounts. Rips (1994, p. 336), for example, argues that it should be possible to frame an account of causal reasoning based on formal rules. Is there any critical test to show that individuals are using models in reasoning about causal relations? The answer is that mental models predict the occurrence of certain systematic errors in reasoning. Consider, first, an inference of the following form:

A causes B.
B prevents C.

What follows?
The premises yield the following mental models (seeTable 1):

a   b   ¬ c
.   .   .

Reasoners should therefore conclude:
A prevents C.

The fully explicit models of the premises also support this conclusion, and so it is valid. Second, consider an inference of the following form:

A prevents B.
B causes C.
What follows?
The premises yield the following mental models:

a   ¬ b
        b   c
.   .   .

Reasoners should therefore conclude:
A prevents C

because A occurs in a model that does not yield C, and C occurs in a model without A. In this case, however, the conclusion is invalid. The fully explicit models of the premises are as follows:

a	¬ b	c
a	¬ b	¬ c
¬ a	b	c
¬ a	¬ b	c
¬ a	¬ b	¬ c

193

Table 2: The 16 causal inferences in Experiment 3, and the conclusions that the mental models of the premises support (invalid conclusions are asterisked). The table also shows the number of participants (n = 20) who drew the predicted conclusions.

FIRST PREMISE

SECOND PREMISE	A causes B	A allows B	A prevents B	Not-A causes B
B causes C	A causes C: 20	*A allows C: 18	*A prevents C: 19	Not-A causes C: 20
B allows C	*A allows C: 19	A allows C: 19	A prevents C: 20	*Not-A allows C: 20
B prevents C	A prevents C: 20	*A allows not-C: 14	*A prevents C: 15	Not-A prevents C: 20
Not-B causes C	*A allows not-C: 8	A allows not-C 12	A causes C: 17	*Not-A prevents C:15

As these models show, there is no causal relation between A and C.

Experiment 3 examined all possible inferences in which the premises were laid out in the following figure:

A - B.
B - C.

and each of the four causal relations was systematically inserted in order to yield the 16 distinct inferences shown in Table 2. The model theory predicts that reasoners should draw a conclusion in all 16 cases, but as the Table shows, half of these conclusions are valid, and half of them are invalid.

Twenty Princeton undergraduates carried out the 16 inferences in random orders, and A, B, and C, referred to abstract topics, e.g. obedience allows motivation to increase; increased motivation causes eccentricity . These contents were rotated over the set of inferences so that each content occurred equally often with each sort of inference in the experiment as a whole. The participants drew the predicted conclusions whether they were valid (93% of conclusions as predicted) or invalid (80% of conclusions as predicted). Each participant drew more predicted than unpredicted conclusions (p = 0.5^{20}) and 15 out of the 16 inferences (Sign test, p < 0.00).

General Discussion

The experiments corroborated the model theory's account of causal relations. Individuals envisage the possibilities corresponding to causal relations. They can distinguish between enabling conditions and causes, even though, according to the model theory, neither relation necessarily depends on the constant presence of an antecedent (pace Cheng, 1997). Likewise, as the theory predicts, reasoners drew systematically invalid conclusions supported by the mental models of the premises. We have carried out other studies that also support the theory. We

conclude that individuals represent the meanings of causal relations in mental models. These models denote possibilities and constrain the order of antecedents and effects: effects do not precede their antecedents. The interpretation of causal relations depends on how individuals represent the circumstances , i.e., the models that they envisage of enabling conditions, causes, and effects. Because mental models do not make the possibilities fully explicit, reasoners should be misled in the case of certain inferences. Experiment 3 corroborated this prediction.

Is there any alternative explanation of our results? One possibility is that people use formal rules of inference to make causal deductions. No such account currently exists, and it is hard to see that it could account, say, for the results of Experiment 1, which depend on a semantic interpretation of causal verbs. Another class of psychological theories postulates that cause is a probabilistic notion (see e.g. Suppes, 1970; Cheng, 1997):

A cause B $=_{def}$ P(B | A) > P(B | not A)

In our view, probabilities enter into the induction of causal relations from empirical observations, but not their meaning. Thus, for example, naïve reasoners distinguish between the assertion:

Smoking causes lung cancer

and:

Smoking causes lung cancer with a certain probability.

The latter claim would be superfluous if causal claims tacitly embodied probabilities. Likewise, the definition above applies equally to causes and enabling conditions. In fact, cause may appear to be probabilistic because of the role of enabling conditions. The circumstance often allow cases in which the cause does not lead to the effect, because the enabling condition fails to hold.

Is there anything more to the meaning of causal relations apart from possibilities and the temporal constraint? We have yet to discern any such missing element.

Acknowledgments

We are grateful to the following colleagues for their helpful advice: Victoria Bell, Zachary Estes, Hansjoerg Neth, Mary Newsome, Sergio Moreno Rios, Vladimir Sloutsky, Jean-Baptiste van der Henst, and Yingrui Yang.

References

Bell, V., and Johnson-Laird, P.N. (1998) A model theory of modal reasoning. Cognitive Science, 22, 25-51.

Cheng, P.W. (1997) From covariation to causation: A causal power theory. Psychological Review, 104, 367-405.

Cheng, P.W., and Novick, L.R. (1991) Causes versus enabling conditions. Cognition, 40, 83-120.

Einhorn, H.J., and Hogarth, R.M. (1986) Judging probable cause. Psychological Bulletin, 99, 3-19.

Hart, H.L.A., and Honoré, A.M. (1985) Causation in the Law. Second Edition. Oxford: Clarendon Press.

Hesslow, G. (1988) The problem of causal selection. In Hilton, D.J. (Ed.) Contemporary Science and Natural Explanation: Commonsense Conceptions of Causality. pp. 11-32. Brighton, Sussex: Harvester Press.

Hilton, D.J., and Erb, H-P. (1996) Mental models and causal explanation: Judgements of probable cause and explanatory relevance. Thinking & Reasoning, 2, 273-308.

Johnson-Laird, P.N & Byrne, R.M.J.(1991?). Deduction Hillsdale, NJ: Lawrence Erlbaum Associates.

Johnson-Laird, P.N., and Goldvarg, Y. (1997). How to make the impossible seem possible. Proceedings of the Nineteenth Annual Conference of the Cognitive Science Society, 354-357.

Johnson-Laird, P.N., Legrenzi, P., Girotto, P., Legrenzi, M.S., and Caverni, J-P. (1998) Naive probability: A mental model theory of extensional reasoning. Psychological Review, 106, 62-88.

Kahneman, D., and Miller, D.T. (1986) Norm theory: Comparing reality to its alternative. Psychological Review, 93, 75-88.

Kahneman, D., and Tversky, A. (1982) The simulation heuristic. In Kahneman, D., Slovic, P., and Tversky, A. (Eds.) Judgment under Uncertainty: Heuristics and Biases. Cambridge: Cambridge University Press.

Mackie, J.L. (1980) The Cement of the Universe: A Study in Causation. Second edition. Oxford: Oxford University Press.

Mill, J.S. (1843/1973) A System of Logic Ratiocinative and Inductive. Toronto, Ontario: University of Toronto Press.

Rips, L.J. (1994) The Psychology of Proof. Cambridge, MA: MIT Press.

Suppes, P. (1970). A Probabilistic Theory of Causality. Amsterdam: North-Holland.

Turnbull, W., and Slugoski, B.R. (1988) Conversational and linguistic processes in causal attribution. In Hilton, D. (Ed.) Contemporary Science and Natural Explanation: Commonsense Conceptions of Causality. pp. 66-93. Brighton, Sussex: Harvester Press.

Cognition and the Computational Power
of Connectionist Networks

Robert F. Hadley
School of Computing Science
and Cognitive Science Program
Simon Fraser University
Burnaby, B.C., V5A 1S6
hadley@cs.sfu.ca

Abstract

This paper examines certain claims of "cognitive significance" which (wisely or not) have been based upon the theoretical powers of two distinct classes of connectionist networks, namely, the "universal function approximators", and recurrent finite-state simulation networks. Each class will be considered with respect to its potential in the realm of cognitive modeling. Regarding the first class, I argue that, contrary to the claims of some influential connectionists, feed-forward networks do *not* possess the theoretical capacity to approximate all *functions of interest to cognitive scientists*. By contrast, I argue that a certain class of recurrent networks (i.e., those which closely approximate *deterministic finite automata*, DFA) shows considerably greater promise in some domains. However, serious difficulties arise when we consider how the relevant recurrent networks (RNNs) could acquire the *weight vectors* needed to support DFA simulations.

Introduction.

Do connectionist networks have the computational power, in theory, to provide successful explanatory models for *high-level* human cognition? Many connectionists believe so, and their confidence stems from a number of formal results that have appeared in the last decade. These 'computability' results pertain to network architectures which, on the face of it, avoid the shortcomings of the oft-cited McCulloch & Pitts (1943) implementation of conventional 'logic gates'. It is now widely recognized that McCulloch-Pitts 'neural circuitry' designs are radically unlike the kinds of neural structures found in actual brains, and offer no advantage over standard digital circuitry. By contrast, some network architectures involved in recent theoretical results bear at least a superficial resemblance to biological neural structures. This fact has inspired hope, in some quarters, that connectionist networks (abbreviated here as 'c-nets') may both (a) match the power of high-level, symbolic AI programs for explaining *higher cognitive functions* (such as abstract planning, mathematical reasoning, and theory formation), and (b) provide insight into how these higher functions could be instantiated in living brains. Among those who have appealed to the 'theoretical power' of multilayer c-nets (especially their purported 'universal'

function approximation capacity) *within a cognitive context* are Churchland (1990), Elman et al (1996), and Niklasson & van Gelder (1994).

This paper will examine several frequently cited theoretical results with regard to their potential for satisfying both conditions (a) and (b) above. In particular, I focus upon whether the relevant proofs appeal to *processing models* which present a genuine alternative to conventional high-level symbolic processing. In addition, I explore whether the proposed models possess some measure of cognitive/biological plausibility. For even if a given c-net architecture essentially *implements* a classical machine, at the computational level, such an implementation would still be vastly important if it revealed how high-level classical processes could be realized in brain-like systems.

In what follows, I consider 'computability results' which fall into two major categories. These are: (1) 'universal' function approximators (emphatic scare quotes), (2) deterministic finite automata simulators. Papers belonging these categories are: **(1)** Hartman, Keeler, & Kowalski (1990); Hornik, Stinchcombe, & White (1989). **(2)** Casey, 1996; Cleerman, Servan-Schreiber, & McClelland (1989); Omlin & Giles (1996); Sontag (1995).

I shall argue that results in the first of the above categories *hold no promise* for explaining the abstract planning and reasoning capacities of humans. ('Universal' function approximators lack the requisite power – they are not universal.) By contrast, the second category (deterministic finite automata, DFA) offers some measure of hope provided the c-nets are designed in a fashion that ensures close mimicry of the state transitions of some DFA. However, even here, the existence of cognitively/biologically defensible training methods is highly problematic.

(Apart from the foregoing, a third class of computability results exists. This concerns *Turing-equivalent networks*. [See Siegelmann (1996); Siegelmann & Sontag, 1994] These are examined in detail in my extended technical report (Hadley, 1999), where I argue that Turing-equivalent networks *not only* require weights and nodes of infinite precision, but are at least as "brittle" and "hand-crafted" as classical symbolic algorithms. It should be noted that the "infinite precision" difficulties have also been explored by Sontag [1995]. Due to space constraints, I must here omit further discussion of

Turing-equivalent networks.)

2. Universal Function Approximators.

The thesis that multi-layer, feed-forward neural networks can be universal function approximators, derives primarily from two influential publications; Hornik, Stinchcombe & White (1989) and Hartman, Keeler & Kowalski (1990). Although the titles of both these publications contain the phrases 'universal approximators' and 'universal approximations', respectively, the text of each paper makes it clear that only a certain class of functions is being addressed. In both cases, only the class of measurable Borel functions is discussed, but this includes continuously valued functions. Hornik et al contend that this class covers 'virtually any function of interest'. Hartman et al do not address this issue, perhaps because they believed the cited work of Hornik et al had sufficiently explored the question. In any case, the difference between the proofs provided by these two groups of researchers concerns only the activation functions applied to hidden layer units, and does not concern the input-output mapping functions which are to be approximated. For this reason, my discussion will center upon Hornik et al, since theirs is the earlier work. My conclusions, however, apply equally well to any claim or 'proof' that finite multi-layer, feed-forward networks can be universal function approximators.

2.1 Problems with Universality.

Let us note at the outset that any given *measurable Borel function* is a set of ordered pairs, where each element of each ordered pair is a single *real* number. Since the output layer of any *actual* neural network can contain only a finite number of units, each having finite precision, any given output value produced for a given input can have only finite precision. Such an output value cannot *uniquely* approximate, in the general case, a specific real number, no matter how large the output layer may be. For, an infinity of real numbers will lie arbitrarily close to any fixed precision output value. In purely numerical domains, this may often be a negligible issue. However, when numerical output is being used to encode symbolic formulae or sentences, serious problems can arise. Admittedly, such difficulties can at times be obviated by a prudent choice of encoding schemes. Unfortunately, this is often not possible in domains associated with higher cognition.

For example, the realms of logic, mathematics, and linguistic theory each involve infinite sets of formulae or sentences. In addition, within these realms there are important functions which can map a given sentence (selected from an infinite set) onto another sentence (or even an infinite set of sentences). That is, such functions have infinite ranges and domains. Yet, the input/output layers of any physically realizable c-net can contain only a finite number of units, each having finite precision. Thus, these layers can accurately encode values spanning only a finite interval. If an infinite number of *formulae encodings* are to be 'packed' into a finite numerical span, there can be no lower bound on how close together a pair of encodings can be.[1]

Now, since a given output value will seldom *precisely* represent the target encoding, one will usually be forced to rely upon some *approximate* result to determine which formula was being output. However, this would commonly lead to disaster, since, in the present context, two numbers which encode distinct formulae should not be viewed as approximations of each other, no matter how numerically close the encodings are. For example, a given numerical code, C, may be very close to the encoding of some theorem, and yet C may encode some other formula which is logically invalid. This difficulty cannot be circumvented by taking some actually generated output value and searching for the *nearest* number which encodes a well-formed formula. For there simply is no "nearest formula encoding". Given any pair of formulae encodings, some other, distinct formula encoding would always lie between them. The brute fact is that formulae are discrete objects which can seldom be viewed as approximations of each other.

It might now be objected that the problem just described is not unique to neural networks. After all, no physically realizable computer could actually derive an infinity of symbolic theorems, or process arbitrarily long symbolic strings. However, this objection misses the *theoretical* point I am making. The point is that there are important 'functions of interest' which cannot be approximated by any feed-forward network, *even though* such networks can approximate many continuously valued functions. Functions which map formulae/sentences into other formulae/sentences are both interesting and important, as I explain below. Many of these functions cannot be approximated in any sense relevant to the foregoing discussion. Yet, as defined by automata theory, they are *computable* functions (i.e., recursive). We can write useful symbolic programs which *intensionally* embody these functions, and the programs will halt, though not always rapidly. This is true even though the ranges and domains of the computable functions may be infinite. An example of such a function, used by some automated theorem provers, is the clause-form conversion algorithm. It takes any sentence of first-order logic (FOL) and generates a set of *clauses*. This set can be written as a simple conjunction of formula.

Now, of course, a given computation involving a computable function can exceed a computer's memory resources. However, as Fodor and Pylyshyn observe (1988), one can always add more memory without having to alter the program, or the function being computed. This cannot be said, in general, of c nets, where increasing the amount of memory (or nodes) will create a different weight configuration, thereby altering the function being computed.

Returning to the main thread, it might be objected that the foregoing discussion is misguided, since Hornik et al were addressing *measurable* nu-

---

[1] To see this, suppose there existed such a finite lower bound. Since there are an infinity of formulae encodings, and each pair of encodings is separated by at least this lower bound, then the entire spanned region would have to be infinitely broad.

197

merical functions. Functions which map symbolic expressions into symbolic expressions are not measurable in the relevant sense. In reply, I would emphasize two points. First, Hornik et al identify the class of functions they address with 'virtually any function of interest'. They even say that 'failures in application can be attributed to ... the presence of a stochastic rather than a deterministic relation between input and target'. Secondly, regardless of the original intent, Hornik et al's results have been construed by many cognitively motivated connectionists as a clear demonstration that, in principle, multilayer feed-forward c-nets can match the power of conventional AI programs. It is this construal (or misconstrual) which concerns me most.

My critique thus far has centered upon problems that arise when interpreting finite numerical output as approximate encodings of discrete symbolic strings. However, significantly deeper problems are encountered when we consider the kinds of (potentially non-halting) computations required by theorem provers in the realm of FOL and higher mathematics. For these realms, there exist working programs which prove interesting theorems. Moreover, standard AI textbooks present various means whereby theorem provers for FOL can be applied to practical problems such as planning, natural language interpretation, and even programming.

The theorem proving programs just alluded to are not guaranteed to halt (complete their computations) in general, though for many input values, they will halt and produce interesting output. However, these programs are usually deterministic and instantiate partial recursive functions (the relevant sets of theorems are recursively enumerable). Such functions comprise a major focus of recursive function theory. Furthermore, as far as we presently know, humans may at times employ high-level mental processes which are best modeled by non-halting programs of the kind being considered. One obvious example would involve mathematicians who discover complex proofs for existing conjectures. The most successful cognitive models in this area are high-level programs which employ both heuristic rules and subprograms that halt for some inputs but not others.

Now, programs which implement theorem provers for standard first-order logic and higher-order mathematical domains involve iterative processes. Their computational complexity is infinite, since no *a priori* limit can be set on the number of iterations involved. By contrast, non-recurrent, feed-forward networks have computational complexity which is linear. Such networks (implemented in parallel, as connectionist theory assumes) always complete their computations in time which is a linear function of the number of layers present. Moreover, unlike partial recursive functions (or their corresponding programs), these networks always produce an output for any given input. In light of this, it is extremely doubtful that multi-layer feed-forward networks could approximate the partial recursive functions under consideration. For example, consider the partial function which, given any suspected theorem of standard FOL, yields a formal proof for that sentence if and only if a proof exists. Certainly, no

feed-forward network could reliably produce an 'approximate output' which could then be used to mechanically infer the desired theorem-proof, if one exists. For this supposition would entail that there exists an effective decision procedure for ascertaining theoremhood in the full first-order calculus. Such a decision procedure has been proven to be impossible (Church, 1956).

Moreover, nothing in Hornik et al (or in Hartman et al) would suggest that a feed-forward network could even *frequently* produce output that approximated the encoding of a target proof for a suspected theorem. Their proofs simply do not address the class of partial recursive functions. We know also, from the renowned work of Church (1956), Gödel (1931) and others, that a function which relates arbitrary sentences of FOL (or number theory) to their *supposed* proofs is not reducible to any arithmetic (or algebraic) equation. To be sure, the 'proof predicates' employed by Gödel in his famous incompleteness theorems correspond to algebraic formulae. However, these predicates apply (or 'hold') only in cases where a proof actually exists. Thus, these 'proof predicates' could not be embedded in a c-net's weight configuration in order to determine *whether* a suspected theorem had a proof. Any purported feed-forward net of this kind would produce output for theorems and non-theorems alike, and this output could not be used to distinguish the two cases.

It might now be objected that the difficulties just considered are innocuous, since it may appear that *partial recursive functions* (PRFs) are not *really* functions at all. After all, these "functions" are only partially defined; they fail to produce output for certain input values.

In reply, two points are relevant. First, as previously noted, PRFs form a major topic in recursive function theory. Researchers in this realm (including Alan Turing, Alonzo Church, and other giants) certainly have regarded PRFs as an important type of function, and recursive function theory is an essential field within both Mathematics and Computer Science. However, setting aside quibbles about the semantics of 'function', there remains the crucial point that theoretical proofs about "universal" function approximation have been cited by many connectionists as conclusive evidence that any mental *computation* could, in principle, be closely approximated by feed-forward networks. Here lies the crux of the matter. Now, certain of the PRFs cited above *have been used* to model high-level cognitive processes, and it is beyond dispute that computer programs embodying these PRFs actually perform computations. Moreover, it would be entirely question-begging for any connectionist to insist that the mental computations of interest could never be accurately modeled by PRFs. So, regardless of how narrowly we choose to construe the sense of 'function', the theoretical proofs in question simply have not established that all *mental computations of interest* can be approximated by feed-forward networks.

3. Recurrent Neural Networks and Deterministic Finite Automata.

In a 1989 paper, Cleermans et al offered a demonstration that simple recurrent networks "can learn to mimic closely a finite-state automaton (FSA) both in its behavior and in its state representations". Their demonstration was experimental, rather than formal. Using backpropagation, they successfully trained recurrent networks to induce comparatively simple deterministic finite automata (DFA). Experimental evidence of this nature could not, of course, establish any general equivalence between recurrent neural networks (RNNs) and DFA. Indeed, Elman's work with RNNs (1990, 1993) illustrated that simple recurrent nets, trained via backpropagation, may provide only limited approximations to DFA, in that network performance can significantly degrade when even moderately deep recursion is present within input strings. On a more encouraging note, the capacity of RNNs to simulate the *general class* of DFAs has been formally proven (see Sontag, 1995; Casey, 1996; Omlin & Giles, 1996). These proofs are significant because many powerful computational processes can be modeled by DFAs. (Whether high-level human cognition can, in general, be modeled by DFAs is less clear, however. We shall return to that issue below.)

Now, it is noteworthy that RNNs and DFAs are not equivalent classes. Given a particular DFA, there does exist some RNN whose I/O behaviour is equivalent to the DFA. However, there are many RNNs which fail to approximate any DFA to a degree sufficient to enable successful automaton simulation on long input strings. Omlin & Giles (1996) are emphatic on this point. They also provide an algorithm which partially *pre-determines* a network's weights, prior to training, in order to ensure that subsequent learning yields weight vectors that guarantee successful simulation. The resulting networks can achieve very close simulations of DFA, but the question naturally arises, do these simulations possess any advantage over classical DFA? For example, will the network possess a tolerance to 'noise' which would be absent in an entirely *precise* DFA simulation?

In principle, some degree of noise tolerance could be present. Indeed, Casey (1996) has proven that RNN simulations of DFA can, in general, be noise tolerant (within narrow bounds) provided the requisite weight vectors are assumed to be present. However, Casey does not offer an algorithm for generating the required weights. His concerns were of a different order.

The overriding goal of Casey's work (1996) was demonstrated in his first theorem, which states that "if an RNN performs an FSM [finite state machine] computation, then it must organize itself so that it models the minimal DFA which performs the same FSM computation." The minimal DFA is one that achieves the given task with the least number of states (and state nodes). Elsewhere, Casey asserts that the RNN "mimics" the organization of the minimal DFA.

It is important to appreciate the generality of Casey's results. For we now know that *every* deterministic RNN which successfully matches the input-output behaviour of a given DFA must *implement* some particular DFA, viz., the minimal DFA that

performs the given computation. To be sure, the limited noise tolerance of the RNN implementation may bring advantages in some domains. Nevertheless, a precise correspondence will exist between the RNN state representations and those of the minimal DFA in question. Any attempt to expand the noise tolerance of the RNN can introduce errors on each iterative state transition. Such errors are rapidly compounded and lead to increasingly degraded performance.

Now, on the face of it, Casey's results conflict with an observation of Cleermans, et al, namely that "representations [within the simple recurrent network] correspond to nodes in the FSA only when resources are severely constrained." The conflict dissolves, however, when we recall that Cleermans et al are thinking of the FSA which one is ostensibly modeling, whereas Casey's proof refers to the minimal DFA.

The upshot of the foregoing discussion is that successful RNN simulations of DFA can be viewed as close approximations of DFA. Moreover, every DFA implements some classical algorithm. The question naturally arises, then, whether RNN implementations of DFA can provide important insights into how high-level algorithmic processes could be implemented in the brain. As I will now argue, the answer largely depends upon the cognitive plausibility of several assumptions which are crucial to the computability results cited above.

3.1 Genesis of Weight Vectors.

A key premise, found both in (Casey, 1996) and (Sontag, 1995) is that weights vectors, having suitable topological properties, may be assumed to be present in the RNN. Admittedly, Casey shows concern for how weights are to be acquired. For this reason, his proofs are designed with "robust" RNNs in mind (the weight vectors need fall only within certain tolerance ranges). Nevertheless, neither he nor Sontag proves that the requisite weights can be generated by any feasible method. Let us consider, therefore, how the appropriate weight vectors might come to be present.

There appear to be just three salient possibilities. (A) Some or all of the required weights are innately present. (B) The weights are induced by supervised training, e.g., via the backpropagation algorithm. (C) The weights are induced by unsupervised learning methods.

For each of these three possibilities, there may be cognitive realms where considerable plausibility exists. However, the realm of high-level cognition (abstract reasoning, planning, theory formation) seems to present some difficulty for all three possibilities.

Consider first the innateness hypothesis. Given that most forms of abstract reasoning and theory formation rely, in part, on a variety of specialized mental skills (e.g., the ability to apply acquired verbal rules, the ability to find analogies with existing theories/methods; the ability to mentally entertain a list of alternative plans; etc.), it seems highly likely that some prior learning, at least, is required to support these powerful cognitive processes (cf. Hadley, in press). So, the supposition that we employ an innate, fully weight-configured

199

RNN to achieve all such cognitive processing is problematic at best. Nevertheless, there remains the apparent possibility that we possess certain modular NNs which are trained *by experience*, but that we also possess a high-level, innately-configured, general problem solving RNN, which matches the power of some DFA. This general-purpose RNN might invoke the empirically-trained modules to achieve certain sub-tasks, but the high-level RNN could still be viewed as an innately programmed network.

Unfortunately, a serious difficulty remains. Specifically, the *innate weight-configuration* hypothesis would find little support among some prominent neuro-psychologists and connectionists. For example, Elman, Bates, et al (1996) argue forcefully that detailed, pre-specified weights (supporting specific representations) are *not* innately present in the cerebral cortex. While this proposition is not embraced by all developmentalists, it is worrisome that its dissenters have not produced physiological evidence to the contrary. Note also, that the *partial* weight pre-specification strategy of Omlin & Giles (1996) is seemingly undermined by Elman et al's arguments. For, though Omlin & Giles offer an algorithm for pre-specifying weights, this algorithm is not driven by experiential training. Thus, their *a-priori* weight pre-specification should be viewed as comparable to innate structuring.

Turning now to (B), the supervised learning hypothesis, we are again confronted with a serious obstacle. For, all known supervised learning algorithms require a representative sampling of target output values to be available during the training process. Yet, in domains such as abstract reasoning, planning, and especially theory formation, the overwhelming majority of "output values" are not available beforehand. Rather, they have yet to be discovered by the agents who are undergoing training. Also, in this domain, "interpolation" between known output values often fails to work (as we saw in the case of 'universal' function approximation). Moreover, it would be ludicrous to suppose that humans learn how to devise theories or to prove theorems by a simple form of stimulus-response training. Part of the learning process, at the very least, involves being taught general principles, and then applying these principles in novel combinations (cf. Hadley, in press).

Apart from the above, we must bear in mind that abstract reasoning, and theory formation are highly complex processes. Any *computationally complete and effective* DFA model of these processes will involve hundreds or thousands of distinct states and state transitions, at the least.[2] In light of this, any RNN, corresponding to such a DFA, will involve complex and lengthy recursive processes. Now, as Omlin & Giles have emphasized, there are no known supervised learning algorithms which reliably induce weight configurations in RNNs so as to achieve *successful* simulation of DFAs in the face of lengthy recursive processes. Yet, humans certainly engage in

lengthy internal processing in cases where they are devising a proof strategy or, say, planning their next move in a chess game. So, even ignoring the difficulty of 'unavailable target outputs', we presently have no reason to believe that supervised learning methods could induce the requisite weight vectors.

Admittedly, it is *possible* that an appropriate, supervised learning method will someday be discovered. However, from this it does not follow that a purely agnostic attitude on the issue is reasonable. Any successful, supervised learning algorithm (in this domain) must have highly specific mathematical properties, just as a proof for a mathematical proposition must have very special properties. If one arbitrarily pinpoints some extremely complex mathematical proposition and asserts, 'It is just as likely that this proposition is provable as that it is not', one may expect a (figurative) barrage of rotten tomatoes before an audience of mathematicians. Likewise, it would be imprudent to claim that the existence of the required supervised learning algorithm is just as likely as its non-existence.

There remains, of course, possibility (C) – unsupervised learning methods. Unfortunately, as many experienced connectionists can testify, it is even more difficult to induce appropriate weight vectors in a large, complex RNN via unsupervised methods than by supervised training. This explains why the overwhelming majority of connectionist models published (in various Cognitive Science journals and proceedings) employ backpropagation (or its near kin) rather than unsupervised algorithms.

To be sure, it is widely agreed that much human learning occurs in an unsupervised fashion. Moreover, unsupervised competitive and/or Hebbian learning appears to be the foundation for most or all "weight adjustment" (via synaptic modification) that occurs in the cerebral cortex (cf. Elman et al, 1996). Presumably then, unsupervised learning can achieve astounding results in the cognitive realm. Given this, should we not remain open-minded regarding hypothesis (C)?

Unfortunately, matters are not so simple. Although I happily embrace the view that unsupervised learning may achieve wonderful feats, it is far from clear that it does so via *training RNNs to behave like DFA*. I have argued elsewhere (Hadley, in press) that much of the power of high-level cognition derives from architectures that arise through novel interactions of modules. Unsupervised learning, in conjunction with other forms of plasticity, probably plays a large role in the formation of these modules. However, it is unclear, at best, whether these modules can be modeled as DFA. Among other difficulties, DFA do not even possess sufficient power to parse context-sensitive grammars.

In summary, we have now seen that significant doubts arise, for all three possibilities {(A), (B), and (C)}, with respect to how suitable weight vectors, needed to support DFA simulation, could occur within brain-based RNNs. The innateness hypothesis presents fewer difficulties, at first glance. However, this hypothesis finds little favor in some connectionist quarters, due to the dearth of supporting physiological evidence. Of the learning-based pos-

---

[2] The skeptical reader might attempt to construct an *complete* DFA model for SOAR (Laird, et al, 1987), a well known candidate for a 'general cognitive model'.

sibilities, only supervised algorithms (in particular, variants of backpropagation) have successfully induced even approximately correct weights in comparatively simple RNNs. It is noteworthy, moreover, that backpropagation, and its refinements, have thus far resisted any *reduction* to biologically based weight modification methods. This aspect further clouds the prospect that *trained* RNNs may reveal the nature of *cognitively plausible* implementations of DFA in actual brains.

Conclusions and Summary.

In the foregoing, I have examined claims made for the computational powers of two distinct classes of connectionist networks, namely, the (putative) "universal function approximators", and recurrent finite-state simulation networks. Each class was considered with respect to its potential in the realm of cognitive modeling. Regarding the first class, I argued that, contrary to the claims of some influential connectionists, feed-forward networks do *not* possess the capacity to approximate all functions of interest to cognitive scientists. They clearly do not possess the ability to approximate partial recursive (non-halting) functions, though the latter may very well provide good models of some high-level cognitive processes. Moreover, we saw that feed-forward networks cannot approximate certain important, simple recursive (halting) functions which map symbolic strings onto other symbolic strings.

By contrast, we saw that a particular class of recurrent networks (namely, RNNs that closely approximate DFAs) shows considerably greater promise in some domains. For, many computable functions, which map symbolic strings of arbitrary length onto similar strings can, in principle, be modeled by DFAs and their connectionist simulations. However, serious difficulties emerged when we considered how the relevant recurrent networks could acquire the *weight vectors* needed to support DFA simulations. Indeed, the most widely used method of inducing weight vectors (supervised learning) was seen to be implausible in the realm of several high-level cognitive functions. Furthermore, hypotheses founded upon innate-wiring and self-organizing learning likewise presented serious obstacles. This is not to say that a set of separate RNN modules would present similar obstacles. However, a *set* of RNN modules is not an RNN in the sense assumed by the various theoretical proofs we have considered here. Moreover, as previously mentioned, I offer theoretical arguments in (Hadley, in press) to show that a modular connectionist architecture inevitably leads to *classical* patterns of inter-modular processing.

References.

Casey (1996) The Dynamics of discrete-time computation, with application to recurrent neural networks and finite state machine extraction, *Neural computation*, 8, 1135-1178.

Church, A. (1956) *Introduction to mathematical logic*, Princeton, NJ: Princeton University Press.

Churchland, P. (1990) Cognitive activity in artificial neural networks. In Osherson, D.N. & Smith, E.E. (eds.), *An invitation to cognitive science: Thinking (Vol. 3)*, Cambridge, MA: MIT Press.

Cleeremans, A., Servan-Schreiber, D. & McClelland, J.L. (1989) Finite state automata and simple recurrent networks, *Neural computation*, 1, 372-381.

Elman, J.L. (1993) Learning and development in neural networks: The importance of starting small, *Cognition*, 48, 71-99.

Elman, J.L., Bates, E.A., Johnson, M.H., Karmiloff-Smith, A., Parisi, D., Plunkett, K. (1996), *Rethinking Innateness: A Connectionist Perspective on Development*, Cambridge, MA: MIT Press.

Gödel, K. (1931) Über formal unentscheidbare Sätze der Principia mathematica und verwandter System I, *Monatschefte für Mathematik und Physik*, 38, 173-198 (Translated in van Heijenoort, 1967).

Fodor, J.A. and Pylyshyn, Z.W. (1988), Connectionism and Cognitive Architecture: A Critical Analysis, *Cognition*, Vol. 28, 3-71.

Hadley, R.F. (1999) Cognition and the Computational Power of Connectionist Networks, Technical Report, SFU CMPT TR 1999-01, School of Computing Science, Simon Fraser University, Burnaby, B.C., V5A 1S6.

Hadley, R.F. (in press) Connectionism and novel combinations of skills: implications for cognitive architecture, *Minds and Machines*, pp. (to appear).

Hartman, E.J., Keeler, J.D. & Kowalski, J.M. (1990) Layered neural networks with Gaussian hidden units as universal approximations, *Neural computation*, 2, 210-215.

Hornik, K., Stinchcombe, M., & White, H. (1989) Multilayer feedforward networks are universal approximators, *Neural Networks*, 2, 359-366.

Laird, J.E., Newell, A., Rosenbloom, P.S. (1987) SOAR, An architecture for general intelligence, *Artificial Intelligence*, 33, 1-64.

Niklasson, L.F. and van Gelder, T. 1994: On Being Systematically Connectionist. *Mind and Language*, 9, 288-302.

Omlin, C.W. & Giles, C.L. (1996) Constructing deterministic finite-state automata in recurrent neural networks, *Journal of the ACM*, 43, 937-972.

Siegelmann, H.T. (1996) The simple dynamics of super Turing theories, *Theoretical Computer Science*, 168, 461-472.

Siegelmann, H.T. & Sontag, E.D. (1994) Analog computation via neural networks, *Theoretical Computer Science*, 131, 331-360.

Sontag, E.D. (1995) Automata and neural networks, *The Handbook of brain theory and neural networks*, (ed.) Arbib, M.A., Cambridge, MA: MIT Press.

Incrementality and Locality of Language Comprehension: The Pivotal Role of Semantic Interpretation Schemata

Udo Hahn   **Martin Romacker**

⬚ℒⒾF Text Understanding Lab
Computational Linguistics Division, Freiburg University
D-79085 Freiburg, Germany
`http://www.coling.uni-freiburg.de`

Abstract

We introduce a computational model of language comprehension that combines locality of syntactic and semantic analysis with incrementality of processing. As the model incorporates inheritance-based abstraction mechanisms we are able to specify a parsimonious inventory of abstract, simple and domain-independent semantic interpretation schemata.

Introduction

Despite a large body of experimental evidence for the incrementality of human language comprehension (e.g., Tyler & Marslen-Wilson (1977), Thibadeau et al. (1982), Garrod & Sanford (1985)), this issue has not been given comparable attention in computational approaches to natural language processing. While in the past only few studies were concerned with the incremental aspects of semantic interpretation (for an overview, cf. Haddock (1989)), interest in this area has recently increased, especially in the cognitive science community (Paredes-Frigolett & Strube, 1996; Hahn & Strube, 1996; Lombardo et al., 1998).

Our contribution to this discussion lies in a model of incremental semantic interpretation whose specifications are very compact. This allows us to treat a variety of linguistic phenomena by only few and general interpretation schemata. In essence, these schemata address structural *configurations* within dependency graphs rather than hook on particular language phenomena or single rules. The grammar model, as well as the domain model make extensive use of inheritance-based abstraction mechanisms. By interfacing the description of semantic interpretation schemata to these inheritance hierarchies, we are able to specify a parsimonious and domain-independent semantic interpretation system.

We start from a lexicalized grammar model based on dependency relations. *Locality* in this framework has a dual reading. On the one hand, it refers to the reachability of syntactically related content words within "minimal" dependency graphs. On the other hand, these minimal dependency graphs have a direct conceptual interpretation. *Incrementality* comes in as local readings are combined on the fly to form larger units of analysis as comprehension unfolds.

Framework for Incremental Interpretation

Knowledge Sources. We supply grammar and domain knowledge, as well as interpretation schemata mediating between these two knowledge sources. *Grammatical knowledge* for syntactic analysis is based on a fully lexicalized dependency grammar (Hahn et al., 1994). Lexeme specifications form the leaf nodes of a lexicon tree, which are further abstracted in terms of word class specifications at different levels of generality. This leads to a word class hierarchy, which consists of word class names $\mathcal{W}$:= {VERBTRANS, VERB, ARTICLE, DET, ...} and a subsumption relation $isa_{\mathcal{W}}$ = {(VERBTRANS, VERB), (ARTICLE, DET), ...} $\subset \mathcal{W} \times \mathcal{W}$. Inheritance of grammatical knowledge is based on the idea that constraints are attached to the most general word class to which they apply.

A dependency grammar captures binary valency constraints between a syntactic head (e.g., a noun) and one of its possible modifiers (e.g., a determiner or an adjective). In order to establish a dependency relation $\delta \in \mathcal{D}$:= {*specifier, subject, direct-object,* ...} between a head and a modifier, the corresponding constraints on word order, compatibility of morphosyntactic features, as well as semantic criteria have to be fulfilled. Fig. 1 depicts a sample dependency graph in which word nodes are given in bold face and dependency relations between them are indicated by labeled edges.

Conceptual knowledge about the domain is expressed in a KL-ONE-like representation language (Woods & Schmolze, 1992). It consists of concept names $\mathcal{F}$:= {SELL, HARD-DISK, ...} and a subsumption relation on concepts $isa_{\mathcal{F}}$ = {(SELL, ACTION), (HARD-DISK, PRODUCT), ...} $\subset \mathcal{F} \times \mathcal{F}$. The relation names $\mathcal{R}$:= {SELL-AGENT, HAS-HARD-DISK, ...} denote conceptual relations also organized in a subsumption hierarchy $isa_{\mathcal{R}}$ = {(HAS-HARD-DISK, HAS-PHYSICAL-PART), (HAS-PHYSICAL-PART, HAS-PART), ...}.

Linkages between concepts via conceptual relations are determined by dependency relations linking lexical items in the dependency graph directly or in a mediated way (via a series of dependency relations). *Semantic interpretation rules* mediate between both description levels in a way as abstract and general as possible.

Local Computations: Basic Parsing Protocol. The lexicalized grammar and the associated parser we use are embedded in an object-oriented computation model. Dependency relations are computed by lexical objects, so-called *word actors*, through strictly local message passing, only involving the lexical items they represent. The basic protocol for incremental parsing can be sketched as follows (Hahn et al., 1994):

- After a word has been read from textual input, its associated lexeme (specified in the lexicon tree) is identified and a corresponding word actor gets initialized. As most lexemes (verbs, nouns and adjectives) are directly linked to

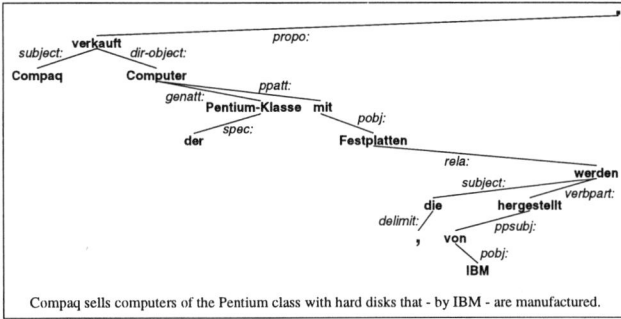

Figure 1: A Sample Dependency Graph

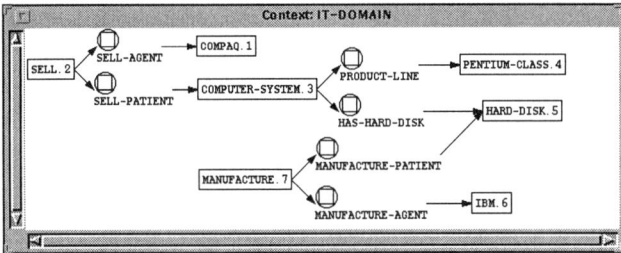

Figure 2: Corresponding Concept Graph

the conceptual system, each lexical item w that has a conceptual correlate C in the domain knowledge base, $w.C \in \mathcal{F}$, gets instantiated in the knowledge base, such that for any instance I_w, initially, $type(I_w) =: w.C$ (e.g., $w = Festplatte$, $w.C = $ HARD-DISK, and $I_w = $ HARD-DISK.5).

- For integration in the parse tree, the newly created word actor searches its head (alternatively, its modifier) by sending a request for dependential government to its left context. The search space is restricted, since this request is only propagated upwards along the "right shoulder" of the dependency graph constructed so far. All word actors which are addressed this way check, in parallel, whether their valency constraints are met by the requesting word actor.

- If all grammatical constraints are fulfilled by one of the targeted word actors, a tentative semantic interpretation is performed incorporating constraints from the currently checked dependency relation. If a valid result is computed, i.e., only if both grammatical and semantic-conceptual integrity are guaranteed, the acknowledged syntactic head h sends an acceptance message to the dependent modifier m and the screened dependency relation is finally established.

Incremental Semantic Interpretation

In the dependency parse tree from Fig. 1, we distinguish lexical nodes that have a conceptual correlate (e.g., *"Compaq"*, *"verkauft" (sells)*) from others that do not have such a correlate (e.g., *"mit" (with)*, *"von" (by)*). This is reflected in the two basic configurational settings for semantic interpretation:

- **Direct Linkage:** If two lexical nodes with conceptual correlates are linked by a *single* dependency relation, a *direct* linkage is given, which can immediately be interpreted in terms of a corresponding conceptual relation. This is illustrated in Fig. 1 by the direct linkage between *"Compaq"*

and *"verkauft" (sells)* via the *subject* relation, which gets mapped to the SELL-AGENT conceptual role linking instances of corresponding conceptual correlates, *viz.* COMPAQ.1 and SELL.2, respectively (cf. Fig. 2). This interpretation uses knowledge about the conceptual correlates and the linking dependency relation, only.

- **Mediated Linkage:** If two lexical nodes with conceptual correlates are linked by a *series* of dependency relations and none of the intervening nodes have a conceptual correlate, a *mediated* linkage is given. Such a "minimal" subgraph can only be interpreted indirectly in terms of a conceptual relation. We include lexical information from intervening nodes in addition to the knowledge about the conceptual correlates and dependency relations. In Fig. 1 this is illustrated by the syntactic linkage between *"Computer"* and *"Festplatten" (hard disks)* via the intervening node *"mit" (with)* and the *ppatt/pobj* relations. This leads to a conceptual linkage between COMPUTER-SYSTEM.3 and HARD-DISK.5 via the relation HAS-HARD-DISK in Fig. 2.

In order to increase the generality and to preserve the simplicity of semantic interpretation we introduce a generalization of the notion of dependency relation such that it incorporates direct as well as mediated linkage: Two content words (nouns, adjectives, adverbs or full verbs) stand in a *mediated syntactic relation*, if one can pass from one word to the other along the connecting edges of the dependency graph without traversing word nodes other than prepositions, modal or auxiliary verbs (i.e., elements of closed word classes). In Fig. 1, e.g., the tuples (*"Compaq"*, *"verkauft"*), (*"verkauft"*, *"Computer"*), (*"Computer"*, *"Festplatten"*), (*"Festplatten"*, *"hergestellt"*) stand in a mediated syntactic relation, whereas, e.g., the tuple (*"verkauft"*, *"Festplatten"*) does not, since *"Computer"* is an intervening content word node.

We then call a series of contiguous words in a sentence S that stand in a mediated syntactic relation a *semantically interpretable subgraph* of the dependency graph of S. Semantic interpretation will be started whenever two word nodes with associated conceptual correlates are dependentially connected so that they form a semantically interpretable subgraph. As we will see, in some cases the dependency structures we encounter will have no constraining effect on the kind of conceptual relations we check (e.g., genitives). There are other cases (e.g., prepositional phrases), however, where constraints on possible interpretations can be derived from dependency structures and the lexical material they embody.

Basic semantic interpretation schema. Semantic interpretation is executed via a search in the domain knowledge base by combining two sorts of knowledge — first, concept pairs for which connecting relation paths have to be determined; second, constraints on the kinds of permitted or excluded conceptual relations when connecting relations are being computed. Concept pairs represent the content words linked at the dependency level within the semantically interpretable subgraph, while constraints on relations account for the dependency relation(s) between them. Schema (1) describes a mapping from the conceptual correlates, $h.C_{from}$ and $m.C_{to}$, in $\mathcal{F}$ of two syntactically linked lexical items, h and m, respectively, to connected relation paths R_{con}. A rela-

203

tion path $rel_{con} \in R_{con}$ composed of n relations, $(r_1, ..., r_n)$, is called *connected*, if for all its n constituent relations the concept type of the domain of relation r_{i+1} subsumes the concept type of the range of relation r_i.

$$si : \begin{cases} \mathcal{F} \times 2^{\mathcal{R}} \times 2^{\mathcal{R}} \times \mathcal{F} & \to & 2^{R_{con}} \\ (C_{from}, R_+, R_-, C_{to}) & \mapsto & \widetilde{R_{con}} \end{cases} \quad (1)$$

As an additional filter, si is constrained by all conceptual relations $R_+ \subset \mathcal{R}$ a priori permitted for semantic interpretation, as well as all relations $R_- \subset \mathcal{R}$ a priori excluded from semantic interpretation (concrete examples will be discussed below). Thus, $rel \in \widetilde{R_{con}}$ holds, if rel is a connected relation path from C_{from} to C_{to}, obeying the restrictions imposed by R_+ and R_-. For ease of specification, R_+ and R_- consist of the most general conceptual relations possible. Prior to semantic processing we expand them into their transitive closures, incorporating all their subrelations in the relation hierarchy. Hence, $R_+^* := \{ r^* \in \mathcal{R} \mid \exists \, r \in R_+ : r^* \, isa_{\mathcal{R}} \, r \}$ (correspondingly, R_-^* is dealt with).

We also define the function $get\text{-}roles(C) =: CR$, which extracts the set of all conceptual relations CR associated with a concept C. Applying $get\text{-}roles$ to C_{from} extracts the roles that are used as starting points for the path search according to the defined restrictions. R_+ restricts the search to relations contained in $CR \cap R_+^*$, iff R_+ is not empty (otherwise, CR is taken as it is), R_- allows only for relations in $CR \setminus R_-^*$.

If the function si returns the empty set (i.e., no valid interpretation can be computed), no dependency relation will be established. Otherwise, for all resulting relations $\text{REL}_i \in \widetilde{R_{con}}$ an assertional axiom is added to the knowledge base by asserting the proposition $(h.C_{from} \, \text{REL}_i \, m.C_{to})$, where REL_i denotes the ith reading.

Integration of Knowledge Levels

In the course of the interpretation process, we apply a number of specializations of the general semantic interpretation schema (1). One major source from which specializations arise are positive lists, D_+^{lexval}, and negative lists, D_-^{lexval}, of syntactic dependency relations, from which corresponding conceptual constraints, R_+ and R_-, can be directly derived. Knowledge about D_+^{lexval} and D_-^{lexval} is encoded at the level of *word classes* $\mathcal{W}$, such that $lexval \in \mathcal{W} \times \mathcal{D}$, and, thereby, inherited by all their subsumed lexemes. For instance, the word class of transitive verbs, VERBTRANS $\in \mathcal{W}$, defines for a subject valency $D_+^{\langle verbtrans, \, subject \rangle}$:= $\{subject\}$ and $D_-^{\langle verbtrans, \, subject \rangle} := \emptyset$. We distinguish three basic cases:

1. Knowledge available from syntax *determines* the semantic interpretation, if $D_+^{lexval} \neq \emptyset$ and $D_-^{lexval} = \emptyset$ (e.g., the subject of a verb).

2. Knowledge available from syntax *restricts* the semantic interpretation, if $D_+^{lexval} = \emptyset$ and $D_-^{lexval} \neq \emptyset$ (e.g., most prepositions).

3. If $D_+^{lexval} = \emptyset$ and $D_-^{lexval} = \emptyset$, no syntactic constraints apply and semantic interpretation proceeds *entirely*

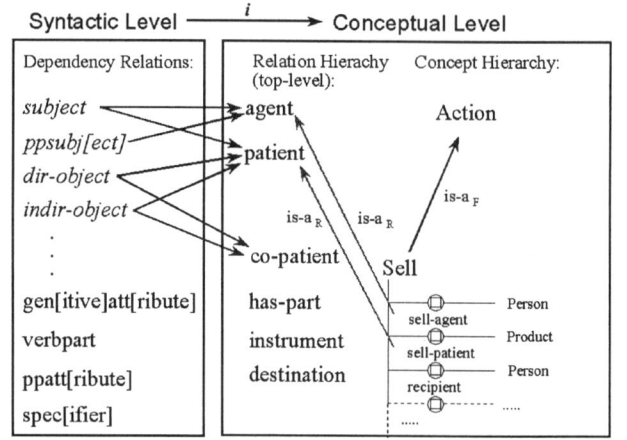

Figure 3: Relations between Knowledge Levels

concept-driven, i.e., it relies on the domain knowledge only (e.g., for genitive attributes).[1]

For syntactic constraints to be propagated to the conceptual level, we define a mapping, $i: \mathcal{D} \to 2^{\mathcal{R}}$, from dependency relations to sets of conceptual relations. R_+ and R_- must be computed from D_+^{lexval} and D_-^{lexval}, respectively, by applying the interpretation function i to each element of D_+^{lexval} and D_-^{lexval}. So, $R_+ := \{y \mid x \in D_+^{lexval} \wedge y \in i(x)\}$ and $R_- := \{y \mid x \in D_-^{lexval} \wedge y \in i(x)\}$. An illustration is given in Fig. 3. On the left side, at the syntactic level proper, a subset of the dependency relations contained in $\mathcal{D}$ are depicted. Those that have associated conceptual relations are shown in italics. *dir[ect]-object*, e.g., must conceptually be interpreted in terms of PATIENT or CO-PATIENT; *gen[itive]att[ribute]*, however, has no direct conceptual correlation as this dependency relation does not restrict conceptual interpretation at all.

At the conceptual level, two orthogonal taxonomic hierarchies exist, one for relations, the other for concepts (cf. Fig. 3, right side). Both are organized in terms of subsumption hierarchies ($isa_{\mathcal{F}}$ and $isa_{\mathcal{R}}$). Also, both hierarchies interact, since relations are used to define concepts. The concept SELL is a subconcept of ACTION. It has a role SELL-PATIENT whose filler's type must be a PRODUCT. SELL-PATIENT itself is subsumed by the more general relation PATIENT.

Sample Analyses

In the examples we discuss now, we start from the interpretation of direct linkage, and then turn to the interpretation of mediated linkages considering increasingly complex configurations in the dependency graph as given by prepositional phrases and passives in relative clauses.

Interpreting direct linkage. When the first word in our sample sentence, *"Compaq"*, is read, its conceptual correlate COMPAQ.1 is instantiated immediately. The next word, *"verkauft"* (*sells*), also leads to the creation of an associated instance (SELL.2). The word actor for *"verkauft"* then attempts to bind *"Compaq"* as its syntactic subject. Since we encounter a direct linkage, the semantic interpretation

---

[1] We have currently no empirical evidence for the fourth possible case, where $D_+^{lexval} \neq \emptyset$ and $D_-^{lexval} \neq \emptyset$.

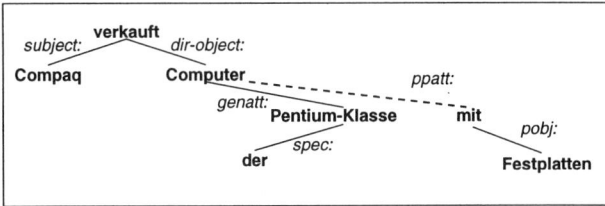

Figure 4: Dependency Graph during PP-Attachment

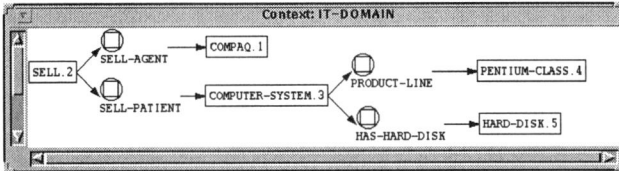

Figure 5: Concept Graph after PP-Attachment

schema si_{dir} is instantiated using the contextual information (the *subject* relation of a transitive verb is to be checked) as actual parameters. We incorporate two types of information — first, grammatical constraints from the word class of the verb *"verkauft"*, viz. $D_+^{\langle verbtrans,\ subject \rangle} := \{subject\}$ and $D_-^{\langle verbtrans,\ subject \rangle} := \emptyset$, which are mapped to {AGENT, PATIENT} by the function i (cf. Fig. 3); second, knowledge about the concept types of COMPAQ.1 and SELL.2 (COMPANY and SELL, respectively). Hence, si_{dir}(SELL, {AGENT, PATIENT}, $\emptyset$, COMPANY). Extracting the role set from SELL (cf. Fig. 3), only SELL-AGENT and SELL-PATIENT are allowed for interpretation as they are included in the transitive closure of AGENT and PATIENT. Checking sortal integrity succeeds only for SELL-AGENT (COMPANY is subsumed by LEGAL-PERSON and by PERSON). In an analogous way, the semantically interpretable subgraph < *"verkauft"* – dir-object – *"Computer"*> is dealt with.

When syntactic information does not constrain the semantic interpretation we have to proceed in an entirely concept-driven way. This holds for the third complete subgraph, < *"Computer"* – genatt – *"Pentium-Klasse"*>. The actual parameters provided lead to si_{dir}(COMPUTER-SYSTEM, $\emptyset$, $\emptyset$, PENTIUM-CLASS). We extract all roles contained in the concept definition of COMPUTER-SYSTEM and iteratively check for each role whether PENTIUM-CLASS is a legal filler. This is only the case for the relation PRODUCT-LINE. Though various linguistic phenomena (subjects, direct objects and genitives) are covered, a single schema, si_{dir}, is sufficient for semantic interpretation of direct linkage configurations.

Interpreting mediated linkage. For the interpretation of mediated linkage, information supplied by the intervening lexical nodes is incorporated. It is contained in *lexeme*-specific lists, $R^{lex} \subset \mathcal{R}$, since specifications at the word-class level ($D_{+/-}^{lexval}$) turn out to be too general here. This applies to specific lexical exemplars from closed word classes encountered in mediated linkages (e.g., prepositions). So, the number of additional specifications required remain fairly small.

Consider Fig. 4 where a semantically interpretable subgraph consisting of three word nodes (*"Computer"*, *"mit"*, *"Festplatten"*) is currently being processed (indicated by the

dashed line). In particular, the word actor for *"mit"* *(with)* tries to determine its syntactic head.[2] We consider prepositions as relators carrying conceptual constraints for the corresponding instances of their syntactic head and modifier. The "meaning" of a preposition is encoded in a set $R^{Prep} \subset \mathcal{R}$, for each preposition in *Prep*, holding all permitted or excluded relations in terms of high-level conceptual relations. For the preposition *"mit"* *(with)*, e.g., we have $R_+^{mit} :=$ {HAS-PART, INSTRUMENT, HAS-PROPERTY, HAS-DEGREE, ...}. If a syntactic dependency relation between a head and a prepositional modifier has to be checked, the corresponding list R^{Prep} has to be matched against the restrictions imposed by the preposition's syntactic head via $D_{+/-}^{lexval}$.

When *"mit"* attempts to be governed by *"Computer"* the mediated linkage results in the instantiation of the specialized interpretation function si_{prep} which is applied exclusively for all occurrences of PP-attachments. The conceptual entities to be related are denoted by the leftmost and the rightmost node in the actual subgraph (i.e., *"Computer"* and *"Festplatten"*). Since the dependency relation between the head *"Computer"* and its modifier *"mit"*, ppatt, does not impose any restrictions at all (cf. Fig. 3), i.e., $D_{+/-}^{lexval} = \emptyset$, semantic interpretation boils down to the evaluation of si_{prep}(COMPUTER-SYSTEM, {HAS-PART, INSTRUMENT, HAS-PROPERTY, HAS-DEGREE, ...}, $\emptyset$, HARD-DISK). By extracting all conceptual roles and checking for sortal consistency, only HAS-HARD-DISK $isa_{\mathcal{R}}$ HAS-PART yields a valid interpretation to directly relate COMPUTER-SYSTEM and HARD-DISK. The state of semantic interpretation after PP-attachment is given in Fig. 5. This also indicates that during each step of the analysis a corresponding conceptual interpretation is already available.

To convey an idea of the generality and flexibility of our approach, we discuss a more complex example. In Fig. 6 the relative clause *"die von IBM hergestellt werden"* *(that are manufactured by IBM)* has already been analyzed and IBM.6 figures as MANUFACTURE-AGENT of MANUFACTURE.7 (cf. Fig. 2). Since the *subject* valency of the passive auxiliary *"werden"* is occupied by a relative pronoun (*"die"*), the interpretation of that structure must be postponed until the pronoun's referent becomes available. Passive interpretation is performed by another specialization of the general interpretation schema, si_{pass}. As with certain prepositions, constraints come directly from a positive list $R_+^{passaux} :=$ {PATIENT, CO-PATIENT}, which resides in the lexeme specification for *"werden"*. The items to be related are contained in the semantically interpretable subgraph spanned by *"die"* *(that)* and *"hergestellt"* *(manufactured)*.[3]

The referent of the relative pronoun *"die"* becomes avail-

---

[2]We do not attach prepositions to possible heads immediately. Rather we have a built-in delay mechanism so that the nominal modifier of a prepositional head has to be determined first. With this content item attached (in our example, HARD-DISK.5), compatibility checks with a potential head, as searched by the prepositional modifier, have a reasonable conceptual grounding.

[3]Pronominal reference is resolved when the head of the relative clause (*"werden"*) has determined its own head. As with prepositions, this delay mechanism is justified by the fact that semantic interpretation is only reasonable when a basic conceptual grounding has already been established (i.e., *"werden"* must have been bound to the content word *"hergestellt"* *(manufactured)*).

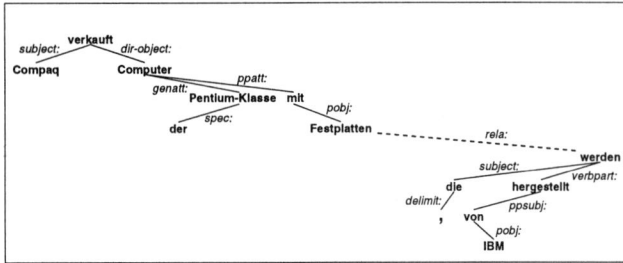

Figure 6: Dependency Graph for Relative Clause

able when the syntactic head of the relative clause (*"werden"*) has determined its head (cf. the dashed line in Fig. 6). Two alternatives arise, namely *"Computer"* and *"Festplatten" (hard disks)*. The choice of the latter leads to those parameters for si_{pass} needed for the *subject* interpretation.[4] si_{pass} inverses the argument structure by taking the leftmost word node (i.e., the pronoun's referent *"Festplatten"*) as C_{to} and the rightmost (*"hergestellt"*) as C_{from}. Hence, si_{pass}(MANUFACTURE, {PATIENT, CO-PATIENT}, ∅, HARD-DISK). The final interpretation is depicted in Fig. 2 linking MANUFACTURE.7 and HARD-DISK.5 via the role MANUFACTURE-PATIENT. Obviously, integrating intra- and extrasentential anaphorical phenomena (Strube & Hahn, 1995) necessitates a slight extension of our notion of a semantically interpretable subgraph. In case pronouns are involved, such a subgraph is interpretable *iff* all referents are made available.

Evaluation of Semantic Interpretation

In a small-scale evaluation study, we started from a domain ontology that is divided into an upper generic part (composed of 1,100 concepts and relations) and various domain-specific parts. In the study we report on two specialized domains were dealt with — an information technology (IT) model and an ontology covering parts of anatomical medicine (MED) (each domain model, in addition, contributes 1,100 concepts and relations). Corresponding lexeme entries in the lexicon provide linkages to the ontology. We also assume a correct parse to be delivered for the semantic interpretation process.

We then took a random selection of 54 texts (comprising 18,500 words) from our two text corpora. For evaluation purposes, we concentrated on the interpretation of genitives (direct linkage), prepositional phrase attachments and auxiliary as well as modal verbs (both variants of mediated linkage). In the following, we will focus on the discussion of the results from the semantic interpretation of genitives (cf. Table 1).

The chosen texts contained a total of almost 250 genitives, from which about 59%/33% (MED/IT) received an automatic interpretation, with correct ones for 57%/31% (recall). An interpretation was considered *correct* when the conceptual correlates of lexical items in a semantically interpretable subgraph were conceptually related in an adequate way.

---

[4]The alternative choice (*"Computer"*) creates a local ambiguity, because COMPUTER-SYSTEM.3 can also be related to MANUFACTURE.7. However, heuristics are applied to select the most plausible reading, which are sensitive to preferential criteria such as the length of role compositions, the types of relations encountered, etc. (Hahn & Markert, 1997).

	MED	IT
# texts	29	25
# words	4,300	14,200
# genitives . . .	100	147
. . . with interpretation	59	49
. . . . . . correct	57	46
. . . . . . incorrect/none	2	3
. . . without interpretation	41	98
recall	57%	31%
precision	97%	94%

Table 1: Empirical Results for the Interpretation of Genitives

Slightly more than half of the loss we encountered can be attributed to insufficient coverage in the two *domain models*. The remaining cases can be explained by insufficient coverage of the *generic model* and reference to *other domains*, e.g., fashion or food. Some minor loss is also due to phrases referring to *time* (e.g., "the beginning of the year"), *space* (e.g., "the surface of the storage medium"), and *abstract* notions (e.g., "the acceptance of IT technology"), as well as *evaluative* expressions (e.g., "the advantages of plasma display") and *figurative* language (e.g., "the heart of the notebook").

The concrete values we found, disappointing as they may be for recall (57%/31%), encouraging, however, for precision (97%/94%), can only be interpreted relative to other data still lacking on a broader scale. Judged from the poor figures of our recall data, there is no doubt whatsoever that conceptual coverage of the domain constitutes *the* bottleneck for any knowledge-based approach to NLP. Sublanguage differences are also mirrored in these data, since medical texts adhere more closely to well-established concept taxonomies than magazine articles in the IT domain.

Related Work

Perhaps the most influential paper that treats incremental semantic interpretation from a modern rule-based perspective is the one by Pereira & Pollack (1991). We share their ideas to preserve as much as possible the principle of *compositionality* (evidenced by the smooth integration of interpreted subgraphs into the already constructed partial dependency graph), and to integrate the *discourse context* into the interpretation process as early as possible (evidenced by the resolution of the relative pronoun, though we have not gone into the details of anaphora resolution; cf. Strube & Hahn (1995)). We differ, however, fundamentally with respect to grammar specification. Pereira and Pollack use a PATR-style unification grammar, which comes without any mechanisms for inheritance to support rule specifications (semantic theories in that area do also not go beyond sortal taxonomies for semantic labels, cf. Creary & Pollard (1985)). This approach suffers from "flat" representations that require to enumerate semantic rules for specific grammatical phenomena and so they tend to proliferate.

Milward (1995) already points out a major advantage of lexicalized over rule-based approaches. In processing a sentence incrementally using a lexicalized grammar, we do not have to look at the grammar as a whole (as with rule-based grammars), but only at the grammatical information indexed for each of the words from the input stream, thus allowing for *efficient processing*. When one takes inheritance mechanisms into account *efficient encoding* becomes another asset to this approach, both for the lexicon and semantic interpretation.

Sondheimer et al. (1984) and Hirst (1988) also propose models of incremental semantic interpretation. Their use of KL-ONE-style representation languages allows them to exploit property inheritance (or typing) built-ins. The main difference to our approach lies in the status of the semantic rules. Sondheimer et al. attach interpretation rules to each *role (filler)* and, hence, have to provide detailed specifications reflecting the idiosyncrasies of each semantically relevant role attachment. Property inheritance comes only into play when the selection of alternative semantic rules is constrained to the one(s) inherited from the most specific case frame. Hirst uses strong typing at the conceptual *object* level only, while we use it simultaneously at the grammar and domain knowledge level for the processing of semantic schemata.

Charniak & Goldman (1988) and Jacobs (1991) specify semantic *rules* in the context of inheritance hierarchies. So, they achieve a similar gain in generality as we do. Unlike our approach, they still provide specific rules for grammatical phenomena (genitives, adjectival noun modifiers, etc.), while we abstract from these phenomena and collapse them in *single* schemata (as with si_{dir}) whenever possible.

The incorporation of inheritance-based abstraction principles within a lexicalized grammar as a basis for a truly cognitive model of on-line comprehension has also been suggested by Jurafsky (1992). While his proposal is focused on architectural issues how to combine linguistic, computational and psychological criteria for efficient comprehension (e.g., by introducing load constraints on STM, lexical salience weights), he is not explicit about the details of grammatical and semantic specifications. The same argument applies to Lombardo et al.'s description of an incremental interpreter based on a lexicalized dependency grammar (Lombardo et al., 1998). Focus is on the procedural aspects of parsing and simultaneous semantic interpretation rather than on a methodology for semantic interpretation. Paredes-Frigolett & Strube (1996) introduce an approach to interfacing an HPSG parser with a powerful logical representation language such that incrementality of parsing and interpretation are preserved. They discuss a sample parse that builds on large amounts of fine-grained knowledge pieces, but they do not provide evidence for the generality and scalability of their approach beyond the example they discuss. Our approach offers descriptional parsimony as required by any reasonable model of cognitively plausible language comprehension. We have also indications that it scales to real-world text understanding applications (Hahn et al., 1999).

Conclusions

We proposed a principled approach to the design of compact, yet highly expressive semantic interpretation schemata. They derive their power from two sources. First, the organization of grammar and domain knowledge, as well as semantic interpretation mechanisms, are based on inheritance principles. Second, grammar and domain knowledge interact closely via a lean interface — the hierarchy of interpretation schemata. So, the incrementality of semantic interpretation can be de-

scribed in a general, parsimonious and, hopefully, plausible way. Also, semantic interpretation which has recently become a somewhat marginalized topic of NLP research is given a focused and self-contained theoretical foundation.

Acknowledgements. M. Romacker is supported by a grant from DFG (Ha 2097/5-1). We owe special thanks to Katja Markert for fruitful discussions.

References

Charniak, E. & R. Goldman (1988). A logic for semantic interpretation. *Proc. of the 26th Annual Meeting of the ACL*, pp. 87–94.

Creary, L. G. & C. J. Pollard (1985). A computational semantics for natural language. In *Proc. of the 23rd Annual Meeting of the ACL*, pp. 172–179.

Garrod, S. & A. J. Sanford (1985). On the real-time character of interpretation during reading. *Language and Cognitive Processes*, 1(1):43–59.

Haddock, N. J. (1989). Computational models of incremental semantic interpretation. *Language and Cognitive Processes*, 4(3/4):337–368.

Hahn, U. & K. Markert (1997). In support of the equal rights movement for literal and figurative language: a parallel search and preferential choice model. In *Proc. of the 19th Annual Conference of the Cognitive Science Soc.*, pp. 289–294.

Hahn, U., M. Romacker & S. Schulz (1999). How knowledge drives understanding — matching medical ontologies with the needs of medical language processing. *Artificial Intelligence in Medicine*, 15(1):25–51.

Hahn, U., S. Schacht & N. Bröker (1994). Concurrent, object-oriented natural language parsing. *International Journal of Human-Computer Studies*, 41(1/2):179–222.

Hahn, U. & M. Strube (1996). Incremental centering and center ambiguity. In *Proc. of the 18th Annual Conference of the Cognitive Science Society*, pp. 568–573.

Hirst, G. (1988). Semantic interpretation and ambiguity. *Artificial Intelligence*, 34(2):131–177.

Jacobs, P. S. (1991). Integrating language and meaning in structured inheritance networks. In J. Sowa (Ed.), *Principles of Semantic Networks*, pp. 527–542. Morgan Kaufmann.

Jurafsky, D. (1992). An on-line computational model of human sentence interpretation. In *Proc. of the 10th National Conference on Artificial Intelligence*, pp. 302–308.

Lombardo, V., L. Lesmo, L. Ferraris & C. Seidenari (1998). Incremental interpretation and lexicalized grammar. In *Proc. of the 20th Annual Conference of the Cognitive Science Society*, pp. 621–626.

Milward, D. (1995). Incremental interpretation of categorial grammar. In *Proc. of the 7th Conference of the European Chapter of the ACL*, pp. 119–126.

Paredes-Frigolett, H. & G. Strube (1996). Integrating world knowledge with cognitive parsing. In *Proc. of the 18th Annual Conference of the Cognitive Science Society*, pp. 92–97.

Pereira, F. & M. Pollack (1991). Incremental interpretation. *Artificial Intelligence*, 50(1):37–82.

Sondheimer, N., R. Weischedel & R. Bobrow (1984). Semantic interpretation using KL-ONE. In *Proc. of the 10th International Conference on Computational Linguistics & 22nd Annual Meeting of the ACL*, pp. 101–107.

Strube, M. & U. Hahn (1995). PARSETALK about sentence- and text-level anaphora. In *Proc. of the 7th Conference of the European Chapter of the ACL*, pp. 237–244.

Thibadeau, R., M. Just & P. Carpenter (1982). A model of the time course and content of reading. *Cognitive Science*, 6:157–203.

Tyler, L. & W. Marslen-Wilson (1977). The on-line effects of semantic context on syntactic processing. *Journal of Verbal Learning and Verbal Behavior*, 16:683–692.

Woods, W. & J. Schmolze (1992). The KL-ONE family. *Computers & Mathematics with Applications*, 23(2/5):133–177.

Structural priming: Purely syntactic?

Mary L. Hare (hare@hare.bgsu.edu)
Department of Psychology; Bowling Green State University
Bowling Green, OH 43403-0228 USA

Adele E. Goldberg (agoldbrg@uiuc.edu)
Department of Linguistics; U. Illinois
Urbana-Champaign Illinois USA

Abstract

In a series of experiments, Bock and colleagues have demonstrated that subjects show a reliable increase in the use of particular syntactic constructions after having heard and repeated that construction in an unrelated sentence. Aspects of the data seem to indicate that it is syntactic constituent structure, independent of meaning, that underlies the facilitation in these situations. In this study we investigate whether more semantic factors might also lead to priming, and specifically whether the assignment of a semantic role to a particular participant in a prime sentence can increase the probability of a target sentence whose structure allows a similar assignment. To test this we replicate Bock's study and include a further set of primes (*provide-with* primes) which have the syntactic constituent structure of the dative, but share semantic role assignment with the ditransitive. If syntactic priming were triggered by constituent structure alone, primes like this would lead to more dative responses, relative to a ditransitive prime. If semantic involvement were crucial, on the other hand, this prime should elicit more ditransitive responses. In this study we find significantly more ditransitive responses following the *provide-with* sentence than following a dative prime, and no difference between the *provide-with* and ditransitive primes, suggesting that semantic factors indeed play a role.

Introduction

Two current views of the mechanics of sentence comprehension posit very different roles for structural information in the comprehension of a sentence. On the one hand are accounts that place primary importance to syntactic considerations (for example, (Frazier, 1979; Frazier & Rayner, 1982; Rayner, Garrod, & Perfetti, 1992), claiming that initial parsing decisions are made on the basis of syntactic information and parsing heuristics alone. On the other hand, a growing body of evidence suggests that even initial hypotheses involve the interaction of multiple sources of information, whether structural, semantic, or pragmatic (Ford, Bresnan, & Kaplan, 1982; MacDonald, 1993; MacDonald, Pearlmutter, & Seidenberg, 1994; MacWhinney & Bates, 1989; St. John & McClelland, 1990; Tanenhaus & Carlson, 1989; Trueswell, Tanenhaus, & Garnsey, 1994).

Despite the data favoring the interactionist approach, there remains evidence that other information can be overridden by syntactic regularities. One set of data that offers strong evidence for the independence of syntactic information has come from the phenomenon of syntactic priming ((Bock, 1986; Bock & Loebell, 1990; Bock, Loebell, & Morey, 1992). Bock and associates have found that subjects show a reliable increase in the use of particular syntactic constructions after having heard and repeated that construction in an unrelated sentence. In these tests, subjects were given a series of spoken sentences interspersed with semantically unrelated simple line drawings, and asked to describe the pictures in a single sentence without using pronouns. For example, when shown a picture that could be described with either ditransitive (e.g. *John gave the dog a biscuit,* with two noun phrases following the verb) or prepositional dative sentences (*John gave a biscuit to the dog,* with one noun phrase followed by a prepositional phrase) subjects were significantly more likely to describe it with a ditransitive if it had been preceded by an unrelated ditransitive.

Aspects of the data suggest that the priming is due entirely to syntactic processes, and independent of semantics. One piece of evidence crucial to this interpretation is the finding that prepositional locatives like (1) below, which have the same syntactic form as prepositional datives, but somewhat different semantics/pragmatics, prime prepositional datives like (2) (Bock & Loebell 1990).

  (1) *The widow drove an old Mercedes to the church.*
(*church* = locative)
  (2) *The widow gave an old Mercedes to the church.*
(*church* = benefactive)

These results can be interpreted as showing that conceptual dissimilarity has no effect on structural priming.

Note, however, that the semantics of the two constructions are not all that dissimilar. Although in (1) the object of *to* is a locative, while in (2) it is a benefactive, there is considerable linguistic evidence showing that both these roles are subsumed under the larger category of *goal* (Jackendoff 1972; Lakoff and Johnson 1980; Goldberg 1995).There are cases such as *The widow sent a package to the church* , where the argument *church* falls somewhere intermediate between a pure locative and a pure benefactive. In addition there is a substantial evidence for a general metaphor, TRANSFER OF POSSESSION as PHYSICAL TRANSFER, that involves transfer of possession without physical transfer, and involves the prepositions *to, from* and *away*. (e.g. *She received the house from her mother's estate. / He gave the house away.*) Thus it is not entirely clear from these data whether the priming is due only to form-based factors, or to a combination of form and meaning.

There is good reason for thinking that semantic factors do play a role in syntactic priming. Bock et al. (1992) found that mapping an inanimate entity onto subject in a prime sentence predisposed mapping an inanimate entity onto the subject in the target. The priming was demonstrated even when the target was active but the prime was passive, indicating that semantic aspects of the priming did not require identical syntactic forms. In the current experiment we follow up on this idea, using different prime and target frames, and controlling for the animacy of the subject argument.

Experiment

In the experiment described below we replicate the Bock and Lobell (1990) ditransitive experiment, but include a third prime condition, the 'fulfilling' frame (Levin, 1993), exemplified in example (3):

(3)   *The officers provided the soldiers with guns.*

Sentences like that in (3) share aspects of both the ditransitive and prepositional dative: the syntax is that of the prepositional dative (NP V NP PP); but the order of semantic roles parallels the ditransitive (<agent, recipient, theme>).

Our hypothesis is that the semantic similarity between the 'provide-with' form and the ditransitive will lead to an increase in ditransitive responses after a 'provide-with' prime. Since the two differ in their syntactic form, the predicted effect would not be attributable to strictly syntactic processes, and would argue that semantic factors as well as form-based factors influence the priming effect.

Methods

Materials for the experiment included 10 pictures of events that could be equally well described with prepositional object constructions or ditransitive paraphrases (a man giving a present to a child, a butler serving a drink to his employer, a woman teaching math to a group of students, and so on) Each picture was matched with three prime sentences: (1) ditransitive, (2) dative (3) provide-with. Intransitive sentences were used as an additional condition in order to assess preferences for the two alternative forms after a minimally related sentence type.

Prime/target pairs were chosen to avoid any semantic relationship between the picture and the prime sentences. The experiment also included 20 each of repeated filler sentences and pictures, and 20 each of non-repeated filler sentences and pictures for a total of 120 filler items. Filler pictures depicted intransitive or simple transitive actions (such as a woman on a swing, or a boy drinking a glass of water). Filler sentences represented a variety of syntactic types. Four experimental lists were constructed. To avoid perseveration of priming effects from earlier sentences, which is known to be long-lasting, prime type was between subjects.

Each list contained 10 priming sentences followed immediately by the target picture from the same stimulus set. In addition to the test and filler items, each list was preceded by a set of 10 practice items, and a number of warm-up trials was inserted before the first trial of the experiment. On each list, test prime/target pairs were preceded by an average of 12 fillers.

Table 1: Example stimulus set

Prime type	Examples
ditransitive	His editor offered Bob the hot story.
dative	His editor promised the hot story to Bob.
provide-with	His editor credited Bob with the hot story.
intransitive	Sasha always dawdles over lunch.
Target Picture	*A man hands a woman a box of candy.*

Procedure. Subjects were tested individually, seated in front of a computer terminal in a sound-attenuated room. On each experimental trial the computer screen displayed either a picture or an empty frame. When the empty frame appeared, the experimenter read a sentence aloud. The cover story used in Bock and Lobell (1990) was adopted here: Subjects were instructed that their task was to remember whether the picture or the sentence was new or repeated in the course of the experiment; as a memory aid, they were to repeat verbatim each sentence that the experimenter read, or to describe each picture in a full sentence.

Picture descriptions were cued by the experimenter, who presented the subject with NP-V sentence fragment (The girl painted ...; The man gave...). The presentation of the subject NP did not affect the choice of the dative or ditransitive form, since both took the same subject. It did, however, constrain the interpretation of the picture toward one that could be described with the dative or ditransitive. This was important in the present context, since each of the events in the test pictures contained two human participants, and could well be described with a simple transitive (e.g. *The girl got a birthday present from her dad*).

Each session was recorded onto audio cassette and the responses of each subject to the target pictures were transcribed following the experimental session.

209

Subjects. 48 subjects participated in the experiment. These were right-handed undergraduates between the ages of 17 and 27 (21 male, 27 female). Subjects were recruited from the Psychology and Cognitive Science department subject pools, and received course credit for participating.

Data from one subject was lost due to experimenter error. Two others were dropped, one for being a non-native speaker of English, and one who failed to complete the task. This left of total of 45 subjects overall, with 11 in each prime condition and 12 on the intransitive list.

Results

Scoring procedure: Adopting the scoring procedure of Bock & Loebell (1990), there were 204 scorable responses among the 330 possible responses to the target pictures (62%). Of the scorable responses, 32% were in the ditransitive condition, 38% in the dative, and 30% in the provide condition.

Analysis: Again following Bock and Loebell (1990), the two dependent variables were the numbers of ditransitives and double-object responses given by each subject (in each cell). Single-factor ANOVAs were performed. In the overall analysis there was a main effect of prime type $(F_1(3,41)=3.131, p<.05; F_2(3,27)=8.991, p<.001)$.

Table 2 gives the mean percentage of ditransitive picture descriptions in each prime condition. Confidence intervals (.05) are shown in parenthesis. There were 32% more ditransitive responses after ditransitive primes than after dative primes (and conversely 32% more dative responses after dative primes than after ditransitive primes). In addition, were 28% more ditransitive responses after provide-with primes than after datives.

Table 2: Subject Means

prime:	% ditransitive responses	%difference from ditransitive prime
ditransitive	87	
dative	55	32 ± (14)
provide-with	83	4 ± (14)

Planned comparisons were run to test the experimental hypothesis. T-tests showed significantly fewer ditransitive responses and more dative responses after dative prime than after either ditransitive (p < .01 for both subjects & items) or *provide-with* primes (p < .05) but no difference in the percentage of ditransitive responses after ditransitive and *provide-with* primes (p > .1).

Finally, we tested whether there was a significant difference between ditransitive and prepostional object responses following an intransitive prime. In the items analysis the ditransitive mean was 56%, while the prepositional object was 44%. This difference is not significant (p > 1) confirming that Bock's finding of no independent preference for either clause type was also true of our items. Although in the subjects analysis the difference was larger, again it did not reach significance (p =.092). We will return to this point in the general discussion.

Discussion

The present experiment indicates that the mapping of semantic features to syntactic positions in a prime sentence has an effect on the production of a target sentence. This finding is especially striking in light of the fact that a purely syntactic account of such priming would have expected target sentences to be produced with a matching syntactic form.

Other explanations for these effects are possible, of course. In earlier work looking at the active/passive alternation, Bock has shown independent effects of both animacy (a semantic factor) and structure on syntactic priming. Given the facts of the ditransitive alternation it is hard to tease apart the effects of animacy order from those of the mapping of conceptual features like thematic roles to syntactic positions – other than in exceptional cases, the object one of a ditransitive must be animate, as is the direct object of the provide-with construction. Given this, it is reasonable to ask if animacy order is what leads to the effects we find here.

In the current experiment there are three prime conditions: One in which both animacy and structure are consistent with the ditransitive, one in which both are consistent with the prepositional dative, and a third which pits the two against each other: animacy is consistent with the ditransitive while structure is consistent with preposional dative. If indeed these should show independent, additive effects, then the third condition should be influenced by both, resulting in behavior somewhere between the other two prime conditions. However, this is not the case. Instead, the third condition patterns closely with the ditransitive, and is significantly different from the prepositional dative.

Given these results, one would have to conclude either that the earlier animacy account is incorrect, or, more plausibly, that what is operative here is a different factor, the order of expression of coarse semantic roles.

It should be noted that in the present experiment the responses in the prepositional dative condition are significantly different from those in the ditransitive, but the datives do not 'prime' dative picture descriptions in the sense that a strong majority of responses after this prime are dative. In fact, only 55% of the responses are. While this does not affect the results, it is an interesting point and worth investigation. One explanation consistent with the current data is that the ditransitive may serve as the default construction for many speakers, from which they are primed away in the prepositional dative condition. If this were true, a purely structural account would predict that subjects should equally be primed by the structure of provide-with condition. However, this was not the case. Even on this account, then, semantics must still be involved: The structural prime only has an effect only when it combines both the syntax and the semantics of the prepositional dative.

The coarse mapping of conceptual features to syntactic positions demonstrated to affect priming in this experiment can account for much of the existing data on "syntactic" priming, including the ability of locative 'to' constructions to prime dative 'to' discussed above: in both cases the theme appears as object and the goal appears in a prepositional phrase marked by *to*. The finding that benefactives (*Pat baked a cake for Chris*) prime datives (*Pat served a cake to Chris*) (Bock 1986), can be accounted for similarly, since in both cases the theme argument appears as direct object. The distinction between benefactives and goals/recipients is apparently not strong enough to prevent the priming.

More work is required to fully account for the priming of passives by intransitive locatives such as *The priest was standing by the stained glass window* (Bock & Loebell 1990). As this example illustrates, the Bock and Loebell study used locative sentences that contained the same *be* auxiliary and preposition *by* that appear in passives. Experiments are being designed to determine whether expressions that have the same syntactic form but different morphemes, such as *The priest might stand near the stained glass window*, actually prime passives as predicted by a purely syntactic account. It may be that some conceptual or morphological overlap is required for structural priming. (See also Dell et al. (in press) for a model of this locative data that in the spirit of the present analysis.).

Thus it seems that "syntactic priming" may be fundamentally influenced by elements of conceptual features or semantic roles. This suggests that rather than being purely syntactic, the phenomenon might be at the level of the mapping between semantics and syntax.

Acknowledgments

This research was supported by NIH grant 5T32 DC0041-06, NSF grant SBR-9873450, and FRC research incentive grant #00009FRCR from Bowling Green State University.

References

Bock, K. (1986). Syntactic persistence in language production. *Cognitive Psychology, 18*, 355-387.

Goldberg, Adele E. 1995. *Constructions: A Construction Grammar Approach to Argument Structure*. Chicago: U of Chicago Press.

Bock, K., & Loebell, H. (1990). Framing sentences. *Cognition, 35*(1), 1-39.

Bock, K., Loebell, H., & Morey, R. (1992). From conceptual roles to structural relations: Bridging the syntactic cleft. *Psychological Review, 99*(1), 150-171.

Ford, M., Bresnan, J., & Kaplan, R.M. (1982). A competence-based theory of syntactic closure. In J. Bresnan (Ed.), *The mental representation of grammatical relations* (pp. 727-796). Cambridge, MA: MIT Press.

Frazier, L. (1979). *On comprehending sentences: Syntactic parsing strategies*. Bloomington, Indiana: Indiana University Linguistics Club.

Frazier, L., & Rayner, K. (1982). Making and correcting errors during sentence comprehension: Eye movements in the analysis of structurally ambiguous sentences. *Cognitive Psychology, 14*, 178-210.

Goldberg, Adele E. 1995. *Constructions: A Construction Grammar Approach to Argument Structure*. Chicago: U of Chicago Press.

Jackendoff, Ray. 1972. *Semantic Interpretation in Generative Grammar*. Cambridge, MAss: MIT Press.

Lakoff, George and Mark Johnson. 1980. *Metaphors We Live By*. Chicago: U of Chicago Press.

Levin, B. (1993). *English verb classes and alternations: A preliminary investigation*. Chicago: University of Chicago Press.

MacDonald, M.C. (1993). The interaction of lexical and syntactic ambiguity. *Journal of Memory and Language, 32*, 692-715.

MacDonald, M.C., Pearlmutter, N.J., & Seidenberg, M.S. (1994). Lexical Nature of Syntactic Ambiguity Resolution. *Psychological Review, 101*(4), 676-703.

MacWhinney, B., & Bates, E. (1989). *The crosslinguistic study of sentence processing*. Cambridge, England: Cambridge University Press.

Rayner, K., Garrod, S., & Perfetti, C.A. (1992). Discourse influences during parsing are delayed. *Cognition, 45*, 109-139.

St. John, M., & McClelland, J.L. (1990). Learning and applying contextual constraints in sentence comprehension. *Artificial Intelligence, 46*, 217-257.

Tanenhaus, M., & Carlson, G. (1989). Lexical structure and language comprehension. In W.D. Marslen-Wilson (Ed.), *Lexical representation and process* (pp. 505-528). Cambridge, MA: MIT Press.

Trueswell, J., Tanenhaus, M., & Garnsey, S. (1994). Semantic influences on parsing: Use of thematic role information in syntactic disambiguation. *Journal of Memory and Language, 33*, 285-318.

Diversity-Based Reasoning in Children Age 5 to 8

Evan Heit (E.Heit@warwick.ac.uk)
Department of Psychology, University of Warwick
Coventry CV4 7AL, United Kingdom

Ulrike Hahn (HahnU@cardiff.ac.uk)
School of Psychology, Cardiff University
Cardiff CF1 3YG, Wales, United Kingdom

Abstract

One of the hallmarks of inductive reasoning by adults is the diversity effect, namely that subjects draw stronger inferences from a diverse set of premise statements than from a homogenous set of premises (Osherson et al., 1990). However, past developmental work (Lopez et al., 1992; Gutheil & Gelman, 1997) has not found diversity effects with children age 9 and younger. In our own experiments, we found robust and appropriate use of diversity information in children as young as 5 years. For stimuli we used pictures of people and their possessions, rather than the stimuli concerning animals and their biological properties in past studies. We discuss implications of these results for models of inductive reasoning.

Introduction

One of the most important functions of categories is that they allow us to make predictions and draw inferences. For example, in seminal work by Rips (1975), subjects drew inferences from one category of animals to another. They were told to imagine an island where all members of one category of animals, such as rabbits, have a particular disease, then they estimated the proportion of another animal category, such as dogs, that would also have this disease. Rips found a predominant tendency towards similarity-based reasoning, namely people were highly sensitive to the similarity between the given, or premise, category, and the target, or conclusion, category. As an example, subjects made stronger inferences from rabbits to dogs than from rabbits to bears. Consistent with proposals from philosophy (e.g., Mill, 1874), it seems that similarity between a premise category and a conclusion category is a crucial determinant of the strength of an inductive inference.

A limitation of this early work is that it only looked at inferences from one category to another. In contrast, people will often face multiple sources of evidence or multiple categories when drawing an inference. Experimental research on inductive inference from multiple categories could be especially revealing about the processes underlying inductive ability. The most extensive and influential work on induction from multiple categories was conducted by Osherson, Smith, Wilkie, Lopez, and Shafir (1990). They reported several phenomena involving reasoning with multiple premise categories, but we will focus on what is perhaps the most basic phenomenon, which we refer to as the diversity effect. This phenomenon is illustrated by the following example. In this notation, the statements above the line are premises, which are assumed to be true, and the task is to assess the strength of the conclusion statement, below the line.

Lions have an ulnar artery.   (1)
Giraffes have an ulnar artery.

Rabbits have an ulnar artery.

Lions have an ulnar artery.   (2)
Tigers have an ulnar artery.

Rabbits have an ulnar artery.

People find arguments like (1) to be stronger than arguments like (2), even though giraffes are very different from rabbits. What is critical is the diversity of the premise categories. For argument (1), lions and giraffes are such a diverse set of premise categories that it seems to license a broad set of inferences, such as that rabbits and many other mammals have an ulnar artery as well. In contrast, for argument (2), lions and tigers are a very non-diverse set, and it seems possible that the property of interest, having an ulnar artery, could be restricted to just these two animals or just to felines. Supported by these diversity effects in adults, Osherson et al. (1990) developed a computational model of induction that includes not only a similarity-based component but also a category-based component, in which people generate a superordinate category and assess how well the premise categories cover this superordinate. In the present example, lions and giraffes would cover the superordinate, mammals, better than lions and tigers, hence a stronger inference to other mammals would be indicated. This account by Osherson et al. not only describes a fairly sophisticated reasoning procedure but also presupposes knowledge of the relevant taxonomic category structure.

Because the diversity effect seems to highly revealing both about reasoning mechanisms as well as categorical knowledge, there has been keen interest among researchers in assessing the generality of this phenomenon. How robust is diversity-based reasoning? Lopez (1993) devised a stricter test of diversity-based reasoning, in which people chose premise categories rather than simply evaluating arguments given a set of premises. In other words, will people's choices of premises reveal that they value diverse evidence? Subjects (American college students, as in Osherson et al.) were given a fact about one mammal category, and they were asked to evaluate whether all mammals have this property. In aid of this task, subjects were allowed to test one other category of mammals. For example, subjects would be told that lions have some property, then they were asked whether they would test leopards or goats as well. The result was that subjects consistently preferred to test the more dissimilar item (e.g., goats rather than leopards). It appears on the basis of Lopez (1993) that for inductive arguments about animals, subjects do make robust use of diversity in not only evaluating evidence but also in seeking evidence. The results are less clear for other subject populations, however. In particular, developmental work has generally failed to find diversity effects in children. (See also Lopez, Atran, Coley, Medin, & Smith, 1997, and Choi, Nisbett, & Smith, 1997, for cross-cultural results.)

The first study of diversity-based reasoning was a developmental one by Carey (1985), comparing 6 year olds and adults. Carey looked at patterns of inductive projection given the premises that two diverse animals, dogs and bees, have some biological property. The purpose of this study was to see whether subjects reason that "if two such disparate animals as dogs and bees" have this property then "all complex animals must" (p. 141). Indeed, adults made broad inferences to all animals, extending the property not only to things that were close to the premises (other mammals and insects) but also to other members of the animal category (such as birds and worms). In contrast, the children seemed to treat each premise separately; they drew inferences to close matches such as other mammals and insects, but they did not use the diversity information to draw a more general conclusion about animals. Therefore in this first attempt there was evidence for effects of diversity in adults but not children. In a follow-up study, Carey looked at diversity effects based on the concept of living thing rather than animal. The results were actually less clear for this study, but again, there was not definitive evidence for mature diversity-based reasoning in children.

Continuing along this line of looking for diversity effects in children, Lopez, Gelman, Gutheil, and Smith (1992) found limited evidence for 9 year olds and no evidence for 5 year olds. For the 5 year olds, choices in a picture-based task did not show any sensitivity to diversity of premise categories, even when the diversity was emphasized by the experimenter. However, 9 year olds did show sensitivity to diversity of premises, but only for arguments with a general conclusion category such as animal. They did not show diversity effects at all for a specific conclusion category such as rabbit. Therefore it seemed that diversity-based reasoning was somewhat shaky in 9 year olds and not at all present in 5 year olds. Lopez et al. interpreted the lack of diversity effects in terms of the Osherson et al. (1990) model, with the reasoning mechanism for generating and using a superordinate category not being well-developed in children.

Gutheil and Gelman (1997) made a further attempt to find evidence of diversity-based reasoning for specific conclusions in 9 year olds, using category members at lower, or more concrete, taxonomic levels which would presumably

make reasoning easier, and also using increased sample size (e.g., three different butterflies rather than two different mammals). However, like Lopez et al. (1992), Gutheil and Gelman did not find diversity effects in 9 year olds, although in a control condition with adults, there was clear evidence for diversity effects with the same stimuli.

We see at least two interesting ways of explaining the lack of diversity effects in children. First, there could be a developmental change in the mechanisms of inductive reasoning; this explanation was elaborated by Lopez et al. (1992). Reasoning in children might not be able to access all the same processes as reasoning by adults. Second, there could be a change in knowledge structures; this explanation was the focus of Carey (1985). It could be the case that children do not have fully developed concepts of animals and the taxonomic structure that relates various animals to each other. Hence it would be difficult to be sensitive to the diversity of a set of animal categories with respect to a superordinate.

Our own experiments were an attempt to distinguish between these two explanations, that the non-existence of diversity effects in children has been due to a lack of reasoning ability or due to lack of knowledge. We attempted to look for diversity-based reasoning in other domains, such as toys and clothing, that should be more familiar to children, compared to animals and their biological properties. If children show diversity-based reasoning for other kinds of categories, there would be evidence that the lack of diversity effects in past studies was due to incomplete knowledge in children rather than differing mechanisms for children's reasoning compared to adult reasoning.

Our procedure was similar to that used by Lopez et al. (1992) and Gutheil and Gelman (1997). The child was given a fact about one set of items, then another fact about another set of items. Then the child was asked which fact was true of a target item. To be more specific, the facts were always related to possession or other interaction with humans. For example, children were shown a set of three different dolls and told that these dolls belong to a girl named Jane. This was a diverse set of dolls, including a china doll, a stuffed doll, and a Cabbage Patch doll. The set was presented as three pictures of Jane playing with the dolls. Then the children were shown another set of dolls, all the same (three pictures of Barbie dolls). The child was shown that these dolls all belong to a girl named Danielle. Then the child was shown a target item, a baby doll. The question was whether this doll belonged to Jane (the diverse choice) or Danielle (the non-diverse choice). Our stimulus design was exactly analogous to past studies of diversity, using everyday objects and properties based on interactions with people. We tested children over a range of ages from 5 to 8 years, with the aim of looking for some evidence of diversity-based reasoning below age 9. Also, in the first experiment, we used two types of instructions, following Lopez et al. For the first four items in each session, the child was given standard instructions that did not refer to diversity. Then for the last four items, the experimenter was emphatic in noting that one set was diverse and the other was not.

213

Experiment 1

Method

Subjects. There were 64 children: 18 in year 1 (mean age 5:7, range 5:3 to 6:0), 19 in year 2 (mean 6:9, range 6:3 to 7:2), 13 in year 3 (mean 7:9, range 7:3 to 8:1, and 14 in year 4 (mean 8:7, range 7:8 to 9:1). All attended St Peter's Primary School in Leamington Spa, England. The experiment was conducted on individual students; each session typically lasted 10 - 15 min.

Materials. There were 8 test questions. For each question, there were two sets of given items as well as a target item, all presented as individual photographs. The given information consisted of a set of 3 non-diverse items and a set of 3 diverse items. Each set was associated with a person. For example, in a non-diverse set there were three photographs of a football (soccer ball), being played with by a boy name Tim. In the corresponding diverse set, there were three photographs of a basketball, a cricket ball, and a tennis ball, each being played with by another boy, named Robby. The target item was a picture of another item from the same general category, such as a photograph of a rugby ball. This photograph was of the item alone, without any person. The test question was to choose whether the target item would go with one person or the other, e.g., Tim or Robby.

The color photographs were mounted on cards approximately 15 cm by 20 cm. The photographs for each test question used a different pair of people. The stimuli are described briefly in Table 1. We tried to choose diverse sets of items that would be as variable as possible, along multiple dimensions, while remaining within the same category. For example, the diverse set of hats varied in terms of color, size, and shape. Likewise, the target item was chosen to be as different as possible from the items in the non-diverse set and the diverse set, in terms of color, size, and shape. Therefore it was not expected that subjects would draw inferences on the basis of simple similarities between pairs of items.

Procedure. The order of test questions was randomized for each subject, and likewise on half the questions the non-diverse pictures were presented before the diverse pictures and half the time presentation was in the opposite order.

Four test questions were given with standard instructions, then four test questions were given with emphatic instructions. The standard instructions involved presenting the 3 non-diverse photographs and 3 diverse photographs, briefly describing each picture. For example, the experimenter would say, "Look, there's my friend Tim. He's playing with a football." The emphatic version of the instructions increased the salience of non-diversity or diversity. For example, the experimenter would say "Look, there's Tim. He's playing with the same thing, another football" for the non-diverse set. Likewise, for the diverse set the experimenter would emphasize the differences between items in the diverse set. The purpose of this manipulation was that for the last 4 test questions, we wanted to be certain that the diversity or non-diversity of each set was highly salient for each subject.

After the 6 given pictures were presented, the child was shown the target photograph and asked who this item would go with, e.g., who would play with this item or who would wear it. The experimenter provided mildly positive feedback after the child's response. Otherwise, the experimenter never gave any reason for the child to favor either the non-diverse set or the diverse set in making predictions, in the standard instructions as well as the emphatic instructions.

Table 1. Stimuli for Experiment 1.

Target Item	Non-Diverse Set	Diverse Set
Rugby ball	Football (soccer ball)	Basketball, Cricket ball, Tennis ball
Baby doll	Barbie doll	China doll, Stuffed doll, Cabbage Patch doll
Purple brimmed hat	White floppy hat	Straw hat, Ski hat, Baseball hat
Red top (shirt)	Green top	Blue top, White top, Gray top
White book	Red book	Black book, Green book, Purple book
Yellow flower	Purple flower	Orange flower, Blue flower, Red flower
Blackcurrant ice cream (cone)	Chocolate ice cream	Strawberry ice cream, Vanilla ice cream, Pistachio ice cream
Horse	Cat	Dog, Guinea pig, Goldfish

Note: These descriptions are simplified. For example, the books varied in size and shape as well as color.

Table 2. Proportion of Diverse Choices, Experiment 1.

Year	Standard (S.E.)	Emphatic (S.E.)
1	.71 (.05)	.78 (.08)
2	.64 (.09)	.86 (.06)
3	.73 (.06)	.88 (.05)
4	.91 (.06)	.93 (.04)

Results and Discussion

Overall, as shown in Table 2, children robustly favored the diverse choice over the non-diverse choice. For example, in the standard condition, the overall proportion of diverse choices was .74, which was significantly greater than a chance level of 50%, $t(63)=6.65$, $p<.001$. Inspection of Table 2 suggests that the emphatic condition led to an even higher level of diverse choice, and that there was a tendency for older children to make more diverse choices. A two-way ANOVA indicated a significant effect of instructions, $F(1,60)=8.66$, $p<.01$. The effect of year was not quite statistically significant, $F(3,60)=2.19$, $p<.10$, and the interaction was not close to the level of significance, $F(3,60)=1.31$. The age-related trend had some further support from a finer-grained analysis, which correlated each child's age with his or her overall proportion of diverse choices. This correlation was .22, $p<.05$, suggesting that older children did indeed make more diverse choices.

The main result, showing diversity effects overall, was consistent across the 8 stimulus sets, with mean proportion of diverse choices ranging from .69 to .90 and no significant item differences found. The consistency across different kinds of stimuli, from toys to clothing to foods, contrasts with the past results showing a lack of diversity effects for biological properties of living things. Indeed, we found diversity effects for flowers and pet animals, using social properties (human interaction or possession) rather than biological properties. (See Heit and Rubinstein, 1994, for further evidence on effects of properties.)

These results represent the first strong evidence for diversity-based reasoning in children under age 9. Indeed, we did not find major age differences in the range of 5 to 8 years. On the first four test questions, which did not emphasize the diversity or non-diversity of given items, children made the diverse choice 74% of the time overall. The proportion was even higher with emphatic instructions that highlighted diversity and non-diversity (but did not indicate which one to choose). This apparent effect of instructions, however, could also be a practice effect because the emphatic instructions were always for the last 4 items. (We could not present the emphatic instructions for the first 4 items because these instructions could carry over to affect subsequent performance on later items.)

Experiment 2

Given the result of Experiment 1, that children as young as age 5 do show diversity effects, we next set out to determine when they *don't* show diversity effects. Having a diverse set of premise categories should license a broad set of inferences compared to a non-diverse set of premises, but it does not license just any inference at all. Do children have a sense of the reasonable scope of inferences, or in Experiment 1 were they simply choosing the more diverse set without a full understanding of the nature of the task? We attempted to address this issue by choosing target stimuli that would not necessarily license strong inferences from a diverse set. In particular, diverse premise categories should have less of an effect on remote conclusion categories, matching the premise categories only at the superordinate level. For example, again the diverse set of dolls belonged to Jane, and the non-diverse set of dolls belonged to Danielle. But sometimes the subjects were asked about a yo-yo rather than another doll (a baby doll). The yo-yo matched the premise categories at a more superordinate level than did the doll, which was a basic-level match. If children have a sophisticated sense of diversity and the scope of inferences, then the diversity effect should be weakened or even eliminated for the more superordinate target items.

This prediction is made by the model of Osherson et al. (1990), because a conclusion item that matches the premise items at a superordinate level would lead the subject to generate a very broad category, such as all toys, for the basis of assessing diversity. Neither set of premise categories, even three different dolls, would seem particularly diverse in terms of the space of all toys. Hence the difference in diversity for the two sets of premise categories would be very minor and less likely to affect choices.

We ran a pilot version of this experiment on 12 adults, comparing responses for 4 basic-level matches (e.g., another kind of doll) and 4 superordinate-level matches (e.g., another kind of toy). The adults chose the diverse set on 93% of the basic-level items, giving results similar to the oldest children in Experiment 1. For the superordinate-level items, the proportion of diverse choices was significantly lower, 69%, $t(11)=2.93$, $p<.05$.

Table 3. Target Items for Experiment 2.

Basic-level	Superordinate-level
Rugby ball	Yo-yo
Baby doll	Yo-yo
Purple brimmed hat	Black shoes
Red top	Black shoes
White book	Newspaper
Yellow flower	Green houseplant
Blackcurrant ice cream	Crisps (potato chips)
Aero chocolate bar	Crisps

Method

Experiment 2 was like Experiment 1, with the following changes. Ninety-two children, who attended Brookhurst Primary School, participated. There were 46 students in year 1 (mean age 5:10, range 5:3 to 6:5) and 46 students in year 4 (mean age 8:10, range 8:3 to 9:5)

The stimuli for one test question from Experiment 1, relating to pets, were replaced because these stimuli belonged to a higher-level taxonomic category than the other stimuli. For example, other stimuli belonged to basic-level categories such as balls or dolls. The photographs for the replacement stimuli were all of chocolate bars. The new basic-level target item was an Aero chocolate bar. The new non-diverse set consisted of three photographs of a man with a Milkybar. The diverse set consisted of three photographs of a man with a Twix, a Mars bar, and a Cadbury's.

In addition to the basic-level target items, each stimulus set was assigned a superordinate-level target item, as shown in Table 3. For example, for the hats, there was a basic-level target (another hat), and a superordinate-level target (a pair of shoes, also in the clothing category). Some pictures, e.g., the shoes, were used as the superordinate target for two stimulus sets. However, any subject only saw a particular picture once. The stimuli were given a random order for each subject, with the constraint that a superordinate target picture could not be used twice for the same subject.

Within each age group, half the students were given four test questions with superordinate-level targets followed by four test questions with basic-level targets. The other half of the students were given four basic-level target questions followed by four superordinate-level questions. We were concerned about possible carry-over effects in which a student might use strategies from one question as the basis for answering a later question. Therefore we considered the first four responses from each subject to be more pure, and we will focus on these responses in this report.

Experiment 2 used the standard form of instructions from Experiment 1.

Results and Discussion

The key result was that children made a lower proportion of diverse choices for superordinate-level target items than for basic-level target items. (See Table 4.) In addition, older children made more diverse choices overall than younger children. A two-way ANOVA supported these observations.

There were main effects of taxonomic level, $F(1,88)=19.35$, $p<.001$, and school year, $F(1,88)=7.41$, $p<.01$. The interaction between these two variables did not approach statistical significance, $F<1$.

These results replicate and extend those of Experiment 1. The overall proportion of diverse choices for basic-level targets, 77%, is similar to that of the standard condition of Experiment 1, 74%. However, children were less likely to make the diverse choice for superordinate-level targets, 54%, apparently chance responding. Clearly children favored the diverse set of premises when this was most appropriate, for basic-level targets, but they did not apply this response strategy in an unconstrained way. Instead they showed the more sophisticated pattern predicted by the Osherson et al. (1990) model and demonstrated in our pilot study by adult subjects. Finally, the results of Experiment 2 supported an age effect, which was also suggested by Experiment 1. However, we would hesitate to over-interpret the developmental trend, because the same pattern was shown by 5 year olds and 8 year olds, with the older children simply showing it more strongly. The age effect could be due to performance differences in, for example, how well children of different ages pay attention to this sort of task.

In both experiments, the children sometimes gave explanations for their choices, and these explanations were recorded when possible. In Experiment 1, it was very obvious from talking with the children that they knew that one set was more diverse than the other, and that they were making inferences on this basis. For Experiment 2, the explanations for the superordinate-level items were much more idiosyncratic, suggesting that children were employing a variety of strategies, such as selecting items based on similarity within a single dimension such as size or color.

Table 4. Proportion of Diverse Choices, Experiment 2.

Year	Superordinate (S.E.)	Basic (S.E.)
1	.46 (.05)	.71 (.05)
4	.62 (.07)	.83 (.04)

General Discussion

In contrast to past studies, our results show clearly that children from age 5 to age 8 can perform diversity-based reasoning, using familiar categories. In addition, this diversity-based reasoning is sophisticated enough to be sensitive to the scope of the inference, with children showing weakened diversity effects for remote inferences about more distantly related target categories. Therefore we would conclude that in terms of diversity, children can assess evidence and reason about categories in a manner similar to adults, if conditions permit. Their reasoning with multiple categories or multiple sources of evidence does not seem to be deficient compared to adults, provided that the materials being reasoned about are familiar enough. The past studies may have used materials that were too difficult or unfamiliar for children to support diversity-based reasoning. For example, Lopez et al. (1992) used properties such as "has leukocytes inside" and "has cellulose inside," rather than simple properties relating to associations with people.

We see this general issue to be important because there is a strong case to be made for the normative status of diversity-based reasoning. The value of diverse evidence for testing a hypothesis has been stressed repeatedly in the philosophical literature on scientific reasoning (e.g., Carnap, 1950; Nagel, 1939; Hempel, 1966). In brief, diversity-based reasoning is related to a falsifying test strategy; testing similar items repeatedly would be seen as a weak, confirmatory strategy. (See also Lopez, 1993.) Likewise diversity-based reasoning can be shown to be compatible with a Bayesian perspective (Heit, 1998; Howson and Urbach, 1993). Having such a powerful tool available early on in development thus seems of great adaptive value.

Acknowledgments

We are grateful to Jane Pollock for assistance in conducting this research. We thank the students and teachers of St Peter's Primary School and Brookhurst Primary School for their participation. This research was supported by a grant from BBSRC.

References

Carey, S. (1985). *Conceptual change in childhood.* Cambridge, MA: Bradford Books.

Carnap, R. (1950). *Logical foundations of probability.* University of Chicago Press.

Choi, I., Nisbett, R. E., & Smith, E. E. (1998). Culture, category salience, and inductive reasoning. *Cognition, 65,* 15-32.

Gutheil, G., & Gelman, S. A. (1997). Children's use of sample size and diversity information within basic-level categories. *Journal of Experimental Child Psychology, 64,* 159-174.

Heit, E. (1998). A Bayesian analysis of some forms of inductive reasoning. In M. Oaksford & N. Chater (Eds.), *Rational models of cognition* (pp. 248-274). Oxford University Press.

Heit, E., & Rubinstein, J. (1994). Similarity and property effects in inductive reasoning. *Journal of Experimental Psychology: Learning, Memory, and Cognition, 20,* 411-422.

Hempel, C.G. (1966) *Philosophy of natural science.* Englewood Cliffs, NJ: Prentice Hall.

Howson, C. and Urbach, P. (1993) *Scientific reasoning: The Bayesian approach.* Chicago: Open Court.

Lopez, A. (1993). The diversity principle in the testing of arguments. *Memory & Cognition, 23,* 374-382.

Lopez, A., Atran, S., Coley, J. D., Medin, D. L., & Smith, E. E. (1997). The tree of life: Universal and cultural features of folkbiological taxonomies and inductions. *Cognitive Psychology, 32,* 251-295.

Lopez, A., Gelman, S. A., Gutheil, G., & Smith, E. E. (1992). The development of category-based induction. *Child Development, 63,* 1070-1090.

Mill, J. S. (1874). *A system of logic.* New York: Harper.

Nagel, E. (1939). *Principles of the theory of probability.* University of Chicago Press.

Osherson, D. N., Smith, E. E., Wilkie, O., Lopez, A., & Shafir, E. (1990). Category-based induction. *Psychological Review, 97,* 185-200.

Rips, L. J. (1975). Inductive judgments about natural categories. *Journal of Verbal Learning and Verbal Behavior, 14,* 665-681.

Selecting Knowledge for Category Learning

Evan Heit (E.Heit@warwick.ac.uk)
Lewis Bott (L.A.Bott@warwick.ac.uk)
Department of Psychology; University of Warwick
Coventry CV4 7AL United Kingdom

Abstract

We present a category learning experiment in which subjects faced the knowledge selection problem, i.e., they needed to use their observations to determine which prior knowledge would be useful for learning. The issue of putting prior knowledge into neural network models is reviewed, and we present a new model which addresses the knowledge selection problem. This model gives a good account of the experimental results.

Introduction

At first glance, categorization would seem to simplify our lives, because a large number of individual observations can be classed together to allow reasoning and communicating about them as a group. But it has been pointed out that categorization itself entails further complexities. Medin and Ross (1997) noted that just 10 objects can be partitioned into categories over 100,000 different ways. So in addressing one computational problem, the high number of unique events, we are led to another computational problem, the high number of possible partitions of events. As a solution to this problem, it has been proposed that, by necessity, category learning is not entirely data driven (e.g., Peirce, 1931-1935). That is, people do not consider all possible partitions of observations when forming a category representation. Instead, we in effect consider a subset of the possibilities, using background knowledge for guidance. Indeed, it has by now been well established empirically that background knowledge has robust effects on facilitating category learning (see Heit, 1997, for a review).

Unfortunately, this solution itself raises yet another problem, namely the problem of selecting prior knowledge. There are many possible sources of background knowledge that could be helpful in learning about a new category. For example, imagine visiting some university campus for the first time and trying to learn about the general layout and architectural styles. Many sources of past knowledge could possibly be helpful, such as memories of other campuses or towns. In fact, it would be easy for the number of past observations to greatly outnumber the number of new observations! In light of this knowledge selection problem, how could background knowledge actually make concept learning easier?

The knowledge selection problem does seem very troublesome for experimental and computational approaches to category learning, but it is important to note that people do manage to solve this problem every day. In addition, it is encouraging to pick up any textbook on Bayesian statistics and find many techniques for combining multiple prior beliefs with observations, and selecting among these beliefs based on the data observed. In Bayesian statistics there is no assumption that a learner starts with optimal or perfectly correct prior beliefs. Instead, the learner begins with a reasonable guess that merely serves as an initial basis for learning, with corrective information then provided by the data. Indeed, it is possible to start with a whole set of different prior beliefs, with a distribution of initial degrees of confidence in each of these. When observations are made, confidence in various prior beliefs can be increased or decreased as appropriate. (See also Heit, 1998.) That is, observations can be used to select from a set of prior hypotheses.

Many previous experiments on knowledge effects on category learning have avoided the knowledge selection problem by more or less telling the subjects which prior knowledge to use. One of the exceptions is a study by Murphy and Allopenna (1994), in which subjects learned about categories of buildings, animals, and vehicles, with labels such as "Category 1" and "Category 2." These category labels did not constrain the knowledge selection problem very much. When a subject learned about a new category of vehicles, for example, there were many known types of vehicles that could be informative. It was impossible to know in advance whether to use prior knowledge about snowmobiles, ice cream vans, heavy trucks, or jeeps. However, the content of the category itself, that is, the descriptions of category members, were helpful in finding useful prior knowledge. For example, when subjects observed a category member with the description "made in Africa, lightly insulated, and drives in jungles," they were able to access knowledge about vehicles used in hot weather such as jeeps, rather than knowledge about other vehicles such as snowmobiles and heavy trucks.

Our own experiment was an attempt to further address the phenomenon of knowledge selection. Like Murphy and Allopenna, we used building categories. (Also see Heit and Bott, 1999, for an experiment with vehicle categories.) Given that people already know about many kinds of buildings, we see these stimuli as encouraging knowledge selection processes. Unlike Murphy and Allopenna, we collected data over the course of learning. One of our goals was to show that in some situations, categorization judgments are not affected early on by prior knowledge, until many observations have been made and relevant prior knowledge can be assembled. Therefore it was necessary to collect categorization judgments after various numbers of category members had been observed. Our general was that in terms of various measures there would be increasing knowledge effects over the course of learning. Another advantage of collecting data along the course of learning was that our data were suitable for developing and testing a computational model of category learning.

We next present an experiment on knowledge selection in category learning, followed by a brief review of computational models that employ prior knowledge and then by the introduction of a computational model that addresses knowledge selection

Method

The 77 subjects learned about two categories of buildings, referred to as Doe buildings and Lee buildings. The subjects were told to imagine that they were reading a book with a series of descriptions of buildings. The stimuli were organized as five blocks, with descriptions of four Doe buildings and four Lee buildings presented in each block. Each description included the category label (Doe or Lee) and a list of featural information. There were two critical features presented in each description and two filler features. The critical features for each category were related to a known type of building (e.g., churches for Doe and office blocks for Lee or vice versa). In contrast, the filler features were general characteristics that could be true of just about any building. Finally, each description contained three pieces of individuating information (name of builder, surveyor, and photographer). The main prediction was that there would be increasing facilitation on critical features over the course of learning, as subjects were increasingly able to select useful prior knowledge.

The critical and filler features were derived from a pretest, which involved a series of sorting tasks in which subjects were asked to place each feature into one of two groups. After a series of iterations, replacing features as necessary, a set of 8 pairs of critical features and 8 pairs of filler features was obtained. A final pre-test group of 20 subjects sorted each of the critical features with at least 90% preferring one group over the other, and for the filler features preference for one group was always less than 75%. In addition, subjects were readily able to describe one sorted pile of features as being related to churches or old buildings, and the other as being related to office buildings or other commercial buildings. The complete list of critical features, as well as sample filler features, are shown in Table 1.

From the 8 pairs of critical features, 4 pairs were randomly assigned to presentation frequency one. Each feature in each pair was presented in one description per block, either Doe or Lee. Two pairs were assigned to presentation frequency two, and each feature presented in two descriptions per block. Finally, 2 pairs of features were not presented at all in the study blocks (but they were tested in test blocks). Likewise, the 8 filler features were assigned to presentation frequencies one, two, and zero.

There was a sequence of 5 study-test blocks. In each study block, the building descriptions, each with a category label, were presented individually, for 6 s each. A sample description would be: {Lee building type, Builder: T Jones, near a river, has gas central heating, Surveyor: R Rawson, Photographer: A Ferraro, has steeply angled roof, has wooden furniture}. Subjects were given memorization instructions. Following each study block was a test block, in which subjects were asked to categorize 40 single features, in the Doe or Lee categories. These test items included 24 individuating features, 8 critical features (4 presented once, 2 presented twice, and 2 not presented), and 8 filler features (same distribution as critical features).

Table 1. Critical and filler features for building stimuli.

Critical Features
has steeply angled roof, has a flat roof
has wooden furniture, has metal furniture
has an interesting structure, has a repetitive structure
old building, new building
quiet building, busy building
lit by candles, lit by fluorescent light
ornately decorated, blandly decorated
built with stone, built with metal and concrete

Sample Filler Features
near a bus station, not near a bus station
designed by a local architect,
designed by an international architect
has gas central heating, has electric central heating

Results and Discussion

Initial analyses did not reveal any significant differences between presentation frequency 1 and presentation frequency 2; therefore the results were pooled over these two presentation frequencies. The average proportions correct are shown in Figure 1. The top panel shows responses to features that had been presented during the study blocks. Overall, there is a trend for performance to improve over blocks. Although there is no difference between critical and filler features in the first block, the difference between the two kinds of features, that is, the gap between lines, widens after the first block, suggesting increased facilitation on critical features over the course of learning. The bottom panel shows responses to the features that had not been presented at all. Responses to filler features essentially represent chance responding. The responses to critical, non-presented features are more interesting. Even though these features were never presented in study blocks, categorization performance clearly improved from the first block to the fifth block, suggesting an increasing influence of knowledge.

The results were analyzed with a three-way ANOVA with block, feature type (critical or filler), and presentation (observed or not observed) entered as variables. Each of the variables had statistically significant main effects, and likewise each of the two-way interactions were significant. Perhaps the most important interaction was the feature type by block interaction, supporting the observation that the difference between critical and filler features increased across blocks.

Finally, performance on the individuating features increased steadily from 51% correct in block 1 to 59% in block 5, suggesting that subjects were devoting increased resources to learning the names on later blocks, as the other features were better learned.

The key result in this experiment was that subjects were increasingly influenced by background knowledge over the course of learning. One source of evidence for increasing influences of knowledge is the results for presented features. There was no difference in classification accuracy for critical and filler features after the first training block, but by the end of the second block subjects had apparently retrieved prior knowledge that facilitated performance on critical features compared to filler features. Realizing that the Doe buildings are church-like and the Lee buildings are like office buildings, for example, would help answer questions about critical features but not filler features. Although performance on critical and filler features continued to improve over the course of learning, the advantage for critical features was persistent. The other source of evidence for changes in knowledge effects is the judgments on non-presented critical features. Subjects were never told the correct category for these features during training blocks. The only way to classify these features correctly was on the basis of general knowledge about buildings. Performance on non-presented critical features improved over the course of learning, suggesting that subjects were increasingly relying on appropriate knowledge for making judgments about these features.

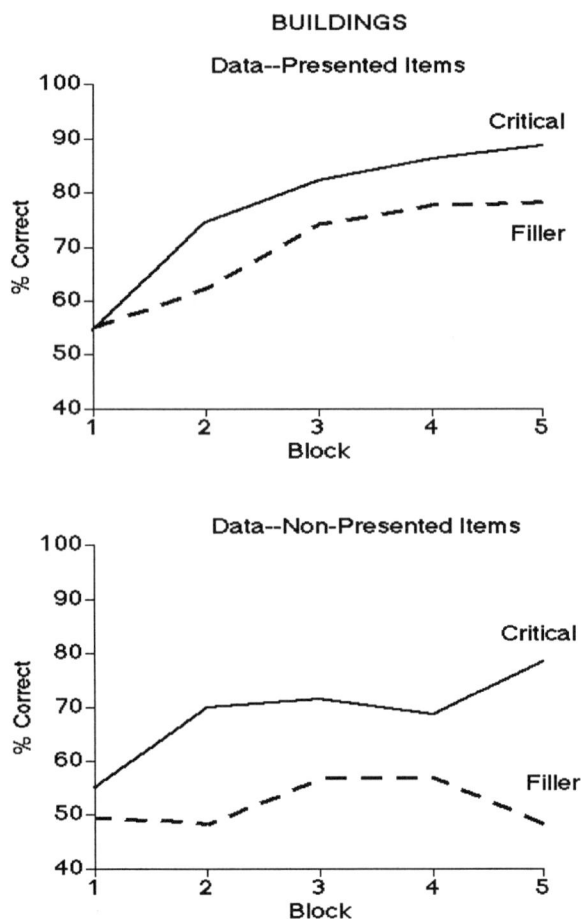

BUILDINGS

Figure 1. Results of experiment.

One surprising result was the lack of difference between features presented once per block and features presented twice per block. For both critical and filler features, we did not find any statistically significant difference in judgments for the two levels of presentation. It is tempting to relate this finding to Murphy and Allopenna (1994), who also found low sensitivity to frequency. Informal debriefing of subjects suggested to us that because each description, containing eight pieces of information, only appeared for 6 s, there may have been some strategic scanning of information. For example, in each block some subjects might have looked for features that had not already been presented in that block, hence overlooking a second presentation.

Putting Knowledge into Neural Networks

Next we set out to develop and apply a computational model that could address knowledge selection. We chose to work within the framework of neural network or connectionist models because they provide such a rich descriptive framework. That is, the complexity of connectionist models provides many opportunities for describing distinctive effects of knowledge on learning, as well as an appropriate framework for describing the dynamics of learning. Also, there has already been a great deal of research on different ways of putting knowledge into neural networks. Before we present our own model, we review some of this past work.

A useful framework for discussing prior knowledge in neural networks has been developed by Geman, Bienenstock and Dourstat (1992), who demonstrated that the generalization error when learning a concept can be broken down into a bias component and a variance component. Models that rely heavily on prior assumptions about the data, e.g., having architectural constraints that favor a particular conceptual structure, can lead to a high bias component, that is the model can persistently fail to capture aspects of the target concept which do not meet its prior assumptions. On the other hand, models that do not make strong assumptions about the concept to be learned can show a high variance component, that is they will be easily swayed by noise in training samples. Therefore a model without many assumptions could require an excessively large training sample to achieve good generalization. Further, reducing one type of error frequently is accompanied by an increase in the other type of error, leading to what Geman at al. referred to as the bias-variance dilemma. To reduce generalization error, both bias and variance must be reduced. We next review a number of learning algorithms that are aimed at reducing generalization error, keeping in mind the need to minimize the number of training examples as well.

One method for reducing the number of examples required for good generalization is to introduce "hints" into neural networks (e.g., Abu-Mostafa, 1995). Hints are general properties of a class of target concepts, independent of the specific details of the training data. Hints are introduced into the network by presenting "virtual examples" of the hint, and altering the error function to incorporate a term for the hint. Another approach to prior knowledge is to insert biases directly into neural networks by artificially setting the weights before learning begins. This approach has been taken by, for example, Giles and Omlin (1993), whose method was to insert transition rules into recurrent neural networks that learned artificial grammars. Known transitions were built into the network and then unknown transitions were learned from the data.

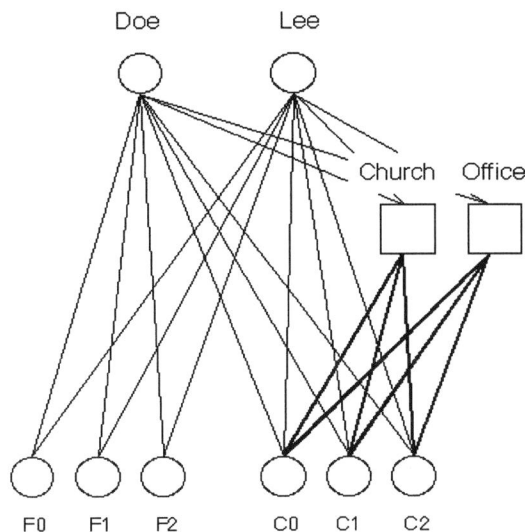

Figure 2. The Baywatch model.

Another way to build in prior knowledge is by varying the network architecture, to allow the network to have sufficient representational power to capture the underlying concept, but also avoid fitting the noise in the data. This goal is another way of looking at the bias-variance dilemma--a network that is too small leads to a high bias, but a network that is too large leads to high variance (and fitting the noise). Constructive networks (e.g., Prechelt, 1997) expand their architecture during learning, allowing the complexity of the network to increase as the data suggests it. Destructive networks, on the other hand, start off with an excess of hidden units and then prune off the hidden units which are not useful (e.g., Mozer & Smolensky 1989).

Rather than varying the network architecture over the course of learning, a different approach is to employ more than one architecture within a mixed network, and allow the network itself to learn which of the architectures is best for a particular problem. An example of this approach is the mixture-of-experts network (e.g., Jacobs, Jordan, & Barto, 1991). Jacobs et al. used a mixed network, with three modules having different structures (no hidden units, medium number of hidden units, and a high number). In effect, each module took a different approach to the bias-variance dilemma, with the simplest network being most constrained in terms of what it could learn and the network with many hidden units being most sensitive to variation in a training sample. The network was trained to perform two tasks, and it learned to allocate the module without hidden units to the simpler task while it allocated one of the modules with hidden units to the more complex task. We see the mixture-of-experts approach as coming close to the Bayesian idea of starting with multiple hypotheses then selecting among them based on the data.

The Baywatch Model

Our own approach to the knowledge selection problem has some parallels to the mixture-of-experts architecture, but instead of using modules with different structures, we used modules with different pools of pre-trained knowledge. Therefore our method also has some relations to techniques that insert prior knowledge directly into networks. Our model, illustrated in Figure 2, can be described as having one module or set of weights for strictly empirical learning. These weights do not get any pre-training. Then the model also has a set of experts which are pre-trained to recognize different known categories. For example, a network for learning about buildings might have experts which can recognize different kinds of buildings such as churches, office blocks, restaurants, and schools. (Only two of these expert modules are illustrated.) We refer to this model as the Baywatch model because it combines a general Bayesian approach to selecting among multiple sources of prior knowledge with an empirical learning component.

The Baywatch model is a feedforward network where the input units represent the individual features and the output units represent the Doe and Lee category nodes. The two hidden units correspond to two expert modules, or prior knowledge category nodes (PK nodes). The input units on the left side of Figure 2 represent filler features, and the input on the right side represent the critical features. The difference between the two types of features is that the filler features are only connected to the output nodes, whereas the critical features are connected both directly to the output nodes and indirectly to the output nodes via the PK nodes. The connections between the critical features and the PK nodes have fixed, pre-learned weights, so that values of critical features of the stimuli that correspond to church features would activate the church PK node, and likewise critical features of the stimuli that correspond to offices would activate the office PK node. The PK nodes have threshold functions, so that if any church feature, say, steeply angled roof, is presented, then the church PK node will be activated. The activation from the PK node would then be propagated to the output units.

In contrast to the connection weights between the critical features and the PK nodes, the other weights in the network are learnable through gradient descent. Adjusting the weights from filler units and the critical units to the output units allows the features to be associated with the category nodes in the empirical learning module. Finally, there are adjustable weights between the PK nodes and the category nodes. These represent the subject's capacity to associate known categories, say churches and office blocks, with the new categories, Doe and Lee buildings. We see this part of the network as addressing (at least in part) the knowledge selection problem, because here the network is learning to select from already known categories and apply this knowledge to judgments about new categories. (See Heit and Bott, 1999, for further details of the model and simulations.)

MODEL PREDICTIONS
Presented Items

Non-Presented Items

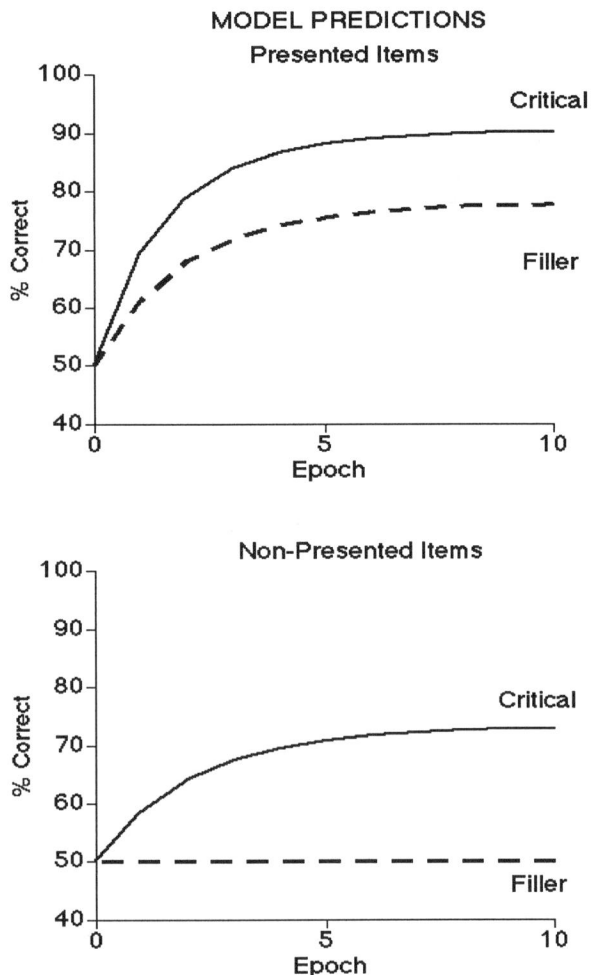

Figure 3. Results of simulations.

Simulations

The network was trained for a total of 10 blocks, with the learning rate in the delta rule set at 0.1 and the activation on each category node converted to a probability using a logistic transformation. The training stimuli consisted of four examples of buildings, two Doe exemplars and two Lee exemplars. Following each training block, the network was tested on the individual features by presenting a vector of all zeroes except for the particular feature of interest, which had a value of either +1 or -1. The results of the simulations are displayed in Figure 3. The top panel shows predictions for presented features, with the predictions for features presented once per block and features presented twice per block pooled together. The bottom panel shows predictions for features that had not been presented during training. The predictions fit well with the main results of the experiment. Critical features were learned more quickly than filler features, and critical features that hadn't been presented were responded to more accurately than chance, whereas filler features which hadn't been presented were at chance level.

To provide a better idea of how the Baywatch model uses prior knowledge, we re-ran the simulations without any PK nodes. In Figure 4, we show predictions on presented items,

comparing versions of the model with and without prior knowledge. For critical features, in the top panel, it can be seen directly that prior knowledge does not have any influence initially on judgments; the model acts the same way with or without PK nodes. However, the beneficial effect of prior knowledge for critical features increases over the course of learning, as the network with PK nodes learns which categories to connect with its prior knowledge. In the bottom panel of Figure 4, there is evidence for a slight detrimental effect of prior knowledge on the learning of filler features. This result can be explained as a kind of overshadowing effect, in which knowledge of some highly predictive cues can reduce learning on other predictive cues.

One difference between the model's predictions and subjects' performance is that the model does predict more accurate judgments for features presented twice per block compared to features presented once per block. In contrast, there was no significant difference between these two levels of presentation in the experiments. This insensitivity to frequency could be an important aspect of concept learning in knowledge-rich domains but on the other hand it could just reflect subjects' reading strategies in this experiment. Therefore further experimental study is required.

Conclusion

How well would the Baywatch model scale up? The simulations were run with just two sources of prior knowledge (i.e., churches and office blocks) and the network was able to link up these two sources with the correct output categories, Doe and Lee. But people would obviously have a much larger number of known categories when facing the knowledge selection problem, due to large numbers of known kinds of buildings. In general, we think the model might scale up well, in terms of adding more prior knowledge nodes. It is useful to distinguish three different classes of PK nodes that might be added to the network in Figure 2, in addition to the church and office nodes.

First, irrelevant prior knowledge nodes might be added, which have little or no connection to the input stimuli. For example, there could be prior knowledge nodes for space stations, igloos, tents, and cave dwellings, added to the network, but these nodes would be hardly activated by the inputs. Therefore, adding PK nodes that are irrelevant to the stimuli would not affect the results of the simulations very much.

Second, additional PK nodes that are similar to the existing PK nodes might be included. For example, a PK node corresponding to cathedrals would entail much of the same connections to inputs as the church node. Putting in additional but similar PK nodes would enhance the prior knowledge effects but it would not really change their nature. Just as adding the PK node for churches helped performance on critical features of churches, relative to a straightforward empirical learning network (see Figure 4), adding another PK node for cathedrals would help even further. Paradoxically, there is no knowledge selection problem here, from adding another similar PK node. To the extent that sources of prior knowledge are mutually supporting, having multiple sources of prior knowledge need not harm performance.

222

MODEL PREDICTIONS
Critical Presented Items

With PK

Without PK

Filler Presented Items

Without PK

With PK

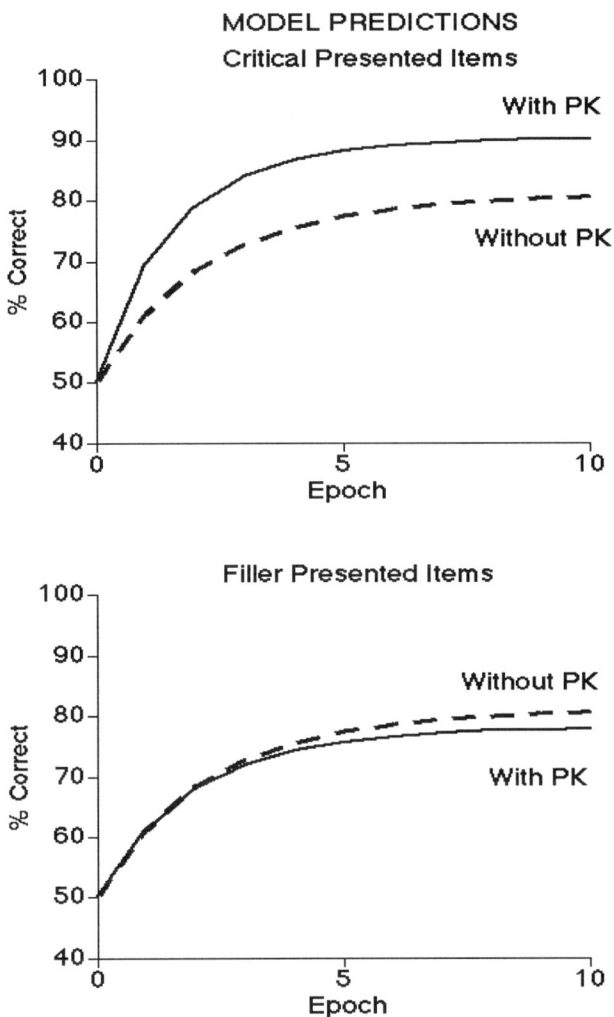

Figure 4. Predictions of model with and without prior knowledge.

Third, "malicious" prior knowledge nodes could be added to the network, for example, prior knowledge about some kind of building that is half-church and half-office block. Such PK nodes that are intermediate between the Doe and Lee categories might reduce the benefits of prior knowledge or even lead to costs due to knowledge, because they could make it more difficult to distinguish between the two categories.

More generally, we see the knowledge selection problem as having many facets. Certainly one of them is that when learning about novel categories, a learner would need to link up knowledge of familiar categories with judgments about the novel categories. The Baywatch model seems to address this aspect of knowledge selection, in terms of the gradual selection of prior knowledge nodes to use for a particular novel output category. In contrast, the prior knowledge in terms of connections from input units to PK nodes is fixed at the start of the simulations. It is assumed that these connections would have been already learned through ordinary associative processes, so that the network can more or less instantly recognize church or office buildings. However,

there could be some gradual aspects of knowledge activation or retrieval that are not captured by the model. It could be the case that somehow the connections between input units and PK nodes would be learned over the course of making observations, so that the recognition of relevant categories in prior knowledge would not be instantaneous when a single observation is made. This aspect of knowledge selection might be studied more directly, for example by showing subjects a series of training examples and asking them to judge directly which familiar categories are related to these stimuli.

Acknowledgments
This research was supported by grants from ESRC and BBSRC.

References

Abu-Mostafa, Y. S. (1995). Hints. *Neural Computation, 7,* 639-671.

Geman, S., Bienenstock, E., & Dourstat, R. (1992). Neural networks and the bias/variance dilemma. *Neural Computation, 4,* 1-58.

Giles, C. L., & Omlin, C. W. (1993). Extraction, insertion and refinement of symbolic rules in dynamically driven recurrent neural networks. *Connection Science, 5,* 307-337.

Heit, E. (1997). Knowledge and concept learning. In K. Lamberts & D. Shanks (Eds.), *Knowledge, concepts, and categories* (pp. 7-41). London: Psychology Press.

Heit, E. (1998). A Bayesian analysis of some forms of inductive reasoning. In M. Oaksford & N. Chater (Eds.), *Rational models of cognition (pp. 248-274).* Oxford: Oxford University Press.

Heit, E., & Bott, L. (1999). Knowledge selection in category learning. In D. L. Medin (Ed.), *Psychology of Learning and Motivation.* San Diego: Academic Press.

Jacobs, R. A., Jordan, M. I., & Barto, A. G. (1991). Task decomposition through competition in a modular connectionist architecture. *Cognitive Science, 15,* 219-250.

Medin, D. L., & Ross, B. H. (1997). *Cognitive Psychology.* (2nd ed.). Fort Worth: Harcourt Brace.

Mozer, M. C., & Smolensky, P. (1989). Using relevance to reduce network size automatically. *Connection Science, 1,* 3-16.

Murphy, G. L., & Allopenna, P. D. (1994). The locus of knowledge effects in concept learning. *Journal of Experimental Psychology: Learning, Memory, and Cognition, 20,* 904-919.

Peirce, C. S. (1931-1935). *Collected papers of Charles Sanders Peirce.* Cambridge: Harvard University Press.

Prechelt, L. (1997). Investigation of the CasCor Family of Learning Algorithms. *Neural Networks, 10,* 885-896.

223

Restricting Working-Memory Capacity Impairs Relational Mapping

Keith J. Holyoak (holyoak@lifesci.ucla.edu)
Department of Psychology, University of California,
Los Angeles, CA 90095-1563

James A. Waltz (waltz@lifesci.ucla.edu)
Department of Psychology, University of California,
Los Angeles, CA 90095-1563

Jean M. Tohill (JeanTohill@aol.com)
Department of Psychology, University of California,
Los Angeles, CA 90095-1563

Albert Lau
Department of Psychology, University of California,
Los Angeles, CA 90095-1563

Sara K. Grewal
Department of Psychology, University of California,
Los Angeles, CA 90095-1563

Abstract

Some theories of analogical mapping predict that finding mappings based on relations between objects requires greater working-memory capacity than finding mappings based on attributes of individual objects. It follows that the ability to make relational mappings will be impaired by any manipulation that constricts available working memory capacity. This prediction was tested in two experiments using a mapping task that required finding correspondences between pairs of pictures in which a critical object was "cross-mapped" (attribute similarity supporting one mapping, relational similarity another). Working memory was constricted in Experiment 1 by requiring participants to maintain a digit load while performing the mapping, and in Experiment 2 by inducing anxiety using a speeded subtraction task administered prior to the analogy task. Both manipulations caused participants to produce fewer relational responses and more attribute responses. The findings support the postulated links among working memory, anxiety, and the ability to perform complex analogical mapping.

Introduction

A basic characteristic of mental representations is that they have a hierarchical form, in which simpler elements are systematically integrated to create more complex structures. Hierarchical representations appear to be used in object and scene perception, in language production and comprehension, and in text comprehension and memory. There is evidence that similarity judgments depend on hierarchical representations based on relations between elements (Medin, Goldstone & Gentner, 1993). Formally, relational representations have a predicate-argument structure, in which one or more elements are bound to distinct roles. Gentner (1983) proposed a taxonomy of representational complexity ranging from attributes (one-place predicates, such as tall (Abe)); to first-order relations (multi-place predicates that take objects as role fillers, such as taller-than (Abe, Bill)); to higher-order relations (multi-place predicates in which at least one role filler is itself a proposition, such as cause (taller-than (Abe, Bill), jealous-of (Bill, Abe))). For the purposes of the present paper it will suffice to distinguish representations based on attributes of individual objects from representations based on relations (either first-order or higher-order) among multiple objects.

Both attributes and relations can provide a basis for finding analogical correspondences, or mappings, between situations. In some cases, correspondences based on information at different levels of complexity may conflict, creating ambiguous cross mappings between elements of the two analogs (e.g., Gentner & Toupin, 1986; Ross, 1987). For example, Markman and Gentner (1993) showed college students pairs of pictures, such as a man bringing groceries from a truck and giving them to a woman, who is thanking him, and a different woman taking food from a bowl and giving it to a squirrel. Participants were asked to indicate which object in the second picture corresponded to the woman in the first picture. Based on attribute mapping, the woman in the first picture would map to the woman in the second picture; but based on relations, the woman in the first picture would map to the squirrel in the second picture because each is a recipient of the food.

Markman and Genter found that different participants gave different responses to such cross-mapped objects, some giving the attribute-based response and some giving the relation-based response. Manipulations that encouraged participants to build an integrated representation of the relations among the objects and of higher-order relations between relations increased the proportion of relational responses. In particular, if participants were asked to match

224

not just one object in the first picture (the woman), but three (the woman, man, and groceries) to objects in the second picture, they were more likely to map the woman to the squirrel on the basis of their similar relational roles than were participants who mapped the woman alone. Active mapping of multiple objects seems to encourage people to process relations, which in turn changes the apparent correspondences between individual objects.

Although people are clearly capable of finding mappings based on multiple levels of representational complexity, there is reason to expect that relational mappings impose greater demands on working memory than do attribute mappings (Halford, 1993; Halford, Wilson & Phillips, 1998; see Baddeley, 1986, 1992, for a discussion of the components of human working memory). For example, mapping the woman in the first picture to the woman in the second picture can be done by focusing on only one object in each picture; whereas mapping the woman to the squirrel requires representing multiple objects and relations in each picture in order to recognize the correspondences between the objects filling matching roles in a system of relations. Some computational models of analogical mapping, such as the STAR model of Halford et al. (1994) and the LISA model of Hummel and Holyoak (1997), postulate inherent limitations on the complexity of possible mappings due to working-memory limits. Such models lead to the general prediction that any manipulation that reduces available working-memory capacity will make it more difficult for reasoners to compute relational mappings, and hence increase the proportion of less-complex attribute mappings in situations in which the mapping is ambiguous.

The results of a number of behavioral studies suggest there is a close relationship between the ability to process multiple relations and the functioning of working memory. Many of these studies have involved the performance of reasoning tasks concurrent with the execution of secondary tasks designed to place demands upon working-memory systems. In a study using a dual-task paradigm to investigate the information processing demands of relational reasoning, Maybery, Bain, and Halford (1986) required subjects to respond to a probe at different stages in the solution of three-term series (transitive inference) problems. Maybery et al. found that reaction times to the probe were longest during the phases of the problems in which premise integration took place: during the presentation of the second premise and during the presentation of the proposition to be verified. Their results were interpreted as indicating that the process of relational premise integration involved greater information-processing requirements than the sequential processing of individual relations, in that premise integration requires multiple relations to be simultaneously active in working memory.

It follows that the ability to make relational mappings will be impaired by any manipulation that constricts available working memory capacity. This prediction was tested in two experiments using a mapping task based on the Markman and Gentner (1993) materials. College students were asked to find correspondences between pairs of pictures in which a critical object was cross-mapped. Working memory was constricted in Experiment 1 by requiring participants to maintain a digit load while performing the mapping, and in Experiment 2 by inducing anxiety using a speeded subtraction task administered prior to the analogy task. The anxiety manipulation was based on Eysenck's (1979, 1985; Eysenck & Calvo, 1992) working-memory restriction theory, according to which the cognitive component of anxiety involves self-preoccupation and concern over performance, which preempt part of the processing and storage resources of the working-memory system. Processes that require working-memory resources will be impaired when anxiety causes the available system capacity to be exceeded, whereas lower-level processes will be relatively unaffected by anxiety. (See Eysenck & Calvo, 1992, for a review of studies that support the working-memory restriction theory of anxiety over alternative accounts.)

Experiment 1

College students performed an analogical mapping task with possible cross-mappings while simultaneously performing a secondary task that would tax working memory. The central prediction was that imposing a load on working memory would selectively impair the generation of correspondences based on relations, so that relational mappings would be made less frequently whereas attribute mappings would be made more frequently.

Method

Participants. Forty-five undergraduate students from the University of California, Los Angeles (UCLA) served in the experiment.

Materials and apparatus. Macintosh microcomputers running the SuperLab software package were used to present the stimuli. The stimuli consisted of the eight pairs of pictures used by Markman and Gentner (1993), as well as eight different 7-digit number strings. Each of the pictures showed a visual scene with three or more objects. One of these objects could be cross-mapped; that is, it could be judged as corresponding to one object on the basis of physical attributes, but to a different object on the basis of the role it plays in a system of relations.

Design and procedure. A 2 x 2 between-subjects design was used, with each participant randomly assigned to one of four conditions. The variables manipulated were the number of objects in each picture for which participants were prompted to identify correspondences (one or three) and the presence or absence of a phonological working memory load. Half of the participants were assigned to the 1-map condition and half were assigned to the 3-map condition. Orthogonally, half of the participants in each group were required to hold digit strings in working memory concurrent with performance of the mapping task, and half were not.

Each participant sat individually at a computer with an experimenter. Written instructions were displayed on the computer screen, explaining the procedure and the tasks to be performed. For each pair of pictures, the experimenter would point to a set of objects in the first picture. The participant's task was to identify the objects in the second picture that corresponded to each of those indicated by the experimenter. In the 1-map condition, participants were prompted to identify the object in the bottom picture corresponding to one object (the cross-mapped one) in the top picture. In the 3-map condition, participants were simultaneously asked to identify the objects in the bottom picture corresponding to three objects in the top picture. The first object to be mapped in the 3-map condition was the same cross-mapped object that was that was the single object to be mapped in the 1-map condition.

The phonological working-memory load required participants to hold a random 7-digit number string in memory while performing the mapping task. After pointing to their answer(s) to the mapping task, participants in this condition were asked to recall the 7-digit number. The experimenter recorded the participants' responses to the mapping questions, as well as their digit recall. All participants viewed each of the eight picture-pairs once, and all participants viewed the pairs in the same order.

Results and Discussion

The dependent variable was the proportion of correspondences identified by the participants on the basis of relational similarity. The results are depicted in Figure 1. The effects of number of mapping questions (1-map versus 3-map) and presence or absence of a working-memory on the percentage of relational mappings were analyzed using a 2 X 2 between-subjects analysis of variance. The analysis revealed a main effect of mapping instruction, such that the mean percentage of relational responses was significantly higher in the 3-map condition than in the 1-map condition ($M = 60\%$ versus 39%), $F (1, 41) = 6.64$, MSE = .0778, $p < .05$. A main effect of working-memory load was also observed, such that the mean percentage of relational responses was significantly lower when a digit string had to be maintained than when there was no memory load, ($M = 35\%$ versus 64%), $F (1, 41) = 12.48$, MSE = .0778, $p < .01$. There was no significant interaction between number of mapping questions and memory load, $F (1, 41) < 1$.

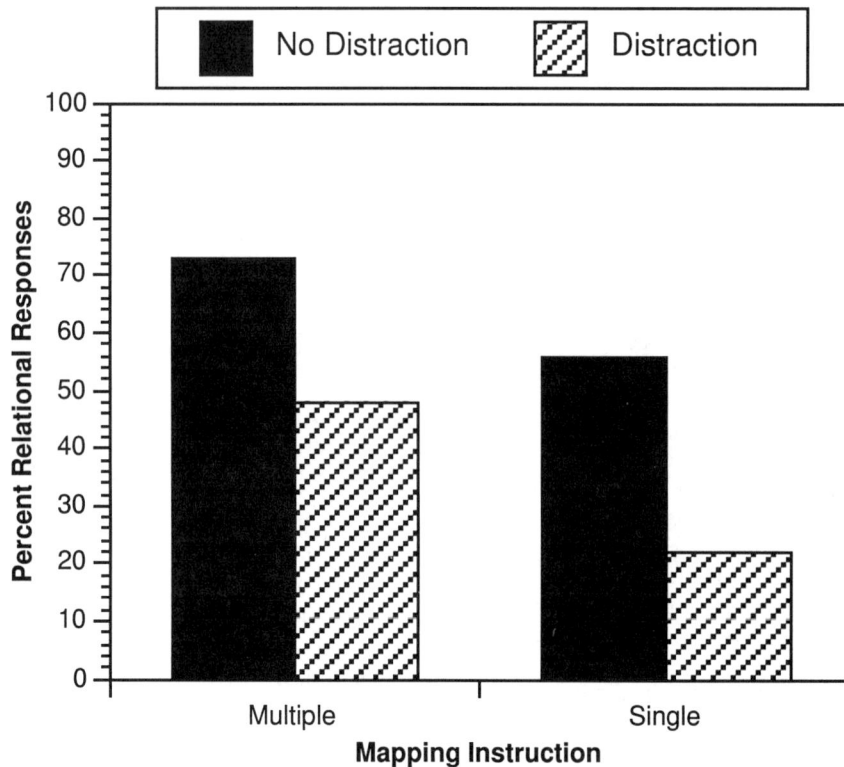

Figure 1. Percent relational responses given by subjects as a function of workinng memory load and mapping instruction.

226

Experiment 2

When Eysenck's (1979) working-memory restriction theory of anxiety is coupled with theories of analogy that specify links between working-memory capacity and the complexity of analogical mappings (Halford et al., 1994; Hummel & Holyoak, 1997), it follows that when objects are cross-mapped, an increase in state anxiety will yield a decrease in relational mappings and a concomitant increase in attribute mappings. Experiment 2 was designed to test this prediction.

Method

Participants. The participants were 22 UCLA undergraduates.

Materials. The stimuli were identical to those used in Experiment 1, with the addition of two extra pairs of pictures that were created by the third author using a computer graphics program.

Design and procedure. Participants were tested individually. They were assigned randomly in equal numbers to one of two conditions, the Anxious and Non-anxious groups. For participants in the Anxious condition, anxiety was induced by the introduction of a stressful task at the beginning of the experimental session. Specifically, their first task was to perform a serial subtraction task aloud. The participant was instructed to count aloud backwards, beginning at 1000 in increments of 13. One experimenter corrected any mistakes, while a second experimenter indicated to the participant at a pre-determined time that their counting speed was too slow. The participant was instructed to stop when 45 seconds had elapsed. The experimenter then informed the participant that they would be asked to repeat the counting task at the end of the experiment. This serial subtraction task has been used successfully to induce anxiety in previous studies (Sgoutas-Emch et al., 1994; White & Yee, 1997).

In the Non-anxious condition, the participant was asked to count aloud (forward) beginning at 1 for 45 seconds. Participants were told that they were not being evaluated in any way and to count at a pace that felt relaxed for them.

For the remainder of the experiment, the procedure was identical for both groups, and was equivalent to the 1-map condition used in Experiment 1 and in Markman and Gentner (1983). Each pair of pictures appeared for a fixed duration before the participant was queried for their response. Specifically, the picture-pair was programmed to appear for 15 seconds, after which the screen flashed, at which point the experimenter pointed to the pre-determined cross-mapped object in the top picture. This procedure was used to avoid participants giving premature responses while allowing ample time to process the pictures relationally. If more than approximately 4 seconds elapsed before the participant responded, the experimenter prompted the participant for an immediate answer. After the participant indicated the object of their choice, the experimenter recorded the response on an answer sheet. The above procedure was repeated for each of the ten picture analogies.

Table 1: Mean STAI Scores and Percentage of Relational Mappings in Experiment 2

Condition	STAI[a]	Percent Relational Mappings
Anxious	43.0 (8.1)	45 .4 (23.4)
Non-anxious	32.7 (8.1)	68 .2 (20.9)

Note. Values enclosed in parentheses represent standard deviations. [a]State score of the Spielberger State Trait Anxiety Inventory (Spielberger et al., 1970).

After the analogical reasoning task, participants completed the state form of the State-Trait Anxiety Inventory (STAI; Spielberger, Gorsuch & Lushene, 1970), and then were debriefed.

Results and Discussion

Mean STAI and mapping scores (percentage of relational mappings) for the Anxious and Non-anxious conditions of Experiment 1 are reported in Table 1. The anxiety manipulation was successful, as the mean STAI state-anxiety rating for the Anxious group was significantly higher than that for the Non-anxious group, $F(1,20) = 8.90$, MSE = 65.21, $p = .008$ (means of 43.0 and 32.7, respectively; maximum = 80). As predicted by the hypothesis that anxiety will restrict working memory, which is required for relational mappings, participants in the Anxious condition produced a significantly lower percentage of relational mappings than did those in the Anxious condition, $F(1,20) = 5.78$, MSE = 4.92, $p = .026$ (means of 45% and 68%, respectively). In both conditions, the remaining responses were primarily attribute mappings (51% and 24%, respectively, for the Anxious and Non-anxious conditions); only about 8% of responses in each condition were neither the relational nor the attribute mapping. Thus the impact of anxiety was not to produce random errors, but rather to systematically shift the dominant basis for mapping from the more complex relational level to the simpler attribute level.

General Discussion

We found that people who were asked to determine correspondences between visual scenes, while either concurrently holding information in phonological working memory or under the burden of high anxiety caused by a preceding difficult task, identified fewer correspondences on the basis of relational similarity than those who were not hindered by working-memory loads. Importantly, people who mapped under working-memory loads did not simply make unsystematic errors; rather, the basis for their mapping responses shifted from relations to attributes. Our results indicate that the process of structural alignment places greater demands upon working memory than does the process of matching attributes.

The finding that the process of structural alignment involves increased working memory demands is consistent with neuropsychological evidence that the capacity to integrate relations is lost as a consequence of damage to prefrontal cortex that reduces working-memory capacity (Waltz et al., 1999). We hypothesize that the representation of a system of relations, which specifies relations between relations, places demands upon working memory because it requires the explicit representation and binding of multiple relations (Robin & Holyoak, 1995). Our findings suggest that one of the primary consequences of factors that reduce available working-memory resources in learning environments is an impairment in the identification and learning of relations.

Acknowledgement

Preparation of this paper was supported by NSF Grant SBR-9729023.

References

Baddeley, A. D. (1986). *Working memory.* Oxford: Clarendon Press.

Baddeley, A. D. (1992). Working memory. *Science, 255,* 556-559.

Eysenck, M. W. (1979). Anxiety, learning, and memory: A reconceptualization. *Journal of Research in Personality, 13,* 363-385.

Eysenck, M. W. (1985). Anxiety and cognitive-task performance. *Personality and Individual Differences, 6,* 579-586.

Eysenck, M. W., & Calvo M. G. (1992). Anxiety and performance: The processing efficiency theory. *Cognition and Emotion, 6,* 409-434.

Gentner, D. (1983). Structure-mapping: A theoretical framework for analogy. *Cognitive Science, 7,* 155-170.

Gentner, D., & Toupin, C. (1986). Systematicity and surface similarity in the development of analogy. *Cognitive science, 10,* 277-300.

Halford, G. S. (1993). Children's understanding: The development of mental models. Hillsdale, NJ: Erlbaum.

Halford, G. S., Wilson, W. H., Guo, J., Gayler, R. W., Wiles, J., & Stewart, J. E. M. (1994). Connectionist implications for processing capacity limitations in analogies. In K. J. Holyoak & J. A. Barnden (Eds.), *Advances in connectionist and neural computation theory, Vol. 2: Analogical connections* (pp. 363-415). Norwood, NJ: Ablex.

Halford, G. S., Wilson, W. H., & Phillips, S. (1998). Processing capacity defined by relational complexity: Implications for comparative, developmental, and cognitive psychology. *Behavioral and Brain Sciences, 21,* 803-864.

Hummel, J. E., & Holyoak, K. J. (1997) Distributed representations of structure: A theory of analogical access and mapping. *Psychological Review, 104,* 427-466.

Markman, A. B., & Gentner, D. (1993). Structural alignment during similarity comparisons. *Cognitive Psychology, 23,* 431-467.

Medin, D., Goldstone, R., & Gentner, D. (1993). Respects for similarity. *Psychological Review, 100,* 254-278.

Premack, D. (1983). The codes of man and beasts. *Behavioral and Brain Sciences, 6,* 125-167.

Robin, N., & Holyoak, K. J. (1995). Relational complexity and the functions of prefrontal cortex. In M. S. Gazzaniga (Ed.), *The cognitive neurosciences* (pp. 987-997).

Ross, B. (1987). This is like that: The use of earlier problems and the separation of similarity effects. *Journal of Experimental Psychology: Learning, Memory, and Cognition, 13,* 629-639.

Sgoutas-Emch, S.A., Cacioppo, J.T., Uchino, B.N., Malarkey, W., Pearl, D., Kiecolt-Glaser, J.K., & Glaser, R. (1994). The effects of an acute psychological stressor on cardiovascular, endocrine, and cellular immune response: A prospective study of individuals high and low in heart rate reactivity. *Psychophysiology, 31,* 264-271.

Spielberger, C. D., Gorsuch, R. L., & Lushene, R. (1970). *STAI manual for the State-Trait Anxiety Inventory.* Palo Alto, CA: Consulting Psychologists Press.

Waltz, J. A., Knowlton, B. J., Holyoak, K. J., Boone, K. B., Mishkin, F. S., de Menezes Santos, M., Thomas, C. R., & Miller, B. L. (1999). A system for relational reasoning in human prefrontal cortex. *Psychological Science, 10,* 119-125.

White, P.M., & Yee, C.M. (1997). Effects of attentional and stressor manipulations on the P50 gating response. *Psychophysiology, 34,* 703-711.

Perceiving Structure in Mathematical Expressions

Anthony R. Jansen (tonyj@csse.monash.edu.au)
School of Computer Science and Software Engineering
Monash University, Clayton, Victoria, Australia

Kim Marriott (marriott@csse.monash.edu.au)
School of Computer Science and Software Engineering
Monash University, Clayton, Victoria, Australia

Greg W. Yelland (Greg.W.Yelland@sci.monash.edu.au)
Department of Psychology
Monash University, Clayton, Victoria, Australia

Abstract

Despite centuries of using mathematical notation, surprisingly little is known about how mathematicians perceive equations. The present experiment provides an initial step in understanding what sort of internal representation is used by experienced mathematicians. In particular, we examined if mathematical syntax plays a role in how mathematicians encode algebraic equations, or if just a simple memory strategy is used. Participants in the experiment performed a memory recognition task that required them to identify both well-formed (syntactically correct) and non-well-formed sub-expressions of equations. As hypothesised, performance was significantly better for well-formed sub-expressions, a result which suggests that mathematicians do indeed use an internal representation based on mathematical syntax to encode equations.

Introduction

Mathematical notation has evolved over hundreds of years. Like natural language, mathematical notation appears to have a well-defined syntax and semantics. It is clear that the expression

$$x(^2 \frac{+}{y8-}$$

is syntactically incorrect, while the equation

$$\frac{7^2 - 9}{5} = (3 - 2)^3$$

is syntactically well-formed but not true.

For decades now, phrase structure grammars have been used to understand how humans parse natural language sentences (for example, see Akmajian, Demers and Harnish, 1984). Such grammars have also been exploited in computer programming languages to process simple linearised mathematical expressions. However, unlike natural language (written or spoken), the syntax of mathematical notation is two-dimensional in nature. For example, the preceding equation relies on both vertical and horizontal adjacency relationships between the symbols to provide the meaning.

Given that mathematical notation has a well-defined syntax and a two-dimensional structure, it is natural to ask how humans comprehend mathematical equations and other notations with similar characteristics. We are especially interested in determining if humans parse mathematical expressions in a manner similar to the way in which they parse natural language. That is, do we assign grammatical structure to equations.

The notion that grammatical structure can be assigned to equations has some support from work on developing computer programs to understand mathematical notation. Since the pioneering work of Anderson (1977), almost all approaches have been grammar based, using some form of context-free attributed multiset grammar to specify the syntax. Attributes are used to capture geometric properties of the symbols, while the grammar works over multisets rather than sequences since there is no single way to sequence an equation. For further information, see the recent survey by Marriott, Meyer and Wittenburg (1998).

When processing an equation, the information needs to be stored in memory. Research has shown that humans have memory procedures for dealing with large amounts of information (Miller, 1956). They encode information into chunks that they can utilise within the limits of working memory. However, chunking is more efficient if it is guided by a structural principle. Research done by Johnson (1968, 1970) has shown that in the context of natural language, chunking is guided by syntax, with individual chunks conforming to grammatically defined units. Our aim is to determine if a similar process is true for mathematical expressions.

This experiment therefore has been designed to examine whether or not experienced users of mathematics use guided encoding when processing equations. That is, do mathematicians use some sort of internal representation that takes syntax into account, or do they use a less guided memory strategy to chunk a mathematical equation.

Surprisingly little work has been done that directly addresses this issue. The two-dimensional nature of mathematical notation seems to place it the same domain as diagrams. There has been much work done in the field of diagrammatic reasoning already (for example, see Glasgow, Narayanan and

229

Table 1: Example equations and sub-expressions.

Equation	Sub-Expression		
	Well-Formed	**Non-Well-Formed**	**Incorrect**
$\dfrac{9}{x(2y-5)} - 4x^3$	$(2y-5)$	$\overline{x(2}$	$x(4y+$
$x = 6yx - \dfrac{2x+2}{x}$	$2x+2$	$= 6yx-$	$\dfrac{6x+2}{y}$

Chandrasekaran, 1995), however this has concentrated on how diagrams are used in reasoning, rather than low-level perception of structure.

Our hypothesis is that experienced mathematicians do use some sort of internal representation to guide their encoding of equations, and that this representation is based on mathematical syntax. To test this hypothesis, we have set up a recognition task to determine if participants can more readily identify syntactically well-formed sub-expressions of an equation, as opposed to non-well-formed sub-expressions. If a simple memory strategy were being used, we would expect chunking in its most basic form, randomly splitting up the equations and providing no advantage for well-formed over non-well-formed sub-expressions. Such a result would mean that our hypothesis has no support. However, we expect that participants will respond significantly faster to the well-formed sub-expressions, indicating that mathematical syntax plays an important role in encoding equations.

Method

Participants Twenty-four participants successfully completed the experiment. All were staff members, graduate or undergraduate students from the Computer Science department, all competent mathematicians who were very familiar with algebra. All participants were volunteers between the ages of 18 and 35 years, with normal or corrected-to-normal vision. Data from an additional 13 participants were not included due to excessive error rates.[1]

Materials and Design One-hundred-and-twenty equations were constructed, all consisting of between twelve and fourteen characters. The equations contained at most one fraction and the variables were x and y, since these are most commonly used. For each equation, sub-expressions of three types were constructed.

a) A *well-formed sub-expression*, which is a component of its equation, and conveys the same meaning on its own that

it conveys in the equation.

b) A *non-well-formed sub-expression*, which is also a component of its equation, but does not convey any coherent mathematical meaning on its own.

c) An *incorrect sub-expression*, which was not part of the original equation. It can be either well-formed or non-well-formed. These act as fillers.

Each of the sub-expressions contained between four and six characters (the average for well-formed sub-expressions was 4.89; for non-well-formed, 4.49; for incorrect, 4.72). See Table 1 for examples of equations and sub-expressions used. As the examples show, a variety of sub-expressions were used, some of which were bracketed, but most of which were not.

In order to present all three sub-expression types for each equation, but ensuring that participants were presented with each equation only once to avoid practice effects, three counterbalanced versions of the experiment were constructed. For each version, there were forty instances of each type of sub-expression. Two additional equations were constructed as practice items. The same practice items were used in each version. The items of each version were presented in a different pseudo-random order for each participant.

Procedure Participants were seated comfortably in an isolated booth. Items were displayed as black text on a white background on a 17" monitor at a resolution of 1024x768, controlled by an IBM compatible computer running a purpose designed computer program. The average width of the equations in pixels was 177 (range 99-220) with an average height of 47 (range 26-61). The average width of the sub-expressions in pixels was 73 (range 39-192) with an average height of 26 (range 16-54).

Participants were given a brief statement of instructions before the experiment began. Practice items preceded the experimental items, and the participants took approximately fifteen minutes to complete the task. Progress was self-paced, with participants pressing the space bar to initiate the presentation of each trial.

Each item was presented in the centre of the monitor in the following sequence. First, a simple algebra equation was shown to the participant for 2500ms. The equation then dis-

---

[1]For the results of a participant to be included in the final analysis, they were required to get at least 70% correct responses overall and at least 50% correct for any given sub-expression type. This resulted in there being twenty-four participants whose data was included, eight for each version of the experiment.

Table 2: Mean correct response times (ms) and error rates (%) as a function of sub-expression type.

Sub-Expression	RT(ms)		%Error	
Well-Formed	1147	(147)	13.1	(6.2)
Non-Well-Formed	1352	(228)	23.5	(8.8)
Incorrect	1429	(213)	32.8	(8.9)

appeared and the screen remained blank for 1000ms. Then the sub-expression was shown. The participant was required to decide whether the sub-expression was in that equation, responding via a timed selective button press. They pressed the green button, (the '/' key on the right side of the keyboard), to indicate that the sub-expression was part of the original equation, and the red button, (the 'Z' key on the left of the keyboard), to indicate that the sub-expression was not part of the original equation. Participants were instructed to respond as quickly as possible, while taking care not to make too many errors. The sub-expression remained on the screen until a response was made.

The response time recorded was the time between the sub-expression first appearing and the participant's response. After the response, the participant was given feedback. If the response was correct then the word "CORRECT" appeared on the screen. Otherwise, the word "INCORRECT" appeared on the screen. In both cases, the participant's response time in milliseconds also appeared on the screen.

Data Treatment Two measures were employed to reduce the unwanted effects of outlying data points. Absolute upper and lower cut-offs were applied to response latencies, such that any response longer than 2500ms or shorter than 500ms was excluded from the response time data analysis and designated as an error. Secondly, standard deviation cut-offs were applied, so that any response time lying more than two standard deviations above or below a participant's overall mean response time was truncated to the value of the cut-off point.

Three items, one from each condition of the experiment, were excluded from the analysis due to error rates in excess of 75%. As a result, the final analyses were over thirty-nine items per condition, not the original forty. Response time and error data were analysed by a series of analyses of variance (ANOVAs), over both participant and item data. Where both the subject-based and item-based analyses were significant they were combined in the *minF'* statistic to ensure the generalisability of results over both these domains (Clark, 1973). Results are reported only where effects are significant.

Results and Discussion

The mean correct response time and error rate for the three sub-expression types are summarised in Table 2, along with the corresponding standard errors (in parentheses). Planned comparisons of the data were conducted using two-way ANOVAs (versions × sub-expression), carried out separately over subject and item data.

The sub-expressions whose content was drawn from their corresponding equations (i.e., both well-formed and non-well-formed sub-expressions), were responded to more rapidly than incorrect sub-expressions (well-formed: $minF'(1, 40) = 57.34$, $p < .01$; non-well-formed: $F_1(1, 21) = 4.33$, $p < .05$, $F_2(1, 114) = 15.73$, $p < .01$, $minF'(1, 34) = 3.40$, $p = .074$).[2] This outcome is reflected in the error rate data also. Fewer errors were made relative to incorrect sub-expressions, on responses to both well-formed ($minF'(1, 105) = 45.15$, $p < .01$) and non-well-formed sub-expressions ($minF'(1, 59) = 5.60$, $p < .05$). While not unexpected, this pattern of outcomes is comforting for it indicates that both the experimental task and the participants are sensitive to the contents of algebraic equations.

More importantly, there is also a 205ms recognition advantage for sub-expressions that are well-formed components of their corresponding equation, over their non-well-formed counterparts ($minF'(1, 46) = 30.70$, $p < .01$). This recognition advantage holds for error rates also, with participants making significantly fewer errors on well-formed than non-well-formed sub-expressions ($minF'(1, 66) = 12.32$, $p < .01$). Clearly, the participants perceive the original equations in a way that allows faster and more accurate recognition of well-formed sub-expressions than non-well-formed sub-expressions.

The results of the experiment give support for the hypothesis stated in the introduction. That is, experienced mathematicians use an internal representation based on mathematical syntax to encode equations. This support comes from the logic of the experiment. Any encoding of an equation that significantly favours recognition of well-formed sub-expressions must rely on an underlying knowledge of mathematics; i.e., on the existence of internal representations of the properties of equations.

An additional source of evidence supporting our hypothesis comes from the fact that performance on non-well-formed sub-expressions provides a recognition advantage over incorrect sub-expressions. This suggests that there is some form

---

[2]The *minF'* is an extremely conservative statistic. It is considered to be significant if (a) it has an alpha level less than or equal to 0.05, or (b) it has an alpha level less than or equal to 0.1, and both subject (F_1) and item (F_2) analyses are significant at an alpha level of less than or equal to 0.05 (Santa, Miller and Shaw, 1979)

of internal mechanism that can rapidly reconstruct the equation from well-formed components, and thus identify sub-expressions that lie across component borders.

General Discussion

The natural question to ask now is what structural principle underlies the encoding of mathematical equations. In the introduction, we considered the possibility that humans might parse mathematical expressions in a manner similar to the way in which they parse natural language. In an experiment conducted by Johnson (1968) it was shown that it was easier for participants to learn sequences of words that conform to grammatical units (phrases), than to learn sequences of words which are equally probable and acceptable, but do not conform to grammatical units. This result is analogous to our own result for equations, in which syntactically well-formed sub-expressions are more readily recognised than non-well-formed sub-expressions, suggesting that experienced mathematicians might encode equations according to the rules of a grammar.

In natural language, the use of a phrase structure grammar allows us to construct *parse trees* for sentences. It is natural to ask then, whether the internal representation used by mathematicians might also utilise a *parse tree* when encoding an equation. For example, consider the following algebraic expression.

$$(x - 3y)^2$$

Figure 1 shows a parse tree of this expression based on mathematical syntax. The dashed boxes in the diagram would be equivalent to phrases in natural language, with the top node of tree (which contains the entire expression) being equivalent to a sentence. Just as natural language consists of noun phrases and verb phrases and so on, the dashed box containing $x - 3y$ might, for example, be considered a subtraction phrase.

While the results of the experiment discussed in this paper do not contradict the possibility of a hierarchical structure such as a parse tree, it will require much further research before such a conjecture can be confirmed. Our research direction for the immediate future therefore is aimed at both reinforcing the results of the experiment conducted, and investigating further the nature of the internal representation used by mathematicians to encode equations. There are two experiments in particular which we plan to conduct in order to meet these aims.

The first of these is essentially the experiment described in this paper, except that rather than using experienced mathematicians as participants, people with very little experience with mathematics will participate. Since the participants have a very weak background in working with mathematical syntax, we would expect no significant advantage in recognising well-formed sub-expressions over non-well-formed sub-expressions. This experiment would be aimed at providing further evidence to support the hypothesis that knowledge of

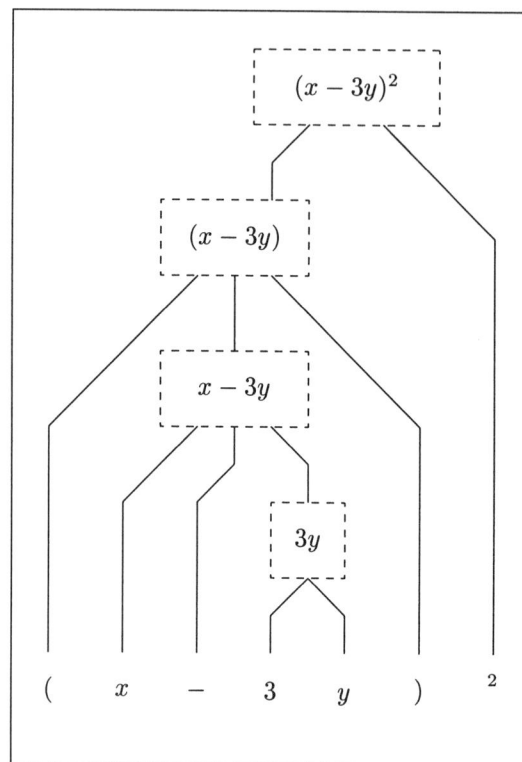

Figure 1: Parse Tree for $(x - 3y)^2$

mathematical syntax is used by experienced mathematicians when encoding equations.

The other experiment is designed to examine the nature of the internal representation used to encode mathematical expressions. The structure of the experiment is similar to the one described in this paper. However, the sub-expressions presented will be of slightly different types. While there will be an incorrect type to again act as fillers, the focus will be on the two types whose content is drawn from their corresponding equations. Consider the following algebraic expression.

$$\frac{8y - 3x^2}{(x - 2y)^3}$$

The sub-expression $x - 2y$ has a valid syntax, and it conveys the same meaning on its own as it conveyed in the equation. It is well-formed, and in a parse tree it would form a phrasal node. However, while the sub-expression $y - 3x$ also has a valid mathematical syntax, it conveys a different meaning on its own than it does in the equation. It would not form a phrasal node on a parse tree.

If the conjecture that mathematicians use a parse tree when encoding an equation is correct, then it would be expected that the first well-formed sub-expression would be recognised significantly faster than the sub-expression which does not form a phrasal node. The result of such an experiment would indicate if the examination of parse trees as an internal representation for encoding equations, is a worthwhile line of investigation.

Finally, it is hoped that this research will not be limited to just mathematical expressions. There are other visual languages with a two-dimensional structure and a well-defined syntax, such as finite state automatas, sheet music and chemical structural formulae. Similar questions need to be asked about these visual languages. A long term aim of this research is to examine the perception of visual languages in general.

References

Akmajian, A., Demers, R.A., & Harnish, R.M. (1984). *Linguistics: An Introduction to Language and Communication* (2nd ed.). Massachusetts: MIT Press.

Anderson, R.H. (1977). Two-dimensional mathematical notation. In K.S. Fu (Ed.), *Syntactic Pattern Recognition Applications*. New York: Springer-Verlag.

Clark, H.H. (1973). The language-as-fixed-effect fallacy: A critique of language statistics in psychological research. *Journal of Verbal Learning and Verbal Behavior, 12,* 335–359.

Glasgow, J., Narayanan, N.H., & Chandrasekaran, B. (Eds.). (1995). *Diagrammatic Reasoning: Cognitive and Computational Perspectives.* AAAI Press.

Johnson, N.F. (1968). The influence of grammatical units on learning. *Journal of Verbal Learning and Verbal Behavior, 7,* 236–240.

Johnson, N.F. (1970). Chunking and organization in the process of recall. In G.H. Bower (Ed.), *The Psychology of Learning and Motivation,* (Vol. 4). New York: Academic Press.

Marriott, K., Meyer, B., & Wittenburg, K. (1998). A survey of visual language specification and recognition. In K. Marriott & B. Meyer (Eds.), *Theory of Visual Languages.* New York: Springer-Verlag.

Miller, G.A. (1956). The magical number seven, plus or minus two: Some limits on our capacity for processing information. *Psychological Review, 63,* 81–97.

Santa, J.L., Miller, J.J., & Shaw, M.L. (1979). Using Quasi *F* to prevent inflation due to stimulus variation. *Psychological Bulletin, 86,* 37–46.

The lexical representation of verbs: The case of the verb "have"

Ian Jantz (ijantz@sfsu.edu)
Psychology Department, San Francisco State University
1600 Holloway Avenue, San Francisco, CA 94132 USA

John J. Kim (johnjkim@sfsu.edu)
Psychology Department, San Francisco State University
1600 Holloway Avenue, San Francisco, CA 94132 USA

Perry Grey (perryg@sfsu.edu)
Psychology Department, San Francisco State University
1600 Holloway Avenue, San Francisco, CA 94132 USA

Abstract

This paper has three goals: (i) to present a partial description of the intricate semantic selectional restrictions on the noun phrases in what we call here the Causal Have Construction (CHC), (ii) to show that four- and five-year old children are sensitive to these selectional restrictions without much exposure to CHCs, and (iii) to discuss some implications of these findings for theories of language and language acquisition. Our interest in this topic derives from the possibilities it opens up for a deeper understanding of the organization of the mental structures that give rise to these semantic selectional facts, an understanding which we believe implicates an intricate and nontrivial interaction between grammatical and conceptual knowledge.

The Causal "Have" Construction (CHC) and its Semantic Restrictions

The main verb "have" has many uses, each with what appears to be its own special idiosyncratic properties. Some uses of main verb "have" follow (from Ritter & Rosen, 1993):

(1) a. John had a good time.　　　　(experience)
　　　b. Harold has a dinner party.　　(cause/creation)
　　　c. John has a new cabinet.　　　(possession)
　　　d. The cabinet has a stereo in it.　(location)
　　　e. John had his students walk out on him.
　　　　　　　　　　　　　　　　　(experience)
　　　f. John had his student go to the principal's office
　　　　　　　　　　　　　　　　　(cause)

We focus here on the causal interpretation of sentences such as (1f), which we will call Causal Have Constructions (CHCs). CHCs include what appear to be two subjects -- the subject of "have", and the subject of the core event (e.g., "his student" in (1f)). The subject of "have" is interpreted as somehow causing the occurrence of the core event.[1] The two subjects in CHCs are subject to an intricate set of restrictions in the way they are interpreted. For instance, the two subjects tend to be restricted to animate noun phrases.[2]

(2) a. In last night's storm, the man had the children cover their heads.
　　　b. #In last night's storm, the lightning had the children cover their heads.
(3) a. The man had the boy break the vase.
　　　b. #The man had the ball break the vase.

The restriction against an inanimate subject of "have" in CHCs does not reduce to a general restriction against inanimate causes because (i) inanimate causes are easily understandable, and (ii) inanimate causes can appear as matrix subjects (underlined in the examples in (4)) in other periphrastic causative constructions:

(4) a. In last night's storm, the lightning made the children cover their heads.
　　　b. In last night's storm, the lightning caused the children to cover their heads.

Similarly, inanimate subjects are acceptable when the core event appears as a simple sentence (as in (5)), and when the core event appears as the core event in other types of periphrastic causative constructions (e.g., with "make" and "cause", as in (6)):

(5)　　The ball broke the vase.
(6) a. The man made the ball break the vase.
　　　b. The man caused the ball to break the vase.

Even so, the selectional restrictions on CHCs are even more complex. The animacy of the subject of "have" and the subject of the core event is not sufficient for the acceptability of CHCs, as the example in (7) shows:

---

[1] Because of this, CHCs are often described as a periphrastic causative construction (e.g., Givon, 1974; Goldsmith, 1984; Pustejovsky, 1995).

[2] We will call these semantic selectional restrictions, though the character and nature of these restrictions tend to differ from the normal use of the term.

(7)   John had Bill drop the ball (#by sneaking up and scaring him).

Again, whatever results in the unacceptability of (7) when it appears with the parenthesized material, it does not transfer to other periphrastic causatives, showing that it too is not a general restriction on sentences which express causation events:

(8) a. John made Bill drop the ball by sneaking up and scaring him.
    b. John caused Bill to drop the ball by sneaking up and scaring him.

Some notion of intention accompanies the sentence in (7), and the unacceptability of the same sentence with the parenthesized material present seems to be related to the lack of intention signaled by that parenthesized material. But it is unclear whether this lack of intention is a lack of intention on the part of the subject of "have" (i.e., John), or on the part of the subject of the core event (i.e., Bill). In fact, it seems to be the case that the intention of both the subject of "have" and the subject of the core event are necessary for the acceptability of the sentence in (7). That is, if we construct contexts which eliminate the intention of only one of the two subjects, the sentence becomes unacceptable:

(9) a. Intention of both subjects: John and Bill are involved in a plot to rob a bank. They develop an intricate system of signals to communicate with one another covertly once in the bank. When John scratches his nose, this is a signal for Bill to distract people by dropping a vase. When the time was right, John gave the signal, and Bill dropped the vase.
    a". John had Bill drop the vase.

    b. Eliminating the intention of the subject of "have": John and Bill are involved in a plot to rob a bank. They develop an intricate system of signals to communicate with one another covertly once in the bank. When John scratches his nose, this is a signal for Bill to distract people by dropping a vase. In the middle of the robbery, John's nose started to itch and he scratched it, forgetting that this was the signal to Bill to drop the vase, and Bill dropped the vase.
    b". #John had Bill drop the vase.

    c. Eliminating the intention of the subject of the core event: John and Bill are involved in a plot to rob a bank. Unbeknownst to Bill, John plans to distract people in the bank by the dropping of a vase which Bill was carrying. When the time was right, John pushed Bill, and Bill dropped the vase.
    c". #John had Bill drop the vase.

The interpretation which emerges, at least with respect to sentences such as that in (7), is that both the subject of "have" and the subject of the core event must be intentional

agents. In fact, the normal interpretation of CHCs involves "a notion of co-agency [between the subject of 'have' and the subject of the core event], brought about by agreement or contractual obligations" (Pustejovsky, 1995; see also Givon, 1974; Goldsmith, 1984).[3]

Experiment One: Adult Knowledge of Constraints on CHCs

We designed an experiment to test whether naive native speakers of English know these intricate selectional restrictions on the subjects of "have" and the core event. In particular, the experiment focused on whether native speakers of English share the intuitions we and our informants have that the subject of the core event must be intentionally involved in the action he/she performs. For purposes of comparison, we collected information about people's interpretations of sentences where the verb "make" replaced "have" as the causative verb.

Method

Participants.   Participants were twelve undergraduate students at San Francisco State University. They were all native speakers of English and participated in the study for course credit.

Design.   The experiment was a single factor, between-subjects design. The independent variable was Question Type. Six participants were asked only "have" questions; six participants were asked only "make" questions.

Materials and Procedure.   Each participant was read four two-part scenarios. Toy figures, representing the characters in the narratives, acted out the story. One of these two-part scenarios follows:

PART A -- Unintentional subject of core event: Sue is in the block area building a large structure with blocks. Her block structure is really nice. She tells Doris, a friend who is close by, that she really likes the building that she is making. Meg comes over. She thinks that it would be fun to watch the large building that Sue is making get knocked over. Sue can't see Meg and she doesn't know that Meg is there. Meg doesn't want to ask Sue to knock over the blocks, so she thinks of another way that she can get Sue to knock the blocks over. Meg goes up to the door to the outside. Meg grabs the door handle and slams the door shut. It makes a loud booming noise. Sue is scared and starts to run away. She runs by the block structure and accidentally gives it a push, knocking it over. The blocks fall all over the ground with a loud crash.

---
[3]There is much more to say about CHCs at the level at which we have been describing them, e.g., CHCs discussed by Ritter and Rosen (1993) which have quasi-arguments/expletive noun phrases as the subject of the core event, etc. For present purposes, the facts outlined here are sufficient.

PART B -- Intentional subject of core event: Later that day, Sue is back in the block area building a large structure with blocks. She has built a structure that is really tall. She tells Doris, a friend who is close by, that she really likes the building that she is making. Cindy comes over. Cindy thinks that it would be fun to watch the large building that Sue is making get knocked over again. Cindy asks Sue nicely if Sue would go up to the building and knock the blocks down. When Sue hears Cindy's suggestion, she smiles and tells Cindy that it would be a great idea to knock over the blocks and watch them fall. Sue goes over to the structure and gives it a big push, knocking it over. The blocks fall all over the ground with a loud crash.

The three figures -- Doris (the friend who is an innocent by-stander), Meg (who scared Sue), and Cindy (who asked Sue to knock over the blocks) in the narrative presented above -- were then placed in random positions in front of the participant. After the figures were placed on the table, the participant was asked one of the following two questions (depending on whether the participant was in the "have" question condition or the "make" question condition):

Have Condition:
Which one had Sue knock over the blocks?

Make Condition:
Which one made Sue knock over the blocks?

An identical procedure was used for the three additional narratives that the each participant heard during a test session. For each participant, the two scenarios within an item were presented in random order (and adjusted for time-relevant linguistic elements, e.g., "later that day....").

Results and Discussion

The dependent measure was the proportion of times each participant chose the cooperative instigator (i.e., who asked the actor to act intentionally) in response to the question. If the participants knew that the subject of the core event in CHCs must be intentionally involved in the event, then they should have picked the cooperative instigator, (i.e., Cindy, the one who asked Sue to knock over the blocks). The subjects that heard the "make" question were not expected to confine their responses to this choice.

The mean proportion of times participants chose the cooperative instigator for each of the two Question Type conditions is displayed in the first column of Table 1. A one-way ANOVA revealed that the difference between the means for the "have" and "make" question conditions was significant ($F(1,10) = 8.033$, $p < .05$).

Table 1: Mean Proportion of Responses Adults Made for Each Character in the Narrative

Condition	cooperative	uncooperative	bystander
Have	.958	.042	.000
Make	.375	.625	.000

Thus, when faced with the question "Who had John drop the ball?", native speakers of English were considerably more likely to choose a character who asked John to perform the action intentionally. They almost never chose a character who scared the actor (John) into performing the action. However, in response to the "make - question" participants did not show this bias. This evidence shows that adult, native speakers of English are sensitive to at least one of the restrictions on the relationship between the subject of "have" and the subject of the core event in CHCs.

Transcript Analysis: Children's Experience with CHCs

The second experiment in this study tested children's knowledge of the same CHC constraint. However, before we discuss the second study, a description of children's exposure to the CHC is useful in providing an idea of what children's direct experience with the CHC is.

We analyzed the language transcripts for the three children -- Adam (ages: 2;3.4-4;10.23), Eve (ages: 1;6-2;3), and Sarah (ages: 2;3.5-5;1.6) -- in the Brown (1973) corpus available in the CHILDES database (MacWhinney & Snow, 1987). From these transcripts we culled out the adult utterances which included "have". We then categorized the utterances based on the following categories of the use of "have": Alienable Possession, Inalienable Possession, Auxiliary, Modal, Location, Nominal (both stative and eventive), the experiencer-have-construction (EHC) (see sentence (1f) and below), and the causal-have-construction (CHC). CHCs and EHCs are the least frequent constructions for each of the three children (proportion of uses of "have" in parentheses):

Table 2: Number of Instances of CHCs and EHCs in Children's Linguistic Input

	CHC	EHC	Total "have"
Adam	4 (.003)	3 (.002)	1148
Eve	2 (.002)	2 (.002)	938
Sarah	11 (.006)	11 (.006)	1175

The relative paucity of CHCs in children's linguistic input is one reason to suspect that the acquisition of this construction is at least in part independent of children's experience with CHCs. However, even if we allow that the minimal experience children gain with the CHC is adequate for them to develop mastery of it, we suggest that EHCs -- which are equally frequent in children's linguistic experience according to these analysis -- potentially confuse the learning situation for the child because CHCs (10a) and EHCs (10b) share the same basic surface form:

(10) a. The teacher had his students walk out (by telling them to.)
b. The teacher had his students walk out (even though he told them not to.)

Though these two constructions have the same surface forms, they do not share the same sorts of interpretive constraint. For instance, it is possible for the subject of the core event to be an inanimate object in EHC's (as in (11)), or that both the subject of the core event and the subject of "have" are unintentionally involved in the event (as in (12)).

(11)   Fred had the hammer fall on his toes.
(12)   Timmy had the teacher step on his toes.

For the child to master the CHC, he/she must determine based on identical surface forms when the construction is a CHC and when the construction is an EHC, and not confuse the properties of what is observed for one construction as being relevant to the other construction. In our transcript searches, we discovered that this is more than a hypothetical problem. In the following EHC, from Eve's mother to Eve, there is no shared intention implied:

(13)   see how frustrating it is # Eve # to have people stomping their feet when you're trying to cook ?

Similarly, in the following EHC, from Sarah's mother to Sarah, the subject of the core event -- i.e., a commercial -- is not animate:

(14)   she can be sound asleep # (a)n(d) have a commercial come on # wake up # look at the commercial (a)n(d) when the commercial's over # right back sound asleep again .

These transcript analyses suggest that children's exposure to CHCs is rather limited, and that learning the semantic restrictions on CHCs by observing the contexts in which they appear is complicated by other factors as well. If children at these ages are sensitive to the semantic restrictions on CHCs that native English-speaking adults showed knowledge of in Experiment One, it is unlikely that they would be able to learn them based on observing the few situations in which CHCs are used in their presence.

Experiment Two: Children's Knowledge of Constraints on CHCs

Experiment Two tested whether children were sensitive to the CHC constraint that adults demonstrated knowledge of in Experiment One.

Method

Subjects. Participants were 20 four and five year old children (mean age of 4;4) at the Child Study Center at San Francisco State University. All spoke English fluently.

Materials and Procedure. Except for the addition of a third question group and a practice phase, the procedure for this experiment was identical to the one used with adult participants. The third condition was a "see - question":

See Condition:
Which one saw Sue knock over the blocks?

Because all three characters in a given narrative sees the actor, we included children's pattern of responses to the "see"

question to obtain a baseline measure of children's preferences for the three characters in the narratives. A practice phase was also added in order to test children's knowledge of other uses of "have" (namely, the locational and the modal uses), and to familiarize children with the experimental procedure. If children failed the practice test, their data were excluded from the analysis; two of the twenty children failed to answer both practice questions correctly. The remaining eighteen participants were evenly distributed across the three Question Type conditions.

Results and Discussion

The dependent measure was the proportion of times each participant chose the cooperative instigator in response to the question. If the participants knew that the subject of the core event in CHCs must be intentionally involved in the event, then they should have picked the cooperative instigator in response to the "have" question. The participants that heard the "make" or "see" questions were not expected to confine their responses to this choice.

The mean proportion of times participants chose the cooperative instigator for each of the three Question Type conditions is displayed in the first column of Table 3. A one-way ANOVA revealed that the difference between the means for the "have", "make", and "see" question conditions was significant ($F(2,15) = 4.687$, $p < .05$).

Table 3: Mean Proportion of Responses Children Made for Each Character in the Narrative

Condition	cooperative	uncooperative	bystander
Have	.763	.180	.055
Make	.333	.625	.042
See[4]	.292	.318	.360

A Tukey-B post hoc analysis showed that the group means differed significantly ($p < .05$) in pairwise comparisons of the mean proportion of cooperative instigator responses between the "have" question group and the "make" question group, and between the "have" question group and the "see" question group. (The pairwise difference between the "make" and "see" question groups was not significant.)

Children in this experiment show the same general pattern of responses as adults did in Experiment One. Thus, these results indicate that children at these ages are sensitive to the constraint on CHCs that the subject of the core event must be intentionally involved in the event. Together with the transcript analysis from the previous section (which shows that children do not receive much, if any, experience with CHCs), these findings cast serious doubt on any account of the acquisition of these constraints which depend on children's exposure to them.

---

[4]The proportions for the See condition do not add up to 1.000 because one participant in that condition did not always respond with one of the three characters we provided.

General Discussion

The results of the experiments and transcript analyses described above place very tight constraints on theories of how the complex set of semantic restrictions on CHCs is acquired. For instance, the entire class of theories of language learning that relies solely on statistical properties of children's linguistic experience as the basis for language acquisition would seem to be unsuited for accounting for the acquisition of such knowledge. Of course, this is not to say that all theories of language learning which depend on at least some exposure to the constructions being learned fail to account for these particular acquisition facts. But these findings do clarify the exact nature of the learning problem facing children acquiring CHCs, and a nontrivial gap between experience and knowledge that such theories must explicitly account for.

Another class of theories which can be (tentatively) eliminated based on these findings are those which suppose that the many uses of "have" have associated with them distinct lexical entries in which learned idiosyncratic information may be stored. If there were many verbs "have", each of which has its own learned idiosyncratic properties, children's little experience with the "have" in CHCs would not be sufficient for accounting for their knowledge of the constraints on CHCs.

This line of argument brings us close to the central assumption in Ritter and Rosen (1993, 1997). Ritter and Rosen show that (much of) the range of interpretations of the subject of "have" across the range of syntactic constructions within which "have" appears may be derived from: (i) the syntactic form and interpretation of the complement of "have", and (ii) principles of Event Structure (e.g., Grimshaw, 1990; Tenny, 1992). Because of the minimal role that the verb "have" plays in assigning an interpretation to its subject, Ritter and Rosen assume that "have" does not assign an interpretation to its subject at all, and that there is a single, semantically unspecified lexical entry for main verb "have".

Ritter and Rosen's theory is broadly consistent with the facts described in this paper in the sense that their theory does not list in the lexical entry for "have" the idiosyncratic interpretations assigned to the subject of "have" for each construction type in which "have" appears (see (1)). Thus, Ritter and Rosen's syntactic analysis lends itself to a theory of the acquisition of "have" in which the single, semantically unspecified lexical entry for "have" may be learned based on children's experience with relatively frequent uses of "have", and in which the idiosyncratic semantic properties of constructions with "have" may be independently derived from the structure and interpretation of the complement of "have" and the principles of Event Structure. Though we see promise in such an approach, we remain agnostic here about Ritter and Rosen's theory because it does not explain the semantic restrictions on the CHC that we have discussed in this paper.

Finally, we would like to note that these facts cut at the heart of claims that CHCs are peripheral to linguistic theory. It is true that CHCs do not occur in all languages and that they are used only infrequently in English (den Dikken, 1997; Ritter & Rosen, 1997), but it is unlikely that CHCs are a class of frozen expressions or slang or idioms specific to English that acquire their meaning by conventional stipulation among language users in a speech community.[5]

But in light of our experimental findings, the relative rareness of CHCs among the world's languages seems not to be cause for dismissal, but cause for serious investigation -- something must account for the systematic knowledge of the subtle semantic restrictions on CHCs that children have acquired in the absence of relevant experience.

In the end, we see these facts as central to a certain sort of interdisciplinary endeavor. On the one hand, it seems that an explanation of these facts will involve linguistic analysis at its core. There is certainly some deep relevance to the fact that CHCs involve only one tense specification (and one event role, if Ritter and Rosen (1993) are correct) -- the verb "have" carries the tense information in the sentence; the verb in the core event is in its infinitival form. There is much to be gained from linguistic analyses regarding the individuation of lexical entries and the structure of the lexicon which may be relevant for the proper analysis of the facts described here. Much of the literature on causation and the semantic composition of predicates is linguistic literature.

On the other hand, it seems unlikely that a solely linguistic analysis will suffice for explaining these facts. In particular, answers to questions like: "Why must the subject of 'have' be an intentional agent in CHCs?" and "Why must the kind of causation describable with CHCs involve the cooperative agency between the subject of 'have' and the subject of the core event?" seem to be questions that will receive (at least part of their) explanations not from linguistic theory, but from a theory of the mental structures responsible for our understanding and conception of agency, causation, intention, and social interaction. Perhaps the semantic restrictions on CHCs are derivable from and will lead to insights about the properties of the interface between the linguistic level of Event Structure that Ritter and Rosen discuss and our capacity for understanding physical causation versus intentional interaction. We leave the explanation of these facts for future research, but hope that our empirical findings may direct the attention of cognitive scientists to develop integrated theories of language and cognition.

Acknowledgments

We would like to thank Sandeep Prasada, Jeffrey Bettger, and Nancy Lee for helpful comments and discussion. We also thank the parents, staff, and most of all the children of the Child Study Center at San Francisco State University.

---

[5]In fact, some of our British informants commented that CHCs are not found spontaneously in British English and that they sound extremely awkward -- "the way Americans talk," as one of our informants put it. But, as with the participants in the experiments we reported in this paper, these informants' intuitions concurred with ours when they were pressed to make judgments regarding the semantic restrictions on CHCs.

References

Brown, R. (1973). A first language. Cambridge, MA: Harvard University Press.

den Dikken, M. (1997). Introduction: The syntax of possession and the verb 'have'. Lingua, 101, 129-150.

Givon, T. (1974). Cause and control: On the semantics of interpersonal manipulation. In J. Kimball (Ed.), Syntax and Semantics 4. San Diego, CA: Academic Press.

Goldsmith, J. (1984). Causative verbs in English. In D. Testen, V. Mishra, & J. Drogo (Eds.), Papers from the Parasession on Lexical Semantics from the Twentieth Regional Meeting, Chicago Linguistics Society. Chicago, IL: Chicago Linguistics Society, University of Chicago.

Grimshaw, J. (1990). Argument structure. Cambridge, MA: MIT Press.

MacWhinney, B., & Snow, C. (1985). The child language data exchange system. Journal of Child Language, 12, 271-296.

Pustejovsky, J. (1995). The generative lexicon. Cambridge, MA: MIT Press.

Ritter, E., & Rosen, S.T. (1993). Deriving causation. Natural Language and Linguistic Theory, 11, 519-555.

Ritter, E., & Rosen, S.T. (1997). The function of 'have'. Lingua, 101, 295-321.

Tenny, C. (1992). The aspectual interface hypothesis. In I. A. Sag & A. Szabolcsi (Eds.), Lexical matters. Stanford, CA: CSLI Publications. Distributed by Chicago, IL: University of Chicago Press.

Rules and Associations

F.W. Jones (FWJ1000@CUS.CAM.AC.UK)
I.P.L. McLaren (IPLM2@CUS.CAM.AC.UK)
University of Cambridge, Psychological Laboratory,
Downing Street, Cambridge CB2 3EB. U.K.

Abstract

Two-process theories of human cognition, that state that learning can occur by both associative and rule-based processes, are currently popular. We report two experiments which support such a view. Both employed a set of six stimuli which varied along a luminance dimension, and followed the same general design. That is, participants were trained to discriminate between the two stimuli in the middle of this set, before being tested on the whole set. In Experiment I, the length of training was varied. Following short training, participants' performance on test exhibited a peak-shift, and therefore may be explained in associative terms. After longer training, however, their behavior was consistent with rule-based learning. In Experiment II, the contingency during the training phase was varied. Participants in the 'Full Contingency' group performed in a manner consistent with rule-learning, while the 'Reduced Contingency' condition produced a peak-shift. These results are discussed in terms of McLaren, Green & Mackintosh's (1994) version of the associative/rule-based distinction.

Introduction

The idea that human cognition comprises both associative and rule-based processes has a long history, that stretches at least as far back as William James. Moreover, its current popularity is illustrated by the volume of literature devoted to the subject (e.g. the entire issue of Cognition 65). One recent incarnation of this 'hybrid' view can be found in McLaren, Green & Mackintosh (1994). Their two process model of human learning comprises: (i) an associative system that is sensitive to the statistical structure present in the surface features of the environment, and operates through the establishment and alteration of connections between representations; and (ii) a 'cognitive' process capable of rule abstraction, whose behavior resembles that of a symbolic logic machine.

There already exists a considerable body of evidence in support of such a dichotomy (see Shanks & St John, 1994; Shanks, 1995; and Sloman, 1996 for reviews). Briefly, in some categorization experiments participants' performance on novel transfer items is dependent on their similarity to the training exemplars (e.g. Perruchet, 1994), which suggests that the knowledge acquired is encoded in terms of surface features. Alternatively, under different conditions, the results from transfer tests and verbal reports are more consistent with participants having abstracted rules (e.g. Regehr & Brooks, 1993). Furthermore, this dissociation can

be observed within the same experiment (e.g. Nosofsky, Clark & Shin, 1989), with some participants responding on the basis of similarity and others abstracting rules.

The aim of the research presented here was to further investigate the viability of this associative/rule-based distinction, and examine some of the conditions under which each process could dominate performance. Like the previous work described above, we differentiated between rule-based and associative learning by examining how participants' performance generalized to novel stimuli. To make this more concrete, consider a set of stimuli that vary along a hypothetical dimension, with a midpoint 'd'. Further, suppose that we create a category structure, such that all those stimuli to the left of d form one group, while those to the right form another; and train participants on one example from each category -namely R_{TRAIN} and L_{TRAIN}. An examination of participants' performance, when they are subsequently required to categorise stimuli spanning the entire dimension, should then allow these two forms of learning to be distinguished.

If their learning is rule-based, then we might expect them to abstract the rule: 'greater than d respond category one, less than d respond category two'. As a result, unless their performance is at asymptote, their accuracy may well be dependent on the stimulus' distance from the category boundary, with more extreme stimuli being classified more accurately. Alternatively, if their learning is just encoded in terms of surface-features, we might expect performance on the transfer stimuli to be dominated by their similarity to the training examples, and to exhibit a pattern known as 'peak-shift'. That is, as we pass along the dimension from the category boundary, we might expect accuracy to increase to a maximum and then drop off, with the peak being positioned further along the dimension than the training stimuli -hence the term peak-shift. Such a pattern of results was first demonstrated by Hanson (1959), in pigeons trained on a wavelength discrimination, and may be explained in associative terms.[1]

Consider the simple associative network illustrated in Figure 1B. The input layer comprises a bank of feature detectors, the pattern of activation across which represents the current stimulus being presented to the network. The activation of each of these detectors is a Gaussian function of stimulus' position on the dimension, with each unit

---

[1] The associative explanation offered here is an extension of the work of Spence (1932), Blough (1975) and Wills & Mackintosh (1998), respectively.

responding maximally to a different point on the dimension (see Figure 1A). The output layer comprises two units, corresponding to the two categories, and the two layers are fully interconnected. During training, on each trial either R_{TRAIN} or L_{TRAIN} is presented as input to the network, and the activation target for the appropriate category unit is set to one, while that for the other unit is zero. The weights are then updated using the delta-rule (e.g. McClelland & Rumelhart, 1985).[2]

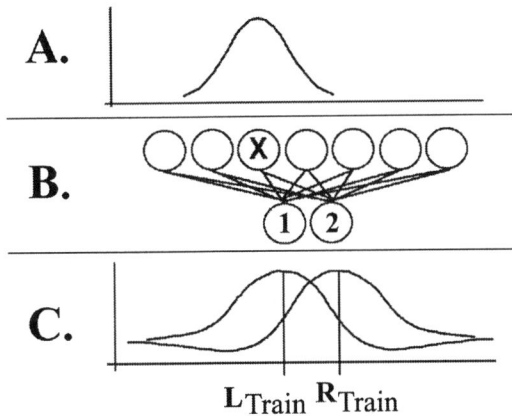

Figure 1: (A): The activation function for feature unit X. (B): A delta-rule network, with feature units on top and category units below. (C): Category units' activations on test.

On test, the network's output activations, in response to stimuli from different points on the dimension, can be measured. The results of just such a simulation are illustrated in Figure 1C. It can be seen that, if participants' responses are dependent on the difference in activation between the two category units, then they will be most accurate at positions outside the training examples, because that is where the difference between the curves is the greatest.[3] Thus, peak-shift can be understood in associative terms.

The differing predictions of associative and simple rule-based accounts are illustrated in Figure 2. Only half the dimension is illustrated, because the pattern should be

---

[2] This may be formally expressed: $dw_{CF} = S\, a_F\, (t_c - a_c)$ where 'dw_{CF}' is the change in the weight connecting feature unit 'F' to category unit 'C', 'a_c' is the activation of category unit c (which is equal to the weighted sum of the activation from the feature units), 'a_F' is the activation of feature unit 'F', 't_c' is the activation target for category unit 'C', and S is a constant that determines the rate of learning.

[3] In fact, a peak-shift may also be obtained if probability of classifying a stimulus into a particular category is dependent upon the ratio of that categories' activation to the total output activation, provided that noise is added to the system. Alternatively, if the output activations are first transformed using an exponential function, then a ratio rule will again produce the desired pattern. In addition, a 'winner-take-all' decision network (e.g. see Jones, Wills & McLaren, 1998) acting on these activations can also produce a peak-shift.

symmetrical either side of the category boundary. As shown, a monotonically increasing trend indicates rule-based learning, and a peak-shift is diagnostic of associatively based performance. Thus, we have a way of distinguishing between rule-based and associative learning.

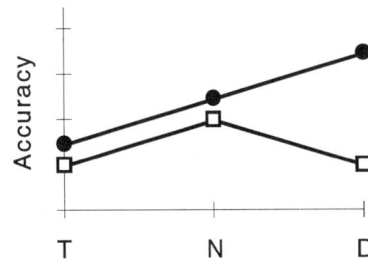

Figure 2: The predictions of associative (open squares) and simple rule-based (filled circles) accounts. T refers to training stimuli, while N and D denote test stimuli that are further out on the dimension.

Wills & Mackintosh (1998, Experiments 3a & 3b) have already performed experiments, on people, following the general design outlined above. For stimuli, they employed green rectangles that varied along a luminance dimension. In total, six different shades of green were used, divided equally between two categories -'dark' (D) and 'light' (L). When arranged in order of increasing brightness these stimuli can be referred to as $D_{DISTANT}$, D_{NEAR}, D_{TRAIN}, L_{TRAIN}, L_{NEAR} and $L_{DISTANT}$, respectively (see Figure 3). During training, on each trial participants were either presented with D_{TRAIN} or L_{TRAIN}, which they had to classify using one of two keys, before receiving feedback. Having learned this discrimination, they were transferred to a test phase. On each trial of this, any one of the six stimuli could appear and participants were again required to categorize them, using the same two keys. No feedback was given.

Figure 3: From left to right: $D_{DISTANT}$, D_{NEAR}, D_{TRAIN}, L_{TRAIN}, L_{NEAR} and $L_{DISTANT}$, respectively.

Wills & Mackintosh found that, during the test phase, participants responded more accurately to the 'Distant' stimuli than to the 'Near' ones, and were more accurate on the 'Near' stimuli than on those presented in training. In other words, moving along the dimension away from the category boundary, participants' performance improved monotonically. This, in conjunction with the fact that all the participants were able to verbalise a variant of the 'bright respond key one, dark respond key two' rule, suggests that their learning was rule-based. Wills & Mackintosh's procedure provides the basis for the two experiments reported here.

Experiment I

The aim of the first experiment was to establish whether varying the amount of training participants received would determine whether associative or rule-based processes dominated performance on test. According to McLaren, Green & Mackintosh (1994), it might be expected that a short training period would produce an associative pattern of results, with rule-based performance only emerging after greater experience of the contingencies. They argue that, initially, the stimuli-category associations will be too weak to support rule abstraction; but that, as training progresses, these traces will become sufficiently strong to enable the 'cognitive' process to use them as the basis for the development of rules.

In order to test this hypothesis, we employed a similar method to Wills & Mackintosh. There was, however, at least one important difference. That is, during both the training and test phases, the green stimuli were only presented on even numbered trials. On odd numbered trials, participants were required to perform a filler task. Specifically, they had to classify stimuli, that comprised a set of colored icons (see Figure 4), using the same two keys. Aside from the stimuli being different, the filler task was identical to that involving the greens. The inclusion of this additional task served to increase the difficulty of the initial discrimination, with the purpose of increasing the likelihood of obtaining an associative pattern of learning after short training. Without the filler task the discrimination would have been far easier, because the next green stimulus would have appeared immediately after the previous one had disappeared, allowing participants to compare them more directly.

Figure 4: An example of an icon stimulus.

Given space constraints, and the fact that the results from the filler task do not bear directly on the question under investigation, details concerning the construction of the filler stimuli, and participants performance on them, will not be reported here. However, aside from making the discrimination harder to learn, we do not believe that the filler task affected the results for the greens, especially since its stimuli were identically generated for both conditions. Suffice to say that, their construction was identical to that in Wills & Mackintosh's Experiments 2a & 2b, and that the results for the filler task do not substantially differ from their findings.

The experiment comprised two conditions, namely, 'Short' and 'Long' training, which differed only in the number of trials of discrimination training the participants received. As already discussed, on the basis of McLaren, Green & Mackintosh we might expect associative learning to occur in the Short Training condition, with rule-based learning being manifest by Long Training participants. If this were the case, then we would expect the results for the

former group to exhibit a peak-shift, compared to a monotonically increasing trend for the latter. Therefore, if these predictions are born out, performance on the Near and Distant test stimuli should differ between the two conditions, with Near being responded to more accurately than Distant after short training and vice versa for long training.

Method

Participants and Apparatus The participants were 58 Cambridge University undergraduates, whose ages ranged between 18 and 35. They were randomly divided equally between the two conditions, and did not receive payment for their help. The experiment was run on a RISC PC 700 computer, situated in a quiet room. Illumination was provided by a small desk lamp. The low light level was employed because pilot work suggested that some participants would be unable to discriminate between some of the shades of green under normal illumination.

Stimuli Both types of stimuli occupied a rectangle measuring 3.6 cm wide by 2.8 high, that was surrounded by a thin grey border. For the greens this rectangle was entirely filled in green. The luminance of these stimuli was determined by the value of a computer parameter, that ranges between 0 and 255. This was set to 50, 108, 137, 166, 195 and 253, for $D_{DISTANT}$, D_{NEAR}, D_{TRAIN}, L_{TRAIN}, L_{NEAR} and $L_{DISTANT}$, respectively. The filler stimuli comprised 12 icons arranged in a 4 by 3 grid, within the rectangle. See Figure 4 for an example, and Wills & Mackintosh (1998), Experiments 2a & 2b, for further details concerning their construction.

Design In the Short Training condition, training lasted for 48 trials, compared to 96 for the Long Training group. The order of stimulus presentation followed a pseudo-random sequence, such that L_{TRAIN} and D_{TRAIN} appeared 3 times during every set of 12 trials and only on even numbered trials. The test phase was identical for both conditions and comprised 120 trials. On test, the greens appeared in a random order, within the constraints that during every batch of 24 trials each of the six stimuli had to appear twice and only on even numbered trials. While the orders of presentation were designed in batches, there was no actual batching of the stimuli. In both training and test, filler stimuli appeared on odd numbered trials. The key assignments were counterbalanced, such that, for a random half of the participants in each condition, the 'x' key equalled 'light' and the '.' key equalled 'dark'. The remaining participants had the mapping reversed.

Procedure Participants sat approximately 1m away from the computer monitor, which was positioned roughly at eye-level. Some general instructions and ones specific to training phase were explained by the experimenter, before he left the room. Participants were informed that they would be performing two unrelated categorization tasks and that the computer would switch between the two on alternate trials. They were told to use the feedback to help them learn which key went with which stimulus. Participants initiated training by pressing the Spacebar. On each trial, the appropriate stimulus appeared in the centre of the screen, and, after 3 sec, the words 'Please respond now' appeared below it. Participants then had to respond using either the 'x' or '.' key, as quickly as possible, whilst avoiding errors. If they pressed the wrong one of these two keys, then the computer beeped. In addition, if they pressed a key other than 'x' or '.', or had not responded within 5 secs of the prompt's onset, then the stimulus was replaced by error messages. Respectively, these read: 'You have pressed an invalid key' and 'You did not respond in time'. Key presses prior to the appearance of the prompt were ignored. The next trial followed immediately after the response.

At the end of training, the computer displayed a message requesting the participants to find the experimenter, who then explained the instructions for the test phase, before again leaving the room. Participants were told to use whatever they had learned in training to classify the new stimuli they would see. The test phase commenced at their initiation. Test trials were identical to those in training, except there was now no prompt. Participants were informed that they could respond as soon as the stimulus appeared, and were again asked to be as fast as possible, whilst avoiding errors. No feedback, concerning the accuracy of their responses, was given. Since there was no prompt, the time-out occurred 5 sec after the beginning of the trial.

Following testing, participants again had to fetch the experimenter, who administered a structured questionnaire. This comprised a series of increasingly specific questions, designed to determine what strategies the participants had employed during the task and whether they could verbalise the underlying rule.

Results and Discussion

During training, unsurprisingly, participants in the Long Training condition responded significantly more accurately than those in the Short Training group (means 71.7 and 55.5 % correct, respectively; $t(56)=2.96$, $p<0.05$). The results for the test phase are shown in Figure 5. It can be seen that the accuracy of the Long Training group follows a monotonically increasing trend, while the means for the Short Training condition exhibit a peak shift. As already discussed, if our predictions are correct, then it is performance on the Near and Distant stimuli that should distinguish between the groups. Therefore, in order to assess whether the group differences were significant, the mean accuracy on Distant stimuli was subtracted from the value for Near stimuli, for each participant, and a planned contrast performed on the resulting scores. This demonstrated that the two conditions did, indeed, significantly differ ($F(1,56)=5.03$, $p<0.05$).

Given this, and in order to allow a more detailed examination, the data from the two groups was then analysed separately. Planned contrasts were used to compare performance on Training and Near stimuli, and performance on Near and Distant stimuli. Since the hypothesis we were testing made clear the directions of the expected effects, these contrasts were one-tailed. For the Short Training group, the contrasts revealed that Near stimuli were responded to

significantly more accurately than to both Training and Distant ones ($F(1,28)=8.05$ and 3.29 respectively, one-tailed $p<0.05$ for both). Therefore, the peak-shift was reliable. With regards to the Long Training condition, accuracy on Near stimuli was significant higher than that on Training ones ($F(1,28)=28.37$, one-tailed $p<0.01$). However, while participants responded more accurately to Distant stimuli than to Near ones, this difference was only marginally significant ($F(1,28)=1.75$, one-tailed $p<0.10$). Nevertheless, the results for this condition are consistent with application of a rule, and clearly do not exhibit a peak-shift. Moreover, when questioned, all the Long Training participants were able to verbalise a variant of the underlying rule, as compared to none in the Short Training group.

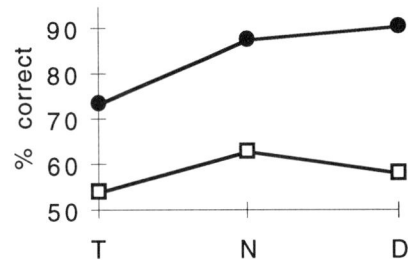

Figure 5: Participants' mean accuracy on test, for Experiment I. Open squares=Short Training, Filled circles= Long Training. T=Training, N=Near, D=Distant.

In summary, as predicted, the short training participants produced a pattern of performance, on test, that is explicable in associative terms. Following more lengthy exposure to the contingencies, participants were able to abstract the underlying rule, and their behavior was consistent with its application.

Experiment II

McLaren, Green & Mackintosh (1994) argue that as we increase the complexity of the contingencies between stimuli, so also we increase the amount of information that needs to be stored in working memory in order for a rule to be abstracted. Thus, as the difficulty of the to-be-learned mappings is increased, so the likelihood that performance will be associatively-based, rather than rule-based, increases.

The second experiment sought to test this prediction, using the same rationale as the first. It comprised two groups, namely 'Full Contingency' and 'Reduced Contingency'. The former was identical to the Long Training condition in the first experiment, while the latter differed only in that, in the training phase, the contingency between the two green stimuli and their respective categories was reduced. The was achieved by reversing the keys assigned to L_{TRAIN} and D_{TRAIN} on 25% of the training trials. Participants were not told about this manipulation.

If McLaren et al.'s predictions proved accurate, we would expect the Full Contingency condition to exhibit rule-based performance, and the Reduced Contingency group to produce a peak-shift. Finally, given that the Full Contingency group was a straight replication of the Long Training condition, and that in the previous experiment this did not produce a completely reliable monotonically increasing trend, a larger number of participants were run in this group. It was hoped that the resulting extra power would lead to a significant monotonically increasing trend.

Method

Participants, Apparatus, Stimuli and Procedure
80 new participants were drawn from the same pool as in Experiment I, split 60:20 between the Full and Reduced Contingency conditions, respectively. The stimuli, apparatus and procedure were identical.

Design The Full Contingency condition was identical to the Long Training condition in Experiment I. The Reduced Contingency condition differed from this only in that during training, on a randomly selected quarter of the L_{TRAIN} trials and arbitrary quarter of the D_{TRAIN} trials, the key assignments were reversed.

Results and Discussion

The data from Experiment II was analysed in exactly same way as that from the first experiment. A comparison of the training scores revealed that participants in the Full Contingency condition responded significantly more accurately than those in the Reduced Contingency group (means 67.4 and 54.9 % correct, respectively; t(78)=2.87, p<0.01). Figure 6 shows the mean accuracy results from the test phase. From this it is clear that, the trend in the Reduced Contingency group follows a peak-shift, while that produced by the Full Contingency condition monotonically increases. As previously, the reliability of these group differences was assessed by performing a planned contrast on the differences between Near and Distant stimuli. This demonstrated that the conditions were significantly different (F(1,78)=5.28, p<0.05).

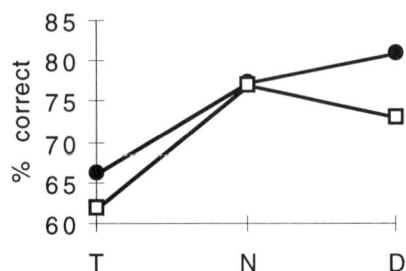

Figure 6: Participants' mean accuracy on test, for Experiment II. Open squares=Reduced Contingency, Filled circles= Full Contingency. T=Training, N=Near, D=Distant.

For each group, planned contrasts were used to compare performance on Training and Near stimuli, and performance on Near and Distant stimuli. Again, since the direction of the predicted effects was pre-specified, these contrasts were one-tailed. For the Full Contingency condition, these revealed that Distant stimuli were responded to significantly

more accurately than to the Near ones, which in turn were responded to more accurately than the Training stimuli (F(1,59)=5.51 and 25.85 respectively, one-tailed p<0.05 for both). Therefore, the monotonically increasing trend was reliable, suggesting rule-based performance. Since this condition was a straight replication of the Long Training condition in Experiment I, this result compensates for the failure to find a completely reliable trend in that experiment, and reinforces the conclusion that Long Training participants were rule-learners. With regards to the Reduced contingency condition, participants were significantly more accurate on Near stimuli than Training ones (F(1,19)=15.69, one-tailed p<0.01), but the difference between Near and Distant stimuli was not significant (F(1,19)=1.51, one-tailed p=0.12). While this means that the peak-shift observed in this experiment was not completely reliable, it is clear that performance in this condition was more consistent with the associative predictions than the rule-based ones.

The structured questionnaire revealed that 63% of the participants in the Full Contingency group reported learning a version of the underlying rule, compared to 40% in the Reduced Contingency condition. A chi-squared performed on these scores demonstrated that, as predicted, this proportion was significantly higher for the Full Contingency group ($\chi^2(1)=$ 3.34, one-tailed p<0.05).

To summarise, the Full Contingency group produced a rule-based pattern of performance, with the majority of participants also being able to verbalise a rule. This reinforces the findings of Experiment I. The results of the Reduced Contingency condition significantly differed from this, and exhibited a peak-shift trend. Moreover, significantly fewer participants in this condition were able to report the rule. This is consistent with the McLaren, Green & Mackintosh's prediction that reducing the contingency decreases the likelihood of rule abstraction, leaving associative learning to dominate performance.

General Discussion

We have argued that associative and rule-based learning, in this type of discrimination task, may be distinguished by examining how participants' performance generalises to test stimuli: rule abstraction being indicated by a monotonically increasing trend and associative learning by a peak-shift. In two experiments, we have shown that, by these criteria, both short training and a reduced training contingency produce associatively based performance, with rule abstraction only emerging after longer training with a 100% contingency.

These findings can be understood in terms of McLaren et al.'s (1994) cognitive/associative dichotomy: After little training, or exposure to a reduced contingency, stimulus-category associations will be too weak to support rule induction, but nevertheless strong enough to produce above chance test performance. With further training, the strength of these associative traces will increase sufficiently to enable rule abstraction, and the resulting rule-based knowledge will then dominate responding on test. Moreover, a similar peak-shift/rule-based dissociation has been found using a different task and type of stimuli (Aitken, McLaren, & Mackintosh, in preparation), suggesting that these findings apply more generally.

We will now address three possible criticisms of this work. First, perhaps apparent rule learning participants correctly guessed the rule on test, rather than learning it during training. This is a real possibility, since a light-dark rule is an obvious way to divide a set of stimuli varying in luminance. However, if the participants had just guessed the rule, then their performance as a group would not have been significantly above chance, because they would not have known which key to assign to bright and which to dark.

Second, maybe participants classified as rule-learners would in fact have shown a peak-shift had we tested them with more extreme stimuli from the dimension. We did not use such stimuli because they no longer appeared green in color, a fact which could have introduced some new artifact into the data. However, we are confident in our conclusion that they were rule learners, because the majority of them were able to verbalise the rule. Moreover, previous work suggests that increasing the amount of training should produce a peak-shift with its peak closer to the training examples (Aitken, et al., in preparation). This makes it unlikely that the Long Training/Full Contingency groups were showing a peak-shift, with its peak further along the dimension than we were testing.

Third, a more complex rule-based account can predict a peak-shift, enabling both patterns of performance to be explained in rule-based terms. Suppose that after short training, or exposure to a reduced contingency, the rule that develops is highly context dependent. If this were the case, then Distant stimuli might evoke the rule less than Near stimuli, resulting in a peak-shift pattern. However, it would be wrong to assume that this 'single' mechanism is more parsimonious than separate cognitive and associative processes, because it also comprises two processes -namely, the similarity based context activation of the rule, and the application of the rule itself. Neither is it clear that such a context sensitive rule-based account can explain the large body of evidence consistent with the cognitive/associative distinction (e.g. Sloman, 1996). Moreover, if we adopted this explanation then we would lose the ability to account for peak-shift in pigeons using the same mechanism, since we probably do not wish to ascribe rule-learning capabilities to them. In short, we believe that considering both pigeons and people together, the cognitive/associative explanation is the more parsimonious.

Finally, it should be made clear that we are not suggesting that learning must always be initially, purely associatively driven. Existing evidence suggests that this would be too simplistic a view for the real world (for a discussion see Keil, Carter Smith, Simons & Levin, 1998). Rather, we would argue that, in everyday life, learning occurs via some complex interaction between cognitive and associative processes. Clearly, attention now needs to be focused on further specifying these two processes, and the way in which they interact.

Acknowledgements

This research was funded by a BBSRC grant awarded to I.P.L. McLaren and an MRC research studentship awarded to F.W. Jones. The authors would thank Chris Ingram, Rob Leech and Daniel Reisel, who ran the participants for Experiment II; and an anonymous reviewer, for their helpful comments.

References

Aitken, M.R.F., McLaren, I.P.L., & Mackintosh, N.J. (In preparation).

Blough, D.S. (1975). Steady state data and a quantitative model of generalization and discrimination. *Journal of Experimental Psychology: Animal Behavior Processes,* 1, 3-21.

Hanson, H.M. (1959). Effects of discrimination training on stimulus generalization. *Journal of Experimental Psychology,* 58, 321-34.

Jones, F.W., Wills, A.J., McLaren, I.P.L. (1998). Perceptual categorization: Connectionist modelling and decision rules. *Quarterly Journal of Experimental Psychology*, 51B (1), 33-58.

Keil, F.C., Carter Smith, W., Simons, D.J., & Levin, D.T. (1998). Two dogmas of conceptual empiricism: Implications for hybrid models of the structure of knowledge. *Cognition*, 65, 103-135.

McClelland, J.L., & Rumelhart, D.E. (1985). Distributed memory and the representation of general and specific information. *Journal of Experimental Psychology: General*, 114, 159-188.

McLaren, I.P.L., Green, R.E.A., & Mackintosh, N.J. (1994). Animal learning and the explicit/implicit distinction: Or why what we think of as explicit for us can be implicit for them. In N. Ellis (Ed.), *Implicit and Explicit Learning of Languages.* Academic Press.

Nosofsky, R.M., Clark, S.E., & Shin, H.J. (1989). Rules and exemplars in categorization, identification, and recognition. *Journal of Experimental Psychology: Learning, Memory, and Cognition*, 15, 282-304.

Perruchet, P. (1994). Learning from complex rule-governed environments: On the proper functions of nonconscious and conscious processes. In C. Umilta & M. Moscovitch (Eds.), *Attention & Performance XV: Conscious and nonconscious information processing* (pp. 811-835). Cambridge, MA: MIT Press.

Regehr, G., & Brooks, L.R. (1993). Perceptual manifestations of an analytic structure: the priority of holistic individuation. *Journal of Experimental Psychology: General*, 122, 92-114.

Shanks, D.R., & St John, M.F. (1994). Characteristics of dissociable human learning systems. *Behavioral and Brain Sciences*, 17, 367-447.

Shanks, D.R. (1995). *The Psychology of Associative Learning. Cambridge*, UK: Cambridge University Press.

Sloman, S.A. (1996). The empirical case for two systems of reasoning. *Psychological Bulletin*, 119 (1), 3-22.

Spence, K.W. (1937). The differential response in animals to stimuli varying within a single dimension. *Psychological Review*, 44, 430-441.

Wills, S., & Mackintosh, N.J. (1998). Peak shift on an artificial dimension. *Quarterly Journal of Experimental Psychology*, 51B (1), 1-31.

An ACT-R Model of Individual Differences in Changes in Adaptivity due to Mental Fatigue

Linda Jongman (linda@tcw3.ppsw.rug.nl)
Experimental and Work Psychology, University of Groningen
Grote Kruisstraat 2/1, 9712 TS Groningen, the Netherlands

Niels Taatgen (niels@tcw3.ppsw.rug.nl)
Cognitive Science and Engineering, University of Groningen
Grote Kruisstraat 2/1, 9712 TS Groningen, the Netherlands

Abstract

In this paper we show that adaptivity is reduced when people become fatigued. Fatigued people adapt worse to changing probability distributions as compared to non-fatigued individuals. In an ACT-R model of the task we show that this decreased adaptivity is due to a decrease in the use of one specific strategy. We argue that the use of this strategy is decreased, because it places high demands on working memory. In previous research we also found indications that mental fatigue is related to changes in working memory functioning. We argue that modeling individual differences in performance will provide better insight in the processes involved in mental fatigue.

Introduction

In this paper, mental fatigue is defined as the subjective feeling of being fatigued, combined which negative changes in performance, apart from the influences of time of day, or investment of physical effort. Many research projects concerning mental fatigue have failed to show decreases in performance as a result from fatigue. It appears that people are able to maintain adequate performance for a substantial amount of time. A growing number of investigations reveal indications that it is the way in which this performance is attained that changes when people become fatigued, as was already suggested by Bartlett (1943) and Broadbent (1979). In the 1970's, Shingledecker and Holding (1974) showed that after 24-32 hours of continuous work on a mentally loading task battery, people changed the order in which they tested possibly defective components on a fault-diagnosis task. The task consisted of finding the defective resistor in three banks of resistors containing one, two and three resistors respectively. All resistors had an equal probability of being defective, so the probabilities for the three banks of containing the defective resistor were respectively 17, 33 and 50 percent. The difficulty of the calculations that had to be made for finding the defective transistor were easiest for the bank with one transistor and most difficult for the bank containing three transistors. It appeared that participants, in the beginning of the experiment, chose to start testing the bank with three transistors which was most likely to contain the defective component. At the end of the experiment, however, they started more often with the bank with only one transistor which was the easiest one to test.

In a more recent article, Schunn and Reder (1998) show for a number of different tasks that people adapt their strategies to changed success rates of these strategies. They also showed that this, what they call extrinsic adaptivity, is a source of differences across individuals and that working-memory capacity and reasoning ability are good predictors of this adaptivity ability.

We developed a task, which is a combination of these two approaches, to investigate whether adaptivity is influenced by changes in mental circumstances, in this case by mental fatigue.

The Coffee Task

In stead of diagnosing transistors as was done in the Shingledecker and Holding experiment, participants have to weigh packets of coffee. On each trial, participants are shown three balances containing a tray with one, two and three packets of coffee respectively. The weights of the six packets and the three trays differs for each trial. The task is to find the one packet that has the same weight as the tray it is on. Participants cannot weigh individual packets, but are only allowed to weigh the whole tray. To find the weight of a specific packet, the balance has to be weighed, the packet must be taken of the balance and the balance has to be weighed again and the difference in weight has to be calculated. Packets cannot be put back on the balance. The task was designed in this way to ensure that calculations for the balance with three packets is hardest, like in the Shingledecker and Holding experiment. Figure 1 shows the interface of the task.

To investigate adaptivity, the probability of success for the three balances is manipulated. At the beginning of the experiment, the probability that the goal packet is on a certain balance is 10% (for the balance with one packet), 20% (for the balance with two packets) and 70% (for the balance with three packets.) So, the probability per packet is highest at the balance with three packets. However, after every five trials, the probabilities are changed according to which balance the participant chooses to weigh first. The balance that is started with most often is reduced in probability. Participants are told that the probability changes in this direction, but not precisely when the probabilities are changed and how big this change in probability is. They are pointed out that it is wise to start with the balance with the highest probability per packet and they are instructed to complete as many trials as possible, making as few mistakes as possible.

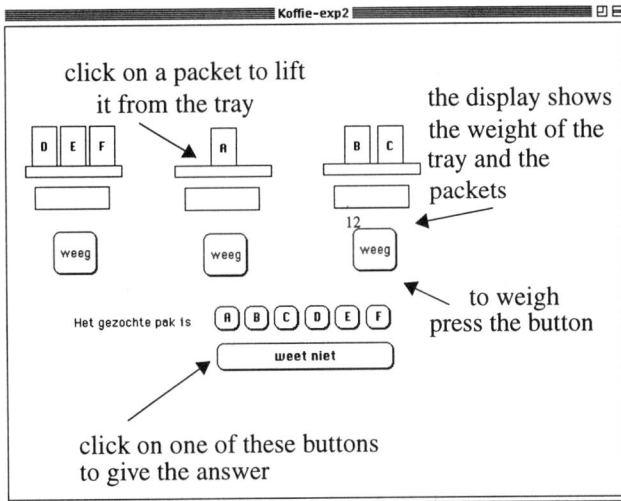

Figure 1: The interface of the coffee task

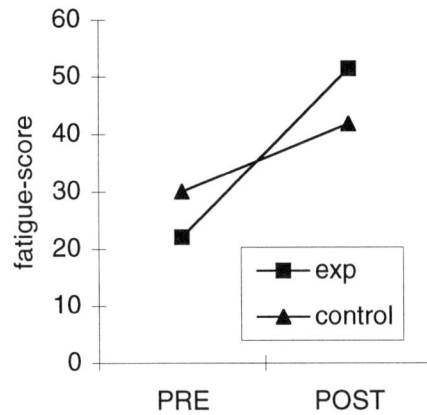

Figure 2: Subjective measurements of fatigue for the PRE- and POST-test.

The Experiment

32 undergraduate students participated in the experiment distributed over two conditions. In both conditions participants performed the coffee task for 25 minutes at the beginning of the experiment (the PRE-test) and at the end of the experiment (the POST-test). Before both tests, participants had to rate how fatigued they felt on a 150-point word-anchored scale. In the time between the PRE-and POST-test, participants of the experimental condition had to continuously solve complex scheduling problems under time-pressure for two hours (for a description of the task see Taatgen, 1997). Participants in the control condition could watch video tapes or read books for two hours. All participants were trained on the task for 3 times 25 minutes on the day preceding the experiment.

Results

Analysis of the reported feelings of fatigue revealed a main effect of session (F(1,30) = 23.937, p<.001) and an interaction of session and condition (F (1,30) = 4.343, p<.05), indicating that the fatigue-manipulation had the intended effect. As a whole group the participants are more fatigued on the POST-test, and this effect is stronger for the participants from the experimental condition. Figure 2 shows the fatigue ratings.

As for the strategy measures, contrary to the findings of Shingledecker and Holding (1974), no difference in global preference for one of the balances could be found on the POST-test as compared to the PRE-test.

More interesting is how well participants adapt their choices according to the changing probability distribution. The neutral probabilities for the three balances are 17, 33 and 50 percent respectively (as used in the Shingledecker and Holding experiment). If a participants always chooses the balance with the highest probability per packet, the distribution will remain close to 17,33 and 50 percent. Large deviations from this neutral probability distribution indicate that the participant often chooses a balance that was not opti-

mal. This deviation for participant i at trial (j) can be calculated according to formula (1).

$$\text{Deviation}_i(j) = 50 - P3 \qquad (1)$$

In this formula, *P3* represents the probability (as a percentage) that the goal-packet is on the balance containing three packets. The deviation is zero when *P3* = 50, as is the case in the neutral distribution. The deviation is plotted positive if the participant chose the optimal balance and negative if the participant chose a non-optimal balance. A deviation close to zero means that the participant adapts to the changing probabilities, whereas a deviation far from zero means he is not. Figure 3 shows an example of a deviation plot, where the participant starts out with a large deviation, but attains a performance close to zero deviation in the second half of the test.

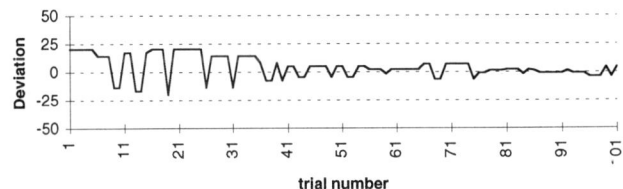

Figure 3: Example of a deviation plot

For each participant, an adaptivity score was calculated for each session (the POST-test and the PRE-test) according to formula (2).

$$\text{Adaptivity}(i) = \frac{\sum_j D_i(j)^2}{n} \qquad (2)$$

$D_i(j)$ is the deviation score for participant i on trial j. This adaptivity is the mean squared deviation score for a whole session. N is the total number of trials the participant completed. We chose to take the squared deviation in order to get rid of the sign and to stress large deviations. Figure 4 shows how this adaptivity changes from the PRE-test to the POST-test.

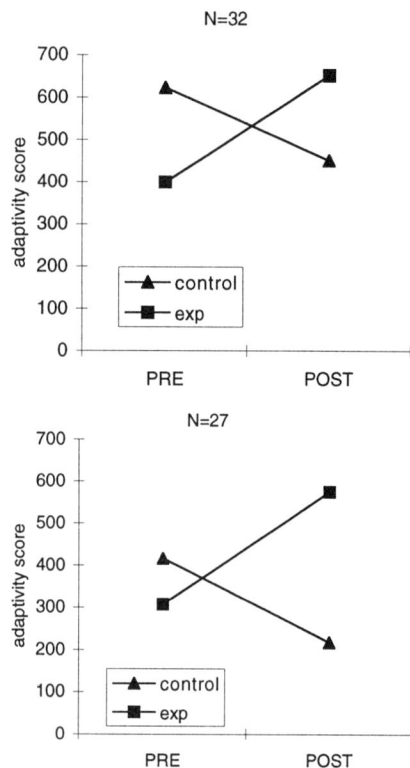

Figure 4: Adaptivity scores for the PRE- and POST-test for all participants (upper figure), and only participants who adapt to the changing probabilities (lower figure)

While no main effect of session (PRE-test, POST-test) could be found, there was a significant interaction of session by condition (F(1,30) = 9.912, p<.01). As can be seen in the upper half of the figure, participants from the control condition adapt better on the POST-test as compared on the PRE-test, while participants from the experimental condition do adapt worse on the POST-test. The difference between the two conditions on the PRE-test was non-significant and could be attributed to five participants who did not adapt at all to the changing probabilities. The lower part of figure 4 shows the adaptivity for the two conditions when these five participants are removed from the set.

These results indicate that strategy adaptivity is reduced when people become fatigued. Moreover, the decrease in performance correlates with the change in reported feelings of fatigue (r=.50, p<.01).

Strategies

An interesting question is how to explain the difference in adaptivity between the two conditions. One approach is to look at which strategies are possible to do the task. We hypothesize that there are two possible strategies: choose the same balance as on the previous trial (P), and choose the balance which contained the answer on the previous trial (A). The latter of the two is an adaptive strategy. Because these strategies overlap in their predicted responses, responses can be categorized into the following four categories:

(1) $P \land A$ (PA)
(2) $P \land \neg A$ (PnA)
(3) $\neg P \land A$ (nPA)
(4) $\neg P \land \neg A$ (nPnA)

Based on the distribution of the responses into these four categories, we estimated the use of the two strategies (A) and (P). Responses in category (3) strongly indicate the use of the adaptive A-strategy. Participants choose the balance that contained the answer on the previous trial. Responses in category (2) indicate the use of the non-adaptive P-strategy. Category (4) is some kind of rest category in which a different balance is chosen, but not the one that contained the answer on the previous trial. Responses in this category can indicate the use of a different strategy as the two mentioned before.

As was described in the introduction, we wanted to see whether adaptivity is influenced by mental fatigue. The results of the experiment indicated that the experimental group had a decreased adaptivity score. An interesting question is whether this reduction in their adaptivity scores could be explained by a reduction in the use of the adaptive A-strategy. If so, this should be visible by a decrease in responses in the nPA-category. We must note that not all participants were fatigued to the same degree by the experimental manipulation. Therefore, we have split the participants in a high-fatigue group and a low-fatigue group, based on the median increase in fatigue scores for the experimental group. Although we did not find a main effect of session in the number of responses in the nPA category, there was a significant interaction of session and the two fatigue groups (F(1,30) = 5.548, p=.025). So, only the high-fatigue group showed a decrease in responses in the nPA category.

Furthermore, the four different categories correlate strongly with the adaptivity scores of the participants as calculated according to formula (2):

	PRE adaptivity	POST adaptivity
PA	-.53**	-.72***
PnA	.90***	.86***
nPA	-.67***	-.61***
nPnA	-.42*	-.44*

* p<.05, ** p<.01, *** p<.001

As this table shows, the A-strategy, indicated by nPA responses, has a strong negative correlation with the adaptivity score, which implies using the A-strategy has a positive effect on performance. The P-strategy on the other hand has a very negative effect on performance, as can be concluded from the positive correlation between PnA and performance.

The ACT-R Model

In order to explore the question whether the proposed strategies fully characterize the behavior of participants on this task, we developed an ACT-R model to simulate the behavior of individual participants.

ACT-R (Anderson & Lebiere, 1998) is a hybrid cognitive architecture based on a production system. It has been used to explain a wide range of cognitive phenomena by producing models that make precise predictions about choices, latencies and errors. The main mechanism we will use is ACT-R's conflict resolution that will be used to choose between strategies.

The basis for this choice between strategies consists of the following three rules:

(1) A rule that proposes to start with the same balance as the previous trial, corresponding to the P-strategy

(2) A rule that proposes to use the answer to the previous trial as the basis for the choice, corresponding to the A-strategy

(3) A rule that picks a random balance to start with, which differs from the balance chosen first in the previous trial. We will call this the rest (R) strategy.

This last rule is used to represent the nPnA cases, for which it is not clear what strategy the participant pursues.

In order to choose between rules, ACT-R (Anderson & Lebiere, 1998) uses a conflict-resolution mechanism based on the expected gain of a rule. The expected gain of a rule is calculated by taking the following factors into account: an estimate of the probability that the rule achieves the current goal, an estimate of the costs that are involved in achieving this goal, and that value of the goal itself. Basically, the rule with the highest expected gain is selected. However, since noise is added to the expected gain, the best rule not always fires, it only has the highest probability of firing, governed by the following equation:

$$\text{Probability of choosing rule } i = \frac{e^{E_i/t}}{\sum_j e^{E_j/t}} \qquad (3)$$

In this equation, E_i represents the expected gain of rule i, and the t parameter determines the level of noise.

From the experiment we have, for each participant, the proportion of times they chose a response in the categories PA, PnA, nPA and nPnA for both the PRE- and the POST-test. These values can be used to estimate the probability the participant uses the P-strategy, the A-strategy, or the R-strategy. Consequently, these estimates can be used to calculate suitable expected gains for the three rules that choose the strategies.

To see how well the model can estimate the adaptivity score for each participant in each test, the model was run for 50 times with the three expected-gain parameters estimated for each participant and each test. The result is shown in figure 5. Each point in the graph corresponds to one test (PRE or POST) of a single participant). The correlation between the data and the model predictions is 0.77, which is not particularly high, although encouraging. The problem is, that there is a lot of randomness involved in the model. Even if the average score for one of the models is 200, values for individual runs may range from 100 to 500. So we decided to see how far apart the experimental score and each model prediction was in terms of the standard deviation of the model (based on the 50 runs for each score). The result was, that 66% of the experimental scores was within one S.D. of the model prediction, and 97% within two S.D.'s, exactly what one would expect in a normal distribution.

To get a better idea of how the model's performance can

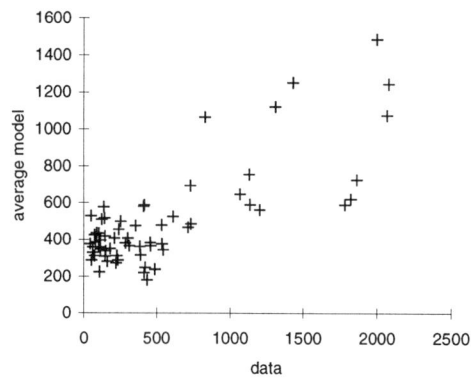

Figure 5: Predictions of the model compared to the data

be compared to what goes within a test, we did a second run of 20 simulations for each participant and each test. In stead of averaging these simulations, we picked the simulation which adaptivity score was closest to the adaptivity score in the experiment. This "best of 20" strategy, nor surprisingly, boosts the correlation between the data and the model to 0.99. Figure 6 shows the match between the model and the data.

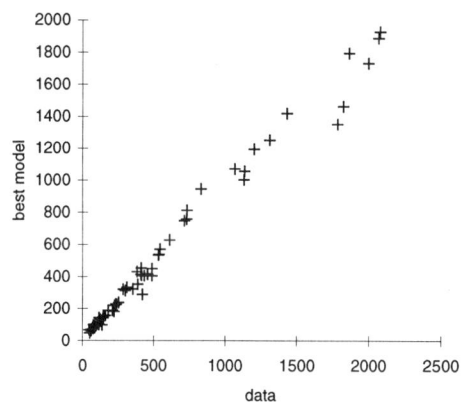

Figure 6: Predictions of the "best of 20" model and the data

Since finding a close match between model and data is not such a big feat if one uses a "best of 20" strategy, we looked at how well this model can predict the details of the experimental data. We plotted the course of the individual deviation scores during the experiment, and compared it to the predictions of the best model. The results for four participants are depicted in figure 7. The left-hand column shows data from the experiment: a PRE- and POST-test for each participant. The right-hand column shows the predictions of the model for each of the individual runs. The four participants shown are all from the high fatigue-group and all performed worse on the POST-test than on the PRE-test, as measured by the adaptivity score. A total of six participants satisfied both of these criteria, so two-third of the "interesting" group is shown in figure 7.

As one can see in figure 7, the plots of the deviation scores show huge individual differences. The model, however, captures these differences quite nicely, especially given the fact

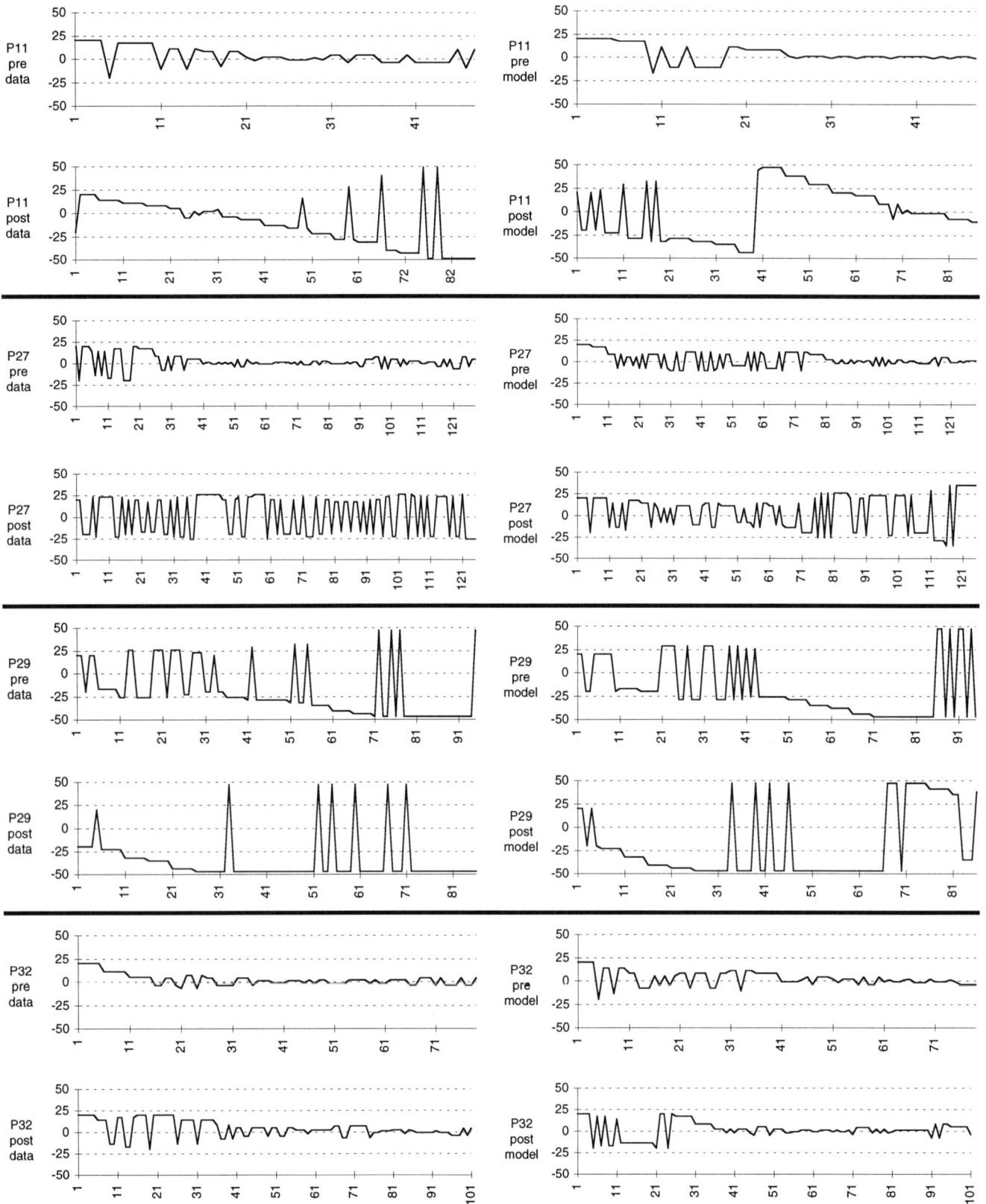

Figure 7: Deviation scores during the course of the experiment for the experiment and the model's predictions. The four participants shown are all from the experimental condition, all reported fatigue at the post-test, and all had a worse performance on the post-test compared to the pre-test.

that no information about the course of the deviations has been put into the model. Basically, each model is based on four parameters from the experimental data: the PC, nPC, PnC, and the adaptivity score.

Each of the four fatigued participants in the figure shows a slightly different pattern of fatigue. Participant 11 does quite well on the PRE-test, and keeps her deviation score quite close to zero. At the POST-test, however, she shows hardly any adaptivity at all anymore. Participant 27 also starts out very well on the PRE-test, but cannot maintain this performance in the POST-test, where he starts oscillating. Participant 29 already starts with poor deviation scores, but gets even worse in the POST-test, where she goes down to a deviation score of -50 with an occasional spike to +50. Participant 32, finally, exhibits a good performance on both the PRE- and the POST-test, but she takes slightly more time to arrive at a deviation of around zero in the post-test.

In all four cases the model shows the same pattern of the effect of fatigue as the data. This is an indication that although the deviation plots of the participants are all quite different, the essence is captured in the four parameters that are put into the model.

Discussion

In the introduction, we hypothesized that mental fatigue would influence adaptivity. The results from the experiment show that adaptivity is reduced when people become fatigued. Moreover, the increase in reported feelings of fatigue strongly correlates with a reduction of adaptivity. If we zoom in on adaptivity in more detail, we see that the reduction in adaptivity for fatigued people could be largely explained by a reduction in the use of the adaptive strategy. In the model, this adaptive strategy was defined as choosing the balance that contained the answer on the previous trial, which is a fairly simple implementation. However, it is possible, people do not realize that such a simple strategy will do the job. It is likely that people will base their decisions on which balance contained the answer on the last two or three trials. In that way, this adaptive strategy will place high demands on working memory. This is consistent with the findings of Schunn and Reder (1998) who report that working-memory capacity is a good predictor of adaptivity. A possible reason that this adaptive strategy is used less when people become fatigued, is that working memory functioning is impaired by mental fatigue. Jongman (1998) also found some indications that working memory functioning could play a role in mental fatigue.

As for the model, we hypothesized that there are two possible strategies to perform the task. Figure 7 shows that the model is able to capture many aspects of individual participants's performance on the task. So, the two strategies gave an adequate representation of people's performance. This was also confirmed by the strong correlations between the different response categories and the adaptivity scores. However, we also found a moderately significant negative correlation between the nPnA category and the adaptivity scores. This may indicate that participant uses a more elaborate adaptive strategy, related tot the A-strategy, but using more trials to base the decision on. Or it may indicate a totally different strategy, meaning people use at least a third strategy as well, which was not captured by our model, but which did have a positive influence on their adaptivity scores.

Overall, the model gave an encouraging fit of the data. Six people from the high-fatigue group showed huge changes in adaptivity from the PRE- to the POST-test. Although the pattern of change was different for the six persons, as shown for four persons in figure 7, the model fitted these different patterns quite nicely. So, the differences in these patterns can be adequately explained by changes in the frequencies these strategies are used.

Many research projects concerning mental fatigue show very specific changes in performance for different individuals. In this paper we showed that different patterns could be explained by changes in the use of a single strategy. We will argue therefore, that fatigue research and related fields will benefit from an approach that focuses on modeling individual differences, thus avoiding the risk of throwing the baby out with the bath water.

Acknowledgments

This research is part of the Netherlands Concerted Research Action "Fatigue at Work" funded by the Netherlands Organization for Scientific Research (NWO).

References

Anderson, J.R. & Lebiere, C. (1998). *The atomic components of thought.* Mahwah, NJ: Erlbaum.

Bartlett, F.R.S. (1943). Fatigue following highly skilled work. *Proceed. Royal Society*, 131, 247-257.

Broadbent, D.E. (1979). Is a fatigue test possible now? The Society Lecture 1979. *Ergonomics, 12*, 1277-1290.

Jongman, L. (1998) How to fatigue ACT-R? Proceedings of the Second European Conference on Cognitive Modelling. Nottingham: Nottingham University Press, 52-57.

Shingledecker, C.A. & Holding, D.H. (1974). Risk and effort measures of fatigue. *Journal of motor behavior, 6*, 17-25.

Schunn, C.D & Reder, L.M. (1998). Strategy adaptivity and individual differences. *The psychology of learning and motivation, 38*, 115-154.

Taatgen, N.A. (1997). A rational analysis of alternating search and reflection strategies in problem solving. *Proceedings of the 19th Annual Conference of the Cognitive Science Society*. Hillsdale, NJ: Erlbaum.

Mirroring the Inverse Base-Rate Effect:
The Novel Symptom Phenomenon

Peter Juslin (Peter.Juslin@psyk.uu.se)
Department of Psychology, Uppsala University
Box 1225, S-751 42, Uppsala, Sweden

Pia Wennerholm (Pia.Wennerholm@psyk.uu.se)
Department of Psychology, Uppsala University
Box 1225, S-751 42, Uppsala, Sweden

Anders Winman (Anders.Winman@psyk.uu.se)
Department of Psychology, Uppsala University
Box 1225, S-751 42, Uppsala, Sweden

Abstract

The *elimination model* is proposed as an account of the *inverse base-rate effect* (D. L. Medin & S. M. Edelson, 1988). A key-assumption is that participants sometimes rely on *eliminative inference* to decide among candidate categories. A new prediction is that there will be an inverse base-rate effect also for an entirely novel symptom presented in the transfer phase—a prediction that contrasts with that by *ADIT* (J. K. Kruschke, 1996). This was tested and confirmed in 2 experiments.

Introduction[1]

In 1988, Medin and Edelson reported an interesting but complex pattern of findings regarding how people utilize base-rates. In their experiments, participants were asked to decide whether patients with ambiguous symptom patterns were suffering from previously learned common or rare diseases. Surprisingly, in some cases participants chose the less frequent of the diseases. A number of explanations of this base-rate inverse (BRI) effect have been proposed (Kruschke, 1996; Medin & Bettger, 1991; Medin & Edelson, 1988; Shanks, 1992).

In this paper, we propose a further mechanism that may contribute to both the BRI effect and the unspecified guessing strategy reported by Kruschke (1996). Basically, the elimination model suggests that the participants eliminate options that are inconsistent with well-supported inference rules, leading to the prediction of an intricate pattern of responses in which participants sometimes favor the common diseases, and sometimes the rare ones. A presentation of the details, and the fit of a quantitative implementation of the elimination model, is provided in Juslin, Wennerholm, and Winman (1999). In this paper we will focus on one prediction by the elimination model that goes beyond what previous models can predict or account for, the prediction of a novel symptom phenomenon.

[1]The research reported here was supported by the Swedish council for Research in the Humanities and Social Sciences.

The Experimental Paradigm

The basic task introduced by Medin and Edelson (1988) involves a training- and a transfer phase. On each training trial a pair of symptoms is presented, and participants are requested to choose which of six fictitious diseases the hypothetical patient is suffering from. After each choice the participant is informed about the proper diagnosis (disease), after which another training trial is presented. The critical manipulation concerns the base-rate of each disease, with the common diseases occurring three times more often than the remaining rare ones (see Table 1).

Table 1: The basic design of the training phase in the Medin and Edelson Experiment 1 (1988).

Base-rate	Symptoms	Disease	Inference rule
3	I_1+PC_1	C_1	$I_1+PC_1 \rightarrow C_1$
1	I_1+PR_1	R_1	$I_1+PR_1 \rightarrow R_1$
3	I_2+PC_2	C_2	$I_2+PC_2 \rightarrow C_2$
1	I_2+PR_2	R_2	$I_2+PR_2 \rightarrow R_2$
3	I_3+PC_3	C_3	$I_3+PC_3 \rightarrow C_3$
1	I_3+PR_3	R_3	$I_3+PR_3 \rightarrow R_3$

During training every instance of a common disease, C, occurs in the presence of two symptoms: One imperfect, I, and one perfect, PC. Similarly, every instance of a rare disease, R, has two symptoms: One imperfect, I, and one perfect, PR. Thus, each imperfect predictor is associated with both a common and a rare disease, and each perfect predictor is uniquely associated with only one disease.

In a succeeding transfer phase participants are tested with previously uncombined symptoms. Medin and Edelson (1988) found that when tested with the imperfect symptom, I, the majority of participants chose the common disease.

When tested with the ambiguous combination, I+PC+PR (the combined probe), the participants again tended to choose the common disease. However, when tested with two perfect predictors, PC+PR (the conflicting probe), the majority of participants chose the rare disease in contrast to the base-rate—the inverse base-rate effect (Figure 2C below).

Accounts of The Inverse Base-rate Effect

Most previous accounts of the BRI effect revolve around a common theme: Because of cue-competition, symptom PR becomes more strongly associated with disease R than symptom PC does with disease C (Gluck & Bower, 1988; Kruschke, 1996; Shanks, 1992).

Kruschke (1996) suggested that ADIT can explain both the inverse base-rate effect and apparent base-rate neglect (Gluck & Bower, 1988). By the application of two separate mechanisms: (a) A base-rate bias, that participants apply consistently on all training trials, and (b) an attention-shifting mechanism that rapidly shifts attention from typical to distinctive features, ADIT provided a good fit to the transfer data. Specifically, because the common disease C, occurs more often than the rare one, R, participants first learn to associate both the imperfect symptom I, and the perfect symptom PC with the common disease. Later in training when they are presented with the symptoms that are associated with the rare disease, R, they focus on the symptom that is perfectly predictive of that disease, PR, and thereby encode it by this single symptom. This explains why participants choose the rare disease on the PC+PR (conflicting) test case. When confronted with the remaining two ambiguous test cases, I and I+PC+PR, people apply both their base-rate knowledge and their associative knowledge, where the base-rate knowledge dominates the responses.

Although ADIT provides a good quantitative fit to transfer data, Kruschke (1996) reported that his participants responded better-than-chance for the rare categories—an effect he attributed to an unspecified non-random guessing strategy (Kruschke, 1996). Likewise, when ADIT was fitted to the training data it performed much worse than human learners on early training trials. Thus, although appealing, ADIT fails to fully account for the complete pattern of data observed with the Medin and Edelson (1988) design.

The Elimination Model

To illustrate the inferential mechanisms of the elimination model, consider the following example: You are told that a friend of yours has bought a pet animal called George who is either a goldfish or a Psittaciformes. Not being a zoologist, you have a pretty good idea of what a goldfish is, but you have no notion whatsoever of what a Psittaciformes is. Your task is to guess what kind of pet animal George is. First, you receive the cue George lives in water. George is thus similar to a goldfish in the sense that he lives in water. In the absence of knowledge about what a Psittaciformes is you might be tempted to guess that George is a goldfish. This illustrates one (weak) form of induction.

Now consider the situation where you instead are given the cue George can fly. In this case you would probably guess that George is a Psittaciformes—he is certainly not a goldfish (in fact a Psittaciformes is a parrot). You would use your knowledge about the category goldfish to eliminate the possibility that George is a goldfish. The elimination model takes this latter kind of inference into account.

A reasonable assumption in the Medin and Edelson (1988) design is that the participants perceive the task as involving a set of perfectly valid inference rules (see Nosofsky, Palmeri, & McKinley, 1994, for similar approaches). If the participants succeed to learn these rules, they will make 100 percent correct classifications at the end of training (see Table 1).

In the quantitative implementation in Juslin et al. (1999), we assume that at each trial the inference rule appropriate for the presented training probe is formed with a rule-activation probability. This probability, that is higher for the early training trials (implementing "freezing" at the initial stages of learning, cf. Medin & Bettger, 1991), is controlled by a single parameter. From these rule-activation probabilities, we can compute the probability c that the rule appropriate for a common disease is active and accessible at the transfer phase, and the corresponding probability r that the rule appropriate for a rare disease is accessible.

In the training phase, every probe precisely matches one of the six inference rules (see Table 1). In the transfer phase, however, the participants' inferences will have to be based on the similarity between the new symptom combinations and the conditions of the inference rules. The elimination model consists of two decision mechanisms that determine how an inference rule is applied to a probe:

(1). The induction mechanism applies when the probe has exactly the symptoms in the condition-part of the inference rule, or when the symptoms of the probe are perceived to be sufficiently similar to the rule conditions. Whenever similarity is larger than a similarity criterion the induction mechanism applies, and the probe is assigned to the category with the most similar rule. If the probe is equally similar to several rules, the participants will decide randomly among the set of equally similar rules.

(2). When a probe is dissimilar to the rule-conditions, as indicated by a similarity smaller than the similarity criterion, the elimination mechanism is used to eliminate the possibility that the probe belongs to the category and the probe is assigned randomly to any category but the dissimilar one. For example, if there is no basis for induction and the probe eliminates one or several of the categories, the participant will have to decide randomly among the still admissible categories—the diseases that are not inconsistent with the symptoms of the probe.

When the elimination model is applied to the Medin and Edelson design, we need to impose a similarity structure on the probes presented in the transfer phase. There are two crucial assumptions: (a) The conflicting probe, PC+PR, is less similar to the inference rules formed in the training phase for C and R than the combined probe, I+PC+PR and the imperfect probe, I. While the combined probe is ambiguous in the sense of being consistent with two inference rules, the conflicting probe actually contradicts both. (b) The similarity criterion for induction versus elimination is located between the similarities of the combined and the conflicting probes implying that the combined probe elicits induction and the conflicting probe elimination. The example of such a similarity structure, derived from the multiplicative similarity rule of the original context model (Medin & Schaffer, 1978), is provided in Juslin et al. (1999).

If we refer to the common-rare disease-pairs relevant to a particular probe (e.g., C_1 and R_1 in Table 1) as the focal disease-pair, a participant may be in one of four knowledge states when entering the transfer phase. State 1: With probability $(1-c)(1-r)$ neither the inference rule for the focal common C_1 nor the focal rare disease R_1 is accessible. State 2: With probability $(c-cr)$ only the inference rule for the focal common disease C_1 is accessible. State 3: With probability $(r-cr)$ only the inference rule for the focal rare disease R_1 is accessible. State 4: With probability (cr) both of the focal inference rules, C_1 and R_1, are accessible. Table 2 shows an example of Knowledge State 1, the state in which neither of the rules for the focal diseases are accessible.

The predicted response patterns may be exemplified by reference to the combined and the conflicting probes. For example, imagine that you are presented with the symptom combination $I_1+PC_1+PR_1$. These three symptoms are consistent with two of the six inference rules, the first and the second in the right-most column of Table 1. If you are in Knowledge state 2 you will only know the rule $I_1+PC_1 \rightarrow C_1$ which is executed by the induction mechanism. If you are in Knowledge state 3 you will only know the rule $I_1+PR_1 \rightarrow R_1$ which is executed by the induction mechanism. Because of the base-rate manipulation the probability of Knowledge state 2 is higher and most responses will thus favor C_1. Knowledge states 1 and 4 are assumed to elicit random decisions favoring neither common nor rare categories.

Now you are presented with the conflicting probe, PC_1+PR_1 that is dissimilar to both the first and the second inference rule.

Table 2: An example of a hypothetical Knowledge State.

Ratio	Symptoms	Disease	Rule
3	I_1: Stomach pain PC_1: Loss of hair	C_1: Coralgia	Unknown
1	I_1: Stomach pain PR_1: Impaired hearing	R_1: Buragamo	Unknown
3	I_2: Epidermophytosis PC_2: Back pain	C_2: Midosis	Known or Unknown
1	I_2: Epidermophytosis PR_2: Loosening of the teeth	R_2: Namitis	Known or Unknown
3	I_3: Visual defect PC_3: Impaired short-term memory	C_3: Terrigitis	Known or Unknown
1	I_3: Visual defect PR_3: Swollen arms	R_3: Althrax	Known or Unknown

If you are in Knowledge state 2 you will eliminate the rule $I_1+PC_1 \rightarrow C_1$; If you are in Knowledge state 3 you will eliminate the rule $I_1+Pr_1 \rightarrow R_1$. Again, because of the base-rate manipulation Knowledge state 2 is more probable, and

most eliminations will concern C_1 and thus favor the choice of a rare disease category. Given the particular knowledge-state and the similarity between the inference rules formed and the presented transfer probe, the decision mechanisms of induction and elimination can be applied in the manner specified above. Although this is straightforward in principle, application to the Medin and Edelson design is complicated by the fact that the number of unknown diseases is a random variable.

This can be illustrated by reference to the example in Table 2. You may be faced with the conflicting probe *loss of hair + impaired hearing*. Because you are in Knowledge state 1 and know neither of the focal rules, there is no possibility for an inductive inference with the focal rules. The choice of an unknown category in this case amounts to an elimination of any of the four non-focal diseases that possibly are known. The number of unknown diseases is a random variable controlled by the same parameter that defines c and r; that is, one to four of the non-focal rules may have been activated the training. Thus, in knowledge State 1, the guessing rate of responses in the focal common category (C_1) is anything between 1/6 (no disease has been learned) to 1/2 (all the four non-focal diseases have been learned).

The predicted response proportions for each probe therefore equal the proportions of inductive inferences that fall in this category plus the expected value of the guessing rates when the participant eliminates, where the probability of induction and elimination is jointly determined by the knowledge-state and the similarity between the probe and the known inference rules. Since both the probabilities of the knowledge states and the expected guessing-rates are controlled by the parameter that defines the rule-activation probability, predicted response proportions are controlled by a single parameter. These computations are detailed in Juslin et al. (1999).

In Juslin et al. (1999) the quantitative predictions were fitted to the data from Experiment 1 by Medin and Edelson (1988), to Kruschke (1996, Exp. 1), and to the data from the two experiments reported below. In all of these data sets, the model reproduced the observed pattern of base-rate findings, in general with an impressive quantitative fit given the use of one single free parameter. Figure 1 illustrates the fit of the model to the data from Kruschke' s (1996) Experiment 1. Although factors such as cue competition probably play an important role in the Medin and Edelson paradigm, the quantitative predictions presented above demonstrate that the elimination mechanism alone has the potential to reproduce the BRI effect.

The Novel Symptom Phenomenon

A straightforward prediction by the model is that the presentation of a novel symptom in the transfer phase will lead to a preponderance of *rare-disease* responses, mirroring the BRI effect for the conflicting symptoms. Participants will notice that the novel symptom is dissimilar to the symptoms of known categories and as a result they will guess on some of the unknown diseases.

By virtue of the base-rate manipulation these *unknown* categories are likely to be rare rather than common ones.

This prediction is important for two reasons: First, it is the most critical test of the presence of eliminative inferences. To the extent that this type of inferences underlies both the responses for the conflicting and novel transfer probes, the response patterns observed for these probes should be similar. Second, this prediction amounts to a response pattern contrary to that by ADIT (Kruschke, 1996). The novel transfer probe has not been affected by any shift of attention during training, thus the only factor at work is the base-rate bias toward common-disease responses. As we have seen, the elimination model predicts rare-disease responses that mirror those observed for the conflicting test probe. Next, we present results from two experiments that confirm this prediction.

A.

B.

C.

Figure 1: The model fitted to Kruschke (1996, Exp.1). A) Predicted/Observed response proportions. B) The pattern of data predicted for the transfer probes, and C) the corresponding observed pattern.

Experiment 1: A Test of the Novel Symptom Phenomenon

The main purpose of Experiment 1 was to test the prediction of a BRI effect for novel symptoms. One hundred and nine participants were divided into a 3:1 base-rate ratio group and a 7:1 base-rate ratio group (cf. Shanks, 1992). The material, stimuli and procedure were more or less identical to those used by previous researchers (see e.g., Medin & Edelson, 1988; Kruschke, 1996; Shanks, 1992). Participants were told that they would be allowed to practice on 168 patients with feedback informing them if they had made a correct or incorrect diagnosis. They were instructed to apply the knowledge they had acquired during the training phase to 24 patients in a transfer phase with no feedback. For each participant nine of the twelve symptoms were randomly selected and matched with the six different diseases (see Tables 1 and 2). The three remaining "novel" symptoms were used in the transfer phase in order to test the novel symptom phenomenon.

On the transfer trials participants were required to make responses to six perfect predictors, PC and PR, three imperfect predictors, I, three combined probes, I+PC+PR, three conflicting probes, PC+PR, and three novel probes, N. The remaining six test trials in which the imperfect predictors were paired with perfect predictors were mainly used to disguise the purpose of the transfer phase. The transfer trials were followed by a short break after which another training - and transfer phase followed. The purpose of this manipulation was to see whether additional training would decrease the BRI effect. The experiment including the break lasted approximately one hour.

Training Results

As in most previous studies, only those participants who had reached asymptotic learning were used in the subsequent analysis. Participants with more than one incorrect answer in the last 24 trials of the first training session were excluded. In the 3:1 ratio group, 40 participants out of 55 (73%) met this criterion. In the 7:1 ratio group, 41 out of 54 (76%) participants met the criterion. Whereas the 7:1 group showed evidence of slightly faster learning, both groups converged on asymptotic learning at the end of the first session and these levels of performance were maintained throughout the second training session.

Transfer Results for the 3:1 Group

For the perfect predictors, PC and PR, in the 3:1 condition, most responses were assigned to the disease that the cue had been a perfect predictor of, both after training session 1 and 2 (all $t(39) = 15$ with $p < .05$, given a null-hypothesis of .5). The use of base-rate information is evident for the imperfect transfer probes, I, both after training sessions 1 and 2 (.67, $t(39) = 5.77$, $p < .05$, and .66, $t(39) = 4.03$, $p < .05$, respectively), and for the combined transfer probe, I+PC+PR, both after training sessions 1 and 2 (.62, $t(39) = 3.60$, $p < .05$, and .62, $t(39) = 2.07$, $p < .05$, respectively).

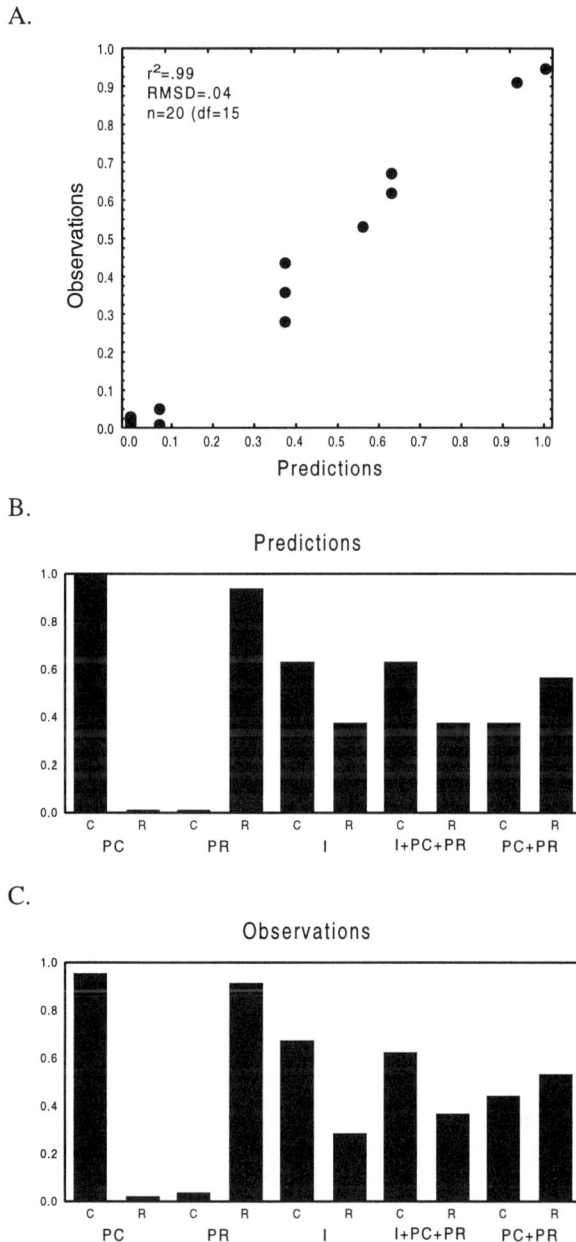

255

The results show a BRI effect for the conflicting transfer probe, PC+PR, although the trend is non-significant. After session 1, the proportion of common category responses, C, was .410, $t(39) = 1.42$, N. S. After training session 2, the proportion increased to .483, $t(39) = .24$, N. S. Finally, the results for novel symptoms, N, mirror the result for the conflicting probe, PC+PR (see Figure 2A). Although slightly less pronounced, the participants favor the rare diseases (response proportion .45, $t(39) = 1.12$, N. S.). After training session 2, the participants have altered into base-rate use (response proportion .58, $t(39) = 1.74$, N. S.).

Transfer Results for the 7:1 Group

Results for the 7:1 group parallel those for the 3:1 group, although the effects were larger, and in contrast to the 3:1 condition there was a significant BRI effect for the conflicting transfer probe (see Figure 2B, session 1: proportion .38, $t(40) = 2.10$, $p < .05$: session 2; proportion .40, $t(40) = 1.65$, N. S.). As described above, the elimination model is supported if the BRI effect on the conflicting and novel probes is similar Figure 2 presents the mean response proportions for the novel and conflicting transfer probes in the 3:1 and 7:1 base-rate ratio conditions of Experiment 1, and of the (single) 3:1 condition of Experiment 2

As can be seen, the response patterns are similar for novel and conflicting probes in both conditions, a result predicted by the elimination model but contrary to ADIT (Kruschke, 1996).

Experiment 2: Does the Inverse Base-Rate Effect Disappear with Additional Training?

Experiment 1 replicated the base-rate effects of the Medin and Edelson (1988) study. Interestingly, however, the BRI effect was diminished after training session 2, and it vanished altogether for the novel transfer probes. This could however be due to the testing between the first and second training phase. When the novel transfer probe has been presented in the transfer phase after training session 1, it will obviously not be "novel" when the retested in the second transfer phase. As a result Experiment 2 was designed to test whether the diminished BRI effect was due to the repeated transfer exposure. Another potential explanation is that the BRI effect disappears after extensive training. Medin and Bettger (1991), for example, discussed the possibility that "with enough experience the BRI effect that we have attributed to competitive learning may be overcome altogether" (p. 328). Thus, the alternative hypothesis was that the diminished effect was due to the higher extent of learning. Experiment 2 involved a prolonged training phase without transfer phases interspersed midways through the training trials. If the BRI effect would persist even with this prolonged training, it would suggest that the diminished BRI effect after training session 2 in Experiment 1 was a consequence of the repeated exposures. On the other hand, if the correct explanation lies in the increased learning the BRI effect should be gone after four times as many trials. Twenty-five students participated in Experiment 2. Procedures were identical to Experiment 1, with the differences that (a) the number of training trials was 672 (168 × 4), and that (b) the base-rate ratio was 3:1 for all

participants. Between the first and second half of learning phase the participants were given a one-hour lunch break. At the end of training session 2 they were presented with the transfer phase (identical to the one in Experiment 1).

A.

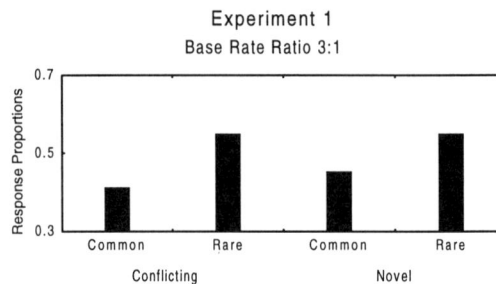

B.

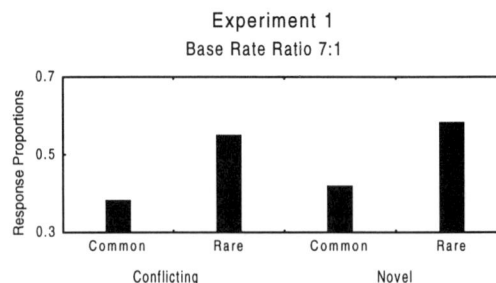

C.

Figure 2: Mean response proportions for the conflicting and novel transfer probes in the 3:1 (Panel A) and the 7:1 (Panel B) base-rate ratio condition of Experiment 1, and in Experiment 2 (Panel C).

Results

As in Experiment 1, the learning criterion was set at 96% correct responses in the last training block of 24 trials. Again, Experiment 2 replicated the standard pattern found by Medin and Edelson: The common disease was chosen more often than the rare one for the imperfect (common response proportion .66, $t(22) = 3.6$, $p < .05$) and the combined probes (common response proportion .46, $t(22) = .38$, N. S), but the rare disease was chosen for the conflicting probe (common response proportion .39, $t(22) = 1.1$, $p < .28$, N. S) - the BRI effect. Although not significant, the effect is not diminished in size after ample learning (see Figure 2C). Finally, a BRI effect for the novel symptoms was observed too (common other response proportion .40, $t(22) = 1.59$, $p < .125$, N. S). We found it hard to obtain statistical significance due to the very high measurement error. Therefore, it should be noted that if all data in Figure 2 are collapsed, there is no doubt a reliable BRI effect (t(103) =2.7, p<.01) as well as a reliable novel symptom phenomenon (t(103) = 2.6, p<.01).)

In sum, it seems that the diminished BRI effect after training session 2 reported in Experiment 1 is due to the repeated exposures with the transfer probes rather than to more extensive training. Similar to Experiment 1, participants guessed on a rare disease category when presented with a novel symptom, as predicted by the elimination model but in contrast to ADIT (Kruschke, 1996).

General Discussion

In this paper, a new mechanism has been proposed that may contribute to the base-rate effects observed with the Medin and Edelson design. The main merits of the elimination model are threefold: First, the model has intuitive appeal in the sense that it seems hard to deny that people at least sometimes rely on eliminative inferences. Second, in terms of the psychological mechanisms involved, the model is simple: The participants either make inductive - or eliminative inferences depending on the similarity of the probe to the known diseases. Finally, the model provides a good quantitative account of the data given the reliance on one single free parameter (Juslin et al., 1999). ADIT is unable to account for the novel symptom phenomenon, and Kruschke (1996, p. 20) noted that ADIT needs "additional mechanisms not implemented in the model" to account for the non-random guessing strategy observed in the early training trials. The elimination model provides such a mechanism.

Nevertheless, we do not suggest that the mechanism proposed by the elimination model is the only factor that contributes to the BRI effect. The ideas of cue-competition have immense support in the literature on animal learning

and Kruschke's (1996) proposal of a rapid attention-shifting mechanism is both reasonable and appealing. We do, however, take the confirmation of the novel symptom phenomenon as fairly strong evidence that eliminative inferences are at work at least to some extent. The quantitative formulation of the elimination model in Juslin et al. (1999) serves to demonstrate that these processes alone have the potential to produce an BRI effect. The relative importance of explanations in terms of attention-shifting mechanisms and eliminative inferences needs to be determined by future research.

References

Gluck, M. A., & Bower, G. H. (1988). From conditioning to category learning: An adaptive network model. *Journal of Experimental Psychology: General, 117*, 227-247.

Juslin, P., Wennerholm, P., & Winman A. (1999). *The elimination model: The inverse base-rate effect as a result of eliminative inferences.* Manuscript submitted for publication. Department of Psychology, Uppsala University, Sweden.

Kruschke, J. K. (1996). Base-rates in category learning. *Journal of Experimental Psychology: Learning, Memory, and Cognition, 1,* 3-26.

Medin D. L., & Bettger, L.G. (1991). Sensitivity to changes in base-rate information. *American Journal of Psychology, 104,* 311-332.

Medin, D. L., & Edelson, S. M. (1988). Problem structure and the use of base-rate information from experience. *Journal of Experimental Psychology: General, 117,* 68-85.

Medin, D. L., & Schaffer, M. M. (1978). Context model of classification learning. *Psychological Review, 85,* 207-238.

Nosofsky, R. M., Palmeri, T. J., & McKinley, S. C. (1994). Rule-plus-exception model of classification learning. *Psychological Review, 101,* 53-79.

Shanks, D. R. (1992). Connectionist accounts of the inverse base-rate effect. *Connection Science, 4,* 3-18.

Changes in Self-Explanation while Learning Vector Arithmetic

Troy D. Kelley (tkelley@arl.mil)
Human Factors & Applied Cognition Program
George Mason University
Fairfax, VA 22030

Irvin R. Katz (ikatz@gmu.edu)
Human Factors/Applied Cognition Program
George Mason University
Fairfax, VA 22030

Abstract

Verbal elaboration of a worked example has been shown to be helpful to learners before attempting to solve similar problems. This has been termed as the self-explanation effect. (Chi, Bassok, Lewis, Reimann & Glaser, 1989). This study examined how self-explanation changes before and after sequential problem solving rounds. We found that changes in self-explanation within an individual may affect individual performance across a series of problem solving episodes. Also, some participants appear to use the worked-out example as a self-generated feedback (SGF) mechanism to help with their problem solving rounds, while other participants do not. Locations or points in a worked-out example where self-explanation (elaboration) is most likely to occur for students with higher performance scores versus those with lower performance scores, is discussed. The implications of these differences for the design of a computational cognitive model are also addressed.

Introduction

Learning from examples has been shown to be an important aid in the learning process (VanLehn, 1986, 1996). Using examples to provide a basis for learning has also been shown to be the preferred way of learning by novices (Anderson, Farell, & Sauers, 1984; Pirolli & Anderson, 1985; Recker & Pirolli, 1995). However, most research has been conducted on worked-out examples which were presented before problem solving episodes (Chi, Bassok, Lewis, Reimann & Glaser, 1989). Chi et. al.'s (1989) original study was limited since the worked examples were only presented prior to problem solving rounds, which did not allow for an examination of the changes in self explanation as learning progressed. In this study, we will look at how learning might proceed if worked-out examples are presented following problem solving episodes instead of prior to problem solving.

In the seminal work on learning from examples (Chi, Bassok, Lewis, Reimann & Glaser, 1989), a *self-explanation effect* was found in effective learners who could, among other things, use a worked-out example to elaborate upon broader principles which they had previously acquired while studying text. The authors also found that effective learners monitored their own performance and knowledge base better than ineffective learners; which has been confirmed by some researchers (Ferguson-Hessler & DeJong, 1990), but questioned by others (Renkl, 1997).

We were interested in how self explanations might change as subjects examined worked examples after problem solving rounds. One hypothesis might be that subjects will have a strategy of using the worked-out example which follows the problem solving differently than they used earlier worked examples. Given that subjects have had the experience of attempting to solve earlier problems, they may choose to use the latter worked examples as a feedback mechanism to the previous problem solving rounds. We characterized this as self generated feedback (SGF). This feedback mechanism would allow subjects to use a strategy of analyzing their previous problem solving rounds (from memory) in order to improve subsequent problem solving. However, other subjects may decide not to use the latter worked examples as a feedback mechanism, instead they may concentrate all of their learning efforts on the first worked example. In which case, these subjects would not show any signs of SGF.

Our research agenda addresses: 1) How do self-explanations change as performance improves? 2) Do subjects use self-explanation strategically and can these strategies be detected? 3) Where in the worked example is a subject most likely to engage in self-explanation behavior; are these locations stable across different worked examples?

We wanted to examine the possibility that subjects might have an identifiable strategic use for the different worked examples. One strategic use could be that subjects would use the later worked examples as a feedback mechanism to their earlier problem solving episodes. If this were occurring, this would change the nature of the latter self-explanation statements. Subjects would begin to make statements which referenced earlier gaps in their knowledge. For example, a subject might say, "Oh, now I see how to use that equation, that is not what I was doing before."

We hypothesized that if participants were using the second or latter worked examples as a SGF mechanism, then they would show fewer self-explanation statements than learners who relied heavily on the first worked-out example. Also, they would utter fewer words on the first worked example and instead concentrate their efforts on the latter worked examples. However, if a subject were relying heavily on the first worked example to establish a foundation for subsequent problem solving, and not relying on latter worked-examples to provide SGF to their problem solving episodes; then the change in explanation statements should be significant. Also, the number of utterances while studying the first worked example would be relatively high. We also hypothesized that if a subject were using a delayed SGF strategy, then their performance would be worse than a subject who was not using the strategy.

Finally, we wanted to identify specific points in the worked example which were likely to trigger a self-explanation episode. This was particularly critical for building any models of the self-explanation process over time. As our model proceeded through the worked example, the model's actions should correlate with what the average subject does at each line of the example. If certain lines in the worked example are more likely to trigger explanation events by the participants, then the model should also respond with explanation events in the same places of the worked example. Furthermore, it was necessary to see if the worked examples differed in the places where explanation statements were likely to occur, as the subjects progressed through the experiment.

Method

Participants

Participants were seven high school juniors and three college students. The high school students were recruited from the same physics class which, a few months earlier, had covered vector arithmetic concepts simpler than those covered in the present experiment. The college students (freshman and two seniors) had no prior physics training beyond a high school course. All were paid for their participation.

Materials and Design

Participants completed four self-explanation tasks conducted at regular intervals throughout the experiment. Each task consisted of the participant talking aloud while studying one of four worked-out examples. All participants were given the same tasks to perform in the same order.

Problem solving tasks.

Participants completed four rounds of problem solving, three sets of 10 problems and a final round of 8 problems. All the problems were similar to the worked examples in that they described two or more vectors (in the context of the story) and asked the participants to find the magnitude and direction of the resultant. Unlike the worked examples, the problem statement was included no diagram.

Procedure

Students participated in individual sessions lasting 2.5 - 3 hours. To complete the experiment, students generally attended 5 sessions, although some students needed fewer sessions. In the first session, students studied a textbook chapter on vector arithmetic and completed the first self-explanation task. For the self-explanation task, students were asked to, "Study this example as if you were studying for a test. Try to understand why each solution step was taken and why the solution correctly answers the question." Students had access to a calculator while performing all tasks. During the study, students were asked to "talk aloud" (Ericsson & Simon, 1993), providing concurrent verbal protocols.

The remaining sessions included both self-explanation tasks and problem solving. In the second session, students completed 10 vector arithmetic problems (referred to as round 1 or R1), without any feedback on the correctness of their responses. The work done by the students during the problem solving episodes was not analyzed in great detail for this paper. We were primarily concerned with the work done by the students while they where studying the examples. Information about the problem solving rounds is presented here for clarity and completeness.

In the third and fourth sessions, students completed a self-explanation task (SE2 and SE3) followed by solving of another set of 10 vector arithmetic problems. In the final sessions, students completed a set of 8 vector arithmetic problems, some of which were easier and some of which were more difficult than problems completed during the previous sessions. After this fourth round of problem solving (R4), students completed the final self-explanation task (SE4). Students were then debriefed.

Results

Background

This study proceeded in two major phases and consequently the results will be presented in two major sections: replication and changes in self-explanation. The replication section addresses the major findings of the Chi et al. (1989) study. The "changes in self-explanation" section includes discussions on the observed changes in self-explanation over time. There are also subsections which include the strategies participants exhibited while using the worked examples and the computational cognitive model which was based on expert performance while studying one worked example.

Replication

Replication of the Chi et al (1989) study was conducted to determine the validity and generalizability of our data and to determine if there were any inconsistencies with the original work of Chi et al. (1989). However, a few problems were encountered. Most of the data which Chi et al. (1989) analyze was split into good versus poor performers, of which they seemed to have a clear delineation. We conducted a similar split (a median split) however, we had only one subject performing above 50 percent.

Chi et al. (1989) first analyze their data determining a count of the number of phrases made by good and poor students during the problem solving episodes. Instead of using a line count, we used word count by each subject while they were studying the example. Because some of the verbal protocol lines were long, whereas other lines were single words or phrases, we feel a word count might be more accurate than a line count. They find the line count to be significant "(142 lines versus 21 lines, t(6) = 1.97, p < .05)." We found that on average the good students uttered more words than the poor performing subjects (1119.33 versus 586.75, t(5) = 1.96, p > .05) performance on the first set of ten problems for 7 subjects in the experiment. A fairly high Pearson's correlation coefficient of (r(6) = .42, p > .05) was obtained but, due to the small sample size, this was not significant.

Next, Chi et al. (1989), found that good students produced significantly more explanations that related to the content of the problem than did the poor students (15.3 versus 2.8). Our data support this result (22.6 versus 12.5). Chi et al. (1989) go on to analyze the number of times the good and poor students refer to the worked-out examples during problem solving rounds. They found the good students referred to the example less often than did the poor students. We looked at the two best performing subjects (with performance scores i.e. questions answered correctly of 80 and 50) in comparison with the two worst performing subjects (with performance scores of 20 and 30) and found that the best performing students refer to the example fewer times than did the worst performing subjects. Furthermore, we found that the amount of explanations while studying an example was correlated to the subsequent performance of the subject during the following problem solving rounds (r(6) = .65).

Our data was consistent with the results presented by Chi et al. (1989), with the exception of one area - the amount of negative versus positive monitoring statements uttered by participants. Positive monitoring statements included statements such as: "OK, I understand this", while negative monitoring statements consisted of statements such as, "What does this mean? I don't understand." Chi et al. (1989) found significant results on the negative monitoring variable. Poor performers averaged 1.1 negative monitoring statements while good performers averaged 9.3 negative monitoring statements. This is where our data is inconsistent with Chi et al.'s (1989) original findings. We found, if anything, that negative monitoring and subsequent performance seemed to be slightly inversely related, however the result was non-significant (r(6) = -.39, p > .05). Good students had an average of 3.3 negative monitoring statements while poor subjects had an average of 4.0 negative monitoring statements. Renkl (1997) also found no relation between the amount of negative monitoring statements during the study of a worked-out example subsequent problem solving performance.

Changes in self-explanation

As an extension to the Chi et al. (1989) data, we were interested in three major points. 1) How do self-explanations change as performance improves during problem solving? 2) Do subjects have a specific strategy of using the latter worked examples as a feedback mechanism to the earlier problem solving rounds and does this affect their subsequent performance on the problem solving rounds? 3) Where in the worked examples is a subject most likely to engage in self-explanation behavior, and how does this likelihood to explain change across problem solving rounds?

Nine out of ten subjects showed an decrease in the amount of explanation statements for the second worked-out example. The total amount of explanation statements for 10 participants for the first worked-out example (SE1) was 163, or an average of 16.3 per subject. For SE2 and SE4 the explanation statements dropped to 96 (9.6 per subject) statements for SE2 and 76 (7.6 per subject) explanation statements for SE4. This change in explanation statements yielded a significant sign test of $(X2(1) = 6.4, p < .025)$. A Wilcoxon matched-pairs signed-ranks test showed a significant difference between SE1 and SE2 on explanation (t(10) = 10, p < .05) as well as between SE1 and SE4 on explanation (t(10) = 0, p < .005). A significant difference was also found for overall word count. The Wilcoxon matched-pairs signed-ranks test yielded a significant effect for number of words for SE1 compared with SE2 (t(7) = 0, p < .001).

In general, while the overall trend in explanation statements decreases across the different worked-out examples as problem solving continues, learning is still continuing as evidenced by the improved performance of each subject across the rounds. Out of ten participants, the average improvement in score from the first problem solving round to the last problem solving round was 77 percent. The most improved subject (JE08) went from a score of 20 on the first round to a score of 100 by the last round of problem solving.

Across all the participants, the percentage of explanation statements, in relation to their total number of statements (which would include the extra categories of "read", "monitor" and "other") did not change across the self-explanation rounds. The percentage of all statements which were explanation statements, for 10 subjects, for SE1, SE2 and SE4 was 27%, 26%, and 27% respectively. However, there were some small and relatively consistent differences when comparing good versus poor participants. On SE1, good performing subjects had 30 percent explanation statements while poor performing subjects had 26 percent. By SE2, the gap widened slightly with the good students having 30 percent explanation statements while the poor students had 24 percent explanation statements. Finally, on SE4, this difference was still fairly consistent with the good students having 30 percent explanation statements and the poor subjects having 25 percent explanation statements.

Self-explanation strategies

In general, some participants appeared to use the first worked example (SE1) to provide a solid foundation for their subsequent problem solving rounds, which we termed the upfront strategy. This was evidenced by an apparent decrease in word count from SE1 to SE2. These learners appeared to expend less effort while examining the second worked example, as compared to the first, and consequently had a reduction in word count. However, other participants used SE2 as more of a SFG mechanism, relying on it to fill in any gaps they may have had in their knowledge which they may have still had after the first round of problem solving. We termed this the catch-up strategy. Again, this was evidenced by the increase in word count from SE1 to SE2. We hypothesized that if participants were using SE2 as a SGF mechanism, then they would show less of a reduction in word count than a subject who relied heavily on SE1. However, if someone is relying heavily on the first worked example to establish a foundation for subsequent problem solving, and not relying on SE2 to provide SGF to their problem solving episodes; then the change in word count from SE1 to SE2 should be large. More importantly, the number of explanation statements should show the same change in direction as was hypothesized for the word counts.

We found that changes in word count from the first worked-out example (SE1) to the second worked-out example (SE2) were correlated with performance in the hypothesized direction ($r(6) = .65, p > .05$). However, the more sensitive count of explanation statements, and using more participants, produced a small negative correlation in the opposite direction ($r(9) = -.13$). Those subjects we had categorized as using the upfront strategy, based on their change in explanation statements, had a total combined score 47.5 questions answered correctly. Those subjects which were categorized as using the catch-up strategy, based on their change in explanation statements, had a total combined score of 51.3 questions answered correctly. So one would have to conclude that even though the changes in word counts were occurring in the hypothesized direction, changes in explanation statements, which must be considered a better indicator, were not occurring in the hypothesized direction.

The subjects which had the smallest absolute changes in word counts from SE1 to SE2 were subjects KB07, JE08, MT16 and MT11 respectively. These protocols were searched for examples of possible SGF examples, and examples were found for subjects KB07, JE08 and subject MT11. These appeared to be instances where the participant was referring back to the previous problem solving rounds while studying a later worked-out example. For this analysis, we concentrated specifically on the latter worked examples after the problem solving episodes (SE2 and SE4) and looked for any statements which made references to earlier problem solving episodes.

While examining the fourth worked example, on lines 30 to 32, subject KB-07 makes these statements:

30) Okay, add 'em up, you get Rx and get Ry
31) Woa, Woa, Woa, Oh.. so that's where you put in
32) but it's still positive 1.8 ft.

This subject is examining where certain positive and negative values came from and appears to realize at what part in the process of solving the equation that the values are actually necessary.

While examining the second worked example, on lines 77-84, subject KB-07 realizes:

77) X squared
78) Rx, Ry squared
79) So we're looking
80) Oh!!
81) So we're looking for this too
82) So this would be equal to 6.25, wait.

Participant KB-07 has realized that part of the process she had previously used did not include a necessary step. Hence the exclamation, "oh, so we are looking for this too".

Participant JE08, on lines 54 to 56 makes the statement:

54) The direction of R may be found using an inverse trigonometric function such as arctan
55) Now here is where I get lost.

Subject JE08 knows from previous problem solving rounds that there is a gap in her declarative knowledge which still has not be resolved, even by the time she gets to the second worked example.

While examining the second worked example, on lines 49 to 53, subject MT-11 makes these statements:

49) Ok, they find out where this was
50) So,..... and use this right angle
51) arc tan
52) adjacent over
53) hypotenuse ... Ah, it doesn't matter

The subject has realized that the example provides an alternative way to approach the problem from what the subject had previously been doing. The subject realizes, from studying the example, that a step the subject had previously taken while solving the problem "doesn't matter", and the example shows how the step can be eliminated.

So while the explanation statements did not decrease in the anticipated direction to show evidence of possible SGF for those subjects who we categorized as using the catch-up strategy, there was some indication within the protocols that SGF was taking place.

Model data

Results from our analysis will be used to build a cognitive model of self-explanation behavior. An expert level model of a subject solving vector arithmetic tasks has already been developed. The model solves a vector arithmetic problem by using the first worked-out example (SE1) as a guide. The model assumes expert performance in that the model knows what each next step is, and the model knows in what order to do each step. The model has four basic decision points as it precedes through the worked-out example. The model uses logical evaluations at each of the four steps to determine the information needed by the model to find any unknown variables, then it proceeds to the next step. The progression is linear, through the worked-out example, toward a solution.

To further analyze the data, and to further help us develop our cognitive model, we were interested in where in the worked-example participants were elaborating or doing self-explanation. Most participants proceeded in a linear fashion through the worked example. As a subject reached each line of the example, the number of explanation statements that occurred at that line were totaled. The highest points of explanation occur at lines 6, 7, 12, 14, 17 and 21. The four major decision points of the previously described expert model occur at lines 6, 7, (one decision) 11, 12, (one decision) 17 and 21. So the model seems to be making critical decisions at the appropriate points in the worked-out example. These points appear to be occurring primarily at mathematical areas of the worked example (formulas) and not textual (written) sections.

The total number of explanation statements for seven subjects were also totaled for each example (not just SE1 on which the model was based upon) and the totals were then compared. As with SE1, subjects appear to be concentrating on the formulas of the worked examples much more than the textual components of the worked examples. Also, the subjects studying SE2 and SE4 tended to do a great deal of explaining near the end of the round. For SE2, which had 10 lines, the largest amount of explanation statements occurred at lines 10, 7 and 6 receptively. For SE4, which had 21 lines, the largest number of explanation statements occurred at lines 11, 14, and 15 (which were tied) and lines 21, 10 and 2 (which were also tied). A direct comparison of SE1 with SE4 was possible since these worked examples were highly similar to each other. Generally, in percentage terms, the amount of explanation statements decreases from SE1 to SE4. But also, there was a tendency for explanation statements to occur later in the worked example for SE4 as opposed to SE1.

While the current version of the model can account for the areas where a participant is most likely to self-explain, it does not account for the tendency of subjects to do most of their explaining near the end of the worked example. While the model does show and increase in explanation statements near the end of the example, it is not proportional to the amount shown by the subjects.

Further development of the model needs to take place in two specific areas. First, in order to account for our empirical data, the model needs to show explanation statements at the very end of the worked example. Secondly, model does not simulate what we have called self-generated feedback, and does not account for the utterances we identified in our protocols as SGF. SGF is an important change in self-explanation behavior that does not occur during the first SE round but rather it is more likely to occur in the latter rounds. This change in self-explanation behavior must be addressed by our model of the self-explanation effect.

Discussion

It is clear that the self-explanation effect is a powerful phenomena in the study of examples. Replication of the self-explanation effect has been conducted by other researchers (Renkl, 1997). The majority of our data shows a clear indication of the self-explanation effect as it was first defined by Chi et al. (1987). However, we did find differences from the Chi et al. (1987) study in the amount of negative monitoring statements and performance during problem solving rounds. Our data is consistent with that of the later Renkl (1997) study, therefore it would be difficult for us to conclude that increased negative monitoring is one of the underlying features of the self-explanation effect.

Beyond replication, we were interested in examining how self-explanations change as performance improves during problem solving. We found that self-explanation decreases as problem-solving performance increases. This would be expected if self-explanation was occurring in order to fill in gaps in their declarative knowledge. If subjects were using self-explanations to fill in gaps in their declarative knowledge base, then as their performance improved, then there should be less need to do any self-explanation.

Next, we addressed the question of whether subjects have a specific strategy of using the latter worked examples as a feedback mechanism to the earlier problem solving rounds and does this affect their subsequent performance on the problem solving rounds. What we found were examples in the protocols of subjects using the latter worked examples as a feedback mechanism to their earlier problem solving episodes. This occurred in subjects who had the smallest change in word count from SE1 to SE2. However, we could not find any reliable changes in performance from the subjects using this strategy.

Finally, we examined areas in the worked example where a subject was most likely to engage in self-explanation behavior, and how does this likelihood to explain change across problem solving rounds. We also found that subjects tended to do a great deal of explaining at the end of the worked-examples, especially for latter worked examples (SE2 and SE4). This was also supported by an analysis of the best performing subject (AM19), who seemed to do most of her explaining at the end of the example, while the worst performing subject (JE08) did not. We construed that participants will frequently reflect at the end of a problem solving episode, even when no gap in their declarative knowledge has been identified. Apparently a significant amount of learning could occur during these reflective, non-impasse periods, and this assumption is supported by the data of our best performing subject (AM 19), as well as our aggregate data. The fact that the best performing subject uses this strategy might reflect differences in good versus poor performance of participants. Perhaps good students tend to be more reflective after problem solving, while poor performing students do most of their problem solving only when they encounter a gap in their knowledge. However, the important assumption here is that gaps are not present when the learner has reached the end of the worked example. Obviously, a counter argument could be made that this has not been proven to be the case, and more research needs to be done to clarify this issue. However, if one accepts the assumption that a gap in knowledge is not likely to be identified after an example has been completed, then this is consistent with VanLehn (1992) findings that not all learning occurs at impasses.

Other models imply that there are strategy differences within participants who self-explain. The Cascade model (VanLehn, 1992) uses strategy differences to distinguish between good and poor learners by forcing the simulation of good learners to rederive an example's solution, while the simulation of poor learners never attempts any new derivations. Their model also found that this strategy caused the good learner model to acquire more knowledge while solving problems than the poor learner model. Our data indicates that participants continue to learn even while self-explanation behavior decreases, and it would seem that the Cascade model can account for this aspect of our data.

Our computational cognitive model simulated subjects' performance during the first worked-example. We found that our model of SE1 was consistent with our empirical data. The model engaged in self-explanation behavior in the same areas where participants were most likely to engage in self explanation behavior. However, our model does not account for the large number of self-explanation statements that occur at the end of the problem solving episodes. The model also needs to incorporate specific instances of SGF, which we had identified in the latter protocols of subjects (SE2 and SE4). These are two important aspects of the change in self-explanation behavior, which will be addressed in future versions of the model.

References

Anderson, J. R., Farrell, R., & Sauers, R. (1984). Learning to program in LISP. *Cognitive Science, 8*, 87-129.

Chi, M. T. H., Bassok, M., Lewis, M. W., Reimann, P., & Glaser, R. (1989). Self-explanations: How students study and use examples in learning to solve problems. *Cognitive Science, 18*, 145-182.

Ferguson-Hessler, M. G. M., & DeJong, T. (1990). Studying physics texts: Differences in study processes between good and poor performers. *Cognition and Instruction, 7*, 41-54.

Pirolli, P., & Anderson, J. R. (1985). The role of learning from examples in the acquisition of recursive programming skills. *Canadian Journal of Psychology, 39*, 240-272.

Recker, M. M., & Pirolli, P (1995). Modelling individual differences in students' learning strategies. *The Journal of the Learning Sciences, 4*, 1-38.

Renkl, A., (1997). Learning from worked-out examples: A study on individual differences. *Cognitive Science, 21*(1), 1-29.

VanLehn, K. (1986). Arthmetic procedures are induced from examples. In J. Hiebert (Ed.), *Conceptual and procedural knowledge: The case of mathematics.* Hillsdale, NJ: Erlbaum.

VanLehn, K. (1992). A model of the self-explanation effect. *The Journal of the Learning Sciences 2*(1), 1-59.

Modeling Perceptual Learning of Abstract Invariants

Philip J. Kellman
<Kellman@cognet.ucla.edu>

Timothy Burke
<MizeRai@ucla.edu>

John E. Hummel
<Jhummel@psych.ucla.edu>

University of California, Los Angeles
405 Hilgard Avenue, Los Angeles, CA 90095-1563

Abstract

We present the beginnings of a model of the human capacity to learn abstract invariants, such as *square*. The model is founded on four primary assumptions, which we believe to be neurally plausible and generic: Metric space, Topology, Comparison operations (subtraction, greater-than/less-than), and Extraction of vertices. The model successfully learns to discriminate simple planar quadrilaterals, and generalizes that learning across variations in viewpoint and modest variations in shape.

Introduction

A hallmark of human information processing is the ability to detect and respond to abstract invariants. Many of the most important outputs of visual perception and cognition involve shape and spatial relations. As the Gestalt psychologists pointed out early in this century, these are relational notions. A square, or a melody, is not definable in terms of any particular elements, but in terms of invariant relationships.

We not only encode certain characteristic relations, but we discover new ones with experience. This ability makes possible high level, sometimes almost magical, expertise. In chess, the best human player can compete with a computer system that examines 250 million moves per second, although plainly we did not evolve to process the specific relations of shape and arrangement that are important in chess. Human perceptual learning allows the discovery of features and relations that make possible efficient performance in almost every domain of expertise (Gibson, 1969; Goldstone, 1998).

Some key aspects of perceptual learning remain deeply mysterious. A crucial one might be called the *discovery problem*: Expertise in a classification task grows as new stimulus relationships, often quite abstract, become the basis of response. How does the visual system discover abstract invariants, such as squareness, roundness, or parallelism? By abstract invariant we mean a visual property that, while *computable from*, is not *definable in* the vocabulary of primitive features from which it is derived. For example, no logical concatenation—conjunctive, disjunctive or otherwise—of neural responses in primary visual cortex (i.e., local visual properties such as edges, vertices or Gabor components) defines the invariant *squareness*. Squareness is both more and less than any finite set of such features. It is more because some new activation pattern might also form a square, and it is less because many of the attributes of any given activation pattern (e.g., its precise location, size and orientation) have nothing to do with their squareness.

Because an abstract invariant is defined less in terms of the visual primitives that compose it than in terms of the relations among those primitives, it is a mystery how the visual system discovers invariants in the outputs of such primitives. It is possible to build a *local* feature detector by systematically combining the outputs of a finite number of simpler feature detectors (e.g., it is possible to define an edge detector based on a weighted sum of the outputs of a finite number of contrast detectors). Learning such a feature is therefore a relatively simple matter of finding an algorithm to discover the appropriate "wiring diagram" (i.e., weighting terms) from the input (a specific set of contrast detectors) to the output (the local edge detector). Many such learning algorithms exist, including supervised learning algorithms such as back-propagation (Rumelhart, Hinton & Williams, 1986), as well as numerous unsupervised learning algorithms. All these algorithms work precisely because (a) they operate by exploiting statistical regularities in their inputs (or, in the case of supervised rules, regularities in the input-output mappings), and (b) the feature to be learned can be defined in terms of the more basic features of which it is composed. By contrast, because squareness does not correspond to any finite collection of local features, there exists no analogous wiring diagram that can detect all (and only) instances of "squareness". As such, there is no straightforward statistical basis for learning "squareness" based on the outputs of local feature detectors.

We assume that an invariant such as "squareness" is not detected simply by constructing a large (potentially infinite) number of detectors for specific squares and then summing their outputs. As the Gestaltists argued against their structuralist predecessors, we can always devise a new square of a different size, made by arranging some new elements, in a new position. It would still be readily detected as a square. An algorithm geared to learning each and every possible instance of a square would be unwieldy to say the least. Humans, by contrast, can learn many invariants in as little as one exposure and transfer that learning to new instances.

How can we account for this performance? Although perceptual learning of abstract relationships is well documented (e.g., Chase & Simon, 1973; Gibson, 1969; for an excellent recent review, see Goldstone, 1998), little modeling has addressed the learning of abstract invariants. Our aim in the present program of research is to develop models of invariant detection and learning in visual shape perception. Here we report initial progress toward those goals.

Mixing Architectures in a Principled Fashion

What appears difficult about the problem of detecting invariants may depend on one's starting point. From the standpoint of symbolic descriptions, squareness does not seem too formidable. Given a closed figure, its vertex locations, edge lengths and angle measurements, we can easily write a mathematical test for squareness. But it would be unwieldy to have such a routine for every spatial invariant. Nor is it clear how such static routines could come to learn *new* invariants. On the other hand, traditional connectionist approaches readily support learning, but will not come to classify (correctly) new squares if their initial inputs are limited to concrete elements such as local features.

Toward a Grammar of Form

What initial recodings of concrete inputs could allow both the extraction of abstract relationships and the learning of new ones? At the root of our approach is the idea that the apparent openendedness of the human ability to learn abstract invariants calls for a system that is formally like a grammar with recursive rules. Natural languages can generate an unlimited number of novel sentences based on a fixed set of words and a recursive rule system for combining them. The same idea has been applied to object recognition: Objects may be decomposed into a finite vocabulary of volumetric primitives and spatial relations connecting them (e.g., Biederman, 1987; Marr & Nishihara, 1978). It seems plausible that human perceptual learning of abstract spatial invariants depends on some basic set of relations and some processes for concatenating them.

In the present work, we pursue this approach in building a prototype *shape network* that extracts a small set of important relations early on and uses them as inputs for learning to classify simple shapes. We choose as our domain the names of simple planar quadrilaterals, including squares, rectangles, parallelograms, trapezoids and rhombuses. This domain is useful for several reasons. Most importantly, the labels refer to abstract entities. Changes in retinal position, scale, and planar orientation (for the most part) of the constituent elements do not affect the correct labels. Second, this kind of classification is arguably natural for humans. Young children readily come to distinguish and name squares, rectangles and so on, and generalize naturally across changes in size, position and orientation. Third, planar shape classification—and the subtleties of quadrilaterals in particular—hinge on interesting spatial relations and comparisons. Ultimately, our aim is to encompass richer domains of shape description, including 3-d form, but the planar shape domain is a reasonable choice for confronting the basic challenges of how encoding and learning might cope with abstract spatial relations.

A key challenge is to postulate only those steps for recoding or finding relations that can be justified on independent grounds. That is, success will not consist of finding a special-purpose processor that responds to squareness, but allowing squareness to emerge naturally from basic operations that we would expect, on independent grounds, to be part of visual processing. By implementing only operations that can be justified independently, we can begin to develop the fundamental grammar of invariant processing and ultimately test whether the scope and limits of what is learnable within that grammar approximate those of human visual cognition.

Based on a small set of early recodings that make sense on general grounds, this first shape network can learn abstract notions—specifically various kinds of quadrilateral (4-vertex, planar) figures. It also classifies new instances, such as a square of a size and position not previously seen.

Modeling Assumptions

The vocabulary of a "visual grammar" consists of a finite set of basic operations corresponding to (presumably innate) assumptions on the part of the visual system about the nature of the visual world. For our current purposes, we assume that the visual system comes into the world equipped with (at least) the following knowledge/capacities:

I. Metric Space: We assume that neurons in early visual processing (e.g., retina, LGN, V1) "know" (perhaps implicitly, in the form of their interconnections) about metric space: They know that their receptive fields correspond to finite regions of larger metric space, and they know (approximately) where in that space their receptive fields are located. We assume that this knowledge manifests itself in a neuron's (or hypercolumn's) ability to signal its location independently of any of its other properties, namely by activating other neurons representing location (e.g., in Euclidean coordinates) independently of the nature of whatever visual features happen to reside there.

II. Adjacency (Topology): Implicit in (I), but deserving of mention, we assume that neurons in early visual processing "know about" their adjacency relations (and possibly other topological relations). This kind of knowledge is manifest in the local lateral connectivity among, for example, neurons in visual area V1.

III. Difference Operations: Third, we assume that, given pairs of numerical quantities (e.g., coordinates in a Euclidean space), the nervous system is equipped with routines for performing various kinds of difference operations, including subtraction, and evaluating greater-than and less-than relations.

IV. Vertex Finding: Finally, we assume that early visual routines can find vertices and other local changes in contour curvature based on the outputs of basic local edge computations.

These four assumptions, along with the way they are embodied and the ways in which they interact, form the theoretical foundation of the current modeling effort. Our goal in this paper is to demonstrate that, embodied in an appropriate architecture, these assumptions are sufficient to "bootstrap" the learning of abstract invariants such as "square."

The Shape Network Model

The assumptions are instantiated in a four-layer "neural"-style network. Units in the first layer represent the retinal coordinates of the vertices in an image of a quadrilateral. We assume that the vertices are detected and their spatial coordinates registered by an early preprocessing stage (Assumptions IV and I, above). (We acknowledge that simply "handing" the model the coordinates of vertices is a strong simplification. We are currently working to relax it and equip the model with a more realistic front-end.) Units in the second layer compute the pair-wise Euclidean distances between the coordinates coded in the first layer (Assumptions I and III). In the current implementation, each unit represents one distance (e.g., there is one unit for the distance between vertex 1 and vertex 2, another for vertices 2 and 3, etc.), and distance is represented as activation. That is, in Layer 2 layer, distance is rate-coded. Coordinates are "read into" the model in a fixed order, starting from some corner on the stimulus, and proceeding around the figure clockwise. This convention is a simplified implementation of our more general assumption that the system knows which vertices are connected to which by virtue of an intervening contour (Assumption II, topological relations). (In the current model, implicit knowledge of sequential order of vertices in a connected figure is important, although the particular starting point in the sequence is not.)

Units in the third layer take their inputs from pairs of distance units (i and j) in the second layer, and compare the distances for their equality (Assumption III). Specifically:

$$e_{ij} = \begin{cases} 1 & \text{if } m|d_i - d_j| < \delta \\ \dfrac{\delta}{\left|m|d_i - d_j|\right|} & \text{otherwise,} \end{cases} \quad (1)$$

where e_{ij} is the activation of equality unit ij, d_i and d_j are distances i and j, δ is a difference threshold that determines the rate at which e drops off as the absolute difference in distances increases, and $m = \max(d_i, d_j)$, serves as a scaling factor. δ was set to .02 in the simulations reported here. e_{ij} takes a value of 1 when d_i and d_j are within δ of being equal, and falls off toward zero as |d_i - d_j| approaches infinity.

The resulting pairwise distance comparisons serve as the input to the fourth (output) layer of units, which learn to classify their inputs as representing various four-sided geometrical figures (squares, parallelograms, trapezoids, etc.). The net input to output node i is simply the dot product:

$$n_i = \Sigma_j \, a_j \, w_{ij} \quad (2)$$

where a_j is the activation of node j in layer 3, and w_{ij} is the weight on the connection from j to i. The activation of output node i is given by the logistic function:

$$a_i = 1/(1 + e^{-n_i}). \quad (3)$$

At the beginning of training, the connections between the third and fourth (output) layers were initialized to zero. On each training trial, one four-sided figure was presented at a time to the network (as detailed below), and layers of units were updated in sequence (layer 1, then layer 2, etc.) until a pattern of activation was generated on the fourth (output) layer. The connections between the third and fourth layers were modified according to the difference between the actual activation of output node i (a_i) and the desired output (d_i) in response to the training pattern (Rumelhart et al., 1986):

$$\Delta w_{ij} = \eta a_j \, (\, 1 - a_i)(a_i - d_i\,), \quad (4)$$

where η is a learning rate.

Simulations

The present simulations were to designed to test the model's ability to learn invariants such as square, rectangle, etc., from a small number of examples (typically one) and to explore its ability to generalize to new instances that differ from the training examples in their location, size and orientation. We also explored the model's ability to generalize across small distortions in the coordinates of a figure's vertices.

Training

We trained the model to classify six types of quadrilaterals. They are shown in Figure 1 along with the coordinates used for training. The model was trained on one example of each, except for the trapezoid, of which there were two examples.

We trained the model to perform two types of classification. *Inclusive* classifications required the model to respond to each stimulus with every label that applied to it (e.g., a square is also a rectangle, a rhombus, etc., so in this condition, the model was trained to respond with all these labels given a square as a stimulus). *Exclusive* classifications required the model to respond only with the most specific label corresponding to a stimulus (e.g., a square would activate only the square unit). The exclusive classification is arguably the more humanly natural, and serves as the basis of the majority of the tests reported here. During training, stimuli were presented one at a time, and the weights at the output layer were corrected in response to the model's output, as described above. The weights were updated "in batch", after all stimuli had been presented. Training proceeded until the mean square error of the model's response at the output layer fell below .01. The classification tasks were trained separately and stored as separate matrices of connection weights. The model's responses to the trained stimuli are shown in the second column of Table 1 (*Training*) for the purposes of comparison with the results of the other simulations. Not surprisingly, the model learned to classify the patterns on which it was trained.

266

Generalization Across Viewpoint

We next tested the model with translated and scaled version of the stimuli on which it had been trained. Table 1 shows the model's responses to scaled, oriented and distorted versions of the training stimuli using the exclusive response criteria.

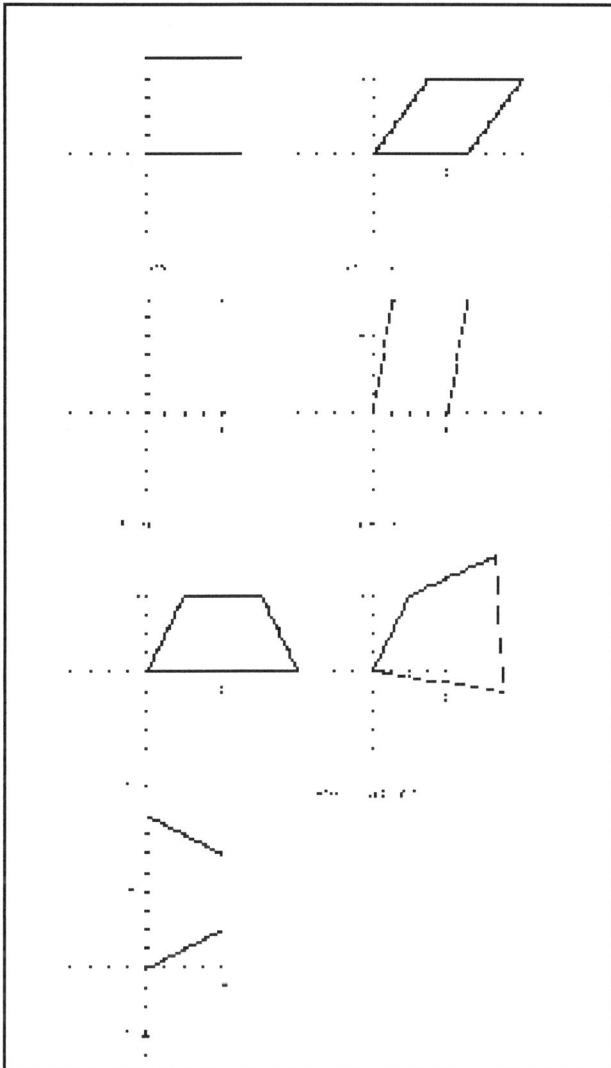

Figure 1. Training Stimuli.

Results were similarly successful in the inclusive labelling condition: generally, the model correctly assigned all relevant labels to a given test display.

These results demonstrate that the model treats the trained figures as invariants, responding equivalently to them regardless of their location in the visual field, size, or orientation. This kind of strong invariance with orientation illustrates a limitation of the current simplified model. To the human visual system, a square rotated 45 in the picture plane may be a diamond, not a square (Mach, 1897).

Tolerance for Distortion

Human shape classification includes some tolerance for distortion. A square with one side slightly too long may still look squarelike. As asymmetry increases, squareness decreases and other classifications become more probable. We do not have quantitative data about human observers' tolerance for distortion, or whether it varies by task, etc. Qualitatively speaking, however, we would expect a plausible model to accept some distortion but not too much in maintaining a shape response.

In the model, this tolerance is controlled by the parameter δ, which modulates the magnitude of response for departures from equality of given length pairs.

Table 2 shows the model's responses with the value of δ used in these simulations (.02). It can be seen that, as would be expected of human observers, small (2%) distortions that technically violate squareness are ignored. Larger deviations do change the response of the model, however. For example, a 12% lengthening of parallel sides leads the network to abandon the "square" response and classify the shape as a rectangle.

Shape	*Training*	*Scale*	*Orientation*	*Proportion*
Square	.94	.96	.94	--
Rhomb.	.93	.97	.93	.93
Rect.	.93	.93	.91	.92, .83
Par.	.91	.92	.92	.93
Trap.	.95, .95	.98	.95	.91, .45[a]
Quad.	.97	.97	.97	.34[b]

Table 1. Classification Results for Exclusive Categorization. Classification scores are shown for transfer tests involving changes in scale, orientation and proportion. Classification scores were calculated from shape network outputs as: c/(c+w), where c is the network's output for the correct response and w is the network's highest response for any incorrect response. Scale changes consisted of a doubling of all lengths. Orientation was changed by 90 deg from the training set except for the square, which was rotated 45 deg. Proportion change displays varied and also included position and orientation changes.

a The network gave a higher score to parallelogram (.51) for this display (vertex coordinates: (2,3), (5,6), (9, 8), (4, 3).)

b The network gave a higher score to parallelogram (.66) for this display (vertex coordinates: (0,0), (5,2), (7, 0), (3, 1).)

Vertex Coord.	Technically Correct Shape		Shape Net's Highest Resp.	
(0.0,0.0)(0.0,5.0) (5.1,5.1)(5.0,0.0)	QUAD	.004	SQU.	.94
(0.0,0.0)(0.0,5.1) (5.0,5.1)(5.0,0.0)	RECT.	.20	SQU.	.80
(0.0,0.0)(0.0,5.0) (5.6,5.6)(5.0,0.0)	QUAD.	.043	RHMB.	.97
(0.0,0.0)(0.0,5.4) (5.0,5.4)(5.0,0.0)	RECT.	.88	RECT.	.88

Table 2. Distortion Tolerance. (See text.)

Discussion

The shape network succeeded in learning to distinguish squares, rectangles, parallelograms and other quadrilaterals based on training with one example of each (two examples of trapezoids). It did so in both inclusive and exclusive response conditions, generally activating strongly all correct shape labels in the former condition and limiting its response to the single most specific (and correct) choice in the latter condition. The model correctly classified new instances that had no overlap in coordinates or edge lengths with the training instances. The basic comparisons built into the early layers of processing made possible learning of abstract invariants involving spatial relationships.

Although the model is simplified in a number of respects and performs a restricted shape classification, it demonstrates how combining early registration of certain relations with learning in a neural network might account for discovery of abstract invariants. The interest of the approach depends on whether the basic operations in the model are *ad hoc* manipulations to perform the task under study or are likely to be part of a basic inventory of relations in a visual "grammar." How plausible are the key ingredients here?

Finding Vertices. The model assumes the ability to locate vertices—points of contour slope discontinuity—in the visual array. This makes sense if the human visual system readily encodes such points and if they are required for a variety of tasks in visual processing. For vertices, this appears to be the case. The mere fact that these points have many names in the literature (e.g., vertices, tangent discontinuities, key points, etc.) is suggestive. Location of vertices appears to be required for many middle and high-level visual processes, such as contour interpolation (Shipley & Kellman, 1990) and object recognition (Hummel & Biederman, 1992). Even so, we do not yet have a completely satisfactory model of vertex finding. One goal of the present research is to develop a suitable method to extract vertices from luminance maps of natural images.

Distance Computations. The model computes distances between vertices. In the present work, retinal extents (rather than real lengths in the physical world) are all that are required. Evidence from a variety of perceptual tasks, including studies of size perception and size constancy, implies that visual space has a metric, and retinal as well as real extents are routinely measured.

Distance Comparisons. Comparing extents would seem to be of the essence of form classification, e.g., it defines the difference between a square and a rectangle. There is also reason to believe that humans are highly sensitive to aspect ratios in shape perception.

In short, each of the three kinds of information extracted by the early stages of the model seems to be not only a type of information potentially available to visual processing, but information that is routinely used for a variety of visual tasks. These are the kinds of information that are plausibly members of a set of basic inputs from which higher-level relationships can be constructed. The model's shape classification abilities emerged more from "off-the-shelf" components than from tools specially engineered for a limited task.

At the same time, there are limitations of the present model, and some aspects of the results reflect arbitrary simplifications. Many limitations involve the domain of shape classification. Additions to the model will be needed to encompass planar shapes of various numbers of vertices, and even more elaboration may required to perform meaningful classification of smooth forms, beginning with the simple circle, which has no vertices. These challenges may help reveal more about the grammar of shape. Attempts to organize the shape domain have a long and continuing history (e.g., Attneave, 1954; Hochberg & Brooks, 1960; Leyton, 1993), yet no system of general utility has emerged. It is possible that further development of the modeling efforts begun here, combined with research on the abstract relationships learnable in human shape classification can help clarify and constrain the grammar of shape.

Another kind of limitation of the present results involves learning. Although we believe the relations extracted in the early layers of the network are generic and not contrived, our current model has the good fortune to include only these several sorts of information. In natural circumstances, human perceptual learners must discover which among routinely computed or potentially computable basic functions will be relevant to a particular classification task. Suppose our stimulus inputs had included many more concrete and relational features, and feedback in our task was supposed to allow the system to converge on the notion of "square." At a minimum, many more examples would be needed by the network to separate the useful invariants from irrelevant variation. Even more may be needed, however. A system that registers lengths and colors, for example, and has comparisons such as equality/difference may have to learn to compare lengths rather than colors, by sampling possible comparisons or by applying previously learned strategies. Even if some comparisons are automatically computed, as in our simple model, it seems unlikely that all learnable comparisons are carried out all the time. If, as we suspect, the most advanced varieties of perceptual learning involve sensitivity to higher-order relations that are synthesized from new combinations of basic relations, then a fundamental problem will be how the search for useful new combinations is guided.

The model's current architecture is unrealistic in the sense that we postulate a separate unit for every vertex-to-vertex distance (layer 2), and every distance comparison (layer 3). (The model's input is similarly unrealistic in the sense that units are dedicated to particular vertices on the quadrilateral, rather than vertices at locations in the visual image.) These representations are spatially multiplexed, in the sense that identical properties of different entities (e.g., different lengths, length comparisons), are represented by completely separate units in the network. This architectural convention cannot be expected to scale to represent figures with arbitrary numbers of vertices. In future incarnations of the system, we intend to replace this spatial multiplexing with temporal multiplexing, allowing separate vertex coordinates, distances, and distance comparisons to be represented by the same units firing at different times (for similar ideas, and for a summary of neurophysiological support for temporal multiplexing in the visual system, see Hummel & Biederman, 1992, and the references therein).

Our prototype shape network can discover, from plausible building blocks, the abstract invariants that determine a simple shape classification. Building on this foundation, we hope to discover the visual grammar and computational processes that make possible and constrain human perceptual learning.

References

Attneave, F. (1954). Some informational aspects of visual perception. *Psychological Review, 61*, 181-193.

Biederman, I. (1987). Recognition-by-components: A theory of human image understanding. *Psychological Review, 94* (2): 115-117.

Chase, W. & Simon, W. (1973). Perception in chess. *Cognitive Psychology, 4*, 55-81.

Gibson, E. J. (1969). *Principles of perceptual learning and development.* New York: Prentice-Hall.

Goldstone, R. L. (1998). Perceptual learning. *Annual Review of Psychology, 49*, 585-612.

Heitger, F. & von der Heydt, R. (1993). A computational model of neural contour processing: figure-ground segregation and illusory contours. *Fourth International Conference on Computer Vision.* Los Alamitos, CA, USA: IEEE Comput. Soc. Press.

Hochberg, J. & Brooks, V. (1961). The psychophysics of forms: Reversible perspective drawings of spatial objects. *American Journal of Psychology, 73*, 337-354.

Hummel, J. & Biederman, I. (1992). Dynamic binding in a neural network for shape recognition. *Psychological Review, 1992 Jul, 99* (3):480-517.

Leyton, M. (1993). Symmetry, causality, mind. Cambridge, MA: MIT Press.

Mach, E. (1897). The analysis of sensations. English translation. New York: Dover, 1959.

Marr, D. & Nishihara, K. (1978).

Rumelhart, D.E., Hinton, G.E. & Williams, R.J. (1986). Learning representations by back-propagating errors. *Nature, 323* (6088): 533-536.

Shipley, T.F. & Kellman, P. J. (1990). The role of discontinuities in the perception of subjective contours. *Perception & Psychophysics*, **48,** (3), 259-270.

Author Note

We thank Randy Gallistel for helpful discussions of perceptual learning and shape representation. This research was supported by NSF SBR 9720410 (Learning and Intelligent Systems Program). Correspondence should be addressed to: Philip J. Kellman, UCLA Department of Psychology, 405 Hilgard Avenue, Los Angeles, CA 90059-1563 or by email to <Kellman@cognet.ucla.edu>.

Short-Term Memory Resonances

Stephen N. Kitzis (SKITZIS@Fhsu.Edu)
Department of Psychology, Fort Hays State University
600 Park Street
Hays, KS 67601 USA

Abstract

A cascading neural loop model is proposed to address the question of how to represent continuous experience. A prediction of the model is that short-term memory decay should exhibit a set of bumps or dips superimposed on a smooth exponential base. The prediction was tested using a Brown-Peterson distractor task, with distractor intervals from 1 to 24 seconds spaced every second apart. In one study with 22 participants, fits of nested regression models indicated that peaking functions with periods near harmonics of 1.6 seconds provided a better description of the data than an exponential function alone. In a replication study with 29 participants, peaking functions with a period of 3.2 seconds provided the best fit. In both studies, 5% rises above an exponential base were evident near 7, 10 to 11, 13 to 14, and 16 seconds. This short-term memory effect has not been reported before and needs further replication.

Introduction

The paradigmatic treatment of short-term memory in experimental psychology has been relatively stable for a long time, though still not resolved. After Miller's (1956) identification of an information processing bottle-neck, Brown (1958) and Peterson and Peterson (1959) established a smooth, rapid decay as the principal empirical characteristic of short-term memory, and Broadbent (1958) and Atkinson and Shiffrin (1968) formalized models in which short-term memory was theoretically separate from other forms of memory. However, while providing a clear focus for empirical and theoretical questions, decay was never universally accepted as the actual mechanism of forgetting. Keppel and Underwood (1962) immediately cast doubt on a simple trace decay interpretation of the Peterson and Peterson (1959) results, and the decay versus interference question is still with us today (Laming, 1992, Crowder, 1993). Similarly, Atkinson and Shiffrin's (1968) model has spawned more debate than consensus. In general, Baddeley's (1992) more complex view of multiple working memory components has become more accepted than their proposal of a single short-term memory store, but even the question of whether separate stores exist is still quite open (Cowan, 1988).

Another, less well recognized facet of this paradigmatic view is how closely short-term memory has been thought to resemble a computer data buffer. Broadbent (1958) was quite explicit in this regard, and most other models have followed in exactly the same tradition. One result of this relatively unquestioned assumption is that no model can adequately address what James (1890) had described as a "stream of consciousness".

Introspection would lead me to believe that both my immediate experience and the recollection of that experience are continuous, yet the typical model of short-term memory would suggest otherwise. That is, short-term memory supposedly can maintain only a small set of information at any given time, some or all of which can be replaced a short time later only by another small set of information. However, if this were the case, then where does that feeling of continuity come from? Any continuous function can be approximated by a sufficiently high enough resolution digital function, but, using neural firing rates as a guide, this implies a digitalization rate on the order of 10 to 100 items per second. Short-term memory may be narrow in width, as suggested by Miller's (1956) seven plus or minus two chunks, but, in effect, must be much longer in length to capture continuity of experience. Even selective attention, unitary store models like Cowan's (1988) do not escape this problem in that recollection of sequences of events somehow must be maintained in memory that can only allow the passage of time to be represented as scanning across different locations in memory.

This admittedly simplistic analysis forces one to ask the question of how any neurologically realistic model can maintain streams of experience instead of merely cross-referencing static "snapshots" of that experience. Neural network models (Rumelhart & McClelland, 1986) provide a good class of candidates, but, by design, the inputs and outputs of these models are restricted to fixed values. Even when the inputs and outputs represent sequential information, as in Jordan (1986), the sequence consists of a series of fixed values. Or, when the time course of processing actually is the subject of interest, as in Kawamoto and Kitzis (1991), the time-varying signals are restricted to approaching fixed asymptotic values. In short, there are no neurologically realistic models available that address the question of maintaining sequences of dynamically varying signals.

On the other hand, dropping the restriction of neurological plausibility would allow all sorts of engineering-like models to be considered. The basis for these would be the equivalent of any sequential recording device, like a tape recorder, or random access computer memory or disc used to store a sequence of data. In essence, storage elements remain empty until filled, maintain perfect data integrity until overwritten, and then are overwritten with new data without any interference from the prior data stored at that location. It seems very unlikely that real neurons would maintain information in such a localist, completely nondistributed, manner. If nothing else, the idea of a neuron "waiting" for that first input, and then maintaining that single fixed value forever after just seems very biologically wasteful.

For these and other reasons, I was led to consider a recurrent neural network approach towards memory in which sensory and cognitive experience is recorded in a series of closed neural network loops. Within a loop, a specific experience at a specific time is recorded as a set of feature values sufficiently rich enough to represent that experience. Parallel to that set of units is another set that records the prior experience in time, followed by another and another. In short, time is represented as distance along a loop "perpendicular" to the units actually maintaining the experience, and experience "flows" along a loop as successive copies of prior experience. Where a loop closes back on itself, current experience coming into a loop is blended with the recording of prior experience from one time cycle before. The loop cycle time would be determined by the physical length of a loop and the time required to copy information from one set of units to the next. Finally, to complete the picture being drawn here, loops are arranged in series such that shorter duration, higher experiential resolution loops provide inputs to longer duration, lower resolution loops. And, some series of loops are dedicated to maintaining specific types of sensory experiences, and others to more progressively integrated sensory or cognitive information.

In many ways, the conceptual basis of this model is that of Jordan's (1986) recurrent network model, but with the emphasis on maintaining potentially longer sequences of information rather than learning a common, efficient set of weights to maintain shorter sequences. If anything, learning is not an immediate consideration, as each individual memory loop acts much like Atkinson and Shiffrin's (1968) conceptualization of sensory memory. Alternatively, each loop could be considered to act like a continuous experiential tape recorder, but with the recording of current experience being affected by prior experiences. However, learning parameters would enter into the model when details of blending information within and between loops are specified. For example, the blending process should allow for different weights to be given to the present and prior experiences, and this could conceivably vary with loop duration and type. Although it is beyond the scope of this paper, it is assumed that as in Cowan's (1988) model, selective attention can focus or amplify information stored at different locations. Here, this would involve focusing on different loops or loop positions and, of course, whatever was recollected and became part of the current cognitive experience would then re-enter the system all over again.

Within this cascading neural loop model, the usual distinctions made between sensory, short-term, and long-term memory (Atkinson and Shiffrin, 1968), or between episodic and semantic memory (Tulving, 1972) have become blurred. What is being proposed, instead, is that memory ranges all the way from highly detailed, short duration sensory memory to semantic memory with little experiential recall but of long duration, and everything in between. Looked at in this way, short-term memory is not a categorically different type of memory but just another set of specializations within the complete system.

Even though the model outlined here is not defined well enough for rigorous testing, there is one prediction that should be amenable to an empirical check. If a loop has the equivalent of a fixed pickup point and information is flowing through it at a constant rate, then recall should be best whenever the relevant information happens to be right at the pickup point or, conversely, worst whenever it just went by and has to go all the way around again. In other words, if there is any reality at all to neural loops, accuracy of recall should rise or fall to some degree with a period equal to a loop cycling time.

This suggests that the decay curve for short-term memory should exhibit a set of nonmonotonic bumps or dips superimposed on the already well established smooth decay curve, resonant with any loops cycling in the range of about 3 to 10 seconds. This constitutes an entirely new testable prediction but, as short-term memory already has been extensively studied, how could such an effect not have been previously observed? The most obvious possibility is that the effect does not exist or is so small as to be lost in the "noise" of interparticipant variability. A second possibility is that loop cycling times vary greatly between participants, and any averaging together of individual results would completely blur away any resonance effects. A third possibility is more interesting and exploits a small methodological oversight.

Short-term memory decay has always been measured in conjunction with some distractor task to prevent item rehearsal (Crowder, 1993). However, as a convenience, researchers never measure every possible distractor interval but only a convenient representative set, such as every 3, 4, or 5 seconds. Peterson and Peterson (1959) and Murdock (1961) used the most complete interval set of every 3 seconds, but researchers thereafter used only every 4, 5, 6, or 9 seconds (see Laming, 1992, for a summary). Apparently, distractor intervals of 7, 11, 13, 14, 17, 19, 21, 22, and 23 seconds have never been sampled, and no study has used distractor interval spacings less than 3 seconds. If the resonance effects in question happen to be relatively narrow in terms of timing width and not fall on any of the typically sampled interval times, then they could have been missed. This is equivalent to fine-grained features in physics or astronomy, such as spectral lines, not being observable until a device with a high enough resolution was built capable of measuring them. In this case, we don't need to build a better device but only to use a more complete, methodologically well controlled sampling interval.

The purpose of the following two studies was to examine the short-term memory decay curve in better detail to determine whether resonance effects might be present. The task used was equivalent to a Brown-Peterson distractor task, with distractor intervals sampled from 1 to 24 seconds, spaced every second apart. Though not exclusive of other possible explanations, short-term memory resonance effects were predicted by, and would be consistent with, the cascading neural loop model outlined above. However, the primary purpose here is not to support that or any other model, but to describe an interesting empirical short-term memory effect that has not otherwise been predicted.

Experiment 1

Ideally, to detect short-term memory resonance effects, a wide range of closely spaced, tightly controlled, repeated measures from the same individual should be taken. The reason for a wide range would be to get as many resonance cycles as possible; closely spaced, to detect narrow resonance effects; tightly controlled, to prevent blurring of those effects; repeated measures, since memory is more likely to be stochastic than deterministic; and the same individual, since different individuals are unlikely to have exactly the same resonance cycle times.

However, people are not physical particles or photons, and psychological research requires compromises. After a number of small pilot studies, it was found both that participants improved with experience across sessions, but also became fatigued or bored if a session went on for too long. The few attempts made to gather extensive data from single individuals quickly led to the conclusion that, assuming rest breaks were sufficient, the quantity of useful data quickly fell off in terms of increasingly lower error rates. And, of course, task vigilance soon became next to impossible without sufficient rest breaks. On the other hand, these same pilot studies unexpectedly also seemed to indicate that resonance effects did occur for many individuals with a period of roughly 6 seconds. Accordingly, the decision was made to follow up this pilot result with a more rigorous study in which the task was restricted to three half-hour sessions, the range of 1 to 24 seconds was sampled with a resolution of 1 second, and to average data across participants.

Method

Participants Twenty-two university undergraduates, mainly psychology majors and approximately two-thirds female, participated in the study in exchange for extra credit. Neither age nor gender was restricted.

Apparatus Micro-computers (286 processor) running programs written by the author were used to present instructions and each trial of the short-term memory task. All times were measured by the computer as differences between successive calls to the onboard clock function. Estimated accuracy was no better than that of the screen refresh rate, approximately 32 milliseconds.

Procedure Up to four participants worked at the same time in the same testing room at divider-separated work stations. Instructions were presented by the computer, but a research assistant was always available to answer questions. Each memory trial consisted of three random but nonrepeated consonants presented on the computer screen for 1 second, followed by a series of random two-digit numbers presented at the rate of two per second for an integer number of seconds between 1 and 24. Participants were to repeat the numbers out loud until a prompt appeared, at which point they were to enter the three consonants. The research assistant first made participants practice the distractor task

until they could comfortably do it, and then remained in the room to ensure that participants continued to do so throughout all three sessions.

Each session lasted about 25 minutes, and participants were encouraged to take a short rest break between them. Sessions began with practicing the distractor task and then doing 6 practice memory trials to gain (or regain) familiarity with the task. Participants then did the actual memory test consisting of three filler trials followed by two blocks of 24 trials in which all 24 intervals were randomly presented. Participants were tested a total of six times at each interval. Participants had an unlimited time to respond on each trial, and would push a key to indicate when they were ready for the next trial. For each trial, a target consonant was scored as remembered correctly if it appeared within the entered response, regardless of position.

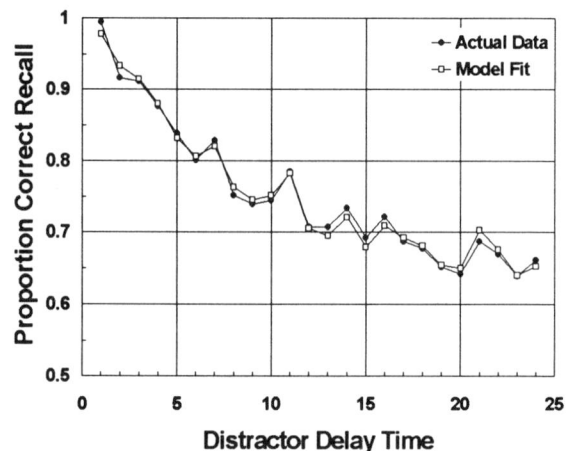

Figure 1: Actual mean proportion correct and best model fit as a function of distractor delay time, experiment 1.

Results

The overall proportion of correct answers was .75, with minimal changes across (in order, .75, .75, and .76) and within sessions (largest difference between blocks occurred in the third session, .77 and .75). Figure 1 provides the mean proportion of correct responses for each of the 24 distractor intervals. As found in earlier studies, such as Peterson and Peterson (1959), the overall shape was that of a descending exponential. In contrast to those studies, however, the points did not form a visually smooth curve but exhibited as many as six peaks: at 3, 7, 11, 14, 16, and 21 seconds. These happened to fall between the usual intervals sampled by previous studies and, in particular, restricting the data set to every 3, 4, 5, or 6 seconds would produce a much smoother curve.

To quantitatively test for the existence of these peaks, three nested regression models were fitted to the data. The base model is an exponential of the form: $y = b0 + b1*\exp(b2*t)$, where y is the proportion correct, t is the distractor interval, and $b0$, $b1$, and $b2$ are free parameters. The next inclusive model adds a three-parameter cyclic peaking function, where $b3$ is the starting time, $b4$ is the period, and $b5$ is the amplitude to be added to the base ex-

272

ponential model. If *b3* plus the next multiple of *b4* does not produce an integer value, then *b5* is split between the two adjacent time points in proportion to the complement of the time difference. For example, if a peak of 10 units occurs at 4.7 seconds, then 3 units would be added to the 4 second interval and 7 units to the 5 second interval. This is equivalent to representing the peaking function as a series of isosceles triangles separated by *b4* seconds, each with a 2-second wide base and a height of *b5* units. The most inclusive model adds a second three-parameter peaking function of the exact same form, but independent of the first series.

The base exponential model accounted for 95% of the total variance. The other models produced a number of different "best" fits which were significantly better than the base exponential. For the single-series model, three fits were found to be significantly better than the base model and accounted for another 2% of the total variance, with $F(3,18) = 5.3$, $p < .01$, $F(3,18) = 5.1$, $p < .01$, and $F(3,18) = 4.9$, $p < .05$ respectively. The first case picked up the peaks at 11, 16, and 21 seconds, with a period of 5.2 seconds; the second, peaks at 3, 7, 11, 14, and 21 seconds, with period 3.5; and the third, peaks at 11, 14, 16, and 21 seconds, with period 1.6. Peaking amplitudes added an extra .05 or .06 to the correct response rate above the base function. The case 1 and 2 periods were close to integer multiples (3.3 and 2.2) of the case 3 period.

For the double-series model, six fits were found to be significantly better than the best single-series fit, all being combinations of single-series fits. The best double-series fit, $F(3,15) = 8.0$, $p < .01$, picked up all six peaks and accounted for 99% of the variance, a 4% improvement over the base model. This particular fit is the one shown along with the actual data in Figure 1. Other types of peaking functions were tested in addition to the triangular, including polynomial power series, sinusoidals, and series of exponential shaped peaks with variable widths. None produced better fits than the simpler triangular shapes.

Discussion

These results imply that short-term memory resonances can be found superimposed on a more basic exponential decay curve. This does not constitute an enormous effect, only about 5%, so it could easily have been previously overlooked.

In a small way, this study does support the cascading neural loop model outlined earlier, as resonance effects were a direct prediction of that model. The fact that the best fitting periods occurred as close harmonics of some base period, 1.6 seconds, also would be very consistent with such a model though not necessarily a hard prediction. That is, the simplest systematic arrangement of a cascade would be a doubling or some other multiple of the smallest loop. As logical as this may sound, however, other more random arrangements can not be excluded on biological or any other grounds. Of course, further tests of the model are necessary before it can be considered more seriously. These might involve looking for similar effects on longer time scales, such as minutes, hours, or even days. Given the fact that distractor tasks are not realistic over such durations,

more subtle procedures would be necessary to minimize the possibility of purposeful practice. Misleading participants about the true target information or perhaps recording the times of spontaneous reminiscences might be techniques by which to accomplish this.

As short-term memory resonance effects have not been reported before, it was felt that replication was essential. If resonance peaks were spurious, then they would not be expected to reappear the same way in a second independent study. On the other hand, the same peaks occurring with an entirely different group of participants and a modified task would be a very convincing argument towards establishing their reality.

Experiment 2

The purpose of this study was to replicate the results of the first. The procedure was modified so that the two blocks of three-consonant trials within each session were replaced by a single block of randomly intermixed three and four-consonant trials. One reason for this variation was to eliminate the small possibility of some unique aspect of the procedure accounting for the resonance peaking. Another was to make the task slightly harder as a few participants in the first study managed to have very low error rates. A third reason, subsequently dropped, was the possibility of investigating whether the resonance effects might systematically vary with task difficulty. This would be consistent with some form of subvocal or subconscious rehearsal strategy rather than an automatic physiological mechanism as proposed in the cascading neural loop model. However, the sample size here was not sufficient to reliably establish or exclude any such variations, and the analysis will collapse the three and four-consonant trials together.

Method

Participants Twenty-nine university undergraduates participated in the study in exchange for extra credit. Neither age nor gender was restricted, but participants of the first study could not also participate in the second.

Apparatus and Procedure Apparatus and procedure were the same as before except that each of the three sessions now consisted of 24 four-consonant trials randomly intermixed with 24 three-consonant trials.

Results

Performance in the second study was slightly lower than in the first due to half the trials having four target items to be recalled instead of three. The overall proportion correct was .72, with only a small increase between the first and second sessions (in order, .69, .73, and .74). Figure 2 provides the mean proportion of correct responses and, again, the overall result was that of a descending exponential with superimposed peaks. As in the first study, three nested regression models were fitted to the data, and the best fit is shown in Figure 2.. This time, the base exponential accounted for only 93% of the variance, and there were three single-series fits that were significantly better than the base model, $F(3,18) = 3.7$, $p < .05$, $F(3,18) = 3.6$, $p < .05$, and $F(3,18) = 3.5$, $p < .05$. These accounted for another 3% of the total variance, and all three fits had essentially the same period of 3.2 seconds. This is exactly twice the smallest period found in the first study (1.6), and very close to the next larger period (3.5). There were no double-series fits that were significantly better than the best single-series fit.

Figure 3 provides an averaging together of results from both studies. Qualitatively, the peaks at 3, 7, 10 or 11, 13 or 14, and 16 seconds seem fairly consistent across the two studies, though whether the peaks near 11 and 14 seconds are broad, shifted, or perhaps closely spaced doubles is unresolved by this data.

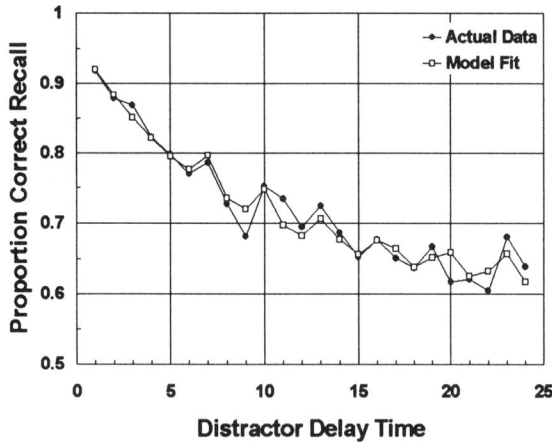

Figure 2: Actual mean proportion correct and best model fit as a function of distractor delay time, experiment 2.

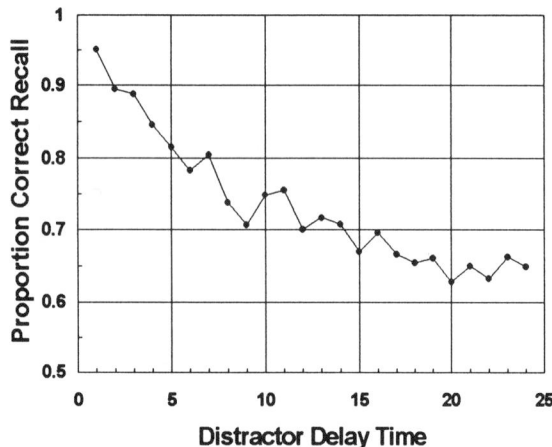

Figure 3: Actual results of experiments 1 and 2 averaged together.

Discussion

Not only did the results of the second study replicate the first, but the peaks appear as even more regular features than before. This strongly suggests that short-term memory resonances do in fact exist and that efforts to replicate and further investigate them would be warranted. Though systematic variations in memory recall on the order of 5% may not have any immediate practical applications, they do imply that short-term memory is a more complicated phenomenon than many theorists might have thought. But one that still is amenable to classic experimental techniques.

These results need further replication and extension. To me, personally, the most unexpected finding was the high degree of consistency in cycle timing between individuals. Though memory itself is a stochastic process, the existence of relatively narrow peaks implies that the underlying physiological parameters may be relatively constant. One extension would be to use essentially the same methodology to expand the resolution of observation in smaller regions to better determine the width or shape of selected peaks and how much variability is normal. For example, sampling every half second between 5 and 16 seconds probably would be sufficient to determine the actual shape of the peaks near 11 and 14 seconds. Another extension would be to extend the range of observations to longer intervals. An example here would be to sample every second between 8 and 31 seconds to determine whether that hint of an upturn in Figure 6 starting at 20 seconds happens to be spurious or not. In either case, increasing the task difficulty might allow single-participant studies to be more feasible, which should provide even more stable results. However, all extensions of the same basic methodology have to balance task difficulty and a reasonable limit on how much time or effort any one participant can be expected to contribute to the task. The critical element always will be to maintain as much consistency as possible in temporal sampling, both between and within participants.

Assuming such replications continue to produce reliable resonance effects, an entirely different level of extension then becomes necessary. This involves the determination of the causes of these cycles, and perhaps even factors that cause consistent variations in their timing. The latter potentially would include any of the factors that now are known to affect attention or vigilance, such as sleep deprivation, drugs, task complexity, prior experience leading to task automaticity, and, of course, time on task without a break. The former would involve looking for the actual physiological correlates of these resonance effects, and determining whether they are innately neural or a matter of experience, such as a learned subvocal rehearsal cycle. If nothing else, if the 1.6 or 3.2 second periods found in these studies were consistent across individuals, it would provide an empirical basis against which EEG or perhaps even functional MRI data could be compared. On the other hand, it would be even more impressive if variations in short-term memory resonances between individuals were matched by related variations in specific neurological cycles. In effect, it would be interesting if individuals could be categorized by relatively unique patterns of neural loop-related "spectral lines" superimposed on the short-term memory decay curve, much as stars can be classified by their spectra.

General Discussion

These studies were motivated by the asking of a simple question: how is continuous experience neurally represented? This led to the conceptual development of a potential answer, the cascading neural loop model, which led in turn to the prediction and observation of a previously unreported memory effect. Though the model is supported by this outcome, it is more important that a new empirical phenomenon may have come to light that ultimately will require some form of theoretical explanation.

In terms of empirical phenomenon that need theoretical explanation, I'd like to point out an interesting coincidence between the results here and from another empirical paper. Kristofferson (1980) found evidence for discrete steps in the discrimination of time durations when participants have had sufficient practice at the task. He determined values for four different "time quantum": 13, 25, 50, and 100 milliseconds. If these temporal discrimination step values were assumed to be related to a cascade of neural loops and the doubling progression continued, the next values would be 0.2, 0.4, 0.8, 1.6, and 3.2 seconds. Those last two, of course, are extremely close to best fitting resonance periods found in this paper. As mentioned in the discussion at the end of the first study, one of the most logical arrangements for a cascading neural loop model would be a simple doubling of duration for each loop in a cascade. This happens to be exactly the pattern found in Kristofferson's (1980) series of temporal quantal steps. It would be truly exciting if this were not a coincidence, but a convergence of evidence from completely different time scales onto a single model of memory.

Acknowledgments

This research was supported in part by grants from the Graduate School, Fort Hays State University. I thank K. Shelly-Carney and M. Reed for their help in working with participants.

References

Atkinson, R. C., & Shiffrin, R. M. (1968). Human memory: A proposed system and its control processes. In K. W. Spence & J. T. Spence (Eds.), *The psychology of learning and motivation: Advances in research and theory* (Vol. 2). New York: Academic Press.

Baddeley, A. (1992). Working memory. *Science, 255,* 556-559.

Broadbent, D. E. (1958). *Perception and communication.* London: Pergamon Press.

Brown, J. (1958). Some tests of the decay theory of immediate memory. *Quarterly Journal of Experimental Psychology, 10,* 12-21.

Cowan, N. (1988). Evolving conceptions of memory storage, selective attention, and their mutual constraints within the human information-processing system. *Psychological Bulletin, 104,* 163-191.

Crowder, R. G. (1993). Short-term memory: Where do we stand? *Memory & Cognition, 21,* 142-145.

James, W. (1890). *Principles of psychology.* New York: Holt.

Jordan, M. I. (1986). Attractor dynamics and parallelism in a connectionist sequential machine. *Proceedings of the Eighth Annual Conference of the Cognitive Science Society,* 531-546.

Kawamoto, A. H., & Kitzis, S. N. (1991). Time course of regular and irregular pronunciations. *Connection Science, 3,* 207-217.

Keppel, G., & Underwood, B. J. (1962). Proactive inhibition in short-term retention of single items. *Journal of Verbal Learning and Verbal Behavior, 1,* 153-161.

Kristofferson, A. B. (1980). A quantal step function in duration discrimination. *Perception & Psychophysics, 27,* 300-306.

Laming, D. (1992). Analysis of short-term memory: Models for Brown-Peterson experiments. Journal of Experimental Psychology: *Learning, Memory, and Cognition, 18,* 1342-1365.

Miller, G. W. (1956). The magical number seven, plus or minus two: Some limits on our capacity for processing information. *Psychological Review, 63,* 81-97.

Murdock, B. B., Jr. (1961). The retention of individual items. *Journal of Experimental Psychology, 62,* 618-625.

Peterson, L. R., & Peterson, M. J. (1959). Short-term retention of individual verbal items. *Journal of Experimental Psychology, 58,* 193-198.

Rumelhart, D. E., & McClelland, J. (Eds.). (1986). *Parallel distributed processing: Explorations in the microstructure of cognition. Volume 1: Foundations.* Cambridge: MIT Press.

Tulving, E. (1972). Episodic and semantic memory. In E. Tulving & W. Donaldson (Eds.), *Organization of Memory.* New York: Academic Press.

Resolving Impasses in Problem Solving:
An Eye Movement Study

Günther Knoblich (knoblich@mpipf-muenchen.mpg.de)
Cognition and Action,
Max-Planck-Insitute for Psychological Research
Leopoldstr. 24, 80802 Munich, Germany

Stellan Ohlsson (stellan@uic.edu) and Gary Raney(geraney@uic.edu)
Department of Psychology
University of Illinois at Chicago
1007 W. Harrison St., Chicago, IL 60607

Abstract

Insight problems cause impasses because they deceive the problem solver into constructing an inappropriate initial representation. The main theoretical problem of explaining insight is to identify the cognitive processes by which impasses are resolved. In past work, we have hypothesized two such processes: constraint relaxation and chunk decomposition. In the study reported here, we derive detailed predictions about the structure of eye movements from these hypotheses. Eye movement data from a study of match stick algebra problems were consistent with the predictions. The results support the view that a key component of creative thinking is to overcome the processing imperatives of past experience.

Impasse Resolution

Some problems are difficult because they require the problem solver to find the right sequence of steps in a vast space of alternative step sequences (Newell & Simon, 1972). The main resources for solving such problems are search heuristics and memories of prior problem solving episodes, both of which help constrain the search. In contrast, other problems, traditionally called *insight problems*, quickly generate the impression that they are unsolvable (Ohlsson, 1984; Seifert, Meyer, Davidson, Patalano, & Yaniv, 1995). Instead of floundering in a myriad of possibilities, problem solvers cannot think of even a single useful step to take: they quickly find themselves at an *impasse*.

Insight problems cause impasses because they deceive the problem solver into constructing an incomplete or overly constrained problem space that does not include the solution (Kaplan & Simon, 1990; Ohlsson, 1984; Wickelgren, 1974). Impasses are encountered after an initial phase in which the procedures activated by the initial representation have been applied to the problem without success (Ohlsson, 1992). During impasses no further ideas about how to proceed come to mind. Feeling-of-knowing judgments show that problem solvers do not feel that they are approaching the solution until the moment immediately before it is attained (Metcalfe & Wiebe, 1987).

If the impasse persists, the problem solver will eventually have to give up and declare failure. To resolve the impasse, the problem solver must revise his or her initial representation of the problem. The new representation might change the problem space by activating previously dormant but task relevant knowledge (operators, procedures, rules, etc.) that allow problem solving to continue. Identifying the cognitive processes involved in resolving impasses is the main theoretical problem in explaining insight problem solving in particular and creative thinking in general.

In past work we have proposed two processes for impasse resolution. First, constraint relaxation changes the representation of the goal (Ohlsson, 1992). Isaak and Just (1995) have suggested that constraint relaxation was involved in many historically important technological innovations. Second, chunk decomposition changes the representation of the problem situation (Knoblich, Ohlsson, Haider, & Rhenius, submitted). The purpose of the study reported here was to develop and test the implications of these process hypotheses for eye movements during problem solving. Before reporting the study, we develop the process hypotheses in more detail vis-a-vis the particular task domain we used.

Process Hypotheses

Constraint Relaxation

The initial representation of the goal to be sought in solving a particular problem is biased by prior knowledge. Such knowledge generates expectation about, among other things, which aspects of the problem situation are invariant and which are variable. For instance, when confronted with an arithmetic problem like $10 + 2 = ?$, the problem solver knows that the arithmetic operation is an invariant aspect of the task; the only thing that can vary is the value of the answer. His or her representation of the goal will naturally constrain the solution to look like the initial state except that the result of the arithmetic operation is filled in.

When encountering an unfamiliar type of problem, it is less obvious which aspects of the initial problem situation should be thought of as invariant and which as variable. Consider the domain of *match stick algebra* (Knoblich et al., submitted). A match stick algebra problem consists of an incorrect arithmetic expression written with Roman numerals and with the operators "+" and "-" and the equal sign (see Figure 1); for brevity, we shall let the term "operator" include the equal sign from this point on. The numerals and operators are constructed out of match sticks. The task is to move exactly one stick so as to change the given expression into a true expression consisting of nothing but Roman numerals in the range of I to XIII and the arithmetic operators.

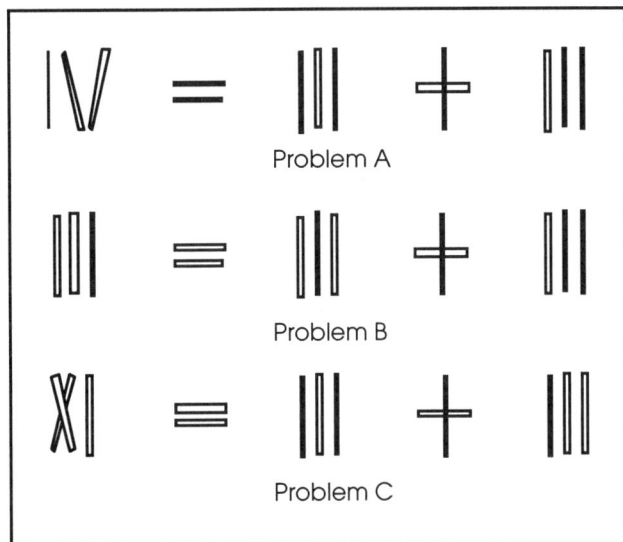

Figure 1. Match stick algebra problems.

Match stick algebra is unfamiliar to many people. However, their similarity to standard arithmetic makes it highly likely that these problems activate prior knowledge of arithmetic. This will bias the problem solver to view the arithmetic operators as constants and only the values as variable. In this representation, the only moves that will be considered are those that transform the values. Problem A in Figure 1 can be solved with a move of this type: pick up the vertical stick (I) in the left hand side of the expression and move it to the other side of the V, thus changing the IV to a VI and the expression as a whole to VI = III + III, a true arithmetic expression.

However, problem B cannot be solved as long as we only consider moves that change values. To find the solution to B, the operators have to be considered variable as well. To solve B, pick up the vertical stick in the plus sign to the right, rotate it 90 degrees and put it down again to make a second equal sign. This transforms the expression into III = III = III, a true arithmetic expression.

In short, constraint relaxation extends the space of possible solutions by replacing constants in the goal structure by variables in response to an impasse.

Chunk Decomposition
Familiarity with certain configurations of perceptual features establish those as patters (chunks) in memory.

Configurations of perceptual features that match a chunk are automatically recognized as an instance of that chunk. When encoding a problem, perceptual chunks that have proven useful in past encounters with superficially similiar problems are automatically applied. In most situations, chunks facilate problem solving (Chase & Simon, 1973; Ericsson & Lehmann, 1996).

Nevertheless, chunks that have proven useful in the past might hinder the solution to an unfamiliar problem. To find a solution, the components of a chunk might have to be separated from each other and reconfigured into a different patterns. We refer to this as *decomposing* the chunk.

The possibility of decomposition may be more or less obvious depending on characteristics of the chunks. To decompose a chunk like "IV" into "I" and "V" is easy, because those componens are themselves meaningful symbols (chunks) within the system of Roman numerals. We call this a *loose* chunk. To solve problem A (see Figure 1), it is sufficient to decompose that chunk.

However, other chunks consists of components that are not meaningful, at least not in the context of match sick algebra. To solve problem B (see Figure 1), the problem solver has to decompose the plus sign into one vertical and one horizontal stick. To solve problem C, the problem solver has to decompose the numeral X into two sticks, one that is slanted to the left and one that is slanted to the right. In neither case are the components of the chunks meaningful within the system of Roman numerals. We call these *tight* chunks.

In short, chunk decomposition splits chunks into their components in response to an impasse. This leads to a more fine grained problem representation which might activate dormant but relevant knowledge (operators, procedurs, rules, etc.) that can be applied to the problem and hence resolve the impasse and enable problem solving to resume.

Eye Movement Predictions
At the surface, the three problems in Figure 1 differ only in the numeral on the left-hand side. However, to solve problem B, the problem solver has to relax constraints on altering operators that do not need to be relaxed to solve problems A or C. To solve problems B and C, the problem solver needs to decompose tight chunks that do not need to be decomposed to solve problem A. Hence, we predicat that problems B and C are more difficult than problem A, in spite of their surface similarities, and that B is more difficult than C.

However, the probability of solution and the time to solution do neither allow us to determine how the problem solving process unfolds nor how succssful and unsuccessful problem solvers differ. Think aloud protocols provide a window onto problem solving (Ericsson & Simon, 1993), but problem solvers do not verbalize during impasses (Duncker, 1945) and verbalizing may interfer with the processes involved in representational change (Schooler,

Ohlsson, & Brooks, 1993). To overcome these problems, we recorded the participants' eye movements.

How do the two processes of constraint relaxation and chunk decomposition influence the movements of the problem solver's eyes? The answer depends on the particular quantity or measure that we derive from the raw eye movement recordings.

Mean *fixation duration* is likely to be affected by impasses, because the duration of fixations indicates how long an information item is processed in working memory (Just & Carpenter, 1976). During impasses, problem solving activity ceases or at least slows down. Hence, processing speed should drop and mean fixation duration should increase. Because we expect impasses in problems B and C but not in problem A, mean fixation duration should be greater in the former two problems. Furthermore, because impasses only occur after the initial exploration of the problem space, mean fixation duration should increase as problem solving continues for problems B and C but not for problem A.

A second measure is the *proportion of fixation time* spent on different elements of the task. This measure indicates which elements were processed most extensively. Differences in the proportion of fixation time devoted to a particular element during successive intervals is a sign that the participants' attention shifted from one element to another during the problem solving process.

In problem B relaxing the constraint to keep operators constant should result in a shift of attention from the result and the operands (i.e., the values) to the operators. Because problem B can only be solved when this constraint is relaxed, this shift should be present or more pronounced in participants who solved problem B, and absent or less pronounced in participant who failed to solve that problem.

In problem C the process of decomposing the tight chunk "X" should result in a shift of attention to that chunk. Because problem C can only be solved when this chunk is decomposed, the shift should be present or more pronounced in participants who solved problem C, and absent or less pronounced in participants who failed to solve that problem.

Our final measure is the *proportion of fixation changes* that moves the point of fixation between a particular pair of problem elements. A high proportion of fixation changes between elements I and J indicates that these two elements were processed together.

Predictions about fixation changes can be derived for problem B. Relaxing the constraint to keep operators constant should allow the problem solver to consider moving sticks between operators as well as between operators and values. Hence, the proportion of fixation changes between operators should increase as problem solving continues. Because problem B can only be solved when such moves are considered, this increase should be present or more pronounced in participants who solved problem B, and absent or less pronounced in participans who did not.

Method

Twentyfour undergraduate students at the University of Illinois at Chicago participated in the study for course credit. All participants attempted the three problems in Figure 1. The problems were presented in random order on a computer screen. Eye movements were recorded concurrently. As soon as a participant announced that she or he had found a solution, the experimenter terminated the eye movement recording and the participant said the solution out aloud. If a problem was not solved within five minutes, the participant was told the solution and the next problem was presented.

Results
Frequency of Solution and Solution Time

Figure 2 shows the solution rate in terms of the percentage of problems solved (panel a), and the mean solution time for the successful solutions (panel b). Consistent with our predictions, problem B, which requires both the relaxation of the constraint to keep operators constant and the decomposition of a tight chunk, was solved least often, and the successful solutions required more time than for the other two problems. Problem C, which requires the decomposition of a tight chunk but not constraint relaxation, was solved more often than B but less often than A, and the mean solution time for the successful solutions was also between the corresponding measures for the other two problems. Problem A, which does not require either constraint relaxation or the decomposition of any tight chunk, was solved most often and fastest. These results are consistent with our hypotheses, but they only provide weak support because many other process hypotheses might predict the same outcome pattern. Stronger support requires more fine grained analyses of the participants' behavior.

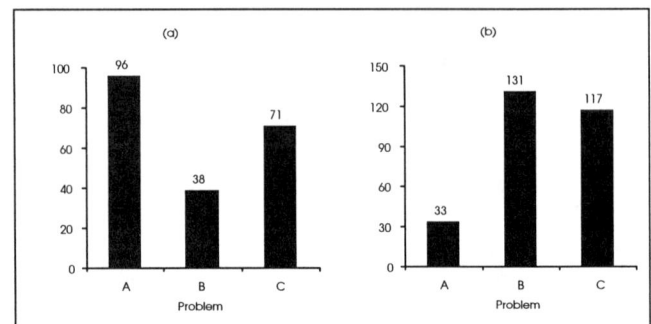

Figure 2. Solution rate (a) and mean solution time (b) for each of three match stick algebra problems.

Eye Movements

We divided each problem solving performance into three invervals, each interval representing one-third of the overall duration of the performance. (Intervals thus were of equal duration within each performance, but varied in duration across performances.) The three eye movement measures were calculated separately for each interval and participant

278

and averaged across participants for each problem. We discuss each measure separately.

Fixation duration. As we predicted, mean fixation duration was greater for problems B and C than for problem A; see Figure 3. Moreover, the fixation duration for problems B and C increased monotonically across the three intervals. In contrast, the fixation duration for problem A increased from interval 1 to interval 2, but did not increase from interval 2 to interval 3. These results support the hypotheses that the participants encountered more impasses on problems B and C, particularly towards the end of the problem solving attempt.

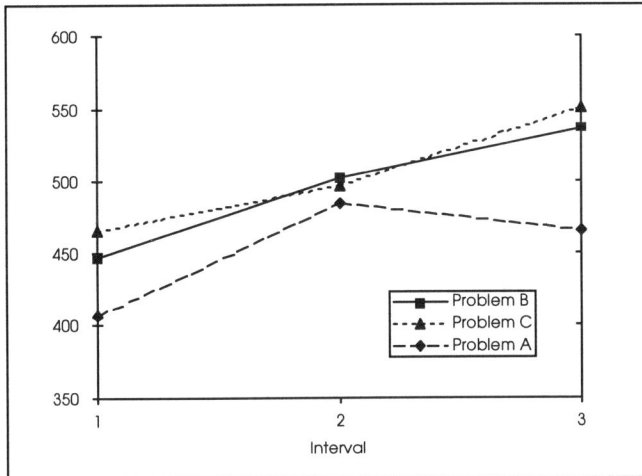

Figure 3. Mean fixation duration as a function of interval and problem

.

Fixation time. We computed the proportion of fixation time allocated to the different problem elements for each interval. Figure 4 shows the result for participants who solved and did not solve problem B. During the first interval there were no differences in attention allocation between those who solved the problem and those who did not. All participants payed more attention to the numerals in the equation than to the operators. This is consistent with the hypothesis that the initial goal representation constrains the problem space to moves that change values but not operators. However, the two groups differ in intervals 2 and 3. For participants who did not solve the problem, the initial pattern of attention allocation did not exhibit any consistent trend across the three invervals. For those who solved the problem, attention gradually migrated from the numerical values to the operators. This is consistent with the hypothesis that a key step in the solution is to relax the constraint that operators must be kept constant. Interestingly, the change in attention allocation is already present between the first and second intervals, i.e., at a time when problem solvers are still at an impasse and still think that the problem is unsolvable.

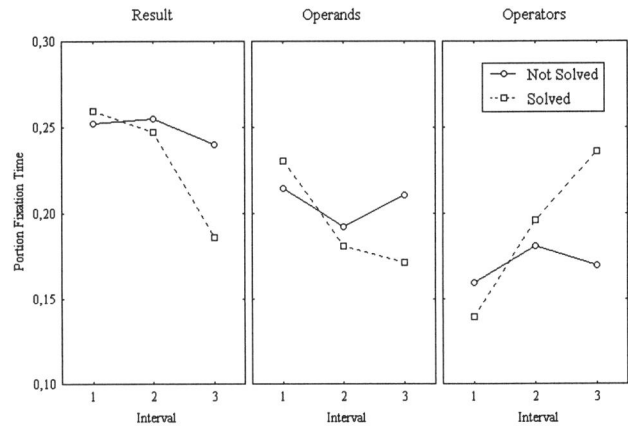

Figure 4. Mean proportion fixation time allocated to results, operands and operators for successful and unsuccessful solutions to problem B.

Figure 5 shows the corresponding results for problem C. During the first interval attention was allocated more to the result than to the operands and operators for both successful and unsuccessful problem solvers. For unsuccessful solvers, the pattern of attention allocation was virtually unchanged in intervals 2 and 3. For the successful solvers, on the other hand, the amount of fixation time allocated to the crucial component -- the numeral X -- increased monotonically across intervals.

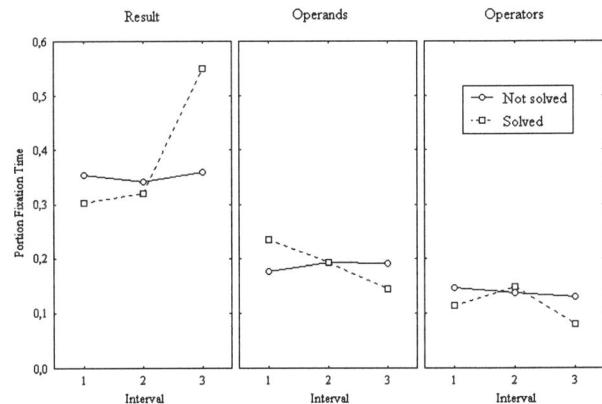

Figure 5. Mean proportion fixation time allocated to results, operands and operators for successful and unsuccessful solutions to problem C.

Moreover, the proportion fixation time allocated to the result almost doubled from interval 2 to interval 3 for the successful solutions. This result is consistent with the hypothesis that problem C can only be solved when the problem solver pays close attention to the tight chunk that need to be decomposed.

Fixation changes. We computed the proportion of fixation changes that moved the fixation point between operators for problem B. Figure 6 shows the results for participants who solved and did not solve problem B.

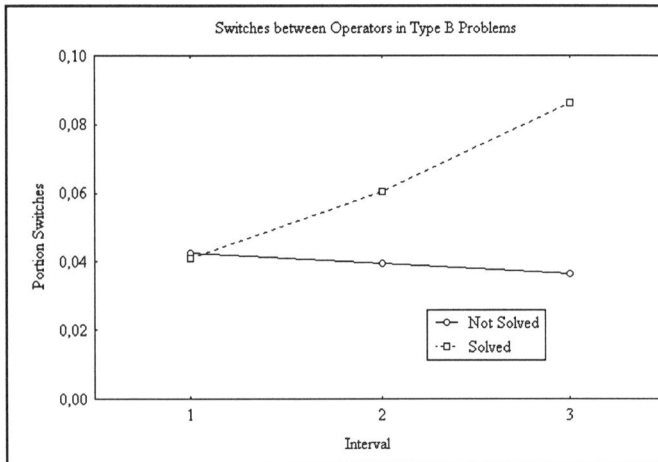

Figure 6. Proportion fixation changes that moved the fixation point between operators for each interval for successful and unsuccessful solutions to problem B.

Consistent with the constraint relaxation hypothesis, the proportion of fixation changes between operators was equal for both groups in the first interval and remained constant in intervals 2 and 3 for unsuccessful participants, but increased monotonically across the three intervals for successful participants. This result is consistent with the hypothesis that participants who solved the problem eventually related the two crucial operators while the unsuccessful participants did not.

Discussion

The fundamental characterisic of insight problem solving is the occurrence of an impasse, i.e., a state of mind in which the problem solver has run out of ideas about what to do next. The main theoretical problem of explaining insight is to specify the cognitive processes by which impasses are resolved (Ohlsson, 1984; Ohlsson, 1992). We propose that constraint relaxation extends the problem space by changing the status of certain problem elements from invariants to variables that can be manipulated, and that chunk decomposition extends the problem space by allowing features or components of the problem situation that are normally perceived as linked in a particular configuration to be separated and reconfigured. When these processes occur, previously unheeded possiblities suddenly come to mind and problem solving can continue.

These hypotheses generate detailed predictions about the expected pattern of eye movements on insight problems. In particular, they generate predictions about differences between superficially similar problems and between successful and unsuccessful problem solvers. The fact that our data were consistent with these predictions lend strong support to the hypotheses. Although there are alternative theories of insight problem solving (Simonton, 1988; Smith, Ward, & Finke, 1995; Sternberg & Davidson, 1995; Weisberg, 1986), they are too vague to generate detailed predictions about the temporal structure of problem solving behavior, about the differential difficulty of individual problems or about differences between problem solvers.

The constraint relaxation and chunk decomposition hypotheses are instances of a more general principle: Although human beings have to base their approach to each new problem or situation on past experience -- there is no other choice -- success vis-a-vis an unfamiliar problem might nevertheless require that the mind overrides the computational imperatives of experience. Automatized encoding rules and habitual response patterns have to be suppressed in order for novel actions to come to mind. Although the particular processes involved might vary across task domains, we suggest that overcoming past experience is a fundamental component of creative thinking in general.

Acknowledgements

This research was conducted during a visit of the first author to the University of Illinois at Chicago, which was made possible by a grant from the German Academic Exchange Service (HSPII/AUFE). It was also supported, in part, by Grant No. N00014-95-1-0748 from the Cognitive Science Program of the Office of Naval Research (ONR) to the second author. The eye movement equipment equipment was purchased with the help of grant *** to the third author. The opinions expressed are not necessarily those of the sponsoring agencies and no endorsement should be inferred. We thank Andrew Halpern and Steven Raminiak for helping to collect the data.

References

Chase, W. G., & Simon, H. A. (1973). Perception in chess. Cognitive Psychology, 4(1), 55-81.

Duncker, K. (1945). On problem-solving. Psychological Monographs, 58(5), ix, 113.

Ericsson, K. A., & Lehmann, A. C. (1996). Expert and exceptional performance: Evidence of maximal adaptation to task constraints. Annual Review of Psychology, 47, 273-305.

Ericsson, K. A., & Simon, H. A. (1993). Protocol analysis: Verbal reports as data (rev. ed.). Cambridge, MA, USA: Mit Press.

Isaak, M. I., & Just, M. A. (1995). Constraints on thinking in insight and invention. In R. J. Sternberg & J. E. Davidson (Eds.), The nature of insight (pp. 281-325). Cambridge, MA, USA: Mit Press.

Just, M. A., & Carpenter, P. A. (1976). Eye fixations and cognitive processes. Cognitive Psychology, 8(4), 441-480.

Kaplan, C. A., & Simon, H. A. (1990). In search of insight. Cognitive Psychology, 22(3), 374-419.

Knoblich, G., Ohlsson, S., Haider, H., & Rhenius, D. (submitted). Constraint relaxation and chunk decomposition in insight.

Metcalfe, J., & Wiebe, D. (1987). Intuition in insight and noninsight problem solving. <u>Memory & Cognition, 15</u>(3), 238-246.

Newell, A., & Simon, H. A. (1972). <u>Human problem solving</u>. Englewood Cliffs, N.J.: Prentice-Hall.

Ohlsson, S. (1984). Restructuring revisited: II. An information processing theory of restructuring and insight. <u>Scandinavian Journal of Psychology, 25</u>(2), 117-129.

Ohlsson, S. (1992). Information-processing explanations of insight and related phenomena. In M. Keane & K. Gilhooly (Eds.), <u>Advances in the psychology of thinking</u> . London: Harvester-Wheatsheaf.

Schooler, J. W., Ohlsson, S., & Brooks, K. (1993). Thoughts beyond words: When language overshadows insight. <u>Journal of Experimental Psychology: General, 122</u>(2), 166-183.

Seifert, C. M., Meyer, D. E., Davidson, N., Patalano, A. L., & Yaniv, I. (1995). Demystification of cognitive insight: Opportunistic assimilation and the prepared-mind perspective. In R. J. Sternberg & J. E. Davidson (Eds.), <u>The nature of insight</u> (pp. 65-124). Cambridge, MA, USA: Mit Press.

Simonton, D. K. (1988). <u>Scientific genius: A psychology of science</u>. New York, NY, USA: Cambridge University Press.

Smith, S. M., Ward, T. B., & Finke, R. A. (Eds.). (1995). <u>The creative cognition approach</u>. Cambridge, MA, USA: Mit Press.

Sternberg, R. J., & Davidson, J. E. (Eds.). (1995). <u>The nature of insight</u>. Cambridge, MA, USA: Mit Press.

Weisberg, R. (1986). <u>Creativity: Genius and other myths</u>. New York, NY, USA: W.

Wickelgren, W. A. (1974). <u>How to solve it</u>. San Francisco: Freeman.

Belief Bias, Logical Reasoning and Presentation Order on the Syllogistic Evaluation Task

Nicola J. Lambell (n.lambell@derby.ac.uk)
Institute of Behavioural Sciences; University of Derby
Mickleover, Derby, DE3 5GX, UK.

Jonathan St. B.T Evans and Simon J. Handley
(j.evans@plymouth.ac.uk, s.handley@plymouth.ac.uk)
Department of Psychology; University of Plymouth
Drake Circus, Plymouth, PL4 8AA, UK.

Abstract

Evans, Barston and Pollard, (1983) found that on the syllogistic evaluation task participants tended to endorse believable conclusions as being valid but reject unbelievable conclusions as invalid. A phenomenon known as "Belief Bias". Additionally, they collected verbal protocols from participants and established that this influence of belief was primarily associated with initial reference to the conclusions of these syllogistic arguments. In contrast, better logical reasoning was associated with initial reference to the premises. This experiment was designed to try to direct participants' attention to either the conclusion or the premises of a syllogistic argument with the intention of manipulating participants' logical reasoning ability and susceptibility to belief. The results reflected an inability to alter the influence of beliefs, but in one condition where the conclusion was presented prior to the premises, there was a successful reduction in participants' reasoning ability. The results are discussed with respect to the current theories of belief bias.

General Introduction

Traditional syllogisms consist of three statements, containing one of the four logical quantifiers *all, some, no and some...not* (see Example 1).The first two statements, the premises, each specify the relationship between an end term (A and C in Example 1) and a middle term (B). The conclusion, on the other hand, specifies a relationship between the two end terms.

All A are B
All B are C
Therefore,
All A are C

Example 1

There are three ways that psychologists utilise syllogisms in order to investigate logical reasoning. Firstly, in the evaluation task, the participant is asked to determine whether a given conclusion necessarily follows from the information given in the premises. Secondly, in the production task, the participant is required to produce a conclusion which necessarily follows from the information given in the premises. Finally, in the multiple choice task, the participant is required to choose a conclusion which necessarily follows from the premises from amongst a set of alternatives.

Early research into the multiple choice paradigm soon established that reasoning performance was poor and that participants demonstrated a systematic pattern of errors (c.f Chapman & Chapman, 1959; Dickstein, 1975; Erickson, 1974). Early research also established that the believability of the conclusion influenced the extent to which people endorsed conclusions (c.f Kaufman & Goldstein, 1967; Revlin, Leirer, Yopp & Yopp, 1980). This phenomenon has been termed belief bias. However, this early research into the influence of belief was heavily criticised on methodological grounds (see Evans, 1989; Evans, Barston & Pollard, 1983).

The first experiments to take account of these possible extraneous influences were the evaluation task studies performed by Evans, Barston and Pollard (1983). In these studies participants were presented with a conclusion for evaluation which was either valid (necessitated by the premises) or invalid (possible but not necessitated by the premises), believable or unbelievable. Their results clearly demonstrated that participants were more willing to accept believable than unbelievable conclusions as being valid. However, the greater acceptance of valid conclusions than invalid conclusions additionally demonstrated clear evidence of deductive capabilities.

What was surprising was that the influence of belief was more marked on invalid problems than valid problems. Whilst there was a small difference in acceptance of believable and unbelievable valid problems, there was a large difference in the acceptance of believable and unbelievable invalid problems. To facilitate understanding of their methodology and to demonstrate the large influence of belief bias observed on invalid problems consider the syllogisms presented in Examples 2 and 3. In Example 2 the conclusion to this syllogism is invalid as the class of highly trained dogs could include all police dogs. This would contradict the given conclusion. Yet a staggering

71% of participants across three experiments erroneously endorsed this conclusion as being valid.

No highly trained dogs are aggressive
Some police dogs are aggressive
Therefore,
Some highly trained dogs are not police dogs.

Example 2

No police dogs are aggressive
Some highly trained dogs are aggressive
Therefore,
Some police dogs are not highly trained.

Example 3

Now consider Example 3; this problem is logically equivalent to the previous example so this presented conclusion is also invalid. The only difference is that the terms have been re-arranged. However, only 10% of participants across three experiments erroneously endorse this conclusion as being valid (Evans et al., 1983). The difference between these two example syllogisms is the believability of the presented conclusion. Example 2 has a believable conclusion which participants tended to erroneously endorse as being invalid, whilst Example 3 has an unbelievable conclusion which participants tended to correctly reject as being invalid.

The pattern of results observed by Evans et al. led them to posit two possible explanations of how beliefs might influence reasoning. The first account, the Selective Scrutiny Model, proposes that participants initially scan the believability of the presented conclusion. When the conclusion is believable the model suggests that participants are likely to accept it without any consideration of its logical validity. On the other hand, when the conclusion is unbelievable the model proposes that some logical analysis takes place to determine whether the conclusion necessarily follows from the premises. The term "Selective Scrutiny" is derived from the fact that only unbelievable conclusions promote any attempts at reasoning. The model explains the belief bias phenomenon by proposing that participants do not reason when presented with believable conclusions. Instead they tend to unequivocally accept them as valid. This accounts for the high erroneous acceptance of invalid-believable conclusions and thus enables the model to explain the belief by logic interaction demonstrated by Evans et al. (1983).

The second account, the Misinterpreted Necessity Model, was motivated by the claims of Dickstein (1981) that participants often misunderstand logical necessity. The model proposes that participants initially engage in logical reasoning but tend to fall back on beliefs when reasoning is inconclusive (in other words, it fails to establish that the conclusion is falsified or necessitated by the premises). The model explains the belief by logic interaction as the invalid problems have conclusions which are consistent, but not necessitated, by the premises. Participants will, therefore, fall back on the believability of the conclusion to determine their response.

The robustness of the belief bias findings was subsequently demonstrated by Newstead, Pollard, Evans and Allen, (1992) and Evans, Newstead, Allen and Pollard (1994). Not only did they successfully replicate the pattern of results observed by Evans et al., they also found that increasing the logical nature of the instructions had little influence on the robustness of the belief bias findings. Their additional inclusion of belief-neutral conclusions also allowed them to investigate the direction of the belief bias effect. They found that only the unbelievable conclusions significantly differed in acceptance from the neutral conclusions, which enabled them to posit that belief bias is primarily associated with the rejection of unbelievable conclusions.

Other research into belief bias has primarily focused on the production task and has demonstrated that beliefs can also influence the production of conclusions (Oakhill & Johnson-Laird, 1985; Oakhill, Johnson-Laird & Garnham, 1989). Oakhill and her colleagues adapted the Mental Model theory of reasoning (Johnson-Laird & Byrne, 1991) to account for their findings. The Mental Model account proposes that participants construct an initial model of the premises and then produce a conclusion which is consistent with this model. The final deductive stage involves participants searching for alternative models of the premises in which this conclusion doesn't hold and positing other potential conclusions for similar evaluation. Oakhill hypothesised that believable conclusions influenced this process by curtailing a participants' willingness to search for alternative models of the premises. The model, therefore, explains the belief bias phenomenon by positing that all believable conclusions are unequivocally accepted without an attempt to search for falsifying models. This would explain the high erroneous acceptance of invalid believable conclusions found by Evans et al. (1983).

Some potentially important data which are often ignored are the concurrent and retrospective verbal protocol reports collected by Evans et al., (1983) during their experiments. Coding of these protocols revealed that when participants initially focused on the conclusion of a syllogistic argument they were more susceptible to the influences of belief. In contrast, when participants initially focused on the premises of a syllogistic argument they tended to show much better levels of logical responding.

This research is controversial because it suggests that belief bias arises due to the initial consideration of the conclusion, yet the production task findings demonstrated that belief bias occurred in the absence of a given conclusion. Thus, these findings appear inconsistent. Perhaps more importantly, these results are only consistent with the explanation of beliefs as proposed by the Selective Scrutiny Model. The other two models propose that participants initially attempt to reason from the premises thus reducing the extent to which belief bias should arise on the evaluation task. Thus, if belief bias on the evaluation task is due to initial focusing on the conclusion the Misinterpreted Necessity Model and Mental Model accounts are not adequate explanations of how beliefs influence performance.

The Experiment

The primary aim of this experiment was to direct participants attention to either the premises or the conclusion of a syllogistic argument in an attempt to encourage participants to adopt methods of responding akin to the verbal protocol findings. By presenting the conclusion of a syllogistic problem prior to its premises (CP condition) initial attention should be focused on the conclusion. Alternatively by presenting the premises of a syllogistic argument prior to its conclusion (PC condition) initial attention should be focused on the premises. However, Evans, et al. (1983) have already established from their protocol work that under this standard presentation condition some participants still give initial consideration to the conclusion. Therefore in order to enhance the focus of attention on the premises or conclusion some participants were given a delay (D condition) after the first piece of information was displayed whilst others received the standard simultaneous presentation of the problem information (N condition).

Two issues can be addressed by using this manipulation. First, if belief bias is a result of the initial consideration of the believability of the conclusion, then encouraging participants to focus on the conclusion should increase the influence that prior beliefs have on performance. Secondly, if logical performance is a result of the initial consideration of the premises, then encouraging participants to focus on the conclusion should reduce logical competence, whilst encouraging participants to focus on the premises should increase logical performance. These issues give rise to a number of testable predictions:

i) There should be more evidence of belief bias in the CP condition than the PC condition.

ii) There should also be less evidence of an effect of validity in the CP condition compared to the PC condition.

iii) There should be more evidence of belief bias in the CPD condition than the CPN condition.

iv) There should be more evidence of an effect of validity in the PCD condition than the PCN condition.

Evans, Handley and Buck, (1998) employed a similar methodology to investigate whether the order of information would affect performance on a conditional inference task. Using the Mental Model account of the reasoning process as a guideline of how participants were responding on the inference task they proposed that presenting the conclusion first should facilitate the building of models in which the conclusion held. They tested this proposal by suggesting that there should be a boosting of acceptance rates in the CP condition, especially for the more difficult Modus Tollens inference. They also explored whether the inclusion of a delay would exaggerate this effect. These hypotheses were not, however, confirmed. The absence of an increase in acceptance rates was explained by the boosting of the acceptance of affirmative conclusions for this condition in the presence of a general reluctance to endorse any conclusion which was presented first. They rejected their initial proposal in favour of the notion that participants' natural mode of reasoning was from the premises to the conclusion. This claim was supported by their additional finding that it took participants longer to respond to problems where the conclusion was presented first.

A secondary aim of this experiment was to compare our findings to the results observed by Evans et al. (1998). In addition to capturing the actual response that participants gave, a second dependant measure of how long participants took to respond to each problem was also included for comparison.

Method

Design. A four-way mixed design was used incorporating two between participant variables and two within participant variables. The first between participants variable was Presentation Order; this was whether they received the conclusion prior to the premises (CP) or the premises prior to the conclusion (PC). The PC condition acted as the control condition as it is the traditional syllogistic format used in the study of belief bias. The second factor was Delay; this was whether participants received the whole syllogism simultaneously (N condition) or whether a three second delay was introduced between the presentation of the first part of the syllogism and the rest of the problem (D condition). Participants were randomly assigned to one of four experimental groups; CPN, PCN, CPD, or PCD.

Each participant received eight syllogisms, half of which were valid (the conclusion presented followed logically from the premises) and half of which were invalid (the conclusion presented did not follow logically from the premises). This was the within participants variable of Validity. Half of these syllogisms had conclusions which were believable and half had conclusions which were unbelievable. This was the within participants variable of Belief. In all participants were given two valid-believable syllogisms, two valid-unbelievable syllogisms, two invalid-believable syllogisms and two invalid-unbelievable syllogisms.

Two dependent measures were taken during the experiment. The first dependent variable was a measure of whether participants accepted a conclusion as being valid or rejected a conclusion as being invalid. The second dependent variable was a measure of the time that participants took to evaluate a syllogistic problem.

Participants. Eighty undergraduate students from the University of Plymouth acted as paid volunteers in this experiment. None of them had any previous experience of syllogistic reasoning or any formal training in logic.

Materials. The EIO-2 form of the syllogisms and the materials employed in this experiment were identical to those employed by Evans et al. (1983). In order to control for any material differences two lists of materials were created. Conclusions which supported valid and believable arguments in list 1 supported invalid and unbelievable arguments in list 2. The syllogisms were presented to participants using a computer program which controlled for the presentation of the premises and the conclusion.

Procedure. Participants were initially presented with a set of instructions to read on the computer screen. The bracketed information denotes additions to the instructions for conditions where the conclusion was presented first and a delay was introduced.

"This experiment is designed to find out how people solve logical problems. On the screen there will be a series of reasoning problems presented one at a time. You will be shown two premises which you should assume to be true and a conclusion which may or may not follow from these premises.

In each case the premises are printed prior to the conclusion **(or the conclusion is printed prior to the premises)**. *You have to evaluate the conclusion in respect to the premises.* **(There will be a short delay after the presentation of the conclusion before the premises are presented).**

A logical conclusion is one which has to be true, if the premises are true. If you believe that the conclusion must follow from the premises answer YES, otherwise NO. You must give your answer to each problem by pressing either the left or right hand mouse button as follows:

LEFT button- answer YES, the conclusion must follow from the premises.

RIGHT button- answer NO, the conclusion need not follow from the premises.

Please take your time and be sure that you have the logically correct answer before deciding.

If you have any questions please ask them now as the experimenter cannot answer any questions once you have begun the experiment. You are free to leave the experiment now or at any time during the presentation of the reasoning problems. Thank you very much for participating."

Participants were then shown the syllogisms one at a time on the computer screen. The latency measure was taken from when the last piece of information was shown on the screen to when the appropriate choice had been made. The program fully randomised the presentation of the eight syllogisms for each participant.

Results

Acceptance Responses. It was initially necessary to try to establish whether, in general, across all conditions there was a replication of the findings of Evans et al. (1983). Table 1 presents the overall percentage acceptance rates for four problem types.

Analysis revealed an effect of believability (sign test, 10/56, with 14 ties; p<.001, one-tailed), with substantially more believable (72%) than unbelievable conclusions (44%) being accepted by participants. A sizeable effect of logic was also established whereby 67% of participants accepted the conclusion of valid problems compared to 49% of participants who erroneously accepted the conclusions of invalid problems (sign test, 14/46, with 20 ties; p<.001, one-

Table 1: The overall percentage acceptance rates for the four problem types, collapsed across the four experimental conditions.

	Believable	Unbelievable	Combined
Valid	74	59	67
Invalid	70	29	49
Combined	72	44	58

tailed). This demonstrated strong evidence of logical competence. Sign test analysis also revealed a significant interaction between Logic and Belief (sign test, 42/17, with 21 ties; p<.001, one-tailed). Comparison of the effects of Belief on the valid and invalid problem types revealed that there was an effect of Belief on valid problems (sign test, 7/29, with 44 ties; p<.001, one-tailed) but the effect of Belief on invalid problems was slightly greater (sign test 6/51, with 23 ties; p<.001, one-tailed). It is interesting to note that like the Evans et al. (1983) study there was an effect of belief bias on valid problems.

Having successfully replicated the belief bias findings of Evans et al. (1983), it was then necessary to consider whether there was any general influence of the variables of Presentation Order and Delay on the acceptance rates of conclusions. Analysis revealed that there was a main effect of Presentation Order (U=1261.0; p<.001, two tailed) such that more conclusions were accepted in the PC condition (68%) than the CP condition (48%). Analysis of the Delay variable (U=1717.0; n.s, two tailed) revealed that there was no difference in the level of acceptance of conclusions in the delay condition (55%) compared to the no delay condition (61%).

Table 2: The percentage acceptance rates for the four problem types, divided according to the variable of Presentation Order and collapsed across Delay.

		Believable	Unbelievable	Combined
CP	Valid	59	50	54
	Invalid	63	23	43
	Combined	61	36	48
PC	Valid	89	69	79
	Invalid	78	35	56
	Combined	83	52	68

Analysis of the data with respect to the four predictions produced the following results:

i) There should be more evidence of belief bias in the CP condition than the PC condition. (see Table 2). For the CP condition there was a greater acceptance of believable (61%) than unbelievable conclusions (36%) (sign test, 7/24, with 9 ties; p<.002, one tailed). However, for the PC condition there was also a greater acceptance of believable (83%) than unbelievable conclusions (52%) (sign test 3/32, with 5 ties; p<.001, one tailed). A comparison of the effects of belief revealed similar levels of acceptance for both conditions. (U=1535.5; n.s, one tailed). Thus, these results reflect that presenting the conclusion prior to the premises does not make participants more susceptible to the influences of belief. It is also interesting to note that significant belief by logic interactions were observed in both conditions.

ii) There should be less evidence of an effect of validity in the CP condition compared to the PC condition. (see Table 2). For the CP condition there was a slightly greater acceptance of valid (54%) than invalid conclusions (43%), however, analysis revealed the difference was not significant (sign test 8/17, with 15 ties; n.s, one tailed). In contrast, the greater acceptance of valid (79%) than invalid conclusions (56%) in the PC condition was significant (sign test, 6/29, with 5 ties; p<.001, one tailed). A comparison of the effects of validity across these two conditions revealed a significant difference (U=435.5; p<.034, one tailed). These findings, are therefore in clear support of prediction ii) and reflect that presenting the conclusion prior to the premises in some way disrupts logical responding.

Table 3: The percentage acceptance rates for the four problem types, divided according to the four experimental conditions.

		Believable	Unbelievable	Combined
CPN	Valid	68	53	60
	Invalid	65	20	43
	Combined	66	36	51
PCN	Valid	95	73	84
	Invalid	80	35	58
	Combined	88	54	71
CPD	Valid	50	48	49
	Invalid	60	25	42
	Combined	55	36	46
PCD	Valid	83	65	74
	Invalid	75	35	55
	Combined	79	50	64

Note : CPN =conclusion first with no delay, PCN = premises first with no delay CPD = conclusion first with delay, PCD = premises first with delay.

iii) There should be more evidence of belief bias in the CPD condition than the CPN condition. (See Table 3). In contrast to this prediction, the acceptance of believable (55%) compared to unbelievable conclusions (36%) in the CPD condition was slightly smaller than the acceptance of believable (66%) compared to unbelievable conclusions (36%) in the CPN condition. These results therefore reflect that the introduction of a delay had no influence on the effect of belief bias.

iv) There should be more evidence of an effect of validity in the PCD condition than the PCN condition. (See Table 3). In contrast to this prediction, the acceptance of valid

(74%) compared to invalid conclusions (55%) in the PCD condition was slightly smaller than the acceptance of valid (84%) compared to invalid conclusions (58%) in the PCN condition. These results therefore reflect that the introduction of a delay had no influence on the effect of logical validity.

In summary, the order of presentation of information had a clear effect on logical performance but not susceptibility to belief bias. Encouraging focus of attention by the introduction of a delay had no effect on logical performance.

Latency Responses. Whilst no predictions were made concerning presentation order and delay variables, inspection of these latencies may provide insight into the possible differences between the presentation conditions. For the Presentation Order variable it was found that participants took significantly longer to respond to problems in the CP condition (M=19.4 seconds) than in the PC condition (M=15.6 seconds), (F(1,76) = 4.39, p<.039). A similar pattern of responding was reported by Evans et al. (1998). It was also found that when participants were given a delay during the presentation of a problem they took significantly less time to reach a decision (M=14.4 seconds) than when participants were given no delay (M=21.0 seconds), (F(1,76) = 12.63, p<.001). Again, this was reported by Evans et al. (1998). This result reflects the fact that some evaluation of the task is occurring during the delay.

In summary, the latency findings are consistent with the findings of Evans et al. (1998) that Presentation Order and Delay have clear influences on how long it takes participants to evaluate a given conclusion.

Discussion

The acceptance results reflect that belief bias does not necessarily arise from initial consideration of a syllogistic conclusion, as clear evidence of belief bias was apparent across all conditions. These findings are in contrast to the verbal protocol findings of Evans et al. (1983) and are more consistent with the findings of Oakhill et al. that belief bias occurs even in the absence of a conclusion for evaluation. It seems that whilst participants might adopt different methods of responding (as demonstrated by Evans et al.'s protocols) it has not been possible either to encourage people to adopt these methods or to demonstrate that focusing on the conclusion is primarily responsible for increasing susceptibility to belief bias. In hindsight, one plausible explanation for the absence of an increase in the effect of beliefs in the CP condition is that belief bias is already a very strong and robust phenomenon. It might not be possible to increase the levels of belief bias as we may have reached a ceiling effect in terms of the influence of beliefs.

What is slightly puzzling is that in the absence of any reduction in belief bias there was a clear reduction in logical performance when the conclusion was presented prior to the premises. This disruption in logical performance was clearly due to the order in which the information was presented as there was no additional effect of introducing a delay. There was also a general suppression of acceptance of conclusions to all problems when the conclusion was presented prior to the premises. Both of these findings are consistent with the theoretical interpretation that Evans et al. (1998) posited to explain their findings.

Using the Mental Model theory as a framework for their explanation they proposed that participants' natural mode of reasoning was from the premises to the conclusion. This is consistent with the superior logical performance observed in the current experiment when the conclusion was presented first. Secondly, they proposed that presenting the conclusion first facilitates its inclusion in the initial model that participants construct. This attempt to include the conclusion as part of an initial model, could increase the difficulty that participants have in constructing any model of the premises. This would account for the finding in the current experiment of a suppression of acceptance of conclusions when they are presented first. Further support for this hypothesis comes from the latency findings which reflect the increased time taken to respond to problems when the conclusion is presented first. Whilst these findings should not be taken as clear support for the Mental Model account, the account does provide a useful framework in which to propose a possible explanation of the logical disruption caused by presenting the conclusion first.

The robust evidence of belief bias on all conditions fails to distinguish between the accounts of how beliefs influence performance on the syllogistic evaluation task. However, the superior reasoning performance demonstrated when the premises are presented first is consistent with the Mental Model theory and the Misinterpreted Necessity Model's notion that participants reason from the premises.

Perhaps the best way in which to view these findings is to adopt the distinction between our belief system and logical reasoning system proposed by Evans and Over (1996). They argue that when presented with a logical reasoning task participants attempt explicitly to comply with the logical instructions of the task but are unable to ignore the implicit influences of our beliefs. This would not only explain why it was possible to alter participants' logical reasoning performance by presenting the conclusion prior to the premises, it also suggests why the manipulations had no influence on the levels of belief bias observed.

Acknowledgments

This work was funded by a studentship grant awarded to Nicola Lambell by the Biotechnology and Biological Sciences Research Council (Reference: 94305619).

References

Chapman, L.J., & Chapman, J.P. (1959). Atmosphere effect re-examined. *Journal of Experimental Psychology, 58,* 220-226.

Dickstein, L.S. (1975). Effects of instructions and premise order on errors in syllogistic reasoning. *Journal of Experimental Psychology: Human Learning and Memory, 104,* 376-384.

Dickstein, L.S. (1981). Conversion and possibility in syllogistic reasoning. *Bulletin of the Psychonomic Society, 18,* 229-232.

Erickson, J.R. (1974). A set analysis theory of behaviour in formal syllogistic reasoning tasks. In R.L. Solso (Ed.), *Theories of cognitive psychology: The Loyola Symposium.* Hillsdale, NJ: Lawrence Erlbaum Associates.

Evans, J.St.B.T. (1989). *Bias in human reasoning: Causes and consequences.* Hove, UK: Lawrence Erlbaum Associates Ltd.

Evans, J.St.B.T., Barston, J.L., & Pollard, P. (1983). On the conflict between logic and belief in syllogistic reasoning. *Memory and Cognition, 11,* 295-306.

Evans, J.St. B. T., Handley, S.J., & Buck, E. (1998). Ordering of information in conditional reasoning. *British Journal of Psychology, 89,* 383-403.

Evans, J.St.B.T., & Over, D.E. (1996). *Rationality and reasoning.* Hove: UK. Psychology Press.

Evans, J.St.B.T., Newstead, S.E., Allen, J.L., & Pollard, P. (1994). Debiasing by instruction: The case of belief bias. *European Journal of Cognitive Psychology, 6* (3) 263-285.

Johnson-Laird, P.N., & Byrne, R.M.J. (1991). *Deduction.* Hove, UK: Lawrence Erlbaum Associates Ltd.

Kaufman, H., & Goldstein, S. (1967). The effects of emotional value of conclusions upon distortions in syllogistic reasoning. *Psychonomic Science, 7,* 367-368.

Newstead, S.E., Pollard, P., Evans, J.St.B.T. & Allen, J.L. (1992). The source of belief bias in syllogistic reasoning. *Cognition, 45,* 257-284.

Oakhill, J., & Johnson-Laird, P.N. (1985). The effect of belief on the spontaneous production of syllogistic conclusions. *Quarterly Journal of Experimental Psychology, 37A,* 553-570.

Oakhill, J., Johnson-Laird, P.N., & Garnham, A. (1989). Believability and syllogistic reasoning. *Cognition, 31,* 117-140.

Revlin, R., Leirer, V.O., Yopp, H. & Yopp, R. (1980). The belief bias effect in formal reasoning: The influence of knowledge on logic. *Memory and Cognition, 8,* 584-592.

Concrete and Abstract Models of Category Learning

Pat Langley[1] (LANGLEY@ISLE.ORG)
Institute for the Study of Learning and Expertise
2164 Staunton Court, Palo Alto, CA 94306 USA

Abstract

In this paper, we compare the rhetoric that sometimes appears in the literature on computational models of category learning with the growing evidence that different theoretical paradigms typically produce similar results. In response, we suggest that concrete computational models, which currently dominate the field, may be less useful than simulations that operate at a more abstract level. We illustrate this point with an abstract simulation that explains a challenging phenomenon in the area of category learning – the effect of consistent contrasts – and we conclude with some general observations about such abstract models.

Introduction and Overview

Learning is one of the ubiquitous aspects of human behavior, so it seems natural that the process of learning has drawn significant attention within both cognitive psychology and artificial intelligence. Over time, different candidate mechanisms have arisen to account for learning phenomena, leading to distinct theoretical camps that have direct analogues across the two disciplines. Another clear parallel lies in the rhetorical stances often taken by authors, which assume that the success of a learning method on a specific problem derives from that method's distinguishing features, rather than from other factors.

In this paper, we review five main paradigms in the computational study of learning, and we consider the mounting evidence that, for purposes of both artifact construction and psychological modeling, these different frameworks typically give equivalent results. Indeed, analysis of successful applications and successful models suggests decisions about how to cast the learning task and how to encode training data are the main source of power in computational learning. This observation leads us to question the usefulness of developing detailed, concrete computational models of human learning.

In response, we draw on the notion of an *abstract* computational model that makes predictions about behavior but that does not actually carry out the task. We discuss

some earlier work in this alternative framework that has focused on skill learning, then apply the approach to a phenomenon from category learning – the effect of consistent contrasts – which poses challenges to most computational accounts. We show that a certain abstract model explains this finding without taking a position on the details of representation or learning, whereas another abstract simulation, which matches the assumptions of most concrete models, does not explain the phenomenon. We close with responses to some natural criticisms of abstract models and with comments on their long-term role in developing theories of human behavior.

Rhetoric and Reality in Learning

Much of the research on mechanisms of learning, both within AI and cognitive psychology, has focused on the acquisition of knowledge for classification or categorization. The performance task here involves assigning a new instance or stimulus, typically described using attribute-value pairs, to some category or class, given a known set of mutually exclusive classes. The associated learning task involves finding some function or mapping that categorizes novel instances, given a set of training instances and their assigned classes. The typical performance measure is classification accuracy or error, though measures of speed and typicality sometimes appear as well.

The machine learning community has explored five main representations of knowledge about categories, each which its associated mechanisms. The first major paradigm represents knowledge as *decision lists*, which consist of rules that specify the logical conditions for membership in a category, typically learned one at a time. A second framework represents category knowledge as a *decision tree* that is acquired through a process of recursive partitioning. A third paradigm represents knowledge as a multilayer *neural network*, often relying on a weight-adjusting method known as *backpropagation*. Yet another framework encodes knowledge about categories as experiences or stimuli stored in long-term memory, using *nearest neighbor* or *case-based* methods for classification. A final paradigm uses training instances to update *probabilistic* descriptions, often using simple methods like naive Bayesian classifiers for categorization.

---

[1] Also affiliated with the DaimlerChrysler Research & Technology Center, Palo Alto, and the Center for the Study of Language and Information at Stanford University.

Superficially, these five paradigms appear quite distinct, and early research in machine learning emphasized differences among them. For example, for many years the common wisdom posited that methods for decision-tree and rule induction were most appropriate for 'symbolic' domains, whereas backpropagation in neural networks was best suited for sensori-motor tasks. Indeed, some felt that such different representations, performance elements, and learning algorithms could not even operate in the same domains. These beliefs were encouraged by the different notations used in various communities, but they were also aided by rhetorical claims, unbacked by evidence, coming from the various camps.

This perception started to change with the first experimental comparisons among different methods for classification learning (e.g., Mooney, Shavlik, Towell, & Gove, 1989). These studies and ensuing ones showed that induction algorithms from separate frameworks, although superficially very different, could operate on the same problems. Their experimental results also suggested that no one induction method was always superior to others, and a decade of experimental comparisons has supported these early results. Although methods for classification learning have steadily improved over time, no one *paradigm* has emerged as superior to others in terms of classification accuracy.

However, contributors to each paradigm have found some quite different factors that affect the success of learning. These include decisions about the formulation of the learning task, the representation or encoding of the stimuli, and the quality of the training cases. Both experimental studies and application efforts suggest that such factors are more important determinants of learning effectiveness than the induction algorithm or the representational formalism itself, although authors seldom emphasize these issues in papers. Langley and Simon (1995) argue that these items – problem formulation, representation engineering, and data collection – are the main sources of explanatory power in machine learning.

Each paradigm in machine learning has a direct analogue in theories of human learning. Techniques for learning decision lists bear a close relation to production-system models of human category learning (e.g., Anderson & Kline, 1979), whereas methods for decision-tree induction are quite similar to psychological models of learning that construct *discrimination networks* (e.g., Richman & Simon, 1989). Backpropagation and its relatives have been used not only for applied problems but also play a role in many models of human learning (e.g., Gluck & Bower, 1988). Case-based methods figure prominently in the papers on human concept learning, where they are known as *exemplar* models (Smith & Medin, 1981), and probabilistic methods have also been proposed as models of human category formation (e.g., Anderson, 1991; Fisher & Langley, 1990).

The literature on computational models of human learning has also seen a period dominated by rhetorical claims. The typical research paper begins by arguing the strengths of connectionism, production systems, or exemplar models, whichever happens to represent the author's paradigm. The text then reviews some psychological phenomena and describes a computational model, cast within this paradigm, that replicates those findings. In closing, the authors conclude that these positive results are evidence for their theoretical framework, ignoring the possibility that the source of explanatory power lies elsewhere, such as in carefully selected stimulus encodings or in a well-crafted training regimen.

The reason for drawing such hasty conclusions are understandable even if the conclusions themselves are questionable. One simply cannot construct a detailed computer simulation of human behavior without making many assumptions, such as representational decisions, that are not central to one's theoretical claims. Naturally, many scientists are tempted to conclude that, when their simulation succeeds at modeling some phenomenon, their core assumptions are responsible rather than the peripheral ones.

Yet not all authors follow this natural inclination, with one revealing counterexample coming from Richman and Simon (1989). They suggest that two alternative accounts of word-recognition findings – connectionist models (which posit parallel processing) and discrimination networks (which posit sequential processing) – are not due to these paradigms' core assumptions. Rather, they argue that a hierarchical representation of words, an auxiliary assumption that both classes of model share, constitute the real source of explanatory power in this domain. We believe many similar examples exist in the literature on computational models of human learning.

Abstract Models of Learning

These observations suggest that traditional computer simulations of human learning, although useful contributions to artificial intelligence and machine learning, may be unnecessary or even misleading in our attempts to explain psychological phenomena. In place of such *concrete* models, we need process models which operate at some more abstract level that lets us make predictions from the central claims of a theory, without needing an overwhelming number of peripheral assumptions.

Of course, there exists a long tradition of such abstract models of learning within mathematical psychology. But many process accounts developed within this framework have drawbacks of their own, in that they usually make constraining assumptions and embody simple theories for the sake of analytical tractability. Such restrictions on analytic models were originally an important factor in the development of computer simulations that actually carry out the task at hand.

However, the decision to work at an abstract level does not mean one must develop an analytic mathematical model; nor does the use of computer simulation mean one's program must accomplish a complete task. Instead, a process-oriented psychologist can develop an *abstract computational model*, a notion championed by Ohlsson and Jewett (1997). In this framework, the scientist still implements a running computer program that predicts behavior, but the system omits details that are not essential to the phenomena it aims to explain. For example, to model learning in problem-solving domains, they retain the idea of search through a problem space, but remove details about the states and operators that define the space. Instead, they specify the structure or connectivity of the space and model the learning process using mechanisms that alter the probability of taking given branches in the future.

Ohlsson and Jewett's goal was to model the power law of learning, in which the rate of improvement decreases with the number of training steps. Simulations on synthetic problems revealed that two learning schemes, involving positive feedback for selecting good branches and negative feedback for bad selections, produced power curves across a broad range of parameter settings. For instance, varying the branching factor, the length of solution, and the probability of feedback did not affect the shape of the learning curve, but extreme parameter settings for success-driven learning gave different simulated behavior. Moreover, failure-driven models that incorporated additive weight reductions in response to negative feedback produced exponential curves, although multiplicative updates gave the power law.

Another abstract computational model of learning comes from Rosenbloom and Newell (1987), who also focused on the power law. Their primary aim was to develop a concrete computer simulation that exhibited this effect on a finger-manipulation task. The key idea in their model is that humans acquire *chunks* which let them link complex perceptual configurations to complex actions, thus reducing the need to carry out multiple reasoning steps at the cognitive level. Rosenbloom and Newell embedded their learning mechanism in a detailed theory of the human cognitive architecture, cast as a production system, and showed that their mechanism for chunk acquisition reduced response time with practice. However, to actually fit the psychological data, they invoked a simple abstract model with four parameters that embodied the core assumptions of their chunking theory.

Shrager, Hogg, and Huberman (1988) present yet another explanation of power-law learning. Like Ohlsson and Jewett's, their abstract model describes a problem space only in terms of nodes and links, along with the probability that a selected branch will lead toward the goal node. Their computer simulations show that power-level behavior can result from two quite different learning

processes. One mechanism (similar to Rosenbloom and Newell's) creates new links from a problem's initial state to its goal state, letting the problem solver make future traversals in one step. Another mechanism (closer to Ohlsson and Jewett's) alters the probability of traversing a link based on whether it led to a solution. Shrager et al. also carried out an average-case analysis of their task, which gave good fits to simulated behaviors.

Langley (1996) reports a rather different abstract model for the task of flying an aircraft simulator through a three-dimensional slalom course. His model's central assumptions are that differences among subjects are due to differences in sensing skills, and that the main form of learning involves improving the ability to focus on relevant features during skill execution. Langley describes an implementation of this abstract model of sensory learning, along with a system that searches the space of parameter settings in order to fit the model to the experimental data. He compares the sensory-learning framework to an alternative model based on the power law, finding that the latter fits the data slightly better but that it requires many more parameters.

There are clear kinships between these abstract simulations and models from mathematical learning theory, such as Estes' stimulus sampling account of learning. Both frameworks typically assume that subjects' decisions are probabilistic in nature and that learning follows from simple changes to probability distributions. As we have noted, the key difference lies in abstract models' reliance on computer simulation rather than detailed analysis, which supports a wider range of process models. A similar relation holds with respect to the average-case analyses occasionally published in machine learning.

Of course, the different approaches to process modeling are not mutually exclusive. The Rosenbloom and Newell work showed that concrete and abstract simulations can coexist, and the Shrager et al. analysis made the same point with respect to abstract simulations and purely analytical models. Ohlsson and Jewett's contribution was the realization that neither mathematical analysis nor the concrete model are really necessary, and that researchers may often find it useful to work entirely at the level of abstract computational models. Nevertheless, research in this paradigm remains rare, especially in the otherwise well-studied topic of category learning. In the remaining pages, we apply the abstract modeling framework to an intriguing phenomenon in this area.

The Effect of Consistent Contrasts

As we have noted, considerable effort has gone into computational models of human category learning, typically using techniques very similar to those from machine learning. For example, Kruschke's (1992) ALCOVE incorporates a variant on the nearest-neighbor method that places weights on attributes, Martin and Billman's

Table 1: Schema for the stimuli used in Billman and Dávila's (1995) experimental study of category learning. 'Consistent contrast' subjects saw instances from categories characterized by the same two attributes, whereas those in the inconsistent contrast condition learned categories characterized by different attributes.

	Cons. contrast	Incons. contrast
Category 1	11 xx xx	11 xx xx
Category 2	22 xx xx	xx 22 xx
Category 3	33 xx xx	xx xx 33

(1992) TWILIX constructs a form of multivariate decision tree, and Anderson's (1991) RA model bears a close relation to the naive Bayesian classifier. All three systems have shown good matches to experimental results on human category learning. However, here we consider an interesting phenomenon that seems difficult to explain within the standard theoretical frameworks.

Billman and Dávila (1995) noted that most psychological studies of concept induction assume that some attributes are relevant and others irrelevant, but that the *same* ones are relevant to each category. They hypothesized that subjects would find concepts easier to learn when such *consistent contrast* occurs than when distinct categories are defined by *different* features. Table 1 shows the structure of the stimuli Billman and Dávila used to test this hypothesis, using a cover story in which subjects classified animals from an alien zoo and received feedback after each guess. Both conditions involve three classes and six attributes with three values each; moreover, all target concepts involve a conjunction of two relevant features. However, in the consistent condition, the same attributes are relevant to recognizing cases from all three classes, whereas in the inconsistent condition, a different pair plays this role for each class.

The learning curves in Figure 1 (a) show a clear difference between the two experimental conditions. Subjects who dealt with consistent contrasts improved very rapidly, achieving over 90 percent predictive accuracy after only ten training stimuli. Subjects in the inconsistent condition hovered around 50 percent during most of the 45 instances, better than the 33 percent that results from random guesses, but far below the accuracy for the consistent subjects. Separate tests on novel stimuli, some that matched the intended category definitions and others that did not, showed that subjects in the consistent condition were much more accurate at this task as well.

Naturally, Billman and Dávila attempted to explain this phenomenon using existing computational models of category induction. However, simulation runs with Kruschke's ALCOVE predicted no differences between the two conditions, and similar studies with Anderson's RA indicated a slight advantage for the *inconsistent* condition. Even runs with Martin and Billman's TWILIX, which they had expected to reflect the observed differences, failed to produce the desired result. Further analysis suggested that all three models lack a strong bias toward category descriptions incorporating fewer attributes overall, which seems the obvious explanation for the large difference in learning rate.

Of course, we could incorporate such an inductive bias into yet another concrete simulation of category learning, based on one of the above models or embedded in a new one entirely. But this would require us to adopt a position on the representation of knowledge, to select a complete performance element, and to propose a detailed learning algorithm. Yet the above account states that none of these factors are important in explaining the consistent contrast phenomenon. Rather, the key issue is whether learners are biased toward category descriptions that, across concept boundaries, require fewer features. Thus, this seems like an ideal context in which to illustrate the notion of abstract models.

An Abstract Model of Contrast Effects

We want our abstract model to make as few assumptions about representation, performance, and learning as necessary to account for the phenomena at hand. However, we can view all induction methods as constructing decision regions that partition a multi-dimensional space of instances or stimuli. Moreover, all basic induction algorithms incorporate some type of *locality* bias, so they are typically more accurate on test cases that fall near to observed training cases in this space. We would like a modeling framework that reflects this bias without committing to a particular encoding of learned knowledge.

For discrete domains like the one in Billman and Dávila's study, we can model conceptual knowledge as a table with c columns (one per attribute) and r rows, with each row specifying a unique combination of attribute values. We also need one extra column to specify the category or class associated with each value combination or to state that the class is unknown for this situation. This notation lets us describe arbitrary contrasting concepts that map from combinations of discrete values to class labels. Given a attributes with v values each, we can have a table with $a + 1$ columns and v^a rows, but many concepts are much simpler in nature. For example, if only c attributes are relevant, we need only include c columns, which means we need at most v^c rows. And not all possible combinations of values may occur in practice, which lets us reduce the number of rows still further.

Our performance and learning elements are similarly abstract. Given a test stimulus, described as a attributes and their values, we assume the subject finds the table's row whose c attribute values match this instance. If the

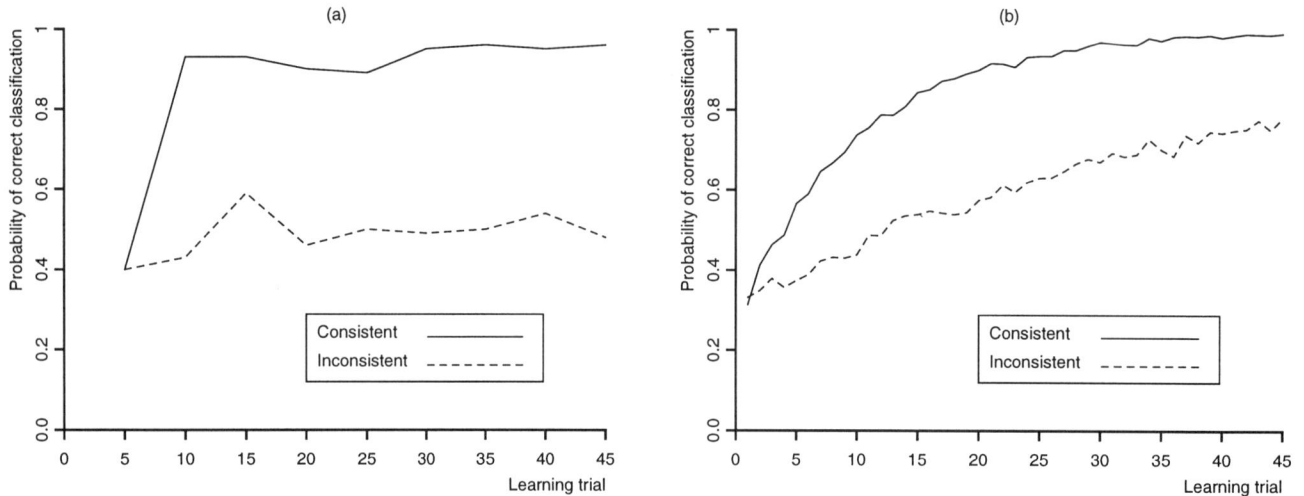

Figure 1: (a) Learning curves that Billman and Dávila's observed for subjects in conditions involving consistent and inconsistent contrast; and (b) learning curves that the abstract model predicts for the same conditions when $p = 0.3$.

row has an associated class, then the subject predicts it; otherwise he selects a class at random from a uniform distribution. We posit two distinct learning mechanisms, one that selects relevant features and another that assigns class labels to rows in the table. We assume that feature selection happens early in the learning process, and thus we model only its result in terms of the number of columns c in the table. For labeling, we assume that each time the subject sees a training instance that matches a given row with an unknown label, he stores, with probability p, the label observed with that stimulus.

When we instantiate this model for Billman and Dávila's two conditions, we see that it should predict quite different behavior. For the situation involving consistent contrast, we have two attributes that are relevant across all categories, giving a table with only two columns. Moreover, since different values on the other four attributes do not matter, we need only three rows in the table, one for each co-occurring pair of relevant values. On the other hand, the table for the inconsistent condition requires six columns, since all six attributes play a role in some concept description; this means we must have 12 rows, one for each combination of values in the training set. Even ignoring the stage of feature selection, which we do not model, subjects should take longer to master categories that require the larger table.

We implemented this abstract model as a simple Lisp program that accepts as input the number of simulated subjects, rows, classes, and training items, along with the probability p of learning on each trial. Figure 1 (b) shows the behaviors that the model generates when we set p to 0.3 and averaged over 1000 simulation runs. As intended, there is a clear difference between simulated subjects under conditions of consistent and inconsistent contrast, with the former learning much more rapidly than the latter. The match to Billman and Dávila's re-

sults is only qualitative, as the simulated learning curve for the consistent condition is slower, and the one for the inconsistent condition higher, than they observed.[2] Altering the parameter p does not help, since this speeds or slows the curves for both conditions. However, Billman (personal communication, 1998) reports that using stimuli with different within-class similarity reduces the separation between the two curves. An extended model might incorporate such additional factors, but the current one still produces the basic effect intended.

We can contrast this qualitative behavior with that for a different abstract model that operates in the same manner but that does not include feature selection. We can simulate this situation by assuming that the tables encoding the learned knowledge have the same number of columns and rows for both the consistent and inconsistent conditions. Thus, they predict identical behavior for subjects in both situations. As such, it constitutes an abstract version of the concrete models developed by Martin and Billman, Anderson, and Kruschke. But, to reiterate, we need not descend to their detailed level to explain the consistent contrast effect.

Closing Remarks

In the preceding pages, we reviewed the main research paradigms in machine learning and their links to computational models of human learning. We also argued that, for purposes of both developing artifacts and matching human behavior, one can usually achieve very similar results with each of the various approaches. Moreover, we claimed that the source of explanatory power often

[2]On novel test items, the model also predicts very high accuracy for the consistent situation and chance for the inconsistent condition. In this case, the experimental differences are smaller than the model predicts, but the behaviors again match at the qualitative level.

lies not in whether one uses rule induction, neural networks, exemplar models, decision trees, or probabilistic schemes, but rather in the features used to describe experience, the formulation of the problem, and the nature of the training items. Our response was to recommend the use of abstract computational models to explain phenomena, rather than the concrete models that have direct analogs in machine learning. We reviewed some examples of abstract models and applied this approach to specific experimental results in category learning.

Before closing, we should examine some likely criticisms of abstract models. For example, one might claim that such models merely 'describe' the data rather than explain them. But the models we have reported all posit explicit (although abstract) processes, and thus embody some form of explanatory structure. A more interesting question concerns whether such models' assumptions are necessary or merely sufficient to explain the phenomena. Since we reviewed three abstract models of the power law, each making somewhat different assumptions, they clearly constitute the sufficient variety, but necessity is a difficult hurdle to leap in any science.

A deeper criticism is that, to date, abstract modeling efforts have focused on explaining isolated phenomena. Clearly, we do not want to develop 20 unrelated models of category learning, one for each robust phenomenon in the literature. A more desirable approach would imitate older sciences like physics, which devise separate models for each phenomenon but constrain them with links to deep theoretical principles. The concrete modeling community has made some progress on this front, as in using discrimination networks to explain diverse memory phenomena (H. A. Simon, personal communication, 1998), but the same strategy should work for abstract models.

In the long term, these two frameworks need not remain antithetical. As we gradually extend abstract models to cover more phenomena, we must place ever more constraints on them to ensure consistency with previous accounts. At some point, we may even have enough constraints to take defensible positions on issues like the underlying representation of knowledge, the performance mechanisms that operate on that knowledge, and the learning processes that generate it. Eventually, we may have enough data to justify the construction of concrete models or even a unified theory of the cognitive architecture that covers behavior in many domains. However, we do not feel the study of human learning has reached that stage, and abstract models, even isolated ones that focus on specific results, seem worthy of increased attention.

Acknowledgements

We owe thanks to Dorrit Billman and Michael Pazzani for discussions that led to many of the ideas in this paper, and to Dorrit Billman for making available the results of her experimental study.

References

Anderson, J. R., & Kline, P. J. (1979). A learning system and its psychological implications. *Proceedings of the Sixth International Joint Conference on Artificial Intelligence* (pp. 16–21). Tokyo: Morgan Kaufmann.

Anderson, J. R. (1991). *The adaptive character of thought*. Hillsdale, NJ: Lawrence Erlbaum.

Billman, D., & Dávila, D. (1995). Consistency is the hobgoblin of human minds: People care but concept learning models do not. *Proceedings of the Seventeenth Conference of the Cognitive Science Society* (pp. 188–193). Pittsburgh: Lawrence Erlbaum.

Fisher, D. H., & Langley, P. (1990). The structure and formation of natural categories. In G. H. Bower (Ed.), *The psychology of learning and motivation: Advances in research and theory* (Vol. 26). Cambridge, MA: Academic Press.

Gluck, M. A., & Bower, G. H. (1988). Evaluating an adaptive network model of human learning. *Journal of Memory and Language, 27*, 166–195.

Kruschke, J. K. (1992). ALCOVE: An exemplar-based connectionist model of category learning. *Psychological Review, 99*, 22–44.

Langley, P. (1996). An abstract computational model of learning selective sensing skills. *Proceedings of the Eighteenth Annual Conference of the Cognitive Science Society* (pp. 385–390). Lawrence Erlbaum.

Langley, P., & Simon, H. A. (1995). Applications of machine learning and rule induction. *Communications of the ACM, 38*, November, 55–64.

Martin, J., & Billman, D. (1991). Variability bias and category learning. *Proceedings of the Eighth International Workshop on Machine Learning* (pp. 90–94). Evanston, IL: Morgan Kaufmann.

Mooney, R., Shavlik, S., Towell, G., & Gove, A. (1989). An experimental comparison of symbolic and connectionist learning algorithms. *Proceedings of the Eleventh International Joint Conference on Artificial Intelligence* (pp. 775–780). Detroit: Morgan Kaufmann.

Ohlsson, S., & Jewett, J. J. (1997). Simulation models and the power law of learning. *Proceedings of the Nineteenth Annual Conference of the Cognitive Science Society* (pp. 584–589). Stanford, CA: Lawrence Erlbaum.

Richman, H. B., & Simon, H. A. (1989). Context effects in letter perception: Comparison of two models. *Psychological Review, 96*, 417–432.

Rosenbloom, P. S., & Newell, A. (1987). Learning by chunking: A production system model of practice. In D. Klahr, P. Langley, & R. Neches (Eds.), *Production system models of learning and development*. Cambridge, MA: MIT Press.

Shrager, J., Hogg, T., & Huberman, B. A. (1988). A graph-dynamic model of the power law of practice and the problem-solving fan effect. *Science, 242*, 414–416.

Attractor Dynamics in Speech Production:
Evidence from List Reading

Adam P. Leary (adamlear@indiana.edu)

CRANIUM
Indiana University
406 Lindley Hall, IN

Abstract

To date, the vast amount of research done on the isochrony of English speech rhythm has not accounted for the emerging organization of rhythmicity. Our observation that speech rhythmicity is naturally occurring and even preferred as a strategy for optimizing the production and perception of a language-related task has been left untested. A set of experiments were devised to simulate list reading, i.e., a finite set of word tokens that a speaker must convey to hearers. Three lists were used that differed in prosodic structure to investigate the effect of stress pattern on isochrony. The results are analyzed as a low-dimensional dynamical system in which stress determines the cycle of an oscillator. The subjects show consistency in their speech rhythm across all list conditions. There is evidence of attractor dynamics in list reading.

Discussion

In many studies of isochrony and speech rhythmicity in English, subjects are asked to listen to, produce, and/or manipulate utterances so as to make their rhythm regular (Rapp, 1971; Allen, 1972; Morton, Marcus, & Frankish 1976; Tuller & Fowler, 1980; Hoequist, 1983; Howell, 1984; Fox & Lehiste, 1987). Despite this vast amount of research done on isochrony in speech, our observation that speech rhythmicity is naturally occurring and even preferred as a strategy for simplifying specific language-related tasks has been left untested. Thus, our question is, "Do subjects naturally, with minimal instruction, fall into a speech rhythm when asked to read a list of randomly ordered items?" Further, "Can this speech rhythm be defined by a low-dimensional dynamical system?"

A set of experiments were carried out on a group of four native speakers (NSs) of American English in which subjects were asked to read three different lists containing letters (i.e., "A" "B" "C" "D") (List 1), monosyllabic words (List 2), and bisyllabic words with alternating stress (List 3). The timing between vowel onsets was measured for each speaker across all three lists using Bex and hand measurements. Phase angles were measured by averaging the three previous inter-stress intervals (ISIs) of any particular ISI and then dividing the current ISI by the average of the previous three. This manner of measurement could be thought of as a simulation of the process by which auditory perceptual oscillators use short-term memory to make phase adjustments in rhythmic speed (McAuley 1995). In strictly-defined, task-specific systems such as the reading of a list for hearer verification, the establishment of isochronous rhythm may act as one oscillator in a coupled system to which the second oscillator, the internal perceptual oscillator of the hearer (McAuley, 1998), may be entrained in a 1:1 ratio.

In the reading of all of these lists, across four NS subjects, we observed an isochronous rhythm. Specifically, we conclude that:

1) the subjects naturally, with minimal instruction, fall into a regular, resting speech rhythm when asked to read a list of randomly ordered items;
2) the subjects showed consistency in their phase (.8-1.1 msec) across all three lists;
3) there is evidence for a low-dimensional dynamical system falling out of list reading;
4) regression plots of phase:phase-1 indicate the presence of a weak attractor for all three lists(List 1: R=.09 R^2=.008; List 2: R=.05 R^2=.002; List 3: R=.20 R^2=.041 respectively) (Figures 1-3.), with the bisyllabic word lists showing the largest difference between mean phase and current phase (F=2.68 p<.11);
5) simple meter is implicated as a temporal object that is used to regulate the ISIs in list reading.

The observed simple meters of 2:1 and 3:1 suggests that perhaps speakers adjust the timing of their list reading to line up onsets of stressed syllables with preferred points in the auditory oscillator. Thus, an effect similar to a Harmonic Timing Effect (Cummins & Port 1996) is suggested. Also, as phase is biased in multiples of 500 and 600 ms units, it is possible that the preferred rhythm of our subjects corresponds to the resting perceptual oscillatory rate suggested by McAuley (1995).

Thus, we find evidence in a simple speech rhythm task for attractor dynamics and view isochrony as a product of a dynamical system and not a segment manipulation on the part of the speakers.

Figure 1.

List 1 regression plot of phase to mean phase

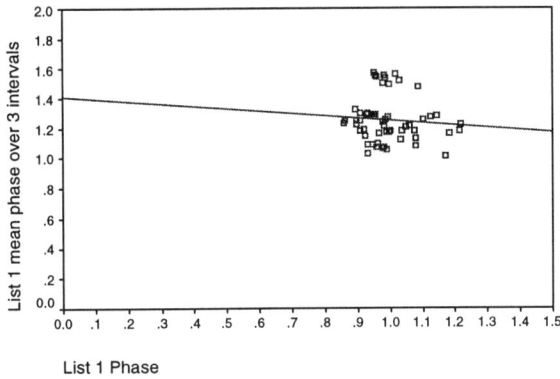

Figure 2.

List 2 regression plot of phase to mean phase

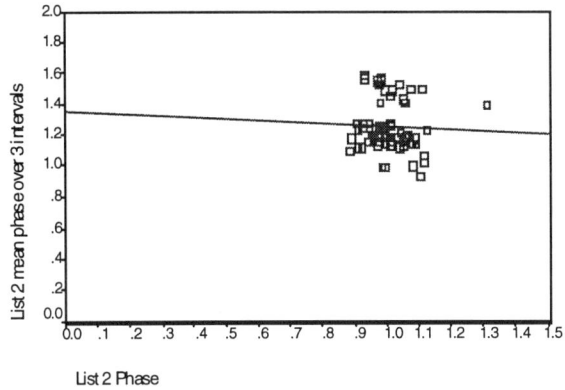

Figure 3.

List 3 regression plot of Phase to Mean Phase

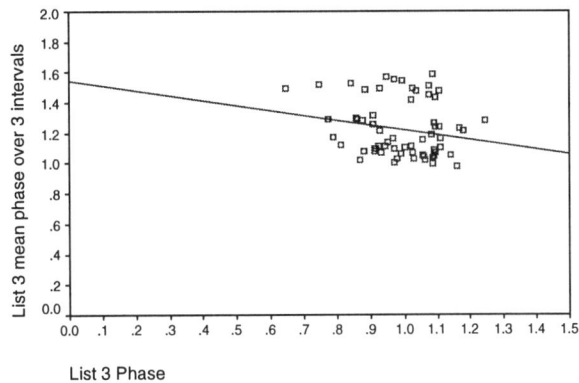

References

Allen, G., *The location of rhythmic stress beats in English: An experimental study I+II*. Language and Speech, 1972. **15**(72-100): p. 179-195.

Cummins, F.and R.F. Port. *Rhythmic constraints on English stress timing*. in *Fourth International Conference on Spoken Language Processing*. 1996: Alfred DuPont Institute.

Fox, R.and I.Lehiste., *The effect of vowel quality variation on stress-beat location*. Journal of Phonetics, 1987. **15**: p. 1-13.

Hocquist, C., *The perceptual centers and rhythm categories*. Language and Speech, 1983. **26**: p. 367-376.

Howell, P. *An acoustic determinant of perceived and produced anisochrony*. in *Xth International Congress of Phonetic Sciences*. 1984.

McAuley, J.D., *Perception of time as phase: Toward an adaptive oscillator model of rhythmic pattern processing.*, . 1995, Indiana University: Bloomington, IN.

McAuley, J.D.and G.R.Kidd, *Effect of deviations from temporal expectations on tempo discrimination of isochronous tone sequences*. JEP: HPP, 1998.

Morton, J., S. Marcus, and C. Frankish, *Perceptual centers (P-centers)*. Psychological Review, 1976. **83**: p. 405-408.

Rapp, K., *A study of syllable timing.*, in *Quarterly Progress and Status Report*. 1971, Speech Transmission Laboratory, Royal Institute of Technology: Stockholm. p. 14-19.

Tuller, B., and C.A. Fowler, *Some articulatory correlates of perceptual isochrony*. Perception and Psychophysics, 1980. **27**: p. 277-283.

A Dynamic ACT-R Model of Simple Games

Christian Lebiere (cl+@cmu.edu)
Psychology Department
Carnegie Mellon University
Pittsburgh, PA 15213

Robert L. West (rwest@hku.hk)
Department of Psychology
Carleton University
Ottawa, Canada

Abstract

A model of humans playing the simple game of Paper Rock Scissors based on the ACT-R architecture (Anderson, 1993; Anderson & Lebiere, 1998) is presented. This model stores in long-term memory sequences of moves and attempts to anticipate the opponent's moves by retrieving from memory the most active sequence. This results in a tightly linked dynamical system in which each player drives the play of its opponent. The performance of this model as a function of the length of the sequences stored and the amount of noise in the system is investigated, and is compared to the performance of human subjects.

Introduction

From the point of view of classical game theory (e.g. von Neumann & Morgenstern, 1944; Nash, 1950; Fudenberg & Tirole, 1991), the simple game of Paper Rock Scissors (PRS) is quite trivial. Each of the three possible moves is as good as the other ones: Paper beats Rock, Rock beats Scissors and Scissors beats Paper. Since the players make their moves simultaneously without any a priori knowledge of each other's move, the optimal strategy is to play randomly and thus guarantee the expected outcome of a tie. However, it is generally accepted that game theory's optimally rational strategies often do not accurately describe human behavior due to the fact that human rationality is bounded (Simon, 1972). Also, game theory does not provide an account of how human players learn. Instead, human game players are best viewed as cognitively limited learners (Erev & Roth, 1998).

As Bracht, Lebiere and Wallach (1998) have demonstrated, ACT-R can be used to model how strategies are applied by conceptualizing the possible moves as productions. There are two advantages to this approach. The first is that ACT-R has been used to model many behavioral phenomena and thus it integrates game playing into the larger context of human cognition. The second is that the method for selecting between productions is consistent with the way game playing is understood in game theory and in Experimental Economics. That is, each move is associated with a probability that reflects its utility. Thus,

while game theory can provide the optimal distribution of the probabilities, ACT-R can provide a cognitively justifiable account of how the actual probabilities are learned.

However, recently West (1998a, 1998b, 1999) has provided an alternative, dynamic systems account of how simple games are played, based on the principle of *reciprocal causation*. Reciprocal causation refers to a state in which two systems are coupled together so that each system's outputs are affected by the other system's behavior (Clark, 1997, 1998). The importance of reciprocal causation is that it is often associated with "emergent behaviors whose quality and complexity far exceeds that which either subsystem could display in isolation" (Clark, 1998). The approach of West (1998a, 1998b, 1999) is based on the findings that humans are quite bad at generating random outputs (see Tune, 1964, and Wagenaar, 1972 for reviews), but quite good at detecting sequential dependencies (e.g. Ward, 1973; Ward, Livingston, & Li, 1988). West (1998a, 1998b, 1999) assumed that players attempt to predict their opponent's next move by detecting sequential dependencies in their opponent's past moves and modeled the process using neural networks. The result was that the modeled players were in a state of reciprocal causation, i.e. each player's moves were determined by their opponent's previous moves.

The reciprocal causation resulted in a chaos-like process that caused both players to generate outputs that appeared random. This result was consistent with the game theory prediction but it was contingent on the players being evenly matched in terms of how many previous moves (lags) they could remember (it was assumed that the players could only remember a limited number of lags back on each trial). When the players were unequally matched in terms of how many lags back they could remember, the player who could remember more lags enjoyed a systematic advantage. Importantly, this was also found to be the case for human subjects (West, 1998a, 1998b, 1999).

This phenomena, which can be considered an emergent property of the dynamic interaction between the players, is very difficult to account for by treating the moves as productions with associated, learned utility values. However, unlike the various specialized models in Experimental Economics, ACT-R is not limited to learning in this way. As Lebiere and Wallach (1998) have demonstrated, the declarative memory system of ACT-R can be used to account for implicit learning tasks without relying on production-based learning. In this paper we demonstrate how the neural network-like qualities of the ACT-R declarative memory system can produce the same phenomena found by West (1998a, 1998b, 1999) in a very straightforward manner.

Model

To emulate the neural network model of West (1998a, 1998b, 1999), we used a simplified version of the ACT-R Sequence Learning Model (Lebiere & Wallach, 1998) that operated by building chunks encoding short sequences of stimuli. For clarity, if the model builds sequences of moves of length 3, we will call it a lag2 model because it remembers the previous two moves of the opponent in addition to its current move. Similarly, a lag1 model refers to sequences of length 2. We will describe below a lag2 model, but we will also report results for a lag1 model.

ACT-R is a goal-directed architecture. At all times, the system focuses on a single goal, and any production must first match that goal before firing. In this model, the current goal can be understood as the player's working memory (Lovett, Reder & Lebiere, in press). It holds a number of the opponent's previous moves in a chunk such as:

Goal
    isa   PRS
    lag2   Paper
    lag1   Rock
    lag0   nil

PRS is the *type* of the goal, and its *slots* are lag2, lag1 and lag0[1]. Lag0 holds the opponent's current move (a value of nil indicates that that move has not yet been played), lag1 holds the opponent's previous move (Rock) and lag2 holds the opponent's move before that (Paper). After a move is made and the lag0 value is filled in, the goal is popped and becomes a chunk in declarative memory. If an identical chunk already exists, then that chunk is reinforced instead of creating a copy.

The model is composed of three productions. The main production, **Sequence Prediction**, attempts to retrieve from memory a chunk that encodes a sequence of three moves (L2, L1, L) played by the opponent, the first two of which match the opponent's last two moves (L2, L1). Then given the third move of that sequence (L), it retrieves the move that beats it (M) and plays that move (M).

Sequence Prediction
IF no move has been played
    and the opponent last played moves L2 and L1
    and moves L2 and L1 are usually followed by move L
    and move L is beaten by move M
THEN play move M

This corresponds to trying to anticipate the move that the opponent is going to make given his most recent moves and making the move that defeats it. If no such sequence of the opponent's moves can be retrieved from memory (for example, at the start of the game), then the second production, **Random Guess**, applies. It simply selects a move (L) at random and plays the move that defeats it (M).

Random Guess
IF no move has been played
    and move L is beaten by move M
THEN play move M

Finally, after the players have each made their move, the third production, **Next Move**, applies. It records the opponent's move (L) in the current goal, thus completing the opponent's most recent three-move sequence (L2, L1, L). It then pops that goal, which becomes a chunk in declarative memory (or reinforces an identical chunk if it already exists), and focuses on a new goal which contains the opponent's two most recent moves (L1, L).

Next Move
IF the opponent has played move L after moves L2 and L1
THEN note move L in the current goal, pop that goal and
        focus on a new goal holding previous moves L1 and L

The production cycle can then start anew. The crucial part of this model is the retrieval from long-term declarative memory in production **Sequence Prediction** of a chunk holding the opponent's move sequence matching the current situation. Retrieval from memory depends upon a chunk's activation. Anderson and Schooler (1991) reported that the odds of an item in the environment being needed decrease as a power function of its past uses. In ACT-R, the activation of a chunk[2] is interpreted as the logarithm of the odds of that chunk being needed from memory, and thus will be defined as:

$$A_i = \ln \sum_{j=1}^{n} t_j^{-d} \qquad (1)$$

A_i is the activation of chunk i, n is the total number of past references to that chunk, t_j is the time since the jth reference and d is the decay rate. This activation equation incorporates both the power law of practice (through the summation) and power law decay (of each reference). Past references refer both to chunk creation (and re-creations) and to retrievals from memory. If the references are assumed to be evenly distributed over the chunk's past history, then the activation of the chunk can be simplified to be[3]:

---

[1] PRS, lag2, lag1 and lag0 are arbitrary names to designate the goal type and its slots.

[2] Strictly speaking, this is only the base-level activation. Additional components of activation include spreading activation and mismatch penalties, but neither is relevant to this model.

[3] For efficiency reasons, the results reported in the next section correspond to models for which Equation (2) is used instead of

$$A_i = \ln \frac{n \cdot L^{-d}}{1-d} \qquad (2)$$

L is the life of the chunk, i.e. the length of time since its creation (t_j). If several chunks satisfy the condition, then the one with the highest activation is retrieved. Zero-mean Gaussian noise is added to the activations, which makes retrieval a probabilistic process. The probability of retrieving chunk i among all the alternatives j is a function of their respective activations and the magnitude of the noise:

$$p(i) = \frac{e^{A_i/t}}{\sum_j e^{A_j/t}} \qquad (3)$$

t is a measure of the noise proportional to its standard deviation[4]. Assuming that all the chunks were created around the same time, i.e. have a similar L, then Equations (2) ad (3) can be simplified to yield:

$$p(i) = \frac{n_i^{1/t}}{\sum_j n_j^{1/t}} \qquad (4)$$

A noise value t of 1 would yield Luce's linear choice rule (Luce, 1959). As Lebiere (1998) established, the noise magnitude t is the crucial parameter that determines the dynamics of the retrieval process. When $t=1$, the probabilities of retrieval match the distribution of past references[5], and retrieval leaves the statistics of occurrence unchanged. For values of t larger than 1, the differences in past references are reduced, and retrieval becomes increasingly random. For values of t smaller than 1, the system becomes increasingly deterministic in selecting the most active chunk. A rich-get-richer dynamics develops, in which the most active chunks become even more so and the less active ones gradually decay away.

Essentially, the model uses the declarative memory system of ACT-R to detect sequential dependencies. Of course, this is only the behavior of a single cognitive system in isolation. Similar to West (1998a, 1998b, 1999), when two of these systems are coupled together the result is a state of reciprocal causation. Thus the important question was whether this particular coupled system would produce the same emergent pattern of behavior that West (1998a, 1998b, 1999) found in his models and human subjects.

Equation (1). There was little indication however that the simplification altered in any way the behavior of the model.

[4] Formally, $t = \sqrt{6}\sigma/\pi$ where σ is the standard deviation.

[5] The phenomenon of reproducing in one's choices the probabilities of occurrence of events in the environment is known as probability-matching (Friedman et al., 1964; Myers, Fort, Katz, and Suydam, 1963).

Results

We will first describe the behavior of the model playing against an identical copy of itself. The model is a lag2 model as described in the previous section. The only parameter is the noise magnitude of 0.25. This parameter is taken from the model of (Lebiere, 1998), which reflected stochasticity in the learning of arithmetic. Examining the difference in score across trials, the output resembles a random walk, with possible fractal properties.

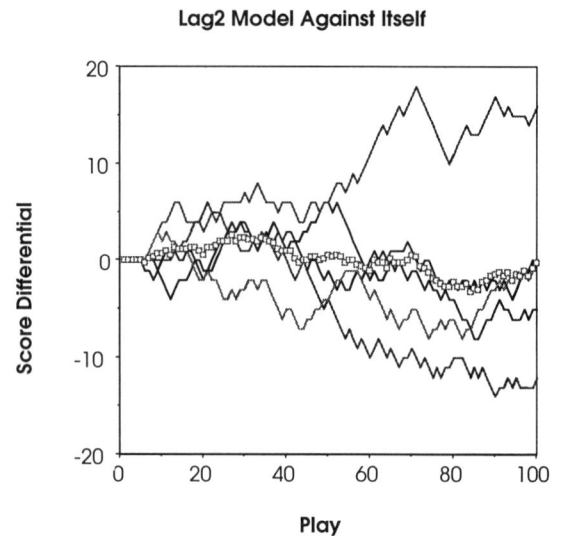

Figure 1: Score differential of lag2 model vs. itself. 5 sample runs of 100 plays. Mean of the 5 runs in squares.

The next question was whether an imbalance between the players in terms of working memory would produce a bias in favor of the player who processed more lags.

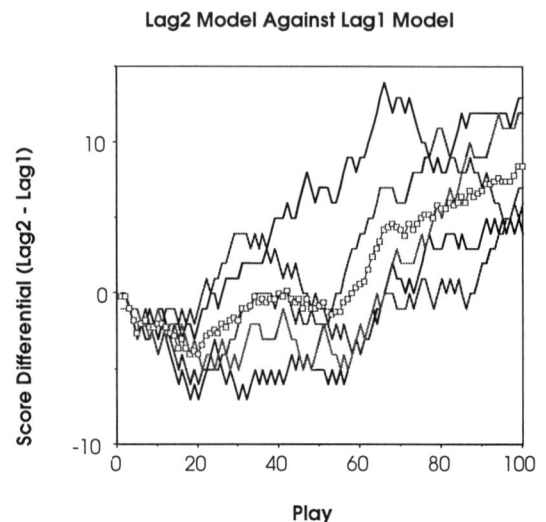

Figure 2: Score differential of lag2 model vs. lag1 model. 5 sample runs of 100 plays. Mean of the 5 runs in squares.

298

While the differential in score between the lag 2 and lag1 models fluctuates as it did between evenly matched models, the long-term trend is clearly in favor of the more powerful lag2 model. But how do these models compare to humans? West (1998a, 1998b, 1999) found that humans play similarly to a lag2 model in that they are able to beat a lag1 model. Following this approach we had human subjects play against the ACT-R lag1 model. The subjects were five participants in the ACT-R summer school.[6]

Human Against Lag1 Model

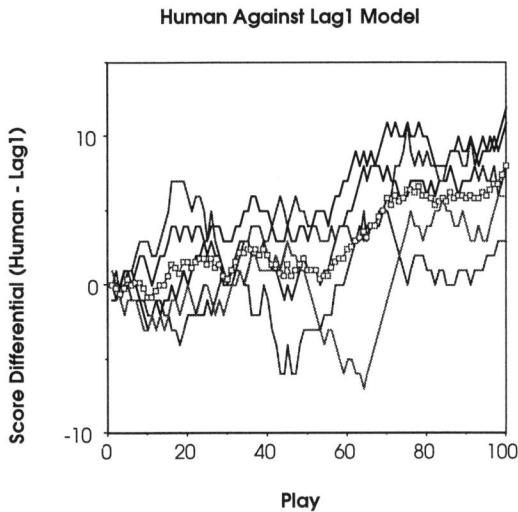

Figure 3: Score differential of humans vs. lag1 model. 5 sample runs of 100 plays. Mean of the 5 runs in squares.

The results were very similar to West (1998a, 1998b, 1999) and also to the performance of the lag2 model playing against the lag1 model (Figure 2), including the fluctuations in the score differential and the average winning margin against the lag1 model. An intriguing feature is that in both Figures 2 and 3 the superior (lag2) player initially loses against the lag1 model, then somewhere between approximately 20 and 30 trials begins to win. This is consistent with the fact that the lag2 model builds longer chunks than the lag1 model, and thus takes longer to accumulate the proper set of sequences. Thus in this range the prediction is reversed and the lag1 model should perform better than the lag2 model. To test this prediction we had 8 human subjects from the University of Hong Kong play short games of 30 trials each against both a lag1 model and a lag2 model. The results, displayed in Figure 4, show that early on the lag1 model is indeed more difficult to beat than the lag2 model. A paired t-test on the score differences revealed that this difference was significant at P<.001.

Human Against Lag1 and Lag2 Models

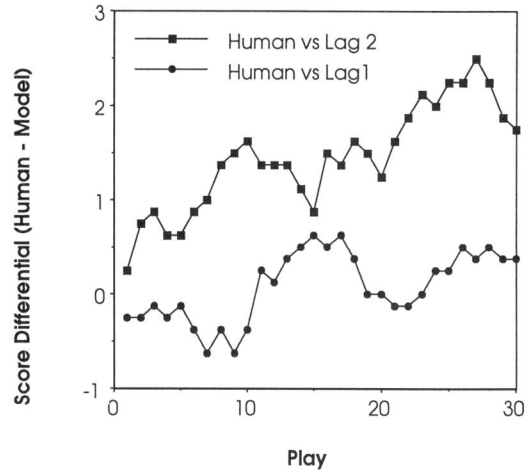

Figure 4: Score differential of humans vs. lag1 and lag2 models for 30 plays. Mean of eight subject runs.

While a larger lag is clearly an advantage, what about the other variable characteristic of our model, the noise? If two models have the same lag and identical noise levels, then as we have seen they will play evenly in the long run. If one model has a very high noise level, it will play randomly (the game theory solution) and will also draw in the long run. This can be a good thing if a player is intrinsically at a disadvantage, as when a lag1 model plays a lag2 model. But is randomness simply a way for a player to limit its losses against a superior opponent? What if both networks have the same lag but different limited noise levels? Is noise an advantage or a disadvantage? Obviously the noisier model is less predictable, but it is also a less powerful learner, slower to pick up on existing sequential dependencies.

Effect of Noise (Lag2 Against Lag2)

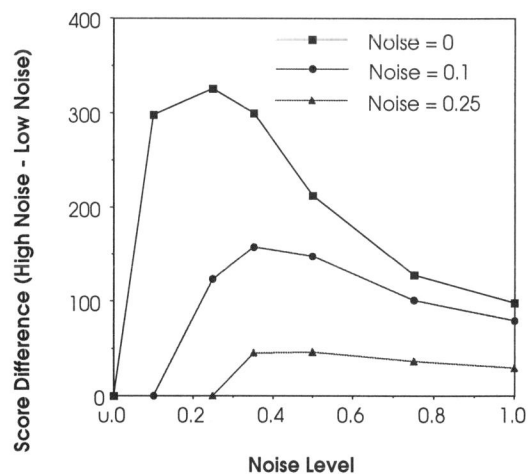

Figure 5: Average over 200 runs of 1000 plays of the final difference in score between two lag2 models with different noise levels.

[6] The model is available for playing on the world-wide web at http://bk1.psy.cmu.edu/inter/models?

We see that all things being equal a higher noise level is indeed an advantage. The advantage in differential score increases for a while as the difference in noise levels increases, then declines because the whole system just becomes increasingly random. To further investigate, we ran a lag2 model against a lag1 model at various noise levels.

Effect of Noise (Lag2 Against Lag1)

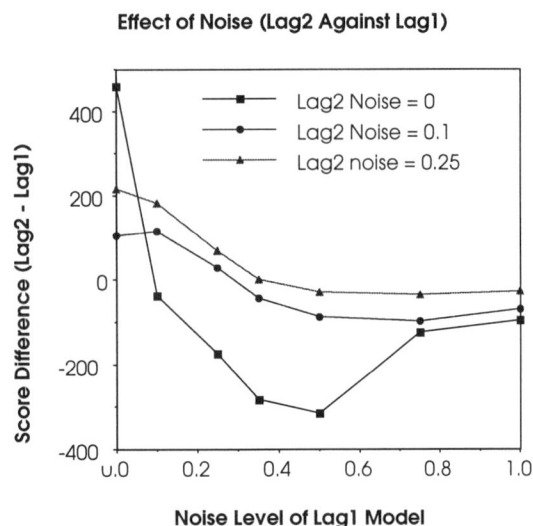

Figure 6: Average over 100 runs of 1000 plays of the final difference in score between a lag2 model and a lag1 model with different noise levels.

The results show that noise can override the lag factor causing a lag1 model to beat a lag2 model. This was particularly true when the noise level of the lag2 model was set to zero or was very low.

Discussion

By manipulating parameters of the model we were able to recreate the phenomena found by West (1998a, 1998b, 1999) and generate further predictions (some of which have yet to be confirmed with human subjects). It is tempting to interpret our results in terms of the individual models. For example, in terms of our findings for number of lags processed and the amount of noise, one interpretation is that the beneficial effect of behaving less predictably (more noise) can outweigh the effects of being a more powerful learner (more lags). However, the picture is potentially more complex. When a model wins it does so by predicting its opponent's moves from sequential dependencies present in past behavior. The past behavior is generated by a reciprocal causation process between the models that results in a chaos-like process. This process produces sequential dependencies as well as a random walk quality. The problem is that it is very difficult to disentangle the process that generates the outputs from the ability of the winning network to detect sequential dependencies, since the action of detecting sequential dependencies is also part of the process that generates them. If we change the way the models detect sequential dependencies we also change the

way the sequential dependencies are generated. Thus, we cannot, at this point, rule out the possibility that the noise factor or the lag factor may operate by altering the type of sequential dependencies produced by the system.

The noise factor can also be understood in another way. Lebiere (1998) found that randomness serves a beneficial cognitive function by keeping the system's dynamics fluid and thus preventing errors from becoming entrenched facts. In our model the opponent is always changing in response to the history of the game. Facts are a short-lived phenomenon in this constantly changing environment. Thus an injection of stochasticity may have the effect of optimizing the system for this environment. More generally, this type of environment probably provides a much better replica of the environment in which our cognitive system evolved than a formal system of unchanging facts and rules such as arithmetic.

The ACT-R model that we used has many features in common with the neural network model of West (1998a, 1998b, 1999). They both work by storing sequences of the opponent's moves and using them to anticipate the opponent's next move. West's (1998a, 1998b, 1999) fixed-length two-layer feedforward neural networks, whose inputs are the opponent's past moves and whose output is the opponent's next move, correspond closely to the ACT-R model's chunks, whose slots holding the opponent's past moves are primed during memory retrieval and whose output is the value of the slot holding the next move. Also, both models resort to a random choice in the case of two moves being equally weighted. However, (again in both models) the main source of randomness is the chaos-like effect generated by coupling the networks together in a state of reciprocal causation. This effect is also the source of the sequential dependencies, which would not occur if the process were based on a truly random process.

However, the ACT-R model has several advantages. First, because ACT-R is a unified cognitive architecture, the model is more informative as to the cognitive structures involved in the process. Specifically, the ACT-R model situates the detection of sequential dependencies in declarative memory, while the lag factor can be interpreted in terms of the amount of working memory (Lovett, Reder & Lebiere, in press). ACT-R also allows for a principled investigation of background random noise, which turns out to be an important factor. Also, ACT-R is capable of modeling more complex games, involving knowledge and strategy, through the use of productions. Because these games also often involve an element of guessing, we suggest that a full model of game playing will integrate both processes. ACT-R is important in this respect because it provides a ready-made model of how to structure this integration.

The origins of this ACT-R model should be emphasized to illustrate the lack of degrees of freedom in its conception. The basic idea to play PRS by storing fixed-length sequences of the opponent's moves was adopted from West (1998a, 1998b, 1999), as was the default length of those sequences. The very simple chunks and productions used to implement that idea were taken from an existing ACT-R model of a seemingly very different paradigm from another field. The default value of the only parameter, the magnitude of the noise, came from an ACT-R model of a separate phenomenon. Those elements were assembled almost automatically and provided a very accurate model of human playing. That it happened on the first try, without any engineering or parameter tuning, is a demonstration of the predictive power of unified architectures.

Conclusion

We proposed a model of human playing for Paper Rock Scissors. This model was inspired by known psychological limitations and inclinations instead of the ideal strategies of classical game theory. The players were viewed not as isolated cognitive entities but as part of a dynamical system in which they constantly influence each other's actions. Crucial parameters of this cognitive system are the raw power of the actors in terms of the length of sequences that the players can process and the degree of stochasticity with which they select their actions. This model was found to closely account for human behavior, without the benefit of unexamined degrees of freedom in its knowledge structures or parameters.

Acknowledgements

This research was partially supported by a grant from the Office of Naval Research (N00014-95-10223).

References

Anderson, J. R. (1993). *Rules of the Mind*. Hillsdale, NJ: Lawrence Erlbaum Associates.

Anderson, J. R., & Lebiere, C. (1998). *The Atomic Components of Thought*. Mahwah, NJ: Lawrence Erlbaum Associates.

Anderson, J. R., & Schooler, L. J. (1991). Reflections of the environment in memory. *Psychological Science*, 2, 396-408.

Bracht, J., Lebiere, C., & Wallach, D. (1998). On the need of cognitive game theory: ACT-R in experimental games with unique mixed strategy equilibria. Paper presented at the Joint Meetings of the Public Choice Society and the Economic Science Association, New Orleans, LA.

Clark, A. (1997). *Being there: Putting brain, body and world together again*. Cambridge, MA: MIT Press.

Clark, A. (1998). The dynamic challenge. *Cognitive Science*, 21 (4), 461-481.

Erev, I., & Roth, A. E. (1998). Predicting how people play games: Reinforcement learning in experimental games with unique, mixed strategy equilibria. *American Economic Review*, 88(4), 848-881.

Friedman, M. P., Burke, C. J., Cole, M., Keller, L., Millward, R. B., & Estes, W. K. (1964). Two-choice behavior under extended training with shifting probabilities of reinforcement. In R. C. Atkinson (Ed.), *Studies in Mathematical Psychology* (pp. 250-316). Stanford, CA: Stanford University Press.

Fudenberg, D., & Tirole, J. (1991). *Game Theory*. Cambridge, MA: MIT Press.

Lebiere, C. (1998). The dynamics of cognition: An ACT-R model of cognitive arithmetic. Ph.D. Dissertation. *CMU Computer Science Dept Technical Report* CMU-CS-98-186. Pittsburgh, PA.

Lebiere, C. & Wallach, D. (1998). Implicit does not imply procedural: A declarative theory of sequence learning. Paper presented at the *41st Conference of the German Psychological Association*, Dresden, Germany.

Lovett, M. C., Reder, L. M., & Lebiere, C. (in press). Modeling working memory in a unified architecture: An ACT-R perspective. To appear in Miyake, A. & Shah, P. (Eds.) *Models of Working Memory: Mechanisms of Active Maintenance and Executive Control*. New York: Cambridge University Press.

Luce, R. D. (1959). *Individual Choice Behavior: A Theoretical Analysis*. New York: Wiley.

Myers, J. L., Fort, J. G., Katz, L., & Suydam, M. M. (1963). Differential monetary gains and losses and event probability in a two-choice situation. *Journal of Experimental Psychology*, 66, 521-522.

Nash, J. (1950). Equilibrium points in N-person games. *Proceedings of the National Academy of Sciences,* 36, 48-49.

Simon, H. A. (1972). Theories of bounded rationality. In C. B. Radner & R. Radner (Eds.), *Decision and Organization* (pp. 161-176) Amsterdam: North-Holland.

Tune, G. S. (1964). A brief survey of variables that influence random generation. *Perception and Motor Skills, 18*, 705-710.

von Neumann, J., & Morgenstern, O. (1944). *Theory of Games and Economic Behavior*. Princeton University Press.

Wagenaar, W. A. (1972). Generation of random sequences by human subjects: A critical survey of the literature. *Psychological Bulletin, 77*, 65-72.

Ward, L. M. (1973). Use of markov-encoded sequential information in numerical signal detection. *Perception and Psychophysics, 14,* 337-342.

Ward, L. M., Livingston, J. W., & Li, J. (1988). On probabilistic categorization: The markovian observer. *Perception and Psychophysics, 43,* 125-136.

West, R. L. (1998a). Zero Sum Games as Distributed Cognitive Systems [Summary]. In *Proceedings of the Twentieh Annual Meeting of the Cognitive Science Society*. Mahwah, NJ: Erlbaum.

West, R. L. (1998b). Zero Sum Games as Distributed Cognitive Systems. In *Proceedings of the Complex Games Workshop*. Tsukuba, Japan: Electrotechnical Laboratory Machine Inference Group.

West, R. L. (1999). Simple Games as Dynamic Distributed Systems: A Neural Network Model of the Emergent Properties. Manuscript submitted for publication.

Learning Under High Cognitive Workload

F. Javier Lerch (fl0c@andrew.cmu.edu)
Cleotilde Gonzalez (conzalez@andrew.cmu.edu)
Christian Lebiere (cl@andrew.cmu.edu)

Center for Interactive Simulations
Carnegie Mellon University
5000 Forbes Ave. Pittsburgh PA 15213 USA

Abstract

This research investigates the impact of time pressure and individual differences on learning in a Real-Time Dynamic Decision Making (RTDDM) task. Our empirical results indicate that high time pressure generates high cognitive loads inhibiting learning. The results also show that high time pressure have a differential impact on the learning of individuals with high or low Working Memory (WM) capacity. We present a cognitive model based on ACT-R intended to explain learning in this task. Our cognitive model simulates learning by recognizing regularities in the decision task, and building "chunks" that guide decision making (instance-based learning). We describe how the model will be used to explain the impact of time pressure and WM capacity by varying the number of chunks acquired by the system given alternative time pressure conditions and individual differences.

Introduction

Real-Time Dynamic Decision Making (RTDDM) tasks have three main characteristics: a) the decision maker has to make a series of interdependent decisions; b) the environment changes because of exogenous events and because of prior decisions; and c) the pacing of decisions is dictated by the task rather than by the decision maker (Brehmer, 1990). This research investigates the impact of time pressure and individual differences on learning in a RTDDM task. It attempts to explain these phenomena by building a detailed cognitive model of the decision maker. The rationale for the cognitive model is to have a more in-depth understanding of why time pressure and individual differences foster or inhibit learning. We expect this detailed understanding would help us build better training and decision aids for RTDDM tasks.

Theory

In most RTDDM tasks the rules for making individual decisions are simple. For example, air traffic controllers need to identify if two airplanes are in a collision course. If this is the case, they need to ask one of the airplanes to change direction. But the tasks are rather complex because of the interdependency of decisions and the time pressure to make them. Under these conditions, we expect most learning will be instance-based learning. That is, decision makers will learn chunks expressing under which task conditions specific decisions have the desired effect on the system.

WM is the system for holding and manipulating information during the performance of cognitive tasks (Baddeley, 1990). Limitations in WM capacity have been recognized as a major bottleneck in human cognitive processing. We expect that differences in WM capacity will have a great impact on how individuals perform and learn in RTDDM tasks because these environments impose a high cognitive workload. More specifically, we expect that individuals with high WM resources will learn faster because they have the additional cognitive resources to reflect on the impact of their prior decisions, and to store more and better chunks. Also, since WM capacity is used for both performance and learning, we expect that decision makers will learn faster if they are first trained in a low time pressure environment, and then they are asked to make decisions in the higher time pressure environment. Conversely, individuals that are trained from the beginning in the high time pressure environment should find it harder to learn because all their cognitive resources are devoted to executing the task, and they have less spare resources devoted to learning. This prediction should be mediated by individual differences in WM.

WM is divided into two subsystem: 1) a linguistic sub-system, and 2) a spatial sub-system. In the linguistic sub-system, information is kept in linguistic code, and the processing can be characterized as sequential and propositional. In the spatial sub-system, information is kept in visual code, and the processing can be characterized as more parallel and analogical. There is strong evidence that language processing and spatial thinking are supported by separate pools of WM capacity (Shah and Miyake, 1996). Prior studies have shown that individuals with high linguistic WM capacity perform better than individuals with low linguistic WM capacity in a variety of real-time tasks such as reading comprehension (Just and Carpenter, 1992) and phone-based interaction (Huguenard, Lerch, Junker, Patz and Kass, 1997). Our RTDDM task is highly spatial so we expect that individuals with high spatial WM capacity will perform and learn better than individuals with low spatial WM capacity.

In our research we use traditional measurements of linguistic and spatial WM capacity. We also use the Raven Progressive Matrices Test (Raven, 1962) as an additional measurement of spatial WM. Prior research using detailed eye-tracking analysis has shown that differences in Raven tests can be explained by the ability to induce abstract spatial relations and the ability to dynamically manage a large set of problem-solving goals in WM (Carpenter, Just and Shell, 1990).

Laboratory Study

We used a simulation tool called Pipes. Pipes is an abstraction of a resource management task that can be performed by a single individual or a group. The task is an isomorph of a real-world task in an organization with large-scale logistical operations (the United States Postal Service). We have built a realistic simulation of the task, but this simulation is too complex and takes too long to learn to be practical in laboratory studies (See Lerch, Ballou and Harter, 1997 for a detailed description of the realistic simulation). On the other hand, Pipes can be learned in approximately one hour, and a complete trial can be run in few minutes. Pipes simulates a water distribution system (isomorph to mail sorting) with multiple deadlines for alternative tanks in the system. The whole simulation is spatial. Decision makers have to decide when to activate or de-activate pumps given that the number of pumps working at any given time is restricted (this is isomorphic to having a limited number of sorting machines in the USPS). The task is highly dynamic because water may arrive into a tank at any time, and the level of water in each tank depends on prior decisions (i.e., the pumps that were activated or de-activated by the decision maker in the past). The task is also real-time because pumps are activated or de-activated as the simulation clock is running (See Figure 1 for the main layout of the simulation).

We ran 33 participants using this simulation. Each participant was run in five consecutive days, and paid $50 at the end of the 5 days. In the first two days, each participant completed three psychological tests: the Reading Span Test (Daneman and Carpenter, 1980) that measures WM capacity for language processing; the Spatial Span Test (Shah and Miyake, 1996) that measures WM capacity for spatial thinking; and the Raven Progressive Matrices Test (Raven, 1962).

We manipulated time pressure by changing the speed of events. In the last three days, each participant was randomly assigned to one of two groups: the Fast-Fast (FF) condition and the Very Slow-Fast (VSF) condition. The exogenous events in all trials were identical. The simulation was run either in a Fast mode (8 minutes trials), or in a Very Slow mode (24 minutes trials). In the FF condition, participants ran the simulation 6 times for three days in the Fast mode (18 trials over three days). In the VSF condition, participants ran the simulation in the Very Slow mode for

the first *two* days. In these *two* days, they only ran 2 trials per day, so their total time on task was the same as the time on task for Fast-Fast participants (Very Slow-Fast: 2 trials x 24 minutes = 48 minutes per day; Fast-Fast: 6 trials x 8 minutes = 48 minutes per day). In the third (last) day, the Very Slow-Fast participants ran the simulation 6 times in the Fast mode (8 minutes trials), the same as the Fast-Fast participants. We expect Very Slow-Fast (VSF) participants will exhibit more learning than Fast-Fast (FF) participants.

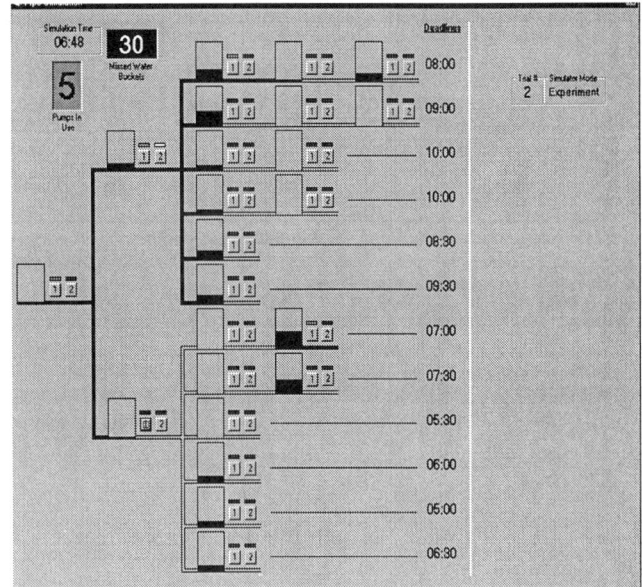

Figure 1. The Pipes simulation

Experimental Results

Figure 2 shows that all three measures of individual differences are correlated. Also, as expected, they show that Raven is more highly correlated to spatial WM capacity than to linguistic WM capacity.

We averaged the results of each participant across trials for each day so each participant had only three repeated performance measures (one for each day). Our performance measure is the number of water buckets that were not pumped in time, therefore the higher the number of water buckets missed, the worse the performance of the decision maker.

We first ran an analysis of variance with three repeated measures with only spatial WM as a covariate. The results show that individuals with high spatial WM capacity performed better than those with low spatial WM capacity [$F(1,29)=4.813$, $p=.036$)]. Second, we ran the same analysis with only linguistic WM capacity as a covariate. Linguistic WM capacity was not significant [$F(1,29)=.341$,ns]. We then ran the same analysis with spatial WM capacity and Raven as covariates. The results are shown in Figure 3.

	Raven	Linguistic WM	Spatial WM
Raven	1		
Linguistic WM	0.391	1	
Spatial WM	0.504	0.545	1

Figure 2. Correlations Among the Three Tests.

	Between-Subjects Effects				
	df	Mean Square	F	Sig	Observed Power
Condition	1	3638.892	0.879	0.357	0.148
Spatial WM	1	2140.571	0.517	0.478	0.107
Condition*Spatial WM	1	7816.475	1.888	0.181	0.263
RAVEN	1	29031.594	7.013	**0.013**	0.723
Condition*Raven	1	6710.012	1.621	0.214	0.233
Error	27	4139.791			
	Within-Subjects Effects				
	df	Mean Square	F	Sig	Observed Power
Trial	2	683.558	1.065	0.352	0.227
Trial*Condition	2	4587.988	7.149	**0.002**	0.919
Trial*Spatial WM	2	312.358	0.487	0.617	0.126
Trial*Condition*Spatial WM	2	515.578	0.803	0.453	0.180
Trial*RAVEN	2	186.382	0.290	0.749	0.094
Trial*Condition*RAVEN	2	4158.304	6.480	**0.003**	0.890
Error (Trial)	54	641.757			

Figure 3. Between and Within Subjects Effects

These results indicate that when both covariates are used only Raven is significant (p=.013). The analysis of variance also shows two significant interaction effects: Trial X Condition interaction (p=.002) and Trial X Condition X Raven interaction (p=.003).

Figure 4 shows the interaction between trial and time pressure. The graph shows that performance was very similar the first day between the FF and VSF participants. But VSF participants improved their performance faster than FF participants. It is important to remember that VSF participants only had 4 trials in the first two days (2 trials per day) while FF participants ran the simulation 12 times (6 times per day). All participants ran the simulation 6 times the last day under high time pressure.

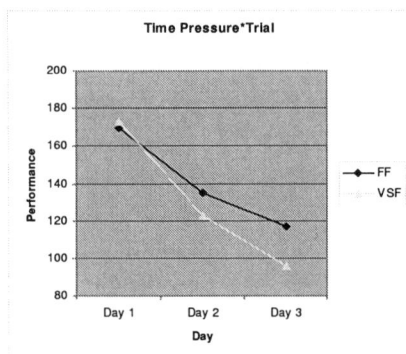

Figure 4. Trial - Time pressure interaction.

Finally, Figure 5 shows the triple interaction. The left panel shows the results of the Low Raven subjects (we divided subjects by using the mean of our sample). The graph shows that Low Raven subjects greatly benefited by first being trained in the low time pressure condition before being exposed to the high time pressure version of the simulation (VSF condition). Although Low Raven subjects in the FF condition performed better the first days than subjects in the VSF condition, they exhibited little learning throughout the three days.

The right panel graph shows the results for the High Raven subjects. In this case, the benefits of the VSF condition on learning and performance are very small throughout the 3 days. It also shows that High Raven subjects had a better performance than Low Raven subjects consistently (compare left and right panels).

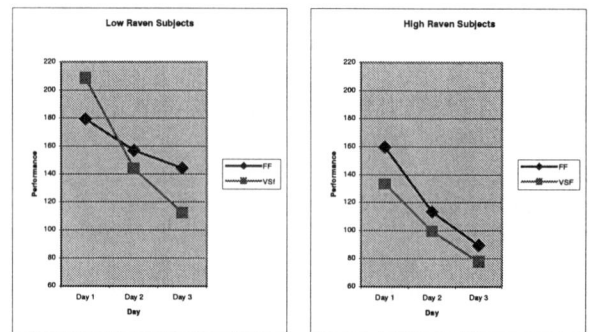

Figure 5. Triple interaction

What Was Learned?

Our next step was to analyze each decision made by each subject in order to figure out what subjects were learning. We hypothesized that subjects would learn chunks representing under which task conditions specific decisions improve performance rather than learning decision rules (or improving their implementation of these rules). In our analysis, we compared each decision in each trial (between 30 and 60 decision per trial) for each day (6 trials in the Fast condition and 2 trials in the Very-Slow condition) for each subject (33 participants) against standard decision rules. These rules were derived from the scheduling literature and the verbal protocols of pilot subjects. For example, a standard rule is the slack rule. In the slack rule you take into account the time left before the deadline and the volume of water in each tank, and select to activate the pump(s) of the tank with the lowest slack (we call this strategy the Time-Volume rule). We did this analysis using several decision rules. For each decision and for each decision rule we calculated a goodness of fit coefficient using the following formula:

Goodness of fit = 1 - ((slack - minimum) / (maximum - minimum))

This coefficient has values between 0 and 1 and represents the similarity between a decision rule and each decision made by the subject. A coefficient fit of 1 means perfect agreement (i.e., the slack of the subject's decision is the same as the minimum slack in the environment). In such a case the subject has chosen the best decision according to the Time-Volume rule. On the other hand, a coefficient fit of 0 is equivalent to the subject selecting the maximum slack in the environment. In this case, the subject has chosen the worst decision according to this rule. In this paper we only report the results of the fit for the Time-Volume rule since the results are similar for the other rules (and because of space constraints).

Figure 6 shows the results of the average fit of all decisions for each day (several trials per day) across all subjects in the FF and VSF conditions. The graph shows that the rule fit *declines* through time, that is, subjects follow the rule less as they are learning. It also shows that VSF subjects (the best learners) had a more pronounced decline in their rule-following fit. Similar declines were found for simpler and for more complex rules. Those subjects that learn the most are those that learn to follow the standard rules less often. These subjects seem to make decisions by being more data driven, that is, by exploiting specific task conditions and making decisions that may have worked in the past.

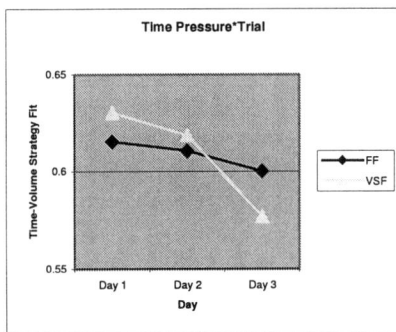

Figure 6. Average fit of all decisions per day

Figure 7 shows the results of the average fit for the Time-Volume decision rule for Low and High Raven subjects. We would expect that Low Raven subjects may be less able to exploit the specific conditions of the task environment because they have less cognitive resources to analyze and store chunks on what decisions worked under which task conditions, especially if they are trained in the high time pressure simulation (FF condition). The left panel shows that Low Raven subjects in the FF condition in fact *increase* their fit to the Time-Volume rule from day 1 to day 3. These are the subjects who exhibited the worst learning. On the other hand, Low Raven subjects in the VSF condition started with a high fit coefficient but lower this coefficient through time. The right panel shows the results for the High Raven subjects. Their rule fit coefficients were lower than for the Low Raven subjects, and decline through time for both experimental conditions.

Our hypothesis here is that subjects that are more data driven should perform better and learn more. If this is true, then we should expect that the best performers were not only

those that followed the rule the least, but are also more adaptive. One measure of adaptation is the standard deviation of the goodness of fit coefficients *within* a trial. Two subjects may have the same average goodness of fit coefficient in a trial, but very different standard deviations. The subject with the high standard deviation follows the rule very closely for some decisions, and not all for others, while the subject with the low standard deviation follows the rule at the same level for all decisions. To test this hypothesis we ran a regression of performance for each trial (450 trials for all subjects)against the following variables:

a)   Raven score
b)   Average rule fit per trial
c)   Standard deviation of rule fit per trial
d)   Two other measurements of how well subjects used the task environments resources (i.e., pump time)

The results were highly significant (Adjusted R^2 = .785). There are no co-linearity problems among the explanatory variables. The highest standardized coefficients were for the two measurements of resource utilization (-.647 and -.320; negative coefficients mean performance improvements). The standardized coefficients for the other three variables are: Raven = -.161 (p<.001), Average fit = .114 (p=.015), and Standard Deviation of fit = -.208 (p<.001). These coefficients indicate that subjects with higher Raven scores have better performance, trials with a higher average rule fit have worse performance (after controlling for Raven score and resource utilization), and finally, trials with higher standard deviation have higher performance. The last coefficient suggests that the less consistent subjects are following the Time-Volume rule (after controlling for average fit), the better their performance. Similar results apply to other decision rules. These results indicate that data driven decision making is beneficial in this task environment.

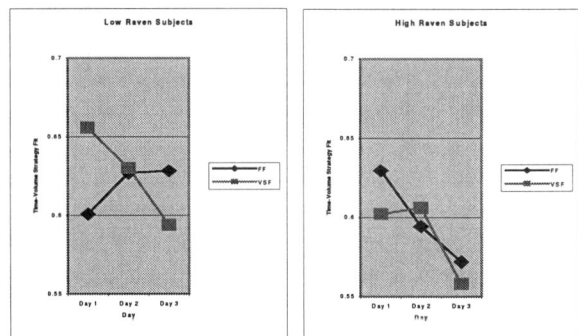

Figure 7. Average fit for Time-Volume Rule for Low and High Raven.

The Act-R Theory

ACT-R is a theory of cognition that has been applied to a wide range of cognitive tasks since its introduction in 1993 (Anderson and Lebiere, 1998). ACT-R assumes two types of memory: procedural and declarative. Procedural memory contains skills in the form of productions or rules of action.

Declarative memory holds explicit knowledge represented as chunks. Production rules specify how the chunks are used to solve problems. ACT-R is a goal-directed architecture. At each cycle, one goal is designated as the top goal or focus of attention. A production is then selected that matches that goal, retrieves a chunk from memory (if necessary), then transforms the goal. This is a symbolic description of ACT-R in terms of how productions and chunks interact.

ACT-R has also a sub-symbolic level. At this level, ACT-R provides real-valued quantities associated with declarative and procedural knowledge to produce a more accurate picture of the graduated nature of human cognition. Their purpose is to resolve conflicts: when several productions match the current goal or several chunks match a production condition, the sub-symbolic parameters associated with those symbolic structures will determine which is selected, and how quickly. Those parameters are learned to optimize the model to the structure of the environment.

In this model, we will concentrate on the acquisition and use of declarative knowledge. We will assume that the productions used to manipulate those chunks, and their parameters, reflect some general, well-established knowledge on how to solve problems of this type.

For declarative knowledge, a chunk is defined as a collection of slots, each of which can hold another chunk as value, and is associated to a quantity called activation which represents the chunk's history of use and determines its availability. Specifically, the activation A_i of chunk i is defined by the formula:

$$A_i = B_i + \sum_j W_j \cdot S_{ji} + N(0, \sigma)$$

B_i is called the base-level activation. It increases with the number of references to the chunk (practice) and decays with time (forgetting). The second term represents the activation spread from each source j according to its source level W_j and its strength of association, S_{ji}, to the chunk. The values of the current goal are the sources of activation, which evenly share a total source amount W. Finally, zero-mean Gaussian noise of standard deviation σ is added to the activation.

In a task such as this featuring continuously evolving quantities such as time and amount of water, no match to declarative memory is ever likely to be perfect because no situation is ever encountered precisely the same way twice. A mechanism called partial matching allows a chunk to be retrieved even if it only partially matches a production condition. A quantity called the match score of chunk i to production p is defined by subtracting from the chunk activation an amount proportional to the degree of mismatch:

$$M_{ip} = A_i - MP \cdot \sum_{v,d} (1 - Sim(v, d))$$

MP is a scaling parameter called the mismatch penalty and Sim(v,d) is the similarity between each production condition d and corresponding chunk value v. The chunk with the highest match score will be retrieved from memory if its score is above the activation threshold τ. Otherwise, the production fails and another is selected. Finally, the latency to retrieve a chunk is inversely proportional to its match score, making more active, better-matching chunks faster to access. The addition of noise to the activation makes declarative retrieval a probabilistic process, and the mechanism of partial matching makes it an approximate process. Stochasticity and generalization are two human qualities in constant display in this task.

These mechanisms of ACT-R can be used to implement a "user model" of the task that will generate detailed predictions for each action and its latency at every step of the problem-solving process.

An ACT-R Model of the Pipes RTDDM Task.

ACT-R has successfully modeled phenomena of memory, problem solving and skill acquisition (Anderson and Lebiere, 1998). However, most of the tasks modeled up to now are static, relatively simple tasks. Recently, there has been more interest in modeling more dynamic and complex tasks in ACT-R. Figure 8 represents our proposed ACT-R model for the Pipes task. The overall goal is represented as a set of deadline chunks. The focus of attention is the deadline closest to the current simulation time.

Declarative memory has two main chunk structures: "the tank" and "the decision." The information provided in the tank chunk corresponds to the physical representation of a tank in the system, and what the user is aware of: the water amount, the deadline, the connections with other tanks, and the status of the pumps in that tank. The chunk called "decision" stores the information on the evaluations performed on the tanks during the course of the learning process, namely water amount, time until the deadline and evaluation. The first time the model is run, no decision chunks exist. Decision chunks are created in the course of solving the current problem: when a goal that was set to evaluate a tank is completed, it becomes a declarative memory chunk holding the information relevant to the evaluation. If the same evaluations are made or retrieved in future trials, the decision chunk increases its activation value, increasing the probability of being retrieved in the future. Decisions are updated according to the feedback provided by the system (i.e., no missed buckets decreases the evaluation because more slack was available whereas a high number of missed buckets increases the evaluation that generated this situation because the tank should have been given a higher priority). Since an identical situation is unlikely to occur, the partial matching mechanism provides a certain amount of generalization in finding the "correct" decision, i.e. a particular decision chunk may be retrieved if the time until deadline and amount of water in the tank is close enough to the current situation.

Procedural memory consists of 5 basic activities: evaluate tanks for which pumps may be turned on or off, turn specific pumps on or off, and review the environment (e.g., re-start the evaluation cycle). When the user evaluates the tanks, the model assumes that the user will keep a value of "urgency" for each tank and type of action (urgency to turn-pumps associated with each tank on or off) in the chunks. Two productions are available to evaluate a tank. The first one will try to retrieve a prior decision closely matching the characteristics of this tank (water, deadline). If no prior decision is sufficiently active and matches closely enough to reach the retrieval threshold, then that production will fail. The second production then will be selected that evaluates the chunk according to some general heuristic function. After all the tanks have been evaluated, the user then decides to turn on the pump associated to the most urgent tank if a pump is available, or to turn off the pump associated to the least urgent tank, thereby freeing a pump, assuming that the urgency of that tank is significantly less than that of the tank that needs to be turned on. Actions in the user model modify the status of the environment, which is updated by the simulation. According to the definition of RTDDM, the environment also changes independently from user's actions. The system adds water to the tanks, pumps water from previous tanks, automatically turns off pumps that correspond to tanks with no more water, and turns on pumps that have been queued by the user. The simulation also verifies the deadlines to provide feedback to the user.

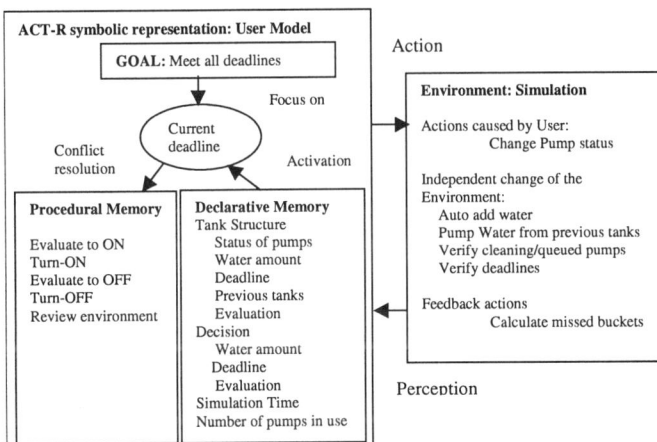

Figure 8. ACT-R Model for Pipes

Although very little work has been done in ACT-R to model individual differences, work by Lovett, Reder, and Lebiere (1999) indicate that the W parameter may be manipulated to capture individual differences in WM. However, that parameter controls the spreading activation component, which is essential to accounts of memory phenomena such as the fan effect, but not particularly relevant in this model. Therefore, we will also investigate if other parameters can account for individual differences, including the decay rate of base-level activation d, the mismatch penalty MP, the retrieval threshold τ and the activation noise magnitude σ. All of these parameters affect the activation that controls the availability of decisions

chunks, but they act on separate components of the activation and thus are expected to exhibit different effects.

Time pressure in the model is implemented by modifying the rate at which the environment changes, and comparing the ACT-R's time to that rate. If the rate of change is very low (no time pressure), the user model will have time to complete more evaluations before the environment changes, and to better reflect and update its evaluations following system feedback. If the rate of change is very high, the user may not have time to evaluate the environment completely before it changes again, or to update its evaluations.

Conclusions

This research suggests that learning in real-time dynamic decision tasks depends on the spare WM resources available during task execution. It also suggests that most learning is based on acquiring relevant decision instances that exploit the task environment.

Acknowledgments

The research reported here was partially supported by a grant from the Air Force Office of Scientific Research (F49620-97-1-0368).

References

Anderson J.R., Lebiere Christian.(1998). *The Atomic Components of Thought*. Hillsdale, NJ: Lawrence Erlbaum Associates.

Baddeley A. D. (1990*). Working Memory: Theory and Practice*. Boston: Allyn and Bacon.

Brehmer, B. (1990). Strategies in Real-Time, Dynamic Decision Making. In R.M. Hogarth (Ed.), *Insights in Decision Making*, University of Chicago Press, 262-279.

Carpenter P.A., Just M.A., Shell P. (1990). What One Intelligence Test Measures: A Theoretical Account of the Processing in the Raven Progressive Matrices Test. *Psychological Review*, 97,3,404-431.

Daneman, M. and Carpenter, P.A. (1980) Individual Differences in Working Memory and Reading. *Journal of Verbal learning and Verbal Behavior*. 19, 450-466.

Huguenard B. R., Lerch F.J., Junker B.W., Patz R.J. and Kass R.E. Working-Memory Failure in Phone-Based Interaction (1997). *ACM Transactions on Computer-Human Interaction*, 4,2, 67-102.

Just, M.A. and Carpenter, P.A. (1992) A Capacity Theory of Comprehension: Individual differences in Working Memory. *Psychology Review*. 99, 1, 122-149.

Lerch, F.J., Ballou, D.J. and Harter, D.E. (1997). Using Simulation Based Experiments for Software Requirements Engineering. *Annals of Software Engineering*, 3, 345-366.

Lovett, M. C., Reder, L. M., & Lebiere, C. (1999). Modeling Working Memory in a Unified Architecture: An ACT-R Perspective. In Miyake, A. and Shah, P. (Eds.) Models of Working Memory: Mechanisms of Active Maintenance and Executive Control. New York: Cambridge University Press, 1999.

Raven J.C. Advanced Progressive Matrices test. (1962) (distributed in the US by the Psychological Corporation).

Shah P. and Miyake A. (1996). The Separability of Working Memory Resources for Spatial Thinking and Language Processing: An Individual Differences Approach. *Journal of Experimental Psychology*, 125, 4-27.

Generalization, Representation, and Recovery in a Self-Organizing Feature-Map Model of Language Acquisition

Ping Li (**Ping@Cogsci.Richmond.Edu**)
Department of Psychology
University of Richmond
Richmond, VA 23173, USA

Abstract

This study explores the self-organizing neural network as a model of lexical and morphological acquisition. We examined issues of generalization, representation, and recovery in a multiple feature-map model. Our results indicate that self-organization and Hebbian learning are two important computational principles that can account for the psycholinguistic processes of semantic representation, morphological generalization, and recovery from generalizations in the acquisition of reversive prefixes such as *un-* and *dis-*. These results attest to the utility of self-organizing neural networks in the study of language acquisition.

Introduction

Language learning is characterized by the learner's ability to generalize beyond what is heard in the input. One current debate on connectionist models of language acquisition concerns the issue of generalization (Elman, 1998). Probably the best-known example in this debate has to do with the acquisition of the English past tense: children generalize *-ed* to irregular verbs, producing errors like *falled*, *breaked*, and *comed*. Connectionist researchers argue that their networks, like human children, display generalizations in a U-shaped pattern of learning (Rumelhart & McClelland, 1986; MacWhinney & Leinbach, 1991; Plunkett & Marchman, 1991). In contrast, symbolic theorists argue that generalization is rule-based (Pinker, 1991; Pinker & Prince, 1988).

Most of this debate has revolved around a specific cluster of connectionist models, the back-propagation network as a model of language acquisition. Several limitations are known to the back-propagation algorithm, especially in the context of language acquisition: in particular, back-propagation relies on a gradient-descent weight adjustment process to reduce the error between desired and actual outputs. According to the well-known "no negative evidence" argument (Baker, 1979; Bowerman, 1988), children do not receive constant feedback about what is incorrect in their speech, or receive the kind of error corrections on a word-by-word basis as provided to the back-propagation network.

In this study, we explore self-organizing neural networks, in particular, the self-organizing feature maps as a potential class of models of language acquisition. In contrast to back propagation, self-organizing networks use unsupervised learning that requires no presence of a supervisor or an explicit teacher; learning is achieved entirely by the system's self-organization in response to the input. The self-organizing process extracts an efficient and compressed internal representation from a high-dimensional input space and expresses this new representation in a map structure (Kohonen, 1989). There are three important properties of self-organizing feature maps that make them particularly well suited to the study of language acquisition.

(1) *Self-organization.* Self-organization in these networks typically occurs in a two-dimensional map, where each unit is a location on the map that can uniquely represent one or several input patterns. At the beginning of learning, an input pattern randomly activates one of the many units on the map, according to how similar by chance the input pattern is to the weight vectors of the units. Once a unit becomes active in response to a given input, the weight vectors of the unit and its neighboring units are adjusted so that they become more similar to the input and will therefore respond to the same or similar inputs more strongly the next time. In this way, every time an input is presented, an area of units will become activated on the map (the activity "bubbles"), and the maximally active units are taken to represent the input. Initially activation occurs in large areas in the map, but gradually learning becomes more focused so that only the maximally responding units are active. This process continues until all the inputs have found some maximally responding units.

(2) *Representation.* As a result of this self-organizing process, the statistical structures implicit in the high-dimensional space of the input are represented as topological structures on a two-dimensional space. Because the network develops activity bubbles to capture the input space, similar inputs will end up activating the same units or units in nearby regions, yielding a new similarity structure that becomes clearly visible on the map. This self-organized representation has clear implications for language acquisition: the formation of activity bubbles may capture critical processes of the emergence of lexical categories in children's acquisition of the lexicon. In particular, the network organizes information first in large areas of the map and gradually zeros in on small areas; this zero-in process is a process from diffuse to focused patterns of activity that leads to continuous adaptation of the network's representation. This process can naturally explain many generalization errors reported in the literature: for example, substitutions of *put* for *give* ("put me the bread") or *fall* for *drop* ("I falled it") reflects the child's recognition of diffuse lexical similarities but not the focused fine distinctions between words (Bowerman, 1982). Miikkulainen (1997) showed that in a lesioned self-organizing feature map, behaviors of dyslexia (e.g., producing *dog* in response to *sheep*) can result from partial damage

308

to the semantic representation (in effect a diffuse representation of meaning).

(3) *Hebbian learning.* Hebbian learning is essentially a co-occurrence learning mechanism, according to which the associative strength between two neurons is increased if the neurons are both active at the same time (Hebb, 1949). The amount of increase is proportionally to the level of activation of the two neurons. Different self-organizing maps can be connected via Hebbian learning, such as in Miikkulainen's (1997) multiple feature-map model: initially all units on one map are connected to all units on the other map; as self-organization takes place, the associations become more focused, so that in the end only the maximally active units on the two (or more) maps are associated. Hebbian learning has strong implications for language acquisition in that it can account for how the child abstracts relationships between phonological, semantic, and morphological properties of words on the basis of how often these properties co-occur and how strongly they are co-activated in the representation.

Because of these properties, self-organizing networks (a) allow us to track the development of the lexicon as an emergent process more clearly in the network's self-organization (from diffuse to focused patterns or from incomplete to complete associative links); (b) allow us to model one-to-many or many-to-many associations between forms and meanings in the development of the lexicon and morphology, and (c) provide us with a set of biologically more plausible and computationally more relevant principles to study language acquisition without relying on negative evidence to learn. They are biologically more plausible because one could conceive of the human cerebral cortex as essentially a self-organizing map (or multiple maps) that compresses information on a two-dimensional space (Spitzer, et al., 1998). They are computationally more relevant because one could argue that child language acquisition in the natural setting (especially organization and reorganization of the lexicon) is largely a self-organizing process that proceeds without explicit teaching (MacWhinney, 1998).

In this paper I focus on the problem of the English reversive prefixes that has been discussed by Whorf (1956) and Bowerman (1982) in the context of morphological generalization. In English, one can use the prefix *un-* to indicate the reversal of an action in verbs like *unbuckle, uncoil, undress, unfasten,* and *untie,* but not **unfill, *unhang, *unkick, *unpush,* or **unsqueeze.* Why is *un-* allowed with some verbs but not with others? Whorf hypothesized that there is some underlying semantic category that licenses the use of *un-* (roughly "a covering, enclosing, and surface-attaching meaning"). Because this category functions only covertly (i.e., by the restrictions it places on *un-*), he called it a "cryptotype". To Whorf, the problem is that the precise meaning of the cryptotype is "subtle" or even "intangible", but the prefix that it licenses is productive. Bowerman argued that the notion of cryptotype, though elusive, might play an important role in children's acquisition of *un-.* Her data showed that children produce generalization errors like **unbury, *unhang, *unhate, *unopen, *unpress, *unspill,* or **unsqueeze* starting from about age 3. She had two hypotheses on the role of cryptotype: (a) "generalization via

cryptotype", i.e., recognition of the cryptotype leads to overly general uses (overgeneralizations); e.g., *bury* fits the cryptotype just as *cover* does, so say **unbury.* (b) "recovery via cryptotype", i.e., children use the cryptotype to recover from overgeneralization errors; e.g., *hate* does not fit the cryptotype meaning and only verbs in the cryptotype can take *un-,* so stop saying **unhate.*

But how could the child extract the cryptotype and use it as a basis for morphological generalization or recovery, when the cryptotype is intangible even to linguists like Whorf? In Li (1993) Li & MacWhinney (1996) we attempted to answer this question by simulating *un-* and its cryptotype in a back-propagation network. We hypothesized that cryptotypes are intangible only because traditional symbolic methods are less effective for analyzing the complex semantic structure: words in a cryptotype vary in the number of relevant semantic features, the strength of activation of each feature, and the degree of overlap of features. These complex structural properties lend themselves naturally to distributed representations and connectionist learning. We trained a network to map semantic features of verbs to three prefixation patterns: *un-,* its competitor *dis-,* and no-prefixation. Our results indicated that (a) the network formed internal representations of semantic categories that corresponded roughly to Whorf's cryptotype, on the basis of learning limited semantic features of verbs and morphological classes; (b) the network produced overgeneralization errors similar to those reported by Bowerman (1982), Clark et al (1995), and those observed in the CHILDES database.

In this study, we examine the representation of cryptotypes, the generalization of prefixes, and the recovery from generalizations in a self-organizing feature-map model. As discussed above, self-organizing feature maps learn on the basis of self-organization, produce representations in a map structure, and form associative connections via Hebbian learning. These properties have recently been implemented in DISLEX, a multiple feature-map model of the lexicon (Miikkulainen, 1997). In this study, we use DISLEX as a basis to simulate generalization, representation, and recovery. We think that the self-organizing and Hebbian learning processes as simulated in DISLEX can help us to understand the representational basis of morphological generalization and the learner's recovery from generalizations.

Method

Network Architecture

DISLEX is a multiple feature-map model of the lexicon, in which different self-organizing maps dedicated to different types of linguistic information (orthography, phonology, or semantics) are connected through associative links via Hebbian learning. During learning, an input pattern activates a unit or a group of units on one of the input maps, and the resulting bubble of activity propagates through the associative links and causes an activity bubble to form in the other map. If the direction of the associative propagation is from phonology or orthography to semantics, comprehension is modeled; production is modeled if it goes from semantics to phonology or orthography. The activation of co-occurring lexical and semantic representations leads to continuous organization in these maps, and to adaptive formations of associative connections between the maps. Figure 1 presents a schematic diagram of the architecture of the model.

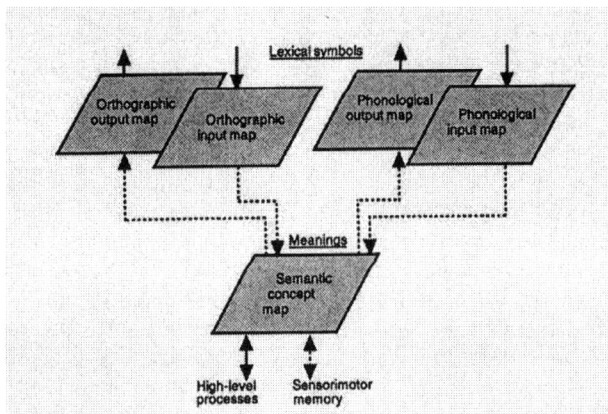

Figure 1: A multiple feature-map model of the lexicon (Miik-kulainen, 1997)

In this study, we applied DISLEX to the examination of lexical and morphological acquisition. We constructed two self-organizing maps, each of the size of 25 x 25 units, one for the organization of phonological input (henceforth the phonological map), and the other for the organization of semantic input (the semantic map). We used no orthographic maps since we were modeling acquisition in young children who are preliterate.

Input Representations

The input data to our network were 228 verbs based on Li (1993) and Li and MacWhinney (1996). Forty-nine of them were verbs with the prefix *un-*, 19 of them were verbs with the competitor prefix *dis-*, and the remaining 160 were verbs with no prefixes (*un-* and *dis-* both indicate the reversal of the action of the verb, as in *untie* and *disassemble*). The relative higher proportion of the last type of verbs (i.e., zero verbs) as compared with *un-* and *dis-* verbs was intended to reflect the distribution of these forms in the input to children.

Previous connectionist models of language acquisition have often relied on the use of artificial input/output representations (e.g., randomly generated patterns of phonological or semantic representations) or representations that are constructed ad hoc by the modeler. For example, in our previous studies we represented each verb as a pattern of 20 semantic features, selected on the basis of our linguistic analyses (see Li, 1993; Li & MacWhinney, 1996). However, the use of this type of representation is subject to the criticism that the model works just because of the presence of these features in the representation. In this study, we wanted to use more linguistically grounded input data to simulate lexical and morphological acquisition. Thus, we represented our inputs as follows.

Phonological representations to our network were based on a syllabic template coding developed by MacWhinney and Leinbach (1991). Instead of a simple phonemic representation, this representation reflects current autosegmental approaches to phonology, according to which the phonology of a word is made up by combinations of syllables in a metrical grid, and the slots in each grid made up by bundles of features that correspond to phonemes, *C*'s (consonants) and *V*'s (vowel). The MacWhinney-Leinbach model used 12 *C*-slots and 6 *V*-slots that allowed for representation of words up to three syllables. For example, the 18-slot template *CCC VV CCC VV CCC VV CCC* represents a full tri-syllabic structure in which each *CCCVV* is a syllable (the last *CCC* represents the consonant endings). Each *C* is represented by a set of 10 feature units, and each *V* by a set of 8 feature units.

Semantic representations to our network were based on the lexical co-occurrence analyses in the Hyperspace Analogue to Language (HAL) model of Burgess and Lund (1997). HAL represents word meanings through multiple lexical co-occurrence constraints in large text corpora. In this representation, the meaning of a word is determined by the word's global lexical co-occurrences in a high-dimensional space: a word is anchored with reference not only to other words immediately preceding or following it, but also to words that are further away from it in a variable co-occurrence window, with each slot (occurrence of a word) in the window acting as a constraint dimension to define the meaning of the target word. Thus, a word is represented as a vector that encodes the multiple constraints (dimensions) in a high-dimensional space of language use. We used 100 dimensions for the unit length of the vectors.

Task and Procedure

Upon training of the network, a phonological input representation of the verb was inputted to the network, and simultaneously, the semantic representation of the same input was also presented to the network. By way of self-organization, the network formed an activity on the phonological map in response to the phonological input, and an activity on the semantic map in response to the semantic input. Depending on whether the verb is prefix-able with *un-* or *dis-*, the phonological representation of *un-* or *dis-* was also co-activated with the phonological and the semantic representations of the verb stems. At the same time, through Hebbian learning the network formed associations between the two maps for all the active units that responded to the input. The network's task was to create new representations in the corresponding maps for all the input words and to be able to map the semantic properties of a verb to its phonological shape and its morphological pattern.

To observe effects of learning on the network's representation, generalization, and recovery, we designed four stages to train the network. (1) A verb's phonological representation was co-activated with its semantic representation on a one-to-one basis, which means that the network saw only the verb's phonological representation and its semantic representation simultaneously. This was done to model the whole-word learning stage, at which children have not analyzed morphological devices as entities separate from the verb stems (Bowerman, 1982). (2) One-to-one mapping was relaxed, so that the phonological and semantic representations of verb stems (e.g., *tie*, *connect*), prefixed verbs (*untie*, *disconnect*), and the prefixes themselves (*un-*, *dis-*) were all co-activated in the network. (3) Twenty-five novel verbs were introduced to the network in order to test whether generalizations would occur in our network as in children's speech. These were verbs on which previous studies have reported children's generalizations (e.g, *ungrip, *unpress, and *untighten; see Bowerman, 1982; Clark et al., 1995). Generalization was tested by inputting the verbs to the network without having the network self-organize the verbs or learn the phonological-semantic associations. (4) Self-organization and Hebbian learning resumed for the novel verbs introduced at Stage 3 in order to test if the network could recover from generalizations.

All simulations were run on a SUN Ultra 1 workstation, using the DISLEX codes configured by Miikkulainen (1999).

Results and Discussion

To analyze the simulation results, we focus here on three levels of analysis: the network's representation of verb semantics, its patterns of morphological generalization, and its ability to recover from generalization errors.

The Representation of Cryptotype

In this study, we wanted to analyze whether our network developed structured representation as a function of the self-organization of verb semantics. In particular, we wanted to see how the patterns of activity formed in the maps can capture Whorf's notion of cryptotype.

As discussed earlier, a distinct property of self-organizing feature maps is that the structures in the network's new representation are clearly visible as activity bubbles or patterns of activity on the two-dimensional map; this property obviates the need of extra steps of mathematical analysis (e.g., cluster analysis or principal component analysis) as required in other connectionist networks. In our network, the self-organization process extracted the semantic structures from the high-dimensional space of the HAL semantic vectors and expressed them on the two-dimensional map as concentrated patterns of activity. Figure 2 presents a snapshot of the network's self-organization of 120 verbs after the network was trained for 600 epochs at Stage 1.

An examination of the semantic map shows that the network has clearly developed forms of representation that correspond to the category of cryptotype that Whorf believed governs the use of un-. In Li and MacWhinney (1996) we suggested that a connectionist model provides a formal mechanism to capture Whorf's cryptotype, in that there can be several "mini-cryptotypes" that work collaboratively as interactive "gangs" (McClelland & Rumelhart, 1981) to support the formation of the larger cryptotype. The idea of "mini-cryptotype" is realized most clearly in the emerging structure of the self-organizing map.

Our network, without the use of ad hoc semantic features, formed clear "mini-cryptotypes" by mapping similar words onto nearby regions of the map. For example, towards the lower right-hand corner, verbs like *lock, clasp, latch, lease,* and *button* are mapped to the same region of the map, and these verbs all share the "binding/locking" meaning. A similar mini-cryptotype also occurs towards the lower left-hand corner, including verbs like *snap, mantle, tangle, ravel, twist, tie,* and *bolt*. Still a third mini-cryptotype can be

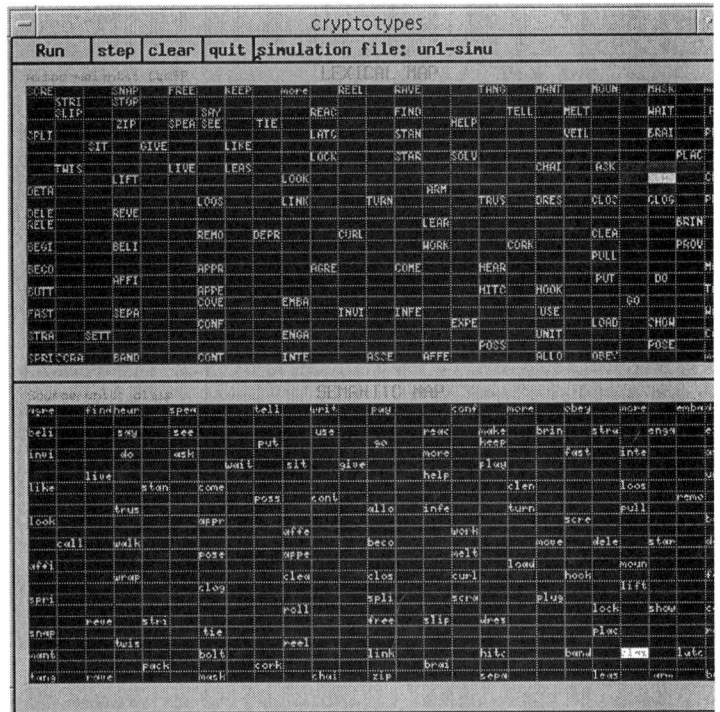

Figure 2: Phonological and semantic representations in DISLEX after the network was trained on 120 verbs for 600 epochs. The upper panel is the phonological map (in capital letters), and the lower panel the semantic map. Words longer than four letters are truncated.

found in the upper left-hand corner, including *hear, say, speak, see,* and *tell,* verbs of perceptions and audition. Finally, one can observe that *embark, engage, integrate, assemble,* and *unite* are being mapped toward the upper right-hand corner of the map, which all seem to share the "connecting" or "putting-together" meaning and interestingly, these are the verbs that can take the prefix *dis-*. Of course, the network's representation at this point is still incomplete, as self-organization is moving from diffuse to more focused patterns of activity; for example, the verb *show,* which shares similarity with none of the above mini-cryptotypes, is grouped with the binding/locking verbs. What is crucial, however, is that these mini-cryptotypes form the semantic basis for the larger cryptotype of *un-* verbs. As shown in the figure, the network has mapped most verbs in the cryptotype to the bottom layer of the semantic map, and these are the verbs that can take the prefix *un-*.

Representation and Generalization

Neural networks are considered to be able to generalize to novel patterns (Elman, 1998). But do they show the same types of generalization as children do? And on what basis do they generalize? Our simulation results indicate that our network was not only able to capture the elusive cryptotype category by way of self-organization, but also able to generalize on the basis of its representation of the cryptotype. For example, the network produced overgeneralization errors that match up with empirical data when tested for generalization at Stage 3, including *unbreak, *uncapture, *unconnect, *unfreeze, *ungrip, *unpeel, *unplant, *unpress, *unspill, *unstick, *untighten,* etc. These overgeneralizations were based both on the network's representation of the meaning of verbs and on the associative connections that the network formed through Hebbian learning in the semantics-phonology mapping process.

First, most of these overgeneralizations involve verbs that fall within the *un-* cryptotype. These verbs (e.g., *connect, freeze, grip, peel, plant, press, spill, stick,* and *tighten*) were mapped to the bottom layer on the semantic map within which we identified the network's representation of the cryptotype. Earlier, we pointed out two hypotheses regarding the role of cryptotype in children's acquisition of un- according to Bowerman: "generalization via cryptotype" and "recovery via cryptotype". Our results here are consistent with the generalization via cryptotype hypothesis, that is, the representation of cryptotype leads to overly general uses of un- (see also discussion of the *clench* example below). Consistent with our previous simulations, we found no flagrant violations of the cryptotype in the network's generalizations such as *unhate* or *untake* (as in Bowerman's data); hence there was no basis for the recovery via cryptotype hypothesis, that is, that the learner can use the representation of cryptotype to recover from overgeneralizations.

Second, all the above generalizations were simulated production errors, in which case patterns of activity in the semantic map were propagated through associative links to the phonological map. The ability to simulate both comprehension and production through associative connections is a distinct property of DISLEX (see Network Architecture). The associative connections formed via Hebbian learning provide the basis for the production of overgeneralization errors. For example, the semantic properties of *tighten* and *clench* are similar and they were mapped onto nearby regions of the semantic map. During learning, the semantics of *clench* and *unclench* were co-activated, and the phonology of *clench, unclench,* and *un-* were also co-activated. When the semantics and the phonology of these items were associated through Hebbian learning, the network can associate the semantics of *tighten* with *un-* because of *clench*, even though the network learned only the association for *unclench* and not *un-tighten* (i.e., at an earlier stage *tighten* was not included in the training). This associative process of correlating semantic features, lexical forms, and morphological devices simulates the process of learning and generalization in children's productive speech, and shows that overgeneralizations can naturally result from the semantic structures in the lexical representation (which in turn is a result of self-organization) and from the associative learning of semantics and phonology.

Finally, generalization errors in our data were not limited only to morphological generalizations. We also found lexical generalizations similar to those reported by Bowerman (1982) and Miikkulainen (1997) (see Introduction). Most important, these generalizations demonstrate further the intimate relationship between representation and generalization. For example, our network produced *see* in response to *say, detach* in response to *delete, begin* in response to *become,* due to its representation of these pairs of words in the same region on the phonological map. These generalizations well resemble lexical errors in surface dyslexia (Miikku-lainen, 1997). Similarly, the network comprehended *see* as *speak, arm* as *clasp,* and *unscrew* as *hook,* due to its representation of these pairs of words in the same region on the semantic map, and these generalizations resemble lexical errors in deep dyslexia. Again, self-organization of lexical information and Hebbian learning of associative connections account for the origin of this type of lexical generalizations.

Mechanisms of Recovery from Generalizations

Our last analysis of the simulation results involves the network's ability to recover from generalization errors. The network in Li (1993) and Li and MacWhinney (1996) suffered, by and large, from the failure to recover from overgeneralization errors. This failure, we hypothesized, was due to the gradient-descent error-adjustment process used in back-propagation. Can our self-organizing network recover from generalizations? If so, what computational mechanisms permit its recovery?

Our network displayed a significant ability of recovery from generalization errors. When tested for generalizations at Stage 3, no learning took place in the network for self-organization or associative connection. When tested for recovery at Stage 4, self-organization and Hebbian learning resumed. Within 200 epochs of new learning during this stage, our network recovered from the majority of the overgeneralizations tested at Stage 3. Recovery in this case is a process of restructuring of the mapping between phonological, semantic, and morphological patterns, and the restructuring is based on the network's ability to reconfigure the associative links through Hebbian learning, in particular, the ability to form new associations between prefixes and verbs and the ability to eliminate old associations that were the basis of erroneous generalizations.

As discussed earlier, adjustment of associative connections via Hebbian learning in DISLEX is proportional to how strongly the units in the associated maps (phonological and semantic maps in this case) are co-activated. When a given phonological unit and a given semantic unit have fewer chances to become co-activated, the strengths of their associative links are correspondingly decreased. For example, un- and *tighten* were co-activated because of *clench* at Stage 3; at Stage 4 un- and *clench* continue to be co-activated, but un- and *tighten* do not get co-activated. Hebbian learning determines that the associative connection between un- and *clench* remains to be strong, but that between un- and *tighten* gets eliminated, thereby simulating what happens at the final phase of the U-shaped learning when errors disappear. This result models the process that children's overgeneralizations are gradually eliminated when there is no auditory support in the input about specific co-occurrences that they expect (MacWhinney, 1997). Of course, in the real learning situation, the strength of the connection between un- and *tighten* may also be reduced by a competing form such as *loosen* that functions to express the meaning of *untighten* (e.g., Clark, 1987, MacWhinney, 1987).

Hebbian learning coupled with self-organization provides a simple but powerful computational principle to account for the recovery process. Restructuring of associative connections often goes hand-in-hand with the reorganization of the corresponding maps. For example, at Stage 4, the network developed finer representations for verbs such as *clench* and *tighten*: as the associative strengths of these verbs to *un-* varied, their representations also became more distinct. This process in our simulation is consistent with the criteria approach of Pinker (1989) which argues that children recover from generalizations by recognizing fine and subtle semantic and phonological properties of verbs (although we do not assume as Pinker does that fine distinctions among verbs rely on the child's innate capacity). Interestingly, in the few cases in which our network did not recover from generalizations, the network was unable to make the fine distinctions between verbs on the basis of meanings; for example, because it was unable to separate on the semantic map *stick* from *screw*, *press* from *zip*, and *freeze* from *bolt*, it continued to produce the erroneous *unstick, *unpress, and *unfreeze. This inability might be due to resource limitations (i.e., size of the map); we are currently investigating this problem using much larger feature maps (e.g., map of 50 x 50 units).

Conclusion

In this paper I showed that self-organizing neural networks can be used successfully to model and provide insights into language acquisition, particularly with respect to issues of generalization, representation, and recovery. Our simulated DISLEX model, without receiving hand-crafted features, was able to capture elusive semantic categories such as Whorf's cryptotype, display overgeneralization errors as children do, and recover significantly from overgeneralizations. Although the simulation results presented here are preliminary, we think that they serve to deepen the link between previous empirical and modeling results and new models in neural networks. Future research in this direction will involve the development of more realistic training schedules (e.g., incremental learning), the use of input representations that are grounded in children's language (e.g., semantic vectors based on lexical co-occurrence analysis of the CHILDES database), and the development of network architecture that is better suited to the task of morphological acquisition (e.g., use of separate morphological maps that allow the interaction among morphemes and verb stems on both phonological and semantic levels).

Acknowledgments

This research was supported by a Faculty Research Grant from the University of Richmond. I am very grateful to Brian MacWhinney and Risto Miikkulainen for their help, comments, and discussions at various stages of the project, and to Curt Burgess and Kevin Lund for making available their HAL semantic vectors to our modeling.

References

Bowerman, M. (1982). Reorganizational processes in lexical and syntactic development. In E. Wanner & L. Gleitman (Eds.), *Language acquisition: the state of the art*. Cambridge, UK: Cambridge University Press.

Bowerman, M. (1988). The "no negative evidence" problem: How do children avoid constructing an overly general grammar? In J. Hawkins (ed.), *Explaining language universals*. New York, NY: Blackwell.

Burgess, C. & Lund, K. (1997). Modelling parsing constraints with high-dimensional context space. *Language and Cognitive Processes, 12*, 1-34.

Clark, E.V. (1987). The principle of contrast: A constraint on language acquisition. In B. MacWhinney (ed.), *Mechanisms of language acquisition*. Hillsdale, NJ: Lawrence Erlbaum.

Clark, E.V., Carpenter, K., & Deutsch, W. (1995). Reference states and reversals: Undoing actions with verbs. *Journal of Child Language, 22*, 633-662.

Elman, J. (1998). Generalization, simple recurrent networks, and the emergence of structure. In M. Gernsbacher & S. Derry (eds.), *Proceedings of the 20th Annual Conference of the Cognitive Science Society*. Mahwah, NJ: Lawrence Erlbaum.

Hebb, D. (1949). *The organization of behavior: A neuropsychological theory*. New York, NY: Wiley.

Kohonen, T. (1989). *Self-organization and associative memory*. Heidelberg: Springer-Verlag.

Li, P. (1993). Cryptotypes, form-meaning mappings, and overgeneralizations. In: E. V. Clark (Ed.), *The Proceedings of the 24th Child Language Research Forum*, Center for the Study of Language and Information, Stanford University, 162-178.

Li, P., & MacWhinney, B. (1996). Cryptotype, overgeneralization, and competition: A connectionist model of the learning of English reversive prefixes. *Connection Science, 8*, 3-30.

MacWhinney, B. (1987). The competition model. In B. MacWhinney (ed.), *Mechanisms of language acquisition*. Hillsdale, NJ: Lawrence Erlbaum.

MacWhinney, B. (1998). Models of the emergence of language. *Annual Review of Psychology, 49*, 199-227.

MacWhinney, B., & Leinbach, J. (1991). Implementations are not conceptualizations: Revising the verb learning model. *Cognition, 40*, 121-157.

Miikkulainen, R. (1997). Dyslexic and category-specific aphasic impairments in a self-organizing feature map model of the lexicon. *Brain and Language, 59*, 334-366.

Pinker, S. (1989). *Learnability and cognition: The acquisition of argument structure*. Cambridge, MA: MIT Press.

Pinker, S. (1991). Rules of language. *Science, 253*, 530-535.

Plunkett, K., & Marchman, V. (1991). U-shaped learning and frequency effects in a multi-layered perceptron: Implications for child language acquisition. *Cognition, 38*, 43-102.

Rumelhart, D. & McClelland, J. (1986). On learning the past tenses of English verbs. In J. McClelland, D. Rumelhart, and the PDP research group (Eds.), *Parallel distributed processing: Explorations in the microstructure of cognition* (Vol. II). Cambridge, MA: MIT Press.

Whorf, B. (1956). Thinking in primitive communities. In J. B. Carroll (Ed.), *Language, thought, and reality*. Cambridge, MA: MIT Press.

The Influence of Verbal Ability on Mediated Priming

Kay Livesay (livesay@cassandra.ucr.edu)
Department of Psychology
University of California, Los Angeles, 90095

Curt Burgess (curt@doumi.ucr.edu)
Department of Psychology
University of California, Riverside, 92521

Abstract

A set of analyses are presented that replicate the mediated priming effect (e.g., *lion-stripes*) using a naming latency task, and demonstrate that the mediated priming effect is influenced by individual differences in sensitivity to this priming effect. Previous research (Livesay & Burgess, 1998) has shown that stimulus differences are a major factor in whether or not mediated priming is obtained. The present research explores the influence of verbal ability on this effect. The primary finding is that individuals with low verbal ability are not sensitive to mediated word relationships, whereas, individuals with high verbal ability manifest a robust mediated priming effect.

Mediated, or two-step priming (MP; Balota & Lorch, 1986; McNamara & Altarriba, 1988), occurs for prime-target pairs which, on the surface, do not appear to be directly related (e.g., *lion-stripes*). Traditionally, the MP effect has been explained using a spreading activation theory of memory. Spreading activation involves of a set of interconnected nodes where activation spreads from one node to related nodes. Thus, according to spreading activation theory, MP occurs because activation spreads from the prime (*lion*) through the mediating item (*tiger*) to the target (*stripes*), thereby facilitating the response to the target word.

An alternative explanation for the MP effect has been suggested by McKoon and Ratcliff (1992), using the compound cue theory of retrieval. According to McKoon and Ratcliff, MP is not "mediated," but, instead, any priming is due to weak, but direct, relationships in memory. Thus, priming between prime-target pairs, either direct or mediated, occurs because of previous experience of the prime-target pair in context.

Livesay and Burgess (1997, 1998) found that stimulus differences contributed to the lability of the MP effect. According to these researchers, there are two kinds of stimuli used to demonstrate mediated priming: contextually consistent stimuli (CC) and contextually inconsistent stimuli (CI). The CC stimuli were mediated prime-target pairs that were judged as likely to be experienced in a common context (e.g., *bat-bounce*, common context is *ball*), CI stimuli were mediated prime-target pairs that were judged as not as likely to be experienced in a common context (e.g., *day-dark*, common context *night*). When stimuli were separated into these categories and the data reanalyzed, the CC items carried the mediated priming effect, while the CI items did not. It is difficult to reconcile these results with a simple localist view of memory (e.g., spreading activation) because there should be priming regardless of contextually consistency, given the clear, direct relationship between the prime and the mediating item and the mediating item and the target.

Contextual consistency with any two items hinges on the likelihood that items consistently occur in similar contexts which is a function of the relationship between objects in the environment and the mapping of those relationships to language use. The role of experience may play another role in whether or not a person experiences mediated priming. Sensitivity to these variations in contextual sensitivity could be a product of a person's verbal ability. If so, contextual consistency and verbal ability could both predict the presence of mediated priming. According to a spreading activation theory of memory both high and low scorers should show mediated priming and direct priming. This is predicted because the connections in two-step priming should be strong, thus facilitating reaction times for both. From a compound cue view only those individuals who have had experience with the mediated prime-target pairs (or direct prime-target pairs) should get priming. And finally, from an experience mediated meaning representation argument, it is more likely that individuals with high verbal ability (high language experience) will display a mediated priming effect because they will have richer representations for the words, while individuals with low verbal ability will be less likely to display a mediated priming effect, because they will have a more impoverished representation (compared to the high ability representations).

We analyzed mediated and direct priming results based on participants' performance on the following three individual differences measures, the Nelson-Denny Reading Comprehension Test, the Print Exposure measure, and Daneman and Carpenter's (1980) working memory span test. For each analysis the participants were separated into high scorers and low scorers based on their scores on the individual difference measure being analyzed.

Methods

Participants. Seventy-three University of California, Riverside undergraduates participated as part of a course requirement. All participants were right-handed, native speakers of English with normal or corrected-to-normal vision.

Materials. Forty-eight prime-target pairs were taken from Balota and Lorch (1986). Each test list consisted of 54 items, 16 mediated trials, 16 unrelated trials, 16 directly related trials and 6 warmup trials. Unrelated prime-target pairs were generated by quasi-randomly pairing targets with primes from the 48 original pairs. For example, the prime *lion* in the mediated pair *lion-stripes* was replaced with *breeze* to form the unrelated pair *breeze-stripes*; *breeze* was originally paired with *blow*. The prime words were counterbalanced; a target preceded by a mediated prime on list 1 would be preceded by an unrelated item on list 2 and a directly related item on list 3.

Procedures. The stimuli were presented on a computer monitor; participant responses were collected via a microphone connected to the computer by a Digitry CTS system. Each trial began with a fixation cross presented for 500 ms. Following the fixation cross, a prime word was presented for 350 ms, immediately followed by a target word; the target word remained on the screen until participant answered or 2500 ms had elapsed. Participants simply had to name the word that appeared on the screen. If a participant failed to respond, or did not speak loud enough for the microphone to detect their voice the computer would beep and the word time-out would appear on the screen. The participants were then instructed to speak louder on the next trial.

Results

For these analyses the participants were separated into high scorers and low scorers based on their scores on the individual difference measure being analyzed; high scorers were defined as the eight highest scores on

each list, low scorers were defined as the lowest eight scores on each list. Participants were chosen this way in order to assure equal representation from each list. The number eight represents the lowest and highest one-third of each list. We analyzed each set of data using the CC/CI categorization. As discussed previously, this categorization has shown a separation of the mediated priming effect in previous research (Livesay & Burgess, 1997, 1998).

Overall Priming Results

One-way ANOVAs were calculated separately for the CC and CI pairs. Both subject (F_1) and item (F_2) analyses were calculated.

Contextually Consistent Pairs. There was a priming effect for both direct and mediated pairs. Mediated pairs were responded to faster (557 ms) than unrelated pairs (570 ms), $F_1(1, 72) = 9.11, p = .0035; F_2(1, 23) = 4.57, p = .043$. Directly related pairs were also responded to faster (546 ms) than unrelated pairs (570 ms), $F_1(1, 72) = 36.79, p = .0001; F_2(1, 23) = 9.73, p = .0048$.

Contextually Inconsistent Pairs. There was not a priming effect for the mediated pairs, $F_s < 1.0$ (mediated, 563 ms, unrelated, 565 ms). However, there was a direct priming effect. Directly related pairs, were responded to faster (550 ms) than unrelated pairs (565 ms), $F_1 (1, 72) = 10.10, p = .0022; F_2 (1, 22) = 5.45, p = .02$.

Individual Differences

For each individual differences measure one-way ANOVAs were calculated for high scorers and low scorers examining priming effects, separated by CC and CI pairs. Figures 1 and 2 illustrate magnitude of mediated and direct priming separated by contextual consistency (CC/CI), group (high/low) and measure.

Nelson-Denny Reading Comprehension Test

The Nelson-Denny reading comprehension test is a two-part test consisting of a reading comprehension test and a vocabulary test; participants receive two scores, a reading comprehension score and a vocabulary score. Therefore, two different analyses will be computed, one for the reading comprehension scores and one for the vocabulary scores.

Nelson-Denny Comprehension
High Scorers

Contextually Consistent Pairs. There was a priming effect for both mediated and directly pairs. Mediated prime-target pairs were responded to faster (533 ms) than unrelated prime-target pairs (550 ms), $F_1(1, 22) = 6.09; p = .022, F_2(1, 23) = 3.13, p = .090$. Directly

Mediated Priming

CC

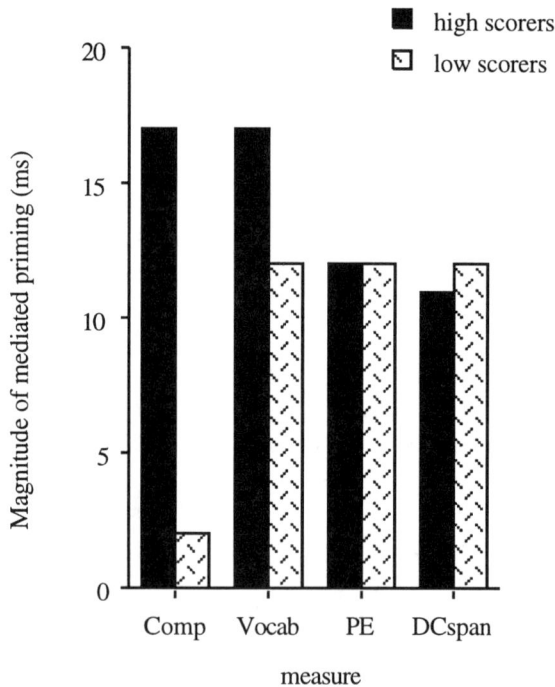

CI

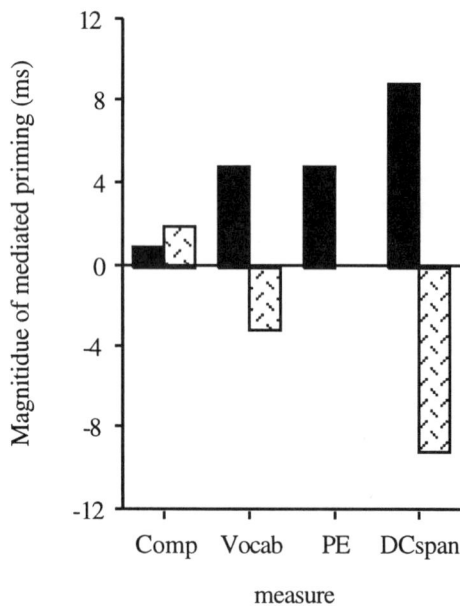

Figure 1. Magnitude (ms) of mediated priming plotted by contextual consistency (CC/CI), high/low scorers and measure.

related pairs (533 ms) were responded to faster than unrelated pairs (550 ms), $F_1(1, 22) = 6.92$, $p = .015$; $F_2(1, 23) = 2.75$, $p = .110$.

Contextually Inconsistent Pairs. There was not a priming effect for the mediated pairs, $F_s < 1.0$ (mediated, 540 ms, unrelated 541 ms). In addition, there was no priming for the directly related pairs, $F_1(1, 22) = 2.39$, $p = .136$; $F_2(1, 22) < 1.0$ (directly related, 533 ms, unrelated 541 ms).

Low Scorers

Contextually Consistent Pairs. There was no priming for mediated pairs, $F_s < 1.0$ (mediated pairs, 577 ms, unrelated pairs, 579 ms). There was, however, direct priming, $F_1(1, 23) = 7.71$, $p = .011$, $F_2(1, 22) = 6.20$, $p = .020$, directly related pairs were responded to faster (562 ms) than unrelated pairs (579 ms),

Contextually Inconsistent Pairs. There was not a priming effect for the mediated pairs, $F_s < 1.0$ (mediated, 569 ms, unrelated pairs, 571 ms). There was also no direct priming effect, $F_s < 1.0$ (directly related pairs, 565 ms, unrelated pairs, 571 ms).

Nelson-Denny Vocabulary

High Scorers

Contextually Consistent Pairs. There was a priming effect for mediated pairs. Mediated pairs were responded to faster (528 ms) than unrelated pairs (545 ms), $F_1(1, 23) = 5.29$, $p = .030$, $F_2(1, 23) = 3.58$, $p = .071$. There was also a priming effect for the directly related pairs, $F_1(1, 23) = 8.51$, $p = .0078$; $F_2(1, 22) = 3.69$, $p = .067$; directly related pairs (525 ms) were responded to faster than unrelated pairs (545 ms).

Contextually Inconsistent Pairs. There was not a priming effect for either mediated or directly related pairs, $F_1 < 1.0$; $F_2(1, 22) = 1.11$, $p = .303$ (mediated pairs, 536 ms, unrelated pairs, 541 ms). Directly related pairs, $F_1(1, 23) = 2.79$, $p = .108$; $F_2(1, 23) = 1.57$, $p = .223$ (direct pairs, 527 ms, unrelated pairs, 541 ms).

Low Scorers

Contextually Consistent Pairs. There was no priming effect for mediated pairs, $F_1(1, 23) = 2.42$, $p = .133$; $F_2(1, 23) = 2.90$, $p = .102$ (mediated 576 ms, unrelated, 588 ms). There was, however, direct priming, $F_1(1, 23) = 11.12$, $p = .0029$; $F_2(1, 23) = 9.92$, $p = .0045$, directly related pairs were responded to faster (564 ms) than unrelated pairs (588 ms).

Contextually Inconsistent Pairs. There was not a priming effect for the mediated pairs, $F_s < 1.0$ (mediated 584 ms, unrelated 581 ms). There was also no direct priming effect, $F_1(1, 23) = 2.07$, $p = .164$;

Direct Priming

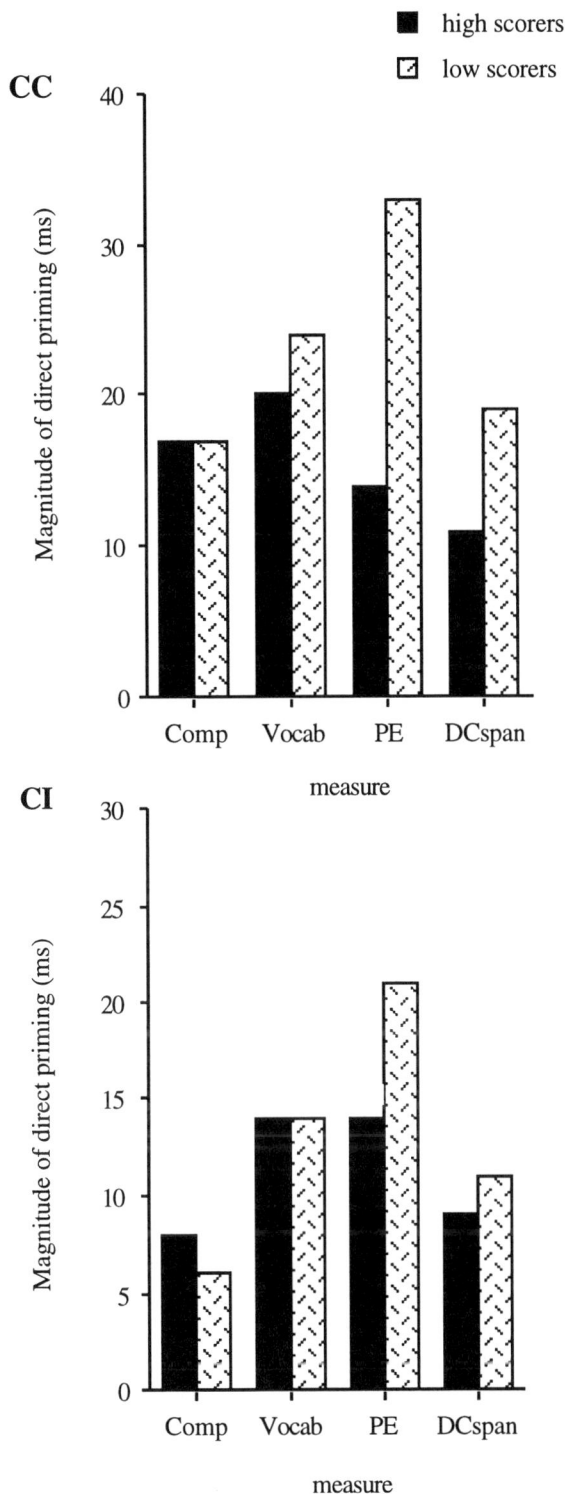

CC

Legend: ■ high scorers ▨ low scorers

Y-axis: Magnitude of direct priming (ms); values 0, 10, 20, 30, 40
X-axis (measure): Comp, Vocab, PE, DCspan

CI

Y-axis: Magnitude of direct priming (ms); values 0, 5, 10, 15, 20, 25, 30
X-axis (measure): Comp, Vocab, PE, DCspan

Figure 2. Magnitude (ms) of direct priming plotted by contextual consistency (CC/CI), high/low scorers and measure.

F_2 (1, 22) = 2.63, p = .119 (directly related, 567 ms, unrelated 581 ms).

Print Exposure

The print exposure measure is composed of three tests, the author recognition test (ART), the magazine recognition test (MRT) and the newspaper recognition test (NRT). The score for each measure is totaled to give an overall index of print exposure. These measures were originally developed by Stanovich and Cunningham (1993; see also, Stanovich & West, 1989) and have been correlated with actual exposure to print and reading comprehension ability. The tests used in this paper were modified versions of the original test, updating the author, magazine and newspaper names. The general procedure is to give a participant a list of names, half real names, half foil names, their task is to circle those names that are either author names (on the ART), magazine names (on the MRT) or newspaper names (on the NRT). A score is calculated as the number hits minus the number of false alarms.

High Scorers

Contextually Consistent Pairs. There was a priming effect for both mediated and directly related pairs. Mediated pairs were responded to faster (535 ms) than unrelated pairs (547 ms), $F_1(1, 23)$ = 5.54, p = .027, $F_2(1, 23)$ = 2.90, p = .102. Directly related pairs (533 ms) were responded to faster than unrelated pairs (547 ms), $F_1(1, 23)$ = 5.09, p = .03; $F_2(1, 23)$ = 3.15, p = .089.

Contextually Inconsistent Pairs. There was not a priming effect for the mediated pairs, F_s < 1.0 (mediated pairs, 541 ms, unrelated pairs, 546 ms). There was, however, a priming effect for the directly related pairs, $F_1(1, 23)$ = 4.74, p = .040; $F_2(1, 23)$ = 4.89, p = .037, directly related pairs (532 ms) were responded to faster than unrelated pairs (546 ms).

Low Scorers

Contextually Consistent Pairs. There was not a priming effect for mediated pairs, $F_1(1, 22)$ = 1.86, p = .186; $F_2(1, 22)$ = 2.62, p = .119 (mediated pairs, 575 ms, unrelated pairs, 587 ms). There was, however, direct priming, $F_1(1, 22)$ = 20.87, p = .0002; $F_2(1, 22)$ = 13.75, p = .0012, directly related pairs (554 ms) were responded to faster than unrelated pairs (587 ms),

Contextually Inconsistent Pairs. There was not a priming effect for mediated pairs, F_s < 1.0 (mediated pairs, 580 ms, unrelated pairs, 580 ms). However, there was a marginal direct priming effect, $F_1(1, 22)$ = 3.88, p = .061; F_2 (1, 22) = 3.75, p = .065, directly related pairs (559 ms) were responded to faster than the unrelated pairs (580 ms).

Daneman and Carpenter Working Memory Span Test

The stimuli for this test were taken from the Daneman and Carpenter (1980). The standard procedures were used. Participants read aloud sentences presented one at a time on a computer screen and were then asked verbally to recall the sentence final word from each sentence read. The number of sentences presented increased from two to six sentences (for a more detailed description of the procedures see, Daneman & Carpenter, 1980). High scores had spans of 3.5-5, low scorers had spans of 1.5-3.0 (span increases occur in steps of .5).

High Scorers

Contextually Consistent Pairs. There was not a priming effect for mediated pairs; $F_1(1, 22) = 2.34$, $p = .141$; $F_2(1, 23) = 2.93$, $p = .100$ (mediated pairs, 550 ms, unrelated pairs, 561 ms). There was marginal priming for directly related pairs, $F_1(1, 22) = 1.38$, $p = .253$; $F_2(1, 23) = 3.67$, $p = .068$ (directly related pairs, 550 ms, unrelated, 561 ms).

Contextually Inconsistent Pairs. There was no priming for either mediated or directly related pairs, $F_1(1, 22) = 1.50$, $p = .234$; $F_2(1, 22) = 1.11$, $p = .303$ (mediated pairs, 553 ms, unrelated pairs, 562 ms); directly related pairs, $F_1 < 1.0$; $F_2(1, 23) = 2.44$, $p = .132$ (directly related pairs, 553 ms, unrelated pairs, 562 ms).

Low Scorers

Contextually Consistent Pairs. There was a marginal priming effect for mediated pairs and a reliable priming effect for directly related pairs. Mediated pairs (558 ms) were responded to faster than unrelated pairs (570 ms); $F_1(1, 23) = 3.69$, $p = .067$; $F_2(1, 23) = 3.47$, $p = .075$. Directly related pairs (551 ms) were responded to faster than unrelated pairs (570 ms); $F_1(1, 23) = 5.20$, $p = .036$; $F_2(1, 23) = 4.16$, $p = .053$.

Contextually Inconsistent Pairs. There was not a priming effect for the mediated pairs, $F_s < 1.0$ (mediated pairs, 584 ms, unrelated pairs, 576 ms). There was not a direct priming effect either, $F_1(1, 23) = 1.08$, $p = .313$; $F_2 < 1.0$ (directly related pairs, 564 ms, unrelated pairs 576 ms).

Discussion

A general pattern of results is apparent. For the most part, both high and low scorers are sensitive to direct priming, however, high and low scoring subjects differ on their sensitivity to subtle, mediated priming effects. With the exception of the working memory span measures, low scorers showed no mediated priming effect for either CC pairs or CI pairs. However, low scorers did show a direct priming effect in the CC condition. These findings suggest, that low scoring participants may not have the representation of the information needed for mediated priming available during processing, unlike the high scorers.

These results support the experience mediated view of representation where differences in experience with language may have structural effects on the lexicon. These results are problematic for the spreading activation model of memory for two reasons. Associative norming results have shown that each step in the "mediated" priming chain are strong associates -- *tiger* is a strong associate of *lion* and *stripes* is a strong associate of *tiger*; thus, according to spreading activation theory, all pairs should get priming. However that is not the case, only the CC items get mediated priming. In addition, all participants should get mediated priming because the spread of activation is considered to occur quickly, and automatically. Even if there was a limited amount of activation available, activation should spread immediately because the prime and mediator and target are all strong associates and should robustly participate in the spreading activation process.

A possible explanation for these results is that low scorers have a slower spread of activation and therefore, given the 350 ms SOA, do not have enough time for activation to spread from prime to target. This has not been empirically tested and remains to examined in a timecourse experiment.

A notable exception that appeared among the individual differences results was the fact that only the DCspan measure showed no mediated priming for high scorers. This result seems unusual since all other measures show a mediated priming effect for the high scorers. Perhaps the DCspan is a measure that is not sensitive to these lower, more lexical aspects of verbal ability. Priming is not usually considered to be a resource demanding task and therefore is not likely to overload an individual's capacity, hence, high and low span participants should show similar performance, which they do. However, the results obtained for high and low span participants appear to be contrary to the other measures of individual differences (Nelson-Denny, print exposure). This is an indication that the DCspan measure is not sensitive to these more lexical level effects, while the other measures of verbal ability are sensitive to these effects.

Another distinction that was expected to appear, but did not, was the difference between the Nelson-Denny vocabulary test and the Nelson-Denny comprehension test. It was assumed that since the vocabulary test is a more lexical level test than the reading comprehension test it would be more sensitive to these lexical level effects. However, both the vocabulary and reading comprehension tests showed the same pattern of results.

As stated earlier, the lack of mediated priming exhibited by low scorers indicates that subtle lexical relationships may not be strongly represented (or represented at all), by low scoring individuals. What is not clear is how this difference in representation would occur.

From a localist, spreading activation view, these low scoring participants appear to be limited in their ability to spread activation, possibly due to an impoverished representation. It is possible this could be attributable to a weak relationship between the directly related and mediated items. This weak relationship between directly related and mediated items could be based on experience with those two words. However, this explanation seems unlikely given that prime, mediators and target words were all rated as strong associates in norming studies.

From a distributed representation view, the differences seen between high and low scorers can be attributed to a difference in contextual experience with these words. Low scorers having less varied contextual experience, thus leading to less enriched representations for those words. The differences seen here could be a consequence of retrieval using semantic representations that are less developed. This is partially suggested given that the print exposure measures appear to be more sensitive to both the mediated and direct priming relationships than the other measures of individual ability.

Previous research (Livesay & Burgess, 1998) has shown that stimuli differences can contribute to differences seen in the mediated priming effect. Present results suggests that verbal ability -- and perhaps more specifically experience -- is also a contributing factor to whether or not the MP effect will be obtained. It appears that print exposure (a measure of language experience) is the most sensitive of the four tests of individual differences to the effects of mediated and direct priming.

Acknowledgments

This research was supported by an NSF Presidential Faculty Fellowship award SBR-9453406 to Curt Burgess.

References

Balota, D. & Lorch, R. (1986). Depth of automatic spreading activation: Mediated priming effects in pronunciation but not in lexical decision. *Journal of Experimental Psychology: Learning, Memory and Cognition, 12,* 336-345.

Daneman, M. & Carpenter, P. (1980). Individual differences in working memory and reading. *Journal of Verbal Learning and Verbal Behavior, 19,* 450-466.

Livesay, K. & Burgess, C. (1997). Mediated priming in high-dimensional meaning space: What is "mediated" in mediated priming? *Proceedings of the Cognitive Science Society* (pp. 436-441). Hillsdale, N.J.: Lawrence Erlbaum Associates.

Livesay, K. & Burgess, C. (1998). Mediated priming does not rely on weak semantic relatedness or local co-occurrence *Proceedings of the Cognitive Science Society* (pp. 609-614). Hillsdale, N.J.: Lawrence Erlbaum Associates.

McKoon, G. & Ratcliff, R. (1992). Spreading activation versus compound cue accounts of priming: Mediated priming revisited. *Journal of Experimental Psychology: Learning, Memory and Cognition, 18,* 1155-1172.

McNamara, T. & Altarriba, J. (1988). Depth of spreading activation revisited: Semantic mediated priming occurs in lexical decisions. *Journal of Memory and Language, 27,* 545-559.

Stanovich, K. & Cunningham, A. (1992). Studying the consequences of literacy within a literate society: The cognitive correlates of print exposure. *Memory and Cognition, 20,* 51-68.

Stanovich, K.& West, R. (1989). Exposure to print and orthographic processing. *Reading Research Quarterly, 24,* 402-433.

Inductive Reasoning Revisited: Children's reliance on category labels and appearances

Jonathan J. Loose (J.J.Loose@exeter.ac.uk)
School of Psychology
University of Exeter,
Perry Rd, Exeter, EX4 4QG, UK

Denis Mareschal (D.Mareschal@bbk.ac.uk)
Center for Brain and Cognitive Development,
Department of Psychology,
Birkbeck College, University of London,
Malet St., London, WC1E 7HX, UK.

Abstract

Previous studies of children's inductive reasoning have attempted to demonstrate that label information is preferred to perceptual similarity as the basis for inductive inference (Gelman and Markman, 1986; Gelman and Markman, 1987; Gelman, 1988). A connectionist model of the development of inductive reasoning predicts that this will only be true when the perceptual variability of category exemplars is high (Loose and Mareschal, 1997). We report three studies investigating the model's predictions. Study 1 demonstrates that patterns of categorization can depend on perceptual variability. In study 2 we develop a set of stimuli with differing variability but equal discriminability. Study 3 demonstrates that young children's patterns of reasoning are more affected by the presence of category labels when the inference is from an exemplar of a more perceptually variable category. This study also demonstrates that the basis of inference is not explicable in terms of the ease of the ability to categorize of the stimuli. Implications for the original model are discussed.

General Introduction

Studies of the basis of children's inductive reasoning have been used to support the notion that even young children's representations are abstract and conceptually sophisticated, as opposed to the historic view that they are perceptually grounded and limited (Inhelder and Piaget, 1964; Gelman and Markman, 1986). A connectionist model of the inductive reasoning paradigm used by Gelman and Markman demonstrates that their results are replicable by a system which does not utilise complex taxonomic representations (Loose and Mareschal, 1997). The model relies on the fact that there is greater variability inherent in perceptual information than in label information—which by its very nature serves to uniquely identify classes of objects. On this basis, learning to make useful inferences in the environment naturally leads to a reliance on category labels. However, this will only be the case if reliable information about the category cannot be extracted simply from the perceptual world. Labels may not be required when category instances are more homogeneous in appearance. Thus, the connectionist model makes a prediction regarding the decision to treat similarity of appearance or shared labels as the basis of inference. The prediction is that the basis of inference is mediated by the variability in appearance of the exemplars which were used in the formation of the category.

An empirical study of the effect of category variability on adult inferences within a taxonomic domain suggests that category variability is important for adults, and also suggests a parallel between the effect of development on conceptual systems and the effect of greater learning within a particular domain (Loose and Mareschal, 1998). This study, and others (e.g. Hampton, 1995) shed some doubt on the assumed target of the developmental process being a global preference for label information (at least within the domain of natural kinds) The question remains as to whether such effects will also be demonstrated by pre-school children. In the studies described here, we investigate the effect of category variability on inferences made by pre-schoolers—the population sampled for the original studies. The questions of interest are (a) whether the basis of young children's inferences is affected by variability, and (b) whether or not performance on inference tasks is best viewed as a simple product of categorization processes.

The rest of the paper unfolds as follows. First, two studies of categorization are described. These studies serve to validate the stimuli used in the final inference study. Different kinds of responses, to categories of different variability, lead to a possible explanation of the effect of category variability in inference. An inference study that investigates the relationship between category variability and the accuracy of children's inductive inference (holding category discriminability constant) is then reported. The results of this study are discussed with respect to the conclusions of the categorization studies. Finally, implications for the development of the connectionist model are briefly discussed.

Experiment 1: Categorization of objects drawn from populations with different perceptual variabilities.

This first study was designed to examine whether or not children could explicitly categorize the algorithmically generated stimuli to be used in the final inference study. Two stimulus categories were used, and are described below. Given that the stimuli were generated algorithmicly, we know what the differences in variability between the different categories used. It is important to know whether participants can explicitly detect the differences between exemplars of different categories. It is also important to know that participants adequately represent the range of the various dimensions along which objects of a particular category can vary.

Because this work deals with category-based inferences,

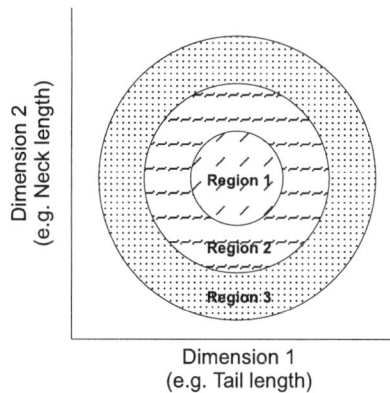

Figure 1: *Regions of feature space from which different 'category exemplars' were drawn.*

Figure 2: *Training stimuli used to familiarize children with the high variability category.*

it is crucial that the participants can categorize the stimuli. This point is made all the stronger by the fact that most other studies of this type have utilized stimuli which are previously known to the child and which can be clearly categorized (e.g. Gelman & Coley, 1990).

Method

Design & Materials The study has a within-subjects, single factor design. The dependent variable is category discriminability—a measure of the likelihood that stimuli will be accurately judged as instances of a particular category, or as being outside of that category. The only factor in this study is category variability. Variability has two levels, high and low, corresponding to categories formed from exemplars constructed across a wider or narrower range of a fixed set of dimensions. Subjects were 10 pupils in a primary reception class[1], six females and four males. Mean age was 4 years 11 months (4;11).

The materials for the study consisted of six color picture sets of artificial animals—two training sets plus four test sets. The training sets consisted of a low variability group and a high variability group. The test sets consisted of three levels of variability in order to provide pictures of animals inside/outside the high/low variability categories. The variability of pictures outside the low variability category is the same as that for pictures inside the high variability category—thus there are three variability levels, and four sets of stimuli.

The stimuli were constructed systematically such that differences between 'animals' consisted of controlled changes along a set of known dimensions. The overall look of the stimuli is similar to those used by Younger (1990) in her studies of categorization in infants and kindergarten children. Examples of the kind of stimuli used are given in figure 2. The reasons for having stimuli that looked like this were firstly so that they would be easily controlled, manipulated and generated, and secondly that they could be proposed as natural kinds from, "another world," which would promote the idea

that children should reason about them as if they were biological kinds—albeit of a fictional nature.

In this study, the term 'category variability' refers to the spread of values across the different dimensions used to derive each category exemplar. The term 'category' is thus used in a non-standard way, since we used a nested category structure—that is, (referring to figure 1), region 1 is one category, while region 2 is a second category. Region 3 is necessary to provide exemplars outside the region 2 category. Contrasting categories with the same central tendency is consistent with a view of category representation which explicitly incorporates information from individual exemplars as well as central tendency (Smith and Medin, 1981).

The actual dimensions that were used for generation of the stimuli were body length, neck length and tail size (for the low variability category only) plus leg length and body lightness (high variability category). The high variability stimuli were also drawn across a broader range of each of these dimensions[2].

Procedure The study utilized familiarization and categorization phases. The familiarization process was motivated by our previous research. It minimizes the time taken with each child to avoid the risk of episodic learning.

Both categories were presented to each subject in random order, with the following procedure:

After spending a short time putting the child at ease, the experimenter presented one of the target sets, chosen at random. The child's familiarization was a guided process. First, the child was asked to point to each animal on the sheet, and then to 'count around' the animals. Finally, the child was asked to find exemplars with particular features—for example, "the animal with the longest neck". These questions focussed attention on the salient dimensions of the stimuli. Having done this, the child was told, "These animals are actually called 'wugs', and on the planet which they come from, there are lots of wugs. *But!...* on the planet there are also some other

[1]The ages of UK primary reception class children are equivalent to US pre-schoolers.

[2]The stimuli were all placed on an (identical) brightly colored "sky" background to make the pictures a little more interesting.

	$\bar{x}$	s
Var:High, Test:Exemplars	3.20	1.03
Var:High, Test:Non-exemplars	1.60	1.07
Var:Low, Test:Exemplars	3.50	0.71
Var:Low, Test:Non-Exemplars	2.60	1.58

Table 1: *Number of correct identifications of objects as either "wugs" or not ($N = 10$ in each case).*

animals which look a bit like wugs—but they're not! They're something else! Those other animals that we saw were like that."

The child was then told that they were going to play a game—they must look at new pictures of animals, and decide which animals actually were 'wugs' and which were impostors. It was expected that by emphasizing that there would be animals in the test sets which would look like wugs but would not actually *be* wugs, participants would be more discriminating in their judgements. Without such instruction, there could be a tendency towards over-generalization, since the child had only seen positive exemplars of 'wugs' during the familiarization process.

The child was then presented with the appropriate test set of eight individual pictures for the target category being familiarized. The pictures were presented to the child in random order. Half of the pictures were of category exemplars, and half were taken from a more variable category with similar prototype. Each time a picture was presented, the child was encouraged to take a quick look at the sheet of known exemplars before deciding on the identity of the new animal. Children were not given time to explicitly compare each test picture with each known exemplar—the purpose of taking a quick look was to reinforce the child's original impression of the kinds of things that could count as category exemplars. This was a continuation of the familiarization process, saving time and allowing the study to be successfully performed with children of the target age.

The child's response was recorded as either correct or incorrect. Once this procedure had been completed for all eight members of the test set, attention switched to the other category which was not originally chosen. The procedure was repeated with the second target category. This new category of animals was given a new name. Exemplars were called 'keeches' rather than 'wugs'. Note that these names are non-words, and have been used previously in similar studies, e.g. Florian (1994). Order effects were removed by presenting each of the target categories first in 50% of cases.

Results

Response accuracy is shown in Table 1. It is easier to detect non-exemplars from a low variance category than from a high variance category. It is possible that children might be biased to give a 'yes' or 'no' response irrespective of the question. This is accounted for by a recoding of the response data. This recoding takes advantage of a measure derived

from signal detection theory (McNicol, 1972 cited in Monk & Eiser, 1980). $P(A)$—an approximation of the area under the ROC curve[3] is a bias free measure of discriminability. Given that there are only two response categories in this study, the measure is easily understood informally. We assume that I_i is the proportion of category exemplars judged as such (hits), O_i is the proportion of non-category exemplars judged as exemplars (false alarms), I_o is the proportion of category exemplars judged as non-exemplars (misses) and O_o is the proportion of non-category exemplars judged as such (all clears)[4]. These assumptions allow us to state $P(A)$ as in equation 1.

$$P(A) = I_i O_o + \frac{I_i O_i + I_o O_o}{2} \qquad (1)$$

This equation has two components which correspond to different aspects of table 2. Correct judgements are found along the I_i and O_o diagonal. If all responses are in these cells, then the first component of $P(A)$ evaluates to 1. The second component reflects response bias. In this case, it will evaluate to 0—however, to the extent that subjects' judgements are the same irrespective of the correct answer, responses will be distributed across a single table row ($I_i O_i / I_o O_o$), leading to an increase in $\frac{I_i O_i + I_o O_o}{2}$, and a corresponding decrease in $I_i O_o$. Given the denominator of $\frac{I_i O_i + I_o O_o}{2}$, it can be seen that response bias will tend to reduce the overall result.

Thus, computing $P(A)$ provides a measure of discriminability ranging from 0—1 which is not subject to response bias.

		Correct Judgment	
		IN	OUT
Subject	IN	I_i	O_i
Judgment	OUT	I_o	O_o

Table 2: *Response bias and correct inference.*

The boundary of the low variability category was discriminated more clearly than the boundary of the high variability category (Mean discriminabilities 0.68/0.33). This difference is reliable ($N = 10$, $t = -5.314$, $p < 0.001$). Thus, despite equivalent objective differences between exemplars/non-exemplars of each category, the boundary of the less variable category is still more discriminable.

Discussion

The results of study 1 suggest that the more perceptually diverse a category is, the more an object must be perceptually different from category members before it is recognized as not being a category member. This is an interesting finding,

---

[3]The 'Receiver Operating Characteristic' curve plots the probability of false-positive identifications against true-positive identifications under different conditions of noise and signal strength.

[4]Thus, for example, for this study I_i is the number of 'wug' responses given to genuine 'wug' stimuli, divided by the total number of genuine 'wug' stimuli presented (always 4 in this study).

suggesting something like a "Weber's law for categorization". The general idea of such a law would be that the required (perceptual) distance between an exemplar of a category and an object which is not an exemplar so that the category distinction is clearly noticed becomes larger with the perceptual variability of the category. It is helpful to define a distinct new here. If we define a 'just noticeable category discrimination', or JNCD, as the distance in feature space between the edge of a category and the nearest point which is reliably judged to be outside of the category, then we can make the simple claim that the JNCD will grow with the perceptual variability of the category.

It may be that the differences in the JNCDs of categories with different variabilities explains why higher variability categories promote more label based inductive reasoning. Therefore, it would be interesting to modify our stimuli to take account of the JNCD, and see if we are then able to find an effect of category variability on inductive reasoning performance.

Experiment 2: Categorization of stimuli with equivalent JNCD

The second categorization study was almost identical to the first, involving ten more pupils from the same primary school class (mean age 5;0). The difference between this study and study 1 is that a new set of stimuli were generated accounting for the effect of differing JNCD with increased variability. Stimulus modifications are described below.

Revised Materials

In order to be able to increase exemplar differences without making some stimuli extremely large, an appropriate extra dimension of variability was added—that of texture. It has been demonstrated that young children's classifications of natural kinds are extremely sensitive to texture. Thus the texture dimension should have a disproportionately large effect on children's judgements (Jones et al., 1991). A texture scale was taken from a computer painting program, and applied to the pictures in the same way that the other dimensions had been. This would serve to place extra 'out of category' information in the non-exemplar pictures.

Further to this, the pictures of animals intended as instances drawn from *outside* the *high* variability category were also modified such that they included more feature information which was further outside the boundaries of anything used in the high variability category. This explicit modification of only the high variability category is required to remove the effect of differing JNCDs.

Results

Discriminability ratings were computed for each subject as in the previous study. Mean discriminability ratings in this study were 0.82/0.73 for the low/high variability categories respectively. The difference in means in this second study is not reliable ($N = 10, t = -1.231, p = 0.25$). Thus, there is no evidence of differing discriminability between the more and less homogeneous categories. This result also demonstrates that the modified stimuli account for the effect of differing JNCDs.

Discussion

The difference in discriminability between the two categories has been removed by the modifications to the stimuli. This is confirmed by a comparison between the results from the two studies.

A two way mixed analysis of variance was performed, comparing the discriminability ratings between the previous studies. The within subjects factor was category variability (two levels, low and high), the between subjects factor was the study from which the results came. The discriminability of exemplars of both categories has improved from the first to the second study. This would be expected from the fact that we have added another dimension to all stimuli (texture). There is also a greater improvement in the high variability category. This would be expected from the extra modifications to stimuli drawn from the high variability category. The main effect of stimulus group is significant ($N = 20, F(1, 18) = 9.311, p = 0.007$), as is the main effect of category variability ($N = 20, F(1, 18) = 17.778, p = 0.001$). Importantly, the interaction is also significant ($N = 20, F(1, 18) = 5.133, p = 0.036$). Thus the effect of category variability is moderated by the stimulus set used. In this case, there is only a reliable difference between the two categories when the original stimuli were used.

These preliminary categorization studies have achieved two things. First, if our model's prediction is correct—that an important factor in the choice to make inferences on the basis of appearance or label information is the perceptual variability of the set of exemplars which form the category— then our first study suggests one possible explanation. It may be that some kind of psychophysical law applies such that the greater the perceptual variability of a category, the larger the "just noticeable category difference." This would lead to a potential re-interpretation of at least some of the comparisons made in previous studies on the basis that the notion of "perceptual similarity" is not an absolute measure, and therefore should not be compared across stimuli without regard for JNCD. Secondly, we now have a set of stimuli which allow us to conduct a study of inductive inference. These stimuli have taken into account the effect of changing JNCD, since categories are not significantly different in their discriminability despite having different levels of perceptual variability.

Experiment 3: Category-based inductive inferences

This study uses the previous stimuli to investigate whether the accuracy of inferences from exemplars of more/less perceptually variable categories will be affected in the same way by the addition of label information. The categories used are equally discriminable, despite having distinct perceptual variabilities.

Method

Subjects & Materials Twenty-eight children participated in the study, taken from three primary school reception classes in Exeter, UK. 14 males and 14 females participated. The mean age of all participants was 4 years 10 months. Participants were chosen at random from school classes. Care was taken to exclude children known to have learning difficulties.

Stimuli were as described above. In the high variability condition, "wugs"/"keeches" were drawn from regions 2/3 of figure 1. In the low variability condition, "wugs"/"keeches" were drawn from regions 1/2. Importantly, the modified stimuli from Experiment 2 were used here so that both low/high variability categories were equally discriminable. Examples of the non-perceptual properties used in the inference questions are, "very quiet and shy of people", and "live only where it is very cold."

Design & Procedure The study utilizes a 2x2 between subjects design. The first factor was the variability of the set of exemplars used to give an impression of the category of animals (two levels, high/low). The second factor was the presence or absence of stimulus labels (two levels, present/absent). The presence of labels was indicated by giving test exemplars different names (wug/keech) depending on whether they were category exemplars or not. When labels were absent, all pictures were described as "animals".

Each participant went through two phases in the study, familiarization and inference. After a short time spent putting the child at ease, s/he began the familiarization phase, and was shown a target category in the form of a single sheet of paper with a set of exemplars printed on it. All exemplars were presented along a single line, and without any white space around them. The variability of the exemplars was varied across conditions. All children were told that these were animals from "another planet" and that they were called "wugs".

The familiarization process was the same across all conditions. It consisted of ensuring that the child looked at each exemplar individually, and also attended to each of the salient dimensions. It also included the use of the terms "wug" and "keech" to describe the stimuli.

Having spent some time examining the target category and learning its name, the children entered the second phase—inference. During this phase, children in the *label* condition were told when they were looking at a wug, and when they were looking at a keech. Children in the *no-label* condition were always told/asked about this or that *animal*. In the rest of this section, the things that the children were told in each condition are contained within a single description, with the differing wording represented as: [label condition/no-label condition].

First, each participant was told that the 'planet' which the [wugs/animals] live on contains some animals that look like [wugs/these animals], but which are not. This was to give the children the idea that what they were about to see *might*

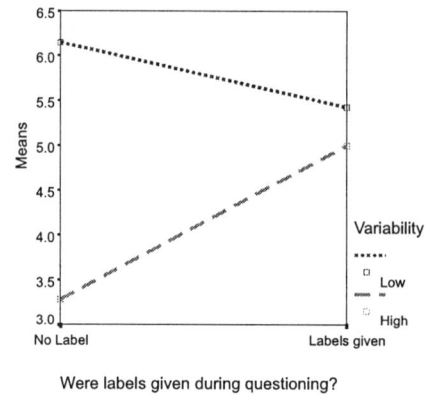

Figure 3: *Number of correct responses (/8) given by children in the different experimental conditions of the inference study.*

be a *different kind* of alien animal. Since the wugs look distinct from anything the children have seen before, it is quite possible that without being told this, they would assume that everything they were shown was a wug.

Next, the children were told that all [wug/animal] pictures on the target sheet had a particular non-perceptual property. For example, "All these [wugs/animals] can see in the dark." The child was then shown a further picture—one that had not been seen before, and was asked, "Do you think that this [(wug/keech)/animal] can [see in the dark]?". We make the assumption that an inference is valid *only* if the source and target objects are drawn from the same category. When children generalized a property *across* categories, or did *not* generalize a property *within* categories, then the response was considered *incorrect*. Responses were recorded as correct/incorrect.

Results

Figure 3 shows the mean number of correct inferences in each of the conditions.

Significant differences were apparent between the accuracies of the different groups ($N = 28$, $F(3, 24) = 4.423$, $p = 0.013$). The data show a significant main effect of category variance ($F(1, 24) = 8.097$, $p = 0.009$). There is no significant *main* effect of adding labels ($F(1, 24) = 1.750$, $p = 0.395$). There is, however, a significant interaction between variance and label ($F(1, 24) = 10.321$, $p = 0.046$)[5]. The effect of labelling on inference therefore depends on the variability of the category used. One-way analyses of the effect of labelling at different levels of category variance demonstrate a significant effect of labelling for high variance categories only ($N = 14$, $F(1, 12) = 5.760$, $p = 0.034$)[6].

Thus, subjects were more likely to make correct inferences when they were making inferences from a low as opposed

---

[5]Note that the interaction is ordinal, but there is a change in direction. The low variance target performance is made worse by the addition of a label—the high variance target performance is improved.

[6]For low variance categories, $N = 14$, $F(1, 12) = 0.620$, $p = 0.446$.

to high variability category. Overall, there is not a significant effect of adding labels. This is explicable in terms of the surprising (and unreliable) reduction in accuracy when labels were provided in the low variability condition, as opposed to the increase in accuracy when labels were provided in the high variability condition.

Discussion

These studies support the predictions of the connectionist model of the development of inductive inference (Loose and Mareschal, 1997). The variability of the set of exemplars from which a category is inferred is important in subsequent inductive reasoning with that category. The more variable the set of exemplars, the greater is the effect of adding category labels.

Study 1 demonstrates that there is a tendency to overgeneralize properties of more variable categories to a greater extent than properties of less variable categories. This might explain the findings of some inference studies, in that it is harder to find perceptual distinctions between categories on which to base inferences when those categories are more variable. However, experiments 2 and 3 demonstrate that there is more to inference than this. Experiment 2 demonstrates that we have produced a set of stimuli taking JNCD into account, since we find no evidence of a difference in discriminability between the two categories used in that study. Experiment 3 investigates inference using these revised stimuli. The study finds that the accuracy of inferences from the high variability category are more affected by the addition of label information than inferences from the low variability category.

The principle that perceptually more heterogeneous categories will be more affected by category labels in inference is predicted by the Loose & Mareschal model. However, depending on the interpretation of these results, a reconsideration of the model may be required.

If it turns out that the two categories are in fact not equally discriminable, then the simplest explanation is that the perceptually more homogeneous category promotes both categorization and inference. Previously reported simulations demonstrate that this interpretation is consistent with the model's performance (Loose and Mareschal, 1997). However, our results suggest that we should treat the two categories as *equally* discriminable. Thus, we are led to argue that the integration of new conceptual information into a perceptually heterogeneous category is actually *more difficult* than the integration of new conceptual information into a perceptually homogeneous category. The previously reported model does not seem to account for this specific finding. The model is currently being developed to investigate what "more difficult" might mean in this context.

References

Florian, J. (1994). Stripes do not a zebra make, or do they - conceptual and perceptual information in inductive inference. *Developmental Psychology*, 30:88–101.

Gelman, S. (1988). The development of induction within natural kind and artifact categories. *Cognitive Psychology*, 20:65 – 95.

Gelman, S. and Coley, J. (1990). The importance of knowing a dodo is a bird: Categories and inferences in 2-yr-old children. *Developmental Psychology*, 26:796–804.

Gelman, S. and Markman, E. M. (1986). Categories and induction in young children. *Cognition*, 23:183—209.

Gelman, S. and Markman, E. M. (1987). Young children's inductions from natural kinds: the role of categories and appearances. *Child Development*, 58:1532—1541.

Hampton, J. A. (1995). Testing prototype theory of concepts. *Journal of Memory and Language*, 34:686—708.

Inhelder, B. and Piaget, J. (1964). *The Early Growth of Logic in the Child: Classification and Seriation*. Routledge & Kegan Paul Ltd, London.

Jones, S., Smith, L., and Landau, B. (1991). Object properties and knowledge in early lexical learning. *Child Development*, 62.

Loose, J. J. and Mareschal, D. (1997). When a word is worth a thousand pictures: A connectionist account of the percept to label shift in children's reasoning. In Shafto, M. and Langley, P., editors, *Proceedings of the Nineteenth Annual Conference of the Cognitive Science Society*, pages 454–459, London. Lawrence Erlbaum Associates.

Loose, J. J. and Mareschal, D. (1998). Inductive reasoning tasks revisited: Adults don't rely on label information when inferring hidden properties. In *Proceedings of the Twentieth Annual Conference of the Cognitive Science Society*.

Monk, A. F. and Eiser, J. R. (1980). A simple, bias-free method for scoring attitude scale responses. *British Journal of Social and Clinical Psychology*, 19:17–22.

Smith, E. and Medin, D. (1981). *Categories and Concepts*. Harvard University Press, Cambridge, MA.

Younger, B. (1990). Infants' detection of correlation among feature categories. *Child Development*, 61:614–620.

Interactive Skill in Scrabble

Paul P. Maglio (pmaglio@almaden.ibm.com)
IBM Almaden Research Center
650 Harry Rd., NWED-B2
San Jose, CA 95120 USA

Teenie Matlock (tmatlock@cats.ucsc.edu)
Psychology Department
University of California, Santa Cruz
Santa Cruz, CA 95064 USA

Dorth Raphaely (sunnyboy@cats.ucsc.edu)
Psychology Department
University of California, Santa Cruz
Santa Cruz, CA 95064 USA

Brian Chernicky (brian@tapestry.net)
Psychology Department
University of California, Santa Cruz
Santa Cruz, CA 95064 USA

David Kirsh (kirsh@ucsd.edu)
Cognitive Science Department
University of California, San Diego
La Jolla, CA 92093 USA

Abstract

An experiment was performed to test the hypothesis that people sometimes take physical actions to make themselves more effective problem solvers. The task was to generate all possible words that could be formed from seven Scrabble letters. In one condition, participants could use their hands to manipulate the letters, and in another condition, they could not. Results show that more words were generated with physical manipulation than without. However, an interaction was obtained between the physical manipulation conditions and the specific letter sets chosen, indicating that physical manipulation helps more for generating words in some circumstances than in others. Overall, our findings can be explained in terms of an interactive search process in which external, physical activity effectively complements internal, cognitive activity. Within this framework, the interaction can be explained in terms of the relative difficulty of generating words from the letters given in the different sets.

Introduction

People often adapt their physical environments to take better advantage of cognitive or perceptual skills (Clark, 1997; Hutchins, 1995a; Kirsh & Maglio, 1994; Kirsh, 1996). For instance, Tetris players take actions to set up their external environment to facilitate perceptual processing (Kirsh & Maglio, 1994; Maglio & Kirsh, 1996); gin rummy players physically organize the cards they have been dealt so as to be able simply to read off what is in a hand (Kirsh, 1995);

and airline pilots place external markers on their controls to help keep track of appropriate speed and flap settings (Hutchins, 1995b). In each of these cases, people take action to set up their external environments so that their mental jobs are easier, faster, or less error-prone.

The key to such processes of *interactive skill* is that the benefits of adapting the physical environment outweigh the costs of taking the physical actions. In the case of Tetris, for example, the cost of rotating a falling Tetris piece too many times is small (because over-rotation can quickly be corrected) compared to the benefit of relying on the visual system to determine whether the piece fits in its visible orientation. Thus, the hypothesis is that it is more efficient to rotate the falling piece in the external environment than to imagine how it would look in a different orientation.

Most people are familiar with the board game Scrabble[1], in which players form words by arranging tiles with letters printed on them. When trying to come up with words in this game, people can either mentally rearrange the letters or physically rearrange the letters. Based on the idea that people routinely set up their environments to make their cognitive jobs easier, it is reasonable to suppose that it is easier to form words by physically moving the tiles than by simply imagining their rearrangement (Kirsh, 1995). But is this really true? If so, is it always the case? Our first objective

---

[1] "Scrabble" is a registered trademark of Hasbro, Inc.

is to test the conjecture that the physical action of moving Scrabble tiles facilitates the discovery of anagram solutions. More precisely, given a sequence of seven letters, such as "RDLOSNA", and the task of calling out all legal words containing at least two letters in five minutes, will more words be formed in the condition in which the tiles can be moved than in the condition in which the tiles cannot be moved? As we will show, the answer is yes.

Overall, we found that participants generated more words when they were allowed to manipulate the Scrabble tiles than when they were not. However, we also found that this was not always the case. In particular, physically arranging the letters led to more words for only one of the two sets of letters tested. Though we attempted to control for productiveness of the letter sets through a norming task, it turned out that the words found in one of the sets are far more frequent in English than the words found in the other. Use of hands facilitated word generation only for the less frequent set. It is reasonable to suppose that less frequent words are harder to generate and so would more likely benefit from any external help.

From a theoretical perspective, it makes sense that physically arranging letters simplifies the task of forming words, as it ought to be easier to see words by looking than to see words by mentally swapping letters around. But there are many possible ways a person might generate words when given a set of letters. Our modeling objective is to discover an underlying process model to explain our finding that people form more words when they can move the Scrabble tiles than they form when they cannot.

Clearly, people engage in a search process of some sort, but we are not sure of the state-space that they are searching, the operators that yield neighboring states, or the subjective metric that is used to judge states. The most obvious state-space search is one in which the states represent letter strings and the transitions between the states represent the operations of adding, deleting, and swapping letters. In such a state-space, for instance, "chat" might be found from "hat" by adding a "c" to the front of the string. An alternative state-space might contain operators that can move from one state to another along a semantic or associative dimension; for instance. In this case, "cat" might be found from "hat" because "cat" is associated with "hat" in the familiar title, "The Cat in the Hat". In any event, the model must account for how the external actions of manipulating letters can have the cognitive effect of improving performance, especially for the set of less frequent words.

In what follows, we first sketch a model of interactive skill in Scrabble, and then describe our empirical study.

Model of Interactive Scrabble Skill

One way to think about the process of generating Scrabble words is in terms of the metaphor of energy landscapes. If we regard the set of legal words created from a letter sequence seven letters long to be a set of attractors distributed in a state-space consisting of letter sequences between two and seven letters long, then we can interpret the search for words to be some sort of stochastic hill-climbing process. The energy metric for this landscape might be determined,

for example, by frequency of bigrams, trigrams, and words, as well as the probability that a bigram such as **br** will be continued with an **e**, or continued with an **o**, and so forth. Given such a landscape, we can then attempt to explain both the timing and sequence of the anagram solutions that participants provide by suggesting that they engage in a particular type of search.

In fact, if we assume some sort of stochastic search, the reason hands help would be the same regardless of the details of the model. Specifically, physical manipulation allows one to instantly jump to new parts of the state space and to begin searching there. In a sense, the mental search for words is hampered by the data-driven nature of looking at the tiles. Consider the analogy of a rubber band: People can generate diverse letter combinations in their heads, but when they re-examine the tiles, they are drawn back to the original arrangement, like a rubber band springs back to its original shape. Thus, it is hard to continue searching from positions in the search space that are distant from the visible arrangement of the tiles. If words are not too difficult to find, they can be discovered quickly without having to look again at the tiles. If words are difficult to find, however, it may be helpful to be reminded of the letters by looking at them.

Nevertheless, to make good on this sort of search-based model, we must specify in detail not only the operators that define the state-space, but also the energy landscape of the space. We are just now beginning to explore such a model, so we can only sketch it in the broadest strokes. For instance, we do not yet know how to combine information about the frequency of words and their parts into a single number that gives the "closeness" – the energy level – of a state. Nor can we make precise the notion of locality; that is, how to decide when two states are neighbors (regardless of how distant they are in energy terms). Then there is the question of defining the state-space itself; for instance, should it be defined over letter sequences, using the operators of "add", "delete", "substitute", "rearrange" as primitives? And there is the problem of words that cannot be formed from the allotted letters but that are attractive because they have a similar sound or meaning to words that can be formed from the allotted letters.

To start, suppose we choose the following operators,

ear $\Rightarrow$ **bear**   **ore** $\Rightarrow$ **ogre** – *arbitrary add*
ago $\Rightarrow$ **age**   **boor** $\Rightarrow$**boer** $\Rightarrow$**boar** – *arbitrary substitute*
bore $\Rightarrow$ **ore**   **brag** $\Rightarrow$ **bag** – *arbitrary delete*
ogre $\Rightarrow$ **gore**   **bear** $\Rightarrow$ **bare** – *arbitrary rearrange*

Taken together, these seem too powerful. After all, there is a cost to mental search, and to performing computations in working memory. Perhaps it would be appropriate to restrict "add", "delete", and "substitute" operators so that they only apply to the beginning or end of a word (i.e., no center embedding in a single step). To achieve arbitrary letter strings, then, these operators would have to be combined with "rearrange". Perhaps we should include "reverse" as a special type of rearrangement that permits more global changes in a single move, such as

gob ⇒ bog   garb ⇒ brag – *reversal, special rearrange*

Of course, people are not dealing with abstract and arbitrary strings when playing Scrabble. Strings form syllables, syllables form words, and so on. In arranging letter strings, people must contend with word-formation constraints of English, including permissible consonant clusters, consonant-vowel sequences, or vowel-vowel sequences. For example, **rd** and **dr** are both permissible consonant clusters in English; however, **rd** is not allowed in word-initial position, and **dr** is not allowed in word-final position. Such orthographic and phonetic constraints are far too numerous and complex to list here.

A similar issue concerns the units the operators act on. These units might be restricted to individual letters, so that only one letter can be altered in a single action, as in **ogre ⇒ gore,** or they may be able to operate on entire letter sequences (**bear ⇒ bare**). Perhaps the operators ought to be restricted to bigrams at the beginning or ending of a string, as in

bar ⇒ **barge** – *append bigram*
rage ⇒ **gear** – *rearrange bigram*

In this first pass analysis, however, such questions about operators are of less consequence than the metric that determines how close (easy to reach) neighbors are. Accordingly, even if we are wrong in supposing that **garb** is an immediate neighbor of **brag** in state-space, the key factor determining whether **brag** is generated soon after **garb**, is principally a function of their closeness in terms of energy – that is, the relative difficulty of making the transition from one state to another. In this way, our modeling approach is similar to Hofstadter's (1983) Jumbo program, which used relatively little knowledge of English to solve anagram puzzles through a stochastic search of the space of letter and syllable clusters. In such a stochastic system, the primary means for assuring that enough of the space is searched (i.e., to downplay the influence of local maxima) is to increase the chance of moving from one arbitrary state to another.

As mentioned, however, given any choice of operators and energy metric, our hypothesis is that people improve when they are permitted to physically move letters because movement enables them to instantly "reset" their position in the landscape. Thus, when they find themselves trapped in a particular region, they can use tile rearrangement as an interactive strategy to assist internal search. In a sense, such physical reorganization provides an element of randomness that supports intelligent behavior (cf. Mitchell & Hofstadter, 1995).

We now turn to our experimental data on the Scrabble task.

Scrabble Experiment

The goal of the experiment was to examine performance in a word formation task using Scrabble letters. Specifically, we hypothesized that people would generate more words with a set of Scrabble tiles when allowed to physically ma-

nipulate the tiles than when not allowed to physically manipulate the tiles. Because we could not test the same person on the same set of tiles in both conditions, we first attempted to establish two sets of letters from which people naturally generate about the same number of words.

Norming Task

Six sets of seven letters each were created by randomly selecting tiles from the Scrabble game. Two sequences for each of the sets were randomly generated (e.g., "RDLOSNA" and "ARLNDOS") to test whether there was an effect of the order of the letters.

Sixteen undergraduates from the University of California, Santa Cruz participated in the norming task to fulfill a requirement in a psychology course. Each participant saw one of the two sequences for each of the six originally chosen sets. The sequences and the order in which the sequences were presented were balanced across participants. Thus, eight participants saw one of the two sequences of each set.

In this pencil-and-paper task, participants were given five minutes to write down as many words as they could by rearranging the letters from each sequence. The participants were informed that words did not have to use all the letters in the sequence but could vary in length between two and seven letters.

For each of the twelve letter sequences, mean number of words generated was calculated. We then compared total number of words generated for each set. A series of *t*-tests between each pair of orders for a given set showed no effect for the number of words generated, so the results for each set were collapsed across the two orderings. A one-way analysis of variance (ANOVA) among the six sets of letters showed a significant effect, $F(1, 5) = 26.2$, $p < 0.001$, indicating that more words were generated for some of the sets than for others. Inspection of the data revealed that about the same number of words were generated for three of the six sets (see Table 1).

Table 1: Mean number of words generated per letter set.

Letter Sequence	Number of Words
"NDRBEOE"	19.88
"ESIFLCE"	12.06
"EMTGPEA"	22.25
"RDLOSNA"	20.81
"IRCDEOE"	16.19
"LNAOIET"	26.07

Another one-way ANOVA indicated that the number of words for "EMTGPEA", "RDLOSNA", and "NDRBEOE" did not differ, $F(1,2) = 1.01$, NS. Thus, we chose the first two of these three sequences as stimuli for our experiment because they shared the fewest letters.

Scrabble Method

Twenty undergraduates from the University of California, Santa Cruz participated in the experiment to meet a requirement in a psychology course. Each participant was a native speaker of English or demonstrated a high proficiency in the language, as determined by responses on a questionnaire and vocabulary test given prior to the experiment.

The experiment was a 2 x 2 mixed design, with physical manipulation (Hands vs. No Hands) as the within-subjects factor, and letter sequence ("RDLOSNA" vs. "EMTGPEA") as the between-subjects factor. The letter sequence and the order in which the Hands or No Hands condition was performed was balanced across participants.

Participants were informed that they would have five minutes to generate as many English words as possible that were at least two letters long. Words were legal only if they were made from the tiles given. Participants were instructed not to use proper names (e.g., "Ron") or acronyms (e.g., "IBM"). They were also instructed to spell out the words as they found them (e.g., "TEAM, T-E-A-M") so that homophones (such as "BE" and "BEE") could be easily distinguished.

The task began with a practice trial. Half the participants were given instructions for the Hands condition, and the other half were given instructions for the No Hands condition. Participants in the Hands condition were told that they could use their hands to physically rearrange the tiles, but that it was not necessary to move the tiles to find and call out words. Participants in the No Hands condition were told that they could not use their hands to physically rearrange the tiles.

The set of Scrabble tiles was laid out on the table in front of the participant and the practice trial began. Practice proceeded for five minutes and then the participant performed a distractor task for five minutes. At this point, the participant performed the test trial in the same condition as the practice trial (i.e., Hands practice followed by Hands test). After five minutes of the word generation task, the participant moved to another distractor task, and then onto the other Hands or No Hands condition in the same way as before: practice followed by distractor followed by test. Throughout the practice and test trials, the experimenter transcribed the words as they were called out and the session was taped.

Results

The number of legitimate, unique words generated by each participant in each condition was calculated. The mean for the Hands condition was 20.70 (SD = 5.00) and for the No Hands condition, 19.30 (SD = 5.58). A two-way repeated measures ANOVA showed a main effect for the within-subjects factor (Hands vs. No Hands), $F(1, 18) = 5.165$, $p < 0.05$, indicating a difference in performance for Hands vs. No Hands. The effect of letter sequence was not significant, $F(1,18) < 1$, indicating there was no difference between "RDLOSNA" and "EMTGPEA". However, an interaction was obtained between manipulation condition and letter sequence, $F(1, 18) = 91.739$, $p < 0.0001$, indicating the use of hands had different effects on the two different sequences (see Table 2).

Table 2: Mean number of words generated.

	Hands	No Hands
"EMTGPEA" (n = 10)	23.30	16.00
"RDLOSNA" (n = 10)	18.10	22.60

To try to make sense of the manipulation by sequence interaction, post hoc comparisons were conducted. For "EMTGPEA", a significant difference was found between the Hands and the No Hands conditions, $t(18) = 4.97$, $p < 0.0001$; but for "RDLOSNA", no such difference was found, $t(18) = 1.87$, $p > 0.05$. Thus, physically moving the tiles improved performance for one letter sequence, but it had marginal and opposite effect on the other letter sequence.

Additional tests were conducted to investigate whether order of presentation (Hands, No Hands vs. No Hands, Hands) had an effect on the number words produced, or whether the average length of words produced varied across conditions. No effect for order was obtained, $t(18) < 1$, NS, indicating that the number of words generated did not depend on which condition (Hands or No Hands) was seen first. Similarly, there was no effect of manipulation condition or letter set on the average length of the words produced per participant, $F(1,18) < 1$, NS, in all cases. Overall, the mean word length was 3.30 (SD = 0.22).

Discussion

Overall, more words were generated when participants were allowed to manipulate the tiles than when they were not allowed to manipulate the tiles. This bears out our initial hypothesis. The interaction between manipulation and letter sequence, however, was unanticipated – the norming data were supposed to assure us that an equivalent number of words would be generated for both letter sequences.

One possible explanation for this interaction concerns the relative difficulty of producing words from the different letter sets. The more difficult it is to generate words from a set of letters, the more we suppose physical rearrangement would help. In terms of the state-space search model outlined previously, use of hands might be more effective if the words are more spread out in the space. In this case, physically rearranging the tiles has the effect of easily resetting the system's position in the energy landscape, enabling wider coverage during search.

One simple measure of word-generation difficulty might be word length. If the state-space search were based primarily on orthographic features and operations, we would expect longer words to be more difficult to generate, as long words must require more operations to compose than short words. Of course, we found no effect of the physical manipulation conditions on the length of words generated, suggesting that word length is not related to difficulty.

A semantic measure of word-generation difficulty relates to the productiveness of the letter strings and the frequency of the words that can be formed. First, 92 words can be generated from the letters "RDLOSNA", whereas only 53 can be generated from "EMTGPEA". Second, 47 of the 92 words in "RDLOSNA" do not appear in the Kucera and

Francis (1967) corpus of written English and the mean frequency of the remaining 45 words is 2735; nineteen of the 58 words in "EMTGPEA" do not appear in Kucera and Francis and the mean frequency of the remaining 39 is 336. In the Brown (1984) corpus of spoken English, the mean frequency for "RDLOSNA" is 1395, and for "EMTGPEA", 221. In both written and spoken English, the words contained in "RDLOSNA" are far more frequent in English than those contained in "EMTGPEA". Thus, it is plausible to suppose that physically arranging the letters would be helpful when trying to produce words from the less productive and less frequent set, as it must be more difficult to produce words in this case.

For the participants in the experiment who chose to use their hands very little or not at all, there was little difference between the Hands and No Hands conditions.[2] Interestingly, when asked why they did not move the letters more, these participants almost universally responded that they thought they could move the letters faster in their heads than they could on the table. As we have argued, this common intuition seems to be false in this case and in many others (Clark, 1997; Hutchins, 1995a, 1995b; Kirsh, 1995; Kirsh & Maglio, 1994).

Conclusion

We tested the hypothesis that physical actions can make problem solving easier. In our study, people were given sets of Scrabble letters and asked to generate words in two conditions: with their hands and without their hands. The results indicated that more words were generated when people used their hands than when they did not, although the story is somewhat more complicated. We argued that in this case, the physical actions of moving the letters allow people to effectively use the external environment as part of an interactive process of searching for words. In future experiments, we hope to control better for the productiveness of the strings and for the frequency of the words that can be produced from them.

Acknowledgments

Thanks to Ray Gibbs for supporting this research in his lab, and for many thoughtful comments and suggestions on this work. Thanks to Jenny Lederer for helping to run subjects. Thanks to Chris Campbell and Rom Brafman for useful discussions on experimental design and data analysis. Thanks to Chris Dryer, Denis Lalanne, and an anonymous reviewer for insightful comments on a draft of the paper.

References

Brown, G. D. A. (1984). A frequency count of 190,000 words in the London-Lund corpus of English conversation, *Behavioural Research Methods Instrumentation and Computers, 16*, 502-532.

Clark, A. (1997). *Being there: Putting body, brain, and world together again.* Cambridge, MA: MIT Press.

Hofstadter, D. R. (1983). The architecture of JUMBO. In *Proceedings of the International Machine Learning Workshop* (pp. 161-170).

Hutchins, E. (1995a). *Cognition in the wild.* Cambridge, MA: MIT Press.

Hutchins, E. (1995b). How a cockpit remembers its speeds. *Cognitive Science, 19*, 265-288.

Kirsh, D. (1995). The intelligent use of space. *Artificial Intelligence, 73*, 31--68.

Kirsh, D. (1996). Adapting the environment instead of oneself. *Adaptive Behavior, 4*, 415-452.

Kirsh, D., & Maglio, P. (1994). On distinguishing epistemic from pragmatic action. *Cognitive Science, 18*, 513--549.

Kucera, H., & Francis, W. N. (1967). *Computational analysis of present-day American English.* Providence, RI: Brown University Press.

Maglio, P. P., & Kirsh, D. (1996). Epistemic action increases with skill. In *Proceedings of the Eighteenth Annual Conference of the Cognitive Science Society* (pp. 391-396). Mahwah, NJ: LEA.

Mitchell, M., & Hofstadter, D. R. (1995). The Copycat project. In D. R. Hofstadter (Ed.) *Fluid concepts and creative analogies: Computer methods of the fundamental mechanisms of thought.* New York: Basic Books.

---

[2] Roughly one third of the participants in the Hands condition chose not to use their hands or used their hands only briefly. The small sample here makes statistical analysis difficult. We note this only as a passing observation.

Spoken Word Recognition in the Visual World Paradigm

Reflects the Structure of the Entire Lexicon

James S. Magnuson (magnuson@bcs.rochester.edu)
Michael K. Tanenhaus (mtan@bcs.rochester.edu)
Richard N. Aslin (aslin@cvs.rochester.edu)
Delphine Dahan (dahan@bcs.rochester.edu)
Brain and Cognitive Sciences, University of Rochester, Meliora Hall, Rochester, NY 14627 USA

Abstract

When subjects are asked to move items in a visual display in response to spoken instructions, their eye movements are closely time-locked to the unfolding speech signal. A recently developed eye-tracking method, the "visual world paradigm", exploits this phenomenon to provide a sensitive, continuous measure of ambiguity resolution in language processing phenomena, including competition effects in spoken word recognition (Tanenhaus, Spivey-Knowlton, Eberhard, & Sedivy, 1995). With this method, competition is typically measured between names of objects which are simultaneously displayed in front of the subject. This means that fixation probabilities may not reflect competition within the entire lexicon, but only that among items which become active because they are displayed simultaneously. To test this, we created a small, artificial lexicon with specific lexical similarity characteristics. Subjects learned novel names for 16 novel geometric objects. Objects were presented with high, medium or low frequency during training. Each lexical item had two potential competitors. The crucial comparison was between high-frequency items which had either high- or low-frequency competitors. In spoken word recognition, performance is correlated with the number of frequency-weighted neighbors (phonologically similar words) a word has, suggesting that neighbors compete for recognition as a function of frequency and similarity (e.g., Luce & Pisoni, 1998). We found that in the visual world paradigm, fixation probabilities for items with high-frequency neighbors were delayed compared to those for items with low-frequency neighbors, even when the items were presented with unrelated items. This indicates that fixation probabilities reflect the internal structure of the lexicon, and not just the characteristics of displayed items.

Introduction

Understanding the structure and role of the lexicon in spoken word recognition has implications at higher and lower levels of processing. Recent theories have placed much syntactic, semantic and pragmatic knowledge in the lexicon (e.g., MacDonald, Pearlmutter & Seidenberg, 1994; Tanenhaus & Trueswell, 1995). Thus, representations activated in the resolution of word recognition may have cascading effects which come into play at higher levels. At the same time, lexical knowledge also has effects at lower levels, such as aspects of speech perception which have often been considered pre-lexical (e.g., Andruski, Blumstein, & Burton, 1994; Marslen-Wilson & Warren, 1994). Understanding the structure of the lexicon and lexical activation patterns in spoken word recognition will clearly provide a vital step towards understanding language processing.

A few key parameters have been identified which account for substantial amounts of variability in spoken word recognition. Luce and colleagues (e.g., Luce and Pisoni, 1998) have shown that *log word frequency* alone can account for 4 to 6% of the variance observed in word identification under noise, whereas 16 to 22% of the variance can be accounted for by the *frequency-weighted neighborhood probability rule* (FWNPR), which is the basis of their *Neighborhood Activation Model* (NAM). The FWNPR estimates the amount of expected competition between a word and its "neighbors" (similar words, often defined as words differing by no more than one phoneme), weighted by their frequencies.

These sorts of results inform us about what Marslen-Wilson (1993) has termed the *macrostructure* of spoken word recognition. However, they provide little information about the *microstructure* of the on-line lexical processing -- e.g., what determines the nature of the competitor set over time and the time course of competition effects. Instead, they provide coarse, indirect information. In a typical experiment, recognition and accuracy are measured, but these are usually all-or-nothing data measures of post-recognition decisions, which do not tell us *directly* about on-line processing. Instead, mechanisms of on-line processing must be inferred *indirectly* by seeing how well different parameters (e.g., frequency) correlate with performance.

The interactive-activation connectionist model, TRACE (McClelland and Elman, 1986), is an example of a class of models which provide a different method of testing predictions (an implementation of the Luce and Pisoni, 1998, NAM would be another example). Given a simulated input, TRACE provides a continuous prediction over time of which words in the lexicon should be active and competing for recognition. In the top panel of Figure 1, we present TRACE activations for a target input, *beaker*, an onset competitor (called a "cohort" item because the Cohort model predicts that mainly items which share onsets compete), *beetle*, a rhyme, *speaker*, and an unrelated item (the activations are scaled; see below). But with these fine-grained predictions in hand, how can we test them? Conventional psycholinguistic tasks cannot provide continuous, on-line measures of activation using continuous speech; tasks which are used to try to make time course measurements, such as gating, require interrupting the speech stream and thus using an unnatural stimulus.

Tanenhaus and his colleagues have developed an eye tracking method for studying spoken language comprehension which provides a sensitive, continuous measure of lexical activation (e.g., Tanenhaus et al., 1995).

In this "visual world paradigm", a subject sees a display containing several objects (either real objects or pictures on a computer display). When subjects are asked to perform an action with one of the objects (e.g., "pick up the beaker" or "click on the beetle"), their eye movements are closely time-locked to the speech stream. For example, subjects might be shown a display containing objects *beaker*, *beetle*, *speaker* and *carriage*. If they are asked to "pick up the beaker", the probabilities of fixating each item over time can be compared directly to TRACE predictions.

In the lower panel of Figure 1 are data from Allopenna, Magnuson and Tanenhaus (1998), who presented many such displays to several subjects (the data shown are averaged over several items and several subjects). As can be seen by comparing the upper and lower panels of Figure 1, the data are very similar to the TRACE predictions. Note that the fixation probabilities do not sum to 1 because subjects begin each trial fixating a central fixation cross, and that the TRACE activations have been transformed to simulate the experimental situation of having limited response possibilities (see Allopenna et al. for details).

Note also that while most of the change in probabilities occur after target offset, the fixation probabilities are more closely time-locked to the spoken input than they appear. In very simple tasks, participants require approximately 150 msec to plan and launch a saccade (e.g., Matin, Shao, & Boff, 1993). Allowing for this planning time, it is clear that the earliest eye movements are being planned approximately 100 msec after target onset.

Some other notable qualities of the paradigm are that it does not require subjects to make explicit decisions about stimuli. Instead, eye movements are monitored as subjects respond naturally to continuous spoken instructions. Given a properly constrained task (one in which visually guided movements are required, which allows a functional interpretation of eye movements and avoids the problems identified by Viviani, 1990; see Allopenna et al., 1998, for further discussion), eye movements provide an incidental measure of moment-to-moment attention.

The results reported by Allopenna et al. demonstrate the sensitivity of the visual world paradigm. While cohort (onset overlap) effects were well-established (e.g., Marslen-Wilson & Zwitserlood, 1989), rhyme effects had proven more elusive. For example, weak rhyme effects had been reported in cross-modal and auditory-auditory priming (Connine, Blasko & Titone, 1993; Andruski et al., 1994) only when the rhymes differed by only one or two phonetic features. Allopenna et al.'s rhymes all differed by more than two features. Thus, in addition to providing information about the time course of activation, the visual world paradigm also proved to be more sensitive than other spoken word recognition paradigms.

However, objects whose names were predicted to compete were displayed at the same time. While this allowed the most direct comparison with, e.g., TRACE predictions, the results are ambiguous in one respect: they may have been due to the use of displays with a restricted set of items. That is, competition may have been limited to the set of displayed items, and may not have reflected the influence of the rest of the lexicon. This is important because the

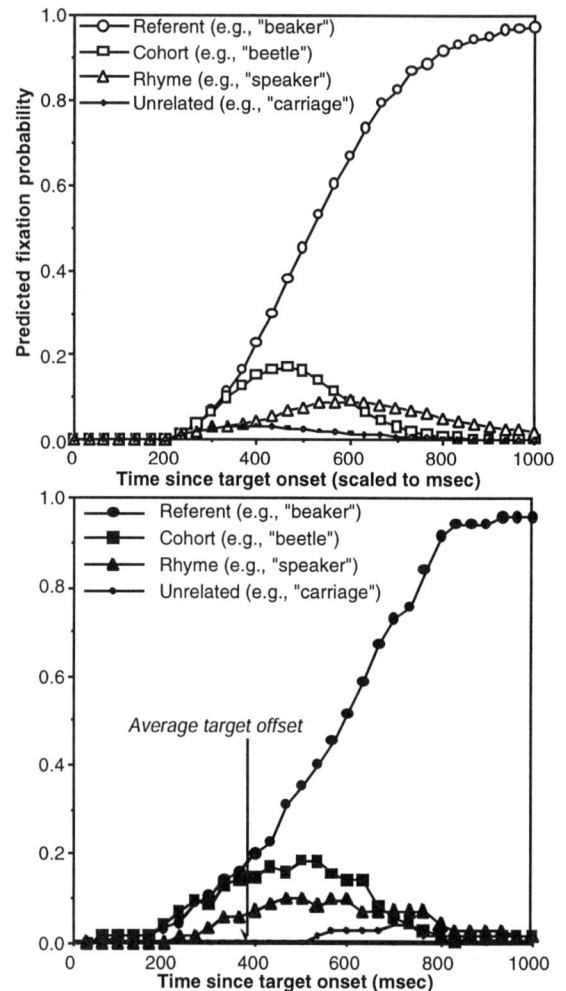

Figure 1: TRACE activations converted to predicted fixation probabilities (top panel) and observed probabilities of fixating a target, a cohort, a rhyme, and an unrelated object from Allopenna et al. (1998).

strength of other paradigms is their ability to inform hypotheses about which lexical items are activated by a given input -- although this is through *indirect* evidence of competition, as reflected in reduced performance. The visual world paradigm provides a relatively *direct* measure of the relative activations of displayed lexical items, but does it indicate the activation of items which are not present?

One way to determine whether competition is limited to the displayed portion of the lexicon is to measure responses to items with different frequency-weighted neighborhood densities (FWNDs). Recall that Luce and colleagues have shown that a word's FWND accounts for much of the variance in spoken word recognition, and consider an example of two words, A and B. If both have 2 neighbors, and all their neighbors have equal occurrence frequencies, recognition times for A and B should be equivalent. If we increase A's FWND by giving it more neighbors, it should take longer to recognize A (because now, given A, more words compete for recognition). If instead we increase the frequency of A's 2 neighbors, it should still take longer to recognize A (since words of higher frequency compete more strongly, and A's FWND has increased). How would such a neighborhood density effect manifest itself under the visual world paradigm? We could present word A among three unrelated items, and present word B among three unrelated items. If A's FWND is higher, the probability of fixating A over time should increase more slowly than for B.

Here, we report an experiment of just this type, which replicates and extends previous work using an artificial lexicon (Magnuson, Dahan, Allopenna, Tanenhaus and Aslin, 1998). The advantage of using an artificial lexicon is that we can carefully control the statistics of the lexicon. A set of stimuli drawn from a natural language will be more variable. The artificial lexicon lets us test our hypothesis with a minimum of potential confounds. Before turning to the current experiment, we will briefly review the artificial lexicon study on which the current study is based.

Artificial Lexicons and the Visual World Paradigm

In the previous artificial lexicon study (Magnuson et al., 1998), we trained subjects to recognize a lexicon of 16 novel words. Each lexical item (e.g., /pibo/) had two potential competitors: a cohort (e.g., /pibu/) and a rhyme (e.g., /dibo/). Each item in the lexicon was randomly associated with a novel geometrical object. Subjects learned the lexicon by learning the names for each object.

Figure 2 shows examples of the sorts of displays subjects saw on a computer screen. Initially, subjects saw pairs of objects, and heard instructions to click on one with the computer mouse (e.g., "click on the pibo"). At first, subjects had to guess. But after they clicked on one object, they received feedback: one object would disappear, and they knew that the remaining one was being named, and then they would hear the name again. Different levels of word frequency were approximated by presenting the items with "high" or "low" frequency (with a ratio of 7:1/high:low during training). Item frequency was crossed with competitor frequency: four items were high frequency (HF) and had HF neighbors (HF/HF); four were low frequency (LF) with low frequency neighbors (LF/LF); four were HF/LF, and four were LF/HF.

Subjects quickly reached ceiling on the 2AFC (alternative forced choice) task, and training continued with a 4AFC task (see Figure 2). After 80 minutes of training on each of two days, we monitored eye movements as subjects performed the basic visual world paradigm task without feedback: given

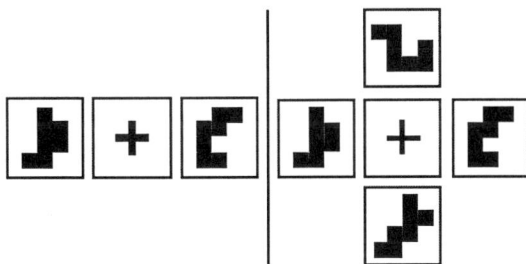

a display containing four objects, subjects were instructed to click on one of the objects. On most trials, the items were all unrelated. On critical trials, a cohort or rhyme competitor was present. The results after two days closely resembled Allopenna et al.'s (1998) results using real words: there was cohort and rhyme activation, and the fixation probabilities for each object varied as a function of the similarity of the stimulus with the object's name. There were also interactions between item frequency and competitor frequency. For example, the cohort effect was stronger -- with a substantial initial advantage for the cohort -- for low-frequency items with high-frequency competitors than for high-frequency items with high-frequency competitors. This indicates that, as with real words, neighbors competed as a function of their similarity and frequency.

The experiment included a FWND manipulation (although all items had the same number of neighbors, FWND varied because competitor frequency varied), and a condition where items were presented along with three unrelated distractors. However, this condition did not provide a complete test of the question at hand: namely, whether changes in fixation probability over time reflect activation of present and absent competitors. This is because there were only two levels of frequency, and target items were presented among unrelated distractors which were matched in frequency with the target's competitors (e.g., for a high-frequency target with low-frequency competitors, low-frequency, unrelated distractors were used). Thus, we cannot be certain that differences in fixation probabilities were due to the frequencies of the absent competitors, or to the frequencies of the simultaneously presented distractors.

In order to have a clean test of the hypothesis, we need a third level of frequency. Then, HF/HF and HF/LF items can both be presented among unrelated distractors of the same frequency, and differences we observe should be due to the frequencies of (absent) competitors, not the characteristics of the unrelated distractors. This was the design we used for the current experiment.

Absent Competitors and the Visual World Paradigm

The design of the current study was similar to that used by Magnuson et al. (1998). We trained subjects to recognize a lexicon of 16 novel words. Each lexical item (e.g., /pibo/) had two potential competitors: a cohort (e.g., /pibu/) and a rhyme (e.g., /dibo/), and was randomly associated with a novel geometrical object. Target and competitor frequency were varied, but with three levels rather than two. The third level (medium) provided distractors of uniform frequency to serve as distractors for the other items.

Method

Participants Seven students at the University of Rochester were paid for their participation. All were native speakers of English with normal or corrected-to-normal vision and normal hearing.

Materials The visual stimuli were simple patterns, formed by filling eight randomly-chosen, contiguous cells of a four-by-four grid (see Figure 2). 10,000 such randomly-generated patterns were randomly ordered, and sixteen were selected

Figure 2: Examples of stimulus displays. The left panel shows a possible display in 2AFC training; the right panel shows a possible 4AFC display.

from the beginning of the set (with two items replaced due to visual similarity with other items).

The novel words consisted of sixteen bisyllabic nonsense words. The sixteen words comprised four four-word sets, such as /pibo/, /pibu/, /dibo/, and /dibu/. Note that for each word, there is an onset ("cohort") competitor which differs only in the final vowel, a rhyme, and a relatively dissimilar item (differing by two phonemes, which would not qualify it as a neighbor using the most standard definition of a word differing by a single phoneme). A small set of phonemes was selected in order to achieve consistent similarity within and between sets. The consonants /p/, /b/, /t/, and /d/ were chosen because they are among the most phonetically similar stop consonants. In each set, rhymes differed by two phonetic features (place and voicing) in the first phoneme. Transitional probabilities were controlled such that all phonemes and combinations of phonemes were equally predictive at each position or combination of positions.

The auditory stimuli were produced by a male, native speaker of English in a sentence context ("click on the ____"). The average duration of the target words was 496 msec. The stimuli were recorded to tape, and then digitized using the standard analog/digital devices on an Apple Macintosh 8500 at 16 bit, 44.1 kHz. The stimuli were converted to 8 bit, 11.127 kHz (SoundEdit format) in order to be used with the experimental control software, PsyScope 1.2 (Cohen, MacWhinney, Flatt & Provost, 1993).

Procedure Participants came to the lab for two 2-hour sessions on consecutive days. Each day consisted of seven training blocks with feedback and a testing block without feedback. We tracked eye movements during the test.

Participants were seated at a comfortable distance from the experimental control computer (an Apple Macintosh 7200 PowerPC). The structure of the training blocks was as follows. First, a fixation cross would appear on the screen. The participant had to click on the cross to begin the trial. After 500 msec, either two shapes (in the first four training blocks) or four shapes (in the rest of the training blocks and the tests) would appear. If only two shapes were presented, they appeared at about 1.5 degrees of visual angle to the left and right of the fixation cross. When four shapes were presented, two would also appear about 1.5 degrees above and below the fixation cross (see Figure 2).

Participants heard the instruction, "Look at the cross" through headphones 750 msec after the objects appeared. Then, they fixated the cross and clicked on it. Participants were instructed at the beginning of the session that they should fixate the cross until they heard the next instruction. 500 msec after clicking on the cross, an instruction to click on one of the items (with the computer's mouse) was presented (e.g., "Click on the pibu").

When participants responded by clicking on one of the items, or at the end of 15 seconds, all of the items disappeared except for the shape that was actually named. The correct shape's name was repeated 500 msec later. The object disappeared 500 msec later, and the subject would click on the cross to begin the next trial. The testing block

was identical to the four-item training, except that no feedback was given.

Shapes were randomly mapped to names, with a different random mapping for each subject. Half the items were medium frequency. Six items were high frequency, and two were low frequency. All of the medium frequency items had medium frequency competitors. The high- and low-frequency items were assigned such that four of the high frequency items had high frequency competitors, and two of the high frequency items had low frequency competitors (and the competitors for the two low frequency items were those two high frequency items).

Each training block consisted of 68 trials. High frequency items appeared 7 times per block, low-frequency items appeared once per block, and medium frequency items appeared 3 times per training block. Across all training blocks, all items appeared as visual distractors approximately equally often. Within training, distractors were randomly assigned to each trial.

The tests consisted of 96 trials. Each item appeared in six trials: one with its onset competitor and two unrelated items, one with its rhyme competitor and two unrelated items, and four with three unrelated items. For the crucial comparisons (HF/HF and HF/LF), medium frequency items were used as unrelated distractors.

We tracked eye-movements with an Applied Scientific Laboratories (E4000) eye tracker. Two cameras mounted on a lightweight helmet provide the input to the tracker. The eye camera provides an infrared image of the eye. The center of the pupil and the first Purkinje corneal reflection are tracked to determine the position of the eye relative to the head. Accuracy is better than 1 degree of arc, with virtually unrestricted head and body movements. A scene camera is aligned with the participant's line of sight. A calibration procedure allows software running on a PC to superimpose crosshairs showing the point of gaze on a HI-8 video tape record of the scene camera. The scene camera samples at a rate of 30 frames per second, and each frame is stamped with a time code. The auditory stimuli were presented binaurally through headphones using the standard digital-to-analog devices provided with the experimental control computer. Audio connections between the computer and HI-8 VCR provided an audio record of each trial. Each trial was analyzed frame-by-frame from stimulus onset to the subject's response (clicking on the appropriate object) by coding fixations from each saccade onset.

Results

Subjects were able to achieve high levels of accuracy relatively quickly; accuracy for high, medium and low frequency items was .83, .89, and .77, respectively, on the 4AFC test without feedback on day 1, and .96, .97, and .94 on day 2. Figure 3 demonstrates that we replicated the basic cohort and rhyme effects reported by Allopenna et al. (1998) and Magnuson et al. (1998). The results are averaged over all conditions in which cohort or rhyme competition was possible.

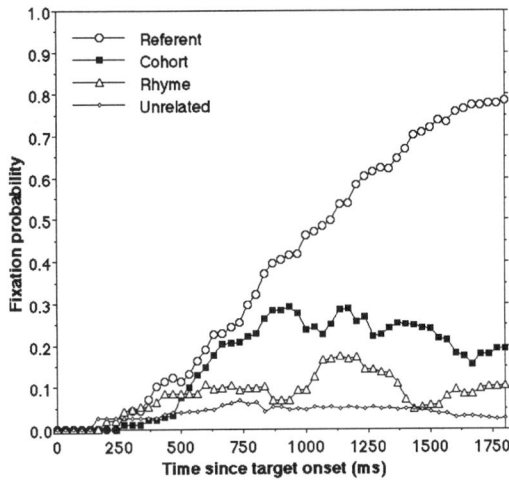

Figure 3: Combined cohort and rhyme effects.

Figure 4 shows how the cohort effect is modulated by target and competitor frequency. In the top panel, the target is low-frequency, and the cohort is high, and there was an initial advantage for the cohort. In the middle, both target and cohort were high frequency, and the result resembles the cohort effect shown in Figure 1, with no advantage for either item initially. On the bottom, the target was high- and the cohort was low-frequency. Although we do not see the expected initial advantage for the target, the cohort clearly is less active than in the other two panels. Thus, the results replicate Magnuson et al. (1998) and show that the degree to which (simultaneously present) items compete depends both on their similarity to the input, and their frequency -- as predicted by models such as NAM and TRACE.

Figure 5 shows the crucial comparison between HF/HF and HF/LF items presented among three unrelated, medium frequency distractors. The frequencies of the targets are equal, and the unrelated distractors are matched in the two cases. The only difference between the items is the frequency of their *absent* competitors. As predicted, the probability of fixating the HF/HF target increases much more slowly than the probability of fixating the HF/LF target. In a 2-way ANOVA (frame x competitor frequency) on the two referent probabilities from frame 10 (333 ms, when the probabilities first diverge) to frame 45 (1500 ms), both main effects were reliable (frame: $F(35, 210) = 30.97$, $p < .001$; competitor frequency: $F(1,6) = 9.00$, $p < .05$). A paired, one-tailed t-test on average fixation probabilities over the same window was also significant ($t(6) = 1.98$, $p < .05$, mean difference = .103).

Discussion

The current results show that fixation probabilities in the visual world paradigm reflect lexical competition which includes items which are not visually present. As in tasks such as lexical decision, recognition time depends on the internal structure of the lexicon. Since the only difference between the HF/HF and HF/LF targets is the frequency of their competitors, we see that lexical items for which no visual referents are available still compete for recognition.

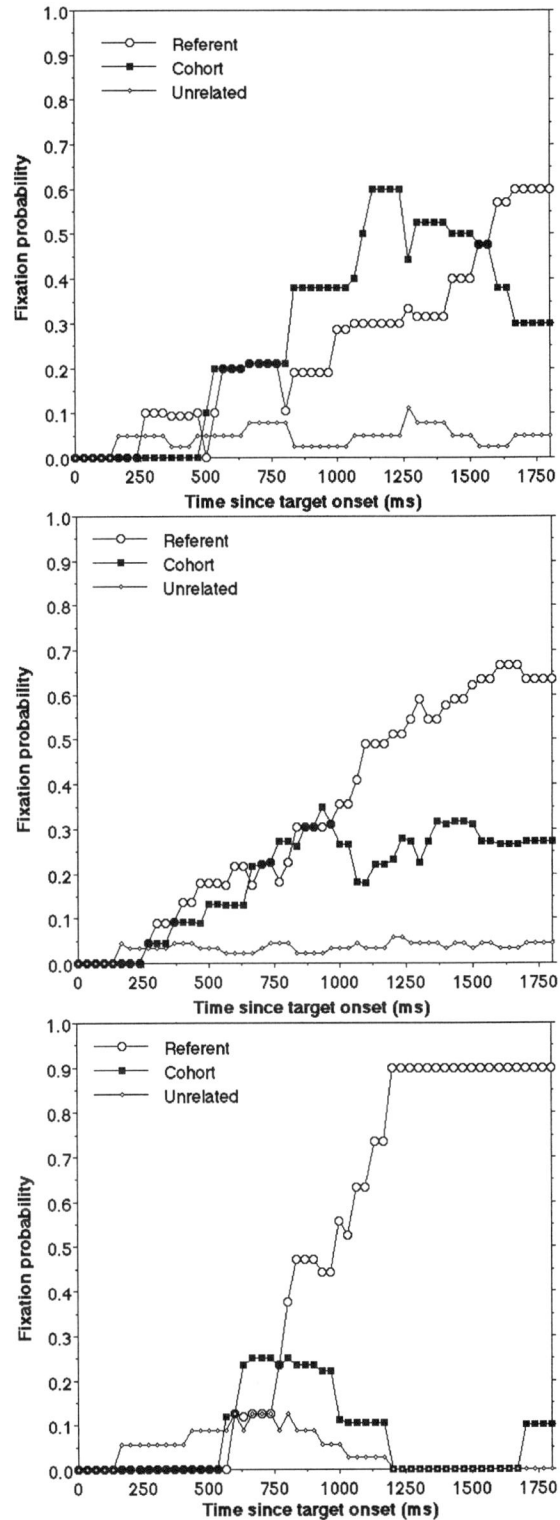

Figure 4: Cohort effects as a function of target and competitor frequency. Top: low-frequency target, high-frequency cohort. Middle: high-frequency target and cohort. Bottom: high-frequency target, low-frequency cohort.

The data in Figure 5 allow us to disregard an alternative interpretation of the results shown in Figure 4, where target

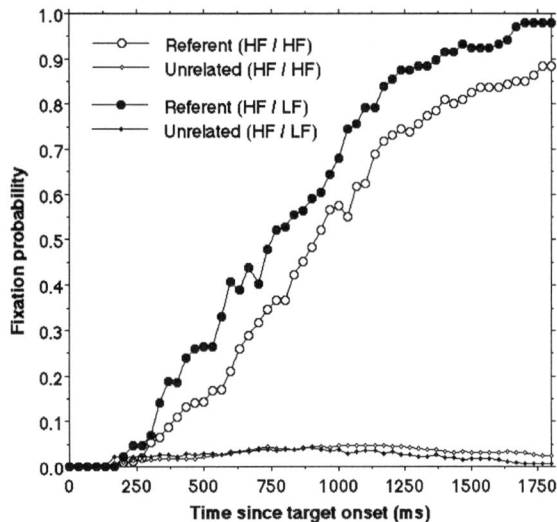

Figure 5: Target fixation probabilities as a function of competitor frequency. Both targets were high frequency, and were presented with 3 unrelated distractors of medium frequency. The only difference between the targets was the frequency of their competitors, *which were not displayed.*

probabilities rise most quickly when the target is high frequency and the competitor is low frequency. We might infer that this indicates competition between the items. However, fixations are necessarily serial. Given what appears to be competition among a set of simultaneously displayed items, one cannot infer competition at the lexical level, rather than simple co-activation (when data is sparse). The site of the competition could be the motor programming to move the eye. Once an item is fixated, the subject cannot simultaneously indicate the activation of other objects. An increased fixation probability for one item may be accompanied by a reduced probability for another, which could lead to the appearance of lexical competition. This problem of interpretation is diminished with sufficiently many data points. Despite having data from relatively few subjects, the current results can be interpreted as indicating lexical competition rather than competition at fixation generation, since the differences in target fixation probabilities shown in Figure 5 are not accompanied by commensurate differences in unrelated fixation probabilities. Therefore, the differences indicate that more time was needed for the activation of the target to become sufficiently large to generate initial eye movements when the target had high-frequency competitors.

In summary, we have replicated the results of Magnuson et al. (1998), showing an interaction of target and competitor frequency. This work also shows that effects in the visual world paradigm are not driven simply by competition among visible referents; changes in fixation probabilities are also driven by competition for recognition with competitors which are not visually available.

Acknowledgments

Supported by NIH HD27206 to MKT, NSF SBR-9729095 to MKT and RNA, and an NSF Graduate Research Fellowship to JSM.

References

Allopenna, P. D., Magnuson, J. S., & Tanenhaus, M. K. (1998). Tracking the time course of spoken word recognition using eye movements: Evidence for continuous mapping models. *Journal of Memory and Language,* 38, 419-439.

Andruski, J. E., Blumstein, S. E., & Burton, M. (1994). The effect of subphonetic differences on lexical access. *Cognition,* 52, 163-187.

Cohen J. D., MacWhinney B., Flatt M. & Provost J. (1993). PsyScope: A new graphic interactive environment for designing psychology experiments. *Behavioral Res. Methods, Instruments & Computers,* 25(2), 257-271.

Connine, C. M., Blasko, D. G., & Titone, D. (1993). Do the beginnings of spoken words have a special status in auditory word recognition? *J. Memory & Language,* 32, 193-210.

Luce, P. A., & Pisoni, D. B. (1998). Recognizing spoken words: The Neighborhood Activation Model. *Ear and Hearing,* 19, 1-36.

MacDonald, M. C, Pearlmutter, N. J., & Seidenberg, M. S. (1994). Lexical nature of syntactic ambiguity resolution. *Psychological Review,* 101, 676-703.

Magnuson, J. S., Dahan, D., Allopenna, P. D., Tanenhaus, M. K., and Aslin, R. N. (1998). Using an artificial lexicon and eye movements to examine the development and microstructure of lexical dynamics. *Proc. 20th Annual Conference of the Cognitive Science Society,* 651-656.

Marslen-Wilson, W. (1993). Issues of process and representation in lexical access. In G. T. M. Altmann & R. Shillcock (Eds.), *Cognitive Models of Speech Processing: The Second Sperlonga Meeting.* Erlbaum.

Marslen-Wilson, W., & Warren, P. (1994). Levels of perceptual representation and process in lexical access: Words, phonemes, and features. *Psych. Rev.,* 101, 653-675.

Marslen-Wilson, W., & Zwitserlood, P. (1989). Accessing spoken words: The importance of word onsets. *Journal of Experimental Psychology: Human Perception and Performance,* 15, 576-585.

Matin, E., Shao, K. C., and Boff, K. R. (1993). Saccadic overhead: Information-processing time with and without saccades. *Perception & Psychophysics,* 53, 372-380.

McClelland, J. L., & Elman, J. L. (1986). The TRACE model of speech perception. *Cognitive Psych.,* 18, 1-86.

Tanenhaus, M. K., and Trueswell, J. C. (1995). Sentence comprehension. In J. L. Miller & P. D. Eimas (Eds.), *Handbook of Perception and Cognition, Volume 11: Speech, Language and Communication.* San Diego: Academic Press.

Tanenhaus, M. K., Spivey-Knowlton, M., Eberhard, K., & Sedivy, J. C. (1995). Integration of visual and linguistic information is spoken-language comprehension. *Science,* 268, 1632-1634.

Viviani, P. (1990). Eye movements in visual search: Cognitive, perceptual, and motor control aspects. In E. Kowler (Ed.), *Eye Movements and Their Role in Visual and Cognitive Processes. Reviews of Oculomotor Research V4.* Amsterdam: Elsevier.

A Connectionist Account of Perceptual Category-Learning in Infants[1]

Denis Mareschal
Centre for Brain and Cognitive Development
Department of Psychology
Birkbeck College
London, WC1E 7HX, UK
d.mareschal@bbk.ac.uk

Robert M. French
Psychology Department, B32
Université de Liège
4000 Liège, Belgium
rfrench@ulg.ac.be

Abstract

This paper presents a connectionist model of correlation-based categorization by 10-month-old infants (Younger, 1985). Simple autoencoder networks were exposed to the same stimuli used to test 10-month-olds. Both infants and networks used co-variation information (when available) to segregate items into separate categories. The model provides a mechanistic account of category learning within a test session. It shows how distinct categories are developed and demonstrates how categorization arises as the product of an inextricable interaction between the subject (the infant) and the environment (the stimuli).

Introduction

The ability to categorize underlies much of cognition. It is a way of reducing the load on memory and other cognitive processes (Rosch, 1975). Because of its fundamental role, any developmental changes in the ability of infants to categorize is likely have a significant impact on subsequent cognitive development as a whole. As a result, categorization is one of the most fertile areas of research in infant cognitive development.

Many studies of infant categorization have relied on visually presented material. The basic idea of these studies is to show infants a series of images that could be construed as forming a category. The infant's subsequent response to a previously unseen image is used to gauge whether the infant has formed a category based on his or her experience with the familiarization exemplars. Generalization to a novel exemplar from the familiar category, coupled with a preference or heightened responsiveness to a novel exemplar from a novel category is taken as evidence of category formation. There is considerable evidence that young infants can form categorical representations of shapes, animals, furniture, faces, etc. (see Quinn & Eimas, 1996, for a recent review).

At first, the categories developed by infants may appear similar to those developed by adults. However, occasionally, infant categories differ dramatically from those of adults. Quinn, Eimas, and Rosenkrantz (1993) report one striking example. These authors found that when 3.5-month-olds were shown a series of cat photographs, the infants would develop a category of CAT that included novel cats and excluded novel dogs (in accordance with the adult category of CAT). However, when 3.5-month-olds were shown a series of dog photographs, they would develop a category of DOG that included novel dogs but

also included novel cats (in contrast to the adult category of DOG). There is an asymmetry in the exclusivity of the CAT and DOG categories developed by 3.5-month-olds.

To understand the source of this asymmetry, one needs to explore the basis on which infants categorize items. While there have been many studies describing infant categorization competence at various ages, there have been few mechanistic accounts of how the underlying categorical representations might emerge. One partial exception is the work by Quinn and Johnson (1997). These authors used a connectionist model to explore the order in which basic and super-ordinate level categories are acquired. Because the model was implemented as a working computer simulation, it is one of the first studies to ask how the mechanisms of learning constrain the nature of the categories that are acquired. Although this work explored how the characteristics of exemplars at different levels might dictate the order in which categories are acquired by infants as a whole, it did not directly address the issue of how categories are learned within the short-term testing sessions characteristic of many published categorization studies.

We believe that the way to a comprehensive synthesis of the numerous competence studies that abound in the infancy literature is to shift the debate to a mechanistic level. If the different studies are tapping into a common categorization ability, then there must exist a common set of mechanisms that can account for the observed behaviors. The search for a common set of mechanisms underlying performance on different tasks has already been successfully applied to explaining the causes of the exclusivity asymmetry mentioned above and a catastrophic interference effect in infant memory studies (Mareschal & French, 1997; Mareschal, French, & Quinn, submitted).

Mareschal & French (1997) and Mareschal et al (submitted) presented connectionist networks with the same cat and dog exemplars used to familiarize infants in the original Quinn et al. (1993) study. The networks developed the same exclusivity asymmetries as had the infants (i.e., the category of CAT excluded novel dogs, whereas the category of DOG did not exclude novel cats). This was accounted for in terms of the distribution of feature values in the familiarization stimuli and the fact that the connectionist networks developed internal representations reflecting the variability of the inputs they experienced. For almost all features, the distribution of

---

[1] A longer version of this paper will appear in Infancy.

CAT values was subsumed within the distribution of DOG values. The same mechanism was used to account for the fact that sometimes (but not always) material presented to infants during a retention interval leads to the catastrophic forgetting of the initial material. The model made the prediction that the subsequent learning of the DOG category would disrupt the prior learning of the CAT category, but that the subsequent learning of the CAT category would not disrupt the prior learning of the DOG category. This prediction was tested and found to be true for 3.5-month-olds (Mareschal & French, 1997; Mareschal, French, & Quinn, submitted). In short, the model demonstrated how the previously unrelated exclusivity asymmetry and interference effects were two sides of the same mechanistic coin.

This previous work establishes that autoassociators provide a good model of how categories overlap. However, one purpose of categorization is to parse the world into distinct units that are then acted on differently. Ultimately infants learn to separate out categories. In this paper, we will extend the previous work by exploring the basis on which distinct categories are developed by infants and connectionist networks given a series of exemplars. Younger (1985) showed that 10-month-olds could use the correlation between feature values to segregate items into separate categories. Although these results are based on presenting infants with line drawings of artificial animals, Younger (1990) found that infants could still use correlation information with natural kind images similar to those used in the Quinn et al. studies. We will explore whether the autoencoder connectionist architecture used to model the Quinn et al. data (Mareschal & French, 1997; Mareschal, French, Quinn, submitted) also responds to correlation information in the same way as infants.

The rest of this paper unfolds as follows. First we will describe in detail Younger's (1985) categorization studies with 10-month-olds. Next we will present connectionist simulations of categorization using the same stimuli as Younger used with her infants. We will present an illustration of the internal representations developed by the networks.

Category formation in 10-month-olds

The two simulations described below are attempts to model the behavior of 10-month-olds reported by Younger (1985). The network training regime is kept as close as possible to the infant familiarization conditions. Younger examined 10-month-olds' abilities to use the correlation between the variation of attributes to segregate items into categories. In the real world certain ranges of attribute values tend to co-occur. Thus, animals with long necks tend to have long legs whereas animals with short necks tend to have short legs. Younger examined whether infants could use these co-variation cues to segment artificial animal line drawings into separate categories.

In a first experiment, infants were familiarized with a set of exemplars. They were then tested with either: (a) an exemplar whose attribute values were the average of all the previously experienced values along each dimension, or (b) an exemplar containing the modal attribute values (i.e., the most frequently experienced values) along each dimension. Based on the finding that infants direct more attention to novel or unfamiliar stimuli, preference for a modal versus the average stimulus was interpreted as evidence that the infants had formed a single category from all the exemplars (as evidenced by the greater familiarity of the average stimulus). Preference for the average stimulus was interpreted as evidence that the infants had formed two categories (as indicated by the lesser familiarity of the average stimulus) since the boundary between correlated clusters lay on the average values. Younger found that 10-month-olds looked more at the modal stimuli when the familiarization set was unconstrained (i.e., all attribute values occurred with every other attribute value) suggesting that they had formed a single representation of the complete set of exemplars. However, the 10-month-olds looked more at the average stimuli when the familiarization set was constrained such that ranges of feature values were correlated suggesting that they had formed two distinct categories.

In a second experiment, Younger (1985) provided a more stringent test of category formation in infancy. In this experiment, the infants were presented with a constrained familiarization set (i.e., ranges of feature values were correlated across dimensions). However, the familiarization set was designed such that the modal stimulus was identical to the average stimulus. Infants were then tested with the modal/average stimulus and two stimuli with previously unseen attribute values but which were prototypical of the two possible categories contained within the familiarization set. Preference for the average/modal stimulus was interpreted as evidence that the infants had formed two categories (as indicated by the greater familiarity of the previously unseen stimuli) since the boundary between correlated clusters lay on the average/modal values. Preference for the previously unseen stimuli was interpreted as evidence that the infants had formed a single category from all the exemplars (as evidenced by the greater familiarity of the average/modal stimulus). Younger found that, under these conditions, 10-month-old infants looked longer at the average/modal stimuli suggesting that they had formed two distinct categories.

To model performance on these two experiments (in simulations 1 and 2 below respectively), the same artificial animal stimuli used by Younger were encoded for presentation to the networks. These animals were defined by their values along 4 dimensions: Leg length (ranging from 1.5 to 3.5 in intervals of 1.0), Neck length (ranging in value from 1.2 to 5.2 in intervals of 1.0), Tail length (ranging in value from 0.5 to 2.3 in intervals of 0.45), and Ear separation (ranging in values from 0.3 to 2.7 in intervals of 0.6). Because none of the attributes are intended to be more salient than any other attribute, each attribute was scaled to range between 0.0 to 1.0. This transformation ensures that the greater magnitude of one dimension (e.g., Ear separation) does not bias the networks to attend preferentially to that dimension. Normalization was achieved by dividing each attribute value by the maximum value along that dimension.

Networks were trained in batch mode. That is, all 8 familiarization items were presented as a batch to the network and the cumulative error was used to update the weights (to drive learning). This ensures that all the items in the familiarization set are weighted equally by the networks and is intended to reflect the fact that there were no significant changes in infant looking times across all familiarization trials. Batch learning also ensures that all order effects are averaged out.

Modeling habituation-dishabituation

Infant categorization tasks rely on preferential looking or habituation techniques based on the finding that infants direct more attention to unfamiliar or unexpected stimuli. The standard interpretation of this behavior is that infants are comparing an input stimulus to an internal representation of the same stimulus (e.g., Solokov, 1963; Charlseworth, 1969; Cohen, 1973). As long as there is a discrepancy between the information stored in the internal representation and the visual input, the infant continues to attend to the stimulus. While attending to the stimulus the infant updates its internal representation. When the information in the internal representation is no longer discrepant with the visual input, attention is directed elsewhere. When a familiar object is presented there is little or no attending because the infant already has a reliable internal representation of that object. In contrast, when an unfamiliar or unexpected object is presented, there is much attending because an internal representation has to be constructed or adjusted. The degree to which a novel object differs from existing internal representations determines the amount of adjusting that has to be done, and hence the duration of attention.

We used a connectionist autoencoder to model the relation between attention and representation construction. An autoencoder is a feedforward connectionist network with a single layer of hidden units. The network learns to reproduce on the output units the pattern of activation across the input units. Thus, the input signal also serves as the training signal for the output units. The number of hidden units must be smaller than the number of input or output units. This produces a bottleneck in the flow of information through the network. Learning in an autoencoder consists in developing a more compact internal representation of the input (at the hidden unit level) that is sufficiently reliable to reproduce all the information in the original input. Information is first compressed into an internal representation and then expanded to reproduce the original input. The successive cycles of training in the autoencoder are an iterative process by which a reliable internal representation of the input is developed. The reliability of the representation is tested by expanding it, and comparing the resulting predictions to the actual stimulus being encoded. Similar networks have been used to produce compressed representations of video images (Cottrell, Munro, & Zipser, 1988).

We suggest that during the period of captured attention infants are actively involved in an iterative process of encoding the visual input into an internal representation and then assessing that representation against the continuing perceptual input. This is accomplished by using the internal representation to predict what the properties of the stimulus are. As long as the representation fails to predict the stimulus properties, the infant continues to fixate the stimulus and to update the internal representations.

This modeling approach has several implications. It suggests that infant looking times are positively correlated with the network error. The greater the error, the longer the looking time. Stimuli presented for a very short time will be encoded less well than those presented for a longer period. However, prolonged exposure after error (attention) has fallen off will not improve memory of the stimulus. The degree to which error (looking time) increases on presentation of a novel object depends on the similarity between the novel object and the familiar object. Presenting a series of similar objects leads to a progressive error drop on future similar objects. All of this is true of both autoassociators (where output error is the measurable quantity) and infants (where looking time is the measurable quantity).

The modeling results reported below are based on the performance of a standard 4-3-4 (4 input units, 3 hidden units, and 4 output units) feedforward backpropagation network. The learning rate was set to 0.1 and momentum to 0.9. Networks were trained for a maximum of 200 epochs or until all output bits were within 0.2 of their targets. This was done to reflect the fact that in the Younger (1985) studies infants were shown pictures for a fixed duration of time rather than using a proportional looking time criterion.

Simulation 1

In this simulation 24 networks were presented with 8 stimuli in which the full range of values in one dimension occurred with the full range of values in the other dimension (the Broad condition). Another 24 networks were presented with the 8 stimuli in which restricted ranges of values were correlated (the Narrow condition). The networks in both conditions were then tested with stimuli made up of the average feature values or the modal feature values. Table 1 shows the normalized values defining the stimuli in the Broad and Narrow familiarization conditions, and the three test stimuli. Figure 1 shows the networks' response to the average and modal test stimuli when familiarized in either the Narrow or Broad conditions. As with the 10-month-olds, networks familiarized in the Narrow condition showed more error (preferred to look) when presented with the average test stimulus than the modal test stimuli. Similarly, as with the 10-month-olds, networks familiarized in the Broad condition showed more error (preferred to look) when presented with the modal test stimuli than the average test stimuli.

Table 1. Normalized familiarization and test stimuli (Exp. 1)

Familiarization Stimuli							
Broad Condition				Narrow Condition			
Legs	Neck	Tail	Ears	Legs	Neck	Tail	Ears
0.27	1.0	0.22	1.0	0.27	1.0	0.8	0.33
0.27	0.23	1.0	1.0	0.27	0.81	1.0	0.33
0.45	0.81	0.41	0.78	0.45	0.81	1.0	0.11
0.45	0.42	0.8	0.78	0.45	1.0	0.8	0.11
0.82	0.42	0.8	0.33	0.82	0.42	0.22	1.0
0.82	0.81	0.41	0.33	0.82	0.23	0.41	1.0
1.0	0.23	1.0	0.11	1.0	0.23	0.41	0.78
1.0	1.0	0.22	0.11	1.0	0.42	0.22	0.78

Test Stimuli					
Average		0.64	0.62	0.61	0.56
Modal1		0.27	1.0	1.0	0.11
Modal2		1.0	0.23	0.22	1.0

<u>Note</u>: Values are scaled to range from 0.0 to 1.0.

This was confirmed by an analysis of variance with one between-subject factor (Conditions: narrow or broad) and one within-subject factor (Stimulus: average or modal) which revealed a significant interaction of Condition x Stimulus ($\underline{F}$(1,46)=752, $\underline{p}$<.0001).

Internal category representation

This section describes the internal representations developed by the networks in Simulation 1and discusses how they lead to the observed preferential looking behaviors described above.

From a behavioral perspective, categorization can be said to have occurred when identifiably different exemplars are treated in the same way. In hidden unit space, members

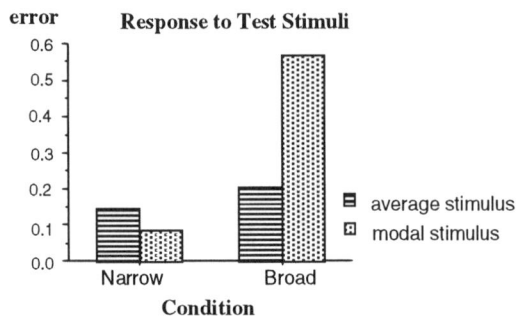

Figure 1. Responses to the average and modal test stimuli for networks familiarized in Broad and Narrow conditions.

of the same category will be mapped to points close together; they will elicit similar activation patterns across the hidden units. Members of different categories will be mapped to points further apart; they will elicit different activation patterns across the hidden units. Because members of a category produce similar hidden unit activation patterns, they will be responded to in a similar fashion by the output units. In contrast, members of a different category that produce different hidden unit activation patterns will be responded to differently by the output units.

Figure 2 shows the distribution of exemplars within the hidden unit space for a representative network trained in the Narrow and Broad conditions of Simulation 1. In the Narrow condition (Figure 2a), exemplars are grouped together in two distinct clusters. One cluster corresponds to those exemplars forming one category and the other cluster correspond to those exemplars forming the second category. The test exemplars are also plotted. Note that the two modal exemplars each fall within (or very close to) one of the category clusters whereas the average exemplar falls between the two clusters. This explains why there is more error (longer looking) to the average exemplar than to either of the modal exemplars. The modal patterns fall within areas that are well covered by the category representations, and hence, for which the network has already learned to make accurate responses. In contrast, the average pattern falls in an area that is not well covered, and hence, for which the network has no experience of making accurate responses.

Figure 2b shows the exemplars within hidden unit space for networks trained in the Broad condition. The internal representations are spread throughout the hidden unit space, reflecting the fact that the exemplars are maximally spread out. Remember that in this condition any feature value can occur with any other feature value. All three of the test stimuli (the average and modal stimuli) project to a similar location at the center of the space. This is because all three have comparable similarities (in terms of feature values) to all of the familiarization exemplars considered individually. There isn't the space in this article to discuss the different ways that similarity can be measured, but by referring to Table 1 we can see intuitively why the test stimuli have comparable similarities to all the familiarization exemplars. Because of the systematic structure of the familiarization set, the average stimulus has feature values that lie mid-way within the range of all possible values. Thus, it is about "half as similar" to any exemplar along any dimension. The modal stimuli have 2 out of the 4 feature values that tend to match the feature values of any particular exemplar. In some cases the match is exact and in others the match is approximate (i.e., both values are high or both values are low). The remaining two values always go in the opposite direction (i.e., the modal value is high when the exemplar value is low or *vice versa*). In short, the three test stimuli are comparably related to the exemplars in the familiarization set: the average stimulus because it has feature values mid-way between the possible range of feature values, and the modal stimuli because they share (approximately) 2 out of 4 feature values with every exemplar.

Finally, because the internal representations are located close to each other in hidden unit space, the network will tend to respond to them in a similar fashion. Since they are in sparsely populated region of the space, the network has little experience with decoding these types of internal

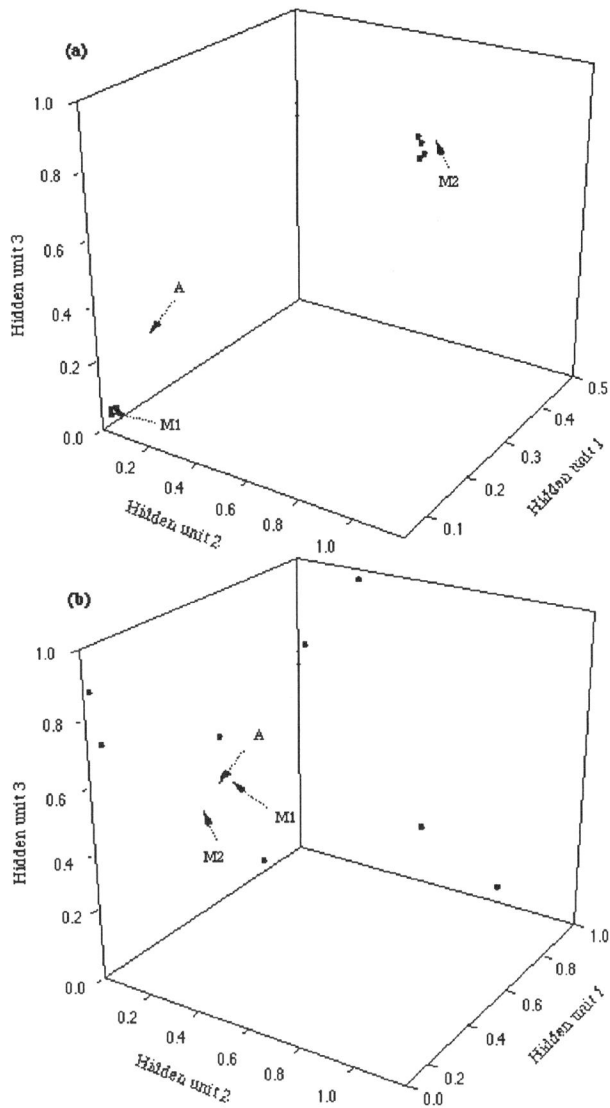

Figure 2. Locations of hidden-unit representations of members of items in the Narrow (2a) and Broad (2b) conditions. M1 and M2 are modal test items, A is an average test item.

representation. As a result, it will output an average of all the outputs it is familiar with. This is fine for the average stimulus since the correct response is precisely the average of all responses (remember that the autoassociation task requires the network to reproduce on the output units the original input values), but it is completely inappropriate for the modal stimuli whose feature values lie at the ends of the possible ranges. Hence, there is more error for the modal stimuli than the average stimulus.

Simulation 2

Younger's (1985) experiment 2 provides a stronger test of category segregation by equating the average and modal values for the full set of familiarization items. In this simulation 24 networks were familiarized with the 10

exemplars designed such that the modal and average values were the same. Under these conditions, the greater familiarity of a stimulus containing previously unseen values (but which are prototypes of two distinct categories) over the average/modal values, would provide strong evidence that the items had been segregated into two distinct categories. As in the Narrow condition of Experiment 1, familiarization stimuli were constructed such that restricted ranges of values were correlated. The networks were then tested with stimuli made up of the average/modal feature values or the previously unseen feature values. Table 2 shows the normalized values defining the stimuli in the Broad and Narrow familiarization phase, and the three test stimuli.

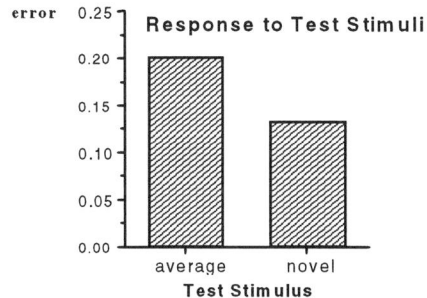

Figure 3. Network response to the average/modal and previously unseen test stimuli.

Figure 3 shows the networks response to the average/modal test stimulus and the previously unseen stimuli. As with 10-month-olds, networks showed more error (longer looking) when presented with the average/modal test stimulus than the stimuli with previously unseen values suggesting that they had formed two distinct categories. A two-way Student t-test revealed this difference was highly significant ($t(23)=6.59$, $p<.001$).

Discussion

This paper presented a model of correlation based categorization by 10-month-old infants. Simple autoencoder networks were exposed to the same stimuli used to test 10-month-olds. The familiarization regime was kept as close as possible to that used with the infants. The model's performance matched that of the infants. Both infants and networks used co-variation information (when available) to segregate items into separate categories.

The model makes the explicit prediction that, in general, looking time to the test stimuli in the Broad condition will be higher than that in the Narrow condition. This can be related to the structure of the internal representations developed by the networks. Encouraging trends that support this prediction can be found in the original Younger (1985) data. Exploration of the model's internal representations suggests that in the Broad condition, looking times are determined by the similarity of the test stimuli to the familiarization stimuli.

This model extends work reported by Mareschal & French (1997) and Mareschal et al. (submitted). It is a

341

model of category learning <u>within</u> a single test session. It leaves open questions of how this categorization ability develops. In other words, how does

Table 2 Normalized familiarization and test stimuli (Exp 2)

Familiarization Stimuli				
	Legs	Neck	Tail	Ears
	0.27	0.62	1.0	0.56
	0.27	1.0	0.61	0.11
	0.27	1.0	0.61	0.56
	0.64	1.0	1.0	0.11
	0.64	0.62	1.0	0.11
	0.64	0.62	0.22	1.0
	0.64	0.23	0.22	1.0
	1.0	0.23	0.61	0.56
	1.0	0.23	0.61	1.0
	1.0	0.62	0.22	0.56
Test Stimuli				
Average/Modal	0.64	0.62	0.61	0.56
Novel1	0.45	0.81	0.80	0.33
Novel2	0.82	0.42	0.41	0.78

<u>Note</u>: Values are scaled to range from 0.0 to 1.0

the developmental time scale interact with the course of learning during a task? Younger & Cohen (1986) describe a sequence of development from no use of correlation information at 4 months of age to the use abstract invariant relations at 10 months. Future modeling needs to explore how the ability to use correlation information comes about.

The complex relationship between the similarity of test stimuli to familiarization stimuli, and relative looking times can be explored through the model before making further empirical predictions. This illustrates the function of a model as a tool for reasoning about untested contexts. In the same way that a model bridge can help engineers reason about a real bridge, a computer model can help experimental psychologists reason about categorization. However, it is also important to understand that in the same way that a model bridge is never meant to embody all the characteristics of the real bridge, the computer model is not meant to capture all the richness of infant behavior.

We do not wish to claim that simple autoassociator networks can capture the full richness of infant categorization. There is far more to an infant than 11 neurons! This model is intended as an illustration of the computational properties of an associative system with distributed representations. There are other such systems that share many of the same computational properties (e.g. Grossberg, 1980; Knapp & Anderson, 1984).

Connectionism has inherited the Hebbian rather than the Hullian tradition of associative learning. What goes in inside the head is crucial for understanding behavior. Connectionist models force us to think about internal representations, to ask how they interact with each other, and to ask how they determine observed behaviors. We argue that connectionist methods are fruitful tools for exploring perceptual and cognitive development.

Finally, we wish to suggest that the observed infant categorization behaviors are inextricably linked to both the categorization mechanisms internal to the infant, and the properties of the external stimuli shown to the infants during the study. Thus, categorization is the product of an inextricable interaction between the infant and its environment. The computational characteristics of both subject and environment must be considered *in conjunction* to explain the observed behaviors.

References

Charlesworth, W. R. (1969). The role of surprise in cognitive development. In D. Elkind & J. Flavell (Eds.), *Studies in cognitive development. Essays in honor of Jean Paiget,* 257-314, Oxford, UK: Oxford University Press.

Cohen, L. B. (1973). A two-process model of infant visual attention. *Merrill-Palmer Quarterly, 19,* 157-180.

Cottrell, G. W., Munro, P., & Zipser, D. (1988). Image compression by backpropagation: an example of extensional programming. In N. E. Sharkey (Ed.), *Advances in cognitive science, Vol. 3.* Norwood, NJ: Ablex.

Grossberg, S. (1982). How does a brain build a cognitive code? *Psychological Review, 87,* 1-51.

Knapp, A. G. & Anderson, J. A. (1984). Theory of categorization based on distributed memory storage. *JEP:LMC, 10,* 616-637.

Mareschal, D. & French, R. M. (1997). A connectionist account of interference effects in early infant memory and categorization. In *Proc. of the 19th annual conference of the Cognitive Science Society* NJ: LEA,484-489.

Mareschal, D., French, R. M., & Quinn, P. C. (submitted). Interference effects in early infant memory and categorization: A connectionist model.

Quinn, P. C., Eimas, P. D., & Rosenkrantz, S. L. (1993). Evidence for representations of perceptually similar natural categories by 3-month-old and 4-month-old infants, <u>Perception, 22,</u> 463-475.

Quinn, P. C., & Johnson, M. H. (1997). The emergence of perceptual category representations in young infants, *J. of Exp. Child Psychology, 66,* 236-263.

Quinn, P. C., & Eimas, P. D. (1996). Perceptual organization and categorization in young infants. *Advances in infancy research, 10,* 1-36.

Rosch, E. (1975). Cognitive representations of semantic categories. *JEP:General 104,* 192-233.

Rosch, E., Mervis, C. B., Gray, W. D., Johnson, D. M., & Boyes-Braem, P. (1976). Basic objects in natural categories. *Cognitive Psychology, 8,* 382-439.

Solokov, E. N. (1963). *Perception and the conditioned reflex.* NJ: LEA.

Younger, B, A. (1985). The segregation of items into categories by ten-month-old infants, *Child Dev., 56,* 1574-1583.

Younger, B. A. (1990). Infants' detection of correlations among feature categories. *Child Dev., 61,* 614-620.

Younger, B. & Cohen, L. B. (1986). Developmental changes in infants' perception of correlation among attributes. *Child Development, 57,* 803-815.

Developmental Mechanisms in the Perception of Object Unity

Denis Mareschal (d.mareschal@bbk.ac.uk)
Centre for Brain and Cognitive Development; Department of Psychology
Birkbeck College; Malet St
London, WC1E 7HX UK

Scott P. Johnson (spj@psyc.tamu.edu)
Department of Psychology; Texas A&M University
College Station, TX 77843-4235 USA

Abstract

Neonates seem to perceive two ends of a partly occluded rod as two separate objects. However, by 4 months of age infants often appear to perceive a similar stimulus as comprised of a single unified object. Little is known about the mechanisms of development underlying this change. We constructed four connectionist models of how perception of object unity might develop in human infants, based on experience with a variety of visual cues known to be important to infants' performance. After exposure to a simulated visual environment, all the models were able to perceive a partly occluded object as unified. A rich perceptual environment and the presence of units for internal representations were found to improve generalization of acquired unity knowledge. These results lend plausibility to mechanistic accounts of human perceptual development, based on learning the statistical regularities inherent in the normal visual environment.

Introduction

Research exploring the development of object perception often employs simple displays while recording young infants' responses to object properties. For example, the display depicted in Figure 1 appears to adults to consist of a center-occluded rod, moving back and forth behind a nearer box. By 4 months of age, infants appear to perceive such a partly occluded rod as consisting of a single unified object. Earlier studies of the cues that support the perception of object unity concluded that common motion of the rod parts was the primary visual cue used by infants in determining that the rod parts belonged to a common object (Kellman & Spelke, 1983; Kellman, Spelke, & Short, 1986).

However, more recent studies have called this finding into question by systematically varying the cues available in occlusion displays. Three-dimensional depth cues were found to be not necessary for the perception of unity, given that 4-month-olds perceived object unity in a two-dimensional (computer generated) rod-and-box display, in which two rod parts moved above and below a stationary box, against a textured background (Johnson & Náñez, 1995). However, in the absence of a textured background, responses of 4-month-olds appeared to reflect ambiguity with respect to object unity (Johnson & Aslin, 1996). The relatability of the two rod segments (the fact that, if extended, they would meet behind the screen) was also found to be important to infants' perception of unity (Johnson & Aslin, 1996).

Currently, there are few accounts of how this fundamental skill develops. Spelke (1990; Spelke & Van de Walle, 1993) has suggested that young infants' object perception is tantamount to reasoning, in accordance with a set of core principles. However, infants' performance on object unity tasks appears to be strongly dependent on the presence or absence of motion, edge alignment, accretion and deletion of background texture, and other cues, implying that low-level perceptual variables influence the development of veridical object perception, rather than reasoning from core principles (Johnson & Aslin, 1996; Kellman & Spelke, 1983). Evidence from younger infants also casts doubt on accounts based on innate reasoning: Neonates appear to perceive the rod stimulus depicted in Figure 1 as arising from two disjoint objects (Slater et al., 1990). Two-month-olds have been found to perceive object unity, but only with additional perceptual support, relative to displays used with older infants (Johnson, 1997; Johnson & Aslin, 1995).

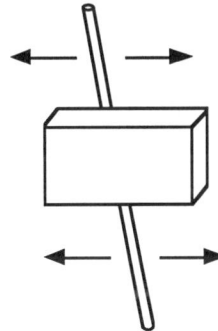

Figure 1. Typical occlusion stimulus

In this paper we explore whether the perception of object unity can be *learned* by experience with objects and events in early infancy. With this goal on mind, we developed a connectionist model that learned to identify a unified partly occluded stimulus from lower-level perceptual cues. The key idea is that when direct perception is not available, a percept of unity can be mediated by an appropriate combination of other supporting cues.

Connectionist models are ideal for modeling learning and development because they develop their own internal representations in response to environmental pressures (Elman et al., 1996). However, they are not simply *tabula rasa* learning machines. The learning that occurs can be strongly determined by innate constraints in the form of specific learning mechanisms or pre-wired connections.

The rest of this paper unfolds as follows. First, we will describe the general model architecture, the input to the model, and what drives learning. Several variations on architecture and training set were tested, but in this paper we report only on the best performing combination. Finally, implications for the development of perception of object unity in human infants will be discussed.

The Basic Architecture

Figure 2 illustrates the basic model architecture. The model receives input via a simple retina. The retinal information is processed by separate encapsulated modules. Each module identifies the presence of one of the following cues: (a) motion, (b) texture deletion and accretion, (c) t-junctions, (d) co-motion (i.e., simultaneous motion) in the upper and lower halves of the retina, (e) common motion in the upper and lower halves of the retina, (f) co-linearity of objects in the upper and lower halves of the retina, (g) the relatability of objects in the upper and lower halves of the retina (i.e., whether the objects' edges would meet if extended behind the occluder).

Unity is also a primitive, like the other cues, in that the network can immediately perceive it (via direct perception). Indeed, when testing the perception of unity in human infants, researchers assume that infants can distinguish single rods from disjoint rod parts. In the absence of direct perception (i.e., when the object(s) are partly occluded) the perception of unity is mediated by its association with other (directly perceivable) cues.

We do not wish to make the claim that a mediated route is unique to the percept of unity. In the brain, there is likely a highly complex and interactive network of connections allowing any number of not-directly-perceivable cues to be indirectly computed from the activation of other directly-perceivable cues. However, in the interest of clarity, we have considered only the one mediated route.

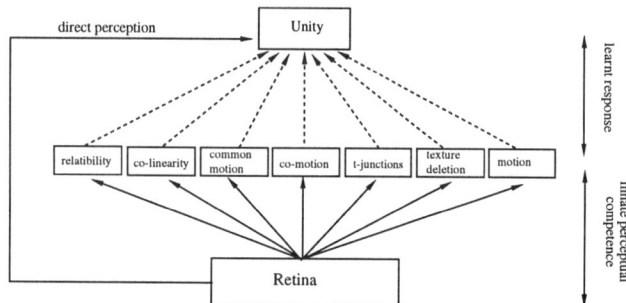

Figure 2. Schema of network architecture. Each module processes the retinal information separately and in parallel.

The bottom half of the network embodies innate abilities. We assume that neonates are able to perceive the components of each of these cues. Indirect evidence suggests that this is the case (Slater, 1995). There is no learning in any of these encapsulated modules. The top half of the network embodies the learning that can occur through interactions with the environment. The models discussed below illustrate several ways in which architectural and environmental constraints can be combined to guide the learning of object unity.

Input to the Model

The network "sees" a series of images from the world and responds with whether a perceived object is unified or not. The response is coded across two output units: (+1, -1) signifies that the object is unified; (-1, +1) signifies that the object is NOT unified. (+1, +1) or (-1, -1) are interpreted as an ambiguous response.

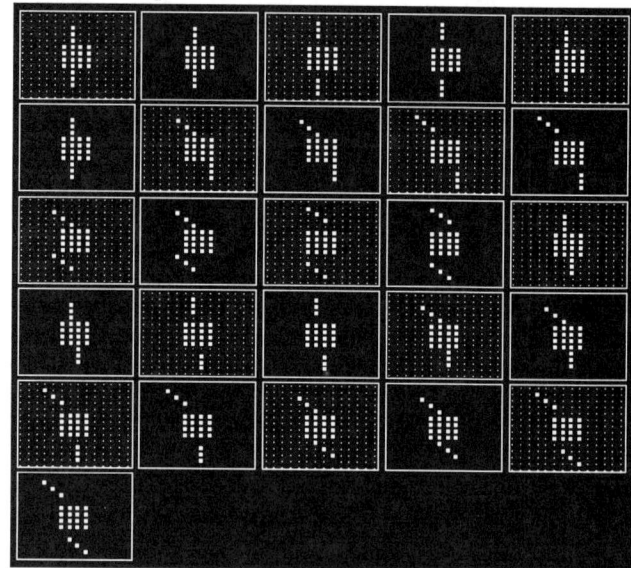

1: motion, texture, t-junction, co-motion, common motion, co-linearity, relatability	2: motion, t-junction, co-motion, common motion, co-linearity, relatability	3: motion, texture, co-motion, common motion, co-linearity, relatability	4: motion, co-motion, common motion, co-linearity, relatability	5: texture, t-junction, co-linearity, relatability
6: t-junction, co-linearity, relatability	7: motion, texture, t-junction, co-motion, common motion, relatability	8: motion, t-junction, co-motion, common motion, relatability	9: motion, texture, co-motion, common motion, relatability	10: motion, co-motion, common motion, relatability
11: motion, texture, t-junction, co-motion, common motion, co-linearity	12: motion, t-junction, co-motion, common motion, co-linearity	13: motion, texture, common motion, co-linearity	14: motion, co-motion, common motion, co-linearity	15: motion, texture, t-junction, co-linearity
16: motion, texture, t-junction, co-linearity	17: motion, texture, co-linearity	18: motion, co-linearity	19: motion, texture, t-junction	20: motion, t-junction
21: motion, texture	22: motion	23: motion, texture, t-junction, co-linearity	24: motion, t-junction, co-linearity	25: motion, texture, co-linearity
26: motion, co-linearity				

Figure 3. Complete set of 26 possible occlusion events. The cues present in the display are listed in the corresponding position of the table.

The input retina consists of a 196-bit vector mapping all the cells on a 14x14-unit grid. In the middle of the grid is a 4x4-unit occluder. All units corresponding to the position of the screen are given a value of 1. When background texture is required, all other units on the retina are given a value of 0.0 or 0.2, depending on the texture pattern. Units with values of 0.2 correspond to position on which there is a texture element (e.g., a dot). Units corresponding to the position of an object (i.e., rod or occluder) are given a value of 1.0. Figure 3 shows a snapshot taken from the "ambiguous" portion of all 26 events in the environment.

An event is made up of a series of snapshots like this one in which the rod moves progressively across the retina. All events begin with the object moving onto the retina from the side. We will call this the *unambiguous* portion of the event. The object then moves across the retina, passing behind the area occupied by the occluding screen. We will call this the *ambiguous* portion of the event. Finally, the object reappears on the other side of the screen and continues off the retina.

All events except 5 and 6 involve motion. The presence of texture, t-junctions, relatability and co-linearity are varied systematically. All events with motion involve motion in the upper and lower half of the retina (co-motion) but only half of those involve common motion. This leads to a total of 26 possible events.

The Perceptual Modules

These modules are not intended to model closely human neurophysiology. Although the modules embody general neural computational principles of summation, excitation, inhibition and local computation, they are also tailored to the specific nature of our networks' visual experience. Thus, they are not general models of the human visual system. However, they do embody some of the basic principles believed to underlie the computation of various visual cues (see Spillman & Werner, 1990 for a review). In essence, they are neurally plausible information processing modules.

All the modules compute the presence or absence of a relevant cue from the retinal image. The principles on which the modules function are as follows:

• *Motion detection module*

Takes the current retinal image and compares it to the previous retinal image. If there is a difference between the images, then there has been motion. If not, then there has not been any motion.

• *Texture module*

Counts the number of texture dots in the input image and compares it to the number of dots in the previous image. If there is a difference in the number of dots in the two images, then there has been deletion and/or accretion of texture elements.

• *T-junction module*

Focuses on the area immediately above and below the edge of the occluding screen and computes whether there is a gap everywhere along the edge.

• *Co-motion module*

Splits the retina into two halves and computes whether there is motion both in the upper half and in the lower half.

• *Common-motion module*

Splits the retina into two halves and computes whether there is the same direction of motion in the upper and the lower halves.

• *Co-linearity module*

Computes the tangent of the angle that the object's axis of principle length makes with the horizontal for both the upper and lower halves of the retina and compares these two.

• *Relatability module*

Computes whether the extension of the axis of principle length for objects in the upper and lower halves of the retina will intersect.

Although alignment has been manipulated as a cue in some infant studies, note that two objects are aligned *if and only if* they are co-linear and relatable. Thus, co-linearity and relatability are more primitive cues than alignment in the sense that the later cannot be computed without computing the former (at least implicitly), whereas the converse is not true: Both co-linearity and relatability can be computed independently of alignment.

As an example, the motion detection module is illustrated in Figure 4. This module takes in the retinal input and returns 1 if there is motion on the retina or 0 if there is no motion on the retina. The basic principle of the module is to take the current retinal image and compare it to the previous retinal image. If there is a difference between the images, then there has been motion. If not, then there has not been any motion.

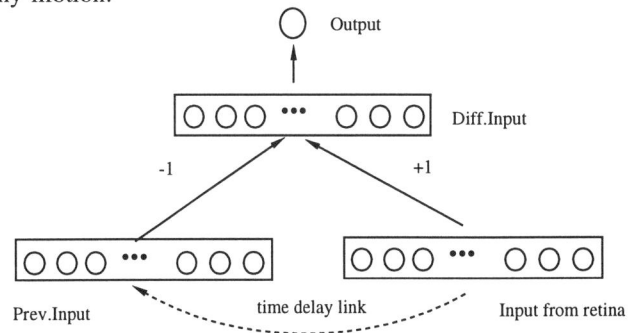

Figure 4. Schema of motion detection module

The Input is copied to a memory buffer (Prev.Input). A layer of hidden units (Diff.Input) computes the unit by unit difference between the current input and the previous input. The output unit then sums the activity across the hidden layer and returns 1 if there are any non-zero values or 0 if there are no non-zero values.

What drives learning?

Learning is partly driven by a feedback signal from the environment and partly driven by memory. When the object is visible, the environment provides immediate feedback about the unity of the object via the direct perception link. When the object is not completely visible, the environment cannot provide feedback about the unity of the object. To overcome this problem, the model has a short-term, rapidly decaying memory that encodes unity information obtained from direct perception (i.e., a kind of visual memory). Immediately following occlusion there is a clear trace of the state of the rod before occlusion. After a short delay, that information disappears and can no longer be used for learning.

This relation between direct perception and memory is embodied in the target signal used for training the weights:

$$T_i(t) = E_i(t) + \mu.T_i(t-1) \qquad (1)$$

with $-1 < T_i < +1$, $0 < \mu < 1$, and $E_i = 0.0$ when the rod is occluded.

E_i is the unity signal obtained from the environment by direct perception for output i, and μ is a parameter controlling the depth of memory. When $E_i = 0.0$ (i.e., there is no direct percept of unity), the target (T_i (t)) is derived entirely from the memory component $\mu.T_i(t-1)$, the second term in the right-hand side of the equation.

An interesting component of the model's performance is the mediated route's ability to predict whether a test event corresponds to a single unified object or to two disjoint objects. Network performance can be assessed either when direct perception is possible, or when it is not possible (i.e., on events 1, 2, 5, 6, 7, and 8 in Figure 3).

When direct perception is not possible, the model's prediction of unity (via the mediated route) can be compared to the modeler's knowledge of whether the test event arises from a unified object or not. When direct perception is possible, the model's prediction of unity (via the mediated route) can be compared to the signal coming from direct perception.

The degree to which the model's prediction is correct when direct perception is NOT possible reflects how well the model is able to respond to incomplete information. This can then be compared to infants' performance when faced with the same ambiguous stimuli. The degree to which the prediction is correct when direct perception IS possible reflects how well the network has extracted general information about objects that applies across its entire learning environment.

Model Performance

Four combinations of architecture and environment were explored. The models either had no hidden units between the output of the modules and the unity response units (see Figure 2), or they had three hidden units between the output of the modules and the unity response units. (The addition of hidden units provides the model with the power to develop internal representations of cue combinations and to represent to non–linearly separable cue relations, such as the exclusive-or, of a set of cues.) In addition, the models were trained either with a *basic* but ecologically plausible set of events (events 1, 2, 3, and 4), or were trained with an *enriched* set of events that sampled more evenly the space of possible object events (events 1, 2, and 17 though 22).

In the interest of brevity we only report the models' performance in the best conditions (complete results will be reported in a future paper). For the moment, it suffices to note that the presence of hidden units and a richly varying environment led to the best generalization performance.

Three hidden units were added between the perception modules and the output response units. The training environment was enriched to capture the fact that there are far more examples of disjoint objects in the real world than unified but occluded objects. Weights were adjusted by applying the backpropagation algorithm with learning rate = 0.5, momentum = 0.03, logistic activation functions, and memory (μ) = 0.4. The results are based on 10 replications with different random initial weights.

The networks very quickly learned (by 10 epochs) to perceive one or two objects during the unambiguous portion of the events. The unambiguous portion of the events corresponds to the time when the rod(s) were moving across the retina and had not yet reached a position of partial occlusion (one rod part above and the other below the occluder). That is, the rod or rod parts were directly visible.

Figure 5 shows the networks' performance when tested with the ambiguous portion of events 1, 3, 5, and 7, events used to test infants (see Johnson, 1997). Note that only event 1 was part of the original training set.

Veridical perception of a single unified object (event 1) was apparently rather difficult to learn (see Figure 5, top left). Over the first 4000 epochs, the networks perceived event 1 as arising from two disjoint objects. Then, from 5000 to 8000 epochs the majority of networks gradually came to perceive the event as arising from a single object. In contrast, the networks very quickly learned (by 100 epochs) to perceive two objects when the event was actually produced by two objects (event 3; see Figure 5, top right).

There was a different developmental profile in the absence of motion (event 5; see Figure 5, bottom left). Up to epoch 500, the networks perceived this event as arising from two disjoint objects. From epoch 1000 onwards, the networks consistently perceived the event as arising from a single unified object. When the object segments were misaligned but relatable (event 7; see Figure 5, bottom right), the pattern of development was rather different. Throughout development, the networks tended to perceive this event as arising from disjoint objects. At different times, only up to 20% of networks perceived a unified object, whereas the rest perceived two disjoint objects.

Network performance on these four events matches human performance very well (see Johnson, 1997 for review). Initially, the single object depicted in Figure 1 (event 1) was perceived as two disjoint objects (similar to human neonates). There was then a transition period in which either of two responses resulted. After more extensive training, the display was perceived as arising from a single object (similar to older infants and adults). Moreover, human neonates, like these networks, have been found to perceive separate rod parts undergoing common motion (event 3) as consisting of separate objects, after little exposure. Finally, the networks tended to perceive objects with misaligned edges as disjoint (event 7), similar to infants and adults.

The networks were effective in generalizing their "knowledge" to the complete set of test events. By the end of training, they responded correctly to 23 of the 26 test events when tested with the ambiguous portion of the event. That is, the mediated route made incorrect predictions on 3 events (events 22, 23 and 24). The response to events 23 and 24 were altogether incorrect, whereas half the networks provided the correct response to event 22 and half provided an incorrect response. Moreover, the networks performed very well on the unambiguous, visible segments of the trajectories. The mediated route produced the correct percept on 24 of the 26 events. In particular, the networks failed to respond appropriately on events 20 and 22. In the former case, four networks correctly predicted two objects while six predicted a single object, and in the latter case six networks correctly predicted two objects while four predicted a single object.

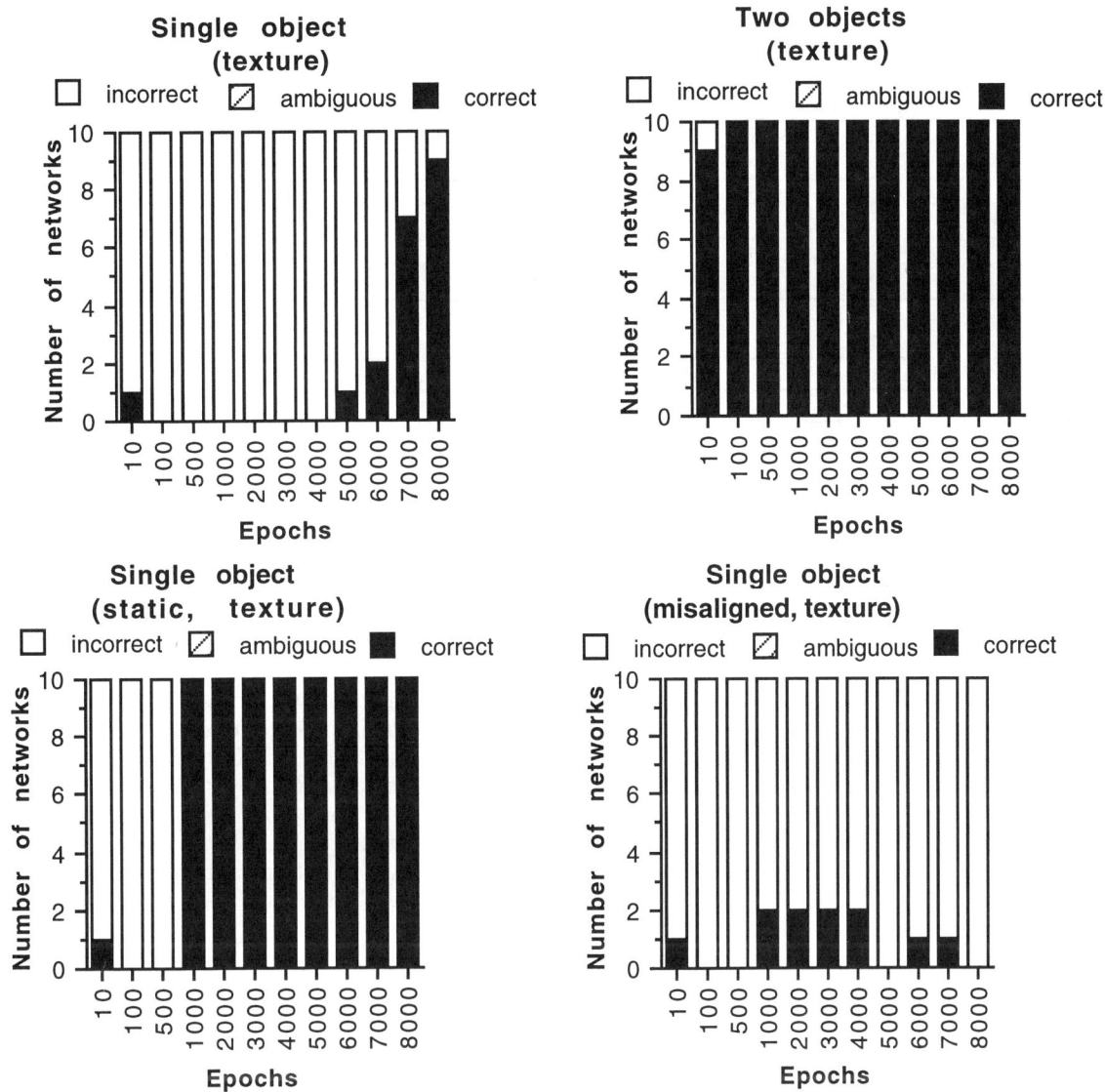

Figure 5. Number of network showing a correct response when tested with the ambiguous segment of events 1, 3, 5, and 7.

An examination of the internal representations developed across the hidden units allows us to explain these responses (Table 1). All three hidden units are used to encode the cue information, but the units have quite different effects on the output responses. Hidden unit 2 is strongly associated with the percept of one object whereas hidden units 1 and 3 are strongly associated with the percept of two objects. Hidden unit 3 has about twice as much impact as hidden unit 1.

The presence of T-junctions is strongly associated with hidden unit 2, and thereby with the percept of unity. Along this dimension, it is the dominant feature. Relatability and common motion are also positively associated with hidden unit 2 (and therefore to the percept of unity), but to a much weaker extent.

Hidden unit 3 is positively associated with all the cues. This means that the unit will almost always be active, whatever the percept. However, because the impact of hidden unit 3 on the output response is less than that of hidden unit 2 (e.g., -2.797 vs. 3.183), the activation of hidden unit 1 will determine which way the network responds. If Hidden unit 1 is active, then the total activation sent to the outputs will produce a "two object" response (activation of the not-unified output node) whereas if the Hidden unit 1 is not active, the total activation sent to the outputs will produce a "one object" response (activation of the unified output node).

Table 1. Connection weights in a representative network

	Relat-ability	Co-linearity	Common-motion	Co-motion	T-junction	Texture deletion	Motion	Bias unit
Hidden unit 1	-0.174	-1.178	-0.659	0.749	-0.237	-0.477	1.513	-0.266
Hidden unit 2	0.431	-0.861	0.555	-1.990	2.358	-0.918	-1.559	3.991
Hidden unit 3	0.825	2.058	1.741	0.208	2.491	0.946	1.010	-2.007

	Hidden unit 1	Hidden unit 2	Hidden unit 3	Bias
unified node	-1.601	3.183	-2.797	-0.101
NOT unified node	1.600	-3.183	2.769	0.101

Hidden unit 1 is positively associated with Motion and Co-motion. Thus if both of these are present, the unit will tend to fire and the response will tend toward the signalling of two objects. If either Motion or Co-motion is absent, hidden unit 1 will be weakened in activity and the network's response will tend toward two objects.

Discussion

Outcomes of these models suggest that perception of object unity can be *learned* rapidly through interaction with the environment. The networks respond to the statistical regularities in the environment: *No prior object representations are required.*

The models can be used to predict the type of percept that will arise from each of the conditions above. Rather than appealing to "core principles" that guide inferences about object unity, the resolution of ambiguous stimuli relies on the previous association of lower-level cues with direct percepts of unity. That is, a strong prediction of the model is that experience viewing unoccluded objects that are progressively occluded and unoccluded lies at the heart of learning to resolve ambiguous stimuli.

In these models, the use of backpropagation is not necessarily crucial; in principle, any algorithm that permits multi-layer learning would suffice (e.g., O'Reilly, 1998). However, there is a need for hidden units (the power for internal re-representation) for proper generalization of knowledge.

Finally, we believe that connectionist models are an effective means of investigating outstanding questions in developmental psychology, in this case by providing a mechanistic account of how learning to perceive object unity could occur.

References

Elman, J. L., Bates, E. A., Johnson, M. H., Karmiloff-Smith, A., Parisi, D., & Plunkett, K. (1996). *Rethinking innateness: A connectionist perspective on development.* Cambridge, MA: MIT Press.

Johnson, S. P. (1997). Young infants' perception of object unity: Implications for development of attentional and cognitive skills. *Current Directions in Psychological Science, 6,* 5-11.

Johnson, S. P., & Aslin, R. N. (1996). Perception of object unity in young infants: The roles of motion, depth, and orientation. *Cognitive Development, 11,* 161-180.

Johnson, S. P., & Aslin, R. N. (1995). Perception of object unity in 2-month-old infants. *Developmental Psychology, 31,* 739-745.

Johnson, S. P. & Náñez, J. E. (1995). Young infants' perception of object unity in two-dimensional displays. *Infant Behavior and Development, 18,* 133-143.

Kellman, P. J., & Spelke, E. S. (1983). Perception of partly occluded objects in infancy. *Cognitive Psychology, 15,* 483-524.

Kellman, P. J., Spelke, E. S., & Short, K. R. (1986). Infant perception of object unity from translatory motion in depth and vertical translation. *Child Development, 57,* 72-86.

O'Reilly, R. C. (1998). Six principles for biologically-based computational models of cortical cognition. *Trends in Cognitive Sciences, 2,* 455-462.

Slater, A. (1995). Visual perception and memory at birth. In C. Rovee-Collier and L. P. Lipsitt (Eds.), *Advances in infancy research* (Vol. 9). Norwood, NJ: Ablex.

Slater, A. M., Morison, V., Somers, M., Mattock, A., Brown, E., & Taylor, D. (1990). Newborn and older infants' perception of partly occluded objects. *Infant Behavior and Development, 13,* 33-49.

Spelke, E. S. (1990). Principles of object perception. *Cognitive Science, 14,* 29-56.

Spelke, E. S., & Van de Walle, G. (1993). Perceiving and reasoning about objects: Insights from infants. In N. Eilan, R. A. McCarthy, & B. Brewer (Eds.), *Spatial representation: Problems in philosophy and psychology.* Oxford: Blackwell.

Spillman, L. & Werner, J. S. (1990). *Visual perception. The neuropsychological foundations.* London, UK: Academic Press.

Activation of Russian and English Cohorts
During Bilingual Spoken Word Recognition

Viorica Marian (vm22@cornell.edu)
Michael Spivey (mjs41@cornell.edu)
Department of Psychology
Cornell University
Ithaca, NY 14853

Abstract

The traditional language switch hypothesis, according to which bilinguals can selectively activate and deactivate either language, has been repeatedly challenged in recent studies. In particular, an eyetracking experiment investigating spoken language processing suggests that bilinguals maintain both languages active in parallel even during monolingual input. The present study extends this finding to circumstances exhibiting between-language competition, within-language competition, or both. In this experiment, we find evidence for lexical items in the first language interfering with processing of the second language. We find that, in addition to competing activation between languages, bilinguals (like monolinguals) encounter competition within languages. Moreover, the results suggest that when simultaneous competition is encountered from items in both languages, within-language competition may be stronger than between-language competition. It appears that a bilingual's irrelevant language continues to be processed even when not actively used. However, this phenomenon is considerably influenced by language mode, even when such variables as word frequency, phonetic overlap, and language preference are taken into account.

Introduction

With the majority of the world's population speaking more than one language (Romaine, 1995), studying bilingualism and polyglotism can provide valuable insights into human cognition and language. The capability of one cognitive system to successfully manage two languages is striking. Do bilinguals use the two languages independently, alternating between them by turning them on and off, or do they constantly keep both languages active and process the two in parallel at all times? The traditional language switch hypothesis, according to which bilinguals are able to selectively activate and deactivate their two languages (Gerard and Scarborough, 1989; McNamara & Kushnir, 1971), has been challenged by a number of recent findings. Parallel activation has been inferred from early studies using the bilingual Stroop task (Chen & Ho, 1986; Preston & Lambert, 1969), from studies using code-switching (Grainger, 1993, Grainger & Dijkstra, 1992; Li, 1996; Soares & Grosjean, 1984), interlingual homographs (Dijkstra, van Jaarsveld, & ten Brinke, 1998; Dijkstra,

Timmermans, and Schriefers, 1997), and cognates (DeGroot & Nas, 1991; Kroll & Stewart, 1994). More recently, evidence for parallel activation comes from masked orthographic priming (Bijeljac-Babic, Biardeau, & Grainger, 1997) and interlingual neighbors (van Heuven, Dijkstra, & Grainger, in press). Attempts to extend the hypothesis of generalized lexical access from the visual to the phonological domain using code-switching were also encouraging (Nas, 1983; Doctor & Klein, 1992).

However, the most compelling evidence supporting automatic activation of both lexicons during monolingual input comes from research investigating spoken language processing in bilinguals using eye-tracking (Spivey & Marian, 1999). The eye-tracking technique, merging input from both the visual and auditory modalities, allows one to index the activation of a second language non-linguistically. It allows testing processing of both languages without compromising a monolingual mode, something that is otherwise difficult. Spivey and Marian (1999) presented subjects with a visual display consisting of four objects and asked them to manipulate a target object. The onset of the name of the target object bore phonetic similarity to the name of one of the other objects in the other language. For example, when instructed "Poloji *marku* nije krestika" ("Put the stamp below the cross"), the visual display in front of the subject contained, among other objects, a *marker*, the name of which shares several phonemes with "marka," the Russian word for stamp. It was found that, while processing the target word "marka," Russian-English bilinguals made eye movements to the between-language competitor "marker," thus suggesting that lexical items in both languages were activated simultaneously, even though only one language was being used.

While this work shows cross-linguistic activation in bilingual spoken language processing, it leaves many questions unanswered. It does not, for example, indicate what happens during bilingual language processing when competition takes place in more realistic visual context. In every-day environments, while processing spoken language, bilinguals are surrounded by a multitude of objects, some of them do indeed compete cross-linguistically, while others compete within the same language, and, finally, in many cases the competition may come simultaneously from both languages. What happens under these circumstances of simultaneous competition from both languages?

A second question arises from the fact that Spivey and Marian (1999) found an asymmetry in their results, with stronger competition from the second language into the first language. We hypothesized that the asymmetry may have been due to the slight preference for English among the bilinguals in that study, or to the subjects' current immersion in an English-speaking environment. We tested both of these hypotheses in the present study, by manipulating the strength of the language mode and by considering language preference as a variable. We aimed to replicate the between-language competition phenomenon observed by Spivey and Marian (1999), while trying to instill a better Russian language mode to examine if competition from the first into the second language can occur.

Finally, we wanted to see if bilinguals would show within-language competition from items whose name bore phonetic overlap in the same language, as did English monolinguals (Allopenna, Magnuson, & Tanenhaus, 1998; Spivey-Knowlton, 1996), thus extending the within-language competition phenomenon to the bilingual domain. In sum, the present study followed three goals: (1) replicate the robustness of the between-language activation phenomenon observed by Spivey and Marian (1999), (2) investigate within-language parallel activation in bilinguals, and, most importantly, (3) investigate language processing in the case of simultaneous competition from both languages. We also took into account such variables as word frequency, amount of phonetic overlap, language preference, and language mode.

Method

Participants

Fifteen Russian-English bilinguals participated in the study, 5 males and 10 females. All were students at Cornell University, their mean age was 22.04 years (SD=3.77). All were Russian-English bilinguals, born in former Soviet Union, who immigrated to the United States at the mean age of 15.62 years (SD=3.65). Six participants indicated that Russian was their preferred language of use at the time when the study was conducted, five participants indicated that English was their preferred language of use, and four indicated no language preference. Informed consent was obtained and participants' rights were protected. All participants were paid for their participation.

Apparatus

A headband-mounted ISCAN eyetracker was used to record the subjects eye movements during the experiment. A scene camera, yoked with the view of the tracked eye, provided an image to the subject's field of view. A second camera, focused on the center of the pupil and the corneal reflection, provided an infrared image of the left eye. Gaze position was indicated by crosshairs superimposed over the image generated by the scene camera. The output was recorded onto a Hi8 VCR with frame-by-frame playback.

All the instructions were pre-recorded by a fluent Russian-English bilingual speaker who acquired both languages in early childhood and had no noticeable accent in either language. SoundEdit was used to record and play the video record for data analysis. the speech files, and the audio record was synchronized with

Design and Stimuli

All participants were tested in two parts, a Russian part and an English part, with order counterbalanced across subjects. Each part consisted of 20 trials, equally distributed across four conditions. In the no-competition condition, of the four objects presented in the display one was the target object and three were control filler objects. The target object was the object actively named in the experiment. The filler objects were objects whose name did not overlap with the name of the target object in either language. This first condition served as the baseline for all analyses.

In the between-language competition condition, of the four objects presented in the display, one was the target object, one was the between-language competitor, and two were filler objects. The between-language competitor was an object whose name in the other language carried phonemic overlap with the name of the target object. For example, during the English part, if "speaker" was the target object, then "speachki," the Russian word for "matches," was the between-language competitor. The name of the between-language competitor was never spoken in either language during the entire experiment.

In the within-language competitor condition, of the four objects presented in the display, one was the target object, one was the within-language competitor, and two were filler objects. The within-language competitor was the object whose name carried phonemic similarity to the target object in the same language. For example, during the English part, if the target object was a "speaker," then the within-language competitor was a "spear." Similarly, in the Russian part, if the target object was "speachki" (matches), then the within-language competitor was "speatsy" (knitting needles). The name of the within-language competitor was also never spoken during the entire experiment.

Finally, in the fourth condition, of the four objects presented in the display, one was a target object, one was a between-language competitor, one was a within-language competitor, and one was a filler object. This fourth condition allowed testing a situation in which simultaneous between-language and within-language competitions take place. The four conditions were intermixed throughout the experiment. A complete list of all target items, between-language competitors, and within-language competitors for both Russian and English can be found in Table 1.

To avoid potential confounds, we considered in each trial such variables as the physical similarity of the items, the word frequency in the two languages, and the amount of phonemic overlap. During the experiment, similarities in the physical properties (size, shape, color) of a target object and one of the filler objects in the display were noticed for one Russian set (a baloon and a pear) and one English set (a greeting card and a napkin). As a result, all trials containing these sets were discarded from analyses.

To compute word frequency, we used three sources. For the English language, we used Zeno et al.'s 1995 Word Frequency Guide based on a corpus of 17,274,580 word tokens. For the Russian words, we used Lenngren's (Ed.) 1993 frequency

Table 1. English and Russian Target and Competitor Words.

Russian Target			English Competitor	Russian Competitor		
Спички	[spichki]	(Matches)	Spear	Спицы	[spitsi]	(Knitting needles)
Бусы	[boosy]	(Beaded necklace)	Book	Бубен	[buben]	(Tambourine)
Черепаха	[cherepaha]	(Turtle)	Chair	Червяк	[chervyak]	(Worm)
Марка	[marka]	(Stamp)	Marker	Марля	[marlya]	(Cheese cloth)
Баран	[baran]	(Ram)	Barbed Wire	Бархат	[barhat]	(Velvet)
Платье	[platye]	(Dress)	Plum	Плащ	[plashch]	(Raincoat)
Гайка	[gaika]	(Nut)	Gun	Галстук	[galstuk]	(Necktie)
Карта	[karta]	(Map)	Car	Картошка	[kartoshka]	(Potato)
Лак для погтей	[Lak dlea nogtei]	(Nailpolish)	Lock	Лопата	[lapata]	(Shovel)

English Target			English Competitor	Russian Competitor		
Speaker			Spear	Спички	(spichki)	Matches
Boot			Book	Бубен	(buben)	Tambourine
Shark			Shovel	Шарик	(sharik)	Balloon
Chair			Chess set	Черепаха	(cherepaha)	Turtle
Marker			Marbles	Марка	(marka)	Stamp
Barbed wire			Bark	Бархат	(barhat)	Velvet
Plug			Plum	Платье	(platye)	Dress
Gun			Gum	Гайка	(gaika)	Nut
Lock			Lobster	Лак для погтей	(Lak dlea nogtei)	Nailpolish

dictionary based on a corpus of 1,000,000 word tokens, as well as Zasorina's (Ed.) 1977 frequency dictionary based on 40,000 word tokens. In addition, we translated all the Russian words and considered the frequency of the translated words in the English language, and translated all the English words into Russian and considered the frequency of the translated words in the Russian language based on the two Russian sources. Both raw word frequency and word frequency adjusted for dispersion were considered. None of the performed analyses showed any statistically significant difference in the frequency of the target and competitor items in either language.

Phonetic overlap was analyzed based on the raw number of overlapping phonemes, as well as the proportion of word overlap across two items. T-tests were performed both across the two languages as well as for each language separately. Overlap within and between languages was considered. No significant difference was found in any of the analyses. The only marginally significant difference was observed for proportion of overlap (but not the raw number of phonemes) in the analysis of English, where overlap with the target word was higher for within-language competitors than for between-language competitors; t=2.26, p<0.1.

Procedure

Upon arrival to the lab, subjects were greeted and interacted with exclusively in the language appropriate for that part of the experiment, including the written consent form and spoken instructions. In addition, to better instill a Russian language mode, popular Russian songs were played via a tape-recorder at the beginning of the Russian session. This manipulation was done to increase the strength of the Russian mode, since Russian was a passive language in the subjects'

overall environment at the time and to test whether significant competition from Russian into English can occur.

Participants were then seated at arm's length from a 61 cm by 61 cm white board set on a table. The board was divided into 9 equal squares and a black cross in the center square served as a neutral fixation point. After the eyetracker was calibrated, each participant was presented with 40 trials, 20 in each language. In each trial, subjects were asked to look at the central cross, followed by instructions to manipulate the target object. An example of verbatim instructions is "Pick up the speaker. Put the speaker below the cross" in English, and "Podnimite speachki. Polojite speachki nije krestika." ("Pick up the matches. Put the matches below the cross") in Russian. Subjects then received two filler instructions to manipulate filler objects in order to prevent them from guessing the hypothesis of the study.

Results

To test between-language competition, we compared the percentage of trials when subjects made eye movements to the between-language competitor with the percentage of trials with looks to the same square when a control filler was present. Analyses of variance were computed by subjects (F1) and by items (F2). As shown in Figure 1, combined across languages, subjects were more likely to make eye movements to the between-language competitor (16%) than to the control filler (7%); $F1(1,14)=7.22$, $p<0.02$; $F2(1,16)=3.48$, p=0.08. Separate by-language analyses indicated that, during English trials, subjects made eye movements to the between-language Russian

351

Between-Language Competition

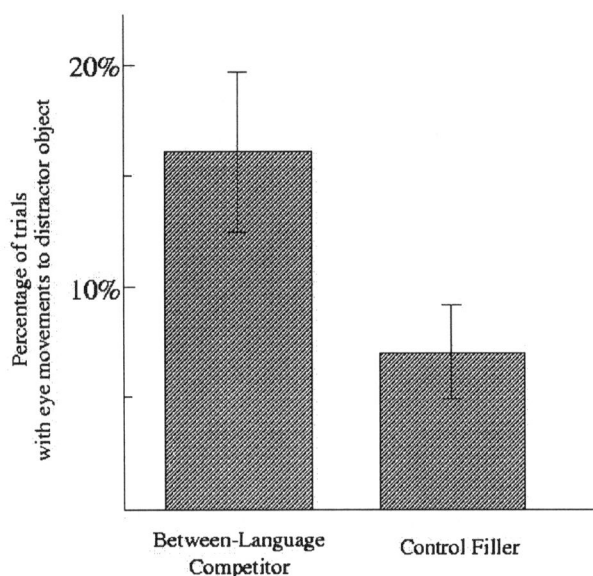

Figure 1. Mean proportion of trials where subjects looked at the between-language competitors in condition 2 compared to the filler object in the same square in condition 1.

competitor 21% of the time and to the control filler 6% of the time, $F(1,14)=12.43$, $p<0.01$; $F2(1,8)=11.85$, $p<0.01$. During Russian trials, subjects made eye movements to the between-language English competitor and the control filler, 11% and 9% of the time, respectively, $F1(1,14)=0.2$, $p>0.1$; $F2(1,8)=0.02$, $p>0.1$. This asymmetry in the results, with stronger competition from Russian items during English trials, follows the mirror reverse pattern to that observed by Spivey and Marian (1999), suggesting that manipulating the strength of the Russian language mode in the present study had a direct influence on the strength of between-language competition[1].

To test within-language competition, the proportion of trials with eye movements to the within-language competitor was compared to the proportion of trials with eye movements to the same square when a control filler was present. As shown in Figure 2, subjects looked significantly more often to the within-language competitor (21% of the time) than to the control filler (9%); $F1(1,14)=7.6$, $p<0.02$; $F2(1,16)=3.74$, $p=0.07$. Analyses by language showed that, during English trials, subjects looked at the within-language competitor 21% of the the time and at the control filler 4% of the time; $F1(1,14)=8.33$, $p<0.02$; $F2(1,8)=4.67$, $p=0.06$. During Russian trials, subjects looked at the within-language competitor 21% of the time and at the control filler 13% of the time, $F1(1, 14)=1.34$, $p>0.1$; $F2<1$.

---

[1] Analyses of language preference showed no significant effect on between-language competition. A closer look revealed, however, that Russian-preferring subjects showed absolutely no interference from English into Russian. Therefore, part of the reason why English did not interfere with Russian is because only the English-preferring subjects contributed to that interference.

Within-Language Competition

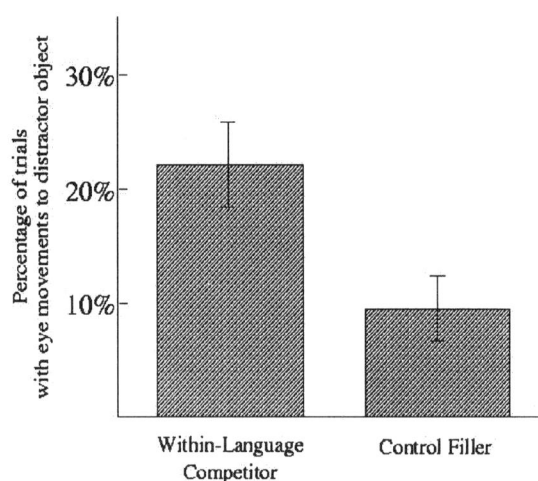

Figure 2. Mean proportion of trials where subjects looked at the within-language competitors in condition 3 compared to the filler object in the same square in condition 1.

The results of simultaneous within-language competition and between-language competition are shown in Figure 3. Combined across both languages, subjects looked at the within-language competitor 19% of the time, compared to the control filler 9%; $F1(1,14)=6.76$, $p<0.05$; $F2(1,16)=3.58$, $p=0.08$. Subjects looked at the between-language competitor 13% of the time, compared to the control filler 7% of the time, $F1(1,14)=2.2$, $p>0.1$; $F2(1,16)=0.58$, $p>0.1$. These results suggest that, in situations of simultaneous within-language and between-language competition, the within-language competition may be stronger.

A comparison of the first half and the second half of the experiment, in order to test for order effects, did not reveal a significant difference in the patterns of competition before the language switch and after it. Within-language competition, between language competition, and simultaneous within- and between-language competition were relatively similar in the two parts.

Discussion

The present study reinforces the hypothesis of parallel activation of two languages during bilingual language processing. While the main effect of between-language competition across languages replicates the basic finding of Spivey and Marian (1999), the asymmetries in the results of by-language analyses in the two studies are in exactly opposite directions. Spivey and Marian found that competition was stronger from the second language into the first, while in the present study competition was stronger from the first language into the second. These differences can be reconciled if we consider more carefully the

352

Simultaneous Competition From Both Languages

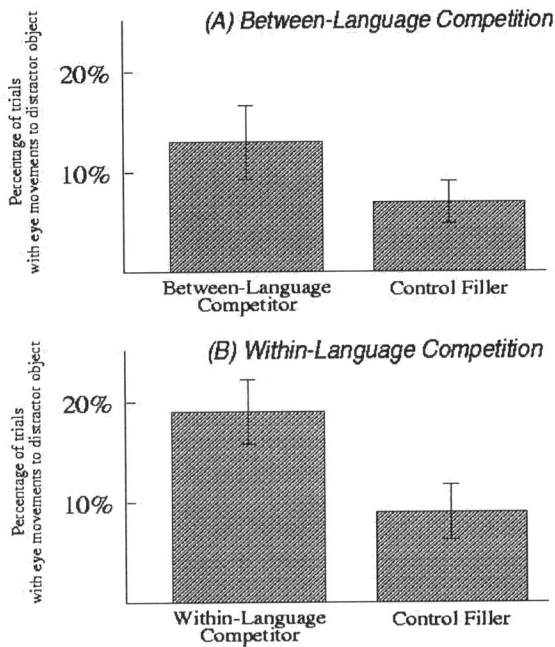

Figure 3. Mean proportion of trials in which subjects looked at the within-language competitors and at the between-language competitors in condition 4 compared to the filler objects in the same squares in condition 1.

participants' language modes in the two studies. In the Spivey and Marian study, it is likely that participants were in a stronger second-language (English) mode, since their everyday surrounding environment was predominantly English (the campus of an American University) and since efforts to instill a Russian mode were minimal. In the present study, however, the strength of the Russian language mode was manipulated by exposing subjects to Russian music that carried strong cultural and linguistic connotations. These were Russian children songs, songs from popular TV shows, and otherwise popular songs that were known to all former Soviet citizens. Subjects systematically commented on the Russian songs they heard. It is possible that such distinct cultural components were strong triggers of a Russian language mode. Thus, it appears that, when greater time and effort is put into instilling a Russian environment in the laboratory, the English session shows reliable interference from Russian. This pattern of results is consistent with the proposed role of language mode in on-line comprehension (Grosjean, in press; Grosjean, 1999; Soares & Grosjean, 1984). Apparently, as the present study suggests, subtle manipulations of the linguistic environment may alter a bilingual's language mode and dramatically influence the pattern of language processing.

Our study was the first to consider eyetracking evidence for within-language competition in bilinguals. The results suggest that, similar to monolingual English speakers, bilingual Russian-English speakers also encounter within-language lexical competition. These findings provide further

support for the robustness of the within-language competition phenomenon. However, analyses by language show that this main effect was driven chiefly by within-language competition in the English language. It is worth noting, though, that, in Russian, the average frequency of the target words was about twice that of the within-language Russian competitors (depending upon the source of the frequency ratings, the respective frequencies are 41.7 (SD=52.7) (Zeno et al., 1995), 44.2 (SD=48.8) (Zasorina, 1977), and 31.9 (SD=28.6) (Lenngren, 1993) words per million for the targets, compared to 17.5 (SD=24.1), 19.9 (SD=22.9), and 21.2 (SD=25.6) words per million for the competitors, respectively). These differences, although not statistically significant, should not be completely discarded as possible factors in contributing to what might be a Type II error in analyses on within-language competition in Russian.

Finally, an investigation of simultaneous within- and between-language competition during bilingual spoken language processing suggests that competition may be stronger within than between languages. The results of the study permit us to conclude that, during spoken language processing, the interplay between visual and auditory information processing can cause bilingual listeners to encounter competition from items between their two languages, as well as within their two languages.

To account for the possibility that the results of the present study were influenced by the linguistic items we chose, word frequency and amount of phonetic overlap were both considered. There was no significant differences in word frequency and amount of phonetic overlap between targets and competitors within or between languages. Similarly, computations on the items in the Spivey & Marian (1999) study did not reveal any significant differences in phonetic overlap between the two languages, or in word frequency between target and competitor items. These results suggest that, while amount of phonetic overlap may play a role in item activation, it is unlikely to have been driving the phenomena reported here. Nevertheless, discarding the two variables as irrelevant to bilingual language processing would be a mistake. A careful study in which word frequency and phonetic overlap were independent variables may prove insightful. For example, if competition occurs from more than one item within the same language, word frequency and amount of phonetic overlap are likely to be important determinants of strength of competition from each item[2].

To conclude, the results of our study provide strong support for parallel spoken language processing in

---

[2] In comparing the present findings with those of Spivey and Marian (1999), we also observed differences in the latencies of eye movements to the target object depending upon the argument structure of the verb in the spoken instructions. Subjects' saccade latencies were shorter for single-argument verbs such as "Pick up" (as in "Pick up the marker. Put it below the cross.") than for double-argument verbs such as "Put" (as in "Put the marker below the cross"). Interestingly, this pattern of findings was also observed with English monolinguals by Spivey & Tanenhaus (submitted).

bilinguals. Together with other findings of parallel processing in bilinguals, these results suggest that bilingual listeners simultaneously accumulate phonemic input into both of their lexicons (with it presumably cascading to higher levels of representation) as a spoken word unfolds in real time, even when in a monolingual situation. Future efforts in investigating bilingual spoken language processing will need to focus on more systematic manipulation of the language mode, as well as on investigating activation of translated item names and their phonetic neighbors. Moreover, explicit tests of competing theories will eventually require computational models of bilingual language processing (e.g., van Heuven, Dijkstra, & Grainger, 1998) with architectures that can accommodate this apparent interaction between the two languages all the way up and down the processing stream.

Acknowledgements

The authors wish to thank Eugene Shildkrot, Marina Basina, and Olga Kats for their help in constructing stimuli, data collection, and coding, and Tobey Doeleman for help with the phonetic count. This work was supported by a Cognitive Studies Dissertation Fellowship (VM) and a Sloan Foundation Research Fellowship in Neuroscience (MS).

References

Allopenna, P., Magnuson, J., & Tanenhaus, M. (1998). Tracking the time course of spoken word recognition using eye movements: Evidence for continuous mapping models. *Journal of Memory and Language, 38,* 419-439.

Bijeljac-Babic, R., Biardeau, A., & Grainger, J. (1997). Masked orthographic priming in bilingual word recognition. *Memory and Cognition, 25,* 447-457.

Chen, H. C., & Ho, C. (1986). Development of Stroop interference in Chinese-English bilinguals. *Journal of Experimental Psychology: Learning, Memory, and Cognition, 12,* 397-401.

DeGroot, A. & Nas, G. (1991). Lexical representation of cognates and noncognates in compound bilinguals. *Journal of Memory and Language, 30,* 90-123.

Dijkstra, A., Timmermans, M., & Schriefers, H. (1997). Cross-language effects on bilingual homograph recognition. Manuscript in preparation.

Dijkstra, A., van Jaarsveld, H., & ten Brinke, S. (1998). Interlingual homograph recognition: Effects of task demands and language intermixing. *Bilingualism, 1* (1).

Doctor, E. A., & Klein, D. (1992). Phonological processing in bilingual word recognition. In R. J. Harris (Ed.), *Cognitive Processing in Bilinguals.* Amsterdam: Elsevier, pp. 237-252.

Gerard, L. D., & Scarborough, D. L. (1989). Language-specific lexical access of homographs by bilinguals. *Journal of Experimental Psychology: Learning, Memory, and Cognition, 15* (2), 305-313.

Grainger, J. (1993). Visual word recognition in bilinguals. In R. Schreuder & B. Weltens (Eds.), *The Bilingual Lexicon.* Amsterdam: John Benjamins.

Grainger, J., & Dijkstra, A. (1992). On the representation and use of language information in bilinguals. In R. J. Harris (Ed.), *Cognitive Processing in Bilinguals.* Amsterdam: Elsevier.

Grosjean, F. (in press). Studying bilinguals: Methodological and conceptual issues. *Bilingualism: Language and Cognition.*

Grosjean, F. (1999). The bilingual's language modes. In J. L. Nicol and T. D. Langendoen (Eds.). *Language Processing in the Bilingual.* Oxford: Blackwell, 1999.

Kroll, J., & Stewart, E. (1994). Category interference in translation and picture naming: Evidence for asymmetric connections between bilingual memory representations. *Journal of Memory and Language, 33,* 149-174.

Lenngren, L. (Ed.). (1993). *Chastotnyi Slovari Sovremennogo Russkogo Yazyka (Frequency Dictionary of Modern Russian Language).* Uppsala.

Li, P. (1996). Spoken word recognition of code-switched words by Chinese-English bilinguals. *Journal of Memory and Language, 35,* 757-774.

MacNamara, J., & Kushnir, S. (1971). Linguistic independence of bilinguals: The input switch. *Journal of Verbal Learning and Verbal Behavior, 10,* 480-487.

Nas, G. (1983). Visual word recognition in bilinguals: Evidence for a cooperation between visual and sound based codes during access to a common lexical store. *Journal of Verbal Learning and Verbal Behavior, 22,* 526-534.

Preston, M., & Lambert, W. (1969). Interlingual interference in a bilingual version of the Stroop Color-Word Task, *Journal of Verbal Learning and Verbal Behavior, 8,* 295-301.

Soares, C., & Grosjean, F. (1984). Bilinguals in a monolingual and a bilingual speech mode: The effect on lexical access. *Memory and Cognition, 12 (4),* 380-386.

Spivey, M., & Marian, V. (1999). Crosstalk between native and second languages: Partial activation of an irrelevant lexicon. *Psychological Science, 10,* 281-284.

Spivey, M., & Tanenhaus, M. (submitted). Integration of visual context and linguistic information in resolving temporary ambiguities in spoken language comprehension.

Tanenhaus, M., Spivey-Knowlton, M., Eberhard, K., & Sedivy, J. (1995). Integration of visual and linguistic information during spoken language comprehension. *Science, 268,* 1632-1634.

van Heuven, W. J. B., Dijkstra, T., & Grainger, J. (1998). Orthographic neighborhood effects in bilingual word recognition. *Journal of Memory and Language, 39,* 458-483.

Zasorina, L. N. (Ed.) (1977). *Chastotnyi Slovari Russkogo Yazyka (Frequency Dictionary of Russian Language).* Moscow: Russkii Yazyk.

Zeno, S., Ivens, S., Millard, R., Duvvuri, R. (1995). *The Educator's Word Frequency Guide.* Touchstone Applied Science Associates.

Language-Dependent Memory

Viorica Marian (vm22@cornell.edu)
Department of Psychology
Cornell University
Ithaca, NY 14853

Abstract

Research with bilinguals may provide insights into the complex relationship between autobiographical memory and language. The present paper suggests existence of language-dependent memory, where linguistic factors at the time of recall influence memory retrieval. In two experiments, Russian-English bilingual immigrants were interviewed using the word-prompt technique. In the first experiment, bilinguals retrieved more autobiographical memories when there was a match between language of recall and language of encoding than when there was a mismatch. More memories from the period before immigration were recalled in Russian than in English and more memories from the United States were recalled in English than in Russian. To examine the mechanisms underlying these results, the ambiance language and the word-prompt language were considered separately in the second experiment. Both the linguistic ambiance and the word prompt were found to influence recall of autobiographical memories. These results, and particularly the effect of linguistic ambiance on recall, suggest language-dependent memory.

Language-Dependent Memory

Since the encoding specificity principle was first introduced a quarter of a century ago (Tulving & Thomson, 1973), cognitive psychologists have investigated context-dependent memory in a number of different domains (see Davies & Thomson, 1988, for reviews). Environmental-context-dependent memory has been found when the encoding and retrieval contexts have been drastically manipulated, as, for example, when encoding took place under water and recall took place on land (Godden & Baddeley, 1975). Context-dependent memory has also been studied with more moderate changes of context, as, for example, in studies where encoding and recall took place in different rooms (e.g., Smith, Glenberg, & Bjork, 1978; Smith, 1979). Similarly, mood-state dependent memory has been supported and challenged by numerous studies with both clinical and normal populations, using both real-life as well as laboratory memory material (for reviews, see Bower, 1981, Christianson, 1992, and Eich, 1995). The main idea behind the work on context-dependent memory is that memories become more accessible when environmental or internal factors at retrieval are similar to circumstances at encoding.

In the present paper, it is proposed that language may lead to similar state-dependency effects. The concept of language-dependent memory is introduced, suggesting that the linguistic ambiance at the time of retrieval influences recall of memories, so that memories become more accessible when language at retrieval matches language at encoding. In its broad application, language-dependent memory may occur both for semantic and for episodic memories. It may be observed with monolingual as well as multilingual speakers, when factors in the linguistic environment undergo significant change.

The phenomenon of language-dependent memory appears to be particularly evident in the memories recalled by bilinguals. Bilinguals have an opportunity to experience life events while using different languages, and the drastic differences between the two linguistic environments may be particularly conducive to studying language dependency. If language is a key factor in memory encoding, then language of retrieval should affect the accessibility of a bilingual's memories. Indeed, anecdotal evidence supporting this hypothesis is abundant. For example, when asked for her apartment number in her native language, a bilingual who has lived in the United States for over a decade has erroneously provided the number of the flat in her native country. Upon quickly correcting herself, she explained the immediate response by saying that the number of the old apartment just popped into her mind because of the way the question was asked. In another case, a bilingual child who learned a French song while on vacation in France could not recall the song upon return to the United States. However, once finding himself in a French-speaking environment again, he remembered the song without any effort. The phenomenon that appears to underlie such effects underlies the present investigation.

In general, most work on the relationship between language and memory in bilinguals has been at the lexical level (see deGroot & Kroll, 1997, for reviews). For autobiographical recall, the idea that bilinguals' memories may be more accessible in the language of origin has been proposed mainly from a clinical psychoanalytic prospective (Aragno and Schlachet, 1996; Javier, 1995; Javier, Barroso and Munoz, 1993). For example, Aragno and Schlachet (1996) review three cases in which using the first language with bilingual clients resulted is more successful psychoanalysis sessions. Similarly, Javier, Barroso and Munoz (1993) found that the memories of five bilingual speakers were richer and more elaborate when accessed in the language in which the events took place than in the other language.

Empirical studies attempting to demonstrate similar effects have traditionally used the word-prompt technique (Galton, 1879, Robinson, 1976), where subjects were presented with a word and asked to produce the first memory that came to mind. Bugelski (1977) found that, when prompted with English words, Spanish-English bilinguals accessed 70% of "thoughts" primarily from the mature life period and 13% of "thoughts" from childhood. When prompted with the same words in Spanish, 43% of the "thoughts" were from maturity and 45% were from childhood.

In a study examining the role of cultural transition and language on memory encoding and recall, Otoya (1987) cued bilinguals with 10 word prompts in each language. Of these, six were translation equivalents of each other. She found that English words triggered later memories than Spanish words, the difference being significant for 3 words.

Contrary to Bugelski (1977) and Otoya (1987), Schrauf and Rubin (1997) found no such tendency to recall earlier memories when prompted in the first language than when prompted in the second language with elderly Spanish-English bilinguals. A possible explanation for Schrauf and Rubin's (1997) failure to find language-dependent memory is that age at which the event took place may not be a sensitive measure of language effects in bilinguals who used both languages concurrently for most of their lives. In Schrauf and Rubin's study, participants' mean age at immigration was about 28 years, and their mean age at the time when the experiment took place was 64.58. During the post-immigration period of almost 40 years, participants used both Spanish and English in parallel. Memories from that post-immigration period may have been encoded at times when English, or Spanish, or both languages were spoken. And, with the mean age from which memories were accessed considerably later than immigration (39.79 years in Spanish and 40.55 years in English), age may indeed not show any language effects. Instead, it may be more useful to establish, for each memory accessed, what language was spoken at the time when the remembered event took place.

The match between encoding and retrieval languages, and its effect on the memories retrieved, became the main focus of the present investigation. The effect of language on recall of autobiographical memories was examined through two experimental studies. We hypothesized that individuals who acquired a second language later in life will exhibit language-dependent memory, eliciting more memories from times when the first language was spoken when interviewed in the first language, and more memories from times when the second language was spoken when interviewed in the second language. The first experiment sought to establish the existence of language-dependent memory. In the second experiment,

the mechanisms underlying this phenomenon were examined more carefully.

One advantage of the present research over previous work is that it actively tried to avoid the demand characteristics that arise when subjects get cues in two different languages while participating in what they know to be a memory experiment. To avoid this problem, the present experiment used a cover task: our Russian-English bilingual subjects were led to believe that we were interested in the characteristic narrative styles of the two languages rather than in the particular memories they recalled. Participants were told that the experiment is part of a larger linguistic study in which properties of narratives in different languages are investigated. They were told that in this case we are examining Germanic and Slavic languages and that we would like them to provide a few narratives for our multi-language database. Participants were also led to believe that the reason for using word prompts was to assist them in coming up with stories.

Another advantage of our study was that subjects were not prompted with the same words in both languages. Instead, the word prompts were presented in the two languages in counterbalanced order across subjects. Finally, in addition to comparing the ages from which memories were accessed in different languages, we also considered the match between language of encoding and language of retrieval. If participants accessed more memories when the language of retrieval matched the language of encoding, then language-dependent memory is concluded.

For purposes of clarity and simplicity, in the present paper, a memory is called a *Russian memory* if it refers to an event that comes from a time when Russian was spoken by, to, or around the participant. A memory is called an *English memory* if it refers to an event that comes from a time when English was spoken by, to, or around the participant. And a memory is called a *Mixed memory* if it comes from a time when a mixture of both Russian and English was spoken by, to, or around the participant.

Experiment 1

The first experiment investigated whether Russian-English bilingual immigrants show language-dependent memory in their recall of autobiographical events. We hypothesized that bilinguals would access more memories when the language of recall matched the language of retrieval.

Methods

Participants

Twenty Cornell students participated in the experiment, 9 females and 11 males. All were Russian-English bilinguals, fluent in both languages, who had immigrated to the U.S. at the mean age of 14.2 years (SD=4.1). Their mean age at the time of the experiment was 21.8 years (SD=2.9). Four subjects indicated that Russian was their preferred language of communication, twelve subjects indicated that English was their preferred language of communication, and four subjects stated no language preference.

Materials

Sixteen Russian/English pairs of cue words were selected such that each member of a pair was the direct translation of the other. Pilot work had shown that these words were effective prompts for autobiographical memories. The following 16 prompt words were used: summer, neighbors, birthday, cat, doctor, getting lost, frightened, bride, snow, friends, holiday, dog, blood, contest, laughing, and newborn. Their Russian translations, respectively, were: лето, соседи, день рождения, кошка, врач, потеряться, испугаться, невеста, снег, друзья, праздник, собака, кровь, конкурс, смеяться, and новорожденный. Each participant received 8 of these prompt words in one language, and 8 in the other language.

Procedure

Each participant was interviewed individually; all interviews were tape-recorded. Upon arrival, participants were welcomed to a study of "storytelling in different languages" and told that we were comparing the psycholinguistic properties of Russian and English narratives. They were asked to tell brief stories of specific events from their lives. They were also told that, since it may be difficult to come up with numerous stories upon request, they will be prompted with some words, to facilitate the storytelling process. In each case, their task was to describe an event from their own life that the prompt had brought to mind. They were encouraged to respond as quickly as possible and to tell the first story that came to mind. Disguising the study as one of "storytelling in different languages" was necessary in order to prevent subjects from guessing the real focus of the experiment. With this and other similar instructions, indeed none of the participants identified the experiment to be a study of memory, as became evident during the post-experimental debriefing.

Participants were interviewed in two parts, an English part and a Russian part. The order of the languages was counterbalanced across subjects. In each part, participants received 8 word prompts in the assigned language. An effort was made to establish a very definite linguistic milieu for each language. Both the experimenter and the participant spoke only in the assigned language. In each part, the participant began by reading and signing a consent form in the appropriate language, followed by a warm-up task. In the warm-up task for the first part, the participant was asked to tell four stories from four specific periods of his or her life (no cue words were given). In the warm-up for the second part, the participant was asked to describe the experience of immigrating to the U.S. in some detail. The warm up tasks were mainly used to get the subject comfortable with the target language, to indicate that code-switching (using words from the other language) is not acceptable, and to make sure that they provide specific events from their lives, rather than loose associations, preferences,

thoughts, or opinions. After all memories were collected, each participant was asked to indicate what language s/he spoke, was spoken to, or was surrounded by at the time when the event took place and to estimate his or her age at the time.

Results

A total of 318 autobiographical memories were collected. Of these, 160 were memories of events that were experienced at a time when Russian was spoken, either before or after immigrating to the U.S. (Russian memories). Ninety-two were memories of events that were experienced at a time when English was spoken after immigrating to the U.S. (English memories). The remaining 66 were memories of events that were experienced at a time when a mixture of Russian and English was spoken; all except one are post-immigration memories (Mixed memories).

Analyses on the ages from which subjects drew their memories showed that participants consistently recalled memories from an earlier age (before emigration) when interviewed in Russian (M=13.1 years) than when interviewed in English (after immigration, M=16.1 years), $t(19)=4.829$, $p<0.001$.

Separate analyses on each type of memories showed that participants accessed more Russian memories when interviewed in Russian (5.15, SD=2.28) than when interviewed in English (2.85, SD=1.69), $t(19)=4.479$, $p<0.001$. Similarly, participants accessed more English memories when interviewed in English (3.35, SD=1.57) than when interviewed in Russian (1.3, SD=1.26), $t(19)=5.712$, $p<0.001$. Mixed Russian-English memories were accessed more or less equally in Russian (1.45, SD=1.32) and in English (1.75, SD=1.44), $t(19)=1.071$, $p>0.1$. No significant effect of order was found. The results of the first experiment are shown in Figure 1.

Finally, the distribution of memories across the lifespan showed a drop in the number of memories accessed from the period immediately following immigration.

Discussion

These results indicate that bilinguals typically access different autobiographical memories in their two languages. Bilingual speakers are more likely to retrieve memories in which the language at the time of retrieval matches the original language of the event. It appears that this effect is not simply a result of demand characteristics, but rather a robust relationship between the language of retrieval and the autobiographical memories retrieved.

Two possible mechanisms may account for the observed language-dependent memory. On the one hand, it is possible that the language milieu at the time of retrieval may establish a mindset that functions like the "states" of state-specific memory. In this case, the important congruity is between the language ambiance at encoding and at retrieval. On the other hand, the effect may depend on the cue word itself, and its specific congruity with words that were spoken at the time of the original event. The second experiment was designed to investigate these two hypotheses.

(a) Russian Memories

*p < 0.001

(b) English Memories

*p < 0.001

(c) Mixed Russian & English Memories

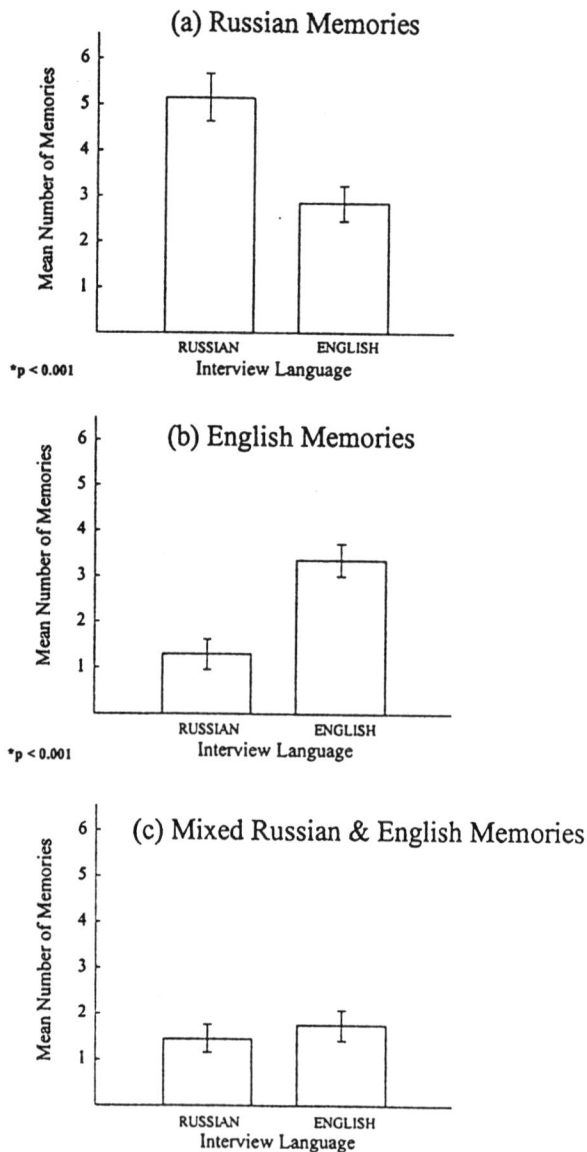

Figure 1: Number of Russian, English and Mixed memories retrieved in each language in Experiment 1.

Experiment 2

The possible mechanisms underlying language-dependent memory were explored by examining the effect of the language ambiance and the effect of the cue words separately from each other. While the interviews continued to be conducted in either the first or the second language, the word prompts did not always correspond to the language of the interview. That is, when interviewed in Russian, half of the word prompts were given in Russian and half of the word prompts were given in English. Regardless of the language of the word-prompt, participants were always to respond and the experimenter was always to interact in the language of the interview. This second experiment was designed to investigate which of the two variables—language ambiance or word-prompt language—was responsible for the language-dependent recall of autobiographical memories.

Methods

Participants

Twenty-four Russian-English bilingual Cornell students participated in the study, 12 males and 12 females. Their mean age at the time of the experiment was 20.2 (SD=2.0), their mean age at the time when they immigrated to the U.S. was 13.4 (SD=2.4). Five indicated that Russian was their preferred language at the time when the experiment was conducted, 13 indicated that English was their preferred language, and 6 indicated no language preference.

Procedure

The procedure used in the second experiment was identical to the one used in experiment 1. Again, the experiment was disguised as a study of storytelling in different languages, and again participants had to undergo a warm-up task in each language. The main difference in this experiment was the separation of ambiance language from word prompt language. The ambiance language refers to the language of the interview, that in which both the experimenter and the participant spoke, and that in which the stories were narrated. However, in each interview, the experimenter pronounced half of the prompt words in the other language. For example, a participant may be interviewed in Russian first and in English second. Of the 8 word prompts used in the Russian interview, only the first, the fourth, the sixth, and the seventh were presented in Russian (ABBA design). The remaining 4 word prompts were presented in English. Thus, autobiographical recall was tested in four conditions: (1) Russian ambiance—Russian word prompt, (2) Russian ambiance—English word prompt, (3) English ambiance—Russian word prompt, and (4) English ambiance—English word-prompt. The same sixteen Russian/English pairs of cue words as in the first experiment were used in experiment two.

Results

A total of 384 memories were collected in the second experiment. Two hundred and forty eight were memories of events from times when Russian was spoken, 91 were memories of events from times when English was spoken, and 45 were memories of events from times when a mixture of Russian and English was spoken. The number of memories accessed in each condition (by ambiance and word prompt language) are shown in Table 1.

A 2x2x2 analysis of variance (ambiance language X word prompt language X order) showed that recall of Russian memories was significantly influenced by the language of the ambiance and by the language of the word prompt. Participants recalled more Russian memories when interviewed in a Russian ambiance (5.9, SD=1.5) than when interviewed in an English ambiance (4.5, SD=1.8), $F(1, 22)=13.306$, $p=0.001$. Participants also recalled more Russian memories when prompted with Russian word prompts (5.6, SD=1.6), than when prompted with English word prompts

Table 1: Total number of Russian, English and Mixed memories retrieved in each condition in Experiment 2. (RA–Russian Ambiance; EA–English Ambiance; RW–Russian Word Prompts; EW–English Word Prompts.)

a. Russian Memories Retrieved

	RA	EA	Total
RW	75	59	134
EW	56	48	114
Total	141	107	N=248

b. English Memories Retrieved

	RA	EA	Total
RW	10	27	37
EW	17	37	54
Total	27	64	N=91

c. Mixed Memories Retrieved

	RA	EA	Total
RW	11	10	21
EW	13	11	24
Total	24	21	N=45

(4.7, SD=1.5), $F(1,22)=6.875$, $p<0.05$. There was no interaction between the effect of ambiance language and the effect of word prompt language, $F(1,22)=0.048$, $p>0.1$. The order in which the two languages were used did not affect recall of Russian memories and did not interact with the ambiance and word prompt effects.

A similar analysis on English memories showed that recall of English memories was also significantly affected by both ambiance language and word prompt language. Participants recalled more English memories when interviewed in an English ambiance (2.7,SD=1.6) than when interviewed in a Russian ambiance (1.1,SD=1.2), $F(1,22)=25.83$, $p<0.001$, and more English memories when prompted with English word prompts (2.3, SD=1.3) than when prompted with Russian word prompts (1.5, SD=1.4), $F(1,22)=8.3$, $p<0.01$. No interaction between the ambiance language and the word prompt language was found, $F(1,22)=0.108$, $p>0.1$. The effect of order on recall of English memories was non-significant and did not interact with ambiance and word-prompt effects.

Analyses on recall of Mixed memories revealed no effect of ambiance language, word prompt language, order, or interactions between any of these factors. Mixed memories were accessed about equally when the ambiance was Russian (1, SD=0.8) as when the ambiance was English (0.9, SD=0.9), $F(1,22)=0.279$, $p>0.1$, and

when the word prompt was Russian (0.9, SD=0.8) as when the word prompt was English (1, SD=0.8), $F(1,22)=0.4$, $p>0.1$.

Since both ambiance language and word prompt language had significant effects on recall of autobiographical memories, analyses on the amplitude of each effect were performed to determine which of the two was stronger. A t-test comparing the word prompt effect versus the ambiance language effect showed that ambiance language had a significantly stronger effect than word prompt language on recall of English memories, $t(23)=2.318$, $p<0.05$. For Russian memories, the effect was in the same direction, but did not reach significance, $t(23)=1.127$, $p>0.05$.

Finally, the distribution of memories accessed in the second experiment replicated the pattern observed in the first experiment, and showed that fewer memories were recalled from the period immediately following immigration.

Discussion

Analyses on the effect of ambiance language on recall showed that more memories were accessed when the language in which the event was experienced matched the language of the ambiance, and analyses on the effect of word prompt language on recall showed that more memories were accessed when the language in which the event was experienced matched the language of the word prompt. No interaction was observed between ambiance language and word prompt language, or between these two factors and order. Instead, the effects of ambiance and word order were cumulative. Thus, both interview language and word prompt language were found to influence recall. The effect of ambiance language tended to be stronger, significantly so for English memories.

An interesting observation emerged from the inspection of the distribution of memories across the lifespan. In both experiments, there is a drop in the number of memories accessed from the period immediately following immigration. A number of possible hypotheses can be advanced to explain why those memories were less accessible. It is possible, for example, that the cognitive load present during the period immediately following immigration is higher, not allowing for memories to be processed in the same manner and to the same depth as during the more usual cognitive load. These differences in cognitive processing at the time of encoding may make the memories less accessible. Another explanation may be that memories encoded during that period are less reliant on language, less language-loaded, and therefore linguistic cues such as word prompts will not trigger recall as successfully. Yet another explanation can be linked to the schema theory, according to which a person seeks to integrate new events into existing frameworks or schemata (Bartlett, 1932). If these frameworks rely on language, however, then the drop in post-immigration memories can be explained with the fact that many of the newly experienced memories do not fit existing schemas. For example, the American experiences of using drive-throughs, writing checks, using credit cards and ATMs, shopping at a mall or for groceries, are only some of the instances that required new or different schemas. Once relatively operational schemas have been created, new events can again be integrated into existing frameworks and are recalled more easily, as indicated by the observed increase in memories after just a few post-immigration years[1].

Conclusions

Bower (1981) proposed that different emotions can create dissociative states in such a way that when a person "feels angry and aggressive, one set of memories, beliefs, and competencies are activated; when he feels romantically or sexually aroused, a different set of memories, beliefs and competencies are activated" (p.147). Perhaps we can think in the same manner about language. Changing the linguistic environment may lead to alterations in recall, whether the change is from language to language, or within the same language, as in child-directed speech and baby-talk, or across domains of use. All these are venues that remain to be pursued in the future. To extend this work to semantic memory and knowledge, future efforts will include investigating whether semantic information learned in one language becomes more accessible when recall takes place in the same language. Studies that experimentally extend the language-dependent memory phenomenon to monolingual speakers may gradually reveal which specific factors in one's linguistic environment can facilitate access of memories. Finally, it is even possible that changes in the linguistic environments may lead to altered self-perception, with language and culture intertwined together in their effect on cognition. Studies of the latter may, for example, consider the content of freely recalled memories in the two languages (e.g., comparing references to self versus others, comparing references to internal states, or comparing references indicative of gender attitudes). With research possibilities looming large, the idea of language-dependent memory may prove to be a useful tool for studying the complex relationship between memory and language.

---

[1] Although the length of time it takes to show an increase in autobiographical memories after immigration varies from subject to subject, it appears that, on average, this increase occurred about 3 to 4 years after immigration. These findings may be useful in thinking about childhood amnesia. The two patterns of recall--immigrants learning a second language upon immersion into a new culture and children acquiring their first language early in life--may exhibit some similar patterns. The schema/framework theory may be one way to explain these results. Future research may show whether these similarities are real and whether they could provide interesting insights into the complex relationship between autobiographical memory and language.

Acknowledgements
The author wishes to thank Dr. Ulric Neisser for support and comments during this research. Partial support for this work was provided by a President's Council of Cornell Women Dissertation Grant and by a Cognitive Studies Dissertation Fellowship to the author.

References

Aragno, A. & Schlachet, P. (1996). Accessibility of early experience through the language of origin: A theoretical integration. *Psychoanalytic Psychology, 13,* 23-34.

Bartlett, F.C. (1932). *Remembering.* Cambridge, England: Cambridge University Press.

Bower, G. H. (1981). Mood and memory. *American Psychologist, 36,* 2, 129-148.

Bugelski, B. R. (1977). Imagery and verbal behavior. *Journal of Mental Imagery, 1,* 39-52.

Christianson. S.A. (Ed.) (1992*). Handbook of emotion and memory.* Hillsdale, N.J.: Erlbaum.

Davies, G.M., & Thomson, D. M. (Eds.). (1988). *Memory in Context: Context in Memory.* Chichester, U.K.: Wiley.

DeGroot, A. M. B., & Kroll, J. F. (1997). *Tutorials in bilingualism: Psycholinguistic Perspectives.* Mahwah, New Jersey: Lawrence Erlbaum.

Eich, E. (1995). Searching for mood dependent memory. *Psychological Science, 6,* 2, 67-75.

Galton, F. (1879). Psychometric experiments. *Brain, 2,* 149-162.

Godden, D.R., & Baddeley, A.D. (1975). Context-dependent memory in two natural environments: On land and underwater. *The British Journal of Psychology, 66,* 325-331.

Javier, R. A. (1995). Vicissitudes of autobiographical memories in a bilingual analysis. *Psychoanalytic Psychology, 12* (3), 429-438.

Javier, R. A., Barroso, F., & Munoz, M. A. (1993). Autobiographical memory in bilinguals. *Journal of Psycholinguistic Research, 18,* 449-472.

Otoya, M. T. (1987). *A study of personal memories of bilinguals: The role of culture and language in memory encoding and recall.* Unpublished doctoral dissertation, Harvard University.

Robinson, J. A. (1976). Sampling autobiographical memory. *Cognition, 8,* 578-595.

Schrauf, R. & Rubin, D. C. (1997). Effects of a major change in language, culture, and environment on autobiographical memory. Poster presented at *the 38th Annual Meeting of the Psychonomic Society.* Philadelphia, PA.

Smith, S. M. (1979). Remembering in and out of context. *Journal of Experimental Psychology: Human Learning and Memory, 5* (5), 460-471.

Smith, S. M., Glenberg, A., & Bjork, R. A. (1978). Environmental context and human memory. *Memory & Cognition, 6,* 4, 342-353.

Tulving, E., & Thomson, D. (1973). Encoding specificity and retrieval processes in episodic memory. *Psychological Review, 80,* 3, 352-373.

Grounding Figurative Language Use in Incompatible Ontological Categorizations

Katja Markert

Language Technology Group
HCRC, Edinburgh University
Edinburgh EH8 9LW, U.K.
markert@cogsci.ed.ac.uk

Udo Hahn

(L)F Text Understanding Lab
Freiburg University
D-79085 Freiburg, Germany
http://www.coling.uni-freiburg.de

Abstract

We propose a formal criterion for delineating literal from figurative speech (metonymies, metaphors, etc.). It is centered around the notion of categorization conflicts that follow from the context of the utterance. In addition, we consider the problem of granularity, which is posed by the dependence of our approach on the underlying ontology, and compare our distinction with alternative reference-based explanations.

Introduction

Figurative language use comes in different varieties (e.g., as metonymy in example (2) and as metaphor in example (3) below), and is typically contrasted with literal language use (e.g., example (1)) on the basis of some notion of deviance.

(1) *"The **man** left without paying."*

(2) *"The **ham sandwich** left without paying."*

(3) *"The Internet is a **gold mine**."*

Cognitive linguists have been struggling for decades to draw a proper distinction between literal and figurative utterances. Their interest derives from the question how a basic, lexical meaning representation must be conceived from which figurative (and possibly literal) readings can be derived. This viewpoint implies to assume a computational process and, hence, requires to be explicit about the representational foundations from which to proceed.

Currently, two approaches prevail, which spell out this distinction. The first one, e.g., Lakoff & Johnson (1980), simply regards *deviation from literal reference* as a sufficient condition for figurativeness. No formal criteria for the nature of such a deviation are given so that the discrimination of literal and figurative meaning rests on subjective ascription.

The second approach (Fass, 1991; Pustejovsky, 1991; Stallard, 1993) introduces such a formal criterion. Each time *selectional restrictions* are violated, e.g., through type conflicts, an instance of figurative speech is encountered. Special reasoning patterns are then activated, like type coercion for metonymies (Pustejovsky, 1991) or analogy-based structure mapping for metaphors (Carbonell, 1982; Gentner et al., 1989), in order to cope with the triggering instance such that a reasonable interpretation can be derived, one that no longer violates the underlying constraints. The proponents of this approach present a lot of supporting evidence for their methodological claims (cf. example (4)) but obviously fail to cover a wide range of residual phenomena (example (5) lacks any violation of selectional restrictions though being figurative, at least, assuming the "writings-by-Chaucer" reading):

(4) *"I read **Chaucer**."*

(5) *"I like **Chaucer**."*

In this paper, we aim at providing a formal framework from which a proper distinction between literal and figurative language use can be made. Rather than formalizing the notion of deviation with recurrence to selectional restrictions, we will base our distinction on conceptual criteria that incorporate the influence from the context of an utterance. These criteria allow us further to focus on the dependence of literal and figurative speech on individual ontologies. Considering granularity issues of ontologies we may even overcome the influence of subjectivity by taking additional formal criteria into account.

Lexical Meaning

We will base our considerations on the notion of *context-independent lexical meaning* of lexemes, from which the notions of literal and figurative meaning in context will be derived. Lexical meaning will be a function from lexemes to categories (concepts) of an ontology.

So, let $\mathcal{L}$ be the set of *lexemes* of a given natural language and let $\mathcal{L}' \subset \mathcal{L}$ be the subset of lexemes containing nouns, main verbs and adjectives only (e.g., *man* or *policeman* are elements of $\mathcal{L}'$). We also assume an ontology composed of a set of concept types $\mathcal{F} := \{$MAN, POLICEMAN, HAM-SANDWICH, ... $\}$, a set of instances $\mathcal{I} := \{$man-1, policeman-2, ... $\}$ related to concept types, and a set of relations $\mathcal{R} := \{$*has-part, part-of, agent,* ... $\}$, which link concept types or instances. We take a set theoretical semantics for granted as is commonly assumed in description logics (Woods & Schmolze, 1992). The *lexical meaning* β_{lex} can then be defined as a relation $\beta_{lex} \subset \mathcal{L}' \times \{\mathcal{F} \cup \mathcal{I}\}$. While we refrain from considering the linkage between lexemes and ontological entities in depth (cf., e.g., Cruse (1986) or Jackendoff (1990)), we require the relation β_{lex} to fulfill the following properties:

1. If $lexeme \in \mathcal{L}'$ is a proper name, then a unique lexeme.i $\in \mathcal{F} \cup \mathcal{I}$ with $(lexeme,$ lexeme.i$) \in \beta_{lex}$ exists such that lexeme.i $\in \mathcal{I}$. Thus, every proper name is linked to a single instance in the domain knowledge base.

361

2. If $lexeme \in \mathcal{L}'$ is not a proper name, then a concept $lexeme.\text{CON} \in \mathcal{F}$ must exist so that $(lexeme, lexeme.\text{CON}) \in \beta_{lex}$. Also, no instance $\texttt{lexeme.i} \in \mathcal{I}$ exists such that $(lexeme, \texttt{lexeme.i}) \in \beta_{lex}$.

3. For reasons of simplicity, we will now restrict β_{lex} appropriately. If $lexeme \in \mathcal{L}'$ is not a proper name, then we require for all $\texttt{i} \in \beta_{lex}(lexeme)$ that $\texttt{i}$ can be referred to by $lexeme$ in a *context-independent* way. Hence, we assume that reference to any $\texttt{i}$ via $lexeme$ is always possible. (We cannot, e.g., relate the lexeme *fool* to MAN as not every man can be referenced by *fool* independent of the context.) The condition of context-independence may, however, still hold for several concepts that stand in a subsumption relation to each other. So, when we regard the lexeme *man*, this condition holds for both the concepts MAN and PO-LICEMAN, as all $\texttt{i} \in$ POLICEMAN and all $\texttt{i} \in$ MAN can be referenced by *man*. We then regard the most general concept to which this unconditioned reference relation applies (here, MAN) as the lexical meaning, and, in general, consider *lexical meaning* as a function[1] $\beta_{lex} : \mathcal{L}' \mapsto \mathcal{F} \cup \mathcal{I}$. By convention, we denote $\beta_{lex}(lexeme)$ by $lexeme.\text{CON}$.

Lexical meaning is thus considered as a context-independent function from lexemes to categories (concepts) of an ontology. As there is no agreement on canonical ontologies, this mapping introduces subjective conceptualizations.

Finally, we extend our definition to words w of a discourse so that their corresponding lexeme be $w.lex \in \mathcal{L}'$. We simply assume $\beta_{lex}(w) := \beta_{lex}(w.lex)$. We distinguish the range of that mapping by $w.\texttt{i}$ for proper names and $w.\text{CON}$ in all other cases. Hence, the lexical meaning of the word *"man"* in example (1) is given by the concept MAN.[2]

Literal *vs.* Figurative Meaning

While in the previous section we have been dealing with the isolated lexical meaning of a word only, in this section we will incorporate the *context of an utterance* in which a word appears. Hence (cf. Fig. 1), we introduce the word w' with respect to which word w is syntactically related — w' is either head or modifier of w. Such a dependency relation (either a direct one or a well-defined series of dependency relations) at the linguistic level induces a corresponding conceptual relation $r \in \mathcal{R}$ at the ontological level (Romacker et al., 1999). The conceptual relation r links the conceptual correlates, $w.\texttt{sf}$ and $w'.\texttt{sf}$, of w and w', respectively. Accordingly, we may now say that w *StandsFor* a corresponding domain entity $w.\texttt{sf}$; alternatively, $w.\texttt{sf}$ is called the (intended) meaning of w. The comparison of $w.\texttt{sf}$ with $w.\text{CON}$ or $w.\texttt{i}$ lies at the heart of the decision criterion we

propose for judging whether a reading is literal or figurative. So, in the well-known example (2), *"ham sandwich"* $(= w)$ *StandsFor* "the man who ordered the ham sandwich" $(= w.\texttt{sf})$, which is distinct from its lexical meaning, HAM-SANDWICH $(= w.\text{CON})$.

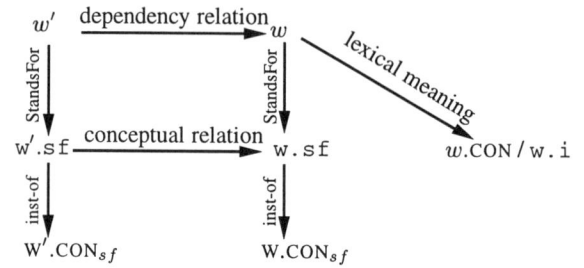

Figure 1: Framework for Contextual Interpretation

We may now consider some examples to distinguish several cases how $w.\texttt{sf}$ can be related to $w.\text{CON}$ or $w.\texttt{i}$. This will also lead us to clarify the notion of distinctiveness between the items involved. Let $w.\texttt{sf}$ be an instance from $\mathcal{I}$, and let $w.\text{CON}_{sf}$ be the least general concept such that $w.\texttt{sf} \in w.\text{CON}_{sf}$.[3] This assumption will be shortcut as $w.\texttt{sf}$ *inst-of* $w.\text{CON}_{sf}$.

In the simplest case, $w.\texttt{sf}$ and $w.\text{CON}$ / $w.\texttt{i}$ are related by an *inst-of* relation. Then $w.\text{CON}_{sf} = w.\text{CON}$ holds. In the utterance

(6) *"A **man** left without paying."*

we have $w =$ *"man"* and $w' =$ *"left"*. Furthermore, $w.\texttt{sf} =$ man-1 *inst-of* MAN $= w.\text{CON} = w.\text{CON}_{sf}$. So, in the example (6), lexical meaning and actual meaning coincide.

If we consider all relations other than equality as deviant, we characterize a class of phenomena that is certainly larger than the one containing figurative speech only. Example

(7) *"A policeman left without paying. The **man** lost his job."*

illustrates an anaphoric, non-figurative relation between *"man"* $(= w)$ and *"policeman"*. A subsumption relation holds between $w.\text{CON}_{sf}$ $(=$ POLICEMAN$)$ and $w.\text{CON}$ $(=$ MAN$)$, which means that $w.\text{CON}$ is either more general than or equal to $w.\text{CON}_{sf}$. In particular, we have (policeman-1 $=$) $w.\texttt{sf} \in w.\text{CON}$, but not $w.\texttt{sf}$ *inst-of* $w.\text{CON}$, in general (as in example (6)).

Loosening ties a bit more, we may abandon the subsumption relation between $w.\text{CON}_{sf}$ and $w.\text{CON}$ as in example

(8) *"A policeman left without paying. The **fool** lost his job."*

We have (policeman-1 $=$) $w.\texttt{sf} \in w.\text{CON}$ $(=$ FOOL$)$, but the specialization relation between $w.\text{CON}_{sf}$ $(=$ POLICE-MAN$)$ and $w.\text{CON}$ $(=$ FOOL$)$ no longer holds. Instead, we are set back to $w.\texttt{sf} \in w.\text{CON}_{sf} \cap w.\text{CON}$ and, therefore, $w.\text{CON}_{sf} \cap w.\text{CON} \neq \emptyset$. We say that $w.\text{CON}_{sf}$ and $w.\text{CON}$

---

[1]In order to make β_{lex} a function we assume in the case of *polysemy* one of several meaning alternatives to be the primary one from which the others can be derived. In the case of *homonymy*, we assume the existence of different lexemes which can be mapped directly to mutually exclusive concepts.

[2]The lexical meaning of a word w must be distinguished from the concrete referent of w in the given discourse.

[3]The least general concept $w.\text{CON}_{sf}$ with $w.\texttt{sf} \in w.\text{CON}_{sf}$ is the intersection of all concepts $C \in \mathcal{F}$ with $w.\texttt{sf} \in C$.

are *compatible*, as no categorization conflict arises. This also holds for all previously discussed examples. As a consequence, the notion of *categorization conflict* turns out to become crucial for our distinction between literalness and figurativeness — the latter being based on an underlying categorization conflict, whereas the former is not. We summarize these observations in the following definition:

Definition 1 (Literalness via Syntactic Constraints)
*A word w in an utterance U is used according to its **literal meaning**, if for every instance* $w.sf \in \mathcal{I}$ *which w StandsFor, one of the following two conditions hold:*

$$w.sf = w.i \qquad \text{if } w \text{ is a proper name} \qquad (1)$$

$$w.sf \in w.\text{CON} \qquad \text{else} \qquad (2)$$

Especially, $w.\text{CON}_{sf} \cap w.\text{CON} \neq \emptyset$ holds for non-proper nouns.

We here restrict the notion of figurative speech to those relationships between $w.sf$ and the lexical meaning of w in terms of $w.\text{CON}$, which are not inclusive ones. A literal use of the word w for an instance $w.sf$ *inst-of* $w.\text{CON}_{sf}$ is only possible, if $w.\text{CON}_{sf} \cap w.\text{CON} \neq \emptyset$. If, however, a categorization conflict occurs, i.e., $w.\text{CON}_{sf} \cap w.\text{CON} = \emptyset$, then we call the use of w *figurative* (as illustrated by *"ham sandwich"* in example (2) or by *"gold mine"* in example (3)). We would like to stress the following implications:

1. We can determine exactly the place where *subjectivity* comes in when a distinction between literalness and figurativeness is made — it is mirrored by subjectivity in categorization. *"fool"* in example (8) can only be considered as literal, if the concepts FOOL and POLICEMAN are considered as being compatible (in the set theoretic sense introduced above). If one does not share this conceptualization, this usage of *"fool"* must be considered as figurative (or even absurd). Thus, we capture the subjectivity of figurativeness formally in the ontological premises, not via intuitive considerations.

2. Definition 1 does not depend on the *violation of selectional restrictions*. The example (5) (*"I like **Chaucer**."*) allows for the same analysis as example (4) (*"I read **Chaucer**."*), because the intended patient of *like* are, in both cases, Writings-by-Chaucer(=$w.sf$), although this is not indicated by selectional restrictions at all. In both cases $w.\text{CON}_{sf} \cap w.\text{CON} = \emptyset$, i.e., figurativeness holds.

Granularity

The (non-)inclusion criterion we have set up for the distinction between literal and figurative speech in Definition 1 introduces a particularly strong tie to the underlying ontology. One of the problems this might cause lies in *granularity* phenomena of domain knowledge bases and their impact on literal/figurative distinctions. Given different levels of granularity, it may well happen that a word w *StandsFor* an instance $w.sf$ *inst-of* $w.\text{CON}_{sf}$ with $w.\text{CON}_{sf} \cap w.\text{CON} = \emptyset$, though,

intuitively, one would rate the usage of w as a literal one. Assume we have a knowledge base KB_1 in which CPU happens to be PART-OF MOTHERBOARD, while MOTHERBOARD itself turns out to be PART-OF COMPUTER. If we analyze the example

(9) *"The CPU of the **computer** ... "*

accordingly, we end up with the determination of a figurative usage for (w =) *"computer"*, since MOTHERBOARD $\cap$ COMPUTER = $\emptyset$ (cf. Fig. 2).

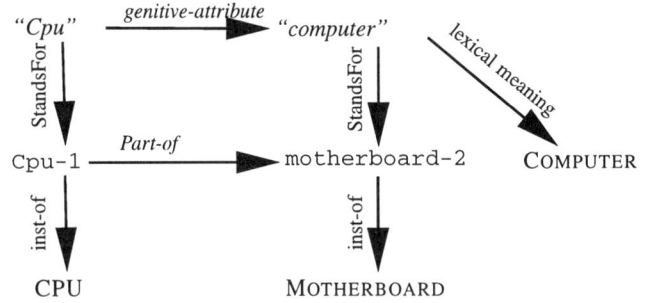

Figure 2: Example (9) Assuming KB_1

If we assume, however, a representation in a knowledge base KB_2 such that CPU is an *immediate* PART-OF COMPUTER, then we derive a literal usage for w (cf. Fig. 3).

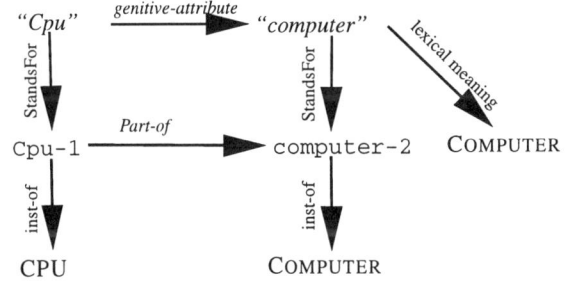

Figure 3: Example (9) Assuming KB_2

In order to lower the dependence on knowledge base granularity we may derive a weaker condition of literalness from Definition 1. Assume $w.sf$ and $w'.sf$ to be related by a conceptual relation r (technically, $w'.sf\ r\ w.sf$). Thus, for literal usage of w the following conditions hold:

$$w'.sf\ r\ w.i \qquad \text{if } w \text{ is a proper name} \qquad (3)$$

$$\exists i \in w.\text{CON}:\quad w'.sf\ r\ i \qquad \text{else} \qquad (4)$$

(3) immediately follows from (1) in Definition 1, since $w'.sf\ r\ w.sf\ (= w.i)$ holds. (4) can be deduced from (2) by defining $i := w.sf$.

Since these conditions provide weaker conditions of literal language use than the ones we have agreed upon in Definition 1, all literal usages determined by the strong condition still remain literal (in particular, example (9) is considered a literal usage of *"computer"* given KB_2). Considering the granularity effects for example (9) with respect to

363

KB_1, we may determine the literal usage of *"computer"* by the following consideration. Since CPU is PART-OF MOTHERBOARD, and MOTHERBOARD is PART-OF COMPUTER, we conclude with the transitivity of the PART-OF relation[4] that CPU is PART-OF COMPUTER. Hence, criterion (4) is fulfilled (assuming i = computer-2, w'.sf = cpu-1, w.sf = motherboard-2, and r = PART-OF). Unlike the examples discussed previously, we do not have w.sf $\in$ w.CON (criterion (2) from Definition 1). So, by moving from the strict criteria in Definition 1 to the weaker ones stated by criteria (3) and (4) we are able to balance granularity phenomena of knowledge bases, to a certain extent at least.

Figurativeness and Reference

One might argue that the problem just discussed, the dependence of the distinction between literal and figurative usage on knowledge base structures, follows from the dependence of *StandsFor* on syntactic context. Accordingly, some researchers, e.g. Lakoff & Johnson (1980), have proposed to build the definition of figurative speech upon the notion of *reference*. The assumption being made is that w uniquely refers to a knowledge base item w.ref *inst-of* w.CON$_{ref}$ and that figurativeness results from the deviation of this reference from literal meaning. Although their notion of deviance is not formalized, referentially-based literalness can now be defined straightforwardly in our approach by proceeding exactly along the lines of Definition 1:

Definition 2 (Literalness in the Referential Approach)
*A word w is called **literal in the referential approach**, if:*

$$w.\text{ref} = w.\text{i} \quad \text{if } w \text{ is a proper name} \quad (5)$$

$$w.\text{ref} \in w.\text{CON} \quad \text{else} \quad (6)$$

Without doubt, we here circumvent the granularity problem, since no change in reference occurs for example (9), no matter whether KB_1 or KB_2 is assumed.[5] But the reference approach runs into severe problems when one considers, e.g., classical examples of metonymies such as

(10) *"I like to read **Chaucer**. **He** was a great writer."*

We have w = *"Chaucer"* as a typical example for a writer-for-writings metonymy (Lakoff & Johnson, 1980).[6] The assumption to link literal/figurative usage to reference relations is flawed by the fact that w = *"Chaucer"* does not refer to the "writings by Chaucer", because in this case the referentially determined anaphor *"He"* could not be resolved. In particular, we have Chaucer.ref = Chaucer, therefore w.ref = w.i. Hence, *"Chaucer"* must be considered,

counterintuitively, as a literal use according to criterion (5) (similar problems have been discussed at length by Stallard (1993)). Given our context-dependent definitions ((1) or (3)), we get w.sf = Writings-by-Chaucer so that w.sf $\neq$ Chaucer. Thus, w = *"Chaucer"* is analyzed figuratively in our approach.

Summarizing our discussion, we combine criteria (1) to (6) by the following conditions:

1. A word w is used in its *literal* meaning in all definitions, if w.sf $\in$ w.CON (analogously, w.sf = w.i) for all w.sf, and w.ref $\in$ w.CON hold (combining Definition 2 and 1 with respect to literal usage).

2. If w.sf $\notin$ w.CON (analogously, w.sf $\neq$ w.i) for some w.sf (with respect to a relation r and another word w'), but w.ref $\in$ w.CON (w.ref = w.i), two cases must be distinguished:

   - In cases of granularity effects criteria (4) or (3) hold. By this, an $i \in w$.CON exists with w'.sf r i (analogously, w'.sf r w.i). We can include this in our definition of literal usage as its analysis is only due to implications a particular ontology design brings to bear.

   - In cases of figurative speech like the one in example (10) the criteria (4) / (3) do not hold. We include these cases in our definition of figurative usage as phrased below.

3. A word w is used in its *figurative* meaning according to the syntactic and the referential definition, if w.ref $\notin$ w.CON holds and there exists a w.sf $\notin$ w.CON. This is the case, e.g., in example (2).

Having only considered the figurative usage of a word w so far, we end up by defining U as a *figurative utterance*, if it contains at least one word w that is used in a figurative way.

Related Work

We consider as the main contribution of this paper the introduction of a *formal notion of deviance* that is both general and simple. To the best of our knowledge, no comparable work has been done so far on this issue. Although there exist formal characterizations of metaphors (Indurkhya, 1988; Gentner et al., 1989) these studies rather account for *structural properties* of metaphors (e.g., constraints on domain mappings, aptness conditions), than they deal with the *distinction* between literal and figurative speech.

Usually, however, utterly vague characterizations of what constitutes figurative language prevail such as the famous and often cited description by Lakoff & Johnson (1980) who characterize a metonymy by the use of "one entity to refer to another that is related to it". There is no restriction on the kind of relatedness between the objects. For example, relatedness might include class inclusion, similarity or part-whole relations. But only the latter are included in metonymy, in general, and the examples Lakoff and Johnson put forward suggest that it is this conventional kind of metonymy they

---

[4]We are aware of empirical observations about the transitivity of PART-WHOLE relations made by Chaffin (1992) and Winston et al. (1987), and take their constrained notion of transitivity for granted.

[5]Note that this definition is, nevertheless, still dependent on the knowledge base and on the lexical meaning of w.

[6]This example is ambiguous in several ways as, e.g., also the "style of Chaucer" could be another reasonable metonymic reading, thus giving rise to a writer-for-style metonymy. The following arguments hold for those alternative metonymic readings as well.

are talking about. The relation of class inclusion even leads to literal meaning as we have shown. Similar criticism applies to Tourangeau & Sternberg (1982), Fauconnier (1984), Kittay (1987), Turner (1988), Nunberg (1995), and many others. The same shadowy definitions of figurative language are then often adopted by computational linguists (Fass, 1991; Martin, 1992). This leads to the fact that it is mostly not clear at all, which phenomena are treated by these approaches.

In addition, a tendency can be observed in more formal approaches — pressed by the need to look for computationally feasible definitions of metaphor or metonymy — to consider figurative language as a violation of selectional restrictions (Carbonell, 1982; Fass, 1991; Hobbs et al., 1993; Pustejovsky, 1991) or communicative norms (Grice, 1975; Searle, 1979). Such an approach *equates* an often used triggering condition, *viz.* constraint violation, with the phenomenon of figurative language (or, subsets, like metonymies). Hence, it confuses the possible, but not necessary effects of a phenomenon with the phenomenon to be explained.

Despite the lack of formal rigor in previous work, it is worth to investigate how our formal criterion is compatible with other views on figurative speech from cognitive science. The tendency to see figurative speech rooted in conceptual categories, as we do, is becoming consensus in cognitive linguistics. The main trend is, e.g., to treat metaphors as a means of categorization by way of similarity (Gibbs, 1992) and to retrace figurative speech to cognitive procedures involving categorization and (subjective) experience (Lakoff & Johnson, 1980; Fauconnier, 1984; Lakoff, 1987). So, Lakoff and Johnson see metaphors rooted in our way of conceptualization via mappings. Kittay (1987) and Turner (1988) regard some kind of conceptual incompatibilities as the basis of metaphorization. Nevertheless, they do neither explicate their theory of categorization and incompatibility, nor do they recognize that these incompatibilities are relevant for other kinds of figurative speech, as well as for metaphors in the strict sense of the word. The dependence of lexical, literal and figurative meaning on ontologies is, therefore, realized, but no explicit formal treatment is given of particular problems this implies.

This is where we see the second major contribution of the paper. Once a formal distinction between literal and figurative meaning is given, it allows us to characterize *subjectivity*, so far an entirely informal notion, by reference to the particular ontology underlying the natural language understanding process. We aim at adapting different ontologies such that by way of abstracting away different *granularities* of representation structures (e.g., by generalizing more fine-grained representations to a coarser grain size, as in criterion (4)) disagreement might turn into consensus (e.g., considering example (9)). Contrary to that, the majority of researchers in our field of study attribute the difference in opinion to the existence of different, incompatible ontologies, and leave it with that explanation without further attempt at smoothing (Lakoff & Johnson, 1980; Turner, 1988).

An exception to this rule is the work by Veale & Keane (1994). While the authors still adopt an entirely informal *definition* of metaphors (close to the one from Turner (1988)), with all its drawbacks, Veale & Keane incorporate a concise, formal explication of how different viewpoints of different speakers influence metaphor interpretation. In contrast to our work, they offer the possibility to accept or reject beliefs in their knowledge base, depending on whether the speaker believes the propositions to be true or not. This then accounts for connotations which might arise in the metaphor interpretation. Instead, we focus on the problem of granularity, offering the possibility to derive coarse-grained views of the ontology from fine-grained views, thus not (de)activating certain propositions but viewing the same propositions in different grain sizes. This is not meant to explain metaphorical effects, but to reconcile different notions of literalness.

The third major proposal we make relates to the contextual embedding of figurative speech. The criterion we formulate is entirely based on syntactic relations that guide conceptual interpretation. In particular, and unlike most algorithmic accounts (Norvig, 1989; Fass, 1991; Pustejovsky, 1991; Hobbs et al., 1993), it does not rely at all upon the violation of selectional restrictions (for a notable exception, cf. Martin (1992)), since this criterion accounts only for a subset of the phenomena naturally recognized as figurative language. In addition, the syntax-based proposal we make avoids to consider reference changes as an indicator of figurativeness as is commonly assumed (e.g. by Lakoff & Johnson (1980)). Instead, our proposal is inspired by Fauconnier's (1984) "connector" function. Though Fauconnier aims at an embedding of figurative language into syntax, there exists no formalization of this notion in relation to an established grammar framework nor is the notion of conceptual incompatibility (and other purely conceptual issues of figurative language) formalized. A more formal criticism of the reference changes proposal was made by Stallard (1993) who, nevertheless, then only dealt with figurative language violating sortal constraints. Another notion of context is again used by Veale & Keane (1994), who do not use syntactic or selectional restriction properties for explaining effects of metaphorical speech, but rely on the speaker's belief system.

Our approach is fully compatible with viewing figurative language as *regular and not violating* linguistic norms. Whereas literal language is grounded in inclusion relations to lexical meaning, figurative language is grounded in other relations to lexical meaning. These can, nonetheless, be systematic and conventionalized relations like part-whole relations. There is no need to claim that inclusion relations are prior or are to be preferred to other (conventional) relations. This is in accordance with the conventional metaphor view first stipulated by Lakoff and Johnson. It is also in accordance with psycholinguistic research showing that figurative speech is in most cases as easily understood as literal speech. This is especially the case when the instance of figurative speech is conventional, i.e., grounded in systematic and pervasive ontological relationships (Blasko & Connine, 1993). The essence of this is that pervasive and structured relations or relations made salient by the context (Inhoff et al., 1984) may be as easily available to comprehension as inclusion relations.

Conclusion

In this paper, we have drawn a distinction between literal and figurative speech which is based on *formal* criteria. These are grounded in the solid framework of description logics, in particular, by relying on its set theoretic semantics. A crucial condition of whether language use is considered literal or figurative is introduced by the particular *ontology* referred to. While earlier formal approaches appeal to semantic types, sortal constraints, etc., this is not fully convincing, since the entire structure and *granularity* of the theory of the domain being talked about contributes to the understanding process, whether literally or figuratively based. In particular, we captured the notion of subjectivity in ontological premises and explained how granularity problems may be overcome.

A *recognition procedure* for figurative language is reported in Hahn & Markert (1997). Contrary to almost all competing approaches, we do <u>not</u> rely on a special triggering mechanism to start figurative interpretation when the literal one has failed. Rather we compute <u>both</u> interpretations in parallel, i.e., without preference for literal interpretations. The distinction between both forms of interpretation (and the need for a corresponding criterion) comes in, finally, when the text understander is required to disambiguate between competing readings. Among the preference criteria we apply are the distinction whether a reading is literal (preferred) or figurative.

The model we have presented does currently not account for neologisms, as those have no *a priori* lexical meaning, and many tricky cases of quantification and the use of proper names. In addition, we considered only rather simple figurative descriptions (words or compounds), not touching the issue of compositionality of figurative speech. From a more technical perspective, we have also not scrutinized the different kinds of relations that are still required to hold between $w.\text{CON}_{sf}$ and $w.\text{CON}$, if $w.\text{CON}_{sf} \cap w.\text{CON} = \emptyset$. So, a necessary condition for figurative speech has been established that needs to be supplemented by sufficient ones. We also have no criteria available right now that lead us to distinguish between various types of figurative speech (e.g., metaphors *vs.* irony). Finally, we stop short of distinguishing between innovative figurative speech (like in the *ham sandwich* example) and conventionalized figurative speech (*systematic polysemy* (Pustejovsky, 1991; Nunberg, 1995)).

Acknowledgements. K. Markert was a member of the Graduate Program *Human and Machine Intelligence* at Freiburg University, funded by DFG.

References

Blasko, D. G. & C. M. Connine (1993). Effects of familiarity and aptness on metaphor processing. *Journal of Experimental Psychology: Learning, Memory and Cognition*, 19(2):295–308.

Carbonell, J. G. (1982). Metaphor: an inescapable phenomenon in natural-language comprehension. In W. G. Lehnert & M. H. Ringle (Eds.), *Strategies for Natural Language Processing*, pp. 415–434. Hillsdale, N.J.: Lawrence Erlbaum.

Chaffin, R. (1992). The concept of a semantic relation. In A. Lehrer & E. F. Kittay (Eds.), *Frames, Fields and Contrasts*, pp. 253–288. Hillsdale, N.J.: Lawrence Erlbaum.

Cruse, D. (1986). *Lexical Semantics*. Cambridge: Cambridge University Press.

Fass, D. C. (1991). met*: A method for discriminating metonymy and metaphor by computer. *Computational Linguistics*, 17(1):49–90.

Fauconnier, G. (1984). *Espace Mentaux*. Paris: Editions de Minuit.

Gentner, D., B. Falkenhainer & J. Skorstad (1989). Viewing metaphors as analogy: the good, the bad, and the ugly. In Y. Wilks (Ed.), *Theoretical Issues in Natural Language Processing*, pp. 171–177. Hillsdale, N.J.: Lawrence Erlbaum.

Gibbs, R. (1992). Categorisation and metaphor understanding. *Psychological Review*, 99(3):572–577.

Grice, H. (1975). Logic and conversation. In P. Cole & J. Morgan (Eds.), *Syntax and semantics*, Vol. 3, pp. 41–58. New York: Academic Press.

Hahn, U. & K. Markert (1997). In support of the equal rights movement for literal and figurative language: A parallel search and preferential choice model. In *Proceedings of the 19th Annual Conference of the Cognitive Science Society*, pp. 609–614.

Hobbs, J. R., M. E. Stickel, D. E. Appelt & P. Martin (1993). Interpretation as abduction. *Artificial Intelligence*, 63(1-2):69–142.

Indurkhya, B. (1988). Constrained semantic transference: a formal theory of metaphors. In A. Prieditis (Ed.), *Analogica*, pp. 129–157. Los Altos, CA: Morgan Kaufmann.

Inhoff, A., S. Lima & P. Carroll (1984). Contextual effects on metaphor comprehension in reading. *Memory and Cognition*, 12:558–567.

Jackendoff, R. (1990). *Semantic Structures*. Cambridge, MA: MIT Press.

Kittay, E. (1987). *Metaphor: Its Cognitive Force and Linguistic Structure*. Oxford: Clarendon Press.

Lakoff, G. (1987). *Women, Fire, and Dangerous Things. What Categories Reveal about the Mind*. Chicago, IL: University of Chicago Press.

Lakoff, G. & M. Johnson (1980). *Metaphors We Live By*. Chicago, IL: University of Chicago Press.

Martin, J. (1992). Computer understanding of conventional metaphoric language. *Cognitive Science*, 16:233–270.

Norvig, P. (1989). Marker passing as a weak method for inferencing. *Cognitive Science*, 13(4):569–620.

Nunberg, G. (1995). Transfers of meaning. *Journal of Semantics*, 12:109–132.

Pustejovsky, J. (1991). The generative lexicon. *Computational Linguistics*, 17(4):409–441.

Romacker, M., K. Markert & U. Hahn (1999). Lean semantic interpretation. In *IJCAI'99 – Proceedings of the 16th International Joint Conference on Artificial Intelligence*.

Searle, J. (1979). Metaphor. In A. Ortony (Ed.), *Metaphor and Thought*, pp. 93–123. Cambridge: Cambridge University Press

Stallard, D. (1993). Two kinds of metonymy. In *Proceedings of the 31st Annual Meeting of the ACL*, pp. 87–94.

Tourangeau, R. & R. Sternberg (1982). Understanding and appreciating metaphors. *Cognition*, 11:203–244.

Turner, M. (1988). Categories and analogies. In D. Helman (Ed.), *Analogical Reasoning*, pp. 3–24. Dordrecht: D. Kluwer.

Veale, T. & M. T. Keane (1994). Belief modelling, intentionality and perlocution in metaphor comprehension. In *Proceedings of the 16th Annual Conference of the Cognitive Science Society*, pp. 910–915.

Winston, M., R. Chaffin & D. Herrmann (1987). A taxonomy of part-whole-relations. *Cognitive Science*, 11:417–444.

Woods, W. A. & J. G. Schmolze (1992). The KL-ONE family. *Computers & Mathematics with Applications*, 23(2-5):133–177.

Using a Sequential SOM to Parse Long-term Dependencies

Marshall R. Mayberry, III (martym@cs.utexas.edu)
Department of Computer Sciences
The University of Texas, Austin, TX 78712

Risto Miikkulainen (risto@cs.utexas.edu)
Department of Computer Sciences
The University of Texas, Austin, TX 78712

Abstract

Simple Recurrent Networks (SRNs) have been widely used in natural language processing tasks. However, their ability to handle long-term dependencies between sentence constituents is somewhat limited. NARX networks have recently been shown to outperform SRNs by preserving past information in explicit delays from the network's prior output. However, it is unclear how the number of delays should be determined. In this study on a shift-reduce parsing task, we demonstrate that comparable performance can be derived more elegantly by using a SARDNET self-organizing map. The resulting architecture can represent arbitrarily long sequences and is cognitively more plausible.

Introduction

The subsymbolic approach (i.e. neural networks with distributed representations) to processing language is attractive for several reasons. First, it is inherently robust: the distributed representations display graceful degradation of performance in the presence of noise, damage, and incomplete or conflicting input (Miikkulainen, 1993; St. John and McClelland, 1990). Second, because computation in these networks is constraint-based, the subsymbolic approach naturally combines syntactic, semantic, and thematic constraints on the interpretation of linguistic data (McClelland and Kawamoto, 1986). Third, subsymbolic systems can be lesioned in various ways and the resulting behavior is often strikingly similar to human impairments (Miikkulainen, 1993, 1996; Plaut, 1991). These properties of subsymbolic systems have attracted many researchers in the hope of accounting for interesting cognitive phenomena, such as role-binding and lexical errors resulting from memory interference and overloading, aphasic and dyslexic impairments resulting from physical damage, and biases, defaults and expectations emerging from training history (Miikkulainen, 1997, 1993; Plaut and Shallice, 1992).

Since its introduction in 1990, the simple recurrent network (SRN) (Elman, 1990) has become a mainstay in connectionist natural language processing tasks such as lexical disambiguation, prepositional phrase attachment, active-passive transformation, anaphora resolution, and translation (Allen, 1987; Chalmers, 1990; Munro et al., 1991; Touretzky, 1991). However, this promising line of research has been hampered by the SRN's inability to handle long-term dependencies, which abound in natural language tasks.

Another class of recurrent neural networks called Nonlinear AutoRegressive models with eXogenous inputs (NARX; Chen et al., 1990; Lin et al., 1996) have been proposed as an alternative to SRNs that can better deal with such long-term dependencies. In NARX networks, previous sequence constituents are explicitly represented in a predetermined number of output delays, thus reducing the effects of vanishing gradients, which is the primary source of memory degradation in recurrent networks (Bengio et al., 1994). The performance of these networks is strongly dependent on the number of output delays, and there are no guidelines on how many are needed.

This paper describes a method of extending recurrent networks such as the SRN and NARX with SARDNET (James and Miikkulainen, 1995), a self-organizing map algorithm designed to represent sequences. SARDNET permits the sequence information to remain explicit, yet generalizable in the sense that similar sequences result in similar patterns on the map. When SARDNET is coupled with SRN or NARX, the resulting networks perform better in the shift-reduce parsing task taken up in this study. Even with no recurrency and no explicit delays, the performance is almost as good. These results show that SARDNET can be used as an effective, concise, and elegant sequence memory in natural language processing tasks, and the approach should also scale up well to realistic language.

The Task: Shift-Reduce Parsing

Shift-reduce (SR) parsing is one of the simplest approaches to sentence processing that has the potential to handle a substantial subset of English (Tomita, 1986). Its basic formulation is based on the pushdown automata for parsing context-free grammars, but it can be extended to context-sensitive grammars as well.

The parser consists of two data structures: the input buffer stores the sequence of words remaining to be read, and the partial parse results are kept on the stack (figure 1). Initially the stack is empty and the entire sentence is in the input buffer. At each step, the parser has to decide whether to shift a word from the buffer to the stack, or to reduce one or more of the top elements of the stack into a new element representing their combination. For example, if the top two elements are currently *NP* and *VP*, the parser reduces them into *S*, corresponding to the grammar rule $S \rightarrow NP\ VP$ (step 17 in figure 1). The process stops when the elements in the stack have been reduced to S, and no more words remain in the input. The reduce actions performed by the parser in this process constitute the parse result, such as the syntactic parse tree (line 18 in figure 1).

The sequential scanning process and incremental forming of partial representations is a plausible cognitive model for language understanding. SR parsing is also very efficient,

Stack	Input Buffer		Action
	the boy who liked the girl chased the cat .	1	Shift
the	boy who liked the girl chased the cat .	2	Shift
the boy	who liked the girl chased the cat .	3	Reduce
NP[the,boy]	who liked the girl chased the cat .	4	Shift
NP[the,boy] who	liked the girl chased the cat .	5	Shift
NP[the,boy] who liked	the girl chased the cat .	6	Shift
NP[the,boy] who liked the	girl chased the cat .	7	Shift
NP[the,boy] who liked the girl	chased the cat .	8	Reduce
NP[the,boy] who liked NP[the,girl]	chased the cat .	9	Reduce
NP[the,boy] who VP[liked,NP[the,girl]]	chased the cat .	10	Reduce
NP[the,boy] RC[who,VP[liked,NP[the,girl]]]	chased the cat .	11	Reduce
NP[NP[the,boy],RC[who,VP[liked,NP[the,girl]]]]	chased the cat .	12	Shift
NP[NP[the,boy],RC[who,VP[liked,NP[the,girl]]]] chased	the cat .	13	Shift
NP[NP[the,boy],RC[who,VP[liked,NP[the,girl]]]] chased the	cat .	14	Shift
NP[NP[the,boy],RC[who,VP[liked,NP[the,girl]]]] chased the cat	.	15	Reduce
NP[NP[the,boy],RC[who,VP[liked,NP[the,girl]]]] chased NP[the,cat]	.	16	Reduce
NP[NP[the,boy],RC[who,VP[liked,NP[the,girl]]]] VP[chased,NP[the,cat]]	.	17	Reduce
S[NP[NP[the,boy],RC[who,VP[liked,NP[the,girl]]]],VP[chased,NP[the,cat]]]		18	Stop

Figure 1: **Shift-Reduce Parsing a Sentence.** Each step in the parse is represented by a line from top to bottom. The current stack is at left, the input buffer in the middle, and the parsing decision in the current situation at right. At each step, the parser either shifts a word onto the stack, or reduces the top elements of the stack into a higher-level representation, such as **the boy** → **NP[the,boy]** (step 3). (Phrase labels such as "NP" and "RC" are only used in this figure to make the process clear.)

and lends itself to many extensions. For example, the parse rules can be made more context sensitive by taking more of the stack and the input buffer into account. Also, the partial parse results may consist of syntactic or semantic structures.

The general SR model can be implemented in many ways. A set of symbolic shift-reduce rules can be written by hand or learned from input examples (Hermjacob and Mooney, 1997; Simmons and Yu, 1991; Zelle and Mooney, 1996). It is also possible to train a neural network to make parsing decisions based on the current stack and the input buffer. If trained properly, the neural network can generalize well to new sentences (Simmons and Yu, 1992). Whatever correlations there exist between the word representations and the appropriate shift/reduce decisions, the network will learn to utilize them.

Another important extension is to implement the stack as a neural network. This way the parser can have access to the entire stack at once, and interesting cognitive phenomena in processing complex sentences can be modeled. The SPEC system (Miikkulainen, 1996) was a first step in this direction. The stack was represented as a compressed distributed representation, formed by a RAAM (Recursive Auto-Associative Memory) auto-encoding network (Pollack, 1990). The SPEC architecture, however, was not a complete implementation of SR parsing; it was designed specifically for embedded relative clauses. For general parsing, the stack needs to be encoded with neural networks to make it possible to parse more varied linguistic structures. We believe that the generalization and robustness of subsymbolic neural networks will result in powerful, cognitively valid performance. However, the main problem of limited memory accuracy of the subsymbolic parsing network must first be solved.

Parser architectures

A subsymbolic parser is a recurrent network such as SRN or NARX. The network reads a sequence of input word representations into output patterns representing the parse results, such as syntactic or case-role assignments for the words. At each time step, a copy of the hidden layer (SRN) or prior outputs (NARX) is saved and used as input during the next step, together with the next word. In this way each new word is interpreted in the context of the entire sequence so far, and the parse result is gradually formed at the output.

Recurrent neural networks can be used to implement a shift-reduce parser in the following way (figure 2: the network is trained to step through the parse (such as that in figure 1), generating a compressed distributed representation of the top element of the stack at each step (formed by a RAAM network). The network reads the sequence of words one word at a time, and each time either shifts the word onto the stack (by passing it through the network, e.g. step 1), or performs one or more reduce operations (by generating a sequence of compressed representations corresponding to the top element of the stack: e.g. steps 8-11). After the whole sequence is input, the final stack representation is decoded into a parse result such as a parse tree. Such an architecture is powerful for two reasons: (1) During the parse, the network does not have to guess what is coming up later in the sentence, as it would if it always had to shoot for the final parse result; its only task is to build a representation of the current stack in its hidden layer and the top element in its output. (2) Instead of having to generate a large number of different stack states at the output, it only needs to output representations for a relatively small number of common substructures. Both of these features make learning and generalization easier. The parser can be implemented with various network architectures; the SRN and NARX networks are compared in this study.

SRN

In the simple recurrent network, the hidden layer is saved and fed back into the network at each step during sentence processing. The network is typically trained using the standard backpropagation algorithm (Rumelhart et al., 1986). A well-known problem with the SRN model is its low memory accuracy. It is difficult for it to remember items that occurred several steps earlier in the input sequence, especially if the network is not required to produce them in the output layer during the intervening steps (Stolcke, 1990; Miikkulainen, 1996). The intervening items are superimposed in the hidden layer, obscuring the traces of earlier items. As a result, parsing with an SRN has been limited to relatively simple sentences with shallow structure.

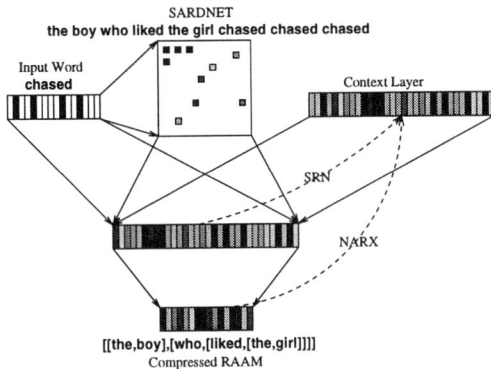

SARDNET
the boy who liked the girl chased chased chased

Input Word
chased

Context Layer

SRN

NARX

[[the,boy],[who,[liked,[the,girl]]]]
Compressed RAAM

Figure 2: **The Parser Network.** This snapshot shows the network during step 11 of figure 1. The representation for the current input word, **chased**, is shown at top left. Each word is input to the SARD-NET map, which builds a representation for the sequence word by word. In the SRN implementation of the parser, the previous activation of the hidden layer is copied (as indicated by the dotted line labelled SRN) to the Context assembly at each step. In the NARX implementation, a predetermined number of previous output representation compose the Context (indicated by the dotted line labelled NARX). The Context, together with the current input word and the current SARDNET pattern, is propagated to the hidden layer of the network. As output, the network generates the compressed RAAM representation of the top element in the shift-reduce stack at this state of the parse (in this case, line 12 in figure 1). SARDNET is a map of word representations, and is trained through the Self-Organizing Map (SOM) algorithm (Kohonen, 1997, 1990). All other connections are trained through backpropagation (for SRN) or BPTT (for NARX) (Rumelhart et al., 1986; Werbos, 1974).

NARX

Nonlinear AutoRegressive models with eXogenous inputs (NARX; Chen et al., 1990; Lin et al., 1996) have been proposed as an alternative to SRNs. They are good at dealing with long-term dependencies that typically arise in nonlinear systems such as system identification (Chen et al., 1990), time series prediction (Connor et al., 1992), and grammatical inference (Horne and Giles, 1995). NARX is a feedforward network with copies of previous outputs called delays fed back into the network during sequence processing. The network is trained via BackPropagation through Time (Rumelhart et al., 1986; Werbos, 1974), which allows "vanishing gradient" information to influence later outputs by unfolding the network in time. The performance of the network improves in an exponentially decreasing manner with the number of output delays provided.

SARDNET

The solution described in this paper is to use an explicit representation of the input sequence on a self-organizing map as additional input to the hidden layer. This representation provides more accurate information about the sequence, such as the relative ordering of the incoming words, and it can be combined with the weak hidden layer representation to generate accurate output that retains all the advantages of distributed representations. The sequence representation is also not limited by length.

The SARDNET (Sequential Activation Retention and Decay Network; James and Miikkulainen, 1995) used for this purpose is based on the Self-Organizing Map neural network (Kohonen, 1990, 1997), and organized to represent the space of all possible word representations. As in a con-

ventional self-organizing map network, each input word is mapped onto a particular map node called the maximally-responding unit, or winner. The weights of the winning unit and all the nodes in its neighborhood are updated according to the standard adaptation rule to better approximate the current input. The size of the neighborhood is set at the beginning of the training and reduced as the map becomes more organized.

In SARDNET, the sentence is represented as a distributed activation pattern on the map (figure 2). For each word, the maximally responding unit is activated to a maximum value of 1.0, and the activations of units representing previous words are decayed according to a specified decay rate (e.g. 0.9). Once a unit is activated, it is removed from competition and cannot represent later words in the sequence. Each unit may then represent different words depending on the context, which allows for an efficient representation of sequences, and also generalizes well to new sequences.

In this parsing task, a SARDNET representation of the input sentence is formed at the same time as the network hidden layer representation, and used together with the previous hidden layer (in the SRN) or output (in NARX) representations and the next word as input to the hidden layer (figure 2). This architecture allows these networks to perform their task with significantly less memory degradation. The sequence information remains accessible in SARDNET, and the network is able to focus on capturing correlations relating to sentence constituent structure during parsing.

Experiments

Input Data

The data used to train and test the SRN and SARDSRN networks was generated from the phrase structure grammar in figure 3, adapted from a grammar that has become common in the literature (Elman, 1991; Miikkulainen, 1996), but limited to a maximum of one relative clause per sentence. From this grammar training targets corresponding to each step in the parsing process were obtained. For shifts, the target is simply the current input. In these cases, the network is trained to auto-associate, which these networks are good at. For reductions, the targets consist of representations of the partial parse trees that result from applying a grammatical rule. For example, the reduction of the sentence fragment **who liked the girl** would produce the partial parse result **[who,[liked,[the,girl]]]**. Two issues arise: how should the parse trees be represented, and how should reductions be processed during sentence parsing?

The approach taken in this paper is the same as in SPEC, as well as in other connectionist parsing systems (Miikkulainen, 1996; Berg, 1992; Sharkey and Sharkey, 1992). Compressed representations of the syntactic parse trees using RAAM are built up through auto-association of the constituents. This training is performed beforehand separately from the parsing task. Once formed, the compressed representations can be decoded into their constituents using just the decoder portion of the RAAM architecture.

Word representations were hand-coded to provide basic part-of-speech information together with a unique ID tag that identified the word within the syntactic category (figure 4). The basic encoding of eight units was repeated eight times to make a redundant 64-unit representation.

369

Rule Schemata	$S \rightarrow NP(n)\ VP(n,m)$	$VP(n,m) \rightarrow \textbf{Verb}(n,m)\ NP(m)$	$NP(n) \rightarrow \textbf{the}\ Noun(n)$
	$RC(n) \rightarrow \textbf{who}\ VP(n,m)$	$NP(n) \rightarrow \textbf{the}\ Noun(n)\ RC(n)$	$RC(n) \rightarrow \textbf{whom}\ NP(m)\ Verb(m,n)$

Nouns	$Noun(0) \rightarrow \textbf{boy}$	$Noun(1) \rightarrow \textbf{girl}$	$Noun(2) \rightarrow \textbf{dog}$	$Noun(3) \rightarrow \textbf{cat}$

Verbs	$Verb(0,0) \rightarrow \textbf{liked, saw}$	$Verb(0,1) \rightarrow \textbf{liked, saw}$	$Verb(0,2) \rightarrow \textbf{liked}$
	$Verb(0,3) \rightarrow \textbf{chased}$	$Verb(1,0) \rightarrow \textbf{liked, saw}$	$Verb(1,1) \rightarrow \textbf{liked, saw}$
	$Verb(1,2) \rightarrow \textbf{liked}$	$Verb(1,3) \rightarrow \textbf{chased}$	$Verb(2,0) \rightarrow \textbf{bit}$
	$Verb(2,1) \rightarrow \textbf{bit}$	$Verb(2,2) \rightarrow \textbf{bit}$	$Verb(2,3) \rightarrow \textbf{bit, chased}$
	$Verb(3,0) \rightarrow \textbf{saw}$	$Verb(3,1) \rightarrow \textbf{saw}$	$Verb(3,3) \rightarrow \textbf{chased}$

Figure 3: Grammar. This phrase structure grammar generates sentences with subject- and object-extracted relative clauses. The rule schemata with noun and verb restrictions ensure semantic agreement between subject and object depending on the verb in the clause. Lexicon items are given in bold face.

the	10000000	who	01010000
whom	01100000	.	11111111
boy	00101000	dog	00100010
girl	00100100	cat	00100001
chased	00011000	saw	00010010
liked	00010100	bit	00010001

Figure 4: Lexicon. Each word representation is put together from a part-of-speech identifier (first four components) and a unique ID tag (last four). This encoding is then repeated eight times to form a 64-unit word representation. Such redundancy makes it easier to identify the word.

System Parameters and Training

SARDNET maps were added to the SRN and NARX networks to yield SARDSRN and SARDNARX parsing architectures. Additionally, a SARDNET map was added to a simple feedforward network (FFN). This (SARDFFN) network provides a baseline for evaluating the map itself in the parsing task. The performances of all the architectures was compared in the shift-reduce parsing task. Additionally, for the NARX and SARDNARX networks, delays of 0, 3, 6, 9, 12, and 15 prior inputs (covering almost the entire sentence) were constructed. The size of the hidden layer for each network was determined so that the total number of weights was as close to 64,000 as the topology would permit (figure 6).

Four data sets of 20%, 40%, 60%, and 80% of the 436 sentences generated by the grammar were randomly selected and each parser was trained on each dataset four times. Training on all 256 runs was stopped when the error on a 22-sentence (5%) validation set began to level off. The same validation set was used for all the simulations and was randomly drawn from a pool of sentences that did not appear in any of the training sets. Testing was then performed on the remaining sentences that were neither in the training set nor in the validation set. All networks were trained with a learning rate of 0.1, and the maps had a decay rate of 0.9. A map of 100 units was pretrained with a learning rate of 0.6, and then used for all of the SARDNARX networks. A slightly larger map with 144 units was used for the SARDFFN and SARDSRN networks since these networks had otherwise much fewer weights. Training took about one day on a 400 MHz Pentium Pro workstation for each network.

Results

Epoch error, the average error per output unit during each epoch, is usually used to gauge the networks' performance in experimental studies like this. It tells us how closely the output representations matched the target representations during parsing. Presumably, if the epoch error is low, the output

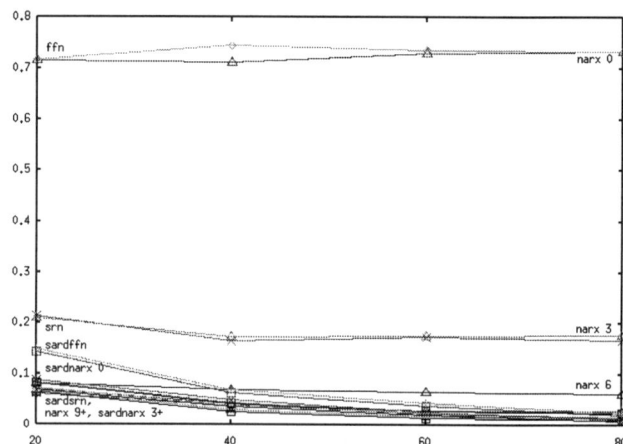

Figure 5: Summary of Parsing Performance. Averages over four simulations each for the fifty two networks tested using the stricter *average mismatches* per sentence measure on the test data. Most of the differences among the SARD networks and NARX networks with nine or more delays were statistically insignificant. SRN, NARX-3, SARDNARX-0, and NARX-6 differ in some of the data points as can be seen is this plot. The FFN and NARX-0 networks provide baselines of how simple feedforward networks would perform on this task (with regular BP and with BPTT), and the SRN shows how much simple recurrency helps. By comparison, the SARDFFN, SARDNARX-0, and SARDSRN graphs demonstrate that storing sequence information on SARDNET can significantly improve performance, while adding delays will improve that performance only minimally.

representations still permit accurate decoding into the correct parse tree. However, because this measure only reports the average performance over an entire epoch, it gives us no sense of the network's performance at each step in the parsing process. For example, there remains the danger that a low epoch error could also be achieved by learning the shift operations very accurately, with lower accuracy on the reductions, resulting in incorrect decoding of the compressed representations of the parse tree.

A more informative measure, *average mismatches*, therefore, was used in the comparisons. This measure reports the average number of leaf representations per sentence that could not be correctly identified by nearest match in Euclidean distance from the lexicon. As an example, if the target is **[who,[liked,[the,girl]]]]** (step 11 of figure 1), but the output is **[who,[saw,[the,girl]]]]**, then a mismatch would occur at the leaf labelled **saw** once the RAAM representation was decoded. Average mismatches provides a measure of the correctness of the information in the RAAM representation, and is a true measure of the utility of the network.

Most of the sentences in the training and test datasets were

370

network	delays	hidden layer	weights	network	delays	hidden layer	map size	weights
FFN		500	64000	SARDFFN		201	144	63888
SRN		197	64025	SARDSRN		134	144	64016
NARX	0	500	64000	SARDNARX	0	252	100	63856
NARX	3	200	64000	SARDNARX	3	137	100	63940
NARX	6	125	64000	SARDNARX	6	94	100	63928
NARX	9	91	64064	SARDNARX	9	71	100	63484
NARX	12	72	64512	SARDNARX	12	58	100	64168
NARX	15	59	64192	SARDNARX	15	48	100	63424

Figure 6: **Network Parameters.** In order to keep the network size as consistent as possible, the number of units in the hidden layers size was varied according to the size of the inputs. Because the SARD networks included a 100-unit map (144 units in the SARDFFN and SARDSRN) that was connected to both the input and hidden layers, the size of the hidden layer was proportionally made smaller.

seventeen words long. The longest long-term dependency the networks had to overcome was at step three in the parsing process where the first reduction occurred, which was part of the final compressed RAAM parse representation for the complete sentence. It was in decoding this final parse representation that even the best networks made errors. The results are summarized in figure 5. The main result is that the performance of all the SARD networks is comparable to NARX with nine or more delays, and clearly superior to SRN and NARX with zero to six delays. These results demonstrate that adding SARDNET to a recurrent network results in very robust performance without the expense of restricting the network to a prespecified number of delays. However, if the domain is well-behaved enough that this number can be determined, adding such delays will improve performance somewhat.

The SARDFFN and SARDNARX results are slightly weaker with the 20% dataset. On closer inspection it turned out that the map was not smooth enough to allow as good generalization as in the larger datasets, where there was sufficient data to overcome the map irregularities. It is also interesting to note that adding even a single delay completely eliminated this problem, bringing the performance in line with the others. We plan to improve generalization on the map further in future work.

Discussion

These results demonstrate a practicable solution to the memory degradation problem of recurrent networks. When prior constituents are explicitly represented at the input, the recurrent network does not have to maintain specific information about the sequence, and can instead focus on what it is best at: *capturing structure*. Although the sentences used in these experiments are still relatively uncomplicated, they do exhibit enough structure to suggest that much more complex sentences could be tackled with recurrent networks augmented with SARDNET or with delays.

These results also show that networks with SARDNET can perform as well as NARX networks with many delays. Why is this a useful result? The point is that it will always be unclear how many delays are needed in a NARX network, whereas SARDNET can accommodate sequences of indefinite length (limited only by the number of nodes in the network). This relieves the designer from having to specify, by trial and error, the appropriate number of delays. It should also lead to more graceful degradation with unexpectedly long sequences, and therefore would allow the system to scale up better and exhibit more plausible cognitive behavior.

The operation of the recurrent networks on the shift-reduce parsing task is a nice demonstration of holistic computation. The network is able to learn how to generate each RAAM parse representation during the course of sentence processing without ever having to decompose and recompose the constituent representations. Partial parse results can be built up incrementally into increasingly complicated structures, which suggests that training could be performed incrementally. Such a training scheme is especially attractive given that training in general is still relatively costly.

The SARDNET idea is not just a way to improve the performance of subsymbolic networks; it is an explicit implementation of the idea that humans can keep track of identities of elements, not just their statistical properties (Miikkulainen, 1993). The subsymbolic networks are very good with statistical associations, but cannot distinguish between representations that have similar statistical properties. People can; whether they use a map-like representation or explicit delays (and how many) is an open question, but we believe the SARDNET representation suggests an elegant way to capture a lot of the resulting behavior. SARDNET is a plausible cognitive approach, and useful for building powerful subsymbolic language understanding systems. SARDNET is also in line with the general neurological evidence for topographical maps in the brain.

Conclusion

We have shown how explicit representation of constituents on a self-organizing map allows recurrent networks to process sequences more effectively. We demonstrated that neural networks equipped with SARDNET sequence memory achieve comparable performance as NARX networks with many delays in a nontrivial shift-reduce parsing task. SARDNET, however, is more elegant and cognitively plausible in that it does not impose limits on the length of the sequences it can process. In future work, we will examine how exactly the networks are using the map information to improve the generalization ability of SARDNET, as well as extending the method to other recurrent neural network architectures.

Acknowledgments

This research was supported in part by the Texas Higher Education Coordinating Board under grant ARP-444.

References

Allen, R. B. (1987). Several studies on natural language and back-propagation. In *Proceedings of the IEEE First In-*

ternational Conference on Neural Networks (San Diego, CA), volume II, pages 335–341, Piscataway, NJ. IEEE.

Bengio, Y., Simard, P., and Frasconi, P. (1994). Learning long-term dependencies with gradient is difficult. *IEEE Transactions on Neural Networks*, 5(2):157–166.

Berg, G. (1992). A connectionist parser with recursive sentence structure and lexical disambiguation. In *Proceedings of the Tenth National Conference on Artificial Intelligence*, pages 32–37, Cambridge, MA. MIT Press.

Chalmers, D. J. (1990). Syntactic transformations on distributed representations. *Connection Science*, 2:53–62.

Chen, S., Billings, S., and Grant, P. (1990). Non-linear system identification using neural networks. In *International Journal of Control*, pages 1191–1214.

Connor, J., Atlas, L., and Martin, D. (1992). Recurrent networks and narma modeling. *Advances in Neural Information Processing Systems*, 4:301–308.

Elman, J. L. (1990). Finding structure in time. *Cognitive Science*, 14:179–211.

Elman, J. L. (1991). Distributed representations, simple recurrent networks, and grammatical structure. *Machine Learning*, 7:195–225.

Hermjacob, U. and Mooney, R. J. (1997). Learning parse and translation decisions from examples with rich context. In *Proceedings of the 35th Annual Meeting of the ACL*.

Horne, B. and Giles, C. (1995). An experimental comparison of recurrent neural networks. *Advances in Neural Information Processing Systems*, 7:697–704.

James, D. L. and Miikkulainen, R. (1995). SARDNET: A self-organizing feature map for sequences. In Tesauro, G., Touretzky, D. S., and Leen, T. K., editors, *Advances in Neural Information Processing Systems 7*, pages 577–584, Cambridge, MA. MIT Press.

Kohonen, T. (1990). The self-organizing map. *Proceedings of the IEEE*, 78:1464–1480.

Kohonen, T. (1997). *Self-Organizing Maps*. Springer, Berlin; New York, second edition.

Lin, T., Horne, B. G., and Giles, C. L. (1996). Learning long-term dependencies in narx recurrent neural networks. *IEEE Transactions on Neural Networks*, 7(6):1329–1338.

McClelland, J. L. and Kawamoto, A. H. (1986). Mechanisms of sentence processing: Assigning roles to constituents. In McClelland, J. L. and Rumelhart, D. E., editors, *Parallel Distributed Processing, Volume 2: Psychological and Biological Models*, pages 272–325. MIT Press, Cambridge, MA.

Miikkulainen, R. (1993). *Subsymbolic Natural Language Processing: An Integrated Model of Scripts, Lexicon, and Memory*. MIT Press, Cambridge, MA.

Miikkulainen, R. (1996). Subsymbolic case-role analysis of sentences with embedded clauses. *Cognitive Science*, 20:47–73.

Miikkulainen, R. (1997). Dyslexic and category-specific impairments in a self-organizing feature map model of the lexicon. *Brain and Language*, 59:334–366.

Munro, P., Cosic, C., and Tabasko, M. (1991). A network for encoding, decoding and translating locative prepositions. *Connection Science*, 3:225–240.

Plaut, D. C. (1991). *Connectionist Neuropsychology: The Breakdown and Recovery of Behavior in Lesioned Attractor Networks*. PhD thesis, Computer Science Department, Carnegie Mellon University, Pittsburgh, PA. Technical Report CMU-CS-91-185.

Plaut, D. C. and Shallice, T. (1992). Perseverative and semantic influences on visual object naming errors in optic aphasia: A connectionist account. Technical Report PDP.CNS.92.1, Parallel Distributed Processing and Cognitive Neuroscience, Department of Psychology, Carnegie Mellon University, Pittsburgh, PA.

Pollack, J. B. (1990). Recursive distributed representations. *Artificial Intelligence*, 46:77–105.

Rumelhart, D. E., Hinton, G. E., and Williams, R. J. (1986). Learning internal representations by error propagation. In Rumelhart, D. E. and McClelland, J. L., editors, *Parallel Distributed Processing, Volume 1: Foundations*, pages 318–362. MIT Press, Cambridge, MA.

Sharkey, N. E. and Sharkey, A. J. C. (1992). A modular design for connectionist parsing. In Drossaers, M. F. J. and Nijholt, A., editors, *Twente Workshop on Language Technology 3: Connectionism and Natural Language Processing*, pages 87–96, Enschede, the Netherlands. Department of Computer Science, University of Twente.

Simmons, R. F. and Yu, Y.-H. (1991). The acquisition and application of context sensitive grammar for English. In *Proceedings of the 29th Annual Meeting of the ACL*, Morristown, NJ. Association for Computational Linguistics.

Simmons, R. F. and Yu, Y.-H. (1992). The acquisition and use of context dependent grammars for English. *Computational Linguistics*, 18:391–418.

St. John, M. F. and McClelland, J. L. (1990). Learning and applying contextual constraints in sentence comprehension. *Artificial Intelligence*, 46:217–258.

Stolcke, A. (1990). Learning feature-based semantics with simple recurrent networks. Technical Report TR-90-015, ICSI, Berkeley, CA.

Tomita, M. (1986). *Efficient Parsing for Natural Language*. Kluwer, Dordrecht; Boston.

Touretzky, D. S. (1991). Connectionism and compositional semantics. In Barnden, J. A. and Pollack, J. B., editors, *High-Level Connectionist Models*, volume 1 of *Advances in Connectionist and Neural Computation Theory*, Barnden, J. A., series editor, pages 17–31. Ablex, Norwood, NJ.

Werbos, P. J. (1974). *Beyond Regression: New Tools for Prediction and Analysis in the Behavioral Sciences*. PhD thesis, Department of Applied Mathematics, Harvard University, Cambridge, MA.

Zelle, J. M. and Mooney, R. J. (1996). Comparative results on using inductive logic programming for corpus-based parser construction. In Wermter, S., Riloff, E., and Scheler, G., editors, *Connectionist, Statistical, and Symbolic Approaches to Learning for Natural Language Processing*, pages 355–369. Springer, Berlin; New York.

Thinking about What Might Have Been: If Only, Even If, Causality and Emotions

Rachel McCloy (mccloyr@tcd.ie)
Department of Psychology, University of Dublin, Trinity College,
Dublin 2, Ireland

Ruth M.J. Byrne (rmbyrne@tcd.ie)
Department of Psychology, University of Dublin, Trinity College,
Dublin 2, Ireland

Abstract

We discuss two different kinds of thinking about what might have been: Counterfactual "if only" thinking about how things might have been different and semifactual "even if" thinking about how things might have turned out the same. We report the results of an experiment that showed that the two kinds of thinking have different effects on cause and emotion judgements. The experiment provides the first demonstration that semifactual "even if" thoughts reduce peoples judgements of causality and their emotional reactions compared to no thoughts about what might have been, and it replicates recent findings that counterfactual "if only" thoughts increase peoples judgements of causality and their emotional reactions.

Counterfactual Thoughts and Emotions

People frequently consider what might have been different in their everyday thinking and these counterfactual thoughts are closely linked with judgements of causality and with a range of emotions, from regret (Byrne & McEleney, 1997; Gilovich & Medvec, 1994; Landman, 1987) to guilt and shame (Niedenthal, Tangney & Gavanski, 1994). The more easily people can imagine a situation turning out differently, the more their emotions about that situation are amplified (Kahneman & Miller, 1986). For example, consider the plight of an Olympic runner who injures herself the day before an important race (Boninger, Gleicher, & Strathman, 1994). The runner must chose between two painkillers, an older drug whose side-effects include nausea and fatigue and a new drug whose side-effects are unknown. She chooses the older drug, experiences the side-effects, and narrowly misses winning a medal. Participants who were told that the new drug turned out to have no side-effects judged that the runner would experience more regret and self-blame compared to participants who were told that the new drug had the same side-effects as the old one. The result suggests that thinking counterfactually about how things could have been different "if only" a different decision had been made can increase emotions such as regret and self-blame (Boninger et al., 1994).

Counterfactuals and Semifactuals

When people think about what might have been, they think not only about how things could have been different *if only* something else had happened (e.g., "if only I had not had an accident, I would have won a medal"); they also think about how things might have turned out the same *even if* something else had happened (e.g., "even if I had taken the other drug, I would still have experienced the side effects"). Philosophers have long recognized the distinction between counterfactual "if only" thinking and semifactual "even if" thinking (e.g., Goodman, 1973), but it has only recently begun to receive attention in psychological research (e.g., Branscombe, Owen, Garstka, & Coleman, 1996; McCloy & Byrne, in press).

Counterfactual and semifactual thoughts focus on different imaginary alternatives to factual events. Counterfactual "if only" thoughts focus on alternative antecedents that would undo an outcome whereas semifactual "even if" thoughts focus on alternative antecedents that would not undo the outcome (McCloy & Byrne, in press). As a result, counterfactual and semifactual thinking have very different effects on people's judgments of cause and blame: counterfactual thoughts increase judgements of cause and blame compared to semifactual thoughts (Branscombe, et al., 1996; McCloy & Byrne, 1999a). Our aim in the experiment we report was to examine more closely the effects of semifactual thinking on judgements of cause and emotion.

Semifactual Thoughts and Emotions

Do semifactual thoughts reduce people's emotional reactions? Semifactual thoughts do not *increase* judgements of cause and blame, compared to counterfactual thoughts, but do they *decrease* judgements of cause and blame? The answer is unknown because in the little available research on semifactual thoughts, their effects have been compared only to the effects of counterfactual thoughts, and not to an appropriate neutral baseline, such as no thoughts about what might have been (e.g., Boninger et al., 1994).

Our first aim in the experiment was to compare three sorts of thoughts about the Olympic scenario: "if only" thoughts, "even if" thoughts, and no thoughts about what might have been. These thoughts may have different consequences for judgements of cause and emotions. Consider, for example, judgements about how much the decision to take the older drug caused the outcome. How people judge how causal an antecedent is has received considerable attention in both philosophy and psychology (e.g., Cheng & Novick, 1991; Hilton & Slugowski, 1986; Mackie, 1974; Mill, 1872). The causal judgement may evoke spontaneously the construction of counterfactual and semifactual alternatives to assess how necessary and sufficient the antecedent is to bring about the outcome (e.g., Johnson-Laird & Byrne, 1991; Mill, 1872; N'Gbala & Branscombe, 1995). An antecedent for which people can readily construct a counterfactual alternative may be judged to be very causal whereas an antecedent for which people can readily construct a semifactual alternative may be judged to be not very causal (Goodman, 1973; Kahneman & Miller, 1986). Judgements of causality in the situation where people have been provided only with information about the factual situation provides a baseline measure of how causal the antecedent is judged on the basis of background knowledge alone.

One way in which counterfactual thinking has been hypothesized to influence people's emotional reactions is by way of causal inferences (e.g., Roese & Olson, 1995). People may, for example, regret events to the extent that they believe them to have caused negative outcomes. As generating counterfactuals about an event increases its perceived causal importance in producing an outcome, it also increases regret for that event. By denying that an antecedent event was causal in producing an outcome, semifactual "even if" thoughts may reduce people's emotional reactions to that event. Our predictions for emotional reactions were therefore the same as those for causal judgements.

Second, orthogonal to this variable, we compared three sorts of information about alternatives: a scenario in which there was an available alternative that would undo an outcome (a counterfactual alternative), one in which there was an available alternative that would not undo the outcome (a semifactual alternative), and one in which there was no information about alternatives. The explicit provision of a different counterfactual alternative may increase judgements of, for example, causality, whereas the explicit provision of a semifactual alternative may decrease judgements of causality, compared to the situation where no alternatives are given. Again, we predicted that people's emotional reactions would follow the same pattern and would increase where causality increases and decrease where causality decreases.

The Experiment

The participants were 367 undergraduates from the University of Dublin, Trinity College who took part in the experiment voluntarily. The 264 women and 101 men (two participants did not record their gender) ranged in age from 17 to 46 years old.

We gave all of them the Olympic scenario described earlier (based on the story in Boninger et al., 1994; see Appendix 1). We manipulated two independent variables in the experiment: the nature of the mutation task following the scenario and the nature of the alternatives described in the scenario. We manipulated the nature of the mutation task by ensuring that one of three mutation tasks followed the scenario: a counterfactual mutation task, in which participants were asked to imagine that in the days and weeks following the race they thought "if only..." and they were asked how they completed this thought; a semifactual mutation task, in which participants were asked to imagine that they thought "even if..." and they were asked how they completed this thought; or no mutation task, for which participants proceeded directly from reading the story to carrying out cause and emotion rating tasks.

We manipulated the nature of the alternatives by ensuring that the scenario had three different endings. For the different alternative scenario, it ended with the information that athletes who used the other, newer drug felt no pain and experienced no side effects. For the same alternative scenario, it ended with the information that those who had taken the newer drug felt no pain, but experienced the same side effects (see Appendix 1). For the no alternative scenario, no information was included about other athletes experiences with the other drug. These two independent variables, each with three levels, resulted in nine different scenario-task combinations. We assigned the participants at random to one of the nine groups, and each group had approximately 40 participants.

The dependent variables were the participants ratings of causes of, and emotional reactions to, the outcome of the scenario. The participants rated on a 9 point scale (where 1 indicated they did not feel the emotion at all and 9 indicated they felt it a great deal), first, how much they regretted taking the well-known drug; second, how bad they felt about what happened; third, how much they blamed themselves for the outcome of the race; and last, how much they thought deciding to take the well-known drug had caused them not to get a medal (see Appendix 2).

Results

We carried out a three (mutation task: if only, even if, none) by three (available alternatives: different, same, none) multivariate analysis of variance on the four dependent rating measures: regret, feeling bad, self-blame and causal ascription. The MANOVA showed that there were main effects for each of the independent variables, the mutation task and the available alternatives (Wilks' lambda = 0.95, $F_{(4, 356)}$ = 2.07, $p < 0.05$, and Wilks' lambda = 0.77, $F_{(4, 356)}$ = 12.26, $p < 0.0001$ respectively), as we describe below, and that there was no interaction between them (Wilks' lambda = 0.93, $F_{(4, 356)}$ = 1.51, $p < 0.87$).[1]

---

[1] Participants' "if only" thoughts tended to make the outcome of the scenario different (90%), whereas their "even if" thoughts tended to leave the outcome of the scenario unchanged (78%; see McCloy & Byrne, 1999b).

"If only" and "Even if" thoughts have different effects

The sort of mutation task participants carried out affected their ratings of emotions and causes, as shown by the main effect of mutation task. The sort of mutation task only affected ratings of feeling bad and ratings of causality, ($F_{(2, 356)} = 4.57$, $p < 0.01$ and $F_{(2, 356)} = 4.77$, $p < 0.01$, as shown by univariate tests) but not ratings of self-blame or regret ($F_{(2, 356)} = 1.55$, $p < 0.21$, and $F_{(2, 356)} = 2.50$, $p < 0.08$), as Table 1 shows. Participants ratings of feeling bad decreased following the generation of "even if" thoughts (mean 6.68) compared to the generation of "if only" thoughts (mean 7.29), or no thoughts (mean 7.30), as shown by post-hoc Student-Neuman-Keuls tests ($p < 0.05$). Participants' ratings of the causal role of the decision to take the well-known drug also decreased following the generation of "even if" thoughts (mean 4.82) compared to no thoughts (mean 5.67), although not reliably compared to "if only" thoughts (mean 5.23), see McCloy & Byrne, 1999b, for further details.

Table 1: The effects of the different mutation tasks (collapsed over different available alternatives) on ratings of emotions and causes

	Regret	Feeling Bad	Self-Blame	Cause
If only	5.36	7.29	5.35	5.23
Even if	5.29	6.68	4.85	4.82
None	5.88	7.30	5.25	5.67

Different available alternatives have different effects

The nature of the available alternatives described in the scenario affected participants ratings of emotions and causes, as shown by the main effect of alternatives. The nature of the available alternative only effected ratings of regret and ratings of causality, ($F_{(2, 356)} = 43.21$, $p < 0.001$ and $F_{(2, 356)} = 27.91$, $p < 0.001$, as shown by univariate tests) but not ratings of feeling bad or self-blame ($F_{(2, 356)} = 0.13$, $p < 0.88$, and $F_{(2, 356)} = 0.46$, $p < 0.63$), as Table 2 shows. Participants' ratings of regret decreased in the same alternative condition (mean 4.08) compared to the no alternative (mean 5.73), and different alternative condition (mean 6.78), as shown by post-hoc Student-Neuman-Keuls tests ($p < 0.05$). Participants' ratings of the causal role of the decision to take the well-known drug also decreased in the same alternative condition (mean 4.26) compared to the no alternative (mean 5.25), and the different alternative condition (mean 6.23), see McCloy & Byrne, 1999b, for further details.

Table 2: The effects of the different available alternatives (collapsed over different mutation tasks) on ratings of emotions and causes

	Regret	Feeling Bad	Self-Blame	Cause
Different	6.78	7.06	5.25	6.23
Same	4.08	7.07	4.99	4.26
None	5.73	7.16	5.24	5.25

Discussion

The results of the experiment provide the first demonstration that semifactual "even if" thoughts reduce peoples judgements of causality and their emotional reactions, at least ratings of feeling bad. The reduction is particularly clear when the effects of semifactual thinking are compared to an appropriate neutral baseline of no thoughts about what might have been, rather than when semifactual thoughts are compared only to counterfactual "if only" thoughts, as in the few previous studies on semifactual thinking (e.g., Branscombe, et al., 1996; McCloy & Byrne, in press). Our experiment also replicates recent studies which show that counterfactual thoughts increase people's judgements of causality and their emotional reactions.

The results of the experiment also provide the first demonstration that the availability of an alternative antecedent which would have led to the same outcome reduces people's judgements of causality and their emotional reactions, at least ratings of regret, compared to a baseline of no alternatives. Our experiment replicates the findings of previous studies that the availability of an alternative antecedent which would have led to a different outcome increases peoples judgements of causality and their emotional reactions, at least ratings of regret.

Something that we did not predict in our results is the divergence between effects of being presented with alternative antecedents and those of explicitly generating "if only" or "even if" thoughts. The explanation for this, we believe, lies in the events that each manipulation causes participants to focus on. When we provided participants with available alternatives, we did so to just one event in the scenario, the choice of drug, whereas our mutation task manipulation was more open ended. People's mutations focus, not only on the choice of drug, but also on the accident and on other events.[2] The manipulation of alternatives is therefore more likely to effect emotions about that one specific event (e.g., regret), whereas the mutation task manipulation could effect reactions to any number of events in the scenario and might be more likely to effect more general measures of negative affect (e.g., feeling bad).

The results suggest that the nature of the mental representations of factual events that people construct affect their construction of alternatives and their subsequent ratings of causality and emotional impact (e.g., Byrne, 1997). When

---

[2] See McCloy & Byrne (1999b) for a breakdown of the content of participants' responses to the two mutation tasks.

people are asked to think "if only", they must undo the outcome and examine how the undone outcome could have come about. They keep in mind two situations, one in which both the antecedent and consequent occurred and another in which neither occurred. As a result the antecedent is represented necessary for the consequent, and hence their rating of causality is increased compared to the baseline. When they are asked to think "even if", they must leave the outcome the same and examine whether the same outcome could have been brought about by different antecedent events. They keep in mind two situations, one in which both the antecedent and consequent occurred and another in which the antecedent did not occur but the consequent did. As a result the antecedent is represented as not necessary for the consequent, and hence their rating of causality is decreased compared to the baseline. The degree to which people think that the antecedent event may have caused the outcome following the generation of counterfactual or semifactual thoughts then effects their emotional reactions to that event.

Our experiment shows that the impact of semifactual thinking on our judgements of causality and emotions may be as important as the impact of counterfactual thinking. Of course, the Olympic scenario that we have examined has a negative outcome, the athlete does not win a medal, and it is well-known that counterfactual thinking is evoked more often following a bad outcome than following a good outcome (e.g., Landman, 1987). Participants in our experiment may have spontaneously thought counterfactually even in situations where they were not asked to (i.e., the no thoughts conditions). Whether semifactual thinking exhibits the same tendencies as counterfactual thinking, such as a prevalence after bad outcomes compared to good outcomes, remains an open research question.

Acknowledgements

We would like to thank David Boninger, Alice McEleney and Clare Walsh for their support and comments. The first author is supported by a postgraduate studentship from the Department of Education for Northern Ireland.

References

Boninger, D.S., Gleicher, F. & Strathman, A. (1994). Counterfactual Thinking: From What Might Have Been to What May Be. *Journal of Personality and Social Psychology, 67 (2),* 297-307.

Branscombe, N.R., Owen, S., Garstka, T.A., & Coleman, J. (1996). Rape and accident counterfactuals: Who might have done otherwise and would it have changed the outcome? *Journal of Applied Social Psychology, 26(12),* 1042-1067.

Byrne, R.M.J. (1997). Cognitive processes in counterfactual thinking about what might have been. In D.L. Medin (Ed.). *The Psychology of Learning and Motivation (Vol. 37).* San Diego, C.A.: Academic Press.

Byrne, R.M.J. & McEleney, A. (1997). Cognitive processes in regret for actions and inactions. In M. Shafto & P. Langley (Eds.). *Proceedings of the 19th Annual Confer-*
ence of the Cognitive Science Society. Hillsdale, N.J.: Erlbaum.

Cheng, P.W. & Novick, L.R. (1991). Causes versus enabling conditions. *Cognition, 40,* 83-120.

Gilovich, T. & Medvec, V.H. (1994). The temporal pattern to the experience of regret. *Journal of Personality and Social Psychology, 67,* 357-65.

Goodman, N. (1973). *Fact, Fiction and Forecast (3rd edition).* Cambridge, M.A.: Harvard University Press.

Hilton, D.J. & Slugowski, B.R. (1986). Knowledge-based causal reasoning: The abnormal conditions focus model. *Psychological Review, 93,* 75-88.

Johnson-Laird, P.N. & Byrne, R.M.J. (1991). *Deduction.* Hove, U.K.: Lawrence Erlbaum Associates Ltd.

Kahneman, D. & Miller, D.T. (1986). Norm theory: Comparing reality to its alternatives. *Psychological Review, 93,* 136-53.

Landman, J. (1987). Regret and elation following action and inaction: Affective responses to positive versus negative outcomes. *Personality and Social Psychology Bulletin, 13,* 524-36.

Mackie, J.L. (1974). *Cement of the universe: A study of causation.* London: Oxford University Press.

McCloy, R. & Byrne, R.M.J. (In press). Thinking about what might have been different and what might have been the same. In Dunnion, J., Smyth, B., O'Hare, G. & O'Nuaillain, S. (Eds). *AICS 98.*

McCloy, R. & Byrne, R.M.J. (1999a). The other side of the story: A comparison of counterfactual and semifactual thinking. Manuscript currently under review.

McCloy, R. & Byrne, R.M.J. (1999b). Affective consequences of counterfactual and semifactual thinking. Manuscript in preparation.

Mill, J.S. (1872). A system of logic, racionative and inductive (8th Ed.). London: Longmans, Green, & Reader.

N'Gbala, A. & Branscombe, N.R. (1995). Mental simulation and causal attribution: When simulating an event does not affect fault assignment. *Journal of Experimental Social Psychology, 31,* 139-62.

Niedenthal, P.M., Tangney, J.P. & Gavanski, I. (1994). If only I weren't versus If only I hadn't: Distinguishing shame and guilt in counterfactual thinking. *Journal of Personality and Social Psychology, 67 (4),* 585-95.

Roese, N.J. & Olson, J.M. (1995). Counterfactual Thinking: A Critical Overview, In N.J. Roese & J.M. Olson (Eds.) *What Might Have Been: The Social Psychology of Counterfactual Thinking.* Mahwah, N.J.: Erlbaum.

Wells, G.L. & Gavanski, I. (1989). Mental simulation of causality. *Journal of Personality and Social Psychology, 56,* 161-9.

Appendices

Appendix 1: The scenario and the tasks used in the experiment

You are a runner and since the age of eight you have competed in the sprint races in local track and field events. Up through school you had won every race in which you had competed. It was at the age of 13 that you began to dream about the Olympics. At the age of 18, before starting college, you decide to give the Olympics one, all out shot. You make the Irish Olympic team for the 400 metre race.

On the day before the 400 metre race, in a freak accident during training, you sprain your left ankle. Although there is no break or fracture, when you try to run, the pain is excruciating. Your trainer tells you about many advances in pain killing medications and assures you that you will still be able to participate. He recommends that you choose between two drugs, both legal according to Olympic guidelines. One is a well-known pain killer that has been proved effective but also has some serious side effects including temporary nausea and drowsiness. The other pain killer is a newer and less well-known drug. Although the research suggests that the newer drug might be a more effective pain killer, its side effects are not yet known because it has not been widely used.

After considerable thought, you elect to go with the more well-known drug. On the day of the race, although there is no pain in your ankle, you already begin to feel the nausea and find yourself fighting off fatigue. You finish in fourth place, only 1 tenth of a second from a Bronze medal, 4 tenths from a silver, and 5 tenths from a gold medal.

Different Alternative:
After the event, you learn that some athletes in other events who were suffering from similar injuries used the other, newer drug. They felt <u>no</u> pain and experienced <u>no</u> side effects.

Same Alternative:
After the event, you learn that some athletes in other events who were suffering from similar injuries used the other, newer drug. They felt <u>no</u> pain but experienced the <u>same</u> side effects.

If Only mutation task:
In the days and weeks following the race you think "if only...". How do you complete this thought?

Even If mutation task:
In the days and weeks following the race you think "even if...". How do you complete this thought?

Appendix 2: The rating tasks used in the experiment

Rating tasks:

1. How much do you regret taking the more well-known drug?

	1	2	3	4	5	6	7	8	9	

I feel no regret at all — I feel a great deal of regret

2. To what extent do you feel bad about how things turned out?

	1	2	3	4	5	6	7	8	9	

I do not feel bad at all — I feel extremely bad

3. How much do you blame yourself for not getting an Olympic medal in the 400 metre race?

	1	2	3	4	5	6	7	8	9	

I do not blame myself at all — I blame myself a great deal

4. To what extent do you think your decision to take the well-known drug led to your failure to obtain an Olympic medal in the 400 metre race?

	1	2	3	4	5	6	7	8	9

Definitely did not lead to my failure — Definitely led to my failure

Taking Time to Structure Discourse: Pronoun Generation Beyond Accessibility

Kathleen F. McCoy
Dept. of Computer and Information Sciences
University of Delaware
103 Smith Hall
Newark, DE 19716, USA
mccoy@cis.udel.edu

Michael Strube
Institute for Research in Cognitive Science
University of Pennsylvania
3401 Walnut Street, Suite 400A
Philadelphia, PA 19104, USA
strube@linc.cis.upenn.edu

Abstract

In order to produce coherent text, natural language generation systems must have the ability to generate pronouns in the appropriate places. In the past, pronoun usage was primarily investigated with respect to the accessibility of referents. That is, it was assumed that a pronoun should be generated whenever the referent was sufficiently accessible so as to make its resolution easy. We found that such an explanation does not seem to account well for the patterns of pronoun usage found in naturally occurring texts. We present an algorithm for generating appropriate anaphoric expressions which takes into account the temporal structure of texts (as a discourse structuring device) and knowledge about ambiguous contexts. Other important factors in our algorithm are sentence boundaries and the distance from the last mention of the anaphor. We back up our hypotheses with some empirical results indicating that our algorithm chooses the right referring expression in 85% of the cases.

Introduction

Anaphoric expressions are an important component to generating coherent discourses. While there has been some work on generating appropriate referring expressions, little attention has been given to the problem of when a pronoun should be used to refer to an object. In most instances the assumption has been that a pronoun should be generated when referring to a discourse entity that is highly prominent (accessible). However, a study of naturally occurring texts reveals that factors beyond accessibility must be taken into account in order to explain the patterns of pronoun use found.

In this paper we attempt to investigate factors that might influence the use of pronominal reference forms. In particular, we attempt to answer the question of when it is appropriate to generate a pronoun versus some other kind of definite description (i.e., a definite noun phrase or proper name) when realizing an anaphoric expression. Pronouns are prevalent in text and play a role in text coherence, yet pronoun generation has gotten very little attention. To date most of the research involving pronouns in text has concentrated on the problem of pronoun resolution (needed for natural language understanding). While it is likely that the work on pronoun resolution may be relevant to the problem of pronoun generation (one would not want to generate a pronoun that the reader could not resolve with reasonable effort), additional explanations are needed.

Our tack has been to study naturally occurring examples and to try to hypothesize rules that explain the pronoun use in those examples. To date we have concentrated our study on New York Times news articles. We hope to generalize some of our findings to other types of text genres as well.

Consider the following passage from the first several lines of one of the stories we analyzed:

Example 1:
When **Kenneth L. Curtis** was wheeled into court nine years ago, mute, dull-eyed and crippled, it seemed clear to nearly everyone involved that it would be pointless to put **him** on trial for the murder of **his** former girlfriend, Donna Kalson, and the wounding of her companion.

It had been a year since **Mr. Curtis** had slammed **his** pickup truck into them, breaking their legs. **He** then shot them both and, finally, fired a bullet into **his** own brain. **Mr. Curtis** lingered in a coma for months, then awoke to a world of paralysis, pain and mental confusion from which psychiatric experts said **he** would never emerge.

One expert calculated **his** I.Q. at 62.....

For convenience, we have indicated all references to the main character in bold.

A surprising thing to note about this passage is that not all of the anaphoric references to *Mr. Curtis* are pronouns even though he is arguably the focus of every sentence included. Previous work on pronoun generation would predict that a pronoun should be used if the same item remains in focus. Thus it appears that something other than a straightforward application of focusing or other pronoun resolution algorithms is necessary.

A second thing to note about this passage is that the sentences are generally long and complex and often contain several references to the same character. These types of sentences are very different from those that have been considered by any generation system that has rules for generating pronouns. In addition, it is not clear how most focusing or pronoun interpretation algorithms would handle them.

Looking for an explanation, one might turn to the underlying structure of the text. However, in doing so, care must be taken to ensure that the chosen structure both affects pronoun generation decisions and is something that is well-defined from a generation perspective. A structuring device such as paragraph breaks fails on both of these counts. At first one might expect to see a definite description at the beginning of a paragraph, and a pronoun inside a paragraph (as long as the item being referred to is in focus). While this correctly predicts the name at the beginning of paragraph two, notice there is a name in the middle of that same paragraph and a pronoun at the beginning of paragraph three. More importantly, from the perspective of a generation system, it is not at all clear how a paragraph break should be defined. Thus positing a rule based on such a structuring device is not helpful. We must search for an answer that (1) explains the patterns of pronouns found in naturally occurring text, and (2) is based on information available to a sentence generation system.

In this paper we present our preliminary work in doing just that. Our work so far has led us to hypothesize several factors that affect pronoun generation and that are available to a sentence generation system. These factors include:

Sentence Boundaries – pronouns appear to be the preferred referring form for subsequent reference to an item within the same sentence.

Distance from Last Mention – when the last mention of an item is several sentences back in the text, a definite description is preferred.

Discourse Structure (as indicated by time change) – if we take time as a discourse structuring device, when the previous reference to an item is at the same time as the current reference, a pronoun is preferred. A definite description is preferred when the time has changed.

Ambiguity – potential ambiguity must be taken into account when choosing an anaphoric expression in that a pronoun should only be generated if it can be resolved correctly.

In the next section we discuss previous research on pronoun generation. This is followed by an introduction of time as a structuring device which affects pronoun generation. Next we investigate ambiguous anaphoric references. We follow this with an algorithm which decides when to use a pronoun versus a definite description when referring to some discourse entity. After this, we report on empirical results of the application of our algorithm to a corpus of New York Times articles.

Previous Work on Pronoun Generation

Few researchers have given serious consideration to the problem of pronoun generation. The most common factor considered has been the accessibility of the referent. For example, a pronoun would be generated if its referent was sufficiently prominent in the preceding text. Some early generation work (e.g., McDonald (1980), McKeown (1985), McKeown (1983), Appelt (1981)) used a simple rule to implement this idea based on focus (Sidner, 1979). The rule roughly stated that if the current sentence is about the same thing that the previous sentence was about, use a pronoun to refer to that thing. As was pointed out above, this rule does not provide a very good match with the referring expressions in our corpus.

In Dale (1992), Dale also discussed the generation of pronouns in the context of work on generating referring expressions (Appelt, 1985; Reiter, 1990). He specified an algorithm that essentially generated the smallest referring expression that distinguished the object in question from all others in the context.[1] He generated a pronoun (or ellipsis) if one were adequate and if the object being referred to was the center of the last utterance. As an example of the kinds of texts he generated consider: "Soak the butterbeans. Drain and rinse *them*."

---

[1]This algorithm was later revised in Dale & Reiter (1995) to more adequately reflect human-generated referring expressions and to be more computationally tractable.

Such an account of pronoun generation, based on center constancy appears to work quite well in the domain studied by Dale, but is unable to account for the patterns in other kinds of naturally occurring text.

The centering model (Grosz et al., 1995) itself makes predictions about pronoun generation only in the instances where Rule 1 is applicable. Recall that centering associates a set of forward looking centers with each utterance. This is an ordered list of the discourse entities introduced in the utterance. While an active area of research, in English the order is generally taken to be subject, object, other arguments to the verb in surface order, adjuncts of the verb in surface order. The backward looking center of an utterance is the highest ranked element of the previous utterance's forward looking centers that is realized in the current utterance. Centering's Rule 1 states that if any element of the previous utterance's forward looking centers list is realized in the current utterance as a pronoun, then the backward looking center must be realized as a pronoun as well (Grosz et al., 1995, p.214). Notice that the *Mr. Curtis* at the beginning of the second sentence in Example 1 is an apparent violation of this rule (because it is the subject of the previous sentence but is not pronominalized while another element mentioned in that sentence is). But, more generally, we must have a theory that is able to handle all cases of pronoun use.

A pronoun interpretation algorithm based on centering which relied on centering transition preferences was developed in Brennan et al. (1987). Using transition preferences in a pronoun generation rule would cover more cases of pronoun use than is covered by Rule 1, but the application of such transition preferences also proved unhelpful in explaining pronoun patterns in our corpus.

Grosz & Sidner (1986) and Reichman (1985) indicate that discourse segmentation has an effect on the linguistic realization of referring expressions. While this is intuitively appealing, it is unclear how to apply this to the generation problem (in part because it is unclear how to define discourse segments to a generation system). Passonneau (1996) argues for the use of the principles of information adequacy and economy in generating anaphoric expressions. Her algorithm takes discourse segmentation into account through the use of focus spaces which are associated with each discourse segment. Thus, Passonneau explains that a fuller description might be used at a discourse segment boundary because the set of accessible objects changes at such boundaries (though she combines this consideration with centering theory which may override the decisions due to segment boundaries). While Passonneau's algorithm seems quite appealing, notice that it provides no explanation of how a discourse segment (boundary) should be defined. The evaluation that she provided used the discourse segments provided by a set of naive subjects who indicated discourse segment boundaries in her texts. Without such boundaries provided, it is impossible to apply her algorithm. In some sense, the work presented here is consistent with Passonneau's theory. What we attempt to add is a definition of discourse segment which is well-defined and can be derived from input that any sentence generator must have in order to generate a sentence. On the other hand, we differ from Passonneau in that we do not attempt to make use of focus spaces in generation. Rather we argue for evaluating informational adequacy on the basis of confusable objects near the current sentence in a discourse.

In the following section we hypothesize that discourse segment boundaries (or, perhaps, setting changes) do have an effect on appropriate anaphoric expression choice and that changes in time are markers of such boundaries (in the stories that we have analyzed). We hypothesize that if the current and previous reference to an entity occur in clauses referencing two different times, a definite description is used (and when the time referenced is the same, a pronoun is used).

Temporal Discourse Structure

In this section, we describe our approach to using discourse structure for choosing the right referring expression. Since we are working with stories from newspapers we were not able to identify the kind of discourse structure as assumed by Grosz & Sidner (1986), whose dialogues are more task-oriented and have clear intentional goals. We needed to find a structuring device that was both recognizable and part of the input to a sentence generation system. After investigating some work on narrative structure (Genette, 1980; Prince, 1982; Vogt, 1990), we concentrated on temporal structure. Temporal structure (an impoverished notion of the deictic center (Nakhimovsky, 1988; Wiebe, 1994) and related to temporal focus (Webber, 1988)) informally relates to the time being referenced in a text. This structure is often indicated by linguistic means and must be part of the input to a sentence generator. Changes in temporal structure or time may require world knowledge reasoning to recognize but are often indicated by either cue words and phrases (e.g., *"nine years ago"*, *"a year"*, *"for months"*, *"several months ago"*), a change in grammatical time of the verb (e.g., past tense versus present tense), or changes in aspect (e.g., atomic versus extended events versus states as defined by Moens & Steedman (1988)).

In considering how time change might affect anaphoric expression choice, we consider the choice for the first mention of a discourse entity in a sentence where that entity has recently been referred to in the discourse. Our hypothesis is that: Changes in time reliably signal discourse segment boundaries in newspaper articles; definite descriptions should appear when the current reference to a discourse entity is in a different discourse segment from the last reference to that entity and pronouns should occur when the previous mention is in the same discourse segment.

Notice that our hypothesis does *not* cover *long distance* situations – where a discourse entity has not been referred to for several sentences; we have found that a definite description is almost always used in long distance situations regardless of time changes. A second place where time does not affect anaphoric expression choice is at very short distances – subsequent reference to an object *within the same sentence* is almost always realized as a pronoun regardless of whether or not the time has changed.

In order to evaluate this hypothesis, we mapped out the time being referenced in our texts on a clause-by-clause basis. For each clause in the texts, we indicated the time which was referred to. We distinguished between events that occurred at a single instance in time (atomic events) and events or states that occurred over a span of time (repeated atomic

events, extended events, and states). For atomic events we allowed for both a specific time at which it occurred and for a non-specific time that indicated the range of uncertainty. We allowed time spans to have both specific end points and unspecified end points as well.

An example from our corpus with its associated temporal structure may illustrate these labels, the complexity of the texts under consideration, and how we propose pronoun generation is affected.

Example 2:

(47a) Questioned about the criminal activities of the football club,

(47b) **Mrs. Mandela** maintained

(47c) that **she** had never had any control over them.

(48a) This despite testimony from a half dozen former members

(48b) that they even had to get permission to go in and out of **her** yard.

(49a) **Mrs. Mandela** also said

(49b) **she** had disbanded the club

(49c) after **her** husband asked **her** to, despite evidence to the contrary.

(50) **Mrs. Mandela** faced questions from more than 10 lawyers representing various victims and the panel of commissioners and their investigators.

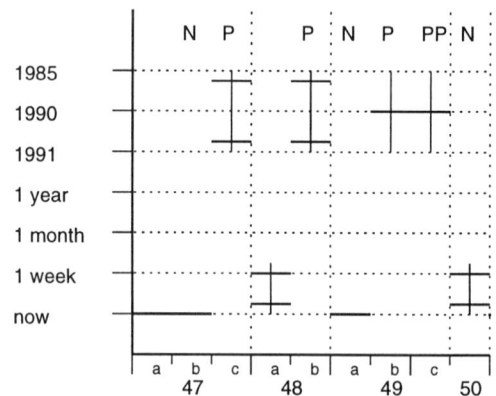

Figure 1: Temporal Structure for Example 2

Figure 1 contains the temporal structure for (each clause) of sentences 47-50 of one of our texts. The corresponding sentences (also broken into finite clauses are contained in Example 2.

Notice that sentence 47 consists of three clauses. The first two (47a and 47b) describe atomic events that are taking place at the "now" time of the story (during the proceedings against Mrs. Mandela). The third clause (47c) refers to an indefinite span of time in the past (roughly from 1985 to 1991, during which Mrs. Mandela's football club existed). Note, the use of the past perfect indicating the change in time and setting.

As Figure 1 illustrates, there is a name (N) reference to Mrs. Mandela in (47b), and a pronominal (P) reference to her in (47c). This pronoun is used even though there is a change in time between (47b) and (47c) and is explained because this is the second reference to Mrs. Mandela within the same sentence (a condition that overrules the time change hypothesis).

(48a) represents a change back to the time of the proceedings (note the discourse-deictic reference (Webber, 1991) "This"). In this case, a time span indicates multiple atomic events occurring during that time. (48b) again points to the time span in the past, though this is not explicitly marked linguistically as it was in (47c). Here world knowledge must be used to understand the time referenced in this clause. Note, however, that the time would have to be part of the input to a generator, and thus our rules are completely well-defined from the generation perspective.

The use of a pronoun to refer to Mrs. Mandela in (48b) is warranted by our hypothesis, because the clause containing the previous mention of Mrs. Mandela, (47c), references the same time as is referenced in (48b).

Because there is a time change between (48b) and (49a), our hypothesis explains the appearance of the proper name in (49a) even though it occurs just after a pronoun (in (48b)) co-specifying with the same character.[2] The pronouns in the remainder of (49) are explained because they are subsequent references within the same sentence (despite the fact that they refer to an unspecified time in the past which is different from the time referenced in (49a)). Finally, the use of a name in (50) is again indicated by the change in time between (49c) and (50).

Ambiguities

Of course, the choice of referring expression is not only guided by discourse structure; there is also an influence due to ambiguities. Dale (1992) generated referring expressions so that they could be distinguished from the other discourse entities mentioned in the context. This strategy can be interpreted as: Generate a pronoun whenever it is not ambiguous. However, how one should define context is not quite clear. For this definition we choose a span of text considered important in our previous work on anaphora resolution (Strube, 1998), and define a referring expression as ambiguous if there is a competing antecedent (i.e. another discourse entity matching in number and gender) mentioned in the previous sentence or to the left of that referring expression in the current sentence. Of the 437 referring expression in the texts we analyzed, 104 were considered ambiguous by this definition. Of these only 51 were realized as a definite description. Thus a rule which says to use a definite description if a pronoun would be ambiguous according to the definition appears to be too strict. Therefore we need to consider ambiguous cases in more detail.

To handle these cases we turned to pronoun resolution algorithms. Our intuition was that a pronoun could be generated to refer to a particular discourse entity if a pronoun resolution algorithm would choose that entity as a referent for the pronoun. To our knowledge, there are only two pronoun resolution algorithms that are specified in enough detail to work on unrestricted naturally occurring text: Brennan et al. (1987) using the definition of utterance according to Kameyama (1998), and Strube (1998). Strube (1998) eval-

uated the effectiveness of these two algorithms on the task of pronoun resolution in some naturally occurring texts. Because Strube's algorithm showed significantly better results, we have turned to it for guidance in pronoun generation.

The idea is that if we want to refer to a discourse entity, E, but there is a competing antecedent, C, we look to Strube's algorithm in the following way. If Strube's algorithm would resolve a pronoun to be E, we use a pronoun. If, instead, Strube's algorithm would prefer C as the referent of the pronoun, we use a definite description to refer to E.

Our analysis showed that the use of Strube's algorithm showed improvement, but it seemed to be too liberal with suggesting pronouns when the competing antecedent was in the previous sentence. Thus, the rule that we settled on is shown in Figure 2.

1. *If* this is the first occurrence of X in the current sentence and
   (a) *if* there is a competing antecedent in the *previous* sentence, use a definite description;
   (b) *if* there is a competing antecedent in the *same* sentence (i.e., to the left) and
      i. *if* Strube's algorithm would resolve a pronoun in this position to be X, use a pronoun;
      ii. *else* use a definite description;

2. *if* this is a subsequent occurrence of X in the current sentence and
   (a) *if* there is an intervening competing antecedent, use a definite description;
   (b) *if* there is no intervening competing antecedent, use a pronoun.

Figure 2: Realization of the Referring Expression X when Competing Antecedents Exist

Anaphoric Expression Generation Algorithm

In the previous sections we have argued that temporal structure should influence pronoun choice and that ambiguous cases need to be handled separately. In addition, we found that if the distance between the current reference and the previous mention of the discourse entity being referred to was far (over two sentences), a definite description was almost always used. Finally, we found that when an entity was referred to multiple times in a sentence, a pronoun was usually used for subsequent references in a sentence.

Based on these findings, we propose the algorithm for realizing anaphoric expressions shown in Figure 3.

Empirical Data

We applied the algorithm described in the previous section to three texts from the *New York Times*. Articles were ranging from a frontpage article to local news. We applied the algorithm to all references to persons in these texts. The algorithm was correct in 370 cases (84.7%), and wrong in 67 cases (15.3%). In Figure 4 we show the distribution over the rules specified in the algorithm.

In order to interpret the results of the algorithm, we must have some comparison. We use the simple scheme shown in Figure 5 for comparison purposes.

---

[2]Note that the use of this definite description cannot be explained by a topic shift since there is no topic shift in between the previous text and (49). At least two discourse entities (*"Mrs. Mandela"* and the *"football club"*) are constant, only Mrs. Mandela's husband does not occur in the immediately preceding sentences.

1. *If* this is a **long distance anaphoric reference** (i.e., if the previous reference to X was more than two sentences prior) use a **definite description**;

2. *else*
   if this is an **unambiguous reference** (i.e., there is no competing antecedent) and this is an **intra-sentential anaphor** (i.e., this is not the first mention of X in the current sentence) use a **pronoun**;

3. *else*
   if this is a **time change** (i.e., there is a difference between the time of the clause with the previous reference to X and the time of the current clause) use a **definite description**;

4. *else*
   if there is a **competing antecedent** (i.e., another object in the previous or current sentence that matches the type and number of X) use the **rule found in Figure 2**;

5. *else*
   for the **remaining cases** use a **pronoun**.

Figure 3: Algorithm for Generating the Appropriate Form for a Referring Expression X

Rule Name		number	percentage
All Rules	correct	370	84.7%
	wrong	67	15.3%
Long Distance Anaph.	correct	46	97.9%
	wrong	1	2.1%
Intra-sentential Anaph.	correct	168	96%
	wrong	7	4%
Time Rules (3 & 5)	correct	116	72.5%
	wrong	44	27.5%
Competing Antecedent	correct	40	72.7%
	wrong	15	27.2%

Figure 4: Results of the Algorithm

The results of applying these rules give 343 correct cases (78.5%) and 94 incorrect ones (21.5%). Hence our algorithm reduces the error rate by 28.9%.

Related Research

A significant amount of work in linguistics has investigated the use of different kinds of anaphoric referring expressions in discourse and their relationship to ease of comprehension. See Arnold (1998) for a discussion of several of the factors involved in referring expression choice. In many cases the various factors seem to affect the *accessibility* of a referent (where accessibility is intended in a broad sense to cover both "topic accessibility" (Givon, 1983) and accessibility due to factors such as recency of mention). Basically, the more accessible a referent the more underspecified a referring expression should be. Accessibility explains the apparent "name-name penalty" as examined in Gordon & Hendrick (1998), for example.

Our work argues that factors beyond accessibility must be considered in anaphoric expression choice. It is consistent with work such as Vonk et al. (1992) whose experiments indicate that a referring expression "... that is more specific than is necessary for the recovery of the intended referent ... marks the beginning of a new theme concerning the same dis-

1. *If* this is a **long distance anaphoric reference** (i.e., if the previous reference to this item is greater than two sentences prior) use a **definite description**;

2. *else*
   if there is a **competing antecedent** (i.e., another object in the previous or current sentences that matches the pronoun which would be used to refer to this entity) use a **definite description**;

3. *else*
   the anaphoric expression would be **unambiguous** so use a **pronoun**.

Figure 5: Simple Algorithm

course referent." (Vonk et al., 1992, page 304). They argue that such overspecified expressions are serving a discourse function of indicating boundaries. This work does not define what a discourse segment boundary actually is. On the other hand, using the definition of time change as a boundary condition, our work is consistent with their hypothesis. Interestingly, Vonk et al. (1992) found that in discourses where a theme change was well marked by other means (e.g., by a preposed adverbial phrase or a subordinate clause indicating time or place) that pronouns were much more common even though a new theme was begun. Presumably such phrases mark the theme change well, and thus it is not necessary to also mark the change via an overspecified description. We are currently reanalyzing our data in light of that finding to see of it provides a fuller account of the naturally occurring data.

Approaches which define discourse segments on the basis of reference resolution (Sidner, 1979; Suri & McCoy, 1994; Strube & Hahn, 1997) are not useful for our purposes because they require referring expressions for recognizing segment boundaries. In contrast to these approaches, we define segment boundaries independently from reference resolution so that in this respect our work is in line with Grosz & Sidner's (1986) definitions.

Future Directions

In analyzing our data, there are several places for further consideration. One problem is that our rule which indicates a definite description should be used in a time change overgenerates definite descriptions. Following Vonk et al. (1992) we plan to investigate whether definite descriptions might best be viewed as boundary markers and whether other markers of discourse boundaries (e.g., preposed adverbial phrases) are found in places where our algorithm suggests a definite description because of a time change but a pronoun appears in the text.

In places where our algorithm overgenerates pronouns, a more sophisticated time analysis may be helpful. Our current analysis distinguishes between four types of time and is driven by both semantic cues in the text (e.g., adverbial time phrases) and changes in tense. Nakhimovsky (1988) also uses changes in "time scale" as a marker for changes in time. We plan to investigate this to see whether it explains more of the examples. Nakhimovsky (1988) also describes several other markers for a setting change, and these will also be investigated to see if they are indicative of definite description use.

Another line of future research involves further investigation of the ambiguous cases. Our current rule was developed by evaluating several different possibilities (e.g., using time change rules, different pronoun resolution algorithms) and selecting a rule that explains most of the cases. Still, the number of ambiguous cases is fairly small and analyzing more texts and concentrating on cases where the current rule makes an incorrect prediction may lead us to a more robust rule.

Conclusions

Pronouns occur frequently in texts and have been hypothesized to play a large role in text coherence. Yet, pronoun generation has not been studied in detail. If future natural language generation systems are to produce coherent, natural texts, they must use rules for generating pronouns that produce pronouns in roughly the same places that human-produced texts do. Moreover, the rules must be based on information that would be available to a sentence generator. At the same time, in order to evaluate rules, they must be based on information that can be gleaned from a text.

In this work we have argued that changes in setting, as indicated by changes in time, provide an explanation for patterns of pronoun use in naturally occurring text. That is, even in places where a pronoun would be unambiguous, a definite description might be used when the time of the sentence is different from the time of the sentence in which the previous mention was made. This hypothesis provides an explanation for many of the uses of definite descriptions found in the studied texts. Other uses of definite descriptions occur because of ambiguities. We have suggested a rule which addresses when such ambiguities should not preclude the generation of a pronoun. Our scheme appears to be a reasonable explanation for the patterns of pronoun use found in our corpus.

Acknowledgments. This work was done while the first author was visiting the Institute for Research in Cognitive Science and while the second author was a post-doctoral fellow there (NSF SBR 8920230). We would like to thank Jennifer Arnold and the centering group at UPenn for helpful discussions. We would also like to thank the anonymous reviewers.

References

Appelt, D. E. (1981). *Planning Natural-Language Utterances to Satisfy Multiple Goals*, (Ph.D. thesis). Stanford University. Also appeared as: SRI International Technical Note 259, March 1982.

Appelt, D. E. (1985). Planning English referring expressions. *Artificial Intelligence*, 26(1):1–33.

Arnold, J. E. (1998). *Reference Form and Discourse Patterns*, (Ph.D. thesis). Stanford University, Department of Linguistics.

Brennan, S. E., M. W. Friedman & C. J. Pollard (1987). A centering approach to pronouns. In *Proceedings of the 25th Annual Meeting of the Association for Computational Linguistics*, Stanford, Cal., 6–9 July 1987, pp. 155–162.

Dale, R. (1992). *Generating Referring Expressions: Constructing Descriptions in a Domain of Objects and Processes.* Cambridge, Mass.: MIT Press.

Dale, R. & E. Reiter (1995). Computational interpretations of the Gricean maxims in the generation of referring expressions. *Cognitive Science*, 18:233–263.

Genette, G. (1980). *Narrative Discourse: An Essay in Method.* Ithaca, N.Y.: Cornell University Press.

Givon, T. (1983). Topic continuity in spoken English. In T. Givon (Ed.), *Topic Continuity in Discourse: A Quantitative Cross-Language Study.* Amsterdam, Philadelphia: John Benjamins.

Gordon, P. C. & R. Hendrick (1998). The representation and processing of coreference in discourse. *Cognitive Science*, 22(4):389–424.

Grosz, B. J., A. K. Joshi & S. Weinstein (1995). Centering: A framework for modeling the local coherence of discourse. *Computational Linguistics*, 21(2):203–225.

Grosz, B. J. & C. L. Sidner (1986). Attention, intentions, and the structure of discourse. *Computational Linguistics*, 12(3):175–204.

Kameyama, M. (1998). Intrasentential centering: A case study. In M. Walker, A. Joshi & E. Prince (Eds.), *Centering Theory in Discourse*, pp. 89–112. Oxford, U.K.: Oxford University Press.

McDonald, D. D. (1980). *Natural Language Production as a Process of Decision Making Under Constraint*, (Ph.D. thesis). MIT.

McKeown, K. R. (1983). Focus constraints on language generation. In *Proceedings of the 8th International Joint Conference on Artificial Intelligence*, Karlsruhe, Germany, August 1983, pp. 582–587.

McKeown, K. R. (1985). *Text Generation: Using Discourse Strategies and Focus Constraints to Generate Natural Language Text.* Cambridge, U.K.: Cambridge University Press.

Moens, M. & M. Steedman (1988). Temporal ontology and temporal reference. *Computational Linguistics*, 14(2):15–28.

Nakhimovsky, A. (1988). Aspect, aspectual class, and the temporal structure of narrative. *Computational Linguistics*, 14(2):29–43.

Passonneau, R. (1996). Using centering to relax Gricean constraints on discourse anaphoric noun ophrases. *Language and Speech*, 39(2):229–264.

Prince, G. (1982). *Narratology: The Form and Functioning of Narrative.* Berlin: Mouton.

Reichman, R. (1985). *Getting Computers to Talk like You and Me.* Cambridge, Mass.: MIT Press.

Reiter, E. (1990). Generating descriptions that exploit a user's domain knowledge. In R. Dale, C. Mellish & M. Zock (Eds.), *Current Research in Natural Language Generation.* London: Academic Press.

Sidner, C. L. (1979). *Towards a Computational Theory of Definite Anaphora Comprehension in English.* Technical Report AI-Memo 537, Cambridge, Mass.: Massachusetts Institute of Technology, AI Lab.

Strube, M. (1998). Never look back: An alternative to centering. In *Proceedings of the 17th International Conference on Computational Linguistics and 36th Annual Meeting of the Association for Computational Linguistics*, Montréal, Québec, Canada, 10–14 August 1998, Vol. 2, pp. 1251–1257.

Strube, M. & U. Hahn (1997). Centered segmentation: Scaling up the centering model to global referential discourse structure. In *Proceedings of the 19th Annual Conference of the Cognitive Science Society*, Palo Alto, Cal., 7–10 August 1997.

Suri, L. Z. & K. F. McCoy (1994). RAFT/RAPR and centering: A comparison and discussion of problems related to processing complex sentences. *Computational Linguistics*, 20(2):301–317.

Vogt, J. (1990). *Aspekte erzählender Prosa: Eine Einführung in Erzähltechnik und Romantheorie* (7th ed.). Opladen: Westdeutscher Verlag.

Vonk, W., L. G. Hustinx & W. H. Simons (1992). The use of referential expressions in structuring discourse. *Language and Cognitive Processes*, 7(3/4):301–333.

Webber, B. L. (1988). Tense as discourse anaphor. *Computational Linguistics*, 14(2):61–73.

Webber, B. L. (1991). Structure and ostension in the interpretation of discourse deixis. *Language and Cognitive Processes*, 6(2):107–135.

Wiebe, J. M. (1994). Tracking point of view in narrative. *Computational Linguistics*, 20(2):233–287.

Holistic and Part-based Processes in Recognition of Upright and Inverted Faces

Margaret C. McKinnon (margaret@psych.utoronto.ca)
Department of Psychology, University of Toronto at Mississauga
3359 Mississauga Rd. North, Mississauga, ON, L5L 1C6

Morris Moscovitch (momos@credit.erin.utoronto.ca)
Department of Psychology, University of Toronto at Mississauga
3359 Mississauga Rd. North, Mississauga, ON, L5L 1C6

Abstract

Participants made a same-different judgment of the internal features of two faces presented simultaneously on screen. Whereas responding to upright faces on "same" trials relied upon holistic processing strategies, responding to upright faces on "different" trials, as well as responding to inverted faces, relied upon part-based processing strategies. Our results are also contrary to earlier reports in that we found that when attention is focused upon the internal features, presentation of these features alone is sufficient to form a discrimination judgment.

Introduction

One line of evidence concerning possible dissociations between face and object processing (e.g., Moscovitch, Winocur, & Behrmann, 1997) concerns the differential effects of inversion on recognition of faces and objects. The greater detrimental effect of inversion upon faces than objects is well-documented (e.g., Bartlett & Searcy, 1993). Specifically, inversion of faces relative to the viewer results in impaired encoding; this effect is greater for faces than for most, but not all (Diamond & Carey, 1986), other objects. Moreover, the available evidence indicates that inversion has a greater effect upon the encoding and subsequent discrimination of the spatial-relational (i.e., configurational) information contained in facial stimuli than it does on discrimination of isolated facial components (e.g., eyes). Indeed, several researchers maintain (e.g., Farah, Tanaka, & Drain, 1995) that whereas recognition of upright faces may rely upon holistic or "configurational" representation schemes, recognition of inverted faces, much like objects, may be reliant upon part-based decompositional strategies. In a pattern of findings similar to the reported dissociation between processing of faces and objects, several researchers report that whereas responding on "same" trials in same-different tasks may rely upon rapid, global modes of processing, responding on "different" trials may instead rely on more analytic or feature-based comparisons (e.g., Taylor, 1976). In fact, this pattern of responding may extend to same-different tasks involving facial stimuli (e.g., Bradshaw & Wallace, 1971).

The present experiment examined these claims in light of one aspect of facial processing that has received some attention in the recent literature: the relationship between the internal (eyes, nose, and mouth) and external (chin, forehead, and hairline) facial features. Claims of differential processing of these features with regards to both familiar and unfamiliar faces are well-documented. For example, whereas some studies (e.g., Nachson, Moscovitch, & Umiltà, 1995) that used different experimental methods provided evidence that external features are more efficacious than internal features in matching tasks for unfamiliar faces, other studies (e.g., Young et al., 1985) have failed to replicate this pattern of findings. By contrast, the results of other researchers (e.g., Ellis, Shepard, & Davies, 1979) converge on the finding that recognition of familiar faces is achieved best through presentation of the internal features.

In the present series of experiments, we proposed to investigate differential processing of the internal and external facial features through a series of manipulations involving same-different judgments of unfamiliar faces. We investigated these claims using an inversion paradigm by initially presenting participants with an upright face and testing their ability to make same-different judgments between this face and a comparison face that was presented either upright or inverted.

Experiment 1

Method

In the first experiment, participants were initially presented with a whole face at study (internal and external features present) and were required to make a same-different judgment of a comparison face. There were four different sets of comparison faces: i) upright faces (whole face); ii) upright faces (internal features only); iii) inverted faces (whole face) and; iv) inverted faces (internal features only). Twenty-four right-handed undergraduate students participated in the experiment; participants made their same-different judgment following simultaneous presentation of a vertically-aligned test and comparison face. "Same" in this experiment referred to full congruency between the internal features of study and comparison faces; the external features were held constant across study and comparison presentations. Each session involved 128 test presentations; sixty-four were "same" judgments and sixty-four "different" judgments.

Because initial study presentations involved *upright* whole faces, we reasoned that participants would rely upon holistic or configurational processing strategies at encoding of these stimuli (e.g., Farah et al., 1995). Indeed, we expected these strategies to result in superior performance for whole (internal and external features) rather than part (internal features only) presentations of comparison faces on upright "same" trials where a match would be provided between holistic processing strategies for study and comparison faces that dictated attention to the entire face. By contrast, presentations of whole comparison faces would no longer be favored on upright "different" trials and on all inverted comparison presentations ("same" or different"), when discrimination performance may be based upon individual (separate) consideration of the internal and external features as the result of viewers' reliance upon analytic or part-based encoding strategies.

Results and Discussion

Indeed, the findings of this experiment were consistent with these predictions. Our results were confirmed using a factorial ANOVA design. Mean accuracy scores were entered into a 2 (Mode of Responding: Same and Different) X 2 (Target Orientation: Upright and Inverted) X 2 (Target Format: Part and Whole) factorial ANOVA with Mode of Responding, Target Orientation, and Target Format treated as within-subjects factors. Median reaction times were calculated for each subject for each condition; only reaction times for accurate responses were included in the analysis. We eliminated from the analysis all reaction times that fell more than two standard deviations from the mean and calculated new medians using the remaining data. Testing revealed a main effect of Target Orientation ($p < .001$); upright faces were processed more efficiently than were inverted faces [REGW-Q ($p < .05$)]. A three-way interaction between Mode of Responding, Target Orientation, and Target Format ($p < .05$) confirmed that whereas responding for upright items on "same" trials varied as a function of part versus whole presentations ($p < .01$), responding did not vary for either upright "different" responses ($p > .05$) or inverted "same' ($p > .05$) and "different" ($p > .05$) responses. Hence, viewers exhibited superior performance for whole, rather than part, comparison faces upon upright "same" presentations only (see Table 1).

Table 1: Experiment 1: Mean accuracy and reaction time by mode of response and target format.

	Same-Upright	Same-Inverted	Different-Upright	Different-Inverted
ACCURACY				
Part	0.94	0.82	0.90	0.76
Whole	0.98	0.81	0.91	0.77
REACTION TIME (in ms)				
Part	1642.94	2154.79	1582.15	2056.13
Whole	1475.29	2146.02	1645.04	2023.52

Analysis of participants' mean accuracy scores revealed a main effect of Target Orientation [$p < .001$; REGW-Q ($p < .05$)]; upright faces were processed more accurately than were inverted faces. The three-way interaction between Mode of Responding, Target Orientation, and Target Form, however, failed to attain significance, ($p > .05$) (see Table 1).

Although analysis of the accuracy scores revealed little evidence of differences in participant responding, viewers nonetheless exhibited more efficient performance for whole

comparison faces on upright "same" trials; this finding may stem from a reliance upon holistic encoding strategies requiring attention to the entire comparison face, as would also be the case at encoding of study faces. We found little evidence of differences in responding to part versus whole faces on upright "different" trials; an initial reliance upon holistic processing strategies at encoding of the study faces may have counteracted more part-based representations of comparison faces that relied upon separate consideration of the internal and external features. Similar processes likely resulted in no differences being observed for part versus whole comparison faces on inverted trials ("same" or "different").

Experiment 2

Method

In a second experiment, we examined whether initial presentation of internal features alone would heighten viewers' reliance upon these features for discrimination. Hence, we conducted a second experiment identical to the first, with the exception that study faces were comprised of the internal features only. Initial encoding of the internal features was expected to result in superior performance being observed for part rather than whole comparison faces, regardless of the mode of responding required, or the target orientation of the stimuli. Although initial presentation of upright study faces may have engaged holistic encoding strategies, viewer representations would nonetheless include the internal features only. On upright "same" trials, such representations would be congruent with holistic representations of part (internal features) comparison faces only. Moreover, because upright "different" and all inverted trials may rely upon part-based encoding strategies, we also expected to observe superior performance for part faces on these trials. Indeed, part-based decompositional strategies may require attention to individual facial features such as the eyes, nose, and mouth contained in the internal features. Under such conditions, the external features may prove distracting at test, when attention may be focused on the internal features alone.

Results and Discussion

Our findings were consistent with these predictions. Analysis of participants' median reaction time scores indicated main effects of Target Orientation ($p < .001$) and Target Format ($p < .001$); upright faces were processed more efficiently than inverted faces and part faces were processed more efficiently than whole faces [REGW-Q ($ps < .05$)]. The three-way interaction between Mode of Responding, Target Orientation, and Target Format also approached significance ($p < .07$); viewers exhibited superior performance for part, rather than whole, presentations of comparison faces upon upright "same" ($p < .001$) and "different" ($p < .001$) trials, and inverted "same" trials ($p < .001$). Such differences were not apparent on inverted "different" trials ($p > .05$) (see Table 2).

Analysis of participants' mean accuracy scores also revealed main effects of Target Orientation ($p < .001$) and Target Format ($p < .01$); upright faces were processed more accurately than were inverted faces and part faces were processed more accurately than were whole faces [REGW-Q (ps < .05)] (see Table 2).

Table 2: Mean accuracy and reaction time by mode of response and target format

		Same-Upright	Same-Inverted	Different-Upright	Different-Inverted
ACCURACY					
	Part	0.98	0.90	0.93	0.75
	Whole	0.94	0.89	0.84	0.77
REACTION TIME (in ms)					
	Part	1230.77	1981.23	1549.10	2351.64
	Whole	1893.10	2578.21	2121.15	2443.56

Thus, the results of Experiment 2 were consistent with our suggestion that initial encoding of the internal features alone would result in superior performance being observed for part rather than whole comparison faces. Although faster responding for part faces was not observed for inverted faces on "different" trials, an inspection of the response data indicated that "different" trials may have engaged time-consuming serial processing strategies not required for "same" responses (e.g., Taylor, 1976). Moreover, in contrast to previous investigations of recognition performance for unfamiliar faces (e.g., Nachson et al., 1995), these results also indicate that attending to the internal features alone is sufficient to form a discrimination judgment. These findings are in line with earlier suggestions by Moscovitch and his colleagues (Moscovitch et al., 1997) that the internal features may carry the burden of information in facial recognition processing. Indeed, the external features may prove an unnecessary distraction at test when attention is focused on the internal features alone. The slower and less accurate responding observed for presentations of comparison faces comprised of both the internal and external features upon all upright presentations of comparison faces, as well as on inverted "same" trials, indicates that these features may add no new or informative information for discrimination.

Discussion

Our experiments provide clear evidence that whereas responding to upright faces on "same" trials may rely upon holistic or configurational processing strategies, responding to upright faces on "different" trials, as well as to inverted faces, relies upon part-based or decompositional processing strategies. In addition, contrary to earlier reports that recognition of unfamiliar faces may be most efficacious following presentation of the external features (e.g., Naschon et al., 1995), we found that presentation of the internal features alone is sufficient to form a discrimination judgment when attention is focused upon these features. Future experiments are planned to determine whether the same pattern of responding will extend to the processing of objects (e.g., houses).

Acknowledgments
The authors would like to thank Giampaolo Moraglia for comments on an earlier version of this paper and Marilyne Zeigler for technical assistance. Funding for this project was provided by a grant awarded to the second author by the Natural Sciences and Engineering Research Council of Canada - Grant A8347.

References
Bartlett, J. C., & Searcy, J. (1993). Inversion and configuration of faces. *Cognitive Psychology, 25,* 281-316.

Bradshaw, J.L., & Wallace, G. (1971). Models for the processing and identification of faces. *Perception and Psychophysics, 9,* 443-448.

Diamond, R., & Carey, S. (1986). Why faces are and are not special: An effect of expertise. *Journal of Experimental Psychology: General, 115,* 107-117.

Ellis, H. D., Shepard, J. W., & Davies, G. M. (1979). Identification of familiar and unfamiliar faces from internal and external features: Source implication for theories of recognition. *Perception, 8,* 431-439.

Moscovitch, M., Behrmann, M., & Winocur, G. (1997). What is special about face recognition? Nineteen experiments on a person with visual object agnosia and dyslexia but normal face recognition. *Journal of Cognitive Neuroscience, 9,* 555-604.

Naschon, I., Moscovitch, M., & Umiltà, C. (1995). The contribution of external and internal features to the matching of unfamiliar faces. *Psychological Research, 58,* 31-37.

Taylor, D. A. (1976a). Holistic and analytic processes in the comparison of letters. *Perception and Psychophysics, 20,* 187-190.

Young, A. W., Hay, D. C., McWeeny, K. H., Flude, B. M., & Ellis, A. W. (1985). Matching familiar and unfamiliar faces on internal and external features. *Perception, 14,* 737-746.

Training Reading Strategies

Danielle S. McNamara (dmcnamar@odu.edu)
Jennifer L. Scott (wecarpedium@earthlink.net)
Old Dominion University; Department of Psychology
Norfolk VA 23529 USA

Abstract

Readers who self-explain texts aloud understand more from a text and construct better mental models of the content. This study examined the effects of providing self-explanation training on text comprehension, as well as course grades. Effects of prior knowledge and reading skill were also examined in relation to the benefits of self-explaining and self-explanation training. In general, low-knowledge readers gained more from training than did high-knowledge readers.

Introduction

We often read in order to learn new information. In these situations, compared to when we read for pleasure, the texts tend to be more difficult to understand and the information tends to be more difficult to learn. This research focuses on providing readers with reading strategies to help them understand more and learn more from difficult expository texts. Thus, the primary purpose of this research is to examine the effectiveness of a training intervention designed to improve learning from texts and to determine whether the success of this intervention depends on readers' individual differences. Specifically, we examine the following individual differences: reading skill, working memory capacity, and prior domain knowledge. Reading skill is considered important to learning from text because it makes available resources for higher-level processing (e.g., Perfetti, 1985). The availability of resources is also assumed to be at least partially determined by the individual's working memory capacity. This latter assumption stems from research showing correlations between reading comprehension measures and working memory tasks (e.g., LaPointe & Engle, 1990; Daneman & Carpenter, 1980; cf. McNamara & Scott, 1999).

Perhaps the most influential factor in learning from texts is the reader's prior domain knowledge (e.g., Chiesi, Spilich, & Voss, 1979). Students with more domain knowledge better understand difficult text material from that domain. However, actually using this prior knowledge is also a key to learning. Readers can be induced to actively engage their prior knowledge by introducing gaps in the text, which require knowledge-based inferences (e.g., McNamara & Kintsch, 1996; McNamara, Kintsch, Songer, & Kintsch, 1996). Alternatively, this active engagement can emerge from reading strategies. Individuals who more actively read text also tend to comprehend more and learn more from text (e.g., Chi & Bassok, 1989). Active processing has long been recognized as critical for ensuring superior and stable retention because it results in the integration of new information with prior knowledge.

However, our theoretical understanding of the relationship and interdependency between prior knowledge and active processing remains incomplete, and we are unable to reliably predict the specific conditions under which prior knowledge and skill will play a role during learning. One premise of this research is that developing a better understanding of the relationship between prior knowledge and instructional techniques is a key to improving training methodologies.

One intervention that has been found to increase the reader's use of prior knowledge, and thus improve comprehension during reading, is called *self-explanation*. Self-explanation is the process of actively explaining the text to yourself while reading (e.g., Chi, de Leeuw, Chiu, & LaVancher, 1994). Chi and her colleagues have found that readers who explain the text, either spontaneously or when prompted to do so, understand more from a text and construct better mental models of the content. However, some readers are better self-explainers than others; less-skilled self-explainers offer little to the text to help them better understand it. One question addressed by this research is whether readers can be trained to become more-skilled self explainers (see also, Bielaczyc, Pirolli, & Brown, 1995). We address this question by providing reading strategy instruction combined with practice using the self-explanation technique. Secondly, we investigate how prior knowledge impacts the benefits of self-explanation. In Experiment 1, we first examine the effects of self-explanation in comparison to reading aloud during a training phase. We then test the benefits of training by having all of the participants self-explain a text aloud and examining their comprehension of the text. Experiment 2 looks further at the effects of this strategy training on students' course grades.

General Method

In each experiment, approximately half of the participants were randomly assigned to a self-explanation training condition, and the other half to a control condition. During two sessions of the training phase, participants in both conditions read four science texts (each concerning a different topic in science). The participants in the self-explanation training condition first received a brief instruction in self-explanation and reading strategies. The instruction focused on the benefits of using logic to understand the text, predicting what the text would say, making bridging inferences, and monitoring comprehension. For each strategy, a description of the strategy and examples were provided. The participants in the training condition then read and self-explained aloud four science texts. During this training phase, the experimenter prompted the participant to provide additional explanation for the text when necessary (e.g., when simply paraphrasing the text). After reading each text, the participants answered three to six open-ended questions about the text. For each text, the participants in the training condition then watched a video of another student self-explaining the text and identified strategies used by the student in the video. Control participants read aloud the same four science texts and answered the same questions about the texts, but did not self-explain the texts and did not watch the videos. The control condition affords the comparison of reading aloud to both reading aloud and self explanation during the training phase. It also allows us to compare individuals with and without reading strategy training.

After the training phase, all of the participants self-explained a low-coherence, difficult text about cell mitosis. The participants were told to self-explain while reading the text, but were not prompted to do so by the experimenter. The participants then answered written questions about the text. The test comprised two types of comprehension questions, text-based and bridging inference questions. To answer text-based questions, the participant must only remember one particular sentence, or idea, from the text. However, on bridging inference questions, the participant must remember separate portions of the text and understand the relationships between those ideas. The participants also answered questions designed to assess their prior knowledge of cells and cell mitosis. These questions were also related to cells and cell division, but the information to answer the questions was not presented in the text (e.g., Name three reasons for, or purposes of, cell reproduction).

Experiment 1

In Experiment 1, the participants were 43 undergraduate psychology and biology students. In addition to the measures described in the General Method section, participants were administered the Nelson Denny reading comprehension test and a working memory (WM) capacity test. The WM capacity test used was a reading span test (e.g., La Pointe & Engle, 1990). This test requires the participant to read two to six sentences, each followed by an unrelated word. After the sentences are presented, the participant is to recall the unrelated words. WM capacity is estimated on the basis of the number of words recalled in order.

Comprehension Performance During Training

An ANOVA was conducted on the proportion of questions answered correctly during training including the between-

subjects variable of condition (self explain, read aloud) and the within-subjects variable of text (i.e., four texts). Participants who self-explained while they read the text answered more questions correctly (M=0.56) than those who simply read aloud (M=0.47), $F(1,41)=5.7$, $p=.021$. There was also a reliable effect of text, $F(3,39)=82.5$, $p<.001$, and a marginal interaction of text and condition, $F(3,39)=2.6$, $p=.067$. As shown in Figure 1, the effect of text was primarily due to low comprehension performance for the second and fourth texts; these texts concerned more difficult topics than did the other two texts. The interaction of text and condition seems to be due to a pronounced effect of condition for the second text. Indeed, post-hoc comparisons confirmed that the only text for which self-explanation had a reliable effect was the second text, $F(1,41)=9.3$, $p<.01$ (all other $p >.10$). It is noteworthy that in addition to its difficulty in terms of content, the second text also seemed to require more causal inferences.

Did individual differences mediate these effects? WM capacity as measured by the reading span test did not; it failed to correlate with comprehension performance during training, and did not interact with any of the variables. However, reading skill, in terms of Nelson Denny performance (M=28.5), reliably correlated with overall comprehension (r=.35, $p=0.02$) during training. Although skilled readers better understood the four texts (n=21; M=0.58) than did less-skilled readers (n=22; M=0.49), $F(1,41)=5.4$, $p<.02$, an analysis of covariance indicated that this effect was marginal, $F(1,39)=4.0$, $p=.06$, when reading condition was also considered, $F(1,39)=8.2$, $p=.04$. Thus, self-explanation seemed to have a stronger effect than reading skill.

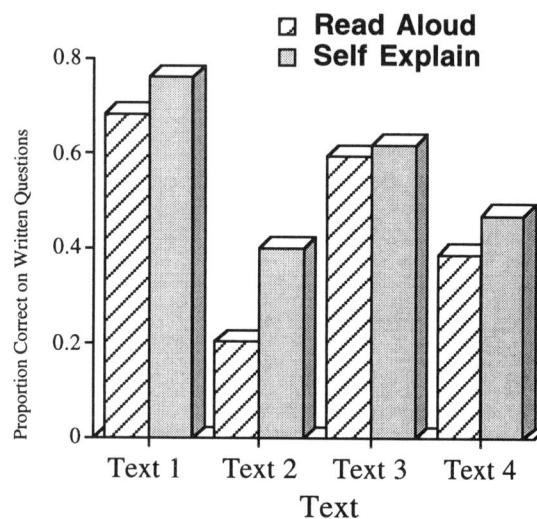

Figure 1: Comprehension performance during training across the four texts. There was little effect of self-explanation in comparison to reading aloud, except for the second text concerning the role of plants in the carbon cycle.

Prior knowledge of cells also correlated with overall comprehension performance during training (r=.44, p<.01). Prior knowledge of cells was measured to determine its effects for the fifth target text that was read after training; however, this specific knowledge also reflected general science knowledge. This correlation reflects the finding that high-knowledge readers better understood the four texts (n=23; M=0.58) than did low-knowledge readers (n=20; M=0.48), F(1,41)=10.0, p<.01. Utilizing an analysis of covariance, there were reliable effects of condition, F(1,39)=4.6, p<.01, and prior knowledge, F(1,39)=14.4, p<.01. Thus, in contrast to the effects of reading skill, the effects of prior knowledge were accentuated by reading condition. This makes sense given that the primary activity during self-explanation is the use of prior knowledge.

In summary, it appears that reading condition and prior knowledge have the strongest effects on comprehension during training. Indeed, regression analyses confirm that when all three variables were considered (i.e., reading condition, prior knowledge, reading skill), only reading condition and prior knowledge reliably predicted performance. That is, reading skill was not a reliable predictor when prior knowledge was also entered into the equation (regardless of text).

Post-training Comprehension Performance

Correlations and regression analyses indicated that neither reading skill nor WM capacity contributed to comprehension of the fifth text, self-explained by all participants (one participant was excluded from these analyses due to ceiling performance on both the post-training comprehension and the prior knowledge questions). On the other hand, prior knowledge correlated highly with comprehension (r=0.63, p<.001). Therefore, an analysis of covariance was conducted including the between-subjects variables of condition (trained, control) and prior knowledge, and the within-subjects variable of question type (text-based, bridging). This analysis yielded a reliable effect of prior knowledge, F(3,38)=29.6, p<.01, and a marginal effect of training condition, F(3,38)=3.7, p=.06. There was also a reliable difference between the two types of questions, F(1,38)=55.5, p<.01.

In addition, it appeared that the effects of prior knowledge and training condition depended somewhat on the type of comprehension question, F(1,38)=3.1, p=.08. As shown in Figure 2a, for bridging inference questions, only prior knowledge affected performance: Low-knowledge readers answered fewer questions correctly (M=0.16) than high-knowledge readers (M=0.38), F(1,38)=30.4, p<.01. As shown in Figure 2b, the results for text-based questions were quite different. For text-based questions, both prior knowledge, F(1,38)=15.5, p<.01, and training condition, F(1,38)=4.1, p=.05, reliably impacted comprehension. There was also a marginal interaction, F(1,38)=3.8, p=.06. This interaction indicated that training had little effect for high-knowledge participants (F<1), but had a large effect for low-

knowledge participants, F(1,18)=9.4, p<.01. Thus, self-explanation training greatly benefited low-knowledge readers, but only in terms of their textbase level of understanding. In contrast, low-knowledge participants who had been in the read aloud, control condition comprehended very little of the text.

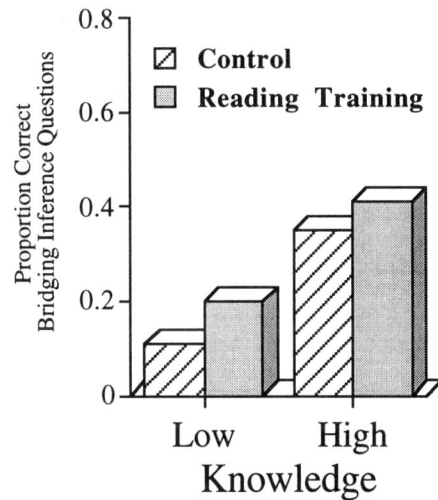

Figure 2a: Post-training comprehension performance on bridging inference questions as a function of condition and prior knowledge in Experiment 1.

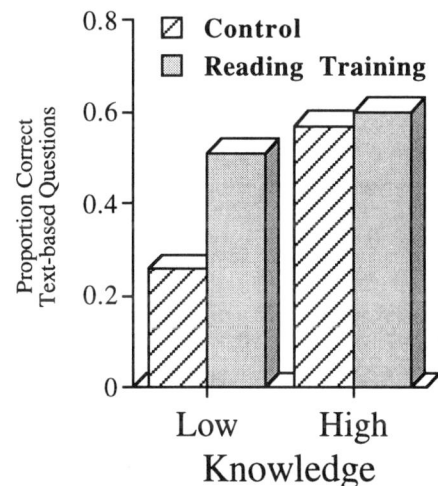

Figure 2b: Post-training comprehension performance on text-based questions as a function of condition and prior knowledge in Experiment 1.

Experiment 2

The primary purpose of Experiment 2 was to examine whether our training program would improve students' course grades in an introductory biology course. This course is a life sciences course in which students learn about living organisms and cells. A total of 360 students in the course were administered reading skill and prior knowledge tests, including the Nelson Denny reading comprehension test and a prior knowledge test including 31 multiple choice questions concerning scientific concepts, and 20 multiple choice questions concerning general knowledge (e.g., humanities, literature, art, etc.). The questions for the prior knowledge test were selected from published test banks based on the experimenters' judgment that skilled (or high-knowledge) freshman college students should be able to answer them. The students completed the tests in small groups. Of the total 360 students, 41 students participated in the training study. The students participated in the training study for extra credit in the course and were randomly assigned to either the self-explanation training condition or the comparison condition. The same training procedure was used as described in the General Method section.

Comprehension Performance During Training

An ANOVA was conducted on proportion correct during training including the between-subjects variable of condition (self explain, read aloud) and the within-subjects variable of text (i.e., the four texts). In contrast to Experiment 1, there was not a reliable effect of self-explanation (M=0.42) in comparison to reading aloud (M=0.46) (all F<2). Once again, the advantage of self-explanation was more pronounced for the text concerning plants (i.e., the second text), but this effect was not reliable. (No individual difference variables interacted with training condition either).

Correlational analyses were preformed to determine the effects of individual differences during training. Performance on the Nelson Denny (M=29.0) correlated with comprehension (r=.45, p<.01), reflecting better overall comprehension for skilled readers (M=0.55) than less-skilled readers (M=0.39). In addition, all three measures of knowledge correlated reliably with comprehension performance: domain knowledge of cells (r=.43, p<.01), general science, (r=.49, p<.01), and general humanities (r=.37, p<.02). In terms of predicting comprehension performance, regression analyses indicated that the best model resulted from entering in the model only two variables, reading comprehension (F(1,38)=9.0) and prior domain knowledge of cells (F(1,38)=8.2) (Model F(2,38)=10.0, p<.001, R^2=.36). However, the effect of knowledge was somewhat different than in Experiment 1. This is because the average level of prior domain knowledge for these participants was considerably greater (M=0.48) than the participants in Experiment 1 (M=0.26). Therefore, knowledge in this case was better characterized in terms of three levels. Accordingly, low-knowledge participants performed less well on the comprehension questions (n=15, M=0.38) than either medium (n=13, M=0.55) or high-knowledge participants (n=13, M=0.56).

Post-training Comprehension Performance

Correlations were examined between individual difference variables and performance on comprehension questions after training (see Table 1). The strongest correlation for both types of questions was with prior knowledge of cells. General Science knowledge also correlated reliably with both types of questions, but more so with text-based questions. Nelson Denny scores correlated only with the text-based questions. These latter two findings support (to some extent) the idea that understanding a text at the textbase level relies more on reading skill. These findings collectively indicate that an individual difference measure correlates with comprehension performance during self-explanation largely to the extent that it taps into the reader's prior knowledge of the domain; and open-ended, domain-specific questions do so better than other types of measures.

Table 1: Correlations between questions and tests in Experiment 2.

| | Question Type | |
	Text-based	Bridging Inf.
Nelson Denny Comprehension	0.329 *	0.172
Prior Knowledge of Cells	0.651 **	0.562 **
Science Knowledge	0.484 **	0.335 *
Humanities Knowledge	0.227	0.081

*p<.05, **p<.01

As in Experiment 1, an analysis of covariance was conducted including the between-subjects variables of condition (trained, control) and prior knowledge (of cells), and the within-subjects variable of question type (text-based, bridging). This analysis yielded a reliable effect of prior knowledge, F(3,37)=30.7, p<.01, and a marginal effect of training condition, F(3,37)=3.0, p=.09. There was also a reliable difference between the two types of questions, F(1,37)=9.0, p<.01. This difference depended on prior knowledge, F(1,37)=4.1, p=.05, but not experimental condition.

□ **Control Condition**
□ **Reading Training Condition**

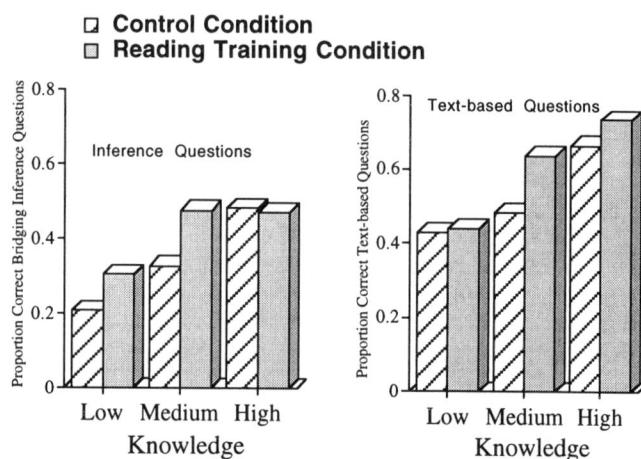

Figures 3a and 3b: Figure 3a, on the left, shows post-training comprehension performance on bridging inference questions. Figure 3b shows post-training comprehension performance on text-based questions.

390

Figures 3a and 3b show participants' performance on comprehension questions after training. These graphs indicate that there is a moderate effect of training for low and medium-knowledge participants. Indeed, further analyses showed that training condition reliably improved performance for these participants, but only for the bridging inference questions, $F(1,24)=4.6$, $p=.04$. Thus, although Experiment 2 is showing an effect of training for participants with relatively less domain knowledge, in contrast to Experiment 1, the effect emerges on the situation model level of understanding rather than the textbase level.

Course Performance

Students in the Biology 108 course took six quizzes, a midterm and a final exam. Thirty-nine of the participants completed the course. A comparison of course scores between control (n=20) and trained (n=19) participants revealed no reliable effects of training. There were no differences on any of the course quizzes or exams between the two groups of students. However, within a questionnaire administered at the end of the course, many of the students in the control condition reported using the self-explanation strategy and that this strategy had been useful in the course. Although they had not been provided with strategy training nor had they extensively practiced the technique, they had been given a brief description of the technique and had used self-explanation while reading the text about cell mitosis. It was hypothesized that the lack of difference between the participants in the training and comparison conditions was because the participants in the comparison condition had been provided sufficient exposure to the technique to use it for their course readings. The control participants' self-reports of using the self-explanation strategy in the course supported that assumption. In contrast, the remaining 319 students in the course who had not participated in the study also had not been introduced to the self-explanation technique. Therefore, 39 non-participant students were identified who matched the 39 participants in terms of both prior knowledge and reading skill. These students are referred to here as *matched controls*.

First, which measures best predicted performance in the course? To answer this question, correlations were computed including all of the students in the course. The three individual difference measures, reading comprehension, science knowledge, and humanities knowledge, reliably correlated with course performance without exception (r=.18 to r=.30). The strongest relationship to course performance was with general science knowledge (r=.30, p<.001). (This result should be of little surprise.) However, the best predictor of performance was found to be a combination of all three variables (r=.35, p<.001). Thus, for further analyses, *skill* was defined in terms of a multiplicative function of the three individual difference variables in order to capture the effects of all three variables. The multiplicative function assumes a network-like relationship between the variables. This definition also assumes that either variable can add to performance, but that they are somewhat interdependent. Hence, if you can read well and have knowledge, you will perform very well in the course; but if you can only read well, or only have knowledge, you will do less well.

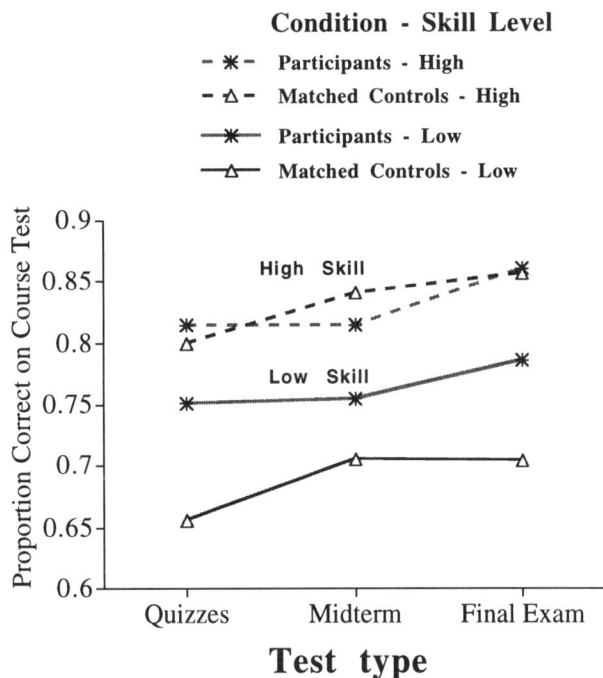

Figure 4. Participants' and matched control students' (who had not participated) course performance as a function of prior skill. Participants with less prior skill particularly benefited from the self-explanation training on course tests and exams.

Based on the three individual difference variables, the students were classified as high, medium, and low skill. This classification resulted in 26 students per skill level, with an equal number of students from each skill level (i.e., n=13) in the participant and matched control group. There was little difference between the high- and medium-skilled students' performance; thus, this analysis focuses on only the high- and low-skill students. Figure 4 shows the participants' and matched control students' performance on three quizzes (i.e., quizzes after training occurred), the midterm, and the final as a function of skill. It is evident that there were no differences during the course between high-skill participants and their matched controls (the result was similar for medium-skill students). However, low-skill participants (who had been trained or introduced to the self-explanation technique) performed better in the course (M=0.76) than did their matched controls (M=0.68). For these students, there was a 10% advantage on the quizzes and a 9% advantage on the final exam. This advantage was reliable for the overall course performance, $F(1,24)=4.3$, $p=.04$, and for quiz performance, $F(1,24)=4.7$, $p=.04$, and marginal in terms of performance on the final exam, $F(1,24)=3.0$, $p=.09$. It was not reliable for performance on the midterm exam, $F<2$.

These results indicate that both self-explanation training and a brief exposure to the self-explanation technique improved course performance for low-knowledge participants in comparison to low-knowledge students who did not learn about the technique.

Conclusions

Is self-explanation beneficial in comparison to reading aloud? Experiment 1 indicated that it was beneficial, but primarily for more difficult text. Experiment 2, however, showed little advantage for self-explanation. This absence of an effect was probably because the students in Experiment 2 possessed more domain knowledge than did those in Experiment 1. Indeed, both experiments indicated that the strongest predictor of comprehension during training was prior knowledge. These findings qualify previous findings showing little effect of knowledge during self-explanation (e.g., Chi et al., 1994). In contrast, here we see a large effect of prior knowledge, compared to moderate effects of self-explanation.

Was self-explanation training effective? We answered this question by having all of the participants self-explain a text after the training phase. In Experiment 1, training had a positive effect on text comprehension but only for low-knowledge participants, and only at the textbase level of understanding. However, in Experiment 2, there was a moderate effect of training for low and medium-knowledge participants, but this time only for bridging questions. Once again, perhaps this finding is a function of the relative level of knowledge of the participants between the two experiments. Indeed, the best predictor of comprehension in both experiments was prior domain knowledge. In any case, both experiments demonstrated that self-explanation training helped the low-knowledge reader to exercise what knowledge they had, such as logic and common sense, to construct a meaningful representation of the text. In comparison, the active processing during self-explanation was not successful for the low-knowledge readers who were not provided with the training.

Experiment 2 further indicated that either training or exposure to self-explanation improved course performance, but only for low-skilled students. Although there were no differences in course performance between participants who received self-explanation training and control participants who only self-explained one text, it was found that exposure to the self-explanation technique in the control condition led those participants to use this method in their course. Thus, the 39 participants were compared to 39 students in the course (matched to each participant in terms of prior knowledge and reading skill). This analysis showed that for less skilled students, but not skilled students, exposure to the self-explanation reading technique led to superior scores (and grades) in the course in comparison to their counterparts.

Why does self-explanation primarily help low-knowledge readers? Primarily because the knowledge is readily available to the high-knowledge reader - there is no need to use strategic processing, particularly when a difficult, low-coherence text (as used here) forces the reader to use prior knowledge (see e.g., McNamara & Kintsch, 1996). On the other hand, the low-knowledge reader cannot readily access knowledge to understand the difficult text. Self-explanation, and self-explanation training teaches the reader in that situation to use what knowledge is available -- that is, logic and common sense. Thus, self-explanation training will be particularly helpful for readers when they encounter a text from an unfamiliar, and difficult domain.

Acknowledgments

This project was funded by a Career Development award from the James S. McDonnell Foundation, and from a Research award from Old Dominion University College of Sciences, to Danielle S. McNamara. We are grateful to a number of students who helped with this project, including Bryan Hayes, Sharon Kruzka, Susan Lee, Ann Rumble, and Quinn Schroeder.

References

Bielaczyc, K., Pirolli, P. L., & Brown, A. L. (1995). Training in self-explanation and self-regulation strategies: Investigating the effects of knowledge acquisition activities on problem solving. *Cognition and Instruction, 13*, 221-252.

Chi, M. T. H., de Leeuw, N., Chiu, M., & LaVancher, C. (1994). Eliciting self-explanations improves understanding. *Cognitive Science, 18*, 439-477.

Chi, M. T. H., & Bassok, M. (1989). Learning from examples via self-explanations. In L. B. Resnick (Ed.), *Knowing, learning, and instruction: Essays in honor of Robert Glaser* (pp. 251-282). Hillsdale, NJ: Erlbaum.

Chiesi, H. I., Spilich, G. J., & Voss, J. F. (1979). Acquisition of domain-related information in relation to high and low domain knowledge. *Journal of Verbal Learning and Verbal Behavior, 18*, 275-290.

Daneman, M., & Carpenter, P. A. (1980). Individual differences in working memory and reading. *Journal of Verbal Learning and Verbal Behavior, 19*, 450-466.

La Pointe, L. B., & Engle, R. W. (1990). Simple and complex word spans as measures of working memory capacity. *Journal of Experimental Psychology: Learning, Memory, and Cognition, 16*, 1118-1133.

McNamara, D. S., & Kintsch, W. (1996). Learning from Text: Effects of prior knowledge and text coherence. *Discourse Processes, 22*, 247-287.

McNamara, D. S., Kintsch, E., Songer, N. B., & Kintsch, W. (1996). Are good texts always better? Text coherence, background knowledge, and levels of understanding in learning from text. *Cognition and Instruction, 14*, 1-43.

McNamara, D. S., & Scott, J. L. (1999, August). Is it memory, or is it metamemory? Paper presented at the Ninth Annual Meeting of the Society for Text and Discourse, Vancouver, Canada.

Perfetti, C. A. (1985). *Reading Ability*. NY: Oxford Univ. Press.

Exploring the Role of Context and Sparse Coding on the Formation of Internal Representations

David A. Medler (medler@cnbc.cmu.edu)
James L. McClelland (jlm@cnbc.cmu.edu)
Center for the Neural Basis of Cognition; Carnegie Mellon University
Pittsburgh, PA 15213 USA

Abstract

Recently, Bayesian principles have been successfully applied to connectionist networks with an eye towards studying the formation of internal representations. Our current work grows out of an unsupervised, generative framework being applied to understand the representations used in visual cortex (Olshausen & Field, 1996) and to discover the underlying structure in hierarchical visual domains (Lewicki & Sejnowski, 1997). We modified Lewicki and Sejnowski's approach to study how incorporating two specific constraints—context and sparse coding—affect the development of internal representations in networks learning a feature based alphabet. Analyses of the trained networks show that (1) the standard framework works well for limited data sets, but tends to poorer performance with larger data sets; (2) context alone improves performance while developing minimalistic internal representations; (3) sparse coding alone improves performance and actually develops internal representations that are somewhat redundant; (4) the combination of context and sparse coding constraints increases network accuracy and forms more robust internal representations, especially for larger data sets. Furthermore, by manipulating the form of the sparse coding constraint, networks can be encouraged to adopt either distributed or local encodings of surface features. Feedback connections in the brain may provide context information to relatively low-level visual areas, thereby informing their ability to discover structure in their inputs.

Introduction

Bayesian principles have been regaining popularity within cognitive science, both in the more traditional approaches to cognitive psychology (e.g., Anderson, 1990) and within the connectionist approach to cognition (e.g., MacKay, 1995; McClelland, 1998). Our current work is a preliminary investigation of incorporating two specific constraints, *context* and *sparse coding*, into an existing Bayesian unsupervised learning paradigm for multilayered architectures (Lewicki & Sejnowski, 1997). The concept underlying the original framework is that higher order internal representations can be formed by exploiting the statistical structure of simple features within an input stream. Indeed, Lewicki and Sejnowski were able to show that their networks could extract hierarchical structure from simple visual domains.

In this paper, we modify and expand the original framework to explore the internal representations of networks trained on feature-based letters. We first modified the framework to directly incorporate contextual information into the deep structure of the network. In the current experiments, "context" is defined as unique information that is presented to the network concurrently with an input pattern. Therefore, context can be used to uniquely identify input patterns and, thus, provide hints about which collection of simple features constitute higher-order representations.

The second—and more substantial—manipulation was to place prior constraints on the base probabilities of unit activations within the networks. This "sparse coding" constraint encourages a network to use relatively few units to represent any specific input pattern. That is, a sparsely coded network uses only a relatively small proportion of units to encode the internal representation for a given pattern. Sparse encoding within neural networks has previously been shown to create more biologically plausible receptive fields (e.g., Olshausen & Field, 1996, 1997).

Three different experiments were carried out. The first two experiments used a reduced stimulus set to study the base effects of independently manipulating the context and sparse coding constraints; Experiment 1 manipulated context with the simplest sparse coding constraint, while the second experiment specifically focused on different forms of the sparse coding constraint. In Experiment 3, the context and sparse coding constraints were investigated using the full alphabet. Networks were analyzed both in terms of their ability to reproduce the training set and in terms of their internal structure via weight analysis.

Networks and Bayesian Theory

It is assumed that the internal representations used by a system must come to represent the external world in some manner. In other words, internal representations could be thought of as hypotheses about the external world. Thus, the problem of defining these internal representations can be reformulated as computing the probability of a given hypothesis (internal representation) given the observed data (external world)—a potentially difficult task.

Fortunately, a relatively simple theoretical framework exists for computing this probability. In its simplest form, *Bayes' theorem* (see Equation 1) states that for a given hypothesis, $\mathcal{H}$, and observed data, $\mathcal{D}$, the posterior probability of $\mathcal{H}$ given $\mathcal{D}$ is computed as

$$P(\mathcal{H}|\mathcal{D}) = \frac{P(\mathcal{D}|\mathcal{H}) \times P(\mathcal{H})}{P(\mathcal{D})} \qquad (1)$$

where $P(\mathcal{H})$ is regarded as the prior probability of $\mathcal{H}$ before observing the data $\mathcal{D}$, $P(\mathcal{D})$ is the probability of the data, and $P(\mathcal{D}|\mathcal{H})$ is the probability of the data given the hypothesis. Thus, by specifying $P(\mathcal{D}|\mathcal{H})$ and $P(\mathcal{D})$, the mechanisms of

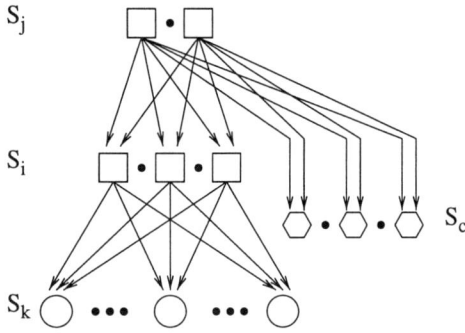

Figure 1: The basic network configuration for a three-layer network. S_k are surface units; S_i are mediating layer units; S_j are deep layer units; S_c are context units.

Bayesian theory provide a solution to the problem of learning from data (Bernardo & Smith, 1994; MacKay, 1995; McClelland, 1998).

We can also rearrange the model to predict the data given the hypothesis; in other words, this framework can be used to construct a *generative* model, such that the higher-order internal representations predict the lower-level simple features.

Network Architecture

To help explain the network dynamics, we will consider the simple case of a three-layer network (see Figure 1) consisting of an surface layer, S_k, a single mediating layer, S_i (in practice, there could be any number of mediating layers), and a deep layer, S_j. It is assumed that connections exist only between adjacent layers; that is, there are no direct connections between the surface and deep structure layers. Furthermore, the generative nature of the model means that connections are uni-directional and flow from the deep to the surface layer as indicated by the directed connections within Figure1. Thus, we can define three different relationships for a given unit; the parents ($pa[S_i]$; units contributing activation), children($ch[S_i]$; units receiving activation), and siblings ($sib[S_i]$; units within the same layer).

Units are assumed to be stochastic and are probabilistically active or inactive as determined by the summed activations being sent by their parents via weighted connections. Consequently, the network weight vector, $\mathbf{W}$, can be interpreted as encoding the underlying probabilities of the generative model. This means that weights are constrained to be zero or positive.

In the present studies, this basic network architecture has been expanded to include a *context*-layer, (S_c), as illustrated by the hexagonal units in Figure 1. This context layer is connected directly to the deep layer and thus provides contextual information to the deep layer only.

Learning Objective

Within this framework, the learning objective is to find the most probable explanation, $\mathcal{H}$, for the input patterns, $\mathcal{D}$, presented to the network. In other words, we wish to develop a generative model that encodes the probabilities of the input data within the network's weight structure. Therefore, the learning objective reduces to adapting the weight vector, $\mathbf{W}$,

to find the most probable explanation for the input patterns.

If we knew the weight vector, we could calculate the probability of the input data as

$$P(\mathcal{D}_{1:N}|\mathbf{W}) = \prod_n P(\mathcal{D}_n|\mathbf{W}) \qquad (2)$$

where

$$P(\mathcal{D}_n|\mathbf{W}) = \sum_m P(\mathcal{D}_n|\mathcal{S}_m, \mathbf{W})P(\mathcal{S}_m|\mathbf{W})$$

is the marginalization of all possible unit states, $\mathcal{S}_m$, of the network.

It should be noted that the number of all possible network states, $\mathcal{S}_m$, increases exponentially with the number of units in the network. Therefore, computing the exact sum becomes intractable as the networks become larger. One desirable property of generative models, however, is for most patterns to have one—or just a few—possible explanations; therefore, only a few terms, $P(\mathcal{D}_n|\mathcal{S}_m, \mathbf{W})$, will be non-zero and it becomes tractable to sample $\mathcal{S}_m$ according to $P(\mathcal{S}_m|\mathcal{D}_n, \mathbf{W})$.

Of course, we do not know the weight vector but must adapt it instead. One way of adapting the weight vector is to use a variation of the *expectation maximization* (EM) algorithm. EM is typically used to find parameter estimates in models where some variables are unknown or unobserved. The algorithm is composed of two steps: (i) an estimation (E) step that samples network states , and (ii) a maximization (M) step that adjusts weights. For our purposes, the E step can be accomplished by Gibbs sampling whereas the M step can use maximum-likelihood (ML) estimation.

Computing Network Probabilities

Being a generative model, the probability of any unit's state is directly computable from the states of its parents:

$$P(S_i = 1|pa[S_i], \mathbf{W}) = h(\sum_j S_j w_{ij}) \qquad (3)$$

where S_j are the parents of S_i and w_{ij} is the weight from unit S_i to S_j.The function h in Equation 3 specifies how these underlying causes are to be combined to produce the probability of $S_i = 1$. One function that can be used for this is the "noisy OR" function:

$$h(u) = 1 - e^{-u} \qquad (4)$$

where $u = \sum_j S_j w_{ij}$ is the causal input to S_i. Note that because weights are constrained to be positive, u is never negative, and therefore $0 \leq h(u) \leq 1$.

Thus, the joint probability density of a such a network can be computed as the product of the conditional probabilities

$$P(S_1 \ldots S_n|\mathbf{W}) = \prod_i P(S_i|pa[S_i], \mathbf{W}). \qquad (5)$$

Sampling Network States

In Lewicki and Sejnowski's (1997) original formulation of the problem, each state of the network, $\mathcal{S}_m$, is updated iteratively according to the probability of each unit state, S_i, given the states of the remaining units in the network. This conditional probability is computed as

$$P(S_i|S_{j;j\neq i}, \mathbf{W}) \propto$$
$$P(S_i|pa[S_i], \mathbf{W}) \prod_{j \in ch[S_i]} P(S_j|pa[S_j], S_i, \mathbf{W}) \qquad (6)$$

394

Thus, the Gibbs equations as used in this framework can be interpreted in terms of a stochastic recurrent neural network, where the feedback from the higher (or deeper) layers influences the states at the lower (or surface) layers. Whereas Lewicki and Sejnowski (1997) computed the probability of a unit *changing* its state, the problem can be reformulated as one where the probability of a unit being active given the remaining states of the network is calculated.

Consequently, one can compute the probability of a unit being active, $S_i = 1$, given the remaining states of the network as

$$P(S_i = 1 | S_{j;j \neq i}, \mathbf{W}) = \frac{1}{1 + e^{-\Delta x_i}} \quad (7)$$

This function will produce a $P(S_i = 1) \approx 0$ for negative evidence, a $P(S_i = 1) \approx 0.5$ for inconclusive evidence, and $P(S_i = 1) \approx 1$ for positive evidence.

The variable Δx_i in Equation 7 indicates how much changing the unit state, S_i, to being active changes the overall probability of the network state. In multilayered networks (where the number of layers is greater than 2), this term will have both a feedback component from the parents in the deeper layers, and a feedforward component from the children in the more surface layers:

$$\Delta x_i = fb(pa[S_i]) + ff(ch[S_i]) \quad (8)$$

In networks with only two layers, or in the deepest layer of a multilayer network, this feedback term will drop out. Typically, the feedforward component of Equation 8 will dominate the term, but if the feedforward input is ambiguous, then the feedback component becomes important as it allows the more surface level units to use information computable only at the deeper layers.

The feedback term in Equation 8 is simply computed as the log probability of the unit being on minus the log probability of the unit being off. This is calculated as:

$$fb(S_i) = \log \frac{h(u)}{1 - h(u)} \quad (9)$$

where the function $h(u)$ is computed as described earlier.

Feedforward is computed as the probability of the unit being on given the activity of its children. Therefore, for a given unit, we want to sum the evidence of the unit being on minus the evidence of the unit being off. We also want to weight the evidence according to the number of other units contributing to the child's activity (the more units contributing, the less effect any one unit will have).

$$ff(S_i) = \sum_{k \in ch[S_i]} S_k \log \frac{h(u - S_i w_{ik} + w_{ik})}{h(u - S_i w_{ik})}$$
$$+ (1 - S_k) \log \frac{1 - h(u - S_i w_{ik} + w_{ik})}{1 - h(u - S_i w_{ik})} \quad (10)$$

Thus, if $S_i = 0$, then the weight from S_i is added to the top portions of Equations 10, whereas if $S_i = 1$ then the weight from S_i is removed from the bottom portion of the above equation. Furthermore, if $S_k = 0$ (indicating that the parent node should be off), then the first term of Equation 10 drops out, whereas if $S_k = 1$ (indicating that the parent node should be on), then the second term of Equation 10 drops out.

Adding Contextual Information

As defined earlier, context is the added information that can provide hints about which collection of simple features constitute higher-order representations, and thus helps constrain the internal representations developed at the deep-layer. Context can be added to the network dynamics simply by directly connecting a set of context units (denoted S_c in Figure 1) to the deep-layer units, S_j, via weighted connections w_{cj}.

$$ff(S_j) = \sum_{c \in cn[S_j]} S_c \log \frac{h(u - S_j w_{cj} + w_{cj})}{h(u - S_j w_{cj})}$$
$$+ (1 - S_c) \log \frac{1 - h(u - S_j w_{cj} + w_{cj})}{1 - h(u - S_j w_{cj})} \quad (11)$$

where $cn[S_j]$ are the context units directly connected to the deep-layer units. Thus, context information is directly added to the activation probabilities of the deep-layer units by summing the contributions of Equations 10 and 11.

Adding Sparse Coding Constraints

A further modification to the original framework is to add a sparse coding constraint on the unit activation probabilities. That is, all things being equal, sparse coding encourages a network to use relatively few units to represent any specific input pattern. In the standard framework, in the absence of any guiding information, a unit will be active with baseline probability $P = 0.5$. Sparse coding can be encouraged within the network by modifying Equation 8 to include a sparcity constraint:

$$\Delta x_i = fb(pa[S_i]) + ff(ch[S_i]) + \lambda \cdot sp(S_i) \quad (12)$$

where λ is equivalent to a gain function which modulates how much effect $sp[S_i]$ exerts on the rest of the equation.

Four sparse coding functions are defined. The first and simplest function is an implicit, independent prior constraint that reduces the baseline probability of a unit being active by a constant amount:

$$\text{Constant}: \quad sp_1[S_i] = log \frac{\phi}{1-\phi}, \quad 0 \leq \phi \leq 1 \quad (13)$$

The three other functions defined encourage sparse coding in an explicit, dependent manner; that is, sparse coding is dependent on the number of sibling units co-active (excluding the current unit):

$$j = \sum_{n \in sib[S_i]} S_n, \quad \text{where } n \neq i \quad (14)$$

The first dependent sparse coding function (*Logistic*), uses a modified logistic function to probabilistically limit the number of units active from 0 to ϕn units.

$$\text{Logistic}: \quad sp_2(S_i) = log \frac{\ell_{\phi,\mu,n}(j+1)}{\ell_{\phi,\mu,n}(j)} \quad (15)$$

where
$$\ell_{\phi,\mu,n}(j) = 1 - \frac{1}{1 + e^{-\frac{(j/n) - \phi}{\mu}}} \quad (16)$$

The second dependent sparse coding function places a prior activation constraint on the units such that probabilistically ϕn units will be active at any given time. This is accomplished by sampling the unit activation states from the binomial distribution:

Binomial : $\quad sp_3(S_i) = log\frac{b_{\phi,n}(j+1)}{b_{\phi,n}(j)}$ (17)

where

$$b_{\phi,n}(j) = \frac{n!}{j!(n-j)!} \cdot \phi^j \cdot (1-\phi)^{n-j}$$ (18)

Finally, the last dependent sparse coding function is a mixture of poisson and binomial distributions.

Pois + Bin : $\quad sp_4(S_i) = log\frac{\pi p_\gamma(j+1)+(1-\pi)b_{\phi,n}(j+1)}{\pi p_\gamma(j)+(1-\pi)b_{\phi,n}(j)}$ (19)

where $b_{\phi,n}(j)$ is defined in Equation 18 and

$$p_\gamma(j) = e^{-\gamma} \cdot \frac{\gamma^j}{j!}$$ (20)

This mixture of distributions has the effect of probabilistically having 0 units active as determined by Equation 20 with probability π, and having ϕn units active with probability $1 - \pi$ as determined by Equation 18.

Weight Estimation

Once we have sampled the activation space, we are in the position to estimate the weights. To control the complexity of the model, a prior is placed on the weights. In using the "noisy OR" function where all weights are constrained to be positive, it is assumed that the weight prior is a product of independent gamma distributions parameterized by α and β. Hence, the objective function we wish the maximize becomes

$$\mathcal{L} = P(D_{1:N}|\mathbf{W})P(\mathbf{W}|\alpha,\beta)$$

Using the maximization step from the EM algorithm, we want to set $\partial\mathcal{L}/w_{ij} = 0$ and solve for w_{ij}. Lewicki and Sejnowski (1997) show this can be accomplished by using the transformations $f_{ij} = 1 - e^{-w_{ij}}$ and $g_i = 1 - e^{-u_i}$ and solving for

$$f_{ij} = \frac{\alpha - 1 + 2f_{ij} + \sum_n S_i^{(n)} S_j^{(n)} f_{ij}/g_j^{(n)}}{\alpha + \beta + \sum_n S_i^{(n)}}$$ (21)

It should be noted that in the equation, S_i is the *cause* of S_j. Furthermore, $S^{(n)}$ is the unit's state obtained via Gibbs sampling for the nth input pattern. The sum in the above equation is simply the weighted average of the number of times unit S_i was active when unit S_j was active. The ratio f_{ij}/g_j weights each term in the sum inversely to the number of causes for S_j; if S_i is the sole cause of S_j (meaning that $f_{ij} = g_j$), then the term has full weight.

Method

The Alphabet We adopted Rumelhart and Siple's (1974) featured based alphabet; Each letter was composed of simple visual features such as horizontal, vertical, and oblique lines. We modified the original alphabet by breaking both the top and bottom horizontal line segments into two segments each in order to equalize all line segment lengths.

Figure 2 shows each of the 26 letters overlayed on the 16 base line segments; a "Space" character (no features active) was also presented to the network. A subset of these letters ('SPC',H, I, N, O, S, X, Z) was used in the first two studies and the full alphabet was used for the third study.

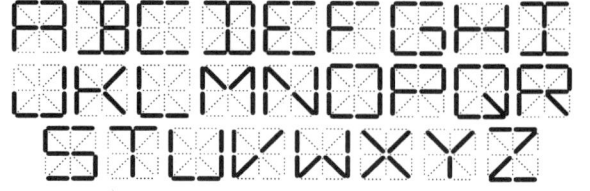

Figure 2: The full alphabet superimposed on the 16 features.

Each line segment was represented by a unary code; therefore, each letter was represented by turning on the appropriate bits in a 16 bit code. Context was also represented using a simple unary scheme; there was one unique unit active for each of the letters within the training set. Thus, there were 8 context representations for the reduced alphabet and 27 context representations for the full alphabet.

Network Architecture and Training The network architecture consisted of 16 surface units, no mediating units, and either 10 or 30 deep units for the reduced and full alphabet training sets respectively. If context was being tested, then the architectures included 8 or 27 context units in accordance with the training set.

For all networks, the weight prior was specified with $\alpha = 1.0$ and $\beta = 1.25$; weights were initialized between 0.05 and 0.15. Internal units were initialized with $P(S_i = 1) = 0.5$. Gibbs sampling was performed 15 times for each input pattern, or until the maximum state change probability was less than 0.05. For the sparse coding experiments, the parameters were set to $\lambda = 1.0$, $\mu = 1.0$, $\pi = 0.5$, and $\gamma = 0.1$; ϕ was set to 0.1 for the first two experiments and 0.05 for the third experiment. It should be noted that the parameters were chosen to maximize network performance (with all things being equal) and a more thorough exploration of parameter space will be required in the future. For each condition, 25 networks were trained with different randomized initial weights.

Results and Discussion

Network performance was analyzed via two methods. First, the generative nature of the models was tested in terms of their ability to reproduce the surface pattern presented. Each pattern was presented to the network and Gibbs sampling was performed to produce an internal pattern of activity at the deep layer. This internal activity was then propagated back to the surface layer units and the number of bits in error—either "Addition" (i.e., 1 instead of 0) or "Omission" (i.e., 0 instead of 1) errors—was calculated. This was performed 100 times for each pattern.

Second, the underlying weight structure of each network was analyzed both qualitatively and quantitatively. The first qualitative measure is based on the visual inspection of the weight matrix as in Lewicki and Sejnowski (1997). The weight for each input feature is passed through Equation 4 to produce a color code fading from "black" to "white" (representing $1 \rightarrow 0$) and then plotted as the appropriate line segment. This type of analysis is shown in Figure 4; unfortunately, it is restricted to single networks. The second is a quantitative measure that can be averaged over runs and is based on the number of weight vectors (i.e., the weights leav-

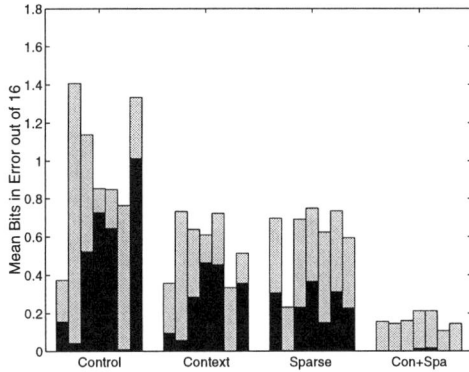

Figure 3: Mean number of errors for the 7 letters across the 4 conditions in Experiment 1. Bottom portion of the bars are 'Addition' errors and upper portions are 'Omission' errors.

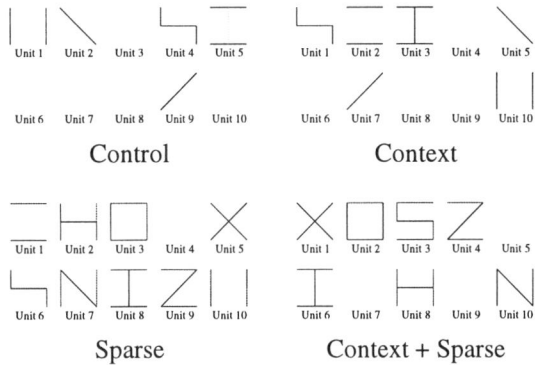

Figure 5: Mean number of errors for the 7 letters across the 4 conditions in Experiment 2. Graphical interpretation is the same as in Figure 3.

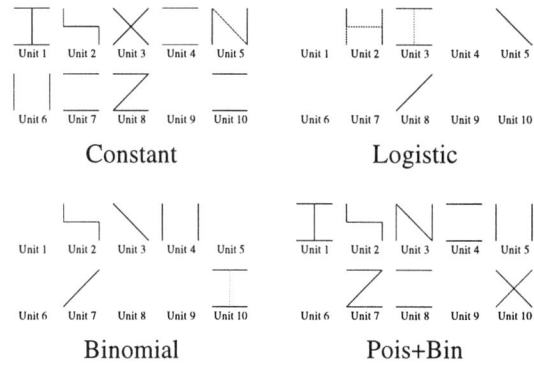

Figure 4: Typical network weights for the context and sparse coding manipulations.

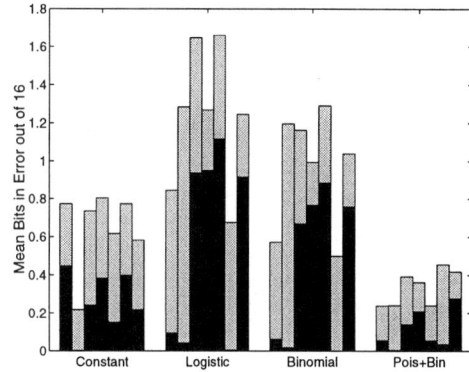

Figure 6: Typical network weights for the sparse coding manipulation.

ing a parent node) that have at least one non-zero element; this measure gives a rough estimation of how many units are being used to represent the data set and therefore local (one unit per pattern) versus distributed encoding.

Experiment 1: Figure 3 shows the mean number of bits in error for the reduced alphabet over the four conditions (only sp_1 with $\lambda = 1.0$, $\phi = 0.1$ was tested in Experiment 1). The mean number of bits in error collapsed over all the letters (excluding 'SPC') for the Control, Context, Sparse, and Context+Sparse conditions are 0.96 ($SD = 0.36$), 0.56 ($SD = 0.16$), 0.62 ($SD = 0.18$), and 0.30 ($SD = 0.09$) respectively.

As can be seen, the standard network performs fairly well on the reduced input set; it averages only 1 bit in error. The addition of the constraints, however, improves the performance of the networks, especially when applied in conjunction. Furthermore, it should be noted that variability in network performance (as indicated by standard deviations) is decreased when constraints are added.

The typical weight structures for the four conditions are shown in Figure 4. As can be seen, the Control, Context, and Sparse conditions tended to extract groups of individual features (indicating a distributed representation) whereas the other condition tends to pick out complete letters (a more lo-

calist representation). It should be noted, however, that the Sparse condition appears to redundantly encode information in terms of replicating line segments. Quantitative analyses show that on average, the number of non-zero weight vectors for each of the four conditions are 4.7 ($SD = 0.8$), 5.4 ($SD = 0.9$), 8.4 ($SD = 1.0$), and 7.32 ($SD = 0.8$) respectively. This analysis confirms that a combination of the context and sparse coding constraints encourages local representations of complete letters to develop.

Experiment 2: Figure 5 shows the mean number of error bits for the four different sparse coding functions (Constant [sp_1], Logistic [sp_2], Binomial [sp_3], and Pois+Bin [sp_4]). The mean error for each of the four conditions are 0.64 ($SD = 0.21$), 1.23 ($SD = 0.37$), 0.96 ($SD = 0.31$), and 0.34 ($SD = 0.09$) respectively. The first thing to note is that two of the sparse coding functions (sp_2 and sp_3) are worse than or equal to the control condition in Experiment 1. The fourth function (sp_4), however, actually improves performance over the simple, independent sparse coding constraint.

Figure 6 show the typical weight structures for the four sparse coding conditions. It is interesting to note that sp_2 and sp_3 have similar structure (i.e., weaker weights pulling out individual lines) to the control condition in Experiment 1. The

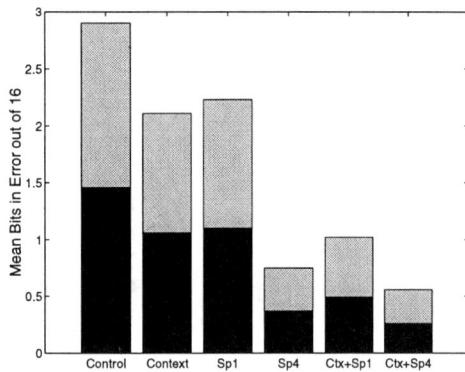

Figure 7: Mean number of bits in error collapsed across letters for the full alphabet plotted for the six conditions.

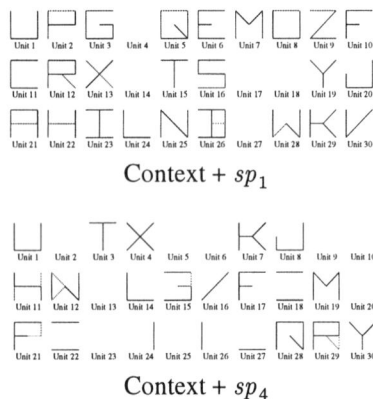

Context + sp_1

Context + sp_4

Figure 8: Weights for the fully constrained networks.

other two functions, sp_1 and sp_4, have again pulled out somewhat redundant codings of line segments. This distinction is supported by the quantitative analysis of the weight structure: the mean number of non-zero weight vectors for the four conditions are 8.7 ($SD = 0.8$), 4.3 ($SD = 0.7$), 4.9 ($SD = 0.9$), and 8.0 ($SD = 0.9$).

Experiment 3: In this final experiment, we tested the context and sparse coding constraints on the full alphabet. Only the sp_1 and sp_4 sparse coding constraints were tested. Furthermore, to improve performance, β was reduced to 1.05 and each pattern was only sampled five times per epoch.

Figure 7 shows the average number of bits in error collapsed across all 26 letters for the six conditions (Control, Context, sp_1, sp_4, Context + sp_1, Context + sp_4). Standard deviations for these six conditions were 0.58, 0.62, 0.42, 0.36, 0.34, and 0.27. Moving to the full data set was detrimental to Control, Context, and sp_1 networks; each network tended to have at least one 'Additive' error bit and one 'Omission' error bit for each letter. This was not the case for the three other conditions, with the best performance being produced by the combination of the Context and the sp_4 constraints.

The average number of non-zero weight vectors for each of the six conditions were 25.5 ($SD = 3.7$), 13.8 ($SD = 6.4$), 30.0 ($SD = 0.0$), 18.5 ($SD = 1.6$), 25.0 ($SD = 1.5$), and 19.5 (SD

= 1.4). The network weight structures for the two fully constrained networks (i.e., Context + Sparse Coding) are shown in Figure 4 (the four other conditions were not graphed as they tended to have smaller weights). As can be seen, combining the context and sparse coding constraints produced networks with distinct weight structures. Once again, it appears that the Context + sp_1 function encourages local encoding. Interestingly, however, the Context + sp_4 function developed a more distributed, redundant encoding.

General Conclusions

The results from these three experiments suggest that a combination of context and sparse coding constraints are required for the formation of adequate internal representations, especially when the data set is large. Moreover, analysis of the weight structure suggests that more accurate performance is due to the development of internal representations that are both distributed and redundant, as opposed to purely local. Although these results are preliminary, they suggest future studies within this generative framework. Specifically, future research will consider networks with mediating layers, and networks trained on words using the feature based letters.

In terms of visual cognition, these results suggest that feedback connections in the brain may provide context information to relatively low-level visual areas, thereby aiding their ability to discover structure in their inputs. Furthermore, sparse coding may be required to create redundant representations that actually increase performance.

References

Anderson, J. R. (1990). *The adaptive character of thought.* Hillsdale, NJ: Lawrence Erlbaum Associates.

Bernardo, J. M., & Smith, A. F. M. (1994). *Bayesian theory.* New York: John Wiley & Sons.

Lewicki, M. S., & Sejnowski, T. J. (1997). Bayesian unsupervised learning of higher order structure. In M. C. Mozer, M. I. Jordan, & T. Petsche (Eds.), *Advances in Neural Information Processing Systems* (Vol. 9, pp. 529–535). Cambridge, MA: MIT Press.

MacKay, D. J. C. (1995). Bayesian methods for supervised neural networks. In M. A. Arbib (Ed.), *The handbook of brain theory and neural networks* (pp. 144–149). Cambridge, MA: MIT Press.

McClelland, J. L. (1998). Connectionist models and Bayesian inference. In M. Oaksford & N. Chater (Eds.), *Rational models of cognition.* Oxford: Oxford University Press.

Olshausen, B. A., & Field, D. J. (1996). Emergence of simple-cell receptive field properties by learning a sparse code for natural images. *Nature, 381,* 607–609.

Olshausen, B. A., & Field, D. J. (1997). Sparse coding with an overcomplete basis set: A strategy employed by V1? *Vision Research, 37,* 3311–3325.

Rumelhart, D. E., & Siple, P. (1974). Process of recognizing tachistoscopically presented words. *Psychological Review, 81,* 99-118.

Language Acquisition and Ambiguity Resolution: The Role of Frequency Distributions

Paola Merlo
LATL-University of Geneva
Department of Linguistics
2 rue de Candolle
1211 Genève 4
Switzerland
merlo@lettres.unige.ch

Suzanne Stevenson
Department of Computer Science
and Center for Cognitive Science (RuCCS)
Rutgers University
110 Frelinghuysen Road
New Brunswick, NJ 08854-8019
suzanne@cs.rutgers.edu

Abstract

This paper proposes that the set of frequencies that the human language processor keeps track of are those that are useful to it in learning. In a computational experimental setting, we investigate four linguistically motivated features which distinguish subclasses of intransitive verbs, and suggest that those features that are the most useful to automatically classify verbs into lexical semantic classes are related to mechanisms used in adult processing to resolve structural ambiguity.

Introduction

Models of human language comprehension have traditionally focused on discrete linguistic properties—structural or interpretive factors—as the guiding influence in determining the preferred interpretation of an ambiguity (e.g., [8, 18]). Theories of human sentence processing that are founded on such linguistic distinctions have generally assumed that their use is an inherent (and universal) property of the language processor. Even if certain distinctions must be learned (such as those that involve parametric variation among languages), the learning process itself has been irrelevant to the later use of those features in resolving ambiguity.

Recently, the experience-based paradigm has shifted the emphasis to the role of frequencies—continuously-valued information that weights the contribution of individual linguistic and contextual features within the ambiguity resolution process [9, 15, 17, 25]. Clearly, frequencies are acquired through on-going exposure to linguistic input, and so these approaches promise to integrate learning and processing more closely. While some work has shown that properties of a learning mechanism may underlie difficulty in processing (e.g., in the case of embedded structures, as in [4]), to our knowledge none has explored the connection between properties of the learning process and the features that play a role in guiding ambiguity resolution.

Here we extend the experience-based point of view by proposing that the frequency distributions that play a guiding role in ambiguity resolution are those that contribute to language acquisition. We assume that the language learner must learn certain fundamental linguistic distinctions. (Not all can be innate, since at least some are lexically specific.) We further assume that distributional data (the frequencies of alternative features and/or constructions) play a role in the learning process. We view the adult language processor as a mature version of the child language processor, with access to the type of knowledge that guides acquisition of language, including the distributional information. Since different resolutions of an ambiguity may be distinguished by fundamental linguistic factors that must be learned, we expect that the frequency distributions of those factors will contribute to the process of selecting the preferred interpretation.

Our general hypothesis then is that the distributional knowledge used by the language processor in guiding interpretation is restricted to specific features that are required to learn necessary linguistic distinctions. Alternatively stated, the frequencies used in language comprehension are a subset (not necessarily a proper subset) of the frequencies used in language acquisition.

Verb Classes and Ambiguity Resolution

We investigate our general hypothesis by exploring the specific instance of the relation between the learning of lexical semantic verb classifications from distributional data, and the use of those frequency distributions in ambiguity resolution. We focus on the main verb/reduced relative (MV/RR) ambiguity [1]:

(1) The horse **raced** past the barn fell.
(2) The boy **washed** in the tub was angry.

In (1) and (2), the boldfaced verb can be interpreted as either a past tense main verb, or as a past participle within a reduced relative clause (e.g., *the horse* [that was] *raced past the barn*). In each case, the reduced relative interpretation is required for a coherent analysis of the complete sentence. The main verb interpretation of *raced* is so strongly preferred that people have great difficulty understanding sentence (1). However, the ease of sentence (2) shows that the reduced relative interpretation is not difficult for all verbs.

The differences in ease of interpreting the resolutions of this ambiguity have been shown to be sensitive both to frequency differentials [16, 26] and to verb class distinctions [24, 7]. Within the context of our general hypothesis above, we relate these two factors by proposing that the frequency measures that influence the ambiguity resolution process in this construction are intimately related to the defining properties of three classes of verbs that present this ambiguity. Building on the idea of syntactic bootstrapping [10], and on statistical approaches to extracting verb information from frequency distributions [12, 2, 23], we follow a computational experimental methodology in which we investigate as indicated each of the following specific hypotheses:

H1: Linguistically motivated features for distinguishing the verb classes are apparent within linguistic experience.

We analyze the three classes to determine potentially relevant distinctive features, and count those features (or approximations to them) in a very large corpus.

H2: The distributional patterns of (some of) those features contribute to learning the classifications of the verbs.

We apply machine learning techniques to determine whether the features support the learning of the classifications. We analyze the contribution of different features to the classification process.

H3: Features informative in learning the verb classes also influence the resolution of the MV/RR ambiguity, and features not helpful in learning do not affect processing of the ambiguity.

We examine whether the features that are found to be informative in classification play a role in resolution of the ambiguity. Conversely, we discuss whether features that are not found to contribute to learning similarly play no role in processing.[1]

To preview, we find that, related to (H1), linguistically motivated features that distinguish the verb classes can be extracted from a corpus with a moderate amount of annotation. We assume that these features are available to the learner once it can make certain fundamental syntactic distinctions (e.g., part of speech, constituency). In relation to (H2), a subset of these features is sufficient to halve the error rate compared to chance in automatic verb classification, suggesting that distributional data does in fact contribute to our knowledge of the classification of verbs. Furthermore, we find that features that are distributionally predictable, because they are highly correlated to other features, contribute little to classification performance. We conclude that the usefulness of distributional features to the learner is determined by their linguistic informativeness. Finally, we find that the results

in (H3) provide important preliminary evidence concerning our general hypothesis that the set of frequency differentials that are available to the learning algorithm and the processing algorithm are the same. Features which contribute most to learning the verb classes have been experimentally demonstrated to influence processing of the MV/RR ambiguity, and one which is not informative for learning has no evidence for its role in MV/RR ambiguity resolution.

Determining the Features

In this section, we present motivation for the features that we investigate in terms of their role in learning the verb classes. We first present the linguistically derived features, then turn to an analysis of the MV/RR ambiguity to extend the set of potentially relevant features.

Features of the Verb Classes

The MV/RR ambiguity involves a choice between a main verb form that is intransitive, and a reduced relative form that is transitive (because the reduced relative is a passive use of the verb). The three verbs classes under study—unergative, unaccusative, and object-drop—were thus chosen because they exhaustively partition the optionally intransitive verbs in English. The three verb classes differ in the properties of their intransitive/transitive alternations, which are exemplified below.

Unergative:
  (3a) The horse raced past the barn.
  (3b) The jockey raced the horse past the barn.
Unaccusative:
  (4a) The butter melted in the pan.
  (4b) The cook melted the butter in the pan.
Object-drop:
  (5a) The boy washed the hall.
  (5b) The boy washed.

The sentences in (3) use an unergative verb, *raced*. Unergatives are intransitive action verbs whose transitive form is the causative counterpart of the intransitive form. Thus, the subject of the intransitive (3a) becomes the object of the transitive (3b) [3, 11, 14]. The sentences in (4) use an unaccusative verb, *melted*. Unaccusatives are intransitive change of state verbs (4a); like unergatives, the transitive counterpart for these verbs is also causative (4b). The sentences in (5) use an object-drop verb, *washed*; these verbs have a non-causative transitive/intransitive alternation, in which the object is simply optional.

Both unergatives and unaccusatives have a causative transitive form, but differ in the semantic roles that they assign to the participants in the event described. In an intransitive unergative, the subject is an Agent (the doer of the event), and in an intransitive unaccusative, the subject is a Theme (something affected by the event).

---

[1]Note that the latter is not directly testable, since it may be that current experiments have simply not shown the effect of the variable even though it does influence processing. Thus any conclusions here are suggestive only.

The role assignments to the corresponding semantic arguments of the transitive forms—i.e., the direct objects—are the same, with the addition of a Causal Agent (the causer of the event) as subject in both cases. This leads to an unusual situation for a transitive unergative, because it assigns two agentive roles—the subject is the agent of causation, and the object is the agent of the action expressed by the verb [24]. Object-drop verbs have a simpler participant/role mapping than either unergatives or unaccusatives, assigning Agent to the subject and Theme to the optional object.

We expect the differing semantic role assignments of the verb classes to be reflected in their syntactic behavior [13, 6], and consequently in the distributional data we collect from a corpus. The three classes can be characterized by their occurrence in two alternations: the intransitive/transitive alternation and the causative alternation. Unergatives are distinguished from the other classes in being rare in the transitive form (due to their "double agentive" nature); we expect this to be reflected in a greater degree of intransitive use in a corpus. Both unergatives and unaccusatives are distinguished from object-drop in being causative in their transitive form, and similarly we expect this to be reflected in amount of detectable causative use. Furthermore, since the causative is a transitive use, and the transitive use of unergatives is expected to be rare, causativity should primarily distinguish unaccusatives from object-drops. In conclusion, we expect the defining features of the verb classes—the intransitive/transitive and causative alternations—to lead to distributional differences in the observed usages of the verbs in these alternations.

Features of the MV/RR Alternatives

We now examine the features that distinguish the two resolutions of the MV/RR ambiguity:

Main Verb: The horse raced past the barn quickly.

Reduced Relative: The horse raced past the barn fell.

In the main verb resolution, the ambiguous verb *raced* is used in its intransitive form, while in the reduced relative, it is used in its transitive, causative form. These features correspond directly to the defining alternations of the three verb classes under study (intransitive/transitive, causative). Additionally, we see that other, related features serve to distinguish the two resolutions of the ambiguity. The main verb form is active and a main verb part-of-speech (tagged as VBD); by contrast, the reduced relative form is passive and a past participle (tagged as VBN). Note that these properties are redundant with the intransitive/transitive distinction, as passive implies transitive use, and necessarily entails the use of a past participle.

We add the VBD/VBN and active/passive distinctions to our features for classification (the transitive and

causative alternations) motivated by two factors. First, recent work in machine learning [21, 22] has argued that using overlapping features can be beneficial for learning. Second, one of the features—the VBD/VBN distinction—has already been shown to influence resolution of the MV/RR ambiguity [26].

In the next section, we describe how we compile the corpus counts for each of the four properties, in order to approximate the distributional information of these alternations available to a human learner.

Frequency Distributions of the Features

We assume that currently available large corpora are a reasonable approximation to the linguistic input that the learner is exposed to [19]. Using a combined corpus of 65-million words, we measured the relative frequency distributions of the linguistic features (VBD/VBN, active/passive, intransitive/transitive, causative) over a sample of verbs from the three lexical semantic classes.

We chose a total of 60 verbs, 20 verbs from each class, based on the classification of verbs in [13] (see Appendix A). Each verb presents the same form in the simple past and in the past participle, as in the MV/RR ambiguity, and most of the verbs can occur in the transitive and in the passive. Most counts were performed on the Linguistic Data Consortium release of the Brown Corpus and of years 1987, 1988, 1989 of the Wall Street Journal, a combined corpus in excess of 65 million words labeled with part-of-speech (POS) tags.[2] Due to the need for additional annotation, the causative feature was counted only for the 1988 year of the WSJ, a parsed corpus of 29 million words also available from the LDC (parsed with the parser from [5]).

The counts for VBD/VBN were automatically extracted for all verbs based on the part-of-speech label according to the tagged corpus. For intransitive/transitive, we searched for the closest nominal group following the verb which was considered to be the object. For active/passive uses, we looked for the preceding auxiliary: *have* indicates an active use and *be* indicates a passive use. The causative feature was approximated by extracting heads of subjects and objects for each verb occurrence, and calculating a token-based percentage of overlap between the two sets of nouns. This captures the property of the causative construction that the subject of the intransitive can occur as the object of the transitive. All counts were normalized, yielding a total of four relative frequency features: VBD (%VBD tag), ACT (%active use), INTR (%intransitive use), CAUS (%causative use). All raw and normalized corpus data are available from

[2] Because we need a very large corpus due to the constraints imposed by the statistical methods used, we are restricted at the present time primarily to newspaper text of this type. Clearly, verificational work across different kinds of corpora, including child-directed speech, would be helpful in elaborating our proposal.

Features	Accuracy
1. VBD ACT INTR CAUS	52%
2. VBD INTR CAUS	66%
3. ACT INTR CAUS	47%
4. VBD ACT CAUS	54%
5. VBD ACT INTR	45%

Table 1: Accuracy of the Verb Clustering Task.

the authors.

Frequencies in Learning and Processing

The frequency distributions of the verb alternation features yield a vector for each verb that represents the relative frequency values for the verb on each dimension; the set of 60 vectors constitute the data for our machine learning experiments.

Vector template: [verb-name, VBD, ACT, INTR, CAUS]

Example: [opened, .793, .910, .308, .158]

It must be determined experimentally which of the distributions actually contribute to learning the verb classifications. First we describe computational experiments in unsupervised learning, using hierarchical clustering, then we turn to supervised learning, using decision tree induction. We conclude with a discussion of the relation of informative features in learning and processing.

Unsupervised Learning

We used the hierarchical clustering algorithm available in SPlus5.0, imposing a cut point that produced three clusters, to correspond to the three verb classes. Table 1 shows the accuracy achieved using the four features that discriminate the resolutions of the MV/RR ambiguity (row 1), and all three-feature subsets of those four features (rows 2–5). Note that chance performance in this task (a three-way classification) is 33% correct.

The highest accuracy in clustering, of 66%—or half the error rate compared to chance—is obtained only by the triple of features in row 2 in the table: VBD, INTR, and CAUS. All other subsets of features yield a much lower accuracy, of 45-54%. We can conclude that some of the features contribute useful information to guide clustering, but the inclusion of ACT actually degrades performance. Clearly, having fewer but more relevant features is important to accuracy in verb classification. We will return to the issue of which features contribute most to learning in our discussion of supervised learning below.

A problem with analyzing the clustering performance is that it is not always clear what counts as a misclassification. We cannot actually know what the identity of the verb class is for each cluster. In the above results, we imposed a classification based on the class of the majority

Features	Decision Trees		Rule Sets	
	A%	SE%	A%	SE%
1. VBD ACT INTR CAUS	64.2	1.7	64.9	1.6
2. VBD INTR CAUS	60.9	1.2	62.3	1.2
3. ACT INTR CAUS	59.8	1.2	58.9	0.9
4. VBD ACT CAUS	55.4	1.5	55.7	1.4
5. VBD ACT INTR	54.4	1.4	56.7	1.5

Table 2: Accuracy (A%) and Standard Error (SE%) in the Verb Classification Task.

of verbs in a cluster, but a problem arises when there is no clear majority. The determination of the number of cut points introduces a further degree of uncertainty in interpreting the results.

To evaluate better the effects of the features in learning, we therefore turned to a supervised learning method, in which the algorithm trains on cases with known classification. We do not believe that supervised learning is realistic for the human learner in this case. Rather, we perform these computational experiments in order to measure, on clear classification results, the contribution in principle of the features to learning.

Supervised Learning

For our supervised learning experiments, we used the publicly available version of the C5.0 machine learning algorithm (http://www.rulequest.com/), a newer version of C4.5 [20]. Given a training set with known classifications, this system applies inductive learning to create a procedure that can be applied to label new cases whose classification is unknown. The output of the system is presented in the form of either decision trees or rule sets, alternative data structures for encoding the use of the features in classification. For all reported experiments, we ran a 10-fold cross-validation repeated ten times, and the numbers reported are averages over all the runs.[3]

Table 2 show the results of our experiments on (different combinations of) the four features we counted in the corpus. Recall that chance performance in this task (a three-way classification) is 33% correct. We attain the best performance, of 64–65% (again, almost halving the chance error rate), by using all four features that discriminate the resolutions of the MV/RR ambiguity: VBD, ACT, INTR, and CAUS (row 1 in the table).

Comparing the accuracy of each of the three-feature subsets to the four-feature subset is very informative.

---

[3] A 10-fold cross-validation means that the system randomly divides the data into ten parts, and runs ten times on a different 90%-training-data/10%-test-data split, yielding an average accuracy and standard error. This procedure is then repeated for 10 different random divisions of the data, and accuracy and standard error are again averaged across the ten runs.

When either the INTR or CAUS feature is removed (rows 4 and 5 respectively), performance degrades considerably, with a decrease in accuracy of 8–10% from the maximum achieved with the four features. However, when the VBD feature is removed (row 3), there is a smaller decrease in accuracy, of 4–6%. When the ACT feature is removed (row 2), there is an even smaller decrease, of 2–4%. In fact, taking the standard error into account, the accuracy using all features (row 1) is not distinct from using only the three features VBD, INTR, and CAUS (row 2)—the same subset of features that yielded the best performance in clustering. We conclude then that INTR and CAUS contribute most to the accuracy of the classification, while ACT seems to contribute little. This shows that not all the linguistically relevant features are equally useful in learning.

We think that this pattern of results is related to the combination of the feature distributions: some distributions are highly correlated, while others are not. According to our calculations, CAUS is not significantly correlated with any other feature. Of the features that are significantly correlated, ACT and VBD are the most highly correlated (R=.67), ACT and INTR the next most highly correlated (R=.44), and VBD and INTR the least correlated (R=.36). Thus, CAUS is uncorrelated with the other features, and VBD and INTR are the least correlated pair of the remaining three.

We expect combinations of features that are less correlated to yield better classification accuracy. If we compare the accuracy of the 3-feature combinations in Table 2 (rows 2–5), this hypothesis is confirmed. The three combinations that contain the feature CAUS (rows 2, 3 and 4)—the uncorrelated feature—have better performance than the combination that does not (row 5), as expected. Now consider the subsets of three features that include CAUS with a pair of the other correlated features. The combination containing VBD and INTR (row 2)—the least correlated pair of features—has the best accuracy, the combination containing ACT and INTR (row 3)—the next least correlated pair—has the next highest accuracy, while the combination containing the most highly correlated ACT and VBD (row 4) has the worst accuracy. Thus, redundancy of features appears to play a clear role in determining their usefulness in learning.[4]

Comparison of Learning and Processing

Recall our motivating hypothesis that only those features that are useful in acquisition are available to influence interpretation. Given our learning results, then, we expect

---

[4]We suspect that another factor comes into play, namely how noisy the feature is. The similarity in performance using INTR or CAUS in combination with VBD and ACT (rows 4 and 5) might also be due in part to the fact that the counts for CAUS are a more noisy approximation of the actual feature distribution than the counts for INTR. We reserve defining a precise model of noise, and its interaction with the other factors, for future research.

to see that INTR, CAUS, and VBD play a role in ambiguity resolution, while ACT does not. Currently available experimental data yield preliminary support for this conclusion. Relative frequency of VBD/VBN use [26] and intransitive/transitive use [16] have been shown to influence resolution of the MV/RR ambiguity.

Demonstrating a processing effect of the frequency of causativity is more problematic. For unergatives, we note that there is an ineliminable confound between verb class and distribution of the CAUS feature. Unergatives never occur as causatives in our very large corpus, thus indicating that the causative use of unergatives is very rare. We interpret this fact as indicating that the causative use of an unergative is difficult, even though linguistically possible, because it entails a double agent structure, as discussed above [24]. Whether caused by knowledge of verb class directly or by causativity, effects of extreme difficulty have been found for unergatives in the MV/RR ambiguity.

On the other hand, unaccusatives occur more often in a causative construction, with a gradient of use. Recently, [7] have manipulated the degree of proto-agent properties, which include causativity, in the MV/RR ambiguity. They collected subject ratings, and showed that the ratings of these properties can influence the ease or difficulty of the reduced relative interpretation. While it remains to be shown that proto-agent ratings reflect the causative frequency distribution in a corpus, this data is highly suggestive.

By contrast to the INTR, CAUS, and VBD features, there is no experimental data indicating that ACT plays a role in MV/RR interpretation (or, in fact, in the resolution of any other ambiguity). Although these results are preliminary, they lend initial support within this constrained domain for our general hypothesis regarding the relation between learning and processing.

Conclusions

In this paper, we propose that the human language processor keeps track of the set of frequencies that are useful to it in learning. To explore this general hypothesis within a specific domain, we investigate how a set of linguistic features that distinguish intransitive verbs perform in a classification task. Linguistic analysis predicts certain features are most relevant (INTR and CAUS), but learning is improved by also using a partially redundant feature (VBD). Experimental evidence in sentence processing shows that some of these features, such as INTR and VBD, are used in resolving ambiguity. Others, such as CAUS, have been indirectly shown to have an influence on ambiguity resolution. On the other hand, a highly redundant feature, ACT, is not informative for learning. Interestingly, there is no experimental data indicating that ACT plays a role in syntactic disambiguation. Our proposal thus receives preliminary support, and also makes a simple prediction which is easily testable, that ACT is not useful in disambiguating verb alternation ambiguities.

Acknowledgments

This research was partly sponsored by the Swiss National Science Foundation, under fellowship 8210-46569 to P. Merlo, and by the US National Science Foundation, under grants #9702331 and #9818322 to S. Stevenson. We thank Martha Palmer for getting us started on this work, and Michael Collins for giving us acces to the output of his parser.

Appendix A

The unergatives are manner of motion verbs: *jumped, rushed, marched, leaped, floated, raced, hurried, wandered, vaulted, paraded, galloped, glided, hiked, hopped, jogged, scooted, scurried, skipped, tiptoed, trotted.*

The unaccusatives are verbs of change of state: *opened, exploded, flooded, dissolved, cracked, hardened, boiled, melted, fractured, solidified, collapsed, cooled, folded, widened, changed, cleared, divided, simmered, stabilized.*

The object-drop verbs are unspecified object alternation verbs: *played, painted, kicked, carved, reaped, washed, danced, yelled, typed, knitted, borrowed, inherited, organized, rented, sketched, cleaned, packed, studied, swallowed, called.*

References

[1] Thomas G. Bever. The cognitive basis for linguistic structure. In J. R. Hayes, editor, *Cognition and the Development of Language*, pages 278–352. John Wiley, NY, 1970.

[2] Michael Brent. From grammar to lexicon: Unsupervised learning of lexical syntax. *Computational Linguistics*, 19(2):243–262, 1993.

[3] Anne-Marie Brousseau and Elizabeth Ritter. A non-unified analysis of agentive verbs. In Dawn Bates, editor, *West Coast Conference on Formal Linguistics*, number 20, pages 53–64, Stanford, CA, 1991. Center for the Study of Language and Information.

[4] Morten H. Christiansen and Maryellen C. MacDonald. Individual differences in sentence comprehension: The importance of experience. Talk at the 11th Annual CUNY Conference on Human Sentence Processing, 1998.

[5] Michael Collins. Three generative, lexicalized models for statistical parsing. In *Proceedings of the 35th Annual Meeting of the Association for Computational Linguistics*, pages 16–23, 1997.

[6] Hoa Trang Dang, Karin Kipper, Martha Palmer, and Joseph Rosenzweig. Investigating regular sense extensions based on interesective Levin classes. In *Proc. of the 36th Annual Meeting of the ACL (COLING-ACL '98)*, pages 293–299, 1998.

[7] Hana Filip, Michael K. Tanenhaus, Gregory N. Carlson, Paul D. Allopenna, and Joshua Blatt. Reduced relatives judged hard require constraint-based analyses. In Paola Merlo and Suzanne Stevenson, editors, *Sentence Processing and the Lexicon: Formal, Computational and Experimental Perspectives*. John Benjamins, Holland, 1999.

[8] Lyn Frazier. *On Comprehending Sentences: Syntactic Parsing Strategies*. PhD thesis, University of Connecticut, 1978. Available through the Indiana University Linguistics Club, Bloomington, IN.

[9] Susan M. Garnsey, Neal J. Pearlmutter, Elizabeth Myers, and Melanie A. Lotocky. The contributions of verb bias and plausibility to the comprehension of temporarily ambiguous sentences. *Journal of Memory and Language*, 37:58–93, 1997.

[10] Lila Gleitman. The structural sources of verb meaning. *Language Acquisition*, 1:3–56, 1990.

[11] Ken Hale and Jay Keyser. On argument structure and the lexical representation of syntactic relations. In K. Hale and J. Keyser, editors, *The View from Building 20*, pages 53–109. MIT Press, Cambridge, MA, 1993.

[12] Judith L. Klavans and Martin Chodorow. Degrees of stativity: The lexical representation of verb aspect. In *Proceedings of the Fourteenth International Conference on Computational Linguistics*, 1992.

[13] Beth Levin. *English Verb Classes and Alternations*. University of Chicago Press, Chicago, IL, 1993.

[14] Beth Levin and Malka Rappaport Hovav. *Unaccusativity*. MIT Press, Cambridge, MA, 1995.

[15] Maryellen MacDonald, Neal Pearlmutter, and Mark Seidenberg. Lexical nature of syntactic ambiguity resolution. *Psychological Review*, 101(4):676–703, 1994.

[16] Maryellen C. MacDonald. Probabilistic constraints and syntactic ambiguity resolution. *Language and Cognitive Processes*, 9(2):157–201, 1994.

[17] D. C. Mitchell, F. Cuetos, M. M. B. Corley, and M. Brysbaert. Exposure-based models of human parsing: Evidence for the use of coarse-grained (non-lexical) statistical records. *J. of Psycholinguistic Res.*, 24:469–488, 1995.

[18] Bradley Pritchett. *Grammatical Competence and Parsing Performance*. University of Chicago Press, 1992.

[19] Geoffrey K. Pullum. Learnability, hyperlearning, and the poverty of the stimulus. In Jan Johnson, Matthew L. Juge, and Jeri L. Moxley, editors, *22nd Annual Meeting of the Berkeley Linguistics Society: General Session and Parasession on the Role of Learnability in Grammatical Theory*, pages 498–513, Berkeley, California, 1996. Berkeley Linguistics Society.

[20] J. Ross Quinlan. *C4.5 : Programs for Machine Learning*. Series in Machine Learning. Morgan Kaufmann, San Mateo, CA, 1992.

[21] Adwait Ratnaparkhi. A linear observed time statistical parser based on maximum entropy models. In *Proceedings of the 2nd Conference on Empirical Methods in NLP*, pages 1–10, 1997. Providence, RI.

[22] Adwait Ratnaparkhi. Statistical models for unsupervised prepositional phrase attachment. In *Proceedings of the 36th Annual Meeting of the Association for Computational Linguistics*, pages 1079–1085, 1998.

[23] Philip Resnik. Selectional constraints: an information-theoretic model and its computational realization. *Cognition*, 61(1–2):127–160, 1996.

[24] Suzanne Stevenson and Paola Merlo. Lexical structure and parsing complexity. *Language and Cognitive Processes*, 12(2/3):349–399, 1997.

[25] John Trueswell and Michael J. Tanenhaus. Toward a lexicalist framework for constraint-based syntactic ambiguity resolution. In Charles Clifton, Lyn Frazier, and Keith Rayner, editors, *Perspectives on Sentence Processing*, pages 155–179. Lawrence Erlbaum, NJ, 1994.

[26] John C. Trueswell. The role of lexical frequency in syntactic ambiguity resolution. *J. of Memory and Language*, 35:566–585, 1996.

Integrating psychometric and computational approaches to individual differences in multimodal reasoning

Padraic Monaghan (Padraic.Monaghan@ed.ac.uk)
Keith Stenning (K.Stenning@ed.ac.uk)
Jon Oberlander (J.Oberlander@ed.ac.uk)
Human Communication Research Centre, Division of Informatics, University of Edinburgh
2 Buccleuch Place, Edinburgh EH8 9LW, UK

Cecilia Sönströd (cecilia@www.phil.gu.se)
Department of Philosophy, Gothenburg University, Gothenberg, Sweden

Abstract

Psychometric measures of ability are unsuited to computational descriptions of tasks, primarily because they cannot take process into account. Studies of aptitude–treatment interactions have often failed to replicate from task to task precisely because of this difficulty. The current study aligns psychometric measures with process accounts in the domain of multimodal reasoning. Learning from multimodal logic courses transfers to other reasoning tasks, and this transfer has been found to relate to differences in strategic use of graphical representations in proof construction. The current study is a replication and an extension of these findings. Different goal types are distinguished in terms of: their modality; whether they involve proofs of consequence or non-consequence; and whether they can be solved by constructing single or multiple cases. We report on the interaction of a range of psychometric measures, and the ways in which they relate to the development and deployment of strategies. In particular, students who develop coping strategies to overcome difficulties with certain problems find that these strategies arise at the expense of appropriate use of a variety of strategies. Our approach, which characterises goals in terms of their logical as well as phenomenal properties, supports a computational perspective on psychometric measures in reasoning tasks.

Introduction

The process of learning to construct formal proofs combining diagrammatic and sentential representations provides a unique microcosm for investigating aptitude–treatment interactions (ATIs) based on different representational behaviours (Stenning, Cox & Oberlander, 1995). The formality of the representations and processes offers the possibility of producing computational models of the mental processes involved. Psychometric approaches which posit no accounts of the mental processes which underly their measurements not only block connections to cognitive accounts of mental process, but frequently lead to failure to replicate ATIs (Cronbach & Snow, 1977). Transfer of scientific theory from situation to situation is dependent on accounts of underlying structure and process—just like transfer of students' learning.

This paper seeks to replicate and extend earlier work on ATIs in learning logic from Hyperproof (HP), a multimodal proof environment due to Barwise and Etchemendy (1994). The HP interface is shown in Figure 1. The top of the main window displays a graphical situation. Below the situation are sentential statements that refer to the graphical situation. The other windows indicate goals for which the student has to construct proofs—in this example, there are four different goals, one sentential and three graphical. Graphical situations can contain abstraction in several ways: size (small,

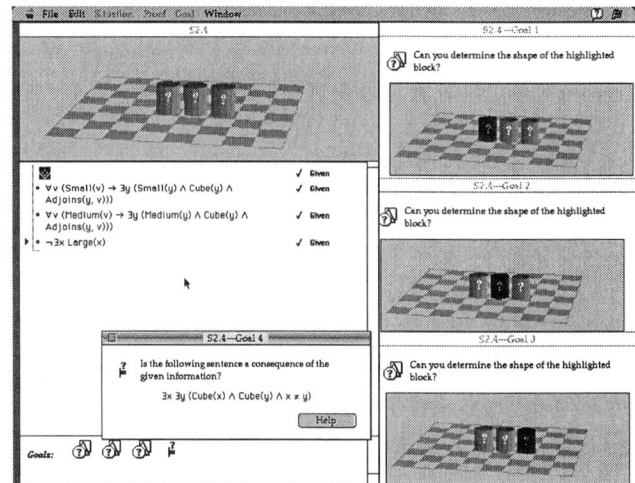

Figure 1: The Hyperproof interface—Problem 4.

medium, large) and shape (tetrahedron, cube, dodecahedron) can be left unspecified via cylinders and bags, respectively; and position of blocks in the situation can be left unspecified by blocks appearing off the board.

An earlier study (Stenning, Cox & Oberlander, 1995) revealed that scores on a subscale of the Graduate Record Exam (GRE) analytical reasoning test predicted outcomes of learning supported by HP with or without its diagrammatic component. Students were classified into high and low scoring on the constraint satisfaction problems of this test. For students taught a 10 week course with HP, those that scored high on the GRE pre-test showed pre- to post-test improvement on a 'blocks-world' (BW) test (see Methods section for a description of this test), whereas those that scored low on the GRE pre-test actually showed decrements on the BW test. These results were reversed when the teaching was a 10 week conventional logic course taught using only the sentential component of HP. Subsequent analysis of the logs of the proofs these students produced in their exams showed that the two groups displayed contrasting proof structures on some problems. These problems were characterised, as predicted, on computational grounds, by the use of a high degree of graphical abstraction (Oberlander et al., 1996).

Monaghan & Stenning (1998) replicated this ATI in the domain of syllogistic reasoning. Subjects who scored high on the GRE constraint satisfaction problems learned a graphical method for solving syllogisms with fewer errors and greater ease than their lower scoring counterparts. Teaching with a method that was based on sentential natural deduction reversed the effect. The study also showed that other tests of spatial ability correlated with difficulties at different stages of syllogistic reasoning. Thus, in these two domains of study, representation and strategy have been shown to inter-relate: the interplay of style and modality is an important factor in learning to solve reasoning tasks.

As well as aiming to replicate these previous studies with a different student population, we had two further goals. The first was to examine the relationship between several established psychometric measures that deal with 'spatial' and 'verbal' processing, and a multimodal reasoning environment. The second was to deepen our understanding of the dimensions of proof strategy which distinguish students with different reasoning styles.

The earlier study revealed systematic differences in the use of graphical abstraction on one indeterminate exam problem. Here we seek to explore the effects of goal-types in HP differing in (i) modality (sentential versus graphical); (ii) consequentiality (versus non-consequentiality); and (iii) multiplicity of required cases (versus the possibility of a single case being sufficient). For a given goal, there is a range of proof methods available. This classification of goal types and proof methods enables a controlled examination of strategic flexibility in choice of proof methods. The relative contributions of representational and strategic differences is an important issue in the psychometrics of reasoning (see for instance Roberts, 1998), as is the issue of flexibility of approach for psychometrics more generally (Guilford, 1980).

Method

84 students registered on a philosophy degree at the University of Gothenberg participated in the experiment. They followed the HP course material (Barwise & Etchemendy, 1994) as part of a course on introductory logic. The HP coursework was done in parallel with the students learning from more traditional sources: in particular, they learned a traditional natural deduction method of proof (Bennet, Haglund, Westerståhl & Sönströd, 1997), which was based on Mates' (1965) natural deduction method.

At the end of the course, students were set six problems in Hyperproof, which they were free to solve in their own time. These problems were designed to be 'indeterminate', containing a high degree of abstraction in the graphics. One exam question (no. 4) was the same as one used in the original HP study. Proofs were computer-logged, providing detailed data on temporal and ordinal aspects of proof construction.

The students sat a range of pre-tests and one post-test voluntarily. The number of students participating in each stage varied (see Results section).

Pre- and post-course tests

Pre- and post-course tests were used in order to replicate and extend the original HP studies. The same tests as were used in the first HP study were administered, but in addition we use other psychometric measures that are relevant to spatial and verbal information processing.

The GRE test was the same as that used in the first HP study. The test has two types of problem: 'analytical' items

(GREA) are those where the construction of a diagram is useful – these are the constraint satisfaction problems where a model can be constructed from the information; and verbal reasoning items (GREV) which require argument analysis, and assessment of the similarity of arguments. For these verbal items, several models may be consistent with the given information. For more details on this test see Cox et al. (1996).

As a measure of transfer of reasoning skills, the Blocks' World test was administered. This is a paper and pencil test which requires students to reason about situations similar to those presented in HP but with natural language descriptions of the conditions on those situations. Different versions of this test were given both before and after the course.

To measure 'spatial ability', the paper folding test (PFT) (French, Ekstrom & Price, 1963) was used. This requires the participant to decide on the array of holes resulting from a piece of paper being folded in various ways, having a hole punched in it, and then unfolded again. This can be interpreted as a measure of strategic flexibility in using spatial information, rather than reflecting spatial ability *per se* (Kyllonen & Lohman, 1983).

Students also took the embedded figures test (EFT). This requires the student to locate geometrical figures in a rectangle of crossing lines. Students who perform well on this task are classified as field-independent, those who perform less well are field-dependent (Witkin et al., 1971). Field-independent students are more likely to process information independently from the context, whereas field-dependent students are more likely to take the context into account.

All tests were set in English, and students were classified according to a median-split on each of the pre-tests.

Results and Discussion

61 students did the EFT and the PFT, 57 students did the pre-course BW test, and 59 students did the GRE test; 72 students completed the HP exam problems; and 27 students did the BW post-course test. Some of the HP records were lost due to bugs in the logging program, so n-values vary throughout the analyses. Full data for all questions and pre-tests exists for 39 students.

Correlations between the pre-test measures (shown in Table 1[1]) were in accord with the literature (Jonassen & Grabowski, 1993): the two subscales of the GRE are correlated; GREA correlates with EFT and PFT scores; and EFT and PFT show a significant, but slight, correlation.

These pre-test measure correlations indicate that the group is representative of a general student population. Analyses of

---

[1]Numbers in parentheses are the degrees of freedom. * indicates two-tailed significance p<0.05.

Table 1: Correlations between pre-tests.

	GREV	EFT	PFT
GREA	0.30* (57)	0.34* (55)	0.27* (55)
GREV		0.22 (55)	0.22 (55)
EFT			0.26* (59)

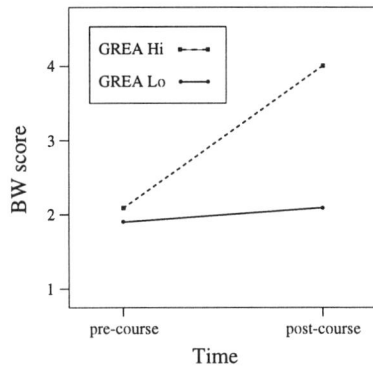

Figure 2: Change in BW score by GREA group.

the differences in generalising *from* and differences in reasoning *within* HP are now reported.

Transfer from HP

To relate the current study to existing results from the original HP experiment, the 'transfer' of reasoning skills from the course to the BW test was measured. Students were divided by a median split into those that scored high and those that scored low on the GREA subscale. Figure 2 illustrates the relationship between pre- and post-test score and GREA group.

As with the original study, those scoring high on the GREA subscale benefitted more from following the HP course (t(23) = 1.80, one-tailed p<0.05).

Strategies within HP

The original HP studies reported two different responses to graphical abstraction in HP, which related to the GREA groups. GREA high scoring students use more abstraction in their proofs, whereas GREA low students utilise highly concrete situations as they solve the problems. These different strategies were found only on HP problems that contained graphical abstraction (Figure 1 shows an indeterminate problem (no. 4)). As the exam in the current study was deliberately designed to consist only of this sort of problem, we expected a similar dichotomy of strategies to emerge.

The strategies observed in the original HP study can be distinguished by the proportion of fully concrete as opposed to abstract graphical situations that the student creates in a proof (Monaghan, 1998). GREA high students' proofs have more abstract situations, GREA low students' proofs have more concrete situations. However, this distinction does not predict different proofs in the current study. The GREA highs use fully concrete situations 40% of the time, GREA lows construct concrete situations 37% of the time. However, the EFT scores did seem to reflect this stylistic difference better: EFT high scoring students used concrete situations 32% of the time, compared to EFT low scoring students' 51% use, which is significantly more (t(51) = 2.29, p<0.05).

It may be that the EFT is a better indicator of this strategic difference for this population – answering the GRE test in a second language may mean that scores also reflect language competence. It might be expected that the EFT and the GREA correlate, and in the current study this is the case. However, in the syllogistic study reported earlier (Monaghan & Stenning, 1998) no correlation was found between these measures for native English speakers (r(20) = 0.03, p>0.8), and in the current study transfer of ability to the BW test was not found to improve more for one EFT group over the other (t(23) = 0.11, p>0.9).

Differences in the observed strategies used in HP may also be due to differences in teaching in the current study, which combined sentential and multimodal teaching. In the original HP study, when sentential materials alone were used to teach logic, the GREA low group transferred reasoning skills better to the BW test.

This complicated mixture of success and failure to replicate at the level of psychometric tests has bedevilled the ATI literature (see Cronbach & Snow, 1977, for a review). Spatial ability, for example, has been variously defined as ability for the "encoding, transformation, and recognition of spatial information" (Salthouse et al., 1990). The measure suffers from even less precision when strategic variation is considered. Kyllonen & Lohman (1983) show that such tests as the PFT are solved by some students with strategies that seem to invoke representations that are not spatial. This has led Roberts (1998) to consider such measures from the perspective of strategic variation and change, arguing that it is these patterns of strategic variation and change which characterise the psychometric measures.

Our response is to look for more principled ways of analysing proof styles which have clearer relations to computational processes. Ultimately, theoretically motivated process accounts of proof styles and an account of how they generalise across tasks should replace unprincipled test scores. For the present, the psychometric scores provide some empirical reassurance of general applicability. We had expanded the indeterminate problems in the exam used in the present study precisely so that we could explore styles more systematically. HP offers an environment where the student's preference for expressing information *within* a particular modality can be assessed; where the transfer of information *between* modalities can be plotted; and where variation of proof method with problem type can be explored.

In order to explore the issue of strategic flexibility within HP we need two classifications: of goal-types and of proof methods. HP problems can pose a number of goals that have to be solved with reference to one situation. These goals vary in terms of *modality*: the goal can be about a graphical state of affairs, or about a sentential statement (G- or S-goal). Within the graphical modality, goals can vary as to whether the requirement is to prove that a particular situation is a consequence of the given information (graphical-consequence: GC-goal), or to prove that the situation is not a consequence of the given information (graphical-non-consequence: GN-goal). Finally, the GC-goals can be distinguished between those that require splitting into *multiple* situations in order to reach the conclusion (GCM-goal), and those that can be achieved by applying sentential information to a *single* graphical situation (GCS-goal). In HP, the former are usually those that require the position of blocks to be decided (though problem 4 illustrated in Figure 1 is an exception). The latter require the shape or size of the shape to be determined. In problem 4 (Figure 1), goal 4 is an S-goal, goals 1 and 3 are GN-goals, and goal 2 is a GCM goal.

Having chosen a classification of goals, we also need a classification of proof-methods so that we can explore how problem-type and learning style combine to determine proof method. HP is designed to teach 'proof-by-cases'. All goal types can be achieved by constructing a number of graphical situations (cases) and using sentential or graphical rules to prove or exclude the goal for each case. Using fewer situations reflects the use of more abstraction in the way the situations are characterised. This suggests the use of number of cases as a general structural index of proof method. We first seek some empirical support for this measure.

When the pattern of situations that the students produced in their HP proofs was examined more closely, one striking feature emerged: some students used very few graphical situations in their proofs. For one problem in particular (in fact no. 4, the question from the original study), all steps in these students' proofs were rules that operated on sentential information in the sentential window of HP. These proofs maintained the maximum level of abstraction, and so can be seen as being the extreme case of the graphical proof that invokes abstract graphical situations. Thus, three types of proof are distinguishable in the current study, and they can be located, not in terms of abstract or concrete graphical situation construction, but in terms of the mean level of concreteness of the proof. For question 4, where instances of the 'sentential' strategy occur, mean concreteness and EFT score are significantly related ($r(49) = 0.35$, $p < 0.02$).

Students were classified as using one of these three proof-types on problem 4, and pre-test scores were compared for these three groups. The results are shown in Table 2.

On semantic grounds, the three strategies constitute points on a continuum—sentential proofs are merely more abstract proofs which happen to be in a different modality. This continuum is reflected in EFT and PFT scores. Those students producing more abstract proofs independent of modality, have high EFT and PFT scores, and they better transfer learned reasoning skills to the post-course BW test. The differences in EFT score between the two extremes of abstraction/concreteness are significantly different ($t(44) = 2.65$, $p < 0.02$). The group including both sentential and abstract graphical strategies scored higher on the EFT than the concrete graphical strategy group ($t(49) = 2.64$, $p < 0.02$). However, the EFT score for the graphical abstract group alone did not differ significantly from either other strategy group. The ordering of EFT scores across the three proof methods helps justify the use of the number of situations generated in a proof as an index of proof methods across problems.

The modality independence constituted by sentential proof is consonant with the EFT being a measure of field-dependence–field-independence. When information is presented graphically, students who are more field independent (scoring high on the EFT) are more able to represent that information abstracted from the context it is presented in (Witkin et al., 1971).

This classification of proof methods applied only to the data of problem 4 does not reveal any relation to GRE scores. However, relations reappear when the two classifications of goal-type and proof method are applied to all the data.

Proof methods analysed by goal-types

For each goal type, the number of situations used to achieve the goal was analysed. Two-way ANOVAs were carried out, with number of situations used for each type of goal as a repeated measure and median splits on the pre-tests (EFT, PFT, GREA, GREV) as between-subjects variables. Using these analyses, strategic approach to the different types of goal can be assessed and related to the pre-tests. Three different ANOVAs were carried out: G-goals compared to S-goals; GC-goals compared to GN-goals; and GCM-goals compared to GCS-goals. Only results reaching significance are reported below. Between-subjects effects indicate whether the pre-test alone distinguishes overall differences in the number of situations used to achieve the goals; within-subjects effects measure the interaction between goal type and the pre-tests.

Sentential & Graphical goals

Between subjects effects that proved significant are shown in the first part of Table 3.

This analysis shows that the number of situations used to solve the goals is sensitive to several of the pre-test measures. Those that score higher on the PFT use fewer situations to solve the goals. Those that score higher on both the EFT and the PFT use fewer situations than those that score lower

Table 3: Comparing S and G goals: between-subjects effects and within-subjects effects.

Pre-test(s)	Between-subjects effects
PFT	F(1, 25) = 8.31, p<0.01
EFT by PFT	F(1, 25) = 13.01, p<0.005
EFT by GRE-A	F(1, 25) = 7.05, p<0.02
PFT by GRE-V	F(1, 25) = 7.49, p<0.02
GRE-A by GRE-V	F(1, 25) = 4.74, p<0.05
EFT by PFT by GRE-V	F(1, 25) = 6.40, p<0.02
EFT by GRE-A by GRE-V	F(1, 25) = 7.61, p<0.02

Pre-test(s)	Within-subjects effects
goal type	F(1, 25) = 176.99, p<0.001
EFT by goal-type	F(1, 25) = 4.24, p<0.05
PFT by goal-type	F(1, 25) = 4.99, p<0.05
EFT by PFT by goal-type	F(1, 25) = 7.04, p<0.02
PFT by GRE-V by goal-type	F(1, 25) = 4.74, p<0.05

Table 2: Pre-test scores for students classified by three strategies for Problem 4.

Pre-test	Strategy		
	sentential	abstract graphical	concrete graphical
GREA	5.00	4.25	4.67
GREV	2.00	1.75	2.97
EFT	19.10	16.40	12.56
PFT	13.50	12.80	12.36
BW improvement	1.50	1.00	0.76

Table 4: Mean number of situations for goal type by PFT and GREV.

S-goal	PFT Lo	PFT Hi
GREV Lo	10.08	4.89
GREV Hi	7.25	7.33

G-goal	PFT Lo	PFT Hi
GREV Lo	37.63	20.88
GREV Hi	25.25	29.00

on one or both of these measures. However, there is not a simple association between high test scores and efficiency in situation use, as measuring the overall proof length alone is unrelated to any of the psychometric scores.

The second part of Table 3 displays within-subjects effects and these results indicate that students that score differently on the pre-test differ in their solutions for each of the goal types. Both EFT Hi students and PFT Hi students use fewer situations for each type of goal. Here high ability on these measures relates to using fewer situations, but the interaction of PFT by GREV on goal-type points to a more stylistic variation, and one that is glossed over when only a single psychometric is used to distinguish response. Table 4 shows the mean number of situations used for each goal type distinguished by PFT and GREV. There are similar interactions for both types of goal: PFT Hi students use fewer situations for each goal type, but this is modulated by GREV group. If the student is in the GREV high group, then it doesn't matter which PFT group they are in: they use the same number of situations for each goal type. If the student scores low on the GREV, then being PFT Lo means a large number of situations are used for each goal, and being PFT Hi means that the fewest situations are used for each goal.

Students who score high on the GREV scale are those who are good at solving problems that do not require breaking into cases. Scoring high on this scale means that flexibility in using graphical representations (measured by the PFT scale) is irrelevant. Scoring low on the GREV means that students who are good at using graphical representations to support reasoning (PFT Hi) utilise the graphical abstraction facilities of HP to their full potential. Those who are low on both scales rely more on concretising the problem's information.

Graphical goals: consequence & non-consequence

For between-subjects effects, the results were identical to the S- and G-goal analysis. EFT Hi and PFT Hi students used fewer situations for both types of problem. However, the only within-subjects effect was for goal-type: in general, students use more situations for GC-goals than for GN-goals, and no interaction between goal type and the pre-tests emerged. This lack of effect is due to the small amount of variation in the strategy used to solve GN-goals. Most students solve them by constructing two situations that differ in terms of the feature in question.

Graphical consequence goals: multiple & single case

When different types of GC-goal are distinguished, different approaches to the goals are highlighted by the pre-tests. Table

5 indicates the between-subjects and within-subjects effects.

Table 5: Comparing GCM and GCS goals: between-subjects effects and within-subjects effects.

Pre-test(s)	Between-subjects effects
PFT	$F_{(1, 25)} = 4.43, p < 0.05$
EFT by PFT	$F_{(1, 25)} = 9.38, p < 0.01$
EFT by GREA	$F_{(1, 25)} = 4.83, p < 0.05$
PFT by GREV	$F_{(1, 25)} = 4.47, p < 0.05$
EFT by PFT by GREV	$F_{(1, 25)} = 6.94, p < 0.02$
EFT by GREA by GREV	$F_{(1, 25)} = 6.55, p < 0.02$

Pre-test(s)	Within-subjects effects
goal-type	$F_{(1, 25)} = 42.98, p < 0.001$
EFT by PFT by goal-type	$F_{(1, 25)} = 8.55, p < 0.01$
PFT by GREA by goal-type	$F_{(1, 25)} = 4.55, p < 0.05$
EFT by PFT by GREV by goal-type	$F_{(1, 25)} = 7.15, p < 0.02$
EFT by GREA by GREV by goal-type	$F_{(1, 25)} = 8.52, p < 0.01$

Again, those with high scores on the EFT and PFT use the fewest number of situations for GC-goals. For within-subjects effects, a different pattern emerges to that found for S- and G-goals. There is an interaction of PFT by GREA by goal-type, and this indicates different approaches to the two types of goal.

For GCM goals, those that score low on the GREA but high on the PFT use the fewest situations (see Table 6). For GCS goals, these are the students that use the most situations. GCS goals can be solved by using one situation that indicates the sentential information being applied to the graphical situation. Alternatively, they can be solved by constructing multiple situations that explore the constraints. Those students that effectively exploit graphical abstractions in solving the GCM problems seem to maintain this strategy when a shorter solution is available. The GREA low scoring subjects are those that are poorer at solving problems where one model can be constructed. These are exactly the GCS-goals. The current analysis suggests that these students have learned a coping strategy for such problems which is not efficient, but it does at least achieve the result, and being flexible in using graphical representations enables the development of this strategy. Students who are GREA low but do not learn

Table 6: Mean number of situations for goal type by PFT and GREA.

GCM-goal	PFT Lo	PFT Hi
GREA Lo	13.40	9.86
GREA Hi	12.33	10.89

GCS-goal	PFT Lo	PFT Hi
GREA Lo	1.15	2.00
GREA Hi	1.71	1.21

this coping strategy—due to being less able to exploit graphical abstraction—do not experience this interference effect on GCS problems. Those students that score highest on both scales seem to choose optimal strategies: their proofs for the GCM goals are not so short, but when they come across GCS goals, they can recognise and use a more appropriate strategy. This reflects flexible use of the graphical abstraction facilities and recognition of differing proof constraints.

Conclusions

The current study replicates the transfer effects of learning logic in a multimodal environment to other reasoning domains, but the strategic differences previously observed do not emerge in the same way. This is due, in part, to differences in the teaching method which provides extra encouragement for using sentential representations to solve logical problems. This gives rise to three types of strategy: sentential, graphical abstract and graphical concrete, which form a continuum in terms of the extent to which they utilise HP's graphical abstraction facilities.

Distinguishing different goal types in HP offers a window into flexible strategic change and the inter-relationship of multiple psychometric measures. No *one* psychometric captures strategic variation in using graphical abstraction in HP, but combinations reflect the options open to students. If psychometrics index cognitive style, then cognitive style dictates the development of strategies in solving multimodal problems. It is this chaining that reflects the observations of strategic variation and change being related to purported measures of 'spatial ability'. A student's propensity for solving problems in a certain way is tempered or licenced by their ability to use representations to achieve the goal. Some students develop strategies that counteract their difficulty with particular problem types, but these can then result in an inflexibility of approach. Those students who seem flexible in their strategies in HP do not necessarily use these strategies optimally, but their ability to switch strategy according to the problem's constraints makes up for this.

The highlighting of strategic variation and change in the current study enables a recharacterisation of the psychometric measures which takes into account the computational features of the task. Thus, HP offers a window into what the psychometrics mean from the computational perspective. Because HP is based on a principled (if highly abstract) theory of reasoning, its categories can be applied to understanding performance on tests such as the GRE. For example, 'splitting into cases' is something that has to be achieved in reasoning, whether in a formal domain, or in more informal problem solving.

Acknowledgements

This research was supported by ESRC research studentship R00429634206, by ESRC's Centre Grant to HCRC, and by a grant from the McDonnell Foundation's CSEP Initiative. The third author is an EPSRC Advanced Fellow.

References

Barwise, J. & Etchemendy, J. (1994). *Hyperproof.* Stanford: CSLI Publications.

Bennet, C., Haglund, B., Westerståhl, D. & Sönströd, C. (1997). *En introduction till första ordningens språk.* University of Gothenberg.

Cronbach, L. J. & Snow, R.E. (1977). Aptitudes and Instructional Methods: A Handbook for Research on Interactions. Irvington: N.Y.

Cox, R., Stenning, K., & Oberlander, J. (1994). Graphical effects in learning logic: reasoning, representation and individual differences. *Proceedings of the 16th Annual Conference of the Cognitive Science Society* (237-242). Georgia: Lawrence Erlbaum Associates.

French, J. W., Ekstrom, R. B., & Price, L. A. (1963). *Kit of reference tests for cognitive factors.* Princeton, NJ: Educational Testing Service.

Guilford, J.P. (1980). Cognitive styles: what are they? *Educational and Psychological Measurement*, 40, 715-735.

Jonassen, D. H. & B. L. Grabowski (1993). *Handbook of Individual Differences, Learning, and Instruction.* Hillsdale: Lawrence Erlbaum Associates.

Kyllonen, P. C. & Lohman, D. F. (1983). Individual differences in solution strategy on spatial tasks. In R. F. Dillon & R. R. Schmeck (Eds.) *Individual Differences in Cognition Volume 1.* New York: Academic Press.

Mates, B. (1965). *Elementary Logic.* Oxford: Oxford University Press.

Monaghan, P. (1998). Holist and serialist strategies in complex reasoning tasks: cognitive style and strategy change. Research Paper EUCCS/RP-73. Edinburgh: University of Edinburgh, Centre for Cognitive Science.

Monaghan, P. & Stenning, K. (1998). Effects of representational modality and thinking style on learning to solve reasoning problems. *Proceedings of the 20th Annual Conference of the Cognitive Science Society of America* (716-721). Madison: Lawrence Erlbaum Associates.

Oberlander, J., Cox, R., Monaghan, P., Stenning, K., & Tobin, R. (1996). Individual differences in proof structures following multimodal logic teaching. *Proceedings of the Eighteenth Annual Conference of the Cognitive Science Society*, (201-206). La Jolla, CA: Lawrence Erlbaum Associates.

Roberts, M. J. (1998). Individual differences in reasoning strategies: a problem to solve or an opportunity to seize? In G. d'Ydewalle, W. Schaeken, A. Vandierendonck & G. De Vooght (Eds.), *Deductive reasoning and strategies.* Mahwah: Lawrence Erlbaum Associates.

Salthouse, T. A., Babcock, R. L., Mitchell, D. R. D., Palmon, R. & Skovronek, E. (1990). Sources of individual differences in spatial visualization ability. *Intelligence*, 14, 187-230.

Stenning, K., Cox, R. & Oberlander, J. (1995). Contrasting the cognitive effects of graphical and sentential logic teaching: reasoning, representation and individual differences. *Language and Cognitive Processes*, 10, 333-354.

Witkin, H., Oltman, P., Raskin, E. & Karp, S. (1971). *A manual for the embedded figures test.* Palo Alto: Consulting Psychologists Press.

410

Feeling Low but Learning Faster: Effects of Emotion on Human Cognition

Simon C. Moore (MooreSC@Cardiff.ac.uk)
School of Psychology, PO Box 901,
Cardiff University, Cardiff, CF1 3YG, UK

Mike Oaksford (Oaksford@Cardiff.ac.uk)
School of Psychology, PO Box 901,
Cardiff University, Cardiff, CF1 3YG, UK

Abstract

This study examined the effects of emotion on the long-term acquisition of a procedural skill over a five-day period. Two tasks were employed: a word association task (WAT) and a visual discrimination task (VDT). Over the initial four days of the study participants went through a mood induction procedure (MIP) then subsequently completed both tasks. Both tasks showed a reduction in reaction time consistent with the power law of learning. No significant change in reaction time between day four and day five (one week later) was noted suggesting the change in reaction time was robust. These data further suggest that emotion modifies the rate at which the VDT is acquired.

Introduction

That emotion modulates human cognition is now well accepted in contemporary psychology (Damasio, 1994; Ekman & Davidson, 1994; Isen, 1987). However, experimental data supporting this view is principally derived from experiments employing a single session protocol. In consequence, the long-term effects of emotion on cognition have not been addressed. In particular, whether emotions modulate the acquisition of a cognitive skill has not been studied. This paper addresses this issue by running participants on two different tasks over a five-day period. Similar tasks demonstrate a steady reduction in the time required for their completion over a series of days (Anderson, 1990; Karni & Sagi, 1991).

The capacity to learn is a central feature of human ability. This ability has been grouped into two forms: the procedural, incidental, acquisition of repetitive skills, and the declarative, explicit, acquisition of memory for places and events (Squire, Knowlton and Musen, 1993). Acquisition in declarative memory generally occurs on a one shot basis: either you know it or you do not. In contrast, acquisition of a procedural skill can take many days showing a steady improvement over time (Fitts & Posner, 1973).

Emotion has been found to affect episodic memory. Bradley, Greenwald, Petry, & Lang (1992) showed that the (subjectively reported) emotional arousal element of an emotion-eliciting picture predicts recall for that picture up to one year later. More recently, Cahill and McGaugh (1995) found that slides of varying emotional arousal, accompanied by a matched narrative, produce more accurate delayed recall when the level of arousal is highest at the encoding stage. However, administering propranolol, a β-adrenergic antagonist, cancels this effect (Cahill, Prins, Weber & McGaugh, 1994). These data suggest that noradrenalin (NA) may play an important role in memory performance under varying emotional states.

If the effects of emotion on memory are as broad as Bradley, et al. (1992) and Cahill, et al. (1994) findings suggest, then we hypothesize that the acquisition of a procedural skill may also be affected. The same underlying plasticity is believed to facilitate both declarative and procedural learning (e.g. Garcia, 1984).

Oatley and Johnson-Laird (1987) argue that emotion differentially modulates cognition, enhancing some processes over others. Consistent with this modulatory role, Davidson (1998) groups emotion into two forms: those that motivate 'approach' behavior, such as love and hunger, and those that motivate 'withdrawal', such as fear. Moreover, there is now a substantial body of evidence that shows that approach emotions are associated with the activation of anterior regions of the left cerebral hemisphere whereas withdrawal emotions are associated with activation of the right anterior hemisphere (Davidson, 1998).

It has consistently been shown that greater left hemisphere arousal (measured using EEG) is associated with increased verbal and reduced spatial ability. Whereas higher right hemisphere arousal is associated with increased spatial ability and reduced verbal ability (e.g. Levy, Heller, Banich, & Burton, 1983). This leads to the prediction that tasks demonstrating greater right hemisphere activation should show better initial performance in a negative emotional state whereas tasks demonstrating left hemisphere activation will show poorer performance. Moreover, this pattern should be reversed for participants in a positive emotional state. Therefore, when assessing the acquisition of a procedural skill it is of interest to use cognitive tasks that depend differentially on the cerebral hemispheres.

To test the hypothesis that emotion may modulate the rate at which a cognitive skill is learnt we used a standard learning experiment where a task is repeated over a series of days and change in reaction time provides the index of learning. Two tasks were selected that had previously been shown to rely differentially on the two cerebral hemispheres: a visual discrimination task (VDT; adapted from Corbetta, et al., 1993) which relies more on the right hemisphere, and a word association task (WAT; adapted from Vandenburghe, et al., 1996) which relies more on the left hemisphere.

Methodology

Participants performed the same tasks on five separate days. Days one to four ran consecutively with each participant going through one mood induction procedure, one VDT and WAT each day. The fifth day, one week later participants performed only the WAT and VDT.

Thirty-six undergraduate psychology students from the University of Wales, Cardiff participated. Each had English as their first language, was right handed, was between the ages of 18 and 25 years and had normal or corrected to normal eyesight. Prior to the experiment potential participants were first screened. Volunteers scoring 12 or above on the Beck Depression Inventory (Beck & Steer, 1987; post hoc analysis revealed no significant interaction between emotion group and BDI score) or undertaking a course of medication that might effect the variables under consideration in this study (such as psychopharmacological drugs) were not allowed to participate. Both factors were viewed as potential confounds.

Materials.

Positive, neutral and negative emotional states were induced through use of an amalgamation of three procedures (Martin, 1990; Westermann, Spies, Stahl & Hesse, 1996). An adapted version of the Velten mood induction procedure (Velten, 1968). Affectively valenced music and requesting participants to enter the required mood state (Slyker & McNally, 1991).

Lykert scales were used as a mood induction check (Isen & Gorgoglione, 1983; Oaksford, Morris, Grainger & Williams, 1996). Each scale was arranged along a numbered line with two, affectively salient, statements at each end. To the right were presented positive adjectives (refreshed, calm, alert, positive and amused) and to the left affectively negative adjectives (tired, anxious, unaware, negative and sober, respectively).

Visual discrimination task (VDT; figure 1, a & b). Sixteen objects were presented in a circular fashion (the circle subtended a visual angle of 40°, presented until participant response), equally spaced around a central fixation point. There were two types of object: a '∨' and a '∧'. Each trial was one of two conditions, either all sixteen objects were of the '∨' shape or, on half of the total trials, one object was replaced with the '∧'. Participants were instructed to look for the '∧' and respond with a 'yes' if it was present, otherwise respond 'no'. For the VDT there were 64 trials with the inverted shape appearing in all 16 positions equally often. Trials were presented in a pseudo-randomized order, not more than three trials of the same type following consecutively.

Word association task (WAT; figure 1, c & d). Each trial consisted of three words: one target word (presented in the center of the screen, using the Geneva font and for a duration of 1,000ms). This was followed by two probe words (one to the left and one to the right of the screen, with the

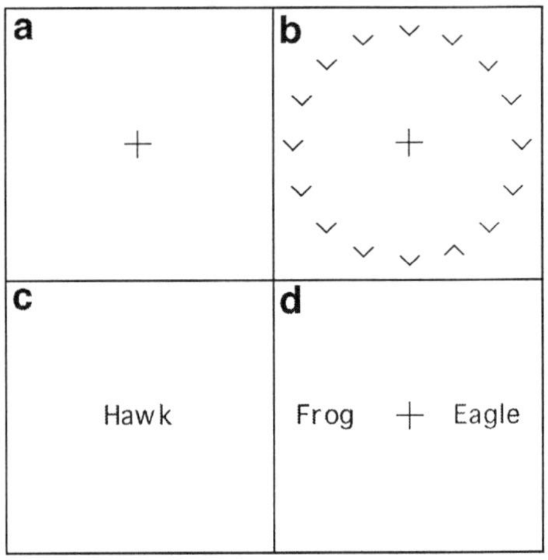

Figure 1: Stimuli. This figure shows example of stimuli from the two tasks used in this study. VDT -- Top row, left to right. Participants were initially presented with a centrally presented fixation point followed by a circular array. Participants were asked to judge whether the '∧' was present or not. WAT -- bottom row left to right, participants were initially presented with a cue word followed by two target words. Participants had to decide which of these two words were most associated with the previously shown target word.

centers of the two words subtending a visual angle of 32°) which were simultaneously presented until participants responded. The words were selected from the Birkbeck Word Association Norms (Moss & Older, 1995). Each days word set was novel to that day.

In each task, trials were presented in a pseudo-randomized order with not more than three responses of the same type (left, right) following consecutively. Four practice trials were constructed to allow participants to become familiarized with the procedure. For each WAT there were 64 trials.

All tasks were presented on a Power Macintosh computer (4400/200) and participants used appropriately marked keys on the keyboard to make responses. The MacLab program (Costin, 1988) controlled stimulus delivery and recorded responses and response times. Music used in the mood induction procedure was presented through stereo headphones attached to a Macintosh Power PC (6500/275) located in a separate control room. For all tasks, other than the mood induction procedure, a headrest was placed 44cm from the screen with participants eyes being level with the centrally presented fixation point.

Procedure.

A statement of informed consent was read and signed by each participant. Participants were then randomly assigned to a condition in the experiment.

Participants went through the appropriate mood induction procedure, each undergoing the same procedure for the first four days of the study. Participants were presented with written instructions that asked them to read the Velten statements aloud (in order to ensure they engaged in the

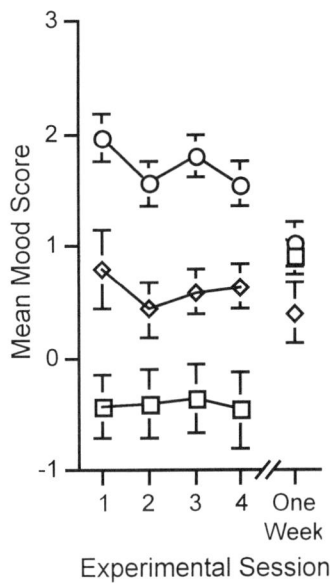

Figure 2: Subjective rating of emotional state. The positive emotion group (circles) rated themselves as more positive than the neutral group (diamonds) who rated themselves as more positive than the negative group (squares). One week later no lasting effects of the MIP were observed.

mood induction procedure). Each statement was presented for 11 seconds followed by a blank screen for 5 seconds. Simultaneously participants listened to music appropriate to their group.

Directly following the mood induction procedure, participants completed the mood induction check. Once the headphones were removed, they performed both the WAT and VDT tasks. Order of task was counterbalanced between participants. Prior to leaving the laboratory, participants were asked to complete a second mood induction check to ensure no one left in an overly negative emotional state. On the fifth day, (one week later) participants performed the same tasks but without the mood induction procedure. Following the tasks participants were debriefed, had all questions regarding the study answered and had the experimental hypotheses explained to them.

Results

Each participant's reaction time (RT) was trimmed to within 2.5 standard deviations from the mean for each task (2.9% of the total data set was removed from analysis). For the VDT, only trials where the inversion was present were analyzed.

MIP. On each day the positive group rated themselves as significantly more positive and the negative group rated themselves as significantly more negative than the neutral group at at least the .05 level (for all eight planned comparisons, $F(1, 33) > 5.0$). Self-report data from the retention session (day 5) where no MIP was performed, revealed no significant emotion state differences between groups (see figure 2).

Learning Rates. Practice and RTs are related as a power function (Fitts & Posner, 1973). We therefore fitted power curves (of the form At^{-k}) to the first four days data: overall, mean $r2 = .974$ (sd = .05). We also fitted power curves to each individual participant's data: overall, mean $r2 = .70$ (sd = .3). The exponent (k) is an index of learning rate. We therefore used the exponents for each participant's learning curve as the dependent variable in our subsequent analyses. In the WAT there were no significant differences in learning rate between emotion conditions. However, in the VDT there was a significant effect of emotional state on learning rate, $F(1, 33) = 16.38$, $p < .0005$ (see figure 3). In pairwise comparisons using Scheffe's S test, the learning rate for the negative emotion condition was significantly higher than the neutral, $p < .05$, and the positive conditions, $p < .01$. Between tasks, in simple effects comparisons, the learning rates were identical for positive emotion, $F(1, 33) = .004$, however, they were significantly higher in the VDT than the WAT for both the neutral condition, $F(1, 33) = 7.28$, $p < .025$, and the negative condition, $F(1, 33) = 18.06$, $p < .0001$. The effects of practice were robust: performance levels one week after the final training session were at similar levels indicating that the improvement over the training phase was retained.

Day One and Day Four Reaction Times. On Day 1 there was an interaction between emotion (positive-negative) and task, $F(1, 22) = 6.00$, $p < .025$. However, between group there was no significant interaction on the WAT ($F < 0.7$).

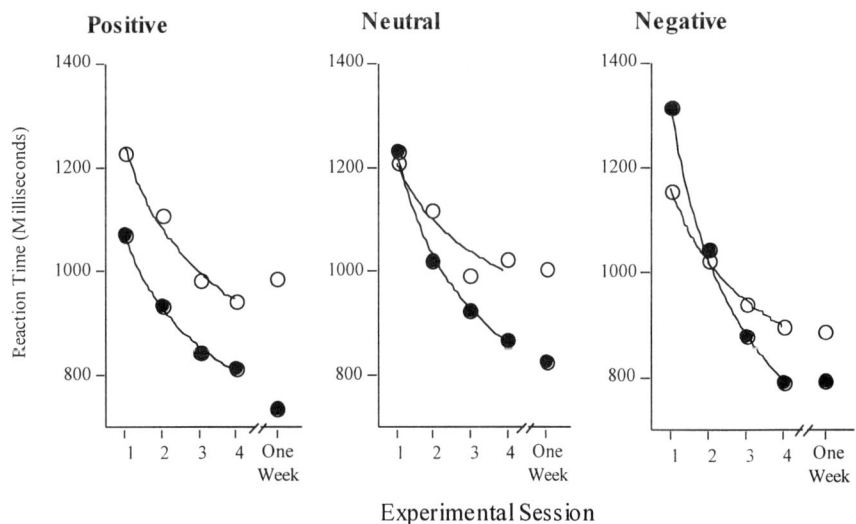

Figure 3: Learning rates. This diagram shows the mean trimmed RTs obtained in each emotion condition, with regression lines fitted, on each day of training (first four consecutive days) for both VDT (filled circles and dashed line) and the WAT (unfilled circles and solid line).

For the VDT participants in a negative emotional state recorded significantly higher RTs than those in the positive emotion state, $F(1, 22) = 5.78$, $p = 0.025$. On Day 4 no interaction was observed, however, there was a significant main effect of task, $F(1, 33) = 12.26$, $p < .0025$. Participants were faster to respond on the VDT than on the WAT. Moreover, no speed accuracy tradeoff was observed across task and across group. 16 trials worth of data were lost on day 5 for one participant in the positive group due to computer error.

Laterality. The RT for associates presented to the left of the fixation point were compared with associates presented to the right of the fixation point on the WAT. Target stimuli presented to the right of the fixation point were compared with target stimuli presented to the left of the fixation point (excluding instances where the target appeared at the top most and lower most positions) in repeated measures ANOVA. We took RTs from the first four days of the study. Participants demonstrated a facilitation effect on RT for associates presented to the left visual field on the WAT and a facilitation effect on RT for targets presented to the right visual field on the VDT ($F(1, 99) = 8.91$, $p < 0.01$).

Discussion

The data presented here show that both tasks (VDT and WAT) demonstrate a reduction in reaction time over the initial four days of the testing period. This reduction is consistent with the power law of learning (Fitts & Posner, 1973). Relative to emotion it is found that the VDT shows a higher rate of learning in the negative group relative to the neutral and the positive groups. Furthermore, the positive group learns both tasks (VDT and WAT) at a comparable rate and the WAT learning rate is consonant between positive, neutral and negative groups. If our assumption that the change in RTs on the VDT is representative of a procedural acquisition of a skill then this data suggests that the rate of learning may be modulated through manipulation of the emotional state of participants.

This view is further consistent with the already discussed research on the effects of emotion on memory. There are data to suggest emotion has long term effects on the encoding of information (Bradley, et al., 1992; Cahill, et al., 1994). Specifically, NA is implicated in the activation of the amygdala. The amygdala is, in turn, implicated in the modulation of long-term memory (Bianchin, Souza, Medina & Izquierdo, 1999; Phelps, La Bar, Anderson, O'Connor, Fulbright & Spencer, 1998); greater stimulation of this region leads to an enhancement. Moreover, heightened activity of the amygdala is associated with negative emotion states, but not positive emotion states (Davidson, 1994).

Triesman and Gormicon (1988) further argue that early visual processes decompose stimuli along a number of dimensions and into a number of separable components. In visual search paradigms 'pop-out', where the target becomes easier to discriminate from its distracters, occurs through the coding of the unique features of that stimulus. Moreover, NA has been implicated in the improvement on visual discrimination tasks across the consolidation period (Karni, et

al., 1994). The improvement demonstrated on the VDT is consistent with this view.

The laterality of emotion and cognition predicted that performance on a spatial task, when participants were placed in a negative emotional state should be facilitated relative to a left hemisphere verbal task. It is clear, from day one RTs that the opposite occurred: participants in a negative emotional state took longer to complete the VDT than those in a positive state. These data may be consistent with the view that NA is involved with human emotion.

Usher, Cohen, Servan-Schreiber, Rajkowski and Aston-Jones (1999) present evidence to suggest activity in locus coeruleus (LC) NA neurons correlates with performance on a visual discrimination task. An instance where LC activity is relatively high correlates with a detriment on a visual discrimination task. First, these data support the view that NA may be implicated in the effects observed on the VDT in this study. Second, it provides a plausible route through which the VDT data may be accounted for.

In consequence, the action of NA provides a plausible explanation for both the day one biases on the VDT and the heightened learning rate in the negative condition.

Conclusion

These results suggest that varying a person's emotional state can modulate the rate of procedural learning. A plausible neuroanatomical basis for this effect is proposed that centers on the role of NA and its involvement in memory consolidation and early visual processing.

Further work is required to substantiate this explanation. We need to delineate what aspects of a task make it susceptible to modulation of the learning rate. The differential effects observed on the VDT and WAT may be explained by the fact that in the WAT novel words were used on each day, whereas the VDT employed the same target shape on each day. Similarly, that the data appears inconsistent with the proposed hemispheric effects of emotion requires more work. It may be that the tasks employed are inappropriate.

Although the findings presented here are from a controlled laboratory setting they may have a broader application. Human emotion is pervasive across the life span and hence it might be expected to affect learning of many important skills such as reading. It might be possible to match emotional state to the to-be-learned task, so as to optimize the learning rate.

Acknowledgments

We would like to thank Janice Muir for offering comments on an earlier draft of this paper; Åse Kvist Innes-Ker for assistance in preparing the mood induction procedures; Allan Jones for technical assistance; Joselyn Sellen for assistance in preparing the manuscript. Simon Moore is funded by a studentship from the Economic and Social Research Council of the United Kingdom.

References

Anderson, J. R. (1990). *The Adaptive Character of Thought*. Hillsdale, NJ: Lawrence Erlbaum Associates.

Aston-Jones, G., Foote, S. L. & Bloom, F. E. (1984) Anatomy and physiology of the locus coeruleus neurons: Functional implications. In Ziegler, M. G. & Lake, C. R. *Norepinephrine: Frontiers of Clinical Neuroscience. Vol. 2*. Baltimore: Williams and Wilkins, 92-116.

Beck, A. T. & Steer, R. A. (1987). *Beck Depression Inventory: Manual*. New York: The Psychological Corporation.

Benloucif, S., Bennett, E. L. & Rosenweig, M.R. (1995). Norepinephrine and neural plasticity: The effects of Xylamine on experience-induced changes in brain weight, memory and behavior. *Neurobiology of Learning and Memory*, 63, 33-42.

Bianchin, M., Souza, T. M., Medina, J. H. & Izquierdo, I. (1999). The amygdala is involved in the modulation of long-term memory, but not in working or short-term memory. *Neurobiology of Learning and Memory*, 71, 127-131.

Bradley, M. M., Greenwald, M. K., Petry, M. C. & Lang, P. J. (1992). Remembering pictures: Pleasure and arousal in memory. *Journal of Experimental Psychology: Learning, Memory and Cognition*, 18, 379-390.

Cahill, L. & McGaugh, J. L. (1995). A novel demonstration of enhanced memory associated with emotional arousal. *Consciousness and Cognition*, 4, 410-421.

Cahill, L., Prins, B., Weber, M. & McGaugh, J. L. (1994). β-adrenergic activation and memory for emotional events. *Nature*, 371, 702-704.

Corbetta, M., Niezen, F. M., Shulman, G. L. & Petersen, S. E. (1993). A PET study of visiospatial attention. *Journal of Neuroscience*, 13, 1202-1226.

Costin, D. (1988). MacLab: A Macintosh system for psychology labs. *Behavior Research, Methods, Instruments, and Computers*, 20, 197-200.

Damasio, A. R. (1994). *Descartes' Error: Emotion, Reason, and the Human Brain*. New York: Avon Books.

Davidson, R. J. (1994). Honoring biology in the study of affective style. In Ekman, P. & Davidson, R. J. [Eds.] *The Nature of Emotion: Fundamental Questions*. Oxford: Oxford University Press, 321-328.

Davidson, R. J. (1998). Affective style and affective disorders: perspectives from affective neuroscience. *Cognition and Emotion*, 12, 307-330.

Ekman, P. & Davidson, R. J. (1994). *The Nature of Emotion: Fundamental Questions*. Oxford: Oxford University Press.

Fitts, P. M. & Posner, M. I. (1973). *Human Performance*. London: Prentice Hall, Inc.

Garcia, J., Quick, D. F. & White, B. (1984). Conditioned disgust and fear from mollusk to monkey. In Alkon, D. L. & Farley, J. [Eds.] *Primary Neural Substrates of Learning and Behavioral Change*. London: Cambridge University Press.

Isen, A. M. & Gorgoglione, J. M. (1983). Some specific effects of four affect-induction procedures. *Personality and Social Psychology Bulletin*, 9, 136-143.

Isen, A. M. (1987). Positive affect, cognitive processes and social behavior. *Advances in Experimental Psychology*, 20, 203-253.

Karni, A. & Sagi, D. (1991). Where practice makes perfect in texture discrimination: Evidence for primary visual cortex plasticity. *Proceedings of the National Academy of Science*, USA, 88, 4966-4970.

Karni, A., Tanne, D., Rubenstein, B. S., Askenasy, J. J. M. & Sagi, D. (1994). Dependence on REM sleep of overnight improvement of a perceptual skill. *Science*, 265, 679-682.

Levy, J., Heller, W., Banich, M. T. & Burton, L. A. (1983). Are variations among right-handed individuals in perceptual asymmetries caused by characteristic arousal differences between hemispheres? *Journal of Experimental Psychology: Human Perception and Performance*, 9(3), 329-359.

Martin, M. (1990). On the induction of mood. *Clinical Psychology Review*. 10, 669-697.

Moss, H. & Older, L. (1996). *Birkbeck Word Association Norms*. London: Psychology Press.

Oaksford, M., Morris, F., Grainger, B. & Williams, J. M. G. (1996). Mood, reasoning, and central executive processes. *Journal of Experimental Psychology: Learning Memory and Cognition*, 22, 476-492.

Oatley, K. & Johnson-Laird, P. N. (1987). Towards a Cognitive Theory of Emotions. *Cognition and Emotion*, 1, 1, 29-50.

Phelps, E. A., La Bar, K. S., Anderson, A. K., O'Connor, K.J., Fulbright, R. K. & Spencer, D. D. (1998). Specifying the contributions of the human amygdala to emotional memory: A case study. *Neurocase*, 4, 6, 527-540.

Slyker, J. P. & McNally, R. J. (1991). Experimental induction of anxious and depressed moods - are Velten and musical procedures necessary? *Cognitive Therapy and Research*, 15, 33-45.

Squire, L. R., Knowlton, B. & Musen, G. (1993). The structure and organization of memory. *Annual Review of Psychology*, 44, 453-495.

Triesman, A. & Gormicon, S. (1988). Feature analysis in early vision: Evidence from search asymmetries. *Psychological Review*, 95, 1, 15-48.

Usher, M., Cohen, J. D., Servan-Schreiber, D., Rajkowski, J. & Aston-Jones, G. (1999). The role of locus coeruleus in the regulation of cognitive performance. *Nature*, 283, 549-554.

Vandenberghe, R., Price, E., Wise, R., Josephs, O., Frackowiak, R. S. J. (1996). Functional-anatomy of a common semantic system for words and pictures. *Nature*, 383, 254-256.

Velten, E. (1968). A laboratory task for the induction of mood states. *Behavior Research and Therapy*, 6, 473-482.

Westermann, R., Spies, K., Stahl, G. & Hesse, F. W. (1996). Relative effectiveness and validity of mood induction procedures: A meta-analysis. *European Journal of Social Psychology*. 26, 557-580.

The Effects Of Age Of Acquisition In Processing Famous Faces And Names: Exploring The Locus And Proposing A Mechanism.

Viv Moore, (V.Moore@gold.ac.uk)
and
Tim Valentine, (T.Valentine@gold.ac.uk)
Goldsmiths College, University of London, New Cross, London. SE14 6NW. UK

Abstract

Information acquired early in life is processed faster than information acquired late in life. Moore and Valentine (1998) report naming celebrities' faces follows the same pattern of results. This is problematic for the account of age of acquisition (AoA) based on language development because knowledge of celebrities is acquired after early representations are formed in the phonological lexicon. Also, the effects of AoA in lexical decision tasks (LDT) are assumed to be the result of automatic activation of phonology from the printed word. Such an account would predict null effects of AoA on face processing tasks not requiring name production (i.e. names are not automatically accessed, Valentine, Hollis & Moore, 1998). Significant effects of AoA were established in three Experiments: reading aloud printed names, making familiarity decisions to celebrities' names and faces. It is argued that temporal order of acquisition rather than age of acquisition may be the chief determinant of processing speed.

Introduction

A number of studies report faster naming for pictures with high frequency names like 'chair' than for low frequency names like 'metronome' (e.g. Oldfield & Wingfield, 1964; Jescheniak & Levelt, 1994). High frequency words are judged to be English words faster than low frequency words and are also read aloud faster (e.g. Gerhand & Barry, 1998). These effects influence the design of current models of lexical processing, with the mechanism postulated as greater connection strengths between levels of representations for frequently encountered items than for less frequently encountered items. This is especially pertinent to connectionist models, for example backward error propagation can simulate word frequency (WF) effects on word naming and lexical decision tasks (LTD). Interactive activation and competition architectures (IAC) can simulate WF effects on picture naming (e.g. Humphreys, Lamote & Lloyd-Jones, 1995). However, it has been demonstrated that the age at which a word was first learned is a powerful determinant of processing speed. A controversy over whether WF or age of acquisition (AoA) is *the* important processing determinant arose because high inter-correlations exist between WF and AoA, and because most WF studies did not control for AoA.

Carroll and White (1973a) reanalysed Oldfield and Wingfield's data and included AoA as a variable, they report that the age at which object names were acquired was the chief determinant of naming speed. They argued that measures of WF only predicted naming latency to the extent that they reflect AoA. Furthermore, when the correlation between WF and AoA was taken into account, WF played no independent role. These findings have been replicated many times (e.g. Morrison, Ellis & Quinlan, 1992).

An effect of AoA, but not of written WF was apparent for word naming speed. Both spoken WF and AoA exerted independent effects in a LDT (e.g. Gerhand & Barry, submitted). However, AoA alone affects auditory lexical decisions (Turner, Valentine & Ellis, 1998). The effects of

AoA in object and word processing tasks have been located at the stage of lexical retrieval (e.g. Ellis & Morrison, 1998). Consistent with these findings, Moore and Valentine (1998) report that early-acquired celebrities' faces are named faster than late-acquired faces. These effects were interpreted in terms of a functional model of face recognition (e.g. Bruce & Young, 1986) which evolved as an analogue of the logogen model of object recognition.

The phonological completeness hypothesis (Brown & Watson, 1987) assumes that AoA effects arise from phonological representations established during language development. Early learned words are stored in a 'more complete form' during language acquisition. A functionally different mechanism occurs for late-acquired words, which are stored in a less complete form and require re-assembly for production. This account of AoA is consistent with a primary locus at the level of phonological representations.

The majority of studies support a single locus for AoA effects at the level of speech output. According to Levelt's (1989) model of lexicalization, two pre-articulatory processing stages for lexical access exist. The first stage of activation is retrieval of an abstract representation of semantic and syntactic information (or lemma selection). The second stage involves activation of the word's phonological representation (or lexeme activation), that will initiate articulatory encoding. If indeed AoA effects are located at speech output, this model allows for three possible loci: lemma selection, links between lemma and lexemes or lexeme selection. Based on an interaction between the effects of WF and AoA, established in a repetition priming paradigm, Barry, Morrison & Ellis (1997) propose the locus of AoA to be at the lexeme. This locus was also proposed for the effects of WF (Jescheniak & Levelt, 1994), but as AoA was not included in that study, it is possible that the effects of WF and AoA were confounded.

However, a single locus for AoA effects is not universally supported, Yamazaki, Ellis, Morrison and Lambon Ralph

(1997) report that the reading speed of Japanese Kanji characters was affected by the age at which words entered the *spoken* vocabulary and the age that children learn the *written* characters. Yamazaki *et al.* argue that AoA affects the quality of lexical representations in the speech output *and* visual input lexicons, requiring at least two loci of AoA.

Moore and Valentine (1998) explored AoA effects on naming famous faces. They report consistently faster naming for early-acquired celebrities (acquired between 6 & 12 years) than late-acquired celebrities (acquired after 18 years). Their study showed that familiarity with celebrities was the major predictor of naming speed. The instructions for rated familiarity explicitly required ratings to reflect the number of times a celebrity had been encountered (in the media, etc.) and was interpreted as an explicit measure of accumulated (lifetime's) frequency of encounter. However, AoA significantly influenced naming speed (when matched on familiarity ratings). While these effects are consistent with a locus at the phonological output level. It is implausible that they are explicable in term of language development, because early-acquired celebrities were rated as acquired between 6 and 12 years of age. In contrast, the majority of early-acquired object names are acquired between 2 and 6 years of age.

A developmental view of language specificity currently proposes that infants are innately equipped to process the tone, stress, vowel length, etc. of any language. Infants become attuned to the phonemic contrasts in their linguistic environment during the first year of life (Werker, 1994). Once established, these representations are used to discover regularities in speech, for example by nine months of age infants show a 'preference' for listening to words rather than non-words (Juscyk, Cutler & Redanz, 1993). Infants also show a 'preference' for listening to phoneme structures conforming to their own language (Juscyk & Aslin, 1995), implying that infants use language regularities to hypothesise word boundaries in speech streams. Thirteen-month old infants can learn novel words from as few as nine presentations, suggesting that a powerful learning mechanism for forming object-label associations already exists (Woodward, Markman & Fitzsimmons, 1994). By around eighteen months of age a 'spurt' of language comprehension and often production (Goldfield & Reznick, 1992) suggests the triggering of a new principle of organisation into the child's understanding of the object-to-label relationship. These features are consistent with AoA effects resulting from a 'critical' period of language development, and a locus for AoA effects to be at the phonological output level. Also it might be predicted that representation of phonological *input* might also be a locus for AoA effects, which is supported by the effects of AoA on auditory LDTs (Turner *et al.*, 1998). Furthermore, language development may explain the absence of AoA effects on semantic classification tasks (e.g. natural or man-made, Morrison *et al.*, 1992). This is because, the stress occurs for associations between an object's appearance and its name. The acquisition of superordinate categories for objects would occur after the formation of object-to-label associations.

The completeness hypothesis argues that AoA effects arise from the development of phonological representations. Alternatively, we argue that these effects arise from the *order* in which information is acquired. Furthermore, we argue that a critical period of language development cannot account for the effects of AoA established in face naming. We propose that an explanation of *temporal order* of acquisition is supported by data from the neuropsychological literature. For example, SS, a 65 year-old man suffering from organic amnesia, evinced an effect of order of acquisition (Verfaellie, Croce & Milberg, 1995). Items were words or concepts, for which entry into the English language were dated into hemi-decades. Pre-morbid items entered the language between 1920 to 1970, post morbid items between 1971 and 1990. SS could recall and recognise the meaning of novel words acquired between 1970 and 1980 significantly better than words acquired in the 1980s, although both had entered the general vocabulary after the onset of his amnesia. SS did recognise a few of the 1980's items, but these were new combinations of old words (e.g. sun-block). Control subjects also showed a non-significant effect of *order* on word recall.

Patient WK, (56 year old) a man with a person-naming deficit (Shallice & Kartsounis, 1993) could name highly familiar personalities famous 20 years ago or more, e.g. Harold Wilson (British prime minister twice between 1964 to 1976) but not Margaret Thatcher (contemporary British prime minister). He could name historical personalities but not contemporary media personalities. However, the effect of temporal order was not specific to peoples' names, but generalised to naming from definitions of words that entered the vocabulary over the past 20 years (e.g. "A device used to record TV programmes so that one can see them at a later date" - *video*).

We propose that the face naming, developmental and neuropsychological literature may offer converging support for an effect of *temporal order of acquisition*. To understand the mechanism(s) giving rise to age or order effects it is necessary to explore these effects in a domain where items are acquired after the period of language development. Therefore, processing famous faces and names provides an ideal domain to determine whether these effects are specific to language acquisition. Recognition of famous faces and names is particularly suitable because current many theories of face and name processing were developed by analogy to theories of object and visual word recognition. The same hierarchy of representations are assumed for naming both faces and objects. Initially visual representation of the stimulus are activated (object or face recognition units) before access to semantic or identity-specific semantic information; finally representations of the name are activated (Bruce & Young, 1986). The major difference between object recognition and face recognition is the assumption that access to semantic information about people and their names is achieved via a Person Identity Node or PIN (e.g. Burton, Bruce & Johnson, 1990). PINs play the role of token markers in memory (denoting an individual), and are assumed to be the critical difference between processing proper name stimuli (e.g. celebrities, landmark names) and common name stimuli (e.g. words & objects).

Having established a robust effect of AoA in face naming paradigms (Moore & Valentine, 1998) the same items (matched on variables other than AoA) are used in the experiments reported below. In Experiment 1 printed names were read aloud (analogous to word reading). Experiments 2 and 3 report face and name familiarity decision tasks that do not require a spoken response. A speech output locus would predict an advantage for reading early-acquired names. Automatic activation of the phonological output lexicon *may* predict and effect of AoA for familiarity decisions to printed names, but not for familiarity decisions to faces.

Methodology.

These details pertain to all of the reported experiments. The same critical items were used in each experiment because AoA effects were established from these items and the selection validated by *post hoc* ratings (Moore & Valentine, 1998). The two groups of celebrities did not differ significantly on: rated familiarity (all were highly familiar); facial distinctiveness; surname frequency; initial phonemic power, name-letter and phoneme length. The groups differed in ratings of AoA (early vs. late) in a one tailed *t*-test ($t(48)$ = 10.20, $p<.0001$). *Post hoc* ratings were collected after each experiment to check the validity of those ratings.

Images of celebrities and filler faces were created by scanning up-to-date, quality photographs or by capturing video stills. Images were monochromatic (256 x 256 pixels in size) and displayed at a resolution of 640 x 480 on a 14 inch screen, using 16 grey levels. Images were edited to obscure background and clothing and displayed individually in the centre of a PC screen. Written names were displayed individually in uppercase Geneva font (20 point).

Each Experiment employed 24 UK university students between the ages of 18 and 25, who had spent the first 18 years of life in the UK. They participated in one experiment only and were paid for their time.

There was one independent variable AoA with two levels (early vs. late). Analyses were performed by related (t_1 participants) and independent (t_2 items) *t* tests. The dependant variable is latency of response.

Apparatus and Procedure. MEL software (Schneider, 1988) controlled the rating tasks and experiments, randomised stimuli for presentation and recorded response latencies (with millisecond accuracy). Ten practice stimuli (not experimental items or analysed) preceded the experiments. Participants focused on a centralised fixation point for 250 ms. the screen cleared, a tone sounded followed by a 250 ms. interval before the stimuli appeared (name or face). Participants' response (via a hand-held response box or by voice-key connected to a computer port) ended the display and logged decision latencies.

Post Hoc **Stimulus Ratings:** Following each experiment, participants gave ratings of the stimuli. The instructions emphasised that *personal opinion was the important factor.*

Participants were given as much time as required. Ratings were entered into the computer by pressing the space-bar to see the appropriate response scale, this remained on the screen until the score was confirmed. The ratings took approximately 25 minutes.

Familiarity: The instructions stressed that *ratings should reflect how many times, prior to the experiment, each celebrity had been encountered, on TV, in films, etc.*, and could be regarded as a measure of cumulative frequency of lifetime encounters. Ratings were made on a 7 point scale (1 = unknown to 7 = very familiar).

Distinctiveness: These ratings were made to faces following Experiment 3 only. Participants were instructed to imagine that they had not seen the faces before and ignore previous *knowledge of characteristics other than those apparent in the images* (e.g. height, hair colour, etc.). They were to imagine how easy identification would be if they were sent to meet each celebrity in a crowded railway station. A score of 1 should be given for faces that would be hard to spot (typical faces), a score of 6 for easy to spot faces (distinctive). Ratings were made on a 6 point scale as 'unknown' would be inappropriate.

Age of Acquisition: Participants estimated when they *first became aware of each celebrity.* Ratings were made on a 7 point scale where 1 = unknown; two for celebrities first acquired under 3 years; three, for a celebrity acquired under 6 years; four, a celebrity acquired under 9 years; five, a celebrity acquired under 12 years; six, acquired under 18 years and seven, acquired over 18 years.

Experiment 1: Reading Celebrities Names.

Participants (9 males & 13 females; mean age = 19.96 years, s.d. = 1.23) *read 50 famous names as quickly and accurately as possible.* Spoken responses triggered a voice key and latencies were recorded. The task took 15 minutes.

Results and Discussion

Incorrect responses were removed (92) and mean scores derived (RT = 563ms, s.d. = 117; accuracy = 23.08; s.d. = 1.98). Significantly faster reading times occurred for early-acquired (mean = 551ms, s.d. = 111) than late-acquired celebrities' names (mean = 575ms, s.d. = 123) by participants ($t_1(23)=35.30$, $p<.0001$) and approached significance by items ($t_2(48)=1.44$, $p<.08$).

The *post hoc* ratings confirmed the validity of selected stimuli. The differences between rated familiarity (early = 5.14 (s.d. = .69) vs. late = 5.32 (s.d. = .63)) was not significant. The difference between rated AoA (early 4.03 (s.d. = .75) vs. late = 6.20 (s.d. = 45) was significant ($t(48)=9.88$, $p<.001$). These AoA effects are analogous to the AoA effects for reading words aloud (e.g. Gerhand & Barry, 1998).

Experiment 2 Name Familiarity Decision Task.

Effects of AoA in LDTs are assumed to occur because visual word recognition automatically activates the phonology of the word (e.g. Morrison & Ellis 1995).

Familiarity decisions are based on activation of the PINs. An effect of AoA on familiarity decisions to printed names is predicted, only if it is assumed that phonology is automatically activated from a printed name.

Participants (12 males & 12 females; mean age = 19.38 years, s.d. = 1.44) were asked to *decide as quickly and accurately as possible* whether each of 100 printed were famous or not (50 celebrities & 50 unknown names). "YES" (famous) or "NO" (not famous) buttons were pressed on a response box. This task took approximately 10 minutes.

Results and Discussion

Incorrect responses were removed (91) and mean scores derived (RT = 646ms, s.d. = 100; accuracy = 23.10; s.d. = 1.80). Significantly faster familiarity decisions were made to early-acquired (mean = 630ms, s.d. = 100) than to late-acquired celebrities' names (mean = 662ms, s.d. = 100), by participants ($t_1(23)=3.09$, $p<.01$) and approached significance by items ($t_2(48)=1.39$, $p<.08$). Participants were faster to make a familiarity decision to famous names rated as early-acquired than to late-acquired names.

Post hoc ratings of printed names confirmed the groups' validity. They significantly differed on measures of AoA (early = 4.23 (s.d. = .58) vs. late = 6.01, (s.d. = .35)), $t(48)=12.09$, $p<.01$. Familiarity ratings did not significantly differ (early = 5.68 (s.d. = .50) vs. late = 5.81, (s.d. = .59)).

It is possible that automatic activation of phonology (e.g. Gerhand & Barry, 1998) or dual loci (at input and output) (Yamazaki, *et al.*, 1997) could explain these effects of AoA.

Experiment 3: Face Familiarity Decision Task.

Bruce (1983) developed the face familiarity task as an analogue of the LDT. By extension we argue that an effect of AoA in face familiarity would be analogous to that established for LDT because activation of a stored representation is required. Face familiarity decisions are assumed to be based on activation of the PINs (Burton, *et al.*, 1990) and do not require phonology. Automatic activation of the phonology of a person's name from seeing their face is most unlikely, naming difficulties are well documented (e.g. Brédart, 1996). Also, Valentine *et al.* (1998) showed that face naming significantly primed a subsequent name familiarity decision, but a face familiarity decision did not prime a name familiarity decision. This result demonstrates that names are not automatically activated by a face familiarity decision. Therefore, an effect of AoA is not predicted for familiarity decisions to faces by all of the theoretical accounts.

Participants (7 males & 17 females; mean age = 22.46 years, s.d. = 1.80) were asked to *decide as quickly and accurately as possible* whether each of 100 faces were famous or not (50 celebrities & 50 unknown faces). "YES" (famous) or "NO" (not famous) buttons were pressed on a hand held response box. This task lasted 10 minutes.

Results and Discussion

Responses errors were removed (125) mean scores and derived (RT = 662ms, s.d. = 81; accuracy = 22.40, s.d. = 2.31). Significantly faster familiarity decisions were made for early-acquired celebrities' faces (mean = 642ms, s.d. = 86) than for late-acquired celebrities' faces (mean = 682ms, s.d. = 76) $t_1(23)=6.29$, $p<.0001$; $t_2(48)=1.72$, $p<.05$.

Analyses of *post hoc* ratings confirmed the selection of items. The differences between rated familiarity (early = 5.85 (s.d. = .63) vs. late = 5.65 (s.d. = .41)) and distinctiveness (early = 3.81 (s.d. = 1) vs. late = 3.35 (s.d. = .86)) were not significant. The difference between rated AoA (early 4.51 (s.d. = .61) vs. late = 6.05 (s.d. = 34)) was significant $t(48)=10.20$, $p<.0001$.

As the result of Experiment 3 is difficult to reconcile with existing accounts of AoA a replication study was conducted to the same result. Significantly faster familiarity decisions were made to early-acquired celebrities' faces (mean = 753ms, s.d. = 162) than late-acquired (mean = 825ms, s.d. = 135) celebrities' faces ($t_1(22)=3.01$, $p<.01$; $t_2(48)=1.71$, $p<.04$).

This result is inconsistent with a single locus of AoA in face naming at the level of name retrieval as no vocal responses were made. Automatic activation of phonology from a face to a name is untenable (Valentine, *et al*, 1998). Taken together these data suggest a locus for AoA effects in processing faces at or before the PINs.

General Discussion.

These Experiments use the same celebrities for whom an effect of AoA occurred in face naming (Moore & Valentine, 1998). The effect is now established on tasks of reading aloud printed names and on face and name familiarity decisions. Celebrities rated as early-acquired were processed significantly faster than late-acquired celebrities.

Early-acquired celebrities' names were read faster than late-acquired names (analogous to reading printed words aloud, (e.g. Gerhand & Barry, 1998). It has been argued that AoA effects have a single locus at the phonological output lexicon. An effect of AoA for lexical decision was attributed to automatic activation of phonology from visual word recognition (Morrison & Ellis 1995; Gerhand & Barry, submitted). A single locus at this level may account for AoA effects on celebrities' printed names, but it is untenable for the effects of AoA established for face familiarity decisions. Experiment 3 (and replication) shows that a single locus at a phonological level for all AoA effects can no longer be maintained. The effect on face familiarity decisions require a locus at, or before the PINs, because familiarity decisions are assumed to be caused by activation of the PINs (Burton et al., 1990). Valentine, *et al.* (1998) demonstrated a face familiarity decision does not automatically activate a phonological representation of the name. Therefore, automatic activation of phonology from face familiarity decisions is implausible. It is possible that representations of all familiar words, faces or objects are organised in a way that produces an effect of AoA, including the representation of lexical items in the semantic lexicon.

As yet there is insufficient converging evidence to enable specific loci of AoA to be identified. This task may be prove to be as difficult as identifying the loci of WF effects. However, three conclusions can be drawn. First, a single locus is no longer adequate to account for AoA effects. Second, AoA effects are widespread. Third, AoA may reflect a general property of the mental representation of perceptual and lexical information. Age of acquisition may be a feature of the representation of information while WF may reflect the strength of connections. The challenge for any cognitive model is to account for the effects of AoA as well as frequency. One such challenge is to account for an effect of AoA in the absence of an effect of WF on *auditory* lexical decision (Turner *et al.,* 1998).

It is obvious that, even when familiarity is controlled, age or order of acquisition significantly affects processing of celebrities' faces and names, and in varying degrees, lexical and object processing tasks too. Equally clear is that frequency of encounter (and especially spoken frequency) affects very similar processing tasks. However, connectionist models fail to account for both of these influences. The effects of AoA present serious problems for current connectionist models of cognition. Connectionist models that use backward error propagation to learn distributed representations, can readily model the effects of frequency (or familiarity). However, these networks suffer from interference of early-learned material by subsequently acquired material. Therefore, it is not clear how such an architecture could model an effect of AoA. Interactive activation models of face recognition and naming do not generally include a learning mechanism. However Burton (1994) has developed an algorithm that enables IAC models to learn localist representations of new stimuli. It can be appreciated how this algorithm can model the effects of frequency (or familiarity) by increasing the weight of connections between nodes. It is not clear how it could model the effect of AoA.

Kohonen (1990), proposed a model based on 'self-organising maps'. This type of network is capable of learning to distinguish between different patterns of input by unsupervised learning. Similar patterns cluster at units in the same area, whereas dissimilar patterns are topologically distant. Therefore, early encountered patterns played a prominent role in the organisation of input representations. However, it is not clear whether this could provide an adequate model of both AoA and cumulative frequency.

The early-acquired celebrities in these studies were rated as acquired much later (between 6 and 12 years of age) than for early-acquired word and object processing studies (between 2 to 6 years of age). This difference is important, because, it has been proposed that AoA effects result from a developmental process where language-specific phonology is established. This is an unlikely candidate for the effects established to famous faces and names because these were acquired after the period of language development. In addition, critical periods of language development cannot account for effects of temporal order from cited patient studies. The evidence from the cognitive, developmental and neuropsychological literature support *temporal order of acquisition*, which provides a plausible explanation for the effects on face and name processing. It is possible that *all* new patterns of information are processed in a fundamentally different way to later-acquired related material. We suggest that initial encounters of exemplars from a new class of information, of any type and at any age, would trigger the setting up of a fundamental organisation of the relevant information. Later acquired related information would be added onto the previous material but would be represented in a different manner, because the early exemplars were actively involved in setting up the dedicated system. Such a mechanism may also serve to clarify the specific roles of WF and AoA using two assumptions. First, that an individual's interaction of initial unique patterns of information is responsible for the set-up of a dedicated processing system. Second, that frequent exposure of appropriate stimuli is required to maintain activity or connection strength. What results is an economical method for dealing with early exemplars of new classes of information, because a unit would be created to meet the demands of processing unique patterns of information. When representations of the same ilk occur it is incorporated into the existing module. The set-up of a systematised processing module is intuitively credible considering the normal temporal patterns of information acquisition. It follows that earlier acquired information should also be more robust to neurological insult. A significant advantage for early-acquired information was reported for aphasic patients (e.g. Hirsh & Funnell, 1995) and dysphasic patients (Rochford & Williams, 1962).

This approach suggests some future lines of research. First, it suggests that it should be possible to demonstrate an effect of order for any modular input system (Fodor, 1983). According to the principle of modularity a variety of cognitive skills are mediated by a number of independent cognitive processes (e.g. face recognition, word recognition). Each module performs a particular type of processing independent of the activity in other modules, although there is obviously communication between the outputs of these systems. Interestingly, Fodor proposed that faces would be candidates for a modular processing system (cf. Experiment 3). Although Fodor proposed that modular systems are innate, processing of written language is a good example of a skill that is only learnt with considerable instruction and effort. Nevertheless, there is considerable evidence of modular organisation of reading skills. Following, this line of thought would suggest that effects of AoA may be found in any area of highly skilled recognition of a stimulus class. The changes in representation that underlie the effects of AoA may underlie expert - novice differences in a wide range of skills.

References

Barry, C., Morrison, C. M., & Ellis, A. W. (1997). Naming the Snodgrass and Vanderwart pictures: Effects of age of acquisition, frequency and name agreement. *Quarterly Journal of Experimental Psychology*, 50A, 560-585.

Brédart, S. (1996). Person familiarity and name-retrieval failures: How are they related? *Current Psychology of*

Cognition, 15, 113-120.

Brown, G. D. A., & Watson, F. L. (1987). First in, first out: Word learning age and spoken word frequency as predictors of word familiarity and word naming latency. Memory and Cognition, 15, 208-216.

Bruce, V. (1983). Recognising faces. Philosophical Transactions of the Royal Society London. B302, 423-436.

Bruce, V. & Young, A. W. (1986). Understanding face recognition. British Journal of Psychology, 77, 305-327.

Burton, A. M. (1994). Learning new faces in an IAC model. Visual Cognition,1, 313-348.

Burton, A. M. (1994). Learning new faces with an IAC model. Visual Cognition 1, 313-348.

Burton, A. M., Bruce, V. & Johnson, R. A. (1990). Understanding face recognition with IAC model. British Journal of Psychology, 81, 361-380.

Carroll, J. B. & White, M. N. (1973a). Word frequency and age of acquisition as determiners of picture-naming latency. Quarterly Journal of Experimental Psychology, 25, 85-95.

Ellis, A. W. & Morrison (1998). Real age of acquisition effects in object naming. Journal of Experimental Psychology: Learning, Memory and Cognition, 24, 515-523.

Fodor, J. A. (1983). The modularity of mind: An essay on faculty psychology. Massachusetts,London, England; MIT Press Cambridge.

Gerhand, S. & Barry, C. (1998). Word frequency effects in oral reading are not merely age of acquisition effects in disguise. Journal of Experimental Psychology: Learning, Memory and Cognition, 24, 267-283.

Gerhand, S. & Barry, C. (submitted). Age of acquisition, frequency and the role of retrieved phonology in the lexical decision task.

Goldfield, B. & Reznick, J. S. (1992). Rapid change in lexical development in comprehension and production. Developmental Psychology, 28, 406-413.

Hirsh, K. W. & Funnell, E. (1995). Those old familiar things: Age of acquisition, familiarity and lexical access in progressive aphasia. Journal of Neurolinguistics. 9, 23-32.

Humphreys, G. W., Lamote, C. & Lloyd-Jones, T. J. (1995). An interactive activation approach to object processing. Effects of structural similarity name frequency, and task in normality and pathology. Memory, 3, 535-586.

Jescheniak, J. D. & Levelt, W. J. M. (1994). Word frequency in speech production: Retrieval of syntactic information and phonological form. Journal of Experimental Psychology: Learning Memory and Cognition, 20, 824-843.

Juscyk, P. W. & Aslin, R. N. (1995). Infant's detection of sound patterns of words in fluent speech. Cognitive Psychology, 29, 1-23.

Juscyk, P. W., Cutler, A., & Redanz, N. J. (1993). Infants' preference for the predominant stress patterns of English words. Child Development, 64, 675- 687.

Kohonen, T. (1990). Self organising map. Proceedings of the IEEE, 78, 1464-1480.

Levelt, W. J. M. (1989). Speaking: From intention to articulation. Cambridge, MA: MIT Press.

Moore, V. & Valentine, T. (1998). Naming faces: The effect of AoA on speed and accuracy of naming famous faces.

Quarterly Journal of Psychology, 51A, 485-513.

Morrison, C. M. & Ellis, A. W. (1995). The roles of word frequency and age of acquisition in word naming and lexical decision. Journal of Experimental Psychology: Learning, Memory & Cognition, 21, 116-133.

Morrison, C. M., Ellis, A. W. & Quinlan, P. T. (1992). Age of acquisition, not word frequency, affect object naming not recognition, Memory and Cognition, 20, 705-714.

Oldfield, R. C. & Wingfield, A. (1964). The time it takes to name an object. Nature, 202, 1031-1032.

Rochford, G. & Williams, M. (1962). Studies in the breakdown of the use of names: The relationship between nominal dysphasia and the acquisition of vocabulary in childhood. Journal of Neurology, Neurosurgery and Psychiatry, 25, 222-233.

Schneider, W. (1988). Micro Experimental Laboratory. An integrated system for IBM-PC. In: Behaviour Research Methods. Instruments and computers, 20. 206-217.

Shallice, T. & Kartsounis, L. D. (1993). Selective memory impairment of retrieving people's names: A category specific disorder? Cortex, 29, 281-291.

Turner, J. E., Valentine, T. & Ellis, A. W. (1998). Contrasting effects of AoA and word frequency on auditory and visual lexical decision. Memory and Cognition. 26, 1282-1291.

Valentine, T., Hollis, J., & Moore, V. (1998). On the relationship between reading, listening and speaking: It's different for people's names. Memory and Cognition. 26, 740-753.

Verfaellie, M., Croce, P., & Milberg, W. P. (1995). The role of episodic memory in semantic learning: An examination of vocabulary acquisition in a patient with amnesia due to encephalitis. Neurocase, 1, 291-304.

Werker, J. F. (1994). Cross language speech perception: developmental change does not involve loss. In Words J. C. Godman and H. C. Nusbaum (Eds.) The Development of Speech Perception: the Transition from Speech Sounds to Spoken, MIT Press. Cambridge.

Woodward, A. L., Markman, E. M. & Fitzsimmons, C. M. (1994). Rapid word learning in 13- and 18-month-olds. Developmental Psychology, 30, 553-566.

Yamazaki, M. Ellis, A. W., Morrison, C. M., & Lambon Ralph, M. A. (1997). Two age of acquisition effects in the reading of Japanese Kanji. British Journal of Psychology, 88, 407-421.

Acknowledgements: We thank Geralda Odinot for running participants for the replication of Experiment 3. Also we would like to thank Chris Barry, Andrew Ellis and an anonymous reviewer for their helpful comments on an earlier draft of this paper.

This research was supported by the Economic and Social Research Council through a project grant awarded to the second author (No. R000 234612) and a postgraduate award to the first author (No. R00 429624208).

The Effects of Multiple Schematic Constraints on the Recall of Limericks

Jason S. Moore (jsmoore@bgnet.bgsu.edu)
Department of Psychology; Bowling Green State University
Bowling Green, OH 43403 USA

Abstract

Traditional theories of text memory and comprehension posit that text is represented and reconstructed based upon its semantic content. In contrast, Rubin (1995) found that poetic materials are remembered based not only on semantic content, but based also on the schematic constraints, such as rhythm and rhyme, present in the surface structure of the verse. Rubin's research has done much to record the phenomenon of memory for poetic, structured materials. The present study is an investigation of the effects of multiple schematic constraints on participants' recall for words in limericks. This study provides support for Rubin's claims that surface structure and schematic constraints facilitate recall for schema-consistent poetic materials. In addition, the present study extends the analysis of the effects of schematic constraints, illustrating that the schematic constraints present in structured verse serve to guide recall for schema-inconsistent material, making the inconsistent material schema-consistent upon recall.

Introduction

Traditional theories of text memory and comprehension posit that text is represented and reconstructed based upon its meaning or gist (Bartlett, 1932; van Dijk & Kintsch, 1983). According to van Dijk and Kintsch (1983) there are three levels of representation for text and discourse in memory. At the surface level, information is represented by the exact words and phrases present in the text. At the second level, it is the semantic content of the text that is represented in memory. At the third level, it is not the text itself that is represented, but the situation described in the narrative. This level of representation is detached from the text structure and is embedded in pre-existing fields of knowledge. At this level, the text is organized and assimilated by the schemas already present in the mind. Thus, according to van Dijk and Kintsch's thesis, it is the semantic content that is extracted from a text and subsequently represented in memory. This theory is based upon the analysis of comprehension for materials that are themselves organized and constrained only by their meaning or narrative content. There is evidence to suggest that memory for structured, highly constrained text is based on more than a macro-level semantic representation (Kintsch, 1998; Rubin, 1995). Traditional theories have not given serious consideration to memory for poetic materials whose surface features and structural constraints may be as important to the reconstruction of such materials as is the gist of the material.

Rubin's (1995) analysis of memory in oral traditions investigated the recall of highly structured material. Rubin's results indicated that poetics aid in the reconstruction of song lyrics (Hyman & Rubin, 1990), epic ballads (Wallace & Rubin, 1991), and counting-out rhymes (Kelly & Rubin, 1988). Rubin viewed the structure of verse as a highly constrained situation in which rhyme, rhythm, and meaning combine to restrict the range of alternatives activated by the stimuli. According to Rubin (1995) these schematic constraints combine to add cue-item discriminability, increasing the amount recalled, and restricting the form of errors and variations in recall.

The constraining effects of schematic structures were illustrated by Hyman and Rubin's (1990) analysis of memory for lyrics from Beatles' songs. Consistent with schema theory, Hyman and Rubin found that recall errors fit the constraints present in the structure of Beatles' song lyrics. Errors in recall adhered to the rhythm, rhyme, and semantic structure of the songs. While this and similar studies provide a good description of the phenomenon, the findings are not free of ambiguity, and the effects of schematic constraints have not been fully investigated. The present study offers support for Rubin's theory, while extending the investigation of the effects of schematic constraints by utilizing materials that are schematically inconsistent.

The ambiguity of Hyman and Rubin's (1990) findings on memory for song lyrics is due to a lack of control over the stimuli. The investigators could not control for subjects' prior exposure to the Beatles' songs, which may have affected both the recall and perception of the song lyrics. Thus, the novelty of the stimuli utilized in the present study allows for clear demonstration that the form of recall errors is not due to errors in encoding, but due to the effects of schematic constraints which combine to guide the reconstruction of the verse in memory.

The present study extends the investigation of the effects of schematic constraints by utilizing materials that are schematically inconsistent. According to Rubin (1995), information is most accurately recalled when it fits the structural constraints, as in Hyman and Rubin's (1990) investigation of memory for song lyrics. Schema theory also predicts that when information violates structural constraints, participants will reconstruct the information to conform to the structure of the context. Thus, by including to-be-remembered materials that do not conform to limerick structure, this study may lead to more understanding of this additional feature of schema-guided recall.

Experiment 1 of the present study investigated participants' ability to predict words missing in limericks given the amount the words were constrained by the structure of the limerick. A limerick, with its set rhythm, rhyme, and thematic structure, is a high-constraint context. The target words in the proposed investigation of recall errors are constrained by the limerick's strict rhyme and rhythm structure, and also by the narrative theme of the verse. According to Holman and Harmon (1992), limericks are made up of five lines of which the first, second, and fifth both rhyme and contain three anapestic feet (the unit of rhythm in verse) and the third and fourth lines, which rhyme and consist of two anapestic feet. The definite pattern of the limerick makes it an ideal medium in which to study the effects of multiple schematic constraints on memory. The consistent rhyme and rhythm pattern and the narrative semantic content combine to provide the maximum constraint for recollection as well as a schema of organization which serves to guide participants' recall of target words in Experiment 2 of this study.

Experiment 1 was a norming study. The data were collected in 1994 and analyzed as part of this investigation. Participants were instructed to generate the missing words for 100 limericks. The study was designed to investigate how multiple constraints aid in the prediction of words missing from limericks. A word was left out of each limerick in either the middle or at the end of one of the five lines. Half of the limericks have a word missing in the middle of a line, and half at the end of a line. There were ten blanks in each position for each of the five lines. Participants filled in the missing words and these responses were analyzed to determine whether there were differences caused by the position of the missing word in the limerick. It was hypothesized that there would be more correct predictions of target words for those limericks where the blank occurred at the end of a line. This hypothesis is based on the theory that constraints combine to restrict the range of response alternatives. The last word in a line of a limerick is constrained by rhyme as well as rhythm and meaning. However, words in the middle of a line in a limerick are constrained only by the meter and meaning of the verse. Thus, it was expected that there would be more accurate prediction for the blanks at the end of lines. The responses participants provided were used as alternates for the study packet in Experiment 2.

In Experiment 2, participants were presented with 50 limericks, some of which have the last word (target word) in one of the five lines altered to violate the rhyme, rhythm, semantic, or all three schematic constraints, and some which were not altered. Participants were asked to read the limericks and then were given a memory test in which they were asked to recall the target words.

Based upon traditional theories of the mental representation of text, one would predict the greatest number of recall errors for those target words that violate the semantic content of the limerick. A nonsense narrative would presumably hinder participants' ability to represent the limerick's gist in memory. Thus, according to the traditional theory, of the one-violation conditions, recall would be lowest for the meaning-violated condition.

It could be argued that the memory test in Experiment 2 amounts to little more than the prediction of the target words, as in Experiment 1. Perhaps the limericks in the study packet are not represented in memory, and

participants merely fill in the blanks with likely words in the test packet. If this were the case, there should be very few schema-inconsistent target words recalled.

Based on the theory of multiple constraints, I hypothesize that the fewest recall errors and fewest schema inconsistent responses will occur in the condition where the limerick structure is not violated. I also predict that the condition where all three structural constraints are violated will produce the least accurate recall and the least structurally consistent responses. I expect that the number of errors and inconsistent responses for the one-constraint-violated conditions will be more than for the no-violation condition and less than for the all-constraints-violated condition. At this time, there is no theoretical basis for predicting differences among the three one-constraint-violated conditions. Based on schema theory, I hypothesize that participants will make recall errors which alter the missing word to fit the structure of the limerick. Recall errors will be schema-consistent.

Experiment 1: Prediction of Missing Words in Limericks

Methods

Participants Twenty-eight undergraduates participated in the study.

Materials and Procedures Participants received a packet containing 100 limericks. Participants then received instruction on limerick structure and an example. Each limerick contained a blank in the middle or the end of one of its five lines. There were ten limericks representing each of the ten positions. Participants were instructed to fill in the blank with the first word that came to mind, keeping in mind the rhyme, rhythm, and meaning patterns.

Results

The analysis of the data consisted of a 5 X 2 ANOVA with Tukey follow-up tests. The independent variables were as follows: Limerick *line* containing the blank (one; two; three; four; or five); and *position* of blank in the line (middle vs end). The dependent variable is the number of correct predictions.

The ANOVA revealed a significant difference in the number of correct predictions, $F(1, 108) = 923.688$, $MSE = 1.18$, $p < .05$, $\omega^2 = .73$, as a function of the position of the blank within the line. There were no significant differences in the number of correct predictions across the line of the limerick in which the blank occurred. There was an interaction between line and position, only in the fourth line, which was considered an artifact.

Experiment 2: Recall of Schema-Inconsistent Limericks

Methods

Participants Twenty (7 males and 13 females, mean age = 23.6) Western Washington University undergraduates participated in the study.

Materials and Procedures Participants were given training in limerick structure and an example. They then received a study packet containing fifty limericks which were read by the participants individually. In the study packet, the last word of a line in each limerick was made to fit the five levels of the independent variable: Constraint Violated (original--no violations; rhyme; rhythm; meaning; and all). There were ten limericks for each level of the independent variable. The alterations were taken from participants' responses in Experiment 1. The alterations for each limerick were chosen at random with the restriction that there would be an equal number of limericks for each level of the independent variable. Participants were given twenty minutes to study this packet.

Participants then received a test packet containing the same limericks with blanks for the target words. The participants were instructed to fill in the blank with the target word for each limerick. No time limit was placed on the completion of the test packet.

Results

The analysis of the data consisted of a one-way ANOVA with Tukey follow-up tests. The independent variable was as follows: constraint violated (none; rhyme; rhythm; meaning; or all). The dependent variables were as follows: number of correct responses and proportion of incorrect responses that were consistent with schema structure. The first analysis tested the effects of the different violations on subjects' ability to recall the target word. As hypothesized, recall was significantly better for the no violation condition than for the others, $F(4, 76) = 103.73$, $\underline{MSE} = 1.60$, $\underline{p} < .05$, $\omega^2 = .84$ (see Figure 1). Follow-up comparisons revealed that, contrary to prediction, there was a significant difference between the one-constraint conditions (M rhythm = 4.05; $\underline{M}$ meaning = 3.55; $\underline{M}$ rhyme = 2.70). Also, the rhyme-violated condition was not significantly different from the all-violations condition, while both the rhythm- and meaning-violated conditions were significantly different from the all-violations condition.

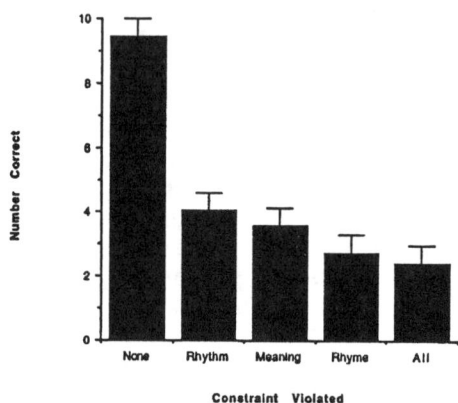

Figure 1: Number of correct responses as a function of the violation condition ($\underline{n}$ = 20). Error bars represent 95% confidence intervals calculated with the method prescribed by Loftus and Masson (1994).

In order to assess the degree to which subjects altered the to-be-remembered word such that it would be consistent with limerick structure, the proportion of incorrect but schema-consistent responses was analyzed across violation conditions. A response was deemed schema-consistent if it fit the rhyme, rhythm, and semantic structures of the given limerick. The results indicate a significant effect across conditions $F(3, 57) = 9.83$, $\underline{MSE} = .021$ $p < .05$, $\omega^2 = .25$ (see Figure 2). Follow-up tests revealed significant differences among the one-violation conditions ($\underline{M}$ rhyme = .61; $\underline{M}$ meaning = .59; $\underline{M}$ rhythm = .50), and all one-violation conditions were significantly different from the all-violation condition ($\underline{M}$ all = .39).

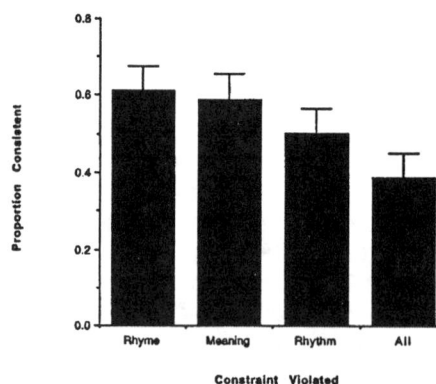

Figure 2: Proportion of errors which are consistent with limerick structure as a function of the violation condition ($\underline{n}$ = 20). Error bars represent 95% confidence intervals calculated with the method prescribed by Loftus and Masson (1994).

The proportion of incorrect-but-consistent responses was taken from the number of incorrect responses that were consistent with limerick structure divided by the total number of incorrect responses. There were so few incorrect responses in the no-violations condition that there were few values for this dependent variable in the no-violations condition. Thus the no-violations condition was not included in this analysis.

Discussion

In general, the results presented above indicate that poetics play a role in guiding and shaping memory for structured verse. The more closely a word fits the structure of the verse, the more accurately it is recalled. And when the word is not correctly recalled, it is often altered to fit the structure of the verse. These findings clearly support Rubin's (1995) claim that poetic structure guides recall. In addition, the experimental manipulation of violating the schematic constraints in the verse has made possible the prediction of where recall errors for words in structured verse will occur. The same constraints that combine to guide recall also shape the reconstruction of the to-be-remembered information. This allows for the prediction that when errors occur, they will be schema consistent. The structure of the verse guides the reconstruction of the missing word.

While the results of the present study support Rubin's (1995) theory of memory and verse, these findings seem at odds with the traditional theory that memory for text and discourse consists of a macro-level semantic representation. Rhyme and rhythm seem to have a central role in memory for structured verse. In the present study, violations of meaning did not hinder recall significantly more than the other one-constraint violation conditions. In fact, rhyme violations did decrease the number of target words recalled by a significant amount over both rhythm and meaning. This suggests that the rhyme schema is an important component organizing memory for limericks.

In his brief treatment of the topic, Kintsch (1998) illustrates that findings for the importance of surface level constraints can be generated utilizing his bottom-up Construction-Integration model. Kintsch claims that given the proper production rules, the model will show the importance of non-semantic components in the comprehension and organization of structured verse. Rubin (1995), however, prefers to view the situation as one in which the participant seeks to satisfy the schematic constraints of the structure, and seems not as concerned with the surface level of representation as is Kintsch in his model.

The results of the present study lend support to the idea that poetics and structure guide recall for verse. On a more general level, these findings support the theory that multiple constraints aid memory by making more stringent the criteria by which an alternative is considered to be the target information. And on an even more general level, these findings support schema theory, specifically, the idea that people's schema for a general event guide their recall for a specific event. However, discussion of the implications of this study must be tempered with the acknowledgment that there are limitations present in methodology, as well as issues concerning the generalizability of the findings.

While the present study has implications for many areas of research and application, it would be problematic, at this time, to generalize the findings to areas beyond the scope of the investigation. It may be that these findings cannot be generalized to all structured verse. It is likely that the relative importance of some constraint varies with its prevalence or complexity in the genre under investigation. For example, in the sonnet, the patterns of rhyme and rhythm are more complex than those in limericks. Perhaps in this condition, people would rely more on the semantic content of the sonnet to guide their recall than they do when provided a sing-song structure of rhythm and rhyme, like that present in limericks. Also, it is likely that the methodology employed in this study has not fully captured the effect of multiple constraints on memory for structured verse.

While it seems clear that the more a target word fits the constraint of the structure the better it will be recalled, it would be beneficial to include conditions in which two of the constraints were violated (i.e. meaning and rhyme violated, meaning and rhythm violated, and rhyme and rhythm violated). In this case, one would expect that recall for words in these conditions would be poorer than for the one-violation conditions and better than for the all-violations condition. Another benefit of including these conditions is that they would provide further indication of which constraints present in structured verse are most important to memory.

Future research should be designed with an eye to examining the effects of schematic structure and constraint in other genre of structured verse. It would also be beneficial to investigate the effects of two-violation conditions on recall for limericks. The inclusion of two-violation conditions should provide a better indication of the relationship between the effects of constraint on recall. It should also provide further indication of the extent to which specific constraints guide the recall for words missing from structured verse, and the degree to which each constraint is implicated in the mental representation of structured text.

The present study is potentially an important step toward understanding the relationship between schematic constraint and memory. It has been shown that schematic and structural constraints combine to aid memory for schema-consistent information. These same constraints guide the reconstruction of schema-inconsistent information, resulting in memory that is schema-consistent. In this way, the present study addresses the general topic of schema-consistent recall error, and more specifically, how schematic constraint influences the way information is recalled in structured verse.

Acknowledgments

I wish to thank my original advisor, Dr. Ira E. Hyman, and my current advisor, Dr. Richard B. Anderson.

References

Bartlett, F. C. (1932). *Remembering: An experimental and social study.* Cambridge: Cambridge University Press.

Holman, H. C., & Harmon, W. (Eds.). (1992). *A handbook to literature.* (6th ed.) New York: Macmillan.

Hyman, I. E., Jr., & Rubin, D. C. (1990). Memorabeatlia: A naturalistic study of long-term memory. *Memory & Cognition, 18,* (2), 205-214.

Kelly, M. H., & Rubin, D. C. (1988). Natural rhythmic patterns in English verse: Evidence from child counting-out rhymes. *Journal of Memory and Language, 27,* 718-740.

Kintsch, W. (1998). *Comprehension.* Cambridge, U. K.: Cambridge University Press.

Loftus, G. R., & Masson, M. E. J. (1994). Using confidence intervals in within-subjects designs. *Psychonomic Bulletin & Review, 1,* (4), 476-490.

Rubin, D. C. (1995). *Memory in oral traditions.* New York: Oxford University Press.

van Dijk, T. A., & Kintsch, W. (1983). *Strategies of discourse comprehension.* New York: Academic Press.

Wallace, W. T., & Rubin, D. C. (1991). Characteristics and constraints in ballads and their effects on memory. *Discourse Processes, 14,* 181-202.

Monitoring the Inner Speech Code

Jane L. Morgan (J.L.Morgan.20@bham.ac.uk)
School of Psychology, University of Birmingham, Edgbaston, Birmingham, B15 2TT, UK

Linda R. Wheeldon (L.R.Wheeldon@bham.ac.uk)
School of Psychology, University of Birmingham, Edgbaston, Birmingham, B15 2TT, UK

Abstract

The aim of this paper is to expand and replicate the findings of Wheeldon and Levelt (1995). They employed an internal speech monitoring task which required Dutch speakers to monitor silently generated words for target syllable or phoneme sequences. On the basis of the obtained data several claims were made concerning the locus, time-course and nature of the internal speech code. The series of experiments reported here examined these predictions using English stimuli. In contrast to the Dutch study, no evidence of any reaction time advantage to syllable over nonsyllable strings was found. A phoneme monitoring experiment replicated the left-to-right pattern of results observed by Wheeldon and Levelt. In addition, a perception version of the task failed to replicate these effects suggesting that they were independent of the position of the target in the speech stream. Implications of the results in terms of the time course of phonological encoding are discussed.

Introduction

In this paper we report a series of experiments which investigate the time course of the generation of an abstract phonological representation during speech production. Current models of speech production postulate a process of phonological encoding which specifies a word's phonemic and prosodic properties. Following the selection of a word based on its semantic and syntactic specifications, it is proposed that a word's constituent phonemes and its metrical frame (e.g., syllable structure and stress pattern) are made available and are subsequently combined during a process termed segment-to-frame association (Levelt and Wheeldon, 1994; Roelofs, 1992, 1997). The output from this process is a syllabified phonological code which forms the input to phonetic encoding processes which retrieve syllabic gestural scores detailed enough to guide articulation. With regard to phonological encoding recent theories propose that the process of segment-to-frame association occurs in sequence from left-to-right across a word and a number of experimental findings exist which suggest that the beginning of a word is phonologically encoded before the end (Meyer, 1990, 1991; Meyer and Schriefers, 1991; Levelt, Roelofs and Meyer, 1999). However, there is a lack of data that would allow the temporal properties of such a process to be specified in any detail.

Our main aim was to replicate and extend the findings of Wheeldon and Levelt (1995) who developed an internal speech monitoring task in order to investigate the time course of the phonological encoding process. The phenomenon of inner speech is one which we all might experience, for instance whilst reading, writing, problem solving or whilst performing short term memory tasks. During such activities we often generate and monitor some form of internal speech code. Moreover, there is evidence to suggest that the generation of this code entails the same mechanisms which are employed in normal speech production (Dell and Repka, 1992; Levelt, 1989; Motley, Camden and Baars, 1982). Wheeldon and Levelt (1995) designed a series of experiments which exploited this ability to monitor one's own inner speech. They adapted a methodology traditionally employed in the field of speech perception in which subjects monitor spoken words for target syllables and phonemes (Cutler, Mehler, Norris and Segui, 1986; Mehler, Dommergues, Frauenfelder and Segui, 1981; Zwisterlood, Schriefers, Lahiri and van Donselaar, 1993). Native Dutch speakers were required to silently generate Dutch translations (e.g., kado) of English prompt words (e.g., gift) and to monitor the Dutch carrier words for a prespecified target sequence (e.g., /k/ or /ka/).

The main findings of these experiments were interpreted as indicating that the code being generated by subjects during the monitoring tasks was an abstract phonological one. This was based on the demonstration that subjects' ability to perform the monitoring task was unimpaired by the addition of a secondary articulatory suppression task. It was also argued that the code could not be phonetic as no relationship was found between the pattern of monitoring latencies and the timing of the same targets in overt speech production. Furthermore it was claimed that subjects were not basing their responses on the initial availability of the word's phonemes but rather on the output of the segment-to-frame association process as it was found that the code being generated was syllabified. When subjects were required to monitor bisyllabic words for target strings which either did or did not correspond to the first syllable in a word (targets /ma/ and /maag/ in the carrier words *ma.ger* - thin, and *maag.den* - virgin) reaction times were faster to the target sequence which was a syllable in the carrier word irrespective of target size. Importantly this syllabification process was argued to be on-line and not merely involve the downloading of stored syllable units.

Subjects were also instructed to monitor CVC.CVC carrier words (e.g., *lif.ter*) for the four constituent consonants and it was seen that the time taken to detect each segment increased in a left-to-right manner. In addition, a 50ms difference was observed between the monitoring latencies for the two medial phonemes which flanked the syllable boundary. This difference was similar in size to that found in

426

monitoring latencies for the first syllable consonants which are separated by an intervening vowel. This would concur with the notion that within syllables phonemes are assigned to their frame in a left-to-right manner. The authors argued that the syllable boundary effect suggests that this assignment process is initiated for the first syllable and the encoding of the second syllable is held up until it is completed. It was also observed that there was a significant difference in the time taken to monitor for the two phonemes in the first syllable but this difference was not seen for the phonemes in final syllable. This was taken as evidence that even though the encoding of the second syllable of the word is delayed the constituent phonemes are still made available. This means that when assignment to the syllable frame can begin it occurs much faster than that of the previous syllable.

While the results of the Wheeldon and Levelt (1995) are intriguing a couple of aspects of their data are problematic. The first problem concerns the locus of the observed syllable monitoring effect. As the authors acknowledge, it is possible that the effects being observed, especially those concerning the structural properties of inner speech, might be a function of the monitor itself rather than the actual speech code being monitored. A number of monitoring devices have been proposed which fall into two broad categories: production and perception based monitors. It is possible that speakers can monitor their speech at every level of the speech production process (from conceptualisation onwards) as proposed in models such as Laver (1980). However, this implies much duplication between the monitor and the actual process itself. A more parsimonious account holds that it is only the output of phonological processing which can be monitored (Motley et al, 1982). Although this model overcomes the duplication issue it still does not satisfactorily explain how speakers are able to monitor their speech on a variety of other linguistic levels (i.e. for semantic, syntactic, social appropriateness, Levelt, 1983; 1989). In order to address the shortcomings of such production based models, Levelt (1989) outlines an device which uses as its input the internal speech code but feeds it through an internal loop into the speech comprehension system. Put simply, Levelt argues that prearticulatory speech is monitored and edited as if it where the speech of an external speaker. If this is the case, then it follows that any observed effects arising from Wheeldon and Levelt's monitoring task could, at least in part, be due to the parsing processes of the speech perception mechanisms.

Zwisterlood et al (1993) observed a syllable match effect when subjects were required to monitor auditorily presented Dutch words. It seems, therefore, that Dutch speakers make use of syllable units during their monitoring of an incoming speech stream. However, this is not universally true of all languages. It was, therefore, thought worthy to investigate whether Wheeldon and Levelt's findings can be replicated in a language which does not appear to employ the syllable in the same way during monitoring. English is the ideal candidate. It is a language which contains many instances of words which are ambisyllabic, has an irregular stress pattern and contains a large range of syllabic structures. This means that employing a syllable-based segmentation strategy would be unproductive (Cutler, 1997; Cutler et al., 1986; Cutler and Norris, 1988; Norris and Cutler, 1985). Indeed, there has been no evidence of a syllable effect using a monitoring task in the perception of English (Cutler, 1997; Cutler et al., 1986; Bradley, Sanchez-Casas and Garcia-Albea, 1993). By contrast there is much evidence to suggest that the syllable is a salient unit in the production of English (Ferrand, Segui and Humphreys, 1997; Sevald, Dell and Cole, 1995). If a syllable effect can be demonstrated in the internal production of English it would add weight to the notion that the properties observed are ones of the production based speech code and not features of the comprehension system by which it is monitored.

The second issue concerns the predictions made regarding the time course of phonological encoding. These were based chiefly on the results of the phoneme monitoring task which has some methodological shortcomings. The left-to-right effect was not robust in that it only reached significance in the subjects analysis of the data and not the items analysis. In addition, the significant monitoring difference which was found between the two phonemes on either side of the syllable boundary was confounded by the fact that for the majority of stimuli this also served as a morphological boundary. As a consequence there is a need not only to replicate this pattern of results but also to employ a more stringent set of stimuli.

Finally, it was thought that a fruitful way to extend the Dutch study would be to directly compare reaction times on the inner speech monitoring task with an identical perceptual version. It still cannot be argued unequivocally from the Wheeldon and Levelt data that the pattern of results obtained from the phoneme monitoring study is one which is specific to speech production. In particular the speeding up of the encoding of the final syllable could be a feature of speech perception mechanisms. For instance, once the uniqueness point of a word has been reached processing could progress at a faster rate. For this reason it is important to establish how closely the monitoring latencies observed in the inner speech experiment correspond to those obtained using a perception task.

The Task
It was not possible to exactly replicate the methodology employed by Wheeldon and Levelt (1995) as they relied upon a population of subjects who were comfortably bilingual and unfortunately such subjects were not available to us. Instead a task was employed which made use of word association pairs. Specifically the word forms to be monitored were elicited using a semantically related prompt words (e.g., prompt - *baboon*, carrier word - *monkey*). The prompt words were chosen carefully to avoid any potential priming effects due to phonological relationships with the associated word. In an attempt to encourage the use of the internal speech code rather that any visual or graphemic representations the target sequences and prompt words were always presented auditorily and the pre-experimental training which the subject undertook in order to learn the word pairs was given verbally.

Experimental trials were structured as follows. First an auditory description of the target sequence was presented (e.g., /mon/). This was followed by the auditory presentation of a prompt word (e.g., rain). Subjects made a push button yes/no response depending upon whether the carrier word (e.g., *monsoon*) contained the target string. Reaction times and errors were recorded. Responses were measured from the onset of the prompt word to the subject's response. All reported statistical analyses were conducted treating subjects (t_1 and F_1) and items (t_2 and F_2) as random factors.

Investigating the Syllable Effect

Experiment 1 - Initial Syllable

Method This experiment investigated whether English speaking subjects would detect strings which corresponded to the first syllable in a word faster than if the sequence constituted more than a syllable. Carrier words were chosen with CVC.C (or VC.C) syllable boundaries. Words were selected for which syllabification was unambiguous according to the phonotactic rules of the language. For instance, the word *tempest* is syllabified *tem.pest* , not *temp.est* because according to the maximisation of onset principle the second syllable onset will attract the /p/ consonant but not the /mp/ cluster which is an illegal syllable onset in English. The target sequences to be monitored for were the initial CVC/VC (syllable - e.g., /tem/) and CVCC/VCC (nonsyllable - e.g., /temp/).

Results and Discussion The mean monitoring latencies for the two target sequence types are given in Table 1. It can be seen that subjects were quicker at detecting a nonsyllable string than a syllable string. A related samples t-test was performed and confirmed that this difference in means was significant in the subjects analysis only, t_1 (23) = 3.0, Standard Error = 16, $p < .01$. The same analyses were conducted using the error rates and no significant differences were observed between the two conditions for either analysis.

Table 1: The Mean Latencies (in ms) for the Two Strings Employed in the Syllable Monitoring Experiments. Mean Percentage Errors are Presented in Brackets.

Experiment	Syllable	Nonsyllable	Difference
(1) Initial CVC	1144 (4.2)	1096 (4.2)	48
(2) Internal CVC	1431 (9.4)	1350 (6.8)	82
(3) Initial CV	1308 (5.5)	1239 (5.5)	70

This experiment, therefore, yielded no evidence of a syllable match effect. Instead there was an insignificant tendency for the longer CVCC target to be monitored for faster than the CVC syllable target. It is possible that a priming effect swamped any syllable effect which might be present; simply the more of the word that is heard the quicker it can be processed. As a consequence the next experiment was designed in order to dissociate the potential syllable effect and the influence of priming by requiring subjects to monitor the carrier word for word internal syllables. If the sequence to be monitored for is no longer coming from the initial portion of the carrier word then hearing the string beforehand should not prime its generation (Meyer, 1990; 1991).

Experiment 2 - Internal Syllable

Method The method was exactly the same as employed in the previous study with the exception that the target strings occurred internally in the carrier word. The syllable sequence to be monitored for was of CVC (and in one instance, CCVC) structure and the nonsyllable sequence CVCC (or CCVCC). For example, the carrier word *romantic* would be preceded by the target strings /man/ (syllable) and /mant/ (nonsyllable).

Results and Discussion As can be seen in Table 1 the time taken to monitor for the nonsyllable string was again faster than for the string which corresponded to a syllable in the carrier word. This difference was significant according to both the subject, t_1 (23) - 4.0, SError = 20, $p < .01$ and the item t_2 (15) = -2.2, SError = 40, $p < .05$ analysis. No significant difference was found between the number of errors made in each condition. Therefore, even when the position of the target sequence is changed to an internal sequence the effect remains constant in that the longer string produces the faster reaction times. These data contradict the notion that any syllable match effect is being masked by a priming effect.

From the data of these first two experiments it must be concluded that there is no evidence that subjects' monitoring latencies are faster when the sequence to be detected in their internal generations corresponds to a syllable in that word. However, there still remains the possibility that the observed effect is a feature of the stimuli being utilised. As outlined in the Introduction the English language is notoriously ambisyllabic and one factor which influences syllabification in English is stress. Treiman and Zukowski (1990) using an oral syllable repetition task observed that for words with a medial consonant cluster (such as those used in Experiment 1 and 2 - for example, tem.pest) the second consonant (so in this example the /p/) is likely to be attached to the first stressed syllable in addition to the onset of the following unstressed syllable; in other words become ambisyllabic. This should not be overstated as the task used did not reflect normal speech production processes and the effect was seen for only certain types of consonant clusters. However, as this ambisyllabic predisposition is one which holds for most of the stimuli described so far a final syllable monitoring experiment was conducted which employed second syllable stress words and avoided medial consonant clusters.

Experiment 3 - Initial Syllable with Noninitial Stress

Method The same design as employed in the previous two experiments was replicated but words of CV.CVC or (V.CVC) structure acted as the carrier words (e.g., *pla.toon*). Gussenhoven (1986) and Treiman and Danis (1988) have identified certain rules which make ambisyllabic intervocalic consonants less prevalent. Specifically, a consonant is more likely to be syllabified with the following vowel when that vowel is stressed; in addition obstruent consonants are more often grouped with the following vowel than sonorant consonants. Following this second syllable stress words were used and it was ensured that the second syllable onset was always a stop consonant or fricative. The syllable sequence which was to be monitored for was of CV structure (e.g., /pla/) and the nonsyllable sequence was of CVC composition (e.g., /plat/).

Results and Discussion As in Experiments 1 and 2 the nonsyllable sequence produced faster monitoring latencies than the syllable sequence. The respective means are detailed in Table 1. This difference was significant by subjects analysis only, t_1 (23) = 3.3, SError = 21, p < .01. The two conditions did not significantly differ in the number of errors which were observed.

Therefore, yet again there is no evidence to suggest that the syllable sequence has a facilitative effect on monitoring latencies. Indeed this is the indisputable finding from the entire series of experiments. It has been consistently found that the trend in subjects' reaction times goes against that which would be predicted by the syllable effect hypothesis. This is a very similar pattern of results as seen for English in the perception studies (Cutler et al 1986; Bradley et al 1993) which casts doubt on the conclusion drawn by Wheeldon and Levelt, namely that their syllable effect reflects properties of the production code. The lack of a syllable effect in the present studies seems to verify the internal loop theory of monitoring (Levelt, 1989) and specifically that it is the properties of the speech comprehension system which are reflected in the syllable monitoring task.

The next two experiments to be reported focused on the monitoring of phonemes in bisyllabic words. This was to establish whether the claims made by Wheeldon and Levelt regarding the time course of phonological encoding could be repeated.

The Time Course of Phonological Encoding

Experiment 4 (a) and (b) - Phoneme Monitoring

Method - Experiment 4 (a) The task employed was similar to that described for the syllable monitoring experiments. The carrier words used were of CVC.CVC structure with a clear, phonotactically correct syllable boundary (e.g., *lit.mus*). The majority of the words were monomorphemic and each contained four different consonants which served as the target phonemes which were to be detected during the task. In this manner the four target positions were situated at

Table 2: The Mean Monitoring Latencies (in ms) for the Four Target Positions In Experiments 4 (a) and (b). The Mean Percentage Errors are Detailed in Parenthesis.

| Experiment | Target Type | | | |
	C1	C2	C3	C4
Internal Production	1392 (3.5)	1501 (6.0)	1564 (4.8)	1589 (5.0)
External Perception	622 (1.6)	886 (1.6)	901 (0.9)	1028 (2.3)

the first syllable onset (C1); first syllable offset (C2); final syllable onset (C3); and final syllable offset (C4). All the intervening vowels (with exception of three) were short.

Monitoring trials were grouped into lists for a given target phoneme. Such lists comprised of between four and twelve prompt words. Subjects heard a description of the target phoneme followed by a series of prompt words. For each of these the subject was required to decide whether the corresponding carrier word contained the sound described to them. If the carrier word did contain the phoneme a, 'yes' response was required, if it did not, no response was necessary.

Results and Discussion As can be seen in Table 2 the mean latencies increase as a function of their position in the word in a left-to-right manner. The differences in monitoring latencies across consonant positions are given in Table 3. The difference between the initial syllable onset and offset is the greatest, followed by that between the consonants which flank the syllable boundary with the difference between the constituent phonemes of the final syllable being the smallest. ANOVAs were performed on the data and a significant difference between the four target positions was demonstrated, F_1 (3, 108) = 30.5, MSError = 9540, p < .01, F_2 (3, 51) = 19.8, MSError = 7350, p < .01. Newman-Keuls pairwise comparisons yielded a significant difference between C1 and C2 and between C2 and C3. The difference between C3 and C4 was not significant. Identical ANOVAs were repeated using the error rates which yielded no significant effects.

This pattern of results is identical to that found by Wheeldon and Levelt (1995). In this way the effects demonstrated by the phoneme monitoring methodology appear to be more robust across languages than those involving syllable monitoring.

Method - Experiment 4 (b) This was an attempt to establish whether the pattern of latencies observed on the production task bore any resemblance to the actual position of the target phonemes in the speech stream. The experiment was repeated but instead of requiring subjects to generate the carrier words internally they were presented to them auditorily.

429

Table 3: A Comparison of the Difference in Monitoring Latencies (in ms) between the Four Target Positions in Experiments 4 (a) and (b). Percentage Error Scores are Given in Brackets.

| | Difference | Experiment | |
		Internal Production	External Perception
(0)	C1-C2	109 (2.5)	264
	C2-C3	63 (1.2)	16 (0.7)
	C3-C4	25 (0.2)	127 (1.4)

Results and Discussion The data for this perception experiment is also shown in Tables 2 and 3. Once again the mean monitoring latencies for the four experimental conditions increases the further along in the word the phoneme is positioned. ANOVAs demonstrated that the difference between these conditions was significant, F_1 (3, 69) = 118.4, MSError = 5893, $p < .01$ and by items, F_2 (3, 51) = 44.8, MSError = 11770, $p < .01$. Pairwise comparisons showed that the difference between the two syllables onsets and offset were significant. In contrast, the difference between the two phonemes which flanked the syllable boundary was not. This reflects a different pattern of monitoring latencies to the ones observed in the production experiment but is consistent with the position of the target segments in the speech stream in that the syllable onsets and offsets are temporally separated by a vowel whereas C2 and C3 are not. This would suggest that the code being monitored in the internal speech task is an abstract one.

General Discussion

In summary, this paper reports a series of experiments which required subjects to monitor their internal productions of English words for target syllables or phonemes. Experiments 1-3 show that there is no evidence of a syllable match effect in English. However, when subjects were required to monitor bisyllabic CVC.CVC words for their constituent phonemes a clear left-to-right effect was observed. In addition, the time taken to detect the phonemes on either side of the syllable boundary was seen to differ and a significant difference was seen between the monitoring latencies for the first syllable onset and offset which was not repeated in the final syllable. These findings relate to those of Wheeldon and Levelt (1995) as follows.

Regarding the locus of the syllable monitoring effect, our results seem to reflect characteristics of the perception rather than the production mechanisms. This supports the claim that the prearticulatory speech code is monitored by the speech comprehension system via an internal loop and that this architecture is not sensitive to syllable structure for English (Cutler, 1997; Cutler, et al., 1986; Cutler and Norris, 1988, Norris and Cutler, 1988).

In contrast, however, the claims regarding the time course of the phoneme monitoring task hold true in that the above results mirror those obtained in the Dutch study. It has again

been shown that speech encoding runs in a left-to-right manner and that the code is syllabified. Importantly, the syllable boundary effect was replicated with words which were not morphologically complex supporting the idea that the syllabification of subsequent syllables is held up until phonemes have been assigned to the initial syllable. These findings compliment current models of speech production such as proposed by Levelt, Roelofs and Meyer (1999) and the computer-based WEAVER model (Roelofs, 1992; 1997) which propose that phonological encoding runs in a strictly serial order and that the syllabification of an utterance is generated on-line in accordance with the rules of the language.

Finally, it was found that when the phoneme monitoring task was repeated using external speech a completely different pattern of results was seen. Specifically, it was seen that the speeding up of encoding in the final syllable which was observed in the inner speech task was not found in the perception version. This confirms that the code being monitored during the inner speech task is phonological and not encumbered by phonetic or articulatory specifications.

In this way the data presented in this paper has been able to resolve some of the problematic aspects of the Wheeldon and Levelt studies. The lack of a syllable matching effect in the above data allows certain aspects of the Dutch data to be attributed to a comprehension not production-based monitor. It has been shown that the left-to-right pattern is robust across different languages and different word eliciting tasks. Finally, it has been confirmed that the time course of internal speech monitoring is different to external speech in theoretically interesting ways. Thus, the internal speech monitoring task can be seen to be a valid methodology for the investigation of the time course of phonological encoding during speech production.

Acknowledgements

This research was conducted by the first author and funded by means of an ESRC research grant awarded to the second author. The authors wish to thank Jan Zandhuis for the design and programming of the experimental set-up.

References

Bradley, D. C., Sanchez-Casas, R. M., & Garcia-Albea, J.E. (1993). The status of the syllable in the perception of Spanish and English. *Language and Cognitive Processes, 8*, 197-233.

Cutler, A. (1997). The syllable's role in the segmentation of stress languages. *Language and Cognitive Processes, 12*, 839-845.

Cutler, A., Mehler, J., Norris, D., & Segui, J. (1986). The syllable's differing role in the segmentation of French and English. *Journal of Memory and Language, 25*, 385-400.

Cutler, A., & Norris, D. (1988). The Role of strong syllables in segmentation for lexical access. *Journal of Experimental Psychology: Human Perception and Performance, 14*, 113-121.

Dell, G. S., & Repka, R. J. (1992). Errors in inner speech. In B. J. Baars (Ed) *Experimental Slips and Human Error: Exploring the Architecture of Volition.* New York: Plenum Press.

Ferrand, L., Segui, J., & Humphreys, G. W. (1997). The syllable's role in word naming. *Memory and Cognition, 25,* 458-470.

Gussenhoven, C. (1986). English plosive allophones and ambisyllabicity. *Gramma, 10,* 119-141.

Laver, J. D. M. (1980). Monitoring systems in the neurolinguistic control of speech production. In V. A. Fomkin (Ed) *Errors in Linguistic Performance.* New York: Academic Press.

Levelt, W. J. M. (1983). Monitoring and self repair. *Cognition, 14,* 41-104.

Levelt, W. J. M. (1989). *Speaking: From intention to articulation.* Cambridge, MA: MIT Press.

Levelt, W. J. M., Roelofs, A., & Meyer, A. S. (1999). A theory of lexical access in speech production. *Behavioural and Brain Sciences, 22,* 1-75.

Levelt, W. J. M., & Wheeldon, L. R. (1994). Do speakers have access to a mental syllabary? *Cognition, 50,* 239-269.

Mehler, J., Dommergues, J., Frauenfelder, U., & Segui, J. (1981). The syllable's role in speech segmentation. *Journal of Verbal Learning and Verbal Behaviour, 20,* 298-305.

Meyer, A. S. (1990). The time course of phonological encoding in language production: The encoding of successive syllables of a word. *Journal of Memory and Language, 29,* 524-545.

Meyer, A. S. (1991). The time course of phonological encoding in language production: Phonological encoding inside a syllable. *Journal of Memory and Language, 30,* 69-89.

Meyer, A. S., & Schriefers, H. (1991). Phonological facilitation in picture-word interference experiments: Effects of stimulus onset asynchrony and types of interfering stimuli. *Journal of Experimental Psychology: Learning, Memory and Cognition, 17,* 1146-1160.

Motley, M. T. , Camden, C. T., & Baars, B. J. (1982). Covert formulation of anomalies in speech production: Evidence from experimentally elicited slips of the tongue. *Journal of Verbal Learning and Verbal Behaviour, 21,* 578-594.

Norris, D., & Cutler, A. (1988). The relative accessibility of phonemes and syllables. *Perception and Psychophysics, 43,* 541-550.

Roelofs, A. (1992). A spread-activation theory of lemma retrieval in speaking. *Cognition, 42,* 107-142.

Roelofs, A. (1997). The WEAVER model of word-form encoding in speech production. *Cognition, 64,* 249-284.

Sevald, C. A., Dell, G. S., & Cole, J. S. (1995). Syllable structure in speech production: Are syllables chunks or schemas? *Journal of Memory and Language, 34,* 807-820.

Treiman, R., & Danis, C. (1988). Syllabification of intervocalic consonants. *Journal of Memory and Language, 27,* 87-104.

Treiman, R., & Zukowski, A. (1990). Toward an understanding of English syllabification. *Journal of Memory and Language, 29,* 66-85.

Wheeldon, L. R., & Levelt, W. J. M. (1995). Monitoring the time course of phonological encoding. *Journal of Memory and Language, 34,* 311-334.

Zwisterlood, P., Schriefers, H., Lahiri, A., & van Donselaar, W. (1993). The role of syllables in the perception of spoken Dutch. *Journal of Experimental Psychology: Learning, Memory and Cognition, 19,* 200-271.

Developmental Differences in Young Children's Solutions of Logical vs. Empirical Problems

Bradley J. Morris (bmorris@andrew.cmu.edu)
Dept. of Psychology, Carnegie Mellon University
5000 Forbes Avenue, Pittsburgh, PA 15213, USA

Vladimir Sloutsky, (Sloutsky.1@osu.edu)
Center for Cognitive Science, The Ohio State University
1945 North High Street, Columbus, OH 43210, USA

Abstract

We examined the development of the ability to differentiate logically determinate from logically indeterminate problems. The results indicated that a) young children tend to reduce the number of empirical possibilities via "cutting" the second half of less informative propositions, b) these errors do not stem from encoding or recall errors, c) from elementary to middle school, children tend to increase their understanding of logical form, and d) this increase corresponds to a decrease in the rate of cuts.

There is a large body of research examining children's understanding of empirical indeterminacy (Fay & Klahr, 1996; Piéraut-Le Bonniec, 1980; Sodian, Zaitchik, & Carey, 1991). A problem is empirically determinate if it corresponds to exactly one empirical possibility; otherwise, it is empirically indeterminate (Piéraut-Le Bonniec, 1980). Previous findings suggest that young children often (1) fail to appreciate empirical indeterminacy, confusing indeterminate problems with determinate ones, but not vice versa, and (2) have less difficulty solving determinate problems than indeterminate problems (Bindra, Clarke, & Schultz, 1980; Byrnes & Overton, 1986; Fay & Klahr, 1996; Pieraut-Le Bonniec, 1980). However, there is another class of problems that should be considered in conjunction with the issue of determinacy- problems that require logical, but not empirical solutions. These problems are logically determinate (LD) if they are solved logically, but they are indeterminate if they are solved empirically.

Researchers have demonstrated that children (and many adolescents) do not fully understand logical determinacy (Byrnes & Overton, 1986; Moshman & Franks, 1986; A. Morris & Sloutsky, 1998) and they often attempt to provide empirical solutions to logically determinate problems (A. Morris & Sloutsky, 1998; Osherson & Markman, 1975). In this article, we examine the development of solution strategies to some logical and empirical problems and possible cognitive mechanisms underlying these strategies.

Information-processing analysis of solving logical vs. empirical problems

Logically determinate problems are those that can be solved *a priori* based on their logical form. Some LD problems yield logically true or necessary conclusions, whereas others yield logically false or impossible conclusions. Problems that yield conclusions that are true in some, but not all states of affairs are defined as logically indeterminate (LI), or empirical.

It has been traditional since the early work of Newell & Simon (1972) to conceptually model problem solving as search through a problem space for a desired goal state. A three-stage model outlines the creation of problem space, search and creating an output. Encoding is the creation of problem space from the information in the environment. The more clearly the problem space represents salient elements in the environment, the more veridical the representation, and the greater opportunity the organism has for solving the problem (Newell & Simon, 1972; Newell, 1990). The second phase is search in which a decision matrix is examined for possible outcomes and actions of the represented problem. Two types of search are utilized: problem search, in which search proceeds through possible outcomes of states and operators, and knowledge search, in which search is through memory (Newell, 1990). Once a goal state, an impasse, or some terminating point in search is reached, an output is then created. Therefore, it seems safe to assume that the mentioned reasoning and problem solving errors may occur due to the following factors (or any of their combinations):

1) Limits on encoding.
2) Poor representation of problem space.
3) Incomplete search through problem or memory space.
4) Inaccurate mapping of a problem solution onto a verbal response.

Evidence from several related domains such as scientific reasoning, logical reasoning, and practical reasoning indicates that children and adults often limit their search in both problem space and memory space (Kuhn et al., 1995; Markovits, 1988; Mynatt, Dohetry, & Tweney, 1977; Tweney et al, 1980). They were also found to exhibit a 'positive capture' strategy failing to consider equally plausible alternatives (Bindra et al., 1980; Fay & Klahr, 1996). The tendency to limit their search has also been found in studies of logical reasoning in that both children and adults tend to not search for counterexamples in forms such as conditional reasoning (Johnson-Laird, 1993; Markovits, 1988; Wason & Johnson-Laird, 1972). In practical reasoning (when the task was to compare size of foreign cities) people often made fast and frugal "Take-the-Best" inferences relying on a small number of the most salient cues (such as the familiarity of the city), while ignoring the rest of the cues (Gigerenzer & Goldstein, 1996).

There is also evidence that young children exhibit difficulties when solving problems corresponding to no empirical possibilities, such as contradictions (B. Morris, 1998; Sloutsky, Rader, & Morris, 1998). In particular, when presented with a contradiction, preschoolers responded as if they ignored (or "cut") the second half of the contradiction thus transforming it into a statement corresponding to one empirical possibility functioning to limit problem space.

Therefore, it seems plausible that young children implicitly assume that propositions correspond to exactly one empirical possibility, thus creating a "defective" or incomplete problem space. If this is the case, then the number of empirical possibilities compatible with the problem could predict the problem difficulty. The easiest problems are those that correspond to <u>exactly one</u> empirical possibility whereas an increase or decrease in the number of empirical possibilities leads to an increase in problem difficulty and subsequently to the number of errors (for a detailed discussion see Sloutsky, Morris, & Rader, in review).

Three groups of children (preschool, elementary and middle school) were presented with reasoning problems. The experiment focused on (a) the ability of children and adolescents to distinguish logical from empirical problems; (b) solution strategies for different types of problems; (c) patterns of errors; (d) the relationship between the number of empirical possibilities and the problem difficulty; and (e) accuracy of encoding and mapping of verbal responses. We deemed it necessary to reduce the number of possible sources of error via eliminating the necessity of search through knowledge space. In so doing, we presented participants with knowledge-lean problems that required "deriving a solution from givens" rather than requiring memory search.

Method

Participants

38 four-and five- year-old children enrolled in three child care centers (average age = 4.3 years; 16 girls and 22 boys), 34 third grade children in three elementary school classrooms (average age =8.4 years; 19 girls and 15 boys), and 35 sixth grade children enrolled in two middle school classrooms (average age = 11.7 years, 16 girls and 19 boys).

Materials

The tasks consisted of a series of predictions by an imaginary character, ZZ, as to the outcome of one of two separate items a) a ball dropped in the Tautology Machine and b) opening or not opening a book. The Tautology machine is a 21" x 24" board with a chute at the top in which a ball dropped will fall to one of two terminating points (in this experiment labeled "Red" and "Green"). The "predictions" took the form of one of four logical forms Tautologies, Contradictions, Disjunctions, and Conjunctions. The Tautology Machine has a switch (occluded from participants) in the back of the machine that moves a lever to one of two sides directing the ball to either the red or the green side. ZZ made eight predictions regarding the outcomes of the game, two predictions for each logical form. The experimenter presented ZZ, the Tautology Machine that has two possible outcomes, and a book that could be either opened or closed. ZZ made predictions pertaining (1) only to the ball's landings (tautologies and contradictions) and (2) to the ball's landing and to whether the book will be opened or closed (conjunctions and disjunctions).

Procedure

In this experiment, there was one within-subject factor, the logical form of the prediction. The experiment was conducted in a single 10-15 minute session that included two phases: warm-up/instruction phase and the experimental phase. In the warm-up phase, each child was read a set of instructions that explained the purpose of the game as evaluating the predications of an imaginary character named "ZZ." Six participants were eliminated from further consideration because they gave "Can't tell" responses to all warm-up questions.

Participants were asked five questions about each prediction. (1) An encoding measure (repeat the prediction). (2) An *a priori* evaluation, of the prediction ("True," "False," and "Can't tell"). (3) Request for empirical verification ("Do we need to drop the ball?"). (4) An *a posteriori* evaluation of the prediction (only if empirical verification was requested). And (5) a measure of encoding (repeat the initial prediction). The recall measure was introduced only with elementary school and middle school.

Results

A priori evaluation

A two way 3 (age group) by 4 (logical form) repeated measures ANOVA was performed on the a priori evaluations with age as a between-participant factor and logical form as a repeated measure. The analysis yielded a significant main effect for age, F (2, 98)= 28.890, p<.001, form, F (2,98)= 5.3171, p<.002, and the interaction between age and logical form, F (2, 98)= 3.171, p<.005. Tukey's HSD post-hoc tests indicated that for conjunctions and tautologies, middle school children made more correct a priori evaluations than both elementary and preschool children, Tukey's HSD, all ps< .01. For contradictions, middle school made more correct a priori evaluations than elementary school children, Tukey's HSD, p< .01. Middle school children made more correct a priori evaluations of disjunctions than preschool children, Tukey's HSD, p< .02. Post-hoc tests by form indicate that conjunctions and disjunctions were evaluated as correct more frequently than tautologies and contradictions more frequently than tautologies , Tukey's HSD, p< .02.

Requests for Empirical Verification

To analyze requests for empirical verification, participants' responses to particular problems were collapsed into two groups, responses to logically determinate problems (i.e., tautologies and contradictions) and to logically indeterminate problems (conjunctions and disjunctions). These collapsed responses were subjected to two-way 3 (age group) by 2 (logical determinacy v. logical indeterminacy) repeated measures ANOVA. The analysis yielded a main effect for age, F (2, 98)= 75.974, p< .0001, form, F (2,98) = 51.061, p< .0001, and the interaction between age and logical forms, F (2, 98)= 44.545, p<.0001. For logically determinate forms, middle school children requested significantly less empirical verification than elementary and preschool children, Tukey's HSD, ps< .001. For logically indeterminate forms, middle school children requested more empirical verification than elementary and preschool children, Tukey's HSD, ps< .01. Additionally, elementary school children requested less empirical verification than preschool children, Tukey's HSD, ps< .01. While younger children equally frequently requested empirical verifications for both logically determinate and logically indeterminate forms, older children less frequently requested empirical verification of logically determinate forms. Within-age group differences were only present in middle school children on logically determinate forms, Tukey's HSD, p< .001, all others were not significantly different.

Cuts

As stated earlier, we hypothesized that the number of empirical possibilities compatible with a proposition will be a predictor of accuracy of processing. It is also predicted that those children who did not recognize logical form should exhibit more distortions of data as a function of the predicted informativeness of each proposition. Furthermore, we expected these distortions to exhibit systematicity. The pattern to be analyzed is a "cut" that functions to reduce the number of empirical possibilities. To analyze patterns of errors, a rigorous decision procedure was introduced into the analysis to distinguish "cuts" from logically appropriate responses in which the second half of propositions was "rigged" so that "cuts" could be distinguished form other responses.

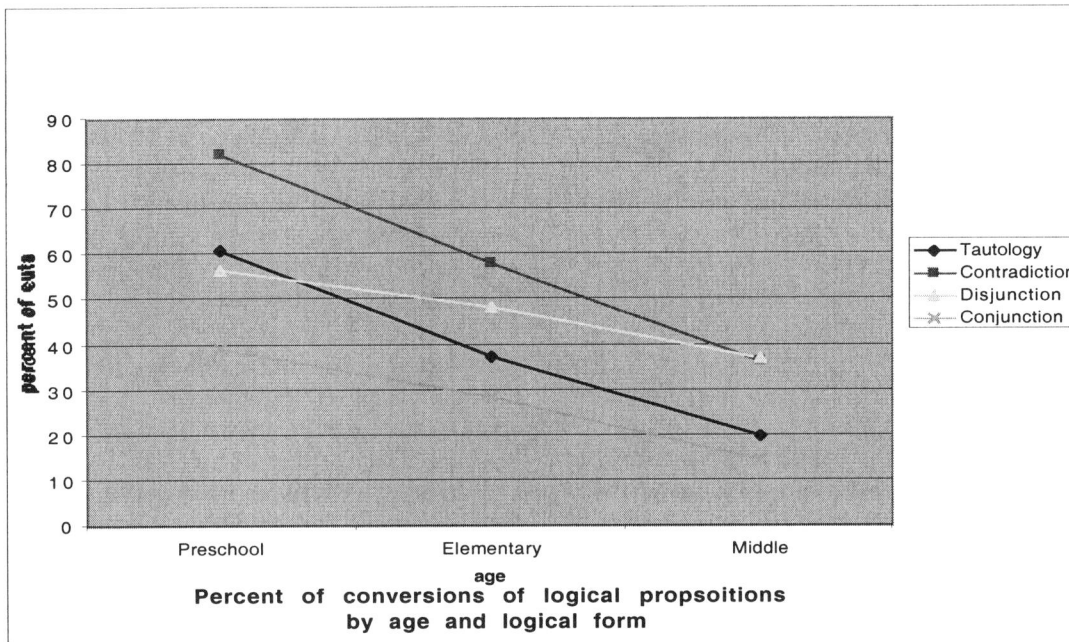

Figure 1- Percent of "cuts" of logical propositions by age and logical form

A series of McNemar and Cochran's chi-squares were conducted to compare conversion levels by logical form. The error rates are depicted in Figure 2. Data in the figure suggest that in the younger groups, conjunctions were least likely to be cut, whereas contradictions and tautologies were most likely to be cut, and disjunctions were in-between the two extremes. In middle school children, disjunctions were cut at the highest rate while low levels of conversions of tautologies and contradictions seem to be due to recognition of logical form and therefore eliminating the need for empirical evidence and its effect on outcome. For preschool age children, a Cochran Q test, with McNemar chi-square tests for post hoc pairwise comparisons, indicated significant differences in conversion rates across the logical forms (Cochran Q (3, 32) = 22.3, p<. 0001). Pairwise comparisons indicate that conjunctions were significantly less probable to be converted than disjunctions (*McNemar* (1, 37) = 4.5, p <. 05), tautologies (*McNemar* (1, 45) = 14.7, p <. 0001), and contradictions (*McNemar* (1, 46) = 28, p <. 0001), whereas disjunctions are less probable to be converted than contradictions (*McNemar* (1, 36) = 10.3, p <. 005). These data suggest that the probability of a conversion increases with an increase in the number of possibilities compatible with the statement.

In elementary school children, a Cochran Q test, with McNemar chi-square tests for post hoc pairwise comparisons, indicated significant differences in conversion rates across the logical forms (Cochran Q (3, 34= 21.7, p<. 0001). Pairwise comparisons indicate that conjunctions were significantly less probable to be

converted than contradictions (*McNemar* (1, 35) = 9.62, p <. 001), whereas contradictions are more probable to be converted than disjunctions (*McNemar* (1, 35) = 6.0, p <. 025), and tautologies (McNemar (1, 35= 8.9, p<.005). These data, like preschool data, suggest that the probability of a conversion increases with an increase in the number of possibilities compatible with the statement.

In middle school children, a Cochran Q test, with McNemar chi-square tests for post hoc pairwise comparisons, indicated significant differences in conversion rates across the logical forms (Cochran Q (3, 35) = 24.6, p<. 0001). Pairwise comparisons indicate that conjunctions were significantly less probable to be converted than contradictions (*McNemar* (1, 34) = 6.36, p<. 025), and disjunctions (*McNemar* (1, 34) = 15.96, p <. 001), whereas disjunctions are more probable to be converted than tautologies (McNemar (1,34)= 6.0, p<.025) and conjunctions (*McNemar* (1, 34) = 9.45, p <. 005). These data indicate that for all age groups, problems corresponding to exactly one empirical possibility (i.e., conjunctions) elicit fewer errors than problems corresponding to more than one or less than one empirical possibility. Data in the figure, also indicate a marked developmental progression with respect to error rates.

Accuracy of encoding & recall

From the very beginning, we have contemplated several possibilities as to where in the course of information processing conversions take place. Two possibilities seemed most plausible: (1) conversions occur in the course of encoding the propositions into working memory or (2) in the course of retrieving the propositions from long-term memory, or (3) they occur in the course of the creation of a problem space.

The results indicate that preschool children encoded about 23% of all predictions incorrectly while elementary and middle school children encoded less than 10% incorrectly. The data demonstrate that the levels of encoding errors cannot account for conversions for two reasons: a) mean encoding levels would have to increase dramatically to equal those levels in conversions and b) the rates of encoding errors are lowest for those forms which are converted at the highest rates. Therefore, the evidence does not suggest that encoding is responsible for conversions.

Recall rates indicate that overall children tend to recall the initial predictions correctly, even those that had high conversion rates. These findings seem to suggest that participants represent a proposition veridically without veridically representing corresponding state of affairs. This dissociation indicates that the reported conversions do not stem from memory limitations. It seems that a likely source of these conversions is an incomplete representation of empirical possibilities or states of affairs compatible with the proposition.

Discussion

The findings could be summarized as follows. (1) As opposed to middle school children, young children do not distinguish between logical and empirical problems, and often attempt to empirically solve logical problems. (2) Participants of all age groups who exhibited errors, exhibited the same pattern of errors — the tendency to represent a problem as if it were compatible with only one empirical possibility. (3) In so doing, across the age groups, participants exhibited one strategy- a "cut" in the second half of logical propositions. (4) Cuts markedly decreased with age and the acquisition of an understanding of logical form.

As in previous experiments (Fay & Klahr, 1996; Sloutsky, Morris, & Rader, in review) preschool and elementary children did not distinguish logically determinate from empirical statements. Two sets of evidence demonstrate a lack of recognition of logical sufficiency of tautologies and contradictions in preschool and elementary school children; (1) no recognition of *a priori* logical form and (2) requests for empirical verification. This evidence also suggests that middle school age children differentiate logically determinate from indeterminate problems.

Cuts were demonstrated by *a posteriori* evaluations of the ball's landing that systematically ignored the second half of the proposition in question. For example, when given "the ball will land on red and not red" and the ball actually landed on red, the child responds that the prediction was correct. Two factors are necessary in order to draw this conclusion: a) inability to distinguish *a priori* logical form, and b) inability to recognize when empirical verification is necessary.

The data also indicate that problems compatible with more than one or less than one empirical possibility elicited more errors than problems compatible with exactly one empirical possibility. Furthermore, participants of all age groups exhibited the same pattern of errors — a "cut" in the second half of propositions.

Developmental trends suggest that the proportion of "cuts" decreases with the acquisition of an understanding of logical form, theoretically a more adaptive strategy for solving these types of problems. While preschool and elementary school children did not recognize a priori logical form and did not differentiate determinate from indeterminate problems in terms of empirical verification, middle school children performed significantly better on both measures. Additionally, levels of cuts were very low for middle school children overall. Encoding and recall rates were not significantly different for all groups and did not occur at levels high enough to account for conversion phenomena. Therefore, developmental changes that seem to be related to the decrease in cuts are an increase in recognition of logical forms and a decrease in requests for empirical verification in logically determinate forms. These changes suggest the acquisition of a strategy of logical reasoning would also function to supplant search through problem space for a solution with a search through knowledge space for the correct solution. This suggests that as children acquire an understanding of logically determinate problems the need for limiting the number of possibilities compatible with the problem decreases.

Conclusion

The presented evidence supports the hypothesis that problems compatible with more than one possibility or less than one possibility elicit more errors than problems compatible with exactly one possibility. Three main findings seem to be particularly important: (a) as opposed to middle school children, preschoolers and elementary school children do not distinguish logically determinate from logically indeterminate forms; (b) cuts seem to stem from the creation of an incomplete problem space, and not from inaccuracies in encoding; and (c) decrease in the rate of cuts corresponds to increases in the ability to distinguish logical from empirical problems. However, additional studies will help establish what else is missing and what else develops in solving logical versus empirical problems.

References

Bindra, D., Clarke, K., & Shultz, T. (1980). Understanding predictive relations of necessity and sufficiency in formally equivalent "causal" and "logical" problems. *Journal of Experimental Psychology: General, 109*(4), 422-443.

Byrnes, J., & Overton, W. (1986). Reasoning about certainty and uncertainty in concrete, causal, and propositional contexts. *Developmental Psychology, 22*(6), 793-799.

Fay, A. L., & Klahr, D. (1996). Knowing about guessing and guessing about knowing: Preschoolers' understanding of indeterminacy. *Child Development, 67*(2), 689-716.

Gigerenzer, G., & Goldstein, D. (1996). Reasoning the fast and frugal way: Models of bounded rationality. *Psychological Review, 103*(4), 650-669.

Johnson-Laird, P. N. (1993). *Human and Machine Thinking*. Hillsdale, NJ: Lawrence Erlbaum.

Kuhn, D., Garcia-Mila, M., Zohar, A., & Anderson, C. (1995). Strategies of knowledge acquisition. *Monographs of the Society for Research in Child Development, 60*(4, Serial No. 245).

Markovits, H. (1988). Conditional reasoning, representation, and empirical evidence on a concrete task. *Quarterly Journal of Experimental Psychology, 40(3-A)*, 483-495.

Morris, A., & Sloutsky, V. (1998). Understanding of logical necessity: Developmental antecedents and cognitive consequences. *Child Development, 69(3)*, 721-741.

Morris, B. J. (1998). Maximizing informativeness: Conversions of logical propositions in preschool children. *Unpublished doctoral dissertation*, The Ohio State University, Columbus, OH.

Moshman, D., & Franks, B. (1986). Development of the concept of inferential validity. *Child Development, 57*, 153-165.

Mynatt, C. R., Doherty, C. E., & Tweney, R. D. (1977). Confirmation bias in a simulated research environment: An experimental study of scientific inference. *Quarterly Journal of Experimental Psychology, 29*, 85-95.

Newell, A. (1990). *Unified theories of cognition*. Cambridge, MA: Harvard Press.

Newell, A., & Simon, H. (1972). *Human problem solving*. Englewood Cliffs, NJ: Prentice-Hall.

Osherson, D., & Markman, E. (1975). Language and the ability to evaluate contradictions and tautologies. *Cognition, 3*(3), 213-226.

Piéraut-LeBonniec, G. (1980). *The development of modal reasoning: The genesis of necessity and possibility notions*. New York, NY: Academic Press.

Sloutsky, V.M., Morris, B.J., & Rader, A. (in review). Increasing informativeness and reducing indeterminacy: an adaptive constraint in human cognition.

Sloutsky, V. M., Rader, A. W., & Morris, B. J. (1998). Increasing informativeness and reducing ambiguities: Adaptive strategies in human information processing. In M. A. Gernsbacher & S. J. Derry (Eds.), *Proceedings of the twentieth annual conference of the Cognitive Science Society (pp. 997-1002)*. Mahwah, New Jersey: Lawrence Erlbaum.

Sodian, B., Zaitchik, D., & Carey, S. (1991). Young children's differentiation of hypothetical beliefs from evidence. *Child Development, 62(4)*, 753-766.

Tweney, R.D., Doherty, M.E., Worner, W.J., Pliske, D.B., Mynatt, C.R., Gross, K.A., & Arkkelin, D.L. (1980). Strategies of rule discovery in an inference task. *Quarterly Journal of Experimental Psychology, 32(1)*, 109-123.

Wason, P. C., & Johnson-Laird, P. N. (1972*). Psychology of reasoning: Structure and content*. Cambridge, Mass: Harvard U. Press.

The Empirical Acquisition of Grammatical Relations

William C. Morris (wmorris@ucsd.edu)
Department of Computer Science and Engineering, University of California, San Diego
9500 Gilman Drive, Dept. 0114; La Jolla CA 92093-0114

Garrison W. Cottrell (gary@cs.ucsd.edu)
Department of Computer Science and Engineering, University of California, San Diego
9500 Gilman Drive, Dept. 0114; La Jolla CA 92093-0114

Abstract

We propose an account for the acquisition of grammatical relations using the concepts of connectionist learning and a construction-based theory of grammar. The proposal is based on the observation that early production of childhood speech is formulaic and the assumption that the purpose of language is communication. If one assumes that children's comprehension of multiword speech is not globally systematic, but based initially on semi-rote knowledge (so-called "pivot grammars"), a pathway through small-scale systematicity to grammatical relations appropriate to the child's target language can be seen. We propose such a system and demonstrate a portion of the emergence of grammatical relations using a connectionist network.

Introduction

Grammatical relations are frequently a problem for language acquisition systems.[1] In one sense they represent the most abstract aspect of language; subjects transcend all semantic restrictions—virtually any semantic role can be a subject. Where semantics is seen as being related to world-knowledge, syntax is seen as existing on a distinct plane. For this reason there are language theories in which grammatical relations are considered theoretical primitives, the most obvious examples of this are Relational Grammar (Perlmutter, 1982; Perlmutter & Postal, 1983) and Arc-Pair Grammar (Johnson & Postal, 1980).

One approach to learning syntax has been to relegate grammatical relations and their behaviors to the "innate endowment" that each child is born with. There are a number of theories of language acquisition, (e.g., Pinker, 1984, 1989; Hyams, 1986; Borer & Wexler, 1987, 1992) which start with the assumption that syntax is a separate component of language, and that the acquisition of syntax is largely independent of semantic considerations. Accordingly, in these theories there is an innate, skeletal syntactic system present from the very beginning of multiword speech. The acquisition of syntax, then, consists of modifying and elaborating the skeletal system to match the target language.

In order to avoid the need for innate knowledge, we propose a language acquisition system that does not rely on innate syntactic knowledge (Morris, 1998). The proposal is based on Construction Grammar (Goldberg, 1995) and on the learning mechanisms of PDP-style connectionism (Rumelhart & McClelland, 1986). We assume that the purpose of language is communication, and that children learn syntax as part of the mediating mechanism between spoken words and their aggregate meaning (cf. Slobin, 1997: 297). We hypothesize that abstractions such as "subject" emerge through rote learning of particular constructions, followed by the merging of these "mini-grammars". The claim is that in using this sort of a language acquisition system it is possible for a child to learn grammatical relations over time, and in the process accommodate to whatever language-specific behaviors his target language exhibits.

We have made a preliminary study showing that a neural net which is trained with its sole task being the assignment of semantic roles to sentence constituents can acquire grammatical relations. We have demonstrated this in two ways: by showing that this network associates particular subjecthood properties with the appropriate verb arguments, and by showing that the network has gone some distance toward abstracting this nominal away from its semantic content.

Theoretical Proposal

The proposal that we are basing our modeling on involves a three-stage process. The first stage involves rote understanding of speech, reflecting the formulaic speech that children exhibit. The second stage involves progressive abstraction over the formulas, based on both semantic and syntactic similarities. The third stage involves associating the resultant abstractions with specific "subjecthood properties".

---

[1] Grammatical relations are the relationships that noun phrases bear with a clause. These include subjects, objects, and indirect objects.

In the first stage a child learns verb argument structures as separate, individual "mini-grammars". This word is used to emphasize that there are no overarching abstractions that link these individual argument structures to other argument structures. Each argument structure is a separate grammar unto itself.

In the second stage the child develops correspondences between the separate mini-grammars; initially the correspondences are based on both semantic and syntactic similarity,[2] later the correspondences are established on purely syntactic criteria. The transition is gradual, with the role that semantics plays decreasing slowly. The result of the correspondences involves the creation of a larger grammar that includes the constituent mini-grammars. These larger grammars, in turn, will with each other.

For example, the verbs *eat* and *drink* are quite similar to each other, and will "merge" quickly into a larger grammar (while retaining their separate identities within that grammar, however). Similarly, the verbs *hit* and *kick* will merge early, since their semantics and syntax are similar. While all four of these verbs have agents and patients as verb arguments, there are many semantic differences between the verbs of ingestion and the verbs of physical assault, therefore the merge between these two verb groups will occur later in development.

Ultimately, these agent-patient verbs will merge with experiencer-percept verbs (e.g., *like, fear, see, remember*), percept-experiencer verbs (e.g., *please, frighten, surprise*), and others, yielding a prototypical transitive construction, with an extremely abstract argument structure. The verb-arguments in these abstract argument structures can be identified as "A", the transitive actor, and "O", transitive patient (or "object"). In addition there is prototypical intransitive argument structure with a single argument, "S", the intransitive "subject". (This schematic description was first put forward by Dixon, 1979.)

In the third stage, the child begins to associate the abstract arguments of the abstract transitive and intransitive constructions with the "bridging constructions" that instantiate the properties of, for example, clause coordination, control structures, and reflexivization. So, for example, an intransitive-to-transitive bridging construction will associate the S of an intransitive first clause to the deleted co-referent A of a transitive second clause. This will enable the understanding of a sentence like *Max arrived and hugged everyone.* Similarly, a transitive-to-intransitive bridging construction will map the A of an initial transitive clause to the S of a following intransitive clause; this will enable the understanding of a sentence like *Max hugged Annie and left.*

From beginning to end this is a usage-based acquisition system. It starts with rote-acquisition of verb-argument structures, and by finding commonalities, it slowly builds levels of abstraction. Through this bottom-up process, it accommodates to the target language. (For other accounts of usage based systems, see also Schlesinger, 1982,1988; Bybee, 1985; Langacker, 1987, 1991a, 1991b; St. John & McClelland, 1990; Tomasello, 1992; Elman et al., 1996.)

Psycholinguistic Evidence

The above proposal is based on the notion that children start with rote behavior, progress through a period in which small-scale systematic behaviors emerge in many very limited arenas, and finally come to a period in which the numerous small-scale systematic behaviors draw together into a small number of large-scale systematic behaviors. There is considerable evidence for this notion.

Child language specialists have noted for years that the earliest multiword utterances that children produce are formulaic. Many of the formulas appear to be "frozen phrases", or unanalyzed sequences that children can treat as individual lexical items (Peters, 1983; Barton & Tomasello, 1994). Other formulas, frequently referred to as "pivot grammars", consist of word combinations in which one element is fixed and the other "open"; examples include *more* + X, as in *more juice, more banana*, etc., and X + *allgone*, as in *juice allgone, milk allgone*, etc. (Braine, 1963, 1976; Bloom, 1973; Brown, 1973; Horgan, 1978). Pine, Lieven, & Rowland (forthcoming) have collected samples of 12 children's speech over six months starting when the children acquired vocabularies of approximately 100 words. They found that each child's five most common pivot formulas could account for between 65 and 90 percent of the child's subject + verb combinations.

Pivot grammars point toward the isolation of early systematic behavior. It is evident that syntactic sophistication likewise appears in isolated small-scale "puddles" of systematic behavior. Tomasello (1992) has shown that when children first begin to vary the types of syntax that they use with a verb, the novel use does *not* extend to other verbs. For a considerable period, children learn syntactic variation on a verb-by-verb basis. Tomasello refers to this analysis as the "Verb-Island Hypothesis".

There appear to be discernable patterns to the ways that children's grammars coalesce. Studies by Bloom, Lifter, & Hafitz (1980) and Clark (1996) examined the semantics behind sets of verbs that shared morphology, i.e., the verbs that appeared with *-ed, -s*, or *-ing*.. Interestingly, they found that these sets of verbs did tend to share semantic features. Pine, Lieven, & Rowland (forthcoming) analyzed the overlap of verb sets with these same verb endings, and found that, by and large, the sets defined by common morphology were distinct from one another. That is, morphological generalization was limited to small groups of semantically similar verbs.

---

[2] For our purposes, "syntactic similarity" refers to similarity of constructions. In English this primarily refers to word order, in other languages case marking performs the same function. In a connectionist network that is sensitive to word order, the instantiation of syntactic similarity is in similarity of trajectories through activation space.

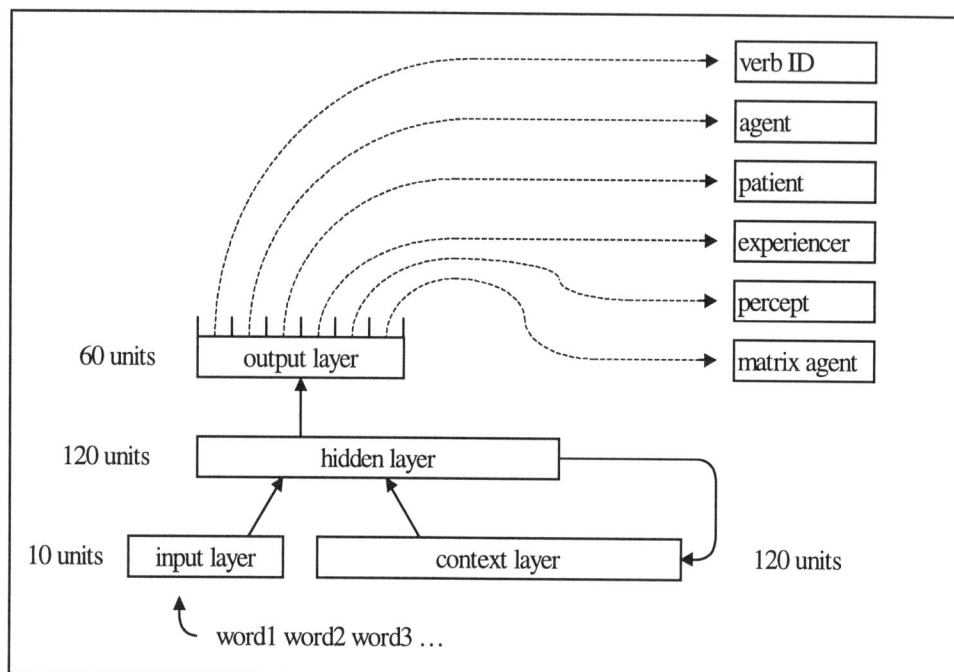

Figure 1: The Network

Finally, we note that studies by Maratsos, Kuczaj, Fox, & Chalkley (1979) and Maratsos, Fox, Becker, & Chalkley (1985), among others, show a surprising differential in children's rate of comprehension of the passive voice depending on which verbs are used. Passive sentences with verbs having agent and patient arguments (as in *Larry was hit by Marvin*) are understood at a higher rate than passive sentences with verbs having experiencer and percept verb arguments (as in *Larry was seen by Marvin*). The difference was dramatic at age 4 (85% comprehension for agent-patient verbs vs. 34% comprehension for experiencer-percept verbs), and did not disappear until the age of ten or eleven (Maratsos et al., 1985). This study demonstrates that even such a thoroughly abstract operation as passivization is learned not as an abstract operation defined in terms of grammatical relations, but as a semantically restricted operation which later "grows" into a greater abstraction. Upon investigation, Maratsos showed that there was a corresponding gap in the parental input to children; in child-directed speech there appear to be few, if any, passive sentences with experiencer-percept verbs.[3]

A Connectionist Simulation

In this section we present a connectionist simulation to test whether a network could build abstract relationships corresponding to "subjects" and "objects" given an English-like language with a variety of grammatical constructions.

This was done in such a way that there is no "innate" knowledge of language in the network. In particular, there are no architectural features that correspond to "syntactic elements", i.e., no grammatical relations, no features that facilitate word displacement, and so forth.

The motivation behind the network is the notion that merely the drive to map input words to output semantics is sufficient to induce the necessary internal abstractions to facilitate the mapping. This is an instantiation of the notion that the sole purpose of language is communication, and syntax is emergent. We were preceded in this approach by St. John & McClelland (1990). Our network and task is a simplified version of the one used by them. We differ from them in that our noun meanings are extremely simplified, while our syntactic constructions are more complex.

Our model uses a Simple Recurrent Network (Elman, 1990) implemented using the Stuttgart Neural Network Simulator (SNNS). The network is shown in Figure 1.

The input layer is ten units wide; each pattern represents one of 56 words or one of 2 punctuations. Each of these is represented by a unique pattern of 5 zeroes and 5 ones. The input consists of sentences drawn from the vocabulary of 56 words and two punctuations. Of these 56 words, 25 are verbs, 25 are nouns, and remaining 6 are a variety of function words. All of the nouns are proper names. Of the verbs, five are unergative (intransitive, with agents as the sole arguments, e.g., *run, sing*), five are unaccusative (intransitive, with patient arguments, e.g., *fall, roll*), ten are "action" transitives (with agent & patient arguments, e.g., *hit, kick, tickle*), and five are "experiential" transitives (with experiencer & percept arguments, e.g., *see, like, remember*).

---

[3] There are, however, a number of percept-experiencer verbs in passive sentences, e.g., *I was surprised by...*, or *He was frightened by ....*

In addition there was a "matrix verb", *persuade,* which was used for embedded sentence structures. The five remaining words were *who, was, by, and,* and *self.* The two punctuations were "period" and "reset". The network was trained to hold the output values during a "period" and to reset the output values at a "reset".

The output layer is 60 units wide. These are divided into 6 fields that are 10 units wide. The first field is the verb identifier, the second through the fifth are the identifiers for the agent, the patient, the experiencer, and the percept. (Note that at most only two of these four fields should be asserted at a single time.) The sixth field is the "matrix agent" field, which will be explained below. The internal identifiers of the nouns are different from (and unrelated to) the external identifiers—the internal identifiers each have only two units asserted.

Using the back-propagation learning procedure (Rumelhart, Hinton, & Williams, 1986) the network was taught to assign the proper noun identifier(s) to the appropriate role(s) for any of a number of sentence structures. Thus for the sentence, *Sandy persuaded Kim to kiss Larry,* the matrix agent role is filled by *Sandy,* the agent role is filled by *Kim,* and the patient role is filled by *Larry.* In the sentence, *Who did Larry see,* the experiencer role is filled by *Larry* and the percept role is filled by *who.*

Training was conducted with 50 epochs, with 10,000 sentences in each epoch. The learning rate was 0.2, initial weights set within a range of ± 1.0. There was no momentum. The learning function was online back-propagation using the mean-square error criterion. The squashing function was the standard 0-1 logistic function.

Examples of the types of sentences in the training set are shown in examples 1-6. The numbers in parentheses indicate the percentage of the total training corpus represented by each type of sentence. Semantic roles (i.e., agent, patient, experiencer, percept, or matrix agent) present in each example sentence are indicated in parentheses after the example.

1. Simple declarative intransitives (18%)
   e.g., *Sandy jumped*   (agent)
   Sandy fell   (patient)
2. Simple declarative transitives (26%)
   e.g., *Sandy kissed Kim*   (agent & patient)
   Sandy saw Kim   (exper. & percept)
3. Simple declarative passives (6%)
   e.g. *Sandy was kissed.*   (patient)
4. Questions (20%)
   e.g., *Who did Sandy kiss?*   (agent & patient)
   Who kissed Sandy?   (agent & patient)
   Who did Sandy see?   (exper. & percept)
   Who saw Sandy?   (exper. & percept)
5. Control (equi-NP) sentences (25%)
   e.g., *Sandy persuaded Kim to run.*
                  (matrix agent & agent)
   Sandy persuaded Kim to fall.

   (matrix agent & patient)
   Sandy persuaded Kim to kiss Max.
                  (matrix agent, agent, & patient)
   Sandy persuaded Kim to see Max.
                  (matrix agent, exper. & percept)
6. Control (equi-NP) sentences with questions (5%)
   e.g., *Who did Sandy persuade to run/fall?*
                  (questioning embedded subject, whether agent or patient, of an intransitive verb)
   Who persuaded Sandy to run/fall?
                  (questioning matrix agent; note embedded intransitive verb)
   Who persuaded Sandy to kiss/see Max?
                  (questioning matrix agent; note embedded transitive verb)
   Who did Sandy persuade to kiss Max?
                  (questioning embedded agent)

The distribution of verb types (agent-only, patient-only, agent-patient, and experiencer-percept types) is intended to be as realistic as possible for such a small vocabulary. The distribution of constructions is intended to provide some syntactic richness and to lay the groundwork for our generalization tests. The richness of the syntax here is out of proportion to the vocabulary size. In future work we hope to reduce the imbalance between vocabulary and syntax.

The test There were two systematic gaps in the data presented to the network; both involved experiential verbs: passive sentences with experiential verbs, e.g., *Sandy was seen by Max,* and questioning embedded subjects in transitive clauses with experiential verbs, e.g., *Who did Sandy persuade to see Max?*

Neither of these sentence types occurred with experiencer-percept verbs; all of the training involving these constructions used only agent-patient verbs. The test involved probing these gaps.

The network was not expected to generalize over these two systematic gaps in the same way. The questioning-of-embedded-subject-sentences gap is part of an interlocking group of constructions which "conspire" to compensate for the gap. The "members of the conspiracy" are the transitive sentences (group 2 above), the questions (group 4), and the control sentences (group 5). These sentences are related to each other, and they should cause the network to treat the agents of action verbs and the experiencers of experiential verbs similarly in the context of embedded clauses. The passive gap has no such compensating group of constructions. Only the transitive sentences (group 2) provides support for the passive generalization; as we shall see, these were insufficient to bridge the gap.

The sentences in groups 5 & 6 involve the "subjecthood property" of control structures (or equi-NP deletion). Part of the point of including these was to show that a network

Table 1: Sentence comprehension using Euclidean distance decisions

Sentence description	Percent correct	
Simple active clauses, action verbs	97.6%	
Simple active clauses, experiential verbs	97.6%	
Simple passive clauses, action verbs	91.8%	
Simple passive clauses, experiential verbs	6.2%	⇦ Failure to generalize
Control (equi-NP) structures	83.6%	
Questioning embedded subjects, action verbs	91.4%	
Questioning embedded subjects, experiential verbs	67.4%	⇦ Successful generalization

can learn to associate subjecthood properties with the appropriate nominal.

The Results

In Table 1 we show the result of testing a variety of constructions, some forms of which were trained, and two were not. Five hundred sentences of each listed type were tested. The results were computed using Euclidean distance decisions—each field in the output vector was compared with all possible field values (including the all-zeroes vector), and the fields assigned the nearest possible correct value. For a sentence to be "correct" all of the output values had to be correct. The two salient lines are for simple passive clauses with experiential verbs, which had a 6.2% success rate, and questioning embedded subjects with experiential verbs, which had a 67.4% success rate. (These lines are italicized in the table below.)

The relationship between these rates is as expected. The near complete failure of generalization for simple passive clauses with experiential verbs shows that the nonappearance of experiential verbs in the passive voice in the training set causes the network to learn the passive voice as a semantically narrow alternation. This is similar to an undergeneralization found by Maratsos et al. (1979; 1985), in which 4- and 5-year old children were shown to not comprehend passive sentences containing experiencer-percept verbs at an age when they could readily understand passive sentences containing agent-patient verbs. This gap has been shown by Maratsos et al. (1985) to be one that actually exists in parental input to children.

On the other hand, the questioning of embedded subjects with experiential verbs, which likewise did not appear in the training set, showed much greater generalization, in all likelihood because there is a "conspiracy of syntactic constructions" surrounding this gap. As a result we are seeing a level of abstraction, with the network able to "define", in some sense, the gap in terms of the embedded subject rather than merely an embedded agent.

This second gap, that of the questioning of embedded subjects of experiential verbs, is an unnatural omission, unlike the passive voice gap with experiential verbs. There does not appear to be a corresponding gap in English. This omission was introduced in the training set precisely because it is unnatural, and the network went a considerable distance in overcoming its effects, thus demonstrating the emergence in the network of the subject, i.e., a syntactic constituent

abstracted away from semantics. This was demonstrated by the fact that the network had begun to define the gap found in the questioning of embedded agents of action verbs in more semantically abstract terms, thus allowing the network to correctly interpret two-thirds of the sentences involving the questioning of embedded experiencers.

Discussion and Conclusions

This simulation and the planned extensions of it are intended to demonstrate that the most abstract syntactic aspects of language are learnable. There are two broad areas in which this is explored: control of "subjecthood" properties and demonstration of abstraction across semantic roles.

In the area of control of properties, this simulation demonstrated that the network was capable of learning equi-NP deletion (also known as "control constructions"). This is shown in the ability of the network to correctly process sentences such as *Sandy persuaded Kim to run* (these are shown in groups 5 & 6, in section 4 above). As was seen above, the network was able to correctly understand these sentences at a rate of 84%.

The network's ability to abstract from semantics was shown in the ability of the network to partially bridge the artificial gap in the training set, that of the questioned embedded subject of experiential verbs. The network was able to define the position in that syntactic construction in terms of a semantically-abstract entity, that is, a subject rather than an agent.

We are currently examining the representation of "subject" in the network's hidden layer. In the examination we are seeing the way that the network has developed a partially-abstract representation of the subject. We are also seeing the limitations of abstraction; the network's representation of the subject of a given sentence is also partially specified in semantically loaded units. And, as we have seen in the Maratsos (1985) study, this appears to be appropriate to the way that humans learn language. (See also Goldberg, 1995, for a theoretical analysis that predicts this semantically-limited scope to certain syntactic constructions.)

We believe that we have shown that syntax as a separate entity from semantic processing is an unnecessary assumption. Rather, what we see in our network is that "syntax", in the usual understanding of that term, is part and parcel of the *processing* required to map from a sequence of input words to a set of semantic roles.

Acknowledgments

We thank Jeff Elman, Adele Goldberg, and Ezra Van Everbroeck for their comments and suggestions in the course of this study.

References

Bloom, Lois. (1973). *One Word at a Time: The Use of Single Word Utterances Before Syntax*. The Hague: Mouton.

Bloom, L., K. Lifter, & J. Hafitz. (1980). Semantics of verbs and the development of verb inflection in child language. *Language 56*, 386-412.

Brown, Roger. (1973). *A First Language/The Early Stages*. Cambridge MA: Harvard University Press.

Bybee, Joan. (1985). *Morphology: A Study of the Relation between Meaning and Form*. Amsterdam: John Benjamins.

Clark, Eve V. (1996). Early verbs, event-types, and inflections. In C. E. Johnson & J. H. V. Gilbert (eds.), *Children's Language*, Volume 9. Mahwah NJ: Lawrence Erlbaum Associates.

Dixon, Robert M. W. (1979). Ergativity. In *Language*, 55. 59-138.

Elman, Jeffrey L. (1990). Finding Structure in Time. In *Cognitive Science* 14: 179-211.

Elman, Jeffrey L., Elizabeth A. Bates, Mark H. Johnson, Annette Karmiloff-Smith, Domenico Parisi, & Kim Plunkett. (1996). *Rethinking Innateness: A Connectionist Perspective on Development*. Cambridge MA: MIT Press.

Goldberg, Adele E. (1995). *A Construction Grammar Approach to Argument Structure*. Chicago: University of Chicago Press.

Horgan, D. (1978). How to answer questions when you've got nothing to say. *Journal of Child Language, 5*. 159-165.

Johnson, David E., & Paul M. Postal. 1980. *Arc pair grammar*. Princeton NJ: Princeton University Press

Keenan, E. L. (1976). Towards a Universal Definition of "Subject." In Li, C. (ed.) *Subject and Topic*. New York: Academic Press.

Langacker, Ronald W. (1987). *Foundations of Cognitive Grammar*, Vol. 1, *Theoretical Prerequisites*. Stanford: Stanford University Press.

Langacker, Ronald W. (1991a). *Foundations of Cognitive Grammar*, Vol. 2, *Descriptive Application*. Stanford: Stanford University Press.

Langacker, Ronald W. (1991b). *Concept, Image, and Symbol: The Cognitive Basis of Grammar*. Berlin: Mouton de Gruyter.

Marantz, Alec P. (1984). *On the nature of grammatical relations*. Cambridge MA: MIT Press.

Maratsos, Michael, Stanley A. Kuczaj II, Dana E. C. Fox, & Mary Anne Chalkley. (1979). Some Empirical Studies in the Acquisition of Transformational Relations: Passives, Negatives, and the Past Tense. In W. A. Collins (ed.) *Children's Language and Communication: The Minnesota Symposia on Child Psychology*, volume 12. Hillsdale NJ: Lawrence Erlbaum Associates.

Maratsos, Michael, Dana E. C. Fox, Judith A. Becker, & Mary Anne Chalkley. (1985). Semantic restrictions on children's passives. *Cognition, 19*. 167-191.

Morris, William C. (1998). *Emergent Grammatical Relations: An Inductive Learning System*. Ph.D. dissertation, University of California, San Diego.

Perlmutter, David. 1982. Syntactic representation, syntactic levels, and the notion of subject, in P. Jacobson and G. Pullum (eds.) *The Nature of Syntactic Representation*, Dordrecht: Reidel.

Perlmutter, David, & Paul Postal. 1983. Toward a universal definition of the passive, in D. Perlmutter (ed.) *Studies in Relational Grammar I*. Chicago: The University of Chicago Press.

Pine, Julian M., Elena V. M. Lieven, & Caroline F. Rowland. (Forthcoming). Comparing different models of the development of the English verb category. MS.

Pinker, Steven. (1984). *Language Learnability and Language Development*. Cambridge MA: Harvard University Press.

Pinker, Steven. (1989). *Learnability and Cognition: The Acquisition of Argument Structure*. Cambridge MA: MIT Press.

Rumelhart, David E., Geoffrey E. Hinton, & Ronald J. Williams. (1986). Learning Internal Representations by Error Propagation. In D. E. Rumelhart & J. L. McClelland. (eds.) *Parallel Distributed Processing: Explorations in the Microstructure of Cognition*. Volume 1: *Foundations*. Cambridge MA: The MIT Press.

Rumelhart, David E., & James L. McClelland. (1986). *Parallel Distributed Processing: Explorations in the Microstructure of Cognition*. Volume 1: *Foundations*. Cambridge MA: The MIT Press.

Schlesinger, Itzchak M. 1982. *Steps to language: Toward a theory of language acquisition*. Hillsdale NJ: Lawrence Erlbaum Associates.

Schlesinger, Itzchak M. 1988. The Origin of Relational Categories. In Y. Levy, I. M. Schlesinger, and M. D. S. Braine (eds.) *Categories and Processes in Language Acquisition*. Hillsdale NJ: Lawrence Erlbaum Associates.

Slobin, Dan I. (1997). The Origins of Grammaticizable Notions: Beyond the Individual Mind. In D.I. Slobin (ed.) *The Crosslinguistic Study of Language Acquisition* Volume 5: *Expanding the Contexts*. Mahwah NJ: Lawrence Erlbaum Associates.

St. John, Mark F., & James F. McClelland. (1990). Learning and Applying Contextual Constraints in sentence Comprehension. *Artificial Intelligence, 46*. 217-257.

Tomasello, Michael. (1992). *First verbs: A case study of early grammatical development*. Cambridge UK: Cambridge University Press.

Age of Acquisition, Lexical Processing and Ageing: Changes Across the Lifespan

Catriona M. Morrison (Morrison@cardiff.ac.uk)
School of Psychology, Cardiff University,
P.O. Box 901, Cardiff, CF11 3YG, Wales, U.K.
Andrew W. Ellis, University of York (A.Ellis@psych.york.ac.uk)
Department of Psychology, University of York,
Heslington, York, YO1 5DD, England, U.K.

Abstract

An important determinant of picture and word naming speed is the age at which the words were learned, that is, their age of acquisition (AoA). Two possible interpretations of these effects are that they reflect differences between words in their cumulative frequency of use, or that they reflect differences in the amount of time early- and late-acquired words have spent in lexical memory. Both theories predict that differences between early- and late-acquired words will be smaller in older than younger adults. We report three experiments in which younger and older adults read words varying in AoA or frequency, or named objects varying in AoA. There was no effect of word frequency when AoA was controlled. In contrast, strong AoA effects which did not diminish with age were found. The implications of these results for theories of how AoA affects lexical processing are discussed.

Introduction

Recent evidence suggests that AoA, and not word frequency, is the most important determinant of lexical processing speed (e.g., Morrison, Ellis & Quinlan, 1992; Morrison & Ellis, 1995). High frequency words tend to be learned earlier in life than low frequency words, so frequency and word learning age correlate highly (typically, around $r = .6$). The consequence of this natural correlation is that word sets matched for frequency are likely to be confounded on AoA. We argue this confounding of frequency with AoA has resulted in an overestimation of the role of frequency in determining word naming speed. Effects of word frequency in object naming have been claimed by a number of authors who have failed to control for differences in AoA (e.g., Oldfield & Wingfield, 1965; Jescheniak & Levelt, 1994), yet when the two variables are controlled statistically the effect of AoA appears robust while the independent effect of frequency fails to achieve significance in most studies (e.g., Morrison et al., 1992; Vitkovitch & Tyrell, 1995).

Cumulative Frequency and Residence Time Accounts of AoA Effects

How could the age at which a word is learned come to affect the speed with which it can be produced in response to a written word or a picture? Gilhooly and Watson (1981) proposed an explanation based on Morton's (1979) logogen model of word recognition and production. According to that model, the spoken forms of words that are produced in word and picture naming tasks are stored in, and retrieved from, the speech output logogen system. They suggested that the thresholds of individual logogens might be determined by AoA, with early-learned words having lower thresholds, and hence being easier to access, than later-learned words. Brown and Watson (1987) and Morrison and Ellis (1995) have proposed somewhat different accounts which nevertheless share Gilhooly and Watson's belief that AoA effects lie in the speed with which spoken word-forms can be accessed.

These proposals all share in common the idea that word learning age is the factor underlying the AoA effect: the accessibility of a word is determined at the time it is acquired and remains more or less unchanged thereafter. Previous studies have tentatively suggested AoA effects are not reducible to cumulative frequency (Carroll and White, 1973) or residence time (Gilhooly, 1984) and have led to the conclusion that whatever determined the accessibility of words in the mental lexicon is more or less fixed at the time the word is learned.

The present experiments take a different approach to evaluating the rival accounts of AoA effects. Imagine two words, one of which is acquired early in childhood at the age of 2 years while the other is acquired later at the age of 10. By the time a person is 20 years old the early-acquired word will have been resident in memory for 18 years while the late-acquired word will have been resident for 10 years. By the time that person has reached the age of 70, the early-acquired word will have been resident in memory for 68 years while the late-acquired word will have been resident for 60 years. The absolute difference is still 8 years, but in proportional terms the difference is residence time between the two words is greater for the 20-year-old that for the 70-year-old. Hence, if AoA effects are due to differences in residence time, then differences between early- and late- acquired words should gradually diminish as a person grows older. Cumulative frequency is just residence time multiplied by the number of times a word is encountered or used each year. If we make the simplifying assumption that the two words are matched in terms of the frequency, then the same prediction holds for the cumulative frequency hypothesis of AoA as for the residence time hypothesis: differences in cumulative frequency which are substantial when a person is young will become less significant as the person grows older. That is, the cumulative frequency hypothesis, like the residence time hypothesis, predicts that AoA effects will diminish with age. In contrast, the theory that word learning age *per se* predicts that AoA effects will be as large in old people as in young people. The present Experiments 1 employs the word naming task to discover whether or not the effect of AoA varies in younger and older participants. Experiment 2 looks for effects of frequency in the two groups. Experiment 3 examines the effect of AoA on object naming speed in groups of participants of different ages.

Experiments 1 and 2

In Experiments 1 and 2, we compared word naming performance in a group of young adult participants and a group of older adults. The word sets used either varied on AoA with word frequency controlled (Experiment 1) or varied on word frequency with AoA controlled (Experiment 2).

Method

Participants. The young adult group comprised 12 undergraduates at the University of York, with a mean age of 20.2 years (range 18-25). They were paid £2 or given a course credit for their participation. The 12 members of the older adult group had a mean age of 44.1 years (range 38-55). They were mature students from the Psychology Department, attendants at an Open University Summer School or Psychology teaching staff.

Stimuli. The stimuli were word sets previously used in a word naming study by Morrison and Ellis (1995). The word sets for Experiment 1 consisted of 24 early and 24 late acquired words matched for frequency and length. Another two sets of 24 words formed the stimuli for Experiment 2. These consisted of high and low frequency words matched for AoA and length.

Procedure. The experiments were conducted using a Macintosh computer and the stimuli were presented via a Hypercard program. A fixation dot appeared in the centre of the screen for 500 milliseconds before each word was presented. Stimuli were positioned such that the initial letter of each word appeared where the fixation dot had been. There was an interstimulus interval of 1000 milliseconds before the next fixation dot appeared. When the word appeared, a square wave signal was sent from the computer to a tape recorder. Naming responses were recorded on tape via a high sensitivity microphone. Reaction times were measured from the recording of the speaker's utterance using the SoundEdit program (see Morrison & Ellis, 1995, for details).

Participants were told that the experiment was aimed at measuring the speed at which people could name words and they were asked to name the words as quickly and accurately as possible. They were instructed to say only the target word, and were warned that mispronunciations or verbal hesitations would invalidate their response. The experiment began with 20 practice items.

Results

All scores representing incorrect responses or verbal hesitations were removed from the analyses. This amounted to 34 responses in total (an error rate of 1.5%).

Experiment 1. Means were calculated for the early and late AoA words and all responses falling more than two and a half standard deviations from the mean were removed. Means were recalculated by subjects and by items and the data were analysed. The by-items data are shown in Table 1.

Two-way analyses of variance were carried out with participant age (young/old) as the between-subjects factor and AoA of the words (early/late) as the within-subjects factor. There was no significant effect of participant age either by subjects, $F_1(1,22) = .02$, $MSE = 244.8$, $p = .90$, or by items, $F_2(1,46) = .42$, $MSE = 467.3$, $p = .50$. The effect of AoA was highly significant both by subjects, $F_1(1,22) = 35.07$, MSE = 20232.7, $p < .0001$, and by items, $F_2(1,46) = 31.3$, $MSE = 41669.6$, $p < .0001$. There was no indication of a significant interaction between participant age and AoA either by subjects, $F_1(1,22) = 1.25$, $MSE = 718.58$, p = .28, or by items, $F_2(1,46) = .35$, $MSE = 465.3$, p = .5687. That is, the AoA effect was of similar magnitude for the younger and older participants.

Experiment 2. Means were calculated for the high and low frequency words and all responses falling more than two and a half standard deviations from the mean were removed. Means were recalculated by subjects and by items and the data were analysed. The by-items data are shown in Table 1.

Two-way analyses of variance were carried out with participant age (young/old) as the between-subjects factor and word frequency (high/low) as the within-subjects factor. Again there was no significant effect of participant age either by subjects, $F_1(1,22) = .003$, $MSE = 29.6$, $p = .96$, or by items, $F_2(1,46) = .05$, $MSE = 29.03$, $p = .82$. The effect of frequency was also nonsignificant in both the by-subjects analysis, $F_1(1,22) = 1.58$, $MSE = 457.6$, $p = .22$, and the by-items analysis, $F_2(1,46) = .88$, $MSE = 634.02$, $p = .35$. The interaction between age and frequency approached significance in the by-subjects analysis, $F_1(1,22) = 3.63$, $MSE = 1049$, $p = .07$, with the young adult group tending to show more of an effect of frequency than the older group, but the interaction was far from being significant in the by-items analysis, $F_2(1,46) = 1.72$, $MSE = 634.02$, $p = .20$.

Discussion

Experiment 1 found a clear effect of AoA on word naming speed which was as large in older as in younger subjects. This pattern is contrary to the prediction of cumulative frequency hypothesis that there will be a reduction in the AoA effect in the older group compared with the younger group. Experiment 2 also replicates Morrison and Ellis (1995) in finding no effect of frequency on word naming once AoA is controlled.

The older participants responded just as quickly as the younger participants in naming the words. The average age of our older participants was over twice that of our younger participants, so the experiment constituted a fair test of the cumulative frequency hypothesis of AoA effects. However, the older participants were only aged between 38 and 55 years so may not have reached the age at which cognitive slowing becomes apparent (although we note that previous studies of the effects of age on word naming failed to find significant slowing with age despite using a wider range of ages than were employed here [Cerella & Fozard, 1984; Waugh & Barr, 1980]).

Experiment 3

Like Experiment 1, Experiment 3 is concerned with whether AoA effects change across the adult life span. It differs from Experiment 1 in three important respects: first, it involves the naming of pictures of objects rather than reading words aloud; second, the participants cover a wider range of ages than those in Experiment 1; and third, the division of words into early and late-acquired is based on normative data on children's naming rather than on adult estimates of AoA.

Because the present investigation is primarily concerned with evaluating alternative theories of AoA effects, Experiment 3 compared the naming of objects with early- and late-acquired names that were matched on frequency and other object and word properties. To the best of our knowledge there have been no previous factorial investigations of differences in object naming latency between early- and late-acquired words.

In Experiment 3, the objects were chosen to be ones which will have been equally commonplace in the childhood experience of people born in the early decades of this century and those born in the 1970s. AoA was determined using normative data from children, taken from Morrison et al. (1997), rather than using adult ratings of word learning age, as all previous studies have done. The young adults were again students. Older participants were drawn from the North East Age Research (NEAR) panel. NEAR is a longitudinal study of cognitive processing in several hundred older adults in Newcastle, England. Most panel members have been involved in research for at least 10 years so there is substantial data on various measures of their language and memory performance, allowing us to select the participants carefully.

Method

Participants. There were three groups of participants. The young adults were 17 psychology students at the University of York who varied in age between 18 and 32 years, with a mean age of 20 years 7 months. They were given a course credit for their participation. The older participants were drawn from the NEAR subject pool. There were 32 participants in each of two age groups - 60-69 year olds and 80+ year olds. They were selected on the basis that they scored highly on four measures of language ability - Mill Hill tests of synonym judgement and word definition (Raven, 1965), and the Alice Heim (AH) tests of general and spatial reasoning (Heim, 1970). All the NEAR panellists reported that they were in good health and had normal or corrected-to-normal vision. They were paid £4 for participation.

Stimuli. The experimental stimuli were 50 black-and-white drawings of objects, taken from the picture set used by Morrison et al. (1997). The AoA scores upon which picture selection was based were the objective norms data reported

by Morrison et al., who obtained objective measures of AoA for the names of 220 pictured objects from children aged from 2:6 up to 10:11. Twenty five pictures were selected that were known to children below the age of 26 months; these were the early-acquired items. The late-acquired items had AoA scores of 50.5 months, or more. The two sets of pictures were matched pairwise on rated visual complexity, name agreement (the degree to which speakers give the target name in response to the picture), Cobuild combined written and spoken frequency (Centre for Lexical Information, 1993), rated imageability and phoneme length.

Procedure. The stimuli were presented on a Macintosh computer, using SuperLab software. Participants wore a set of headphones with a high sensitivity microphone attached. The microphone was linked to a voice key that detected verbal responses and relayed reaction times to the computer. A 350 ms fixation dot in the centre of the screen cued the participant for the appearance of the stimulus which immediately followed the dot. The stimulus remained on screen until the participant made a verbal response or for 4 seconds, whichever was shorter. There was then an interstimulus interval of 2500 ms before the presentation of the next fixation dot.

Participants were told that a picture would appear on the screen and that they had to name the pictured object as quickly and as accurately as possible using a single-word label. They were instructed that the experiment was a test of their reaction time and were encouraged to respond as rapidly as possible, but with an emphasis on accuracy as well as speed. The experiment began with 30 practice items.

Results

Mean naming latencies were calculated by subjects and by items for each of the three age groups. Data from three of the participants in the oldest group were omitted from the analyses because they failed to name, or misnamed, more than 25% of the pictures. Seven items had an error rate of more than 25% across all three groups and these were also removed from the analyses (*beetle, boot, camera, cannon, cowboy, glasses, rocket*) along with their corresponding paired items (*duck, bow, cake, frog, butterfly, chain, rabbit*). This left 18 pairs of early- and late-acquired items for analysis. Thus, the analyses reported here are based on responses to 18 early and 18 late acquired items from 17 young adults, 32 60-69 year olds, and 29 80+ year olds.

The error rates for the young adults (7.9%) and the 60-69 year olds (7.4%) were comparable, though the error rate for the oldest participants was somewhat higher (11.9%). Some of this difference can be accounted for by the fact that older participants tended to make more elaborations (e.g., saying 'crescent moon' for *moon*), and that they used alternative names rarely or never used by younger participants (e.g., 'fiddle' for *violin*; 'keg' for *barrel*). However, older participants also made recognition errors never made by younger participants, (e.g., naming *cake* as 'cheese'), and occasionally gave semantic alternatives, (e.g., naming *violin* as 'banjo'), which younger participants rarely do.

The results of Experiment 3 are illustrated in Figure 1. Analyses of variance were carried out with AoA (early/late)

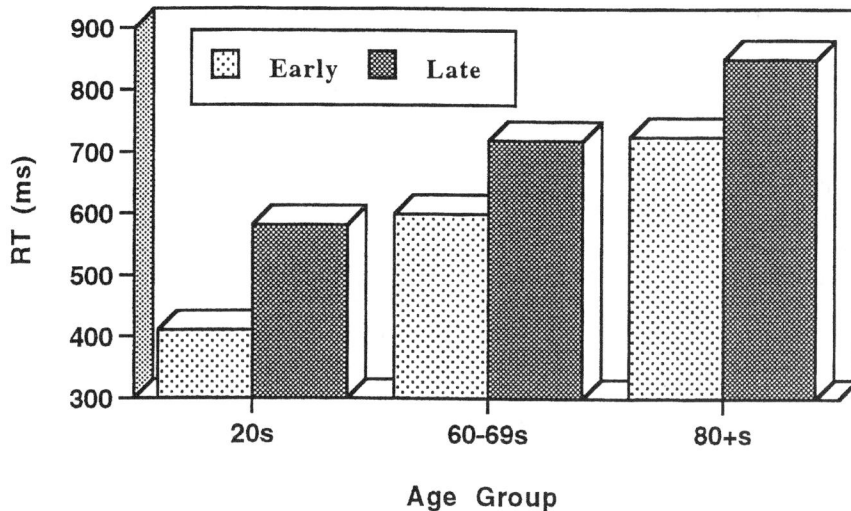

Figure 1. Results of Experiment 3. Mean picture naming RTs for early- and late-acquired items for each age-group.

as a within-subjects factor and group (young adults/60-69s/80+s) as a between-subjects factor. These revealed significant main effects of AoA, $F_1(1, 75) = 103.21$, $MSE = 691093$, $p < .0001$; $F_2(1, 51) = 24.75$, $MSE = 1056879$, $p < .0001$, with naming speeds being much faster for pictures with early-acquired names than for pictures with late-acquired names, and of group, $F_1(1, 75) = 16.51$, $MSE = 917201$, $p < .0001$; $F_2(1, 51) = 17.04$, $MSE = 684570$, $p < .0001$, with naming speeds being progressively slower as age increased. The interaction between AoA and group did not approach significance, $F_1(1, 75) = 1.17$, $MSE = 7865$, $p = .30$; $F_2(1, 51) = .41$, $MSE = 17359$, $p = .70$.

Discussion

AoA exerted a significant effect on picture naming speed in all three age groups. The effect of AoA in the young adult group replicates previous findings on picture naming speed (e.g., Morrison et al., 1992; Vitkovitch & Tyrell, 1995). Naming speed increased with age, which is in line with the results of previous studies (Mitchell, 1989; Thomas, Fozard and Waugh, 1977). Importantly, there was no indication of an interaction between participant age and AoA: indeed the AoA effect was as great in adults over 80 years of age as it was in young adults.

General Discussion

Experiments 1 and 3 found strong effects of AoA on word naming speed and object naming speed respectively using sets of items that were matched on word frequency, length, etc. In Experiment 1 the differentiation of early- from late-acquired words was based on adult ratings of AoA whereas in Experiment 3 it was based on normative data from children and adults. These results replicate the previous reports of effects of AoA on word naming (e.g., Brown & Watson, 1987; Morrison & Ellis, 1995) and object naming (Morrison et al., 1992; Vitkovitch & Tyrell, 1995).

In Experiment 2, there was no effect of word frequency on word naming speed. These results are compatible with the

view that at least a substantial proportion of the so-called 'frequency' effect in word naming is, in fact, due to differences in AoA between high and low frequency words.

The central focus of the present paper is whether AoA effects change across the adult life span. It is clear from our results that AoA effects remain invariant across age for both word naming (Experiment 1) and object naming (Experiment 3). A similar lack of interaction with age has been reported for frequency effects in picture naming (Thomas et al., 1977) and lexical decision (Allen, et al., 1993) in studies which failed to control for the natural correlation between frequency and AoA, and where a proportion of the reported frequency effects is probably due to differences in AoA between the high and low frequency word sets (Morrison & Ellis, 1995).

The lack of any interaction between age and AoA is incompatible with either the cumulative frequency or the residence time hypotheses of AoA effects, both of which predict that the impact of AoA will diminish with chronological age. The results are, however, in accord with the view that AoA effects reflect intrinsic properties of lexical representations which are fixed when those words are first learned and remain unchanged thereafter.

Theoretical models of lexical processing postulate distinct stages of processing in object and word naming, and researchers have used these frameworks in an attempt to trace the locus of frequency and AoA effects in picture and word recognition tasks. Currently, architectures used in the speech production literature are a popular explanatory tool (e.g., Levelt, 1989; Jescheniak & Levelt, 1994). External information maps on to semantic nodes which in turn activate lexical nodes called lemmas - abstract representations of word forms. Lemmas then activate lexemes - phonological nodes specifying a word's spoken form. The lexeme specification maps onto phonetic articulatory units and drives the process of articulation.

Roelofs (1992) extended this framework to account for both picture and word naming. In line with many accounts of word recognition (e.g., Seidenberg & McClelland, 1989),

447

he suggested that word naming by-passes the semantic level in one of two ways. The first is that orthographic representations (visual lexemes) activate lemmas which are involved in both the comprehension and production of words. The lemmas can then activate phonological lexemes and output processes without involving semantic representations. Another possibility is that there is a direct mapping between orthographic and phonological representations. Such mappings are widely proposed in models of word naming (e.g., Seidenberg & McClelland, 1989), and one reason for believing in their existence is to explain effects of the consistency or regularity of spelling-sound correspondences on word naming speed (e.g., Jared, McRae & Seidenberg, 1990). If abstract lemma representations were interposed between orthographic and phonological representations, then effects of sublexical consistency between spelling and sound should be lost.

Morrison et al. (1992) found no effect of AoA on semantic classification time for pictures but an effect on picture naming, and argued that AoA exerts its effect at or beyond the stage of lexical access. Morrison and Ellis (1995) found an effect of AoA on word naming in an immediate naming task when the response was produced as rapidly as possible following a word's appearance on the screen. There was no effect of AoA in a delayed naming task where the word was followed by an unpredictable delay and naming was prompted by the appearance of a visual cue to respond. Morrison and Ellis concluded that AoA did not affect post-lexical articulatory processes. If AoA does not affect semantic activation or output processes in object naming, then by a process of elimination, the effect of age of spoken acquisition on picture naming would seem to arise in the process of lexicalisation. That could be at the lemma stage, the lexeme stage, or both. But if word naming involves direct mappings between visual and phonological lexemes, and if the same locus is to be proposed for AoA effects in both object naming and word naming, then the best candidate would seem to be lexeme activation. Though the terminologies differ, that is roughly where it was placed in the theoretical accounts of Gilhooly and Watson (1981) and Brown and Watson (1987).

We suggest therefore, that the effect of the age at which a spoken word is learned on the speed with which it can be produced in object and word naming tasks have something to do with the organisation of the lexeme layer. And because AoA effects are fixed across the adult life span we suggest that they reflect differences between early- and late-acquired words which are fixed at the time when those words are learned. Few attempts have been made to simulate AoA effects using connectionist models. Morrison & Ellis (1995) suggested that one form of architecture which might provide a plausible account of how AoA effects could arise is provided by self-organising networks of the sort proposed by Kohonen (1990). Self-organising networks learn to discriminate between patterns by organising the output layer in such a way as to represent similar patterns close together and different patterns further apart. Morrison (1993) showed that patterns introduced early into the training cycle are spread across the whole of the output layer. Patterns introduced later have to be fitted in around them in such a way that fewer cells in the network are ever involved in representing them. Although only preliminary, such work indicates how AoA effects which remain invariant across the life span might begin to be understood.

In conclusion, our results indicate that AoA effects are just as strong in older participants as in younger participants for both word and picture naming. We take this as evidence that effects of AoA observed in young adults are a genuine reflection of the age at which words are learned (or the order in which they are learned). AoA effects are not reducible to cumulative frequency or residence time. We propose that AoA effects have their locus at the lexeme level and suggested that learning in a self-organising neural network architecture might provide a useful analogy for the development of the lexeme system.

References

Allen, P.A., Madden, D.J., Weber, T.A. & Groth, K.E. (1993). Influence of age and processing stage on visual word recognition. *Psychology and Aging*, **8**, 274-282.

Brown, G.D.A., & Watson, F.L. (1987). First in, first out: Word learning age and spoken word frequency as predictors of word familiarity and word naming latency. *Memory and Cognition,* **15**, 208-216.

Carroll, J.B. & White, M.N. (1973). Word frequency and age of acquisition as determiners of picture-naming latency. *Quarterly Journal of Experimental Psychology*, **25**, 85-95.

Center for Lexical Information (1993). *The Celex Lexical Database*. Nijmegen.

Cerella, J. & Fozard, J.L. (1984). Lexical access and age. *Developmental Psychology*, **20**, 235-243.

Cirrin, F.M. (1983). Lexical access in children and adults. *Developmental Psychology*, **19**, 452-460.

Gilhooly, K.J. (1984). Word age-of-acquisition and residence time in lexical memory as factors in word naming. *Current Psychological Research and Reviews*, **3**, 24-31.

Gilhooly, K.J., & Watson, F.L. (1981). Word age-of-acquisition effects: a review. *Current Psychological Research*, **1**, 269-286.

Heim, A. (1970). *The Alice Heim Four Group Test of General Intelligence*. Windsor: NFER Nelson.

Jared, D., McRae, K., & Seidenberg, M.S. (1990). The basis of consistency effects in word naming. *Journal of Memory and Language*, **29**, 687-715.

Jescheniak, J.D. & Levelt, W.J.M. (1994). Word frequency effects in speech production: Retrieval of syntactic information and of phonological form. *Journal of Experimental Psychology: Learning, Memory and Cognition*, **20**, 824-843.

Kohonen, T. (1990). The self-organising map. *Proceedings of the IEEE*, **78**, 1464-1480.

Levelt, W.J.M. (1989). *Speaking: from intention to articulation*. Cambridge, MA: MIT Press.

Mitchell, D.B. (1989). How many memory systems? Evidence from aging. *Journal of Experimental Psychology: Learning, Memory & Cognition*, **15**, 31-49.

Morrison, C.M. (1993). *The loci and roles of word age of acquisition and word frequency in lexical processing.* University of York: Unpublished D.Phil. thesis.

Morrison, C.M., Chappell, T.D. & Ellis, A.W. (1997). Age of acquisition norms for a large set of object names and their relation to adult estimates and other variables. *Quarterly Journal of Experimental Psychology,* **50A**, 528-559.

Morrison, C.M. Ellis, A.W. & Quinlan, P.T. (1992). Age of acquisition, not word frequency, affects object naming, not object recognition. *Memory & Cognition,* **20**, 705-714.

Morrison, C.M. & Ellis, A.W. (1995). The roles of word frequency and age of acquisition in word naming and lexical decision. *Journal of Experimental Psychology: Learning, Memory & Cognition,* **21**, 116-133.

Morton, J. (1979). Word recognition. In J. Morton & J.C. Marshall (Eds.), *Psycholinguistics Series, Vol. 2.* London: Elek & Cambridge, Mass: MIT Press.

Oldfield, R.C. & Wingfield, A. (1965). Response latencies in naming objects. *Quarterly Journal of Experimental Psychology,* **17**, 273-281.

Raven, J.C. (1965). *The Mill Hill Vocabulary Scale.* London: H.K. Lewis.

Roelofs, A. (1992). A spreading activation theory of lemma retrieval in speaking. *Cognition,* **42**, 107-142.

Salthouse, T.A. (1985). *A theory of cognitive aging.* Netherlands: Elsevier Science Publishers.

Seidenberg, M.S., & McClelland, J.L. (1989). A distributed, developmental model of word recognition and naming. *Psychological Review,* **96**, 523-568.

Thomas, J.C., Fozard, J.L. & Waugh, N.C. (1977). Age-related differences in naming latency. *American Journal of Psychology,* **90**, 499-509.

Vitkovitch, M. & Tyrell, L. (1995). Sources of disagreement in object naming. *Quarterly Journal of Experimental Psychology,* **48A**, 822-848.

Waugh, N.C. & Barr, R.A. (1980). Memory and mental tempo. In L.W. Poon, J.L. Fozard, L.S. Cermak, D. Arenberg & L.W. Thompson (eds.), *New directions in memory and aging.* Hillsdale, NJ: Erlbaum.

Waugh, N.C., Thomas, J.C. & Fozard, J.L. (1978). Retrieval time from different memory stores. *Journal of Gerontology,* **33**, 718-724.

449

Simulating the Effects of Relational Language in the Development of Spatial Mapping Abilities

Tom Mostek[1], Jeff Loewenstein[2], Ken Forbus[1], Dedre Gentner[2]
Northwestern University
1890 Maple Avenue
Evanston, IL, 60201, USA
{t-mostek, loewenstein, forbus, gentner} @nwu.edu

Abstract

Young children's performance on certain mapping tasks can be improved by introducing relational language (Gentner, 1998). We show that children's performance on a spatial mapping task can be modeled using the Structure-Mapping Engine (SME) to simulate the comparisons involved. To model the effects of relational language in our simulations, we vary the quantity and nature of the spatial relations and object descriptions represented. The results reproduce the trends observed in the developmental studies of Loewenstein & Gentner (1998; in preparation). The results of these simulations are consistent with the claim that gains in relational representation are a major contributor to the development of spatial mapping ability. We further suggest that relational language can promote relational representation.

Introduction

Spatial reasoning is one of the core abilities in human cognition. An important test of spatial reasoning is the *mapping task* (DeLoache, 1987, 1995; Huttenlocher, Newcombe, & Sandberg, 1994; Uttal, Schreiber, & DeLoache, 1995; Uttal, Gregg, Tan, Chamberlin, & Sines, submitted). In a mapping task, the goal is to find a correspondence between two different spatial situations. In DeLoache's classic task, a child is shown two rooms, similar in layout and furniture (though not necessarily in size). A toy is hidden in one room and the child must look for another toy in the corresponding place in the other room (e.g., DeLoache, 1995). It has been proposed (Gentner & Rattermann, 1991) that the same process of structural alignment that is used in analogy and similarity may play a role in spatial mapping tasks. That is, spatial mapping tasks can be viewed as a kind of analogy in which the spatial relationships of the situations involved provide the base and target descriptions for the structural alignment, and the correspondences computed in structural alignment provide the basis for inferring the correct answer.

This paper provides evidence for the role of structural alignment in spatial mapping tasks. We show how the pattern of developmental results found by Loewenstein and Gentner (1998, in preparation) can be modeled using SME (the Structure-Mapping Engine (Falkenhainer, Forbus, & Gentner, 1989; Forbus, Ferguson, & Gentner 1994)) a simulation of Gentner's (1983) structure-mapping theory. We start by describing Loewenstein and Gentner's spatial mapping task. Next we describe how we used SME to model the results.

Spatial Mapping Tasks

Mapping and symbolic reference is ubiquitous in adult daily life, but it develops only gradually in children. Studies by Blades and Cooke (1994), DeLoache (1995), Uttal (Uttal, Schreiber, & DeLoache, 1995), and others have shown that preschool children have great difficulty with the seemingly simple task of finding an object in the 'same place' as an object in an almost identical model, even though they can easily retrieve the original hidden object. Gentner and her colleagues have suggested that one contribution to the great gains children make in their performance on spatial mapping tasks is relational knowledge, and further, that acquiring relational language promotes this relational knowledge (Gentner & Loewenstein, in preparation; Gentner & Rattermann, 1991; Gentner, Rattermann, Kotovsky, & Markman, 1995; Kotovsky & Gentner, 1996; Loewenstein & Gentner, 1998).

Experiment 1. The first study (Loewenstein & Gentner, 1998) used the setup in the left of Figure 1, with neutral appearances for the cards. Three cards are placed *on*, *in*, and *under* the Hiding box. The instructions given in the baseline condition avoided language that used spatial relationships. During the orientation trial, the experimenter said "I'm putting the winner right here" while placing the card in its location at the Hiding box. The instructions in the language condition used spatial relationships during the orientation task to describe where the cards were being placed: The experimenter said, "I'm putting the winner [in/on/under] the box." For both conditions, no language was used in the finding task: The Experimenter gestured generally towards the Finding box, saying "Can you find the

---

[1] Computer Science Department
[2] Psychology Department

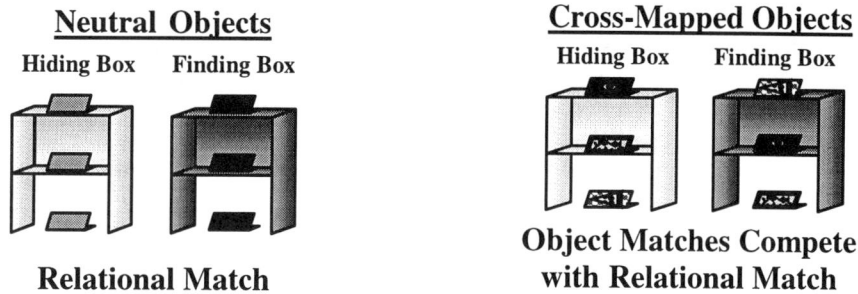

Figure 1 – Experiment Setup

winner here, in the very same place?" In the baseline condition, 44-month-olds found the sticker only 42% of the time, not significantly above chance performance of 33%. However, in the language condition, they performed far better, finding the winner 70% of the time (see Figure 2, left). The 49-month-olds performed fairly well in both the baseline (63%) and language (73%) conditions. It appears that hearing spatial relational terms led the younger children to form stronger representations of the spatial relational structure.

Experiment 2. Gentner and Loewenstein (in preparation) use a similar task as in Experiment 1, but the three cards associated with a box were all distinctive and unique, as shown in Figure 1. While there is an exact object match for each card, the cards that match in appearance have different spatial relationships with the box. For example, the card that is ON the Hiding box matches the card that is IN the Finding box. This is an example of a *cross-mapping task* (Gentner & Toupin, 1986; Gentner & Rattermann, 1991), in which object similarity is pitted against relational similarity. Such tasks are useful in testing for the availability and salience of the child's relational knowledge. The results bore out previous findings that cross-mapping tasks are difficult: 49-month-olds were at chance at finding the winner in both conditions. Even 62-month-olds were correct only 53% of the time in the baseline condition. However, in the language condition, when given the spatial relation during the hiding task, their performance improves to 73% correct (see Figure 2, right).

These results suggest the following conjectures:

- Age-related improvements are largely due to improved understanding of spatial relationships.
- Relational language highlights spatial relationships, supporting children's relational mapping abilities.

This pattern of results is consistent with other findings on the role of relational language in domain learning (e.g., Gentner & Rattermann, 1991; Gentner, Rattermann, Markman & Kotovsky, 1995; Kotovsky & Gentner, 1996), and lends evidence to the position that relational language fosters the development of relational thought (Gentner, 1998).

Modeling Spatial Mapping as Visual Comparisons

The spatial mapping task above involves encoding descriptions of the hiding box and the finding box and comparing these descriptions to predict, based on the location of the winner in the hiding box, where the winner will be in the finding box. Since we are modeling the comparison process via structure-mapping, we first briefly review structure-mapping theory and SME.

Review of Structure-Mapping

According to structure-mapping theory, the process of structural alignment takes as input two structured representations (*base* and *target*) and produces as output a set of

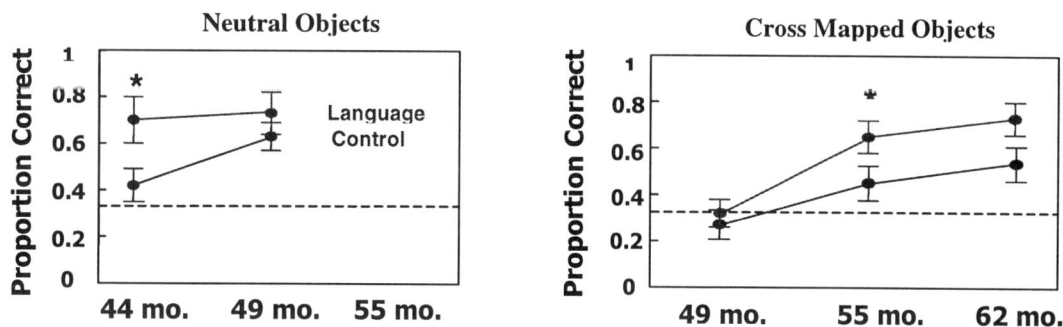

Figure 2 – Experiment Results

mappings. Each mapping consists of a set of *correspondences* that aligns items in the base with items in the target and a set of *candidate inferences*, which are surmises about the target made on the basis of the base representation plus the correspondences. The constraints on the correspondences include *structural consistency*, i.e., that each item in the base maps to at most one item in the target and vice-versa (the *1:1 constraint*), and that if a correspondence between two statements is included in a mapping, then so must correspondences between its arguments (the *parallel connectivity* constraint). Which mapping is chosen is governed by the *systematicity* constraint: Preference is given to mappings that match systems of relations in the base and target. Each of these constraints is motivated by the role analogy plays in cognitive processing. The 1:1 and parallel connectivity constraints ensure that the candidate inferences are well-defined. The systematicity constraint reflects a (tacit) preference for inferential power in analogical arguments.

The Structure-Mapping Engine (SME) (Falkenhainer *et al* 1989; Forbus *et al* 1994) is a cognitive simulation of analogical matching. Given base and target descriptions, SME finds globally consistent interpretations via a local-to-global match process. SME begins by proposing correspondences, called *match hypotheses,* in parallel between statements in the base and target. Then, SME filters out structurally inconsistent match hypotheses. Mutually consistent collections of match hypotheses are gathered into global mappings using a greedy merge algorithm. An evaluation procedure based on the systematicity principle is used to compute the *structural evaluation* for each match hypothesis and mapping. These numerical estimates are used both to guide the merge process and as one component in the evaluation of an analogy. It is important to note that SME can produce multiple mappings for a given pair of base and target descriptions, corresponding to different ways in which they might be aligned.

Using SME to Model the Effects of Relational Knowledge in Spatial Mapping

Our focus is on demonstrating that, if SME is used to model structural alignment, that changes in available relational knowledge can explain the pattern of results. Consequently, we do not model the encoding processes themselves, only their results. Our goal is to show that the assumed outcomes of encoding in the explanations for the experiments do in fact lead to the pattern of results found, given that SME is used to model the structural alignment involved.

We model the role of structural alignment in these tasks using SME by the following assumptions:

- The base is our construal of what the child might have encoded about the hiding box.
- The target is our construal of what the child might have encoded about the finding box.

- The child's response will be based on the mapping they use, specifically, the candidate inference for which card has the sticker behind it (and hence is the "winner").
- The use of relational language leads to increased relational content in the child's representations
- The likelihood that a child uses a particular mapping is a function of its structural evaluation.

This last assumption requires further explanation. Why not just take SME's top-rated mapping as the child's mapping, as was done in Markman & Gentner (1993) or Gentner, Rattermann, Markman, and Kotovsky (1995)? We believe that variability in children's performance is caused by variations in how they encode the situations. There are many sources of encoding variability, including domain knowledge and context. We model the potential variability in these processes by representing what the final result of such processes might be, and taking alternate mappings generated by SME as representative of different possible outcomes of an interleaved encoding/mapping process.

For each experiment, we generated a set of descriptions intended to be representative of the encodings children used at different ages and different conditions. We then ran SME on these descriptions to compute the mappings. The prediction made by a mapping is generated from the candidate inferences for a mapping: The location of the card in the Hiding box is represented by the statement (BEHIND <card> STICKER) in the base, with no such statement appearing in the target. The lack of corresponding statement in the target means that there will be a candidate inference computed from this statement. Consequently, whatever corresponds to the winning card in the hiding box within a mapping will be the predicted winning card in the Finding box. We randomly chose a location for the winner card, and this location had no effect on the results.

As noted above, SME can produce multiple mappings for a given base and target. We used these multiple mappings to estimate the relative likelihood of a particular card being suggested as the winner for each base and target as follows. To eliminate size effects, the score for each mapping is divided by an ideal score (i.e., the score obtained by mapping the target to itself). The relative likelihood of choosing a card is calculated by the sum of the scores of all the mappings that predict it. The relative likelihoods for the cards are scaled so that they sum to 1.

Since the specific values of these numbers depend on the particular choice of representations and processing parameters, they must be interpreted with care. Specifically, we do not consider the particular numerical values produced as the probability that a specific outcome will occur. For robustness, we only consider meaningful the ordinal relationships between these numbers. That is, if one number calculated is larger/smaller than another, then the corresponding frequency in the human data should be larger/smaller.

```
(defdescription exp1-target          (at box card-3)                   (open-object card-2)
  :documentation                                                       (folding-card card-2)
  "neutral object finding task"      ;; unique relations              (horizontally-oriented card-2)
  :entities                          (on box card-1)                   (green-colored card-2)
    (box card-1 card-2 card-3)       (in--cont-open box card-2)
  :expressions                       (under box card-3)                (open-object card-3)
  ((open-object box)                                                   (folding-card card-3)
   (rectangle-volume box)            ;; cards                          (horizontally-oriented card-3)
   (vertically-oriented box)         (open-object card-1)              (green-colored card-3)))
   (touches box ground)              (folding-card card-1)
   (at box card-1)                   (horizontally-oriented card-1)
   (at box card-2)                   (green-colored card-1)
```

Figure 3 – Sample Description

Simulation Experiment 1

To model the effects of relational knowledge in Experiment 1 we used three different base descriptions. All base descriptions used the spatial relation AT (e.g., (AT BOX CARD-1)), descriptions of the box and the three cards, and the location of the winner (i.e., (BEHIND CARD-1 STICKER)). The rest of the base content varied as follows:

- Level 1: No additional information
- Level 2: Includes the spatial relationship between the box and the winner (e.g., (IN CARD-2 BOX)).
- Level 3: Included all spatial relations involving the cards and the boxes.

In other words, with each increase in level is an increase in the number of first-order spatial relations.

The target description included descriptions of the box, the cards, and all of the spatial relationships involving them, but did not mention the location of the sticker. (It is sufficient to vary the base representations without varying the target representations because the failure to include a proposition in either base or target will prevent a correspondence from being formed.)

We then ran SME on each of the three base descriptions and the target. For each pair, SME generates multiple legitimate mappings (e.g. the first mapping has the top card corresponding to the top card, and the second mapping has the top card corresponding to the middle card). We used the procedure outlined above to calculate the accuracy that each level of description predicts.

Results

As expected, increasing the number of unique spatial relations resulted in stronger mappings between the cards 'in the same place' in each representation. This in turn resulted in a greater calculated probability of the correct card being selected as 'the winner'. With no specific spatial relations represented (Level 1), the model gave each possible mapping the same score. (yielding 33% selection probability for the correct relational choice). With one spatial relation representing the location of the winner (Level 2), the model gave the correct choice the greatest score (40%). When all three first-order spatial relations were represented (Level 3), the model's selection probability for the correct choice increased (to 48%). Thus, the model predicts (quite reasonably) that children who did not represent any spatial relations would perform at chance (33%), and that more fully representing the spatial relations in the two scenes would increase performance.

Discussion

In Lowenstein and Gentner's study, the 44 month olds performed at chance in the baseline condition and at 70% in the language condition. The simulation results suggest that in the baseline condition, the children were not encoding any of the spatial relations that uniquely identify the card. The language condition explicitly draws attention to one of the identifying spatial relations. The simulation results suggest that encoding this information is sufficient to raise the performance as seen. In the second simulation experiment we consider the more difficult cross-mapped model task and investigate the effect of encoding higher-order relations.

| | Level 1 | | | Level 2 | | | Level 3 | | |
	Score	% from best	Selection probability	Score	% from best	Selection probability	Score	% from best	Selection probability
'the winner'	0.224	0	33.3%	0.241	N/A	40%	0.266	N/A	48%
Other card 1	0.224	0	33.3%	0.232	3.7	30%	0.252	5.7	26%
Other card 2	0.224	0	33.3%	0.232	3.7	30%	0.252	5.7	26%

Table 1 – Results from Simulation Experiment 1
Note – max score in these cases is 0.302

Simulation Experiment 2

Recall that Experiment 2 pitted relational similarity against object similarity. The rich object matches between the individual cards suggest a different set of correspondences than the parallel spatial relationships that define the correct responses. We suggested that the effects of age and relational language in increasing accuracy both stemmed from greater encoding of relations (thereby preventing strong attribute overlap from overwhelming relational mappings). Our purpose in this simulation was to see if varying the amount of relational structure does indeed lead to these effects.

The method used is similar to that used in modeling Experiment 1 – the target representation is fully specified, and the base representation is varied with respect to its relational specificity. The cards are described at the same level of detail in both target and base descriptions. In contrast to Experiment 1 (for which the object descriptions were all identical and could be represented with only four attributes), the rich cross-mapped objects used in Experiment 2 required nine attributes and two relations. They were represented as uniquely colored and as having a different picture on the front of each object. The base descriptions include different levels of information, as follows:

3. Basic relations – The relations ON, IN and UNDER are included. (the Level 3 case in Experiment 1)
4. Extra binary relations – In addition to ON, IN and UNDER, several binary relations are added, specifically (ABOVE CARD-1 CARD-2) and (ABOVE CARD-2 CARD-3)
5. Extra ternary relation – In addition to ON, IN and UNDER, a relation that ties all three cards together is added: (IN-A-COLUMN CARD-1 CARD-2 CARD-3)
6. One higher-order relation – In addition to ON, IN and UNDER, some of the inferential structure linking relations is added: (IMPLIES (AND (ON CARD-1 BOX) (IN CARD-2 BOX)) (ABOVE CARD-1 CARD-2))
7. Two higher-order relations – Like #4, but with two higher-order relations: (IMPLIES (AND (IN CARD-2 BOX) (UNDER CARD-3 BOX)) (ABOVE CARD-2 CARD-3))

We then ran SME using each of these bases with the same target, as we did in the first model, and generated relative predictions using the same method.

Results

The results range from a below chance selection probability of 27% (always preferring the 'same card' attribute mapping over the 'same place' card) to a strong performance of 60%, as shown in Table 2. The model's selection probability for the correct choice increases with increased relational knowledge. This is consistent with the possibility that children's improved performance between 49 and 62 months of age, and between the baseline and language condition, is due to better encoding of the spatial relations relevant for the task.

In order for the model to consistently make the correct relational matches in the face of richly represented cross-mapped objects, higher-order relations needed to be included in the representations. Even with all the basic first-order relations represented (Level 3) -- the most successful model in Simulation Experiment 1 -- the model still chose the object choice (47% selection probability for the object match). Adding a ternary relation was also not sufficient to enable the model to make the correct relational choice. One higher-order relation was just enough to induce a shift to the relational choice as the mapping with the best score. However, the scores of the relational mapping and the object mapping were extremely close (relational choice: 50%; object match: 47%). Representing two higher-order relations yields a clearly dominant selection of the correct relational choice (60%). Although the precise threshold at which relational overlap will overcome attribute overlap depends on specific modeling parameters (such as the amount of each type of information available), what is clear here is that the addition of higher-order relations contributes to improved performance. Thus, the improved performance of children in Experiment 2 may be due to learning of higher-order relations.

General Discussion

Relational language helped preschool children perform mapping tasks with neutral and cross-mapped objects. However, because the effects came about at different ages, it was possible that language was providing a different kind of

	Level 3			Level 4			Level 5			Level 6			Level 7		
	Score	% from best	Prob.	Score	% from best	Prob.	Score	% from best	Prob.	Score	% from best	Prob.	Score	% from best	Prob.
Winner	4.50	16.5	27%	4.55	16.4	27%	4.52	16.3	27%	7.77	N/A	50%	8.36	N/A	60%
Same	5.39	N/A	47%	5.44	N/A	47%	5.40	N/A	47%	7.67	1.3	47%	7.67	8.2	35%
Other	4.48	16.7	26%	4.53	16.6	26%	4.49	16.7	26%	6.32	18.8	3%	6.85	18.0	5%

Table 2 – Results from Simulation Experiment 2
Note – max score in these cases is 9.67

support for 44- and 62-month-olds. The simulation experiments provide evidence for this suggestion by showing a link between level of relational understanding and mapping task performance. The simulation experiments also provide support for the claim that spatial mapping tasks involve structural alignment. Simulation experiment 1, with no competing object mappings, showed that first order relations are sufficient for success on the relational mapping task. Simulation Experiment 2, with rich cross-mapped objects, suggested that higher-order relations are needed to perform relational mappings under these more challenging circumstances. Further experimentation providing higher-order relations could provide evidence for this hypothesis.

During spatial mapping tasks, as in many everyday tasks, there is a vast array of perceptual information available to the child. Our evidence suggests that knowing what to encode may be in part a learned skill. Learning relational language may contribute to this ability.

Acknowledgements

This research was supported by the Learning and Intelligent System Program of the National Science Foundation and by the Cognitive Science Program of the Office of Naval Research.

References

Blades, M., & Cooke, Z. (1994). Young children's ability to understand a model as a spatial representation. *Journal of Genetic Psychology, 155,* 201-218.

DeLoache, J. S. (1987). Rapid change in the symbolic functioning of very young children. *Science, 238,* 1556-1557.

DeLoache, J. S. (1989). Young children's understanding of the correspondence between a scale model and a larger space. *Cognitive Development, 4,* 121-139.

DeLoache, J. S. (1995). Early symbol understanding and use. In D. Medin (Ed.) *The psychology of learning and motivation* (Vol. 32). New York: Academic Press.

Falkenhainer, B., Forbus, K., & Gentner, D. (1989). The Structure Mapping Engine: Algorithm and Examples. *Artificial Intelligence, 41.* (pp. 1-63).

Forbus, K. D., Ferguson, R. W., & Gentner, D. (1994). Incremental structure mapping. In A. Ram & K. Eiselt (Eds.), *Proceedings of the Sixteenth Annual Conference of the Cognitive Science Society* (pp. 313-318). Atlanta, GA:Lawrence Erlbaum Associates.

Gentner, D. (1983). Structure-mapping: A theoretical framework for analogy. *Cognitive Science. Vol 7(2),* (pp. 155-170).

Gentner, D. (1998). Relational language and relational thought. Plenary paper presented at the Meeting of the Piaget Society, Chicago, IL.

Gentner, D., & Loewenstein, J. (in preparation). Relational language contributes to the development of spatial mapping.

Gentner, D., & Rattermann, M. J. (1991). Language and the career of similarity. In S. A. Gelman & J. P. Byrnes (Eds.), *Perspectives on Language & Thought: Interrelations in Development* (pp. 255-277). Cambridge: Cambridge University Press.

Gentner, D., Rattermann, M. J., Markman, A. B., & Kotovsky, L. (1995). Two forces in the development of relational similarity. In T. J. Simon & G. S. Halford (Eds.), *Developing cognitive competence: New approaches to process modeling* (pp.263-313). Hillsdale, NJ: Erlbaum.

Gentner, D., & Toupin, C. (1986). Systematicity and surface similarity in the development of analogy. *Cognitive Science, 10,* 277-300.

Halford, G. S. (1992). Analogical reasoning and conceptual complexity in cognitive development. *Human Development. Vol 35(4),* (pp. 193-217).

Huttenlocher, J., Newcombe, N., & Sandberg, E. H. (1994). The coding of spatial location in young children. *Cognitive Psychology, 27,* 115-147.

Kotovsky, L., & Gentner, D. (1996). Comparison and categorization in thedevelopment of relational similarity. *Child Development, 67,* (pp. 2797-2822).

Loewenstein, J., & Gentner, D. (1998). Relational language facilitates analogy in children. *Proceedings from the 20th Annual Conference of the Cognitive Science Society,* (pp. 615-619).

Markman, A. B. and Gentner, D. (1993). Structural alignment during similarity comparisons. Cognitive Psychology, 25, 431-467.

Uttal, D. H., & Schreiber, J. C., & DeLoache, J. S. (1995). Waiting to use a symbol: The effects of delay on children's use of models. *Child Development, 66,* (pp. 1875-1891).

Uttal, D. H., Gregg, V., Tan, L. S., Chamberlin, M., & Sines, A. (submitted). Connecting the dots: Children's use of a figure to facilitate mapping and search.

Do Visual Attention and Perception Require Multiple Reference Frames?
Evidence from a Computational Model of Unilateral Neglect

Michael C. Mozer

Department of Computer Science and Institute of Cognitive Science
University of Colorado
Boulder, CO 80309–0430

Abstract

A key question motivating research in perception and attention is how the brain represents visual information. One aspect of this representation is the coordinate or *reference* frame with respect to which visual features are encoded. To determine the frames of reference involved in human vision and attention, neurological patients with unilateral neglect have been extensively studied. Neglect patients often fail to orient toward, explore, and respond to stimuli on the left. The interesting question is: with respect to what frame of reference is neglect of the left manifested? When a neglect patient shows a deficit in attentional allocation that depends not merely on the location of an object with respect to the viewer but on the extent, shape, or movement of the object itself, the inference is often made that attentional allocation must be operating in an object-based frame of reference. Via simulations of an existing connectionist model of spatial attention (Mozer, 1991; Mozer & Sitton, 1998), we argue that this inference is not logically necessary: object-based attentional effects in neglect can be obtained without object-based frames of reference.

Introduction

A key question motivating research in perception and attention is how the brain represents visual information. One aspect of this representation is the reference frame with respect to which visual features are encoded. The reference frame specifies the center location, the up-down, left-right, and front-back directions, and the relative scale of each axis. Reference frames can be prescribed by the viewer, objects, or the environment. *Viewer-based* frames are determined by the gaze, head orientation, and/or torso position of the viewer. *Object-based* frames are determined by intrinsic characteristics of an object, such as axes of symmetry or elongation, or knowledge of the object's standard orientation.

Determining which reference frame or frames are used by the brain to encode visual features is a key step to understanding the mechanisms of visual cognition and attention. For this reason, there has been intense interest in neurological patients with unilateral neglect, who provide a rich source of data diagnostic of the reference frames involved in human perception.

Unilateral neglect

Damage to parietal cortex can cause patients to fail to orient toward, explore, and respond to stimuli on the contralesional side of space (Farah, 1990; Heilman, Watson, & Valenstein, 1993). Unilateral neglect is more frequent, longer lasting, and severe following lesions to the right hemisphere than to the left. Consequently, all descriptions in this paper will refer to right-hemisphere damage and neglect of stimuli on the left. The interesting question surrounding unilateral visual neglect is: With respect to what reference frame is left neglect manifested? Clever behavioral experiments have been designed to dissociate various reference frames and determine the contribution of each to neglect. In several experiments, patients show a deficit in attentional allocation that depends not merely on the location of an object with respect to the viewer, but on the extent, shape, or movement of the object itself. From this finding of *object-based attentional effects*, the inference is often made that attentional allocation must be operating in an object-based frame of reference, and consequently, object-based representations are key to visual information processing. The point of this paper is to show that this inference is not logically necessary: *Object-based attentional effects can be obtained without object-based reference frames.* We argue this point via a computational model that utilizes only viewer-based frames, yet can account for data from experimental studies that were interpreted as supporting object-based frames.

MORSEL

MORSEL (Mozer, 1991; Mozer & Sitton, 1998) is a connectionist model of visual perception and attention, which has previously been used to explain a large corpus of experimental data, including reading deficits in neglect dyslexia (Mozer & Behrmann, 1992), and line bisection performance in neglect (Mozer, Halligan, & Marshall, 1997). MORSEL (Figure 1) includes a *recognition network* that can identify multiple shapes in parallel and in arbitrary locations of the visual field, but has capacity limitations. MORSEL also includes an *attentional mechanism* that determines where in the visual field to focus processing resources.

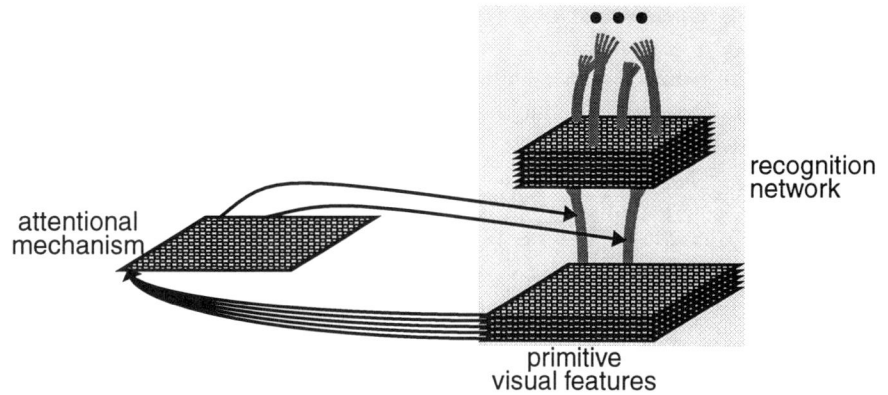

FIGURE 1. Key components of MORSEL (Mozer, 1991) include a recognition network, the first stages of which are depicted against a grey background, and an attentional mechanism.

Visual input presented to MORSEL is encoded by a set of feature detectors arrayed on a topographic map. Activity from the topographic map propagates through both the recognition network and the attentional mechanism. The topographic map is in a viewer-based reference frame, meaning that the input representation changes as the viewer moves through the world.

The attentional mechanism

In the present work, only the attentional mechanism, or *AM* for short, is required to account for data from unilateral neglect. The AM is a set of processing units in one-to-one correspondence with the locations in the topographic map. Activity in an AM unit indicates the salience of the corresponding location, and serves to gate the flow of activity from feature detectors at that location in the topographic map into the recognition network (indicated in Figure 1 by the connections from the AM into the recognition network); the more active an AM unit is, the more likely that features in the corresponding location of the topographic map will be detected and analyzed by the recognition network.

Each unit in the AM receives bottom-up or *exogenous* input from the detectors in the corresponding location of the topographic map (indicated in Figure 1 by the connections from the primitive features to the AM). Given the exogenous input, cooperative and competitive dynamics within the AM cause a subset of locations to be activated.

Figure 2 shows an example of the AM in operation. Although the model appears to have formed a spotlight of attention, the dynamics of the model do not mandate the selection of a contiguous or convex region. Typically, however, a single region is selected, and the selected region conforms to the shape of objects in the visual input, tapering off at object boundaries.

The operation of the AM is based on three principles concerning the allocation of spatial attention, which most would view as noncontroversial: (1) Attention is directed to locations in the visual field where objects appear, as well as to other task-relevant locations. (2) Attention is directed to contiguous regions of the visual field. (3) Attention has a selective function; it should choose some regions of the visual field over others.

These abstract principles concerning the direction of attention can be incorporated into a computational model like the AM by translating them into rules of activation, such as the following:

(1) *Locations containing visual features should be activated.* This rule provides a *bias* on unit activity (i.e., all else being equal, the principle indicates whether a unit should be on or off). One can see this rule at work in Figure 2, where the initial activity of the AM (upper-middle frame) is based on the exogenous input (upper-left frame).

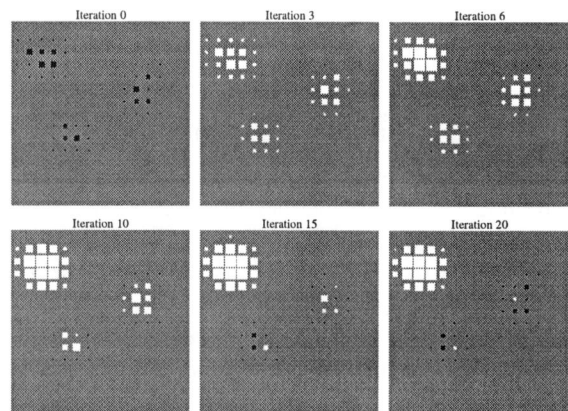

FIGURE 2. Example of the operation of the AM. Each panel contains a 15×15 topographic map depicting the state of the AM at a particular processing iteration The area of a black square is proportional to the exogenous input at that location. The area of a white square is proportional to the AM activity. The white squares are superimposed on top of the black; consequently, the exogenous input is not visible at locations with AM activity. The exogenous input pattern indicates three objects, the largest one—the one producing the strongest input—is in the upper left portion of the field. By iteration 20, the AM has reached equilibrium and has selected the region surrounding the largest object.

457

(2) *Locations adjacent to activated locations should also be activated.* This rule results in *cooperation* between neighboring units, and is manifested in Figure 2 by the increase in activity over time for the blob in the upper left portion of the field.

(3) *Locations whose activity grows the slowest should be suppressed.* This rule results in *competition* between units, and is manifested in Figure 2 by the decrease in activity for the two lower blobs once the upper-left blob begins to dominate in activity. This rule allows a large region to become activated, if the activity of all units in the region rises at more-or-less the same rate.

These three rules qualitatively describe the operation of the model. The model can be characterized quantitatively through an update equation, which expresses the activity of a processing unit in the AM as a function of the input to the AM and the activities of other AM units (Mozer, 1991, 1999).

Lesioning the AM to produce neglect

In our earlier modeling of neglect, we proposed a particular form of lesion to the model—damaging the connections from the primitive feature maps to the AM. The damage is graded monotonically, most severe at the left extreme of the topographic map and least severe at the right (assuming a right hemisphere lesion, as we will throughout this article). The graded damage we propose is inspired by Kinsbourne's (1987) orientational bias account of neglect. The damage affects the probability that primitive visual features are detected by the AM. The specifics of the damage are described in Mozer (1999). We emphasize that the damage is to a viewer-centered representation of space.

Simulations 1 and 2

When an experimental stimulus is presented upright and centered on the fixation point, viewer-centered and object-centered reference frames are confounded. To dissociate the two frames, Behrmann and Tipper (1994) rotated a display containing a *barbell*—two disks, one colored red and the other blue, connected by a solid bar. The barbell first appeared with, say, the red disk on the left and the blue disk on the

right. It remained stationary for one second, allowing subjects to establish an object-based frame of reference. In the *moving* condition, the barbell then rotated 180° (Figure 3a) such that the two disks exchanged places (Figure 3b). Following the rotation, the red disk appears on the left with respect to the object-based frame, but on the right with respect to the viewer-based frame. The subjects' task was to detect a target appearing on either the red or the blue disk. A *static* condition, in which the barbell did not rotate, was used as a baseline (Figure 3b). Left-neglect subjects showed facilitation for targets appearing on the blue disk in the moving condition relative to the static condition, and showed inhibition for targets appearing on the red disk. Essentially, the laterality of neglect reversed with reversal of the barbell. Results were therefore consistent with object-based, not viewer-based, neglect.

Tipper and Behrmann (1996) showed that the phenomenon appeared to depend on the disks being encoded as one object: in contrast to the condition depicted in Figure 3 in which the two disks are *connected*, when the bar between the disks is removed—the *disconnected* condition—the reversal of neglect no longer occurred when the disks rotate. This finding is what one would expect if neglect occurred in an object-based frame, because rotation of the display no longer corresponds to rotation of a single object.

General simulation methodology

The AM as described is identical to the model used in our earlier simulation studies of neglect. Because our earlier simulations involved static displays and the present simulations involve dynamic displays, one minor technical change was made to the nature of input noise, as described in Mozer (1999).

To simulate an experimental task, the experimental stimuli are mapped to a pattern of exogenous input to the AM. As we have done in earlier work, the mapping was accomplished by laying a silhouette of the stimulus over the topographic map and generating a pattern of activity based on the silhouette, emphasizing the object borders.

The experimental task has as its dependent variable the response time to detect a target. Rather than running the full MORSEL model and using the object-recognition network to determine detection or identification responses, we make a simple readout assumption that allows us to perform a simulation using only the AM. The assumption is that *the reaction time to detect a target is inversely proportional to the attentional activation in locations that correspond to the target.* This assumption is justified by earlier simulations of MORSEL (Mozer, 1991), in which output activity of the recognition network was found to be monotonically related to the allocation of attention to locations of a target. In all results reported, we average activity over a window of 20 iterations following target onset, over all locations of the

FIGURE 3. Barbell stimulus used in the Behrmann and Tipper experiment. The disk labeled "R" is colored red, the disk labeled "B" blue. In the moving condition, the initial display (panel a) was rotated 180°, resulting in the left and right disks exchanging places (panel b). In the static condition, no rotation occurred (panel b).

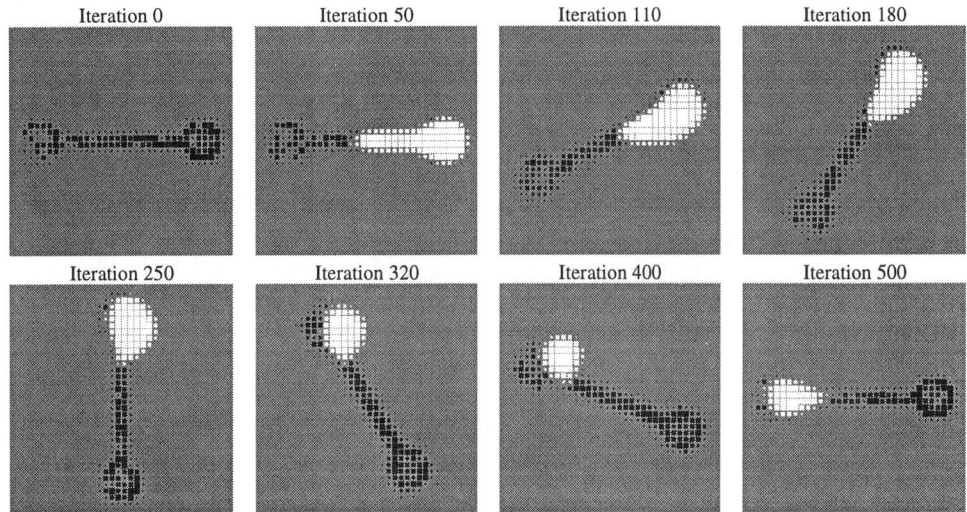

| Iteration 0 | Iteration 50 | Iteration 110 | Iteration 180 |
| Iteration 250 | Iteration 320 | Iteration 400 | Iteration 500 |

FIGURE 4. One trial of the lesioned model on the Behrmann and Tipper (1994) rotating-barbell stimulus.

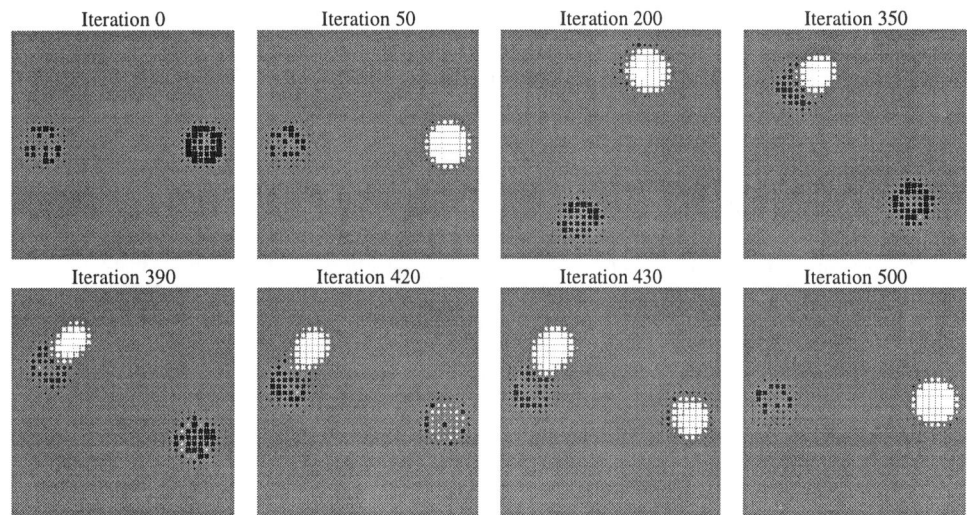

| Iteration 0 | Iteration 50 | Iteration 200 | Iteration 350 |
| Iteration 390 | Iteration 420 | Iteration 430 | Iteration 500 |

FIGURE 5. One trial of the lesioned model on the Tipper and Behrmann (1996) rotating disconnected disks.

disk on which a target appears, and over 200 trials. We call this the *read-out activity*. Greater read-out activity for a disk indicates a shorter response time to the target appearing in that disk.

Results

For the unlesioned model, the read-out activity on left and right disks and in the static and moving conditions is the same. The lesioned AM shows quite different behavior (Figure 4). A relative degradation to the exogenous input on the left side of the barbell can be observed due to the damage, causing the right half of the barbell to be selected initially. As the barbell begins to rotate, the focus of attention narrows further to just the disk. As rotation continues, attentional activity lags slightly behind the exogenous input, but catches up when the rotation is completed. Given the final distribution of attention in the moving condition, the model will be faster to respond to a target on the left than on the right. This

reversal does not occur in the static condition, as suggested by the AM state at iteration 50. The trial depicted in Figure 4 is representative; it is consistent with the more quantitative measure of read-out activity (Table 1, connected condition) which indicates greater activity for the left disk in the moving versus the static condition, and less activity for the right disk.

When the disks are disconnected, attention jumps from the disk that started off on the left to the disk that ends up on the left (Figure 5). After the disks cross the midline, the disk rotating into the right field begins to receive more support from the exogenous input than the disk rotating into the left field. Eventually this exogenous support is sufficient to activate the right disk, and competition kicks in to suppress the left disk. This pattern is observed reliably, as indicated by the measure of read-out activity (Table 1, disconnected condition). The read-out activity shows nearly full activity to the right disk and none to the left disk, and no difference between moving and static conditions.

To summarize, the AM simulation replicates the primary findings of Behrmann and Tipper (1994; Tipper & Behrmann, 1996): (1) For normals, no reliable differences are obtained across conditions. (2) For patients shown connected disks, left-sided facilitation and right-sided inhibition is obtained in the moving condition relative to the static. (3) For patients shown disconnected disks, left-sided facilitation and right-sided inhibition are not observed. (4) For patients, there is a main effect of target side: left is slower than right.

To understand the simulation results, consider first the moving connected-disk trials. The model appears to track the right disk into the left field. Because attentional activity in the model corresponds to covert attention, this tracking is not necessarily overt and is therefore consistent with the finding of Tipper and Behrmann (1996) that eye movements are not critical to the phenomenon. Tracking occurs because the attentional state has hysteresis: the state at some iteration is a function of both the exogenous input and the state at the previous iteration. Attention is not ordinarily drawn to a disk on the left given a competing disk on the right because the exogenous input to the left disk is weaker. Nonetheless, if attention is already focused on the disk on the left, even a weak exogenous input may be sufficient to maintain attention on the disk. Returning to the rules of activation of the model described earlier, the disk that has moved into the left field has support via the bias and cooperation rules, whereas the disk that has moved into the right field has support only via the bias rule.

However, the winner is not determined simply by the number of activation rules that support it. Key to the model's behavior is the total *quantitative* support provided to each of the disks. If the total support is greater for the right disk, then attention will flip to the right. This flipping occurs on the disconnected-disk trials. Based on an exploration of alternative stimuli, it appears that the flipping occurs for the disconnected but not connected trials due to the presence of the *neck* of the barbell on connected trials—the region where the disk makes contact with the bar. The neck provides an region of exogenous input adjacent to the disk, and by the cooperation rule, therefore provides an environment that supports attentional activity. Figure 4 clearly shows that activation is centered on the neck as the disk rotates into the left field. Without the neck to "hook" activity in place, activity drops to the point that the left disk cannot fend off attack from the right disk. Although this account is not entirely satisfactory, in that we have not explained the phenomena in linguistically simple, qualitative terms, it is sometimes the best one

can hope for in characterizing the behavior of a complex, dynamical system such as the AM.

Simulation 3

Recently, Behrmann and Tipper (1999) have explored an intriguing variation of the rotating-barbell experiment in which the display contains, in addition to the rotating barbell, two elements—squares which remain stationary during the trial (Figure 6). Subjects were asked to detect a target that could appear either on one of the disks or on one of the squares. As in the earlier studies, facilitation is observed for targets appearing on the left (blue) disk in the moving condition relative to the static condition, and inhibition is observed for targets appearing on the right (red) disk, consistent with neglect in the object-based frame of the barbell. Simultaneously, however, neglect is observed in the viewer-based frame for the squares: target detection in the left square is slower than in the right square. The finding of neglect in both viewer- and object-based reference frames suggests that attention can select and access information encoded with respect to multiple reference frames.

Without delving into the details of the simulation, Figure 7 presents a single trial of the lesioned AM which gives an intuition of how the model can explain the data. Initially, attention is drawn to the right side of the display, which includes the right disk and right square. As the barbell begins to rotate, attention is stretched to span the disk and the square, but as the disk and square separate, the attentional blob connecting them is broken into two blobs. One might expect one blob to be suppressed due to competition between the blobs, but the competition is weak, for the following reason. The competition rule depends on the *rate* of activity growth in one blob versus another, not the total activity in one blob versus another. By the time the two blobs are formed, the activities of individual locations in the two blobs are comparable and near asymptote. Consequently, the competition rule does not produce significant inhibition of either blob. Quantitative measures of read-out activity are consistent with the example presented in Figure 7 and with the

TABLE 1

condition		left disk	right disk
connected	moving	0.22	0.04
	static	0.00	0.99
disconnected	moving	0.00	0.93
	static	0.00	0.99

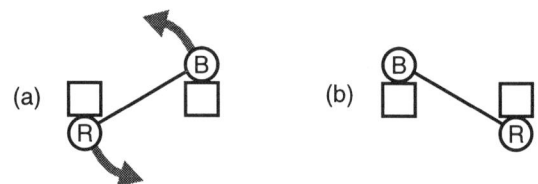

FIGURE 6. The multiple-object display studied by Behrmann and Tipper (1999). In the moving condition, the initial display (panel a) consists of two stationary squares and a barbell, which—as in the earlier studies—rotates such that its two disks exchange horizontal positions (panel b). Subjects were asked to detect a target that could appear on either disk or either square. In the static condition, no rotation occurred (panel b).

460

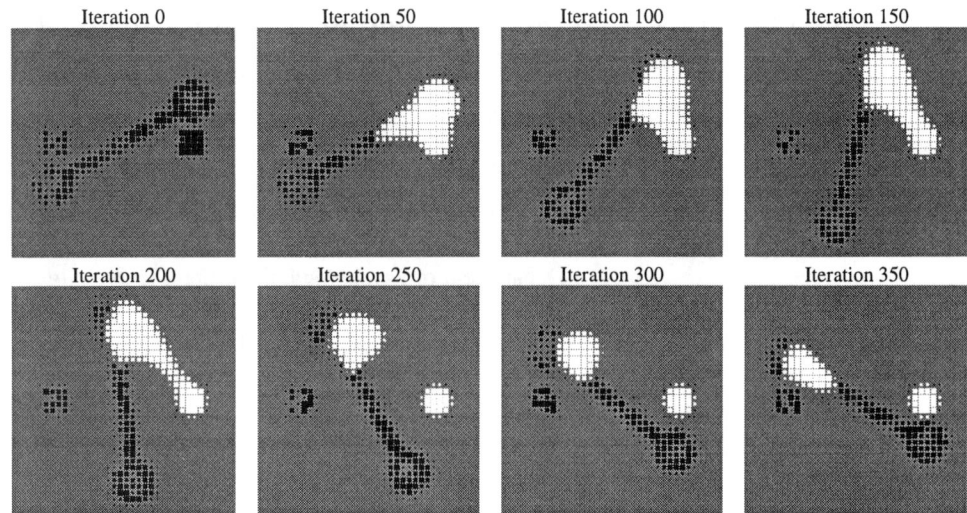

FIGURE 7. One trial of the lesioned model on the Behrmann and Tipper (1999) barbell-square experiment.

results of Behrmann and Tipper (see Mozer, 1999, for details).

Discussion

The neuropsychological studies of Behrmann and Tipper have been taken as support for the hypothesis that neglect, and hence attention, can operate in object-based coordinates. However, simulations of the AM provide an alternative explanation, because the AM operates only in a viewer-based frame of reference, yet can account for the data. The AM's account involves covert attentional tracking. But the AM's account is more complex, because the AM can in addition explain the lack of neglect reversal for disconnected displays or for the stationary squares.

The AM has also been successful in explaining a variety of other neglect studies that have been interpreted to support the psychological reality of object-based frames of reference (Mozer, 1999). The accumulation of such studies has caused many researchers to believe in the existence of object-based reference frames and object-based representations in the brain. But the AM strikes a blow against this interpretation. In the absence of strong empirical support for object-based frames, it would seem more parsimonious to suppose that object-based frames of reference play little or no role in the course of ordinary perception.

Acknowledgments

This work greatly benefited from the comments and insights of Paul Smolensky, Brenda Rapp, John McGoldrick, and Yuko Munakata. This research was supported by Grant 97-18 from the McDonnell-Pew Program in Cognitive Neuroscience, and by NSF award IBN-9873492.

References

Behrmann, M., & Tipper, S. P. (1994). Object-based attentional mechanisms: Evidence from patients with unilateral neglect. In C. Umilta & M. Moscovitch (Eds.), *Attention and performance XV: Conscious and nonconscious processing and cognitive functioning* (pp. 351–375). Cambridge, MA: MIT Press.

Behrmann, M. & Tipper, S. P. (1999). Attention accesses multiple reference frames: Evidence from visual neglect. *Journal of Experimental Psychology: Human Perception and Performance, 25,* 83–101.

Farah, M. J. (1990). *Visual agnosia.* Cambridge, MA: MIT Press/ Bradford Books.

Heilman, K. M., Watson, R. T., & Valenstein, E. (1993). Neglect and related disorders. *Clinical neuropsychology* (2nd edition) (pp. 279–336). New York: Oxford University Press.

Kinsbourne, M. (1987). Mechanisms of unilateral neglect. In M. Jeannerod, *Neurophysiological and neuropsychological aspects of spatial neglect* (pp. 69–86). Amsterdam: North Holland.

Mozer, M. C. (1991). *The perception of multiple objects: A connectionist approach.* Cambridge, MA: MIT Press/Bradford Books.

Mozer, M. C. (1999). Frames of reference in unilateral neglect and spatial attention: A computational perspective. *Submitted for publication.*

Mozer, M. C., & Behrmann, M. (1992). Reading with attentional impairments: A brain-damaged model of neglect and attentional dyslexias. In R. G. Reilly & N. E. Sharkey (Eds.), *Connectionist approaches to natural language processing* (pp. 409–460). Hillsdale, NJ: Erlbaum Associates.

Mozer, M. C., Halligan, P. W., & Marshall, J. C. (1997). The end of the line for a brain-damaged model of unilateral neglect. *Journal of Cognitive Neuroscience, 9,* 171–190.

Mozer, M. C., & Sitton, M. (1998). Computational modeling of spatial attention. In H. Pashler (Ed.), *Attention* (pp. 341–393). London: UCL Press.

Tipper, S. P., & Behrmann, M. (1996). Object-centered not scene-based visual neglect. *Journal of Experimental Psychology: Human Perception and Performance, 22,* 1261–1278.

True to Thyself: Assessing Whether Computational Models of Cognition Remain Faithful to Their Theoretical Principles

In Jae Myung (MYUNG.1@Osu.Edu)
August E. Brunsman IV (BRUNSMAN.3@Osu.Edu)
Mark A. Pitt (PITT.2@Osu.Edu)
Department of Psychology, Ohio State University
1885 Neil Avenue
Columbus, OH 43210 USA

Abstract

This study investigated the model selection problem in cognitive psychology: How should one decide between two computational models of cognition? The focus was on model "faithfulness," which refers to the degree to which a model's behavior originates from the theoretical principles that it embodies. The guiding principle is that among a set of models that simulate human performance equally well, the model whose behavior is most stable or robust with variation in parameter values should be favored. This is because such a model is likely to have captured the underlying mental process in the least complex way while at the same time being faithful to the theoretical principles that guided the model's development. Sensitivity analysis is introduced as a tool for assessing model faithfulness. Its application is demonstrated in the context of two localist connectionist models of speech perception, TRACE and MERGE.

Introduction

One of the most challenging tasks for researchers interested in modeling human cognition is developing techniques for choosing among a set of computational models (e.g., Grossberg, 1987; Grainger & Jacobs, 1998). The goal is to choose the model that best captures the underlying cognitive process. It is standard practice to select the model that most accurately simulates or fits data generated by humans. Justification for using this procedure, termed descriptive adequacy, is that the best-fitting model most closely approximates the mental process being modeled. The adequacy of this model selection criterion is limited to cases in which the models do not capture the underlying process equally well. When they do, how should one decide among models? There are at least two important issues to consider.

The first issue is *model complexity*, which refers to the flexibility inherent in a model (i.e., how the parameters are combined mathematically) that enables it to fit diverse patterns of data. A model may describe data well, but may not do so in a parsimonious manner. It is well established (e.g., Linhart & Zucchini, 1986; Myung, in press) that model selection based solely on descriptive adequacy will result in the choice of an unnecessarily complex model that over-fits the data, and therefore fails to capture the true regularities of the underlying mental process. To avoid this mistake, both descriptive adequacy and model complexity must be taken into account in model selection (Myung & Pitt, 1997). These two criteria embody the principle of Occam's razor in model selection: The model that fits data sufficiently well in the least complex way should be preferred.

The second, equally critical, issue is determining the cause of a model's behavior. Is a model's success in mimicking human behavior due to the theoretical principles embodied in the model or due to other aspects of its computational instantiation? Put another way, is the computational instantiation faithful to its theoretical principles? These are not one in the same, as the latter can take on many forms (Uttal, 1990). Even if a model provides an excellent description of human data in the simplest manner possible, it is often difficult to determine what properties of the model are critical for explaining human performance and what aspects are not. Ideally, theoretical principles from which the model was developed must be clearly identifiable and their contribution to determining model behavior clearly demonstrated. In other words, the behavior of the model must originate from the theoretical ideas that motivated its creation, not from the computational choices made in its instantiation. Failure to make this distinction runs the risk of erroneously attributing a model's behavior to its underlying theoretical principles: computational complexity is mistaken for theoretical accuracy.

In this paper, we undertake an investigation of model faithfulness. The behaviors of two localist models of phonemic perception, TRACE (McClelland & Elman, 1986; McClelland, 1991) and MERGE (Norris et al, 1998) were compared to determine which architectural properties are most responsible for their behavior. Sensitivity analysis, a measure of how sensitive the behavior of a model is to variation in the values of its parameters, is introduced as a tool to assess model faithfulness. An additional attractive property of sensitivity analysis is that the results reveal the relative complexities of the models. A highly sensitive model is complex. A small change in the value of a parameter can change the model's behavior drastically. This property makes the model very powerful and adaptable, being able to fit a wide range of data patterns, perhaps many more than are necessary to model the mental process of interest. On the other hand, a model whose behavior changes minimally with variation in parameter values is far less flexible, but behaviorally more stable (i.e., robust). If such a model happens to simulate human data well, then it is an indication that the model may have captured the regularities of interest in the data and little else. Following Occam's razor, such a model should be favored. Thus, model selection using sensitivity analysis favors the model that is least sensitive to parameter variation, and as a consequence captures the data the best under the widest range of parameter variation.

Connectionist Models of Speech Perception

Model development in speech perception is divided on the issue of how prior information from different sources is integrated during recognition (see Frauenfelder & Tyler, 1987). A wide range of experimental results has demonstrated that a listener's knowledge about a word can influence how the phonemes (i.e., speech sounds) of that word are perceived. The theoretical debate in the literature has focused on determining how these two forms of information (lexical and phonemic) are combined during perception. In most computation models of word recognition, there exist at least two levels of processing, a phonemic level and a lexical (i.e., word) level. Information flow from the phoneme to the lexical level is common among models. The theoretical distinction of primary importance is how lexical information is integrated with phonemic information. In Figure 1, TRACE and MERGE illustrate the two positions architecturally. In TRACE (McClelland & Elman, 1986), activation of phonemes is influenced by bottom-up sensory input from the speech signal itself *and* from top-down connections to the lexical level. In MERGE (Norris et al, 1998), there are no top-down connections from the lexical level that *directly* affect phoneme activation. Rather, phonemic processing is split in two, with an activation/input stage and a phonemic decision stage. Lexical processes affect only phonemic decision making; they cannot directly influence phoneme activation. In MERGE, phonemic and lexical influences on phonemic decision making are independent of each other, being integrated only at the decision stage.

Norris et al (in press) showed that the models are fairly comparable in their ability to simulate two sets of human data. But what properties of the models are most responsible for their similar behavior? MERGE was proposed as a non-interactive alternative to TRACE, with no top-down feedback directly to the phonemic input stage. However, the models

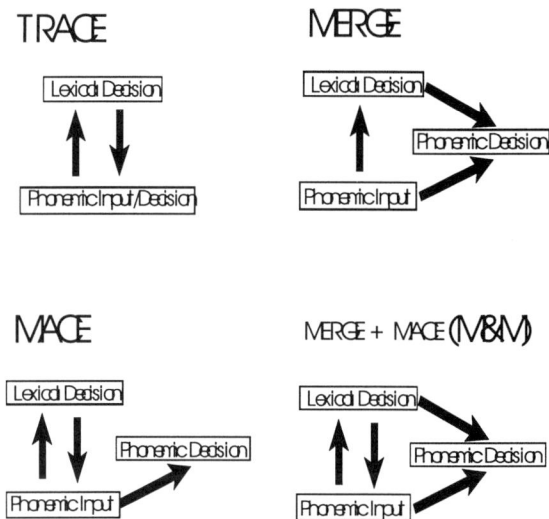

Figure 1. Network architectures of the four models tested.

also differ in the number of phonemic stages, with TRACE possessing one and MERGE possessing two. Which of these properties is most responsible for MERGE's behavior, direction of information flow or an additional stage? Additionally, are the two models equally complex? That is, we know their architectures are sufficient to capture the test data, but are they also necessary? Or are there redundancies in their design that make them overly complex?

To address these questions, sensitivity analyses were carried out on TRACE and MERGE and two other models (shown in the bottom of Figure 1), which were included to understand better the implications of model design on model performance. MACE is a hybrid of TRACE and MERGE that was intended to assess the consequences of independently integrating lexical and phonemic input at a separate decision stage. Like TRACE, lexical information flows back to the phonemic activation level, but like MERGE, phoneme decisions are made separately. If MERGE derives its descriptive power solely from its non-interactive architecture, then MACE's performance should be significantly inferior to that of MERGE. Similar performance would suggest that splitting phoneme processing into two stages is more important than whether lexical information flows directly to the phoneme activation level or instead to the phoneme decision level. MERGE+MACE (M&M) is a combination of MERGE and MACE. Lexical connections are redundant, feeding to both the phoneme activation level and the decision making level. It is included for completeness and to serve as a check on the predictability of models with different configurations of information flow.

Table 1. Experimental Conditions and Human Data.

| Experimental Condition | | Human Data | | | | | |
Condition	Example	Phonemic /b/	/g/	/z/	/v/	Lexical "job"	"jog"
W1W1 (Word 1 + Word 1)	JOb + joB (JOB)	*				*	
W2W1 (Word 2 + Word 1)	JOg + joB (JOB)	*				*	
N2W1 (Nonword 2 + Word 1)	JOv + joB (JOB)	*				*	
N1N1 (Nonword 1 + Nonword 1)	JOz + joZ (JOZ)			*			
W2N1 (Word 2 + Nonword 1)	JOg + joZ (JOZ)			*			
N2N1 (Nonword 2 + Nonword 1)	JOv + joZ (JOZ)			*			

Note: For each condition, the star '*' indicates the phoneme or word that is recognized by listeners.

Method

Overview

Norris et al (in press) evaluated TRACE and MERGE on their ability to simulate data showing listeners' sensitivity to mismatching phonemic information at the end of an utterance (Whalen, 1984; Marslen-Wilson & Warren, 1994). This same data set was used in the evaluation of the four models in Figure 1. First, the ability of the models to simulate the human data was assessed to replicate Norris et al and demonstrate that all models were comparable in descriptive adequacy. Second, a sensitivity analysis was performed on the models by systematically varying the parameter values around the optimal parameter settings that provided the best fit to the data in the first analysis. As mentioned above, the sensitivity analysis assessed the robustness of a model's behavior in the face of parameter variation. It enabled us to ascertain the degree to which performance arises from theoretical principles that the model purports to implement. The more the behavior of a model changes over the range of parameter values, the less likely the model derives its power from its theoretical design principles, in this case how lexical and phonemic information are integrated, than from idiosyncratic choices of parameter values.

Data That Were Modeled

Following Norris et al (in press), the four models were compared in their ability to simulate data from Marslen-Wilson and Warren (1994; McQueen, Norris & Cutter, in press), in which listeners were shown to be sensitive to conflicting phonemic input in both phonemic decision making and lexical decision making. When listening to speech, listeners exhibit considerable sensitivity to deviations from the natural production of an utterance. For example, if the portion of the phoneme (i.e., letter sound) /g/ in the word "jog" is spliced off and replaced with a token of the phone /b/ from the word "job," listeners are slower to identify the final phoneme, /b/, in the newly created cross-spliced word "job" than in the original, unspliced token of "job." This is because the acoustic information signaling the identity of the final phoneme cannot be fully removed, as it blends into the immediately preceding vowel, creating conflicting information about the identity of the final phone. The acoustic information at the end of the vowel specifies /g/ whereas the subsequent information specifies /b/.

By varying the original source of the two parts of a cross-spliced token (i.e., whether they came from words or nonwords), lexical influences on phoneme and word processing can be explored. The six conditions shown in Table 1 were used. The alpha-numeric condition names (column 1) refer to the composition of the cross-spliced stimulus. For example, W1W1 refers to the stimulus described above in which the cross-spliced stimulus was created from two words. In the examples, the capital letters refer to the portions of the two utterances that formed part of the cross-spliced utterance, with the resulting cross-spliced stimulus in parentheses. Both identification of the final phoneme and recognition of the cross-spliced word were simulated. For each condition, the asterisk '*' indicates the phoneme or word that was recognized by listeners.

Phonemic Decision Making Data

W1W1 and N1N1 were two control conditions, included only to demonstrate that when no conflicting phonemic cues are present in the stimulus, recognition of the final phoneme is not impeded. In the four remaining conditions, there is conflicting phonemic information in the cross-spliced tokens because stimuli with different final phonemes were cross-spliced. The lexical status of the two "source" utterances influences identification of the final phoneme. When the initial item is a word (e.g., "jog"), as in conditions W2W1 and W2N1, recognition of the final phoneme, /b/ or /z/, are comparatively slower, presumably because the lexical entry for "jog" affects phonemic processing in some manner. Further, the amount of the slowdown is less in W2W1 than in W2N1, which is thought to be due to lexical competition between "job" and "jog" diminishing lexical influences in processing the final phoneme. If the initial portion of the cross-spliced stimulus comes from a nonword (e.g., "jov'), as in conditions N2W1 and N2N1, there is no slowdown in recognition of /b/ or /z/. Although there is conflicting phonemic information in the cross-spliced stimulus, the use of a nonword effectively shuts down lexical influences.

Table 2. Description of Model Parameters

Parameter	TRACE	MERGE	MACE	M&M
PE (phoneme excitation)	✔	✔	✔	✔
PWE (phoneme to word excitation)	✔	✔	✔	✔
PWI (phone to word inhibition)	-	✔	✔	✔
PTE (phoneme to target excitation)	-	✔	✔	✔
PD (phoneme decay)	✔	✔	✔	✔
WTE (word to target excitation)	-	✔	-	✔
WPE (word to phoneme excitation)	✔	-	✔	✔
WWI (word to word inhibition)	✔	✔	✔	✔
WD (word decay)	✔	✔	✔	✔
TTI (target to target inhibition)	-	✔	✔	✔
TD (target decay)	-	✔	✔	✔
TM (target momentum)	-	✔	✔	✔
CPS (word/target cycles per input slice)	✔	✔	✔	✔
PPI (phoneme to phoneme inhibition)	✔	-	-	-

Note: For each parameter, the check '✔' indicates the models that adopt the parameter.

The bottom-up information specifying /v/ is too weak to affect phonemic decision making.

Lexical Decision Making Data

Lexical decision making with the cross-spliced stimuli was straightforward. Between the two possible word responses (e.g., "job" and "jog"), if the final item in the cross-splice is a word (e.g., "job") as in conditions W1W1, W2W1 and N2W1, then only the word "job" (i.e., final item) should be recognized, regardless of whether the initial portion originally came from a word or a nonword. On the other hand, if the second portion came from a nonword (e.g., /z/ in "joz") as in conditions N1N1, W2N1, and N2N1, then word recognition depends on whether the initial item in the cross-splice is a word or nonword. If it is a word (e.g., "jog") as in W2N1, then the word "jog" should be activated, but not substantially to be recognized because of the following mismatching information (/z/). If it is a nonword (e.g., "joz" or "jov") as in N1N1 and N2N1, then both "jog" and "job" will be activated too weakly to be recognized.

Model Implementation and Simulation Procedure

The four models in Figure 1 were constructed by modifying the architecture of MERGE, which is a localist network consisting of six input nodes corresponding to the phonemes /dʒ/ "j", /o/, /b/, /g/, /v/ and /z/, four lexical decision nodes representing two words ("job", "jog") and two nonwords ("jov", "joz"), and finally, four phonemic decision nodes representing the target phonemes /b/, /g/, /v/ and /z/. In MERGE, each lexical decision node receives inputs from the phonemic input nodes through excitatory connections and also receives activations from other lexical decision nodes through lateral inhibitory connections. Similarly, the phonemic decision nodes are linked to the phonemic input

nodes as well as to the lexical decision nodes through excitatory connections. Lateral inhibitory connections are assumed among phonemic decision nodes whereas no such connections are assumed among phonemic input nodes. TRACE, MACE, and M&M, were created either by pruning existing connections and/or adding new connections to MERGE. The TRACE model had 8 parameters, MERGE and MACE models had 12, M&M model had 13 parameters. The parameters used in the models are cross-tabulated in Table 2 to provide one view of their similarities and differences.

In simulating the human data, each model was presented with the same input, six numerically represented cross-spliced tokens that were all three phonemes long and were either words or nonwords. All six of the tokens, one for each condition, were identical to the ones used by Norris, McQueen, and Cutler (1998). Each token was represented by six Mx1 vectors (one vector for each phoneme) where M is the number of time slices or iterations. The first vector gave activation to the /dʒ/ node in the phoneme input layer, the next to /o/, and so on for /b/, /g/, /v/, and /d/. Each vector began at zero except for /dʒ/, which was .25 in the first time slice, .5 in the second, and then 1 in the third (maintained for the rest of the iterations). In the fourth time slice, /o/ went to .25, .5 in the fifth, and 1 in the sixth (maintained for the rest of the trail). In a similar fashion, the final phoneme was constructed, depending upon the condition. For a given input stimulus, the activation profiles of the phonemic decision nodes and the lexical decision nodes were obtained and then compared to the predictions from human data to evaluate the model's performance.

Each of the four models was qualitatively fit to the human data, and a set of optimal parameter values was obtained by a hand-done parameter search, relying on initial estimates reported in Norris, McQueen, & Cutler (1998). A model's behavior was judged to be either "human-like" or "not human-like"in each condition by determining whether the model's output matched predictions in the corresponding condition. Twelve judgements were made for each model, six phoneme decisions and six lexical decisions (Table 1). In the sensitivity analysis, the parameter values of each model were systematically varied ±75% from the optimum value. At each value of the parameter, the model's behavior (phonemic decision making and lexical decision making) was re-assessed in the twelve judgements.

Results and Discussion
Simulating Human Data

All four models produced "human-like" results in every condition in both phonemic decision making and lexical decision making, including the all-important slowdown in phonemic processing in conditions W2W1 and W2N1. Thus in terms of descriptive adequacy, all models were functionally equivalent in their ability to simulate this set of data. This finding suggests that the two ways in which lexical and phonemic information are integrated does not matter: Direct top-down feedback (TRACE, MACE) simulates human performance just as well as integrating the two sources of information independently (MERGE) or a combination of the two methods (M&M).

Given the similar behavior of the four models, how should we choose among them? Overly complex models should be avoided. Recall that MERGE, MACE and M&M assume that phoneme activation is separate from phoneme decision making. Relative to TRACE, which makes no such distinction, these models require extra parameters (4 for MERGE and MACE, 5 for M&M). The finding of virtually no difference in descriptive adequacy between any of the models suggests that splitting phonemic processing across levels is a redundant property of these models, one unnecessary to simulate human behavior. Instead, splitting phonemic processing in two may introduce unnecessary complexity that only reduces generalizability of the models, making them less stable amidst parameter variation. The sensitivity analysis explores this possibility.

Sensitivity Analysis

Figure 2 shows the proportion of non-human-like data patterns (i.e., errors) generated by each model when the model's parameters were systematically varied around the optimum values. The proportions were averaged over all parameters and are shown separately for each of the 12 testing conditions. TRACE was the least error prone, with a 2.1% error rate, whereas the other three models made considerably more errors (5 - 7 %). This result is clear confirmation that splitting phonemic processing into two stages (MERGE, MACE, and M&M) does more than reflect the regularities of human speech processing. It introduces

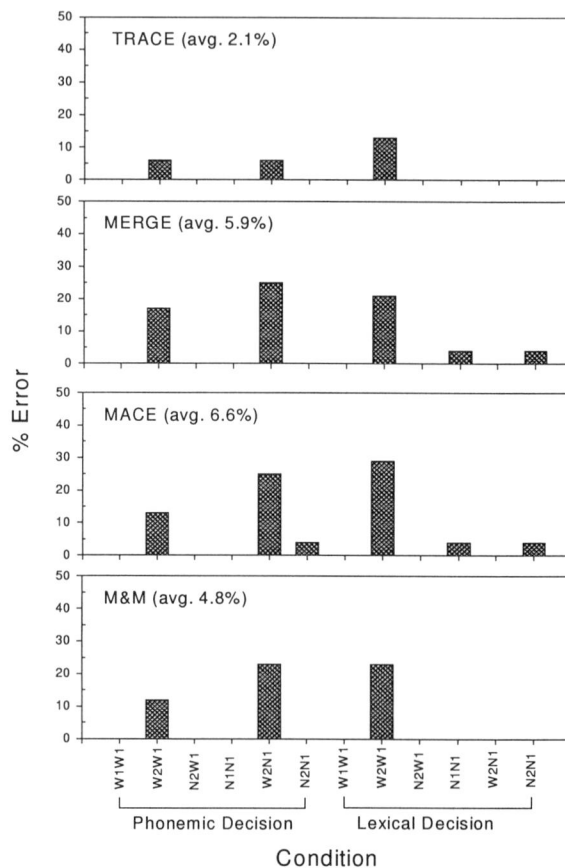

Figure 2. Results of sensitivity analysis..

unnecessary flexibility not needed to capture the phenomena of interest.

Examination of Figure 2 reveals that errors occurred most frequently in three conditions: W2W1, W2N1 in phonemic decision making and W2W1 in lexical decision making. In the first two, phonemic and lexical information must be integrated to simulate accurately human data. In the third, the effects of lexical inhibition must be simulated. The three models with independent lexical and phonemic influences on phonemic decision making (MERGE, MACE, M&M) were very sensitive to parameter variation in these conditions, producing many patterns that were not human-like. TRACE, on the other hand, was able to exhibit human-like performance under a wider range of parameter values.

Note also in Figure 2 the similar performance profiles of MERGE and MACE across the 12 conditions. Recall that MACE, like TRACE, contains top-down flow of lexical information directly to the phonemic input level, but like MERGE, phonemic processing is split into two stages. The fact that this hybrid model behaves so similarly to MERGE suggests that MERGE's behavior is determined more by the separation of phonemic processing into two stages than by lexical information affecting phonemic decision making rather than phonemic activation. In other words, what differentiates MERGE from TRACE is not so much how information flows between processing stages, but the number of processing stages. This result suggests that the current implementation of MERGE is only partially faithful to the theoretical principle that motivated its development.

The sensitivity analysis suggests that the extra parameters that MERGE, MACE, and M&M require as a result of separating phonemic processing into two stages, and thus making the models non-interactive, increases the complexity of the models. This design characteristic has the detrimental side effect of decreasing model robustness. TRACE explains human data sufficiently well in the least complex manner.

Summary and Conclusion

The purpose of this preliminary investigation was to explore the model selection problem in cognitive psychology: How should one decide between two computational models of cognition? The particular focus of the study has been on assessing model faithfulness, which refers to the degree to which a model's behavior originates from the theoretical principles that it embodies. The idea is that among a set of models that simulate human performance equally well, the model whose behavior is most stable with variation of parameter values should be favored. This is because such a model is likely to be most faithful to the theoretical principles that guided the model's development; it is also likely to have captured the underlying mental process in the least complex way. Sensitivity analysis was introduced as a tool for assessing model faithfulness. An application of the method was demonstrated for comparing the behaviors of four connectionist models of speech perception, TRACE, MERGE, MACE and M&M.

All four models were functionally indistinguishable in their ability to simulate human data. Sensitivity analysis, however, revealed that TRACE was the most stable model, suggesting that it best reflects the underlying regularities of human behavior and therefore should be preferred. An important implication of these results for modeling speech perception is that the separation of phonemic decision making from phonemic activation, as assumed in MERGE, MACE, and M&M, may be an overly complex architectural design that is not necessary to capture the phenomenon of interest (i.e., lexical and phonemic interaction).

Acknowledgments

The authors wish to thank Dennis Norris for kindly answering many questions we had about implementation of MERGE. This research was supported by NIMH Grant MH57472 to I.J.M. and M.A.P., and by the Ohio State University Colleges of Arts and Sciences Honors Award to A.E.B.

References

Frauenfelder, U.H. & Tyler, L.K. (1987) The process of spoken word recognition: An Introduction. *Cognition* 25:1-20.

Grainger, J., & Jacobs, A. M. (1998). On localist connectionism and psychological science. In J. Grainger & A. M. Jacobs (eds.), *Localist Connectionist Approaches to Human Cognition*. Lawrence Erlbaum Associates.

Grossberg, S. (1987). Competitive learning: From interactive activation and adaptive resonance. *Cognitive Science*, 11, 23-63.

Linhart, H., & Zucchini, W. (1986). *Model Selection*. New York: Wiley.

Marslen-Wilson, W. & Warren, P. (1994) Levels of perceptual representation and process in lexical access: words, phonemes, and features. *Psychological Review* 101:653-675.

McClelland, J. L., & Elman, J. L. (1986). The TRACE model of speech perception. *Cognitive Psychology*, 18, 1-86.

McClelland, J. L. (1991). Stochastic interactive processes and the effect of context on perception. *Cognitive Psychology*, 23, 1-44.

McQueen, J. M., Norris, D. & Cutler, A. (in press) Lexical influence in phonetic decision making: Evidence from subcategorical mismatches. *Journal of Experimental Psychology: Human Perception & Performance*.

Myung, I. J. (in press). The importance of complexity in model selection. *Journal of Mathematical Psychology*.

Myung, I. J., & Pitt, M. A. (1997). Applying the Occam's razor in modeling cognition: A Bayesian approach. *Psychonomic Bulletin & Review*, 4, 79-95.

Norris, D., McQueen, J.M. & Cutler, A. (in press) Merging phonetic and lexical information in phonetic decision-making. *Behavioral and Brain Sciences*.

Uttal, W. R. (1990). On some two-way barriers between models and mechanisms. *Perception & Psychophysics*, 48, 188-203.

Whalen, D. H. (1984). Subcategorical phonetic mismatches slow phonetic judgments. *Perception & Psychophysics*, 35, 49-64.

How Knowledge Interferes with Reasoning – Suppression Effects by Content and Context

Hansjörg Neth (neth@psychologie.uni-freiburg.de)
Sieghard Beller (beller@psychologie.uni-freiburg.de)
Department of Psychology; University of Freiburg
79 085 Freiburg, Germany

Abstract

The suppression of logically valid inferences by the content or context of premises can be seen as an instance of knowledge having a detrimental influence on reasoning. Although Henle (1962) has claimed that invalid deductions are due to additional premises drawn from background knowledge, current research on content effects ignores the methodological implications of this claim. Elaborating on the suppression effect in conditional reasoning (Byrne, 1989), we present a knowledge-based approach that makes relevant features of background knowledge an integral part of the analysis. After identifying the sufficiency and necessity of conditions as the type of knowledge mediating the effect, we construct and validate task materials independently from any assessment of reasoning (Experiment 1). We then replicate and extend suppression effects in syllogism tasks (Experiment 2) and show that participants are able to couch their background knowledge in formally correct wordings (Experiment 3).

Suppose you were presented with the following premises:

(1) If Ann is in New York, she visits the Guggenheim.
(2) If it is still open, she visits the Guggenheim.
(3) Ann is in New York.

Would you conclude that Ann visits the Guggenheim? If not, you fell prey to the *suppression effect* (Byrne, 1989) by not drawing the valid inference of Modus Ponens, warranted by premises (1) and (3). The demonstration that the valid inferences of Modus Ponens (MP) and Modus Tollens (MT) can be suppressed by an additional premise (2) was used against an earlier result that additional premises can also prevent the fallacious inferences 'Affirmation of the Consequent' (AC) and 'Denial of the Antecedent' (DA), thereby *facilitating* logical performance (Rumain, Connell, & Braine, 1983).

The ambiguous nature of this phenomenon provoked a vigorous debate about contemporary theories of human reasoning. Proponents of a mental models theory (Byrne, 1991; Johnson-Laird & Byrne, 1991) focus on semantic procedures of integrating formally identical premises and argue that contextual information facilitates the search for counterexamples to putative conclusions (Byrne, Espino, & Santamaria, 1998). Theorists who conceptualize the mind as equipped with tacit rules of inference claim that premise (2) leads participants to question the truth of premise (1), and to combine the two antecedent propositions in a conjunctive way

(Politzer & Braine, 1991). This conjunction yields "If Ann is in New York *and* it is still open, she visits the Guggenheim" and blocks the MP in question. As both theories refer to interpretive processes that are not within their scope, they suffer from the same shortcoming (Fillenbaum, 1993): neither can explain *why* premises of one and the same syntactic form are interpreted differently based on their content. To understand this process, it is necessary to embed the study of deductive reasoning in a knowledge-based approach (Chan & Chua, 1994; Beller, 1997; Beller & Spada, 1998).

A Knowledge-Based Approach

To introduce our point of view, let us analyze another version of the above example. Based on the premises

(1) If Ann is in New York, she visits the Guggenheim.
(4) Ann is *not* in New York.

many certainly would conclude that Ann does *not* visit the Guggenheim. But this conclusion seems logically unjustified, for concluding 'not q' from 'if p then q' and 'not p' means to commit the fallacy of DA. Yet, again an extra premise might be to blame for this apparent error in reasoning. For if someone *knows* that the Guggenheim is actually located in New York, he or she might introduce this spatial information as

(5) If Ann visits the Guggenheim, she is in New York.

Based on this implicit premise it follows by MT that Ann does not visit the Guggenheim if she is not in New York. But even this explanation depends on background knowledge. For believing (5) means to be negligent of the fact that there is another Guggenheim museum in Bilbao, Spain.

The relation between knowledge and reasoning has puzzled philosophers ever since Plato. Experimental evidence indicating that additional information can suppress logical thinking seems to refute the naïve equation "the more knowledge the better the reasoning." Henle (1962) sought to reestablish universal deductive competence by claiming that all inferences are perfectly valid but occasionally based on additional or distorted premises. Yet dissolving the issue of faulty reasoning by mere reference to interpretive processes introduces a second black box to account for the first. Even if additional or misinterpreted premises may account for a variety of apparent errors, the claim that eventually all fallacies can be reduced to this mechanism could still be premature.

A crucial point in the debate on content effects is the dependence of logical validity on the premises *as understood by the reasoner*. But despite its appeal, the methodological consequences of this idea have widely been ignored. To account for content effects, at least four levels of analysis have to be distinguished: (a) the linguistic framing of the premises, (b) their formal properties as defined by the semantics of logic, (c) the inferential properties of background knowledge that is triggered by the premises' content, and finally, (d) the mental integration of all this in the reasoner's mind.

Within our knowledge-based framework we focus on the complementary levels (b) and (c): Separate formalizations of form and content yield two predictive models whose combination predicts the *facilitation* of formally correct inferences whenever the conclusions warranted by both models coincide. Likewise, *suppression* of formally correct inferences is expected to occur whenever both predictions diverge (as it was the case in both examples above). However, when form and content predict inconsistent conclusions, evaluation of logical validity does not depend on logic, but on the preferred point of view.

Methodologically, the relativity of conclusions to different predictive models suggests experimental research on content effects to proceed in a two step strategy:

1. When background knowledge is claimed to have effects on reasoning, it must not be consulted only *post hoc* to justify otherwise invalid answers. To avoid circular argumentation (Smedslund, 1970), relevant inferential features of content have to be specified in advance and tested independently from any judgment about the validity of inferences.

2. Once assumptions about the relevant type of background knowledge have been secured, its influence on reasoning can be addressed, e.g., by assessing the extent to which changes in content or context invite formally invalid inferences.

General Design of the Experiments

To apply this framework to account for effects of suppression and facilitation in conditional reasoning, we follow Thompson (1994) by taking knowledge about the *sufficiency* and *necessity* of conditions as key variables to capture the relevant features of content. Combining both dimensions results in four *patterns of dependence* between a condition p and a consequence q:

S+N+: p is both sufficient and necessary for q.
S+N−: p is sufficient but not necessary for q.
S−N+: p is not sufficient but necessary for q.
S−N−: p is neither suffient nor necessary for q.

Sometimes it is only known that a condition p correlates with a consequence q but the precise nature of the relation is unknown. This can be represented as a fifth pattern (S?N?). Furthermore, several conditions may act additively or alternatively. Additive conditions can be understood as conjunctions of single factors (e.g. $p \land p_{add}$), and alternative conditions as disjunctions (e.g. $p \lor p_{alt}$). For instance, in the

Table 1: An example of a S−N+ scenario: According to general social knowledge p is necessary but insufficient for q.

primary condition p:	She has enough money.
consequent q:	She buys herself a dress.
alternative condition p_{alt}:	She has a credit card.
additional condition p_{add}:	The shops are open.

example at the very beginning, the two conditions ("Ann is in New York" and "It is still open") may be interpreted as conjunctive factors for her visit to the Guggenheim.

Having described how relevant background knowledge can be conceptualized, we were able to develop concrete task scenarios. The core of each scenario consists of a primary condition p and a consequence q, whose relation is one of the five patterns of dependence defined above. For each scenario we introduced an *alternative* condition p_{alt} and an *additional* condition p_{add}. As sufficiency and necessity are abstract concepts we did not expect effects due to highly domain-specific reasoning schemas (Cheng & Holyoak, 1985). But to allow for their possibility, each pattern of dependence was instantiated in three knowledge domains (*causal, social,* and *conceptual* relations). Combining both factors resulted in fifteen scenarios of which Table 1 presents an example.

In line with the methodological program outlined in the previous section, our first experiment assures that participants actually perceive our scenarios' presumed patterns of dependence as intended. In the following two experiments the influence of background knowledge on two different tasks is analyzed: Experiment 2 shows that suppression and facilitation in conditional syllogisms depends systematically on the premises' content and context. Experiment 3 demonstrates peoples' linguistic competence to frame their knowledge in formally correct wordings.

Experiment 1:
Rating Patterns of Dependence

The goal of this experiment was to validate our materials and to test a prediction about the mechanism of context effects. If different contexts trigger different aspects of background knowledge, the introduction of an extra condition should systematically alter the perceived sufficiency and necessity of the primary condition p: Mentioning an alternative condition p_{alt}, should specifically reduce the perceived necessity $\hat{n}$ of p, but not its perceived sufficiency $\hat{s}$. Likewise, we expect an additive condition p_{add} to reduce $\hat{s}$, but not $\hat{n}$.

Method

To assess perceived sufficiency $\hat{s}$ and necessity $\hat{n}$ we devised a *rating task*. Despite encouraging evidence by Thompson (1995), a pilot study showed that participants' conceptions of these notions varied considerably. The natural language usage of "sufficiency" in particular seemed susceptible to deviations from its logical semantics. In logic, the presence of a sufficient condition *necessitates* its consequence; in everyday-contexts, however, possession of a \$100 bill might be considered to be "sufficient" to buy a beer, yet does not necessitate its purchase. To avoid these linguistic issues, we asked for expected contingencies: If p is sufficient for q, it is *not possible* that p is the case without q being the case as well; therefore, whenever one knows that p, one would expect q. Similarly, when p is necessary for q, p *must* be the case whenever q is the case. Thus, assessing contingencies is logically equivalent to asking about sufficiency and necessity, but can be done in a more manageable format.

Materials To create a situation similar to the syllogism tasks to be used in Experiment 2, we wanted participants to notice conditional relations without actually asserting them. Therefore, each scenario introduced a conditional as the statement of an alter ego: "Suppose someone stated that '*if p, then q*'." We then pointed out that the actual truth of the conditional was irrelevant to the task, which was to rate – based on background knowledge – the degree of confidence that q is the case, provided that p, and vice versa. Both ratings were performed on a scale from 0–100%. Context was introduced by adding a statement like "No information about p_{alt} is available." Thus, an extra condition was mentioned, but the rating task nonetheless addressed the relation between p and q.

Procedure Thirty volunteers were recruited on the campus of University of Freiburg and randomly assigned to one of three groups. Each participant rated all 15 scenarios in a randomized order. Whereas the first group received only scenarios without context (only p/q), in the other groups an extra condition was always present (either p_{alt} or p_{add}, counterbalanced across groups).

Results and Discussion

As the manipulation of specific knowledge domains yielded no significant results it will be excluded from this and all subsequent analyses.

Figure 1 shows the mean ratings on both dimensions in the absence of any contextual condition. Overall ratings of perceived sufficiency $\hat{s}$ were significantly higher for S+ scenarios (with conditions that were meant to be sufficient) than for those of type S– (91.8%>21.3%; Wilcoxon signed ranks: z=6.7, n=60, p<.001). Likewise, perceived necessity $\hat{n}$ was larger for scenarios of type N+ than of type N– (93.3%>27.0%; z=6.6, n=60, p<.001).

Similar analyses of perceived sufficiencies and necessities within each single pattern of dependence indicated a deviation from our predictions only in one case: A significantly higher rating of $\hat{n}$ than of $\hat{s}$ within scenarios of type S–N– suggested that they do not correspond as close to our intentions as the others (34.3%>21.3%; z=3.16, n=30, p=0.002). But as both means still were within the negative range, this slight deviation is acceptable.

The overall effect of "context" can be seen in Table 2, aggregated over all five patterns of dependence. Whereas there was no difference in perceived sufficiency and necessity without extra conditions (53.2%≈58.3%; Wilcoxon: z=.97, n=150, p=0.33), the introduction of an alternative condition p_{alt} selectively reduced perceived necessity $\hat{n}$ (58.3%>46.5%; Mann-Whitney-U: z=3.15, n=150, p=0.002), but not perceived sufficiency $\hat{s}$ (53.2%≈53.7%; U: z=0.05, n=150, p=0.96). Mentioning an additional condition p_{add} instead had just the opposite effect of selectively

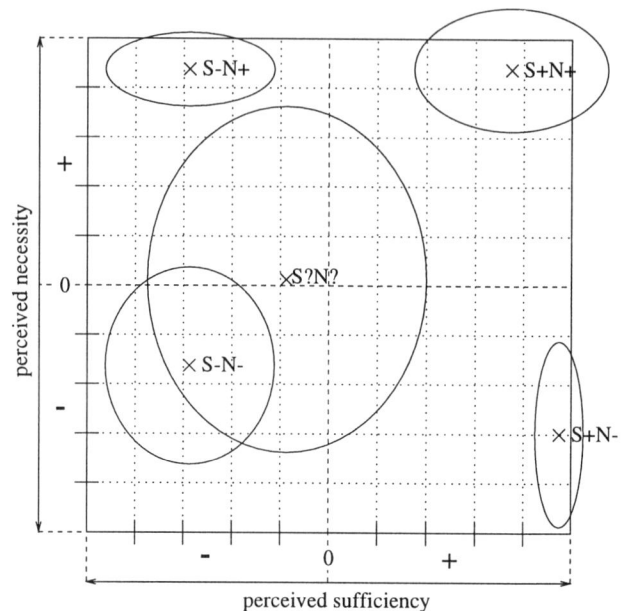

Figure 1: The mean perceived sufficiency and necessity of p for the five patterns of dependence in absence of any contextual condition. The ellipses extend one standard deviation in either dimension. Each mean represents 60 ratings.

reducing $\hat{s}$ (53.2%>40.4%; U: z=3.42, n=150, p=0.001) but not $\hat{n}$ (58.3%≈53.6%; U: z=0.98, n=150, p=0.33). As a whole, the pattern illustrated by Table 2 is a rating equivalent of Byrne's (1989) original suppression effect.

Both goals of Experiment 1 could be achieved: Participants' knowledge led them to perceive the scenarios as theoretically predicted, and the perceived patterns of dependence were selectively modulated by the introduction of extra conditions. This validates our materials and will be the basis to interpret the results of the following experiments.

Experiment 2:
Drawing Conditional Inferences

Correct conditional reasoning, that is, drawing the valid inferences of MP and MT while abstaining from the invalid ones of AC and DA, is one of the hallmarks to assess human logical competence. In our second experiment, we combined the methodologies employed by Byrne (1989) and Thompson (1994, 1995) to replicate and generalize their effects and explain them within our knowledge-based framework. With regard to the two predictive models (based on form vs. con-

Table 2: The effects of context (p_{alt}/p_{add}) on the ratings of sufficiency and necessity aggregated over patterns of dependence. (Standard deviations in brackets.) Each mean represents 150 ratings on a 0–100 scale.

Rated	Context		
dimension	only p/q	$+p_{alt}$	$+p_{add}$
Sufficiency $\hat{s}$	53.2 (37.7)	53.7 (40.5)	40.4 (33.7)
Necessity $\hat{n}$	58.3 (37.1)	46.5 (33.1)	53.6 (37.4)

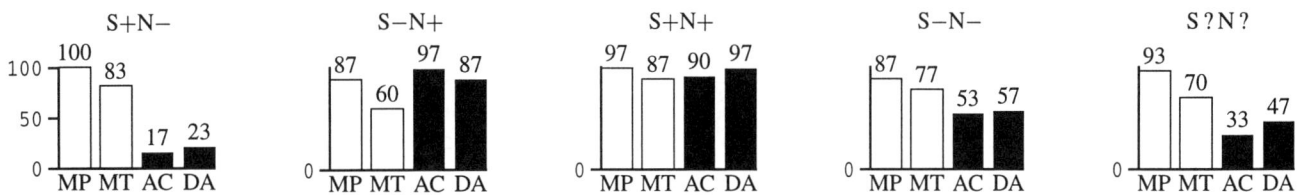

Figure 2: The percentages of inferences based on a single conditional 'If p then q' without an extra conditional as its context. High white and low black bars signify high percentages of formally correct inferences. Every bar represents the mean percentage over 30 trials.

tent) we predicted an increase in formally justified inferences when the predictions of both models coincide, and a corresponding decrease in the case of diverging predictions. As we understand this effect to be mediated by background-knowledge, we expect that it can be experimentally evoked in the same fashion by variations of content and context.

Method

Materials The same fifteen scenarios validated in Experiment 1 were now used to develop conditional syllogisms. Each item consisted of a primary conditional 'If p then q', a categorical premise (e.g. 'p' for MP) and a choice of three possible conclusions (e.g. 'q', 'not q' and 'nothing follows'). For each scenario all four conditional inferences were used (MP, MT, AC, and DA). In the context conditions, an extra conditional with an alternative or additive antecedent ('If p_{alt}/p_{add} then q') was inserted after the primary one.

As the propositions p and q of the primary conditional still exemplified each of the five patterns of dependence, a note on possibly "odd" premises is in place here: According to textbook logic, relations between a necessary and insufficient condition p and a consequence q (S−N+) have to be formalized as '$q \rightarrow p$' (e.g. "If Jack goes skiing, there's snow" – at least as long as alpine skiing is meant). In a natural language context, however, the inversion "If there's snow, Jack goes skiing" does not seem peculiar at all. On the contrary: Whereas the first sentence only states a notorious fact of common knowledge, its inversion conveys new information about Jack.

Procedure 50 volunteers from the University of Freiburg participated in the experiment. Each completed two booklets. The first contained a random sequence of all 24 conditional syllogisms that can be constructed from the primary conditional of the six scenarios of two different patterns of dependence.

After finishing the first booklet, the volunteers received another 24 items made up by the six scenarios of two other patterns of dependence. Now each syllogism was presented in the context of an extra conditonal, half of which had alternative, and half of which had additive antecedents. Thus, every participant took part in a "simple" and a "context" condition and responded to total of 12 scenarios from four different patterns of dependence with four different categorical premises. Position of patterns of dependence was counterbalanced across five groups, to which participants were randomly assigned.

Results and Discussion

The percentages of different conditional inferences with only a single conditional premise, 'If p then q', are represented in Figure 2.

As scenarios of type S?N? yield no knowledge-based prediction we used the inferences drawn under this condition as a baseline to assess the influence of background knowledge by other patterns of dependence. The last bar chart in Figure 2 shows that this baseline consists of a standard result: MP is endorsed nearly universally and more often than MT, and both valid inferences are drawn more often than the two fallacies of AC and DA.

Whenever it follows from background knowledge that a condition p is sufficient for a consequence q to occur (S+), this content prediction coincides with the formal interpretation of the conditional '$p \rightarrow q$', so that an overall increase of MP and MT inferences should result. Vice versa, if the content model tags a condition p as being insufficient for a consequence q (S−), we expected a decrease. However, only the first prediction was supported by our data: Whereas the mean frequency of MP and MT inferences was significantly higher for S+ scenarios than for our reference S?N? ($91.7\% > 81.7\%$; $\chi^2 = 4.01$, df=1, p<.05), they were not significantly reduced by S− scenarios ($77.5\% \approx 81.7\%$; $\chi^2 = .70$, df=1, n.s.).

Complementarily to the influence of perceived sufficiency on MP and MT, perceived necessity should influence the inferences of AC and DA. If, as in the sample scenario shown in Table 1, a condition p is known to be necessary for a consequence q (N+), the content-based prediction contradicts the formal properties of a conditional '$p \rightarrow q$', which defines p as being sufficient for q. As the content corresponds to a reversed formalization '$q \rightarrow p$', we expected an increase of inferences based on categorical premises 'q' and 'not p', i.e., AC and DA with respect to '$p \rightarrow q$'. Similarly, if condition p is known to be unnecessary (N−) a corresponding decrease should result. Again, only the first prediction was supported: The mean frequency of AC and DA inferences of both N+ scenarios was significantly higher than the corresponding frequency of scenario S?N? ($92.5\% > 40.0\%$; $\chi^2 = 68.91$, df=1, p<.01), but scenarios of type S− did not yield the opposite effect ($37.5\% \approx 40.0\%$; $\chi^2 = .16$, df=1, n.s.).

These analyses show a twofold result: Whenever the content predicts a definite conclusion (as in the S+ and N+ cases) this conclusion is likely to be drawn, whether it coincides with the formal prediction (as in the case of MP and MT) or not (as in the case of AC and DA). However, a content

Table 3: The percentages of inferences (out of 150) in three different contexts, aggregated over patterns of dependence.

Context	MP	MT	AC	DA
only p/q	92.7	75.3	58.0	62.0
$+p_{alt}$	92.7	78.0	16.0	14.0
$+p_{add}$	64.0	46.7	50.0	47.3

model that predicts the indefinite answer "nothing follows" does not significantly affect reasoning.

Effects of *context*, i.e., the introduction of an extra conditional premise with an alternative or additive antecedent, are depicted in Table 3. Compared to the frequencies in the condition without an extra conditional (only p/q), the predicted effects of facilitation and suppression were found: The frequency of AC and DA inferences was significantly reduced by the introduction of an alternative condition p_{alt} ($15.0\% < 60.0\%$; $\chi^2 = 253.13$, df=1, p<.01), and also by an additive condition p_{add} ($48.7\% < 60.0\%$; $\chi^2 = 16.06$, df=1, p<.01), though not to the same extent. Vice versa, the mean frequency of MP and MT inferences was significantly reduced in the additive context $+p_{add}$ ($55.3\% < 84.0\%$; $\chi^2 = 183.43$, df=1, p<.01), but not in the alternative context $+p_{alt}$ ($85.3\% \approx 84.0\%$; $\chi^2 = .40$, df=1, n.s.).

To sum up, the main predictions of this experiment were empirically supported: Through variations of content and context we were able to replicate and combine the results of Byrne (1989) and Thompson (1994, 1995), i.e., we found effects of suppression and facilitation in conditional syllogisms. By theoretically accounting for them within our general framework, we demonstrated that performance in conditional reasoning is modulated systematically by background knowledge.

Experiment 3: Selecting Appropriate Wordings

Psychological assessment of logical competence traditionally presupposes participants' proper understanding of semiformal premises, as used in verbal tasks of conditional reasoning. On the other hand, fallacious responses are frequently attributed to misinterpreted logical notions, as, for example, the confusion of conditional and biconditional relations (see Evans, Newstead, & Byrne, 1993, for an overview.) In a third experiment we shifted our focus from processes of language comprehension and reasoning to the selection of logically adequate formulations. If participants' linguistic competence indeed includes a basic mastery of the conditional connective, we would expect them to select different wordings for different patterns of dependence. Moreover, their selections should be altered systematically by the introduction of extra conditions p_{alt} and p_{add}.

Method

Materials To limit the range of possible answers we preferred a *multiple choice* format to a free formulation task. For each scenario, we first presented two conditionals, which, as in Experiment 1, were introduced by an alter ego: "When asked for the proper relation between p and q, someone proposed 'If p then q', and someone else 'If q then p' ". We em-

phasized that both, either one, or none of these propositions could express the actual relationship between p and q, and prompted participants to indicate which of these four possibilities was most appropriate according to their background knowledge. Due to the construction of the task, the four options corresponded to a biconditional ($p \leftrightarrow q$), conditional ($p \rightarrow q$ vs. $q \rightarrow p$), or non-conditional ($p \sim q$) interpretation. Manipulations of context were accomplished by mentioning an extra condition, as in "No information about p_{alt} is available."

Procedure Identical to Experiment 1.

Results and Discussion

As indicated in Table 4, participants were very sensitive to variations in patterns of dependence. Overall, 68.2% of all responses matched the predicted category. This percentage was highest for context 'only p/q' (79.2%), and lower for contexts '$+p_{alt}$' (64.0%) and '$+p_{add}$' (61.3%).

The correspondence between participants' choices and the predictions is best illustrated by an example. For the sample scenario of type S−N+ presented in Table 1, the two proposed conditionals were "If she has enough money, she buys herself a dress" and "If she buys herself a dress, she has enough money." Because having money (p) is a necessary but insufficient prerequisite for buying a dress (q), only the second sentence expresses the general relation between p and q appropriately ('$q \rightarrow p$'). 28 out of 30 participants had this intuition.

What should happen if extra conditions are mentioned? Stating that it is unknown whether or not "she has a credit card" provides participants with an *alternative* condition p_{alt}. The formally appropriate representation of this new situation '$q \rightarrow p \vee p_{alt}$' does not correspond to either conditional formulation. Therefore, if this new information is taken into account, the non-conditional category '$p \sim q$' ought to be

Table 4: The number of selected conditional relations by context and pattern of dependence. Each row represents a total of 30 choices. Predicted choices are **bold**-faced.

Context	Pattern of dependence	'$p \rightarrow q$'	'$q \rightarrow p$'	'$p \leftrightarrow q$'	'$p \sim q$'
only p/q	S+N−	**29**[**]	0	1	0
	S−N+	0	**28**[**]	1	1
	S+N+	1	4	**25**[**]	0
	S−N−[a]	1	4	1	**23**[**]
	S?N?	5	7	5	**13**[*]
$+p_{alt}$	S+N−	**28**[**]	0	0	2
	S−N+	4	**13**[*]	2	11
	S+N+	**15**[**]	3	9	3
	S−N−	1	2	2	**25**[**]
	S?N?	3	3	7	**17**[**]
$+p_{add}$	S+N−	**15**[**]	2	1	12
	S−N+	1	**27**[**]	0	2
	S+N+	2	**12**	10	6
	S−N−	1	5	2	**22**[**]
	S?N?	2	8	1	**19**[**]

*: $P(X \geq 13, 30, .25) < .05$. **: $P(X \geq 14, 30, .25) < .01$.
[a]: one missing value.

selected. As Table 4 ('$+p_{alt}$; S−N+') shows, 11 people followed this prediction. Vice versa, mere mentioning of the additional condition p_{add} "The shops are open" should not induce the same shift in responses, as the resulting formal representation '$q \rightarrow p \wedge p_{add}$' still implies '$q \rightarrow p$', and 27 participants indeed selected the latter category.

In general, participants selected formally appropriate wordings for the basic patterns of dependence, and tended to choose the predicted categories when extra conditions were introduced. Such specific shifts of preferred verbal expressions can be seen as another instance of the suppression effect. The fact that many participants selected the original option even when a change in context seemed to necessitate a shift suggests that some subjects chose to ignore the additional information.

Conclusions

The results of all three experiments back up our general approach to content effects that distinguishes between inferences based on the form of premises and inferences based on their semantic content. First, we confirmed that our experimental scenarios were perceived as theoretically intended. Second, both types of syllogism tasks (with and without an extra conditional premise) yielded the predicted effects of facilitation and suppression, which shows that reasoning performance depends systematically on the specific background knowledge triggered. Finally, a third experiment demonstrated that participants were able to select appropriate wordings for conditional relations although the syllogism tasks seemed to compromise their logical competence.

What do our results mean for the theoretical debate? Our analysis could be adapted to either a mental models framework or a mental rule theory. Presently, our knowledge representation is more similar to the mental models point of view, but a computer implementation uses formal rules to derive inferences from these representations (see Beller, 1997, for details). Thus, our account cannot resolve the debate between mental logic and mental models theories, but by addressing interpretive processes that are ubiquitous in human reasoning it fills a gap that has been neglected by both.

Suppression effects of content and context are multifaceted, knowledge-based phenomena. In order to explain their underlying mechanisms, syntactic and semantic accounts must be integrated rather than pitted against each other. A knowledge-based approach addresses questions of content without having to assert content-specific rules of inference or being commited exclusively to one theoretical framework. It adds predictive power to general theories of reasoning and helps to clarify the complex interplay of knowledge, reasoning, language, and logic.

Acknowledgements

We thank Hans Spada, Michael Scheuermann, and Josef Nerb for many helpful discussions on the topic of this paper. We are also grateful to Phil Johnson-Laird, Zachary Estes, Mary Newsome, and Yingrui Yang for their encouraging criticisms and suggestions.

References

Beller, S. (1997). *Inhaltseffekte beim logischen Denken – Der Fall der Wason'schen Wahlaufgabe. [Content effects in deductive reasoning: The case of Wason's selection task.].* Lengerich: Pabst Science Publishers.

Beller, S., & Spada, H. (1998). Conditional reasoning with a point of view: the logic of perspective change. In M. Gernsbacher & S. Derry (Eds.), *Proceedings of the Twentieth Annual Conference of the Cognitive Science Society* (pp. 138–143). Mahwah, NJ: Lawrence Erlbaum.

Byrne, R. M. J. (1989). Suppressing valid inferences with conditionals. *Cognition, 31,* 61–83.

Byrne, R. M. J. (1991). Can valid inferences be suppressed? *Cognition, 39,* 71–78.

Byrne, R. M. J., Espino, O., & Santamaria, C. (1998). Context can suppress inferences. In A. C. Quelhas & F. Pereira (Eds.), *Cognition and context* (pp. 201–214). Lisboa: Instituto Superior de Psicologia Aplicada.

Chan, D., & Chua, F. (1994). Suppression of valid inferences: Syntactic views, mental models, and relative salience. *Cognition, 53,* 217–238.

Cheng, P. W., & Holyoak, K. J. (1985). Pragmatic reasoning schemas. *Cognitive Psychology, 17*(4), 391-416.

Evans, J. St. B. T., Newstead, S. E., & Byrne, R. M. J. (1993). *Human reasoning: The psychology of deduction.* Hove: Lawrence Erlbaum Associates.

Fillenbaum, S. (1993). Deductive reasoning: What are taken to be the premises and how are they interpreted? *The Behavioral and Brain Sciences, 16*(2), 348–349.

Henle, M. (1962). On the relation between logic and thinking. *Psychological Review, 69*(4), 366–378.

Johnson-Laird, P. N., & Byrne, R. M. J. (1991). *Deduction.* Hove, England: Lawrence Erlbaum Associates.

Politzer, G., & Braine, M. D. (1991). Responses to inconsistent premises cannot count as suppression of valid inferences. *Cognition, 38,* 103-108.

Rumain, B., Connell, J., & Braine, M. D. (1983). Conversational comprehension processes are responsible for reasoning fallacies in children as well as adults: *If* is not the biconditional. *Developmental Psychology, 19*(4), 471-481.

Smedslund, J. (1970). On the circular relation between logic and understanding. *Scandinavian Journal of Psychology, 11,* 217–219.

Thompson, V. A. (1994). Interpretational factors in conditional reasoning. *Memory and Cognition, 22*(6), 742–758.

Thompson, V. A. (1995). Conditional reasoning: The necessary and sufficient conditions. *Canadian Journal of Experimental Psychology, 49*(1), 1–60.

Content, Context and Connectionist Networks

Lars Niklasson

Department of Computer Science
University of Skövde, S-54128, SWEDEN
lars@ida.his.se

Mikael Bodén

Department of CS and EE, University of Queensland,
4072, AUSTRALIA/
Department of Computer Science
University of Skövde, S-54128, SWEDEN
mikael@ida.his.se

Abstract

The question whether connectionism offers a new way of looking at the cognitive architecture, or if its main contribution is as an implementational account of the classical (symbol) view, has been extensively debated for the last decade. Of special interest in this debate has been to achieve tasks which easily can be explained within the symbolic framework, i.e., tasks which seemingly require the possession of a systematicity of representation and process, in a novel way in connectionist systems. In this paper we argue that connectionism can offer a new explanational framework for aspects of cognition. Specifically, we argue that connectionism can offer new notions of compositionality, content and context-dependence based on connectionist primitives, i.e., architectures, learning, weights and internal activations, which open up for new variations of systematicity.

Introduction

Ever since Fodor and Pylyshyn (1988) published their seminal paper in which they defined the relation between systematicity (i.e., the systematic structure of mental representations and the structure-sensitivity of mental processes), compositionality (i.e., the method of composing/decomposing structured mental representations) and the cognitive architecture, the debate has been intense. It has had two main research agendas; i) to exhibit and explain systematicity in connectionist systems (Smolensky, 1990; van Gelder, 1990; Pollack, 1990; Chalmers, 1990; Niklasson and Sharkey, 1992, Niklasson van Gelder, 1994, Phillips, 1994) and ii) to question the relevance of the systematicity and compositionality phenomenon altogether (Goschke and Koppelberg, 1991; van Gelder and Niklasson, 1994; Matthews, 1994).

The success of the early connectionist counter examples was questioned by Hadley (1992, 1994a). He noted that in many of these examples the success might have been due to the constitution of the training set. He therefore re-formulated systematicity in a learning-based fashion, defining different levels of systematicity depending on the content of the training set. He identified three levels of systematicity:

- weak systematicity (concerned with generalization to novel sentences in which tokens appear in syntactic positions in which they have appeared during training),
- quasi systematicity (requires weak systematicity and embedded structures),
- strong systematicity (requires quasi systematicity and generalization across syntactic positioning of tokens).

Hadley argued that no counter-example had achieved the strongest form of systematicity, with the possible exception of Niklasson and van Gelder (1994). Hadley was, however, concerned about the approach adopted by Niklasson and van Gelder for generating representations. They used a separate network which encoded syntactic information in order to generate similar distributed representations (i.e., close in the representational space) for tokens of similar types, which caused Hadley (1994b) to classify their result as a 'border line' case.

Recently, Phillips (1998) pointed out that connectionist architectures using localistic representations, by themselves cannot account for strong systematicity. But also, that this restriction does not preclude separate mechanisms for generating similarity based representations, which could be used in subsequent systematicity tasks. He outlined two research directions; develop architectures which could support systematicity under localistic input/output representations, or justify similarity-based distributed representations sufficient for allowing systematicity.

The former of these directions is exemplified by Hadley and Hayward (1997) when they showed that a network could achieve an even stronger form of systematicity; semantic systematicity, defined as:

> A system possesses semantic systematicity if it is strongly systematic and it assigns appropriate meanings to all words occurring in novel test sentences which (would or could) demonstrate strong systematicity of the network (Hadley, 1994b, p. 434).

The intention of this paper is to take the latter research direction pointed out by Phillips (1998), i.e., to justify similarity-based representations sufficient for systematicity. Two forms of justifications can be identified; i) empirical justification of the exact boundaries of the systematicity phenomenon (an analytic approach), or ii) a technical justification related to the representational primitives of connectionist architectures (a synthetic approach).

We will, in the following, take the latter of these and define meaning (i.e., content) in relation to connectionist architectures, learning, weights and internal activations, and show that the implications of this approach are rather different compared to the notions within classicism. We argue that our view allows systematic processes which are sensitive not to the syntax of the representation (which is a cornerstone of the traditional definition) but instead to the *context* in which the *content* of the representations is defined. This context could naturally also include processing of expressions based on their syntactic structure.

To substantiate our arguments, we will present some examples and performance results which assign the appropriate meaning (admittedly, somewhat different than defined by Hadley) to novel test cases. The examples can also be used to indicate how our approach can be empirically validated or refuted.

If our approach is accepted, that would allow connectionists to go beyond traditional symbol processing and account for context-dependent semantic systematicity.

Content of connectionist representations

In a classical system appropriate meaning can be assigned to words depending the structural positioning of their representational tokens. This is due to the definition of a classical system which hinges on the possession of a combinatorial syntax and semantics for mental expressions. The definition states that the content of a complex representation is a function of the meaning of the constituents, together with the constituent structure of the representation.

The kind of connectionist system we have in mind, does not possess *syntactically* structured representations. Instead it relies on the possession of *spatially* structured representations, formed as a result of an individual learning situation.

The main difference between the two approaches is what can be assumed when trying to extract content from the representations. In a connectionist system, the spatial structure is the result of a specific learning situation. It is generally not, contrary to the classical approach, possible to assume a surrounding context when constructing representations.

In order to objectively compare the two approaches we argue that they must be allowed to make the same assumptions about the context when constructing systematicity examples. Rather than using natural language examples (where it is difficult to define the complete contextual framework), we will here use a different domain including reasoning with both defaults and exceptions.

We propose the following understanding of content and context within the connectionist framework:

- The only *context* supplied to a network is defined in terms of its training set. Any *content* found in the representations depends on this context.
- The organization of the representations is not arbitrary, rather it depends on the *context* expressed in the training set.

- The weights operating on the representations extract the *content* expressed in them. Therefore, the weights and the internal dynamics of the receiving units can be used to define *content* which entails that an explanation to systematic processes working on the representations, is possible.

One possible objection to this view could be the definition of content, but we argue that it is in line with Palmer's (1978) definition of information in cognitive representation:

> The only information contained in a representation is that for which operations are defined to obtain it (Palmer, 1978, p. 266)

Figure 1 exemplifies some of the above points. Two input units (x and y) are connected to a logistic output unit (z) with weights -3.7 and 9.4 respectively. In addition to this, a -3.2 bias weight is connected to the output unit. The figure shows how a particular weight configuration partitions the input space, allowing the extraction of the content of three sample input (A, B and C in Figure 1) representations to the network.

The context (i.e., the relations expressed in the training set) in this example is that representations B and C belong to the same category ($z=0$), and A to a different one ($z=1$). As can be seen in the figure, the network has in this particular case learned this classification. It is also clear that we now can use the weights in the trained network to extract the content in the input representations, even for novel ones, and identify spatial regions for the different classes.

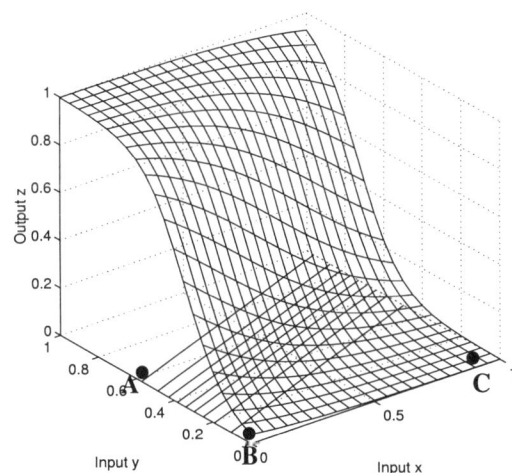

Figure 1: The result of a particular learning situation. The lines represent different values for z (0.9 to 0.1)

What this simple example does not show is how the organization of the representations (i.e., the locations to points in the n-dimensional representational space) can be made sensitive to the particular context expressed in the training set. For this, we need to extend the architecture with features allowing learnable representations for tokens. Here

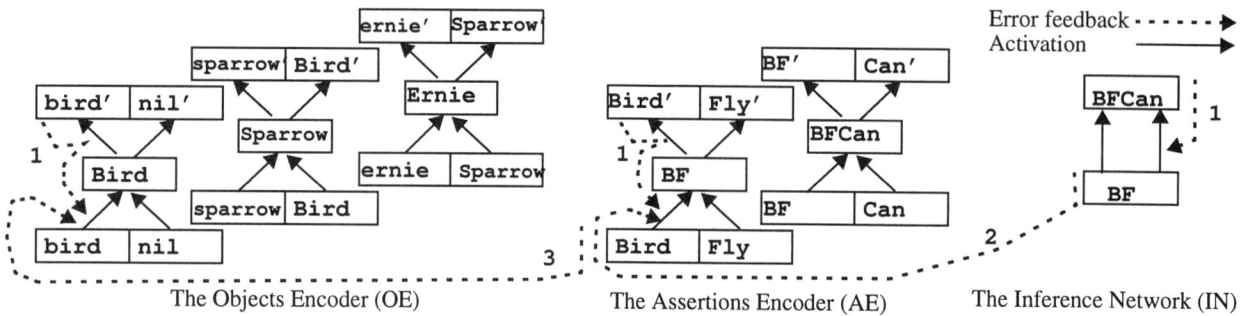

Figure 2: The complete architecture. The OE and the AE are standard RAAM networks. It should be noted that the figure show the same network but at different stages of the encoding process. Numbers indicate different error feedback; 1 is the feedback within a network, 2 is feedback between the IN and the AE and 3 is feedback from the AE to the OE. The between-network feedback is possible since a separate dictionary is used to store all representations (e.g., **Bird**) and the relation to its constituents (e.g., **bird** and **nil**).

we use the same architecture as used by Bodén and Niklasson (1999), see Figure 2. This architecture is intended to incorporate different kinds of context. Here we will use a simple hierarchical taxonomy (e.g., that **Sparrows** are **Birds** and that **Ernie** is a **Sparrow**) and, in the Objects Encoder (OE), generate compositional representations based on this context. In addition to this, the representations needed to train and test the Inference Network (IN) need to be generated (i.e., (**Bird Fly**), ((**Bird Fly**) **Can**)). This is done by the Assertions Encoder (AE). Finally the particular inferences valid in a particular context (e.g., that birds in fact can fly) are trained in the IN. The encoders are standard RAAM networks (Pollack, 1990) and the inference network is related to Chalmers' (1990) transformation network. The main difference from Chalmers is the use of between-network error feedback (which is an extension to Chrisman's (1991) confluent representations).

The within- and between-network error feedback (see Figure 2) allows that the representation for an object (e.g., **Ernie**) is affected by its relation to other objects in the domain, the assertions it appears in and the valid inferences it is part of. It will therefore in the following be referred to as contextual feedback.

An illustrative example

The OE was trained to encode the following:

```
Node:                   Denoted by:
OE(bird nil)            Bird
OE(sparrow Bird)        Sparrow
OE(ernie Sparrow)       Ernie
OE(penguin Bird)        Penguin
OE(tweety Penguin)      Tweety
```

The AE was trained to encode the following assertions:

```
AE(Bird Fly)     AE((Bird Fly) Can)
AE(Sparrow Fly)  AE((Sparrow Fly) Can)
AE(Penguin Fly)  AE((Penguin Fly) Cannot)
AE(Ernie Fly)    AE((Ernie Fly) Can)
                 AE((Ernie Fly) Cannot)
AE(Tweety Fly)   AE((Tweety Fly) Can)
```

```
AE((Tweety Fly) Cannot)
```

For **Ernie** and **Tweety** both possible inferences (i.e., **Can** and **Cannot**) were generated for test purposes. The IN was trained to do the inferences:

```
AE(Bird Fly)    -> AE((Bird Fly) Can)
AE(Penguin Fly) -> AE((Penguin Fly) Cannot)
AE(Sparrow Fly) -> AE((Sparrow Fly) Can)
```

The relations encoded in the OE and the valid inferences in the IN can be visualized as (see Figure 3):

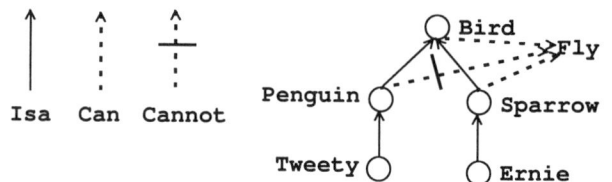

Figure 3: Graphical representation of the domain.

The main purpose of this simplified example is not to show that the architecture can handle both defaults and exceptions, but rather to relate the points made about context and content to a specific example. It, however, shows why the this particular network can generalize to the novel situations:

```
AE(Ernie Fly)    AE(Tweety Fly)
```

For visualization purposes the dimensionality of the hidden layer of the encoders was reduced to two units. The OE was a 12-2-12 sequential RAAM (i.e., the left input slot, in Figure 2, had a size of 10 units and the right had the same as the hidden layer). The representations for the atomic objects (i.e., **bird**, **sparrow**, etc.) were assigned a 10-element localistic non-overlapping representation. The AE was a 4-2-4 sequential RAAM, and the IN a 2-2 feed-forward network.

The hidden space for the encoded objects is shown in Figure 4(a). In this diagram, the hyperplanes for weights connected to the two output units (i.e., units 11-12, represented in the figure by 1 and 2 respectively) representing classes (i.e., **Bird**, **Sparrow** and **Penguin**) are

476

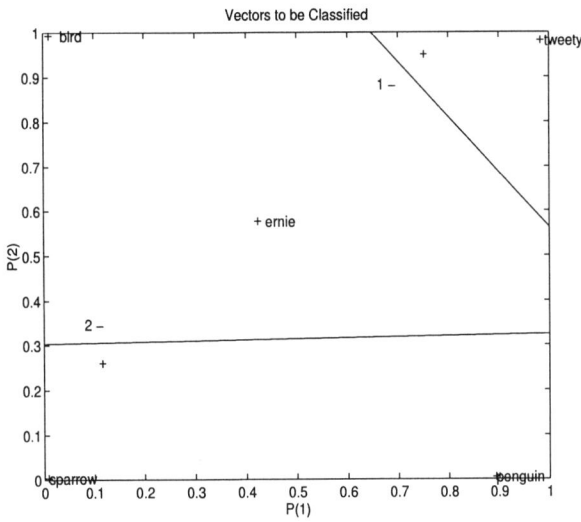

(a)

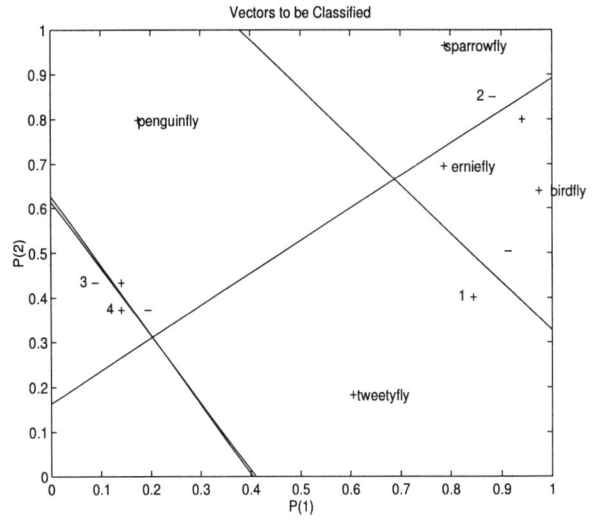

(b)

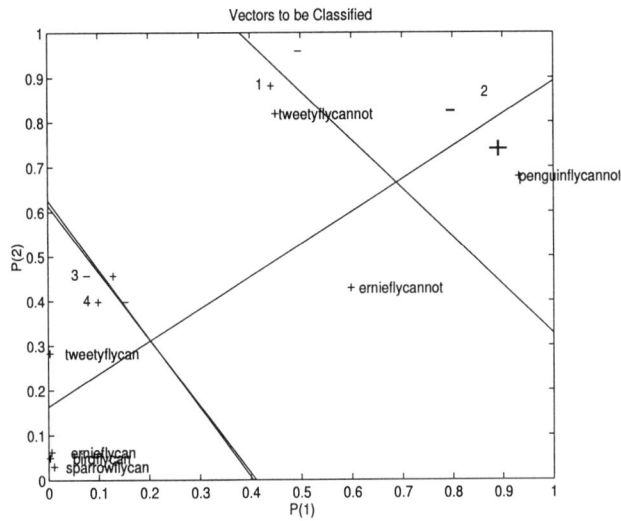

(c)

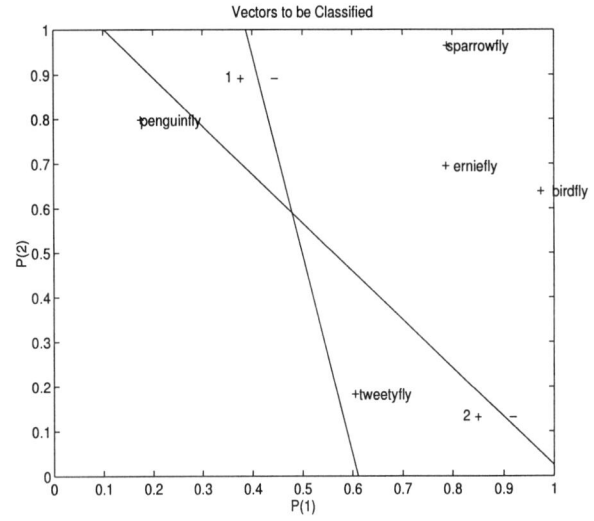

(d)

Figure 4: The hidden space of the OE (a), the AE (b and c) and the IN (d). Please note, that all the locations in the space (e.g., bird) actually are the generated representation after training (e.g., **Bird**).

also included. The first thing to note is that the representation for **Bird** (top left corner) has become close to [0 1]. This means that all members of the class **Bird** (i.e., **Penguin** and **Sparrow**) must end up on the negative side of the hyperplane enforced by the first class unit (i.e., unit 11), and the positive side of the other (i.e., unit 12). The representational region in the OE for members of the **Bird** class therefore becomes the region between the positive side of the second hyperplane and the x-axis. Similarly, the region for members of the **Sparrow** class (represented by [0 0]) becomes the region between the two hyperplanes, and the **Penguin** class the top-right region in the diagram. The reason that **Bird** ([0 1]) ends up in the **Sparrow** class region, is that the representation chosen for **nil** is [0 0] which is the same as the one developed for the **Sparrow** class. That, however, is not

essential for the current purposes.

In the assertion space (Figure 4b and 4c) the representational regions for **Fly**, **Can** and **Cannot** are easy to identify. These regions are defined by output units 3 and 4 in the AE. **Fly** and **Cannot** are located to the positive side of 3 and negative side of 4, and **Can** vice versa. Moreover, the findings for the objects space are useful also in the assertion space, which in turn is the space used by the IN. It is possible to identify the region of the assertion space in which, for instance, new members of the **Penguin** class will end up. We can note that both units of the OE (the x and y axis in Figure 4a) will receive an activation above 0.5 for members of the **Penguin** class. This means that all new members of the class will be on the positive side of

477

the units 1 and 2 in the AE when combined with `Fly` (Figure 4b), i.e., the region in which `TweetyFly` now appears. Combining this with the hyperplanes of the IN (Figure 4d) it is possible to define the region for which the IN will make 'correct' inferences concerning `Penguin`. A location on the positive side of both hyperplanes means that the inference network will transform the location to one above [0.5 0.5] in the assertion space (Figure 4c), which always is classified as a no-`Fly` zone, by hyperplanes 3 and 4. One could also note that not all members of the `Penguin` class are guaranteed to actually end up the positive side of these hyperplanes. Some (e.g., those which receive an activation in the AE of about [1.0 0.3] which are likely to be transformed in the IN to a position close to 0 for the x axis and definitely below 0.5 for the y axis) novel member of the `Penguin` class will be classified as flying. We will in the following simulation see examples of this.

Context-dependent processing

Let us now turn to the remaining issue to be resolved; i.e., the impact of the context for solving practical problems. We will here refer to simulations reported elsewhere (cf. Bodén and Niklasson, 1999). In a series of simulations we examined the performance of the architecture on problems involving defaults and exceptions. Two contexts for some test objects (see D1 and D2 Figure 5) were used to evaluate the performance of the architecture. Of special interest was to evaluate the effect of the contextual feedback between the different sub-networks, allowing fully context-dependent representations.

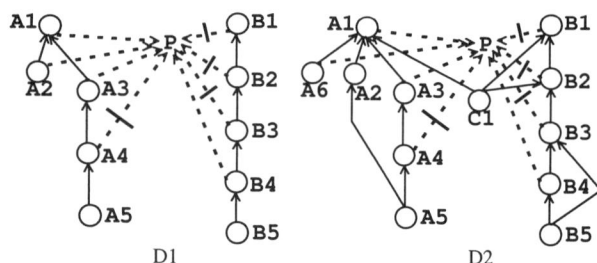

Figure 5: Two sample contexts, D1 and D2.

The architecture was trained on the two contexts (D1 and D2). After training, it was tested which content (`P+` or `P-`) was assigned to the test objects (`A5` and `B5` for both D1 and D2, and `C1` for D2). The content assigned by the IN was compared to the representations formed in the AE, and the one with the closest Euclidean distance was chosen. This can, for reasons explained earlier, be somewhat misleading but this approach do not favor one outcome over the other, which means that an average over several runs will give an objective result. In the first run on D1 the inference on `A5P` (i.e., the output of the IN) was 1.300 from `A5P+` and 0.116 from `A5P-`.

For D1, the size of the RAAMs used were OE 10-3-10, AE 6-3-6 and IN 3-3. For D2, the size were 12-3-12, 6-3-6 and

3-3. For each experiment 30 runs were conducted with contextual feedback enabled, and 30 with it disabled. Training was conducted for 10000 epochs with learning rate 0.1 and momentum 0.9. The results are listed in Table 1.

Table 1: Results from the D1 and D2 data sets

Context	Object	% P+	% P-	Contextual feedback
D1	A5	30	70	Yes
D1	A5	37	63	No
D1	B5	77	23	Yes
D1	B5	62	38	No
D2	A5	10	90	Yes
D2	A5	20	80	No
D2	B5	43	57	Yes
D2	B5	13	87	No
D2	C1	20	80	Yes
D2	C1	47	53	No

Some interesting observations can be made. Generally the architecture supports shortest path reasoning, e.g., for `A5` and `B5` in D1. Contextual feedback accentuates this preference, which is most obvious for `B5` in D2, where the path `B5->B3->P+` is as long as `B5->B4->P-`. The results show that architecture with feedback assigns positive or negative content with almost equal probability (without feedback 13% vs. 87%, and with feedback 43% vs. 57%). Compare this to `A5->A4->P-` and `A5->A2->A1->P+`, where the feedback has increased the bias for the shorter of the paths.

The most obvious reflection one can make, is that the effect on `C1` in D2 is rather dramatic. Without feedback the two outcomes occur with almost equal probability. With feedback the preference is for `P-`. This example can be compared to an extension of the famous Nixon diamond, i.e., Nixon is a quaker, republican and colonel. Quakers are pacifists, but republicans and colonels are not. One way of reasoning is that since the majority of categories of which Nixon is a member are non-pacifists, he is too.

Conclusion

We have argued that connectionism can offer alternatives to classical explanations for cognitive phenomena provided that content and context are defined in terms more natural to connectionist architectures, learning, weights and internal activations. Such definitions were provided and connected to an example.

If the approach we suggest is accepted, it is possible to explain not only context-independent reasoning (see Niklasson and van Gelder (1994) who used a related architecture for syntactic transformations), but also *context-dependent reasoning*, by referring the performance exhibited on data sets like D1 and D2.

Phillips (1998) noted that the networks used by Niklasson and van Gelder (1994) could not support systematicity at the compositional level, only at the component level. The approach used in this paper shows that connectionist architectures can support systematicity at both levels, by incorporating contextual feedback.

We have shown how compositionality and context-dependence can co-exist within the same framework. The explanation we supply is based on weight regions expressing spatial structure which mirror contextual similarities among representations.

We also argued that connectionist and classicist systems should be allowed to make the same assumption about the example domain. Here two rather small data sets were used and cannot give the complete story but they can serve as useful indicators of what to look for. It would be quite easy to define an empirical investigation of how humans perform on D1 and D2. If the performances of humans differ significantly from the performance of our architecture, this would be quite damaging for our argument. If not, our view would be justified both on technical and empirical grounds.

Acknowledgment

This paper was made possible by a grant from The Foundation for Knowledge and Competence Development (1507/97), Sweden, to the first author, an Australian Research Council grant to the second author and a grant to both authors from the University of Skövde, Sweden.

References

Bodén, M. and Niklasson, L., (1999), Semantic systematicity and context in connectionist networks, (submitted to *Connection Science*, Carfax Publishers Ltd.).

Chalmers, D. J., (1990), Syntactic Transformation on Distributed Representations, *Connection Science*, **Vol. 2**, Nos 1 & 2, pp 53 - 62.

Chrisman, L., (1991), Learning Recursive Distributed Representation for Holistic Computation, In *Connection Science*, **Vol. 3**, No. 4, pp. 345 - 366.

Fodor, J. A. & Pylyshyn Z. W., (1988), Connectionism and cognitive architecture: A critical analysis, In *Connections and symbols*, Pinker, S. and Mehler, J., (eds.), MIT Press, pp. 3 - 71.

Goschke, T. and Koppelberg, D., (1991), The Concept of Representation and the Representation of Concepts in Connectionist Models, In Ramsey, W., Stich, S. and Rumelhart, D. E., (eds.), *Philosophy and Connectionist Theory*, LEA, pp. 129 - 162.

Hadley, R. F., (1992), Compositionality and Systematicity in Connectionist Language Learning, *Proceedings of the Fourteenth Annual Conference of the Cognitive Science Society*, pp. 659 - 664.

Hadley, R. F., (1994a), Systematicity in Connectionist Language Learning, *Mind and Language*, vol. 9, no. 3.

Hadley, R. F., (1994b), Systematicity Revisited, Mind and Language, vol. 9, no. 4, Blackwell Publ., pp. 431 - 443.

Hadley, R. F. and Hayward, M. B., (1997) Strong Semantic Systematicity from Hebbian Connectionist Learning, *Mind and Machines*, 7, pp. 1 - 37.

Matthews, R. F., (1994), Three-Concept Monte: Explanation, Implementation, and Systematicity, *Synthese*, 101, Kluwer Academic Publishers, pp. 347 - 363.

Niklasson, L. F. and Sharkey N. E., (1992), Connectionism and the Issues of Compositionality and Systematicity, *Cybernetics and Systems Research*, Trappl (ed.), World Scientific, pp. 1367 - 1374.

Niklasson, L. F. and van Gelder, T., (1994), Can Connectionist Models Exhibit and Explain Non-Classical Structure Sensitivity, *Proceedings of the Sixteenth Annual Conference of the Cognitive Science Society*, LEA, pp. 664 - 669.

Palmer, S. E., (1987), Fundamental Aspects of Cognitive Representation, *Cognition and Categorization*, Rosch, E. and Lloyd, B. B., (eds.), LEA, Hillsdale, NJ, pp. 259 - 303.

Phillips, S., (1994), Strong Systematicity within Connectionism - The Tensor Recurrent Network, *Proceedings of the Sixteenth Annual Conference of the Cognitive Science Society*, LEA, pp. 723 - 727.

Phillips, S., (1998), Are Feedforward and Recurrent Networks Systematic? - Analysis and Implications for a Connectionist Cognitive Architecture, *Connection Science*, Carfax Publishing Ltd., vol 10, no 2, pp. 137 - 160.

Pollack, J. B., (1990), Recursive Distributed Representations, *Artificial Intelligence*, **46**, pp 77 - 105.

Smolensky, P., (1990), Tensor Product Variable Binding and the Representation of Symbolic Structures in Connectionist Systems, *Artificial Intelligence*, **46**, pp 159 - 216.

van Gelder, T., (1990), Compositionality: A Connectionist Variation on a Classical Theme, *Cognitive Science*, **Vol. 14**, pp. 355 - 364.

van Gelder, T. and Niklasson L., (1994), Classicalism and Cognitive Architecture, *The Proceedings of the Sixteenth Annual Conference of the Cognitive Science Society*, LEA, pp. 959 - 964.

Methods for Learning Articulated Attractors over Internal Representations

David C. Noelle
(NOELLE@CNBC.CMU.EDU)
Center for the Neural Basis of Cognition
Carnegie Mellon University
Pittsburgh, PA 15213 USA

Andrew L. Zimdars
(ZIMDARS@ANDREW.CMU.EDU)
School of Computer Science
Carnegie Mellon University
Pittsburgh, PA 15213 USA

Abstract

Recurrent attractor networks have many virtues which have prompted their use in a wide variety of connectionist cognitive models. One of these virtues is the ability of these networks to learn *articulated attractors* — meaningful basins of attraction arising from the systematic interaction of explicitly trained patterns. Such attractors can improve generalization by enforcing "well formedness" constraints on representations, massaging noisy and ill formed patterns of activity into clean and useful patterns. This paper investigates methods for learning articulated attractors at the hidden layers of recurrent backpropagation networks. It has previously been shown that standard connectionist learning techniques fail to form such structured attractors over internal representations. To address this problem, this paper presents two unsupervised learning rules that give rise to componential attractor structures over hidden units. The performance of these learning methods on a simple structured memory task is analyzed.

Introduction

Connectionist attractor networks have been used to model many aspects of cognitive performance, involving domains as diverse as word naming (Plaut and McClelland, 1993), cognitive control (Cohen et al., 1996), and conscious awareness (Mathis and Mozer, 1995). Such networks have many virtues. Processing element activity evolves over time in complex ways in such networks, allowing them to capture various aspects of the dynamics of cognition. Learned recurrent connection weights often lend themselves to interpretation as soft constraints between representational elements, facilitating analysis of such models. One of the most interesting advantages of these networks, however, is the manner in which the learning of attractor basins can aid generalization of performance.

Attractor networks can learn to enforce "well formedness" constraints on representations, and this process of "cleaning up" patterns of activity can facilitate generalization (Mathis and Mozer, 1995). Such enforcement is implemented by the instantiation of a distinct stable fixed-point attractor for every possible well formed representation. It is important to note that this potentially combinatoric space of valid attractor basins need not be explicitly trained, but may arise in the interaction between trained patterns (Plaut and McClelland, 1993). When the dynamics of a network includes such a compositional space of meaningful attractors, arising from the interplay of trained patterns, we refer to the network as possessing *articulated attractors*.

Previous work has shown that standard connectionist learning techniques spontaneously give rise to such structured attractors when recurrent connections are present at the output layer of the network, but they *fail* to learn such attractors over units which do not receive a direct teaching signal (Noelle and Cottrell, 1996). In other words, networks with recurrent connections only at a hidden layer cannot learn articulated attractors from backpropagated error. This means that recurrent backpropagation networks cannot learn to actively maintain a componential representation without having the structure of that representation *explicitly* specified by a teacher or by the environment.

This result poses a problem for cognitive models which employ attractor networks as a form of working memory, since it shows that such networks cannot learn internal representations for a task and simultaneously learn to robustly remember those representations over time. Even models which involve the learning of attractors at an output layer may be challenged by this result, as the output representation of such models is often conceptualized as a learned internal representation in a larger cognitive system. For example, the recurrent phonological output layer of some word naming models (Plaut and McClelland, 1993) is trained with an explicitly structured teaching signal, despite the fact that phonology is thought to involve a learned internal coding scheme.

This paper discusses some preliminary efforts to discover connectionist learning methods which will give rise to articulated attractors over learned internal representations. Specifically, two unsupervised learning rules are presented, simulated, and analyzed.

A Structured Memory Task

To facilitate analysis, we focused on an extremely simple task. Each attractor network was to learn to act as a kind of working memory, maintaining a presented pattern of activation indefinitely, given only a brief initial exposure to that pattern. Furthermore, the networks were to discover regularities in the corpus of presented patterns, identify the structure of well formed patterns, and use that knowledge to "clean up" noisy patterns, thereby enforcing "well formedness" constraints. This task is depicted schematically in Figure 1. Note that the input pattern is made available to the network for the first few time steps only, requiring the network to both "clean up" and remember the pattern over time.

Specifically, each network was briefly presented with an encoding of a simple slot-filler structure. The network was to filter out any noise in this representation and continuously present the resulting clean pattern at the network's output, even after the input was removed. Thus, the network needed

480

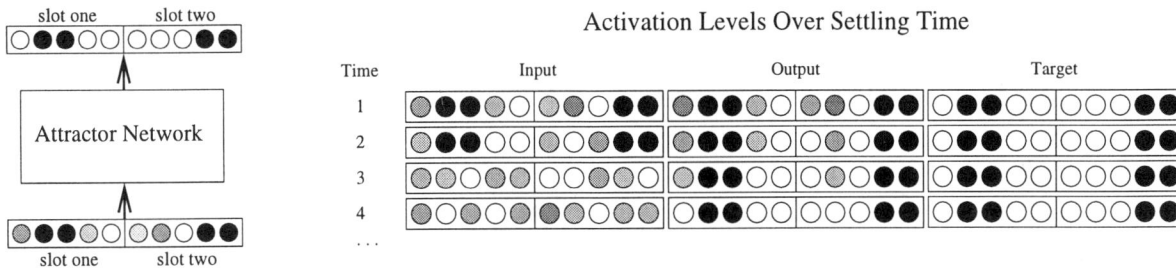

Figure 1: The Slot-Filler Structured Memory Task

to learn a distinct stable attractor for each valid slot-filler structure. Each input pattern represented a structure containing two slots, each holding exactly one of five distinct fillers. The contents of the slots were considered independent, with the specific filler in one slot in no way constraining the filler for the other. The fillers for each slot were encoded over a 5 element binary vector, resulting in a 10 element vector for the entire structure. Each of the five fillers was encoded as a unique pair of adjacent "on" elements.[1] Thus, well formed patterns included exactly two of the first five elements "on" and exactly two of the last five elements "on". With five possibilities for each slot, there were only $5^2 = 25$ "well formed" patterns out of the $2^{10} = 1024$ possible binary input vectors.

These simulation experiments focused on systematic generalization. The networks were to learn an attractor for every valid slot-filler structure, given training on only a fraction of the valid patterns. In order to investigate such generalization, each network was explicitly trained on some subset of the well formed input patterns. Once trained, each network was then tested on *all* valid slot-filler representations, and the number of attractors corresponding to these valid patterns was determined. The dynamic behavior of each trained network was also examined to locate any spurious attractors corresponding to ill formed patterns.

In order to generalize in a systematic way, the network needed to learn two interrelated properties of these input patterns. First, it needed to recognize that the patterns consisted of two *independent* slots. Second, the network needed to identify the structure of the filler patterns, constructing attractors for patterns involving two adjacent active units but not for other cases (e.g., 3 of the 5 units in a group active). It is important to note that a localist code (i.e., 1 of 5 units "on" for each filler) was *not* used here. Such a localist code would make the identification of the two independent slots very difficult. If such a code were used and the training set of the network left out but a single valid pattern — a single pair of slot fillers — the network would discover that the two units for that pair were perfectly anticorrelated in the training set and would hinder the formation of an attractor for the novel pattern involving that pair of fillers. Thus, for a network to discover the independence of the two slots by attending to the pairwise statistics of pattern element values, some form of coarse coding of fillers was needed.

Even with such coarse coding of fillers, a training set consisting of only a few valid patterns displays very little inher-

ent structure. As training sets get larger, however, the underlying slot-filler structure of the input patterns becomes evident. In order to examine this dependence on training set size, networks were trained with varying sized collections of well formed patterns. Each training set contained at least five patterns, as this was the minimum number needed to present each filler pattern to the network at least once. The largest training set consisted of all 25 well formed patterns. The frequency of each filler value in each training set was balanced as much as was possible given the small size of the training sets. During training, zero mean independent Gaussian noise with 0.025 variance was added to each input element. Noise was resampled on every time step, and it persisted even after the input pattern was removed. Network output targets consisted of the "clean" patterns over the entire course of network settling, as shown in Figure 1. A settling period of 10 time steps was used during training, and 100 time steps were used during testing. In our initial simulations, the input pattern was presented for only a single time step.

Problems With Internal Representations

It has previously been shown that network learning techniques based on the backward propagation of error, such as *backpropagation through time (BPTT)* (Rumelhart et al., 1986), can learn appropriate articulated attractors for this structured memory task, but only if an external teaching signal is provided directly to the processing elements which are recurrently connected (Noelle and Cottrell, 1996). Consider the simulation results plotted on the left side of Figure 2. In this graph, the size of the training set is displayed along the horizontal axis, and the resulting number of valid attractors learned is specified vertically. If a given network failed to generalize, only learning attractors for the training set patterns, then data should fall along the displayed unit slope reference line. But this graph displays substantial generalization. Training sets consisting of 15 or more of the 25 valid slot-filler patterns resulted in networks which generalized to all 25 of them. The attractor network exhibiting this good generalization performance contained recurrent connections between the units of its output layer and was trained using BPTT. This means that the teaching signal strictly specified the structure of the activation vectors over which attractors were to form. This external structuring signal is not directly available to recurrent weights, however, when only hidden units are recurrently connected. This results in a failure to generalize when attractors are required to form over an internal representation — over the hidden layer of the network —

---
[1] Each group of five elements was conceptualized as forming a closed loop for the purposes of determining adjacency.

481

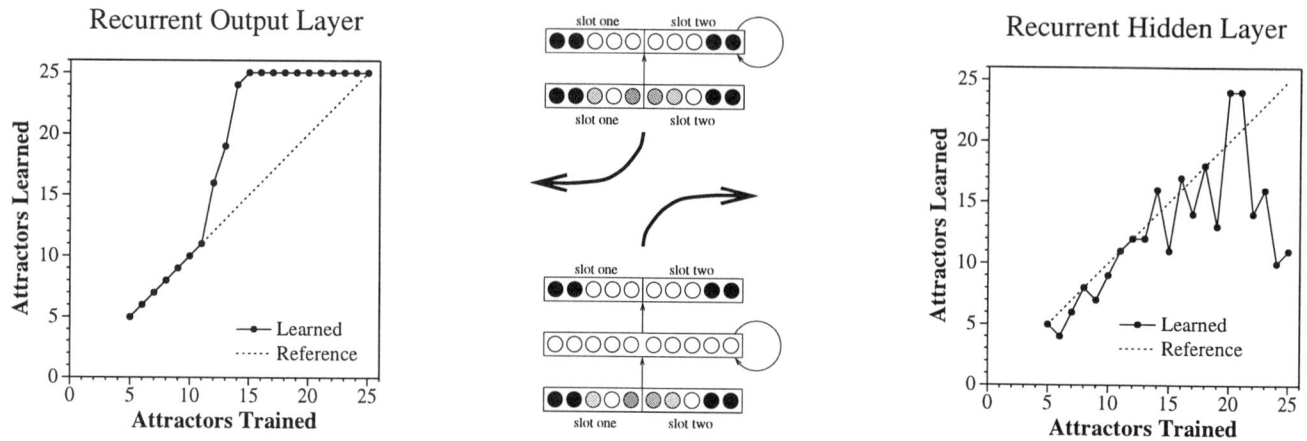

Figure 2: Generalization Performance of BPTT With Recurrence at the Output Layer vs. the Hidden Layer

as shown in the graph on the right side of Figure 2. Without the teaching signal directly enforcing a compositional structure on the recurrently connected layer of units, the network can barely construct stable attractors for the training patterns.

Detailed analyses of these poorly generalizing networks suggested four main reasons for their failures:

(1) *Localist Coding.* The network sometimes developed hidden layer representations for individual slot fillers which were approximately localist in nature. Even though the inputs and output targets were coarse coded, the hidden layer used a single unit to represent a given filler value. This introduced the previously mentioned problem of localist coding to the recurrent weights. The recurrent connections would learn to inhibit novel, yet valid, slot value pairs.

(2) *Small Weights Have Large Effects Over Time.* In typical training sets, each slot filler is somewhat anticorrelated with all of the fillers for the other slot, because individual fillers appear relatively rarely. This leads to small amounts of inhibition between the representations for fillers in opposite slots. This slight inhibition typically poses no problem for the maintenance of a pattern over the course of the 10 time steps that the network is allowed to settle during training, but extending the settling time beyond 10 steps (as is done during testing) allows the network to drift, slowly losing the activation pattern that it had maintained. In other words, the network does not actually learn stable attractors for patterns, but, instead, learns to drift away from well formed patterns slowly enough so as to not impact network error noticeably during the 10 settling time steps of training.

(3) *The Error Gradient Disappears Near Solutions.* Another reason that these networks fail to form stable attractors is that such attractors typically require a particular configuration of large weight values. Starting with small random weights, the connection strengths grow towards the needed values, but network error decreases as those weights are approached, causing the weights to change less and less. Thus, the network remains in a region of weight space corresponding to the "slow drift" strategy, never quite arriving at the magnitude of weights required for stable attractors.

(4) *Polarization.* The first behavior that these networks acquire is the reproduction of the input pattern across the output units during the initial time period when the input is still

directly available. This occurs without much use of the recurrent weights. The hidden layer representation that forms early in training tends to be distributed and graded. The graded nature of this representation makes it hard to form the needed attractors. It is much easier to learn stable fixed-point attractors involving extreme activation values rather than involving activity levels in the linear range of the unit activation function (Noelle et al., 1997).

We sought to modify the learning procedures of our networks so as to alleviate these problems, and we found two very distinct and somewhat successful mechanisms for learning articulated attractors over internal representations.

Asymmetric Hebbian Learning

Our approach involves learning the recurrent connection weights using an unsupervised learning method. The idea is to restrict the pattern of connectivity at the hidden layer so as to promote the formation of attractors, and then learn the recurrent weights in a manner which is independent of the teaching signal being provided at the output of the network. A sketch of the network architecture used appears in Figure 3. Notice that the hidden layer in this architecture contained 20 units, and a new layer of 20 "interneurons" was added. In the network diagram, the thin solid arrows represent sparse random connectivity between the layers, with connection weights bound to be non-negative. The dashed lines specify one-to-one patterns of connectivity, with each hidden unit possessing a non-negative connection to itself and a fixed positive connection to its "partner" unit in the interneurons layer. The bold line from the interneurons layer to the hidden layer represents nearly complete interconnectivity, with weights *bound to be non-positive*. Each interneuron projected to all hidden units except for its hidden layer "partner". Thus, the architecture provided each hidden unit with a positive self connection to maintain it's own activity and a partner interneuron to inhibit other hidden units.

The feedforward connections, from input to hidden and hidden to output, were trained using the standard BPTT algorithm. The self connections on each hidden unit were also learned using the backpropagated error signal, as were the unit biases. The one-to-one connections between the hidden units and the interneurons had fixed weights, forcing

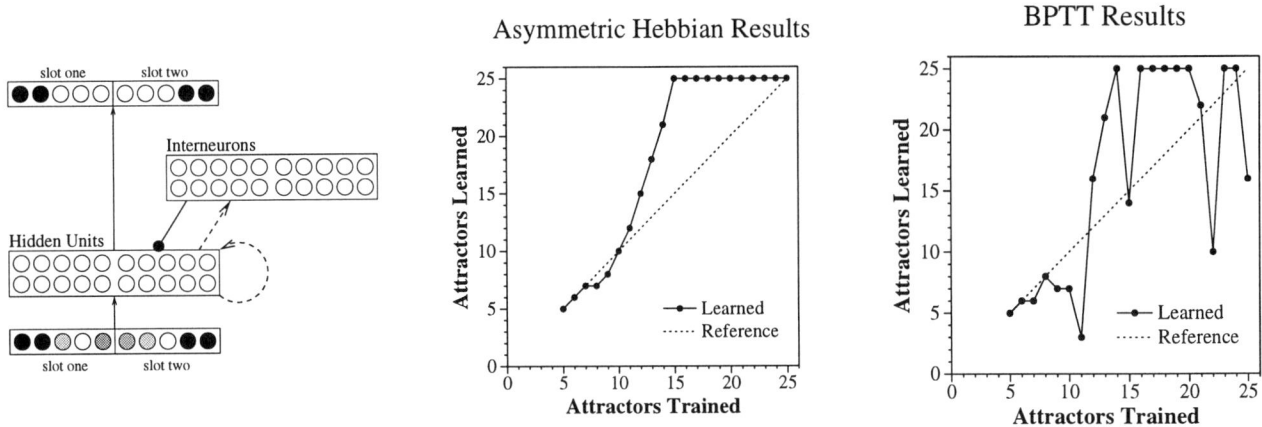

Figure 3: Network Architecture & Results for Asymmetric Hebbian Learning Technique Along With BPTT Control

the partner interneuron to come on whenever its hidden unit became active. The inhibitory connections projecting from the interneurons used the following new asymmetric Hebbian learning procedure:

- If the interneuron is in the upper half of its activation range, and the hidden unit it projects to is also highly active, then decrease the magnitude of the connecting weight by a large amount, proportional to the product of their activities.

- If the interneuron is highly active, and the hidden unit it projects to is not, then lower the connecting weight value slightly, proportional to the product of the interneuron activity and one minus the hidden unit activity.

This learning rule tends to zero out weights associated with pairs of hidden units that have been active together, while allowing hidden units that are never co-active to acquire negative connections between them, mediated by an interneuron.

This network architecture and learning technique avoids the problem of localist codes at the hidden layer (problem 1, above) by using sparse random patterns of connectivity between the input and hidden layers and between the hidden and output layers. Each unit in these layers received input from only five randomly selected units from the preceding layer. Since the inputs and targets were coarse coded, this pattern of connectivity encouraged the appearance of a similar coarse code at the hidden layer. By pushing weights strongly towards zero if a co-occurrence of activity is detected, this scheme also avoids the problem of small inhibitory weights (problem 2, above). The problem of the disappearing gradient (problem 3, above) is alleviated by using a Hebbian scheme which is insensitive to the backpropagated error. Lastly, the problem of developing polarized hidden layer representations (problem 4, above) is handled by initializing the hidden unit self weights to large values, encouraging the network to make use of the extreme hidden unit activations which naturally arise from such positive feedback.

Simulations of this learning mechanism were conducted with the self connection weights initialized to values uniformly sampled from [5.5, 6.5], feedforward weights sampled from [0.0, 0.2], the weights leading to the interneurons fixed at 6, and the weights from the interneurons initialized to

zero. The unit biases were initialized to values sampled from [−2.5, −3.5] and were bound to be non-positive. The bias values on the interneurons, however, were fixed at −3. All weights trained using BPTT, including the bias weights, used a learning rate of 0.1. The weights trained using the asymmetric Hebbian rule used a learning rate of 0.002 when raising weight values towards zero and a rate of 0.0001 when making weights more negative. A weight decay regularization, reducing each weight magnitude by 0.001% on each weight update, kept the Hebbian weights from growing without bound. All processing elements used the logistic activation function, bounding output activity between 0 and 1.

The results of these simulations are shown in the middle of Figure 3. Notice that the network successfully generalized to all 25 patterns for training sets of size 15 and larger. An analysis of the dynamics of these networks revealed more than these 25 stable fixed-point attractors, however. A total of 96 attractors appeared for ill formed input vectors, as well. These spurious attractors had an interesting structure, however. They all consisted of cases in which a group of five units encoding a slot had either one element active or no elements active. In other words, the network successfully learned that the two slots were independent, but came to treat both "no units on" and "one unit on" as valid filler values. In essence, the trained network treated the two adjacent active elements in every valid slot filler as independently modifiable features. This makes sense considering that the recurrent connections in this network could only encode independence (zero weight) or anticorrelation (negative weight).

One might wonder how much of the success of this network was driven by the asymmetric Hebbian learning algorithm and how much relied on the restricted architecture and weight initializations. To investigate this, networks which were identical in architecture and initial weight configuration were trained using BPTT. The weights from the hidden layer to the interneurons remained fixed, but the connections projecting from the interneurons were trained used a backpropagated error signal. The results of these simulations are shown in Figure 3, on the right. While this diagram appears to display good generalization for the larger training set sizes, it hides a multitude of sins. Every training set which resulted in attractors for all 25 well formed patterns also resulted in a stable attractor for *any* of the 1024 possible binary vectors. It appears as if the network architecture and weight initialization scheme provided the means for retaining patterns over time, but the asymmetric Hebbian learning rule provided the means for discerning well formed patterns.

While this learning method generalizes much more effectively than any investigated method involving backpropagated error, it still has a number of problems. In an effort to avoid small negative weights, the learning rule ignores less than perfect correlations between unit activation levels. This forces the network to treat the components of coarse coded filler values as independent (zero weight between them) when they are not. One might imagine augmenting these networks with separate recurrent weights which are bound to be *non-negative*, hoping that these weights would capture the positive correlations between the components of filler representations. Unfortunately, the inclusion of such weights reintroduces the problem posed by small negative weights — such positive weights interfere with the formation of stable attractors. As settling time increases, such weights tend to result in virtually *all* of the hidden units becoming active. Another weakness of this asymmetric Hebbian learning strategy is that it requires the formation of an appropriate coarse code at the hidden layer. The feedforward connections must be tightly restricted in order to ensure that the elements of the input filler representation remain separated in the hidden layer representation. Thus, while this learning method allows articulated attractors to form over internal representations, it only works for a restricted class of such representations.

Competitive Inhibition Learning

In hopes of overcoming the need for tight restrictions on the feedforward weights, our attention shifted toward finding a method for learning the recurrent weights which could effectively handle localist encodings of filler values. As previously discussed, learning to dissociate the two slots is impossible with a localist code *if pairwise correlations in unit activity are used to make the dissociation*. A training set communicates more information than just pairwise statistics, however. If an input unit, A, has been seen co-active with another input element, B, then, under a localist coding scheme, these two elements must code for filler values in separate slots. If a third input element, C, is never seen co-active with either A or B, standard learning mechanisms would come to build inhibition between C and the other two elements. But we know, given a localist code, that C only participates in one slot, so it should only inhibit either A or B, but *not both*. Thus, by attending to these ternary relationships, we can start to identify alternative fillers for a single slot as opposed to novel pairs of fillers across slots.

For this new learning approach we used the same network architecture as before, with one exception. Since localist encodings of filler values were actually desired at the hidden layer in this case, sparse random connectivity in the feedforward connections was not needed. Instead, each unit in the input layer was connected to every hidden unit, and each hidden unit was connected to every output. Once again, the feedforward weights were bound to be non-negative and were trained using standard BPTT. As before, the one-to-one weights projecting to the interneurons were of fixed positive

values. The inhibitory weights projecting from the interneurons were modified according to a competitive learning rule:

- If a hidden unit is in the upper half of its activation range, and an interneuron projecting to it is also highly active, then decrease the magnitude of the connecting weight by a large amount, proportional to the product of their activities.

- If a hidden unit is in the lower half of its activation range, find all highly active interneurons projecting to it and identify the interneuron which inhibits this hidden unit most strongly. This interneuron is the *winner* of the competition. Its weight is made more negative by a moderate amount, proportional to the product of the activity levels of the two units. Active interneurons which lose this competition have the magnitude of their weights decreased by a large amount, proportional to the product of activations.

- If a hidden unit is inactive, but no interneurons are strongly inhibiting it, it needs to join a new group of mutually inhibiting units. This is done by identifying the strongest weight to this hidden unit from the interneurons, and raising this weight towards zero while making all other weights slightly more negative.

The first part of this rule makes sure that only one hidden unit is active in any group of mutually inhibitory units. The second part further separates the slots by ensuring that a unit is never inhibited by units belonging to different slots. The third part keeps the network from creating too many separate groups of mutually inhibitory units by insisting that at least one unit be active in each group.

We found that standard BPTT training did not produce sufficiently strong positive self connections, so a different learning algorithm was used on the self weights. Whenever a hidden unit was in the upper half of its activation range and experienced a negative change in activity level, it effectively received an error signal driving the unit towards its maximum activation level, resulting in an increase in the self weight.

Simulations involving this learning mechanism initialized hidden unit self weights to values uniformly sampled from $[3.5, 4.5]$ and feedforward weights sampled from $[0.0, 0.2]$. The weights leading to the interneurons were fixed at 8, and the weights from the interneurons were initialized to zero. All unit biases were initialized to -2, though these biases were adapted via BPTT in all units except the interneurons. All weight values trained using BPTT used a learning rate of 0.1. The hidden unit self connections used a learning rate of 0.01. The competitive inhibition rule essentially set weights mediating between co-active hidden units to zero, used a learning rate of 0.01 for "winning" interneurons and a rate of 0.1 to reduce the magnitude of weights from "losing" units. When an inactive hidden unit failed to be inhibited by any interneuron, its strongest weight was raised with a learning rate of 0.1 while other weights were made more negative with a learning rate of 0.001. The weights projecting from the interneurons were bound to be no lower than -4. Lastly, to promote the use of a localist code at the hidden layer, the feedforward weights were subjected to the "weight elimination" procedure (Weigand et al., 1991) with a decay rate of 0.001.

Since this learning method was expected to thrive on localist encodings, the simpler localist encoding of slot fillers

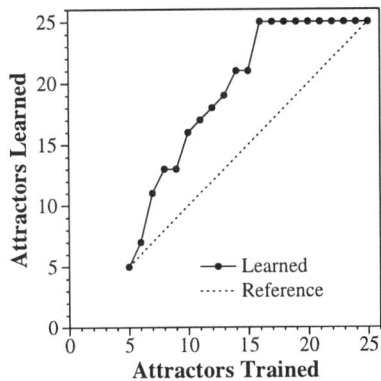

Figure 4: Results for Competitive Inhibition Learning

was used in the inputs and targets for these simulations. Each filler was encoded as one unit active out of the group of five. Also, the input pattern was presented for 4 initial time steps in these simulations rather than one. This aided the learning of strong self weights at the hidden layer.

The results of these simulations are shown in Figure 4. Notice that the networks generalized perfectly for training sets of size 16 and greater. As before, the networks overgeneralized, producing attractors for 11 ill formed patterns. Once again, these spurious attractors had an interesting structure, always involving a slot with no units active. The networks were able to identify the two independent slots, but came to treat "no units on" as a valid filler value.

While this learning method overcame the requirement for a coarse coding of slot fillers in the hidden layer, it introduced its own constraints on the internal representation in use. Specifically, the hidden layer representation of a slot filler had to be localist in nature. If there were multiple hidden units encoding a particular filler, only one of these units would be recruited to participate in the mutual inhibition between fillers for this slot. The unrecruited units would either come to participate in other groups of mutually inhibiting units, making generalization to novel combinations of slot fillers less likely, or they would "stand on their own", potentially encouraging overgeneralization.

Conclusions

We have presented two new methods for the learning of articulated attractors over internal representations. These techniques gain their strength in large part from the use of a restricted network architecture. These learning methods also benefit from being unsupervised — from basing weight updates on an inherent inductive bias rather than on a backpropagated error signal. Such a bias appears necessary for the formation of true stable attractors at hidden layers.

These two learning rules do not completely resolve the problem of hidden layer articulated attractors. Both methods are only successful when the network is constrained to develop certain kinds of internal representations. The asymmetric Hebbian learning method requires a coarse coding of fillers at the hidden layer, while the competitive inhibition method works best with a localist code. Future work will focus on modifying these learning rules so as to allow for more

general distributed representations at the hidden layer.

It might be the case, however, that active maintenance of a pattern of activation requires some restricted coding scheme. Indeed, some researchers have argued that the working memory functions of dorsolateral prefrontal cortex require representations involving isolated components (Cohen et al., 1996). Perhaps the key to learning articulated attractors at a hidden layer, then, is learning an internal representation with an appropriate componential structure.

Acknowledgements

This work was supported, in part, by the NIH through a National Research Service Award (# 1 F32 MH11957-01) from the National Institute of Mental Health, awarded to the first author. We extend our thanks to James L. McClelland, Randy O'Reilly, and the members of the CMU *PDP Research Group* for their comments and suggestions. Thanks are also due to three anonymous reviewers for their helpful advice concerning the clear presentation of this research.

References

Cohen, J. D., Braver, T. S., and O'Reilly, R. C. (1996). A computational approach to prefrontal cortex, cognitive control and schizophenia: Recent developments and current challenges. *Philosophical Transactions of the Royal Society of London B*, 351:1515–1527.

Mathis, D. W. and Mozer, M. C. (1995). On the computational utility of consciousness. In Tesauro, G., Touretzky, D. S., and Leen, T. K., editors, *Advances In Neural Information Processing Systems 7*, pages 11–18, Denver. MIT Press.

Noelle, D. C. and Cottrell, G. W. (1996). In search of articulated attractors. In Cottrell, G. W., editor, *Proceedings of the 18th Annual Conference of the Cognitive Science Society*, pages 329–334, La Jolla. Lawrence Erlbaum.

Noelle, D. C., Cottrell, G. W., and Wilms, F. R. (1997). Extreme attraction: On the discrete representation preference of attractor networks. In Shafto, M. G. and Langley, P., editors, *Preceedings of the 19th Annual Conference of the Cognitive Science Society*, page 1000, Stanford. Lawrence Erlbaum.

Plaut, D. C. and McClelland, J. L. (1993). Generalization with componential attractors: Word and nonword reading in an attractor network. In *Proceedings of the 15th Annual Conference of the Cognitive Science Society*, pages 824–829, Boulder. Lawrence Erlbaum.

Rumelhart, D. E., Hinton, G. E., and Williams, R. J. (1986). Learning internal representations by error propagation. In *Parallel Distributed Processing: Explorations in the Microstructure of Cognition*, volume 1, chapter 8, pages 318–362. MIT Press, Cambridge, Massachusetts.

Weigand, A. S., Rumelhart, D. E., and Huberman, B. A. (1991). Generalization by weight-elimination with application to forcasting. In Lippman, R. P., Moody, J., and Touretzky, D. S., editors, *Advances In Neural Information Processing Systems 3*, pages 875–882, Denver. Morgan Kaufmann.

Procedures are Only Skin Deep: The Effects of Surface Content and Surface Appearance on the Transfer of Prior Knowledge in Complex Device Operation

Tenaha O'Reilly (toreilly@ualberta.ca)
Department of Psychology, University of Alberta
Edmonton, Alberta, Canada T6G 2E9

Peter Dixon (peter.dixon@ualberta.ca)
Department of Psychology, University of Alberta
Edmonton, Alberta, Canada T6G 2E9

Abstract

In this research, we investigated the factors that mediate the use of prior knowledge in learning new procedures. Participants learned to operate two different versions of four tasks on a hypothetical device interface. At a conceptual level, all devices were operated in the same way. However, in some conditions, the appearance of the two versions was manipulated by changing the graphical appearance of the interface. A second manipulation concerned the physical layout: The position of the device controls, graphics, and gauges was either the same or different from one version to the next. Providing the same appearance and providing the same physical layout both increased the amount of transfer. These effects were additive, suggesting that the factors contribute independently to learning. Our interpretation is that appearance affects the use of semantic constraint, while layout affects the use of structural analogy.

Introduction

The aim of the present investigation was to discover the effect of prior knowledge on the acquisition of novel procedures. More specifically, we were interested in determining how superficial task features are used to activate relevant prior knowledge. Ross (1987, 1989) demonstrated that in solving probability problems, superficial similarity in content has a large effect on performance. Performance was relatively poor, for example, when a study problem involved salesmen choosing mechanics to work on a car and the test problem involved mechanics choosing which salesman's car to work on. Learners were affected by these superficial correspondences even though, as far as the solution of the problem was concerned, these characteristics were entirely arbitrary. Furthermore, when participants were explicitly cued to the corresponding aspects of previously learned problems (e.g., "This is like the earlier problem with the golfers getting prizes"), they show increased transfer compared to those who were not so cued (Ross & Kennedy, 1990). Taken together, the results suggest that domain novices are often affected by the irrelevant (superficial) aspects of the problem when they should be more concerned by the deeper (structural) components.

In a similar vein, we investigated the effect of superficial features in the procedural domain of device operation. We propose that the use prior procedural knowledge is mediated by two processes: semantic constraint and structural analogy. Constraint operates by restricting the nature of the actions that are deemed to be appropriate at each step of the procedure. For instance, familiarity with other electronic devices might dictate that an initial step should be some form of power-up or reset operation. Structural analogy fosters transfer by mapping steps from a familiar procedure onto the steps required in a novel procedure. For example, a novel telephone might be operated by mapping the previously learned steps needed to make a phone call onto the corresponding steps needed for the new phone. In previous research, we have found that both processes contribute to transfer when a person is explicitly cued to use their prior knowledge (O'Reilly and Dixon,1999). However, superficial similarity (such as a label applied to the subprocedure) seemed to affect only the semantic constraint process, not the use of structural analogy.

The goal of the present research was to further investigate the nature of the superficial similarity cues that might be involved in the use of prior knowledge in general and semantic constraint specifically. To address this issue, we asked participants to learn eight different tasks on different device consoles, with the last four tasks constituting different versions of the first four. The different versions of a task could differ in physical appearance (with different logos and graphical feedback) or layout (with different control positions and groupings). Despite the different superficial appearances between the versions, the operating procedures for both could be interpreted as having the same semantic context. For example, if the steps for one version included pressing buttons labeled STORE, TRANSFER, INNER, the steps for the second version would have the semantically similar labels MEMORY, MOVE, and CENTRAL. By manipulating both the appearance and layout, we hoped to identify some of the factors that contribute to superficial similarity.

Method

Participants

The sample consisted of 24 introductory psychology students from the University of Alberta who were given extra credit for their participation.

Materials

Seventeen hypothetical device consoles were simulated on a computer screen. The device consoles consisted of a distinctive dialog box with labeled buttons; subjects operated the devices by using a computer mouse to click on the buttons in a particular order. All device consoles contained a set of ten buttons. Six of the buttons were used to carry out the procedure, and four served as distracters to make the task more difficult. One device was used as practice; the remaining 16 consisted of 4 different versions of 4 test procedures on a hypothetical weather monitoring device. These procedures were Start-up, Sensor Activation, Data Location, and System Cleaning. For each procedure, there were two different appearances consisting a distinctive font, manufacturer name and logo, and style of meter (used for feedback). In addition, for each device, appearance was crossed with two alternative spatial arrangements of the buttons, meter, and graphics. Participants were provided with a two-page instruction booklet which described the nature of a device console, the number of consoles they were required to learn, and the consequences of making both correct and incorrect button sequences. The instructions also explained that each console was operated by pressing 6 buttons in sequence. Finally, it was mentioned that the first four consoles were different and unrelated to one another while the second set of four consoles *could* be either similar or dissimilar to the previous four device consoles.

Design and Procedure

Participants were 24 University of Alberta undergraduates. Participants learned 9 tasks to a criterion of two correct trials in a row. First, participants learned a task using the practice device console. This was followed by four training and then four transfer tasks. The corresponding training and transfer devices always used the same procedure but the appearance could be either the same or different, and the physical layout could be either the same or different from training. Corresponding button labels between the two versions of a device were roughly synonymous. All four combinations of same or different appearance and layout were used for each participant. The order of these four transfer conditions as well as the particular devices and versions used in condition were counterbalanced across subjects.

Before beginning each task, all participants were asked to carefully read the two page instruction booklet. After reading the instructional materials participants were encouraged to operate the device. When a participant made an error, the task was aborted by the computer, feedback was given regarding the correct alternative that should have been pressed, and the participant was required to start the task

from the beginning. When a participant made a correct button press, the reading of the feedback meter increased to a value corresponding to the total number of correct steps completed.

Results and Discussion

For each participant, four relative error rates were obtained by dividing the total number of errors on a training task by the total number of errors on the corresponding transfer task. Due to the variability of this measure, the resulting error rate was transformed logarithmically. That is, transfer was defined as:

$$T = \ln \frac{N_2}{N_1} = \ln N_2 - \ln N_1 \qquad (1)$$

where N_1 is the number of errors on the training task, and N_2 is the number of errors on the transfer task. Figure 1 shows the amount of transfer as a function of version and layout; higher scores indicate more transfer. The results can be summarized as follows: Providing the same physical layout increased the amount of transfer from training, and providing the same appearance aided the transfer process but to a somewhat larger extent. These effects were almost perfectly additive.

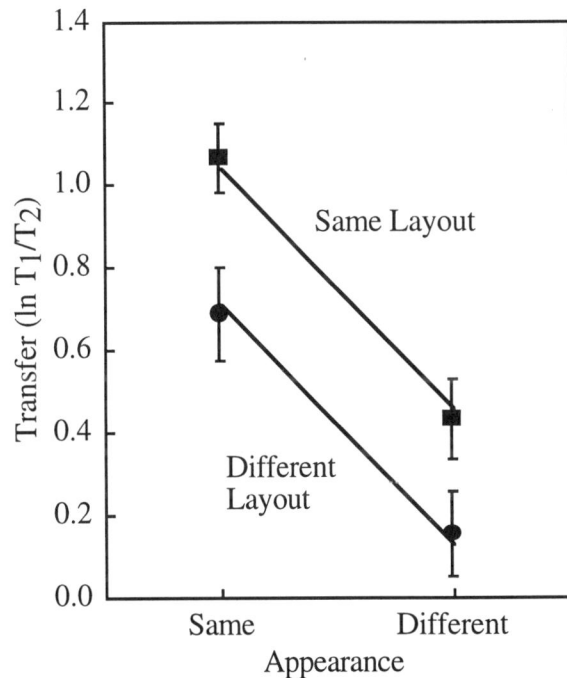

Figure 1. Transfer as a function of condition. Points on the graph represent the mean of the log relative errors in each condition, while the lines represent the predictions generated by a purely additive model. Error bars indicate standard errors of the mean.

This pattern of results can be readily interpreted in terms of the processes of semantic constraint and structural analogy. For simplicity, we assume that each step is learned independently in an all-or-none manner. Under this assumption, the total number of errors on a task is given by:

$$N = \frac{1-r}{s} \qquad (2)$$

where r is the probability of guessing the step correctly and s is the probability of learning the step on a given trial. This learning model is essentially a variation of the all-or-none (one element model) described by Bower (1961). Further, we assume that semantic constraint limits the number of possible buttons that might be considered on each step and hence is reflected in the parameter r. In contrast, the successful formation of a structural analogy would mean faster learning of the steps of the procedure, and hence would be reflected in the parameter s. Combining Equations 1 and 2 allows one to rewrite transfer as the sum of two components:

$$
\begin{aligned}
T &= \ln N_1 - \ln N_2 \\
&= \left[\ln(1-r_1) - \ln s_1 \right] - \left[\ln(1-r_2) - \ln s_2 \right] \\
&= \left[\ln(1-r_1) - \ln(1-r_2) \right] + \left[\ln s_2 - \ln s_1 \right] \\
&= R + S
\end{aligned}
\qquad (3)
$$

where R reflects the increment from training to transfer in the guessing rate (based on semantic constraint) and S reflects the change in how quickly individual steps are learned (based on structural analogy processes). In other words, factors having selective effects on semantic constraint and structural analogy should be additive.

Based on this analysis, our interpretation of the present results is that superficial appearance affects semantic constraint, while layout affects the formation of an analogy. For example, participants might use the graphical logo or the manufacturer's name as a cue for retrieving information about a previously learned procedure; when these are the same from training to transfer, participants may be able to recall the semantic content of the steps that are needed for the new task. In contrast, the results for the layout factor suggest that the formation of an analogy with an earlier task is based on the physical location of the controls. Consequently, when a new layout is used, participants have difficulty mapping the current steps onto those learned previously. The major implication of the present pattern of results is that these processes of constraint and analogy can go on separately and independently, at least in tasks like the present one.

A similar patter of results was obtained by O'Reilly and Dixon (1999) using a different set of tasks and different manipulations. In that study, participants learned two longer tasks that were divided into labeled subprocedures. Performance on the second (transfer) task was examined as a function of whether or not the labels were the same as those used in the first (training) task and whether or not the steps in each subprocedure were in the same order. Analogous to the present results, labels and step order had additive effects on transfer. The interpretation is that presence of identical labels fostered the use of semantic constraint, while having the steps in the same order made the process of mapping the steps from the current task to the earlier one easier. The fact that these effects were additive suggests, as here, that constraint and analogy are separate and independent processes.

Our approach to understanding the present results is based on the assumption that transfer in situations such as the present one is primarily mediated by relatively specific memory representations. An alternative orientation is to explain the operation of complex devices is to assume that participants form a mental model of the internal workings of a device (e.g., Kieras & Bovair, 1984) and use this information to infer task goals and procedures (e.g., Kamouri & Kamouri, 1986). For instance, Kieras and Bovair provided one group of participants a detailed mental model on the internal workings of a hypothetical device and another group of participants with no such model. Participants who were provided with the mental model were more likely to take shortcuts in the procedures and execute the procedures faster than the no-model group. The authors concluded that the provision of a mental model facilitated the participant's ability to draw inferences about the internal working of the device and its operating sequence.

Results such as these may reflect the same mechanisms as the effects on semantic constraint observed in the present study. For example, it might be hypothesized that the effect of superficial appearance here arises because the graphics remind users of a conceptual model they derived during the corresponding training task. Such a conceptual model in turn may allow users to make reasonably accurate guesses about which button to press when. However, the effect of layout suggests that there is another mechanism contributing to transfer that is independent of this form of conceptual knowledge. In particular, we argue that this effect is produced because subjects can map the sequence of physical button locations from the training task to the transfer task.

As alluded to earlier, the additive nature of the present results suggest that both appearance and layout contribute independently to procedural transfer. Previous investigations suggested that the *content* of superficial features affects prior knowledge use (Ross, 1987, 1989). The present findings build on this research research by showing that participants use the superficial *appearance* of a device interface to access previous experience. Further, the spatial layout of the interface may make applying analogical processes either easy or difficult. This poses special concerns for interface designers: Not only are the physical features (i.e, graphics, meters, and the appearance of controls) important in designing user-friendly interfaces, but the spatial organization of the steps in the procedure matters as well. However, our interpretation is that these variables have their effects for different reasons: Appearance is important for using semantic constraint, but spatial layout is important for mapping steps across procedures in an effective way.

Acknowledgments

This research was supported by the Natural Science and Engineering Research Council of Canada through a graduate fellowship to the first author and a research grant to the second.

References

Bower, G.H. (1961). Application of a model to paired-associate learning. *Psychometrika, 26*, 255-280.

Kamouri, A. L., Kamouri, J., & Smith, K, (1986). Training by exploration: facilitating the transfer of procedural knowledge through analogical reasoning. *International Journal of Man and Machine Studies, 24*, 171-192.

Kieras, D. E., & Bovair, S. (1984). The role of a mental model in learning to operate a device. *Cognitive Science, 8*, 255-273.

O'Reilly, T., Dixon, P. (1999). The transfer of prior knowledge in complex procedural domains: Evidence for routine availability (Unpublished manuscript).

Ross, B. (1984). Remindings and their effect on learning a cognitive skill. *Cognitive Psychology, 16*, 371-416.

Ross, B. (1987). This like that: The use of earlier problems and the separation of similarity effects. *Journal of Experimental Psychology: Learning, Memory and Cognition,14*, 629-639.

Ross, B. (1989). Distinguishing types of superficial similarities: Different effects on the access and use of earlier problems *Journal of Experimental Psychology: Learning, Memory and Cognition, 15*, 456-468.

Ross, B. & Kennedy, P.T. (1990). Generalizing from the use of earlier examples in problem solving *Journal of Experimental Psychology: Learning, Memory and Cognition, 16*, 42-55.

Articulating an Explanatory Schema:
A Preliminary Model and Supporting Data

Stellan Ohlsson (stellan@uic.edu)
Joshua Hemmerich (joshh@uic.edu)
Department of Psychology
The University of Illinois at Chicago
1007 West Harrison Street (M/C 285)
Chicago, IL 60607, U.S.A.

Abstract

The schema repertoire model claims that an explanation is constructed by selecting and articulating a schema. Novice evolutionary explanations are analyzed to identify the relevant schemas and to demonstrate competition among schemas. An intervention study shows that a newly acquired schema does not necessarily win the competition against previously acquired schemas. The difference between schemas and beliefs is emphasized.

Explanation

Explanation is a central topic in the cognitive sciences. Systematic analysis of what it means to explain began with Hempel and Oppenheimer's (1948) now classical claim that an explanation is a deductive argument, a claim that philosophers have since replaced with an emphasis on semantic models (Thompson, 1989) and causal relations (Salmon, 1984). Artificial Intelligence researchers realized early that expert systems ought to explain their reasoning to users (Clancey & Shortliffe, 1984) and much research in computational linguistics is aimed at providing computer systems with such explanatory capabilities (e.g., Moore & Paris, 1993). To cognitive psychologists, explanation is an intellectual performance with measurable effects on learning (Chi, DeLeeuw, Chiu & LaVancher, 1994) and a close relation to the elusive concept of understanding (Schank, 1986). In educational research, the ability to construct an explanation is often used as a criterion of successful learning.

In spite of the effort allocated to the topic, central questions about explanation remain without widely accepted answers. What is an explanation? What are the knowledge structures that underpin explanatory competence? How are such structures acquired and applied?

We propose a preliminary model of the cognitive processes involved explanation. Like the prior formulation by Schank (1986), the model puts the concept of an explanation pattern at the center. The model is as yet informal, but makes qualitative predictions. Two corpora of students' explanations of biological adaptations are analyzed to identify their explanatory schemas and to demonstrate intra-individual competition among schemas. Two attempts to teach the Darwinian explanation schema show that successful schema acquisition does not have the effect on the students' explanations that our common sense concept of learning would lead us to expect.

The Schema Repertoire Model

Consider an everyday *explanation question* like, "Why is flight X delayed?", where X is some particular flight. Answers that come to mind include equipment malfunction, bad weather, a tardy pilot, delays elsewhere in the air traffic system, airline management problems and so on.

Several observations are pertinent. First, because we can access these answers in the absence of specifics (which flight? when? where?), we must be accessing *explanation types* instead of particular explanations. Second, each explanation type captures a prototypical *genesis* for flight delays. Equipment problems, bad weather, etc. signify typical ways in which a flight comes to be delayed. Third, the explanation types are relatively *abstract*. To explain a particular flight delay, one must supply the specifics (which malfunction? what weather? delays where?). Fourth, there are *multiple* explanation types, so the process of explaining a particular delay requires selecting one schema over others.

Figure 1 depicts a model that incorporates these observations. The long horizontal line separates the observables (below the line) from the hypothesized cognitive processes (above the line). Time runs from left to right, roughly speaking. The iconography is of course arbitrary, but the claims expressed by it are not.

The central claim, adopted from Schank (1986), is that explanation types are not only in the eye of the beholder. Explanations fall into types because they are generated from explanatory *schemas* in the knowledge base of the person doing the explaining. Schemas are abstract (Ohlsson, 1993; Ohlsson & Lehtinen, 1997).

The ellipses in the middle of Figure 1 depict the explainer's repertoire of schemas. A question *activates* multiple schemas, to varying degrees. A second claim of the model is that there is a process of *selection* among those schemas that compete for control of the explanatory discourse (depicted in Figure 1 as a repetition of the outline of the winning schema to the right.)

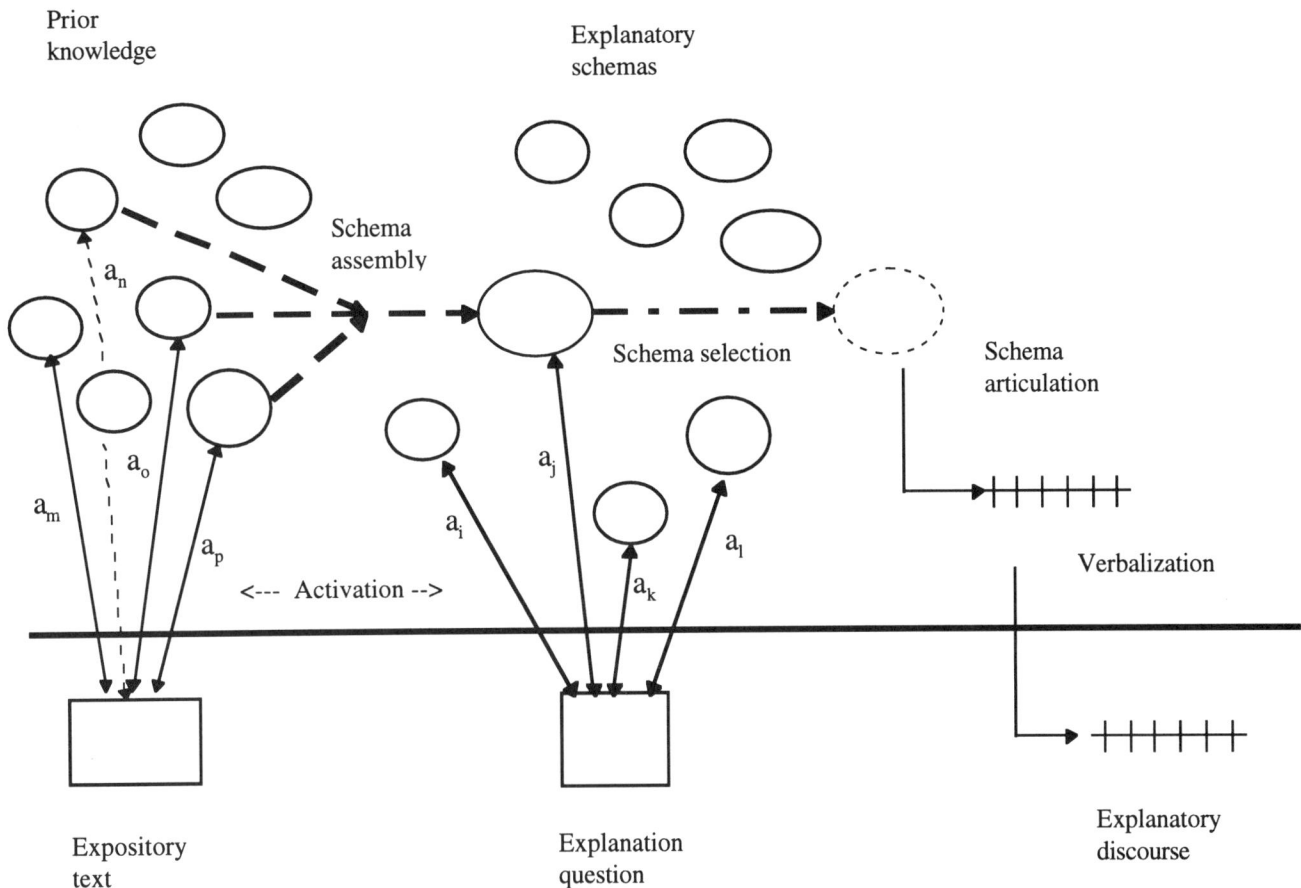

Figure 1: A model of the acquisition, selection and articulation of an explanatory schema.

Because an explanatory schema is abstract, there must also be a process *schema articulation* (Ohlsson, 1992) in which the winning schema is applied to generate a particular explanation (depicted by the upper hashed line to the right), which may or may not be *verbalized* (lower hashed line) depending on context and conditions.

The articulation process constructs a story conforming to the type specified by the schema about the event mentioned in the question. To tell a *delays-elsewhere* story about a flight out of Chicago is to spell out the current weather conditions in another locale, the connection between that locale and the flight (e.g., the aircraft is there) and the consequences of the weather conditions (e.g., the aircraft cannot take off). Different *delays-elsewhere* stories differ in locales, type of weather, impact on the relevant flight and so on. To articulate the *delays-elsewhere* schema is to flush it out with the facts about a particular target event.

The left part of Figure 1 depicts the process of acquiring a new schema in response to instructional input, most typically an expository text. During reading, the text is assimilated to prior knowledge; for the sake of brevity, let us conceptualize prior knowledge as consisting of nothing but schemas. The result of assimilation is then a new knowledge structure that both instantiates an existing schema and represents the message of the text. (Low level reading processes are deliberately ignored here.)

If the text describes an explanatory schema or theory (as is the case in, for example, science education), then two consequences follow. First, successful comprehension should produce a new explanatory schema, to be inserted into the repertoire and so be subject to competition when an explanation question is encountered. Second, the prior schema that guides the assimilation of such a text must be more abstract than the schema communicated by the text.

Finally, consider what happens when there is no prior schema to which the text can be assimilated. Either the text is not understood properly or else a new abstract schema is constructed during reading. The model claims that schema construction is not a matter of induction but of *assembling* prior schemas into a new configuration (Ohlsson & Lehtinen, 1997). This process is depicted by the converging arrows to the left in Figure 1.

The schema repertoire model is as yet too informal to generate quantitative predictions. However, it suggests new ways to analyze data and it does predict some qualitative properties of explanations. We report two empirical studies of novice explanations in biology which illustrate and support those predictions.

491

A Schema Repertoire For Biology

Evolutionary biology is a rich source of examples of explanations and explanatory knowledge. The basic explanation question in this domain is, "Why did species X evolve trait Y?" or, "How did species X acquire trait Y?". I will refer to this as *the phylogenetic question*. There are other types of questions, e.g., "Why is species X distributed geographically in the way it is?", but they will not be dealt with in this paper (see Kitcher, 1993, for a discussion of question types in evolutionary theory). What types of explanations do students construct in response to the phylogenetic question? Our theory implies that their explanations should fall into a set of types, corresponding to schemas acquired in other contexts.

Method Two groups of psychology students, 50 from the University of Pittsburgh and 95 from the University of Illinois at Chicago, participated in return for course credit. The two sets of explanations they generated will be referred to as *the Pittsburgh corpus* and *the Chicago corpus*.

The participants were given sheets of paper with a version of the phylogenetic question written across the top and asked to write down their answers. The Pittsburgh participants were asked why dinosaurs became so large and how birds developed flight, while the Chicago participants were asked those two questions, plus how tigers got their stripes. Both groups were encouraged to ignore factual issues (e.g., the climate millions of years ago) and to invent an explanation that seemed plausible to themselves. They were given no help or instruction.

Results The answers fell into recognizable types, including the following (paraphrases of observed explanations in parentheses):

(a) *Environmentalism*. Traits develop when the environment provides a demand or an opportunity. (Dolphins needed to reach food in the water, so they became aquatic.) (b) *Survival*. Both the relevant trait and its opposite were once present in the species, but all members without the trait died. (There were once both large and small dinosaurs, but all the small ones were eaten.) (c) *Creationism*. Animals were created by a deity with the characteristics they have today. (Dinosaurs were created large so as to flatten the Earth in preparation for the coming of humans.) (d) *Training* (Lamarckianism). Traits are caused by the activity of the organism. (Birds flapped their proto-wings until they grew large enough to support flight.) (e) *Mutationism*. The trait suddenly appeared in a small number of organisms. (One day a bird was born with wings.) (f) *Mentalism*. Animals decide, discover, learn or are taught new behaviors (traits?). (A bird discovered that it could fly and taught its offspring.) (g) *Crossbreeding*. Traits arise via interbreeding between species. (A black panther and a tiger without stripes mated and produced a tiger with stripes.) (h) *Dissemination*. Organisms with the trait gradually increased in numbers until they replaced those without. (In every generation there were more and more tigers with stripes.)

Once the eight explanation types had been identified, the data were analyzed by two coders. The coders were given a definition and examples of each explanation type. They went through cycles of coding examples, discussing disagreements and coding additional examples until 85% of their codes were identical. They then coded the target data set independently of each other. The first author arbitrated any disagreements. The coders were instructed to look for expressions of the eight explanation types, as opposed to classify each answer into a single type. Consequently, an answer could be scored as providing evidence for more than one type.

The frequency of each explanation type in each group is shown in Figure 2. To facilitate comparison between the two unequal-sized corpora, the raw frequencies have been converted to proportions. That is, a value of .40 for explanation type X in corpus Y means that 40% of the explanations in corpus Y were coded as expressing ideas that are parts of explanation type X.

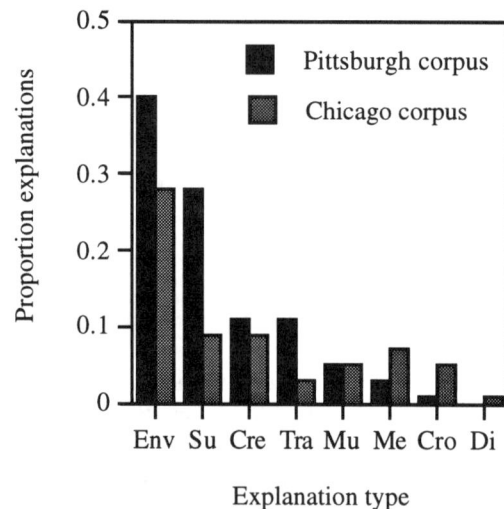

Figure 2: Frequency of eight explanation types in two novice corpora, expressed as proportions of the total number of explanations in each corpus.

Seven of the eight explanation types occurred in both groups. (There were no dissemination explanations in the Chicago corpus.) The explanation types were applied with varying frequency, but there is a rough correspondence between the two corpora. The participants regarded the environment and individual survival as key factors in evolution (as do biologists). Creationism and Lamarckianism--often claimed to be the most common ideas among biology novices (Ferrari & Chi, in press; Samarapungavan & Wiers, 1997)--are not dominant in these data. In the Pittsburgh corpus, training (Lamarckianism) is fourth from the top in frequency; in the Chicago corpus, seventh. There are other non-Darwinian explanation types (e.g., mutationism, mentalism, crossbreeding) that are as frequent or more frequent.

492

Evidence For Schema Competition

The schema repertoire model claims that in the typical case, a person is in possession of a repertoire of qualitatively different explanatory schemas which are more or less relevant for a given explanation question, and that these schemas compete for control of the answer to the question. If so, we would expect single individuals to give one type of explanation in response to question X and a different type in response to question Y.

Single cases Compare the two following two answers to the dinosaur and tiger questions, produced by the same student in the same one-hour session:

Dinosaurs who existed at the period of time had no competition for food source, therefore some species of dinosaur who eat as much food as they want to without any other animal stopping it. This led to the gigantic size of some dinosaurs.
[Chicago corpus, S32, Dinosaur problem]

The tiger was originally all black. After thousand of years, the black color of the tiger's outer body began to fade and continued to fade for generations. The black faded into the stripes we see on the tiger today. Generations from now, we may see no stripes on the tiger. [Chicago corpus, S32, Tiger problem]

S32 explains the size of dinosaurs by referring to their eating habits. However, the explanation for the tiger's stripe is qualitatively different: It refers to a natural tendency (a type of explanation we did not yet code for) and it contains no reference to the tiger's behavior.

A similar diversity of explanations is evident in the following pair of explanations from student S55:

I think that some species of dinosaurs became so gigantic because of the warm climate. Many other animals back then were gigantic too. I think because of the pleasant temperatures dinosaurs could develop very well. It was the temperature, very cold temperature, that cause dinosaurs to go extinct. Their bodies were mainly dependent of high, warm temperatures.
[Chicago corpus, S55, Dinosaur problem]

I would assume that the tiger got its black stripes from some kind of biological cross between lion and black panther. Because of the genetic combination some of the characteristics of black panther blended with some characteristics of the lion's genes.
[Chicago corpus, S55, Tiger problem]

The first explanation is a pure environmentalist explanation; the climate was warm and this apparently constitutes an opportunity to grow in the opinion of this student, perhaps in analogy with the fast growth of plants in warm climates. The second explanation is a crossbreeding explanation that does not mention the environment.

Broader view Table 1 shows quantitative evidence for between-question shifts among explanation schemas. We identified all pairs of successive explanations for which both explanations could be classified in one of the eight explanation types listed above. Next, we calculated the proportion of such pairs in which the second explanation was of the same type as the first. This quantity is an estimate of the conditional probability that a student answers question N+1 with an explanation of type X, given that he or she answered question N with an explanation of type X.

As Table 1 shows, the probability of switching between explanation types was .43 in the Pittsburgh sample and .59 in the Chicago sample. Hence, the students switched explanation types as often as not. Because the explanation types are qualitatively different, these data provide no support for the idea that students operate with consistent explanatory frameworks (Ferrari & Chi, in press; Samarapungavan & Wiers, 1997; Tamir & Zohar, 1991).

Table 1. The probability that a student's answer to question N+1 is of the same (or different) type as that student's answer to question N.

Corpus	Probability	
	Same	Different
Pittsburgh	.57	.43
Chicago	.41	.59

Even stronger evidence for schema competition is provided by cases in which a single answer contains within itself evidence of multiple explanation types. The following example contains at least three distinct explanations:

The stripes are just a variation of nature. They help the tiger stand out from all the rest. Its also a warning signal for other animals to stay away. It could also be used in mating. They could use their stripes to show off for the women. It also could have been that they were a genetic defect that got amplified and stayed through the years.
[S9, Chicago corpus, Tiger problem]

This student thus proposed three qualitatively different explanations: that tiger stripes relate tigers to other species; play a role in courtship and mating; and constitute a genetic defect. There are no transitions in the student's text. It is as if each schema automatically took over as soon as the previous one had spent its potential to control verbal output.

Acquiring A New Schema

A biologist's answer to the phylogenetic question draws upon five distinct ideas: (a) Variation. The members of a species are not all alike. (b) Selectivity. Some species members are better adapted than others. (c) Replication rate. Better adapted members reproduce more. (d) Inheritance. Reproducing members of the species pass on their traits to the next generation. (e) Accumulation. Small changes from generation to generation eventually add up to a large change. Kitcher (1993) refers to the explanation schema built out of these five ideas as the *simple individual selection* schema (p. 28).

The schema repertoire theory makes a counterintuitive prediction about what a student's behavior should look like after acquiring the Darwinian schema: Very little should change. The reason is that there is no process in the model that erases old schemas. New schemas are added to the repertoire, but they do not replace previously acquired schemas. Hence, contrary to the wishes of a hopeful teacher, there is no reason to expect a newly acquired schema to dominate a student's explanatory discourse, *even if that schema has been successfully acquired.* Instead, the new schema competes on equal terms with all the other schemas in the repertoire, winning sometimes and loosing sometimes. Hence, the acquisition of a new explanation schema should increase the diversity of a person's explanations, but not make their discourse conform exclusively to the newly acquired schema.

To investigate this prediction, we conducted an intervention study. As in the base line study, the data collected consisted of written answers to phylogenetic questions. The answers were coded for presence or absence of the five Darwinian ideas described above. (The coding method was the same as in the base line study.) Each explanation was assigned a score between 0 (no trace of any Darwinian idea in the explanation) and 5 (all five Darwinian ideas are expressed in the explanation). This *Darwin score* is a rough measure of how closely an explanation conforms to the Darwinian schema.

Method Two groups of students were taught the Darwinian explanation schema with two different instructional methods. (a) Text only. Twenty undergraduate psychology students from the University of Pittsburgh read a two-page exposition of Darwin's' theory that emphasized the five ideas described above. They were given no other help, instruction or preparation. After reading, they were asked to explain the size of the dinosaurs and the tiger's stripes. (b) Text plus feedback. Twenty psychology undergraduate students read the same expository text as the students in the text only group. They then answered five phylogenetic questions, pertaining to horses, tigers, ducks, polar bears and dinosaurs. After each answer, they received feedback in the form of an expert answer to that question. That is, the participants read a phylogenetic question, wrote down their own answers, turned the page and read a Darwinian answer to that same question.

Results Given the famously poor science knowledge of U.S. students and the religious opposition to the theory of evolution, one would not expect young adults in the U. S. to spontaneously generate Darwinian explanations. The data confirm this sad expectation. Figure 3 shows the frequency distribution of Darwin scores in the base line study. (The Chicago and Pittsburgh corpora were combined in this analysis.) For ease of comparison, the frequencies displayed in Figure 3 are expressed as proportions of the number of explanations produced by each group. That is, a value of .20 for a Darwin score of 2 in study X means that 20% of the explanations produced in study X contained 2 out of the 5 Darwinian ideas.

Without instruction, zero on the Darwin scale was the modal value. Approximately 35% of the explanations in the base line group contained no trace of the five Darwinian ideas, and an additional 30% expressed no more than one of those ideas. None of the 385 explanations from the base line groups contained traces of all five Darwinian ideas.

The distributions for the two intervention groups are very different. A Darwin score of five is the most frequent value. Over half the explanations contain either four or five Darwinian ideas, while less then 10% have zero such ideas. In short, the interventions worked in the sense that they increased the students' use of the target ideas.

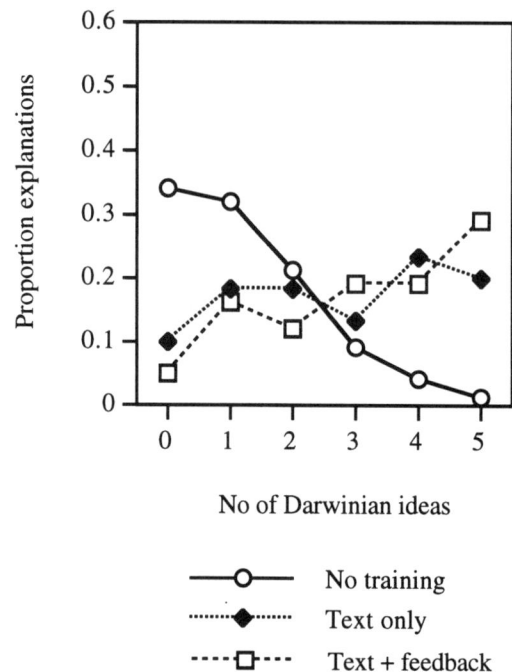

Figure 3: Frequency distributions for the number of Darwinian ideas per explanation (range 0-5) in each of four studies. The frequencies are expressed as proportions of the total number of explanations produced in each study.

It is correct but misleading to summarize Figure 3 by saying that the students learned something about Darwinism. The schema repertoire theory warns that the acquisition of the Darwinian schema does not cause prior, non-Darwinian schemas to disappear. Figure 4 shows that this warning is indeed warranted. In this analysis, we asked how many different explanation types each participant used in each of the three studies. Once again, we show the entire distributions instead of the merely the means and we have converted frequencies to proportion for ease of comparison.

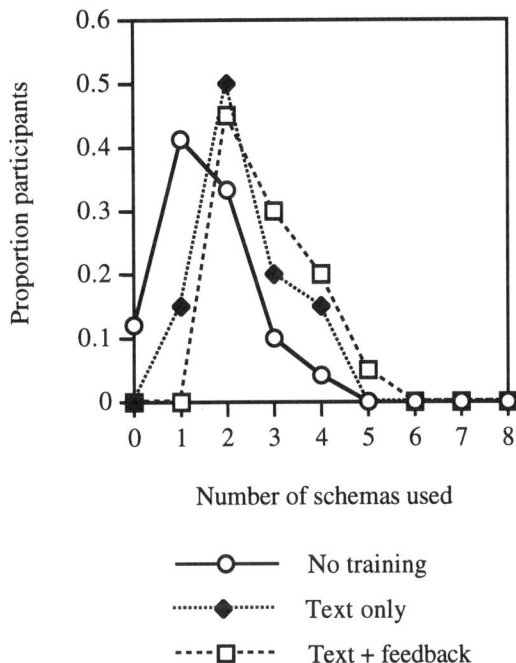

Figure 4. The frequency distribution of participants who used N of the 8 explanation types (N = 0, 1, ..., 8), expressed as the proportion of all participants in each study.

The result is clear: Successful acquisition of the Darwinian schema did *not* lower the tendency to use the eight non-Darwinian explanation types. This is what the schema repertoire theory predicts. There are no processes that erase prior schemas, and there is no guarantee that a newly schema will win over the prior ones on every relevant explanation question. Hence *successful* schema acquisition will not necessarily affect a student's explanations.

Discussion

The theory presented here differs from alternative theories by insisting on the distinction between schemas and beliefs or explanatory frameworks (Ferrari & Chi, in press; Samarapungavan & Wiers, 1997; Tamir & Zohar, 1991). It is highly unlikely that the explanations we recorded expressed entrenched beliefs or consistent frameworks. The participants in our studies had probably never thought about how this or that animal acquired such-and-such a trait. They made up their answers in response to our questions by articulating schemas acquired in other contexts.

Where did these schemas come from? Consider crossbreeding. Many students have come across the concept of crossbreeding in the context of animal breeding. Also, crossbreeding is only a specialized version of a yet more abstract schema of blending or mixing: Combine X and Y and you get something that is intermediate between X and Y. The latter schema applies to colors, spices, decorating styles and many other everyday phenomena. Similarly, the training schema (Lamarckianism) is strongly present in everyday culture: Exercise and your limbs will grow bigger. The origins of the other schemas in everyday experience is no more mysterious.

References

Chi, M. T. H., DeLeeuw, N., Chiu, M.-H., & LaVancher, C. (1994). Eliciting self-explanations improves understanding. *Cognitive Science, 18*, 439-477.

Clancey, W. J., & Shortliffe, E. H. (1984). *Readings in medical artificial intelligence: The first decade.* Reading, MA: Addison-Wesley.

Ferrari, M., & Chi, M. T. H. (in press). The nature of naive explanations of natural selection. *International Journal of Science Education.*

Hempel, C. G., & Oppenheimer, P. (1948). Studies in the logic of explanation. *Philosophy of Science, 15*, 135-175.

Kitcher, P. (1993). *The advancement of science.* New York: Oxford University Press.

Moore, J. D., & Paris, C. L. (1993). Planning text for advisory dialogues: capturing intentional and rhetorical information. *Computational Linguistics, 19*, 651-695.

Ohlsson, S. (1992a). The cognitive skill of theory articulation: A neglected aspect of science education? *Science & Education, 1*, 181-192.

Ohlsson, S. (1993). Abstract schemas. *Educational Psychologist, 28*, 51 66.

Ohlsson, S. & Lehtinen, E. (1997). Abstraction and the acquisition of complex ideas. *International Journal of Educational Research, 27*, 37-48.

Salmon, W. C. (1984). *Scientific explanation and the causal structure of the world.* Princeton, NJ: Princeton University Press.

Samarapungavan, A., & Wiers, R. (1997). Children's thoughts on the origin of species: A study of explanatory coherence. *Cognitive Science, 21*, 147-177.

Schank, R. C. (1986). *Explanation patterns.* Hillsdale, NJ: Erlbaum.

Tamir, P, & Zohar, A. (1991). Anthropomorphism and teleology in reasoning about biological phenomena. *Science Education, 75*, 57-67.

Thompson, P. (1989). *The structure of biological theories.* New York: State University of New York Press.

When Learning is Detrimental: SESAM and Outcome Feedback

Henrik Olsson (henrik.olsson@psyk.uu.se)
Department of Psychology, Uppsala University
Box 1225, SE-751 42 Uppsala, Sweden

Peter Juslin (peter.juslin@psyk.uu.se)
Department of Psychology, Uppsala University
Box 1225, SE-751 42 Uppsala, Sweden

Abstract

The *sensory sampling model* (*SESAM*; P. Juslin & H. Olsson, 1997) accounts for the underconfidence observed in sensory discriminations with pair-comparisons. In the present study the model is applied to a single-stimulus task and a comparison is made with pair-comparisons. The model predicts that in the single-stimulus condition training with feedback should lead to *poorer* calibration with more underconfidence. In pair-comparison the feedback should have little or no effect on calibration. The results confirm these predictions.

Introduction

A common presumption is that experience should improve the quality of our judgments and foster insights into the limitations of our knowledge. Indeed, in cognitive tasks there is ample of evidence that experts' confidence judgments are more realistic than those of novices (for a review, see Yates, 1990). The evidence on sensory discrimination is less clear-cut. Some studies report little or no improvement in realism of confidence when participants are provided with outcome feedback (Winman & Juslin, 1993), while other studies report improvement for difficult task sets and worse calibration in easy task sets (Petrusic & Baranski, 1997).

In Juslin and Olsson (1997), it was suggested that this and other discrepancies between inferential and sensory discrimination tasks arise because two different sources of uncertainty are involved. The *sensory sampling model* (*SESAM*) was developed to elucidate confidence in sensory discrimination. This paper extends the work to the case of a single-stimulus task where the participant is to decide whether a presented line is longer than a specified but not seen reference length (e.g., a Swedish twenty-kronor note). We will concentrate on two counter-intuitive predictions by *SESAM*: (a) The underconfidence observed with pair-comparisons will be *unaffected even by prolonged sessions of outcome feedback.* (b) The realism of confidence in a single-stimulus task will deteriorate with outcome feedback, leading to *poorer calibration with more underconfidence.* The experiment reported below provides a test of these two predictions.

Realism of Confidence in Sensory Discrimination

Realism of confidence, or *calibration*, is commonly investigated by presenting participants with a large set of two-alternative decision tasks. For each task-item, the participant decides on one of the two alternatives and assesses his or her confidence in the correctness of this decision as a subjective probability between .5 (random choice) and 1.0 (certainty). Participants are said to be realistic, or well calibrated, to the extent that items assigned subjective probability .xx are correct with relative frequency .xx. An index of *over/underconfidence* is obtained by subtracting the overall proportion of correct decisions from the mean subjective probability, where a positive difference is overconfidence.

In a review of early psychophysical studies, Björkman, Juslin, and Winman (1993) observed that these studies often suggest *underconfidence* (although interpretations in terms of calibration are problematic, because confidence was not assessed as subjective probabilities). For one-hundred years, results such as these has led researchers to speculate about *subconscious mental processes* (Fullerton & Cattell, 1892), or *implicit perception* (Kihlstrom, Barnhardt, & Tataryn, 1992). More recently, underconfidence in sensory discrimination has been replicated in studies with the modern calibration paradigm (for a review, see Juslin & Olsson, 1997). In a meta-analysis (Juslin, Olsson, & Winman, 1998), which aggregated data from 21 sensory discrimination tasks and 44 inferential tasks, a clear main effect of sensory versus inferential tasks revealed more underconfidence for sensory discrimination. This was true even when the effect of proportion correct was removed as a co-variate. These results refute the claim that there is no difference between confidence in inferential and sensory-discrimination tasks (Baranski & Petrusic, 1994; Ferrell, 1995).

The Sensory Sampling Model (SESAM)

Consider yourself as a participant in a difficult pair-comparison task, such as deciding which of two almost equivalent lines is the longest. It may take some time before you reach a decision. The presence of neural noise makes computation of an error-free estimate of the 'true' difference μ between the two lines impossible, where the *true difference* refers to the estimate that would result if there was no noise in the processing of the sensory information. *SESAM* proposes that the nervous system repeatedly computes new estimated values X_i of μ that vary from moment to moment due to the neural noise. The X_i are assumed to be a Normally and Independently Distributed (NID) random variable with mean μ and variance σ_X^2, where the parameter μ is defined in the unit variance of X_i. *Sensory discrimination* refers to a perceptual task where erroneous decisions emanate primarily from neural noise (or approximations to these situations).

In *SESAM*, the decision about the difference μ is based on the participant's overall impression as modeled by a statistical aggregate; the mean sensation $\bar{X}_i$ (where the index i denotes that the mean is computed after sensation X_i). However, this mean is computed only from the n last estimates X_i that are still contained in a memory window that represents a limitation in the processing capacity of the nervous system. The number of sensations n contained in the memory window at any moment is the *sample size parameter*. When a new sensation enters, the oldest one is pushed out and for every new X_i a new mean $\bar{X}_i$ is computed from the last n sensations. Thus, the process is modeled as a capacity limited sequential sampling process. A decision cannot be made unless the absolute value of the mean $\bar{X}_i$ exceeds a *response threshold* $\phi_{\bar{X}}$, where the threshold corresponds to a definite experience of a difference between the stimuli. When the estimation of μ is that $\mu > 0$ then the decision will be that the left line is longer, and vice versa.

Note two consequences: First, the number of sensations X_i sampled before a decision is made provides predictions of response times (after a linear transformation). Second, occasionally there will be an erroneous decision due to the sampling error in the estimates X_i. The probability of a correct decision is determined by the sampling error variance $\sigma_{\bar{X}}^-$ of the mean used to make the decision. In the case of non-sequential or static sampling of n sensations (i.e., obtained by setting the response threshold $\phi_{\bar{X}}$ at 0), this variance takes the familiar form of $\sigma_{\bar{X}}^2 = \sigma_X^2 / n$. Finally, the *stopping parameter* N_{max} is the maximum number of iterated computations of X_i before the process is ended. If no noticeable difference is detected within the allotted time, a random decision with subjective probability .5 is made.

When making a conditional probability assessment, the participant first makes a decision and then an assessment of the probability that the decision is correct. Say that the participant decides that $\mu > 0$, that is, the left line is longer. According to *SESAM*, confidence reflects the consistency of the information contained in the sample used for the decision. It is proposed that the subjective probability is computed as the proportion of the last n sensations that support the decision. If, for instance, the decision is $\mu > 0$ and 70% of the X_i are larger than zero, the subjective probability is .7 (and vice versa when the decision is $\mu < 0$). This means that subjective probability is based on a proportion defined by the variance σ_X^2 of the sensations.

While confidence reflects the variability of single sensations (σ_X^2), decisions benefit from the greater precision of a statistical aggregate obtained across n sensations (defined by $\sigma_{\bar{X}}^2$). The most immediate test of these assumptions is that when both proportions of decisions $\mu > 0$ and mean subjective probability that $\mu > 0$ is plotted against (negative and positive) physical stimulus differences, both functions should approximate normal ogive functions, although the ogive for subjective probability should be less steep (larger variance) than the ogive for decisions. This, indeed, is the common finding (e.g., Johnson, 1939, Juslin & Olsson, 1997). A second implication is that in studies with conditional probability assessment, there will be a disposition towards underconfidence, in particular, for stimulus differences with moderate and high proportions correct. *SESAM* provides a good account of subjective probability distributions, hit-rates, and response times in a line discrimination

task (Juslin & Olsson, 1997).

The disposition towards underconfidence is alleviated in three circumstances: First, when the physical stimulus difference is zero or close to zero; Second, when the sequential sampling process is constrained by time pressure (Baranski & Petrusic, 1994; Olsson & Juslin, 1998); Third, when the processing is dominated by a perceptual bias.

SESAM with Perceptual Bias

One of the most striking demonstrations of perceptual bias in a calibration task of the kind modeled by *SESAM* is Experiment 2 in Baranski and Petrusic (1994). The participants' task was to indicate which of two vertical lines was located farther from a vertical central referent line. In this task, Baranski and Petrusic observed a "left-looks-farther" effect: When the left line was closer to the central referent line (i.e., the left line was located 296 pixels to the right of the central referent and the right line was located 300 pixels to the right of the central referent) it was perceived as being farther away. The proportion of correct decisions in the 296, 300 order was only .26 and the proportion correct in the 300, 296 order was .87.

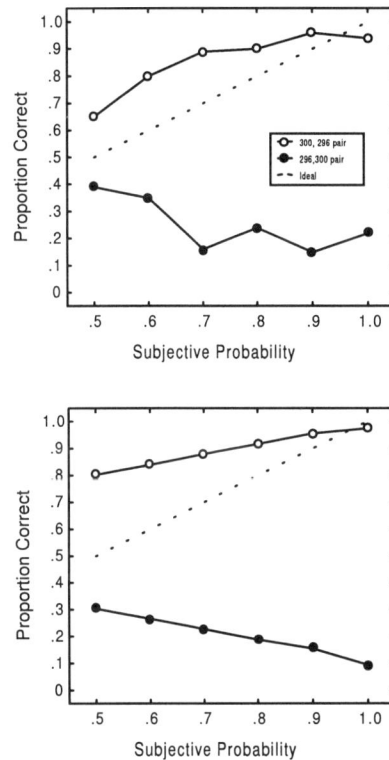

Figure 1: Top panel: Empirical calibration curves from Experiment 2 in Baranski and Petrusic (1994). Lower panel: Calibration curves simulated with *SESAM*.

In realism of confidence studies, data are sometimes presented in calibration curves where the proportion of correct decisions are plotted against the levels of subjective probability (in a conditional probability task with two alternatives these are .5, .6, .7, .8, .9, and 1.0). The effect of perceptual bias is dramatically illustrated in the calibration curve in the top panel of Figure 1. For the 296, 300 order, the participants in Baranski and Petrusic's experiment had a lower proportion of correct decisions when they were absolutely certain (1.0) than when they were guessing (.5).

The effects of perceptual bias can be accounted for by adding a perceptual bias parameter b (with a permissible range of $-\infty$ to ∞) to μ in *SESAM* (Olsson, 1999). For example, consider a line discrimination task with two lines A and B. If we assume that $\mu = .1$ and $b = -.2$, the participant has a tendency to falsely perceive line B to be longer than line A. This means that compared to an unbiased version of the same task, the proportion of correct answers will be lower. If the bias is large the proportion of correct answers can approach 0. The lower panel in Figure 1 shows simulated calibration curves obtained with perceptual bias. The only parameter difference between the two calibration curves is the sign of the bias parameter (.17 and -.17). The other parameters were: $\mu = .06$, $\phi_{\bar{x}} = .35$, and n = 10. The stopping parameter N_{max} was not used in this simulation. It can be seen that the model is successful in accounting for the empirical calibration curves from Baranski and Petrusic (1994) (see Olsson 1999, for a detailed account).

Pair-Comparison and Single-Stimulus

The idea of a distorted perceptual representation can not only be applied to a pair-comparison task but also to a single-stimulus task where the reference is a memory representation. Consider a variation of the pair-comparison task used by Juslin and Olsson (1997). Instead of two lines you are looking at a single line and you are to decide if this line is longer or shorter than a stated reference length, say the diameter of a compact disc. Even if you do not know the exact length of this reference, you will have some apprehension of it, presumably by retrieval of a memory representation. When *SESAM* is applied to this single-stimulus task the nervous system computes the subjective difference between the memory representation and the stimulus line. Whereas in pair-comparison, all error is assumed to stem from neural noise, in the single stimulus case there will be one additional source of error because your memory representation of the reference length may be biased.

Effects of Feedback In a pair-comparison task dominated by neural noise, *SESAM* suggests that feedback should have little or no effect. Both decisions and probabilities fundamentally arise from the finite precision of the sensory system, and this precision is not likely to be affected by the amount of feedback provided in a single experimental session. Aside from the stimulus units with very low proportion correct, we thus expect underconfidence both before and after a long feedback session. This is illustrated in the top panel of Figure 2 (see Winman & Juslin, 1993, for empirical evidence).

For single-stimulus, on the other hand, we expect the provision of feedback to allow the participants to detect the bias in their memory representation of the reference. Initially, there will be small underconfidence or overconfidence depending on the size of the bias. The reduction of

bias will make the proportion correct rise and subjective probability will not be much affected. These predictions are illustrated in the lower panel of Figure 2, where it is assumed that the bias is decreased by a constant fraction. These two effects taken together imply that there will be a change towards underconfidence. If the bias is completely eliminated, the task becomes identical to the pair-comparison task and the prediction is thus the same: Underconfidence.

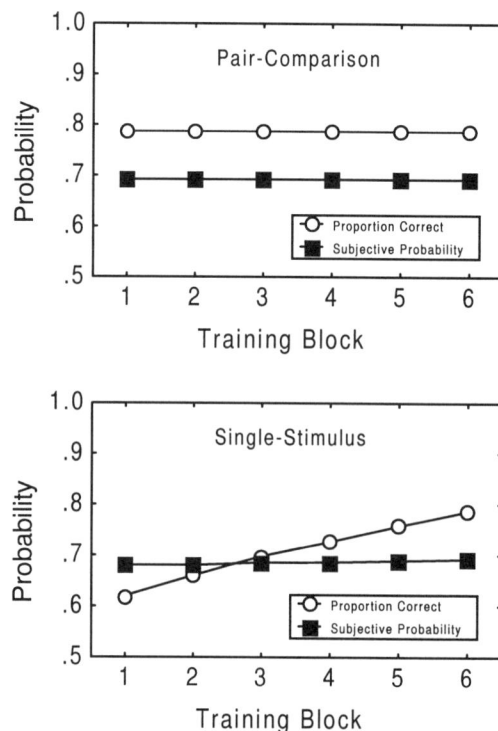

Figure 2: The predicted interaction between provision of outcome feedback and task. Top panel: Expected proportion correct and subjective probability as a function of feedback for pair-comparison. Lower panel: Simulated proportion correct and subjective probability as a function of feedback (or bias b) for single-stimulus. The values of b was: -.15, -.12, -.09, -.06, -.03, and 0. The response threshold $\phi_{\bar{x}}$ was .35, the stimulus difficulty μ was .25, and the sample window parameter n was 7 for all simulations.

The Experiment In one condition participants were to decide which of two lines that was the longer. In the second condition participants had to decide whether an observed line was longer or shorter than a given but not seen reference length, the length of a Swedish twenty-kronor note. A pilot test indicated that the participants did have a biased representation of this reference length (i.e., they guess that a presented length equal to a Swedish twenty-kronor note is longer than the note in 74% of the trials). In pair-comparisons there will be initial underconfidence that prevails in the face of feedback. With single stimulus, the provision of outcome feedback should make the participants underconfident. To the extent that there is no strong initial

498

over- or underconfidence bias, outcome feedback will make these participants *more poorly calibrated*.

Method

Participants

Twenty-four participants, aged between 19 and 32, attended. Thirteen were males and eleven were females. Most were undergraduate psychology students at Uppsala University who participated in exchange for credit for a course requirement. All participants had normal vision or corrected to normal vision.

Apparatus and Stimuli

All participants responded with a mouse. In the pair-comparison condition the stimulus display consisted of two vertical lines. The standard stimulus that was 130 mm (the same length as a Swedish 20 kronor note—the reference in the single-stimulus condition). There were three levels of difficulty, L1 (hardest), L2, and L3 (easiest). The difficulty of the standard-variable combination is expressed in terms of the ratio r of the longer line to the shorter line in the pair: 1.01, 1.02, and 1.025. Each stimulus unit consisted of two pairs of stimuli; in one pair the left line was longer; in the other pair the right line was the longer. Thus, in half of the presentations the standard was on the left, in the other half the standard was on the right. The order of the standard-variable pairs was determined randomly with a new permutation for each participant.

The lines were black, 1 mm wide and appeared on a white background. All of the standard-variable combinations were centered both horizontally and vertically. To ascertain that the stimuli combinations were of correct length and properly centered, all combinations were measured directly on the computer screen. This procedure was repeated on several occasions.

In the single-stimulus condition, the participant was presented with a single line. Half of the lines with stimulus difficulties L1, L2, and L3 were longer than the standard reference and the other half were shorter. The stimuli were the same as the variable stimulus in the pair comparison condition. The line appeared in equal proportion in the right and the left position with a random order of stimulus presentation.

Design and Procedure

Two independent variables were investigated; pair-comparison versus single-stimulus tasks (between-subjects), and performance before and after training with outcome feedback (within-subjects). Dependent measures were decisions and confidence, refined into measures of over/underconfidence. The effect of feedback was studied by first presenting participants with a pretest block without feedback. Then in four blocks, outcome feedback was given after each trial. Finally, a posttest without feedback was administered.

Each block consisted of 120 judgments. In a block the participant made 40 judgments of 3 different stimulus units. In the pair-comparison condition, a standard-variable pair of lines was shown on the screen and the question "Which line is the longest?" appeared below the stimulus unit. Beneath the question there were two small buttons, one labeled "Left" and one labeled "Right". When participants had decided which of the lines they thought was the longest, they clicked on the appropriate button with the mouse. Then the screen was cleared and the participants assessed how certain they were that they had made the correct decision. The subjective probability scale appeared in the middle of the screen and consisted of six buttons labeled "50%, "60%", "70%", "80%", "90%", and "100%". The written instructions briefly introduced the notion of calibration and the scale was anchored with "random choice" at "50%" and "certainty" at "100%". After the participants had selected a subjective probability level with the mouse, the screen was cleared and the next trial began.

In the single-stimulus condition, the question was "Is this line longer or shorter than a 20-kronor note?" and the two buttons were labeled "shorter" and "longer". In other respects the conditions were identical. The session was interrupted by a 20-minutes break and the whole session took between 2.5 and 3.5 hours.

Results and Discussion

In Figure 3 proportion correct, mean subjective probability and over/underconfidence scores are plotted for the pretest, the four training blocks, and the posttest for pair-comparison and single stimuli.

In the pair-comparison condition training had no (positive) effect on proportion correct. In fact, the proportion correct was higher in pretest than in posttest. The proportions correct were .79 (95 % confidence interval, CI, $\pm$ .05)[1] and .76 (95 % CI, $\pm$ .05) respectively). The prediction was that feedback would have no (positive) effect and this is the case. The confidence intervals overlap, so we can not be sure that the slight decrease in proportion correct constitutes a real effect.

In the single-stimulus condition, training had a positive effect on proportion correct as predicted. In Pretest the proportion correct was .64 (95% CI, $\pm$ .05). When feedback was given this figure grew steadily with every block until it peaked at .76 (95 % CI, $\pm$ .05) in training block 4. When feedback was withdrawn the proportion correct fell to .72 (95 % CI, $\pm$ .05) in the posttest. The drop in proportion correct from training block 4 to the posttest (.76 to .72) in the single-stimulus condition could be an effect of fatigue. One additional hypothesis is that the memory representation of the reference length may not be very stable and once the feedback is withdrawn, the older and more biased memory representation once again comes to dominate the responses. This latter hypothesis is not supported by the data, however. The bias towards responding "longer" was .74 in the pretest of the single-stimulus condition, decreased to .57 in the training block 4, but was still no more than .56 in the post-test. Fatigue seems to be a more viable alternative.

---

[1] All within group comparisons have a standard error contrast based on the condition × subject in groups mean square; see Estes, 1997, for details.

Therefore, we decided to test the effects of training on training block 4 rather than the posttest. This can be motivated on two grounds: First, the proportion correct in training block 4 of the single-stimulus condition (.76) is similar to the proportion correct in training block 4 for the pair-comparison condition (.77), and not much lower than the pretest result of .79 for pair comparisons. The small difference between the two conditions in training block 4 indicates that nearly all of the bias is eliminated. The structure of the single-stimulus condition is thus almost equivalent to the pair-comparison task, both of which can be modeled by the bias-free version of *SESAM*. Second, the signs of fatigue in the posttest of both conditions indicate that these data may not be representative of the true performance level of the participants. It is important to note, though, that all qualitative results are the same if the posttest data are used (see Figure 3).

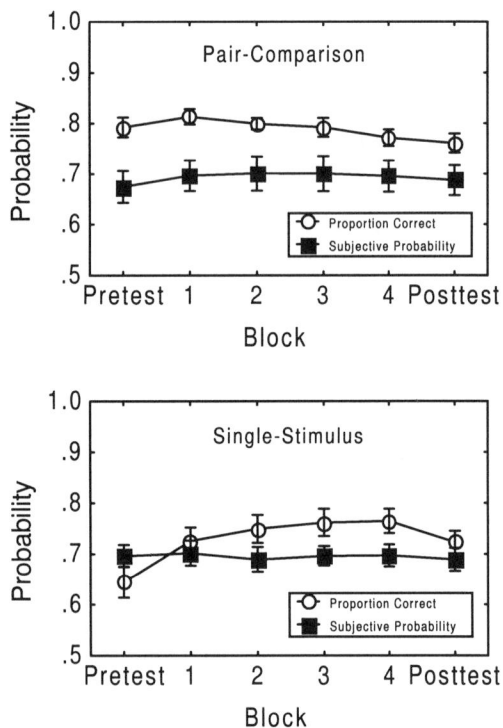

Figure 3: Proportion correct and mean subjective probability for pair-comparisons and single-stimulus as a function of training-block. The error bars are standard errors.

In Figure 3 we see that mean subjective probability is roughly the same in both conditions and constant across training blocks. In the pair-comparison condition, the pretest mean confidence is .67 (95 % CI, ± .03) and in training block 4 .70 (95 % CI, ± .03). In the single-stimulus condition, mean confidence is .70 (95 % CI, ± .03) both in the pretest and in training block 4. This means that there was underconfidence -.12 (95 % CI, ± .06) in the pretest of the pair-comparison condition. In the pretest of the single-

stimulus condition there was a moderate overconfidence of .05 (95 % CI, ± .06). For the pair-comparison condition there was a clear underconfidence bias also in training block 4, -.08 (95 % CI, ± .06). As is evident from Figure 3, the minor decrease in underconfidence in the pair-comparison condition, is wholly explained by a decrease in the proportion correct, perhaps due to fatigue, rather than to accommodation of subjective probability judgments to feedback. In the single-stimulus condition there was also underconfidence in training block 4, -.07 (95 % CI, ± .06). this confirms two predictions. First, it was predicted that the reduction of bias should lead to a change in the direction of underconfidence and, second, that the single-stimulus condition should give the same result as the pair-comparison condition after training.

Note, finally, that the proportions correct in all blocks except the pretest of the single-stimulus condition (.64) are between .7 and .8, whereas mean subjective probability never exceeds .7. Across all blocks in the pair-comparison condition, underconfidence is -.10 at a proportion correct of .78. This pattern of uniform underconfidence for proportions correct between .7 and .8 deviates from the pattern in cognitive or inferential tasks, where there is generally close to zero over/underconfidence at this level of difficulty (Juslin, Olsson, & Björkman, 1997).

Conclusion

As predicted by *SESAM*, feedback produced different results in a pair-comparison and a single-stimulus task with no improvement in the former and deteriorating realism in the latter. These results illustrate that to predict the effect of outcome feedback on realism of confidence one needs to take into account how the specific task relates to the cognitive processes and representations that underlie performance.

The results from the conditions where the role of bias was minimized, pretest and training block 4 of the pair-comparison condition, and training block 4 of the single-stimulus condition, replicates the finding of underconfidence in sensory discrimination (for a review, see Juslin & Olsson, 1997). The results also indicate that bias is an important limiting condition for underconfidence to occur, a conclusion that is consistent with the data reported by Baranski and Petrusic (1994).

Acknowledgments

The research reported in this paper was supported by the Swedish Council for Research in Humanities and Social Sciences. We are indebted to Tomas Foucard for running the experiment. We thank Magnus Persson, Pia Wennerholm, and Anders Winman for helpful comments.

References

Baranski, J. V., & Petrusic, W. M. (1994). The calibration and resolution of confidence in perceptual judgments. *Perception & Psychophysics, 55*, 412-428.

Björkman, M., Juslin, P., & Winman, A. (1993). Realism of confidence in sensory discrimination: The underconfidence phenomenon. *Perception & Psychophysics, 54*, 75-81.

Estes, W. K. (1997). On the communication of information by displays of standard errors and confidence intervals. *Psychonomic Bulletin & Review, 4*, 330-341.

Ferrell, W. R. (1995). A model for realism of confidence judgments: Implications for under-confidence in sensory discrimination. *Perception & Psychophysics, 57*, 246-254.

Fullerton, G. S., Cattell, J. M. (1892). *On the perception of small differences.* Philadelphia: University of Pennsylvania Press.

Johnson, D. M. (1939). Confidence and speed in two-category judgment. *Archives of Psychology, 34*, 1-53.

Juslin, P., & Olsson, H. (1997). Thurstonian and Brunswikian origins of uncertainty in judgment: a sampling model of confidence in sensory discrimination. *Psychological Review, 104*, 344-366.

Juslin, P., Olsson, H., & Winman, A. (1998). The calibration issue: theoretical comments on Suantak, Bolger, and Ferrell (1996). *Organizational Behavior and Human Decision Processes, 73,* 3-26.

Juslin, P., Olsson, H., & Björkman, M. (1997). Brunswikian and Thurstonian origins of bias in probability assessment: on the interpretation of stochastic components of judgment. *Journal of Behavioral Decision Making, 10*, 189-209.

Kihlstrom, J. F., Barnhardt, T. M., & Tataryn, D. J. (1992). Implicit Perception. In R. F. Bornstein & T. D. Pittman (Eds.), *Perception without awareness* (pp. 17-54). New York: Guilford Press.

Olsson, H. (1999). *Explorations of the sensory sampling model: Sampling mechanisms, response time distributions, and bias.* Manuscript in preparation.

Olsson, H., & Winman, A. (1996). Underconfidence in sensory discrimination: The interaction between experimental setting and response strategies. *Perception & Psychophysics, 58*, 374-382.

Petrusic, W. M., & Baranski, J. V. (1997). Context, feedback, and the calibration and resolution of confidence in perceptual judgments. *American Journal of Psychology, 110*, 543-572.

Winman, A., & Juslin, P. (1993). Calibration of sensory and cognitive judgment: Two different accounts. *Scandinavian Journal of Psychology, 34*, 135-148.

Yates, J. F. (1990). *Judgment and decision making.* Englewood Cliffs, NJ: Prentice Hall.

501

From deep to superficial categorization with increasing expertise

Thomas C. Ormerod
Lancaster University, UK.
T.Ormerod@lancaster.ac.uk

Catherine O. Fritz
Bolton Institute, UK.
cof1@bolton.ac.uk

James Ridgway
University of Durham, UK.
jimridgway@durham.ac.uk

Abstract

An experimental study of task design expertise is reported wherein a set of 12 mathematics tasks were sorted by specialist designers of mathematics tasks and by experienced mathematics teachers without specialist design experience. Contrary to the frequent finding of increasing conceptual depth with increasing expertise, conceptual depth did not differ between groups. Teachers sorted on the basis of mathematical content earlier than designers, and were more specific in their content-based categories. Designers produced more sorts than teachers and were more individualistic in their sorting. These findings suggest that domain expertise does not necessarily impair creative problem solving, as has been suggested in other studies. Instead, expertise includes the ability to shift perspectives with respect to the domain.

One of the basic phenomena of skilled performance is an increasing conceptual depth at which domain knowledge is mentally represented. Experts are able to call upon abstract and generalizable representations, such as schemata, which they subsequently adapt to meet current task demands. Typically, these representations embody fundamental principles that capture significant and useful commonalities among domain problems. In contrast, novices rely upon shallow representations that focus upon superficial features of the domain or task. For example, McKeithen, Reitman, Rueter & Hirtle (1981) investigated recall by intermediate and novice programmers. Intermediates recalled programming terms in an order that suggested organisation by algorithm or function, whereas novices' recall orders reflected superficial relations, nicely illustrated by the recall chunk of the terms "bits", "of" and "string". Similar expert/novice differences have been found in many domains, such as computing systems (Doane, Pellegrino & Klatzky, 1990), physics (Chi, Feltovitch & Glaser, 1981), geometry (Koedinger & Anderson, 1990) and experimental design (Schraagen, 1993). Where expert performance is based on deep representations, it is characterised by the rapid recognition of problem states and the structured development of solutions following a predetermined pattern.

While a deep conceptual representation confers many advantages, there may be situations where the re-use of established domain knowledge is insufficient or inappropriate. For example, Adelson (1984) presented expert and novice programmers with abstract (output-oriented) or concrete (step-by-step) program flowcharts prior to participants answering abstract or concrete questions about program code. She found that experts made fewer errors than novices when the level of abstraction of flowchart and question matched. However, where an abstract flowchart primed a concrete question, novices outperformed experts. The source of this effect appears to be the misapplication by experts of conceptual knowledge primed by the abstract flowcharts. More recently, Wiley (1998) has shown that priming of domain knowledge can impair performance in a remote associates task, in which participants are presented with three words, such as *plate*, *broken* and *shot* , and are required to generate a fourth word, such as *glass*, that forms a familiar phrase with each of the three presented words.

Baseball primes impaired performance on trials where only the first word fits a baseball theme (e.g., *home plate*).

These studies, combined with other demonstrations of impaired expert performance, encourage a view that experts are unable to 'turn off' their deep domain knowledge when it is inappropriate for task performance. However, it might be argued that these demonstrations are artefactual. In these studies, experts' skills are systematically undermined, either by domain priming, as in Wiley's (1998) study, or by having them perform a task that typifies novice problem-solving behaviour, as in Adelson's (1984) study. One might argue that the message of these studies is simply that experts make poor novices. Whether there are cases of realistic domain activities where the presence of deep conceptual knowledge impairs expert performance remains to be demonstrated.

Of particular interest to the current research is Wiley's (1998) suggestion that domain knowledge can sometimes act to inhibit creative problem-solving. Design is a creative problem-solving activity where a case might be made for expertise involving more than re-use of conceptual knowledge. Design has been studied extensively (e.g., Goel & Pirrolli, 1989), and evidence has accumulated showing the same kinds of conceptual representation underlying expert design that are found in other domains of expertise (e.g., Visser, 1991). However, the application domains of these studies (architecture, engineering and software design) are constrained, either by technology or by the design brief or context, such that highly original solutions are the exception rather than the rule (see Goel, 1994, for a useful exposition on the nature of design constraints). When originality is the primary concern, prior knowledge may be less efficacious, perhaps leading to design fixation (Jansson & Smith, 1991).

The development of instructional tasks presents an interesting test case of creative design, and is the focus of the current study, conducted as part of a wider investigation into the nature of task design expertise funded by the UK Economic and Social Research Council. Changes in educational practice, such as increasing use of problem-based teaching, place an emphasis on creativity in task design. This is magnified by the need for tasks that are motivating for students, and that address curriculum and assessment goals without disenfranchising minority groups.

We have recently carried out a study (Ormerod and Fritz, 1999) in which we analysed verbal protocols of designers developing novel tasks to appear in English as a Foreign Language (EFL) textbooks The protocols of specialist designers, experienced authors of EFL textbooks, were typified by the early depth-first development of multiple task ideas, prior to a phase in which a single idea was developed in breadth. In contrast, the protocols of non-specialists, experienced EFL teachers without specialist task design experience, reveal an initial phase in which a single generic task was developed in breadth and subsequently instantiated in depth. The early depth-first work of specialists allowed them to generate and test alternative task ideas to a point where task feasibility could be evaluated before choosing one to develop more completely. The task generation of non-specialists appears constrained by their application of a pre-determined 'schema' embodying a generic task structure. Although it is difficult to assess objectively, the specialists' tasks appear more original than those of the non-specialists, whose tasks were strongly based on popular ESL textbooks. This is supported by protocols in which non-specialists refer to common ESL task types (e.g., 'information-gap').

What is the source of creativity in task design shown by specialists? The non-specialists' behavior appears consistent with Wiley's (1998) suggestion that conceptual knowledge can impair creative problem-solving. So, why does it not impair that of specialist task designers? It is possible, though unlikely, that specialists do not have the deep conceptual representations of tasks that ESL teachers develop. Alternatively, it may be that specialists acquire strategic knowledge that enables them to bypass the application of conceptual domain knowledge where necessary, or that they acquire alternative conceptual structures that take precedence over principles-based conceptual knowledge in design contexts. The study reported in the remainder of this paper set out to explore these hypotheses.

The study

One approach to exploring expertise is the sort method (e.g., Hoffman, Shadbolt, Burton & Klein, 1995). In this method, participants are given sets of domain item descriptions, and are required to sort these into categories according to one or more dimensions that are significant to them. By examining the nature of the sorts produced (e.g., the dimensions used to define categories, assignment to categories and the order in which dimensions are produced), one can infer something about participants' mental representations of conceptual knowledge. Chi et al (1981) used the sort method to explore expert/novice differences in conceptual representation of physics knowledge. They found that sorts produced reflected a deep/superficial distinction, with experts sorting according to underlying principles and novices sorting according to surface features of physics problems. Similarly, Schoenfield & Herman (1982) used the sort method to investigate mathematics expertise, again replicating the deep/superficial distinction. The sort method has been used to explore other forms of expertise such as programming (Davies, Gilmore and Green, 1995), Archeology (Burton, Shadbolt, Rugg & Hedgecock, 1990). and engineering design (Ormerod, Rummer & Ball, in press).

The present study used the sort method to investigate expertise in the design of mathematics tasks. Because we were not interested in studying mathematical expertise per

se, which generally occurs alongside expertise in the design of mathematical tasks, it was important that our expert and non-expert groups be well matched with respect to their education and experience with mathematics. We selected specialist designers of assessment items for English exam boards as our expert group. Mathematics teachers, with equivalent educational and teaching backgrounds, served as our non-expert group. Because domain content varies considerably depending upon school year, we targeted GCSE-level mathematics (equivalent to the middle of high school). Both the designers and the teachers worked primarily at the GCSE level. The cards to be sorted each contained a task from the prior year's GCSE exams (e.g., Figure 1), so as to be realistic and familiar to both groups.

We were interested to see whether teachers and designers, being well matched on most dimensions other than actual design expertise, would perform the card sorts differently. If designers, more than teachers, are fixed in their concepts regarding tasks, then we would expect to find designers producing fewer sorts than teachers. Unless task design is idiosyncratic, designers might also be expected to be more similar to one another than teachers. On the other hand, if designers benefit from greater flexibility in their approach to tasks, then they should produce more sorts and be less fixed in their assignment of tasks to conceptual categories. Unlike other sort studies (e.g., Chi et al, 1981; Schoenfeld & Herrmann, 1982) we did not specify the sort dimensions or the pre-classify tasks according to conceptual level. Our interest lies in the sorts that participants produce spontaneously to reflect their own choice of dimensions.

Method

Participants

Participants included 20 GCSE-level math teachers from Northwest England and 14 GCSE task designers from 4 different exam boards. The designers were also experienced in teaching mathematics with 4-10 years experience in designing tasks for GCSE examinations.

Materials

Twelve GCSE tasks were selected from the MEG and NEAB 1996 exams, and were reduced to fit on A5 card (as in Figure 1). The tasks were selected to be representative of the exams as a whole, while still being reasonably related to other tasks in the set. Tasks were selected from all three exam levels (lower, intermediate, and higher).

Design and Procedure

Expertise was a between-participants factor with two levels, designers and teachers. The study was conducted in the form of an interview between experimenter and participant. Each participant was interviewed individually for approximately one hour. All participants received the tasks approximately one week prior to their interviews so that they could look them over at their leisure. The interviews began with participants giving an account of their teaching or design training and experience. The sort and another related activity were counter-balanced with respect to order between participants. For the sort, the experimenter

503

then demonstrated the sort activity, sorting a set of mammal names twice under example dimensions such as ferocity and attractiveness, while giving a verbal account of her reasons for choosing each dimension, category and assignment.

Participants were instructed "I'd like for you to organise these tasks into groups, more or less as I've just done with the animals. You can make as many or as few groups as you choose. Sort in ways that are useful and meaningful to you as a professional. All of the tasks are different, but sort them based upon the commonalities that you identify, that is, how they fall into different categories for a particular dimension of your choice. " Participants' verbalizations were recorded, along with a record of the task groupings derived from each sort. Where the participant's verbal report had not already revealed sufficient information concerning the dimensions and categories of a sort, the experimenter indicated each of the groups in turn, asking "What makes these form a group?". When all categories were described by the participant, the experimenter asked "Is there some overall theme or explanation to the way these have been grouped?" After each sort the tasks were shuffled, and the participant was asked to produce another sort under a different dimension. Participants were encouraged to continue with further sorts for as many dimensions as they could reasonably form.

Results

Designers produced reliably more sorts than teachers. (Designers mean = 4.2 , sd=0.77, teachers = 3.5, sd=1.05), t(33)=2.17, p=.037. A hierarchical cluster analysis of the participants was run using Euclidean distances and a complete linkage procedure. For each participant, each task pair was assigned a score calculated as the percentage of times that the pair was assigned to the same group by that participant. Thus, if two tasks were always grouped together by one participant, then the score for that pair for that participant would be 100%. If two tasks were never grouped together, the score would be zero. If the participant produced four sorts and the two tasks were grouped together in one sort but were not together in any of the others, the score would be 25%. The clusters that emerged from the analysis, using a reasonable cutoff, included six pairs of designers, 1 pair of teachers, and 4 larger groups (Ns=5,4,3,3) containing teachers with a single designer. There were no instances of designers forming clusters larger than a pair.

MEG Paper 5, 1996

For examiner's use only

10. Two photographs of a yacht are pictured to the right.

Photograph B is an enlargement of photograph A.
Photograph A has width 5.5 cm and photograph B has width 8.8 cm.

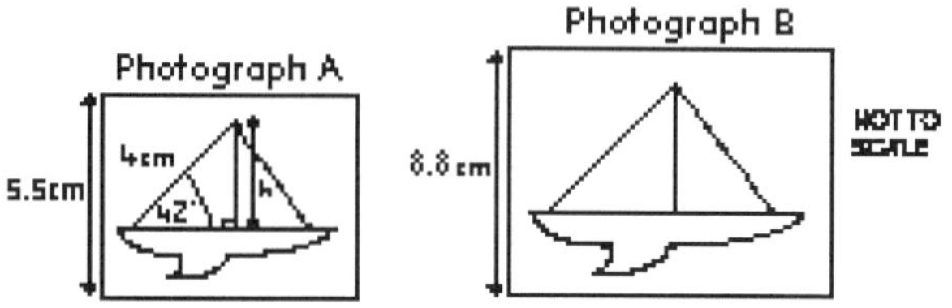

(a) (i) Find the scale factor of the enlargement. Give your answer in form $\frac{p}{q}$ where p and q are whole numbers.

Answer (a) (i) _____ (2)

In photograph A the sail of the yacht is a triangle with one side 4 cm and one angle 42 °.
(ii) Find the length of the corresponding side of the sail on photograph B.

Answer (a) (ii) _____ cm (3)

(iii) Write down the size of the corresponding angle on photograph B.

Answer (a) (iii) _____ (1)

(b) calculate the height, h, of the mast of the yacht in photograph A.

Answer (b) _____ cm (3)

Figure 1. An example of a GCSE Mathematics task used in the study (©MEG examination board, UK).

Designers and teachers used many of the same dimensions, but one interesting difference was that many teachers identified 'thinking' tasks as a category whereas designers identified 'open' tasks as a category. (See Table 1.) In addition, designers were more likely to sort on the basis of the level of the tasks than were teachers, and produced 'open', 'thinking' and 'level' sorts earlier than the teachers.

All participants except one teacher included at least one sort based on the mathematical content of the tasks. All 19 teachers who produced a content-based sort, produced it as their first sort. Nine of the 15 designers led with a content-based sort, but many designers began by sorting on the basis of level and referred to math content in the later sorts. This difference is reliable, U=85.5, p=.048. The number of math content sorts did not differ reliably. Designers produced an average of 4.8 groups (sd=1.01) and teachers produced 5.3 groups on average (sd=1.06), t(32)=1.43, p>.05.

Nevertheless, it was apparent from cluster analysis that there were differences in the ways that designers and teachers sorted the tasks. (See Figure 2.) This cluster analysis, again using hierarchical clustering, Euclidean distances, and a complete linkage procedure, was run using only the participants' math content-based sort. (Four teachers provided more than one content-based sort; we used only the first content-based sort from each of these participants.) The linkage distances do not show a marked increase, used to suggest a cutoff for accepting clusters, until near the end of the run, at approximately 4.8. Whether using that cutoff, or no cutoff at all, it is evident that there are two main clusters forming, and that those clusters are specialist designers on the one hand, and teachers on the other. Designers and teachers were sorting the tasks differently.

Detailed examination of the categories and category members assigned by the two groups provides some explanation. Teachers often produced more specific categories, such as 'linear inequalities', 'fractions', and 'number patterns' as compared to the more general category of 'Algebra and number' which was more often adopted by designers. When more specific categories were collapsed to form the four Attainment Targets defined for the English mathematics curriculum (Applying & using math; Algebra & number; Shape & space; and Data handling), designers' and teachers' sorts were very similar with the only notable difference being in the use of the Data handling category, which designers used far more than did teachers. Otherwise, it was clear that teachers' and designers' perceptions of the tasks in terms of gross mathematical content were not distinguishable; differences were primarily due to the greater specificity on the part of the teachers.

Table 1. % of designers (D) and teachers (T) producing dimensions and mean position in which the dimension. Other dimensions produced by < 10% participants were Complexity, Exam board, Wordiness, Response type, and Mark.

Participant	Math topic	Level	Openness	Thinking vs rote	Difficulty	Context	Prefs / turn-offs	Graphics	Ramping
% D	100	60	50	14	14	33	29	14	21
% T	95	20	10	37	35	15	20	25	0
Position D	1.6	1.6	2.7	2.0	2.5	3.2	3.5	3.0	3.7
Position T	1.0	2.5	3.5	3.3	2.8	2.7	2.8	2.4	0

Discussion

Although designers produced reliably more sorts than teachers, it is possible that the difference was due to the different structures of their work days. Most teachers were scheduled to teach shortly after the interview whereas designers were less rigidly scheduled. However, teachers did not appear distracted and seemed to be as fully occupied by the task at hand as the designers. Furthermore, a related activity (not reported in this paper) was also scheduled during the hour; for half of the participants the sort occurred first and for the other half, the sort occurred last. If teachers limited their responses, then a difference between those who sorted first and second would be predicted. No difference was found (Teachers = 3.6 and 3.4 sorts, designers = 4.1 and 4.3 sorts first and second, respectively.)

The results suggest that different kinds of domain role invoke different kinds of conceptual representation. Teachers appear primarily to use the kinds of conceptual representations found in other studies of expert knowledge (e.g., Chi et al, 1981). This knowledge is precisely what is needed for the task of selecting appropriate Mathematics exercises for a particular stage of the curriculum. Designers, on the other hand, appear to use a wider range of conceptual representations, of which principles-based deep conceptual representations are not always primary. There are two potential explanations for this. The first is that designers have lost or under-rehearsed their principles-based representations. This seems unlikely given that all the designers used Math content for one of their sorts. The second is that design requires different kinds of knowledge to teaching. 'Superficial' dimensions may reflect the very things that make tasks interesting, original and practicable. A similar finding of distinct types of conceptual representation underlying different forms of expertise in the same domain

was made by Weiser & Shertz (1983), again using a sort paradigm to explore conceptual representation. They found that expert computer programmers sorted problems by algorithm type while novices sorted by application area. In contrast, programming managers, all formerly experienced programmers, sorted by 'kinds of programmer' needed to solve each problem.

In much of the expertise research reported in the literature, there is an assumption that what is elicited through methods such as recall and sorting is relatively static. This assumption uynderlies reports of impaired performance resulting from inappropriate application of expert domain knowledge. We argue that to characterise experts' conceptual knowledge in this way is to miss an essential aspect: Experts have many layers of domain knowledge, and, when they are given realistic domain roles and contexts in which to perform, they know when and how to use it <u>and</u> when not to use it.

The notion of static conceptual representations is further challenged by Barsalou's (1985) distinction between taxonomic and goal-directed categories. In this view, goal-directed categories and their members are not fixed, but are determined by the task faced by the individual at any one time. The distinction between goal-directed and taxonomic categories has important methodological implications for the use of the sort method in studies of expertise. It has been suggested by some authors (e.g., Burton et al, 1991) that the sort method provides an equally informative but more cost-effective method for knowledge elicitation than traditional methods such as the analysis of verbal protocols. However, studies that restrict participants to a single sort or that impose pre-specified dimensions may limit elicitation to the kinds of taxonomic knowledge that underlie routine expertise, and fail to capture the sorts of goal-directed categories that may underlie highly skilled performance in non-routine activities.

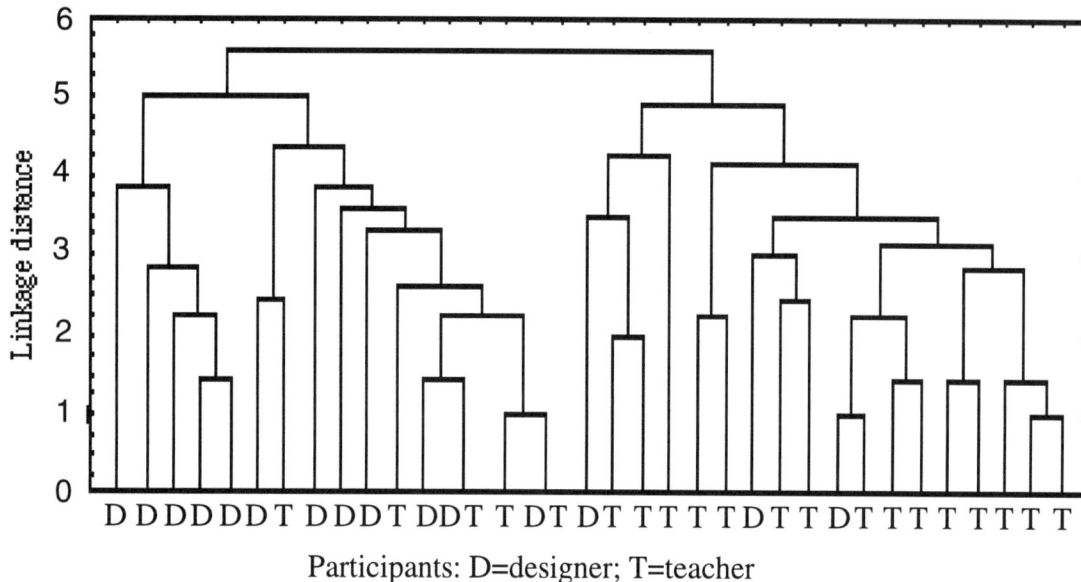

Participants: D=designer; T=teacher

Figure 2. Cluster analysis of designer and teacher groupings based upon sorts under a Mathematics Topic dimension

Acknowledgements

The research reported in this paper was supported by a grant from the Economic and Social Research Council's (UK) Cognitive Engineering initiative, No. L127251031. We thank all the designers and teachers who took part in the study.

References

Adelson, B. (1984). When novices surpass experts : the difficulty of a task may increase with expertise. Journal of Experimental Psychology : L,M&C, 10, 483-495.

Barsalou, L. W. (1985). Ideas, central tendency, and frequency of instantiation as determinants of graded structure in categories. Journal of Experimental Psychology: L,M&C, 11, 629-654.

Burton, A. M., Shadbolt, N. R., Rugg, G. & Hedgecock,A.P. (1990). The efficacy of knowledge acquisition techniques, Knowledge Acquisition, 2, 167-178.

Chi, M. T. H., Feltovich, P. J., & Glaser, R. (1981). Categorization and representation in physics problems by experts and novices. Cognitive Science, 5, 121-152.

Davies, S. P., Gilmore, D. J., & Green, T. R. G. (1995). Are objects that important? Effects of expertise and familiarity on classification of object-oriented code. Human-Computer Interaction, 10, 227-248.

Doane, S. M., Pellegrino, J. W., & Klatzky, R. L. (1990). Expertise in a computer operating system: conceptualization and performance. Human-Computer Interaction, 5, 267-304.

Goel, V. (1994). A comparison of design and non-design problem spaces. AI in Engineering, 9, 53-72.

Hoffman, R. R., Shadbolt, N. R., Burton, A. M., & Klein, G. (1995). Eliciting knowledge from experts: A methodological analysis. Organizational behavior and human decision processes, 62, 129-158.

Jansson, D. G., & Smith, S. M. (1991). Design fixation. Design Studies, 12, 3-11.

Koedinger, K. R., & Anderson, J. R. (1990). Abstract planning and perceptual chunks: Elements of expertise in geometry. Cognitive Science, 14, 511-550.

McKeithen, K., Reitman, J.., Rueter, H., & Hirtle, S. (1981). Knowledge organization and skill differences in computer programmers. Canadian Jnl of Psychology, 13, 307-325.

Ormerod, T. C., Rummer, R., & Ball, L. J. (in press). An ecologically valid study of categorisation by designers. D.Harris (Ed.), Cognitive Ergonomics & Engineering Psychology Hampshire: Ashgate.

Ormerod, T. C. &. Fritz., C.O. (1999). Strategy changes across phase in problem solving: the case of task design Unpublished MS. Lancaster University.

Schoenfeld, A. H., & Herrmann, D. J. (1982). Problem perception and knowledge structure in expert and novice mathematical problem-solvers. Journal of Experimental Psychology: L,M&C, 8, 484-494.

Schraagen, J. M. (1993). How experts solve a novel problem in experimental design. Cognitive Science, 17, 285-309.

Visser, W. (1990). More or less following a plan during design: opportunistic deviations in specification. Int. Journal of Man - Machine Studies, 33, 247-278.

Weiser, M., & Shertz, J. (1983). Programming problem representation in novice and expert programmers. Int. Journal of Man - Machine Studies, 19, 391-398.

Wiley, J. (1998). Expertise as mental set: The effects of domain knowledge increative problem-solving. Memory & Cognition, 26, 716-730.

Expressing manner and path in English and Turkish:
Differences in speech, gesture, and conceptualization

Asli Özyürek (asliozu@mpi.nl)
Max Planck Institute for Psycholinguistics, Wundtlaan 1
6500 AH Nijmegen, The Netherlands

Sotaro Kita (kita@mpi.nl)
Max Planck Institute for Psycholinguistics, Wundtlaan 1
6500 AH Nijmegen, The Netherlands

Abstract

This study investigates how speakers of typologically different languages, Turkish (verb-framed) and English (satellite-framed) express motion events in their speech and accompanying gestures. 14 English and 16 Turkish speakers narrated an animated cartoon and one motion event scene was selected for analysis. English speakers depicted this scene with one verb with a satellite "the cat *rolls down*", combining manner and path of the motion in one clause. Whereas Turkish speakers used two verbal clauses (e.g., *yuvarlanarak iniyor* (rolling descends)), separating manner from path. Gestures showed a similar pattern. Turkish speakers compared to English were more likely to use a) pure rotation gestures (representing manner only) and b) pure trajectory gestures (representing path only). These findings support the claim that speakers of typologically different languages conceptualize motion events in different ways during on-line speaking. While more Turkish speakers represent two components of a motion event as separate, English speakers represent them as one unit.

Introduction

Languages vary typologically in terms of how they map lexical and syntactic elements onto semantic domains (Bowerman, 1996; Slobin 1996; Talmy, 1985). This variation has been most prominent in the domain of spatial relations, especially in expressions of motion events. The typological differences among languages have recently inspired other questions: Do speakers of typologically different languages also have different ways of thinking and conceptualizing spatial relations? In this paper we will address this question by comparing how speakers of two typologically different languages, Turkish and English, use their speech as well as spontaneous gestures to express motion events in narrative discourse.

Turkish and English belong to two typologically different group of languages in terms of how they map lexical elements onto semantic elements of motion events (Talmy, 1985). Turkish prefers to encode path of motion in a verb (e.g., *gir* 'enter', *cik* 'exit', *in*' descend'), whereas English encodes path of motion by verb particles or satellites (e.g., go *in*, *out*, *down*). In this regard Turkish belongs to the so called verb-framed languages (e.g., Semitic, and Romance languages as well as Japanese and Korean) and English belongs to satellite-framed languages (e.g., most Indo-European languages except Romance) since the core semantic domain, the path, is mapped onto the verb in Turkish but onto the satellite or verb particle in English.

Recent research by Slobin et al. (e.g., Berman and Slobin, 1996; Slobin, 1987; Özcaliskan & Slobin, in press) has shown that differences in how semantic elements are mapped onto lexical elements influence speakers to conceptualize motion events in different ways. For example, it has been found that speakers of verb-framed languages pay less narrative attention to manner of motion than speakers of satellite-framed languages unless manner is the salient information in the discourse context. This difference is mainly due to how manner is lexicalized in the two types of languages. Since satellite-framed languages do not prefer to encode path in the main verb, this slot is available for manner verbs (e.g., *roll* down the hill). On the other hand verb-framed languages tend to use the main verb to encode path, that is, this slot is generally reserved for path verbs and manner tends to be encoded as subordinated to the main verb (e.g., *yuvarlanarak indi* 'descended rolling'). Since satellite-framed languages encode manner in the main verb they have been shown to use manner verbs more frequently than speakers of verb-framed languages. However, since speakers of verb-framed languages such as Turkish have to encode manner as subordinated to the main verb, they express manner less frequently and usually omit it unless it is salient or foregrounded information in the narrative context. Slobin et al. have concluded from findings of this sort that speakers of satellite-framed languages devote a great amount of narrative attention to details of manner of movement compared to speakers of verb-framed languages. They have also reported similar results about speakers of verb-framed languages applying more attention to scene setting than speakers of satellite-framed languages due to typological differences.

Slobin (1987) calls this way of conceptualizing events for purposes of speaking "thinking-for-speaking" hypothesis. According to this hypothesis, the preferred typological options in one's language tune speakers to deal with experience in different ways and these subjective orientations affect speakers to organize their thinking to meet the demands of linguistic encoding during acts of speaking.

In these studies spoken linguistic patterns have been the only information used to show the differences in conceptualization of motion events in narrative. However, linguistic expreessions are limited in terms of informing us about the underlying on-line thinking processes during speaking. Therefore we need a measure additional to the linguistic expressions and yet that will still provide an insight into speakers' thinking and conceptualization processes used during narrative communication.

One other way that speakers externalize their on-line representations during speaking is the spontaneous gestures they use accompanying their speech. After detailed observation of speakers' spontaneous gestures used in narratives, McNeill (1985; 1992) has shown that gestures and speech are systematically organized in relation to one another. Gestures together with the speech segments they synchronize with form meaningful combinations. That is, gestures in form and manner of execution exhibit a meaning relevant to the simultaneously expressed linguistic meaning.

The meaning relations between the gesture and the linguistic segment can vary. For example the content of gesture and the content of the linguistic segment can parallel each other. Consider this example from a narrator telling a cartoon story:

(1) 'and he climbs up the drainpipe'
[hands go up in a climbing manner]

In this example both gesture and speech represent the manner (i.e., climbing motion of the hands expressing manner of motion and the verb *climb*) and path of the moving figure (i.e., the rising hand expressing upward path and the verb particle *up*). The contents of gesture and speech can parallel each other as in this example but need not be. Consider this other example from a narrator telling the same cartoon story (McNeill, 1985).

(2) 'she chases him out again'
[hand as if gripping an object swings from left to right]

Here speech conveys the concepts of pursuit (chases) and recurrence (again) and gesture conveys the means of pursuit that is, swinging an umbrella. Thus the content in gesture and content in speech might complement as well as be parallel to each other.

This kind of gestures cited above is called 'iconic' or 'representational' gestures. That is, they have a formal resemblance to the referents they represent[1]. In this study these kind of spontaneous gestures will be the focus of study and considered as an enhanced index of the speakers' thinking processes in addition to linguistic expressions.

McNeill (1985, 1992; McNeill & Duncan, in press) and Kita (in press) have proposed the following relations between gestural and verbal expressions and thinking processes. According to McNeill representational gestures and speech reflect different parts of one underlying unit of cognitive representation. The thrust of McNeill's argument is that "to get the full representation that the speaker had in mind both the sentence and the gesture must be taken into account" (McNeill 1985; 353). For example, in the above utterance (2), the speaker had the chasing, and the recurrence aspects of the action expressed by speech and the pursuit of chasing expressed by gesture as one unit of cognitive representation underlying the utterance. This mental unit consists of both visuo-spatial cognition as manifested in gestures and linguistic content as manifested by the structural and lexical possibilities of languages. Kita (in press) further proposes that, one of the functions of gestures is to help the "organization of complex information into a message that can be verbalized". According to this hypothesis, visual thinking adapts to the specific linguistic system in which gesture is performing its organizing function in utterances: the gesture is shaped "so as to make the informational content as compatible as possible to linguistic encoding possibilities." In both views the gesture content and the linguistic content are integrated to each other and the joint representation of the two informs us about the underlying cognitive representation of the speaker rather than one or the other alone[2].

The aim of this paper is to investigate whether speakers of typologically different languages differ in the way they use their gestures as well as their speaking patterns in expressing motion events. If both speech and gesture index the full representation of the speaker during speaking, then differences in gestures will provide additional insight about the differences in spatial conceptualization among speakers of different languages. McNeill and Duncan (in press) have provided evidence for this claim by comparing English, Spanish and Chinese speakers' gestures accompanying motion event expressions and have shown that the content in speech and gesture are combined in different ways suggesting different conceptualizations among speakers. Here we investigate a similar question by comparing speakers of Turkish and English and focusing on the differences in verbal packaging of manner and path elements of a motion event.

---

[1] Iconic gestures differ from the conventional or emblematic ones such as "OK" gesture which can be used in the absence of speech or hand signs used in sign languages (Kendon, 1997; McNeill, 1992). In the conventional gestures the form-meaning relationship is arbitrary and the form of the gesture does not bear an iconic relation to the referents represented.

---

[2] This hypothesis argues against the other assumptions that speech and gesture are independent communicative channels (Sanders, 1987) or that the function of gestures is merely lexical retrieval (Butterworth and Hadar, 1989).

Linguistic Differences in Motion Event Representations in Turkish and English

A motion event consists of 4 semantic components: path, manner, ground and figure. Path refers to the translational motion of a figure (a moving entity) which, in the most elaborated sense, moves from a source to a goal through some medium, passing one or more milestones. Ground refers to an explicit feature of the physical environment serving as source, medium, milestone, or goal. Manner refers to factors such as the motor pattern of the movement of the figure, rate, and degree of effort (Slobin, 1996). In this paper the focus will be on manner and path components of a motion event.

As mentioned above, Talmy (1985) has grouped world's languages into two according to the way lexical and syntactic structures are mapped onto these semantic elements of motion events. In this categorization Turkish belongs to verb-framed languages whereas English belongs to satellite-framed languages. Here we will outline how English and Turkish differ in the way they lexicalize the following semantic elements of a motion event: a) path, and b) both manner and path with regard to Talmy's typology.

a) expressing path: Turkish, as a verb-framed language encodes path of motion in a verb (e.g., *gir* 'enter', *cik* 'exit', *in* 'descend'), whereas English, as a satellite-framed language encodes path of motion with a particle or satellite (e.g., go *in, out, up, down*).

b) expressing both manner and path: If speakers of English and Turkish have to encode manner of motion in addition to path the following differences arise. Since in English path is encoded by a verb particle but not in the verb, the manner can be encoded in the main verb. Therefore English speakers can easily encode both manner and path within one verbal clause, that is, manner in the verb and path in the satellite. However in Turkish since the main verb is used to encode path, manner tends to be encoded as subordinated to the main verb (e.g., *yuvarlanarak iniyor* 'descends rolling'). Thus Turkish speakers have to use two verbal clauses to express both manner and path components of the motion event (Figure 1).

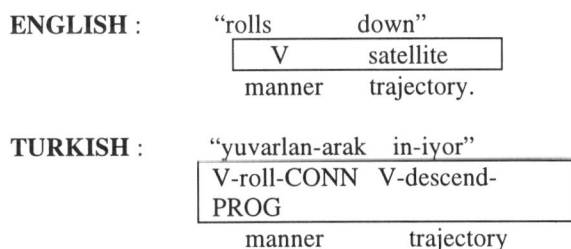

ENGLISH :	"rolls	down"
	V	satellite
	manner	trajectory.

TURKISH :	"yuvarlan-arak	in-iyor"
	V-roll-CONN	V-descend-PROG
	manner	trajectory

Figure 1. Differences in the mapping of manner and path components of a motion event onto syntactic elements in English and Turkish (CONN = connective marker, PROG = progressive tense marker)

We want to focus on possible consequences of these differences between English and Turkish, for the conceptualization of these components for the speakers of both languages. The prediction is that since the verb + satellite construction in English allows speakers to package manner and path components within one verbal clause, it will be easier for English speakers to process both manner and path components within one mental processing unit. However, since Turkish does not allow expressing both manner and path within one verbal clause but instead needs two, Turkish speakers might have to conceptualize manner and path as separated components that is, in two mental processing units. Therefore it will not be as easy for Turkish speakers as it is for English speakers to package both components in one conceptual unit during on-line speaking.

In order to test this prediction we will compare Turkish and English speakers' on-line speaking patterns as well as the gestural expressions of a given motion event. With regard to verbal expressions we expect English speakers to combine both components using one verbal unit whereas Turkish speakers to use separate verbs to do so. We also expect similar patterning with regard to gestures. If this is the case, English speakers will be more likely represent both manner and path components within one gesture (e.g., gesture 1: hands rising up in an imaginary climbing motion to depict a character climbing up a tree). However, Turkish speakers will prefer to represent manner and path components in separate gestures (e.g., gesture 1: hands representing an imaginary climbing motion (but not rising up); gesture 2: hands rising up (without the climbing manner of the hands). If the gestural representations of motion event components parallel differences in speaking patterns as predicted, this will provide further evidence that speakers of different languages differ in their spatial conceptualization as well.

Method

Subjects

14 American English and 16 Turkish speakers participated in this study. All subjects were monolingual speakers. English speaking subjects were undergraduate students at the University of Chicago, USA and Turkish speaking subjects attended Yeditepe and Marmara Universities in Istanbul, Turkey.

Procedure

Each subject was asked to see an animated cartoon 'Canary Row' (8 minutes). In the cartoon Sylvester the Cat attempts to catch Tweety Bird in different ways, each including a series of motion events. Each subject was asked to narrate the cartoon story to an addressee who has not seen it. The narratives were videotaped but subjects were not told that the focus of the study was on gestures.

Coding

Scene sampling: One scene from the cartoon was selected for detailed analysis of speech and gesture. In this scene Sylvester swallows a bowling ball that Tweety Bird throws into his mouth and with the force of this bowling ball he rolls down the street and ends up in a bowling alley. The aim of picking descriptions of this scene is that in this scene both manner and path components are represented in a simultaneous way (i.e., Sylvester goes down the hill as he rolls).

Coding speech: Each subject's verbal descriptions of this scene were coded in terms of whether each speaker used a) verb + satellite construction or b) separate verbs to describe the manner and path components of the cat's rolling down the hill.

Coding gestures: Speakers' gestures[3] that accompanied verbal expressions of this scene were categorized into 3 types:

a) Manner-only gestures: Represent the manner of the motion event only (i.e., hand(s) or finger(s) rotate/wiggle without any trajectory component)

b) Trajectory-only gestures: Represent the path of the motion event only (i.e., hand(s) move along a lateral or sagittal trajectory without any rotation/wiggling of the hand(s) or finger(s))

c) Manner-Trajectory Conflated gestures: Represent both the path and the manner of motion simultaneously (i.e., hand(s) move along a lateral or sagittal trajectory while the hand(s) or the finger(s) rotate/wiggle).

All gestures were initially coded by a single coder and reliability was established by having a second coder code 25% of the data. Agreement between coders was 100 % for categorizing gesture phrases.

Results

Motion Event Descriptions in Speech

Do the differences in lexicalization patterns between English and Turkish have an effect on speakers' on-line verbalizations of this event? If this is the case, then English speakers are expected to use one verbal clause whereas Turkish speakers are expected to use separate verbs to express both manner and path components to express the cat's rolling down the hill in the cartoon event. Here note that even though English allows speakers to express both manner and path in one verbal clause, speakers are not constrained to do so. That is, they can also use separate verbs such as "he went down the street rolling" to depict the scene. However, if the typological pattern has an influence

on on-line verbalization of this scene then English speakers will use one verbal clause to do so.

The results showed in line with the expectations that all English speakers used verb+satellite construction (i.e., *rolled down*) to express both components. On the other hand, all but one Turkish speaker used separate verbs to talk about the motion event (e.g., *yuvarlanarak iniyor* 'descended rolling') to express both manner and path. Only one Turkish speaker expressed both manner and path in one verbal clause: *yokus asagi kayiyo* 'slides down hill'. This one example also shows that it is possible for Turkish speakers to use one verbal clause instead of two to express both components but most of them did not prefer to do so. Thus typological differences influence Turkish and English speakers' on-line ways of speaking about manner and path in different ways.

It was also predicted that since English easily allows to encode both manner and path within one verbal clause, speakers would be more likely to mention both in their description. But since Turkish does not allow this possibility, speakers would be more likely to omit one or the other and let the other be inferred from the discourse context. To answer this question, percentage of subjects in each language who expressed both manner and path versus either manner or path were calculated (Table 1).

Table 1. Percentage of subjects who mentioned either Path or Manner versus both of them in the Turkish and English sample

Language	Either Path or Manner	Both Path and Manner
English	0	100% (N=14)
Turkish	56% (N=9)	43% (N=7)

The results were in line with the initial prediction. More Turkish speakers mentioned either one of the motion event components in their description than English speakers (Chi square = 11.1, df=1, p<.001). On the other hand all English speakers mentioned both components[4].

These differences in ways of speaking about manner and path support the idea that there might be differences between the speakers of both languages in conceptualizing motion event components. Since it is easier to combine the two components in one clause English speakers conceptualize both components together all the time. However, since packaging of both components in one

---

[3] According to McNeill (1992) conventions, gestures have three phases in their production: a preparation phase, a stroke, and a retraction phase. The three phases together constitute a *gesture phrase* . In the present study the gesture phrase was the basic unit of analysis. See McNeill (1992) for more information on how to define a gesture phrase.

[4] As mentioned before Slobin et al. has found that Turkish speakers omit manner more frequently than English speakers due to the differences in lexicalization patterns. However, in the present data set Turkish speakers did not omit manner more frequently than the path. That is both components were likely to be omitted (55% of all the omitted verbs were Manner and 45% were Path). This might be due to the fact that both manner and path were salient components for the description of this scene in the narrative discourse.

verbal unit is not lexically easy for Turkish speakers, they conceptualize manner and path separately and thus easily omit one or the other.

Motion Event Descriptions in Gesture

We expected speakers' gestural representations to parallel differences in their on-line speaking patterns. That is, more Turkish speakers were expected to represent manner and path components in separate gestures than English speakers, whereas English speakers were expected to use one gesture to express both components simultaneously.

For this analysis the percentage number of subjects who used either one of the representational strategies in gestures, that is a) Manner-only, b) Trajectory-only, and c) Manner-Trajectory Conflated gestures were calculated. Many speakers used more than one gesture of different types to depict this scene. Thus percentage subjects who used either one of the 3 types at least once in their descriptions were calculated.

Since the stimulus event in the cartoon scene included both manner and trajectory represented simultaneously, we expected similar amount of English and Turkish speakers to use Manner-Trajectory Conflated gestures at least once in their repertoire. In this type of gesture (e.g., the hand moves along a trajectory while the hand or the fingers rotate/wiggle), manner is represented simultaneously with path within one gesture. This would be an isomorphic representation of this scene and speakers of both languages would be as likely to use this type of gesture. However, if Turkish speakers also have cognitive representations in which manner and path are encoded as separate units, we expected more Turkish speakers to use Manner-only and Trajectory-only gestures in their repertoire. On the other hand, English speakers were expected to have fewer Manner-only and Trajectory-only gestures but rather have mostly Manner-Trajectory Conflated gestures where manner and path are represented as one unit.

First, percentage of subjects who used Manner-only and Trajectory-only gestures in their repertoire at least once in both languages was calculated (Tables 2 and 3).

Table 2. Percentage of subjects who used Manner-only gestures at least once in their repertoire of gestures

Language	Used at least once	Never used
English	14 %(N=2)	86% (N=12)
Turkish	62 %(N=10)	38% (N=6)

Table 3. Percentage of subjects who used Trajectory-only gestures at least once in their repertoire of gestures

Language	Used at least once	Never used
English	35% (N=5)	65% (N=9)
Turkish	75% (N=12)	25% (N=4)

As expected more Turkish speakers than English speakers used Manner-only and Trajectory-only gestures (for Manner-only: Chi square = 7.18, df=1, p<.001; for Trajectory-only: Chi square = 4.58, df=1, p<.05). These findings support the idea that Turkish speakers conceptualize both components of this motion event separately. This pattern was more visible in Manner-only gestures than in Trajectory-only gestures. However not as many English speakers had these types of gestures where manner and path were represented as separate units.

On the other hand there was no difference found between language samples in the percentage of speakers who used Manner-Trajectory Conflated gestures at least once in their repertoire. This was also in line with the inital predictions since Manner-Trajectory Conflated represented the stimulus as perceived. Thus equal number of speakers in both languages had Manner-Trajectory Conflated gestures in their descriptions.

In sum the repertoire of English speakers' gestures can be characterized mostly with Manner-Trajectory Conflated gestures where manner and path are represented simultaneously within one gesture. However, Turkish speakers also had Manner-only and Trajectory-only gestures in their descriptions more than English speakers did. The additional use of Manner-only and Trajectory-only gestures by Turkish speakers parallel their verbalization patterns about this motion event. On the other hand, since English speakers could verbalize both components of this motion event within one verbal clause, they used Manner-only and Trajectory-only gestures less frequently than Turkish speakers did.

Conclusion and Discussion

This study showed that typological differences among languages, that is, how they map lexical and syntactic elements onto semantic domains influence speakers' on-line speaking and gestural patterns that describe a motion event. Differences in both verbal and gestural patterns give support to the idea that there might also be differences in the conceptualization of motion events among speakers of typologically different languages as also have been suggested by McNeill and Duncan (in press).

Here we demonstrated further evidence for this claim by comparing how Turkish (verb-framed) versus English (satellite-framed) speakers express manner and path of a motion event both in their linguistic expressions and gestures. Results on linguistic patterns showed that Turkish speakers typically used separate clauses to express both manner and path and also half of the speakers omitted one or the other in their descriptions. However, English speakers used one verbal clause to express both manner and path and all the speakers mentioned both components in their descriptions. With regard to gestures, English speakers mostly used one type of gesture where both manner and path components are represented as one unit. Even though English speakers' repertoire also included gestures where manner and path are presented separately, fewer speakers preferred to do so than Turkish speakers. Turkish speakers also had gestures where they represented manner and path within one gesture in ways similar to English speakers. However, in addition to this, more Turkish speakers than English speakers represented manner and path in separate gestures.

These findings support the claim about differences in conceptualization of motion events. While more Turkish speakers are likely to conceptualize manner and path of a motion event as separate components, English speakers conceptualize the two components as one unit during on-line speaking.

If Turkish speakers are more likely to conceptualize manner and path as separate components, one would also expect them to use speech and gesture combinations where speech expresses Manner-only but gesture represents Trajectory-only or vice-versa. In fact out of all speech and gesture combinations where Turkish speakers used Manner-only and Trajectory–only gestures, gestures complemented speech 30% of the time. This is further evidence for the claim that Turkish speakers can decompose representations about manner and path and express them as a combined unit of representation through speech and gesture together.

The findings presented in this paper are in line with "thinking-for-speaking" hypothesis outlined by Slobin (1987) and show that speakers' gestures as well as their speaking patterns index different ways of conceptualizing motion events for purposes of speaking. The findings also support the idea that the speech content and gestural content are integrated to each other as proposed by McNeill (1985; 1992). They further show that speakers organize their gestures in order to make the informational content as compatible as possible to the linguistic encoding possibilities of their language (Kita, in press).

Even though this study has shown that there are differences in gestural patterns in line with differences in typological patterns, there might be other reasons for the differences between the gestural patterns of English and Turkish speakers, such as cultural factors. In order to attribute the differences in gestures to their correlation with the lexicalization patterns for sure one needs to test these predictions in another language and culture. In fact, Kita (1996) has conducted the same study with Japanese speakers. Japanese belongs to the same typological group as Turkish, that is it is a verb-framed language and yet the Japanese cultural context is quite different than that of Turkish. The gestural patterns of Japanese speakers paralleled the gestural patterns of Turkish speakers and thus differed than those of English speakers. This finding is evidence for the claim that differences in the gestural patterns observed with regard to path and manner representations are due to typological differences among languages rather than due to cultural factors[5]

---

[5] This claim does not deny that there might be other differences between gestures that might be due to cultural factors such as the size of the gestures.

Acknowledgements

Data from American English speakers was collected using the resources of the David McNeill lab at the University of Chicago. We would like to thank McNeill for providing us with a portion of the data, resources, and intellectual support. The Turkish part of the data was collected by a grant given to the first author by the Max Planck Institute for Psycholinguistics. We would like to thank members of the Gesture Project at the MPI for their valuable comments.

References

Berman, R. A. & Slobin, D. I. 1994. *Relating events in narrative: A cross-linguistic developmental study.* Hillsdale, NJ: Lawrence Erlbaum Associates.

Butterworth, B. & Hadar, U. 1987. Gesture, speech, and computational stages: A reply to McNeill. *Psychological Review, 96, 168-174.*

Kendon, A. 1997. Gesture. *Annual Review of Anthropology*, 26: pg. 109-28.

Kita, S. 1996. "Linguistic effects on spatial thinking as probed by spontaneous speech-accompanying gestures". Paper presented at the 5th International Pragmatics Conference, Mexico City.

Kita, S. in press. How representational gestures help speaking. *Speech and gesture: Window into thought and action*, in D. McNeill ed. Cambridge: Cambridge University Press.

McNeill, D. & Duncan, S. in press. Growth points in thinking-for-speaking. In D. McNeill ed. *Speech and gesture: Window into thought and action.* Cambridge: Cambridge University Press.

McNeill, D. 1985. So you think gestures are nonverbal? *Psychological Review*, 92, pg. 350-71.

McNeill, D. 1992. *Hand and mind: What gestures reveal about thought.* Chicago: University of Chicago Press.

Özcaliskan, S. & Slobin, D. I. (in press). Learning how to search for the frog: Expression of manner of motion in English, Spanish, and Turkish. *Proceedings of the 26th Annual Boston University Confrence on Language Development*, Boston, MA: Cascadilla Press.

Sanders, R.E. 1987. The interconnection of utterances and non-verbal displays. *Research on Language and Social Interaction,* 20, pg. 141-170.

Slobin, D. 1987. Thinking for speaking. *Proceedings of the Thirteenth Annual Meeting of the Berkeley Linguistics Society*, pg. 435-444.

Slobin, D. 1996. Two ways to travel: Verbs of motion in English and Spanish. In M. Shibatani & S. A. Thompson ed.. *Essays in Semantics.* pg. 195-217. Oxford: Oxford University Press.

Talmy, L. 1985. Lexicalization patterns: Semantic structure in lexical forms. *Language typology and syntactic description. Vol. III. Grammatical categories and the lexicon.* In T. Shopen ed., pg. 57-149. Cambridge: Cambridge University Press.

Investigating the Relationship Between Perceptual Categorization and Recognition Memory Through Induced Profound Amnesia

Thomas J. Palmeri (thomas.j.palmeri@vanderbilt.edu)
Department of Psychology; Vanderbilt University
Nashville, TN 37240 USA

Marci A. Flanery (marci.flanery@vanderbilt.edu)
Department of Psychology; Vanderbilt University
Nashville, TN 37240 USA

Abstract

Are perceptual categorization and recognition memory sub-served by a single memory system or by separate memory systems? A critical piece of evidence for multiple memory systems is that amnesics can categorize stimuli as well as normals but recognize those same stimuli significantly worse than normals (Knowlton & Squire, 1993). An extreme case is E.P., a profound amnesic who can categorize as well as normals but cannot recognize better than chance. This paper demonstrates that the paradigm used to test E.P. and other amnesics may be fundamentally flawed in that memory may not even be necessary to categorize the test stimuli in their paradigm. We "induced" profound amnesia in normals by telling them they had viewed subliminally presented stimuli that were never actually presented. Without any prior exposure to training stimuli, subjects' recognition performance was completely at chance, as expected, yet their categorization performance was quite good.

Single Versus Multiple Memory Systems

What processes are involved in judging whether an object belongs in a particular category (a categorization decision) and in judging whether an object is something that has been seen before (a recognition decision)? Formal theoretical accounts have suggested that both of these fundamental types of cognitive judgments are subserved by a single memory system. By contrast, many neuropsychological accounts have suggested that there are separate neural systems subserving categorization and recognition memory. This paper will briefly review the evidence for single memory systems and for multiple memory systems and then present recent experimental work that may challenge some of the critical evidence used to support the multiple memory systems view.

Exemplar-based models, such as the Generalized Context Model (GCM; Nosofsky, 1986; see also Nosofsky & Palmeri, 1997; Palmeri, 1997), assume that both categorization and recognition rely on memory for stored instances but differ in the way that memory is probed. According to the GCM, categorization is based on the relative summed similarity of a probe item to the stored instances of the possible category responses whereas recognition is based on the absolute summed similarity of a probe item to all inst-

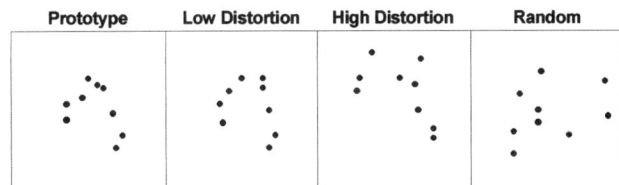

Figure 1: Examples of a prototype, a low-distortion, a high-distortion, and a random pattern.

ances stored in memory. In other words, categorization and recognition decisions rely on the same memories, but differ in the decision rules they use. Nosofsky (1988, 1991) has shown the GCM to provide excellent accounts of observed categorization and recognition data in a variety of experimental paradigms using normal individuals.

Knowlton and Squire (1993) provided evidence for multiple memory systems by contrasting performance of amnesics and normal individuals on categorization and recognition memory tasks. They used a variant of the well known dot pattern classification and recognition paradigm (Posner & Keele, 1968). In a categorization task, amnesics and normals were initially exposed to forty high-level distortions of a prototype pattern (see Figure 1) without being told that the patterns belonged to the same category. At test, subjects were told that the patterns they had just seen were all members of the same category and were then asked to judge whether a new set of patterns were members or non-members of that category. Subjects were tested on the prototype, low distortions of the prototype, and high distortions of the prototype, which were all to be judged as members, and new random patterns, which were all to be judged as nonmembers. In a recognition task, amnesics and normals were initially exposed to five random patterns repeated eight times each without being told that they would later be tested on their memory for those patterns. At test, subjects were shown the five training patterns and five new random patterns and were asked to judged which were old and which were new.

As shown in the top panel of Figure 2, Knowlton and Squire (1993) found that recognition memory was significantly impaired for amnesics compared to normals. However, no significant difference was observed between amne-

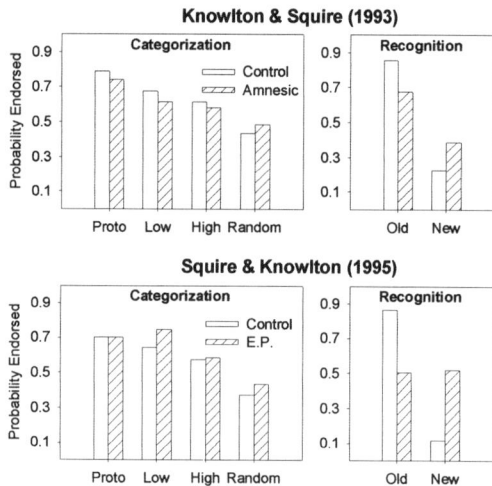

Figure 2: Top panel displays data from Knowlton and Squire (1993) comparing amnesics with normals. Bottom panel displays data from Squire and Knowlton (1995) comparing E.P. with normals. Each panel shows probability of categorization and recognition for each type of pattern.

sics and normals at categorization. This pattern of results was used as evidence for two separate, independent memory systems: an explicit system subserving recognition memory, which is impaired in amnesia, and a separate implicit categorization system, which is spared in amnesia. Knowlton and Squire conclude that *"single-factor models in which classification judgments derive from, or in any way depend on, long-term declarative memory do not account for the finding that amnesic patients perform well on the classification tasks"* (p. 1748).

While these results seemed to suggest the existence of multiple memory systems, Nosofsky and Zaki (1998) recently reported theoretical analyses that showed the GCM capable of accounting for this apparent task dissociation in a fairly straightforward manner. By simply assuming that amnesics had poorly discriminated memory traces compared to normals, which was instantiated by variation in a single parameter of the model, the GCM was able to account for the observed difference in recognition and categorization performance. In addition, by experimentally producing poor memory discrimination in normal individuals through the use of a delay between study and test, Nosofsky and Zaki were able to reproduce the exact pattern of categorization and recognition results observed with amnesics.

One of the reasons for the success of the GCM in accounting for the Knowlton and Squire (1993) results is that amnesics had poor but above chance recognition memory. More recent evidence reported by Squire and Knowlton (1995) may be more challenging. They tested E.P., a profoundly amnesic individual, on a task similar to that used by Knowlton and Squire (1993). As shown in the lower panel of Figure 2, E.P. was able to categorize as well as normals, but was completely unable to recognize above chance levels. It may prove impossible for a single-system model, such as the GCM, to account for this extreme pattern of results (see Nosofsky & Zaki, 1998). In summariz-

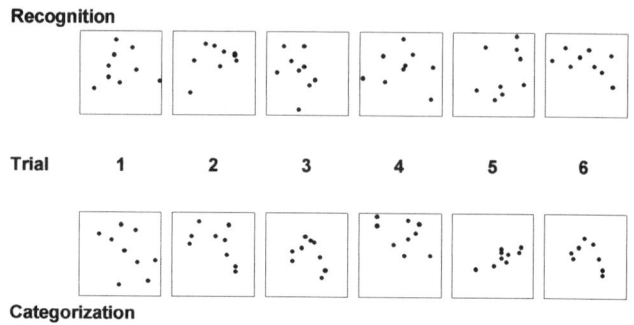

Figure 3: Example sequences of recognition (top row) and categorization (bottom row) using stimuli from Knowlton and Squire (1993). For the recognition trials, correct answers are: (1) Old, (2) New, (3) Old, (4) New, (5) Old, (6) New. For the categorization trials, correct answers are: (1) Nonmember, (2) Member (high), (3) Member (prototype), (4) Nonmember, (5) Nonmember, (6) Member (low).

ing these results, Squire and Zola (1996) concluded that *"these results suggest that category knowledge can develop independently of and in the absence of normal declarative memory ... the information supporting classification learning must be distinct from declarative knowledge about the specific items presented for training. Models in which classification judgments derive from, or in any way depend on, long-term declarative memory do not account for the finding that amnesic patients can acquire category knowledge as well as normal subjects"* (pp. 13517-13518).

The Squire and Knowlton (1995) findings may appear to be devastating to the single system models. However, in this paper, we will suggest that the experimental procedures used to test E.P. and other amnesics may be fundamentally flawed in that prior exposure to training stimuli may be unnecessary to perform the categorization task they used. To illustrate, the top row of Figure 3 displays a sequence of recognition test trials from Knowlton and Squire (1993). Clearly, if asked to judge which of these patterns were old or new without ever having been shown any training patterns, it would be impossible to perform better than chance. The bottom row of Figure 3 displays a sequence of categorization trials. Recall that subjects were required to judge as members the prototype, low distortions of the prototype, and high distortions of the prototype, and to judge as nonmembers a set of completely random patterns. Without previous exposure to any training patterns, it may be quite easy to discover that the set of very similar patterns all belong to the same category and that the set of very dissimilar patterns are all nonmembers of that category. This determination can possibly be made after only a few test stimuli have been shown. Thus, a profound amnesic, such as E.P., who has relatively intact working memory and other cognitive functions, may be able to accurately judge category membership without any memory for the studied patterns.

Experiment 1

Our basic claim is that even without memory for having seen any category members, it may be possible to correctly categorize members versus nonmembers in the particular type of categorization task used by Knowlton and Squire (1993; Squire & Knowlton, 1995). By contrast, it is simply impossible to judge old from new members without remembering which stimuli had been presented before.

Our first goal was to demonstrate that information about the category structure can be extracted from the sequence of patterns used in the categorization test. As a way of maximally assessing how much information about the category structure could possibly be extracted from the test sequence, a particularly well-motivated subject (the second author) participated in ten categorization test sessions. These categorization tests had the exact same structure as those used by Knowlton and Squire (1993); however, in our experiment, the subject did not receive any prior exposure to category members. Although she was aware of how the category members and nonmembers were defined abstractly, she had absolutely no prior knowledge of the particular prototypes and distortions that were used within a given test session – she needed to discover the category structure (judging members versus nonmembers) without the benefit of any prior exposure and without the benefit of any corrective feedback.

It is important to emphasize that even with a complete understanding of the procedure for how old and new patterns in a recognition test were generated, without any prior exposure to study patterns it would be absolutely impossible to recognize old from new patterns better than chance.

Method

Subject. The second author (M.A.F.) completed ten categorization sessions over a two week period.

Procedure. On each trial of the categorization task, a dot pattern was displayed and the subject was asked to judge whether it was a member or nonmember of a category; no prior exposure to category members had been provided. The subject judged four instances of the prototype, 20 low-level distortions, 20 high-level distortions, and 40 random patterns (identical to procedures used by Knowlton and Squire, 1993). Order of stimuli was randomized and no corrective feedback was supplied.

Stimuli. Stimuli were patterns of nine dots. At the start of each session, the computer randomly generated a pattern and designated it the prototype of the category. Distortions of the prototype were created with a commonly used statistical distortion algorithm (Posner, Goldsmith, & Welton, 1967). Random nonmember patterns were also newly created at the start of each session. The subject was completely unaware of the particular set of dot patterns that had been created until they were presented during the experiment.

Results

Figure 4 displays the probability of endorsing each type of stimulus as a member of the category. Without any prior

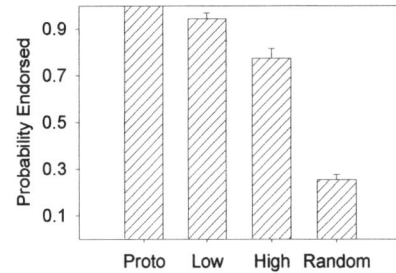

Figure 4 : Proportions of categorization decisions for each type of pattern from Experiment 1.

exposure to category members, M.A.F. was able to categorize the prototypes perfectly, the low distortions nearly perfectly, and the high distortions and random patterns extremely well. Overall, M.A.F. was 81.3% correct at classifying members and nonmembers. A statistically significant effect of stimulus type was observed, $F(3,39)=153.28$, $MS_e=0.008$, and planned comparisons confirmed the observed ordering of levels.

Discussion

As we predicted, there was sufficient information in the sequence of categorization trials at least for a particularly well-motivated subject to accurately categorize the test stimuli in the Knowlton and Squire (1993) paradigm without any prior exposure to category members. It is now necessary to demonstrate that naïve subjects can also categorize without any prior exposure to category members.

Experiment 2

In the previous experiment, the subject knew that category members were distortions of a prototype pattern whereas category nonmembers were random patterns. Obviously, the amnesic subjects tested by Knowlton and Squire had no such intimate knowledge of the experimental procedures. Could naïve individuals categorize the test patterns without prior exposure to category members? If so, then it is quite possible that amnesic individuals, without memory for training patterns, could do the same thing.

Without access to an amnesic population, we wanted to test normal individuals under conditions that closely mimicked those used to test amnesics. As a way of inducing profound amnesia in college undergraduates, Palmeri and Flanery (in press) eliminated the study session altogether but led subjects to believe they had seen a set of patterns. In their experiment, subjects first completed a simple word identification task. As a ruse, after the word identification task was completed, subjects were informed that dot patterns had been flashed on the computer screen during the task so quickly that they could only be perceived subliminally. In fact, no dot patterns had ever been presented. Subjects were then given the exact same categorization and recognition memory tasks used by Knowlton and Squire (1993). The expectation was that subjects would be completely unable to recognize the old patterns but would be able to categorize members versus nonmembers.

Not surprisingly, Palmeri and Flanery (in press) found that subjects were completely at chance at recognizing old versus new patterns. However, subjects were 60.4% correct at categorizing members versus nonmembers (endorsing as members 70.9% of the prototypes, 61.3% of the low distortions, 51.4% of the high distortions, and 36.6% of the random patterns). These data are in close correspondence to categorization performance by amnesics (59.9%, Knowlton & Squire, 1993), by E.P (61.1%, Squire & Knowlton, 1995), and by college students after a one week delay (57%, Nosofsky & Zaki, 1998).

Palmeri and Flanery (in press) used the same testing procedures used by Knowlton and Squire (1993), which involved presenting all subjects with one particular set of dot patterns. It is important to rule out the possibility that the ability to categorize without prior exposure is limited to a small subset of dot patterns, such as were used in those experiments. Therefore, in the present experiment, every subject was tested on a different set of dot patterns, randomly generated by the computer.

In the present experiment, we also included a group of subjects that had received prior exposure to dot patterns (thereby replicating the original Knowlton & Squire experiment). This allowed us to explicitly measure the relative effects of prior exposure on categorization and recognition performance. Although Palmeri and Flanery (in press) did test a group of individuals on categorization with prior exposure, finding that they did not perform significantly better than the nonexposed "amnesic" group, these two groups of individuals were not tested at the same time.

Method

Subjects. Subjects were 88 undergraduates students from Vanderbilt University who received course credit for their participation. Subjects were randomly assigned to the categorization-exposure, categorization-nonexposure, recognition-exposure, and recognition-nonexposure conditions.

Procedure. The exposure conditions were replications of Knowlton and Squire (1993). In the exposure phase, subjects viewed dot patterns and were asked to point to the center dot of each pattern. In the categorization condition, the dot patterns were forty high-level distortions of a prototype patterns. In the recognition condition, the dot patterns were five random patterns repeated eight times each.

The nonexposure conditions were replications of Palmeri and Flanery (in press). In the "study" task, subjects were asked to identify words rapidly flashed on a computer screen. On each trial, a pair of words differing by one letter was selected and one word from the pair was designated the target. A crosshairs appeared at the center of the screen, the target word was displayed for 25ms, and then the pair of words was displayed side by side. Subjects judged which of the two words has been flashed.

Following this task, subjects were informed that we were not really interested in word identification after all. Rather, they were told that during the word identification task, patterns of dots had been flashed on the computer screen, centered at the crosshairs, so quickly that they could only be perceived subliminally. The reason for doing the word identification task, they were told, was so that they would attend to the location on the screen where the dot patterns

were being subliminally presented. Extensive pilot work was conducted to construct a believable cover story.

Subjects in the categorization task were provided the following instructions (adapted from instructions used by Reber, Stark, & Squire, 1998, and Squire & Knowlton, 1995):

"During the previous word identification task, patterns with nine dots were quickly flashed on the computer screen so as to be perceived subliminally (without conscious awareness). All of these patterns belonged to a single category of patterns in the same sense that, if a series of dogs had been presented, they would all belong to the category dog. While you probably have no conscious recollection of the patterns, we would like you to try as hard as possible to figure out which of the following patterns are members of the same category which was displayed earlier and which are not."

If subjects claimed not to completely understand what we meant by a "category" they were then shown a picture of some dogs and asked to think about what it would mean for a new animal to belong to the same category as these animals. Subjects in the recognition condition were provided similar instructions, except they were asked to decide which patterns were old or new.

On each trial of the categorization task (for subjects exposed and nonexposed to previous patterns), a dot pattern was displayed and subjects were asked to judge whether it was a member or nonmember of the previously exposed category (for which they had been exposed in the exposure condition but for which they had never actually been exposed in the nonexposure condition). Subjects judged four instances of the prototype, 20 low-level distortions, 20 high-level distortions, and 40 random patterns (identical to procedures used by Knowlton and Squire, 1993). On each trial of the recognition task (for subjects exposed and nonexposed to previous patterns), a dot patterns was displayed and subjects were asked to judge whether it was old or new. Subjects judged five patterns old and five new patterns; four blocks of ten recognition trials were presented.

In both recognition and categorization, order of stimuli was randomized for every subject, no corrective feedback was supplied, and subjects were informed that approximately equal numbers of members/nonmembers or old/new stimuli would be presented.

Stimuli. In the "study" task of the nonexposure conditions, stimuli were forty pairs of four-letter words that differed by a single letter (e.g., WORD vs. WORK). The location of the changed letter was roughly equated across positions. Because this task was just a ruse, we did not systematically control other aspects of the word pairs.

In the categorization conditions (both exposure and nonexposure), the computer randomly generated a pattern and designated it the prototype of the category. Distortions of the prototype were created with a commonly used statistical distortion algorithm (Posner, Goldsmith, & Welton, 1967). Random nonmember patterns were also newly created at the start of each session. The subject was completely unaware of the particular set of dot patterns that had been created until they were presented during the experiment.

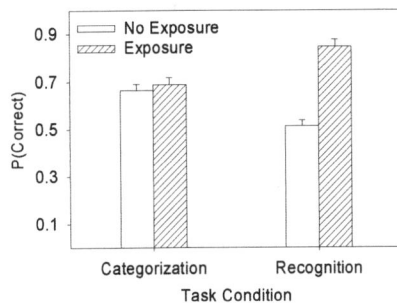

Figure 5: Categorization and recognition accuracy as a function of exposure versus nonexposure to prior patterns from Experiment 2.

In the recognition conditions (exposure and nonexposure), the computer randomly generated ten dot patterns, designating five as old and five as new. In the exposure condition, subjects saw the five old patterns.

Results

Figure 5 displays probability correct as a function of task (categorization versus recognition) and as a function of exposure (exposure versus no exposure). A two (task) x two (exposure) between-subjects ANOVA was conducted on the accuracy data revealing a significant main effect of exposure, $F(1,84)=40.635$, $MS_e=0.018$, and a significant two-way task x exposure interaction, $F(1,84)=30.403$, $MS_e=0.018$. Planned comparisons revealed no significant difference between exposed and nonexposed subjects at categorization but revealed a significant difference between exposed and nonexposed subjects at recognition. Recognition accuracy for nonexposed subjects was not different from chance.

Discussion

Replicating and extending Palmeri and Flanery (in press), we induced a state of profound amnesia in a group of normal college undergraduates by convincing them that they had been subliminally exposed to a series of dot patterns that were never actually presented. We compared performance of this group of "profound amnesics" to performance of a group who received normal exposure to stimuli in a study task. Not surprisingly, without exposure to any training patterns, recognition was at chance, but with exposure to training patterns, recognition was quite good. Yet, categorization by the "profound amnesics" was quite good, and was statistically indistinguishable from performance by an exposed group of subjects. Moreover, performance of both groups was comparable to what had been observed with amnesics and normals in studies by Knowlton and Squire (1993; Squire & Knowlton, 1995).

The apparent dissociation between categorization and recognition reported by Knowlton and Squire had been taken as evidence for multiple memory systems. However, our results suggest that their findings could simply reflect intact cognitive abilities that amnesics might have for detecting categories of similar patterns presented within a relatively short period of time, without any need to rely on long-term memory for those patterns. Unlike the recognition memory test, prior exposure to category members is unnecessary to perform the categorization test.

It should be noted that Knowlton and Squire did conduct a control condition similar to the nonexposed categorization condition reported here. They instructed subjects to imagine that they had seen a series of dot patterns but were never presented any training patterns, and then were given the same instructions and test stimuli as were given to a group of experimental subjects (amnesics and normal individuals in the second experiment of Knowlton and Squire, 1993). They reported that these subjects performed at chance on the classification test. So what explains the difference between their findings and ours? One potentially important difference is that we led our subjects to believe that they had actually seen patterns and that this subliminal exposure should be sufficient for them to perform the categorization task. By contrast, Knowlton and Squire (1993) simply told their subjects to imagine that they had seen patterns, probably leaving subjects uncertain as to whether it would be possible to perform the categorization task (consistent with this hypothesis, data from these subjects revealed an overall bias to classify every pattern as a nonmember). Even with little or no recollection for the training patterns, their amnesics subjects, like our induced amnesics, probably believed that it would be possible to perform the categorization task.

Many amnesics, including E.P., are elderly. One potential criticism of this experiment (and of Palmeri & Flanery, in press) is that we tested young college students, thereby raising questions about the validity of directly comparing our work with that of Knowlton and Squire. The main defense to this criticism is to note that performance of our undergraduates was not much different from that of the elderly individuals that had been previously studied; in addition, Nosofsky and Zaki (1998) also tested young college students and found similar results as we report.

General Discussion

Are various fundamental cognitive processes, such as categorization and recognition, subserved by a common memory system or by independent memory systems? While a dissociation between categorization and recognition in amnesics and normals has been taken as evidence for multiple memory systems (Knowlton & Squire, 1993), this dissociation is apparently consistent with a single memory system as well (Nosofsky & Zaki, 1998). However, the results from E.P., who has no detectable recognition memory yet categorizes nearly as well as normals, have been taken as powerful evidence against this single-system view.

While this remains a viable possibility, our results suggest that the evidence from E.P. may not be as compelling as once believed. We induced profound amnesia in undergraduates by telling them they had seen patterns which we never presented. While completely unable to recognize, they categorized as well as amnesics and normals who had prior exposure to training patterns. We have shown that the categorization task used by Squire and Knowlton allows subjects to discover which clusters of patterns are likely to be members simply because all members are similar to one

517

another and all nonmembers are dissimilar from one another.

Our results emphasize the importance of equating categorization and recognition studies prior to testing memory-impaired individuals. Without prior exposure to any patterns, subjects should be entirely at chance at recognizing old versus new stimuli and at categorizing members versus nonmembers. One simple way of improving the present categorization paradigm is to have nonmembers be distortions of a different prototype; induced profound amnesics might notice that there are two clusters of patterns, but without prior exposure to one of the categories, it would be impossible to correctly label the members versus the nonmembers.

Finally, these results also point out some important oversights by formal theories of categorization and recognition (e.g., Nosofsky & Zaki, 1998). Our subjects categorized quite well without memories for training exemplars. Yet, exemplar models account for the observed dissociations by relying entirely on such memories. Although exemplar models can be augmented to allow test items to become part of the category representations (e.g., Nosofsky, 1986), the mechanisms by which this kind of unsupervised category learning takes place have yet to be fully investigated.

Acknowledgments

This work was supported by Vanderbilt University Research Council Direct Research Support grants to T.J.P. We thank Allison Bell, Kinna Patel, and Talia Day for testing participants in this experiment. We also thank Woo-kyoung Ahn, Randolph Blake, Keith Clayton, Jeffrey Franks, Isabel Gauthier, Larry Jacoby, Joseph Lappin, Arthur Markman, Timothy McNamara, and Robert Nosofsky for their comments on early versions of this work.

References

Knowlton, B.J., & Squire, L.R. (1993). The learning of categories: Parallel brain systems for item memory and category knowledge. *Science, 262,* 1747-1749.

Nosofsky, R.M. (1986). Attention, similarity, and the identification-categorization relationship. *Journal of Experimental Psychology: General, 115,* 39-57.

Nosofsky, R.M. (1988). Exemplar-based accounts of relations between classification, recognition, and typicality. *Journal of Experimental Psychology: Learning, Memory, and Cognition, 14,* 700-708.

Nosofsky, R.M. (1991). Tests of an exemplar model for relating perceptual classification and recognition memory. *Journal of Experimental Psychology: Human Perception and Performance, 17,* 3-27.

Nosofsky, R.M., & Palmeri, T.J. (1997). An exemplar-based random walk model of speeded classification. *Psychological Review, 104,* 266-300.

Nosofsky, R.M., & Zaki, S.R. (1998). Dissociations between categorization and recognition in amnesic and normal individuals: An exemplar-based interpretation. *Psychological Science, 9,* 247-255.

Palmeri, T.J. (1997). Exemplar similarity and the development of automaticity. *Journal of Experimental Psychology: Learning, Memory, and Cognition, 23,* 324-354.

Palmeri, T.J., & Flanery, M.A. (in press). Learning about categories in the absence of training: Profound amnesia and the relationship between perceptual categorization and recognition memory. *Psychological Science.*

Posner, M.I., Goldsmith, R., & Welton, K.E. Jr. (1967). Perceived distance and the classification of distorted patterns. *Journal of Experimental Psychology, 73,* 28-38.

Posner, M.I., & Keele, S.W. (1968). On the genesis of abstract ideas. *Journal of Experimental Psychology, 77,* 353-363.

Reber, P.J., Stark, C.E.L., & Squire, L.R. (1998). Cortical areas supporting category learning identified using functional MRI. *Proceedings of the National Academy of Sciences USA, 95,* 747-750.

Squire, L.R., & Knowlton, B.J. (1995). Learning about categories in the absence of memory. *Proceedings of the National Academy of Sciences USA, 92,* 12470-12474.

Squire, L.R., & Zola, S.M. (1996). Structure and function of declarative and nondeclarative memory systems. *Proceedings of the National Academy of Sciences USA, 93,* 13515-13522.

The Time-Course of the Use of Background Knowledge in Perceptual Categorization

Thomas J. Palmeri (thomas.j.palmeri@vanderbilt.edu)
Department of Psychology; Vanderbilt University
Nashville, TN 37240 USA

Celina Blalock (ladybug531@juno.com)
Department of Child and Family Development; University of Georgia
Athens, GA 30604 USA

Abstract

We examined the time-course of the utilization of background knowledge in perceptual categorization by manipulating the meaningfulness of labels associated with categories and by manipulating the amount of time given to subjects to make a categorization decision. Extending a paradigm originally reported by Wisniewski and Medin (1994), subjects learned two categories of children's drawings that either were given standard labels (drawing by children from group 1 or group 2) or were given theory-based labels (drawings by creative or noncreative children); meaningfulness of the label had a profound effect on how new drawings were categorized. Half of the subjects were given unlimited time to respond, the other half of the subjects were given a quick response deadline; speeded response conditions had a relatively large effect on categorization decisions by subjects given the standard labels but had a relatively small effect on categorization decisions by subjects given the theory-based labels. These results suggest that background knowledge may have its influence at relatively early stages in the time-course of a categorization decision.

Introduction

In cognitive science, two distinct approaches to research on categorization and concept formation have been considered. One approach focuses on the statistical aspects of learning categories via induction over a series of instances. A main theme of this work involves developing and testing formal mathematical models of categorization that embody various assumptions about the kinds of representations thought to be formed during category learning. Examples include rule-based models (e.g., Nosofsky & Palmeri, 1998; Nosofsky, Palmeri, & McKinley, 1994; Palmeri & Nosofsky, 1995), exemplar-based models (Estes, 1994; Hintzman, 1986; Medin & Schaffer, 1978; Nosofsky, 1986; Nosofsky & Palmeri, 1997), various clustering models (e.g., Anderson, 1990; Fisher, 1987), and numerous others. What these theoretical approaches have in common is that the category representations that are formed during learning are entirely dependent on the empirical regularities in the particular set of category instances that have been experienced.

A contrasting approach focuses on how background knowledge or theories might influence what is learned about a category from experience with particular instances (Murphy & Medin, 1985). For example, background knowledge can influence the ease of learning linearly separable versus nonlinearly separable categories (Wattenmaker, Dewey, Murphy, & Medin, 1986) and can influence the ease of learning conjunctive versus disjunctive rules (Pazzani, 1991). In addition, a number of studies have found a facilitative effect of prior background knowledge on learning new categories (e.g., Heit, 1994, 1998; Murphy & Allopenna, 1994; Murphy & Wisniewski, 1989).

We will describe the results of one particularly illuminating study by Wisniewski and Medin (1994), both to illustrate how background knowledge can influence categorization and because the present experiment builds on this particular study. Subjects learned two categories of drawings, shown in Figure 1. One group of subjects was provided standard labels (i.e., children from "group 1" drew the pictures in category A, children from "group 2" drew the pictures in category B), while another group of subjects was provided theory-based labels (i.e., "creative" children drew the pictures in category A, "noncreative" children drew the pictures in category B). Subjects were required to develop a set of rules that could be used to partition those drawings as well as any new drawings into the two distinct categories. When Wisniewski and Medin analyzed the rules subjects formed, they found that those subjects given the standard labels had generated rules based on fairly concrete perceptual features (such as "... all of the characters have their arms out straight from their bodies and they're also standing very straight, facing the front") whereas those subjects given the theory-based labels had generated rules based on fairly abstract features (such as "much more attention was given to the clothing ..." or "make drawings that show more positive emotional expression ...").

Wisniewski and Medin next tested their subjects on new drawings that factorially combined the perceptual features and the abstract features of the two categories, as shown in Figure 2. For example, stimulus T5 had the abstract features of category A but had the concrete perceptual features of category B, whereas stimulus T15 had the abstract features of category B but had the concrete perceptual features of category A. As shown in Figure 3, Wisniewski and Medin observed that subjects given different labels classified these conflicting stimuli in markedly different ways; subjects provided the theory-based labels tended to classify according to the abstract features and subjects provided the

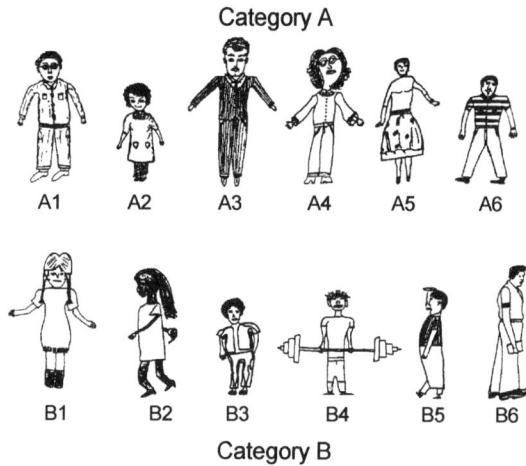

Figure 1. Training drawings used in the experiment (from Wisniewski & Medin, 1994).

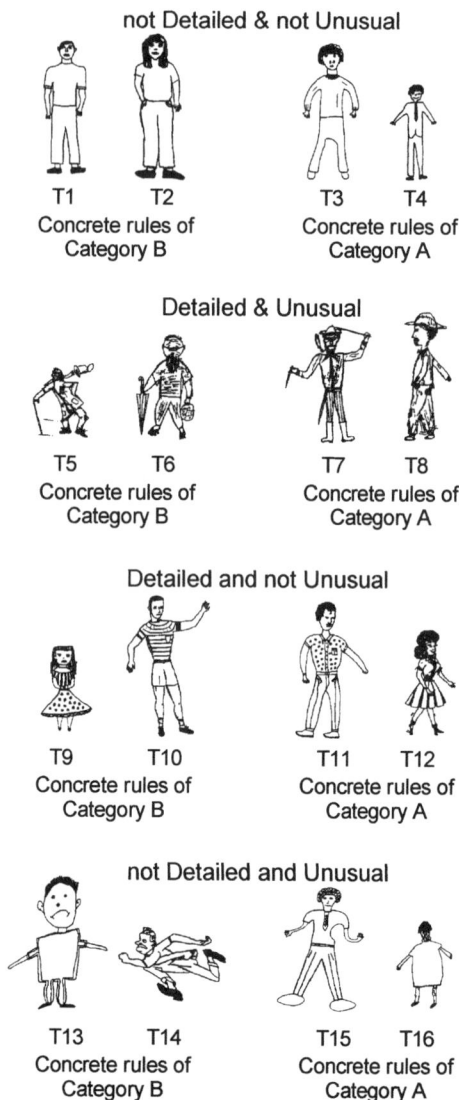

Figure 2. Transfer drawings used in the experiment (from Wisniewski & Medin, 1994).

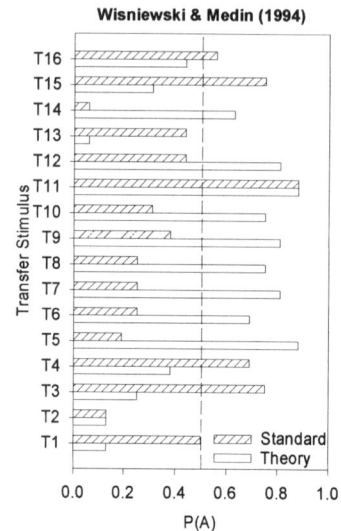

Figure 3. Probability of categorizing each of the transfer drawings into category A as a function of the type of label provided to subjects (data from Experiment 1A of Wisniewski and Medin, 1994).

standard labels tended to classify according to the concrete perceptual features. For example, note in Figure 3 how stimuli T5 and T15 were classified into different categories depending on what kind of label subjects were provided.

These results may be difficult to explain by theories of categorization that assume nothing more than some form of an abstraction of the statistical structure of the training patterns. In this experiment, both groups of subjects observed the identical set of training drawings. Yet, by simply manipulating the meaningfulness of the labels applied to those drawings, subjects categorized new transfer drawings in markedly different ways. To explain these results, Wisniewski and Medin (1994) suggested that background knowledge and empirical information about instances closely interact during categorization. They contended that background knowledge does not just weight the features of an object that are extracted during early perceptual processing. Rather, background knowledge may influence what features are actually extracted from an object for purposes of categorization (see also Wisniewski & Medin, 1991). The suggestion is that knowledge may have an influence on categorization at the very earliest stages of processing, not just after the objects have been analyzed into their constituent parts.

We chose to investigate this claim by manipulating the amount of time provided to subjects for making a categorization decision. We chose to introduce a response deadline in which one group of subjects was signaled to response very soon after stimulus onset while another group of subjects was given unlimited time to respond. Such response signal paradigms can be quite useful for determining what information is available at various points within the time-course of a categorization decision (e.g., Lamberts, 1998; see also Meyer, Irwin, Osman, & Kounios, 1988).

For example, it is possible that background knowledge might influence categorization at relatively late stages of processing, contrary to the claims made by Wisniewski and Medin (1994) – knowledge might be used to interpret features after they have been extracted, or knowledge might be used to guide the combination of a number of simple perceptual features into something more meaningful and useful for determining category membership. Introducing a response deadline might force subjects to make their categorization response before background knowledge has yet had any time to perform these interpretive operations. Therefore, under a response deadline, subjects given the theory-based labels might have much more difficulty categorizing both training and transfer drawings than subjects given the standard labels.

In this experiment, we extended the paradigm developed by Wisniewski and Medin (1994). Subjects learned the categories of drawings shown in Figure 1 either using standard labels or using theory-based labels. Subjects were then asked to classify the training drawings and new transfer drawings either under speeded conditions (issuing a response signal just 200ms after stimulus onset) or under unspeeded conditions (unlimited decision time).

Method

Subjects. Sixty Vanderbilt University undergraduates participated for partial course credit.

Stimuli. The stimuli were drawings by children who were given a "draw-a-person" test and were the same as those used in Experiment 1 of Wisniewski and Medin (1994; taken from Koppitz, 1984, and Harris, 1963). The drawings are displayed in Figures 1 and 2. For the first part of the experiment, the drawings were individually mounted and covered in clear-coat plastic. For the remaining parts of the experiment, scanned drawings were individually displayed on a computer monitor.

Following Wisniewski and Medin (1994), the training set consisted of two categories of six drawings each, as shown in Figure 1. The first set (category A) were deemed to be relatively detailed and the second set (category B) were deemed to be relatively unusual. In pilot studies Wisniewski and Medin conducted, people expected drawings made by creative children to be more detailed or more unusual than drawings made by noncreative children.

In addition to these abstract properties, the drawings in the two categories could be distinguished on the basis of at least three simple rules based on concrete perceptual features (Wisniewski and Medin modified the drawings to conform to these rules). For category A (top row of Figure 1), the drawings had "curly hair and arms not at the sides," "light-colored shoes or not smiling," and "wearing a collar or tie." For category B (bottom row of Figure 1), the drawings had "straight hair or arms at the sides," "dark-colored shoes and not smiling," or "ears and short sleeves."

Sixteen test drawings, shown in Figure 2, were selected and appropriately modified to factorially combine the abstract and concrete features of the training drawings. The first set of four drawings, shown in the top row of Figure 2, were not detailed and not unusual; the second set of four drawings, shown in the second row of Figure 2, were both detailed and unusual; the third set of four drawings, shown

in the third row of Figure 2, were detailed and not unusual; the fourth set of four drawings, shown in the bottom row of Figure 2, were not detailed and unusual. For each set of four drawings, two were best described by the concrete rules of the category A drawings and the other two were best described by the concrete rules of the category B drawings. For example, the first two drawings of the first set, test drawings T1 and T2, had the concrete rules of the category B, whereas the second two drawings, test drawings T3 and T4, had the concrete rules of category A. See Figure 2 for more information.

Procedure. Half of the subjects were instructed that they would be studying drawings by "creative" or "noncreative" children, the other half of the subjects were instructed that they would be studying drawings by children from "group 1" or "group 2." In the initial study phase, subjects were shown the mounted drawings simultaneously, laid out on the table in front of them, and separated into two labeled groups (with theory-based labels of "creative" or "noncreative", or standard labels of "group 1" or "group 2"). The instructions for the "creative" / "noncreative" subjects were as follows (adapted from Wisniewski and Medin, 1994):

In this experiment, you will be shown two groups of children's drawings. One group was done by creative children. The other group was done by noncreative children. Your task is to come up with a rule that someone could use to decide whether a new drawing belongs to the group drawn by the creative children. In writing down a rule, it is important that the rule "works" for all of the drawings in the creative group and none of the drawings in the noncreative group. That is, if someone were given your rule and the drawings from the two groups (all mixed up), they should be able to use it to divide the drawings into those that belong in one group and those that belong in the other group. Also, please write your rule clearly and describe the rule so that the experimenter will be able to understand what your mean.

The instructions for the "group 1" / "group 2" subjects were virtually the same, except that "creative" was replaced by "group 1" and "noncreative" was replaced by "group 2." Whereas Wisniewski and Medin (1994) randomized the assignment of category labels to sets of drawings, we instead chose to have the "creative" and "group 1" labels apply to the category A drawings and to have the "noncreative" and "group 2" labels apply to category B drawings.

After developing a set of rules to classify the drawings into the two categories, the mounted drawings were removed but subjects were allowed to keep their rules at hand. Subjects were then shown each of the training drawings one at a time on the computer, were asked to classify them as "creative" / "noncreative" or "group 1" / "group 2", were supplied corrective feedback, and were permitted to modify their rules if necessary. After classifying each of the drawings twice, their rules were removed. Order of drawings was randomized for every subject.

In the next phase of the experiment, subjects were asked to classify the training drawings and some new test drawings using the rules they had developed in the first part of the experiment. They were also instructed that they would receive no corrective feedback following their response. Half of the subjects in each group were allowed to classify the drawings without any time limit (unspeeded condition), the other half of the subjects in each group were instructed that they would be required to make their responses according to a response signal (speeded condition).

In the speeded condition, a tone was presented 200ms after the onset of the drawing. Subjects were required to make their classification response within 300ms of hearing the tone. These subjects were provided a series of practice trials to familiarize themselves with the demands of the speeded condition. On the practice trials, a category label (either "creative" or "noncreative", or "group 1" or "group 2") was displayed at the center of the screen. After 200ms, the tone sounded and subjects were required to press the response key associated with the displayed category label. If the subject responded before the tone sounded, they were informed to wait until the tone sounded; if the subject responded more than 300ms after the tone sounded, they were informed to respond more quickly; if the subject made an incorrect response, they were informed not to make errors. The practice trials continued until the subject was able to make 15 valid responses in a row. Once the subject had achieved this criterion for responding appropriately to the signal, they were moved on to the test phase.

In the test phase, a crosshairs appeared at the center of the screen for 1 sec. Then one of the 28 drawings (12 training and 16 test) was displayed at the location of the crosshairs and the subject was required to classify it as "creative" / "noncreative" or "group 1" / "group 2". In the unspeeded condition, the subject could take as much time as they needed to make their response. In the speeded condition, a tone sounded 200ms after the drawing was displayed and subjects were required to make their response within 300ms of the tone; subjects were informed if they had made a response prior to the tone or had made a response more than 300ms after the tone. In both the speeded and unspeeded conditions, no other corrective feedback was provided. After the response was made, the drawing was erased from the screen. After a one second interval, the next drawing was presented. Subjects classified each of the 28 drawings eight times. Each stimulus was presented once per block in a randomized order for every subject.

To summarize the design of the experiment, each subject could be classified into one of four groups: theory-based labels (creative versus noncreative) with unspeeded responses, theory-based labels with speeded responses, standard labels (group 1 versus group 2) with unspeeded responses, and standard labels with speeded responses.

Results and Discussion

The primary data of interest were the probabilities of categorizing each stimulus as a member of category A in the testing phase as a function of category label (standard versus theory-based) and as a function of response deadline (speeded versus unspeeded). Figure 4 displays the probability of categorizing each of the original training drawings into category A. Figures 5 and 6 both display the probability of categorizing each of the new transfer drawings into

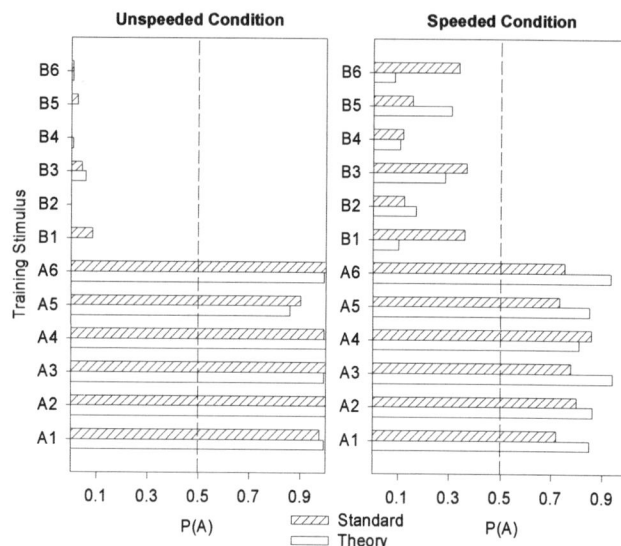

Figure 4. Probability of categorizing each of the training drawings into category A as a function of response deadline (unspeeded in the left panel, speeded in the right panel) and as a function of label (hatched bars for standard labels, open bars for theory-based labels).

category A (Figure 5 plots the two response deadline conditions separately, Figure 6 plots the two category label conditions separately). In the speeded condition, categorization responses that were made prior to the response signal or more than 300ms after the response signal were excluded from the analyses (12.8% of the observations); further analyses in which we excluded none of the categorization responses were essentially identical to those presented here.

A 28 (stimulus) x 2 (category label) x 2 (response deadline) mixed analysis of variance was conducted on the probability data with stimulus (12 training drawings and 16 transfer drawings) as a within-subjects variable, and with label (standard versus theory-based) and deadline (speeded versus unspeeded) as between-subjects variables. A significant main effects of stimulus, $F(27,1512)=56.454$, and significant interactions of stimulus x label, stimulus x deadline, and stimulus x label x deadline were observed, $F(27, 512)=18.282$, $F(27,1512)=5.787$, and $F(27,1512)=1.636$, respectively (an alpha level of .05 was established for all statistical tests reported in this article). No main effects of label or deadline were observed.

Not surprisingly, for the training drawings, subjects were significantly more accurate in the unspeeded condition than the speeded condition. Somewhat surprisingly, for the training drawings, subjects were significantly more accurate in the speeded condition when applying theory-based labels (84.1% correct) than when applying standard labels (75.6% correct). This provides our first piece of evidence inconsistent with the hypothesis that background knowledge might influence categorization only at relatively late stages in the decision making process.

Turning to the categorization of the transfer drawings, let us first note the fairly close correspondence between the findings we obtained in the unspeeded condition, shown in

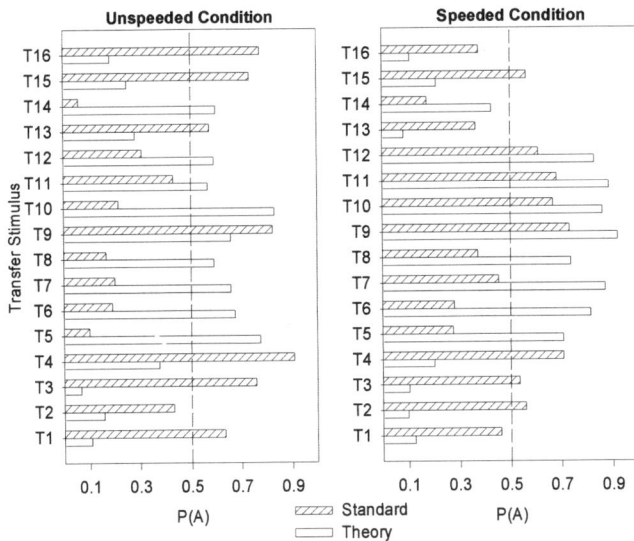

Figure 5. Probability of categorizing each of the transfer drawings into category A as a function of response deadline (unspeeded in the left panel, speeded in the right panel) and as a function of label (hatched bars for standard labels, open bars for theory-based labels).

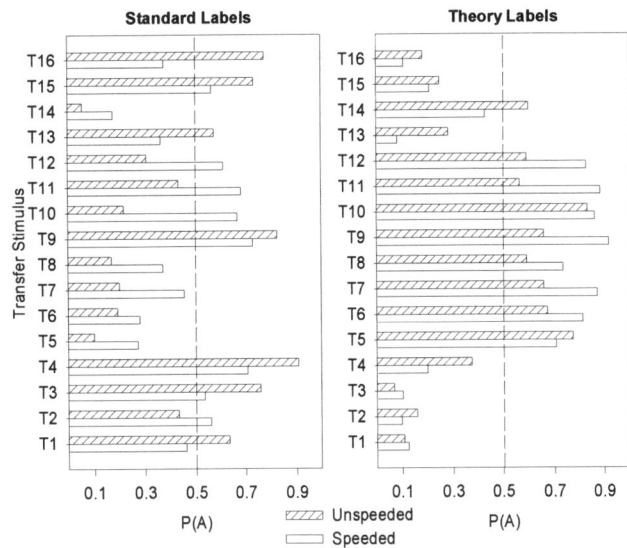

Figure 6. Probability of categorizing each of the transfer drawings into category A as a function of label (standard labels in the left panel, theory-based labels in the right panel) and as a function of response deadline (hatched bars for unspeeded, open bars for speeded).

the left panel of Figure 5, and the findings obtained by Wisniewski and Medin, shown in Figure 3. Our results provide an important replication of their study by again demonstrating that the meaningfulness of a label can have a striking effect on how subjects categorize new transfer drawings. The general pattern of how the transfer drawings were categorized can probably best be explained by assuming that subjects given theory-based labels tended to categorize on the basis of abstract features whereas subjects given standard labels tended to categorize on the basis of concrete perceptual features (see Wisniewski & Medin, 1994).

The novel aspect of the present work is a finding that response deadlines had quite different effects depending on the type of label that was provided. To see this, compare categorization probabilities under speeded and unspeeded conditions for subjects given standard labels, shown in the left panel of Figure 6, with those for subjects given theory-based labels, shown in the right panel of Figure 6. One useful way of summarizing this data is as follows: With standard labels, categorization probabilities tended to be closer to 50% or even crossed over to the opposite category in the speeded condition compared to the unspeeded condition. By contrast, with theory-based labels, categorization probabilities actually tended to be further from 50% in the speeded condition compared to the unspeeded condition. A simple explanation for this difference solely in terms of the amount of time needed to apply the rules in the two conditions is unlikely; in the unspeeded conditions, there was no significant difference in response times for subjects given standard (1420ms) and theory-based labels (1601ms).

Another useful way of summarizing this data requires slightly recoding the categorization responses as follows. To develop a measure analogous to categorization accuracy for the new transfer drawings, for each stimulus, we simply define the modal categorization response in the unspeeded

condition as the "correct" category response for that stimulus (e.g., for subjects given standard labels, the "correct" response for T5 is category B, but for subjects given the theory-based labels, the "correct" response for T5 is category A). For each stimulus, we then recode each subjects' category responses with respect to this nominally "correct" response. Figure 7 displays categorization "accuracy" averaged across all the transfer stimuli. With standard labels, subjects were significantly less "accurate" in the speeded condition than the unspeeded condition; with theory-based labels, subjects actually tended to be more "accurate" in the speeded condition than the unspeeded condition, although this difference did not reach significance.

Either way we examine the data, we find that introducing a very rapid response deadline, causing subjects to make fairly complex categorization responses in an average of just 370ms, had surprisingly little effect on categorization performance when subjects were provided meaningful labels that tapped into background knowledge. Subjects

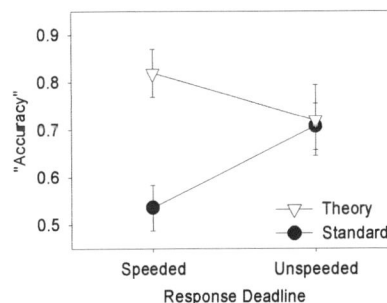

Figure 7. Categorization "accuracy" averaged across the transfer drawing as a function of response deadline (x-axis) and as a function of label (closed circles for standard labels, open triangles for theory-based labels).

appeared to be able to utilize this background knowledge at the earliest stages of a categorization decision, consistent with the hypothesis that the joint influences of background knowledge and empirical evidence are tightly coupled in the service of making a categorization decision (see Wisniewski, 1995). This is one of the first fairly direct demonstrations that background knowledge can have a very early influence in the time course of a perceptual categorization decision (see also Lin & Murphy, 1997).

It was somewhat surprising that introducing a very rapid response deadline had a quite a detrimental effect on categorization performance when subjects were provided standard labels. Although the rules subjects developed in this condition were based on simple perceptual features, testing these rules on a given drawing may have required a serial evaluation of the various component features making up that categorization rule (see Palmeri & Nosofsky, 1995). Under a response deadline, subjects simply did not have sufficient time to check all of the various features of the rule they formed, causing them to respond in a haphazard fashion.

Acknowledgments

The research described in this paper is based on work completed in partial fulfillment of an undergraduate honors thesis submitted to Vanderbilt University by the second author. This work was supported by Vanderbilt University Research Council Direct Research Support grants to the first author. We thank Douglas Medin and Evan Heit for their comments on this work.

References

Anderson, J.R. (1990). *The adaptive character of thought.* Hillsdale, NJ: Erlbaum.

Estes, W.K. (1994). *Classification and cognition.* Oxford, England: Oxford University Press.

Fisher, D.H. (1987). Knowledge acquisition via incremental conceptual clustering. *Machine Learning, 2,* 139-172.

Gluck, M.A., & Bower, G.H. (1988). Evaluating an adaptive network model of human learning. *Journal of Memory and Language, 27,* 166-195.

Harris, D.B. (1963). *Children's drawings as measures of intellectual maturity.* New York: Harcourt Brace & World.

Heit, E. (1994). Models of the effects of prior knowledge on category learning. *Journal of Experimental Psychology: Learning, Memory, and Cognition, 20,* 1264-1282.

Heit, E. (1998) Influences of prior knowledge on selective weighting of category members. *Journal of Experimental Psychology: Learning, Memory, and Cognition, 24,* 712-731.

Koppitz, E.M. (1984). *Psychological evaluation of human figure drawings by middle school pupils.* Orlando, FL: Grune & Stratton.

Lamberts, K. (1998). The time course of categorization. *Journal of Experimental Psychology: Learning, Memory, and Cognition, 24,* 695-711.

Lin, E.L., & Murphy, G.L. (1997) Effects of background knowledge on object categorization and part detection. *Journal of Experimental Psychology: Human Perception and Performance, 23,* 1153-1169.

Medin, D.L., & Schaffer, M.M. (1978). A context theory of classification learning. *Psychological Review, 85,* 207-238.

Meyer, D.E., Irwin, D.E., Osman, A.M., Kounios, J. (1988). The dynamics of cognition: Mental processes inferred from a speed-accuracy decomposition technique. *Psychological Review, 95,* 183-237.

Murphy, G.L., & Allopenna, P.D. (1994). The locus of knowledge effects in concept learning. *Journal of Experimental Psychology: Learning, Memory, and Cognition, 20,* 904-919.

Murphy, G.L., & Medin, D.L. (1985). The role of theories in conceptual coherence. *Psychological Review, 92,* 289-316.

Murphy, G.L., & Wisniewski, E.J. (1989). Feature correlations in conceptual representations. In G. Tiberghien (Ed.), *Advances in cognitive science: Vol. 2. Theory and applications.* Chichester, England: Ellis Horwood.

Nosofsky, R.M. (1986). Attention, similarity, and the identification-categorization relationship. *Journal of Experimental Psychology: General, 115,* 39-57.

Nosofsky, R.M., Palmeri, T.J., & McKinley, S.C. (1994). Rule-plus-exception model of classification learning. *Psychological Review, 101,* 53-79.

Nosofsky, R.M., & Palmeri, T.J. (1997). An exemplar-based random walk model of speeded classification. *Psychological Review, 104,* 266-300.

Nosofsky, R.M., & Palmeri, T.J. (1998). A rule-plus-exception model for classify objects in continuous-dimension spaces. *Psychonomic Bulletin & Review, 5,* 345-369.

Palmeri, T.J., & Nosofsky, R.M. (1995). Recognition memory for exceptions to the category rule. *Journal of Experimental Psychology: Learning, Memory, and Cognition, 21,* 548-568.

Pazzani, M.J. (1991). Influence of prior knowledge on concept acquisition: Experimental and computational results. *Journal of Experimental Psychology: Learning, Memory, and Cognition, 15,* 416-432.

Wattenmaker, W.D., Dewey, G.I., Murphy, T.D., & Medin, D.L. (1986). Linear separability and concept learning: Context, relational properties, and concept naturalness. *Cognitive Psychology, 18,* 158-194.

Wisniewski, E.J. (1995). Prior knowledge and functionally relevant features in concept learning. *Journal of Experimental Psychology: Learning, Memory, and Cognition, 21,* 449-468.

Wisniewski, E.J., & Medin, D.L. (1991). Harpoons and long sticks: The interaction of theory and similarity in rule induction. In D.H. Fisher, M.J. Pazzani, & P. Langley (Eds.), *Concept formation: Knowledge and experience in unsupervised learning.* San Mateo, CA: Morgan Kaufmann.

Wisniewski, E.J., & Medin, D.L. (1994). On the interaction of theory and data in concept learning. *Cognitive Science, 18,* 221-281.

The Independent Sign Bias: Gaining Insight from Multiple Linear Regression

Michael J. Pazzani (pazzani@ics.uci.edu)
Stephen D. Bay (sbay@ics.uci.edu)
Department of Information and Computer Science
University of California, Irvine
Irvine, CA 92697

Abstract

As electronic data becomes widely available, the need for tools that help people gain insight from data has arisen. A variety of techniques from statistics, machine learning, and neural networks have been applied to databases in the hopes of mining knowledge from data. Multiple regression is one such method for modeling the relationship between a set of explanatory variables and a dependent variable by fitting a linear equation to observed data. Here, we investigate and discuss some factors that influence whether the resulting regression equation is a credible model of the data.

Introduction

Multiple linear regression (e.g., Draper and Smith, 1981) is a technique for finding a linear relationship between a set of explanatory variables (x_i) and a dependent variable (y): $y = b_0 + b_1 x_1 + b_2 x_2 + \ldots + b_n x_n$. The coefficients, ($b_i$) provide some indication of the explanatory variables effect on the dependent variable. With the wide availability of personal computers and the inclusion of regression routines in commonly available statistics or spreadsheet software such as Microsoft Excel©, there is an increased recognition of the value of gaining insight from data. There are also free web servers (e.g., Autofit http://www.lava.net/~seekjc) for fitting data to linear models. As a consequence, multiple linear regression is being applied to a wide array of problems ranging from business to agriculture. The goal of this application is to 'convert data to information'. Such information might help guide future decision-making. For example, many lenders use a credit score to help determine whether to make a loan. This score is a combination of many factors such as income, debt, and past payment history which positively or negatively affect the credit risk of a borrower. In this paper, we show that multiple regression as used in practice can produce models that are unacceptable to experts and laypeople because factors that should positively affect a decision may have negative coefficients and vice versa. We introduce a constrained form of regression that produces regression models that are more acceptable.

To illustrate the problem we are addressing, consider forming a model of professional baseball players' salaries as a function of statistics describing the players' performance. An agent representing a player might use the model as part of an argument that the player is underpaid. A player could use the model to determine how to improve certain aspects of his game to increase his salary. We have created a model of baseball players' salaries in 1992 as a function of the players' performance in 1991 using data on 270 players collected by CNN/Sports Illustrated and The Society for American Baseball Research. The model is listed below:

```
s=-180+10r+5hit+0.9obp+15hr+14rbi-0.8ave-18db-39tr
```

where s is the salary (in thousands), r is the number of runs scored, h is the number of hits, obp is the on base percentage (between 0 and 1000), rbi is the number of runs batted in, ave is the batting average (between 0 and 1000), db is the number of doubles and tr is the number of triples. Most people knowledgeable about baseball are confused by the negative coefficients for ave, db, and tr. It is unlikely that someone familiar with the sport would consider this insightful or advise anyone to act upon this model. If a baseball player interested in maximizing his income were to follow this equation literally, he would always stop at first base when hitting, rather than trying for a double. In this paper, we discuss why incorrect signs occur in multiple linear regression and present some alternative means of inferring linear models from data that do not suffer from this problem.

In multiple linear regression, the best equation fitting the data is found by choosing the coefficients (b_i) so that the sum of the squared error for the training data points is minimized. The coefficients, b_i, can be found with matrix manipulations, also known as a least squares approach (Draper and Smith, 1981). Mullet (1976) discusses a variety of reasons that multiple linear regression produces the "wrong sign" for some coefficients:

- Computational Error. Some computational procedures for computing least squares have problems with precision when the magnitudes of variables differ drastically. To avoid this problem, we internally convert variables to standard form (i.e., 0 mean, and unit variance) for calculations and convert back to the original form for displaying the coefficients.

- Coefficients that don't significantly differ from zero. In this case, the sign of the coefficient does not matter because it is small enough so that it does not significantly affect the equation. One recommended way to avoid this problem is to eliminate these irrelevant variables. Forward stepwise regression methods (Draper and Smith, 1981) do not include a variable in the model unless the variable significantly improves the fit of the model to the data. We used this method (with an alpha of 0.05) to model the baseball data and obtained the following equation:

`s=-114+16r+17rbi-59tr`

Although this equation reduces the number of violations, it still has the wrong sign for the variable `tr`.

- Multicollinearity. When two or more explanatory variables are not independent, the sign of the coefficient of one of the variables may differ from the sign of that coefficient if the variable were the only explanatory variable. One approach to deal with this problem is manually eliminating some of the variables from the analysis.

In this paper, we consider an alternative approach to address the "wrong sign" problem in multiple regression. The goal is to produce linear models that are as accurate predictors of the dependent variable as the least-squares model but are more acceptable to people knowledgeable in the domain. Following the methodology commonly used in machine learning, we evaluate accuracy not by goodness of fit to a collection of data but by the ability to generalize to unseen data. Furthermore, we report on an experiment that evaluates what types of linear models are more acceptable to people in the baseball salary domain.

Constrained Regression

We hypothesize that a linear model is more acceptable to people knowledgeable in a domain when the effect of each variable in the regression equation in combination with the other variables is the same as the effect of each variable in isolation. That is, if in general, baseball salaries increase as the number of doubles increases, we would like the sign of the coefficient of this variable to be positive in the full linear model. Here, we propose and evaluate three methods to constrain regression to make this true. We call this constraint the *independent sign bias*.

Independent Sign Regression (ISR) treats the problem of fitting the linear model to the data as a constrained optimization problem: i.e., find the regression coefficients (b_i) that minimize the squared error on the training data subject to the constraint that all coefficients must have the same sign as they would in isolation (simple regression). There are many numerical algorithms for performing constrained optimization, and Lawson and Hanson (1974) present a comprehensive set for this case. The new contribution in ISR is that the constraints (i.e., the sign of the coefficient) are determined automatically by analysis of the data. Explanatory variables positively correlated with the dependent variable have a positive sign, while those negatively correlated have a negative sign.

We used ISR to create the following model of the baseball salary data:

`s=-207+15r+0.8hit+11hr+11rbi+0.33ave+5db`

In the next sections, we evaluate how well a constrained form of regression fits the data and whether people prefer regression equations with this constraint. Here, we note that the signs agree with our intuition. However, it does eliminate some variables such as the on base percentage (obp). This occurs because the best fit to the data subject to the constraint that obp ≥ 0 is that obp = 0. This occurs because obp is correlated with other variables such as ave (r = 0.81).

Next, we consider how to modify forward stepwise regression to constrain the sign of the variable. In forward stepwise regression (Draper and Smith, 1981), we start with an empty set of variables and then add the single variable that improves the model's fit to the training data the most. We continue this process of adding variables to those present until we have either included all variables or the remaining variables do not significantly improve the fit based on the partial F-test (Draper and Smith, 1981). *Independent sign forward regression* (ISFR) modifies this procedure by adding the constraint that the entering variable must also not result in sign violations (i.e., we add the variable that improves model fit the most subject to the constraint of no sign violations in the fitted equation). ISFR produces the following equation on the baseball data:

`s=-148+15r+15rbi`

The previous two constrained forms of regression both may eliminate variables. Here, we introduce a form of regression we call *Mean Coefficient Regression* (MCR) that does not eliminate variables but ensures that the signs agree with the sign in isolation. MCR finds the regression coefficients for each of the variables in isolation and then simply uses those values (dividing by the number of variables) for the multiple regression case. This is equivalent to treating each variable as a predictor and then averaging the results. The intercept is found automatically through the conversion of coefficients from standard form to the original scaling and minimizes the mean squared error of the linear equation with those coefficients. If all of the explanatory variables are uncorrected, MCR would produce the same equation as multiple linear regression. MCR produces the following equation on the baseball data:

`s=-162+4r+2hit+1.1obp+10hr+3rbi+1.2ave+9db+16tr`

Accuracy and the Independent Sign Bias

In this section, we evaluate the five regression algorithms on the several data sets. In each case, we report the squared multiple correlation coefficient (R^2), which is the percent of the total variance explained by the regression equation, and the descriptive mean squared error (MSE) of the regression routines on the entire data set. Both of these statistics measure how well the algorithms fit the data. Note that multiple linear regression always has the best fit to the training data, because it by definition minimizes squared training error. The more constrained forms of regression are limited in their ability to fit the data. We also report on the predictive mean squared error, which measures the ability of the regression algorithm to produce models that generalize

526

to unseen data. The predictive mean squared error is found by 5-fold cross-validation: The entire data set is randomly divided into five equal sized partitions. The data from four of the partitions is used to form a linear model that is evaluated on the fifth partition. This is repeated five times with each partition used exactly once for evaluation. The predictive MSE is almost always higher than the descriptive MSE. However, the algorithm with the lowest descriptive MSE does not necessarily have the lowest predictive MSE because the less constrained algorithms can overfit the data.

When evaluating the five regression algorithms, we also report on the number of sign violations where a sign violation occurs if the sign of the coefficient in the equation differs from the sign of the coefficient in the simple regression case. In this work, the principle goal is not to find regression routines that generalize better than multiple linear regression, but to find routines that generalize equally well and produce equations that people would be more willing to use.

We ran each of the 5 regression approaches on six data sets available from either the Statlib repository (http://www.stat.cmu.edu) or the UCI archive of databases (http://www.ics.uci.edu/~mlearn). The databases Autompg, Housing, and Pollution deal with automobile mileage, housing costs, and mortality rates respectively. CS Dept is available from the Computing Research Association (http://www.cra.org) and involves computer science department quality ratings. The Alzheimer's database was collected by UCI's Institute of Brain Aging and Dementia and involves predicting the level of dementia from the results of tests that screen for dementia.

In all of these domains, a sign violation could cause credibility problems. For example, in the CS Dept domain, linear regression indicated that the more publications per faculty member the lower the quality of the program, while in isolation this variable has the opposite effect.

Table 1. Summary of five approaches to creating linear models on six data sets

Database	Multiple Linear Regression	Independent Sign Regression	Mean Coefficient Regression	Stepwise Forward Regression	Independent Sign Forward Regression
Alzheimer					
R^2	0.750	0.743	0.420	0.727	0.727
Descriptive MSE	0.124	0.127	0.287	0.135	0.135
Predictive MSE	0.184	**0.166**	0.297	**0.166**	**0.166**
Violations	7	0	0	0	0
Autompg					
R^2	0.849	0.844	0.500	0.844	0.844
Descriptive MSE	9.3	9.47	30.5	9.4	9.55
Predictive MSE	10.6	**10.5**	30.8	10.7	10.6
Violations	4	0	0	1	0
Baseball					
R^2	0.478	0.472	0.375	0.476	0.470
Descriptive MSE	8e+5	8.09e+5	9.57e+5	8.02e+5	8.11e+5
Predictive MSE	8.74e+5	8.55e+5	9.66e+5	8.37e+5	**8.33e+5**
Violations	3	0	0	1	0
CS Dept					
R^2	0.859	0.858	0.414	0.844	0.844
Descriptive MSE	0.135	0.136	0.559	0.148	0.148
Predictive MSE	0.244	**0.213**	0.605	0.24	0.24
Violations	1	0	0	0	0
Housing					
R^2	0.740	0.698	0.328	0.740	0.696
Descriptive MSE	21.9	25.6	56.7	21.9	25.7
Predictive MSE	**23.7**	27.6	57.2	24.4	27.7
Violations	3	0	0	0	0
Pollution					
R^2	0.768	0.728	0.224	0.719	0.719
Descriptive MSE	895	1.04e+3	2.96e+3	1.08e+3	1.08e+3
Predictive MSE	3.53e+3	**1.6e+3**	3.3e+3	1.82e+3	1.78e+3
Violations	5	0	0	2	0

In Table 1, we show the results of the five regression approaches on the six data sets. The best (lowest) predictive MSE is shown in bold to allow simple comparison of the predictive ability. It is typical in such a simulation that no algorithm stands out as uniformly superior on all problems. However, the results indicate that independent sign regression is usually at least as accurate as multiple linear regression. Due to the sign violations, one might prefer to use independent sign regression. Mean coefficient regression does not fit the data nor generalize as well as the other regression algorithms. Independent sign forward regression is usually at least as accurate as stepwise forward regression.

Note that multiple linear regression has at least one sign violation on every data set. This shows that correlated variables frequently occur in naturally occurring databases and that the techniques designed to correct for sign violations may be applicable to a broad range of problems. Although stepwise forward regression mitigates the problem of sign violations, it does not eliminate it entirely.

In the next section, we report on the results of an experiment in which subjects indicate their willingness to use regression equations to make predictions. The goal of the study is to determine whether subjects have a preference for the independent sign bias: i.e., equations in which the sign of each coefficient is the same as the sign in isolation.

Baseball Salary Experiment

In this experiment, subjects are asked to imagine that they are an agent representing a baseball player. Subjects are shown various linear equations and told that they "might be used as a starting point to get a rough estimate of what a player should be paid." For each equation, they were asked to indicate on a [-3,+3] scale "How willing would you be to use this equation as a rough estimate of a baseball player's salary?" We are interested in exploring whether subjects have a preference for regression equations without sign violations.

We hypothesized that subjects would give higher ratings to equations formed with Independent Sign Regression and Mean Coefficient Regression to equations formed with Multiple Linear Regression because such equations do not contain sign violations. Note that Independent Sign Regression does not necessarily use all of the variables, and on the baseball data it typically uses 4-6 of the 8 variables.

We also hypothesized that subjects would give higher ratings to equations found with Independent Sign Forward Regression than Stepwise Forward Regression because they also did not contain sign violations.

Subjects. The subjects were 47 male and female undergraduates attending the University of California, Irvine who indicated that they were somewhat or very familiar with baseball. The subjects participated in this experiment to receive extra credit in an artificial intelligence course. We did not enroll subjects with little or no familiarity with baseball in the study.

Stimuli. The stimuli consisted of 15 linear equations that were displayed to the user in a web browser. Figure 1 contains an example of the type of stimuli used. Three equations were generated by each of five different regression routines:

- Multiple Linear Regression
- Independent Sign Regression
- Mean Coefficient Regression
- Stepwise Forward Regression
- Independent Sign Forward Regression

Three different equations for each algorithm were formed on different random subsets of the baseball data resulting in different coefficients. The coefficients of the equations were rounded to two significant digits. The stimuli were presented in random order for each subject.

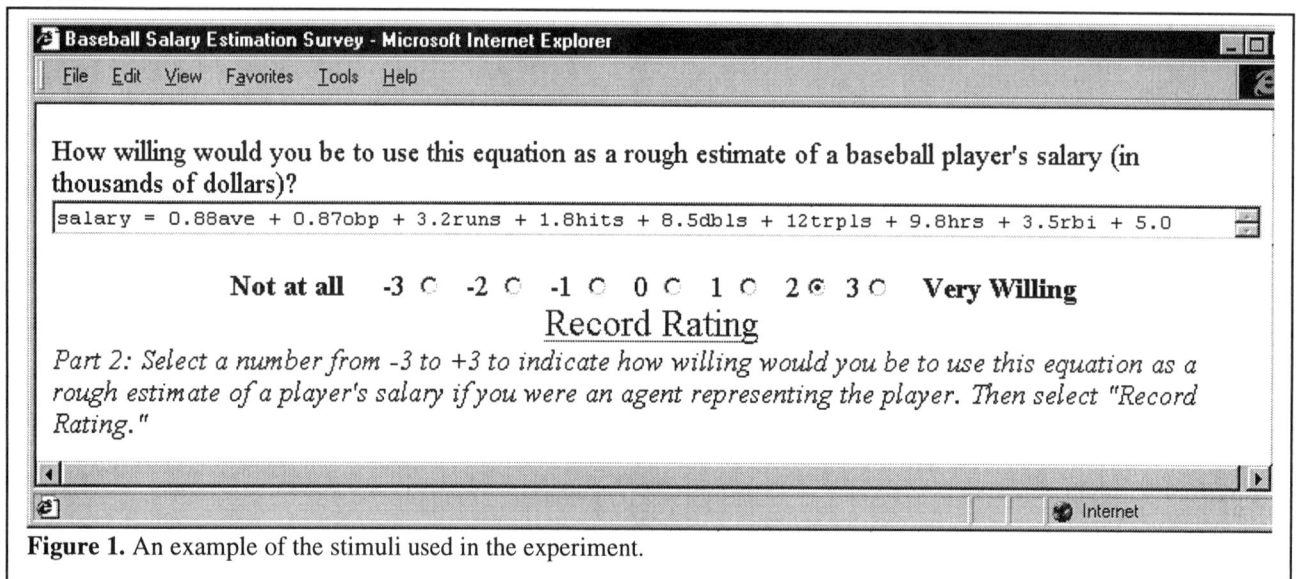

Figure 1. An example of the stimuli used in the experiment.

Procedures. Each subject was shown a single equation at a time, in random order, and asked to indicate on a scale from -3 to +3 how willing they would be to use the equation as a rough estimate of a player's salary by clicking on a radio button. Next they clicked on "Record rating" and were shown another equation. The radio button was reset to 0 before displaying the next equation. This continued until the subject rated all 15 equations.

Results. The average rating of all subjects for each type of equation is shown in Table 2.

Table 2. Average Subjects Ratings for Linear Equations.

Regression Algorithm	Mean Rating
Multiple Linear Regression	-0.816
Independent Sign Regression	0.603
Mean Coefficient Regression	0.851
Stepwise Forward Regression	-1.09
Independent Sign Forward Regression	-0.113

We analyzed the results of the experiment as follows. For each subject, we found the mean rating of the three equations generated by each of the five algorithms. An analysis of variance showed that the algorithm had a significant effect on the rating $F(4,184) = 22.11$, $p < .0001$. A Tukey-Kramer test at the .05 level was used to evaluate the three hypotheses. The critical difference is 0.706 so all three differences are significant:

- Subjects gave significantly higher ratings to equations found by independent sign regression than equations found by multiple linear regression.
- Subjects gave significantly higher ratings to equations found by mean coefficient regression than equations found by multiple linear regression.
- Subjects gave significantly higher ratings to equations found by independent sign forward regression than equations found by stepwise forward regression.

Discussion. The results support the notions that subjects have a preference for linear models that conform to the independent sign bias, i.e., those in which the sign of the coefficients of each explanatory variable agrees with the effect that the explanatory variable has in isolation.

The independent sign regression routine introduced in this paper automatically determines the sign of the coefficient of explanatory variables, produces models that have similar predictive accuracy as those produced by multiple linear regression, and produces linear models that subjects would be more willing to use. A possible disadvantage of the independent sign regression routine is that it may eliminate some variables from the linear equation. This may be a benefit in some cases (e.g., if it was expensive to collect some variables) or if simplicity is a consideration. However, the results of the experiment in this paper suggest that ignoring many explanatory variables may reduce the willingness of subjects to use a linear model. Although this is not a focus of this study, it appears that both independent sign forward regression and stepwise forward regression received relatively low rankings by subjects.

If one is interested in eliminating variables, the simulations showed that independent sign forward regression produces equations with similar predictive accuracy to stepwise forward regression and the experiment showed that users preferred equations created by independent sign forward regression.

We proposed mean coefficient regression as a means of eliminating sign violations while using all explanatory variables. Although it received the highest average ranking by subjects in our experiments, it does not fit the data nor generalize as well as the other regression routines. We suspect that our subjects are sensitive to the sign and perhaps order of magnitude of the coefficients, but there's no reason to believe they'd be able to determine whether two similar equations with slightly different coefficients are a better fit to the data. The inferior accuracy of mean coefficient regression is a result of correlations between the explanatory variables. These same correlations result in multiple linear regression getting the "wrong sign" on the coefficients. It remains an open question whether a linear model can be found that has similar accuracy to multiple linear regression, gets the signs right, and uses all of the variables when there are correlations among the explanatory variables. Because there is a relationship between averaging multiple linear models and mean coefficient regression, it is possible that some of the methods for correcting for correlations in linear models (e.g., in Leblanc and Tibshirani, 1993; Merz and Pazzani, in press) may be useful in this case.

Related Work

The purpose of the independent sign bias is to produce linear models that are as accurate as those produced by multiple linear regression, yet are more acceptable to users because they do not violate the users' understanding of the effect that each explanatory variable has on the dependent variable. If knowledge-discovery in database systems is to produce insightful models that are deployed in practice, it is important that users be willing to accept the models. One implication of this bias is that as additional explanatory variables are added to a model, the magnitude of the effect of the other variables may be changed but not the direction of the effect (cf. Kelley, 1971).

Credit scoring is one important application that may benefit from the global sign bias. In this application, the risk that a borrower may not pay back a loan is assessed as a function of a number of factors such as income, debt, payment history, etc. If a potential borrower is turned down for a loan, it is necessary to explain why. It is important to get the signs of the coefficients right on the models so that the explanation makes sense to the lender and the borrower.

Here, we have introduced constrained regression routines that produced linear models conforming to the independent sign bias. Monotone regression (Lawson and Hanson, 1974) is a related type of constrained regression in which the user indicates the sign constraint on the variables. In contrast, in independent sign regression, the sign is determined automatically.

Having the wrong sign in the regression equation results from having correlated explanatory variables. One way to deal with this is to introduce additional variables to represent the interaction between two explanatory variables. The focus of such work has been to produce models that improve the fit of the data to the model and not to improve the comprehensibility or acceptance of the learned models.

In training artificial neural networks, weight decay (Krogh,. & Hertz, 1995) Sill, & Abu-Mostafa,, 1997) have been proposed as techniques for constraining models. However, the focus has been on improving generalization ability and not improving the user acceptance of the learned models.

Causal Models (Spirtes, Glymour, and Scheines, 1993) and Belief Networks (Pearl, 1988) also explicitly represent the dependencies among variables. The resulting models are more complex than linear models. In this work, we have adopted a fixed representation (linear equations) and addressed what constraints can be imposed upon this representation to improve user acceptance.

In previous work (Pazzani, Mani & Shankle, 1997), we addressed a related problem of rule learning algorithms including counterintuitive tests in rules by having an expert provide "monotonicity constraints." For nominal variables, a monotonicity constraint is expert knowledge that indicates that a particular value makes class membership more likely. For numeric variables, a monotonicity constraint indicates whether increasing or decreasing the value of the variable makes class membership more likely. By showing neurologists rules learned with and without these constraints, we showed that monotonicity constraints biased the rule learning system to produce rules that were more acceptable to experiments.

Pazzani (1998) extended the work on monotonicity constraints by introducing the globally predictive test bias. In this bias, every test in a rule must be independently predictive of the predicted outcome of the rule. Such a bias eliminated the need for a user to specify monotonicity constraints but provided the same advantages. The globally predictive test bias in rule learners is analogous to the independent sign bias in linear models in that the effect of a variable in combination with other variables is constrained to be the same as the effect of that variable in isolation. Here we have shown that such a constraint improves the willingness of people to use linear models without harming the predictive power of the models.

Conclusions

People are not computers and cannot easily find a linear equation that best fits a data set with 10 variables and 500 examples. However, we argue that people have certain constraints on the qualitative properties of the linear equations. We have shown that one factor that influences the willingness of subjects to use linear models is the independent sign bias. By creating regression routines that conform to this bias, we constrain the computer to produce results that are more acceptable to people.

New regression routines were produced that implement the independent sign bias. Experiments and simulations showed that independent sign regression produces linear equations that are approximately as accurate as multiple linear regression and that are more acceptable to users. Independent sign forward regression is similarly preferable to forward stepwise regression.

Acknowledgements

This research was funded in part by the National Science Foundation grant IRI-9713990. Comments by Dorrit Billman and Susan Craw on an earlier draft of this paper help to clarify some issues and their presentation.

References

Draper, N. and Smith, H. (1981). *Applied Regression Analysis.* John Wiley & Sons.

Kelley, H. (1971). Causal schemata and the attribution process. In E. Jones, D. Kanouse, H. Kelley, N. Nisbett, S. Valins, & B. Weiner (Eds.), *Attribution: Perceiving the causes of behavior* (pp 151-174). Morristown, NJ: General Learning Press.

Krogh, A. & Hertz, J. (1995). A Simple Weight Decay Can Improve Generalization Advances in Neural Information Processing Systems 4, Morgan Kauffmann Publishers, San Mateo CA, 950-957.

Lawson, C. L. & Hanson, R. J. (1974). *Solving least squares problems.* Prentice-Hall.

Leblanc, M. & Tibshirani, R. (1993). Combining estimates in regression and classification. Dept. of Statistics, University of Toronto, Technical Report.

Merz, C. & Pazzani, M. (in press). *A Principal Components Approach to Combining Regression Estimates.* Machine Learning.

Mullet, G. (1976). *Why Regression Coefficients Have the Wrong Sign.* Journal of Quality Technology. 8(3), 121-126.

Pazzani, M., Mani, S., & Shankle, W. R. (1997). *Comprehensible Knowledge-Discovery in Databases.* In M. G. Shafto and P. Langley (Ed.), Proceedings of the Nineteenth Annual Conference of the Cognitive Science Society (pp. 596-601). Mahwah, NJ: Lawrence Erlbaum.

Pazzani, M. (1998). Learning with Globally Predictive Tests. The First International Conference on Discovery Science Fukuoka, Japan.

Pearl, J. (1988). *Probabilistic Reasoning in Intelligent Systems: Networks of Plausible Inference.* Morgan Kaufmann Publishers, Palo Alto.

Sill, J. & Abu-Mostafa, Y. (1997). Monotonicity Hints. Advances in Neural Information Processing Systems 9, Morgan Kauffmann Publishers, San Mateo CA, 634-640

Spiegelhalter, D., Dawid, P., Lauritzen, S. and Cowell, R. (1993). *Bayesian Analysis in Expert Systems.* Statistical Science, 8, 219-283.

Spirtes, P., Glymour, C. and Scheines, R. (1993). Causation, Prediction, and Search, New York, N.Y.: Springer-Verlag.

Multiple Processes in Graph-based Reasoning

David Peebles (djp@psychology.nottingham.ac.uk)[1]
Peter C-H. Cheng (peter.cheng@nottingham.ac.uk)[2]
Nigel Shadbolt (nigel.shadbolt@nottingham.ac.uk)[3]

Department of Psychology, University of Nottingham, Nottingham, NG7 2RD, U.K.

Abstract

Current models of graph understanding typically address the encoding and interpretive processes involved during the course of comprehension and largely focus on the visual properties of the graph. An experiment comparing reasoning with two types of graph is presented. On the basis and scope of existing models, performance with the two graphs would not be predicted to differ substantially. There are substantial computational differences between the graphs, however. It is suggested, therefore, that an adequate model of graph use must incorporate different combinations of visual properties of the graphs, levels of graph complexity, interpretive schemas and task requirements.

Introduction

Graphs are widely used to represent quantitative information in many spheres of human activity. They are commonly encountered in business and the media and are ubiquitous in science and engineering. The essential property common to all graphs is a mapping between the quantitative information being represented and visual dimensions (such as length, size or colour) of specific graphical elements. This visuo-spatial representation of numerical information allows perceptual inferences to be made rather than often more difficult and time-consuming logical reasoning or numerical computation.

Research into graph comprehension has largely focussed on specifying the visual and structural properties of particular types of graph and on providing analyses of the various tasks involved in graph understanding (Bertin, 1983; Cleveland & McGill, 1985; Kosslyn, 1989; Pinker, 1990). Researchers have also developed accounts of various perceptual and cognitive processes involved in graph interpretation (Carpenter & Shah, 1998; Gattis & Holyoak, 1996; Lohse, 1993; Tabachneck, Leonardo, & Simon, 1994).

Although researchers are producing ever more detailed and sophisticated models of graphical perception, the scope of much of this research remains rather narrow. Each study typically examines the properties of only one particular class of graph (e.g. Gattis & Holyoak, 1996; Kosslyn, 1989). In addition, most of these investigations have concentrated on the processes of graph comprehension (e.g. Carpenter & Shah, 1998; Pinker, 1990; Shah & Carpenter, 1995) while less research has been carried out on the actual *use* of graphs for reasoning and problem solving.

There are different classes and subtypes of graph, each with characteristic features which may be more or less appropriate for representing certain types of information or facilitating different perceptual and cognitive operations. A case may be made both for conducting comparative analyses of different graph types and for studying graph-based reasoning. Firstly, because different types of graph may be used to represent the same information, two graphs may be created which are *informationally* equivalent without necessarily being *computationally* equivalent (for discussions of these issues see Simon, 1978 and Larkin & Simon, 1987). Comparing the behaviour of subjects carrying out the same task using these two graphs may shed light both on subjects' mental representations of particular graphs and the representational and computational properties of the graphs. In addition, by using a wide range of problems, one can produce a rich and varied set of behavioural data which can reveal complex interactions between the graph user's knowledge, the visual properties of the graph and the different perceptual and cognitive operations afforded by the type or class of graph. Graphs and questions may also be produced specifically to test hypotheses about the effect of particular factors. Results of these tests may then be used to specify the factors and components necessary for an adequate model of graph use. In this paper we present results of an experiment designed to identify some of these factors using the methodology outlined above. In a previous study, the class of line graphs was delineated and analysed in terms of an ontology of graphical components shared by the constituent graph types (Cheng, Cupit & Shadbolt, 1998). In this experiment, the properties of two types of line graph identified by Cheng *et al.* are compared. In the

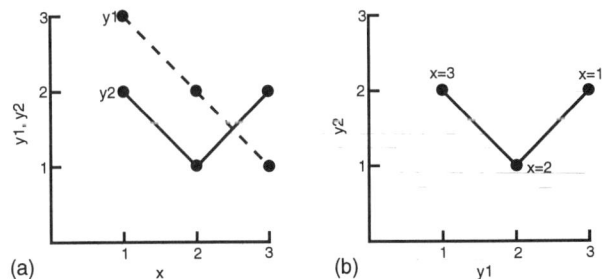

Figure 1: Graph types used in the experiment. (a) *Combined Single Unary* (CSU) graph; (b) *Double Unary Parametric* (DUP) graph

[1] Artificial Intelligence Group
[2] ESRC Centre for Research in Development, Instruction and Training
[3] Artificial Intelligence Group

first, called a *Combined Single Unary* (CSU) graph (see Figure 1a), two dependent variables, each of which represents a single unary function of an independent variable plotted on the horizontal x axis, are represented as separate lines superimposed on a single co-ordinate space, the values of which are plotted on the vertical y axis. In the second, named a *Double Unary Parametric* (DUP) graph (Figure 1b), the two dependent variables are represented on the x and y axes and the values of the independent variable are represented by points on the plotted line.

These graphs would be considered similar in various important ways identified in the literature. Firstly, they are informationally equivalent, having been generated from the same data set (Larkin & Simon, 1987). Secondly, they are both diagrams using locational indexing of information (Larkin & Simon, 1987). Thirdly, they are both Cartesian graphs using a two dimensional co-ordinate system to relate quantities and represent magnitudes, so presumably would invoke similar general schemas and interpretive processes (Pinker, 1990; Kosslyn, 1989). Fourthly, they are both simple line graphs so the same set of general heuristics can be used in their interpretation and they would also be affected by the same set of biases (Carpenter & Shah, 1998; Shah & Carpenter, 1995; Gattis & Holyoak, 1996).

As will be demonstrated in the experiment, CSU and DUP graphs differ substantially in computational terms. But it is unclear, given the basis and scope of any one current model of graph comprehension, how the differences in the graphs may be explained. The detailed analysis of subjects' behaviours on selected problems suggests that an adequate model of graph use must integrate different permutations of visual properties of the graphs, levels of graph complexity, knowledge of appropriate interpretive schemas and task requirements.

Experiment

A set of informationally equivalent CSU and DUP graphs was created and a number of questions about the data represented by the graphs were constructed. An additional factor was included in the experiment design—that of graph *complexity*. As a working hypothesis, we have adopted Carpenter and Shah's (1998) equation of the graphical complexity of a graph with the number of plotted lines which can be regarded as distinct functions. Carpenter and Shah have found that the time taken to interpret a graph is strongly related to the complexity of the graph (as they define it). The experiment was also designed, therefore, to determine what effect, if any, increased graphical complexity has on problem solving.

Method

Participants Sixty-four second and third year psychology students from the University of Nottingham were paid to participate in the experiment. Participants were informed that the two people who achieved the most correct answers in the shortest time would also receive an additional payment.

Apparatus and Stimuli The experiment was carried out on an Apple Macintosh computer with a 17 in colour monitor. The primary stimuli used were six line graphs depicting amounts of gold and silver produced by a fictitious mine each month for two consecutive years. Two *simple* graphs of both graph types were created, each presenting the data for one year. A *complex* graph of each type was then created by superimposing the two simple graphs onto a single set of axes. Thus four experiment conditions were produced—simple CSU, simple DUP, complex CSU and complex DUP (henceforth referred to as CSU(s), DUP(s), CSU(c) and DUP(c) respectively). The six graphs are shown in Figure 2.

In order to construct a set of graphs that shared certain visual characteristics to allow them to be used for a single set of problems, the data were generated so that the graphs for each year had a number of similar properties and all graphs satisfied several constraints. For example, each year contained two months with exactly the same amounts of silver and gold production. A set of sixteen questions were produced, eleven of which were seen by all four conditions, the other five being presented after the eleven general questions only to subjects in the complex graph conditions.

Design and Procedure The experiment was a mixed design involving two between-subjects conditions—graph type (CSU or DUP), graph complexity (simple or complex), and one within-subject condition—the year of the graph (1996 or 1997). The 64 participants were randomly allocated to one of the four conditions.

Participants were instructed to answer the questions as accurately and as rapidly as possible and to continuously point to the part of the computer screen at which they were currently looking with a pointer while thinking aloud at all times. Verbal protocols and pointing movements were recorded on audio and video tape respectively. Before starting the experiment, subjects were given a practice trial with six simple questions using graphs of a similar type to their condition in order to familiarise them with the graph type, the equipment and the procedure for answering questions and thinking aloud.

Before answering any questions, participants were shown the graphs and were asked to try to understand what they was about while thinking aloud and pointing to the part of the graph that they were looking at. Subjects in the simple conditions were presented with the graph for each year one after the other in random order. When subjects were familiar with the graphs, they were presented with the questions. In each trial, a question for a particular year together with the graph for that year was presented on the screen. As soon as the subject indicated that s/he had an answer by pressing a key on the keyboard, the graph was removed from the screen and a prompt for the subject to enter a response appeared. Subjects were encouraged to answer all of the questions, entering a "best guess" if they were unsure of the correct answer. The response time (RT) for the initial key press and the subject's response were recorded for each trial. The presentation order of graph year and the questions were randomised. Subjects in the simple conditions saw 22 trials (11 questions for both years) whereas those in the complex conditions received a total of 27 trials (11 questions for both years plus an additional 5 questions).

Results

From a total of 1568 responses, 9 (from different participants) were not accurately recorded. For the purposes of analysis of variance (ANOVA) therefore, each missing response was replaced by a value estimated by computing the cell mean.

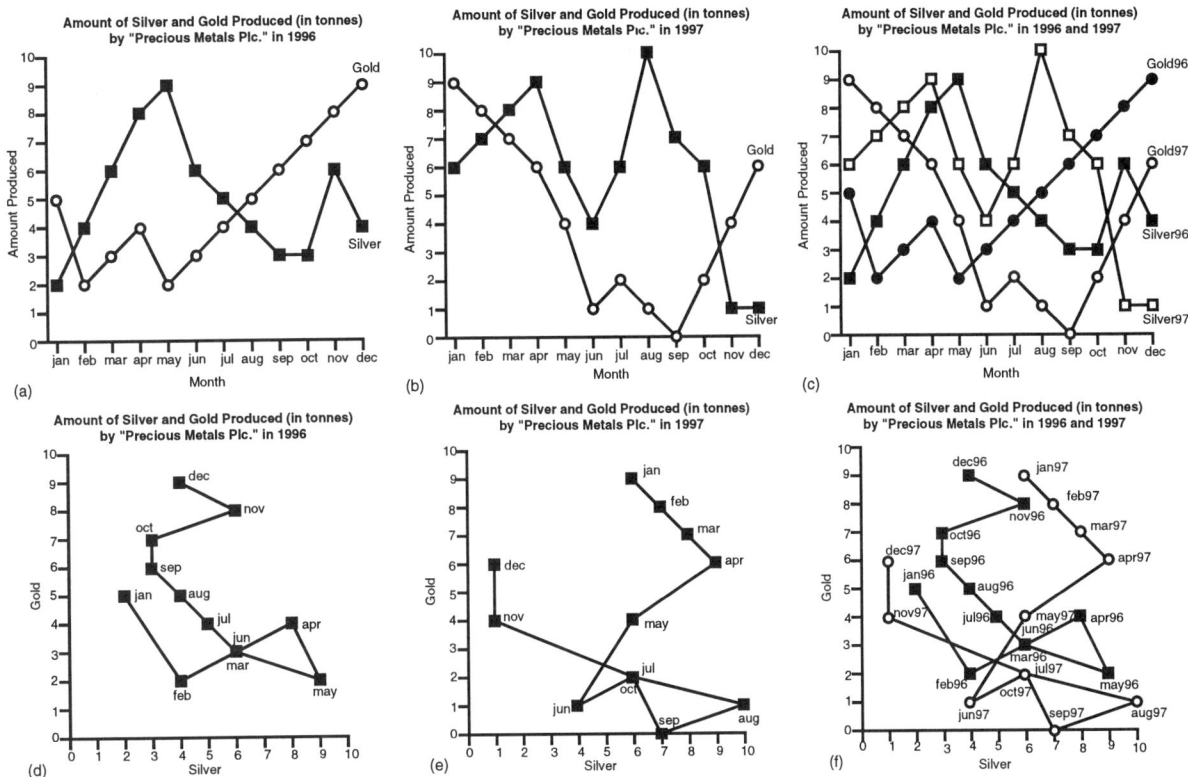

Figure 2: Graphs used in the experiment. (a & b) CSU(s) graphs for 1996 and 1997 respectively, (c) CSU(c) graph for 1996 & 1997, (d & e) DUP(s) graphs for 1996 and 1997 respectively, (f) DUP(c) graph for 1996 & 1997

An ANOVA on the proportions of correct responses for the 11 general questions produced significant main effects of graph type, $F(1, 10) = 19.90, p < .01, MS_e = 0.098$, graph complexity, $F(1, 10) = 31.36, p < .01, MS_e = 0.155$, and question number, $F(10, 10) = 74.71, p < .01, MS_e = 0.368$. Significant interactions were also found between graph type and graph complexity, $F(1, 10) = 19.90, p < .01, MS_e = 0.098$, graph type and question number, $F(10, 10) = 20.10, p < .01, MS_e = 0.099$, and graph complexity and question number, $F(10, 10) = 4.01, p < .05, MS_e = 0.020$. A significant three-way interaction between graph type, graph complexity and question number $F(10, 10) = 3.73, p < .05, MS_e = 0.083$ was also produced. No significant main effect or interaction was found involving the year condition.

An ANOVA on the mean RTs for the 11 general questions produced significant main effects of graph type, $F(1, 10) = 10.75, p < .01, MS_e = 201032528$, graph complexity $F(11, 10) = 11.83, p < 0.01, MS_e = 221300928$, and question, $F(10, 10) = 121.25, p < 0.01, MS_e = 2267433216$, and also yielded significant interactions between graph type and graph complexity, $F(1, 10) = 18.21, p < .01, MS_e = 340499488$, and graph type and question, $F(10, 10) = 12.65, p < .01, MS_e = 236482048$. As with the response accuracy data, no significant main effect or interaction involving the year condition was found.

The ANOVAs indicate a complex interaction between the various factors in the experiment. It is clear that both the accuracy of the responses and the time taken to respond is affected by both the type and complexity of the graph being used and the type of problem being solved.

Subjects' graph familiarity After taking part in the experiment, participants were asked to rate the frequency with which they normally encountered information presented in the form just seen by choosing between *High*, *Medium*, or *Low* frequency, and also to rate on a scale of 1 to 9, how familiar they considered themselves to be with the type of graph they had just encountered, where 1 represented *Very Unfamiliar* and 9 indicated *Very Familiar*.

There were considerable differences in the ratings of exposure frequency between the graph types. Only 19.4% of subjects in the CSU conditions rated their exposure frequency to the graph type as low whereas in the DUP conditions, 71.9% of subjects did so.

A similar situation was found in the ratings of graph familiarity. Whereas 74.2% of subjects in the CSU conditions rated their familiarity with the graph type as being 7 or over and none of them rated themselves in the least familiar third of the range, only 28.1% of subjects in the DUP conditions rated their familiarity as being 7 or over while 40.6% of them rated themselves in the least familiar third of the range.

Individual Question Analysis

In this section, we demonstrate the effects of three different factors on problem solving behaviour by analysing the responses to four questions in detail. Each of the questions was selected to highlight the effect of a particular factor.

Question 4: This question illustrates how different graphical representations of the same information can require different procedures to access certain information and how this can result not only in a considerable difference in the distribution of resulting answers but also in a marked difference in the time taken to retrieve the information. Question 4 had the form: *In [1996/1997], when the amount of Gold production is 4, what is the amount of Silver production?*

The graphs were constructed so that two months in each year had amounts of gold production equal to 4 so that at least three answers to the question could be expected—the amount of silver production for either of these months individually or that for both months. Based on a previous analysis of graph-based reasoning protocols, a set of basic perceptual and cognitive operations has been identified which subjects commonly use when solving graph-based problems. Examples of these basic operations include: (a) identifying the axis associated with a given variable, (b) locating a point on an axis corresponding to a given value of the variable, (c) tracing a straight line from a point on an axis until a plotted line is reached, and (d) measuring the distance between two points on a line relative to a given axis. With this set of basic operations, a model of the procedures used to solve specific problems using different graphs can be constructed which can then be used to make predictions about the relative time taken and the probability of different responses being made using a particular graph type. A detailed discussion of this set of basic operations and an outline of such a model will be reported elsewhere. However, to give some idea of the type of analysis which can be carried out and the explanations which can result from this analysis, an outline of two procedures carried out by subjects using the different graph types for 1996 are presented below. Subjects using CSU graphs who give only one answer to this question generally follow this procedure:

1. Identify the axis (a_1) which represents the values of the variable given in the question, (*Amount Produced* axis).

2. Identify the specific point (a_1p_1) on axis (a_1) corresponding to the specific value of the variable given in the question, (*Amount Produced* = 4).

3. Trace a straight line from axis point (a_1p_1) across the graph until the line (l_1) corresponding to the variable and year given in the question is reached at point (l_1p_1), (Point on *Gold* line where *Amount Produced* = 4, [*Month* = 'apr'])

4. Trace a straight line from line point (l_1p_1) until the line (l_2) on the line corresponding to the variable and year required by the question is reached at point (l_2p_1) so that $(l_1p_1 = l_2p_1)$ on the month axis, (Point on *Silver* line where *Month* = 'apr').

5. Trace a straight line from point (l_2p_1) until the axis (a_1) represents the values of the variable required variable in the question is reached at point (a_1p_2), (Point on *Amount Produced* axis from point on *Silver* line where *Month* = 'apr').

6. Read off value of axis location (a_1p_2) and return as answer, (Answer: "Amount of *Silver* produced = 8").

Table 1: Number of responses to each of three answers to Question 4

Answer	Simple		Complex	
	CSU	DUP	CSU	DUP
		1996		
5 (jul)	1	8	6	9
8 (apr)	9	1	5	3
5 & 8	6	7	4	4
		1997		
1 (nov)	1	8	3	11
6 (may)	4	0	8	2
1 & 6	10	8	3	2

In this procedure, as soon as the first point on the plotted line (l_1) is encountered, the corresponding point on the other line (l_2) is identified the value of that point on the (a_1) axis is returned as the answer. To discover the second month with a gold production value of 4, the straight line traced from (a_1p_1) must be extended to reach the second point on the same line. Subjects using DUP graphs who give only one answer to this question generally follow this procedure:

1. Identify the axis (a_1) corresponding to the variable given in the question, (*Gold* axis).

2. Identify the specific point (a_1p_1) on axis (a_1) corresponding to the specific value of the variable given in the question, (*Gold* = 4).

3. Trace a straight line from axis point (a_1p_1) across the graph until the line (l_1) corresponding to the year given in the question is reached at point (l_1p_1), (Point on *Year* line where *Gold* = 4, [*Month* = 'jul']).

4. Trace a straight line from the point (l_1p_1) until the axis (a_2) corresponding to the variable required by the question is reached at point (a_2p_1), (Point on *Silver* axis where *Month* = 'jul').

5. Read off value of axis location (a_2p_1) and return as answer, (Answer: "Amount of *Silver* produced = 5").

Even from this relatively coarse-grained analysis, one can see that the procedure used for the CSU graphs requires an additional step and involves the identification of more lines and points than that used for the DUP graphs. One might expect, therefore, that the time taken by subjects in the CSU conditions to solve this problem will be greater than that of the DUP subjects. More importantly, the graphs in this experiment were constructed so that the first point reached by the procedures (i.e. the month closest to the starting axis) would be different for each condition. Therefore, if the above procedures are followed, the profile of responses given by subjects in the two conditions will differ. The numbers of responses corresponding to the three answers given by subjects in the four conditions are shown in Table 1. The mean RT for the DUP conditions (14.4 s and 16.4 s for the DUP(s) and DUP(c) conditions respectively) was on average 8.04 s faster than that of the CSU conditions (21.8 s and 25.0 s for the CSU(s) and

CSU(c) conditions respectively). In both the simple and complex conditions the differences between the graph types were significant, $t(30) = 2.55, p < .05$ and $t(30) = 2.94, p < .01$ respectively. In the two CSU conditions, 35.9% of the participants gave the answer of two values, 40.6% gave the answer corresponding to the month nearest the starting axis and 17.2% gave the answer corresponding to the alternative month. In the DUP conditions, 32.8% of the subjects gave the answer of two values, 56.3% gave the answer corresponding to the month nearest the starting axis and 9.4% gave the answer corresponding to the alternative month. In both CSU and DUP conditions, therefore, more subjects gave one answer corresponding to the month nearest the starting axis than to the alternative answer or both answers.

This pattern of results, together with an analysis of the video-taped protocols, provides evidence that the procedures outlined above were carried out by the majority of subjects using the two graph types. Question 4 demonstrates that both the speed and nature of a problem solution can be determined by the graphical representation because of the different procedures which users are required to follow in order to access the same information.

Question 5: This question demonstrates how graph-based reasoning processes involving perceptual components can require less time and be less error-prone than those involving numerical computation. Question 5 had the form: *How many months in [1996/1997] is there a difference between Silver and Gold production of exactly 1 tonne?*

Because the two dependent variables are plotted as lines in CSU graphs, differences between the two variables for a given month are represented by a vertical distance between the two lines at a specific value of the x axis. In the CSU conditions, therefore, subjects are simply required to scan the graph to find points on the x axis where the distance between the two plotted lines for a given year are separated by one unit on the y axis. In DUP graphs, however, this spatial representation is not available as the value of each variable for a given month is represented by the co-ordinate location of one particular point. In the DUP conditions, therefore, subjects are required to attend to each point on the plotted line in turn, obtain the gold and silver value for that point from the two axes, and then compute the difference between the two values. One may expect, therefore, that the more elaborate and error-prone procedure used in the DUP conditions would be reflected in lower response accuracy and longer RT scores from subjects in the DUP conditions.

The proportion of correct responses and mean response times for each of the four conditions are shown in Table 2. The average response time for the CSU(s) condition was 17.55 s faster than that of the DUP(s) condition $t(30) = 6.56, p < .001$ while the mean RT of the CSU(c) condition was 15.00 s faster than that of the DUP(c) condition $t(30) = 4.14, p < .001$. The accuracy of the responses from the CSU conditions was also greater than those from the DUP conditions. The average RT for the CSU(c) condition was only 5.41 s slower than that for the CSU(s) condition and the correct response scores for the two conditions was virtually identical.

A study of the video-taped protocols supported the analysis above, revealing that a large proportion of subjects solved the

Table 2: Proportion of correct responses (CR) and mean response time (RT in seconds) for the four conditions to Question 5

	Simple		Complex	
	CSU	DUP	CSU	DUP
CR	.938	.844	.969	.625
RT	18.9	36.5	24.3	39.2

problem using the procedures appropriate for the particular graph representation, whether perceptual or conceptual.

Questions 12 and 13: The purpose of questions 12 and 13 taken together was to investigate the combined effect of two factors—graph complexity and graphical representation—and to determine whether the effects of the former may be mitigated by the latter. The two questions illustrate how the retrieval of certain information (e.g. the maximum and minimum sum, difference and product of the two dependent variables), can be facilitated by a representation in which the information can be found by searching a local region for a specific point, compared to a representation in which the same information can only be retrieved by a process of search and computation.

Question 12 had the form: *In which month in 1996 was the greatest total amount of metal (i.e. both Silver and Gold) produced?* The form of question 13 was: *In which year was the most metal (i.e. both Silver and Gold) produced in any month?*

Questions 12 and 13 were only presented to subjects in the complex conditions. In the DUP(c) condition, the item of information required by both questions (the maximum sum of the two metals) corresponds to a point on a line, the location of which is the furthest to the top right corner of the graph. In the CSU(c) condition, however, the same information is derived by identifying a point on the x axis at which the points on two specific lines have positions on the y axis which result in a combined value greater than any other. Therefore, whereas the correct answer may be found in the DUP(c) condition using a visual search procedure which identifies the location of a single point, the search procedure in the CSU(c) condition requires the additional computation of the sum of the two y axis values.

Question 13 differs from question 12 in that it requires the maximum sum of the two metals to be found across both years. If subjects in the DUP(c) condition identify the correct year by locating a point on one of two lines which is nearest to the top right corner of the graph, they should take approximately the same time to carry out this procedure as for question 12. In the CSU(c) condition, however, the procedure required to answer question 13 is more demanding than that for question 12 because it requires subjects to take all four lines into account as they must find the maximum sum of two metals from two pairs of lines rather than from only one pair in question 12. Therefore, if the procedures outlined above are followed, one may expect not only that subjects in the CSU(c) condition would take longer to produce an answer than those in the DUP(c) condition, but also that a greater discrepancy should be found between the average RTs between the conditions for question 13.

Table 3: Proportion of correct responses (CR) and mean response time (RT in seconds) for the two conditions to Question 12 and 13

	Question 12		Question 13	
	CSU	DUP	CSU	DUP
CR	.938	.875	.800	1.00
RT	36.8	27.0	54.5	25.4

The response accuracy and RT data for questions 12 and 13 are shown in Table 3. Although there was very little difference in the accuracy of the responses between the conditions for question 12, those from the DUP(c) condition were on average 9.86 s faster than those from the CSU(c) condition, $t(30) = 2.33, p < .05$. Mean RTs of subjects in the DUP(c) condition for question 13 were similar to those from question 12. Responses of subjects in the CSU(c) condition, however, were on average 17.62 s slower than those from question 12, $t(15) = 3.63, p < .01$. Subjects in the DUP(c) condition were on average 29.03 s faster and 20% more accurate in their responses to question 13 than those in the CSU(c) condition, $t(30) = 3.40, p < .01$.

Analysis of the video-taped protocols revealed that many subjects in both conditions initialised a search by attending to months at which one of the values was particularly high (May and December). Several subjects in the CSU(c) condition scanned along the x axis from January to December, computing possible candidate solutions until the highest sum was found. Many subjects in the DUP(c) condition, however, rapidly located the correct point without attending to other points on the line, suggesting that they were aware of the general region at which the required information must lie.

Discussion

The results of the experiment show that a number of interrelated factors can significantly affect graph-based reasoning and that these factors cannot be accounted for simply by an analysis of the visual properties of the graph. These results support the claim that an adequate model of graph use must also take into account the specific representational properties of a graph type and computational procedures which are facilitated by particular graphical representations. Question 4 illustrated how different graphical representations of the same data can require different procedures to access the same information and that the time taken to access the information, and even what information is retrieved, can be affected by the procedure followed. Question 5 demonstrated that representing an item of information in a graph as a visual feature such as distance can result in a more rapid and accurate retrieval of that information than when the information must be computed from the visual information. Questions 12 and 13 illustrated three main points. Firstly, both questions showed individually that accessing certain types of information can be facilitated by a graphical representation in which the information is represented as a specific location compared to a representation in which extraction of the information requires computational effort. Secondly, the two graphs taken together showed that graph complexity can have an affect on problem solving performance but that this effect can be mitigated by

the type of representation used. Thirdly, Questions 12 and 13 showed that the effect of a particular representation can be large, even when users of the particular graph are relatively unfamiliar with its form.

Acknowledgements

This research is funded by a grant from the UK Engineering and Physical Sciences Research Council and by the UK Economic and Social Research Council through the Centre for Research in Development, Instruction and Training. The authors wish to thank the fourth member of the project, James Cupit, for his comments on earlier drafts of this paper.

References

Bertin, J. (1983). *Semiology of graphics: Diagrams, networks, maps.* Madison, WI: The University of Wisconsin Press.

Carpenter, P. A., & Shah, P. (1998). A model of the perceptual and conceptual processes in graph comprehension. *Journal of Experimental Psychology: Applied, 4,* 75–100.

Cheng, P. C-H., Cupit, J., & Shadbolt, N. R. (1998). Knowledge acquisition from graphs: An ontological framework. In B. R. Gaines, & M. Musen (Eds.), *Knowledge Acquisition Workshop, KAW98, 1,* (pp. 1–19). Banff, Alberta: University of Calgary.

Cleveland, W. S., & McGill, R. (1985). Graphical perception and graphical methods for analysing scientific data. *Science, 229,* 828–833.

Gattis, M., & Holyoak, K. (1996). Mapping conceptual to spatial relations in visual reasoning. *Journal of Experimental Psychology: Learning, Memory, and Cognition, 22,* 231–239.

Kosslyn, S. M. (1989). Understanding charts and graphs. *Applied Cognitive Psychology, 3,* 185–226.

Larkin, J. H., & Simon, H. A. (1987). Why a diagram is (sometimes) worth ten thousand words. *Cognitive Science, 11,* 65–100.

Lohse, G. L. (1993). A cognitive model for understanding graphical perception. *Human-Computer Interaction, 8,* 353–388.

Pinker, S. (1990). A theory of graph comprehension. In R. Freedle, *Artificial Intelligence and the Future of Testing.* Hillsdale, NJ: Lawrence Erlbaum Associates.

Shah, P., & Carpenter, P. A. (1995). Conceptual limitations in comprehending line graphs. *Journal of Experimental Psychology: General, 124,* 43–62.

Simon, H. A. (1978). On the forms of mental representation. In C. W. Savage (Ed.), *Minnesota Studies in the Philosophy of Science. Vol. IX: Perception and Cognition: Issues in the Foundations of Psychology.* Minneanapolis: University of Minnesota Press.

Tabachneck, H. J. M., Leonardo, A. M., & Simon, H. A. (1988). How does an expert use a graph? A model of visual and verbal inferencing in economics. *Proceedings of the Sixteenth Annual Conference of the Cognitive Science Society* (pp. 842–846). Hillsdale, NJ: Lawrence Erlbaum Associates.

Coarse Coding In Value Unit Networks:

Subsymbolic Implications Of Nonmonotonic PDP Networks

C. Darren Piercey (dpiercey@bcp.psych.ualberta.ca)
Department of Psychology; University of Alberta
Edmonton, Alberta, CANADA T6G 2E9

Michael R.W. Dawson (mike@bcp.psych.ualberta.ca)
Department of Psychology, University of Alberta
Edmonton, Alberta, CANADA T6G 2E9

Abstract

PDP networks that use nonmonotonic activation functions often produce hidden unit regularities that permit the internal structure of these networks to be interpreted (Berkeley et al, 1995; Dawson, 1998; McCaughan, 1997). In some cases, these regularities are associated with local interpretations (Dawson, Medler & Berkeley, 1997). Berkeley has used this observation to suggest that there are fewer differences between symbols and subsymbols than one might expect (Berkeley, 1997). We suggest below that this kind of conclusion is premature, because it ignores the fact that regardless of their content, the local features of these networks are not combined symbolically. We illustrate this point with the interpretation of a network trained on a variant of Hinton's (1986) kinship problem, and show how the network's behavior depends on the coarse coding of information represented by hidden unit bands, even when these bands have local interpretations. We conclude that nonmonotonic PDP networks actually provide an excellent example of the differences between symbolic and subsymbolic processing.

Introduction

Networks of value units are a PDP architecture whose processors use a Gaussian activation function, and whose connection weights are trained using a variation of the generalized delta rule (Dawson & Schopflocher, 1992).

One property that emerges from this PDP architecture is a marked "banding" of its hidden unit activities (Berkeley et al., 1995; Dawson, 1998; Dawson et al., 1997). This banding is revealed when the responses of hidden units to each of a set of training patterns are plotted in a type of one-dimensional scatter plot called a jittered density plot (Chambers, Cleveland, Kleiner, & Tukey, 1983). One jittered density plot is drawn for each hidden unit in a network. For each pattern in a training set, a dot is added to the density plot. The x-position of the dot indicates the activity produced in that hidden unit by an input pattern. The y-position of the dot is randomly selected to reduce the overlap of different points. For the hidden units of a value unit network, the dots in a jittered density plot are not "smeared" uniformly across the graph. Instead, the plot is typically organized into a set of distinct bands or stripes (see Figure 1).

This banding phenomenon is important, because the bands often enable a researcher to determine the algorithm that is used by a trained network to accomplish a particular pattern recognition task. Training patterns that fall into the same band in a hidden unit do so because they share one or more properties, called *definite features* (Berkeley et al., 1995). By identifying the definite features in a layer of hidden units, and by determining how they are combined by a layer of output units, one can specify in great detail how a network of value units accomplishes a mapping from inputs to outputs.

For example, a network of value units has been trained on a set of logical problems devised by Bechtel and Abrahamsen (1991). When this network was analyzed, its hidden units were highly banded, and bands were associated with specific local features (e.g., type of logical connective, relations among variables in the logic problems). The network combined these local features in such a way that its internal structure represented many of the traditional rules of logic, such as *modus ponens* (Dawson et al., 1997).

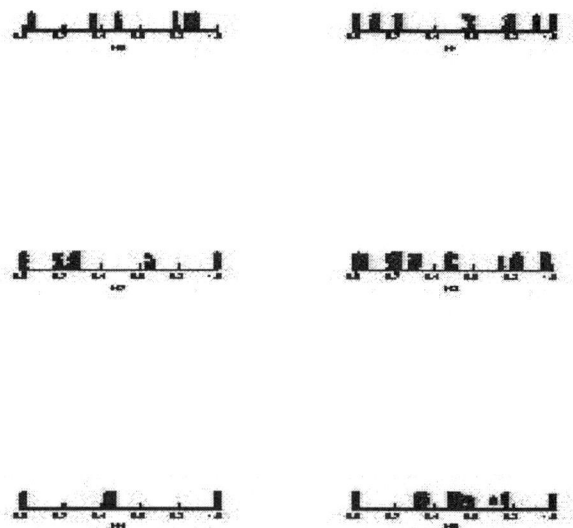

Figure 1. Jittered density plots for the kinship network described below. Each plot is for one of the 6 hidden units in that network. Each plot is comprised from 312 different points, which are organized into distinct bands for each hidden unit

Bands, Symbols, And Subsymbols

One major debate in cognitive science concerns potential differences (and similarities) between symbolic models and connectionist networks (Dawson, 1998). For example, Smolensky has argued that, in contrast to symbolic theories, PDP networks are *subsymbolic* (Smolensky, 1988). To say that a network is subsymbolic is to say that the activation values of its individual hidden units do not represent interpretable features that could be represented as individual symbols. Instead, each hidden unit is viewed as indicating the presence of a *microfeature*. Individually, a microfeature is unintelligible, because its "interpretation" depends crucially upon its context (i.e., the set of other microfeatures which are simultaneously present (Clark, 1993)). However, a collection of microfeatures represented by a number of different hidden units can represent a concept that could be represented by a symbol in a classical model.

It has recently been argued that the banding phenomenon found in value units is relevant to understanding the subsymbolic nature of PDP networks (Berkeley, 1997). This argument is based on an interpretation of a network trained to solve the logic problem (see also Dawson et al., 1997). Berkeley noted that the bands in the logic network are associated with a local interpretation (e.g., some bands represent which connective is present in a stimulus problem, while other bands represent relationships between specific variables in a stimulus problem, such as "Sentence 1 variable 2 is equal to the variable in the conclusion"). Berkeley also noted how such local features become interpretable (as symbols) only after considering a collection of individual hidden unit activations (i.e., a collection of individual dots which in turn produce a band of the sort depicted in Figure 1). Berkeley concluded that the fact that the bands in this network could be construed as being symbols under a liberal interpretation of the term (Vera & Simon, 1993), and suggested that the differences between symbols and subsymbols was smaller than one would believe from the extant literature.

Unfortunately, this conclusion is premature. This is because the "symbolic" nature of the bands in the logic network (i.e., a local interpretation denoting or representing a

specific component of the logic problem, along the lines explored by Vera and Simon, 1993) are actually rarely seen in value unit networks. When most other examples of such networks are interpreted with the banding technique, we find that individual bands do not typically denote entities that would be represented as symbols in a classical theory. Instead, the bands themselves seem much more akin to subsymbols, and the "symbolic" interpretation of a network's internal structure only emerges after considering combinations of bands distributed over a number of different hidden units. Furthermore, even when the representations at the level of hidden units are local (e.g., in the logic network, when individual bands of activity could be assigned local interpretation, and as a result each hidden unit represented a collection of different local features), these local features are not combined into a more global response using symbolic operations.

To illustrate these points, let us consider the interpretation of a different value unit network, one which has been trained to solve a variation of a kinship problem originally reported by Hinton (1986) in the context of interpreting internal network representations.

Simulation

Problem Representation

In Hinton's kinship problem (Hinton, 1986), a network was given an individual's name and a relationship (e.g., "James, father"). This input represented a question about a person (i.e., "Who is James' father?"). The network's task was to generate the name or names representing the correct answer to the question (i.e., "Andrew").

In Hinton's original version of the problem, a network was trained on 100 of the 104 possible relationships in two different family trees of identical structure (i.e., the structure illustrated in Figure 2). In our version of this problem, we used six different versions of this family tree (i.e., six different families with the identical family tree structure), training the network on 52 relationships in each tree, for a total of 312 instances.

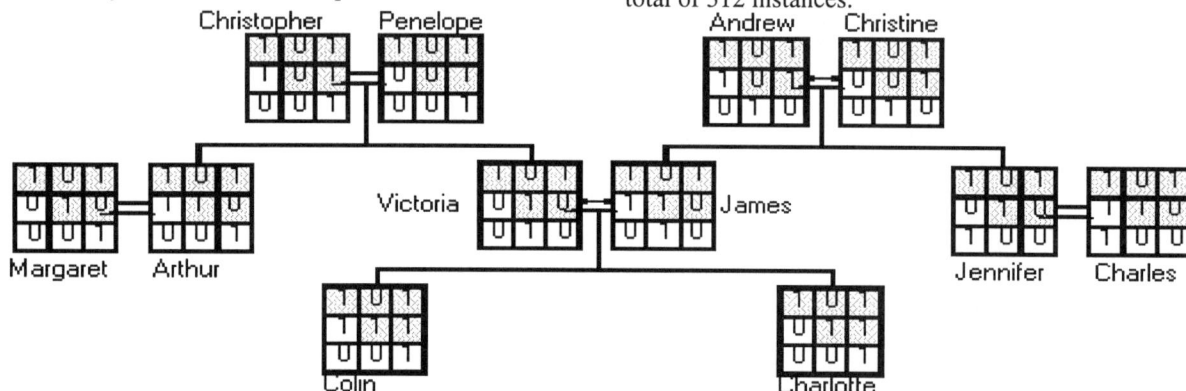

Figure 2. One of the 6 family trees modeled after Hinton (1986). The 9 bits at each node represent the binary code used to represent an individual in the network. The top 3 represent the family, the next three represent gender (white bit) and generation (gray bits), and the bottom 3 represent a code to distinguish individuals of the same generation (see text for more details).

cause the "symbolic" nature of the bands in the logic network (i.e., a local interpretation denoting or representing a

The network had 21 input units. The first 9 represented a person's name using the following coding scheme: The first three bits indicated which of the six families the individual belonged to (001 = family 1, 010 = family 2, 011 = family 3, 100 = family 4, 101 = family 5, 110 = family 6). The fourth bit indicated whether the individual was male (activity = 1) or female (activity = 0). The fifth and sixth bits indicated the generation within the family tree to which the person belonged (01 = first generation, 10 = second generation, 11 = third generation). The seventh, eighth, and ninth bits were local codes that, in combination with gender bit 4, individuated different people belonging to the same generation of the family tree (see Figure 2). The advantage of the local code in these final bits is that the network could generate two names by turning two of these bits on, which is necessary when asked to name the aunts or uncles of Generation 3 children.

The remaining 12 input units of the network represented a relationship using Hinton's local coding scheme (Hinton, 1986). A relationship was encoded by turning one of these 12 units on and by turning the other 11 off. In order from input unit 10 to input unit 21 the represented relations were nephew, niece, aunt, uncle, brother, sister, father, mother, daughter, son, wife, and husband.

The network had 6 hidden units and 9 output units, all of which were value units. The 9 output units encoded an individual's name using the same coding scheme that was used to represent names in the input units.

In each family tree, there is a total of 52 different relationships that can be queried (4 nephew, 4 niece, 2 aunt, 2 uncle, 3 brother, 3 sister, 6 father, 6 mother, 6 daughter, 6 son, 5 wife, 5 husband). Note that there are only 2 aunt and 2 uncle queries because each of these queries results in the network generating a name output that represents two different individuals by turning two of the "local bits" on. Because we trained the network on these 52 relationships for 6 different family trees there was a total of 312 patterns in the training set.

Network Training

The network biases and connections were randomly selected from the range [-0.1,0.1], and the network was trained using a variation of the generalized delta rule developed for value unit networks (Dawson & Schopflocher, 1992) with a learning rate of 0.001 and a momentum of 0. Connection weights and biases were updated after every pattern presentation. During one sweep of training, each of the 312 training patterns was presented to the network. The order of pattern presentation was randomized before every sweep.

The network was said to have converged on a solution to the problem when a "hit" was recorded for the output unit for every pattern presented during the epoch. A "hit" was defined as output unit activity of 0.9 or greater when the desired output was 1.0, or as output unit activity of 0.1 or less when the desired output was 0.0. Convergence was achieved after 2734 sweeps.

Results

Network Interpretation

The jittered density plots that were presented in Figure 1 were actually plots for each of the 6 hidden units in the converged kinship network. It is apparent from these diagrams that there is marked banding in all six of these units. The interpretation of these bands was accomplished by using descriptive statistics to identify the definite unary and binary features in each of these bands in accordance with previously published methods (Berkeley et al., 1995). The interpretations of the definite features that were found are presented in Table 1.

From Table 1, it can be seen that two of the hidden units are completely devoted to representing which of the six possible family trees is being queried. Each of the six bands observed in hidden unit 0 is composed of stimulus questions about only one of the six families. For example, Band A contains all of the questions about family 3 (see Table 1 for more details). Similarly, each non-zero band in Hidden unit 4 contains questions about a specific family.

The network's discovery that some of the input bits correspond to family name is important, because the remaining hidden units can be used to represent regularities *within* the family tree structure. These regularities can be applied to all six of the family trees. Therefore, the regularities represented in the bands of the remaining four hidden units ignore the first three bits of any input name. Table 1 indicates that all four of the remaining hidden units have bands associated with specific definite features, all of which pertain to structure within the family tree, and which ignore the family feature.

Given the Table 1 account of the bands for the hidden units in this network, how does it solve the kinship problem? Qualitatively speaking, the network's algorithm appears to involve two different tasks. When asked a question like *"Who is person X's mother?"*, the network uses two of its hidden units (i.e., units 0 and 4) to identify the family name that is required in the answer, and to write this family name into the first three output units by activating them appropriately. There does not appear to be much of a mystery about how this "writing" is done: hidden units 0 and 4 act as the bottleneck in a 3-2-3 encoder network. In such a network, the values of 3 input units are compressed into a 2-hidden unit representation; the hidden unit activity is then uncompressed to produce a copy of the input bits into the 3 output units.

The second task for the network is to identify the individual's name, and to "write" this into the remaining six output units. How this task is accomplished is much more mysterious, though, because the kind of definite features listed in Table 1 appear to refer to groups of people, and do *not* refer to individuals. How does the network utilise these general features to represent the identity of the individual whose name is to be "written" into the output units?

The answer to this question is that the network uses *coarse coding* to represent individuals (or more specifically, particular nodes in the family tree) using the Table 1 features. In general, coarse coding means that an individual processor is sensitive to a broad range of features, or at least

539

Table 1: Definite features for each band in each hidden unit. Beside each band label is the number of patterns that belong to that band. Key for definite features: F = father, M = mother, B = brother, Sr = sister, Sn = son, D = Daughter, W = wife, H = husband, Nc = niece, Np = nephew, U = uncle, A = aunt, G = generation, P = person, FG = female of generation, MG = male of generation.

UNIT	BAND	DEFINITE FEATURES
Hidden Unit 0	A N=52	Family 3
	B N=52	Family 1
	C N=52	Family 2
	D N=52	Family 5
	E N=52	Family 6
	F N=52	Family 4
Hidden Unit 1	A N=156	Not A and Not U
	B N=36	(Sn of G01 P001) or (A or U of G11 P001)
	C N=18	(H of FG10 P001) or (W of MG10 P001) or (B of F G10 P010)
	D N=24	(D of G01 P001) or (Sr of G10 P001) or (B of G10 P100)
	E N=30	(D of G01 P010) or (Sr or W or H of G10 P010) or (Sr or W or H of G10 P100)
	F N=12	Sn of G01 P010
	G N=36	(F or M or W or H of G10 P010) or (F or M of G11 P001
Hidden Unit 2	A N=240	No definite features
	B N=12	(M of FG10 P010) or (F of FG10 P010)
	C N=24	(H of FG01 P001) or (W of MG01 P001) or (M or F of MG10 P001)
	D N=12	(F of FG10 P100) or (M of FG10 P100)
	E N=24	(H or W of G01 P010) or (F or M of G10 P010)
Hidden Unit 3	A N=66	Not Np and Not Nc and Not U and Not Sn
	B N=96	Np or Nc or B or Sr or D
	C N=24	(Sn of G01 P001) or (Sn of G10 P010)
	D N=36	(W or H of G01 P010) or (F or M of G10 P010)
	E N=6	B of FG10 P010
	F N=24	(Sn of G01 P010) or (W or H of G10 P001)
	G N=60	(F or M or W or H of P001) or (F or M or W or H of P010)
Hidden Unit 4	A N=156	Family 2 or Family 3 or Family 4
	B N=52	Family 1
	C N=52	Family 6
	D N=52	Family 5
Hidden Unit 5	A N=156	Np or U or B or F or Sn or H
	B N=72	Nc or Sr or D or W
	C N=12	D of G01 P010
	D N=6	Sr of MG10 P010
	E N=12	(W of MG01 P001) or (W of MG P010)
	F N=24	(M of G10 P001) or (M of G11 P001) or (M of G01 P100)
	G N=6	W of MG01 P010
	H N=12	M of G01 P010
	I N=12	A of G11 P001

540

to a broad range of values of an individual feature (e.g., Churchland & Sejnowski, 1992, pp. 178-179). As a result, individual processors are not particularly useful or accurate feature detectors. However, if different processors have overlapping sensitivities, then their outputs can be pooled, which can result in a highly useful and accurate representation of a specific feature. Indeed, the pooling of activities of coarse-coded neurons is the generally accepted account of hyperacuity, in which the accuracy of a perceptual system is substantially greater than the accuracy of any of its individual components (e.g., Churchland & Sejnowski, 1992, pp. 221-233).

In the trained kinship network, each of the four hidden units that is not involved in representing a particular family tree is instead involved with the coarse coding of a particular node within a family tree. The network can pick out an individual node in the family tree by pooling (or combining, or intersecting) the coarse coded representation of the four hidden units.

To illustrate this, let us imagine that for any one of the family trees, we asked the network *"Who is the father of the female Person 2 Generation 2?"* (e.g., for the family tree given in Figure 2, the network would be asked "Who is Victoria's father?"). Ignoring hidden units 0 and 4 (which are concerned with picking out family trees, and not concerned with picking out relations within the tree structure), this query will produce activity that falls in Band A of hidden unit 1, Band B of hidden unit 2, Band D of hidden unit 3, and Band A of hidden unit 5.

Importantly, none of these bands picks out an individual node in the family tree by itself, as is revealed in Table 1. Hidden unit 1 Band A picks out 156 different individuals (across family trees) who are not aunts and not uncles. Hidden unit 2 Band B picks out 12 different individuals who are either the mother or the father of the female person 010 in the second generation. Hidden unit 3 Band D picks out the 36 different individuals who are the wife or husband of person 010 in generation 1, or who are the father or mother of person 010 in generation 2. Band A of hidden unit 5 picks out the 156 different individuals who are either nephews, uncles, brothers, fathers, sons, or husbands (i.e., any individual who is male).

While none of the bands by themselves pick out an individual, the *intersection* of the nodes picked out by each of these four bands selects the appropriate individual within the family tree: the only node pointed to by every one of these bands is the male Person 1 in Generation 1. This is the essence of coarse coding -- the overlap of the receptive fields of broadly tuned detectors can be used to represent finely detailed information.

Likewise, we could ask the network a very similar question: *"Who is the mother of the female Person 2 Generation 2?"* This question will produce the identical band activity in the network as was produced in the previous example, with one exception: it will produce activity in hidden unit 5 that falls in Band H, and not in Band A. Because of this change, the result of intersecting the subsets of nodes pointed to by all the bands changes: now, the only node pointed to by all of the bands is the female Person 1 in Generation 1.

Finally, let us consider the two hidden units that detect which of the 6 family trees is being queried. As was noted earlier, and as can be observed in Table 1, the bands for both of these units have very specific local interpretations. However, it is important to realize that their activities must also be pooled in order to "write" the correct family name into the appropriate output units. For instance, when a network is asked about a relationship for a person in Family 5, this will produce activity that falls in Band D of hidden unit 0 and that falls in Band D of hidden unit 4. Both of these bands must be active for the correct family output to be generated. For instance, if hidden unit 0 was ablated from the network, and the network was asked a question about Family 5, the activity of hidden unit 4 by itself would not produce the correct output in the network, even though the local interpretation of hidden unit 4's activity is "Family 5". For the network, the complete representation of family is a result of a distributed representation -- a combination of hidden unit 0 and hidden unit 4 activities.

Discussion

According to Smolensky (1988), subsymbols are constituents of traditional symbols. "Entities that are typically represented in the symbolic paradigm are typically represented in the subsymbolic paradigm by a large number of subsymbols" (p. 3). As a result, " it is often important to analyze connectionist models at a higher level; to amalgamate, so to speak, the subsymbols into symbols".

The analysis of the kinship network that was reported above is completely consistent with this view. To summarize this analysis, the following discoveries were made. First, the jittered density plots revealed a great deal of structure (i.e., bands). Second, the definite features of most of these bands did *not* correspond to a particular local concept (e.g., an individual's name, or the name of a particular relationship). Instead, the bands usually corresponded to disjunctions of general features that picked out sets of individuals (e.g., Hidden Unit 3 Band D), or in some cases a single feature shared by a large number of individuals (e.g., Hidden Unit 5 Band A's detection of "male"). Third, an account of how the network uses such broadly tuned representations to identify particular individuals relies on the notion of coarse coding. Specifically, the intersection of the sets of individuals represented in all of the bands in which the activity of an input pattern falls picks out a single individual, permitting the network to correctly respond to an input question. In short, the bands illustrated in Figure 1 appear to be acting as subsymbols, and the "symbolic" behavior of the network (i.e., its generation of an individual's name in its output units) depends upon the ability of the output units to combine -- to intersect -- the subsymbolic representations.

Results like these are relevant to the comparison between classical and connectionist architectures. Consider a recent attempt to incorporate situated action theories (including connectionism) into classical cognitive science. Vera and Simon (1993) argued that any situation-action pairing can be represented either as a single production in a production system, or (for complicated situations) as a set of productions. "Productions provide an essentially neutral language for describing the linkages between information and action at any desired (sufficiently high) level of aggregation" (p. 42).

Greeno and Moore (1993) take the middle road in their analysis of ALVINN, suggesting that "some of the processes are symbolic and some are not" (p. 54). Disagreements about what counts as a symbol are clearly at the heart of the debate that Vera and Simon initiated (Vera & Simon, 1994).

The problem with Vera and Simon's notion of what defines a symbol is that it focuses exclusively on the content that the symbol represents, and ignores the operations that are used to manipulate this information (e.g. symbolic concatenation, or the parsing of a string into symbolic constituents). The definition of a subsymbol in Smolensky's (1988) terms not only depends on content (i.e., what subsymbols might represent), but also upon the mechanisms for processing this content. Smolensky (p.3) notes that networks "participate in numerical - not symbolic - computation." Similarly, Fodor and Pylyshyn (1988) have pointed out that "even on the assumption that concepts are distributed over microfeatures, '+ has-a-handle' is not a constituent of CUP in anything like the sense that 'Mary' (the word) is a constituent of (the sentence) 'John loves Mary'" (p. 21). This is exactly the position of connectionist critics who believe that Vera and Simon's (1993) definition of "symbol" is too liberal. For example, Touretzky and Pomerleau (1994) argue against Vera and Simon's symbolic reconstrual of a particular network, ALVINN, by noting that its internal features "are not arbitrarily shaped symbols, and they are not combinatorial. Its hidden unit feature detectors are tuned filters" (p. 348). (But for responses to this view, see also Greeno & Moore, 1993; Vera & Simon, 1994).

The coarse coding interpretation of the kinship network is a case study in the nonsymbolic processing of subsymbols, and thus illustrates an important difference between subsymbolic and symbolic accounts. Importantly, this processing difference holds true for value unit networks even when the content associated with bands is local (i.e., the family units discussed above, or the units of the logic network discussed by (Dawson et al., 1997)). When Berkeley used value unit bands to argue for similarities between symbols and subsymbols, he mistakenly focussed on the content of the bands themselves (Berkeley, 1997). As we have shown above, when one considers value unit banding in terms of represented content as well as the processes required to exploit this content, value unit banding provides an excellent example of Smolensky's (1988) subsymbolic level.

Acknowledgments

This work was supported by an NSERC Research Grant awarded to the second author.

References

Bechtel, W., & Abrahamsen, A. (1991). *Connectionism and the mind.* Cambridge, MA: Basil Blackwell.

Berkeley, I. S. N. (1997). What the #$*%! is a subsymbol? . Paper presented at the 1997 meeting of the Society For Exact Philosophy: Web version available at http://www.ucs.usl.edu/~isb9112/dept/phil341/subsymbol/subsymbol.html.

Berkeley, I. S. N., Dawson, M. R. W., Medler, D. A., Schopflocher, D. P., & Hornsby, L. (1995). Density plots of hidden value unit activations reveal interpretable bands. *Connection Science, 7,* 167-186.

Chambers, J. M., Cleveland, W. S., Kleiner, B., & Tukey, P. A. (1983). *Graphic methods for data analysis.* Belmont, CA: Wadsworth International Group.

Churchland, P. S., & Sejnowski, T. J. (1992). *The computational brain.* Cambridge, MA: MIT Press.

Clark, A. (1993). *Associative engines.* Cambridge, MA: MIT Press.

Dawson, M. R. W. (1998). *Understanding Cognitive Science.* Oxford, UK: Blackwell.

Dawson, M. R. W., Medler, D. A., & Berkeley, I. S. N. (1997). PDP networks can provide models that are not mere implementations of classical theories. *Philosophical Psychology, 10,* 25-40.

Dawson, M. R. W., & Schopflocher, D. P. (1992). Modifying the generalized delta rule to train networks of nonmonotonic processors for pattern classification. *Connection Science, 4,* 19-31.

Fodor, J. A., & Pylyshyn, Z. W. (1988). Connectionism and cognitive architecture. *Cognition, 28,* 3-71.

Greeno, J. G., & Moore, J. L. (1993). Situativity and symbols: Response to Vera and Simon. *Cognitive Science, 17,* 49-59.

Hinton, G. E. (1986). *Learning distributed representations of concepts.* Paper presented at the The 8th Annual Meeting of the Cognitive Science Society, Ann Arbor, MI.

McCaughan, D. B. (1997, June 9-12). *On the properties of periodic perceptrons.* Paper presented at the IEEE/INNS International Conference on Neural Networks (ICNN'97), Houston, TX.

Smolensky, P. (1988). On the proper treatment of connectionism. *Behavioural and Brain Sciences, 11,* 1-74.

Touretzky, D. S., & Pomerleau, D. A. (1994). Reconstructing physical symbol systems. *Cognitive Science, 18,* 345-353.

Vera, A. H., & Simon, H. A. (1993). Situated action: A symbolic interpretation. *Cognitive Science, 17,* 7-48.

Vera, A. H., & Simon, H. A. (1994). Reply to Touretzky and Pomerlau: Reconstructing physical symbol systems. *Cognitive Science, 18,* 355-360.

A Three-Level Model of Comparative Visual Search

Marc Pomplun (marc@psych.utoronto.ca)
University of Toronto, Department of Psychology
100 St. George Street, Toronto, Ontario, Canada M5S 3G3

Helge Ritter (helge@techfak.uni-bielefeld.de)
University of Bielefeld, Collaborative Research Center 360
P.O.Box 10 01 31, 33501 Bielefeld, Germany

Abstract

In the experiments of comparative visual search reported here, each half of a display contains simple geometrical objects of three different colors and forms. The two hemifields are identical except for one mismatch either in color or form. The subject's task is to find this difference. Eye-movement recording yields insight into the interaction of mental processes involved in the completion of this demanding task. We present a hierarchical model of comparative visual search and its implementation as a computer simulation. The evaluation of simulation data shows that this Three-Level Model is able to explain about 98% of the empirical data collected in six different experiments.

Comparative Visual Search

Comparative visual search can be considered a complex variant of the picture-matching paradigm (Humphrey & Lupker, 1993). In picture-matching experiments, subjects are typically presented with pairs of images and have to indicate whether or not they show the same object. In comparative visual search, however, pairs of almost identical item *distributions* are to be compared, requiring subjects to switch between the two images several times before detecting a possible mismatch.

The stimuli in the experiments reported here showed patterns of simple geometrical items on a black background. The items appeared in three different forms (triangles, squares, and circles) and three different colors (fully saturated blue, green, and yellow), each of them covering about 0.7 degrees of visual angle in diameter. The item locations were randomly generated, but avoiding item contiguity as well as item overlap. Each stimulus picture consisted of two hemifields (size 11x16 degrees each) separated by a vertical white line. There were 30 items in each hemifield, which were equally balanced for color and form. The hemifields were translationally identical in the color, the form, and the spatial arrangement of the 30 items - with one exception: There was always a single item that differed from its "twin" in the other hemifield, either in color or in form. The subjects' task was to find this single mismatch. They were to press a mouse button as soon as they detected the mismatch. Eye movements during comparative visual search were measured with the OMNITRACK1 system, which has a temporal resolution of 60 Hz and a spatial precision of about 0.6 degrees.

Sixteen subjects participated in Experiment 1, each of them viewing 50 pictures. Subjects knew that the critical mismatch would be either in form or in color, they did not know, however, when to expect what kind of mismatch. In fact, 25 of the 50 trials contained a difference in form and 25 contained a difference in color. Experiments 2 to 6 differed from Experiment 1 in specific aspects (see Table 1) in order to provide comprehensive data on comparative visual search (cf. Pomplun, 1998; Pomplun et al., to appear).

Table 1: Six different experiments of comparative visual search

Experiment	Subjects	Trials per subject	Description
1	16	50	No information about dimension of mismatch
2	20	60	Subjects know dimension of mismatch in advance
3	16	60	No entropy in irrelevant dimension
4	14	60	Search for a match instead of a mismatch
5	16	50	Mirror symmetry between hemifields
6	16	60	Comparison of item groups of varying size

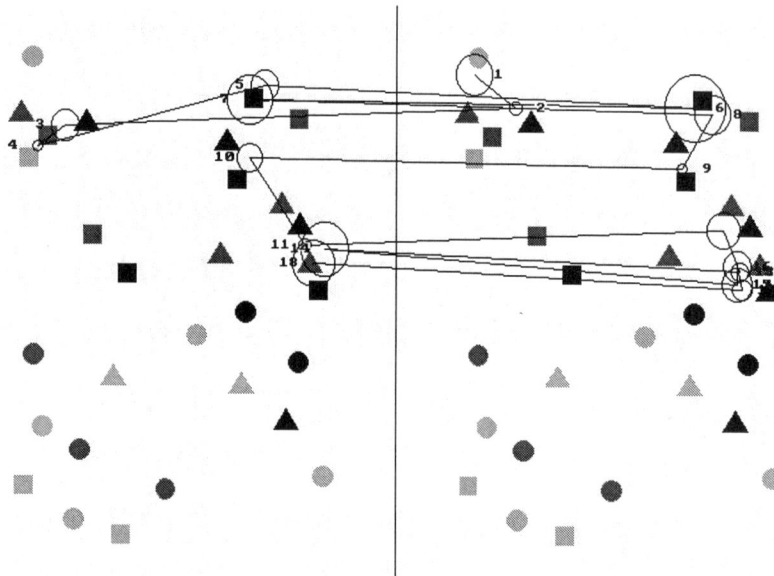

Figure 1: Example stimulus with the plotted visual scan path chosen by one of the subjects. Fixations are numbered; circle size signifies fixation duration.

Figure 1 shows an example stimulus for Experiment 1 with a subject's gaze trajectory superimposed on it. As the example suggests, subjects switch between the hemifields very often and tend to fixate groups of items rather than single items. Moreover, they prefer to use exhaustive, self-avoiding scan paths for optimal search efficiency. For the quantitative analysis of eye movements, the independent variables local item density, local color entropy, and local form entropy were introduced. While local density indicates how closely packed the items are in a certain region of the stimulus, local entropy tells us to what extent different item features are mixed.

There are nine different dependent variables, for example fixation duration (FD) and number of successive fixations within the same hemifield (SF). These variables make it possible to investigate the influence of the local information content on a subject's eye movements. FD, for instance, increases with the local item density at the fixation point, but not with the local entropy values, indicating that short processes like single fixations are controlled by localization processes rather than by identification processes. SF, however, depends on both density and entropy: It decreases with increasing density or entropy, i.e. increasing amount of information, at the first fixation point after switching between hemifields. The quantity of this effect yields data about the capacity of visual working memory.

Taken together, eye-movement analysis in comparative visual search allows us to investigate the interaction of several perceptive and cognitive processes during the completion of a demanding task. In order to test the hypotheses derived from empirical results, a comprehensive model and its computer simulation are required.

The Three-Level Model

The Three-Level Model is not the first attempt to reproduce eye-movement patterns in comparative visual search. A simpler predecessor, the Random-Walk Model (Pomplun, 1998), directly incorporated several empirical eye-movement parameters (e.g. FD and saccade length) and their dependence on local stimulus features. The main shortcoming of the Random-Walk Model turned out to be the exclusion of higher cognitive levels, leading to unstructured search behavior. In contrast, subjects tend to structure their search, e.g. by favoring self-avoiding global scan paths.

Another problem was the direct implementation of statistical properties of empirical eye-movement variables into the model. Although this can tell us to what extent these variables determine the subjects' gaze trajectories, it does not allow us to test our interpretations of empirical findings. It is clearly more comprehensible to use a model that incorporates only these interpretations, i.e. the assumed interaction of several perceptive and cognitive components, instead of the raw empirical data. This model should generate fixations and saccades on the basis of assumed mental processes and their parameters derived from the empirical research. If the model is able to replicate the empirical eye-movement patterns, it supports our interpretations.

The Three-Level Model is a phenomenological approach meeting these requirements. Its structure is essentially motivated by the inadequacy of its predecessor, showing that different levels of processing during comparative visual search have to be distinguished. In addition to the rather schematic processes of perception, memorization, comparison etc., a higher cognitive level must be taken into account, which is responsible for global planning processes.

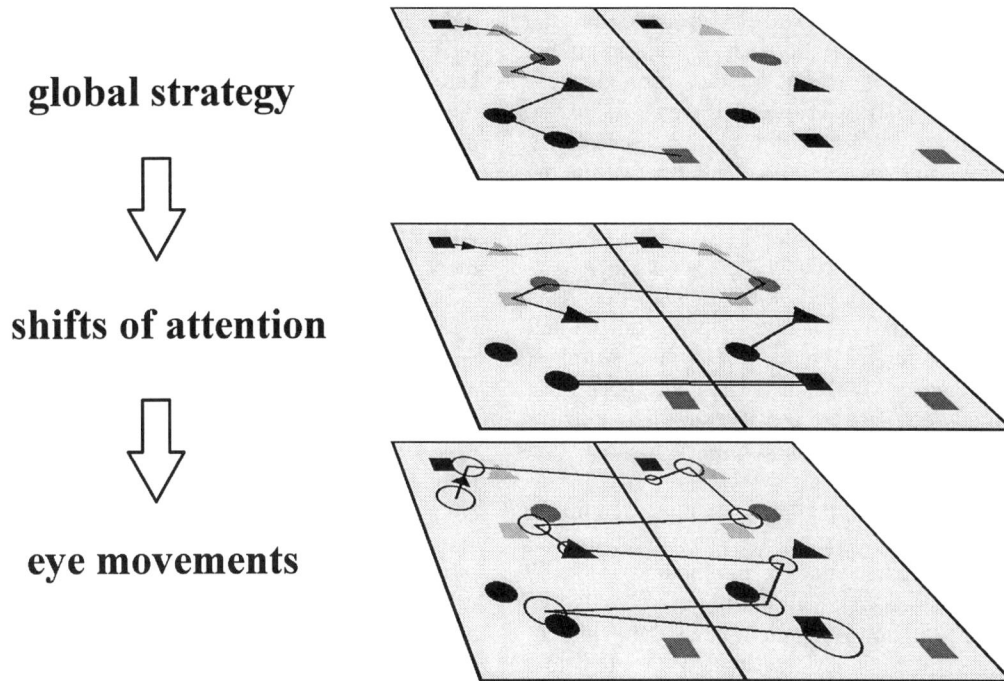

Figure 2: Scheme of the Three-Level Model. The example stimulus contains only eight items per hemifield for the sake of clarity.

Consequently, the Three-Level Model incorporates a vertical organization of mental processes, i.e. a hierarchical scheme of functional modules, better in line with current views about human brain architecture (see, e.g., Velichkovsky, 1990; Gazzaniga, 1997).

A further aspect of the model's vertical organization is the dissociation of eye movements and attention. It is a well-known fact that shifts of attention can be performed without moving the eyes (Wright & Ward, 1994; Tsal, 1983). Accordingly, the finding that subjects fixate groups of items rather than single items might be due to "invisible" shifts of attention: While attention is successively spent to all items in the display, it is not necessary to readjust the foveal gaze position for each of these steps to provide sufficient acuity. This assumption is supported by the results of studies (Pomplun, 1998) investigating the discriminability of the items used in comparative visual search: Reaction time and error rate for detecting color and form features do not vary significantly with retinal eccentricity between 0 and 10 degrees.

Figure 2 presents the structure of the Three-Level Model. On the upper level, the global strategy is planned and realized. Presumably, one of the hemifields is used as a reference with respect to this purpose; hence, the global scan path is plotted only in the left hemifield. The intermediate level is concerned with shifts of attention and processes of memorization and comparison. While the global course of processing is determined by the upper level, the local attentional shifts within and between the hemifields, needed for memorization and comparison of item features, are conducted at this intermediate level. Finally, the lower level is responsible for actually executing eye movements. The gaze position follows the attentional focus, if necessary, to provide appropriate visual acuity for the processing of information. Fixation points are chosen in such a way that the next group of items to be inspected can be memorized or compared employing as few fixations as possible. The integration of the three individual levels into a single model is described in the following sections.

The Upper Level: Global Strategy

The model's global scanning strategy is based on the Color TSP Scanning Model developed in previous research (Pomplun, 1998). It was found that subjects tend to scan a display of randomly distributed items in a "traveling salesman" fashion, i.e. using scan paths of minimal length. Moreover, subjects can take advantage of the items' colors. If the colors are clustered within the display, i.e. if there are separate areas of blue, yellow, and green items, subjects tend to completely scan each of these areas before proceeding to the next one. This strategy reduces their memory load, because keeping in memory the distinction between the items already visited and those still to be processed is easier to achieve on the basis of item clusters than on the basis of single items. No such influence was found for the items' forms, at least for the geometrical shapes used in the present context.

The Color TSP Scanning Model accounts for the influence of both the items' locations and colors on human scanning strategies, and is able to predict approximately 67.5% of gaze transitions between items, if the task is just to look once at each item. This performance is acceptable, given the fact that the maximal predictability within the analyzed scan paths was found to be 71.2% due to the observed, high variability of the data.

Therefore, the Three-Level Model's global strategy for a particular stimulus is determined as the item-by-item path yielded by the Color TSP Scanning Model for the left hemifield. Since most subjects tend to start their search at the top of the display, the items with the uppermost positions in the left hemifield are possible starting points for the path. Among them, the one that allows the construction of the shortest scan path is chosen.

It is implausible, however, that subjects plan a complete item-by-item scan path in advance. They are likely to start searching with a very coarse strategy in mind and to locally refine it to an item-based scan path during task completion. The resulting scan path, however, might be the same in both cases. We do not completely understand the dynamic development of global scan paths so far, but we know some features of the static results. This knowledge constitutes the basis for the Color TSP Scanning Model and we assume it to be transferable to the global strategy level of comparative search.

The Intermediate Level: Shifts of Attention

Attention is modeled in such a way that the sequence of items specified by the global strategy is strictly followed. Starting in a randomly chosen hemifield, attention is shifted between the hemifields during search in order to reproduce processes of memorization and comparison (see below). When the focus of attention reaches the last item in this sequence without the target items being detected, the strategy level calculates a new global scan path starting at the item in focus. A new search cycle begins, guided by the new scan path. This procedure is repeated until the detection of the target.

First, the model memorizes a number of items. This number is limited by the capacity of working memory. As suggested by empirical data, the maximum number of objects to be memorized at a time is set to three for Experiment 1. Moreover, the data show that subjects generally memorize spatially small groups of items at a time. Thus, the Three-Level Model assumes a specific radius of attention. All items to be memorized must fit into a circle of this radius. According to an estimation based on the empirical distance between neighboring fixations, the radius was set to 1.5 degrees of visual angle.

After memorization, the model ideally shifts its attention to the other hemifield, compares the stored information with the corresponding items in the same order, and starts memorizing a new group of items unless the target has been detected (see below). In most cases, however, more than one saccade between the hemifields is necessary to accomplish the comparison of two corresponding sets of items. The results of Experiment 6 indicate that the number of required between-hemifield saccades (BS) strongly depends on the number of memorized features (MF), i.e. the number of different colors plus the number of different forms contained in the set of items that are currently stored in memory. The following equation is a good linear approximation of the empirical findings and is thus incorporated into the model: $BS = 0.23 + 0.39 \cdot MF$.

In accordance with Tsal's (1983) results, the speed of the model's attentional shifts was set to one degree of visual angle per eight milliseconds. As to the dwell time of attention on the items, empirical data suggest that the processing of color may be accomplished faster than the processing of form. Consequently, we can assume attention to be focused on an item for a shorter span during specific color search than during form search or unspecified search. We adjusted the model's span in such a way that the resulting average FD corresponds to the empirical one, which is about 200 ms for all six experiments and can thus be considered a "landmark" of comparative visual search. According to this adjustment, the processing of an item's color requires 70 ms, while the processing of an item's form - or its form and color at the same time - requires 85 ms.

The attentional level is also responsible for target detection. As indicated by the empirical results, the probability of target detection is inversely proportional to working memory load, i.e. the number of item features that are memorized at the same time. In the model, the probability of target detection is defined as an experiment-specific detectability constant divided by the number of memorized features at the moment of comparing the target items to each other. For Experiment 1, the detectability constant equals 2.7.

The Lower Level: Eye Movements

As explained above, a subject's saccades are assumed to follow the shifts of attention in order to provide sufficient visual acuity in the currently attended region of the display. In the simulation, the maximum distance between the gaze position and an item to be processed is given by the radius of attention (see above). If the model directs its attention to an item outside this radius, a saccade is initiated. The target of a saccade is the next item to be inspected, if the next item but one cannot be processed within the same fixation due to a long distance separating the two items. Otherwise, the center point between these two items is chosen as the target of the saccade. Such a behavior is qualitatively indicated by the empirical eye-movement patterns; it is a reasonable way to increase search efficiency.

The model also reproduces saccadic error, i.e. a certain imprecision of saccades, mainly depending on saccade length. Its implementation follows the values reported in literature (e.g. Boff, Kaufman & Thomas, 1986). If, due to saccadic error, a saccade generated by the model does not move the intended item or any of the items to be compared into the radius of attention, another, corrective saccade is executed aiming at the same target position. Between the saccades of this kind, fixations with random durations between 90 and 110 ms are executed. The duration of saccades is modeled as a function of the respective saccade length as it can be found in literature (e.g. Boff, Kaufman & Thomas, 1986).

Finally, the empirical error in the spatial eye-movement measurement is simulated as well. The simulated eye-movement data are randomly shifted in accordance with the average error of the eye-tracker system. This feature of the Three-Level Model improves the comparability between empirical and simulated gaze trajectories.

Results and Discussion

Figure 3a shows an example scan path generated by the model on a stimulus of Experiment 1. A qualitative resemblance to empirical scan paths is obvious: Both a structured search strategy (top-down scanning) and grouping processes (e.g. within fixations number 14 to 16) can be observed. In order to quantitatively compare the simulated with the empirical data, the model was

"presented" with 10.000 randomly generated stimuli for each of the six experiments.

Figure 3b presents the distribution of reaction time (RT) in Experiment 1 for both the subjects and the Three-Level Model. As can clearly be seen, the model correctly replicates a conspicuous plateau in the data between three and ten seconds. While an unstructured (random) search would lead to an exponential decay law, the plateau is indicative of a more structured search strategy. Structured search on a self-avoiding scan path leads to an RT plateau because the number of item pairs to be processed during a search cycle before encountering the target varies homogeneously randomly between 1 and 30. Neither the empirical nor the simulated data show a second plateau corresponding to the second search cycle, which has to be initiated after an unsuccessful first cycle. This result might be due to the variable duration of search cycles.

The diagram in Figure 3c presents FD as a function of local item density at the fixation point in Experiment 1. Both the empirical and the simulated FD increase approximately linearly with item density.

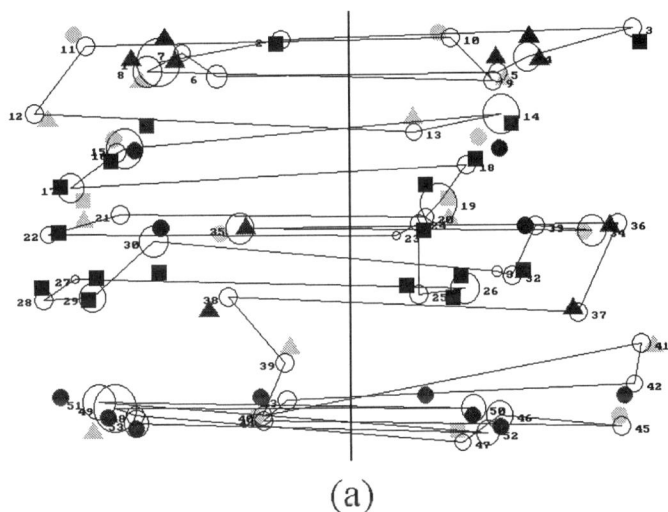

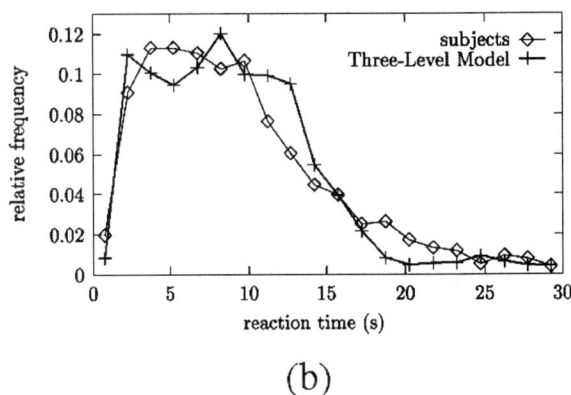

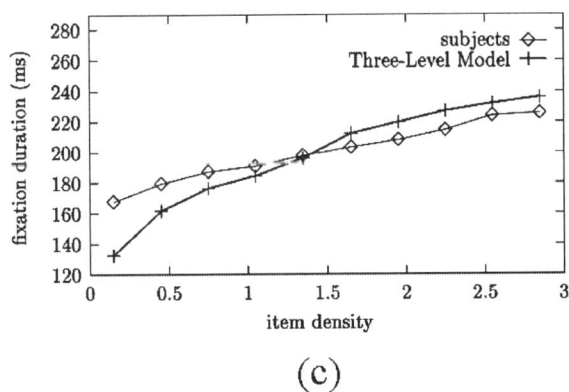

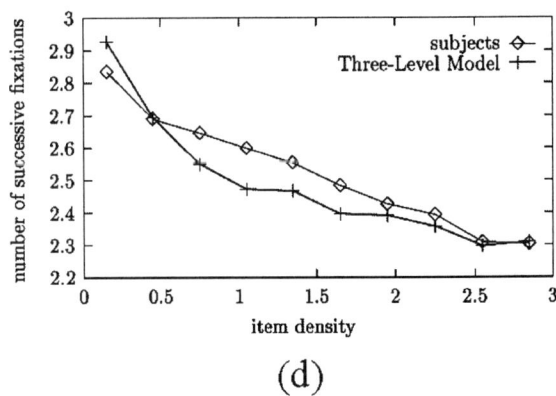

Figure 3: Results of the computer simulation: (a) example scan path, (b) RT histogram,
(c) FD as a function of item density, (d) SF as a function of item density

547

While the two functions match well for density values above 1.5, the model's curve is steeper than the empirical one for lower density. Hence, we can assume a certain minimum duration for empirical fixations in comparative visual search even in stimulus areas providing only little information. This would explain why the duration of empirical fixations does not diminish as strongly at low local information content as could be expected regarding the results of the Three-Level Model.

Figure 3d shows SF for different item densities, measured in Experiment 1. Again, item density was calculated at the first fixation point after every transition between the hemifields. The effect of item density on the model's SF is slightly weaker than on the empirical SF for density values above 0.5, but the situation is reversed for lower density: Here, the effect on the model data is stronger. As observed in the analysis of FD, the empirical effect of low item density is weaker than it could be expected on the basis of the Three-Level Model.

All in all, the data of the Three-Level Model present for most variables a remarkably good correspondence to the empirical data. The model achieves this without directly accessing empirical eye-movement data or using freely adjustable parameters. There are only a few adjustable parameters that were set in accordance with empirical data. The analysis included three independent and nine dependent variables for six different experiments, leading to a total of 54 data distributions and 162 functional relationships to be compared, three of which are outlined above. Only 4 out of these 216 empirical data are not qualitatively reproduced by the model.

In some of the eye-movement variables, however, there are slight deviations of simulated from empirical data for regions of very low local item density. Here, item density exerts a stronger influence on the simulated than on the empirical data. On the one hand, this finding suggests the participation of inhibitory processes in comparative search, which prevent subjects from losing their search strategy, e.g. confusing processed items with unprocessed ones. On the other hand, this discrepancy could be caused by high variability in the empirical eye-movement data, which reduces the strength of measured effects. Future research will investigate to what extent the addition of inhibitory processes and higher variability at different levels can further improve the model.

Summarizing, the results are well in line with the assumed vertical organization of mental processes involved in the completion of comparative visual search. The conclusions drawn from the empirical data are strongly supported, since it is possible to build a model incorporating these conclusions and producing eye-movement patterns that are remarkably similar to the empirical ones. Thus, the Three-Level Model successfully manages to integrate a considerable number of distinct aspects investigated in different experiments into a coherent framework of mental processes and factors underlying comparative visual search.

Acknowledgments

The authors would like to thank Elena Carbone, Hendrik Koesling and Lorenz Sichelschmidt for their contribution to the present work. This research was funded by a grant from the German Science Foundation (DFG CRC 360).

References

Boff, K.R., Kaufman, L., & Thomas, J.P. (Eds.) (1986). *Handbook of Perception and Human Performance, Vol.1: Sensory Processes and Perception.* New York: John Wiley and Sons.

Gazzaniga, M.S. (1997). *The Cognitive Neurosciences.* Cambridge, Mass.: MIT Press.

Humphrey, G.K., & Lupker, S.J. (1993). Codes and operations in picture matching. *Psychological Research, 55,* 237-247.

Pomplun, M. (1998). *Analysis and Models of Eye Movements in Comparative Visual Search.* Göttingen: Cuvillier.

Pomplun, M., Sichelschmidt, L., Wagner, K., Clermont, T., Rickheit, G., & Ritter, H. (to appear). Comparative visual search: A difference that makes a difference. *Cognitive Science.*

Tsal, Y. (1983). Movements of attention across the visual field. *Journal of Experimental Psychology: Human Perception and Performance, 9,* 523-530.

Velichkovsky, B.M. (1990). The vertical dimension of mental functioning. *Psychological Research, 52,* 282-289.

Wright, R.D., & Ward, L.M. (1994). Shifts of visual attention: An historical and methodological overview. *Canadian Journal of Experimental Psychology, 48,* 151-166.

An Entropy Model of Artificial Grammar Learning

Emmanuel M. Pothos (e.pothos@bangor.ac.uk)
School of Psychology, 39 College Road
Bangor, LL57 2DG, UK

Todd M. Bailey (todd@psy.ox.ac.uk)
Department of Experimental Psychology; South Parks Road
Oxford, OX1 3UD, UK

Abstract

We propose a model to characterize the type of knowledge acquired in Artificial Grammar Learning (AGL). In particular, we suggest a way to compute the complexity of different test items in an AGL task, *relative* to the training items, based on the notion of Shannon entropy: The more predictable a test item is from training items, the higher the likelihood that it will be selected as compatible to the training items. Our model is an attempt to formalize some aspects of inductive inference by providing a quantitative measure of the knowledge abstracted by experience. We motivate our particular approach from research in reasoning and categorization, where reduction of entropy has also been seen as a plausible cognitive objective. This may suggest that reducing (Shannon) uncertainty may provide a single explanatory framework for modeling as diverse aspects of cognition, as learning, reasoning, and categorization.

Introduction

Artificial Grammar Learning (henceforth AGL; Reber, 1989; Redington & Chater, 1996) is an experimental paradigm to study inductive inference. An artificial grammar is a set of rules that can be used to generate sequences of symbols. These sequences are labeled grammatical (G) to distinguish them from ungrammatical sequences (NG), which are sequences that violate the rules of the finite state language. Figure 1 shows an example of one such grammar. With this set of rules, while the string MSSV is legal, this would not be the case for string MMSV.

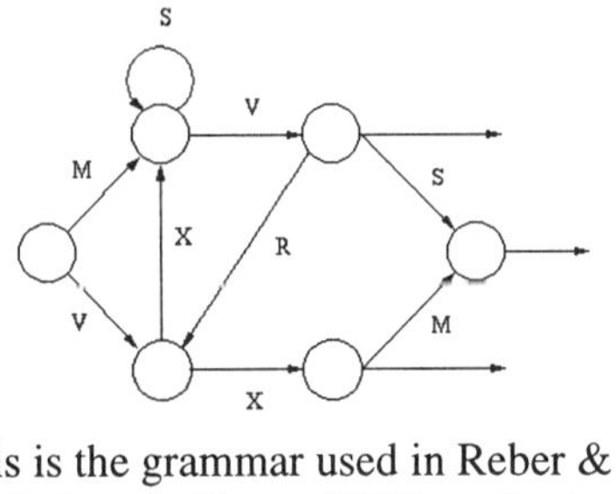

Figure 1: This is the grammar used in Reber & Allen, 1978, as well as Pothos & Chater (1998a, submitted), whose results are analyzed in this work.

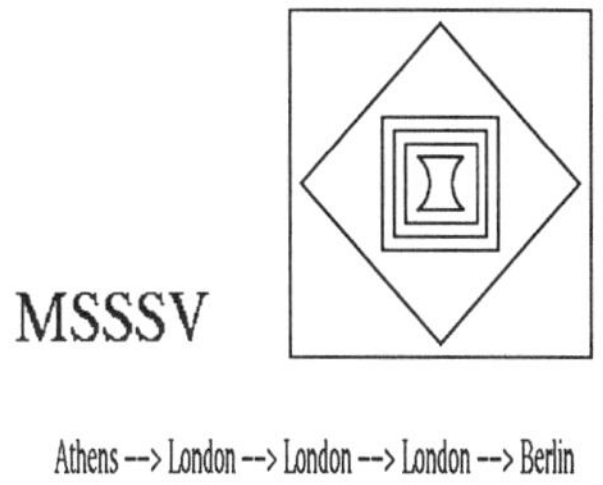

MSSV

Athens --> London --> London --> London --> Berlin

Figure 2: Examples of the types of stimuli used in Pothos & Chater (1998a, submitted). Letter strings were used in Experiment 1, arrangements of shapes in Experiment 2, and city sequences in Experiment 3.

In a typical AGL experiment the sequences of symbols are presented as letter strings, such as MSSV or MSSSV (but these sequences can also be, for example, graphical symbols or musical tones; see Figure 2). Participants are presented with a subset of the G strings in a training phase, and asked to observe them, but no other information is provided either about the nature of the strings, or about the subsequent test phase. After training, they are told that the strings they saw all complied to a set of rules and are then asked to identify the novel G strings in a set that contains both G and NG ones. A robust finding in the literature is that participants can identify the new G strings with above chance accuracy, while in many circumstances they are unable to fully articulate the basis on which they made their decisions.

Pothos and Chater (1998, submitted) provided results indicating that overall performance in AGL does not vary, regardless of whether the stimuli are letter strings (as is the standard condition), embedded shapes, or sequences of cities that correspond to the routes of an airline company (see also Pothos & Bailey, 1997; Bailey & Pothos, 1998).[1]

---

[1] Altmann, Dienes, and Goode (1995), as well as Whittlesea & Wright (1997) also present evidence that AGL-type of learning is possible with stimuli other than the standard AGL strings, but these investigators have not attempted a direct comparison of performance across the different conditions.

Performance was investigated in terms of overall accuracy in detecting G strings, as opposed to NG ones, and also patterns of error across the different sets the test items could be divided into.

The fact that performance does not appear to be different in conditions as different as the ones used by Pothos & Chater (1998a, submitted), appears to suggest that the type of learning observed with AGL reflects general properties of the learning process (that is, properties that do not depend on the particular experimental format used in different situations). Thus, the project of identifying an adequate theory of how participants generalize in an AGL task from training items to the test ones is an important one.

Investigators have proposed accounts of AGL performance in the context of rules, stimulus fragments (parts), or similarity. The original claim by Reber and his colleagues (see Reber, 1989, for a review) has been that participants learn in training something of the abstract, rule structure of the finite state language used to create the stimuli. This view has been corroborated by "transfer" experiments, where the symbols used in training were different from the symbols used in test. However, one has to observe that the actual artificial grammar used in different experiments is an object defined entirely by the experimenter; there is no reason to expect a priori that it will be psychologically relevant. In this sense, Dulany, Carlson, & Dewey's (1984) theory would appear more realistic. These investigators have instead argued that participants acquired "correlated grammars," that is a set of "microrules" which generally approximated the true grammar, but might at the same time include unrepresentative or even wrong rules.

Perruchet and Pacteau (1990) asked what is the minimal type of knowledge that could be used by participants and lead to the observed levels of accuracy. They suggested that all that is learned is information about the legal bigrams, that is which pairs of symbols have been observed in the training items (see Gomez and Schvaneveldt, 1994, and Johnstone & Shanks, in press, for an extension of this approach; Redington & Chater, 1996, for a re-evaluation of these results).

Other theorists suggested that an important factor is similarity: That is, whether a test item is selected as G or NG will depend, to some extent, on how similar it is to training items. Similarity has been operationalized in different ways, for example, as symbol differences between test and training items (Brooks and Vokey, 1991), or empirically computed on the basis of direct similarity data from participants (Bailey & Pothos, 1998; Pothos & Bailey, 1997). A similar approach by Knowlton and Squire (1996) differs in that these investigators computed item similarity in the context of an instantiation of Servan-Schreiber's (1991; see also Servan-Schreiber & Anderson, 1990) "chunking-hypothesis," which is a general theory of learning; the main finding of previous investigations, that similarity is an important predictor of grammaticality performance, has been replicated.

What do all the above theories share? Unfortunately very little. Theories such as the above can be used to make predictions as to which items would be more likely to be selected as G in the test part of an AGL task (that is, predictions about "grammaticality endorsements"). While the actual predictions made by different models in practice often correlate very highly (e.g., see Johnstone & Shanks, in press), there is little theoretical insight as to the extent to which these models are supposed to be mutually exclusive (in terms of representing different hypotheses about learning processes) or not. For example, is the microrules approach (Dulany et al., 1984) the same as Perruchet & Pacteau's (1990) bigram proposal? They both suggest that the knowledge acquired in an AGL task is of the form: If there is an M then a V must follow; although there are some qualifications, these theories would still probably be compatible in terms of their predictions. However, one theory is in terms of rules, while the other might be more reminiscent of exemplar models of classification. Furthermore, if one accepts that AGL is supposed to be a small scale, experimental version of real-life learning tasks (and the utility of investigating AGL would be arguable otherwise), in most of the above cases one cannot readily see how the explanations proposed could generalize to other learning situations, or relate to existing accounts of other aspects of cognition.

Motivation

Our aim in this research is to derive a model of AGL performance from the same *computational* principles that have been seen as relevant in research in reasoning and categorization, namely the assumption that uncertainty is quantified via Shannon entropy and that the cognitive system operates in a way to reduce this uncertainty. Whether the same principles underlie cognitive performance in areas as diverse as reasoning, categorization, and learning is arguable; however, here we suggest as a plausible hypothesis that these processes reflect the same, basic, problem of inductive inference (that is the successful generalization from previously seen instances to future events). We begin with a brief presentation of the models in reasoning and categorization that motivate the present work.

For several years investigators assumed that human reasoning is mediated by the rules of classical logic (e.g., Braine et al., 1995; Evans, 1991). The observation that people often fall prey to an alarmingly large number of logical and probabilistic fallacies, and recent theoretical investigations criticizing the appropriateness of logic for everyday reasoning (Chater & Oaksford, 1993), have led theorists to pursue alternative approaches. The Wason selection task is a simple problem where people are asked to examine whether a conditional rule is true or false, by selecting among a set of cards (the cards are labeled with one clause of the conditional, and contain hidden information about the other clause of the conditional). Oaksford & Chater (1994) suggest that people select these cards that minimize the expected *uncertainty* in deciding whether the rule is true or not. Uncertainty is quantified using the notion of Shannon entropy: If there are N events, that can occur with probability p_i, then the entropy in trying to guess which one will actually occur is given by

$$entropy = -\sum_{i=1}^{N} p_1 \log(p_i).$$ As will become clear in the

presentation of our model of AGL performance to follow, we suggest that the items people will be selecting as G in the test part of an AGL task, are the ones that are most predictable in terms of the training items. That is, we predict that the strings selected as G are the ones that minimize the entropy of specifying them, relative to the training items. This is a strategy very similar to Oaksford and Chater's (1994), who claimed that people select these cards that minimize the entropy of selecting the right hypothesis.

In categorization, Pothos & Chater (1998b) suggested a model whereby people's classifications on a set of items were such so as to reduce the *description length* (used here in a technical sense) of the items as much as possible. In other words, categories were seen as a means to simplify the description of a set of items as much as possible. Pothos (1998) illustrated that the mathematical framework of Pothos & Chater (1998) is equivalent to an entropy minimization one: That is, the preferred classification of a set of items is assumed to be the one that reduces the uncertainty in predicting the similarity structure of these items.

The very brief exposition above can only provide a presentation of the models in question at a very crude, qualitative level. At such a level, it might appear that the theoretical coherence afforded by terms like "reduction of uncertainty," or "entropy minimization," is only an artifact of the fact that the mathematical specification of models based on such notions is relatively loose; so that conceptually different models can still be instantiated in a way that would appear consistent with an entropy maximization process. This is far from true. Although there can be several different entropy maximization procedures to address the same cognitive problem, such alternatives still need to share the same foundation (a specific use of probabilities, quantifying uncertainty in a certain way, etc.), that would make them much more similar, as a class of models, compared to others.

An entropy model of AGL

This model is an attempt to quantify what exactly is learned in training in an AGL task. In such a task the test items are evaluated in terms of whether they are compatible with the training items or not. What can this mean? We suggest that each test item is given a complexity measure according to how "specifiable" it is from training items. This complexity measure is computed by dividing the item into parts, and seeing how "determinable" the continuation to each of these parts can be on the basis of information from training.

First, each test string is broken down into all constituent fragments, "anchored" at the beginning or end of the string. Letting symbols "b" and "e" stand for the beginning and the end of a string, test string MSV would be broken into [b, bM, bMS, bMSV] in the forward direction, and fragments [e, Ve, SVe, MSVe] in the reverse. We consider these fragments as relevant, on the simple assumption that symbol sequences are likely to be parsed/ encoded by the cognitive system in a simple forward and reverse direction. For a given test item, we ask what is the expected difficulty of specifying a continuation, given what one has seen in training, and in this way we compute the S-measures for each fragment (for the reverse chunks, we ask how likely a given symbol is to precede a particular fragment; for simplicity, we use continuation to refer to both, when discussing general properties of the S-measures). In particular, if there are N possibilities for a continuation, each occurring with a probability p_i from training, then the entropy associated with specifying the next symbol in the string is given by $S(fragment) = -\sum_{i=1}^{N} p_1 \log(p_i).$

For example, suppose that the training items consist only of strings MSSV, MSSSSX, and MSVRV. When we see MSV in test, then to compute the overall complexity of this string, relative to the training items, we need to consider, first, $S(b)$: How hard is it to guess what the next symbol is, for the first symbol in a string, from training? All training strings start with an M, thus, we have $S(b) = 0$. Likewise, $S(bM) = 0$. To compute $S(bMS)$, note that after fragment bMS in training, we have an "S" continuation with a probability of 2/3 and a "V" one (the observed continuation in test) with probability 1/3. Thus, $S(bMS)$ would be -1/3log(1/3)-2/3log(2/3). Taking an example for the reverse S measures, $S(e)$ would be computed by noting that in the training items an end symbol is preceded by a V symbol with a probability 2/3 and an X one with probability 1/3.

In cases where there is a novel symbol in a test item, or a fragment not seen in training, we compute S by assuming that all the possible symbols are equiprobable. In the above example, if we had a test string QM, then forward $S(bQ)$ would be given by $-5 \times \frac{1}{5} \log_2 \frac{1}{5}$ (we have five possible symbols, M, S, V, Q, and "e," and since the fragment bQ has not been observed in training, all possible continuations are equiprobable). The underlying hypothesis is that if there is no information from training about a given sequence, the cognitive system will operate as if all possible continuations were equiprobable.

How would the S-measures corresponding to the different fragments lead to an overall complexity measure for a string? We suggest that all forward and reverse S-measures are averaged, and that the resulting number reflects how familiar a given test string is relative to the training items. Without going into too much detail here, using an average, instead of some other way of combining S-measures, has the advantage that the computation of the overall complexity of an item is more balanced across items, and also for individual items that contain a single violation of regularity relative to the training items (e.g., in the above example, think of an item like VMSV), the effect of this single violation is moderated by the presence of regular fragments (regular with respect to the training items).

Investigation of the model

The limited exposition of our model cannot address all the relevant theoretical issues. In this section, we aim to alleviate some concerns by fitting the model to results from three AGL experiments reported in Pothos & Chater (1998a, submitted). They utilized the Reber & Allen (1967) grammar to create AGL stimuli that were standard letter strings (Experiment 1), nested arrangements of geometric shapes (Experiment 2), and sequences of cities (Experiment 3). The average S-measure reflects how specifiable each test item is on the basis of training items. That is, a high S-measure indicates that a particular string is not "intuitive" relative to the training items. Thus, we predicted that the extent to which different test items would be selected as G would negatively correlate with the average S-measure of these items.

Table 1 shows the correlation of the average S-measure for the test strings of the Reber and Allen (1967) grammar, with the probability that these strings would be selected as G in each of the three experiments of Pothos & Chater (1998a, submitted). That is, for each of these experiments, we averaged the total number of times (across participants) each test items was selected as G. Considering the low number of participants in these studies (ten participants each) compared to the high number of endorsements we are trying to predict (50 items), the fact that all the correlations are in the predicted direction and highly significant provides important support for our model.

Table 1: The correlation of the average S-measure of individual test items complexity, with the number of times each test item (50 test items, in total) was selected as G in the three experiments of Pothos & Chater (1998a, submitted).

	Letters	Shapes	Cities
Average S-measures			
Pearson Correlation	-0.577	-0.613	-0.381
p-value	0.000	0.000	0.006

This work is not meant to disconfirm any of the existing models. Our objective has been to propose a computation of "what is learned in AGL," in a wider theoretical context, that is, in a way that relates to research in other areas of cognition. That is, we hope to provide a model of AGL that would still capture as much of the intuitions seen as relevant in other AGL accounts, while the model itself would be inspired from more general computational principles. In this respect, finding that the average S-measure correlates with the predictions from other models of AGL performance, would provide important support for our approach.

Table 2 shows the correlation of the S-measure with grammaticality, global associative chunk strength (Knowlton and Squire, 1996), associative chunk strength at anchor positions (same as before, but computed only for fragments at the beginning or end of a string, that is the anchor positions), novel chunk strength (the fraction of novel fragments in a string; see Meulemans & Van der Linden, 1997), and anchor novel chunk strength (same as before, but one looks at the proportion of novel fragments at the anchor positions). Table 2 reveals that our measure correlates very highly with almost all the above measures.

Table 2: The number in each of the rows represents the correlation of the Average S-measure with the AGL performance measure in each row. The "*" flags correlations significant at the 0.01 level or less. All correlations were computed over the 50 test items in the Pothos & Chater (1998a, submitted), work.

	Average S-measure	
grammaticality	-.630	*
global chunk strength	-.436	*
anchor chunk strength	-.515	*
novel chunk strength	-.378	*
anchor novel chunk strength	-0.022	

Discussion and future direction

We have tried to quantify the amount of information that is available in the test part on an AGL task from training. In this respect, we proposed the average S-measure which, for each string, provides us with a number reflecting how specifiable the string is (that is, how easily it can be determined), given a particular set of training items.

To examine our model, we computed average S-measures for all the strings of the Reber and Allen (1978) grammar and showed that these correlated significantly with grammaticality endorsements from three AGL experiments reported by Pothos & Chater (1998a, submitted). Moreover, we showed that the average S-measure captures many aspects of previously proposed measures of AGL performance, that made different assumptions about what is learned in an AGL task.

The underlying motivation for the present work was to provide a quantitative measure of knowledge acquired in an AGL task, in a broad theoretical context. Thus, our model can be seen as having a foundation very similar to Oaksford & Chater's (1994) model of reasoning in the selection task, and Pothos & Chater's (1998) model of categorization. In all these cases, it is assumed that reduction in uncertainty (quantified via Shannon's entropy) is the objective for the cognitive system in processing information about the world. Further fleshing out the formal relation between such seemingly diverse aspects of cognition is an important future objective.

The average S-measure can be employed to model on-line generalization, that is generalization patterns from individual items, in the sense that the first item can be used to compute the complexity of the second one, the first and second together, the complexity of the third one, and so forth. Moreover, one can manipulate the "total information available from training," and thus make predictions about the overall level of grammaticality accuracy in an AGL task, when the actual number of training items presented is always the same. Both the above considerations represent possible simple extensions of the present work, to further test the psychological plausibility of the average S-measure of generalization.

Acknowledgments

We would like to thank Nick Chater for valuable comments on this work. Emmanuel Pothos was supported by ERSC grant ref. R000222655 to Emmanuel Pothos and Nick Chater. Todd Bailey was supported by grants from the McDonnell-Pew Center for Cognitive Neuroscience, Oxford, and a grant to Kim Plunkett from the Biotechnology and Biological Sciences Research Council, UK

References

Altmann, G. T. M., Dienes, Z. & Goode, A. (1995). Modality Independence of Implicitly Learned Grammatical Knowledge. *Journal of Experimental Psychology: Learning, Memory and Cognition, 21*, 899-912.

Bailey, T. M. & Pothos, E. M. (1998). Unconfounding similarity and rules in artificial grammar learning. In *Proceedings of the Twentieth Annual Conference of the Cognitive Science Society*, 96-101, LEA: Mahwah, NJ.

Braine, M. D. S., O'Brien, D. P., Noveck, I. A., Samuels, M. C., Lea, B. L., Fisch, S. M., Yang Y. (1995). Predicting Intermediate and Multiple Conclusions in Propositional Logic Inference Problems: Further Evidence for a Mental Logic. *Journal of Experimental Psychology: General, 124*, 263-292.

Brooks, L. R. & Vokey, J. R. (1991). Abstract Analogies and Abstracted Grammars: Comments on Reber (1989) and Mathews et al. (1989). *Journal of Experimental Psychology: Learning, Memory and Cognition, 120*, 316-323.

Chater, N., & Oaksford, M. (1993). Logicism, mental models and everyday reasoning. *Mind and Language, 8*, 72-89.

Evans, St B. T. J. (1991). Theories of Human Reasoning: The Fragmented State of the Art. *Theory & Psychology, 1*, 83-105.

Gomez, R. L., Schvaneveldt, R. W. (1994). What Is Learned From Artificial Grammars? Transfer Tests of Simple Association. *Journal of Experimental Psychology: Learning, Memory and Cognition, 20*, 396-410.

Johnstone, T. & Shanks, D. (submitted). Two Mechanisms in Implicit Grammar Learning? Comment on Meulemans and Van der Linden (1997).

Knowlton, B. J., Squire, L. R. (1996), Artificial Grammar Learning Depends on Implicit Acquisition of Both Abstract and Exemplar-Specific Information. *Journal of Experimental Psychology: Learning, Memory and Cognition, 22*, 169-181.

Meulemans, T., Van der Linden, M. (1997). Associative Chunk Strength in Artificial Grammar Learning. *Journal of Experimental Psychology: Learning, Memory, and Cognition, 23*, 1007-1028.

Oaksford, M. & Chater, N. (1994). A Rational Analysis of the Selection Task as Optimal Data Selection. *Psychological Review, 101*, 608-631.

Perruchet, P. & Pacteau, C. (1990). Synthetic Grammar Learning, Implicit Rule Abstraction or Explicit Fragmentary Knowledge?. *Journal of Experimental Psychology, General, 119*, 264-275.

Pothos, E. M. (1998) Aspects of Generalisation. Unpublished D.Phil thesis, University of Oxford.

Pothos, E. M. & Bailey, T. M. (1997). Rules vs. Similarity in Artificial Grammar Learning. In *Proceedings of the Similarity and Categorisation Workshop 97*, 197-203, University of Edinburgh.

Pothos, E. M. & Chater, N. (1998a). Generality of the Abstraction Mechanisms in Artificial Grammar Learning. In *Proceedings of the Twentieth Annual Conference of the Cognitive Science Society*, 854-858, LEA: Mahwah, NJ.

Pothos, E. M. & Chater, N. (1998b). Rational Categories. In *Proceedings of the Twentieth Annual Conference of the Cognitive Science Society*, 848-853, LEA: Mahwah, NJ.

Pothos, E. M. & Chater, N. (submitted). Generality of the Abstraction Mechanisms in Artificial Grammar Learning. *European Journal of Cognitive Psychology*.

Reber, A. S. (1989). Implicit Learning and Tacit Knowledge. *Journal of Experimental Psychology, General, 118*, 219-235.

Reber, A. S., Allen R. (1978). Analogic and abstraction strategies in synthetic grammar learning, A functional interpretation. *Cognition, 6*, 189-221.

Redington, M. & Chater, N. (1996). Transfer in Artificial Grammar Learning, Methodological Issues and Theoretical Implications. *Journal of Experimental Psychology, General, 125*, 123-138.

Servan-Schreiber, E. (1991). *The Competitive Chunking Theory: Models of Perception, Learning, and Memory*. Doctoral Dissertation, Department of Psychology, Carnegie-Mellon University.

Servan-Schreiber, E., Anderson, J. R. (1990). Learning Artificial Grammars With Competitive Chunking. *Journal of Experimental Psychology: Learning Memory and Cognition, 16,* 592-608.

Whittlesea, B. W. & Wright, R. L. (1997). Implicit (and explicit) learning, Acting adaptively without knowing the consequences. *Journal of Experimental Psychology, Learning, Memory and Cognition, 23,* 181-200.

Conceptual representations of exceptions and atypical exemplars: They're not the same thing.

Sandeep Prasada
Dartmouth College

Introduction

Counterexamples play a central role in psychological and philosophical research on concepts. The main argument against the classical theory of concepts is that it is not possible to come up with a set of necessary and sufficient properties that are true of all exemplars of a given kind of thing. Thus, the fact that a dog with three legs is still a dog is taken to be a crucial counterexample to any proposal that states that our concept of dogs is such that having 4 legs is considered to be a necessary property of dogs. Similarly, the fact that a dog that doesn't bark (because he has laryngitis or defective vocal chords) is nevertheless conceived of as being a dog, is taken to be a crucial counterexample to any proposal that states that our concept of dogs is such that the ability to bark is considered to be a necessary property of dogs. As is well known, we can continue in this manner to provide counterexamples against just about any property that is proposed to be conceived of as being a necessary property of a given kind of thing. This type of reasoning constitutes the central argument against the classical theory of concepts, and an important motivation for considering concepts to be prototypes. Counterexamples of the sort considered above have been interpreted as demonstrating that our concept of the kind dog could not represent having four legs as a necessary property of being a dog. Instead, it is posited that four-leggedness is represented as merely a typical property of dogs.

There is an alternative to this conclusion, however. The alternative is that our conceptual systems do, in fact, represent a necessary relationship as holding between being a dog and being four-legged, and that our conceptual systems treat instances of three-legged dogs as exceptions. It is an *empirical* question whether our conceptual systems treat three-legged dogs as evidence that having four legs is merely a typical property of dogs and thus do not represent a necessary connection as holding between the property of being a dog and having four legs, or if our conceptual systems represent a necessary connection between being a dog and having four legs and treat three-legged dogs as exceptions. In this paper, I present evidence that counter to the assumption implicit in all previous research on conceptual representation, our conceptual systems do, in fact, treat certain entities as exceptional exemplars. The notion of an exception requires, (i) that we represent a rule-like or necessary connections between being a certain kind of thing and some property, and (ii) that an exemplar can be considered to be that kind of thing, but nevertheless lack the property in question.

Evidence for necessary connections

In this section, I provide three types of evidence that show that we represent necessary connections between the property of being a dog and having four legs, and the property of being a chair and for being for sitting on, despite the fact that we acknowledge that there are dogs that have three legs and that there are chairs that are not for sitting on. As will be obvious, this type of evidence is available for indefinitely many types of things and properties and I've chosen these examples simply for the purpose of illustration.

Generic statements

Some evidence that our common sense conception of things like dogs and chairs contain necessary connections to certain properties comes from generic sentences such as (1) and (2).[1] Sentences such as these can be used to express thoughts that have contents such as those in (1') and (2'), which clearly display the necessary connection we conceive of as holding between being a dog and having four legs, or being a chair and being for sitting on. These properties are seen as being present simply in virtue of the things being the kinds of things they are. This contrasts sharply with cases in which the relationship between being an instance of a certain kind and a property is conceived of as being contingent, even if the property is typical of the kind of thing in question. Thus, we cannot express the latter type of relationship in a parallel form (3)-(4), and insofar as we may be able to use such sentences, they cannot be understood as expressing the thoughts given in (3') and (4'), which we regard to be false.

(1) Chairs are for sitting on.
(1') Chairs, in virtue of being chairs, are for sitting on.
(2) Dogs are four-legged.
(2') Dogs, in virtue of being dogs, have four legs.
(3) *Chairs are wooden.
(3') *Chairs, in virtue of being chairs, are wooden.
(4) *Dogs are brown.
(4') *Dogs, in virtue of being dogs, are brown.

Thus, generic sentences such as (1) and (2) provide evidence that our conceptual systems represent a rule-like or

---

[1] It is important to note that not all sentences with a surface form of this sort can express the type of relationship discussed here. It is not important for our purposes that there is not a one-to-one correspondence between surface forms of this sort and the types of meanings they express.

555

necessary relation between being a certain kind of thing and having certain properties.

Explanation of one property in terms of the other

A second type of evidence that shows that we conceive of there being necessary relations between being a certain kind of thing and having certain properties is that we can explain the presence of certain properties in an entity by citing the kind of thing the entity is. For example, if someone points at an entity and asks a question such as (5) or (6), we find that an answer such as (7) or (8) is appropriate and explanatory. Note that though the questions may strike us as unlikely, or a little odd, given that they are asked, there are certain answers that we find appropriate and explanatory, and others not. Only those answers that cite properties or kinds which are conceived of as being in a necessary relationship to the explananda are judged to be appropriate or explanatory, whereas those answers that cite properties or kinds that are in a contingent relation to the explananda are not judged to be appropriate or explanatory (7a-b) (8a-b), even if they happen to mention a property that is typically true of the kind of thing in question.

(5) Why can you sit on that ? (pointing to a chair)
(6) Why does that have four legs? (pointing to a dog)
(7) Because it is a chair.
(7a) Because it is brown.
(7b) Because it is furniture.
(8) Because it is a dog.
(8a) Because it is brown.
(8b) Because it is an animal.

These facts show that we conceive of there being necessary relations between being a certain kind of thing and having certain properties and thus can explain the presence of one property by citing the other property. In cases in which we don't understand there to be a necessary relation between properties, the presence of one property cannot be accounted for by reference to the other property.

Redundancy effects in prenominal modification

A third type of evidence that shows that our conceptual systems represent certain necessary relations between being a certain kind of thing and having certain properties comes from redundancy effects in prenominal modification. In cases in which we conceive of there being a necessary connection between a certain kind of thing and a certain property, prenominal modification of the kind by the property engenders a feeling of redundancy (9). In contrast, if the modification is by a property that is conceived of as being only contingently related to the kind, then no redundancy is engendered, even if the property is typical of the kind in question (10).

(9) A four-legged dog is barking.
(10) A red apple is on the table.

There is a sense in which our conceptual systems assume that dogs have four legs, and therefore, when *dog* is modified by *four-legged*, there is a feeling of redundancy. In contrast, our conceptual systems do not assume that being an apple implies being red, even if most apples are red, and therefore, there is no feeling of redundancy when *apple* is modified by *red*. These facts provide further evidence that our conceptual systems distinguish properties that are conceived of as being necessary consequences of being a certain kind of thing from properties which are conceived of as merely being typical of that kind of thing.

The need for exceptions

We seem to have arrived at a paradox. On the one hand, there is evidence that we conceive of there being necessary connections between things like being a dog and being four-legged, while on the other hand, we find no problem in conceiving of the existence of three-legged dogs. There are only three ways out of the paradox. We could reject the validity of the first conclusion, the validity of the second conclusion, or find some way in which we can hold onto both conclusions. There does not seem to be any compelling reason to question the validity of either conclusion. They both rest on strong intuitions. Thus, what is called for is a way of thinking about the data that would allow us to hold onto both conclusions while avoiding the contradiction that seems inherent in them.

One way in which this may be possible is to regard cases in which a putatively necessary property is lacking as being *exceptions* in some sense. Crucially, the exceptional status of these cases, and thus the lack of the putatively necessary property, must derive from something *other* than their being the kind of thing they are. If this were the case, one could maintain that we do conceive of there being a necessary connection between being a dog and having four legs, but that we are willing to allow that dogs might not be four-legged for reasons other than their being dogs. As such, no contradiction would arise as we would not be predicating incompatible properties of dogs. Four-leggedness would be predicated of dogs in virtue of their being dogs, whereas three-leggedness would be predicated of dogs in virtue of something other than their being dogs.

This is, in fact, the solution that our cognitive systems seem to embody. Evidence for this solution comes from sentences such as (11) and (12).

(11) Dogs are four legged, but/however/though these dogs have only three legs.
(11') # Dogs are four legged, and these dogs have only three legs.
(12) Chairs are for sitting on, but/however/though you can't sit on these chairs.
(12') # Chairs are for sitting on, and you can't sit on these chairs.

These sentences transparently express the fact that we conceive of certain properties as following as a consequence of being a certain kind of thing while recognizing that there could be some instances of the kind in question that are exceptional in that they fail to have the property in question.

The fact that the instances that lack the necessary property are conceived of as being exceptional is highlighted by the fact that the use of a conjunction such as *and*, which does not confer exceptional status to that which follows, engenders a sense of oddness and contradictoriness to the sentences (11') - (12'). Notice further that the existence of the exceptional exemplars cries out for explanation. We feel that sentences such as (11) and (12) should be followed by an explanation which gives the reason why the exceptional exemplars lack the property they should have in virtue of being the kinds of things they are. Importantly, neither three-leggedness, nor lack of sitability, unlike the four-leggedness of a dog or sitability of a chair, can be explained by citing the kind of thing the entity is. Thus, whereas (7) and (8) are appropriate answers to (5) and (6), (7) and (8) are not appropriate answers to (13) and (14). Instead, the lack of sitability in a chair or lack of four-leggedness in a dog must be accounted for by citing a reason other than the entity being a chair or a dog.

(13) Why can't you sit on that? (pointing to a chair you can't sit on)
(14) Why does that have three legs? (pointing to a dog with three legs)

Because the lack of four-leggedness of a dog, or the lack of sitability of a chair, are not conceived of as being due to the entity being a dog or a chair, but are conceived of as being due to something other than the entity being a dog or a chair, the three-leggedness of some dogs, or the lack of sitability of some chairs is not treated by our conceptual systems as providing evidence against the conception that dogs, in virtue of being dogs, have four legs, or that chairs, in virtue of being chairs, are for sitting on.

These facts show that the apparent paradox that we started with is exactly that. Our conceptual systems do, in fact, represent necessary connections between being a certain kind of thing and having some property, however, they are also capable of treating instances which lack the necessary property in question as exceptions because the lack of the necessary property in these exemplars is conceived of as being due to factors other than the exemplars being the kinds of things they are.

Implications

Methodological lesson

The foregoing suggests that the common method of trying to determine the necessary and contingent properties of our conceptual representations on the basis of the conceivability of some property being true and not true of exemplars in all possible worlds is flawed because it assumes, wrongly, that there are not other factors that might be responsible for the lack of a property that is conceived of, and represented as, being a necessary consequence of being an instance of that kind of thing. What such intuitions show is that there are relatively few limits on what is conceivable, however, such intuitions do not display the manner in which things becomes conceivable. The four-

leggedness of dogs is conceived of as simply being true in virtue of our conception of dogs, whereas a property such as the three-leggedness of dogs is conceiv*able* only by imagining some factor or set of factors other than the entity being a dog in virtue of which the dog is three-legged. Conceivability, as such, is too weak a tool to investigate the structure of our conceptual representations because it does not distinguish what is conceived *simpliciter* from that which is conceivable but instead involves combination. Conceivability does, of course, put important constraints on theories of conceptual combination.

Necessary connections are not established on purely empirical grounds

If, as has been argued, we do establish necessary connections between concepts, it raises the important question of how it is that we do this. Clearly such connections cannot be established on empirical or statistical grounds because, as is well known, empirical and statistical evidence can never underwrite conclusions of necessity. Before speculating about how we do establish necessary connections, I want to present explicit evidence that the necessary connections that are established between concepts are not established upon the basis of solely statistical or empirical means.

The contrast between sentences such as (2) (reproduced here as (15)) and (16) helps make this point. As discussed earlier, (15) expresses what we conceive of as being a true generic fact about dogs. Because the statement expresses what we conceive of as a necessary connection between the property of being a dog and being four-legged, the fact that there exist some dogs that are three-legged for reasons other than being a dog does not render the sentence false. In contrast, (16) is an extensional statement that expresses a particular fact about dogs which is false because not all dogs are four legged, some are three-legged. Furthermore, even if there did not exist any three-legged dogs, it would be impossible to make a statement such as (16) without qualifying it in some way, for example, as in (16'). This is because (16), unlike, (15) is a statement about the extension of the concept DOG, and one could never make a statement about every instance of a given kind as there are, in principle, indefinitely many instances of any kind.

(15) Dogs are four-legged.
(16) All dogs are four-legged.
(16') All the dogs, observed up to this point in history, have been four-legged.
(17) Ninety percent of dogs are four-legged.

Similarly, all explicitly statistical quantified statements such as (17), can only be made with qualification, as they can only be truthfully asserted of some particular set of exemplars which exist at some particular sets of times and places. In contrast, (15) can be said without qualification, because it expresses a fact that we conceive of being true of dogs in simply in virtue of being dogs, rather than a fact about some particular set of exemplars which we conceive of as having the properties they do in virtue of being dogs as well as all kinds of contingent facts that are particular to the instances in question. This contrast between our understanding of generic and extensional sentences provides explicit evidence that the basis upon which we establish the truth of the necessary connections expressed in generic sentences cannot be derived solely by statistical and empirical means.

A formal system of knowledge for common sense conception?

So, how is it that necessary connections between concepts are established? It seems that there are two options left open. The first is that the content of all of these connections (e.g. between dogs and four-leggedness, chairs and sitabiliity, etc....) are innately specified and somehow triggered given appropriate experience. This, however, strikes me as extremely implausible. Alternatively, what may be innate is a *formal* system of knowledge in terms of which we form common sense concepts by analyzing the things around us in terms of this formal system of knowledge. For the purposes of illustrating how such a system might work, let us consider how the formal system of knowledge that constitutes knowledge of Euclidean geometry is put to use in reasoning about actual things in the world. If we are standing at the corner of a rectangular park, we know that it will *necessarily* take us less time to traverse the diagonal to get to the opposite corner than walking along two sides of the park (assuming that we walk at a constant speed and the park is relatively even). Here we have an example of how formal knowledge, which is not about parks or walking, can nevertheless be applied to establish a fact about parks and walking. Furthermore, the necessity that we know to hold concerning the relative sizes of the sides of geometrical figures is transferred to our reasonings about actual distances of paths in parks as long as we are justified in applying the formal system of knowledge to these aspects of the real world. Thus we see that it is possible for us to make statements about what is *necessarily* the case, either when we are making statements within a purely formal system of knowledge, or when we are using that system to reason about certain aspects of the world. This suggests that just as there is a formal system of knowledge that underlies our intuitions and reasonings about space, there may also be a formal system of knowledge that underlies our abilities to form common sense conceptions of kinds of things and allows us to come to conclusions about what kinds of connections between concepts are necessarily rather than contingently true. Furthermore, if it is the case that there is a formal system of knowledge that underlies our common sense conception of things and our conceptions of necessary connections between concepts, then we also have a way of accounting for the normative aspect of the necessary connections -- (e.g. dogs, in virtue of being dogs, *should* have four legs). Formal systems of knowledge such as those that constitute our knowledge of logic or language specify how we *should* reason or structure our sentences, not how we actually do, as many other factors affects performance. For a sketch of

what this formal system of knowledge might look like see Prasada (1998a; 1998b).

Representing necessary connections/exceptions.

The facts discussed above suggest that we represent a rule-like necessary relation between being a dog and having four legs, but allow for there to be exceptions in particular cases. This suggests that the four-leggedness of dogs should be represented by a *default* process which can be overridden if knowledge about a particular dog dictates otherwise. Defaults are commonly used in psychological research, however, they are usually used to represent probabilistic information. Thus, the default value for the colour of an apple might be "red" based on the fact that most apples are red. While this is a perfectly legitimate thing to do, there are important qualitative differences between representing probabilistic defaults, and rule-based defaults that involve necessary connections. First, as discussed above, probabilistic defaults are updatable on the basis of further experience, however, necessary connections are not revised in the face of evidence that suggests that they the property in question is not displayed by some instances of the kind in question. Second, probabilistic defaults need to be the most frequent value of the property in question, however, this need not be the case for rule-based defaults. It is even possible to have the majority of instances not display the default value. For example, (18) which has the meaning in (18') is considered to be true even though we know that the vast majority of chickens nver live to the age of x because they are raised and killed on poultry farms.

(18) Chickens live to be x years of age.
(18') Chickens, in virtue of being chickens, live to be x years of age.

Third, as the difference in redundancy effects of (9) and (10) show, rule-based defaults can be, and are, assumed to be true with perfect certainty unless evidence for a different value is available, whereas, probabilistic default can never be assumed with perfect certainty. Fourth, it seems that rule-based defaults can only be established through the use of a formal system of knowledge, however, probabilistic defaults can be acquired on the basis of statistical information in the absence of a formal system of knowledge.

There are close parallels between the organization of conceptual knowledge and the organization of aspects of linguistic knowledge. For example, inflectional morphology also makes use of default rules that are formed through the use of a formal system of knowledge, and exceptions which are particular cases which are not formed through the use of a formal system of knowledge but are stored in memory and can block the application of the default rule (Pinker & Prince, 1988; Kim, Pinker, Prince, & Prasada, 1991; Marcus, Brinkmann, Clahsen, Wiest, & Pinker, 1995). Thus, there is independent evidence that the mind has the representational resources being proposed for

558

the representation of conceptual knowledge (see also, Pinker & Prince, 1996).

Conclusion: Typicality, atypicality, necessary connections, and exceptions.

We are now in a position to explicate the title of this paper. Typicality and atypicality are extensional notions. What is typical or atypical can only be established by observing the frequencies with which various properties of a given kind of thing actually appear in its extension. In contrast, exceptions are instances of a given kind that lack a property that is conceived of as being a necessary consequence of being that kind of thing. As such, typicality and exceptionhood are orthogonal to one another -- exceptions need not be atypical, and atypical instances need not be exceptional. However, since an exceptional property such as the three-legged of dogs, is not due simply to the entity being a dog, but in virtue of its being a dog *and* certain other particular properties that happen to be true of the instance, it is usually the case that the rule-based default property is most frequent because there is usually no reason why other circumstances that prevent expression of this property should be systematically present for instances of the kind in question. That is why we usually assume generics such as (18) to be true of the majority of instances unless we are explicitly made aware of a systematic reason why the majority of instances would not display the rule-based default property (Prasada, 1999).

References

Kim, J.J., Pinker, S., Prince, A., & Prasada, S. (1991) Why no mere mortal has flown out to center field. *Cognitive Science, 15*, 173-218.

Marcus, G. F., Brinkmann, U., Clahsen, H., Wiest, R., & Pinker, S. (1995). German inflection: The exception that proves the rule. Cognitive Psychology, 29, 189-256.

Osherson, D. N., Stern, J., Wilkie, O., Stob, M., & Smith, E.E. (1991). Default probability. Cognitive Science, 15, 251-269.

Pinker, S. & Prince, A. (1988). On language and connectionism: Analysis of a parallel distributed processing model of language acquisition. *Cognition, 28,* 73-193.

Pinker, S., & Prince, A. (1996). The nature of human concepts/evidence from an unusual source.*Communication & Cognition, 29*, 307-362.

Prasada, S. (1998a). Formal aspects of common sense conception. http://mitpress.mit.edu/celebration

Prasada, S. (1998b). How to form stable concepts: The formal aspect of common sense conception. Unpublished manuscript, Dartmouth College, Hanover, NH.

Prasada, S. (1999). Names for things and stuff: An Aristotelian perspective. In R. Jackendoff, K.Wynn, and P. Bloom (Eds.), *Language, Logic, and Conceptual Representation: Essays in honor of John Macnamara*. Cambridge, MA: MIT Press.

Representation of Logical Form in Memory

Aaron W. Rader (rader.34@osu.edu)
Center for Cognitive Science and Department of Psychology
The Ohio State University, 208 Stadium East
1961 Tuttle Park Place, Columbus, OH 43210, USA

Vladimir M. Sloutsky (sloutsky.1@osu.edu)
Center for Cognitive Science, and School of Teaching & Learning
The Ohio State University, 208 Stadium East
1961 Tuttle Park Place, Columbus, OH 43210, USA

Abstract

Current theories of human deductive reasoning make different claims about the representation of logical statements in memory. Syntactically-based theories claim that abstract logical forms are represented veridically in memory, separate from content, whereas semantic theories propose that naïve reasoners represent combinations of possibilities that are based on the content of statements. We tested these predictions in two experiments in which participants had to recall and recognize statements of different logical forms. Results indicate that memory for logical form is not veridical, thus failing to support the syntactic view. In particular, results suggest that naïve participants tend, whenever possible, to represent only a single possibility for a statement of any logical form. These findings are consistent with semantic theories of human deductive reasoning and have significant implications for all theories of reasoning.

Introduction

The ability to reason and derive conclusions is ubiquitous in human life; however, naïve reasoning is error-prone (see Evans, Newstead, & Byrne, 1993; Evans & Over, 1996; Johnson-Laird, in press; Johnson-Laird & Byrne, 1991, for reviews). Because these errors exhibit systematic and predictable patterns, we believe that the analysis of these errors can shed light on the mechanism underlying human reasoning.

There are two major theoretical approaches to propositional reasoning, the syntax-based approach and the semantics-based approach. According to the syntactic approach, reasoners extract the syntactic form of the argument and apply certain formal rules of inference, or inferential schemata, to the extracted form (Braine & O'Brien, 1991; Rips, 1994). For example, reasoners easily conclude that B is the case, using the "modus ponens schema," when presented with the following premises:

$A \rightarrow B$ (If A then B)
A.

The syntactic approach thus hinges on assumptions that reasoners (a) veridically represent information in the premises and (b) apply inferential schemata to these representations. However, both assumptions are not uncontroversial. For example, according to the semantic approach, the untrained mind is not equipped with formal rules of inference. Furthermore, reasoning, to a large extent, is a function of representations of information in the premises. In turn, these representations are not veridical but are often incomplete or defective (Johnson-Laird & Byrne, 1991; Evans & Over, 1996; Sloutsky, Rader, & Morris, 1998).

In particular, when premises are compatible with multiple possibilities (e.g., A or B, or both), people tend to reduce the number of represented possibilities in accordance with the "principle of truth." The principle claims that people normally represent only true possibilities, and within these possibilities they represent only those propositions that are true (Johnson-Laird & Byrne, 1991; Yang & Johnson-Laird, in press). It has been also argued that people construe the "minimalist representation," discounting all possibilities except one (Sloutsky, et al. 1998; Sloutsky & Goldvarg, 1999).

If the assumptions of the syntax-based approach are true, then people should be able to extract logical form of a proposition and to construe a veridical representation of this form. If they do construe such veridical representations of the logical form, then, when the task is to remember these propositions, different logical forms should generate comparable error rates. Furthermore, because people tend to represent form, memory errors should be at least as likely to preserve the form, but not the content of propositions, as they should be to preserve the content but not the form.

However, if people do not represent propositions veridically, construing instead the "minimalist representation," then propositions that are compatible with one possibility (such as conjunctions) should generate fewer errors than propositions that are compatible with multiple possibilities (e.g., disjunctions, conditionals, or tautologies) or to no possibilities (such as contradictions). In addition, if they do not extract the logical form of the proposition, then, when committing errors, they should prefer foils that preserve the content, not the form of propositions. The two reported experiments were designed to test these predictions via examining recall and recognition of propositions varying in their logical form.

Experiment 1

Method

Participants Forty-nine undergraduate participants (35 females and 14 males, mean age = 23.0 years) from a large Midwestern university took part in the study. Some volunteered in return for extra credit, whereas others were paid a small cash amount for participation.

Materials A set of 18 pictures and 18 sentences were used in the experiment. Each picture, printed on plain white paper and laminated, measured approximately 4" by 5" and depicted a black-and-white line drawing of a face. The pictures were made to be visibly and distinctly different. Each sentence was printed in 14-point type on a slip of white paper, measured approximately 1.5" by 5", and was laminated. Each sentence was a description of a person; the descriptions always started with the phrase "This professor…" and then continued with either one or two clauses that used simple syntax (i.e., verb and direct object). The 6 one-clause sentences included 3 affirmations and 3 negations, and the 12 two-clause sentences included 3 each of the following logical forms: conjunction, disjunction, tautology and contradiction. In the learning phase (see below) each description was paired with one picture.

The other important materials included 18 lists of sentences, used in the recognition test, with one list corresponding to each description. Each list included that original description as well as 5 critical foils in which the contents of that description was phrased, as closely as possible, in the other five logical forms used in the study. For example, if the original description were a tautology, the list would include that description as well as its content presented as a contradiction, conjunction, disjunction, negation and affirmation. Note that in some cases, changing the form necessarily requires adding or deleting some content. For example, to represent a conjunction as an affirmation, we dropped the second proposition from each conjunction. The other 12 sentences on each list included six that used the same six forms, but content that appeared in none of the original 18 descriptions. Finally, six sentences with different content and forms were included on each list, as checks on guessing. The procedure also used a stopwatch to keep time.

Design and Procedure The experiment included two within-subject factors, Learning Trial and Logical Form of the sentence (affirmation, negation, conjunction, disjunction, tautology, and contradiction). Also, for the recognition test only, participants were divided into two groups, with one group tested on affirmations, conjunctions, and tautologies; and the other on negations, disjunctions, and contradictions. For each participant, the picture-sentence pairs were randomly determined, so that every participant received a unique pairing of pictures and descriptions. All participants were tested individually, seated at a small table across from the experimenter. The procedure was videotaped for subsequent analyses.

The experiment comprised three phases, a learning phase, a distracter phase, and a recognition test. Before the learning phase, the experimenter told each participant that she would be asked to try to remember a series of descriptions of people in response to pictures of those people. Participants also were told to pay attention to the exact wording of the statements and that some of them might sound odd. The learning phase consisted of five learning trials. On each trial, the participant first studied each of the picture-sentence pairs. A picture and its associated sentence were placed on the table in front of the participant, and the participant read the sentence aloud while studying the picture. Once the participant finished reading one description, the experimenter removed that picture-description pair and placed the next pair on the table. This procedure was repeated until all pairs had been presented, with the order of presentation randomized. The experimenter then placed one picture at a time on the table and asked the participant to recall its associated sentence. Again, order of presentation was random. After attempting to recall each picture, the participant repeated the learning trial sequence, until 5 trials were complete.

The next phase was a 5-minute distracter task, during which the participant completed math word problems. Following the distracter task, the participant was presented with the recognition test. Each participant was presented with 6 of the recognition lists, with two lists for each of the three forms in that participant's block (either affirmations, conjunctions, and tautologies; or negations, disjunctions, and contradictions). Every participant received a unique combination of lists, so that each description's list was given to a roughly equal number of participants, but no two participants received the same combination of lists. The experimenter informed the participant that she should decide, for each sentence on the list, whether it had been paired with a picture or not, and to place a check next to those sentences believed to have been presented earlier. Participants could take as much time as desired to finish this task, which was always completed in 3-5 minutes. Following this task the experimenter debriefed the participant.

Results and Discussion

Dependent variables for these analyses included participants' recalls in the learning phase and their choices in the recognition test. In all repeated measures analyses of variance to be reported in this and Experiment 2, only effects that were significant after applying the Geisser-Greenhouse correction to degrees of freedom in the F tests are reported. Additionally, all pairwise comparisons reported in both experiments were computed with the Bonferroni correction to the overall alpha rate of .05. Participants' gender is omitted because no significant differences emerged in preliminary analyses.

We first examined numbers of correct responses across trials and forms in the learning phase. A 5 (Trial) X 6 (Form) ANOVA, with repeated measures on both factors, yielded main effects of Trial, $F(4, 192) = 428.245, p < .001$; of Form, $F(5, 240) = 47.311, p < .001$; and a significant interaction, $F(20, 960) = 7.951, p < .001$. We also performed simple effects analyses of each Trial, following these with pairwise contrasts examining mean differences among Forms. These analyses revealed that, on Trial 1, atomic propositions and logical constants were recalled at higher rates than conjunctions and disjunctions, with the first four forms not differing from one another. On subsequent trials, however, the overall mean differences by form reported below were largely established. Consequently, and for the sake of brevity, we report differences among the overall means, shown in Figure 1.

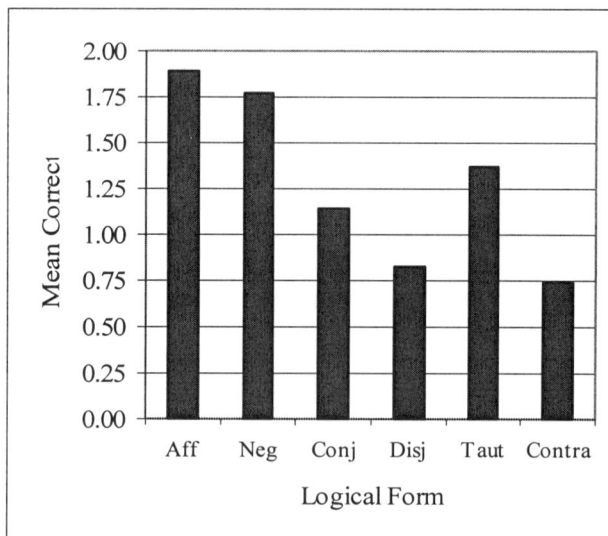

Figure 1: Mean correct recall by logical form, collapsed over trials.

Note that atomic statements (affirmations and negations) were recalled better than composite statements, which is not surprising given that they involve one proposition and no connective. Pairwise comparisons showed that affirmations ($M = 1.89$, $SD = .50$) were recalled significantly better than all other forms except negations ($M = 1.78$, $SD = .45$), all significant $Fs(1,48) > 25$, $ps < .001$. Negations, in turn, were also recalled at significantly higher rates than conjunctions ($M = 1.14$, $SD = .51$), disjunctions ($M = .82$, $SD = .50$), tautologies ($M = 1.38$, $SD = .55$) and contradictions ($M = .74$, $SD = .64$), all $Fs(1, 48) > 50$, $ps < .001$. Conjunctions were recalled at significantly higher rates than disjunctions, $F(1, 48) = 4.84$, $p = .002$; and contradictions, $F(1, 48) = 8.00$, $p < .001$. Recall rates for conjunctions and tautologies did not differ, $F(1, 48) < 1$, NS. Tautologies were recalled significantly more often than disjunctions and contradictions, $Fs(1, 48) = 14.44$ and 19.70, respectively, $ps < .001$. Finally, recall rates of disjunctions and contradictions did not differ significantly, $F(1, 48) < 1$, NS.

A second analysis focused upon recall errors involving substitutions of the logical connectives "and" and "or;" for example, recalling a conjunction as a disjunction by using "or." We refer to such errors as *conversions* because they convert a statement's logical form. Conversions occur when the content is correctly recalled, but the connective is not. This analysis used only recall attempts for the compound propositions, because such errors are not possible with

atomic statements. A 5 (Trial) X 4 (Form) ANOVA, with repeated measures on both factors, revealed significant effects of Trial, $F(4, 192) = 30.09$, $p < .001$; of Form, $F(3, 144) = 37.10$, $p < .001$; and a Form X Trial interaction, $F(12, 576) = 4.54$, $p < .001$. We will focus on the Form effect, because inspection of the interaction showed it to be a function of contradictions' increasing likelihood of being converted over succeeding trials. Conversions for conjunctions, disjunctions, and tautologies did not increase nearly as much over trials, and the relative ordering of conversion rates of these forms was consistent; simple effects analyses of conversion rates by form on each trial revealed no significant cross-overs among these three forms. Means for each form are presented in Figure 2.

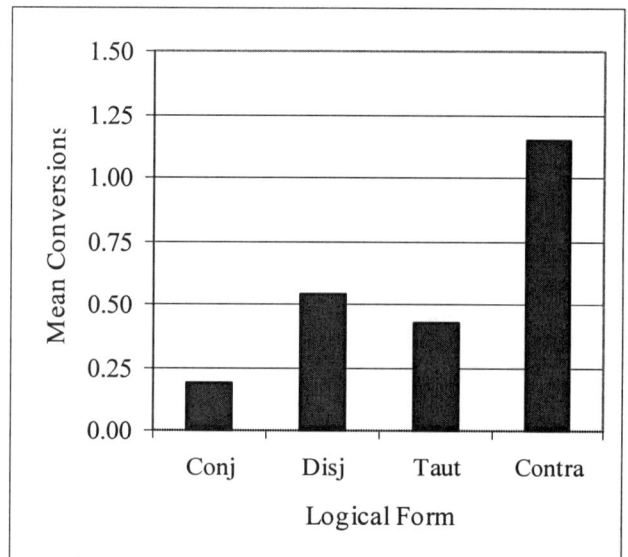

Figure 2: Mean conversions by logical form, collapsed over trials.

Pairwise comparisons following up the main effect of Form revealed that conjunctions ($M = .19$, $SD = .25$) were significantly less likely to be converted than disjunctions ($M = .54$, $SD = .43$), tautologies ($M = .42$, $SD = .48$), and contradictions ($M = 1.15$, $SD = .61$), all $Fs(1, 48) > 8.2$, $ps < .006$. Disjunctions and tautologies were both less likely to be converted than contradictions, $Fs(1, 48) = 31.88$ and 28.69, respectively, $ps < .001$. Finally, disjunctions and tautologies did not differ in their conversion rates.

We now turn to the recognition test results. In contrast to the recall data, these data indicate a substantial ability to remember the form and content of a statement. Separate binomial tests comparing the numbers of hits and false alarms for each form revealed that subjects' hit rates for affirmations (93%), negations (91%), conjunctions (71%), disjunctions (78%), and contradictions (64%) were significantly greater than false alarm rates, all $zs > 2.21$, $ps < .02$. The hit rate for tautologies (55%) did not differ from the false-alarm rate, however, $z = .80$, $p > .20$. Inspections of false alarms revealed that 96% of all false alarms were those in which the content exactly matched the original statement but the connective was wrong (e.g., accepting a disjunction when the original statement was a conjunction). Participants' ability to select statements with precisely the original content was thus very good, and ability to select correct forms was also notable. These results disagree sharply with the learning phase data in two ways. First, no particular advantage exists for conjunctions relative to the other compound forms. Also, tautologies could not be discriminated from contradictions with the same content, whereas tautologies were recalled as well as conjunctions in the learning phase.

The learning phase data largely support our hypotheses and replicate earlier work with compound statements (Sloutsky, et al., 1998), but the recognition data raise questions. In recall, conjunctions were significantly more likely to be recalled than disjunctions and contradictions, and were less likely to be converted than all other compound forms. The lack of a difference between tautologies and conjunctions in recall was not predicted. Possibly, some participants may have been unsure of the connective used but were averse to producing contradictions (see Lakoff, 1971). The finding that atomic statements were easier to recall than all compound statements is not surprising given that atomic statements do not involve a connective and are shorter.

The recognition data not only indicated overall better memory (this finding is trivial on its own) but also indicated that the relative accuracies in memory for tautologies and contradictions switched, compared to recall. We suspected that the overall improvement and perplexing changes with tautologies and contradictions partially stemmed from the recognition test itself (i.e., presenting separate lists for each statement, with only one correct answer per list). In Experiment 2, we made the recognition test somewhat more complex by testing participants for all original statements. We also assessed recognition performance after varying numbers of learning trials because five trials allow many opportunities for rote memorization. With these modifications we could better test our original hypothesis with respect to recognition memory.

Experiment 2

Method

Participants Forty-five undergraduate participants (25 female and 20 male, mean age = 19.3 years) from a large Midwestern university volunteered for the experiment in partial fulfillment of a course requirement. Fifteen participants were assigned to each Learning Trial condition.

Materials All materials were the same as Experiment 1 except the recognition test. The new test included the original 18 descriptions and the five foil statements associated with each description from the lists of Experiment 1. The test also included the six foil statements associated with each description that involved the same logical forms but different content. Overall, the test consisted of 216 items. The different-content, different-form distracters from Experiment 1 were not included

because results of that study and a pilot study of this experiment, using only one learning trial, showed that these items were never chosen. The test was presented as an eight-page typewritten packet, and sentences were randomly ordered.

Design and Procedure The experiment included one within-subject factor, Logical Form, and one between-subject factor, Number of Learning Trials (1, 3, or 5). Instructions to participants were the same as Experiment 1. The learning phase also followed the same procedure, with each participant receiving either one, three, or five trials. The distracter and recognition phases were the same as Experiment 1.

Results

Gender is again omitted because no such differences emerged. We initially conducted a 2 (Number of Trials: 3 or 5) X 6 (Form) mixed ANOVA on number of correct recalls during the learning phase on the last trial for each group. The 1-trial group was omitted because, perhaps owing to a smaller sample size, it contributed virtually no variance (participants in this group largely recalled nothing, and in the recognition test they chose many of the different-content foils, indicating that the task was perhaps too difficult for them). The recall data parallel those in Experiment 1, so we do not report them here.

For the recognition results, we analyzed accuracy for each logical form, operationalized for each participant as (Hits - False alarms)/3. Such a computation puts accuracy on a scale ranging from -1 to $+1$, which was appropriate because overall frequencies of acceptances for each form varied somewhat. Accuracy varies within the range $[-1, +1]$ because every participant chose only some combination of correct choices and same-content, wrong-connective foils; no other distracters were chosen. Accuracy data are presented in Figure 3.

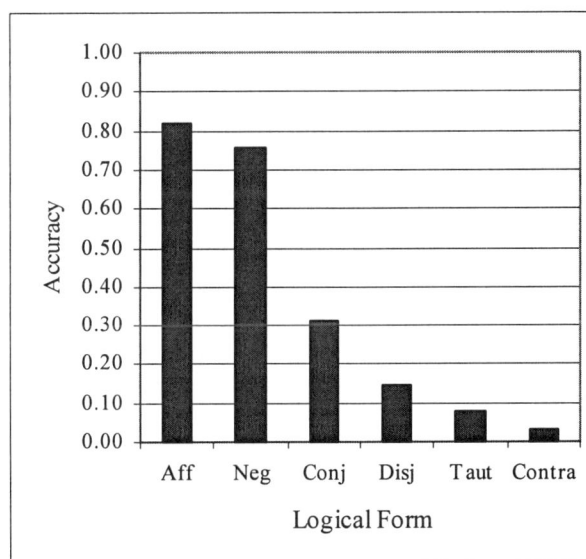

Figure 3: Mean recognition accuracy by form.

A 2 (Number of Trials) X 6 (Form) ANOVA on accuracy yielded only a main effect of Form, $F(5, 140) = 16.65$, $p < .001$. Number of trials and the interaction were not significant, $F(1, 28) = 1.21$ and $F(5, 140) = 1.02$, respectively, $ps > .25$. Pairwise comparisons of these means revealed higher accuracies for affirmations ($M = .82$, $SD = .29$) than for all compound forms, all $Fs(1, 28) > 37$, $ps < .001$. Negations ($M = .75$, $SD = .30$) also had higher accuracies than all compound forms, all $Fs(1, 28) > 30$, $ps < .001$. Affirmations and negations did not differ significantly. No mean differences among conjunctions ($M = .31$, $SD = .47$), disjunctions ($M = .14$, $SD = .61$), tautologies ($M = .08$, $SD = .49$), or contradictions ($M = .03$, $SD = .65$) were significant. For each form, t-tests comparing mean accuracy to 0 showed that accuracy was significantly greater than zero for affirmations and negations, $ts(29) > 13$, $ps < .001$. Mean accuracy for conjunctions was also significantly greater than zero, $t(29) = 3.64$, $p = .001$. For the other three forms, accuracy did not differ significantly from zero, all $ts < 1.28$, $ps > .2$.

General Discussion

Taken together, both experiments offer some support to the hypothesis that reasoners tend to construct a single, conjunctive representation of possibilities for compound propositions. In Experiment 1, conjunctions were more accurately recalled than all other compound forms except tautologies, and participants converted conjunctions less often than all other compound forms. The advantage for conjunction in recall also supports the idea that conjunctions, being more informative than other compound forms, should be easiest to learn (see Sloutsky, et al., 1998). Tautologies were recalled as accurately as conjunctions, but were nevertheless more likely to be converted. As discussed earlier, the high recall rate for tautologies (and the lower rate for contradictions) may partially have resulted from an aversion to producing contradictory statements when the correct connective was in doubt. Recognition of, and aversion to, contradiction is believed to be a basic part of logical competence (Macnamara, 1986). In Experiment 2, although mean recognition accuracies for all compound forms were lower than those for atomic forms, and did not differ from one another, only accuracy for conjunctions was significantly greater than zero. These data suggest that participants may have been less likely to false alarm on same-content, wrong-connective foils with conjunctions than with other compound forms. Such a pattern would be expected if conjunctions are easier to commit to memory and involve less uncertainty in a recognition task.

Our data do not support the claim that naïve reasoners possess a stock of syntactic inference schemata in long-term memory (Rips, 1994). Such a theory would have to predict, for these tasks, that memory errors altering a statement's logical form should be randomly distributed over forms because all forms (except possibly tautologies and contradictions) should be available. The pattern of memory errors, at least in recall, also questions the claim that people automatically parse sentences for their logical forms (Braine, 1990), again because conjunctions seem to be privileged with respect to other compound forms. A critic might claim that disjunctions, tautologies, and contradictions are less acceptable than conjunctions and may be subject to "normalization" in comprehension; that is, they may be altered into the more accepted format (Fillenbaum, 1977). This view cannot predict, however, differences in recall among these forms, which we found. Also, even some conjunctions have no "common topic" that makes them sensible (Lakoff, 1971, p. 116). Because disjunctions and conjunctions in these experiments were intentionally constructed to contain two unrelated propositions, pragmatic explanations cannot easily account for the recall results.

These results do not necessarily impugn the claim that parsing into logical form occurs in an autonomous, on-line fashion (Lea, 1995); possibly, syntactic abstraction does occur but logical syntax is quickly discarded in favor of a semantic representation. However, our experiments make the claim for mental logic more elusive. Investigations of recognition at varying intervals after presentation, from immediate test to substantial delays, will help settle the issue, as will investigations into whether logical inferences putatively represented as procedural rules respond to experimental manipulations in ways consistent with rule-based, autonomous processes (Smith, Langston, & Nisbett, 1992). We are currently conducting both types of investigations.

Acknowledgements

This research was supported by a Summer Fellowship awarded to the first author by the Center for Cognitive Science at the Ohio State University during the summer of 1998.

References

Braine, M. D. S. (1990). The "natural logic" approach to reasoning. In W. F. Overton (Ed.), *Reasoning, necessity, and logic: Developmental perspectives*. Hillsdale, NJ: Erlbaum.

Braine, M. D. S., & O'Brien, D. P. (1991). A theory of *if*: A lexical entry, reasoning program, and pragmatic principles. *Psychological Review, 98,* 182-203.

Evans, J. St. B. T., Newstead, S. E., & Byrne, R. M. J. (1993). *Human reasoning: The psychology of deduction.* Hove, UK: Erlbaum.

Evans, J. St. B. T., & Over, D. E. (1996). *Rationality and reasoning.* Hove, UK: Psychology Press.

Fillenbaum, S. (1977). Mind your *p*'s and *q*'s: The role of content and context in some uses of *and*, *or*, and *if*. In G. H. Bower (Ed.), *The psychology of learning and motivation: Vol. 11.* San Diego: Academic Press.

Johnson-Laird, P. N. (in press). Deductive reasoning. *Annual Review of Psychology.*

Johnson-Laird, P. N., & Byrne, R. M. J. (1991). *Deduction.* Hillsdale, NJ: Erlbaum.

Lakoff, R. (1971). If's, and's, and but's about conjunction. In C. J. Fillmore & D. T. Langendoen (Eds.), *Studies in linguistic semantics*. New York: Holt, Rinehart, and Winston.

Lea, R. B. (1995). On-line evidence for elaborative logical inferences in text. *Journal of Experimental Psychology: Learning, Memory, and Cognition, 21,* 1469-1482.

Macnamara, J. (1986). *A border dispute: The place of logic in psychology.* Cambridge, MA: MIT Press.

Rips, L. J. (1994). *The psychology of proof: Deductive reasoning in human thinking.* Cambridge, MA: MIT Press.

Sloutsky, V. M., & Goldvarg, Y. (1999). *Representation and recall of determinate and indeterminate problems.* Manuscript submitted for publication.

Sloutsky, V. M., Rader, A. W., & Morris, B. J. (1998). Increasing informativeness and reducing ambiguities: Adaptive strategies in human information processing. *Proceedings of the Twentieth Annual Conference of the Cognitive Science Society.* Hillsdale, NJ: Erlbaum.

Smith, E. E., Langston, C., & Nisbett, R. E. (1992). The case for rules in reasoning. *Cognitive Science, 16,* 1-40.

Yang, Y., & Johnson-Laird, P. N. (in press). Illusions in quantified reasoning: How to make the impossible seem possible, and vice versa. *Memory & Cognition.*

Towards Exemplar-based Polysemy

Mohsen Rais-Ghasem (mohsen@cyberus.ca)
Jean-Pierre Corriveau (jeanpier@scs.carleton.ca)

School of Computer Science, Carleton University
Ottawa, ON K1S 5B6 Canada

Abstract

In this paper we criticize existing computational models of lexicon for assuming that for every word there is a fixed number of word sense that must be searched for the proper meaning of that word in a context. We reject this sense enumerative view and argue for a different model of lexicon in which the effects of context are not limited to selecting a word sense, and selected senses can be contextually modulated. We also explain how patterns of contextual effects could evolve in an exemplar-based fashion. A prototype implementation of this model is also discussed.

Introduction

This paper proposes a computational model of lexicon. What distinguishes the proposed model from other such models is the treatment of polysemy. Most existing models of lexicon presuppose a list of possible word senses for each word, and a selection process that searches such a list for the proper meaning of that word in a context. We question this sense enumerative approach and argue for an alternative model in which the ultimate semantic characteristics of a word not only depend on the meaning of the word itself but on the interactions that the word has with other words in its contexts. We will also show how the model can gradually extract patterns of contextual effects from a number of exemplars.

We will first review computational models of lexicon and argue that they essentially present an enumeration of senses in one form or another. We will then question the validity and viability of such an approach. Following this section, we will present our model and explain how word senses in our model are determined. Finally, a prototype implementation of the proposed model and results of some sample runs will be presented.

Overview of Computational Lexicons

Computational models of lexicon can be divided into three categories: symbolic lexicons, structured connectionist lexicons, and distributed connectionist lexicons. In symbolic lexicons, word senses are generally represented by means of *ad hoc* and complicated structures. Typically, such structures are predefined, i.e. hand-coded, and static. Lexicons of most early NLP models (e.g. BORIS, Dyer 1983) fall under this category.

In structured connectionist lexicons (e.g.; Bookman, 1994; Lange and Dyer, 1989; Waltz and Pollack, 1985) each word is represented by a node in a structured network. Word nodes become activated when their corresponding words are presented at the input. Each word node is connected to some sense nodes, each representing one of the word's senses. For example, in the network described in (Waltz and Pollack, 1985) the word *star* is connected to three senses: *movie star*, *celestial body*, and *geometric figure*.

Distributed connectionist networks are the basis of the last category of computational lexicons. Unlike structured connectionist lexicons where every concept is assigned to a node, here a single set of nodes is used to represent all word senses. Examples of such lexicons are reported in McClelland and Kawamoto (1986), Miikkulainen (1993), and Veronis and Ide (1995). For example, in McClelland and Kawamoto (1986), each word is associated with one or more word senses. Word senses are encoded along a number of semantic microfeatures. Examples of these microfeatures are softness (soft/hard), gender (male/female/neuter), volume (small/medium/large), and form (compact/1-D/2-D/3-D).

Despite the difference in their adopted concept representation approach, all three categories of computational lexicons are committed to the same principle of enumerating word senses[1]. This sense enumerative view postulates that, upon hearing a word, listeners access a mental dictionary containing an exhaustive list of potential senses for that word, from which, the proper sense for the word is selected.

On Sense Enumeration

Sense enumeration seems very intuitive. After all, it is how all dictionaries are organized. The question seems to be about the validity of accepting this familiar assumption as a meaning theory. Ruhl remarks:

> I claim that current linguistic theories accept too uncritically the conclusion of dictionaries that words in general have multiple meanings (Ruhl, 1989:vii).

Kilgarriff (1997) contends that standard dictionaries specify the range of meanings of a word in a list merely in response

---

[1] While in most symbolic and structured connectionist model this commitment is made at the architectural level, in distributed connectionist model it is mostly a matter of representation.

to "constraints imposed by tradition, the printed page, compactness, a single, simple method of access, and resolving disputes about what a word does and does not mean." He criticizes word sense disambiguation (WSD) approaches for committing themselves to such lists:

> Much WSD word proceeds on the basis of there being a computationally relevant, or useful, or interesting, set of word senses in the language, approximating to those stated in a dictionary... Meanwhile, the theoreticians provide various kinds of reasons to believe there is no such set of senses (Kilgarriff, 1998:95).

Moreover, it is not always possible to enumerate all potential senses of a word. The use of language, even non-figuratively, by its very nature is creative. That is, "words can take on an infinite number of meanings in novel contexts" (Pustejovsky, 1995: 42). No matter how comprehensive a list can be, it is generally possible to find a new sense for some words. *Eponymous* expressions, expressions built around references to people, provide a good example. Clark and Gerrig (1983) showed how understanding eponymous expressions such as *Please do a Napoleon for the camera* requires sense generation rather sense selection.

Another problem facing sense enumeration is *semantic flexibility*. Many psycholinguistic findings suggest that context highlights (or obscures) certain properties of a single concept as it appears in different contexts. For example, in an early experiment, Barclay et al. (1974) demonstrated how the interpretations of familiar, unambiguous words vary with context. For instance, they argued that the choice of attributes for *piano* is affected by the verb selection in *The man (lifted) (tuned) (smashed) (sat on) (photographed) the piano*. They then provided evidence that the prior acquisition of a sentence like *The man lifted the piano* (vs. *The man tuned the piano*) influences the effectiveness of cues like "something heavy" (vs. "something with a nice sound") in recall. They concluded that context can affect the encoding of concepts in memory. Similar results have been reported in Anderson et al. (1976), Barsalou (1982), Greenspan (1986), and Witney et al. (1985).

Having subscribed to the sense enumeration principle, the effect of context in existing models of lexicon is limited to selecting a sense among a closed number of alternatives. Any changes in the characteristics of a selected sense either has to come in the form of a new sense or is ignored.

Dealing with contextual effects is particularly problematic for symbolic and structured connectionist lexicons. This is due to the fact that commitment to sense enumeration is made at the structural level, making word senses discrete entities. To the contrary, word senses in distributed connectionist models are continuous. Thus, such models are potentially capable of dealing with contextual effects. In fact McClelland and Kawamoto (1986) reported an unintended yet interesting result. They had presented their model with *The ball broke the vase*. Although throughout the training phase *ball* was always associated with the microfeature *soft*, in the output it was associated with the microfeature *hard*. They attributed this result to the fact that *breakers* in their experiment were all *hard* and the model had shaded the

meaning of ball accordingly. Kawamoto later remarked:

> [T]he flexibility of a distributed representation allows a natural account of polysemy and homonymy, and provides a mechanism for new senses to be learned or generated on-line (1993: 510).

However, distributed connectionist approach also has some disadvantages. First of all, such lexicons presuppose a set of universal and fixed microfeatures and demand every sense to be characterized in terms of such a set *in advance*. But what is even more important is the difficulty to separate patterns of contextual effects from the representation of a word sense. For instance, consider *breakers* in McClelland and Kawamoto (1986). It is impossible to isolate *breakers* (the category) from breakers (the instances).

We believe this separation is useful. Firstly, such patterns can be thought of as *ad hoc* categories: categories built by people to achieve goals (Barsalou, 1983). For instance, *breakers* can be instrumental in achieving the goal of "breaking a window". Secondly, from a learning point of view, such patterns can be very useful. Rais-Ghasem (1998) has shown how a concept can evolve (i.e. acquire new properties) from such patterns as it becomes associated with in different contexts. And finally, Rais-Ghasem (*Ibid.*) has employed such patters to implement a metaphor understanding theory, in which metaphors are interpreted as class inclusion assertions (Gluksberg & Keysar, 1990).

A Lexicon for Sense Modulation

In this section we will discuss our proposed model of lexicon. We begin by introducing our two-tiered word senses. Later in this section, we explain how this two-tiered structure allows us to account for contextual effects in an exemplar-based fashion. The structure and overall behavior of the model are described in the last two subsections.

Two-Tiered Word Senses

Word senses in the proposed model consist of two components: **sense-concept** and **sense-view**. Given a word sense, its sense-concept will determine the concept it represents, whereas, the sense-view describes how this sense-concept is to be viewed in its surrounding context. For example, the word sense for *book* in *The book broke the window* will consist of the sense-concept BOOK, representing the concept book, and a sense-view which portrays *book* as an instrument of breaking.

This separation parallels the two different roles that Franks (1995) proposed for concepts when he distinguished between their *representational* and *classificatory* functions. While the former is used to discern instances of one concept from instances of others, the latter specifies how an instance of a concept should be classified. For example, he argues that, depending on context, *fake gun* could be classified as a *gun*, a *toy*, a *replica*, and a *model*.

We contend that this two-tiered structure allows us to account for various contextual effects. In general, Cruse (1986) specifies two ways in which context affects the semantic contribution of a word: *sense selection* and *sense modulation*. Sense selection happens in cases of lexical ambiguity where one sense is chosen among a number of distinct senses. Sense selection is followed by sense modulation in which the semantic characteristics of selected senses are modulated or become specified according to their surrounding contexts.

It is our intention to show that unlike conventional lexicons, which do not go beyond sense selection, using the proposed two-tiered structure for word senses we can account for both processes. Below we will show that sense selection and sense modulation, respectively, lead to the selection of a sense-concept and a sense-view. As an example, consider the word *book*. Being unambiguous, the sense-concept BOOK can be easily selected. However, BOOK must be modulated according to its surrounding context. For example, in a context such as *The book broke the window* it will be associated with a sense-view that portrays it as breaker. In other contexts such as *Many books were burnt in the fire* and *I read the book* the same concept BOOK would be associated with different sense-views namely a flammable object and text.

Development of Sense-Views

Concepts associated with a word, to a large extent are conventionalized. Traditionally, concepts are represented by means of a number of weighted properties (for a review see Barsalou & Hale, 1993). Such properties explain similarities and facilitate comparisons. While representation of concepts and word-concept mapping are pre-defined in the proposed model, sense-views are developed incrementally in an exemplar-based fashion.

This is achieved by defining the **alike** relationship. Essentially, this relationship states that two sense-concepts become alike if they appear in a similar context and they share the same thematic role. For example, *book* in *The book broke the window* and *bat* in *The man smashed the windshield with a bat* are alike. That is to say that, *book* in this context is best classified with *bat* and not with *book* in a context such as *I read the book*, despite the fact that representationally it should be the other way around.

Now considering the two-tiered word senses, we can say that sense-concepts associated with a sense-view are alike. Thus, a sense-view can be considered as a generalization of its associated sense-concepts. Therefore, as new sense-concepts become associated with a sense-view, its characteristics will evolve to become a better representative of all its associated sense-views (see Sample Runs below for examples).

A Lexicon with Exemplars

The proposed model is structured in four levels. Words appear at the bottom level. Concepts associated with words appear at the second level. Unlike other models, the proposed model also maintains examples of how each concept is typically used. Such usage examples, or **exemplars**, appear at the third level. Each exemplar consists of a number of sense-concepts, each representing the occurrence of a concept in the context given by the exemplar. Sense-views constitute the fourth level. Each sense-concept is connected to a sense-view.

The following picture illustrates an example of two words, *bat* and *book* (represented as rectangles) and their associated concepts (represented a rounded rectangles). Concepts are further specified by exemplars (represented as double-lined rectangles) each representing a use-case for its associated concepts.

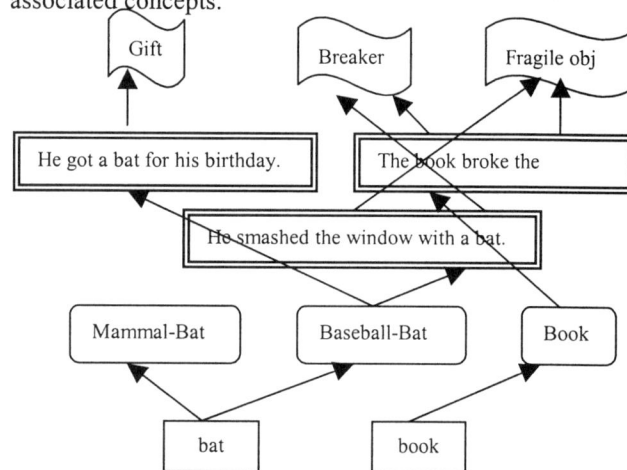

For the sake of simplicity, sense-concepts constituting each exemplar are not displayed. An example of sense-concept is the occurrence of BASEBALL-BAT in *He got a bat for his birthday*. Each sense-concept is connected to a sense-view. Three sense-views are shown: breaker, gift and fragile object. Note that both sense-views and sense-concepts are in fact represented by a set of properties, and that the names used in this figure are only labels used for the sake of clarity and brevity.

Finally, here is the output generated by the implemented prototype (see Implementation) for *book* in *The book broke the window*.

> **SENSE** Generated for Input Word Book
> **[Sense-Concept(s)]** BOOK
> **[Sense View]**
> Thematic Role: Instrument - No. of Exemplars: 4 - Marker(s): with-pp, p-subj,
> STATE-OF-MATTER(0.73305)---->:SOLID,
> MADE-OF(0.82805)---->: MATERIAL,

As shown, the sense-view breaker consists of two properties with strong presence (enclosed number represents weight, between 0 and 1, for each property). Thus, this output should be interpreted to indicate that, in the given context, BOOK is to be viewed as a solid object made of material. In other words, other properties of BOOK such as author or title, as well as the fact that book is made of paper, ink etc., are all irrelevant in this context.

From Words to Senses

Criticizing Lakoff for concluding that *open* is polysemyous because of expressions such as *open the door* and *open the present*, Ruhl writes:

> Admittedly, the phrase *open the door* and *open the present* evoke quite different images, but why is the

difference attributed to *open*, which does not differ, and not also *the*? Why isn't the difference located solely where there is difference: in the door-present distinction and in the knowledge that people have about these two activities (1989:x).

This notion that the meaning of a word is being dynamically made up, at least in part, of meanings of other words it occurs with in a context constitutes the basis of the process that maps input words to word senses in the proposed model[2].

The process begins by presenting an input context, made up of input words to the model. The goal is to find pairs of sense-concept/sense-view for input words. Details of this process are beyond the scope of this paper (see Rais-Ghasem, 1998). What follows is a brief overview of the process.

Word nodes corresponding to the input words are activated. Activated word nodes, in turn, activate their related concepts, which in turn activate their related exemplars. The set of activated exemplars represents the model's knowledge, up to that point, of various ways that input words can interact with other words.

Having located all such possible interactions for every input word, the next step is to find a set of senses for input words that can **interact** with each other. A number of word senses can interact with each other only if they either appear in one of the exemplars, or they can be used in one.

The decision on whether input words can be interpreted based on one of the exemplars depends on the possibility of classifying the input words along that exemplar. In other words, the question is whether a word sense can be found for each input word so that it could be associated with the same sense-view that its corresponding sense-concept in the exemplar is associated with.

For example, assume that *The book broke the window* is presented to the model and two exemplars are activated for this input: *John broke the law* and *The man smashed the windshield with a bat*. The first exemplar is rejected because none of the concepts associated with the word *window* could be classified as law and regulation (the sense-view related to sense-concept LAW). Because senses can be found for input words that are compatible with the sense-views associated with the second exemplar, the input sentence can be interpreted after it. For example, *book* (in fact its related concept BOOK) can be classified as a breaker. Note that finding such set of senses implies finding a pair of sense-concept/sense-view for each input word.

Finally, the examination of activated exemplars can be carried out in parallel. In fact, in the next section, we discuss a prototype implementation in which each activated exemplar measures its adaptability to the input.

Implementation

This section reports on a prototype implementation of the proposed model for sentential contexts. For this

---

[2] Also see *semantic traits* in Cruse, 1986.

implementation, we used syntactic case markers (Delisle et al., 1993) as indications of thematic roles that noun phrases play in a context. Examples of such case markers are p-subj (positional subject) and with-pp (prepositional phrase with).

Architecture

The implemented prototype is a hybrid system. Architecturally, the system is grounded in two marker passing networks. The bottom network, the ontology network, serves as the system's knowledge base to define concepts in the second network. The network is arranged based on the Mikrokosmos ontology (Mahesh & Nirenburg, 1995). The second network, the lexicon network, consists of four layers of nodes: words, concepts, exemplars comprising sense-concepts, and sense-views (as discussed earlier).

However, despite the fact that it is built on top of two structured networks, the behavior of the system is not completely accomplished through a spreading activation process. Instead, its functionality to a large extent arises from the execution of a number of processing elements called *agents*. Attached to network nodes, agents are autonomous processing elements capable of performing a sequence of instructions. Once a node is activated, its attached agent "fires up" and individually and concurrently starts carrying out its instructions. Details of types and responsibilities of agents employed in this implementation could be found in Rais-Ghasem (1998).

Because of its hybrid nature, the implemented system supports parallelism at the knowledge level, the ability to apply symbolic rules and controls, and taking advantage of statistical similarities in developing sense-views.

Sample Runs

Results of some of the test cases conducted on the implemented system are presented in this subsection.

Sense Modulation: This test demonstrates the system's capability to modulate semantic features of one single sense-concept in different contexts. Here is the first input context:

The musician moved the piano.

The output sense generated for piano is displayed below. Note how only those properties of PIANO that portray it as a heavy physical object are highlighted by the sense-view.

```
SENSE Generated for Input Word Piano
[Sense-Concept(s)] PIANO
[Sense View]
Thematic Role: Object - No. of Exemplars: 3  -  Marker(s):
p-obj,
WEIGHT(0.666667)---->:heavy,
IsKindOf-DEVICE(0.513)---->:
STATE-OF-MATTER(0.756)---->:SOLID,
MADE-OF(0.513)---->:PLASTIC,METAL,
IsKindOf-ARTIFACT(0.7047)---->:
COLOR(0.7047)---->:
AGE(0.7047)---->:
OPERATED-BY(0.7047)---->:HUMAN,
```

It is also interesting to look at the selected sense-view for *musician* in which all properties specific to musician are suppressed, since they are irrelevant in the present context.

```
SENSE Generated for Input Word Musician
[Sense-Concept(s)] MUSICIAN
[Sense View]
Thematic Role: Agent - No. of Exemplars: 3 - Marker(s):
p-subj,
GENDER(0.885367)---->:MALE,
IsKindOf-HUMAN(0.8187)---->:
IsKindOf-PRIMATE(0.54)---->:
```

Here is the second input context:

The musician played the piano.

In contrast, here the properties portraying *piano* as a musical instrument and musician are relevant.

```
SENSE Generated for Input Word Piano
[Sense-Concept(s)] PIANO
[Sense View]
Thematic Role: Object - No. of Exemplars: 2 - Marker(s):
p-obj,
IsKindOf-MUSICAL-INSTRUMENT(0.81)---->:
WORK-EQUIPMENT-OF(0.81)---->:MUSICIAN,
IsKindOf-ARTIFACT(0.729)---->:
STATE-OF-MATTER(0.729)---->:SOLID,
COLOR(0.729)---->:
AGE(0.729)---->:
OPERATED-BY(0.729)---->:HUMAN,
IsKindOf-INANIMATE(0.6561)---->:
```

Here is the last input context and generated sense:

The man broke the piano.

```
SENSE Generated for Input Word Piano
[Sense-Concept(s)] PIANO
[Sense View]
Thematic Role: Object - No. of Exemplars: 4 - Marker(s):
p-obj,
IsKindOf-ARTIFACT(0.54675)---->:
STATE-OF-MATTER(0.567)---->:SOLID,
COLOR(0.54675)---->:
AGE(0.54675)---->:
OPERATED-BY(0.54675)---->:HUMAN,
MADE-OF(0.6775)---->:PLASTIC,GLASS,
```

Sense-View Development: This test demonstrates gradual development of sense-views. The destination sense-view is initially exemplified by only one exemplar:

Mary went to the office.

```
[Sense View]
Thematic Role: Destination - Marker(s): to-pp,
IsKindOf-BUILDING(0.9)---->:
IsKindOf-PLACE(0.81)---->:
IsKindOf-PHYSICAL-OBJECT(0.729)---->:
MADE-OF(0.729)---->:MATERIAL,
WEIGHT(0.729)---->:
SIZE(0.729)---->:
IsKindOf-OBJECT(0.6561)---->:
IsKindOf-BUILDING-ARTIFACT(0.81)---->:
IsKindOf-ARTIFACT(0.729)---->:
STATE-OF-MATTER(0.729)---->:SOLID,
COLOR(0.729)---->:
AGE(0.729)---->:
OPERATED-BY(0.729)---->:HUMAN,
IsKindOf-INANIMATE(0.6561)---->:
```

Notice both IsKindOf-Building and IsKindOf-Place are relatively central to office and therefore to this sense-view.

The set properties shrinks rapidly after processing the following exemplar:

The student went to the stadium.

```
SENSE Generated for Input Word Stadium
[Sense-Concept(s)] STADIUM
[Sense View]
Thematic Role: Destination - Marker(s): to-pp,
IsKindOf-BUILDING(0.8145)---->:
IsKindOf-PLACE(0.73305)---->:
IsKindOf-BUILDING-ARTIFACT(0.73305)---->:
```

This trend continues with another input:

John went to the park.

```
SENSE Generated for Input Word Park
[Sense-Concept(s)] PARK
[Sense View]
Thematic Role: Destination - Marker(s): to-pp,
IsKindOf-BUILDING(0.543)---->:
IsKindOf-PLACE(0.7074)---->:
```

Here, unlike previous case, the property IsKindOf-Place is more prominent than IsKindOf-Building. This is because *park* is not a building, nonetheless, its effect is not enough to completely eliminate IsKindOf-Building from the sense-view.

Multiple Word Senses: There are cases in which context does not favor any of the alternative readings of a word, and therefore the ambiguity must be maintained in the output. This test demonstrates the system's ability to handle such cases. In this example, both readings of *bank* are compatible, to some degree, with the destination sense-view.

John went to the bank.

Here is the output word sense for *bank*, with two sense-concepts, both linked to the same sense-view.

```
SENSE Generated for Input Word Bank
[Sense-Concept(s)] RIVER-BANK, BANK-BRANCH
[Sense View]
Thematic Role: Destination - Marker(s): to-pp,
IsKindOf-BUILDING(0.51585)---->:
IsKindOf-PLACE(0.5967)---->:
```

Unknown Words: Here is an example of how sense-views can be used to establish some properties about unknown words. Here is the input:

Mary went to the palladium.

The word *palladium* is not defined in the lexicon. Nevertheless, the system associates it with the proper sense-view. Through this sense-view, some initial properties for *palladium* can be inferred.

```
SENSE Generated for Input Word Palladium
[Sense-Concept(s)] *** unknown ***
[Sense View]
Thematic Role: Destination - Marker(s): to-pp,
IsKindOf-BUILDING(0.51585)---->:
IsKindOf-PLACE(0.5967)---->:
```

Instantiation of General Terms: This test is inspired by the experiment reported by Anderson et al. (1976). They found that *shark* was a better cue than *fish* for subjects in remembering a sentence like the following:

The fish attacked the man.

They concluded that *fish* was instantiated to, and encoded accordingly as, *shark* in their subjects' memory. Notice how in the output, *fish* is associated with properties specific to shark (*aggressiveness* and *black color*).

```
SENSE Generated for Input Word Fish
[Sense-Concept(s)] FISH
[Sense View]
Thematic Role: Agent - Marker(s): p-subj,
COLOR(1)---->:BLACK,
AGGRESSIVE(1)---->:
IsKindOf-FISH(0.9)---->:
IsKindOf-VERTEBRATE(0.81)---->:
IsKindOf-ANIMAL(0.729)---->:
GENDER(0.729)---->:
IsKindOf-ANIMATE(0.6561)---->:
```

Conclusion

In this paper we discussed a lexicon model in which the role of context is not limited to sense selection. Selected senses are also modulated according to their surrounding context. We also described how patterns of contextual effects could be learned by the model. A prototype implementation of the model was also discussed.

Acknowledgments

Support from the Natural Science and Engineering Research Council of Canada is gratefully acknowledged.

References

Anderson, R., Pichert, J., Goetz, E., Schallert, D., Stevens, K., & Trollip, S. (1976) Instantiation of general terms. *Journal of Verbal Learning and Verbal Behavior* 15:667-679.

Barclay, J., Bransford, J., Franks, J., McCarrell, N. & Nitsch, K. (1974) Comprehension and semantic Flexibility. *Journal of Verbal Learning and Verbal Behavior* 13:471-481.

Barsalou L.W.(1982). Context-independent and context-dependent information in concepts. *Memory and Cognition* 10(11):82-93

Barsalou L.W.(1983). Ad hoc categories. *Memory and Cognition* 11(3):211-227.

Barsalou, L.W. & Hale C.R. (1993). Components of conceptual representation: from feature lists to recursive frames. In Mechelen I.V. et al, A. (Eds.), *Categories and concepts* Academic Press.

Clark, H. & Gerrig, R. (1983). Understanding old words with new meanings. *Journal of Verbal Learning and Verbal Behavior* 22:591-608.

Cruse, D. (1986). *Lexical Semantics*, Cambridge University Press.

Delisle S., Copeck, T., Szpakowicz, S. & Barker, K. (1993). Pattern matching for case analysis: A computational definition of closeness. ICCL, 310-315.

Dyer, M. (1983). *In-depth Understanding: A computer model of integrated processing for narrative comprehension.* Cambridge, MA: MIT Press.

Franks, B. (1995). Sense Generation: A "Quasi-Classical" Approach to Concepts and Concept Combination. *Cognitive Science* 19:441-505.

Gluksberg, S. & Keysar, B. (1990). Understanding metaphorical comparisons: Beyond literal similarity. *Psychological Review* 97(1): 3-18.

Greenspan, S. (1986). Semantic flexibility and referential specificity of concrete nouns. *Journal of Memory and Language* 25:539-557.

Kawamoto, A.H. 1993. Nonlinear dynamics in the resolution of lexical ambiguity: a parallel distributed processing account. *Journal of Memory and Language* 32:474-516.

Kilgarriff, A. 1997. What is word sense disambiguation good for?. Proceedings of Natural Language Processing in the Pacific Rim(NLPRS-97) Phuket, Thailand.

Kilgarriff, A. 1998. "I do not believe in word senses". *Computers and Humanities* 31(2): 91-113.

Lakoff, G. (1987). *Women, Fire and Dangerous Things.* The University of Chicago Press.

Lange, T. & Dyer, M. (1989). Frame selection in a connectionist model of high-level inferencing. *Proceedings of the 11th Conference of the Cognitive Science Society.*

Mahesh K. & Nirenburg, S.(1995). A situated ontology for practical NLP. *Proceedings of the Workshop on Basic Ontological Issues in Knowledge Sharing.* IJCAI 95. Montreal, Canada.

McClelland, J. & Kawamoto, A. (1986). Mechanisms of sentence processing. In McClelland J. and Rumelhaurt, D. (Eds.) *Parallel Distributed Processing: Explorations in the Microstructure of Cognition.* Vol. 2. MIT press.

Miikkulainen, R. 1993. *Subsymbolic natural language processing: An integrated model of scripts, lexicon, and memory.* Cambridge, MA: MIT Press.

Pustejovsky, J. (1995). *The generative lexicon.* MA, MIT Press.

Rais-Ghasem,, M. (1998) *An exemplar-based account of contextual effects* (Ph.D. Thesis) Ottawa, ON: Carleton University, School of Comp. Sc.

Ruhl, C. (1989) *On monosemy: A study in linguistic semantics* Albany, N.Y: State University of New York Press.

Veronis, J. & Ide, N. (1995). Large neural networks for the resolution of lexical ambiguity. In Saint-Dizier, P. & Vicgas, E. (Eds.) *Computational Lexical Semantics* NY, Cambridge University Press.

Waltz, D. & Pollack, J. (1985). Massively parallel parsing: A strongly interactive model of natural language interpretation. *Cognitive Science* 9:51-74.

Witney, P., McKay, T., & Kellas, G. (1985). Semantic activation of noun concepts in context. *Journal of Experimental Psychology: Learning, Memory, and Cognition* 11:126-135.

Modeling Cognitive Flexibility of Super Experts in Radiological Diagnosis

Eric Raufaste (raufaste@univ-tlse2.fr)
Laboratoire Travail et Cognition; Université de Toulouse le Mirail
31058 Toulouse Cedex; France

Abstract

The paper presents theoretical propositions for modeling the expert radiologist. The propositions are twofold. First, a *basic model* is given to complement a recent connectionist symbolic framework (Raufaste, Eyrolle, & Mariné, 1998). Empirical data have showed dissociation between two kinds of experts ("basic" and "super") with regard to cognitive flexibility. The difference is conceived as a kind of perseveration in basic experts. Hence, the basic model was combined with a Supervisory Attentional System (Norman & Shallice, 1986) into an "extended model". An analysis of cognitive activity is then presented within this framework, along with a new theoretical explanation of cognitive flexibility.

Modeling the expert radiologist

The Need for a Connectionist-Symbolic Approach

A well-documented ability in expert physicians is early selection of pertinent diagnostic hypotheses (Elstein, Schulman & Sprafka, 1978). In radiology, perceptual processes have a dramatic importance (Lesgold et al., 1981). However, a search time study suggested the existence of two components, the earlier rapid, and the latter slow (Christensen et al., 1981). It has been proposed that a "visual concept" shapes perception (Kundel & Nodine, 1983). Hence, the first component might be more plausibly described by a connectionist approach. But medical diagnosis is a complex task that requires a lot of deliberate reasoning, so it seems also necessary to have a symbolic layer. Empirically, Lesgold et al. (1988) found a nonmonotonic performance curve on some films: Novices (1- and 2-year residents in radiology) sometimes performed better than intermediates (3- and 4-year residents in radiology). Experts performed the best. To explain this nonmonotonicity, the authors proposed a three-stage framework. In the first stage, novices would acquire basic subsymbolic abilities such as recognizing the normal anatomy on the film. On a second stage, intermediates would develop "cognitive" (i.e., symbolic) abilities. For some cases, however, the cognitive processing would conflict with previously developed perceptual processing, resulting in a decreased performance. In experts, cognitive processing would have reached its plain development so that conflict no longer spoils performance. Thus, a connectionist-symbolic approach (Holyoak, 1991) seemed to be appropriate for modeling diagnosis.

Basic and super experts

In their experiments, Lesgold et al. used expert radiologists who were recognized as "outstanding" by peers. But only a few radiologists become outstanding. Moreover, such radiologists often have professional attributions that are substantially different from "normal" radiologists. These attributions induce more "deliberate practice" (Ericsson, Krampe, & Tesh-Römer, 1993) and more symbolic reasoning. In a recent study of expertise in radiological diagnosis, Raufaste, Eyrolle, and Mariné (1998) called "basic experts" the common radiology practitioners, and "super experts" the outstanding radiologists. The distinction appeared to be empirically fruitful and allowed us to reinterpret classical results by Lesgold et al. (1988). Raufaste et al. tested a framework that was initially devised to account for both subsymbolic and symbolic aspects of medical reasoning. Since subsymbolic and symbolic processes are integrated, they should not generate a conflict. Hence, there should not be nonmonotonicity. We found that performance curves on typical features was monotonically increasing from novices to basic experts. In contrast, performance curves on atypical features was monotonically decreasing from novices to basic experts. Such a result was in accord with our framework. However, super experts exhibited several features that could not stem from our model. In particular, they always had a better performance than the other groups, *even on atypical features*. This replicated the apparent non-monotonicity in the results of Lesgold et al. (1988) and showed a dramatic distinction between basic- and super-expertise: the former is accompanied with growing dependence on automatic processes whereas the latter allows a better independence from automatisms. A key point here is the fact that super experts were better than basic experts at detecting an inconspicuous feature *that could not be awaited* from the hypotheses associated with salient features. Such an effect could not be attributed to better perceptual abilities but rather to a better cognitive flexibility in super experts.

The present paper proposes an extension of the initial framework in order to model cognitive flexibility in super experts. After a new formulation of the model, a new level is added in the model. An original analysis of cognitive activity is conducted and explanatory mechanisms are proposed.

The basic model: a new formulation

The initial model had to account for previously known results about general expertise (e.g., Lesgold et al, 1988) and for what we called *pertinence generation*, that is the acquisition through experience –and the use, of the ability to select rapidly pertinent hypotheses. We present here a refined version of the model: The *basic model*. The basic model is grounded on the concept of Long-Term Working-Memory, that is, W-M is viewed as a more activated part of LTM (Anderson, 1983, Ericsson & Kintsch, 1995).

Schemas. Categories, called schemata, may be represented at two levels: They may be symbols (in the sense of Hinton, 1990) and/or they may be patterns that are distributed within subsymbolic networks. A symbol may or not be associated with a subsymbolic pattern. A subsymbolic pattern may or not be associated with a symbol.

Activating Attention postulate: any symbol in Working Memory is a source of activation. Although restricted to a single symbol, a similar postulate can be found in Collins and Loftus (1975).

Inferences and Reasoning. Inferences in the basic model may occur through two distinct processes. The first is spreading activation from a node to another node. For example, if a radiologist detects a cue, the corresponding visual pattern lends activation to its symbol (e.g., a specific syndrome) and activation can spread towards the symbols of the pathologies that are associated with the syndrome. The second process is activation of a production rule in procedural memory. We define a *focal threshold* as the quantity of activation that a symbol must reach for being consciously processed: A category can be symbolically processed (e.g. verbally reported) only if it is associated with a symbol whose activation is above the focal threshold.

As conscious attention works with limited resources, a plausible mechanism for conflict resolution is a competition based on the level of activation (e.g., "contention scheduling", Norman & Shallice, 1986). Thus, we add a new   postulate:
Captivating Activation postulate: the most activated symbols tend to obtain the focus of attention.

Those premises entail several interesting consequences: (1) a distributed pattern that was implicitly acquired cannot be symbolically processed until it has been associated with a symbol. (2) When the activation of a category increases, its probability to be symbolically processed also increases. (3) If the symbol of a category is inhibited, its probability of being symbolically processed decreases. (4) Through conscious call, activation may spread from a symbol to an associated subsymbolic pattern. Thus, some "substance" can be given to abstract concepts. Reciprocally, (5) activation may spread to a symbol from a distributed pattern that was activated by environmental stimuli. (6) The conscious representation can be defined as the set of categories having a symbol whose activation is above the focal threshold.

Initial Learning and Effects of Experience. The basic model is essentially a spreading-activation model of memory. Conscious processing takes place at the symbolic level and can initially generate theoretical knowledge by creating nodes (*canonical schemata*) in the network, and links (*canonical links*) between the nodes. In our view, experience has two main effects. The first effect is the classical view of connectionist networks: acquiring new nodes and/or new links; modifying the strengths of the links and the base-levels of the nodes. The second effect is to complement the knowledge base with examples that are encoded in episodic memory. In other words, declarative knowledge acquired through University learning, constitutes a pre-structured network of canonical schemata and canonical links around which further subsymbolic acquisitions will be arranged. Indeed, symbolic reasoning may still create new nodes and links after the medical degree course is over, through further reading, reflection on the results of actions, and so on...

Low accessibility postulate: without reinforcements from experience, the weak accessibility of symbols only allows a deliberate access to canonical schemata.

This postulate explains why sometimes novices were found not to use knowledge which they have (Custers, Boshuizen, & Schmidt, 1996). Until the link between symbols and subsymbolic background becomes enough strengthened by experience, only deliberate reasoning may trigger a schema. For example, a novice detects an abnormal feature on a film but does not know how to evoke pathology from the feature. The pathology, however, may be activated by a procedural rule whose action part provides activation to its symbol. The more the subject encounters simultaneously a context and a symbol, the more the link between both is strengthened and the more further encountering of the context will automatically activate the symbol. Thus, with experience, rule-based reasoning is replaced by spreading activation. From the same context, the expert will be able to activate more symbols with more links between them, that is, to generate a richer representation. Due to lateral inhibition effects, pertinent symbols will receive more activation from the context and, therefore, are more likely to win the competition for attentional resources. Hence, because they are more integrated, expert representations should also be more pertinent. Because the representation is richer, complex productions rules may also be triggered and so complex reasoning also becomes available in experts.

Basically, our results as well as the literature fit the model. However, one result was clearly not in agreement with the basic model, even in its current form.

A reduced SOS phenomenon in Super Experts

The Satisfaction of Search phenomenon (SOS, Berbaum et al., 1990) is a well-known effect in the literature about observer performance in radiology: an inconspicuous feature (e.g., a lung nodule) has a lower probability to be detected when the film also presents with a more salient unrelated feature. By itself, this phenomenon is clearly "predicted" by the basic model: (1) the most salient features naturally receive first the focus of attention. (2) Due to spreading activation, the symbols that are strongly associated with this initial context receive more activation other symbols; (3) Due to the captivating attention postulate, those symbols are reinforced by conscious attention and (4) they become activation sources. From this moment, they control behavior, and rules that relate to those symbols are more likely to fire. In particular, rules for hypothesis testing will lead further exploration of the film, and features that are non-directly relevant to those hypotheses may be missed. As the model predicted, the SOS phenomenon increased with basic expertise, from novices to basic experts. In novices, SOS is reduced by the lower strength of the links between subsymbolic context and symbols. Indeed, other causes for rigidifying effects may be found (e.g., Feltovich, Spiro & Coulson, 1997). On some cases, however, super experts avoided the SOS phenomenon. This is clearly not compatible with the basic model because they should have been even more subject to SOS than basic experts. One might suggest that they might have a better visual ability but an eye-recording study showed that cognitive processes were responsible for the SOS phenomenon (Samuel et al., 1995): most missed nodules were fixated and erroneously categorized as variants of normal. Structural properties of specific knowledge cannot account for the SOS phenomenon because the same diagnosticians do detect the same nodules when no independent salient feature is present. Because neither perceptual abilities nor specific knowledge can be responsible for the better performance we observed in super experts, we need to turn to general mechanisms of control. Thus, we need to extend the basic model.

An Extended Model of Expertise

From a neuropsychological standpoint, it has been argued that an activation-based competition between schemata is not sufficient to model a normal human subject. It can only model a patient with prefrontal lesions (e.g., Shallice & Burgess, 1993). In our attempt to model basic and super expertise, it seems interesting to view the SOS phenomenon as a particular case of perseveration. In such a view, a major difference between basic- and super-expertise is the relative weight of the specific-knowledge base. Basic experts may be modeled by a contention-scheduling process in a knowledge base whereas modeling super experts requires, in addition, the existence of an instance that modulates contention-scheduling. We adopt here, the concept of *Supervisory Attentional System* (*SAS*, Norman & Shallice, 1986).

Adding a modulation process to the basic model is not sufficient by itself to explain super-expert flexibility: We also need to model cognitive activity. First, we call *mental state* (Smolensky, 1988) a particular pattern of activation in the network. We call *cognitive flow* the sequence of mental states. Now, attentional control has a heavy cognitive cost. Therefore, the SAS is not expected to function actively

without a good reason. We consider that the SAS may be in two states: (1) We call *intervention* the state where the SAS actively modifies the course of contention scheduling. (2) In contrast, we call *survey* the state during which the SAS is not active. We call *natural flow* the cognitive flow when the SAS is in a survey position. Thus, the cognitive flow can be analyzed as a sequence of natural flows which, sometimes, is interrupted by SAS interventions (Figure 1). Natural flow sequences are guided by the procedural schemata associated with the dominant declarative schema (i.e., only the rules that are associated with this schema are activated enough to fire).

Within such an analysis, two questions must be solved: (1) How are the SAS interventions triggered? and (2) How do interventions work? To answer those questions, and in addition to the previous descriptive approach to cognitive activity, we need a general principle that orients the activity:

Principle of coherence maximizing: Cognitive activity is intended to maximize the overall coherence of the cognitive system. At a symbolic level, the principle gives the orientation of cognitive activity. At a subsymbolic level, the principle enables computation in neural nets (for justifications, see Thagard, 1989 and related works).

How are interventions triggered? Because of their importance in daily activity, for both experts and ordinary people, we expect interventions to be a very low-level process. Because cognitive flexibility seems to differ from basic to super experts, and because the two kinds of experts mainly differ by their daily activity, we assume that the low-level procedure can be triggered by procedural schemata which can be learned:

We define a *ruptor* as a schema that operates a categorization on mental states and that activates a low-level procedure of intervention on the cognitive flow. Because of the maximizing coherence principle, ruptors should take as input a kind of information that embodies an estimate of the overall (and/or local) coherence of the network. For example, detecting a deadlock is a good reason for triggering an intervention (See Holyoak, 1991).

With regard to the specific problem of basic and super expertise, we must also justify how interventions explain why super experts are less prone to the SOS phenomenon. The concept of ruptor provides a simple explanation in terms of differential intervention triggering.

_ Because ruptors are schemas, they are subject to the contention-scheduling process. This explains why a basic expert becomes more and more dependent on automatisms. As the automatisms become more efficient, they are more likely to win the competition. With experience they are more and more refined so deadlocks become sparse. Finally, because basic expert daily activity does not include much deliberate practice, ruptors are not systematically trained. Thus, with experience, Basic expertise tends to be more and more in agreement with the basic model because the weight of the SAS is continuously decreasing.

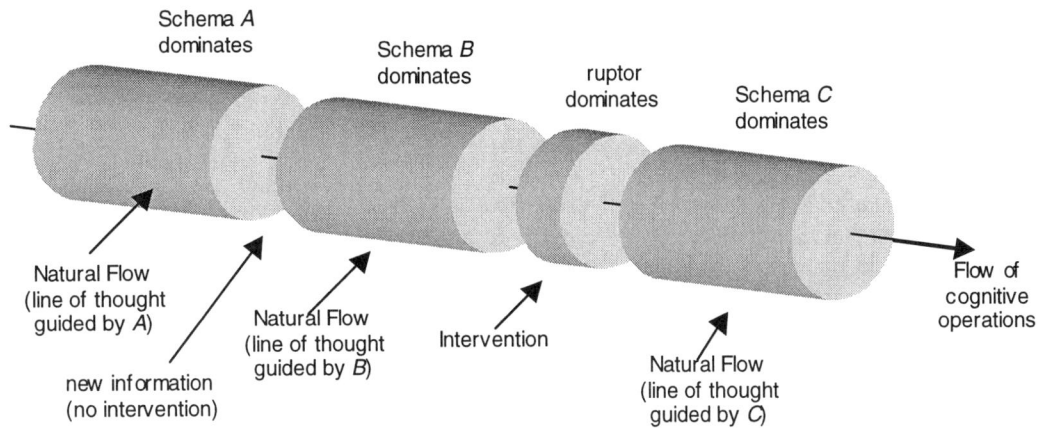

Figure 1 : An analysis of the flow of cognitive operations

_ In super experts, daily activity is accompanied with much more explicit reasoning. As researchers, they have to make their reasoning explicit in order to publish, whereas basic experts only have to provide a diagnosis. As teachers of internists, they have to justify their reasoning, to make their implicit inferences explicit. In addition, super experts are often called for diagnosing those cases that were deemed too much difficult by other radiologists. For all those reasons, they are trained to commit high levels of attention in diagnosis. They are more familiar with the fact that automatisms may lead to wrong solutions. The factors that favor cognitive flexibility (For a review, see Feltovich et al., 1997) are construed in the extended model as generating and training new ruptors. Super expert's SASs are more trained to intervene and their ruptors probably have a higher base-level than basic expert ruptors. Thus we may expect them to be more independent from automatisms than basic experts.

Now, we have a basis for an explanation of the difference between basic- and super-experts with regard to cognitive flexibility. Nevertheless, for the explanation to be complete, we should be able to explain how an intervention may reduce the likelihood of the SOS phenomenon.

How interventions work? The SOS phenomenon can be regarded as a kind of perseveration induced by a positive feedback loop. This loop results from to the combination of captivating activation and activating attention postulates. Then, to avoid the SOS, interventions must be able to break the loop. As we stated earlier, interventions are expected to be a low-level process. In our analysis, interventions are short actions from the SAS. After the intervention, new schemata can gain the control over the cognitive flow. Hence, a minimal action of the SAS is to inhibit the current dominating schema (see McCarthy & Warrington, 1990) so that a new sequence of natural flow can begin. More sophisticated explanations based on concepts like Harmony Optimization (Smolensky, 1986) can be found in the theory of stochastic neural nets. However, they are beyond the scope of the present paper.

A Preliminary Test of the Extended Model

A complete test of the extended model is not out of reach, but it requires such methods as eye-movement recording because in order to observe a real SOS phenomenon, one has to ensure that the critical cue was actually seen. We just want to verify the plausibility of the main idea of the model—the SOS phenomenon might be related to a slight form of perseveration, which accompanies basic expertise and can be avoided in super experts by SAS interventions.

If the explanation we proposed is correct, we should be able to find some traces of interventions in verbal protocols. In particular, we should be able to find more interventions in super experts than in basic experts.

With regard to novices and intermediates, a lot of explicit reasoning, and even of deliberate practice (Ericsson et al., 1993) is likely because they have to acquire a vast specific knowledge-base within few years. Moreover, they are trained to learn because before being internists in a specialty, like radiology, they were selected among the best students in general medicine courses. Nevertheless, as their specific knowledge-base grows, they should depend more on automatisms and less on general mechanisms such as weak heuristics and SAS interventions. Only the few who will some day become super experts can be expected to maintain a high level of deliberate activity. As a consequence, we can expect a monotonely decreasing curve in the number of interventions from novices to basic experts, and a higher number of interventions in super experts than in basic experts.

The next question is how can we measure interventions? The main consequence of interventions is to change the schema that guides the reasoning process. Therefore, we can trace those changes in verbal protocols. We call *line of thought* the verbal trace of a sequence of natural flow. A sequence of natural flow is not observable, whereas a line of thought can be traced in the verbal protocols. The basic idea behind the test is that interventions change line of thoughts and, therefore, tracing the changes in the line of thoughts give some indications on the number of interventions. However, not only interventions change lines of thought: The dominating schema can be inhibited because a crucial new information gives a strong positive support to a concurrent hypothesis or a strong negative support to the dominant schema. In other words, we should not count as interventions the changes that can be attributed to a new information arrival. Another important factor in the changes is the use of a systematic strategy of exploration. Radiology residents are taught to explore the films according to a topographical schema that enables them to explore systematically every important zone. Therefore, when a diagnostician uses such a strategy, the dominant schema is not pathology but the topographic exploration schema. Hence we should not count changes in the dominant schema that can be attributed to the use of such a strategy.

In order to give a preliminary testing of these ideas, the verbal protocols that were used in Raufaste et al. (1998) were coded. It should be noted that the new coding was completely independent and different from the original coding. Thus, the number of interventions could be obtained (Figure 2). The test involved 8 novices, 6 intermediates, 4 basic experts, and 4 super experts.

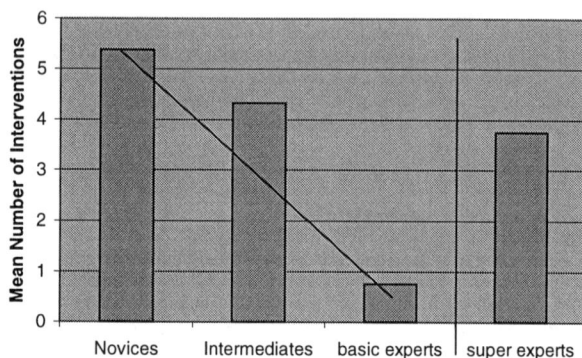

Figure 2: Interventions as a function of expertise

As expected, one can observe a monotonic decrease in the curve: The mean number of interventions is maximal in novices (5.38, $s_d = 3.16$), it decreases down to 4.33 ($s_d = 2.07$) in intermediates, to 0.75 ($s_d = 0.96$) in basic experts. Moreover, the relation is significant ($F_{(2,17)} = 4.63$; $p = .0135$). Also expected, with 3.75 interventions on average ($s_d = 2.22$), the mean number of interventions is significantly higher in super experts than in basic experts ($t_{(6)} = -2,48$; $p = .024$). The graph presented in Figure 2 is typically of the same kind as the graph one can draw with regard to performance on atypical cases.

When examining the whole curve of planed-triggering, which can be produced from the four groups, we obtain a significant decreasing monotonic relation ($F_{(2,17)} = 5.684$; $p = .006$) ranging from 3.63 ($sd = 1.85$) in novices down to 0.50 ($sd = 0.58$) in super experts. The latter result confirms that the better performance of super experts in the detection of an inconspicuous feature cannot be explained by the use of a more systematic exploration procedure.

Discussion

The model of cognitive flexibility proposed here should be considered as complementary to the prescriptions of Cognitive Flexibility Theory (e.g., Feltovich et al., 1997). Explaining differences between basic and super experts might also have been approached through Rasmussen's model of control (e.g., Olsen & Rasmussen, 1989) which construes cognitive flexibility in terms of the ability to adapt the control mode (skill-, rule-, or knowledge-based) to the specificity of the situation. However, if a radiologist devotes most of his attentional resources to an abstract reflection, the control is knowledge-based with regard to the reflection while, at the same time, there is place for a skill-based control of the film exploration. More generally, automated processes can operate in parallel to attentional processes. Hence, the control is not skill-based only, rule-based only, or knowledge-based only. To the contrary, our model uses two mutually exclusive categories: a state of survey and a state of intervention. In many respects, our model could also be compared to a hybrid model of abduction like UEcho (e.g. Wang, Johnson, & Zhang, 1997). In a recent study (Raufaste & Da Silva-Neves, 1998), basic expert radiologists were found to conform to Possibility Theory (Zadeh, 1978; Dubois & Prade, 1988). UEcho, however, is not compatible with nonstandard approaches, which require two measures of uncertainty. Our model, in contrast, is compatible with these results as well as with results in the literature where subjects tend to conform to bayesian rules of reasoning (Raufaste, Da Silva-Neves, & Mariné, submitted manuscript). Moreover, UEcho combines ECHO and the SOAR architecture whereas our model resembles a hybrid form of the ACT-R architecture and Norman & Shallice' theory of action.

This paper presented a refined version of a model of expertise in radiological diagnosis (Raufaste et al., 1998). The model is now twofold. The "basic model" embodies the previous model as well as two new postulates that relate to attention. In its current version, it is sufficient to model the basic expert. The paper also presented an "extended model". In addition to the basic model, it includes a Supervisory Attentional System (Norman & Shallice, 1986) that accounts for super expert's behavior. The extended model can account for a wide range of results about radiological expertise. Being within the frame of symbolic connectionism, it has the potential to deal with purely perceptual aspects of diagnosis as well as attentional deliberate reasoning. Indeed, much work will now be necessary to test the model. It is expressed in a general form, and might serve for many other domains of expertise. Therefore, it could be useful to any researcher who studies abductive reasoning and/or reflective experts.

Acknowledgments

This research was supported by Grant n° 99N35/0005 of the CNRS – "GIS Sciences de la Cognition".

References

Anderson, J.R. (1983). *The Architecture of cognition*, Cambridge Massachusetts, Harvard University Press.

Anderson, J.R. (1993). *Rules of the mind*. NJ: LEA.

Berbaum, K.S., Franken, E.A., Dorfman, D.D., Rooholamini, S.A., Kathol, M.H., Barloon, T.J, Behlke, F.M., Sato, Y., Lu, C.H., El-Khoury, G.Y., Flickinger, F.W., & Montgomery, W.J. (1990): Satisfaction of search in diagnostic radiology. *Investigative Radiology, 25,* 133-140.

Christensen, E.E., Murry R.C., Holland, K., Reynolds, J., Labday, M.J., & Moore, J.G. (1981). The effect of search time on perception. *Radiology, 138,* 361-365.

Collins, A.M., & Loftus, E.F. (1975). A spreading-activation theory of semantic processing. *Psychological Review, 82,* 6, 407-428.

Custers, E.J.F.M., Boshuizen, H.P.A., & Schmidt, H.G. (1996). The influence of medical expertise, case typicality, and illness script component on case processing and disease probability estimates. *Memory & Cognition, 24,* 3, 384-499.

Darling, S., Della Sala, S., Gray, C., & Trivelli, C. (1998). Putative functions of the prefrontal cortex: Historical perspectives and new Horizons. In G. Mazzoni & T.O. Nelson (Eds.), *Metacognition and cognitive neuropsychology: Monitoring and control processes.* Mahwah,NJ: LEA.

Dubois, D., & Prade, H. (1988). *Possibility Theory*, New York: Plenum Press.

Elstein, A.S., Shulman, L. S., & Sprafka, S.A. (1978). *Medical Problem Solving: An analysis of clinical reasoning.* Cambridge, Ma: Harvard University Press.

Ericsson, K.A., Krampe, R.T., & Tesh-Römer, C. (1993). The role of deliberate practice in the acquisition of expert performance. *Psychological Review, 100,* 3, 363-406.

Hinton, G.E. (1990). Mapping part-whole hierarchies into connectionist networks. *Artificial Intelligence, 46,* 47-75.

Holyoak, K.J. (1991). Symbolic connectionism: Toward third-generation theories of expertise. In K.A. Ericsson & J. Smith (Eds.), *Toward a general theory of expertise.* Cambridge University Press.

Kundel, H.L., & Nodine, C.F. (1983). A visual concept shapes image perception. *Radiology, 146,* 2, 363-368.

Lesgold, A.M., Feltovich, P.J., Glaser, R., & Wang, Y. (1981, September). *The acquisition of perceptual diagnostic skill in radiology (Tech. Rep. No. PDS-1).* University of Pittsburgh, Learning Research and Development Center.

Lesgold, A.M., Rubinson, H., Feltovich, P., Glaser, R., Klopfer, D., & Wang, Y. (1988). Expertise in a complex skill: Diagnosing X-Ray Pictures. in M.T.H. Chi, R. Glaser, & M.J. Farr (Eds.), *The Nature of Expertise.* Hillsdale, NJ: LEA.

Feltovich, P. J., Spiro R. J., & Coulson, R. L. (1997). Issues of expert flexibility in contexts characterized by complexity and changes. In P.J. Feltovich, K.M. Ford, & R.R. Hoffman. (Eds.), *Expertise in context. Human and machine.* Cambridge, Mass: MIT Press.

McCarthy, R.A., & Warrington, E.K. (1990). *Cognitive Neuropsychology: A clinical introduction.* San Diego, CA: Academic Press.

Norman, D.A., & Shallice, T. (1986). Attention to action : Willed and automatic control of behavior. in R.J. Davidson, G.E. Schwartz, & D. Shapiro (Eds.), *Consciousness and self-regulation* (vol. 4). New-York: Plenum Press.

Olsen, S.E., & Rasmussen, J. (1989). The Reflective Expert and the Prenovice: Notes on Skill-, Rule- and Knowledge-base Performance in the Setting of instruction and Training. In L. Bainbridge & S.A. Ruiz-Quintanilla (Eds.), *Developing skills with information technology.* Chichester : Wiley.

Patel, V.L., & Groen, G.J. (1986). Knowledge base solution strategies in medical reasoning. *Cognitive Science, 10,* 91-116.

Raufaste, E., & Da Silva Neves, R. (1998). Empirical evaluation of possibility theory in human radiological diagnosis. In H. Prade (Ed.), *Proceedings of the 13th Biennial Conference on Artificial Intelligence,* ECAI'98 (pp. 124-128). London: John Wiley & Sons, Ltd.

Raufaste, E., Da Silva Neves, R.M., & Mariné, C. (submitted manuscript). Consonant Support Theory: A psychological theory of confidence judgements in diagnostic reasoning.

Raufaste, E., Eyrolle, H., & Mariné, C. (1998). Pertinence generation in radiological diagnosis: Spreading activation and the nature of expertise. *Cognitive Science, 22,* 4, 517-546.

Samuel, S., Kundel, H.L., Nodine, C.F., & Toto, L.C. (1995). Mechanism of satisfaction of search: eye-position recordings in the reading of chest radiographs. *Radiology, 194,* 3, 895-902.

Shallice, T., & Burgess, P.W. (1993). Supervisory control of action and thought selection. In A. Baddeley & L. Weiskrantz (Eds.), *Attention: Selection, awareness, and control.* Oxford: Clarendon Press.

Smolensky, P. (1986). Information processing in dynamical systems: Foundations of Harmony Theory. In D.E Rumelhart, J.L. McClelland J.L., & The PDP Research Group. *Parallel Distributed Processing. Explorations in the microstructure of cognition. Volume 1: Foundations.* Cambridge, Mass: MIT Press.

Smolensky, P. (1988). On the proper treatment of connectionism. *Behavioral and Brain Sciences, 11,* 1-23.

Thagard, P. (1989). Explanatory coherence. *Behavioral and Brain Sciences, 12,* 435-502.

Wang, H., Johnson, T. R., & Zhang, J. (1997). UEcho: A model of uncertainty management in human abductive reasoning. Hybrid Technical Report No. 4 (TR-97/ONR-HYBRID-04). The Ohio State University.

Zadeh, L.A. (1978). Fuzzy sets as a basis for a theory of possibility. *Fuzzy Sets and Systems, 1,* 3-28.

A Feedback Neural Network Model of Causal Learning and Causal Reasoning

Stephen J. Read (read@rcf.usc.edu)
Department of Psychology, University of Southern California
Los Angeles, CA 90089-1061

Jorge A. Montoya (gmontoya@rcf.usc.edu)
Department of Psychology, University of Southern California
Los Angeles, CA 90089-1061

ABSTRACT

We present a feedback or recurrent, auto-associative model that captures several important aspects of causal learning and causal reasoning that cannot be handled by feedforward models. First, our model learns asymmetric relations between cause and effect, and can reason in both directions between cause and effect. As a result it can represent an important distinction in causal reasoning, that between necessary and sufficient causes. Second, it predicts cue competition among effects and provides a mechanism for them, something which can only be done with feedforward models by assuming that two separate networks are learned, a highly non parsimonious assumption. Finally, we show that contrary to previous claims, a feedforward model cannot handle Discounting and Augmenting in causal reasoning, although a feedback model can. The success of our feedback model argues for a greater focus on such models of causal learning and reasoning.

Introduction

Connectionist models of causal learning and reasoning have relied on feedforward networks (e.g., Gluck & Bower, 1988; Shanks, 1991; Van Overwalle, 1998). However, as we have recently shown, feedforward networks have serious limitations as models of causal learning and reasoning (Read & Montoya, in press). In that paper, we outlined an alternative, a feedback or recurrent model, that can handle phenomena that a feedforward model cannot. In the current paper we examine further implications of this kind of model for phenomena that feedforward models cannot handle, such as asymmetries in causal learning and reasoning, and cue competition for consequences or effects.

In previous work we have examined how this kind of model can handle a number of phenomena in causal learning and causal reasoning. Read and Montoya (in press) have demonstrated that it can successfully simulate many of the classic phenomena from the animal and human causal learning literature, such as blocking and conditioned inhibition, to which the Rescorla-Wagner model (Rescorla & Wagner, 1972) and feedforward models with delta-rule learning (e.g., Gluck & Bower, 1988; Shanks, 1991), have been applied. Read and Montoya also demonstrated that this auto-associative model, which is a parallel constraint satisfaction model, deals with the principles of explanatory coherence discussed by Thagard (1989, 1992) and

experimentally demonstrated by Read and Marcus-Newhall (1993) and Read and Lincer-Hill (1998) (see also Ranney, in press; Schank & Ranney, 1991, 1992). Finally, several papers (Montoya & Read, 1998; Read & Miller, 1993) have shown that this kind of model can simulate the Discounting and Augmenting principles in causal reasoning (Kelley, 1971), as well as the role of factors, such as construct accessibility and causal strength, that may underlie the closely related Correspondence Bias or Fundamental Attribution Error (Jones, 1990; Ross, 1977).

In the current paper, we focus on the implications for causal learning and reasoning of a central aspect of this model: all nodes are completely interconnected, with an independent link going in each direction between each pair of nodes. This has three implications which we will examine. First, because each pair of nodes is joined by two links, one in each direction, it is possible to reason both from cause to effect and from effect to cause. In contrast, with the feedforward models previously investigated in causal learning and reasoning, it is only possible to learn and reason in one direction, typically from cause to effect. Second, because each member of the pair of links can have different strengths, the link from cause to effect can have a different strength than the link from effect to cause. As a result, with this model one can learn asymmetric relations between cause and effect, and use these asymmetric relations in causal reasoning. Third, because the network is totally interconnected, it can learn relations among possible causes of an event. In contrast, in the feedforward networks used in this domain the only links are forward, from cause to effect. It is not possible to learn links among causes. One implication of this, we will argue, is that the standard feedforward model is incapable of handling either discounting or augmenting in causal reasoning, whereas our model can handle both phenomena.

An Auto associative Model.

Our model is based on McClelland and Rumelhart's (1988) auto-associator, which is a single layer auto-associative network with all units completely interconnected. Each unit receives input from other nodes and simultaneously sends activation to other nodes. Because of the feedback relations, this network functions as a parallel constraint satisfaction system, acting to satisfy multiple simultaneous constraints among elements in the network. Links are modified by delta-rule learning and each link in a pair can end up with a different weight. All of the nodes can receive input from both the environment and other nodes. Thus, both cause and effect nodes can be activated by environmental cues. (Although Thagard's ECHO model is also a feedback model, it assumes that both links between pairs of nodes have identical weights. Thus, there is no way to represent asymmetric causal relations in ECHO and no way to examine the role of differences in links from cause to effect and effect to cause. Further, because ECHO has no learning mechanism, it cannot learn causal links (however, see Wang, Johnson, and Zhang (1997) who have recently added delta rule learning to ECHO).)

This network can learn associations among all the elements that co-occur. That is, not only can it learn the relation between the effect X and potential causes A and B, it can also learn the association between the two potential causes. In contrast, in feedforward networks, there are links in only one direction, from input nodes to output nodes. Output nodes only receive activation from the input nodes, and cannot be directly activated by the environment. Also, there are no links among the nodes in a layer; the only links are between layers. Thus, it cannot learn associations between causes.

Processing in the auto associative network proceeds as follows. After input is received, all the units in the network are synchronously updated at each cycle by an activation function that is essentially the same as that employed in ECHO (Thagard, 1989; 1992) and in Rumelhart and McClelland's (1986) interactive-activation and competition model, as well as in a handful of other models they have explored. This activation function is:

$$a_j(t+1) = a_j(t)(1-d) + \begin{cases} net_j(max-a_j(t)) & \text{if } net_j > 0 \\ net_j(a_j(t)-min) & \text{if } net_j \le 0 \end{cases}$$
$$\text{where } net_j = (istr)[\Sigma w_{ji}a_i] + (estr)ext$$

The only minor difference in this activation function for the auto-associative architecture, compared to other models in which it has been used, is that the total input net_j is now determined by external input from the pattern vector *ext*, as well as the sum of weighted inputs from other units within the network with activations from the previous cycle, $\Sigma w_{ji}a_i$. Note that the internal input and the external input are scaled, by *istr* and *estr*, respectively.

After the system completes a number of processing cycles (defined by the user), the delta rule (or Widrow-Hoff rule) (Widrow & Hoff, 1960) is applied to the network to compare the external input pattern to the internal inputs to units. This learning regime reduces the difference between internal and external inputs to units, by modifying the weights among the nodes, so that the internal input comes to reproduce or match the external input to the units. Hence, the *desired* activation of a unit is determined by the set of external inputs to that unit. The discrepancy between the *desired* and *actual* activation of a unit is the measure of error used in delta rule learning. Weight change is given by:

$$\Delta weight_{ji} = lrate(t-a_j)a_i,$$

where lrate is the learning rate, t is the target or external activation, a_j is the internal or actual activation, and a_i is the activation of the node sending activation to a_j.

Learns and Uses Asymmetries in Causal Relations.

One advantage of this model is that separate links exist from cause to effect and from effect to cause. As a result, this model is able to learn any asymmetries that might exist in these relationships. Further, having learned these asymmetries, they can be used in causal reasoning.

In contrast, neither current associative models (e.g., Gluck & Bower, 1988; Shanks, 1991; Van Overwalle, 1998) nor Cheng's (Cheng & Novick, 1990, 1992) probabilistic contrast model can learn separate relations for cause to effect and effect to cause. In fact, both capture the relationship from cause to effect, but not the reverse relationship. Thus, these models cannot learn asymmetries in cause-effect and effect-cause relations. Further, these models do not allow for reasoning in both directions.

Several authors (e.g., Shanks, Lopez, Darby, & Dickinson, 1996) suggest that one could capture the two different directions of causal learning by using two feedforward networks, one with causes as inputs and the other with effects as inputs. However, with recurrent networks, such as the present model, only one network is required. This is much more parsimonious than assuming that an individual would require two separate networks to capture bi-directionality in causal learning and reasoning.

Table 1 gives a set of learning trials that result in asymmetric learning of links, such that cause A has a stronger forward link to X than does cause B, whereas effect X has a stronger backward link to cause B than to cause A. In this example, assume that we are learning and reasoning about possible causes of a forest fire (X). One possibility is lightning (A) while another is a campfire (B).

Table 1: Learning History for Asymmetry in Causes

Simulation	Unit	Learning history	Epochs
Asymmetry	A	+ + ...	20
	B	. . + + + + + + + +	
	X	+ + + + + + ...	

Because of the pattern of covariation, asymmetric causal relations are learned. The model learns that if it occurs, lightning is more likely to cause a forest fire, than is a campfire. However, it also learns that if there is a forest fire it was more likely preceded by a campfire then by lightning. This asymmetry is apparent in both the activations when causes and effects are separately tested and in the patterns of weights that are learned.

When we separately activate the two causes, A (lightning) alone leads to a higher activation for X (forest fire) than does B (campfire), .35 versus .16. However, if effect X alone (forest fire) is activated then cause B (campfire) is more highly activated, .37, than is cause A (lightning), .27.

The connection strengths leads to the same conclusion. The connection from A (lightning) to X (forest fire) is stronger than the connection from B (campfire) to X (forest fire), 1.58 versus .74. However, the connection from

X(forest fire) to B(campfire) is stronger than the connection from X(forest fire) to A (lightning), 1.88 versus .94. The model has learned that the occurrence of lightning is more likely to cause a forest fire than is the occurrence of a campfire. However, it has also learned that if a forest fire occurs that it is more likely to be caused by a campfire.

Such asymmetries seem to be an important part of human causal reasoning, and our model easily captures them. Yet a feedforward model, because links only go from input to output, is completely unable to learn such asymmetries and thus is unable to reason asymmetrically.

Captures the Distinction between Necessary and Sufficient Causes

Our ability to model asymmetries in causal learning and reasoning also allows us to capture what has been identified as a central distinction in causal reasoning, the difference between necessary and sufficient causes. For instance, a lit match is sufficient to set gasoline on fire, but it is not necessary because there are other ways in which the gasoline can be ignited. This can be captured in our network by assuming that the strength of a link from cause to effect captures the sufficiency of a cause; the stronger this link the more likely the cause is to bring the effect about. In contrast, the link from effect to cause captures the necessity of a cause; the stronger the link, the more likely it is that the cause preceded the effect. A very strong link from effect to cause suggests that the effect is almost always preceded by that cause, suggesting that the cause is necessary for the effect to come about. Because it cannot learn such asymmetries, having links that only run from cause to effect, a feedforward model cannot learn or use information about this fundamental distinction between necessary and sufficient causes.

Cue competition among effects

In the human and animal causal learning literature, there is considerable evidence for cue competition among causes. One example of such cue competition is Blocking, where first learning that cue A strongly predicts an effect prevents the later learning of the relation between cue B and the effect, even if cue B is highly predictive of the effect. The standard explanation is that cues compete for predictive strength and that when cue A is learned to strongly predict the effect, this essentially captures all the available predictive strength, leaving none for B. Both the Rescorla-Wagner rule and feedforward networks with delta rule learning can capture such cue competition for causes.

But does cue competition for effects also occur, when a single cause predicts multiple effects? Waldmann (1996) points out that the Rescorla-Wagner model strongly predicts such effects and that their absence would create serious problems for this model. However, Waldmann argues that his causal-model theory predicts that cue competition for effects should not occur. And across several studies, he found no evidence for cue competition for effects.

However, other researchers (e.g., Chapman, 1991; Shanks, 1991; Shanks, Lopez, Darby, & Dickinson, 1996) have provided evidence for cue competition for effects. Miller and Matute (1996) have argued that the discrepancy in results among various researchers might be attributable to differences in the questions used to assess causal strength.

This possibility is particularly clear in terms of our model, which suggests that whether one gets cue competition for effects may depend strongly on the type of question that is asked to assess causal strength. That is, does the question ask subjects to assess the strength from cause to effect or from effect to cause? Waldmann (1996), among others, has characterized this difference as between asking predictive questions and asking diagnostic questions. A predictive question asks subjects to assess the extent to which the cause predicts potential effects. In terms of our model, such a question asks subjects to assess the strength of the link from the causes forward to the effect. In contrast, a diagnostic question asks subjects to assess the extent to which the effect is diagnostic of the cause, that is, to what extent the existence of the effect provides evidence for the cause. In terms of our model, this question asks subjects to assess the strength of the link from the effect to the cause.

Cue competition for <u>causes</u> is typically demonstrated when two or more causes predict a single effect, and researchers ask a predictive question about the extent to which the causes predict the effects. Our model suggests that cue competition for <u>effects</u> should be demonstrated when a single cause predicts two or more effects, and subjects are asked a diagnostic question, for which they must assess the strength of the link from the effect back to the cause.

One obvious implication of this is that the learner must be able to separately encode the link from cause to effect, and the link from the effect back to the cause. Several researchers (e.g., Shanks, Lopez, Darby, & Dickinson, 1996) have suggested that this can be captured by assuming two feedforward networks, one that learns the relations from causes to effects and the other which learns the relations from effects back to cause. However, such a solution seems inelegant. With the current model, such an assumption is unnecessary, as a basic part of its architecture is that it can learn separate weights for the two links from cause to effect, and from effect to cause.

Table 2: Learning History for Simulation of Cue Competition

Simulation	Unit	Learning history		Epochs
Phase I	A	+ + + + +		10
	X	. . . . .		
	Y	+ + + + +		
Phase II	A		+ + + + +	
	X		+ + + + +	10
	Y		+ + + + +	

We have successfully simulated cue competition for effects when a single cause predicts multiple effects and the right question is asked. In the simulation, there are two alternative stimulus presentations (See Table 2). In the first, a single cause (A) is presented that predicts two effects (X and Y) (Phase II alone). In the second, the network is first presented with a number of instances of one effect (Y) predicted by a single cause (A) (Phase I), followed by two effects (X and Y) predicted by the same cause (A) (Phase II).

In this model, separate links are learned from cause to effect, and from effect to cause. And as can be seen in Figure 1, there is an asymmetry in the learned links for the learning sequence of Phase I followed by Phase II. Moreover, it is clear from the links that in this model whether one should expect to get cue competition for effects, depends upon the direction of reasoning. For Phase II alone, equal weights are learned among all the causes and effects (.76). However, when Phase I is presented first, followed by Phase II, the results are quite different, predicting a cue competition effect for effects or consequences. First, there are strong weights from the cause A to both effects X and Y, although the weight is twice as strong from A to Y (1.53 vs. .76). However, in the reverse direction, the weight from X to A is 0, while the weight from Y to A is 1.53. And when we examine the resulting activations (See Table 3) when each of the causes and effects are tested, we get strong evidence for cue competition in backward reasoning from effects to causes, but not in forward reasoning from causes to effects. When effect X is turned on, neither cause A nor effect Y is activated at all. In contrast, when effect Y is turned on, both cause A and effect X are activated. Further, when cause A is activated, effects X and Y have almost identical activations, although Y is slightly higher. Thus, there is strong evidence for cue competition when reasoning backward, from effect to cause, but not when reasoning forward, from cause to effect. Thus, this model suggests that whether one gets cue competition for effects will depend on the direction of reasoning.

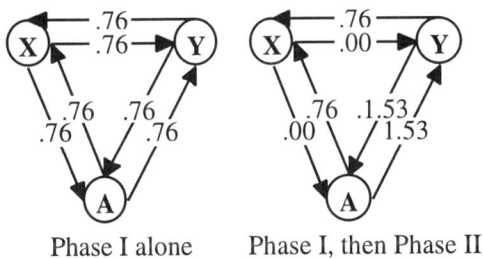

Phase I alone   Phase I, then Phase II

Figure 1: Weights for Cue Competition Simulation

Table 3: Output Activations for Cue Competition Simulation

Simulation	Units Tested	Resulting activations for:		
		A	X	Y
Phase II alone	A	.55	.29	.29
	X	.29	.55	.29
	Y	.29	.29	.55
Phase I followed by Phase II	A	.58	.32	.39
	X	.00	.48	.00
	Y	.39	.32	.58

Represents Learning of and Reasoning with Relations between Causes.

As we noted above, Rescorla-Wagner and feedforward models cannot directly capture relations between causes, but only relations between cause and effect. As a result, we argue that feedforward models are unable to capture Discounting and Augmenting (Jones & Davis, 1965; Kelley, 1971), although feedback models can.

A number of authors (e.g., Baker, Mercier, Vallée-Tourangeau, Frank, & Pan, 1993; Shanks, 1985, 1991; Vallée-Tourangeau, Baker, & Mercier, 1994; Van Overwalle & Van Rooy, 1998) have suggested that a feedforward model with delta rule learning can handle the Discounting and Augmenting principles identified by Kelley (1971) and Jones and Davis (1965). The claim is that Discounting is the same as Blocking found in studies of animal learning, and Augmenting is the same as Super-conditioning. However, despite their apparent similarity the underlying processing mechanisms for the two sets of phenomena are quite different. Moreover, feedforward models lack the necessary mechanism for capturing discounting and augmenting, as they lack the ability to represent relations among causes, which we argue is critical for capturing these effects..

In Blocking, if the organism first learns that A is strongly associated with effect X, when it is later presented examples of B and A covarying with X, the organism fails to learn the new association between B and X. In terms of error correcting learning, such as the delta-rule, once X is strongly predicted by A, when A and B are subsequently paired with X, there is little discrepancy between the actual and predicted value of X (no error) and therefore little change is made in the weight from B to X.

Thus, Blocking clearly deals with competition in the initial learning of the causal links. In contrast, Discounting in the human literature clearly deals with competition among already learned causal explanations. Kelley (1971) and Jones and Davis (1965) were considering adults who were relying on already learned and activated knowledge. For instance, consider adults who are told that a woman wrote a pro-abortion essay after being assigned to the position by her debate coach. Because of the assignment, they should discount a pro-abortion attitude as a cause of her behavior. These adults already know that both a pro-abortion attitude and the assignment by the coach are possible explanations for the behavior. They are not learning these relationships for the first time. Thus, in contrast to Blocking, Discounting does not refer to the failure to learn a causal link, but rather reasoning on the basis of already learned causal knowledge.

What changes in the typical Discounting situation is information about the availability or presence of a potential cause in a particular situation. Both McClure (1998) and Morris and Larrick (1996) have argued that the degree of discounting between two causes is a function of the extent to which they are positively or negatively related. Discounting can be handled in an auto-associative model by assuming that there is an inhibitory link between competing explanations (Read & Miller, 1993; Read & Marcus-Newhall, 1993) (This cannot be done in a feedforward model). Because of the inhibitory link, increased availability of a plausible alternative will reduce the activation of the other explanation. Thus, we aren't looking at competition for learning of links, but rather at competition for the activation of concepts with previously learned causal links.

Now consider Super conditioning and its relation to Augmenting. If the organism learns that A is followed by X, but A and B together are not followed by X, then B develops a negative or inhibitory relationship with X. If the organism then learns that D and B together are followed by X, then the relationship between D and X becomes stronger than it would have been if B had not first developed a negative relationship with X. Again, although this phenomena is similar to Augmenting, it is not the same thing. Augmenting deals with inhibition between an already learned cause and effect, whereas Super-conditioning is based on inhibition in the initial learning of causal relationships. For example, suppose we are told that someone got an A on an extremely difficult exam. We use our preexisting causal knowledge to infer that the individual must be quite smart. We are clearly not learning for the first time that someone who can overcome a major barrier must possess a considerable amount of the relevant ability, which is what Super-conditioning would be concerned with. Clearly, there is a critical distinction between the initial acquisition of information and the ways in which it is later used.

Morris and Larrick (1996) make a similar distinction. They note that in models of causal reasoning, there is a distinction between induction or the initial acquisition of causal knowledge, and reasoning or attribution, the actual use of that knowledge. For instance, Kelley's (1971) ANOVA cube model is a model of the acquisition of causal knowledge, whereas his causal schema model is a model of the use of pre-existing knowledge for reasoning.

Thus, the two types of phenomena are fundamentally different in terms of the underlying processing mechanisms. Blocking and Super-conditioning deal with competition for weight strength in the learning of new causal relations, whereas Discounting and Augmenting deal with competition for activation in the use of already learned causal relations. These are quite different processes. And as we noted, a feedforward model is unable to capture a situation in which the causal mechanism depends on links among causes.

Summary

In this paper we have demonstrated that a feedback or recurrent, auto-associative model can capture several important aspects of causal learning and causal reasoning that cannot be handled by the feedforward models that have been the typical focus of investigation. First, our model can learn asymmetric relations between cause and effect. Second, it can reason in both directions between cause and effect. As a result it can represent an important distinction in causal reasoning, the difference between necessary and sufficient causes. Third, because the nodes in the network are totally interconnected, it can represent cue competition among effects, something which can only be done with feedforward models by assuming that two separate networks are learned, a highly non parsimonious assumption. Finally, we argue that contrary to previous claims, a feedforward model cannot handle Discounting and Augmenting in causal reasoning. However, a feedforward model can. The success of our feedback model suggests that researchers should focus more energy on the capabilities of such models of causal learning and reasoning.

References

Baker, A. G., Mercier, P., Vallée-Tourangeau, F., Frank, R., & Pan, M. (1993). Selective associations and causality judgments: Presence of a strong causal factor may reduce judgments of a weaker one. Journal of Experimental Psychology: Learning, Memory, and Cognition, 19, 414-432.

Chapman, G. B. (1991). Trial order affects cue interaction in contingency judgment. Journal of Experimental Psychology: Learning, Memory, and Cognition, 17, 837-854.

Chapman, G. B., & Robbins, S. J. (1990). Cue interaction in human contingency judgment. Memory and Cognition, 18, 537-545.

Cheng, P. W., & Novick, L. R. (1990). A probabilistic contrast model of causal induction. Journal of Personality and Social Psychology, 58, 545-567.

Cheng, P. W., & Novick, L. R. (1992). Covariation in natural causal induction. Psychological Review, 99, 365-382.

Gluck, M. A., & Bower, G. H. (1988). Evaluating an adaptive network model of human learning. Journal of Memory and Language, 27, 166-195.

Jones, E. E. (1990). Interpersonal perception. New York: W. H. Freeman.

Jones, E. E., & Davis, K. E. (1965) From acts to dispositions: The attribution process in person perception. In L. Berkowitz (Ed.), Advances in experimental social psychology (Vol. 2). New York: Academic Press.

Kelley, H. H. (1971). Attribution in social interaction. In E. E. Jones, D. Kanouse, H. H. Kelley, R. E. Nisbett, S. Valins, & B. Weiner (Eds.), Attribution: Perceiving the causes of behavior. Morristown, NJ: General Learning Press.

McClelland, J. L., & Rumelhart, D. E. (1986). (Eds.). Parallel Distributed Processing: Explorations in the microstructure of cognition. Vol. 2: Psychological and Biological Models. Cambridge, MA: MIT Press/Bradford Books.

McClelland, J. L., & Rumelhart, D. E. (1988). Explorations in parallel distributed processing: A handbook of models, programs, and exercises. Cambridge, MA: MIT Press/Bradford Books.

McClure, J. (1998). Discounting causes of behavior: Are two reasons better than one? Journal of Personality and Social Psychology, 74, 7-20.

Miller, R. R., & Matute, H. (1996). Animal analogues of causal judgment. In D. R. Shanks, K. J. Holyoak, & D. L. Medin (Eds.), The Psychology of Learning and Motivation, Vol. 34: Causal learning. San Diego, CA: Academic Press.

Montoya, J. A., & Read, S. J. (1998). A constraint satisfaction model of the correspondence bias: The role of accessibility and applicability of explanations. In M. A. Gernsbacher & S. J. Derry (Eds.), Proceedings of the Twentieth Annual Conference of the Cognitive Science Society. Mahwah, NJ; Erlbaum.

Morris, M. W., & Larrick, R. P. (1996). When one cause casts doubt on another: A normative analysis of discounting in causal attribution. Psychological Review,

102, 331-355.

Ranney, M. (in press). Explorations in explanatory coherence. In E. Bar-On, B. Eylon, & Z. Schertz (Eds.)., Designing intelligent learning environments: From cognitive analysis to computer implementation. Ablex: Norwood, NJ.

Read, S. J., Lincer-Hill, H.(1999). Principles of explanatory coherence in trait inferences. Unpublished manuscript, University of Southern California, Los Angeles, CA.

Read, S. J., & Marcus-Newhall, A. (1993). Explanatory coherence in social explanations: A parallel distributed processing account. Journal of Personality and Social Psychology, 65, 429-447.

Read, S. J., & Miller, L.C. (1993). Rapist or "regular guy": Explanatory coherence in the construction of mental models of others. Personality and Social Psychology Bulletin, 19, 526-540.

Read, S. J., & Miller, L. C. (1994). Dissonance and balance in belief systems: The promise of parallel constraint satisfaction processes and connectionist modeling approaches. In R. C. Schank & E. Langer (Eds.), Beliefs, reasoning, and decision making: Psycho-logic in honor of Bob Abelson). Hillsdale, NJ: Lawrence Erlbaum Associates.

Read, S. J., & Miller, L. C. (1998). On the dynamic construction of meaning: An interactive activation and competition model of social perception. In S. J. Read & L. C. Miller (Eds.) Connectionist models of social reasoning and behavior.(pp. 27-68). Mahwah, NJ: Erlbaum.

Read, S. J., & Montoya, J. A. (in press). An autoassociative model of causal reasoning and causal learning: Response to Van Overwalle's critique of Read and Marcus-Newhall (1993). Journal of Personality and Social Psychology.

Read, S. J., Vanman, E. J., & Miller, L. C. (1997). Connectionism, parallel constraint satisfaction processes, and gestalt principles: (Re)introducing cognitive dynamics to social psychology. Personality and Social Psychology Review, 1(1), 26-53.

Rescorla, R. A., & Wagner, A. R. (1972). A theory of Pavlovian conditioning: Variations in the effectiveness of reinforcement and non-reinforcement. In A. H. Black & W. F. Prokasy (Eds.), Classical conditioning II: Current research and theory. New York: Appleton-Century-Crofts.

Ross, L. (1977). The intuitive psychologist and his shortcomings: Distortions in the attribution process. In L. Berkowitz (Ed.), Advances in experimental social psychology. (Vol. 10). New York: Academic Press.

Rumelhart, D. E., & McClelland, J. L. (1986). Parallel distributed processing: Explorations in the microstructure of cognition: Vol. 1. Foundations. Cambridge, MA: MIT Press/Bradford Books.

Schank, P.K., & Ranney, M. (1991). An empirical investigation of the psychological fidelity of ECHO: Modeling and experimental study of explanatory coherence. Proceedings of the Thirteenth Annual Conference of the Cognitive Science Society. Hillsdale, NJ: Erlbaum.

Schank, P. K., & Ranney, M. (1992).Assessing explanatory coherence: A new method for integrating verbal data with models of on-line belief revision. Proceedings of the Fourteenth Annual Conference of the Cognitive Science Society. . Hillsdale, NJ: Erlbaum.

Shanks, D. R. (1985). Forward and backward blocking in human contingency judgment. Quarterly Journal of Experimental Psychology, 37B, 1-21.

Shanks, D. R. (1991). Categorization by a connectionist network. Journal of Experimental Psychology: Learning, Memory, and Cognition. 17, 433-443.

Shanks, D. R., Lopez, F. J., Darby, R. J., & Dickinson, A. (1996). Distinguishing associative and probabilistic contrast theories of human contingency judgment. In D. R. Shanks, K. J. Holyoak, & D. L. Medin (Eds.), The Psychology of Learning and Motivation, Vol. 34: Causal learning. San Diego, CA: Academic Press.

Thagard, P. (1989). Explanatory coherence. Behavioral and Brain Sciences, 12, 435-467.

Thagard, P. (1992). Conceptual revolutions. Princeton: Princeton University Press.

Vallée-Tourangeau, F., Baker, A. G., & Mercier, P. (1994). Discounting in causality and covariation judgments. Quarterly Journal of Experimental Psychology, 47B, 151-171.

Van Overwalle, F. (1998). Causal explanation as constraint satisfaction: A critique and a feedforward connectionist alternative. Journal of Personality and Social Psychology, 74, 312-328.

Van Overwalle, F., & Van Rooy, D. (1998). A connectionist approach to causal attribution. In S. J. Read & L. C. Miller (Eds.) Connectionist models of social reasoning and behavior. Mahwah, NJ: Erlbaum.

Waldmann, (1996). Knowledge-based causal induction. In D. R. Shanks, K. J. Holyoak, & D. L. Medin (Eds.), The Psychology of Learning and Motivation, Vol. 34: Causal learning. San Diego, CA: Academic Press.

Wang, H., Johnson, T. R., & Zhang, J. (1997). UEcho: A model of uncertainty management in human abductive reasoning. In M. G. Shafto & P. Langley (Eds.). Proceedings of the Nineteenth Annual Conference of the Cognitive Science Society. Mahwah, NJ: Erlbaum.

Widrow, G., & Hoff, M. E. (1960). Adaptive switching circuits. Institute of Radio Engineers, Western Electronic Show and Convention, Convention Record, Part 4, 96-104

The Development of Explicit Rule-Learning

Martin Redington (m.redington@ucl.ac.uk)
Department of Psychology, University College London,
26, Bedford Way, London, WC1E 6BT, UK

Elliot Ronald (elliot.ronald@corpus-christi.oxford.ac.uk)
Department of Experimental Psychology, University of Oxford,
South Parks Road Oxford, OX1 3UD, UK

Abstract

Implicit and explicit learning were originally distinguished in terms of accessibility to verbal report. We identify evidence for the proposal that the implicit/explicit contrast corresponds to a divide between connectionist and symbolic representations. We show that explicit learning shows marked improvement between 4 and 8 years of age. This finding contrasts against very early implicit learning abilities, and concurs with other evidence on the progressive development of symbolic reasoning abilities.

The identification and study of human learning mechanisms is a central concern of psychology and cognitive science. An important contemporary debate in this area concerns the distinction between implicit and explicit learning. Generally, implicit and explicit learning mechanisms have been distinguished in terms of the accessibility of the knowledge acquired to conscious awareness, as assessed by verbal report (e.g., Reber, 1967).

At first this division appears to be relatively convincing: In implicit learning, by definition, participants' verbally reported knowledge is insufficient to account for their performance on some task. For example, participants who memorise strings generated by a finite state grammar are subsequently able to classify strings as obeying or violating the grammar to a significant degree, despite being unable to report the rules of the grammar verbally (e.g., Reber, 1967; Reber & Allen, 1978).

By implication, explicit learning is defined as those cases where participants are able to verbally report sufficient knowledge to account for their performance. For example, if a person can play a legal game of chess, and can also report the rules of chess, then one might conclude that their ability to play a legal chess game was based on this explicit knowledge.

However, the claim that dissociations between verbal report and performance mark the boundary between two distinct learning mechanisms, differing in the accessibility of their knowledge to conscious awareness, has proved problematic. The strongest critics, Shanks and St. John (1994), argue that the insensitivity of verbal report, and the problems of relating participants' reports to the knowledge representations underlying their performance render apparent dissociations between performance and verbal report suspect. Shank and St. John do however endorse the view that learning mechanisms can be distinguished in terms of the representational form of the knowledge acquired, contrasting exemplar or instance-based learning mechanisms, for which connectionism provides a natural model, against processes of hypothesis testing and rule-discovery, best described by symbolic mechanisms.

In connectionist learning mechanisms, knowledge and cognition are embodied in patterns of activation of many simple units, and in the flow of activations between those units. Learning is the modification of the strengths of the connections between units. In symbolic mechanisms, knowledge is embodied by statements or rules composed of arbitrary symbols, and interpreted according to a consistent syntax and semantics. Learning is the addition of new statements or rules.

Dienes and Perner (1996) suggest that viewing implicit and explicit learning in terms of a divide between connectionist and symbolic mechanisms explains the differing availability of implicit and explicit knowledge to verbal report. The form of knowledge in a connectionist network—the strengths of interunit connections—does not lend itself to verbal communication. In contrast, symbolic knowledge can be easily communicated, and utilised by the receiver.

In this paper we present evidence in support of an implicit/explicit divide based on representational form. We first describe a recent study which dramatically contrasts explicit and implicit learning. We then present evidence on the development of explicit learning, showing marked developmental changes between four and eight years of age. This contrasts against developmental evidence on implicit learning, which appears to function in mature form from the first year of life. We discuss similar findings on the development of symbolic reasoning abilities from other paradigms.

Two dissociable human learning systems

Shanks, Johnstone and Staggs (1997, Experiment 4) report a study where, as in implicit learning studies, they presented participants with a set of rule-governed training strings, and subsequently tested their ability to distinguish between test strings which obeyed or violated the rules underlying the training strings.

However, the materials used by Shanks et al. (1997)

584

were very different to those of implicit learning studies.[1] Generally in artificial grammar learning studies, materials are drawn from complex grammars, with many (e.g., ten) rules, which specify relationships between adjacent letters. The distinction between grammatical and nongrammatical strings is usually correlated with simple local distributional properties of the training materials, such as the frequency of letter pairs and triples.

Shanks et al. (1997) drew their training and test strings from a crypto-grammar: Grammatical strings possessed the structure 1234.1234, with each number being replaced by the two halves of a pair of letters according to the following three rules: D ↔ F, G ↔ L, K ↔ X. Thus a typical string might be DFGK.FDLX. Nongrammatical strings violated one or more of these rules. Additionally, the training and test items were painstakingly constructed so that only conformance to the rules distinguished grammatical and nongrammatical strings: Local distributional properties provided no useful information.

Shanks et al. (1997) utilised two different training conditions. In the *match* condition, participants were shown a single grammatical training item, and then had to match that example to one of a display of five training items. This is akin to the memorisation training usually used in studies of implicit learning, with participants uninformed that the training stimuli were rule-governed. In the *edit* training condition, participants were informed of the rule-governed nature of the materials. They were shown items which violated the rules, and were required to indicate which elements (letters) were correct and which were not. After each item they were shown the correct string and given feedback as to the actual errors present in the string. This training was intended to facilitate rule-discovery processes.

At test, the match group showed no ability to appropriately classify the test items as obeying or violating the rules. Participants in the edit group fell into two distinct subgroups. One subgroup scored at or around chance on the grammaticality judgment test, while, the other subgroup scored at or near 100% of classifications correct. The manipulation of materials and training conditions appears to flip participants between implicit and explicit learning modes, with the latter resulting in a distinctive bimodal pattern of performance.[2]

The Shanks et al. (1997) results also provide support for the view that the differences between implicit and explicit knowledge are best explained in terms of the contrast between connectionist and symbolic representations.

The failure of the match group, who were presented with a typical implicit learning paradigm, to make accurate grammaticality judgments concurs with accounts of implicit learning which stress the learning of local distributional properties (e.g., Perruchet & Pacteau, 1990; Redington & Chater, 1996). In the Shanks et al. (1997) materials, by design, local distributional properties of the materials give no cue to grammaticality. However, in most artificial grammar learning studies, such distributional properties are strong predictors of grammaticality, and sensitivity to such properties is sufficient to account for human performance. Connectionist models, such as the simple recurrent network (Elman, 1990), provide a natural framework for learning of this kind, and are able to capture much of the data on artificial grammar learning (Redington, 1996).

As for the the edit group, three features of their performance contrast clearly against implicit learning, and suggest a process of symbolic, rule-based learning:

1. The step function of participants' performance: Participants either discover the correct rules, or they do not. With "typical" artificial grammar learning materials (e.g., Reber, 1967, or the commonly used set from Reber & Allen, 1978), participants performance is unimodal, and imperfect: Participants' classification scores exceed chance and untrained controls, but never approach 100% (60–70% correct is a typical score).

2. The ability to capture relationships between "arbitrary" (non-adjacent) elements is consonant with a rule-based representation. Evidence from both the Shanks et al. (1997, Experiment 4) study and St. John & Shanks (1997) suggests that implicit learning is limited to local dependencies (between adjacent or near-adjacent letters).

3. A hitherto unmentioned manipulation in the Shanks et al. (1997) crypto-grammar study was that test items were either similar specific training items (two letters different to), or dissimilar (at least four letters different from) to all of the training items. The edit group were equally likely to classify both kinds of items as grammatical. The absence of effects of surface similarity is often proposed as an indicator of rule-learning, and contrasts against studies such as Vokey and Brooks (1992), where under implicit conditions, participants showed clear effects for the similarity of training and test items (which can generally be explained in terms of similarity in terms of distributional properties).

The Shanks et al. (1997) effects appear to be robust: We replicated the crypto-grammar study, using only the edit training condition. Using the exact same procedure and stimuli, six of our participants ($n = 12$) clearly showed evidence of rule-learning (near-ceiling performance), while the remainder scored at or around chance. Verbal reports and a post-task questionnaire provide convergent evidence that this study captures the implicit/explicit distinction: Participants who showed near-ceiling classification were

---

[1]The materials and training conditions used by Shanks et al. (1997) were based on those of a similar study by Mathews, Buss, Stanley, Blanchard-Fields, Cho and Druhan (1989, Experiment 4). However, the contrast between implicit and explicit learning is much clearer in the Shanks et al. study.

[2]It is the interaction of materials and training conditions that is important: In an earlier study Shanks et al. (1997, Experiment 3) used the same training conditions with typical artificial grammar learning materials (with many complex rules, governing local dependencies). Both match and edit groups showed typical implicit learning effects (i.e., performance was unimodal and imperfect).

able to report the rules of the crypto-grammar without error, whereas those who scored near chance were unable to report the rules.

Alternative hypotheses

Although the Shanks et al. (1997) effects do point towards the operation of two distinct learning mechanisms, and two different forms of representation, the possibility that a single learning mechanism (and representational form) underlies performance on both explicit and implicit tasks remains.

To sketch one possible alternative hypothesis, match and edit training might encourage the consideration of different hypotheses, or different orderings of hypotheses, by a single, symbolic, learning mechanism. Edit training might encourage the initial consideration of nonlocal relationships, while memorisation training might limit hypotheses to local dependencies. With a small number of rules, participants might be able to discover and report them all (as for the learners in the edit condition). When the number of rules is large (as in the relatively complex artificial grammars), participants may well fail to discover every rule, and the sheer number of rules might preclude accurate reporting of every rule that has been learnt. These additional assumptions about the effect of training conditions and the relationship between the complexity of the knowledge base and accessibility to verbal report allow the evidence to be reconciled with a single symbolic learning mechanism.

In general, the problem of distinguishing between different kinds of representational form is very difficult (e.g., see Barsalou, 1990). Additional assumptions will always permit apparent dissociations to be reconciled with a single learning mechanism or representational form. However, it may be possible (and possibly necessary) to support the case for distinct learning mechanisms by appealing to multiple lines of converging evidence. For example, if implicit and explicit learning mechanisms show different profiles of development, this would reinforce the case that there are two distinct mechanisms. Below we present some evidence from a developmental study of the explicit learning effects found by Shanks et al. (1997).

Developmental evidence

As far as the development of implicit learning is concerned, preliminary evidence suggests that implicit learning mechanisms are functioning to a considerable degree within the first year of life. For instance, Saffran, Newport, and Aslin (1996) found that 8-month-old infants exposed to a stream of phonemes of an artificial language subsequently exhibited sensitivity to the sequential structure of the sequence in a preferential listening paradigm. Gomez and Gerken (1997) have observed artificial grammar learning effects in 11- to 13-month old infants, again using a preferential listening paradigm. In both of these studies, performance was unimodal and far from ceiling. Although cross-sectional studies remain to be performed, the indication is that implicit learning mechanism functions in essentially the adult form from very early on.

The question we investigated here was the developmental profile of the explicit rule-learning effects found by Shanks et al. (1997). We assessed the performance of four, six, and eight-year-olds on a variation of the edit training condition. We predicted that if explicit learning reflected the the action of a separate learning mechanism, then we would observe a substantial increase in the proportion of learners (participants showing near-perfect classification performance) with increasing age. As well as the classification task, we also administered the Test for Reception of Grammar (TROG test, Bishop, 1981), in order to provide some measure of the cognitive development of each child.

The participants ($n = 36$) were all students at a Hertfordshire primary school. There were 12 participants for each age group, with mean ages of 61, 82, and 102 months.

The materials were based on those used by Shanks et al. (1997, Experiment 4). We simplified the material in order to reduce the effect of differences in cognitive ability or memory capacity due to age. The materials were expressed in terms of animals and fruit in order to provide an engaging task for four- to eight-year-olds.

Grammatical strings all followed the pattern 123.123, with numbers being replaced by elements of the following rules: elephant → orange, lion → apple, rabbit → banana. Note that unlike the Shanks et al. materials, the rules here are unidirectional: Grammatical strings always follow the pattern "animal animal animal fruit fruit fruit" as in the sample test item shown in Figure 1. The divider between the animals and fruits in Figure 1 was present in both training and test stimuli, in order to make the division more salient for participants.

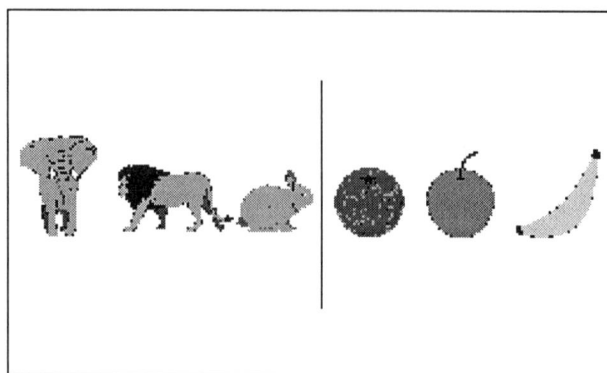

Figure 1: A sample test item (27.5% of actual size).

We constructed two sets of materials (shown in Table 1). Each test set consisted of 12 grammatical and 12 non-grammatical strings. The grammatical and nongrammatical strings had the same combinations of animals, but the combinations of fruit in the nongrammatical strings violated one, two, or three of the above rules. The distributional properties (letter pairs and triples) of the grammatical and nongrammatical items were roughly balanced, and all letter pairs or triples in the nongrammatical strings were also present in the grammatical strings, and vice versa. The nongrammatical items from each test set served as the training items for participants tested on the opposite set. Training and test items were presented to participants on individual sheets of paper.

Table 1: The two sets of test items. A, B, and C refer to elephant, lion and rabbit, and 1, 2, and 3 refer to orange, apple and banana respectively . Underlining indicates a violation of the rules. The training items for each test set were the nongrammatical items from the opposite set.

Set 1		Set 2	
AAB.112	AA<u>B</u>.11<u>3</u>	AAC.113	AA<u>C</u>.11<u>2</u>
ACA.131	AC<u>A</u>.13<u>3</u>	ABA.121	AB<u>A</u>.12<u>2</u>
BAA.211	BA<u>A</u>.21<u>3</u>	CAA.311	CA<u>A</u>.31<u>2</u>
BBC.223	BBC.11<u>3</u>	BBA.221	BB<u>A</u>.33<u>1</u>
BAB.212	BA<u>B</u>.33<u>2</u>	BCB.232	BCB.11<u>2</u>
CBB.322	CBB.13<u>2</u>	ABB.122	AB<u>B</u>.32<u>1</u>
CCA.331	CC<u>A</u>.22<u>3</u>	CCB.332	CCB.11<u>3</u>
CBC.323	CBC.23<u>1</u>	CAC.313	CAC.23<u>1</u>
ACC.133	ACC.32<u>1</u>	BCC.233	BCC.31<u>2</u>
ABC.123	AB<u>C</u>.32<u>3</u>	ACB.132	AC<u>B</u>.23<u>2</u>
BCA.231	BCA.32<u>1</u>	BAC.213	BAC.23<u>1</u>
CAB.312	CA<u>B</u>.12<u>3</u>	CBA.321	CBA.13<u>2</u>

The procedure of the study was as follows: Participants were tested individually in a quiet room, performing the TROG test first, and then performing the edit training and classification test. Prior to edit training, the experimenter informed the child about the task, using the following words:

I am now going to show you some pictures. Each animal likes only one kind of fruit. I will show you three animals and three fruit. In each of these pictures some of the animals don't get the fruit they like. Can you tell me which animals get the fruits they like and which don't?

If you think the animal gets the fruit they like then tick the box under the animal and the fruit. If you think they don't then cross the box under the animal and the fruit.

The order of the animals and the fruit is important. They must be in the right place to get the fruit they

like. At first you will be guessing. I will put the correct answers in the grey boxes to help you learn the rules.

The two different sets of materials shown in Table 1 were counterbalanced within each age group. The training items were presented in random order. Each training item consisted of a nongrammatical item, with a white and grey box under each animal/fruit, for the child's response and feedback from the experimenter. After the child had indicated which animals and fruits they thought were correct, the experimenter would write the correct sequence of ticks and crosses. This is very similar to the edit task used by Shanks et al. (1997), with the exception that the corrected sequences were not presented with the feedback. Participants also received verbal feedback intended to encourage rule-discovery. If the child correctly identified all of errors, they were told "Well done. You got all of them right." In items containing two errors, the experimenter would draw the child's attention to the correct animal-fruit pair, by asking "What do you think the X likes?" In items containing only one error, the experimenter would draw the child's attention to the two "satisfied" animals and their food, by stating "The X gets what they like and the Y gets what they like. The A is eaten, and so is the B." The set of 12 nongrammatical training items was presented twice, for a total of 24 training trials. After training, participants received the following instructions before proceeding to the test phase.

Now you will see some more pictures of the same animals and fruit. I want you to tell me if each page is right. It is right if each animal gets the fruit they like, and wrong if any of the animals don't get the fruit they like.

The 24 test items were then presented in random order. The experimenter recorded the children's responses, but gave no feedback as to whether they were correct or not.

Results

We first analysed the training data. For each training item, the child's response was correct if they correctly identified all rule violations, and incorrect otherwise. The 24 training trials were divided into four blocks of six items each. A $2 \times 3 \times 4$ (Training Set $\times$ Age Group $\times$ Block) mixed model ANOVA revealed effects for Training Set, $F(1, 30) = 11.27, p = 0.002$, Age Group, $F(2, 30) = 13.04, p = 0.0001$, and Block, $F(3, 90) = 5.12, p = 0.0026$. The Training Set $\times$ Age interaction was also reliable, $F(2, 30) = 5.72, p = 0.0079$. The effect of Training Set, and the reliable interaction suggest that Training Set 2 was more likely to encourage rule-learning than Training Set 1, moreso with increasing age, despite the identical nature of the two sets of materials. In fact, it appears that despite random allocation of participants to materials within each age group, participants trained and tested on Set 2 were significantly more advanced, as measured by their TROG scores, than those tested on Set 1,

$t(35) = 14.84, p = 0.0001$, and this difference may well account for the apparent effect of training materials.

The effect of Block is consistent with a process of hypothesis-testing, and the eventual discovery of the correct rules (similar effects during learning were observed by Shanks et al., 1997, for participants trained on the edit, but not the match task). The effects for Age Group suggests that the ability to discover the correct rules increases with age, although we also expected to observe an interaction between Age Group and Block, which was not reliable.

A notable feature of the training data (also observed in our replication of the Shanks et al. 1997 study), was that some participants reached ceiling performance on the training task early on in training (by the third block). These four participants were all eight years of age, and performed perfectly on the subsequent classification task.

Overall, the mean proportion of classifications correct for the four, six, and eight year-olds were, respectively, .49 (.10), .69 (.18), and .82 (.21), (standard deviations in parentheses). A 2×3 Test Set $\times$ Age Group ANOVA, with proportion of classifications correct as the dependent variable, revealed reliable effects for both Test Set, $F(1, 30) = 8.61, p = 0.006$, and Age Group $F(1, 30) = 16.27, p = 0.0001$. There was also a reliable interaction effect, $F(2, 30) = 4.63, p = 0.018$.

Once again, the effect of materials, and the interaction with Age Group is probably due to differing levels of development in the participants, despite random allocation of participants to materials. The effect of Age Group on the classification task provides clear evidence of improvement in explicit rule-learning between the ages of four and eight.

Examination of the raw data (shown in Figure 2) shows that participants' performance exhibited the bimodal pattern observed in the edit group of the Shanks et al. study. Although some participants' performance was intermediate, most children showed either chance or perfect performance (i.e., no learning, or correct learning of all rules). The prevalence of intermediate performance may be due to the fact that some test strings could be rejected merely on the basis of the training instructions. For example, given that "each animal likes only one kind of fruit" in items like "lion lion rabbit apple orange banana" one of the lions will clearly not be happy. However, this possibility cannot account for scores of .80 or more.

Informal questioning of the participants, and a questionaire administered to a subset of participants both confirmed our findings in the adult version of the task: Participants who scored at or near-ceiling were able to accurately report all of the rules, whilst those who scored below ceiling were not.

Discussion

The results reported above suggest that, in contrast to implicit learning, the ability to acquire and utilise a system of explicit rules does show clear developmental effects in early childhood.

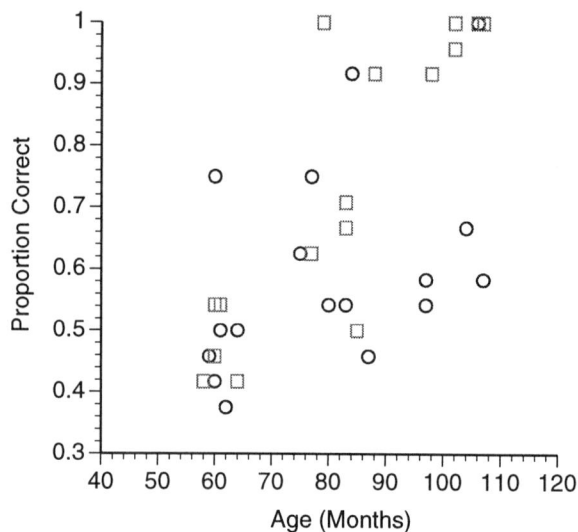

Figure 2: Classification score (proportion of judgments correct) as a function of age (months) and test set. Circles indicate participants tested on Test Set 1, and squares indicate participants tested on Test Set 2.

The improvement over age does not appear to be directly due to improvements in memory capacity or general cognitive ability: While piloting the materials, we tested children's ability to follow the rules (as opposed to having to both discover and follow the rules). When told the rules, children of four, six, and eight years were all able to make appropriate grammaticality judgements, and scored near ceiling.

As well as providing converging evidence in favour of two distinct learning mechanisms; an early functioning implicit system, and a gradually developing explicit system, this study reinforces other evidence on the development of apparently symbolic reasoning abilities. Raijmakers and Molenaar (1995, 1996) have recently drawn attention to the discrimination-shift task, which they argue provides a strong test for distinguishing between purely associative representations, and mechanisms possessing representations of "mediating concepts."

In the discrimination-shift task (Kendler, 1995), participants are reinforced for distinguishing between four stimuli presented in two distinct pairs. The stimuli are distinguishable on two dimensions (e.g., shape [round/triangle] and colour [white/black]), one of which is reinforced. After participants make the appropriate distinction to a given criterion, the shift phase of the experiment begins (without informing the participant). The reinforcement contingencies undergo either a reverse shift (RS), where the previously reinforced stimuli get negative reinforcement and vice versa, or an extradimensional shift (EDS), where the dimension on which reinforcement is based (e.g., shape or color) is shifted. According to Kendler (1995), animals learn by an associative mechanism, and learn an EDS faster than an RS. Adult humans learn by forming "mediated concepts" and learn a RS faster than an EDS.

As for the rule-discovery task reported here, child performance on the discrimination-shift task shows a clear developmental progression, with the proportion of children showing adult-like performance in the majority only after around 6 years of age. Prior to this, most children show discrimination-shift performance characteristic of animals, learning a EDS faster than an RS.

Raijmakers and Molenaar (1996) report that feedforward connectionist networks perform like animals or young children on the discrimination-shift task. In contrast, symbolic representations provide a natural metaphor for mediating concepts. Future studies might assess children's performance in both the rule-discovery and discrimination-shift tasks, in order to see if the apparent correspondence between development in these two tasks is present within particular individuals.

Another interesting line of potential research concerns the performance of amnesic and elderly patients on the explicit rule-discovery task. Implicit learning appears relatively unimpaired in amensic patients (Knowlton, Ramus & Squire, 1992), despite their known deficits in declarative memory. This tallies neatly with the properties of connectionist and symbolic representations. Connectionist representations degrade gracefully in the face of damage, because their knowledge is distributed over many interunit connections. Symbolic mechanisms are brittle, because their knowledge is concentrated in discrete statements, interpreted by a single processing mechanism. Damage to either rules or processor can have drastic consequences. The finding of impaired explicit rule-learning in amnesic patients would further reinforce the case for representationally distinct implicit and explicit learning mechanisms.

Acknowledgements

Thanks to David Shanks and Theresa Johnstone, for their advice and generous provision of their materials and experimental software. We are grateful to all of the children and teachers for their cooperation during this study. Thanks also to Hilary Leevers for advice on appropriate materials for use with children, and to Giles Shilson, Annette Karmiloff-Smith, and three anonymous reviewers for comments on an earlier version of this paper. This research was supported in part by the U.K. Economic and Social Research Council (ESRC) Research Grant number R000236214.

References

Barsalou, L. W. (1990). On the indistinguishability of exemplar memory and abstraction in category representation. In T. K. Srull & R. S. Wyer, Jr. (Eds.) *Advances in Social Cognition*, Vol. III. Hillsdale, NJ: LEA.

Bishop, D. V. M. (1983). *Test for Reception of Grammar.* Medical Research Council, U.K.

Dienes & Perner (1996). Implicit knowledge in people and connectionist networks. In G. Underwood (Ed.), *Implicit Cognition*, (pp. 227–256). Oxford, England: Oxford University Press.

Elman, J. L. (1990). Finding structure in time. *Cognitive Science*, *14*, 179–211.

Gomez, R. L. & Gerken, L. A. (1997). Artificial grammar learning in one-year-olds: Evidence for generalization to new structure. In E. Hughes, M. Hughes, A. Greenhill (Eds.), *Boston University Conference on Language Development 21*, p. 194. Somerville, MA: Cascadilla Press.

Kendler, T. S. (1995). *Levels of Cognitive Development.* Mawah, NJ: Erlbaum.

Knowlton, B. J., Ramus, S. J. & Squire, L. R. (1992). Intact artificial grammar learning in amnesia: Dissociation of classification learning and explicit memory for specific instances. *Psychological Science*, *3*, 172–179.

Mathews, R. C., Buss, R. R., Stanley, W. B., Blanchard-Fields, F., Cho, J. R. & Druhan, B. (1989). Role of implicit and explicit processes in learning from examples: A synergistic effect. *Journal of Experimental Psychology: Learning, Memory, and Cognition*, *15*, 1083–1100.

Perruchet, P., & Pacteau, C. (1990). Synthetic grammar learning: Implicit rule abstraction or explicit fragmentary knowledge? *Journal of Experimental Psychology: General* , *119*, 264–275.

Raijmakers, M. E. J. & Molenaar, P. C. M. (1995). How to decide whether a neural representation is a cognitive concept? *Behavioral and Brain Sciences*, *18*, 641–642.

Raijmakers, M. E. J. & Molenaar, P. C. M. (1995). An experimental test of rule-like network performance. In G. Cottrell (Ed.), *Proceedings of the Eighteenth Annual Conference of the Cognitive Science Society* (p. 827). Mawah, NJ: Erlbaum.

Reber, A. S. (1967). Implicit learning of artificial grammars. *Journal of Verbal Learning and Verbal Behaviour*, *5*, 855–863.

Reber, A. S. & Allen, R. (1978). Analogy and abstraction strategies in synthetic grammar learning: A functional interpretation. *Cognition*, *6*, 189–221.

Redington, M. (1996). *What is learnt in artificial grammar learning.* Unpublished doctoral dissertation, Department of Experimental Psychology, University of Oxford.

Redington, M. & Chater, N. (1996). Transfer in artificial grammar learning: A Reevaluation. *Journal of Experimental Psychology: General*, *125*, 123–138.

St. John, M. F. & Shanks, D. R. (1997). *Implicit Learning from an Information Processing Standpoint.* In D. Berry (Ed.), *How implicit is implicit learning?*, (pp. 162–194). Oxford, England: Oxford University Press.

Saffran, J. R., Aslin, R. N. & Newport, E. L. (1996). Statistical cues in language acquisition: Word segmentation by infants. In G. W. Cottrell (Ed.), *Proceedings of the Eighteenth Annual Conference of the Cognitive Science Society*, (pp. 376–380). Mawah, NJ: Lawrence Erlbaum Associates.

Shanks, D. R., & St. John, M. F. (1994). Characteristics of dissociable human learning systems. *Behavioral and Brain Sciences*, *17*, 367–447.

Shanks, D. R., Johnstone, T. & Staggs, L. (1997). Abstraction processes in artificial grammar learning. *Quarterly Journal of Experimental Psychology*, *50A*, 216–252.

Vokey, J. R., & Brooks, L. R. (1992). Salience of item knowledge in learning artificial grammar. *Journal of Experimental Psychology: Learning, Memory, and Cognition*, *20*, 328–344.

The Impact of Abstract Ideas on Discovery and Comprehension in Scientific Domains

Shamus Regan (sregan@uic.edu)
Stellan Ohlsson (stellan@uic.edu)
Department of Psychology
The University of Illinois at Chicago
1007 West Harrison Street (M/C 285)
Chicago, IL 60607, U.S.A.

Abstract

The domain-specificity principle implies that domain-specific knowledge is the main determinant of scientific discovery. An alternative view is that scientists make discoveries by assembling and articulating abstract schemas. If so, prior activation of the relevant abstractions should facilitate discovery and comprehension. Two *in vitro* studies showed that abstract information can have as much or larger impact on scientific thinking as domain-specific information.

Abstraction and Discovery

The strongest determinant of a person's performance on a cognitive task is his or her task relevant knowledge. Recently, cognitive scientists have emphasized *domain-specific knowledge*, i.e., facts, principles and procedures that apply in one domain but are of limited usefulness in any other. What matters when playing chess is how much, and what, the player knows about the game of chess (Charness, 1989), but chess knowledge is not very useful outside that game; when diagnosing a patient, what matters is what, and how much, the physician knows about the relevant disease (Patel, Evans & Groen, 1989), but knowledge about diseases is not very helpful in non-medical tasks; and so on. The domain-specificity hypothesis implies that the factor that enables a scientist to make a discovery overlooked by others is his or her superior knowledge of the relevant facts.

The domain-specificity view contrasts with the long standing idea that human intelligence is a function of our ability to reason with *abstract knowledge*. Philosophers from Aristotle to Bertrand Russel have assumed that concepts like symmetry, subset, variation, probability, rate of change, complementarity, sequence, etc.,. confer cognitive power precisely because they apply across content domains (Ohlsson & Lehtinen, 1997). Outside cognitive science, it is still widely assumed that abstract thinking plays an essential role in complex cognitive tasks in general and scientific discovery in particular.

Most scholarly discussions of abstraction focus on the question of how abstractions are formed, but the question of interest here is how an abstraction, once formed, functions. One hypothesis is that an abstract concept, idea or schema can serve as a template for a domain-specific one (Ohlsson, 1993; Schank, 1986). By articulating a prior abstraction vis-à-vis a new situation or task, a person can generate a novel domain-specific structure (explanation, mental model, problem solution) for that situation or task (Ohlsson, 1992b).

This hypothesis is intuitively plausible but empirical research tends to support the domain-specificity view. There are numerous studies in which the participants apparently failed to apply a relevant abstraction in what we as observers think is the obvious way (e.g., Chen, Yanowitz & Daehler, 1995). A common weakness of such studies is that they do not compare the effect of abstract concepts, ideas or schemas with the effect of domain-specific ones. A second weakness from the present point of view is that the tasks used in those studies tend to be unrelated to the kinds of tasks involved in scientific reasoning and discovery.

In this paper we report two *in vitro* studies of the impact of abstractions on scientific thinking. First, we compared the impact of domain-specific and abstract information on students' performance on a laboratory version of the discovery of the structure of DNA. Second, we compared the effects of domain-specific and abstract information on students' ability to comprehend the theory of biological evolution.

Study 1: Discovering DNA

James Watson and Francis Crick discovered the structure of DNA in 1953, after several months of problem solving (Olby, 1974). Although prior research had identified the six types of molecules that made up DNA (sugar, phosphate and four nucleotide bases called adenine, thymine, guanine and cytosine), it was not known how those molecular building blocks fit together.

On their path to the correct structure, Watson and Crick identified the following eight properties of DNA: (a) The DNA molecule is an *elongated* structure. (b) The DNA molecule is *symmetric*. (c) The DNA molecule is held together by *two* (rather than one or three) sugar-phosphate strands, so-called backbones. (d) Due to asymmetries in the sugar molecule, each backbone has a direction; the two backbones in DNA are oriented in *inverse* directions. (e) The nucleotide bases are located *inside* the DNA molecule, with the two backbones twirled around them. (f) The backbones are connected via *pairs* of nucleotide bases, each pair analogous to the rung connecting the two sides of a ladder. (g) Every base pair consists of two *different* molecules. (h) More precisely, each base pair is *complementary* in that adenine is always paired with thymine and guanine with cytosine The first two properties were suggested by prior research by others. Properties c through h were *partial insights* (Ohlsson, 1992a) that Watson and Crick had to attain on the path to discovery.

Watson and Crick's main problem solving method was to build a physical model of the DNA molecule. They manufactured metal pieces that represented the various molecules and tried to put these together in a way that was consistent with quantitative measures (primarily X-ray data) of DNA as well as with its function as carrier of genetic information.

In searching the space of possible structures for DNA, Watson and Crick presumably drew upon their prior knowledge. However, Watson and Crick were not, in fact, experts in the relevant domain. Watson had recently been awarded his Ph.D. and Crick had yet to be awarded his; neither had ever resolved the structure of a large biological molecule. Furthermore, several of their insights had no basis in the chemistry of that time. In particular, there were no prior examples of double helixes, nor of molecules held together by complementary base pairs. As one would expect, the solution was unfamiliar; that is why its attainment was a major scientific advance. These observations suggest that the impact of domain-specific knowledge might have been limited.

A possible alternative is that Watson and Crick, while building their model of DNA, drew upon abstract concepts. After all, concepts like chain, symmetry, parallel sides, direction, inverse, pairing, unlike pairs and complementarity must have been in the conceptual repertoire of these two well-educated scientists. This view of the original discovery implies that prior activation of the relevant abstractions would have simplified the discovery. To evaluate this implication, we conducted what Dunbar (1995) calls an *in vitro* study of the discovery of DNA.

Materials In our laboratory version of the DNA problem, the six types of molecules (sugar, phosphate, four types of nucleotides) were modeled by foam board pieces of different shapes and colors. The pieces were shaped in such a way that they fit together in accordance with the possible chemical bonds between the corresponding molecules. There were 32 pieces. The task of putting this jigsaw puzzle together closely resembles the model-building task that occupied Watson and Crick in the summer of '53 (Olby, 1974).

To make the task performance more practical and easier to record, we reduced the target structure from a double helix to a ladder. This also simplified the problem so that the

participants could solve it in a single sitting.(In a pilot study, 80% of the 30 participants solved it in 50 minutes and the median time to solution for those who succeeded was 26 mins. 8 secs.) Notice that all eight partial insights described above have to be attained to solve this two-dimensional version of the DNA problem.

Because many participants might have encountered the DNA molecule in chemistry courses, we removed the problem from its chemical context by describing the puzzle pieces as finds from an archaeological excavation of a fictitious pre-Egyptian civilization. The students were told that the pieces had been used to carry messages from the king to the villages throughout his land, and that they had originally been put together in such a way that the messages could be duplicated with high accuracy by blind priests. The students' task was to figure out the original arrangement of the pieces.

The participants were seated at a table. Their performances were video taped with a camera placed across and slightly above the table.

Design and procedure The 120 participants were randomly divided into seven groups. All groups solved the problem, but under different information conditions.

(a) *No information.* There were 30 participants in this group.

(b) *Domain-embedded information.* Three groups of 15 participants each were given hints formulated within the context of the archaeological cover story. The 2-hint group received hints about properties a and b only, the 5-hint group received hints about properties a through e and the 8-hint group about all eight properties of the target structure. A domain-specific hint presented one of the key concepts embedded within the archaeological cover story. For example, the idea of two parallel strands was presented by saying that ancient sources indicated that the structure "resembled the kingdom itself with its two long and straight eastern and western borders."

(c) *Abstract information.* Three groups of 15 participants each were given 2, 5 or 8 hints formulated abstractly. In a training phase, the experimenter gave the participants a statement of the abstraction, a single example that did not belong to either chemistry or to the archaeological cover story and asked them to generate an example of their own. For example, for the idea of two parallel strands the statement was "things with two parallel parts", the example was "a railroad track" and the instruction was to generate another example of something that has two parallel parts.

Results The main measure derived from the videotapes was the overall time to solve the problem. Figure 1 shows the result. Because the distribution of solution times was skewed, we report medians rather than means. The open points show the effect of the domain-specific hints. Participants who received five hints did better than the participants who received two hints, but the 8-hint group did not do better than the 5-hint group. The difference between two and eight domain-specific hints is 1357 - 1179 = 178 seconds, or 2 minutes 58 seconds. The solid points show the effect of the abstract hints. Although the 2-hint abstract group was slower than the corresponding domain-specific group, the decrease in solution time with number of hints is more systematic. The difference between two and eight abstract hints is 1677 - 1050 = 627 seconds, or 10 minutes and 27 seconds, a huge effect in terms of the reduction in the cognitive processing needed to reach the solution.

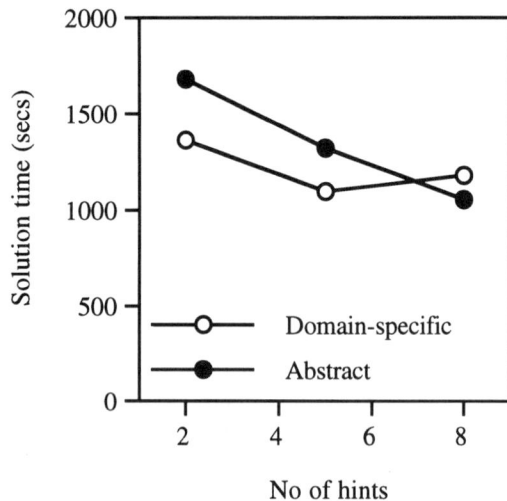

Figure 1: The median time to solution as a function of type of hint and number of hints.

The DNA molecule poses a complex and unfamiliar search problem. Figure 1 suggests that the eight partial insights that we extracted from the historical record functioned as heuristics that guided the search of our participants. The hints did not need to be domain-specific to be helpful. Indeed, the effect of the hints is larger in magnitude and more systematic when the hints were formulated abstractly. This result is consistent with our explanation for how Watson and Crick could solve a major scientific problem without superior domain-specific expertise: The problem is solvable on the basis of abstract concepts that Watson and Crick certainly acquired during their intellectual training.

Study 2: Comprehending Evolution

The elementary version of Darwin's theory of evolution explains particular biological adaptations with five component ideas: variation, inheritance, survival, reproduction rate and the accumulation of small changes over many generations. This biology-specific schema can be articulated vis-à-vis a particular adaptation. ("Polar bears vary in the thickness of their furs; thicker fur enables survival in a cold climate; ... ; over time, they evolved thicker fur.") Evolutionary theory has advanced beyond

Darwin, but in this paper we consider his theory in its original version.

Less is known, at the level of day-to-day details, about how Darwin hit upon natural selection than how Watson and Crick discovered the structure of DNA, but Darwin's autobiography and his notebooks have been interpreted to imply that the idea of natural selection was discovered via an analogy with either the Malthussian theory of population growth or with artificial selection in animal breeding (Gruber, 1974; Holyoak & Thagard, 1995).

However, the five key concepts of the original theory were not themselves novel, and they would certainly have been known by Darwin in abstract form. Hence, an alternative hypothesis is that Darwin assembled these abstractions into a new abstract schema and constructed the biological theory by articulating that schema. This hypothesis implies that prior activation of those abstractions should have made that schema easier to construct.

To evaluate this implication, we conducted a second *in vitro* study in which students encountered biological evolution, not as a discovery problem but in the form of a textbook exposition that was 1-2 pages long and contained one illustrative example (the neck of the giraffe). Pilot studies (Ohlsson & Bee, 1991) had shown that constructing Darwinian explanations on the basis of such an exposition is a challenging intellectual task for young adults. We compared the effects of domain-specific information in the form of expert feedback on the participants' own explanations with the effects of abstract information in the form of training intended to establish an abstract variation-and-selection schema prior to reading.

Design and procedure The participants answered evolutionary explanation questions under three different information conditions.

(a) *No information.* Fifty undergraduate psychology students from the University of Pittsburgh and 95 from the University of Illinois at Chicago were asked questions of the form, "How did X evolve Y?" but they were not given any form of preparation, instruction or help. These two base line groups will be reported separately.

(b) *Domain-specific information.* The 20 participants in this group read an expository text that stated Darwin's theory and demonstrated how it can be applied to explain the long neck of the giraffe. They then answered a series of questions of the form, "How did X evolve Y?" (e.g., How did polar bears evolve thick fur?). They were given feedback in the form of expert answers. For each question, the participants first wrote down their own answer and then turned the page and read an expert Darwinian answer to that question. The data from the groups a and b are analyzed in more detail in Ohlsson and Hemmerich (1999).

(c) *Abstract information.* The 38 participants in this group were given pairs of descriptive texts from other domains than biology and were asked to state what the two texts in each pair had in common. Each text pair was designed to instantiate one of the five Darwinian ideas mentioned above or some combination of them. The last pair in the training sequence instantiated the complete Darwinian schema.

Results The participants' written explanations were coded in two ways: First, all explanations were coded with respect to which of the five Darwinian ideas that were expressed in the explanation. This *Darwin score* varied between 0 and 5. Second, all explanations were coded with respect to eight types of non-Darwinian explanation types that we had identified in a pilot study (Ohlsson & Bee, 1991); these explanation types are described elsewhere in this volume (Ohlsson & Hemmerich, 1999). This *misconception score* varied between 0 and 8. Both dimensions were scored by two independent coders.

Figure 2 shows the result in terms of a two-dimensional outcome space. The mean Darwin score is plotted on the vertical axis, the mean misconception score on the horizontal axis. The four data points represent the four groups of subjects. The two base line studies are shown separately, within the ring.

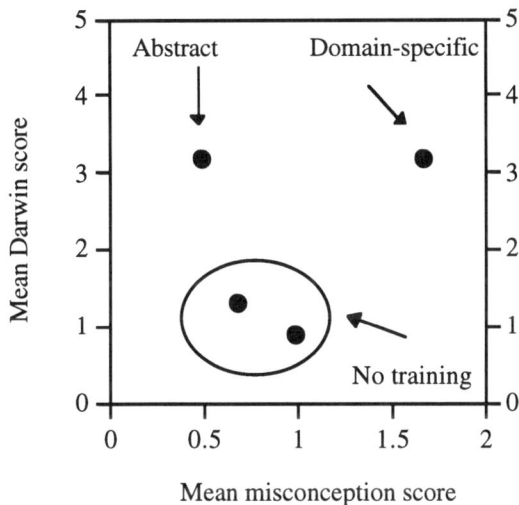

Figure 2: The results of four groups, displayed in a two-dimensional outcome space.

There are two main observations. First, both abstract and domain-specific information increased the use of the five Darwinian ideas in the participants' explanations, compared to the two no-information groups. The abstract information (which did not provide any knowledge about biology) had as strong an effect on the participants' tendency to use Darwinian ideas in their explanations as did the domain-specific feedback. Second, the two types of information had qualitatively different effects when viewed through the lens of the misconception scale. The domain-specific information *increased* the participants tendency to use the eight non-Darwinian ideas (e.g., crossbreeding) that we had identified in pilot work. The abstract information did not have this effect.

These results are consistent with the hypothesis that human beings, Mr. Darwin included, can construct the Darwinian theory by assembling a small group of abstract concepts into a new configuration and then articulating that configuration vis-à-vis biological evolution.

Discussion

The cognitive processes involved in discovery, problem solving, text comprehension, explanation and other cognitive performances must draw upon prior knowledge; there is no other resource for dealing with a situation or a task. The issue is how prior knowledge should be conceived.

There is no doubt that task relevant knowledge impacts on the performance of a task; this is mere common sense. However, the domain-specific knowledge principle might not be the whole story about how people accomplish unfamiliar tasks. In two studies, we showed that domain-specific information had a measurable impact on subsequent task performance, *but so did abstract information that had no relation to the target domain.* Information that aimed to activate certain abstract ideas in the participants' heads produced as large, or larger, effects on subsequent problem solving and comprehension as did domain-specific or domain-embedded information.

These findings are consistent with the idea that prior knowledge should be conceptualized as a repertoire of previously acquired structures (categories, schemas, strategies, rules, etc.) at different levels of abstraction. When a person encounters an unfamiliar task, domain-specific knowledge structures are unlikely to have a high level of fit (or else the task is not unfamiliar after all). In such cases, people respond to the task by articulating the best-fitting abstraction, or assembling a new abstraction which, in turn, is articulated (Ohlsson, 1993; Ohlsson & Lehtinen, 1997).

In both studies, the effects of both domain-specific and abstract information were small in magnitude. This is to be expected with short-term interventions. Future work should compare the effects of training that extends over longer periods of time. Another limitation is that although the quantitative measures verify that the information had an effect, they tell us little about how it worked. Future work will use more detailed data such as think-aloud protocols to elucidate the assembly and articulation of abstractions.

In summary, people in general and scientists in particular cannot operate on the basis of domain-specific knowledge vis-à-vis a task for which they lack such knowledge. The fact that we can solve an unfamiliar problem, understand an unfamiliar text, explain an unfamiliar phenomenon and make unexpected discoveries is due to the fact that we can operate with abstractions. Theories of human cognition that ignore this fact are incomplete.

References

Charness, N. (1989). Expertise in chess and bridge. In D. Klahr & K. Kotovsky (Eds.), *Complex information processing* (pp. 183-208). Hillsdale, NJ: Erlbaum.

Chen, Z., Yanowitz, K.L., & Daehler, M.W. (1995). Constraints on accessing abstract source information: Instantiation of principles facilitates children's analogical transfer. *Journal of Educational Psychology*, 87(3), 445-454.

Dunbar, K. (1995). How scientists really reason: Scientific reasoning in real-world laboratories. In R.J. Sternberg & J.E. Davidson (Eds.), *The nature of insight* (pp. 365-395). Cambridge, Massachusetts: MIT Press.

Gruber, H. E. (1974). *Darwin on man*. Chicago, IL: University of Chicago Press.

Holyoak, K., & Thagard, P. (1995). *Mental leaps*. Cambridge, MA: MIT Press.

Ohlsson, S. (1992a) Information processing explanations of insight and related phenomena. In M. Keane and K. Gilhooly, (Eds.), *Advances in the Psychology of Thinking* (Vol. 1, pp. 1-44) London, UK: Harvester-Wheatsheaf.

Ohlsson, S. (1992b). The cognitive skill of theory articulation: A neglected aspect of science education? *Science & Education, 1*, 181-192.

Ohlsson, S. (1993). Abstract schemas. *Educational Psychologist, 28*, 51-66.

Ohlsson, S. & Bee, N. (1991). *Young adults' understanding of evolutionary explanations: Preliminary observations* (Technical report, November). Pittsburgh, PA: Learning Research and Development Center

Ohlsson, S., & Hemmerich, J. (1999). Articulating an explanation schema: A preliminary model and supporting data. *Proceedings of the 21st Annual Meeting of the Cognitive Science Society*, Vancouver, Canada, August 19-21, 1999.

Ohlsson, S. & Lehtinen, E. (1997). Abstraction and the acquisition of complex ideas. *International Journal of Educational Research, 27*, 37-48.

Olby, R. (1974). *The path to the double helix: The discovery of DNA*. Seattle, Washington: University of Washington Press.

Patel, V., Evans, D., & Groen, G. (1989). Biomedical knowledge and clinical reasoning. In D. Evans & V. Patel (Eds.), *Cognitive science in medicine* (pp. 57-112). Cambridge, MA: MIT Press.

Schank, R. C. (1986). *Explanation patterns*. Hillsdale, NJ: Erlbaum.

A Causal-Model Theory of Categorization

Bob Rehder (brehder@uiuc.edu)

Department of Psychology, University of Illinois, Urbana-Champaign, IL 61801

Abstract

In this article I propose that categorization decisions are often made relative to *causal models* of categories that people possess. According to this *causal-model theory of categorization*, evidence of an exemplar's membership in a category consists of the likelihood that such an exemplar can be generated by the category's causal model. *Bayesian networks* are proposed as a representation of these causal models. Causal-model theory was fit to categorization data from a recent study, and yielded better fits than either the prototype model or the exemplar-based context model, by accounting, for example, for the confirmation and violation of causal relationships and the asymmetries inherent in such relationships.

Several investigators have argued that category learning and categorization are strongly influenced by the theoretical, explanatory, and causal knowledge that people bring to bear (Murphy & Medin, 1985; Murphy, 1993; Heit, 1998). For example, manipulations of stimulus materials affect category learning by eliciting different aspects of people's background knowledge (e.g., Pazzani, 1989; Murphy & Allopenna, 1994). Performance on a variety of tasks has been correlated with the amount of relevant domain knowledge individuals possess (Keil, 1989; Medin, Lynch, Coley, & Atran, 1997). However, there has been relatively little development of this "theory-based" view of categories in terms of detailed theory and computational models (c.f. Heit, 1994). This state of affairs arises in part because of the uncertainty surrounding exactly what knowledge participants deploy in an experimental task. A few recent studies have addressed this problem by employing novel domains and teaching participants "background" knowledge as part of the experimental session (e.g., Ahn & Lassaline, 1996; Rehder & Hastie, 1999; Sloman, et al. 1998). For example, Rehder and Hastie taught participants about fictitious categories described as possessing causal relationships between binary-valued category attributes, and manipulated experimentally whether those causal relationships formed a common-cause or a common-effect causal schema (Figure 1). In the common-cause schema, one attribute (A1) is described as causing the three other attributes, whereas in the common-effect schema one attribute (A4) is caused by three other attributes. For example, one of the fictitious categories was Kehoe Ants, a species of ants described as living on an island in the Pacific Ocean, and one of that category's causal relationships was "Blood high in iron sulfate causes a hyperactive immune system. The iron sulfate molecules are detected as foreign by the immune system, and the immune system is highly active as a result." After learning about such categories and their causal relationships participants performed a transfer categorization task. Rehder and Hastie found that the presence of both a cause and its effect in an instance (e.g., an ant with iron sulfate blood and a hyperactive immune system) led to the instance receiving a higher category membership rating compared to control categories with no causal relationships. Because ratings were also higher when both the cause and effect were *absent* (normal blood and normal immune system), and *lower* when either the cause or the effect was present and the other absent (iron sulfate blood and normal immune system, or normal blood and hyperactive immune system), Rehder and Hastie concluded that participants were attending not merely to the presence of the cause/effect configuration, but rather to whether instances *confirmed* or *violated* causal relationships. Category membership ratings also reflected the asymmetries inherent in causal relationships. For example, a distinct characteristic of common-cause causal networks is that the effect attributes (e.g., A2, A3, and A4 in Figure 1) will be correlated, and indeed the categorization ratings of substantial numbers of common-cause participants were sensitive to whether those correlations were preserved or violated.

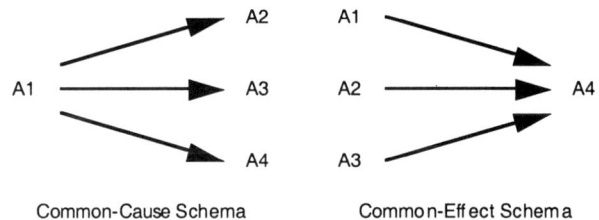

Common-Cause Schema   Common-Effect Schema

Figure 1

Although these results are suggestive of explicit causal reasoning in participants, it is important to consider whether they can be accounted for by the well-known *similarity-based* categorization models, such as the prototype model and the context model (Medin & Shaffer, 1978; Nosofsky, 1986). Similarity-based models are able to accommodate seemingly disparate categorization strategies by adjusting similarity parameters to differentially shrink or expand the dimensions of the stimulus space. In fact, Rehder and Hastie fitted these models to their transfer categorization data, and found that the models yielded only moderate-quality fits. The fits of instances that possessed many confirmations or many violations of causal relationships were particularly poor.

The failure of the similarity-based models to account for these data leads to a search for alternative categorization models that can account for people's apparent ability to reason causally while categorizing. In this article I propose that categorization decisions are often made relative to *causal models* of categories that people possess, and test *Bayesian networks* as a candidate representation of such models.

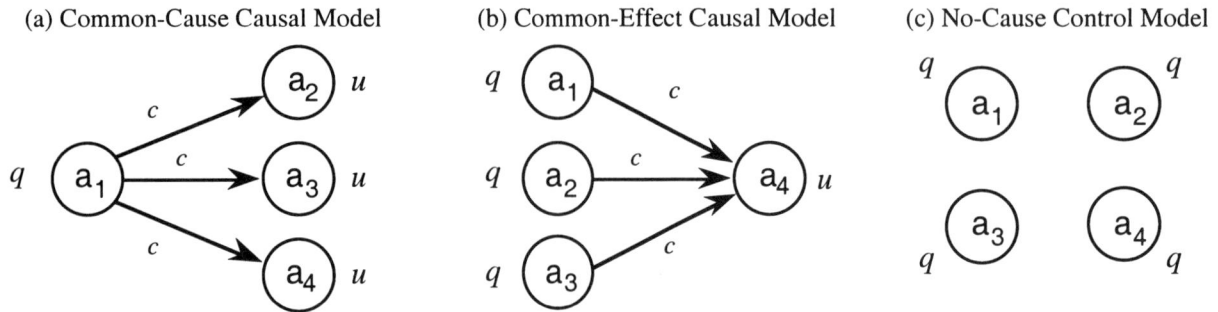

(a) Common-Cause Causal Model   (b) Common-Effect Causal Model   (c) No-Cause Control Model

Figure 2

Below I present this *causal-model theory of categorization*, review the procedures employed in the Rehder and Hastie study, and compare data fits produced by the similarity-based models and the causal-model approach. When causal knowledge was present, causal models produced better fits than the similarity-based models by accounting for the categorization ratings of those instances especially affected by the confirmation or violation of causal relationships, and for the asymmetries inherent in causal relationships.

A Causal-Model Theory of Categorization

As in other categorization models, I assume that the classification process consists of both an evidence stage and a decision stage, and that the decision stage is given by a relative-ratio rule (Luce, 1963),

$$P(C_A|t) = E_A(t)/\Sigma_i E_i(t) \qquad (1)$$

where $P(C_A|t)$ is the probability that instance t is classified into Category A, i ranges over the set of categories that t may belong to, and $E_i(t)$ is the evidence in favor of t belonging to category C_i. The core of the current proposal is that evidence of t's membership in category C_i consists of the likelihood that t could have been generated by C_i's causal model. Figure 2 presents Bayesian networks that serve as common-cause and common-effect causal models. Bayesian networks are directed acyclic graphs in which nodes represent variables (in this work, binary variables whose values are referred to as "present" and "absent", or "1" and "0"), and edges represent direct dependency relations among variables. In particular, Bayesian networks can be used to represent causal dependencies among variables in which the only direct causes of a variable are its immediate parents (Pearl, 1988). For example, the common-cause network (Figure 2a) has one variable (a_1) that directly causes three other variables (a_2, a_3, a_4). The common-effect network (Figure 2b) has three variables (a_1, a_2, a_3) that each directly causes one other variable (a_4). The no-cause control network (Figure 2c) specifies no causal relationships among variables.

For each variable with no edge into it (i.e., for each "exogenous variable"), Bayesian networks require specification of the probability that its value will be present. These probabilities are referred to as q. In addition, each edge requires the probability that the effect is caused when the cause is present, referred to as c. Finally, u is the probability that a caused (endogenous) variable is present even when its causes are absent. (u can be interpreted as the probability that an effect is brought about by some unspecified cause.) These probabilities can be treated as the parameters of the

causal model that the Bayesian networks represents[1].

Given a causal model and its parameters, it is possible to calculate the likelihood that it will generate a particular value on a variable, or a particular set of values. Table 1 presents the equations for computing the likelihood for any settings of the four binary variables in the common-cause, common-effect, and no-cause control models. For example, the probability that the common-cause model will generate the values 1000 (that is, a_1 present, a_{2-4} absent) is the probability that a_1 is present (q) times the probability that all three causal mechanisms fail to operate ($(1-c)^3$) times the probability that the effects are not otherwise caused ($(1-u)^3$).

When the variables of a causal model are interpreted as attributes of a category, and the edges are interpreted as causal relationships between category attributes, the equations of Table 1 can be used to compute the likelihood that the category will generate a particular exemplar, that is, a particular set of attribute values.

The causal-model approach to categorization assumes that categorizers estimate for each candidate category the likelihood that the category generated the exemplar, and then combine this evidence in accord with the relative-ratio rule (Eq. 1) in order to reach a categorization decision.

The Experiments

In addition to Kehoe Ants, Rehder and Hastie (1999) employed five other fictitious categories: one other biological kind, two nonliving natural kinds, and two artifacts. For each of the four binary attributes the base rates of its two values was stated to be 75% and 25%. To relate these attribute values to the networks in Figure 2, the 75% and 25% values are henceforth referred to as "present" and "absent" (or "1" and "0"), respectively. Causal relationships were written such that the "presence" of one attribute caused the presence of another.

---

[1]The networks shown in Figure 2 exhibit a number of restrictions, any of which can be lifted. First, all causal links are assumed to have the equal strength (i.e., the same c) because in the Rehder and Hastie study all relationships were pretested to equate their plausibility. Second, exogenous variables are assumed to have the same base rate (q), because all participants were explicitly told that attributes exhibited equal base rates (see description of the experiments below). Third, endogenous variables are assumed to have the same u for the same reason. Fourth, in the common-effect network I assume that the probabilities that each cause will bring about its effects are independent, that is, the common-effect causal model is related to a "fuzzy or gate" (Pearl, 1988). These restrictions have the advantage of reducing the number of free parameters, facilitating comparison with similarity-based models (see below).

Table 1

Exemplar	Common Cause Causal Model	Common Effect Causal Model	No-Cause Control Model
0000	$q'u'^3$	$q'^3 u'$	q'^4
0001	$q'u'^2 u$	$q'^3 u$	qq'^3
0010	$q'u'^2 u$	$qq'^2 c'u'$	qq'^3
0100	$q'u'^2 u$	$qq'^2 c'u'$	qq'^3
1000	$qc'^3 u'^3$	$qq'^2 c'u'$	qq'^3
0011	$q'u'u^2$	$qq'^2(1-c'u')$	$q^2 q'^2$
0101	$q'u'u^2$	$qq'^2(1-c'u')$	$q^2 q'^2$
0110	$q'u'u^2$	$q^2 q'c'^2 u'$	$q^2 q'^2$
1001	$qc'^2 u'^2(1-c'u')$	$qq'^2(1-c'u')$	$q^2 q'^2$
1010	$qc'^2 u'^2(1-c'u')$	$q^2 q'c'^2 u'$	$q^2 q'^2$
1100	$qc'^2 u'^2(1-c'u')$	$q^2 q'c'^2 u'$	$q^2 q'^2$
0111	$q'u^3$	$q^2 q'(1-c'^2 u')$	$q^3 q'$
1011	$qc'u'(1-c'u')^2$	$q^2 q'(1-c'^2 u')$	$q^3 q'$
1101	$qc'u'(1-c'u')^2$	$q^2 q'(1-c'^2 u')$	$q^3 q'$
1110	$qc'u'(1-c'u')^2$	$q^3 c'^3 u'$	$q^3 q'$
1111	$q(1-c'u')^3$	$q^3(1-c'^3 u')$	q^4

Note. $q'=(1-q)$. $r'=(1-r)$. $s'=(1-s)$.

Table 2

Exemplar	Neutral-Data Experiment Target	Contrast	Congruent-Data Experiment Target CC	CE	Control	Contrast
0000	0	11	4	1	0	15
0001	0	3	2	0	1	5
0010	0	4	2	1	0	5
0100	0	4	2	1	0	5
1000	0	4	0	1	1	5
0011	1	1	1	1	2	2
0101	1	1	1	1	2	1
0110	1	1	1	2	2	2
1001	1	1	0	1	2	2
1010	1	1	0	2	1	2
1100	1	1	0	2	2	2
0111	4	0	0	5	5	1
1011	4	0	3	5	5	0
1101	4	0	3	5	5	0
1110	3	0	3	2	5	1
1111	11	0	26	18	15	0

Note. CC=Common-Cause Schema. CE=Common-Effect Schema. Control=No-Cause Control Schema.

In each of three experiments, each participant learned of one category, and the category's causal schema was manipulated as a between-subjects factor: Participants were given either a common-cause or a common-effect set of causal relationships, or no causal relationships.

After learning about a category and its causal schema, participants were exposed to exemplars of the target category (e.g., Kehoe Ants) in the guise of a training classification task. Participants classified a series of exemplars into the target category or a contrast category labeled "other", with feedback provided on every trial. In the *Neutral-Data Experiment*, target-category exemplars exhibited no correlations between attributes. That is, the inter-attribute correlations that might be expected from the causal relationships learned by the common-cause or common-effect participants were *not* reflected in the target category exemplars (as a result participants in all conditions observed the same training exemplars). In the *Congruent-Data Experiment* target-category exemplars exhibited the inter-attribute correlations implied by the causal relationships: the common-cause participants saw common-cause correlations and the common-effect participants saw common-effect correlations (and the no-cause control participants saw no inter-attribute correlations). Finally, in the *No-Data Experiment* participants were presented with no training exemplars and hence received no empirical information about the target category. Table 2 presents the number of instances of each exemplar presented to participants as members of the target and contrast categories in the Neutral-Data and Congruent-Data experiments. Note that the target category samples exhibited the 75%/25% attribute base rates that participants were told the category possessed. The attribute base rates in the contrast category samples were 25%/75%.

All three experiments concluded with participants performing three transfer tasks: a categorization task, a similarity rating task, and a property induction task. (The similarity and induction results are not discussed further.) During the categorization task, participants rated on a 100-point scale the category membership of 32 exemplars, consisting of all possible 16 examples that could be formed from four binary attributes, each presented twice. No feedback was provided.

216, 234, and 180 University of Colorado undergraduates participated in the Neutral-Data, No-Data, and Congruent-Data experiments, respectively. Average category membership ratings for each exemplar in each condition of the Neutral-Data, No-Data, and Congruent-Data experiments are presented in the Appendix.

Model Fitting Procedure

To fit the transfer categorization data from Rehder and Hastie (1999), causal models must be assumed for both the target category (e.g., Kehoe Ants) and the contrast category. It was assumed that for the target category participants employed a causal model appropriate to the causal schema they learned: a common-cause model for the common-cause schema, a common-effect model for the common-effect schema, and a no-cause control model when they learned of no causal relationships. It was also assumed that all participants employed a no-cause control model for the contrast category, because no causal knowledge was provided about that category. As a result, fits of data from the common-cause and the common-effect schema conditions involve four parameters (q, c, and u for the target category, and q for the contrast category), and fits to the no-cause control condition involve two (one q each for the target and contrast category). Data fitting involved finding the set of parameters that minimized squared error, with predictions being given by the equations in Table 1.

The transfer categorization data were also fitted to the prototype and context model (Nosofsky, 1992). The prototype model assumes that evidence of category membership consists of the (additive) similarity of a stimulus to the category's prototype. In fitting the prototype

Table 3

Model Parameters	Common Cause Schema			Common Effect Schema			No-Cause Control Schema		
	No Data	Neutral Data	Congruent Data	No Data	Neutral Data	Congruent Data	No Data	Neutral Data	Congruent Data
					Prototype Model				
s_1	0.03	0.00	0.00	0.25	0.04	0.11	0.10	0.00	0.00
s_2	0.57	0.00	0.19	0.41	0.00	0.21	0.32	0.00	0.10
s_3	0.53	0.01	0.28	0.53	0.05	0.25	0.33	0.00	0.08
s_4	0.57	0.00	0.33	0.16	0.00	0.00	0.35	0.00	0.05
RMSE	0.097	0.080	0.071	0.077	0.090	0.078	0.023	0.071	0.033
					Context Model				
s_1	–	0.21	0.29	–	0.36	0.37	–	0.22	0.26
s_2	–	0.31	0.44	–	0.33	0.42	–	0.29	0.33
s_3	–	0.32	0.44	–	0.35	0.41	–	0.27	0.29
s_4	–	0.34	0.52	–	0.26	0.24	–	0.30	0.37
RMSE	–	0.046	0.045	–	0.067	0.051	–	0.029	0.025

	Causal Models								
	Common Cause Causal Model			Common Effect Causal Model			No-Cause Control Model		
Target Category									
q	0.52	0.74	0.63	0.47	0.65	0.60	0.56	0.66	0.63
c	0.30	0.31	0.29	0.22	0.40	0.35	–	–	–
u	0.36	0.55	0.47	0.31	0.38	0.39	–	–	–
Contrast Category									
q	0.37	0.30	0.39	0.35	0.32	0.38	0.41	0.31	0.36
RMSE	0.039	0.025	0.038	0.037	0.018	0.024	0.030	0.025	0.030

model, the prototypes of the target and contrast categories were assumed to be 1111 and 0000, respectively. The context model assumes that evidence of category membership is the total (multiplicative) similarity of a stimulus to all stored category exemplars. In fitting the context model, it was assumed that all exemplars presented during training classification were stored in memory with their category membership. Because of the absence of training exemplars in the No-Data experiment, the context model was not fit to that experiment's data. Both the prototype and context model produce four parameters (s_1, s_2, s_3, and s_4) in the range [0,1] representing the saliency of each dimension (where smaller numbers mean more salient).

Although the relative-ratio rule (Eq. 1) is typically used to predict the probability of category membership, in the Rehder and Hastie study participants produced continuously-valued category membership ratings rather than binary-valued categorization decisions. Accordingly, the rule is used to predict category membership ratings, which are divided by 100 to bring them into the range [0-1].

Results

Fits of the prototype model, the context model, and causal-model theory to the Rehder and Hastie (1999) transfer categorization data are presented in Table 3 as a function of causal schema and experiment. In all common-cause and common-effect conditions, causal models produced better fits than either the prototype model or the context model in terms of residual error variance (RMSE), and did so with the same number of free parameters as the similarity-based models (four). In the no-cause control experimental conditions, participants were not instructed on the presence of inter-attribute causal relationships, and the prototype

model and the context model produced adequate data fits in those conditions (with one exception: the prototype model fit in the Neutral-Data experiment). However, causal models produced equally good fits in those conditions, and did so with two fewer free parameters[2].

One reason for the poorer fits of the prototype and context models in the common-cause and common-effect conditions is their inability to account for the category membership ratings of those exemplars especially sensitive to the confirmation or violation of causal relationships. To illustrate, Figure 3 presents context model and causal model fits in the Neutral-Data experiment for such exemplars. As is apparent from the figure, in the common-cause condition the context model *underestimates* the category membership ratings of exemplars that possess three confirmations of causal relationships: 1111 (the common-cause, a_1, and its effects all present) and 0000 (common-cause and its effects all absent). It also *overestimates* the ratings of exemplars that possess three violations of causal relationships: 1000 (the common-cause is present but its effects are absent) and 0111 (the common-cause absent, but its effects are present). Analogously, in the common-effect condition the context model underestimates the ratings of exemplars that possess three confirmations of causal relationships (1111 and 0000), and overestimates the ratings of exemplars that possess three violations (0001 and 1110).

In comparison, in both the common-cause and common-effect conditions causal model theory yielded quite good fits of these exemplars (Figure 3). The causal models' parameter

---

[2] Participants in the No-Cause Control condition of the No-Data Experiment exhibited a substantial response bias in favor of the target category, and hence data fits in that condition include the addition of a bias parameter $b=.58$ ($b=.50$ means zero bias).

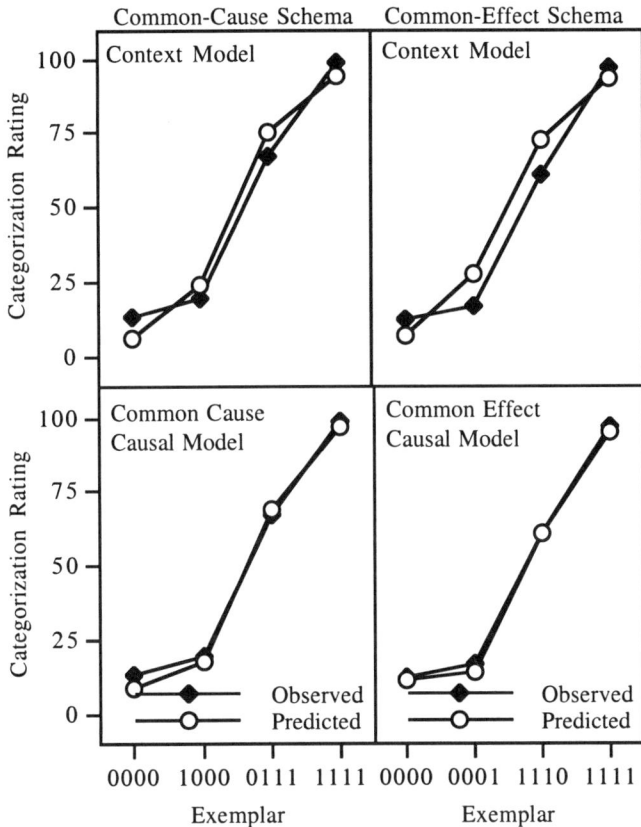

Common-Cause Schema   Common-Effect Schema

Context Model   Context Model

Common Cause   Common Effect
Causal Model   Causal Model

Observed
Predicted

Observed
Predicted

0000 1000 0111 1111 0000 0001 1110 1111

Exemplar   Exemplar

Categorization Rating

Figure 3

values support the view that their superior data fits were due to participants' use of causal relationships when generating categorization ratings. For example, parameter c, which reflects the strength of the causal relationship between a cause and an effect attribute, was large and positive in all common-cause and common-effect schema conditions (Table 3). One effect of those parameter values is to make the generation of exemplars that confirm causal relationships (1111 and 0000 for both the common-cause and common-effect models) more likely, and the generation of exemplars that violate causal relationships (1000 and 0111 for the common-cause model, and 0001 and 1110 for the common-effect model) less likely. Because the likelihood of generation controls category membership ratings (Eq. 1), causal models account for the sensitivity to the correlations between causally-connected features shown in Figure 3.

The context model can also exhibit sensitivity to inter-attribute correlations, but only when category exemplars are present in memory that manifest those correlations. Such exemplars were absent in both the Neutral-Data and No-Data experiments, and hence participants' sensitivity to correlations between causally-connected features in those experiments must be attributed to the causal knowledge they learned. However, exemplars that manifested the appropriate correlations were present in the Congruent-Data experiment and hence that experiment established the most favorable conditions for the context model. Nevertheless, even in the Congruent-Data experiment causal model theory yielded the best fits, apparently because participants weighed the critical inter-attribute correlations more heavily than is predicted by the context model's multiplicative-similarity rule.

Discussion

The failure of the prototype and context models to account for the Rehder and Hastie (1999) categorization data implies that participants were not rating the category membership of exemplars on the basis of (only) similarity, a result that led Rehder and Hastie to suggest that their participants were engaging in explicit causal reasoning while categorizing. In this article "causal reasoning" is rendered computationally explicit in the form of causal-model theory. In fact, the good fits of the categorization data produced by causal models, together with the specific parameter values responsible for those fits, support the claim that people can engage in causal reasoning while categorizing when causal knowledge is present that enables that reasoning.

An alternative way to account for the current data in terms of similarity is to argue that the feature space is expanded to include higher-order features encoding the confirmation or violation of causal relationships, and that similarity is computed in that expanded space. However, Rehder and Hastie also collected ratings of the similarity of pairs of category members, and found that such ratings were insensitive to whether exemplars matched on those higher-order features (also see Rehder & Hastie, 1998), suggesting that the presence of causal knowledge did not result in an expansion of the feature space. Such higher-order features also fail to account for the asymmetries inherent in causal relationships. For example, when the direction of causality in the common-cause and common-effect models (Figure 2) is reversed, the result is substantially worse fits of the data of many common-cause and common-effect participants, respectively. In other words, when classifying exemplars many participants did not just evaluate each causal link in isolation but rather considered interactions among links produced by the entire *network*, where the nature of those interactions is determined by the links' direction of causality.

Despite differing assumptions regarding the form of category representations (prototypes, exemplars, rules, etc.), traditional similarity-based models assume that such representations are built with information taken from the *data* people observe (i.e., exemplars). Causal model theory diverges sharply from this approach by assuming instead that category representations are formed from the category *knowledge* people possess. The superior fits of causal model theory reported here for the Neutral- and Congruent-Data experiments reflects the greater importance of category *knowledge* versus category *data* on categorization in those experiments. Causal model theory also applies when no exemplars of the category have been observed at all (e.g., the No-Data experiment), a domain beyond the purview of the traditional models. In contrast to the traditional models, causal model theory is thus applicable to the many real-world categories about which people know far more than they have observed first hand.

Causal models have been implicated in other domains. For example, Glymour (1998) argues that people's ability to estimate the strength of causal influences controlling for other causes (i.e., Cheng's, 1997, causal power theory) is equivalent to estimating the conditional probability associated with an edge in a Bayesian network (in this article, parameter c). Waldmann, Holyoak, and Fratianne

(1995) demonstrated that the speed of category learning depends on the match between the correlational structure of the learning data and the learner's causal model of the category. An important area of development for causal model theory is to specify the learning algorithm by which learners *integrate* their category knowledge (in the form of causal models) with data (i.e., observations of the category).

In this article I have demonstrated how the claim that causal knowledge affects categorization can be formalized as an explicit computational model, how it can be fitted to empirical data, and how it can be rigorously tested against other models. Bayesian networks were utilized as a device with which causal knowledge was represented and evidence in favor of category membership was calculated. Future work may advance causal model theory by specifying the processes by which likelihood functions (e.g., Table 1) are computed (or approximated). The success of parallel network algorithms in implementing complex reasoning processes in other domains (Pearl, 1988, Thagard, 1989) make the prospect for this development promising.

Acknowledgements

This work was conducted while the author was supported by NSF Grant SBR 97-20304.

References

Ahn, W., & Lassaline, M. E. (1996). *Causal structure in categorization: A test of the causal status hypothesis (Part I).* Unpublished manuscript.

Cheng, P. (1997). From covariation to causation: A causal power theory. *Psychological Review*, 104, 367-405.

Glymour, C. (1998). Learning causes: Psychological explanations of causal explanation. *Minds and Machines*, 8, 39-60.

Heit, E. (1994). Models of the effects of prior knowledge on category learning. *Journal of Experimental Psychology: Learning, Memory, and Cognition, 20*, 1264-1282.

Keil, F. C. (1989). *Concepts, kinds, and cognitive development.* Cambridge, MA: MIT Press.

Luce, R. D. (1963). Detection and recognition. In R. D. Luce, R. R. Bush, & E. Galanter (Eds.), *Handbook of mathematical psychology.* (pp. 103-189). NY:Wiley.

Medin, D. L., & Schaffer, M. M. (1978). Context theory of classification learning. *Psychological Review*, 85,207-38.

Medin, D. L., Lynch, E. B., Coley, J. D., & Atran, S. (1997). Categorization and reasoning among tree experts: Do all roads lead to Rome? *Cognitive Psychology*, 32, 49-96.

Murphy, G. L., & Allopenna, P. D. (1994). The locus of knowledge effects in concept learning. *Journal of Experimental Psychology: Learning, Memory, and Cognition*, 20, 904-919.

Murphy, G. L., & Medin, D. L. (1985). The role of theories in conceptual coherence. *Psychological Review, 92*, 289.

Nosofsky, R. M. (1992). Exemplars, prototypes, and similarity rules. In A. F. Healy, S. M. Kosslyn, & R. M. Shiffrin (Eds.), *From learning processes to cognitive processes: Essays in honor of William K. Estes.* (pp. 149-167). Hillsdale, NJ: Erlbaum.

Pazzani, M. J. (1991). Influence of prior knowledge on concept acquisition: Experimental and computational results. *Journal of Experimental Psychology: Learning, Memory, and Cognition, 17*, 416-432.

Pearl, J. (1988). *Probabilistic reasoning in intelligent systems: Networks of plausible inference.* San Mateo, CA: Morgan Kaufman.

Sloman, S., Love, B. C., & Ahn, W. (1998). Feature centrality and conceptual coherence. *Cognitive Science, 22*, 189-228.

Rehder, B., & Hastie, R. (1998). The differential effects of causes on categorization and similarity. In *The Proceedings of the Twentieth Annual Conference of the Cognitive Science Society* (pp. 893-898). Madison, WI.

Rehder, B., & Hastie, R. (1999). *The essence of categories: The effect of underlying causal mechanism on induction, categorization, and similarity.* Submitted for publication.

Thagard, P. (1989). Explanatory coherence. *Behavioral and Brain Sciences*, 12, 435-502.

Waldmann, M. R., Holyoak, K. J., & Fratianne, A. (1995). Causal models and the acquisition of category structure. *Journal of Experimental Psychology: General, 124*, 181-206.

Appendix – Transfer Categorization Ratings from Rehder & Hastie (1999)

Exemplar	Common Cause Schema			Common Effect Schema			No-Cause Control Schema		
	No Data	Neutral Data	Congruent Data	No Data	Neutral Data	Congruent Data	No Data	Neutral Data	Congruent Data
0000	50.2	13.4	27.1	40.7	12.1	18.1	33.1	9.9	8.8
0001	41.4	19.9	32.8	34.6	16.8	24.1	41.4	20.4	26.8
0010	41.7	20.1	33.0	38.9	22.5	30.3	42.8	20.1	28.8
0100	39.9	19.9	36.6	41.1	22.6	30.9	43.4	17.2	23.2
1000	37.5	19.9	29.8	44.4	25.4	32.0	46.7	19.2	29.1
0011	43.0	47.3	43.4	51.9	50.4	49.1	56.5	50.5	46.3
0101	40.9	46.7	44.5	52.5	52.5	48.5	54.1	51.1	49.0
0110	40.7	43.8	47.2	46.2	40.8	38.7	55.2	49.2	47.3
1001	50.8	51.1	48.9	56.5	47.8	52.7	59.7	56.8	53.3
1010	51.4	50.5	50.8	48.6	38.9	40.0	60.3	56.2	49.1
1100	52.4	54.4	50.4	52.4	41.3	44.5	60.4	54.2	55.0
0111	45.5	67.2	50.0	66.1	84.5	76.1	69.6	80.7	72.4
1011	65.3	86.2	67.9	68.1	84.5	78.2	72.8	84.2	79.5
1101	65.2	84.2	70.9	71.0	84.8	76.8	74.6	83.4	74.7
1110	67.2	86.5	70.7	53.1	61.3	49.8	73.4	82.6	76.1
1111	90.0	98.6	97.8	90.0	97.2	92.3	88.6	95.0	92.2

Argument Detection and Rebuttal in Dialog[1]

Angelo Restificar[1]   Syed S. Ali[2]   Susan W. McRoy[1]
angelo@uwm.edu   syali@uwm.edu   mcroy@uwm.edu
[1]Electrical Engineering and Computer Science
[2]Mathematical Sciences
University of Wisconsin-Milwaukee

Abstract

A method is proposed for argumentation on the basis of information that characterizes the structure of arguments. The proposed method can be used both to detect arguments and to generate candidate arguments for rebuttal. No assumption of *a priori* knowledge about *attack* and *support* relations between propositions, advanced by the agents participating in a dialog, is made. More importantly, by using the method, the relations are dynamically established while the dialog is taking place. This allows incremental processing since the agent need only consider the current utterance advanced by the dialog participant, not necessarily the entire argument, to be able to continue processing.

Introduction and Motivation

Argument detection is an important task in building an intelligent system that can understand and engage in an argument. In an intelligent dialog system (IDS)[6], an interactive system that tailors its response according to the user's needs and intentions, it is necessary to detect whether an utterance given by the user is an argument against an utterance advanced by the system. Two agents, *e.g.* the system and the user, may engage in a conversation and may not always agree. Each of them may attempt to resolve issues either by attacking an agent's claim or by defending its position. Thus, an IDS must be able to determine whether a proposition advanced by an agent in a dialog attacks a claim currently held by the other agent, supports it, or does neither.

The method we propose here, used by our system ARGUER, finds an argument schema which matches the deep meaning representation of a proposition. The schemata characterize general patterns of argument, similar to those studied by [16, 3]. These types of arguments are reasonable but defeasible—they incorporate plausible assumptions in the absence of complete information. ARGUER uses a truth-maintenance system to revise its beliefs. These schemata are used to establish *support* or *attack* relations among the propositions expressed by agents over the course of a dialog. The schemata allow ARGUER to recognize attack or support relations between an utterance and any prior utterance, not just an immediately preceding one. Separate models of the agent's beliefs are maintained, both for the system and the user. Hence, a proposition believed by the system may not be necessarily believed by the user. The system generates a response by taking into consideration the beliefs held by the user.

The work that we describe here focuses on the problem of relating a proposition to an existing knowledge base for argument detection and rebuttal. By using argument schemas, we will show how a system can detect and rebut arguments, automatically. The relation of the input proposition to existing knowledge is established while the dialog is ongoing. This allows a dialog participant to consider only the current utterance advanced by the other participant, not necessarily the entire argument, to continue processing.

Background

Argument and Argument Relations

During a dialog, whenever a participant challenges an utterance of another participant, we say that an *argument* has started. The propositions that are usually advanced by the participants during the exchange are either an *attack* to a proposition (*i.e.,* by attacking the merits of the opponent's claim), or a *support* to a previously held claim. An argument relation is either an *attack* or a *support*. We use rules, called *argument schema rules*, that characterize the structure of arguments and are used to determine whether an input proposition attacks or supports existing knowledge (or does neither). Thus during the exchange, it is possible to keep track of the relationships between utterances by maintaining a labeled directed acyclic graph whose nodes are the propositions corresponding to the utterances and whose arcs are labeled as either *attack* or *support*. This labeled directed graph is called an *argument graph* [4]. In this paper, we refer to an argument as any node in the argument graph.

[1]This work was supported by the National Science Foundation, under grants IRI-9701617 and IRI-9523666.

601

Relation to Prior Work

Argumentation has attracted the interest of researchers in philosophy [16], law [10] and artificial intelligence. Inside the field of artificial intelligence, investigations ranged from understanding arguments [1, 4, 11], building interactive systems [15, 17] and to the use of argumentation to deal with issues in default logic and nonmonotonic reasoning [8, 5].

Our current work deals with issues closely related to understanding arguments in an interactive environment, more specifically, in a dialog. Recent work in interactive argumentation systems include IACAS [15] and NAG [17]. IACAS allows the users to start a dispute and find arguments. However, it does not address the issue of maintaining separate belief models for each participant in a dialog. NAG, on the other hand, uses tagged words from the input and uses Bayesian networks to find the best set of nodes that can be formed as an argument for a proposition. Neither system, however, addresses argument detection; they deal with issues different from the ones that concern us. [7] focus on a different, but related, problem within dialog, the collaborative construction of a plan. This work considers the problem of deciding whether the system should accept or reject a user's proposal, on the basis of its consistency with the system's beliefs and the evidence offered by the user. Unlike ARGUER's approach to deciding whether some evidence supports a belief, which relies on domain-independent schemata, their system's approach relies on domain-specific patterns of evidence. Other systems like ABDUL/ILANA [4] and HERMES [9] do not address the issue of how attack and support relations between propositions may be established computationally. Alvarado's OpEd [1] although designed to understand arguments is limited to processing editorial text and does not address the issues we are concerned in dialog processing.

In ARGUER, the relations between propositions are dynamically established while the dialog is taking place. The use of argument schemas allows *incremental* processing of arguments. Since relations between propositions can be established dynamically, the system may not always have to wait until the complete argument is presented. Moreover, the method is *symmetrical* because it can be used for interpretation or generation of arguments. This is important because the system can have the role of observer or participant. Currently, ARGUER does not address the problem of choosing among a set of possible arguments when generating an utterance; this has been the topic of work by [17], for example. We are currently working on a model of preference that will address this concern.

Argument Detection and Rebuttal

The underlying principle for detecting arguments in ARGUER is to find a general case of an argument schema into which the meaning representation of an utterance can be matched. In ARGUER, argument detection and rebuttal are established via argument schema rules. An argument schema rule characterizes the structure of an argument. The method finds all the schemata into which the deep meaning representation of a proposition fits. If a match is found, the corresponding variables in the argument schema rules are instantiated, thereby establishing attack or support relation. For example, given a proposition α corresponding to `Tweety is a bird`, the schema, NOT α, corresponding to `Tweety is not a bird`, characterizes an argument against α. A simple argument schema rule that allows ARGUER to establish an attack relation between the two propositions is: α *attacks* NOT α.

The sample dialog of Figure 1 will be used throughout the discussion in this paper. The domain knowledge used in the dialog is extracted from the American Heart Association Screener Technician Manual[2], a manual for teaching people how to measure blood pressure and what to say to patients. Two participating agents in the dialog, the system and the user, argue about the need for a blood pressure check. S1, S2 and S3 are the system's utterances while U1 and U2 are those of the user.

```
S1 Have your blood pressure checked.
U1 There is no need.
S2 Uncontrolled high blood pressure can lead
   to heart attack, heart failure, stroke or
   kidney failure.
U2 But I feel healthy.
S3 Unfortunately, there are no signs or
   symptoms that tell whether your blood
   pressure is elevated.
```

Figure 1: Sample blood pressure dialog

We refer to an argument as any proposition corresponding to a node in the argument graph (see Section), e.g. in Figure 1, S1, U1, S2, U2 and S3 are possible arguments. S1 tells the user that he/she must have a blood pressure check. The user responds by telling the system that there is no need (U1). The system then tells the user of the possible consequences if the user does not have a blood pressure check, *i.e.*, uncontrolled high blood pressure can lead to heart attack, heart failure, stroke or kidney failure (S2). Then the user responds that there is nothing wrong with himself/herself: 'But I feel healthy' (U2). The system then responds using S3 that, unfortunately, a person might have a high blood pressure even if no symptoms are felt.

The Types of Knowledge

ARGUER's knowledge base contains at least three types of information: domain knowledge, common-sense knowledge and argumentation knowledge. This information is needed to detect arguments and generate possible responses to them. ARGUER needs domain knowledge to reason about information specific to the topic being talked about. For example, the knowledge associating signs or symptoms with an illness is domain knowledge essential in understanding and responding during dialogs about health. General facts are considered as common-sense knowledge, *e.g.* requiring x from A may imply a need of x by A. In addition, information about the structure of arguments and their various forms is used to detect arguments advanced by agents during a dialog. The argumentation knowledge consists of the argument schema rules.

The Knowledge Representation

ARGUER represents all its knowledge in a uniform framework known as a propositional semantic network. A propositional semantic network is a framework for representing the concepts of a cognitive agent who is capable of using language (hence the term *semantic*). The particular knowledge representation system that is used by ARGUER is SNePS [14] which provide facilities for building and finding nodes as well as for (first- and second-order) reasoning, truth-maintenance, and knowledge partitioning (for user- and system-models). Reasoning in ARGUER is computationally tractable because of the knowledge partitioning; only relevant portions of the knowledge base(s) are used for argumentation.

For brevity and clarity, in this paper, we shall use equivalent logic form representations in our sample interactions.

Models of Belief

ARGUER uses a separate model of belief for each participant in the dialog. In Figure 1, the system plays the role of the medical expert and the user is the patient. The propositions that the system believes may not be necessarily believed by the user. The system, in responding to propositions advanced by the user, takes into account the beliefs currently held in the (system's) user model (as well as its own beliefs). We define a belief model M of an agent A as follows:

Definition 1 (Belief Model)
Let $T = \{\sigma_i \mid \sigma_i$ is a proposition believed by an agent

*A, $i = 1 \ldots n\}$ and $T' = \{\sigma_j \mid \sigma_j$ can be logically derived from T, iteratively}. The belief model of agent A, denoted M_A, is $T \cup T'$. Furthermore, if a proposition $\sigma_k \notin M_A$ then we say that agent A is **ignorant** about σ_k.*

According to Definition 1, the belief model of an agent is the set containing all the propositions believed by the agent together with all its logical consequences[2]. If a proposition does not belong to an agent's model, it is not necessarily false. It may just be the case that the agent is ignorant about it.

During the dialog, as the utterance's meaning representation becomes available, general cases (corresponding to argument schema rules) are used to establish the argument relations. In our method, this process is dynamic and no *a priori* knowledge of argument relations between propositions is assumed. For brevity, we consider only two argument schemas in this paper that are necessary for the examples that we discuss. However, our model is extensible and does support other argument schemas.

Argument Schemas

Suppose someone makes a claim. A common way of challenging that claim is by saying 'No, it is not the case that …'. Let us refer to the first speaker as the *proponent* and the second speaker, *opponent*. The proponent then is obliged to prove or show evidence that would support the claim, i.e. the burden of proof lies with the proponent [16]. In Fragment 1, we intuitively know that U1 is a challenge to S1. By having a general case of argument schema that captures this situation, it is possible to establish computationally that U1 is a challenge to a preceding claim made by the other agent, *i.e.*, S1. The same argument schema could also be used for rebuttal.

Fragment 1
```
S1 Have your blood pressure checked.
U1 There is no need.
```

In Fragment 1, S1 tells the user that there is a need for a blood pressure check. This first utterance makes the *claim* (by implication). The user however responds by challenging the claim and saying, U1, that there is no need. By using an argument schema rule, in this case the rule R1b below, our system can establish that U1 is an *attack* to S1.

---

[2]This is a simplifying assumption. We are working on a resource-limited model of belief.

Argument Schema 1 (Negation)

Let α and β be propositions, then:

R1a: If α is an utterance then NOT α is an attack to α.

R1b: If α is an utterance implying β, then NOT β is an attack to α.

In Fragment 1, U1 attacks S1. The utterance 'Have your blood pressure checked' implies that 'There is a need for a blood pressure check.' We assume that there exists common-sense and domain knowledge that allows us to derive this knowledge. Rules in the common-sense and domain knowledge allow us to derive that 'requiring x of A' (which follows from the imperative form of S1) implies 'the need of A for x'. The rule R1b in Argument Schema 1 allows us to establish that the utterance U1: There is no need, is a proposition that attacks the utterance S1: Have your blood pressure checked. In this case α is S1 and NOT β is U1.

Conversely, suppose the user said S1 and the system wants to generate an attack. Using either R1a or R1b allows us to generate a response that attacks the preceding claim. If α is S1, then there are two possible attacks to α: (1) NOT α via R1a and (2) NOT β via R1b. The best possible response could then be selected using some preference criteria. The method of selecting the best response is a part of our ongoing work (see Section).

We consider a more complex case in the next fragment of the dialog of Figure 1.

Fragment 2

```
U1 There is no need.
S2 Uncontrolled high blood pressure can lead
   to heart attack, heart failure, stroke or
   kidney failure.
```

In Fragment 2, one way to detect that S2 is an attack to U1 is to consider the consequences of U1. Not having a blood pressure check can lead to uncontrolled high blood pressure which can lead to events like heart attack or stroke. These events in turn can lead to a fatal event, *i.e.*, death, which is possibly unacceptable to the user.

There are two different types of unacceptability that we consider here. One type is subjective, *i.e.*, an agent subjectively determines a proposition to be unacceptable (such as the possibility of his/her death). The other type is motivated by reason, *i.e.*, a proposition is unacceptable to the agent because the agent believes its negation is true. For example, the statement It is unacceptable to me that Tweety can fly because I believe that Tweety can not fly is of the second type. We define the notion of unacceptability as follows:

Definition 2 (Unacceptable)

*Let M_A be a belief model of agent A, δ be a proposition, and C,D be the set of facts and rules of the common-sense and domain knowledge, respectively. Let the symbol, $\vdash$, denote logical derivation. Furthermore let $uc(\delta)$ be a proposition equivalent to the statement "δ is not acceptable". The proposition δ is **unacceptable** with respect to M_A if and only if (1) $(M_A \cup C \cup D) \vdash NOT\ \delta$, or (2) $(M_A \cup C \cup D) \vdash uc(\delta)$.*

Definition 2 allows two types of unacceptability. The condition in (1) means we do not accept the proposition because we believe its negation. The condition in (2) means we do not accept the proposition because we simply believe it is subjectively unacceptable. In Fragment 2, α is U1 and δ, the state which is unacceptable to the user, is implied by S2. Any proposition leads to δ from α, attacks α. Our model uses both types of unacceptability[3].

Argument Schema 2

Let α, β, γ and δ be propositions and let the symbol, $\Rightarrow$, denote logical implication, then:

R2: If $\alpha \Rightarrow (\beta \Rightarrow \gamma) \Rightarrow \delta$, where δ is unacceptable to the agent uttering α then $(\beta \Rightarrow \gamma)$ is an attack to the utterance α.

Argument Schema 2 means that any set of rules or propositions that leads to a state that is unacceptable to the speaker is an attack to what has just been uttered. The flow of reasoning is the following: Suppose that there is no need for a blood pressure check. If blood pressure is not checked, blood pressure may become high. High blood pressure can lead to heart attack, heart failure, stroke or kidney failure. Having a heart attack, heart failure, stroke or kidney failure is unacceptable to the agent.

Using R2, the system detects that the proposition 'Uncontrolled high blood pressure can lead to heart attack, heart failure, stroke or kidney failure' is an attack to the utterance 'There is no need'. Moreover, R2 can be used to find possible responses to U1. Possible propositions that attack U1 are (1) 'If a blood pressure is not checked, blood pressure may become high' and (2) 'High blood pressure can lead to heart attack, heart failure, stroke or kidney failure'. These propositions enable the flow of reasoning to reach a state that is unacceptable to the speaker.

A Sample Interaction

In this section, we describe an implementation of the method, used in ARGUER, that detects arguments

---

[3]We are investigating argumentation issues when there is an inconsistency between the user model and the domain and common sense knowledge.

and also finds possible ways to respond. The problem that we are attempting to address here is a part of a larger project to understand various issues involved in robust human-machine communication [12]. We will use a portion of Figure 1 to show how the proposed method works.

ARGUER's knowledge base consists of commonsense knowledge, domain knowledge and argumentation knowledge. The knowledge base and the utterance of the participating agents in the dialog are represented uniformly. On the implementation level, the uniform representation allows a common method of accessing and processing information of different types, which in this case include processing information from domain knowledge, argumentation knowledge and common-sense knowledge. Advantages of using uniform representation are discussed in [12].

We are currently extending ARGUER to process natural language (using the general-purpose grammar of B2 and a template-based generation system [12, 13]). For clarity, the example(s) below are shown in English, ARGUER assumes the interpreted logic-based representations shown for input (and outputs similar representations).

An Example In the following example, S1, U1 and S2 are utterances of the agents participating in the dialog. S1 and S2 are utterances of the system while U1 is an utterance of the user. We show how AR-GUER detects correctly that U1 attacks S1, and S2 attacks U1. Moreover, after each utterance (U1 and S1) we show how ARGUER can use the same method to generate possible rebuttals.

Fragment 3

S1 Have your blood pressure checked.
U1 There is no need.
S2 Uncontrolled high blood pressure can lead to heart attack, heart failure, stroke or kidney failure.

Utterance S1 is interpreted as: agent 'system' requiring agent 'user' to have a blood pressure check. (We assume that U1 has been modified to the form 'There is no need for a blood pressure check'.) We represent the utterance S1 as:

$$\text{S1:} \quad Require(system, user, *check\text{-}BP)$$

The variable *check-BP represents an unknown agent performing a blood pressure check on the agent *user*. The utterance U1 is true in the user's model, *i.e.*, as far as the user is concerned, the user does not need a blood pressure check. This is represented as:

$$\text{U1:} \quad \neg Need(user, *check\text{-}BP)$$

Since models for the user and the system are separately maintained, this proposition is not believed by the system, but can be used to detect an argument.

Detecting an argument: U1 attacks S1 To detect that U1 attacks S1, ARGUER asks the question

$$Attacks(Require(system, user, *check\text{-}BP), \\ \neg Need(user, *check\text{-}BP))$$

(*i.e.*, is U1 an attack on S1?) and uses two rules to answer:

1. Rule R1b: If α is an utterance implying β, then NOT β is an attack to α.

$$\text{R1b:} \quad \forall \alpha, \beta \ Derived(\alpha, \beta) \rightarrow Attacks(\alpha, \neg\beta)$$

2. a common-sense rule, C1, that states that if the system requires some act of an user, then the user needs the required act.

$$\text{C1:} \quad \forall u, a \ Require(system, u, a) \rightarrow Need(u, a)$$

When rule C1 is instantiated and used (with S1), ARGUER deduces the consequent

$$Need(user, *check\text{-}BP)$$

and a meta-level proposition that represents the reasoning path of a ground rule from the antecedent to the consequent:

$$Derived(Require(system, user, *check\text{-}BP), \\ Need(user, *check\text{-}BP))$$

This meta-level proposition is then used in argument schema R1b (with $\alpha = Require(system, user, *check\text{-}BP)$ and $\beta = Need(user, *check\text{-}BP)$) to conclude that the user is attacking the system's utterance:

$$Attacks(Require(system, user, *check\text{-}BP), \\ \neg Need(user, *check\text{-}BP))$$

Generating an argument U1 that attacks S1 To generate an argument that attacks S1 (possibly to preempt a user argument), ARGUER might ask the question

$$Attacks(Require(system, user, *check\text{-}BP), *X)$$

(where *X is a free unbound variable) and deduces possible attacking propositions to S1. As before, this uses rules R1b and C1 to find possible values for *X. In this example, it once again finds:

$$Attacks(Require(system, user, *check\text{-}BP), \\ \neg Need(user, *check\text{-}BP))$$

We note that this method can find *all* arguments that attack S1 (*i.e.*, not just U1), dynamically and with the currently available information and state of the user model.

Generating a rebuttal: S2 attacks U1 To generate a rebuttal to U1, ARGUER asks the question:

$$Attacks(\neg Need(user, \,*check\text{-}BP), \,*X)$$

namely, what attacks U1. In this example, these rules are used:

1. R2: $\forall \, \alpha, \beta, \gamma, \delta \; (Derived(\alpha, \, \beta) \, \wedge$
   $Derived(\beta, \, \gamma) \wedge Derived(\gamma, \, \delta) \wedge uc(\delta))$
   $\rightarrow Attacks(\alpha, \, Derived(\beta, \, \gamma))$

2. C2: $\forall a, x \; \neg Need(a, x) \rightarrow \neg Do(a, x)$

3. D1: $\forall a \; \neg Do(a, \,*check\text{-}BP)$
   $\rightarrow Possible(attribute(a, \, BP, \, high))$

4. D2: $\forall \, a \; Possible(attribute(a, \, BP, \, high))$
   $\rightarrow Possible(event(a, \, \{stroke, \, heart\text{-}failure, \, heart\text{-}attack, \, kidney\text{-}failure\}))$

5. D3: $uc(Possible(event(a, \, \{stroke, \, heart\text{-}failure, \, heart\text{-}attack, \, kidney\text{-}failure\})))$

Note that the simplest case of R2 is when $\alpha = \beta$ and $\gamma = \delta$ (the user's utterance immediately leads to a consequent that is unacceptable). C2 is a common sense rule that states that if a agent does not need to do an act, the agent will not do it. D1 is a domain rule that states that if an agent does not have their blood pressure checked then high blood pressure is possible. D2 is a domain rule that states that an agents having possible high blood pressure can lead to a possible stroke, heart attack or failure, or kidney failure. D3 is a domain fact that states that these outcomes are unacceptable to the user.

To answer its query, ARGUER reasons as follows:

- $Derived(\neg Need(user, \,*check\text{-}BP), \, \neg Do(a, \,*check\text{-}BP))$ From C2 and U1.

- $Derived(\neg Do(user, \,*check\text{-}BP), \, Possible(attribute(user, \, BP, \, high)))$ From D1 and C2.

- $Derived(Possible(attribute(user, \, BP, \, high)), \, Possible(event(user, \, \{stroke, \, heart\text{-}failure, \, heart\text{-}attack, \, kidney\text{-}failure\})))$ From D2, D1 and C2.

Using R2 and D3 with this final derivation, ARGUER can conclude:

$$Attacks(\neg Need(user, \,*check\text{-}BP),$$
$$Derived(Possible(attribute(user, \, BP, \, high)),$$
$$Possible(event(user, \{stroke, \, heart\text{-}failure,$$
$$heart\text{-}attack, \, kidney\text{-}failure\}))))$$

From this, ARGUER decides that S2 is an attack to U1:

S2: $Derived(Possible(attribute(user, \, BP, \, high)),$
$Possible(event(user, \{stroke, heart\text{-}failure,$
$heart\text{-}attack, \, kidney\text{-}failure\})))$

Moreover, because the method is symmetrical, it could be used to detect that S2 attacks U1, to predict possible rebuttals to S2, or to answer the question "What attacks what?"

Summary

We have shown the use of argument schema both for detecting arguments and generating possible rebuttals in dialogs. The method is implemented as ARGUER and allows us to establish argument relations between propositions while the dialog is ongoing. Moreover, the method we have proposed here is *incremental* in that it allows processing of each piece of the utterance and selects only a part of the argument to continue.

References

[1] S. Alvarado. *Understanding Editorial Text: A Computer Model of Argument Comprehension.* Kluwer Academic, 1990.

[2] American Heart Association, Milwaukee, WI. *Blood Pressure Measurement Education Program Screener Technician Manual*, 1998.

[3] K. D. Ashley. *Modeling Legal Argument: Reasoning with Cases and Hypotheticals.* MIT Press, Cambridge, MA, 1990.

[4] L. Birnbaum, M. Flowers, and R. McGuire. Towards an AI Model of Argumentation. In *Proceedings of the AAAI-80*, pages 313–315, Stanford, CA, 1980.

[5] A. Bondarenko, P. M. Dung, R. A. Kowalski, and F. Toni. An Abstract, Argumentation-Theoretic Approach to Default Reasoning. *Artificial Intelligence*, 93:63–101, 1997.

[6] M. Bordegoni, G. Faconti, T. Y. Maybury, T. Rist, S. Ruggieri, P. Trahanias, and M. Wilson. A Standard Reference Model for Intelligent Multimedia Representation Systems. *The International Journal on the Development and Applications of Standards for Computers, Data Communications and Interfaces*, 1997.

[7] Jennifer Chu-Caroll and Sandra Carberry. Generating Information-Sharing Subdialogues in Expert-User Consultation. In *Proceedings of the 14th IJCAI*, pages 1243–1250, 1995.

[8] P. M. Dung. The Acceptability of Arguments and its Fundamental Role in Non-Monotonic Reasoning, Logic Programming and N-Person Games. *Artificial Intelligence*, pages 321–357, 1995.

[9] N. Karacapilidis and D. Papadias. Hermes: Supporting Argumentative Discourse in Multi-Agent Decision Making. In *Proceedings of the AAAI-98*, pages 827–832, Madison, WI 1998.

[10] R. P. Loui and J. Norman. Rationales and Argument Moves. *Artificial Intelligence and Law*, 1993.

[11] R. McGuire, L. Birnbaum, and M. Flowers. Opportunistic Processing in Arguments. In *Proceedings of the 7th International Joint Conference on Artificial Intelligence (IJCAI)*, pages 58–60, Vancouver, B.C., 1981.

[12] Susan McRoy, Syed S. Ali, and Susan Haller. Uniform Knowledge Representation for NLP in the B2 System. *Journal of Natural Language Engineering*, 3(2):123–145, 1997.

[13] Susan W. McRoy, Syed S. Ali, and Susan M. Haller. Mixed Depth Representations for Dialog Processing. In *Proceedings of the 20th Annual Conference of the Cognitive Science Society (CogSci '98)*, pages 687–692. Lawrence Erlbaum Associates, 1998.

[14] Stuart C. Shapiro and William J. Rapaport. The SNePS family. *Computers & Mathematics with Applications*, 23(2–5), 1992.

[15] G. Vreeswijk. IACAS: An Implementation of Chisholm's Principles of Knowledge. In *Proceedings of the 2nd Dutch/German Workshop on Nonmonotonic Reasoning*, pages 225 234, 1995.

[16] D. Walton. *Argument Structure: A Pragmatic Theory*. University of Toronto Press, 1996.

[17] I. Zukerman, R. McConachy, and K. Korb. Bayesian Reasoning in an Abductive Mechanism for Argument Generation and Analysis. In *Proceedings of the AAAI-98*, pages 833–838, Madison, Wisconsin, July 1998.

607

Semantic Competition and the Ambiguity Disadvantage

Jennifer Rodd (jenni.rodd@mrc-cbu.cam.ac.uk)

Gareth Gaskell (gareth.gaskell@mrc-cbu.cam.ac.uk)

William Marslen-Wilson (william.marslen-wilson@mrc-cbu.cam.ac.uk)

MRC Cognition and Brain Sciences Unit
15 Chaucer Road, Cambridge, UK

Abstract

In many recent models of word recognition, words compete to activate distributed semantic representations. Reports of faster visual lexical decisions for ambiguous words compared with unambiguous words are problematic for such models; why does increased semantic competition between different meanings not slow the recognition of ambiguous words? This study challenges these findings by showing that visual lexical decisions to ambiguous words whose meanings were judged to be unrelated were *slower* than either unambiguous words or ambiguous words whose meanings were judged to be related. We suggest that previous reports of an ambiguity advantage are due to the use of ambiguous words with highly related meanings.

Introduction

Many recent connectionist models of written and spoken word recognition use distributed lexical representations; words are represented as a pattern of activation across a set of units. In most of these models, words are represented at an explicitly semantic level (Gaskell & Marslen-Wilson, 1997; Hinton & Shallice, 1991; Joordens & Besner, 1994; Plaut, 1997). Words compete for activation of these semantic representations, and for a word to be recognised, it must produce a familiar pattern of semantic activation. This paper reviews the existing evidence for such semantic competition effects, and reports a visual lexical decision experiment that suggests that semantic competition does play an important role in visual word recognition.

Ambiguous Words

One situation where semantic competition will occur is the case of ambiguous words. Homographs such as *bank* are words where a single orthographic form has more that one meaning. In the above models, the orthographic pattern *bank* must be associated with two different semantic patterns of activation corresponding to its two meanings. When the orthographic pattern is presented to a network model, these two semantic patterns will both be partially activated as the network tries to simultaneously instantiate both patterns across the semantic units. The likely consequence of this is that the two patterns of activation will interfere with each other, and delay the recognition of the word. This prediction of a processing disadvantage for ambiguous words is an example of the more general principle that one-to-one mappings are more easily learned and processed than mappings where a single input pattern corresponds to more than one output pattern. (For a review of such efforts see Joordens & Besner, 1994.)

However, there have been several attempts to show that, given the right assumptions, this class of model can show the opposite effect, an advantage for ambiguous words. The approach taken by Joordens and Besner (1994) relies on the assumption that when the orthography of a word is presented to the network, the initial state of the semantic units is randomly determined. Further, the closer to its target state the network starts, the quicker it will settle into that state. For ambiguous words, there is more than one valid finishing state, and the network is likely to settle into the target state closest to its initial state. On average, the initial state of the network will be closer to one of these states than for an unambiguous word where there is only one valid finishing state. Their network did show an ambiguity advantage, but only when the error rate was unrealistically high.

Kawamoto, Farrar, and Kello (1994) took a different approach to modelling an ambiguity advantage. They assumed that lexical decisions are made on the basis of orthographic, not semantic, representations. Using the least mean square (LMS) algorithm during training, an ambiguity advantage was shown. This arises because of the error-correcting nature of the learning algorithm; in order to compensate for the increased error produced by the ambiguous words in the semantic units, stronger connections are formed between the orthographic units, which are being used as the index of performance.

Therefore, successful demonstration of an ambiguity advantage relies on the assumption that the lexical decision task is primarily orthographic. This assumption has problems in accounting for the range of semantic effects on lexical decision. The simulations reported by Joordens and Besner (1994) make it clear that if the activation of the semantic units is used to model lexical decision in such networks, semantic competition makes it difficult for such networks to produce an ambiguity advantage.

Joordens and Besner (1994) argue that for this class of model "increased competition must occur when an ambiguous word is processed and that this competition must act to reduce processing efficiency". This gives us a testable prediction of models where words compete for activation of semantic representations: semantic competition should act to slow the recognition of ambiguous words relative to words with only one meaning.

The second semantic competition prediction relies on the observation that within the class of ambiguous words there are words that differ widely in the relatedness of their meanings. The linguistic literature has distinguished between two types of ambiguity: homonymy and polysemy. Traditionally homonyms are said to be different words that happen to have the same form, whereas a polysemous word is a single word that has more than one sense (Lyons, 1977). For example, the orthographic form *bank* is assumed to be a homograph, and to and correspond to two separate lexical items that share the same orthography purely by chance. Conversely, the use of the word *mouth* to describe the entrance to a cave as well as a part of the body is assumed to reflect the fact that *mouth* is a word with more than one sense (Lyons, 1977). This distinction between word senses and word meanings is respected by all standard dictionaries; lexicographers decide whether different usages of the same orthography correspond to different lexical entries or different senses within a single entry.

However, while this homonymy/polysemy distinction is easy to formulate, it is difficult to apply with consistency and reliability. There is not always a clear distinction between these two types of ambiguity, and people will often disagree about whether two word meanings are sufficiently related to be classed as senses of a single meaning. Despite this, it is important to remember that words that are described as ambiguous can vary between these two extremes.

This distinction is important when looking at the issue of semantic competition. Models in which words compete for activation of distributed semantic representations predict that recognition of ambiguous words will be slowed by competition between semantic representations. However, as noted by Rueckl (1995), competition between two meanings can be reduced by increasing the similarity of their representations. This means that the situation is different for ambiguous words with semantically related senses. Often these senses relate to a single core literal meaning of the word (Lyons, 1977). For such words there will be a high level of overlap between the competing semantic representations, and so the interference between them will be greatly reduced. We predict that this reduction in interference will result in ambiguous words being recognised faster when their different meanings are related than when they are highly unrelated. This is a second testable prediction of models where words compete for activation of distributed semantic representations.

The following two sections will review the literature that is relevant to these two predictions. We will then go on to report a lexical decision experiment that supports these two predictions.

The Ambiguity Advantage

The first prediction, that semantic competition will slow recognition of ambiguous words, is apparently contradicted by the available experimental evidence. The *ambiguity advantage* is the reported finding that words with many meanings are responded to *faster* in visual lexical decision experiments than words with fewer meanings. Therefore, rather than semantic competition slowing recognition of ambiguous words, there appears to be a processing advantage for such words.

Early reports of an ambiguity advantage (e.g. Rubenstein, Garfield, & Millikan, 1970) have been criticised for not controlling for subjective familiarity; words with more than one meaning are typically more familiar. Since then, Kellas, Ferraro, and Simpson (1988) report faster lexical decision times

for ambiguous words than for unambiguous words, but they do not show that this effect reaches statistical significance. Hino and Lupker (1996) and Borowsky and Masson (1996) also report finding an ambiguity advantage in a lexical decision experiment, but the effect of ambiguity was not significant in the items analysis in either study.

Millis and Button (1989) and Azuma and Van Orden (1997) do show a statistically significant advantage in visual lexical decision experiments for words with many meanings when compared with words with few meanings. However, both these studies used procedures that counted highly related word meanings that correspond to different senses of a single dictionary entry as distinct meanings. For example, Millis and Button (1989) provide the word *tell* as an example of a word that has many meanings. Participants produced up to four meanings for this word; these meanings were: *to inform, to explain, to understand* and *to relate in detail*. Although there are important differences between these four definitions, they all relate to a single core meaning of the word, and are given as different senses within a single entry in the Wordsmyth Dictionary (Parks, Ray, & Bland, 1998).

Further, neither study compared ambiguous words with words with only one meaning. Rather, they compared words with few meanings with words with many meanings. This had the result that many of the low-ambiguity stimuli were themselves highly ambiguous, e.g. *spoke, staff, bark* and *seal* which all have more than one entry in the Wordsmyth Dictionary. We carried out an analysis of the stimuli used in the two experiments, and found that in both experiments, the low- and high-ambiguity words did not differ in their number of different dictionary entries; both groups had a similar number of words with multiple entries in the Wordsmyth Dictionary. Instead, the groups differed in their number of dictionary *senses*, with the high-ambiguity words having far more senses. This suggests that the ambiguity advantage they found was not the result of a benefit for ambiguity of the kind seen in the word *bank*, between two highly unrelated meanings. Rather, they have shown a benefit for ambiguous words that, in addition to any unrelated meanings, have clusters of related word senses.

As discussed earlier, ambiguity between related word senses would not necessarily produce the kind of semantic competition that we would expect for unrelated word meanings. For a reported ambiguity advantage to be problematic for models that incorporate semantic competition, the high-ambiguity words must have fewer semantically unrelated meanings than the low-ambiguity words. We believe that neither of these studies has shown such an effect. Therefore, the question of what effect the presence of semantically unrelated meanings has on word recognition has not yet been resolved.

Relatedness of Word Meanings

Two studies have investigated whether the relationship between the meanings of an ambiguous word can affect its recognition. Azuma and Van Orden (1997) report that for homographs with two to four meanings, visual lexical decision times are slower when their meanings were rated as being unrelated. This is consistent with our prediction that when the meanings of an ambiguous word are highly unrelated, there is greater competition between their semantic representations, and that this increased competition produces slower lexical decision times than when the meanings are highly related. However, as mentioned earlier, the procedure used by Azuma and Van Orden (1997) means that highly related word senses were counted as separate meanings. They were comparing words like *hide* and *park*, whose meanings are completely unrelated with words like *drink* and *rake* which are ambiguous only in that they can be used as either a noun or a verb. It is not clear that words with such highly related meanings should really be considered ambiguous. This study will attempt to replicate their finding using stimuli with greater ambiguity. We will compare words like *cricket*, whose meanings are semantically unrelated, to words like *letter*. *Letter* has two meanings, but they are semantically related in that they both refer to written language.

The second relevant experiment was performed by Rubenstein et al. (1970). They investigated the effects of ambiguity and concreteness on visual lexical decision times, and report faster lexical decision times for homographs with one concrete meaning and one abstract meaning than for homographs with two concrete meanings. If we accept that concreteness is an important semantic dimension, we would expect that on average two concrete meanings will be more similar than one concrete meaning and one abstract meaning. This makes Rubenstein et al.'s (1970) result somewhat counterintuitive; the homographs with less similar meanings were responded to faster. This is inconsistent with the finding of Azuma and Van Orden (1997), and conflicts with the view that increased competition between semantically dissimilar homographs should slow their recognition.

Therefore the two reports in the literature of semantic competition effects give somewhat contradictory findings. The experiment reported here attempts to resolve the issue. Its starting point is the design used by Rubenstein et al. (1970), since this is a result that is problematic for all classes of models.

Aims of the Experiment

The visual lexical decision experiment reported here has three main aims.

Firstly it will investigate the effect of ambiguity on word recognition by comparing words that have more that one meaning with words that are not ambiguous. Evidence of a disadvantage for ambiguous words would contradict previous reports of an ambiguity advantage, but support our first prediction, that semantic competition will slow recognition of ambiguous words. To avoid the problems associated with earlier studies, it will directly compare ambiguous words with unambiguous words that have only one dictionary entry. The ambiguous words will have two meanings with a range of relatedness.

The experiment will also address the prediction of that any slowing of the recognition of ambiguous words will be greater when the ambiguity is between meanings that are semantically highly unrelated. It will use a design that will allow us to investigate the finding by Rubenstein et al. (1970) that ambiguous words were responded to faster when they had one abstract and one concrete meanings than when both meanings are concrete. Finally, it will also look for direct effects of meaning relatedness.

Lexical Decision Experiment

Experimental Design   A summary of the experimental design is given in Table 1. Three groups of homographs are used. Group CC contains homographs with two concrete meanings, homographs in Group CA have one concrete meaning and one abstract meaning, and Group AA contains homographs with two abstract meanings. The control stimuli are two groups of non-homographs. Group C consists of concrete non-homographs, matched for concreteness with the more concrete meaning of the homograph Groups CC and CA. Group A consists of abstract words, matched for concreteness with the abstract meanings of the words in Group CA.

Table 1: Experimental Design

	Homographs			Non-Homographs	
	CC	CA	AA	C	A
Group					
Example	calf	seal	lean	goat	sane

In addition to these between-group comparisons, it was intended to use correlation analyses to investigate the effects of the concreteness and relatedness of the two meanings of the homographs. To increase the number of words in this analysis all homographs that were pre-tested but eliminated during the matching of the groups were also included in the experiment.

Method

Stimuli   The main selection criterion for the homographs was that they should have only two meanings. 130 homographs taken from a range of homograph norms were pre-tested to obtain ratings for the concreteness of each of their two meanings, and for their overall familiarity. Two groups of subjects not included in the lexical decision experiment made the familiarity and concreteness ratings, on a 7-point scale. The pre-tested stimuli were assigned to the five stimulus groups according to these concreteness scores. Concrete meanings had a minimum concreteness rating of 5.22; abstract meanings had a maximum concreteness of 4.96.

The five groups each contained 30 words, and were matched for length, frequency, familiarity, and neighbourhood density. The frequency measure used was the log-transformed frequency of the word in the spoken section of the British National Corpus (LnBNC). The neighbourhood density measure was the number of words in CELEX that differed from each word by only one letter.

These five groups can be collapsed to give a group of homographs (N=90) and a group of non-homographs (N=60); these two groups were also matched on all the above variables, as well as for mean concreteness. For a homograph, Mean Concreteness was the mean of the concreteness rating for its two meanings; for non-homographs it was simply the concreteness score for the word.

Thirty four homographs that had been pre-tested but were not included in one of the five groups were included in the experiment.

The non-word stimuli were pseudohomophones with a similar distribution of word lengths to the word stimuli.

Procedure The participants were 25 members of the MRC Cognition and Brain Sciences Unit subject panel. All the stimulus items were pseudo-randomly divided into four groups which participants saw in a pseudo-random order such that no two participants saw the lists in the same order. Within the lists, the order in which stimulus items were presented was randomised for each participant. All the participants saw all of the stimulus materials. The participants' task was to decide whether each item was a word or a non-word. Once a participant had responded, the next stimulus appeared after 500 ms.

Results

The latencies for responses to the word and non-word stimuli were recorded, and the inverse of these response times (1/RT) were used in the analyses.

ANOVA Analyses Mean values were calculated separately across subjects and items. The subject means were subjected to an analysis of variance (ANOVA), and the item means were subjected to an analysis of covariance (ANCOVA) with frequency, familiarity and length entered as covariates.

Incorrect responses were not included in the analysis. All responses greater than two seconds were also removed from the analysis; participants made between zero and six such responses, nearly all on the non-words; in total, only two such responses were made for the real words. The overall error rate for responses was 5.5%.

Response latencies for each cell of the design, averaged over subjects and items, are presented in Figure 1.

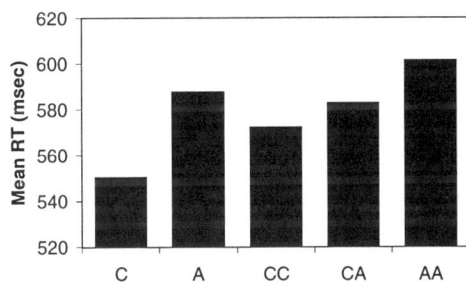

Figure 1: Mean lexical decision times

The analyses of variance using the five groups of words revealed a main effect of stimulus type, $F_s(4, 88) = 13.27$, $p < .001$, $F_i(4, 142) = 6.47$, $p < .001$. Multiple comparisons were made between the individual groups using the Newman–Keuls procedure.

For the non-homographs, there was a significant effect of concreteness, with response times for concrete words being significantly shorter than for abstract words; $q_s(3, 88) = 7.39$, $p < .001$, $q_i(4, 144) = 4.99$, $p < .001$.

For the homographs, although there was a trend towards faster responses for the more concrete groups, there were no significant differences between the groups of homographs.

In order to look at the effect of ambiguity, the three groups of ambiguous words (Groups AA, CA and CC) were col-

lapsed and compared with the unambiguous words (Groups A and C). In the items analyses frequency, familiarity, length and mean concreteness were entered as covariates. The analyses revealed that the ambiguous words were responded to more slowly than the unambiguous words. The 16.5 ms difference was significant in both the subjects and the items analyses; $F_s(1, 23) = 21.88$, $p < .001$, $F_i(1, 144) = 4.93$, $p < .05$.

Analyses of the error data showed a similar pattern to the response time data: slower responses patterned with more errors.

Correlation Analyses The inverse response times for the homographs were also entered in a simple correlation analysis. Included in this were the 90 homographs used in the ANOVA analyses, as well as the 35 additional homographs.

Frequency (*LnBNC*) significantly predicted response times; $p < .001, r = .54$. Familiarity also significantly predicted response times; $p < .001, r = .48$. Partial correlations demonstrated that familiarity did not account for a significant proportion of variance in response times, when frequency effects were partialled out; $p > .3$. Frequency remained a significant predictor when familiarity was partialled out; $r = .31, p < .001$.

Three concreteness-related measures were entered in a correlation analysis of the inverse response times to the homographs in which frequency effects were partialled out: *mean concreteness*, *concreteness difference* and *dominant concreteness*. *Mean concreteness* is the average of the two concreteness scores, and *concreteness difference* is the difference between them. *Dominant concreteness* is the concreteness rating for the more dominant meaning of each homograph; the dominant meaning was determined on the basis of the frequency norms from which the homographs were taken.

Significant correlations were obtained between response times and *mean concreteness r = .20, p < .05*. This indicates a reliable effect of the concreteness trend, which was non-significant in the ANOVA analyses. The correlations between *concreteness difference, dominant concreteness* and response times were not significant.

Discussion

Three interesting results have emerged from this experiment.

Firstly, both the homographs and non-homographs showed a significant effect of concreteness; concrete words were responded to faster than abstract words. This serves as a useful replication of the concreteness effect, using stimuli that were controlled for subjective familiarity as well as frequency.

Secondly, the difference between the groups of homographs reported by Rubenstein et al. (1970) is likely to have been an artefact of their poorly-controlled stimuli. They reported that the homographs with two concrete meanings were responded to more slowly than those with one concrete meaning and one abstract meaning. We found the opposite effect, although this difference was non-significant.

Finally, and most importantly, response times showed an *ambiguity disadvantage*. Using stimuli that are matched for frequency, familiarity, length and orthographic neighbourhood, we have found an ambiguity disadvantage; words with one meaning were responded to significantly *faster* than words with two meanings. This apparently contradicts previous findings of an advantage for ambiguous words on visual lexical decision tasks.

In this experiment we used concreteness as the semantic dimension along which the words varied. This was in response to the finding of Rubenstein et al. (1970). The finding that concreteness difference had no significant effect on response times means that we have no evidence that homographs are recognised faster when their meanings have similar concreteness scores. Therefore, this analysis of the experiment provides no evidence that the relationship between the semantics of the different meanings of ambiguous words affects their recognition.

However, concreteness is only one aspect of the semantic properties of a word; comparing the concreteness of a word's meanings is a rather indirect way of measuring the semantic relationship between its different meanings. The following analyses will address the issue of the semantic relationship between the meanings of ambiguous words, by focusing more directly on semantic similarity. Using participants' judgements of how related the meanings of ambiguous words are, we will look for an effect of relatedness of lexical decision times.

Relatedness Ratings Analysis of Lexical Decision Experiment

Method

Relatedness ratings were obtained for 124 of the homographs used in the lexical decision experiment. Raters were given each homograph, together with short definitions of its two meanings, and asked to rate how related the two meanings were, on a 7-point scale.

Results and Discussion

The mean relatedness rating across all participants and items was 2.64. This low value reflects the fact that participants saw many of the pairs as completely unrelated; a rating of 1 was used more than any other rating. In all the following analyses, rather than using the mean relatedness ratings, the inverse of these values was used. This made the measure more sensitive to small changes at the lower end of the scale.

Correlation Analyses The relatedness values for the homographs were entered into a correlation analysis with the other predictor values used in the earlier analysis. Relatedness was not significantly correlated with frequency ($r = -.14, p = .13$), familiarity ($r = -.16, p = .07$), or mean concreteness ($r = .07, p = .48$). The only variable that relatedness was correlated with was *concreteness difference*, the difference between the concreteness values of the two meanings; $r = .24, p < .01$. This shows that the difference between the concreteness of the two meanings is moderately related to the overall relatedness of the meanings.

The real question of interest is whether the relatedness scores for the homographs significantly predict the lexical decision times. In a correlation analysis in which the effects of frequency, mean concreteness and concreteness difference were partialled out, relatedness was a significant predictor of response times ($r = -.19, p < .05$); homographs were responded to faster when their meanings were judged to be semantically related.

ANOVA Analyses To provide further evidence for the effects of relatedness, ANOVA/ANCOVA analyses were also performed. From the set of 124 homographs, two sets of 27 homographs were selected, containing related and unrelated homographs respectively. They were selected by using only those homographs with a relatedness score of either less than 1.9 or greater than 3.4. A few homographs were then removed so that the two groups were matched for frequency, mean concreteness, length, familiarity and neighbourhood density. A group of 43 non-homographs was selected to match the two homograph groups according to the same variables. The response times for these three groups of words were submitted to ANOVA/ANCOVA analyses. The mean response times are given in Figure 2.

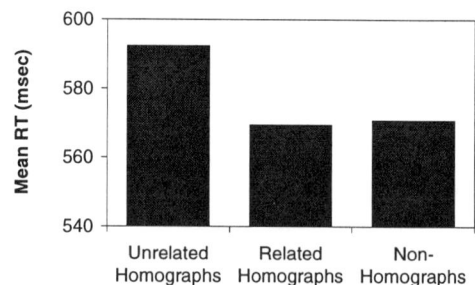

Figure 2: Mean lexical decision times

In the subjects analysis, the effect of group was significant; $F_s(2, 44) = 6.74, p < .003$. In the items analysis, using the log-transformed frequency, familiarity, mean concreteness and length as covariates, the effect of group was marginal; $F_i(2, 90) = 2.80, p < .07$.

Multiple comparisons were made between the individual groups, using the Newman–Keuls procedure. Responses to the group of non-homographs were faster than to the group of homographs with unrelated meanings; this difference was significant in the subjects analysis and marginal in the items analysis; $q_s(3, 44) = 4.89, p < .005, q_i(3, 90) = 3.23, p < .06$. The related homographs were significantly faster than the unrelated homographs in the subjects analysis; again, the difference was marginal in the items analysis; $q_s(2, 44) = 3.95, p < .01, q_i(2, 90) = 2.72, p < .06$

These results confirm the findings of the regression analysis; homographs with related meanings are responded to more quickly than homographs with highly unrelated meanings. Further, they show that homographs are responded to more slowly than non-homographs only when their meanings are unrelated.

In summary, post-hoc analyses using relatedness ratings for the homographs used in the lexical decision experiment times were performed using regression and ANOVA analyses. These showed that responses were quicker for homographs with related meanings (e.g. *letter*) than for those with unrelated meanings(e.g. *cricket*). This result fits well with the idea that words compete at a semantic level of representation, and that recognition of the homographs with unrelated meanings is slowed by increased competition between the competing semantic representations.

General Discussion

The lexical decision experiment reported here confirms predictions made by models of visual word recognition where words compete to activate distributed semantic representations and are recognised as familiar patterns of activation. Homographs with two meanings were responded to more slowly than unambiguous words; we found a significant ambiguity *disadvantage* for a group of 90 homographs, compared with 60 non-homographs that were controlled for frequency, familiarity, length, neighbourhood size, and concreteness. This difference completely disappeared for ambiguous words whose two meanings were judged to be semantically related. These findings are explained by assuming that recognition of an ambiguous word is slowed by competition between the semantic representations corresponding to its different meanings. This semantic competition arises only for words whose different meanings are highly unrelated.

The finding of a disadvantage for ambiguous words clearly contradicts previous reports of an ambiguity advantage. We have suggested that this apparent discrepancy may be explained by noting that all the studies that report a reliable ambiguity advantage count highly related senses as distinct word meanings. Their findings can therefore be explained as a benefit for words with clusters of related meanings rather that a benefit for ambiguity.

Reports of an ambiguity advantage have presented a serious problem for models where words are recognised as familiar patterns of semantic activation, and there have been several attempts to model the effect using such models (Joordens & Besner, 1994; Kawamoto et al., 1994). We have argued that the ambiguity advantage is an artefact of the way in which ambiguous words have been selected for experiments. The significant *disadvantage* we found for ambiguous words is a result straightforwardly predicted by this type of model.

The lexical decision experiment reported here provides important evidence that processes of semantic competition are crucially involved in word recognition, and supports the trend in the literature towards models where semantic representations actively participate in word recognition.

References

Azuma, T., & Van Orden, G. C. (1997). Why safe is better than fast: The relatedness of a word's meanings affects lexical decision times. *Journal of Memory and Language, 36,* 484–504.

Borowsky, R., & Masson, M. E. J. (1996). Semantic ambiguity effects in word identification. *Journal of Experimental Psychology: Learning Memory and Cognition, 22,* 63 85.

Gaskell, M. G., & Marslen-Wilson, W. D. (1997). Integrating form and meaning: A distributed model of speech perception. *Language and Cognitive Processes, 12,* 613–656.

Hino, Y., & Lupker, S. J. (1996). Effects of polysemy in lexical decision and naming - an alternative to lexical access accounts. *Journal of Experimental Psychology: Human Perception and Performance, 22,* 1331–1356.

Hinton, G. E., & Shallice, T. (1991). Lesioning an attractor network: Investigations of acquired dyslexia. *Psychological Review, 98,* 74–95.

Joordens, S., & Besner, D. (1994). When banking on meaning is not (yet) money in the bank - explorations in connectionist modeling. *Journal of Experimental Psychology: Learning Memory and Cognition, 20,* 1051.

Kawamoto, A. H., Farrar, W. T., & Kello, C. T. (1994). When two meanings are better than one: Modeling the ambiguity advantage using a recurrent distributed network. *Journal of Experimental Psychology: Human Perception and Performance, 20,* 1233–1247.

Kellas, G., Ferraro, F. R., & Simpson, G. B. (1988). Lexical ambiguity and the timecourse of attentional allocation in word recognition. *Journal of Experimental Psychology: Human Perception and Performance, 14,* 601–609.

Lyons, J. (1977). *Semantics.* Cambridge, England: Cambridge University Press.

Millis, M. L., & Button, S. B. (1989). The effect of polysemy on lexical decision time: Now you see it, now you don't. *Memory & Cognition, 17,* 141–147.

Parks, R., Ray, J., & Bland, S. (1998). *Wordsmyth english dictionary-thesaurus.* [ONLINE]. Available: http://www.wordsmyth.net [1999, February 1], University of Chicago.

Plaut, D. C. (1997). Structure and function in the lexical system: insights from distributed models of word reading and lexical decision. *Language and Cognitive Processes, 12,* 765–805.

Rubenstein, H., Garfield, L., & Millikan, J. A. (1970). Homographic entries in the internal lexicon. *Journal of Verbal Learning and Verbal Behavior, 9,* 487–494.

Rueckl, J. G. (1995). Ambiguity and connectionist networks - still settling into a solution - comment. *Journal of Experimental Psychology: Learning Memory and Cognition, 21,* 501-508.

A Dynamic Neural Network Model of Multiple Choice Decision-Making

Robert M. Roe (rmroe@indiana.edu)
Indiana University, Department of Psychology, 1101 E. 10th Street
Bloomington, IN 47405 USA

Abstract

A neural network instantiation of Decision Field Theory (Busemeyer & Townsend, 1993) for multiple choice decision tasks is presented. First it is shown how under certain situations this dynamic model reduces to two well-known static models of choice. Next, model simulations of two well-known findings in multiple choice decision literature are presented. The first is the effect of similarity (Tversky, 1972). Several choice models also predict this effect. However, a more challenging effect, which is not predicted by numerous static choice models is the decoy effect (Huber, Payne, & Puto, 1982). Simulations show that the current model predicts this finding by using the concept of lateral inhibition. Finally, predictions of the model are made about the dynamic nature of the deliberation process in the decoy effect. If empirical results are found to be in agreement with this prediction, it would be a strong test of the model

Introduction

Preferential choice is a very complex topic that needs to be studied from many different perspectives. Take for example, the relatively simple task of buying a used car. From one point of view, this is a search problem in which a very large set of options is winnowed down to a much smaller set of satisfactory options (Simon, 1955). From another point of view, this is an evaluation problem requiring tradeoffs among multiple conflicting attributes such as safety, quality, performance, and cost (Keeney & Raiffa, 1976). From a third point of view, this is a choice problem which the candidates engage in a competition for the purpose of identifying a winning or best alternative (Thurstone, 1959).

The purpose of this article is to present a general decision theory that encompasses all of these points of view within a single processing framework. The present theory is based on an earlier theory known as decision field theory. Decision field theory was originally developed to explain choice behavior for decision making under uncertainty by Busemeyer & Townsend (1993). Later it was extended to explain the relation between choice, selling prices, and certainty equivalents by Townsend & Busemeyer (1996). More recently, it was extended to account for multi-attribute decision making by Diederich (1997). However, all of these previous developments were limited to choice situations involving only two choice options. This simplification was initially necessary to focus on other issues in more depth such as multi-attribute outcomes and multiple uncertain outcomes. The purpose of this article is to relax this restriction and present an extension of decision field theory to multiple (more than two) preferential choice problems. Many new and complex issues arise with multiple alternative choice problems that do not appear in the simpler binary choice task -- for example, a winnowing search process is unnecessary in the binary choice task.

A very large literature already exists on the topic of preferential choice with multiple options. What is unique about decision field theory is that it provides a detailed description of the dynamic process that ensues between the onset of the choice task and the final selection. This dynamic description permits the theory to explain the systematic relations between choice probability and decision time, and the important effects of time pressure on choice probability.

A second purpose of this article is to build formal connections between decision field theory and other neurally inspired models of decision processes (e.g. Grossberg & Gutowski, 1987, Usher & Zakay, 1993, & Levin & Levin, 1996). More specifically, decision field theory is recast or reinterpreted in terms of a neural network formulation. One key idea borrowed from neural network theorists (e.g., Grossberg, 1988) is the principle of lateral inhibition among competing nodes. This idea turns out to play a critical role in explaining paradoxical findings that have posed serious challenges to a large class of static choice models.

The remainder of this article is organized as follows. First we introduce the basic ideas of decision field theory. In order to do this, a specification of how this theory operates under two different types of task constraints, one called the experimenter controlled choice task, and the other called the subject controlled choice task is needed. Second, it is shown how some earlier static theories of choice can be viewed as special cases of decision field theory. In particular, it is shown how decision field theory can be used to derive a dynamic version of the classic Thurstone choice model for the experimenter controlled task, and it is shown the well known elimination by aspects model can be mimicked for the subject controlled task. Next some basic findings are reviewed from multiple alternative choice including the effects of similarity on choice and the effects of adding asymmetrically dominated alternatives. The latter result is particularly important because it violates a principle of choice called regularity that is satisfied by a large class of previous choice models. Then it is shown how the multiple choice version of decision field theory provides a simple and natural explanation for these paradoxical results. Finally new predictions from the theory are derived for the effects of deadline time pressure on multiple alternative choice. At this point in time, these predictions are unique to newly developed version of decision field theory and provide a strong test of the theory.

614

Multiple Cue Decision Field Theory

In order to make the description of the model more concrete, an example of a three option choice task we will be presented and referred to throughout the paper. Consider the case of a new car purchase where after some deliberation the choice set has been reduced to three cars. Also consider that the number of dimensions used to deliberate about these choices has been reduced to two, performance and gas mileage.

Below is presented a neural network interpretation of Multiple Choice Decision Field Theory (MCDFT) for this choice task. The model is expressed with the following linear difference equation:

$$P(t+1) = S * P(t) + V(t)$$

where **P** is a 3x1 vector representing the preferences for the three alternatives, **V** is a 3x1 vector of valences which represent the momentary anticipated value of each option, and **S** is a 3x3 constant matrix called the stability matrix that controls the rate of growth of the preferences.

The current model is expressed within a neural network framework as shown in Figure 1. There are three nodes in the system, labeled A, B, and C that represent three options. The nodes in the network are fully connected and have a self-feedback loop. The weights of the connections between these nodes are given in **S** matrix. By choosing the appropriate values in the **S** matrix, the principle of lateral inhibition can be implemented. It turns out that lateral inhibition is a key issue in predicting known findings in the area of multiple choice decision tasks.

The input that drives the system are valences, represented in Figure 1 by **V**'s. Values of the valences for each option change moment by moment as attention is randomly shifted from dimension to dimension in the deliberation process. For example, while deliberating on a new car purchase, attention may be shifting between gas mileage and performance. Therefore, the momentary anticipated value of each option will change depending on what dimension is being attended to. It has been shown that subjects do tend to use a dimension-wise process in many choice tasks (Russo & Dosher, 1983). Because attention is randomly shifting and the value of the **V**'s are fluctuation, they are random variables and are assumed to be independent and identically distributed. The output of the system is the preference, **P**, of each option at time t+1. The option selected in a particular choice task depends on the values of these preferences and the type of choice task.

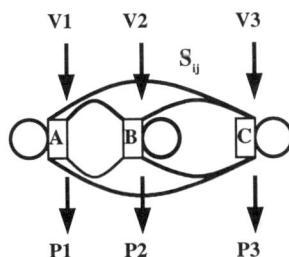

Figure 1: A neural network representation of MCDFT

Given this brief explanation of the model it can now be shown how it maps onto two types of choice tasks, experimenter controlled and subject controlled. In describing these mappings, it will shown how MCDFT can be viewed as special cases of two well-known static theories of choice.

Experimenter Controlled Choice Tasks

Experimenter controlled choice tasks involve placing a deadline on the choice deliberation time. For example, while deliberating on the purchase of a new car, the dealer may interrupt the deliberation process by forcing an immediate decision. Part of Figure 2 represents this type of choice task for the new car example. The abscissa represents time while the ordinate represents the level of preference of each option. The three lines labeled A, B, and C represent the preferences of each option as they evolve over time. The vertical line to the right represents the choice deadline. The option with the highest preference at deadline is the one chosen. In this case, option C was chosen because it had the highest preference at deadline (the meaning of the upper bound is discussed below).

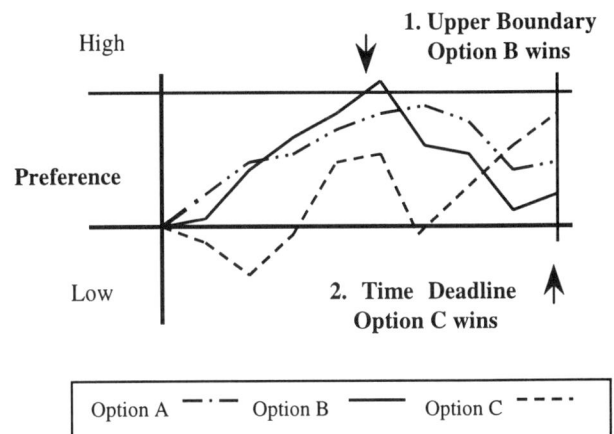

Figure 2: Experimenter and subject controlled tasks

The probability that a particular choice is made at time t, for example option C, is:

$$P(\text{Choose C, AB, at time t}) =$$
$$P (P_C > P_A \text{ at time t, } P_C > P_B \text{ at time t}).$$

To find the mean and the variance of preferences, we look at the expansion of the model equation given earlier: With $E(V) = \mu$ and $Var(V) = \Phi$ and the fact that the V's are i i d. (Therefore their sum becomes multivariate normal), MCDFT reduces to a Multivariate Dynamic Thurstone choice Model with:

$$P(t+1) = \sum_{j=0}^{t-1} S^j V(t-j) + S^t P(0)$$

$$E[P(t)] = (I - S)^{-1} (I - S^t)\mu$$

$$V[P(t)] = \sum S^j \Phi (S^j)^T$$

As can be seen in the above equations, means evolve over time such that at any point in time t, the means can change leading to preference reversals.

By stopping the deliberation process and obtaining a choice response at various points during deliberation, it is possible to study how the deliberation process evolves. A paradigm such as this, known as the response signal method (Reed, 1973), had been used to study many cognitive processes such as recognition memory and lexical decisions (Hintzman & Curran, 1997) and discriminating semantic from episodic associations (Dosher, 1984).

Presently there are no known decision-making studies using experimenter-controlled tasks. However, a model prediction based on this type of task makes a strong test of the model and will be presented below. A second and more frequently used type of choice task is the subject controlled choice task.

Subject Controlled Choice Task

Subject controlled differ from experimenter controlled choice tasks in that no deadline is placed on the deliberation process. Instead, a choice is made when the preference for an option crosses some threshold. Figure 2 also represents this type of choice task for the new car example. The abscissa represents time while the ordinate represents the level of preference of each option. The three lines labeled A, B, and C represent the preferences of each option as they evolve over time. The horizontal line at the top represents the choice threshold. A choice is made when the level of preference for any option crosses the threshold. In this case, option B is chosen.

Within this type of choice task, MCDFT mimics a well-known model of multiple choice decision, Elimination by Aspects (Tversky, 1972). According to this model, options are eliminated from the choice set based on aspects (or dimensions). However, unlike the subject controlled task presented above, a different choice boundary is needed, a lower boundary to discard options. In Figure 3, the results of a computer simulation of a choice task with five options along three aspects (or dimensions) are shown. The abscissa represents time and the ordinate represents preferences. The lines labeled A through E represent each option. Each vertical line represents a shift of attention from one dimension to another in the deliberation process. As can be seen, while focusing on the first dimension two options, A and E, were eliminated from the choice process. After the shift to the second dimension, no items were eliminated, and while focusing on the third dimension, items B and D were eliminated leaving option C as the option selected.

MCDFT mimics well the Elimination by Aspects model of multiple choice decision. In the sections that follow it sill be shown how MCDFT can qualitatively account for two salient findings in the literature, the effect of similarity and violations of regularity.

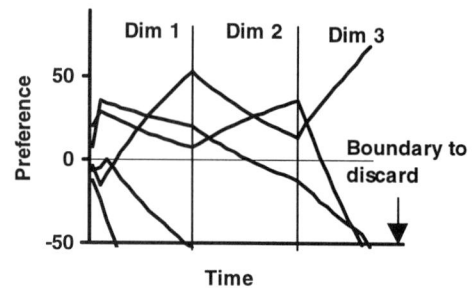

Figure 3: Results of EBA simulation

Effects of Similarity

The similarity effect is a well-known phenomenon in the area multiple-choice decision-making, (Tversky, 1972). It states that by adding a new item to a choice set that is similar to one already in the set, the probability that the original similar item chosen is lowered relative to the other items in the choice set. In other words, the similar alternative takes more of the market share from items similar to it than dissimilar to it. Part of Figure 4 shows an example of this situation for three options. The three options are represented in a two dimensional space of performance and gas mileage. Options A and B are located such that option A has better performance but poorer gas mileage than B, and option B has worse performance but better gas mileage than A. By adding the third option S which is similar to A, the probability that B is chosen relative to A is increased.

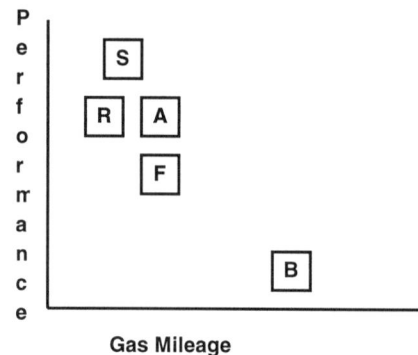

Figure 4: Similarity and Decoy Effects

The similarity effect leads to violations of strong stochastic transitivity. Many static choice models can handle these violations including Elimination by Aspects (Tversky, 1972) and the Edgell-Geisler choice model (Edgell &Geisler, 1980). The dynamic model proposed here can also handle these violations. Computer simulations were run to test model predictions for this finding. Simulations were first run to obtain preferences for two options then a third similar option was added. Figure 5 shows the prediction of the model. The abscissa shows the two possible options A and B while the ordinate shows the probability that an item is chosen. The line connected with the diamonds reflects the probability of each choice when only two options are available. In this case, each option is equally likely to be chosen. The line connected by the squares indicates the effect of adding the similar alternative. As you can see, the probability that A is chosen relative to B is lower when the similar item is added.

This simulation shows that MCDFT can qualitatively reproduce the similarity effect. However, as mentioned above, several static choice models do also. An often more difficult finding to explain is violations of regularity.

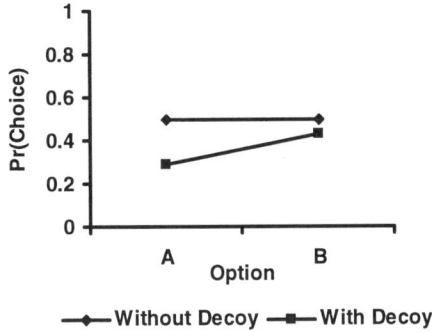

Figure 5: Results of Similarity Effect Simulation

Violations of Regularity

Violations of regularity, called the decoy effect, have been shown in several studies (Huber, Payne, & Puto, 1982 & Ariely &Wallsten, 1995). Regularity implies that by adding an item to a choice set, the probability of choosing an item already in the set cannot be increased. Formally, for any option x that is an element of set A which is in turn a subset of set B, $x \in A \subseteq B$, the probability of choosing x from A must be greater than or equal to choosing x from B $Pr(x;A) \geq Pr(x;B)$.

These violations occur when a new item is added to the choice set which is asymmetrically dominated. An item is asymmetrically dominated if it is dominated by at least one alternative in the choice set but not dominated by at least one other. Figure 4 shows two types of dominated alternatives, or decoys, the range decoy and the frequency decoy. The range decoy R is dominated by A because it has the same performance but worse gas mileage than A. It is called a range decoy because adding it to the choice set increases the range on the gas mileage dimension. The frequency decoy F is dominated by A in that it has the same gas mileage as A but worse performance. It is called a frequency decoy because adding it to the choice set increases the frequency of items below A. Using range and frequency decoys, Huber, Payne, & Puto (1983) found subjects violate regularity and they do more so with range decoys than with frequency decoys.

Computer simulations were run to test model predictions for this finding. Simulations were first run to obtain preferences with two options and then a third option (either a Range or Frequency decoy) was added and another simulation run. Figure 6 shows the prediction of the model. The abscissa gives the probability that the dominating choice is picked (choice A) in the binary condition. The ordinate gives the probability that A is chosen for both two and three option conditions. The line connected by the squares represents the predictions for the probability that A will be chosen without the decoy. The line connected by the circles represents the probability that A will be chosen when the frequency decoy is present. As can be seen, the model predicts that adding the frequency decoy leads to higher probabilities of choosing A across a wide initial range probabilities of choosing A in the binary case. Also notice that the effect is stronger for the range decoy than the frequency decoy

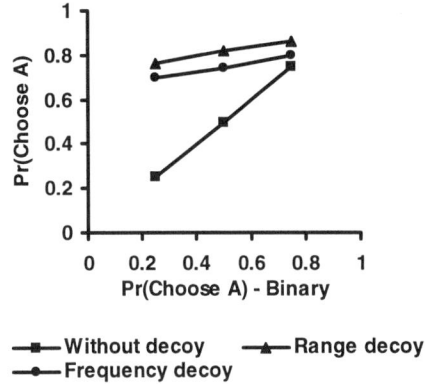

Figure 6: Results of Decoy Effect Simulations

A critical issue in producing this result was the use of lateral inhibition. In the model described here, the S matrix reflects lateral inhibition in which options close together in decision space inhibit each other and this inhibition lessens as distance increases. The S matrix shown below was used in the simulations of the decoy effect for the frequency decoy:

$$S = \begin{matrix} & A & F & B \\ & \begin{bmatrix} .95 & -.09 & -.001 \\ -.09 & .95 & -.02 \\ -.001 & -.02 & .95 \end{bmatrix} & \begin{matrix} A \\ F \\ B \end{matrix} \end{matrix}$$

Referring to Figure 4, and this S matrix, we can see that options A and F are closer to together relative to B. Therefore options A and F inhibit each other more relative to B and F. In much the same way as edge enhancement effects can be produced with lateral inhibition, the closeness of A to F enhances the probability that A is chosen. To produce the larger effect for the range decoy, the matrix **S** is simply altered to reduce the inhibition between R and B (because R is farther away from B than F). This allows a greater effect of the inhibition between A and R.

Effects of Decision Deadline

Because of the dynamic nature of this model, predictions can be made about the effects of deadline time pressure and other time related ideas on the choice process. One prediction of the model is the effect of a time deadline on violations of regularity. As mentioned above, by using the experimenter control method, we can look at the deliberation process at specific instances over time. By stopping the process at various times, the evolution of the process can be studied.

Model simulations of an experimenter controlled choice task were conducted by stopping the choice process at various points in time. The same simulation for the effect of the frequency decoy presented above was used except the time delay was varied.. Figure 7 gives the predictions of the model. The abscissa represents the time deadline with smaller numbers meaning shorter deadline. The ordinate represents the probability of choosing A, the dominating alternative. The line connected with squares represents the probability that A is chosen in the binary choice conditions. Here, no matter what the time cutoff, there is an equal probability that A or B will be chosen. The line connected by diamonds reflects the probability that A is chosen with the decoy present. Notice that with short deadlines, the model predicts that regularity will be satisfied. This can be seen in that by adding the decoy the probability that A is chosen decreases instead of increases. However, as the as the deadline time increases, the model predicts regularity will be violated. This can be seen in that adding the decoy increases the probability that A is chosen. Recall that most static models of choice predict that regularity will always be satisfied even though empirical studies show this is not true. If it is found that with short deadlines regularity is satisfied, it would be a very strong test of the model.

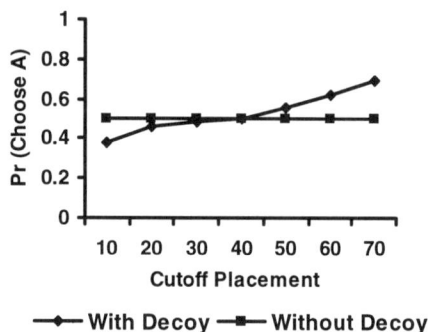

Figure 6: MCDFT predictions for the effect of time pressure

Conclusion

Under certain choice tasks this neural network instantiation of MCDFT has been shown to reduce to special cases of some well-known static models of choice (e.g., Thurstone's (1959) choice model, and Elimination by Aspects (Tversky, 1972)). Also, MCDFT can account for both the similarity and decoy effects found in the literature on multiple choice decision making. Further, the dynamic nature of this model allows for predictions about time deadlines on the choice process. Specifically it predicts that with very short deadlines, subjects will not violate regularity although it has been found that they do. Due to the fact that most models of choice are static, they make no prediction about this. If this result can be found empirically, it would be a strong test of the model. This is currently being tested in our lab.

Acknowledgements
This research was supported by an NIMH Post-Doctoral Training Grant F321MH11988-02. For helpful comments on a previous version of this paper, I would like to thank Jerome Busemeyer and Rachel Barkan.

References

Ariely, D., & Wallsten, T. S. (1995). Seeking subjective dominance in multidimensional space: An exploration of the asymmetric dominance effect. *Organizational Behavior and Human Decision Processes,* 63(3), 223-232.

Busemeyer, J. R., & Townsend, J. T. (1993). Decision Field Theory: A dynamic cognition approach to decision making. *Psychological Review,* 100, 432-459.

Dosher, B. A. (1984). Discriminating pre-experimental (semantic) from learned (episodic) associations: A speed accuracy study. *Cognitive Psychology,* 16(4), 519-555.

Diederich, A. (1997). Dynamic stochastic models for decision making under time constraints. *Journal of Mathematical Psychology,* 41(3), 260-274.

Edgell, S. E., & Geisler W. S. (1980). A set-theoretic random utility model of choice behavior. *Journal of Mathematical Psychology,* 21(3), 265-278.

Grossberg, S. (1988). *Neural Networks and Natural Intelligence.* Cambridge, M.A.:Mit Press.

Grossberg, S., & Gutowski, W. E., (1987). Neural dynamics of decision making under risk: Affective balance and cognitive-emotional interactions. *Psychological Review,* 94(3), 300-318.

Hintzman D. L., & Curran, T. (1997). Comparing retrieval dynamics in recognition memory and lexical decision. *Journal of Experimental Psychology: General,* 126(3), 228-247.

Huber, J., Payne, J. W., & Puto, C. (1982). Adding asymmetrically dominated alternatives: Violations of regularity and the similarity hypothesis. *Journal of Consumer Research,* 9(1), 90-98.

Keeney, R. L, & Raiffa, H. (1976). *Decisions with multiple objectives: Preference and value tradeoffs.* New York: John Wiley & Sons.

Levin, S.J., & Levin, D.S. (1980). Multiattribute decision making in context: A dynamic neural network methodology. *Cognitive Science,* 20, 271-299.

Reed, A. V. (1973). Speed-accuracy tradeoff in recognition memory. *Science(Washington, D.C.),* 181, 574-576.

Russo, J. E. & Dosher, B. A. (1983). Strategies for multiattribute binary choice. Journal of Experimental Psychology: Learning, Memory, and Cognition, 9(4), 676-696.

Thurstone, L. L. (1959). *The measurement of values.* Chicago: University of Chicago Press.

Townsend, J. T., & Busemeyer, J. R. (1995). Dynamic representation of decision making. In R. F. Port and T. van Gelder (Eds.) Mind as Motion. MIT press.

Tversky, A. (1972). Elimination by aspects: A theory of choice. *Psychological Review,* 79(4), 281-299.

Simon, H. A. (1955). A behavioral model of rational choice. Quarterly Journal of Economics, 69, 99-118.

Usher, M., & Zakay, D. (1993). A neural network model for attribute-based decision processes. Cognitive Science, 17(3), 349-396.

Analogies Out of the Blue: When History Seems to Retell Itself

Lelyn Saner (lsaner@gmu.edu)
Department of Psychology, Applied Cognition Program
MSN 3F5; George Mason University
Fairfax, VA 22030-4444

Christian Schunn (cschunn@gmu.edu)
Department of Psychology, Applied Cognition Program
MSN 3F5; George Mason University
Fairfax, VA 22030-4444

Abstract

To explain the origins of new scientific ideas, historians and philosophers of science point to examples where scientists appear to have drawn analogies between their scientific domain and some very different domain. By contrast, research from the psychology lab suggests that those kinds of analogies are very difficult to obtain in even the simplest situations. To resolve this potential conflict, we examine the analogies that occur in psychology lab group and formal colloquium settings. This approach can be viewed as a cross-sectional approximation of an historical analysis. We find that as the setting moves further away from the original discovery, the way different types of analogies appear to be used changes. In particular, analogies between very different domains are never used in reasoning in the lab group, whereas they are frequently used in reasoning in formal colloquium presentations. Yet, we find that analogy between very similar domains remains an important source of new ideas and a method for solving problems in scientific settings.

Introduction

The research reported here has been done in response to a long-standing question in philosophy and psychology of science. From where do scientists get their ideas? Previous research on this question has suggested that analogy can be an important source of scientific ideas (e.g., Gentner, Brem, Ferguson, Markman, Levidow, Wolff, & Forbus, 1997; Gentner & Jeziorski, 1989; Thagard, 1988). For example, Kekulé is said to have developed the idea of a circular molecular structure for benzene from a dream of a snake swallowing its tail.

In addition, there is already much research on how analogies are used by people in various domains to explain, explore, infer, and persuade others with respect to a particular idea or approach to understanding. Some of these domains include politics (Blanchette & Dunbar, 1997), math education (VanLehn, 1998), history education (Young & Leinhardt, in press), narrative interpretation (Holyoak, 1985), and molecular biology (Dunbar, 1995).

A paradox arises from this previous work on analogy. On the one hand, when philosophers and historians of science have discussed the role of analogy in science, they have referred primarily to what one might call long-distance analogies— analogies between two highly distinct domains

(e.g., Boden, 1993; Holyoak & Thagard, 1995; Koestler, 1964; Nersessian, 1985; Rhodes, 1986; Ueda, 1997). For example, ideas about the structure of the atom were said to derive from ideas about the structure of the solar system (i.e., several smaller bodies orbiting one larger body at the center, with much space between). Here, one is connecting a *source* in the domain of astronomy to a *target* in the domain of nuclear physics. Another example is the case of Kekulé and the analogy between benzene and a snake. Here the source domain is either mythology, everyday experience, or artistic imagery, and the target domain is chemistry. In each of these kinds of famous examples that are much discussed in history and philosophy of science, the source and target domains are almost completely unrelated (e.g., Boden, 1993; Koestler, 1964; Nersessian, 1985; Rhodes, 1986).

On the other hand, psychological research on analogy has found that even when cued, subjects will fail to connect two very analogous scenarios when they are from different domains. One of the most famous, and often cited, demonstrations of this result comes from Gick and Holyoak (1983). They showed that, after reading a story of how a general takes a large army across multiple bridges, subjects were still unable to see the application of it to an analogous medical problem. This result was initially quite surprising and has been frequently replicated: subjects usually do not spontaneously note long distance analogies (Holyoak & Koh, 1987; Mandler & Orlich, 1993; Ross, 1989).

Thus we have a paradox: If people do not spontaneously see long-distance analogies in such simple situations, how can scientists see long-distance analogies in the complexities of their research?

There are at least three simple resolutions to this problem. First, it is possible that scientists are more intelligent, creative, or insightful than the average psychology experiment subject, and thus are able to find and use long distance analogies. Certainly, individuals are thought to vary along many cognitive dimensions including creativity (Gardner, 1983).

Second, scientists have much deeper knowledge of their subject domain and this deeper knowledge base may be more supportive of long-distance analogies. Along those lines Novick (1988) found that students with greater math expertise were better able to make analogies involving math problems and other domains.

Third, it is possible that scientists do not make much use of long-distance analogies, and the historical record is misleading. Under this explanation, the psychological research is correct even for scientists: long-distance analogies are not typically used to solve problems. To account for long distance analogies described by historians of science, this explanation assumes that the historical record is misleading. In particular, it may be that the historical record contains descriptions of long distance analogies that were developed by the discoverers <u>after</u> the discovery was made. These analogies, while not used as the source of the discovery, may have been persuasive to the discoverer and were used to explain the discovery to others. In some cases, the scientists themselves have pointed to these long distance analogies as the source of their ideas. However, these accounts were typically long after the discovery, and psychological research has demonstrated that retrospective accounts of insights and analogies are frequently very inaccurate (Dunbar, 1996; Nisbett & Wilson, 1977; Schunn & Dunbar, 1996). This last possibility does not suggest that long-distance analogies are never used by scientists. Instead, it merely argues that long-distance analogies may not be as common or important as the previous accounts would suggest.

To resolve these issues, one can observe scientists as they are conducting their research, before the historical record can be rewritten. One such approach, developed by Kevin Dunbar (1995, 1996; Dunbar & Baker, 1994), is to analyze the discussions that occur in lab groups. Increasingly, science is being conducted in the context of a lab group, consisting of a head researcher, postdocs, graduate students, and research assistants. These lab groups meet on a regular basis and serve a very important function for the discovery process: new experiments are developed; data is interpreted; and alternative hypotheses are proposed.

Dunbar (1996) used this methodology to examine the kinds of analogies used by leading molecular biology lab groups as they made important discoveries. Analogies were coded along two dimensions: whether the analogy was within-organism, other-organism, or non-biological (or distant), and what the goal of the analogy was (form a hypothesis, design an experiment, fix an experiment, or explain a concept). Dunbar found that, of the 99 analogies identified, only two were from distant sources, and both were used to explain a concept rather than form a hypothesis or design or fix an experiment. What he concluded from his study was that distant analogies may not have a role in the making of discoveries, and that scientists use them more to explain a new concept to an audience. He also noted that scientists did not often remember analogies generated during meeting (including the within-organism analogies). This suggests that analogies may be used as scaffolding for the construction of new ideas, and are then discarded once they have served their purpose. This amnesia for the analogies that were used may explain the conflicting view of analogy from history of science. That is, the analogies that were actually used in making the discoveries are simply missing from the historical record, leaving only the analogies used in papers and presentations to convince others (i.e., the distant analogies).

To explore this resolution of the analogy paradox, we use a modification of Dunbar's methodology. We contrast the discussions that occur in lab groups with those that occur in colloquia. The assumption is that research is at an early stage when it is being addressed in lab groups. At this point, ideas are formulated and focused, preliminary studies are planned or data from them is analyzed to determine next steps. Since ideas are formulated and developed in this context, the kinds of analogies used here provides insight into whether long distance analogies are used as sources of ideas. In other words, will Dunbar's results generalize to other domains, in this case, that of psychology?

Colloquia, by contrast, present later-stage or completed studies. Conclusions and theories are typically well developed. Highlights from a sequence of studies rather than all the details of a single study are presented. Well-formulated ideas and completed discoveries are being communicated to others, both within and outside the domain. Here the emphasis is on rhetorical goals rather than the researcher seeking new ideas. The kinds of analogies that occur in this setting will provide insight into whether long distance analogies occur in the descriptions of research long after it has been completed.

Thus, the lab group/colloquium distinction captures much of the life cycle from ongoing discoveries to final presentations of discoveries, while holding several other features constant (e.g., verbal format and length of presentation). By contrasting the analogies used in each context, we hope to provide some insight into the paradox of long-distance analogies: are they used as sources of ideas by scientists, or do they simple serve rhetorical functions, appearing primarily after the discoveries have been made?

In this paper, we will analyze discussions that occur in lab group and colloquia in various psychology domains. Of particular focus will be the frequency, kind, and use of analogies in each setting. Do different kinds of analogies occur in each setting, and are different kinds of analogies used differently in each setting?

Methods

For this research, the sources of data were a series of video and audio-taped recordings of the proceedings of two psychology lab groups and one colloquium series, all from within one psychology department. The psychology department is part of a major university in a large Eastern US city. The department has over 100 faculty, students, postdocs, and research assistants. Each lab group presentation involved one researcher presenting research that was about to be conducted or had just been conducted but not fully analyzed. The colloquium series presenters were from around the world and were invited as colloquium speakers because of the established excellence of their work.

The data set included three colloquia sessions, which were composed of the primary speaker and 30 to 60 audience members (a mix of faculty, postdocs, graduate students, and research assistants). Two of the speakers were well-known, senior researchers from other universities. The third speaker was a senior, internal faculty member. The colloquia range from 70-90 minutes in length, and topics included social psychology, cognitive psychology, and

Table 1: Coding dimensions and example analogies for each code.

Dimension	Codes	Examples
Source	Within Domain	Other than that, the structure of the study is just like the first one.
	Between Domain	Well, a cognitive resource consumption is really bearing uh the analogy that we use to go in and look at the brain activation. That is, in the model you're using up more and more resources, [that is] you're using up a higher proportion of resources in more difficult tasks.
Use	Used	…there's a kind of linear relationship between the two…, but only when you get up close to the limits do you vastly increase the difficulty, [which is] measured by brain activity. It suggests a somewhat different underlying model. I think disk drive, [with a] disk drive you get similar effects. It's not a linear…
	Mentioned	I'm doing a similar study [to the Siegler and Jenkins study] right now at the ** school
Similarity Base	Similarity	Um, but our production system architecture has some particular properties that are connectionist like… representational elements are conceptualized as having faded levels of activation. They can be more or less active depending on how accessible they are to the system.
	Dissimilarity	And, as you can see, what's different about the forking moves, as opposed to the wins and blocks…
Mode	Asserted	*** All the Above are Asserted Analogies ***
	Anticipated	Well, what about the remember/know paradigm that's gotten so much press lately? Have you tried looking at that?
Analogizer	Presenter	Other than that, the structure of the study is just like the first one.
	Audience	Well, what about the remember/know paradigm that's gotten so much press lately? Have you tried looking at that?

cognitive neuropsychology. None of the participants were aware of the goals of the study.

The five lab group recordings were of the meetings of two different lab groups. One was a developmental psychology lab group comprised of two faculty, three postdocs, three graduate students, and two research assistants. The other was a cognitive psychology lab group comprised of three faculty, six postdocs, five graduate students, and five research assistants. The exact number of participants varied from meeting to meeting. These lab group meetings also were approximately 70-90 minutes long. Both lab groups were run by well-established researchers with research records similar to those of the colloquium speakers.

All speech, including all questions and comments made by audience members were transcribed, producing a total data source of almost 95,000 words. From these transcripts, analogies were identified and coded along five dimensions by two independent coders. Based on the recoding of a subset of the data, there was greater than 90% inter-rater reliability. Differences were easily resolved through discussion.

For the coding, an analogy was defined to occur when existing knowledge from a source object or domain was compared with a target object or domain in order to increase understanding of that target. Any time that a reference was made to the similarity or dissimilarity of two things, in a structural or functional sense, it was counted as an analogy. If something was said to be an attribute of something else or an instance of a category, it was not counted as an analogy. For example, if the statement were to be made that the model "is a connectionist system," it would not be considered an analogy. In addition, if one thing was being simply being re-described using new terminology, it was not considered analogy. For example, if one were to re-describe participant accuracy in terms of d' rather than hit rate, this would not be considered an analogy. Once all the analogies were located and agreed upon, they were coded along several dimensions. These are listed on Table 1, with examples for each.

There were two dimensions that were most critical with respect to our research questions. The first was the "Source" dimension. Here we coded for the distance between source and target according to whether the target came from within the same domain as the source or from a different domain altogether. This dimension was taken from Dunbar (1996) and modified to suit the current domain of psychology. Dunbar distinguished between within organism, between organism, and non-biological analogies. While these distinctions are important and appropriate for molecular biology, they are not applied so easily to psychology. For example, why should analogies between social and cognitive phenomena in humans be less distant than analogies between memory phenomena in rats and humans? Thus, we collapsed all within psychology analogies in a single Within-domain category. Analogies between psychology and other domains (e.g., between psychology and economics) or between two similarly different domains were coded as Between-domain analogies.

The other critical factor was the "Use" dimension: whether or not the analogy was used in reasoning. Analogies could simply be mentioned in the course of the dialogue, just to point out an additional connection that may

621

or may not be central to the discussion. Such analogies were coded as "Mentioned". In the Mentioned example presented in Table 1, the speaker simply mentioned that they were doing a study analogous to a previous study, without using that analogy to justify a previous point or to motivate a new point. If the speaker made an analogy and used it to illustrate or substantiate a point, or to draw a further inference on the topic at hand, it was coded as "Used". In the Used example presented in Table 1, the speaker brings up the analogy to argue for an alternative conception of cognitive resource consumption. This new conception is then used in the discussion.

Emphasis in previous research has been on similarity analogies as sources of ideas. However, analogies can also be based on dissimilarities between entities. To examine whether such analogies occurred and what their relative frequencies were, we coded for the "Similarity Base" dimension according to whether analogies were based on similarity or dissimilarity.

Another dimension was that of mode. Not all analogies are fully thought through when first proposed. These new, incomplete analogies are interesting because the may track the introduction of new ideas. So, each analogy was coded as "Asserted" if the relation is observed and stated to be present by the analogizer. If the analogizer was asking if a particular relation exists between two systems, or suggesting that one might exist, it was coded as "Anticipated".

A final dimenion of interest is whether the analogies were produced by the presenter or the audience member. This is the "Analogizer" dimension. The ones produced by the audience members are of particular interest because they are potentially bringing new insights to the presenter's topic. Thus, we examined whether between domain analogies were more likely to be given by the presenter (as a rhetorical function) or by the audience member (as a source of ideas).

Results & Discussion

There were a total of 67 analogies identified. Of these 67, 37 came from the 5 lab group transcripts (with between 3 and 15 analogies in each transcript) and 30 from the 3 colloquia transcripts (with between 4 and 15 analogies in each). Thus, analogies did occur with some regularity in each setting, although some presentations in each setting involved more analogies than others.

Because there were more lab group transcripts and because there were roughly a similar number of analogies in each setting, the comparisons between the two settings will focus on proportions rather than absolute frequencies. The proportion of codes on each of the dimension in each setting are presented in Table 2. Since the Ns are relatively low throughout, the statistical tests were done using Fisher Exact tests (which is a very conservative statistic).

For both colloquium and lab group settings, approximately 80% of the analogies were within-domain analogies. This overall high frequency of within-domain analogies support Dunbar's (1996) suggestion that within-domain analogies are the more common occurrence in scientific settings. However, that the proportions did not differ across the two settings was surprising. We had expected a higher proportion of between domain analogies in the colloquium setting.

By contrast, there was a marginally significant difference in the propoportion of Used analogies in each setting. Analogies were almost twice as likely to be used in colloquium setting. Interestingly, the majority of analogies were simply mentioned in both settings.

These overall weak main effects of Setting on Used and Mentioned hid a very important interaction: Used X Mentioned X Setting. Table 3 presents the proportion of analogies used in reasoning separately for within and between domain analogies. Within domain analogies were just as likely to be used in reasoning in colloquia as in lab groups. By contrast, between domain analogies were very likely to be used in reasoning in colloquia but were never used in reasoning in lab groups. Thus, there appears to be good evidence for a difference in how analogies appear to be used in lab groups and colloquia.

To make this effect concrete, consider the following two examples. From the colloquium setting:

…striving is real hard to study with the nature of this construct because these are not by definition something that's out there, we achieve it once and you've accomplished it. Like to be a helpful person, you know it's not like, "Well I'll open the door for someone - I'm helpful. I can move on to my next goal." It's kind of like always there, you know, always recurring. So it makes it a little harder to study process questions like…

Here the colloquium speaker is making a between domain analogy (from everyday life to social psychology). Although the analogy is used in reasoning to support his arguments, it is an analogy that was previously thought through and there is no indication that it was used in discovery.

From the lab group setting, an audience member is comparing the results from the current study on children learning addition to a previous study of children learning subtraction:

Table 2: Percent of analogies categorized as within domain, used, similarity based, asserted, and presenter given (separately by setting).

Category	Coll.	Lab Group	p
Within Domain	77%	81%	n.s.
Used	43%	27%	.2
Similarity Based	70%	68%	n.s.
Asserted	93%	97%	n.s.
Presenter	90%	70%	.1

Table 3: Percent of between and within domain analogies used in reasoning (separately by setting).

Type of Analogy	Colloquia	Lab Group	p
Between Domain	71%	0%	.02
Within Domain	35%	33%	n.s.

Like [with] addition, if you were looking at conception, seems to be more of a conceptual guide, since they don't ever have a procedure, the min strategy, demonstrated for them. Versus in buggy subtraction they're basing it on the fact that they know the correct, they've seen the correct procedure before and that's what they're using. That is a very different kind of thing.

Here, both source and target are from within psychology (both cognitive/mathematical development). Despite the short distance between the source and target, the analogy is an important one, highlighting a potentially crucial factor in the development of new strategies (the topic of the presentation). As one can see by the verbal disfluencies, the analogy is much less well-formed, providing some evidence that it is a new idea. Moreover, the analogy is provided by an audience member rather than by the speaker.

In general, analogies were three times as likely to be generated by an audience member in the lab group setting than in the colloquium setting (although this difference was only marginally significant). However, this did not explain the difference in use of analogies in the two settings: audience members were just as likely to produce between domain analogies as were presenters (21% vs. 21%), and there was no three way interaction with setting. There was also no difference in how likely it was that analogies generated by audience member versus presenter analogies would be used (57% vs. 57%), nor was there a three way interaction with setting.

The frequencies of analogies coded along the Similarity Base dimension follow well with the common understanding of analogies as generally being put in terms of similarity relations. However, there was a significant proportion of dissimilarity based analogies in both settings. There were no differences in the use of similarity vs. dissimilarity based analogies in the two settings.

In both lab group and colloquium settings, almost all of the analogies were asserted rather than anticipated. As one might expect, all three of the anticipated analogies were produced by audience members, and two of those three were in colloquia, where many audience members have relatively little knowledge of the presenter's research topic.

General Discussion

The primary result of this study is that while the relative frequency of analogies held no large differences between presentation settings along any single dimension, the use of analogy did differ significantly as a function of the distance between the source and the target. Specifically, the analogies drawn between domains were only used in the reasoning processes in colloquia. The between domain analogies, while they did occur in lab group settings, were mentioned and not used in reasoning. By contrast, within domain analogies were used in both settings.

These findings support one resolution of the paradox between psychology lab findings and history and philosophy of science claims: long distance analogies appear to be rarely used in the actual discovery processes and, instead, appear to be rhetorical additions added after the fact. Of course, we do not wish to claim that long distance analogies are never used in science. Rather, this study simply suggests: 1) that long distance analogies do not

occur frequently as sources of new ideas in scientific settings; 2) within domain analogies serve a much larger role in these settings; and 3) the historical record may be misleading with respect to the role that long distance analogies played in previous discoveries.

Our findings have both some important similarities and differences between those found by Dunbar (1996). Like Dunbar, we found that long-distance analogies are not used in reasoning in lab group settings. This similarity across research groups from very different sciences supports the generality of these findings. However, unlike Dunbar, we did find many (just mentioned) long-distance analogies in the lab group settings. It remains to be seen whether these are differences between psychology and biology: research problems in psychology may be easier to relate to everyday life experiences. Alternatively, it is possible that differences were due to differences in coding analogy: Dunbar's definition of analogy included that it must be used in reasoning, and thus there may have been many mentioned long-distance analogies that were excluded.

If the results of the current study and Dunbar's study are correct, then the consequences for historical studies are quite severe: writings or oral reports made after the discovery may not be trusted regarding the role that analogies played in the discoveries. Of course, psychologists have been long suspicious of the accuracy of retrospective reports (Nisbett & Wilson, 1977; Ericsson & Simon, 1993). Moreover, people have been shown to be very inaccurate regarding the origins of their scientific hypotheses (Dunbar, 1996; Schunn & Dunbar, 1996). Not all historical analyses fall prey to this problem equally: lab books will continue to be a good source of data. Unfortunately, not all scientists are good about keeping organized and complete lab books.

It is also important to note that historical and philosophical analyses will always play an important role in the cognitive science of science, regardless of the accuracy of the current study's results. At the very least, they provide descriptions of important macro-level phenomena in the history of science that need to be further explained (e.g., Boden, 1983; Gentner et al., 1997; Nersessian, 1985; Schunn, Crowley, & Okada, 1998; Thagard, 1988).

The methodology used here represents clear tradeoffs in experimental design: as we increase external validity we lose the benefits of studies conducted in the psychology lab. A few particular issues deserve mention. First, there were only two lab groups and only 8 total sessions across both settings. This issue of small Ns is a common problem with detailed process analyses of real world cognition. However, despite the small N, the key findings involved very large effects that were statistically reliable using conservative statistical tests. To examine how stable the findings are within each setting, we are currently extending the analyses to a larger set of colloquia and lab group presentations.

Second, it is possible that important new ideas are developed or discoveries are made outside of the lab group setting and are not discussed until much later presentations (e.g., a colloquium talk). However, the five lab group presentations were of research at various phases of progress (from just started to almost completed), and in no case was there use of a between domain analogy in reasoning. It is unclear why important between domain analogies would remain hidden in each of these cases.

Another potential problem is that there were different topics in the different settings. For example, the topic of one of the colloquia was social psychology, whereas none of the lab groups involved social psychology. However, there was a variety of topics in both settings, with some overlap, and the main findings appeared consistent across topics within settings. Thus, it is likely that these findings will generalize across other topics as well.

It is also true that there were different people present in each of the settings, both as presenters and audience members. More crucially, some of the lab group presenters were graduate students and postdocs—a very different level from the colloquium presenters. However, all of the research presented in the lab groups was designed and conducted in close consultation with the primary faculty researcher. Moreover, there was consistency in the main findings across the different presentations, suggesting that the main results generalize across different presenter levels.

Finally, to anticipate a suggestion, it is clear that a longitudinal design should be the next step. Fortunately, the lab groups that appeared in this study were quite strong, and it is likely that the results from their studies will appear in the literature in the near future. Observation of more formal presentations of these same projects would be a logical follow-up project, and the current study suggests that it will be a worthwhile undertaking.

Acknowledgments

We would like to thank Greg Trafton, Erik Altmann, Susan Trickett, Sheryl Miller, and Audrey Lipps for comments made on earlier drafts of this paper. Work on this paper was supported by a grant from the Mitsubishi Bank Foundation.

References

Blanchette, I. & Dunbar, K. (1997). Constraints Underlying Analogy Use in a Real-World Context: Politics. Poster presented at the *19th Annual Meeting of the Cognitive Science Society*, Stanford, CA.

Boden, M. (1993). *The creative mind: Myths & mechanisms*. New York, NY: BasicBooks, Inc.

Dunbar, K. (1995). How scientists really reason: Scientific reasoning in real-world laboratories. In R.J. Sternberg & J. Davidson (Eds.), *Mechanisms of insight*. Cambridge MA: MIT press. pp 365-395.

Dunbar, K. (1996). How scientists think: Online creativity and conceptual change in science. In T.B. Ward, S.M. Smith, & S.Vaid (Eds.), *Conceptual structures and processes: Emergence, discovery and Change*. Washington, DC: APA Press.

Dunbar, K., & Baker, L. M. (1994). Goals, analogy, and the social constraints of scientific discovery. *Behavioral and Brain Sciences, 17*, 538-539.

Ericsson, K. A., & Simon H. A. (1993). *Protocol analysis: Verbal reports as data* (Revised Edition). Cambridge, MA: MIT Press.

Gardner, H. (1983). *Frames of mind: The theory of multiple intelligences*. New York: Basic Books.

Gentner, D., Brem, S., Ferguson, R. W., Markman, A., Levidow, B. B., Wolff, P., & Forbus, K. (1997). Analogical reasoning and conceptual change: A case study of Johannes Kepler. *Journal of the Learning Sciences, 6*(1), 3-40.

Gentner, D., & Jeziorski, M. (1989). Historical shifts in the use of analogy in science. In W. R. S. B. Gholson, Jr., R. A. Beimeyer, & A. Houts (Eds.), *The psychology of science: Contributions to metascience* (pp. 296-325): Cambridge University Press.

Gick, M. L., & Holyoak, K. J. (1983). Schema induction and analogical transfer. *Cognitive Psychology, 15*, 1-38.

Holyoak, K. J. (1985). The pragmatics of analogical transfer. In G. H. Bower (Ed.), *The psychology of learning and motivation* (Vol. 19) (pp. 59-87). New York: Academic Press.

Holyoak, K. J., & Koh, K. (1987). Surface and structural similarity in analogical transfer. *Memory & Cognition, 15*, 332-340.

Holyoak, K. J., & Thagard, P. (1995). *Mental leaps: Analogy in creative thought*. Cambridge, MA: MIT Press.

Koestler, A. (1964). *The act of creation*. London: Macmillian.

Mandler, J. M., & Orlich, F. (1993). Analogical transfer: The roles of schema abstraction and awareness. *Bulletin of the Psychonomic Society, 31*(5), 485-487.

Nersessian, N. J. (1985). Faraday's field concept. In D. Gooding & F. James (Eds.), *Faraday rediscovered* (pp. 175-187). London: Macmillan.

Nisbett, R. E., & Wilson, T. D. (1977). Telling more than we can know: Verbal reports on mental processes. Psychological Review, 84, 231-259.

Novick, L. R. (1988). Analogical transfer, problem similarity, and expertise. *Journal of Experimental Psychology: Learning, Memory, & Cognition, 14*(3), 510-520.

Rhodes, R. (1986). *The making of the atomic bomb*. New York: Simon & Schuster.

Ross, B. H. (1989). Distinguishing types of superficial similarities: different effects on the access and use of earlier problems. *Journal of Experimental Psychology: Learning, Memory, and Cognition, 15*, 456-468.

Schunn, C. D., & Dunbar, K. (1996). Priming, analogy, and awareness in complex reasoning. *Memory & Cognition, 24*(3), 271-284.

Schunn, C. D., Crowley, K., & Okada, T. (1998). The growth of multidisciplinarity in the Cognitive Science Society. *Cognitive Science*, 22(1), 107-130.

Thagard, P. (1988). *Computational philosophy of science*. Cambridge, MA: MIT Press.

Ueda, K. (1997). Actual use of analogies in remarkable scientific discoveries. Paper presented at the *19th Annual Meeting of the Cognitive Science Society*, Stanford, CA.

VanLehn, K (1998). Analogy events: How examples are used during problem solving. *Cognitive Science*. 22(3). 347-388

Young, K.M., & Leinhardt, G. (in press) *Wildflowers, sheep, and democracy: The role of analogy in the teaching and learning of history*.

Optimal Control Methods for Simulating the Perception of Causality in Young Infants

Matthew Schlesinger (matthew@cs.umass.edu)
Department of Computer Science; University of Massachusetts
Amherst, MA 01003 USA

Andrew Barto (barto@cs.umass.edu)
Department of Computer Science; University of Massachusetts
Amherst, MA 01003 USA

Abstract

There is a growing debate among developmental theorists concerning the perception of causality in young infants. Some theorists advocate a top-down view, e.g., that infants reason about causal events on the basis of intuitive physical principles. Others argue instead for a bottom-up view of infant causal knowledge, in which causal perception emerges from a simple set of associative learning rules. In order to test the limits of the bottom-up view, we propose an optimal control model (OCM) of infant causal perception. OCM is trained to find an optimal pattern of eye movements for maintaining sight of a target object. We first present a series of simulations which illustrate OCM's ability to anticipate the outcome of novel, occluded causal events, and then compare OCM's performance with that of 9-month-old infants. The implications for developmental theory and research are discussed.

Introduction

How does the perception of causality develop? Do we perceive cause-and-effect relations at birth, or are months of experience necessary? Developmental researchers have approached these questions by studying infants' perceptual reactions to causal events (e.g., Baillargeon, 1986; Keil, 1979; Leslie, 1982; Oakes & Cohen, 1990). Much of this research depends on the tendency for infants to anticipate the outcomes of causal events, often showing surprise to unexpected outcomes (as inferred by measures of attention).

Consider the pair of causal events presented in Figure 1. The first (1a) is a simple, occluded movement display; by age 6 months, infants will quickly learn to anticipate the block's reappearance (Bower, Broughton, & Moore, 1971; Rutkowska, 1993). The second event (1b), however, is more complex. A wall obstructs the path of the block; note that the wall is partially occluded by the screen, revealing only

the upper and lower portions of the wall. While both events begin in a similar manner, they end differently, depending on the presence of the wall.

Two broad theoretical views have been proposed to explain infants' reactions to events like those in Figure 1. First, several researchers advocate a top-down view of infant causal knowledge (Baillargeon, 1994; Spelke, 1998). According to this view, infants use naive or intuitive physical principles to predict, reason about, or deduce the outcomes of occluded causal events. Two recent computational models help illustrate how the representations underlying this type of prediction system might develop (Mareschal, Plunkett, & Harris, in press; Munakata, McClelland, Johnson, & Siegler, 1997).

Alternatively, several infant causal perception studies have drawn attention to the role of simple perceptual preferences and associative learning rules (Bogartz & Shinskey, 1998; Rivera, Wakeley, & Langer, in press; Schilling, 1997). These researchers argue for a bottom-up view of causal perception. According to this approach, prediction is not an *a priori* goal, nor is representation of hidden objects necessary for the perception of causality in occluded events.

It is theoretically possible, if not likely, that both top-down and bottom-up factors play a role in infants' causal perception. How should the two views be reconciled? The strategy that we propose is to construct a model based on the bottom-up view, and then to test the extent of its perceptual "abilities" when presented with causal events like those shown to young infants. Any gaps or limitations in the performance of the model could then be addressed, we assume, by using the top-down approach.

Rather than simulating causal perception as a *representational* task (cf., Mareschal et al., in press; Munakata et al., 1997), we model the phenomenon as an optimal control problem. The optimal control model (OCM) is a *sensorimotor* model of infant causal perception. Unlike human infants, OCM: (1) has no intuitive knowledge, (2) cannot generate predictions, and (3) learns only by trial-and-error. OCM's objective is to learn a sequence of eye movements that best maintain a target object in view. After training OCM to track a target, we then test OCM's reactions to novel, occluded causal events like the one presented in Figure 1b. We next briefly describe OCM.

The Optimal Control Model

The Tracking Display

Figure 2a presents a snapshot of the 2-dimensional tracking display used to train OCM. During each trial, the block (rep-

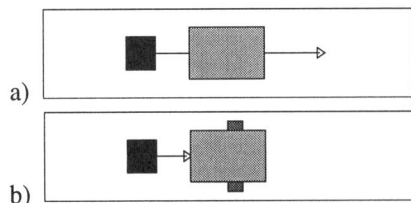

Figure 1: Occluded causal events. In (a), the block passes behind the occluding screen and reappears on the opposite side. In (b), a partially-visible wall obstructs the path of the block; after passing behind the screen, the block fails to reappear.

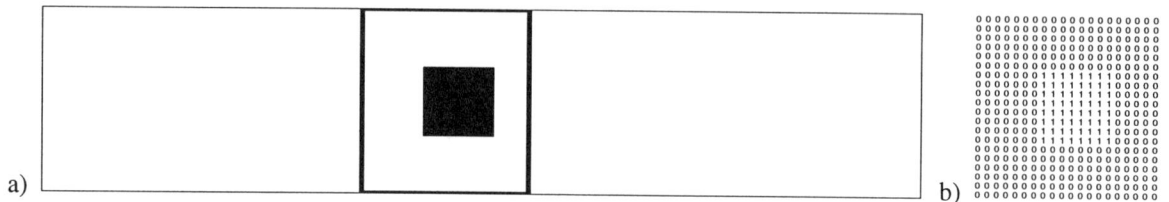

Figure 2: The OCM: (a) the tracking display (the target is represented by the solid black square, while the visual field is indicated by the black frame); (b) OCM's visual input for the corresponding display.

resented by the black square) moves from left to right, at the rate of 1 unit per timestep. At the start of the trial, OCM's visual field (the large, square frame) is positioned with the block in the center of the field. Each trial lasts 50 timesteps.

The display is 100 units wide and 20 units high. OCM's visual field covers a 20-unit square region of the display, permitting only lateral eye movements. The block is an 8-unit square, while the screen (when present) is 20 units wide and 12 units high.

Objects in OCM's visual field activate corresponding input units on its retina (see Figure 2b). Because the block is OCM's target object, its activation on the retina is 1 (i.e., maximum salience). The background has an activation of 0, while the wall and screen's activation levels are 0.6 and 0.2, respectively.

Model Architecture

OCM's sensorimotor "knowledge" is represented by a multi-layer, artificial neural network. There are two input systems. First, OCM receives visual input from its 20-by-20 unit retina. Figure 2b illustrates a typical visual input pattern. OCM also receives an additional input indicating the position of the visual field with respect to the display, normalized from 0 to 1.

There are 20 hidden units, and 5 output (motor) units. The network is fully connected, with only feedforward connections. Each of the motor units controls one of 5 possible eye movements: <-4, -1, 0, 1, 4>. On each timestep, the movement corresponding to the most active motor unit is performed.

Learning Algorithm

OCM is rewarded for generating eye movements which keep the block within the visual field; OCM learns by trial-and-error to find a pattern of eye movements which optimize sight of the block (i.e., maximize the total reward). Any movement which is followed by sight of the block is rewarded; the reward ranges from 0 to 1, as a function of the proportion of the block in the visual field after the eye movement (e.g., 1 when fully visible, 0.5 when half visible, etc.).

The output of each motor unit is an estimate of the value (i.e., probability of reward) for performing the corresponding eye movement. We employed the Sarsa learning algorithm, an unsupervised, online version of reinforcement learning methods (see Sutton & Barto, 1997) to train OCM. Using standard gradient descent methods, the Sarsa algorithm attempts to minimize the difference between the estimated and observed rewards after each eye movement.

Consequently, the direction and magnitude of the weight changes for the output layer depend on the eye movement chosen, and the corresponding reward, during a given timestep. These weight changes are then propagated backwards to the hidden layer (see Lin, 1991, for a discussion of reinforcement learning and back-prop hybrid models).

Simulation Overview

We conducted a series of simulation studies which assess OCM's ability to learn to track visible and occluded targets. In each study, OCM was first trained to track a target during two types of events. In the occluded event, the block passed behind a screen and reappeared on the other side. In the other (fully visible) event, the block encountered a wall and then remained in place. After OCM learned to optimally track the block during these events, we then tested OCM's tracking during a novel, occluded causal event which included both the screen and the wall. In Studies 1 and 2, the wall was partially occluded by the screen, while it was completely occluded in Study 3.

Study 1: Tall Wall

Study 1 addresses the question of how OCM will respond to a partially occluded causal event. Figure 3 displays the events used to train and then test OCM.

Method

Training. During training, OCM was presented with two causal events. On Screen trials, a screen occluded the central portion of the display. On Wall trials, an obstacle was positioned in the center of the display; the block remained in place after making contact with the wall.

Screen and Wall trials alternated randomly. Training continued until OCM's total rewards during both Screen and Wall trials were at least 95% optimal over 10 consecutive trials (i.e., maximum total rewards were 30 and 50 points for Screen and Wall trials, respectively). If criterion was not reached by 300 trials, the run was terminated, the data were discarded, and a new set of random initial weights were generated.

Testing. After training, all weights in the network were frozen (i.e., learning was turned off[1]). OCM was then presented with 10 Wall-Screen trials. During Wall-Screen trials, the wall was positioned behind the screen; when the block passed behind the screen, its path was obstructed by the wall

---
[1]This was done to prevent OCM's responses during early test trials from contaminating later trials.

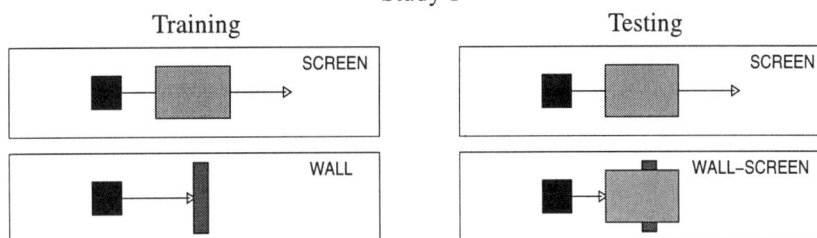

Figure 3: Training and test events presented to OCM in Study 1. Note that the Screen trial type was identical during training and testing.

(as during Wall trials), and consequently did not reappear. In addition, OCM was also presented with 10 Screen trials, in order to assess OCM's ability to track an occluded, unobstructed object.

Results

Training. Figure 4a presents the average number of trials to criterion in Study 1, averaged across 50 runs (36 additional runs were discarded). OCM reached criterion on Wall and Screen trials after 65.5 and 93.2 training trials, respectively. The difference is statistically significant ($t(98) = 1.78$, $p < .05$). Like human infants, OCM learns to track a fully visible target before it learns to track an occluded target. However, an average of 148.8 trials were necessary before reaching criterion on *both* trial types concurrently.

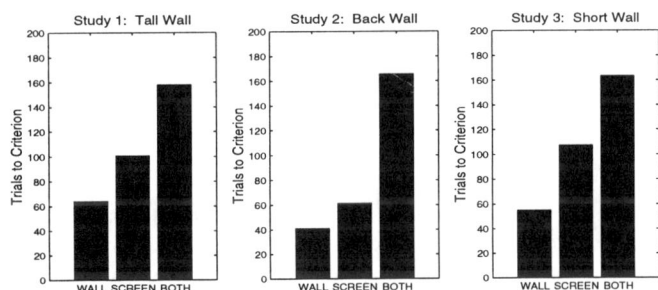

Figure 4: Trials to criterion during training in Studies 1, 2, and 3. See text for details.

Testing. Our analyses of the test trials focus on OCM's tracking behavior once the block disappears behind the screen, and how the presence or absence of the wall affects this behavior. In particular, we are interested in whether or not OCM moves its visual field to the right edge of the screen *before* or *after* the block reappears, during Screen trials. Consequently, we define *tracking latency* as the difference in time between OCM's first fixation of the right edge of the screen, and the block's reappearance at the right edge during Screen trials. Although the block does not reappear during Wall-Screen trials, we can use the same temporal index to compute OCM's tracking latency (i.e., assuming reappearance of the block, had it not been obstructed). A positive latency (or delay) means that OCM fixates the right edge of the screen after the block has (or would have) reappeared, while a negative latency means that OCM anticipates the reappearance of the block.

Figure 5 presents OCM's tracking latencies during the test phase of Study 1. During Screen trials, OCM anticipated the block's reappearance, fixating the right edge of the screen 8.9 timesteps sooner than the reappearance of the block ($t(49) = 7.03$, $p < .01$). In contrast, OCM's average tracking latency was significantly delayed by the presence of the tall wall during Wall-Screen trials; on average, OCM fixated the right edge of the screen 18.9 timesteps after an unobstructed block would have reappeared ($t(49) = 4.21$, $p < .01$).

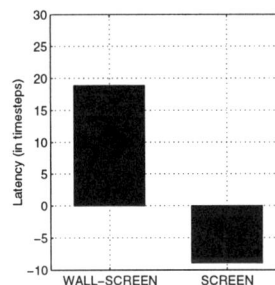

Figure 5: Mean tracking latencies in the test phase of Study 1, for Wall-Screen and Screen trials. OCM anticipated the reappearance of the block during Screen trials, while tracking was delayed during Wall-Screen trials.

Discussion

OCM's learning trajectory parallels that of human infants. OCM learns to track a fully visible object before it learns to track the movements of an occluded object. After training, OCM appears to react as if it "knows" when the occluded path of the block will, or will not, be obstructed. OCM anticipates the reappearance of the occluded object during Screen trials, but not Wall-Screen trials.

It is tempting to conclude that OCM learns to use the presence of the wall as a cue for tracking the occluded block. However, there is more than one way to explain OCM's behavior. One explanation is that OCM learns nothing about the wall when training on Wall trials; rather, it only learns to hold the visual field in place when the block stops moving. According to this explanation, the presence of the partially visible wall, during Wall-Screen trials, simply disrupts the tracking pattern learned during Screen trials. Alternatively, we might argue that OCM learns to associate the sight of the wall with its effect on the block.

These two explanations can be tested by placing the wall

Study 2

Training

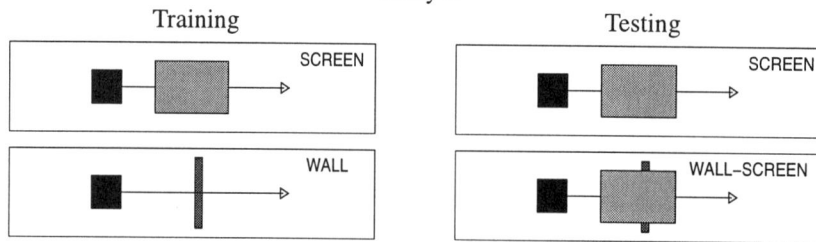

Testing

Figure 6: Training and test events presented to OCM in Study 2. Unlike Study 1, a thinner wall was included in the display, representing a wall which has been moved back relative to its position in Study 1. During Wall and Wall-Screen trials, the block passed in front of the wall.

"back" (from the perspective of OCM), beyond the path of the block. Thus, the movement of the block is identical during Screen and Wall-Screen trials. If sight of the wall is used as a cue, OCM should anticipate the block on *both* Screen and Wall-Screen trials; otherwise, if the wall disrupts tracking during Wall-Screen trials, then OCM should only anticipate the block during Screen trials. Study 2 tests these alternative hypotheses.

Study 2: Back Wall

Study 1 was repeated, replacing the tall wall which obstructs the block's movement with a wall placed "back" (i.e., farther from the observer's point of view), beyond the path of the block. Because the display is a 2-dimensional projection of a 3-dimensional world, we represented the perceptual effect of moving the wall back by decreasing its width (from 4 to 2 units). Consequently, the block passed in front of the wall during both Wall and Wall-Screen trials (see Figure 6).

Results

Training. Figure 4b presents the mean number of trials to criterion, during training in Study 2, across 50 runs (28 additional runs were discarded). Compared to Study 1, fewer trials were needed to independently reach criterion on Wall and Screen trials (41.0 and 61.3, respectively).

Testing. Figure 7 presents OCM's mean tracking latencies, during testing, for the Wall-Screen and Wall events. Placing the wall back significantly reduced OCM's tracking latency during Wall-Screen trials, compared to the tall-wall condition in Study 1 (-3.12 versus 18.9 timesteps; $t(98) = 4.11, p < .01$). However, OCM's anticipatory tracking was slightly slower on Wall-Screen trials, than during Screen trials (see Figure 7).

A closer analysis revealed that during 6 of the 50 runs, tracking of the block was in fact completely interrupted by the partially visible back wall, during Wall-Screen trials. However, when the remaining 44 runs are analyzed, OCM's average tracking latencies during Wall-Screen and Screen trials are -10.37 and -10.5 timesteps, respectively. During the majority of the runs in Study 2, therefore, sight of the wall did not disrupt OCM's anticipatory tracking.

Discussion

Study 2 replicates and extends the findings of Study 1. In both studies, OCM spontaneously learns to anticipate the reappearance of the occluded block. Further, when the wall is

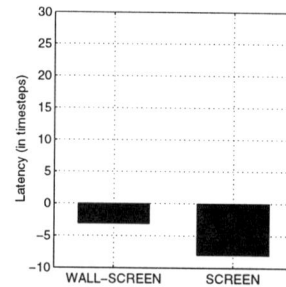

Figure 7: Mean tracking latencies in the test phase of Study 2, for Wall-Screen and Screen trials. OCM anticipated the reappearance of the block during both Screen and Wall-Screen trials.

positioned so as to have no effect on the movement of the block, it does not disrupt OCM's anticipatory tracking. Taken together, the results of Studies 1 and 2 support the conclusion that OCM learns to use both the screen and the wall as cues for perceptual action.

In contrast to Studies 1 and 2, a number of infant causal perception studies present perceptual cues to infants *prior to, rather than during* the occlusion event (e.g., occluded collision events, studied by Baillargeon, 1986; Lucksinger, Cohen, & Madole, 1992). Because the pairs of test events are identical in these studies, infants must *remember and recruit* information made available to them *before* each occluded event is presented.

We can simulate this type of causal event by reducing the height of the wall; when occluded, a short wall is no longer visible. While Studies 1 and 2 presented OCM with *partially* occluded causal events, Study 3 simulates OCM's reaction to a *completely* occluded causal event.

Study 3: Short Wall

Figure 8 presents a display of the training and test events used in Study 3. Three modifications were made to the method employed in Study 1. First, the height of the wall was reduced from 16 to 10 units. Second, 20 new input units were added to OCM's neural network. These "context" units were activated via recurrent connections from OCM's hidden layer, providing a functional memory of past internal states (Elman, 1990).

Third, each trial was preceded by a preview. During the

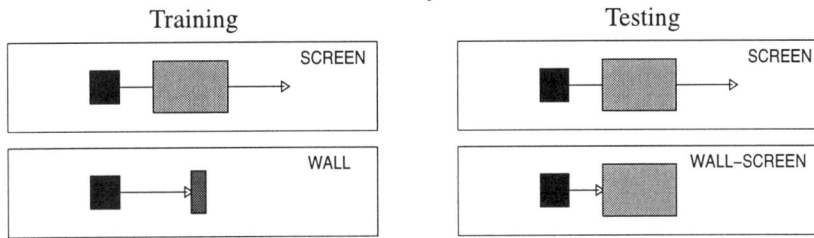

Figure 8: Training and test events presented to OCM in Study 3. Unlike Study 1, a short wall was included in the display, which was fully occluded by the screen during Wall-Screen trials. Note that after the preview, Screen and Wall-Screen trials are perceptually identical; past state information is necessary to differentiate these two trial types.

preview, OCM's visual field was held at the center of the display for 10 timesteps. During Screen and Wall-Screen trials, the screen was not included in the preview (i.e., OCM saw what was "behind" the screen). Learning was turned off during the preview. After the preview, each trial proceeded as in Studies 1 and 2.

Results

Training. Figure 4c presents the mean trials to criterion, during training in Study 3, across 50 runs (21 additional runs were discarded). When compared with Study 1, there were no significant differences in training time after changing the tall wall to the short wall.

Testing. OCM appears to "forget" about the short wall once it is occluded by the screen. As Figure 9 indicates, OCM's tracking latencies during Screen and Wall-Screen trials were identical; regardless of whether or not the short wall was present, OCM anticipated the reappearance of the block by 7.34 timesteps ($t(49) = 4.12, p < .01$).

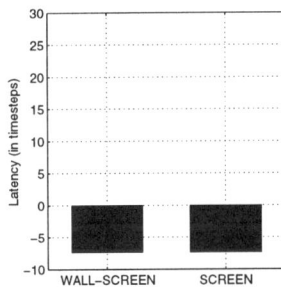

Figure 9: Mean tracking latencies in the test phase of Study 3, for Wall-Screen and Screen trials. Unlike Study 1, OCM anticipated the reappearance of the block during both Screen and Wall-Screen trials.

Discussion

In contrast to the results of Study 1, OCM's tracking behavior was not affected by the presence of a short wall. The findings from Study 3 suggest that OCM relied on its immediate perceptual input, while ignoring or failing to use its memory of the short wall.

However, it is important to remember that there was no pressure on OCM during training to learn to use memory.

First, during the fully visible wall trials, memory is unnecessary. Second, during the Screen trials, OCM learns to use the sight of the screen (rather than an internal representation of the occluded block) as a perceptual cue for anticipating the block's reappearance. Thus, the task constraints operating during training make the use of memory redundant.

General Discussion

The results from the three sets of simulations highlight both the strengths and limitations of the optimal control model of infant causal perception. There are two major findings. First, OCM quickly learns a set of optimal tracking strategies for following a moving object. Second, when presented with a novel causal event, OCM appropriately anticipates the outcome of partially occluded, but not fully occluded, versions of the event.

We can evaluate the performance of the model by directly comparing the results with data obtained from young infants. Berthier et al. (in preparation) conducted a series of experiments with 9-month-old infants, corresponding to Studies 1 through 3. Figure 10 presents a summary of the test results for OCM, and the comparable average tracking latencies (in seconds) for three groups of 9-month-olds. Across all three

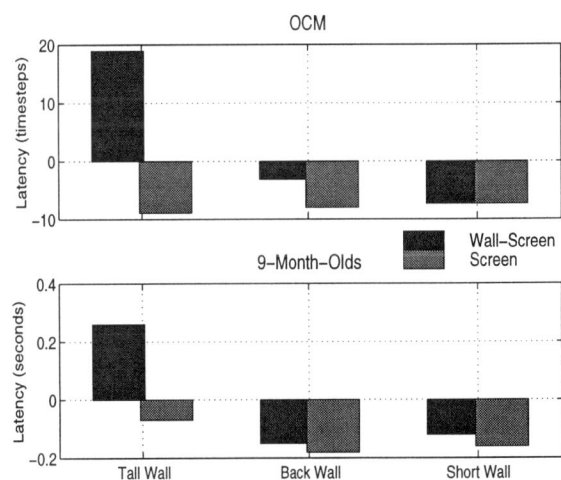

Figure 10: Mean tracking latencies in the test phase of Studies 1-3 for the OCM (top panel) and 9-month-old infants (bottom panel; from Berthier et al., in preparation).

studies, the performance of OCM provides a close qualitative fit to the performance of human infants. Like OCM, 9-month-old infants also use the partially visible wall as a cue, but not the fully occluded wall, to guide their tracking of the occluded target.

When taken together, the human and simulation findings carry a number of implications for developmental theory and research on the perception of causality in young infants. First, many causal perception researchers: (1) assume that infants *explicitly predict* the outcomes of the events they watch, and (2) infer, on the basis of looking-time measures, when infants' predictions are confirmed or violated. While the results from OCM are not intended to provide a reductionist account, they suggest that learning during the habituation or familiarization phase may drive the process of anticipation, helping to shape infants' causal expectations (see Rivera et al., in press; Schilling, 1997).

A second implication concerns the role of internal representations of occluded objects and events. Again, it is often assumed that infants must operate on mental representations, rather than direct perceptions, when the critical objects are occluded or out of sight. However, when tracking an occluded target, OCM relies on *sensorimotor* rather than *representational* strategies for anticipating the target.

For example, the results from Study 1 demonstrate that anticipatory behavior can emerge as a consequence of learning an optimal tracking strategy, without the need for memory or prediction. Indeed, having memory does not seem to facilitate OCM's learning to track the block during Screen training trials (compare Figures 4a and 4c), although there were fewer discarded runs when OCM was trained with a recurrent network (i.e., in Study 3). We suspect that in many causal perception studies, infants employ some combination of sensorimotor and representational strategies. Simulations with models like OCM help to determine if and when the sensorimotor strategies are sufficient to account for the perceptual phenomenon.

This point echoes a question raised in the introduction: what are the performance limits of OCM? On the one hand, there is surprisingly close agreement between the performance of OCM and the recent findings of Berthier et al. Nevertheless, this fit may in part be due to the specific constraints of learning to track, and the way in which this task favors an optimal control solution (e.g., like learning to reach or generate saccades). Thus, two current weaknesses of a bottom-up view in general, and an optimal control approach in particular are: (1) that some tasks may necessarily require predictive, representational strategies, and (2) that OCM may not be able to account for infants' perceptual behavior in other contexts (e.g., preference for "surprising" or impossible events). We are currently exploring an elaborated version of the model which addresses these issues.

Acknowledgments

This work was supported by NSF IRI-9720345 and NSF CDA-9703217.

References

Baillargeon, R. (1986). Representing the existence and the location of hidden objects: Object permanence in 6- and 8-month-old infants. *Cognition, 23*, 21-41.

Baillargeon, R. (1994). Physical reasoning in young infants: Seeking explanations for impossible events. *British Journal of Developmental Psychology, 12*, 9-33.

Berthier, N., Bertenthal, B., Seaks, J., Sylvia, M., Johnson, R., & Clifton, R. (in preparation). Catching a moving object: Infant reaching when trajectory information is removed.

Bogartz, R.S., & Shinskey, J.L. (1998). On perception of a partially occluded object in 6-month-olds. *Cognitive Development, 13*, 141-163.

Bower, T.G.R., Broughton, J.M., & Moore, M.K. (1971). Development of the object concept as manifested in changes in the tracking behavior of infants between 7 and 20 weeks. *Journal of Experimental Child Psychology, 11*, 182-193.

Elman, J.L. (1990). Finding structure in time. *Cognitive Science, 14*, 179-211.

Keil, F. (1979). The development of the young child's ability to anticipate the outcomes of simple causal events. *Child Development, 50*, 455-462.

Leslie, A.M. (1982). The perception of causality in infants. *Perception, 11*, 173-186.

Lin, L.J. (1991). Self-improvement based on reinforcement learning, planning, and teaching. In L.A. Birnbaum, and G.C. Collins (Eds.), *Machine learning: Proceedings of the eighth international workshop on machine learning,* (pp. 323-327). San Mateo, CA: Morgan Kaufmann.

Lucksinger, K.L., Cohen, L.B., & Madole, K.L. (1992, May). What infants infer about hidden objects and events. Presented at the *International Conference for Infant Studies*, Miami, FL.

Mareschal, D., Plunkett, K., & Harris, P. (in press). A computational and neuropsychological account of object-oriented behaviours in infancy. *Developmental Science*.

Munakata, Y., McClelland, J.L., Johnson, M.H., & Siegler, R.S. (1997). Rethinking infant knowledge: Toward an adaptive process account of successes and failures in object permanence tasks. *Psychological Review, 104*, 686-713.

Oakes, L.M., & Cohen, L.B. (1990). Infant perception of a causal event. *Cognitive Development, 5*, 193-207.

Rivera, S.M., Wakeley, A., Langer, J. (1999). The draw-bridge phenomenon: Representational reasoning or perceptual preference? *Developmental Psychology, 35*, 427-435.

Rutkowska, J.C. (1993). *The computational infant: Looking for developmental cognitive science*. London: Harvester Wheatsheaf.

Schilling, T.H. (1997, April). Infants' understanding of physical phenomena: A perceptual hypothesis. Presented at the meeting of the *Society for Research in Child Development*, Washington, D.C.

Spelke, E.S. (1998). Nativism, empiricism, and the origins of knowledge. *Infant Behavior and Development, 21*, 181-200.

Sutton, R.S., & Barto, A.G. (1998). *Reinforcement learning: An introduction*. Cambridge, MA: MIT Press.

Analogical Transfer of Non-Isomorphic Source Problems

Ute Schmid (schmid@cs.tu-berlin.de)
Department of Computer Science, Technische Universität Berlin, Franklinstr. 28, 10587 Berlin, Germany
Joachim Wirth (wirth@mpib-berlin.mpg.de)
Center for Educational Research, Max Planck Institute for Human Development, 14195 Berlin, Germany
Knut Polkehn (knut.polkehn@rz.hu-berlin.de)
Institute of Psychology, Humboldt Universität, 10099 Berlin, Germany

Abstract

In analogical problem solving, non-isomorphic source/target relations are typically only investigated in contrast to the ideal case of isomorphism. We propose to give a closer look to different types of non-isomorphic source/target relations and varying degrees of structural overlap. We introduce a measure of graph distance which captures the "size" of partial isomorphism between two structures and we present two experiments investigating the influence of different non-isomorphic relations on analogical transfer. In the first experiment we contrast transfer performance for isomorphic vs. source inclusive problems with high vs. low superficial similarity. In the second experiment we explore different types of partial isomorphisms: source inclusiveness, target exhaustiveness, and different degrees of source/target overlap. The results indicate that (1) transfer of isomorphs is not significantly influenced by superficial similarity but transfer of partial isomorphs is, and (2) partial isomorphs can be transferred successfully if the amount of structural overlap is at least as high as structurally differences. The experiments were inspired by some open design questions for the analogy module of IPAL (a computational model integrating problem solving and learning).

Introduction

Analogical problem solving is commonly described by the component processes retrieval, mapping and transfer. The work presented in this paper focusses on *analogical transfer*. Transfer can be faulty or incomplete, even if retrieval and mapping were successful (Novick & Holyoak, 1991). We are especially interested in transfer of *non-isomorphic* sources – the standard case in everyday problem solving. Several studies (cf. Reed, Ackinclose, & Voss, 1990; Novick & Hmelo, 1994; Spellman & Holyoak, 1996; Gholson, Smither, Buhrman, Duncan, & Pierce, 1996) show that subjects *can* transfer non-isomorphic sources successfully – at least when retrieval and mapping information is given explicitly. In our experiments, we want to look closer at the influence of different types and degrees of structural source/target similarities on transfer success.

This question is interesting for several reasons: (1) In an educational context (cf. tutoring systems) the provided examples have to be carefully balanced to allow for generalization (learning). Presenting only isomorphs restricts learning to small problem classes, while too large a degree of structural dissimilarity can result in failure of transfer and thereby obstructs learning (Pirolli & Anderson, 1985). (2) A plausible cognitive model of analogical problem solving (cf. Falkenhainer, Forbus, & Gentner, 1989; Holyoak & Thagard, 1989; Hummel & Holyoak, 1997) should generate correct transfer only for such source/target relations where human subjects perform successfully. (3) Computer systems which employ analogical or case-based reasoning techniques (Carbonell, 1986; Schmid & Wysotzki, 1998) should refrain from analogical transfer when there is a high probability of constructing faulty solutions. Thus, it can be avoided that system users have to check – and possibly debug – generated solutions. Information about conditions for successful transfer in human analogical problem solving can provide guidelines for implementing criteria for when analogical reasoning should be rejected in favor of other problem solving strategies.

Our experiments were mainly motivated by this last reason (Schmid, Mercy, & Wysotzki, 1998). We are well aware that analogical problem solving is strongly influenced by semantic and pragmatic aspects of the involved problems (Hummel & Holyoak, 1997). But we believe that there are still open questions with respect to the structural basis (Falkenhainer et al., 1989) of analogical transfer which are worthwhile to investigate (see also results on dominance of systematicity over pragmatic relevance in Markman & Sanchez, 1998).

In the following, we introduce our problem domain and describe how we constructed problems with different types and varying degrees of structural similarity. Afterwards, we first present an experiment contrasting the effect of superficial and structural similarity on transfer success; second we present an experiment contrasting target exhaustiveness, source inclusiveness, and different degrees of structural overlap between problems. Finally, we will describe how the experimental findings can be used to improve the performance of the analogy module of our problem solving and learning system IPAL.

Non-Isomorphic Variants in a Water Redistribution Domain

Water redistribution problems

Our material is based on a modification of the water jug domain (Luchins & Luchins, 1950). In contrast to the classical problems we investigate *redistribution* problems (Atwood & Polson, 1976), for example: Given three jugs with capacities $A = 36$, $B = 45$ and $C = 54$ liters and initial quantities $A = 16$, $B = 27$ and $C = 34$ liters, find a (minimal) sequence of operations *pour from jug x to y* so that the jugs contain $A = 25$, $B = 0$ and $C = 52$ liters. An example problem is given in figure 1.

The *pour*-operator is defined in the following way:

IF not(empty(x)) and not(filled(y)) THEN pour(x,y) resulting

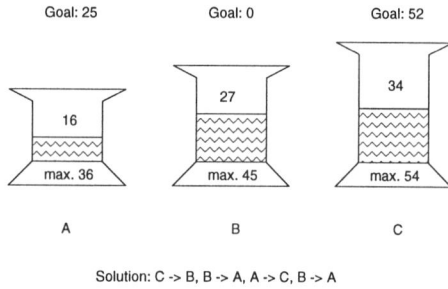

Figure 1: A water redistribution problem

in:

IF current(x) ≤ max(y) - current(y)

THEN current(y) := current(y) + current(x), current(x) := 0

ELSE current(x) := current(x) - (max(y) - current(y)), current(y) := max(y).

Water can be poured only from a non-empty jug and only into a jug which is not completely filled. Pouring results in filling y up to its capacity (possibly leaving a rest of the water in x) or in emptying x (possibly leaving a free capacity in y).

Problem Analysis

We are especially interested in the influence of *structural* similarity between source and target on transfer success. For this reason, we use only problems from the same domain – water redistribution problems – and with identical goals – find a sequence of *pour* operations so that each jug contains the desired amount of water. That is, we keep semantic and pragmatic aspects (Holyoak & Thagard, 1989) constant.

To investigate analogical transfer, we want to make sure that subjects really refer to the source for solving the target problem. Therefore, the problems should be complex enough that the correct solution cannot be found by trial and error, and difficult enough that the abstract solution principle is not immediately inferable. We constructed redistribution problems for which exists only a single (for two problems two) shortest operation sequence – in problem spaces with over 1000 states and more than 50 cycle-free solution paths[1]. The strategy for solving redistribution problems is to express goal quantities in terms of relations between initial and maximum quantities. For example, the goal quantity (25 liters) of the small jug (A) in figure 1 can be obtained in the medium jug by filling it up to its capacity (45) and pouring in the small jug (25 = 45 - (36 - 16) = 45 - 36 + 16). Abstracting from the given values, the relation between the goal quantity of jug A and given initial quantities (*start*) and capacities (*max*) is *goal(small-jug) = max(medium-jug) - max(small-jug) + start(small-jug)* (see jug j_3 in fig. 2a). Even for three-jug problems, calculating the desired redistribution is quite complex[2].

Redistribution problems can be described by the following attributes, operations and relations:

- superficial features: names (A, B, C ...) and positions (left, right, middle ...) of jugs,
- relevant features: jug capacities ($max(j)$), initial quantities ($start(j)$), and goal quantities ($goal(j)$),
- relevant operations and relations: ordinal difference between jug capacities ($j_i < j_j$), relative differences between quantities (e.g. $goal(j_1) = (start(j_1) + start(j_2)) - (max(j_2) - max(j_3))$) for the largest jug, j_1 in fig. 2a).

The structure of the problem given in figure 1 is presented in figure 2a. Note, that we represent only the aspects of the problem structure which are *relevant* for calculating the solution, e.g. we do not represent the relation between $start(j_1)$ and $start(j_3)$ or $max(j_2)$ and $goal(j_2)$.

We do neither claim that human problem solvers without experience with this problem domain represent the relevant problem structure completely and correctly, nor do we make assumptions whether analogical problem solving is better modelled on a symbolic (Falkenhainer et al., 1989) or a subsymbolic (Hummel & Holyoak, 1997) level. We constructed this "normatively complete" symbolic graph representation to explore the impact of different analytically given structural source/target relations on empirical observable transfer success.

Non-Isomorphic Source/Target Relations

In the following experiments, we are interested in a special kind of non-isomorphism – **partial isomorphism**. That is, we do not consider many-to-one (Spellman & Holyoak, 1996; i.e. epimorphisms, see Schmid et al., 1998) or one-to-many (Spellman & Holyoak, 1996; i.e. no morphism, see Schmid et al., 1998) mappings. Instead we investigate source/target relations which share a common substructure. There are different kinds of partial isomorphic source/target relations:

- **target exhaustiveness:** the target problem is completely contained in the source (Gentner, 1980),
- **source inclusiveness:** the source problem is completely contained in the target (cf. Reed et al., 1990),
- **source/target overlap:** source and target share a common part, but both problems have additional aspects (cf. Carbonell, 1986).

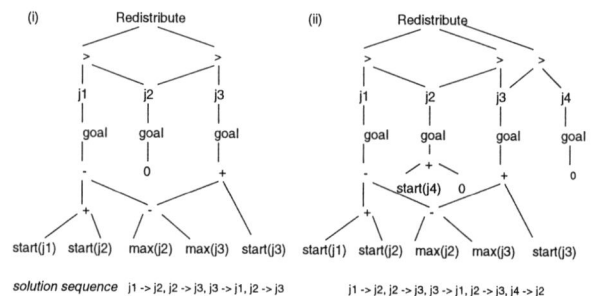

Figure 2: Structure of the source problem (a) and a partial isomorph (b)

---

[1]The algorithm for generating the set of all solutions can be obtained from the authors.

[2]The rules for calculating redistribution sequences can be obtained from the authors.

For all three types of partial isomorphs, the *degree* of structural overlap can vary. For example, the common substructure can consist of only five nodes and their interrelations, while the target problem consists of twenty nodes vs. six nodes.

The degree of structural overlap is captured in measures of graph similarity, as for example:

$$d(G, H) = 1 - \frac{V_{GH} + N_{GH}}{max(V_G, V_H) + max(N_G, N_H)}. \quad (1)$$

The distance between two graphs G and H is defined as the number of common arcs (V_{GH}) and nodes (N_{GH}) in relation to the number of nodes and arcs of the larger graph. This simple measure captures all information relevant for characterizing relations between redistribution problems. To capture mappings between different concepts (as *heat/water*; Falkenhainer et al., 1989) or relations (as $+/-$; Anderson & Thompson, 1989), the definition can be relaxed to bijective mappings between (similar) node and/or arc labels (Schädler & Wysotzki, 1998).

In the following experiments, we investigate a small subset of the possible variations of types and degrees of partial source/target isomorphisms.

Experiments

Experiment 1

In a first experiment, we explored the suitability of our domain for studying transfer of non-isomorphic sources. That is, we investigated (1) whether subjects can solve water redistribution problems by analogical transfer (problems are neither too difficult, resulting in a failure even to solve isomorphical problems, nor too easy resulting in ignoring the source problem for generating a correct solution), and (2) whether at least partial isomorphs with a moderate degree of structural dissimilarity can be transferred successfully, and (3) whether superficial similarity has an influence on transfer success.

The problem given in figure 1 was used as source problem. We investigated two structural variations: one target problem which is isomorphic to the source (deviating only in the absolute numbers for initial, maximum and goal quantities) and one which is a source inclusive partially isomorph with a high degree of structural overlap (one additional jug and one additional solution step, see fig. 2b and table 2). Additionally, we varied superficial similarity between source and target by renaming jugs and switching their positions. All variations (see table 1) were realized between subjects.

Subjects Subjects were 60 pupils of a Berlin gymnasium, aged between fourteen and nineteen (average 17.4), 31 male and 29 female.

Procedure The experiment was fully computer based. The overall time of an experimental session was about 45 minutes.

After general instruction, subjects learned how to use the program and were introduced to the problem domain by solving an initial problem with tutorial guidance. Jugs were represented graphically (see fig. 1); redistributions $pour(x, y)$ were performed by clicking first on x and then on y; impossible moves (from empty or into full jugs) were rejected; all operations could be redone and the subjects could cancel their current solution and start again. The tutor-module intervened if the solution path was longer than four, if the problem was two times restarted without solving it correctly, or after two minutes. The program only proceeded if the problem was correctly solved twice without tutorial help.

Next, subjects were informed about the general principle for finding a shortest solution path (i.e. by thinking about the goal quantities in terms of relations to initial and maximum quantities). Note, that the problems were too difficult to simply apply the general concept given in this instruction. Afterwards, subjects received the source problem – which was isomorphic to the initial problem – and were given the hint that the initial problem was similar to the current problem and that referring to its solution might help solving the new problem. Similarity was pointed out by explicitly presenting the mapping relations between the jugs of the problems (Novick & Holyoak, 1991; Novick & Hmelo, 1994). The initial solution could be retrieved by mouse click. To proceed with the current problem this window had to be closed. Tutorial support was identical to that for the initial problem. The program proceeded after the source problem was correctly solved.

We introduced an initial problem before the source problem for three reasons: Solving the source problem should not be disturbed by difficulties in interacting with the program; the handling of the recall of a prior solution could be introduced; and the subjects were "primed" to use analogical reasoning as solution strategy (in contrast to solving the problem by trial and error or guided search in the problem space).

After solving the source problem, each subject received one of a set of five versions of the target problem (see table 1). All problems had different absolute numbers for jug capacities, start and goal quantities than the source problem. Again, the similarity to the last problem (source) was pointed out by explicitly presenting the mapping relations. The target problem had to be solved without tutorial help, only by referring to the solution of the source (by the same procedure as given above). Time was restricted to ten minutes.

Finally, subjects were asked to give the mapping between the source and target jugs.

Results and Discussion To make sure that we investigate analogical transfer, we restricted the criterium "succesfully solved" to subjects, who produced the correct shortest target solution in a single trial and gave the correct mapping between source and target when questioned afterwards. The main results are given in table 1.

Table 1: Target problems and results for experiment 1

	Target Problems		Results		
	structure	surface	success	no success	χ^2
1	isomorphism	same names	8	4	
2	isomorphism	one renaming (A ↔ B)	10	2	1 vs 2 0.67
3	isomorphism	two ren.s (A → B, B → C, C → A)	8	4	1+2 vs 3 0.22
4	partial isom. (additional jug)	no renaming	8	4	1+2+3 vs 4 0.11
5	partial isom.	one renaming (A ↔ B)	3	9	1+2+3+4 vs 5 8.09*

Kimball's $k \times 2$ test, df = 1, $\alpha > 0.4$ for constrasts 1 vs 2, 1+2 vs 3, 1+2+3 vs 4,
    $\alpha = 0.005$ for constrast 1+2+3+4 vs 5 (Bonferoni adjusted $\alpha = 0.025$)

There is a significant relation between source/target similarity and solution success (5×2 contingence table with $\chi^2 = 9.58$, df = 4, $\alpha < 0.05$). The crucial (only significant, see χ^2 in table 1) difference is between conditions 4 and 5. That is, partial isomorphism is a sufficient condition for successful analogical transfer *if* the surface similarity of problems is high. If partial isomorphs differ in surface features (naming and positioning of jugs), it seems too difficult to transfer the source solution, even if the mapping is given explicitly. The results are in correspondence with the findings of (Reed et al., 1990) in the domain of algebra word problems: In their second experiment they could show that transfer success was low for source inclusive problems sharing a common domain with a target (cf. travel rate) and high for isomorphic problems even if they differ in their domain (cf. travel rate vs. interest rate).

Experiment 1 suggest the following consequences for our further investigation of partial source/target isomorphs: (1) to investigate the influence of *structural* similarity, superficial similarity between source and target should be as high as possible, and (2) to demonstrate that there exists a degree of structural dissimilarity where a source is no longer relevant for generating a target solution we have to construct source/target relations with less structural overlap than problem 4.

Experiment 2

In the second experiment, we used again the problem given in figure 1 as the source, that is, a three jug problem which can be solved by a minimal sequence of four *pour* operations. We investigated the following source/target relations:

- Target exhaustiveness: a target problem in which the last operation of the source solution is not needed (i.e. a three jug, three operations problem; problem 1 in table 2),

- Source inclusiveness (2): a target problem, in which an additional operation is needed (i.e. a three jug problem solvable with five operations; problem 2 in table 2),

- Three degrees of source/target overlap: target problems consisting of four jugs and are solvable with five operations

  - the partial isomorphic target problem used in experiment 1 (called problem 4 in exp. 1, see fig. 2b; problem D1 in table 2),

  - target problems with progressively decreasing structural overlap to the source (problems D2 and D3 in table 2).

To control the degree of structural overlap, we represented all problem structures as graphs – using an extended version of the representation given in figure 2 (additionally, the solution sequences are explicitly coded[3]). We calculated the distances between source and target problems using formula (1). Problems 1 and 2 differ from the source only in the number of necessary operations. This high degree of structural overlap is reflected in the low source/target distances of 0.16 and 0.17 (see table 3).

---

[3]The complete representations for all problems can be obtained from the authors.

Table 2: Target problems for experiment 2

jug:	small	medium-small	medium-large	large
Problem 1 (target exhaustive)				
	–	j_3	j_2	j_1
max	–	48	60	72
start	–	21	36	45
goal	–	0	33	69
solution:	$j_1 \to j_2, j_2 \to j_3, j_3 \to j_1$			
Problem 2 (source inclusive)				
	–	j_3	j_2	j_1
max	–	48	60	72
start	–	21	36	45
goal	–	33	60	9
solution:	$j_1 \to j_2, j_2 \to j_3, j_3 \to j_1, j_2 \to j_3, j_1 \to j_2$			
Problem D1 (high overlap), see problem 4 in exp. 1				
	j_4	j_3	j_2	j_1
max	16	20	25	31
start	3	8	15	18
goal	0	13	3	28
solution:	$j_1 \to j_2, j_2 \to j_3, j_3 \to j_1, j_2 \to j_3, j_4 \to j_2$			
Problem D2 (medium overlap)				
	j_4	j_3	j_2	j_1
max	16	20	25	31
start	6	9	15	18
goal	14	14	0	20
solution:	$j_1 \to j_2, j_2 \to j_3, j_1 \to j_4, j_3 \to j_1, j_2 \to j_3$			
or	$j_1 \to j_2, j_1 \to j_4, j_2 \to j_3, j_3 \to j_1, j_2 \to j_3$			
Problem D3 (less overlap)				
	j_4	j_3	j_2	j_1
max	17	20	25	31
start	7	9	15	18
goal	16	5	0	28
solution:	$j_3 \to j_4, j_1 \to j_2, j_2 \to j_3, j_3 \to j_1, j_2 \to j_3$			
or	$j_1 \to j_2, j_3 \to j_4, j_2 \to j_3, j_3 \to j_1, j_2 \to j_3$			

Problems D1, D2 and D3 were constructed by changing and/or introducing relations between *goal*, *max* and *initial* quantity. On the operational level the modifications result in an additional operator after the solution sequence of the source problem (problem D1), in the middle (problem D2) and at the beginning (problem D3). The values for the source/target distances decrease from 0.37 over 0.55 to 0.59 (see table 3). Note, that the absolute values for graph distances are to some extent dependent on the way in which the graph representation is realized. If all problems are transformed in the same way into graphs, the *relative* source/target distances reflect the varying degrees of structural overlap in an uniform way.

The experimental comparison of the realized degrees of structural overlap was not guided by a specific hypothesis. Our general assumption was, that there exists a degree of structural overlap between source and target (smaller than for the source/D1 relation) which is insufficient for analogical transfer.

Subjects Subjects were 70 pupils of a Berlin gymnasium, aged between 16 and 17 (average 16.3), 18 male, 52 female.

Procedure The procedure was identical to experiment 1. Subjects were presented with one of the five target problems.

Results and Discussion To make sure that we really investigate the impact of source/target relations on analogical transfer, we excluded all subjects from the analysis who did not give the correct mapping of jugs after solving the target.

634

Table 3: Results for experiment 2

	1	2	D1	D2	D3
	target exhaus.	source incl.	decreasing overlap		
distance	0.16	0.17	0.37	0.55	0.59
success	6	7	10	5	1
no success	1	1	3	4	11

The main results are given in table 3.

Interestingly, there are no performance differences between source inclusive and target exhaustive source/target pairs, if problems differ only in the number of operations (problems 1 and 2, exact binomial test, $\alpha = 0.601$). Both problems could be succesfully solved by most of the subjects.

The finding of experiment 1, that a source inclusive partial isomorph can be succesfully solved, could be replicated (problem 4 in exp. 1 and problem D1 in exp. 2). There is a significant relation between structural source/target similarity and solution success (conditions D1, D2 and D3; exact 3×2 test, $\alpha = 0.002$). The crucial difference is between conditions D2 and D3 (exact binomial tests: D1 vs. D2 $\alpha = 0.1003$, D2 vs. D3 $\alpha = 0.0004$, D1 vs. D3 $\alpha = 0.0001$).

Partial isomorphism exists between a lot of problems, although most of them might not be in a source/target relation usually considered in analogical problem solving: Even if two problems share only a single node, formally there exists a partial isomorphism between them. Our results suggest that there is a degree of source/target dissimilarity when the source can be no longer considered as relevant for solving the target. Note, that we are not discussing retrieval of a source – which is much more restricted by semantic similarity (Reed et al., 1990) – but analogical transfer.

For the given problem domain and representation of problem graphs the results show, that source inclusive partial isomorphs can be good candidates for analogical problem solving as long as the structurally identical part of the problems (i.e. the common subgraph) is greater than the structural differences. This is reflected by a distance smaller than 0.5 calculated with the similarity metric given in formula (1). We hope to continue our experiments on conditions for source/target relations for transfer success, exploring different problem domains (cf. algebra word problems or geometry proofs) and further structural source/target relations (cf. one-to-many mappings).

Modeling Analogical Transfer in IPAL

IPAL is a prototype for integrating problem solving and learning based on the machine learning approach of inductive program synthesis (Schmid & Wysotzki, 1998). IPAL is primarily intended as an AI application and not as a cognitive model. It deals with programming problems (such as sorting of lists) and blocksworld problems or puzzles (such as Tower of Hanoi) in an uniform way. Currently, IPAL receives a problem description (initial states, goal, operators) as input, generates problem solutions by planning and generalizes these solutions to cyclic macro-operations (Shell & Carbonell, 1989). Cyclic macros represent solution strategies for problem classes; for example, experience with sorting lists of three numbers can be generalized to a recursive program for sorting lists of arbitraty length, experience with solving a Tower of Hanoi problem with three discs can be generalized to a solution strategy for n disc problems.

We plan to integrate an analogy module in IPAL as a possible way to circumvent macro-generation from scratch. Currently our analogy module operates stand-alone and we are using it to explore conditions for successful analogical transfer based on structural information *alone*. Of course, for many real-world applications it is necessary to consider semantical aspects of problems. But we want to develop a (generic) adaptation algorithm which works context-free as far as possible. Thus, our strategy is, to try to extract as much information as possible from structural source/target relations (Schmid et al., 1998).

The analogy module works in the following way: When a new problem is solved, we check whether there already exists a cyclic macro under which the new solution can be subsumed. If a macro (source) can generate a solution sequence which is isomorphic to the current solution (target), the macro is transferred to the target domain and two new knowledge structures are committed to memory – the macro for the target domain and a macro generalizing over source and target. If there is no isomorphic source/target relation, IPAL has to decide whether (1) to generate a re-representation of the target which might result in an isomorphic source/target relation (currently done on the basis of rewrite-rules provided by the user), (2) to try adaptation nevertheless, or, (3) switch to macro-generation from scratch. If source and target are structurally too dissimilar, analogical transfer might require more effort than inductive inference and additionally has the danger of generating inadequate or erroneous solutions. This is the reason why we are investigating structural criteria for successful analogical transfer.

In our psychological experiments, we investigated the transfer of *problem solutions*; in IPAL we want to employ analogical transfer on the level of macros, i.e. *problem solving strategies*. But the decision *whether* a source-macro can be transferred to the new domain is determined on the basis of problem solutions (the new problem solving trace and a trace generated from the candidate source macro). Thus, information about conditions for successful transfer of problem solutions can give us valuable design hints for IPAL. Up to now, our analogy module eagerly adapts each source to the target but generates more than 50% erroneous solutions for non-isomorphic source/target relations (see Schmid et al., 1998 for the adaptation algorithm and test results for a variety of source/target pairs). On the basis of the experimental results, we plan to run new trials, comparing IPAL's performance when adapting sources with more vs. less than fifty percent overlap. Also based on the experimental results, we will not prefer target exhaustiveness (which involves deleting of information from the source) to source inclusiveness (which involves inserting additional information). In our current implementation deletion is preferred. To introduce new information we currently rely on structural constraints given by the partial source solution only. Another possibility might be to introduce a mechanism of internal analogy (Hickman & Lovett, 1991).

Conclusions

We reported two experiments investigating the influence of different types and degrees of non-isomorphic source/target relations on transfer success. For the water-jug redistribution domain we could show that partial isomorphic sources can transferred successfully if source and target do not differ in surface features. Furthermore, we could demonstrate that there exists a degree of structural overlap between source and target where the source is no longer helpful for constructing a solution.

Cognitive models and other systems using analogical reasoning techniques usually make no restrictions with respect to the *structural* overlap between problems when retrieving a source. Retrieval (or at least its first stage) is usually guided by feature-based (i.e. superficial) similarity alone. Thus, a source which shares only a very small isomorphic substructure with the target will be treated in exactly the same way as a source with a high structural overlap. In this case, transfer might result in meaningless inferences or erroneous solutions. Of course, problems which share a greater amount of similar attributes might often also share a greater structural overlap, but this must not be true in general. For example, in the domains we are exploring with IPAL, there might be the following source candidates for solving the blockworld problem *"build a tower of alphabetically ordered blocks"*: (1) another blocksworld problem – *"unstack a tower of alphabetically ordered blocks"* – which shares a lot of attributes with the target, and (2) a list sorting problem which shares no attributes but can be solved with the same underlying strategy. We would prefer to retrieve the structurally rather than the superficially similar source problem. That is, we propose to consider not only attribute-based but also structural similarity between problems to minimize the risk of erroneous transfer.

Acknowledgements

This manuscript was written while the first author was on leave at the School of Computer Science, Carnegie Mellon University, supported by a DFG research scholarship. Thanks to Klaus Eyferth, Bruce Burns, Alex Petrov and Dario Salvucci for helpful discussions and again to Bruce Burns for comments on the manuscript.

References

Anderson, J., & Thompson, R. (1989). Use of analogy in a production system architecture. In S. Vosniadou & A. Ortony (Eds.), *Similarity and analogical reasoning.* Cambridge University Press.

Atwood, M. E., & Polson, P. G. (1976). A process model for water jug problems. *Cognitive Psychology, 8,* 191-216.

Carbonell, J. (1986). Derivational analogy: A theory of reconstructive problem solving and expertise acquistion. In R. Michalski, J. Carbonell, & T. Mitchell (Eds.), *Machine learning - an artificial intelligence approach* (Vol. 2). Los Altos, CA: Morgan Kaufmann Pub.

Falkenhainer, B., Forbus, K., & Gentner, D. (1989). The structure mapping engine: Algorithm and example. *Artificial Intelligence, 41,* 1-63.

Gentner, D. (1980). *The structure of analogical models in science.* Cambridge, MA: Bolt Beranek and Newman Inc.

Gholson, B., Smither, D., Buhrman, A., Duncan, M. K., & Pierce, K. (1996). The sources of children's reasoning errors during analogical problem solving. *Applied Cognitive Psychology, Special Issue: "Reasoning Processes", 10,* 85-97.

Hickman, A. K., & Lovett, M. C. (1991). Partial match and search control via internal analogy. In *Proceedings of the 13th Annual Conference of the Cognitive Science Society* (p. 744-749). Hillsdale, NJ: Lawrence Erlbaum Ass.

Holyoak, K. J., & Thagard, P. (1989). Analogical mapping by constraint satisfaction. *Cognitive Science, 13,* 295-355.

Hummel, J., & Holyoak, K. (1997). Distributed representation of structure: A theory of analogical access and mapping. *Psychological Review, 104*(3), 427-466.

Luchins, A., & Luchins, E. (1950). New experimental attempts at preventing mechanization in problem solving. *Journal of General Psychology, 42,* 279-297.

Markman, A., & Sanchez, A. (1998). Structure and pragmatics in analogical inference. In *Advances in analogy research: Integration of theory and data from the cognitive, computational, and neural sciences* (p. 191-200). Sofia, Bulgaria.

Novick, L. R., & Hmelo, C. E. (1994). Transferring symbolic representations across nonisomorphic problems. *Journal of Experimental Psychology: Learning, Memory, and Cognition, 20*(6), 1296-1321.

Novick, L. R., & Holyoak, K. J. (1991). Mathematical problem solving by analogy. *Journal of Experimental Psychology: Learning, Memory, and Cognition, 17*(3), 398-415.

Pirolli, P., & Anderson, J. (1985). The role of learning from examples in the acquisition of recursive programming skills. *Canadian Journal of Psychology, 39,* 240-272.

Reed, S. K., Ackinclose, C. C., & Voss, A. A. (1990). Selecting analogous problems: similarity versus inclusiveness. *Memory & Cognition, 18*(1), 83-98.

Schädler, K., & Wysotzki, F. (1998). Application of a neural net in classification and knowledge discovery. In *Neural Networks in Applications NN'98, Proceedings of the Third International Workshop* (p. 219-226). Magdeburg.

Schmid, U., Mercy, R., & Wysotzki, F. (1998). Programming by analogy: Retrieval, mapping, adaptation and generalization of recursive program schemes. In *Proc. of the Annual Meeting of the GI Machine Learning Group, FGML-98* (pp. 140–147). TU Berlin.

Schmid, U., & Wysotzki, F. (1998). Induction of recursive program schemes. In *Proceedings of the 10th European Conference on Machine Learning (ECML-98)* (p. 228-240). Springer.

Shell, P., & Carbonell, J. (1989). Towards a general framework for composing disjunctive and iterative macrooperators. In *Proceedings of the 11th IJCAI-89.* Detroit, MI.

Spellman, B. A., & Holyoak, K. (1996). Pragmatics in analogical mapping. *Cognitive Psychology, 31,* 307-346.

636

The Production of Noun Phrases:
A Cross-linguistic Comparison of French and German

Herbert Schriefers (Schriefers@nici.kun.nl)
NICI, University of Nijmegen
P.O. Box 9104, 6500 HE Nijmegen, The Netherlands

Encarna Teruel (encarnat@t-online.de)
Median Klinik II fuer Neurologie
Parkstrasse, D-39345 Flechtingen, Germany

Abstract

Two experiments investigated the grammatical encoding processes during the production of noun phrases consisting of an article, an adjective, and a noun. Experiment 1 shows that for noun phrases in German, with the adjective in prenominal position, the lemmas of the noun and the adjective, and the noun's grammatical gender are selected before utterance onset. Experiment 2 shows that for noun phrases in French, with the adjective in postnominal position, only the noun lemma and its grammatical gender are selected. This suggests that grammatical advance planning at the level of grammatical encoding can operate with the smallest full phrase which can be expanded rightwards during articulation. Furthermore, the data show that gender is selected irrespective of whether it surfaces in the eventual phonological form of the noun phrase or not. This result is in line with the assumption that the grammatical encoder operates independently of the phonological encoder.

Introduction

Psycholinguistic models of language production assume that the production of an utterance occurs in several processing steps. The major processing stages are conceptualization, grammatical encoding, phonological encoding, and articulation (e.g., Levelt, 1989; Levelt et al., 1999). For the description of a colored line drawing by means of a noun phrase like "the red table", the main processing steps are as follows. On the basis of the pictorial input, the abstract lexical entities (so-called lemmas) for the adjective and the noun have to be retrieved from the mental lexicon. For languages marking the grammatical gender of nouns by gender marked articles and / or gender marking inflections on the adjective (e.g., German, Dutch, and French), the noun lemma activates the lexical-syntactic information about the noun's grammatical gender. This information is also assumed to be stored in the mental lexicon and is used to select so-called agreement targets. In the case of definite determiner noun phrases in German, the gender marked definite determiner is the agreement target. For the corresponding noun phrases in French, the definite determiner and the inflectional ending of the color adjective are agreement targets. The result of these grammatical encoding processes is a syntactic representation. In the next step, this representation is passed to the phonological encoder, resulting in a phonetic plan which guides the articulation of the noun phrase.

The present study reports experiments on the grammatical encoding processes for the production of French and German noun phrases consisting of a definite determiner, a color adjective and a noun. Although simple in syntactic structure, these noun phrases require the retrieval of a lexical-syntactic feature of the noun lemma, i.e. its grammatical gender. The grammatical gender is used to determine specific gender agreement targets like gender marked definite determiners and gender marked inflections of the adjective. As these noun phrases can easily be elicited in an experimental setting by asking subjects to name colored line drawings of common objects, they provide an interesting window for the experimental study of grammatical encoding processes in language production.

Noun phrases in French and German differ in two aspects that are of central interest for the present study. First, the noun phrases in the two languages differ in word order. Whereas in German, the color adjective occurs in prenominal position (e.g., der gruene Tisch - the (masc) green table (masc)), it occurs in postnominal position in French (e.g., la table verte - the (fem) table green (fem.)). In French, the first part of the noun phrase ("la table") is a perfectly grammatical and complete phrase which can be expanded rightwards by the color adjective. By contrast, in German, the first part of the noun phrase (e.g., "der gruene") does not constitute a complete phrase (though it can occur in certain elliptical utterances). Thus, it appears that in German, the complete noun phrase has to be grammatically encoded before phonological encoding and articulation can be initiated, i.e. the noun lemma, the adjective lemma, and the noun's grammatical gender have to be retrieved from the mental lexicon. By contrast, in French, it might be possible that phonological encoding and articulation can already be initiated right after completion of the grammatical encoding of the first complete phrase ("la table"). The grammatical encoding for the remainder of the noun phrase, i.e. the retrieval of the adjective lemma, could be carried out while the first part of the noun phrase is processed at the phonological and articulatory levels.

Second, in German, the noun's grammatical gender always surfaces in the eventual phonological form of a noun phrase in the form of different definite determiners for the different types of grammatical gender. In French, by contrast, the grammatical gender does not always surface in the eventual phonological form of the noun phrase. For example, in the noun phrase "la table verte" the gender marking surfaces in the phonological form of the determiner and of the adjective. However, in a noun phrase with a noun starting with a vowel and an adjective which does not change its phonological form as a function of grammatical gender (e.g., "rouge" - red), the phonological form of the noun phrase does not contain any reflection of the grammatical gender of the noun (e.g., "l'assiette rouge" - the red plate).

The critical question is whether the grammatical gender of the target noun is also selected if this lexical-syntactic property of the noun has no consequences for the eventual phonological form of the noun phrase. According to the model proposed by Levelt (1989; Levelt et al., 1999) this should be the case because the grammatical encoding processes (lemma retrieval and gender selection) are assumed to strictly precede phonological encoding. That is, they are assumed to be modular and completely blind with respect to the noun phrase's eventual phonological shape. However, recently, Caramazza (1997, see also Caramazza & Miozzo, 1997) has proposed an alternative model of lexical processing in language production. At least in principle, this model allows for gender selection to be skipped if the grammatical gender of the noun has no consequences for the phonological form of the noun phrase. According to this independent network model (IN-model), the phonological form and the lexical-syntactic properties of words (like a noun's grammatical gender) can be looked up independently from each other and in parallel. This has important potential implications for noun phrases for which the phonological form of the definite determiner and the adjective does not depend on the noun's gender. If the retrieval of the phonological forms can be completed before selection of the grammatical gender has been completed, the selection of grammatical gender may not occur before utterance onset or may even be skipped completely. If this is the case, we should not obtain empirical evidence for gender selection for this type of noun phrases. Barbaud et al. (1982) have provided evidence which could be interpreted in this way. In a corpus of spoken Canadian-French, gender agreement errors were more frequent for agreement targets (like gender marked adjectives) agreeing with a noun starting in a vowel than for agreement targets agreeing with a noun starting in a consonant. This could suggest that a noun's gender is sometimes not selected if it is not needed for the determination of the phonological form of the noun's local syntactic context.

In summary, we will investigate two questions. First, does the amount of advance planning in grammatical encoding vary as a function of word order differences between French and German? More specifically, are the noun lemma, its grammatical gender, and the adjective lemma retrieved before initiation of phonological encoding in both languages, or do speakers of French only retrieve the noun lemma and its grammatical gender before initiating articulation? Second, is grammatical gender always selected, irrespective of whether it surfaces in the phonological form of a gender agreement target in the noun phrase (e.g., a definite determiner), or can gender selection be skipped if the gender marking does not surface in the phonological form of the utterance.

These questions were addressed by means of an extension of the picture-word interference paradigm that allows us to obtain empirical evidence for lemma selection and for the selection of the grammatical gender of a noun.

With respect to gender selection, the present experiments extend previous research on the production of noun phrases in Dutch (Schriefers,1993; see also van Berkum, 1997; LaHeij et al. 1998). In these experiments, subjects were asked to name colored line drawings as quickly as possible by means of noun phrases consisting of a gender marked determiner, a prenominal color adjective and a noun. In addition, they were presented (visually or auditorily) with so-called distractor words which they were instructed to ignore. The results showed that a distractor word which had a different grammatical gender than the to-be-produced target noun (hereafter gender-incongruent distractor, INC) prolonged utterance onset latencies relative to a distractor word which had the same gender as the target noun (hereafter gender-congruent distractors, CON). This gender interference effect is assumed to be due to a competition between the gender of the distractor and the gender of the target noun in the INC condition and leads to a prolongation of the selection of the gender of the target noun relative to the CON condition.

Lemma selection in the picture-word interference paradigm is reflected in a so-called semantic interference effect. The naming latency for the picture of an object (e.g., a chair) is prolonged in the presence of a semantically related distractor word (e.g. table) relative to a condition with unrelated distractor words (e.g., car). This effect is assumed to be due to the fact that a semantically related distractor (hereafter SEM) introduces an additional lexical competition in the selection of the to-be-produced target lemma relative to an unrelated distractor (hereafter UNR, e.g., Roelofs, 1992; Schriefers et al., 1990).

Experiment 1:
Production of noun phrases in German

In this experiment, native speakers of German named colored line-drawings by noun phrases with a definite determiner (e.g., der gruene Tisch - the (masc.) green table (masc.)). In addition, they were auditorily presented with distractor words. There were 18 critical target objects from three semantic categories, with 6 exemplars per category. Within each category, each of the three grammatical genders (feminine, masculine, neuter) was represented by two object names. Each target object could occur in one of five different colors. In four distractor conditions, the distractor words were nouns. They were either semantically related to the target noun (SEM), or unrelated (UNR), and they had either the same gender as the target noun (CON) or a different gender (INC). The crossing of the factors semantic relatedness (SEM, UNR) and gender relation

(CON, INC) yields four distractor conditions. The distractors were assigned to the target objects such that each specific distractor contributed equally to each of the four distractor conditions.

In two additional distractor conditions, the distractors were adjectives. In one of these conditions, the distractor was a color adjective which was different from the one to-be-produced in the target utterance (hereafter SEM-A). In the other condition, the adjective did not have any semantic or phonological relation with any of the words in the target utterance. These UNR-A distractors were matched with the color adjectives for word length and frequency. Finally, in the so-called NONE condition, no distractor was presented. In addition, the point in time at which the distractor was presented (the stimulus onset asynchrony, SOA) was systematically varied across four levels, ranging from distractor presentation preceding picture onset by 150 ms (SOA -150 ms) to distractor presentation following picture onset by 300 ms (SOA +300 ms, steps of 150 ms). The critical dependent variable was utterance onset latency, measured from the onset of the colored line drawing to the beginning of the articulation of the noun phrase. Sixteen native speakers of German participated in the experiment.

Tables 1 and 2 give the mean utterance onset latencies as a function of distractor conditions and SOA.

Table 1: Mean utterance onset latencies (in ms) as a function of conditions with nouns as distractors and SOA in Experiment 1 (percentage of errors in parentheses)

	SOA			
	-150	0	+150	+300
NONE	713	710	708	697
	(2.8)	(4.5)	(5.2)	(6.2)
SEM-CON	776	785	776	723
	(3.8)	(4.9)	(4.2)	(5.9)
SEM-INC	771	804	795	725
	(4.9)	(8.7)	(6.9)	(5.2)
UNR-CON	756	784	764	719
	(5.9)	(3.1)	(4.5)	(2.4)
UNR-INC	747	766	793	718
	(6.2)	(3.1)	(5.9)	(3.5)

The results for the conditions with noun distractors were analyzed in separate ANOVAs per SOA with the crossed factors semantic relatedness (SEM, UNR) and gender relation (CON, INC). At SOA -150 ms, there was a significant main effect of semantic relatedness, i.e. a semantic interference effect ($F1(1,15) = 8.35$, $p < .05$, $F2(1,17) = 5.4$, $p = .05$). Neither the main effect of gender relatedness nor the interaction were significant, showing that the semantic interference effect was independent of the gender relation between target noun and distractor.

At SOA +150 ms, we obtained a gender-interference effect, ($F1(1,15) = 12.2$, p , .01, $F2(1,17) = 17.2$, $p < .001$). This gender-interference effect was independent of the factor semantic relatedness as indicated by the absence of an effect of this factor and an interaction. At an SOA of 0 ms,

we only obtained a significant interaction of the two factors ($F1(1,15) = 10.6$, $p < .01$, $F2(1,17) = 4.5$, $p < .05$). At SOA +300, neither the two main effects nor their interaction were significant. The comparison of the utterance onset latencies for the conditions SEM-A and UNR-A was significant at SOA +150 ms ($F1(1,15) = 9.0$, $p < .01$, $F2(1,17) = 6.0$, $p < .05$).

Table 2: Mean utterance onset latencies (in ms) of distractor conditions SEM-A and UNR-A in Experiment 1 (percentage of errors in parentheses)

	SOA			
	-150	0	+150	+300
SEM-A	740	779	787	726
	(6.9)	(3.8)	(4.2)	(2.8)
UNR-A	747	765	756	719
	(5.2)	(2.1)	(5.2)	(3.8)

These results provide empirical evidence for the three main processes of grammatical encoding in the production of noun phrases, retrieval of the noun lemma, selection of the corresponding grammatical gender, and retrieval of the adjective lemma. All three processes appear to be completed before utterance onset. The ordering of the effects along the SOA dimension furthermore suggests that the retrieval of the noun lemma precedes the selection of its corresponding gender and the retrieval of the adjective lemma. As the retrieval times for noun lemma and gender information are presumably not deterministic but rather variable between trials, the interaction of semantic relatedness and gender relation at SOA 0 ms presumably reflects the fact that we are here dealing with a mixture of trials. Some of them are still in the stage of lemma retrieval, and some of them are already in the stage of gender selection. Finally, at an SOA of 300 ms, distractors are presented too late to have any systematic influence on grammatical encoding.

Experiment 2: Production of noun phrases in French

Experiment 2 contained the same distractor conditions and SOA manipulations as the experiment on the production of German noun phrases. However, in contrast to Experiment 1, two different types of noun phrases were used as target utterances, noun phrases with an explicit reflection of the noun's grammatical gender in the eventual phonological form, and noun phrases without a reflection of the noun's gender in the eventual phonological form of the noun phrase.

There are two critical issues here. First, does the fact that the color adjective occurs in postnominal position in French lead to a situation in which only the first part of the noun phrase (i.e., determiner and noun) is planned before utterance onset? If this is the case, we should only obtain a semantic interference effect for the noun (SEM vs UNR), but not for the adjective (SEM-A vs UNR-A). Second, can the grammatical encoding processes bypass the selection of grammatical gender if grammatical gender does not surface in the phonological form of the eventual utterance?

Eight objects with names starting in a consonant (C-nouns) and another eight objects with names starting in a vowel (V-nouns) were selected. Each object could occur in four different colors. Two of the color adjectives require an explicit phonologically and orthographically realized inflection in dependence of the noun's gender (vert (masc) / verte (fem) - green; blanc (masc) / blanche (fem) - white; inflected adjectives hereinafter), whereas the other two did not (rouge (masc or fem) - red; jaune (masc or fem) - yellow; uninflected adjectives hereinafter). The noun types and adjective types were combined such that they yielded two different types of noun phrases. Type 1 noun phrases consist of C-nouns and an inflected or uninflected adjective. Thus, they require an explicit phonological realization of the noun's gender in the definite determiner (le = masc., la = fem.). It should already be mentioned here that the type of adjective (inflected vs uninflected) did not modulate the pattern of results for these noun phrases. Type 2 noun phrases consist of a V-noun and an uninflected adjective. For these noun phrases, grammatical gender does not surface in the phonological form. Sixteen native speakers of French participated in the experiment.

Utterance onset latencies for type 1 noun phrases (C-nouns) are given in Tables 3 and 4.

Table 3: Mean utterance onset latencies (in ms)
as a function of conditions with nouns as distractors and SOA
in Experiment 2, noun phrases of type 1
(percentage of errors in parentheses)

	SOA			
	-150	0	+150	+300
NONE	699	670	679	651
	(5.5)	(8.2)	(5.5)	(4.7)
SEM-CON	800	783	746	662
	(6.3)	(6.3)	(6.7)	(7.4)
SEM-INC	794	781	779	689
	(6.3)	(6.7)	(6.7)	(7.0)
UNR-CON	749	744	764	658
	(3.9)	(5.1)	(6.2)	(7.4)
UNR-INC	763	757	750	689
	(5.1)	(5.5)	(6.7)	(5.9)

Table 4: Mean utterance onset latencies (in ms)
of distractor conditions SEM-A and UNR-A in Experiment 2,
noun phrases of type 1 (percentage of errors in parentheses)

	SOA			
	-150	0	+150	+300
SEM-A	744	726	720	662
	(8.6)	(6.2)	(5.1)	(5.9)
UNR-A	744	734	745	666
	(3.9)	(6.2)	(5.5)	(3.1)

The results were analyzed in the same way as in Experiment 1. For the conditions with nouns as distractors, there was a significant effect of semantic relatedness at SOA -150 (F1(1,15) = 27.55, p < .001, F2(1,7) = 9.6, p < .05). The main effect of gender relation and the interaction of the two factors was not significant. The same pattern obtained for SOA 0 (semantic relatedness: F1(1,15) = 16.5, p < .01, F2(1,7) = 6.5, p < .05). At SOA +150 ms, only the interaction between the two factors was significant in the analysis by subjects (F1(1,15) = 7.1, p < .05, F2(1,7) = 3.5, p = .10). At SOA +300 ms, the effect of gender relation was significant (F1(1,15) = 4.6, p < .05, F2(1,7) = 6.5, p < .05), with longer utterance onset latencies in the INC-condition than in the CON-condition. Neither the main effect of semantic relatedness nor the interaction of gender relation and semantic relatedness were significant. The difference between the conditions SEM-A and UNR-A was not significant at any SOA.

Tables 5 and 6 give the mean utterance onset latencies for type 2 noun phrases (V-nouns with uninflected adjectives).

Table 5: Mean utterance onset latencies (in ms)
as a function of conditions with nouns as distractors and SOA
in Experiment 2 for noun phrases of type 2
(percentage of errors in parentheses)

	SOA			
	-150	0	+150	+300
NONE	694	663	672	628
	(3.9)	(7.0)	(6.2)	(6.2)
SEM-CON	754	770	750	687
	(4.7)	(4.7)	(10.2)	(3.9)
SEM-INC	758	806	780	685
	(7.8)	(5.5)	(7.8)	(5.5)
UNR-CON	716	742	748	689
	(3.1)	(5.5)	(7.8)	(3.9)
UNR-INC	716	780	766	694
	(3.9)	(3.9)	(6.2)	(5.5)

Table 6: Mean utterance onset latencies (in ms)
of distractor conditions SEM-A and UNR-A in Experiment 2,
noun phrases of type 2 (percentage of errors in parentheses)

	SOA			
	-150	0	+150	+300
SEM-A	712	717	721	680
	(3.9)	(5.1)	(6.3)	(6.3)
UNR-A	741	728	738	684
	(4.7)	(5.5)	(0.8)	(4.7)

For the conditions with nouns as distractors, we obtained the following statistical results. For SOA -150 ms, the effect of semantic relatedness was significant in the subject analysis and marginally significant in the item analysis (F1(1,15) = 13.1, p < .01, F2(1,7) = 4.1, p = .08). The effect of gender relation and the interaction of the two factors was not significant. For SOA 0 ms, both main effects were significant or marginally significant (semantic relatedness: F1(1,15) = 4.2, p = .08, F2(1,7) = 6.4, p < .05); gender relation: F1(1,15) = 6.7, p < .05, F2(1,7) = 3.3, p = .10). The interaction of the two factors was not significant. At the remaining two SOAs (+150, +300) neither the main effects nor their interaction reached significance, although there was a trend towards a gender interference effect at SOA +150 of 24 ms across the levels of semantic relatedness ($p =$ .10 in subject and item analysis). The comparison of the SEM-A and UNR-A conditions did not yield a significant difference at any of the four SOAs.

In summary, for both types of noun phrases we obtain evidence for the retrieval of the noun lemma, as indicated by the effect of semantic relatedness, and for gender selection, as indicated by the gender interference effect. However, the time courses of these effects are different for the two different types of noun phrases. In particular, the gender interference effect for type 1 noun phrases (C-nouns) is obtained at the longest SOA whereas for type 2 noun phrases (V-nouns) it already occurs at SOA 0, and in combination with an effect of semantic relatedness (though at SOA +150 ms, there is also some indication of a pure gender interference effect). As the two types of noun phrases are necessarily based on different objects, it is not clear what the reasons for these differences in time course are. Nevertheless, it is clear that gender selection occurs irrespective of whether it is reflected in the phonological form of the noun phrase or not. Furthermore, neither for type 1 nor for type 2 noun phrases is there any indication of a semantic interference effect from the SEM-A condition relative to the UNR-A condition. This is in clear contrast to the experiment on noun phrases in German.

Discussion and Conclusion

In contrast to German noun phrases with prenominal color adjectives, the production of French noun phrases with postnominal color adjectives does not require the selection of the adjective lemma before utterance onset. Rather, it appears that speakers of French start phonological encoding and articulation as soon as they have completed the grammatical encoding of article and noun which, in French, already delivers a complete noun phrase. The further rightward expansion of this noun phrase with a postnominal color adjective takes place while the speaker is about to start articulation or during articulation of the phrase consisting of determiner and noun.

These results provide evidence for the cross-linguistic variation of grammatical advance planning. It appears that the grammatical encoder constructs one full noun phrase (in French, determiner and noun; in German, determiner, adjective and noun) before passing the result to the next processing stages. If the beginning part of the to be planned utterance is a full noun phrase on its own, as in French, the further rightward expansion of this phrase by a postnominal color adjective can be done incrementally. That is, while the first part is already processed at the phonological and articulatory level, the adjective is still being processed at the level of grammatical encoding. The results for German converge with corresponding results for noun phrases with prenominal adjectives in Dutch (Schriefers, 1993) as well as in German (Schriefers et al., in press). The results for French show that the scope of advance planning is different for noun phrases with postnominal color adjectives.

The second main issue concerned the retrieval of lexical-syntactic features like the grammatical gender of nouns. A gender interference effect was obtained irrespective of whether the noun's gender did or did not appear overtly in the eventual phonological form of the noun phrase. The gender interference effect for French noun phrases without an explicit reflection of gender in the phonological form (i.e. noun phrases of type 2, V-nouns) was somewhat weaker than the corresponding effect in German and in French noun phrases of type 1 (C-nouns), and the time course of the gender interference effects over SOAs shows some as yet unexplained variability. Nevertheless, it appears to be clear that gender selection occurs in all noun phrases investigated in the present experiments. This conclusion is further supported by results for French noun phrases consisting of a V-noun and an inflected (i.e. gender marked) postnominal adjectives. The results for these noun phrases were not presented here for reasons of space. So we can only mention that also for these noun phrases a gender interference effect and a semantic interference effect for noun distractors were obtained. Again, no semantic interference effect was obtained for the color adjectives. Thus, although the grammatical advance planning does not comprise the gender marked adjective in this latter type of noun phrases, grammatical gender is again selected despite the fact that it does not surface in the first part of the noun phrase. However, this latter type of noun phrases had a higher proportion of gender agreement errors on the postnominal inflected adjective than noun phrases consisting of a C-noun and an inflected adjective. Although the number of gender agreement errors was generally low, this result is in agreement with Barbaud et al's (1982) data on gender errors in spontaneous speech.

Overall, the results are in line with Levelt's (1989; Levelt et al. 1999) model of language production. The grammatical encoding processes appear to be blind with respect to the eventual phonological shape of an utterance; grammatical gender is selected irrespective of whether it surfaces in the phonological form of the noun phrase or not. This result can also been seen as compatible with the IN-model (Caramazza, 1997), but it gives an important constraint on the model. Even if lexical-syntactic properties of a word and its phonological form can be retrieved independently and in parallel, the lexical-syntactic properties of a word are selected if there is an agreement target in the utterance (like a definite determiner), irrespective of whether gender agreement has a reflection in the eventual phonological form of the noun phrase or not. However, the pattern of gender errors suggests that very occasionally, gender selection can be bypassed if the gender does not surface in the phonological form of the actual grammatical planning unit (i.e. determiner and noun, in French). This leads to a somewhat higher proportion of gender errors on agreement targets later on in the utterance, like inflected postnominal adjectives.

A final point concerns the relation of the present data with recent experiments on noun phrase production in Italian by Miozzo and Caramazza (in press). In Italian, the definite determiner of masculine nouns not only depends on the noun's grammatical gender, but also on the phonology of the next word in the noun phrase. More specifically, before words starting with a "z", a "s + consonant", or an affricate, the masculine determiner is "lo", in all other cases "il" (e.g., "lo sgabello" - the stool, "il grande sgabello" - the big stool, "il treno" - the train, "il grande treno" - the big train). Miozzo and Caramazza propose that due to these properties Italian is a so-called "late selection language"; the determiner can only be selected in a late stage, when the phonological form of the to-be-produced noun phrase becomes available. By contrast, German is an early selection language because selection of the definite determiner exclusively depends on the noun's grammatical gender, and not on the eventual phonological form of the noun phrase. Miozzo and Caramazza (in press) propose, on the basis of a repeated failure to obtain a gender interference effect in the production of noun phrases in Italian, that the late selection status of Italian might render the gender interference effect invisible. Given this proposal, the question arises of whether French would also qualify as a late selection language. If one assumes that, for example, the two forms of the masculine determiner ("le" for C-nouns, and "l'" for V-nouns) are treated as two different determiners (just as "il" and "lo" are in the case of Italian), this might be the case. On the other hand, one could assume that French qualifies as an early selection language and that the reduced form of the determiner ("l'") is a late phonetic accomodation of the determiner "le". If the absence of a gender interference effect is taken as a clear diagnostic tool for identifying late selection languages, then, given the present results, French would not be a late selection language. However, at present it is not yet clear whether the late selection languages necessarily imply the absence of a gender interference effect. Nevertheless, the proposal of Miozzo and Caramazza (in press) is in line with the conclusions from the present experiments in so far as there appear to be clear cross-linguistic differences in the grammatical encoding processes involved in the production of noun phrases.

References

Barbaud, Ph., Ducharme, Ch. & Valois, D. (1982). D'un usage particulier du genre en canadien-francais: la feminisation des noms a initiale vocalique. *Canadian Journal of Linguistics, 27 (2),* 103-133.

Caramazza, A. (1997). How many levels of processing are there in lexical access? *Cognitive Neuropsychology, 14,* 177-208.

Caramazza, A., & Miozzo, M. (1997). The relation between syntactic and phonological knowledge in lexical access: evidence from the 'tip-of-the-tongue' phenomenon. *Cognition, 64,* 309-343

LaHeij, W., Mak, P., Sander, J., & Willeboordse, E. (1998). The gender congruency effect in picture-word tasks. *Psychological Research, 61,* 209-219.

Levelt, W.J.M. (1989). *Speaking. From Intention to articulation.* Cambridge, MIT Press.

Levelt, W.J.M., Roelofs, A. & Meyer, A.S. (1999). A theory of lexical access in speech production. *Behavioral and Brain Sciences,* in press.

Miozzo, M. & Caramazza, A. (in press). The selection of determiners in noun phrase production. *Journal of Experimental Psychology: Learning, Memory and Cognition.*

Roelofs, A. (1992). A spreading activation theory of lemma retrieval in speaking. *Cognition, 42,* 107-142.

Schriefers, H. (1993). Syntactic processes in the production of noun phrases. *Journal of Experimental Psychology: Learning, Memory, and Cognition, 19,* 841-850.

Schriefers, H., Meyer, A.S. & Levelt, W.J.M. (1990). Exploring the time course of lexical access in language production: Picture-word interference studies. *Journal of Memory and Language, 29,* 86-102.

Schriefers, H., deRuiter, J.P. & Steigerwald, M. (in press): Parallelism in the production of noun phrases: Experiments and reaction time models. *Journal of Experimental Psychology: Learning, Memory and Cognition.*

van Berkum, J. J. A. (1997). Syntactic processes in speech production: The retrieval of grammatical gender. *Cognition, 64,* 115-152.

The Presence and Absence of Category Knowledge in LSA

Christian D. Schunn (cschunn@gmu.edu)

Department of Psychology; MSN 3F5
George Mason University
Fairfax, VA 22030-4444

Abstract

How much information about meaning is contained in the statistical structure of the environment? LSA is a theoretical and practical tool that is challenging previous notions about what is contained in the statistical structure of the environment. This paper examines what kind of category knowledge can be obtained from the environment using LSA. In particular, two experiments are conducted with LSA to test what kind of category structure it embodies. LSA ratings about the relatedness of categories to their properties are compared with human judgments regarding the centrality of properties to the categories. LSA is found to capture aspects of property centrality for some object and event categories. However, it is found to only capture those aspects related to typicality: how often do members of the category have that property? LSA fails to capture other aspects of centrality that can be found in human category judgments. Thus, it appears that humans do bring other constraints to bear in shaping their categories.

Introduction

Latent Semantic Analysis

Latent Semantic Analysis (LSA) is a technique developed by Landauer and Dumais (1997) for automatically constructing a semantic representation of terms based on how they co-occur in a large corpus of texts. In the last several years, LSA has become an exciting tool for cognitive science, both as a psychological model and as a practical tool.

As a psychological theory, LSA claims that people learn the meaning of words not from learning definitions or from innate constraints, but instead by simply observing the contexts under which terms are used. In other words, the claim is that the statistical structure of the environment contains all the information that is needed to determine the meanings of words. Within this theoretical framework, LSA has been used as a model of child word acquisition (Landauer & Dumais, 1997), subject matter knowledge (Landauer, Foltz, & Laham, 1998), and semantic priming (Landuaer & Dumais, 1997). In each case, it has provided powerful demonstrations regarding how much information can be derived purely from the statistical structure of the environment.

As a practical tool, LSA has also been used for grading essays (Landauer, Foltz, & Laham, 1998), text research (Foltz, 1996), information retrieval (Dumais, 1994), and selecting reviewers for papers (Dumais & Neilson, 1992). A longer list of applications of LSA can be found in Landauer, Foltz, and Laham (1998). Again, in each of these cases, the statistical structure of the environment was proven to be surprisingly rich.

How does LSA work?

LSA begins with a very large text corpus (e.g., an encyclopedia, a series of textbooks in an area, or a large set of smaller documents). Next, all words that occur in the corpus are found. Those that occur with extremely high frequency (e.g., 'the', 'of', 'a') are removed. For all the remaining words (potentially tens of thousands), the frequency with which they co-occur in a given context (usually defined as a paragraph) are counted. For example, how often does the word 'beak' occur in paragraphs that have the word 'tail'? This produces an enormous matrix of co-occurrence information of size N by N, where N is the number of words being examined (which can be as high as 100,000).

LSA then reduces this huge frequency matrix into a much smaller matrix by a process called singular value decomposition (SVD). This new matrix is now of size N by M, where M is some number usually between 5 and 500. This reduction in size is conceptually similar to factor analysis or multidimensional scaling: it produces a more compact representation of the important statistical regularities in the larger matrix.

In the reduced matrix, each word can be thought of as a vector or point in an M-dimensional space. Since the reduced matrix is derived from frequency co-occurrence information, it represents words that occur in similar contexts with similar representations. This process forces synonyms to have very similar representations because, even though they rarely both occur in the same context, they co-occur with the same other words.

In LSA, one can also represent groups of words, sentences and paragraphs as a point in this M-dimensional space. By adding context words, the meanings of polysemous words are disambiguated. In any case, the relative similarity in meaning of two items (two words, two sentences, two paragraphs, a word and a sentences, etc.) is defined by their proximity in the M-dimensional space (in particular, the cosine of the angle between each vector). The closer the two items, the more similar their meanings.

On the surface, this complex process may seem psychologically implausible. However, some classes of neural nets do produce an approximation of SVD. Thus, the LSA claim is that humans may be doing something computationally equivalent to SVD. A more precise description of the mechanics of LSA and SVD can be found in Deerwester, Dumais, Furnas, Landauer, and Harshman (1990).

In sum, LSA develops a semantic representation of terms based on underlying (or latent) information found in the statistical regularities in the environment. While the success of its past applications to a variety of domains is impressive, the nature of the representation that it develops requires further investigation. Of particular focus in this paper is the following question: what kind of category knowledge does LSA develop? If LSA develops a human-like semantic representation of terms, then categorical structure of human knowledge should also be found in its representations.

This research question about LSA is also a more general question about how much information can be derived from the statistical structure of the environment (specifically of text and speech input that humans receive). If LSA can produce a good account of category structure, then the statistical structure of the environment may include more information than we may have originally thought.

Property centrality

Almost all categories have a structure such that some instances of the category are considered more central than other instances. For example, a robin is considered a more central instance of the category of birds than is an ostrich. Moreover, some properties of the category are considered more central than other properties. For example, having wings is considered more central to the category of birds than having a beak. In this paper, I will focus on the centrality of properties.

First, it is important to distinguish between property typicality and property centrality. Property typicality is how often members of the category have that property (e.g., what proportion of birds have wings?). By contrast, property centrality is the importance of the property to a person's concept. For example, imagine something that was like a bird, but didn't have wings. How good an example of the concept of birds would that be? How much would it change your concept? How likely would you be to call it a bird? All of these questions address the centrality of the property to the category, or how important the property is to the category.

What makes a property central? Certainly the property's typicality is very important (Rosch, 1975). However, centrality is not synonymous with typicality. Consider the case of curvedness and bananas and boomerangs. All bananas and all boomerangs that people will have seen are curved (i.e., curvedness is equally typical of both boomerangs and bananas). Yet, curvedness is considered more central to the category of boomerangs than it is to the category of bananas (Murphy & Medin, 1985). It was argued that curvedness is especially important to boomerangs because people have a theory about the role that curvedness plays to the fundamental activities of a boomerang.

In a more recent study, Schunn and Vera (1995) found that it is specifically causal factors that play a very important role in determining property centrality. For both objects (e.g., pigeons and cars) and events (e.g., birthday parties and elections), the properties that were considered to play an important causal role in the functioning of the object/event were considered central to the category. In fact, they were considered more central than properties considered part of a formal definition or properties used to recognize the object/event. Of course, property typicality was also an important predictor of property centrality.

In sum, property centrality in humans appears to be a combination of property typicality and causal theories (see also Ahn and Lassaline, 1995). What kind of account of categorization does LSA provide? If LSA is deriving true word meaning, then it should understand the relationship between a category term and its properties. Since LSA derives its semantic structure from frequency information, it should at least be able to provide a good account of the typicality effects in category structure.

Yet, it is unclear whether LSA will be able to account for more of centrality (meaning) than just typicality. Or, asked another way, what information is found in the statistical structure of the environment? Certainly first order statistics (as measured by typicality) do not fully capture centrality. However, the complete statistical information found in textual corpuses may produce patterns in LSA that mimic what appears to be the role of causal theories in human categorization. This added information may occur either because textual corpuses contain a bias towards mentioning causal information or because higher-order statistics contain additional information.

To examine these questions, I compared categorization data from Schunn and Vera (1995) (henceforth called SV) against the responses produced by LSA to the exact same categories and properties. In Experiment 1, I report the results for analyses of object categories (both artifacts and biological kinds). In Experiment 2, I report the results for analyses of event categories.

Experiment 1: Objects

Methods

Human Data Source The human data for Experiment 1 were taken from SV Experiments 1b, 1c, and 1d. In particular, the data from the six familiar objects were used. These objects included three familiar biological kinds (cats, pigeons, and mice) and three artifacts (cars, toilets, and power lawn mowers). Each object had 6 properties. These properties were the ones most commonly generated by people as properties for each of these categories (SV, Experiment 1a). Table 1 presents an example of one of the categories and its properties. The full list of properties can be found in the original paper.[1]

Table 1: An example object (pigeons) and its properties and z-scored property centrality ratings.

Centrality	Properties
0.628	Pigeons have wings.
0.612	Pigeons have feathers.
0.392	Pigeons fly.
-0.429	Pigeons are gray.
-0.555	Pigeons coo.
-0.648	Pigeons live in big cities.

For all six properties of all six objects, SV obtained five different ratings from five different groups of subjects. First, property centrality ratings (the importance of the property to the category) were obtained by asking subjects how much their concepts were changed by negating each property. For example, subjects were asked how much does your concept of pigeons change if a particular instance did not have wings. Each subject's ratings were z-score transformed. Then a mean normalized rating for each property was determined. This mean for each property served as the input for the current analyses. Table 1 presents the centrality ratings for each property of pigeons. Higher ratings represent more central properties.

Second, property typicality ratings (the frequency with which category members have the property) were obtained by asking subjects either what proportion of those objects in the world had each property (or what proportion of objects that they had seen had each property). The two variations produced highly correlated responses, and the average frequency rating across both variants was computed for each property.

Third, subjects were asked to rate each property of each object in terms of how important it was for a scientific or expert definition of the object category. Fourth, subjects were asked to rate each property of each object in terms of how important it was for recognizing members of the object category. Fifth, subjects were asked to rate each property in terms of how important the property was for the object to be successful at what it did. Mean ratings were determined on each scale for each property. These last three ratings were called definition, recognition, and cause, respectively.

In sum, the human data consisted of ratings of each property of each object on five dimensions: centrality, typicality, definition, recognition, and cause. SV determined that centrality judgments were best predicted by typicality and cause. At issue now is to what LSA judgments correspond.

[1] An online version of the paper can be found at http://hfac.gmu.edu/~schunn.

LSA Data The LSA data was gathered using the web version of LSA available at http://lsa.colorado.edu.[2] As previously used in many tests of LSA, a data space derived from the TASA college corpus was selected. Of the available data spaces, this one best represents general knowledge and experiences of a college student (who was the source of the human data). The TASA college corpus uses a variety of texts, novels, newspaper articles, and other information. It includes 37,651 documents and 92,409 terms.[3] The default setting of 300 dimensions was used and those results are reported below. However, all analyses were also redone using only 100 dimensions to examine the role of number of dimensions.

To get judgments of semantic similarity or relatedness from LSA, LSA was asked to compare each category name with its properties. For example, LSA was asked to compare "pigeon" with "has wings". The same object and property names were given to LSA as were given to the human subjects. For each of these comparisons, LSA produced a number between -1 and 1 representing the perceived similarity (with 1 being perfectly similar and -1 being perfectly dissimilar).

Results

How Well Does LSA Predict Property Centrality?

To see how well LSA values predicted property similarity, the 36 LSA values were correlated against the 36 human centrality means (6 objects and 6 properties per object). Across the 36 properties, LSA correlated $r=.26$, $p<.1$ with property centrality. The correlations were slightly larger when done separately for artifacts ($r=.29$) and biological kinds ($r=.32$). Computing the correlations separately for each object, two of the three artifacts (cars $r=.42$, toilets $r=.91$) and one of the three biological kinds (pigeons $r=.99$) had noticeably positive correlations. Correlations for the other three objects were: cats $r=-.37$, mice $r=.04$, and power lawn mowers $r=-.46$. While some of these correlations are based on extremely low Ns and are correspondingly noisy, it appears that LSA did predict some aspects of property centrality, but quite poorly for some artifacts and biological kinds.

What Predicts Which Objects Will Have Centrality Well Predicted By LSA?

Why did LSA predict centrality quite well for some objects but terribly for others? In particular, what generally determined whether LSA could predict property centrality for a particular object? The most important determinant was how well LSA predicted property typicality. For several of the objects, LSA did not predict property typicality

[2] In particular, the matrix comparisons application was used, with the "term to term" comparison method

[3] At the web location, this topic space is called "General reading up to 1st year college"

(i.e., how often members of the category have the given property). This can be readily seen by comparing the correlation between LSA and typicality for each object to the correlation between LSA and centrality for each object. The correlation among these correlation values was r=.97 (N=6, p<.01). In other words, if LSA did not account for typicality effects, it did not account for centrality effects. But, this explanation does not tell us whether there was something inherently different between the three objects LSA did predict well and the three objects LSA did not predict well.

Another explanation that focuses on properties of the objects themselves is how well the different components of centrality correlated with one another. There were four measures (from SV) that all played some role in centrality to varying degrees (in decreasing order of importance): typicality, cause, recognition, and definition. For each object, the correlations were computed between each of these measures. Then the average of the 6 pairwise correlations was computed for each object. This average correlation represents how consistent the various components of centrality were with one another. For example, for some objects, properties high in typicality were also high in causal importance, high in recognitional importance, and high in definitional importance, whereas for other objects, properties high in typicality may not have been high in causal importance or not high in recognitional importance. This average correlation is purely a property of the objects and is not logically tied to LSA. It is truly an independent predictor of LSA's quality of fits to centrality. Thus, it is impressive that it did predict well how LSA was related to property centrality for each object: the correlation among correlations (average intercorrelation vs. LSA to centrality correlation) was r=.82 (N=6, p<.02). Moreover, since each individual correlation was based on such a small N and so is likely to be quite noisy, the strength of this correlation is noteworthy.

To make this last analysis more concrete, let us consider the examples of objects with centrality well and poorly predicted by LSA. The six properties of pigeons are listed in Table 1. The two most central properties of the category are having feathers and wings. Those properties are also in the top three for ratings of typicality, definition, recognition, and cause. Thus, LSA could receive information about any of these factors from the text corpus structure and be able to predict centrality well.

By contrast, consider the case of cats. For the category of cats, having fur and four legs was considered central. While these two properties were also rated as most typical of cats, the relatively less typical property of having claws and meowing were viewed as very important in ratings of definition, recognition, and cause (although also tied with having four legs). Thus, the various sources of information that LSA may be abstracting conflict with one another, and thus produce poor judgments of centrality.

In sum, one can say that LSA did a good job of predicting centrality when it did a good job of predicting typicality, or one can say that LSA was able to predict centrality for objects that had a simple and consistent property structure.

Does LSA Capture More Than Typicality? For humans, centrality is more than just typicality (Murphy & Medin, 1985; Schunn & Vera, 1995). In this human data set (as SV showed), cause was a very important predictor of centrality above and beyond typicality. For example, partialling out the correlation of typicality with centrality, cause still has a high partial correlation with centrality (partial r=.48, N=36, p<.01).

Does LSA also have more to it than typicality? In particular, can it predict aspects of centrality above and beyond typicality effects? To examine this issue, LSA predictions were correlated with centrality after partialling out the correlation of typicality with centrality. Unfortunately, LSA predicted very little of centrality beyond typicality (partial r=.06, N=36, p>.5).

Discussion

At least for objects, it appears that LSA can only account for typicality aspects of category structure. It does not predict aspects of centrality beyond the effects of typicality. These results can be viewed in a positive or negative light. On the positive side, LSA can predict property centrality quite well for some objects. Give that is simply uses word co-occurrence from a textual corpus, this result is not trivial. On the negative side, its correlations overall are quite weak, and for some objects, it does not at all predict property centrality.

The analyses provided some insight into the circumstances under which LSA can predict centrality. When LSA does not predict typicality well, it does a terrible job of predicting centrality. Moreover, those objects that have complex property structure appear to be the most difficult for LSA to capture.

In testing LSA against human judgments, the default (and recommended) value of 300 dimensions was used. To examine whether the results depended on the number of dimensions, the analyses were redone LSA values based on 100 dimensions. The new values produced slightly larger correlations for some objects and slightly smaller correlations for others. Overall, results remained the same. Thus, the results appear to generalize across a range of dimension settings.

The results also do not appear to be very specific to minor variations in how the text is presented to LSA. LSA produces similar numbers when the singular form of the category names were used (e.g., mouse versus mice or car versus cars). Correlations between singular and plural ranged between .72 and .98.

Experiment 1 focused on object categories. Events are another important class of categories, and have a wide variety of subtypes. For example, there are socially defined events (weddings, parties) that are similar to artifacts. There are also more naturally defined events (thunderstorms, car accidents) that are similar to biological kinds. Experiment 2 examines how well LSA can predict category structure for events.

Experiment 2: Events

Methods

Human Data Source The human data for Experiment 2 were taken from SV Experiments 2b, 2c, and 2d. The data included 16 familiar events, including social, formal, and physical events of varying durations: birthday party, birth, breakfast at diner, car accident, getting dressed to go out, elections, getting a haircut, grocery shopping, having a cold, using an ATM machine, making coffee, making photocopies, making a phone call, taking a final exam, thunderstorm, and wedding. There were between 6 and 10 properties for each event. As with Experiment 1, these properties were the ones most commonly generated by people (SV, Experiment 2a). Table 2 presents an example of one the categories and its properties.

Table 2: An example event (weddings) and its properties and z-scored property centrality ratings.

Centrality	Property
1.006	groom
.913	bride
.303	relatives
.150	rings
.078	cake
.009	priest
-.093	gifts
-.426	throw bouquet
-.440	drinking

As with Experiment 1, the human data consists of norms on the same five dimensions using similar procedures: centrality, typicality, definition, recognition, and cause. Again, mean subject ratings on each dimension will be used as the input for the current analyses.

LSA Data The LSA was gathered in the same manner as for Experiment 1: the web implementation of LSA using the semantic space derived from the TASA database with 300 dimensions. Of the 118 properties, LSA did not know two of the property terms: photocopier (for making photocopies) and mudslinging (for elections), and thus those two properties were deleted from the analyses.

Results

How Well Does LSA Predict Property Centrality?
To see how well LSA values predicted property similarity, the 116 LSA values were correlated against the 116 human centrality means (16 events and 6–10 properties per event). Across the 116 properties, LSA cor-

related r=.21, p<.05 with property centrality. To see how well it predicted property centrality within each of the events, the correlations were computed separately for each event. Eight of the events produced noticeably positive correlations with centrality (in decreasing order of predictiveness): weddings r=.88, phone call r=.88, thunderstorm r=.85, births r=.74, using ATM machine r=.44, making coffee r=.34, having a cold r=.26, and dressed to go out r=.25. The other eight events produce very small or negative correlations with centrality (in decreasing order of predictiveness): car accident r=.18, elections r=.11, getting a haircut r=-.12, breakfast at diner r=-.23, making photocopies r=-.39, birthday parties r=-.48, taking final exam r=-.53, grocery shopping r=-.54. In sum, LSA did not predict centrality well overall, but did predict centrality for half the events.

What Predicts Which Objects Will Have Centrality Well Predicted By LSA? As with the objects, one can examine what determines when LSA will predict event property centrality. The same two factors proved to be important. First, when LSA did not predict property typicality, it did not predict centrality. The correlation of correlations (LSA vs. typicality and LSA vs. centrality) was r=.86, (N=16, p<.01). Second, when the events had a complex property structure, LSA did not predict centrality. The correlation between the average intercorrelation for each event (mean pairwise intercorrelation among typicality, cause, recognition, and definition) and the LSA-centrality correlation for each event was r=.47 (N=16, p<.05).

Does LSA Capture More Than Typicality? As with objects, for humans cause is a very important predictor of centrality. In this SV data set, cause predicts centrality significantly above the contributions of typicality (partial r=.46, N=118, p<.001). By contrast, LSA only predicts a very small amount of centrality beyond typicality (partial r=.18, N=116, p<.1).

Discussion

As with objects, it appears that LSA can only account for typicality aspects of category structure. When LSA does not predict typicality, it does not predict centrality. Again, when objects had complex property structure, LSA also did not predict centrality well. However, LSA did predict centrality relatively well for at least half of the events, and this accomplishment should not be minimized.

As in Experiment 1, LSA was tested using the default value of 300 dimensions. To examine whether the results depended on the number of dimensions, the analyses were redone using LSA values based on 100 dimensions. As with objects, the overall, results remained the same. Thus, the results continue to generalize across a range of dimension settings.

Once again, the results also do not appear to be very specific to minor variations in how the text is presented to LSA. For example, correlations between the singular and plural (e.g., birth vs. births, birthday vs. birthdays) were usually in the .9 range.

General Discussion

One goal of the research presented here was to determine whether and under which circumstances LSA has a human-like category structure to its semantic space. In general, the results were a story of the glass half-full and the glass half-empty. For the glass half-full, LSA was able to produce modest correlations with centrality overall, and reasonably strong correlations with centrality for at least half the objects and events that were examined. Thus, there are some important human-like properties to the semantic space produced by LSA. For the glass half-empty, LSA was primarily restricted to typicality effects. LSA could not account for aspects of property centrality beyond typicality. This is in sharp contrast to human categories, for which causal aspects play a very important role in both objects and events. When categories had complex structure, in which typicality, definitional, recognitional, and causal factors did not all correlate highly with one another, LSA did not predict property centrality well.

The implications of these findings extend beyond LSA. They suggest that the textual environment appears not to have aspects of centrality beyond typicality hidden in simple statistical structure. If this implication is correct, then this provides further evidence that if human categories have aspects of centrality beyond typicality, these aspects must come purely from a top-down bias. Of course, it is possible that other methods for retrieving information from statistical structure may have found other aspects of property centrality.

There are important limitations of the work presented here that must be acknowledged. First, there are other kinds of category knowledge than property centrality. For example, there is also instance centrality—how central an instance is of a category (e.g., that a robin is more central an instance of the category birds than is an ostrich). Preliminary work by Laham (1998) suggests that LSA can capture some aspects of this category knowledge. Moreover, there are other ways of defining property centrality.

Second, it is possible that using another textual corpus to construct the semantic space with LSA or using different similarity metrics between items in LSA would have produced better category knowledge. Similarly, alternative schemes to LSA might provide a better extraction of information from the environment. For example, the HAL model (Burgess, 1998; Burgess & Conley, 1998) uses a different scheme that takes into account how close the words appear within a context. However, HAL can only represent the similarity between words, not words and phrases as in LSA.

Third, there are many possible variations that could have been used in presenting the properties to LSA. LSA was given the variations most similar to what the human participants saw. However, it is possible that other variations (e.g., adding the object and event names to the property sentences) might have changed the results, and this needs to be explored further.

As a final note, the research presented here illustrates the advantages of the world wide web as a research tool. Complex computational engines can be made available to other researchers without requiring the overhead of maintaining the software or specialized computers to run the software.

Acknowledgements

I would like to thank Greg Trafton, Erik Altmann, Lelyn Saner, Sherryl Miller, Audrey Lipps, and Susan Trickett for comments made on drafts of this paper.

References

Ahn, W., & Lassaline, M. (1995). Causal structure in categorization. In the *Proceedings of the 17th Annual Conference of the Cognitive Science Society*.

Burgess, C. (1998). From simple associations to the building blocks of language: Modeling meaning in memory with the HAL model. *Behavior Research Methods, Instruments, & Computers*, 30, 188-198.

Burgess, C., & Conley, P. (1998). Representing proper names and objects in a common semantic space: A computational model. *Brain and Cognition*.

Deerwester, S., Dumais, S. T., Furnas, G. W., Landauer, T. K., & Harshman, R. (1990). Indexing by latent semantic analysis. *Journal of the American Society for Information Science*, 41, 391-407.

Dumais, S. T., & Nielson, J. (1992). Automating the assignment of submitted manuscripts to reviewers. *Proceedings of the Fifteenth Annual International ACM SIGIR Conference on Research and Development in Information Retrieval*.

Foltz, P. W. (1996). Latent Semantic Analysis for text-based research. *Behavioral Research Methods, Instruments and Computers*, 28(2), 197-202.

Laham, D. (1998). Latent Semantic Analysis approaches to categorization. Poster presented at the *19th Annual Meeting of the Cognitive Science Society*.

Landauer, T. K. & Dumais, S. T. (1997). A solution to Plato's problem: The Latent Semantic Analysis theory of the acquisition, induction, and representation of knowledge. P*sychological Review*, 104, 211-240.

Landauer, T. K., Foltz, P. W., & Laham, D. (1998). Introduction to Latent Semantic *Analysis. Discourse Processes*, 25, 259-284.

Murphy, G. L. & Medin, D. L. (1985). The role of theories in conceptual coherence. *Psychological Review*, 92, 289-316.

Schunn, C. D., & Vera, A. H. (1995). Causality and the categorization of objects and events. *Thinking & Reasoning, 1*(3), 237-284.

Saccadic selectivity during visual search: The effects of shape and stimulus familiarity

Jiye Shen (jiye@psych.utoronto.ca)
Eyal M. Reingold (reingold@psych.utoronto.ca)
Department of Psychology; 100 St. George Street
Toronto, ON M5S 3G3, Canada

Abstract

Three experiments were designed to examine the influence of shape feature and stimulus familiarity on saccadic selectivity during visual search. Robust shape feature based guidance was found in Experiment 1. In contrast, familiarity-based guidance was much smaller in magnitude and was observed with an unfamiliar target (Experiments 2 & 3) but not with a familiar target (Experiments 1, 2 & 3). Results from the current study suggest that there are qualitative and quantitative differences between the saccadic selectivity produced by stimulus familiarity and that produced by low-level features.

The Guided Search Model proposed by Wolfe, Cave and Franzel (1989) and Wolfe (1994) argues that information extracted preattentively could guide the shifts of attention during visual search. One potential prediction from this theory is that if a particular feature or stimulus dimension guides visual search, distractors which share that feature or dimension with the target will be fixated on more often than those distractors which do not. Studies monitoring eye movements have produced ample evidence that is consistent with this prediction. Stimulus dimensions such as color, orientation, shape and size (e.g., Findlay, 1997; Motter & Belky, 1998; Williams, 1967; Williams & Reingold, 1999; but see Zelinsky, 1996) have been shown to bias the distribution of the saccadic endpoints.

The current study examined whether participants could use learned stimulus properties, such as the familiarity of the stimuli, to guide eye movements during the search process. Stimulus familiarity has been shown to strongly influence visual search efficiency in several studies (e.g., Frith, 1974; Krueger, 1984; Reicher, Snyder & Richards, 1976). Wang, Cavanagh and Green (1994) further claimed that stimulus familiarity behaves like a primitive feature that could be processed preattentively. If this were the case, stimulus familiarity should bias saccadic endpoints in a manner similar to low-level features such as color and shape. In the current study, three experiments were conducted to examine whether stimulus familiarity could produce saccadic selectivity during visual search.

General Method

The eyetracker employed in this research was the SR Research Ltd. EyeLink system. This system has high spatial resolution (0.005°), and a sampling rate of 250 Hz (4 ms temporal resolution). By default, only the subject's dominant eye was tracked in our studies. In the present study, the configurable acceleration and velocity thresholds were set to detect saccades of 0.5° or greater.

Stimulus displays were presented on two monitors, one for the participant (a 17-inch Viewsonic 17PS) and one for the experimenter. The experimenter monitor was used to give feedback in real-time about the participant's computed gaze position. In general, the average error in the computation of gaze position was less than 0.5° of visual angle.

Participants were presented with a number of visual search displays. At the beginning of each trial, a fixation dot was displayed in the center of the computer screen in order to correct for drift in gaze position. Participants were asked to fixate on the dot and then press a start button to initiate a search display in the center of the screen. They were asked to decide quickly and accurately whether the target was in the display or not. The trial terminated if parti-

Table 1: Search targets and Distractors used in the current study

Experiment	Targets	Familiar Distractors	Unfamiliar Distractors
Expt 1	P F	B D E T	ꓭ ᗡ Ǝ ⊥
Expt2	A ∀	F N Y	⊣ ⋊ ⅄
Expt 3 (Group 1)	OF ꟻO	GO TO OK OR	Oᘓ O⊥ ꓘO ꓤO
Expt 3 (Group 2)	FIT TIꟻ	TIN TIP SIT BIT	ИIT ꟼIT TIƧ TIꓭ

cipants pressed one of the response buttons or if no response was made within 20 seconds. The particular buttons used to indicate target presence or absence were counterbalanced across participants.

Experiment 1

The goal of the first experiment was to examine whether shape feature and stimulus familiarity would produce saccadic selectivity. Two different search targets were used in the current experiment: **P**, which has curvature and closure, and **F**, which does not. Eight distractors (**B**, **D**, **E**, **T** and their 180° degree rotated form) were used (see Table 1). These distractors could be categorized into four groups: familiar distractors with curvature and closure (**B** and **D**); familiar distractors without curvature or closure (**E** and **T**); unfamiliar distractors with curvature and closure (rotated **B** and **D**) and unfamiliar distractors without curvature or closure (rotated **E** and **T**). All stimuli subtended 1.6 degree vertically and 1.3 degree horizontally. Displays consisted of 24 stimuli and were created using an imaginary 6 × 6 grid of stimulus positions, which subtends 12 × 12 degrees (see Figure 1 for an example).

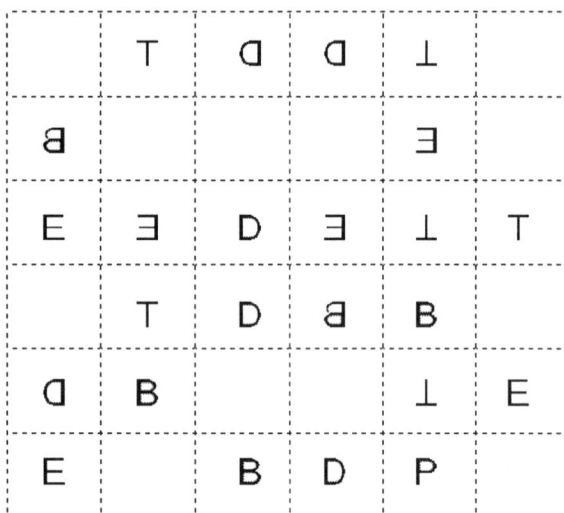

Figure 1: A sample search display used in Experiment 1 (target was a **P**). The dotted grid was not shown to the participants.

Twelve participants were tested individually in a single one-hour session. Each participant received six blocks of 96 trials (three blocks for **F** and three blocks for **P**). At the beginning of the experiment, participants received two practice blocks of 18 trials, one for each search target. Each participant searched both targets with the order of target presentation counterbalanced across individuals.

Trials in which participants responded incorrectly were excluded from analysis (3.4% of all test trials). Following

Zelinsky (1996), only target-absent trials were included in the current analysis. The fixations were assigned to the nearest distractors and then proportions of fixations to each type (similar vs. dissimilar feature) were calculated. When **F** was the search target, participants made 56.0 % fixations to those distractors without curvature and closure (**E**, **T**, and rotated **E** and **T**) and 44.0 % fixations to the distractors with curvature and closure (**B**, **D**, and rotated **B** and **D**). Similarly, when **P** was the search target, 58.1% of the fixations were directed to the distractors with curvature and closure whereas 41.9% fixations to were directed to the distractors without curvature and closure. The overall shape-based guidance, calculated by subtracting the proportion of fixations made to the distractors with dissimilar shape-feature from the proportion of fixations made to the distractors with similar shape-feature, was 14.1%, t (11) = 7.75, p < .01. This effect was quite robust and observed across all 12 participants and across saccades of different amplitude (see Figure 2). The magnitude of the saccadic guidance, however, was stronger for the saccades within 2 degrees than for those above 6 degrees. In addition, the guidance for **P** was slightly stronger than that for **F** (16.2% vs. 12.0%, t (11) = 2.2, p < .05).

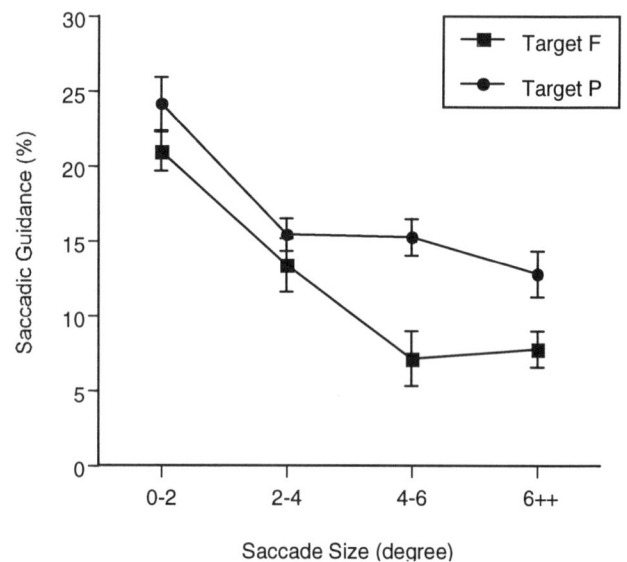

Figure 2: Percentage of saccadic guidance as a function of preceding saccadic amplitude. Saccadic guidance was calculated as the difference in the proportion of saccades between the similar and different shape-feature distractors.

Following the same procedure, proportions of fixations to the familiar versus unfamiliar distractors were calculated. Participants were equally likely to make fixations to the familiar and unfamiliar distractors (49.8% to the familiar distractors vs. 50.2% to the unfamiliar distractors). Thus, in the present experiment, when a familiar target was used, there was no significant guidance by familiarity (the difference was – 0.4 %, t < 1).

Experiment 2

To examine the generality of the findings from Experiment 1, this experiment employed a more powerful manipulation of the stimulus familiarity. In addition, both familiar and unfamiliar targets were included.

The search targets used in the current experiment were **A** and **Λ**, which was derived by moving the middle bar of **A** to the top. The familiar distractors were **F**, **N** and **Y** whereas the unfamiliar distractors were formed by recombining the same set of features from the familiar distractors (Reingold & Jolicoeur, 1993; see Table 1). All stimuli subtended 1.6 degree vertically and 1.3 degree horizontally. Similar display composition as in the previous experiment was used except that the display size was kept at 18 (see Figure 3 for an example). Each of twelve participants received 6 blocks of 96 trials preceded by two blocks of 24 practice trials representing each search target.

Figure 3: A sample search display used in Experiment 2 (Search target was a **Λ**). The dotted grid was not shown to the participants.

Similar to the previous experiment, fixations were assigned to the nearest distractors. For both the familiar and unfamiliar target, proportions of fixations made to the familiar versus unfamiliar distractors were calculated. When participants were searching for a familiar target, there was no difference in the distribution of fixations between the familiar and unfamiliar distractors (48.3 % of the fixations to the familiar distractors vs. 51.7 % to the unfamiliar distractors; the difference was – 3.4%, t (11) = 1.75, p > .05). In marked contrast, when the target was unfamiliar, more fixations were made to the unfamiliar distractors than to the familiar ones (45.4 % to the familiar distractors vs. 54.6% to the unfamiliar distractors; the difference was about 9.2 %, t (11) = 8.01, p < .01). This effect was observed when the

size of the preceding saccade was within 6 degrees but disappeared at larger amplitudes (see Figure 4).

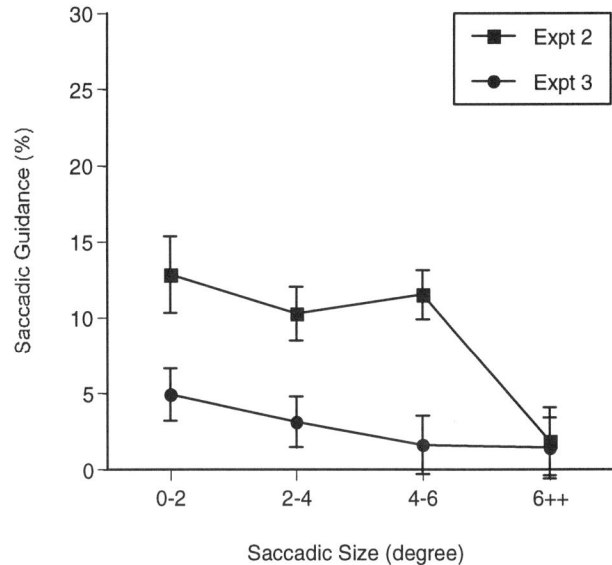

Figure 4: Percentage of saccadic guidance as a function of preceding saccadic amplitude (degree) for the unfamiliar target in Experiment 2 (in square) and 3 (in circle). Saccadic guidance was calculated as the difference in the proportion of saccades between the familiar and unfamiliar distractors.

Experiment 3

This experiment was designed to replicate the findings from Experiment 2, with words used as familiar stimuli and rotated or reflected words as unfamiliar stimuli. Two groups of six participants were tested. For one group, the targets were **OF** and its 180° rotation whereas the distractors were **GO**, **TO**, **OK** and **OR**, and their 180° rotated forms. Each stimulus subtended 1.3 degree vertically and 1.6 degree horizontally. For the other group, the targets were **FIT** and its left-right reflection, with distractors being **TIN**, **TIP**, **SIT** and **BIT**, and their left-right reflection (see Table 1). Each stimulus subtended 1.1 degree vertically and 1.6 degree horizontally. For both groups, the display size was fixed at 16. Each of 12 participants received six blocks of 96 test trials with two practice blocks of 24 trials, representing each of the search targets.

The results from both groups were identical and thus were reported together. Just as in previous two experiments, there was no guidance with a familiar target (49.5 % of the fixations made to the familiar distractors vs. 50.5% to the unfamiliar distractors; the difference was -1.0%, t (11) < 1). When the search target was unfamiliar, participants made more fixations to the unfamiliar distractors than to the familiar distractors (51.6% vs. 48.4%; the difference was 3.2%, t (11) = 3.43, p < .01). This effect was much smaller than that observed in the previous experiment (9.1%) and was evident only when the preceding saccade size was below 4 degrees (See Figure 4).

General Discussion

The current study examined whether shape feature (curvature and closure) and stimulus familiarity could guide visual search effectively. Results from Experiment 1 revealed a robust shape feature based guidance. Both the presence and the absence of the shape feature (curvature and closure) biased saccadic endpoints, though stronger bias was observed for the presence of the curvature and closure. Furthermore, the shaped-based saccadic selectivity was observed across all saccadic amplitude. Thus, the current study provided further evidence for guidance by low-level features during visual search (e.g., Findlay, 1997; Motter & Belky, 1998; Williams, 1967; Williams & Reingold, 1999; but see Zelinsky, 1996).

Another finding from the current study is that stimulus familiarity biased saccadic endpoints in a different manner than the shape feature. Across all three experiments, when a familiar search target was involved, there was no bias in the distribution of saccadic endpoints. In Experiments 2 and 3, a small but consistent familiarity-based guidance was observed with an unfamiliar target. The familiarity-based saccadic selectivity was only observed when the preceding saccade was small in amplitude (no more than 6 degrees).

Why was saccadic selectivity only observed with an unfamiliar target but not with a familiar target? There are many potential differences between searching for a familiar versus an unfamiliar target. Such differences may include the nature of target representation as well as comparison processing efficiency (Reingold & Jolicoeur, 1993). Therefore, a strong interpretation of the current data is premature. Nevertheless, one possible explanation can be based on the interaction of the bottom-up activation and top-down activations as postulated by the guided search theory (Wolfe, 1994; Wolfe et. al., 1989). It has been speculated that the unfamiliar stimuli elicit more activity and constitute larger bottom-up activation during the search process (Treisman & Gormican, 1988; Wang et. al., 1994). When an unfamiliar item is specified as the search target, both the bottom-up activation and top-down activation guide attention towards the unfamiliar distractors. Accordingly, participants tend to direct more saccades towards the unfamiliar distractors. On the other hand, when a familiar item is designated as the search target, the goal-directed top-down activation guides attention towards the familiar distractors whereas the bottom-up activation guides attention towards the unfamiliar distractors. In this case, the bottom-up and top-down activation may largely cancel out each other.

In summary, the current study demonstrated that the guidance with presence or absence of a shape feature (curvature and closure) was quite robust whereas familiarity-based guidance was much smaller in magnitude and only observed with an unfamiliar target but not with a familiar target. This suggests that the guidance by stimulus familiarity is qualitatively and quantitatively different from that by low-level features.

Acknowledgements

Preparation of this paper was supported by a grant to Eyal M. Reingold from the Natural Science and Engineering Research Council of Canada.

References

Findlay, J. M. (1997). Saccade target selection during visual search. *Vision Research, 37*, 617-631.

Frith, U. (1974). A curious effect with reversed letters explained by a theory of schema. *Perception & Psychophysics, 16*, 113-116.

Krueger, L. E. (1984). The category effect in visual search depends on physical rather than conceptual differences. *Perception & Psychophysics, 35*, 558-564.

Motter, B., & Belky, E. (1998). The guidance of eye movements during active visual search. *Vision Research, 38*, 1805-1815.

Reicher, G. M., Snyder, C. R. R., & Richards, J. T. (1976). Familiarity of background characters in visual scanning. *Journal of Experimental Psychology: Human Perception and Performance, 2*, 522-530.

Reingold, E. M., & Jolicoeur, P. (1993). Perceptual versus postperceptual mediation of visual context effects: Evidence from the letter-superior effect. *Perception & Psychophysics, 53*, 166-178.

Treisman, A., & Gormican, S. (1988). Feature analysis in early vision: Evidence from search asymmetries. *Psychological Review, 95*, 15-48.

Wang, Q., Cavanagh, P., & Green, M. (1994). Familiarity and pop-out in visual search. *Perception & Psychophysics, 56*, 495-500.

Williams, D. E., & Reingold, E. M. (1999). Preattentive guidance of eye movements during triple conjunction search tasks. Manuscript submitted for publication.

Williams, L. G. (1967). The effects of target specification on objects fixated during visual search. *Acta Psychologica, 27*, 355-360.

Wolfe, J. M. (1994). Guided search 2.0: A revised model of visual search. *Psychonomic Bulletin & Review, 1*, 202-238.

Wolfe, J. M., Cave, K. R., & Franzel, S. L. (1989). Guided search: An alternative to the feature integration model for visual search. *Journal of Experimental Psychology: Human Perception and Performance, 15*, 419-433.

Zelinsky, G. J. (1996). Using eye saccades to assess the selectivity of search movements. *Vision Research, 36*, 2177-2187.

The Optimal Behaviour of a Split Model of Word Recognition Resembles Observed Fixation Behaviour

Richard Shillcock (rcs@cogsci.ed.ac.uk)
T. Mark Ellison (marke@cogsci.ed.ac.uk)
Padraic Monaghan (pmon@cogsci.ed.ac.uk)
Institute for Adaptive and Neural Computation,
Division of Informatics, University of Edinburgh,
2 Buccleuch Place, Edinburgh, EH8 9LW, U.K.

Abstract

We expand upon the case for believing that the initial precise splitting of the foveal projection to the visual cortex fundamentally conditions the whole process of visual word recognition. We explore the optimal behaviour of a split architecture that attempts to divide its processing load equally between its two halves. We successfully model three aspects of fixation behaviour in human readers: (a) the positioning of the optimal viewing position to the left of the midpoint of the word, (b) a displaced Gaussian curve of letter-report accuracy resembling an RVF advantage, (c) the tendency for shorter words not to be directly fixated.

Modelling visual word recognition

The connectionist modelling of visual word recognition in the tradition of the Seidenberg and McClelland developmental model (1989) has led to a richer and deeper understanding of both normal and impaired visual word recognition over the last decade. However, such modelling has not consistently addressed issues of fixation behaviour in reading, for the reason that this tradition of modelling has essentially conceived of word recognition as the abstract mapping between orthographic, phonological and semantic representations, with all of the orthographic input being simultaneously and completely available to the model when any particular word is presented for processing. As with much other research in visual word recognition, this research begins with the abstract problem of distinguishing a target word from 50,000 other words in the lexicon, and the problem is not constrained by what is known about the human visual system.

In contrast, we have argued that the cognitive modelling of word recognition may be advanced by beginning the modelling at an earlier, anatomically constrained stage – that of the precise splitting of a word when it is initially presented to the fovea (Shillcock & Monaghan, 1998). We have demonstrated that "vertically" split, small-scale feedforward networks exhibit behaviour relevant to word recognition when such networks are required to co-ordinate input that straddles the split in the architecture. For instance, such a model automatically prioritises the processing of the end-letters of words (cf. Jordan, 1990), the model's learning and final performance exhibit superadditivity in regard to the separate and combined functioning of the two halves of the model (cf. Banich & Belger, 1990), and its learning evidences a bilateral effect when the same word is presented to each hemifield (see Mohr, Pulvermüller & Zaidel, 1994) (For further discussion, see Shillcock & Monaghan, *in press*; Shillcock & Monaghan, *submitted*). Below we rehearse some of the arguments in favour of this claimed role for foveal splitting in word recognition and we then compare the optimal behaviour of such a model with the observed behaviours of human readers.

A split fovea conditions word recognition

Research with commissurotomy patients reveals that the human fovea is precisely vertically split (Fendrich & Gazzaniga, 1989; Fendrich, Wessinger & Gazzaniga, 1996; Sugishita, Hamilton, Sakuma & Hemmi, 1994). When a word is fixated the part of the word in the left visual field (LVF) goes initially to the right hemisphere (RH) and the part in the right visual field (RVF) goes to the left hemisphere (LH). After this initial projection we claim that substantial processing directly relevant to word recognition occurs *intra*hemispherically before any putative complete sharing of information from the two hemifields. The argument for this claim can only be summarised here, but includes the following points.

(a) There are robust hemispheric differences in word recognition, which would not be apparent if all word recognition occurred after complete sharing of information (Hellige, 1995; Mohr, Pulvermüller & Zaidel, 1994).

(b) Cortical processing involves orchestrations of activity, replete with recurrent connectivity, rather than unidirectional cascades of processing towards a single abstract goal (see, e.g., VanEssen & Felleman, 1991). Activity in the (neatly split) primary visual cortex seems to be crucial for visual awareness (see, e.g., Cowey & Stoerig, 1992).

(c) Cells in the visual cortex require their receptive fields to contain the vertical midline in order to have a direct callosal connection (Whitteridge, 1965).

(d) Even very high level lexical processing (reading, oral spelling, mental rotation of words) is frequently impaired by unilateral neglect in a predictable spatial manner (Hillis & Caramazza, 1990).

(e) There is direct connectivity between the RH's inferior-temporal cortex and the LH's language regions (DiVirgilio & Clarke, 1998).

(f) The word-beginning superiority effect interacts with hemispheric dominance for language for foveally presented stimuli (Brysbaert, 1994). In particular, see Brysbaert (1994) for further discussion of the issue of foveal splitting.

The evidence cited above indicates that researchers in visual word recognition should explore the cognitive consequences of foveal splitting. As we will demonstrate below, the parsimony of the cognitive modelling accounts based on a generic "Split Model" bears out the claim that foveal splitting fundamentally conditions even the high-level cognitive aspects of visual word recognition.

A split word is already informative

Let us begin by assuming that a word is fixated somewhere close to its midpoint. We have seen that the two halves of the word are then projected to different hemispheres. Conventional modelling of word recognition assumes two partly contradictory next steps: first, that the information should all be moved into one hemisphere, and second, that detailed information is represented about the location of all the individual letters. We can assess these two proposals by considering just how much information is in our split starting position, which in effect resembles a model of word recognition with just two letter locations, the RVF and the LVF. If each word is split as close to its midpoint as possible, then *carpet* is split as $[a, c, r]$ and $[e, p, t]$, ignoring the precise order of the letters. This already gives us enough letter and location information to identity *carpet* uniquely. In contrast, $[i, t]$ and $[e, m]$ is ambiguous between *item* and *time*, and we need more letter-location information to distinguish between these two possibilities. What proportion of the lexicon is unambiguously specified, like *carpet*, by such a two-slot model?

The answer is that 98.6% of English words in the CELEX database (Baayen, Pipenbrock & Gulikers, 1995) are like *carpet*. This result is intuitively plausible for longer words, so let us consider four-letter words only; four-letter words are the most labile case in this analysis. When no information is given about letter position in four-letter words, then 34% are ambiguous (i.e. they are anagrams of other words). When four-letter words are split, the RVF *versus* LVF information disambiguates all but 4.7% of words (like *item*). These data constitute an empirical result. If the processor can fixate near the mid-point of a word, the split information is already very informative of word identity. A split word is an attractive goal rather than an inconvenience caused by the anatomy and something to be transcended.

An ambiguity level of 4.7% is still not ideal and more information is required. Happily this information is readily available from an aspect of split architectures that we have described elsewhere (Shillcock & Monaghan, 1998; Shillcock & Monaghan, *submitted*): split architectures naturally prioritise the processing of the outside letters of words when they are trying to co-ordinate inputs that are typically spread across the two halves of the model. Identifying the outer letters disambiguates all of the four-letter words and leaves the greatest ambiguity at five-letter words, of which only 0.62% remain ambiguous like *trial* and *trail*.

Thus, we see that splitting words is an informationally attractive precursor to identification, and that fast normal visual word recognition may not need exhaustive information about letter location. We now consider the further processing implications of split architectures by exploring their optimal behaviour.

Optimal behaviour of a split model

One of the factors affecting the behaviour of a split network concerns the balancing of the independent activity in its two halves. *Ceteris paribus*, such a network might be expected to function optimally if labour is evenly balanced between its two halves. A simple approximation to this state of affairs would emerge from fixating each word at its physical midpoint; each half of the model would have to cope with the same quantity of orthographic input We term this fixation strategy the *middle-justified strategy*. Alternatively. the model could arrange to fixate each word so that the left end of each word appears at the left end of a putative foveal window. This *left-justified strategy* is actually suggested by Legge et al. (1997) as an approximation of the behaviour of their model of fixation behaviour. A further (hypothetical) model might be a *right-justified strategy* .

Let us assume complete control over word fixation, so that each word is individually fixated at the optimal location (Although this strategy presupposes prior knowledge of the word's identity, such an analysis gives us an illuminating picture of the optimal behaviour of a split model.) In this case, the optimal fixation point for each word in the lexicon will be one that creates the minimum competition in the right and in the left half of the model. We developed a splitting algorithm that iterated through all the words in the lexicon, in random order each time, moving the split-point (which equals the fixation point) so as to minimise l/lr, in which l and r are respectively the numbers of occurrences of the left and right half of the word in the left and right halves of the split words in the lexicon[1]. The algorithm begins with a random placement of the split-point in every word, and converges after many iterations on one of a number of qualitatively closely comparable outcomes. Figure 1 shows the mean of ten different runs of the algorithm.

Each line of the graph shows the distribution of split-points for words of a particular length. The graph shows that the mean split-point is to the left of the physical midpoint of the word, and that this split-point predominates for the longer words. (We discuss the shorter words below.) This result matches the observation that the starts of English words are typically more informative than their ends. It confirms that a principle of dividing the processing equally between the two halves of the model produces a psychologically realistic outcome for English: word recognition is better for most words when fixated to the left of the midpoint, at an *optimal viewing position* (OVP) (O'Regan, 1990). One heuristic that approximates this optimal behaviour is to fixate words slightly to the left of their midpoints. The orthodox interpretation of the OVP is that it is the result of the positioning of a limited, high resolution foveal window over the most informative part of the word. Our results demonstrate that the OVP may equally be interpreted as a split processor attempting to divide the processing load evenly between its two halves.

The splitting algorithm produces detailed behaviour that further approximates the observed performance of human readers. The algorithm is free to choose a fixation point adjacent to any letter in the current word, the choice being determined solely by the reduction of l/lr. On some occasions the algorithm places the split-point to the left of the first letter or to the right of the last letter of the word: ‖*rip* or *the*‖. In such a case, the algorithm has placed the

---

[1] Both l and r can only fall to 1, representing their occurrence in the current word.

word exclusively in one hemifield, balancing the uninformativeness of an empty hemifield against the increased informativeness of, for instance, three letters rather than two in the other hemifield. As Figure 2 shows, this aspect of the optimal behaviour of the split model resembles the observed human failure to fixate short words, in inverse proportion to word-length (Rayner & McConkie, 1976).

This analysis provides a novel explanation for failure to directly fixate words in text: it is inappropriate to fixate on a word that is so short that its division gives the two halves of the processor only very short, multiply ambiguous parts

of words. We propose that this explanation contributes to the observed human behaviour along with other accounts such as the increased predictability of the (predominantly short) function words and the parafoveal preview of short words.

We can see some of the computational implications of the different fixation strategies, introduced above, by arranging every word in the lexicon "vertically", one above the other, in the way suggested by each strategy and then calculating the entropy of the distribution of letters in each

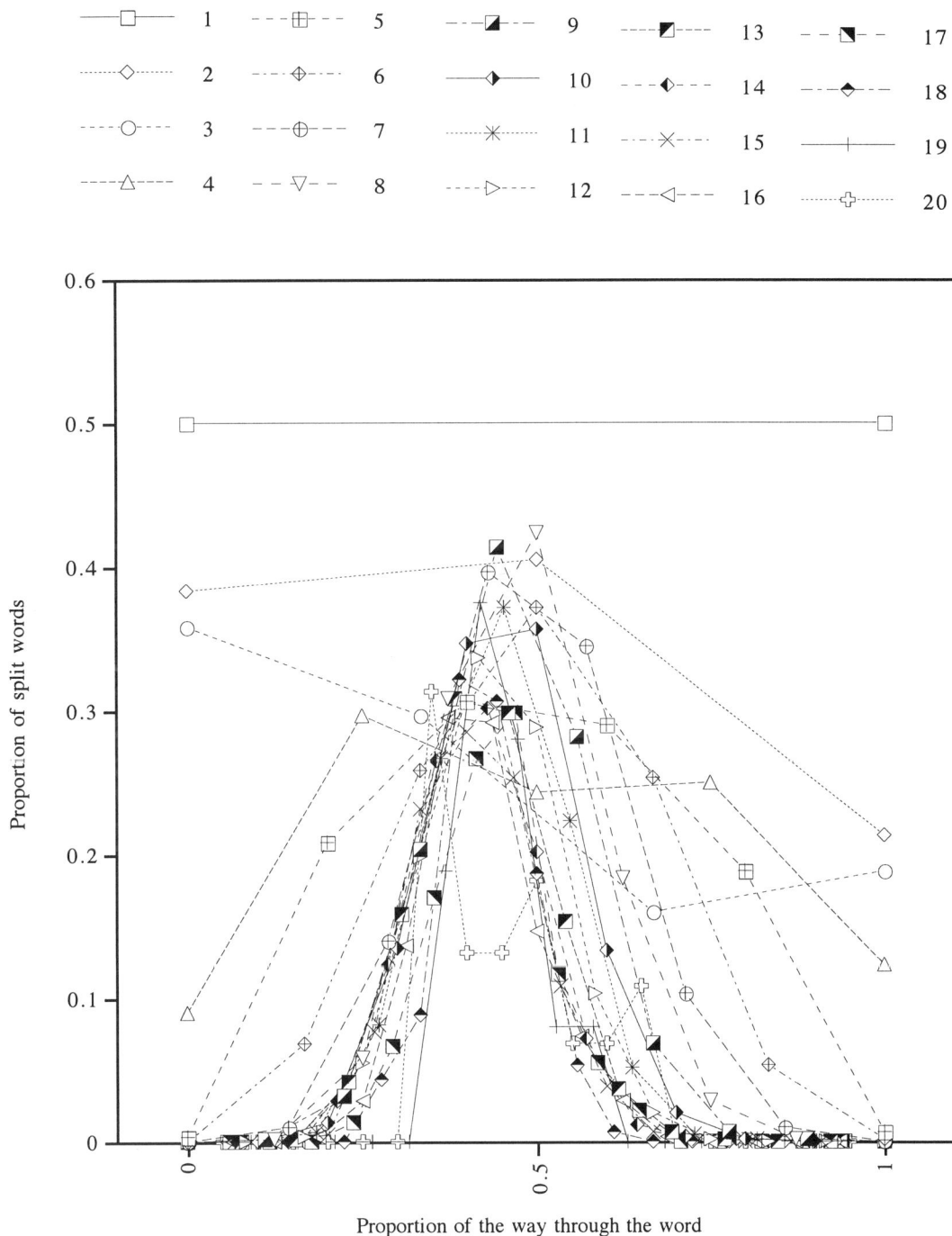

Figure 1: The curves, for each word-length, of the proportion of words optimally split at different proportions of their total length.

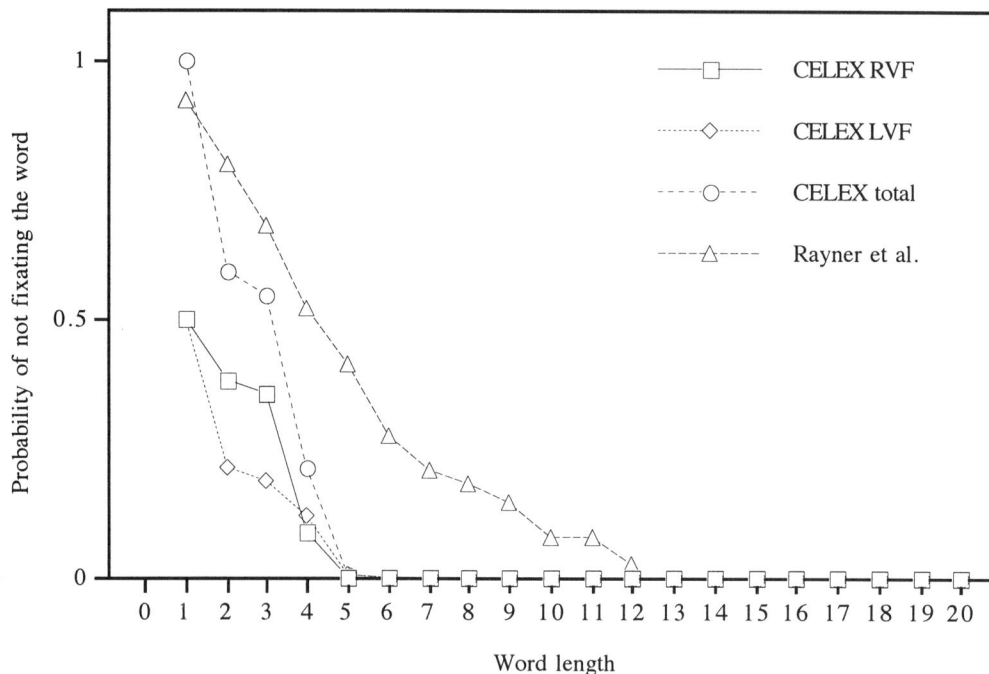

Figure 2. The probability of a word not receiving a direct fixation, as a function of word length. The "CELEX" curves show the probability that the splitting algorithm positions the word completely in the LVF or RVF. The human data are from Rayner and McConkie (1976).

vertically aligned letter location. Thus, for the left-justified strategy, the leftmost column of letters is the first letter of each word in the lexicon, and the entropy of this distribution reflects the skewedness of this distribution. For the middle-justified strategy and for the optimal strategy produced by the splitting algorithm, the fixation point in each word is vertically aligned. We present the results of this analysis in Figure 3. The entropy curves allow us to compare the different fixation strategies. A consistently high curve means that there is high quality information spread across the relevant part of the visual field.

First, note that in Figure 3 the left- and right-justified curves are arbitrarily positioned and may be moved along the *x*-axis to align them with the respective ends of any foveal window. As Figure 3 shows, the middle- and optimally-split curves provide the most symmetrical spread of information across the visual field, although there is no significant difference between the total entropy of the left- and middle-justified strategies. The optimal strategy, as might be expected from the nature of the algorithm, produces the most sustained spread of high quality information across the visual field. Note that the optimal strategy does not need to imply that reliable letter recognition occurs in a window as wide as 26 letters, only that high quality information can be effectively spread across the visual field. Such a spread of information maximises the utility of any parafoveal processing. It naturally emerges from the splitting algorithm that a span of 26 letters is an appropriate span; this span matches observed human behaviour (McConkie & Rayner, 1975). Finally, we might note that the entropy curve from the optimal strategy is asymmetric, in that the curve in the

RVF falls off less sharply than that in the LVF. Brysbaert, Vitu and Schroyens (1996) claim that letter-report accuracy (in the form of an extended OVP curve) follows a Gaussian distribution across both foveal and parafoveal parts of the visual field; they claim that the reported RVF advantage for word recognition may be subsumed into this EOVP curve. The entropy curve for the optimal strategy in Figure 3 is in agreement with this EOVP curve: letters are reported at different locations with respect to the fixation point with an accuracy in proportion to the probability that useful information may be found at that position. The optimal-strategy curve in Figure 3 indicates this probability. Thus, the EOVP curve may be accounted for without any necessary role for LH dominance for language.

Discussion

We have shown that a word split between the two hemifields and between the two hemispheres is potentially a very informative starting point for visual word recognition. A midpoint split is a simple fixation strategy but it provides an effective starting point for word recognition. A more sophisticated, optimal split-point for each word illustrates the upper bound for a split processor that is attempting an equal division of labour. This optimal strategy has many features that closely correspond to reading behaviour in human subjects, suggesting that skilled readers have developed fixation behaviour that is perhaps closer to the optimal behaviour than is the simple midpoint split strategy.

One possible way in which this distributional analysis of a full-sized lexicon might be related to the connectionist

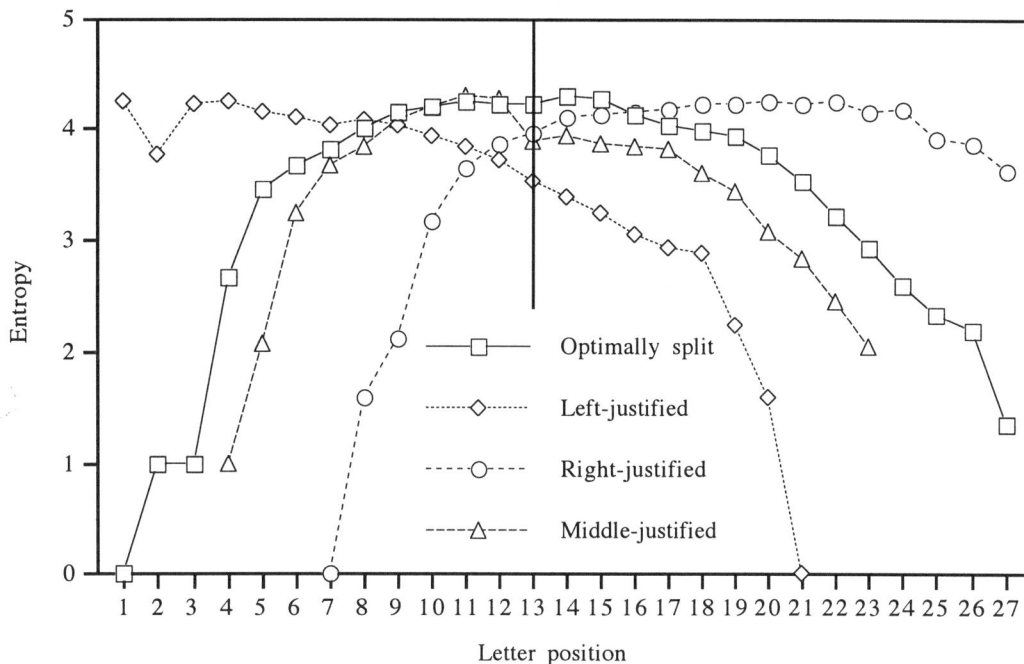

Figure 3: Entropy curves for the optimal splitting strategy, versus other fixation strategies.

modelling with small-scale networks we describe elsewhere is for us to assume that the processor can indeed deal with any word presented at any position relative to the fixation point, but that for each word type there is a strongest, best template situated in one particular position, contingently defined by the rest of the words that must be stored.

Conclusions

We have shown that if we follow through the implications of the initial precise splitting of the foveal projection, to consider the effects on the later stages of visual word recognition, then psychologically realistic behaviours result. These behaviours emerge naturally from the proposed Split Model. We have provided accounts for a number of different phenomena in visual word recognition, for which alternative yet separate accounts previously existed. The coherence of the accounts based on foveal splitting, and their grounding in known anatomy are arguments for favouring these accounts of the phenomena and for further exploring the claim that foveal splitting fundamentally determines visual word recognition at all levels.

References

Baayen, R.H., Pipenbrock, R. & Gulikers, L. (1995). The CELEX Lexical Database (CD-ROM). Linguistic Data Consortium, University of Pennsylvania, Philadelphia, PA.

Banich, M.T. & Belger, A. (1990). Interhemispheric interaction: How do the hemispheres divide and conquer a task? *Cortex*, **26**, 77–94.

Brysbaert, M. (1994). Interhemispheric transfer and the processing of foveally presented stimuli. *Behavioural Brain Research*, **64**, 151–161.

Brysbaert, M., Vitu, F. & Schroyens, W. (1996). The right visual field advantage and the optimal viewing position effect: On the relation between foveal and parafoveal word recognition. *Neuropsychology*, **10**, 385– 395.

Cowey, A. & Stoerig, P. (1992). Reflections on blindsight. In *The neuropsychology of consciousness* (Eds. A.D. Milner & M.D. Rugg), Academic Press, London.

DiVirgilio, G. & Clarke, S. (1997). Direct interhemispheric visual input to human speech areas. *Human Brain Mapping*, *5*, 347–354.

Fendrich, R. & Gazzaniga, M.S. (1989). Evidence of foveal splitting in a commissurotomy patient. *Neuropsychologia*, Vol. 27, No. 3, pp. 273-281.

Fendrich, R., Wessinger & Gazzaniga, M.S. (1996). Nasotemporal overlap at the retinal vertical meridian – investigations with a callosotomy patient. *Neuropsychologia*, **34**, 637–646

Hellige (1995). Coordinating the different processing biases of the left and right cerebral hemispheres. In F. Kitterle, (Ed.) *Hemispheric communication: Mechanisms and models*. Law. Erl. Assoc., Hove.

Hillis, A.E. & Caramazza, A. (1990). The effects of attentional deficits on reading and spelling. In A. Caramazza (Ed.) *Cognitive Neuropsychology and Neurolinguistics: Advances in Models of Cognitive Function and Impairment*. Law. Erl. Assoc., Hove, UK.

Jordan, T.R. (1990). Presenting words without interior letters: Superiority over single letters and influence of postmark boundaries. *Journal of Experimental Psychology: Human Perception and Performance*, *16*, 893–909.

Legge, G.E., Klitz, T.S. & Tjan, B.S. (1997). Mr. Chips: An ideal-observer model of reading. *Psychological Review*, *104*, 524–553.

McConkie, G.W. & Rayner, K. (1975). The span of the effective stimulus during a fixation in reading. *Perception & Psychophysics*, **17**, 578–586

Mohr, B., Pulvermüller, F. & Zaidel, E. (1994). Lexical decision after left, right and bilateral presentation of function words, content words and non-words: Evidence for interhemispheric interaction. *Neuropsychologia, 32,* 105–124.

O'Regan, J.K. (1990). Eye movements and reading. In E. Kowler (Ed.) *Eye movements and their role in visual and cognitive processes,* North Holland: Elsevier Science Pub.

Rayner, K. & McConkie, G.W. (1976). What guides a reader's eye movements? *Vision Research,* **16,** 829–837.

Seidenberg, M. S. & McClelland, J. L. (1989). A distributed, developmental model of word recognition and naming. *Psychological Review,* 96, 523-568.

Shillcock, R.C. & Monaghan, P. (*submitted*). The computational exploration of visual word recognition in a split model.

Shillcock, R.C. & Monaghan, P. (1998). Using anatomical information to enrich the connectionist modelling of normal and impaired visual word recognition. *Proceedings of the 1998 Cognitive Science Society Conference,* Wisconsin, 945–950.

Shillcock, R., Monaghan, P. & Ellison, T.M. (*in press*). The SPLIT model of visual word recognition: Complementary connectionist and statistical cognitive modelling. Proceedings of the 5th Neurocomputation and Psychology Workshop (eds. G. Humphreys et al.), Springer-Verlag.

Sugishita, M., Hamilton, C.R., Sakuma, I. & Hemmi, I. (1994). Hemispheric representation of the central retina of commissurotomized subjects. *Neuropsychologia, 32,* 399–415.

VanEssen, D.C. & Felleman, D.J. (1991). Distributed hierarchical processing in the primate cerebral cortex. *Cerebral Cortex,* **1,** 1–47.

Whitteridge, D. (1965). Area 18 and the vertical meridian of the visual field. In (ed. E.G. Ettlinger) *Functions of the Corpus Callosum.,* pp. 115–120. Churchill, London.

Consonance Network Simulations of Arousal Phenomena in Cognitive Dissonance

Thomas R. Shultz (shultz@psych.mcgill.ca)
Department of Psychology; McGill University
Montreal, QC H3A1B1 Canada

Mark R. Lepper (lepper@psych.stanford.edu)
Department of Psychology; Stanford University
Stanford, CA 94305-2130 USA

Abstract

The consonance constraint satisfaction model, recently used to simulate the major paradigms of cognitive dissonance theory, is extended to deal with emotional arousal phenomena in dissonance. The impact of arousing drugs is implemented in the simulations by a scalar that modulates the intensity of unit activations representing the relevant cognitions and the connection weights representing their implications. The simulations show that even exotic dissonance phenomena can be explained in terms of the relatively common process of constraint satisfaction.

For centuries, cognitive consistency has been a battleground in the continuing philosophical debates over human rationality. Consistency was often seen as a hallmark of reason, or as a necessary condition for a set of beliefs or arguments to be considered logical. Alternatively, consistency was dismissed as wooden-headed conservatism—the "hobgoblin of small minds" or the "last refuge of the unimaginative."

Within psychology, cognitive consistency has similarly played a major role in debates about human reason (Abelson, 1971; Lepper, 1994). In many of the early consistency models, pressures toward balance (Heider, 1946, 1958), congruity (Osgood & Tannenbaum, 1955), or symmetry (Newcomb, 1953) in a person's cognitive system were viewed as expressions of a logical need to achieve a coherent understanding of a sometimes contradictory world (McGuire, 1960). In contrast, other consistency models, especially Festinger's (1957) theory of cognitive dissonance, focused on the ways in which the human desire to avoid inconsistency (dissonance) could lead to behavior that appeared fairly irrational. In this latter tradition, people were seen more as "rationalizing," than as "rational," creatures (Aronson, 1969). This emphasis on irrationality in dissonance theory contributed to the view that dissonance phenomena were noticeably different, even somewhat exotic, compared to other everyday psychological phenomena.

After a considerable period of relative quiescence, interest in cognitive consistency in general, and cognitive dissonance in particular, has recently been rekindled, from two quite disparate directions (e.g., Harmon-Jones & Mills, 1999).

One source of renewed interest in cognitive consistency has come from attempts to model relationships among social attitudes and beliefs in computational terms (e.g., Read & Miller, 1994, 1998; Shultz & Lepper, 1996, 1998; Spellman, Ullman, & Holyoak, 1993). Shultz and Lepper (1996, 1998), for example, modeled a variety of phenomena from some of the central and most robust cognitive dissonance paradigms. Their so-called "consonance" model uses connectionist neural networks operating by the principle of constraint satisfaction to simulate traditional findings in cognitive dissonance. Their model captured the results of a number of classic "insufficient justification" phenomena, including the effects of threats of punishment on liking for a forbidden toy (e.g., Freedman, 1965), the consequences of "forced compliance" with a request to engage in counter-attitudinal activities (e.g., Linder, Cooper, & Jones, 1967), and the psychological effects of initiations (e.g., Gerard & Mathewson, 1966). It also captured basic phenomena concerning the consequences of making a free choice (e.g., Brehm, 1956) and predicted new free-choice effects that were subsequently confirmed in further psychological experimentation (Shultz, Léveillé, & Lepper, 1999).

These simulations suggested that even the apparently exotic, counter-intuitive, and irrational effects emphasized by early dissonance theorists could be interpreted in terms of considerably more mundane cognitive processes. In several cases, psychological phenomena were covered more accurately in the simulations than they were by classical dissonance theory. The superior coverage of the consonance model was due to the inclusion of constraints not present in dissonance theory and to the increased precision inherent to a computational formulation. The success of the consonance model allows a reinterpretation of cognitive dissonance and its reduction that emphasizes what it has in common with many other psychological phenomena operating according to constraint satisfaction principles (Holyoak & Thagard, 1989; Kintsch, 1988; Read & Miller, 1994, 1998; Rumelhart, Smolensky, McClelland, & Hinton, 1986; Sloman, 1990; Spellman & Holyoak, 1992; Spellman et al., 1993; Thagard, 1989).

A second main source of the recent resurgence of interest in consistency phenomena has come from social psychologists wanting to highlight and investigate the affective and motivational properties of cognitive dissonance (Elliot & Devine, 1994; Harmon-Jones & Mills, 1999). Some of these efforts involved attempts to study the critical role of actual physiological arousal in the production and reduction of dissonance (e.g., Cooper & Fazio, 1984).

The present paper attempts to integrate these two newer directions in cognitive dissonance research by presenting simulations, based on the consonance model, of key phenomena in the more recent dissonance literature concerning the emotionally-arousing properties of cognitive dissonance. Simulations of this sort can provide a relatively precise and reliable extension of theory to explain empirical psychological findings (Smith, 1996).

Arousal and Dissonance

Even in the earliest dissonance research, it was postulated that dissonance involves a state of aversive affective arousal (Festinger, 1957). In more recent years, a number of experiments supported this notion (Cooper & Fazio, 1984). Thus, dissonance energizes dominant responses, just as other arousal states do (Pallak & Pittman, 1972), and produces actual physiological arousal (Croyle & Cooper, 1983). Even more dramatically, the presence and size of dissonance effects vary with independent manipulations of arousal.

For example, Cooper, Zanna, and Taves (1978) directly manipulated arousal with drugs and showed the necessity of arousal for subsequent attitude change. Cooper et al.'s participants were asked to write an essay that went against their own attitudes. The essay favored pardoning former President Richard Nixon—a policy with which virtually all participants strongly disagreed. The participants were given either high or low choice to write these counter-attitudinal essays, under three different drug conditions. In the context of an earlier experiment, however, the participants had each just taken a pill that they had been told was a harmless placebo. Actually, however, the drug was either a placebo, a "downer" (Phenobarbital), or an "upper" (amphetamine).

Results from the Cooper et al. study are shown in Figure 1. In the placebo condition, there was the usual dissonance effect, i.e., more attitude change in the direction of the essay under high choice than under low choice. This effect of choice was eliminated in the "downer" condition, where there was very little attitude change regardless of choice, and enhanced in the "upper" condition, producing a significant dissonance effect even under low choice. These results became the focus of our simulations.

The Consonance Model

The consonance model is based on the idea that dissonance reduction can be interpreted as a constraint satisfaction problem. The motivation to seek cognitive consistency that is postulated by dissonance theory and related theories can

be viewed as imposing constraints on the beliefs and attitudes that an individual holds at any given moment (e.g., Abelson, Aronson, McGuire, Newcomb, Rosenberg, & Tannenbaum, 1968; Abelson & Rosenberg, 1958; Feldman, 1966). Such consistency problems can be solved by satisfying multiple soft constraints, conditions that are desirable, but not essential, to satisfy and which may vary in their relative importance to the individual.

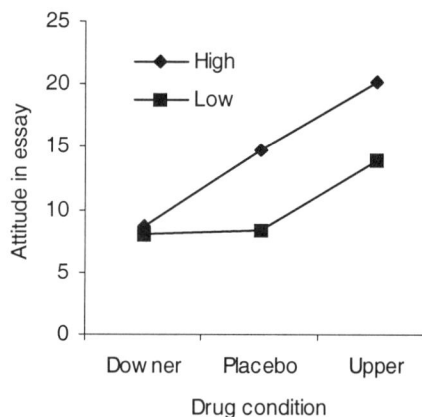

Figure 1: Mean attitude as a function of choice and drug (from Cooper et al., 1978).

In this model, consonance networks correspond to an individual's representation of the situation created in the various conditions of a cognitive dissonance experiment. Unit activations represent the direction and strength of the person's attitudes and beliefs. Units can also differ in their resistance to activation change, reflecting differences in the extent to which particular cognitions are supported by other cognitions or are clearly anchored in reality. Connection weights between different cognitions represent psychological implications among the person's attitudes and beliefs. The connections between any two units can be excitatory, inhibitory, or nonexistent. Both unit activations and connection weights can vary across the different conditions of a single experiment.

Consonance is roughly the degree to which similarly evaluated units are linked by excitatory weights and oppositely valued units are linked by inhibitory weights. More formally, the consonance contributed by a particular unit i is:

$$consonance_i = \sum_j w_{ij} a_i a_j \qquad (1)$$

where w_{ij} is the weight between units i and j, a_i is the activation of the receiving unit i, and a_j is the activation of the sending unit j. The overall consonance in a given network is the sum of the values given by Equation 1 over all receiving units in that network:

$$consonance_n = \sum_i \sum_j w_{ij} a_i a_j \qquad (2)$$

Activations change over time cycles in order to satisfy constraints and increase consonance. Activation spreads over time cycles by two update rules:

$$a_i(t+1) = a_i(t) + net_i(ceiling - a_i(t)) \text{, when } net_i \geq 0 \quad (3)$$

$$a_i(t+1) = a_i(t) + net_i(a_i(t) - floor) \text{, when } net_i < 0 \quad (4)$$

where $a_i(t+1)$ is the activation of unit i at time $t + 1$, $a_i(t)$ is the activation of unit i at time t, $ceiling$ is the maximal level of activation, $floor$ is the minimal activation, and net_i is the net input to unit i, defined as:

$$net_i = resist_i \sum_j w_{ij} a_j \quad (5)$$

The parameter $resist_i$ indexes the resistance of receiving unit i to having its activation changed.

At each time cycle during the simulation, n units are randomly selected and updated according to Equations 3 and 4. For most of our simulations, n is the number of units in the network. The update rules described in Equations 3-5 ensure that consonance increases or stays the same across time cycles. When consonance reaches an asymptote, the updating process is stopped.

The design of consonance networks to implement particular dissonance experiments follows a set of five principles that map cognitive dissonance theory to the consonance model. Principle 1 specifies that a cognition is implemented by the net activation of a pair of negatively connected units, one representing the positive pole and the other representing the negative pole. This permits the model to deal with both conflict and ambivalence. Net activation for the cognition is the difference between activation of the positive unit and activation of the negative unit. Activations range from a floor to a ceiling. In our dissonance simulations, the floor parameter has a default value of 0. The ceiling parameter is 1 for positive poles, and 0.5 for negative poles.[1]

Principle 2 specifies that connections between cognitions are based on inferred causal implications between those cognitions (Abelson, 1968). Connection weights range from -1 to +1, with 0 representing a lack of causal relation. When two cognitions are positively related, their positive poles are connected with excitatory weights, as are their negative poles. Inhibitory weights connect the positive pole of one cognition with the negative pole of the other cognition. These connections are exactly reversed for cognitions that are negatively related. Each unit has an inhibitory self-connection specified by the *cap* parameter, and all connection weights are bi-directional. Connection weights have a default value of 0.5.

Principle 3 specifies that the total amount of dissonance in a network is the negative of total consonance divided by r, which is the number of nonzero relations among cognitions:

$$dissonance = \frac{-\sum_i \sum_j w_{ij} a_i a_j}{r} \quad (6)$$

Dividing by r standardizes dissonance across networks by controlling for the number of relevant relations. Self-connections are excluded from this computation of dissonance. This definition of dissonance differs from Festinger's (1957) definition, not only because it is formalized, but also because it measures the amount of dissonance within each inter-cognition relation, includes within-cognition ambivalence, and varies even when all relations are dissonant or all relations are consonant.

Principle 4 maps dissonance reduction to activation updates by specifying that networks settle into more stable, less dissonant states as unit activations are updated with Equations 3, 4, and 5. The *cap* parameter with a default value of -0.5, corresponding to the value of the connection between each unit and itself (w_{ii}), prevents activations from reaching the ceiling.

Principle 5 specifies that cognition unit activations, but not connection weights, are allowed to change, and that some cognitions are more resistant to change than others, as implemented in Equation 5. Typically, beliefs, behaviors, and justifications are more resistant to change than are evaluations or attitudes. The *resist* parameter has default values of 0.5 for low resistance and 0.01 for high resistance. As specified in Equation 5, the larger the resistance multiplier, the more readily the unit changes its activation.

In order to assess the robustness of simulation results across variability in the specific parameters, weights, resistances, caps, and initial activations are all randomized by adding or subtracting a random proportion of their initial amounts. The *rand%* parameter specifies the proportion in which additions or subtractions are randomly selected with a uniform distribution. Randomization increases psychological realism in the sense that not everyone can be expected to have precisely the same parameter values. Randomization of weight values violates connection weight symmetry, such that $w_{ij} \neq w_{ji}$, and increases instability of network solutions (Hopfield, 1982, 1984). We often compare low (0.1), medium (0.5), and high (1.0) levels of *rand%* in order to assess the stability of network solutions (Shultz & Lepper, 1996). All of the default values mentioned above are consistent across all of our simulations. Additional details about the consonance model and discussions of its assumptions are available in other sources (Shultz & Lepper, 1996, 1998).

The Present Simulations

Network Design

The present simulations focused on the Cooper et al. (1978) experiment described earlier. Because there were no differential payments, as is typical in forced compliance experiments (e.g., Collins, 1973; Linder et al., 1967), there are only two cognitions to model: the attitude and the

---

[1] When neurons are organized into excitatory and inhibitory groups that respond in opposite ways to the same input, the activation range for excitatory neurons is usually greater for positive than for inhibitory neurons (Anderson, 1995, pp. 150-152).

counter-attitudinal essay. Initially, attitude is given a high negative activation (-0.5) because it is strongly against pardoning Nixon. The essay cognition starts with a high positive activation (0.5) because it is in favor of pardoning Nixon. Using Principle 2, we implement degree of choice by varying the strength of the connection between attitude and essay: high (0.5) vs. low (0.1). Both relations are positive because the more favorable one's attitude, the more likely it would be for one to support this position in writing.[2] Following Principle 5, writing the essay is given high resistance to change because it is a public and irrevocable behavior; attitude is given low resistance to change because it is a subjective evaluation.

The various drug conditions are implemented with a scalar value that multiplies the initial values of connection weights and cognitions: 1.0 for the placebo, 0.5 for the "downer," and 1.5 for the "upper" condition. The basic idea is that these arousal-modulating drugs dampen or boost everything in the system, connection weights and initial activations. This interpretation is consistent with the energizing properties of dissonance (Cooper & Fazio, 1984). The fact that even the low choice/"upper" participants said in manipulation checks that they felt a high level of choice (Cooper et al., 1978) underscores the importance of scaling the connection weights. This scaling is also consistent with evidence that phenobarbital depresses, whereas amphetamine increases, neural firing rates and synaptic transmission (Quastel, 1975). Initial network values for the placebo condition obviously do not change with a multiplier of 1.0.

Twenty networks were run in each of the six experimental conditions, each network having somewhat different parameter settings and a different randomly determined pattern of activation updates. Each network ran for 30 cycles, because dissonance and unit activation values typically reached asymptote by that time.

Results

Mean attitudes towards the view espoused in the essay are presented in Figure 2 at the lowest level of parameter randomization (0.1). The drug x choice interaction resembles that produced in Cooper et al.'s (1978) study. We performed an ANOVA in which drug, with three levels, and choice, with two levels, served as factors. As predicted, there was an interaction between drug and choice, $F(2, 114) > 900$, $p < .001$. The precise nature of this interaction was assessed using a contrast F with weights suggested by Cooper et al.'s human data. The contrast weights were -1, 0, and +3 for the high choice "downer," placebo, and "upper" conditions, respectively; and -1, -1, and 0 for the low choice "downer," placebo, and "upper" conditions, respectively. The contrast was highly significant $F(1, 114) > 12000$, $p <$

.001, accounting for 89% of the between-conditions variance.

To test for a significant dissonance effect, even under low choice, when amphetamines had been administered, we computed a regression F for only the three low choice cells, using weights of 0 for the high choice cells, and weights of -1, -1, and 2 for the low choice "downer," placebo, and "upper" conditions, respectively, $F(1, 114) > 500$, $p < .001$. Just as with Cooper et al.'s participants, a dissonance effect was found in the low choice, "upper" condition.

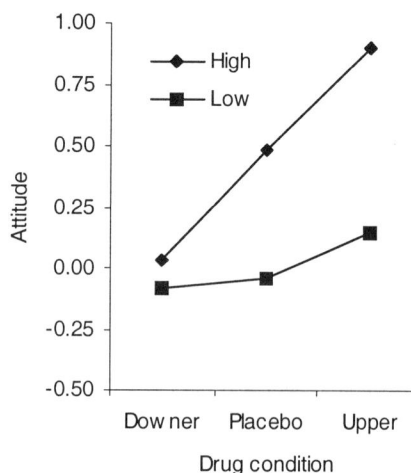

Figure 2: Mean simulated attitude as a function of choice and drug condition.

Mean simulated dissonance scores for the six conditions of the Cooper et al. experiment are plotted in Figure 3 over time cycles. A substantial amount of dissonance in the high choice "upper" condition is strongly reduced, a moderate amount of dissonance in the high choice placebo condition is moderately reduced, and a small amount of dissonance in the low choice "upper" condition is slightly reduced. The other three conditions show almost no dissonance and almost no dissonance reduction. All of these results held up at medium (0.5) and even at high (1.0) levels of parameter randomization.

Discussion

The consonance network simulations fit the human data from Cooper et al. quite precisely. There was the typical dissonance effect in the placebo condition (i.e., more attitude change in the direction of the views expressed in the essay under high choice than under low choice), very little attitude change in the "downer" condition, and enhanced attitude change in the "upper" condition. The only required change from our previous forced-compliance simulations was the inclusion of a scalar parameter to multiply initial values of activations and weights. Consistent with psychological and neurophysiological evidence, this scalar enhanced activations and weights in the "upper" condition

---

[2] In our previous simulation of the Linder et al. (1967) forced compliance experiment, which had no choice rather than low choice, we cut this link between attitude and essay to zero (Shultz & Lepper, 1996).

and dampened them in the "downer" condition, relative to the placebo control condition. Plots of dissonance reduction suggest that the amount of attitude change is a direct function of the amount of dissonance reduced in the networks.

It appears that as though the basic phenomena on arousal in dissonance can be captured with constraint satisfaction consonance networks, again suggesting that dissonance arousal and reduction has much in common with other constraint satisfaction processes.

Acknowledgments

This research was supported by a grant to the first author from the Social Sciences and Humanities Research Council of Canada and by grant MH-44321 to the second author from the U.S. National Institute of Mental Health. We are grateful for comments on an earlier draft by Sylvain Sirois and Dave Buckingham.

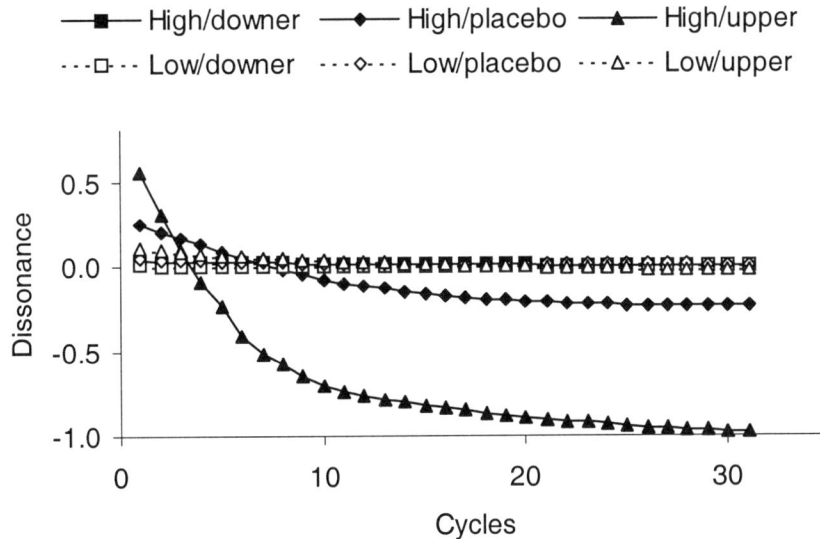

Figure 3: Simulated dissonance over time cycles.

References

Abelson, R. P. (1968). Psychological implication. In R. P. Abelson, E. Aronson, W. J. McGuire, T. M. Newcomb, M. J. Rosenberg, & P. H. Tannenbaum (Eds.), *Theories of cognitive consistency: A sourcebook*. Chicago: Rand McNally.

Abelson, R. P. (1971). Are attitudes necessary? In B. T. King & E. McGinnies (Eds.), *Attitudes, conflict, and social change*. New York: American Marketing Association.

Abelson, R. P., Aronson, E., McGuire, W. J., Newcomb, T. M., Rosenberg, M. J., & Tannenbaum, P. H. (Eds.) (1968). *Theories of cognitive consistency: A sourcebook*. Chicago: Rand McNally.

Abelson, R. P., & Rosenberg, M. J. (1958). Symbolic psycho-logic: A model of attitudinal cognition. *Behavioral Science, 3*, 1-13.

Anderson, J. A. (1995). *An introduction to neural networks*. Cambridge, MA: MIT Press.

Aronson, E. (1969). The theory of cognitive dissonance: A current perspective. In L. Berkowitz (Ed.), *Advances in experimental social psychology* (Vol. 4). New York: Academic Press.

Brehm, J. W. (1956). Post-decision changes in the desirability of choice alternatives. *Journal of Abnormal and Social Psychology, 52*, 384-389.

Collins, B. E. (1973). *Public and private conformity: Competing explanations by improvisation, cognitive dissonance, and attribution theories*. Andover, MA: Warner Modular Publications.

Cooper, J., & Fazio, R. H. (1984). A new look at dissonance theory. In L. Berkowitz (Ed.), *Advances in experimental social psychology* (Vol. 17). New York: Academic Press.

Cooper, J., Zanna, M. P., & Taves, P. A. (1978). Arousal as a necessary condition for attitude change following forced compliance. *Journal of Personality and Social Psychology, 36*, 1101-1106.

Croyle, R., & Cooper, J. (1983). Dissonance arousal: Physiological evidence. *Journal of Personality and Social Psychology, 45*, 782-791.

Elliot, A. J., & Devine, P. G. (1994). On the motivational nature of cognitive dissonance: Dissonance as psychological discomfort. *Journal of Personality and Social Psychology, 67*, 382-394.

Feldman, S. (Ed.) (1966). *Cognitive consistency*. New York: Academic Press.

Festinger, L. (1957). *A theory of cognitive dissonance.* Evanston, IL: Row, Peterson.

Freedman, J. L. (1965). Long-term behavioral effects of cognitive dissonance. *Journal of Experimental Social Psychology, 1,* 145-155.

Gerard, H. B., & Mathewson, G. C. (1966). The effects of severity of initiation on liking for a group: A replication. *Journal of Experimental Social Psychology, 2,* 278-287.

Harmon-Jones, E., & Mills, J. (Eds.). (1999). *Cognitive dissonance: Progress on a pivotal theory in social psychology.* Washington, DC: American Psychological Association.

Heider, F. (1946). Attitudes and cognitive organization. *Journal of Psychology, 21,* 107-112.

Heider, F. (1958). The psychology of interpersonal relations. New York: Wiley.

Holyoak, K. J., & Thagard, P. (1989). Analogical mapping by constraint satisfaction. *Cognitive Science, 13,* 295-355.

Hopfield, J. J. (1982). Neural networks and physical systems with emergent collective computational abilities. *Proceedings of the National Academy of Sciences, USA, 79,* 2554-2558.

Hopfield, J. J. (1984). Neurons with graded responses have collective computational properties like those of two-state neurons. *Proceedings of the National Academy of Sciences, USA, 81,* 3088-3092.

Kintsch, W. (1988). The role of knowledge in discourse comprehension: A construction-integration model. *Psychological Review, 95,* 163-182.

Lepper, M. R. (1994). "Hot" vs. "cold" cognition: An Abelsonian journey. In R. Schank & E. Langer (Eds.), *Beliefs, reasoning, and decision making: Psycho-logic in honor of Bob Abelson.* Hillsdale, NJ: Erlbaum.

Linder, D. E., Cooper, J., & Jones, E. E. (1967). Decision freedom as a determinant of the role of incentive magnitude in attitude change. *Journal of Personality and Social Psychology, 6,* 245-254.

McGuire, W. J. (1960). A syllogistic analysis of cognitive relationships. In W. J. Rosenberg, C. I. Hovland, W. J. McGuire, R. P. Abelson, & J. W. Brehm (Eds.), *Attitude organization and change.* New Haven, CT: Yale University Press.

Newcomb, T. M. (1953). An approach to the study of communicative acts. *Psychological Review, 60,* 393-404.

Osgood, C. E. & Tannenbaum, P. H. (1955). The principle of congruity in the prediction of attitude change. *Psychological Review, 62,* 42-55.

Pallak, M. S., & Pittman, T. S. (1972). General motivation effects of dissonance arousal. *Journal of Personality and Social Psychology, 21,* 349-358.

Quastel, J. H. (1975). Effects of drugs on energy metabolism of the brain and on cerebral transport. In L. L. Iverson, S. D. Iverson, & S. H. Snyder (Eds.), *Handbook of psychopharmacology: Vol. 5. Synaptic modulators.* New York: Plenum.

Read, S. J., & Miller, L. C. (1994). Dissonance and balance in belief systems: The promise of parallel constraint satisfaction processes and connectionist modeling approaches. In R. C. Schank & E. Langer (Eds.), *Beliefs, reasoning, and decision making: Psycho-logic in honor of Bob Abelson.* Hillsdale, NJ: Erlbaum.

Read, S. J., & Miller, L. C. (Eds.). (1998). *Connectionist models of social reasoning and social behavior.* Hillsdale, NJ: Erlbaum.

Rumelhart, D. E., Smolensky, P., McClelland, J. L., & Hinton, G. (1986). Schemata and sequential thought processes in PDP models. In D. E. Rumelhart & J. L. McClelland (Eds.), *Parallel distributed processing: Explorations in the microstructure of cognition* (Vol. 2). Cambridge, MA: MIT Press.

Shultz, T. R., & Lepper, M. R. (1996). Constraint satisfaction modeling of cognitive dissonance phenomena. *Psychological Review, 103,* 219-240.

Shultz, T. R., & Lepper, M. R. (1998). The consonance model of dissonance reduction. In S. J. Read & L. C. Miller (Eds.), *Connectionist models of social reasoning and social behavior.* Hillsdale, NJ: Erlbaum.

Shultz, T. R., Léveillé, E., & Lepper, M. R. (1999). Free-choice and cognitive dissonance revisited: Choosing "lesser evils" vs. "greater goods." *Personality and Social Psychology Bulletin, 25,* 40-48.

Sloman, S. (1990). *Persistence in memory and judgment: Part-set inhibition and primacy.* Unpublished Doctoral Dissertation, Stanford University.

Smith, E. R. (1996). What do connectionism and social psychology offer each other? *Journal of Personality and Social Psychology, 70,* 893-912.

Spellman, B. A., & Holyoak, K. J. (1992). If Saddam is Hitler, then who is George Bush? Analogical mapping between systems of social roles. *Journal of Personality and Social Psychology, 62,* 913-933.

Spellman, B. A., Ullman, J. B., & Holyoak, K. J. (1993). A coherence model of cognitive consistency: Dynamics of attitude change during the Persian Gulf War. *Journal of Social Issues, 49,* 147-165.

Thagard, P. (1989). Explanatory coherence. *Behavioral and Brain Sciences, 12,* 435-502.

Rule learning by Habituation can be Simulated in Neural Networks

Thomas R. Shultz (shultz@psych.mcgill.ca)
Department of Psychology; McGill University
Montreal, QC H3A 1B1 Canada

Abstract

Contrary to a recent claim that neural network models are unable to account for data on infant habituation to artificial language sentences, the present simulations show successful coverage with cascade-correlation networks using analog encoding. The results demonstrate that a symbolic rule-based account is not required by the infant data.

One of the fundamental issues of cognitive science continues to revolve around which type of theoretical model better accounts for human cognition -- a symbolic rule-based account or a sub-symbolic neural network account. A recent study of infant habituation to expressions in an artificial language claims to have struck a damaging blow to the neural network approach (Marcus, Vijayan, Rao, & Vishton, 1999). The results of their study show that 7-month-old infants attend longer to sentences with unfamiliar structures than to sentences with familiar structures.

Because of certain features of their experimental design and their own unsuccessful neural network models, Marcus et al. conclude that neural networks cannot simulate these results and that infants possess a rule-learning capability unavailable to neural networks. A companion article suggests that rule learning is an innately provided capacity of the human mind, distinct from associative learning mechanisms like those in neural networks (Pinker, 1999).

My paper presents neural network simulations of the key features of the Marcus et al. (1999) experiment, thus showing that their infant data do not uniquely support a rule-based account.

Psychological Evidence and One Interpretation

Marcus et al. (1999) present experiments in which 7-month-old infants habituate to three-word sentences in an artificial language and are then tested on novel sentences that are either consistent or inconsistent with those to which the infant has habituated. In one experiment, illustrated in the first three columns of Table 1, infants habituated to sentences exhibiting an ABA pattern, for example, *ga ti ga* or *li na li*. There were 16 of these ABA sentences, created by combining four A words (*ga, li, ni,* and *ta*) with four B words (*ti, na, gi,* and *la*). Then the infants were presented with two novel sentences that were consistent with the ABA pattern (*wo fe wo,* and *de ko de*) and two novel sentences that were inconsistent with ABA because they followed an ABB pattern (*wo fe fe,* and *de ko ko*). A second, control condition habituated infants to sentences with an ABB pattern, for example, *ga ti ti* and *ga na na*. Again, 16 such sentences were created by combining the four A words with the four B words. The test sentences were the same in this second condition, but here the novel ABB sentences were consistent and the novel ABA sentences were inconsistent with the habituated ABB pattern.

Table 1: Conditions and error in simulation of Experiment 1

Procedure	Condition 1	Condition 2	Mean	SE
Habituate	ABA	ABB		
Consistent	ABA	ABB	0.649	0.107
Inconsistent	ABB	ABA	1.577	0.088

The dependent measure was looking time. During the test phase, if the infant looked at a flashing light to her left or right, a test sentence was played from a speaker near that light. A test sentence was played over and over until the infant either looked away or until 15 s elapsed. Infants attended more to inconsistent novel sentences than to consistent novel sentences, indicating that they were sensitive to grammatical differences between the sentences.

Marcus et al. designed another experiment, described in the first three columns of Table 2, that contrasted habituation to ABB sentences with AAB sentences. The idea was to rule out the possibility that infants might have used the presence or absence of duplicated words to distinguish grammatical types in their other experiments. For example, ABA sentences duplicate no words, but ABB sentences do (by duplicating B). In this Experiment 3, both grammatical sequences have duplicated words.

Table 2: Conditions and error in simulation of Experiment 3

Procedure	Condition 1	Condition 2	Mean	SE
Habituate	ABB	AAB		
Consistent	ABB	AAB	0.570	0.100
Inconsistent	AAB	ABB	1.491	0.072

Infants performed in a similar fashion in both experiments, i.e., they attended more to inconsistent than to consistent novel sentences. All infants except one showed the predicted preference for inconsistent over consistent test sentences. The issue is the proper theoretical account of this grammatical knowledge -- is it based on rules or on connections?

Marcus et al. argue that these simple grammars could not be learned by a computational system that is sensitive only to transitional probabilities or event frequencies.

Transitional probabilities would not work because the transitional probabilities for novel words would be 0. Counting the numbers of duplicated words might work for Experiment 1, but not for Experiment 3, where both grammars had duplicate words. Nonetheless, Marcus et al. briefly mention their unsuccessful attempts to simulate these habituation data with simple recurrent networks such as those used by Elman (1990).

No details of the course of habituation of attention or the extent of recovery were reported by Marcus et al. Nor was there an implementation of a rule-based model to account for the habituation data or a theoretical analysis of how rule learning might be used in the computation of habituation. In any case, the challenge raised by Marcus et al. is interesting and worthwhile. It is interesting because habituation is important and still poorly understood, and worthwhile because of the implications for the fundamental debate on rules vs. connections.

Habituation as Encoding and Decoding in Neural Networks

One computational account of habituation has been in terms of the encoding and decoding processes involved in so-called encoder networks (Mareschal & French, 1997). Encoder networks have output units identical to their input units. Their task is to reproduce their inputs on their output units. With layer-to-layer connectivity, an encoder network must encode input signals onto a layer of hidden units and then decode the hidden unit representations onto the output units. If the number of hidden units is less than the number of input or output units, then the encoder network learns to abstract a compact representation of the problem on its hidden units. Such compact abstractions generalize to novel inputs and enable prototype phenomena and pattern completion skills (Hertz, Krogh, & Palmer, 1991).

How might encoder networks be related to habituation? The habituation technique is arguably the most important methodological advance in developmental psychology in this century. The reason for this is that habituation enables the systematic study of perceptual and conceptual abilities in non-verbal, response-impoverished infants (Cohen, 1979). Unlike the study of mere preferences, habituation can be used even when no preferences exist. Even if the infant exhibits no natural preferences between stimulus categories, such preferences can be experimentally introduced by habituating the infant to one category and measuring dishabituation to another contrasting category. The responses required to show habituation and dishabituation are available at birth (Slater, 1995). All that is required is visual attention, head turning, or something as passive as heart rate. These advantages have enabled dozens of discoveries of perceptual and cognitive abilities in young infants over the past 30 years using habituation methodology. Infants have been demonstrated to perceive color, form, complex patterns, faces, and intricate relations, to learn categories and prototypes, to perceive perceptual constancies, to know about object permanence and

causality, to identify objects, and to form both short-term and long-term memories for objects and events (e.g., Cohen, 1979; Haith, 1990; Oakes & Cohen, 1990; Quinn & Eimas, 1996). The memories identified in habituation studies are essentially recognition memories -- the recognition of a stimulus as being a member of a previously habituated category.

What is going on during the processes of habituation and dishabituation? The standard view is that infants gradually construct representational categories for stimuli that they encounter (Cohen, 1973; Sokolov, 1963). This category building is enabled by visual attention, as well as by other sensory modalities. Once a representational category is constructed via attention and processing, the infant no longer needs to attend so much to stimuli of that category. When the infant encounters a new stimulus, he compares it to stored representations of existing stimulus categories. If the new stimulus matches a stored category, then it will likewise elicit little or no attention. But if the new stimulus is not recognized as a member of an existing category, then it receives additional attention and processing. This is a system that seems adaptive in encouraging the infant to expend cognitive resources on novel information and thus continue to learn about the world.

There are many interesting aspects to the habituation literature. Among them is a tendency for attention to habituate gradually in a negatively accelerated fashion -- fast at first and then slowing down to an asymptote of no attention. The gradual decrease is perhaps a natural consequence of the fact that building representations in relatively naive infants takes time and effort. The negative acceleration is a natural consequence of the fact that attention to a stimulus may start at a high level and is bounded at none.

The basic idea enabling a link between habituation and encoder networks is that encoder networks model how one might learn about stimuli from attending to them. Relations among stimulus features are abstracted in the hidden unit representations as connection weights are adjusted. New stimuli that produce similar representations and little or no error are in a sense recognized as familiar. Those stimuli that produce different representations and large error are essentially unrecognized and considered as novel. The key assumption in this modeling of habituation with encoder networks is that network error corresponds to the need to direct current attentional resources (Mareschal & French, 1997). The theoretical contribution of this analysis is a computationally precise implementation of the representation and processing involved in habituation as presently understood.

The Current Model

Cascade-correlation

The proposed model of habituation uses an encoder version of the cascade-correlation learning algorithm. Cascade-correlation is an algorithm for learning in feed-forward neural networks (Fahlman & Lebiere, 1990). Unlike standard back-propagation networks, whose topologies are designed by hand and remain static as connection weights are adjusted, cascade-correlation networks grow as well as learn. They grow by recruiting new hidden units into the network as required to reduce error at the output units. New hidden units are recruited one at a time and installed each on a separate layer with input connections from the input units and from any existing hidden units. The candidate hidden unit that gets recruited is the one whose activations correlate most highly with the network's current error as the input weights to the candidates are adjusted.

Cascade-correlation also differs from standard back-propagation by using curvature as well as slope information from the error surface in making weight adjustments. This additional information about the error surface, which is approximated in a computationally efficient way, enables more decisive and effective weight adjustments.

Cascade-correlation was designed to solve two of the major problems with back-propagation -- slow learning and inability to learn some difficult problems. On average, it learns about 10-50 times faster than standard back-propagation, and it learns problems that are too difficult for standard back-propagation networks (Fahlman & Lebiere, 1990). Some of the neurological justification for generative networks such as cascade-correlation is reviewed by Quartz and Sejnowski (1997).

Cascade-correlation has proved useful in simulating many aspects of cognitive development, including the balance scale (Shultz, Mareschal, & Schmidt, 1994), conservation (Shultz, 1998), seriation (Mareschal & Shultz, in press), pronoun semantics (Takane, Oshima-Takane, & Shultz, 1995), number comparison (Hashmi & Shultz, 1998), discrimination shift learning (Sirois & Shultz, 1998), and integration of velocity, time, and distance cues (Buckingham & Shultz, 1994, 1996). In these models, network behavior becomes rule-like with learning, but rules are not the actual representations of knowledge and rule firing is not the mechanism for cognitive processing. Rules are instead high-level, epi-phenomenal characterizations of what is happening at the sub-symbolic level of unit activations and connection weights. Among the many advantages of implementing rule-like behavior in neural activity are acquisition of non-normative rules, natural variation across problems and individuals, theoretical integration of perceptual and cognitive phenomena, and achievement of the right degree of crispness in knowledge representations. Several network predictions were confirmed in subsequent psychological studies.

An Encoder Version of Cascade-correlation

An apparent problem for using standard cascade-correlation in encoder problems is that it creates many cross-connections that bypass hidden units. The most troublesome of these for encoder simulations are the direct connections from input to output units, which could solve any encoder problem in a trivial way by rapidly learning weights of 1 between an input unit and its corresponding output unit. The solution is to freeze these direct input-to-output links to have values of 0, not modifiable by subsequent learning. As with back-propagation encoder networks, all of the computation must then employ hidden units.

Coding the Marcus et al. Experiments

The coding scheme for simulation of these experiments is a straightforward translation of words into an analog representation of real numbers. In such analog representations, degree of activation encodes distinct inputs and outputs. The assignment of words to numbers is arbitrary but consistent. In the training patterns, the four levels of A (*ga, li, ni, ta*) are represented by the numbers 1, 3, 5, and 7, respectively, and the four levels of B (*ti, na, gi,* and *la*) by the numbers 2, 4, 6, and 8, respectively. Hence, the ABA sentences *ga ti ga* and *li na li* are represented by 1 2 1 and 3 4 3, respectively. The ABB sentences *ga ti ti* and *ga na na* are represented by 1 2 2 and 1 4 4, respectively. The test patterns have values not used in training, but are interpolated within the training values: 2.5 for *wo*, 3.5 for *fe*, 5.5 for *de*, and 6.5 for *ko*. Thus, the ABA test sentence *wo fe wo* is represented by 2.5 3.5 2.5; and the ABB test sentence *de ko ko* is represented by 5.5 6.5 6.5.

In previous simulations with cascade-correlation networks, we have found that analog coding schemes often enable excellent learning and generalization. The use of analog representations is also supported by many psychological studies, particularly on numerical operations (e.g., Gelman & Gallistel, 1978, 1992).

Procedure

Sixteen cascade-correlation networks were run in each of the conditions of Marcus et al.'s Experiments 1 and 3. Each network, starting with its own randomly determined connection weights, including those initial weights used for candidate hidden units, corresponds to a unique infant. In each network, there were three input units to represent each of the three words in a sentence, and three output units to represent the target response, that is, the same three-word sentence. The output units had linear activation functions to enable their approximation of real numbers. The encoder option ensured that direct input-to-output connections were frozen at 0, so hidden units, with sigmoid activation functions would have to be recruited.

All cascade-correlation parameters were equal to Fahlman's default values with the following exceptions. Score-threshold, the tolerated difference between target and actual outputs was raised from the default of 0.4 to 1.0 in order to reduce the crispness of the rules learned by the network. Without this increased sloppiness, networks would never reverse the difference between consistent and inconsistent test sentences as did one of Marcus et al.'s infants. Training continued until all output units produced activations within score-threshold of their targets. There are also parameters for input-patience and output-patience with default settings of 8. They represent the number of epochs allowed to pass with little or no increase in correlation or reduction in error, respectively, before shifting phase.[1] Cascade-correlation alternates between input and output phases, depending on whether a hidden unit is being recruited or weights going into output units are being adjusted, respectively. I changed these two patience values to 1, partly to increase sloppiness in network learning and partly because, on this problem, performance did not improve much after it failed to improve on a single epoch.

Results

Results in terms of error on the two types of test patterns are presented in Table 1 for the simulation of Experiment 1. Error on the test patterns was subjected to a repeated measures ANOVA in which condition (1 vs. 2) served as a between network factor and test pattern (consistent vs. inconsistent) served as a repeated measure. Neither the main effect of condition or the condition x test pattern interaction was significant. However, there was a substantial main effect of test pattern, $F(1,30) = 228$, $p < .0001$. As revealed in Table 1, there was more error to the inconsistent test patterns than to the consistent test patterns. With error considered to be equivalent to the need for further cognitive processing, this result mirrors that found with Marcus et al.'s infant participants. In a further parallel to the infant study, one network produced a reversal of the general trend, i.e., it showed (slightly) more error to the consistent test patterns than to the inconsistent test patterns.

Analogous results for the simulation of Experiment 3 are presented in Table 2. A similar ANOVA yielded only a substantial main effect of test pattern, $F(1,30) = 356$, $p < .0001$. Again, as revealed in Table 2, there was more error to inconsistent test patterns than to consistent test patterns. And again, there was one network with a reversal of the general trend, i.e., it showed (slightly) more error to the consistent test patterns than to the inconsistent test patterns.

Apart from needing a score-threshold of at least 1.0 to produce any reversals on the test patterns, other simulations showed that results were robust against systematic variation in the score-threshold and patience parameters.

A plot of results for one representative network is presented in Figure 1. It shows a negatively accelerated decrease in error over output epochs on the training patterns, much like the shape of declining attention in infant habituation experiments. After complete success with the training patterns, the consistent test patterns likewise show very little error, but the inconsistent test patterns show considerable error recovery, much like dishabituation of attention in infants. The epochs at which hidden units are recruited are marked with diamonds just above the training errors. As in other cascade-correlation simulations, it is noteworthy that error often decreases sharply after a new hidden unit is recruited.

Preliminary analyses of the knowledge representations learned by these networks suggest that the hidden units cluster on two fundamental components, each of which is sensitive to variation in both the A and B categories of words.[2] The two-dimensional nature of this problem was further verified by PCAs of the raw training data in each experimental condition.

Discussion

The simulation results show that a neural network model without variable-laden symbolic rules can indeed simulate the results of Marcus et al.'s (1999) infant habituation experiments. Like the infants, the networks showed gradual habituation to a repeated syntactic form, and recovery of interest to an inconsistent novel form but not to a consistent novel form. Even the occasional reversal preference by a single individual was captured. These results show that Marcus et al.'s findings with infants do not uniquely require a symbolic rule-based account. It may well turn out that some computation in humans is based on explicit symbolic rules, but the Marcus et al. data do not provide definitive proof for this claim as it applies to infants. Pinker's (1999) argument that the Marcus et al. data suggest an innate rule-learning capacity seems, at best, premature.

The key feature of the present simulations would appear to be the use of analog encoding for the input and output words composing a sentence. Generalization to novel items is known to be facilitated by analog coding schemes in which activation intensity corresponds to particular representations (Jackson, 1997).[3] Although the details of Marcus et al.'s (1999) unsuccessful simulations were not published, it might be speculated that they employed non-analog binary codes. The reason such codes might not generalize is that novel items are coded on units with untrained connection weights. Such failures are analogous to expecting that I can speak Spanish just because someone I never interact with has learned to do so.

The use of analog coding is not merely a way of smuggling in variable binding. Analog coding by itself does not implement variable binding because assignments of values to input units are lost as activation is propagated forward onto non-linear hidden units. In explicit variable-binding schemes, assignments of values to variables are preserved for use in later computation.

It is possible that other neural network algorithms, such as back-propagation or auto-association, would be able to simulate these habituation data, using analog or some other coding scheme. If so, differences between successful algorithms could be explored for relative accuracy in accounting for the psychological data.

Other future work on this model might profitably address the issue of whether a successful neural network model could use a more realistic phonetic coding of the input. Very few neural network models attempt to cover everything from raw stimuli to high-level cognitive manipulations. But

---

[1] An epoch is a presentation of all of the training patterns.

[2] These analyses are based on PCA of network contributions (Shultz, Oshima-Takane, & Takane, 1995).

[3] The present networks generalize to A and B syntactic categories even outside of the range of the training patterns.

such extensions would generate more complete understanding of psychological phenomena. The finding that phonemes vary continuously on sonority suggests that a more realistic analog encoding might be feasible (Vroomen, van den Bosch, & de Gelder, 1998). Research on such realistic coding schemes would be necessary to simulate Marcus et al.'s Experiment 2, which was designed with particular phonetic properties in mind.

Another useful extension might involve the use of recurrent networks. Non-recurrent networks and recurrent networks can both learn temporal problems. The essential difference between them is a trading of space for time in the input coding (Hertz et al., 1991). Recurrent networks process inputs in sequence over time, allowing for sequences of indeterminate length; non-recurrent networks represent inputs on different input units simultaneously. There is a recurrent version of cascade-correlation that might be interesting to try on these sentence habituation problems.

It is worth stressing that the present model is not the definitive treatment of habituation. The process of habituation is still poorly understood and there are many phenomena in the habituation literature that would need to be accounted for by any comprehensive model. This study is essentially a demonstration that the Marcus et al. (1999) data can be covered by a neural network model.

Because a turnabout is often considered fair play, I would like to issue a reciprocal challenge to those favoring symbolic rule-based models of human cognition to implement serious models of habituation phenomena. The habituation literature is extensive and contains some of the most important discoveries in developmental psychology. They are largely untouched by computational modeling. It does not suffice to merely re-describe psychological phenomena in terms of a few symbolic rules. It is critically important to implement working models that include not only knowledge representation and processing but also learning and development as appropriate. Among the significant challenges for rule-based models of habituation are clear links with new or standard theories of the habituation process, the gradual negatively accelerated shape of habituation curves, individual differences in habituation rates and occasional reversals of general trends, and the dishabituation differences reported by Marcus et al. (1999). One of the quickest cures for incorrect or inappropriate theoretical statements and models is the discipline of actually trying to implement a model that covers phenomena in a principled way.

Another set of infant habituation data that has been successfully simulated by an encoder neural network model concerns asymmetric exclusivity effects in infant memory and categorization (Mareschal & French, 1997). Infants learn the categories of *dog* and *cat*, but with some interesting asymmetries (Quinn, Eimas, & Rosenkrantz, 1993). Essentially, dogs are not included in the category of cats, but cats are included in the category of dogs.

At this point, the only successful models of infant habituation employ feed-forward connectionist models without explicit variable binding.

Acknowledgments

This research was supported by a grant from the Natural Sciences and Engineering Research Council of Canada. Comments on an earlier draft by Alan Bale, Dave Buckingham, Yasser Hashmi, Yuriko Oshima-Takane, Sylvain Sirois, and Yoshio Takane were gratefully received.

References

Buckingham, D., & Shultz, T. R. (1994). A connectionist model of the development of velocity, time, and distance concepts. *Proceedings of the Sixteenth Annual Conference of the Cognitive Science Society* (pp. 72-77). Hillsdale, NJ: Lawrence Erlbaum.

Buckingham, D., & Shultz, T. R. (1996). Computational power and realistic cognitive development. *Proceedings of the Eighteenth Annual Conference of the Cognitive Science Society* (pp. 507-511). Hillsdale, NJ: Lawrence Erlbaum.

Cohen, L. B. (1973). A two-process model of infant visual attention. *Merrill-Palmer Quarterly, 19,* 157-180.

Cohen, L. B. (1979). Our developing knowledge of infant perception and cognition. *American Psychologist, 34,* 894-899.

Elman, J. L. (1990). Finding structure in time. *Cognitive Science, 14,* 179-211.

Fahlman, S. E., & Lebiere, C. (1990). The Cascade-correlation learning architecture. In D. S. Touretzky (Ed.), *Advances in Neural Information Processing Systems 2.* Los Altos, CA: Morgan Kaufmann.

Gelman, R., & Gallistel, C. R. (1978). *The child's understanding of number.* Cambridge, MA: Harvard University Press.

Gelman, R., & Gallistel, C. R. (1992). Preverbal and verbal counting and computation. *Cognition, 44,* 43-74.

Haith, M. M. (1990). Progress in the understanding of sensory and perceptual processes in early infancy. *Merrill-Palmer Quarterly, 36,* 1-26.

Hashmi, Y., & Shultz, T. R. (1998). A neural network model of number comparison. (Submitted for publication).

Hertz, J., Krogh, A., & Palmer, R. G. (1991). *Introduction to the theory of neural computation.* Reading, MA: Addison Wesley.

Jackson, T. O. (1997). Data input and output representations. In E. Fiesler & R. Beale (Eds.), *Handbook of neural computation.* Oxford: Oxford University Press.

Marcus, G. F., Vijayan, S., Rao, S. B., & Vishton, P. M. (1999). Rule learning by seven-month-old infants. *Science, 283,* 77-80.

Mareschal, D. & French, R. M. (1997). A connectionist account of interference effects in early infant memory and categorization. In *Proceedings of the 19th annual conference of the Cognitive Science Society* (pp. 484-489). Mahwah, NJ: LEA.

Mareschal, D., & Shultz, T. R. (in press). Development of children's seriation: A connectionist approach. *Connection Science*.

Oakes, L. M., & Cohen, L. B. (1990). Infant perception of a causal event. *Cognitive Development, 5*, 193-207.

Pinker, S. (1999). Out of the minds of babes. *Science, 283*, 40-41.

Quartz, S. R, & Sejnowski, T. J. (1997). The neural basis of cognitive development: A constructivist manifesto. *Behavioural and Brain Sciences, 20*, 537-596.

Quinn, P. C., & Eimas, P. D. (1996). Perceptual organization and categorization in young infants. *Advances in Infancy Research, 10*, 1-36.

Quinn, P. C., Eimas, P. D., & Rosenkrantz, S. L. (1993). Evidence for representations of perceptually similar natural categories by 3-month-old and 4-month-old infants. *Perception, 22*, 463-475.

Shultz, T. R. (1998). A computational analysis of conservation. *Developmental Science, 1*, 103-126.

Shultz, T. R., Mareschal, D., & Schmidt, W. C. (1994). Modeling cognitive development on balance scale phenomena. *Machine Learning, 16*, 57-86.

Shultz, T. R., Oshima-Takane, Y., & Takane, Y. (1995). Analysis of unstandardized contributions in cross connected networks. In D. Touretzky, G. Tesauro, & T. K. Leen, (Eds). *Advances in Neural Information Processing Systems 7* (pp. 601-608). Cambridge, MA: MIT Press.

Sirois, S., & Shultz, T. R. (1998). Neural network modeling of developmental effects in discrimination shifts. *Journal of Experimental Child Psychology, 71*, 235-274.

Slater, A. (1995). Visual perception and memory at birth. *Advances in Infancy Research, 9*, 107-125.

Sokolov, E. N. (1963). *Perception and the conditioned reflex.* Hillsdale, NJ: Erlbaum.

Takane, Y., Oshima-Takane, Y., & Shultz, T. R. (1995). Network analyses: The case of first and second person pronouns. *Proceedings of the 1995 IEEE International Conference on Systems, Man and Cybernetics* (pp. 3594-3599).

Vroomen, J., van den Bosch, A., & de Gelder, B. (1998). A connectionist model for bootstrap learning of syllabic structure. *Language and Cognitive Processes, 13*, 193-220.

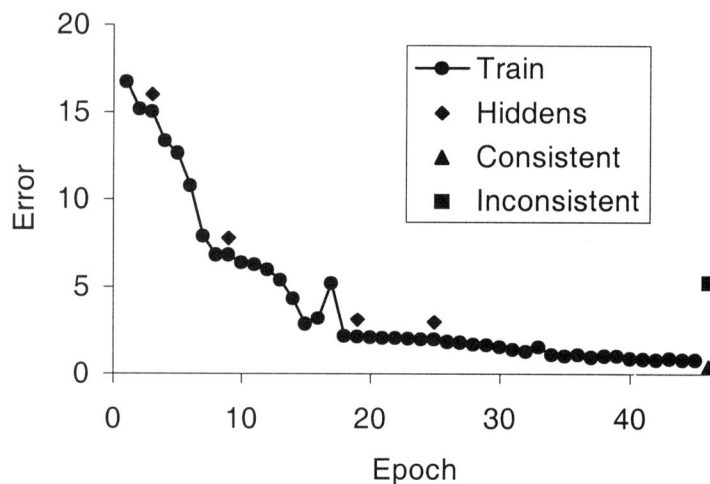

Figure 1: Results for a representative network in the ABA condition of the simulation of Experiment 1. Error on the training patterns decreases in a negatively accelerated fashion over time, representing habituation. Error remains low for the consistent test, but increases for the inconsistent test, demonstrating dishabituation to novelty. The diamond shapes represent the epochs at which hidden units were recruited.[4]

---

[4] Error is divided by the number of patterns and plotted over output-phase epochs. Input-phase epochs are not included in such plots because there is no change in network performance during input phases. The first three output epochs are omitted from this plot to improve clarity because error starts quite high and drops dramatically by virtue of adjustment of weights from the bias unit. The bias unit is always on, with an activation of 1, regardless of the input pattern. It has trainable connection weights to all non-input units, specifying their resting levels of activation.

Changes in Student Decisions with *Convince Me*: Using Evidence and Making Tradeoffs

Marcelle A. Siegel (mcgull@socrates.berkeley.edu)
Science and Math Education (SESAME); 4533 Tolman Hall
University of California at Berkeley
Berkeley, California 94720-1670 USA

Abstract

This study examined the cognitive processes of decision making in an urban high school classroom in which tenth graders analyzed scientific evidence about current issues of technology and society. A computer program, called *Convince Me* (Schank, Ranney & Hoadley, 1996), provided scaffolding for making evidence-based decisions for the experimental group. During the course of instruction, both the control and experimental classes completed open-ended assessments. Student progress, in using evidence to support claims and in weighing benefits and drawbacks, was mixed. Reasons for the changes in decision making are offered.

Coherent Reasoning about Evidence

This research emphasizes reasoning skills in using evidence. Other studies have examined these skills as well. For example, in the Knowledge Integration Environment (KIE) project, students interpret and critique scientific information garnered via the internet and make conjectures about it, forming a scientific argument. KIE researchers have hypothesized that engaging students in the creation of an argument facilitates conceptual change (e.g., Bell & Linn, 1997).

The program used in this project, *Convince Me* (CM), possesses a connectionist network (called ECHO) that simulates human reasoning. Using ECHO, CM offers feedback as to whether the student's evaluation of each proposition matches the values that are simulated by the computer. ECHO's principles of reasoning are based on the Theory of Explanatory Coherence, established by the philosopher, Thagard (1989). The theory assumes that the plausibility of a belief increases with, for instance: a) the simplicity with which it is explained, b) increasing breadth of evidential coverage, and c) decreasing competition with alternative beliefs (Ranney & Schank, 1998). Table 1 lists these principles in more detail.

While using the program, a student enters alternatives, beliefs, and evidence about an issue and then evaluates the plausibility of his decision. CM's interface provides scaffolding for making a decision through prompts for entering hypotheses and evidence. Students are also asked to make links between and among hypotheses and evidence and must choose whether each link they make is supportive ("explain") or contradictory ("conflict"). Students rate the reliability of each piece of evidence, as well as how much they believe each statement that they have entered. Next, they run the ECHO simulation, and then they contrast their ratings with ECHO's activations by pressing the Model's Fit button that calculates a correlation score and responds with, for instance "The correlation between your ratings and ECHO's evaluations is: 0.29 (mildly related)...."

Table 1. Some of ECHO's Principles for a Coherent Argument

1)	Plausibility increases with more support from explanatory statements.
2)	Plausibility increases with less competition from contradictory statements.
3)	Simplicity: The plausibility of a belief is inversely related to the number of cohypotheses it needs to explain a proposition.
4)	Data priority: Results of observations, such as evidence and acknowledged facts, have a degree of acceptability on their own.

Prior studies with CM indicate that it is a useful tool for learning about reasoning. Students using the program performed better than students doing similar pen and paper exercises, perhaps because of the computer's feedback (Schank, 1995). Also, undergraduates working with CM improved at distinguishing between hypotheses and evidence (Ranney, Schank, Hoadley & Neff, 1994). High school students using CM supported their beliefs with objective evidence and generated more than one alternative while making complex decisions (Siegel, 1997).

Research Focus

This study investigated whether using *Convince Me* with high school students in an issue-oriented biology class helped the students become better at using scientific evidence to support their decisions and to weigh tradeoffs in their choices. The hypothesis was that CM's principles of coherence (Table 1) would help engender better decision-making skills.

To test whether CM activities significantly improved students' uses of evidence and the weighing of tradeoffs, two classes were examined for several months. The researcher observed, taught, participated with, and tested the classes daily and equally from January through June of 1998. The same teacher led both advanced Biology classes. The main curriculum was *Science and Sustainability,* a new course developed by the Science Education for Public Understanding

Program (SEPUP) at the University of California at Berkeley's Lawrence Hall of Science. SEPUP's courses include written materials for students and teachers, as well as laboratory equipment. Students learned science by studying, discussing, debating, and experimenting, based on issues relevant to society.

Assessing Student Reasoning

In addition to SEPUP instructional materials, this study utilized an assessment system developed by SEPUP. The system was developed for SEPUP's middle school course using Rasch measurement techniques (e.g., Masters, 1982).

The SEPUP assessment system consists of *variables* that are a set of scientific concepts, processes and skills that are central to the course (Roberts, Wilson & Draney, 1997; Sloane, Wilson & Samson, 1996). *Understanding Concepts* and *Evidence and Tradeoffs* were the variables used in this study. Items for measuring these variables were embedded in tasks throughout the curriculum. Students wrote short essays or sentences in response to open-ended questions. Both Evidence and Tradeoffs and Understanding Concepts questions were scored on a criterion-referenced, 0-4 scale according to the SEPUP rubric (0 is low, 4 is high: see Roberts, et al., 1997).

Students were assessed on two elements of the variable Evidence and Tradeoffs:

- *Using Evidence* (student supports claims with relevant evidence)
- *Using Evidence to Make Tradeoffs* (student sees drawbacks as well as benefits in choice and supports these tradeoffs with evidence)

For example, a student who provided the major objective reasons for her decision and supported each reason with relevant and accurate evidence would receive a score of 3, on the 0-4 scale for Using Evidence.

Students were also assessed on two elements of the Understanding Concepts variable to determine how well they grasped the principles of coherent decisions:

- *Recognizing Relevant Content* (identify and describe the principles of coherence used in *Convince Me)*
- *Applying Relevant Content* (use the principles of coherence in new situations)

In accordance with the SEPUP method (Sloane, et al., 1996), students were introduced to the assessment system before the evaluation took place. They completed practice questions and received feedback. They also used the scoring guides while constructing their responses in order to learn to distinguish the qualitative differences between score levels.

Participants

The school involved in the project faced socioeconomic challenges typical of the inner city. According to the school district's data, 50% of students qualified for Aid for Families with Dependent Children (AFDC); 42% of students were identified as Limited English Proficient. The most recent SAT scores reported for the approximately 41% of school seniors who took the test were far below the national aver-

age: 321 on the verbal portion (national average is 423) and 437 on mathematics (national average is 479).

Before the study began, the two classes were compared to see if there were initial differences in ability. Scores on the standardized Terra Nova test (CTB, 1997) were obtained. The SEPUP control class had an average reading performance level on a 1-5 scale (5="advanced" and 3="nearing proficient") of 2.34 (average percentile of 60.25), while the experimental class had an average reading performance level of 2.56 (average percentile of 67.25). However, a t-test revealed that the difference was not significant (p=.35).

Procedure

Both of the two tenth-grade classes participated in the *Science and Sustainability* course activities. For a period of two months, the experimental class used both the course and the CM computer activities. Due to scarcity of computer facilities, half of the class would use CM one day while the other half completed SEPUP activities; the next day they would switch. The control class engaged in SEPUP decision-making activities during the time the experimental class used CM in this manner. The timing of the activities, the evaluations and the topics covered are summarized in Figure 1.

Figure 1: Timeline

The 2nd column is separated into rows with each representing 1 week. The light areas represent weeks when both classes were working on the same activities. The darkly shaded area represents the experimental period. The asterisks indicate times of testing (except that the Evidence and Tradeoffs tests for Rasch analysis lasted three weeks).

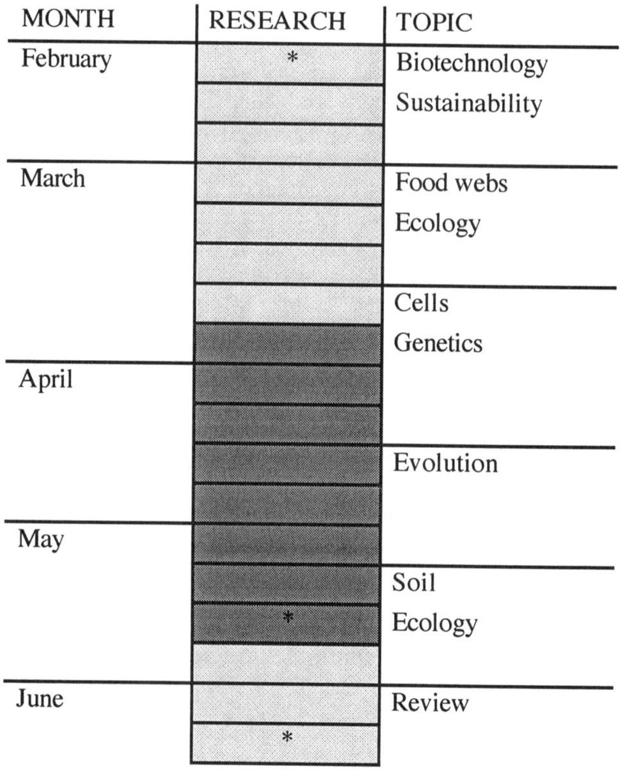

MONTH	RESEARCH	TOPIC
February	*	Biotechnology
		Sustainability
March		Food webs
		Ecology
		Cells
		Genetics
April		
		Evolution
May		
		Soil
	*	Ecology
June		Review
	*	

672

Some of the questions used on the Evidence and Tradeoffs evaluations are shown in Table 2. Ten open-ended items were taken by 56 students in the two classes during three weeks at the beginning of the study, and ten open-ended items were taken over three weeks after the decision-making activities at the end of the study. In addition, five Understanding Concepts questions (examples in Table 2) were given to the experimental class immediately after the *Convince Me* activities.

Table 2. Examples of Evidence and Tradeoffs and Understanding Concepts Questions

Evidence and Tradeoffs Items	Question #
Do you think humans should be included in the Antarctic ecosystem? Why or why not? What role, if any, do you think humans play in the Antarctic ecosystem?	3, 4
Imagine that you are the principal of a school that is having a problem with broken windows. Your choice is to replace the broken windows with either glass or plexiglass (plastic). Glass and plexiglass have different properties, and the plexiglass costs about 25% more. What material would you use and why? Be sure to describe the trade-offs involved in your decision. A complete answer will discuss the advantages and disadvantages of both materials.	7, 8 also 19,20
...Table of Final Radish Heights (Calculate the average height for each treatment.) a) Would you add fertilizer to soil to increase agricultural output? Give reasons for your answer. b) Do you think that adding fertilizer to soil to increase agricultural output is a sustainable process? Explain.	13, 14

Understanding Concepts Questions	
H1: Diazinon should not be used because it is dangerous to animals E1: It causes birth defects in chickens (Reliability=3) E2: It poisons and kills birds, bees, and fish (Reliability=3) v s. H2: Diazinon may be used because it is safe E3: It does not cause birth defects in rats or rabbits QUESTIONS ABOUT THE ABOVE ARGUMENT: 1. Which side of the argument, H1 or H2, would ECHO think is stronger? Why? 2. How would you change the argument to make H1 stronger?	4,5

Analyses

Two types of analyses of the data were carried out. The first analysis (*basic*) included data in raw form, and looked at particular points in time (*tests*). In this basic analysis, the first question (**pretest**, week 1) was compared to the question given first after the treatment period (**posttest**, week 16) to the last question given (**delayed posttest,** week 18). While this type of testing might be more familiar to researchers and classroom teachers, this type of analysis is not fully sound according to theories of measurement.

For example, the researcher has no way to tell which items are more difficult than other items. To address such measurement issues, a second analysis was carried out using Rasch modeling (e.g., Masters, 1982; Wright & Masters, 1982). Using this advanced statistical technique, one can compare the difficulty of items in detail, rather than giving questions without knowing how they differ, because "item difficulty" and "person ability" are estimated on the same scale. In addition, Rasch modeling is not only norm-referenced (i.e. comparing a student to other students), but also criterion-referenced (i.e. comparing a student's work to content standards or criteria) which offers more avenues for interpreting the results.

In the Evidence and Tradeoffs Rasch analysis, two sets of data were modeled: all the items answered before the treatment (**preexam**) and all the items answered after the treatment (**postexam**). Because the questions were open-ended and were often embedded in a laboratory activity, multiple questions could not be given on the same day; the preexam and postexam both took three weeks to complete. Students' progress over time and difficulty of items were thus confounded. In order to control for the difficulty of the items, a simulation was run with items anchored to an analysis of 830 students in another SEPUP course (which was possible because there were common items) (Wilson, Sloane & Roberts, 1995). Four preexam items and two postexam items were anchored at their appropriate difficulty levels from this previous study. In the Understanding Concepts analysis, the five items were modeled as one (post) exam. All the student scores were analyzed using *Quest* software (Adams & Khoo, 1993).

The ET and UC questions had been pilot tested for reliability and validity in a previous study. Using this information, ET items were assigned to the two exams.

Rasch Results: UC Correlation

Results from the Understanding Concepts evaluation suggested that understanding the principles of *Convince Me* was helpful in building better decisions in Evidence and Tradeoffs. Students who did better on the Evidence and Tradeoffs postexam also did better on the Understanding Concepts exam. The correlation between ET and UC was .50 for the basic analysis and .61 for the Rasch analysis.

Basic Results: Significant Improvement on Evidence and Tradeoffs

Student work revealed higher scores over time on measures of using evidence and making tradeoffs. The results from comparing the basic Evidence and Tradeoffs scores, including both Using Evidence and Using Evidence to Make Tradeoffs, portray progress for both classes. Note that throughout this time both classes were using SEPUP activities and so would be expected to improve on Evidence and Tradeoffs measures. It appears this was true, in addition to the CM group showing marked improvement on the delayed posttest. Their average scores (on a 0-4 scale, 4 is high) are shown in Table 3.

Table 3. Average Evidence and Tradeoffs scores.

The units are from the Evidence and Tradeoffs 0-4 scale. P values are shown in parentheses.

Class	Pretest	Posttest	Delayed Posttest
Control	1.87	2.35	2.92
Experimental	1.52	1.98	3.00
Difference	.35 (<.001)	.37 (<.05)	-.08 (>.05)

Table 4. Average Evidence and Tradeoffs Gains.

The units are from the Evidence and Tradeoffs 0-4 scale. P values are shown in parentheses.

Class	Pre to Posttest	Post to Delayed Posttest	Total: Pre to Delayed Posttest
Control	.48 (<.001)	.57 (<.001)	1.05 (<.001)
Experimental	.46 (<.001)	1.02 (<.001)	1.48 (<.001)

All of these differences were significant, except for the difference between the control and experimental class on the delayed posttest (p values shown in Table 3). The improvement for the experimental group was numerically larger than for the control group; the gain scores are shown in Table 4. The experimental group shows the most improvement, not immediately after using *Convince Me*, but on the delayed posttest. A sample of work from one of the students (alias "Jamie") who improved over time follows. Jamie's answer on the pretest received a low score because it did not employ scientific evidence according to the question.

Question: Should we allow human cloning for research or medicine? Should any cloning experiments be done? Give evidence from the articles to support your view. Think in terms of both advantages and disadvantages.

Response: *Jamie's answer does not include scientific evidence from the activity and received low scores:*
I think that there should not be human cloning because if you clone a person you would have same DNA and everything and if the identical clone go and do something bad, like killing somebody, and when the police go catch the person, they might find the wrong person and they can not say anything because the clone and the person have everything the same. In a way I think that allowing human cloning is good because the clone could be much healthier and everything.

Score: Using Evidence: 1 Using Evidence to Make Tradeoffs: 1

Jamie was in the control class and did not take part in the decision-making instructional treatment. Three months and eight items later, Jamie used more evidence on the posttest:

Question: Pretend you are in charge of transporting things by car two hours away. You need to pack the fragile items carefully so they are not harmed. You may pick from the following...etc.

Response: I think I would use a card board box because it can hold the skull perfectly and that we are driving a car so we will not be scared that the skull will fall on the water and the recyclable and it will break the skull. If we use newspaper the skull will roll around the car and will break. The plastic bucket can not be recycled and the can make the skull float around so it might break. Water can make the dirty stuff away, but it may be important.

Score: Using Evidence: 3

After another month and 6 questions, Jamie's delayed posttest answer received the highest marks. Recall that the score only reflects the criteria described in the scoring guide. Jamie's answer is not "well written," but shows "higher-level reasoning" because of the way the evidence is weighed on both sides of the issue:

Question: (see Table 2, #19,20)

Response: If I was the principal of the school I would use the plexiglass because the plexiglass it can not be break easily and if they are having so much problem and they still use the same kind of material I think they should change it. The plexiglass cost 25% more, but think of the safety of the student and not replacing it so much time. If you replace it a lot it will cost you more and it might hurt a student and for putting it up you will be interuppting the student and the worker will need you to pay for them then you will use more because every time they put it up that is hundreds of dollar already, with plexiglass it will not break easily and there will not be a lot of interruptness and hurt and you don't have to pay as much time to the workers as the glass window.

Score: Using Evidence: 4 Using Evidence to Make Tradeoffs: 4

Rasch Results: No Progess on Evidence and Tradeoffs

The Rasch estimates of students' abilities did not indicate improvement for the two classes. The ability estimate represents the Rasch model's prediction of each student's ability on Evidence and Tradeoffs. The units are expressed in logits

$(log[\pi/(1-\pi)]$ where π = response probability according to the model). The Rasch estimates of students' abilities that were anchored to a previous analysis as described above, showed that the classes' average estimates were worse on the final ten questions than on the first ten questions (see Table 5). However, the changes were not significant. The standard deviation on the preexam was 1.44 and on the postexam was .79 putting the two scores within reach of each other.

Table 5. Anchored Rasch Estimates of Evidence and Tradeoffs

Units are expressed in logits.

Class	Preexam (1-10)	Postexam (11-20)
Control	-.76	-1.66
Experimental	-.70	-1.55

Also, when the logit estimates are plotted onto the map of performance levels, the preexam and postexam levels are both at 2 (see Figure 2). This indicates that when item difficulty is controlled, neither class improves, but both remain at the same level. (Level 2 is similar to a score of 2 on the Evidence and Tradeoffs scoring guide.)

Figure 2. Evidence and Tradeoffs Levels

EXP=Experimental Class
CRL=Control Class

Logits	Average Preexam	Average Postexam	Levels of Performance
3.2 1.6			**Level 3** Provides relevant and accurate evidence for each claim or for at least two perspectives on the issue.
0 -1.6	EXP CRL	EXP CRL	**Level 2** Some reasons offered but evidence is incomplete or only one perspective is provided.

The two class's average pre and post responses were not qualitatively different according to the chart. Both classes began with an average ability level of 2, meaning that they offered reasons for their decision, but did not provide sufficient relevant evidence. Both classes ended up with an average ability level of 2 as well.

Discussion

The results demonstrated that some students from both classes became more sophisticated at answering questions that required them to use evidence and make tradeoffs. The basic analysis showed that, as expected, both groups using SEPUP activities scored higher on Evidence and Tradeoffs over time. The group using CM had significantly better posttest scores than the control group, indicating that CM was a useful tool for helping students learn to make evidence-based decisions. The experimental group's average gain from the pretest to the delayed posttest was 1.48, while the control group's was 1.08--both representing qualitative differences of at least one score level. Interestingly, the experimental group's greatest gain was between the posttest and delayed posttest. This result might imply that students benefited from integrating their experience with CM with additional SEPUP activities and evaluations. However, the Rasch estimates did not indicate improvement for the average of either class.

There are two reasons why the Rasch analysis might not have showed improvement. One possibility is that with moderate improvement (as shown in the basic analysis), one could not expect significance without a larger sample of students. A second possible reason is that the Rasch analysis compared two long time periods. The preexam took place over three weeks and there was improvement during that time; the postexam also consisted of several questions over three weeks during which there was improvement (e.g., shown by the basic analysis gains). The improvement during the lengthy preexam and postexam could mask the general improvement overall from preexam to postexam. The Rasch analysis had practical constraints preventing the administration of multiple questions on the same day: the analysis required several items in order to model the results; the items were complex questions, often associated with a laboratory activity, and took at least half an hour to answer. Thus, it was not possible to give more than one question per day.

The results suggest that *Convince Me* helped students use evidence and weigh tradeoffs in their argument. From the researcher's daily experience in the classroom, it appeared that students learned that to build an argument in CM, one was **obligated** to include evidence and not just opinions. The correlational data support, but do not prove or negate, the original hypothesis that CM's principles of coherence would enhance students' use of evidence. Still, other aspects of interacting with CM could have been beneficial beyond the principles of coherence. For instance, CM scaffolded students in connecting supporting statements and linking conflicting statements into a web. This type of reasoning experience might have been helpful when answering the Evidence and Tradeoffs questions. The act of checking one's decision after receiving feedback from the computer, then revising it, may have also been useful for the students. Further studies are necessary before a full assessment of CM's assets and flaws can be made. Previous studies have targeted parts of the program that assist learning in particular ways (e.g., Ranney et al., 1994). Current analyses of students' use of *Convince Me* will provide further evidence about students' development of decision-making skills. One reason this research is essential is that developing ways of enhancing these decision-making skills is vital for educating students as critically-thinking citizens, as noted in many national proposals for science education reform (e.g., NRC, 1996).

References

Adams, R.J., & Khoo, S-T. (1993). *Quest: The Interactive Test Analysis System.* Hawthorn, VIC: Australian Council for Educational Research.

CTB (1997). *TerraNova.* CTB/McGraw-Hill.

Bell, P. & Linn, M.C. (1997). Scientific arguments as learning artifacts: Designing for learning on the web. Paper presented at the Annual Meeting of the American Educational Research Association, Chicago.

Masters, G.N. (1982). A Rasch model for partial credit scoring. *Psychometrika, 47, 149-174.*

National Research Council (1996). *National Science Education Standards.* Washington, D.C.: National Academy Press.

Ranney, M. & Schank, P. (1998). Toward an integration of the social and the scientific: Observing, modeling and promoting the explanatory coherence of reasoning. In S. Read & L. Miller (eds.)*Connectionist and PDP models of social reasoning.* Hillsdale, NJ: Lawrence Erlbaum.

Ranney, M., Schank, P., Hoadley, C. &Neff, J. (1994). 'I know one when I see one:' How (much) do hypotheses differ from evidence? In: *Proceedings of the Fifth Annual American Society for Information Science Workshop on Classification Research,* 139-156.

Roberts, L., Wilson, M. & Draney, K. (1997). The SEPUP assessment system: An overview. *BEAR Report Series SA-91-1.* Berkeley, CA: University of California.

Schank, P. K. (1995). *Computational Tools for Modeling and Aiding Reasoning: Assessing and Applying the Theory of Explanatory Coherence.* Unpublished Doctoral Dissertation, University of California at Berkeley.

Schank, P., Ranney, M. & Hoadley, C. (1996). *Convince Me* [Revised computer program (on CD, etc.) and manual]. In: J.R. Jungck, V. Vaughan, J.N. Calley, N.S. Peterson, P. Soderberg, & J. Stewart, (Eds.), The 1996-1997 BioQUEST Library (fourth edition). College Park, MD: Academic Software Development Group, University of Maryland.

Siegel, M.A. (1997). Developing decision-making skills with *Convince Me.* In: Proceedings of the Nineteenth Annual Conference of the Cognitive Science Society, 1049. Mahwah, NJ: Erlbaum.

Sloane, K., Wilson, M. & Samson, S. (1996). Designing an embedded assessment system: From principles to practice. *BEAR Report Series-96-1.* Berkeley, CA: University of California.

Thagard, P. (1989). Explanatory coherence. *Behavioral and Brain Sciences, 12,* 435-502.

Wilson, M., Sloane, K., Roberts, L. & Henke, R. (1995). SEPUP Course I, *Issues, Evidence and You:* Achievement evidence from the pilot implementation. *BEAR Report Series-SA-95-2.* Berkeley, CA: University of California.

Wright, B.D. & Masters, G.N. (1982). *Rating Scale Analysis: Rasch Measurement.* Chicago: Mesa Press.

Acknowledgements

I would like to thank Michael Ranney, Mark Wilson, Christine Diehl, and the Reasoning Group for comments on earlier drafts of this paper. I also extend much gratitude to the teacher and students who participated in my study. This research was supported by a traineeship from the National Science Foundation, Reforming Education through Science and Design, and other support from the University of California at Berkeley.

Faces are Different Than Words: Evidence from Associative Priming Studies

Amy L. Siegenthaler (amy@psych.utoronto.ca)
Department of Psychology, University of Toronto at Mississauga
3359 Mississauga Rd. North, Mississauga, ON, L5L 1C6

Morris Moscovitch (momos@credit.erin.utoronto.ca)
Department of Psychology, University of Toronto at Mississauga
3359 Mississauga Rd. North, Mississauga, ON, L5L 1C6

Abstract

Associative memory for familiar faces was investigated in two experiments. Pairs of familiar faces were presented for deep or shallow encoding; memory for these pairs was tested by presenting old-intact pairs, old-recombined pairs, and pairs consisting of one or two new faces. In Experiment 1, pairs consisted of two different individuals whereas in Experiment 2, pairs consisted of different views of the same individual. In both experiments, explicit recognition was best for old-intact pairs under deep encoding conditions. No associative priming effects were obtained in either experiment despite using a simultaneous familiarity-judgment task, similar to one that has produced associative priming effects with words (e.g., Goshen-Gottstein & Moscovitch, 1995a). It is proposed that the different associative priming effects obtained with the two types of stimuli may arise from differences in the modular perceptual representation systems for faces and words.

Introduction

Learning someone's name, the names of objects, which groups of people belong together or the context in which they are known all require forming arbitrary associations. To date, the majority of the theories concerned with associative memory have focussed on verbal associations. It is unclear whether these same processes apply to non-verbal associations as well. The two experiments presented here, using paradigms similar to those used with words, focus on associations between pairs of faces. Results similar to those obtained with words would suggest that the same associative processes apply across different classes of stimuli. Dissimilar results, on the other hand, would imply that the type of processes needed for forming new associations may be determined by the class of stimuli.

These two experiments investigate the associative process in terms of explicit and implicit memory (Graf & Schacter, 1985). Whether implicit memories can be formed for associative material has been a question of considerable debate in the literature. Using words as stimuli, some studies claim to have found convincing evidence for associative priming (e.g., Graf & Schacter, 1985; Schacter & Graf, 1986; Moscovitch, Winocur & McLachlan, 1986; McKoon & Ratcliff, 1979; Goshen-Gottstein & Moscovitch, 1995a, 1995b, 1995c) whereas other studies claim to have found no convincing evidence for associative priming (e.g., Carroll & Kirsner, 1982; Mayes & Gooding, 1989; Smith, MacLeod, Bain & Hoppe, 1989).

In the present set of experiments, we use a procedure, a simultaneous two-item task, that has produced reliable verbal associative priming effects in both normal and amnesic subjects (Goshen-Gottstein & Moscovitch, 1995a, 1995b, 1995c) and adapt it slightly for use with faces. In the original task, participants studied unrelated pairs of words such as *pause-weird* and *slope-plate*. At test, participants saw words in intact pairings (e.g., *pause-weird*), in recombined pairings (e.g., *pause-plate*) or new pairings (e.g., *soldier-apple*) and were asked to judge the lexical status (e.g., word or non-word) of the two words presented. Accordingly, some pairs were presented which consisted of one or two non-words. In this paradigm, repetition (or item) priming effects are measured by comparing reaction times to new pairs with reaction times to old or recombined pairs. Associative priming effects are measured by comparing reaction times to the recombined pairs with reaction times to the old pairs.

Theoretically, there is good reason to suppose that faces may give very different results from words in associative priming studies. First, theories of memory such as Schacter's (1990) and Moscovitch's (1992) contend that priming is mediated through modular perceptual representation systems which represent the form and structure of stimuli. Words are often combined into different pairs and sequences, both meaningful and non-meaningful, which create strong perceptual associations. The emphasis on *perceptual* association is important because Goshen-Gottstein and Moscovitch (1995a, 1995b, see also Light, La Voie & Kennison, 1995) concluded that two items *must* be perceived as a coherent perceptual gestalt in order for associative priming to occur. Faces, on the other hand, are not often combined into different combinations. It is thus likely that the modular perceptual representation system for faces would not possess the capability for creating strong perceptual associations (or gestalts) with other faces. Second, Farah (1991) contends that complex stimuli such as faces are recognized as single units whereas words are recognized by decomposition into multiple parts. It is very possible that these fundamental differences in how the two classes of stimuli are perceived and recognized may lead to differences in their ability to support associative priming.

Experiment 1

The aims of this experiment were two-fold: First, to demonstrate that explicit associative memory effects could be demonstrated with our famous face stimuli and encoding tasks (e.g., Winograd & Rivers-Bulkeley, 1977); and second, to investigate whether implicit memory for new associations could be demonstrated using pairs comprised of different famous individuals.

Methods

Sixty-four undergraduates participated in this experiment, half in the explicit memory task and half in the implicit memory task. Of these, half were instructed were to perform a deep encoding task (e.g., how likely are they to be friends?) and half were instructed to perform a shallow encoding task (e.g., how similar are the two skin tones?). Participants performed these encoding tasks with 36 pairs of familiar faces, each pair consisting of two different individuals. In all cases, encoding was incidental as participants were not informed that they would be tested later on their memory for these pairs. Of the 36 pairs presented, thirty pairs were critical pairs which in the test phase would form the Old-Old, Old-Recombined and Old-New pairs and three pairs were presented at both the beginning and end of the list to minimize primacy and recency effects.

The test phase differed for the explicit and implicit versions of the experiment. For the explicit version, participants were presented with 40 pairs of faces pairs which were presented in four kinds of pairings: Old-Old (same pairs as at study), Old-Recombined (same faces as at study but in new pairs), Old-New (faces from study paired with new faces not seen before) and New-New (faces not seen before). Participants were instructed to respond "Old" if both members of the pair were Old (the Old-Old and Old-Recombined pairs) and "New" if at least one member of the pair was New (the Old-New and New-New pairs).

For the implicit version, participants saw the 40 pairs of faces described above as well as 40 pairs containing unfamiliar faces to permit them to perform the familiarity judgment task. These pairs consisted of unfamiliar faces paired with old familiar faces, unfamiliar faces paired with new familiar faces and unfamiliar faces paired with other unfamiliar faces. Participants were instructed to respond "Familiar" if both members of the pair were Familiar (the Old-Old, Old-Recombined, Old-New, and New-New pairs) and "Unfamiliar" if at least one member of the pair was Unfamiliar (these consisted of Unfamiliar-Old, Unfamiliar-New and Unfamiliar-Unfamiliar pairs). This task is called a simultaneous familiarity-judgment task because participants are required to make a single response to two stimuli rather than responding to each stimulus individually. For

both explicit and implicit versions of the task, participants were instructed to respond as quickly and accurately as possible.

Results and Discussion

Accuracy and reaction time means are presented below in Table 1 (Explicit) and Table 2 (Implicit).

Table 1: Explicit memory: Accuracy and reaction time means to judge whether associated pairs of faces in different conditions were "Old" or "New".

	Old-Old	Old-Recomb.	Old-New	New-New
ACCURACY (H-FA)[1]				
Deep	0.75	0.70	0.67	0.83
Shallow	0.63	0.55	0.50	0.74
REACTION TIME (in ms)[2]				
Deep	1342	1379	1435	1182
Shallow	1492	1590	1796	1457

Table 2: Implicit memory: Accuracy and reaction time means to judge whether associated pairs of faces in different conditions were "Familiar" or "Unfamiliar."

	Old-Old	Old-Recomb.	Old-New	New-New	Un-familiar
ACCURACY (H-FA)					
Deep	0.83	0.80	0.70	0.68	0.71
Shallow	0.80	0.80	0.64	0.70	0.69
REACTION TIME (in ms)					
Deep	963	953	1117	1168	1130
Shallow	1108	1073	1244	1275	1283

Accuracy data for the explicit task (Table 1) were entered into a 2 (Study Condition: Deep and Shallow) x 4 (Test Condition: Old-Old, Old-Recombined, Old-New and New-New) repeated measures ANOVA with Study Condition treated as a between-subjects factor and Test Condition treated as a within-subjects factor. Study Condition was significant (F (1,30) = 6.27, MSE = 0.122, p < 0.02), indicating that deep encoding led to more accurate responding than did shallow encoding. Test Condition was also significant (F (4,120) = 23.70, MSE = 0.0095, p < 0.0001). Post-hoc pairwise contrasts between the four Test Condi-

---

[1] In all tables, accuracy scores are Hits – FA.

[2] In all tables, reaction time outliers were removed by calculating the means in each condition for each subject and eliminating responses that were more than 2 standard deviations from these means; new means were then calculated.

678

tions, using the REGW multiple range q-test, confirmed that recognition accuracy in the explicit task was better for old-old pairs than for old-recombined pairs ($p<0.05$). This indicates an associative memory effect.

Reaction time data were analyzed in the same manner. Results of the ANOVA revealed a significant effect of Study Condition ($F_{(1,30)} = 6.10$, $MSE = 482076$, $p < 0.02$) indicating that reaction times for pairs encoded deeply were faster than for pairs encoded shallowly. Test Condition was also significant ($F_{(4,120)} = 11.01$, $MSE = 50593$, $p < 0.0001$). Unlike the accuracy data, post-hoc testing did not reveal significant differences between the old-old and old-recombined pairs but the difference between the old-old and new-new condition was significant ($p < 0.05$). Thus although two faces are recognized more accurately when they are in the same pair as opposed to a different pair, they are not necessarily recognized more quickly.

Accuracy data for the implicit task must be interpreted with caution due to the subjective nature of the familiarity-judgment task. That is, a score of "incorrect" may have been obtained because a participant was genuinely unfamiliar with a particular face. The overall error rates were 13% for the deep condition and 15% for the shallow condition. These data were not analyzed further but a breakdown of the accuracy rates across conditions appears in Table 2.

Reaction time data for the implicit task (Table 2), were analyzed in the same way as for the explicit task. Results of the ANOVA revealed a nonsignificant effect of Study Condition ($F_{(1,30)} = 2.77$, $MSE = 450019$, $p > 0.1$) but a significant effect of Test Condition ($F_{(4,120)} = 12.36$, $MSE = 27929$, $p < 0.0001$). Post-hoc tests revealed a large item priming effect between the reaction times to the new-new pairs and both old-old and old-recombined pairs ($p < 0.05$). No associative priming effect was observed, however, as reaction times to the old-old and old-recombined pairs did not differ.

These results occurred despite using a simultaneous familiarity-judgment task that has produced reliable associative priming effects with words (e.g., Goshen-Gottstein & Moscovitch, 1995a). Because reaction times in the implicit task were on average about 400 ms faster for each pair condition than in the explicit task, we were confident that our task was measuring *priming* rather than some form of conscious recollective process. On the basis on these results, we concluded that faces do not lend themselves to associative priming in the same manner that words do. One question we posed, however, was whether the face-processing system was incapable only of forming new associations between two *different* people. We hypothesized that the face-processing system instead may be adapted for forming associations between different views of the same individual. It is important, after all, to be able to integrate different views of the same individual into a single perceptual repre-

sentation. This is the question we attempted to answer in Experiment 2.

Experiment 2

Methods

Sixty-four undergraduates participated in this experiment, half in the explicit condition and half in the implicit condition. The pairs of faces in this experiment all consisted of two different views of the same individual. For this reason, the encoding tasks had to be varied as the friendship-judgment task would not make sense with two different pictures of the same person. Thus, the deep encoding task became an honesty-judgment task (e.g., which picture looks the most honest?) and the shallow encoding task became a picture-shading task (e.g., which picture is the darkest in shading?).

The pairing conditions for the explicit task remained the same. Participants viewed 10 pairs in each condition (Old-Old, Old-Recombined, Old-New, and New-New) for a total of 40 pairs. As in Experiment 1, participants were instructed to respond "Old " if both member of the pair were old and "New" if at least one member of the pair was new. The Old-New condition became slightly more difficult, however, as participants were required to recognize which particular *picture* of an individual had been presented previously, rather than recognizing which individual had been presented previously.

For the implicit task, we changed the familiarity-judgment task to a person-identity task in which participants were required to judge whether the pair consisted of two pictures of the same person or of two different people. Participants viewed 10 pairs in each of the Old-Old, Old-Recombined and New-New conditions and 30 pairs in the Different condition (pairs consisting of two different individuals) for a total of 60 pairs. The faces making up the Different pairs were taken from 10 of the study pairs (the faces which made up the Old-New pairs in Experiment 1). These 20 faces were combined into 30 pairs. Participants were instructed to respond "Same" if both the pictures were of the same person and "Different" if the two pictures were of different people. As in Experiment 1, participants were asked to respond as quickly and as accurately as possible.

Results and Discussion

Accuracy and reaction time means are presented below in Table 3 (Explicit) and Table 4 (Implicit).

Data were analyzed in the same manner as Experiment 1. Data were entered into a 2 (Study Condition: Deep and Shallow) x 4 (Test Condition: Old-Old, Old-Recombined, Old-New and New-New) repeated measures ANOVA with Study Condition treated as a between-subjects factor and Test condition treated as a within-subjects factor. For the accuracy data in the explicit task (Table 3), significant effects were found for both Study Condition ($F_{(1,30)} =$

14.17, $\underline{MSE}$ = 0.7813, $\underline{p}$ < 0.001) and Test Condition ($\underline{F}$ (3,90) = 49.89, $\underline{MSE}$ = 0.5484, $\underline{p}$ < 0.0001). Thus, deep encoding led to more accurate responding than did shallow encoding. Post-hoc pairwise contrasts between the four Test Conditions, using the REGW multiple range q-test, confirmed that recognition accuracy was better for old-old pairs than for old-recombined pairs ($\underline{p}$ < 0.05), indicating an associative memory effect.

Table 3: Explicit memory: Accuracy and reaction time means to judge whether associated pairs of faces in different conditions were "Old" or "New".

		Old-Old	Old-Recomb	Old-New	New-New
ACCURACY (H-FA)					
	Deep	0.75	0.67	0.58	0.84
	Shallow	0.59	0.51	0.37	0.73
REACTION TIME (in ms)					
	Deep	1360	1406	1785	1114
	Shallow	1329	1394	1613	1028

Table 4: Implicit memory: Accuracy and reaction time means to judge whether associated pairs of faces in different conditions consisted of views of the "Same" or of "Different" individuals.

		Old-Old	Old-Re-comb.	New-New	Different
ACCURACY (H-FA)					
	Deep	0.98	0.94	0.96	0.96
	Shallow	0.96	0.95	0.92	0.94
REACTION TIME (in ms)					
	Deep	751	741	889	707
	Shallow	749	751	904	667

For the reaction time data (Table 3), results of the ANOVA revealed a nonsignificant effect of Study Condition ($\underline{F}$ (1,30) = 0.61, $\underline{MSE}$ = 179957, $\underline{p}$ > 0.4) but a significant effect of Test Condition ($\underline{F}$ (1,30) = 29.65, $\underline{MSE}$ = 2118586, $\underline{p}$ < 0.0001). Thus, unlike Experiment 1, deep encoding did not lead to faster performance than shallow encoding. Post-hoc testing on Test Condition revealed that, similar to Experiment 1, reaction times to old-old pairs were not faster than reaction times to old-recombined pairs but were faster than reaction times to old-new pairs. Thus, like Experiment 1, the pairs of faces are recognized more accurately when in the same pair as opposed to a recombined pair, but not necessarily more quickly. It is not clear why deep encoding did not lead to faster reaction times than shallow encoding, as in Experiment 1, but differences in the encoding tasks may have been a factor. Judging

which view of an individual is more honest may not elaborate the association as much as judging how likely two different people would be to be friends.

For the implicit task, a full analysis of the accuracy data could be performed due to the change in the task requirements. Judging "Same" versus "Different" is much more objective than judging "Familiar" versus "Unfamiliar" and has definite correct and incorrect responses. It also appears to be an easier task as overall accuracy rates were much higher for the implicit task in Experiment 2 compared with Experiment 1 (see Tables 2 and 4, respectively). Results of the ANOVA revealed no significant differences for either Study or Test Condition. This is desirable as any differences in reaction time cannot then be attributed to variations in task difficulty.

The reaction time data (Table 4) were analyzed in the same manner as in Experiment 1. Results of the ANOVA revealed a nonsignificant effect of Study Condition ($\underline{F}$ (1,30) = 0.01, $\underline{MSE}$ = 545.6, $\underline{p}$ > 0.9) but a significant effect of Test Condition ($\underline{F}$ (1,30) = 71.21, $\underline{MSE}$ = 255565, $\underline{p}$ < 0.0001). Similar to Experiment 1, post-hoc testing on Test Condition revealed a large item priming effect between the reaction times to the new-new pairs and both old-old and old-recombined pairs ($\underline{p}$ < 0.05) but no associative priming effect, as reaction times to the old-old pairs and old-recombined pairs did not differ significantly. Once again, reaction times in the implicit task were much faster than reaction times in the explicit task, suggesting that the person-identity task is measuring priming rather than some conscious recollective process.

Contrary to our hypothesis, it does not appear that pairs consisting of two views of the same individual support associative priming any more than pairs consisting of two pictures of different individuals. Thus, the inability to form perceptual associations is not restricted to new associations between faces which share no relation (e.g., Boris Yeltsin and Suzanne Sommers) but also applies to faces which share a direct relation (e.g., two different views of Harrison Ford).

Discussion

The results of these two experiments led us to conclude that the face-processing system does not lend itself to form perceptual associations in the same manner that the word-processing system does. This provides partial support for Farah's (1991) theory that faces and words are perceived and recognized in fundamentally different ways. Because words are recognized by decomposition into multiple parts, meaningless words like "housefrog" form a perceptual gestalt just as coherent as meaningful words like "houseboat." The individual letters that make up a word can be mixed and matched and no detriment is observed on recognition performance for the individual letters. Faces, on the other hand, do not share this property. As Tanaka and Farah (1993) have shown, recognition of the individual elements

680

that make up a face (e.g., the nose, the eyes) is drastically reduced when these elements are taken out of the context of the whole face. Because faces do not share the "mix-and-match" property that words have, it is likely that two individual faces, side by side, would not form a coherent perceptual gestalt in the same way that "housefrog" does. For this reason, it is not likely that pairs of faces would support associative priming as Goshen-Gottstein and Moscovitch (1995a, 1995b) concluded that perceptual associations are a necessary condition in order to demonstrate associative priming.

It is possible, however, that these results apply only to familiar faces. With a familiar face, the different viewpoints are already represented and thus forming associations between them may be unnecessary. With unfamiliar faces, however, each new view is unique and it is important to be able to associate these various views into a single perceptual representation. We have been investigating this possibility in the lab and preliminary data seem to suggest that this is indeed the case.

It is important to note that our conclusions apply only to implicit memory for associations between faces and not to explicit memory. As our results showed, it is possible to form associations between unrelated or related faces and to recollect them as long as they are recollected with an explicit test of memory. Explicit memory is not restricted to modular perceptual representation systems but has access to higher-level central systems (see Moscovitch, 1992). Therefore, for explicit recollection, it does not matter that pairs of faces do not form a coherent perceptual gestalt; this only becomes important when one wishes to demonstrate implicit recollection. Further, the disparity between implicit and explicit recollection reveals that the inability to form associations is not an inherent property of face stimuli themselves, but rather a property of the way faces are represented in their perceptual representation system (e.g. Farah, 1991). Future research in this area could use stimuli that vary in the extent to which they are processed holistically or by parts to determine precisely what are the necessary and sufficient conditions to demonstrate associative priming.

Acknowledgments

The authors would like to thank Gordon Winocur for comments on an earlier version of this paper and Marilyne Zeigler and Brenda Hannon for technical assistance. Funding for this project was provided by a grant awarded to the second author by the Natural Sciences and Engineering Research Council of Canada, Grant A8347.

References

Carroll, M., & Kirsner, K. (1982). Context and repetition effects in lexical decision and recognition memory. *Journal of Verbal Learning and Verbal Behavior, 21*, 55-69.

Goshen-Gottstein, Y. & Moscovitch, M. (1995a). Repetition priming for newly formed and preexisting associations: Perceptual and conceptual influences. *Journal of Experimental Psychology: Learning, Memory and Cognition, 21(5)*, 1229-1248.

Goshen-Gottstein, Y. & Moscovitch, M. (1995b). Repetition priming effects for newly formed associations are perceptually based: Evidence from shallow encoding and format specificity. *Journal of Experimental Psychology: Learning, Memory and Cognition, 21(5)*, 1249-1262.

Goshen-Gottstein, Y. & Moscovitch, M. (1995c). Intact implicit memory for newly-formed verbal associations in amnesic patients. Paper presented at Society for Neuroscience, Nov. 11-16, 1995.

Graf, P. & Schacter, D.L. (1985). Implicit and explicit memory for new associations in normal and amnesic subjects. *Journal of Experimental Psychology: Learning, Memory and Cognition, 11(3)*, 501-518.

Farah, M. (1991). Patterns of co-occurrence among the associative agnosias: Implications for visual object representation. *Cognitive Neuropsychology, 8*, 1-19.

Light, L.L., La Voie, D. & Kennison, R. (1995). Repetition priming of nonwords in young and older adults. *Journal of Experimental Psychology: Learning, Memory and Cognition, 21(2)*, 327-346.

Mayes, A.R. & Gooding, P. (1989). Enhancement of word completion priming in amnesics by cueing with previously novel associates. *Neuropsychologia, 27(8)*, 1057-1072.

McKoon, G. & Ratcliff, R. (1979). Priming in episodic and semantic memory. *Journal of Verbal Learning and Verbal Behavior, 18*, 463-480.

Moscovitch, M. (1992). A neuropsychological model of memory and consciousness. In L.R. Squire & N. Butters (eds.) *Neuropsychology of Memory, 2nd ed.* (pp. 5-22). New York: Guildford Press.

Moscovitch, M., Winocur, G. & McLachlan, D. (1986). Memory as assessed by recognition and reading time in normal and memory-impaired people with Alzheimer's disease and other neurological disorders. *Journal of Experimental Psychology: General, 115(4)*, 331-347.

Schacter, D.L. (1990). Perceptual representation systems and implicit memory: Toward a resolution of the multiple memory systems debate. *Annals of the New York Academy of Sciences, 608*, 543-571.

Schacter, D.L. & Graf, P. (1986). Effects of elaborative processing on implicit and explicit memory for new associations. *Journal of Experimental Psychology: Learning, Memory and Cognition, 12*, 432-444.

Smith, M.C., MacLeod, C.M., Bain, J.D. & Hoppe, R.B. (1989). Lexical decision as an indirect test of memory: Repetition and list-wide priming as a function of type of encoding. *Journal of Experimental Psychology: Learning, Memory and Cognition, 15*, 1109-1118.

Tanaka, J.W. & Farah, M.J. (1993). Parts and wholes in face recognition. *Quarterly Journal of Experimental Psychology, 46A(2)*, 225-245.

Winograd, E. & Rivers-Bulkeley, N.T. (1977). Effects of changing context on remembering faces. *Journal of Experimental Psychology: Human Learning and Memory, 3(4)*, 397-405.

Perceptual Learning in Mathematics: The Algebra-Geometry Connection

Ana Beatriz V. e Silva (Silva@psych.ucla.edu)
Department of Psychology, University of California, Los Angeles
Los Angeles, CA 90095

Philip J. Kellman (Kellman@cognet.ucla.edu)
Department of Psychology, University of California, Los Angeles
Los Angeles, CA 90095

Abstract

An important component of expertise is the rapid pickup of complex, task-relevant pattern structure, yet such skills are seldom trained explicitly. We report initial results applying principles of perceptual learning to the processing of structure in mathematics, specifically the connection between graphed functions and their symbolic expressions. Subjects in two experiments viewed graphs of functions and made a speeded, forced choice match from several equations. Training consisted of many short trials of this active classification task involving examples of a function (e.g., sine) subjected to various transformations (e.g., scaling, shifting, reflection). Experiment 1 used *contrastive feedback* -- the graph for a trial was shown superimposed on the canonical function to accentuate transformations. Subjects showed substantial performance gains from 45 minutes of training and transferred to new instances, new function families and a new task. In Experiment 2, with contrastive feedback removed, subjects showed no transfer to new functions. The results indicate the value of perceptual training in producing mathematical expertise and the value of contrastive feedback in particular.

What does it mean to attain mathematical expertise? Instruction most often emphasizes declarative knowledge -- facts and concepts. A student may learn, for example, that the function $y = \text{Sin } x$ can be generated by a certain construction involving a triangle. Having learned about the sine function, the student may be able to answer certain factual questions and work out problems on a test.

There is more to expertise, however, than facts, concepts and inferences. Suppose we ask the student who is familiar with $y = \text{Sin } x$ what the graph of $y = \text{Sin } (x-2)$ would look like. Chances are the student will not know immediately. Getting the answer may be an inference process requiring several steps: We can substitute 2 for x to find that the function now crosses the x axis at $(2,0)$ instead of $(0,0)$. If we check a few more points, perhaps the answer would become clear.

How would an expert respond to this query? At a glance, it is intuitive that the "-2" in the function $y = \text{Sin } (x-2)$ shifts the whole function rightward on the x axis by two units. If it were $y = \text{Sin } (x+2)$, the shift would be to the left. Likewise, $y = 3 \text{ Sin } x$ amplifies the function along the y axis; $y = \text{Sin } (4x)$ compresses along the x; and $y = 2 - \text{Sin}$

x causes a reflection around the x axis and shifts the whole function upward by two units. The expert detects at a glance the structural relationships in each equation and intuitively knows their meaning in the spatial representation (the graph in Cartesian coordinates).

This kind of expertise is important, and in some tasks, decisive. To be able to look at a plot of data and recognize how it could be approximated by an equation, or to visualize the consequences of changes in an equation for the shape of the function, would seem to be basic to the use of mathematics in science. The student who can work out these connections through factual knowledge and reasoning lags behind one who intuits the relevant patterns at a glance. The scientist's ability to extract relevant structures in both equations and graphs allows her attention and effort to be allocated to the scientific problem at hand -- without having to pause to work through what e^{-x} looks like. These aspects of skilled performance are our concern in this paper.

How these skills arise, in mathematics or in other domains, may appear mysterious. In the examples of sine functions above, we could specifically state to the student, in a lecture or text, all of the transformations mentioned. We might find that even after the student learned to state the facts, classification of new examples would be slow and arduous. Attaining the fluent pattern classification skills of the expert may not come from this kind of instruction. We would say that the students "need experience" and that they will attain greater fluency with time. The same is said to beginning radiologists, instrument pilots, accountants and novices in other domains about the structures and patterns they work with.

The passage of time is not very satisfactory as an explanatory notion. A specific hypothesis about the development of such skills is that they involve perceptual learning (Gibson, 1969). Broadly, perceptual learning is defined as "an increase in the ability to extract information from the environment, as a result of experience and practice with stimulation coming from it" (Gibson, 1969). Perceptual learning is a cornerstone of advanced human performance. In some domains, it leads to competence that appears nearly magical. The magic comes from processes that allow for continuing improvement in the extraction of pattern structure with practice. For example, in 1996, Garry Kasparov, the world champion chess player, defeated Deep Blue, a chess-playing computer that examined 125 million possible moves per second. In 1997, Kasparov lost a close match to an improved Deep Blue that examined approximately 250 million moves per second. How can a human, who examines a smaller number of possible moves on each turn (about 4) even begin to compete with a machine that computes all the possibilities for exactly what the board will look like many moves later? The grandmaster has developed pattern pickup skills, specifically relevant to the game of chess (Chase & Simon, 1973). These allow efficient processing of the board structures that will be relevant to the outcome of the game. Much of this knowledge is not accessible to the player. If it could be clearly articulated, the grandmaster's strategy could be implemented in a computer chess-playing program, allowing computers that look at mere thousands or even hundreds of moves to defeat the best humans.

We do not yet have good process models of perceptual learning. Evidence indicates, however, that the performance of certain acts of information processing, not the passage of time, lead to advances in perceptual learning (Gibson, 1969; Hock, 1987; Karni & Sagi, 1993; Kellman & Kaiser, 1994; Lewicki, 1992; Pick 1965; for a review see Goldstone, 1998). Impressive changes in detection and pattern classification have often been obtained, even using relatively brief training procedures in laboratory experiments. Such experiments shed light on the conditions that lead to perceptual learning, and they raise the possibility of systematizing procedures that might accelerate the development of pattern extraction skills in educational and training settings. In short, even while we lack complete process models, we know quite a bit and can learn more about how to produce perceptual learning. Expert pattern processing skills represent a component of expertise that differs from declarative knowledge and must be trained differently. In the present research, we seek to apply perceptual learning principles to mathematics and investigate how they can be optimized.

What are the ingredients for obtaining perceptual learning? Our answer is tentative, but a number of ideas have received support. In the first place, information pickup is a skill that is not much exercised by hearing a recitation of facts. Training using many short trials requiring a speeded response may optimize perceptual learning. In these trials, the learner must be exposed to a range of variation in a stimulus set that contains the invariants that support some discrimination or classification (Gibson, 1969). Often it is suggested that the learner must be actively involved in a classification task, i.e., must attend and respond on a number of trials. Where discovery of differences is primary, such trials may allow learning without feedback (Gibson, 1969), whereas for other classifications feedback may be crucial.

The Algebra – Geometry Connection

In the present research, we developed a perceptual learning module (PLM) to advance subjects' abilities to relate graphs of mathematical functions to their symbolic expressions. Obviously, connecting graphs and equations is a complex

task, one that no doubt has conceptual and perceptual components.[1] We do not attempt to separate these components here. We chose the task, however, because subjects appear to be quite poor at it initially, despite having satisfactorily completed relevant coursework in mathematics. It seemed plausible that their difficulties were due in part to the limitations of traditional instruction and might be overcome by perceptual learning.

We have three ultimate goals in this research. One is to test whether a brief period of perceptual training can improve subject's performance in interpreting graphs and equations. If so, a second goal is to determine what variables are important in producing and optimizing perceptual learning. These include the type of feedback and the role of active classification. Finally, we are interested in how acquired pattern processing skills transfer to new stimuli and more complex tasks. In this paper, we report some initial findings related to all three of these goals.

Experiment 1

In Experiment 1, we tested a perceptual learning procedure for developing pattern processing skills in matching graphs to equations. The procedure incorporated a number of ingredients that may be important in accelerating perceptual classification skills. First, subjects performed an active classification task. They performed many short trials, each requiring a speeded, perceptual classification response. On these trials, a graph appeared, followed two seconds later by 3 choices of equations, one of which matched the graph. Functions spanned a range of variation appropriate for subjects to extract the invariant patterns specifying various aspects. Finally, a particular kind of feedback display --

---

[1] We should make clear what we mean by "perceptual." It is common to think of the senses as providing low-level, concrete, sensory information, such as color, and invoking higher cognitive processes, such as inference processes, to account for more abstract descriptions of reality. Our view of *perception* is a much more inclusive one: Perceptual mechanisms respond to abstract patterns of information and produce abstract and meaningful descriptions of reality (c.f., J. Gibson, 1966, 1979; Marr, 1982; for recent discussions see Kellman & Arterberry, 1998; Barsalou, in press). Although this is not the place to attempt to find a clear boundary between perception and conception, our working hypothesis is that any potentially detectable pattern in the stimulus is a candidate for perceptual learning. The shape of a sine function in a graph, and the difference in patterns between $y = \text{Sin } x$ and $y = \text{Sin } 2x$ are candidates for perceptual learning, as are aspects of the elements, positions and sequencing of symbols in the equations. On the other hand, knowledge that the sine function is derived from a certain construction involving a triangle is not potentially discernible from looking at the graph of $y = \text{Sin } x$; contributions of that knowledge to performance are therefore not perceptual. The reader who worries about the boundaries of perceptual learning may feel more at home with our occasional substitution of the more neutral phrase "pattern learning."

what we call *contrastive feedback* -- was used. This display showed the canonical function (e.g., y = Sin x) as a dotted line in the background, and the particular function for the trial (e.g., y = 4 Sin (x+2)) in front. Contrastive feedback may highlight the particular pattern transformations relating the basic function to its variants. It is a form of augmented feedback (e.g., Lintern, 1980).

Method

Participants. Participants were 20 undergraduate students at the University of California, Los Angeles who received credit units for participation in the one-hour experiment.

Materials and Apparatus. Stimuli were designed with *Mathematica*, version 2.2.1, and consisted of graphs of four types of mathematical functions: Cosines, Logarithms, Sines, and Exponentials. They were displayed in a Power Macintosh 7100/66.

Design and Procedure. Subjects were randomly assigned to one of 2 training groups: Sines and Exponentials, or Cosines and Logarithms.

Training consisted of 8 blocks with 20 trials each. At the beginning of each block, the basic function on which the subject was to be trained was displayed on the screen (i. e., the graph and equation for y = Sin x). Then, for each trial, a graph with some transformation from the basic function was presented (i. e., Sin 3x) along with three equations.

Variations within function families were created using 6 transformations. These included:

1)   <u>X-shifting</u>: Adding some integer to the variable x within the scope of the basic function (e.g., y = Sin (x+4))

2)   <u>Y-shifting</u>: Adding some integer outside the scope of the basic function (e.g., y = 4+Sin x)

3)   <u>X-scaling</u> (e.g., Sin x/4)

4)   <u>Y-scaling</u> (e.g., y = 4 Sin x)

5)   <u>X-reflection</u> (e.g., y = Sin (-x))

6)   <u>Y-reflection</u> (e.g., x = - Sin x)

Many problems included combinations of these (e.g., y = 2-Sin(5x)). The materials given to the 2 training groups were matched regarding the types of transformations seen during training. However, matching of specific equations were not necessarily made. For example, if y = Sin 2x was given to one group, the other group would not necessarily be given y = Cos 2x. The second group may have received Cos 4x, for example.

The subject had to choose which of the possible answers corresponded to the graph. Responses were entered on a keyboard. A trial feedback screen reported whether the subject's answer was right or wrong. This screen also showed a display with both the tested graph (depicted with a thick blue line) and the basic one (depicted with a dotted line) superimposed, along with a label indicating the correct equation. This design for the feedback screen aimed to highlight the relevant transformations relating the tested expression and graph to the basic function type. At the end of each block, average accuracy and reaction time for the previous 20 trials was reported. The training phase of the experiment lasted 40-45 minutes.

Dependent Measures. Accuracy and reaction times for the first 10 trials of the first 2 blocks were taken as a pre-test measure, and scores for the last 10 trials of the last 2 blocks were taken as a measure of the subjects' end-of-training performance (EOT). Three kinds of posttests were administered to both groups:

1)   A *Familiar Functions Posttest* (FFP) presented in the same format as training, composed of 10 *new* instances in the function families the subjects saw in training.

2)   An *Unfamiliar Functions Posttest* (UFP) in the same format, consisting of 10 trials from the function families they had not seen (i. e., subjects trained on Sines and Exponentials were tested on Cosines and Logarithms).

3)   A Remote Transfer Posttest (RTP) assessing transfer to a different task. In this task, subjects tried to make sense of complicated "combination" functions (such as y = - Cos (2x) * Log (-x) or y = Exp (-x/3)–5 Sin (5-x)). Eight such combination functions were generated for both studied and not studied pairs of function families. This yielded two written forms with 8 graphs and 8 functions. Subjects' task was to perform a matching test, indicating the correct equation for each graph. Total time and accuracy to complete each form was measured. Half of the subjects did one form first, and the other half did the other form first.

A control group of 20 subjects received no training and was given the same tests as in the experimental condition: the post-test with both sets of basic functions and the RTP with combination functions. In the regular training, the pretest (initial learning trials) formed a within-subjects baseline. The control group served as an additional baseline group for assessing effects in FFP, UFP, and RTP. Because subjects were not given any feedback in the control group, their performance allowed a check on possible rapid learning effects that might have elevated performance in the pretest.

Results

Training, FFP and UFP Results. Training had clear and highly reliable effects on accuracy (shown in Figure 1) and RT. Subjects improved from about 50% correct in the pretest to about 70% in the final block of training. Response times decreased about 40% during training. Accuracy at EOT and in the FFP were both higher than in the pretest and did not differ from each other. In the UFP, in which subjects saw new instances, accuracy was slightly higher than the pretest but not as high as in FFP. Accuracy in the control group was similar to the pretest and worse than the EOT and FFP. Reaction times were negatively correlated with accuracy.

These patterns were confirmed by the analyses. Accuracy and RTs were analyzed using a 2 (functions trained: sines & exponentials or cosines & logarithmics) X 4 (phase: pretest, EOT, FFP, UFP) analysis of variance (ANOVA), with repeated measures on the latter factor. There was a significant main effect of phase, for accuracy $F(3,54) =$

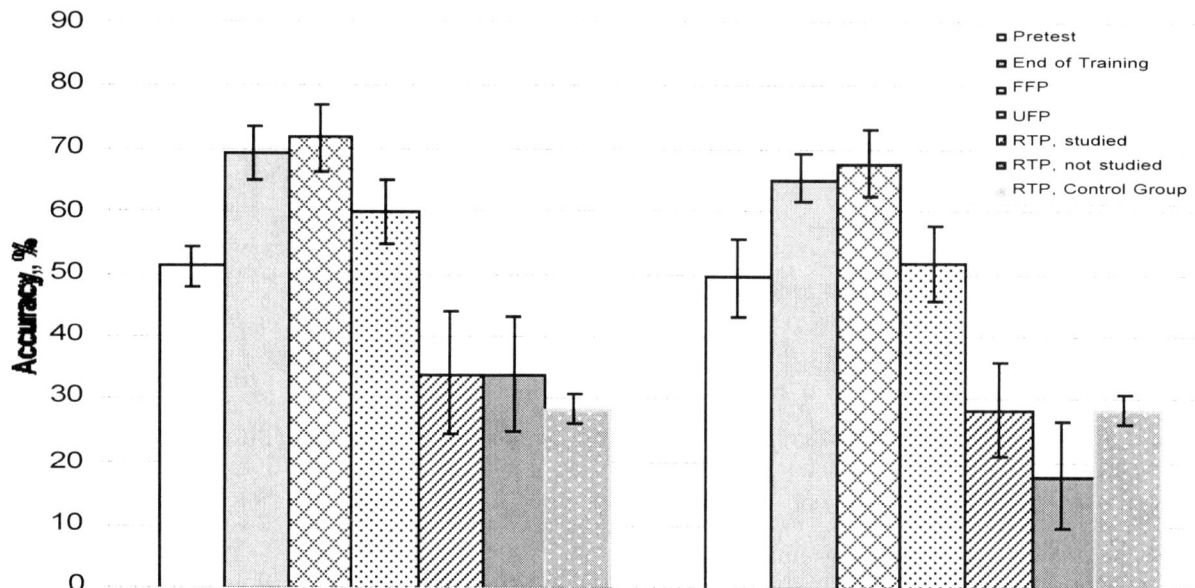

Figure 1. Results for Experiments 1 and 2. Mean accuracy for the different stages of training are shown, including pre-test, end of training, familiar functions post-test (FFP), unfamiliar functions post-test (UFP), remote transfer post-test (RTP) for studied and non-studied functions, and remote transfer post-test for the control group.

12.669, p<0.001, and RT, F(3,54) =18.596, p<0.001. Individual comparisons showed that participants were significantly better and faster at the end of the training than at the beginning (accuracy: t(19) = 7.241, p<0.001; reaction time: t(19) = 7.486, p<0.001), and in the UFP (accuracy: t(19) = 2.582, p<0.02; reaction time: t(19) = 4.174, p<0.001). Accuracy was higher in FFP than in both the pre-test, t(19) = 4.547, p<0.001, and the UFP, t(19) = 2.239, p<0.05. RTs in the FFP were faster than in the pretest, t(19) = 4.098, p<0.001, but not reliably different from UFP. Subjects' UFP performance was superior to pretest performance (accuracy: t(19) = -2.281, p< 0.05; reaction time: t(19) = 2.666, p<0.015). Accuracy did not differ between EOT and FFP, but the former phase showed better reaction time, t(19) = 4.198, p<.001. There was a reliable interaction between function trained and condition, F(3,54)=3.226, p<0.03. For reasons that are unclear, a somewhat larger training effect was found with sines and exponentials than with cosines and logarithmic functions.

The control group did not differ from the pretest group in accuracy, t(38)=-0.453, n.s. UFP was faster than the control group (t(38)=-3.551, p<.001) but not more accurate (t(38)=1.425, n.s.). Reaction times in the control group were reliably slower than in the pretest, mean difference = 4.6 sec, t(38)=-2.661, p<0.011. This result suggests some rapid improvements in response time during the early training trials. (Note that in order to get 10 pretest trials with each function type studied, the pretest comprised the first 10 trials of each of the first two trial blocks, i.e., trials 1-10 and trials 21-30).

Remote Transfer Post-test Results. Performance on the two forms containing combination functions did not differ depending on the function families seen during training.

This observation was verified by the analyses. Two 2 X 2 (functions trained by functions tested) ANOVAs, one for accuracy and another for reaction time, yielded no significant main effects or interactions. Comparisons with the control group revealed no reliable differences in terms of accuracy. Response times were quite long for the RTP, on the order of three minutes per form. Control subjects were reliably slower than Experiment 1 subjects for non-studied functions, t(38)=-2.19, p<0.035, and marginally slower for the studied functions post-test, t(38) =-1.802, P<0.05 (one tailed).

Discussion

The results of Experiment 1 support several conclusions. First, training designed specifically to foster perceptual learning can improve subjects' performance in relating graphs and equations. Although our subjects had previously learned about the relevant functions in mathematics classes, they had not become skilled in classifying patterns and recognizing transformations. A relatively brief intervention substantially improved both accuracy and speed. A second finding was that training generalized to new instances in the same function families and also to similar transformations deployed in new function families. Training in this experiment did not lead to reliable effects on our remote transfer (combination function) test, which proved quite difficult for subjects.

It does not appear that the 45 minutes of training were sufficient to achieve automatic pattern recognition. At the end of training, accuracy had not reached ceiling. Still, response times had fallen to about 5.5 sec on average. Response times on this order suggest that subjects were becoming more automatic in their pattern classification rather than using an elaborate reasoning process. Further training might lead to greater automaticity in classification.

What did subjects learn? The transfer results suggest that the training effects were not merely about the trained functions, such as sines. Transformations such as compression, scaling and shifting have the same sorts of effects across function classes. Initial performance levels of subjects suggested that these generalities have not been well learned from conventional mathematics instruction. Our results showed enhancement of performance after training even with new function families, suggesting that the symbolic and graphical "meanings" of basic transformation patterns were learned to some degree. The results suggest that PLMs may have great promise for developing fluent pattern processing in mathematics.

Experiment 2: Contrastive Feedback

Experiment 1 indicated the efficacy of perceptual classification training, using several ingredients suggested by earlier research and by intuition. Which ingredients are crucial to the usefulness of this kind of training? These questions have hardly begun to be addressed, especially in the application of perceptual learning to complex skills, such as doing mathematics. A goal of our research is to determine systematically the effects of particular aspects of training in order to optimize PLMs in education.

In Experiment 2, we examined whether the particular kind of contrastive feedback used in Experiment 1 had important effects on learning. Recall that after each problem, participants viewed the problem function superimposed on the canonical function. If this particular type of feedback facilitated discovery of relevant pattern transformations, then eliminating it might reduce the success of training. In particular, we hypothesized that this type of feedback might have been especially helpful in producing transfer to new function families. Eliminating it might therefore be expected to reduce or eliminate transfer of learning to new functions.

Method

Participants. Twenty undergraduate students at the University of California, Los Angeles, received credit units for participation.

All aspects of the method were identical to Experiment 1 except that contrastive feedback was eliminated. Instead, feedback screens indicated whether the response on the trial had been correct and displayed the graph and correct equation for that problem.

Results

Training, FFP and UFP Results. As in experiment 1, training produced highly reliable effects on accuracy (shown in Figure 1) and RT. Subjects improved from about 50% correct in the pretest to about 65% in the final block of training. Response times decreased about 40% during training. Both the accuracy and speed at the end of training were maintained in the FFP. In contrast to the results of Experiment 1, there was little or no transfer of training to

the UFP; training effects were largely confined to the familiar (studied) function types. The control group, pretest and UFP performance of the training group were very similar.

These observations were confirmed by the analyses. Accuracy and RTs were each analyzed using a 2 (functions trained) X 4 (phase: Pretest, EOT, FFP, UFP) ANOVA, with repeated measures on the latter factor. There were reliable main effects of phase, for accuracy $F(3,54) = 8.892$, $p<0.001$, and for RT, $F(3,54) =8.405$, $p<0.001$. Individual comparisons showed that participants were significantly better and faster at EOT than at the beginning (for accuracy: $t(19) = 3.507$, $p<0.002$; for reaction time: $t(19) = 3.816$, $p<0.001$). Accuracy and reaction time were better in FFP than in the pre-test (accuracy: $t(19) = 3.789$, $p<0.001$; reaction time: $t(19) = 3.165$, $p<0.005$) and also better than in the UFP (accuracy: $t(19) = 3.47$, $p<0.003$; reaction time: $t(19) = 2.687$, $p<0.015$). Accuracy did not reliably differ between EOT and FFP, $t(19) = 0.76$, n.s. For reaction time, there was a marginal advantage for the EOT problems over FFP, $t(19) = 1.96$, $.05<p<0.10$. Subjects' UFP performance did not differ significantly from the pretest (accuracy: $t(19) = 0.453$, n. s.; reaction time: $t(19) = 0.805$, n. s.).

Accuracy in the control group did not differ from either the pretest ($t(38) = -0.59$, n.s.) or the UFP ($t(38) = -0.192$, n.s). Reaction times in the control group were significantly slower than in both the pretest, $t(38) = -2.706$, $p<0.01$, and UFP, $t(38) = -3.359$, $p<.002$.

Remote Transfer Post-test Results. As in Experiment 1, performance on the two forms containing combination functions did not differ depending on the trained function families. Two 2 X 2 (functions trained by functions tested) ANOVAs, one for accuracy the other for reaction time, yielded no significant main effects or interactions.

Accuracy and reaction times on both forms (studied and unstudied functions) of the RTP did not differ reliably from a control group that received no training.

Cross-Experiment Comparisons

To more carefully assess the effect of contrastive feedback, we performed several analyses including the data from both Experiments 1 and 2.

Training, FFP and UFP Results. ANOVAs were performed on accuracy and reaction time data. Each was a 2 X 2 X 4 (experiment by functions trained by condition) design with repeated measures on the condition factor. There was a strong main effect of condition (accuracy: $F(3, 108) - 20.56$, $p <.001$, RT: $F(3,108) - 21.99$, $p <.001$), indicating the effects of training in both experiments. There was also a condition by functions trained interaction for accuracy, $F(3, 108) = 4.188$, $p <.008$, indicating that in both experiments training effects were greater for subjects trained on sines and exponentials. No other main effects or interactions reached significance in the accuracy or RT analyses.

The lack of a reliable experiment by condition interaction for accuracy contrasts somewhat with a difference between

Experiment 1 and 2 observed above. Specifically, in Experiment 1, participants showed better accuracy on unfamiliar functions (UFP) than in the pretest, $t(19) = -2.281$, $p < 0.05$, whereas no such difference appeared in Experiment 2, $t(19) = 0.453$, n.s.

This difference suggests that contrastive feedback produced better learning of transformations that could be applied to new function families. The lack of direct support for this interaction in the combined analysis is probably due to the greater variability present in between-subjects comparisons. Nevertheless, the mixed results suggest caution. Fortunately, the cross-experiment comparisons of RTP performance (below) provide some confirmation of the difference in training effects between the two experiments.

Remote Transfer Post-test Results. Remote transfer effects of training appeared to be somewhat larger in Experiment 1, which used contrastive feedback, than in Experiment 2, which did not. Even though no reliable differences were found between the studied and non-studied functions within each experimental condition, a difference was found in terms of accuracy when comparing experiments 1 and 2. More specifically, subjects in experiment 1 performed better than subjects in experiment 2 for the non-studied functions ($t(38)=1.906$, $p<.05$, one tailed). The superior performance of experiment 1 subjects' in the non-studied functions attests for the effect of the contrastive feedback in producing better learning of transformations that can be applied to new function families. No other comparisons reached significance.

General Discussion

Taken together, the results of experiments 1 and 2 indicate a clear training effect. Subjects in both experiments substantially improved their accuracy and speed in relating graphs and equations, as compared to a control group. Subjects' improvement extended beyond the specific examples on which they were trained. They showed equally good performance on new examples from the function families on which they had trained, indicating that learning did not depend merely on memorizing specific instances.

Subjects in Experiment 1 also transferred their learning to *new* function families. This result did not appear in Experiment 2, however. These outcomes suggest that the *contrastive feedback* used in Experiment 1 may have laid the foundation for transfer by directing attention to the relevant *transformations* in the stimulus patterns. The results are consistent with the idea that perceptual learning might be accelerated by augmented feedback that helps to direct attention to relevant features and dimensions.

Our remote transfer test was difficult and showed modest effects. Here again, however, a comparison between the two experiments revealed that Experiment 1 produced more gains in accuracy for the non-studied functions than Experiment 2, which did not have contrastive feedback.

The contributions of other procedural ingredients remain to be assessed. For example, it is often asserted that perceptual learning depends on an active classification task (e.g., Karni & Sagi, 1993), but there have been few careful

tests of this idea. In the present work, both groups were given active classification tasks. Currently we are studying whether mere exposure in short episodes to corresponding graphs and equations produces learning and transfer.

In summary, our results indicate that a brief period of perceptual training can substantially improve subjects' performance in processing mathematical structures expressed in graphs and equations. Contrastive feedback – highlighting the dimensions of difference between a basic function and transformations of it -- enhances the learning of relationships that transfer to novel function families. Perceptual learning modules may have great promise in accelerating the development of components of expertise that do not arise easily from traditional instruction.

Acknowledgments

This work was supported by NSF Grant SBR 9720410. We thank Robert Bjork, Orville Chapman, Randy Gallistel, Rochel Gelman, James Stigler and Arlene Russell for helpful discussions of perceptual learning and its applications. Address correspondence to A. B. V. e Silva.

References

Barsalou, L.W. (in press). Perceptual symbol systems, *Behavioral and Brain Sciences.*

Chase, W. & Simon, W. (1973). Perception in chess. *Cognitive Psychology, 4,* 55-81.

Gibson, E. J. (1969). *Principles of perceptual learning and development.* New York: Prentice-Hall.

Gibson, J. J. (1966). *The senses considered as perceptual systems.* Boston: Houghton Mifflin.

Gibson, J. J. (1979). *The ecological approach to visual perception.* Boston: Houghton Mifflin.

Goldstone, R. L. (1998). Perceptual Learning. *Annual Review of Psychology*, 49, 585-612.

Hock, H. S., Webb, E., & Cavedo, L. C. (1987). Perceptual learning in visual category acquisition. *Memory & Cognition*, 15, 544-556.

Karni, A. & Sagi, D. (1993). The time course of learning a visual skill. *Nature*, 365, 250-252.

Kellman, P. J. & Arterberry, M. E. (1998). *The cradle of knowledge – Development of perception in infancy.* Cambridge, MA: The MIT Press.

Kellman, P. J. & Kaiser, M. K. (1994). Perceptual learning modules in flight training. *Proceedings of the Human Factors and Ergonomics Society 38th Annual Meeting*, 1183-1187.

Lewicki,, P., Hill, T, & Czyzewska, M. (1992). Nonconscious acquisition of information. *American Psychologist*, 47, 796-801.

Lintern, G. (1980). Transfer of landing skill after training with supplementary visual cues. *Human Factors*, 22 (1) : 81-88.

Marr, D. (1982). Vision. San Francisco: Freeman.

Pick, A. D. (1965). Improvement of visual and tactual form discrimination. *Journal of Experimental Psychology*, 69, 331-339.

Learning, Development, and Nativism: Connectionist Implications

Sylvain Sirois (sirois@psych.mcgill.ca)
Department of Psychology, McGill University
1205 Penfield Avenue
Montréal, Qc H3A 1B1 CANADA

Thomas R. Shultz (shultz@psych.mcgill.ca)
Department of Psychology, McGill University
1205 Penfield Avenue
Montréal, Qc H3A 1B1 CANADA

Abstract

Feedforward neural network models of cognitive development are reviewed within the framework of a functional distinction between learning and development. This analysis suggests that static architecture networks implement a learning theory, whereas generative architecture networks combine learning and development. Both types of networks are then evaluated in terms of genetic costs. Within a levels-of-innateness framework, generative architectures are viewed as more plausible than static ones. Static architecture networks appear to implement a form of nativistic elicitation.

Introduction

Feedforward neural networks process information through brain-inspired principles: excitatory and inhibitory stimulation, activation summation, activation threshold, unit activation, and massively parallel and distributed processing. Although much simpler than neural tissue found in most (if not all) animals, these networks can process complex information and provide an alternative framework to rule-based symbolic approaches to the study of cognition. The fact that such networks can learn also provides researchers with powerful tools for the study of human learning and development (Elman, Bates, Johnson, Karmiloff-Smith, Parisi, & Plunkett, 1996; Shultz & Mareschal, 1997). This paper is an expansion on work by Quartz (1993), who studied assumptions of PDP models within the framework of Valiant's probably approximately correct (PAC) model of learning. The first section proposes a formal distinction between learning and development, which serves to evaluate the theoretical implications of developmental work using different neural network architectures. The second section evaluates underlying assumptions of different types of neural network algorithms within the levels-of-innateness framework outlined by Elman and his colleagues (1996). Each of the two arguments that can be formulated to support static neural networks as models of human cognition are inconsistent with the neurological evidence. The discussion stresses that static networks implement a form of nativistic elicitation, as suggested by their inability to escape Fodor's paradox. Overall, static networks do not seem to be good candidates for modeling cognitive development.

A Distinction Between Learning and Development

In order to evaluate the possible contribution of neural network modeling to our understanding of human cognitive development, it is important to distinguish it from learning. As Carey (1985) suggested, attempts to differentiate between learning and development often confound two distinctions: whether knowledge acquisition requires restructuring, and whether changes are domain-general or domain-specific. This paper focuses on the former.

How knowledge acquisition may or may not require restructuring is a question that was not directly addressed by Elman and his colleagues in their landmark book on the connectionist perspective of cognitive development (1996). Although they do provide justification for the study of development above and beyond learning, they do not commit to a formal distinction between these two processes. Because of the substantial impact their book has on the study of cognitive development, a clarification appears timely as it bears on the theoretical implications of different types of neural network research.

In this paper, the following functional distinction is proposed. Learning is defined as a change *within* an existing processing structure in order to adapt to information from the environment. This broad description is compatible with general statements about learning, such as found in nativistic accounts (e.g., Fodor, 1980) and developmental models (e.g., Carey 1985; Piaget, 1982). In contrast, development is defined as change *of* an existing structure to enable more complex and adaptive cognitive activity. This general statement highlights the key idea underlying most theories of development; that is, a qualitative change in the structure supporting cognition. Such a general definition of a developmental mechanism can be found explicitly or implicitly in Piaget's (1982) abstraction, Karmiloff-Smith's (1992) representational redescription, and Carey's (1985) conceptual change, for example. These functional definitions of learning and development allow a distinction between the two processes, removing overlap between them and constraining their individual contribution to cognitive change. Learning is viewed as parameter adjustment within a given structure; development as change of structure within which learning (as well as other cognitive processes) takes place.

The outlined distinction between learning and development is also useful in the evaluation of neural network models of cognitive development. A simple example that illustrates this point is the XOR Boolean operator. Implemented as a function, XOR takes two arguments that can be either true or false and returns *true* if one and only one argument is true, otherwise it returns *false*. Figure 1 presents two different networks that solve the XOR problem. Given that the network on the left is a typical static feedforward network and the one on the right is a generative cascade-correlation network, how they achieve these indistinguishable solutions at the output level from initially random weights requires two different stories.

At the onset of training, the static network of Figure 1 has the representational power to solve the problem. Within the multidimensional weight space determined by its topology, there is a region that will produce the correct output. Only quantitative changes are required to move the network from its initially random position in weight space to the region that solves the problem. Because the learning algorithm capitalizes on the nonlinear properties of hidden and output units, these gradual weight changes will not be linear (i.e., the delta value for a weight is a function of the receiving unit's nonlinear activation, which changes from epoch to epoch). Although these nonlinear changes qualify as learning, they do not qualify as development, because only parameters of the current structure are changed.

The learning history in the cascade-correlation network from Figure 1 would be different. The initial two-layer architecture of the network does not allow it to properly solve the XOR problem. There is no region in its weight space that will produce the correct output. Therefore error reduction in output training will stagnate above a satisfactory level (typically, the network will reduce all weights to near-zero values in order to minimize error on all patterns). This stagnation in error reduction spurs the recruitment of a hidden unit that is then used for further output training. This new unit increases the dimensionality of the weight space, in which a solution region now exists. For this network, both learning and development combine to achieve a solution to the XOR problem. Learning (i.e., parameter adjustment) within the initial structure is unsuccessful, prompting a modification of the architecture. This qualifies as development because the structure within which learning takes place is changed. Further learning within the new structure finally solves the problem.

We are not arguing that a generic XOR function is a human developmental problem. However, this basic example highlights how static and generative networks tell a different story about learning and development for nonlinear problems. Namely, that the former network type is a learning model, and that the latter is a developmental model (that also incorporates learning).

Elman and his colleagues (1996), by not providing a clear distinction between learning and development, make an equivocal statement about connectionism as a model of development. Even though they make a good case that there is development and not just learning in human cognitive change, the simulations they report consist of learning models that capture developmental data. A good example is their discussion of a balance-scale model by McClelland (1989).

The balance-scale task consists of a beam with a series of unit-spaced pegs on both of its sides, centered on a fulcrum. On a given trial, a number of unit weights are placed on one peg on each side. The participant is required to predict which side of the beam (if any) would go down, provided that the balance would be free to move. Robust and replicable developmental effects have been observed in children of various ages performing this task (Siegler, 1981).

Initially, younger children perform at chance level (stage 0). As they grow older, they begin to use weight information in their predictions (stage 1). They predict that one side will go down if it has more weight. At the next stage, they begin to use distance information, only if there is equal weight on both arms of the scale (stage 2). Stage 3 children use both types of information yet fail to integrate them, so when weight and distance conflict, they perform at chance level. This is associated with a U-shaped effect on conflict problems in which the side with larger weight would go down. Stage 2 children make a correct prediction on these problems whereas older, stage 3 children perform at chance level. Finally, stage 4 is the level where performance is correct on all problem types. At stage 4, children's answers appear to follow the torque-rule solution of the problem, which states that if the products of weight and distance on each side of the beam are different, the beam will not balance but tip to the side with the larger product.

The balance-scale task provides modelers with a robust target of stage-wise developmental data. Mathematical models based on catastrophe theory maintain that such a

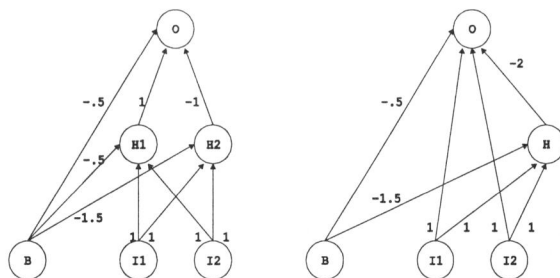

Figure 1: Two different solutions of the XOR problem. The network on the left represents a static network. All the weights and units depicted are there at the onset of training. The solution of the XOR problem required only gradual changes of the initially random weights. The cascade-correlation network on the right differed at the onset of training, with only bias, input, and output units. The recruitment of a hidden unit being necessary at the end of the first output training phase, the network increased its representational power in order to solve the problem. Whereas the output of both networks will be indistinguishable on all four instances of the problem after training, the static network learned and the generative network both learned and developed.

developmental profile is better described as discontinuous (van der Maas & Molenaar, 1992). A neural network that claims to be an adequate model of human development should replicate this ordered progression through the 5 stages of performance. According to Elman and colleagues (1996), this is exactly what was accomplished by McClelland's (1989) model.

McClelland (1989) used static backpropagation networks to model the balance-scale task. His results show that the networks progress through all stages, without reliably settling into stage 4 performance. At the end of training, network behavior is between stage 3 and stage 4 levels of performance. Because not all human adults spontaneously reach stage 4 performance, the results were considered satisfactory (McClelland, 1989).

Irrespective of the fact that some humans do reach stage 4, the networks were viewed as apt developmental models, exhibiting competence acquisition that follows qualitatively distinct stages. However, Raijmakers, van Koten, & Molenaar (1996) suggest that it is the evaluation method and not network learning that is responsible for this stage-wise progression. In McClelland's simulations, network performance was assessed at every training epoch. The performance categories were mutually exclusive, so at any point in training, network performance could be associated with only one stage. Because of the binary decision involved in determining whether a network is at one stage or not, Raijmakers and colleagues (1996) suggest that stage-like discontinuous progression in these networks is an artifact of the evaluation procedure that was applied to the continuous, gradual learning taking place in the networks.

Elman and colleagues did not discuss a cascade-correlation model of the balance-scale task (Shultz, Mareschal, & Schmidt, 1994). This model progresses through all stages in an orderly fashion, and performs at stage 4 by the end of training. Moreover, there are genuine discontinuities in network learning, as it needs to alter its topology to solve the task.

Although it is possible that the discontinuities observed in child performance are due to measurement intervals too broad to assess continuous change taking place at a smaller intervals (Siegler, 1998), this appears unlikely (van der Maas & Molenaar, 1992). If this were the case, the static network might be a better model. What is relevant for this paper, though, is that a learning model may be construed as capturing developmental phenomena if learning and development are not formally distinguished. The ability of a network to mimic developmental data does not, in itself, make it a developmental model. Static backprop models only implement learning, even when one observes nonlinear changes during learning. As we argue in the next section, they also offer a different view of innateness than generative networks.

Elman (1993) reports an interesting simulation using the simple recurrent network architecture (SRN). SRN networks are similar to static feedforward networks, with the addition of a bank of context units. These units take as activation values the activations of the hidden units at one time step, and are fed back to the hidden units at the following time step. Such context units provide the network with a working memory, essential for sequential problems such as language. Elman (1993) found that he could improve the performance of SRN networks on complex problems that they failed to learn by enabling the context units to deal with progressively longer strings. Elman (1993) actually implemented a sort of sequential generative architecture. However, the rest of the static architecture of the SRN is problematic from a nativistic perspective, as is the case for all static models.

Levels of Innateness

One important contribution of the Elman and colleagues' (1996) book is their review of innateness. They identify three levels at which concepts could have an innate basis: representations, architectures, and timing. The focus is not as much towards identifying what is or is not innate, but rather towards defining how things could be innate.

Hardwired concepts or knowledge would be at the representational level. This level of innateness implies that specific synapses in the brain must be designed in order to represent concepts before experience could have shaped such connections. Spelke's (1994) suggestion of innately specified core theories in infants would fall in this category of innateness, for example. So would a language acquisition device for a universal grammar (Chomsky, 1975). For nature to implement such information prior to any experience, a large amount of precisely designed connections need to be made.

Elman and his colleagues reject such a level of cognitive innateness based on two observations: human DNA cannot encode such a large amount of precise information, and the human cortex exhibits a significant equipotentiality that is incompatible with pre-specified representations. Both observations also have implications for the evaluation of neural network models.

Human DNA is found on twenty-three pairs of chromosomes. It is estimated that it can carry up to 10^9 bits of information on base pairs (Elman et al., 1996). This is not enough data to specify the specific location and configuration of each cell in the body (i.e., mosaic development), so nature must rely on heuristics to generate a viable being from minimal information (i.e., regulatory development). Elman and colleagues (1996) provide an extensive review of known mechanisms through which cells organize themselves functionally and spatially through an interaction between their DNA and the environment.

Their conclusion is that the different processes through which nature makes use of minimal DNA to build humans argues against representational innateness. The information in genes operates at a more abstract level than is required for innate representations, so other constraints are suggested to account for species specific stereotypical behavior.

The plasticity or equipotentiality of the cortex also argues against representational innateness (Elman et al., 1996). Compelling evidence from studies of brain rewiring in small mammals is reported to highlight the plasticity of neural tissue, even in species for which behavior is typically considered more rigid (i.e., innately specified). First, redirecting the visual input of mice to the auditory cortex and vice-versa has the resulting effect that the auditory cortex will process visual signals and the visual cortex will process auditory signals. Moreover, the auditory cortex will develop receptive fields and ocular dominance columns as would normally take place in the visual cortex. Second, transplanting cortical columns from one cortical area to another has a similar effect on the transplanted tissue. Rather than processing the type of information it would have initially, it will process information from its new location and develop to be indistinguishable from its neighbor columns. Both these effects argue against innate constraints on representation, because the cortex will learn to process whatever it is fed and will be strongly influenced by its neighbors.

One justification that is invoked to sustain the idea of innate knowledge in humans is that it would be unreasonable to assume it in all animals but humans (e.g., Karmiloff-Smith, 1992). Quartz and Sejnowski (1997) reviewed the literature on brain development, and concluded that plasticity is most often found in species that are phylogenetically recent and proximal to humans. They suggest that it may be more appropriate to speak of human evolution as moving towards maximal plasticity rather than towards hyperspecialization. Like Elman and colleagues (1996), they suggest that plasticity may be the more adaptive solution, that it is more compact than innately specified knowledge, and that it is sustained by comparative data (Quartz & Sejnowski, 1997).

The next level of innateness is architectural. According to Elman and his colleagues (1996), this is the level where innate constraints on the brain can have a plausible effect on knowledge. Architectural constraints themselves can be divided into three levels. Unit level architectural constraints deal with the specific properties of neurons (e.g., neuron types, response characteristics, type of transmitter). Local architectural constraints, as the name implies, deal with the local organization of neural tissue (e.g., layers, density, degree and nature of connectivity). Finally, global architectural constraints concern the global organization of the local areas. According to the view of the brain as a network of networks, global constraints specify how networks are interconnected.

In these architectural forms of innateness, knowledge is not innate, but the overall structure of the brain constrains how, where, and what information will be processed. This embodies species specific aspects of cognition, without requiring representational innateness. The almost universal specific localization of many important cognitive processes (e.g., Broca's area) is guaranteed by specifying what input is sent to different areas. For Elman et al. (1996), architectural constraints are not only a reasonable way through which genes may constrain cognition: connectionist models implement this level of innateness.

The final level at which something cognitive can be innate is with chronotropic constraints. These affect the timing of maturational events, from cell division in neurogenesis to waves of synaptic growth and pruning, as well as the temporal development of different cortical areas (Elman et al., 1996).

Whereas architectural constraints deal with *what* and *where* information is processed, chronotopic constraints add a *when* dimension to the equation. The order in which information can be processed and integrated over development will have an important impact on the nature of cognitive processes (Elman et al., 1996).

This revised interpretation of innateness leaves one important question unanswered, though: What is implied by different neural network algorithms with respect to innate specification? Elman et al. (1996) took the opposite perspective: What do innate specifications imply? Their answer is that only architectural and chronotopic constraints are reasonable forms of innateness, and consequently so are neural networks with unspecified weights. However, we argue that different architectures have different genetic costs and theoretical implications.

Neural Networks and Innateness

Quartz (1993) has shown that a static network cannot learn what is beyond its representational power (defined by the number of weights, the activation functions of units, and the topology of the network). However, too powerful a network may correctly produce output without having abstracted any relevant information through training. What is implied for the human brain by these results? The exact topology of a network is crucial for its learning behavior. Therefore an important issue, in order to consider static architecture networks as models of human cognition, is sustaining that the brain would have the appropriate topology beforehand as an experience-independent given.

Using static networks as models of human cognitive development would require one of the following two assumptions. The first assumption would be that the large number of neurons in the brain are highly interconnected in such a way that for anything humans need to (and can) learn, the probability of an appropriately connected network existing in advance is extremely high. We call this the *probable-network* assumption. The second assumption is that the brain is provided a priori knowledge of what will have to be learned, and has the appropriately connected networks before any experience, which would be a form of representational innateness (Quartz, 1993). We refer to this as the *knowledgeable-network* assumption, because networks are specified from a priori knowledge about tasks. Let us consider each of these assumptions.

The Probable-Network

There is indeed a very large number of neurons in the brain, estimated at about 10^{11}, and with an average of 10^3 connections per neuron, it is safe to say that it is a powerful computer (Churchland, 1989). But like a supercomputer with an operating system that would allow only single digit arithmetic, such power is useless if not properly wired. According to the probable-network assumption, a great deal of computational power in the human brain is wasted. Only those neurons organized in the appropriate topology for a given task will be of use, the others discarded. Elman and colleagues (1996) refer to the observation that there is an initial proliferation of neurons in young children, followed by substantial pruning. The problems with this approach are three-fold.

First, the simulations reported by Elman and his colleagues should not be understood as making a one-to-one correspondence between neurons in the brain and units in the networks. Rather, units in neural network models should be viewed as analogous to groups of neurons in the brain (columns, modules, regions...), and weights as pathways. This implies that the unit cost of a given network is greater at the level that implements it, namely neurons. In order for the brain to make sure that there is somewhere a network appropriate for the task at hand, a huge number of neurons must go to waste. Such a costly solution does not appear adaptive (Quartz & Sejnowski, 1997).

Second, given that learning and development occur across the lifespan and that most of the neural pruning takes place in childhood (Elman et al., 1996), the implication is that in order to learn a new task in later life, the brain will have kept the appropriate network from childhood. This would be at best odd, because a) pruning was understood as an experience-driven process by which useless neurons are removed whereas their usefulness has not yet been evaluated, and b) the brain would require some mysterious access to the solutions of yet to be encountered problems to keep suitable solutions for future learning.

Finally, support for the probable-network assumption would go against recent findings in neuroscience research (Quartz & Sejnowski, 1997). It is suggested that the overproduction/pruning model of development is overstated in the literature, and that flexibility in the brain provided by synaptogenesis, the generation of new synapses, has a greater role in cognitive change. Moreover, this flexibility available through synaptogenesis is more often found in species that are close to humans phylogenetically.

For these reasons, static neural networks do not appear a tenable approach to modeling human cognition, unless one commits to the second assumption, that networks are already properly connected.

The Knowledgeable-Network

The problem with this suggestion is that it implies more than architectural innateness, unlike what is suggested by Elman and colleagues (1996). They argue that representations need not be innate because a network with initially random connections will find its way to the appropriate representation through weight adjustment. Representational innateness would imply that the weights, and not just the topology, would be pre-specified. This suggestion masks the fact that in order to generate the appropriate static topology, the brain would still require some a priori representation of the problem it will come to learn, because network topology determines what can be learned, and how. If an appropriate topology is not available beforehand, the organism could fail to learn, or learn inappropriately. Given that the appropriate topology is defined as a function of the problem, networks with initially random weights still imply representational innateness (if only a relaxed version). In which case the networks implement elicitation of knowledge through parameter adjustment. Elicitation is a nativistic synonym of learning, where only parameter values need

to be derived through experience, because the organism was provided with the required parameters. So the suggestion the authors raise about the implausibility of genetically specified representations, presented earlier, should be taken a step further and would argue against genetically specified topologies for each problem humans will come to learn.

Conclusion

Overall, static neural networks do not fare well in light of the reasonable objections to some forms of innateness highlighted by Elman and his colleagues (1996). Either they imply a costly and mysterious use of neurons that clashes with data from developmental neuroscience (the probable-network assumption), or they require a disguised form of representational innateness for their implementation (the knowledgeable-network assumption). This is not the case with generative networks such as cascade-correlation. These are cost efficient, because they will develop the architecture required as they learn. Because of this generative property, there is no requirement, through either overproduction of neurons or innately specified architectural maps, for an appropriate topology to be present prior to learning. And generative networks are more consistent with developmental observations in neuroscience (Quartz & Sejnowski, 1997).

As a final note, generative networks can escape the nativistic paradox of development formulated by Fodor (1980), whereas static networks cannot (Mareschal & Shultz, 1996; Shultz & Mareschal, 1997). Fodor's paradox states that a system with a given level of logical power will be unable to generate a logical system at a higher level (Fodor, 1980). This implies that it is impossible to represent something for which one does not already have the representational power. Formulated in the context of computational approaches to learning, this is a strong argument, and it has been used to argue against development and in favor of nativistic ideas.

Cascade-correlation escapes Fodor's paradox through its principled recruitment of additional units during the learning process (Mareschal & Shultz, 1996). Recall the XOR example from the first section. Initially, with its two-layer topology, a cascade-correlation network cannot represent a logical operator of the XOR level. Only linear functions like OR and AND could be learned. XOR, a combination of AND and OR, can only be represented in the network through recruiting an additional, hidden unit. The algorithm does just that when error reduction stagnates, and the new unit is trained to track the network's residual error. The solution is not given to the network, it develops and learns one. Static networks, at the onset of training, must have an appropriate topology in order to succeed. To some extent, the solution is thus built in because networks are powerful enough to represent it. No matter what the initial weight values are, the number and arrangement of these weights determines what the networks can and cannot learn (Quartz, 1993). As such, they fail to escape Fodor's paradox. Moreover, given the implications at the implementational level discussed in the previous section, the learning model implied by static networks may not only fail to realize development, but may very well succeed at implementing nativism as elicitation. Elicitation implies that mere exposure to stimuli will produce the predetermined behavior through parameter tweaking. A priori topologies do exactly that by constraining the representations of a network.

The ideas presented in this paper are not meant to be definitive. The purpose is to raise awareness to a common confound between learning and development, as well as to the biological implications associated with different neural network models. Because of the immense potential of neural network tools for the study of human development, it would be of great disservice to ignore the basic questions that pertain to theoretical assumptions and implications associated with these models.

Acknowledgments

This research was supported in part by a Natural Sciences and Engineering Research Council of Canada (NSERC) graduate fellowship to the first author, and an NSERC operating grant to the second author. The authors thank Isabelle Blanchette, David Buckingham, and Yoshio Takane for their comments on an earlier version.

References

Carey, S. (1985). *Conceptual Change in Childhood*. Cambridge, MA: MIT Press.

Chomsky, N. (1975). *Reflections on Language*. New-York, NY: Parthenon Press.

Churchland, P. M. (1989). Cognitive activity in artificial neural networks. In D.N. Osherson & E.E. Smith (Eds.), *Thinking: An Invitation to Cognitive Science*, Vol. 3 (pp.199-227). Cambridge, MA: MIT Press.

Elman, J. L. (1993). Learning and development in neural networks: The importance of starting small. *Cognition*, *48*, 71-99.

Elman, J. L., Bates, E. A., Johnson, M. H., Karmiloff-Smith, A., Parisi, D., & Plunkett, K. (1996). *Rethinking Innateness: A Connectionist Perspective on Development*. Cambridge, MA: MIT Press.

Fodor, J. (1980). Fixation of belief and concept acquisition. In Piatelli-Palmarini (Ed.), *Language and Learning: The Debate Between Chomsky and Piaget* (pp. 143-149). Cambridge, MA: Harvard Press.

Karmiloff-Smith, A. (1992). *Beyond Modularity: A Developmental Perspective on Cognitive Science*. Cambridge, MA: MIT Press / Bradford Books.

Mareschal, D., & Shultz, T. R. (1996). Generative connectionist networks and constructivist cognitive development. *Cognitive Development*, *11*, 571-603.

McClelland, J. L. (1989). Parallel distributed processing: Implications for cognition and development. In R.G.M. Morris (Ed.), *Parallel Distributed Processing: Implications for Psychology and Neurobiology* (pp. 8-45). Oxford: Oxford University Press.

Piaget, J. (1982). La psychogenèse des connaissances et sa signification épistémologique. In M. Piattelli-Palmarini (Ed.), *Théories du Language, Théories de l'Apprentissage* (pp. 53-64). Paris: Éditions du Seuil.

Quartz, S. R. (1993) Neural networks, nativism, and the plausibility of constructivism. *Cognition*, *48*, 223-242.

Quartz, S. R., & Sejnowski, T. (1997). The neural basis of cognitive development: A constructivist manifesto. *Behavioural and Brain Sciences*, *20*, 537-596.

Raijmakers, M. E. J., van Koten, S., & Molenaar, P. C. M. (1996). On the validity of simulating stagewise development by means of PDP networks: Application of catastrophe analysis and an experimental test of rule-like network performance. *Cognitive Science*, *20*, 101-136.

Shultz, T. R., & Mareschal, D. (1997). Rethinking innateness, learning, and constructivism: connectionist perspectives on development. *Cognitive Development*, 12, 563-586.

Shultz, T. R., Mareschal, D., & Schmidt, W. C. (1994). Modeling cognitive development on balance scale phenomena. *Machine Learning*, *16*, 57-86.

Siegler, R. S. (1981). Developmental sequences between and within concepts. *Monographs of the Society for Research in Child Development*, *46* (Whole No. 189).

Spelke, E. S. (1994). Initial knowledge: Six suggestions. *Cognition*, *50*, 431-445.

van der Maas, H., & Molenaar, P. (1992). Stagewise cognitive development: An application of catastrophe theory. *Psychological Review*, *99*, 395-417.

Effects of externalization on representation and recall of indeterminate problems

Vladimir M. Sloutsky (sloutsky.1@osu.edu)
Center for Cognitive Science & School of Teaching & Learning; 1945 N. High Street
Columbus, OH 43210, USA

Yevgeniya Goldvarg (goldvarg@phoenix.princeton.edu)
Department of Psychology, Princeton Univeristy
Green Hall, Princeton, NJ 08554 USA

Abstract

Naïve reasoning and problem solving is error-prone. One such pattern is manifested in that people err more often when problems are indeterminate than when problems are determinate We suggest that an incomplete problem representation could account for the observed pattern of errors. We further contend that in verbal reasoning such incomplete representation stems from a lack of systematic representations of connectives (e.g., *and*, *or*, *if*, etc.), and, therefore, externalization of relations denoted by sentential connectives should improve people's representations of multiple possibilities. These predictions were tested in three reported experiments. Results indicate that determinate problems were easier to represent and recall than indeterminate problems. Furthermore, there was a tendency to represent and recall indeterminate problems as if they were determinate ones by truncating the number of possibilities compatible with the problem. Finally, external aids dramatically improved representation and recall of indeterminate problems. These results are discussed in relation to theories of representation and reasoning.

Introduction

Naïve reasoning and problem solving is error-prone (see Evans & Over, 1996; Johnson-Laird & Byrne, 1991; for reviews) and errors exhibit systematic and predictable patterns. One particular pattern of errors is especially robust: people err more often if problems are indeterminate (i.e., compatible with multiple possibilities) than if problems are determinate (i.e., compatible with a single possibility). For example, the problem is compatible with multiple possibilities if an observed outcome could be caused by several factors, an observed symptom could be indicative of several conditions, or an observed pattern could be generated by several rules.

When problems are indeterminate, people (including expert scientists, politicians, lawyers, and physicians) tend to attribute observed outcomes to one or few factors, overlooking other plausible possibilities. Such a pattern was founds in the study of scientific thinking and problem solving (Kuhn, Garcia-Mila, Zohar, & Andersen, 1995; Mynatt, Dohetry, & Tweney, 1977), verbal reasoning (Evans & Over, 1996; Johnson-Laird, 1998), and in learning and recall (Sloutsky, Rader, & Morris, 1998).

The most obvious theoretical explanation of the reviewed difficulties is that indeterminate problems require more working memory resources than determinate problems. As a result, the overall error rate in indeterminate problems is greater than that in determinate ones. Such an interpretation assumes that people attempt to construe veridical problem representations, but they err because they cannot keep all these possibilities in their working memory. Although cognitive overload is a plausible factor, it is likely to result in an increase of unsystematic errors, whereas the observed patterns of errors are quite systematic. Therefore, it seems necessary to carry out a more detailed cognitive analysis and to consider other factors.

It seems that deriving a problem solution requires at least five steps. These include: (1) encoding of a verbal description or perceptual arrangement of the problem; (2) construing a problem representation that includes various elements of the problem and relations among them; (3) search through memory space for a putative solution; (4) manipulation of components of the problem for a putative solution; and (5) mapping a solution onto a verbal response. Therefore, the described difficulties may arise at each of these steps.

We believe that, when other factors are controlled or eliminated via simplifying the task, an incomplete problem representation could account for the observed pattern of errors, although with more demanding tasks other factors may also add to the problem difficulty. The idea stems from prior findings that children do create an incomplete problem representation by truncating the number of possibilities compatible with the problem and often considering just one possibility (Sloutsky, et al., 1998). We call such problem representation the "minimalist" representation and suggest that untrained adults may also tend to construe such problem representations.

One possible mechanism underlying the "minimalist" representation could be a lack of representations of relations among the alternative possibilities. Typically, in the case of verbal descriptions, these relations are denoted by sentential, or logical, connectives, such as *and*, *or*, *or else*, *if*, etc. In the case of *and* the considered possibilities (*A and B*) must co-occur, whereas in the case of *or*, the considered possibilities (*A or B*) may or may not co-occur. However, if people do not have consistent representations of sentential connectives, they would create identical (or similar) problem representations for conjunctive (those connected with *and*) and disjunctive (those connected with *or*) problem statements. In fact, it has been demonstrated (Sloutsky, et al., 1998) that people tend to recall disjunctions as conjunctions, but not vice versa.

It is known that external representations are capable of improving the process of inference and problem solving (Bauer & Johnson-Laird, 1993; Larkin & Simon, 1987; Zhang, 1997; Zhang & Norman, 1994). It seems plausible that if people do construe defective or "minimalist" representations of possibilities denoted by different sentential connectives, failing to represent multiple possibilities compatible with some of the connectives (e.g., disjunction or conditional), then helping them to externally represent multiple possibilities should reduce errors in representation and recall. On the other hand, if externalization of possibilities corresponding to sentential connectives fail to improve representation and recall of propositions containing these connectives, then it is likely that difficulties in recall stem from other reasons than the "minimalist" representation.

If these theoretical considerations are true, then people should err when problems are indeterminate, corresponding to multiple possibilities, and these errors should exhibit a particular pattern -- people construe "defective" problem representations, truncating the number of alternatives compatible with a problem. Therefore, fully determinate problems, those that correspond to exactly one possibility, should be the easiest ones. Furthermore, in many problems, such as propositions with connectives, the total number of possibilities is fixed ($\Sigma P_i = 2^n$, where ΣP_i is the total number of possibilities, and n is the number of atomic statements in the connective). Therefore, TRUE and FALSE possibilities in such problems are related inversely. As a result, the proportion of errors when the task is to represent what is TRUE, should be a mirror image of the proportion of errors when the task is to represent what is FALSE. Finally, if participants are assisted only in externalizing connectives, the predicted effects should decrease. The following critical predictions were tested in the three reported experiments:

1. People err more often when problems correspond to multiple possibilities rather than a single possibility: they construe "defective" or "minimalist" problem representations, truncating the number of alternatives compatible with a problem.
2. Because the total number of TRUE and FALSE possibilities are related inversely, proportions of errors when the task is to represent what is TRUE, should be a mirror image of the proportion of errors when the task is to represent what is FALSE.
3. Externalization of relations denoted by sentential connectives should improve people's representations of multiple possibilities.

Experiment 1

The goal of this experiment was to test hypotheses 1-2 via investigating people's representation and recall of propositions with various logical connectives. Participants were presented with content-based propositions having different connectives (e.g., *A and B*; *A or B*; *If A then B*) and cutout cards containing each of the atomic statements and their negations (e.g., *A*, *not-A*; *B*, *not-B*). Note that there were no cutout cards corresponding to the connectives. The task was to select those cutouts that would (a) communicate that the proposition is true (True) or (b) communicate that the proposition is false (False).

Method

Participants The sample consisted of 26 Ohio State University post-baccalaureate students majoring in education (8 men and 18 women; Mean age = 24.9 years). They received extra credit for participation.

Materials The experiment included five logical forms: conjunctions (*A and B*), inclusive disjunctions (*A or B, or both*), exclusive disjunctions (*A or B, but not both*), conditionals (*If A then B*), and bi-conditionals (*If and only if A then B*). Each logical form appeared three times with different neutral content. The content was rotated across the connectives. Below are the examples of propositions:

Conjunction: *This person drinks orange juice in the morning and watches the history channel.*

Inclusive Disjunction: *This person likes fishing or volunteers in a public school, or both.*

Exclusive Disjunction: *This person either collects stamps or teaches classes on Thursdays, but not both.*

Conditional: *If this person works on weekends, then he supports scientific research.*

Bi-Conditional: *If and only if this person is honest he drives a blue minivan.*

Each proposition was accompanied by several cutout cards. For all logical forms, cutout cards stated atomic propositions in the sentences, negations of atomic propositions, and unrelated filler statements. For each statement, there were two cards that stated each atomic proposition in the sentence, two cards that stated the negation of each atomic sentence, and two unrelated filler items. For example the sentence *If and only if this person is honest he drives a blue minivan* was presented with the following cutout cards: *This person is honest* (two cards*)*, *This person is not honest* (two cards), *This person drives a blue minivan* (two cards), *This person does not drive a blue minivan* (two cards), and two unrelated cards: *This person likes hamburgers,* and *This person plays tennis.* Such an arrangement was created to allow participants to veridically represent the logical forms. For example, to veridically represent conjunctions, they had to select just two *cards (A & B)*, whereas to veridically represent inclusive disjunctions, they had to select six cards (*A & not-B*; *not-A & B*, and *A & B*).

The experiment had a mixed design with the Truth condition as a between-subject factor and Logical form as a within-subject factor. The Truth condition had two levels: (a) communicate that a proposition is true (True) and (b) communicate that a proposition is false (False).

Procedure Participants were tested individually in a quiet room. The experiment consisted of four phases: warm-up, selection, distraction, and cued recall. During the warm-up phase participants read instructions and completed two practice trials: one in which they represented true meaning of a sentence and another in which they represent false meaning of a sentence.

During the selection phase the participants were presented

with one sentence at a time, and, to control for encoding, were asked to repeat the sentence. If encoding was incorrect the sentence was reread and participants repeated it again. This procedure continued until the sentence was repeated correctly. After that, the experimenter laid down 11 cards in front of the participant. One card had the whole sentence printed on it, whereas ten cards had atomic propositions (2 propositions * 2 cards = 4 cards), negations of these propositions (2 negated propositions * 2 cards = 4 cards), and filler items printed on them. Then, depending on what condition they were in, participants were asked to select those cards that made the sentence true or false. For each sentence all selected cards corresponding to each proposition were placed into separate envelopes for future cued recall.

The selection phase was followed by the distraction phase, during which the participants solved simple numerical problems. This phase continued for seven minutes.

During the cued recall phase, the participants were presented with the cards that they selected and asked to recall the entire original sentence. Note that the cards stated only atomic statements and filler items, but they did not state connectives. Participants recalled one sentence at a time.

If participants recalled an entire sentence, including the connective, the answer was scored as correct. If participants recalled a sentence correctly, but substituted one connective with another, the answer was scored as a substitution error. If they recalled an affirmative proposition as a negation or a negation as an affirmative statement, if they forgot a proposition in a compound statement, or if they included one that originally was not there, the answer was scored as incorrect.

Results and Discussion

In this section, we consider how participants represented and recalled propositions. The correct (truth table) representations for the True and False conditions are presented in Table 1. The table presents the five logical forms used in the experiment, and possible card arrangements representing correct and incorrect choices by logical forms and truth conditions. A representation was coded as correct in a given condition, if a participant selected all and only card arrangements marked by the plus sign.

Proportions of correct representations are depicted in Figure 1. These data clearly indicate that participants tended to represent correctly conjunctions in the True condition (Percent Correct = 100%, Chance = 6.25%, 95% Confidence Interval = 91.4% to 100%) and disjunctions in the False condition (Percent Correct = 100%, Chance = 6.25%, 95% Confidence Interval = 78.5% to 100%). In other words, they tended to represent correctly only those forms that were compatible with exactly one possibility. In the True condition, all other logical forms generated low correct response rates that did not surpass 5%. Similarly,

low correct response rates were observed in the False condition, except for the conditional, which was still not significantly different from chance (Percent Correct = 37%, Chance = 6.25%, 95% Confidence Interval = 40% to 81%).

Because the proportion of correct responses was surprisingly low, we deemed it necessary to analyze patterns of responses. In the True condition, these patterns fell into three categories: (1) correct representations (selection of cards corresponding to <u>all</u> true possibilities, (2) "conversion-to-conjunction" (a tendency consider only one true possibility, A & B), and (3) "other" errors (those that did not fall in the first two categories). In the False condition, we also identified three response categories: (1) correct representations (selection of cards corresponding to <u>all</u> false possibilities), (2) "conjunction of negations," (a tendency to consider only one false possibility, $\neg A$ & $\neg B$), and (3) "other" errors (those that did not fall in the first two categories).

In the True condition, the dominant pattern of responses was the "conversion-to-conjunction" (more than 90% of all responses), whereas in the False condition, the dominant pattern of response was the "conjunction of negations" (around 70% of all responses). To establish significance of the prevalence these patterns of responses, within each Truth condition and for each logical form, the number of responses conforming to the patterns was cross-tabulated with the total number of "other" and correct responses and subjected to 2 by 2 chi-square analyses. In the True condition, the prevalence of conversion-to-conjunctions was significant for the bi-conditional, $\chi^2 (1, 96) = 80.67$, $p < .0001$, for the conditional, $\chi^2 (1, 96) = 96$, $p < .0001$, for the exclusive disjunction, $\chi^2 (1, 96) = 66.7$, $p < .0001$, and for the inclusive disjunction, $\chi^2 (1, 96) = 80.67$, $p < .0001$. In the False condition, the "conjunction-of-negations" was prevalent for the conjunction, $\chi^2 (1, 60) = 45$, $p < .0001$, and for the exclusive disjunction, $\chi^2 (1, 60) = 9.6$, $p < .003$. For the conditional and the bi-conditional this pattern, although not reaching significance, accounted for more than a half of all responses.

Recall rates across the Truth conditions are presented on Figure 2. Note that while there were significant differences in correct representations of true and false possibilities of conjunctions and of disjunctions, recall rates for these logical forms did not differ across the Truth conditions (F (1, 24) = 1.7, p = 0.2, for conjunctions and F (1, 24) = 1.6, p = .22, for disjunctions). Overall, conjunctions were more likely to be recalled than all other logical forms, F (1, 25) = 74.8, $p < .0001$. Recall rates of the other logical forms did not differ significantly from each other.

Pattern of responses was similar to that in the representation phase. In both conditions, the conjunction generated significantly more correct responses than incorrect responses. In the True condition, there were 85% correct responses, $\chi^2 (2, 144) = 90.2$, $z = 6.3$, $p < .0001$. In the False condition, there were 77% correct responses, $\chi^2 (2, 90) = 41.7$, $z = 4.1$ $p < .0001$.

Table 1: Truth table representations by the logical forms.

Logical forms	Truth conditions and possible card arrangements							
	True condition				False condition			
	A & B	*¬A & B*	*A & ¬B*	*¬A & ¬B*	*A & B*	*¬A & B*	*A & ¬B*	*¬A & ¬B*
Conjunction (AND)	+	-	-	-	-	+	+	+
Inclusive Disjunction (XOR)	+	+	+	-	-	-	-	+
Exclusive Disjunction (OR)	-	+	+	-	+	-	-	+
Conditional (IF)	+	+	-	+	-	-	+	-
Bi-Conditional (IFF)	+	-	-	+	-	+	+	-

Note: "+" indicates a card arrangement that correctly represents a choice in a given condition, whereas "-" indicates an incorrectly selected card arrangement.

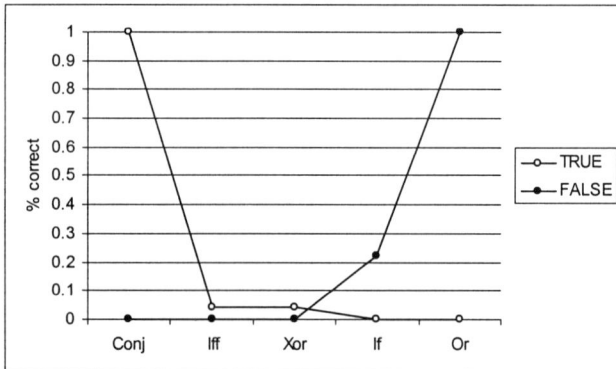

Figure 1: Percent of correct representations in the "TRUE" vs. "FALSE" conditions.

For the other logical forms, in the True condition the dominant pattern of recall was the "conversion-to-conjunction" (recall of a proposition as if it had a form *A and B*), whereas in the False condition, the dominant pattern of recall was the conjunction of negations. Significance of the prevalence of conversion-to-conjunction and conjunction of negations responses was established in the same manner as it was established for representations. In the True condition, the bi-conditional conversion-to-conjunctions accounted for 41% of all responses, *NS*, for 54% of exclusive disjunctions, χ^2 (2, 144) = 15.8, z = 2.5, $p <$ .001, for 67% of inclusive disjunctions, χ^2 (2, 144) = 36.8, z = 4.0, $p <$ .0001, and for 44% of conditionals, *NS*.

In the False condition, in addition to conjunctions, bi-conditionals generated mostly correct responses, 57%, χ^2 (2, 90) = 11.8, z = 2.2, $p <$ .005. At the same time, the conjunction of negations accounted for 57% of exclusive disjunctions, χ^2 (2, 90) = 11.1, z = 2.1, $p <$ .005, for 63% of inclusive disjunctions, χ^2 (2, 90) = 18.9, z = 2.8, $p <$ .0001, and for 40% of conditionals, *NS*.

In short, in the True condition participants tended to represent and recall different logical forms as conjunctions, whereas in the False condition, they tended to represent propositions of all logical forms as the "conjunction of negations," while recalling them (except for the bi-conditional) as conjunctions of true possibilities. Across the conditions, conjunctions were likely to be recalled correctly. These findings support our hypothesis that people tend to construe the "minimalist" representation, one that is compatible with a single possibility.

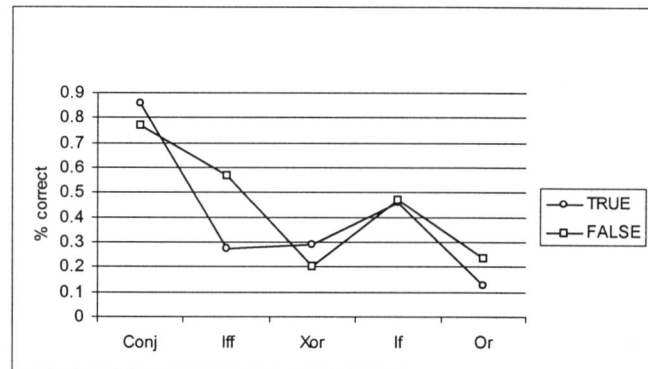

Figure 2: Percent of correct recall in the "TRUE" vs. "FALSE" conditions.

However, the experiment left a number of issues unresolved. The strongest argument against the reported findings could be that participants might have misinterpreted the task. In particular, they may have thought that the task was to select possibilities merely exemplifying that the proposition is true or false rather than selecting all possibilities that are compatible with the true or false state of affairs. It could be also argued that people poorly understand meanings of all logical connectives, except for conjunctions.

To address these arguments, we conducted the Experiments 2 and 3. In Experiment 2, we clarified the instructions, specifically asking participants to select ALL possibilities corresponding to true or false states of affairs respectively. In Experiment 3, we provided them with simple representational aids that allowed them to externalize possibilities denoted by a connective.

Experiment 2

Method

Participants Participants were 109 undergraduate students at the Ohio State University (Mean Age = 22.1 years; 15 men and 94 women). Students were recruited through psychology classes and were given extra credit for participating in the study.

Materials The problem set was identical to that in the first experiment. There were, however a number of important differences. First, each problem was followed by a full ("truth table" type) list of possibilities. For example when a problem consisted of two atomic statements *A* and *B* connected by a sentential connective, a full list of possibilities included the following: *A & B*; *not-A & B*; *A & not B*; and *not-A & not-B*. These possibilities were randomized within each problem and accompanied by two filler items. An example of an item and subsequent choice options is as follows: *This person drinks orange juice in the morning and watches history channel*.

A. This person drinks orange juice in the morning and does not watch history channel
B. This person has running shoes and does not smoke cigars
C. This person does not drink orange juice in the morning and does not watch history channel
D. This person drinks orange juice in the morning and watches history channel
E. This person does not drink orange juice in the morning and watches history channel
F. This person does not have running shoes and smokes cigars

The participants were presented with booklets with experimental tasks. The instruction asked them to encircle ALL choices that correspond to the proposition if it was TRUE (True condition) or if it was FALSE (False condition).

Procedure Participants were tested in groups ranging from 20 to 60 participants. The experiment was conducted in a single 20-minute session.

Results and Discussion

In the True condition, the proportion of correct representations was above chance for the conjunction (Chance = 0.167; Confidence Interval = .88 to 1, $p <$ .001) and the exclusive disjunction (Chance = 0.03, Confidence Interval = .2 to .37, $p <$ .01). Correct performance was at the chance level for the inclusive disjunction (Chance = 0.008, Confidence Interval = 0 to .16) and for the bi-conditional (Chance = 0.03, Confidence Interval = 0 to 0.29). Finally, the proportion of correct performance was below chance for the conditional (Chance = 0.008; Confidence Interval = 0, $p <$ .05). In the False condition, the proportion of correct representations was above chance for the exclusive disjunction (Chance = 0.167, Confidence Interval = .5 to .9, $p<$ .01), and at the chance level for the other forms. Overall patterns of responses were similar to those in Experiment 1. In the True condition, the majority of errors (more than 75%) were conversion-to-conjunction responses, whereas in the False condition a large number of propositions were represented as the conjunction of negations (about 55%).

Experiment 3

The goal of this experiment was to find a remedy for "conversions-to-conjunctions" and "conjunction-of-negations." If reported errors stem from the "minimalist" representation of sentential connectives, then providing participants with external tools to represent these connectives, should allow for more complete representations, thus decreasing the proportion of errors.

Method

Participants, Materials, and Procedure The sample was selected from the same population of Ohio State University post-baccalaureate students majoring in education, as the sample for Experiment 1. It consisted of 23 Ohio State University students (5 men and 18 women; Mean age = 23.8 years). They received extra credit for participation. The experiment had the same materials, design, and procedure as Experiment 1, except that participants were given sheets of paper to externally represent sentential connectives.

Results and Discussion

Aggregated effects of externalization on correctness of representation and recall are presented on Figures 3 and 4 respectively.

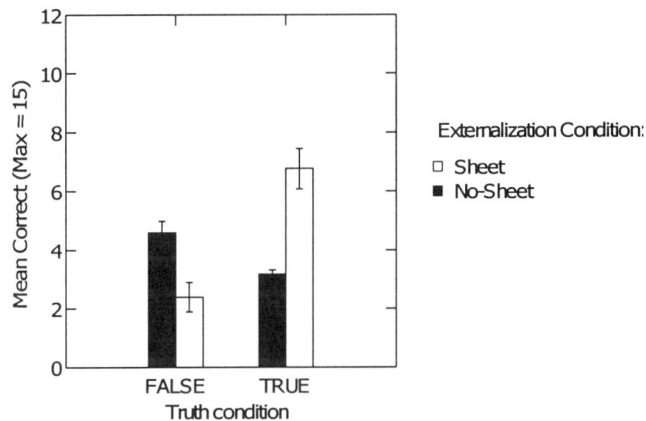

Figure 3: Effects of externalization on representation aggregated across logical forms.

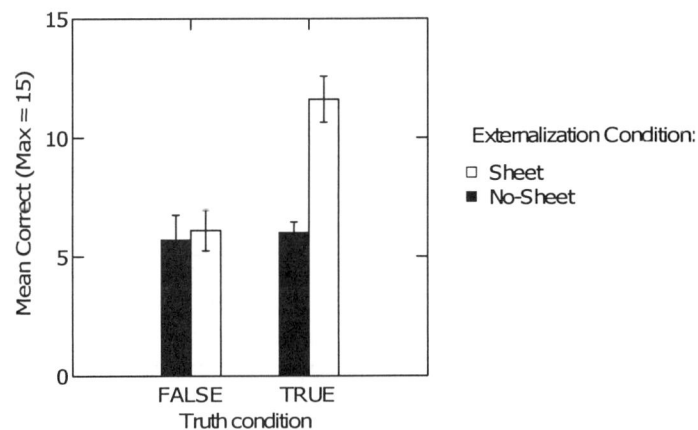

Figure 4. Effects of externalization on recall aggregated across logical forms.

As depicted in Figure 3, externalization had positive effects in the True condition ($M_{external\ rep}$ = 6.8; $M_{no-extenal\ rep}$ = 3.2) and negative effects in the False condition ($M_{external\ rep}$ = 2.4; $M_{no-external\ rep}$ = 4.6). A two (Truth condition) by two (Externalization condition) between-subjects ANOVA indicated that there was indeed a significant Truth condition * Externalization interaction, F (1, 45) = 42, p < .0001. At the same time, the main effect of Externalization was non-significant, F (1, 45) = 2.4, p = .128.

Data in Figure 4 indicate that in the true condition, recall drastically increased with the introduction of external representation, whereas in the false condition, recall rates remained unchanged. A 2 (Truth condition) by 2 (Externalization condition) between-subjects ANOVA indicated that the main effect of Externalization was significant, with recall rates with external representations (M = 9.2, SD = 4.1) higher than recall rates without external representations (M = 5.9, SD = 2.3), F (1, 45) = 14.6, p < .0001. The main effect of the Truth condition was also significant, F (1, 45) = 13.7, p = .001 and there was also a significant Truth condition * Externalization interaction, F (1, 45) = 11, p < .003. While effects of Externalization were pronounced in the True condition ($M_{external\ rep}$ = 11.6; $M_{no-external\ rep}$ = 6.0), there were no such effects in the False condition ($M_{external\ rep}$ = 6.1; $M_{no-external\ rep}$ = 5.7).

In short, in Experiment 3 it was found that in the True condition, external representations lead to (a) a decrease of conversion-to-conjunction responses; (b) an increase in the number of represented possibilities; (c) an increase in the percent of correct representation; and (d) an increase of correct recall. The results were more mixed for the False condition.

General Discussion

As predicted, people err more often when problems are indeterminate. These errors exhibit a particular pattern: people tend to construe "defective" problem representations, truncating the number of alternatives compatible with a problem. Second, proportions of errors when the task is to represent what is TRUE, were a mirror image of the proportion of errors when the task is to represent what is FALSE. Finally, as predicted, externalization of relations denoted by sentential connectives improved people's representations of multiple possibilities.

These findings indicate that participants tended to minimize the number of represented possibilities, often creating a representation with just one possibility compatible with the problem. We define such representation a "minimalist" representation. One possible mechanism underlying the "minimalist" representation could be a lack of consistent representations of logical connectives. In this case they would create identical (or similar) problem representations for conjunctive and disjunctive problems.

Findings support our contention that reasoning errors often stem from an incomplete or "defective" problem

representation. We took special care to eliminate alternative sources of problem difficulty. First, we constructed sufficiently simple tasks where no solutions were required and where there was no need to search through memory or problem space or to simultaneously manipulate many items in working memory. Furthermore, because the tasks have a very simple linguistic form and encoding was controlled, we could assume that errors are unlikely to stem from an inaccurate mapping. Therefore, the likely remaining candidate is the tendency to create an incomplete problem representation, although when problems are more difficult, other factors may also affect error rates.

References

Bauer, M., & Johnson-Laird, P. N. (1993). How diagrams can improve reasoning. *Psychological Science, 4*, 372-378.

Evans, J. S. B., & Over, D. (1996). *Rationality and reasoning*. Hove, UK: Psychology Press.

Johnson-Laird, P. N., & Byrne, R. (1991). *Deduction*. Hove, UK: Lawrence Erlbaum.

Kuhn, D., Garcia-Mila, M., Zohar, A., Andersen, C. (1995). Strategies of knowledge acquisition. *Monographs of the Society for Research in Child Development, 60*(4).

Larkin, J. H., & Simon, H. A. (1987). Why a diagram is (sometimes) worth ten thousand words. *Cognitive Science , 11*, 65-99.

Morris, A. K., & Sloutsky, V. M. (1998). Understanding of logical necessity: Developmental antecedents and cognitive consequences. *Child Development, 69*(3), 721-741.

Morris, B., & Sloutsky, V. M. (1998). Children's solutions of logical vs. empirical problems: What's missing and what develop. Submitted for publication.

Mynatt, C. R., Dohetry, C. E., & Tweney, R. D. (1977). Confirmation bias in a simulated research environment: An experimental study of scientific inference. *Quarterly Journal of Experimental Psychology, 29,* 85-95.

Sloutsky, V. M., Rader, A., & Morris, B. (1998). Increasing informativeness and reducing ambiguities: Adaptive strategies in human information processing. In M. A. Gernsbacher & S. J. Derry (Eds.), *Proceedings of the XX Annual Conference of the Cognitive Science Society* (pp. 997-999). Hillsdale, NJ: Erlbaum.

Zhang, J., & Norman, D. A. (1994). Representations in distributed cognitive tasks. *Cognitive Science, 18*, 87-122.

Zhang, J. (1997). The nature of external representations in problem solving. *Cognitive Science, 21*(2), 179-217.

Acknowledgements

This research has been supported by a grant from the James S. McDonnell Foundation awarded to the first author.

Problem representations and illusions in reasoning

Vladimir M. Sloutsky (sloutsky.1@osu.edu)
Center for Cognitive Science & School of Teaching & Learning; 1945 N. High Street
Columbus, OH 43210, USA

Philip N. Johsnon-Laird (phil@clarity.princeton.edu)
Department of Psychology, Princeton Univeristy
Green Hall, Princeton, NJ 08554 USA

Abstract

The mental model theory of reasoning postulates that reasoners build models of the situations described in premises, and that these models normally make explicit only what is true. The theory has an unexpected consequence: it predicts the occurrence of inferences that are systematically invalid. These inferences should arise from reasoners failing to take into account what is false. We report an experiment that corroborated the occurrence of these illusory inferences, and that eliminated a number of alternative explanations for them. Results illuminate the controversy among various current theories of reasoning.

Introduction

While reasoning is ubiquitous in human life, its underlying mechanisms are a matter of controversy. Although there are many theories of reasoning with conditionals (e.g., Cheng & Holyoak, 1985; Cosmides, 1989), there are two major general approaches to reasoning: the syntactic and the semantic. According to the syntactic approach, reasoning hinges on a set of formal rules of inference or "inferential schemata" (Braine & O'Brien, 1998; Rips, 1994). For instance, when individuals are presented with premises of the following form:

> *A* or *B*
> not-*A*

they draw the conclusion:

> *B*

using the "disjunction elimination schema", which is a formal rule that has precisely the form of this inference.

According to the semantic approach, the untrained mind is not equipped with formal rules of inference, but relies instead on the validity of arguments. The principle states that an inference is valid if its conclusion must be true given that its premises are true (e.g. Johnson-Laird & Byrne, 1991; Polk & Newell, 1995). Among other theories within the semantic approach, the mental logic theory (Johnson-Laird & Byrne, 1991) attempts to explain both correct and erroneous reasoning using a small number of fundamental assumptions. These include: (1) people represent information in premises in a systematic manner and (2) these representations are constructed in accordance with a principle of truth:

- Individuals normally represent only true possibilities.
- Within these possibilities they represent only those literal propositions in the premises (affirmative or negative) that are true.

Given the example above, reasoners therefore construct the following mental models, where each row denotes a mental model of a possibility, and "¬" denotes negation:

First premise	Second premise
A B A B	¬A

They can infer that *B* is the case, because the second premise rules out those models of the first premise in which *A* occurs, i.e., the first and third models of possibilities.

Although the syntactic and semantic approaches postulate different mechanisms underlying reasoning, they often yield similar predictions. Yet, there are inferences for which their predictions diverge. In particular, the principle of truth leads to the prediction that certain inferences should lead reasoners systematically astray. For example, consider the following inference in which you must assume that only one of the two conditional premises is true:

If George was on the team then Dan was on the team.

If Dan wasn't on the team than George was on the team.
Assuming that only one of the above statements is true, is it possible that George was on the team, and Dan was on the team?

The mental model theory predicts that reasoners should envisage the following mental models of possibilities for the first premise:

> George   Dan
>
>   ...

where the first model makes explicit the possibility in which both George and Dan are on the team, and the second model (ellipses) corresponds to those possibilities in which the antecedent of the conditional is false. The theory accordingly assumes that individuals do not normally make these possibilities explicit (Johnson-Laird & Byrne, 1991). Likewise, reasoners should envisage the following mental models for the second premise:

> ¬ Dan   George
>
>   ...

According to the principle of truth, when reasoners think about the truth of one premise, they will fail to think about the falsity of the other premise. The question posed in the problem is whether it is possible that George was on the team and Dan was on the team. This situation corresponds to a possibility in the mental models of the first premise, and so reasoners should respond "yes". The response is an illusion, however. Indeed, the fully explicit models of the premises are respectively:

First premise		Second premise	
George	Dan	¬ Dan	George
¬ George	Dan	Dan	George
¬ George	¬Dan	¬ Dan	¬George

It follows that the question *Is it possible that George was on the team, and Dan was on the team* has the correct answer, "no", because the case in which both George and Dan are members of the team is true in <u>both</u> premises. We label this inference as a "yes/no" problem, where the first word ("yes") is the predicted answer and the second word ("no") is the correct answer. The same premises support a control inference to which reasoners should get the correct answer even though they fail to represent what is false: *Is it possible that George was on the team, but Dan wasn't on the team?* The situation corresponds to a mental model of the second premise, but it is also correct because it occurs only in the fully explicit models of the second premise. We define such control inferences as "yes/yes" problems.

By pairing the same premises with two other sorts of question, we can create inferences to which reasoners should fall into the trap of responding "no" incorrectly ("no/yes" problems): *Is it possible that George was not on the team, and Dan was not on the team?* We can also create a corresponding control problem to which reasoners should respond "no" correctly ("no/no" problems): *Is it possible that George wasn't on the team, but Dan was on the team?*

The source of the illusions according to the model theory is the failure to represent what is false. This failure is likely to be more pronounced among those who have little or no training in reasoning, and among those who score less well on SAT tests, which Keith Stanovich (personal communication) has shown to correlate significantly with logical performance.

It is also known from previous work on reasoning about possibilities (Bell & Johnson-Laird, 1998) that it is easier for people to establish that a conclusion is possible than it is to establish that the conclusion is impossible. However, it is easier for them to establish that a conclusion is not necessary than that it is necessary. The reason for this divergence is that to establish that a conclusion is impossible or necessary, one needs to search through all models, whereas to establish that a conclusion is possible or unnecessary, it is sufficient to find just one example or counterexample. Therefore, we can predict that both illusory and control problems that require participants to answer "No" (impossible), should be harder than respective problems requiring them to answer "Yes" (possible). As a result, Yes/No illusion should be harder than No/Yes illusions, and No/No controls should be harder than Yes/Yes controls.

In contrast to the model theory, current theories based on formal rules of inference (e.g. Rips, 1994; Braine & O'Brien, 1998) do not predict the occurrence of the illusory inferences. These theories rely solely on valid rules of inference, and so in principle they cannot predict the occurrence of systematically erroneous conclusions.

Because both illusory and control problems are based on the same premises, the formal rules theorist should predict no systematic differences between illusory and control problems.

An experiment comparing illusory inferences and control inferences

In order to test the model theory's predictions, we carried out an experiment in which we examined performance on four types of inference problems. They were illusory inferences with a predicted response of "yes" (yes/no problems), illusory inferences with a predicted response of "no" (no/yes problems), and their respective controls (yes/yes problems) and (no/no problems). The four types of problems were based on the same premises so that no differences in the premises could be responsible for the results. Similarly, all the questions following the premises were in the form of conjunctions. The form of the problems is summarized in Table 1. The problems in set A are those that we described in the Introduction. Those in set B are comparable with the four types of problems all based on the same premises.

Method

Participants Two groups of undergraduate students participated in the experiment. One group of 18 were recruited from a private (highly selective) university ($M = 20.6$ years, $SD = 1.4$; 11 men and 7 women) and 20 from a large public (mainly non-selective) university ($M = 20.0$ years, $SD = 1.5$; 9 men and 11 women). Hence, the two groups were drawn from two populations, which differ in their required SAT scores and in the emphasis in their curricula on mathematical training.

Materials Each participant carried out 16 problems (the eight sorts of problems in Table 1 and eight filler items) in one of two random orders. The content of the problems concerned team memberships and each of the 16 problems was about a different pair of individuals, i.e. frequent one- or two-syllable first names of males and females.

Table 1: The inferences in the experiment, their mental models, and their fully explicit models.
In each set, only one of the premises is true.

Problem set A	Mental models		Fully explicit models	
	Premise 1	Premise 2	Premise 1	Premise 2
If B then A If not A then B	B A …	¬A B …	B A ¬B A ¬B ¬A	¬A B A ¬B A B
1. Is B & A possible? (yes/no)	B & A are in the first premise		B & A are in both premises	
2. Is B & not-A possible? (yes/yes)	B & ¬ A are in the second premise		B & ¬A are in the second premise	
3. Is not B & not-A possible? (no/yes)	¬B & ¬A are not in the premises		¬B & ¬A are in the first premise	
4. Is not B & A possible? (no/no)	¬B & A are not in the premises		¬B & A are not in the premises	
Problem set B	Mental models		Fully explicit models	
	Premise 1	Premise 2	Premise 1	Premise 2
If B then not A If A then B	B ¬A …	A B …	B ¬A ¬B A ¬B ¬A	A B ¬A B ¬A ¬B
1. Is B & not A Possible? (yes/no)	B & ¬A are in the first premise		B & ¬A are in both premises	
2. Is A & B Possible? (yes/yes)	A & B are in the second premise		A & B are in the second premise	
3. Is not B & A possible? (no/yes)	B & ¬A are not in the premises		¬B & A are in the first premise	
4. Is not A & not B possible? (no/no)	¬A & ¬B are not in the premises		¬A & ¬B are in both premises	

Procedure The participants were tested individually. The experimenter read them the instructions and presented them with a warm-up problem. The key component of the instructions was as follows: *Imagine that there is a meeting of two old coaches who coached together two competing teams, the Bulls and the Wildcats. They started talking about the good old days when their teams competed. But it soon turned out that as in the good old days they could not agree on anything. In particular, they weren't even able to agree on who was on each team. So they might need your help. But before helping them, I want you to know that in every argued case, they cannot be both right or both wrong. In other words, in each case <u>one of them is right and the other is wrong</u>. Sometimes it is the coach of the Bulls (Coach Bull) who is right, sometimes it is the coach of the Wildcats (Coach Wildcat), but it is always the case that <u>only one is right and another is wrong</u>. Your goal is to decide whether or not some of the things they say are possible, given that one is right and another is wrong.*

The participants then carried out a simple warm up problem:

Coach Bull: *Sara wasn't on the team*
Coach Wildcat: *Megan wasn't on the team*

Is it possible that Sara and Megan were on the team? Why?

If a participant gave an incorrect answer (i.e., if they asserted that it was possible for Sara and Megan to be on the team), the experimenter provided an explanation and offered another warm-up problem. The experimenter also took special care to explain to the participants that the coaches were always arguing about the same team. Once the participants had understood the instruction, they proceeded to the experiment proper, which lasted for approximately half an hour. The problems were in booklets, and the participants were told to respond "yes" or "no" to each question.

Results

Table 2 presents the results of the experiment for the two groups of participants. The table shows that both groups of participants tended to succumb to the illusions (only 31% correct Yes/No inferences and 50% correct No/Yes inferences), but performed better on the control problems (79% correct Yes/Yes inferences and 66% correct No/No inferences). Across groups, participants performed significantly better on control problems than on illusory problems (Wilcoxon $z = 3.59$, $p<.0001$).

In the Table, the percentages for the group of the selective university students are on the left of each cell, and those for the non-selective students are in parentheses on the right of each cell. All eight percentages depart significantly from chance (p ranging from $< .0001$ to $< .05$).

Table 2: The percentages of correct responses to the four sorts of problems for the two groups of participants.

Illusory inferences		Control inferences	
1. Yes/No	33 (27)	2. Yes/Yes	86 (73)
3. No/Yes	67 (35)	4. No/No	70 (65)

Note that students in the non-selective university were more likely to succumb to illusions, and their results conformed better to the mental model theory's predictions than did results of students in the selective university. Both groups responded "yes" to the Yes/No illusions significantly more often than a chance rate (all $ps < .05$ or better), but only the public university students succumbed to the "No/Yes" illusions significantly more often than a chance rate ($p < .05$). Both groups performed better than chance on the control problems (ps ranged from $< .0001$ to $< .05$), and there were no differences on these problems between the selective university the non-selective university students. At the same time, the two groups differed on Yes/No (Mann-Whitney $U = 190$, $p < .05$) and No/Yes (No (Mann-Whitney $U = 255$, $p < .03$) illusory problems. Finally, as predicted the yes/no illusory problems were harder than the no/yes illusions, Wilcoxon $z = 2.1$, $p < .05$, whereas the yes/yes controls were somewhat easier than the no/no controls, Wilcoxon $z = 1.8$, $p = .07$.

Discussion

A surprising consequence of the mental model theory of reasoning is its prediction of illusory inferences, that is, inferences that lead to systematic but fallacious conclusions. The results of the experiment corroborated this prediction of the theory. For example, given a problem of the form below, most of our participants wrongly concluded that the answer was "yes".

Assuming that only one of the above statements is true:

If A then B.
If ¬ B then A.
Is A & B possible?

Is there any plausible alternative explanation of our results apart from the failure to represent what is false? One alternative hypothesis is that the task, instructions, or premises of the inferences are so complex, ambiguous, or pragmatically odd, that they confused the experimental participants, thus adversely affecting their performance. However, this hypothesis fails to explain the systematicity and predictability of errors, as well as the correct performance of the control problems. Another alternative hypothesis is that the participants failed to notice that one premise is true and the other premise is false. Again, this idea is most likely implausible given the framing of the problems as disagreements between two coaches, and the participants' practice with such a problem.

A more plausible alternative hypothesis is that that conditionals have an interpretation distinct from the one that we have proposed. There are various versions of this hypothesis, e.g., conditionals are interpreted as having a "defective" truth table (e.g. Wason & Johnson-Laird, 1972), or as having some other, as yet unknown,

sophisticated meaning. The hypothesis of a "defective" truth table treats conditionals as having no truth value whenever their antecedents are false. The idea, however, runs into insuperable difficulties with biconditionals, such as:
If, and only if, Dan was in the game, then George was in the game.

People judge that this assertion is true when both Dan and George were in the game or not in the game; but when one was in the game and the other was not, they judge it to be false. Hence, the biconditional has a complete truth table. Yet, it can be paraphrased by the following conjunction of two conditionals:
If Dan was in the game then George was in the game, and if George was in the game then Dan was in the game.

Consider the case where neither Dan nor George was in the game. Neither of the two conditionals has a truth value, yet the biconditional is true. How can the conjunction of two assertions lacking a truth value yield an assertion that is true? The answer is: it cannot.

As for an unknown sophisticated meaning, our analysis offers a parsimonious explanations that rests only on two simple and testable assumptions, such as (1) the mental model representation of the conditional and (2) the principle of truth. Recall that the first states that people construe only two models to represent the conditional, such as *If A then B*:
  A B
    …

and the second states that people represent only true possibilities.

Robert Mackiewicz and Walter Schaeken (personal communication) have drawn our attention to an interesting possibility. Perhaps, if reasoners think about one proposition in a disjunction then they <u>forget</u> about the other. However, if reasoners merely forgot one of the clauses, then they should be liable to forget A or to forget B in dealing with a <u>conjunction</u> of A and B (cf. Johnson-Laird & Savary, 1996). However, neither adults nor children usually forget the constituents of conjunctions when they reason (Sloutsky, Rader, & Morris, 1998; Morris & Sloutsky, 1999). On the other hand, for exclusive disjunctions, such as the one in our experiment, it seems that people think about the truth of one proposition while forgetting about the falsity of the other proposition. Such dissociation between not forgetting propositions in conjunctions and forgetting them in exclusive disjunctions, if demonstrated empirically, would strongly support the principle of truth.

Unlike these rival hypotheses, the mental model theory of reasoning predicts that any manipulation that emphasizes falsity should reduce the illusions. Recent studies have corroborated this prediction. For example, the rubric, "Only one of the following two premises is <u>false</u>," reliably reduced the illusions (Tabossi, Bell, & Johnson-Laird, 1998). They were reduced when the participants had to generate false instances of the premises before they carried out the inferential task (Newsome & Johnson-Laird, 1996). Likewise, they were reduced when the participants had to check whether the conclusions were consistent with the relations between the premises (Goldvarg & Johnson-Laird,

704

1999; Yang & Johnson-Laird, 1998a; 1998b). A final advantage of the model theory is that it bases its predictions on a single principle -- reasoners take into account truth, not falsity.

There is, however, another potential source of difficulty. Reasoners cannot cope very well with inferences that call for multiple models of the premises. Indeed, they often appear to construct only a single model. Bauer & Johnson-Laird (1993) reported that the most frequent error in a study of disjunctive inferences was that the participants constructed only one model of the premises. Likewise, children often appear to construct only a single model. Hence, a more radical source of error than the principle of truth is that reasoners may sometimes construct a "minimalist" representation of just a single possibility – just a single mental model of the premises (Sloutsky, Rader, & Morris, 1998; Sloutsky & Goldvarg, 1999).

The illusions are an important shortcoming in human reasoning, and they are worth investigating further for their own intrinsic interest. The neglect of falsity, however, appears to underlie a number of other well-established inferential phenomena, such as Wason's selection task and the difficulty of *modus tollens* inferences (for a review, see Evans, Newstead, & Byrne, 1993). The illusions contravene all current formal rule theories (e.g. Braine & O'Brien, 1998; Rips, 1994). These theories rely solely on valid rules of inference, and so the only systematic conclusions that they can account for are valid ones. These theories therefore need to be amended -- either in their implementation or in a more radical way in order to account for the illusions. Our study shows that naïve reasoning depends crucially on how individuals represent the premises.

References

Bauer, M. I., & Johnson-Laird, P. N. (1993). How diagrams can improve reasoning. *Psychological Science, 4*, 372-378.

Bell, V. A., & Johnson-Laird, P. N. (1998). A model theory of modal reasoning. *Cognitive Science, 1*(22), 25-51.

Braine, M. D. S., & O'Brien, D. P., Eds. (1998). *Mental logic*. Mahwah, NJ: Erlbaum.

Cheng, P. N., & Holyoak, K. J. (1985). Pragmatic reasoning schemas. *Cognitive Psychology, 17*, 391-416.

Cosmides, L. (1989) The logic of social exchange: Has natural selection shaped how humans reason? Studies with the Wason selection task. *Cognition, 31*, 187-276.

Evans, J. St. B. T., Newstead, S. E., & Byrne, R. M. J. (1993). *Human Reasoning: The Psychology of deduction*. Hillsdale, NJ: Lawrence Erlbaum Associates.

Goldvarg, Y., & Johnson-Laird, P. N. (1999). *Memory & Cognition*, in press.

Johnson-Laird, P. N., & Byrne, R. M. J. (1991). *Deduction*. Hillsdale, NJ: Erlbaum.

Johnson-Laird, P. N., & Savary, F. (1996). Illusory inferences about probabilities. *Acta Psychologica, 93*, 69-90.

Morris, B. J., & Sloutsky, V. M. (1998). Developmental differences in young children's solutions of logical vs. empirical problems. In *Proceedings of the Twenty First Annual Conference of the Cognitive Science Society*. (pp. 000-000). Mahwah, NJ: Erlbaum.

Newsome, M. R., & Johnson-Laird, P. N. (1996). An antidote to illusory inferences? In Cottrell, G.W. (Ed.) *Proceedings of the Eighteenth Annual Conference of the Cognitive Science Society*. Mahwah, NJ: Lawrence Erlbaum Associates, p. 820.

Oaksford, M., & Stenning, K. (1992). Reasoning with conditionals containing negated constituents. *Journal of Experimental Psychology: Learning, Memory, & Cognition, 18*, 835-854.

Polk, T. A., & Newell, A. (1995). Deduction as verbal reasoning. *Psychological Review, 102*, 533-566.

Rips, L. J. (1994). *The Psychology of Proof*. Cambridge, MA: MIT Press.

Sloutsky, V. M., & Goldvarg, Y. (1999). Effects of externalization on representation and recall of indeterminate problems. In *Proceedings of the Twenty First Annual Conference of the Cognitive Science Society*. (pp. 000-000). Mahwah, NJ: Erlbaum.

Sloutsky, V. M., Rader, A., & Morris, B. (1998). Increasing informativeness and reducing ambiguities: Adaptive strategies in human information processing. In Gernsbacher, M.A., & Derry, S.J. (Eds.) *Proceedings of the Twentieth Annual Conference of the Cognitive Science Society*. (pp. 997-999). Mahwah, NJ: Erlbaum.

Tabossi, P., Bell, V. A., & Johnson-Laird, P. N. (1998). Mental models in deductive, modal, and probabilistic reasoning. In Habel, C., & Rickheit, G. (Eds.) *Mental Models in Discourse Processing & Reasoning*. Amsterdam: North-Holl&, in press.

Wason, P. C., & Johnson-Laird, P. N. (1972). *The Psychology of Deduction: Structure & Content*. Cambridge, MA: Harvard University Press. London: Batsford.

Yang, Y., & Johnson-Laird, P. N. (1998a). Illusions in quantified reasoning: How to make the impossible seem possible, and *vice versa*. *Memory & Cognition*, in press.

Yang, Y., & Johnson-Laird, P. N. (1998b) Systematic fallacies in quantified reasoning and how to eliminate them. Under submission.

Acknowledgements

This research has been supported by a grant from the James S. McDonnell Foundation awarded to the first author.

Connectionist Learning to Read Aloud and Comparison to Human Data

Ivelin Stoianov (stoianov@let.rug.nl)
Dept. Alfa-Informatica, Faculty of Arts, University of Groningen, POBox 716, 9700 AS, The Netherlands

Laurie Stowe (l.a.stowe@let.rug.nl)
Dept. of Linguistics, Faculty of Arts, University of Groningen, POBox 716, 9700 AS, The Netherlands

John Nerbonne (nerbonne@let.rug.nl)
Dept. Alfa-Informatica, Faculty of Arts, University of Groningen, POBox 716, 9700 AS, The Netherlands

Abstract

Research on connectionist mapping from written to spoken forms in natural language is presented. For this task, the more plausible Simple Recurrent Networks were used instead of static Neural Networks. The model was trained on a Dutch monosyllabic corpus. The effects of frequency, length and consistency were examined and were found similar to reported data in psycholinguistic experiments.

Introduction

Among a number of linguistic problems attracting attention in cognitive science is the word reading task, in particular the mapping from written to spoken forms in natural language. Connectionist models, if successfully trained on this problem and if their performance correlates to human performance in reading, can supply a framework for lexical processing. For example, Seidenberg & McClelland (1989, hence SM89) and Plaut, McClelland, Seidenberg & Patterson (1996) suggest that a *single-route distributed* process performs this transformation, as opposed to the symbolic *dual-route* model which claims that the reader must also have a lexical route that handles exceptions, or irregular words(Coltheart, 1980, 1993). The dual-route model has a connectionist implementation too: Zorzi (1998) proposed a Multilayered Perceptron (MLP) with an alternative structure to handle both easier rule-based mappings and the more difficult, exceptional words.

Although these connectionist models are reported to perform well, they employ static lexical encoding, which imposes constraints and might be considered a theoretical drawback. A better account for the variable and sequential nature of words would be sequential processing, which produces single phoneme at a time, as in the Sejnowski & Rosenberg's NETtalk model (1987). Simple Recurrent Networks (SRN) by Elman (1990) fit even better in this lexical representation scheme, with the advantage of gradual left context dependence, as opposed to the window context dependence in the NETtalk model. This capacity is due to the distributed contextual memory in SRNs, gradually evolving in time, while in the NETtalk the temporal information is encoded in a fixed-size window.

SRNs are reported to be successful in other difficult lexical tasks, e.g., learning the phonotactics of monosyllabic Dutch corpus (Stoianov, Nerbonne & Bouma, 1998, hence SNB98), which raises hopes for success in the reading task. Also, Plaut (*in press*) employs an extended SRN model for this task. In the current paper we propose further exploitation of the SRN model on this challenging problem.

To accomplish this, we trained a SRN to map the orthographic to phonologic representations of all 6100 Dutch monosyllables, as found in the CELEX lexical database. This difficult data contain also rare and foreign words. The lexical encoding that was used in phonotactics learning (SNB98), in which the left context only is presented, is not enough for this task, because only the proper output phoneme should be activated, as opposed to predicting all successors in the phonotactics problem. Therefore, a more specific data presentation was applied. The learning was successful and the network generalized well too. In addition to learning, we examined the SRN's errors for various effects found in previous psycholinguistic experiments, such as word frequency and grapheme-to-phoneme mapping consistency. The sequential data representation allowed us to observe other effects too, such as word length and error positioning. The SRN error profile matched closely to the human performance in reading, which supports the suggestion that SRN models can be used as a basic sequential processing module in a larger cognitive framework, explaining our cognitive linguistic capacity.

Mapping from orthography to phonology & the evolution of connectionism.

The reading process is complex. It involves acquiring visual input; transformation to abstract graphemic representations; further mapping to abstract auditory representations and finally, production of motor commands that cause sounds. If the visual data has semantic meaning, it is accessed too. Simultaneous modeling of all these stages is difficult, so assuming that the input and output steps are done, one can work on the intermediate level. Our work involves only the mapping from abstract graphemic representations to abstract phonetic representations. Therefore, hereafter by 'reading', we will mean this mapping.

Since the first successful connectionist system that transformed text to phonemes – the NETtalk (Sejnowski & Rosenberg, 1987) – a number of other connectionist architectures that model the human reading process have been suggested. Among these, Seidenberg & McClelland (1989) and Plaut, McClelland, Seidenberg & Patterson (1996), have been influential on connectionist NLP research (e.g., Milostan & Cottrell, 1998; Harm, 1998). A connectionist model that takes the opposite side of the *single route / dual-route* controversy was proposed by Zorzi (1998), who suggests that the MLP might benefit from an extra set of connections, from the input to the output layer. This extra set of connections is interpreted as a second "route", similar to the Grapheme-to-Phoneme-Conversion (GPC) rules (Coltheart, 1978). The information flow through the standard hidden layer is expected to represent the more complex mappings (exceptional words).

All these models are based on the general PDP postulates of using distributed representations and distributed knowledge, principles that stem from our neural system. Connectionist models benefit other useful properties too – generalization for unseen data, noise resistance, etc. – which are considered hard for symbolic systems.

An important divergence between NETtalk and the latter models is the employed lexical representation. In NETtalk, which is a static feedforward MLP, words are represented sequentially to the network, one letter at a time, together with a context of a few letters surrounding the letter to be pronounced. The network is trained to produce the phoneme that corresponds to the current letter and context. In contrast, the models in SM89, Plaut et al. (1996) and Zorzi (1998) explore static lexical representations, where words are presented to the NN at once and the corresponding phonologic representations are produces at once, too.

The SM89 model uses representation based on triples of graphemes and phonemes: "Wickelfeatures". In the input pattern correspondent to a given word, active orthographic Wickelfeatures will be those, which are parts of the input word. The output phonetic encoding is similar. The model used 400 input orthographic units and 460 output phonetic units. This representation raises the necessity of a complex feature encoder and decoder.

The connectionist models proposed in Plaut et al. (1996) are a feedforward MLP and an attractor NN (an extension of MLP with a recurrent layer at the output, which aimed at more precise target identification). They explore an alternative static data representation, which accounts for the spelling-to-sound regularities of English monosyllabic words. In this representation, there are slots for the onset and coda consonants and the vowels of the nucleus. By observing the existing graphotactic and phonotactic restrictions in the orthographic and phonetic onset, nucleus and coda representations, the authors manually constructed reduced input and output representations with 105 input

grapheme and 61 output phoneme units. These units stand for a limited number of orthographic and phonetic onsets, nuclei and codas. After a number of successful experiments on learning a orthography to phonology mapping, their claim is that this connectionist model and lexical representation can account for the basic abilities of skilled readers to pronounce correctly both regular and exceptional items, while still generalizing to novel items.

In addition, network error profile with respect to word frequency and grapheme-to-phoneme mapping consistency were tested against human performance in reading. For this purpose, reading latencies were considered to correspond to network error (in MLP) or to the time required for the network to settle to a stable output pattern (in attractor NN model). The experiments and mathematical analysis explain how the networks succeeded in handling quasi-regular domains (both regular and exception words) and producing frequency and consistency interactions exhibited by humans. In spite of this, in natural languages language objects are dynamic, including the words in their orthographic and phonetic representations, which has been dismissed by unjustified application of one of the Hinton's principles about connectionist models:

> For processing to be fast, the major constituents of an item should be processed in parallel. (Hinton, 1990)

Words have constituents (graphemes or phonemes) that are inevitably encountered in a strictly sequential fashion, therefore, words should be processed sequentially. Phonemes span time too, but dealing with words, we can consider the phonemes as static objects and process them in parallel. As far as the graphemes are concerned, one might argue that in reading we perceive visual objects larger than single graphemes, e.g., words. But this is done by skilled readers, possibly by use of some extra mechanisms. Beginners initially read one letter at a time (or group of letters) therefore the models should account for this.

Given this fact, there is nothing wrong with the NETtalk model, which was criticized in Plaut et al. (1996) because of the Hinton's principle. Nevertheless, NETtalk has another problem, based on the fixed limited context. The network would not map correctly two words with different phonetic representations, which differ somewhere beyond the temporally shifting graphemic context scope. A model that is theoretically able to handle such dependencies is Simple Recurrent Networks (Elman, 1990), where the output of the network depends on the whole left context, and which we explore in the following section.

Learning to Read Aloud with SRN

The experimental setting in our research is based on a standard SRN, trained to learn sequential association in orthography-to-phonology conversion (Fig.1). The network performance was measured on the training and unseen words. Further, the performance was analyzed for different variables, such as word frequency, length, grapheme-to-phoneme mapping consistency and error positioning, from which we drew some conclusions about the syllabic structure in Dutch. Preliminary analysis of a damaged network is provided, too.

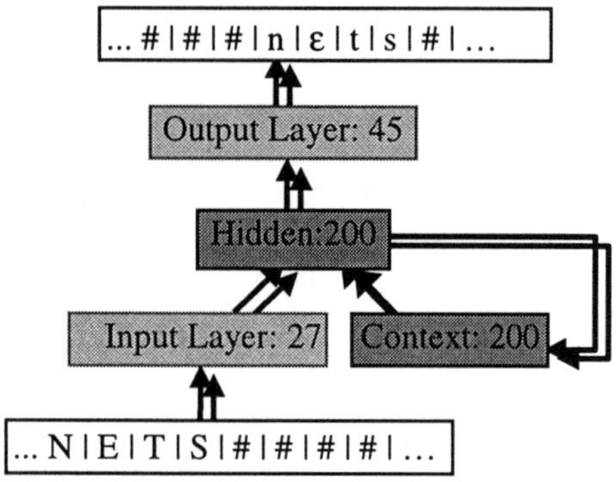

Fig.1 Simple Recurrent Networks and mapping from orthography to phonology.

Method

A Simple Recurrent Network (Elman, 1990) was trained to learn a sequential mapping: the orthography-to-phonology conversion of all 6100 monosyllabic Dutch words, as extracted from CELEX lexical database. The same corpus was used in our previous studies on graphotactics and phonotactics (SNB98). The data set contains orthographic and phonetic word representations and the frequencies of word occurrence in the Dutch language. This corpus was split into a training set (5100 words) and a test set (1000 words). The orthographic and phonetic word representations have mean length 4.53 ($\sigma=1.08$, $min=3$, $max=9$), and 3.92 ($\sigma = 0.94$, $min=2$, $max=8$) respectively. The word representations are built of 26 graphemes and 44 phonemes, plus one extra symbol representing space ('#'), used as a filler specifing end-of-word. The graphemes and phonemes were encoded orthogonally, that is, for each grapheme and phoneme, there is one input or output neuron respectively. One might speculate that with regard to the network associative capacity, this representation is equivalent to a distributed, feature-based representation, because we can always add two more static layers that decode and encode such feature-based representations to the orthogonal ones. During training and testing, the orthographic and phonetic word representations were given to the network sequentially, one symbol at a time. In order for the network to be able to correctly reproduce different phonological representations for words with identical beginnings (left context), phonological production is delayed for three steps. In this manner, the network receives partial right context as well. Therefore, the network decision at each moment is based on the full left and partial right context (initially 3 graphemes), simultaneously and distributively encoded in the context layer.

Experiments were conducted with different number of hidden neurons, ranging from 100 to 400. Best performance was found with the largest network. In this report, we present results for a network with 200 hidden neurons, resulting in about 55,000 weights. The network architecture

is given in Fig.1, where an example mapping from the orthographic to the phonetic representation of the word 'nets' is shown as well. The delay is implemented technically by targeting 3 starting filling symbols (end-of-word – '#') at the output layer. The input sequence is completed with the same symbol until the full phonetic representation is generated at the output layer (Fig.1).

Training

The training process is organized in epochs, in the course of which the whole training data set (5100 words) is presented to the network in accordance with word distribution, that is, word frequencies (SNB98). In order to reduce the learning time, the actual word frequencies were shrunk by applying a logarithm function, resulting in about 12,500 training sequences per session. Such an approach has been used by other authors as well (e.g., Plaut et al., 1996; Zorzi, 1998). Next, for each word, the sequence of graphemes is presented to the input, one by one, followed by end-of-word symbols. Each time step is completed by copying the hidden layer activations to the context layer, which are used in the next step (Elman, 1990). At the same time, after the network generates its expectations for the phonemes at the output layer, the representation of the true phoneme is used to compute an error for the current time step. This error is used by the Backpropagation Through Time (BPTT) learning algorithm (see for details Haykin, 1994; SNB98), which includes a forward move where errors are collected and a backward move, during which global error is back-propagated through time until the beginning of the current training sequence. This process is completed by updating the network weights with values, accumulated during the backward move. The state of the network (i.e., the context memory) is reset after processing one word.

The network was trained on 18 epochs, resulting in approximately 200,000 word presentations. The total number of individual word presentations ranged from 18 to 200, according to the individual word frequencies. The network started with a sharp error drop to about 4%, slowly decreasing down to 1.2% (see Table 1).

Table 1. Dynamics of the SRN error during the training.

Epoch	1	2-4	4-10	10-17	18
Error (%)	4.1%	2%	1.6%	1.4%	1.2%

We expect further error decrease with longer training, although this would need much more training time, because the learning coefficient η decreases by 30% after each epoch, starting from 0.3 and restricted with a bottom limit at 0.001. Learning grapheme to phoneme conversion is quite a difficult task, so we had to apply some other special techniques to improve it. First, instead of the standard Backpropagation algorithm, we used a BPTT learning scheme as described above. Next, standard momentum $\alpha=0.5$ term was applied. Further, the training process was supervised by an evolutionary algorithm that trains a pool of networks on the same problem and after each training epoch, it eliminates the neural network with the worst performance, keeping clones of the networks that performs better. This training method was developed in our previous studies on phonotactics (SNB98) and was found to perform better than the standard single-network training.

Performance

A procedure that examined all training examples followed each training epoch, in which we distinguished phonemic and network errors. *Phonemic* error occurred if during the network processing, the most active neuron didn't correspond to the expected phoneme. *Network* error estimated the percent of all mispronounced phonemes, weighted by the frequency of the word the phonemes belong to. This procedure results in fair estimation of the network performance, accounting for the distribution of the words in natural language. As we mentioned earlier, the final network error estimated during the training was 1.2%. We analyzed the type of the incorrect productions the network has made and found that 75% of them were substitutions between close phonemes, mainly vowels, which might be used as an argument to reduce error.

The generalization capabilities of the network were tested on a test set, which contained the orthographic and phonetic representations of 1000 unseen during training words. For the sake of comparison to Plaut et al. (1996), we should note, that the test words should be interpreted as *nonwords* because they have not been used for training. We could use them for testing, because we had their correct phonetic representations. Still, there might be words that are exceptional with regard to the reading, therefore we expected higher error. The performance on this test set was 1.4%, which confirmed that the network learned the Dutch GPC rules for monosyllables. As we predicted, error increase was primary due to the exceptional words.

The overall performance is similar to Zorzi (1998) and worse than SM89, and Plaut et al. (1996), which we attribute to the twice larger data set and incomplete training. Obviously, the 18 training epochs resulting in 18 up to 200 exposures for a single word were not enough to achieve perfect performance, especially for the exceptional words. Also, networks with larger hidden layers tend to learn better, however at the cost of longer training time.

Error profile analysis

In addition to overall network accuracy, connectionist systems that model lexical tasks also aim at approximating the correspondent human performance with regard to different variables such as word frequency. This aspect in connectionist modeling is important, because it contributes to verifying whether the suggested models can be used for modeling the correspondent human processes. For this purpose, the model performance is compared with reaction time or error in reading.

The variables we examined were word *frequency*, word *length*, *consistency* of the grapheme-to-phoneme mapping and *error positioning*. Previous reports (Plaut et al., 1996;

Zorzi, 1998) deal mainly with word frequency and consistency, unable to exhibit significant length effects. The sequential nature of SRNs and structure of the training process naturally involve these characteristics, so we were able to test them, as does Plaut (*in press*).

Consistency in orthography to phonology conversion measures how much the pronunciation of a given item is coherent to the pronunciation of orthographically similar items. An interesting issue is how to measure consistency. Plaut et al. (1996) used the similarity in spelling of *rhymes* (see below) in order to estimate it. We adopted another definition, suggested by Jared et al. (1990), according to which consistency depends on the summed frequency of the word's *friends* and word's *enemies*. "Friends" are words with similar spelling and similar pronunciation, while "enemies" are words with similar spelling, but distinct pronunciation. We categorized consistency into four categories (similarly to Plaut et al., 1996). Words with many more friends than enemies are named *regular*, as opposed to words with many more enemies than friends, which are named *exceptions*. There are two intermediate categories – *ambiguous* – with as many friends as enemies – and *semi-regular* – with somewhat more friends than enemies. Error with regard to consistency is given in Table 2. The strong interaction between error and consistency is in parallel to the observed effect of reduced naming latencies and greater accuracy in pronunciation for regular words and increased latencies and lower pronunciation accuracy for irregular ones (Coltheart, 1978; Glusko, 1979). Still, we see in Table 2 much higher error for exceptional words, which we attribute to the insufficient training.

Table 2. SRN performance against word consistency.

Consistency	Exception	Ambiguous	Semi-Regular	Regular
Error (%)	30 %	5 %	0.8 %	0.1 %
Entropy	1.4	1.2	1.05	1.0

The next important effect we verified was the network performance for different word frequencies (Table 3). This effect was observed in most of the models (Plaut et al., 1996; Plaut, 1998; Zorzi, 1998) and psycholinguistic studies; The SRN also exhibited good frequency effects with up to five times better performance for high-frequency words as compared to the low-frequency items.

Table 3. SRN performance against word frequency.

Frequency	Low	Mid-Low	Mid	Mid-High	High
Error (%)	4.1%	1.5%	1.0%	0.7%	0.8%

In previous studies (Plaut et al., 1996 for review) important frequency-consistency interaction was found, where the frequency effect almost disappears for consistent words. To test for this effect, we studied the neural network performance varying both consistency and frequency (Fig.2) and found the pattern exhibited in human reading studies. Frequency is unimportant for consistent words, somewhat important for ambiguous words and crucial for exceptional words. We should note that the significant error for very exceptional words doesn't influence much the overall error, due to the very small number of words from that category.

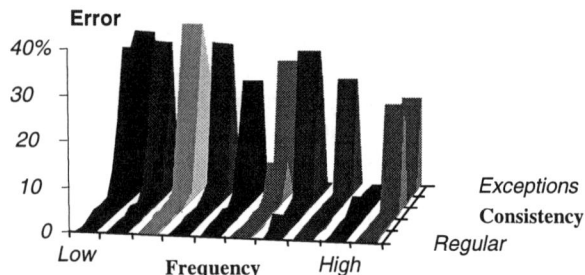

Fig.2 Network error for various degrees of grapheme-to-phoneme consistency as a function of frequency & consistency. The frequency influence on error almost disappears for regular words, where error is very small.

Further, we analyzed the network performance with regard to word length. Connectionist models based on static word representations didn't show apparent interaction between error and length, because static word processing can not be affected by word length. This was the reason Plaut to turn to a sequential connectionist model in his recent paper (*in press*). As we see in Table 4, there is a specific U-shaped dependency between error and word length. For short and long words, error is higher than for words of size 4 to 7. Our explanation for the higher error of the short words is that the network doesn't have enough information to produce the correct pronunciation. On the other hand, the higher error for longer words is easy to explain, as error accumulates during left context processing.

Table 4. SRN performance against word length.

Length	Short	Mid-Short	Mid	Mid-Long	Long
Error (%)	11%	3.1%	2.5%	3.0%	5.6%

And finally, we present an error analysis with regard to the position at which it appears during the process of phoneme producing (Table 5). Given the higher error for longer words one might expect that higher error at the final positions, but it is not so clear why it is higher at the 2nd and 3rd positions. In order to answer to this question, we conducted further finer-grained analyze of the position of the error concentrating on sub-lexical units.

Table 5. SRN performance against error position.

Position	1	2	3	4-5	6-7
Error (%)	1.9%	4.8%	3.7%	1.7%	2.5%

Syllabic structure

Most linguists divide syllables into *onset* (which contains the initial consonants), followed by *nucleus* (intermediate vowels) and ending with *coda* (the final consonants). Also, they look for a more complex internal structure of the syllable, e.g., (1) or (2) (Kessler & Treiman, 1997), most theoretical linguists preferring (1).

(1)   (onset – rhyme (nucleus – coda))

(2)   (body (onset – nucleus) – coda)

Observing the non-uniform error distribution in the words (see above), we were interested how the error was distributed into the above sub-syllabic units. The results showed that the higher error in the middle of the words was due to higher error at the vowel positions (which includes positions 2 and 3 in Table 5 due to variable onset length). The error at the vowel was 8%, while the error in the coda was about 1.9%. There was lower error in the onset as well (1.5% – 1.8%). This means that there is a specific error break at the transition onset-nucleus. This error peak is in parallel to another interesting fact that we observed, namely that the mean entropy in the body was 3.37, σ=0.55, while the mean entropy in the rhyme was 1.90, σ=1.21. This supports statistically the position that the body is less coherent, and the rhyme more.

Therefore, we can conclude that the syllabic structure in Dutch is not plain, but follows the onset-rhyme division (1). The same structure was found in English as well (Kessler & Treiman, 1997). With a non-sequential model this would be difficult to measure.

Network damaging and dyslexia

The PDP models are well known for their damage resistance due to their distributed way of representing data and knowledge. A network damage might consist of loss of neurons or memory distortion. We experimented with adding noise to weights and neuron removal. The noise was represented as random numbers, uniformly distributed in the range [-p ... p]. There was no apparent effect of 1-2% hidden neurons removal, which was similar to adding slight noise (p=0.10) to 90% of the neurons, while removing 5% of the hidden neurons resulted in 20-25% error, which was similar to adding noise with p=0.30. A milder effect was observed with noise p=0.20, where the error jumped to 6%.

We investigated also how damages influenced network performance for various frequency and consistency levels, searching for processes similar to those found in dyslexics. There are two main categories of dyslexia resulting from brain damage (Coltheart, 1993; Plaut et al, 1996). In *surface* dyslexia, patients can read non-words but have problems with exceptional words – they regularize them by using Grapheme-to-Phoneme-Conversion (GPC) rules (Coltheart, 1993). The other pattern, *phonological* dyslexia is characterized by difficulties in pronunciation of non-words, although familiar words can be read, i.e., patients seem to have lost GPC rules.

The pattern of errors we observed after network damage did not correlate to any specific type of phonological dyslexia. Both exceptional and regular words were affected. Mild damage (noise, p=0.20 or discarding 2-3% of the hidden neurons) affected regular (from 0.2-1% to 2-10% error), ambiguous (from 4-5% to 15% err) and exceptional words (from 20-30% to 35-50% error). At the same time, the frequency effect was reduced significantly. More severe damage (removing 5% neurons or noise, p=0.30) resulted in even larger error for regular words and a fading of consistency effect. Also, increasing noise affected almost twice as many test words (i.e., *nonwords*) with exceptional pronunciations as training words from the same category; that is, some test words with exceptional pronunciations perhaps assimilate to the reading of similar words, but a slight amount of noise easily disrupts this relation. Next, weight damage was more likely to affect the end of the words, which we attribute to corrupting the rules that the network has built in order to encode the context in memory.

The fail to find a specific type of dyslexia we attribute to the general weight damaging. We expect more specific damaging to affect more specific groups of words.

Discussion

In this paper we presented our initial experiments on learning grapheme-to-phoneme mapping in Dutch with Simple Recurrent Networks. The main goal in the study was to test the ability of SRNs to learn such a complex mapping employing simple data encoding – sequential presentation of a single grapheme at a time – as opposed to static connectionist models (Plaut et al., 1996; Zorzi, 1998) and the more complex sequential mapping scheme in Plaut (*in press*). In order to test how well SRNs approximate human performance in reading, we studied the influence of word frequency and consistency, as well as word length and error-position, which static connectionist models could not observe. Further, trained networks were deliberately damaged with an aim to model dyslexia. SRN performed well on training and unseen test data sets even after very limited number of training epochs. Also, there were significant consistency and frequency effects on error. The error interacted with word length as well. The observed pattern of error positioning suggested a specific non-symmetric syllabic structure for Dutch, which was found in English as well. The reported data on damaging did not show specific dyslexic patterns, but we should note that the experiments are still in progress and the data is suggestive.

How does this study contribute to the dialog between single- and dual-route models? We consider SRNs as a single-route model, where a single, although complex mapping produces all outputs in contrast to the "dual-route" network by Zorzi (1998). However, is Zorzi's model dual-route? We claim that it can be simulated by a single-route MLP with restrictions on the weights during training. Zorzi's network structure – with connections that map directly from orthography to phonology (standing for GPC rules) and another set of connections that maps through a standard hidden layer (supposed to maintain a lexicon) – can emerge in training a general MLP. Still, because it is difficult to analyze the way a uniform network does its job, Zorzi's model contributes to our understanding how neural networks can handle such a difficult problem. Maybe, if we structure SRNs in a similar way, we might achieve even better performance, by minimizing the complexity of the learning task. In addition, this would help to model surface and phonological dyslexia in a more direct way.

References

Christiansen, M. & Nick Chater (1998). Toward a connectionist model of recursion in human linguistic performance. *Cognitive Science* (in press).

Coltheart, M. (1978). Lexical Access in simple reading task. In Underwood (ed.) Strategies of information processing, pp. 151-216. London, Acad. Press.

Coltheart, Max, K. Petterson & J.C. Marshall (1980). *Deep Dyslexia. London*, Boston: Routledge & Kegan Paul.

Coltheart, Max, B. Curtis, P. Atkins & M. Haller (1993). Models of Reading Aloud: Dual-Route and Parallel-Distributed-Processing Approach. In *Psychological Review*, Vol.100, N.4, 589-608.

Elman J.L. (1990). Finding structure in time. *Cognitive Science*, 14, 213-252.

Glushko, R.J. (1979). The organisation and activation of orthographic knowledge in reading aloud. Journal of Experimental Psychology: Human Perc&Perf.,5,674-691.

Harm, M. (1998). Division of Labor in a Computational Model of Visual Word Recognition. Ph.D. thesis at UCS.

Haykin, Simon (1994). *Neural Networks*. Macmillan College Publications.

Hinton, G.E. (1990). Mapping part-whole hierarchies into connectionist networks. *Artificial Intelligence*,46(1),47-76

Jared, D., McRae, K. & M. S. Seidenberg (1990). The basis of consistency effects in word naming. Journal of Memory and Language, 29, 687-715.

Kessler, Brett & Rebecca Treiman (1997). Syllable Structure and the Distribution of Phonemes in English Syllables. *Journal of Memory and Language*, 37,295-311.

Milostan, J.C. & G.W. Cottrell, (1998). Serial Order in Reading Aloud: Connectionist Models and Neighborhood Structure. In *Advances in Neural Information Processing Systems*, 10, MIT, Cambridge, MA.

Plaut, D.C., J McClelland, M Seidenberg & K Patterson (1996). Understanding Normal and Impaired Word Reading: Computational Principles in Quasi-Regular Domains. *Psychological Review*, 103, pp.56-115.

Plaut, David C. (*in press*). A Connectionist Approach to Word Reading and Acquired Dyslexia: Extension to Sequential Processing. *Cognitive Science*.

Seidenberg, M.S. & J.L. McClelland (1989). A distributed, developmental model of word recognition & naming. *Psychological Review*, 96, 523-568.

Sejnowski, T.J. & C.R. Rosenberg (1987). Parallel networks that learn to pronounce English text. *Complex Systems*, 1, 145-168

Stoianov, Ivelin P., John Nerbonne & Huub Bouma (1998). Modeling the phonotactic structure of natural language words with Simple Recurrent Networks. In Coppen, van Halteren & Teunissen, (eds.) *Computational Linguistics in the Netherlands 1997*, Rodopi, Amsterdam, Netherlands, pp. 77-95.

Zorzi, Marco (1998). Two Routes or One in Reading Aloud? A Connectionist Dual-Process Model. In *Journal of Experimental Psychology: Human Perception and Performance*. Vol. 24, N.4, pp. 1131-1161.

Investigating Language Change:
A Multi-Agent Neural-Network Based Simulation

Scott C. Stoness (scs@sfu.ca)
School of Computing Science
Simon Fraser University

Christopher Dircks (chrisd@sfu.ca)
School of Computing Science
Simon Fraser University

Abstract

Multiple agents, equipped with a feature-based phonetic model and a connectionist cognitive model, interact via the *naming game*, with lexicon formation and change as emergent properties of this complex adaptive system. We present a new description of the *naming game*, situating it as a general, implementation-independent paradigm. Our addition of richer phonetic and cognitive models provides the agents with a greater degree of cognitive validity than does earlier work, while enhancing the flexibility of the system and reproducing empirical results. Feature-based phonetics, piecewise reinforcement learning, and a connectionist architecture with local representation allows language discrimination based on schemata instead of entire utterances.

Introduction

All things change. Despite societal and personal predispositions towards stability, constant modification is an incontrovertible fact, irrespective of the observed impact on our quotidian existence. We may not be cognizant, as individuals, of the process of change as it occurs, because the brief time span of a human life does not permit the recognition of developments preceding great change; events do not always match the meteoric pace set by the limits of our frail biology, but rather transpire with a certain glacial implacability. One phenomenon with which we are all intimately acquainted that undergoes just such an incremental process of adjustment is natural language; the idiolect of any single individual remains relatively fixed subsequent to the initial acquisition of the mother tongue, yet the language as a whole clearly experiences periodic alteration.

It is no coincidence that the description above of a gradual, gradational modification also applies to evolution by natural selection, since language change shares many of the basic attributes of its biological parallel. The deliberate nature of this type of process can obfuscate the situation from an individual perspective. This tendency was exacerbated in the late 19th century by the conflation of language change with an incomplete comprehension of the biological principle – obsession with progress, and the usurption for this goal of the phrase 'survival of the fittest' – leading to the simultaneous adoption of both historical reconstruction, which would map out the myriad developments in the evolution of language *and* prescriptive grammar, where the stated goal is to maintain the purity of the current linguistic norms.

20th century linguists have relinquished their grip on the reins and recognized the inevitability of language change, but have been unable to converge upon a single theory, or more accurately, have been unable to produce a universally compelling explanatory mechanism for language change. It puts modern linguists in much the same position as biologists before the advent of Darwin's theory of natural selection, or perhaps before the discovery of DNA by Watson & Crick; there is a universal acceptance of a general process, but the details are not known.

Unfortunately, the data available for historical research is limited since the vast majority of the world's languages spoken to date had no written form, and much of what was committed to paper(or other appropriate media) has been lost. Controlled linguistic experimentation is extraordinarily difficult, the more so when the phenomenon we are examining would require investigations which would not begin to bear fruit for several generations. The linguistic analogue to *Drosophilia* is not obvious; there simply are no biological entities which exhibit sufficient similarities to human linguistic communication to make experimentation worthwhile *and* are also short-lived enough for such experiments to be feasible. Fortunately, modern computers are sufficiently powerful to enable us to produce simulations of language change which are, to a certain degree, cognitively and linguistically accurate, yet simplistic enough to allow controlled investigation of this phenomenon.

Much recent work has been done in this vein, in particular by the members of the Sony CSL Paris, examining language as a complex adaptive system which produces language change as emergent behaviour. [9] postulates that four factors are necessary for linguistic variation and evolution to occur: self-organization, stochasticity in transmission and production, tolerance of minor linguistic variation, and a certain rate of population change. The authors first produced completely deterministic agents, showing that linguistic variation was not tolerated. As a result, the final model (reported in [8], [9], and prototypically in [7]) of agent-internal linguistic processing is something of a hybrid, with probabilistic measures grafted onto what is essentially a deterministic backbone, while linguistic utterances in the system are represented by a random sequence of characters. We present a new system for investigating lexicon change which follows a framework for agent interaction similar to the Steels and Kaplan model; however, we have made our system more valid from a cognitive perspective by using artificial neural networks for the agent-internal cognitive model and a phonetic model based on Chomsky-Halle binary features for utterances.

Agent Interactions: The Naming Game

It is foolhardy to undertake research involving multiple software agents without a principled framework within which the agents can interact; that is, the precise details of all possible interactions between agents and their environment (and / or each other) must be specified in an unambiguous fashion. To this end, [7] introduces the *naming game*, an austere paradigm for interaction tailored especially for the development, transmission, and evolution of a lexicon in either a static or dynamic population of agents.

The *naming game* is appropriate for a population of agents and a number objects[1]; an interaction proceeds as follows:

1) Two agents are selected from the population; one is designated as the Speaker, the other as the Hearer.
2) The Speaker chooses an object, possibly at random.
3) The Speaker, through whatever process encoded in the agent model, names the object; that is, accesses the appropriate form-meaning pair, and produces the form.
4) The Speaker (virtually) points at the object.
5) The Hearer interprets this combination of linguistic and extra-linguistic information produced by the speaker to be a reference to some particular object.
6) The game succeeds if the Hearer correctly interprets the information provided by the Speaker; if the agents do not agree on the object referenced, then the game fails.

Presumably, upon completion of a *naming game* interaction, learning occurs; while the bulk of research which follows this paradigm uses some variation of the adaptive rules outlined in [9], there is no reason to suppose that this framework must be coupled with that particular set of learning rules. In fact, we demonstrate that the *naming game* provides an excellent paradigm for use with agents possessing different internal mechanisms and learning procedures.

The procedure outlined above is the *naming game* in its simplest incarnation; many enhancements can, and have been, made, including noisy channels for both linguistic and 'visual' communication, and changing populations of both agents and objects.

One might imagine that the *naming game* is not a valid model of language acquisition, since the Hearer has no way of

knowing that the spoken utterance names the object; it could conceivably be any form of communication, or even none at all. However, the *naming game* is intended to model neither language acquisition nor the origins of language, but rather language coalescence and change through highly constrained interactions between adult speakers.

The Phonetic Model

Rather than retain the character-based randomly-generated words of earlier systems, we have chosen to move towards linguistic validity by the inclusion of a rudimentary feature-based phonetic model.

Our agents communicate by means of single-syllable utterances consisting of a consonant followed by a vowel. Each phoneme is represented by a set of binary features loosely based on Chomsky-Halle features and the cardinal vowel system (see Table 1)

Table 1: Binary Feature Matrix for Phonemes

	Consonants																Vowels									
IPA	p	b	f	v	t	d	s	z	c	j	ʃ	ʒ	k	g	χ	ɣ	i	u	e	o	ɛ	ʌ	a	ɑ		IPA
Ant	+		+		+		-			-							+	+	-	-			-			Closed
Cor	-				+		+		+			-					-	+	+	-						Mid
Vcd	-	+	-	+	-	+	-	+	-	+	-	+	-	+	-	+	-	+	-	+	-	+	-	+		Back
Cont	-	-	+	+	-	-	+	+	-	-	+	+	-	-	+	+										

There appears to be little to differentiate a model consisting of a sequence of characters from one which consists of a sequence of abstract phonemes represented by binary feature sets, but the phonetic model we introduce does in fact provide at least one major advantage other than the semblance of cognitive validity. Rather than forcing each of the features to have a discrete binary value of either zero or one, we allow values across the real interval (0,1). Not only does this allow a much more flexible connectionist implementation than the equivalent using binary features, but it is in fact phonetically justified. The cardinal vowel system is little more than a set of standard reference locations for the infinitely variable tongue position observed in vowel production in the real world. Similarly, voicing delays on consonants vary from speaker to speaker and context to context, as do tongue positions in consonant.

By moving away from the character-based model, which can only encode a fixed amount of information depending on the character set, we arrive at a representation which allows us to encode a much higher degree of variability in the utterances, modelling crudely the acoustic signals received by the human ear.

Agents' Internal Neural Nets

Each agent is furnished with two completely separate neural networks, one for determining the utterance from precise object information (henceforth the S-Net, or Speech Network), and one for settling upon a particular object given

---

[1]Usually these are software agents and virtual objects, but some work has been done with autonomous robots. [9]

713

an utterance and (possibly) some non-linguistic information (the H-Net, or Hearing Network).

The S-Net

The neural net used for the production of utterances is of extraordinarily simple design. It is a two-layer, fully connected network, with an input node for each distinct object in the simulation and one node in the output layer for each phonetic feature in a word. Once the Speaker has randomly chosen an object as the topic, the activation of the corresponding input node is set to 1.0, while that of all other nodes becomes 0.0. The activation level on an output feature node is simply the sum of all inputs to the node, with no use of a threshold or normalization. Thus, the weights on the links from the active object node appear directly on the output layer; the range of values for the weights is the continuous interval [0,1], as this also defines the values desired for our features. The activations of the output nodes can therefore be interpreted directly as values for the corresponding features.

The H-Net

The H-Net is also a two-layer, fully connected network, with an output layer which is virtually competitive; inhibitory links are not implemented directly, but rather through a winner-take-all choice.

The output of the Speaker's S-Net is placed directly on nodes in the input layer of the Hearer's H-Net, modelling the reception of auditory information. There is a further subset of the H-Net input layer which is dedicated to extra-linguistic information.

This extra-linguistic information is meant to correspond vaguely to the real-world visual cues experienced by the Hearer in a *naming game* where the Speaker points at an object. Accordingly, the topic receives the highest score of all objects: not a perfect score, but rather a random number between ⅔ and 1. Four or five other potential objects are assigned smaller scores between 0 and ½ to represent physical proximity to the topic, the main source of ambiguity in pointing. The random choice of error-objects and the high degree of variability in the object scores is an attempt to crudely model a wide variety of pointing situations, where the topic will be surrounded by different objects in different configurations in every interaction. This differs significantly from a fixed object layout, pointed to in every interaction, since in that restricted instance, certain objects will never need to be disambiguated by phonetic information.

Once the input layer is fully initialized, activation levels for the output layer are calculated with a straight sum of products rule. The scores of the object nodes are compared, and the node with the highest score is chosen as the eventual winner of the virtual competition within this layer.

This object chosen by the H-Net is compared with the original topic, and the success of the game determined.

Training Regimen

Individual speakers are unlikely to drastically change their speech patterns when they are being understood; similarly, if a listener is able to comprehend a speaker, there is little reason to adapt one's model of the language to their accent. Accordingly, in our model, learning only occurs when the *naming game* fails.

In the real world, communication occurs for a purpose, and in the case of a misunderstanding about the topic of a conversation, it is unlikely that the participants will simply give up; the speaker will repeat the word, and perhaps even identify the object physically in an unambiguous manner (i.e. by picking it up). It is therefore reasonable to suppose that the Hearer agents are familiar with both the utterance produced by the Speaker and with the intended topic, even when the *naming game* does not succeed. We have arbitrarily chosen to have the Hearer adapt its behaviour to match the Speaker; when discussing this object in the future, the Hearer's speech will more closely resemble that of the Speaker, and the Hearer will also be more accepting of utterances similar to the Speaker's designation for that topic.

Initialization

All phoneme-object weights in both the S-Net and the H-Net are initialized to random values between 0 and 1, representing in the first instance phoneme values, and in the latter relative contribution of features to the object score.

Since the weights in the H-Net between the input and output object nodes undergo no training, their initialization must be performed more carefully. Weights between input and output nodes which represent the same object are set to 0.6, while all other weights in this set are given random values uniformly distributed over the interval [0,0.5]. This approach attempts to model in a simple way similarities between objects, while avoiding the undesirable extremes where object information either overpowers the contribution of the object's name or cannot affect the result.

S-Net Training

One of the tasks of the Hearer is to interpret the continuous phonetic output of the Speaker in terms of idealized binary features. This is implicit in the normal actions of the H-Net, but explicit during S-Net training; rather than adapting its speech towards the actual output of the Speaker, the Hearer moves its speech towards an idealized binary feature set. Because we only train the Hearer when the *naming game* fails, its speech will never reach this ideal, but will only move in that direction as far as is necessary for effective communication to occur.

$$w' = \sqrt[3]{\frac{w - 1/2}{4}} + \frac{1}{2} \qquad (1)$$

$$w' = 4(w - 1/2)^3 + 1/2 \qquad (2)$$

Each feature in the S-Net of the Hearer is examined independently to determine if its idealized value is the same as that of the corresponding feature in the Speaker's utterance. If so, its value is reinforced (see equation 1); if not, it is punished (see equation 2). The punishment equation moves values towards 0.5, while the reinforcement function moves values towards (but not beyond) 1 or 0, depending on the polarity of the weight. A random number between -0.05w′ and 0.05w′ is then generated and added to this new value, and this 'fuzzy' result is forced within the interval [0,1]. This last step is required, else punishment will set the weight on an exponential growth pattern. The random fuzz is also necessary, since the punishment function, on its own, will never force a weight across the fixed point of 0.5.

H-Net Training

The only weights in the H-Net which are trained are those between the phonetic input nodes and the output nodes (as discussed above, the object weights remain fixed at their initial values). Again, training only occurs when the Hearer has chosen the wrong object; the goal of this weight modification is to make the H-Net more likely to settle on the correct topic when given similar phonetic input in the future.

There are a number of ways to accomplish this result, but we settled on decreasing the score of the false positive, and increasing the score of the correct answer. This is a very straightforward procedure which does not overly complicate the dynamics of the network, and tends to restrict the weights to a reasonable range.

$$w' = w - \delta \cdot w \cdot p \qquad (3)$$

$$w' = w + \delta \cdot w \cdot p \qquad (4)$$

At the implementation level, we apply equation (3) to the false positives, and equation (4) to the missed answer; in these equations, w' is the new weight, w is the old weight, δ is the learning rate, and p is the value on the phonetic input node. Essentially, we modify the weight by a certain percentage of its contribution to the activation of the object node in question. The current value of δ in the system is 0.05, and since the value of p ranges from 0 to 1, in practice, the weight is modified by an average of 2.5% of its own value.

The Simulation World

We have tested and run our simulation with up to 50 agents and 20 objects, but for the most part we have kept to 20 agents and 10 objects, so that our results are comparable to those of [9]; these numbers seem to produce interesting results, yet have simulation run times which are reasonable.

Each agent has its H-Net and S-Net randomly initialized as described above. There is no internal communication between the nets, and we do not explicitly train the agents to understand their own utterances; the eventual consistency exhibited by the system is a result of self-organization (at a societal, rather than agent level).

We conduct instances of the *naming game* in groups of 20;

the agents speak in order, and a random partner is chosen to be the Hearer. This approach has the advantage that speech starvation will not occur – every agent gets its chance to speak – but it is theoretically possible that an agent could survive a simulation completely unchanged, never being selected as the Hearer. However, the probabilities involved are so small that it is not an issue at present, and starvation-avoidance techniques could be easily added it if became a problem.

In the next section, we present results from four different types of simulation runs: with and without population flux, with either 5000 or 20,000 groups of *naming game* interactions. Since each group consists of 20 *naming games*, altogether the simulations consist of 100,000 and 400,000 instances of the *naming game*. In the simulations with population flux, a random individual is removed every 2000 games and a new, randomly initialized agent takes its place; in the longest simulations with population flux, there have been 200 new individuals inserted in the population.

Experimental Results

In some initial experiments with a learning rate (δ) of 0.20, we achieved an 86% average naming game success rate in interactions without object information given to the Hearers, and a 95% average success rate when such information was provided. This latter figure rose to 98% when the learning rate was changed to 0.05. We recently ran several simulations using only the object information, which resulted in a success rate of around 50%.

In our long-term trials, we achieved success rates around 95% when the population was stable. Certain periods of the simulation exhibited success rates around 98%, but the global average was lower because of language change and periods of instability. With a dynamic population, the success rate hovers around 80%.

Form Distributions

In order to quantify the development of our agents we plotted, for each object, the number of speakers for each linguistic variant as a function of time. Since we ran 15 simulations with a static population and had 10 objects in each, this gives a total of 150 separate graphs over the length of 100,000 games, 60 over 400,000 games. Examining these graphs, we recognized that they fell into several different patterns, based both on the overall appearance of the graph, and some underlying statistics.

In *dominance*, one form (almost) completely dominates the phonetic space quickly, and retains control for the duration of the simulation (See Fig 1a). In *70-30* graphs, two forms exist in the population, one spoken by about 70% of agents, the other by about 30% (See Fig 1b). *Parity* graphs have two common forms, splitting the majority of speakers between them (Fig 1c). When a large number of competing forms arise (usually 4), none spoken by more than 40% of the population (See Fig 1d), we call this a *mush* graph. In the *step-up* pattern, one form appears destined for dominance, but it is overtaken

(a) (b)

(c) (d)

(e) (f)

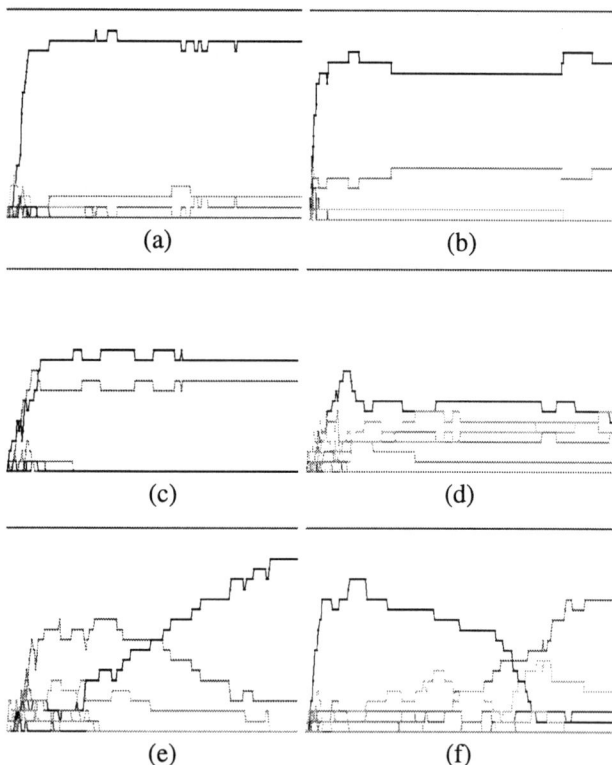

Figure 1: The graphs above represent the number of speakers of linguistic variants as a function of time for a particular object. The vertical axis counts the number of agents speaking a form, while the horizontal axis represents time. Each shade of gray shows the frequency fluctuations over time for a particular linguistic form. Each graph is a canonical example for its category: (a) dominance (b) 70-30 (c) parity (d) mush (e) step-up (f) switch. As an example, graph (c) shows a simulation where two forms, after an initial learning period, achieve a steady state where each is spoken by about half the population.

Table 2: Relative Frequency of Form Patterns

Form Patterns	% of Static Pop.		% of Dynamic Pop.	
	100k games	400k games	100k games	400k games
Dom.	30	38	30	21
70-30	9	17	9	2
Parity	16	12	9	3
Mush	30	7	12	3
Step-Up	9	17	26	24
Switch	6	10	13	47
Total # of Graphs	150	60	160	70

by a second form, which goes on to dominance (See Fig 1e). Finally, some of the graphs exhibit a form *switch*, where one form dominates with over 60% of the speakers for a period, and then is replaced by a different form, which then dominates (see Fig 1f).

Table 2 outlines the relative frequency (as a percentage) for each category of graph. We show statistics for both our short and long simulations, separated into runs with and without population flux.

Obviously, these categories have very fuzzy boundaries, and some graphs simply do not fit particularly well into any category. However, there are clear examples of each group, (including the *switch* form representing lexicon change) and the fact that these particular patterns are the most common provides insight into the learning processes of our agents.

Discussion

We are not willing to claim that coherence in language **must** be due to self-organization, but our simulation (along with [8] and [9]) makes it clear that extremely simple self-organizing systems **can** achieve a coherent lexicon. Our results reinforce the idea (presented in [9]) that population flux increases the incidence of lexicon change, but we also show significant change even in a static population (see [3] for a discussion).

When we examined the graphs of our simulations, we at first had a difficult time reconciling a 95% success rate with the fairly high frequency (30%) in short simulations of the *mush* graph, where there were multiple competing forms. However, an examination of the lexical forms in these mush patterns revealed that all forms were fairly similar. The combination of these forms and the weights on the network showed that our H-Nets were acquiring schemata, which may include one or more □ (don't care) values, to use the notation of [5]. For example, a particular H-Net recognized a voiced anterior consonant followed by an 'i' as Object 8, regardless of whether the input utterance was 'di', 'zi', 'vi', or 'bi'; the agent has internalized a schema which includes a □ value for the coronal and continuant features.

Schemata allow us to explain not only the high success rate in the face of many variants, but also the stability shown by the variants themselves. No matter which of the four forms above is heard, the H-Net will settle on Object 8; this allows all four variants to flourish, since in the absence of misunderstanding, no learning occurs, and the forms are stable.

With our schemata firmly in hand, we can also investigate the high degree of stability demonstrated in the *parity* and *70-30* graphs. Both patterns exhibit two forms which together dominate the population of speakers, which can be nicely explained by the acquisition of a single □ value by the H-Nets of the population. The two patterns in fact represent different facets of the same underlying situation: in the *parity* case, the single-□ schema first dominates the population at a time when the two forms have approximately equal shares of the speakers, whereas in the *70-30* pattern, this stability is achieved when the distribution is somewhat lopsided. The *dominance* graphs, of course, are a result of H-Nets learning fully specified schemata (i.e. no □ values).

Even the *step-up* graphs rely to some extent on the existence of schemata in the population of H-Nets. In this pattern, one form achieves a certain degree of initial prominence amongst the speakers, but is quickly overtaken by a similar form. In fact, this occurs when part of the H-Net population settles on a schema having a □ value which allows the initial form. If at some point before this schema dominates the population (which would result in parity), a significant portion of the H-Nets learn a fully specified schema for a different form which is also allowed by the □-schema, this new form will eventually take over. Even if the usurper's frequency is initially low, the new form will dominate, as it is meaningful to all agents, whereas the first form is understood only by those with the □-schema.

Another interesting phenomenon is the near-disappearance of the *mush* pattern in longer simulations, especially those involving population change. This is a more drastic example of the process explained in the previous paragraph. Longer simulation runs involve more lexicon change, and this tendency is exacerbated by a changing population. The phonetic space in our model is limited; since a *mush* form occupies far more than its share of this limited resource, there is a certain degree of pressure to reduce the number of □ features. If for some reason a form for another object moves into this space, object-confusion will result. This will apply pressure to differentiate the two schemata; the simplest adaptation is simply for the *mush* pattern to lose one of its □ values, resulting in a *70-30* or a *parity*. Of course, this same process is moving these single-□ patterns into fully specified *dominance*.

Although less probable, the reverse operation also occurs in our simulation, with *parity* patterns (one □) becoming *mush* patterns (two □s). Over very long simulations, one would expect these two forces to come into balance, resulting in a relatively stable distribution of □ features over the population.

Conclusions and Future Work

While the results reported in [8] and [9] are exciting, the authors make little attempt to exhibit any sort of low-level cognitive validity. Our approach recasts this earlier work in a more natural form, introducing a connectionist cognitive model for the agents and a much richer phonetic model. We have also refigured the *naming game* paradigm as implementation-independent, divorcing its description from the details of the accompanying model, a characteristic which is distinctly lacking in other definitions.

Our most significant result is not merely that language emerges from our system, which we've taken care to provide with a cognitively valid base, but rather that the linguistic systems which our agents learn are themselves cognitively valid. The schemata learned by the agents do not just provide

an explanation of their behaviour, but represent valid phonetic generalizations in their own right. Human speakers do not learn fully specified feature sets, but rather schemata with one or more □ values. For example, English speakers do not differentiate phonemically between aspirated and non-aspirated consonants, whereas this has been constructed as a distinctive difference in proto-Indo-European, a distant ancestor.

Clearly, some of the most exciting future work involves following up the notion of the schemata which our agents learn, determining how the distribution of schemata affects the evolution of the system, and to what degree future behaviour of the system can be predicted. These schemata should also prove crucial in planned investigations of the complex interactions at the border of two stable languages.

Both the phonetic model and the object model used in our simulation could be improved. We plan to model physical constraints of the vocal tract so as to have the agents produce even more realistic sound combinations, which will allow us to expand the feature set and thus the number of phonemes. We hope to introduce an object model where objects are represented by feature vectors rather than simply atomic nodes, to see if hierarchical concepts might be instantiated as lexical items under these conditions.

Our results build on other recent work, demonstrating not only that modelling language as the emergent behaviour of a complex adaptive system can be a valuable tool for linguistic investigation, but that these systems can be created in a cognitively valid manner.

References
[1] Carr, Philip. *Phonology.* London: The Macmillan Press, Ltd, 1993.
[2] Caudill, Maureen and Charles Butler. *Naturally Intelligent Systems.* Cambridge, MA: The MIT Press, 1991.
[3] Dircks, Christopher, and Stoness, Scott C. "Lexicon Change in the Absence of Population Flux", Proceedings from ECAL 99, Springer-Verlag, forthcoming.
[4] Gurney, Kevin. *An Introduction to Neural Networks.* London: UCL Press Ltd., 1997.
[5] Holland, John H. *Adaptation in Natural and Artificial Systems.* Cambridge: MIT Press, 1995.
[6] Rogers, Henry. *Theoretical and Practical Phonetics.* Toronto: Copp Clark Pitman, Ltd., 1991.
[7] Steels, Luc. "Self-organizing vocabularies" In: Langton, C. (ed.) (1996) *Proceedings of Alife V*, Nara Japan.
[8] Steels, L. and F. Kaplan. "Stochasticity as a source of innovation in language games," *Proceedings of Artificial Life VI*, Adami et al eds. Los Angeles: MIT Press, 1998.
[9] Steels, L. and F. Kaplan. "Spontaneous Lexicon Change," *Proceedings from COLING-ACL 98.* Montreal: Université de Montréal, 1998.
[10] Steels, L and P. Vogt. "Grounding adaptive language games in robotic agents." In Harvey, I. et.al. (Eds.) *Proceedings of ECAL 97*, Brighton UK, July 1997. The MIT Press, Cambridge Ma., 1997.

The Effect of Clausal and Thematic Domains on Left Branching Attachment Ambiguities

Patrick Sturt (patrick@psy.gla.ac.uk)
Human Communication Research Centre; 53 Hillhead Street
Glasgow G12 8QF, UK

Holly P. Branigan (holly@psy.gla.ac.uk)
Human Communication Research Centre; 53 Hillhead Street
Glasgow G12 8QF, UK

Yoko Matsumoto-Sturt (Y.M.Sturt@ed.ac.uk)
Centre for Japanese Studies; George Square, Edinburgh
Edinburgh EH8 9JX, UK

Abstract

Recent work has emphasised the importance of thematic domains in sentence processing. Two questionnaire studies examined whether thematic domains influence attachment of relative clauses to complex NPs in Japanese. The results suggest that definitions of thematic domains should be revised to cover left-branching structures, but do not support a distinction between domains associated with clauses and adpositional phrases.

Introduction

In this paper, we present two experiments which examine how postpositional and clausal domains affect the resolution of relative clause ambiguities in Japanese. Our two concerns are firstly to establish whether the notion of thematic domain is relevant to left-branching structures; and secondly to compare the effects of clausal and postpositional phrase thematic domains on relative clause attachments. Our results will clarify whether existing definitions of thematic domain must be modified to cover left-branching structures, and will also provide evidence to distinguish between current theories of thematic domain effects in parsing.

Relative Clause Attachment and Thematic Domains

This paper considers constructions that involve a relative clause which can attach to either one of two possible sites in a complex noun phrase, as in the English example shown in (1), from Cuetos and Mitchell (1988):

(1) Somebody shot the servant of the actress who was on the balcony.

Several studies have shown that the interpretation of such ambiguities depends partly on the thematic status of the preposition in the complex noun phrase. In (1), the preposition *of* is commonly assumed not to assign a theta role, but appears for purely syntactic reasons (perhaps to assign case to the lower NP, or as a "reflex" of the lower NP's inherent case (Chomsky, 1986)), mediating the assignment of a theta role from *servant* to *actress*. Cross-linguistically,

the eventual interpretation of ambiguities such as (1) tends to favour high attachment of the relative clause, where it is the *servant* rather than the *actress* who is interpreted as being on the balcony. This high attachment preference has been established in Spanish (Cuetos & Mitchell, 1988), French (Zagar, Pynte, & Rativeau, 1997), Dutch (Brysbaert & Mitchell, 1996) and Italian (De Vincenzi & Job, 1995), though in English the preference seems to be much weaker, and if anything favours low attachment (i.e., with the *actress* on the balcony) (Cuetos & Mitchell, 1988). These findings contrast with similar examples where the preposition assigns a thematic role in its own right. For example, consider the following sentence (from De Vincenzi & Job, 1995).

(2) Everybody admires the man with the daughter who began to sing an opera.

In (2), *with* is commonly assumed to assign a thematic role to *daughter*. Several studies have found that the high attachment preference is reduced for complex NPs containing a thematic preposition such as *with*, compared to similar examples containing a non-thematic preposition such as *of*, often resulting in a low attachment preference when a thematic preposition is used. This has been found in English (Traxler, Pickering, & Clifton, 1998), Italian (De Vincenzi & Job, 1995) and Spanish (Gilboy, Sopena, Clifton, Jr, & Frazier, 1995).

These facts can be explained by adopting the notion of a *thematic processing domain*. Construal Theory (Frazier & Clifton, 1996) defines the *Current Thematic Processing Domain* as "the extended maximal projection of the last theta assigner". Construal Theory claims that modifiers, such as the relative clauses in (1) and (2), are associated with the current thematic processing domain. Thus any potential attachment site within the current thematic processing domain is accessible, but sites lying beyond the current thematic processing domain are inaccessible. Let us apply this definition to (1) and (2) at the point where the relative clause is first encountered. In (1), assuming that *servant* assigns a thematic role to *actress*, then the last theta assigner is *servant*. Hence the current thematic processing domain corresponds to the complex NP *the servant of the actress*. Therefore, both *servant* and *actress* are available as heads of attachment sites for the relative clause. In (2), by contrast, the current thematic processing domain corresponds to the PP *with the daughter*, since *with* assigns a theta role, and is therefore the last theta assigner at the point that the relative clause is encountered. Therefore, the higher noun phrase headed by *man* is inaccessible as an attachment site of the relative clause, resulting in a low attachment preference.

Relative Clause Attachment and Left-Branching Structure

In head-final languages, where recursive structure is predominantly left-branching, the relative clause *precedes* both of the nouns which it may eventually modify. Consider the Japanese translation of (1), given in (3):

(3) barukonii ni   iru joyuu   no   mesitukai wo
    balcony   LOC is   actress GEN servant   ACC
    dareka   ga   utta.
    somebody NOM shot.

This contrasts with predominantly right branching languages such as English or Italian, where both possible attachment sites have been read in the input at the point where the relative clause is attached.

The postposition *no* in (3) is similar to the English preposition *of*. We assume that, like *of*, it "transmits" a thematic role from the higher noun to the lower NP. Henceforth, we will call such postpositions *non-thematic* postpositions. Now consider a pair of complex NPs which differ only in their postposition. In (4a) below, the postposition is again the non-thematic postposition *no*; but in (4b) it is another postposition, *kara-no*, which we will assume *assigns* rather than *transmits* a thematic role to its complement. Henceforth, we will call such postpositions *thematic* postpositions.[1] Figure 1 shows the thematic structure of the two complex NPs in (4):

(4) a. genjuumin ga   odosita   tankentai   no
       natives   NOM threatened expedition-force of
       taichou
       commander
       "The commander of the expedition force that the natives threatened"

    b. genjuumin ga   odosita   tankentai
       natives   NOM threatened expedition-force
       kara-no taichou
       from   commander
       "The commander from the expedition force that the natives threatened"

---

[1] *Kara-no* is actually composed of two postpositions, *kara* and *no*.

Consider (4a) and (4b) in the light of the Construal definition of thematic domains. An on-line study of constructions similar to (4a) by Kamide and Mitchell (1997) suggests that the relative clause is initially attached to the first noun that becomes available (i.e. the *low* site). In (4), this noun is *tankentai* ("expedition force"). However, under the definition above, the Construal notion of thematic domain corresponds to the maximal projection of the last theta assigner. At the point where the relative clause is initially attached, the last theta assigner is actually the verb inside the relative clause itself. Hence Construal would predict the domain for attachment of the relative clause to be the relative clause itself, which clearly makes no sense. Therefore, the Construal definition cannot be applied to left-branching constructions such as (4). This suggests that the Construal notion of thematic domains would not apply in such cases.

However, in Sturt (1997) and Sturt and Crocker (1997), we develop a definition of thematic domains which is applicable in such cases. These domains are defined in terms of thematic nodes, which for the purposes of this paper can be assumed to correspond to the (extended) maximal projections of theta-assigners. The thematic domain for a node N corresponds to the subtree rooted at the thematic node which most immediately dominates N if such a dominating thematic node exists, and corresponds to the root of the entire tree otherwise (see Sturt (1997) and Sturt and Crocker (1997) for more precise definitions). In left-branching structures, the thematic domain for a node N may be established after N has been processed, as subsequent theta assigners are read in the input. Once established, such a thematic domain defines the set of alternative attachment sites to which N may be easily reanalyzed (see Figure 1).

Once the relative clause has been attached in the low site (Kamide & Mitchell, 1997), the maximal projection of a *thematic* postposition will demarcate the thematic domain for the relative clause. Hence reanalyzing the initially low-attached relative clause to the high site is predicted to be dispreferred in such cases, since it entails moving the relative clause outside its domain (4b). However, if the postposition is *non-thematic*, then the thematic domain for the relative clause will be established when the higher of the two nouns is read. In the case of (4a), this noun is *taichou* ("leader"). This noun will assign a theta role to the lower NP headed by *taikentai* ("expedition force"). Hence, the thematic domain will correspond to the entire complex NP headed by *taichou*, and it is predicted that, even if the relative clause is initially attached to the NP headed by the lower noun, it can subsequently be freely reanalyzed to the other possible attachment site, since it is within the same thematic domain. In fact, there is evidence that such reanalysis takes place when the postposition is non-thematic. In an off-line questionnaire, Kamide and Mitchell (1997)

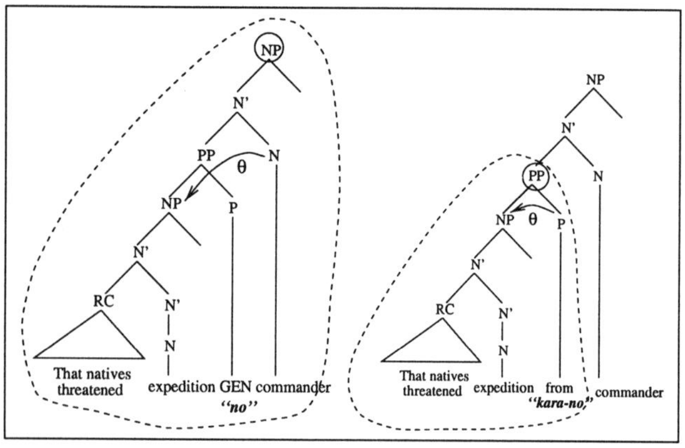

Figure 1: Syntactic structure of (4a) and (4b). The circles indicate thematic assigner nodes, and the dotted loop indicates the domain within which the relative clause can be easily reanalyzed

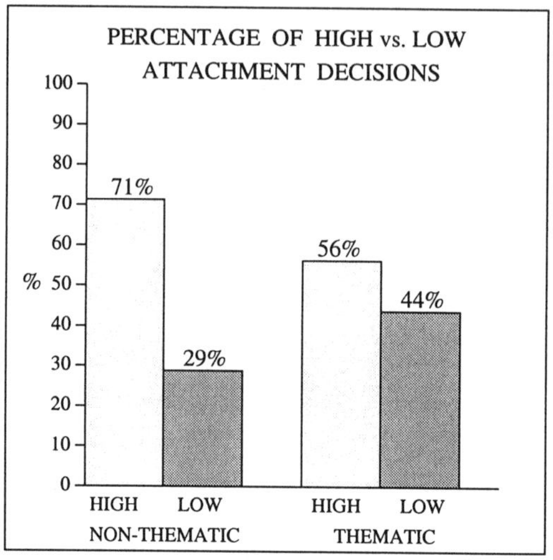

Figure 2: Results of Experiment 1

found that the final attachment preference for constructions similar to (4a) is for the high site, despite the fact that an on-line self paced reading study had found evidence for an initial attachment to the low site.

If thematic domains do have an effect on these ambiguities, then clearly the definition given in Construal should be revised along the lines of Sturt (1997) and Sturt and Crocker (1997), where thematic domains can be applied to left-branching structures. The purpose of Experiment 1 was to establish whether such an effect could be found.

Experiment 1

Method

Participants   The participants were 24 native speakers of Japanese attending a course at the Park Language School in Sheffield.

Materials and Design   The materials were 24 pairs of complex NPs involving a relative clause and two NPs, similar to the example given in (4). In each pair, the two possible attachment sites for the relative clause were separated by either a non-thematic postposition (4a) or a thematic postposition (4b); the two members of the pair differed only in the postposition separating the two noun attachment sites. To control for possible effects of animacy, half of the 24 items included an animate noun in the high site and an inanimate noun in the low site, and half involved the reverse configuration. We will call the former configuration "animate-high" and the latter "animate-low". We used 26 fillers, which consisted of NPs with various types of internal structure, and various types of ambiguity.

Procedure   Participants saw individually randomized lists of materials, which were presented in printed booklets. The materials were rotated in a Latin Square design, so that each participant saw only one version of each item. The fillers were interspersed among the materials in a random order, and in such a way that no two experimental items appeared adjacent to each other. Underneath each experimental or filler item, two sentences were included, which indicated the two possible interpretations of the item. For example, in the item given in (4), the two possible interpretations were indicated by (Japanese translations of) the following:

a. The natives threatened the commander.

b. The natives threatened the expedition force.

Participants had to indicate whether a. or b. corresponded to their first interpretation. The sentences were balanced so that the a. sentence indicated the high attachment interpretation half the time and vice versa.

Results

The results are summarized in Figure 2. Overall, participants chose the high attachment interpretation more often than the low attachment interpretation. Collapsing over the two postposition types, we compared the proportion of high attachment decisions with 50% (i.e. the population mean expected on the null hypothesis). This revealed a significant preference ($t_1(23) = 3.67$, $p < .01$; $t_2(23) = 3.93$, $p < .001$). However, there were significantly more high-attachment decisions in the non-thematic condition than in the thematic condition ($t_1(23) = 3.03$, $p < .01$; $t_2(23) = 3.76$, $p < .01$).[2] The proportion of high

---

[2]The Wilcoxon Matched-Pairs Signed-Ranks Test yielded equivalent results for all pairwise comparisons in this paper.

attachment decisions for the non-thematic condition was significantly greater than 50% ($t_1(23) = 4.96, p < .001$; $t_2(23) = 6.66, p < .001$). In contrast, the proportion of high attachment decisions for the thematic condition was not significantly greater than 50% ($t_1(23) = 1.24, p > .1$; $t_2(23) = 1.24, p > .1$). This indicates that the overall high attachment preference is driven by the non-thematic condition.

Discussion

The study replicated the off-line high-attachment preference found by Kamide and Mitchell (1997) for non-thematic postpositions, but also demonstrated that postposition type has a reliable effect on attachment decisions in Japanese. This demonstrates that thematic domains are relevant to processing in left-branching constructions, and therefore the definition of thematic domain in Construal theory should be revised.

Our findings are explicable if we assume that reanalyzing a relative clause to a position outside its thematic domain is dispreferred in relation to a reanalysis which does not cross a domain boundary in this way. However, the fact that no low-attachment preference was observed for the thematic condition indicates that the processor does sometimes reanalyze across a thematic domain boundary, even in the absence of a syntactic or semantic cue.

Experiment 2

Experiment 1 demonstrated what we can call a "containing effect" for postpositional thematic domains: Assuming an initial low attachment, the presence of a thematic domain boundary reduces the chances of the relative clause being re-attached high. In Experiment 2, we compared such domains with clausal thematic domains. We might expect this containing effect to be "stronger" in cases where the thematic domain boundary is clausal than in cases where it is merely postpositional. This is the prediction of Gibson et al's (1996) *Predicate Proximity* principle, which specifies that an attachment is costly if it is not as close as possible to a *predicate phrase* in the tree. Gibson et al intend *predicate phrases* to correspond to verbal projections, but not prepositional or postpositional projections.[3]

Gibson et al also postulate a pure recency preference, under which an attachment is costly to the extent that it is made to non-recent material. An attachment which is favoured by both recency and predicate proximity will be preferred over an attachment that is favoured by only one of these principles. Now consider (4b) which includes a postpositional thematic domain boundary (between *kara-*

no and *taichou*), and compare it with with (5), which includes a clausal thematic domain boundary (marked $_S$]). Note that (5) involves two relative clauses; the initial relative clause, whose attachment is ambiguous, and a second relative clause, which modifies the higher of the two attachment sites (see also Figure 3).

(5) [$_{RC}$ genjuumin ga odosita] tankentai wo
      natives   NOM threatened expedition ACC
   hikiita $_S$] taichou
   led   commander

"The commander who led the expedition force that the natives threatened"

In (4b), there is no predicate phrase outside the relative clause itself. In (5), by contrast, there is a predicate phrase headed by the verb *hikiita* ("led"). Therefore, the lower noun *tankentai* would be favoured as an attachment site for the relative clause by both recency and predicate proximity in (5), while this noun would be favoured only by recency in (4b). Hence, predicate proximity would predict more low attachments in (5) than in (4b).

Thus in the type of ambiguity under discussion here, predicate proximity effectively predicts that clausal domains, rather than thematic domains in general, will affect the attachment of the relative clause. Of course, this also means that predicate proximity cannot account for the results of Experiment 1, since the relevant domain there was postpositional rather than clausal. This could perhaps be remedied by invoking an extra principle, for example, *PP-proximity*. This extra principle would have to be weaker than predicate proximity, however, to account for the findings presented in Gibson et al. (1996). Hence, whether or not an extra principle is invoked, clausal domains would be predicted to have a stronger containing effect than postpositional thematic domains.

We can contrast this view with the Construal notion of thematic domains, which does not differentiate between clausal and pre/postpositional domains. Given that the Construal definitions could be amended to allow for left branching constructions, as discussed above, then clausal domains would be predicted to have a containing effect of equal strength to postpositional domains.

Experiment 2 aimed to distinguish between these two alternative accounts of domains. If the clause-based notion of domains implied by predicate proximity is correct, then there should be fewer high attachment conditions where a clause boundary separates the two possible noun attachment sites, as in (5), than where a postpositional thematic domain separates these nouns, as in (4b). However, if the more general notion of thematic domains implied by Construal is correct, there should be no difference between (5) and (4b); but both of these should differ from cases where no thematic domain boundary intervenes between the two

---

[3]This is clearly presupposed in their explanation for attachment preferences in Spanish and English 3-site relative clause ambiguities.

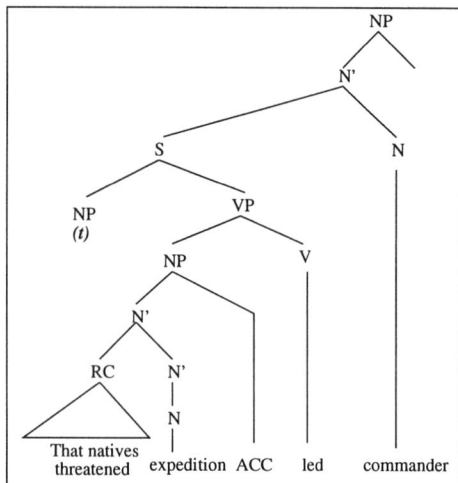

Figure 3: Syntactic structure of (5)

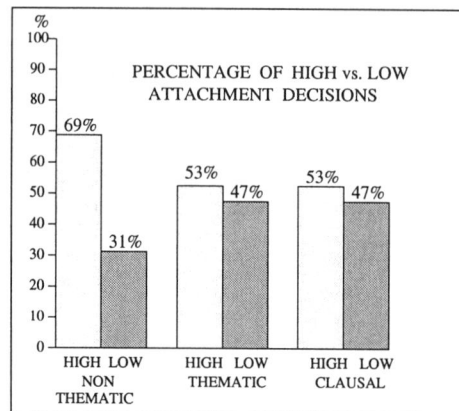

Figure 4: Results of Experiment 2

possible noun attachment sites, as in (4a). The use of clausal domains (and thus verbal theta assigners) also allowed us to generalize the findings of Experiment 1 to domains involving a wider range of thematic roles.

Method

Participants Participants were 24 Japanese native speakers who were attending a course at the Institute of Applied Language Studies at the University of Edinburgh.

Materials and Design The materials consisted of 24 triples of complex NPs. Each triple consisted of three conditions. The first two were a *non-thematic* condition and a *thematic condition*, which together correspond to the two conditions of Experiment 1; the third was a *clausal* condition, as in (5) (see also Figure 3).

Twelve items were animate-high and 12 were animate-low. In all the animate-high items in the clausal condition, the second relative clause was a *subject relative* (i.e., included a subject gap, coindexed with the (animate) higher noun). In all the animate-low items, the second relative clause was a *non-subject* relative (i.e., included a non-subject gap, coindexed with the (inanimate) higher noun). This was necessary because of the strong preference for transitive verbs to have animate subjects in Japanese.

The 24 experimental items were selected from a larger group of materials (including the materials from Experiment 1) in a pretest run at the University of Surugadai. This pretest ensured that the high and low interpretations were matched for plausibility (both t's < 1). The materials for Experiment 1 were found to have had a slight plausibility bias towards the low attachment reading (the opposite direction to the high attachment preference actually found in experiment 1). Experiment 2 used the same fillers as Experiment 1, and the procedure was identical.

Results

The results are summarized in Figure 4. Overall there was a preference for high attachment. t-tests comparing the overall proportion of high attachments with 50% revealed that this preference was significant by subjects but not by items ($t_1(23) = 2.5, p < .05; t_2(23) = 1.47, p = 0.16$). Planned contrasts revealed that there were significantly more high-attachment decisions in the non-thematic condition than in the thematic condition ($t_1(23) = 4.53, p < .001; t_2(23) = 3.82, p < .001$), and also that there were significantly more high attachment decisions in the non-thematic condition than in the clausal condition ($t_1(23) = 4.17, p < .001; t_2(23) = 2.81, p < .02$). However, there was no difference between the thematic and the clausal conditions—in fact the means were numerically identical ($t_1 = t_2 = 0, p = 1$).

We also found a significant effect of animacy in this experiment, with an overall high attachment preference for the animate-low items and an overall low attachment preference for the animate-high items. As this factor was not the focus of the experiment, and did not interact with the thematicity effect, we will not discuss it further here (though see Sturt (1997) for details).

Discussion

Experiment 2 replicated the finding of Experiment 1 that a thematic postposition neutralizes the high attachment preference found with the non-thematic postposition. However, Experiment 2 shows no evidence that a clausal thematic domain is treated any differently from a postpositional one. These results support (a left-branching version of) the Construal notion of thematic domains, but do not support the clause-based notion of domain implied by Predicate Proximity.

722

General Discussion

Our experiments demonstrate that altering thematic characteristics of a complex NP influences the eventual attachment preferences for relative clauses in Japanese. The effect is identical, whether the domain in question is clausal or postpositional. However, the presence of a thematic domain boundary does not *block* the high attachment of the relative clause.

We can exclude an alternative explanation of our results in terms of the relative phonological weight of the attaching constituent and its potential sister (Fodor, 1998; Hirose, Inoue, Fodor, & Bradley, 1997). The length of the complex NP to which the relative clause may attach differs between conditions in our materials (e.g. *N kara-no N* is longer than *N no N*). According to the phonological weight theory, this could have affected the attachment preferences in our materials. However, post-hoc analyses correlating the number of high attachment decisions with the length (in terms of the number of morae) of a. the relative clause, b. the complex NP, and c. the length of the relative clause minus the length of the complex NP, showed no significant correlations.

It is not clear how current theories can account for the overall high attachment preference that was found in both Experiments 1 and 2. One possible explanation could be some version of *Relativised Relevance* (Frazier, 1990), which predicts a preference for associating a modifier with the "main assertion of the sentence". We note that, as our study examined NPs, which are not sentences and do not denote "assertions", this principle would have to apply either to non-overt clausal structure or to the "main content" of the utterance, whatever its categorial expression or semantic type.

In conclusion, our results support the general notion of thematic domains given in Construal (Frazier & Clifton, 1996); but any such definition must be altered to account for left-branching structures such as those considered here. Our data argue against theories in which clausal domains are seen as stronger than pre/postpositional domains.

Acknowledgements

This research was supported by British Academy Postdoctoral Fellowships awarded to HB and PS, and an ESRC studentship awarded to PS. We thank Kuabrat Branigan, Masayo Iida, Ruth Linden, Jim Hutton, Peter Sells and Toyomi Takahashi.

References

Brysbaert, M., & Mitchell, D. C. (1996). Modifier attachment in sentence parsing: Evidence from dutch. *Quarterly Journal of Experimental Psychology, Section A—Human Experimental Psychology, 49*(3), 664–695.

Chomsky, N. (1986). *Knowledge of language: its nature, origin and use.* New York: Praeger.

Cuetos, F., & Mitchell, D. C. (1988). Cross-linguistic differences in parsing: Restrictions on the use of the late closure strategy in Spanish. *Cognition, 30*, 72–105.

De Vincenzi, M., & Job, R. (1995). An investigation of late closure: the role of syntax, thematic structure and pragmatics in initial and final interpretation. *Journal of Experimental Psychology. Learning, Memory and Cognition, 21*(5), 1303–1321.

Fodor, J. D. (1998). Learning to parse? *Journal of Psycholinguistic Research, 27*(2), 285–319.

Frazier, L. (1990). Parsing modifiers: Special purpose routines in the HPSM? In D. A. Balota, G. B. F. d'Arcais, & K. Rayner (Eds.), *Comprehension processes in reading* (pp. 303–331). Hillsdale, New Jersey: Lawrence Erlbaum Associates.

Frazier, L., & Clifton, C. (1996). *Construal.* Cambridge MA: MIT Press.

Gibson, E., Pearlmutter, N., Canesco-Gonzalez, E., & Hickok, G. (1996). Recency preference in the human sentence processing mechanism. *Cognition, 59*(1), 23–59.

Gilboy, E., Sopena, J. M., Clifton, Jr, C., & Frazier, L. (1995). Argument structure and association preferences in Spanish and English compound NPs. *Cognition, 54*, 131–167.

Hirose, Y., Inoue, A., Fodor, J. D., & Bradley, D. (1997, March). *Adjunct ambiguity in Japanese: The role of constituent weight.* Poster presented at the *11th Annual CUNY Conference on Human Sentence Processing, New Brunswick, NJ.*

Kamide, Y., & Mitchell, D. C. (1997). Relative clause attachment: Nondeterminism in Japanese parsing. *Journal of Psycholinguistic Research, 26*(2), 247–254.

Sturt, P. (1997). *Syntactic reanalysis in human language processing.* Unpublished doctoral dissertation, Centre for Cognitive Science, University of Edinburgh, Edinburgh, Scotland.

Sturt, P., & Crocker, M. W. (1997). Thematic monotonicity. *Journal of Psycholinguistic Research, 26*(3), 297–322.

Traxler, M. J., Pickering, M. J., & Clifton, C. (1998). Adjunct attachment is not a form of lexical ambiguity resolution. *Journal of Memory and Language, 39*, 558–592.

Zagar, D., Pynte, J., & Rativeau, S. (1997). Evidence for early-closure attachment on first-pass reading times in french. *Quarterly Journal of Experimental Psychology, Section A—Human Experimental Psychology, 50*(2), 421–438.

Conditional Probability and Word Discovery:
A Corpus Analysis of Speech to Infants

Daniel Swingley (swingley@bcs.rochester.edu)
Department of Brain and Cognitive Sciences
Meliora Hall; University of Rochester
Rochester, NY 14627 USA

Abstract

Analyses of an idealized corpus of English speech to infants revealed that simple conditional decision rules can separate frequent bisyllabic words from bisyllables not corresponding to words. If infants accurately represent speech in terms of syllables, and compute conditional statistics over these syllables, such computations have the potential to inform infants of likely English words.

Introduction

Researchers have shown an increasing interest in the possible use of statistical regularities by language learners. Recent experiments have shown that infants are sensitive to certain distributional properties of syllable sequences (e.g. Aslin, Saffran, & Newport, 1998; Goodsitt, Morgan, & Kuhl, 1993; Morgan, 1994; Saffran, Aslin, & Newport, 1996). If the statistical properties to which infants' sensitivity has been demonstrated are in fact present in speech to infants, distributional regularities could serve as valuable cues for word discovery.

Several researchers have shown that the segments comprising words in speech or in text exhibit distributional regularities that can be computationally exploited for word boundary detection (Aslin, Woodward, LaMendola, & Bever, 1996; Cairns, Shillcock, Chater, & Levy, 1997; Christiansen, Allen, & Seidenberg, 1998; Elman, 1990; see also Brent & Cartwright, 1996; de Marcken, 1996). However, corpus analyses have yet to assess the value of distributional regularities for clustering *syllables* into words in speech. Given that infants spontaneously cluster syllables in the laboratory, determining whether this ability could aid in identifying words in speech is of interest.

Syllables are widely considered to be a unit of speech infants are capable of processing and representing. Several experiments have shown that infants categorize varied sets of words by their number of syllables, but not their number of segments, suggesting that syllables are crucial units in infants' representation of speech (Bertoncini, Floccia, Nazzi, & Mehler, 1995; Bijeljac-Babic, Bertoncini, & Mehler, 1993). Furthermore, Bertoncini, Bijeljac-Babic, Jusczyk, Kennedy, and Mehler (1988) found that when infants were habituated to sets of CV syllables containing a common onset (such as /b/), infants dishabituated equally after the introduction of novel /b/-initial CVs, and after the introduction of novel CVs containing a new C, suggesting that infants did not consider the /b/-initial syllables as similar. Though these results do not necessarily indicate that infants fail to represent segments as units, much evidence now suggests that the syllabic level of representation is significant in infancy, and that syllables may be relevant as units over which statistical computations may be done.

In the present study, a corpus of speech directed to children under 18 months was used to evaluate the utility of transitional probability information for grouping syllables into words. The problem may be stated as follows: given that over 80% of the words children hear are monosyllabic (Aslin et al., 1996; Christiansen et al., 1998), would infants' tendencies to cluster syllables according to conditional probability criteria lead infants to inappropriately conflate monosyllabic words? Or, alternatively, would these clustering mechanisms help lead infants to discover words in speech? Answering these questions requires a statistical analysis of large samples of speech directed to infants. Below, we demonstrate that given certain assumptions about the mechanisms that underlie infants' sensitivities to statistical structure, American English speech does contain statistical regularities that could be used for word discovery.

Methods

Corpora of 15 parents' speech to American infants under 18 months (CHILDES; MacWhinney, 1995; see Bernstein-Ratner, 1984; Bloom, 1973; Hayes & Ahrens, 1988; Higginson, 1986; Sachs, 1983; Warren-Leubecker & Bohannon, 1983) were combined to form a 50,000–word corpus. Spelling of words throughout was regularized by hand, and pronunciations of the resulting words were estimated using the CMU pronouncing dictionary (v. 0.4, 1995). This phonemic corpus was syllabified using an implementation of Kahn's (1980) formalism for slow speech (essentially maximal onset). The syllabification algorithm was run over words, not over utterances; thus, no segments were syllabified across word boundaries (a point we return to below). Over the resulting corpus of syllables, three metrics were calculated for each consecutive pair of syllables AB (bigrams): predictive transitional probability , or $p(B|A)$; reverse transitional probability, or $p(A|B)$, and mutual information, or $log_2[p(AB)/p(A)p(B)]$. (Mutual information is a measure of how much the occurrence of one syllable is informative about the other syllable; cf. Charniak, 1993).

Predictive transitional probability is high when one syllable makes the following one predictable. This metric would be useful if it tended to be higher in words (such as "pretty": $p(B|A) = 1$) than in other sequences (such as "thank you": $p(B|A) = 1$). Reverse transitional probability is high when one syllable makes the previous one predictable. This metric would be useful if sequences like "little" ($p(A|B) = .81$) were more common than sequences like "the door" ($p(A|B) = .81$). Mutual information is high when both syllables tend to co-occur, and would be useful if sequences like "daddy" (m.i. 7.7) were more common than sequences like "sit down" (m.i. 7.5).

Because the vast majority of the word types in speech to children are monosyllabic (in the present corpus, about 55%) or bisyllabic (about 42%), the syllable-grouping problem in English reduces largely to a problem of deciding whether two syllables form a bisyllabic word. If this distinction could be made accurately, 97% of words in the corpus would be correctly identified. Thus, the current analyses examined whether conditional-probability metrics could be used to distinguish the bigrams that are words, from those that are not.

In a series of analyses, a threshold value of one of these three conditional probability metrics was set, and the bigrams above that threshold were identified as words. The question asked was whether thresholds for the three metrics could be set to produce a favorable ratio of hits to false alarms (*precision, or accuracy*), and of hits to misses (*completeness;* cf. Brent & Cartwright, 1996). Precision indicates whether "yes, it's a word" responses tend to be correct; completeness measures the proportion of words that are identified. At the same time, the frequency of bigrams considered as possible words was varied, to evaluate possible interactions between frequency and conditional-probability information.

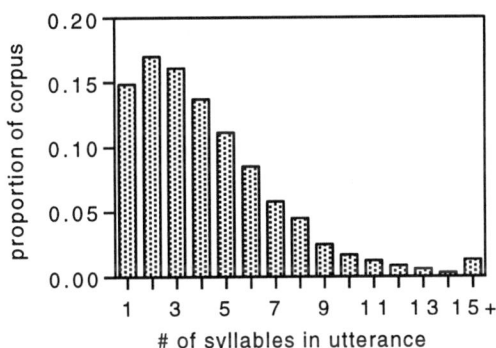

Figure 1: Sentence lengths in syllables.

For computational convenience, analyses included only the utterances containing 2–14 syllables. (The upper bound excluded about 1.3% of the utterances.) The remaining corpus included approximately 41,000 bigram tokens, and about 13,000 bigram types. Among these types were 761 different bisyllabic words, the primary targets of the analyses. As shown in Figure 1, the majority of utterances directed to infants contained only a few syllables. This illustrates a common observation about speech to infants (e.g. Snow, 1972).

Results and Discussion

As a baseline for comparison, we first consider the precision and completeness scores that would be expected from simple guessing, an estimate of "chance" in finding bisyllables. A guessing-based decision rule could say "yes" or "no" equally often (p=0.5), or could say "no" most of the time (say, p=0.8).

Considered in Figure 2 are precision and completeness (by types) for a range of guessing rates, from 0 (calling all bigrams words) to 1.0 (calling all bigrams nonwords). The results make clear the fact that bisyllables cannot be located effectively by guessing; they are too rare. Without more information differentiating word and nonword bigrams, the infant would be considerably better off simply assuming that all syllables are monosyllabic words.

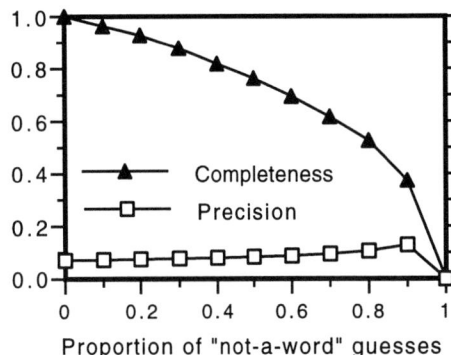

Figure 2: Results for guessing, shown by types (i.e. not frequency weighted).

Next, we consider the possibility that infants cluster syllables in natural speech primarily on the basis of frequency. This has a certain intuitive appeal–one might imagine that, upon hearing a given pair of syllables many times, this pair might come to be considered a linguistic unit. The following decision rule was evaluated: if a bigram appeared *x* or more times in the corpus, it was considered to be a word. Figure 3 shows the results over types (i.e. displayed are the number of different words found, not weighted by their frequency of occurrence). Clearly, many common bigrams are not bisyllabic words; in fact, even if we admit only the bigrams that occur 50 or more times (which is true of 108 bigrams), there are still more than twice as many nonwords as words. Examples of hits include very frequent words like "baby," "doing," and "very." But a frequency criterion false-alarms to sequences like "good girl," "is it," "that's right," and "we put."

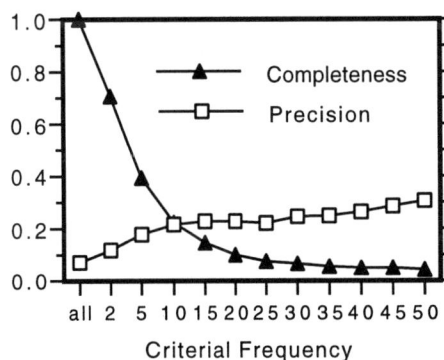

Figure 3: Results for frequency criterion, by types.

The experimental results of Aslin, Saffran, and Newport

(1998), however, show that infants' clustering of syllables is not mediated only by frequency. In that study, the frequency of occurrence of words and nonwords in an artificial language was controlled, and differed only in the conditional probabilities with which syllables followed one another. Eight-month-olds discriminated word and nonword lists, demonstrating sensitivity to conditional probabilities independent of frequency. These results suggest that infants do not simply assume that common bisyllables are words. Given the statistics of the language, this is fortuitous; as shown above, such a mechanism would usually be wrong.

Subsequent analyses consider decision rules based on conditional probability metrics. In Figure 4, precision and completeness scores are shown for decision rules using a mutual information threshold; when the threshold was equaled or surpassed, the bigram was considered a word. As Figure 4 demonstrates, performance using mutual information was better than performance using frequency. However, even the best precision scores were never above about 45%. Similar results were obtained using predictive and reverse transitional probability.

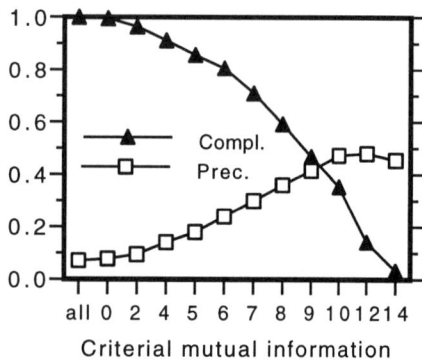

Figure 4: Mutual information criterion with all words counted. Results by types.

However, this analysis included many words of very low frequency. The clustering mechanism that is operating here is likely to be one of implicit memory (Saffran et al., 1997). Implicit learning of sequential stimuli typically involves considerable repetition of the training materials. Thus, it seems reasonable to suppose that the learning mechanism would only make assumptions about bigrams after several exposures to them. Exactly how many exposures we would expect the system to require is not clear; in the subsequent analyses we will examine the degree to which simple decision rules detect words that occurred 5 or more times in the corpus. (The results are similar if we exclude only the bigrams occurring 1–2 times, 1–3 times, etc.) Because (as in any natural corpus) a large proportion of the types occurred infrequently, even a small frequency criterion excludes many types. In the present corpus, a frequency criterion of 5 excludes 32% of the bigram tokens, and excludes 84% of the bigram types; at the same time, this criterion excludes only 11% of the bisyllabic word tokens, and 61% of the bisyllabic word types. Of the 761 bisyllabic words in the corpus, 299 meet the frequency criterion. The following analyses consider whether conditional proba-

bility information can help identify the 299 most common bisyllabic words, out of the 1676 most common bigrams. (Note that this is still a nontrivial task, because word types make up less than 18% of the common bigram types.)

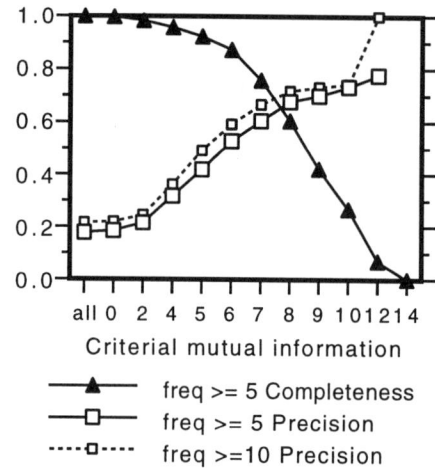

Figure 5: Mutual information criterion with frequency thresholds of 5 and 10.

Figure 5 shows precision and completeness results using mutual information and a frequency criterion of 5 (solid lines). The dashed line shows precision scores when a frequency criterion of 10 is applied. Two results are evident here. First, precision is quite high: in the higher mutual information range, precision varies from about 60% to 80%. Where the lines cross, both precision and completeness are above 60%. Second, precision is only marginally improved by increasing the frequency criterion to 10 occurrences. This small improvement is offset by substantial decreases in completeness (not shown here), reflecting the fact that only 170 bisyllabic words occurred 10 times or more.

Although performance is much better than chance, even in the best case the decision rule makes many errors. An analysis of the false alarms, however, suggests that many of the errors are not pernicious ones: often the "false alarms" are instances in which the decision rule groups together two syllables from a trisyllabic word. Figure 6 shows the proportion of false alarms that are clusters of syllables that form 2/3 of a trisyllabic word, over a range of mutual information criteria.

At the higher mutual information criteria, the proportion of "false alarms" within trisyllabic words is quite high—over 50% in some cases. In some of these, there are two "false alarms" within a trisyllable. For example, at a mutual information threshold of 7, the decision rule clusters "kanga" and "garoo." If all false alarms that actually cluster bigrams within trisyllabic words are counted as hits, rather than false alarms, precision at the higher thresholds reaches 85%–90%. Some persistent errors remain: among them are "I'm so(rry)," "much fun," and "(pe)ter pi(per)." However, in spite of these errors, it is clear that infants computing conditional probabilities are likely to group together syllables that belong within words.

How robust are these results? Changing the frequency

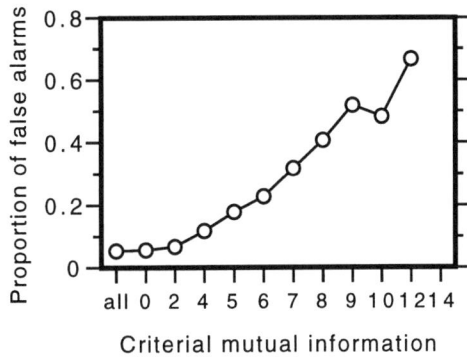

Figure 6: Proportion of false alarm types that form part of a trisyllabic word.

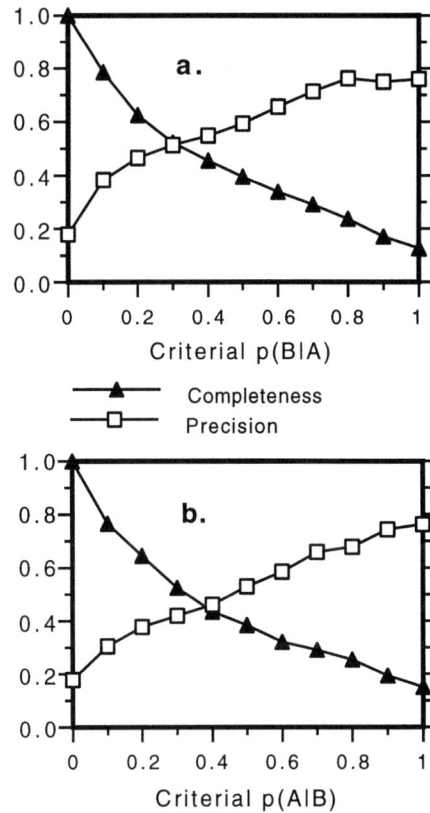

Figure 7: Transitional probability used in decision rule. Graph a, predictive probability; b, reverse probability.

threshold does not have much impact on precision, as long as bigrams of very low frequency are not included as possible words. Furthermore, similar results are obtained when using predictive and reverse transitional probability rather than mutual information. Figure 7 shows precision and completeness at probabilities ranging from 0 to 1, using a frequency criterion of 5.

Another sort of decision rule compatible with current experimental results would not use an absolute thresholding mechanism to cluster syllables. Rather, two syllables might be grouped together if their conditional probabilities were higher than those of the neighboring bigrams. Consider the utterance ABCDEF, with each letter a syllable. Suppose the mutual information values between syllables are as follows: $A_2B_5C_1D_6E_6F$. On a "neighbor-comparison" rule, BC would be grouped together, but DE would not, because the value for DE does not exceed the value for EF. Several varieties of this decision rule are possible. For example, a bigram might have to exceed its neighbors by 1, or 2, or 6 (and in this last case, BC would not be considered a word, because the difference between 5 and 2 is less than 6). Rules of this sort amount to attempts to find peaks in the mutual information function across the sentence. Figure 8 shows results from this decision rule. The x-axis represents the number of mutual information units by which a given bigram must be greater than its neighbors, to be considered as a word. As the figure shows, performance using such a rule is comparable to performance using absolute thresholds.

These results suggest that a variety of decision rules, conditional statistics, and threshold values might lead an infant to correctly identify frequent bisyllabic words with a high success rate. However, the current results must be qualified by three important considerations. First, the syllabification algorithm did not permit consonants to be resyllabified to adjacent words. This sort of resyllabification does occur in English, although it is not clear how often it occurs in speech to infants, or under what circumstances. It is unlikely that all consonantal codas preceding vowel onsets are resyllabified. For example, upon hearing a sentence like "I like that one," infants probably do not group the /t/ of "that" with "one." In the absence of a model of resyllabification in infant-directed speech, and without experimental data on infants' assignment of con-

sonants to syllables, this remains an open question. However, preliminary analyses based on a "worst-case scenario" in which coda consonants are always resyllabified to the following syllable, if thereby creating a legal onset cluster, show that the use of conditional statistics in decision rules still substantially improves performance. Under these conditions, as expected, precision and completeness scores are lower; but baselines produced by guessing or frequency criteria are also lower. Thus, regardless of our particular assumptions about the transparency of syllable boundaries, information about words is still present in conditional statistics.

Second, the current corpus is idealized in the sense that it assumes a fixed pronunciation for each orthographic word. The truth was certainly messier, although without the recordings themselves we cannot attempt a precise reconstruction. In principle, some of the variability in actual pronunciations could be modeled using probabilistic rules (e.g. de Marcken, 1996), yielding a noisier (but more veracious) corpus. This procedure would be particularly useful if more were known about infants' compensation for the processes that lead to variable realizations of words. Our implicit assumption in the current work is that this compensation is perfect, but we recognize that this is an idealization.

Third, it is possible that the structure of English favors the success of decision rules of the sort employed here, and that these decision rules would prove ineffective in the analysis of other languages. This outcome would not necessarily indicate

727

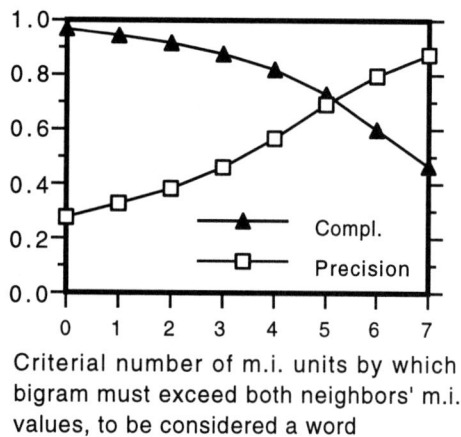

Figure 8: Results with decision rule based on comparison with neighboring bigrams' mutual information values.

that infants do not perform computations like those we have evaluated. It would indicate that if infants do, it will not lead them to discover words.

The current results are not themselves a model of the infant learner. Presumably infants do not store a corpus of the speech they hear for several months, and then (implicitly) group together cohesive units. Rather, infants' mental computations occur incrementally over time, as more and more speech is heard. A model of the infant learner would take this into account by simultaneously calculating conditional statistics and applying decision rules; such a model is currently under development. The analyses presented here suggest that such a model would show that infants' ability to cluster syllables based on statistical characteristics would result in the identification of words more often than not, perhaps with very high precision.

Several researchers have proposed that infants might use prosodic information, such as lexical stress, to help identify words in speech (e.g. Cutler, 1994; Gleitman, Gleitman, Landau, & Wanner, 1988). In fact, evidence from infant experiments suggests that English-learning infants tend to extract (from continuous speech) words with strong-weak stress patterns more readily than words with weak-strong patterns (Newsome & Jusczyk, 1995). Because English content words tend to begin with strong syllables, this tendency may well help English-learning infants to discover words. It is not clear at present whether these tendencies hold for all infants (in which case infants learning some other languages will be disadvantaged by this decision rule), or only for infants in certain language environments (in which case an account of how this prosodic knowledge is acquired will be necessary). In either case, however, there is no reason to suppose that a prosodic strategy and a statistical-learning strategy are incompatible. Although prosodic cues to word boundaries vary with different languages, it may be that statistical cues of the sort examined here are true of most languages. If so, statistical cues might help "bootstrap" a prosodic (or any other) strategy. This is obviously an important area for future cross-linguistic research.

Acknowledgment

This research was supported by NIH grant F32–HD08307.

References

Aslin, R. N., Saffran, J. R., & Newport, E. L. (1998). Computation of conditional probability statistics by 8–month–old infants. *Psychological Science, 9*, 321–324.

Aslin, R. N., Woodward, J. C., LaMendola, N. P., & Bever, T. G. (1996) Models of word segmentation in fluent maternal speech to infants. In Morgan & Demuth (Eds). *Signal to syntax: Bootstrapping from speech to grammar in early acquisition*. Mahwah, NJ: LEA.

Bernstein Ratner, N. (1984). Patterns of vowel modification in mother-child speech. *Journal of Child Language, 11*, 557–578.

Bertoncini, J., Bijeljac-Babic, R., Jusczyk, P. W., Kennedy, L. J., and Mehler, J. (1988). An investigation of young infants' perceptual representations of speech sounds. *Journal of Experimental Psychology: General, 117*, 21–33.

Bertoncini, J., Floccia, C., Nazzi, T., & Mehler, J. (1995). Morae and syllables: Rhythmical basis of speech representations in neonates. *Language & Speech, 38*, 311–329.

Bijeljac-Babic, R., Bertoncini, J., & Mehler, J. (1993). How do 4–day–old infants categorize multisyllabic utterances? *Developmental Psychology, 29*, 711–721.

Bloom, L. (1973). *One word at a time: the use of single word utterances before syntax*. The Hague: Mouton.

Brent, M. R., & Cartwright, T. A. (1996). Distributional regularity and phonotactic constraints are useful for segmentation. *Cognition, 61*, 93–125.

Cairns, P., Shillcock, R., Chater, N., & Levy, J. P. (1997). Bootstrapping word boundaries: A bottom-up corpus-based approach to speech segmentation. *Cognitive Psychology, 33*, 111–153.

Charniak, E. (1993). *Statistical language learning*. Cambridge, MA: MIT Press.

Christiansen, M. H., Allen, J., & Seidenberg, M. S. (1998). Learning to segment speech using multiple cues: A connectionist model. *Language & Cognitive Processes, 13*, 221–268.

Cutler, A. (1994). Segmentation problems, rhythmic solutions. *Lingua, 92*, 81–104.

Elman, J. L. (1990). Finding structure in time. *Cognitive Science, 14*, 179–211.

Gleitman, L., Gleitman, H., Landau, B., & Wanner, E. (1988). Where learning begins: Initial representations for language learning. In Newmeyer, et al. (Eds). *Language: Psychological and biological aspects*. Cambridge, UK: CUP.

Goodsitt, J., Morgan, J. L., & Kuhl, P. K. (1993). Perceptual strategies in prelingual speech segmentation. *Journal of Child Language, 20*, 229–252.

Hayes, D. P., & Ahrens, M. G. (1988). Vocabulary simplification for children: A special case of "motherese?" *Journal of Child Language, 15*, 395–410.

Higginson, R. P. (1986). *Fixing: Assimilation in language acquisition.* Doctoral dissertation, Washington State University.

Kahn, D. (1980). *Syllable-based generalizations in English phonology.* New York: Garland.

MacWhinney, B. (1995) *The CHILDES project: Tools for analyzing talk (2nd ed.).* Hillsdale, NJ: LEA.

de Marcken, C. (1996). *Unsupervised acquisition of a lexicon from continuous speech.* MIT AI Lab Memo 1558; CBCL memo 129.

Morgan, J. L. (1994). Converging measures of speech segmentation in preverbal infants. *Infant Behavior & Development, 17,* 389–403.

Newsome, M. R., & Jusczyk, P. W. (1995). Do infants use stress as a cue in segmenting fluent speech? In *Proceedings of the 19th Boston University Conference on Language Development* (pp. 415–426). Boston, MA: Cascadilla.

Sachs, J. (1983). Talking about the There and Then: The emergence of displaced reference in parent-child discourse. In K. E. Nelson (ed.) *Children's Language.* Hillsdale, NJ: LEA.

Saffran, J. R., Aslin, R. N., & Newport, E. L. (1996). Statistical learning by 8–month–old infants. *Science, 274,* 1926–1928.

Saffran, J. R., Newport, E. L., Aslin, R. N., Tunick, R. A., & Barrueco, S. (1997). Incidental language learning: Listening (and learning) out of the corner of your ear. *Psychological Science, 8,* 101–105.

Snow, C. (1972). Mothers' speech to children learning language. *Child Development, 43,* 549–565.

Warren-Leubecker, A., & Bohannon, J. (1983). The effects of verbal feedback and listener type on the speech of preschool children. *Journal of Experimental Child Psychology, 35,* 540–548.

A Model of Learning Task-specific Knowledge for a new Task

Niels A. Taatgen (niels@tcw3.ppsw.rug.nl)

Cognitive Science and Engineering, University of Groningen

Grote Kruisstraat 2/1, 9712 TS Groningen, the Netherlands

Abstract

In this paper I will present a detailed ACT-R model of how the task-specific knowledge for a new, complex task is learned. The model is capable of acquiring its knowledge through experience, using a declarative representation that is gradually compiled into a procedural representation. The model exhibits several characteristics that concur with Fitts' and Anderson's theories of skill learning, and can be used to show that individual differences in working-memory capacity initially have a large impact on performance, but that this impact diminished after sufficient experience, which is consistent with Ackermans's theory of skill learning. Some preliminary experimental data support these findings.

Introduction

From the viewpoint of cognitive modeling it has always been hard to explain why people can learn new skills as fast as they usually do. In a typical psychological experiment, participanta are told to do something they have never done before. Nevertheless, only some verbal instructions and maybe one or two practice trials are enough to get them started. Initial performance is characterized by the fact that it is slow, and that many errors are made. Once a participant gains some practice, speed goes up, and the number of errors decreases. Fitts (1964) has described this process in three stages: a cognitive stage, in which processing is conscious, deliberate, slow, and requires full attention, an associative stage, in which processing gradually speeds up and less attention is needed, and finally an autonomous stage, in which a skill is performed very fast and requires very little attention. The autonomous stage is also characterized by the fact that deliberate control is gradually lost: it is hard for an expert in some domain to explain exactly what he does.

Ackerman (e.g., 1988) has investigated these three stages by looking at individual differences in skill learning. He found differences in the cognitive stage can be attributed mainly to figural, numerical and verbal skills, as well as working-memory capacity, differences in the associative stage to perceptual speed differences, and differences in the autonomous stage to psychomotor abilities.

Models of skill learning based on rules start out with a set of production rules that fully implement the skill. Several explanations for the speed-up produced by learning are offered. For example, in Soar (Newell, 1990) the speed-up is explained by chunking: reasoning steps that previously required multiple rules are carried out by a single rule after learning. Another explanation is that the efficiency of a rule itself is improved. For example, in ACT-R (Anderson & Lebiere, 1998) strength parameters are maintained for each rule.

As a rule is practiced, its strength value increases, and the time it takes to use it decreases. Although these models often predict the data very well, a conceptual problem remains: where do these initial production rules come from? The general critique is that these models are "preprogrammed": they already contain the information they are supposed to learn. Also the more qualitative aspects of the stages in skill learning, like the requirement for attention at the start and the lack of conscious access in the end, remain largely unexplained.

One of the reasons why the initial stage of skill learning is so hard to model, is the fact that the participant's general common sense knowledge comes into play. This knowledge is necessary to interpret the instructions, and to fill in the gaps in these instructions. For example, if the instruction is "push the button when an X appears on the screen", it is assumed to participant knows what an X is, what the screen is, and how to push a button. In other cases, for example in the scheduling task I will discuss later on, participants have to discover for themselves how to perform the task.

A theory that can explain the transition from the cognitive to the autonomous stage is proposed by Anderson (1982). This theory is based on the distinction between declarative and procedural memory. Declarative memory contains factual knowledge, and is available to conscious access. Procedural memory on the other hand contains production rules. These rules can only act, and cannot be inspected themselves. The idea is that in the cognitive stage the task-specific knowledge is represented declaratively. Declarative knowledge cannot act by itself, so it has to be interpreted by production rules. This explains why processing in the cognitive stage is slow. It also explains the fact that it is a conscious process, since declarative knowledge is available to consciousness. Gradually, during the associative stage, this declarative knowledge is compiled in production rules. These rules can act much faster than declarative knowledge, but are not available to consciousness. When all declarative knowledge is compiled, the autonomous stage is reached. Since the declarative knowledge is no longer needed, it is gradually forgotten and conscious control of task performance is lost.

Although this theory of skill learning is specified in terms that can be implemented in a production system, this is hardly ever done. In this paper I will discuss a model that does acquire its skills in this fashion. The model is implemented in ACT-R, and the task it models is scheduling.

The General Skill-learning Model

ACT-R (Anderson & Lebiere, 1998) is a cognitive architecture based on a production system. It has two long-term

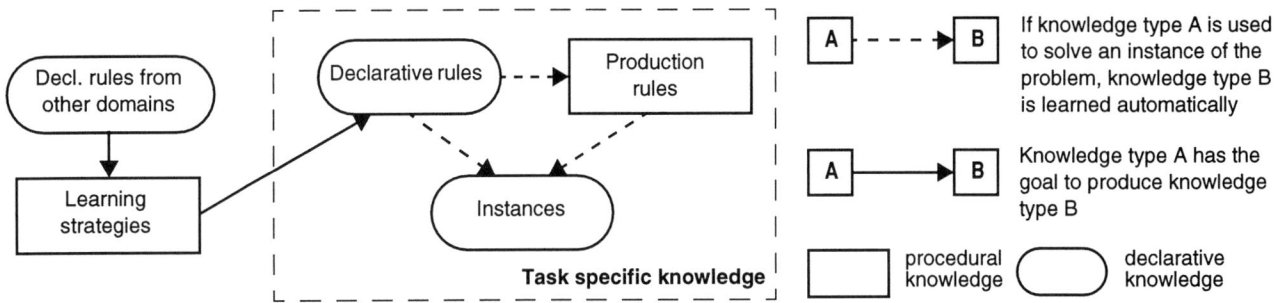

Figure 1: Overview of the proposed skill-learning model in ACT-R

memory stores: a declarative memory and a procedural memory. Although knowledge is represented by symbols in each of these memories (chunks and productions, respectively), it has a rich underlying sub-symbolic layer of representation that handles aspects like choice, errors, reaction times and forgetting.

Figure 1 shows an outline of the general skill-learning model. Each of the boxes in the diagram represents a type of knowledge: rectangles represent procedural knowledge, rounded boxes represent declarative knowledge. The dashed boundary indicates which of these knowledge types are task specific. The three types of knowledge that are part of the task-specific knowledge each have a different representation. Declarative rules are rules that are stored as a fact. An example of such a rule is:

```
Example-Declarative-Rule
    Isa Declarative-Rule
    Goal Count
    Retrieve Number-order
    Test "current count is equal to first number
        in number-order"
    Action "set the count to the second number
        in the number-order"
```

This rule is part of a counting procedure. It specifies that it is applicable to a goal of type count. When it is applied, a number-order fact (e.g., "Two is followed by Three") has to be retrieved of which the first number matches the current count. The second number in this fact should be stored in the goal. The procedural counterpart of this rule is:

```
IF      the goal is to count and the current count is num1
        AND num1 is followed by num2
THEN    set the current count to num2
```

Each time a declarative rule is used, there is a small probability that it will be compiled into a production rule. Although both representations will lead to the same results, there are differences. The declarative rule cannot act by itself. It needs to be interpreted by production rules. Figure 2 shows the process: given a certain goal, a suitable declarative rule is retrieved first, than the fact specified in the rule is retrieved while checking whether the test is satisfied at the same time. Finally the action is carried out and the goal is cleaned up. The procedural representation can do this in a single step, so

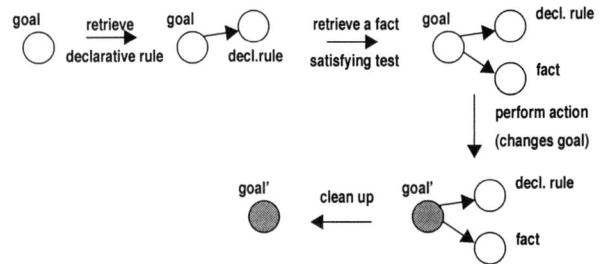

Figure 2: Interpreting declarative rules

is much faster. A declarative rule has its own advantages, however. Since it is declarative, it can be inspected by other rules than the rules that are used to carry it out. The counting rule, for example, can be modified slightly to count letters instead of digits. Production rules do not offer this flexibility, since they cannot be inspected themselves. If a declarative rule leads to an error, a modified version can be created to try something else. A final aspect of using a declarative rule is that it uses working-memory capacity. Although ACT-R has no separate workin memory, the function normally attributed to working memory, keeping track of currently relevant task knowledge, is related to the spread of activation from the goal. Due to the interpretation process of the declarative rule, the activation that originates from the goal has a larger fan: it is spread out over more chunks and becomes "thinner". In terms of working-memory capacity: the declarative rule consumes some of the working-memory capacity that is available for normal processing.

Instead of using a rule to solve a problem, an example or instance can be retrieved that immediately contains the answer, in a fashion that is comparable to Logan's (1988) instance theory. In ACT-R, achieved goals are kept in declarative memory automatically, an serve as instances that can be retrieved later (provided they have not been forgotten).

Since there is no initial task-specific knowledge, except for some uninterpreted instructions and possible biases, general "common sense" knowledge is needed to make a start. This knowledge is indicated in figure 1 by the term *learning strategies*. Learning strategies interpret instructions or try to modify declarative rules for other, similar tasks, or use other strategies to come up with methods to do the new task. The general idea is that this set of learning strategies is not fixed, but may be a source of individual differences. Taatgen (1997), for example, describes how different learning strate-

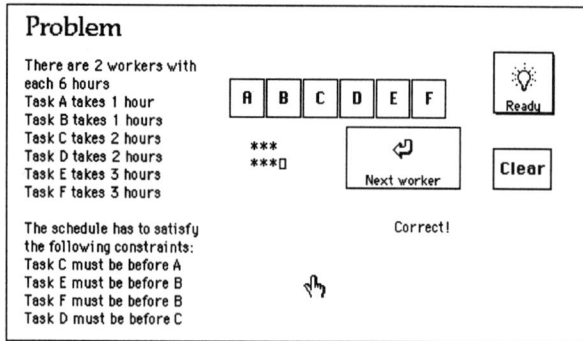

Figure 3: Example of a scheduling experiment

gies can explain the difference in behavior between adults and children in the discrimination-shift task. In the example model analogy will be used as a learning strategy.

At each stage in the reasoning process, there are multiple possible strategies to try. According to ACT-R, the strategy with the highest expected gain will win the competition. Retrieving an instance or using a production rule is generally the fastest strategy, followed by using a declarative rule. Using a learning strategy is generally only a good idea if the existing knowledge is insufficient or incorrect.

The Scheduling Task

An example of a complex task that is easy to explain but which is new to most people is scheduling. Figure 3 shows an example of a scheduling task used in our experiments. The goal is to assign a number of tasks (6 in the example) to a number of workers (2 in the example), satisfying a number of order constraints. A solution to the example in figure 3 is to assign DEA to the first worker, and FCB to the second. The participants have to solve the problem entirely by heard: the interface only allows them to type in the answer, which is represented by asterisks on the screen. An example fragment of verbal protocol is as follows:

> Yes. There are two workers with each six hours. Two. Task A, task B, task C. The schedule has to satisfy the following constraints... Task C before A, C before A, E before B, F before B and D before C. [..unintelligible..] First now D. D.. D..C..A..B.., D..C..A..B.., D.C.A.B., DCAB, and then, DCAB, [keys in DCAB] and then E... E..F, E..F. [keys in EF] [Receives feedback] Oh, task F is not before B. C.., D has to be before C. D.. No, C..D.., D has to be before C. C.. D.., C.. D.., A...B [keys in CDAB] that's one worker. E..F..., [keys in EF]. [receives feedback] Huh?! Task F is not before B and task D is not before C? Oh wait. D has to be before C, so first D... D...C..AB..AB [keys in DCAB]. Next worker, F.. yes, F..E., ready. [keys in FE] [receives feedback] Task E is not before B? Isn't it? Yes? [Emphasizing, keys in] D..C..A..B..E..E..F...ready. [receives feedback]. Well! Ehmm.. Task D takes two hours. [Silence] Task F is not before B, so F should be before B. Task E before... E should be before B, so E and F shouldn't be done by.... by the same worker. So we will, let's see. Task C before A, so we will first.... E before B, so we will first E..E..E..B..C. E...E..B..C.., EBC, no that's not right. EBC..F..A..B.. Ah.. start again. The D should be before C. [silence]. E... Ehm... The D should be before the C, so we put the D with worker one, and C with worker two. So we start with E with worker one... E..C..A.. E.C.A. ECA.. E.C.A. No, I don't get it... E..C..A..D..F.. Oh.. wrong again.

As can be seen is this small fragment of protocol, the partici-

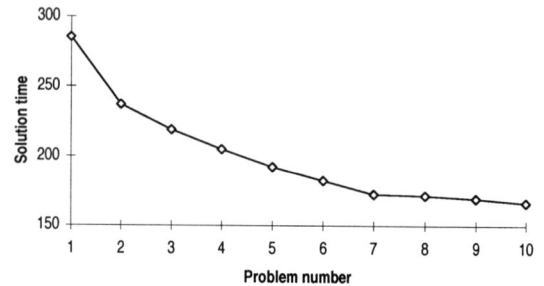

Figure 4: Learning curve of the first model

pant has a hard time memorizing partial solutions and deciding what to do next. This often leads to forgetting or mixing up letters, signs of an overloaded working memory.

I will explore this task in two models, the second of which is an extension of the first. The first will only model the proceduralization aspect of the general model from figure 1, while the second model will use a learning strategy to produce new task-specific declarative rules. The second model will also suffer from working memory limitations, as will be evident when I discuss its performance.

The First Model

A first approximation of a model of scheduling has the following components:

1. Production rules that interpret and proceduralize declarative rules as outlined in the previous section

2. A top-goal that reads the task and pushes a task subgoal and types out the answer of the subgoal has been reached

3. Productions that store elements in a list, and implement rehearsal, both maintenance and elaborate.

4. A set of declarative rules that implements a simple strategy for scheduling.

5. Productions that produce some sort of verbal protocol.

Results of the First Model

The model was tested using a set of ten example problems, all of which consisted of two workers and six or seven tasks. Although the problems are not particularly hard, this is not yet important since the answer given by the model is not checked. The model uses only symbolic learning, and has all subsymbolic learning turned off. New chunks in declarative memory do not have a role in the problem solving process yet. Improvements in performance can therefore be attributed to production compilation. Figure 4 shows the learning curve of the model. To get some idea of the rate of learning, the growth in the number of productions is plotted in figure 5. The more interesting part is the pseudo verbal protocol produced by the model. To see the impact of proceduralization, examples of the output of the first and the tenth problem have been printed in below.

Protocol of the First Problem

> There are two workers. Each of the workers has seven hours. Task A takes two hours. Task B takes two hours. Task C takes two hours. Task D takes two hours. Task E takes three hours. Task F takes three hours.

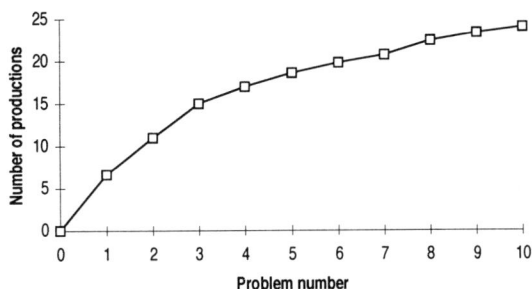

Figure 5: Growth in the number of production rules

Task B before F. Task F before A. First I will find a task to begin with. Let's look at an order constraint. B before F. Let's see if there is no earlier task. There is no earlier task. Begin with B. B Can we find a next task just by looking at the order? B before F. B.. F.. Can we find a next task just by looking at the order? F before A. B.. F.. A.. Can we find a next task just by looking at the order? Is this a schedule for one worker or for more? Now I am going to count how many hours we already have B.. How long does this one take? Task B takes two hours. Add this to what we have. nothing plus two equals two. F.. How long does this one take? Task F takes three hours. Add this to what we have. Two plus three equals five. A.. How long does this one take? Task A takes two hours. Do we have enough for one worker? Each worker has seven hours. We can move to the next worker.. B.. F.. A.. next.. Let's do the rest Now we are going to look at all the tasks, and see which ones are not yet in the schedule. Let's start with A. Task A takes two hours. Let's try to put it in the schedule. A is already in the schedule. OK, what is the next letter? B comes after A. Task B takes two hours. Let's try to put it in the schedule. B comes after A. B is already in the schedule. OK, what is the next letter? C comes after B. Task C takes two hours. Let's try to put it in the schedule. C comes after B. B.. F.. A.. next.. C.. OK, what is the next letter? D comes after C. Task D takes two hours. Let's try to put it in the schedule. D comes after C. B.. F.. A.. next.. C.. D.. OK, what is the next letter? E comes after D. Task E takes three hours. Let's try to put it in the schedule. E comes after D. B.. F.. A.. next.. C.. D.. E. Task F takes three hours. Let's try to put it in the schedule. F comes after E. F is already in the schedule. OK, that was the last task, we're done! The answer is B F A next C D E

Protocol of the Tenth Problem

There are two workers. Each of the workers has six hours. Task A takes one hours. Task B takes one hours. Task C takes two hours. Task D takes two hours. Task E takes three hours. Task F takes three hours. Task D before E. Task E before A. First I will find a task to begin with. Let's see if there is no earlier task. Begin with D. D.. D.. E.. D.. E.. A.. Can we find a next task just by looking at the order? Is this a schedule for one worker or for more? Now I am going to count how many hours we already have D..E..A.. D..E..A.. next Now we are going to look at all the tasks, and see which ones are not yet in the schedule. Let's start with A. A is already in the schedule. D..E..A.. next.. B D..E..A..next..B..C D is already in the schedule. E is already in the schedule. D..E..A.. next.. B..C..F. OK, that was the last task, we're done! The answer is D E A next B C F

Clearly, the protocol of the first problem is a protocol analyst's dream, because participants are hardly ever that precise. But the tenth protocol looks more familiar: many steps in the process are omitted, and we can only guess why some decisions have been made. This concurs with the general idea that proceduralized skills produce no verbal protocol (Ericsson & Simon, 1984).

Although this first model shows some interesting proper-ties similar to real problem-solving behavior, it is far from complete. The current model just takes a single shot at the solution, and does not retry if it is incorrect. Only ACT-R's production compilation had been turned on, so the model will never forget any intermediate results it has found. And finally, the model starts out with a set of task-specific declarative rules. One of the desired properties of the model was to start without any task-specific knowledge. These issues will be addressed in the second version of the model.

Learning Scheduling

People may not know anything about schedules, but they do know something about lists, and how to construct them. Suppose we need to make a schedule. We may use knowledge about lists to start with. How do we make a list? First we have to find a first item for the list, a beginning. Once we have a beginning, we find a next task until we are done. But how do we find something to begin with, and how do find a next task? We may choose to handle these problems making them subgoals, or we may try to find mappings between 'beginning' and 'next' and terms in the scheduling problem. For example, a mapping can be made between 'next' and an order-constraint in the scheduling problem. The result is a modified version of the list-building declarative rules, with 'list' substituted by 'schedule' and 'next' substituted by 'order'. Note that for sake of the explanation, the terms 'list', 'beginning', 'before' and 'next' will be used to refer to general terms, and 'schedule' and 'order' to refer to task-specific terms. Except for knowledge on how to build a list, the analogy between a schedule and a list may also offer knowledge on how to retain a list in memory by rehearsal.

Although these new declarative rules may find a partial schedule, they are insufficient to build a complete schedule, mainly because the mapping between 'next' and 'order' is inadequate. When this declarative rule fails to make a complete schedule, another plan may take over and contribute to the schedule.

An idea that may take over if the list-building plan fails to add any more tasks to the schedule is the plan that tries to complete the first worker. A useful general plan may state that whenever something has to be completed, the difference between the desired size and the current size has to be calculated, after which an object had to found with a size equal to this difference.

The central emerging idea is therefore that several strategies from similar domains are adopted and patched together. This method of adapting old strategies for new purposes is similar to *script* and *schema* theories. Traditional script and schema theories assume that a complete script is first adapted to fit the current task, and then carried out. The ACT-R model uses a more on-demand style of adaptation: a new declarative rule is created at the moment it is needed.

The second model

The second model solves some of the shortcomings of the first. It learns new declarative rules as outlined in the previous section. Furthermore, the following aspects have been added to the model:

1. The model has to come up with a correct solution within 300 seconds. If no solution has been found in this period, the attempt is counted as a failure.

2. ACT-R's activation learning is turned on. As a consequence, the model can forget all kinds of partial results it derives, most notably the list that contains the partial solution, but also read constraints (which have to be reread in that case), newly derived declarative rules, etc.

3. Several extra declarative rules have been added to ensure that correct solutions are eventually found by the model. The model now tries to satisfy the order constraints for the second worker as well, and uses the feedback it gets when it makes an error as a starting constraint for the next try.

Results of the second model

The following protocol fragment, produced by the model, gives an impression of the additional aspects of the model:

> There are two workers. Each of the workers has six hours. Task A takes one hours. Task B takes one hours. Task C takes two hours. Task D takes two hours. Task E takes three hours. Task F takes three hours. Task B before C. Task F before A. I have to think of some new way to find a schedule. Let's use what I know about lists. First I will find something to begin with. Let's look at a before constraint. I have to think of some new way to find a before. Let's use what I know about order. Let's use a order fact as a before fact. F before A. I have to think of some new way to find a before following a failed declarative rule 12. F before A. I have to think of some new way to find a before following declarative rule 12. Let's look at a before constraint. Let's see if there is no earlier element. Let's use a order fact as a before fact. There is no earlier element.

> [three failed episodes removed]

> First I will find something to begin with. Begin with F. F.. Now I have to find the next thing. F before A. A.. Now I have to find the next thing. No more items for the list, let's check whether we're done. F.. A.. Is this a schedule for one worker or for more? Now I am going to count how many hours we already have F.. Add this to what we have. Nothing plus three equals three. A.. Add this to what we have. Three plus one equals four. Do we have enough for one worker? No, the schedule is not full, yet. F.. A.. Now find the task that fits in. Task C takes two hours. C.. We can move to the next worker.. NEXT-WORKER Let's do the rest F.. A.. C.. NEXT-WORKER.. I now try to find any unused order constraints. B before C. B before C. This one hasn't been used, so the constraint has been found. B before C. B before C. B.. Now we are going to look at all the tasks, and see which ones are not yet in the schedule. Let's start with A. Task A takes one hours. A is already in the schedule. OK, what is the next task? Task B takes one hours. B is already in the schedule. Let's move on to the next task. OK, what is the next task? Task C takes two hours. C is already in the schedule. Let's move on to the next task. OK, what is the next task? Task D takes two hours. D.. Let's move on to the next task. OK, what is the next task? Task E takes three hours. E.. Let's move on to the next task. OK, what is the next task? Task F takes three hours. F is already in the schedule. Let's move on to the next task. OK, what is the next task? OK, that was the last task, we're done! F.. A.. C.. NEXT-WORKER.. B.. D.. E.. The answer is F A C NEXT-WORKER B D E

This particular problem required five attempts, only two of which are shown: the first and the final, successful episode. In the first fragment, the model is busy figuring out how

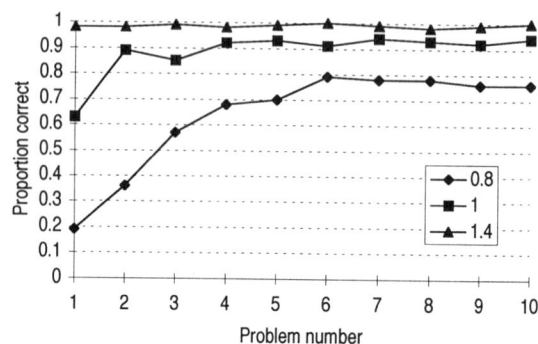

Figure 6: Accuracies of the second model for a low (0.8), average (1) and high working memory (1.4) capacity

aspects of the problem can be mapped onto things it knows something about. Unfortunately, the primitive protocol generating part of the model produces some awkward sentences with references to internal symbols. Somewhere along the line the model gets stuck, because it cannot keep track of all the constraints in the task and all the newly derived declarative rules. The third search episode (not shown) is slightly more successful: it can use the declarative rules derived in the first two episodes. Unfortunately, it fails just when it is done, because it cannot retrieve the start of the list anymore for typing in the answer. In the fifth, successful episode some of the earlier derived results can be retrieved, which is evidence that the model uses the instance strategy. For example, the model immediately starts with "Begin with F" instead of deriving this fact.

Individual differences

The basic performance results of the second model have roughly the same shape as the results from the first model. The second model is, however, sensitive to working memory limitations. According to Lovett, Reder and Lebiere (1997), individual differences in working-memory capacity can be modeled by variations in the spread of activation in ACT-R, as mediated by the source activation (W) parameter. When this parameter is varied between 0.8 and 1.4, it produces the accuracy rations depicted in figure 6. Accuracy means in this case: the proportion of problems solved correctly within 300 seconds. The interesting aspect of different working memory capacities is that at the start of the experiment the differences are very high: the low capacity model hardly gets any problem done, and the high capacity model is successful almost all the time. At the tenth problem however, these differences have diminished: the low capacity model has proceduralized enough knowledge to be able to reach a correct solution most of the time. This finding is consistent with Ackerman's theory that cognitive abilities like working-memory capacity are mainly important in the cognitive stage of skill acquisition.

Preliminary Empirical Evidence for the Model

As part of a study of mental fatigue (Jongman & Taatgen, 1999), we have used the scheduling task and the digit working memory task that has been modeled in ACT-R by Lovett, Reder and Lebiere (1997). The digit working memory task

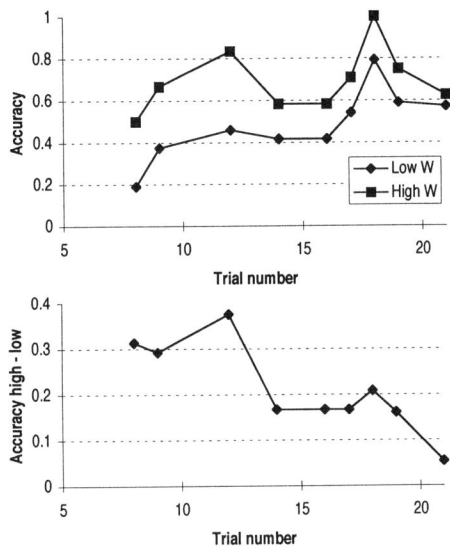

Figure 7: Proportion of correctly solved schedules for the 8 highest source activations and the 8 lowest source activations. The top graph shows the accuracy curves for both groups, the bottom graph shows the difference between the two.

was used to make an estimate of the working-memory capacity of a participant, expressed in the ACT-R source activation parameter. This working-memory capacity was related to the performance on the scheduling task. Unfortunately, the scheduling task as it was used in this particular experiment was a mixture of problems with two and three workers with varying difficulty and varying time limitations. It is therefore hard to directly compare the results to the model predictions. Nevertheless some of the more qualitative predictions of the model can be tested with respect to individual differences. The model predicts a strong correlation between working-memory capacity and the performance on the scheduling task. This proved to be the case in the experiment: the correlation between the estimated source activation and the number of successfully solved schedules is 0.56 (with n=16). This correlation increases to 0.66 if the analysis is restricted to the three-worker schedules, the schedules that require most working-memory capacity. A more specific prediction of the model is that the effect of working-memory capacity on performance will diminish due to proceduralization. To investigate this prediction, the group of participants is split into eight low source-activation participants (W<0.95) and eight high source-activation participants (W>0.95). The proportions of correct solutions for each of the groups is plotted in figure 7. In this graph only three-workers problems are shown, and to average out part of the noise each data point is averaged with its predecessor and its successor. There is a clear convergence between the two curves, as can be seen in the bottom graph that depicts the difference.

Discussion

The skill-learning model discussed above exhibits many of the characteristics normally attributed to different stages in problem solving. In the cognitive stage, declarative rules are used. While these rules are interpreted, they are part of the current goal context and are available to consciousness. The interpretation process is slow, and susceptible to errors, especially if the task-demands with respect to working memory are high. On the other hand declarative rules may be manipulated by learning strategies, offering flexibility. In the autonomous stage, all processing is handled by production rules, which are fast and less demanding for working memory. Moreover, once the declarative rule itself is forgotten, conscious access to the skill is lost.

Although the model discussed above implements the skill-learning process for the scheduling task, the basic framework can be used for learning other tasks as well. The prior knowledge of the model can to be changed, and maybe also the learning strategies, in order to enable it to learn a different task.

Availability of Models

Both models can be retrieved from: http://tcw2.ppsw.rug.nl/~niels/thesis. They are listed under the "chapter 7" caption.

References

Ackerman, P. L. (1988). Determinants of individual differences during skill acquisition: cognitive abilities and information processing. *Journal of Experimental Psychology: General, 117*, 288-318

Anderson, J. R. & Lebiere, C. (1998). *The atomic components of thought*. Mahwah, NJ: Erlbaum.

Anderson, J. R. (1982). Acquisition of cognitive skill. *Psychological Review, 89*, 369-406.

Ericsson, K. A. & Simon, H. A. (1984). *Protocol analysis. Verbal reports as data*. Cambridge, MA: The MIT Press.

Fitts, P. M. (1964). Perceptual-motor skill learning. In A. W. Melton (Eds.), *Categories of human learning*. New York: Academic Press.

Jongman, L. & Taatgen, N.A. (1999). An ACT-R model of individual differences in changes in adaptivity due to mental fatigue. *Proceedings of the 21th Annual Conference of the Cognitive Science Society*. Mahwah, NJ: Erlbaum

Logan, G. D. (1988). Toward an instance theory of automization. *Psychological Review, 22*, 1-35.

Lovett, M. C., Reder, L. M., & Lebiere, C. (1997). Modeling individual differences in a digit working memory task. *Proceedings of the Nineteenth Annual Conference of the Cognitive Science Society* (pp. 460-465). Hillsdale, NJ: Erlbaum.

Newell, A. (1990). *Unified theories of cognition*. Cambridge, MA: Harvard university press.

Taatgen, N.A. (1997). A rational analysis of alternating search and reflection in problem solving. *Proceedings of the 19th Annual Conference of the Cognitive Science Society*. Mahwah, NJ: Erlbaum

Incremental Grammatical Encoding in Event Descriptions

Mark Timmermans (M.Timmermans@nici.kun.nl)
NICI, University of Nijmegen
P.O. Box 9104, 6500 HE Nijmegen, The Netherlands

Herbert Schriefers (Schriefers@nici.kun.nl)
NICI, University of Nijmegen
P.O. Box 9104, 6500 HE Nijmegen, The Netherlands

Simone Sprenger (Simone.Sprenger@mpi.nl)
MPI for Psycholinguistics, Nijmegen
P.O. Box 310, 6500 AH Nijmegen, The Netherlands

Ton Dijkstra (Dijkstra@nici.kun.nl)
NICI, University of Nijmegen
P.O. Box 9104, 6500 HE Nijmegen, The Netherlands

Abstract

Speech is produced incrementally. The Incremental Parallel Formulator (De Smedt, 1996) is a computational model of grammatical encoding that takes this notion of incrementality into account. It predicts that the order and time-scale with which conceptual fragments activate lexical segments affect the syntactic shape of an utterance. We derived predictions from this model and tested these in two online experiments. In these experiments, participants described computer animations in which two objects moved in upward or downward directions. We manipulated the availability of pieces of the conceptual input by withholding either the information about the movement direction, or about the identity of one of the objects for various amounts of time. The experiments showed that both the type and the temporal availability of conceptual information strongly affect the syntactic shape of an utterance.

Introduction

According to Levelt (1989) three processing modules for fluent speech production can be distinguished: the Conceptualizer, the Formulator, and the Articulator. These modules work in an incremental fashion on different parts of an utterance simultaneously. The view that speakers produce spoken language incrementally has now become common ground in the psycholinguistic community (e.g., Levelt, 1989). On the modeling side, this has resulted in the formulation of several computational models of grammatical encoding, one of which is the Incremental Parallel Formulator (IPF; De Smedt, 1996). A central prediction of this model is that the order and time-scale with which conceptual fragments become available to the speaker exert an influence on the syntactic format of an utterance. We tested this prediction in two experiments in which we asked participants to describe simple computer animations.

The Incremental Parallel Formulator has its roots in Segment Grammar (e.g., Kempen, in preparation). The building blocks of Segment Grammar are segments: small structures of lexical origin that have a root node and a foot node with an arc in between. The arc specifies the syntactic function of a segment, the root and the foot node represent syntactic categories. The construction of syntactic structures takes place in the so-called Unification Space (see also Kempen & Vosse, 1989), where the roots of segments can unify with root nodes (by bifurcation) or with foot nodes (by concatenation) of other segments.

An important general principle in incremental language production has been formulated by Hoenkamp (1983, p. 18) as follows: "What can be uttered, must be uttered immediately". If we extend this principle to the process of grammatical encoding, we can formulate the following processing assumption for unification: "What can be unified, must be unified immediately".

Consider the application of this processing assumption to the following example. Imagine a speaker who is asked to describe a scene in which a circle moves in upward direction. The conceptual input will activate the following lexical segments in the Unification Space:

(1)	NP_1	----	head	----	n	*(circle)*
(2)	NP_1	----	det	----	art	*(the)*
(3)	S	----	HEAD	----	verb	*(move)*
(4)	ADVP	----	head	----	adv	*(up)*
(5)	S	----	SUBJ	----	NP	
(6)	S	----	MOD	----	ADVP	

By unification of the segments (i.e., by bifurcation of segments (1) & (2), and of segments (3), (5) & (6), and by concatenation of segments (5) & (1,2), and of segments (6) & (4)), the speaker can utter the simple sentence: *"The circle moves up"* (denoted as S_1 hereafter). Now suppose that the speaker sees an animation displaying a circle and a triangle both moving upward simultaneously. In this case, the speaker can either say *"The circle moves up and the triangle moves up"* (a Sentence Coordination or SC), or the speaker can say *"The circle and the triangle move up"* (a Noun Phrase Coordination or NPC). To describe the scene by means of an NPC, the conceptualizer will activate the following additional segments:

(7)	NP_2	----	head	----	n	*(triangle)*
(8)	NP_2	----	det	----	art	*(the)*
(9)	S/NP	----	OP	----	conj	*(and)*

Segment (9) is a special kind of segment that requires some further comment. As Kempen (in preparation) proposes, a few segments have the special status of Operator. These special segments accommodate for conjunctions (*and* and *but*) and negations (*not* and *nor*). As the double category of the root (S/NP) of segment (9) indicates, we introduce a special syntactic segment here that can be regarded as a 'generic syntactic conjunctor'. That is, it can unify two sentences or two noun phrases. In the former case, the operator is nothing more than just an ordinary conjunction between two full sentences. In the latter, it combines two noun phrases. The result of grammatical encoding will then be an NPC. We further assume that this special segment should not be allowed to unify with only one other segment at the time. That is, for a coordination to take place, segment (9) requires two roots of the same category to be available at the same time (i.e., two S-nodes or two NP-nodes).[1]

So far, we have only considered a situation in which all segments are available at the same time. What would happen if we systematically delay the availability of some parts of the conceptual input? This was tested in our first experiment.

Experiment 1

Participants were asked to describe simple computer animations as fast and as accurately as possible. The animations displayed two geometrical objects making movements in either upward or downward directions. The objects either moved in the same direction (conjunct movements) or in opposite directions (disjunct movements). Both objects appeared on the screen simultaneously. Immediately following their appearance, one of the objects (e.g., a circle) started to move, whereas the other one (e.g., a triangle) started to move with some delay. As a result of this Movement Onset Asynchrony (henceforth MOA), the conceptual information associated with the movement direction of the object moving second was withheld from the grammatical encoder by MOA milliseconds relative to that of the object moving first. In terms of IPF, the Conceptualizer will activate lexical

segments (1) - (6) to verbalize the event in which a circle moves upward. In the mean time, segments (7) and (8) are also activated. The movement of the triangle cannot yet be conceptualized because it is still stationary at this point. If we now apply the principle "what can be unified, must be unified immediately", then unification of segments (1) - (6) yields a full sentence: *"The circle moves up"*. At the same time, segments (7) and (8) are also unified, resulting in a noun phrase: *"the triangle"*. After MOA milliseconds, the triangle has also started to move upward. The conceptualizer can now activate the conjunctor and (re)activate segments (3) - (6), thereby unifying two full sentences, yielding an SC: *"The circle moves up and the triangle moves up"*. Had there been no delay between onset of the movements (i.e., MOA = 0 ms), a situation as described in our earlier example would have resulted. The grammatical encoder might have had segment (9) available before unification of the first sentence had been completed and the speaker would have uttered an NPC. As a consequence, we expect a decrease in the proportion of NPCs with increasing MOA.

Until now, we have argued as if the processes of conceptualization and unification are deterministic. That is, we assumed that not only the speed with which the event is conceptually encoded is constant, but also the speed with which lexical and syntactic segments are activated. Thus, going from an MOA of 0 milliseconds up to a certain MOA of *n* milliseconds, speakers would produce only NPCs, and at MOAs longer than *n* milliseconds only SCs. However, this is not a realistic assumption. Because both the time needed to conceptualize objects and their movements, and the time needed for segments to be unified will probably vary around some mean value, one would not expect a sudden transition from NPCs to SCs at a certain MOA. Rather, one would expect a gradual decrease in the proportion of NPCs for describing conjunct movements with increasing MOA. This prediction presupposes that speakers do not wait until the movement direction of the second object becomes available too, but start uttering as soon as possible. In other words, we expect that the grammatical encoder does not have to wait until all conceptual information about the to-be-described event is processed before a (partial) result is passed on to phonological encoding and articulation. If this is indeed the case, then utterance onset latencies should not be affected by MOA.

Thus far, we have only discussed the effect of the temporal availability of conceptual fragments on utterance format (either SC or NPC) and neglected utterance onset latencies (i.e., the time between onset of the animation and onset of the utterance). However, utterance onset latencies will allow us to look at the temporal development of incremental grammatical encoding also at a more fine-grained level. Let us assume that the grammatical encoder starts constructing the first noun phrase (by bifurcation of segments (1) & (2)) as soon as the corresponding object becomes available. The time needed for unification of this noun phrase (NP_1) will vary around some mean value. If, on a given trial, NP_1 is completed early, then the likelihood that the information about the movement direction of the object moving second is

---

[1] In addition, there is a second constraint. The conceptualizer has to indicate that the two objects triggering their respective segments move in the same direction. Otherwise, the two NPs might, erroneously, also be unified during the grammatical encoding of a description of an event in which two objects move in different directions. This, of course, would result in an inappropriate structure (an NPC), because only an SC is an adequate description for such an event.

737

already available should be rather low. As a consequence, NP_1 will probably be unified with the other available segments in the Unification Space, thus yielding the first simple sentence (S_1) in an SC. If, by contrast, NP_1 is completed relatively late, then the likelihood that information about the movement direction of the object moving second should be higher. As a result, the probability that an NPC is constructed will be higher. Thus, taken together, within a given MOA, the proportions of NPCs should be higher for a subset of responses that have relatively long utterance onset latencies compared to a subset of responses with relatively short utterance onset latencies. This would also show that the eventual utterance format (SC or NPC) is determined by both MOA and utterance onset. In order to quantify this prediction we will divide all descriptions (irrespective of type of utterance) of conjunct movements for a given participant and MOA into two equally large sets: one set containing all responses with utterance onset latencies that were shorter than the median latency for a given participant and MOA (fast responses), and one set with responses with latencies slower than the median (slow responses). We will refer to this division of responses according to their latencies as the factor Speed. Thus, we would expect that the resulting factor Speed yields higher proportions of NPCs in the set of slow responses than in the set of fast responses. Furthermore, this difference should become larger with increasing MOA.

In sum, we hypothesize that with increasing MOA, the chance that a speaker describes a conjunct movement by means of an NPC will gradually decrease. Furthermore, the proportion of NPCs will be lower for fast than for slow responses.

Method

Participants Twenty undergraduate students at the University of Nijmegen participated in the experiment.

Design Participants were asked to describe 120 short animations in which two geometrical objects, positioned to the left and the right of the center of a computer screen, made phi-movements in either upward or downward directions. On every trial, participants saw a combination of two objects selected from a set containing a circle, a triangle, a square, and a cross. All possible combinations of two objects occurred equally often, excluding pairs of identical objects. There were 48 critical trials in which the objects made conjunct movements (24 upward and 24 downward). In addition, there were 48 distractor trials in which the objects made disjunct movements (24 trials with the left object moving upward and the right downward and 24 trials with reversed directions). Movement Type (conjunct or disjunct) was crossed with MOA. There were four different MOAs: 0 ms, 300 ms, 500 ms, and 700 ms, which were equally distributed over left and right objects. In addition to these 96 trials, 24 filler trials (12 upward and 12 downward) were added in which the objects started to move in the same direction without delay (i.e., the MOA was 0 ms). These trials were added to induce participants to use NPCs in their descriptions where applicable.

Procedure Participants were instructed to describe the animations on the screen as fast and accurately as possible. This instruction provided two examples possible animations. For conjunct movements, the example was: *If you see a cross moving up and a triangle moving up, you say: "the cross and the triangle move up"*. For disjunct movements, the example was: *If you see a circle moving up and a square moving down, you say: "the circle moves up and the square moves down"*. The instruction did not explicitly state to use only NPCs for describing conjunct movements and only SCs for disjunct movements. Participants were also instructed to complete their utterance as naturally as possible, without making corrections or re-starting their utterances. Participants were free to choose the object they wanted to mention first (either the left or the right object).

Results and Discussion

As valid NPC utterances we scored all responses like *"The circle and the triangle (both) move up"*. As valid SC utterances we scored all responses like *"The circle moves up (and) the triangle (also) moves up (too)"*. Moreover, SC utterances containing ellipses (e.g., *"The circle moves up and the triangle too"* or *"The circle moves up and the triangle down"*) were also treated as valid responses. Apart from the type of utterance, we also scored hesitations, self-corrections and errors. All responses containing any of these dysfluencies (8.5% in total) were excluded from further analyses.

To determine the effect of the factor Speed, we computed the median of the utterance onset latencies for descriptions of conjunct movements for each MOA and subject separately. Next, the proportions of NPCs below (fast responses) and above (slow responses) the median were computed.

Table 1: Proportions of NPCs
on Critical Trials by MOA and Speed in Experiment 1.

Speed	MOA in ms			
	0	300	500	700
Fast	.950 (.134)	.597 (.401)	.354 (.322)	.236 (.271)
Slow	.975 (.082)	.633 (.408)	.498 (.385)	.412 (.384)

Note. Standard deviations in brackets.

The proportions of NPCs on the critical trials (see Table 1) were subjected to a 4 (MOA) × 2 (Speed) analysis of variance. The main effect of MOA was significant for the linear contrast, $F(1,19) = 122.16$; $p < 0.01$, and for the quadratic contrast, $F(1,19) = 6.08$; $p < 0.05$.[2] In addition, the main effect of Speed and the interaction effect between

---

[2] We tested the linear, quadratic, and cubic polynomial contrasts. Therefore, the degree of freedom of the numerator is always equal to 1, even if a factor has more than two levels. In the remainder, we will only report the F-ratios associated with the linear contrasts. Unless explicitly stated, the tests on the quadratic and cubic contrasts were not significant.

MOA and Speed were significant, $F(1,19) = 18.571$; $p < 0.01$ and $F(1,19) = 4.61$; $p < 0.05$, respectively. An analysis of variance on utterance onset latencies of NPCs on conjunct movements (see Table 2) yielded no significant effect of MOA, $F(1,19) < 1$, showing that participants did not systematically postpone their utterances with increasing MOA.

Table 2: Average Utterance Onset Latencies (UOL) for NPCs on Critical Trials by MOA in Experiment 1.

	MOA in ms			
	0	300	500	700
UOL in ms	862 (148)	886 (153)	901 (143)	862 (143)

Note. Standard deviations in brackets.

The results of the experiment are in line with the predictions outlined above. They stated that there should be a systematic decrease in proportions of NPCs with increasing MOA, and that the proportion of NPCs would be smaller in the set of fast responses than in the set of slow responses. The results strongly suggest that the temporal relation between the point in time at which conceptual fragments are activated and utterance onset affects utterance format. The results further show that across MOAs, 70% of all utterance onset latencies were longer than the longest MOA of 700 ms. This implies that the average utterance onset latencies in the set of slow responses are presumably longer than their respective MOA. Nevertheless, we see a decrease in proportions of NPCs also for the slow set of responses at MOA = 300 ms and MOA = 500 ms. This implies that even when utterances were initiated after onset of the second movement, the grammatical encoder has on a relatively substantial proportion of trials committed itself to an SC. This commitment, moreover, appears not to be revised anymore within the time frame between onset of the second movement and initiation of the utterance. Thus, a possible revision of the syntactic format of the utterance under construction seems highly unlikely, perhaps even impossible, even when additional conceptual information becomes available before speech onset.

This conclusion is also in line with the observation that the proportion of NPCs for the utterances in the set of fast responses was lower than in the set of slow responses. One possible account of this effect is as follows: The grammatical encoder passes information to phonological encoding and articulation as soon as a commitment has been made for an SC or an NPC. As a result, the proportion of NPCs should be lower for fast than for slow responses. Under this view, the grammatical encoder forwards information to later processing stages as soon as a commitment to an SC or an NPC has been made. From that point onward, the encoded structure can only be expanded further to the right, but it cannot be revised anymore.

It should be noted, however, that the results do not exclude an alternative option. According to this option, the decision to construct an SC or an NPC might be taken while phonological encoding and/or articulation of the first noun phrase is carried out. Nevertheless, the large proportion of SCs in responses where utterance onset follows the onset of the second movement strongly suggests that once a commitment for an SC is taken, the probability that this commitment is revised to an NPC appears to be low (or even zero).

In sum, it appears that the temporal availability of conceptual fragments and, consequently, of the corresponding segments for grammatical encoding, plays a crucial role for the eventual syntactic format of an utterance. More specifically, the later the information with respect to the movement direction of the object moving second comes into play relative to speech onset, the lower the probability that speakers produce an NPC.

This conclusion was further tested in Experiment 2. In this experiment, the identity of only one object, but the movement directions of both objects, were available from the very beginning of the animation. There are two possible hypothesis in this situation. First, the grammatical encoder might use the movement information about both objects to take an 'early decision' on whether to encode an SC or an NPC in absence of information about the identity of the object moving second. In this case, one would expect the proportions of NPCs in this experiment not to be affected by the onset asynchrony manipulation. Second, if the coordinating segment (9) can only yield an NPC if both to-be-coordinated segments are available (as assumed by IPF), we should obtain the same pattern of proportions of NPCs as in Experiment 1.

Experiment 2

Method

Participants Sixteen undergraduate students from the same pool as in Experiment 1 took part in this experiment.

Design The experiment was identical to Experiment 1 except for two changes. First, instead of having an MOA, we introduced an Object Onset Asynchrony (OOA). At the start of the animation one object and a cloud of random dots started to move immediately. After OOA ms, the moving cloud of dots was replaced by the target object. Second, in order to obtain a more fine-grained picture of the development of the proportions of NPCs across asynchronies, we replaced the 700 ms delay by a 400 ms delay. Thus, the factor OOA had four levels: 0 ms, 300 ms, 400 ms, and 500 ms.

Results and Discussion

All utterances were scored in the same fashion as in Experiment 1. All responses containing dysfluencies (12% in total) were excluded from further analyses. The proportions of NPCs on the critical trials (see Table 3) were subjected to a 4 (OOA) × 2 (Speed) analysis of variance, which yielded a significant main effect for OOA, $F(1,15) = 8.93$, $p < 0.01$. The main effect of Speed and the interaction effect between OOA and Speed were not significant, $F(1,15) = 2.95$, $p > 0.10$ and $F(1,15) < 1$, respectively. Average utterance onset latencies on NPCs for descriptions of conjunct movements (see Table 4) showed no significant effect of OOA on utterance onset latencies, $F(1,15) = 1.49$, $p > 0.10$. It appears that participants did not postpone their utterances with increasing OOA. Had this been the case, this could have implied that they waited for the information about the identity of the object moving second to be able to produce an NPC.

Table 3: Proportions of NPCs
on Critical Trials by OOA and Speed in Experiment 2.

Speed	OOA in ms			
	0	300	400	500
Fast	.948 (.146)	.833 (.184)	.777 (.332)	.814 (.221)
Slow	.967 (.094)	.938 (.103)	.881 (.159)	.801 (.288)

Note. Standard deviations in brackets.

Thus far, we have proposed that speakers do not revise syntactic commitments made once articulation is initiated. This conclusion is based on the substantial proportions of SCs in Experiment 1, even in trials where the movement of the object moving second starts before speech onset. By introducing an OOA in this experiment instead of an MOA (as in Experiment 1), participants are, in principle, provided with sufficient information at the start of the animation to encode an NPC. One could argue, therefore, that participants might have used a strategy such as prolonging the first part of an utterance to be able to produce an NPC with increasing delay. Obviously, such a prolongation would not be reflected in utterance onset latencies, but in the duration of the first part of the utterance.

To investigate this possibility, we measured the duration of the first NP (NP_1) and the conjunctor up to the beginning of the second NP (NP_2) for NPCs describing conjunct events (see Table 4). An analysis of variance yielded no significant effects for the linear and cubic contrasts, $F(1,15) < 1$ and $F(1,15) = 2.92$, $p > 0.10$, respectively. The quadratic contrast was only marginally significant, $F(1,15) = 3.85$, $p < 0.10$. In sum, although there was a slight tendency to prolong the first part of the utterance, this prolongation did by far not compensate for the longer OOA (see Table 4).

Table 4: Average Utterance Onset Latencies (UOL)
and Duration of the First Part of the Utterance (DUR)
for NPCs on Critical Trials by OOA in Experiment 2.

	OOA in ms			
	0	300	400	500
UOL in ms	928 (166)	896 (155)	934 (132)	876 (120)
DUR in ms	673 (111)	671 (106)	701 (142)	667 (125)

Note. Standard deviations in brackets.

The results provide only partial support for our predictions with respect to the effect of OOA. Although the proportions of NPCs for OOAs > 0 ms were considerably higher than in Experiment 1, there still was a significant decrease in the proportions of NPCs with increasing OOA. This was the case even though the information with respect to movement direction of both objects was available from the beginning of the animation. However, this decrease appeared to be considerably smaller than in the first experiment. An analysis of variance comparing Experiment 1 and 2 with respect to the

proportions of NPCs for the MOAs/OOAs of 0 ms, 300 ms, and 500 ms supported this observation. This analysis yielded significant main effects of the factors Experiment, $F(1,34) = 10.31$, $p < 0.01$, MOA/OOA, $F(1,34) = 56.21$, $p < 0.01$, and Speed, $F(1,34) = 6.76$, $p < 0.05$. The interaction between Experiment and MOA/OOA was also significant, $F(1,34) = 17.86$, $p < 0.01$. No other interactions reached significance.

These results imply that the availability of information with respect to the movement direction of the objects is not the only information that determines the eventual syntactic format of the descriptions. Rather, the tendency to describe conjunct movements by means of an NPC seems to benefit from early availability of the identity of the object moving second. Had there been no such benefit, Experiment 2 should have yielded the same proportions of NPCs for all OOAs.

General Discussion

In sum, we have seen the following. First, the eventual syntactic structure of a description of a conjunct movement crucially depends on the point in time at which information about both movements becomes available, as indicated by the strong effect of MOA on the proportions of NPCs in Experiment 1. Second, the effect of Speed in Experiment 1 further qualifies this conclusion. It is not simply the absolute timing of the availability of conceptual information about the movement direction of the object moving second (MOA), that determines the commitment for an SC or an NPC, but rather the temporal relation between MOA and utterance onset. Third, as soon as the grammatical encoder has made a commitment to an SC, the chance that this commitment is revised during the articulation of the first part of the description appears to be low. Fourth, although there are indications that speakers prolonged the articulation of the first part of their descriptions with increasing OOA (Experiment 2), this prolongation was by far insufficient to compensate for the increase in OOA. Fifth, the immediate availability of information with respect to both movements does not appear to be the only force driving the encoding of an NPC, because we found a significant effect of OOA on the proportions of NPCs in Experiment 2. This result can be interpreted in two ways.

First, one could assume that availability of the identity of the object moving second is an independent force favoring the construction of NPCs. For example, one could assume that with the availability of two NPs, the grammatical encoder has a tendency to unify them in an NPC even if it is unknown whether the movements to be described are conjunct or disjunct. Note that this is in fact the state of affairs in Experiment 1, where at the start of an animation participants saw two objects, but only one movement.

Second, one could assume that information with respect to the identity of the object moving second is an additional force favoring the construction of NPCs which comes into play only when it is known that one is dealing with a conjunct movement. In other words, the grammatical encoding process leading to an NPC is started by information about the presence of a conjunct movement, but can be facilitated

740

by the early availability of the identity of the object moving second.[3]

To conclude, although there is considerable consensus on the assumption that the grammatical structure of utterances is generated in a piecemeal fashion, there is almost no experimental research on the precise (temporal) properties of incremental grammatical encoding. This is due to the fact that such investigations would require the simultaneous measurement of the flow of thought forming the to-be-formulated conceptual input to grammatical encoding and the development of the corresponding syntactic structures over time. Obviously, this creates a very difficult, if not unsolvable empirical problem. Therefore, we instead developed an experimental paradigm that enables us to manipulate the temporal availability of pieces of conceptual information and thus to exert experimental control on the conceptual input. With this new paradigm, we were able to show that withholding information about either the movement direction of one of two objects (Experiment 1) or its identity (Experiment 2) for a short period of time, produces systematic effects on grammatical encoding. Furthermore, at some point in planning an utterance, which presumably lies before speech onset, there seems a point of no return in grammatical encoding. More specifically, the probability that a commitment for a specific syntactic structure (e.g., an SC) is revised appears to be low, even in situations in which the conceptual information that could trigger such a revision becomes available before utterance onset.

We believe that the experimental paradigm applied in this study has potential for the future: One no longer needs to rely solely on off-line data such as speech errors. Rather, this approach enables researchers to investigate the claim that language is produced incrementally under full experimental control in online tasks.

References

De Smedt, K. (1996). Computational models of incremental grammatical encoding. In T. Dijkstra & K. De Smedt (Eds.), *Computational psycholinguistics: AI and connectionist models of human language processing.* (pp. 279-307). London, UK: Taylor & Francis.

Hoenkamp, E. (1983). *Een computermodel van de spreker: psychologische en linguistische aspecten.* [A computer model of the speaker: psychological and linguistic aspects.] Ph.D. dissertation, University of Nijmegen.

Kempen, G. (in preparation). *Human Grammatical Coding.* [working title]. Manuscript, Leiden University.

Kempen, G., & Vosse, T. (1989). Incremental syntactic tree formation in human sentence processing: A cognitive architecture based on activation decay and simulated annealing. *Connection Science, 1(3),* 273-290.

Levelt, W. (1989). *Speaking: from intention to articulation.* Cambridge, MA: MIT Press.

---

[3] At present, we are planning experiments that allow to differentiate between these two interpretations.

Note-taking as a Strategy for Learning

Susan B. Trickett (stricket@osf1.gmu.edu)
George Mason University, Fairfax, VA 22030

J. Gregory Trafton (trafton@itd.nrl.navy.mil)
NRL, Code 5513, Washington, DC 20375

Abstract

We explore the effects of taking notes on problem-solving and learning in a scientific discovery domain. Participants solved a series of five scientific reasoning problems in a computer environment in which they had access to an on-line, unstructured notepad. The results show that participants who used the notepad performed better than those who did not use it. This improvement held even when these participants no longer used the notepad on subsequent tasks. However, not all uses of the notepad were equally effective; only those that involved deeper levels of processing were related to improved performance.

Introduction

At the heart of much scientific endeavor lies the scientific method—the systematic design of experiments to test hypotheses and the interpretation of the results of the experiments to assess the validity of those hypotheses. Mastery of the "scientific method" is considered crucial to the enterprise of science, because it applies across scientific domains.

However, many studies show that although some people conform to a normative model of scientific reasoning, many do not (e.g., Klahr & Dunbar, 1988; Trickett, Trafton, & Raymond, 1998). Instead, people frequently adopt other, less optimal strategies, such as conducting experiments without a hypothesis (Klahr & Dunbar) or even generating all possible experiments (Trickett et al.). It is a consistent result of such studies that people in general—children, college students, and adults alike—find scientific reasoning tasks hard and may fail to solve them altogether (e.g. Kuhn, 1989).

What can be done to support students, both as they engage in scientific reasoning tasks (performance) and as they learn to solve them without scaffolding (learning)? One might take a "systems" approach to bolstering performance. Several options come to mind: an intelligent tutoring system, partial scaffolding, a complex help system, to name a few. However, these are expensive, complex and time-consuming to build (e.g., Anderson, Corbett, Koedinger, & Pelletier, 1995). Another, less costly option is to focus on strategies rather than systems. For example, students who are taught strategies of self-explanation when studying problem examples have been shown to outperform those who do not (Chi, de Leeuw, Chiu, & LaVancer, 1994).

One general strategy that may help students learn is taking notes. Many studies have shown that students who take notes perform better than those who do not; however other studies have found no advantage for students who take notes (see Kiewra, 1985 for a review). Taken as a whole, the literature on note-taking shows mixed results.

In reviewing the findings of the note-taking literature, Kiewra (1985) suggests that the mixed results are due to the *kind* of note-taking participants engaged in. He argues that note-taking studies should focus on levels of processing during note-taking. Different note-taking strategies may vary considerably in the level of processing participants must engage in. Sometimes participants merely copy verbatim what is read or heard, involving only transcription (e.g., Laidlaw, Skok, & McLaughlin, 1993). At other times, participants engage in "conceptual note-taking" (e.g., Rickards & McCormick, 1988) or summarizing material (e.g., King, 1992), requiring some kind of filtering. Or participants may base their notes on some form of self-questioning, which involves some level of synthesis (e.g., Spires, 1993).

These different levels of engagement—"transcription," "filtering," and "synthesis"—can be understood in terms of theories of levels of processing within the psychological literature (Craik & Lockhart, 1972). Levels of processing research suggests that participants recall items better when they process material more elaboratively. Techniques that bring about deeper processing include generating elaborations (Bobrow & Bower, 1969), and using advance organizers and generating questions (Frase, 1975).

Viewed from the levels of processing perspective, we see that more elaborative note-taking strategies lead to better performance than more shallow strategies. The "transcription" level of note-taking corresponds to levels of processing which involve no elaboration. Indeed, when participants simply copy material from a text or lecture, note-taking does not result in better learning (Laidlaw, Skok, & McLaughlin, 1993). In fact, such note-taking is no more effective than underlining (Ayer & Milson, 1993). "Conceptual note-taking" and summarizing (the "filtering" level of note-taking) involve deeper levels of processing, and are more effective than merely copying material (Rickards & McCormick, 1988; King, 1992). Note-taking that involves self-questioning or reorganizing material (the "synthesis" level) maps directly to the elaborative self-questioning strategies implicated in superior performance on memory tasks. These note-taking strategies result in better performance than either copying verbatim (Spires, 1993; Shimmerlik & Nolan, 1976) or summarizing (King, 1992). In summary, the deeper the level of processing involved in the note-taking strategy, the greater and more stable the learning that results.

Typically, studies of note-taking have been conducted in traditional learning environments, such as classrooms and lectures. Little research has been done on the effects of note-taking on synthesizing information in problem-solving environments. But combining the note-taking and depth-of-processing approaches suggests that some kinds of note-taking might indeed be helpful in problem-solving, and that the strategy of taking notes—particularly certain kinds of notes—might help performance in ways other than simply improving memory for information. If so, providing a note-taking facility in computer-learning environments might be a relatively straightforward and inexpensive means of improving performance and learning on problem-solving tasks.

One computer-based problem-solving environment in which note-taking is supported is Smithtown, an economics microworld that is considered a scientific discovery learning environment (Shute & Glaser, 1990). Empirical studies of students using Smithtown have found that successful students made more notebook entries, overall, than less effective students (Shute & Glaser, 1990). However, Smithtown's notepad is highly structured. It contains both implicit and explicit instruction that not only prompts the student to take notes but also suggests how to set about doing so.

The type of note-taking supported by Smithtown does not appear to involve the deeper levels of processing discussed above. It is not known what effect, if any, more elaborative note-taking has on problem-solving performance. Perhaps more importantly for a learning system, it is not known how using such tools affects students' learning, that is, how well students perform on subsequent tasks when they might no longer have access to such a note-taking tool.

In this paper, we present a re-analysis of 3 studies in which students solved some simple scientific reasoning tasks in an environment which provided access to a note-taking facility that allowed them to take free-form notes. This re-analysis focuses on the note-taking behaviors of the participants and the relationship of this behavior to performance on the tasks. It provides important insights into the relationships among note-taking, performance, and learning.

Method

Three separate studies were conducted to investigate different issues in scientific reasoning. Two of the studies were carried out at the same time; the third was conducted the following semester of the same school year. In all 3 studies, participants performed the same tasks and used an identical interface. Because studies of students engaged in problem-solving tasks typically involve relatively small numbers of participants, and because there were only minor procedural differences among the studies, we combine data from all 3 studies in order to increase the power of our analyses.

Participants

Participants in all three studies were George Mason University undergraduates, who received course credit for their participation. There were 30 participants in each study—a total of 90 participants (42 males and 48 females).

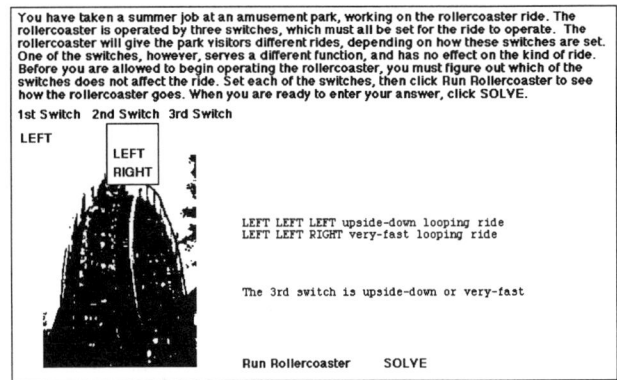

Figure 1: Screen snapshot of roller-coaster task

Materials

Five isomorphic scientific-reasoning tasks were developed, based on an adaptation of a task from Siegler and Atlas, (1976). For example, in one of the tasks, participants were told they were running a roller-coaster that was operated by three switches. The roller-coaster gave a different ride, depending on how the switches were set. All three switches had to be set for the roller-coaster to work; however, one switch did not affect the kind of ride. Participants had to identify the switch that did not affect the roller-coaster ride.

Each switch had two possible settings. Participants manipulated the setting of each switch and clicked on a "Run Roller-coaster" button to learn the kind of ride produced by that combination of settings. We refer to each new setting of the 3 variables, followed by clicking on the "Run Roller-coaster" button, as an *experiment*. Participants could run as many experiments as they wished before entering their solution. A record of each experiment and its results was displayed in a text box that remained visible throughout the task. If the text box became full, participants could use a scroll bar to view the results of their earlier experiments.

The interface also included a notepad, consisting of a blank text box, on which participants could enter information or comments if they chose. Figure 1 shows a screen snapshot of the interface for the roller-coaster task. The interface was the same for each task; only the instructions, variables, and answer were different across tasks.

Different cover stories were developed for four additional tasks. Instead of a roller-coaster, these tasks involved a musical instrument, a catapult, chemicals, or genetics (no domain knowledge was required for any of the tasks). The tasks were isomorphic in that they shared the same deep structure and could be solved by applying identical procedures (Simon & Hayes, 1976). For each task, there were three possible causal variables, each with two levels. One variable had no effect, and the goal was identify that variable.

Analysis of the problem space identified several different strategies by which the tasks could be solved. Participants could test each variable in turn by changing its setting while holding the other two constant—the optimal vary-one-thing-at-a-time strategy, or VOTAT (Tschirgi, 1980). They could identify the effect of each level of each variable (e.g., the third switch in the left position makes the roller-coaster go upside-down). They could identify the effect of each variable (e.g., the third switch makes it go upside-down or fast). They could also find two different experiments that yielded the same result and deduce that the variable whose setting was different in this pair of experiments must not affect the ride (e.g., LEFT, LEFT, LEFT: looping, upside-down ride; RIGHT, LEFT, LEFT: looping, upside-down ride).

Design

There were five different tasks, as described above. In studies 1 and 2, there were two conditions, a "same task" condition and an "isomoporh" condition. In the "same task" condition, participants were asked to solve the same task five times. In the "isomorph" condition, participants were asked to solve the series of five different, isomorphic tasks. In Study 3, all participants solved the series of five different tasks, i.e., were in the "isomorph" condition. In *both* conditions, the correct solution for a task was randomly generated for each task. In all three studies, the interface for the tasks was identical. All participants had access to the notepad, but were neither encouraged to use it nor discouraged from doing so.

Measures

Keystroke data, including entries participants made on the notepad, were collected as participants solved the tasks. In addition, in Study 1, verbal protocols were collected.

We used keystroke data to determine the accuracy of each participant's solution for each task. In order to investigate the use of the notepad, we identified three patterns of notepad use and coded each participant on each task as follows. Each task on which the participant made an entry on the notepad was coded as a *use* of the notepad. We were also interested in any possible carry-over effect of using the notepad. Consequently, after a participant used the notepad on one or more of the five tasks, each subsequent task on which they did *not* use it was coded as a *scaffolded non-use* of the notepad. All other tasks on which the participant did not use the notepad were coded as *non-use* of the notepad (i.e., tasks for which participants did not use the notepad or had not used it on a previous task). Table 1 illustrates this coding scheme.

Subj	Task	Use Notepad?	Notepad Code
100	1	No	Non-use
100	2	No	Non-use
100	3	Yes	Use
100	4	Yes	Use
100	5	No	Scaffolded non-use

Table 1: Codes for whether notepad was used

In addition to coding whether participants used the notepad, we coded *how* they used it. Several types of notepad entry were identified, as follows. Some participants identified what the variables did (IV). This category includes identifying the effect of each *level* of a variable and identifying the effect of the *whole* variable. Some participants noted that two different experiments yielded the same result (2-same). These two uses of the notepad map directly to the strategies outlined above by which participants could successfully solve the task. Some participants used the notepad to represent or re-represent the experiments they had run (RE). Some participants stated a hypothesis (SH). Entries on the notepad that engaged the task but did not fit any of these categories, that merely typed text that was visible on the screen, and/or were used by only one participant were coded as "Other" uses. Notepad entries that demonstrated more than one category of use were coded as "Mixed" entries.

There were nine entries on the notepad that did not fit any of the categories described above. These entries did not pertain to the task participants were asked to solve; for example, they were comments to the experimenter. Because there was no connection between these entries and the problem-solving task, these uses were discarded and recoded as *non-use* of the notepad. Table 2 summarizes the coding scheme for types of entry and gives examples of each category.

Type	Example
IV	Yerk-bubbly; Anjo-green; Ilop-hot Second switch affects looping or backwards
2-same	RRR & RRL same ride; LLL & LLR same ride
RE	1+1+1 = green glowing 1+2+2 = bubbly green
SH	It may be the second switch.
Other	Green-HIG BIG; Clear-BYA HYA
Mixed	Ivory piece is there or not Plastic mouthpiece makes it treble
Discard	If the answer is not chromosome 5 then these surveys should be deemed ineffective …

Table 2: Codes for type of notepad entry

Finally, among participants who used the notepad, we identified the number of experiments they had conducted when they first used the notepad for a task. This number could range from 0 (if a participant used the notepad before beginning any experimentation) to the total number of experiments run (if they used it at the end of experimentation).

Procedure

Participants were trained on the interface. They practiced designing and running experiments, viewing the results, and using the notepad. They were told that they did not have to use the notepad, but should do so if they wished.

Results and Discussion

Although all 3 studies were experimental, each was designed to explore different issues in scientific reasoning. Thus the results presented in this paper are correlational in nature and are subject to the usual caveats in interpreting correlational data. However, we find evidence within the studies that supports a broader interpretation, as we discuss below.

We first analyzed the extent to which participants used the notepad in each of the three studies. In each study, there were 30 participants, who each performed 5 tasks, i.e., there were 150 tasks or opportunities for notepad use. In Study 1, there were 26 uses of the notepad (17% of tasks); in Study 2, there were 16 uses (11% of tasks); and in study 3, there were 40 uses (27% of tasks). In all 3 studies combined, there were 82 uses of the notepad over 450 tasks (18% of tasks).

In order to ascertain that there were no quantitative differences between conditions in terms of notepad use, we performed an ANOVA comparing use of the notepad by condition (same-task, isomorph). The results of this analysis were non-significant, $F(1, 448) = 1.23$, $MSE = .54$, $p = .26$, with means of .42 and .5, respectively. This suggests that participants in the same-task condition were neither more nor less likely to use the notepad than those in the isomorph condition. Therefore, because the number of notepad uses in each individual study was rather small, we combine the results across the three studies in all subsequent analyses.

Use vs. Non-Use of Notepad

In order to investigate whether using the notepad had an effect on problem-solving performance, we performed an ANOVA comparing participants' correct solutions in *non-use* of the notepad with correct solutions in *use* of the notepad. The result was significant, $F(1, 382) = 3.99$, $MSE = .22$, $p < .05$. (Means were for .7 *non-use* and .86 for *use*). This result suggests that participants who used the notepad, regardless of how they used it, were more likely to solve the problem correctly than participants who did not use the notepad. It appears that using the notepad was associated with better performance in this type of problem-solving.

Scaffolded Non-Use of Notepad

Next, in order to explore the relationship between use of the notepad and learning, we investigated whether there was an effect of the *scaffolded non-use* of the notepad. Recall that *scaffolded non-use* of the notepad refers to those tasks on which participants did not use the notepad but had used it on prior tasks; that is, it refers to tasks on which they *no longer* used the notepad. *Scaffolded non-use* of the notepad could not occur on the first task in the series of five tasks that participants performed and was more likely to occur later in the series. Because in general performance improved as participants proceeded through the series of tasks (Trickett, Trafton, & Raymond, 1998), we performed an ANCOVA comparing correct solutions in *non-use* with correct solutions in *scaffolded non-use* of the notepad, with task as covariate. The result of this ANCOVA was significant, $F(1, 365) = 4.04$, $MSE = .21$, $p < .05$. (Means were .65 for *non-use* and .82 for *scaffolded non-use*).

The result of this analysis suggests that participants who did not use the notepad but had used it earlier were more likely to solve the problem correctly than participants who never used the notepad. Thus, using the notepad can be seen to serve a scaffolding function, as the benefit of using the notepad carries over to later tasks even when a participant no longer uses it. This result shows an association between using the notepad and improved learning on these tasks.

Issues of Self-Selection

As discussed above, these data are correlational. One plausible interpretation of the results is therefore that participants who used the notepad were simply more likely to be successful for reasons unrelated to use of the notepad. Another possibility is that some participants were inherently "note-takers" and therefore more likely to do better on the task. While we cannot entirely reject these possible explanations, we believe that there is evidence against both of them.

First, recall that there were a number of strategies by which participants could solve these problems. As an explicit hypothesis-testing strategy, the varying-one-thing-at-a-time (VOTAT) strategy is considered the optimal strategy in scientific reasoning tasks (Vollmeyer, Burns, & Holyoak, 1996). If participants were "naturally good" at scientific reasoning tasks, we would expect them to use the VOTAT strategy. Contrary to this expectation, however, in Study 1, at least, there were very few instances of VOTAT (Trickett, Trafton, & Raymond, 1998)[1]. Instead, in that study, accuracy was highly correlated with systematicity.[2]

Second, given that systematicity was strongly related to accuracy on these tasks, we might expect participants using the notepad to be more systematic than those who did not. However, this was not the case; there was no significant correlation between systematicity and notepad use in Study 1. Thus we find two distinct strategies (systematicity and notepad use), both of which correlate with successful performance, but which do not correlate with each other. This result, coupled with the general lack of the VOTAT strategy, suggests that notepad users were not just better students.

Third, if some participants were simply more inclined to take notes, we would expect the majority of entries on the notepad to occur very early in the problem-solving process, that is, after participants had run only a very few experiments. However, again, this was not the case. Across all 5 tasks, only 16 (19.5%) of the 82 notepad entries were made in the initial stages of problem-solving (after 0 to 2 experiments). Participants used the notepad for the first time in the middle stages of problem-solving (after 3 to 5 experiments) on 30 tasks (36.5%), and in the late stages of problem-solving (after 6 or more experiments) on 36 tasks (44%). Table 3 shows a complete breakdown of these results.

# experiments	# first uses	% first uses
0	7	8.5%

---

[1] Use of the VOTAT strategy could only be determined in Study 1, because verbal protocols are needed in order to ascertain that this strategy is being used. We do not expect, however, that use of the VOTAT strategy would have been greater in studies 2 and 3; note, for example, that there were no uses of VOTAT recorded on the notepad in any of the studies, whereas all the other strategies were represented on the notepad.

[2] Systematicity refers to the entire set of experiments a participant generated for a task. Participants' data collection was coded as systematic if at least 75% of their experiments conformed to a discernible pattern.

1	5	6%
2	4	5%
3	13	16%
4	10	12%
5	7	8.5%
6	7	8.5%
7	5	6%
8	19	19%
9+	5	6%

Table 3: Experiments run at first use of notepad

These results show that by far the majority of participants did not approach these tasks already equipped with the strategy of using the notepad. Rather, they appear to have run several experiments and *then* turned to the notepad. Particularly interesting is the fact that the largest number of first uses occurred after participants had run 8 experiments. There were 8 possible different experiments that could be run. 19% of first notepad uses occurred when participants had run 8 experiments. It seems likely that, having exhausted all possibilities for collecting data, participants used the notepad to try to make sense of these data. Certainly, it does not appear that participants were *a priori* inclined to take notes.

How the Notepad Was Used

The analyses described above show that participants were more accurate if they used the notepad and that this advantage was maintained even if they subsequently stopped using it. However, the analyses do not differentiate among different types of notepad use. They do not address the question of whether some uses of the notepad were more effective than other uses. To investigate this question, we examined type of notepad entry in relation to the accuracy of the solution.

Because the number of notepad entries for some categories was quite small, it was not appropriate to perform statistical tests to determine the effectiveness of each of the different ways of using the notepad. Instead, we report the percentages of tasks with correct solutions for each type of entry. Table 4 summarizes the results of this analysis. The findings show that if participants did not use the notepad, they were correct 65% of the time (baseline). Four types of entry on the notepad (identify variable, note 2-same, re-represent experiments, and state hypothesis) were at least 18% above baseline. Two types of entry, "Other" and "Mixed," were below baseline.

Entry type	#correct	total	% correct
None	197	302	65%
Identify effect of variable	35	38	92%
Note 2 results the same	5	6	83%
Re-represent experiments	10	11	91%
State hypothesis	7	8	88%
Other	1	10	10%

Mixed	5	9	56%

Table 4: Percent correct per type of entry

Thus it would appear that use of the notepad *per se* is not linked to more successful performance. Instead, some uses of the notepad seem to be more helpful than others. Specifically, using the notepad to identify the effect of a variable, note two identical results, re-represent experiments, or state a hypothesis seem related to successful performance. We thus reclassified these four uses as "good" uses of the notepad. On the other hand, using the notepad for uses coded "Other" or mixing uses of the notepad seems to be associated with unsuccessful performance. We thus reclassified these two uses as "poor" uses of the notepad.

In order to test whether "good" uses of the notepad were more likely to lead to successful performance than "poor" uses, we conducted a Fisher's exact test. In doing so, we violate one assumption of this test, namely that the observations were independent (they were not independent, because each participant performed 5 tasks). However, such violations usually lead to a more conservative test. The results of the Fisher's exact test (two-tailed) were significant, $p < .001$.

Thus it seems that the four "good" uses are more likely to be associated with successful performance than the two "poor" uses of the notepad. Why might this be the case? We believe it is because the four good uses all involve a deeper level of processing. These helpful uses—identifying what a variable does, identifying 2 identical results, re-representing experiments and stating a hypothesis—all require reorganization or synthesis, both of which involve deeper processing than simply copying what is on the screen. This deeper level of processing would not be involved in the "Other" uses of the notepad, nor perhaps in the "Mixed" uses, depending on the combination of uses involved. Furthermore, identifying what a variable does and identifying 2 identical results correspond to strategies by which the tasks could be solved. Possibly, if participants were having difficulty solving the task, using the notepad might have been instrumental in helping them to develop a good strategy for the tasks.

Concerning the "poor" uses, several observations can be made. First, recall that the "Other" category included instances where the participant just re-typed some text that was visible on the screen. The low success-rate for participants using the notepad for "Other" uses (10%) suggests that using the notepad in ways that involve shallow processing did not help. Second, some "Other" uses of the notepad were idiosyncratic, in that they could not be categorized according to the general types of entry identified among the majority of participants. Thus it may be that these "Other" uses indicate confusion on the part of the participants. It seems plausible to think that they turned to the notepad in an effort to do something to move their problem-solving forward. However, because they had no clear idea about how to use the notepad effectively, it was of no benefit in these cases.

Finally, in 8 out of the 9 cases of "Mixed" entry, one of those uses was "Other." On 5 of those 8 tasks, the "Other" use came first and was followed by the helpful use. Participants were correct on 4 of those 5 tasks. If, the "Other" use is evidence of floundering, as suggested above, this shift in type of notepad entry might be an indication of a transition point in their problem-solving. Although there are too few instances to allow us to draw firm conclusions, the data and the trace of problem-solving activity provided by the notepad entries suggest that these participants were able to recover from their confusion and move to a good solution strategy.

General Discussion

The results described above show that participants who used the notepad were more likely in general to solve the problem correctly than those who did not use it. Moreover, this advantage was maintained even if participants stopped using the notepad on subsequent tasks. Although these data are correlational, there is some evidence to suggest that participants who used the notepad were neither "better" students nor for some reason intrinsically inclined to use the notepad. In addition, only uses of the notepad that involve deeper levels of processing were associated with successful performance; those that involve shallow levels were not.

These findings extend the current research on note-taking during problem-solving in a number of ways. First, the general finding that participants who used the notepad were more successful than those who did not suggests that taking notes can indeed be a helpful strategy in a problem-solving domain. It appears that the benefits of taking notes extend beyond boosting the simple recall of learned material, to helping students make sense of data they have generated and possibly to helping them to develop good problem-solving strategies.

Second, we suggest that the reason some uses of the notepad were related to better performance while others were not is that the "good" uses engaged the participants in deeper, more elaborative processing than the poor uses. We further specify those kinds of note-taking that involve deeper processing.

Third, the finding that the benefit of using the notepad carried over to tasks on which participants no longer used it suggests that taking notes can actually help students *learn*. This finding further supports our interpretation that using the notepad helped participants develop robust problem-solving strategies that they continued to apply across tasks. Having developed a sound strategy, participants most likely no longer needed the scaffolding provided by the notepad but could continue their problem-solving efforts independently.

Taken together, the results suggest that having students take notes that involve deeper processing as they solve scientific reasoning tasks is one means by which their performance might be improved. Providing a note-taking facility for students engaged in these types of problems is an inexpensive and comparatively straightforward form of scaffolding that may have the further advantage of helping students learn general strategies by which these problems can be solved.

As we have mentioned, these conclusions remain tentative, because of the correlational nature of the data. We cannot decisively reject other possible explanations, such as that the notepad users were "better" problem-solvers, although we find evidence to suggest that this is not the case. Clearly, our next step is to test these results experimentally, by manipulating use of the notepad and providing the appropriate controls. This experiment is currently being run.

Acknowledgments

This research was supported in part by a student fellowship from George Mason University to the first author and by grant number 55-7294-A8 from the Office of Naval Research to the Naval Research Laboratory. We wish to thank Erik M. Altmann, Robert W. Holt, Irvin R. Katz, Audrey W. Lipps, Sheryl L. Miller, Llelyn D. Saner, and Christian D. Schunn for their comments. We give special thanks to Paula D. Raymond for her continuing comments on this research.

References

Anderson, J. R., Corbett, A. T., Koedinger, K. R. & Pelletier, R. (1995). Cognitive tutors: Lessons learned. *Journal of the Learning Sciences, 4*(2), 167-207.

Ayer, W. W., & Milson, J. L. (1993). The effect of notetaking and underlining on achievement in middle school life science. *Journal of Instructional Psychology,*] *20*(2), 91-95.

Bobrow, S. A. , & Bower, G. H. (1969). Comprehension and recall of sentences. *Journal of Experimental Psychology. 80*(3, Pt. 1), 455-461.

Chi, M. T. H., de Leeuw, N., Chiu, M., & LaVancher, C. (1994). Eliciting self-explanations improves understanding. *Cognitive Science, 18,* 439-477.

Craik, F. I. M., & Lockhart, R. S. (1972). Levels of processing: A framework for memory research. *Journal of Verbal Learning and Verbal Behavior, 11,* 671-684.

Frase, L.T. (1975). Prose processing. In G. H. Bower (Ed.), *The psychology of learning and motivation,* (Vol. 9). New York: Academic Press.

Kiewra, K. A. (1985). Investigating notetaking and review: A depth of processing alternative. *Educational Psychologist, 20*(1), 23-32.

King, A. (1992). Comparison of self-questioning, summarizing, and notetaking-review as strategies for learning from lectures. *American Educational research Journal, 29*(2), 303-323.

Klahr, D., & Dunbar, K. (1988). Dual search space during scientific reasoning. *Cognitive Science, 12,* 1-48.

Kuhn, D. (1989). Children and adults as intuitive scientists. *Psychological Review, 96*(4), 674-689.

Laidlaw, E. N. Skok, R. L., & McLaughlin, T. F. The effects of notetaking and self-questioning on quiz performance. *Science & Education.. 77*(1), 75-82.

Rickards, J. P. , & McCormick, C. B. (1988). Effect of interspersed conceptual prequestions on note-taking in listening comprehension. *Journal of Educational Psychology, 80*(40), 592-594.

Shimmerlik, S. M., & Nolan, J. D. (1976). Reorganization and the recall of prose. *Journal of Educational Psychology, 68*(6), 779-786.

747

Shute, V. J., & Glaser, R. (1990). A large-scale evaluation of an intelligent discover world: Smithtown. *Interactive Learning Environments, 1*, 51-77.

Siegler, R. S. & Atlas, M. (1976). Acquisition of formal scientific reasoning by 10- and 13-year-olds: Detecting interactive patterns in data. *Journal of Educational Psychology, 68*(3), 360-370.

Simon, H. A., & Hayes, J. R. (1976). The understanding process: problem isomorphs. *Cognitive Psychology, 8,* 165-190.

Spires, H. A. (1993). Learning from a lecture: Effects of comprehension monitoring. *Reading Research & Instruction, 32*(2(, 19-30.

Trickett, S. B., Trafton, J. G., & Raymond, P. D. (1998). Exploration in the experiment space: The relationship between systematicity and performance. *Proceedings of the 20th Annual Meeting of the Cognitive Science Society* (pp. 1067-1072). Mahwah, NJ: Erlbaum.

Tschirgi, J. E. (1980). Sensible reasoning: A hypothesis about hypotheses. *Child Development, 51,* 1-10.

Vollmeyer, R., Burns, B. D. & Holyoak, K. J. (1996). The impact of goal specificity on strategy use and the acquisition of problem structure. *Cognitive Science, 20,* 75-100.

The Nominal Competitor Effect:
When One Name Is Better Than Two.

Tim Valentine (T.Valentine@gold.ac.uk)
Jarrod Hollis (J.Hollis@gold.ac.uk)
and
Viv Moore (V.Moore@gold.ac.uk)
Goldsmiths College, University of London, New Cross, London. SE14 6NW. UK

Abstract

Brédart, Valentine, Calder and Gassi (1995) described an interactive activation and competition (IAC) model in which the lexical representations of people's names have inhibitory connections between each other, but do not receive inhibition from the representation of biographical properties. The model predicts that people would be slower to name a celebrity for whom two names are equally available than they would be to name an equally familiar celebrity for whom only one name is available. However, naming should only be slowed by competition from a competing *name*; a highly available biographical property should not increase face naming latency. These predictions were confirmed in a simulation of the model. The effect is referred to as the *nominal competitor effect*. Experiment 1 showed that participants who had practiced naming actors using both the actor's name (e.g. John Cleese) and the character's name (e.g. Basil Fawlty) were slower to produce the actor's name at test than were participants who had practiced producing only the actor's name. However, practice in naming the relevant television series (e.g. Fawlty Towers) did not inhibit subsequent production of the actor's name. In contrast to the semantic competitor effect in picture naming, the effect reported here was found to be long-lasting (Experiment 2).

Introduction

A challenge addressed in the face processing literature has been to account for findings that the recall of identity-specific semantic information (biographic information; e.g. occupation) is far more robust than recall of a familiar person's name. In a diary study Young, Hay and Ellis (1985) found many errors when people could recall biographical details of somebody they encountered but were unable to recall the person's name. However, they found *no* occurrences of successful name retrieval when the diarist was unable to recall *any* identity-specific semantics. Furthermore, recall of biographical information is faster than recall of people's names (Young, Ellis and Flude, 1988). Any model of the normal process of face naming must be able to account for the contingency of name retrieval upon retrieval of biographical information and the longer latency of retrieval of name information than of identity-specific semantic information.

Burton and Bruce (1992) introduced an interactive activation and competition (IAC) model of face naming in which faces, names and biographical properties of familiar people were represented by localist representations. In the model biographical properties and names are represented within a single pool of units referred to as semantic information units (SIUs). It is assumed that when a familiar face is seen activation is passed via a face recognition unit (FRU) to a person identity node (PIN). The PIN is a localist representation of a specific familiar person activated by a stimulus from any domain (e.g. by seeing the face or name of that individual, or by hearing their name or their voice). A person is recognised when the activation of the relevant PIN passes a threshold. Familiarity decisions to a face or name are assumed to be based on the activation of PINs.

The structure of the Burton and Bruce (1992) model appears to allow for names to be retrieved as rapidly as other semantic information and for recall of a name not to be contingent on retrieval of other semantic properties. However, this is not the case because biographical properties are often shared between individuals but people's full names are usually unique to one person. All links between units in different pools are assumed to be excitatory and bi-directional. All units within a pool are connected by inhibitory links of equal connection strength to all other units within the pool. As the activation of a PIN rises it sends activation to the SIUs to which it is connected. For example, the PIN for Bill Clinton would activate units representing the fact that he is a politician, American, president, a Democrat etc. As between-pool links are bi-directional activation will be passed back to units in the PIN layer which are connected to any of these properties. All of the SIUs which have become activated will pass activation back to the Bill Clinton's PIN. This mutual support speeds up the rise in activation of the SIUs that represent properties which are shared by other familiar people. In contrast the unit representing Bill Clinton's name does not benefit from any such mutual support because his name is unique and therefore connected only to Bill Clinton's PIN. (i.e. It is assumed that most people know only one 'Bill Clinton'.)

Burton and Bruce showed that in their model 'unique' SIUs reach their threshold of activation more slowly than units representing shared properties. They propose that this effect is the basis of slower retrieval of names than of biographical information. Furthermore, attenuation of the links between PINs and SIUs will cause the activation of unique SIUs to fail to reach threshold before the SIUs representing shared biographical properties. If attenuation of

PIN - SIU links is assumed to model difficulties in accessing identity-specific semantic information (either in brain-injured patients or temporary failures in neurologically intact individuals), the 'uniqueness' of names explains why biographical information may be recalled when a person's name is inaccessible, but it is never the case that a person's name can be recalled in the absence of any biographical details.

Some difficulties for the Burton and Bruce (1992) model have recently become apparent. The model predicts that knowing many biographical details about somebody should increase the time taken to name their face. SIUs that become activated as the PIN is activated will inhibit the activation of the SIU that represents the person's name. Therefore, the name will be recalled more quickly if few biographical details are known. Brédart et al. (1995) tested this prediction empirically. Participants named two groups of celebrities' faces. The groups were matched on rated familiarity but differed in the number of biographical facts about the celebrities that could be produced. Brédart et al. found that the faces of celebrities about whom many facts were known were named more quickly than were the celebrities about whom few facts were known.

Another prediction of the Burton and Bruce model is that patients who cannot name famous faces should be unable to retrieve other identity-specific semantic information that is unique to the individual. This prediction arises because the model makes no distinction at all between unique biographical information and a person's name. However, two patients have been reported who can retrieve unique information about celebrities they cannot name (Hanley, 1995; Harris and Kay, 1995). Patient BG reported by Harris and Kay (1995) was able to produce the catch-phrases of several celebrities she could not name.

Brédart et al. (1995) noted that the Burton and Bruce (1992) model is incompatible with models of speech production because it does not draw a distinction between non-lexical conceptual representations and lexical representations. This distinction is found in most models of speech production (e.g. Butterworth, 1989; Dell, 1986; Garrett, 1980; Harley, 1993; Kempen & Huijbers, 1983; Levelt, 1989; Levelt, Roelofs & Meyer, in press; Roelofs, 1992). If SIUs are assumed to represent non-lexical conceptual knowledge, what is the conceptual knowledge of a person's name? Proper names are, by definition, lexical units. If this is the case either all SIUs are lexical units, which implies the representations in semantic memory are lexical, or a justification needs to be made for storing lexical and conceptual units in a common pool.

Brédart et al. (1995) showed that all of these issues can be addressed by slightly modifying the architecture of the interactive activation model. First, they argued that names must be represented in a pool of Lexical Output Units (LOUs) which are separate from the semantic information units that represent biographical information. These pools represent different types of information: LOUs are lexical units, SIUs represent non-lexical, conceptual knowledge.

LOUs represent the first stage of lexical access (equivalent to the lexical representations termed *lemmas* by Levelt, 1989). Second, different types of biographical information are represented in separate pools of SIUs. Thus there might be a pool of SIUs representing different nationalities; another pool may represent political opinions and yet another may represent occupations. All other aspects of the architecture and the processing assumptions are the same as the Burton and Bruce (1992) IAC model. All units receive inhibitory connections of the same strength from all other units within the same pool. All connections between units in different pools are excitatory and bi-directional with a single uniform weight (except where stated in the simulation reported below).

Brédart et al. (1995) demonstrated that the revised architecture showed the same properties which Burton and Bruce (1992) had set out to explain: no naming without recall of some biographical details was possible and retrieval of names was slower than recall of biographical details. Although the structure of the model allows for the possibility that names might be recalled in the absence of recall of any biographical details, Brédart et al. showed that if PIN - SIU links were attenuated sufficiently to prevent any SIU reaching threshold, no lexical output unit could reach threshold. Although the PIN - LOU links were intact, attenuation of the PIN - SIU links depressed the activity in the network to the extent that LOUs could not exceed their threshold. The effect arises because the activity of PINs was receiving less support from feedback from SIUs. The revised model relied on the uniqueness of names (i.e. the one-to-one mapping between PINs and LOUs) to account for the slower retrieval of names than of semantic properties. Therefore, the account of this effect is identical to that proposed by Burton and Bruce (1992).

Furthermore, Brédart et al. (1995) showed that the revised model correctly simulated the finding that knowing many biographical details about a person reduces the time taken to recall their name: LOUs associated with PINs connected to many SIUs reached their threshold activation faster than LOUs associated with PINs connected to few SIUs. Another advantage of the revised architecture is that it allows for the possibility that patients with an anomia for people's names will be able to recall unique biographical information.

The aim of the present study is to provide a further empirical test of the Brédart et al. model. More specifically, we describe a direct test of the assertion that people's names are represented separately from other biographical details.

The first experiment reported makes use of actors who could be identified by either their own name or the name of a character with whom they have become closely associated. If both names are equally available (i.e. the links from the PIN have equal connection strengths), the Brédart et al. model predicts that participants should be slower to name the actor's face than they would be if only one name was available (*the nominal competitor effect*). If two LOUs receive activation from a PIN they will start to mutually inhibit one another within the LOU pool. This mutual inhibition would slow down retrieval of the name. If only one LOU received activation from a PIN, there would be no inhibition from other LOUs, so name retrieval would be relatively fast. It is important to note that the maximum effect of inhibition will be observed when the two competing names are equally available. If one LOU becomes more activated than the other it will send more inhibition to the competing LOU than it receives from its competitor. The effect of inhibition will be to increase the difference in activation between the two LOUs so that the winner will rapidly suppress the activation of the other. In order to make the two names equally available one group of participants were given extensive and equal practice producing both the actor's and the character's name to a small set of faces of famous actors. Another group of participants practiced producing only the actor's name to provide a baseline condition for when only one name is much more available than the other. The model also makes the clear prediction that practice accessing a biographical property cannot inhibit recall of a person's name. Biographical properties and people's names are stored in separate pools and therefore do not share any inhibitory connection.

Before reporting the experiment, we demonstrate the behaviour of the Brédart *et al.* model by simulation.

Simulation 1

The architecture of the network was the same as the 90-unit network described by Brédart *et al.* (1995) and was implemented using the McClelland and Rumelhart (1988) software package. There were 24 person identity nodes, 24 lexical output units and 42 semantic information units arranged in 4 separate pools. The pools might represent biographical properties such as occupation, nationality, political opinion etc. One pool of SIUs included 24 units, each was connected to a different PIN. There were 8 units in another SIU pool and 6 units in third, each unit was connected to between 1 and 4 PINs. Finally, there was a pool of 4 units each connected to between 2 and 6 PINs. The values of the model parameters are listed in the Appendix. In the Brédart *et al.* simulations each PIN had been connected to one LOU (or name). These connections were altered such that 8 PINs were connected to 2 LOUs, 8 PINs were to 1 LOU and 8 PINs were not connected to any LOUs. All links between units in different pools were set to 1.0 unless otherwise stated. The effect of practicing a name was simulated by increasing the weight of the PIN - LOU link from 1.0 to 3.0. Following the procedure adopted by Brédart *et al.* the threshold value of the LOUs was set at 80% of the mean of their maximum level of activation. The threshold in this simulation was 0.52. The simulation focused on the 8 PINs that were connected to two LOUs. All simulations were repeated for all of these 8 PINs, by setting the initial activation of each PIN to 1.0 in turn.

The mean rise in activation of LOUs was recorded as a function of the number of processing cycles under the following experimental conditions: 1) The activation of an LOU connected to a PIN by a link with connection strength 3.0. The PIN was connected to another LOU with connection strength 1.0. This simulated the condition in which participants practiced producing the actor's name only, so that the desired name was much more available than the competitor name: 2) This simulation was identical to (a) except that the connection strength to the 'competitor' LOU was increased to 3.0. This simulated practice in producing both the actor's and the character's name: 3) This simulation was identical to (a) except that the weight of a link from the relevant PIN to an SIU in a pool of non-unique biographical properties was increased from 1.0 to 3.0. The SIUs were connected to a mean of 2.7 PINS. This simulated practice in producing the actor's name and the name of the TV series or film.

It is assumed that naming latency is simulated in an IAC model by a monotonic function of the number of cycles taken for the relevant LOU to reach threshold. The rise in activation of the LOU was slower and the maximum level of activation (MLA) achieved in condition 2 (threshold reached after 17.0 cycles; MLA = 0.62) than in conditions 1 and 3 (threshold reached after 15.7 and 15.0 cycles respectively; MLA = 0.67 for both conditions). Note that there is no random factor in the interactive activation model. Thus, simulation 1 confirmed the prediction derived from the model. The aim of Experiment 1 was to provide an empirical test of the predictions.

Experiment 1

Method

Participants: Fifty-four students took part (34 male and 20 female).

Stimuli: The stimuli consisted of a set of twelve pictures of famous actors playing characters with which they have become strongly associated Therefore the pictures could be named with either the actor's name (e.g. John Cleese) or with the character's name (e.g. Basil Faulty).

The pictures were digitized as 16 grey level monochrome images 256 x 256 pixels. They showed head and shoulders views of the actor in approximately a full-face view.

Apparatus: Images were displayed in the centre of a PC screen using the MEL software package at a screen resolution of 640 x 480 pixels on a 14" screen. This software controlled the display of stimuli and logged vocal naming latency (with millisecond accuracy) from a voice key. A throat microphone was used to detect the participants' vocal responses. The experimenter coded each response recorded as acceptable or unacceptable. This enabled data from trials on which there was an accidental triggering of the voice key, a failure to trigger the voice key or an incorrect naming response to be excluded from the analysis.

Design: The experiment consisted of a between-participants design incorporating two phases; a practice phase and a test phase. There was one independent variable; the nature of the naming tasks in which the participant was practiced prior to the test phase. There were three levels of this variable. Either participants only practiced naming the pictures with the actor's name (condition 1); or they practiced naming the picture with both the actor's and the character's name (condition 2); or they practiced producing the actor's name and naming the TV series or film in which the character appeared (condition 3). In the test phase all participants were required to produce the actor's name only. The dependent variable was the mean naming latency of correct responses in the test phase.

Procedure: PRACTICE PHASE. The procedure for the practice phase of the experimental condition in which participants practised producing the actor's name only was as follows. Instructions were presented on the computer screen. Participants were informed that the face of a famous actor would appear in the centre of the PC screen after a brief warning tone. Their task was to say aloud the name of the actor as quickly as possible. Participants were instructed to give the full name of the actor. The image disappeared when the voice key was triggered by the participant's response. If the participant did not know the name or gave an incorrect response, the experimenter said the correct name and the participant was required to repeat it. The experimenter entered a response using the keyboard to code each response as correct or incorrect. Entering this response triggered the next trial. The warning tone at the start of each trial lasted for 250ms. There was a 500ms interval between the end of the tone and the display of the stimulus face. All 12 stimuli were presented once in a random order before any stimulus was repeated. The stimulus set was presented 4 times in a different random order each time. Although the presentation was organised into four 'blocks', the stimuli were presented as one list of 48 trials in which each stimulus appeared four times.

In the two other experimental conditions the procedure differed in only two respects. The group of participants who practiced producing both the actor's name and the character's name received 48 trials practice in producing the actor's name as described above. These trials were mixed in a pseudo-random order with 48 trials practice in producing the character's name, making a total of 96 trials. Each stimulus was presented once for each naming response before any of the stimuli were repeated. Immediately after the warning tone, the instruction 'ACTOR'S NAME' or 'CHARACTER'S NAME' was presented in the centre of the screen for a 2 second interval. The stimulus face then replaced the instruction.

The procedure for the participants who practiced production of the actor's name and the name of the TV series or film was the same as described above for the 'actor's name and character's name' condition, except that the instructions were modified appropriately. 'TV SERIES OR FILM' was presented in the centre of the screen in the place of 'CHARACTER'S NAME'.

TEST PHASE: The procedure for the test phase was identical for all participants. Instructions were presented on the computer screen. Each of the twelve stimulus faces was presented once in a random order. Participants were required to give the actor's name only as quickly as possible. In the test phase the latency of the vocal response (the time between the presentation of the stimulus and the triggering of the voice key) were logged by the computer. The logging of the accuracy of the response by the experimenter and the triggering of the presentation of the next image was as described for the practice phase.

Results

The accuracy of participants in producing the correct actor's name in the test phase was close to ceiling. The number of correct responses was as follows (maximum = 12): 11.67 (Actor's name only); 10.94 (Actor's and character's name) and 11.17 (actor's name and TV series or film). In view of the restricted variability in the data no further analysis was carried out.

The mean naming latency of correct responses in each experimental condition were as follows (standard errors are given in parentheses): actor's name only = 969ms (41ms); actor's and character's name = 1312ms (51ms) and actor's name and TV series or film = 939ms (35ms). These data were subjected to two separate one-way analyses of variance taking participants (F_1) or items (F_2) as the random factor, with repeated measures in the items analysis. The main effect of condition was significant $F_1(2,51)=9.66$, p<.001; $F_2(2,22)=32.03$, p<.001 Planned comparisons showed that naming latencies in the test phase following practice with the actor's name and the character's name were significantly slower than latencies following either practice with the actor's name alone $t_1(51)=3.64$, p<.002; $t_2(11)=5.15$, p<.001 or with the actor's name and the TV series or film $t_1(51)=3.95$, p<.001; $t_2(11)=7.14$, p<.001. Naming latencies in the test phase following practice with actors name and the TV series or film did not differ from naming latencies following practice with the actors name alone $t_1(51)=0.30$; $t_2(11)=1.01$.

Discussion

Practice naming the face of an actor using both the actor's name and the name of the character who (s)he portrays increased the time taken to subsequently name the face using the actor's name compared to the naming latency following practice in producing the actor's name only. Practice in producing the name of the TV series did not slow the speed of face naming. The predictions derived from the Brédart et al. (1995) model, were clearly supported by the data.

A comparison between priming a competitor name in picture naming and face naming.

Wheeldon and Monsell (1994) found that priming a semantic competitor increased the time taken to name a picture. For example, having recently produced the word 'shark' in response to a definition increased the naming latency of a picture of a whale. The effect was found to be short lived: it was reliable after an interval of approximately 12 seconds during which two trials had intervened but no effect was found after 4 - 8 minutes during which 38 - 100 names were produced. This is much shorter-lived than the facilitation that results from repetition priming. They suggest that the difference in time scale can be explained by attributing repetition priming to an increase in connection strength between the lemma and phonology but the semantic competitor effect to greater availability of the lemma of a semantic competitor following recent activation.

The results reported above raise the following question: Is the inhibition from having practiced an alternative name of a celebrity analogous to the semantic competitor effect in speech production? The present procedure is somewhat different to that used by Wheeldon and Monsell (1994). First, in Experiment 1 the participants were highly practiced in naming a small set of items. This was not the case in Wheeldon and Momsell's experiments. Second, the semantic competitor effect arises from previously naming a closely related but *different* concept (e.g. whale vs. shark), but the nominal competitor effect arises from having named the *same* person with a different name. Naming a whale as a shark is an error, but naming John Cleese as Basil Fawlty is not incorrect.

If the mechanism for the nominal competitor effect is the same as that which produces the semantic competitor effect both phenomena should be short-lived. In contrast, the Brédart *et al.* model uses the same mechanism to explain the nominal competitor effect and repetition priming. Therefore, this model predicts that the nominal competitor effect will be long-lived.

Experiment 2

In Experiment 1 the test phase followed the practice phase immediately. The number of intervening items between the last practice trial and the test trial of an item was between 0 and 12. The delay was between approximately 30 seconds and 2 minutes. These parameters include the number of intervening items and the delay after which the semantic competitor effect has been observed. Experiment 2 replicated conditions 1 and 2 of Experiment 1 but the test phase following after a minimum delay of 5 minutes and 46 intervening face naming trials.

Method

Participants: Twenty-four students (23 female and 1 male) took part for a course credit.

Stimuli and Apparatus: The stimuli and apparatus were the same as described above for Experiment 1.

Design: The design was the same as Experiment 1 except that there were only two experimental conditions: Either participants only practiced naming the pictures with the actor's name (condition 1) or they practiced naming the picture with both the actor's and the character's name (condition 2).

Procedure: The procedure was identical to that for conditions 1 and 2 of experiment 1, except for the intervening period between the practice phase and the test phase. After completing the practice phase, participants carried out a face naming task in which 46 famous faces were presented on the same PC screen for the participant to name as quickly as possible This task took approximately

five minutes. The interval between the final practice trial and the first trial of the test phase was a minimum of 5 minutes and a maximum of 10 minutes.

Results

The number of correct responses in the test phase was as follows (maximum = 12): 10.17 (Actor's name only); 9.58 (Actor's and character's name). An independent t-test taking participants as the random factor and a related t-test taking items as the random factor confirmed that there was no reliable difference in the accuracy of face naming between the two experimental conditions (both t's < 1).

The mean naming latency of correct responses in each condition were as follows (standard errors are given in parentheses): actor's name only = 1178ms (64ms); actor's and character's name = 1920ms (268ms). An independent t-test (not assuming equal variances) was used to analyse the data by participant and a related t-tests were used for the by-items analysis. Naming latencies in the test phase following practice with the actor's name and the character's name were significantly slower than latencies following practice with the actor's name alone $t_1(12)=2.70$, $p<.02$; $t_2(11)=4.02$, $p<.01$.

Discussion

The results of Experiment 2 show that the nominal competitor effect survives over 46 intervening trials and a period of approximately 5 minutes. Therefore the effect is longer lasting than the semantic competitor effect. These data suggest different mechanisms underlie the two effects. The long-lasting nature of the nominal competitor effect suggests that it could arise from the same mechanism as that which gives rise to repetition priming. Therefore, the data are consistent with the interpretation in terms of an increase in the weight of the PIN – lexical unit link postulated by the Brédart *et al.* (1995) model.

Processing assumptions required to model inhibition of naming by priming a competitor name.

The architecture of the Brédart *et al.* model is broadly compatible with current models of speech production, in particular it has similarities with spreading activation models such as Dell (1986), Harley (1993), Roloefs (1992) and Levelt, Roloefs & Meyer (in press). However, only the Harley (1993) model and the Brédart *et al.* model include inhibitory connections in the lemma or morphological layer. The simulation reported above demonstrated that the Brédart *et al.* model can simulate the experimental result successfully. The question, therefore, arises of whether the models that do not postulate inhibitory links between lemmas can simulate the slowing of face naming by increased availability of a competitor. Wheeldon and Monsell (1994) point out that such slowing can be produced only in models which have either inhibitory links between lemmas or which have a differential activity criterion for lemma selection. Roloefs (1992) and Levelt *et al.* (in press) use a ratio rule which takes the activity of other lemmas into account. As practicing the character's name will result in the lemma (or LOU) for a competing name becoming more activated than otherwise would be the case, these models can simulate the experimental effect. However, Dell's (1986) model neither has inhibitory connections between lemmas nor uses a differential criterion in lemma selection. In Dell's model the most activated lemma is selected after a period determined by the rate of speech. In conclusion, Dell's model does not have a mechanism that could account for the nominal competitor effect.

How do speech production models account for the lack of inhibition following naming of a TV series? The Brédart *et al.* model does not include lemmas for biographical properties. However, a simulation of an extended network, in which lemmas for biographical properties were included in a separate pool produced that same result as Simulation 1 above. In order to account for the data, the selection rule in the speech production models must be restricted to the relative activation of people's names. Names of TV series or films must not enter the selection process either because they are not in the response set used (Levelt *et al.*, in press) or because the selection can be limited to only lemmas marked as representing only the class of people's names.

References

Brédart, S., Valentine, T., Calder, A. & Gassi, L. (1995). An interactive activation model of face naming. *Quarterly Journal of Experimental Psychology*, 48A, 466-486.

Burton, A.M. & Bruce, V. (1992). I recognize your face but I can't remember your name: a simple explanation? *British Journal of Psychology*, 83, 45-60.

Butterworth, B. (1989). Lexical access in speech production. In: Marslen-Wilson, W. (ed.) *Lexical representations and process.* Cambridge, MA: MIT Press.

Dell, G. S. (1986). A spreading activation theory of retrieval in sentence production. *Psychological Review,* 93, 283-321.

Garrett, M. F. (1980). Levels of processing in sentence production. In: Butterworth, B. (ed.) *Language production, volume 1, Speech and talk.* London: Academic Press.

Hanley, J. R. (1995). Are names difficult to recall because they are unique? A case study of a patient with anomia. *Quarterly Journal of Experimental Psychology*, 48A, 487-506.

Harley, T. A. (1993). Phonological activation of semantic competitors during lexical access in speech production. *Language and Cognitive Processes,* 8, 291-309.

Harris, D. M. & Kay, J. (1995). I recognize your face but I can't remember your name: Is it because names are unique? *British Journal of Psychology*, 86, 345-358.

Kempen, G. & Huijbers, P. (1983). The lexicalisation process in sentence production and naming: Indirect election of words. *Cognition,* 14, 185-209.

Levelt, W. J. M. (1989). *Speaking: From intention to articulation.* Cambridge, MA: MIT Press.

Levelt, W.J.M., Roelofs, A., & Meyer, A.S. (in press). A theory of lexical access in speech production. *Behavioral and Brain Sciences*

McClelland, J.L., & Rumelhart, D.E. (1988). *Explorations in parallel distributed processing: A handbook of models, programs and exercises.* Cambridge, MA: MIT Press.

Roelofs, A. (1992). A spreading-activation theory of lemma retrieval in speaking. *Cognition,* 42, 107-142.

Wheeldon, L. R. & Monsell, S. (1994). Inhibition of spoken word production by priming a semantic conpetitor. *Journal of Memory and Language,* 33, 332-356.

Young, A.W., Ellis, A.W. & Flude, M. (1988). Accessing stored information about familiar people. *Psychological Research,* 50, 111-115.

Young, A.W., Hay, D.C. & Ellis, A.W. (1985). The faces that launched a thousand slips: Everyday difficulties and errors in recognizing people. *British Journal of Psychology*, 76, 495-523.

Appendix

Parameters used in the interactive activation and competition network.

The simulation reported here was run using McClelland and Rumelhart iac program (McClelland & Rumelhart, 1988). The global parameters were set as follows:

Maximum activation	1.0
Minimum activation	-0.2
Resting activation	-0.1
Decay rate	0.1
Alpha (strength of excitatory input)	0.1
Gamma (strength of inhibitory input)	0.1
Estr (strength of external input)	0.1

All excitatory connections have a weight of 1.0 except for 'practiced' connections which had a weight of 3.0 (see text). All inhibitory connections have a weight of -0.5.

Acknowledgements: We thank Victoria Lay, Geralda Odinot, Julie Lanham-Jackson and Angela Taylor for running the some of the participants in this research project.

Language Type Frequency and Learnability. A Connectionist Appraisal.

Ezra Van Everbroeck (ezra@ucsd.edu)
UCSD Department of Linguistics; 9500 Gilman Drive
La Jolla, CA 92093 USA

Abstract

In this paper, I present experimental data bearing on the controversial issue of the possible relationship between the frequency of language types and how easily they can be learnt. Using simple, artificial languages which only differ with respect to the properties we are interested in, I show that there does appear to be a relationship of some kind, although not as strong as one might have hoped. In particular, if a language type can be learnt relatively easily, then the models fail to predict its actual frequency in the real world. On the other hand, the connectionist models provide evidence that the language types which are unattested or highly infrequent are also impossible or hard to learn.

Introduction

It has been known for a long time that languages differ in the ways in which they express 'who did what to whom?'. The three most important dimensions of this variability are: first, the word order of Subject, Object and Verb; second, the presence or absence of markers on the Verb; and, third, the presence or absence of markers on the Subject and/or Object (Nichols, 1986). The sentences below illustrate how the three strategies work. All (a) examples mean the same thing, but they get the message across in a different way — notice that the word order of S, O and V in (2) and (3) remains constant from the (a) to the (b) sentence.

1. (a)   The matador killed the bulls.
   (b)   The bulls killed the matador.

2. (a)   The matador-he the bulls-them killed.
   (b)   The matador-him the bulls-they killed.

3. (a)   Killed-he-them the matador the bulls.
   (b)   Killed-they-him the matador the bulls.

The three dimensions are also theoretically independent from each other, so we can imagine languages which (redundantly) combine a fixed word order with verbal and nominal marking, languages with none of the three, as well as the six other logical possibilities.

Things get even more complex, however, because the three dimensions are not necessarily binary. For example, in word order alone there are already six possible orders of S, O and V, next to a 'free' word order type in which many different combinations of the words are usually possible, albeit normally with pragmatically distinct meanings (Payne, 1992). And when case markers are used on the Subject and Object, we can at least distinguish between accusative strategies, in which (nominative) S is always marked differently from (accusative) O, ergative strategies, in which the (absolutive) S of the intransitive clause is marked similarly to the (absolutive) O of the transitive clause but the transitive S has a different (ergative) marker (Van Valin, 1992), and unmarked (or null) strategies.

So, we actually have at least 42 different language types (i.e. 7 word orders * 3 types of nominal marking * 2 types of verbal marking). As careful study of this many real languages would threaten to take up a lifetime of research, the route taken here has been to create artificial languages instead, and to use connectionist models to investigate the latter in a systematic manner.

Given the plethora of possible language types, it is not surprising that they are not found with equal frequency in the world. For example, the types with a fixed word order of SOV account for about half of the world's languages, whereas the types with OSV may even be unattested (see also below). In general, there are also no known languages which consistently fall on the extremes of the three dimensions: i.e. which either don't have any such mechanism for signaling 'who did what to whom', or which simultaneously employ all three mechanisms in a single sentence. While there are good common-sense reasons for these last two facts — i.e. there has to be some strategy or all communication would fail, on the one hand, and useless redundancy is not likely when it wastes resources, on the other hand — the connectionist simulations to be presented below are an attempt to explain the data by bringing learnability issues into the picture (cf. Christiansen & Devlin 1997). The important questions are:

- Can a neural network learn the attested language types?

- Will a neural network fail to learn the unattested language types?

- Will a neural network learn the more frequent language types faster/better?

If the answer to all three were to be 'yes', we could claim that there is a strong causal relationship between frequency and learnability of language types. However, the experimental evidence presented here only warrants a weaker conclusion.

The structure of this paper is as follows: in the next section, I will present the available linguistic frequency data in more detail, so that the phenomena to be accounted for are clear. In section 3, I briefly go over the setup of the simulations. Section 4 presents the results of the various simulations. The last section, then, wraps up the paper and provides

755

some language acquisition evidence supporting the posited connection between learnability and frequency.

Linguistic Frequency Data

Historically, the focus in language typological research has been on finding correlations between the word orders of various pairs like adposition — NP, genitive — noun, or NP — relative clause (see e.g. Greenberg 1963; Hawkins 1988; Dryer 1992). However, such correlations do not play a role in the simulations reported in this paper (but see Christiansen & Devlin (1997) and Van Everbroeck (in prep) for related work in which they do), because the sentences used only contain a Subject, Verb, and possibly an Object.

The information summarized in Table 1 below is much more relevant, then, as it shows the frequencies of the 6 possible fixed word orders of S, O and V (Tomlin, 1986; Dryer, 1989). One should keep in mind that these numbers control for historical and geographical biases, so that only unrelated languages are taken into consideration.[1]

Table 1: Language type frequencies.

SOV	SVO	VSO	VOS	OVS	OSV
51%	23%	10%	9%	.75%	.25%

It is not hard to see that SOV is by far the most frequent word order, with OVS and OSV being extremely rare — the latter may even be absent completely (Polinskaja, 1989). The percentages in Table 1 also show that Subject-before-Object languages are much more common than their Object-before-Subject counterparts — compare Greenberg's (1963: 77) Universal 1: "In declarative sentences with nominal subject and object, the dominant order is almost always one in which the subject precedes the object." Moreover, the Subject-initial SOV and SVO are noticeably more frequent than either the Verb-initial or the Object-initial language types. Finally, the missing 6% in Table 1 accounts for free word order languages like Native American Klamath (Barker, 1964).

With regards to the other two dimensions, nominal marking

---

[1] We are not interested in the raw number of individual languages which exhibit a certain word order, partly because there is still no clear definition of what makes a language as opposed to a dialect, and partly because we do not want closely related languages to count individually: e.g. German and Dutch might as well have ended up as a single language if history had taken a different turn; similarly, due to their geographical proximity to one another, all the languages spoken in the Balkan have some linguistic features in common though they belong historically to different groups. If we only look at a single representative language from each larger family instead of counting individual languages, we have a much better chance of capturing a universal phenomenon which is independent of where a language is spoken.

and verbal marking, only limited frequency information is available (see Nichols 1986 for the best summary to date). In general, language types with redundant nominal and verbal marking, and language types with neither kind of marking (as in English), are less common than the other two possibilities. It also appears that most languages use some form of verbal marking, though usually with supplementary information being provided by either word order or nominal marking. With respect to the latter, it seems safe to say that accusative systems outnumber ergative systems by a considerable margin, though the presence of mixed systems (e.g. accusative for pronouns, but ergative for other nominals — see Morris 1998) again complicates matters considerably.

In summary, then, at least some of the frequency patterns which one would expect to find mirrored in the learnability findings below — if there is indeed a connection between frequency and learnability — are at various levels of abstractness. First, any of the three strategies should be sufficient in at least some cases. Second, the language types which do not make use of word order or any kind of marking should be unlearnable. Third, the ones that are highly redundant should be learnable, but should not offer much of an improvement over their less verbose learnable counterparts — there is a production and processing cost to verbosity (Kirby, 1997). Fourth, Subject-initial languages should be more easily learnable than Object-initial languages. Fifth, accusative language types should present fewer problems for the neural networks than ergative language types.

Experimental Setup

The network model used to test the hypotheses just mentioned follows the following steps:

- Generate 42 artificial languages which only differ with respect to the three dimensions;
- Train an identical network on a corpus of sentences of each language;
- Compare 1) how well each language type is mastered, and 2) how well the network can generalize each time.

The artificial languages have been generated using simple context-free grammars which produce sentences appropriate to each language type. For example, the doubly redundant accusative SOV/VN grammar generates transitive SOV or intransitive SV sentences in which there are also verbal markers (indicated by the /V) as well as nominal markers (/N). (If a strategy is not used in a language type, X's are used instead of the other letters. So, VOS/XN is a VOS language with only nominal marking; XXX/VX is the free word order language type with only verbal marking. These mnemonic references will be used constantly below.)

The neural network architecture used in these simulations is illustrated in Figure 1. It is basically a simple recurrent neural network (Elman, 1992) with an extra recurrent layer at the output to provide it with more memory capacity. The task of this network, which sees one word per time step at the input layer, is to construct a representation at the output layer which shows which words it parses to be the Subject, Object or Verb.

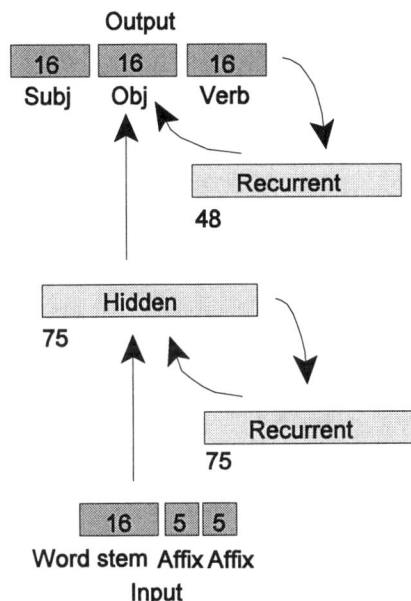

Figure 1: The network used in the simulations.

At the input layer, the 26 units encode a single lexical item (16 units with always exactly 6 units asserted) and up to two verbal or nominal suffixes (5 units each; 3 units asserted for a suffix). The corpus for each language contains 600 nouns and 100 verbs, with half of each category being used in the training set and the other half in the test set; the suffixes remain constant across the two sets.

At the output layer, there are three slots with 16 units — one slot each for S, O and V. Hence, as the network sees each of the words in a sentence similar to English 'They see me', it has to put the lexical item representation for 'they' in the first slot, then add that of 'see' to the third slot, and finally that of 'me' in the second. Each sentence is followed by a time step in which the pattern for the entire sentence has to be maintained, and then a reset signal is sent to the network. It is at the maintenance time step that the performance of the network on each sentence is calculated. For a sentence to be correct, the pattern of activation in each of the three output slots has to be closest in Euclidean space to the target word. As soon as a single word does not match the correct output, the entire sentence is considered incorrect.

For training, each network sees a corpus of 3,000 sentences of the relevant language (e.g. SVO/XN) for 10 epochs. The backpropagation algorithm is used to adjust the weights,

with a learning rate of 0.15. For generalization testing, then, another corpus of 3,000 sentences (but with completely different nouns and verbs) is presented to the network and the performance determined. In the next section, the numbers reflect the percentages of sentences correctly processed on the training and the test set.

Results and Discussion

Fixed Word Order and Accusative Marking

The results of the simulations with the fixed word order language types with accusative N-marking are shown in Table 2 below. If we just look at the Subject-before-Object languages, we find excellent performance on the training set (>97%) as well as the test set (>93%, with one exception).

The 23.7% test set performance for SOV/XX, however, is actually a positive result, because it meshes well with Greenberg's (1963: 96) Language Universal 41, which specifies: "If in a language the verb follows both the nominal subject and the nominal object as the dominant order, the language almost always has a case system". Hence, SOV/XX is a language type which is extremely rare among the languages of the world, and the fact that the network has problems with the test set is compatible with the posited relationship between frequency and learnability.

Table 2: Percentages of sentences correctly analyzed (accusative N-marking; fixed word order languages).

		Train	Test		Train	Test
VSO	XX	100	99.0	VOS	75.9	28.0
	XN	100	99.6		99.9	98.0
	VX	100	98.1		80.0	50.0
	VN	100	98.7		100	98.3
SVO	XX	99.9	95.5	OVS	81.2	12.4
	XN	99.9	95.9		93.3	87.5
	VX	99.9	97.0		96.9	91.9
	VN	99.9	97.9		99.9	99.2
SOV	XX	84.4	23.7	OSV	24.9	16.3
	XN	98.4	97.3		99.9	99.1
	VX	97.4	93.9		27.2	17.1
	VN	100	98.8		99.0	99.0

It turns out that there is a very good reason for the bad performance on the SOV/XX test set: when one 'hears' a sentence with all new words, and the words are also unmarked (i.e. there is no marker distinguishing even the nouns from the verbs), it is impossible to know whether the second word in the sentence is the Object of a transitive clause (in SOV), or the Verb of an intransitive clause (in SV) — all one can know is that it is the second word in the sentence. Hence, the network hedges its bets and spreads partial activation over both the Object and the Verb units at the output. Though in theory the presence/absence of a third word in the sentence can disambiguate between the two options — if another word follows, it must be the Verb, so the second word must have been the Object — the network fails to recover sufficiently (i.e. remove activation from the incorrect units and fully activate the correct ones) by the time performance is determined.[2]

If we now turn our attention to the twelve Object-before-Subject language types, we find that the networks in general have a harder time learning them — especially generalization is often problematic. As with SOV/XX, the language types without any kind of marking are very bad performers, and they are also infrequent or unattested. If verbal marking is added, only OVS benefits enough to produce a useful communicative system (>90% on training and test set), because the Verb now separates the two ambiguous nominals. With nominal case-marking, on the other hand, all the resulting systems appear learnable, be they /XN or /VN. Still, as with the Subject-before-Object language types, we again find that the /VN types tend to have little or no advantage over their /XN counterparts — a finding which is expected if we assume that speakers/hearers would prefer to avoid overly redundant and verbose systems.

It is also worth pointing out that OSV/XX and OSV/VX are the only two language types for which not even the training set is learnt successfully. OSV/VN is unrealistic in that is highly redundant. This leaves us with OSV/XN as an apparently viable type for the network, though it may be unattested among natural languages. A possible explanation here may be that case systems tend to degrade over time (Venneman, 1975), which would result in unlearnable languages with OSV. Hence, this basic word order would automatically become extinct over time.

Finally, I want to return to the general frequency hierarchy given earlier in Table 1, and its counterpart in Table 2. It turns out that one can find interesting parallels between this hierarchy on the one hand, and a comparison of the absolute order of Subject, Object and Verb in transitive and intransitive sentences, on the other hand. This is diagramed in Table 3 below.

---

2. This finding demonstrates that the simple recurrent network architecture used here does not perform well when there is a heavy memory load — i.e. it has to store the exact identity of the second word in the recurrent layer when it waits for the third word. An architecture in which such a word could be stored in a dedicated buffer would obviously have a better chance of dealing with this task. Alternatively, the types SOV/VX and SOV/XN solve the problem by disambiguating the word forms through the use of morphological markers. One should also keep in mind that the generalization task which the network faces is unrealistically hard in that it has to process sentences in which *all* the words are completely unknown.

Table 3: Relationship between transitive and intransitive clauses in terms of absolute word order.

VSO	V	S	O	VOS	V	O	S
	V	S			V	S	
	✓	✓			✓	✗	
SVO	S	V	O	OVS	O	V	S
	S	V			V	S	
	✓	✓			✗	✗	
SOV	S	O	V	OSV	O	S	V
	S	V			S	V	
	✓	✗			✗	✗	

Notice that with VSO and SVO, the fixed word order is actually most useful in that it guarantees that the Verb and Subject will always appear in the same slot — thereby facilitating processing. With SOV and VOS, only the first word remains constant; with the infrequent OVS and OSV types, there is no overlap between transitive and intransitive sentences at all. Table 3, however, does not explain at all why SOV actually accounts for 51% of the world's language types, and VSO for only 10%. I will return to this quandary in the general discussion, but let me simply mention that the psychological primacy of agentive/animate Subjects is likely involved.

Fixed Word Order and Ergative Marking

Let us now take a look at the results for the ergative language types in Table 4. Recall that in an ergative language, the O of the transitive clauses is treated similarly to the S of the intransitive clause (both are semantical Patients most of the time), whereas the S of the transitive clause (usually the Agent) receives different marking. For the Subject-before-Object language types, we find an almost identical picture to the one in Table 2 above. All language types are learnable and can be generalized from quite well, except for the rare language type SOV/XX. (Obviously, the /XX language types are really identical to the ones from the accusative set. They are repeated here for completeness, and to give some idea of the range of variation between different simulations of the same type.)

The fact that the Subject-before-Object ergative language types can be learnt easily is an encouraging finding, because such languages do certainly exist. When we turn to the Object-before-Subject languages, however, we find a different picture — even a cursory glance suffices to see that the percentages are generally much lower than those for the accusative counterparts. But for three OVS types, performance on the training set is down, and generalization also turns out to be quite problematic in almost all cases. For OVS, the redeeming feature appears to be that the Verb separates the nouns, which makes it easier to tell S, O and V apart. With the VOS types, the networks are rote learning the training set; a strategy which fails miserably when the new words in the test set are presented. And things are even worse for the — luckily unattested — OSV language types, because these networks fail to even pick up anything useful in the training set.

Table 4: Percentages of sentences correctly analyzed (ergative N-marking; fixed word order languages).

		Train	Test		Train	Test
VSO	XX	99.9	98.0	VOS	84.8	33.1
	XN	99.9	98.6		66.9	40.5
	VX	99.9	98.6		76.9	42.0
	VN	99.9	98.6		79.2	49.5
SVO	XX	99.9	96.8	OVS	73.2	14.4
	XN	99.9	96.1		98.5	96.6
	VX	100	96.7		97.9	92.3
	VN	99.9	97.5		99.8	99.0
SOV	XX	80.5	21.7	OSV	12.9	4.2
	XN	98.3	98.3		12.5	4.4
	VX	98.2	95.4		11.8	4.1
	VN	99.9	99.4		12.2	3.4

What is so difficult about ergative nominal marking and Object-before-Subject languages? In short, the case markers are no longer helpful for telling the Subject and Object apart. With an accusative system, an accusative always signals an Object; the nominative always a Subject. But in an ergative system, Subjects can be marked either with the absolutive (intransitive) or the ergative (transitive) marker. Objects will always appear with the absolutive case, but the ambiguity has already been introduced into the system. And as before, the networks fail to recover from such an initially ambiguous input sequence: in theory, the presence or absence of an ergative form should allow the network to figure out the grammatical role of the absolutive marked noun, but this does not seem to happen.[3]

Free Word Order

The final set of results are shown in Table 5. They are for the free word order languages, both accusative and ergative.

[3.] Additional training with the current network does not alleviate this problem, but Bill Morris (personal communication) has found that a much larger Elman net can be taught to handle such patterns much better. Still, one would expect language types which require more resources than others to be at a competitive disadvantage when children have to acquire them, and, therefore, to be either absent, or very rare — cf. Kirby 1997.

Table 5: Percentages of sentences correctly analyzed (accusative/ergative marking; free word order languages).

Acc		Train	Test	Erg	Train	Test
XXX	XX	15.0	2.3	XXX	11.5	1.5
	XN	84.7	75.1		32.2	25.8
	VX	45.3	33.8		40.5	29.1
	VN	99.0	94.0		38.3	19.2

It is easy to see that free word order languages as implemented here — i.e. with a completely random word order — are generally harder to learn than their fixed word order counterparts. Only the accusative XXX/VN type performs adequately. However, ergative XXX/VN is actually also attested, though the simulations would not predict this. Although the answer may lie in the overly random nature of the free word order simulated here, this issue does require further attention.

General Discussion and Conclusion

The motivation of this paper has been to explore the possible relationship between frequency of language types, and their learnability. We have seen that four of the five predictions formulated earlier find support in the connectionist simulations: first, the language types which do not use any of the three (i.e. XXX/XX) are unlearnable. Second, language types which are highly redundant (e.g. SVO/VN) do not have a significant advantage — if at all — over the simpler SVO/VX or SVO/XN, so the latter should be preferred by processing cost sensitive mechanisms. Third, the Subject-before-Object language types are indeed much easier to learn than their less frequent Object-before-Subject counterparts. Fourth, accusative language types as a group are apparently easier for a simple recurrent neural network than similar ergative types. With respect to the fifth prediction, namely that all three strategies (word order, verb marking, nominal marking) should be independently capable of expressing 'who did what to whom', we have found that this bit of conventional linguistic wisdom is not supported by the simulations: XXX/VX is not a learnable type. On the other hand, adding verbal marking to a fixed word order or nominal marking does make the task of the network noticeably easier in a number of the simulations.

Other findings which are compatible with the typological data are the bad performance on SOV/XX (Greenberg, 1963); the fact that V-initial languages prefer verbal marking (Nichols, 1986); and the fact that free word order languages generally occur with case systems (Payne, 1992).

However, we have also seen that the network results do not map perfectly onto the frequencies observed in the real world: e.g. the VSO networks generally perform much better than their frequency would lead one to expect. Similarly, some of the Object-before-Subject language types, or the ergative types are not as frequent as the modeling results predict. Also, there is at least one language type, ergative XXX/VN, which is attested but on which the model performs badly.

There are two different approaches to these criticisms: one is to make the models more complex — e.g. by adding semantics and perceptual salience for Subjects (Langacker, 1993), one could build in a psychologically plausible bias for SVO and SOV. The other one is to take the network results at face value and use them as a heuristic for further investigation: if a language type appears to be unlearnable, then it may employ other, compensatory mechanisms (e.g. tone) to express 'who did what to whom'. The simulations tell us where to look for such mechanisms.

So, how tight is the relationship between frequency and learnability? The simulations presented here do not justify a strong causal connection between the two, but they do lend support to a weaker position. Namely, they can tell us which language types are unattested because they are impossible to learn and which ones are unattested because they are overly redundant. If a language type is attested, however, then the networks fail to predict its frequency. It seems safe to assume that other factors, more historical and geographical in nature, determine the actual frequency of a language type, just as they determine the number of the speakers of any given language.

This conclusion should not really come as a surprise. Christiansen & Devlin (1997) found that their neural networks had a harder time learning language types with head-dependent word order inconsistencies — exactly the types which are also less frequent in the real world. And some time ago, Slobin & Bever (1982) reported that children acquiring different language types do indeed differ significantly in how fast they learn to understand 'who did what to whom' in their native languages. For example, infants learning Turkish (a language with abundant and unambiguous nominal marking) are much better at this task than infants acquiring Serbo-Croatian (which requires attention to word order and often ambiguous case markers). Children are somewhat like neural networks then, at least with respect to this task.

Acknowledgments

I would like to thank Chris Barker, Gary Cottrell, Ron Langacker, Bill Morris and Masha Polinsky, as well as an anonymous reviewer for their comments on an earlier version of this paper. Numerous useful suggestions were also made by the audiences who sat through talks at the Center for Research in Language and the AI Research Group. The early stages of the research presented here were performed when the author was a fellow of the Belgian American Educational Foundation.

References

Barker, M. (1964), *Klamath Grammar*, Berkeley: UC Press

Christiansen, M. and J. Devlin (1997), Recursive Inconsistencies Are Hard to Learn: A Connectionist Perspective on Universal Word Order Correlations, in *Proceedings of the 19th Annual Cognitive Science Society Conference*, Mahwah, NJ: Lawrence Erlbaum, 113-118

Dryer, M. (1989), Large Linguistic Areas and Language Sampling, in *Studies in Language* 13.2, 257-292

Dryer, M. (1992), The Greenbergian Word Order Correlations, in *Language* 68.1, 81-138

Elman, J.L. (1992), Grammatical Structure and Distributed Representations, in S. Davis (Ed.) *Connectionism: Theory and Practice*. New York: Oxford University Press

Greenberg, J. (1963), Some Universals of Grammar with Particular Reference to the Order of Meaningful Elements, in J. Greenberg (Ed.) *Universals of Language*. Cambridge, MA: MIT Press

Hawkins, J. (1988), Explaining Language Universals, in J. Hawkins (Ed.) *Explaining Language Universals*. Oxford: Basil Blackwell

Kirby, S. (1997), Competing motivations and emergence: explaining implicational hierarchies, in *Language Typology* 1, 5-32

Langacker, R. (1993), Reference-point constructions, in *Cognitive Linguistics* 4.1, 1-38

Morris, W. (1998), *Emergent Grammatical Relations: An Inductive Learning System*, unpublished doctoral dissertation, UC San Diego

Nichols, J. (1986), Head-marking and dependent-marking grammar, *Language* 62.1, 56-119

Payne, D. (1992), Introduction, in D. Payne (Ed.) *Pragmatics of Word Order Flexibility*. Amsterdam: John Benjamins

Polinskaja, M. (1989), Object initiality: OSV, in *Linguistics* 27, 257-303

Slobin, D. and T. Bever (1982), Children use canonical sentence schemas: A crosslinguistic study of word order and inflections, in *Cognition* 12, 229-265

Tomlin, R. (1986), *Basic Word Order. Functional Principles*, London: Croom Helm

Van Everbroeck (in prep.), Syntax vs. Morphology

Van Valin, R. (1992), An Overview of Ergative Phenomena and Their Implications for Language Acquisition, in D. Slobin (Ed.) *The Crosslinguistic Study of Language Acquisition. Volume 3*. Hillsdale, NJ: Lawrence Erlbaum

Vennemann, T. (1975), An Explanation of Drift, in C. Li (Ed.) *Word Order and Word Order Change*. Austin, TX: University of Texas Press

How Categories Shape Causality

Michael R. Waldmann (michael.waldmann@bio.uni-goettingen.de)
Department of Psychology, University of Göttingen,
Gosslerstr. 14, 37073 Göttingen, Germany

York Hagmayer (york.hagmayer@bio.uni-goettingen.de)
Department of Psychology, University of Göttingen,
Gosslerstr. 14, 37073 Göttingen, Germany

Abstract

The standard approach guiding research on the relationship between categories and causality views categories as reflecting causal relations in the world. We provide evidence that the opposite direction also holds: Categories that have been acquired in previous learning contexts may influence subsequent causal learning. In three experiments we show that identical causal learning experiences yield different attributions of causal capacity depending on the pre-existing categories that the learning exemplars are assigned to. There is a strong tendency to continue to use old conceptual schemes rather than switch to new ones even when the old categories are not optimal for predicting the new effect. This tendency is particularly strong when there is a plausible semantic link between the categories and the new causal hypothesis under investigation.

Introduction

The Standard View: Causality Shapes Categories

The standard view guiding research on causality presupposes the existence of networks of causes and effects in the world that cognitive systems try to mirror. Regardless of whether causal learning is viewed as the attempt to induce causality on the basis of statistical information or on the basis of mechanism information it is typically assumed that the goal of causal learning is to form adequate representations of the texture of the causal world (see Shanks, Holyoak, & Medin, 1996, for an overview of recent research). This view also underlies research on the relationship between categories and causality. According to the view that categorization is theory-based traditional similarity-based accounts of categorization are deficient because they ignore the fact that many categories are grounded in knowledge about causal structures (Murphy & Medin, 1985). As Murphy and Medin pointed out, natural kind categories (e.g., animals) are not adequately represented as lists of features because this format excludes functional and causal relations that also are part of our category knowledge. Categories should rather be seen as embodying intuitive theories of the target domain. One example of the standard view is Waldmann, Holyoak, and Fratianne's (1995) work on causal categories (also see Waldmann, 1996). In a series of experiments they have shown that category learning is affected by assumptions about the causal structure underlying the categories (e.g., disease categories).

The Neglected Direction: Categories Shape Causality

Even though it is certainly true that in many cases knowledge of causal structures influences the way categories are formed, the opposite may also hold true: The categories that have been acquired in previous learning contexts may have a crucial influence on subsequent causal learning. This direction has typically been neglected in research on the relationship between categories and causality.

The basis of the potential influence of categories on causal induction lies in the fact that the acquisition and use of causal knowledge is based on categorized events. Regardless of whether causal relations are viewed as statistical relations (probabilistic causality view) or as mechanisms (mechanism view), both accounts postulate causal regularities that refer to *types* of events. Causal laws, such as the fact that smoking causes heart disease, can only be noticed on the basis of events that are categorized (e.g., events of smoking and cases of heart disease). Without such categories causal laws neither could be detected nor could causal knowledge be applied to new cases. Thus, causal knowledge not only affects the creation of categories, it also presupposes already existing categories for the description of causes and effects.

Given that the induction of new causal knowledge is based on already existing categories the question arises whether the outcome of causal learning may be influenced by the categories that are used. The potential influence of categories is due to the fact that one of the most important cues to causality is statistical covariation between causes and effects. Many (otherwise conflicting) views agree that causal induction is based on the observation of causes altering the probability of effects (e.g., contingency view; associationist theories)(see Shanks et al., 1996). However, statistical regularities are not invariant across different categorial segmentations of domains. This can easily be shown with a simple example. Let us assume, for example, a world with four different (uncategorized) event tokens, A, B, C, and D, that represent potential causes. It has been observed that A and C are followed by a specific effect but B and D are not. Now the statistical regularities that are observed in this mini-world are crucially dependent on how these four events are categorized. If A and B are exemplars of Category 1, and C and D exemplars of Category 2, no causal regularity would be observed. Within this conceptual framework the effect has a base rate of 0.5 that is invariant across the two categories. By contrast, categorizing A and C (Category 3), and B and D (Category 4) together would lead to the induction of a deterministic causal law. Events that belong to Category 3 always produce the effect, whereas Category 4 is never associated with the effect. Thus, the causal regularities observed in a domain are dependent on the way the domain is categorized. In fact, as pointed out by Clark and Thornton (1997) in an example with (non-causal) continuous features, there is an infinite number of descriptions of the world with a potentially infinite number of statistical regularities entailed by these descriptions.

At this point it could be argued that the dependence of causal knowledge on pre-existing categories is a philosophical rather than a psychological problem as long as it has not been shown that there is evidence for the possibility of different categorizations of the same domains. Following the work of Rosch on natural categories many psychologists have assumed that natural categories are relatively stable since they are reflecting the correlational structure in the world (see Rosch, 1978). However, recently it became clear that this assumption is too strong. For example, Medin, Lynch, Coley, and Atran (1997) have shown that the way natural objects (e.g., trees) are categorized is dependent on pragmatic factors such as the profession of the categorizer (also see Barsalou, 1983). Schyns, Goldstone, and Thibaut (1998) have demonstrated that not even the object features used in categorizations are invariant. Their work demonstrates that the way the world is perceived may be influenced by the categories that are being used.

Another reason for the potential bi-directional interaction of categories and causality is the dynamic character of knowledge acquisition. Causal knowledge, in everyday life as well as in science, is typically not acquired at one point in time after which it remains stable but is rather the result of a long process in which it undergoes dynamic changes such as continuous modifications or even paradigm shifts (see Carey, 1991; Horwich, 1993). Categories acquired in specific contexts may not always be optimal for the new learning task at hand. For example, a learner who is equipped with Categories 1 and 2 in our example may be better off abandoning the old conceptual scheme altogether and instead forming Categories 3 and 4 that allow her to optimize predictability. On the other hand, switching to a novel conceptual scheme or keeping two different schemes in parallel incurs a cost that is computationally demanding. Therefore, there is a possible trade-off between sticking to old conceptual schemes that may not be currently optimal and switching to a new paradigm.

The hypothesized impact of pre-existing categories on causal learning constitutes a new type of transfer effect. Unlike in research on analogical transfer (see Holyoak & Thagard, 1995), no specific relational knowledge is transferred. The transfer effect is rather based on the indirect influence pre-existing categories may have on the statistical regularities observed in a domain. For example, traditionally, psychiatric diseases were classified on the basis of a taxonomy of symptoms, whereas today many researcher are more interested in neuropsychological analyses for which the old categories may not be optimal anymore. The original categories have never been created with the new research questions in mind. Nevertheless, it is possible that the continued use of the old categories in the new context may seriously bias the outcome of the causal investigations.

The following three experiments demonstrate how causal induction is affected by the way a novel domain is categorized. It will be shown that participants tend to use category knowledge acquired in a different context in a subsequent causal learning task.

Experiment 1

The goal of this experiment was to demonstrate how the way exemplars in a domain are categorized affects causal learning. The experiment consisted of three phases: In Phase 1, the *category learning* phase, participants were told that scientists had discovered new types of viruses that vary in the dimensions brightness, size, shape, and number of molecules on the surface. Cytophysiological investigations had revealed two types of viruses which can be distinguished on the basis of their appearance, *allovedic* and *hemovedic* viruses. After these instructions participants received index cards with pictures of viruses one after another, and had to judge whether the respective exemplar represented a hemovedic or an allovedic virus. After each judgment feedback was given. Learning proceeded until participants met a learning criterion, 10 correct classifications in a row. The exemplars varied continuously in the four features. The two relevant features were size and brightness. The diameter of the viruses varied between 30 and 48mm (Levels 1 to 4 in Table 1), and brightness was manipulated by using four equally spaced levels of grayness (20% to 80%)(Levels 1 to 4 in Table 1). The two irrelevant features also came in four levels which allowed us to create 256 different items. Our goal was to discourage exemplar learning. Table 1 shows examples of the 16 crucial types of viruses than can be created by factorially combining four values of size and brightness.

Two conditions were compared: In the size condition participants learned, for example, that the bigger viruses were allovedic, and the smaller ones hemovedic, in the orthogonal brightness condition they learned, for example, that the darker exemplars were allovedic and the lighter ones hemovedic. In Phase 1 128 different exemplars were presented to the participants.

While Phase 1 differed between conditions, the subsequent Phases 2 and 3 were *identical* across conditions. In Phase 2, the *causal learning* phase, participants were told that physicians were interested in exploring the relationship between the newly discovered viruses and diseases in animals. In particular, they wanted to find out whether the viruses cause splenomegaly, that is a swelling of the spleen. Therefore they studied animals that were infected with the new viruses. It was pointed out that any outcome of this study was possible including the possibility that there was no causal relationship between the viruses and splenomegaly. In Phase 2 participants saw a new set of 32 viruses one after another representing single instances of the viruses. On the backside of each card information was given on whether the respective virus had caused splenomegaly or not. In all conditions the same items with identical associations with the effect were presented to participants.

Table 1: Structure of learning items in Experiments 1 and 3 (see text for explanations).

	1 (A)	2 (B)	3 (B)	4 (A)
1		~ effect	~ effect	
	5 (B)	6 (A)	7 (A)	8 (B)
2	effect			~ effect
	9 (B)	10 (A)	11 (A)	12 (B)
3	effect			~ effect
	13 (A)	14 (B)	15 (B)	16 (A)
4		effect	effect	
Size	1	2	3	4
		Brightness		

Table 1 displays the structure of the items with respect to the two relevant dimensions size and brightness. In the brightness condition the left half of the table (Levels 1 and 2 of the brightness dimension) may represent allovedic viruses and the right half hemovedic viruses (Levels 3 and 4). By contrast, in the size condition the upper half represented one category (Levels 1 and 2 of the size dimension), and the lower half the other category (Levels 3 and 4). Half of the items, indicated by an A, were shown in the category learning phase (Phase 1), the other half (indicated by a B) in the causal learning phase (Phase 2). (Again, our goal was to discourage exemplar encoding by presenting items that differed in their appearance.) Table 1 also shows which of the B-items caused the effect splenomegaly (*effect*), and which did not (*~effect*). The assignment of exemplars to the learning phases (A, B), and the association of dimensional attributes (dark, light, big, small) with the effects was counterbalanced.

In Phase 3, the *test* phase, we switched back to exemplars corresponding to the A-items from the category learning phase. These items had not been presented in the causal learning phase. Participants received eight exemplars. Their task was to express their assessment of the likelihood that the respective virus causes splenomegaly by using a rating

scale that ranged from 0 ("never") to 100 ("always"). After these ratings participants also gave a general assessment of the likelihood that the two virus types, allovedic and hemovedic viruses, caused the effect. These ratings allowed us to check whether participants encoded the causal relation on the category level.

Despite the fact that participants in the two conditions received identical learning inputs in the causal learning phase (Phase 2), and were confronted with identical test items (Phase 3) we expected that the different categories learned in Phase 1 would influence the causal judgments. Item 1 (A) in Table 1 may exemplify our predictions: We expected that participants in the size condition would group this item along with the other items in the category of relatively small viruses (Items 1 to 8). Since only one out of four items (Items 2, 3, 5, 8) from the group presented in the causal learning phase produced the effect, it seems reasonable to infer that Item 1 would have a relatively low likelihood of causing splenomegaly. By contrast, participants in the brightness condition were expected to classify this item along with the other relatively light viruses. Within this group three out of four viruses (Items 2, 5, 9, 14) caused splenomegaly which should lead participants in this condition to give relatively high ratings. A similar prediction can be derived for Item 6, whereas for Items 11 and 16 the inverse pattern is predicted (i.e., high ratings in the size condition, and low ratings in the brightness condition). The remaining four test items (4, 7, 10, 13) should not yield any differences across the category conditions because they were associated with the same number of effects regardless of which category structure was used.

Alternatively, participants could ignore the category-level information from Phase 1 in the causal learning phase, and compare the test exemplar with the causal pattern of adjacent exemplars (exemplar-based learning), or they could create new categories that are more optimal for the causal task at hand. Both strategies should not lead to any differences in the ratings of the items across the two category conditions.

Results and Discussion

The results are based on 48 participants, 24 in the size condition and 24 in the brightness condition. (Two participants were excluded because they did not meet the learning criterion in Phase 1.) The most interesting analyses involved the test items whose ratings should differ across conditions (e.g., Items 1, 6, 11, 16 in Table 1). To make these items comparable, the ratings of Items 11 and 16 were assigned the inverse rating (e.g., 80 was recoded as 20). For the statistical analyses the average of these four items was used.

The results clearly confirm the predictions. The mean ratings of the four critical items clearly differed depending on which categories had been learned in the prior category learning phase, $F(1,46)=11.8$, $p<.001$, $MSE=350.9$. The mean ratings for these critical items were 43.5 versus 62.1 in the two contrasting category learning conditions. By contrast, no reliable difference was obtained for the test items which were not expected to differ across conditions ($M=62.7$ vs. $M=65.8$). A 2 (category conditions) × 2 (critical vs. non-critical items) analysis of variance yielded a significant interaction, $F(1,46)=5.62$, $p<.05$, $MSE=253.8$. These results were mirrored in the final category-effect ratings. The mean ratings for the category with strong associations with the effect was 70.0, the contrast category with weak associations yielded a mean value of 24.6. All but four participants who rated both categories equally gave ratings consistent with this difference. An analysis of variance with categories as a within-subject factor yielded a clearly significant effect, $F(1,47)=153.5$, $p<.001$, $MSE=299.2$. This effect was independent of whether the categories were based on the size or the brightness of the items. Thus, participants not only registered the association of the individual exemplars with the effect but they also encoded the causal relations on the category level.

In summary, despite the fact that participants in all conditions received identical learning inputs in the causal learning phase, and had to make predictions about identical exemplars in the test phase, the attribution of causal capacity clearly differed depending on the categories the exemplars were assigned to in Phase 1. The same virus exemplars were either seen as causally effective or ineffective with respect to splenomegaly.

Although it is certainly true that viruses may generally be viewed as categories that are potentially responsible for symptoms such as splenomegaly, it is by no means certain that viruses that have been classified on the basis of their *appearance* represent classes that are optimal with respect to all kinds of effects that later may be studied. Potential causal links between the viruses and specific symptoms never were an issue when the rationale for the classifications in Phase 1 was introduced. In fact, other classifications (e.g., segmenting the exemplars in Table 1 using a diagonal boundary) are better able to capture the observed causal regularities. Nevertheless, Experiment 1 shows that participants rather continued to use categories acquired in a different learning context than create new categories for the induction task at hand.

Experiment 2

Experiment 1 used relatively simple category structures that were based on one-dimensional rules. By contrast, the rules underlying the causal regularities in Phase 2 were quite complex relative to the categorization rules. In Experiment 2 we investigated a task with more realistic, complex category structures, and with a comparably complex causal structure. In this experiment we used linearly separable, family resemblance categories that were based on four binary features. None of these features was individually sufficient for achieving correct classifications. However, correct classifications could be learned by an additive integration of the four features.

As learning exemplars we used items similar to the ones in Experiment 1. In the present experiment the exemplars varied on *four binary* dimensions, however: brightness (20% vs. 60%), size (30mm vs. 42mm), number of corners (5 vs. 7), and number of molecules on the surface (2 vs. 4). Table 2 displays the structures of the learning items with the feature value 1 representing high values and the value 0 low values.

Again the cover stories from Experiment 1 were used so that participants' task was to categorize the items into allovedic and hemovedic viruses in Phase 1 of the experiment. Two categorization conditions were compared that manipulated the location of the category boundaries (see Table 2). In *Condition A* hemovedic viruses had at least two high values on the four dimensions (Items 1 to 11), whereas allovedic viruses (Items 12 to 16) only had one or no high value. (The category labels were counterbalanced.) By contrast, in *Condition B* hemovedic viruses (Items 1 to 5) had at least three high values, whereas allovedic viruses (Items 6 to 16) only had two high values or less. Again we used a learning criterion in Phase 1. Learning proceeded until participants managed to correctly classify one block of 16 items (maximum of 8 blocks). The items were presented in random orders within blocks.

Table 2: Structure of learning items in Experiment 2

Items	Features				Effect	Categorization A	B
1	1	1	1	1	E	Hemovedic viruses	Hemovedic viruses
2	1	1	1	0	E		
3	1	1	0	1	E		
4	1	0	1	1	E		
5	0	1	1	1	E		
6	1	1	0	0	E		Allovedic viruses
7	0	0	1	1	E		
8	0	1	1	0	-		
9	1	0	0	1	-		
10	1	0	1	0	~E		
11	0	1	0	1	~E		
12	0	0	0	1	~E	Allovedic viruses	
13	0	0	1	0	~E		
14	0	1	0	0	~E		
15	1	0	0	0	~E		
16	0	0	0	0	~E		

Whereas Phase 1 differed across the two conditions, the subsequent causal learning phase and the test phase were *identical* for all participants. Again participants received index cards that depicted exemplars of the viruses with information on the backside on whether the respective virus causes splenomegaly (E) or not ($\sim E$). To avoid an unequal association of individual features with the effect Items 8 and 9 were not presented in this phase. In the particular counterbalancing condition shown in Table 1 the upper half of the items (1-7) caused the effect, whereas the lower half (10-16) did not cause it.

In Phase 3, the test phase, participants received ten exemplars (1, 3, 6 to 11, 14, 16) and had to assess their likelihood of producing splenomegaly using the rating scale from

Experiment 1. The most important results involved the six items (6 to 11) lying between the category boundaries of the two conditions. In Condition A these items should be viewed as being members of the hemovedic virus type. Since within this group seven out of nine viruses caused splenomegaly, high ratings are to be expected. By contrast, the very same items should yield low ratings in Condition B. In this condition the six items belong to the allovedic viruses which cause the effect in only two out of nine cases.

Results and Discussion

The analyses are based on 32 participants (16 in Condition A and 16 in Condition B). All participants met the learning criterion. The most important analysis involved the test items between the two category boundaries (Items 6-11). The mean ratings for these six items clearly differed across the two category conditions A and B, $F(1,30)=14.7$, $p<.01$, $MSE=224.3$. The two contrasting mean values (averaged over the six items) were 65.9 versus 45.6. No significant differences were obtained for the (non-critical) test items that were not expected to differ across conditions. The average ratings of these items (with the items expected to yield low ratings being recoded to the corresponding high values) were 82.7 and 82.6. A 2 (category conditions) × 2 (critical vs. non-critical items) analysis of variance revealed a significant interaction, $F(1,30)=9.19$, $p<.01$, $MSE=178.3$.

Interestingly, the influence of the categories was strongest for the exemplars participants had seen in the causal learning phase (Phase 2). The mean ratings for these items (6, 7, 10, 11) were 70.8 versus 43.4 which was, of course, highly reliable, $F(1,30)=25.1$, $p<.01$, $MSE=238.3$. Thus, even though all participants had, for example, seen Item 6 as causing splenomegaly, they nevertheless gave this exemplar a lower rating in the test phase when it was categorized as an allovedic virus in Condition B than when it belonged to the hemovedic category in Condition A. (In this experiment the appearance of the items did not vary across phases.) By contrast, the two non-presented items 8 and 9 did not significantly differ across category conditions ($M=56.3$ vs. $M=50.0$). This rather surprising result seems to indicate that category level information is only used when it is actively encoded along with the item in the causal learning phase. Since these two items were not presented during this phase the relation between these items and the effects were apparently not encoded on the category level.

Again the final ratings showed that participants generally encoded the relationship between categories and the effect. They rated the causal efficacy of the two categories clearly different regardless of the location of the category boundary, $F(1,31)=80.5$, $p<.01$, $MSE=509.5$ ($M=74.8$ vs. $M=24.2$). All but five participants gave ratings consistent with this trend.

In summary, Experiment 2 confirms the results of Experiment 1 with family resemblance category structures. Despite the fact that all participants received identical cause-effect information in the causal learning phase, the ratings of the causal efficacy of the exemplars seen in this phase were moderated by the categories to which they belonged.

Experiment 3

The two previous experiments have shown that participants tended to use category knowledge that they had acquired in a previous learning context when learning about a new causal relation. Even though there was no reason to believe that the appearance-based categories learned in Phase 1 would provide useful classifications for the induction of the cause-effect relations in Phase 2, participants rather continued to use these categories than switch to a new conceptual scheme. Experiment 3 aimed at exploring the boundary conditions for this effect. In Experiments 1 and 2 category labels were used in Phase 1 (types of viruses) that seem to be plausible candidates for being causes of the target effect in Phase 2 (splenomegaly). Thus, despite the fact that the classification of the virus types was based on a rationale which was conceptually independent of the causal hypothesis in Phase 2 it may still be plausible to assume that viruses are generally useful categories for predicting health-related symptoms. It is possible that in a situation in which the semantic relatedness between categories and target effect is reduced fewer participants would continue to use the old categories.

To test the relevance of the semantic relation between categories and effect we designed cover stories that attempted to exclude all possible associations between categories and effects. Thus, our goal was to present participants with a learning situation in which there was no a priori reason to transfer the category knowledge from Phase 1 to the causal learning situation in Phase 2.

We used the same learning exemplars and the same learning procedure as in Experiment 1 but changed the cover stories. In Experiment 3 we introduced the items displayed in Table 1 as belonging to two types of objects, *Alpha-Objects* and *Beta-Objects*, that were distinguished on the basis of their appearance. In Phase 1 participants learned to classify the items into these two classes. In Phase 2, the causal learning phase, it was mentioned that these objects may be the causes of a novel *Effect*. No further semantic characterization of the kind of effect was given. In Phase 3 participants gave ratings of the likelihood that the test items caused this unknown effect. After these ratings we also required participants to assess the causal efficacy of the two contrasting categories, Alpha- and Beta-Objects.

Results and Discussion

The results are based on 48 participants (24 in the size condition and 24 in the brightness condition). Three further participants were excluded because they did not meet the learning criterion. Again the most interesting result involved the critical test items whose ratings should differ in case category level information was used. In this experiment the effect was again in the right direction but clearly weaker than in Experiment 1. The mean values of the averaged four critical items (two of them were recoded) were $M=59.6$ versus $M=49.6$ in the contrasting category conditions, $F(1,46)=3.92$, $p<.06$, $MSE=306.3$. As in the previous experiments the uncritical items whose ratings were not expected to differ across conditions yielded similar ratings ($M=64.3$ vs. $M=63.6$). The 2 (category conditions) × 2 (critical vs. non-critical items) analysis of variance did not show a significant interaction ($p=.16$) in this experiment. The ratings of the category-effect relations indicated that overall the differential relation of the two categories and the effect was learned, $F(1,47)=52.4$, $p<.01$, $MSE=595.2$. The mean ratings were 68.8 versus 32.7. However, a closer inspection of the data revealed that unlike in the previous experiments a considerable number of participants did not encode the differences on the category level. 14 out of the 48 participants gave equal ratings to the two categories. An analysis in which these cases were excluded showed that these 14 participants were mainly responsible for the decrease of the size of the effect for the critical items. The remaining 34 participants (14 in the size and 20 in the brightness condition) gave mean ratings of 43.6 versus 62.1 for the critical items, $F(1,32)=11.6$, $p<.01$, $MSE=245.3$, which closely corresponds to the results of Experiment 1. An analysis of variance that only included the data from these remaining participants yielded a significant interaction between category conditions, and type of items (critical vs. non-critical), $F(1,32)=7.55$, $p<.01$, $MSE=211.5$.

In summary, once again a considerable number of participants continued to use the old categorial scheme even though there was *no* semantic link between categories and the effect that suggested the usefulness of the categories for the causal context. However, we also found evidence that the semantic relatedness between categories and effect affects the likelihood of transfer. Unlike in the previous experiments a relatively large number of participants (ca 30%) seemed to have concluded that the category level information was not useful for the subsequent causal learning task, and therefore did not encode the relation between category level and effect.

Conclusions

Overall the three experiments provide clear evidence for the tendency to continue to use categories that have been acquired in previous learning contexts when learning about new causal relations. Identical learning experiences yielded different attributions of causal capacity depending on the categories that the learning exemplars were assigned to. This holds true even though there was no compelling reason that the old categories were useful, and in fact other possible category structures yielded stronger statistical relations between categories and the effect. In our view, these findings show that the relation between categories and causality is bi-directional. Categories not only reflect pre-existing knowledge of causal structures they also affect the acquisition of new causal knowledge. Like in scientific paradigms there is a tendency to continue to use old conceptual schemes at the potential cost of suboptimal predictability but with the computational gain of not having to use many categorization schemes in parallel. As in science there seem to be conditions, however, in which old paradigms tend to be abandoned. The results of Experiment 3 suggest that there is a tendency to acquire new knowledge from scratch when the semantic link between old categories and the new causal hypotheses is weak.

The inherent bi-directionality of the relation between categories and causality may, of course, extend to multiple steps in a dynamic process of theory revisions. Prior categories affect causal induction which in turn may create new causal categories that influence what kind of statistical structure new data exhibit. The present results show that the outcome of this dynamic process of theory development may be crucially dependent on how it started.

References

Barsalou, L. W. (1983). Ad-hoc categories. *Memory & Cognition, 11*, 211-227.

Carey, S. (1991). Knowledge acquisition: Enrichment or conceptual change? In S. Carey & R. Gelman (Eds.), *The epigenesis of mind: Essays on biology and cognition.* Hillsdale, NJ: Erlbaum.

Clark, A., & Thornton, C. (1997). Trading spaces: Computation, representation, and the limits of uninformed learning. *Behavioral and Brain Sciences, 20*, 57-90.

Holyoak, K. J., & Thagard, P. (1995). *Mental leaps. Analogy in creative thought.* Cambridge, MA: MIT Press.

Horwich, P. (Ed.)(1993). *World changes: Thomas Kuhn and the nature of science.* Cambridge, MA: MIT Press.

Medin, D. L., Lynch, E. B., Coley, J. D., & Atran, S. (1997). Categorization and reasoning among tree experts: Do all roads lead to Rome? *Cognitive Psychology, 32*, 49-96.

Murphy, G. L., & Medin, D. L. (1985). The role of theories in conceptual coherence. *Psychological Review, 92*, 289-316.

Rosch, E. (1978). Principles of categorization. In E. Rosch & B. B. Lloyd (Eds.), *Cognition and categorization.* Hillsdale, NJ: Erlbaum.

Schyns, P. G., Goldstone, R. L., & Thibaut, J.-P. (1998). The development of features in object concepts. *Behavioral and Brain Sciences, 21*, 1-54.

Shanks, D. R., Holyoak, K. J., & Medin, D. L. (Eds.)(1996). *The psychology of learning and motivation, Vol. 34: Causal learning.* San Diego: Academic Press.

Waldmann, M. R. (1996). Knowledge-based causal induction. In D. R. Shanks, K. J. Holyoak & D. L. Medin (Eds.), *The psychology of learning and motivation, Vol. 34: Causal learning.* San Diego: Academic Press.

Waldmann, M. R., Holyoak, K. J., & Fratianne, A. (1995). Causal models and the acquisition of category structure. *Journal of Experimental Psychology: General, 124*, 181-206.

A study of complex reasoning:
The case of GRE 'logical' problems

Yingrui Yang (yingruiy@phoenix.princeton.edu)
Department of Psychology, Princeton University,
Green Hall, Princeton, NJ 08544, USA

P.N. Johnson-Laird (phil@clarity.princeton.edu)
Department of Psychology, Princeton University,
Green Hall, Princeton, NJ 08544, USA

Abstract

Complex reasoning, such as that elicited by GRE 'logical' reasoning problems, is demanding for human reasoners and beyond the competence of any existing computer program. We report four experiments carried out to investigate the question of what makes these problems difficult. The experiments established three causes of difficulty: the nature of the logical task (Experiment 1), the nature of the incorrect foils (Experiment 2), and the nature of the correct conclusions (Experiments 3 and 4).

Introduction

Most psychological studies of reasoning concern deductions that are logically straightforward. Even those deductions that are difficult for human reasoners are easy for computer reasoning programs. Certain inferential problems, however, are demanding for human reasoners and impossible for all current computer programs. They include a class of problems in the Graduate Record Examination (GRE) developed by Educational Testing Service to predict performance in graduate school. The examination has sections measuring mathematical, verbal, and analytical ability. The analytical section contains two sorts of problem, which are known respectively as analytical problems and logical problems. Here is an example of a GRE logical reasoning problem:

> Children born blind or deaf and blind begin social smiling on roughly the same schedule as most children, by about three months of age.
> The information above provides evidence to support which of the following hypotheses:

A. For babies the survival advantage of smiling consists in bonding the caregiver to the infant.
B. Babies do not smile when no one else is present.
C. The smiling response depends on inborn trait determining a certain pattern of development.
D. Smiling between persons basically signals a mutual lack of aggressive intent.
E. When a baby begins smiling, its caregivers begin responding to it as they would to a person in conversation.

This problem is easy, as readers may check for themselves. (The correct answer is below). Yet, no current computer program can take such problems verbatim and reason its way through to the correct conclusion. There are at least three difficulties: the extraction of the logical structure of the problems, the variety of the inferential tasks posed by the problems, and their reliance - often in subtle ways - on general knowledge. Some GRE logical reasoning problems are much harder than this example, but what all the problems have in common is a basis in real life examples and a three-part structure: an initial text, a sentence that poses the task, and a set of five options containing one correct answer and four incorrect foils.

No-one knows what makes an GRE logical reasoning problem difficult for human reasoners. Our aim in the present paper is to begin to answer this question. Our studies relied on the 120 GRE logical problems that are in the public domain. The correct answer to the problem above is option C. It depends on the following argument: If blind children start to smile at the same point in their development as sighted children, then smiling is not learned, because it could be learned only from seeing a caregiver smile. If smiling is not learned, it must depend on an inborn disposition. This problem calls for an inference, but other problems pose other tasks. We begin our investigation with a study of these tasks.

The nature of the inferential task: Experiment 1

GRE logical problems fall into four principal categories in terms of the tasks they pose: 1. Problems of identifying which option can be inferred from the text, 2. Problems of identifying a missing premise in the argument of the text, 3. Problems of identifying a weakness in the argument in the text, and 4. Problems of identifying the logical relation between two propositions in the text. These tasks probably differ in their intrinsic difficulty, but it is impossible to make a general comparison, because the nature of the task is inevitably confounded with its content. One comparison, however, is feasible, and it concerns a difference that the mental model theory predicts. It should be easier to identify a conclusion than to identify a missing premise. Consider a simple illustrative example, such as an inference of the following form:

A or B, or both.

Not A.

∴ B.

According to the theory of mental models (see e.g. Johnson-Laird and Byrne, 1991), the task of drawing an inference calls for constructing models of each of the possibilities consistent with the premises. In fact, these premises have only a single model:

¬ A   B

where "¬" denotes negation. Reasoners can then determine which option is true in the models, i.e. an option of the form, B. In contrast, a missing-premise problem has a text which is an inference with a missing premise:

A or B, or both.

∴ B.

Reasoners can construct the models of the disjunctive premise:

A   ¬ B

¬ A   B

A   B

where each row is a model of a different possibility. Reasoners can now try to work out how the models should be modified in order to

yield the conclusion: B. One such modification is to eliminate the first model. The next step is to formulate a premise that will do the job, that is, a premise that negates the model but leaves the other models intact, e.g.: If A then B. The options in the problem, however, may not contain this premise. Another modification to the models is to eliminate the first and third models. Again, this step calls for working out a premise that will do the job, i.e., not A, and checking it against the options. Hence, the task of identifying a missing premise is more complicated than the task of identifying a conclusion. Reasoners must examine the relation between the premises and the conclusion, try to figure out what information is needed for an inference from the premises to the conclusion, and then check whether this information is among the options.

Experiment 1 tested the prediction that inferential problems should be easier than missing-premise problems. It used problems with an identical content, consisting of a text followed by a single test item for the participants to evaluate. The problems were based on six inferential problems and six missing-premise problems from the sample of 120 GRE problems. All 12 problems were difficult, as performance with them in the GRE showed. We constructed four versions of each of the problems: a valid inference, an invalid inference, a correct missing premise, and an incorrect missing premise. The valid conclusion was the original correct option, the invalid conclusion was the most frequently chosen foil in the original item, and so on. The resulting 48 experimental problems were divided into four sets, each including three different problems the four sorts, and the participants carried out all the problems in a set. Thus, the participants encountered a particular content only once, and carried out three problems of the four different sorts, but each content occurred equally often in the three sorts of problem in the experiment as a whole. The participants were allowed to use paper and pencil, and were encouraged to write or draw whatever they had in mind on the problem page during the course of solving a problem. After they had evaluated the putative conclusion, they rated its difficulty on a 7-point scale. We tested 20 Princeton undergraduates individually.

768

Table 1: The percentages of correct responses, the mean latencies (in minutes), and the means of the rated difficulties to the four sorts of problems in Experiment 1.

	Inferential problems		
	% Correct	Latency	Rating
Valid conclusion	51	2.75	3.20
Invalid conclusion	63	2.84	3.35
	Missing-premise problems		
	% Correct	Latency	Rating
Valid conclusion	61	3.21	3.82
Invalid conclusion	65	3.31	3.49

The results are shown in Table 1. There was no reliable difference in accuracy between the problems, but the participants responded to the inferential problems significantly faster than to the missing-premise problems (z=2.88, p<.002), and they also rated them as significantly easier (z=1.66, p<.05). (Here and throughout the paper, the tests of significance were Wilcoxon signed-ranks matched-pairs tests.) No other differences were reliable. The results accordingly corroborated our prediction that it should be easier to evaluate a putative conclusion than a putative missing premise.

The effect of foils:
Experiment 2

Experiment 2 concerned only those GRE logical reasoning problems in which the task was to select the option that could be inferred from the text. A salient source of difficulty should be the set of five options from which the testees select their choice. We can classify options into those that are: 1. valid, i.e. they follow from the text (and general knowledge); 2. consistent with the text, i.e. they may be true given the text, but they do not follow from it; 3. inconsistent, i.e. they are false given the text; and 4. irrelevant, i.e. whether true or false, they are consistent with the text. Readers should note that all the foils in the sample problem in the Introduction are irrelevant. Indeed, if all the foils are irrelevant or inconsistent, a problem is likely to be easy. But, the presence of a consistent foil should render a problem more difficult. Reasoners often fail to construct all the models of the premises, and in this case they may easily confuse a consistent conclusion, which is true in some of the models of the premises, with a

necessary conclusion, which must be true in all the models of the premises (Johnson-Laird and Byrne, 1991). We therefore used six difficult inferential problems (those in the previous experiment) and six easy inferential problems from the pool of 120 GRE problems. For each of the difficult problems, we changed the most seductive foil (as shown by the results from the GRE test) to make it inconsistent with the text. Likewise, for each of the easy problems, we changed an inconsistent foil to make it consistent with the text. We divided the resulting 24 problems into two sets, each consisting of the original versions of three difficult and three easy problems, and modified versions of the other six problems. We assigned a set to each participant at random, so that he or she saw a text only once. We tested 32 Princeton undergraduates individually. They were allowed to take as much time as they needed for each problem.

Table 2: The percentages of correct responses in Experiment 2

	Easy problems / Difficult problems	
Original version	99	53
Modified version	78	70

Table 2 presents the percentages of correct responses to the four sorts of problem. As predicted, the modified easy problems were made harder whereas the modified difficult problems were made easier, and the interaction was highly reliable (z = 4.25, p < .0001). Hence, it is simple to increase the difficulty of an easy problem by introducing a consistent foil and to decrease the difficulty of a hard problem by introducing an inconsistent foil. As it happens, the modified problems no longer differed reliably in difficulty (z = .21, p > .4). But, the manipulation of the foils was unable to make the difficult problems as easy as the original versions of the easy problems, or to make the easy problems as hard as the original versions of the difficulty problems. Hence, other factors must influence difficulty. Indeed, Experiment 1 showed that the difficult problems were just as difficult when there were no foils. We examined the effect of conclusions in the next experiment.

The nature of conclusions:
Experiment 3

In this experiment, each text was presented with only a single putative conclusion, and the task was to determine whether or not it followed from the text. The materials were based on the same set of problems as those in the previous experiment. The easy problems had as their putative conclusion either the original conclusion (valid) or the original inconsistent foil (invalid), and the difficult problems had as their putative conclusion either their original conclusion (valid) or the original consistent foil (invalid). The participants were allowed to use paper and pencil, and the procedure was the same as Experiment 1. We tested 20 Princeton undergraduates individually.

Table 3: The percentages of correct responses, the mean latencies (in minutes), and the means of the rated difficulties to the four sorts of problems in Experiment 3.

	Easy problems		
	% Correct	Latency	Rating
Valid conclusions	100	1.78	2.08
Invalid conclusions	83	1.82	2.82
	Difficult problems		
	% Correct	Latency	Rating
Valid conclusions	75	3.22	3.93
Invalid conclusions	58	3.20	4.28

Table 3 presents the results of the experiment. Overall, the easy problems were reliably easier than the difficult problems on all three measures ($z \geq 3.0$, $p \leq .002$, in all three cases). The valid problems yielded a greater percentage of correct responses than the invalid ones ($n = 15$, $c > 116$, $p < .0002$) and were rated as more difficult ($z = 2.6$, $p < .005$). The invalid easy problems depended on a conclusion that was inconsistent with the text, and the invalid difficult problems depended on a conclusion that was consistent with the text. Hence, the invalid difficult problems should show a greater increase in difficulty than the invalid easy problems. This prediction was not reliable, perhaps because the invalid difficult problems sank to a level of accuracy that was no better than chance. This 'floor' effect may have vitiated the latency and rating measures.

Experiment 4

In this experiment, each text was again presented with only a single putative conclusion, and the task was to determine whether or not it followed from the text. The materials were

based on the same set of 24 problems as Experiment 2. The easy problems had either the original conclusion (valid) or the modified foil that was consistent with the text (invalid), and the difficult problems had either the original conclusion (valid) or the modified foil that was inconsistent with the text. The procedure was the same as the previous experiment. We tested 20 Princeton undergraduates individually.

Table 4: The percentages of correct responses, the mean latencies (in minutes), and the means of the rated difficulties to the four sorts of problems in Experiment 4.

	Easy problems		
	% Correct	Latency	Rating
Valid conclusion	83	2.04	2.93
Invalid conclusion	53	2.27	3.47
	Difficult Problems		
	% Correct	Latency	Rating
Valid conclusion	92	2.94	3.67
Invalid conclusion	90	2.77	3.50

Table 4 presents the results of the experiment. At first sight, the effect of the experimental manipulation was extraordinary. Overall, the difference between the easy and difficult problems almost disappeared: if anything, the accuracies switched around, though not reliably, perhaps because performance with the invalid easy problems was at chance. Only the latencies show a significant advantage for the easy problems ($z = 3.85$, $p < .0001$). What is striking is the interaction: on every measure, the easy problems show an advantage for the valid conclusion, whereas the difference disappears for the difficult problems (the interaction was reliable, $z \geq 2.03$, $p \leq .05$ on all three measures).

General Discussion

Our experiments have shown the existence of three factors that affect the difficulty of GRE logical problems. The first factor is the nature of the task. Experiment 1 corroborated the model theory's prediction that inferential problems are easier than missing-premise problems. Hence, the nature of the logical task affects performance. A second factor is the nature of the foils. Experiment 2 showed that an easy problem can be made harder by introducing a foil that is consistent with the text, and a difficult problem can be made easier by introducing a foil that is inconsistent with the text. A third factor is the relation between the text and the correct conclusion. Experiment 3 showed that easy problems are easier than difficult problems even when there are no foils and the task is merely to evaluate a single putative conclusion. When we consider the results of Experiment 3 and 4 together, a striking and unexpected phenomenon emerges. The participants developed a strategy based on problems with invalid conclusions that were inconsistent with the texts. They showed a bias towards responding "no" to any such problem, which is the correct response. This bias, however, often led them to often respond "yes" to any other problem. The result in Experiment 3 was that performance with the invalid difficult problems, which have conclusions that are consistent with the texts, fell to a chance level. The effect was more dramatic in Experiment 4. The difficult problems became easy, because it was simple to determine that a conclusion was invalid -- it was inconsistent with the premises. But, performance with the invalid conclusions to the easy problems, which are consistent with the text, dropped to chance – the participants responded 'yes', because the conclusion was not inconsistent with the text. The moral of these studies is that reasoners develop particular strategies during the course of experiments (see also Johnson-Laird, Savary, and Bucciarelli, 1999), and these strategies can be exquisitely tuned to the exigencies of the problems.

Complex reasoning problems, such as the logical problems in the GRE examination, vary in their difficulty. We have succeeded in isolating three causes of difficulty – the nature of the task, the nature of the foils, and the nature of the conclusions. A task for the future is to determine how the logical structure of the text, taken in conjunction with the correct response, influences difficulty.

Acknowledgements
Yingrui Yang is a visiting fellow at Princeton University and an E.T.S. post-doctoral fellow. We are grateful to colleagues in the Logical Reasoning group at E.T.S. for their advice and help: Malcolm Bauer, Charles Davis, Larry Frase, Karen Kukich, Carol Tucker, and Ming-Mei Wang. We also thank our colleagues at Princeton and elsewhere for their useful suggestions: Victoria Bell, Zachary Estes, Yevgeniya Goldvarg, Hansjoerg Neth, Mary Newsome, Sergio Moreno Rios, Vladimir Sloutsky, and Jean-Baptiste van der Henst.

References
Johnson-Laird, P.N., and Byrne, R.M.J. (1991) Deduction. Hillsdale, NJ: Lawrence Erlbaum Associates.

Johnson-Laird, P.N., Savary, F., and Bucciarelli, M. (1999) Strategies and tactics in reasoning. In W.S. Schaeken, A. Vandierendonck, G. De Vooght, & G. d'Ydewalle (Eds.) Deductive Reasoning and Strategies. Mahwah, NJ: Lawrence Erlbaum Associates. In press.

A Computational Model of Number Comparison

Marco Zorzi (mzorzi@psico.unipd.it)
Dipartimento di Psicologia Generale, Università di Padova
via Venezia 8, 35131 Padova, Italy

Brian Butterworth (b.butterworth@ucl.ac.uk)
Institute of Cognitive Neuroscience, University College London
17, Queen Square, London WC1N 3AR, UK

Abstract

Number comparison is a task that has been widely used to investigate the mental representation of number magnitudes. It is frequently assumed that the mapping from numerals to a "mental number line" is compressive (i.e., logarithmic) or that magnitude representations have the property of scalar variability. In this study, we simulate the process of selecting the larger of two numbers in a neural network model. We show that it is possible to account for the main experimental effects (e.g., the distance effect and the number size effect) with a simple architecture using a linear representation of numerical magnitudes. The compressive effects that are found in the reaction times emerge from the non-linear interactions that are intrinsic to the decision process.

Introduction

Number comparison is one of the fundamental numerical abilities. McCloskey (1992) takes the ability to select the larger of two numbers to be the criterion of understanding of numbers. Neurological patients who perform abnormally on this task, turn out to be profoundly acalculic (Delazer & Butterworth, 1997). Recent research on infant numerosity discrimination is consistent with the idea that infants can recognise which of two visual arrays contains the more objects (e.g., Wynn, 1992). It is also known that some primates, when presented with two visual arrays, can reliably select the array with more objects, for small numbers (Brannon & Terrace, 1998; Washburn & Rumbaugh, 1991).

As with many stimulus dimensions, number size shows a symbolic *distance effect*. That is, it is easier and quicker to select the larger of two numbers when they are numerically dissimilar than when they are similar (Moyer & Landauer, 1967). This "distance effect" has been found in young children (Sekuler & Mierkiewicz, 1977) and even in primates (Washburn & Rumbaugh, 1991; Brannon & Terrace, 1998). Thus, the ability to compare numerosities could be ontogenetically and phylogenetically basic.

According to McCloskey, Caramazza, and Basili's (1985) model, comparing the magnitude of two numerals requires the generation of an abstract representation corresponding to each numeral. However, the way in which the magnitudes of these abstract representations are compared is an unspecified process. In Dehaene's (1992) triple-code model, number comparison is performed on the basis of an *analogue magnitude* code (Dehaene & Cohen, 1995). Again, the details of the comparison process operating on these codes is largely unspecified. In effect, number comparison has not been simulated to our knowledge within a computational framework (but see Dehaene & Changeaux, 1993, for a simulation of the preverbal elementary ability to compare small sets of up to 3-4 objects). The serious limit of any verbal model of number comparison is that the comparison process itself and nature of the input-output representations on which this processes operates are not made explicit; the corollary to this fact is that the origin of the classic effects found in magnitude comparison tasks, such as the symbolic-distance and the number size effects, are still poorly understood.

The symbolic-distance effect refers to the finding that the latency of the comparative judgement is an inverse function of the numerical distance between the two numerals. That is to say, it is easier (i.e., faster) to compare 3 and 5 than 3 and 4. This classic result, indexed by the "split" or difference between the two numbers, has been found with arabic numerals (Banks, Fujii, & Kayra-Stuart, 1976; Buckley & Gillman, 1974; Duncan & MacFarland, 1980; Moyer & Landauer, 1976; Parkman, 1971; Sekuler & Mierkiewicz, 1977), patterns of dots (Buckley & Gillman, 1974), and written-word numerals (Foltz, Poltrock, & Potts, 1984). The number size (or serial-position) effect refers to the fact that, for a given symbolic distance, pairs of small numbers are compared faster than pairs of large numbers. Again, this is a classic result which has been observed with arabic numerals (Buckley & Gillman, 1974; Parkman, 1971), with patterns of dots (Buckley & Gillman, 1974), and with written-word numerals (Foltz et al., 1984).

We present below a connectionist model of number comparison, which can account for the main findings with a very simple processing architecture and a very limited set of basic assumptions.

Representation of number magnitudes

What representations are used in the comparison task to select the larger? The discoverers of the distance effect in numerical judgments, Moyer and Landauer (1967) state that "the decision process ... is one in which the displayed numerals are converted to analogue magnitudes, and a comparison is then made between these magnitudes in much the same way that comparisons are made between physical stimuli such as loudness or length of line." (p. 1520). They carefully separate the process of *conversion* of symbols to analogue representations from the process of *deciding* which is the larger. It is frequently assumed, however, that the key parametric findings should attributed to the process of conversion alone. The rationale for this position is very clearly stated by Dehaene who, in a series of papers, argues that the mapping from numerals to the number line is non-linear. This is because the line is held to be compressive, that is obeying Weber-Fechner logarithmic law. Accordingly, the subjective difference between two numbers will depend on their positions on the line, that is, the subjective difference between N and N+1 will be smaller as N increases (Dehaene, 1992; Dehaene, Dupoux & Mehler, 1992).

The similarity between numerical judgments and physical judgments has struck many other authors including Gallistel and Gelman (1992), who argue that the *mechanism* of comparing the magnitudes, again conceptualised as analogue, of two numbers is equivalent to comparing the lengths of two lines. However, their conception of analogue magnitude is subtly but importantly different from Dehaene's. Gallistel and Gelman (1992) propose that the mapping from number symbol (word or numeral) to the magnitude representation is linear, not compressive, but the variability of the mapping increases in proportion to the magnitude. For this reason, "the discriminability of the two numbers decreases as their mean numerical value increases, not because they are subjectively closer together, but because the variability (noise) in the mapping is scalar." (p. 57).

Analogue representations, however, fail to capture our intuitive notion of whole numbers, and whole-number arithmetic. Perhaps because of our early experience with counting, we intuitively think of whole numbers as meaning not approximate analogue magnitudes, but discrete numerosities.

In particular, we think of a whole number as denoting the numerosity (or cardinality) of a set with discrete members. Intuitively, we think of arithmetical operations on whole numbers in terms of sets and numerosities. For example, we think of the addition of x and y as being the numerosity of the union of two disjoint sets whose numerosities are x and y (Giaquinto, 1995).

Our working hypothesis is, then, that number representations are ordered by numerosity: smaller numbers denote proper subsets of the sets denoted by bigger numbers. Notice that this hypothesis is not trivial. If we conceptualise numbers essentially as words, then they will not be intrinsically ordered by numerosity magnitude; they will be instead be intrinsically ordered along some verbal dimension, such as the alphabet. Even ordinal numbers are not ordered by magnitude - the first past the post is not smaller than the second past the post, even though 1 is smaller than 2.

Our principal question is which aspects of the comparison phenomena should be attributed to the representation of numerical magnitudes and which to the implementation of a decision process. The reaction time data for the judgement of physical magnitudes across a wide range of domains (e.g., line length, pitch, weight) are well represented by the equation proposed by Welford (1960):

$$RT = a + k \log[L/(L-S)]$$

where L and S are the larger and the smaller physical magnitudes, respectively, and a and k are constants. The same equation has been found to be the best predictor of number comparison reaction time data, accounting for about 50% of the variance (e.g., Moyer & Landauer, 1973; Dehaene, 1989). Given this striking similarity, it is unclear why the experimental effects found in number comparison should be attributed to the *representation* of numerical magnitudes rather than to the decision process *per se*.

This lead us to the issue of *implementation*. Within a connectionist framework, a two-choice decision process can be implemented by two nodes that compete with each other for responding to the input (e.g., Zorzi & Umiltà, 1995). What is less straightforward, however, is how to represent numbers as activation patterns over a set of processing units. The analogue **"number line" hypothesis** represents number magnitudes as points or regions on a continuous psychological dimension. In one of the few attempts to model numerical processes in a neural network, that is McCloskey and Lindemann's (1989) model of multiplication facts retrieval, numbers were encoded over an ordered sequence of input nodes, where each node stood for a particular number. Moreover, the two immediate neighbours of the number were activated as well: thus 5 was represented as the activation of the node labelled "5" plus (lesser) activation of "4" and "6". Although this provides some ordering of numbers, "8" and "4", with no overlapping neighbours, would activate orthogonal representations (i.e., nodes 7-8-9 for "8" and nodes 3-4-5 for "4"). McCloskey and Lindeman (1989) did not however attempt to model number comparison, and it is not clear how it would succeed in capturing the distance effect. In any event, this representation is very different from, and incompatible with, a numerosity representation (see Figure 1).

Our approach is to represent **numerosity magnitude** straightforwardly as the number of units activated, such that bigger numbers include smaller numbers; therefore, for N>M, a set with M members can be put in 1-1 correspondence with a proper subset of the set with N members. This representational scheme is also known as a "thermometer" representation (see Figure 1, right panel).

| McCloskey-Lindemann | Numerosity magnitudes |

Figure 1: Alternative schemes for representing numbers (top row: 3; bottom row: 7). On the left is the McCloskey-Lindemann scheme in which an input number activates its own representation and, to a lesser extent, its immediate neighbours. On the right is our numerosity (i.e. the cardinality) representation, where each number is represented by a set of activated units corresponding to its numerosity.

The numerosity representation just described has several advantages. First, it readily maps onto lower level perceptual processes (e.g., object identification) and enumeration procedures (e.g., subitizing, counting). That is, each magnitude increment in our numerosity representation corresponds to the enumeration of a further element in the to-be-counted set. Second, it entails that larger numbers are more similar to each other than smaller numbers, without assuming a logarithmic compression, since large numbers share more active nodes. For example, 9 and 8 would share 8 nodes, whereas 1 and 2 would share only 1 node. This can also be formalized in terms of the cosine of the angle formed by the vectors coding the two numbers. Finally, we do not assume that the variability of the mapping from symbols to magnitude representation increases with size as Gallistel and Gelman (1992) proposed. Rather, the mapping in our scheme is linear and not noisy.

Model of number comparison

Architecture

The model is implemented in a network, in which each node is associated with an activation value. Nodes are connected by weighted links, which may be excitatory (positive) or inhibitory (negative).

We assume that number comparison is performed on the basis of magnitude, "semantic" codes (e.g., McCloskey et al., 1985; Dehaene, 1992). Therefore, this level of representation is used as input data for the model. The numbers to be compared are each represented by a set of 9 nodes, which are activated according to the "numerosity magnitude" scheme discussed above. The representation of the two possible responses (left or right button-press, to indicate which of the numbers is the larger) consists of two nodes, which we call the "response system" (see Zorzi & Umiltà, 1995). Activation values of the input nodes (number magnitudes) are in the [0,1] range, whereas in the response system suppressed states (modelled as negative activations) are permitted. In this case the range of activation is [-1,1], but only positive activations propagate through the connections. At stimulus onset, the relevant input nodes are clamped to the "1" value.

The response system incorporates a competitive mechanism, that, via lateral inhibition, implements response competition (e.g., Zorzi & Umiltà, 1995). Thus, the response system can be represented as a dipole, where two mutually exclusive responses compete: each response node has an inhibitory connection to the other node. The state of each response node changes smoothly over time in response to influence (of both excitatory and inhibitory kind) from the other nodes of the network (magnitude nodes for the two numbers, and the other response node). For simulation purposes, continuous time units can be approximated with discrete time units, in which time is discretized into ticks of some duration τ. Therefore, the new state of each response node at time $t+\tau$ is a weighted average of its current state at time t and the state dictated by its external input, according to the following equations:

$$a_j^{[t+\tau]} = \tau\sigma(net_j) + (1-\tau)a_j^{[t]} + \eta \qquad (1)$$

where a_j is the activation level of the response node j, and net_j is the external input to j, and τ is a parameter defining the weighting proportion that determines how gradually the state of the node changes over time. Note that η represents a small random noise, which gives the model non-deterministic (stochastic) properties.

To bound the activation values in the range [-1,1], the function $\sigma(x)$ in (2) is a S-shaped squashing function (hyperbolic tangent):

$$\sigma(x) = \frac{2}{1 + e^{-\lambda x}} - 1 \qquad (2)$$

where λ is a "temperature" parameter defining the sigmoidal shape of the function. The net input (external input) to the response nodes is given by:

$$net_j = (\sum_i w_{ij}o_i) - w^- a_k \qquad (3)$$

where w_{ij} is the weight of the connection from the input node i to the response node j, o_i is the output of the input node i, a_k is the activation of the other response node, and

w^- is the weight of the inhibitory link from the other response node. Free parameters for all simulations reported in this paper: $\lambda=4$, $\tau=0.01$, η=random gaussian noise (mean = 0, standard deviation = 0.01).

After stimulus onset (i.e., activation of the relevant numerical magnitudes), the system is allowed to cycle until there is a winning node. We simply assume that a response occurs when the difference between the activations of the two response nodes exceeds a certain threshold. The number of cycles required by the system to settle represents a measure of the reaction time (RT) in responding to the stimulus.

The connections linking the magnitude nodes for the two numbers with the response system are learnt in the model by simple association of the input with the required response. This is done in the model by simple "hebbian learning": a connection is strengthened if the activation of the nodes that it connects are correlated (e.g., both nodes are active). Formally:

$$w_{ij} = \varepsilon \, a_i a_j \qquad (4)$$

where w_{ij} is the weight of the connection between the nodes i and j, and a_i and a_j are the activation values of the two nodes, and ε is a small learning rate.

Learning is done in a "one-shot" fashion, in the sense that the 72 possible input patterns (i.e., all combinations of two 1-9 digits, excluding the ties) are presented just once, simultaneously with the required response. The connections involved in the learning phase are those from the magnitude nodes to the response nodes. The model's architecture, with its nodes and connections, is depicted in Figure 2. Not surprisingly, the connections linking uncorrelated nodes are not strengthened.

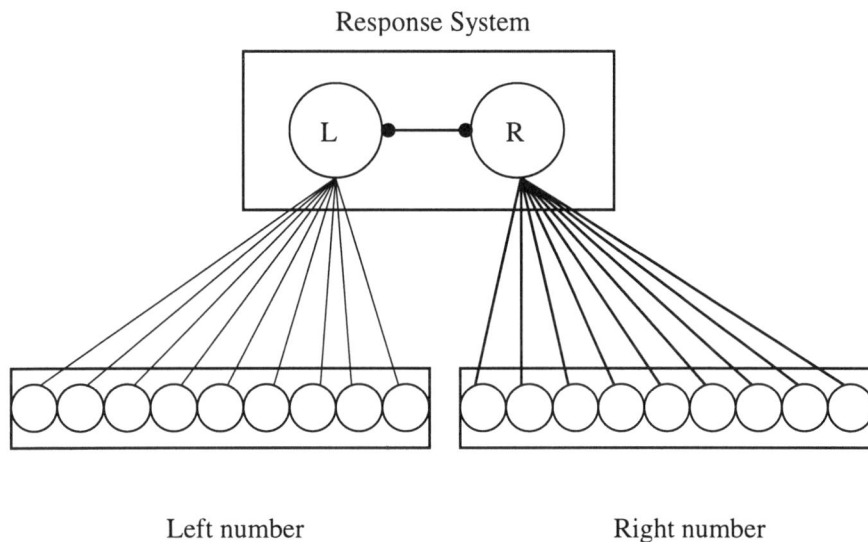

Response System

Left number

Right number

Figure 2: Architecture of the model. The connection linking the two response nodes is an inhibitory dipole implementing response competition. All connections from magnitude representations to response system are excitatory.

Results

The model was presented with the numerosity representations of all possible pairs of single digit numbers (1-9). Activation of the magnitude nodes propagates gradually to the response nodes, and the model is allowed to cycle until response criterion is reached, which consists in the difference of the activations of the two response nodes; we assume that a response can be unambiguously selected when this difference becomes equal to [0.5] or bigger. At that point, the number of cycles needed by the system to reach response threshold is taken as a measure of the reaction time. Crucially, this will in turn depend on the amount of competition between the two nodes. Note that response competition is what accounts in general for the relevant part of empirical RTs, across domains as different as attention (e.g., Cohen & Huston, 1994; Houghton & Tipper, 1994; Zorzi & Umiltà, 1995) and reading aloud (e.g., Zorzi, Houghton, & Butterworth, 1998).

The presence of noise in the activation function of the response nodes implies that the model can exhibit a relative variability in the response times. Therefore, each pair of numbers is presented 100 times to the model, and a mean RT is computed for each pair. The mean RTs produced by the model are analysed by regressing the standard structural variables (the magnitudes of the two numbers, and their difference) onto them. The two main effects that are usually found in a number comparison task are the distance or "split" effect (i.e., RTs increase as the difference between the two numbers becomes smaller) and the number size effect (i.e., RTs increase as the size of the two numbers increases). A variable that is standardly used to index the latter effect is the sum of the two numbers. A linear regression onto the model's RTs showed that the split accounts for 40.3% of the variance (p<.001) and the sum accounts for 57.6% of the variance. A different variables that has been often reported as a good predictor of comparison times is the Welford function, i.e. Log (Larger-Smaller)/Larger)). Used as a predictor of the model's RTs, the Welford functions accounts for 88.3% of the variance. As with human performance, the reaction times produced by the model are sensitive to the difference between the two numbers (i.e., split) and to the overall size of the two numbers. The effect of the split can be seen more clearly in Figure 3.

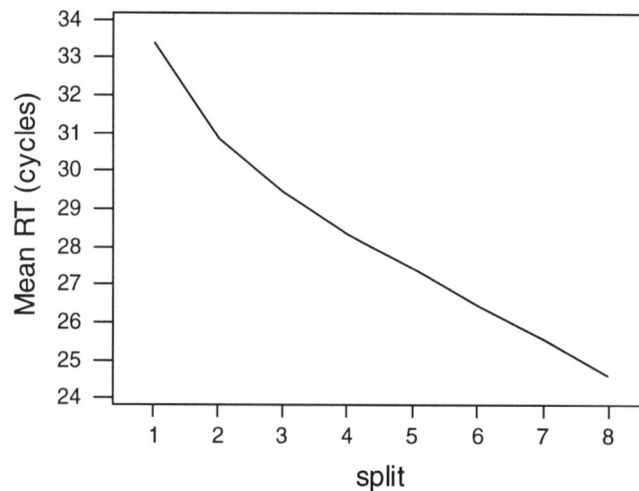

Figure 3: Graph shows the effect of the split (difference between the two numbers) on the number of cycles that the system needs to select a response.

General Discussion

We have modelled number comparison for the first time, and have done it in such a way that seems to capture out intuitive understanding of whole numbers. This shows that analogue representations of number magnitudes are not necessary to fit the data from comparison tasks, as has been often claimed. The crucial point, however, is that magnitude representations need not be compressed in order to observe a Weber-Fechner logarithmic effect in number comparison, contrary to the claims of Dehaene and his colleagues. In our simulation, numerals were mapped linearly on to magnitude representations, and the compressive effect on the comparison times emerges by virtue of the non-linear interactions that are intrinsic to the decision process itself. The non-linear squashing function in the *response units* produces a compression of the input magnitudes which is larger for bigger numbers. It is also not necessary to postulate that magnitude representations have the property of scalar variability, that is, that the standard deviation of mapping from numerals to magnitudes increases with the mean magnitudes of the numbers, as claimed by Gallistel and Gelman (1992).

There are two main psychological advantages of linear mapping from number symbols to number magnitudes. For one thing, it corresponds to our intuitive notion that each counting increment when enumerating a set of objects is equivalent, regardless of the size of the set. Secondly, these magnitudes are appropriate for arithmetical operations on whole numbers, which are linear, whereas compressive representations would not be. As Gallistel and Gelman note, "the concatenation of mental magnitudes is isomorphic to addition of the corresponding values." (1992, p. 57).

Acknowledgments

This research was supported by Wellcome Biomedical Research Collaboration Grant 048004 and Wellcome Grant 045013.

References

Banks, W. P., Fujii, M., & Kayra-Stuart, F. (1976). Semantic congruity effects in comparative judgements of magnitudes of digits. *Journal of Experimental Psychology: Human Perception and Performance, 2,* 435-447.

Brannon, E. M., & Terrace, H. S. (1998). Ordering of the numerosities 1 to 9 by monkeys. *Science, 282,* 746-749.

Buckley, P. B., & Gillman, C. B. (1974). Comparisons of digits and dot patterns. *Journal of Experimental Psychology, 103,* 1131-1136.

Cohen, J. D., & Huston, T. A. (1994). Progress in the use of interactive models for understanding attention and performance. In: C. Umiltà & M. Moscovitch (Eds.), *Attention & Performance XV* (pp. 453-475). Cambridge (MA): MIT Press.

Dehaene, S. (1989). The psychophysics of numerical comparison: A reexamination of apparently incompatible data. *Perception & Psychophysics, 45,* 557-566

Dehaene, S. (1992). Varieties of numerical abilities. *Cognition, 44,* 1-42

Dehaene, S., & Changeaux, J. (1993). Development of elementary numerical abilities: A neuronal model. *Journal of Cognitive Neuroscience, 5,* 390-407.

Dehaene, S., & Cohen, L. (1995). Towards and anatomical and functional model of number processing. *Mathematical Cognition, 1,* 83-120.

Dehaene, S., Dupoux, E., & Mehler, J. (1992). Is numerical comparison digital: Analogical and symbolic effects in two-digit number comparison. *Jorrnal of Experimental Psychology: Human Perception and Performance, 16,* 626-641

Delazer, M., & Butterworth, B. (1997). A dissociation of number meanings. *Cognitive Neuropsychology, 14,* 613-636.

Duncan, E. M., & McFarland, C. E. (1980). Isolating the effects of symbolic distance and semantic congruity in comparative judgments: an additive factors analysis. *Memory and Cognition, 8,* 612-622.

Foltz, G. S., Poltrock, S. E., & Potts, G. R. (1984). Mental comparison of size and magnitude: Size congruity effects. *Journal of Experimental Psychology: Learning, Memory and Cognition, 10,* 442-453.

Houghton, G., & Tipper, S. P. (1994). A model of inhibitory mechanisms in selective attention. In D. Dagenbach & T. Carr (Eds), *Inhibitory Mechanisms in Attention, Memory and Language*. San Diego, CA: Academic Press.

Gallistel, C. R., & Gelman, R. (1992). Preverbal and verbal counting and computation. *Cognition, 44,* 43-74.

Giaquinto, M. (1995). Concepts and calculation. *Mathematical Cognition, 1,* 61-81.

McCloskey, M. (1992) Cognitive mechanisms in number processing: Evidence from acquired dyscalculia. *Cognition, 44,* 107-157 .

McCloskey, M., Caramazza, A., & Basili, A. (1985). Cognitive mechanisms in number processing and calculation: Evidence from dyscalculia. *Brain and Cognition, 4,* 171-196.

McCloskey, M., & Lindemann, A. (1992). Mathnet: preliminary results from a distributed model of arithmetic fact retrieval. In J.I.D. Campbell (Ed.), *The nature and origins of mathematical skill* (p. 365-409). Amsterdam: Elsevier.

Moyer, R. S., & Landauer, T. K. (1967). Time required for judgments of numerical inequality. *Nature, 215,* 1519-1520.

Parkman, J. M. (1971). Temporal aspects of digit and letter inequality judgments. *Journal of Experimental Psychology, 91,* 191-205.

Sekuler, R., & Mierkiewicz, D. (1977). Children's judgements of numerical inequality. *Child Development, 48,* 630-633.

Washburn, D. A., & Rumbaugh, D. M. (1991). Ordinal judgments of numerical symbols by macaques (*Macaca Mulatta*). *Psychological Science, 2,* 190-193.

Wynn, K. (1992). Addition and subtraction by human infants. *Nature, 358,* 749-751.

Zorzi, M., Houghton, G., & Butterworth, B. (1998). Two routes or one in reading aloud? A connectionist dual-process model. *Journal of Experimental Psychology: Human Perception and Performance, 24,* 1131-1161.

Zorzi, M. & Umiltà, C. (1995). A computational model of the Simon effect. *Psychological Research, 58,* 193-205.

Routes, Races, and Attentional Demands in Reading: Insights from Computational Models

Marco Zorzi (**mzorzi@psico.unipd.it**)
Dipartimento di Psicologia Generale, Università di Padova
via Venezia 8, 35131 Padova, Italy
Institute of Cognitive Neuroscience, University College London
17, Queen Square, London WC1N 3AR, UK

Abstract

One influential view about the attentional demands of the reading processes maintains that phonological assembly is less automatic and more attention-demanding than phonological retrieval. The strongest evidence is this respect is the release-from-competition (RFC) effect (Paap & Noel, 1991), in which the pronunciation of low frequency exception words is speeded when participants have to perform a concurrent memory task. However, the results of follow-up investigations have led to a sharp controversy regarding whether the phenomenon is real and whether it can be replicated or not. The debate has reached stalemate, partly because the discussion about architectural and processing assumptions has been carried out only in verbal terms. This paper investigates the RFC phenomenon through simulations with two computational models of reading, the Connectionist Dual-Process model (Zorzi et al., 1998) and the DRC model (Coltheart et al., 1993). Both models failed to reproduce the RFC effect, even when the specific assumptions made by Paap and Noel were accurately implemented in the simulations. This finding casts further doubts about the reality of the phenomenon.

Introduction

Models of reading aloud distinguish between at least two different ways to derive the phonological form of written words. One route, usually referred to as lexical route, is thought to operate by retrieving the phonology of known words (*addressed or retrieved phonology*) from the visual word form through a word-specific association mechanism. The second route, named the assembly or phonological route, is conceptualized as a spelling to sound mapping process which allows the computation of the phonology (*assembled phonology*) for any (legal) string of letters (Carr & Pollatsek, 1985; Patterson & V. Coltheart, 1987, for reviews). This fairly general "two-process" architecture is supported by a large body of empirical data, and by converging evidence coming from the neuropsychological studies of reading disorders (acquired dyslexia; Denes, Cipolotti, & Zorzi, 1998; Shallice, 1980, for reviews).

Although most of the researchers would probably agree with the broad definition provided above, the details of the models in the literature vary widely, and the different claims about the nature of the computations underlying the reading system have sparked a vigorous debate (see, e.g., Besner, Twilley, McCann, & Seergobin, 1990; Coltheart, Curtis, Atkins, & Haller, 1993; Plaut, McClelland, Seidenberg, & Patterson, 1996; Seidenberg & McCelland, 1989; Van Orden, Pennington, & Stone, 1990; Zorzi, in press; Zorzi, Houghton, & Butterworth, 1998a, 1998b).

A disputed issue regards the nature of the assembly mechanism and the role of phonology in lexical access from print. One view is that phonological assembly consists of grapheme to phoneme correspondence (GPC) rules, that is, a system based upon explicit rules specifying the dominant (e.g., most frequent) relationships between letters and sounds (Coltheart, 1978; Coltheart et al., 1993, for a computational version of the GPC system). Furthermore, the GPC route is held to be slow and serial (i.e., it delivers one phoneme at a time), and is therefore regarded as a controlled process, which is resource-demanding and subject to strategic control (e.g., Monsell, Patterson, Graham, Hughes, & Milroy, 1992; Paap & Noel, 1991). More generally, the dual-route model (Baron, 1973; Coltheart, 1978; Meyer, Schvaneveldt, & Ruddy, 1974) gives a predominant role to the visual route, because the assembly of phonology is believed to be too slow to affect lexical access. A sublexical (or prelexical) activation of phonology (coming from the assembly route) can make some contribution to word recognition only for very low frequency words, that is, those that are too slowly dealt with by the lexical route (e.g., Seidenberg, Waters, Barnes, & Tanenhaus, 1984). Furthermore, because the GPC route delivers the regular pronunciation of a letter string, the assembled phonology is beneficial only in the case of regular words, while it is detrimental in the case of words with irregular spelling-sound correspondences (Coltheart et al., 1993; Coltheart & Rastle, 1994). On this account, written word recognition in real circumstances is largely a matter of direct visual access from print.

A radically different view is that phonological assembly is a fast and automatic process (e.g., Perfetti, Bell, & Delaney, 1988; Perfetti & Bell, 1991), that plays an important role in written word recognition (for reviews see Berent & Perfetti, 1995; Van Orden et al., 1990). Phonological properties of printed words can affect performance in a variety of reading tasks (see Frost, 1998, for review) and the results of several studies suggest a fast and automatic assembly process, preceding word identification. A fast and parallel process of phonological assembly is implemented in the recent connectionist "dual-process" model developed by Zorzi and colleagues (1998a, 1998b; Zorzi, in press).

The Release-From-Competition Effect

In 1991 Paap and Noel reported a new experimental phenomenon which seemed to provide strong evidence in favor of a "classic" dual-route model of reading. The participants in Paap and Noel experiments had to maintain in memory a number of digits while reading words aloud. The surprising finding was that a memory load of five digits speeded the pronunciation of low frequency exception words in the reading task compared to a memory load of one digit. By contrast, high frequency exceptions and both high and low frequency regular words slowed down in the same comparison. The speeding of low frequency exception was in fact predicted by Paap and Noel (although a concurrent task usually has a detrimental effect on performance of the primary task). They made specific assumptions (within the dual-route framework) about the attentional demands of the two reading routes. Paap and Noel proposed that phonological assembly is less automatic and more attention-demanding than phonological retrieval; therefore, the former would be more susceptible to interference in the case of a concurrent task. In the case of substantial slowing of the assembly route (as in the case of the five digits load), the lexical retrieval process would not be affected by the competing candidate that are usually assembled in reading exception words. Thus, low frequency exception words would not suffer from the competition derived from the existence of conflicting pronunciations. Bernstein and Carr (1996) named this effect "release from competition" (or RFC). The effect of the load manipulation on the four types of words can be seen in Figure 1.

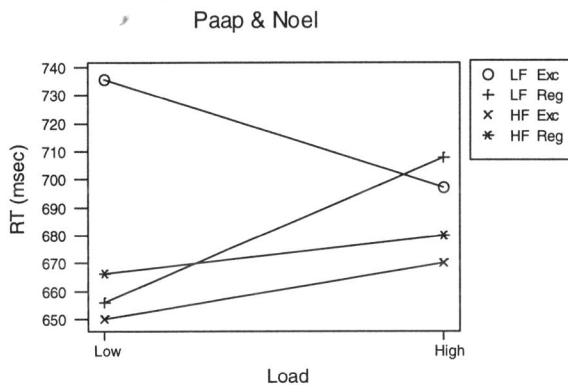

Figure 1: Data from Paap and Noel's (1991) study. The graph shows the effect of a digit memory load (low = one digit; high = five digits) on the naming latencies in the reading task.

The striking results of Paap and Noel (1991) prompted some researcher to carry out follow-up investigations. The results of these studies have led to a sharp controversy regarding whether the phenomenon reported by Paap and Noel is real and whether it can be replicated or not. Only one study by Herdman and Beckett (1996) replicated the RFC effect. Pexman and Lupker (1995) reported a complete failure to replicate the RFC effect in a series of five experiments. In their experiments, all word types slowed down under the high load condition (see Figure 2). Bernstein and

Carr (1996) succeeded in replicating the RFC effect, but only in a subset of readers, suggesting individual differences in the architecture of the reading system. However, Pexman and Lupker (1998) criticized Bernstein and Carr's finding as artifactual and failed to replicate it with a different way of selecting readers. In a follow-up study of the individual differences account, Bernstein, DeShon, and Carr (1998) found little evidence of systematic individual differences in the occurrence of the RFC effect. Thus, Bernstein et al. concluded that the effect cannot be replicated (but see Paap and Herdman, 1998, for opposing views).

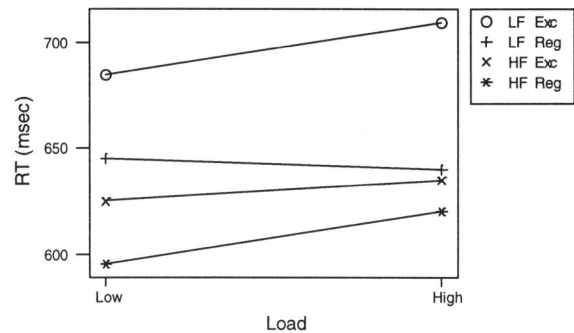

Figure 2: Data from Pexman and Lupker's (1995) experiment 1. Note that the increasing digit memory load (low = one digit; high = five digits) slowed down all word types in the same way

Simulations of the RFC Effect

It is clear that the debate concerning the Paap and Noel (1991) phenomenon has reached stalemate. The investigators involved in this debate have discussed at length why the effect should or should not exist, focusing on the predictions derived from different models and/or different architectural assumption. However, these detailed analyses and discussions have been carried out only in verbal terms, that is without actually testing any of the available computational models. Pexman and Lupker (1995) have even hypothesized how memory load might affect a single route connectionist model. In the limit, this enterprise might be misleading, because the behavior of a complex computational model can be neither predicted nor inferred without actually running the simulations.

In this article, I investigate the Paap and Noel (1991) phenomenon through simulations with the Connectionist Dual-Process model (Zorzi et al., 1998a) and with Coltheart et al.'s (1993) DRC (Dual-Route Cascade) model. Starting with simple computational assumptions regarding the effect of memory load (which follow directly the assumptions made by Paap and Noel), I assess whether the RFC effect can be observed in the models in any of a range of tested conditions.

Simulations with the Dual-Process Model

Zorzi et al. (1998a) developed a connectionist model of reading where a dual-route processing system emerges from the interaction of task demands and initial network architecture in the course of reading acquisition. In this model, the distinction between phonological assembly and lexical retrieval is realized in the form of connectivity (either direct of mediated) between orthographic input and phonological output patterns (see Houghton & Zorzi, 1998, for similar treatment of the problem of learning the sound-spelling mapping in writing). The model thus maintains the uniform computational style of the PDP models, but makes a clear distinction between lexical and sublexical processes in reading. The model has been shown to account for a number of empirical results, including the interaction between frequency and regularity, the effects of consistency, the effect of the position of irregularity, the interaction between word length and lexicality, and the impaired performance of patients with neuropsychological disorders (Zorzi et al., 1998a; 1998b; Zorzi, in press).

The two processing pathways of the Dual-Process model are activated in parallel and their phonological output builds up over time in a cascaded fashion. However, in contrast to a classic dual-route horserace model, the processing rates of the two routes are very similar (the assembly route is actually faster than the lexical route). Following Paap and Noel's (1991) assumption that a digit memory load slows down processing in the assembled route, the effect of memory load was simulated in the Dual-Process model by manipulating the ramp parameter of the phonological assembly route. This parameter dictates the number of processing cycles that are necessary for the assembly route to reach its asymptotic (i.e., maximum) output value. Therefore, an increase in the value of ramp corresponds to a slower production of the assembled phonology in the model.

Method The low load condition was simulated by holding the ramp parameter unchanged (8 cycles). That is to say, the low load condition was a no-load condition in the model. For the high load condition, the model was tested with two different values of the ramp parameter, that is 15 and 20 cycles. The word lists used by Paap and Noel (1991) were submitted to the model. Of the 40 words used in the Paap and Noel study, 8 were not in the model's lexicon (two of these were bisyllabic, and four other were inflected words) and were therefore excluded. Because the words are matched in quadruplets, the three corresponding words for

each of the excluded word were removed from the analyses. This left with 14 words in each list.

RT data were collected from the model by first running the word lists with the original model's parameter (low load condition). The same lists were presented to the model two other times with the assembly ramp parameter changed to 15 and then to 20 cycles.

Results - Ramp 15 The RT data obtained from the model were analyzed in a 2 (Frequency: high vs. low) X 2 (Regularity: regular vs. exception) X 2 (Load: low vs. high) repeated measures analysis of variance. The effects of frequency ($F_{1,13}=64.72$, Mse=91.08, p<.001), regularity ($F_{1,13}=39.45$, Mse=23.22, p<.001), as well as their interaction ($F_{2,13}=17.67$, Mse=9.72, p=.001) were significant. Naming latencies for high frequency words (2.62 cycles) were faster compared to low frequency words (4.43 cycles), and regular words (3.07) were faster compared to exception words (3.98). The effect of load was significant ($F_{1,13}=162.87$, Mse=33.22, p<.001), showing that naming latencies under high load (4.07) were longer than under low load (2.98). However, load interacted with both regularity ($F_{2,13}=36.38$, Mse=3.22, p<.001) and frequency ($F_{2,13}=33.85$, Mse=8.58, p<.001). That is, load was more affecting low frequency words (3.6 vs. 5.24) than high frequency words (2.35 vs. 2.89), and regular words (2.35 vs. 3.78) more than exception words (3.60 vs. 4.35). The three-way interaction was not significant. The results are shown in Figure 3, panel A.

Results - Ramp 20 The RT data obtained from the model were analyzed in a 2 (Frequency: high vs. low) X 2 (Regularity: regular vs. exception) X 2 (Load: low vs. high) repeated measures analysis of variance. The effects of frequency ($F_{1,13}=114.49$, Mse=106.08, p<.001), regularity ($F_{1,13}=38.21$, Mse=16.51, p<.001), as well as their interaction ($F_{2,13}=13.72$, Mse=7.51, p<.01) were significant. The effect of load was significant ($F_{1,13}=250.13$, Mse=58.58, p<.001), showing that naming latencies under high load (4.43) were longer than under low load (2.98). However, load interacted with both regularity ($F_{2,13}=52.36$, Mse=6.51, p<.001) and frequency ($F_{2,13}=49.81$, Mse=13.58, p<.001). The three-way interaction was significant ($F_{3,13}=4.94$, Mse=.72, p<.05. The results are shown in Figure 3, panel B.

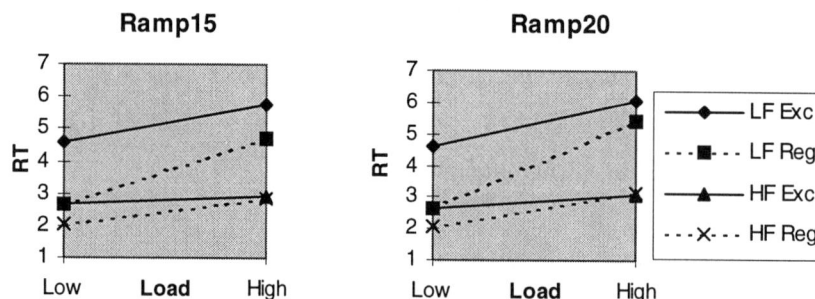

Figure 3: Results of the load manipulation with the Dual-Process model. Panel A: ramp parameter was set to 15 cycles (Ramp15). Panel B: ramp parameter was set to 20 cycles (Ramp20).

Discussion Inspecting the graphs in Figure 3, it can be noted that the results are quite similar for both values of ramp. More important, however, is that in both cases the model does not show a RFC effect. That is, the significant interactions in the two anova qualify a pattern that is very different from that found by Paap and Noel (1991). The load manipulation does not speed up any of the word types. Load has more impact on low frequency words; within low frequency words, the detrimental effect of load is maximal for the regular words. This pattern of results can be ascribed to the reduced contribution of phonological assembly in the model. Note, however, that load has some effect on the high frequency words as well. This is because the assembly procedure in the Dual- Process model is a fast process, interacting with the lexically-derived phonology even in the case of high frequency words.

The pattern produced by the Dual-Process model resembles rather closely that found by Pexman and Lupker (1995) in their Experiment 1, which attempted to replicate the Paap and Noel (1991) phenomenon using the same experimental material.

Simulations with the DRC model

The DRC model (Coltheart et al., 1993; Coltheart & Rastle, 1994) is a traditional dual-route model, where the assembly procedure is conceptualized (and implemented) as a slow, serial route, operating on grapheme-phoneme correspondences. Note that this characterization perfectly fits the dual-route horse-race model described by Paap and Noel (1991). Therefore, it might be anticipated that DRC is the model that would most likely show a pattern similar to the Paap and Noel phenomenon. In contrast to the Dual-Process model, the processing rates of the two routes are quite different. In this regard, the key feature of DRC is that the lexical route operates in parallel on the letter constituents of the input word, whereas the assembly (GPC) route operates serially on the individual letters. The letters are therefore submitted to the GPC route one at a time, with the speed of serial processing determined by a parameter specifying the time interval between submission of each letter. Such parameter is tied to the processing in the lexical route, and is therefore measured in processing cycles. The parameter is normally set to 17; that is, each letter is submitted to the GPC algorithm every 17 processing cycles in the lexical route. An increase of the interval would slow down the assembly procedure and as such is a good candidate for simulating a memory load manipulation. A second parameter of DRC which can be useful to simulate a memory load condition is the strength of the GPC procedure, that is, the excitation received by the phoneme system when a given phoneme is produced by the GPC rules.

Method A procedure similar to that used with the Dual-Process model was employed to investigate the effect of memory load in DRC (I am grateful to Kathy Rastle for running the simulations). The low load condition was simulated in DRC by holding all GPC parameters unchanged (inter-letter interval=17 cycles; GPC excitation=.055). Thus, similarly to the simulations with the Dual-Process model, the low load condition was actually a no-

load condition. For the high load condition, the model was tested with two different values of the inter-letter interval parameter, which was increased to 25 and to 50 cycles. In addition, a third simulation of the high load condition was obtained by reducing the strength of GPC excitation from .055 to .03.

The word lists used by Paap and Noel (1991) were submitted to the model. Of the 40 words used in the Paap and Noel study, 4 were not in the model's lexicon (two of these were bisyllabic) and one was wrong in the database. These words were therefore excluded, and the corresponding matched words in the other three lists were removed from the analyses. This left with 15 words in each list.

RT data were collected from the model by first running the word lists with the original model's parameter (low load condition). The same lists were presented to the model three other times with the GPC parameters altered as to simulate the high load condition.

Results - Inter-letter interval 25 The RT data obtained from the model were analyzed in a 2 (Frequency: high vs. low) X 2 (Regularity: regular vs. exception) X 2 (Load: low vs. high) repeated measures analysis of variance. The effects of frequency ($F_{1,14}$=42.58, Mse=1209.68, p<.001) and regularity ($F_{1,14}$=21.17, Mse=880.21, p<.001) were significant. Their interaction, however, was not significant ($F_{2,14}$= .46, Mse= 20.01, p>.5). Naming latencies for high frequency words (76.9 cycles) were faster compared to low frequency words (83.2 cycles), and regular words (77.3 cycles) were faster compared to exception words (82.7 cycles). The effect of load was not significant ($F_{1,14}$=1.16, Mse=8.01, p>.1). However, load interacted with regularity ($F_{2,14}$=14.91, Mse=81.68, p=.002). That is, regular words slowed down under high load (76.8 vs. 77.9 cycles), whereas exception words sped up (83.8 vs. 81.7 cycles). The load by frequency and the three-way interactions were not significant. The results are shown in Figure 5, panel A.

Results - Inter-letter interval 50 The RT data obtained from the model were analyzed in a 2 (Frequency: high vs. low) X 2 (Regularity: regular vs. exception) X 2 (Load: low vs. high) repeated measures analysis of variance. The effects of frequency ($F_{1,14}$=46.56, Mse=974.70, p<.001) and regularity ($F_{1,14}$=12.76, Mse=410.70, p<.01) were significant. Their interaction, however, was not significant ($F_{2,14}$= .09, Mse= 2.13, p>.7). Naming latencies for high frequency words (75.2 cycles) were faster compared to low frequency words (80.2 cycles), and regular words (76.2 cycles) were faster compared to exception words (79.9 cycles). The effect of load was significant ($F_{1,14}$=72.10, Mse=616.53, p<.001), showing that naming latencies under low load (80.3 cycles) were longer than under high load (75.8 cycles). Load interacted with regularity ($F_{2,14}$=50.80, Mse=340.03, p<.001). That is, both regular and exception words were faster under high load compared to low load, but exception words sped up (83.8 vs. 75.9 cycles) more than regular words (76.7 vs. 75.6 cycles). The load by frequency interaction was not significant, but the three-way interaction was reliable ($F_{3,14}$=5.86, Mse=38.53, p<.05. The results are shown in Figure 5, panel B.

Results - GPC Excitation .03 The RT data obtained from the model were analyzed in a 2 (Frequency: high vs. low) X 2 (Regularity: regular vs. exception) X 2 (Load: low vs. high) repeated measures analysis of variance. The effects of frequency ($F_{1,14}=47.19$, Mse=1098.08, p<.001) and regularity ($F_{1,14}=20.48$, Mse=658.01, p<.001) were significant. Their interaction, however, was not significant ($F_{2,14}=.94$, Mse= 20.01, p>.3). Naming latencies for high frequency words (75.7 cycles) were faster compared to low frequency words (81.8 cycles), and regular words (76.4) were faster compared to exception words (81.1). The effect of load was significant ($F_{1,14}=45.55$, Mse=285.21, p<.001), showing that naming latencies under low load (80.3 cycles) were longer than under high load (77.2 cycles). Load interacted with regularity ($F_{2,14}=32.92$, Mse=170.41, p<.001). That is, exception words were faster under high load compared to low load (83.8 vs. 78.4 cycles), whereas the effect of load for the regular words was quite small (76.8 vs. 76.1 cycles). The load by frequency interaction was not reliable, but there was a trend towards a three-way interaction ($F_{3,14}=3.27$, Mse=10.21, p=.09. The results are shown in Figure 5, panel C.

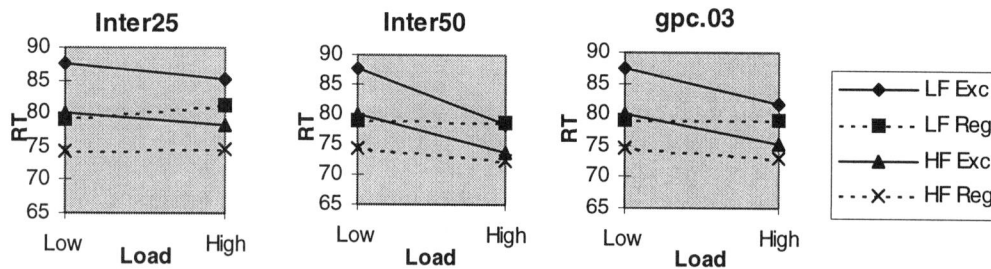

Figure 5: Results of the load manipulation with the DRC model. Panel A: the inter-letter interval was set to 25 cycles (Inter25). Panel B: the inter-letter interval was set to 50 cycles (Inter50). Panel C: GPC excitation was set to 0.03 (gpc.03).

Discussion Inspecting the graphs in Figure 5, it can be seen that the results produced by the three different high load conditions are similar, in particular for Inter50 and gpc.03. The crucial finding, however, is that in all cases the low frequency exception words speed up under high load. Thus, it would seem that the DRC model produces the RFC effect. However, other word types speed up under high load. For instance, high frequency exception words benefit of the high load as much as the low frequency exceptions. What is more striking, however, is that even high frequency regular words speed up under high load in two of the simulations, whereas low frequency regular words are virtually unaffected by the load manipulation. It should be clear that this pattern does not match that found by Paap and Noel (1991). This becomes more evident by visually comparing the original data from Paap and Noel in Figure 1 with the graphs of Figure 5.

General Discussion

Paap and Noel's (1991) RFC phenomenon is potentially very important because, if real, it poses a number of constraints on the architectural and processing assumptions of reading models. This is why the controversy about its replicability has been relatively sharp (e.g., Bernstein & Carr, 1996; Bernstein et al., 1998; Paap & Herdman, 1998; Pexman & Lupker, 1995, 1998). However, the theoretical debate has been carried out on purely verbal grounds. For example, Paap and Noel predicted that slowing down the nonlexical route (through a digit memory load) would facilitate processing of low frequency exception words. Pexman and Lupker (1995) argued that the same result could be explained within a single-route PDP account: they speculated that a memory load would hinder the ability of the competitors of the correct word pronunciation to compete during a phonological cleanup process. The debate has reached stalemate for two main reasons: first, several studies failed to replicate the RFC phenomenon (even when possible individual differences among readers where considered; Pexman & Lupker, 1995, 1998; Bernstein et al., 1998), but at least one study provided a complete replication of the effect (Herdman & Beckett, 1996); second, the various predictions and/or accounts have not been tested computationally.

The basic question addressed by the present study is not that of replicability of the effect, but rather that of whether a dual-route processing system does actually predict the Paap and Noel phenomenon. The RFC effect was investigated using two computational models of reading: the DRC model and the Dual-Process model. I assumed, following Paap and Noel (1991), that a memory load slows down the nonlexical route. However, the simulations with both models failed to produce the complete Paap and Noel phenomenon. The Dual-Process model showed a pattern of results that resembled the findings of Pexman and Lupker (1995). The simulations with the DRC model revealed a more complex pattern, but no simulation resembled the overall pattern observed by Paap and Noel (1991). This might be surprising, given that the architectural and processing assumptions of the DRC model appear to largely overlap with Paap and Noel's own characterization of the dual-route model.

In summary, the present study has important implications for the debate concerning the Paap and Noel (1991) phenomenon. Indeed, the predictions made by Paap and Noel were not confirmed in either of two different computational implementations of dual-route theories (the Dual-Process model and the DRC model). The models fail to produce the effect even when Paap and Noel's assumptions are accurately implemented in the simulations. This finding casts further doubts about the reality of the phenomenon.

References

Baron, J., & Strawson, C. (1976). Use of orthographic and word-specific knowledge in reading words aloud. *Journal of Experimental Psychology: Human Perception and Performance, 2,* 386-392.

Berent, I., & Perfetti. C.A. (1995). A rose is a reez: The two-cycles model of phonology asssembly in reading English. *Psychological Review, 102,* 146-184.

Bernstein, S.E., & Carr, T.H (1996). Dual-route theories of pronouncing printed words: What can be learned from concurrent task performance? *Journal of Experimental Psychology: Learning, Memory, and Cognition, 22,* 86-116.

Bernstein, S.E., DeShon, R.P., & Carr, T.H (1998). Concurrent task demands and individual differences in reading: How to discriminate artifacts from real McCoys. *Journal of Experimental Psychology: Learning, Memory, and Cognition, 24,* 822-844.

Besner, D., Twilley, L., McCann, R.S., & Seergobin, K. (1990). On the connection between connectionism and data: Are a few words necessary? *Psychological Review, 97,* 432-446.

Carr, T.H., & Pollatsek, A. (1985). Recognizing printed words: A look at current models. In D. Besner, T.G. Waller, & G.E. MacKinnon (Eds.), *Reading research: Advances in theory and practice* (Vol. 5, pp. 1-82). San Diego, CA: Academic Press.

Coltheart, M. (1978). Lexical access in simple reading tasks. In G. Underwood (Ed.), *Strategies of information processing* (pp. 151-216). London: Academic Press.

Coltheart, M., Curtis, B., Atkins, P., & Haller, M. (1993). Models of reading aloud: Dual-route and parallel-distributed-processing approaches. *Psychological Review, 100,* 589-608.

Coltheart, M., & Rastle, K. (1994). Serial processing in reading aloud: Evidence for dual-route models of reading. *Journal of Experimental Psychology: Human Perception and Performance, 20,* 1197-1211.

Denes, F., Cipolotti, L., & Zorzi, M. (1998). Acquired dyslexias and dysgraphias. In G. Denes & L. Pizzamiglio (Eds.), *Handbook of Clinical and Experimental Neuropsychology.* Hove: Psychology Press.

Frost, R. (1998). Toward a strong phonological theory of visual word recognition: True issues and false trails. *Psychological Bulletin, 123,* 71-99.

Herdman, C.M., & Beckett, B.L. (1996). Code specific processes in word naming: Evidence supporting a dual-route model of word recognition. *Journal of Experimental Psychology: Human Perception and Performance, 22,* 1149-1165.

Houghton, G., & Zorzi, M. (1998). A model of the sound-spelling mapping in English and its role in word and nonword spelling. In M.A. Gernsbacher & S.J. Derry (Eds.), *Proceedings of the Twentieth Annual Conference of the Cognitive Science Society* (p. 490-495). Mahwah (NJ): Erlbaum.

Meyer, D.E., Schvaneveldt, R.W., & Ruddy, M.G. (1974). Functions of graphemic and phonemic codes in visual-word recognition. *Memory & Cognition, 2,* 309-321.

Monsell, S., Patterson, K.E., Graham, A., Hughes, C.H., & Milroy, R. (1992). Lexical and sub-lexical translation of spelling to sound: Strategic anticipation of lexical status. *Journal of Experimental Psychology: Learning, Memory and Cognition, 18,* 452-467.

Paap, K.R., & Herdman, C.M. (1998). Highly skilled participants and failures to redirect attention: Two plausible reasons for failing to replicate Paap and Noel's effect. *Journal of Experimental Psychology: Learning, Memory, and Cognition, 24,* 845-861.

Paap, K.R., & Noel, R.W. (1991). Dual route models of print to sound: Still a good horse race. *Psychological Research, 53,* 13-24.

Patterson, K.E., & Coltheart, V. (1987). Phonological processes in reading: A tutorial review. In M. Coltheart (Ed.), *Attention and Performance XII: The psychology of reading* (pp. 421-447). Hillsdale, NJ: Erlbaum.

Perfetti, C.A., & Bell, L.C. (1991). Phonemic activation during the first 40 ms of word identification: Evidence from backward masking and masked priming. *Journal of Memory and Language, 30,* 473-485.

Perfetti, C.A., & Bell, L.C.,& Delaney, S.M. (1988). Automatic (prelexical) phonemic activation in silent word reading: Evidence from backward masking. *Journal of Memory and Language, 27,* 59-70.

Pexman, P.M., & Lupker, S.J. (1995). Effects of memory load in a word naming task: Five failures to replicate. *Memory & Cognition, 23,* 581-595.

Pexman, P.M., & Lupker, S.J. (1988). Word naming and memory load: Still searching for an individual differences explanation. *Journal of Experimental Psychology: Learning, Memory, and Cognition, 24,* 803-821.

Seidenberg, M.S., & McClelland, J.L. (1989). A distributed, developmental model of word recognition and naming. *Psychological Review, 96,* 523-568.

Seidenberg, M.S., Waters, G.S., Barnes, M.A., & Tanenhaus, M.K. (1984). When does irregular spelling or pronunciation influence word recognition? *Journal of Verbal Learning and Verbal Behaviour, 23,* 383-404.

Shallice, T. (1988). *From neuropsychology to mental structure.* Cambridge, England: Cambridge University Press.

Van Orden, G.C., Pennington, B.F., & Stone, G.O. (1990). Word identification in reading and the promise of sub-symbolic psycholinguistics. *Psychological Review, 97,* 488-522.

Zorzi, M. (in press). Serial processing in reading aloud: No challenge for a parallel model. *Journal of Experimental Psychology: Human Perception and Performance.*

Zorzi, M., Houghton, G., & Butterworth, B. (1998a). Two routes or one in reading aloud? A connectionist dual-process model. *Journal of Experimental Psychology: Human Perception and Performance, 24,* p. 1131-1161.

Zorzi, M., Houghton, G., & Butterworth, B. (1998b). The development of spelling-sound relationships in a model of phonological reading. *Language and Cognitive Processes, 13,* 337-371.

Towards Including Simple Emotions in a Cognitive Architecture in Order to Better Fit Children's Behavior

Roman V. Belavkin[1,2] (rvb@cs.nott.ac.uk)
Frank E. Ritter[2] (frank.ritter@nottingham.ac.uk)
David G. Elliman[1] (dge@cs.nott.ac.uk)

1 School of Computer Science,
2 School of Psychology,
U. of Nottingham; Nottingham, NG7 2RD UK

Introduction

Emotions are an important aspect of human behavior that should be included in cognitive models. Including them in a cognitive architecture is the safest way to ensure consistency, and to get reuse. We demonstrate here how interest, distress, and pleasure, three primary emotions shown by children when problem solving, can be implemented in a specific ACT-R (Anderson & Lebiere, 1998) model. These emotions can be initially implemented by modifying some of the ACT-R problem solving mechanisms and our reusable model of vision.

While emotions can be a result of information and knowledge processing, they are not purely cognitive processes. They may result in bodily changes (hormones etc.) that influence behavior, perception and decision making. Our task is to implement a simple theory of emotions into the ACT-R cognitive architecture and test it on an existing model of human behavior.

The Task and the Model

The model is a simulation of Tower of Nottingham problem (Wood & Middleton, 1975), which consists of several blocks of different shapes and sizes that have to be assembled to form a pyramid. The ACT-R model can interact with a simulation of the blocks through the Nottingham eye and hands model. The model predicts well how adults solve the puzzle. The model is accurate enough to test theories of development (Jones & Ritter, 1998) through varying parameters of the model (working memory, Expected Gain Noise, etc.). While there is good match to most of the data, there is still is room for improving the model's fit to some of the children's data (e.g., time per layer, errors per layer, strategies used). Providing this model with emotions could improve that fit, as emotions seem to play a greater role in children problem solving, particularly in this task.

The Architectural Changes

We consider three basic emotions: interest, distress, and pleasure. These emotions can be initially implemented relating interest to the goal value variable in ACT-R and by introducing two new variables representing amounts of two synthetic hormones, each produced by "positive" or "negative" emotions. The amounts will decay over time. The goal value will also decay over time and may depend on the amount of hormones. This will allow the model to abandon tasks after some time, when its interest decreases and no reinforcement by positive emotions has occurred. Successful productions will activate positive emotions, which change the amount of "positive" hormones, and associate the successful rule with that emotion as well as the knowledge facts used by the rule. A similar mechanism will support negative emotions (distress). Associations with emotions are realized through additional coefficients in expected gains of rules and base level activations of chunks. These coefficients amplify the chunks associated with a particular emotion according to the amount of the hormone corresponding to it. This will increase the possibility of retrieveing a positively associated chunk in good moods and visa versa. The coefficients in the expected gain of production rules will make the choice of actions affectively dependent. Some performance parameters, like speed of rules firing, expected gain noise, or perceptual characteristics in the model (sizes of fovea and parafovea) depend on the amount of hormones as well.

What We Expect to Find

Emotions seem to play a greater role in and sometimes take control of children's problem solving on this task. These changes will allow the existing ToN model to better match children's behavior by (a) slowing down performance in general, (b) slowing down initial performance as the child explores the puzzle driven by interest, (c) making performance sensitive to, and influenced by the success of previous actions, and (d) abandoning the task if performance is not successful. The mechanisms also can be reused to study more complex emotional behaviors using cognitive architecture where the task involves interaction.

Acknowledgments

This work was supported by DERA and the ESRC Centre for Research on Development, Instruction, and Training at the U. of Nottingham. Gary Jones and David Wood have provided useful comments.

References

Jones, G., & Ritter, F. E. (1998). Initial explorations of simulating cognitive and perceptual development by modifying architectures. In *Proceedings of the 20th Annual Conference of the Cognitive Science Society.* 543-548. Mahwah, NJ: LEA.

Wood, D., D. Middleton (1975). A study of assisted problem-solving. *Br. J. Psychol.*, 66, 2, pp. 181-191

Anderson, J. R., & Lebiere, C. (1998). *The atomic components of thought.* Mahwah, NJ: LEA.

Constructions as the Main Determinants of Sentence Meaning

Giulia Bencini (bencini@uiuc.edu) Adele E. Goldberg (agoldbrg@uiuc.edu)
Department of Linguistics
University of Illinois

Introduction

What types of linguistic information do people use to construct the meaning of a sentence? Most linguistic theories and psycholinguistic models of sentence comprehension assume that the main determinant of sentence meaning is the verb. In this study, a sorting paradigm (Healy and Miller, 1970) was used to explore the possibility that the main determinant of sentence meaning is the argument structure construction: a pairing of form and meaning at the clausal level (Goldberg, 1995; Jackendoff, 1997; Kay & Fillmore, 1999; Rappaport Hovav & Levin, 1998). Examples of English argument structure constructions with their form and meaning are shown in Table 1. Healy and Miller (1970) asked participants to sort sentences according to their meaning and found that participants were more likely to sort sentences according to the main verb in the sentence than according to the subject argument. On the basis of these results the authors concluded that the verb is the main determinant of sentence meaning. The study reported here was aimed at determining whether the morphological form of the verb or the argument structure provides the better indicator of overall sentence meaning.

The Experiment

Sixteen English sentences were used, obtained by crossing 4 verbs and 4 constructions. Participants sorted the sentences into four piles, each pile consisting of four sentences. They were asked to place sentences with the same overall meaning into the same pile.

The results showed that participants were more likely to perform a constructional sort than a verb-based sort: 41% of the participants sorted entirely by construction, no one sorted entirely by verb. Participants' sorts were overall closer to a constructional sort than to a verb sort. The average number of changes required for a sort to be entirely by construction was significantly smaller than the average number of changes required for the sort to be entirely by verb ($p < .0002$).

Conclusion

The results of this study suggest that argument structure constructions are better predictors of overall sentence meaning than the morphological form of the verb. Participants in this study frequently sorted entirely by construction and never wholly by the morphological form of the verb; averaging across all subjects, sorts were significantly closer to a constructional sort than to a verb sort.

The most important contribution of this study is that it provides a sufficiency proof for the fact that people recognize abstract relationships between formal phrasal patterns and meaning: i.e. constructions. The fact that people in our experiment sorted by construction suggests that argument structure constructions may be 'natural' linguistic categories that speakers have access to in comprehension.

Acknowledgements

The work reported in this article was supported by NSF Grant SBR-9873450 to the second author.

References

Goldberg, A. (1995). *Constructions: A Construction Grammar Approach to Argument Structure.* Chicago: University of Chicago Press.

Healy, A., & Miller, G. (1970). The verb as the main determinant of sentence meaning. *Psychonomic Science, 20,* 372.

Kay, P., & Fillmore, C. (1999). Grammatical Constructions and Linguistic Generalizations. *Language* 75 1. 1-33.

Rappaport Hovav, M., & Levin, B. (1998). Building verb meanings. In M. Butt & W. Geuder (Eds.), *The Projection of Arguments: Lexical and Compositional Factors.* Stanford: CSLI Publications.

Construction	Form	Meaning	Example
Transitive	Subject Verb Object	X act on Y	*Anita threw the hammer.*
Ditransitive	Subject Verb Object1 Object2	X causes Y to receive Z	*Beth got Liz an invitation.*
Resultative	Subject Verb Object Complement	X causes Y to become Z	*Nancy sliced the tire open.*
Caused motion	Subject Verb Object Oblique	X causes Y to move Z	*Kim took the cat into the barn.*

Table 1: Examples of English Argument Structure Constructions

Systematicity and the Cognition of Structured Domain

Jim Blackmon
Daivd Byrd
Robert Cummins
Pierre Poirier
Martin Roth

University of California at Davis

Abstract

The current debate over what conditions a scheme of mental representation needs to satisfy in order to explain the systematicity of thought is characterized in such a way that (contrary to Fodor, Pylyshyn, and McLaughlin) any *complete* representational scheme (whether classical or non-classical) can explain the systematicity of thought. Though FPM might reply that non-classical schemes only satisfy these conditions in an unprincipled fashion, this shifts the discussion to less empirical considerations.

Recasting the debate, we show that FPM can maintain their objection of unprincipledness only at the price of representational pluralism. Our thesis is that one can maintain representational monism if one uses what we call structured encodings. This will be accomplished by spelling out a representational taxonomy that makes evident what properties need obtain for a given representational scheme to exhibit systematicity effects.

Study of the Time Course of the Updating Process

Nathalie Blanc (Nathalie.Blanc@univ-lyon2.fr)
Isabelle Tapiero (Isabelle.Tapiero@univ-lyon2.fr)
University of Lyon II (France)
Laboratory of Cognitive Psychology
5, avenue Pierre Mendès-France, C.P. 11
69676 BRON cedex (France)

This study focuses on the way readers integrate new incoming textual information into the mental representation they initially constructed. According to Morrow, Bower and Greenspan (1989), the mental representation activated at the beginning of the reading is on-line updated by the integration of new incoming information. More specifically, their findings based on on-line probed task are consistent with the assumption that readers make on-line inferences to update their mental representation. To the opposite, de Vega (1995) postulates that the integration of new incoming information is a backward process. By using an inference judgment task, he provided experimental supports to this assumption. Thus, models of comprehension differ in their hypothesis on the occurrence of the updating (i.e., integration) process. Issued from these opposite results, our intention was to demonstrate that the updating process is composed of two temporal components, a first one occurring on-line and a second one intervening with a certain delay. According to our view, each incoming information may be mapped on-line into the existing mental representation, but its final integration to the complete model depends on a backward updating which occurrence is function of readers' prior knowledge. In line with Ericsson and Kintsch (1995), we proposed that the expansion of short-term working memory due to a high level of prior knowledge allows to maintain more new incoming information and to delay their integration in long-term memory as a function of their relevance for the rest of the text. On the contrary, if subjects do not have enough knowledge about the situation, they should use an on-line process to integrate new incoming information. This assumption also leads to rise questions on the duration of the integration process (temporary or definitive) and how this duration interacts with readers' prior knowledge.

Experiment

Method

72 students from the University of Lyon II (France) were instructed to read a text that described a play with two topics, the scenery of the stage and the characters' goal and actions. The text was divided into two equal parts, the first one described the spatial arrangement of objects in the layout and the location of three characters, and the second one mentioned the movement of the characters with some of the objects previously located. The scenery of the stage was composed of a living and a dining room that contained 8 objects each. Prior to the reading of the first part of the text, we provided subjects with prior information on the described situation. They received a short text that described either all the spatial arrangement of objects in the scenery (high specific condition), or the location of fixed objects only (low specific condition), or general information about the layout of a theatre (general condition). The situation model built was tested after the reading of each part of the text via an inference judgment task that contained 8 spatial inferences for each topic. Participants judged whether each statement was true or false in relation to the part of the text previously read. The first task dealt with location information only and the second one concerned only motion information that were also probed during the reading of the second part of the text (Morrow et al., 1989). After each sentence describing a character's movement, 6 pairs of words naming objects of the scenery were presented one after another. Participants judged whether they were located in the character's goal room. The objects were either located in the same or in different rooms and were either fixed or moved by a character in the second part of the text.

Results and Discussion

From the results obtained to these different tasks, we draw four main conclusions. First, the updating process can occur either in an on-line or in a backward way. Second, readers' prior knowledge determine the time course of the updating process. Readers with high specific prior knowledge used the two temporal components whereas subjects in the low specific and general conditions preferentially used the on-line component. Third, the duration of the integration varies as a function of prior knowledge: Whereas performances increased between on-line and backward tasks for readers with high specific prior knowledge, the reverse pattern was observed for readers of the two other conditions. In other words, these latters appear to temporarily update their situation model whereas the former seems to definitively integrate new incoming information. Finally, these results highlighted the limits of on-line tasks that mainly assess the temporary integration whereas backward tasks evaluate the permanent integration. Thus, we underlined that the effects of readers' prior knowledge as well as the task bias have to be taken into account in the study of the updating process.

References

de Vega, M. (1995). Backward updating of mental models during continuous reading of narratives. Journal of Experimental Psychology: Learning, Memory and Cognition, 21, 373-385.

Ericsson, K. A., & Kintsch, W. (1995). Long-term working memory. Psychological Review, 102, 211-245.

Morrow, D. G., Bower, G. H., & Greenspan, S. L. (1989). Updating situation models during narrative comprehension. Journal of Memory and Language, 28, 292-312.

The Effect of Explanation and Alternative Hypotheses on Information-Seeking Strategies: Implications for Science Literacy

Sarah K. Brem (sbrem@soe.berkeley.edu)
University of California, Berkeley
Berkeley, CA 94720-1670

Recent assessments of science education emphasize science literacy—the ability to use scientific information in everyday life (National Research Council, 1995). To do this, students must exhibit self-reliance in seeking out new information. Brem and Rips (in press) showed that people use speculative explanations to fill gaps in their knowledge. Explanation improves comprehension and performance (e.g., Chi, deLeeuw, Chiu & LaVancher, 1994), but also leads to overconfidence (Koehler, 1991). Given that explaining has both positive and negative consequences, I examine how it affects students' testing of scientific claims.

First, explaining may affect student goals. I focus on two: Determining whether a relationship exists (Existence), and determining why the relationship exists (Mechanism). Suppose we test the claim: "Redwood harvesting is causing a decline in the hawk population." Existence questions address whether harvesting reliably causes a decline. Mechanism questions focus on how harvesting affects the hawks (e.g., destroying nesting sites). Because explanations focus on mechanisms, explaining should increase the number of Mechanism questions. This is a potentially undesirable shift—better to establish that there is a relationship before trying to determine how it works.

Second, explaining may affect the kind of information sought. Previous studies focused on the covariational (CV) and noncovariational (nonCV) distinction (e.g., Ahn, Kalish, Medin & Gelman, 1995). NonCV questions—"Do hawks nest in dwarf redwoods?"—can be made CV by explicitly addressing whether changes in cause result in changes in effect: "Do declines in dwarf redwoods coincide with declines in nesting sites?" Both CV and nonCV questions can serve Existence and Mechanism goals. What the distinction can tell us is how students approach a problem. CV questions specify explicit comparisons and measures; nonCV questions do not. Given the strong relationship between nonCV and mechanistic questions (Ahn et al., 1995), explaining should cause an undesirable shift to nonCV questions.

Given these opportunities for error, I consider a well-documented antidote for problems induced by explanation—considering alternative claims. Many studies show that considering alternatives reduces overconfidence (Koehler, 1991). However, do alternatives simply induce a lack of confidence, or do they encourage more rigorous testing? If alternatives have a positive effect, their presence should increase the number of Existence and CV questions.

Method

The design was a 2X2 between-participants factorial, varying Explanation (Explain vs. Don't Explain) and Alternate (Present vs. Absent). Participants read about 8 ecological problems and saw a primary claim as to the cause of each problem. In the Alternative Present condition, they also read an assertion regarding an alternative cause for each problem. Participants in the Explain condition speculated how the primary cause could lead to the problem, then generated three questions to ask an expert in assessing the validity of the primary claim. In the Don't Explain condition, participants posed questions without speculating.

The participants were 53 novice undergraduates.

Results

Participants' questions were coded by a blind rater as CV or nonCV, and as Existence or Mechanism. Only significant results are reported.

As predicted, with no alternatives present, the percentage of CV questions declined (47.9% vs. 27.8% ($t(7)$ = 2.75, p < 0.05). With alternatives present, the drop is not significant (48.9% vs. 43.5%; $t(7)$ = 1.49, p > 0.10). Again as predicted, most CV questions were of the Existence type (66%), and most nonCV questions were of the Mechanism type (79%). However, regardless of the presence of alternatives, explaining increased students' focus on mechanisms (52.5% vs. 30.9%; $t(7)$ = 3.02, p< 0.05).

Discussion

The results do not recommend explaining as a gap-filling strategy. It produced less rigorous queries and assumptions regarding causal relationships. Although students still made assumptions, alternatives did encourage more specific tests. Using alternatives in concert with interventions to discourage unwarranted assumptions may help students achieve science literacy.

Acknowledgments

The author was supported by NSF (DGE-9843256).

References

Ahn, W., Kalish, C., Medin, D., & Gelman, S. (1995). The role of covariation versus mechanism information in causal attribution. Cognition, 54, 299-352.

Brem, S. & Rips, L. J. (in press). Explanation and evidence in informal argument. Cognitive Science.

Chi, M. T. H., deLeeuw, N., Chiu, M., & LaVancher, C. (1994). Eliciting self-explanations improves understanding. Cognitive Science, 18, 439-477.

Koehler, D. (1991). Explanation, imagination, and confidence in judgment. Psych. Bulletin, 110, 499-519.

National Research Council Staff (1995). National Science Education Standards: Observe, Interact, Change, Learn. Washington: National Academy Press.

Distributed Cognition of a Navigational Instrument Display Task

Johnny Chuah (chuah.5@osu.edu)
The Ohio State University
204 Lazenby Hall, 1827 Neil Avenue, Columbus, OH 43210, USA

Jiajie Zhang (jiajie.zhang@uth.tmc.edu)
Todd R. Johnson (todd.r.johnson@uth.tmc.edu)
University of Texas, Houston
7000 Fannin, Suite 600, Houston, TX 77030, USA

The information necessary for the performance of almost any everyday task is distributed across information perceived from the external world and information retrieved from the internal mind. These tasks are known as distributed cognitive tasks (Zhang & Norman, 1994). The external representations constructed from the information extracted from external objects (such as written symbols) and the internal representations in the mind (such as schemas) dynamically integrate and interweave to result in a rich pattern of cognitive behavior. The principle of distributed representations is that a distributed cognitive task involves a system of distributed representations that consists of internal and external representations (Zhang & Norman, 1994, 1995). The task is neither exclusively dependent on internally nor exclusively dependent on externally processed information, but rather on the interaction of the two information spaces formed by the internal and external representations.

In the aviation industry, there are a wide variety of navigational systems. However, there exists a set of very basic navigational instruments. These instruments are selectively tuned to transmitting radio stations on the ground. The received signals are then presented onto a display in the cockpit for the navigator to interpret. There is only so much information that a navigation instrument needs to display: azimuth or directional information, and distance information. However, the various instruments present these information differently and result in varying degrees of precision and efficiency as interpreted by the navigator.

Cockpit information displays are examples of distributed representation systems. Navigational information in a cockpit information system can and is represented through a variety of isomorphic navigation instruments. Although these instruments are isomorphic and provide similar necessary information, they vary in their relative degree of directness and efficiency in their representation of scale information (Zhang & Norman, 1995). The scale information of the orientation and distance dimensions in a cockpit information display is represented across internal and external representations and can dramatically affect the representational efficiency of the display and the navigator's behavior (Zhang, 1997). This research seeks to study the varying cognitive properties of the representations that such instruments produce. The specific assumption to be tested is that with the most direct system, scale information is maximally represented externally, resulting in higher efficiency, faster and more direct responses. An experiment was carried out to test this hypothesis on four sets of navigation instruments which are isomorphic to each other but have different degrees of directness.

The resulting behavior variance from the experiment indicates that some representations are more 'efficient' in extending the necessary information for a task. Although the different isomorphic representations result in varying initial levels of performance and learning curves, performances appear to converge after a sufficient period of learning.

An argument could be made for learning and practice to eliminate such a representational effect. However, further research need to be done in more complex and dynamic settings. The experimental task was a simple position-fixing task in a very controlled and calm environment. In an unpredictable and complex environment such as that of the cockpit of an aircraft, the representational effect could be more pronounced and a possible regression to initial performance levels should be studied. Another issue that is worth of further study is whether the converged performance after learning for different representations will diverge again under extreme conditions such as high cognitive workload and time pressure.

Acknowledgements

This research was in part supported by Grant N00014-96-1-0472 from the Office of Naval Research, and a Summer Fellowship Research Award from the Center for Cognitive Science, The Ohio State University.

References

Zhang, J., & Norman, D. A. (1994). Representations in Distributed Cognitive Tasks. *Cognition Science, 18*, 87-122.

Zhang, J., & Norman, D. A. (1995). A representational analysis of numeration systems. *Cognition, 57*, 271-295.

Zhang, J., (1997). Distributed representation as a principle for the analysis of cockpit information displays. *International Journal of Aviation Psychology, 7(2)*, 105-121.

PSI plays „island": Comparison of the PSI-theory with human behavior

Frank Detje (frank.detje@ppp.uni-bamberg.de)
Lehrstuhl Psychologie II, Universität Bamberg, Markusplatz 3
D-96047 Bamberg, Germany

During the last two decades amazing progress has been made in cognitive science and related fields in explaining and predicting human behavior (especially in cognitive modeling). Best known and most elaborated are probably the two "cognitive architectures" that "put it all together" as "unified theories of cognition": ACT (Anderson & Lebiere, 1998) and Soar (Newell, 1990; Rosenbloom, Laird & Newell, 1993). But though the term "cognition" should not be understood too narrowly (Newell, 1990), their architectural conception does not cover too many psychological phenomena beside cognitive ones (Cooper & Shallice, 1995; Detje, 1999). A theory that explicitly tries to cover more psychological relevant phenomena such as emotional and motivational processes is the PSI-theory of action regulation (Dörner & Wearing, 1995).

Intention Regulation

Central to the PSI-theory is the concept of intentions. Every state of deprivation will lead to a need as soon as internal regulation cannot reach equilibrium. But intentions are more than merely the goal to satisfy the needs, they also include a number of elements that contain context information. The memory contains of three network-like hierarchies, a sensory network, a motor network and a motivational network as an internal sensor for needs. A lot of processes work on these data structures to show action and intention regulating behavior and to perform in complex problem-solving tasks (see Dörner & Wearing, 1985). The PSI theory and its precursers were tested in several environments and compared with human behavior. The results showed sufficient similarities (e.g. Dörner & Wearing, 1995).

„Insel" [„island"]: A short description

„Insel" consists of a number of localities that are linked by paths. Each locality consists of objects that can be manipulated. These manipulations can have effects on the manipulating agent and / or the environment. Human subjects are asked to control a robot on a distant island. Their task is to use the robot in order to look for "nucleotides" (that are fictional lumps of fuel) and to collect as many as possible. At the same time they need to satisfy the robot's existential needs for food and water. They also have to prevent the robot from damage. When PSI plays "island" it is the autonomous agent "James", it "is" the robot, by means of the needs it has.

Results of the comparison

In an experiment 29 subjects participated in the "island" setting (1 hour) and 19 simulation runs for PSI were completed (300 cycles). We find that PSI and subjects do not differ much (see table 1) with respect to the main achievement indicators a) the number of locations visited (degree of exploration of the island; 45 existing) and b) the number of nucleotides collected (subjects were paid according to this amount; 80 nucleotides existing). Even the number of objects approached is quite similar. But if we take a look at other behavioral data we see that the PSI-simulation and the subjects differ in their tactical means or strategies. PSI shows in general much more changes of locations but much less manipulations of objects than the subjects show.

Table 1: Results for achievement indicators.

Achievement indicator	Mean (sd)
Locations visited (PSI)	29.37 (5.92)
Locations visited (subjects)	31.07 (7.28)
Nucleotides collected (PSI)	41.76 (17.70)
Nucleotides (subjects)	39.00 (9.98)
Objects approached (PSI)	150.00 (44.27)
Objects (humans)	161.24 (28.87)

Conclusion

First comparisons show that PSI is already able to deal with the "island"-game with the same success as human beings do. The island is neither explored completely nor are all nucleotides found by both groups. Still both groups deal with the simulation quite successfully. But they do it in different ways. PSI is more willing to change locations than to explore a given location in greater detail. Humans are more willing to explore what they already see than to figure out what they do not yet see.

References

Anderson J. R. & Lebiere, C. (1998): *Atomic Components of Thought*. Hillsdale, NJ: Erlbaum.

Cooper, R. & Shallice, T. (1995): Soar and the Case for Unified Theories of Cognition. *Cognition, 55,* 115-149.

Detje, F. (1999): *Handeln erklären* ["explaining acting"]. Wiesbaden: Deutscher UniversitätsVerlag.

Dörner, D. & Wearing, A. J. (1995): Complex Problem Solving: Toward a (Computer-simulated) Theory. In P. A. Frensch & J. Funke (eds.): *Complex Problem Solving. The European Perspective*. Hillsdale, NJ; Hove, UK: Erlbaum.

Newell, A. (1990): *Unified Theories of Cognition*. Cambridge, Mass.: Harvard University Press.

Rosenbloom, P. S., Laird, J. E. & Newell, A. (eds.) (1993): *The Soar Papers: Research on Integrated Intelligence*. 2 vols. Cambridge, Mass.: MIT Press.

Hypotheses and Evidence About Evidence and Hypotheses

Christine Diehl (CDiehl@Socrates.Berkeley.Edu)
Michael Ranney (Ranney@Cogsci.Berkeley.Edu)
Grace Lan (Graslawn@Uclink4.Berkeley.Edu)
Sergio Castro (Floyd@Uclink4.Berkeley.Edu)
Graduate School of Education; 4533 Tolman Hall
University of California, Berkeley, CA 94720 USA

Introduction

We report on two studies that investigate students' understandings of the relationship between hypothesis and evidence. The *Convince Me* software and associated reasoning curriculum developed by the ECHO Educational Project aid students in generating and analyzing scientific arguments, requiring students to identify propositions that fill the roles of evidence and hypothesis (Ranney, Schank & Diehl, 1996). The distinction between evidence (or data) and hypothesis (or theory) appears to be fundamental in scientific reasoning, yet research shows that students seem to differ from scientists in their process of differentiating hypothesis and evidence. However, even experts do not always exhibit good agreement regarding this distinction. Although training with *Convince Me* lends sophistication to students' epistemic criteria, making their categorization appear more expert-like, the categorization of individual propositions may still be disputed (Ranney, Schank, Hoadley, & Neff, 1994). We hypothesize that students rely on prototypical models of hypothesis and evidence in structuring their arguments. Prior analysis of students' *Convince Me* arguments points to linguistic markings of this prototypic representation, which we investigate with paper-and-pencil surveys.

Study 1

Sixty-three undergraduate students completed a survey asking them to list five to seven very good examples each of hypotheses and evidence. We coded the surveys for the presence of fourteen linguistic categorical features, then used a stepwise multiple regression analysis to identify the best ("prototypical") models for propositional categorization. The model for evidence explained more variance than the model for hypothesis and included only one linguistic marker, the specifier category (e.g., when, which), as well as past tense and non-sentential phrases referencing objects or actions (e.g., fingerprints, data analysis). The predictor model for hypothesis included the modal (e.g., may, can) and causal (e.g., because, results in) linguistic marker, as well as references to a particular scientific theory.

Study 2

Sixty-two undergraduate students completed a survey with twenty-one propositions for which they were asked to indicate how good an example of a *hypothesis* or of a piece of *evidence* the statement/phrase is. One-third of the proposi-

tions contained only linguistic features for the hypothesis model identified in Study 1; one-third contained only features for the evidence model; and the last third were intended to be ambiguous with respect to categorization and contained either no such features or conflicting pairs of features. A multiple regression analysis shows that proposition type is a significant predictor of both hypothesis-ness and evidence-ness.

Discussion

Our results support the claim that the presence of certain linguistic features can predict the epistemic categorization of a proposition. Propositions that contained linguistic features representative of "prototypical" hypothesis were rated as being more like hypothesis than evidence. Propositions that contained linguistic features representative of "prototypical" evidence were rated as being more like evidence than hypothesis. Subjects considered ambiguous propositions with conflicting linguistic features to be more like hypothesis and ambiguous propositions with no obvious epistemic features to be more like evidence. This result indicates that evidence may be the "default" categorization of an epistemic statement, where as a single "hypothetical" feature biases an entire statement toward seeming hypothetical in nature. Extensive use of evidentials (especially modals) to express the strength and confidence of an assertion is evident in scientific discourse and writing; therefore, it seems reasonable that students should use these linguistic features in evaluating a proposition's epistemic status and determining its role assignment in an argument's structure (Crismore & Farnsworth, 1990).

References

Crismore, A. & Farnsworth, R. (1990). Metadiscourse in popular and professional science discourse. In W. Nash (Ed.), *The writing scholar: Studies in academic discourse*. Newbury Park, CA: Sage Publications.

Ranney, M., Schank, P., & Diehl, C. (1996). Competence versus performance in critical reasoning: Reducing the gap using *Convince Me. Psychology Teaching Review,4*(2), 153-166.

Ranney, M., Schank, P., Hoadley, C., & Neff, J. (1994). "I know one when I see one": How (much) do hypotheses differ from evidence? In the *Proceedings of the Fifth Annual American Society for Information Science Workshop on Classification Research*.

A Framework for Modeling Representational Change in Scientific Communities

Sean C. Duncan and **Ryan D. Tweney**
Department of Psychology
Bowling Green State University
Bowling Green, OH 43403
{seand,tweney}@bgnet.bgsu.edu

Though representational change has traditionally been a central issue for cognitive studies of science (e.g., Duncan & Tweney, 1997), little research has focused on how these changes occur in scientific communities. A framework is elaborated which may lead to the successful computational modeling of group representational change in science. Of particular interest is how group-level representational dynamics develop from the interaction of individual cognitive processes.

Constraint satisfaction

The framework is connectionist in structure, with sub-elements of cognitive representations encoded as nodes, and their excitatory or inhibitory relationships represented as weights between nodes (à la the schema models of Rumelhart, et al, 1986). Thagard (1989) described a theory of "explanatory coherence" which assessed theories according to the degree that internal and external constraints were satisfied. Using ECHO (a computational model of the theory), Thagard simulated several scientific controversies (e.g., Nowak & Thagard, 1992). However, while ECHO is a useful tool for judging the consistency of a set of subtheoretic propositions with itself and external evidence, it is rarely used in modeling the individual cognitive agent.

Hutchins (1995) used similar architectures for studying the interaction of multiple cognitive agents. He studied a series of linked constraint satisfaction models and showed that differences in organizational structure affected a simple measure of confirmation bias in the networks. The models were relatively abstract, though, and have not yet been expanded into complex domains. The present framework applies an ECHO-like structure to Hutchins' linked networks.

A framework

This framework postulates that the interaction of individual constraint satisfaction processes may be behind complex representational change at the level of groups. Thus, theories are represented as patterns of activation over the set of nodes within several linked networks, each network representing an individual. Like ECHO, component elements of the representation of a given phenomenon are encoded as nodes in each network, with weighted connections between the nodes. As with Hutchins' linked networks, patterns of interconnectedness between networks represents "who talks to whom" while the strength of the connections between networks represents the persuasiveness between individuals.

The networks are interpreted such that the path each takes to a stable goodness-of-fit maximum is analogous to the process of representational change in a real individual. Other modifications include the introduction of external evidence at different times during processing and weight modification during the network runs. It is theorized that these changes will increase the fit between a given set of networks and the data being simulated.

The framework will be applied to the modeling of an important historical case, namely the "Great Devonian Controversy" of early nineteenth-century British geology (Rudwick, 1985). The controversy, which involved several dozen professional and amateur geologists, centered about how to classify a range of rock strata found in Devon, England and was eventually resolved with the development of a new representation of the geologic strata (which has become known as the "Devonian system"). Rudwick's narrative analysis is quite detailed with respect to the representational dynamics at play during the controversy and thus provides a true test of our framework.

References

Duncan, S. & Tweney, R. (1997). The problem-behavior map as cognitive-historical analysis. In M. Shafto & P. Langley (eds.), *Proceedings of the Nineteenth Annual Conference of the Cognitive Science Society*, (p. 901). Hillsdale, NJ: Lawrence Erlbaum Associates.

Hutchins, E. (1995). *Cognition in the wild.* Cambridge, MA: MIT Press.

Nowak, G. & Thagard, P. (1992). Copernicus, Ptolemy, and explanatory coherence. In R. N. Giere (ed.) *Cognitive models of science.* Minneapolis, University of Minnesota Press, 274-309.

Rudwick, M. (1985). *The great Devonian controversy.* Chicago : University of Chicago Press.

Rumelhart, D., Smolensky, P., McClelland, J. & Hinton, G. (1986). Schemata and sequential thought processes in PDP models. In J. L. McLelland, D. E. Rumelhart, and PDP research group (eds.). *Parallel distributed processing. Vol. 2. Psychological and biological models.* Cambridge, MA: MIT Press, 7-57.

Thagard, P. (1989). Explanatory coherence. *Behavioral and Brain Sciences*, 12, 435-467.

Modelling Cognitive Dynamic Units of Analysis

Vaccari Erminia (email: vaccari@di.uniba.it)

Dipartimento di Informatica, Universita' di Bari, Bari,Italy

D'Amato Maria (email: damato@di.uniba.it)

Dipartimento di Informatica, Universita' di Bari, Bari,Italy

Abstract Poster

We outline a new approach (Vaccari, 1998; Vaccari, 1998a) to cognitive modelling developed in the Dynamic System Theory (DST) paradigm. It is based on structured models representing a set of interacting functional systems (Vaccari, 1998) amenable to be formally represented by independent generative models which are solved simultaneously taking into account their interactions (Delaney & Vaccari 1989). These generative models may be specified at a 'mathematical level' with different (continuous/discrete) formalisms, while at the implementation level we use the discrete event formalism, which allows different time scales and can represent practically any continuous time/discrete time model. Conceptualising systems as structured models refers to the process of structuring models (i.e. using composition) from a conceptual point of view and not simply at the implementation level by means of mathematical artifices. A global functional system (whose invariant activity emerges from the interplay of the activities of its functional subsystems) is formally represented by a structured DST model

In our approach the problem of descriptive complexity affecting the dynamic approach to cognitive modelling (Port & Van Gelder, 1995) is overcome by adopting complementarity of holism and reductionism in the formulation of the structured model. It is also possible to represent in the model hierarchical relations not foreseen by classical DST formalisms.

We show how the above mentioned notion of functional system, characterized by its invariant activity, is conceptually equivalent to a Cognitive Dynamic Unit of Analysis as defined by Mandelblit and Zachar (1998). We also point out that some of the foreseen (type 3 and type 4) forms of unity are not representable by classical dynamic models while they can be easily accomodated in our approach. which foresees also more sophisticated forms of unity. This is possible because our modelling framework foresees representing various different types of hierarchical systems including anticipatory systems governed by feed-forward mechanisms, systems characterized by constitutive hierarchy i.e. connections between a functional system and its constitutive lower level functional systems and by authority hierarchy. i.e. connections which represent authority whereby certain subordinate functional systems are controlled by a higher level functional system.

The generative models constituting a DST structured model can be structured models themselves and it is possible to recursively define higher order structured models. This possibility to represent a sub-model, in turn, as a structured model allows hierarchical representations in a well established theoretical framework such as systems theory. This issue is very important because it assures a correct time synchronization among system activities/events, very difficult to implement in hybrid dynamic approaches based on heuristics.

Our approach has been implemented in a software environment for the formulation and solution of structured dynamic models characterized by modularity and hierarchy (Vaccari, D'Amato & Delaney,1998).

References

Delaney,W. & Vaccari,E. (1989). *Dynamic Models and Discrete Event Simulation*, Marcel Dekker, N.Y.

Mandelblit, N. and Zachar, O. (1998). The Notion of Dynamic Unit: Conceptual Developments in Cognitive Science, *Cognitive Science, 22*, 229-268.

Port, R. F. and Van Gelder, T.(1995). It's about time: an overview of the dynamical approach to cognition. In Port, R.F. and Van Gelder, T.(eds.), *Mind as Motion,* 1-43,MIT Press, Cambridge, MA.

Vaccari, E.(1998). Knowing as Modelling, *Cybernetics and Human Knowing, 5*(2),59-72.

Vaccari, E.(1998a). Some Considerations on Cognitive Systems Modelling. In Trappl, R.(ed.), *Cybernetics and Systems'98,* vol.2, 663-668,Austrian Society for Cybernetic Studies, Vienna.

Vaccari, E., D'Amato, M., Delaney, W.(1998). Hierarchical Modelling: a systemic framework, *Third International Conference on Emergence, August 98, Helsinki.* In Farre, G. and Oksala, T. (eds.),*Acta Polytechnica Scandinavica,*.179-184.

Babies, Variables, and Connectionist Networks

Michael Gasser (GASSER@CS.INDIANA.EDU)
Eliana Colunga (ECOLUNGA@CS.INDIANA.EDU)
Computer Science Department, Cognitive Science Program
Indiana University
Bloomington, IN 47405

Two recent papers in *Science* have demonstrated the remarkable language learning abilities that are possessed by infants. In both cases the infants were presented with sequences of syllables embodying some sort of regularity and later tested with sequences that agreed or disagreed in certain ways with the training set. In the experiments of Saffran, Aslin, and Newport (1996), eight-month-olds were able to distinguish three-syllable "words" that they had heard previously in a stream of syllables from those that they had not. In the experiments of Marcus, Vijayan, Bandi Rao, and Vishton (1999), seven-month-olds were able to distinguish three-syllable "sentences" that followed a training pattern, either AAB, ABB, or ABA (where A and B are variables representing syllables), from those that did not. What is striking about the latter experiments is that the test sentences consisted of syllables not heard at all during training. That is, the pattern that is learned seems to be independent of specific syllable content and to require variables for its representation.

Marcus et al. (1999) and Pinker (1999) argue that these experiments demonstrate two distinct learning mechanisms, which will later prove useful for learning language: a statistical/associationist mechanism, illustrated in Saffran et al.'s experiments, and a symbolic mechanism, illustrated in Marcus et al.'s experiments. In particular, because variables are excluded from simple associationist accounts, these accounts fail to handle Marcus et al.'s results.

We focus on Marcus et al.'s experiments, showing how an associationist device, a particular neural network architecture, can learn the patterns in the experiments, generalizing to novel sequences, and how this account, rather than being simply an uninteresting implementation of a symbolic model, makes novel predictions about the learning of sequences.

A neural network needs two features to generalize over "grammatical" patterns like those in Marcus et al.'s experiments. First, it needs a means of distinguishing particular objects or sets of objects from each other in short-term memory. This allows it to *segment* input sequences on the basis of inter-item similarity, for example, for the sequence *le le di*, to treat the first and second syllables as belonging to a group distinct from the third. Second, it needs a means of handling relational knowledge. This allows it to detect and store the similarity relations that characterize segmented sequences, for example, to remember that the first and second syllables in a sequence resembled each other but differed from the third.

Playpen (Gasser & Colunga, 1998) is a neural network architecture which is designed to represent and learn relational knowledge and to deal with simple sequential patterns. Seg-mentation is handled through the use a second dimension, in addition to activation, along which units in the network can vary. Units which are synchronized along this dimension are treated as belonging together. Relational knowledge in Playpen is handled with special-purpose *micro-relation units* (MRUs), which are activated to the extent that input to their two *micro-roles* is out of synchronization, that is, that it represents distinct objects or groups.

To simulate Marcus et al.'s task, we used an instantiation of Playpen in which similarity relations between the syllable input units were represented by hard-wired connections. These connections cause similar syllables in a sequence to be synchronized and different syllables to be desynchronized. MRUs in the network represented possible binary relations between the syllables in a sequence, and connections between the MRUs represented correlations between these simple binary relations.

The network was trained on sequences of syllables adhering to one or another of the patterns used in Marcus et al.'s experiments, and it successfully generalized to sequences of novel syllables by producing more activation across the MRU layer for "grammatical" than for "ungrammatical" sequences. But in the model this rule-like behavior is just one extreme of a continuum along which sensitivity to item content varies. Specifically, in the network the similarity of the test sequences to the training sequences had an effect on generalization. This behavior contrasts with that of symbolic accounts, where the variables in the learned rule would be oblivious to the specific content of the items in the test patterns.

References

Gasser, M. & Colunga, E. (1998). Where do relations come from?. Tech. rep. 221, Indiana University, Cognitive Science Program, Bloomington, IN.

Marcus, G. F., Vijayan, S., Bandi Rao, S., & Vishton, P. M. (1999). Rule learning by seven-month-old infants. *Science*, *283*, 77–80.

Pinker, S. (1999). Out of the minds of babes. *Science*, *283*, 40–41.

Saffran, J., Aslin, R., & Newport, E. (1996). Statistical learning by eight-month-old infants. *Science*, *274*, 1926–1928.

Centrality and Property Induction

Constantinos Hadjichristidis
(constantinos.hadjichristidis@durham.ac.uk)
Dept. of Psychology, University of Durham,
Durham, DH1 3LE, UK
Rosemary J. Stevenson
(rosemary.stevenson@durham.ac.uk)
Human Communication Research Center,
University of Durham, Durham, DH1 3LE, UK

Steven A. Sloman
(steven_sloman@brown.edu)
Dept. of Cognitive & Linguistic Sciences, Brown University,
Box 1978, Providence, RI 02912, USA
David E. Over
(david.over@sunderland.ac.uk.)
School of Social Sciences, University of Sunderland,
Sunderland, SR1 3SD, UK

Introduction

We address why some properties are more generalizable than others. In search for a determinant of projectibility, we make the general and weak assumption that concepts involve features embedded in networks of asymmetric relations. We take mutability to be a structural aspect of representations that measures the extent to which a feature is integral to the coherence of a concept. Following Sloman, Love, and Ahn (1998) we take a feature to be central to the extent that other (central) features depend on it. We note that centrality is concept-relative. "Roundness", for instance, is central for Basketballs but not for Cantaloupes (cf. Medin & Shoben, 1988). We thus hypothesize that the more central a feature in a category's representation, the higher its projectibility among concepts that share common structure.

Methods. Participants (N=24) were informed that an animal (base) had two properties: one upon which lots of its functions depend (central property), and one upon which few of its functions depend (non-central property). Participants were then presented with a new animal (target) and had to estimate the likelihood of that animal having each of the two properties. As a surrogate of the extent to which two animals share common structure, we manipulated the physiological similarity between the base and target animals. Out of 18 items, 6 involved animals from the same superordinate and highly similar (SS-HS), 6 from the same superordinate but lowly similar (SS-LS), and 6 from a different superordinate and lowly similar (DS-LS). The assignment of animal pairs to similarity conditions was controlled by a separate group of participants. Consider a sample item from the SS-HS condition:

Many of a squirrel's physiological functions depend on the enzyme amylase, but only a few on the enzyme streptokinase.
Please rate the likelihood of the following statements:
A. Mice have amylase. _____%
B. Mice have streptokinase. _____%

Results. Table 1 summarizes the results. Central properties were more projectible than non-central ones. This effect was proportional to the base-target similarity.

Table 1. Mean inductive strength estimates. Underneath each column is the two-tailed level of significance.

	SS-HS	SS-LS	DS-LS
Central	75	53	44
Non-central	55	47	47
	$p<.001$	$p<.07$	$p>.40$

Discussion

The results confirm both parts of our hypothesis since: (i) central features were more projectible than non-central ones, and (ii) the more structure the base and target categories shared, the higher the preference to project the central feature.

There is much evidence that can be said to corroborate our hypothesis. Gelman (1988) found that young children prefer to project properties that are intrinsic/stable (e.g., "has a spleen") rather than extrinsic/unstable (e.g., "is cold"). To the extent that people believe that lots of properties depend on intrinsic features (Medin & Ortony, 1989), but only a few on extrinsic ones, the former are more central than the latter. The advantage of our theory is that it predicts violations of such general biases; e.g., even an enzyme that seems pretty intrinsic and stable is not highly projectible when it is stipulated that only few of the animal's functions depend on it.

Importantly, our results cannot be accounted for by current models of categorical inference. Models based on feature-similarity (e.g., Osherson, Smith, Wilkie, Lopez, & Shafir, 1990) appeal only to the relations among categories to predict inductive strength; centrality has no place in their equations. Also, models based on structural alignment (e.g., Lassaline, 1996) fail to account for such effects because for such models to work predicates, as well as their relation to other predicates, must all be clearly specified. Since the dependencies of the predicates in the current study were left vague, it is unclear how such models could apply.

References

Gelman, S.A. (1988). The development of induction within natural kind and artifact categories. *Cognitive Psychology, 20*, 65-95.

Lassaline, M.E. (1996). Structural alignment in induction and similarity. *Journal of Experimental Psychology: Learning, Memory, and Cognition, 22*, 754-770.

Medin, D.L. & Ortony, A. (1989). Psychological essentialism. In S. Vosniadou & A. Ortony (eds), *Similarity and analogical reasoning* (pp. 179-196). New York: Cambridge University Press.

Medin, D.L. & Shoben, E.J. (1988). Context and structure in conceptual combination. *Cognitive Psychology, 20*, 158-190.

Osherson, D.N., Smith, E.E., Wilkie, O., Lopez, A., & Shafir, E. (1990). Category based induction. *Psychological Review, 97*, 185-200.

Sloman, S.A., Love, B., & Ahn, W. (1998). Feature centrality and conceptual coherence. *Cognitive Science, 22*, 189-228.

Using 'basic level categories' to retrieve multimedia from the world-wide-web

E. Hoenkamp (HOENKAMP@ACM.ORG)

Nijmegen Institute for Cognition and Information (NICI)

Search engines today allow a user to quickly find relevant documents among the hundreds of millions of documents on the world-wide-web. These search engines use keywords as the primary way to index into documents. Most users learn to live with this, as keywords are still an approximation to textual material. However, it becomes a serious contraint for searching other media such as images. Search engines that provide facilities to search pictures (e.g. AltaVista) usually link to specialized closed image databases. The results, however, in no way parallel the success of text retrieval. This is deplorable, given the vast amount of multimedia on the web and their significance for communication. In our research we begin

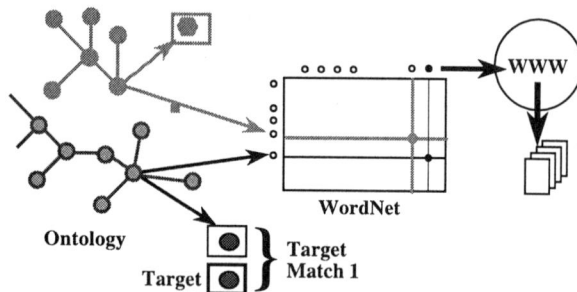

Figure 1: Retrieving an image step one (schematic): (1) match the target with basic level images, (2) look up the corresponding basic level word, (3) retrieve documents from WWW.

to see how the cognitive status of 'basic level' categories can improve this situation. In a taxonomy of concepts, a concept at the *basic level* adds many attributes to the level above (e.g. from 'chair' to 'furniture'), but very few are added to the level below (e.g. from 'chair' to 'kitchen chair'). At the basic level, many remarkable connections exists between imagery and language (Rosch, Mervis, Gray, Johnson, & Boyes-Braem, 1976). For example:

- the basic level appears the most abstract level for which an image can represent a class as a whole,
- when people have to name a picture of an object at

the subordinate level, they almost always use the word for the basic level,

- For words that stand for physical objects and organisms, parts notably proliferate at the basic level.

We built a system that uses the basic level to retrieve images through linguistic means (keywords), as depicted in the figures. The results are promising. To further assess the role of the basic level we investigated the following conjectures:

- Documents about parts of a basic level category can be retrieved by searching for the basic level word,
- Images of a basic level category can be retrieved by searching for the basic level word,
- The previous two cases should show notably higher precision for unambiguous basic level words than for polysemous basic level words.

These conjectures were confirmed. For an impression of the actual data, see:

http://ontolingua.nici.kun.nl/~stanf4/images.html

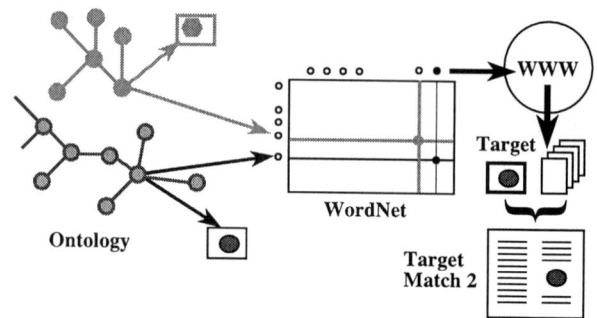

Figure 2: Retrieving an image step two: (1) select retrieved documents that contain images, (2) match images to target image, (3) collect matching images.

References

Rosch, E., Mervis, C. B., Gray, W. E., Johnson, E. M., & Boyes-Braem, P. (1976). Basic objects in natural categories. *Cognitive Psychology*, *8*, 382–439.

Implicit memory in children: Are there age-related improvements in a conceptual test of implicit memory?

Almut Hupbach (hupbach@cogpsy.uni-trier.de)
Silvia Mecklenbräuker (mecklen@cogpsy.uni-trier.de)
Werner Wippich (wippich@cogpsy.uni-trier.de)
University of Trier; Department of Psychology
54286 Trier, Germany

In recent years, implicit tests of memory have become very popular. In an implicit test, participants are not asked to report on memory for an event as they would in an explicit test; rather, they engage in some task that can indirectly reflect memory for the occurrence of that event. For instance, they are presented with category names and are asked to produce the first exemplars of that category that comes to mind. The finding is that prior presentation of a category exemplar increases the likelihood of that word being named as a category instance. This facilitation of performance is a phenomenon known as priming. Priming effects have attracted considerable interest because experimental manipulations (e.g., varying depth of processing) and subject variables can dissociate them from explicit test performance (for a review, see Roediger & McDermott, 1993). For instance, age-related dissociations have been found. Age-related differences between young and older children typically observed on explicit memory tests are largely reduced with implicit memory testings (for a review, see, e.g., Parkin, 1993). This conclusion, however, is based on relatively few studies that have mainly used picture identification tasks. Yet to draw conclusions about the developmental invariance of priming from findings on one single type of implicit memory test seems problematic because of dissociations observed between different implicit tests. These dissociations have led researchers to classify memory tasks according to the forms of information they access or the type of processing they require (see Roediger & McDermott, 1993). So-called perceptual tests are distinguished from so-called conceptual tests. Perceptual implicit tests tap the perceptual record of past experience. The perceptual similarity between study and test events affects performance, but manipulations of levels of encoding usually do not. The reverse is true for most conceptual tests (e.g., general knowledge questions or category production). These tests emphasize the semantic relationships between the studied and tested item. They require conceptually driven processing by relying on the encoded meaning of concepts, or on semantic processing, elaborative encoding, and the like.

Results from the few studies that examine developmental trends in priming using conceptual measures of implicit memory also usually indicate that priming is age-invariant (e.g., Anooshian, 1997). These findings are inconsistent with theoretical accounts of implicit memory and knowledge development. Developmental research has extensively demonstrated that existing semantic knowledge is a crucial factor in accounting for age-related differences in explicit memory performance (e.g., Bjorklund, 1987). We expect that the developing knowledge base has similar effects on conceptual priming.

Two experiments examining the development of conceptual priming in childhood using a category-production task were conducted. Four- and 6-year-old preschoolers and 8- and 10-year-old elementary school children studied exemplars (pictures) of different categories and were required to spontaneously produce exemplars of studied and unstudied categories at test. In Experiment 1, a significant but rather small age-related improvement from preschool to school age was demonstrated which was neither affected by study tasks (with or without reference to category information) nor by category-exemplar typicality. However, both variables were important inasmuch as preschoolers only showed significant priming of atypical exemplars when the category names were available at study. In Experiment 2, when using a between-subjects manipulation of item typicality and giving the category names at study, reliable age-related improvement in priming of atypical exemplars could be observed. It is concluded that the notion of developmental invariance of priming should be modified. The important research question is not whether there is age constancy of conceptual priming in childhood. Rather, the conditions should be studied in which age-related differences can be observed as well as what kind of conceptual knowledge and conceptual processes are responsible for these differences. For the category-production task, (1) identifying the category names, (2) strengthening the associations between category and exemplar or, for atypical exemplars, forming new associations, and (3) organizing the exemplars according to conceptual categories all seem to be relevant processes at encoding.

Anooshian, L. J. (1997). Distinctions between implicit and explicit memory: Significance for understanding cognitive development. *International Journal of Behavioral Development, 21,* 453-478.

Bjorklund, D. F. (1987). How age changes in knowledge base contribute to the development of children's memory: An interpretive review. *Developmental Review, 7,* 93-130.

Parkin, A. J. (1993). Implicit memory across the lifespan. In P. Graf & M. E. J. Masson (Eds.), *Implicit memory: New directions in cognition, development, and neuropsychology* (pp. 17-48). Hillsdale: Erlbaum.

Roediger, H. L., & McDermott, K. B. (1993). Implicit memory in normal human subjects. In H. Spinnler & F. Boller (Eds.), *Handbook of neuropsychology* (Vol. 8, pp. 63-131). Amsterdam: Elsevier.

Selecting Evidence to Limit Hypotheses

Alexandra Kincannon (kincannon@virginia.edu)
Barbara A. Spellman (spellman@virginia.edu)
Department of Psychology; University of Virginia
Charlottesville, VA 22903 USA

Introduction

When people want to *generalize* a hypothesis, they recognize the importance of getting evidence from different sources. This preference, the "diversity principle", is both assumed as normative by philosophers of science and is a strategy used by participants in scientific reasoning tasks (Popper, 1962; Spellman, López, & Smith, 1999). To *limit* a hypothesis, however, it seems that evidence from similar sources would be more informative. Using an example from our first experiment, "Suppose you know for a fact that elephants have an ostic vesicle. What organism would you examine to test whether or not ONLY mammals have an ostic vesicle?" The response choices were hippopotamus (inside category and similar), fox (inside category and dissimilar), crocodile (outside category and similar), and snake (outside category and dissimilar). It seems to us that the best answer is crocodile—the non-mammal that is the most similar to mammals. Most people agree that no information about hippos or foxes is relevant to the question of whether only mammals have this property, but the choice of crocodile over snake is less obvious. Crocodiles are more similar to elephants than snakes are, and perhaps the property in question (ostic vesicle) has something to do with that similarity (e.g., four-leggedness). Thus, although finding that crocodiles or snakes do have the property is equally conclusory (i.e., that the property is not limited to only mammals), finding that crocodiles don't have the property is more informative than finding that snakes don't have the property (because if snakes don't have it, crocodiles still might).

Experiments

In the first experiment, we tested participants' ability to limit hypotheses using stimuli like that from the elephant example above. Similarity was established by the results of an earlier experiment in which 80 participants made pairwise ratings on a 6-point scale. As expected, participants were more likely to test an animal outside the category of the target than inside the category, $X^2(1,N=63)=21.7$, $p<0.001$ for the elephant question. Contrary to our prediction, the dissimilar animal was chosen more often than the similar one, $X^2(1,N=50)=6.5$, $p=0.01$ for the elephant question. Similar results were found with the other target (robin).

In the next experiment, the test questions had a premise/conclusion format like that used by Osherson, et al. (1990). Participants saw one first premise (Robins have a condyloid canal) and one conclusion (Therefore ONLY birds have a condyloid canal) and two different second premises (Bats DO NOT have a condyloid canal. / Gorillas DO NOT have a condyloid canal.). Participants chose which second premise gave stronger support to the conclusion. Here, they tended to choose the similar animal premise (bats) more than the dissimilar animal premise (gorillas); however, this preference was not significant. When we compared across experiments, a 2x2 contingency table revealed a statistically significant difference between the frequencies of similar versus dissimilar animal choices depending on the test question format, $X^2(1,N=128)=10.24$, $p=0.001$ for the robin question. The frequency of choosing the outside-category similar animal, compared to the frequency of choosing the outside-category dissimilar animal, was greater in the premise choice task than in the evidence choice task.

Discussion

We have found that some conditions seem to promote better reasoning in hypothesis limitation tasks; what needs more exploration is the process behind that reasoning. In ongoing experiments, we are delving further into how people select evidence for limiting hypotheses. Participants are conducting multiple tests of their hypothesis, given several animals in several different categories to choose from. One factor that could play a role in our participants' behavior is the desire for a quick confirmation of their hypothesis (as in Wason, 1960). Perhaps people start by picking the most dissimilar animal, but if given the option to continue testing, they may see the value of working their way towards the more similar animal on subsequent tests. Under many real-life circumstances, such a strategy might be viewed as quite rational.

References

Osherson, D. N., Smith, E. E., Wilkie, O., López, A., & Shafir, E. (1990). Category-based induction. *Psychological Review, 97*, 185-200.

Popper, K. R. (1962). *Conjectures and refutations: The growth of scientific knowledge*. New York: Basic Books.

Spellman, B. A., López, A., & Smith, E. E. (1999). Hypothesis testing: Strategy selection for generalising versus limiting hypotheses. *Thinking & Reasoning, 5*, 67-91.

Wason, P. C. (1960). On the failure to eliminate hypotheses in a conceptual task. *The Quarterly Journal of Experimental Psychology, 12*, 129-140.

Rethinking the Role of External Representation in Problem solving

Simon M. K. Lai (h9408159@hkusua.hku.hk)
Department of Psychology
University of Hong Kong, Pokfulam Road, HK

Alonso H. Vera (vera@hku.hk)
Department of Psychology
University of Hong Kong, Pokfulam Road, HK

The so-called "cognitive revolution" of the mid-1950s has resulted in extensive research on *internal* cognitive processes. Researchers studying different areas of human cognitive functions were typically focused on both the contents and processes that occur inside the head. Newer approaches, like situated action and distributed cognition, question the information-processing approach to the understanding of human problem solving and knowledge. They emphasize external representation, both in the physical and social world, over internal representation in various kinds of human activities. Under the influence of the two newer approaches, studies demonstrating the importance and facilitation effects of external representation have increased in number (Gero, 1998; Scaife & Rogers, 1996; Suwa & Tversky, 1997; Zhang, 1997; Zhang & Norman, 1994). They focus on the study of the tasks and situations in which the activities are performed.

Studies by Zhang and his colleagues (Zhang, 1997; Zhang & Norman, 1994) showed the positive effect of externalizing rules in solving traditional cognitive problems. Architectural design research also emphasizes the interaction between external representation and internal representation (Gero, 1998; Suwa & Tversky, 1997). However, observation from a study by Vera, Kvan, West, and Lai (1998) on architectural design showed similar design process by participants using distinct ways of representing the problem.

In this study, eighteen pairs of participants worked on the Twelve Balls Problem[1], a cognitive problem not like other traditional ones such as tic-tac-toe isomorphy or missionaries and cannibals, for forty-five minutes under chat line, audio video, or face to face condition. They were allowed to freely represent the problem on a shared electronic whiteboard. Some pairs solved the problem through representing the problem solving in tree diagrams, while others used production rules, or actual manipulation of drawn balls.

Verbal protocols between pairs of participants were coded into six categories for the analysis of the quality of their problem solving. Three of the categories are good indicators of problem solving, i.e. *Meta-rule, Both Possible Outcomes Evaluated*, and *Path Ruled Out*, that help the participants move toward the solution. Three other categories, *Improper Move, Unbalanced Strategy*, and *Guess* are of bad problem solving in that they don't move participants toward the solution. Analyses were done comparing differences between groups with and without using a particular representation on the indicators mentioned above. No significant difference was found. The number of ideas was also counted in order to compare differences between using and not using a particular representation and the result was also not significant.

There was no dramatic impact of external representation found in this study. Even pairs using very distinct modes of representation did not differ much from each other in their problem space search. In general, tree diagram representation and forms of production rules are abstract ways of exploring possibilities and relations of different tasks. They differ significantly from the creation and manipulation of explicit objects and manipulating them. The similar problem solving behaviors displayed by pairs in this study suggest that the link between external representation and better performance in various tasks is not as simple and direct as it appears at first glance. With the different types of representation used, participant pairs did not differ in generating more ideas, having more reflective thoughts on the problem, or in better avoiding useless moves. Aside from being more complex, the Twelve Balls Problem is not inherently different from the kinds of problems and isomorphs that have been studied in recent distributed cognition work. There is no *a priori* reason to expect that representation would have no effect in this task.

References

Gero, J. S. (1998). Conceptual designing as a sequence of situated acts. In I. Smith (Ed.), *Artificial intelligence in structural engineering: Information technology for design, collaboration, maintenance, and monitoring.* Berlin: Springer.

Scaife, M. & Rogers, Y. (1996). External cognition: how do graphical representations work? *International journal of human-computer studies, 45,* 185-213.

Suwa, M. & Tversky, B. (1997). What do architects and students perceive in their design sketches? A protocol analysis. *Design studies, 18*(4), 385-403.

Vera, A. H., Kvan, T., West, R. L., & Lai, S. (1998). Expertise and collaborative design. *Proceedings of CHI'98* (pp. 502-510). LA: ACM.

Zhang, J. (1997). The nature of external representations in problem solving *Cognitive Science, 21*(2), 179-217.

Zhang, J. & Norman, D. A. (1994). Representations in distributed cognitive tasks. *Cognitive Science, 18,* 87-122.

---

[1] We are grateful to Chris Schunn and David Klahr for introducing us to this problem.

Problem Solving with Diagrams :

Modelling the Learning of Perceptual Information

Peter C.R. Lane, Peter C-H Cheng and Fernand Gobet
ESRC Centre for Research in Development, Instruction and Training,
Department of Psychology, University of Nottingham,
University Park, Nottingham NG7 2RD, UK
{pcl,pcc,frg}@psychology.nottingham.ac.uk

This project attempts to model the process by which humans improve their ability to solve problems using diagrams. In order to achieve this, we have chosen an established model for human perceptual memory, and extended it with an ability to learn and manipulate plans for drawing schematic diagrams. This model is applied to the learning of planning information in a specific problem-solving domain: the construction of a diagrammatic representation for electric circuits.

The Computer Model The general form of a model for problem solving with diagrams has been established in earlier work on reasoning and inferencing with external representations (e.g. Tabachneck-Schijf, Leonardo & Simon, 1997). In brief, the model must interact with its external environment (e.g. the sheet of paper) using an eye for input and a pen for output. The model will require both short-term and long-term memories for visuo-spatial, verbal and planning information. However, this earlier work has not addressed the question of learning information about external representations, which is the focus of our project.

The long-term memory in our model is based on a model for the recall of chess positions, CHREST (Gobet & Simon, in press). CHREST is a development of EPAM (Feigenbaum & Simon, 1984) and uses a discrimination network to index chunks of perceptual information. However, CHREST is unable to learn complex problem-solving behaviour, because it lacks the ability to form and represent plans. Accordingly, we have adapted CHREST to handle multiple external representations: one set of representations corresponding to the problem space, one set corresponding to the solution space. A discrimination network is learnt which uses perceptual information to index each space. When a solution is given for a specific problem, an *equivalence link* is formed, linking the two representations. When confronted with novel problems, this new model will attempt to match the problem with a previous entry in its memory. If such a match exists, any equivalence link will suggest a possible solution to the current problem based on what was learnt earlier.

Problem Solving with Diagrams The work in this project attempts to simulate a specific example of problem-solving behaviour: the construction of a diagrammatic representation for a given electric circuit. We use AVOW diagrams as a representation for electric circuits (Cheng, 1998). An AVOW diagram represents each load in a circuit as a separate rectangle, known as an AVOW box. The dimensions of the AVOW boxes represent various electrical properies in their loads. Simple composition rules enable separate AVOW boxes to be combined to form a representation for a complete circuit. The visual nature of the representation makes it suitable for use by a model of perceptual memory, such as CHREST.

We can compare problem-solving behaviour with electric circuits with that in other domains, e.g. the construction of geometry proofs studied by Koedinger and Anderson (1990). This and related work has suggested that experts use *schemas* when solving problems, and it is the propagation of domain specific information in these schemas which explains the superiority of experts over novices.

However, our proposal extends such work by attempting to explain how schemas may be acquired whilst the learner is being taught to solve problems. In brief, schemas may be identified in our model with the chunks of diagrammatic information stored in the discrimination networks. Perceptual similarities ensure that each chunk may be generalised to a range of future examples, showing how our novel version of CHREST learns to improve its problem-solving ability. Further details of the project may be found in Lane, Cheng and Gobet (1999) and at our website:

http://www.psychology.nottingham.ac.uk/research/
    credit/projects/problem_solving/

The authors wish to thank the Economic and Social Research Council for funding this project.

References

Cheng, P. C-H. (1998). A framework for scientific reasoning with law encoding diagrams: Analysing protocols to assess its utility. In M. A. Gernsbacher & S. J. Derry (Eds.) *Proceedings of the Twentieth Annual Conference of the Cognitive Science Society* (pp. 232-235). Mahwah, NJ: Erlbaum.

Feigenbaum, E. A., & Simon, H. A. (1984). EPAM-like models of recognition and learning. *Cognitive Science*, 8, 305-336.

Gobet, F. & Simon, H. A. (in press). Five seconds or sixty? Presentation time in expert memory. *Cognitive Science*.

Koedinger, K. R., & Anderson, J. R. (1990). Abstract planning and perceptual chunks: Elements of expertise in geometry. *Cognitive Science*, 14, 511-550.

Lane, P. C. R., Cheng, P. C-H., & Gobet, F. (1999). Problem solving with diagrams: Modelling the learning of perceptual information. *Technical Report No.59. ESRC CREDIT, University of Nottingham.*

Tabachneck-Schijf, H. J. M., Leonardo, A. M., & Simon, H. A. (1997). CaMeRa: A computational model of multiple representations. *Cognitive Science*, 21, 305-350.

ASPECTS OF INFORMATION STRUCTURE:
CROSS-LINGUISTIC EVIDENCE OF CONTRASTIVE TOPIC

Chungmin Lee clee@snu.ac.kr
Seoul National University

I. Introduction

This paper addresses characterizing Contrastive Topic by critically examining in what sense it is both topical and focal as claimed by Krifka (1991), and supported by means of alternative semantics (Buering 1997, 1998), and a discourse model (Roberts 1996), thus clarifying aspects of information structure -- i.e., Topic-Focus structure -- and by explaining why scope inversion occurs and how reversed polarity or contrast implicature occurs cross-linguistically. A Contrastive Topic (CT), marked by either some high-toned prosodic feature such as a fall-rise contour or by some morphological marking such as a CT marker accompanied by some H-toned prosodic feature (in Korean and Japanese), induces a contrast set in the speaker's mind. The set is scalar in quantification and event-contrast in terms of affectedness or goal accessibility. Verbs are newly claimed to be included in CTs via event-contrast, contra others.

II. Event-Contrast

CTs are underlyingly based on concessive admission of an event/proposition with regard to a cell of a partition of the referent set denoted by a given or accommodated Topic in contrast with the rest of the alternatives in the Contrastive Set (Cset). Concessive admission, however, is for evocation of an implicature normally in the reversed polarity. Verbs/adjectives and other event-denoting predicates, contrary to what is commonly believed, can also be topical (occurring in the previous question) and thus contrastive. For the question 'Did she arrive already?,' the answer can be (1) with the relevant implicature from the contextually salient ordered scalar Cset:

(1) Arrive she did. *or* She arrived. ---
Contour: LH*LH%

Implicature: (But she is *not* ready for the performance.)

Cset on the scale: *<be ready for the performance, arrive>* The concessive affirmative admission of her arrival evokes implicating the denial or the reversed poalarity of the stronger alternative on the scale of event-expressions in different degrees of goal accessibility in the Cset by the Gricean maxim of quantity, Horn's (1972) and Gazdar's (1977) scales. Such effects occur in English, with B accent (Jackendoff 1972), virtually same as LH*LH%, in predicates *in-situ* or in VP preposing constructions (Ward 1985). The verb *arrive* in the answer is not new and can be topical, in an otherwise default (wide) focus *in-situ* position or in preposed/'topicalized' position. In this situation, Korean similarly shows a H-toned CT marker attached to the main verb stem (nominalized when necessary), which is followed by a light V, as in *tochak* 'arrival'-*UN* 'CT' *ha-yess-e* 'did.' The question can be directly whether she is ready for the performance and (1) can serve as an answer to the new question with a stronger negative implicature. Similarly, *didn't kill him* in the fall-rise contour or CT-marking as in Korean, implicates a weaker affirmative alternative such as *beat him* or *pushed him* from the scalar Cset of event expressions in different degrees of affectedness such as *<kill, beat, push>*.

III. Contrast in Quantificational Expressions

If a universal quantifier, numeral, modal operator or 'because' clause, being CT-marked, interacts with negation, it gets a narrow interpretation, because of the nature of the contrastive qualifying denial, e.g. in (2) but note (3) and (4):

(2) *All didn't come* H*LH% Interpretation: ¬∀

(3) **All came* H*LH% (in CT intonation)

(4) **motu -NUN o -ass eo* (Korean)
all CT come Past
Dec(larative)

In (2), negation is concessively partly admitted, naturally evoking a polarity-reversed affirmative implicature. On the contrary, the utterance (3) in the contrastive contour and its equivalent in Korean (4) are anomalous because of the lack of any stronger (higher) alternative to negate on the quantificational scale of relevant quantificational expressions. Numeral expressions in CT also evoke polarity-reversed implicatures. In this connection, denotational implicatures vs. meta-linguistic negation, CT of nominal referents and in embedded sentences, and Topic vs. Focus are explored.

Rethinking the Consistency Assumptions of the Process-Dissociation Procedure

Yuh-shiow Lee (psyysl@ccunix.ccu.edu.tw)
Department of Psychology; National Chung-Cheng University
Chiayi, 621, Taiwan, ROC
Chun-lei Fan and Ying Zhu
Department of Psychology; Peking University
Beijing, China

According to the process-dissociation procedure (Jacoby, 1991), both conscious *(R)* and unconscious *(A)* components of memory can be estimated by contrasting the performance in an inclusion condition with that in an exclusion condition. In formal terms, the probabilities of completing a stem with a studied word in the inclusion test *(I)* and in the exclusion test *(E)* condition are

$$P(I) = R + A (1 - R); \quad P(E) = A (1 - R)$$

Given these two conditions, the probabilities of conscious recollection *(R)* and automatic influence *(A)* will be

$$R = P(I) - P(E); \quad A = P(E) / (1-R)$$

The above calculations are based on three critical assumptions; (1). The criterion for responding on the basis of automatic influences are equivalent in the inclusion and exclusion conditions (*Ain = Aex*); (2). Participants are equally likely to attempt to recollect previously studied items in the inclusion and exclusion conditions (*Rin = Rex*), (3). Automatic memory processes and recollection are independent, that is, the probability of *A* does not depend on the probability of *R*. (e.g., Jacoby, 1991; Jacoby et al, 1993). The first two assumptions have been referred to as the consistency assumption (Dodson & Johnson, 1996).

To deal with the problem of violating consistency assumptions, we make a distinction between the estimated probabilities, P'(I) and P'(E), which are data from experiments, and the true probabilities, P(I) and P(E), which are values derived from the logic of the process dissociation model given no violation of consistency assumptions. We first derived mathematically the true probabilities of I and E. These two values were theoretical probabilities assuming that the inclusion and exclusion performance are measured concurrently and thus the underlying assumptions are always met. We then found the relationship between the estimated probabilities and the true probabilities: P(I) = P'(I);   P(E) = P'(I) P'(E)

Analyses of Curran and Hintzman (1995) Based on the Concurrent Measurement

This study reports the results from reanalysing Curran & Hintzman's (1995) data to demonstrate that our approach is feasible and in a way solves the problem of the violation of consistency assumptions in the process dissociation procedure. The original results and the new calculated R and A are presented in Table 1.

Curran & Hintzman (1995) manipulated the presentation duration in five experiments. Previous studies have suggested that recollection increases with the presentation duration, but automatic priming does not. However, since there was a violation of the assumptions, they found that the estimate of A' decreased as the presentation duration (and thus R') increased based on both the participant means and the item means. This was particularly evident in the full sample case in which significant negative duration effects of A were revealed in all five experiments except for Experiment 4. Once again, when the assumptions were violated, there was an underestimate of A' as R' increases.

On the other hand, in both Tables 1 and 2, based on the logic of the concurrent measurement, the new calculated A remained constant across presentation durations in Experiments 1, 2, and 3. As for Experiment 4, the presentation duration may have had some real, positive effect on automatic influences, thus the true A was larger for the long duration condition than for the short duration condition. Once again, our new approach enabled us to provide a more reasonable estimate of A even from a full sample.

Table 1 Means of R and A Computed from Participant Means
from Curran & Hintzman (1995)

		R			A	
Exp.	new	1s	10s	new	1s	10s
1	.01	.17	.32	.12	.16	.12
2	.02	.20	.40	.11	.18	.13
3	.12	.32	.50	.30	.31	.23
4	.00	.19	.35	.30	.36	.35
5	.05	.33	.47	.32	.32	.20

from Concurrent Measurement

		R			A	
Exp.	new	1s	10s	new	1s	10s
1	.11	.26	.37	.02	.06	.06
2	.12	.30	.45	.02	.08	.08
3	.30	.42	.55	.11	.19	.18
4	.21	.34	.45	.12	.23	.27
5	.25	.43	.52	.15	.21	.15

References

Curran, T., & Hintzman, D. L. (1997). Consequences and causes of correlations in process dissociation. Journal of Experimental Psychology: Learning, Memory and Cognition, 23, 496-504.

Dodson, C. S., & Johnson, M. K. (1996). Some problems with the process-dissociation approach to memory. Journal of Experimental Psychology: General, 125, 181-194.

Jacoby, L. L. , Toth, J. P. , & Yonelinas, A. P. (1993). Separating conscious and unconscious influences of memory: Measuring recollection. Journal of Experimental Psychology: General, 122, 139-154.

Jacoby, L. L. (1991). A process dissociation framework: Separating automatic from intentional uses of memory. Journal of Memory and Language, 30, 513-541.

The Syntax-Semantics Interface and the Innateness of Scope

John D. Lewis (jlewis5@po-box.mcgill.ca)
Department of Linguistics, McGill University, 1001 Sherbrooke St. West
Montreal, Quebec H3A1G5 Canada

On the standard conception of the syntax-semantics interface *wh*-phrases and quantifiers take scope in a clause initial position. Due to the absence of overt evidence for such movement it is then necessary to hypothesize that Universal Grammar specifies these scopal requirements. In principle, this is not a problem: the language learner can acquire overt movement where there is evidence for it, and utilize covert movement where there is not (as per Chomsky, 1986). The acquisition facts, however, raise difficulties for this model. Children both hypothesize movement in the absence of any apparent evidence, and also develop adult interpretations with substantial individual variation. These facts indicate that this knowledge is acquired, and thus that the stimulus is richer than is standardly assumed.

The functional approach to *wh*-questions (Chierchia, 1991,1993; Lewis,1999), provides a way to answer to these facts. Through the assumption of a functional semantics tightly reflected in the syntax, the functional *wh* approach identifies a relevant source of evidence.

On the functional *wh* approach the pair-list interpretation of a question like (1a) is represented semantically as (1b), or equivalently (1c), and this semantics is then mapped to the syntax in (1d). The mapping is achieved through the assumption that a wh-trace is a complex structure containing two empty categories: one with a function-index bound by the fronted wh-element, and the other with an anaphoric index bound to the function's argument.

(1) a) *Who does everyone love?*

b) which f is such that for everyone$_x$, x loves $f(x)$

c) for which f : everyone$_x$ [x loves $f(x)$]

d) [$_{CP}$ who$_j$ [$_{IP}$ everyone$_i$ [$_{IP}$ τ_i love [τ_j e$^{+pro}_i$]$_j$]]]

On this approach the crucial outcome of quantifier raising is the existence of a trace rather than the final location of the moved element. Since traces are the syntactic correlates of variables, semantic relationships involving variable binding will necessitate syntactic traces — and hence movement. Stimuli for the acquisition of quantifier raising are thus available in *wh*-quantifier interactions.

Virtually the same analysis can be given for covert *wh*-movement, though in this case multiple *wh*-questions present the evidence to the learner.[1] A question like (2a),

when assigned the semantics in (2b), requires a syntactic form in which the subject wh-phrase has undergone movement; the trace left by the movement binds the anaphoric component of the complex functional wh-trace, and thus maps to the variable in the semantics. In the absence of this movement no trace is created, and the syntax-semantics mapping fails.

(2) a) *Who brought what?*

b) which f is such that who$_x$, x brought $f(x)$

Several facts speak in support of this account of acquisition. Critically, there is evidence that children determine the binding possibilities for pronouns based on distributivity, allowing a pronoun to take a plural antecedent in a collective, but not a distributive context (Avrutin and Thornton, 1994). This indicates that the pronoun-antecedent relation utilized by the functional *wh* account can play the hypothesized role in acquisition. And Crain *et al* (1996) found that children who exhibited deviant behavior with respect to quantification in declarative sentences, gave only collective readings for quantified *wh*-questions; thus there is a plausible link between the issues. Also, based on the order of acquisition facts, and on erroneous overt *wh*-movement, multi-clause multiple *wh*-questions can be argued to trigger *wh*-movement (de Villiers, 1991; Thornton, 1990).

Such evidence is clearly not irrefutable, but I believe that it does indicate that the idea is worth pursuing.

References

Avrutin, S. & Thornton, R. (1994). Distributivity and binding in child grammar. *Linguistic Inquiry*, 25,1.

Chierchia, G. (1991). Functional wh and weak crossover. In D. Bates (ed.) *Proceedings of WCCFL 10*, CSLI, Stanford, CA

Chierchia, G. (1993). Questions with quantifiers. *Natural Language Semantics* 1, 181-234.

Chomsky, N. (1986). *Barriers*, MIT Press, Cambridge, MA

Crain, S., Thornton, R., Boster, C., Conway, L., Lillo-Martin, D. & Woodams, E. (1996). Quantification without Qualification. *Language Acquisition* 5(2), 83-153.

de Villiers, J. (1991). Why questions? In T. Maxfield & B. Plunkett (eds.), *UMOP: Papers in the acquisition of WH*. Amherst: University of Massachusetts, GLSA.

Lewis, J. (1999). On multiple wh-questions: Weak crossover, D-linking and the third wh-phrase effect. To appear in *Proceedings of WCCFL 18*.

Thornton, R. (1990). *Adventures in long-distance moving: The acquisition of complex wh-questions*. Doctoral dissertation, University of Connecticut, Storrs.

---

[1] The ambiguity of multiple *wh*-questions — they may be interpreted as distributive over either the subject or the object *wh*-phrase — allows evidence both for *wh*-subject movement, and for covert movement of the *wh-in-situ*.

Modelling human performance on the travelling salesperson problem

James N. MacGregor (jmacgreg@HSD.UVic.CA)
Department of Public Administration; University of Victoria
Victoria, BC, Canada V8W 2Y2

Thomas C. Ormerod & Edward P. Chronicle (t.ormerod,e.chronicle@lancaster.ac.uk)
Department of Psychology; Lancaster University
Lancaster LA1 4YF, UK

Abstract

The travelling salesperson problem (TSP) is a classic problem in combinatorial optimisation which is of considerable theoretical and practical importance. Here, we propose and test a computational model of how humans solve TSPs. Model behaviour was highly consistent with human solutions. Human TSP performance depends on boundary formation. A human-emulating model may allow better solutions than conventional algorithms.

Introduction

The Euclidean form of the Travelling Salesperson Problem (TSP) requires finding the shortest path ("tour") through a set of points in the plane and returning to the origin. Although it has attracted a great deal of research, the general problem remains unsolved, in that no practical algorithm has been discovered that guarantees an optimal solution. MacGregor & Ormerod (1996) reported, however, that with a range of TSPs, humans are able to draw solutions that are significantly shorter than those provided by the best known algorithms. Human performance appears to depend upon the ability of visual perception to identify a global figure in the TSP point array. This figure is then utilised to construct a tour (Ormerod & Chronicle, 1999).

The research reported here developed and tested a mathematical model of human performance, with two aims: first, to provide support for the account of human performance suggested by Ormerod & Chronicle (1999), and second to investigate the performance of human-derived heuristics against conventional algorithms.

Method

Three psychologically plausible features were incorporated into the model: (i) a convex hull basis, (ii) a sequential procedure, (iii) capability of producing a variety of solutions. The resultant algorithm is described in general terms below. Note that "closest" can be defined in a number of possible, ways.

Step 1: sketch arcs between adjacent boundary points, select one randomly (the "current arc") and choose a tour direction

Step 2: find the closest interior point to the current arc, and check that no other boundary arcs are closer to that point. If so, the next arc in the direction of travel becomes the current arc and Step 2 is repeated. If not, retain the current arc.

Step 3: replace the current arc with two arcs that include the interior point identified in Step 2. Move to the next arc in the direction of travel, and return to Step 2.

Tour lengths generated by the model were first tested against those produced by MacGregor & Ormerod (1996)'s participants. Second, the optimality of those tours was assessed against known or predicted optimal tour lengths for the 13 problems concerned.

Results

Regression through the origin was used to find the slopes of the lines predicting the experimental data from the model results. In all cases the results supported the goodness-of fit of the model, with r-squared values in excess of 0.99 for all problems.

The best solutions of the model were 1% above optimality across the same 13 test problems. The experiment also employed the highly-structured benchmark 10-node problem described by Dantzig *et al* (1959), and for this problem the best human solutions were 2.7 and 3.0 percent above optimal. By comparison, the present model found the optimal solution.

Discussion

We proposed and tested a model of human performance on TSPs. Because of empirical evidence that people are influenced by the convex hull in generating solutions to TSPs the model was designed to conform in a general way to a convex hull approach. However, it differs from conventional convex hull heuristics by generating solutions from a given starting point and progressing in a specified direction (clockwise or counterclockwise). The results of the model conformed closely to those of the human subjects, both quantitatively and qualitatively. The model also found tours that were nearer to optimal than commonly-used heuristics. Human performance may be capitalised upon in this novel domain for cognitive science.

References

Dantzig, G.B., Fulkerson, D.R. & Johnson, S.M. (1959) On a linear-programming, combinatorial approach to the traveling salesman problem. *Operations Research*, 7, 58-66

MacGregor, J.N. & Ormerod, T.C. (1996) Human performance on the traveling salesman problem. *Perception and Psychophysics*, 58, 527-539

Ormerod, T.C. & Chronicle, E.P (1999) Global perceptual processes in problem-solving: the case of the traveling salesperson. *Perception and Psychophysics*, in press

The Role of Motion and Category Label in Preschoolers' Categorization of Animals

Benise S.K. Mak (h9313988@hkusua.hku.hk)
Alonso H. Vera (vera@hkucc.hku.hk)
Department of Psychology, The University of Hong Kong
Pokfulam Road, Hong Kong

Dynamic perceptual cues have recently been found to be more important than static perceptual cues for young children to determine category membership (Mak & Vera, in press). Four-year-olds tended to categorize animals and geometric figures based on motion similarity rather than on appearance similarity. For instance, they were more likely to judge a donkey and an antelope, rather than a donkey and a horse, as sharing a given property when they were shown to jump in the same way. Recent studies by Gelman and her colleagues have found that 2-year-old children are able to override perceptual similarity and use category label to make inferences about natural kinds (e.g., Gelman & Coley, 1990). Since motion and category label have individually been demonstrated to have an overriding role in children's categorization, the question then becomes which type of features, motion or category label, is more important.

Research by Barbara Tversky (1985) has shown that although preschool children are able to group objects by category name, their knowledge about categories is rather shallow and, at times, may simply be limited to perceptual information. For instance, 3- and 4-year-olds tended to group a plate with a teapot, instead of a clock, because they were white or they shared similar round shapes, rather than because they were tableware. This seems to suggest that category name is less effective than perceptual information as a cue for children to make categorical choice. Although category labels have been shown to be more important than static perceptual cues, it may not be able to override dynamic perceptual cues, such as motion.

Motion information, among other perceptual cues, has been demonstrated to be important not only in children's object categorization (Mak & Vera, in press) but also in some animals' categorization of potential predators (Evans & Marler, 1995). Infant vervets (a species of small gray African monkey) were found to use motion exclusively to make alarm calls, for example, giving eagle alarms not only to eagles but also to various other birds in motion, and even to leaves falling from trees. Young children, like infant vervets, may also be initially guided by motion to categorize objects.

This study, therefore, was an attempt to determine preschool children's development in the use of motion and category label as well as static appearance to draw inferences about the category membership of animals. We expected that preschool children, particularly 3- and 4-year-olds who have been shown to have limited knowledge about categories, would be more likely to use motion cues over category labels to draw inferences about the category membership of animals.

An inductive methodology was used to test four hundred and eighty children, between 3 and 5 years old. In the experiment, children were required to consider two pairs of animals, each of which consisted of a target and a test (e.g., a horse and a donkey). We showed children one pair of animals at a time. They were first taught a new property (e.g., having good vision) about the target animal (the horse) and were then asked to infer if the property of the target animal applied to the test animal (the donkey).

To determine children's development in the use of appearance, motion, and category label to indicate category membership, we adopted a 2 (similar & different appearance) x 2 (same & different motion) x 2 (same & different label) x 3 (3-, 4- & 5-year-olds) between-subject factorial design.

Results showed that 3-year-old children tended to ignore category label and use motion to indicate category membership. The data also showed a motion-label developmental shift in children's use of these features. Three-year-olds used motion to make judgments more often than 4- and 5-year-olds, whereas 4- and 5-year-olds used label more often than 3-year-olds. All of these effects were statistically significant at the .01 level. This clearly provides support for our main hypothesis: Children are initially guided by motion in object categorization.

Although Gelman et al.'s and our studies have consistently shown that static perceptual appearance may very well be the least important feature for children to indicate category membership, it is important to note that appearance is not at all irrelevant. There are some instances in which it does play a part. For example, appearance can help 2-year-old children to determine category membership when no label is given (Gelman & Coley, 1990). It can also help 7-year-olds and adults to make categorical judgments about geometric figures when motion cues become relatively irrelevant (Mak & Vera, in press). The significance of appearance is also found in adult vervets' martial eagle (a dangerous predator) alarms. Adult monkeys sometimes make martial eagle alarms mistakenly in response to other species that share not only the general silhouette but also aspects of the ventral marking with martial eagles (Evans & Marler, 1995). These results are not intended to minimize the role of appearance but rather to stress the role of motion that has been given short shrift in the general literature on children's categorization.

References

Evans, C. S., & Marler, P. (1995). Language and animal communication: Parallels and constrasts. In H. L. Roitblat & J. A. Meyer (Eds.), *Comparative Approaches to Cognitive Science*. Cambridge, MA: MIT Press.

Gelman, S. A., & Coley, J. D. (1990). The importance of knowing a dodo is a bird: Categories and inferences in 2-year-old children. *Developmental Psychology, 26 (5)*, 796-804.

Mak, B. S. K., & Vera, A. H. (in press). The role of motion in children's categorization of objects. *Cognition*.

Tversky, B. (1985). Development of taxonomic organization of named and pictured categories. *Developmental Psychology, 21(6)*, 1111-1119.

The Search for Counterexamples in Human Reasoning

Hansjörg Neth (hneth@princeton.edu)
Philip N. Johnson-Laird (phil@clarity.princeton.edu)
Department of Psychology; Princeton University
Princeton, NJ 08544, USA

A major point of contention about human reasoning is whether or not individuals search for counterexamples to conclusions. According to theorists who argue that the mind is equipped with tacit rules of inference, the decision that an argument is invalid depends on a failure to find a derivation leading from the premises to the conclusion (see e.g. Rips, 1994). However, with this procedure, one can never be certain that the space of possible derivations has been searched exhaustively.

Alternatively, reasoners may base their inferences on mental models (Johnson-Laird and Byrne, 1991). This theoretical account rests on the semantic principle of validity: a conclusion is valid if and only if it allows for no counterexamples, i.e., possibilities in which the premises are true but the conclusion is false. Hence, by constructing a counterexample, reasoners are able to *know* that an inference is invalid.

So, how do logically naïve individuals establish invalidity? Surprisingly for so central a question, there is a dearth of evidence. In the case of syllogisms, Polk and Newell (1995) have defended an account in terms of models, but argued that their explanation of individual differences gains little, if anything, by postulating a search for counterexamples. Bucciarelli and Johnson-Laird (1999) have shown that people are able to search for counterexamples if prompted to do so. Whether individuals spontaneously engage in this search is still unknown.

An experiment on reasoning with non-standard quantifiers

To search for a search for counterexamples, we carried out an experiment in which the participants had to evaluate eight inferences based on *non-standard* quantifiers. Quite common in everyday life, such inferences call for a higher-order predicate calculus, which is incomplete. Each problem had one, two, or three premises, each based on the quantifier "more than half", and a putative conclusion based on "at least one" or "more than half", e.g.:

> More than half of the visitors speak English.
> More than half of the visitors speak French.
> More than half of the visitors speak German.
> Therefore, at least one visitor is trilingual.

Because our main interest was the search for counterexamples, five of the eight inferences were invalid. 20 Princeton undergraduates were instructed to think aloud as they tackled the inferences in a random order. Each problem was presented on a separate sheet of paper on which the participants were encouraged to write or draw whatever would help them.

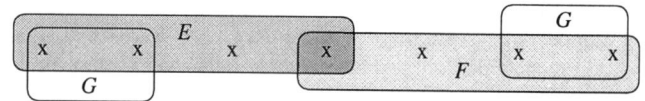

Figure 1: A visual counterexample in which each x represents an individual, more than half of them have features E, F, and G, but it is false that at least one x has all three features.

Overall, participants responded correctly in 72% of all trials. The two-premise problems were easiest (95% correct), the one-premise problems were intermediate in difficulty (73%), and the three premise problems were hardest (48%), and this trend was reliable (Kendall's W=.56, p<.01).

A striking phenomenon was the participants' spontaneous use of a variety of strategies. In 80% of all trials, they constructed a single specific instance of the premises. Typically, the participants constructed a diagram in which they tried to *minimize* the overlap between the given sets. Indeed, every single participant came up with at least one counterexample (see Figure 1). The participants evaluated inferences as invalid on 69 trials and were correct on 62 of them. They produced a counterexample (58% of these trials), claimed that a counterexample was possible (10%), or used some method that the protocols did not reveal (32%). Likewise, they frequently used an unsuccessful search for counterexamples as a basis for concluding that an inference was valid. Where the protocols revealed the nature of the participants' strategies, they relied on counterexamples more often than not (60% of all trials, Wilcoxon test, N=13, T^+=83.5, p<.003).

We still know very little about what governs the selection of strategies. In general, one domain of reasoning may elicit a search for counterexamples more readily than another. Also, the evaluation of a given conclusion may be more likely to trigger a search than the formulation of one's own conclusions. Our study has shown, however, that logically naïve individuals do spontaneously search for counterexamples for at least one sort of deduction.

References

Bucciarelli, M., & Johnson-Laird, P. N. (in press). Strategies in syllogistic reasoning. *Cognitive Science*.

Johnson-Laird, P. N., & Byrne, R. M. J. (1991). *Deduction*. Hove, England: Lawrence Erlbaum Associates.

Polk, T. A., & Newell, A. (1995). Deduction as verbal reasoning. *Psychological Review, 102*, 533–566.

Rips, L. J. (1994). *The psychology of proof: Deductive reasoning in human thinking*. Cambridge, MA: MIT Press.

Implicit Learning and Deliberate Problem Solving: What is the Connection?

Timothy Nokes (tnokes@uic.edu)
Andrew Corrigan-Halpern (ahalpe1@uic.edu)
Stellan Ohlsson (stellan@uic.edu)
Department of Psychology, University of Illinois at Chicago
Chicago, IL 60607-7513, U.S.A.

Introduction

Many theoretical proposals conceptualize the acquisition of deep knowledge as a deliberate, effortful and constructive process. In contrast, research on implicit learning of artificial grammars (Reber, 1989) suggests that learning is a passive, inductive process which is independent of any intention to learn and which creates knowledge not accessible to the learner. In the traditional artificial grammar paradigm participants first memorize letter strings that are generated according to a particular set of relational rules (the training phase). Participants are then given a new list of letter strings and are asked to identify those that are similar to the strings previously memorized (the test phase). Participants perform better than chance in the test phase, implying that an abstract representation is extracted from the training phase and used in the recognition task of the test phase (Reber, 1989).

What is the nature of the knowledge generated by the string memorization procedure? How does that knowledge function in subsequent processing? Can it support problem solving and other higher-order cognitive processes?

To investigate these questions, we revised the standard artificial grammar learning paradigm by replacing the string classification task typically used in the test phase with a letter sequence extrapolation problem (Simon, 1972). If what is learned in implicit pattern learning is available for deliberate problem solving, prior implicit learning of the relevant pattern should facilitate performance on sequence extrapolation.

Method

Participants. Eighty-four students from the University of Illinois at Chicago participated in return for course credit.

Materials. The target tasks were three letter sequence extrapolation problems; see Table 1 for an example.

Table 1: Letter sequence extrapolation problem 1.
For example, given the string
B D X E C Z E G X H F Z
Infer the 8-step extrapolation
H J X K I Z K M

There were 18 training strings consisting of 12 double-digit numbers, six for each of the three problems; see Table 2 for an example. The six number strings followed the exact same pattern as the associated letter sequence extrapolation problem. In addition, there were 18 strings of random double-digit numbers used in the control condition.

Table 2: Relevant training string for Problem 1.
For example,
13 15 35 16 14 37 16 18 35 19 17 37

Design and procedure. The participants were randomly assigned to either the *relevant training* group or to the *irrelevant training* group. Both groups solved the three sequence extrapolation problems. In the relevant training condition, the participants memorized number strings that embodied the same patterns as those in the extrapolation problems. In the irrelevant condition, the participants memorized the random number sequences.

Results and Discussion

Training. As expected, the relevant training group performed significantly better than the irrelevant training group on the memorization task [p<.01].

Problem Solving. The relevant group was slightly better than the irrelevant group on the problem solving tasks, but the difference was small in magnitude and it did not reach statistical significance [p>.06]. Also, performance on the memorization task did not correlate significantly with problem solving performance for two out of the three extrapolation problems.

There are at least two possible explanations. It is possible that although the participants did acquire the pattern underlying the number strings they memorized, their representation of that pattern was not abstract enough to transfer to letter sequences. A second explanation is that the pattern representation learned during string memorization is abstract but not generative. It can support familiarity judgments, but it cannot be equated with the abstract concepts, ideas and schemas that support higher-order thinking (Ohlsson & Lethtinen, 1997). Studies currently under way aim to resolve these issues.

References

Ohlsson, S. & Lehtinen, E. (1997). Abstraction and the acquisition of complex ideas. *International Journal of Educational Research, 27,* 37-48.

Reber, S., R. (1989). Implicit learning and tacit knowledge. *Journal of Experimental Psychology: General, 118,* 219-235.

Simon, H. (1972). Complexity and the representation of patterned sequences of symbols. *Psychological Review, 79,* 369-382.

Attention Is Automatically Allocated To Negative Emotional Stimuli

Clark Ohnesorge (cohnesor@carleton.edu)
Department of Psychology; Carleton College
Northfield MN. 55057

Simine Vazire
vazires@carleton.edu

Introduction

Two recent papers have demonstrated surprisingly large effects of emotionally valenced stimuli on low level perceptual processes and argued that the effect occurred through a differential influence of emotionally valenced stimuli on the allocation of attention (Pratto & John, 1991; Ohnesorge & Bierman, 1998). The results of both these studies supported a similar conclusion; negative emotion stimuli demand more processing capacity than do positive stimuli. This result accords well with data from behavioral and physiological studies. One important issue, however, remains unresolved. The argument for differential attentional allocation is carried through an indirect inference rather than a direct demonstration of the effect. The present paper demonstrates that negative emotion stimuli strongly influence the automatic allocation of visual attention.

Experiment One

Subjects
27 Carleton undergraduates participated.

Design
The experiment was conducted within subjects There were three levels of Emotional Focus stimuli (Negative, Neutral, Positive), and five levels of Target Location.

Stimuli
There were two groups of stimuli in the study; Emotion Focus stimuli and single letter Target stimuli. The Focus stimuli were sets of Negative, Neutral and Positive Emotion words. Target stimuli were curved or angular letters.

Procedure
Following a fixation cross subjects viewed an emotion word presented for 300 ms. After a 100 ms ISI a target was presented at one of 5 positions and remained visible until the subject's response. There were 480 experimental trials.

Result
The planned comparison of Emotional Valence at Fixation (position 0) was significant, $\underline{f}$ (1,192) = 10.0, p < .01. Target Classification time following a negative focus stimulus was 580 ms. Vs 608 ms. if the target followed a Neutral or Positive stimulus.

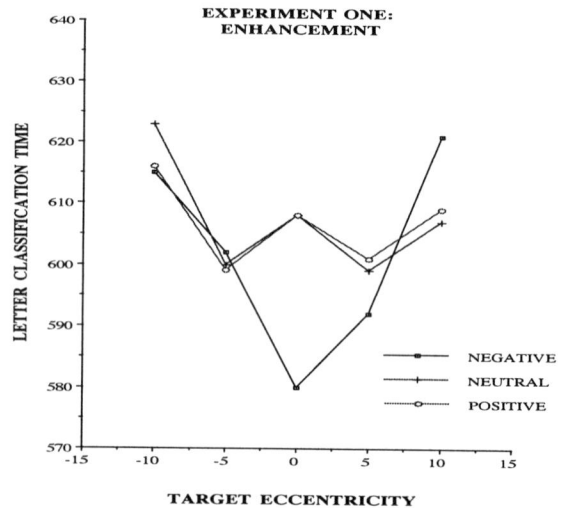

Figure 1; Letter Classification Following Presentation of Negative, Neutral, and Positive Emotion Words.

Discussion

The data reveal that a high efficiency attentional channel was opened at the spatial location occupied by a negative stimulus and it remained open for at least 100 milliseconds following the offset of a negative focus stimulus disconfirming the Disruption hypothesis. No such facilitation was observed for targets that followed a focusing stimulus from the positive or neutral set.

References

Ohnesorge, C.G., & Bierman, C. (1998). The Influence Of Emotional Valence In Backward: Evidence For Early Appraisal. *Proceedings of the 20th Annual Meeting of the Cognitive Science Society.*

Pratto, F., & John, O. (1991). Automatic Vigilance: The Attention-Grabbing Power of Negative Social Information. *Journal Personality & Social Psychology*, 61, 380-391.

Effects of Music Expertise on Evaluative Judgments

Mark G. Orr (morr@uic.edu)
Stellan Ohlsson (stellan@uic.edu)
University of Illinois, 1007 West Harrison Street
Chicago, Illinois 60607 USA

Performers Versus Connoisseurs

Expertise research has to date focused on the type of expertise that is expressed in temporally extended, complex performances. However, there is a class of experts -- commonly referred to as connoisseurs -- whose expertise is not expressed in complex, temporally extended performances but in evaluative judgments.

Do the cognitive mechanisms that are responsible for performance expertise -- e.g., chunking -- also suffice to explain connoisseurship. In one of the few systematic studies of connoisseurship, Solomon (1997) found that experts' descriptions of wines were more complex than those of novices. This finding suggests that wine tasting practice triggers an increase in perceptual subtlety, as common sense would suggest, rather than a decrease in subjective complexity via chunking.

Music straddles the distinction between performance and judgment. On the one hand, musicians are performance experts. On the other hand, the quality of music can only be measured through an evaluative judgment. We are interested in the connection between performance expertise and evaluative judgment in music? Performance expertise should lead to reduced stimulus complexity. However, it is unclear how this will effect evaluative liking judgments.

The purpose of this study was to determine how musical expertise influences liking of music within and across styles. We asked novices, jazz musicians and bluegrass musicians to judge short jazz improvisations with respect to both complexity and liking. We also asked participants to describe the reasons for their judgments.

Method

Sixty-four undergraduate students (musical novices), 12 jazz musicians and 8 bluegrass musicians rated 40 jazz improvisations for complexity and likability on 7-point scales in two sessions. At the end of the second session the subjects were asked to describe what makes one piece of music more or less likable than another.

Results

Liking ratings were regressed onto the complexity ratings for each group of listeners. As shown in Table 1, complexity predicted liking for the novices but not for either expert group.

Table 1: Regression analyses (liking regressed onto complexity) for novices and expert listeners.

Listeners	F	R^2_{Linear}	$R^2_{Quadratic}$
Novices	35.02	--	.65
Jazz	3.31	--	.15
Bluegrass	ns	--	--

Further analyses indicated that experts (jazz and bluegrass) did not perceive the improvisations as less complex compared to novices, $F = .94$, $p > .30$. However, there was a marked increase of liking with increased expertise, $M = 3.4$, 4.0, and 4.8 for the novices, bluegrass and jazz listeners, respectively, $F = 15.65$, $p < .05$.

We coded both novice and expert responses to the question of what makes one piece of music more or less likable than another. As Table 2 shows, novices and experts are both concerned with whether they understood the music, but otherwise they differed in that the novices focused primarily on the music itself, while the experts focused primarily on how well it was performed.

Table 2: Verbal responses of novices and experts.

Category	Novices[a]	Experts[b]
Listener	115	59
Performer	2	72
Misc.	19	0

Discussion

The data do show that liking was systematically related to complexity for our novice listeners but not for experts. Furthermore, musical training does not lower the subjective complexity of jazz improvisations, but it changes the relationship between complexity and liking. Liking increases, but it becomes more independent of complexity. Furthermore, these effects transfer from one style of music to another. Although chunking certainly has to be a part of any complete theory of music expertise, the chunking hypothesis cannot, by itself, explain these training effects.

References

Solomon, G. E. A. (1997). Conceptual change and wine expertise. *Journal of the Learning Sciences*, **6**(1), 41-60.

An Evaluation of the Weight of Evidence Theory of Figural Goodness

Emmanuel M. Pothos (e.pothos@bangor.ac.uk) and **Rob Ward** (r.ward@bangor.ac.uk)

School of Psychology, 39 College Road
Bangor, LL57 2DG, UK

The Weight of Evidence theory of figural goodness (Helm and Leeuwenberg,1996; H&L) has been proposed to explain why certain types of regularity are psychologically more relevant than others. This research is the first attempt to directly empirically investigate some predictions of the theory.

Theories of figural goodness explain why some objects appear to be more well-formed than others. For example, consider the upper two patterns in Figure 1. The two patterns are composed of the same segments, but in one case they are organized to form a perceptual grouping of high regularity, and in the other case they are not. The effects of figural goodness can be operationalized in several ways. For example, a random noise element has been introduced into the lower two patterns in Figure 1. The noise element is readily noticeable in the "good" left pattern but less so on the "bad" right one. Intuitively, the good figure segments away from the random noise in the left pattern, but there is no such basis for segmentation of figure and noise in the right pattern.

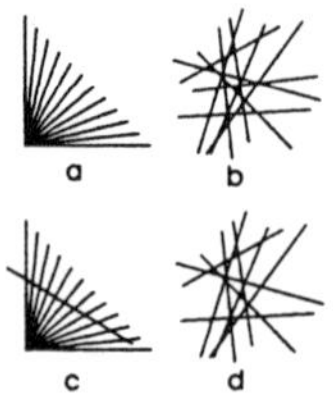

Figure 1: Noise elements are more easily noticed in a good figure than a poor one: The (a), (c) difference is more readily detectable than between (b) and (d).

The effects of regularity on figural goodness have been examined in numerous studies, and the recent Weight of Evidence model by Helm and Leeuwenberg (1996) is an important attempt to explain these effects. The H&L theory has been successfully applied to a range of existing results in the perception literature, but many of the theory's predictions have not yet been subject to direct testing. For example, consider the following symbolic sequences, and corresponding predictions of H&L's theory:

Pair of sequences:	Predicted figural goodness
1) abccbabccbaabccbabccba	high
ababababababababababab	low
2) aaabbabaaaabbaba	high

abababababababab	low
3) abcdfghhgfdcba	high
ababababababab	low
4) abababababababababababbababababababababababa	high
abababababababababababababababababababababab	low

In one line, the H&L theory would always favor more intricate patterns, in contrast to intuition which would suggest that simpler patterns are actually better-formed. To evaluate this intuition, we conducted two experiments where the above sequences were mapped onto simple arrangements of dot patterns (Figures 2 and 3). In all cases, the results were inconsistent with the predictions of the H&L theory, while the intuitive simplicity of the stimuli used was a much better indicator of their figural goodness. This suggests that H&L's theory may not be an accurate account of figural goodness.

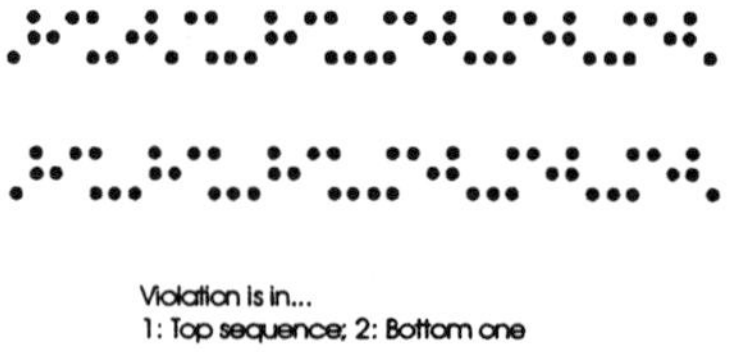

Violation is in...
1: Top sequence; 2: Bottom one

Figure 2: An example of a trial in Experiment 1. The sequences above were mapped onto arrangements of dots, as shown. In each pair, the two sequences were identical, but for one violation of regularity in one of the sequences. We assume that the speed in identifying the violation would be a measure of figural goodness.

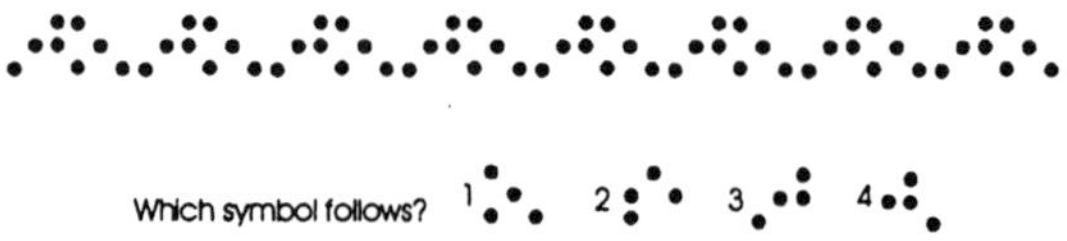

Which symbol follows?

Figure 3 A typical trial in Experiment 2. The sequences were shown individually and the task was to predict the next "dot-symbol" in the sequence. The assumption here is that the faster participants can parse the regularity in the sequence, the higher its figural goodness.

References

Helm P A van der, Leeuwenberg E L J, 1996 "Goodness of visual regularities: A non-transformational approach." *Psychological Review, 103,* 429–456.

Linguistic Structure and Short Term Memory

Emmanuel M. Pothos* (e.pothos@bangor.ac.uk), **Patrick Juola****, (juola@mathcs.duq.edu), & **Nick C. Ellis***
(n.ellis@bangor.ac.uk)

*School of Psychology, 39 College Road
Bangor, LL57 2DG, UK

**Department of Mathematics and Computer Science,
Duquesne University, Pittsburgh, PA 15282

Introduction

The induction of linguistic knowledge displayed by children is perhaps one of the most puzzling aspects of learning performance, because of the intricacies and complexities of language and also because of the unerring accuracy with which (apparently) most people seem to approximate the same linguistic structure (within a society). The latter observation suggests that language learning depends on universal features of the learning process. The feature we examine here is the short term memory (STM) span. If STM is relevant in language acquisition then we expect language structure to reflect the STM span. In other words, the STM span will be reflected in language only insofar that it is a relevant aspect of the language learning problem.

Mutual information and linguistic structure

All the analyses presented are based on the notion of "mutual information," a measure of relatedness between different probability distributions. Let "range" be the number of words between two given words, x, and y plus one. For instance, a range of 1 will indicate that words x and y are separated by only 1 other word. We are asking whether our expectation of obtaining word y at a particular location is affected by the knowledge that we have word x in an earlier location. A measure of this is the mutual information (MI) between P(x) and P(y), the probabilities of obtaining word x and word y respectively. Mutual information indicates how much the uncertainty involved in expecting y is reduced by knowledge that we have x, and is

given by $\sum_{x,y} P(x,y) \log \frac{P(x,y)}{P(x)P(y)}$. For different

ranges, P(x,y) is the probability of having both words x and y, separated by a number of words equal to the range. By MI profile, we mean the way MI varies with increasing range. This we take to be an indicator of statistical structure in language.

Analyses

We investigated samples from seven different languages, all from the CD-ROM database of the European Corpus Initiative Multilingual Corpus 1 (ECI/MC1), distributed by the Association for Computational Linguistics. Table 1 shows the number of samples and average number of words in each language.

Table 1: "SE" is the standard error of the mean sample size, for each language.

language	mean words	SE	samples
Bulgarian	1468	256	4
Czechoslovakian	27591	860	29
Dutch	181407	33483	35
English	97272	11805	12
Estonian	19944	15043	2
French	166620	202	26
Gae	200239	.	1
German	129270	96532	8

In this research our objective is to examine the evidence that different languages display a similar MI profile, regardless of sample differences. Therefore, our approach has been to standardize the average mutual information values for each language, for the different ranges. Standardization converts a set of variables so that the mean of each variable is 0 and the standard deviation 1, so that essentially all the variables are on the same scale. This means that the same differences in each of the variables are now directly comparable. Figure 1 shows the results of this calculation. While there is considerable noise, one can see that the mutual information dependence "elbows" at about four items for all the languages. This we take to indicate that STM is indeed a relevant aspect of language learning. With future work, we aim to extend our analyses so as to more accurately examine MI profile similarities.

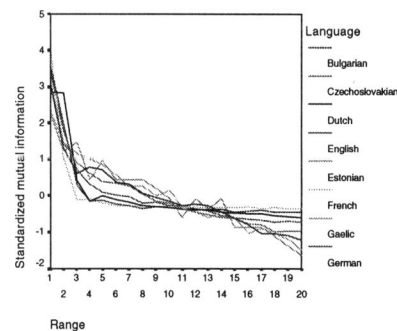

Figure 1: MI profile when MI values were standardized for the different languages.

Inductive inferences about disease: The effect of shared causal features

Tanja Rapus (rapus@ego.psych.mcgill.ca)
Department of Psychology; McGill University, 1205 Dr. Penfield Av.
Montreal, PQ, Canada, H3A 1B1

Introduction

Causal or *deep* features of concepts have been found to play an important role in categorization judgments (e.g.,Ahn, 1998). Individuals are more likely to categorize based on a matching cause than matching effects (Ahn & Dennis, 1997). Causal features can be characterized as those that give rise to or explain the surface features of a concept, such as perceptual or behavioral attributes.

The present research explored how the presence of shared deep and surface features affects other types of induction, specifically, the projection of new attributes or information to a novel case. The key issue examined was whether shared deep-features have a greater impact than shared surface-features of a concept on the mapping of information from a known case to a novel one. Specifically, does a disease which shares a deep feature with a novel disease (i.e., a feature that is more central to the concept of disease) serve as a better case or analogue from which to infer new information about a novel disease?

Experiment 1

Method and Results

Subjects received descriptions of a new, unclassified disease (target) and a known disease (source). Further, the source disease contained additional information about an attribute that could potentially be mapped to or inferred in the target disease. The source and target diseases shared either the same deep feature (same underlying cause) or the same surface feature (same symptom).

People were more likely to map additional information to the target disease when diseases shared a deep rather than a surface feature (p<.0167). Further, deep-feature matches were judged to be more similar, and had a greater impact on categorization judgments, in that a novel disease was more likely to be classified as an instance of the known disease when it had a matching deep versus a matching surface feature.

Experiment 2 was conducted to determine what effect the number of shared features has on inductive judgments. Specifically, diseases either shared a single deep feature or shared multiple surface features.

Experiment 2

Method and Results

Subjects received descriptions of target and source diseases, where deep-feature matches shared the same underlying cause, and surface-feature matches shared multiple symptoms. Again, source disease descriptions contained additional information about an attribute that could potentially be mapped to the new (target) disease.

In contrast to the results of Experiment 1, it was found that people were more likely to project or map new information to the novel disease when diseases shared multiple surface features rather than a single deep feature (p<.0167). Additionally, surface-feature matches were also rated more similar than deep-feature matches, but surface-feature matches did not have a greater impact on categorization judgments than deep-feature matches.

General Discussion

When diseases shared a cause versus a symptom, the likelihood that new information would be mapped to a novel disease increased. However, when diseases shared multiple symptoms the likelihood of projecting new information was greater than when diseases just shared a cause. The present results suggest that several surface features can be as powerful as one shared deep feature in mapping new information to a novel case. It is possible that shared multiple symptoms were viewed as implying the existence of a common cause. Multiple shared surface features, however, did not drive categorical judgments to a greater extent than a single shared deep feature. In general, the causal status of features plays an important role in inductive processes.

Acknowledgments

This research was supported by an NSERC postgraduate fellowship.

References

Ahn, W. (1998). Why are different features central for natural kinds and artifacts?: The role of causal status in determining feature centrality. *Cognition, 69,* 135-178.

Ahn, W. & Dennis, M. (1997). Deep versus surface features in categorization and similarity judgment. *Proceedings of the Nineteenth Annual Conference of the Cognitive Science Society* (pp. 850). Mahwah: Erlbaum.

Developing a Theory of Mind: A Connectionist Investigation

Philip J. Rudling*, J. Richard Eiser* and Denis Mareschal**

*Department Of Psychology, University Of Exeter, Perry Rd, Exeter EX4 5QR, UK.
{p.j.rudling, j.r.eiser}@exeter.ac.uk

**Centre For Brain And Cognitive Development, Department Of Psychology, Birkbeck College, University Of London, London, WC1E 7HX, UK.
d.mareschal@bkk.ac.uk

Introduction

It is generally accepted that by 4 or 5 years of age, children have normally developed a theory of mind (Mitchell, 1996). One of the key signs that a child is acknowledging another's mental states as a causal factor in their behavior is to pass the false belief test. The aim of this abstract is to present a connectionist model that accounts for this change in behavior without necessarily positing an explicit theory of mind.

The particular task modeled here is the deceptive box or "Smarties" test, in which a child is shown a Smarties tube (well known British confectionery) containing some pencils and then asked what another child might think is inside (Perner, Leekam and Wimmer, 1987). The younger child answers "Pencils", supplying their own knowledge, whereas the older says "Smarties", as another would not know the unusual contents of the box.

The Network

By using a connectionist approach, one can show a smooth and seamless change in performance that is accounted for simply by the training set presented to the network. The model is similar in some respects to the Cohen, Dunbar and McClelland (1991) model of the conflicting Stroop task: in this case the two conflicting channels are what is known (reality) and what is seen (appearance).

The network consists of three layers; an input bank, a partially connected hidden layer and an output. In this simplified case, the network is trained to answer the question "what is in the box?". The model is given three types of information; the agent in question (self or other), what is seen (e.g. a Smarties box), and what is known to be inside (e.g. pencils). The network is trained with situations such as "I see Smarties box and I know it contains Smarties". The network computes an output which is compared to target (c.f. opening a box to check) and the connection weights are updated according to the backward propagation rule. One crucial point in the training is that, as in the real world, the network/child does not have direct access to others' knowledge and has to use their own knowledge to predict the behavior of others. The fact that one's own knowledge is a poor predictor of other's behavior is reflected in the training set by adding noise in these situations.

Conclusions

The results of the network match the developmental progression of children, passing from an early "Pencils" stage to a "Smarties" stage, with a smooth developmental progression. The network is able to replicate other aspects of this development, for example the fact that a "noisier" environment (i.e. lots of siblings) causes change in performance to occur earlier (Perner, Ruffman and Leekam, 1994). Such an approach has qualities of both "Theory" Theory and Simulation Theory (Stich and Nichols, 1995) and also suggests that (given an initial innate endowment), a simple constructivist model can arrange its representations by a process of development though experience. The model also gives useful predictions about children's performance under certain conditions. These are currently being followed up.

References

Cohen, J. D., Dunbar, K. & McClelland, J. L. (1991) "On the control of automatic processes: a parallel distributed processing account of the Stroop effect ", Psychological Review, 97, 332-361

Mitchell, P. (1996) "Acquiring a conception of mind", Hove: Psychology Press

Perner, J., Leekam, S., and Wimmer, H. (1987) "Three year olds' difficulty with false belief: the case for conceptual deficit", British Journal of Developmental Psychology. 5. 125-37

Perner, J., Ruffman, T. and Leekam, S. R. (1994) "Theory of mind is contagious: you catch it from your sibs", Child Development, 65, 1228-1238

Stich, S. P., and Nichols, S., (1992) "Folk Psychology: simulation or tacit theory?", Mind and Language, 7, 35-71

Empirical Evidence for Derivational Analogy

Ute Schmid (schmid@cs.tu-berlin.de)
Department of Computer Science, Technische Universität Berlin
Franklinstr. 28, 10587 Berlin, Germany

Jaime Carbonell (jgc@cs.cmu.edu)
School of Computer Science, Carnegie Mellon University
5000 Forbes Avenue, Pittsburgh, PA 15213, USA

Analogical problem solving is mostly described as transfer of a source solution to a target problem based on the structural correspondences (mapping) between source and target. Derivational analogy (Carbonell, 1986) proposes an alternative view: A target problem is solved by replaying a remembered problem solving episode. Thus, experience with the source is used to guide the search for the target solution by applying the same solution technique rather than by a transferring the complete solution. We postulate that both transformational (TA) and derivational analogy (DA) are problem solving strategies realized by human problem solvers. Which strategy is evoked in a given problem solving context depends among other factors on the problem domain. A problem can be solved easily by TA, if the domain involves few objects and relations and/or there are strong (semantic) constraints for mapping; otherwise using DA is more efficient.

We investigated this hypothesis in an empirical study. An example for a domain where TA on an analytical level is inefficient are the network problems presented by Novick and Hmelo (1994). The authors presented (1) an island problem, where a travel route has to be calculated such that each bridge connecting pairs of islands is travelled exactly once, and (2) a handshaking problem, where the goal is to figure out who shook hands with whom at a cocktail party. The first problem corresponds to finding an Eulerian trail in a graph, the second to finding a Hamiltonian cycle. We modified these problems such that the underlying graphs are isomorphic (have the same number of nodes and arcs and identical structure). For each problem class we constructed an additional problem, embedding it into a different context. We used the following problems as source candidates: Euler/travel (*islands*), Hamilton/party (*cocktail*), Euler/party (*birthday*), and Hamilton/travel (*castles*). As target problem we used another Eulerian trail problem from the travel domain (fig. 1) which again could be represented by the same graph.

Because there are no constraints for mapping (unknown) cities to other cities or names of (unknown) persons, node-to-node mapping has exponential effort – that is, TA is a highly inefficient strategy. In our study we wanted to explore if people prefer DA to TA to solve such graph problems and which variables are suitable to discriminate between DA and TA.

Method. Ten subjects were presented first with the four source candidates (in one of two permutations). They had 5 *min* to try to solve each problem and were presented with the solution afterwards. Then, they read the target problem, were asked to select one of the source candidates, and to solve the target problem. To discriminate between DA and TA we (1) obtained thinking aloud protocols dur-

---

> **Boat**. [...] The eight [river] locks are located between the following pairs of cities: Schwetzingen and Blaubeuren, Schwetzingen and Ludwigsburg, Schwetzingen and Marbach, Marbach and Ludwigsburg, Marbach and Blaubeuren, Blaubeuren and Ludwigsburg, Blaubeuren and Ulm, Ulm and Ludwigsburg. The Hamiltons plan to start their trip in Schwetzingen. From Schwetzingen, they wish to travel along a route on the rivers that will enable them to go through each of the eight locks exactly once. Note that their desire to travel through every lock once necessarily means that they will visit some of the cities more than once. Plan a route for the Hamiltons so that they travel over through every lock exactly once and visit each city as many times as necessary.

Figure 1: Target problem (abridged)

ing problem solving, and afterwards (2) asked subjects to map (a) global concepts of source and target (boat, rivers, river locks, cities), and (b) the concrete objects (the cities Schwetzingen, Blaubeuren, etc.) which correspond to the nodes in the graph. For both mappings subjects were instructed to answer each question only if they have the source correspondence available and otherwise skip the question. Finally, (3) subjects were asked to select one from seven predefined strategies. Among these strategies were (a) abstraction, (b) TA (*Because I think that the example problem and the boat problem are very similar, I just tried to replace the node labels in the graph of the example solution with the corresponding names of the boat problem*), and (c) DA (*I remembered how I solved the example problem, that is, how I constructed the graph and how I found the required path in the graph and did it in the same way for the boat problem*).

Our results are promising: Six subjects reported DA, two TA, and two abstraction. Verbal protocols correspond to the strategy reports. Results for node-to-node mapping are not conclusive, however. Our hypothesis was that only subjects using TA should be able to give these mappings easily. But two subjects reporting DA gave the correct mapping and one subject reporting TA did not give the correct mapping. But this subject was the only of the ten which could not produce a correct solution of the target: she tried to draw the graph in exactly the same layout which was given for the source solution and thereby missed one arc.

The results suggest that DA – although proposed in an AI context – is a strategy employed by human problem solvers. To strengthen our claim that humans use TA when mapping effort is low and DA otherwise, we plan follow-up studies contrasting domains with few vs many constraints for node-to-node mapping.

References

Carbonell, J. (1986). Derivational analogy: A theory of reconstructive problem solving and expertise acquistion. In R. Michalski et al., *Machine learning – An AI approach*. Morgan Kaufmann.

Novick, L. R., & Hmelo, C. E. (1994). Transferring symbolic representations across nonisomorphic problems. *Journal of Experimental Psychology: Learning, Memory, and Cognition*, 20(6), 1296-1321.

Have You Read this Before?

Evaluating Familiarity Using Response Times

Travis L. Seymour (NOGARD@Umich.Edu)

Shana R. Pallota (SPALLOT@Umich.Edu)

Colleen M. Seifert (SEIFERT@Umich.Edu)

University of Michigan
Department of Psychology; 525 East University
Ann Arbor, MI 48109-1109 USA

Eyewitnesses are often asked to recognize faces from mugshots or in police lineups. But how good is our ability to recognize faces with minimal prior exposure, and in the absence of the appropriate context? Studies suggest eyewitnesses may be much less accurate than performance in laboratory tests of recognition memory, even though great weight is given to eyewitness identification (Shapiro & Penrod, 1986; Watkins, Ho & Tulving, 1976). How can recognition memory be accurately evaluated?

One method requires witnesses recall the face of a suspect, either by describing it to a police sketch artist or by building a face from a set of features (e.g., using an "Identi-kit"). However, several studies have shown that, rather than eliminating context effects, such methods may require participants to transform visual memories into verbal form, or choose similar features when no exact match is offered.

Further, because these methods take far longer than memory recognition itself, they may introduce bias to the selection of subsequent features (e.g., Mauldin & Laughery, 1981). For example, while considering whether a face looks "familiar," a witness may recall the crime circumstances, and reason that the face should have a particular feature (e.g., a teenager is more likely to have been at the schoolgrounds). The long time periods involved in these methods may allow other thoughts to influence the recognition process.

An alternative method is suggested by Seymour, Seifert, Shafto and Mosmann (1999), who showed that response times (RT] can detect whether participants are familiar with verbal stimuli. Participants were asked to study a set of phrases, and then to perform an Old/New recognition task. Some of the "new" items in this task were already familiar from an earlier task. Participants had to reject the new items within a strict response deadline, too fast to allow strategic processing (< 750 ms). The results showed participants were slower to reject familiar non-target items compared to unfamiliar non-target items. This method reliably detected whether participants had seen the "new" items before, even though that familiarity was unrelated to the test task.

In the present study, we replicate Seymour et al. (1999) using faces as stimuli instead of words. Our hypothesis was that non-target faces familiar from a prior task will interfere with responses in a new task. In Part 1, participants studied pictures of six individual faces ("Probes"), and then had to identify whether a picture was the same as studied, or mirror-reversed. In Part 2, participants were told to study six more faces, the "Target" faces. Then, participants completed an Old/New judgment task, and were told to respond "old" only to the Target pictures. They were instructed to respond "new" to both new faces ("Fillers") and to any of the earlier mirror test faces (Probes). The results show that participants' familiarity with the Probe faces interfered with fast and accurate responses. Probe faces were more often incorrectly called "Old" (M = .57; SD = .21) than were Filler faces (M = .18; SD = .06), t (23) = 9.14, p < .001. Also, RTs for Probe faces correctly called "New" (M = 670ms; SD = 124) were reliably slower than for Filler items (M = 597ms; SD = 85), t (23) = 2.42, p < .03.

We conclude that participants are unable to identify the source of a feeling of familiarity within the short time frame required by this task. Therefore, they are more likely to initially falsely recognize a familiar item as a Target item, thus affecting their response. This paradigm may hold promise for evaluating visual recognition memory in other settings. To the extent that items interfere with fast and accurate responses, this method easily identifies them as familiar to the participant. Although further studies are needed, these results suggest that the RT-based "Guilty Knowledge" test reported by Seymour et al. (1999) may offer a robust paradigm with which to evaluate recognition memory for faces.

References

Watkins, M. J., Ho, E., & Tulving, E. (1976). Context effects in recognition memory for faces. *Journal of Verbal Learning and Verbal Behavior, 15*, 505-517.

Seymour, T. L., Seifert, C. M., Shafto, M. G., & Mosmann, A. L. (1999). Using response times to assess "Guilty Knowledge." *Journal of Applied Psychology*, In Press.

Shapiro, P. N., & Pendrod, S. (1986). Meta-analysis of facial identification studies. *Psychological Bulletin, 100(2)*, 139-156.

Mauldin, M. A., & Laughery, K. R. (1981). Composite production effects on subsequent facial recognition. *Journal of Applied Psychology, 66(3)*, 351-357.

Is the same name like the same color? The role of linguistic labels in similarity judgment

Vladimir M. Sloutsky (sloutsky.1@osu.edu)
Center for Cognitive Science & School of Teaching & Learning; 1945 N. High Street
Columbus, OH 43210, USA

Ya-Fen Lo (Lo.37@osu.edu)
School of Teaching & Learning & Center for Cognitive Science; 1945 N. High Street
Columbus, OH 43210, USA

Abstract

We propose a model of the label as a discrete attribute of an object. According to the model, a relative weight of the label decreases with the child's age. Predictions derived from the model were tested in two experiments. In Experiment 1, children aged 6 to 12 years were presented with triads of schematic faces and were asked to make similarity judgments. These triads were administered under the label (members of the triads were labeled) and no-label conditions. In both conditions, similarity of faces within the triads was manipulated via systematic variation of distinct facial features. In Experiment 2, labels were substituted with colored dots. It was found that (1) labels could be considered as attributes of object that affect similarity judgment in a quantifiable manner, (2) labels' weight decreased with age, and (3) effects of labels do not stem from children's inability to ignore task-irrelevant information. These results have implications for theories of categorization.

Experiments 1 and 2

A total of 107 children aged 6 to 12 years participated in the study. The participants represented three age groups: (1) 34 five-to-seven year-olds, 41 seven-to-nine year-olds, and 32 nine-to-eleven year-olds. The design included an experimental (label) and a control (non-label) condition. The conditions varied across participants. In both conditions, participants were presented with triads of 2" by 2" schematic faces, two of which were Backgrounds and one was a Target. The participants had to select which of the Background faces was more similar to the Target. Each schematic face had three distinct attributes (shape of head, shape of ears, and shape of nose), and each attribute had three values (e.g., "curve-lined" nose, "straight-lined" nose, and "angled" nose). A Target stimulus could share zero, one, or two attribute values with the Background stimuli. In the experimental condition, one Background stimulus (Background A) shared attributes with the Target, whereas another Background stimulus (Background B) always shared the category label (an artificial word) with the Target. No labels were introduced in the control condition. The design also included six within-subject stimulus pattern conditions. (1) Pattern T-0-0 — the Target stimulus shared zero attributes with each of the Background stimuli. (2) Pattern T-1-0 — the Target shared one attribute with Background A. (3) Pattern T-1-1 — the Target shared one attribute with each of the Background stimuli. (4) Pattern T-2-1 — the Target shared two attributes with Background A and one attribute with Background B. (5) Pattern T-2-0 — the Target shared two attributes with Background A. And (6) Pattern T-2-2 — the Target shared two attributes with both Background stimuli.

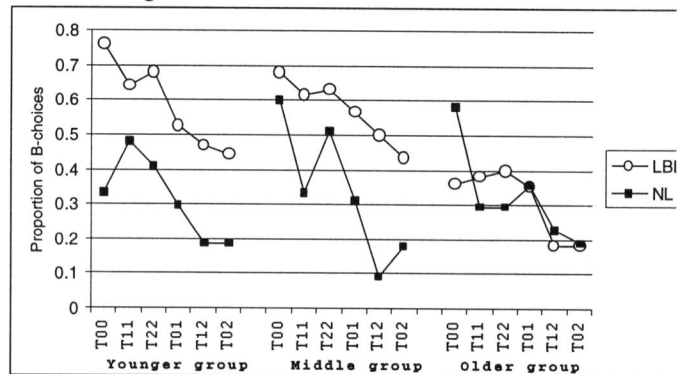

Figure 1: Proportions of B-choices by age and stimulus pattern condition.

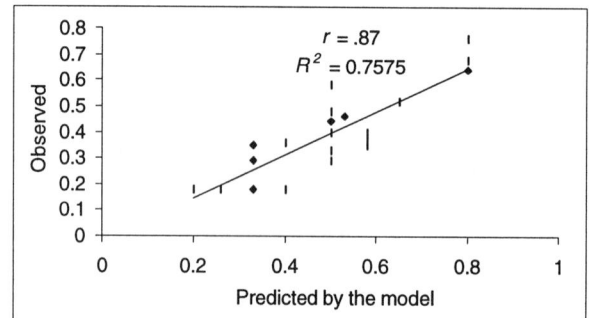

Figure 2: Predicted and observed probabilities across stimulus pattern conditions and age groups

However, these data do not rule out an alternative explanation that the effects stem from inability or unwillingness of young children to ignore task-irrelevant information. To test this alternative, in Experiment 2 labels were substituted with colored dots (task-irrelevant features). Results of Experiment 2 indicate that the contribution of labels in children's similarity judgment is significantly greater than the contribution of dots that were task-irrelevant stimuli.

Acknowledgements

This research has been supported by a grant from the James S. McDonnell Foundation awarded to the first author.

Levels of Competence in Procedural Skills

Jon R. Star (jonstar@umich.edu)

Combined Program in Education and Psychology; 1406 School of Education;
610 E. University; Ann Arbor, MI 48109-1259 USA

Introduction

How do we master procedural skills in domains such as mathematics? According to Anderson's model of developing procedural skills, learners progress through a series of three stages -- a declarative stage, a knowledge compilation stage, and a tuning stage (Anderson, 1996).

One criticism of Anderson's model is its inability to distinguish among levels of "skilled" performance. Anderson's theory proposes that the achievement of automaticity is the endpoint in the acquisition of a skill, where a learner's knowledge is fully compiled and s/he is able to fluently execute a skill. But automaticity may not be the only endpoint of procedural skill development. Karmiloff-Smith has found that competent, automatic performance can be followed by a period of meta-procedural representational redescription, where procedural knowledge becomes increasingly explicit, flexible, manipulatable, and available to conscious access (Karmiloff-Smith, 1992).

In this study, I explore the issues of expertise, automaticity, and "skilled" performance through a close analysis of the development of one child's competence in the domain of linear equation solving.

Method

The subject of this study was a 14-year old, male, 8th grade student, Noah. The experimenter (JS) worked with Noah for 4 sessions over a period of three weeks. Each session lasted approximately 45 minutes. All sessions were videotaped. At the time of the tutoring sessions, Noah had not covered the sections of his symbolic algebra text which dealt with equation solving.

Instruction in equation solving procedures was done by pattern recognition on portions of problems. For example, JS gave Noah the problem step $x+7=12$ and told him a cue to the problem pattern (e.g., "When you see a problem like this,") and then the operator action (e.g., "subtract a 7 from both sides"). Noah was not given instruction which mentioned operator subgoals nor on the order that operators should be applied.

Noah engaged in a two-step verbal protocol during the equation solving portion of the sessions -- planning and solving. In the first step (planning), Noah was asked to verbally describe or plan all problem-solving steps that he thought would be needed to solve a particular problem. Planning occurred before Noah began any written work on the problem. In the second step (solving), Noah attempted to solve the problem. Noah and JS had very little verbal interaction while Noah was engaged in solving.

Results and Discussion

Three main claims are based on a close examination of the video records of Noah's problem-solving. First, the development of algorithms was an important achievement for Noah, however it was not the culmination of his learning. In fact, the abandonment of algorithms in favor of heuristics represents a much more advanced way of approaching equations.

Second, the ability to execute operators individually, which was achieved in the very early stages of Noah's learning, did not enable Noah to use operators in the context of equation solving. Noah had to learn about operators in context, including possible orders of operators and subgoals of operators, before he was able to exhibit knowledge beyond mere operator competence. Similarly, the ability to solve equations of the form $ax+b=c$ using an algorithm did not mean that Noah was necessarily able to solve complex equations such as $5x+4(x+1)+3=2x+3(x+4)-1$. Despite the fact that he had achieved fluency on all sub-skills necessary for the solution of such a problem, he was not able to utilize this knowledge to solve more complex problems.

Third, the instruction given to Noah in operator use was purposefully "procedural," in that JS chose not to provide Noah with any "concepts" which underlie the equation solving operators. In the absence of conceptual instruction, Noah was nevertheless able to develop very robust and useful heuristics for solving many different kinds of equations. Noah's understanding of these heuristics involved knowing such things as the order of steps, the subgoals of steps, the types of problems for which certain heuristics may or may not be useful, and the ability to plan to the solution state. Expertise in algebra equation solving would require an even more thorough procedural understanding.

I conclude that a learner can become quite accomplished and competent at a procedural skill, and yet fail to display some qualities of expert behavior. In addition to fluency in performance, a successful account of procedural learning must include a way to incorporate these more general characteristics of expertise.

References

Anderson, J. R. (1996). *The architecture of cognition*. Cambridge, MA: Harvard University Press.

Karmiloff-Smith, A. (1992). *Beyond modularity: A developmental perspective on cognitive science*. Cambridge, MA: MIT Press.

On the Relationship Between Knowing and Doing in Procedural Learning

Jon R. Star (jonstar@umich.edu)
Combined Program in Education and Psychology; 1406 School of Education;
610 E. University; Ann Arbor, MI 48109-1259 USA

What is the relationship between learners' knowledge of concepts and their ability to execute procedural skills? In this paper, I draw from research in mathematics education, cognitive psychology, and developmental psychology in order to examine how knowledge of procedures and concepts has been studied and what conclusions have been reached about the relationship between them.

Three observations can be made from a review of the existing literature on the relationship between concepts and procedures in children's mathematics learning. First, most research in this area has tried to determine the optimum developmental relationship between concept and procedure learning. In other words, most studies have sought to answer the question, "Which comes first?" Second, almost all research in this area has been limited to the learning of topics in elementary school mathematics. Notably absent are studies of the development of procedural and conceptual knowledge in algebra, geometry, and calculus. Third, knowledge of concepts and knowledge of procedures are assessed in very different ways. Knowledge of concepts is often assessed verbally and through a variety of tasks. It appears that conceptual knowledge is viewed to be complex and multi-faceted. By contrast, procedural knowledge is assessed non-verbally by observing the execution of a procedure. Procedural knowledge is viewed as an entity that a student either has or does not have.

It is this final point that I wish to explore in more depth. The current assumption in the field is that the endpoint of acquisition for concepts is when they are "understood". By contrast, the endpoint of acquisition of procedures is when skills become routine and can be executed with fluency; in other words, when such knowledge has become automatized. I suggest that this portrayal of procedural knowledge does not adequately reflect the complex ways in which procedures can be known or even understood.

What might it mean to have understanding of a procedure? I mention three descriptions of what "procedural understanding" might look like. First, Davis (1983) writes about the process that a student goes through in planning how to approach an unfamiliar problem. Such planning requires that the student have knowledge of a range of necessary techniques, each with an appropriate cognitive label or "tag" which specifies what the technique can accomplish and its relevant goals and subgoals.

Second, Ohlsson and Rees (1991) propose that a procedure is executed with understanding when the "problem solver monitors his or her performance on the problem by comparing the successive states of the problem with what he or she knows about the task environment" (p. 108). They propose that there are two types of knowledge of the task environment: One is knowledge of the principles which guide events and objects in the domain and the other is knowledge of the purposes of each step in a procedure. This second type of knowledge about the procedural task environment -- knowing the purposes of each step in a procedure -- is very similar to Davis' (1983) planning knowledge.

Third, VanLehn and Brown have written about teleological semantics (VanLehn & Brown, 1980). The teleological semantics of a procedure is "knowledge about [the] purposes of each of its parts and how they fit together. ... Teleological semantics is the meaning possessed by one who knows not only the surface structure of a procedure but also the details of its design" (p. 95). VanLehn and Brown (1980) note that a procedure can be cognitively represented on multiple levels. On a very superficial level, a procedure may be represented simply as a chronological list of actions or steps; on a more abstract level, a procedure can include planning knowledge in its representation. Planning knowledge includes not only the surface structure (the sequential series of steps) but also "the reasoning that was used to transform the goals and constraints that define the intent of the procedure into its actual surface structure" (p. 107).

According to these three views, to understand a procedure is to have planning knowledge -- knowledge of such things as the order of steps, the goals and subgoals of steps, the environment or type of situation in which the procedure is used, constraints imposed upon the procedure by the environment or situation, and any heuristics or common sense knowledge which are inherent in the environment or situation. I suggest that this expanded view of procedural knowledge can lead to better theory about the development of procedural and conceptual knowledge.

References

Davis, R. B. (1983). Complex mathematical cognition. In H. P. Ginsburg (Ed.), *The development of mathematical thinking* (pp. 253-290). New York: Academic Press.

Ohlsson, S., & Rees, E. (1991). The function of conceptual understanding in the learning of arithmetic procedures. *Cognition and Instruction, 8*(2), 103-179.

VanLehn, K., & Brown, J. S. (1980). Planning nets: A representation for formalizing analogies and semantic models of procedural skills. In R. E. Snow, P. A. Federico, & W. E. Montague (Eds.), *Aptitude, learning, and instruction* (Vol. 2, pp. 95-137). Hillsdale, NJ: Lawrence Erlbaum

The Effect of Visuo-spatial Ability on the Selection of Route-Learning Strategies within Virtual Environments

Jonathan R. Sykes (j.sykes@dcs.napier.ac.uk)
School of Computing; Napier University, 219 Colinton Road
Edinburgh, EH14 1DJ UK

Introduction

Siegel and White (1975) posit four, necessary stages in the development of the mental representation of a novel environment, starting with landmark identification, and culminating in the building of survey knowledge. However, recent studies suggest that there may be various different route-learning strategies, and that individual cognitive ability may determine the strategy employed. Studies investigating sex differences suggest that females adopt a landmark strategy whereas males prefer a geometric-based strategy (e.g. Holding & Holding, 1989). This is thought to reflect well-documented gender differences in visuo-spatial ability (Kimura, 1992).

To determine the influence of individuals' cognitive skills upon the route-learning process, subjects were pre-tested for visuo-spatial ability, and then asked to navigate through a computer simulated maze. If different learning strategies are indeed used and also dependant upon an individual's cognitive strengths, it is predicted that a participant with high visuo-spatial ability will adopt a geometric-based strategy regardless of the presence of landmarks. Therefore, even when landmarks are present, participants' visuo-spatial ability should correlate with a measure of maze completion time (CT) such that those with high ability complete the maze more quickly.

Method

31 Edinburgh University undergraduates (16 female, 15 male) aimed to find the exit in a computer simulated maze. Participants' CT was measured on each of five successive trials, in a virtual environment where landmarks were either present or not present. As a pre-test measure of spatial ability, a form of Cooper's 2-Dimensional mental rotation task was used, where participants judge whether two presented shapes (one of which is rotated) are the same or mirror-images (Cooper, 1975). Maze CT was then correlated with the participant's mean response time for images presented at 180 degrees in the cognitive pre-test; this was found to be a difficult task and should therefore make considerable use of the participant's visuo-spatial ability.

Results

It was found that for those adept at the visuo-spatial task, correlation between the cognitive test score and maze CT during the latter part of the learning process reached significance ($p < 0.05$) in the non-landmark condition, and tended towards significance in the landmark condition. No significant correlation was found during the early phases of the learning process. Additionally, no significant correlation was identified for those subjects with poor visuo-spatial ability during any part of the learning process.

Discussion

It was hypothesised that participants who were particularly proficient at visuo-spatial tasks would adopt a geometric-based learning strategy, regardless of the presence or absence of landmarks, and that this would be demonstrated by a significant positive correlation between their navigation and visuo-spatial task scores. For both landmark and non-landmark conditions, the correlation tended towards significance during the later sessions within the virtual environment. If the visuo-spatially 'talented' adopted a geometric strategy it would be expected that the correlation would be stronger during the early part of the learning phase, during which time they would be busy identifying the environment's geometric properties. The correlation demonstrated in this paper, between visuo-spatial skills and navigation task performance during the *later* stages of the spatial orientation process, fits more comfortably with the Siegel and White model, where the agent's spatial ability is thought to aid the development of survey knowledge in the later stages of the learning curve.

References

Cooper, L.A. (1975). Mental rotation of random two-dimensional shapes. *Cognitive Psychology, 7,* 20-43.

Holding, C.S., & Holding, D.H. (1989). Acquisition of route knowledge by males and females. *The Journal of General Psychology, 116,* 29-41.

Kimura, D. (1992). Sex differences in the brain. *Scientific American, 267,* 81-87.

Siegel, A.W., & White, S.H. (1975). The development of spatial representations of large-scale environments. *Advances in Child Development and Behaviour, 10,* 9-55.

Cognition and History: Toward a Cognitive Understanding of Science

Ryan D. Tweney (tweney@opie.bgsu.edu)
Department of Psychology, Bowling Green State University
Bowling Green, OH 43403

This symposium reports recent research using a cognitive-historical approach to understand scientific thinking. Because of the richness of the historical record in particular cases, it is possible to achieve a depth of analysis that extends beyond laboratory studies of "science-like" thinking or "in vitro" studies of real-world science.

ALEC: A Computational Simulation of the Invention of the Telephone

Marin Simina (marin@cc.gatech.edu)
College of Computing, Georgia Institute of Technology
Michael E. Gorman (meg3c@virginia.edu)
TCC & Systems Engineering, University of Virginia

This paper investigates the role of historical cases in developing computational simulations of technoscientific thinking by focusing on Alexander Graham Bell's invention of the telephone. Bell's cognitive processes, as described in his notebooks and other materials, have been analyzed using methods similar to those used by Tweney and by Gooding to study Faraday. From these, Gorman derived a series of generalizations about scientific discovery and invention.

Independently, Simina analyzed Bell's invention of the telephone using case-based reasoning as an investigation tool. He then integrated Gorman's work into a program called ALEC that simulates Bell's problem-solving processes. ALEC helped identify the limitations of the traditional case-based reasoning paradigm for addressing scientific thinking and suggested ways of overcoming these limitations at a computational level. ALEC also allowed us to consider whether Gorman's generalizations can be converted into a computationally adequate account of technoscientific thinking. We conclude that historical data can be used to advance cognitive and computational theories of technoscientific thinking.

A Simulation of Multi-Agent Reasoning about Disparate Phenomena

D. C. Gooding (hssdcg@bath.ac.uk)
Department of Psychology, University of Bath

This paper describes a computer model which originated in cognitive-historical analysis of the diaries of Michael Faraday, and has now been extended to represent belief revision in a community of scientists. The formation and revision of beliefs is modeled as a process mediated both by observation and experimentation and by communication between individuals within groups and between groups. Beliefs are represented with varying degrees of generality, from those which can be fully expressed by logical models to those requiring some interpretation of, and negotiation about, qualitative descriptions. Agents are defined as having a number of attributes, including varying confidence in a range of hypotheses or models and variable sensitivity to the opinions and findings of other observers, and they have the ability to make decisions about whether and how to make new experiments, or consult other actors.

The simulations can be used to explore such factors in scientific discovery as: (i) the consequences of situations in which agents exchange information about logically well defined experiments that produce unambiguous results, versus situations more like that of real science, in which agents exchange information drawn from results of variable precision and ambiguity, and (ii) the consequences of variability of agents' access to information produced or held by others and to particular experiments.

Conceptual Change: Development, Learning, and Science

Nancy J. Nersessian (nancyn@cc.gatech.edu)
Cognitive Science, Georgia Institute of Technology

The "cognitive-historical" method is reflexive in application: While it attempts to integrate findings from research on cognition and findings from historical research into models of actual scientific practices, assessments of the fit between cognitive findings and historical practices are fed back to aid in developing richer and more realistic models of cognition.

This paper focuses on what cognitive-historical analyses can contribute to a central issue in cognitive science: conceptual change. An extensive literature in cognitive development claims that there are significant parallels between conceptual change in development and in science. Most earlier work focused on similarities of the products of conceptual change. Thus, salient differences between the child's conceptual structure and the adult's are claimed to be like those between the beginning and end points of conceptual change in a "scientific revolution".

Recently, attention in these areas has shifted to the nature of the mechanisms or processes of conceptual change, especially in the debate between the "neo-nativist" notion of conceptual enrichment and the "theory-formation theory" notion of conceptual change. I focus on the "mechanisms" issue and evaluate the cognitive science claims in light of cognitive-historical analyses of scientific practice. My verdict is that there are indeed significant parallels that can be exploited by cognitive scientists. However, current arguments in favor of this position are weakened by inadequate understanding of the practices leading to conceptual change in science.

Action-Effect Contingency Judgment Tasks
Foster Normative Causal Reasoning

Frédéric Vallée-Tourangeau and **Robin A. Murphy**

Department of Psychology, University of Hertfordshire
Hatfield, Hertfordshire, UNITED KINGDOM, AL10 9AB
`psyqfv, psyqram@herts.ac.uk`

Abstract

We report two experiments using an action-effect causal inference task in which subjects were asked to evaluate the importance of an action they performed in producing the effect. In Experiment 1, judgments of positive and zero contingencies were a function of the actual action-effect contingency and they were not influenced by the effect base rate. Experiment 2 replicated this finding but did record a significant influence of the effect base rate on ratings of negative contingencies. We identify a number of research avenues that may elucidate why an inferential context involving instrumental learning fosters causal inferences that approximate so closely the actual degree of contingency.

A reasoner who aims to infer the causal importance of a candidate cause in producing a target effect in a novel domain may be unable to recruit prior knowledge to help her formulate an initial hypothesis as to the causal importance of the candidate cause. In this situation the contingency between the causal candidate and the effect is an informative cue. Research has shown that causal inferences in such situations are significantly determined by the actual level of cause-effect contingency, but they are also significantly influenced by the overall probability of the effect. Thus, if the target effect frequently occurs then subjects are more likely to attribute greater importance to the causal candidate than if the target effect seldom occurs, even if in both cases the actual cause-effect contingency is held constant. In contrast, prior research using an action-effect contingency judgment task has revealed that subjects attribute less causal importance to their action at high levels of the effect base rate. Paradoxically, then, a passive learning procedure fosters inflated causal ratings at high levels of the effect base rate whereas an active learning procedure yields deflated causal ratings at high levels of the effect base rate.

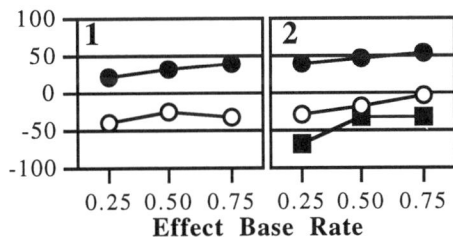

Figure 1. Mean causal ratings in each of the six action - effect contingencies of Experiment 1, and in each of the nine action-effect contingencies of Experiment 2. Black circles correspond to the positive contingency conditions, white circles to the zero contingency conditions, and black squares to the negative contingency conditions.

A careful examination of the previous experimental procedures for action-effect contingency judgment tasks revealed that the contiguity between the action and the effect varied across time intervals. That is, during an interval in which the effect was programmed to occur following an action, the presentation of the effect was delayed until the end of the one-second interval. As a consequence, if a response was recorded at the beginning of the time-interval then there could be nearly up to a one-second delay before the appearance of the figure. At high levels of the effect base rate this degraded contiguity could suggest to the participants that something other than their behavior is causing the effect to occur and hence attenuate their perception of the causal effectiveness of their action. Hence, high effect base rates might lead to weaker causal ratings. In the experiments reported here we have removed this variability in contiguity by ensuring that during the time intervals in which the effect followed the action, it did so immediately after the action was performed.

In Experiment 1, 26 subjects were exposed to 6 conditions reflecting the factorial combination of two levels of the action-effect contingency (0, 0.5) and three levels of the overall density of the effect (0.25, 0.5, 0.75). Ratings are plotted in the left panel of Figure 1. Subjects easily discriminated between the two levels of contingency, but their ratings seemed generally uninfluenced by the base rate of the effect. A 2-factor repeated measures ANOVA confirmed these impressions: The main effect of contingency was reliable, $F(1, 25) = 65.2$, but neither the main effect of the effect base rate, nor the interaction were reliable, both $Fs < 1$. Experiment 2 replicated these 6 conditions but also included three new conditions in which the actual action-effect contingencies were negative (-0.5). The mean ratings of the 34 participants are plotted in the right panel of Figure 1. Exposure to the negative contingencies helped subjects calibrate their estimates of the positive and zero contingencies better. The effect base rate seemed to have influenced the ratings especially of the negative contingencies. The main effect of contingency was reliable, $F(2, 66) = 95$, as well as the main effect of the base rate of the effect, $F(2, 66) = 6.86$; the interaction was not reliable.

Our procedure maximized the contiguity between the action and the effect. In this respect one could argue that in the positive and zero contingency conditions better contiguity would catalyze the influence of the effect base rate since with high base rates, subjects were exposed to more perfectly contiguous pairings of their action and the effect. Yet causal ratings in conditions with high and low base rates did not differ significantly, a result observed in both Experiments 1 and 2. The stronger contiguity between an action and the effect encouraged better contingency judgments.

Acknowledgements

This research was supported by grant R000222542 from the Economic and Social Research Council (UK).

Explaining Success in Discovery Learning.
Analyses Leading Towards a Computational Model.

Hedderik van Rijn and **Maarten van Someren** (rijn,maarten@swi.psy.uva.nl)

Department of Social Science Informatics; University of Amsterdam;

Roetersstraat 15; 1018 WB Amsterdam; The Netherlands

Abstract

This paper describes the results both of a rational analysis of discovery learning and of an analysis of think aloud data. These data were gathered while subjects learn the effects of different nominal factors in a simulated domain. The reported analyses are conducted to guide the modeling of discovery learning in Act-R.

Introduction

The project reported here is about the construction of an Act-R model (Anderson & Lebiere, 1998) of discovery learning. Learners in our task (see Wilhelm et al.) had to discover what factors determine whether a boy (Peter) comes late at school. Learners construct instances of five given factors with 2x2x3x2x2 levels (e.g., the type of bike used). After constructing an instance, the outcome (in number of minutes too late) for that combination is given. Below we describe the analyses which serve as basis for the modeling.

Rational Discovery Learning

The most economical approach to discovery learning in a domain as sketched above, is to apply a "Vary One Thing At a Time strategy" (VOTAT, Schauble, 1996). A learner systematically using VOTAT in the Peter domain constructs a base instance and six instances in which each time one factor has another level. Given these seven instances, all main effects can be derived by comparing these.

But by using this strategy, a learner does not discover possible higher level effects. If a learner suspects that the domain might contain these effects, there are a three different approaches to test for higher level effects. (1) If something is known about the possible outcomes, the learner can check whether the found outcomes fit the possible outcomes. (2) If (domain related) prior-knowledge does not align with the found effects, this might be an indication of higher level effects. (3) A learner might construct some "random" instances to test by trial whether the found effects are generalizable. Although this last approach might not seem to rational, it is a useful technique because it does not require a lot of effort for a relative high change of discovering higher order effects. These three approaches are identified in the current data.

Protocol Analysis of Discovery Learning

Besides a rational task analysis of discovery learning, we also examined the think aloud protocols of learners working in the Peter-task. As initial framework for the analysis of think aloud protocols, we used the SDDS theory (Scientific Discovery as Dual Space Search, Klahr & Dunbar, 1988). However, although the SDDS theory can be used to classify learners at a higher grained level, it is not specific enough to classify the motives for individual actions.

For example, a lot of subjects try to construct an instance that makes Peter arrive on time. This kind of behavior, often called "engineering", is located in the SDDS Experiment Space. That is, no concrete hypothesis is tested when constructing these instances. Another strategy mainly concerned with actions in the Experiment Space is one in which a learner systematically varies one or more factors, without explicitly mentioning a hypothesis. According to the SDDS theory, both strategies are typical strategies of "Experimenters", learners who derive knowledge from experiments, without using hypotheses. To be able to categorize learners at a more fine-grained level, we used the following categories: (1) **Empiricist**. A learner who systematically changes one or more factors per instance, not necessarily with mentioning a hypothesis, to induce the effect of the varied factors. (2) **Theorist**. A learner who constructs instances to test a theory about the domain, not necessarily using sound experimenting techniques. (3) **Engineer**. A learner who focuses on the outcomes of instances instead of on the underlying effects causing the particular outcomes. The difference between these strategies is the motive of the actions, instead of focussing on the actions themselves as in the SDDS theory. During the discovery process, learners often shift from engineering or theorizing strategies to a more empiricist approach.

Conclusion

Currently, we are expanding the implemented VOTAT strategies to incorporate the different motives learners have for constructing their instances. By incorporating the motives and modeling the shift from one strategy to another, we aim at explaining why some learners perform better than others.

References

Anderson, J., & Lebiere, C. (1998). The Atomic Components of Thought. Hillsdale, NJ: Lawrence Erlbaum.

Klahr, D., & Dunbar, K. (1988). Dual Space Search During Scientific Reasoning. *Cognitive Science*, 12, 1-48.

Schauble, L. (1996). The Development of Scientific Reasoning in Knowledge-Rich Contexts. *Developmental Psychology*, 32(1), 102-119.

Wilhelm, P., van Rijn, H., Beishuizen, J.J., Niewold, P. (submitted). *Studying Self-Directed Inductive Learning: The Peter-task.*

Order Effects in Human Belief Revision

Hongbin Wang (Hongbin.Wang@uth.tmc.edu)
Jiajie Zhang (Jiajie.Zhang@uth.tmc.edu)
Todd R. Johnson (Todd.R.Johnson@uth.tmc.edu)
Department of Health Informatics
University of Texas - Houston Health Science Center
7000 Fannin, Suite 600,Houston, TX 77030

Psychological investigations of human belief revision have revealed a robust finding – *the order effect* (e.g., Hogarth & Einhorn, 1992; Zhang, Johnson, & Wang, 1997). Generally speaking, the order effect refers to the phenomenon in which the final belief is significantly affected by the temporal order of information presentation. Normative theories, such as Bayes' Theorem, has no room for the order effect since it simply violates commutativity. Consequently, the order effect is generally regarded as yet another bias in human reasoning.

Several theories have been proposed to explain the order effect (see Wang, 1998 for a review). They are all based on a serial weight-assignment mechanism. According to this mechanism, evidence items are weighted differently based on their positions in the presentation sequence. A primacy effect occurs when earlier evidence item(s) are weighted more heavily, and a recency effect occurs when later evidence item(s) are weighed more heavily. Different theories differ in terms of what factors are important to determine the weight assignment. For some, the difference in weights results from memory decay so earlier items are weighted less than later items (e.g., Miller and Campbell, 1959). For some others, attention is more critical – attention decrement makes later items to be less weighted (e.g., Anderson 1981).

Wang (1998) proposes a different approach to the order effect in human belief revision. This approach is based on two important findings in the fields of epistemology and uncertainty management. First, human belief has a coherence foundation. Beliefs hold each other as a coherent system. As a result, in terms of belief revision, new positive evidence does not necessarily reinforce a belief, and new negative evidence does not necessarily discredit a belief. Second, human belief has a multi-component structure. A single probability number cannot capture all the important aspects of a belief. A confidence dimension, measured in terms of the amount of previous experience a belief is based on, determines how easily a belief can be revised.

The current study aims to investigate this approach, both empirically and computationally.

The probability/confidence distinction predicts that the order effect pattern may change with different levels of experience. Specifically, when one gains more experience about the environment, one's confidence increases. As a result, the order effect tends to diminish and disappear. The experiment, using a serial tactical decision making task, is designed to test this prediction. It is found that participants showed significant recency effects at the beginning of the experiment when their experience about the environment is little. Recency effects disappeared as more training trials were performed. The disappearance of the recency effect suggests that instead of over-reacting in the light of new evidence, participants made more confident belief judgments, which eliminated over-reaction.

UEcho, first proposed in Wang, Johnson, and Zhang (1998) as an extension to Echo (Thagard 1989), is a coherence-based model of belief revision. It is further developed to model the experimental results. The modeling results show that UEcho is able to capture the changes of order effect patterns – the order effect occurs when confidence is low and it tends to disappear when confidence increases.

It is suggested that the fact that UEcho, constructed based on rational postulates and intended to prescribe what people should do, naturally shows order effects (when the confidence is low) convincingly "debiases" order effects. The ecological implications of the order effect are discussed.

Acknowledgements

This work is funded by Office of Naval Research Grant No. N00014-95-1-0241.

References

Anderson, N. H. (1981). *Foundations of information integration theory.* New York, NY: Academic Press.

Hogarth, R.M. & Einhorn, H.J. (1992). Order effects in belief updating: The belief-adjustment model. *Cognitive Psychology, 24,* 1-55.

Miller, N., & Campbell, D. T. (1959). Recency and primacy in persuasion as a function of the timing of speeches and measurement. *Journal of Abnormal and Social Psychology, 59,* 1-9.

Thagard, P. (1989). Explanatory Coherence. Behavioral and Brain Sciences, 12(3), 435-502.

Wang, H. (1998). *Order effects in human belief revision.* Ph.D. Dissertation, The Ohio State University.

Wang, H., Johnson, T. R., & Zhang, J., (1998). UECHO: A model of uncertainty management in human abductive reasoning. In *Proceedings of the Twentieth Annual Conference of the Cognitive Science Society.* Hillsdale, NJ: Erlbaum.

Zhang, J., Johnson, T.R., & Wang, H. (1997). The relation between order effects and frequency learning in tactical decision making. *Thinking and Reasoning, 4(2),* 123-145.

What's in a word?: A sublexical effect in a lexical decision task

Chris Westbury
Lori Buchanan
Department of Psychology
P220 Biological Sciences Building
University Of Alberta
Edmonton, Alberta T6G 2E5

One of the most robust findings in the single word reading literature is that high frequency words are processed as a unitary whole. The cognitive process by which such words are read has proven to be relatively impervious to the disparate influences which can systematically affect reading of other word types, such as regularity (Andrews, 1982; Jared & Seidenberg, 1990) and pronunciation consistency (Glushko, 1979, Jared et al., 1990- but see also Jared, 1997). In this paper we present evidence that the process by which high frequency words are read can be affected by a systematic manipulation of a sublexical feature of those words. Sublexical orthographic features are of particular theoretical interest because they can be controlled in the same manner for both words and nonwords. Having such control may make it possible to make new inferences about the timing and structure of single word processing. If we can identify features that affect different sets of stimulus categories, it will be possible to infer that certain types of cognitive operations must operate only across certain types of stimuli. With a sufficient number of differential effects, inferences about which categories of words and nonwords were processed together, and for how long, may become possible. Such inferences require that several sublexical features with differentiable effects be identified. In this study we examine the role the sublexical feature of minimal bigram frequency.

We examined the effect the frequencies of two-letter pairs within letter strings (bigram frequencies) on a lexical decision task. The stimuli we used were selected on the basis of the frequency of the least-frequent bigram in each (word or nonword) letter string. We hypothesized that words with high minimal bigram frequencies would bias the word reading system towards using subprocesses specialized for high frequency words. We therefore expected faster lexical decision times among high frequency words with high minimal bigrams, and slower times among low frequency words and nonwords with high frequency minimal bigrams, as compared to stimuli closely matched on all characteristics except minimal bigram frequency.

Method/Subjects

79 native English undergraduate subjects participated.

The stimuli we used were selected from a database of 4251 words and 2946 nonwords for which we have computed a wide range of lexical and sublexical measures. The nonwords were a subset (selected for phonological consistency and length 12) from a larger set of nonwords randomly generated using pair-wise Markov chaining of the words. This stochastic computational method of generating nonwords guarantees that every nonword contains only bi-grams among the nonwords is roughly identical to the distribution among the words. The 240 stimuli used in this experiment were comprised of 30 high and low frequency words with high minimal bigram frequencies, 30 high and low frequency words with low minimal bigram frequencies, 60 nonwords with high minimal bigram frequencies, and 60 nonwords with low minimal bigram frequencies.

Results

High frequency words with high minimal bigrams are recognized more slowly (average = 596.9 msecs.) than high frequency words with low minimal bigrams (average = 576.6 msecs.). In contrast, there is no significant difference between words with high (average = 695.1 msecs) and low (average = 701.4) minimal bigrams within the low frequency category of words (t(78) = -0.9; p > 0.05). Subjects were significantly slower (t(78)=6.84; p < 0.001) in correctly classifying nonwords with low minimal bigrams (average RT = 837.1 msecs.) than they were at correctly classifying nonwords with high minimal bigrams (average RT = 804.4 msecs).

Conclusion

The reading system is sensitive to minimal bigram frequency; however, our findings were in the opposite direction of the predicted pattern. Instead of specifically facilitating lexical decision among high frequency words, a high frequency minimal bigram within a word slows down reaction times for high frequency words only. This effect of minimal bigram frequency is reversed among nonwords. Correct reaction times to nonwords are significantly faster when the nonword contains a high frequency minimal bigram.

References

Andrews, S. (1982). Phonological recoding: Is the regularity effect consistent? *Memory And Cognition, 10*, 565-575.

Glushko, R.J. (1979). *The organization and activation of orthographic knowledge in reading aloud.* Journal Of Experimental Psychology: Human Perception and Performance, 5, 674-691.

Jared, F., McRae, K., & Seidenberg, M.S. (1990). *The basis of consistency effects in word naming.* Journal of Memory and Language, 29, 687-715.

Jared, D. & Seidenberg, M. (1990). *Naming multi-syllabic words.* Journal Of Experimental Psychology: Haumn Perception & Performance, 16, 92-105.

Jared, D. (1997). *Consistency effects in high-frequency words.* Journal Of Memory And Language, 36:505-529.

Assessing student contributions in a simulated human tutor with Latent Semantic Analysis

Peter Wiemer-Hastings (PWMRHSTN@MEMPHIS.EDU)
Katja Wiemer-Hastings (KWIEMER@CC.MEMPHIS.EDU)
Arthur C. Graesser (A-GRAESSER@MEMPHIS.EDU)
Department of Psychology; Campus Box 526400
Memphis, TN 38152-6400 USA

Introduction

One-on-one tutoring is a highly-effective means of education compared to classroom instruction. But what accounts for its learning gains? We are developing an intelligent tutoring system called AutoTutor which is based on studies of human tutors. This paper describes the findings from these studies that form the foundation of our tutor. We also show how a corpus-based, statistical natural language understanding technique called Latent Semantic Analysis (LSA) allows AutoTutor to understand student responses and respond appropriately. Finally, we describe analyses of the performance of LSA with respect to human raters.

Psychological foundations of AutoTutor

A fairly complete description of the architecture of AutoTutor has been given elsewhere [Wiemer-Hastings et al., 1999]. Here, we give a brief description of the tutorial discourse foundations of the system in order to highlight how LSA is used by it.

Graesser et al (1997) compared the frequency of a number of features between one-on-one tutoring and classroom education. They found that the following types of activities occur rarely, or at least not more often than in the classroom: active student learning, convergence toward shared meanings, error diagnosis, anchored learning, and sophisticated pedagogical strategies. The following types of activities were significantly more prevalent in tutoring situations than in classroom teaching: use of examples, curriculum scripts, explanatory reasoning by the student and tutor, and collaborative question answering and problem solving.

Assessing student contributions with LSA

LSA relies on a statistical technique that reduces the co-occurrence information in a corpus to a high-dimensional space in which meanings of words and sentences are represented as vectors. The similarity between any two meaning vectors can be computed by calculating the cosine between them. Previous work has shown human-like performance by LSA on variety of tasks.

We trained LSA on two textbooks, 30 articles, and the items of our curriculum script: the questions, expected good answers, and responses that AutoTutor uses. We used a 200-dimensional LSA space to compare student contributions with expected good answers, and calculated a compatibility score that reflected the extent to which the student contribution matched part of the good answer.

We tested LSA's performance by comparing its ratings to those of four human raters: two subject-area experts, and two with intermediate domain knowledge. The correlation between the two intermediate-knowledge raters was $r=0.52$. The correlation between the two expert raters was $r=0.78$. The correlation between the average human rating and the LSA rating was $r=0.47$, almost equalling the interrater reliability for the intermediate-knowledge human raters.

Effects of student text attributes on LSA

We performed an ANOVA with the LSA compatibility scores as the dependent variable, and these three independent variables measuring attributes of the student contributions: (1) *quality*: measured by the compatibility score from the human raters, (2) *length*: the number of words in the student speech contribution, and (3) *information content*: the number of glossary terms in the contribution divided by the number of words. As expected, there was a main effect of the quality of the student contributions. If the absolute LSA score is caused simply by longer contributions, there would be a main effect of the number of words. There was not a significant effect here however. There was a main effect of the information content of the student contributions. If the density of glossary terms in the student contributions is indicative of their quality, this suggests that LSA is measuring the right thing. However, because this effect was independent of the effect of the human quality judgment, it also suggests that there is something to the LSA judgments which is independent of the quality of the contribution alone, at least as human raters judge it.

Acknowledgments

This project is supported by grant number SBR 9720314 from the National Science Foundation's Learning and Intelligent Systems Unit.

References

[Graesser et al., 1997] Graesser, A., Millis, K., and Zwaan, R. (1997). Discourse comprehension. In Spence, J., Darley, J., and Foss, D., editors, *Annual Review of Psychology*, volume 48. Annual Reviews Inc., Palo Alto, CA.

[Wiemer-Hastings et al., 1999] Wiemer-Hastings, P., Wiemer-Hastings, K., and Graesser, A. (1999). Improving an intelligent tutor's comprehension of students with Latent Semantic Analysis. In *Proceedings of Artificial Intelligence in Education*, Amsterdam. IOS Press.

Knowledge structure and type of explanation in the domain of bodily functioning

Reinout W. Wiers (R.Wiers@psychology.unimaas.nl)
Faculty of Psychology, University of Maastricht, PO BOX 616,
6200 MD, Maastricht, The Netherlands

Cindy van de Velde
(former student) Clinical Psychology; University of Amsterdam
Roetersstraat 15, 1018 WB Amsterdam, The Netherlands

Baukje Hemmes (B.Hemmes@psychology.unimaas.nl)
(student) Faculty of Psychology University of Maastricht,
PO BOX 616, 6200 MD, Maastricht, The Netherlands

This study addresses two related issues of current debate: the coherence of intuitive knowledge structures and the use of "vitalistic" explanations in the domain of biology. Hatano and Inagaki (1994) proposed that early theories in this domain are characterized by a unique type of explanation: vitalistic explanation (VE). With VE, a biological phenomenon is explained by the activity of an internal organ as if the organ is a (semi-) autonomous agent, <u>functioning independently of the person's intentions</u>. VE is fundamentally different from intentional explanation (IE) and mechanistic explanation (ME). The type of explanation used is important with respect to the nature of cognitive change (e.g. Gutheil, Vera & Keil, 1998). The biological function that was primarily targeted in this study was digestion. We were interested in the coherence of knowledge when explaining the digestion of "good stuff" (milk and bread) and "bad stuff" (alcohol, generally regarded "bad", Wiers, Gunning, & Sergeant, 1998) and the types of explanations used.

Methods

Participants Ten girls and ten boys from a primary school (7-12 years old), eleven girls and eight boys from a secondary school (12-18 years) and four female and eight male psychology students (18-25 years old) participated. **Materials** A semi-structured interview was developed consisting of factual and generative questions. **Procedure-Scoring** Participants were interviewed individually. After scoring the transcribed interviews at the question level, the overall frameworks used and the types of explanation used were scored (as in Samarapungavan & Wiers, 1997). Protocols were scored by two independent judges, with 92% agreement. Disagreements were resolved through discussion.

Results

Three theories were found (framework level): 1. foods and drinks remain in the alimentary canal; 2. only good stuff enters the body; 3. good and bad stuff enters the body. Adherence shifted with age from theory 1 to 3. Counter intuitively, primary school children responded significantly more consistent when compared with secondary school children and students, $\chi2$ (1) = 7.0, $\underline{p}$ < .01 (Table 1).

Table 1: Consistent and inconsistent use of a theory

Group	Theory 1		Theory 2		Theory 3	
	Consis	Incons	Consis	Incons	Consis	Incons
Primar	2	2	12	4		
Secon			8	11		
Adult			1	10	1	

We hypothesized that young children remained more consistent due to their more frequent use of VE when confronted with anomalies such as: how does alcohol influence behavior when it does not leave the alimentary canal (as a bad stuff)? Indeed, young children used more VE, and older participants more ME, $\chi2$ (4) = 11.6, $\underline{p}$ < .05 (Table 2).

Table 2: Types of explanations used

Group	Intentional - IE	Vatalistic - VE	Mechanistic-ME
Primar	3	16	1
Secon	2	11	6
Adult		5	7

Discussion

When confronted with an anomaly in the domain of biology, young children remain more coherent than older children. This is probably due to their more frequent use of VE, which was higher here than in Hatano & Inagaki (1994).

References

Gutheil, G., Vera, A. & Keil, F. C. (1998). Do houseflies think? Patterns of induction and biological beliefs in development. *Cognition, 66,* 33-49.

Hatano, G. & Inagaki, K. (1994). Young children's naive theory of biology. *Cognition, 50,* 171-188.

Samarapungavan, A. & Wiers, RW (1997). Children's thoughts on the origin of species. *Cognitive Science, 21,* 147-177.

Wiers, R.W., Gunning, W.B. & Sergeant, J.A. (1998). Do young children of alcoholics hold more positive or negative alcohol-related expectancies than controls? *Alcoholism: Clinical and Experimental Research, 22,* 1855-1863.

Verbal Agreement in Sign Language of the Netherlands (SLN)

Inge Zwitserlood (inge.zwitserlood@let.uu.nl)
Utrecht institute for Linguistics, Universiteit Utrecht
Trans 10, NL-3512 JK UTRECHT, The Netherlands

Introduction

Languages over the world employ several ways to express the relationship between a verb and its arguments. SLN (like all other sign languages investigated) has two kinds of verbs: non-agreement and agreement verbs. The arguments of non-agreement verbs are expressed by lexical NPs or pronouns. Agreement verbs mark their arguments spatially. However, there are also verbs that carry a different agreement morpheme: a meaningful *handshape*.

Agreement systems

Languages employ various ways to mark the arguments of verbs. Firstly, they differ in the way they are expressed. In languages like English and Dutch they are expressed by overt NPs and pronouns. These languages also have a system of subject agreement: a verbal suffix. These systems are very poor. Many languages, however, like Spanish and Italian, have *rich* agreement systems. In general, the arguments do not have to be expressed overtly (unless the speaker wants to stress them), since the syntactic relations between verb and arguments are clear enough.

In the second place, languages vary in the number of arguments with which a verb can agree. Some languages not only have subject agreement, but also object agreement, like Choctaw. Languages like Georgian and Basque agree with direct and indirect objects.

Thirdly, languages vary widely in regard of the features the agreement affix contains. Person, gender and number are the best known. The Indo-European languages employ person and number agreement systems. Gender agreement is found in Bantu languages. Recently another way of verbal agreement marking is recognised. In many Amerindian languages verbs carry morphemes that do not have person or gender features, but refer to some salient characteristic of the argument, e.g. its shape or semantic class. These agreement markers are called verbal classifiers.

Agreement in SLN

SLN non-agreement verbs do not undergo systematic changes to express the arguments. Many verbs, however, show spatial agreement markers. At first sight it may appear that this marking concerns person agreement. However, the agreement markers do not have person features, but rather they agree with the *locations* of the referents in signing space. An example is given in (1). Subscript numbers refer to the locations in signing space.

(1)

| father | INDEX$_3$ | INDEX$_1$ | $_1$look-at$_3$ |

"I look at my father."

Classifiers in SLN

SLN also has verbal classifiers. In sign languages, a verbal classifier consists of a handshape, combined with a certain orientation of the palm of the hand and the fingers. Classifiers are bound morphemes, occurring on verbs of movement and location. They typically refer to entities that move or are being moved and, thus, working in a GB-model, can be argued to carry the semantic role of *Theme*. In discourse, as soon as a referent is introduced, it need not be expressed lexically anymore: the use of a verbal classifier suffices to make clear what the argument of the verbs is. An example is given in (2). Mind that the sign for "girl" can be left out.

(2)

| street | girl | LONG-THIN-ENTITYMOVE |

"The girl went along the street."

Conclusion

The following pattern occurs in SLN: next to spatial agreement markers, there are handshapes that refer to arguments of verbs of movement or location. In case such a handshape is used, the arguments can be left implicit. SLN has about fifteen verbal classifiers. These considerations lead to the conclusion that the SLN-classifiers form a way of agreement marking, strongly resembling the verbal classifiers in spoken languages.

Author Index

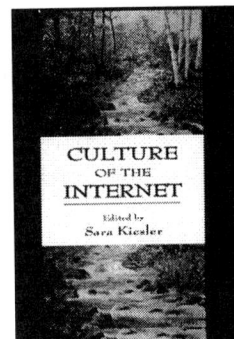

More Leading Titles in Cognitive Science From LEA

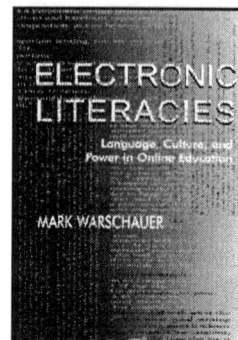